Plumbing Costs with RSMeans data

Brian Adams, Senior Editor

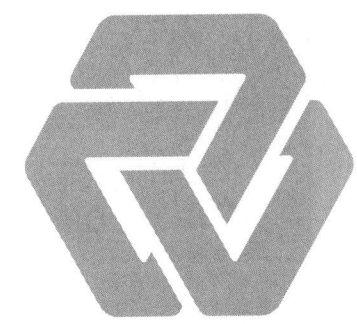

2021
44th annual edition

Vice President, Data
Tim Duggan

Vice President, Product
Ted Kail

Product Manager
Jason Jordan

Principal Engineer
Bob Mewis

Engineers: Architectural Divisions
1, 3, 4, 5, 6, 7, 8, 9, 10, 11, 12, 13, 41
Matthew Doheny, Manager
Sam Babbitt
Sergeleen Edouard
Richard Goldin

Scott Keller
Thomas Lane

Engineers: Civil Divisions and Wages
2, 31, 32, 33, 34, 35, 44, 46
Derrick Hale, PE, Manager
Christopher Babbitt
Stephen Bell
Michael Lynch
Elisa Mello
David Yazbek

Engineers: Mechanical, Electrical, Plumbing & Conveying Divisions
14, 21, 22, 23, 25, 26, 27, 28, 48
Joseph Kelble, Manager
Brian Adams

Michelle Curran
Antonio D'Aulerio
Thomas Lyons
Jake MacDonald

Contributing Engineers
John Melin, PE, Manager
Paul Cowan
Barry Hutchinson
Gerard Lafond, PE
Matthew Sorrentino

Technical Project Lead
Kevin Souza

Production
Debra Panarelli, Manager
Jonathan Forgit
Sharon Larsen
Sheryl Rose
Janice Thalin

Data Quality
Joseph Ingargiola, Manager
David Byars
Audrey Considine
Ellen D'Amico

Cover Design
Blaire Collins

Innovation
Ray Diwakar, Vice President
Kedar Gaikwad
Todd Glowac
Srini Narla
Joseph Woughter

RSMeans data from Gordian
Construction Publishers & Consultants
1099 Hingham Street, Suite 201
Rockland, MA 02370
United States of America
1.800.448.8182
RSMeans.com

Copyright 2020 by The Gordian Group Inc.
All rights reserved.
Cover photo © iStock/DenBoma

Printed in the United States of America
ISSN 1537-8411
ISBN 978-1-950656-65-3

0211 $335.00 per copy (in United States)
Price is subject to change without prior notice.

Related Data and Services

Our engineers recommend the following products and services to complement *Plumbing Costs with RSMeans data*:

Annual Cost Data Books
2021 Assemblies Costs with RSMeans data
2021 Square Foot Costs with RSMeans data

Reference Books
Estimating Building Costs
RSMeans Estimating Handbook
Green Building: Project Planning & Estimating
How to Estimate with RSMeans data
Plan Reading & Material Takeoff
Project Scheduling & Management for Construction
Universal Design Ideas for Style, Comfort & Safety

Virtual, Instructor-led & On-site Training Offerings
Unit Price Estimating
Training for our online estimating solution
Practical Project Management for Construction Professionals
Mechanical & Electrical Estimating

RSMeans data
For access to the latest cost data, an intuitive search, and an easy-to-use estimate builder, take advantage of the time savings available from our online application.

To learn more visit: **RSMeans.com/online**

Enterprise Solutions
Building owners, facility managers, building product manufacturers, and attorneys across the public and private sectors engage with RSMeans data Enterprise to solve unique challenges where trusted construction cost data is critical.

To learn more visit: **RSMeans.com/Enterprise**

Custom Built Data Sets
Building and Space Models: Quickly plan construction costs across multiple locations based on geography, project size, building system component, product options, and other variables for precise budgeting and cost control.

Predictive Analytics: Accurately plan future builds with custom graphical interactive dashboards, negotiate future costs of tenant build-outs, and identify and compare national account pricing.

Consulting
Building Product Manufacturing Analytics: Validate your claims and assist with new product launches.

Third-Party Legal Resources: Used in cases of construction cost or estimate disputes, construction product failure vs. installation failure, eminent domain, class action construction product liability, and more.

API
For resellers or internal application integration, RSMeans data is offered via API. Deliver Unit, Assembly, and Square Foot Model data within your interface. To learn more about how you can provide your customers with the latest in localized construction cost data visit: **RSMeans.com/API**

Table of Contents

Foreword

The Value of RSMeans data from Gordian

Since 1942, RSMeans data has been the industry–standard materials, labor, and equipment cost information database for contractors, facility owners and managers, architects, engineers, and anyone else that requires the latest localized construction cost information. More than 75 years later, the objective remains the same: to provide facility and construction professionals with the most current and comprehensive construction cost database possible.

With the constant influx of new construction methods and materials, in addition to ever-changing labor and material costs, last year's cost data is not reliable for today's designs, estimates, or budgets. Gordian's cost engineers apply real-world construction experience to identify and quantify new building products and methodologies, adjust productivity rates, and adjust costs to local market conditions across the nation. This adds up to more than 22,000 hours in cost research annually. This unparalleled construction cost expertise is why so many facility and construction professionals rely on RSMeans data year over year.

About Gordian

Gordian originated in the spirit of innovation and a strong commitment to helping clients reach and exceed their construction goals. In 1982, Gordian's chairman and founder, Harry H. Mellon, created Job Order Contracting while serving as chief engineer at the Supreme Headquarters Allied Powers Europe. Job Order Contracting is a unique indefinite delivery/indefinite quantity (IDIQ) process, which enables facility owners to complete a substantial number of repair, maintenance, and construction projects with a single, competitively awarded contract. Realizing facility and infrastructure owners across various industries could greatly benefit from the time and cost saving advantages of this innovative construction procurement solution, he established Gordian in 1990.

Continuing the commitment to provide the most relevant and accurate facility and construction data, software, and expertise in the industry, Gordian enhanced the fortitude of its data with the acquisition of RSMeans in 2014. And in an effort to expand its facility management capabilities, Gordian acquired Sightlines, the leading provider of facilities benchmarking data and analysis, in 2015.

Our Offerings

Gordian is the leader in facility and construction cost data, software, and expertise for all phases of the building life cycle. From planning to design, procurement, construction, and operations, Gordian's solutions help clients maximize efficiency, optimize cost savings, and increase building quality with its highly specialized data engineers, software, and unique proprietary data sets.

Our Commitment

At Gordian, we do more than talk about the quality of our data and the usefulness of its application. We stand behind all of our RSMeans data—from historical cost indexes to construction materials and techniques—to craft current costs and predict future trends. If you have any questions about our products or services, please call us toll-free at 800.448.8182 or visit our website at gordian.com.

How the Cost Data Is Built: An Overview

Unit Prices*

All cost data have been divided into 50 divisions according to the MasterFormat® system of classification and numbering.

Assemblies*

The cost data in this section have been organized in an "Assemblies" format. These assemblies are the functional elements of a building and are arranged according to the 7 elements of the UNIFORMAT II classification system. For a complete explanation of a typical "Assembly", see "RSMeans data: Assemblies—How They Work."

Residential Models*

Model buildings for four classes of construction—economy, average, custom, and luxury—are developed and shown with complete costs per square foot.

Commercial/Industrial/ Institutional Models*

This section contains complete costs for 77 typical model buildings expressed as costs per square foot.

Green Commercial/Industrial/ Institutional Models*

This section contains complete costs for 25 green model buildings expressed as costs per square foot.

References*

This section includes information on Equipment Rental Costs, Crew Listings, Historical Cost Indexes, City Cost Indexes, Location Factors, Reference Tables, and Change Orders, as well as a listing of abbreviations.

- **Equipment Rental Costs:** Included are the average costs to rent and operate hundreds of pieces of construction equipment.
- **Crew Listings:** This section lists all the crews referenced in the cost data. A crew is composed of more than one trade classification and/or the addition of power equipment to any trade classification. Power equipment is included in the cost of the crew. Costs are shown both with bare labor rates and with the installing contractor's overhead and profit added. For each, the total crew cost per eight-hour day and the composite cost per labor-hour are listed.

Unit Cost data

Assembly Cost data

Square Foot Models

- **Historical Cost Indexes**: These indexes provide you with data to adjust construction costs over time.
- **City Cost Indexes**: All costs in this data set are U.S. national averages. Costs vary by region. You can adjust for this by CSI Division to over 730 cities in 900+ 3-digit zip codes throughout the U.S. and Canada by using this data.
- **Location Factors**: You can adjust total project costs to over 730 cities in 900+ 3-digit zip codes throughout the U.S. and Canada by using the weighted number, which applies across all divisions.
- **Reference Tables**: At the beginning of selected major classifications in the Unit Prices are reference numbers indicators. These numbers refer you to related information in the Reference Section. In this section, you'll find reference tables, explanations, and estimating information that support how we develop the unit price data, technical data, and estimating procedures.
- **Change Orders**: This section includes information on the factors that influence the pricing of change orders.

- **Abbreviations**: A listing of abbreviations used throughout this information, along with the terms they represent, is included.

Index (printed versions only)

A comprehensive listing of all terms and subjects will help you quickly find what you need when you are not sure where it occurs in MasterFormat®.

Conclusion

This information is designed to be as comprehensive and easy to use as possible.

The Construction Specifications Institute (CSI) and Construction Specifications Canada (CSC) have produced the 2018 edition of MasterFormat®, a system of titles and numbers used extensively to organize construction information.

All unit prices in the RSMeans cost data are now arranged in the 50-division MasterFormat® 2018 system.

---

* Not all information is available in all data sets

Note: The material prices in RSMeans cost data are "contractor's prices." They are the prices that contractors can expect to pay at the lumberyards, suppliers'/distributors' warehouses, etc. Small orders of specialty items would be higher than the costs shown, while very large orders, such as truckload lots, would be less. The variation would depend on the size, timing, and negotiating power of the contractor. The labor costs are primarily for new construction or major renovation rather than repairs or minor alterations. With reasonable exercise of judgment, the figures can be used for any building work.

Estimating with RSMeans data: Unit Prices

Following these steps will allow you to complete an accurate estimate using RSMeans data: Unit Prices.

1. Scope Out the Project

- Think through the project and identify the CSI divisions needed in your estimate.
- Identify the individual work tasks that will need to be covered in your estimate.
- The Unit Price data have been divided into 50 divisions according to CSI MasterFormat® 2018.
- In printed versions, the Unit Price Section Table of Contents on page 1 may also be helpful when scoping out your project.
- Experienced estimators find it helpful to begin with Division 2 and continue through completion. Division 1 can be estimated after the full project scope is known.

2. Quantify

- Determine the number of units required for each work task that you identified.
- Experienced estimators include an allowance for waste in their quantities. (Waste is not included in our Unit Price line items unless otherwise stated.)

3. Price the Quantities

- Use the search tools available to locate individual Unit Price line items for your estimate.
- Reference Numbers indicated within a Unit Price section refer to additional information that you may find useful.
- The crew indicates who is performing the work for that task. Crew codes are expanded in the Crew Listings in the Reference Section to include all trades and equipment that comprise the crew.
- The Daily Output is the amount of work the crew is expected to complete in one day.
- The Labor-Hours value is the amount of time it will take for the crew to install one unit of work.
- The abbreviated Unit designation indicates the unit of measure upon which the crew, productivity, and prices are based.
- Bare Costs are shown for materials, labor, and equipment needed to complete the Unit Price line item. Bare costs do not include waste, project overhead, payroll insurance, payroll taxes, main office overhead, or profit.
- The Total Incl O&P cost is the billing rate or invoice amount of the installing contractor or subcontractor who performs the work for the Unit Price line item.

4. Multiply

- Multiply the total number of units needed for your project by the Total Incl O&P cost for each Unit Price line item.
- Be careful that your take off unit of measure matches the unit of measure in the Unit column.
- The price you calculate is an estimate for a completed item of work.
- Keep scoping individual tasks, determining the number of units required for those tasks, matching each task with individual Unit Price line items, and multiplying quantities by Total Incl O&P costs.
- An estimate completed in this manner is priced as if a subcontractor, or set of subcontractors, is performing the work. The estimate does not yet include Project Overhead or Estimate Summary components such as general contractor markups on subcontracted work, general contractor office overhead and profit, contingency, and location factors.

5. Project Overhead

- Include project overhead items from Division 1–General Requirements.
- These items are needed to make the job run. They are typically, but not always, provided by the general contractor. Items include, but are not limited to, field personnel, insurance, performance bond, permits, testing, temporary utilities, field office and storage facilities, temporary scaffolding and platforms, equipment mobilization and demobilization, temporary roads and sidewalks, winter protection, temporary barricades and fencing, temporary security, temporary signs, field engineering and layout, final cleaning, and commissioning.
- Each item should be quantified and matched to individual Unit Price line items in Division 1, then priced and added to your estimate.
- An alternate method of estimating project overhead costs is to apply a percentage of the total project cost—usually 5% to 15% with an average of 10% (see General Conditions).
- Include other project related expenses in your estimate such as:
  - Rented equipment not itemized in the Crew Listings
  - Rubbish handling throughout the project (see section 02 41 19.19)

6. Estimate Summary

- Include sales tax as required by laws of your state or county.
- Include the general contractor's markup on self-performed work, usually 5% to 15% with an average of 10%.
- Include the general contractor's markup on subcontracted work, usually 5% to 15% with an average of 10%.
- Include the general contractor's main office overhead and profit:
  - RSMeans data provides general guidelines on the general contractor's main office overhead (see section 01 31 13.60 and Reference Number R013113-50).
  - Markups will depend on the size of the general contractor's operations, projected annual revenue, the level of risk, and the level of competition in the local area and for this project in particular.
- Include a contingency, usually 3% to 5%, if appropriate.
- Adjust your estimate to the project's location by using the City Cost Indexes or the Location Factors in the Reference Section:
  - Look at the rules in "How to Use the City Cost Indexes" to see how to apply the Indexes for your location.
  - When the proper Index or Factor has been identified for the project's location, convert it to a multiplier by dividing it by 100, then multiply that multiplier by your estimated total cost. The original estimated total cost will now be adjusted up or down from the national average to a total that is appropriate for your location.

Editors' Note:
We urge you to spend time reading and understanding the supporting material. An accurate estimate requires experience, knowledge, and careful calculation. The more you know about how we at RSMeans developed the data, the more accurate your estimate will be. In addition, it is important to take into consideration the reference material such as Equipment Listings, Crew Listings, City Cost Indexes, Location Factors, and Reference Tables.

How to Use the Cost Data: The Details

What's Behind the Numbers? The Development of Cost Data

RSMeans data engineers continually monitor developments in the construction industry in order to ensure reliable, thorough, and up-to-date cost information. While overall construction costs may vary relative to general economic conditions, price fluctuations within the industry are dependent upon many factors. Individual price variations may, in fact, be opposite to overall economic trends. Therefore, costs are constantly tracked and complete updates are performed yearly. Also, new items are frequently added in response to changes in materials and methods.

Costs in U.S. Dollars

All costs represent U.S. national averages and are given in U.S. dollars. The City Cost Index (CCI) with RSMeans data can be used to adjust costs to a particular location. The CCI for Canada can be used to adjust U.S. national averages to local costs in Canadian dollars. No exchange rate conversion is necessary because it has already been factored in.

[G] The processes or products identified by the green symbol in our publications have been determined to be environmentally responsible and/or resource-efficient solely by RSMeans data engineering staff. The inclusion of the green symbol does not represent compliance with any specific industry association or standard.

Material Costs

RSMeans data engineers contact manufacturers, dealers, distributors, and contractors all across the U.S. and Canada to determine national average material costs. If you have access to current material costs for your specific location, you may wish to make adjustments to reflect differences from the national average. Included within material costs are fasteners for a normal installation. RSMeans data engineers use manufacturers' recommendations, written specifications, and/or standard construction practices for the sizing and spacing of fasteners. Adjustments to material costs may be required for your specific application or location. The manufacturer's warranty is assumed. Extended warranties are not included in the material costs. **Material costs do not include sales tax.**

Labor Costs

Labor costs are based upon a mathematical average of trade-specific wages in 30 major U.S. cities. The type of wage (union, open shop, or residential) is identified on the inside back cover of printed publications or selected by the estimator when using the electronic products. Markups for the wages can also be found on the inside back cover of printed publications and/or under the labor references found in the electronic products.

- If wage rates in your area vary from those used, or if rate increases are expected within a given year, labor costs should be adjusted accordingly.

Labor costs reflect productivity based on actual working conditions. In addition to actual installation, these figures include time spent during a normal weekday on tasks, such as material receiving and handling, mobilization at the site, site movement, breaks, and cleanup.

Productivity data is developed over an extended period so as not to be influenced by abnormal variations and reflects a typical average.

Equipment Costs

Equipment costs include not only rental but also operating costs for equipment under normal use. The operating costs include parts and labor for routine servicing, such as the repair and replacement of pumps, filters, and worn lines. Normal operating expendables, such as fuel, lubricants, tires, and electricity (where applicable), are also included. Extraordinary operating expendables with highly variable wear patterns, such as diamond bits and blades, are excluded. These costs are included under materials. Equipment rental rates are obtained from industry sources throughout North America—contractors, suppliers, dealers, manufacturers, and distributors.

Rental rates can also be treated as reimbursement costs for contractor-owned equipment. Owned equipment costs include depreciation, loan payments, interest, taxes, insurance, storage, and major repairs.

Equipment costs do not include operators' wages.

Equipment Cost/Day—The cost of equipment required for each crew is included in the Crew Listings in the Reference Section (small tools that are considered essential everyday tools are not listed out separately). The Crew Listings itemize specialized tools and heavy equipment along with labor trades. The daily cost of itemized equipment included in a crew is based on dividing the weekly bare rental rate by 5 (number of working days per week), then adding the hourly operating cost times 8 (the number of hours per day). This Equipment Cost/Day is shown in the last column of the Equipment Rental Costs in the Reference Section.

Mobilization, Demobilization—The cost to move construction equipment from an equipment yard or rental company to the job site and back again is not included in equipment costs. Mobilization (to the site) and demobilization (from the site) costs can be found in the Unit Price Section. If a piece of equipment is already at the job site, it is not appropriate to utilize mobilization or demobilization costs again in an estimate.

Overhead and Profit

Total Cost including O&P for the installing contractor is shown in the last column of the Unit Price and/or Assemblies. This figure is the sum of the bare material cost plus 10% for profit, the bare labor cost plus total overhead and profit, and the bare equipment cost plus 10% for profit. Details for the calculation of overhead and profit on labor are shown on the inside back cover of the printed product and in the Reference Section of the electronic product.

General Conditions

Cost data in this data set are presented in two ways: Bare Costs and Total Cost including O&P (Overhead and Profit). General Conditions, or General Requirements, of the contract should also be added to the Total Cost including O&P when applicable. Costs for General Conditions are listed in Division 1 of the Unit Price Section and in the Reference Section.

General Conditions for the installing contractor may range from 0% to 10% of the Total Cost including O&P. For the general or prime contractor, costs for General Conditions may range from 5% to 15% of the Total Cost including O&P, with a figure of 10% as the most typical allowance. If applicable, the Assemblies and Models sections use costs that include the installing contractor's overhead and profit (O&P).

Factors Affecting Costs

Costs can vary depending upon a number of variables. Here's a listing of some factors that affect costs and points to consider.

Quality—The prices for materials and the workmanship upon which productivity is based represent sound construction work. They are also in line with industry standard and manufacturer specifications and are frequently used by federal, state, and local governments.

Overtime—We have made no allowance for overtime. If you anticipate premium time or work beyond normal working hours, be sure to make an appropriate adjustment to your labor costs.

Productivity—The productivity, daily output, and labor-hour figures for each line item are based on an eight-hour work day in daylight hours in moderate temperatures and up to a 14' working height unless otherwise indicated. For work that extends beyond normal work hours or is performed under adverse conditions, productivity may decrease.

Size of Project—The size, scope of work, and type of construction project will have a significant impact on cost. Economies of scale can reduce costs for large projects. Unit costs can often run higher for small projects.

Location—Material prices are for metropolitan areas. However, in dense urban areas, traffic and site storage limitations may increase costs. Beyond a 20-mile radius of metropolitan areas, extra trucking or transportation charges may also increase the material costs slightly. On the other hand, lower wage rates may be in effect. Be sure to consider both of these factors when preparing an estimate, particularly if the job site is located in a central city or remote rural location. In addition, highly specialized subcontract items may require travel and per-diem expenses for mechanics.

Other Factors—

- season of year
- contractor management
- weather conditions
- local union restrictions
- building code requirements
- availability of:
  - adequate energy
  - skilled labor
  - building materials
- owner's special requirements/restrictions
- safety requirements
- environmental considerations
- access

Unpredictable Factors—General business conditions influence "in-place" costs of all items. Substitute materials and construction methods may have to be employed. These may affect the installed cost and/or life cycle costs. Such factors may be difficult to evaluate and cannot necessarily be predicted on the basis of the job's location in a particular section of the country. Thus, where these factors apply, you may find significant but unavoidable cost variations for which you will have to apply a measure of judgment to your estimate.

Rounding of Costs

In printed publications only, all unit prices in excess of $5.00 have been rounded to make them easier to use and still maintain adequate precision of the results.

How Subcontracted Items Affect Costs

A considerable portion of all large construction jobs is usually subcontracted. In fact, the percentage done by subcontractors is constantly increasing and may run over 90%. Since the workers employed by these companies do nothing else but install their particular products, they soon become experts in that line. As a result, installation by these firms is accomplished so efficiently that the total in-place cost, even with the general contractor's overhead and profit, is no more, and often less, than if the principal contractor had handled the installation. Companies that deal with construction specialties are anxious to have their products perform well and, consequently, the installation will be the best possible.

Contingencies

The allowance for contingencies generally provides for unforeseen construction difficulties. On alterations or repair jobs, 20% is not too much. If drawings are final and only field contingencies are being considered, 2% or 3% is probably sufficient and often nothing needs to be added. Contractually, changes in plans will be covered by extras. The contractor should consider inflationary price trends and possible material shortages during the course of the job. These escalation factors are dependent upon both economic conditions and the anticipated time between the estimate and actual construction. If drawings are not complete or approved, or a budget cost is wanted, it is wise to add 5% to 10%. Contingencies, then, are a matter of judgment.

Important Estimating Considerations

The productivity, or daily output, of each craftsman or crew assumes a well-managed job where tradesmen with the proper tools and equipment, along with the appropriate construction materials, are present. Included are daily set-up and cleanup time, break time, and plan layout time. Unless otherwise indicated, time for material movement on site (for items that can be transported by hand) of up to 200' into the building and to the first or second floor is also included. If material has to be transported by other means, over greater distances, or to higher floors, an additional allowance should be considered by the estimator.

While horizontal movement is typically a sole function of distances, vertical transport introduces other variables that can significantly impact productivity. In an occupied building, the use of elevators (assuming access, size, and required protective measures are acceptable) must be understood at the time of the estimate. For new construction, hoist wait and cycle times can easily be 15 minutes and may result in scheduled access extending beyond the normal work day. Finally, all vertical transport will impose strict weight limits likely to preclude the use of any motorized material handling.

The productivity, or daily output, also assumes installation that meets manufacturer/designer/standard specifications. A time allowance for quality control checks, minor adjustments, and any task required to ensure proper function or operation is also included. For items that require connections to services, time is included for positioning, leveling, securing the unit, and making all the necessary connections (and start up where applicable) to ensure a complete installation. Estimating of the services themselves (electrical, plumbing, water, steam, hydraulics, dust collection, etc.) is separate.

In some cases, the estimator must consider the use of a crane and an appropriate crew for the installation of large or heavy items. For those situations where a crane is not included in the assigned crew and as part of the line item cost, then equipment rental costs, mobilization and demobilization costs, and operator and support personnel costs must be considered.

Labor-Hours

The labor-hours expressed in this publication are derived by dividing the total daily labor-hours for the crew by the daily output. Based on average installation time and the assumptions listed above, the labor-hours include: direct labor, indirect labor, and nonproductive time. A typical day for a craftsman might include but is not limited to:

- Direct Work
  - ☐ Measuring and layout
  - ☐ Preparing materials
  - ☐ Actual installation
  - ☐ Quality assurance/quality control
- Indirect Work
  - ☐ Reading plans or specifications
  - ☐ Preparing space
  - ☐ Receiving materials
  - ☐ Material movement
  - ☐ Giving or receiving instruction
  - ☐ Miscellaneous
- Non-Work
  - ☐ Chatting
  - ☐ Personal issues
  - ☐ Breaks
  - ☐ Interruptions (i.e. sickness, weather, material or equipment shortages, etc.)

If any of the items for a typical day do not apply to the particular work or project situation, the estimator should make any necessary adjustments.

Final Checklist

Estimating can be a straightforward process provided you remember the basics. Here's a checklist of some of the steps you should remember to complete before finalizing your estimate.

Did you remember to:

- factor in the City Cost Index for your locale?
- take into consideration which items have been marked up and by how much?
- mark up the entire estimate sufficiently for your purposes?
- read the background information on techniques and technical matters that could impact your project time span and cost?
- include all components of your project in the final estimate?
- double check your figures for accuracy?
- call RSMeans data engineers if you have any questions about your estimate or the data you've used? Remember, Gordian stands behind all of our products, including our extensive RSMeans data solutions. If you have any questions about your estimate, about the costs you've used from our data, or even about the technical aspects of the job that may affect your estimate, feel free to call the Gordian RSMeans editors at 1.800.448.8182.

Unit Price Section

Table of Contents

RSMeans data: Unit Prices— How They Work

All RSMeans data: Unit Prices are organized in the same way.

03 30 Cast-In-Place Concrete

03 30 53 – Miscellaneous Cast-In-Place Concrete

03 30 53.40 Concrete In Place	Crew	Daily Output	Labor-Hours	Unit	Material	2021 Bare Costs Labor	Equipment	Total	Total Incl O&P
0010 **CONCRETE IN PLACE**									
0020 Including forms (4 uses), Grade 60 rebar, concrete (Portland cement									
0050 Type I), placement and finishing unless otherwise indicated									
3540 Equipment pad (3000 psi), 3' x 3' x 6" thick	C-14H	45	1.067	Ea.	55.50	57	.61	113.11	147
3550 4' x 4' x 6" thick	"	30	1.600	"	84	85.50	.91	170.41	220
3560 5' x 5' x 8" thick	C-14H	18	2.667	Ea.	150	143	1.52	294.52	380
3570 6' x 6' x 8" thick		14	3.429		204	184	1.95	389.95	500
3580 8' x 8' x 10" thick		8	6		415	320	3.42	738.42	945
3590 10' x 10' x 10" thick		5	9.600		720	515	5.45	1,240.45	1,550
3800 Footings (3000 psi), spread under 1 C.Y.	C-14C	28	4	C.Y.	208	209	.96	417.96	540
3825 1 C.Y. to 5 C.Y.		43	2.605		266	136	.63	402.63	495
3850 Over 5 C.Y.	R033053-60	75	1.493		245	78	.36	323.36	385
3900 Footings, strip (3000 psi), 18" x 9", unreinforced	C-14L	40	2.400		159	123	.67	282.67	360
3920 18" x 9", reinforced	C-14C	35	3.200		189	167	.77	356.77	460

It is important to understand the structure of RSMeans data: Unit Prices so that you can find information easily and use it correctly.

---

① Line Numbers

Line Numbers consist of 12 characters, which identify a unique location in the database for each task. The first 6 or 8 digits conform to the Construction Specifications Institute MasterFormat® 2018. The remainder of the digits are a further breakdown in order to arrange items in understandable groups of similar tasks. Line numbers are consistent across all of our publications, so a line number in any of our products will always refer to the same item of work.

② Descriptions

Descriptions are shown in a hierarchical structure to make them readable. In order to read a complete description, read up through the indents to the top of the section. Include everything that is above and to the left that is not contradicted by information below. For instance, the complete description for line 03 30 53.40 3550 is "Concrete in place, including forms (4 uses), Grade 60 rebar, concrete (Portland cement Type 1), placement and finishing unless otherwise indicated; Equipment pad (3,000 psi), 4' × 4' × 6" thick."

③ RSMeans data

When using **RSMeans data**, it is important to read through an entire section to ensure that you use the data that most closely matches your work. Note that sometimes there is additional information shown in the section that may improve your price. There are frequently lines that further describe, add to, or adjust data for specific situations.

④ Reference Information

Gordian's RSMeans engineers have created **reference** information to assist you in your estimate. **If** there is information that applies to a section, it will be indicated at the start of the section. The Reference Section is located in the back of the data set.

⑤ Crews

Crews include labor and/or equipment necessary to accomplish each task. In this case, Crew C-14H is used. Gordian's RSMeans staff selects a crew to represent the workers and equipment that are typically used for that task. In this case, Crew C-14H consists of one carpenter foreman (outside), two carpenters, one rodman, one laborer, one cement finisher, and one gas engine vibrator. Details of all crews can be found in the Reference Section.

Crews - Standard

Crew No.	Bare Costs		Incl. Subs O & P		Cost Per Labor-Hour	
Crew C-14H	Hr.	Daily	Hr.	Daily	Bare Costs	Incl. O&P
1 Carpenter Foreman (outside)	$56.70	$453.60	$84.60	$676.80	$53.53	$79.68
2 Carpenters	54.70	875.20	81.65	1306.40		
1 Rodman (reinf.)	58.90	471.20	88.05	704.40		
1 Laborer	44.40	355.20	66.25	530.00		
1 Cement Finisher	51.80	414.40	75.90	607.20		
1 Gas Engine Vibrator		27.15		29.86	.57	.62
48 L.H., Daily Totals		$2596.75		$3854.67	$54.10	$80.31

6. Daily Output

The **Daily Output** is the amount of work that the crew can do in a normal 8-hour workday, including mobilization, layout, movement of materials, and cleanup. In this case, crew C-14H can install thirty 4' × 4' × 6" thick concrete pads in a day. Daily output is variable and based on many factors, including the size of the job, location, and environmental conditions. RSMeans data represents work done in daylight (or adequate lighting) and temperate conditions.

7. Labor-Hours

The figure in the **Labor-Hours** column is the amount of labor required to perform one unit of work–in this case the amount of labor required to construct one 4' × 4' equipment pad. This figure is calculated by dividing the number of hours of labor in the crew by the daily output (48 labor-hours divided by 30 pads = 1.6 hours of labor per pad). Multiply 1.6 times 60 to see the value in minutes: 60 × 1.6 = 96

minutes. Note: the labor-hour figure is not dependent on the crew size. A change in crew size will result in a corresponding change in daily output, but the labor-hours per unit of work will not change.

8. Unit of Measure

All RSMeans data: Unit Prices include the typical **Unit of Measure** used for estimating that item. For concrete-in-place the typical unit is cubic yards (C.Y.) or each (Ea.). For installing broadloom carpet it is square yard and for gypsum board it is square foot. The estimator needs to take special care that the unit in the data matches the unit in the take-off. Unit conversions may be found in the Reference Section.

9. Bare Costs

Bare Costs are the costs of materials, labor, and equipment that the installing contractor pays. They represent the cost, in U.S. dollars, for one unit of work. They do not include any markups for profit or labor burden.

10. Bare Total

The **Total column** represents the total bare cost for the installing contractor in U.S. dollars. In this case, the sum of $84 for material + $85.50 for labor + $.91 for equipment is $170.41.

11. Total Incl O&P

The **Total Incl O&P column** is the total cost, including overhead and profit, that the installing contractor will charge the customer. This represents the cost of materials plus 10% profit, the cost of labor plus labor burden and 10% profit, and the cost of equipment plus 10% profit. It does not include the general contractor's overhead and profit. Note: See the inside back cover of the printed product or the Reference Section of the electronic product for details on how the labor burden is calculated.

National Average

The RSMeans data in our print publications represent a "national average" cost. This data should be modified to the project location using the City Cost Indexes or Location Factors tables found in the Reference Section. Use the Location Factors to adjust estimate totals if the project covers multiple trades. Use the City Cost Indexes (CCI) for single trade

projects or projects where a more detailed analysis is required. All figures in the two tables are derived from the same research. The last row of data in the CCI—the weighted average—is the same as the numbers reported for each location in the location factor table.

Project Name: Pre-Engineered Steel Building | **Architect: As Shown**

Location:	Anywhere, USA						01/01/21	STD
Line Number	Description	Qty	Unit	Material	Labor	Equipment	SubContract	Estimate Total
03 30 53.40 3940	Strip footing, 12" x 24", reinforced	15	C.Y.	$2,625.00	$1,830.00	$8.40	$0.00	
03 30 53.40 3950	Strip footing, 12" x 36", reinforced	34	C.Y.	$5,644.00	$3,315.00	$15.30	$0.00	
03 11 13.65 3000	Concrete slab edge forms	500	L.F.	$225.00	$1,390.00	$0.00	$0.00	
03 22 11.10 0200	Welded wire fabric reinforcing	150	C.S.F.	$3,150.00	$4,575.00	$0.00	$0.00	
03 31 13.35 0300	Ready mix concrete, 4000 psi for slab on grade	278	C.Y.	$35,862.00	$0.00	$0.00	$0.00	
03 31 13.70 4300	Place, strike off & consolidate concrete slab	278	C.Y.	$0.00	$5,560.00	$136.22	$0.00	
03 35 13.30 0250	Machine float & trowel concrete slab	15,000	S.F.	$0.00	$10,350.00	$750.00	$0.00	
03 15 16.20 0140	Cut control joints in concrete slab	950	L.F.	$38.00	$437.00	$57.00	$0.00	
03 39 23.13 0300	Sprayed concrete curing membrane	150	C.S.F.	$1,830.00	$1,125.00	$0.00	$0.00	
Division 03	**Subtotal**			**$49,374.00**	**$28,582.00**	**$966.92**	**$0.00**	**$78,922.92**
08 36 13.10 2650	Manual 10' x 10' steel sectional overhead door	8	Ea.	$9,800.00	$3,880.00	$0.00	$0.00	
08 36 13.10 2860	Insulation and steel back panel for OH door	800	S.F.	$4,760.00	$0.00	$0.00	$0.00	
Division 08	**Subtotal**			**$14,560.00**	**$3,880.00**	**$0.00**	**$0.00**	**$18,440.00**
13 34 19.50 1100	Pre-Engineered Steel Building, 100' x 150' x 24'	15,000	SF Flr.	$0.00	$0.00	$0.00	$367,500.00	
13 34 19.50 6050	Framing for PESB door opening, 3' x 7'	4	Opng.	$0.00	$0.00	$0.00	$2,300.00	
13 34 19.50 6100	Framing for PESB door opening, 10' x 10'	8	Opng.	$0.00	$0.00	$0.00	$9,400.00	
13 34 19.50 6200	Framing for PESB window opening, 4' x 3'	6	Opng.	$0.00	$0.00	$0.00	$3,360.00	
13 34 19.50 5750	PESB door, 3' x 7', single leaf	4	Opng.	$2,740.00	$772.00	$0.00	$0.00	
13 34 19.50 7750	PESB sliding window, 4' x 3' with screen	6	Opng.	$2,940.00	$660.00	$68.70	$0.00	
13 34 19.50 6550	PESB gutter, eave type, 26 ga., painted	300	L.F.	$2,595.00	$906.00	$0.00	$0.00	
13 34 19.50 8650	PESB roof vent, 12" wide x 10' long	15	Ea.	$705.00	$3,615.00	$0.00	$0.00	
13 34 19.50 6900	PESB insulation, vinyl faced, 4" thick	27,400	S.F.	$14,522.00	$10,412.00	$0.00	$0.00	
Division 13	**Subtotal**			**$23,502.00**	**$16,365.00**	**$68.70**	**$382,560.00**	**$422,495.70**
			Subtotal	$87,436.00	$48,827.00	$1,035.62	$382,560.00	$519,858.62
Division 01	General Requirements @ 7%			6,120.52	3,417.89	72.49	26,779.20	
			Estimate Subtotal	$93,556.52	$52,244.89	$1,108.11	$409,339.20	$519,858.62
			Sales Tax @ 5%	4,677.83		55.41	10,233.48	
			Subtotal A	98,234.35	52,244.89	1,163.52	419,572.68	
			GC O & P	9,823.43	26,331.42	116.35	41,957.27	
			Subtotal B	108,057.78	78,576.31	1,279.87	461,529.95	$649,443.91
			Contingency @ 5%					32,472.20
			Subtotal C					$681,916.11
			Bond @ $12/1000 +10% O&P					9,001.29
			Subtotal D					$690,917.40
			Location Adjustment Factor		114.20			98,110.27
			Grand Total					**$789,027.67**

This estimate is based on an interactive spreadsheet. You are free to download it and adjust it to your methodology.
A copy of this spreadsheet is available at **RSMeans.com/2021books.**

Sample Estimate

This sample demonstrates the elements of an estimate, including a tally of the RSMeans data lines and a summary of the markups on a contractor's work to arrive at a total cost to the owner. The Location Factor with RSMeans data is added at the bottom of the estimate to adjust the cost of the work to a specific location.

---

❶ Work Performed

The body of the estimate shows the RSMeans data selected, including the line number, a brief description of each item, its take-off unit and quantity, and the bare costs of materials, labor, and equipment. This estimate also includes a column titled "SubContract." This data is taken from the column "Total Incl O&P" and represents the total that a subcontractor would charge a general contractor for the work, including the sub's markup for overhead and profit.

❷ Division 1, General Requirements

This is the first division numerically but the last division estimated. Division 1 includes project-wide needs provided by the general contractor. These requirements vary by project but may include temporary facilities and utilities, security, testing, project cleanup, etc. For small projects a percentage can be used—typically between 5% and 15% of project cost. For large projects the costs may be itemized and priced individually.

❸ Sales Tax

If the work is subject to state or local sales taxes, the amount must be added to the estimate. Sales tax may be added to material costs, equipment costs, and subcontracted work. In this case, sales tax was added in all three categories. It was assumed that approximately half the subcontracted work would be material cost, so the tax was applied to 50% of the subcontract total.

❹ GC O&P

This entry represents the general contractor's markup on material, labor, equipment, and subcontractor costs. Our standard markup on materials, equipment, and subcontracted work is 10%. In this estimate, the markup on the labor performed by the GC's workers uses "Skilled Workers Average" shown in Column F on the table "Installing Contractor's Overhead & Profit," which can be found on the inside back cover of the printed product or in the Reference Section of the electronic product.

❺ Contingency

A factor for contingency may be added to any estimate to represent the cost of unknowns that may occur between the time that the estimate is performed and the time the project is constructed. The amount of the allowance will depend on the stage of design at which the estimate is done and the contractor's assessment of the risk involved. Refer to section 01 21 16.50 for contingency allowances.

❻ Bonds

Bond costs should be added to the estimate. The figures here represent a typical performance bond, ensuring the owner that if the general contractor does not complete the obligations in the construction contract the bonding company will pay the cost for completion of the work.

❼ Location Adjustment

Published prices are based on national average costs. If necessary, adjust the total cost of the project using a location factor from the "Location Factor" table or the "City Cost Index" table. Use location factors if the work is general, covering multiple trades. If the work is by a single trade (e.g., masonry) use the more specific data found in the "City Cost Indexes."

Estimating Tips
01 20 00 Price and Payment Procedures

- Allowances that should be added to estimates to cover contingencies and job conditions that are not included in the national average material and labor costs are shown in Section 01 21.
- When estimating historic preservation projects (depending on the condition of the existing structure and the owner's requirements), a 15–20% contingency or allowance is recommended, regardless of the stage of the drawings.

01 30 00 Administrative Requirements

- Before determining a final cost estimate, it is good practice to review all the items listed in Subdivisions 01 31 and 01 32 to make final adjustments for items that may need customizing to specific job conditions.
- Requirements for initial and periodic submittals can represent a significant cost to the General Requirements of a job. Thoroughly check the submittal specifications when estimating a project to determine any costs that should be included.

01 40 00 Quality Requirements

- All projects will require some degree of quality control. This cost is not included in the unit cost of construction listed in each division. Depending upon the terms of the contract, the various costs of inspection and testing can be the responsibility of either the owner or the contractor. Be sure to include the required costs in your estimate.

01 50 00 Temporary Facilities and Controls

- Barricades, access roads, safety nets, scaffolding, security, and many more requirements for the execution of a safe project are elements of direct cost. These costs can easily be overlooked when preparing an estimate. When looking through the major classifications of this subdivision, determine which items apply to each division in your estimate.
- Construction equipment rental costs can be found in the Reference Section in Section 01 54 33. Operators' wages are not included in equipment rental costs.
- Equipment mobilization and demobilization costs are not included in equipment rental costs and must be considered separately.

- The cost of small tools provided by the installing contractor for his workers is covered in the "Overhead" column on the "Installing Contractor's Overhead and Profit" table that lists labor trades, base rates, and markups. Therefore, it is included in the "Total Incl. O&P" cost of any unit price line item.

01 70 00 Execution and Closeout Requirements

- When preparing an estimate, thoroughly read the specifications to determine the requirements for Contract Closeout. Final cleaning, record documentation, operation and maintenance data, warranties and bonds, and spare parts and maintenance materials can all be elements of cost for the completion of a contract. Do not overlook these in your estimate.

Reference Numbers

Reference numbers are shown at the beginning of some major classifications. These numbers refer to related items in the Reference Section. The reference information may be an estimating procedure, an alternate pricing method, or technical information.

Note: Not all subdivisions listed here necessarily appear. ■

Same Data. Simplified.

Enjoy the convenience and efficiency of accessing your costs anywhere:

- **Skip the multiplier** by setting your location
- **Quickly search,** edit, favorite and share costs
- **Stay on top of price changes** with automatic updates

Discover more at rsmeans.com/online

01 11 Summary of Work

01 11 31 – Professional Consultants

01 11 31.10 Architectural Fees

		Crew	Daily Output	Labor-Hours	Unit	Material	2021 Bare Costs Labor	Equipment	Total	Total Incl O&P
0010	**ARCHITECTURAL FEES**									
0020	For new construction									
0060	Minimum				Project				4.90%	4.90%
0090	Maximum								16%	16%
0100	For alteration work, to $500,000, minimum								7.35%	7.35%
0125	Maximum								24%	24%
0150	For alteration work, over $500,000, minumum								6.13%	6.13%
0175	Maximum								20%	20%

01 11 31.20 Construction Management Fees

		Crew	Daily Output	Labor-Hours	Unit	Material	2021 Bare Costs Labor	Equipment	Total	Total Incl O&P
0010	**CONSTRUCTION MANAGEMENT FEES**									
0020	$1,000,000 job, minimum				Project				4.50%	4.50%
0050	Maximum								7.50%	7.50%
0300	$50,000,000 job, minimum								2.50%	2.50%
0350	Maximum								4%	4%

01 11 31.30 Engineering Fees

		Crew	Daily Output	Labor-Hours	Unit	Material	2021 Bare Costs Labor	Equipment	Total	Total Incl O&P
0010	**ENGINEERING FEES** R011110-30									
0020	Educational planning consultant, minimum				Project				.50%	.50%
0100	Maximum				"				2.50%	2.50%
0200	Electrical, minimum				Contrct				4.10%	4.10%
0300	Maximum								10.10%	10.10%
0400	Elevator & conveying systems, minimum								2.50%	2.50%
0500	Maximum								5%	5%
0600	Food service & kitchen equipment, minimum								8%	8%
0700	Maximum								12%	12%
0800	Landscaping & site development, minimum								2.50%	2.50%
0900	Maximum								6%	6%
1000	Mechanical (plumbing & HVAC), minimum								4.10%	4.10%
1100	Maximum								10.10%	10.10%
1200	Structural, minimum				Project				1%	1%
1300	Maximum				"				2.50%	2.50%

01 21 Allowances

01 21 16 – Contingency Allowances

01 21 16.50 Contingencies

		Crew	Daily Output	Labor-Hours	Unit	Material	2021 Bare Costs Labor	Equipment	Total	Total Incl O&P
0010	**CONTINGENCIES**, Add to estimate									
0020	Conceptual stage				Project				20%	20%
0050	Schematic stage								15%	15%
0100	Preliminary working drawing stage (Design Dev.)								10%	10%
0150	Final working drawing stage								3%	3%

01 21 53 – Factors Allowance

01 21 53.50 Factors

		Crew	Daily Output	Labor-Hours	Unit	Material	2021 Bare Costs Labor	Equipment	Total	Total Incl O&P
0010	**FACTORS** Cost adjustments R012153-10									
0100	Add to construction costs for particular job requirements									
0500	Cut & patch to match existing construction, add, minimum				Costs	2%	3%			
0550	Maximum					5%	9%			
0800	Dust protection, add, minimum					1%	2%			
0850	Maximum					4%	11%			
1100	Equipment usage curtailment, add, minimum					1%	1%			
1150	Maximum					3%	10%			
1400	Material handling & storage limitation, add, minimum					1%	1%			

For customer support on your Plumbing Costs with RSMeans Data, call 800.448.8182.

01 21 Allowances

01 21 53 – Factors Allowance

01 21 53.50 Factors	Crew	Daily Output	Labor-Hours	Unit	Material	2021 Bare Costs Labor	Equipment	Total	Total Incl O&P
1450	Maximum				Costs	6%	7%			
1700	Protection of existing work, add, minimum					2%	2%			
1750	Maximum					5%	7%			
2000	Shift work requirements, add, minimum						5%			
2050	Maximum						30%			
2300	Temporary shoring and bracing, add, minimum				↓	2%	5%			
2350	Maximum					5%	12%			

01 21 53.60 Security Factors

		Crew	Daily Output	Labor-Hours	Unit	Material	2021 Bare Costs Labor	Equipment	Total	Total Incl O&P
0010	**SECURITY FACTORS** R012153-60									
0100	Additional costs due to security requirements									
0110	Daily search of personnel, supplies, equipment and vehicles									
0120	Physical search, inventory and doc of assets, at entry				Costs		30%			
0130	At entry and exit						50%			
0140	Physical search, at entry						6.25%			
0150	At entry and exit						12.50%			
0160	Electronic scan search, at entry						2%			
0170	At entry and exit						4%			
0180	Visual inspection only, at entry						.25%			
0190	At entry and exit						.50%			
0200	ID card or display sticker only, at entry						.12%			
0210	At entry and exit				↓		.25%			
0220	Day 1 as described below, then visual only for up to 5 day job duration									
0230	Physical search, inventory and doc of assets, at entry				Costs		5%			
0240	At entry and exit						10%			
0250	Physical search, at entry						1.25%			
0260	At entry and exit						2.50%			
0270	Electronic scan search, at entry						.42%			
0280	At entry and exit				↓		.83%			
0290	Day 1 as described below, then visual only for 6-10 day job duration									
0300	Physical search, inventory and doc of assets, at entry				Costs		2.50%			
0310	At entry and exit				"		5%			
0320	Physical search, at entry				Costs		.63%			
0330	At entry and exit						1.25%			
0340	Electronic scan search, at entry						.21%			
0350	At entry and exit				↓		.42%			
0360	Day 1 as described below, then visual only for 11-20 day job duration									
0370	Physical search, inventory and doc of assets, at entry				Costs		1.25%			
0380	At entry and exit						2.50%			
0390	Physical search, at entry						.31%			
0400	At entry and exit						.63%			
0410	Electronic scan search, at entry						.10%			
0420	At entry and exit				↓		.21%			
0430	Beyond 20 days, costs are negligible									
0440	Escort required to be with tradesperson during work effort				Costs		6.25%			

01 21 53.65 Infectious Disease Precautions

		Crew	Daily Output	Labor-Hours	Unit	Material	2021 Bare Costs Labor	Equipment	Total	Total Incl O&P
0010	**INFECTIOUS DISEASE PRECAUTIONS** R012153-65									
0100	Additional costs due to infectious disease precautions									
0120	Daily temperature checks				Costs		1%			
0130	Donning & doffing masks and gloves						1%			
0140	Washing hands						1%			
0150	Informational meetings						2%			
0160	Maintaining social distance				↓		1%			

For customer support on your Plumbing Costs with RSMeans Data, call 800.448.8182.

9

01 21 Allowances

01 21 53 – Factors Allowance

01 21 53.65 Infectious Disease Precautions	Crew	Daily Output	Labor-Hours	Unit	Material	2021 Bare Costs Labor	Equipment	Total	Total Incl O&P
0170 Disinfecting tools or equipment				Costs		3%			
0200 N95 rated masks				Ea.	1.40			1.40	1.54
0210 Surgical masks				"	.16			.16	.18
0220 Black nitrile disposable gloves				Pair	.18			.18	.20
0250 Hand washing station & 2 x weekly service				Week				158	174

01 21 55 – Job Conditions Allowance

01 21 55.50 Job Conditions	Crew	Daily Output	Labor-Hours	Unit	Material	2021 Bare Costs Labor	Equipment	Total	Total Incl O&P
0010 **JOB CONDITIONS** Modifications to applicable									
0020 cost summaries									
0100 Economic conditions, favorable, deduct				Project				2%	2%
0200 Unfavorable, add								5%	5%
0300 Hoisting conditions, favorable, deduct								2%	2%
0400 Unfavorable, add								5%	5%
0700 Labor availability, surplus, deduct								1%	1%
0800 Shortage, add								10%	10%
0900 Material storage area, available, deduct								1%	1%
1000 Not available, add								2%	2%
1100 Subcontractor availability, surplus, deduct								5%	5%
1200 Shortage, add								12%	12%
1300 Work space, available, deduct								2%	2%
1400 Not available, add				Project				5%	5%

01 21 57 – Overtime Allowance

01 21 57.50 Overtime	Crew	Daily Output	Labor-Hours	Unit	Material	2021 Bare Costs Labor	Equipment	Total	Total Incl O&P
0010 **OVERTIME** for early completion of projects or where	R012909-90								
0020 labor shortages exist, add to usual labor, up to				Costs		100%			

01 21 63 – Taxes

01 21 63.10 Taxes	Crew	Daily Output	Labor-Hours	Unit	Material	2021 Bare Costs Labor	Equipment	Total	Total Incl O&P
0010 **TAXES**	R012909-80								
0020 Sales tax, State, average				%	5.08%				
0050 Maximum	R012909-85				7.50%				
0200 Social Security, on first $118,500 of wages						7.65%			
0300 Unemployment, combined Federal and State, minimum	R012909-86					.60%			
0350 Average						9.60%			
0400 Maximum						12%			

01 31 Project Management and Coordination

01 31 13 – Project Coordination

01 31 13.20 Field Personnel	Crew	Daily Output	Labor-Hours	Unit	Material	2021 Bare Costs Labor	Equipment	Total	Total Incl O&P
0010 **FIELD PERSONNEL**									
0020 Clerk, average				Week		495		495	750
0100 Field engineer, junior engineer						1,241		1,241	1,877
0120 Engineer						1,825		1,825	2,775
0140 Senior engineer						2,400		2,400	3,625
0160 General purpose laborer, average	1 Clab	.20	40			1,775		1,775	2,650
0180 Project manager, minimum						2,175		2,175	3,300
0200 Average						2,500		2,500	3,800
0220 Maximum						2,850		2,850	4,325
0240 Superintendent, minimum						2,125		2,125	3,225
0260 Average						2,325		2,325	3,525

For customer support on your Plumbing Costs with RSMeans Data, call 800.448.8182.

01 31 Project Management and Coordination

01 31 13 – Project Coordination

01 31 13.20 Field Personnel

		Crew	Daily Output	Labor-Hours	Unit	Material	2021 Bare Costs Labor	2021 Bare Costs Equipment	Total	Total Incl O&P
0280	Maximum				Week		2,650		2,650	4,025
0290	Timekeeper, average				▼		1,350		1,350	2,050

01 31 13.30 Insurance

			Crew	Daily Output	Labor-Hours	Unit	Material	2021 Bare Costs Labor	2021 Bare Costs Equipment	Total	Total Incl O&P
0010	**INSURANCE**	R013113-40									
0020	Builders risk, standard, minimum					Job				.24%	.24%
0050	Maximum	R013113-50								.80%	.80%
0200	All-risk type, minimum									.25%	.25%
0250	Maximum	R013113-60				▼				.62%	.62%
0400	Contractor's equipment floater, minimum					Value				.50%	.50%
0450	Maximum					"				1.50%	1.50%
0600	Public liability, average					Job				2.02%	2.02%
0800	Workers' compensation & employer's liability, average										
0850	by trade, carpentry, general					Payroll		11.97%			
1000	Electrical							4.91%			
1150	Insulation					▼		10.07%			
1450	Plumbing					Payroll		5.77%			
1550	Sheet metal work (HVAC)					"		7.56%			

01 31 13.50 General Contractor's Mark-Up

			Crew	Daily Output	Labor-Hours	Unit	Material	2021 Bare Costs Labor	2021 Bare Costs Equipment	Total	Total Incl O&P
0010	**GENERAL CONTRACTOR'S MARK-UP** on Change Orders										
0200	Extra work, by subcontractors, add					%				10%	10%
0250	By General Contractor, add									15%	15%
0400	Omitted work, by subcontractors, deduct all but									5%	5%
0450	By General Contractor, deduct all but									7.50%	7.50%
0600	Overtime work, by subcontractors, add									15%	15%
0650	By General Contractor, add					▼				10%	10%

01 31 13.80 Overhead and Profit

			Crew	Daily Output	Labor-Hours	Unit	Material	2021 Bare Costs Labor	2021 Bare Costs Equipment	Total	Total Incl O&P
0010	**OVERHEAD & PROFIT** Allowance to add to items in this	R013113-50									
0020	book that do not include Subs O&P, average					%				25%	
0100	Allowance to add to items in this book that	R013113-55									
0110	do include Subs O&P, minimum					%				5%	5%
0150	Average									10%	10%
0200	Maximum									15%	15%
0300	Typical, by size of project, under $100,000									30%	
0350	$500,000 project									25%	
0400	$2,000,000 project									20%	
0450	Over $10,000,000 project					▼				15%	

01 31 13.90 Performance Bond

			Crew	Daily Output	Labor-Hours	Unit	Material	2021 Bare Costs Labor	2021 Bare Costs Equipment	Total	Total Incl O&P
0010	**PERFORMANCE BOND**	R013113-80									
0020	For buildings, minimum					Job				.60%	.60%
0100	Maximum					"				2.50%	2.50%

01 31 14 – Facilities Services Coordination

01 31 14.20 Lock Out/Tag Out

		Crew	Daily Output	Labor-Hours	Unit	Material	2021 Bare Costs Labor	2021 Bare Costs Equipment	Total	Total Incl O&P
0010	**LOCK OUT/TAG OUT**										
0020	Miniature circuit breaker lock out device	1 Elec	220	.036	Ea.	24	2.32		26.32	30
0030	Miniature pin circuit breaker lock out device		220	.036		19.90	2.32		22.22	25.50
0040	Single circuit breaker lock out device		220	.036		21	2.32		23.32	26.50
0050	Multi-pole circuit breaker lock out device (15 to 225 Amp)		210	.038		19.90	2.43		22.33	25.50
0060	Large 3 pole circuit breaker lock out device (over 225 Amp)		210	.038		21	2.43		23.43	27
0080	Square D I-Line circuit breaker lock out device		210	.038		33	2.43		35.43	39.50
0090	Lock out disconnect switch, 30 to 100 Amp		330	.024		11	1.54		12.54	14.40
0100	100 to 400 Amp		330	.024	▼	11	1.54		12.54	14.40

01 31 Project Management and Coordination

01 31 14 – Facilities Services Coordination

01 31 14.20 Lock Out/Tag Out

		Crew	Daily Output	Labor-Hours	Unit	Material	2021 Bare Costs Labor	Equipment	Total	Total Incl O&P
0110	Over 400 Amp	1 Elec	330	.024	Ea.	11	1.54		12.54	14.40
0120	Lock out hasp for multiple lockout tags		200	.040		5.60	2.55		8.15	9.95
0130	Electrical cord plug lock out device		220	.036		10.50	2.32		12.82	15
0140	Electrical plug prong lock out device (3-wire grounding plug)		220	.036		6.95	2.32		9.27	11.10
0150	Wall switch lock out		200	.040		25	2.55		27.55	32
0160	Fire alarm pull station lock out	1 Stpi	200	.040		18.75	2.73		21.48	24.50
0170	Sprinkler valve tamper and flow switch lock out device	1 Skwk	220	.036	Ea.	16.60	2.08		18.68	21.50
0180	Lock out sign		330	.024		18.90	1.38		20.28	23
0190	Lock out tag		440	.018		5	1.04		6.04	7.05

01 32 Construction Progress Documentation

01 32 33 – Photographic Documentation

01 32 33.50 Photographs

					Unit	Material			Total	Total Incl O&P
0010	**PHOTOGRAPHS**									
0020	8" x 10", 4 shots, 2 prints ea., std. mounting				Set	545			545	600
0100	Hinged linen mounts					550			550	605
0200	8" x 10", 4 shots, 2 prints each, in color					575			575	630
0300	For I.D. slugs, add to all above					5.10			5.10	5.60
1500	Time lapse equipment, camera and projector, buy				Ea.	2,600			2,600	2,850
1550	Rent per month				"	1,325			1,325	1,450
1700	Cameraman and processing, black & white				Day	1,300			1,300	1,425
1720	Color				"	1,525			1,525	1,675

01 41 Regulatory Requirements

01 41 26 – Permit Requirements

01 41 26.50 Permits

					Unit				Total	Total Incl O&P
0010	**PERMITS**									
0020	Rule of thumb, most cities, minimum				Job				.50%	.50%
0100	Maximum				"				2%	2%

01 51 Temporary Utilities

01 51 13 – Temporary Electricity

01 51 13.80 Temporary Utilities

		Crew	Daily Output	Labor-Hours	Unit	Material	Labor	Equipment	Total	Total Incl O&P
0010	**TEMPORARY UTILITIES**									
0350	Lighting, lamps, wiring, outlets, 40,000 S.F. building, 8 strings	1 Elec	34	.235	CSF Flr	3.55	15		18.55	26.50
0360	16 strings	"	17	.471		7.10	30		37.10	52.50
0400	Power for temp lighting only, 6.6 KWH, per month								.92	1.01
0430	11.8 KWH, per month								1.65	1.82
0450	23.6 KWH, per month								3.30	3.63
0600	Power for job duration incl. elevator, etc., minimum								53	58
0650	Maximum								110	121
0675	Temporary cooling				Ea.	1,050			1,050	1,150

12

01 52 Construction Facilities

01 52 13 – Field Offices and Sheds

01 52 13.20 Office and Storage Space

		Crew	Daily Output	Labor-Hours	Unit	Material	2021 Bare Costs Labor	Equipment	Total	Total Incl O&P
0010	**OFFICE AND STORAGE SPACE**									
0020	Office trailer, furnished, no hookups, 20' x 8', buy	2 Skwk	1	16	Ea.	10,000	915		10,915	12,400
0250	Rent per month					192			192	211
0300	32' x 8', buy	2 Skwk	.70	22.857		15,000	1,300		16,300	18,500
0350	Rent per month					243			243	267
0400	50' x 10', buy	2 Skwk	.60	26.667		30,500	1,525		32,025	35,800
0450	Rent per month				Ea.	355			355	390
0500	50' x 12', buy	2 Skwk	.50	32		25,800	1,825		27,625	31,200
0550	Rent per month					450			450	500
0700	For air conditioning, rent per month, add					53			53	58.50
0800	For delivery, add per mile				Mile	11.95			11.95	13.15
0900	Bunk house trailer, 8' x 40' duplex dorm with kitchen, no hookups, buy	2 Carp	1	16	Ea.	89,000	875		89,875	99,500
0910	9 man with kitchen and bath, no hookups, buy		1	16		91,500	875		92,375	102,000
0920	18 man sleeper with bath, no hookups, buy		1	16		98,500	875		99,375	110,000
1000	Portable buildings, prefab, on skids, economy, 8' x 8'		265	.060	S.F.	29.50	3.30		32.80	37.50
1100	Deluxe, 8' x 12'		150	.107	"	33.50	5.85		39.35	45.50
1200	Storage boxes, 20' x 8', buy	2 Skwk	1.80	8.889	Ea.	3,100	510		3,610	4,200
1250	Rent per month					88			88	96.50
1300	40' x 8', buy	2 Skwk	1.40	11.429		4,600	655		5,255	6,050
1350	Rent per month					135			135	148

01 54 Construction Aids

01 54 16 – Temporary Hoists

01 54 16.50 Weekly Forklift Crew

		Crew	Daily Output	Labor-Hours	Unit	Material	2021 Bare Costs Labor	Equipment	Total	Total Incl O&P
0010	**WEEKLY FORKLIFT CREW**									
0100	All-terrain forklift, 45' lift, 35' reach, 9000 lb. capacity	A-3P	.20	40	Week		2,225	1,725	3,950	5,200

01 54 19 – Temporary Cranes

01 54 19.50 Daily Crane Crews

		Crew	Daily Output	Labor-Hours	Unit	Material	2021 Bare Costs Labor	Equipment	Total	Total Incl O&P
0010	**DAILY CRANE CREWS** for small jobs, portal to portal R015433-15									
0100	12-ton truck-mounted hydraulic crane	A-3H	1	8	Day		490	735	1,225	1,525
0200	25-ton	A-3I	1	8			490	810	1,300	1,625
0300	40-ton	A-3J	1	8			490	1,275	1,765	2,150
0400	55-ton	A-3K	1	16			910	1,525	2,435	3,025
0500	80-ton	A-3L	1	16			910	2,200	3,110	3,775
0900	If crane is needed on a Saturday, Sunday or Holiday									
0910	At time-and-a-half, add				Day		50%			
0920	At double time, add				"		100%			

01 54 19.60 Monthly Tower Crane Crew

		Crew	Daily Output	Labor-Hours	Unit	Material	2021 Bare Costs Labor	Equipment	Total	Total Incl O&P
0010	**MONTHLY TOWER CRANE CREW**, excludes concrete footing									
0100	Static tower crane, 130' high, 106' jib, 6200 lb. capacity	A-3N	.05	176	Month	10,800	38,200		49,000	58,000

01 54 23 – Temporary Scaffolding and Platforms

01 54 23.70 Scaffolding

		Crew	Daily Output	Labor-Hours	Unit	Material	2021 Bare Costs Labor	Equipment	Total	Total Incl O&P
0010	**SCAFFOLDING** R015423-10									
0015	Steel tube, regular, no plank, labor only to erect & dismantle									
0090	Building exterior, wall face, 1 to 5 stories, 6'-4" x 5' frames	3 Carp	8	3	C.S.F.		164		164	245
0200	6 to 12 stories	4 Carp	8	4			219		219	325
0301	13 to 20 stories	5 Clab	8	5			222		222	330
0460	Building interior, wall face area, up to 16' high	3 Carp	12	2			109		109	163
0560	16' to 40' high	"	10	2.400			131		131	196
0800	Building interior floor area, up to 30' high	3 Carp	150	.160	C.C.F.		8.75		8.75	13.05

01 54 Construction Aids

01 54 23 – Temporary Scaffolding and Platforms

01 54 23.70 Scaffolding	Crew	Daily Output	Labor-Hours	Unit	Material	2021 Bare Costs Labor	Equipment	Total	Total Incl O&P
0900 Over 30' high	4 Carp	160	.200	C.C.F.		10.95		10.95	16.35
0906 Complete system for face of walls, no plank, material only rent/mo				C.S.F.	81.50			81.50	90
0908 Interior spaces, no plank, material only rent/mo				C.C.F.	5.60			5.60	6.15
0910 Steel tubular, heavy duty shoring, buy									
0920 Frames 5' high 2' wide				Ea.	92.50			92.50	102
0925 5' high 4' wide					118			118	130
0930 6' high 2' wide					120			120	132
0935 6' high 4' wide					119			119	131
0940 Accessories									
0945 Cross braces				Ea.	19.20			19.20	21
0950 U-head, 8" x 8"					23			23	25
0955 J-head, 4" x 8"					16.70			16.70	18.35
0960 Base plate, 8" x 8"					18.05			18.05	19.90
0965 Leveling jack					36			36	39.50
1000 Steel tubular, regular, buy									
1100 Frames 3' high 5' wide				Ea.	54			54	59.50
1150 5' high 5' wide					71			71	78
1200 6'-4" high 5' wide					83.50			83.50	92
1350 7'-6" high 6' wide					156			156	171
1500 Accessories, cross braces					19.20			19.20	21
1550 Guardrail post					24			24	26.50
1600 Guardrail 7' section					8.15			8.15	9
1650 Screw jacks & plates					26			26	28.50
1700 Sidearm brackets					26			26	29
1750 8" casters					37.50			37.50	41
1800 Plank 2" x 10" x 16'-0"					66.50			66.50	73.50
1900 Stairway section					298			298	330
1910 Stairway starter bar					32.50			32.50	35.50
1920 Stairway inside handrail					57.50			57.50	63.50
1930 Stairway outside handrail					91.50			91.50	101
1940 Walk-thru frame guardrail					42			42	46.50
2000 Steel tubular, regular, rent/mo.									
2100 Frames 3' high 5' wide				Ea.	9.90			9.90	10.90
2150 5' high 5' wide					9.90			9.90	10.90
2200 6'-4" high 5' wide					5.75			5.75	6.35
2250 7'-6" high 6' wide					9.90			9.90	10.90
2500 Accessories, cross braces					3.35			3.35	3.69
2550 Guardrail post					3.25			3.25	3.58
2600 Guardrail 7' section				Ea.	11.10			11.10	12.20
2650 Screw jacks & plates					4.35			4.35	4.79
2700 Sidearm brackets					5.20			5.20	5.70
2750 8" casters					7.85			7.85	8.65
2800 Outrigger for rolling tower					9.10			9.10	10
2850 Plank 2" x 10" x 16'-0"					9.90			9.90	10.90
2900 Stairway section					32			32	35.50
2940 Walk-thru frame guardrail					8.25			8.25	9.10
3000 Steel tubular, heavy duty shoring, rent/mo.									
3250 5' high 2' & 4' wide				Ea.	8.45			8.45	9.25
3300 6' high 2' & 4' wide					8.45			8.45	9.25
3500 Accessories, cross braces					.91			.91	1
3600 U-head, 8" x 8"					2.49			2.49	2.74
3650 J-head, 4" x 8"					2.49			2.49	2.74
3700 Base plate, 8" x 8"					.91			.91	1

For customer support on your Plumbing Costs with RSMeans Data, call 800.448.8182.

01 54 Construction Aids

01 54 23 – Temporary Scaffolding and Platforms

01 54 23.70 Scaffolding

		Crew	Daily Output	Labor-Hours	Unit	Material	2021 Bare Costs Labor	Equipment	Total	Total Incl O&P
3750	Leveling jack				Ea.	2.47			2.47	2.72
5700	Planks, 2" x 10" x 16'-0", labor only to erect & remove to 50' H	3 Carp	72	.333			18.25		18.25	27
5800	Over 50' high	4 Carp	80	.400	↓		22		22	32.50
6000	Heavy duty shoring for elevated slab forms to 8'-2" high, floor area									
6100	Labor only to erect & dismantle	4 Carp	16	2	C.S.F.		109		109	163
6110	Materials only, rent/mo.				"	43.50			43.50	48
6500	To 14'-8" high									
6600	Labor only to erect & dismantle	4 Carp	10	3.200	C.S.F.		175		175	261
6610	Materials only, rent/mo				"	64			64	70

01 54 23.75 Scaffolding Specialties

		Crew	Daily Output	Labor-Hours	Unit	Material	2021 Bare Costs Labor	Equipment	Total	Total Incl O&P
0010	**SCAFFOLDING SPECIALTIES**									
1200	Sidewalk bridge, heavy duty steel posts & beams, including									
1210	parapet protection & waterproofing (material cost is rent/month)									
1220	8' to 10' wide, 2 posts	3 Carp	15	1.600	L.F.	65.50	87.50		153	203
1230	3 posts	"	10	2.400	"	96.50	131		227.50	300
1500	Sidewalk bridge using tubular steel scaffold frames including									
1510	planking (material cost is rent/month)	3 Carp	45	.533	L.F.	9.80	29		38.80	54.50
1600	For 2 uses per month, deduct from all above					50%				
1700	For 1 use every 2 months, add to all above					100%				
1900	Catwalks, 20" wide, no guardrails, 7' span, buy				Ea.	148			148	163
2000	10' span, buy					207			207	228
3720	Putlog, standard, 8' span, with hangers, buy					78.50			78.50	86
3730	Rent per month					16.30			16.30	17.90
3750	12' span, buy				↓	100			100	110
3755	Rent per month				Ea.	20.50			20.50	22.50
3760	Trussed type, 16' span, buy					254			254	280
3770	Rent per month					24.50			24.50	27
3790	22' span, buy					287			287	315
3795	Rent per month					32.50			32.50	35.50
3800	Rolling ladders with handrails, 30" wide, buy, 2 step					289			289	320
4000	7 step					1,125			1,125	1,225
4050	10 step					1,375			1,375	1,500
4100	Rolling towers, buy, 5' wide, 7' long, 10' high					1,250			1,250	1,375
4200	For additional 5' high sections, to buy				↓	180			180	198
4300	Complete incl. wheels, railings, outriggers,									
4350	21' high, to buy				Ea.	2,050			2,050	2,250
4400	Rent/month = 5% of purchase cost				"	103			103	113

01 54 36 – Equipment Mobilization

01 54 36.50 Mobilization

		Crew	Daily Output	Labor-Hours	Unit	Material	2021 Bare Costs Labor	Equipment	Total	Total Incl O&P
0010	**MOBILIZATION** (Use line item again for demobilization) R015436-50									
0015	Up to 25 mi. haul dist. (50 mi. RT for mob/demob crew)									
1200	Small equipment, placed in rear of, or towed by pickup truck	A-3A	4	2	Ea.		111	44	155	214
1300	Equipment hauled on 3-ton capacity towed trailer	A-3Q	2.67	3			167	93	260	350
1400	20-ton capacity	B-34U	2	8			425	224	649	885
1500	40-ton capacity	B-34N	2	8			440	345	785	1,050
1600	50-ton capacity	B-34V	1	24			1,350	995	2,345	3,100
1700	Crane, truck-mounted, up to 75 ton (driver only)	1 Eqhv	4	2			123		123	183
1800	Over 75 ton (with chase vehicle)	A-3E	2.50	6.400			360	70.50	430.50	620
2400	Crane, large lattice boom, requiring assembly	B-34W	.50	144	↓		7,775	7,075	14,850	19,400
2500	For each additional 5 miles haul distance, add						10%	10%		
3000	For large pieces of equipment, allow for assembly/knockdown									
3100	For mob/demob of micro-tunneling equip, see Section 33 05 07.36									

For customer support on your Plumbing Costs with RSMeans Data, call 800.448.8182.

01 55 Vehicular Access and Parking

01 55 23 – Temporary Roads

01 55 23.50 Roads and Sidewalks		Crew	Daily Output	Labor-Hours	Unit	Material	2021 Bare Costs Labor	Equipment	Total	Total Incl O&P
0010	ROADS AND SIDEWALKS Temporary									
0050	Roads, gravel fill, no surfacing, 4" gravel depth	B-14	715	.067	S.Y.	5.35	3.13	.30	8.78	10.85
0100	8" gravel depth	"	615	.078	"	10.65	3.64	.35	14.64	17.55
1000	Ramp, 3/4" plywood on 2" x 6" joists, 16" OC	2 Carp	300	.053	S.F.	2.44	2.92		5.36	7.05
1100	On 2" x 10" joists, 16" OC	"	275	.058	"	3.39	3.18		6.57	8.50

01 56 Temporary Barriers and Enclosures

01 56 13 – Temporary Air Barriers

01 56 13.60 Tarpaulins

	01 56 13.60 Tarpaulins	Crew	Daily Output	Labor-Hours	Unit	Material	2021 Bare Costs Labor	Equipment	Total	Total Incl O&P
0010	TARPAULINS									
0020	Cotton duck, 10-13.13 oz./S.Y., 6' x 8'				S.F.	.74			.74	.81
0050	30' x 30'					.62			.62	.68
0100	Polyvinyl coated nylon, 14-18 oz., minimum					1.46			1.46	1.61
0150	Maximum					1.46			1.46	1.61
0200	Reinforced polyethylene 3 mils thick, white					.06			.06	.07
0300	4 mils thick, white, clear or black					.17			.17	.19
0400	5.5 mils thick, clear					.16			.16	.18
0500	White, fire retardant					.56			.56	.62
0600	12 mils, oil resistant, fire retardant					.51			.51	.56
0700	8.5 mils, black					.31			.31	.34
0710	Woven polyethylene, 6 mils thick					.06			.06	.07
0730	Polyester reinforced w/integral fastening system, 11 mils thick					.23			.23	.25
0740	Polyethylene, reflective, 23 mils thick					1.34			1.34	1.47

01 56 13.90 Winter Protection

	01 56 13.90 Winter Protection	Crew	Daily Output	Labor-Hours	Unit	Material	2021 Bare Costs Labor	Equipment	Total	Total Incl O&P
0010	WINTER PROTECTION									
0100	Framing to close openings	2 Clab	500	.032	S.F.	.77	1.42		2.19	2.97
0200	Tarpaulins hung over scaffolding, 8 uses, not incl. scaffolding		1500	.011		.25	.47		.72	.99
0250	Tarpaulin polyester reinf. w/integral fastening system, 11 mils thick		1600	.010		.25	.44		.69	.94
0300	Prefab fiberglass panels, steel frame, 8 uses		1200	.013		3.06	.59		3.65	4.25

01 56 16 – Temporary Dust Barriers

01 56 16.10 Dust Barriers, Temporary

	01 56 16.10 Dust Barriers, Temporary	Crew	Daily Output	Labor-Hours	Unit	Material	2021 Bare Costs Labor	Equipment	Total	Total Incl O&P
0010	DUST BARRIERS, TEMPORARY, erect and dismantle									
0020	Spring loaded telescoping pole & head, to 12', erect and dismantle	1 Clab	240	.033	Ea.		1.48		1.48	2.21
0025	Cost per day (based upon 250 days)				Day	.22			.22	.25
0030	To 21', erect and dismantle	1 Clab	240	.033	Ea.		1.48		1.48	2.21
0035	Cost per day (based upon 250 days)				Day	.73			.73	.81
0040	Accessories, caution tape reel, erect and dismantle	1 Clab	480	.017	Ea.		.74		.74	1.10
0045	Cost per day (based upon 250 days)				Day	.34			.34	.38
0060	Foam rail and connector, erect and dismantle	1 Clab	240	.033	Ea.		1.48		1.48	2.21
0065	Cost per day (based upon 250 days)				Day	.10			.10	.11
0070	Caution tape	1 Clab	384	.021	C.L.F.	2.55	.93		3.48	4.18
0080	Zipper, standard duty		60	.133	Ea.	10.75	5.90		16.65	20.50
0090	Heavy duty		48	.167	"	10.85	7.40		18.25	23
0100	Polyethylene sheet, 4 mil		37	.216	Sq.	2.70	9.60		12.30	17.25
0110	6 mil		37	.216	"	7.65	9.60		17.25	23
1000	Dust partition, 6 mil polyethylene, 1" x 3" frame	2 Carp	2000	.008	S.F.	.49	.44		.93	1.19
1080	2" x 4" frame	"	2000	.008	"	.60	.44		1.04	1.31
1085	Negative air machine, 1800 CFM				Ea.	900			900	990
1090	Adhesive strip application, 2" width	1 Clab	192	.042	C.L.F.	10.65	1.85		12.50	14.50
4000	Dust & infectious control partition, adj. to 10' high, obscured, 4' panel	2 Carp	90	.178	Ea.	580	9.70		589.70	655

For customer support on your Plumbing Costs with RSMeans Data, call 800.448.8182.

01 56 Temporary Barriers and Enclosures

01 56 16 – Temporary Dust Barriers

01 56 16.10 Dust Barriers, Temporary

		Crew	Daily Output	Labor-Hours	Unit	Material	2021 Bare Costs Labor	2021 Bare Costs Equipment	Total	Total Incl O&P
4010	3' panel	2 Carp	90	.178	Ea.	550	9.70		559.70	620
4020	2' panel	↓	90	.178		430	9.70		439.70	490
4030	1' panel	1 Carp	90	.089		285	4.86		289.86	320
4040	6" panel	"	90	.089		265	4.86		269.86	299
4050	2' panel with HEPA filtered discharge port	2 Carp	90	.178		500	9.70		509.70	565
4060	3' panel with 32" door		90	.178		895	9.70		904.70	1,000
4070	4' panel with 36" door		90	.178		995	9.70		1,004.70	1,125
4080	4'-6" panel with 44" door		90	.178		1,200	9.70		1,209.70	1,350
4090	Hinged corner		80	.200		185	10.95		195.95	220
4100	Outside corner		80	.200		150	10.95		160.95	181
4110	T post		80	.200		150	10.95		160.95	181
4120	Accessories, ceiling grid clip		360	.044		7.45	2.43		9.88	11.85
4130	Panel locking clip		360	.044		5.15	2.43		7.58	9.30
4140	Panel joint closure strip		360	.044		8	2.43		10.43	12.45
4150	Screw jack	↓	360	.044		6.65	2.43		9.08	10.95
4160	Digital pressure difference gauge					275			275	305
4180	Combination lockset	1 Carp	13	.615	↓	200	33.50		233.50	271
4185	Sealant tape, 2" wide	1 Clab	192	.042	C.L.F.	10.65	1.85		12.50	14.50
4190	System in place, including door and accessories									
4200	Based upon 25 uses	2 Carp	51	.314	L.F.	9.85	17.15		27	36.50
4210	Based upon 50 uses	↓	51	.314		4.94	17.15		22.09	31
4230	Based upon 100 uses		51	.314		2.47	17.15		19.62	28

01 56 23 – Temporary Barricades

01 56 23.10 Barricades

		Crew	Daily Output	Labor-Hours	Unit	Material	2021 Bare Costs Labor	2021 Bare Costs Equipment	Total	Total Incl O&P
0010	**BARRICADES**									
0020	5' high, 3 rail @ 2" x 8", fixed	2 Carp	20	.800	L.F.	9.45	44		53.45	76
0150	Movable		30	.533		8.15	29		37.15	52.50
1000	Guardrail, wooden, 3' high, 1" x 6" on 2" x 4" posts		200	.080		2.05	4.38		6.43	8.80
1100	2" x 6" on 4" x 4" posts	↓	165	.097		3.46	5.30		8.76	11.70
1200	Portable metal with base pads, buy					14.35			14.35	15.80
1250	Typical installation, assume 10 reuses	2 Carp	600	.027	↓	2.19	1.46		3.65	4.59
1300	Barricade tape, polyethylene, 7 mil, 3" wide x 300' long roll	1 Clab	128	.063	Ea.	7.65	2.78		10.43	12.55
3000	Detour signs, set up and remove									
3010	Reflective aluminum, MUTCD, 24" x 24", post mounted	1 Clab	20	.400	Ea.	3.46	17.75		21.21	30.50
4000	Roof edge portable barrier stands and warning flags, 50 uses	1 Rohe	9100	.001	L.F.	.08	.03		.11	.14
4010	100 uses	"	9100	.001	"	.04	.03		.07	.09

01 56 26 – Temporary Fencing

01 56 26.50 Temporary Fencing

		Crew	Daily Output	Labor-Hours	Unit	Material	2021 Bare Costs Labor	2021 Bare Costs Equipment	Total	Total Incl O&P
0010	**TEMPORARY FENCING**									
0020	Chain link, 11 ga., 4' high	2 Clab	400	.040	L.F.	3.05	1.78		4.83	6
0100	6' high	2 Clab	300	.053	L.F.	5.80	2.37		8.17	9.90
0200	Rented chain link, 6' high, to 1000' (up to 12 mo.)		400	.040		3.44	1.78		5.22	6.45
0250	Over 1000' (up to 12 mo.)	↓	300	.053		4.07	2.37		6.44	8
0350	Plywood, painted, 2" x 4" frame, 4' high	A-4	135	.178		9.45	9.25		18.70	24
0400	4" x 4" frame, 8' high	"	110	.218		17.05	11.35		28.40	35.50
0500	Wire mesh on 4" x 4" posts, 4' high	2 Carp	100	.160		12.25	8.75		21	26.50
0550	8' high	"	80	.200	↓	18	10.95		28.95	36

01 56 Temporary Barriers and Enclosures

01 56 29 – Temporary Protective Walkways

01 56 29.50 Protection	Crew	Daily Output	Labor-Hours	Unit	Material	2021 Bare Costs Labor	2021 Bare Costs Equipment	Total	Total Incl O&P
0010 **PROTECTION**									
0020 Stair tread, 2" x 12" planks, 1 use	1 Carp	75	.107	Tread	7.50	5.85		13.35	16.95
0100 Exterior plywood, 1/2" thick, 1 use		65	.123		3.01	6.75		9.76	13.35
0200 3/4" thick, 1 use		60	.133	↓	4.15	7.30		11.45	15.45
2200 Sidewalks, 2" x 12" planks, 2 uses		350	.023	S.F.	1.25	1.25		2.50	3.24
2300 Exterior plywood, 2 uses, 1/2" thick		750	.011		.50	.58		1.08	1.42
2400 5/8" thick		650	.012		.58	.67		1.25	1.64
2500 3/4" thick	↓	600	.013	↓	.69	.73		1.42	1.85

01 57 Temporary Controls

01 57 33 – Temporary Security

01 57 33.50 Watchman

01 57 33.50 Watchman	Crew	Daily Output	Labor-Hours	Unit	Material	2021 Bare Costs Labor	2021 Bare Costs Equipment	Total	Total Incl O&P
0010 **WATCHMAN**									
0020 Service, monthly basis, uniformed person, minimum				Hr.				27.65	30.40
0100 Maximum								56	61.62
0200 Person and command dog, minimum								31	34
0300 Maximum				↓				60	65

01 58 Project Identification

01 58 13 – Temporary Project Signage

01 58 13.50 Signs

01 58 13.50 Signs	Crew	Daily Output	Labor-Hours	Unit	Material	2021 Bare Costs Labor	2021 Bare Costs Equipment	Total	Total Incl O&P
0010 **SIGNS**									
0020 High intensity reflectorized, no posts, buy				Ea.	22			22	24

01 66 Product Storage and Handling Requirements

01 66 19 – Material Handling

01 66 19.10 Material Handling

01 66 19.10 Material Handling	Crew	Daily Output	Labor-Hours	Unit	Material	2021 Bare Costs Labor	2021 Bare Costs Equipment	Total	Total Incl O&P
0010 **MATERIAL HANDLING**									
0020 Above 2nd story, via stairs, per C.Y. of material per floor	2 Clab	145	.110	C.Y.		4.90		4.90	7.30
0030 Via elevator, per C.Y. of material		240	.067			2.96		2.96	4.42
0050 Distances greater than 200', per C.Y. of material per each addl 200'	↓	300	.053	↓		2.37		2.37	3.53

01 74 Cleaning and Waste Management

01 74 13 – Progress Cleaning

01 74 13.20 Cleaning Up

01 74 13.20 Cleaning Up	Crew	Daily Output	Labor-Hours	Unit	Material	2021 Bare Costs Labor	2021 Bare Costs Equipment	Total	Total Incl O&P
0010 **CLEANING UP**									
0020 After job completion, allow, minimum				Job				.30%	.30%
0040 Maximum				"				1%	1%
0050 Cleanup of floor area, continuous, per day, during const.	A-5	24	.750	M.S.F.	2.77	33.50	2.07	38.34	56
0100 Final by GC at end of job	"	11.50	1.565	"	2.93	70.50	4.32	77.75	113

01 91 Commissioning

01 91 13 – General Commissioning Requirements

01 91 13.50 Building Commissioning	Crew	Daily Output	Labor-Hours	Unit	Material	2021 Bare Costs Labor	Equipment	Total	Total Incl O&P
0010 **BUILDING COMMISSIONING**									
0100 Systems operation and verification during turnover				%				.25%	.25%
0150 Including all systems subcontractors								.50%	.50%
0200 Systems design assistance, operation, verification and training								.50%	.50%
0250 Including all systems subcontractors				▼				1%	1%

01 93 Facility Maintenance

01 93 13 – Facility Maintenance Procedures

01 93 13.15 Mechanical Facilities Maintenance

		Crew	Daily Output	Labor-Hours	Unit	Material	2021 Bare Costs Labor	Equipment	Total	Total Incl O&P
0010	**MECHANICAL FACILITIES MAINTENANCE**									
0100	Air conditioning system maintenance									
0130	Belt, replace	1 Stpi	15	.533	Ea.		36.50		36.50	54.50
0170	Fan, clean		16	.500			34		34	51
0180	Filter, remove, clean, replace		12	.667			45.50		45.50	68
0190	Flexible coupling alignment, inspect		40	.200			13.65		13.65	20.50
0200	Gas leak, locate and repair		4	2			137		137	204
0250	Pump packing gland, remove and replace		11	.727			49.50		49.50	74
0270	Tighten		32	.250			17.10		17.10	25.50
0290	Pump, disassemble and assemble	▼	4	2			137		137	204
0300	Air pressure regulator, disassemble, clean, assemble	1 Skwk	4	2			114		114	172
0310	Repair or replace part	"	6	1.333			76		76	115
0320	Purging system	1 Stpi	16	.500			34		34	51
0400	Compressor, air, remove or install fan wheel	1 Skwk	20	.400			23		23	34.50
0410	Disassemble or assemble 2 cylinder, 2 stage		4	2			114		114	172
0420	4 cylinder, 4 stage		1	8			455		455	685
0430	Repair or replace part	▼	2	4	▼		228		228	345
0700	Demolition, for mech. demolition see Section 23 05 05.10 or 22 05 05.10									
0800	Ductwork, clean									
0810	Rectangular									
0820	6" G	1 Shee	187.50	.043	L.F.		2.79		2.79	4.21
0830	8" G		140.63	.057			3.72		3.72	5.60
0840	10" G		112.50	.071			4.65		4.65	7
0850	12" G		93.75	.085			5.60		5.60	8.40
0860	14" G		80.36	.100			6.50		6.50	9.85
0870	16" G	▼	70.31	.114	▼		7.45		7.45	11.25
0900	Round									
0910	4" G	1 Shee	358.10	.022	L.F.		1.46		1.46	2.20
0920	6" G	1 Shee	238.73	.034	L.F.		2.19		2.19	3.31
0930	8" G		179.05	.045			2.92		2.92	4.41
0940	10" G		143.24	.056			3.66		3.66	5.50
0950	12" G		119.37	.067			4.39		4.39	6.60
0960	16" G	▼	89.52	.089	▼		5.85		5.85	8.80
1000	Expansion joint, not screwed, install or remove	1 Stpi	3	2.667	Ea.		182		182	272
1010	Repack	"	6	1.333	"		91		91	136
1200	Fire protection equipment									
1220	Fire hydrant, replace	Q-1	3	5.333	Ea.		325		325	485
1230	Service, lubricate, inspect, flush, clean	1 Plum	7	1.143			77.50		77.50	115
1240	Test	"	11	.727	▼		49		49	73.50
1310	Inspect valves, pressure, nozzle	1 Spri	4	2	System		133		133	199
1800	Plumbing fixtures, for installation see Section 22 41 00									
1801	Plumbing fixtures									

01 93 13.15 Mechanical Facilities Maintenance		Crew	Daily Output	Labor-Hours	Unit	Material	2021 Bare Costs Labor	Equipment	Total	Total Incl O&P
1850	Open drain with toilet auger	1 Plum	16	.500	Ea.		34		34	50.50
1870	Plaster trap, clean	"	6	1.333			90.50		90.50	135
1881	Clean commode	1 Clab	20	.400		.14	17.75		17.89	26.50
1882	Clean commode seat		48	.167		.07	7.40		7.47	11.15
1884	Clean double sink		20	.400		.14	17.75		17.89	26.50
1886	Clean bathtub		12	.667		.27	29.50		29.77	44.50
1888	Clean fiberglass tub/shower		8	1		.34	44.50		44.84	67
1889	Clean faucet set		32	.250		.05	11.10		11.15	16.60
1891	Clean shower head		80	.100		.01	4.44		4.45	6.65
1893	Clean water heater		16	.500		.48	22		22.48	33.50
1900	Relief valve, test and adjust	1 Stpi	20	.400			27.50		27.50	41
1910	Clean pump, heater, or motor for whirlpool	1 Clab	16	.500		.07	22		22.07	33
1920	Clean thermal cover for whirlpool	"	16	.500		.68	22		22.68	34
2000	Repair or replace, steam trap	1 Stpi	8	1			68.50		68.50	102
2020	Y-type or bell strainer		6	1.333			91		91	136
2040	Water trap or vacuum breaker, screwed joints		13	.615			42		42	63
2100	Steam specialties, clean									
2120	Air separator with automatic trap, 1" fittings	1 Stpi	12	.667	Ea.		45.50		45.50	68
2130	Bucket trap, 2" pipe		7	1.143			78		78	117
2150	Drip leg, 2" fitting		45	.178			12.15		12.15	18.15
2200	Thermodynamic trap, 1" fittings		50	.160			10.95		10.95	16.30
2210	Thermostatic		65	.123			8.40		8.40	12.55
2240	Screen and seat in Y-type strainer, plug type		25	.320			22		22	32.50
2242	Screen and seat in Y-type strainer, flange type		12	.667			45.50		45.50	68
2500	Valve, replace broken handwheel		24	.333			23		23	34
3000	Valve, overhaul, regulator, relief, flushometer, mixing									
3040	Cold water, gas	1 Stpi	5	1.600	Ea.		109		109	163
3050	Hot water, steam		3	2.667			182		182	272
3080	Globe, gate, check up to 4" cold water, gas		10	.800			54.50		54.50	81.50
3090	Hot water, steam		5	1.600			109		109	163
3100	Over 4" ID hot or cold line		1.40	5.714			390		390	585
3120	Remove and replace, gate, globe or check up to 4"	Q-5	6	2.667			164		164	245
3130	Over 4"	"	2	8			490		490	735
3150	Repack up to 4"	1 Stpi	13	.615			42		42	63
3160	Over 4"	"	4	2			137		137	204

Estimating Tips

02 30 00 Subsurface Investigation

In preparing estimates on structures involving earthwork or foundations, all information concerning soil characteristics should be obtained. Look particularly for hazardous waste, evidence of prior dumping of debris, and previous stream beds.

02 40 00 Demolition and Structure Moving

The costs shown for selective demolition do not include rubbish handling or disposal. These items should be estimated separately using RSMeans data or other sources.

- Historic preservation often requires that the contractor remove materials from the existing structure, rehab them, and replace them. The estimator must be aware of any related measures and precautions that must be taken when doing selective demolition and cutting and patching. Requirements may include special handling and storage, as well as security.

- In addition to Subdivision 02 41 00, you can find selective demolition items in each division. Example: Roofing demolition is in Division 7.

- Absent of any other specific reference, an approximate demolish-in-place cost can be obtained by halving the new-install labor cost. To remove for reuse, allow the entire new-install labor figure.

02 40 00 Building Deconstruction

This section provides costs for the careful dismantling and recycling of most low-rise building materials.

02 50 00 Containment of Hazardous Waste

This section addresses on-site hazardous waste disposal costs.

02 80 00 Hazardous Material Disposal/Remediation

This subdivision includes information on hazardous waste handling, asbestos remediation, lead remediation, and mold remediation. See reference numbers RO28213-20 and RO28319-60 for further guidance in using these unit price lines.

02 90 00 Monitoring Chemical Sampling, Testing Analysis

This section provides costs for on-site sampling and testing hazardous waste.

Reference Numbers

Reference numbers are shown at the beginning of some major classifications. These numbers refer to related items in the Reference Section. The reference information may be an estimating procedure, an alternate pricing method, or technical information.

Note: Not all subdivisions listed here necessarily appear. ■

Same Data. Simplified.

Enjoy the convenience and efficiency of accessing your costs anywhere:

- **Skip the multiplier** by setting your location
- **Quickly search,** edit, favorite and share costs
- **Stay on top of price changes** with automatic updates

Discover more at rsmeans.com/online

02 21 Surveys

02 21 13 – Site Surveys

02 21 13.09 Topographical Surveys

		Crew	Daily Output	Labor-Hours	Unit	Material	2021 Bare Costs Labor	Equipment	Total	Total Incl O&P
0010	**TOPOGRAPHICAL SURVEYS**									
0020	Topographical surveying, conventional, minimum	A-7	3.30	7.273	Acre	35	430	10.40	475.40	695
0050	Average	"	1.95	12.308		52.50	730	17.60	800.10	1,175
0100	Maximum	A-8	.60	53.333	↓	70	3,075	57.50	3,202.50	4,750

02 21 13.13 Boundary and Survey Markers

		Crew	Daily Output	Labor-Hours	Unit	Material	2021 Bare Costs Labor	Equipment	Total	Total Incl O&P
0010	**BOUNDARY AND SURVEY MARKERS**									
0300	Lot location and lines, large quantities, minimum	A-7	2	12	Acre	34.50	715	17.15	766.65	1,125
0320	Average	"	1.25	19.200		61	1,150	27.50	1,238.50	1,800
0400	Small quantities, maximum	A-8	1	32	↓	77	1,850	34.50	1,961.50	2,875
0600	Monuments, 3' long	A-7	10	2.400	Ea.	37.50	143	3.43	183.93	258
0800	Property lines, perimeter, cleared land	"	1000	.024	L.F.	.08	1.43	.03	1.54	2.26
0900	Wooded land	A-8	875	.037	"	.10	2.11	.04	2.25	3.30

02 21 13.16 Aerial Surveys

		Crew	Daily Output	Labor-Hours	Unit	Material	2021 Bare Costs Labor	Equipment	Total	Total Incl O&P
0010	**AERIAL SURVEYS**									
1500	Aerial surveying, including ground control, minimum fee, 10 acres				Total				4,700	4,700
1510	100 acres								9,400	9,400
1550	From existing photography, deduct				↓				1,625	1,625
1600	2' contours, 10 acres				Acre				470	470
1850	100 acres								94	94
2000	1000 acres								90	90
2050	10,000 acres				↓				85	85

02 41 Demolition

02 41 13 – Selective Site Demolition

02 41 13.17 Demolish, Remove Pavement and Curb

		Crew	Daily Output	Labor-Hours	Unit	Material	2021 Bare Costs Labor	Equipment	Total	Total Incl O&P
0010	**DEMOLISH, REMOVE PAVEMENT AND CURB** R024119-10									
5010	Pavement removal, bituminous roads, up to 3" thick	B-38	690	.058	S.Y.		2.90	1.76	4.66	6.25
5050	4"-6" thick	"	420	.095	"		4.76	2.90	7.66	10.30

02 41 13.23 Utility Line Removal

		Crew	Daily Output	Labor-Hours	Unit	Material	2021 Bare Costs Labor	Equipment	Total	Total Incl O&P
0010	**UTILITY LINE REMOVAL**									
0015	No hauling, abandon catch basin or manhole	B-6	7	3.429	Ea.		165	31	196	280
0020	Remove existing catch basin or manhole, masonry	"	4	6	"		289	54	343	490
0030	Catch basin or manhole frames and covers, stored	B-6	13	1.846	Ea.		89	16.65	105.65	150
0040	Remove and reset	"	7	3.429			165	31	196	280
0900	Hydrants, fire, remove only	B-21A	5	8			440	86.50	526.50	750
0950	Remove and reset	"	2	20	↓		1,100	216	1,316	1,875
2900	Pipe removal, sewer/water, no excavation, 12" diameter	B-6	175	.137	L.F.		6.60	1.24	7.84	11.20
2930	15"-18" diameter	B-12Z	150	.160			8	10.45	18.45	23.50
2960	21"-24" diameter		120	.200			10	13.05	23.05	29.50
3000	27"-36" diameter	↓	90	.267			13.35	17.40	30.75	39
3200	Steel, welded connections, 4" diameter	B-6	160	.150			7.20	1.35	8.55	12.25
3300	10" diameter	"	80	.300	↓		14.45	2.70	17.15	24.50

02 41 13.30 Minor Site Demolition

		Crew	Daily Output	Labor-Hours	Unit	Material	2021 Bare Costs Labor	Equipment	Total	Total Incl O&P
0010	**MINOR SITE DEMOLITION** R024119-10									
4000	Sidewalk removal, bituminous, 2" thick	B-6	350	.069	S.Y.		3.30	.62	3.92	5.60
4010	2-1/2" thick		325	.074			3.55	.67	4.22	6.05
4100	Concrete, plain, 4"	↓	160	.150	↓		7.20	1.35	8.55	12.25

02 41 Demolition

02 41 13 – Selective Site Demolition

02 41 13.30 Minor Site Demolition	Crew	Daily Output	Labor-Hours	Unit	Material	2021 Bare Costs Labor	2021 Bare Costs Equipment	Total	Total Incl O&P
4110	Plain, 5"	B-6	140	.171	S.Y.		8.25	1.54	9.79	14
4120	Plain, 6"	↓	120	.200	↓		9.60	1.80	11.40	16.35

02 41 19 – Selective Demolition

02 41 19.19 Selective Demolition

		Crew	Daily Output	Labor-Hours	Unit	Material	2021 Bare Costs Labor	2021 Bare Costs Equipment	Total	Total Incl O&P
0010	**SELECTIVE DEMOLITION**, Rubbish Handling	R024119-10								
0020	The following are to be added to the demolition prices									
0600	Dumpster, weekly rental, 1 dump/week, 6 C.Y. capacity (2 tons)				Week	415			415	455
0700	10 C.Y. capacity (3 tons)					480			480	530
0725	20 C.Y. capacity (5 tons)	R024119-20				565			565	625
0800	30 C.Y. capacity (7 tons)					730			730	800
0840	40 C.Y. capacity (10 tons)				↓	775			775	850
2000	Load, haul, dump and return, 0'-50' haul, hand carried	2 Clab	24	.667	C.Y.		29.50		29.50	44
2005	Wheeled		37	.432			19.20		19.20	28.50
2040	0'-100' haul, hand carried		16.50	.970			43		43	64
2045	Wheeled		25	.640			28.50		28.50	42.50
2050	Forklift	A-3R	25	.320			17.75	11.35	29.10	39
2080	Haul and return, add per each extra 100' haul, hand carried	2 Clab	35.50	.451			20		20	30
2085	Wheeled		54	.296			13.15		13.15	19.65
2120	For travel in elevators, up to 10 floors, add		140	.114			5.05		5.05	7.55
2130	0'-50' haul, incl. up to 5 riser stairs, hand carried		23	.696			31		31	46
2135	Wheeled		35	.457			20.50		20.50	30.50
2140	6-10 riser stairs, hand carried		22	.727			32.50		32.50	48
2145	Wheeled		34	.471			21		21	31
2150	11-20 riser stairs, hand carried		20	.800			35.50		35.50	53
2155	Wheeled	↓	31	.516	↓		23		23	34
2160	21-40 riser stairs, hand carried	2 Clab	16	1	C.Y.		44.50		44.50	66.50
2165	Wheeled		24	.667			29.50		29.50	44
2170	0'-100' haul, incl. 5 riser stairs, hand carried		15	1.067			47.50		47.50	70.50
2175	Wheeled		23	.696			31		31	46
2180	6-10 riser stairs, hand carried		14	1.143			50.50		50.50	75.50
2185	Wheeled		21	.762			34		34	50.50
2190	11-20 riser stairs, hand carried		12	1.333			59		59	88.50
2195	Wheeled		18	.889			39.50		39.50	59
2200	21-40 riser stairs, hand carried		8	2			89		89	133
2205	Wheeled		12	1.333			59		59	88.50
2210	Haul and return, add per each extra 100' haul, hand carried		35.50	.451			20		20	30
2215	Wheeled		54	.296	↓		13.15		13.15	19.65
2220	For each additional flight of stairs, up to 5 risers, add		550	.029	Flight		1.29		1.29	1.93
2225	6-10 risers, add		275	.058			2.58		2.58	3.85
2230	11-20 risers, add		138	.116			5.15		5.15	7.70
2235	21-40 risers, add	↓	69	.232	↓		10.30		10.30	15.35
3000	Loading & trucking, including 2 mile haul, chute loaded	B-16	45	.711	C.Y.		33	12.85	45.85	63.50
3040	Hand loading truck, 50' haul	"	48	.667			31	12.05	43.05	60
3080	Machine loading truck	B-17	120	.267			13.05	5.20	18.25	25
5000	Haul, per mile, up to 8 C.Y. truck	B-34B	1165	.007			.35	.50	.85	1.08
5100	Over 8 C.Y. truck	"	1550	.005			.26	.37	.63	.81

02 41 19.20 Selective Demolition, Dump Charges

		Crew	Daily Output	Labor-Hours	Unit	Material	2021 Bare Costs Labor	2021 Bare Costs Equipment	Total	Total Incl O&P
0010	**SELECTIVE DEMOLITION, DUMP CHARGES**	R024119-10								
0100	Building construction materials				Ton	74			74	81
0300	Rubbish only					63			63	69.50
0500	Reclamation station, usual charge				↓	74			74	81

For customer support on your Plumbing Costs with RSMeans Data, call 800.448.8182.

23

02 41 Demolition

02 41 19 – Selective Demolition

02 41 19.27 Selective Demolition, Torch Cutting		Crew	Daily Output	Labor-Hours	Unit	Material	2021 Bare Costs Labor	Equipment	Total	Total Incl O&P
0010	**SELECTIVE DEMOLITION, TORCH CUTTING** R024119-10									
0020	Steel, 1" thick plate	E-25	333	.024	L.F.	2.23	1.50	.04	3.77	4.81
0040	1" diameter bar	"	600	.013	Ea.	.37	.83	.02	1.22	1.72
1000	Oxygen lance cutting, reinforced concrete walls									
1040	12"-16" thick walls	1 Clab	10	.800	L.F.		35.50		35.50	53
1080	24" thick walls	"	6	1.333	"		59		59	88.50

02 65 Underground Storage Tank Removal

02 65 10 – Underground Tank and Contaminated Soil Removal

02 65 10.30 Removal of Underground Storage Tanks

			Crew	Daily Output	Labor-Hours	Unit	Material	2021 Bare Costs Labor	Equipment	Total	Total Incl O&P
0010	**REMOVAL OF UNDERGROUND STORAGE TANKS** R026510-20										
0011	Petroleum storage tanks, non-leaking										
0100	Excavate & load onto trailer										
0110	3,000 gal. to 5,000 gal. tank	G	B-14	4	12	Ea.		560	54	614	895
0120	6,000 gal. to 8,000 gal. tank	G	B-3A	3	13.333			630	232	862	1,200
0130	9,000 gal. to 12,000 gal. tank	G	"	2	20	↓		945	350	1,295	1,775
0190	Known leaking tank, add					%				100%	100%
0200	Remove sludge, water and remaining product from bottom										
0201	of tank with vacuum truck										
0300	3,000 gal. to 5,000 gal. tank	G	A-13	5	1.600	Ea.		89	149	238	296
0310	6,000 gal. to 8,000 gal. tank	G		4	2			111	187	298	370
0320	9,000 gal. to 12,000 gal. tank	G	↓	3	2.667	↓		148	249	397	495
0390	Dispose of sludge off-site, average					Gal.				6.25	6.80
0400	Insert inert solid CO_2 "dry ice" into tank										
0401	For cleaning/transporting tanks (1.5 lb./100 gal. cap)	G	1 Clab	500	.016	Lb.	1	.71		1.71	2.16
1020	Haul tank to certified salvage dump, 100 miles round trip										
1023	3,000 gal. to 5,000 gal. tank					Ea.				760	830
1026	6,000 gal. to 8,000 gal. tank									880	960
1029	9,000 gal. to 12,000 gal. tank					↓				1,050	1,150
1100	Disposal of contaminated soil to landfill										
1110	Minimum					C.Y.				145	160
1111	Maximum					"				400	440
1120	Disposal of contaminated soil to										
1121	bituminous concrete batch plant										
1130	Minimum					C.Y.				80	88
1131	Maximum					"				115	125
2010	Decontamination of soil on site incl poly tarp on top/bottom										
2011	Soil containment berm and chemical treatment										
2020	Minimum	G	B-11C	100	.160	C.Y.	6.25	8.25	2.16	16.66	21.50
2021	Maximum	G	"	100	.160	↓	8.10	8.25	2.16	18.51	23.50
2050	Disposal of decontaminated soil, minimum									135	150
2055	Maximum					↓				400	440

02 81 Transportation and Disposal of Hazardous Materials

02 81 20 – Hazardous Waste Handling

02 81 20.10 Hazardous Waste Cleanup/Pickup/Disposal	Crew	Daily Output	Labor-Hours	Unit	Material	2021 Bare Costs Labor	Equipment	Total	Total Incl O&P
0010 **HAZARDOUS WASTE CLEANUP/PICKUP/DISPOSAL**									
0100 For contractor rental equipment, i.e., dozer,									
0110 Front end loader, dump truck, etc., see 01 54 33 Reference Section									
1000 Solid pickup									
1100 55 gal. drums				Ea.				240	265
1120 Bulk material, minimum				Ton				190	210
1130 Maximum				"				595	655
1200 Transportation to disposal site									
1220 Truckload = 80 drums or 25 C.Y. or 18 tons									
1260 Minimum				Mile				3.95	4.45
1270 Maximum				Mile				7.25	7.98
3000 Liquid pickup, vacuum truck, stainless steel tank									
3100 Minimum charge, 4 hours									
3110 1 compartment, 2200 gallon				Hr.				140	155
3120 2 compartment, 5000 gallon				"				200	225
3400 Transportation in 6900 gallon bulk truck				Mile				7.95	8.75
3410 In teflon lined truck				"				10.20	11.25
5000 Heavy sludge or dry vacuumable material				Hr.				140	155
6000 Dumpsite disposal charge, minimum				Ton				140	155
6020 Maximum				"				415	455

02 82 Asbestos Remediation

02 82 13 – Asbestos Abatement

02 82 13.39 Asbestos Remediation Plans and Methods

	Crew	Daily Output	Labor-Hours	Unit	Material	Labor	Equipment	Total	Total Incl O&P
0010 **ASBESTOS REMEDIATION PLANS AND METHODS**									
0100 Building Survey-Commercial Building				Ea.				2,200	2,400
0200 Asbestos Abatement Remediation Plan				"				1,350	1,475

02 82 13.41 Asbestos Abatement Equipment

	Crew	Daily Output	Labor-Hours	Unit	Material	Labor	Equipment	Total	Total Incl O&P
0010 **ASBESTOS ABATEMENT EQUIPMENT** R028213-20									
0011 Equipment and supplies, buy									
0200 Air filtration device, 2000 CFM				Ea.	965			965	1,050
0250 Large volume air sampling pump, minimum					350			350	385
0260 Maximum					320			320	350
0300 Airless sprayer unit, 2 gun					2,100			2,100	2,325
0350 Light stand, 500 watt					34			34	37.50
0400 Personal respirators									
0410 Negative pressure, 1/2 face, dual operation, minimum				Ea.	28.50			28.50	31.50
0420 Maximum					26.50			26.50	29
0450 P.A.P.R., full face, minimum					680			680	745
0460 Maximum					1,225			1,225	1,325
0470 Supplied air, full face, including air line, minimum					289			289	320
0480 Maximum					470			470	515
0500 Personnel sampling pump					210			210	231
1500 Power panel, 20 unit, including GFI					500			500	550
1600 Shower unit, including pump and filters					1,225			1,225	1,350
1700 Supplied air system (type C)					4,125			4,125	4,525
1750 Vacuum cleaner, HEPA, 16 gal., stainless steel, wet/dry					1,300			1,300	1,450
1760 55 gallon					1,500			1,500	1,650
1800 Vacuum loader, 9-18 ton/hr.					100,000			100,000	110,000
1900 Water atomizer unit, including 55 gal. drum					310			310	340
2000 Worker protection, whole body, foot, head cover & gloves, plastic					8.90			8.90	9.80

For customer support on your Plumbing Costs with RSMeans Data, call 800.448.8182.

25

02 82 13 – Asbestos Abatement

02 82 13.41 Asbestos Abatement Equipment

		Crew	Daily Output	Labor-Hours	Unit	Material	2021 Bare Costs Labor	Equipment	Total	Total Incl O&P
2500	Respirator, single use				Ea.	27.50			27.50	30.50
2550	Cartridge for respirator					4.20			4.20	4.62
2570	Glove bag, 7 mil, 50" x 64"					8.75			8.75	9.60
2580	10 mil, 44" x 60"					6.85			6.85	7.50
2590	6 mil, 44" x 60"					4.19			4.19	4.61
6000	Disposable polyethylene bags, 6 mil, 3 C.F.					1.23			1.23	1.35
6300	Disposable fiber drums, 3 C.F.					19.80			19.80	22
6400	Pressure sensitive caution labels, 3" x 5"					3.46			3.46	3.81
6450	11" x 17"					7.80			7.80	8.55
6500	Negative air machine, 1800 CFM					900			900	990

02 82 13.42 Preparation of Asbestos Containment Area

		Crew	Daily Output	Labor-Hours	Unit	Material	2021 Bare Costs Labor	Equipment	Total	Total Incl O&P
0010	PREPARATION OF ASBESTOS CONTAINMENT AREA R028213-20									
0100	Pre-cleaning, HEPA vacuum and wet wipe, flat surfaces	A-9	12000	.005	S.F.	.01	.33		.34	.52
0200	Protect carpeted area, 2 layers 6 mil poly on 3/4" plywood	"	1000	.064		2.16	3.90		6.06	8.35
0300	Separation barrier, 2" x 4" @ 16", 1/2" plywood ea. side, 8' high	2 Carp	400	.040		4.07	2.19		6.26	7.75
0310	12' high		320	.050		3.98	2.74		6.72	8.45
0320	16' high		200	.080		3.94	4.38		8.32	10.90
0400	Personnel decontam. chamber, 2" x 4" @ 16", 3/4" ply ea. side		280	.057		4.74	3.13		7.87	9.85
0450	Waste decontam. chamber, 2" x 4" studs @ 16", 3/4" ply ea. side		360	.044		4.74	2.43		7.17	8.85
0500	Cover surfaces with polyethylene sheeting									
0501	Including glue and tape									
0550	Floors, each layer, 6 mil	A-9	8000	.008	S.F.	.08	.49		.57	.84
0551	4 mil		9000	.007		.03	.43		.46	.69
0560	Walls, each layer, 6 mil		6000	.011		.08	.65		.73	1.09
0561	4 mil		7000	.009		.03	.56		.59	.88
0570	For heights above 14', add						20%			
0575	For heights above 20', add						30%			
0580	For fire retardant poly, add					100%				
0590	For large open areas, deduct					10%	20%			
0600	Seal floor penetrations with foam firestop to 36 sq. in.	2 Carp	200	.080	Ea.	13.45	4.38		17.83	21.50
0610	36 sq. in. to 72 sq. in.		125	.128		27	7		34	40
0615	72 sq. in. to 144 sq. in.		80	.200		54	10.95		64.95	75.50
0620	Wall penetrations, to 36 sq. in.		180	.089		13.45	4.86		18.31	22
0630	36 sq. in. to 72 sq. in.		100	.160		27	8.75		35.75	42.50
0640	72 sq. in. to 144 sq. in.		60	.267		54	14.60		68.60	81
0800	Caulk seams with latex	1 Carp	230	.035	L.F.	.19	1.90		2.09	3.05
0900	Set up neg. air machine, 1-2k CFM/25 M.C.F. volume	1 Asbe	4.30	1.860	Ea.		113		113	173
0950	Set up and remove portable shower unit	2 Asbe	4	4	"		244		244	375

02 82 13.43 Bulk Asbestos Removal

		Crew	Daily Output	Labor-Hours	Unit	Material	2021 Bare Costs Labor	Equipment	Total	Total Incl O&P
0010	BULK ASBESTOS REMOVAL									
0020	Includes disposable tools and 2 suits and 1 respirator filter/day/worker									
0200	Boiler insulation	A-9	480	.133	S.F.	.43	8.15		8.58	12.90
0210	With metal lath, add				%				50%	50%
0300	Boiler breeching or flue insulation	A-9	520	.123	S.F.	.34	7.50		7.84	11.85
0310	For active boiler, add				%				100%	100%
0400	Duct or AHU insulation	A-10B	440	.073	S.F.	.20	4.44		4.64	7
0500	Duct vibration isolation joints, up to 24 sq. in. duct	A-9	56	1.143	Ea.	3.14	69.50		72.64	110
0520	25 sq. in. to 48 sq. in. duct		48	1.333		3.67	81.50		85.17	128
0530	49 sq. in. to 76 sq. in. duct		40	1.600		4.40	97.50		101.90	154
0600	Pipe insulation, air cell type, up to 4" diameter pipe		900	.071	L.F.	.20	4.34		4.54	6.85
0610	4" to 8" diameter pipe		800	.080		.22	4.88		5.10	7.70
0620	10" to 12" diameter pipe		700	.091		.25	5.60		5.85	8.85

02 82 Asbestos Remediation

02 82 13 – Asbestos Abatement

02 82 13.43 Bulk Asbestos Removal	Crew	Daily Output	Labor-Hours	Unit	Material	2021 Bare Costs Labor	Equipment	Total	Total Incl O&P
0630	14" to 16" diameter pipe	A-9	550	.116	L.F.	.32	7.10		7.42	11.20
0650	Over 16" diameter pipe		650	.098	S.F.	.27	6		6.27	9.50
0700	With glove bag up to 3" diameter pipe		200	.320	L.F.	9.95	19.50		29.45	41
1000	Pipe fitting insulation up to 4" diameter pipe		320	.200	Ea.	.55	12.20		12.75	19.25
1100	6" to 8" diameter pipe		304	.211		.58	12.85		13.43	20.50
1110	10" to 12" diameter pipe		192	.333		.92	20.50		21.42	32
1120	14" to 16" diameter pipe		128	.500	▼	1.38	30.50		31.88	48
1130	Over 16" diameter pipe		176	.364	S.F.	1	22		23	35
1200	With glove bag, up to 8" diameter pipe		75	.853	L.F.	6.55	52		58.55	86.50
2000	Scrape foam fireproofing from flat surface		2400	.027	S.F.	.07	1.63		1.70	2.57
2100	Irregular surfaces		1200	.053		.15	3.25		3.40	5.15
3000	Remove cementitious material from flat surface		1800	.036		.10	2.17		2.27	3.43
3100	Irregular surface		1000	.064	▼	.13	3.90		4.03	6.10
6000	Remove contaminated soil from crawl space by hand	▼	400	.160	C.F.	.44	9.75		10.19	15.45
6100	With large production vacuum loader	A-12	700	.091	"	.25	5.60	1.07	6.92	10
7000	Radiator backing, not including radiator removal	A-9	1200	.053	S.F.	.15	3.25		3.40	5.15
9000	For type B (supplied air) respirator equipment, add				%		10%		10%	

02 82 13.44 Demolition In Asbestos Contaminated Area

		Crew	Daily Output	Labor-Hours	Unit	Material	2021 Bare Costs Labor	Equipment	Total	Total Incl O&P
0010	**DEMOLITION IN ASBESTOS CONTAMINATED AREA**									
0200	Ceiling, including suspension system, plaster and lath	A-9	2100	.030	S.F.	.08	1.86		1.94	2.93
0210	Finished plaster, leaving wire lath		585	.109		.30	6.65		6.95	10.55
0220	Suspended acoustical tile		3500	.018		.05	1.12		1.17	1.77
0230	Concealed tile grid system		3000	.021		.06	1.30		1.36	2.05
0240	Metal pan grid system		1500	.043		.12	2.60		2.72	4.11
0250	Gypsum board		2500	.026	▼	.07	1.56		1.63	2.47
0260	Lighting fixtures up to 2' x 4'	▼	72	.889	Ea.	2.44	54		56.44	85.50
0400	Partitions, non load bearing									
0410	Plaster, lath, and studs	A-9	690	.093	S.F.	.86	5.65		6.51	9.60
0450	Gypsum board and studs	"	1390	.046	"	.13	2.81		2.94	4.44
9000	For type B (supplied air) respirator equipment, add				%		10%		10%	

02 82 13.45 OSHA Testing

		Crew	Daily Output	Labor-Hours	Unit	Material	2021 Bare Costs Labor	Equipment	Total	Total Incl O&P
0010	**OSHA TESTING**									
0100	Certified technician, minimum				Day		340		340	340
0110	Maximum						670		670	670
0121	Industrial hygienist, minimum						385		385	385
0130	Maximum				▼		755		755	755
0200	Asbestos sampling and PCM analysis, NIOSH 7400, minimum	1 Asbe	8	1	Ea.	15.85	61		76.85	111
0210	Maximum		4	2		27	122		149	217
1000	Cleaned area samples		8	1		185	61		246	297
1100	PCM air sample analysis, NIOSH 7400, minimum		8	1		17.05	61		78.05	112
1110	Maximum	▼	4	2		20	122		142	209
1200	TEM air sample analysis, NIOSH 7402, minimum								80	106
1210	Maximum				▼				360	450

02 82 13.46 Decontamination of Asbestos Containment Area

		Crew	Daily Output	Labor-Hours	Unit	Material	2021 Bare Costs Labor	Equipment	Total	Total Incl O&P
0010	**DECONTAMINATION OF ASBESTOS CONTAINMENT AREA**									
0100	Spray exposed substrate with surfactant (bridging)									
0200	Flat surfaces	A-9	6000	.011	S.F.	.42	.65		1.07	1.46
0250	Irregular surfaces		4000	.016	"	.45	.98		1.43	1.99
0300	Pipes, beams, and columns		2000	.032	L.F.	.61	1.95		2.56	3.66
1000	Spray encapsulate polyethylene sheeting		8000	.008	S.F.	.42	.49		.91	1.21
1100	Roll down polyethylene sheeting		8000	.008	"		.49		.49	.75
1500	Bag polyethylene sheeting	▼	400	.160	Ea.	1.05	9.75		10.80	16.10

27

02 82 Asbestos Remediation

02 82 13 – Asbestos Abatement

02 82 13.46 Decontamination of Asbestos Containment Area

		Crew	Daily Output	Labor-Hours	Unit	Material	2021 Bare Costs Labor	Equipment	Total	Total Incl O&P
2000	Fine clean exposed substrate, with nylon brush	A-9	2400	.027	S.F.		1.63		1.63	2.49
2500	Wet wipe substrate		4800	.013			.81		.81	1.24
2600	Vacuum surfaces, fine brush	↓	6400	.010	↓		.61		.61	.93
3000	Structural demolition									
3100	Wood stud walls	A-9	2800	.023	S.F.		1.39		1.39	2.13
3500	Window manifolds, not incl. window replacement		4200	.015			.93		.93	1.42
3600	Plywood carpet protection	↓	2000	.032	↓		1.95		1.95	2.99
4000	Remove custom decontamination facility	A-10A	8	3	Ea.	15.45	183		198.45	298
4100	Remove portable decontamination facility	3 Asbe	12	2	"	13.25	122		135.25	202
5000	HEPA vacuum, shampoo carpeting	A-9	4800	.013	S.F.	.11	.81		.92	1.36
9000	Final cleaning of protected surfaces	A-10A	8000	.003	"		.18		.18	.28

02 82 13.47 Asbestos Waste Pkg., Handling, and Disp.

		Crew	Daily Output	Labor-Hours	Unit	Material	2021 Bare Costs Labor	Equipment	Total	Total Incl O&P
0010	**ASBESTOS WASTE PACKAGING, HANDLING, AND DISPOSAL**									
0100	Collect and bag bulk material, 3 C.F. bags, by hand	A-9	400	.160	Ea.	1.23	9.75		10.98	16.30
0200	Large production vacuum loader	A-12	880	.073		1.04	4.44	.85	6.33	8.85
1000	Double bag and decontaminate	A-9	960	.067	↓	1.23	4.07		5.30	7.55
2000	Containerize bagged material in drums, per 3 C.F. drum	A-9	800	.080	Ea.	19.80	4.88		24.68	29.50
3000	Cart bags 50' to dumpster	2 Asbe	400	.040	"		2.44		2.44	3.73
5000	Disposal charges, not including haul, minimum				C.Y.				61	67
5020	Maximum				"				355	395
9000	For type B (supplied air) respirator equipment, add				%				10%	10%

02 82 13.48 Asbestos Encapsulation With Sealants

		Crew	Daily Output	Labor-Hours	Unit	Material	2021 Bare Costs Labor	Equipment	Total	Total Incl O&P
0010	**ASBESTOS ENCAPSULATION WITH SEALANTS**									
0100	Ceilings and walls, minimum	A-9	21000	.003	S.F.	.44	.19		.63	.76
0110	Maximum		10600	.006	"	.53	.37		.90	1.14
0300	Pipes to 12" diameter including minor repairs, minimum		800	.080	L.F.	.59	4.88		5.47	8.10
0310	Maximum	↓	400	.160	"	1.27	9.75		11.02	16.35

02 87 Biohazard Remediation

02 87 13 – Mold Remediation

02 87 13.16 Mold Remediation Preparation and Containment

		Crew	Daily Output	Labor-Hours	Unit	Material	2021 Bare Costs Labor	Equipment	Total	Total Incl O&P
0010	**MOLD REMEDIATION PREPARATION AND CONTAINMENT**									
6010	Preparation of mold containment area									
6100	Pre-cleaning, HEPA vacuum and wet wipe, flat surfaces	A-9	12000	.005	S.F.	.01	.33		.34	.52
6300	Separation barrier, 2" x 4" @ 16", 1/2" plywood ea. side, 8' high	2 Carp	400	.040		3.11	2.19		5.30	6.70
6310	12' high		320	.050		3.52	2.74		6.26	7.95
6320	16' high		200	.080		2.64	4.38		7.02	9.45
6400	Personnel decontam. chamber, 2" x 4" @ 16", 3/4" ply ea. side		280	.057		5.40	3.13		8.53	10.55
6450	Waste decontam. chamber, 2" x 4" studs @ 16", 3/4" ply each side	↓	360	.044	↓	5.30	2.43		7.73	9.50
6500	Cover surfaces with polyethylene sheeting									
6501	Including glue and tape									
6550	Floors, each layer, 6 mil	A-9	8000	.008	S.F.	.08	.49		.57	.84
6551	4 mil		9000	.007		.03	.43		.46	.69
6560	Walls, each layer, 6 mil		6000	.011		.08	.65		.73	1.09
6561	4 mil	↓	7000	.009	↓	.03	.56		.59	.88
6570	For heights above 14', add						20%			
6575	For heights above 20', add						30%			
6580	For fire retardant poly, add					100%				
6590	For large open areas, deduct					10%	20%			
6600	Seal floor penetrations with foam firestop to 36 sq. in.	2 Carp	200	.080	Ea.	13.45	4.38		17.83	21.50

02 87 Biohazard Remediation

02 87 13 – Mold Remediation

02 87 13.16 Mold Remediation Preparation and Containment	Crew	Daily Output	Labor-Hours	Unit	Material	2021 Bare Costs Labor	Equipment	Total	Total Incl O&P
6610	36 sq. in. to 72 sq. in.	2 Carp	125	.128	Ea.	27	7		34	40
6615	72 sq. in. to 144 sq. in.		80	.200		54	10.95		64.95	75.50
6620	Wall penetrations, to 36 sq. in.		180	.089		13.45	4.86		18.31	22
6630	36 sq. in. to 72 sq. in.		100	.160		27	8.75		35.75	42.50
6640	72 sq. in. to 144 sq. in.		60	.267		54	14.60		68.60	81
6800	Caulk seams with latex caulk	1 Carp	230	.035	L.F.	.19	1.90		2.09	3.05
6900	Set up neg. air machine, 1-2k CFM/25 M.C.F. volume	1 Asbe	4.30	1.860	Ea.		113		113	173

02 87 13.33 Removal and Disposal of Materials With Mold

		Crew	Daily Output	Labor-Hours	Unit	Material	2021 Bare Costs Labor	Equipment	Total	Total Incl O&P
0010	**REMOVAL AND DISPOSAL OF MATERIALS WITH MOLD**									
0015	Demolition in mold contaminated area									
0200	Ceiling, including suspension system, plaster and lath	A-9	2100	.030	S.F.	.08	1.86		1.94	2.93
0210	Finished plaster, leaving wire lath		585	.109		.30	6.65		6.95	10.55
0220	Suspended acoustical tile		3500	.018		.05	1.12		1.17	1.77
0230	Concealed tile grid system		3000	.021		.06	1.30		1.36	2.05
0240	Metal pan grid system		1500	.043		.12	2.60		2.72	4.11
0250	Gypsum board		2500	.026		.07	1.56		1.63	2.47
0255	Plywood		2500	.026		.07	1.56		1.63	2.47
0260	Lighting fixtures up to 2' x 4'		72	.889	Ea.	2.44	54		56.44	85.50
0400	Partitions, non load bearing									
0410	Plaster, lath, and studs	A-9	690	.093	S.F.	.86	5.65		6.51	9.60
0450	Gypsum board and studs		1390	.046		.13	2.81		2.94	4.44
0465	Carpet & pad		1390	.046		.13	2.81		2.94	4.44
0600	Pipe insulation, air cell type, up to 4" diameter pipe		900	.071	L.F.	.20	4.34		4.54	6.85
0610	4" to 8" diameter pipe		800	.080		.22	4.88		5.10	7.70
0620	10" to 12" diameter pipe		700	.091		.25	5.60		5.85	8.85
0630	14" to 16" diameter pipe		550	.116		.32	7.10		7.42	11.20
0650	Over 16" diameter pipe		650	.098	S.F.	.27	6		6.27	9.50
9000	For type B (supplied air) respirator equipment, add				%				10%	10%

02 87 19 – Bio-Hazard and Infectious Disease Control Remediation

02 87 19.10 Temporary Barriers, Including Ppe

		Crew	Daily Output	Labor-Hours	Unit	Material	2021 Bare Costs Labor	Equipment	Total	Total Incl O&P
0010	**TEMPORARY BARRIERS, including PPE**									
0100	Infectious control partition, adjustable to 10' high, obscured, 4' panel	2 Carp	75	.213	Ea.	580	11.65		591.65	655
0110	3' panel		75	.213		550	11.65		561.65	620
0120	2' panel		75	.213		430	11.65		441.65	490
0130	1' panel	1 Carp	75	.107		285	5.85		290.85	325
0140	6" panel	"	75	.107		265	5.85		270.85	300
0150	2' panel with HEPA filtered discharge port	2 Carp	75	.213		500	11.65		511.65	565
0160	3' panel with 32" door		75	.213		895	11.65		906.65	1,000
0170	4' panel with 36" door		75	.213		995	11.65		1,006.65	1,125
0180	4'-6" panel with 44" door		75	.213		1,200	11.65		1,211.65	1,350
0190	Hinged corner		68	.235		186	12.85		198.85	223
0200	Outside corner		68	.235		151	12.85		163.85	185
0210	T post		68	.235		151	12.85		163.85	185
0220	Accessories, ceiling grid clip		300	.053		7.60	2.92		10.52	12.75
0230	Panel locking clip		300	.053		5.30	2.92		8.22	10.20
0240	Panel joint closure strip		300	.053		8.15	2.92		11.07	13.35
0250	Screw jack		300	.053		6.80	2.92		9.72	11.85
0260	Digital pressure difference guage					275			275	305
0280	Combination lockset	1 Carp	11	.727	Ea.	202	40		242	283
0285	Sealant tape, 2" wide	1 Clab	165	.048	C.L.F.	10.85	2.15		13	15.10
0290	System in place, including door and accessories									
0300	Based upon 25 uses	2 Carp	43	.372	L.F.	11.10	20.50		31.60	42.50

02 87 19 – Bio-Hazard and Infectious Disease Control Remediation

02 87 19.10 Temporary Barriers, Including Ppe

		Crew	Daily Output	Labor-Hours	Unit	Material	2021 Bare Costs Labor	Equipment	Total	Total Incl O&P
0310	Based upon 50 uses	2 Carp	43	.372	L.F.	6.15	20.50		26.65	37.50
0330	Based upon 100 uses		43	.372	↓	3.69	20.50		24.19	34.50
0400	Separation barrier, 2" x 4" @ 16" O.C., 1/2" plywood each side, 8' high		340	.047	S.F.	4.22	2.57		6.79	8.50
0410	12' high		275	.058		4.17	3.18		7.35	9.35
0420	16' high		170	.094		4.25	5.15		9.40	12.35
0430	5/8" FR GWB each side, 8' high		290	.055		2.68	3.02		5.70	7.45
0440	12' high		235	.068		2.60	3.72		6.32	8.40
0450	16' high	↓	144	.111	↓	2.55	6.10		8.65	11.85

02 87 19.20 Temporary Protection, Including Ppe

		Crew	Daily Output	Labor-Hours	Unit	Material	2021 Bare Costs Labor	Equipment	Total	Total Incl O&P
0010	**TEMPORARY PROTECTION, including PPE**									
0100	Infectious control protection, 6 mil plastic sheeting, applied to floors	1 Skwk	1900	.004	S.F.	.08	.24		.32	.44
0110	Applied to walls		1400	.006		.08	.33		.41	.57
0120	Applied to ceilings	↓	1000	.008	↓	.08	.46		.54	.77
0130	Sealant tape, 2" wide	1 Clab	165	.048	C.L.F.	10.85	2.15		13	15.10
0140	Floor covering, 1/8" tempered hardboard	2 Carp	878	.018	S.F.	.69	1		1.69	2.25
0150	1/2" CDX		878	.018		1.18	1		2.18	2.79
0160	3/4" CDX	↓	755	.021	↓	1.56	1.16		2.72	3.45

02 87 19.30 Infectious Control Equipment and Supplies

		Crew	Daily Output	Labor-Hours	Unit	Material	2021 Bare Costs Labor	Equipment	Total	Total Incl O&P
0010	**INFECTIOUS CONTROL EQUIPMENT AND SUPPLIES**									
0100	Entry/exit decontamination chamber, portable, 10 uses	A-10C	12	2	Ea.	228	122	16.55	366.55	455
0105	Negative air machine, 1800 CFM					900			900	990
0110	Disposable polyethylene bags, 6 mil, 3 C.F.					1.23			1.23	1.35
0130	Paper towels, carton of 12, 350' rolls					65.50			65.50	72.50
0140	Heavy duty wipes, 160 per box					38			38	42
0200	Tent, including frame, weights, roof, walls, 5' x 5'	2 Skwk	32	.500		2,150	28.50		2,178.50	2,425
0205	5' x 10'		24	.667		2,500	38		2,538	2,800
0210	10' x 10'		24	.667		3,125	38		3,163	3,475
0220	13' x 26'		12	1.333		9,350	76		9,426	10,400
0230	13' x 26', flat roof infirmary		12	1.333		7,600	76		7,676	8,475
0240	13' x 26', drive through screening	↓	12	1.333		8,800	76		8,876	9,800
0242	LED light kit for tents, 15 watt	1 Elec	48	.167		199	10.60		209.60	235
0244	Halogen heating kit, 1500 watt	"	48	.167		500	10.60		510.60	565
0250	Medical partition system, including frame, drapes, 5' x 5'	2 Skwk	48	.333	↓	1,300	19.05		1,319.05	1,450

02 87 19.40 Infectious Control Cleaning, Includes Ppe

		Crew	Daily Output	Labor-Hours	Unit	Material	2021 Bare Costs Labor	Equipment	Total	Total Incl O&P
0010	**INFECTIOUS CONTROL CLEANING, Includes PPE**									
0100	Infectious control cleaning, including PPE and antimicrobial spray, floors	1 Skwk	3600	.002	S.F.	.01	.13		.14	.20
0110	Walls	"	3000	.003	"	.01	.15		.16	.24
0120	Ceilings	1 Skwk	2700	.003	S.F.	.01	.17		.18	.26
0130	Countertops		3300	.002		.01	.14		.15	.22
0140	Other surfaces		2600	.003	↓	.01	.18		.19	.27
0150	Plumbing fixtures		14	.571	Ea.	1.87	32.50		34.37	51
0160	Bathroom accessories		30	.267		.87	15.25		16.12	24
0170	Kitchen appliance		9	.889		2.92	51		53.92	79.50
0180	Electrical switches and receptacles		140	.057		.19	3.26		3.45	5.10
0190	Work station		9	.889		2.92	51		53.92	79.50
0200	Door and window hardware		88	.091	↓	.30	5.20		5.50	8.15
0250	Supply / return ductwork, up to 2 S.F. cross section		240	.033	L.F.	.11	1.90		2.01	2.98
0260	2 to 4 S.F. cross section		192	.042		.14	2.38		2.52	3.73
0270	4 to 8 S.F. cross section		144	.056		.19	3.17		3.36	4.97
0280	8 SF cross section		102	.078	↓	.26	4.48		4.74	7.05
0300	Duct coil		5	1.600	Ea.	5.25	91.50		96.75	143
0310	Duct damper and vane set	↓	77	.104	↓	.35	5.95		6.30	9.30

02 87 Biohazard Remediation

02 87 19 – Bio-Hazard and Infectious Disease Control Remediation

02 87 19.40 Infectious Control Cleaning, Includes Ppe	Crew	Daily Output	Labor-Hours	Unit	Material	2021 Bare Costs Labor	Equipment	Total	Total Incl O&P
0320	Fan coil unit or VAV box	1 Skwk	8	1	Ea.	3.28	57		60.28	89.50
0330	Air handling unit and coil		2	4		13.10	228		241.10	360
0340	Grill / diffuser or register		32	.250		.82	14.30		15.12	22.50

For customer support on your Plumbing Costs with RSMeans Data, call 800.448.8182.

31

Division Notes

	CREW	DAILY OUTPUT	LABOR-HOURS	UNIT	BARE COSTS				TOTAL INCL O&P
					MAT.	LABOR	EQUIP.	TOTAL	

Estimating Tips
General

- Carefully check all the plans and specifications. Concrete often appears on drawings other than structural drawings, including mechanical and electrical drawings for equipment pads. The cost of cutting and patching is often difficult to estimate. See Subdivision 03 81 for Concrete Cutting, Subdivision 02 41 19.16 for Cutout Demolition, Subdivision 03 05 05.10 for Concrete Demolition, and Subdivision 02 41 19.19 for Rubbish Handling (handling, loading, and hauling of debris).

- Always obtain concrete prices from suppliers near the job site. A volume discount can often be negotiated, depending upon competition in the area. Remember to add for waste, particularly for slabs and footings on grade.

03 10 00 Concrete Forming and Accessories

- A primary cost for concrete construction is forming. Most jobs today are constructed with prefabricated forms. The selection of the forms best suited for the job and the total square feet of forms required for efficient concrete forming and placing are key elements in estimating concrete construction. Enough forms must be available for erection to make efficient use of the concrete placing equipment and crew.

- Concrete accessories for forming and placing depend upon the systems used. Study the plans and specifications to ensure that all special accessory requirements have been included in the cost estimate, such as anchor bolts, inserts, and hangers.

- Included within costs for forms-in-place are all necessary bracing and shoring.

03 20 00 Concrete Reinforcing

- Ascertain that the reinforcing steel supplier has included all accessories, cutting, bending, and an allowance for lapping, splicing, and waste. A good rule of thumb is 10% for lapping, splicing, and waste. Also, 10% waste should be allowed for welded wire fabric.

- The unit price items in the subdivisions for Reinforcing In Place, Glass Fiber Reinforcing, and Welded Wire Fabric include the labor to install accessories such as beam and slab bolsters, high chairs, and bar ties and tie wire. The material cost for these accessories is not included; they may be obtained from the Accessories Subdivisions.

03 30 00 Cast-In-Place Concrete

- When estimating structural concrete, pay particular attention to requirements for concrete additives, curing methods, and surface treatments. Special consideration for climate, hot or cold, must be included in your estimate. Be sure to include requirements for concrete placing equipment and concrete finishing.

- For accurate concrete estimating, the estimator must consider each of the following major components individually: forms, reinforcing steel, ready-mix concrete, placement of the concrete, and finishing of the top surface. For faster estimating, Subdivision 03 30 53.40 for Concrete-In-Place can be used; here, various items of concrete work are presented that include the costs of all five major components (unless specifically stated otherwise).

03 40 00 Precast Concrete
03 50 00 Cast Decks and Underlayment

- The cost of hauling precast concrete structural members is often an important factor. For this reason, it is important to get a quote from the nearest supplier. It may become economically feasible to set up precasting beds on the site if the hauling costs are prohibitive.

Reference Numbers

Reference numbers are shown at the beginning of some major classifications. These numbers refer to related items in the Reference Section. The reference information may be an estimating procedure, an alternate pricing method, or technical information.

Note: Not all subdivisions listed here necessarily appear. ■

Same Data. Simplified.

Enjoy the convenience and efficiency of accessing your costs anywhere:

- **Skip the multiplier** by setting your location
- **Quickly search,** edit, favorite and share costs
- **Stay on top of price changes** with automatic updates

Discover more at rsmeans.com/online

03 11 Concrete Forming

03 11 13 – Structural Cast-In-Place Concrete Forming

03 11 13.40 Forms In Place, Equipment Foundations

		Crew	Daily Output	Labor-Hours	Unit	Material	2021 Bare Costs Labor	Equipment	Total	Total Incl O&P
0010	**FORMS IN PLACE, EQUIPMENT FOUNDATIONS**									
0020	1 use	C-2	160	.300	SFCA	3.96	16		19.96	28.50
0050	2 use		190	.253		2.18	13.45		15.63	22.50
0100	3 use		200	.240		1.59	12.80		14.39	21
0150	4 use	↓	205	.234	↓	1.30	12.50		13.80	20

03 11 13.45 Forms In Place, Footings

		Crew	Daily Output	Labor-Hours	Unit	Material	2021 Bare Costs Labor	Equipment	Total	Total Incl O&P
0010	**FORMS IN PLACE, FOOTINGS**									
0020	Continuous wall, plywood, 1 use	C-1	375	.085	SFCA	7.85	4.45		12.30	15.30
0050	2 use		440	.073		4.32	3.79		8.11	10.40
0100	3 use		470	.068		3.14	3.55		6.69	8.75
0150	4 use		485	.066		2.56	3.44		6	7.95
5000	Spread footings, job-built lumber, 1 use		305	.105		3.28	5.45		8.73	11.75
5050	2 use		371	.086		1.82	4.50		6.32	8.70
5100	3 use		401	.080		1.31	4.16		5.47	7.65
5150	4 use		414	.077		1.06	4.03		5.09	7.15

03 11 13.65 Forms In Place, Slab On Grade

		Crew	Daily Output	Labor-Hours	Unit	Material	2021 Bare Costs Labor	Equipment	Total	Total Incl O&P
0010	**FORMS IN PLACE, SLAB ON GRADE**									
3000	Edge forms, wood, 4 use, on grade, to 6" high	C-1	600	.053	L.F.	.45	2.78		3.23	4.65
6000	Trench forms in floor, wood, 1 use		160	.200	SFCA	3.15	10.45		13.60	19
6050	2 use		175	.183		1.73	9.55		11.28	16.15
6100	3 use		180	.178		1.26	9.25		10.51	15.25
6150	4 use	↓	185	.173	↓	1.02	9		10.02	14.60

03 15 Concrete Accessories

03 15 05 – Concrete Forming Accessories

03 15 05.75 Sleeves and Chases

			Crew	Daily Output	Labor-Hours	Unit	Material	2021 Bare Costs Labor	Equipment	Total	Total Incl O&P
0010	**SLEEVES AND CHASES**										
0100	Plastic, 1 use, 12" long, 2" diameter		1 Carp	100	.080	Ea.	1.27	4.38		5.65	7.95
0150	4" diameter			90	.089		2.99	4.86		7.85	10.55
0200	6" diameter			75	.107		11.30	5.85		17.15	21
0250	12" diameter			60	.133		18.05	7.30		25.35	31
5000	Sheet metal, 2" diameter	G		100	.080		1.61	4.38		5.99	8.30
5100	4" diameter	G		90	.089		2.01	4.86		6.87	9.45
5150	6" diameter	G		75	.107		1.74	5.85		7.59	10.60
5200	12" diameter	G		60	.133		3.67	7.30		10.97	14.95
6000	Steel pipe, 2" diameter	G		100	.080		4.35	4.38		8.73	11.35
6100	4" diameter	G		90	.089		17.20	4.86		22.06	26
6150	6" diameter	G		75	.107		19.45	5.85		25.30	30
6200	12" diameter	G		60	.133	↓	87	7.30		94.30	107

03 15 16 – Concrete Construction Joints

03 15 16.20 Control Joints, Saw Cut

		Crew	Daily Output	Labor-Hours	Unit	Material	2021 Bare Costs Labor	Equipment	Total	Total Incl O&P
0010	**CONTROL JOINTS, SAW CUT**									
0100	Sawcut control joints in green concrete									
0120	1" depth	C-27	2000	.008	L.F.	.03	.41	.06	.50	.70
0140	1-1/2" depth		1800	.009		.04	.46	.06	.56	.79
0160	2" depth	↓	1600	.010	↓	.06	.52	.07	.65	.90
0180	Sawcut joint reservoir in cured concrete									
0182	3/8" wide x 3/4" deep, with single saw blade	C-27	1000	.016	L.F.	.04	.83	.11	.98	1.38
0184	1/2" wide x 1" deep, with double saw blades		900	.018		.08	.92	.13	1.13	1.58
0186	3/4" wide x 1-1/2" deep, with double saw blades	↓	800	.020	↓	.17	1.04	.14	1.35	1.87

03 15 Concrete Accessories

03 15 16 – Concrete Construction Joints

03 15 16.20 Control Joints, Saw Cut		Crew	Daily Output	Labor-Hours	Unit	Material	2021 Bare Costs Labor	Equipment	Total	Total Incl O&P
0190	Water blast joint to wash away laitance, 2 passes	C-29	2500	.003	L.F.		.14	.04	.18	.25
0200	Air blast joint to blow out debris and air dry, 2 passes	C-28	2000	.004	↓		.21	.02	.23	.32
0300	For backer rod, see Section 07 91 23.10									
0342	For joint sealant, see Section 07 92 13.20									

03 15 19 – Cast-In Concrete Anchors

03 15 19.10 Anchor Bolts

			Crew	Daily Output	Labor-Hours	Unit	Material	2021 Bare Costs Labor	Equipment	Total	Total Incl O&P
0010	**ANCHOR BOLTS**										
0015	Made from recycled materials										
0025	Single bolts installed in fresh concrete, no templates										
0030	Hooked w/nut and washer, 1/2" diameter, 8" long	G	1 Carp	132	.061	Ea.	1.55	3.32		4.87	6.65
0040	12" long	G		131	.061		1.72	3.34		5.06	6.90
0050	5/8" diameter, 8" long	G		129	.062		3.40	3.39		6.79	8.80
0060	12" long	G		127	.063		4.18	3.45		7.63	9.75
0070	3/4" diameter, 8" long	G		127	.063		4.18	3.45		7.63	9.75
0080	12" long	G	↓	125	.064	↓	5.25	3.50		8.75	11
0090	2-bolt pattern, including job-built 2-hole template, per set										
0100	J-type, incl. hex nut & washer, 1/2" diameter x 6" long	G	1 Carp	21	.381	Set	7.20	21		28.20	39
0110	12" long	G		21	.381		7.90	21		28.90	39.50
0120	18" long	G		21	.381		8.90	21		29.90	41
0130	3/4" diameter x 8" long	G		20	.400		12.80	22		34.80	46.50
0140	12" long	G		20	.400		14.90	22		36.90	49
0150	18" long	G		20	.400		18.05	22		40.05	52.50
0160	1" diameter x 12" long	G		19	.421		28	23		51	65.50
0170	18" long	G		19	.421		33	23		56	70.50
0180	24" long	G		19	.421		39	23		62	77.50
0190	36" long	G		18	.444		52	24.50		76.50	93.50
0200	1-1/2" diameter x 18" long	G		17	.471		49	25.50		74.50	92.50
0210	24" long	G		16	.500		57.50	27.50		85	105
0300	L-type, incl. hex nut & washer, 3/4" diameter x 12" long	G		20	.400		21.50	22		43.50	56
0310	18" long	G		20	.400		26.50	22		48.50	61.50
0320	24" long	G		20	.400		31	22		53	67
0330	30" long	G		20	.400		38.50	22		60.50	74.50
0340	36" long	G		20	.400		43	22		65	80
0350	1" diameter x 12" long	G		19	.421		34	23		57	72
0360	18" long	G		19	.421		41.50	23		64.50	80
0370	24" long	G		19	.421		50	23		73	89.50
0380	30" long	G		19	.421		58.50	23		81.50	98.50
0390	36" long	G		18	.444		66	24.50		90.50	109
0400	42" long	G		18	.444		79	24.50		103.50	124
0410	48" long	G		18	.444		88.50	24.50		113	134
0420	1-1/4" diameter x 18" long	G		18	.444		43	24.50		67.50	84
0430	24" long	G		18	.444		50	24.50		74.50	92
0440	30" long	G		17	.471		57.50	25.50		83	102
0450	36" long	G		17	.471		64.50	25.50		90	110
0460	42" long	G	2 Carp	32	.500		72.50	27.50		100	121
0470	48" long	G		32	.500		82	27.50		109.50	132
0480	54" long	G		31	.516		96	28		124	148
0490	60" long	G		31	.516		105	28		133	158
0500	1-1/2" diameter x 18" long	G		33	.485		48	26.50		74.50	92.50
0510	24" long	G		32	.500		55.50	27.50		83	102
0520	30" long	G		31	.516		62	28		90	110
0530	36" long	G	↓	30	.533	↓	70.50	29		99.50	121

For customer support on your Plumbing Costs with RSMeans Data, call 800.448.8182.

03 15 19.10 Anchor Bolts		Crew	Daily Output	Labor-Hours	Unit	Material	2021 Bare Costs Labor	Equipment	Total	Total Incl O&P
0540	42" long	[G] 2 Carp	30	.533	Set	79.50	29		108.50	131
0550	48" long	[G]	29	.552		89	30		119	143
0560	54" long	[G]	28	.571		107	31.50		138.50	165
0570	60" long	[G]	28	.571		117	31.50		148.50	176
0580	1-3/4" diameter x 18" long	[G]	31	.516		80.50	28		108.50	131
0590	24" long	[G] 2 Carp	30	.533	Set	93.50	29		122.50	147
0600	30" long	[G]	29	.552		108	30		138	164
0610	36" long	[G]	28	.571		123	31.50		154.50	182
0620	42" long	[G]	27	.593		137	32.50		169.50	200
0630	48" long	[G]	26	.615		150	33.50		183.50	217
0640	54" long	[G]	26	.615		186	33.50		219.50	255
0650	60" long	[G]	25	.640		200	35		235	273
0660	2" diameter x 24" long	[G]	27	.593		147	32.50		179.50	211
0670	30" long	[G]	27	.593		166	32.50		198.50	231
0680	36" long	[G]	26	.615		181	33.50		214.50	251
0690	42" long	[G]	25	.640		202	35		237	275
0700	48" long	[G]	24	.667		231	36.50		267.50	310
0710	54" long	[G]	23	.696		274	38		312	355
0720	60" long	[G]	23	.696		295	38		333	380
0730	66" long	[G]	22	.727		315	40		355	405
0740	72" long	[G]	21	.762		345	41.50		386.50	440
1000	4-bolt pattern, including job-built 4-hole template, per set									
1100	J-type, incl. hex nut & washer, 1/2" diameter x 6" long	[G] 1 Carp	19	.421	Set	9.95	23		32.95	45.50
1110	12" long	[G]	19	.421		11.30	23		34.30	47
1120	18" long	[G]	18	.444		13.40	24.50		37.90	51
1130	3/4" diameter x 8" long	[G]	17	.471		21	25.50		46.50	62
1140	12" long	[G]	17	.471		25.50	25.50		51	66.50
1150	18" long	[G]	17	.471		31.50	25.50		57	73.50
1160	1" diameter x 12" long	[G]	16	.500		51.50	27.50		79	97.50
1170	18" long	[G]	15	.533		61	29		90	111
1180	24" long	[G]	15	.533		73.50	29		102.50	125
1190	36" long	[G]	15	.533		99	29		128	153
1200	1-1/2" diameter x 18" long	[G]	13	.615		94	33.50		127.50	154
1210	24" long	[G]	12	.667		111	36.50		147.50	177
1300	L-type, incl. hex nut & washer, 3/4" diameter x 12" long	[G]	17	.471		38.50	25.50		64	81
1310	18" long	[G]	17	.471		48.50	25.50		74	91.50
1320	24" long	[G]	17	.471		58	25.50		83.50	102
1330	30" long	[G]	16	.500		72.50	27.50		100	121
1340	36" long	[G]	16	.500		82	27.50		109.50	131
1350	1" diameter x 12" long	[G]	16	.500		63.50	27.50		91	111
1360	18" long	[G]	15	.533		78.50	29		107.50	130
1370	24" long	[G]	15	.533		95.50	29		124.50	149
1380	30" long	[G]	15	.533		112	29		141	167
1390	36" long	[G]	15	.533		128	29		157	184
1400	42" long	[G]	14	.571		154	31.50		185.50	216
1410	48" long	[G]	14	.571		172	31.50		203.50	236
1420	1-1/4" diameter x 18" long	[G]	14	.571		81.50	31.50		113	136
1430	24" long	[G] 1 Carp	14	.571	Set	96	31.50		127.50	153
1440	30" long	[G]	13	.615		110	33.50		143.50	173
1450	36" long	[G]	13	.615		125	33.50		158.50	188
1460	42" long	[G] 2 Carp	25	.640		141	35		176	208
1470	48" long	[G]	24	.667		160	36.50		196.50	231
1480	54" long	[G]	23	.696		188	38		226	263

For customer support on your Plumbing Costs with RSMeans Data, call 800.448.8182.

03 15 Concrete Accessories

03 15 19 – Cast-In Concrete Anchors

03 15 19.10 Anchor Bolts

		Crew	Daily Output	Labor-Hours	Unit	Material	2021 Bare Costs Labor	Equipment	Total	Total Incl O&P
1490	60" long	G 2 Carp	23	.696	Set	206	38		244	283
1500	1-1/2" diameter x 18" long	G	25	.640		91.50	35		126.50	154
1510	24" long	G	24	.667		106	36.50		142.50	172
1520	30" long	G	23	.696		119	38		157	188
1530	36" long	G	22	.727		136	40		176	210
1540	42" long	G	22	.727		155	40		195	230
1550	48" long	G	21	.762		173	41.50		214.50	253
1560	54" long	G	20	.800		210	44		254	297
1570	60" long	G	20	.800		230	44		274	320
1580	1-3/4" diameter x 18" long	G	22	.727		156	40		196	232
1590	24" long	G	21	.762		183	41.50		224.50	263
1600	30" long	G	21	.762		212	41.50		253.50	295
1610	36" long	G	20	.800		241	44		285	330
1620	42" long	G	19	.842		270	46		316	365
1630	48" long	G	18	.889		297	48.50		345.50	400
1640	54" long	G	18	.889		365	48.50		413.50	480
1650	60" long	G	17	.941		395	51.50		446.50	510
1660	2" diameter x 24" long	G	19	.842		290	46		336	390
1670	30" long	G	18	.889		325	48.50		373.50	435
1680	36" long	G	18	.889		360	48.50		408.50	470
1690	42" long	G	17	.941		400	51.50		451.50	515
1700	48" long	G	16	1		460	54.50		514.50	585
1710	54" long	G	15	1.067		545	58.50		603.50	685
1720	60" long	G	15	1.067		585	58.50		643.50	730
1730	66" long	G	14	1.143		625	62.50		687.50	785
1740	72" long	G	14	1.143		685	62.50		747.50	850
1990	For galvanized, add				Ea.	75%				

03 15 19.45 Machinery Anchors

		Crew	Daily Output	Labor-Hours	Unit	Material	2021 Bare Costs Labor	Equipment	Total	Total Incl O&P
0010	**MACHINERY ANCHORS**, heavy duty, incl. sleeve, floating base nut,									
0020	lower stud & coupling nut, fiber plug, connecting stud, washer & nut.									
0030	For flush mounted embedment in poured concrete heavy equip. pads.									
0200	Stud & bolt, 1/2" diameter	G E-16	40	.400	Ea.	60.50	24.50	3.72	88.72	109
0300	5/8" diameter	G	35	.457		67	28	4.25	99.25	122
0500	3/4" diameter	G	30	.533		80.50	32.50	4.96	117.96	145
0600	7/8" diameter	G	25	.640		95	39	5.95	139.95	171
0800	1" diameter	G	20	.800		105	49	7.45	161.45	199
0900	1-1/4" diameter	G	15	1.067		135	65.50	9.90	210.40	260

03 21 Reinforcement Bars

03 21 11 – Plain Steel Reinforcement Bars

03 21 11.60 Reinforcing In Place

		Crew	Daily Output	Labor-Hours	Unit	Material	2021 Bare Costs Labor	Equipment	Total	Total Incl O&P
0010	**REINFORCING IN PLACE**, 50-60 ton lots, A615 Grade 60									
0020	Includes labor, but not material cost, to install accessories									
0030	Made from recycled materials									
0502	Footings, #4 to #7	G 4 Rodm	4200	.008	Lb.	.62	.45		1.07	1.35
0552	#8 to #18	G	7200	.004		.62	.26		.88	1.07
0602	Slab on grade, #3 to #7	G	4200	.008		.62	.45		1.07	1.35
0900	For other than 50-60 ton lots									
1000	Under 10 ton job, #3 to #7, add					25%	10%			
1010	#8 to #18, add					20%	10%			
1050	10-50 ton job, #3 to #7, add					10%				

03 21 Reinforcement Bars

03 21 11 – Plain Steel Reinforcement Bars

03 21 11.60 Reinforcing In Place	Crew	Daily Output	Labor-Hours	Unit	Material	2021 Bare Costs Labor	Equipment	Total	Total Incl O&P
1060	#8 to #18, add					5%				
1100	60-100 ton job, #3 to #7, deduct					5%				
1110	#8 to #18, deduct					10%				
1150	Over 100 ton job, #3 to #7, deduct					10%				
1160	#8 to #18, deduct					15%				

03 22 Fabric and Grid Reinforcing

03 22 11 – Plain Welded Wire Fabric Reinforcing

03 22 11.10 Plain Welded Wire Fabric

		Crew	Daily Output	Labor-Hours	Unit	Material	Labor	Equipment	Total	Total Incl O&P
0010	**PLAIN WELDED WIRE FABRIC** ASTM A185									
0020	Includes labor, but not material cost, to install accessories									
0030	Made from recycled materials									
0050	Sheets									
0100	6 x 6 - W1.4 x W1.4 (10 x 10) 21 lb./C.S.F.	2 Rodm	35	.457	C.S.F.	14.95	27		41.95	57

03 22 13 – Galvanized Welded Wire Fabric Reinforcing

03 22 13.10 Galvanized Welded Wire Fabric

		Crew	Daily Output	Labor-Hours	Unit	Material	Labor	Equipment	Total	Total Incl O&P
0010	**GALVANIZED WELDED WIRE FABRIC**									
0100	Add to plain welded wire pricing for galvanized welded wire				Lb.	.24			.24	.27

03 22 16 – Epoxy-Coated Welded Wire Fabric Reinforcing

03 22 16.10 Epoxy-Coated Welded Wire Fabric

		Crew	Daily Output	Labor-Hours	Unit	Material	Labor	Equipment	Total	Total Incl O&P
0010	**EPOXY-COATED WELDED WIRE FABRIC**									
0100	Add to plain welded wire pricing for epoxy-coated welded wire				Lb.	.49			.49	.53

03 30 Cast-In-Place Concrete

03 30 53 – Miscellaneous Cast-In-Place Concrete

03 30 53.40 Concrete In Place

		Crew	Daily Output	Labor-Hours	Unit	Material	Labor	Equipment	Total	Total Incl O&P
0010	**CONCRETE IN PLACE**									
0020	Including forms (4 uses), Grade 60 rebar, concrete (Portland cement									
0050	Type I), placement and finishing unless otherwise indicated									
3540	Equipment pad (3000 psi), 3' x 3' x 6" thick	C-14H	45	1.067	Ea.	55.50	57	.61	113.11	147
3550	4' x 4' x 6" thick	"	30	1.600	"	84	85.50	.91	170.41	220
3560	5' x 5' x 8" thick	C-14H	18	2.667	Ea.	150	143	1.52	294.52	380
3570	6' x 6' x 8" thick		14	3.429		204	184	1.95	389.95	500
3580	8' x 8' x 10" thick		8	6		415	320	3.42	738.42	945
3590	10' x 10' x 12" thick		5	9.600		720	515	5.45	1,240.45	1,550
3800	Footings (3000 psi), spread under 1 C.Y.	C-14C	28	4	C.Y.	208	209	.96	417.96	540
3825	1 C.Y. to 5 C.Y.		43	2.605		266	136	.63	402.63	495
3850	Over 5 C.Y.	R033053-60	75	1.493		245	78	.36	323.36	385
3900	Footings, strip (3000 psi), 18" x 9", unreinforced	C-14L	40	2.400		159	123	.67	282.67	360
3920	18" x 9", reinforced	C-14C	35	3.200		189	167	.77	356.77	460
3925	20" x 10", unreinforced	C-14L	45	2.133		153	109	.60	262.60	335
3930	20" x 10", reinforced	C-14C	40	2.800		178	146	.67	324.67	415
3935	24" x 12", unreinforced	C-14L	55	1.745		150	89.50	.49	239.99	299
3940	24" x 12", reinforced	C-14C	48	2.333		175	122	.56	297.56	375
3945	36" x 12", unreinforced	C-14L	70	1.371		144	70	.38	214.38	263
3950	36" x 12", reinforced	C-14C	60	1.867		166	97.50	.45	263.95	330
4000	Foundation mat (3000 psi), under 10 C.Y.		38.67	2.896		259	151	.70	410.70	510
4050	Over 20 C.Y.		56.40	1.986		225	104	.48	329.48	405

03 30 Cast-In-Place Concrete

03 30 53 – Miscellaneous Cast-In-Place Concrete

03 30 53.40 Concrete In Place	Crew	Daily Output	Labor-Hours	Unit	Material	2021 Bare Costs Labor	Equipment	Total	Total Incl O&P
4650	Slab on grade (3500 psi), not including finish, 4" thick	C-14E	60.75	1.449	C.Y.	149	77.50	.45	226.95	279
4700	6" thick	"	92	.957	↓	144	51	.30	195.30	234
4701	Thickened slab edge (3500 psi), for slab on grade poured									
4702	monolithically with slab; depth is in addition to slab thickness;									
4703	formed vertical outside edge, earthen bottom and inside slope									
4705	8" deep x 8" wide bottom, unreinforced	C-14L	2190	.044	L.F.	4.58	2.24	.01	6.83	8.40
4710	8" x 8", reinforced	C-14C	1670	.067		7.50	3.51	.02	11.03	13.50
4715	12" deep x 12" wide bottom, unreinforced	C-14L	1800	.053		9.05	2.73	.01	11.79	14.05
4720	12" x 12", reinforced	C-14C	1310	.086		14.40	4.47	.02	18.89	22.50
4725	16" deep x 16" wide bottom, unreinforced	C-14L	1440	.067		15	3.41	.02	18.43	21.50
4730	16" x 16", reinforced	C-14C	1120	.100		21.50	5.25	.02	26.77	31.50
4735	20" deep x 20" wide bottom, unreinforced	C-14L	1150	.083		22.50	4.27	.02	26.79	31
4740	20" x 20", reinforced	C-14C	920	.122		30.50	6.35	.03	36.88	43
4745	24" deep x 24" wide bottom, unreinforced	C-14L	930	.103		31.50	5.30	.03	36.83	42.50
4750	24" x 24", reinforced	C-14C	740	.151	↓	42	7.90	.04	49.94	58

03 31 Structural Concrete

03 31 13 – Heavyweight Structural Concrete

03 31 13.35 Heavyweight Concrete, Ready Mix

		Crew	Daily Output	Labor-Hours	Unit	Material	2021 Bare Costs Labor	Equipment	Total	Total Incl O&P
0010	**HEAVYWEIGHT CONCRETE, READY MIX**, delivered									
0012	Includes local aggregate, sand, Portland cement (Type I) and water									
0015	Excludes all additives and treatments									
0020	2000 psi				C.Y.	109			109	120
0100	2500 psi				C.Y.	112			112	124
0150	3000 psi					124			124	137
0200	3500 psi					127			127	139
0300	4000 psi					129			129	142
1000	For high early strength (Portland cement Type III), add					10%				
1300	For winter concrete (hot water), add					5.85			5.85	6.40
1410	For mid-range water reducer, add					4.22			4.22	4.64
1420	For high-range water reducer/superplasticizer, add					6.65			6.65	7.30
1430	For retarder, add					4.22			4.22	4.64
1440	For non-Chloride accelerator, add					8.70			8.70	9.60
1450	For Chloride accelerator, per 1%, add					3.69			3.69	4.06
1460	For fiber reinforcing, synthetic (1 lb./C.Y.), add					8.30			8.30	9.15
1500	For Saturday delivery, add				↓	7.70			7.70	8.45
1510	For truck holding/waiting time past 1st hour per load, add				Hr.	114			114	125
1520	For short load (less than 4 C.Y.), add per load				Ea.	97			97	107
2000	For all lightweight aggregate, add				C.Y.	45%				

03 31 13.70 Placing Concrete

		Crew	Daily Output	Labor-Hours	Unit	Material	2021 Bare Costs Labor	Equipment	Total	Total Incl O&P
0010	**PLACING CONCRETE**									
0020	Includes labor and equipment to place, level (strike off) and consolidate									
1200	Duct bank, direct chute	C-6	155	.310	C.Y.		14.25	.35	14.60	21.50
1900	Footings, continuous, shallow, direct chute	"	120	.400			18.40	.45	18.85	28
1950	Pumped	C-20	150	.427			20	3.19	23.19	33.50
2000	With crane and bucket	C-7	90	.800			38.50	12.10	50.60	70.50
2100	Footings, continuous, deep, direct chute	C-6	140	.343			15.75	.39	16.14	24
2150	Pumped	C-20	160	.400			18.95	2.99	21.94	31.50
2200	With crane and bucket	C-7	110	.655			31.50	9.90	41.40	57.50
2400	Footings, spread, under 1 C.Y., direct chute	C-6	55	.873			40	.99	40.99	60.50
2450	Pumped	C-20	65	.985	↓		46.50	7.35	53.85	77.50

39

03 31 Structural Concrete

03 31 13 – Heavyweight Structural Concrete

03 31 13.70 Placing Concrete

		Crew	Daily Output	Labor-Hours	Unit	Material	2021 Bare Costs Labor	2021 Bare Costs Equipment	Total	Total Incl O&P
2500	With crane and bucket	C-7	45	1.600	C.Y.		77	24	101	141
2600	Over 5 C.Y., direct chute	C-6	120	.400			18.40	.45	18.85	28
2650	Pumped	C-20	150	.427			20	3.19	23.19	33.50
2700	With crane and bucket	C-7	100	.720			34.50	10.90	45.40	63.50
2900	Foundation mats, over 20 C.Y., direct chute	C-6	350	.137			6.30	.15	6.45	9.55
2950	Pumped	C-20	400	.160			7.60	1.20	8.80	12.60
3000	With crane and bucket	C-7	300	.240			11.50	3.63	15.13	21

03 35 Concrete Finishing

03 35 13 – High-Tolerance Concrete Floor Finishing

03 35 13.30 Finishing Floors, High Tolerance

		Crew	Daily Output	Labor-Hours	Unit	Material	2021 Bare Costs Labor	2021 Bare Costs Equipment	Total	Total Incl O&P
0010	**FINISHING FLOORS, HIGH TOLERANCE**									
0012	Finishing of fresh concrete flatwork requires that concrete									
0013	first be placed, struck off & consolidated									
0015	Basic finishing for various unspecified flatwork									
0100	Bull float only	C-10	4000	.006	S.F.		.30		.30	.44
0125	Bull float & manual float		2000	.012			.59		.59	.87
0150	Bull float, manual float & broom finish, w/edging & joints		1850	.013			.64		.64	.94
0200	Bull float, manual float & manual steel trowel		1265	.019			.94		.94	1.38
0210	For specified Random Access Floors in ACI Classes 1, 2, 3 and 4 to achieve									
0215	Composite Overall Floor Flatness and Levelness values up to FF35/FL25									
0250	Bull float, machine float & machine trowel (walk-behind)	C-10C	1715	.014	S.F.		.69	.05	.74	1.08
0300	Power screed, bull float, machine float & trowel (walk-behind)	C-10D	2400	.010			.49	.08	.57	.81
0350	Power screed, bull float, machine float & trowel (ride-on)	C-10E	4000	.006			.30	.06	.36	.51
0352	For specified Random Access Floors in ACI Classes 5, 6, 7 and 8 to achieve									
0354	Composite Overall Floor Flatness and Levelness values up to FF50/FL50									
0356	Add for two-dimensional restraightening after power float	C-10	6000	.004	S.F.		.20		.20	.29
0358	For specified Random or Defined Access Floors in ACI Class 9 to achieve									
0360	Composite Overall Floor Flatness and Levelness values up to FF100/FL100									
0362	Add for two-dimensional restraightening after bull float & power float	C-10	3000	.008	S.F.		.39		.39	.58
0364	For specified Superflat Defined Access Floors in ACI Class 9 to achieve									
0366	Minimum Floor Flatness and Levelness values of FF100/FL100									
0368	Add for 2-dim'l restraightening after bull float, power float, power trowel	C-10	2000	.012	S.F.		.59		.59	.87

03 63 Epoxy Grouting

03 63 05 – Grouting of Dowels and Fasteners

03 63 05.10 Epoxy Only

		Crew	Daily Output	Labor-Hours	Unit	Material	2021 Bare Costs Labor	2021 Bare Costs Equipment	Total	Total Incl O&P
0010	**EPOXY ONLY**									
1500	Chemical anchoring, epoxy cartridge, excludes layout, drilling, fastener									
1530	For fastener 3/4" diam. x 6" embedment	2 Skwk	72	.222	Ea.	4.89	12.70		17.59	24.50
1535	1" diam. x 8" embedment		66	.242		7.35	13.85		21.20	29
1540	1-1/4" diam. x 10" embedment		60	.267		14.70	15.25		29.95	39
1545	1-3/4" diam. x 12" embedment		54	.296		24.50	16.90		41.40	52.50
1550	14" embedment		48	.333		29.50	19.05		48.55	61
1555	2" diam. x 12" embedment		42	.381		39	22		61	75.50
1560	18" embedment		32	.500		49	28.50		77.50	97

03 82 Concrete Boring

03 82 13 – Concrete Core Drilling

03 82 13.10 Core Drilling

	Crew	Daily Output	Labor-Hours	Unit	Material	2021 Bare Costs Labor	Equipment	Total	Total Incl O&P
0010 **CORE DRILLING**									
0015 Includes bit cost, layout and set-up time									
0020 Reinforced concrete slab, up to 6" thick									
0100 1" diameter core	B-89A	17	.941	Ea.	.20	48	7.15	55.35	79.50
0150 For each additional inch of slab thickness in same hole, add		1440	.011		.03	.56	.08	.67	.98
0200 2" diameter core		16.50	.970		.27	49	7.35	56.62	82.50
0250 For each additional inch of slab thickness in same hole, add		1080	.015		.04	.75	.11	.90	1.30
0300 3" diameter core		16	1		.41	51	7.60	59.01	85
0350 For each additional inch of slab thickness in same hole, add		720	.022		.07	1.13	.17	1.37	1.96
0500 4" diameter core		15	1.067		.50	54	8.10	62.60	90.50
0550 For each additional inch of slab thickness in same hole, add		480	.033		.08	1.69	.25	2.02	2.91
0700 6" diameter core		14	1.143		.80	58	8.70	67.50	97.50
0750 For each additional inch of slab thickness in same hole, add		360	.044		.13	2.26	.34	2.73	3.90
0900 8" diameter core		13	1.231		1.16	62.50	9.35	73.01	105
0950 For each additional inch of slab thickness in same hole, add		288	.056		.19	2.82	.42	3.43	4.90
1100 10" diameter core		12	1.333		1.43	67.50	10.15	79.08	114
1150 For each additional inch of slab thickness in same hole, add		240	.067		.24	3.38	.51	4.13	5.85
1300 12" diameter core		11	1.455		1.63	74	11.05	86.68	125
1350 For each additional inch of slab thickness in same hole, add		206	.078		.27	3.94	.59	4.80	6.85
1500 14" diameter core		10	1.600		2.02	81	12.15	95.17	138
1550 For each additional inch of slab thickness in same hole, add		180	.089		.34	4.51	.68	5.53	7.85
1700 18" diameter core		9	1.778		2.80	90	13.50	106.30	153
1750 For each additional inch of slab thickness in same hole, add		144	.111		.47	5.65	.84	6.96	9.90
1754 24" diameter core		8	2		4.07	102	15.20	121.27	173
1756 For each additional inch of slab thickness in same hole, add		120	.133		.68	6.75	1.01	8.44	12
1760 For horizontal holes, add to above						20%	20%		
1770 Prestressed hollow core plank, 8" thick									
1780 1" diameter core	B-89A	17.50	.914	Ea.	.26	46.50	6.95	53.71	77.50
1790 For each additional inch of plank thickness in same hole, add		3840	.004		.03	.21	.03	.27	.39
1794 2" diameter core		17.25	.928		.36	47	7.05	54.41	78.50
1796 For each additional inch of plank thickness in same hole, add		2880	.006		.04	.28	.04	.36	.52
1800 3" diameter core		17	.941		.55	48	7.15	55.70	80
1810 For each additional inch of plank thickness in same hole, add		1920	.008		.07	.42	.06	.55	.78
1820 4" diameter core		16.50	.970		.67	49	7.35	57.02	83
1830 For each additional inch of plank thickness in same hole, add		1280	.013		.08	.63	.10	.81	1.14
1840 6" diameter core		15.50	1.032		1.06	52.50	7.85	61.41	88.50
1850 For each additional inch of plank thickness in same hole, add		960	.017		.13	.85	.13	1.11	1.56
1860 8" diameter core	B-89A	15	1.067	Ea.	1.55	54	8.10	63.65	91.50
1870 For each additional inch of plank thickness in same hole, add		768	.021		.19	1.06	.16	1.41	1.97
1880 10" diameter core		14	1.143		1.90	58	8.70	68.60	98.50
1890 For each additional inch of plank thickness in same hole, add		640	.025		.24	1.27	.19	1.70	2.37
1900 12" diameter core		13.50	1.185		2.18	60	9	71.18	102
1910 For each additional inch of plank thickness in same hole, add		548	.029		.27	1.48	.22	1.97	2.76

03 82 16 – Concrete Drilling

03 82 16.10 Concrete Impact Drilling

	Crew	Daily Output	Labor-Hours	Unit	Material	2021 Bare Costs Labor	Equipment	Total	Total Incl O&P
0010 **CONCRETE IMPACT DRILLING**									
0020 Includes bit cost, layout and set-up time, no anchors									
0050 Up to 4" deep in concrete/brick floors/walls									
0100 Holes, 1/4" diameter	1 Carp	75	.107	Ea.	.05	5.85		5.90	8.75
0150 For each additional inch of depth in same hole, add		430	.019		.01	1.02		1.03	1.53
0200 3/8" diameter		63	.127		.05	6.95		7	10.40
0250 For each additional inch of depth in same hole, add		340	.024		.01	1.29		1.30	1.93

For customer support on your Plumbing Costs with RSMeans Data, call 800.448.8182.

41

03 82 Concrete Boring

03 82 16 – Concrete Drilling

03 82 16.10 Concrete Impact Drilling		Crew	Daily Output	Labor-Hours	Unit	Material	2021 Bare Costs Labor	Equipment	Total	Total Incl O&P
0300	1/2" diameter	1 Carp	50	.160	Ea.	.05	8.75		8.80	13.10
0350	For each additional inch of depth in same hole, add		250	.032		.01	1.75		1.76	2.62
0400	5/8" diameter		48	.167		.08	9.10		9.18	13.70
0450	For each additional inch of depth in same hole, add		240	.033		.02	1.82		1.84	2.74
0500	3/4" diameter		45	.178		.11	9.70		9.81	14.60
0550	For each additional inch of depth in same hole, add		220	.036		.03	1.99		2.02	3
0600	7/8" diameter		43	.186		.13	10.20		10.33	15.35
0650	For each additional inch of depth in same hole, add		210	.038		.03	2.08		2.11	3.15
0700	1" diameter		40	.200		.21	10.95		11.16	16.60
0750	For each additional inch of depth in same hole, add		190	.042		.05	2.30		2.35	3.50
0800	1-1/4" diameter		38	.211		.35	11.50		11.85	17.60
0850	For each additional inch of depth in same hole, add		180	.044		.09	2.43		2.52	3.73
0900	1-1/2" diameter		35	.229		.37	12.50		12.87	19.05
0950	For each additional inch of depth in same hole, add	▼	165	.048	▼	.09	2.65		2.74	4.06
1000	For ceiling installations, add						40%			

Estimating Tips
05 05 00 Common Work Results for Metals

- Nuts, bolts, washers, connection angles, and plates can add a significant amount to both the tonnage of a structural steel job and the estimated cost. As a rule of thumb, add 10% to the total weight to account for these accessories.

- Type 2 steel construction, commonly referred to as "simple construction," consists generally of field-bolted connections with lateral bracing supplied by other elements of the building, such as masonry walls or x-bracing. The estimator should be aware, however, that shop connections may be accomplished by welding or bolting. The method may be particular to the fabrication shop and may have an impact on the estimated cost.

05 10 00 Structural Steel

- Steel items can be obtained from two sources: a fabrication shop or a metals service center. Fabrication shops can fabricate items under more controlled conditions than crews in the field can. They are also more efficient and can produce items more economically. Metal service centers serve as a source of long mill shapes to both fabrication shops and contractors.

- Most line items in this structural steel subdivision, and most items in 05 50 00 Metal Fabrications, are indicated as being shop fabricated. The bare material cost for these shop fabricated items is the "Invoice Cost" from the shop and includes the mill base price of steel plus mill extras, transportation to the shop, shop drawings and detailing where warranted, shop fabrication and handling, sandblasting and a shop coat of primer paint, all necessary structural bolts, and delivery to the job site. The bare labor cost and bare equipment cost for these shop fabricated items are for field installation or erection.

- Line items in Subdivision 05 12 23.40 Lightweight Framing, and other items scattered in Division 5, are indicated as being field fabricated. The bare material cost for these field fabricated items is the "Invoice Cost" from the metals service center and includes the mill base price of steel plus mill extras, transportation to the metals service center, material handling, and delivery of long lengths of mill shapes to the job site. Material costs for structural bolts and welding rods should be added to the estimate. The bare labor cost and bare equipment cost for these items are for both field fabrication and field installation or erection, and include time for cutting, welding, and drilling in the fabricated metal items. Drilling into concrete and fasteners to fasten field fabricated items to other work is not included and should be added to the estimate.

05 20 00 Steel Joist Framing

- In any given project the total weight of open web steel joists is determined by the loads to be supported and the design. However, economies can be realized in minimizing the amount of labor used to place the joists. This is done by maximizing the joist spacing and therefore minimizing the number of joists required to be installed on the job. Certain spacings and locations may be required by the design, but in other cases maximizing the spacing and keeping it as uniform as possible will keep the costs down.

05 30 00 Steel Decking

- The takeoff and estimating of a metal deck involve more than the area of the floor or roof and the type of deck specified or shown on the drawings. Many different sizes and types of openings may exist. Small openings for individual pipes or conduits may be drilled after the floor/roof is installed, but larger openings may require special deck lengths as well as reinforcing or structural support. The estimator should determine who will be supplying this reinforcing. Additionally, some deck terminations are part of the deck package, such as screed angles and pour stops, and others will be part of the steel contract, such as angles attached to structural members and cast-in-place angles and plates. The estimator must ensure that all pieces are accounted for in the complete estimate.

05 50 00 Metal Fabrications

- The most economical steel stairs are those that use common materials, standard details, and most importantly, a uniform and relatively simple method of field assembly. Commonly available A36/A992 channels and plates are very good choices for the main stringers of the stairs, as are angles and tees for the carrier members. Risers and treads are usually made by specialty shops, and it is most economical to use a typical detail in as many places as possible. The stairs should be pre-assembled and shipped directly to the site. The field connections should be simple and straightforward enough to be accomplished efficiently, and with minimum equipment and labor.

Reference Numbers

Reference numbers are shown at the beginning of some major classifications. These numbers refer to related items in the Reference Section. The reference information may be an estimating procedure, an alternate pricing method, or technical information.

Note: Not all subdivisions listed here necessarily appear. ∎

Same Data. Simplified.

Enjoy the convenience and efficiency of accessing your costs anywhere:

- **Skip the multiplier** by setting your location
- **Quickly search,** edit, favorite and share costs
- **Stay on top of price changes** with automatic updates

Discover more at rsmeans.com/online

05 05 19 – Post-Installed Concrete Anchors

05 05 19.10 Chemical Anchors		Crew	Daily Output	Labor-Hours	Unit	Material	2021 Bare Costs Labor	Equipment	Total	Total Incl O&P
0010	**CHEMICAL ANCHORS**									
0020	Includes layout & drilling									
1430	Chemical anchor, w/rod & epoxy cartridge, 3/4" diameter x 9-1/2" long	B-89A	27	.593	Ea.	11.90	30	4.50	46.40	63
1435	1" diameter x 11-3/4" long		24	.667		24	34	5.05	63.05	82.50
1440	1-1/4" diameter x 14" long		21	.762		38.50	38.50	5.80	82.80	106
1445	1-3/4" diameter x 15" long		20	.800		70.50	40.50	6.10	117.10	145
1450	18" long		17	.941		84.50	48	7.15	139.65	172
1455	2" diameter x 18" long		16	1		148	51	7.60	206.60	247
1460	24" long	B-89A	15	1.067	Ea.	194	54	8.10	256.10	305

05 05 19.20 Expansion Anchors			Crew	Daily Output	Labor-Hours	Unit	Material	2021 Bare Costs Labor	Equipment	Total	Total Incl O&P
0010	**EXPANSION ANCHORS**										
0100	Anchors for concrete, brick or stone, no layout and drilling										
0200	Expansion shields, zinc, 1/4" diameter, 1-5/16" long, single	G	1 Carp	90	.089	Ea.	.31	4.86		5.17	7.60
0300	1-3/8" long, double	G		85	.094		.42	5.15		5.57	8.15
0400	3/8" diameter, 1-1/2" long, single	G		85	.094		.51	5.15		5.66	8.25
0500	2" long, double	G		80	.100		1.22	5.45		6.67	9.50
0600	1/2" diameter, 2-1/16" long, single	G		80	.100		1.23	5.45		6.68	9.50
0700	2-1/2" long, double	G		75	.107		2.25	5.85		8.10	11.20
0800	5/8" diameter, 2-5/8" long, single	G		75	.107		1.97	5.85		7.82	10.85
0900	2-3/4" long, double	G		70	.114		3.80	6.25		10.05	13.55
1000	3/4" diameter, 2-3/4" long, single	G		70	.114		3.75	6.25		10	13.50
1100	3-15/16" long, double	G		65	.123		5.55	6.75		12.30	16.15
2100	Hollow wall anchors for gypsum wall board, plaster or tile										
2300	1/8" diameter, short	G	1 Carp	160	.050	Ea.	.31	2.74		3.05	4.42
2400	Long	G		150	.053		.30	2.92		3.22	4.68
2500	3/16" diameter, short	G		150	.053		.61	2.92		3.53	5
2600	Long	G		140	.057		.71	3.13		3.84	5.45
2700	1/4" diameter, short	G		140	.057		.58	3.13		3.71	5.30
2800	Long	G		130	.062		1.04	3.37		4.41	6.15
3000	Toggle bolts, bright steel, 1/8" diameter, 2" long	G		85	.094		.33	5.15		5.48	8.05
3100	4" long	G		80	.100		.38	5.45		5.83	8.55
3200	3/16" diameter, 3" long	G		80	.100		.40	5.45		5.85	8.60
3300	6" long	G		75	.107		.62	5.85		6.47	9.40
3400	1/4" diameter, 3" long	G		75	.107		.44	5.85		6.29	9.20
3500	6" long	G		70	.114		.59	6.25		6.84	10
3600	3/8" diameter, 3" long	G		70	.114		.73	6.25		6.98	10.15
3700	6" long	G		60	.133		1.62	7.30		8.92	12.70
3800	1/2" diameter, 4" long	G		60	.133		2.66	7.30		9.96	13.85
3900	6" long	G		50	.160		3.05	8.75		11.80	16.40
4000	Nailing anchors										
4100	Nylon nailing anchor, 1/4" diameter, 1" long		1 Carp	3.20	2.500	C	19.40	137		156.40	226
4200	1-1/2" long			2.80	2.857		22	156		178	258
4300	2" long			2.40	3.333		24	182		206	299
4400	Metal nailing anchor, 1/4" diameter, 1" long	G		3.20	2.500		20.50	137		157.50	227
4500	1-1/2" long	G		2.80	2.857		21	156		177	256
4600	2" long	G		2.40	3.333		44	182		226	320
5000	Screw anchors for concrete, masonry,										
5100	stone & tile, no layout or drilling included										
5700	Lag screw shields, 1/4" diameter, short	G	1 Carp	90	.089	Ea.	.28	4.86		5.14	7.55
5800	Long	G		85	.094		.56	5.15		5.71	8.30
5900	3/8" diameter, short	G		85	.094		.55	5.15		5.70	8.30
6000	Long	G		80	.100		1.06	5.45		6.51	9.30

For customer support on your Plumbing Costs with RSMeans Data, call 800.448.8182.

05 05 19 – Post-Installed Concrete Anchors

05 05 19.20 Expansion Anchors		Crew	Daily Output	Labor-Hours	Unit	Material	2021 Bare Costs Labor	Equipment	Total	Total Incl O&P
6100	1/2" diameter, short	G 1 Carp	80	.100	Ea.	1.02	5.45		6.47	9.25
6200	Long	G	75	.107		1.53	5.85		7.38	10.40
6300	5/8" diameter, short	G	70	.114		1.60	6.25		7.85	11.10
6400	Long	G	65	.123		2.32	6.75		9.07	12.60
6600	Lead, #6 & #8, 3/4" long	G	260	.031		.22	1.68		1.90	2.75
6700	#10 - #14, 1-1/2" long	G	200	.040		.52	2.19		2.71	3.84
6800	#16 & #18, 1-1/2" long		160	.050		.70	2.74		3.44	4.85
6900	Plastic, #6 & #8, 3/4" long		260	.031		.05	1.68		1.73	2.57
7000	#8 & #10, 7/8" long		240	.033		.04	1.82		1.86	2.76
7100	#10 & #12, 1" long		220	.036		.07	1.99		2.06	3.05
7200	#14 & #16, 1-1/2" long	↓	160	.050	↓	.04	2.74		2.78	4.12
8000	Wedge anchors, not including layout or drilling									
8050	Carbon steel, 1/4" diameter, 1-3/4" long	G 1 Carp	150	.053	Ea.	.76	2.92		3.68	5.20
8100	3-1/4" long	G	140	.057		1	3.13		4.13	5.75
8150	3/8" diameter, 2-1/4" long	G	145	.055		.43	3.02		3.45	4.98
8200	5" long	G	140	.057		.76	3.13		3.89	5.50
8250	1/2" diameter, 2-3/4" long	G	140	.057		1.14	3.13		4.27	5.95
8300	7" long	G	125	.064		1.96	3.50		5.46	7.40
8350	5/8" diameter, 3-1/2" long	G	130	.062		2.13	3.37		5.50	7.35
8400	8-1/2" long	G	115	.070		4.54	3.81		8.35	10.70
8450	3/4" diameter, 4-1/4" long	G	115	.070		3.45	3.81		7.26	9.50
8500	10" long	G	95	.084		7.85	4.61		12.46	15.55
8550	1" diameter, 6" long	G	100	.080		5.65	4.38		10.03	12.75
8575	9" long	G	85	.094		7.35	5.15		12.50	15.75
8600	12" long	G	75	.107		7.95	5.85		13.80	17.40
8650	1-1/4" diameter, 9" long	G	70	.114		33.50	6.25		39.75	46.50
8700	12" long	G ↓	60	.133	↓	43	7.30		50.30	58.50
8750	For type 303 stainless steel, add					350%				
8800	For type 316 stainless steel, add					450%				
8950	Self-drilling concrete screw, hex washer head, 3/16" diam. x 1-3/4" long	G 1 Carp	300	.027	Ea.	.21	1.46		1.67	2.41
8960	2-1/4" long	G	250	.032		.29	1.75		2.04	2.93
8970	Phillips flat head, 3/16" diam. x 1-3/4" long	G	300	.027		.20	1.46		1.66	2.40
8980	2-1/4" long	G ↓	250	.032		.31	1.75		2.06	2.95

05 05 21 – Fastening Methods for Metal

05 05 21.15 Drilling Steel

0010	**DRILLING STEEL**									
1910	Drilling & layout for steel, up to 1/4" deep, no anchor									
1920	Holes, 1/4" diameter	1 Sswk	112	.071	Ea.	.06	4.31		4.37	6.70
1925	For each additional 1/4" depth, add		336	.024		.06	1.44		1.50	2.29
1930	3/8" diameter		104	.077		.08	4.64		4.72	7.30
1935	For each additional 1/4" depth, add		312	.026		.08	1.55		1.63	2.48
1940	1/2" diameter		96	.083		.08	5.05		5.13	7.90
1945	For each additional 1/4" depth, add		288	.028		.08	1.68		1.76	2.68
1950	5/8" diameter		88	.091		.12	5.50		5.62	8.65
1955	For each additional 1/4" depth, add		264	.030		.12	1.83		1.95	2.96
1960	3/4" diameter		80	.100		.16	6.05		6.21	9.55
1965	For each additional 1/4" depth, add		240	.033		.16	2.01		2.17	3.29
1970	7/8" diameter		72	.111		.19	6.70		6.89	10.55
1975	For each additional 1/4" depth, add		216	.037		.19	2.23		2.42	3.67
1980	1" diameter		64	.125		.30	7.55		7.85	12
1985	For each additional 1/4" depth, add		192	.042		.30	2.51		2.81	4.22
1990	For drilling up, add				↓		40%			

05 05 21.90 Welding Steel

		Crew	Daily Output	Labor-Hours	Unit	Material	2021 Bare Costs Labor	2021 Bare Costs Equipment	Total	Total Incl O&P
0010	**WELDING STEEL**, Structural	R050521-20								
0020	Field welding, 1/8" E6011, cost per welder, no operating engineer	E-14	8	1	Hr.	5.45	62.50	18.60	86.55	123
0200	With 1/2 operating engineer	E-13	8	1.500		5.45	90	18.60	114.05	165
0300	With 1 operating engineer	E-12	8	2	↓	5.45	118	18.60	142.05	206
0500	With no operating engineer, 2# weld rod per ton	E-14	8	1	Ton	5.45	62.50	18.60	86.55	123
0600	8# E6011 per ton	"	2	4		22	249	74.50	345.50	490
0800	With one operating engineer per welder, 2# E6011 per ton	E-12	8	2		5.45	118	18.60	142.05	206
0900	8# E6011 per ton	"	2	8	↓	22	470	74.50	566.50	820
1200	Continuous fillet, down welding									
1300	Single pass, 1/8" thick, 0.1#/L.F.	E-14	150	.053	L.F.	.27	3.32	.99	4.58	6.55
1400	3/16" thick, 0.2#/L.F.		75	.107		.54	6.65	1.98	9.17	13.10
1500	1/4" thick, 0.3#/L.F.		50	.160		.82	9.95	2.97	13.74	19.55
1610	5/16" thick, 0.4#/L.F.		38	.211		1.09	13.10	3.91	18.10	26
1800	3 passes, 3/8" thick, 0.5#/L.F.		30	.267		1.36	16.60	4.96	22.92	32.50
2010	4 passes, 1/2" thick, 0.7#/L.F.		22	.364		1.90	22.50	6.75	31.15	44.50
2200	5 to 6 passes, 3/4" thick, 1.3#/L.F.		12	.667		3.54	41.50	12.40	57.44	82
2400	8 to 11 passes, 1" thick, 2.4#/L.F.	↓	6	1.333		6.55	83	25	114.55	164
2600	For vertical joint welding, add						20%			
2700	Overhead joint welding, add						300%			
2900	For semi-automatic welding, obstructed joints, deduct						5%			
3000	Exposed joints, deduct				↓		15%			
4000	Cleaning and welding plates, bars, or rods									
4010	to existing beams, columns, or trusses	E-14	12	.667	L.F.	1.36	41.50	12.40	55.26	79.50

05 05 23.30 Lag Screws

			Crew	Daily Output	Labor-Hours	Unit	Material	Labor	Equipment	Total	Total Incl O&P
0010	**LAG SCREWS**										
0020	Steel, 1/4" diameter, 2" long	G	1 Carp	200	.040	Ea.	.13	2.19		2.32	3.41
0100	3/8" diameter, 3" long	G		150	.053		.44	2.92		3.36	4.83
0200	1/2" diameter, 3" long	G		130	.062		1.08	3.37		4.45	6.20
0300	5/8" diameter, 3" long	G	↓	120	.067	↓	1.72	3.65		5.37	7.35

05 05 23.35 Machine Screws

			Crew	Daily Output	Labor-Hours	Unit	Material	Labor	Equipment	Total	Total Incl O&P
0010	**MACHINE SCREWS**										
0020	Steel, round head, #8 x 1" long	G	1 Carp	4.80	1.667	C	3.98	91		94.98	140
0110	#8 x 2" long	G		2.40	3.333		14.70	182		196.70	288
0200	#10 x 1" long	G		4	2		6.25	109		115.25	170
0300	#10 x 2" long	G	↓	2	4	↓	10.90	219		229.90	335

05 05 23.50 Powder Actuated Tools and Fasteners

			Crew	Daily Output	Labor-Hours	Unit	Material	Labor	Equipment	Total	Total Incl O&P
0010	**POWDER ACTUATED TOOLS & FASTENERS**										
0020	Stud driver, .22 caliber, single shot					Ea.	146			146	160
0100	.27 caliber, semi automatic, strip					"	625			625	685
0300	Powder load, single shot, .22 cal, power level 2, brown					C	7.85			7.85	8.60
0400	Strip, .27 cal, power level 4, red						12			12	13.20
0600	Drive pin, .300 x 3/4" long	G	1 Carp	4.80	1.667		16.95	91		107.95	155
0700	.300 x 3" long with washer	G	"	4	2	↓	22	109		131	187

05 05 23.55 Rivets

			Crew	Daily Output	Labor-Hours	Unit	Material	Labor	Equipment	Total	Total Incl O&P
0010	**RIVETS**										
0100	Aluminum rivet & mandrel, 1/2" grip length x 1/8" diameter	G	1 Carp	4.80	1.667	C	2.17	91		93.17	138
0200	3/16" diameter	G		4	2		5.15	109		114.15	169
0300	Aluminum rivet, steel mandrel, 1/8" diameter	G		4.80	1.667		1.71	91		92.71	138
0400	3/16" diameter	G		4	2		18.45	109		127.45	184
0500	Copper rivet, steel mandrel, 1/8" diameter	G		4.80	1.667		9.55	91		100.55	147

05 05 Common Work Results for Metals

05 05 23 – Metal Fastenings

05 05 23.55 Rivets			Crew	Daily Output	Labor-Hours	Unit	Material	2021 Bare Costs Labor	Equipment	Total	Total Incl O&P
0800	Stainless rivet & mandrel, 1/8" diameter	G	1 Carp	4.80	1.667	C	21	91		112	159
0900	3/16" diameter	G		4	2		29.50	109		138.50	196
1000	Stainless rivet, steel mandrel, 1/8" diameter	G		4.80	1.667		8.05	91		99.05	145
1100	3/16" diameter	G		4	2		15.85	109		124.85	180
1200	Steel rivet and mandrel, 1/8" diameter	G		4.80	1.667		11.55	91		102.55	149
1300	3/16" diameter	G		4	2		11.45	109		120.45	176
1400	Hand riveting tool, standard					Ea.	71			71	78
1500	Deluxe						415			415	455
1600	Power riveting tool, standard						540			540	595
1700	Deluxe						1,700			1,700	1,875

05 05 23.70 Structural Blind Bolts			Crew	Daily Output	Labor-Hours	Unit	Material	2021 Bare Costs Labor	Equipment	Total	Total Incl O&P
0010	**STRUCTURAL BLIND BOLTS**										
0100	1/4" diameter x 1/4" grip	G	1 Sswk	240	.033	Ea.	1.27	2.01		3.28	4.51
0150	1/2" grip	G		216	.037		1.01	2.23		3.24	4.57
0200	3/8" diameter x 1/2" grip	G		232	.034		1.69	2.08		3.77	5.10
0250	3/4" grip	G		208	.038		1.96	2.32		4.28	5.75
0300	1/2" diameter x 1/2" grip	G		224	.036		3.82	2.15		5.97	7.55
0350	3/4" grip	G	1 Sswk	200	.040	Ea.	3.82	2.41		6.23	7.95
0400	5/8" diameter x 3/4" grip	G		216	.037		6	2.23		8.23	10.05
0450	1" grip	G		192	.042		6	2.51		8.51	10.50

05 05 23.90 Welding Rod			Crew	Daily Output	Labor-Hours	Unit	Material	2021 Bare Costs Labor	Equipment	Total	Total Incl O&P
0010	**WELDING ROD**										
0020	Steel, type 6011, 1/8" diam., less than 500#					Lb.	2.72			2.72	2.99
0100	500# to 2,000#						2.45			2.45	2.70
0200	2,000# to 5,000#						2.30			2.30	2.53
0300	5/32" diam., less than 500#						3.94			3.94	4.33
0310	500# to 2,000#						3.55			3.55	3.91
0320	2,000# to 5,000#						3.34			3.34	3.67
0400	3/16" diam., less than 500#						2.87			2.87	3.16
0500	500# to 2,000#						2.59			2.59	2.85
0600	2,000# to 5,000#						2.43			2.43	2.68
0620	Steel, type 6010, 1/8" diam., less than 500#						5.70			5.70	6.30
0630	500# to 2,000#						5.15			5.15	5.65
0640	2,000# to 5,000#						4.83			4.83	5.30
0650	Steel, type 7018 Low Hydrogen, 1/8" diam., less than 500#						2.99			2.99	3.28
0660	500# to 2,000#						2.69			2.69	2.96
0670	2,000# to 5,000#						2.53			2.53	2.78
0700	Steel, type 7024 Jet Weld, 1/8" diam., less than 500#						3.72			3.72	4.09
0710	500# to 2,000#						3.35			3.35	3.69
0720	2,000# to 5,000#						3.15			3.15	3.46
1550	Aluminum, type 4043 TIG, 1/8" diam., less than 10#						7			7	7.70
1560	10# to 60#						6.30			6.30	6.90
1570	Over 60#						5.90			5.90	6.50
1600	Aluminum, type 5356 TIG, 1/8" diam., less than 10#						6.55			6.55	7.20
1610	10# to 60#						5.90			5.90	6.50
1620	Over 60#						5.55			5.55	6.10
1900	Cast iron, type 8 Nickel, 1/8" diam., less than 500#						46.50			46.50	51.50
1910	500# to 1,000#						42			42	46
1920	Over 1,000#						39.50			39.50	43.50
2000	Stainless steel, type 316/316L, 1/8" diam., less than 500#						7.20			7.20	7.90
2100	500# to 1,000#						6.50			6.50	7.15
2220	Over 1,000#						6.10			6.10	6.70

05 12 Structural Steel Framing

05 12 23 – Structural Steel for Buildings

05 12 23.40 Lightweight Framing		Crew	Daily Output	Labor-Hours	Unit	Material	2021 Bare Costs Labor	Equipment	Total	Total Incl O&P
0010	**LIGHTWEIGHT FRAMING**									
0015	Made from recycled materials									
0400	Angle framing, field fabricated, 4" and larger	G	E-3	440	.055	Lb.	.77	3.33	.34	4.44	6.35
0450	Less than 4" angles	G	E-3	265	.091	Lb.	.79	5.50	.56	6.85	10.05
0600	Channel framing, field fabricated, 8" and larger	G		500	.048		.79	2.93	.30	4.02	5.75
0650	Less than 8" channels	G		335	.072		.79	4.37	.44	5.60	8.10
1000	Continuous slotted channel framing system, shop fab, simple framing	G	2 Sswk	2400	.007		4.09	.40		4.49	5.10
1200	Complex framing	G	"	1600	.010		4.62	.60		5.22	6.05
1250	Plate & bar stock for reinforcing beams and trusses	G					1.45			1.45	1.60
1300	Cross bracing, rods, shop fabricated, 3/4" diameter	G	E-3	700	.034		1.58	2.09	.21	3.88	5.20
1310	7/8" diameter	G		850	.028		1.58	1.72	.18	3.48	4.59
1320	1" diameter	G		1000	.024		1.58	1.46	.15	3.19	4.16
1330	Angle, 5" x 5" x 3/8"	G		2800	.009		1.58	.52	.05	2.15	2.61
1350	Hanging lintels, shop fabricated	G		850	.028		1.58	1.72	.18	3.48	4.59
1380	Roof frames, shop fabricated, 3'-0" square, 5' span	G	E-2	4200	.013		1.58	.80	.41	2.79	3.41
1400	Tie rod, not upset, 1-1/2" to 4" diameter, with turnbuckle	G	2 Sswk	800	.020		1.71	1.21		2.92	3.76
1420	No turnbuckle	G		700	.023		1.65	1.38		3.03	3.94
1500	Upset, 1-3/4" to 4" diameter, with turnbuckle	G		800	.020		1.71	1.21		2.92	3.76
1520	No turnbuckle	G		700	.023		1.65	1.38		3.03	3.94

05 12 23.60 Pipe Support Framing

		Crew	Daily Output	Labor-Hours	Unit	Material	Labor	Equipment	Total	Total Incl O&P
0010	**PIPE SUPPORT FRAMING**									
0020	Under 10#/L.F., shop fabricated	G	E-4	3900	.008	Lb.	1.77	.50	.04	2.31	2.75
0200	10.1 to 15#/L.F.	G		4300	.007		1.74	.45	.03	2.22	2.66
0400	15.1 to 20#/L.F.	G		4800	.007		1.71	.41	.03	2.15	2.55
0600	Over 20#/L.F.	G		5400	.006		1.69	.36	.03	2.08	2.45

05 54 Metal Floor Plates

05 54 13 – Floor Plates

05 54 13.20 Checkered Plates

		Crew	Daily Output	Labor-Hours	Unit	Material	Labor	Equipment	Total	Total Incl O&P
0010	**CHECKERED PLATES**, steel, field fabricated									
0015	Made from recycled materials									
0020	1/4" & 3/8", 2000 to 5000 S.F., bolted	G	E-4	2900	.011	Lb.	1.05	.67	.05	1.77	2.26
0100	Welded	G		4400	.007	"	1.02	.44	.03	1.49	1.84
0300	Pit or trench cover and frame, 1/4" plate, 2' to 3' wide	G		100	.320	S.F.	12.65	19.45	1.49	33.59	45.50
0400	For galvanizing, add	G				Lb.	.30			.30	.33
0500	Platforms, 1/4" plate, no handrails included, rectangular	G	E-4	4200	.008		3.49	.46	.04	3.99	4.60
0600	Circular	G	"	2500	.013		4.37	.78	.06	5.21	6.10

Estimating Tips
06 05 00 Common Work Results for Wood, Plastics, and Composites

- Common to any wood-framed structure are the accessory connector items such as screws, nails, adhesives, hangers, connector plates, straps, angles, and hold-downs. For typical wood-framed buildings, such as residential projects, the aggregate total for these items can be significant, especially in areas where seismic loading is a concern. For floor and wall framing, the material cost is based on 10 to 25 lbs. of accessory connectors per MBF. Hold-downs, hangers, and other connectors should be taken off by the piece.

  Included with material costs are fasteners for a normal installation. Gordian's RSMeans engineers use manufacturers' recommendations, written specifications, and/or standard construction practice for the sizing and spacing of fasteners. Prices for various fasteners are shown for informational purposes only. Adjustments should be made if unusual fastening conditions exist.

06 10 00 Carpentry

- Lumber is a traded commodity and therefore sensitive to supply and demand in the marketplace. Even with "budgetary" estimating of wood-framed projects, it is advisable to call local suppliers for the latest market pricing.

- The common quantity unit for wood-framed projects is "thousand board feet" (MBF). A board foot is a volume of wood—1" x 1' x 1' or 144 cubic inches. Board-foot quantities are generally calculated using nominal material dimensions—dressed sizes are ignored. Board foot per lineal foot of any stick of lumber can be calculated by dividing the nominal cross-sectional area by 12. As an example, 2,000 lineal feet of 2 x 12 equates to 4 MBF by dividing the nominal area, 2 x 12, by 12, which equals 2, and multiplying that by 2,000 to give 4,000 board feet. This simple rule applies to all nominal dimensioned lumber.

- Waste is an issue of concern at the quantity takeoff for any area of construction. Framing lumber is sold in even foot lengths, i.e., 8', 10', 12', 14', 16', and depending on spans, wall heights, and the grade of lumber, waste is inevitable. A rule of thumb for lumber waste is 5–10% depending on material quality and the complexity of the framing.

- Wood in various forms and shapes is used in many projects, even where the main structural framing is steel, concrete, or masonry. Plywood as a back-up partition material and 2x boards used as blocking and cant strips around roof edges are two common examples. The estimator should ensure that the costs of all wood materials are included in the final estimate.

06 20 00 Finish Carpentry

- It is necessary to consider the grade of workmanship when estimating labor costs for erecting millwork and an interior finish. In practice, there are three grades: premium, custom, and economy. The RSMeans daily output for base and case moldings is in the range of 200 to 250 L.F. per carpenter per day. This is appropriate for most average custom-grade projects. For premium projects, an adjustment to productivity of 25–50% should be made, depending on the complexity of the job.

Reference Numbers

Reference numbers are shown at the beginning of some major classifications. These numbers refer to related items in the Reference Section. The reference information may be an estimating procedure, an alternate pricing method, or technical information.

Note: Not all subdivisions listed here necessarily appear. ∎

---

06 11 10 – Framing with Dimensional, Engineered or Composite Lumber

06 11 10.24 Miscellaneous Framing		Crew	Daily Output	Labor-Hours	Unit	Material	2021 Bare Costs Labor	Equipment	Total	Total Incl O&P
0010	MISCELLANEOUS FRAMING									
8500	Firestops, 2" x 4"	2 Carp	.51	31.373	M.B.F.	1,050	1,725		2,775	3,700
8505	Pneumatic nailed		.62	25.806		1,050	1,400		2,450	3,275
8520	2" x 6"		.60	26.667		1,075	1,450		2,525	3,350
8525	Pneumatic nailed		.73	21.858		1,075	1,200		2,275	2,950
8540	2" x 8"	2 Carp	.60	26.667	M.B.F.	1,200	1,450		2,650	3,500
8560	2" x 12"		.70	22.857		1,250	1,250		2,500	3,250
8600	Nailers, treated, wood construction, 2" x 4"		.53	30.189		1,350	1,650		3,000	3,950
8605	Pneumatic nailed		.64	25.157		1,350	1,375		2,725	3,550
8620	2" x 6"		.75	21.333		1,250	1,175		2,425	3,150
8625	Pneumatic nailed		.90	17.778		1,275	970		2,245	2,850

06 16 36 – Wood Panel Product Sheathing

06 16 36.10 Sheathing		Crew	Daily Output	Labor-Hours	Unit	Material	2021 Bare Costs Labor	Equipment	Total	Total Incl O&P
0010	SHEATHING									
0012	Plywood on roofs, CDX									
0050	3/8" thick	2 Carp	1525	.010	S.F.	.78	.57		1.35	1.72
0055	Pneumatic nailed		1860	.009		.78	.47		1.25	1.56
0200	5/8" thick		1300	.012		1.16	.67		1.83	2.27
0205	Pneumatic nailed		1586	.010		1.16	.55		1.71	2.09
0300	3/4" thick		1200	.013		1.38	.73		2.11	2.61
0305	Pneumatic nailed		1464	.011		1.38	.60		1.98	2.41
0500	Plywood on walls, with exterior CDX, 3/8" thick		1200	.013		.78	.73		1.51	1.95
0505	Pneumatic nailed		1488	.011		.78	.59		1.37	1.74
0700	5/8" thick		1050	.015		1.16	.83		1.99	2.51
0705	Pneumatic nailed		1302	.012		1.16	.67		1.83	2.27
0800	3/4" thick		975	.016		1.38	.90		2.28	2.86
0805	Pneumatic nailed		1209	.013		1.38	.72		2.10	2.60

Estimating Tips
07 10 00 Dampproofing and Waterproofing

- Be sure of the job specifications before pricing this subdivision. The difference in cost between waterproofing and dampproofing can be great. Waterproofing will hold back standing water. Dampproofing prevents the transmission of water vapor. Also included in this section are vapor retarding membranes.

07 20 00 Thermal Protection

- Insulation and fireproofing products are measured by area, thickness, volume, or R-value. Specifications may give only what the specific R-value should be in a certain situation. The estimator may need to choose the type of insulation to meet that R-value.

07 30 00 Steep Slope Roofing
07 40 00 Roofing and Siding Panels

- Many roofing and siding products are bought and sold by the square. One square is equal to an area that measures 100 square feet.

  This simple change in unit of measure could create a large error if the estimator is not observant. Accessories necessary for a complete installation must be figured into any calculations for both material and labor.

07 50 00 Membrane Roofing
07 60 00 Flashing and Sheet Metal
07 70 00 Roofing and Wall Specialties and Accessories

- The items in these subdivisions compose a roofing system. No one component completes the installation, and all must be estimated. Built-up or single-ply membrane roofing systems are made up of many products and installation trades. Wood blocking at roof perimeters or penetrations, parapet coverings, reglets, roof drains, gutters, downspouts, sheet metal flashing, skylights, smoke vents, and roof hatches all need to be considered along with the roofing material. Several different installation trades will need to work together on the roofing system. Inherent difficulties in the scheduling and coordination of various trades must be accounted for when estimating labor costs.

07 90 00 Joint Protection

- To complete the weather-tight shell, the sealants and caulkings must be estimated. Where different materials meet—at expansion joints, at flashing penetrations, and at hundreds of other locations throughout a construction project—caulking and sealants provide another line of defense against water penetration. Often, an entire system is based on the proper location and placement of caulking or sealants. The detailed drawings that are included as part of a set of architectural plans show typical locations for these materials. When caulking or sealants are shown at typical locations, this means the estimator must include them for all the locations where this detail is applicable. Be careful to keep different types of sealants separate, and remember to consider backer rods and primers if necessary.

Reference Numbers

Reference numbers are shown at the beginning of some major classifications. These numbers refer to related items in the Reference Section. The reference information may be an estimating procedure, an alternate pricing method, or technical information.

Note: Not all subdivisions listed here necessarily appear. ■

---

07 65 10 – Sheet Metal Flashing

07 65 10.10 Sheet Metal Flashing and Counter Flashing	Crew	Daily Output	Labor-Hours	Unit	Material	2021 Bare Costs Labor	Equipment	Total	Total Incl O&P
0010 **SHEET METAL FLASHING AND COUNTER FLASHING**									
0011 Including up to 4 bends									
0020 Aluminum, mill finish, .013" thick	1 Rofc	145	.055	S.F.	1.30	2.66		3.96	5.75
0030 .016" thick		145	.055		1.16	2.66		3.82	5.60
0060 .019" thick		145	.055		1.56	2.66		4.22	6.05
0100 .032" thick		145	.055		1.53	2.66		4.19	6
0200 .040" thick		145	.055		2.38	2.66		5.04	6.95
0300 .050" thick		145	.055		3.25	2.66		5.91	7.90
0325 Mill finish 5" x 7" step flashing, .016" thick		1920	.004	Ea.	.17	.20		.37	.52
0350 Mill finish 12" x 12" step flashing, .016" thick		1600	.005	"	.61	.24		.85	1.06
0400 Painted finish, add				S.F.	.31			.31	.34
1000 Mastic-coated 2 sides, .005" thick	1 Rofc	330	.024		1.87	1.17		3.04	3.95
1100 .016" thick		330	.024		2.06	1.17		3.23	4.16
1600 Copper, 16 oz. sheets, under 1000 lb.		115	.070		7.90	3.35		11.25	14.15
1700 Over 4000 lb.		155	.052		7.90	2.49		10.39	12.75
1900 20 oz. sheets, under 1000 lb.		110	.073		11.20	3.51		14.71	18.05
2000 Over 4000 lb.	1 Rofc	145	.055	S.F.	10.65	2.66		13.31	16
2200 24 oz. sheets, under 1000 lb.		105	.076		15.60	3.67		19.27	23
2300 Over 4000 lb.		135	.059		14.80	2.86		17.66	21
2500 32 oz. sheets, under 1000 lb.		100	.080		22.50	3.86		26.36	31.50
2600 Over 4000 lb.		130	.062		21.50	2.97		24.47	29
5800 Lead, 2.5 lb./S.F., up to 12" wide		135	.059		6.30	2.86		9.16	11.60
5900 Over 12" wide		135	.059		4.34	2.86		7.20	9.40
8900 Stainless steel sheets, 32 ga.		155	.052		3.55	2.49		6.04	7.95
9000 28 ga.		155	.052		5.10	2.49		7.59	9.65
9100 26 ga.		155	.052		5.20	2.49		7.69	9.80
9200 24 ga.		155	.052		5.95	2.49		8.44	10.60
9400 Terne coated stainless steel, .015" thick, 28 ga.		155	.052		8.15	2.49		10.64	13
9500 .018" thick, 26 ga.		155	.052		9.10	2.49		11.59	14.05
9600 Zinc and copper alloy (brass), .020" thick		155	.052		9.05	2.49		11.54	14
9700 .027" thick		155	.052		10.65	2.49		13.14	15.75
9800 .032" thick		155	.052		13.30	2.49		15.79	18.70
9900 .040" thick		155	.052		19.45	2.49		21.94	25.50

07 65 13 – Laminated Sheet Flashing

07 65 13.10 Laminated Sheet Flashing

	Crew	Daily Output	Labor-Hours	Unit	Material	2021 Bare Costs Labor	Equipment	Total	Total Incl O&P
0010 **LAMINATED SHEET FLASHING**, Including up to 4 bends									
0500 Aluminum, fabric-backed 2 sides, mill finish, .004" thick	1 Rofc	330	.024	S.F.	1.58	1.17		2.75	3.63
0700 .005" thick		330	.024		1.86	1.17		3.03	3.94
0750 Mastic-backed, self adhesive		460	.017		3.41	.84		4.25	5.10
0800 Mastic-coated 2 sides, .004" thick		330	.024		1.58	1.17		2.75	3.63
2800 Copper, paperbacked 1 side, 2 oz.		330	.024		2.15	1.17		3.32	4.26
2900 3 oz.		330	.024		3.14	1.17		4.31	5.35
3100 Paperbacked 2 sides, 2 oz.		330	.024		2.38	1.17		3.55	4.51
3150 3 oz.		330	.024		2.23	1.17		3.40	4.34
3200 5 oz.		330	.024		3.88	1.17		5.05	6.15
3250 7 oz.		330	.024		6.10	1.17		7.27	8.65
3400 Mastic-backed 2 sides, copper, 2 oz.		330	.024		2.27	1.17		3.44	4.39
3500 3 oz.		330	.024		2.57	1.17		3.74	4.72
3700 5 oz.		330	.024		3.91	1.17		5.08	6.20
3800 Fabric-backed 2 sides, copper, 2 oz.		330	.024		2.04	1.17		3.21	4.13
4000 3 oz.		330	.024		2.88	1.17		4.05	5.05
4100 5 oz.		330	.024		4.04	1.17		5.21	6.35

For customer support on your Plumbing Costs with RSMeans Data, call 800.448.8182.

07 65 Flexible Flashing

07 65 13 – Laminated Sheet Flashing

07 65 13.10 Laminated Sheet Flashing

	07 65 13.10 Laminated Sheet Flashing	Crew	Daily Output	Labor-Hours	Unit	Material	2021 Bare Costs Labor	Equipment	Total	Total Incl O&P
4300	Copper-clad stainless steel, .015" thick, under 500 lb.	1 Rofc	115	.070	S.F.	6.90	3.35		10.25	13.05
4400	Over 2000 lb.		155	.052		6.80	2.49		9.29	11.50
4600	.018" thick, under 500 lb.		100	.080		7.90	3.86		11.76	14.95
4700	Over 2000 lb.	↓	145	.055	↓	8.15	2.66		10.81	13.25
8550	Shower pan, 3 ply copper and fabric, 3 oz.	1 Rofc	155	.052	S.F.	4.12	2.49		6.61	8.55
8600	7 oz.		155	.052		4.98	2.49		7.47	9.55
9300	Stainless steel, paperbacked 2 sides, .005" thick	↓	330	.024	↓	4.05	1.17		5.22	6.35

07 65 19 – Plastic Sheet Flashing

07 65 19.10 Plastic Sheet Flashing and Counter Flashing

		Crew	Daily Output	Labor-Hours	Unit	Material	2021 Bare Costs Labor	Equipment	Total	Total Incl O&P
0010	**PLASTIC SHEET FLASHING AND COUNTER FLASHING**									
7300	Polyvinyl chloride, black, 10 mil	1 Rofc	285	.028	S.F.	.27	1.35		1.62	2.49
7400	20 mil		285	.028		.26	1.35		1.61	2.48
7600	30 mil		285	.028		.33	1.35		1.68	2.55
7700	60 mil		285	.028		.88	1.35		2.23	3.16
7900	Black or white for exposed roofs, 60 mil	↓	285	.028	↓	4.78	1.35		6.13	7.45
8060	PVC tape, 5" x 45 mils, for joint covers, 100 L.F./roll				Ea.	185			185	203
8850	Polyvinyl chloride, 30 mil	1 Rofc	160	.050	S.F.	1.90	2.41		4.31	6

07 71 Roof Specialties

07 71 16 – Manufactured Counterflashing Systems

07 71 16.20 Pitch Pockets, Variable Sizes

		Crew	Daily Output	Labor-Hours	Unit	Material	2021 Bare Costs Labor	Equipment	Total	Total Incl O&P
0010	**PITCH POCKETS, VARIABLE SIZES**									
0100	Adjustable, 4" to 7", welded corners, 4" deep	1 Rofc	48	.167	Ea.	23	8.05		31.05	38.50
0200	Side extenders, 6"	"	240	.033	"	3.16	1.61		4.77	6.10

07 72 Roof Accessories

07 72 53 – Snow Guards

07 72 53.10 Snow Guard Options

		Crew	Daily Output	Labor-Hours	Unit	Material	2021 Bare Costs Labor	Equipment	Total	Total Incl O&P
0010	**SNOW GUARD OPTIONS**									
0100	Slate & asphalt shingle roofs, fastened with nails	1 Rofc	160	.050	Ea.	12.45	2.41		14.86	17.60
0200	Standing seam metal roofs, fastened with set screws		48	.167		17.80	8.05		25.85	32.50
0300	Surface mount for metal roofs, fastened with solder		48	.167	↓	7.75	8.05		15.80	21.50
0400	Double rail pipe type, including pipe	↓	130	.062	L.F.	36	2.97		38.97	44.50

07 76 Roof Pavers

07 76 16 – Roof Decking Pavers

07 76 16.10 Roof Pavers and Supports

		Crew	Daily Output	Labor-Hours	Unit	Material	2021 Bare Costs Labor	Equipment	Total	Total Incl O&P
0010	**ROOF PAVERS AND SUPPORTS**									
1000	Roof decking pavers, concrete blocks, 2" thick, natural	1 Clab	115	.070	S.F.	3.57	3.09		6.66	8.55
1100	Colors		115	.070	"	3.56	3.09		6.65	8.55
1200	Support pedestal, bottom cap		960	.008	Ea.	2.65	.37		3.02	3.47
1300	Top cap		960	.008		4.86	.37		5.23	5.90
1400	Leveling shims, 1/16"		1920	.004		1.21	.19		1.40	1.61
1500	1/8"		1920	.004		1.99	.19		2.18	2.47
1600	Buffer pad		960	.008	↓	4.50	.37		4.87	5.50
1700	PVC legs (4" SDR 35)		2880	.003	Inch	.18	.12		.30	.38
2000	Alternate pricing method, system in place		101	.079	S.F.	8	3.52		11.52	14

07 84 13 – Penetration Firestopping

07 84 13.10 Firestopping		Crew	Daily Output	Labor-Hours	Unit	Material	2021 Bare Costs Labor	Equipment	Total	Total Incl O&P
0010	**FIRESTOPPING**									
0100	Metallic piping, non insulated									
0110	Through walls, 2" diameter	1 Carp	16	.500	Ea.	3.70	27.50		31.20	45
0120	4" diameter		14	.571		8.30	31.50		39.80	55.50
0130	6" diameter		12	.667		12.70	36.50		49.20	68.50
0140	12" diameter		10	.800		25	44		69	93
0150	Through floors, 2" diameter		32	.250		1.88	13.70		15.58	22.50
0160	4" diameter		28	.286		3.31	15.65		18.96	27
0170	6" diameter		24	.333		6.50	18.25		24.75	34
0180	12" diameter		20	.400		12.90	22		34.90	46.50
0190	Metallic piping, insulated									
0200	Through walls, 2" diameter	1 Carp	16	.500	Ea.	8.65	27.50		36.15	50.50
0210	4" diameter		14	.571		12.65	31.50		44.15	60.50
0220	6" diameter		12	.667		16.90	36.50		53.40	73
0230	12" diameter		10	.800		28.50	44		72.50	97
0240	Through floors, 2" diameter		32	.250		3.30	13.70		17	24
0250	4" diameter		28	.286		6.35	15.65		22	30.50
0260	6" diameter		24	.333		8.75	18.25		27	36.50
0270	12" diameter		20	.400		13.05	22		35.05	47
0280	Non metallic piping, non insulated									
0290	Through walls, 2" diameter	1 Carp	12	.667	Ea.	54.50	36.50		91	115
0300	4" diameter		10	.800		109	44		153	186
0310	6" diameter		8	1		156	54.50		210.50	254
0330	Through floors, 2" diameter		16	.500		28	27.50		55.50	72
0340	4" diameter		6	1.333		64	73		137	180
0350	6" diameter		6	1.333		120	73		193	241
0370	Ductwork, insulated & non insulated, round									
0380	Through walls, 6" diameter	1 Carp	12	.667	Ea.	16.65	36.50		53.15	73
0390	12" diameter		10	.800		29	44		73	97
0400	18" diameter		8	1		34	54.50		88.50	119
0410	Through floors, 6" diameter		16	.500		8.55	27.50		36.05	50.50
0420	12" diameter		14	.571		14.70	31.50		46.20	62.50
0430	18" diameter		12	.667		17.75	36.50		54.25	74
0440	Ductwork, insulated & non insulated, rectangular									
0450	With stiffener/closure angle, through walls, 6" x 12"	1 Carp	8	1	Ea.	22	54.50		76.50	106
0460	12" x 24"		6	1.333		37	73		110	150
0470	24" x 48"		4	2		93	109		202	265
0480	With stiffener/closure angle, through floors, 6" x 12"		10	.800		10.50	44		54.50	77
0490	12" x 24"		8	1		15.85	54.50		70.35	99
0500	24" x 48"		6	1.333		44	73		117	157
0510	Multi trade openings									
0520	Through walls, 6" x 12"	1 Carp	2	4	Ea.	56	219		275	385
0530	12" x 24"	"	1	8		210	440		650	885
0540	24" x 48"	2 Carp	1	16		675	875		1,550	2,050
0550	48" x 96"	"	.75	21.333		2,550	1,175		3,725	4,550
0560	Through floors, 6" x 12"	1 Carp	2	4		45	219		264	375
0570	12" x 24"	"	1	8		64	440		504	725
0580	24" x 48"	2 Carp	.75	21.333		138	1,175		1,313	1,900
0590	48" x 96"	"	.50	32		325	1,750		2,075	2,975
0600	Structural penetrations, through walls									
0610	Steel beams, W8 x 10	1 Carp	8	1	Ea.	74	54.50		128.50	163
0620	W12 x 14		6	1.333		108	73		181	228
0630	W21 x 44		5	1.600		165	87.50		252.50	315

07 84 Firestopping

07 84 13 – Penetration Firestopping

07 84 13.10 Firestopping

		Crew	Daily Output	Labor-Hours	Unit	Material	2021 Bare Costs Labor	2021 Bare Costs Equipment	Total	Total Incl O&P
0640	W36 x 135	1 Carp	3	2.667	Ea.	234	146		380	475
0650	Bar joists, 18" deep		6	1.333		43	73		116	157
0660	24" deep		6	1.333		61	73		134	176
0670	36" deep		5	1.600		91.50	87.50		179	232
0680	48" deep	↓	4	2	↓	132	109		241	310
0690	Construction joints, floor slab at exterior wall									
0700	Precast, brick, block or drywall exterior									
0710	2" wide joint	1 Carp	125	.064	L.F.	11.05	3.50		14.55	17.40
0720	4" wide joint	"	75	.107	"	17.05	5.85		22.90	27.50
0730	Metal panel, glass or curtain wall exterior									
0740	2" wide joint	1 Carp	40	.200	L.F.	12.55	10.95		23.50	30
0750	4" wide joint	"	25	.320	"	14.25	17.50		31.75	41.50
0760	Floor slab to drywall partition									
0770	Flat joint	1 Carp	100	.080	L.F.	14.45	4.38		18.83	22.50
0780	Fluted joint		50	.160		16.80	8.75		25.55	31.50
0790	Etched fluted joint	↓	75	.107	↓	20	5.85		25.85	30.50
0800	Floor slab to concrete/masonry partition									
0810	Flat joint	1 Carp	75	.107	L.F.	16.85	5.85		22.70	27
0820	Fluted joint	"	50	.160	"	18.35	8.75		27.10	33
0830	Concrete/CMU wall joints									
0840	1" wide	1 Carp	100	.080	L.F.	20.50	4.38		24.88	29
0850	2" wide		75	.107		32	5.85		37.85	43.50
0860	4" wide	↓	50	.160		36	8.75		44.75	52.50
0870	Concrete/CMU floor joints									
0880	1" wide	1 Carp	200	.040	L.F.	21.50	2.19		23.69	27
0890	2" wide		150	.053		24.50	2.92		27.42	31.50
0900	4" wide	↓	100	.080	↓	36	4.38		40.38	46

07 91 Preformed Joint Seals

07 91 13 – Compression Seals

07 91 13.10 Compression Seals

		Crew	Daily Output	Labor-Hours	Unit	Material	2021 Bare Costs Labor	2021 Bare Costs Equipment	Total	Total Incl O&P
0010	**COMPRESSION SEALS**									
4900	O-ring type cord, 1/4"	1 Bric	472	.017	L.F.	.42	.91		1.33	1.83
4910	1/2"		440	.018		1.11	.98		2.09	2.69
4920	3/4"		424	.019		2.23	1.01		3.24	3.98
4930	1"		408	.020		4.48	1.05		5.53	6.50
4940	1-1/4"		384	.021		10.80	1.12		11.92	13.55
4950	1-1/2"		368	.022		13.25	1.17		14.42	16.30
4960	1-3/4"		352	.023		15.45	1.22		16.67	18.85
4970	2"	↓	344	.023	↓	27	1.25		28.25	31.50

07 91 16 – Joint Gaskets

07 91 16.10 Joint Gaskets

		Crew	Daily Output	Labor-Hours	Unit	Material	2021 Bare Costs Labor	2021 Bare Costs Equipment	Total	Total Incl O&P
0010	**JOINT GASKETS**									
4400	Joint gaskets, neoprene, closed cell w/adh, 1/8" x 3/8"	1 Bric	240	.033	L.F.	.35	1.79		2.14	3.09
4500	1/4" x 3/4"		215	.037		.74	2		2.74	3.82
4700	1/2" x 1"		200	.040		1.79	2.15		3.94	5.20
4800	3/4" x 1-1/2"	↓	165	.048	↓	1.88	2.60		4.48	6

For customer support on your Plumbing Costs with RSMeans Data, call 800.448.8182.

55

07 91 Preformed Joint Seals

07 91 23 – Backer Rods

07 91 23.10 Backer Rods		Crew	Daily Output	Labor-Hours	Unit	Material	2021 Bare Costs Labor	Equipment	Total	Total Incl O&P
0010	**BACKER RODS**									
0030	Backer rod, polyethylene, 1/4" diameter	1 Bric	4.60	1.739	C.L.F.	2.77	93.50		96.27	144
0050	1/2" diameter		4.60	1.739		19.90	93.50		113.40	163
0070	3/4" diameter		4.60	1.739		7.70	93.50		101.20	149
0090	1" diameter		4.60	1.739		15.45	93.50		108.95	158

07 91 26 – Joint Fillers

07 91 26.10 Joint Fillers		Crew	Daily Output	Labor-Hours	Unit	Material	2021 Bare Costs Labor	Equipment	Total	Total Incl O&P
0010	**JOINT FILLERS**									
4360	Butyl rubber filler, 1/4" x 1/4"	1 Bric	290	.028	L.F.	.23	1.48		1.71	2.48
4365	1/2" x 1/2"		250	.032		.92	1.72		2.64	3.60
4370	1/2" x 3/4"		210	.038		1.38	2.05		3.43	4.60
4375	3/4" x 3/4"		230	.035		2.07	1.87		3.94	5.10
4380	1" x 1"		180	.044		2.76	2.39		5.15	6.65
4390	For coloring, add					12%				
4980	Polyethylene joint backing, 1/4" x 2"	1 Bric	2.08	3.846	C.L.F.	14.85	207		221.85	325
4990	1/4" x 6"		1.28	6.250	"	33.50	335		368.50	540
5600	Silicone, room temp vulcanizing foam seal, 1/4" x 1/2"		1312	.006	L.F.	.45	.33		.78	.99
5610	1/2" x 1/2"		656	.012		.91	.65		1.56	1.99
5620	1/2" x 3/4"		442	.018		1.36	.97		2.33	2.95
5630	3/4" x 3/4"		328	.024		2.04	1.31		3.35	4.21
5640	1/8" x 1"		1312	.006		.45	.33		.78	.99
5650	1/8" x 3"		442	.018		1.36	.97		2.33	2.95
5670	1/4" x 3"		295	.027		2.72	1.46		4.18	5.20
5680	1/4" x 6"		148	.054		5.45	2.90		8.35	10.35
5690	1/2" x 6"		82	.098		10.85	5.25		16.10	19.85
5700	1/2" x 9"		52.50	.152		16.30	8.20		24.50	30.50
5710	1/2" x 12"		33	.242		21.50	13		34.50	43.50

07 92 Joint Sealants

07 92 13 – Elastomeric Joint Sealants

07 92 13.20 Caulking and Sealant Options		Crew	Daily Output	Labor-Hours	Unit	Material	2021 Bare Costs Labor	Equipment	Total	Total Incl O&P
0010	**CAULKING AND SEALANT OPTIONS**									
0050	Latex acrylic based, bulk				Gal.	30.50			30.50	33.50
0055	Bulk in place 1/4" x 1/4" bead	1 Bric	300	.027	L.F.	.10	1.43		1.53	2.27
0060	1/4" x 3/8"		294	.027		.16	1.46		1.62	2.38
0065	1/4" x 1/2"		288	.028		.21	1.49		1.70	2.48
0075	3/8" x 3/8"		284	.028		.24	1.51		1.75	2.54
0080	3/8" x 1/2"		280	.029		.32	1.53		1.85	2.66
0085	3/8" x 5/8"		276	.029		.40	1.56		1.96	2.78
0095	3/8" x 3/4"		272	.029		.48	1.58		2.06	2.90
0100	1/2" x 1/2"		275	.029		.42	1.56		1.98	2.82
0105	1/2" x 5/8"		269	.030		.53	1.60		2.13	2.99
0110	1/2" x 3/4"		263	.030		.64	1.63		2.27	3.16
0115	1/2" x 7/8"		256	.031		.74	1.68		2.42	3.35
0120	1/2" x 1"		250	.032		.85	1.72		2.57	3.52
0125	3/4" x 3/4"		244	.033		.96	1.76		2.72	3.70
0130	3/4" x 1"		225	.036		1.28	1.91		3.19	4.28
0135	1" x 1"		200	.040		1.70	2.15		3.85	5.10
0190	Cartridges				Gal.	41.50			41.50	45.50
0200	11 fl. oz. cartridge				Ea.	3.57			3.57	3.93

07 92 Joint Sealants

07 92 13 – Elastomeric Joint Sealants

07 92 13.20 Caulking and Sealant Options

		Crew	Daily Output	Labor-Hours	Unit	Material	2021 Bare Costs Labor	Equipment	Total	Total Incl O&P
0500	1/4" x 1/2"	1 Bric	288	.028	L.F.	.29	1.49		1.78	2.57
0600	1/2" x 1/2"		275	.029		.58	1.56		2.14	2.99
0800	3/4" x 3/4"		244	.033		1.31	1.76		3.07	4.09
0900	3/4" x 1"		225	.036		1.75	1.91		3.66	4.81
1000	1" x 1"	↓	200	.040	↓	2.19	2.15		4.34	5.65
1400	Butyl based, bulk				Gal.	41			41	45.50
1500	Cartridges				"	41			41	45.50
1700	1/4" x 1/2", 154 L.F./gal.	1 Bric	288	.028	L.F.	.27	1.49		1.76	2.54
1800	1/2" x 1/2", 77 L.F./gal.	"	275	.029	"	.53	1.56		2.09	2.94
2300	Polysulfide compounds, 1 component, bulk				Gal.	84			84	92
2600	1 or 2 component, in place, 1/4" x 1/4", 308 L.F./gal.	1 Bric	300	.027	L.F.	.27	1.43		1.70	2.46
2700	1/2" x 1/4", 154 L.F./gal.		288	.028		.54	1.49		2.03	2.85
2900	3/4" x 3/8", 68 L.F./gal.		272	.029		1.23	1.58		2.81	3.74
3000	1" x 1/2", 38 L.F./gal.	↓	250	.032		2.21	1.72		3.93	5
3200	Polyurethane, 1 or 2 component				Gal.	54.50			54.50	60
3300	Cartridges				"	65			65	71.50
3500	Bulk, in place, 1/4" x 1/4"	1 Bric	300	.027	L.F.	.18	1.43		1.61	2.36
3655	1/2" x 1/4"		288	.028		.35	1.49		1.84	2.64
3800	3/4" x 3/8"		272	.029		.80	1.58		2.38	3.26
3900	1" x 1/2"	↓	250	.032	↓	1.42	1.72		3.14	4.15
4100	Silicone rubber, bulk				Gal.	65.50			65.50	72.50
4200	Cartridges				"	44.50			44.50	49

07 92 16 – Rigid Joint Sealants

07 92 16.10 Rigid Joint Sealants

		Crew	Daily Output	Labor-Hours	Unit	Material	2021 Bare Costs Labor	Equipment	Total	Total Incl O&P
0010	**RIGID JOINT SEALANTS**									
5800	Tapes, sealant, PVC foam adhesive, 1/16" x 1/4"				C.L.F.	8.95			8.95	9.85
5900	1/16" x 1/2"					9.05			9.05	9.95
5950	1/16" x 1"					15.05			15.05	16.55
6000	1/8" x 1/2"				↓	9.20			9.20	10.10

07 92 19 – Acoustical Joint Sealants

07 92 19.10 Acoustical Sealant

		Crew	Daily Output	Labor-Hours	Unit	Material	2021 Bare Costs Labor	Equipment	Total	Total Incl O&P
0010	**ACOUSTICAL SEALANT**									
0020	Acoustical sealant, elastomeric, cartridges				Ea.	8.05			8.05	8.85
0025	In place, 1/4" x 1/4"	1 Bric	300	.027	L.F.	.33	1.43		1.76	2.52
0030	1/4" x 1/2"		288	.028		.66	1.49		2.15	2.97
0035	1/2" x 1/2"		275	.029		1.31	1.56		2.87	3.80
0040	1/2" x 3/4"		263	.030		1.97	1.63		3.60	4.63
0045	3/4" x 3/4"		244	.033		2.96	1.76		4.72	5.90
0050	1" x 1"	↓	200	.040	↓	5.25	2.15		7.40	9.05

For customer support on your Plumbing Costs with RSMeans Data, call 800.448.8182.

57

Division Notes

	CREW	DAILY OUTPUT	LABOR-HOURS	UNIT	BARE COSTS				TOTAL INCL O&P
					MAT.	LABOR	EQUIP.	TOTAL	

Estimating Tips

General

- Room Finish Schedule: A complete set of plans should contain a room finish schedule. If one is not available, it would be well worth the time and effort to obtain one.

09 20 00 Plaster and Gypsum Board

- Lath is estimated by the square yard plus a 5% allowance for waste. Furring, channels, and accessories are measured by the linear foot. An extra foot should be allowed for each accessory miter or stop.

- Plaster is also estimated by the square yard. Deductions for openings vary by preference, from zero deduction to 50% of all openings over 2 feet in width. The estimator should allow one extra square foot for each linear foot of horizontal interior or exterior angle located below the ceiling level. Also, double the areas of small radius work.

- Drywall accessories, studs, track, and acoustical caulking are all measured by the linear foot. Drywall taping is figured by the square foot. Gypsum wallboard is estimated by the square foot. No material deductions should be made for door or window openings under 32 S.F.

09 60 00 Flooring

- Tile and terrazzo areas are taken off on a square foot basis. Trim and base materials are measured by the linear foot. Accent tiles are listed per each. Two basic methods of installation are used. Mud set is approximately 30% more expensive than thin set.

The cost of grout is included with tile unit price lines unless otherwise noted. In terrazzo work, be sure to include the linear footage of embedded decorative strips, grounds, machine rubbing, and power cleanup.

- Wood flooring is available in strip, parquet, or block configuration. The latter two types are set in adhesives with quantities estimated by the square foot. The laying pattern will influence labor costs and material waste. In addition to the material and labor for laying wood floors, the estimator must make allowances for sanding and finishing these areas, unless the flooring is prefinished.

- Sheet flooring is measured by the square yard. Roll widths vary, so consideration should be given to use the most economical width, as waste must be figured into the total quantity. Consider also the installation methods available—direct glue down or stretched. Direct glue-down installation is assumed with sheet carpet unit price lines unless otherwise noted.

09 70 00 Wall Finishes

- Wall coverings are estimated by the square foot. The area to be covered is measured—length by height of the wall above the baseboards—to calculate the square footage of each wall. This figure is divided by the number of square feet in the single roll which is being used. Deduct, in full, the areas of openings such as doors and windows. Where a pattern match is required allow 25–30% waste.

09 80 00 Acoustic Treatment

- Acoustical systems fall into several categories. The takeoff of these materials should be by the square foot of area with a 5% allowance for waste. Do not forget about scaffolding, if applicable, when estimating these systems.

09 90 00 Painting and Coating

- New line items created for cut-ins with reference diagram.

- A major portion of the work in painting involves surface preparation. Be sure to include cleaning, sanding, filling, and masking costs in the estimate.

- Protection of adjacent surfaces is not included in painting costs. When considering the method of paint application, an important factor is the amount of protection and masking required. These must be estimated separately and may be the determining factor in choosing the method of application.

Reference Numbers

Reference numbers are shown at the beginning of some major classifications. These numbers refer to related items in the Reference Section. The reference information may be an estimating procedure, an alternate pricing method, or technical information.

Note: Not all subdivisions listed here necessarily appear. ∎

Same Data. Simplified.

Enjoy the convenience and efficiency of accessing your costs anywhere:

- **Skip the multiplier** by setting your location
- **Quickly search,** edit, favorite and share costs
- **Stay on top of price changes** with automatic updates

Discover more at rsmeans.com/online

09 22 Supports for Plaster and Gypsum Board

09 22 03 – Fastening Methods for Finishes

09 22 03.20 Drilling Plaster/Drywall		Crew	Daily Output	Labor-Hours	Unit	Material	2021 Bare Costs Labor	Equipment	Total	Total Incl O&P
0010	**DRILLING PLASTER/DRYWALL**									
1100	Drilling & layout for drywall/plaster walls, up to 1" deep, no anchor									
1200	Holes, 1/4" diameter	1 Carp	150	.053	Ea.	.01	2.92		2.93	4.36
1300	3/8" diameter		140	.057		.01	3.13		3.14	4.68
1400	1/2" diameter		130	.062		.01	3.37		3.38	5
1500	3/4" diameter		120	.067		.01	3.65		3.66	5.45
1600	1" diameter		110	.073		.03	3.98		4.01	6
1700	1-1/4" diameter		100	.080		.04	4.38		4.42	6.60
1800	1-1/2" diameter		90	.089		.05	4.86		4.91	7.30
1900	For ceiling installations, add						40%			

09 91 Painting

09 91 23 – Interior Painting

09 91 23.52 Miscellaneous, Interior		Crew	Daily Output	Labor-Hours	Unit	Material	2021 Bare Costs Labor	Equipment	Total	Total Incl O&P
0010	**MISCELLANEOUS, INTERIOR**									
3800	Grilles, per side, oil base, primer coat, brushwork	1 Pord	520	.015	S.F.	.15	.71		.86	1.23
3850	Spray		1140	.007		.16	.33		.49	.66
3880	Paint 1 coat, brushwork		520	.015		.25	.71		.96	1.34
3900	Spray		1140	.007		.28	.33		.61	.79
3920	Paint 2 coats, brushwork		325	.025		.49	1.14		1.63	2.24
3940	Spray		650	.012		.56	.57		1.13	1.47
3950	Prime & paint 1 coat		325	.025		.40	1.14		1.54	2.14
3960	Prime & paint 2 coats		270	.030		.40	1.38		1.78	2.48
4500	Louvers, 1 side, primer, brushwork		524	.015		.09	.71		.80	1.14
4520	Paint 1 coat, brushwork		520	.015		.13	.71		.84	1.20
4530	Spray		1140	.007		.15	.33		.48	.64
4540	Paint 2 coats, brushwork		325	.025		.26	1.14		1.40	1.98
4550	Spray		650	.012		.28	.57		.85	1.16
4560	Paint 3 coats, brushwork		270	.030		.38	1.38		1.76	2.46
4570	Spray		500	.016		.42	.74		1.16	1.56
5000	Pipe, 1"-4" diameter, primer or sealer coat, oil base, brushwork	2 Pord	1250	.013	L.F.	.09	.59		.68	.98
5100	Spray		2165	.007		.08	.34		.42	.60
5200	Paint 1 coat, brushwork		1250	.013		.15	.59		.74	1.05
5300	Spray		2165	.007		.13	.34		.47	.66
5350	Paint 2 coats, brushwork		775	.021		.27	.96		1.23	1.71
5400	Spray		1240	.013		.30	.60		.90	1.22
5420	Paint 3 coats, brushwork		775	.021		.40	.96		1.36	1.86
5450	5"-8" diameter, primer or sealer coat, brushwork		620	.026		.18	1.20		1.38	1.98
5500	Spray		1085	.015		.29	.68		.97	1.34
5550	Paint 1 coat, brushwork		620	.026		.41	1.20		1.61	2.23
5600	Spray		1085	.015		.45	.68		1.13	1.52
5650	Paint 2 coats, brushwork		385	.042		.54	1.93		2.47	3.45
5700	Spray		620	.026		.60	1.20		1.80	2.44
5720	Paint 3 coats, brushwork		385	.042		.80	1.93		2.73	3.74
5750	9"-12" diameter, primer or sealer coat, brushwork		415	.039		.27	1.79		2.06	2.95
5800	Spray		725	.022		.49	1.03		1.52	2.06
5850	Paint 1 coat, brushwork		415	.039		.41	1.79		2.20	3.11
6000	Spray		725	.022		.46	1.03		1.49	2.02
6200	Paint 2 coats, brushwork		260	.062		.80	2.86		3.66	5.10
6250	Spray		415	.039		.89	1.79		2.68	3.64
6270	Paint 3 coats, brushwork		260	.062		1.19	2.86		4.05	5.55

09 91 Painting

09 91 23 – Interior Painting

09 91 23.52 Miscellaneous, Interior		Crew	Daily Output	Labor-Hours	Unit	Material	2021 Bare Costs Labor	Equipment	Total	Total Incl O&P
6300	13"-16" diameter, primer or sealer coat, brushwork	2 Pord	310	.052	L.F.	.36	2.40		2.76	3.95
6350	Spray		540	.030		.40	1.38		1.78	2.48
6400	Paint 1 coat, brushwork		310	.052		.55	2.40		2.95	4.17
6450	Spray		540	.030		.61	1.38		1.99	2.71
6500	Paint 2 coats, brushwork		195	.082		1.07	3.81		4.88	6.85
6550	Spray		310	.052		1.19	2.40		3.59	4.87
6600	Radiators, per side, primer, brushwork	1 Pord	520	.015	S.F.	.09	.71		.80	1.15
6620	Paint, 1 coat		520	.015		.08	.71		.79	1.15
6640	2 coats		340	.024		.26	1.09		1.35	1.90
6660	3 coats		283	.028		.38	1.31		1.69	2.37

Division Notes

	CREW	DAILY OUTPUT	LABOR-HOURS	UNIT	BARE COSTS				TOTAL INCL O&P
					MAT.	LABOR	EQUIP.	TOTAL	

Estimating Tips
General

- The items in this division are usually priced per square foot or each.

- Many items in Division 10 require some type of support system or special anchors that are not usually furnished with the item. The required anchors must be added to the estimate in the appropriate division.

- Some items in Division 10, such as lockers, may require assembly before installation. Verify the amount of assembly required. Assembly can often exceed installation time.

10 20 00 Interior Specialties

- Support angles and blocking are not included in the installation of toilet compartments, shower/dressing compartments, or cubicles. Appropriate line items from Division 5 or 6 may need to be added to support the installations.

- Toilet partitions are priced by the stall. A stall consists of a side wall, pilaster, and door with hardware. Toilet tissue holders and grab bars are extra.

- The required acoustical rating of a folding partition can have a significant impact on costs. Verify the sound transmission coefficient rating of the panel priced against the specification requirements.

- Grab bar installation does not include supplemental blocking or backing to support the required load. When grab bars are installed at an existing facility, provisions must be made to attach the grab bars to a solid structure.

Reference Numbers

Reference numbers are shown at the beginning of some major classifications. These numbers refer to related items in the Reference Section. The reference information may be an estimating procedure, an alternate pricing method, or technical information.

Note: Not all subdivisions listed here necessarily appear. ■

Same Data. Simplified.

Enjoy the convenience and efficiency of accessing your costs anywhere:

- **Skip the multiplier** by setting your location
- **Quickly search,** edit, favorite and share costs
- **Stay on top of price changes** with automatic updates

Discover more at rsmeans.com/online

10 21 13.13 Metal Toilet Compartments

		Crew	Daily Output	Labor-Hours	Unit	Material	2021 Bare Costs Labor	Equipment	Total	Total Incl O&P
0010	**METAL TOILET COMPARTMENTS**									
0110	Cubicles, ceiling hung									
0200	Powder coated steel	2 Carp	4	4	Ea.	540	219		759	920
0500	Stainless steel	"	4	4		890	219		1,109	1,300
0600	For handicap units, add				↓	480			480	530
0900	Floor and ceiling anchored									
1000	Powder coated steel	2 Carp	5	3.200	Ea.	710	175		885	1,050
1300	Stainless steel	"	5	3.200		1	175		176	262
1400	For handicap units, add				↓	380			380	415
1610	Floor anchored									
1700	Powder coated steel	2 Carp	7	2.286	Ea.	920	125		1,045	1,200
2000	Stainless steel	"	7	2.286		1,425	125		1,550	1,725
2100	For handicap units, add					360			360	395
2200	For juvenile units, deduct				↓	45			45	49.50
2450	Floor anchored, headrail braced									
2500	Powder coated steel	2 Carp	6	2.667	Ea.	350	146		496	605
2804	Stainless steel	"	6	2.667		1,200	146		1,346	1,525
2900	For handicap units, add					330			330	360
3000	Wall hung partitions, powder coated steel	2 Carp	7	2.286		690	125		815	945
3300	Stainless steel	"	7	2.286		1,800	125		1,925	2,150
3400	For handicap units, add				↓	330			330	360
4000	Screens, entrance, floor mounted, 58" high, 48" wide									
4200	Powder coated steel	2 Carp	15	1.067	Ea.	243	58.50		301.50	355
4500	Stainless steel	"	15	1.067	"	990	58.50		1,048.50	1,175
4650	Urinal screen, 18" wide									
4704	Powder coated steel	2 Carp	6.15	2.602	Ea.	229	142		371	465
5004	Stainless steel	"	6.15	2.602	"	530	142		672	795
5100	Floor mounted, headrail braced									
5300	Powder coated steel	2 Carp	8	2	Ea.	240	109		349	425
5600	Stainless steel	"	8	2	"	615	109		724	845
5750	Pilaster, flush									
5800	Powder coated steel	2 Carp	10	1.600	Ea.	268	87.50		355.50	425
6100	Stainless steel		10	1.600		615	87.50		702.50	805
6300	Post braced, powder coated steel		10	1.600		169	87.50		256.50	315
6800	Powder coated steel		10	1.600		104	87.50		191.50	245
7800	Wedge type, powder coated steel		10	1.600		137	87.50		224.50	282
8100	Stainless steel	↓	10	1.600	↓	625	87.50		712.50	820

10 21 13.16 Plastic-Laminate-Clad Toilet Compartments

		Crew	Daily Output	Labor-Hours	Unit	Material	2021 Bare Costs Labor	Equipment	Total	Total Incl O&P
0010	**PLASTIC-LAMINATE-CLAD TOILET COMPARTMENTS**									
0110	Cubicles, ceiling hung									
0300	Plastic laminate on particle board	2 Carp	4	4	Ea.	745	219		964	1,150
0600	For handicap units, add				"	480			480	530
0900	Floor and ceiling anchored									
1100	Plastic laminate on particle board	2 Carp	5	3.200	Ea.	810	175		985	1,150
1400	For handicap units, add				"	380			380	415
1610	Floor mounted									
1800	Plastic laminate on particle board	2 Carp	7	2.286	Ea.	540	125		665	780
2450	Floor mounted, headrail braced									
2600	Plastic laminate on particle board	2 Carp	6	2.667	Ea.	805	146		951	1,100
3400	For handicap units, add					330			330	360
4300	Entrance screen, floor mtd., plas. lam., 58" high, 48" wide	2 Carp	15	1.067		525	58.50		583.50	665
4800	Urinal screen, 18" wide, ceiling braced, plastic laminate	↓	8	2		210	109		319	395

For customer support on your Plumbing Costs with RSMeans Data, call 800.448.8182.

10 21 Compartments and Cubicles

10 21 13 – Toilet Compartments

10 21 13.16 Plastic-Laminate-Clad Toilet Compartments

		Crew	Daily Output	Labor-Hours	Unit	Material	2021 Bare Costs Labor	Equipment	Total	Total Incl O&P
5400	Floor mounted, headrail braced	2 Carp	8	2	Ea.	212	109		321	395
5900	Pilaster, flush, plastic laminate		10	1.600		435	87.50		522.50	610
6400	Post braced, plastic laminate		10	1.600		235	87.50		322.50	390
6700	Wall hung, bracket supported									
6900	Plastic laminate on particle board	2 Carp	10	1.600	Ea.	97	87.50		184.50	237
7450	Flange supported									
7500	Plastic laminate on particle board	2 Carp	10	1.600	Ea.	263	87.50		350.50	420

10 21 13.19 Plastic Toilet Compartments

		Crew	Daily Output	Labor-Hours	Unit	Material	2021 Bare Costs Labor	Equipment	Total	Total Incl O&P
0010	**PLASTIC TOILET COMPARTMENTS**									
0110	Cubicles, ceiling hung									
0250	Phenolic	2 Carp	4	4	Ea.	920	219		1,139	1,325
0260	Polymer plastic	"	4	4		1,025	219		1,244	1,450
0600	For handicap units, add					480			480	530
0900	Floor and ceiling anchored									
1050	Phenolic	2 Carp	5	3.200	Ea.	970	175		1,145	1,325
1060	Polymer plastic	"	5	3.200		1,200	175		1,375	1,575
1400	For handicap units, add					380			380	415
1610	Floor mounted									
1750	Phenolic	2 Carp	7	2.286	Ea.	935	125		1,060	1,200
1760	Polymer plastic	"	7	2.286		845	125		970	1,125
2100	For handicap units, add					360			360	395
2200	For juvenile units, deduct					45			45	49.50
2450	Floor mounted, headrail braced									
2550	Phenolic	2 Carp	6	2.667	Ea.	865	146		1,011	1,175
3600	Polymer plastic		6	2.667		890	146		1,036	1,200
3810	Entrance screen, polymer plastic, flr. mtd., 48"x58"		6	2.667		252	146		398	495
3820	Entrance screen, polymer plastic, flr. to clg pilaster, 48"x58"		6	2.667		252	146		398	495
6110	Urinal screen, polymer plastic, pilaster flush, 18" w		6	2.667		244	146		390	485
7110	Wall hung		6	2.667		166	146		312	400
7710	Flange mounted		6	2.667		198	146		344	435

10 21 13.40 Stone Toilet Compartments

		Crew	Daily Output	Labor-Hours	Unit	Material	2021 Bare Costs Labor	Equipment	Total	Total Incl O&P
0010	**STONE TOILET COMPARTMENTS**									
0100	Cubicles, ceiling hung, marble	2 Marb	2	8	Ea.	1,650	425		2,075	2,475
0600	For handicap units, add					480			480	530
0800	Floor & ceiling anchored, marble	2 Marb	2.50	6.400		1,675	340		2,015	2,375
1400	For handicap units, add					380			380	415
1600	Floor mounted, marble	2 Marb	3	5.333		1,050	284		1,334	1,575
2400	Floor mounted, headrail braced, marble	"	3	5.333		1,200	284		1,484	1,750
2900	For handicap units, add					330			330	360
4100	Entrance screen, floor mounted marble, 58" high, 48" wide	2 Marb	9	1.778		725	94.50		819.50	935
4600	Urinal screen, 18" wide, ceiling braced, marble	D-1	6	2.667		760	130		890	1,025
5100	Floor mounted, headrail braced									
5200	Marble	D-1	6	2.667	Ea.	650	130		780	910
5700	Pilaster, flush, marble		9	1.778		850	86.50		936.50	1,075
6200	Post braced, marble		9	1.778		830	86.50		916.50	1,050

For customer support on your Plumbing Costs with RSMeans Data, call 800.448.8182.

65

10 28 13.13 Commercial Toilet Accessories	Crew	Daily Output	Labor-Hours	Unit	Material	2021 Bare Costs Labor	Equipment	Total	Total Incl O&P
0010 COMMERCIAL TOILET ACCESSORIES									
0200 Curtain rod, stainless steel, 5' long, 1" diameter	1 Carp	13	.615	Ea.	26.50	33.50		60	79.50
0300 1-1/4" diameter	"	13	.615	"	18.30	33.50		51.80	70.50
0500 Dispenser units, combined soap & towel dispensers,									
0510 Mirror and shelf, flush mounted	1 Carp	10	.800	Ea.	355	44		399	455
0600 Towel dispenser and waste receptacle,									
0610 18 gallon capacity	1 Carp	10	.800	Ea.	385	44		429	490
0800 Grab bar, straight, 1-1/4" diameter, stainless steel, 18" long		24	.333		10.90	18.25		29.15	39
0900 24" long		23	.348		13.30	19.05		32.35	43
1000 30" long		22	.364		27.50	19.90		47.40	59.50
1100 36" long		20	.400		13.30	22		35.30	47
1105 42" long		20	.400		41.50	22		63.50	78
1120 Corner, 36" long		20	.400		95	22		117	137
1200 1-1/2" diameter, 24" long		23	.348		37.50	19.05		56.55	70
1300 36" long		20	.400		43	22		65	79.50
1310 42" long		18	.444		34	24.50		58.50	74
1500 Tub bar, 1-1/4" diameter, 24" x 36"		14	.571		110	31.50		141.50	168
1600 Plus vertical arm		12	.667		108	36.50		144.50	174
1900 End tub bar, 1" diameter, 90° angle, 16" x 32"		12	.667		102	36.50		138.50	167
2300 Hand dryer, surface mounted, electric, 115 volt, 20 amp		4	2		425	109		534	630
2400 230 volt, 10 amp		4	2		585	109		694	810
2450 Hand dryer, touch free, 1400 watt, 81,000 rpm		4	2		1,350	109		1,459	1,675
2600 Hat and coat strip, stainless steel, 4 hook, 36" long		24	.333		36.50	18.25		54.75	67
2700 6 hook, 60" long		20	.400		106	22		128	149
3000 Mirror, with stainless steel 3/4" square frame, 18" x 24"		20	.400		50	22		72	87.50
3100 36" x 24"		15	.533		63.50	29		92.50	114
3200 48" x 24"		10	.800		168	44		212	251
3300 72" x 24"		6	1.333		385	73		458	535
3500 With 5" stainless steel shelf, 18" x 24"		20	.400		83	22		105	124
3600 36" x 24"		15	.533		290	29		319	365
3700 48" x 24"		10	.800		285	44		329	380
3800 72" x 24"		6	1.333		241	73		314	375
4100 Mop holder strip, stainless steel, 5 holders, 48" long		20	.400		72	22		94	112
4200 Napkin/tampon dispenser, recessed		15	.533		635	29		664	745
4220 Semi-recessed		6.50	1.231		345	67.50		412.50	480
4250 Napkin receptacle, recessed		6.50	1.231		170	67.50		237.50	287
4300 Robe hook, single, regular		96	.083		15.30	4.56		19.86	23.50
4400 Heavy duty, concealed mounting		56	.143		25.50	7.80		33.30	39.50
4600 Soap dispenser, chrome, surface mounted, liquid		20	.400		67.50	22		89.50	107
5000 Recessed stainless steel, liquid		10	.800		172	44		216	255
5600 Shelf, stainless steel, 5" wide, 18 ga., 24" long		24	.333		66.50	18.25		84.75	100
5700 48" long		16	.500		212	27.50		239.50	274
5800 8" wide shelf, 18 ga., 24" long		22	.364		85	19.90		104.90	123
5900 48" long		14	.571		146	31.50		177.50	207
6000 Toilet seat cover dispenser, stainless steel, recessed		20	.400		181	22		203	232
6050 Surface mounted		15	.533		34	29		63	81
6200 Double roll		24	.333		23.50	18.25		41.75	53
6240 Plastic, twin/jumbo dbl. roll		24	.333		30.50	18.25		48.75	60.50
6400 Towel bar, stainless steel, 18" long		23	.348		38.50	19.05		57.55	70.50
6500 30" long		21	.381		46	21		67	82
6700 Towel dispenser, stainless steel, surface mounted		16	.500		38.50	27.50		66	83.50
6800 Flush mounted, recessed		10	.800		67.50	44		111.50	140
6900 Plastic, touchless, battery operated		16	.500		79	27.50		106.50	128

10 28 Toilet, Bath, and Laundry Accessories

10 28 13 - Toilet Accessories

10 28 13.13 Commercial Toilet Accessories

		Crew	Daily Output	Labor-Hours	Unit	Material	2021 Bare Costs Labor	Equipment	Total	Total Incl O&P
7000	Towel holder, hotel type, 2 guest size	1 Carp	20	.400	Ea.	54	22		76	92
7200	Towel shelf, stainless steel, 24" long, 8" wide		20	.400		85	22		107	126
7400	Tumbler holder, for tumbler only		30	.267		37.50	14.60		52.10	63
7410	Tumbler holder, recessed		20	.400		8.25	22		30.25	41.50
7500	Soap, tumbler & toothbrush		30	.267		17.40	14.60		32	41
7510	Tumbler & toothbrush holder		20	.400		11.15	22		33.15	45
8000	Waste receptacles, stainless steel, with top, 13 gallon		10	.800		315	44		359	415
8100	36 gallon		8	1		535	54.50		589.50	670
9996	Bathroom access., grab bar, straight, 1-1/2" diam., SS, 42" L install only		18	.444			24.50		24.50	36.50

10 28 16 - Bath Accessories

10 28 16.20 Medicine Cabinets

		Crew	Daily Output	Labor-Hours	Unit	Material	2021 Bare Costs Labor	Equipment	Total	Total Incl O&P
0010	**MEDICINE CABINETS**									
0020	With mirror, sst frame, 16" x 22", unlighted	1 Carp	14	.571	Ea.	102	31.50		133.50	159
0100	Wood frame		14	.571		128	31.50		159.50	188
0300	Sliding mirror doors, 20" x 16" x 4-3/4", unlighted		7	1.143		136	62.50		198.50	244
0400	24" x 19" x 8-1/2", lighted		5	1.600		220	87.50		307.50	375
0600	Triple door, 30" x 32", unlighted, plywood body		7	1.143		271	62.50		333.50	390
0700	Steel body		7	1.143		355	62.50		417.50	485
0900	Oak door, wood body, beveled mirror, single door		7	1.143		119	62.50		181.50	225
1000	Double door		6	1.333		375	73		448	520
1200	Hotel cabinets, stainless, with lower shelf, unlighted		10	.800		296	44		340	390
1300	Lighted		5	1.600		310	87.50		397.50	475

10 28 19 - Tub and Shower Enclosures

10 28 19.10 Partitions, Shower

		Crew	Daily Output	Labor-Hours	Unit	Material	2021 Bare Costs Labor	Equipment	Total	Total Incl O&P
0010	**PARTITIONS, SHOWER** floor mounted, no plumbing									
0400	Cabinet, one piece, fiberglass, 32" x 32"	2 Carp	5	3.200	Ea.	625	175		800	945
0420	36" x 36"		5	3.200		545	175		720	860
0440	36" x 48"		5	3.200		1,375	175		1,550	1,775
0460	Acrylic, 32" x 32"		5	3.200		225	175		400	510
0480	36" x 36"		5	3.200		1,050	175		1,225	1,400
0500	36" x 48"		5	3.200		815	175		990	1,150
0520	Shower door for above, clear plastic, 24" wide	1 Carp	8	1		150	54.50		204.50	247
0540	28" wide		8	1		190	54.50		244.50	291
0560	Tempered glass, 24" wide		8	1		135	54.50		189.50	231
0580	28" wide		8	1		160	54.50		214.50	258
2400	Glass stalls, with doors, no receptors, chrome on brass	2 Shee	3	5.333		1,625	350		1,975	2,300
2700	Anodized aluminum	"	4	4		1,500	262		1,762	2,050
2900	Marble shower stall, stock design, with shower door	2 Marb	1.20	13.333		2,275	710		2,985	3,600
3000	With curtain		1.30	12.308		2,025	655		2,680	3,200
3200	Receptors, precast terrazzo, 32" x 32"		14	1.143		288	61		349	405
3300	48" x 34"		9.50	1.684		410	89.50		499.50	585
3500	Plastic, simulated terrazzo receptor, 32" x 32"		14	1.143		171	61		232	280
3600	32" x 48"		12	1.333		305	71		376	440
3800	Precast concrete, colors, 32" x 32"		14	1.143		235	61		296	350
3900	48" x 48"		8	2		325	106		431	515
4100	Shower doors, economy plastic, 24" wide	1 Shee	9	.889		133	58		191	234
4200	Tempered glass door, economy		8	1		278	65.50		343.50	405
4400	Folding, tempered glass, aluminum frame		6	1.333		425	87.50		512.50	600
4500	Sliding, tempered glass, 48" opening		6	1.333		615	87.50		702.50	805
4700	Deluxe, tempered glass, chrome on brass frame, 42" to 44"		8	1		320	65.50		385.50	450
4800	39" to 48" wide		1	8		725	525		1,250	1,575
4850	On anodized aluminum frame, obscure glass		2	4		570	262		832	1,025

For customer support on your Plumbing Costs with RSMeans Data, call 800.448.8182.

67

10 28 19.10 Partitions, Shower		Crew	Daily Output	Labor-Hours	Unit	Material	2021 Bare Costs Labor	Equipment	Total	Total Incl O&P
4900	Clear glass	1 Shee	1	8	Ea.	710	525		1,235	1,575
5100	Shower enclosure, tempered glass, anodized alum. frame									
5120	2 panel & door, corner unit, 32" x 32"	1 Shee	2	4	Ea.	1,025	262		1,287	1,525
5140	Neo-angle corner unit, 16" x 24" x 16"	"	2	4		990	262		1,252	1,500
5200	Shower surround, 3 wall, polypropylene, 32" x 32"	1 Carp	4	2		460	109		569	670
5220	PVC, 32" x 32"		4	2		470	109		579	680
5240	Fiberglass		4	2		390	109		499	590
5250	2 wall, polypropylene, 32" x 32"		4	2		310	109		419	505
5270	PVC		4	2		370	109		479	570
5290	Fiberglass		4	2		370	109		479	575
5300	Tub doors, tempered glass & frame, obscure glass	1 Shee	8	1		218	65.50		283.50	340
5400	Clear glass		6	1.333		510	87.50		597.50	690
5600	Chrome plated, brass frame, obscure glass		8	1		287	65.50		352.50	415
5700	Clear glass		6	1.333		705	87.50		792.50	905
5900	Tub/shower enclosure, temp. glass, alum. frame, obscure glass		2	4		390	262		652	825
6200	Clear glass		1.50	5.333		815	350		1,165	1,425
6500	On chrome-plated brass frame, obscure glass		2	4		535	262		797	985
6600	Clear glass		1.50	5.333		1,150	350		1,500	1,800
6800	Tub surround, 3 wall, polypropylene	1 Carp	4	2		249	109		358	435
6900	PVC		4	2		365	109		474	565
7000	Fiberglass, obscure glass		4	2		380	109		489	585
7100	Clear glass		3	2.667		645	146		791	930

10 44 Fire Protection Specialties
10 44 13 – Fire Protection Cabinets

10 44 13.53 Fire Equipment Cabinets

		Crew	Daily Output	Labor-Hours	Unit	Material	2021 Bare Costs Labor	Equipment	Total	Total Incl O&P
0010	**FIRE EQUIPMENT CABINETS**, not equipped, 20 ga. steel box D4020–310									
0040	Recessed, D.S. glass in door, box size given									
1000	Portable extinguisher, single, 8" x 12" x 27", alum. door & frame	Q-12	8	2	Ea.	155	120		275	350
1100	Steel door and frame		8	2		183	120		303	380
1200	Stainless steel door and frame		8	2		238	120		358	440
2000	Portable extinguisher, large, 8" x 12" x 36", alum. door & frame		8	2		286	120		406	495
2100	Steel door and frame D4020–330		8	2		251	120		371	455
2200	Stainless steel door and frame		8	2		415	120		535	640
2500	8" x 16" x 38", aluminum door & frame		8	2		231	120		351	435
2600	Steel door and frame		8	2		240	120		360	445
2700	Fire blanket & extinguisher cab, inc blanket, rec stl., 14" x 40" x 8"		7	2.286		223	137		360	450
2800	Fire blanket cab, inc blanket, surf mtd, stl, 15"x10"x5", w/pwdr coat fin		8	2		95	120		215	283
3000	Hose rack assy., 1-1/2" valve & 100' hose, 24" x 40" x 5-1/2"									
3100	Aluminum door and frame	Q-12	6	2.667	Ea.	570	160		730	870
3200	Steel door and frame D4020–410		6	2.667		229	160		389	490
3300	Stainless steel door and frame		6	2.667		600	160		760	900
4000	Hose rack assy., 2-1/2" x 1-1/2" valve, 100' hose, 24" x 40" x 8"									
4100	Aluminum door and frame	Q-12	6	2.667	Ea.	555	160		715	850
4200	Steel door and frame R211226-10		6	2.667		390	160		550	670
4300	Stainless steel door and frame		6	2.667		765	160		925	1,075
5000	Hose rack assy., 2-1/2" x 1-1/2" valve, 100' hose									
5010	and extinguisher, 30" x 40" x 8"									
5100	Aluminum door and frame	Q-12	5	3.200	Ea.	735	192		927	1,100
5200	Steel door and frame		5	3.200		282	192		474	595
5300	Stainless steel door and frame		5	3.200		640	192		832	990

For customer support on your Plumbing Costs with RSMeans Data, call 800.448.8182.

10 44 Fire Protection Specialties

10 44 13 – Fire Protection Cabinets

10 44 13.53 Fire Equipment Cabinets

		Crew	Daily Output	Labor-Hours	Unit	Material	2021 Bare Costs Labor	2021 Bare Costs Equipment	Total	Total Incl O&P
6000	Hose rack assy., 1-1/2" valve, 100' hose									
6010	and 2-1/2" FD valve, 24" x 44" x 8"									
6100	Aluminum door and frame	Q-12	5	3.200	Ea.	275	192		467	585
6200	Steel door and frame		5	3.200		575	192		767	915
6300	Stainless steel door and frame		5	3.200		865	192		1,057	1,250
7000	Hose rack assy., 1-1/2" valve & 100' hose, 2-1/2" FD valve R211226-20									
7010	and extinguisher, 30" x 44" x 8"									
7100	Aluminum door and frame	Q-12	5	3.200	Ea.	770	192		962	1,125
7200	Steel door and frame		5	3.200		470	192		662	805
7300	Stainless steel door and frame		5	3.200		725	192		917	1,075
8000	Valve cabinet for 2-1/2" FD angle valve, 18" x 18" x 8"									
8100	Aluminum door and frame	Q-12	12	1.333	Ea.	253	80		333	400
8200	Steel door and frame		12	1.333		209	80		289	350
8300	Stainless steel door and frame		12	1.333		340	80		420	495

10 44 16 – Fire Extinguishers

10 44 16.13 Portable Fire Extinguishers

		Crew	Daily Output	Labor-Hours	Unit	Material	2021 Bare Costs Labor	2021 Bare Costs Equipment	Total	Total Incl O&P
0010	**PORTABLE FIRE EXTINGUISHERS**									
0140	CO_2, with hose and "H" horn, 10 lb.				Ea.	335			335	370
0160	15 lb.					395			395	435
0180	20 lb.					475			475	525
1000	Dry chemical, pressurized									
1040	Standard type, portable, painted, 2-1/2 lb.				Ea.	43.50			43.50	48
1060	5 lb.					64			64	70.50
1080	10 lb.					98.50			98.50	109
1100	20 lb.					230			230	253
1120	30 lb.					430			430	475
1300	Standard type, wheeled, 150 lb.					1,750			1,750	1,925
2000	ABC all purpose type, portable, 2-1/2 lb.					40.50			40.50	44.50
2060	5 lb.					48.50			48.50	53.50
2080	9-1/2 lb.					73			73	80.50
2100	20 lb.					170			170	187
5000	Pressurized water, 2-1/2 gallon, stainless steel					98.50			98.50	108
5060	With anti-freeze					117			117	128
9400	Installation of extinguishers, 12 or more, on nailable surface	1 Carp	30	.267			14.60		14.60	22
9420	On masonry or concrete	"	15	.533			29		29	43.50

10 44 16.16 Wheeled Fire Extinguisher Units

		Crew	Daily Output	Labor-Hours	Unit	Material	2021 Bare Costs Labor	2021 Bare Costs Equipment	Total	Total Incl O&P
0010	**WHEELED FIRE EXTINGUISHER UNITS**									
0350	CO_2, portable, with swivel horn									
0360	Wheeled type, cart mounted, 50 lb.				Ea.	1,425			1,425	1,575
0400	100 lb.				"	3,650			3,650	4,000
2200	ABC all purpose type									
2300	Wheeled, 45 lb.				Ea.	770			770	845
2360	150 lb.				"	1,950			1,950	2,150

For customer support on your Plumbing Costs with RSMeans Data, call 800.448.8182.

69

Division Notes

	CREW	DAILY OUTPUT	LABOR-HOURS	UNIT	BARE COSTS				TOTAL INCL O&P
					MAT.	LABOR	EQUIP.	TOTAL	

Estimating Tips
General

- The items in this division are usually priced per square foot or each. Many of these items are purchased by the owner for installation by the contractor. Check the specifications for responsibilities and include time for receiving, storage, installation, and mechanical and electrical hookups in the appropriate divisions.

- Many items in Division 11 require some type of support system that is not usually furnished with the item. Examples of these systems include blocking for the attachment of casework and support angles for ceiling-hung projection screens. The required blocking or supports must be added to the estimate in the appropriate division.

- Some items in Division 11 may require assembly or electrical hookups. Verify the amount of assembly required or the need for a hard electrical connection and add the appropriate costs.

Reference Numbers

Reference numbers are shown at the beginning of some major classifications. These numbers refer to related items in the Reference Section. The reference information may be an estimating procedure, an alternate pricing method, or technical information.

Same Data. Simplified.

Enjoy the convenience and efficiency of accessing your costs anywhere:

- **Skip the multiplier** by setting your location
- **Quickly search,** edit, favorite and share costs
- **Stay on top of price changes** with automatic updates

Discover more at rsmeans.com/online

11 11 Vehicle Service Equipment

11 11 13 – Compressed-Air Vehicle Service Equipment

11 11 13.10 Compressed Air Equipment	Crew	Daily Output	Labor-Hours	Unit	Material	2021 Bare Costs Labor	Equipment	Total	Total Incl O&P
0010 **COMPRESSED AIR EQUIPMENT**									
0030 Compressors, electric, 1-1/2 HP, standard controls	L-4	1.50	16	Ea.	1,125	835		1,960	2,475
0550 Dual controls		1.50	16		1,050	835		1,885	2,400
0600 5 HP, 115/230 volt, standard controls		1	24		1,000	1,250		2,250	2,975
0650 Dual controls	↓	1	24	↓	1,325	1,250		2,575	3,325

11 11 19 – Vehicle Lubrication Equipment

11 11 19.10 Lubrication Equipment

	Crew	Daily Output	Labor-Hours	Unit	Material	2021 Bare Costs Labor	Equipment	Total	Total Incl O&P
0010 **LUBRICATION EQUIPMENT**									
3000 Lube equipment, 3 reel type, with pumps, not including piping	L-4	.50	48	Set	10,400	2,500		12,900	15,300
3700 Pump lubrication, pneumatic, not incl. air compressor									
3710 Oil/gear lube	Q-1	9.60	1.667	Ea.	850	102		952	1,075
3720 Grease	"	9.60	1.667	"	1,250	102		1,352	1,525

11 11 33 – Vehicle Spray Painting Equipment

11 11 33.10 Spray Painting Equipment

	Crew	Daily Output	Labor-Hours	Unit	Material	2021 Bare Costs Labor	Equipment	Total	Total Incl O&P
0010 **SPRAY PAINTING EQUIPMENT**									
4000 Spray painting booth, 26' long, complete	L-4	.40	60	Ea.	18,100	3,125		21,225	24,600

11 21 Retail and Service Equipment

11 21 53 – Barber and Beauty Shop Equipment

11 21 53.10 Barber Equipment

	Crew	Daily Output	Labor-Hours	Unit	Material	2021 Bare Costs Labor	Equipment	Total	Total Incl O&P
0010 **BARBER EQUIPMENT**									
0020 Chair, hydraulic, movable, minimum	1 Carp	24	.333	Ea.	635	18.25		653.25	725
0050 Maximum	"	16	.500		3,950	27.50		3,977.50	4,375
0500 Sink, hair washing basin, rough plumbing not incl.	1 Plum	8	1		505	67.50		572.50	655
1000 Sterilizer, liquid solution for tools				↓	163			163	179

11 21 73 – Commercial Laundry and Dry Cleaning Equipment

11 21 73.16 Drying and Conditioning Equipment

	Crew	Daily Output	Labor-Hours	Unit	Material	2021 Bare Costs Labor	Equipment	Total	Total Incl O&P
0010 **DRYING AND CONDITIONING EQUIPMENT**									
0100 Dryers, not including rough-in									
1500 Industrial, 30 lb. capacity	1 Plum	2	4	Ea.	1,625	271		1,896	2,175
1600 50 lb. capacity	"	1.70	4.706	"	4,575	320		4,895	5,500

11 21 73.26 Commercial Washers and Extractors

	Crew	Daily Output	Labor-Hours	Unit	Material	2021 Bare Costs Labor	Equipment	Total	Total Incl O&P
0010 **COMMERCIAL WASHERS AND EXTRACTORS**, not including rough-in									
6000 Combination washer/extractor, 20 lb. capacity	L-6	1.50	8	Ea.	7,175	530		7,705	8,700
6100 30 lb. capacity		.80	15		10,900	995		11,895	13,500
6200 50 lb. capacity		.68	17.647		13,800	1,175		14,975	17,000
6300 75 lb. capacity		.30	40		19,200	2,650		21,850	25,100
6350 125 lb. capacity	↓	.16	75	↓	30,100	4,975		35,075	40,500

11 21 73.33 Coin-Operated Laundry Equipment

	Crew	Daily Output	Labor-Hours	Unit	Material	2021 Bare Costs Labor	Equipment	Total	Total Incl O&P
0010 **COIN-OPERATED LAUNDRY EQUIPMENT**									
0990 Dryer, gas fired									
1000 Commercial, 30 lb. capacity, coin operated, single	1 Plum	3	2.667	Ea.	1,150	181		1,331	1,525
1100 Double stacked	"	2	4	"	2,925	271		3,196	3,600
5290 Clothes washer									
5300 Commercial, coin operated, average	1 Plum	3	2.667	Ea.	1,350	181		1,531	1,775

11 21 Retail and Service Equipment

11 21 83 – Photo Processing Equipment

11 21 83.13 Darkroom Equipment		Crew	Daily Output	Labor-Hours	Unit	Material	2021 Bare Costs Labor	Equipment	Total	Total Incl O&P
0010	DARKROOM EQUIPMENT									
0020	Developing sink, 5" deep, 24" x 48"	Q-1	2	8	Ea.	630	485		1,115	1,425
0050	48" x 52"		1.70	9.412		1,500	575		2,075	2,500
0200	10" deep, 24" x 48"		1.70	9.412		1,650	575		2,225	2,675
0250	24" x 108"	↓	1.50	10.667	↓	4,050	650		4,700	5,425

11 30 Residential Equipment

11 30 13 – Residential Appliances

11 30 13.15 Cooking Equipment

		Crew	Daily Output	Labor-Hours	Unit	Material	Labor	Equipment	Total	Total Incl O&P
0010	COOKING EQUIPMENT									
0020	Cooking range, 30" free standing, 1 oven, minimum	2 Clab	10	1.600	Ea.	460	71		531	610
0050	Maximum		4	4		2,300	178		2,478	2,800
0150	2 oven, minimum		10	1.600		1,075	71		1,146	1,300
0200	Maximum	↓	10	1.600	↓	3,400	71		3,471	3,825

11 30 13.16 Refrigeration Equipment

		Crew	Daily Output	Labor-Hours	Unit	Material	Labor	Equipment	Total	Total Incl O&P
0010	REFRIGERATION EQUIPMENT									
5200	Icemaker, automatic, 20 lbs./day	1 Plum	7	1.143	Ea.	1,375	77.50		1,452.50	1,625
5350	51 lbs./day	"	2	4	"	1,625	271		1,896	2,175

11 30 13.17 Kitchen Cleaning Equipment

		Crew	Daily Output	Labor-Hours	Unit	Material	Labor	Equipment	Total	Total Incl O&P
0010	KITCHEN CLEANING EQUIPMENT									
2750	Dishwasher, built-in, 2 cycles, minimum	L-1	4	4	Ea.	310	263		573	730
2800	Maximum		2	8		445	525		970	1,275
2950	4 or more cycles, minimum		4	4		420	263		683	850
2960	Average		4	4		560	263		823	1,000
3000	Maximum	↓	2	8	↓	1,700	525		2,225	2,650

11 30 13.18 Waste Disposal Equipment

		Crew	Daily Output	Labor-Hours	Unit	Material	Labor	Equipment	Total	Total Incl O&P
0010	WASTE DISPOSAL EQUIPMENT									
3300	Garbage disposal, sink type, minimum	L-1	10	1.600	Ea.	109	105		214	277
3350	Maximum	"	10	1.600	"	242	105		347	425

11 30 13.19 Kitchen Ventilation Equipment

		Crew	Daily Output	Labor-Hours	Unit	Material	Labor	Equipment	Total	Total Incl O&P
0010	KITCHEN VENTILATION EQUIPMENT									
4150	Hood for range, 2 speed, vented, 30" wide, minimum	L-3	5	3.200	Ea.	107	191		298	400
4200	Maximum		3	5.333		1,050	320		1,370	1,625
4300	42" wide, minimum		5	3.200		152	191		343	450
4330	Custom		5	3.200		1,975	191		2,166	2,450
4350	Maximum	↓	3	5.333		2,400	320		2,720	3,125
4500	For ventless hood, 2 speed, add					18.45			18.45	20.50
4650	For vented 1 speed, deduct from maximum				↓	70			70	77

11 30 13.24 Washers

		Crew	Daily Output	Labor-Hours	Unit	Material	Labor	Equipment	Total	Total Incl O&P
0010	WASHERS									
5000	Residential, 4 cycle, average	1 Plum	3	2.667	Ea.	995	181		1,176	1,375
6650	Washing machine, automatic, minimum		3	2.667		575	181		756	900
6700	Maximum	↓	1	8	↓	1,425	540		1,965	2,350

11 30 13.25 Dryers

		Crew	Daily Output	Labor-Hours	Unit	Material	Labor	Equipment	Total	Total Incl O&P
0010	DRYERS									
0500	Gas fired residential, 16 lb. capacity, average	1 Plum	3	2.667	Ea.	950	181		1,131	1,325
7450	Vent kits for dryers	1 Carp	10	.800	"	44.50	44		88.50	115

11 30 Residential Equipment

11 30 15 – Miscellaneous Residential Appliances

11 30 15.13 Sump Pumps

		Crew	Daily Output	Labor-Hours	Unit	Material	2021 Bare Costs Labor	Equipment	Total	Total Incl O&P
0010	**SUMP PUMPS**									
6400	Cellar drainer, pedestal, 1/3 HP, molded PVC base	1 Plum	3	2.667	Ea.	138	181		319	420
6450	Solid brass	"	2	4	"	250	271		521	680
6460	Sump pump, see also Section 22 14 29.16									

11 30 15.23 Water Heaters

		Crew	Daily Output	Labor-Hours	Unit	Material	2021 Bare Costs Labor	Equipment	Total	Total Incl O&P
0010	**WATER HEATERS**									
6900	Electric, glass lined, 30 gallon, minimum	L-1	5	3.200	Ea.	1,100	210		1,310	1,525
6950	Maximum		3	5.333		1,525	350		1,875	2,200
7100	80 gallon, minimum		2	8		2,200	525		2,725	3,175
7150	Maximum		1	16		3,050	1,050		4,100	4,925
7180	Gas, glass lined, 30 gallon, minimum	2 Plum	5	3.200		2,000	217		2,217	2,525
7220	Maximum		3	5.333		2,775	360		3,135	3,600
7260	50 gallon, minimum		2.50	6.400		1,775	435		2,210	2,600
7300	Maximum		1.50	10.667		2,475	720		3,195	3,800
7310	Water heater, see also Section 22 33 30.13									

11 30 15.43 Air Quality

		Crew	Daily Output	Labor-Hours	Unit	Material	2021 Bare Costs Labor	Equipment	Total	Total Incl O&P
0010	**AIR QUALITY**									
2450	Dehumidifier, portable, automatic, 15 pint	1 Elec	4	2	Ea.	219	127		346	430
2550	40 pint		3.75	2.133		250	136		386	475
3550	Heater, electric, built-in, 1250 watt, ceiling type, minimum		4	2		103	127		230	300
3600	Maximum		3	2.667		196	170		366	470
3700	Wall type, minimum		4	2		235	127		362	450
3750	Maximum		3	2.667		102	170		272	365
3900	1500 watt wall type, with blower		4	2		172	127		299	380
3950	3000 watt		3	2.667		515	170		685	815
4850	Humidifier, portable, 8 gallons/day					132			132	146
5000	15 gallons/day					209			209	230

11 32 Unit Kitchens

11 32 13 – Metal Unit Kitchens

11 32 13.10 Commercial Unit Kitchens

		Crew	Daily Output	Labor-Hours	Unit	Material	2021 Bare Costs Labor	Equipment	Total	Total Incl O&P
0010	**COMMERCIAL UNIT KITCHENS**									
1500	Combination range, refrigerator and sink, 30" wide, minimum	L-1	2	8	Ea.	635	525		1,160	1,475
1550	Maximum		1	16		1,125	1,050		2,175	2,825
1570	60" wide, average		1.40	11.429		1,150	750		1,900	2,400
1590	72" wide, average		1.20	13.333		2,100	875		2,975	3,600
1600	Office model, 48" wide		2	8		1,275	525		1,800	2,175
1620	Refrigerator and sink only		2.40	6.667		2,600	440		3,040	3,500
1640	Combination range, refrigerator, sink, microwave									
1660	Oven and ice maker	L-1	.80	20	Ea.	4,875	1,325		6,200	7,300

11 41 Foodservice Storage Equipment

11 41 13 – Refrigerated Food Storage Cases

11 41 13.20 Refrigerated Food Storage Equipment

		Crew	Daily Output	Labor-Hours	Unit	Material	2021 Bare Costs Labor	Equipment	Total	Total Incl O&P
0010	**REFRIGERATED FOOD STORAGE EQUIPMENT**									
2350	Cooler, reach-in, beverage, 6' long	Q-1	6	2.667	Ea.	2,875	162		3,037	3,425
4300	Freezers, reach-in, 44 C.F.		4	4		4,425	244		4,669	5,250
4500	68 C.F.		3	5.333		5,450	325		5,775	6,450

11 44 Food Cooking Equipment

11 44 13 – Commercial Ranges

11 44 13.10 Cooking Equipment

		Crew	Daily Output	Labor-Hours	Unit	Material	2021 Bare Costs Labor	Equipment	Total	Total Incl O&P
0010	**COOKING EQUIPMENT**									
0020	Bake oven, gas, one section	Q-1	8	2	Ea.	6,250	122		6,372	7,050
0300	Two sections		7	2.286		10,000	139		10,139	11,200
0600	Three sections		6	2.667		12,400	162		12,562	13,800
0900	Electric convection, single deck	L-7	4	7		5,500	370		5,870	6,600
6350	Kettle, w/steam jacket, tilting, w/positive lock, SS, 20 gallons		7	4		10,800	212		11,012	12,200
6600	60 gallons		6	4.667		14,600	248		14,848	16,500

11 46 Food Dispensing Equipment

11 46 83 – Ice Machines

11 46 83.10 Commercial Ice Equipment

		Crew	Daily Output	Labor-Hours	Unit	Material	2021 Bare Costs Labor	Equipment	Total	Total Incl O&P
0010	**COMMERCIAL ICE EQUIPMENT**									
5800	Ice cube maker, 50 lbs./day	Q-1	6	2.667	Ea.	1,825	162		1,987	2,250
6050	500 lbs./day	"	4	4	"	2,625	244		2,869	3,250

11 48 Foodservice Cleaning and Disposal Equipment

11 48 13 – Commercial Dishwashers

11 48 13.10 Dishwashers

			Crew	Daily Output	Labor-Hours	Unit	Material	2021 Bare Costs Labor	Equipment	Total	Total Incl O&P
0010	**DISHWASHERS**										
2700	Dishwasher, commercial, rack type										
2720	10 to 12 racks/hour		Q-1	3.20	5	Ea.	3,500	305		3,805	4,300
2730	Energy star rated, 35 to 40 racks/hour	G		1.30	12.308		5,650	750		6,400	7,325
2740	50 to 60 racks/hour	G		1.30	12.308		10,800	750		11,550	13,000
2800	Automatic, 190 to 230 racks/hour		L-6	.35	34.286		16,600	2,275		18,875	21,700
2820	235 to 275 racks/hour			.25	48		29,000	3,175		32,175	36,700
2840	8,750 to 12,500 dishes/hour			.10	120		56,500	7,975		64,475	74,000

11 53 Laboratory Equipment

11 53 13 – Laboratory Fume Hoods

11 53 13.13 Recirculating Laboratory Fume Hoods

		Crew	Daily Output	Labor-Hours	Unit	Material	2021 Bare Costs Labor	Equipment	Total	Total Incl O&P
0010	**RECIRCULATING LABORATORY FUME HOODS**									
0670	Service fixtures, average				Ea.	375			375	415
0680	For sink assembly with hot and cold water, add	1 Plum	1.40	5.714	"	805	385		1,190	1,450

11 53 Laboratory Equipment

11 53 19 – Laboratory Sterilizers

11 53 19.13 Sterilizers	Crew	Daily Output	Labor-Hours	Unit	Material	2021 Bare Costs Labor	2021 Bare Costs Equipment	Total	Total Incl O&P
0010 **STERILIZERS**									
0700 Glassware washer, undercounter, minimum	L-1	1.80	8.889	Ea.	6,950	585		7,535	8,525
0710 Maximum	"	1	16		15,200	1,050		16,250	18,400
1850 Utensil washer-sanitizer	1 Plum	2	4	↓	8,675	271		8,946	9,950

11 53 23 – Laboratory Refrigerators

11 53 23.13 Refrigerators

	Crew	Daily Output	Labor-Hours	Unit	Material	Labor	Equipment	Total	Total Incl O&P
0010 **REFRIGERATORS**									
1200 Blood bank, 28.6 C.F. emergency signal				Ea.	14,100			14,100	15,600
1210 Reach-in, 16.9 C.F.				"	9,550			9,550	10,500

11 53 33 – Emergency Safety Appliances

11 53 33.13 Emergency Equipment

	Crew	Daily Output	Labor-Hours	Unit	Material	Labor	Equipment	Total	Total Incl O&P
0010 **EMERGENCY EQUIPMENT**									
1400 Safety equipment, eye wash, hand held				Ea.	370			370	410
1450 Deluge shower				"	825			825	905

11 53 43 – Service Fittings and Accessories

11 53 43.13 Fittings

	Crew	Daily Output	Labor-Hours	Unit	Material	Labor	Equipment	Total	Total Incl O&P
0010 **FITTINGS**									
1600 Sink, one piece plastic, flask wash, hose, free standing	1 Plum	1.60	5	Ea.	2,075	340		2,415	2,775
1610 Epoxy resin sink, 25" x 16" x 10"	"	2	4	"	228	271		499	655
8000 Alternate pricing method: as percent of lab furniture									
8050 Installation, not incl. plumbing & duct work				% Furn.				22%	22%
8100 Plumbing, final connections, simple system								10%	10%
8110 Moderately complex system								15%	15%
8120 Complex system								20%	20%
8150 Electrical, simple system								10%	10%
8160 Moderately complex system								20%	20%
8170 Complex system				↓				35%	35%

11 71 Medical Sterilizing Equipment

11 71 10 – Medical Sterilizers & Distillers

11 71 10.10 Sterilizers and Distillers

	Crew	Daily Output	Labor-Hours	Unit	Material	Labor	Equipment	Total	Total Incl O&P
0010 **STERILIZERS AND DISTILLERS**									
0700 Distiller, water, steam heated, 50 gal. capacity	1 Plum	1.40	5.714	Ea.	22,500	385		22,885	25,300
5600 Sterilizers, floor loading, 26" x 62" x 42", single door, steam					162,000			162,000	178,000
5650 Double door, steam					253,000			253,000	278,500
5800 General purpose, 20" x 20" x 38", single door					11,700			11,700	12,900
6000 Portable, counter top, steam, minimum					3,300			3,300	3,625
6020 Maximum					5,125			5,125	5,650
6050 Portable, counter top, gas, 17" x 15" x 32-1/2"					35,600			35,600	39,200
6150 Manual washer/sterilizer, 16" x 16" x 26"	1 Plum	2	4	↓	48,800	271		49,071	54,000
6200 Steam generators, electric 10 kW to 180 kW, freestanding									
6250 Minimum	1 Elec	3	2.667	Ea.	12,800	170		12,970	14,400
6300 Maximum	"	.70	11.429		22,600	730		23,330	25,900
8200 Bed pan washer-sanitizer	1 Plum	2	4	↓	9,875	271		10,146	11,300

11 73 Patient Care Equipment

11 73 10 – Patient Treatment Equipment

11 73 10.10 Treatment Equipment	Crew	Daily Output	Labor-Hours	Unit	Material	2021 Bare Costs Labor	Equipment	Total	Total Incl O&P
0010 **TREATMENT EQUIPMENT**									
1800 Heat therapy unit, humidified, 26" x 78" x 28"				Ea.	3,525			3,525	3,875
2100 Hubbard tank with accessories, stainless steel,									
2110 125 GPM at 45 psi water pressure				Ea.	14,900			14,900	16,400
2150 For electric overhead hoist, add					3,250			3,250	3,575
3600 Paraffin bath, 126°F, auto controlled					270			270	297
4600 Station, dietary, medium, with ice					19,500			19,500	21,500
8400 Whirlpool bath, mobile, sst, 18" x 24" x 60"					6,200			6,200	6,825
8450 Fixed, incl. mixing valves	1 Plum	2	4	▼	4,775	271		5,046	5,650

11 74 Dental Equipment

11 74 10 – Dental Office Equipment

11 74 10.10 Diagnostic and Treatment Equipment

	Crew	Daily Output	Labor-Hours	Unit	Material	2021 Bare Costs Labor	Equipment	Total	Total Incl O&P
0010 **DIAGNOSTIC AND TREATMENT EQUIPMENT**									
0020 Central suction system, minimum	1 Plum	1.20	6.667	Ea.	1,275	450		1,725	2,075
0100 Maximum	"	.90	8.889		6,000	600		6,600	7,500
0300 Air compressor, minimum	1 Skwk	.80	10		2,875	570		3,445	4,000
0400 Maximum		.50	16		9,175	915		10,090	11,500
0600 Chair, electric or hydraulic, minimum		.50	16		2,700	915		3,615	4,325
0700 Maximum		.25	32		4,250	1,825		6,075	7,425
2000 Light, ceiling mounted, minimum		8	1		840	57		897	1,000
2100 Maximum		8	1		2,050	57		2,107	2,325
2200 Unit light, minimum	2 Skwk	5.33	3.002		755	171		926	1,100
2210 Maximum		5.33	3.002		1,575	171		1,746	1,975
2220 Track light, minimum		3.20	5		1,750	286		2,036	2,350
2230 Maximum	▼	3.20	5		2,400	286		2,686	3,050
2300 Sterilizers, steam portable, minimum					1,775			1,775	1,950
2350 Maximum					5,350			5,350	5,900
2600 Steam, institutional					2,725			2,725	3,000
2650 Dry heat, electric, portable, 3 trays				▼	1,400			1,400	1,525

11 76 Operating Room Equipment

11 76 10 – Equipment for Operating Rooms

11 76 10.10 Surgical Equipment

	Crew	Daily Output	Labor-Hours	Unit	Material	2021 Bare Costs Labor	Equipment	Total	Total Incl O&P
0010 **SURGICAL EQUIPMENT**									
5000 Scrub, surgical, stainless steel, single station, minimum	1 Plum	3	2.667	Ea.	6,575	181		6,756	7,525
5100 Maximum				"	9,625			9,625	10,600

11 78 Mortuary Equipment

11 78 13 – Mortuary Refrigerators

11 78 13.10 Mortuary and Autopsy Equipment	Crew	Daily Output	Labor-Hours	Unit	Material	2021 Bare Costs Labor	2021 Bare Costs Equipment	Total	Total Incl O&P
0010 **MORTUARY AND AUTOPSY EQUIPMENT**									
0015 Autopsy table, standard	1 Plum	1	8	Ea.	9,900	540		10,440	11,700
0020 Deluxe	"	.60	13.333	"	16,100	905		17,005	19,100

11 81 Facility Maintenance Equipment

11 81 19 – Vacuum Cleaning Systems

11 81 19.10 Vacuum Cleaning

	Crew	Daily Output	Labor-Hours	Unit	Material	2021 Bare Costs Labor	2021 Bare Costs Equipment	Total	Total Incl O&P
0010 **VACUUM CLEANING**									
0020 Central, 3 inlet, residential	1 Skwk	.90	8.889	Total	1,175	510		1,685	2,075
0200 Commercial		.70	11.429		1,375	655		2,030	2,500
0400 5 inlet system, residential		.50	16		1,850	915		2,765	3,400
0600 7 inlet system, commercial		.40	20		2,400	1,150		3,550	4,375
0800 9 inlet system, residential		.30	26.667		4,275	1,525		5,800	7,000

11 91 Religious Equipment

11 91 13 – Baptisteries

11 91 13.10 Baptistry

	Crew	Daily Output	Labor-Hours	Unit	Material	2021 Bare Costs Labor	2021 Bare Costs Equipment	Total	Total Incl O&P
0010 **BAPTISTRY**									
0150 Fiberglass, 3'-6" deep, x 13'-7" long,									
0160 steps at both ends, incl. plumbing, minimum	L-8	1	20	Ea.	7,575	1,150		8,725	10,000
0200 Maximum	"	.70	28.571		12,400	1,625		14,025	16,100
0250 Add for filter, heater and lights					1,700			1,700	1,875

11 91 23 – Sanctuary Equipment

11 91 23.10 Sanctuary Furnishings

	Crew	Daily Output	Labor-Hours	Unit	Material	2021 Bare Costs Labor	2021 Bare Costs Equipment	Total	Total Incl O&P
0010 **SANCTUARY FURNISHINGS**									
0020 Altar, wood, custom design, plain	1 Carp	1.40	5.714	Ea.	2,775	315		3,090	3,550
0050 Deluxe	"	.20	40	"	13,400	2,200		15,600	18,000

11 98 Detention Equipment

11 98 30 – Detention Cell Equipment

11 98 30.10 Cell Equipment

	Crew	Daily Output	Labor-Hours	Unit	Material	2021 Bare Costs Labor	2021 Bare Costs Equipment	Total	Total Incl O&P
0010 **CELL EQUIPMENT**									
3000 Toilet apparatus including wash basin, average	L-8	1.50	13.333	Ea.	3,525	765		4,290	5,025

For customer support on your Plumbing Costs with RSMeans Data, call 800.448.8182.

Estimating Tips
General
- The items in this division are usually priced per square foot or each. Most of these items are purchased by the owner and installed by the contractor. Do not assume the items in Division 12 will be purchased and installed by the contractor. Check the specifications for responsibilities and include receiving, storage, installation, and mechanical and electrical hookups in the appropriate divisions.
- Some items in this division require some type of support system that is not usually furnished with the item. Examples of these systems include blocking for the attachment of casework and heavy drapery rods. The required blocking must be added to the estimate in the appropriate division.

Reference Numbers
Reference numbers are shown at the beginning of some major classifications. These numbers refer to related items in the Reference Section. The reference information may be an estimating procedure, an alternate pricing method, or technical information.

Same Data. Simplified.

Enjoy the convenience and efficiency of accessing your costs anywhere:

- **Skip the multiplier** by setting your location
- **Quickly search,** edit, favorite and share costs
- **Stay on top of price changes** with automatic updates

Discover more at rsmeans.com/online

12 32 Manufactured Wood Casework

12 32 23 – Hardwood Casework

12 32 23.10 Manufactured Wood Casework, Stock Units	Crew	Daily Output	Labor-Hours	Unit	Material	2021 Bare Costs Labor	Equipment	Total	Total Incl O&P
0010 **MANUFACTURED WOOD CASEWORK, STOCK UNITS**									
0700 Kitchen base cabinets, hardwood, not incl. counter tops,									
0710 24" deep, 35" high, prefinished									
0800 One top drawer, one door below, 12" wide	2 Carp	24.80	.645	Ea.	320	35.50		355.50	405
0840 18" wide		23.30	.687		215	37.50		252.50	293
0880 24" wide		22.30	.717		415	39.50		454.50	520
1000 Four drawers, 12" wide		24.80	.645		296	35.50		331.50	380
1040 18" wide		23.30	.687		330	37.50		367.50	415
1060 24" wide		22.30	.717		375	39.50		414.50	475
1200 Two top drawers, two doors below, 27" wide		22	.727		440	40		480	540
1260 36" wide		20.30	.788		485	43		528	600
1300 48" wide		18.90	.847		585	46.50		631.50	710
1500 Range or sink base, two doors below, 30" wide		21.40	.748		360	41		401	460
1540 36" wide		20.30	.788		420	43		463	525
1580 48" wide		18.90	.847		465	46.50		511.50	580
1800 For sink front units, deduct					172			172	190
9000 For deluxe models of all cabinets, add					40%				
9500 For custom built in place, add					25%	10%			
9558 Rule of thumb, kitchen cabinets not including									
9560 appliances & counter top, minimum	2 Carp	30	.533	L.F.	202	29		231	267
9600 Maximum	"	25	.640	"	450	35		485	550

12 32 23.30 Manufactured Wood Casework Vanities

12 32 23.30 Manufactured Wood Casework Vanities	Crew	Daily Output	Labor-Hours	Unit	Material	2021 Bare Costs Labor	Equipment	Total	Total Incl O&P
0010 **MANUFACTURED WOOD CASEWORK VANITIES**									
8000 Vanity bases, 2 doors, 30" high, 21" deep, 24" wide	2 Carp	20	.800	Ea.	440	44		484	550
8050 30" wide		16	1		465	54.50		519.50	590
8100 36" wide		13.33	1.200		395	65.50		460.50	535
8150 48" wide		11.43	1.400		515	76.50		591.50	680
9000 For deluxe models of all vanities, add to above					40%				
9500 For custom built in place, add to above					25%	10%			

Estimating Tips
General

- The items and systems in this division are usually estimated, purchased, supplied, and installed as a unit by one or more subcontractors. The estimator must ensure that all parties are operating from the same set of specifications and assumptions, and that all necessary items are estimated and will be provided. Many times the complex items and systems are covered, but the more common ones, such as excavation or a crane, are overlooked for the very reason that everyone assumes nobody could miss them. The estimator should be the central focus and be able to ensure that all systems are complete.

- It is important to consider factors such as site conditions, weather, shape and size of building, as well as labor availability as they may impact the overall cost of erecting special structures and systems included in this division.

- Another area where problems can develop in this division is at the interface between systems.

The estimator must ensure, for instance, that anchor bolts, nuts, and washers are estimated and included for the air-supported structures and pre-engineered buildings to be bolted to their foundations. Utility supply is a common area where essential items or pieces of equipment can be missed or overlooked because each subcontractor may feel it is another's responsibility. The estimator should also be aware of certain items which may be supplied as part of a package but installed by others, and ensure that the installing contractor's estimate includes the cost of installation. Conversely, the estimator must also ensure that items are not costed by two different subcontractors, resulting in an inflated overall estimate.

13 30 00 Special Structures

- The foundations and floor slab, as well as rough mechanical and electrical, should be estimated, as this work is required for the assembly and erection of the structure. Generally, as noted in the data set, the pre-engineered building comes

as a shell. Pricing is based on the size and structural design parameters stated in the reference section. Additional features, such as windows and doors with their related structural framing, must also be included by the estimator. Here again, the estimator must have a clear understanding of the scope of each portion of the work and all the necessary interfaces.

Reference Numbers

Reference numbers are shown at the beginning of some major classifications. These numbers refer to related items in the Reference Section. The reference information may be an estimating procedure, an alternate pricing method, or technical information.

Note: Not all subdivisions listed here necessarily appear. ■

Same Data. Simplified.

Enjoy the convenience and efficiency of accessing your costs anywhere:

- **Skip the multiplier** by setting your location
- **Quickly search,** edit, favorite and share costs
- **Stay on top of price changes** with automatic updates

Discover more at rsmeans.com/online

13 11 Swimming Pools

13 11 13 – Below-Grade Swimming Pools

13 11 13.50 Swimming Pools

		Crew	Daily Output	Labor-Hours	Unit	Material	2021 Bare Costs Labor	Equipment	Total	Total Incl O&P
0010	**SWIMMING POOLS** Residential in-ground, vinyl lined									
0020	Concrete sides, w/equip, sand bottom	B-52	300	.187	SF Surf	29.50	9.50	1.91	40.91	48.50
0100	Metal or polystyrene sides R131113-20	B-14	410	.117		24.50	5.45	.53	30.48	35.50
0200	Add for vermiculite bottom					1.87			1.87	2.06
0500	Gunite bottom and sides, white plaster finish									
0600	12' x 30' pool	B-52	145	.386	SF Surf	54.50	19.60	3.95	78.05	93.50
0720	16' x 32' pool		155	.361		49	18.35	3.70	71.05	85.50
0750	20' x 40' pool		250	.224		44	11.35	2.29	57.64	68
0810	Concrete bottom and sides, tile finish									
0820	12' x 30' pool	B-52	80	.700	SF Surf	55	35.50	7.15	97.65	121
0830	16' x 32' pool		95	.589		45.50	30	6.05	81.55	101
0840	20' x 40' pool		130	.431		36.50	22	4.41	62.91	77.50
1100	Motel, gunite with plaster finish, incl. medium									
1150	capacity filtration & chlorination	B-52	115	.487	SF Surf	67.50	24.50	4.98	96.98	117
1200	Municipal, gunite with plaster finish, incl. high									
1250	capacity filtration & chlorination	B-52	100	.560	SF Surf	87	28.50	5.75	121.25	144
1350	Add for formed gutters				L.F.	128			128	141
1360	Add for stainless steel gutters				"	380			380	415
1600	For water heating system, see Section 23 52 28.10									
1700	Filtration and deck equipment only, as % of total				Total				20%	20%
1800	Automatic vacuum, hand tools, etc., 20' x 40' pool				SF Pool				.56	.62
1900	5,000 S.F. pool				"				.11	.12
3000	Painting pools, preparation + 3 coats, 20' x 40' pool, epoxy	2 Pord	.33	48.485	Total	1,300	2,250		3,550	4,775
3100	Rubber base paint, 18 gallons	"	.33	48.485		1,250	2,250		3,500	4,725
3500	42' x 82' pool, 75 gallons, epoxy paint	3 Pord	.14	171		5,500	7,975		13,475	17,900
3600	Rubber base paint	"	.14	171		5,150	7,975		13,125	17,500

13 11 23 – On-Grade Swimming Pools

13 11 23.50 Swimming Pools

		Crew	Daily Output	Labor-Hours	Unit	Material	2021 Bare Costs Labor	Equipment	Total	Total Incl O&P
0010	**SWIMMING POOLS** Residential above ground, steel construction									
0100	Round, 15' diam.	B-80A	3	8	Ea.	880	355	283	1,518	1,800
0120	18' diam.		2.50	9.600		935	425	340	1,700	2,025
0140	21' diam.		2	12		1,050	535	425	2,010	2,425
0160	24' diam.		1.80	13.333		1,200	590	470	2,260	2,700
0180	27' diam.		1.50	16		1,500	710	565	2,775	3,325
0200	30' diam.		1	24		1,725	1,075	850	3,650	4,425
0220	Oval, 12' x 24'		2.30	10.435		1,525	465	370	2,360	2,775
0240	15' x 30'		1.80	13.333		1,700	590	470	2,760	3,275
0260	18' x 33'		1	24		1,925	1,075	850	3,850	4,650

13 11 46 – Swimming Pool Accessories

13 11 46.50 Swimming Pool Equipment

		Crew	Daily Output	Labor-Hours	Unit	Material	2021 Bare Costs Labor	Equipment	Total	Total Incl O&P
0010	**SWIMMING POOL EQUIPMENT**									
0020	Diving stand, stainless steel, 3 meter	2 Carp	.40	40	Ea.	19,800	2,200		22,000	25,100
0300	1 meter	"	2.70	5.926		11,400	325		11,725	13,000
2100	Lights, underwater, 12 volt, with transformer, 300 watt	1 Elec	1	8		465	510		975	1,275
2200	110 volt, 500 watt, standard		1	8		291	510		801	1,075
2400	Low water cutoff type		1	8		315	510		825	1,100
2800	Heaters, see Section 23 52 28.10									

13 12 Fountains

13 12 13 - Exterior Fountains

13 12 13.10 Outdoor Fountains

		Crew	Daily Output	Labor-Hours	Unit	Material	2021 Bare Costs Labor	2021 Bare Costs Equipment	Total	Total Incl O&P
0010	**OUTDOOR FOUNTAINS**									
0100	Outdoor fountain, 48" high with bowl and figures	2 Clab	2	8	Ea.	385	355		740	955
0200	Commercial, concrete or cast stone, 40-60" H, simple		2	8		975	355		1,330	1,600
0220	Average		2	8		1,900	355		2,255	2,600
0240	Ornate		2	8		4,300	355		4,655	5,275
0260	Metal, 72" high		2	8		1,525	355		1,880	2,200
0280	90" high		2	8		2,200	355		2,555	2,950
0300	120" high		2	8		5,275	355		5,630	6,325
0320	Resin or fiberglass, 40-60" H, wall type		2	8		690	355		1,045	1,300
0340	Waterfall type		2	8		1,175	355		1,530	1,800

13 12 23 - Interior Fountains

13 12 23.10 Indoor Fountains

		Crew	Daily Output	Labor-Hours	Unit	Material	2021 Bare Costs Labor	2021 Bare Costs Equipment	Total	Total Incl O&P
0010	**INDOOR FOUNTAINS**									
0100	Commercial, floor type, resin or fiberglass, lighted, cascade type	2 Clab	2	8	Ea.	435	355		790	1,000
0120	Tiered type		2	8		680	355		1,035	1,275
0140	Waterfall type		2	8		485	355		840	1,050

13 17 Tubs and Pools

13 17 13 - Hot Tubs

13 17 13.10 Redwood Hot Tub System

		Crew	Daily Output	Labor-Hours	Unit	Material	2021 Bare Costs Labor	2021 Bare Costs Equipment	Total	Total Incl O&P
0010	**REDWOOD HOT TUB SYSTEM**									
7050	4' diameter x 4' deep	Q-1	1	16	Ea.	4,650	975		5,625	6,575
7100	5' diameter x 4' deep		1	16		4,950	975		5,925	6,900
7150	6' diameter x 4' deep		.80	20		6,300	1,225		7,525	8,750
7200	8' diameter x 4' deep		.80	20		9,000	1,225		10,225	11,700

13 17 33 - Whirlpool Tubs

13 17 33.10 Whirlpool Bath

		Crew	Daily Output	Labor-Hours	Unit	Material	2021 Bare Costs Labor	2021 Bare Costs Equipment	Total	Total Incl O&P
0010	**WHIRLPOOL BATH**									
6000	Whirlpool, bath with vented overflow, molded fiberglass									
6100	66" x 36" x 24"	Q-1	1	16	Ea.	1,075	975		2,050	2,625
6400	72" x 36" x 21"		1	16		1,550	975		2,525	3,150
6500	60" x 34" x 21"		1	16		1,150	975		2,125	2,725
6600	72" x 42" x 23"		1	16		1,450	975		2,425	3,050
6710	For color, add					10%				
6711	For designer colors and trim, add					25%				

13 24 Special Activity Rooms

13 24 16 - Saunas

13 24 16.50 Saunas and Heaters

		Crew	Daily Output	Labor-Hours	Unit	Material	2021 Bare Costs Labor	2021 Bare Costs Equipment	Total	Total Incl O&P
0010	**SAUNAS AND HEATERS**									
0020	Prefabricated, incl. heater & controls, 7' high, 6' x 4', C/C	L-7	2.20	12.727	Ea.	6,425	675		7,100	8,050
0050	6' x 4', C/P		2	14		4,300	745		5,045	5,825
0400	6' x 5', C/C		2	14		5,350	745		6,095	6,975
0450	6' x 5', C/P		2	14		5,150	745		5,895	6,775
0600	6' x 6', C/C		1.80	15.556		6,850	825		7,675	8,750
0650	6' x 6', C/P		1.80	15.556		5,725	825		6,550	7,525
0800	6' x 9', C/C		1.60	17.500		7,400	930		8,330	9,525
0850	6' x 9', C/P		1.60	17.500		5,825	930		6,755	7,775

83

For customer support on your Plumbing Costs with RSMeans Data, call 800.448.8182.

13 24 Special Activity Rooms

13 24 16 – Saunas

13 24 16.50 Saunas and Heaters

	13 24 16.50 Saunas and Heaters	Crew	Daily Output	Labor-Hours	Unit	Material	2021 Bare Costs Labor	Equipment	Total	Total Incl O&P
1000	8' x 12', C/C	L-7	1.10	25.455	Ea.	13,900	1,350		15,250	17,300
1050	8' x 12', C/P		1.10	25.455		7,500	1,350		8,850	10,300
1200	8' x 8', C/C		1.40	20		9,750	1,050		10,800	12,300
1250	8' x 8', C/P		1.40	20		7,100	1,050		8,150	9,400
1400	8' x 10', C/C		1.20	23.333		9,050	1,250		10,300	11,800
1450	8' x 10', C/P		1.20	23.333		8,250	1,250		9,500	10,900
1600	10' x 12', C/C		1	28		14,200	1,475		15,675	17,900
1650	10' x 12', C/P	↓	1	28		14,300	1,475		15,775	17,900
2500	Heaters only (incl. above), wall mounted, to 200 C.F.					1,025			1,025	1,125
2750	To 300 C.F.					1,150			1,150	1,275
3000	Floor standing, to 720 C.F., 10,000 watts, w/controls	1 Elec	3	2.667		2,950	170		3,120	3,475
3250	To 1,000 C.F., 16,000 watts	"	3	2.667	↓	4,050	170		4,220	4,700

13 24 26 – Steam Baths

13 24 26.50 Steam Baths and Components

	13 24 26.50 Steam Baths and Components	Crew	Daily Output	Labor-Hours	Unit	Material	2021 Bare Costs Labor	Equipment	Total	Total Incl O&P
0010	**STEAM BATHS AND COMPONENTS**									
0020	Heater, timer & head, single, to 140 C.F.	1 Plum	1.20	6.667	Ea.	2,150	450		2,600	3,025
0500	To 300 C.F.		1.10	7.273		2,425	490		2,915	3,400
1000	Commercial size, with blow-down assembly, to 800 C.F.		.90	8.889		6,125	600		6,725	7,625
1500	To 2,500 C.F.	↓	.80	10		7,225	675		7,900	8,950
2000	Multiple, motels, apts., 2 baths, w/blow-down assm., 500 C.F.	Q-1	1.30	12.308		6,750	750		7,500	8,550
2500	4 baths	"	.70	22.857		10,200	1,400		11,600	13,300
2700	Conversion unit for residential tub, including door				↓	4,150			4,150	4,550

13 34 Fabricated Engineered Structures

13 34 23 – Fabricated Structures

13 34 23.10 Comfort Stations

	13 34 23.10 Comfort Stations	Crew	Daily Output	Labor-Hours	Unit	Material	2021 Bare Costs Labor	Equipment	Total	Total Incl O&P
0010	**COMFORT STATIONS** Prefab., stock, w/doors, windows & fixt.									
0100	Not incl. interior finish or electrical									
0300	Mobile, on steel frame, 2 unit				S.F.	195			195	215
0350	7 unit					310			310	340
0400	Permanent, including concrete slab, 2 unit	B-12J	50	.320		251	16.95	17	284.95	320
0500	6 unit	"	43	.372	↓	214	19.70	19.80	253.50	287
0600	Alternate pricing method, mobile, 2 fixture				Fixture	6,675			6,675	7,350
0650	7 fixture					11,600			11,600	12,800
0700	Permanent, 2 unit	B-12J	.70	22.857		20,700	1,200	1,225	23,125	25,900
0750	6 unit	"	.50	32	↓	17,800	1,700	1,700	21,200	24,000

13 42 Building Modules

13 42 63 – Detention Cell Modules

13 42 63.16 Steel Detention Cell Modules

	13 42 63.16 Steel Detention Cell Modules	Crew	Daily Output	Labor-Hours	Unit	Material	2021 Bare Costs Labor	Equipment	Total	Total Incl O&P
0010	**STEEL DETENTION CELL MODULES**									
2000	Cells, prefab., 5' to 6' wide, 7' to 8' high, 7' to 8' deep,									
2010	bar front, cot, not incl. plumbing	E-4	1.50	21.333	Ea.	9,650	1,300	99	11,049	12,700

13 47 13 – Cathodic Protection

13 47 13.16 Cathodic Prot. for Underground Storage Tanks	Crew	Daily Output	Labor-Hours	Unit	Material	2021 Bare Costs Labor	Equipment	Total	Total Incl O&P
0010 **CATHODIC PROTECTION FOR UNDERGROUND STORAGE TANKS**									
1000 Anodes, magnesium type, 9 #	R-15	18.50	2.595	Ea.	39.50	162	15.25	216.75	300
1010 17 #		13	3.692		81.50	230	21.50	333	460
1020 32 #		10	4.800		120	300	28	448	610
1030 48 #	↓	7.20	6.667		154	415	39	608	830
1100 Graphite type w/epoxy cap, 3" x 60" (32 #)	R-22	8.40	4.438		142	259		401	540
1110 4" x 80" (68 #)		6	6.213		245	360		605	810
1120 6" x 72" (80 #)		5.20	7.169		1,500	420		1,920	2,275
1130 6" x 36" (45 #)	↓	9.60	3.883		750	226		976	1,150
2000 Rectifiers, silicon type, air cooled, 28 V/10 A	R-19	3.50	5.714		2,350	365		2,715	3,150
2010 20 V/20 A		3.50	5.714		2,375	365		2,740	3,150
2100 Oil immersed, 28 V/10 A		3	6.667		2,400	425		2,825	3,275
2110 20 V/20 A	↓	3	6.667	↓	3,175	425		3,600	4,125
3000 Anode backfill, coke breeze	R-22	3850	.010	Lb.	.27	.56		.83	1.14
4000 Cable, HMWPE, No. 8		2.40	15.533	M.L.F.	435	905		1,340	1,825
4010 No. 6		2.40	15.533		625	905		1,530	2,050
4020 No. 4		2.40	15.533		950	905		1,855	2,400
4030 No. 2		2.40	15.533		1,700	905		2,605	3,225
4040 No. 1		2.20	16.945		2,000	990		2,990	3,675
4050 No. 1/0		2.20	16.945		2,925	990		3,915	4,675
4060 No. 2/0		2.20	16.945		4,800	990		5,790	6,750
4070 No. 4/0	↓	2	18.640	↓	8,500	1,075		9,575	11,000
5000 Test station, 7 terminal box, flush curb type w/lockable cover	R-19	12	1.667	Ea.	79.50	106		185.50	246
5010 Reference cell, 2" diam. PVC conduit, cplg., plug, set flush	"	4.80	4.167	"	158	266		424	570

Division Notes

	CREW	DAILY OUTPUT	LABOR-HOURS	UNIT	BARE COSTS				TOTAL INCL O&P
					MAT.	LABOR	EQUIP.	TOTAL	

Estimating Tips
General
- Many products in Division 14 will require some type of support or blocking for installation not included with the item itself. Examples are supports for conveyors or tube systems, attachment points for lifts, and footings for hoists or cranes. Add these supports in the appropriate division.

14 10 00 Dumbwaiters
14 20 00 Elevators
- Dumbwaiters and elevators are estimated and purchased in a method similar to buying a car. The manufacturer has a base unit with standard features. Added to this base unit price will be whatever options the owner or specifications require. Increased load capacity, additional vertical travel, additional stops, higher speed, and cab finish options are items to be considered. When developing an estimate for dumbwaiters and elevators, remember that some items needed by the installers may have to be included as part of the general contract.

Examples are:
- ☐ shaftway
- ☐ rail support brackets
- ☐ machine room
- ☐ electrical supply
- ☐ sill angles
- ☐ electrical connections
- ☐ pits
- ☐ roof penthouses
- ☐ pit ladders

Check the job specifications and drawings before pricing.

- Installation of elevators and handicapped lifts in historic structures can require significant additional costs. The associated structural requirements may involve cutting into and repairing finishes, moldings, flooring, etc. The estimator must account for these special conditions.

14 30 00 Escalators and Moving Walks
- Escalators and moving walks are specialty items installed by specialty contractors. There are numerous options associated with these items. For specific options, contact a manufacturer or contractor. In a method similar to estimating dumbwaiters and elevators, you should verify the extent of general contract work and add items as necessary.

14 40 00 Lifts
14 90 00 Other Conveying Equipment
- Products such as correspondence lifts, chutes, and pneumatic tube systems, as well as other items specified in this subdivision, may require trained installers. The general contractor might not have any choice as to who will perform the installation or when it will be performed. Long lead times are often required for these products, making early decisions in scheduling necessary.

Reference Numbers
Reference numbers are shown at the beginning of some major classifications. These numbers refer to related items in the Reference Section. The reference information may be an estimating procedure, an alternate pricing method, or technical information.

Note: Not all subdivisions listed here necessarily appear. ■

Same Data. Simplified.

Enjoy the convenience and efficiency of accessing your costs anywhere:

- **Skip the multiplier** by setting your location
- **Quickly search,** edit, favorite and share costs
- **Stay on top of price changes** with automatic updates

Discover more at rsmeans.com/online

14 45 10 – Hydraulic Vehicle Lifts

14 45 10.10 Hydraulic Lifts		Crew	Daily Output	Labor-Hours	Unit	Material	2021 Bare Costs Labor	Equipment	Total	Total Incl O&P
0010	**HYDRAULIC LIFTS**									
2200	Single post, 8,000 lb. capacity	L-4	.40	60	Ea.	7,600	3,125		10,725	13,100
2810	Double post, 6,000 lb. capacity		2.67	8.989		2,375	470		2,845	3,325
2815	9,000 lb. capacity		2.29	10.480		3,125	545		3,670	4,275
2820	15,000 lb. capacity		2	12		6,525	625		7,150	8,125
2822	Four post, 26,000 lb. capacity		1.80	13.333		15,900	695		16,595	18,600
2825	30,000 lb. capacity		1.60	15		62,000	780		62,780	69,000
2830	Ramp style, 4 post, 25,000 lb. capacity		2	12		15,400	625		16,025	17,900
2835	35,000 lb. capacity		1	24		19,400	1,250		20,650	23,300
2840	50,000 lb. capacity		1	24		129,500	1,250		130,750	144,500
2845	75,000 lb. capacity		1	24		150,500	1,250		151,750	167,500
2850	For drive thru tracks, add, minimum					1,100			1,100	1,200
2855	Maximum					1,875			1,875	2,075
2860	Ramp extensions, 3' (set of 2)					445			445	490
2865	Rolling jack platform					1,575			1,575	1,725
2870	Electric/hydraulic jacking beam					12,200			12,200	13,400
2880	Scissor lift, portable, 6,000 lb. capacity					1,700			1,700	1,875

Estimating Tips

Pipe for fire protection and all uses is located in Subdivisions 21 11 13 and 22 11 13.

The labor adjustment factors listed in Subdivision 22 01 02.20 also apply to Division 21.

Many, but not all, areas in the U.S. require backflow protection in the fire system. Insurance underwriters may have specific requirements for the type of materials to be installed or design requirements based on the hazard to be protected. Local jurisdictions may have requirements not covered by code. It is advisable to be aware of any special conditions.

For your reference, the following is a list of the most applicable Fire Codes and Standards, which may be purchased from the NFPA, 1 Batterymarch Park, Quincy, MA 02169-7471.

- NFPA 1: Uniform Fire Code
- NFPA 10: Portable Fire Extinguishers
- NFPA 11: Low-, Medium-, and High-Expansion Foam
- NFPA 12: Carbon Dioxide Extinguishing Systems (Also companion 12A)
- NFPA 13: Installation of Sprinkler Systems (Also companion 13D, 13E, and 13R)
- NFPA 14: Installation of Standpipe and Hose Systems
- NFPA 15: Water Spray Fixed Systems for Fire Protection
- NFPA 16: Installation of Foam-Water Sprinkler and Foam-Water Spray Systems
- NFPA 17: Dry Chemical Extinguishing Systems (Also companion 17A)
- NFPA 18: Wetting Agents
- NFPA 20: Installation of Stationary Pumps for Fire Protection
- NFPA 22: Water Tanks for Private Fire Protection
- NFPA 24: Installation of Private Fire Service Mains and their Appurtenances
- NFPA 25: Inspection, Testing and Maintenance of Water-Based Fire Protection

Reference Numbers

Reference numbers are shown at the beginning of some major classifications. These numbers refer to related items in the Reference Section. The reference information may be an estimating procedure, an alternate pricing method, or technical information.

Same Data. Simplified.

Enjoy the convenience and efficiency of accessing your costs anywhere:

- **Skip the multiplier** by setting your location
- **Quickly search,** edit, favorite and share costs
- **Stay on top of price changes** with automatic updates

Discover more at rsmeans.com/online

21 05 23 – General-Duty Valves for Water-Based Fire-Suppression Piping

21 05 23.50 General-Duty Valves	Crew	Daily Output	Labor-Hours	Unit	Material	2021 Bare Costs Labor	Equipment	Total	Total Incl O&P
0010 **GENERAL-DUTY VALVES**, for water-based fire suppression									
6200 Valves and components									
6210 Wet alarm, includes									
6220 retard chamber, trim, gauges, alarm line strainer									
6260 3" size	Q-12	3	5.333	Ea.	1,625	320		1,945	2,275
6280 4" size	"	2	8		2,475	480		2,955	3,450
6300 6" size	Q-13	4	8		2,375	505		2,880	3,375
6320 8" size	"	3	10.667		2,850	675		3,525	4,150
6400 Dry alarm, includes									
6405 retard chamber, trim, gauges, alarm line strainer									
6410 1-1/2" size	Q-12	3	5.333	Ea.	3,175	320		3,495	3,950
6420 2" size		3	5.333		3,175	320		3,495	3,950
6430 3" size		3	5.333		3,225	320		3,545	4,025
6440 4" size		2	8		3,425	480		3,905	4,500
6450 6" size	Q-13	3	10.667		3,950	675		4,625	5,350
6460 8" size	"	3	10.667		5,825	675		6,500	7,400
6500 Check, swing, C.I. body, brass fittings, auto. ball drip									
6520 4" size	Q-12	3	5.333	Ea.	435	320		755	955
6540 6" size	Q-13	4	8		805	505		1,310	1,650
6580 8" size	"	3	10.667		1,150	675		1,825	2,250
6800 Check, wafer, butterfly type, C.I. body, bronze fittings									
6820 4" size	Q-12	4	4	Ea.	1,925	239		2,164	2,475
6840 6" size	Q-13	5.50	5.818		4,050	370		4,420	5,000
6860 8" size		5	6.400		3,775	405		4,180	4,750
6880 10" size		4.50	7.111		5,750	450		6,200	7,000
8700 Floor control valve, includes trim and gauges, 2" size	Q-12	6	2.667		1,050	160		1,210	1,400
8710 2-1/2" size		6	2.667		815	160		975	1,125
8720 3" size		6	2.667		840	160		1,000	1,175
8730 4" size		6	2.667		840	160		1,000	1,175
8740 6" size		5	3.200		840	192		1,032	1,200
8800 Flow control valve, includes trim and gauges, 2" size		2	8		2,225	480		2,705	3,175
8820 3" size		1.50	10.667		6,850	640		7,490	8,500
8840 4" size	Q-13	2.80	11.429		3,575	725		4,300	5,000
8860 6" size	"	2	16		4,750	1,025		5,775	6,750
9200 Pressure operated relief valve, brass body	1 Spri	18	.444		750	29.50		779.50	870
9600 Waterflow indicator, vane type, with recycling retard and									
9610 two single pole retard switches, 2" thru 6" pipe size	1 Spri	8	1	Ea.	180	66.50		246.50	298

21 05 53 – Identification For Fire-Suppression Piping and Equipment

21 05 53.50 Identification

21 05 53.50 Identification	Crew	Daily Output	Labor-Hours	Unit	Material	2021 Bare Costs Labor	Equipment	Total	Total Incl O&P
0010 **IDENTIFICATION**, for fire suppression piping and equipment									
3010 Plates and escutcheons for identification of fire dept. service/connections									
3100 Wall mount, round, aluminum									
3110 4"	1 Plum	96	.083	Ea.	9.50	5.65		15.15	18.85
3120 6"	"	96	.083	"	46	5.65		51.65	59
3200 Wall mount, round, cast brass									
3210 2-1/2"	1 Plum	70	.114	Ea.	23	7.75		30.75	36.50
3220 3"		70	.114		26.50	7.75		34.25	41
3230 4"		70	.114		45.50	7.75		53.25	61.50
3240 6"		70	.114		53	7.75		60.75	70
3250 For polished brass, add					25%				
3260 For rough chrome, add					33%				
3270 For polished chrome, add					55%				

For customer support on your Plumbing Costs with RSMeans Data, call 800.448.8182.

21 05 Common Work Results for Fire Suppression

21 05 53 – Identification For Fire-Suppression Piping and Equipment

21 05 53.50 Identification

		Crew	Daily Output	Labor-Hours	Unit	Material	2021 Bare Costs Labor	Equipment	Total	Total Incl O&P
3300	Wall mount, square, cast brass									
3310	2-1/2"	1 Plum	70	.114	Ea.	61	7.75		68.75	78.50
3320	3"	"	70	.114	"	64.50	7.75		72.25	82.50
3330	For polished brass, add					15%				
3340	For rough chrome, add					25%				
3350	For polished chrome, add					33%				
3400	Wall mount, cast brass, multiple outlets									
3410	rect. 2 way	Q-1	5	3.200	Ea.	83.50	195		278.50	385
3420	rect. 3 way		4	4		198	244		442	580
3430	rect. 4 way		4	4		247	244		491	635
3440	square 4 way		4	4		247	244		491	635
3450	rect. 6 way		3	5.333		340	325		665	855
3460	For polished brass, add					10%				
3470	For rough chrome, add					20%				
3480	For polished chrome, add					25%				
3500	Base mount, free standing fdc, cast brass									
3510	4"	1 Plum	60	.133	Ea.	45.50	9.05		54.55	63.50
3520	6"	"	60	.133	"	61	9.05		70.05	80.50
3530	For polished brass, add					25%				
3540	For rough chrome, add					30%				
3550	For polished chrome, add					45%				

21 11 Facility Fire-Suppression Water-Service Piping

21 11 11 – Fire-Suppression, Pipe Fittings, Grooved Joint

21 11 11.05 Corrosion Monitors

		Crew	Daily Output	Labor-Hours	Unit	Material	2021 Bare Costs Labor	Equipment	Total	Total Incl O&P
0010	**CORROSION MONITORS**, pipe, in line spool									
1100	Powder coated, schedule 10									
1140	2"	1 Plum	17	.471	Ea.	230	32		262	300
1150	2-1/2"	Q-1	27	.593		236	36		272	315
1160	3"		22	.727		242	44.50		286.50	335
1170	4"		17	.941		255	57.50		312.50	365
1180	6"	Q-2	17	1.412		274	89		363	435
1190	8"	"	14	1.714		293	108		401	485
1200	Powder coated, schedule 40									
1240	2"	1 Plum	17	.471	Ea.	230	32		262	300
1250	2-1/2"	Q-1	27	.593		236	36		272	315
1260	3"		22	.727		243	44.50		287.50	335
1270	4"		17	.941		268	57.50		325.50	380
1280	6"	Q-2	17	1.412		294	89		383	460
1290	8"	"	14	1.714		320	108		428	510
1300	Galvanized, schedule 10									
1340	2"	1 Plum	17	.471	Ea.	249	32		281	320
1350	2-1/2"	Q-1	27	.593		262	36		298	340
1360	3"		22	.727		274	44.50		318.50	365
1370	4"		17	.941		287	57.50		344.50	400
1380	6"	Q-2	17	1.412		305	89		394	470
1390	8"	"	14	1.714		325	108		433	520
1400	Galvanized, schedule 40									
1440	2"	1 Plum	17	.471	Ea.	249	32		281	320
1450	2-1/2"	Q-1	27	.593		262	36		298	340
1460	3"		22	.727		274	44.50		318.50	365

21 11 11.05 Corrosion Monitors

		Crew	Daily Output	Labor-Hours	Unit	Material	2021 Bare Costs Labor	2021 Bare Costs Equipment	Total	Total Incl O&P
1470	4"	Q-1	17	.941	Ea.	300	57.50		357.50	415
1480	6"	Q-2	17	1.412		325	89		414	495
1490	8"	"	14	1.714	↓	350	108		458	545
1500	Mechanical tee, painted									
1540	2"	1 Plum	50	.160	Ea.	191	10.85		201.85	226
1550	2-1/2"	Q-1	80	.200		198	12.20		210.20	235
1560	3"		67	.239		204	14.55		218.55	246
1570	4"		50	.320		210	19.50		229.50	260
1580	6"	Q-2	50	.480		217	30.50		247.50	284
1590	8"	"	42	.571	↓	293	36		329	380
1600	Mechanical tee, galvanized									
1640	2"	1 Plum	50	.160	Ea.	204	10.85		214.85	240
1650	2-1/2"	Q-1	80	.200		210	12.20		222.20	249
1660	3"		67	.239		217	14.55		231.55	260
1670	4"		50	.320		223	19.50		242.50	274
1680	6"	Q-2	50	.480		230	30.50		260.50	298
1690	8"	"	42	.571	↓	330	36		366	420

21 11 11.16 Pipe Fittings, Grooved Joint

		Crew	Daily Output	Labor-Hours	Unit	Material	2021 Bare Costs Labor	2021 Bare Costs Equipment	Total	Total Incl O&P
0010	**PIPE FITTINGS, GROOVED JOINT,** For fire-suppression									
0020	Fittings, ductile iron									
0030	Coupling required at joints not incl. in fitting price.									
0034	Add 1 coupling, material only, per joint for installed price.									
0038	For standard grooved joint materials see Div. 22 11 13.48									
0040	90° elbow									
0110	2"	1 Plum	25	.320	Ea.	25.50	21.50		47	60.50
0120	2-1/2"	Q-1	40	.400		34	24.50		58.50	74
0130	3"		33	.485		44	29.50		73.50	92.50
0140	4"		25	.640		63	39		102	128
0150	5"		20	.800		147	48.50		195.50	235
0160	6"	Q-2	25	.960		173	60.50		233.50	281
0170	8"	"	21	1.143	↓	335	72		407	480
0200	45° elbow									
0210	2"	1 Plum	25	.320	Ea.	25.50	21.50		47	60.50
0220	2-1/2"	Q-1	40	.400		34	24.50		58.50	74
0230	3"		33	.485		44	29.50		73.50	92.50
0240	4"		25	.640		63	39		102	128
0250	5"		20	.800		147	48.50		195.50	235
0260	6"	Q-2	25	.960		173	60.50		233.50	281
0270	8"	"	21	1.143	↓	335	72		407	480
0300	Tee									
0310	2"	1 Plum	17	.471	Ea.	39	32		71	90
0320	2-1/2"	Q-1	27	.593		50.50	36		86.50	110
0330	3"		22	.727		70	44.50		114.50	143
0340	4"		17	.941		105	57.50		162.50	202
0350	5"		13	1.231		241	75		316	380
0360	6"	Q-2	17	1.412		278	89		367	440
0370	8"	"	14	1.714	↓	590	108		698	810
0400	Cap									
0410	1-1/4"	1 Plum	76	.105	Ea.	14	7.15		21.15	26
0420	1-1/2"		63	.127		14.80	8.60		23.40	29
0430	2"	↓	47	.170		17.60	11.50		29.10	36.50
0440	2-1/2"	Q-1	76	.211	↓	20.50	12.85		33.35	41.50

21 11 11.16 Pipe Fittings, Grooved Joint

		Crew	Daily Output	Labor-Hours	Unit	Material	2021 Bare Costs Labor	Equipment	Total	Total Incl O&P
0450	3"	Q-1	63	.254	Ea.	24	15.45		39.45	49.50
0460	4"		47	.340		35	20.50		55.50	69.50
0470	5"	↓	37	.432		70	26.50		96.50	117
0480	6"	Q-2	47	.511		77.50	32.50		110	134
0490	8"	"	39	.615	↓	147	39		186	220
0500	Coupling, rigid									
0510	1-1/4"	1 Plum	100	.080	Ea.	39	5.40		44.40	51
0520	1-1/2"		67	.119		40	8.10		48.10	56
0530	2"	↓	50	.160		44	10.85		54.85	64.50
0540	2-1/2"	Q-1	80	.200		51	12.20		63.20	74
0550	3"		67	.239		59.50	14.55		74.05	87
0560	4"		50	.320		81.50	19.50		101	119
0570	5"	↓	40	.400		104	24.50		128.50	151
0580	6"	Q-2	50	.480		120	30.50		150.50	178
0590	8"	"	42	.571	↓	199	36		235	273
0700	End of run fitting									
0710	1-1/4" x 1/2" NPT	1 Plum	76	.105	Ea.	47.50	7.15		54.65	62.50
0714	1-1/4" x 3/4" NPT		76	.105		47.50	7.15		54.65	62.50
0718	1-1/4" x 1" NPT		76	.105		47.50	7.15		54.65	62.50
0722	1-1/2" x 1/2" NPT		63	.127		47.50	8.60		56.10	65
0726	1-1/2" x 3/4" NPT		63	.127		47.50	8.60		56.10	65
0730	1-1/2" x 1" NPT		63	.127		47.50	8.60		56.10	65
0734	2" x 1/2" NPT		47	.170		50.50	11.50		62	72.50
0738	2" x 3/4" NPT		47	.170		50.50	11.50		62	72.50
0742	2" x 1" NPT		47	.170		50.50	11.50		62	72.50
0746	2-1/2" x 1/2" NPT		40	.200		60	13.55		73.55	86
0750	2-1/2" x 3/4" NPT		40	.200		60	13.55		73.55	86
0754	2-1/2" x 1" NPT	↓	40	.200	↓	60	13.55		73.55	86
0800	Drain elbow									
0820	2-1/2"	Q-1	40	.400	Ea.	118	24.50		142.50	167
0830	3"		33	.485		151	29.50		180.50	210
0840	4"	↓	25	.640		153	39		192	226
0850	6"	Q-2	25	.960	↓	267	60.50		327.50	385
1000	Valves, grooved joint									
1002	Coupling required at joints not incl. in fitting price.									
1004	Add 1 coupling, material only, per joint for installed price.									
1010	Ball valve with weatherproof actuator									
1020	1-1/4"	1 Plum	39	.205	Ea.	178	13.90		191.90	217
1030	1-1/2"		31	.258		211	17.45		228.45	258
1040	2"	↓	24	.333	↓	254	22.50		276.50	315
1100	Butterfly valve, high pressure, with actuator									
1110	Supervised open									
1120	2"	1 Plum	23	.348	Ea.	475	23.50		498.50	555
1130	2-1/2"	Q-1	38	.421		490	25.50		515.50	580
1140	3"		30	.533		530	32.50		562.50	635
1150	4"		22	.727		555	44.50		599.50	675
1160	5"	↓	19	.842		830	51.50		881.50	990
1170	6"	Q-2	23	1.043		855	66		921	1,050
1180	8"		18	1.333		1,200	84		1,284	1,450
1190	10"		15	1.600		2,925	101		3,026	3,375
1200	12"	↓	12	2	↓	4,175	126		4,301	4,800
1300	Gate valve, OS&Y									
1310	2-1/2"	Q-1	35	.457	Ea.	520	28		548	615

For customer support on your Plumbing Costs with RSMeans Data, call 800.448.8182.

93

21 11 11.16 Pipe Fittings, Grooved Joint		Crew	Daily Output	Labor-Hours	Unit	Material	2021 Bare Costs Labor	Equipment	Total	Total Incl O&P
1320	3"	Q-1	28	.571	Ea.	680	35		715	795
1330	4"	↓	20	.800		795	48.50		843.50	950
1340	6"	Q-2	21	1.143		1,075	72		1,147	1,275
1350	8"		16	1.500		1,675	95		1,770	2,000
1360	10"		13	1.846		2,250	117		2,367	2,650
1370	12"	↓	10	2.400	↓	3,525	152		3,677	4,100
1400	Gate valve, non-rising stem									
1410	2-1/2"	Q-1	36	.444	Ea.	530	27		557	620
1420	3"		29	.552		650	33.50		683.50	765
1430	4"	↓	21	.762		770	46.50		816.50	920
1440	6"	Q-2	22	1.091		990	69		1,059	1,200
1450	8"		17	1.412		1,450	89		1,539	1,725
1460	10"		14	1.714		1,950	108		2,058	2,300
1470	12"	↓	11	2.182	↓	2,500	138		2,638	2,950
2000	Alarm check valve, pre-trimmed									
2010	1-1/2"	Q-1	6	2.667	Ea.	119	162		281	370
2020	2"	"	5	3.200		119	195		314	420
2030	2-1/2"	Q-2	7	3.429		139	217		356	480
2040	3"		6	4		139	253		392	530
2050	4"		5	4.800		178	305		483	650
2060	6"		4	6		178	380		558	760
2070	8"	↓	2	12	↓	178	760		938	1,325
2200	Dry valve, pre-trimmed									
2210	1-1/2"	Q-1	5	3.200	Ea.	3,175	195		3,370	3,775
2220	2"	"	4	4		3,175	244		3,419	3,850
2230	2-1/2"	Q-2	6	4		3,225	253		3,478	3,925
2240	3"		5	4.800		3,225	305		3,530	4,000
2250	4"		4	6		3,425	380		3,805	4,350
2260	6"		2	12		3,950	760		4,710	5,475
2270	8"	↓	1	24	↓	5,825	1,525		7,350	8,675
2300	Deluge valve, pre-trimmed, with electric solenoid									
2310	1-1/2"	Q-1	5	3.200	Ea.	4,525	195		4,720	5,275
2320	2"	"	4	4		4,525	244		4,769	5,350
2330	2-1/2"	Q-2	6	4		4,575	253		4,828	5,425
2340	3"		5	4.800		4,575	305		4,880	5,500
2350	4"		4	6		5,475	380		5,855	6,575
2360	6"		2	12		6,525	760		7,285	8,300
2370	8"	↓	1	24	↓	7,375	1,525		8,900	10,400
2400	Preaction valve									
2410	Valve has double interlock,									
2420	pneumatic/electric actuation and trim.									
2430	1-1/2"	Q-1	5	3.200	Ea.	2,875	195		3,070	3,450
2440	2"	"	4	4		2,875	244		3,119	3,525
2450	2-1/2"	Q-2	6	4		2,875	253		3,128	3,525
2460	3"		5	4.800		2,875	305		3,180	3,600
2470	4"		4	6		3,175	380		3,555	4,050
2480	6"		2	12		3,175	760		3,935	4,600
2490	8"	↓	1	24	↓	3,475	1,525		5,000	6,100
3000	Fittings, ductile iron, ready to install									
3010	Includes bolts & grade "E" gaskets									
3012	Add 1 coupling, material only, per joint for installed price.									
3200	90° elbow									
3210	1-1/4"	1 Plum	88	.091	Ea.	69.50	6.15		75.65	85.50

For customer support on your Plumbing Costs with RSMeans Data, call 800.448.8182.

21 11 11 – Fire-Suppression, Pipe Fittings, Grooved Joint

21 11 11.16 Pipe Fittings, Grooved Joint

		Crew	Daily Output	Labor-Hours	Unit	Material	2021 Bare Costs Labor	Equipment	Total	Total Incl O&P
3220	1-1/2"	1 Plum	72	.111	Ea.	71.50	7.50		79	90.50
3230	2"	↓	60	.133		72	9.05		81.05	92.50
3240	2-1/2"	Q-1	85	.188	↓	87	11.45		98.45	113
3300	45° elbow									
3310	1-1/4"	1 Plum	88	.091	Ea.	69.50	6.15		75.65	85.50
3320	1-1/2"	↓	72	.111		71.50	7.50		79	90.50
3330	2"	↓	60	.133		72	9.05		81.05	92.50
3340	2-1/2"	Q-1	85	.188	↓	87	11.45		98.45	113
3400	Tee									
3410	1-1/4"	1 Plum	60	.133	Ea.	107	9.05		116.05	130
3420	1-1/2"		48	.167		110	11.30		121.30	138
3430	2"	↓	38	.211		112	14.25		126.25	145
3440	2-1/2"	Q-1	60	.267	↓	129	16.25		145.25	167
3500	Coupling									
3510	1-1/4"	1 Plum	200	.040	Ea.	29	2.71		31.71	36
3520	1-1/2"		134	.060		29.50	4.04		33.54	38.50
3530	2"	↓	100	.080		33.50	5.40		38.90	44.50
3540	2-1/2"	Q-1	160	.100		38.50	6.10		44.60	51
3550	3"		134	.119		43.50	7.30		50.80	58.50
3560	4"		100	.160		60.50	9.75		70.25	81
3570	5"		80	.200		80	12.20		92.20	106
3580	6"	Q-2	100	.240		95.50	15.15		110.65	128
3590	8"	"	84	.286	↓	95.50	18.05		113.55	132
4000	For seismic bracing, see Section 22 05 48.40									
5000	For hangers and supports, see Section 22 05 29.10									

21 11 13 – Facility Fire Suppression Piping

21 11 13.16 Pipe, Plastic

		Crew	Daily Output	Labor-Hours	Unit	Material	2021 Bare Costs Labor	Equipment	Total	Total Incl O&P
0010	**PIPE, PLASTIC**									
0020	CPVC, fire suppression (C-UL-S, FM, NFPA 13, 13D & 13R)									
0030	Socket joint, no couplings or hangers									
0100	SDR 13.5 (ASTM F442)									
0120	3/4" diameter	Q-12	420	.038	L.F.	1.17	2.28		3.45	4.70
0130	1" diameter		340	.047		1.81	2.82		4.63	6.20
0140	1-1/4" diameter		260	.062		2.85	3.68		6.53	8.65
0150	1-1/2" diameter		190	.084		3.94	5.05		8.99	11.90
0160	2" diameter		140	.114		6.10	6.85		12.95	16.90
0170	2-1/2" diameter		130	.123		10.30	7.35		17.65	22.50
0180	3" diameter	↓	120	.133	↓	15.70	8		23.70	29

21 11 13.18 Pipe Fittings, Plastic

		Crew	Daily Output	Labor-Hours	Unit	Material	2021 Bare Costs Labor	Equipment	Total	Total Incl O&P
0010	**PIPE FITTINGS, PLASTIC**									
0020	CPVC, fire suppression (C-UL-S, FM, NFPA 13, 13D & 13R)									
0030	Socket joint									
0100	90° elbow									
0120	3/4"	1 Plum	26	.308	Ea.	1.87	21		22.87	33
0130	1"		22.70	.352		4.11	24		28.11	40
0140	1-1/4"		20.20	.396		5.20	27		32.20	45.50
0150	1-1/2"	↓	18.20	.440		7.40	30		37.40	52.50
0160	2"	Q-1	33.10	.483		9.20	29.50		38.70	54
0170	2-1/2"		24.20	.661		17.65	40.50		58.15	79.50
0180	3"	↓	20.80	.769		24	47		71	96.50
0200	45° elbow									
0210	3/4"	1 Plum	26	.308	Ea.	2.57	21		23.57	34

21 11 13.18 Pipe Fittings, Plastic		Crew	Daily Output	Labor-Hours	Unit	Material	2021 Bare Costs Labor	Equipment	Total	Total Incl O&P
0220	1"	1 Plum	22.70	.352	Ea.	3.02	24		27.02	39
0230	1-1/4"		20.20	.396		4.37	27		31.37	45
0240	1-1/2"		18.20	.440		6.10	30		36.10	51
0250	2"	Q-1	33.10	.483		7.60	29.50		37.10	52.50
0260	2-1/2"		24.20	.661		13.60	40.50		54.10	75
0270	3"		20.80	.769		19.50	47		66.50	91.50
0300	Tee									
0310	3/4"	1 Plum	17.30	.462	Ea.	2.57	31.50		34.07	49.50
0320	1"		15.20	.526		5.05	35.50		40.55	58.50
0330	1-1/4"		13.50	.593		7.65	40		47.65	68.50
0340	1-1/2"		12.10	.661		11.25	45		56.25	79.50
0350	2"	Q-1	20	.800		16.60	48.50		65.10	91
0360	2-1/2"		16.20	.988		27	60		87	120
0370	3"		13.90	1.151		42	70		112	152
0400	Tee, reducing x any size									
0420	1"	1 Plum	15.20	.526	Ea.	4.31	35.50		39.81	57.50
0430	1-1/4"		13.50	.593		7.90	40		47.90	68.50
0440	1-1/2"		12.10	.661		9.55	45		54.55	77.50
0450	2"	Q-1	20	.800		18.15	48.50		66.65	92.50
0460	2-1/2"		16.20	.988		21	60		81	113
0470	3"		13.90	1.151		24.50	70		94.50	132
0500	Coupling									
0510	3/4"	1 Plum	26	.308	Ea.	1.80	21		22.80	33
0520	1"		22.70	.352		2.38	24		26.38	38
0530	1-1/4"		20.20	.396		3.47	27		30.47	44
0540	1-1/2"		18.20	.440		4.94	30		34.94	50
0550	2"	Q-1	33.10	.483		6.70	29.50		36.20	51.50
0560	2-1/2"		24.20	.661		10.20	40.50		50.70	71.50
0570	3"		20.80	.769		13.30	47		60.30	84.50
0600	Coupling, reducing									
0610	1" x 3/4"	1 Plum	22.70	.352	Ea.	2.38	24		26.38	38
0620	1-1/4" x 1"		20.20	.396		3.60	27		30.60	44
0630	1-1/2" x 3/4"		18.20	.440		5.40	30		35.40	50.50
0640	1-1/2" x 1"		18.20	.440		5.20	30		35.20	50
0650	1-1/2" x 1-1/4"		18.20	.440		4.94	30		34.94	50
0660	2" x 1"	Q-1	33.10	.483		6.95	29.50		36.45	51.50
0670	2" x 1-1/2"	"	33.10	.483		6.70	29.50		36.20	51.50
0700	Cross									
0720	3/4"	1 Plum	13	.615	Ea.	4.05	41.50		45.55	66.50
0730	1"		11.30	.708		5.05	48		53.05	77
0740	1-1/4"		10.10	.792		7	53.50		60.50	87.50
0750	1-1/2"		9.10	.879		9.70	59.50		69.20	99.50
0760	2"	Q-1	16.60	.964		15.80	58.50		74.30	105
0770	2-1/2"	"	12.10	1.322		34.50	80.50		115	158
0800	Cap									
0820	3/4"	1 Plum	52	.154	Ea.	1.09	10.40		11.49	16.75
0830	1"		45	.178		1.55	12.05		13.60	19.65
0840	1-1/4"		40	.200		2.51	13.55		16.06	23
0850	1-1/2"		36.40	.220		3.47	14.90		18.37	26
0860	2"	Q-1	66	.242		5.20	14.75		19.95	27.50
0870	2-1/2"		48.40	.331		7.50	20		27.50	38.50
0880	3"		41.60	.385		12.15	23.50		35.65	48.50
0900	Adapter, sprinkler head, female w/metal thd. insert (s x FNPT)									

21 11 13 – Facility Fire Suppression Piping

21 11 13.18 Pipe Fittings, Plastic

		Crew	Daily Output	Labor-Hours	Unit	Material	2021 Bare Costs Labor	2021 Bare Costs Equipment	Total	Total Incl O&P
0920	3/4" x 1/2"	1 Plum	52	.154	Ea.	4.96	10.40		15.36	21
0930	1" x 1/2"		45	.178		5.25	12.05		17.30	23.50
0940	1" x 3/4"	↓	45	.178	↓	8.25	12.05		20.30	27

21 11 16 – Facility Fire Hydrants

21 11 16.50 Fire Hydrants for Buildings

		Crew	Daily Output	Labor-Hours	Unit	Material	2021 Bare Costs Labor	2021 Bare Costs Equipment	Total	Total Incl O&P
0010	**FIRE HYDRANTS FOR BUILDINGS**									
3750	Hydrants, wall, w/caps, single, flush, polished brass									
3800	2-1/2" x 2-1/2"	Q-12	5	3.200	Ea.	233	192		425	545
3840	2-1/2" x 3"		5	3.200		420	192		612	750
3860	3" x 3"	↓	4.80	3.333		340	200		540	675
3900	For polished chrome, add				↓	20%				
3950	Double, flush, polished brass									
4000	2-1/2" x 2-1/2" x 4"	Q-12	5	3.200	Ea.	795	192		987	1,150
4040	2-1/2" x 2-1/2" x 6"		4.60	3.478		1,300	208		1,508	1,725
4080	3" x 3" x 4"		4.90	3.265		1,175	195		1,370	1,600
4120	3" x 3" x 6"	↓	4.50	3.556		1,175	213		1,388	1,625
4200	For polished chrome, add				↓	10%				
4350	Double, projecting, polished brass									
4400	2-1/2" x 2-1/2" x 4"	Q-12	5	3.200	Ea.	248	192		440	560
4450	2-1/2" x 2-1/2" x 6"	"	4.60	3.478	"	505	208		713	865
4460	Valve control, dbl. flush/projecting hydrant, cap &									
4470	chain, extension rod & cplg., escutcheon, polished brass	Q-12	8	2	Ea.	257	120		377	460
4480	Four-way square, flush, polished brass									
4540	2-1/2" (4) x 6"	Q-12	3.60	4.444	Ea.	3,900	266		4,166	4,700

21 11 19 – Fire-Department Connections

21 11 19.50 Connections for the Fire-Department

		Crew	Daily Output	Labor-Hours	Unit	Material	2021 Bare Costs Labor	2021 Bare Costs Equipment	Total	Total Incl O&P
0010	**CONNECTIONS FOR THE FIRE-DEPARTMENT**									
0020	For fire pro. cabinets, see Section 10 44 13.53									
4000	Storz type, with cap and chain									
4100	2-1/2" Storz x 2-1/2" F NPT, silver powder coat	Q-12	4.80	3.333	Ea.	71	200		271	375
4200	4" Storz x 1-1/2" F NPT, silver powder coat		4.80	3.333		137	200		337	450
4300	4" Storz x 2-1/2" F NPT, silver powder coat		4.80	3.333		143	200		343	455
4400	4" Storz x 3" F NPT, silver powder coat		4.80	3.333		153	200		353	465
4500	4" Storz x 4" F NPT, red powder coat		4.80	3.333		171	200		371	485
4600	4" Storz x 4" F NPT, silver powder coat		4.80	3.333		167	200		367	480
4700	4" Storz x 6" F NPT, red powder coat		4.80	3.333		235	200		435	555
4800	4" Storz x 6" F NPT, silver powder coat		4.80	3.333		234	200		434	555
4900	5" Storz x 2-1/2" F NPT, silver powder coat		4.80	3.333		204	200		404	520
5000	5" Storz x 3" F NPT, silver powder coat		4.80	3.333		215	200		415	535
5100	5" Storz x 4" F NPT, red powder coat		4.80	3.333		188	200		388	505
5200	5" Storz x 4" F NPT, silver powder coat		4.80	3.333		1,200	200		1,400	1,625
5300	5" Storz x 6" F NPT, red powder coat		4.80	3.333		282	200		482	610
5400	5" Storz x 6" F NPT, silver powder coat	↓	4.80	3.333	↓	1,700	200		1,900	2,175
6000	Roof manifold, horiz., brass, without valves & caps									
6040	2-1/2" x 2-1/2" x 4"	Q-12	4.80	3.333	Ea.	252	200		452	575
6060	2-1/2" x 2-1/2" x 6"		4.60	3.478		585	208		793	950
6080	2-1/2" x 2-1/2" x 2-1/2" x 4"		4.60	3.478		410	208		618	760
6090	2-1/2" x 2-1/2" x 2-1/2" x 6"	↓	4.60	3.478		445	208		653	800
7000	Sprinkler line tester, cast brass				↓	38			38	42
7140	Standpipe connections, wall, w/plugs & chains									
7160	Single, flush, brass, 2-1/2" x 2-1/2", Fire Dept Conn.	Q-12	5	3.200	Ea.	152	192		344	455
7180	2-1/2" x 3"	"	5	3.200	"	160	192		352	460

21 11 19 – Fire-Department Connections

21 11 19.50 Connections for the Fire-Department	Crew	Daily Output	Labor-Hours	Unit	Material	2021 Bare Costs Labor	Equipment	Total	Total Incl O&P
7240	For polished chrome, add					15%				
7280	Double, flush, polished brass									
7300	2-1/2" x 2-1/2" x 4"	Q-12	5	3.200	Ea.	585	192		777	930
7330	2-1/2" x 2-1/2" x 6"		4.60	3.478		865	208		1,073	1,275
7340	3" x 3" x 4"		4.90	3.265		1,150	195		1,345	1,575
7370	3" x 3" x 6"		4.50	3.556		1,175	213		1,388	1,625
7400	For polished chrome, add					15%				
7440	For sill cock combination, add				Ea.	101			101	111
7580	Double projecting, polished brass									
7600	2-1/2" x 2-1/2" x 4"	Q-12	5	3.200	Ea.	535	192		727	875
7630	2-1/2" x 2-1/2" x 6"	"	4.60	3.478	"	895	208		1,103	1,300
7680	For polished chrome, add					15%				
7900	Three way, flush, polished brass									
7920	2-1/2" (3) x 4"	Q-12	4.80	3.333	Ea.	1,900	200		2,100	2,375
7930	2-1/2" (3) x 6"	"	4.60	3.478		2,800	208		3,008	3,375
8000	For polished chrome, add					9%				
8020	Three way, projecting, polished brass									
8040	2-1/2" (3) x 4"	Q-12	4.80	3.333	Ea.	1,150	200		1,350	1,550
8070	2-1/2" (3) x 6"	"	4.60	3.478		1,925	208		2,133	2,400
8100	For polished chrome, add					12%				
8200	Four way, square, flush, polished brass,									
8240	2-1/2" (4) x 6"	Q-12	3.60	4.444	Ea.	1,875	266		2,141	2,475
8300	For polished chrome, add				"	10%				
8550	Wall, vertical, flush, cast brass									
8600	Two way, 2-1/2" x 2-1/2" x 4"	Q-12	5	3.200	Ea.	400	192		592	725
8660	Four way, 2-1/2" (4) x 6"		3.80	4.211		1,350	252		1,602	1,850
8680	Six way, 2-1/2" (6) x 6"		3.40	4.706		1,600	282		1,882	2,175
8700	For polished chrome, add					10%				
8800	Free standing siamese unit, polished brass, two way									
8820	2-1/2" x 2-1/2" x 4"	Q-12	2.50	6.400	Ea.	350	385		735	955
8850	2-1/2" x 2-1/2" x 6"		2	8		845	480		1,325	1,650
8860	3" x 3" x 4"		2.50	6.400		640	385		1,025	1,275
8890	3" x 3" x 6"		2	8		1,475	480		1,955	2,325
8940	For polished chrome, add					12%				
9100	Free standing siamese unit, polished brass, three way									
9120	2-1/2" x 2-1/2" x 2-1/2" x 6"	Q-12	2	8	Ea.	990	480		1,470	1,800
9160	For polished chrome, add				"	15%				

21 12 Fire-Suppression Standpipes

21 12 13 – Fire-Suppression Hoses and Nozzles

21 12 13.50 Fire Hoses and Nozzles

0010	**FIRE HOSES AND NOZZLES**									
0200	Adapters, rough brass, straight hose threads	R211226-10								
0220	One piece, female to male, rocker lugs									
0240	1" x 1"	R211226-20			Ea.	47			47	52
0260	1-1/2" x 1"					46			46	50.50
0280	1-1/2" x 1-1/2"					13.10			13.10	14.40
0300	2" x 1-1/2"					79.50			79.50	87.50
0320	2" x 2"					54.50			54.50	60
0340	2-1/2" x 1-1/2"					35			35	38.50
0360	3" x 1-1/2"					136			136	150

For customer support on your Plumbing Costs with RSMeans Data, call 800.448.8182.

21 12 13.50 Fire Hoses and Nozzles	Crew	Daily Output	Labor-Hours	Unit	Material	2021 Bare Costs Labor	Equipment	Total	Total Incl O&P
0380 2-1/2" x 2-1/2"				Ea.	25			25	27.50
0400 3" x 2-1/2"					122			122	134
0420 3" x 3"					109			109	120
0500 For polished brass, add					50%				
0520 For polished chrome, add					75%				
0700 One piece, female to male, hexagon									
0740 1-1/2" x 3/4"				Ea.	49.50			49.50	54.50
0760 2" x 1-1/2"					125			125	138
0780 2-1/2" x 1"					194			194	213
0800 2-1/2" x 1-1/2"					54.50			54.50	60
0820 2-1/2" x 2"					35.50			35.50	39
0840 3" x 2-1/2"					62			62	68
0900 For polished chrome, add					75%				
1100 Swivel, female to female, pin lugs									
1120 1-1/2" x 1-1/2"				Ea.	55			55	60.50
1200 2-1/2" x 2-1/2"					62.50			62.50	69
1260 For polished brass, add					50%				
1280 For polished chrome, add					75%				
1400 Couplings, sngl. & dbl. jacket, pin lug or rocker lug, cast brass									
1410 1-1/2"				Ea.	63.50			63.50	70
1420 2-1/2"				"	49.50			49.50	54.50
1500 For polished brass, add					20%				
1520 For polished chrome, add					40%				
1580 Reducing, F x M, interior installation, cast brass									
1590 2" x 1-1/2"				Ea.	23			23	25
1600 2-1/2" x 1-1/2"					16.70			16.70	18.35
1680 For polished brass, add					50%				
1720 For polished chrome, add					75%				
2200 Hose, less couplings									
2260 Synthetic jacket, lined, 300 lb. test, 1-1/2" diameter	Q-12	2600	.006	L.F.	3.31	.37		3.68	4.19
2270 2" diameter		2200	.007		3.61	.44		4.05	4.62
2280 2-1/2" diameter		2200	.007		5.80	.44		6.24	7.05
2290 3" diameter		2200	.007		7.10	.44		7.54	8.45
2360 High strength, 500 lb. test, 1-1/2" diameter		2600	.006		3.62	.37		3.99	4.53
2380 2-1/2" diameter		2200	.007		6.10	.44		6.54	7.35
5000 Nipples, straight hose to tapered iron pipe, brass									
5060 Female to female, 1-1/2" x 1-1/2"				Ea.	26			26	28.50
5100 2-1/2" x 2-1/2"					44			44	48.50
5190 For polished chrome, add					75%				
5200 Double male or male to female, 1" x 1"					34			34	37.50
5220 1-1/2" x 1"					17.40			17.40	19.15
5230 1-1/2" x 1-1/2"					22			22	24
5260 2" x 1-1/2"					83			83	91.50
5270 2" x 2"					105			105	115
5280 2-1/2" x 1-1/2"					62			62	68.50
5300 2-1/2" x 2"					48.50			48.50	53.50
5310 2-1/2" x 2-1/2"					27.50			27.50	30.50
5340 For polished chrome, add					75%				
5600 Nozzles, brass									
5620 Adjustable fog, 3/4" booster line				Ea.	154			154	170
5630 1" booster line					184			184	202
5640 1-1/2" leader line					120			120	132
5660 2-1/2" direct connection					280			280	310

21 12 Fire-Suppression Standpipes

21 12 13 – Fire-Suppression Hoses and Nozzles

21 12 13.50 Fire Hoses and Nozzles		Crew	Daily Output	Labor-Hours	Unit	Material	2021 Bare Costs Labor	Equipment	Total	Total Incl O&P
5680	2-1/2" playpipe nozzle				Ea.	261			261	287
5780	For chrome plated, add					8%				
5850	Electrical fire, adjustable fog, no shock									
5900	1-1/2"				Ea.	194			194	213
5920	2-1/2"					1,150			1,150	1,275
5980	For polished chrome, add					6%				
6200	Heavy duty, comb. adj. fog and str. stream, with handle									
6210	1" booster line				Ea.	232			232	255
6240	1-1/2"					445			445	490
6260	2-1/2", for playpipe					665			665	735
6280	2-1/2" direct connection					470			470	515
6300	2-1/2" playpipe combination					535			535	585
6480	For polished chrome, add					7%				
6500	Plain fog, polished brass, 1-1/2"					190			190	209
6540	Chrome plated, 1-1/2"					132			132	145
6700	Plain stream, polished brass, 1-1/2" x 10"					58.50			58.50	64.50
6760	2-1/2" x 15" x 7/8" or 1-1/2"					109			109	119
6860	For polished chrome, add					20%				
7000	Underwriters playpipe, 2-1/2" x 30" with 1-1/8" tip				Ea.	675			675	745
9200	Storage house, hose only, primed steel					2,025			2,025	2,225
9220	Aluminum					2,575			2,575	2,850
9280	Hose and hydrant house, primed steel					1,975			1,975	2,175
9300	Aluminum					1,650			1,650	1,800
9340	Tools, crowbar and brackets	1 Carp	12	.667		97.50	36.50		134	162
9360	Combination hydrant wrench and spanner					89			89	98
9380	Fire axe and brackets									
9400	6 lb.	1 Carp	12	.667	Ea.	141	36.50		177.50	210
9500	For fire equipment cabinets, Section 10 44 13.53									

21 12 16 – Fire-Suppression Hose Reels

21 12 16.50 Fire-Suppression Hose Reels

		Crew	Daily Output	Labor-Hours	Unit	Material	2021 Bare Costs Labor	Equipment	Total	Total Incl O&P
0010	**FIRE-SUPPRESSION HOSE REELS**									
2990	Hose reel, swinging, for 1-1/2" polyester neoprene lined hose									
3000	50' long	Q-12	14	1.143	Ea.	157	68.50		225.50	275
3020	100' long		14	1.143		251	68.50		319.50	380
3060	For 2-1/2" cotton rubber hose, 75' long		14	1.143		266	68.50		334.50	395
3100	150' long		14	1.143		300	68.50		368.50	430

21 12 19 – Fire-Suppression Hose Racks

21 12 19.50 Fire Hose Racks

		Crew	Daily Output	Labor-Hours	Unit	Material	2021 Bare Costs Labor	Equipment	Total	Total Incl O&P
0010	**FIRE HOSE RACKS**									
2600	Hose rack, swinging, for 1-1/2" diameter hose,									
2620	Enameled steel, 50' and 75' lengths of hose	Q-12	20	.800	Ea.	132	48		180	217
2640	100' and 125' lengths of hose		20	.800		90	48		138	171
2680	Chrome plated, 50' and 75' lengths of hose		20	.800		66	48		114	145
2700	100' and 125' lengths of hose		20	.800		137	48		185	222
2750	2-1/2" diameter, 100' hose		20	.800		104	48		152	187
2780	For hose rack nipple, 1-1/2" polished brass, add					33.50			33.50	37
2820	2-1/2" polished brass, add					107			107	118
2840	1-1/2" polished chrome, add					39			39	43
2860	2-1/2" polished chrome, add					81.50			81.50	89.50

21 12 23.70 Fire Hose Valves

	Crew	Daily Output	Labor-Hours	Unit	Material	2021 Bare Costs Labor	Equipment	Total	Total Incl O&P
0010 FIRE HOSE VALVES									
0020 Angle, combination pressure adjust/restricting, rough brass R211226-10									
0030 1-1/2"	1 Spri	12	.667	Ea.	96	44.50		140.50	172
0040 2-1/2" R211226-20	"	7	1.143	"	170	76		246	300
0042 Nonpressure adjustable/restricting, rough brass									
0044 1-1/2"	1 Spri	12	.667	Ea.	113	44.50		157.50	190
0046 2-1/2"	"	7	1.143	"	146	76		222	275
0050 For polished brass, add					30%				
0060 For polished chrome, add					40%				
0080 Wheel handle, 300 lb., 1-1/2"	1 Spri	12	.667	Ea.	101	44.50		145.50	177
0090 2-1/2"	"	7	1.143	"	194	76		270	330
0100 For polished brass, add					35%				
0110 For polished chrome, add					50%				
1000 Ball drip, automatic, rough brass, 1/2"	1 Spri	20	.400	Ea.	18.30	26.50		44.80	59.50
1010 3/4"	"	20	.400	"	16.30	26.50		42.80	57.50
1100 Ball, 175 lb., sprinkler system, FM/UL, threaded, bronze									
1120 Slow close									
1150 1" size	1 Spri	19	.421	Ea.	305	28		333	375
1160 1-1/4" size		15	.533		330	35.50		365.50	420
1170 1-1/2" size		13	.615		435	41		476	540
1180 2" size		11	.727		530	48.50		578.50	655
1190 2-1/2" size	Q-12	15	1.067		715	64		779	880
1230 For supervisory switch kit, all sizes									
1240 One circuit, add	1 Spri	48	.167	Ea.	144	11.10		155.10	175
1280 Quarter turn for trim									
1300 1/2" size	1 Spri	22	.364	Ea.	42	24		66	82.50
1310 3/4" size		20	.400		45.50	26.50		72	89.50
1320 1" size		19	.421		50.50	28		78.50	97.50
1330 1-1/4" size		15	.533		82.50	35.50		118	144
1340 1-1/2" size		13	.615		104	41		145	175
1350 2" size		11	.727		123	48.50		171.50	209
1400 Caps, polished brass with chain, 3/4"					67			67	74
1420 1"					84.50			84.50	92.50
1440 1-1/2"					21			21	23.50
1460 2-1/2"					30.50			30.50	33.50
1480 3"					99			99	109
1900 Escutcheon plate, for angle valves, polished brass, 1-1/2"					15.70			15.70	17.30
1920 2-1/2"					24.50			24.50	27
1940 3"					32			32	35
1980 For polished chrome, add					15%				
2000 Foam, control valve, 3"	1 Spri	6	1.333		2,400	88.50		2,488.50	2,750
2020 Supply valve, 2-1/2"		7	1.143		159	76		235	288
2040 Proportioner, 8"		2	4		3,900	266		4,166	4,675
2060 Oscillating foam monitor with electric remote control	Q-12	5.33	3.002		17,800	180		17,980	19,800
3000 Gate, hose, wheel handle, N.R.S., rough brass, 1-1/2"	1 Spri	12	.667		144	44.50		188.50	224
3040 2-1/2", 300 lb.	"	7	1.143		184	76		260	315
3080 For polished brass, add					40%				
3090 For polished chrome, add					50%				
3800 Hydrant, screw type, crank handle, brass									
3840 2-1/2" size	Q-12	11	1.455	Ea.	535	87		622	720
3880 For chrome, same price									
4200 Hydrolator, vent and draining, rough brass, 1-1/2"	1 Spri	12	.667	Ea.	112	44.50		156.50	189

21 12 Fire-Suppression Standpipes

21 12 23 – Fire-Suppression Hose Valves

21 12 23.70 Fire Hose Valves

21 12 23.70 Fire Hose Valves		Crew	Daily Output	Labor-Hours	Unit	Material	2021 Bare Costs Labor	Equipment	Total	Total Incl O&P
4280	For polished brass, add				Ea.	50%				
4290	For polished chrome, add					90%				
5000	Pressure reducing rough brass, 1-1/2"	1 Spri	12	.667		350	44.50		394.50	450
5020	2-1/2"	"	7	1.143		530	76		606	700
5080	For polished brass, add					105%				
5090	For polished chrome, add					140%				
8000	Wye, leader line, ball type, swivel female x male x male									
8040	2-1/2" x 1-1/2" x 1-1/2" polished brass				Ea.	405			405	445
8060	2-1/2" x 1-1/2" x 1-1/2" polished chrome				"	283			283	310

21 13 Fire-Suppression Sprinkler Systems

21 13 13 – Wet-Pipe Sprinkler Systems

21 13 13.50 Wet-Pipe Sprinkler System Components

21 13 13.50 Wet-Pipe Sprinkler System Components			Crew	Daily Output	Labor-Hours	Unit	Material	2021 Bare Costs Labor	Equipment	Total	Total Incl O&P
0010	**WET-PIPE SPRINKLER SYSTEM COMPONENTS**										
1100	Alarm, electric pressure switch (circuit closer)	R211313-10	1 Spri	26	.308	Ea.	116	20.50		136.50	159
1140	For explosion proof, max 20 psi, contacts close or open			26	.308		420	20.50		440.50	490
1220	Water motor gong			4	2		325	133		458	555
1900	Flexible sprinkler head connectors										
1910	Braided stainless steel hose with mounting bracket										
1920	1/2" and 3/4" outlet size										
1940	40" length		1 Spri	30	.267	Ea.	62	17.75		79.75	94.50
1960	60" length		"	22	.364	"	72	24		96	116
1982	May replace hard-pipe armovers										
1984	for wet and pre-action systems.										
2000	Release, emergency, manual, for hydraulic or pneumatic system		1 Spri	12	.667	Ea.	192	44.50		236.50	278
2060	Release, thermostatic, for hydraulic or pneumatic release line			20	.400		850	26.50		876.50	975
2200	Sprinkler cabinets, 6 head capacity			16	.500		75.50	33.50		109	133
2260	12 head capacity			16	.500		87.50	33.50		121	146
2340	Sprinkler head escutcheons, standard, brass tone, 1" size			40	.200		3.80	13.30		17.10	24
2360	Chrome, 1" size			40	.200		5.60	13.30		18.90	26
2400	Recessed type, bright brass			40	.200		22.50	13.30		35.80	44.50
2440	Chrome or white enamel			40	.200		22.50	13.30		35.80	44.50
2600	Sprinkler heads, not including supply piping										
3700	Standard spray, pendent or upright, brass, 135°F to 286°F										
3720	1/2" NPT, K2.8		1 Spri	16	.500	Ea.	19.40	33.50		52.90	71
3730	1/2" NPT, 7/16" orifice			16	.500		17.40	33.50		50.90	68.50
3740	1/2" NPT, K5.6			16	.500		14.90	33.50		48.40	66
3760	1/2" NPT, 17/32" orifice			16	.500		14.25	33.50		47.75	65
3780	3/4" NPT, 17/32" orifice			16	.500		13.90	33.50		47.40	65
3800	For open sprinklers, deduct						15%				
3840	For chrome, add					Ea.	4.25			4.25	4.68
3920	For 360°F, same cost										
3930	For 400°F		1 Spri	16	.500	Ea.	114	33.50		147.50	175
3940	For 500°F		"	16	.500	"	114	33.50		147.50	175
4200	Sidewall, vertical brass, 135°F to 286°F										
4240	1/2" NPT, 1/2" orifice		1 Spri	16	.500	Ea.	31	33.50		64.50	83.50
4280	3/4" NPT, 17/32" orifice		"	16	.500		85	33.50		118.50	143
4360	For satin chrome, add						5.05			5.05	5.55
4400	For 360°F, same cost										
4500	Sidewall, horizontal, brass, 135°F to 286°F										
4520	1/2" NPT, 1/2" orifice		1 Spri	16	.500	Ea.	28	33.50		61.50	80.50

21 13 Fire-Suppression Sprinkler Systems

21 13 13 – Wet-Pipe Sprinkler Systems

21 13 13.50 Wet-Pipe Sprinkler System Components

		Crew	Daily Output	Labor-Hours	Unit	Material	2021 Bare Costs Labor	Equipment	Total	Total Incl O&P
4540	For 360°F, same cost									
4800	Recessed pendent, brass, 135°F to 286°F									
4820	1/2" NPT, K2.8	1 Spri	10	.800	Ea.	52.50	53		105.50	137
4830	1/2" NPT, 7/16" orifice		10	.800		22.50	53		75.50	105
4840	1/2" NPT, K5.6		10	.800		17.65	53		70.65	99
4860	1/2" NPT, 17/32" orifice		10	.800		52.50	53		105.50	137
4900	For satin chrome, add					6.30			6.30	6.95
5000	Recessed-vertical sidewall, brass, 135°F to 286°F									
5020	1/2" NPT, K2.8	1 Spri	10	.800	Ea.	38	53		91	121
5030	1/2" NPT, 7/16" orifice		10	.800		38	53		91	121
5040	1/2" NPT, K5.6		10	.800		38	53		91	121
5100	For bright nickel, same cost									
5600	Concealed, complete with cover plate									
5620	1/2" NPT, 1/2" orifice, 135°F to 212°F	1 Spri	9	.889	Ea.	29	59		88	121
5800	Window, brass, 1/2" NPT, 1/4" orifice		16	.500		42.50	33.50		76	96.50
5810	1/2" NPT, 5/16" orifice		16	.500		42.50	33.50		76	96.50
5820	1/2" NPT, 3/8" orifice		16	.500		42.50	33.50		76	96.50
5830	1/2" NPT, 7/16" orifice		16	.500		42.50	33.50		76	96.50
5840	1/2" NPT, 1/2" orifice		16	.500		45.50	33.50		79	99.50
5860	For polished chrome, add					49			49	54
5880	3/4" NPT, 5/8" orifice	1 Spri	16	.500		47.50	33.50		81	102
5890	3/4" NPT, 3/4" orifice	"	16	.500		47.50	33.50		81	102
6000	Sprinkler head guards, bright zinc, 1/2" NPT					5.35			5.35	5.85
6020	Bright zinc, 3/4" NPT					5.40			5.40	5.95
6025	Residential sprinkler components (one and two family)									
6026	Water motor alarm with strainer	1 Spri	4	2	Ea.	325	133		458	555
6027	Fast response, glass bulb, 135°F to 155°F									
6028	1/2" NPT, pendent, brass	1 Spri	16	.500	Ea.	37	33.50		70.50	90
6029	1/2" NPT, sidewall, brass		16	.500		40	33.50		73.50	93.50
6030	1/2" NPT, pendent, brass, extended coverage		16	.500		30	33.50		63.50	82.50
6031	1/2" NPT, sidewall, brass, extended coverage		16	.500		27	33.50		60.50	79.50
6032	3/4" NPT sidewall, brass, extended coverage		16	.500		31	33.50		64.50	83.50
6033	For chrome, add					15%				
6034	For polyester/teflon coating, add					20%				
6100	Sprinkler head wrenches, standard head				Ea.	28			28	31
6120	Recessed head					43			43	47
6160	Tamper switch (valve supervisory switch)	1 Spri	16	.500		267	33.50		300.50	345
6165	Flow switch (valve supervisory switch)	"	16	.500		267	33.50		300.50	345

21 13 16 – Dry-Pipe Sprinkler Systems

21 13 16.50 Dry-Pipe Sprinkler System Components

		Crew	Daily Output	Labor-Hours	Unit	Material	2021 Bare Costs Labor	Equipment	Total	Total Incl O&P
0010	**DRY-PIPE SPRINKLER SYSTEM COMPONENTS**									
0600	Accelerator	1 Spri	8	1	Ea.	955	66.50		1,021.50	1,150
0800	Air compressor for dry pipe system, automatic, complete									
0820	30 gal. system capacity, 3/4 HP	1 Spri	1.30	6.154	Ea.	1,400	410		1,810	2,150
0860	30 gal. system capacity, 1 HP		1.30	6.154		1,625	410		2,035	2,400
0910	30 gal. system capacity, 1-1/2 HP		1.30	6.154		1,375	410		1,785	2,125
0920	30 gal. system capacity, 2 HP		1.30	6.154		1,350	410		1,760	2,075
0960	Air pressure maintenance control		24	.333		380	22		402	455
1600	Dehydrator package, incl. valves and nipples R211313-20		12	.667		900	44.50		944.50	1,050
2600	Sprinkler heads, not including supply piping									
2640	Dry, pendent, 1/2" orifice, 3/4" or 1" NPT									
2660	3" to 6" length	1 Spri	14	.571	Ea.	197	38		235	273

For customer support on your Plumbing Costs with RSMeans Data, call 800.448.8182.

103

21 13 Fire-Suppression Sprinkler Systems

21 13 16 – Dry-Pipe Sprinkler Systems

21 13 16.50 Dry-Pipe Sprinkler System Components

		Crew	Daily Output	Labor-Hours	Unit	Material	2021 Bare Costs Labor	Equipment	Total	Total Incl O&P
2670	6-1/4" to 8" length	1 Spri	14	.571	Ea.	209	38		247	287
2680	8-1/4" to 12" length		14	.571		221	38		259	300
2690	12-1/4" to 15" length		14	.571		245	38		283	325
2700	15-1/4" to 18" length		14	.571		260	38		298	340
2710	18-1/4" to 21" length		13	.615		271	41		312	360
2720	21-1/4" to 24" length		13	.615		285	41		326	375
2730	24-1/4" to 27" length		13	.615		299	41		340	390
2740	27-1/4" to 30" length		13	.615		310	41		351	400
2750	30-1/4" to 33" length		13	.615		320	41		361	415
2760	33-1/4" to 36" length		13	.615		335	41		376	430
2780	36-1/4" to 39" length		12	.667		350	44.50		394.50	445
2790	39-1/4" to 42" length		12	.667		360	44.50		404.50	460
2800	For each inch or fraction, add					4.37			4.37	4.81
6330	Valves and components									
6340	Alarm test/shut off valve, 1/2"	1 Spri	20	.400	Ea.	27.50	26.50		54	69.50
8000	Dry pipe air check valve, 3" size	Q-12	2	8		2,200	480		2,680	3,150
8200	Dry pipe valve, incl. trim and gauges, 3" size		2	8		3,200	480		3,680	4,250
8220	4" size		1	16		3,425	960		4,385	5,200
8240	6" size	Q-13	2	16		4,050	1,025		5,075	5,975
8280	For accelerator trim with gauges, add	1 Spri	8	1		296	66.50		362.50	425

21 13 19 – Preaction Sprinkler Systems

21 13 19.50 Preaction Sprinkler System Components

		Crew	Daily Output	Labor-Hours	Unit	Material	2021 Bare Costs Labor	Equipment	Total	Total Incl O&P
0010	**PREACTION SPRINKLER SYSTEM COMPONENTS**									
3000	Preaction valve cabinet									
3100	Single interlock, pneum. release, panel, 1/2 HP comp. regul. air trim									
3110	1-1/2"	Q-12	3	5.333	Ea.	26,000	320		26,320	29,100
3120	2"		3	5.333		26,000	320		26,320	29,100
3130	2-1/2"		3	5.333		26,000	320		26,320	29,100
3140	3"		3	5.333		26,000	320		26,320	29,100
3150	4"		2	8		27,000	480		27,480	30,400
3160	6"	Q-13	4	8		28,300	505		28,805	32,000
3200	Double interlock, pneum. release, panel, 1/2 HP comp. regul. air trim									
3210	1-1/2"	Q-12	3	5.333	Ea.	26,800	320		27,120	29,900
3220	2"		3	5.333		26,800	320		27,120	29,900
3230	2-1/2"		3	5.333		26,800	320		27,120	29,900
3240	3"		3	5.333		26,800	320		27,120	30,000
3250	4"		2	8		28,100	480		28,580	31,700
3260	6"	Q-13	4	8		29,300	505		29,805	33,000

21 13 20 – On-Off Multicycle Sprinkler System

21 13 20.50 On-Off Multicycle Fire-Suppression Sprinkler Systems

		Crew	Daily Output	Labor-Hours	Unit	Material	2021 Bare Costs Labor	Equipment	Total	Total Incl O&P
0010	**ON-OFF MULTICYCLE FIRE-SUPPRESSION SPRINKLER SYSTEMS**									
8400	On-off multicycle package, includes swing check									
8420	and flow control valves with required trim									
8440	2" size	Q-12	2	8	Ea.	5,775	480		6,255	7,075
8460	3" size		1.50	10.667		6,350	640		6,990	7,925
8480	4" size		1	16		7,100	960		8,060	9,225
8500	6" size	Q-13	1.40	22.857		8,175	1,450		9,625	11,200

21 13 Fire-Suppression Sprinkler Systems

21 13 26 – Deluge Fire-Suppression Sprinkler Systems

21 13 26.50 Deluge Fire-Suppression Sprinkler Sys. Comp.	Crew	Daily Output	Labor-Hours	Unit	Material	2021 Bare Costs Labor	2021 Bare Costs Equipment	Total	Total Incl O&P
0010 **DELUGE FIRE-SUPPRESSION SPRINKLER SYSTEM COMPONENTS**									
1400 Deluge system, monitoring panel w/deluge valve & trim	1 Spri	18	.444	Ea.	7,650	29.50		7,679.50	8,475
6200 Valves and components									
7000 Deluge, assembly, incl. trim, pressure									
7020 operated relief, emergency release, gauges									
7040 2" size	Q-12	2	8	Ea.	4,675	480		5,155	5,850
7060 3" size		1.50	10.667		5,250	640		5,890	6,725
7080 4" size		1	16		6,025	960		6,985	8,050
7100 6" size	Q-13	1.80	17.778		7,125	1,125		8,250	9,500
7800 Pneumatic actuator, bronze, required on all									
7820 pneumatic release systems, any size deluge	1 Spri	18	.444	Ea.	320	29.50		349.50	400

21 13 39 – Foam-Water Systems

21 13 39.50 Foam-Water System Components

	Crew	Daily Output	Labor-Hours	Unit	Material	2021 Bare Costs Labor	2021 Bare Costs Equipment	Total	Total Incl O&P
0010 **FOAM-WATER SYSTEM COMPONENTS**									
2600 Sprinkler heads, not including supply piping									
3600 Foam-water, pendent or upright, 1/2" NPT	1 Spri	12	.667	Ea.	248	44.50		292.50	340

21 21 Carbon-Dioxide Fire-Extinguishing Systems

21 21 16 – Carbon-Dioxide Fire-Extinguishing Equipment

21 21 16.50 CO2 Fire Extinguishing System

	Crew	Daily Output	Labor-Hours	Unit	Material	2021 Bare Costs Labor	2021 Bare Costs Equipment	Total	Total Incl O&P
0010 **CO$_2$ FIRE EXTINGUISHING SYSTEM**									
0042 For detectors and control stations, see Section 28 31 23.50									
0100 Control panel, single zone with batteries (2 zones det., 1 suppr.)	1 Elec	1	8	Ea.	1,075	510		1,585	1,950
0150 Multizone (4) with batteries (8 zones det., 4 suppr.)	"	.50	16		3,150	1,025		4,175	5,000
1000 Dispersion nozzle, CO$_2$, 3" x 5"	1 Plum	18	.444		163	30		193	224
2000 Extinguisher, CO$_2$ system, high pressure, 75 lb. cylinder	Q-1	6	2.667		1,450	162		1,612	1,850
2100 100 lb. cylinder	"	5	3.200		1,925	195		2,120	2,400
3000 Electro/mechanical release	L-1	4	4		1,075	263		1,338	1,600
3400 Manual pull station	1 Plum	6	1.333		90.50	90.50		181	235
4000 Pneumatic damper release	"	8	1		189	67.50		256.50	310

21 22 Clean-Agent Fire-Extinguishing Systems

21 22 16 – Clean-Agent Fire-Extinguishing Equipment

21 22 16.50 Clean-Agent Extinguishing Systems

	Crew	Daily Output	Labor-Hours	Unit	Material	2021 Bare Costs Labor	2021 Bare Costs Equipment	Total	Total Incl O&P
0010 **CLEAN-AGENT EXTINGUISHING SYSTEMS**									
0020 FM200 fire extinguishing system									
1100 Dispersion nozzle FM200, 1-1/2"	1 Plum	14	.571	Ea.	259	38.50		297.50	345
2400 Extinguisher, FM200 system, filled, with mounting bracket									
2460 26 lb. container	Q-1	8	2	Ea.	2,150	122		2,272	2,525
2480 44 lb. container		7	2.286		2,775	139		2,914	3,250
2500 63 lb. container		6	2.667		3,950	162		4,112	4,600
2520 101 lb. container		5	3.200		5,700	195		5,895	6,575
2540 196 lb. container		4	4		6,875	244		7,119	7,950
6000 FM200 system, simple nozzle layout, with broad dispersion				C.F.	1.87			1.87	2.06
6010 Extinguisher, FM200 system, filled, with mounting bracket									
6020 Complex nozzle layout and/or including underfloor dispersion				C.F.	3.71			3.71	4.08
6100 20,000 C.F. 2 exits, 8' clng					2.08			2.08	2.29
6200 100,000 C.F. 4 exits, 8' clng					1.88			1.88	2.07

21 22 16.50 Clean-Agent Extinguishing Systems	Crew	Daily Output	Labor-Hours	Unit	Material	2021 Bare Costs Labor	Equipment	Total	Total Incl O&P
6300　250,000 C.F. 6 exits, 8' clng				C.F.	1.59			1.59	1.75
7010　HFC-227ea fire extinguishing system									
7100　Cylinders with clean-agent									
7110　Does not include pallet jack/fork lift rental fees									
7120　70 lb. cyl. w/35 lb. agent, no solenoid	Q-12	14	1.143	Ea.	3,100	68.50		3,168.50	3,525
7130　70 lb. cyl w/70 lb. agent, no solenoid		10	1.600		4,100	96		4,196	4,650
7140　70 lb. cyl w/35 lb. agent, w/solenoid		14	1.143		4,650	68.50		4,718.50	5,200
7150　70 lb. cyl w/70 lb. agent, w/solenoid		10	1.600		5,875	96		5,971	6,600
7220　250 lb. cyl, w/125 lb. agent, no solenoid		8	2		7,200	120		7,320	8,100
7230　250 lb. cyl, w/250 lb. agent, no solenoid		5	3.200		7,200	192		7,392	8,200
7240　250 lb. cyl, w/125 lb. agent, w/solenoid		8	2		10,000	120		10,120	11,300
7250　250 lb. cyl, w/250 lb. agent, w/solenoid		5	3.200		14,400	192		14,592	16,200
7320　560 lb. cyl, w/300 lb. agent, no solenoid		4	4		13,800	239		14,039	15,600
7330　560 lb. cyl, w/560 lb. agent, no solenoid		2.50	6.400		21,000	385		21,385	23,700
7340　560 lb. cyl, w/300 lb. agent, w/solenoid		4	4		18,600	239		18,839	20,900
7350　560 lb. cyl, w/560 lb. agent, w/solenoid		2.50	6.400		27,700	385		28,085	31,100
7420　1,200 lb. cyl, w/600 lb. agent, no solenoid	Q-13	4	8		26,200	505		26,705	29,700
7430　1,200 lb. cyl, w/1,200 lb. agent, no solenoid		3	10.667		43,000	675		43,675	48,300
7440　1,200 lb. cyl, w/600 lb. agent, w/solenoid		4	8		64,500	505		65,005	72,000
7450　1,200 lb. cyl, w/1,200 lb. agent, w/solenoid		3	10.667		101,500	675		102,175	112,500
7500　Accessories									
7510　Dispersion nozzle	1 Spri	16	.500	Ea.	125	33.50		158.50	187
7520　Agent release panel	1 Elec	4	2		72	127		199	268
7530　Maintenance switch	"	6	1.333		39	85		124	169
7540　Solenoid valve, 12v dc	1 Spri	8	1		273	66.50		339.50	400
7550　12v ac		8	1		670	66.50		736.50	835
7560　12v dc, explosion proof		8	1		705	66.50		771.50	875

21 31 Centrifugal Fire Pumps
21 31 13 – Electric-Drive, Centrifugal Fire Pumps
21 31 13.50 Electric-Drive Fire Pumps

	Crew	Daily Output	Labor-Hours	Unit	Material	2021 Bare Costs Labor	Equipment	Total	Total Incl O&P
0010　**ELECTRIC-DRIVE FIRE PUMPS** Including controller, fittings and relief valve									
3100　250 GPM, 55 psi, 15 HP, 3550 RPM, 2" pump	Q-13	.70	45.714	Ea.	16,600	2,900		19,500	22,600
3200　500 GPM, 50 psi, 27 HP, 1770 RPM, 4" pump		.68	47.059		20,600	2,975		23,575	27,200
3250　500 GPM, 100 psi, 47 HP, 3550 RPM, 3" pump		.66	48.485		25,100	3,075		28,175	32,200
3300　500 GPM, 125 psi, 64 HP, 3550 RPM, 3" pump		.62	51.613		27,400	3,275		30,675	35,100
3350　750 GPM, 50 psi, 44 HP, 1770 RPM, 5" pump		.64	50		30,100	3,175		33,275	37,900
3400　750 GPM, 100 psi, 66 HP, 3550 RPM, 4" pump		.58	55.172		26,700	3,500		30,200	34,600
3450　750 GPM, 165 psi, 120 HP, 3550 RPM, 4" pump		.56	57.143		34,800	3,625		38,425	43,700
3500　1000 GPM, 50 psi, 48 HP, 1770 RPM, 5" pump		.60	53.333		30,200	3,375		33,575	38,300
3550　1000 GPM, 100 psi, 86 HP, 3550 RPM, 5" pump		.54	59.259		33,800	3,750		37,550	42,800
3600　1000 GPM, 150 psi, 142 HP, 3550 RPM, 5" pump		.50	64		38,300	4,050		42,350	48,200
3650　1000 GPM, 200 psi, 245 HP, 1770 RPM, 6" pump		.36	88.889		50,500	5,625		56,125	64,500
3660　1250 GPM, 75 psi, 75 HP, 1770 RPM, 5" pump		.55	58.182		31,000	3,675		34,675	39,600
3700　1500 GPM, 50 psi, 66 HP, 1770 RPM, 6" pump		.50	64		32,500	4,050		36,550	41,900
3750　1500 GPM, 100 psi, 139 HP, 1770 RPM, 6" pump		.46	69.565		36,700	4,400		41,100	47,000
3800　1500 GPM, 150 psi, 200 HP, 1770 RPM, 6" pump		.36	88.889		45,100	5,625		50,725	58,000
3850　1500 GPM, 200 psi, 279 HP, 1770 RPM, 6" pump		.32	100		48,500	6,325		54,825	63,000
3900　2000 GPM, 100 psi, 167 HP, 1770 RPM, 6" pump		.34	94.118		44,000	5,950		49,950	57,500
3950　2000 GPM, 150 psi, 292 HP, 1770 RPM, 6" pump		.28	114		55,000	7,225		62,225	71,500
4000　2500 GPM, 100 psi, 213 HP, 1770 RPM, 8" pump		.30	107		51,500	6,750		58,250	66,500

For customer support on your Plumbing Costs with RSMeans Data, call 800.448.8182.

21 31 Centrifugal Fire Pumps

21 31 13 – Electric-Drive, Centrifugal Fire Pumps

21 31 13.50 Electric-Drive Fire Pumps

		Crew	Daily Output	Labor-Hours	Unit	Material	2021 Bare Costs Labor	Equipment	Total	Total Incl O&P
4040	2500 GPM, 135 psi, 339 HP, 1770 RPM, 8" pump	Q-13	.26	123	Ea.	73,500	7,800		81,300	92,500
4100	3000 GPM, 100 psi, 250 HP, 1770 RPM, 8" pump		.28	114		61,500	7,225		68,725	78,500
4150	3000 GPM, 140 psi, 428 HP, 1770 RPM, 10" pump		.24	133		98,000	8,450		106,450	120,000
4200	3500 GPM, 100 psi, 300 HP, 1770 RPM, 10" pump		.26	123		76,500	7,800		84,300	95,500
4250	3500 GPM, 140 psi, 450 HP, 1770 RPM, 10" pump		.24	133		98,000	8,450		106,450	120,000
5000	For jockey pump 1", 3 HP, with control, add	Q-12	2	8		3,900	480		4,380	5,025

21 31 16 – Diesel-Drive, Centrifugal Fire Pumps

21 31 16.50 Diesel-Drive Fire Pumps

		Crew	Daily Output	Labor-Hours	Unit	Material	2021 Bare Costs Labor	Equipment	Total	Total Incl O&P
0010	**DIESEL-DRIVE FIRE PUMPS** Including controller, fittings and relief valve									
0050	500 GPM, 50 psi, 27 HP, 4" pump	Q-13	.64	50	Ea.	66,500	3,175		69,675	77,500
0100	500 GPM, 100 psi, 62 HP, 4" pump		.60	53.333		70,500	3,375		73,875	82,500
0150	500 GPM, 125 psi, 78 HP, 4" pump		.56	57.143		71,000	3,625		74,625	84,000
0200	750 GPM, 50 psi, 44 HP, 5" pump		.60	53.333		71,000	3,375		74,375	83,000
0250	750 GPM, 100 psi, 80 HP, 4" pump		.56	57.143		74,000	3,625		77,625	87,000
0300	750 GPM, 165 psi, 203 HP, 5" pump		.52	61.538		76,500	3,900		80,400	90,000
0350	1000 GPM, 50 psi, 48 HP, 5" pump		.58	55.172		71,500	3,500		75,000	83,500
0400	1000 GPM, 100 psi, 89 HP, 4" pump		.56	57.143		76,000	3,625		79,625	89,000
0450	1000 GPM, 150 psi, 148 HP, 4" pump		.48	66.667		75,000	4,225		79,225	89,000
0470	1000 GPM, 200 psi, 280 HP, 5" pump		.40	80		94,000	5,075		99,075	111,000
0480	1250 GPM, 75 psi, 75 HP, 5" pump		.54	59.259		72,000	3,750		75,750	85,000
0500	1500 GPM, 50 psi, 66 HP, 6" pump		.50	64		77,000	4,050		81,050	90,500
0550	1500 GPM, 100 psi, 140 HP, 6" pump		.46	69.565		78,500	4,400		82,900	93,000
0600	1500 GPM, 150 psi, 228 HP, 6" pump		.42	76.190		88,500	4,825		93,325	104,500
0650	1500 GPM, 200 psi, 279 HP, 6" pump		.38	84.211		139,000	5,325		144,325	161,000
0700	2000 GPM, 100 psi, 167 HP, 6" pump		.34	94.118		80,500	5,950		86,450	97,500
0750	2000 GPM, 150 psi, 284 HP, 6" pump		.30	107		109,000	6,750		115,750	130,000
0800	2500 GPM, 100 psi, 213 HP, 8" pump		.32	100		103,000	6,325		109,325	122,500
0820	2500 GPM, 150 psi, 365 HP, 8" pump		.26	123		120,500	7,800		128,300	144,000
0850	3000 GPM, 100 psi, 250 HP, 8" pump		.28	114		118,500	7,225		125,725	141,500
0900	3000 GPM, 150 psi, 384 HP, 10" pump		.20	160		138,000	10,100		148,100	167,000
0950	3500 GPM, 100 psi, 300 HP, 10" pump		.24	133		141,500	8,450		149,950	168,500
1000	3500 GPM, 150 psi, 518 HP, 10" pump		.20	160		200,500	10,100		210,600	235,500

Division Notes

	CREW	DAILY OUTPUT	LABOR-HOURS	UNIT	BARE COSTS				TOTAL INCL O&P
					MAT.	LABOR	EQUIP.	TOTAL	

Estimating Tips
22 10 00 Plumbing Piping and Pumps

This subdivision is primarily basic pipe and related materials. The pipe may be used by any of the mechanical disciplines, i.e., plumbing, fire protection, heating, and air conditioning.

Note: CPVC plastic piping approved for fire protection is located in 21 11 13.

- The labor adjustment factors listed in Subdivision 22 01 02.20 apply throughout Divisions 21, 22, and 23. CAUTION: the correct percentage may vary for the same items. For example, the percentage add for the basic pipe installation should be based on the maximum height that the installer must install for that particular section. If the pipe is to be located 14' above the floor but it is suspended on threaded rod from beams, the bottom flange of which is 18' high (4' rods), then the height is actually 18' and the add is 20%. The pipe cover, however, does not have to go above the 14' and so the add should be 10%.

- Most pipe is priced first as straight pipe with a joint (coupling, weld, etc.) every 10' and a hanger usually every 10'. There are exceptions with hanger spacing such as for cast iron pipe (5')

and plastic pipe (3 per 10'). Following each type of pipe there are several lines listing sizes and the amount to be subtracted to delete couplings and hangers. This is for pipe that is to be buried or supported together on trapeze hangers. The reason that the couplings are deleted is that these runs are usually long, and frequently longer lengths of pipe are used. By deleting the couplings, the estimator is expected to look up and add back the correct reduced number of couplings.

- When preparing an estimate, it may be necessary to approximate the fittings. Fittings usually run between 25% and 50% of the cost of the pipe. The lower percentage is for simpler runs, and the higher number is for complex areas, such as mechanical rooms.

- For historic restoration projects, the systems must be as invisible as possible, and pathways must be sought for pipes, conduit, and ductwork. While installations in accessible spaces (such as basements and attics) are relatively straightforward to estimate, labor costs may be more difficult to determine when delivery systems must be concealed.

22 40 00 Plumbing Fixtures

- Plumbing fixture costs usually require two lines: the fixture itself and its "rough-in, supply, and waste."

- In the Assemblies Section (Plumbing D2010) for the desired fixture, the System Components Group at the center of the page shows the fixture on the first line. The rest of the list (fittings, pipe, tubing, etc.) will total up to what we refer to in the Unit Price section as "Rough-in, supply, waste, and vent." Note that for most fixtures we allow a nominal 5' of tubing to reach from the fixture to a main or riser.

- Remember that gas- and oil-fired units need venting.

Reference Numbers

Reference numbers are shown at the beginning of some major classifications. These numbers refer to related items in the Reference Section. The reference information may be an estimating procedure, an alternate pricing method, or technical information.

Note: Not all subdivisions listed here necessarily appear. ∎

Same Data. Simplified.

Enjoy the convenience and efficiency of accessing your costs anywhere:

- **Skip the multiplier** by setting your location
- **Quickly search,** edit, favorite and share costs
- **Stay on top of price changes** with automatic updates

Discover more at rsmeans.com/online

22 01 02.10 Boilers, General	Crew	Daily Output	Labor-Hours	Unit	Material	2021 Bare Costs Labor	Equipment	Total	Total Incl O&P
0010 **BOILERS, GENERAL**, Prices do not include flue piping, elec. wiring,									
0020 gas or oil piping, boiler base, pad, or tankless unless noted									
0100 Boiler H.P.: 10 KW = 34 lb./steam/hr. = 33,475 BTU/hr.									
0150 To convert SFR to BTU rating: Hot water, 150 x SFR;									
0160 Forced hot water, 180 x SFR; steam, 240 x SFR									

22 01 02.20 Labor Adjustment Factors

	Crew	Daily Output	Labor-Hours	Unit	Material	2021 Bare Costs Labor	Equipment	Total	Total Incl O&P
0010 **LABOR ADJUSTMENT FACTORS** (For Div. 21, 22 and 23) R220102-20									
0100 Labor factors: The below are reasonable suggestions, but									
0110 each project must be evaluated for its own peculiarities, and									
0120 the adjustments be increased or decreased depending on the									
0130 severity of the special conditions.									
1000 Add to labor for elevated installation (Above floor level)									
1080 10' to 14.5' high R221113-70						10%			
1100 15' to 19.5' high						20%			
1120 20' to 24.5' high						25%			
1140 25' to 29.5' high						35%			
1160 30' to 34.5' high						40%			
1180 35' to 39.5' high						50%			
1200 40' and higher						55%			
2000 Add to labor for crawl space									
2100 3' high						40%			
2140 4' high						30%			
3000 Add to labor for multi-story building									
3010 For new construction (No elevator available)									
3100 Add for floors 3 thru 10						5%			
3110 Add for floors 11 thru 15						10%			
3120 Add for floors 16 thru 20						15%			
3130 Add for floors 21 thru 30						20%			
3140 Add for floors 31 and up						30%			
3170 For existing structure (Elevator available)									
3180 Add for work on floor 3 and above						2%			
4000 Add to labor for working in existing occupied buildings									
4100 Hospital						35%			
4140 Office building						25%			
4180 School						20%			
4220 Factory or warehouse						15%			
4260 Multi dwelling						15%			
5000 Add to labor, miscellaneous									
5100 Cramped shaft						35%			
5140 Congested area						15%			
5180 Excessive heat or cold						30%			
9000 Labor factors: The above are reasonable suggestions, but									
9010 each project should be evaluated for its own peculiarities.									
9100 Other factors to be considered are:									
9140 Movement of material and equipment through finished areas									
9180 Equipment room									
9220 Attic space									
9260 No service road									
9300 Poor unloading/storage area									
9340 Congested site area/heavy traffic									

22 05 05.10 Plumbing Demolition		Crew	Daily Output	Labor-Hours	Unit	Material	2021 Bare Costs Labor	Equipment	Total	Total Incl O&P
0010	**PLUMBING DEMOLITION**	R220105-10								
0400	Air compressor, up thru 2 HP	Q-1	10	1.600	Ea.		97.50		97.50	145
0410	3 HP thru 7-1/2 HP	R024119-10	5.60	2.857			174		174	260
0420	10 HP thru 15 HP		1.40	11.429			695		695	1,050
0430	20 HP thru 30 HP	Q-2	1.30	18.462			1,175		1,175	1,750
0500	Backflow preventer, up thru 2" diameter	1 Plum	17	.471			32		32	47.50
0510	2-1/2" thru 3" diameter	Q-1	10	1.600			97.50		97.50	145
0520	4" thru 6" diameter	"	5	3.200			195		195	291
0530	8" thru 10" diameter	Q-2	3	8			505		505	755
0700	Carriers and supports									
0710	Fountains, sinks, lavatories and urinals	1 Plum	14	.571	Ea.		38.50		38.50	57.50
0720	Water closets	"	12	.667	"		45		45	67.50
0730	Grinder pump or sewage ejector system									
0732	Simplex	Q-1	7	2.286	Ea.		139		139	208
0734	Duplex	"	2.80	5.714			350		350	520
0738	Hot water dispenser	1 Plum	36	.222			15.05		15.05	22.50
0740	Hydrant, wall		26	.308			21		21	31
0744	Ground		12	.667			45		45	67.50
0760	Cleanouts and drains, up thru 4" pipe diameter		10	.800			54		54	81
0764	5" thru 8" pipe diameter	Q-1	10	1.600			97.50		97.50	145
0780	Industrial safety fixtures	1 Plum	8	1			67.50		67.50	101
1020	Fixtures, including 10' piping									
1100	Bathtubs, cast iron	1 Plum	4	2	Ea.		135		135	202
1120	Fiberglass		6	1.333			90.50		90.50	135
1140	Steel		5	1.600			108		108	162
1150	Bidet	Q-1	7	2.286			139		139	208
1200	Lavatory, wall hung	1 Plum	10	.800			54		54	81
1220	Counter top		8	1			67.50		67.50	101
1300	Sink, single compartment		8	1			67.50		67.50	101
1320	Double compartment		7	1.143			77.50		77.50	115
1340	Shower, stall and receptor	Q-1	6	2.667			162		162	242
1350	Group	"	7	2.286			139		139	208
1400	Water closet, floor mounted	1 Plum	8	1			67.50		67.50	101
1420	Wall mounted	"	7	1.143			77.50		77.50	115
1440	Wash fountain, 36" diameter	Q-2	8	3			190		190	283
1442	54" diameter	"	7	3.429			217		217	325
1500	Urinal, floor mounted	1 Plum	4	2			135		135	202
1520	Wall mounted	"	7	1.143			77.50		77.50	115
1590	Whirl pool or hot tub	Q-1	2.60	6.154			375		375	560
1600	Water fountains, free standing	1 Plum	8	1			67.50		67.50	101
1620	Wall or deck mounted		6	1.333			90.50		90.50	135
1800	Medical gas specialties		8	1			67.50		67.50	101
1810	Plumbing demo, floor drain, remove		12	.667			45		45	67.50
1820	Plumbing demo, roof drain, remove		9	.889			60		60	90
1900	Piping fittings, single connection, up thru 1-1/2" diameter		30	.267			18.05		18.05	27
1910	2" thru 4" diameter		14	.571			38.50		38.50	57.50
1980	Pipe hanger/support removal		80	.100			6.75		6.75	10.10
1990	Glass pipe with fittings, 1" thru 3" diameter		200	.040	L.F.		2.71		2.71	4.04
1992	4" thru 6" diameter		150	.053			3.61		3.61	5.40
2000	Piping, metal, up thru 1-1/2" diameter		200	.040			2.71		2.71	4.04
2050	2" thru 3-1/2" diameter		150	.053			3.61		3.61	5.40
2100	4" thru 6" diameter	2 Plum	100	.160			10.85		10.85	16.15
2150	8" thru 14" diameter	"	60	.267			18.05		18.05	27

22 05 05 – Selective Demolition for Plumbing

22 05 05.10 Plumbing Demolition		Crew	Daily Output	Labor-Hours	Unit	Material	2021 Bare Costs Labor	Equipment	Total	Total Incl O&P
2153	16" thru 20" diameter	Q-18	70	.343	L.F.		22	1.54	23.54	34
2155	24" thru 26" diameter		55	.436			28	1.95	29.95	43.50
2156	30" thru 36" diameter		40	.600			38.50	2.69	41.19	60
2160	Plastic pipe with fittings, up thru 1-1/2" diameter	1 Plum	250	.032			2.17		2.17	3.23
2162	2" thru 3" diameter	"	200	.040			2.71		2.71	4.04
2164	4" thru 6" diameter	Q-1	200	.080			4.87		4.87	7.25
2166	8" thru 14" diameter		150	.107			6.50		6.50	9.70
2168	16" diameter		100	.160			9.75		9.75	14.55
2170	Prison fixtures, lavatory or sink		18	.889	Ea.		54		54	81
2172	Shower		5.60	2.857			174		174	260
2174	Urinal or water closet		13	1.231			75		75	112
2180	Pumps, all fractional horse-power		12	1.333			81		81	121
2184	1 HP thru 5 HP		6	2.667			162		162	242
2186	7-1/2 HP thru 15 HP		2.50	6.400			390		390	580
2188	20 HP thru 25 HP	Q-2	4	6			380		380	565
2190	30 HP thru 60 HP		.80	30			1,900		1,900	2,825
2192	75 HP thru 100 HP		.60	40			2,525		2,525	3,775
2194	150 HP		.50	48			3,025		3,025	4,525
2198	Pump, sump or submersible	1 Plum	12	.667			45		45	67.50
2200	Receptors and interceptors, up thru 20 GPM	"	8	1			67.50		67.50	101
2204	25 thru 100 GPM	Q-1	6	2.667			162		162	242
2208	125 thru 300 GPM	"	2.40	6.667			405		405	605
2211	325 thru 500 GPM	Q-2	2.60	9.231			585		585	870
2212	Deduct for salvage, aluminum scrap				Ton	455			455	500
2214	Brass scrap					2,175			2,175	2,400
2216	Copper scrap					3,725			3,725	4,100
2218	Lead scrap					910			910	1,000
2220	Steel scrap								204	224
2230	Temperature maintenance cable	1 Plum	1200	.007	L.F.		.45		.45	.67
2240	Toilet partitions, see Section 10 21 13									
2250	Water heater, 40 gal.	1 Plum	6	1.333	Ea.		90.50		90.50	135
3100	Tanks, water heaters and liquid containers									
3110	Up thru 45 gallons	Q-1	22	.727	Ea.		44.50		44.50	66
3120	50 thru 120 gallons		14	1.143			69.50		69.50	104
3130	130 thru 240 gallons		7.60	2.105			128		128	191
3140	250 thru 500 gallons		5.40	2.963			181		181	269
3150	600 thru 1,000 gallons	Q-2	1.60	15			950		950	1,425
3160	1,100 thru 2,000 gallons		.70	34.286			2,175		2,175	3,225
3170	2,100 thru 4,000 gallons		.50	48			3,025		3,025	4,525
6000	Remove and reset fixtures, easy access	1 Plum	6	1.333			90.50		90.50	135
6100	Difficult access	"	4	2			135		135	202
6500	Solar heating system									
6510	Solar panel, hot water system	1 Plum	304	.026	S.F.		1.78		1.78	2.66
7910	Solar panel for pool	"	300	.027	"		1.81		1.81	2.69
8000	Sprinkler system									
8100	Exposed wet/dry	1 Plum	2500	.003	SF Flr.		.22		.22	.32
8200	Concealed wet/dry	"	1250	.006	"		.43		.43	.65
9100	Valve, metal valves or strainers and similar, up thru 1-1/2" diameter	1 Stpi	28	.286	Ea.		19.55		19.55	29
9110	2" thru 3" diameter	Q-1	11	1.455			88.50		88.50	132
9120	4" thru 6" diameter	"	8	2			122		122	182
9130	8" thru 14" diameter	Q-2	8	3			190		190	283
9140	16" thru 20" diameter		2	12			760		760	1,125
9150	24" diameter		1.20	20			1,275		1,275	1,875

For customer support on your Plumbing Costs with RSMeans Data, call 800.448.8182.

22 05 Common Work Results for Plumbing

22 05 05 – Selective Demolition for Plumbing

22 05 05.10 Plumbing Demolition

		Crew	Daily Output	Labor-Hours	Unit	Material	2021 Bare Costs Labor	Equipment	Total	Total Incl O&P
9200	Valve, plastic, up thru 1-1/2" diameter	1 Plum	42	.190	Ea.		12.90		12.90	19.25
9210	2" thru 3" diameter		15	.533			36		36	54
9220	4" thru 6" diameter		12	.667			45		45	67.50
9300	Vent flashing and caps		55	.145			9.85		9.85	14.70
9350	Water filter, commercial, 1" thru 1-1/2"	Q-1	2	8			485		485	725
9360	2" thru 2-1/2"	"	1.60	10			610		610	910
9400	Water heaters									
9410	Up thru 245 GPH	Q-1	2.40	6.667	Ea.		405		405	605
9420	250 thru 756 GPH		1.60	10			610		610	910
9430	775 thru 1,640 GPH		.80	20			1,225		1,225	1,825
9440	1,650 thru 4,000 GPH	Q-2	.50	48			3,025		3,025	4,525
9470	Water softener	Q-1	2	8			485		485	725

22 05 23 – General-Duty Valves for Plumbing Piping

22 05 23.10 Valves, Brass

		Crew	Daily Output	Labor-Hours	Unit	Material	2021 Bare Costs Labor	Equipment	Total	Total Incl O&P
0010	**VALVES, BRASS**									
0032	For motorized valves, see Section 23 09 53.10									
0500	Gas cocks, threaded									
0510	1/4"	1 Plum	26	.308	Ea.	21.50	21		42.50	54.50
0520	3/8"		24	.333		21.50	22.50		44	57
0530	1/2"		24	.333		21	22.50		43.50	56.50
0540	3/4"		22	.364		24.50	24.50		49	64
0550	1"		19	.421		42	28.50		70.50	88.50
0560	1-1/4"		15	.533		55.50	36		91.50	115
0570	1-1/2"		13	.615		82.50	41.50		124	153
0580	2"		11	.727		135	49		184	222
0672	For larger sizes use lubricated plug valve, Section 23 05 23.70									

22 05 23.20 Valves, Bronze

			Crew	Daily Output	Labor-Hours	Unit	Material	2021 Bare Costs Labor	Equipment	Total	Total Incl O&P
0010	**VALVES, BRONZE**	R220523-80									
1020	Angle, 150 lb., rising stem, threaded										
1030	1/8"	R220523-90	1 Plum	24	.333	Ea.	172	22.50		194.50	224
1040	1/4"			24	.333		172	22.50		194.50	224
1050	3/8"			24	.333		171	22.50		193.50	222
1060	1/2"			22	.364		171	24.50		195.50	225
1070	3/4"			20	.400		235	27		262	300
1080	1"			19	.421		325	28.50		353.50	405
1090	1-1/4"			15	.533		395	36		431	490
1100	1-1/2"			13	.615		550	41.50		591.50	665
1102	Soldered same price as threaded										
1110	2"		1 Plum	11	.727	Ea.	865	49		914	1,025
1300	Ball										
1304	Soldered										
1312	3/8"		1 Plum	21	.381	Ea.	19.90	26		45.90	60.50
1316	1/2"			18	.444		19.90	30		49.90	67
1320	3/4"			17	.471		36.50	32		68.50	87.50
1324	1"			15	.533		49	36		85	108
1328	1-1/4"			13	.615		74.50	41.50		116	144
1332	1-1/2"			11	.727		94.50	49		143.50	178
1336	2"			9	.889		154	60		214	259
1340	2-1/2"			7	1.143		505	77.50		582.50	670
1344	3"			5	1.600		620	108		728	845
1350	Single union end										
1358	3/8"		1 Plum	21	.381	Ea.	27.50	26		53.50	69

113

22 05 23.20 Valves, Bronze		Crew	Daily Output	Labor-Hours	Unit	Material	2021 Bare Costs Labor	Equipment	Total	Total Incl O&P
1362	1/2"	1 Plum	18	.444	Ea.	30.50	30		60.50	78.50
1366	3/4"		17	.471		53.50	32		85.50	107
1370	1"		15	.533		70.50	36		106.50	132
1374	1-1/4"		13	.615		104	41.50		145.50	176
1378	1-1/2"		11	.727		125	49		174	212
1382	2"	↓	9	.889	↓	204	60		264	315
1398	Threaded, 150 psi									
1400	1/4"	1 Plum	24	.333	Ea.	19.05	22.50		41.55	54.50
1430	3/8"		24	.333		19.05	22.50		41.55	54.50
1450	1/2"		22	.364		19.05	24.50		43.55	58
1460	3/4"		20	.400		30	27		57	73.50
1470	1"		19	.421		40.50	28.50		69	87
1480	1-1/4"		15	.533		51	36		87	111
1490	1-1/2"		13	.615		81.50	41.50		123	152
1500	2"		11	.727		125	49		174	211
1510	2-1/2"		9	.889		269	60		329	385
1520	3"	↓	8	1	↓	625	67.50		692.50	785
1522	Solder the same price as threaded									
1600	Butterfly, 175 psi, full port, solder or threaded ends									
1610	Stainless steel disc and stem									
1620	1/4"	1 Plum	24	.333	Ea.	15.40	22.50		37.90	50.50
1630	3/8"		24	.333		26	22.50		48.50	62.50
1640	1/2"		22	.364		27.50	24.50		52	67.50
1650	3/4"		20	.400		44.50	27		71.50	89.50
1660	1"		19	.421		54.50	28.50		83	102
1670	1-1/4"		15	.533		87.50	36		123.50	151
1680	1-1/2"		13	.615		110	41.50		151.50	183
1690	2"	↓	11	.727	↓	141	49		190	229
1750	Check, swing, class 150, regrinding disc, threaded									
1800	1/8"	1 Plum	24	.333	Ea.	79.50	22.50		102	121
1830	1/4"		24	.333		81.50	22.50		104	124
1840	3/8"		24	.333		87.50	22.50		110	130
1850	1/2"		24	.333		89.50	22.50		112	132
1860	3/4"		20	.400		124	27		151	177
1870	1"		19	.421		176	28.50		204.50	237
1880	1-1/4"		15	.533		257	36		293	335
1890	1-1/2"		13	.615		294	41.50		335.50	385
1900	2"	↓	11	.727		425	49		474	540
1910	2-1/2"	Q-1	15	1.067		985	65		1,050	1,175
1920	3"	"	13	1.231	↓	1,325	75		1,400	1,550
2000	For 200 lb., add						5%	10%		
2040	For 300 lb., add						15%	15%		
2060	Check swing, 300 lb., lead free unless noted, sweat, 3/8" size	1 Plum	24	.333	Ea.	170	22.50		192.50	222
2070	1/2"		24	.333		170	22.50		192.50	222
2080	3/4"		20	.400		229	27		256	293
2090	1"		19	.421		340	28.50		368.50	415
2100	1-1/4"		15	.533		480	36		516	580
2110	1-1/2"		13	.615		560	41.50		601.50	675
2120	2"	↓	11	.727		825	49		874	980
2130	2-1/2", not lead free	Q-1	15	1.067		1,175	65		1,240	1,400
2140	3", not lead free	"	13	1.231	↓	1,650	75		1,725	1,900
2850	Gate, N.R.S., soldered, 125 psi									
2900	3/8"	1 Plum	24	.333	Ea.	76.50	22.50		99	118

For customer support on your Plumbing Costs with RSMeans Data, call 800.448.8182.

22 05 23 – General-Duty Valves for Plumbing Piping

22 05 23.20 Valves, Bronze		Crew	Daily Output	Labor-Hours	Unit	Material	2021 Bare Costs Labor	Equipment	Total	Total Incl O&P
2920	1/2"	1 Plum	24	.333	Ea.	76.50	22.50		99	118
2940	3/4"		20	.400		84.50	27		111.50	134
2950	1"		19	.421		122	28.50		150.50	177
2960	1-1/4"		15	.533		186	36		222	259
2970	1-1/2"		13	.615		207	41.50		248.50	290
2980	2"		11	.727		293	49		342	395
2990	2-1/2"	Q-1	15	1.067		640	65		705	800
3000	3"	"	13	1.231		780	75		855	965
3350	Threaded, class 150									
3410	1/4"	1 Plum	24	.333	Ea.	106	22.50		128.50	151
3420	3/8"		24	.333		106	22.50		128.50	151
3430	1/2"		24	.333		100	22.50		122.50	144
3440	3/4"		20	.400		110	27		137	162
3450	1"		19	.421		139	28.50		167.50	195
3460	1-1/4"		15	.533		187	36		223	260
3470	1-1/2"		13	.615		267	41.50		308.50	355
3480	2"		11	.727		320	49		369	425
3490	2-1/2"	Q-1	15	1.067		745	65		810	915
3500	3"	"	13	1.231		1,050	75		1,125	1,275
3600	Gate, flanged, 150 lb.									
3610	1"	1 Plum	7	1.143	Ea.	1,650	77.50		1,727.50	1,950
3620	1-1/2"		6	1.333		2,050	90.50		2,140.50	2,400
3630	2"		5	1.600		3,025	108		3,133	3,500
3634	2-1/2"	Q-1	5	3.200		4,675	195		4,870	5,450
3640	3"	"	4.50	3.556		5,475	217		5,692	6,325
3850	Rising stem, soldered, 300 psi									
3900	3/8"	1 Plum	24	.333	Ea.	160	22.50		182.50	210
3920	1/2"		24	.333		157	22.50		179.50	206
3940	3/4"		20	.400		179	27		206	237
3950	1"		19	.421		242	28.50		270.50	310
3960	1-1/4"		15	.533		335	36		371	425
3970	1-1/2"		13	.615		405	41.50		446.50	505
3980	2"		11	.727		650	49		699	790
3990	2-1/2"	Q-1	15	1.067		1,375	65		1,440	1,625
4000	3"	"	13	1.231		2,150	75		2,225	2,450
4250	Threaded, class 150									
4310	1/4"	1 Plum	24	.333	Ea.	97.50	22.50		120	141
4320	3/8"		24	.333		97.50	22.50		120	141
4330	1/2"		24	.333		88	22.50		110.50	131
4340	3/4"		20	.400		99.50	27		126.50	150
4350	1"		19	.421		133	28.50		161.50	190
4360	1-1/4"		15	.533		181	36		217	253
4370	1-1/2"		13	.615		228	41.50		269.50	315
4380	2"		11	.727		305	49		354	410
4390	2-1/2"	Q-1	15	1.067		715	65		780	880
4400	3"	"	13	1.231		995	75		1,070	1,200
4500	For 300 psi, threaded, add					100%	15%			
4540	For chain operated type, add					15%				
4850	Globe, class 150, rising stem, threaded									
4920	1/4"	1 Plum	24	.333	Ea.	131	22.50		153.50	179
4940	3/8"		24	.333		129	22.50		151.50	176
4950	1/2"		24	.333		129	22.50		151.50	176
4960	3/4"		20	.400		172	27		199	231

115

22 05 23.20 Valves, Bronze		Crew	Daily Output	Labor-Hours	Unit	Material	2021 Bare Costs Labor	Equipment	Total	Total Incl O&P
4970	1"	1 Plum	19	.421	Ea.	271	28.50		299.50	340
4980	1-1/4"		15	.533		430	36		466	530
4990	1-1/2"		13	.615		525	41.50		566.50	635
5000	2"		11	.727		785	49		834	940
5010	2-1/2"	Q-1	15	1.067		1,575	65		1,640	1,825
5020	3"	"	13	1.231		2,250	75		2,325	2,575
5120	For 300 lb. threaded, add					50%	15%			
5600	Relief, pressure & temperature, self-closing, ASME, threaded									
5640	3/4"	1 Plum	28	.286	Ea.	267	19.35		286.35	320
5650	1"		24	.333		425	22.50		447.50	505
5660	1-1/4"		20	.400		855	27		882	985
5670	1-1/2"		18	.444		1,650	30		1,680	1,850
5680	2"		16	.500		1,800	34		1,834	2,025
5950	Pressure, poppet type, threaded									
6000	1/2"	1 Plum	30	.267	Ea.	95.50	18.05		113.55	132
6040	3/4"	"	28	.286	"	111	19.35		130.35	151
6400	Pressure, water, ASME, threaded									
6440	3/4"	1 Plum	28	.286	Ea.	58	19.35		77.35	93
6450	1"		24	.333		360	22.50		382.50	430
6460	1-1/4"		20	.400		565	27		592	660
6470	1-1/2"		18	.444		780	30		810	905
6480	2"		16	.500		1,125	34		1,159	1,300
6490	2-1/2"		15	.533		4,425	36		4,461	4,925
6900	Reducing, water pressure									
6920	300 psi to 25-75 psi, threaded or sweat									
6940	1/2"	1 Plum	24	.333	Ea.	560	22.50		582.50	650
6950	3/4"		20	.400		585	27		612	685
6960	1"		19	.421		905	28.50		933.50	1,050
6970	1-1/4"		15	.533		1,550	36		1,586	1,750
6980	1-1/2"		13	.615		2,325	41.50		2,366.50	2,625
6990	2"		11	.727		3,525	49		3,574	3,975
7100	For built-in by-pass or 10-35 psi, add					54.50			54.50	59.50
7700	High capacity, 250 psi to 25-75 psi, threaded									
7740	1/2"	1 Plum	24	.333	Ea.	935	22.50		957.50	1,050
7780	3/4"		20	.400		760	27		787	875
7790	1"		19	.421		1,100	28.50		1,128.50	1,250
7800	1-1/4"		15	.533		1,900	36		1,936	2,125
7810	1-1/2"		13	.615		2,775	41.50		2,816.50	3,100
7820	2"		11	.727		4,025	49		4,074	4,500
7830	2-1/2"		9	.889		5,750	60		5,810	6,450
7840	3"		8	1		6,900	67.50		6,967.50	7,700
7850	3" flanged (iron body)	Q-1	10	1.600		6,775	97.50		6,872.50	7,600
7860	4" flanged (iron body)	"	8	2		8,850	122		8,972	9,900
7920	For higher pressure, add					25%				
8000	Silent check, bronze trim									
8010	Compact wafer type, for 125 or 150 lb. flanges									
8020	1-1/2"	1 Plum	11	.727	Ea.	460	49		509	580
8021	2"	"	9	.889		485	60		545	625
8022	2-1/2"	Q-1	9	1.778		525	108		633	735
8023	3"		8	2		585	122		707	825
8024	4"		5	3.200		975	195		1,170	1,375
8025	5"	Q-2	6	4		1,050	253		1,303	1,525
8026	6"	"	5	4.800		1,125	305		1,430	1,700

22 05 23 – General-Duty Valves for Plumbing Piping

22 05 23.20 Valves, Bronze	Crew	Daily Output	Labor-Hours	Unit	Material	2021 Bare Costs Labor	Equipment	Total	Total Incl O&P
8050	For 250 or 300 lb. flanges, thru 6" no change								
8060	Full flange wafer type, 150 lb.								
8063	1-1/2"	1 Plum	11	.727	Ea.	470	49		519	590
8064	2"	"	9	.889		630	60		690	785
8065	2-1/2"	Q-1	9	1.778		710	108		818	940
8066	3"		8	2		785	122		907	1,050
8067	4"	↓	5	3.200		1,150	195		1,345	1,575
8068	5"	Q-2	6	4		1,500	253		1,753	2,025
8069	6"	"	5	4.800		1,975	305		2,280	2,625
8080	For 300 lb., add				↓	40%	10%			
8100	Globe type, 150 lb.								
8110	2"	1 Plum	9	.889	Ea.	620	60		680	770
8111	2-1/2"	Q-1	9	1.778		775	108		883	1,025
8112	3"		8	2		925	122		1,047	1,200
8113	4"	↓	5	3.200		1,275	195		1,470	1,700
8114	5"	Q-2	6	4		1,575	253		1,828	2,100
8115	6"	"	5	4.800	↓	2,175	305		2,480	2,850
8130	For 300 lb., add					20%	10%			
8140	Screwed end type, 250 lb.								
8141	1/2"	1 Plum	24	.333	Ea.	56.50	22.50		79	95.50
8142	3/4"		20	.400		56.50	27		83.50	103
8143	1"		19	.421		64.50	28.50		93	114
8144	1-1/4"		15	.533		89	36		125	152
8145	1-1/2"		13	.615		100	41.50		141.50	172
8146	2"	↓	11	.727	↓	135	49		184	222
8350	Tempering, water, sweat connections								
8400	1/2"	1 Plum	24	.333	Ea.	119	22.50		141.50	165
8440	3/4"	"	20	.400	"	175	27		202	234
8650	Threaded connections								
8700	1/2"	1 Plum	24	.333	Ea.	167	22.50		189.50	218
8740	3/4"		20	.400		1,050	27		1,077	1,200
8750	1"		19	.421		1,200	28.50		1,228.50	1,350
8760	1-1/4"		15	.533		1,825	36		1,861	2,075
8770	1-1/2"		13	.615		2,000	41.50		2,041.50	2,250
8780	2"	↓	11	.727	↓	3,000	49		3,049	3,375
8800	Water heater water & gas safety shut off								
8810	Protection against a leaking water heater								
8814	Shut off valve	1 Plum	16	.500	Ea.	196	34		230	266
8818	Water heater dam		32	.250		52.50	16.95		69.45	83.50
8822	Gas control wiring harness	↓	32	.250	↓	25.50	16.95		42.45	53.50
8830	Whole house flood safety shut off								
8834	Connections								
8838	3/4" NPT	1 Plum	12	.667	Ea.	1,075	45		1,120	1,250
8842	1" NPT		11	.727		1,100	49		1,149	1,300
8846	1-1/4" NPT	↓	10	.800	↓	1,150	54		1,204	1,325

22 05 23.40 Valves, Lined, Corrosion Resistant/High Purity

		Crew	Daily Output	Labor-Hours	Unit	Material	2021 Bare Costs Labor	Equipment	Total	Total Incl O&P
0010	**VALVES, LINED, CORROSION RESISTANT/HIGH PURITY**									
3500	Check lift, 125 lb., cast iron flanged									
3510	Horizontal PPL or SL lined									
3530	1"	1 Plum	14	.571	Ea.	655	38.50		693.50	780
3540	1-1/2"		11	.727		790	49		839	945
3550	2"	↓	8	1		925	67.50		992.50	1,125

For customer support on your Plumbing Costs with RSMeans Data, call 800.448.8182.

117

22 05 Common Work Results for Plumbing

22 05 23 – General-Duty Valves for Plumbing Piping

22 05 23.40 Valves, Lined, Corrosion Resistant/High Purity		Crew	Daily Output	Labor-Hours	Unit	Material	2021 Bare Costs Labor	Equipment	Total	Total Incl O&P
3560	2-1/2"	Q-1	5	3.200	Ea.	1,200	195		1,395	1,600
3570	3"		4.50	3.556		1,500	217		1,717	1,975
3590	4"		3	5.333		1,975	325		2,300	2,650
3610	6"	Q-2	3	8		3,375	505		3,880	4,475
3620	8"	"	2.50	9.600		7,425	605		8,030	9,075
4250	Vertical PPL or SL lined									
4270	1"	1 Plum	14	.571	Ea.	620	38.50		658.50	745
4290	1-1/2"		11	.727		775	49		824	930
4300	2"		8	1		910	67.50		977.50	1,100
4310	2-1/2"	Q-1	5	3.200		1,350	195		1,545	1,800
4320	3"		4.50	3.556		1,375	217		1,592	1,825
4340	4"		3	5.333		1,900	325		2,225	2,550
4360	6"	Q-2	3	8		2,825	505		3,330	3,850
4370	8"	"	2.50	9.600		5,525	605		6,130	6,975
5000	Clamp type, ductile iron, 150 lb. flanged									
5010	TFE lined									
5030	1" size, lever handle	1 Plum	9	.889	Ea.	1,300	60		1,360	1,525
5050	1-1/2" size, lever handle		6	1.333		1,700	90.50		1,790.50	2,000
5060	2" size, lever handle		5	1.600		2,150	108		2,258	2,500
5080	3" size, lever handle	Q-1	4.50	3.556		2,725	217		2,942	3,300
5100	4" size, gear operated	"	3	5.333		3,725	325		4,050	4,575
5120	6" size, gear operated	Q-2	3	8		13,100	505		13,605	15,200
5130	8" size, gear operated	"	2.50	9.600		15,500	605		16,105	17,900
6000	Diaphragm type, cast iron, 125 lb. flanged									
6010	PTFE or VITON, lined									
6030	1" size, handwheel operated	1 Plum	9	.889	Ea.	325	60		385	450
6050	1-1/2" size, handwheel operated		6	1.333		385	90.50		475.50	560
6060	2" size, handwheel operated		5	1.600		440	108		548	645
6080	3" size, handwheel operated	Q-1	4.50	3.556		730	217		947	1,125
6100	4" size, handwheel operated	"	3	5.333		1,450	325		1,775	2,075
6120	6" size, handwheel operated	Q-2	3	8		2,275	505		2,780	3,250
6130	8" size, handwheel operated	"	2.50	9.600		4,575	605		5,180	5,925

22 05 23.60 Valves, Plastic

		Crew	Daily Output	Labor-Hours	Unit	Material	Labor	Equipment	Total	Total Incl O&P
0010	**VALVES, PLASTIC** R220523-90									
1100	Angle, PVC, threaded									
1110	1/4"	1 Plum	26	.308	Ea.	49.50	21		70.50	85.50
1120	1/2"		26	.308		70.50	21		91.50	109
1130	3/4"		25	.320		83.50	21.50		105	125
1140	1"		23	.348		102	23.50		125.50	147
1150	Ball, PVC, socket or threaded, true union									
1230	1/2"	1 Plum	26	.308	Ea.	38	21		59	73
1240	3/4"		25	.320		45	21.50		66.50	82
1250	1"		23	.348		54	23.50		77.50	94
1260	1-1/4"		21	.381		85	26		111	132
1270	1-1/2"		20	.400		85	27		112	134
1280	2"		17	.471		134	32		166	196
1290	2-1/2"	Q-1	26	.615		189	37.50		226.50	263
1300	3"		24	.667		272	40.50		312.50	360
1310	4"		20	.800		460	48.50		508.50	580
1360	For PVC, flanged, add					100%	15%			
1450	Double union 1/2"	1 Plum	26	.308		37	21		58	71.50
1460	3/4"		25	.320		43.50	21.50		65	80.50

118

For customer support on your Plumbing Costs with RSMeans Data, call 800.448.8182.

22 05 23.60 Valves, Plastic		Crew	Daily Output	Labor-Hours	Unit	Material	2021 Bare Costs Labor	Equipment	Total	Total Incl O&P
1470	1"	1 Plum	23	.348	Ea.	53.50	23.50		77	93.50
1480	1-1/4"		21	.381		72	26		98	118
1490	1-1/2"		20	.400		86.50	27		113.50	136
1500	2"		17	.471		114	32		146	174
1650	CPVC, socket or threaded, single union									
1700	1/2"	1 Plum	26	.308	Ea.	59	21		80	96
1720	3/4"		25	.320		74.50	21.50		96	115
1730	1"		23	.348		89.50	23.50		113	133
1750	1-1/4"		21	.381		142	26		168	195
1760	1-1/2"		20	.400		142	27		169	197
1770	2"		17	.471		196	32		228	264
1780	3"	Q-1	24	.667		640	40.50		680.50	765
1840	For CPVC, flanged, add					65%	15%			
1880	For true union, socket or threaded, add					50%	5%			
2050	Polypropylene, threaded									
2100	1/4"	1 Plum	26	.308	Ea.	44	21		65	79.50
2120	3/8"		26	.308		45.50	21		66.50	81
2130	1/2"		26	.308		39.50	21		60.50	74.50
2140	3/4"		25	.320		48	21.50		69.50	85
2150	1"		23	.348		56	23.50		79.50	96.50
2160	1-1/4"		21	.381		74	26		100	120
2170	1-1/2"		20	.400		92	27		119	142
2180	2"		17	.471		123	32		155	184
2190	3"	Q-1	24	.667		280	40.50		320.50	370
2200	4"	"	20	.800		495	48.50		543.50	620
2550	PVC, three way, socket or threaded									
2600	1/2"	1 Plum	26	.308	Ea.	63.50	21		84.50	101
2640	3/4"		25	.320		75	21.50		96.50	115
2650	1"		23	.348		91.50	23.50		115	136
2660	1-1/2"		20	.400		129	27		156	182
2670	2"		17	.471		165	32		197	229
2680	3"	Q-1	24	.667		520	40.50		560.50	635
2740	For flanged, add					60%	15%			
3150	Ball check, PVC, socket or threaded									
3200	1/4"	1 Plum	26	.308	Ea.	47.50	21		68.50	83.50
3220	3/8"		26	.308		47.50	21		68.50	83.50
3240	1/2"		26	.308		46.50	21		67.50	82
3250	3/4"		25	.320		52	21.50		73.50	89.50
3260	1"		23	.348		65	23.50		88.50	107
3270	1-1/4"		21	.381		109	26		135	159
3280	1-1/2"		20	.400		109	27		136	161
3290	2"		17	.471		149	32		181	211
3310	3"	Q-1	24	.667		415	40.50		455.50	515
3320	4"	"	20	.800		585	48.50		633.50	715
3360	For PVC, flanged, add					50%	15%			
3750	CPVC, socket or threaded									
3800	1/2"	1 Plum	26	.308	Ea.	64.50	21		85.50	102
3840	3/4"		25	.320		85	21.50		106.50	126
3850	1"		23	.348		101	23.50		124.50	146
3860	1-1/2"		20	.400		174	27		201	232
3870	2"		17	.471		234	32		266	305
3880	3"	Q-1	24	.667		630	40.50		670.50	755
3920	4"	"	20	.800		855	48.50		903.50	1,025

For customer support on your Plumbing Costs with RSMeans Data, call 800.448.8182.

119

22 05 23 – General-Duty Valves for Plumbing Piping

22 05 23.60 Valves, Plastic	Crew	Daily Output	Labor-Hours	Unit	Material	2021 Bare Costs Labor	Equipment	Total	Total Incl O&P
3930	For CPVC, flanged, add				Ea.	40%	15%			
4340	Polypropylene, threaded									
4360	1/2"	1 Plum	26	.308	Ea.	51	21		72	87
4400	3/4"		25	.320		72	21.50		93.50	112
4440	1"		23	.348		76.50	23.50		100	119
4450	1-1/2"		20	.400		148	27		175	204
4460	2"		17	.471		187	32		219	253
4500	For polypropylene flanged, add					200%	15%			
4850	Foot valve, PVC, socket or threaded									
4900	1/2"	1 Plum	34	.235	Ea.	64.50	15.95		80.45	94.50
4930	3/4"		32	.250		72.50	16.95		89.45	106
4940	1"		28	.286		95	19.35		114.35	133
4950	1-1/4"		27	.296		179	20		199	227
4960	1-1/2"		26	.308		182	21		203	232
4970	2"		24	.333		212	22.50		234.50	267
4980	3"		20	.400		505	27		532	595
4990	4"		18	.444		885	30		915	1,025
5000	For flanged, add					25%	10%			
5050	CPVC, socket or threaded									
5060	1/2"	1 Plum	34	.235	Ea.	75	15.95		90.95	107
5070	3/4"		32	.250		104	16.95		120.95	141
5080	1"		28	.286		128	19.35		147.35	170
5090	1-1/4"		27	.296		195	20		215	244
5100	1-1/2"		26	.308		195	21		216	245
5110	2"		24	.333		244	22.50		266.50	305
5120	3"		20	.400		495	27		522	585
5130	4"		18	.444		900	30		930	1,025
5140	For flanged, add					25%	10%			
5280	Needle valve, PVC, threaded									
5300	1/4"	1 Plum	26	.308	Ea.	69.50	21		90.50	108
5340	3/8"		26	.308		79.50	21		100.50	119
5360	1/2"		26	.308		79	21		100	118
5380	For polypropylene, add					10%				
5800	Y check, PVC, socket or threaded									
5820	1/2"	1 Plum	26	.308	Ea.	72.50	21		93.50	111
5840	3/4"		25	.320		78.50	21.50		100	119
5850	1"		23	.348		85	23.50		108.50	129
5860	1-1/4"		21	.381		134	26		160	186
5870	1-1/2"		20	.400		146	27		173	202
5880	2"		17	.471		181	32		213	248
5890	2-1/2"		15	.533		385	36		421	480
5900	3"	Q-1	24	.667		385	40.50		425.50	485
5910	4"	"	20	.800		630	48.50		678.50	770
5960	For PVC flanged, add					45%	15%			
6350	Y sediment strainer, PVC, socket or threaded									
6400	1/2"	1 Plum	26	.308	Ea.	57	21		78	93.50
6440	3/4"		24	.333		60.50	22.50		83	100
6450	1"		23	.348		71.50	23.50		95	114
6460	1-1/4"		21	.381		172	26		198	228
6470	1-1/2"		20	.400		148	27		175	204
6480	2"		17	.471		148	32		180	211
6490	2-1/2"		15	.533		360	36		396	450
6500	3"	Q-1	24	.667		360	40.50		400.50	455

22 05 23 – General-Duty Valves for Plumbing Piping

22 05 23.60 Valves, Plastic

		Crew	Daily Output	Labor-Hours	Unit	Material	2021 Bare Costs Labor	Equipment	Total	Total Incl O&P
6510	4"	Q-1	20	.800	Ea.	600	48.50		648.50	735
6560	For PVC, flanged, add					55%	15%			

22 05 29 – Hangers and Supports for Plumbing Piping and Equipment

22 05 29.10 Hangers & Supp. for Plumb'g/HVAC Pipe/Equip.

		Crew	Daily Output	Labor-Hours	Unit	Material	2021 Bare Costs Labor	Equipment	Total	Total Incl O&P
0010	**HANGERS AND SUPPORTS FOR PLUMB'G/HVAC PIPE/EQUIP.**									
0011	TYPE numbers per MSS-SP58									
0050	Brackets									
0060	Beam side or wall, malleable iron, TYPE 34									
0070	3/8" threaded rod size	1 Plum	48	.167	Ea.	5.20	11.30		16.50	22.50
0080	1/2" threaded rod size		48	.167		8.85	11.30		20.15	26.50
0090	5/8" threaded rod size		48	.167		15.85	11.30		27.15	34.50
0100	3/4" threaded rod size		48	.167		21.50	11.30		32.80	40.50
0110	7/8" threaded rod size		48	.167		14.90	11.30		26.20	33.50
0120	For concrete installation, add						30%			
0150	Wall, welded steel, medium, TYPE 32									
0160	0 size, 12" wide, 18" deep	1 Plum	34	.235	Ea.	375	15.95		390.95	440
0170	1 size, 18" wide, 24" deep		34	.235		445	15.95		460.95	515
0180	2 size, 24" wide, 30" deep		34	.235		590	15.95		605.95	675
0200	Beam attachment, welded, TYPE 22									
0202	3/8"	Q-15	80	.200	Ea.	14.30	12.20	1.34	27.84	35.50
0203	1/2"		76	.211		14.85	12.85	1.41	29.11	37
0204	5/8"		72	.222		15.75	13.55	1.49	30.79	39
0205	3/4"		68	.235		20.50	14.35	1.58	36.43	45.50
0206	7/8"		64	.250		28	15.25	1.68	44.93	55.50
0207	1"		56	.286		45.50	17.40	1.92	64.82	78
0300	Clamps									
0310	C-clamp, for mounting on steel beam flange, w/locknut, TYPE 23									
0320	3/8" threaded rod size	1 Plum	160	.050	Ea.	3.68	3.39		7.07	9.10
0330	1/2" threaded rod size		160	.050		4.57	3.39		7.96	10.10
0340	5/8" threaded rod size		160	.050		6.15	3.39		9.54	11.80
0350	3/4" threaded rod size		160	.050		8.05	3.39		11.44	13.90
0352	7/8" threaded rod size		140	.057		15.15	3.87		19.02	22.50
0400	High temperature to 1050°F, alloy steel									
0410	4" pipe size	Q-1	106	.151	Ea.	27.50	9.20		36.70	44
0420	6" pipe size		106	.151		40	9.20		49.20	57.50
0430	8" pipe size		97	.165		45.50	10.05		55.55	65
0440	10" pipe size		84	.190		58	11.60		69.60	81.50
0450	12" pipe size		72	.222		76.50	13.55		90.05	104
0460	14" pipe size		64	.250		134	15.25		149.25	170
0470	16" pipe size		56	.286		149	17.40		166.40	190
0480	Beam clamp, flange type, TYPE 25									
0482	For 3/8" bolt	1 Plum	48	.167	Ea.	7.80	11.30		19.10	25.50
0483	For 1/2" bolt		44	.182		6.20	12.30		18.50	25
0484	For 5/8" bolt		40	.200		6.20	13.55		19.75	27
0485	For 3/4" bolt		36	.222		6.20	15.05		21.25	29.50
0486	For 1" bolt		32	.250		6.20	16.95		23.15	32.50
0500	I-beam, for mounting on bottom flange, strap iron, TYPE 21									
0530	4" flange size	1 Plum	93	.086	Ea.	10.05	5.80		15.85	19.75
0540	5" flange size		92	.087		10.95	5.90		16.85	21
0550	6" flange size		90	.089		12.30	6		18.30	22.50
0560	7" flange size		88	.091		13.65	6.15		19.80	24.50
0570	8" flange size		86	.093		14.55	6.30		20.85	25.50

121

For customer support on your Plumbing Costs with RSMeans Data, call 800.448.8182.

22 05 29.10 Hangers & Supp. for Plumb'g/HVAC Pipe/Equip.	Crew	Daily Output	Labor-Hours	Unit	Material	2021 Bare Costs Labor	Equipment	Total	Total Incl O&P
0600	One hole, vertical mounting, malleable iron									
0610	1/2" pipe size	1 Plum	160	.050	Ea.	1.89	3.39		5.28	7.15
0620	3/4" pipe size		145	.055		2.42	3.74		6.16	8.25
0630	1" pipe size		136	.059		2.57	3.98		6.55	8.80
0640	1-1/4" pipe size		128	.063		3.79	4.23		8.02	10.45
0650	1-1/2" pipe size		120	.067		4.44	4.51		8.95	11.65
0660	2" pipe size		112	.071		5.80	4.84		10.64	13.60
0670	2-1/2" pipe size		104	.077		11.35	5.20		16.55	20.50
0680	3" pipe size		96	.083		15.25	5.65		20.90	25
0690	3-1/2" pipe size		90	.089		9.50	6		15.50	19.45
0700	4" pipe size		84	.095		13.45	6.45		19.90	24.50
0750	Riser or extension pipe, carbon steel, TYPE 8									
0756	1/2" pipe size	1 Plum	52	.154	Ea.	3.58	10.40		13.98	19.50
0760	3/4" pipe size		48	.167		4.45	11.30		15.75	22
0770	1" pipe size		47	.170		4.37	11.50		15.87	22
0780	1-1/4" pipe size		46	.174		4.04	11.75		15.79	22
0790	1-1/2" pipe size		45	.178		6	12.05		18.05	24.50
0800	2" pipe size		43	.186		6	12.60		18.60	25.50
0810	2-1/2" pipe size		41	.195		6.30	13.20		19.50	26.50
0820	3" pipe size		40	.200		7.25	13.55		20.80	28
0830	3-1/2" pipe size		39	.205		8.55	13.90		22.45	30
0840	4" pipe size		38	.211		9.30	14.25		23.55	32
0850	5" pipe size		37	.216		12.40	14.65		27.05	35.50
0860	6" pipe size		36	.222		15.55	15.05		30.60	39.50
0870	8" pipe size		34	.235		20.50	15.95		36.45	46.50
0880	10" pipe size		32	.250		39.50	16.95		56.45	69
0890	12" pipe size		28	.286		56.50	19.35		75.85	91
0900	For plastic coating 3/4" to 4", add					190%				
0910	For copper plating 3/4" to 4", add					58%				
0950	Two piece, complete, carbon steel, medium weight, TYPE 4									
0960	1/2" pipe size	Q-1	137	.117	Ea.	4.52	7.10		11.62	15.55
0970	3/4" pipe size		134	.119		4.52	7.30		11.82	15.80
0980	1" pipe size		132	.121		4.56	7.40		11.96	16
0990	1-1/4" pipe size		130	.123		5.20	7.50		12.70	16.95
1000	1-1/2" pipe size		126	.127		5.20	7.75		12.95	17.30
1010	2" pipe size		124	.129		7.10	7.85		14.95	19.60
1020	2-1/2" pipe size		120	.133		7.60	8.10		15.70	20.50
1030	3" pipe size		117	.137		8.45	8.35		16.80	22
1040	3-1/2" pipe size		114	.140		13.90	8.55		22.45	28
1050	4" pipe size		110	.145		11.70	8.85		20.55	26
1060	5" pipe size		106	.151		26	9.20		35.20	43
1070	6" pipe size		104	.154		29	9.35		38.35	46
1080	8" pipe size		100	.160		35	9.75		44.75	53
1090	10" pipe size		96	.167		66	10.15		76.15	87.50
1100	12" pipe size		89	.180		80.50	10.95		91.45	105
1110	14" pipe size		82	.195		132	11.90		143.90	163
1120	16" pipe size		68	.235		142	14.35		156.35	178
1130	For galvanized, add					45%				
1150	Insert, concrete									
1160	Wedge type, carbon steel body, malleable iron nut, galvanized									
1170	1/4" threaded rod size	1 Plum	96	.083	Ea.	8.90	5.65		14.55	18.20
1180	3/8" threaded rod size		96	.083		2.85	5.65		8.50	11.55
1190	1/2" threaded rod size		96	.083		1.47	5.65		7.12	10

22 05 29.10 Hangers & Supp. for Plumb'g/HVAC Pipe/Equip.	Crew	Daily Output	Labor-Hours	Unit	Material	2021 Bare Costs Labor	Equipment	Total	Total Incl O&P
1200	5/8" threaded rod size	1 Plum	96	.083	Ea.	2.02	5.65		7.67	10.60
1210	3/4" threaded rod size		96	.083		2.22	5.65		7.87	10.85
1220	7/8" threaded rod size		96	.083		12.30	5.65		17.95	22
1250	Pipe guide sized for insulation									
1260	No. 1, 1" pipe size, 1" thick insulation	1 Stpi	26	.308	Ea.	269	21		290	330
1270	No. 2, 1-1/4"-2" pipe size, 1" thick insulation		23	.348		269	24		293	330
1280	No. 3, 1-1/4"-2" pipe size, 1-1/2" thick insulation		21	.381		166	26		192	222
1290	No. 4, 2-1/2"-3-1/2" pipe size, 1-1/2" thick insulation		18	.444		186	30.50		216.50	251
1300	No. 5, 4"-5" pipe size, 1-1/2" thick insulation		16	.500		196	34		230	266
1310	No. 6, 5"-6" pipe size, 2" thick insulation	Q-5	21	.762		226	47		273	320
1320	No. 7, 8" pipe size, 2" thick insulation		16	1		294	61.50		355.50	415
1330	No. 8, 10" pipe size, 2" thick insulation		12	1.333		440	82		522	605
1340	No. 9, 12" pipe size, 2" thick insulation	Q-6	17	1.412		440	90		530	620
1350	No. 10, 12"-14" pipe size, 2-1/2" thick insulation		16	1.500		525	95.50		620.50	725
1360	No. 11, 16" pipe size, 2-1/2" thick insulation		10.50	2.286		525	146		671	800
1370	No. 12, 16"-18" pipe size, 3" thick insulation		9	2.667		740	170		910	1,075
1380	No. 13, 20" pipe size, 3" thick insulation		7.50	3.200		740	204		944	1,125
1390	No. 14, 24" pipe size, 3" thick insulation		7	3.429		1,050	219		1,269	1,475
1400	Bands									
1410	Adjustable band, carbon steel, for non-insulated pipe, TYPE 7									
1420	1/2" pipe size	Q-1	142	.113	Ea.	.84	6.85		7.69	11.15
1430	3/4" pipe size		140	.114		.84	6.95		7.79	11.30
1440	1" pipe size		137	.117		.84	7.10		7.94	11.50
1450	1-1/4" pipe size		134	.119		.48	7.30		7.78	11.40
1460	1-1/2" pipe size		131	.122		.91	7.45		8.36	12.10
1470	2" pipe size		129	.124		.91	7.55		8.46	12.30
1480	2-1/2" pipe size		125	.128		2.20	7.80		10	14.05
1490	3" pipe size		122	.131		2.64	8		10.64	14.85
1500	3-1/2" pipe size		119	.134		.98	8.20		9.18	13.35
1510	4" pipe size		114	.140		3.62	8.55		12.17	16.75
1520	5" pipe size		110	.145		2.34	8.85		11.19	15.80
1530	6" pipe size		108	.148		2.80	9.05		11.85	16.55
1540	8" pipe size		104	.154		4.29	9.35		13.64	18.70
1550	For copper plated, add					50%				
1560	For galvanized, add					30%				
1570	For plastic coating, add					30%				
1600	Adjusting nut malleable iron, steel band, TYPE 9									
1610	1/2" pipe size, galvanized band	Q-1	137	.117	Ea.	7.55	7.10		14.65	18.90
1620	3/4" pipe size, galvanized band		135	.119		7.55	7.20		14.75	19.10
1630	1" pipe size, galvanized band		132	.121		7.75	7.40		15.15	19.55
1640	1-1/4" pipe size, galvanized band		129	.124		7.80	7.55		15.35	19.90
1650	1-1/2" pipe size, galvanized band		126	.127		8.05	7.75		15.80	20.50
1660	2" pipe size, galvanized band		124	.129		8.40	7.85		16.25	21
1670	2-1/2" pipe size, galvanized band		120	.133		13.30	8.10		21.40	26.50
1680	3" pipe size, galvanized band		117	.137		13.90	8.35		22.25	27.50
1690	3-1/2" pipe size, galvanized band		114	.140		25.50	8.55		34.05	41
1700	4" pipe size, cadmium plated band		110	.145		35.50	8.85		44.35	52.50
1740	For plastic coated band, add					35%				
1750	For completely copper coated, add					45%				
1800	Clevis, adjustable, carbon steel, for non-insulated pipe, TYPE 1									
1810	1/2" pipe size	Q-1	137	.117	Ea.	1.65	7.10		8.75	12.40
1820	3/4" pipe size		135	.119		1.53	7.20		8.73	12.50
1830	1" pipe size		132	.121		1.58	7.40		8.98	12.75

For customer support on your Plumbing Costs with RSMeans Data, call 800.448.8182.

123

22 05 29.10 Hangers & Supp. for Plumb'g/HVAC Pipe/Equip.	Crew	Daily Output	Labor-Hours	Unit	Material	2021 Bare Costs Labor	Equipment	Total	Total Incl O&P
1840	1-1/4" pipe size	Q-1	129	.124	Ea.	1.74	7.55		9.29	13.20
1850	1-1/2" pipe size		126	.127		1.83	7.75		9.58	13.55
1860	2" pipe size		124	.129		2.19	7.85		10.04	14.15
1870	2-1/2" pipe size		120	.133		3.48	8.10		11.58	15.95
1880	3" pipe size		117	.137		4.25	8.35		12.60	17.15
1890	3-1/2" pipe size		114	.140		4.47	8.55		13.02	17.65
1900	4" pipe size		110	.145		5.25	8.85		14.10	19.05
1910	5" pipe size		106	.151		6.60	9.20		15.80	21
1920	6" pipe size		104	.154		9.30	9.35		18.65	24
1930	8" pipe size		100	.160		13.65	9.75		23.40	29.50
1940	10" pipe size		96	.167		25.50	10.15		35.65	43.50
1950	12" pipe size		89	.180		34	10.95		44.95	54
1960	14" pipe size		82	.195		50	11.90		61.90	73
1970	16" pipe size		68	.235		72.50	14.35		86.85	102
1971	18" pipe size		54	.296		91	18.05		109.05	127
1972	20" pipe size		38	.421		193	25.50		218.50	252
1980	For galvanized, add					66%				
1990	For copper plated 1/2" to 4", add					77%				
2000	For light weight 1/2" to 4", deduct					13%				
2010	Insulated pipe type, 3/4" to 12" pipe, add					180%				
2020	Insulated pipe type, chrome-moly U-strap, add					530%				
2250	Split ring, malleable iron, for non-insulated pipe, TYPE 11									
2260	1/2" pipe size	Q-1	137	.117	Ea.	9.95	7.10		17.05	21.50
2270	3/4" pipe size		135	.119		10.05	7.20		17.25	22
2280	1" pipe size		132	.121		10.65	7.40		18.05	22.50
2290	1-1/4" pipe size		129	.124		13.20	7.55		20.75	26
2300	1-1/2" pipe size		126	.127		16.05	7.75		23.80	29
2310	2" pipe size		124	.129		18.15	7.85		26	31.50
2320	2-1/2" pipe size		120	.133		26	8.10		34.10	41
2330	3" pipe size		117	.137		32.50	8.35		40.85	48.50
2340	3-1/2" pipe size		114	.140		16.55	8.55		25.10	31
2350	4" pipe size		110	.145		34	8.85		42.85	51
2360	5" pipe size		106	.151		45.50	9.20		54.70	64
2370	6" pipe size		104	.154		101	9.35		110.35	125
2380	8" pipe size		100	.160		149	9.75		158.75	179
2390	For copper plated, add					8%				
2400	Channels, steel, 3/4" x 1-1/2"	1 Plum	80	.100	L.F.	3.74	6.75		10.49	14.20
2404	1-1/2" x 1-1/2"		70	.114		4.25	7.75		12	16.25
2408	1-7/8" x 1-1/2"		60	.133		24.50	9.05		33.55	40.50
2412	3" x 1-1/2"		50	.160		23.50	10.85		34.35	42
2416	Hangers, trapeze channel support, 12 ga. 1-1/2" x 1-1/2", 12" wide, steel		8.80	.909	Ea.	36.50	61.50		98	132
2418	18" wide, steel		8.30	.964		38.50	65.50		104	140
2430	Spring nuts, long, 1/4"		120	.067		.50	4.51		5.01	7.30
2432	3/8"		100	.080		.64	5.40		6.04	8.80
2434	1/2"		80	.100		.72	6.75		7.47	10.90
2436	5/8"		80	.100		3.75	6.75		10.50	14.25
2438	3/4"		75	.107		5.90	7.20		13.10	17.25
2440	Spring nuts, short, 1/4"		120	.067		1.27	4.51		5.78	8.15
2442	3/8"		100	.080		1.40	5.40		6.80	9.65
2444	1/2"		80	.100		1.67	6.75		8.42	11.95
2500	Washer, flat steel									
2502	3/8"	1 Plum	240	.033	Ea.	.07	2.26		2.33	3.45
2503	1/2"		220	.036		.16	2.46		2.62	3.85

22 05 29.10 Hangers & Supp. for Plumb'g/HVAC Pipe/Equip.	Crew	Daily Output	Labor-Hours	Unit	Material	2021 Bare Costs Labor	2021 Bare Costs Equipment	Total	Total Incl O&P
2504	5/8"	1 Plum	200	.040	Ea.	.31	2.71		3.02	4.38
2505	3/4"		180	.044		.27	3.01		3.28	4.79
2506	7/8"		160	.050		.70	3.39		4.09	5.80
2507	1"		140	.057		.75	3.87		4.62	6.60
2508	1-1/4"		120	.067		1.17	4.51		5.68	8.05
2520	Nut, steel, hex									
2522	3/8"	1 Plum	200	.040	Ea.	.22	2.71		2.93	4.28
2523	1/2"		180	.044		.53	3.01		3.54	5.05
2524	5/8"		160	.050		.85	3.39		4.24	6
2525	3/4"		140	.057		1.42	3.87		5.29	7.30
2526	7/8"		120	.067		2.28	4.51		6.79	9.25
2527	1"		100	.080		3.15	5.40		8.55	11.55
2528	1-1/4"		80	.100		7.85	6.75		14.60	18.75
2532	Turnbuckle, TYPE 13									
2534	3/8"	1 Plum	80	.100	Ea.	14.45	6.75		21.20	26
2535	1/2"		72	.111		18.75	7.50		26.25	32
2536	5/8"		64	.125		19.05	8.45		27.50	33.50
2537	3/4"		56	.143		25	9.65		34.65	42
2538	7/8"		48	.167		34.50	11.30		45.80	54.50
2539	1"		40	.200		46	13.55		59.55	70.50
2540	1-1/4"		32	.250		85	16.95		101.95	119
2650	Rods, carbon steel									
2660	Continuous thread									
2670	1/4" thread size	1 Plum	144	.056	L.F.	.43	3.76		4.19	6.05
2680	3/8" thread size		144	.056		.46	3.76		4.22	6.10
2690	1/2" thread size		144	.056		.72	3.76		4.48	6.40
2700	5/8" thread size		144	.056		1.03	3.76		4.79	6.75
2710	3/4" thread size		144	.056		1.81	3.76		5.57	7.60
2720	7/8" thread size		144	.056		2.27	3.76		6.03	8.10
2721	1" thread size	Q-1	160	.100		23	6.10		29.10	34.50
2722	1-1/8" thread size	"	120	.133		21	8.10		29.10	35
2725	1/4" thread size, bright finish	1 Plum	144	.056		1.26	3.76		5.02	7
2726	1/2" thread size, bright finish	"	144	.056		6.50	3.76		10.26	12.75
2730	For galvanized, add					40%				
2820	Rod couplings									
2821	3/8"	1 Plum	60	.133	Ea.	2.92	9.05		11.97	16.65
2822	1/2"		54	.148		3.87	10.05		13.92	19.20
2823	5/8"		48	.167		4.50	11.30		15.80	22
2824	3/4"		44	.182		6.15	12.30		18.45	25
2825	7/8"		40	.200		10.30	13.55		23.85	31.50
2826	1"		34	.235		13.80	15.95		29.75	39
2827	1-1/8"		30	.267		15.60	18.05		33.65	44
2860	Pipe hanger assy, adj. clevis, saddle, rod, clamp, insul. allowance									
2864	1/2" pipe size	Q-5	35	.457	Ea.	50	28		78	97.50
2866	3/4" pipe size		34.80	.460		49	28.50		77.50	96
2868	1" pipe size		34.60	.462		45	28.50		73.50	92
2869	1-1/4" pipe size		34.30	.466		45.50	28.50		74	93
2870	1-1/2" pipe size		33.90	.472		52.50	29		81.50	102
2872	2" pipe size		33.30	.480		51	29.50		80.50	100
2874	2-1/2" pipe size		32.30	.495		62.50	30.50		93	114
2876	3" pipe size		31.20	.513		65	31.50		96.50	119
2880	4" pipe size		30.70	.521		80	32		112	136
2884	6" pipe size		29.80	.537		80.50	33		113.50	138

125

For customer support on your Plumbing Costs with RSMeans Data, call 800.448.8182.

22 05 29.10 Hangers & Supp. for Plumb'g/HVAC Pipe/Equip.		Crew	Daily Output	Labor-Hours	Unit	Material	2021 Bare Costs Labor	Equipment	Total	Total Incl O&P
2888	8" pipe size	Q-5	28	.571	Ea.	98.50	35		133.50	161
2892	10" pipe size		25.20	.635		150	39		189	224
2896	12" pipe size		23.20	.690		175	42.50		217.50	257
2900	Rolls									
2910	Adjustable yoke, carbon steel with CI roll, TYPE 43									
2918	2" pipe size	Q-1	140	.114	Ea.	11.30	6.95		18.25	23
2920	2-1/2" pipe size		137	.117		18.05	7.10		25.15	30.50
2930	3" pipe size		131	.122		20	7.45		27.45	33
2940	3-1/2" pipe size		124	.129		27.50	7.85		35.35	42
2950	4" pipe size		117	.137		27	8.35		35.35	42
2960	5" pipe size		110	.145		30.50	8.85		39.35	47
2970	6" pipe size		104	.154		41.50	9.35		50.85	59.50
2980	8" pipe size		96	.167		58	10.15		68.15	79
2990	10" pipe size		80	.200		77.50	12.20		89.70	104
3000	12" pipe size		68	.235		111	14.35		125.35	144
3010	14" pipe size		56	.286		264	17.40		281.40	315
3020	16" pipe size		48	.333		325	20.50		345.50	385
3050	Chair, carbon steel with CI roll									
3060	2" pipe size	1 Plum	68	.118	Ea.	23.50	7.95		31.45	38
3070	2-1/2" pipe size		65	.123		25.50	8.35		33.85	41
3080	3" pipe size		62	.129		26.50	8.75		35.25	42
3090	3-1/2" pipe size		60	.133		20	9.05		29.05	35.50
3100	4" pipe size		58	.138		33.50	9.35		42.85	51
3110	5" pipe size		56	.143		37	9.65		46.65	55.50
3120	6" pipe size		53	.151		52	10.20		62.20	72.50
3130	8" pipe size		50	.160		70	10.85		80.85	93
3140	10" pipe size		48	.167		87.50	11.30		98.80	113
3150	12" pipe size		46	.174		131	11.75		142.75	162
3170	Single pipe roll (see line 2650 for rods), TYPE 41, 1" pipe size	Q-1	137	.117		16.25	7.10		23.35	28.50
3180	1-1/4" pipe size		131	.122		10.55	7.45		18	22.50
3190	1-1/2" pipe size		129	.124		17.10	7.55		24.65	30
3200	2" pipe size		124	.129		17.80	7.85		25.65	31.50
3210	2-1/2" pipe size		118	.136		19.05	8.25		27.30	33.50
3220	3" pipe size		115	.139		20	8.50		28.50	34.50
3230	3-1/2" pipe size		113	.142		21.50	8.65		30.15	36.50
3240	4" pipe size		112	.143		25	8.70		33.70	40.50
3250	5" pipe size		110	.145		25.50	8.85		34.35	41.50
3260	6" pipe size		101	.158		42	9.65		51.65	61
3270	8" pipe size		90	.178		56.50	10.85		67.35	78
3280	10" pipe size		80	.200		69	12.20		81.20	94
3290	12" pipe size		68	.235		103	14.35		117.35	135
3300	Saddles (add vertical pipe riser, usually 3" diameter)									
3310	Pipe support, complete, adjust., CI saddle, TYPE 36									
3320	2-1/2" pipe size	1 Plum	96	.083	Ea.	191	5.65		196.65	218
3330	3" pipe size		88	.091		154	6.15		160.15	179
3340	3-1/2" pipe size		79	.101		206	6.85		212.85	236
3350	4" pipe size		68	.118		207	7.95		214.95	240
3360	5" pipe size		64	.125		172	8.45		180.45	202
3370	6" pipe size		59	.136		223	9.20		232.20	259
3380	8" pipe size		53	.151		243	10.20		253.20	282
3390	10" pipe size		50	.160		276	10.85		286.85	320
3400	12" pipe size		48	.167		288	11.30		299.30	330
3450	For standard pipe support, one piece, CI, deduct					34%				

126

22 05 29.10 Hangers & Supp. for Plumb'g/HVAC Pipe/Equip.	Crew	Daily Output	Labor-Hours	Unit	Material	2021 Bare Costs Labor	Equipment	Total	Total Incl O&P
3460	For stanchion support, CI with steel yoke, deduct					60%				
3550	Insulation shield 1" thick, 1/2" pipe size, TYPE 40	1 Asbe	100	.080	Ea.	10.35	4.88		15.23	18.85
3560	3/4" pipe size		100	.080		10.60	4.88		15.48	19.10
3570	1" pipe size		98	.082		11.80	4.98		16.78	20.50
3580	1-1/4" pipe size		98	.082		12.65	4.98		17.63	21.50
3590	1-1/2" pipe size		96	.083		11.50	5.10		16.60	20.50
3600	2" pipe size		96	.083		13.55	5.10		18.65	22.50
3610	2-1/2" pipe size		94	.085		22.50	5.20		27.70	33
3620	3" pipe size		94	.085		20.50	5.20		25.70	30.50
3630	2" thick, 3-1/2" pipe size		92	.087		28	5.30		33.30	38.50
3640	4" pipe size		92	.087		25	5.30		30.30	35.50
3650	5" pipe size		90	.089		35.50	5.40		40.90	47.50
3660	6" pipe size		90	.089		47.50	5.40		52.90	61
3670	8" pipe size		88	.091		44	5.55		49.55	57
3680	10" pipe size		88	.091		50	5.55		55.55	63.50
3690	12" pipe size		86	.093		93	5.65		98.65	111
3700	14" pipe size		86	.093		102	5.65		107.65	122
3710	16" pipe size		84	.095		128	5.80		133.80	150
3720	18" pipe size		84	.095		153	5.80		158.80	178
3730	20" pipe size		82	.098		160	5.95		165.95	185
3732	24" pipe size		80	.100		175	6.10		181.10	201
3750	Covering protection saddle, TYPE 39									
3760	1" covering size									
3770	3/4" pipe size	1 Plum	68	.118	Ea.	11.95	7.95		19.90	25
3780	1" pipe size		68	.118		11.95	7.95		19.90	25
3790	1-1/4" pipe size		68	.118		9.90	7.95		17.85	23
3800	1-1/2" pipe size		66	.121		12.90	8.20		21.10	26.50
3810	2" pipe size		66	.121		10.40	8.20		18.60	23.50
3820	2-1/2" pipe size		64	.125		12.90	8.45		21.35	27
3830	3" pipe size		64	.125		12.70	8.45		21.15	26.50
3840	3-1/2" pipe size		62	.129		19.65	8.75		28.40	34.50
3850	4" pipe size		62	.129		14	8.75		22.75	28.50
3860	5" pipe size		60	.133		19.65	9.05		28.70	35
3870	6" pipe size		60	.133		23.50	9.05		32.55	39.50
3900	1-1/2" covering size									
3910	3/4" pipe size	1 Plum	68	.118	Ea.	7.45	7.95		15.40	20
3920	1" pipe size		68	.118		15.50	7.95		23.45	29
3930	1-1/4" pipe size		68	.118		7.45	7.95		15.40	20
3940	1-1/2" pipe size		66	.121		17.65	8.20		25.85	31.50
3950	2" pipe size		66	.121		51.50	8.20		59.70	69.50
3960	2-1/2" pipe size		64	.125		23	8.45		31.45	38
3970	3" pipe size		64	.125		67.50	8.45		75.95	87
3980	3-1/2" pipe size		62	.129		20.50	8.75		29.25	35.50
3990	4" pipe size		62	.129		72	8.75		80.75	92
4000	5" pipe size		60	.133		72	9.05		81.05	92.50
4010	6" pipe size		60	.133		34.50	9.05		43.55	51.50
4020	8" pipe size		58	.138		39.50	9.35		48.85	57.50
4022	10" pipe size		56	.143		36.50	9.65		46.15	54.50
4024	12" pipe size		54	.148		62	10.05		72.05	83
4028	2" covering size									
4029	2-1/2" pipe size	1 Plum	62	.129	Ea.	19.20	8.75		27.95	34
4032	3" pipe size		60	.133		69.50	9.05		78.55	90
4033	4" pipe size		58	.138		11.40	9.35		20.75	26.50

127

22 05 29.10 Hangers & Supp. for Plumb'g/HVAC Pipe/Equip.		Crew	Daily Output	Labor-Hours	Unit	Material	2021 Bare Costs Labor	Equipment	Total	Total Incl O&P
4034	6" pipe size	1 Plum	56	.143	Ea.	32.50	9.65		42.15	50
4035	8" pipe size		54	.148		40.50	10.05		50.55	59.50
4080	10" pipe size		58	.138		40.50	9.35		49.85	58.50
4090	12" pipe size		56	.143		69.50	9.65		79.15	90.50
4100	14" pipe size		56	.143		69.50	9.65		79.15	90.50
4110	16" pipe size		54	.148		94.50	10.05		104.55	119
4120	18" pipe size		54	.148		94.50	10.05		104.55	119
4130	20" pipe size		52	.154		70.50	10.40		80.90	93
4150	24" pipe size		50	.160		123	10.85		133.85	151
4160	30" pipe size		48	.167		135	11.30		146.30	166
4180	36" pipe size		45	.178		105	12.05		117.05	133
4186	2-1/2" covering size									
4187	3" pipe size	1 Plum	58	.138	Ea.	24.50	9.35		33.85	41
4188	4" pipe size		56	.143		26	9.65		35.65	43
4189	6" pipe size		52	.154		35.50	10.40		45.90	54.50
4190	8" pipe size		48	.167		42.50	11.30		53.80	63.50
4191	10" pipe size		44	.182		46.50	12.30		58.80	69.50
4192	12" pipe size		40	.200		72	13.55		85.55	99.50
4193	14" pipe size		36	.222		52	15.05		67.05	80
4194	16" pipe size		32	.250		66.50	16.95		83.45	98.50
4195	18" pipe size		28	.286		103	19.35		122.35	143
4200	Sockets									
4210	Rod end, malleable iron, TYPE 16									
4220	1/4" thread size	1 Plum	240	.033	Ea.	3.24	2.26		5.50	6.95
4230	3/8" thread size		240	.033		3.25	2.26		5.51	6.95
4240	1/2" thread size		230	.035		4.29	2.35		6.64	8.25
4250	5/8" thread size		225	.036		8	2.41		10.41	12.40
4260	3/4" thread size		220	.036		11.25	2.46		13.71	16.05
4270	7/8" thread size		210	.038		15.80	2.58		18.38	21
4290	Strap, 1/2" pipe size, TYPE 26	Q-1	142	.113		2.95	6.85		9.80	13.50
4300	3/4" pipe size		140	.114		3.64	6.95		10.59	14.40
4310	1" pipe size		137	.117		4.56	7.10		11.66	15.60
4320	1-1/4" pipe size		134	.119		4.56	7.30		11.86	15.85
4330	1-1/2" pipe size		131	.122		4.83	7.45		12.28	16.40
4340	2" pipe size		129	.124		5.30	7.55		12.85	17.10
4350	2-1/2" pipe size		125	.128		8.15	7.80		15.95	20.50
4360	3" pipe size		122	.131		10.60	8		18.60	23.50
4370	3-1/2" pipe size		119	.134		11.60	8.20		19.80	25
4380	4" pipe size		114	.140		10.85	8.55		19.40	24.50
4400	U-bolt, carbon steel									
4410	Standard, with nuts, TYPE 42									
4420	1/2" pipe size	1 Plum	160	.050	Ea.	1.21	3.39		4.60	6.40
4430	3/4" pipe size		158	.051		1.73	3.43		5.16	7
4450	1" pipe size		152	.053		1.77	3.56		5.33	7.25
4460	1-1/4" pipe size		148	.054		2.09	3.66		5.75	7.75
4470	1-1/2" pipe size		143	.056		2.25	3.79		6.04	8.15
4480	2" pipe size		139	.058		2.43	3.90		6.33	8.45
4490	2-1/2" pipe size		134	.060		4.06	4.04		8.10	10.50
4500	3" pipe size		128	.063		4.27	4.23		8.50	11
4510	3-1/2" pipe size		122	.066		4.47	4.44		8.91	11.55
4520	4" pipe size		117	.068		4.59	4.63		9.22	11.95
4530	5" pipe size		114	.070		3.41	4.75		8.16	10.85
4540	6" pipe size		111	.072		7.05	4.88		11.93	15.05

22 05 29 – Hangers and Supports for Plumbing Piping and Equipment

22 05 29.10 Hangers & Supp. for Plumb'g/HVAC Pipe/Equip.		Crew	Daily Output	Labor-Hours	Unit	Material	2021 Bare Costs Labor	2021 Bare Costs Equipment	Total	Total Incl O&P
4550	8" pipe size	1 Plum	109	.073	Ea.	8.40	4.97		13.37	16.60
4560	10" pipe size		107	.075		15	5.05		20.05	24
4570	12" pipe size	↓	104	.077	↓	20	5.20		25.20	30
4580	For plastic coating on 1/2" thru 6" size, add					150%				
4700	U-hook, carbon steel, requires mounting screws or bolts									
4710	3/4" thru 2" pipe size									
4720	6" long	1 Plum	96	.083	Ea.	1.15	5.65		6.80	9.65
4730	8" long		96	.083		1.24	5.65		6.89	9.75
4740	10" long		96	.083		1.47	5.65		7.12	10
4750	12" long	↓	96	.083	↓	1.94	5.65		7.59	10.55
4760	For copper plated, add					50%				
7000	Roof supports									
7006	Duct									
7010	Rectangular, open, 12" off roof									
7020	To 18" wide	Q-9	26	.615	Ea.	171	36.50		207.50	243
7030	To 24" wide	"	22	.727		199	43		242	284
7040	To 36" wide	Q-10	30	.800		228	49		277	325
7050	To 48" wide		28	.857		256	52.50		308.50	360
7060	To 60" wide	↓	24	1	↓	285	61		346	405
7100	Equipment									
7120	Equipment support	Q-5	20	.800	Ea.	75	49		124	156
7300	Pipe									
7310	Roller type									
7320	Up to 2-1/2" diam. pipe									
7324	3-1/2" off roof	Q-5	24	.667	Ea.	23.50	41		64.50	86.50
7326	Up to 10" off roof	"	20	.800	"	29	49		78	106
7340	2-1/2" to 3-1/2" diam. pipe									
7342	Up to 16" off roof	Q-5	18	.889	Ea.	52	54.50		106.50	139
7360	4" to 5" diam. pipe									
7362	Up to 12" off roof	Q-5	16	1	Ea.	74	61.50		135.50	174
7400	Strut/channel type									
7410	Up to 2-1/2" diam. pipe									
7424	3-1/2" off roof	Q-5	24	.667	Ea.	21	41		62	84
7426	Up to 10" off roof	"	20	.800	"	27	49		76	103
7440	Strut and roller type									
7452	2-1/2" to 3-1/2" diam. pipe									
7454	Up to 16" off roof	Q-5	18	.889	Ea.	74.50	54.50		129	164
7460	Strut and hanger type									
7470	Up to 3" diam. pipe									
7474	Up to 8" off roof	Q-5	19	.842	Ea.	47.50	52		99.50	130
8000	Pipe clamp, plastic, 1/2" CTS	1 Plum	80	.100		.24	6.75		6.99	10.35
8010	3/4" CTS		73	.110		.23	7.40		7.63	11.30
8020	1" CTS		68	.118		.60	7.95		8.55	12.55
8080	Economy clamp, 1/4" CTS		175	.046		.05	3.09		3.14	4.68
8090	3/8" CTS		168	.048		.05	3.22		3.27	4.87
8100	1/2" CTS		160	.050		.05	3.39		3.44	5.10
8110	3/4" CTS		145	.055		.05	3.74		3.79	5.65
8200	Half clamp, 1/2" CTS		80	.100		.08	6.75		6.83	10.20
8210	3/4" CTS		73	.110		.13	7.40		7.53	11.20
8300	Suspension clamp, 1/2" CTS		80	.100		.24	6.75		6.99	10.35
8310	3/4" CTS		73	.110		.24	7.40		7.64	11.30
8320	1" CTS		68	.118		.54	7.95		8.49	12.50
8400	Insulator, 1/2" CTS	↓	80	.100	↓	.35	6.75		7.10	10.50

129

22 05 29.10 Hangers & Supp. for Plumb'g/HVAC Pipe/Equip.	Crew	Daily Output	Labor-Hours	Unit	Material	2021 Bare Costs Labor	Equipment	Total	Total Incl O&P
8410	3/4" CTS	1 Plum	73	.110	Ea.	.37	7.40		7.77	11.45
8420	1" CTS		68	.118		.39	7.95		8.34	12.35
8500	J hook clamp with nail, 1/2" CTS		240	.033		.11	2.26		2.37	3.49
8501	3/4" CTS	↓	240	.033	↓	.13	2.26		2.39	3.51
8800	Wire cable support system									
8810	Cable with hook terminal and locking device									
8830	2 mm (.079") diam. cable (100 lb. cap.)									
8840	1 m (3.3') length, with hook	1 Shee	96	.083	Ea.	3.83	5.45		9.28	12.45
8850	2 m (6.6') length, with hook		84	.095		4.53	6.25		10.78	14.40
8860	3 m (9.9') length, with hook	↓	72	.111		5.05	7.25		12.30	16.50
8870	5 m (16.4') length, with hook	Q-9	60	.267		6.20	15.70		21.90	30.50
8880	10 m (32.8') length, with hook	"	30	.533	↓	8.90	31.50		40.40	57.50
8900	3 mm (.118") diam. cable (200 lb. cap.)									
8910	1 m (3.3') length, with hook	1 Shee	96	.083	Ea.	11.15	5.45		16.60	20.50
8920	2 m (6.6') length, with hook		84	.095		.55	6.25		6.80	10
8930	3 m (9.9') length, with hook	↓	72	.111		13.70	7.25		20.95	26
8940	5 m (16.4') length, with hook	Q-9	60	.267		11.80	15.70		27.50	36.50
8950	10 m (32.8') length, with hook	"	30	.533	↓	10.55	31.50		42.05	59
9000	Cable system accessories									
9010	Anchor bolt, 3/8", with nut	1 Shee	140	.057	Ea.	2.67	3.74		6.41	8.60
9020	Air duct corner protector		160	.050		.89	3.27		4.16	5.90
9030	Air duct support attachment	↓	140	.057	↓	2.31	3.74		6.05	8.20
9040	Flange clip, hammer-on style									
9044	For flange thickness 3/32"-9/64", 160 lb. cap.	1 Shee	180	.044	Ea.	.61	2.91		3.52	5.05
9048	For flange thickness 1/8"-1/4", 200 lb. cap.		160	.050		.69	3.27		3.96	5.70
9052	For flange thickness 5/16"-1/2", 200 lb. cap.		150	.053		1.01	3.49		4.50	6.35
9056	For flange thickness 9/16"-3/4", 200 lb. cap.		140	.057	↓	1.22	3.74		4.96	7
9060	Wire insulation protection tube		180	.044	L.F.	.58	2.91		3.49	5.05
9070	Wire cutter				Ea.	45			45	49.50

22 05 33.20 Temperature Maintenance Cable

		Crew	Daily Output	Labor-Hours	Unit	Material	2021 Bare Costs Labor	Equipment	Total	Total Incl O&P
0010	**TEMPERATURE MAINTENANCE CABLE**									
0040	Components									
0080	Heating cable									
0100	208 V									
0150	140°F	Q-1	1060.80	.015	L.F.	10.40	.92		11.32	12.80
0200	120 V									
0220	125°F	Q-1	1060.80	.015	L.F.	9.20	.92		10.12	11.50
0300	Power kit w/1 end seal	1 Elec	48.80	.164	Ea.	146	10.45		156.45	176
0310	Splice kit		35.50	.225		172	14.35		186.35	212
0320	End seal		160	.050		12.95	3.19		16.14	19
0330	Tee kit w/1 end seal		26.80	.299		178	19		197	225
0340	Powered splice w/2 end seals		20	.400		147	25.50		172.50	200
0350	Powered tee kit w/3 end seals		18	.444		189	28.50		217.50	250
0360	Cross kit w/2 end seals	↓	18.60	.430	↓	195	27.50		222.50	255
0500	Recommended thickness of fiberglass insulation									
0510	Pipe size									
0520	1/2" to 1" use 1" insulation									
0530	1-1/4" to 2" use 1-1/2" insulation									
0540	2-1/2" to 6" use 2" insulation									
0560	NOTE: For pipe sizes 1-1/4" and smaller use 1/4" larger diameter									
0570	insulation to allow room for installation over cable.									

22 05 48 – Vibration and Seismic Controls for Plumbing Piping and Equipment

22 05 48.40 Vibration Absorbers	Crew	Daily Output	Labor-Hours	Unit	Material	2021 Bare Costs Labor	Equipment	Total	Total Incl O&P
0010 **VIBRATION ABSORBERS**									
0100 Hangers, neoprene flex									
0200 10-120 lb. capacity				Ea.	25.50			25.50	28
0220 75-550 lb. capacity					42.50			42.50	46.50
0240 250-1,100 lb. capacity					86			86	94.50
0260 1,000-4,000 lb. capacity					133			133	146
0500 Spring flex, 60 lb. capacity					38			38	42
0520 450 lb. capacity					59			59	65
0540 900 lb. capacity					86			86	94.50
0560 1,100-1,300 lb. capacity					86			86	94.50
0600 Rubber in shear									
0610 45-340 lb., up to 1/2" rod size	1 Stpi	22	.364	Ea.	23	25		48	62
0620 130-700 lb., up to 3/4" rod size		20	.400		49.50	27.50		77	95.50
0630 50-1,000 lb., up to 3/4" rod size		18	.444		65	30.50		95.50	118
1000 Mounts, neoprene, 45-380 lb. capacity					14.10			14.10	15.50
1020 250-1,100 lb. capacity					59.50			59.50	65.50
1040 1,000-4,000 lb. capacity					117			117	129
1100 Spring flex, 60 lb. capacity					71.50			71.50	78.50
1120 165 lb. capacity					73.50			73.50	81
1140 260 lb. capacity					90.50			90.50	99.50
1160 450 lb. capacity					95.50			95.50	105
1180 600 lb. capacity					99			99	109
1200 750 lb. capacity					151			151	166
1220 900 lb. capacity					148			148	163
1240 1,100 lb. capacity					182			182	200
1260 1,300 lb. capacity					182			182	200
1280 1,500 lb. capacity					203			203	223
1300 1,800 lb. capacity					246			246	271
1320 2,200 lb. capacity					305			305	335
1340 2,600 lb. capacity					305			305	335
1399 Spring type									
1400 2 piece									
1410 50-1,000 lb.	1 Stpi	12	.667	Ea.	208	45.50		253.50	296
1420 1,100-1,600 lb.	"	12	.667	"	247	45.50		292.50	340
1500 Double spring open									
1510 150-450 lb.	1 Stpi	24	.333	Ea.	142	23		165	191
1520 500-1,000 lb.		24	.333		158	23		181	208
1530 1,100-1,600 lb.		24	.333		158	23		181	208
1540 1,700-2,400 lb.		24	.333		155	23		178	204
1550 2,500-3,400 lb.		24	.333		264	23		287	325
2000 Pads, cork rib, 18" x 18" x 1", 10-50 psi					152			152	167
2020 18" x 36" x 1", 10-50 psi					305			305	335
2100 Shear flexible pads, 18" x 18" x 3/8", 20-70 psi					63.50			63.50	70
2120 18" x 36" x 3/8", 20-70 psi					153			153	168
2150 Laminated neoprene and cork									
2160 1" thick	1 Stpi	16	.500	S.F.	74	34		108	133
2200 Neoprene elastomer isolation bearing pad									
2230 464 psi, 5/8" x 19-11/16" x 39-3/8"	1 Stpi	16	.500	Ea.	62	34		96	119
3000 Note overlap in capacities due to deflections									

For customer support on your Plumbing Costs with RSMeans Data, call 800.448.8182.

131

22 05 53.10 Piping System Identification Labels

	Crew	Daily Output	Labor-Hours	Unit	Material	2021 Bare Costs Labor	2021 Bare Costs Equipment	Total	Total Incl O&P
0010 PIPING SYSTEM IDENTIFICATION LABELS									
0100 Indicate contents and flow direction									
0106 Pipe markers									
0110 Plastic snap around									
0114 1/2" pipe	1 Plum	80	.100	Ea.	3.58	6.75		10.33	14.05
0116 3/4" pipe		80	.100		3.78	6.75		10.53	14.25
0118 1" pipe		80	.100		3.78	6.75		10.53	14.25
0120 2" pipe		75	.107		5.20	7.20		12.40	16.50
0122 3" pipe		70	.114		9.60	7.75		17.35	22
0124 4" pipe		60	.133		9.75	9.05		18.80	24
0126 6" pipe		60	.133		10.40	9.05		19.45	25
0128 8" pipe		56	.143		15.05	9.65		24.70	31
0130 10" pipe		56	.143		15.05	9.65		24.70	31
0200 Over 10" pipe size	↓	50	.160	↓	18.65	10.85		29.50	36.50
1110 Self adhesive									
1114 1" pipe	1 Plum	80	.100	Ea.	1.95	6.75		8.70	12.25
1116 2" pipe		75	.107		1.95	7.20		9.15	12.95
1118 3" pipe		70	.114		3.25	7.75		11	15.15
1120 4" pipe		60	.133		3.25	9.05		12.30	17.05
1122 6" pipe		60	.133		3.25	9.05		12.30	17.05
1124 8" pipe		56	.143		5.15	9.65		14.80	20
1126 10" pipe		56	.143		6.55	9.65		16.20	21.50
1200 Over 10" pipe size	↓	50	.160	↓	6.55	10.85		17.40	23.50
2000 Valve tags									
2010 Numbered plus identifying legend									
2100 Brass, 2" diameter	1 Plum	40	.200	Ea.	2.37	13.55		15.92	22.50
2200 Plastic, 1-1/2" diameter	"	40	.200	"	3.53	13.55		17.08	24

22 05 76.10 Cleanouts

	Crew	Daily Output	Labor-Hours	Unit	Material	2021 Bare Costs Labor	2021 Bare Costs Equipment	Total	Total Incl O&P
0010 CLEANOUTS									
0060 Floor type									
0080 Round or square, scoriated nickel bronze top									
0100 2" pipe size	1 Plum	10	.800	Ea.	440	54		494	560
0120 3" pipe size		8	1		460	67.50		527.50	605
0140 4" pipe size		6	1.333		615	90.50		705.50	815
0160 5" pipe size	↓	4	2		1,000	135		1,135	1,300
0180 6" pipe size	Q-1	6	2.667		1,000	162		1,162	1,350
0200 8" pipe size	"	4	4	↓	975	244		1,219	1,450
0340 Recessed for tile, same price									
0980 Round top, recessed for terrazzo									
1000 2" pipe size	1 Plum	9	.889	Ea.	630	60		690	785
1080 3" pipe size		6	1.333		675	90.50		765.50	875
1100 4" pipe size	↓	4	2		805	135		940	1,075
1120 5" pipe size	Q-1	6	2.667		1,275	162		1,437	1,650
1140 6" pipe size		5	3.200		1,275	195		1,470	1,700
1160 8" pipe size	↓	4	4	↓	1,275	244		1,519	1,775
2000 Round scoriated nickel bronze top, extra heavy duty									
2060 2" pipe size	1 Plum	9	.889	Ea.	335	60		395	455
2080 3" pipe size		6	1.333		440	90.50		530.50	620
2100 4" pipe size	↓	4	2		545	135		680	795
2120 5" pipe size	Q-1	6	2.667		880	162		1,042	1,200
2140 6" pipe size	↓	5	3.200	↓	880	195		1,075	1,250

132

For customer support on your Plumbing Costs with RSMeans Data, call 800.448.8182.

22 05 76 – Facility Drainage Piping Cleanouts

22 05 76.10 Cleanouts

		Crew	Daily Output	Labor-Hours	Unit	Material	2021 Bare Costs Labor	Equipment	Total	Total Incl O&P
2160	8" pipe size	Q-1	4	4	Ea.	1,025	244		1,269	1,500
4000	Wall type, square smooth cover, over wall frame									
4060	2" pipe size	1 Plum	14	.571	Ea.	430	38.50		468.50	530
4080	3" pipe size		12	.667		530	45		575	655
4100	4" pipe size		10	.800		440	54		494	565
4120	5" pipe size		9	.889		730	60		790	895
4140	6" pipe size		8	1		820	67.50		887.50	1,000
4160	8" pipe size	Q-1	11	1.455		1,100	88.50		1,188.50	1,325
5000	Extension, CI; bronze countersunk plug, 8" long									
5040	2" pipe size	1 Plum	16	.500	Ea.	192	34		226	262
5060	3" pipe size		14	.571		299	38.50		337.50	390
5080	4" pipe size		13	.615		226	41.50		267.50	310
5100	5" pipe size		12	.667		400	45		445	510
5120	6" pipe size		11	.727		710	49		759	855

22 05 76.20 Cleanout Tees

		Crew	Daily Output	Labor-Hours	Unit	Material	2021 Bare Costs Labor	Equipment	Total	Total Incl O&P
0010	**CLEANOUT TEES**									
0100	Cast iron, B&S, with countersunk plug									
0200	2" pipe size	1 Plum	4	2	Ea.	133	135		268	350
0220	3" pipe size		3.60	2.222		205	150		355	450
0240	4" pipe size		3.30	2.424		310	164		474	585
0260	5" pipe size	Q-1	5.50	2.909		610	177		787	935
0280	6" pipe size	"	5	3.200		895	195		1,090	1,275
0300	8" pipe size	Q-3	5	6.400		1,225	410		1,635	1,975
0500	For round smooth access cover, same price									
0600	For round scoriated access cover, same price									
0700	For square smooth access cover, add				Ea.	60%				
2000	Cast iron, no hub									
2010	Cleanout tee, with 2 couplings									
2012	2"	Q-1	22	.727	Ea.	45	44.50		89.50	116
2014	3"		19	.842		59.50	51.50		111	142
2016	4"		16.50	.970		88	59		147	185
2018	6"	Q-2	20	1.200		211	76		287	345
2040	Cleanout plug, no hub, with 1 coupling									
2042	2"	Q-1	44	.364	Ea.	14.75	22		36.75	49.50
2046	3"		38	.421		23	25.50		48.50	63.50
2048	4"		33	.485		38	29.50		67.50	85.50
2050	6"	Q-2	40	.600		136	38		174	207
2052	8"	"	33	.727		190	46		236	278
4000	Plastic, tees and adapters. Add plugs									
4010	ABS, DWV									
4020	Cleanout tee, 1-1/2" pipe size	1 Plum	15	.533	Ea.	17.30	36		53.30	73
4030	2" pipe size	Q-1	27	.593		24	36		60	80.50
4040	3" pipe size		21	.762		35.50	46.50		82	109
4050	4" pipe size		16	1		77.50	61		138.50	176
4100	Cleanout plug, 1-1/2" pipe size	1 Plum	32	.250		3.59	16.95		20.54	29.50
4110	2" pipe size	Q-1	56	.286		3.93	17.40		21.33	30.50
4120	3" pipe size		36	.444		6.40	27		33.40	47.50
4130	4" pipe size		30	.533		11.05	32.50		43.55	60.50
4180	Cleanout adapter fitting, 1-1/2" pipe size	1 Plum	32	.250		5.30	16.95		22.25	31.50
4190	2" pipe size	Q-1	56	.286		7.35	17.40		24.75	34
4200	3" pipe size		36	.444		18.55	27		45.55	61
4210	4" pipe size		30	.533		35	32.50		67.50	87

22 05 76.20 Cleanout Tees	Crew	Daily Output	Labor-Hours	Unit	Material	2021 Bare Costs Labor	Equipment	Total	Total Incl O&P
5000	PVC, DWV									
5010	Cleanout tee, 1-1/2" pipe size	1 Plum	15	.533	Ea.	14.20	36		50.20	69.50
5020	2" pipe size	Q-1	27	.593		16.55	36		52.55	72
5030	3" pipe size		21	.762		32	46.50		78.50	105
5040	4" pipe size		16	1		57	61		118	154
5090	Cleanout plug, 1-1/2" pipe size	1 Plum	32	.250		3.32	16.95		20.27	29
5100	2" pipe size	Q-1	56	.286		3.73	17.40		21.13	30
5110	3" pipe size		36	.444		6.65	27		33.65	48
5120	4" pipe size		30	.533		9.85	32.50		42.35	59.50
5130	6" pipe size		24	.667		31.50	40.50		72	95
5170	Cleanout adapter fitting, 1-1/2" pipe size	1 Plum	32	.250		4.47	16.95		21.42	30.50
5180	2" pipe size	Q-1	56	.286		6	17.40		23.40	32.50
5190	3" pipe size		36	.444		16.10	27		43.10	58
5200	4" pipe size		30	.533		26.50	32.50		59	77.50
5210	6" pipe size		24	.667		90.50	40.50		131	160
5300	Cleanout tee, with plug									
5310	4"	Q-1	14	1.143	Ea.	71	69.50		140.50	182
5320	6"	"	8	2		80.50	122		202.50	271
5330	8"	Q-2	11	2.182		135	138		273	355
5340	12"	"	10	2.400		725	152		877	1,025
5400	Polypropylene, Schedule 40									
5410	Cleanout tee with plug									
5420	1-1/2"	1 Plum	10	.800	Ea.	36.50	54		90.50	122
5430	2"	Q-1	17	.941		37.50	57.50		95	127
5440	3"		11	1.455		81	88.50		169.50	222
5450	4"		9	1.778		110	108		218	283

22 07 Plumbing Insulation

22 07 16 – Plumbing Equipment Insulation

22 07 16.10 Insulation for Plumbing Equipment

		Crew	Daily Output	Labor-Hours	Unit	Material	2021 Bare Costs Labor	Equipment	Total	Total Incl O&P
0010	**INSULATION FOR PLUMBING EQUIPMENT**									
2900	Domestic water heater wrap kit									
2920	1-1/2" with vinyl jacket, 20 to 60 gal. [G]	1 Plum	8	1	Ea.	19.70	67.50		87.20	123
2925	50 to 80 gal. [G]	"	8	1	"	30.50	67.50		98	135

22 07 19 – Plumbing Piping Insulation

22 07 19.10 Piping Insulation

		Crew	Daily Output	Labor-Hours	Unit	Material	2021 Bare Costs Labor	Equipment	Total	Total Incl O&P
0010	**PIPING INSULATION**									
0110	Insulation req'd. is based on the surface size/area to be covered									
0230	Insulated protectors (ADA)									
0235	For exposed piping under sinks or lavatories									
0240	Vinyl coated foam, velcro tabs									
0245	P Trap, 1-1/4" or 1-1/2"	1 Plum	32	.250	Ea.	17.10	16.95		34.05	44.50
0260	Valve and supply cover									
0265	1/2", 3/8", and 7/16" pipe size	1 Plum	32	.250	Ea.	16.80	16.95		33.75	44
0280	Tailpiece offset (wheelchair)									
0285	1-1/4" pipe size	1 Plum	32	.250	Ea.	13.95	16.95		30.90	41
0600	Pipe covering (price copper tube one size less than IPS)									
1000	Mineral wool									
1010	Preformed, 1200°F, plain									
1014	1" wall									

22 07 19 – Plumbing Piping Insulation

22 07 19.10 Piping Insulation		Crew	Daily Output	Labor-Hours	Unit	Material	2021 Bare Costs Labor	2021 Bare Costs Equipment	Total	Total Incl O&P
1016	1/2" iron pipe size	G Q-14	230	.070	L.F.	1.93	3.82		5.75	7.95
1018	3/4" iron pipe size	G	220	.073		1.96	3.99		5.95	8.25
1022	1" iron pipe size	G	210	.076		2.06	4.18		6.24	8.65
1024	1-1/4" iron pipe size	G	205	.078		2.12	4.28		6.40	8.90
1026	1-1/2" iron pipe size	G	205	.078		2	4.28		6.28	8.75
1028	2" iron pipe size	G	200	.080		2.71	4.39		7.10	9.70
1030	2-1/2" iron pipe size	G	190	.084		2.78	4.62		7.40	10.10
1032	3" iron pipe size	G	180	.089		2.97	4.88		7.85	10.70
1034	4" iron pipe size	G	150	.107		3.66	5.85		9.51	13
1036	5" iron pipe size	G	140	.114		4.14	6.25		10.39	14.15
1038	6" iron pipe size	G	120	.133		4.42	7.30		11.72	16.05
1040	7" iron pipe size	G	110	.145		5.15	8		13.15	17.90
1042	8" iron pipe size	G	100	.160		7.15	8.80		15.95	21.50
1044	9" iron pipe size	G	90	.178		7.50	9.75		17.25	23
1046	10" iron pipe size	G	90	.178		8.50	9.75		18.25	24.50
1050	1-1/2" wall									
1052	1/2" iron pipe size	G Q-14	225	.071	L.F.	3.29	3.90		7.19	9.55
1054	3/4" iron pipe size	G	215	.074		3.38	4.08		7.46	9.95
1056	1" iron pipe size	G	205	.078		3.42	4.28		7.70	10.30
1058	1-1/4" iron pipe size	G	200	.080		3.49	4.39		7.88	10.55
1060	1-1/2" iron pipe size	G	200	.080		3.66	4.39		8.05	10.75
1062	2" iron pipe size	G	190	.084		3.88	4.62		8.50	11.30
1064	2-1/2" iron pipe size	G	180	.089		4.22	4.88		9.10	12.10
1066	3" iron pipe size	G	165	.097		4.40	5.30		9.70	13
1068	4" iron pipe size	G	140	.114		5.30	6.25		11.55	15.40
1070	5" iron pipe size	G	130	.123		5.65	6.75		12.40	16.55
1072	6" iron pipe size	G	110	.145		5.80	8		13.80	18.60
1074	7" iron pipe size	G	100	.160		6.85	8.80		15.65	21
1076	8" iron pipe size	G	90	.178		7.65	9.75		17.40	23.50
1078	9" iron pipe size	G	85	.188		8.80	10.30		19.10	25.50
1080	10" iron pipe size	G	80	.200		9.65	10.95		20.60	27.50
1082	12" iron pipe size	G	75	.213		11.05	11.70		22.75	30
1084	14" iron pipe size	G	70	.229		12.60	12.55		25.15	33
1086	16" iron pipe size	G	65	.246		13.90	13.50		27.40	36
1088	18" iron pipe size	G	60	.267		16.15	14.65		30.80	40.50
1090	20" iron pipe size	G	55	.291		17.25	15.95		33.20	43.50
1092	22" iron pipe size	G	50	.320		19.70	17.55		37.25	48.50
1094	24" iron pipe size	G	45	.356		21	19.50		40.50	53
1100	2" wall									
1102	1/2" iron pipe size	G Q-14	220	.073	L.F.	4.25	3.99		8.24	10.80
1104	3/4" iron pipe size	G	210	.076		4.41	4.18		8.59	11.25
1106	1" iron pipe size	G	200	.080		4.68	4.39		9.07	11.85
1108	1-1/4" iron pipe size	G	190	.084		5	4.62		9.62	12.55
1110	1-1/2" iron pipe size	G	190	.084		5.20	4.62		9.82	12.80
1112	2" iron pipe size	G	180	.089		5.50	4.88		10.38	13.50
1114	2-1/2" iron pipe size	G	170	.094		6.20	5.15		11.35	14.70
1116	3" iron pipe size	G	160	.100		6.60	5.50		12.10	15.70
1118	4" iron pipe size	G	130	.123		7.70	6.75		14.45	18.85
1120	5" iron pipe size	G	120	.133		8.55	7.30		15.85	20.50
1122	6" iron pipe size	G	100	.160		8.90	8.80		17.70	23
1124	8" iron pipe size	G	80	.200		10.80	10.95		21.75	28.50
1126	10" iron pipe size	G	70	.229		13.20	12.55		25.75	34
1128	12" iron pipe size	G	65	.246		14.65	13.50		28.15	36.50

For customer support on your Plumbing Costs with RSMeans Data, call 800.448.8182.

135

22 07 19.10 Piping Insulation		Crew	Daily Output	Labor-Hours	Unit	Material	2021 Bare Costs Labor	Equipment	Total	Total Incl O&P
1130	14" iron pipe size	**G** Q-14	60	.267	L.F.	16.40	14.65		31.05	40.50
1132	16" iron pipe size	**G**	55	.291		18.55	15.95		34.50	45
1134	18" iron pipe size	**G**	50	.320		20	17.55		37.55	49
1136	20" iron pipe size	**G**	45	.356		23	19.50		42.50	55.50
1138	22" iron pipe size	**G**	45	.356		25	19.50		44.50	57.50
1140	24" iron pipe size	**G**	40	.400		26.50	22		48.50	62.50
1150	4" wall									
1152	1-1/4" iron pipe size	**G** Q-14	170	.094	L.F.	13.50	5.15		18.65	23
1154	1-1/2" iron pipe size	**G**	165	.097		14.05	5.30		19.35	23.50
1156	2" iron pipe size	**G**	155	.103		14.30	5.65		19.95	24.50
1158	4" iron pipe size	**G**	105	.152		18.10	8.35		26.45	32.50
1160	6" iron pipe size	**G**	75	.213		21.50	11.70		33.20	41.50
1162	8" iron pipe size	**G**	60	.267		24	14.65		38.65	49
1164	10" iron pipe size	**G**	50	.320		29	17.55		46.55	59
1166	12" iron pipe size	**G**	45	.356		32	19.50		51.50	65.50
1168	14" iron pipe size	**G**	40	.400		34.50	22		56.50	71
1170	16" iron pipe size	**G**	35	.457		38.50	25		63.50	81
1172	18" iron pipe size	**G**	32	.500		41.50	27.50		69	88
1174	20" iron pipe size	**G**	30	.533		45	29.50		74.50	94.50
1176	22" iron pipe size	**G**	28	.571		50.50	31.50		82	104
1178	24" iron pipe size	**G**	26	.615		53.50	34		87.50	110
4280	Cellular glass, closed cell foam, all service jacket, sealant,									
4281	working temp. (-450°F to +900°F), 0 water vapor transmission									
4284	1" wall									
4286	1/2" iron pipe size	**G** Q-14	120	.133	L.F.	7.90	7.30		15.20	19.90
4300	1-1/2" wall									
4301	1" iron pipe size	**G** Q-14	105	.152	L.F.	8.30	8.35		16.65	22
4304	2-1/2" iron pipe size	**G**	90	.178		13.70	9.75		23.45	30
4306	3" iron pipe size	**G**	85	.188		19.65	10.30		29.95	37.50
4308	4" iron pipe size	**G**	70	.229		21.50	12.55		34.05	42.50
4310	5" iron pipe size	**G**	65	.246		21.50	13.50		35	44.50
4320	2" wall									
4322	1" iron pipe size	**G** Q-14	100	.160	L.F.	13.65	8.80		22.45	28.50
4324	2-1/2" iron pipe size	**G**	85	.188		17.15	10.30		27.45	34.50
4326	3" iron pipe size	**G**	80	.200		17.30	10.95		28.25	36
4328	4" iron pipe size	**G**	65	.246		24	13.50		37.50	47
4330	5" iron pipe size	**G**	60	.267		26.50	14.65		41.15	51.50
4332	6" iron pipe size	**G**	50	.320		30.50	17.55		48.05	60.50
4336	8" iron pipe size	**G**	40	.400		36.50	22		58.50	73.50
4338	10" iron pipe size	**G**	35	.457		37	25		62	79
4350	2-1/2" wall									
4360	12" iron pipe size	**G** Q-14	32	.500	L.F.	64.50	27.50		92	113
4362	14" iron pipe size	**G** "	28	.571	"	71	31.50		102.50	127
4370	3" wall									
4378	6" iron pipe size	**G** Q-14	48	.333	L.F.	54.50	18.30		72.80	87.50
4380	8" iron pipe size	**G**	38	.421		59.50	23		82.50	101
4382	10" iron pipe size	**G**	33	.485		77.50	26.50		104	126
4384	16" iron pipe size	**G**	25	.640		83	35		118	145
4386	18" iron pipe size	**G**	22	.727		108	40		148	179
4388	20" iron pipe size	**G**	20	.800		117	44		161	196
4400	3-1/2" wall									
4412	12" iron pipe size	**G** Q-14	27	.593	L.F.	69	32.50		101.50	126
4414	14" piron pipe size	**G** "	25	.640	"	88.50	35		123.50	151

22 07 19 – Plumbing Piping Insulation

22 07 19.10 Piping Insulation		Crew	Daily Output	Labor-Hours	Unit	Material	2021 Bare Costs Labor	Equipment	Total	Total Incl O&P
4430	4" wall									
4446	16" iron pipe size	G	Q-14	22	.727	L.F.	102	40		142	173
4448	18" iron pipe size	G		20	.800		101	44		145	178
4450	20" iron pipe size	G	↓	18	.889	↓	146	49		195	236
4480	Fittings, average with fabric and mastic										
4484	1" wall										
4486	1/2" iron pipe size	G	1 Asbe	40	.200	Ea.	7.85	12.20		20.05	27.50
4500	1-1/2" wall										
4502	1" iron pipe size	G	1 Asbe	38	.211	Ea.	7.40	12.85		20.25	28
4504	2-1/2" iron pipe size	G		32	.250		17.10	15.25		32.35	42.50
4506	3" iron pipe size	G		30	.267		19.35	16.25		35.60	46.50
4508	4" iron pipe size	G		28	.286		19.95	17.40		37.35	48.50
4510	5" iron pipe size	G	↓	24	.333	↓	26	20.50		46.50	60
4520	2" wall										
4522	1" iron pipe size	G	1 Asbe	36	.222	Ea.	13.80	13.55		27.35	35.50
4524	2-1/2" iron pipe size	G		30	.267		17.20	16.25		33.45	44
4526	3" iron pipe size	G		28	.286		24	17.40		41.40	53
4528	4" iron pipe size	G		24	.333		27	20.50		47.50	60.50
4530	5" iron pipe size	G		22	.364		39.50	22		61.50	77.50
4532	6" iron pipe size	G		20	.400		43	24.50		67.50	85
4536	8" iron pipe size	G		12	.667		97.50	40.50		138	169
4538	10" iron pipe size	G	↓	8	1	↓	110	61		171	215
4550	2-1/2" wall										
4560	12" iron pipe size	G	1 Asbe	6	1.333	Ea.	208	81.50		289.50	355
4562	14" iron pipe size	G	"	4	2	"	232	122		354	440
4570	3" wall										
4578	6" iron pipe size	G	1 Asbe	16	.500	Ea.	71	30.50		101.50	125
4580	8" iron pipe size	G		10	.800		106	49		155	192
4582	10" iron pipe size	G	↓	6	1.333	↓	142	81.50		223.50	281
4900	Calcium silicate, with 8 oz. canvas cover										
5100	1" wall, 1/2" iron pipe size	G	Q-14	170	.094	L.F.	4.71	5.15		9.86	13.10
5130	3/4" iron pipe size	G		170	.094		4.74	5.15		9.89	13.10
5140	1" iron pipe size	G		170	.094		4.64	5.15		9.79	13
5150	1-1/4" iron pipe size	G		165	.097		4.71	5.30		10.01	13.35
5160	1-1/2" iron pipe size	G		165	.097		4.76	5.30		10.06	13.40
5170	2" iron pipe size	G		160	.100		5.50	5.50		11	14.45
5180	2-1/2" iron pipe size	G		160	.100		5.85	5.50		11.35	14.80
5190	3" iron pipe size	G		150	.107		6.35	5.85		12.20	15.90
5200	4" iron pipe size	G		140	.114		7.80	6.25		14.05	18.20
5210	5" iron pipe size	G		135	.119		8.25	6.50		14.75	19.05
5220	6" iron pipe size	G		130	.123		8.75	6.75		15.50	20
5280	1-1/2" wall, 1/2" iron pipe size	G		150	.107		5.10	5.85		10.95	14.55
5310	3/4" iron pipe size	G		150	.107		5.20	5.85		11.05	14.65
5320	1" iron pipe size	G		150	.107		5.65	5.85		11.50	15.15
5330	1-1/4" iron pipe size	G		145	.110		6.10	6.05		12.15	15.95
5340	1-1/2" iron pipe size	G		145	.110		6.50	6.05		12.55	16.40
5350	2" iron pipe size	G		140	.114		7.20	6.25		13.45	17.50
5360	2-1/2" iron pipe size	G		140	.114		7.80	6.25		14.05	18.20
5370	3" iron pipe size	G		135	.119		8.25	6.50		14.75	19
5380	4" iron pipe size	G		125	.128		9.50	7		16.50	21
5390	5" iron pipe size	G		120	.133		10.75	7.30		18.05	23
5400	6" iron pipe size	G		110	.145		11.10	8		19.10	24.50
5460	2" wall, 1/2" iron pipe size	G	↓	135	.119	↓	7.90	6.50		14.40	18.60

137

22 07 19.10 Piping Insulation		Crew	Daily Output	Labor-Hours	Unit	Material	2021 Bare Costs Labor	2021 Bare Costs Equipment	Total	Total Incl O&P
5490	3/4" iron pipe size	G	Q-14	135	.119	L.F.	8.30	6.50		14.80	19.05
5500	1" iron pipe size	G		135	.119		8.70	6.50		15.20	19.55
5510	1-1/4" iron pipe size	G		130	.123		9.35	6.75		16.10	20.50
5520	1-1/2" iron pipe size	G		130	.123		9.75	6.75		16.50	21
5530	2" iron pipe size	G		125	.128		10.30	7		17.30	22
5540	2-1/2" iron pipe size	G		125	.128		12.20	7		19.20	24
5550	3" iron pipe size	G		120	.133		12.30	7.30		19.60	25
5560	4" iron pipe size	G		115	.139		14.25	7.65		21.90	27.50
5570	5" iron pipe size	G		110	.145		16.20	8		24.20	30
5580	6" iron pipe size	G		105	.152		17.70	8.35		26.05	32.50
5600	Calcium silicate, no cover										
5720	1" wall, 1/2" iron pipe size	G	Q-14	180	.089	L.F.	4.29	4.88		9.17	12.15
5740	3/4" iron pipe size	G		180	.089		4.29	4.88		9.17	12.15
5750	1" iron pipe size	G		180	.089		4.10	4.88		8.98	11.95
5760	1-1/4" iron pipe size	G		175	.091		4.19	5		9.19	12.25
5770	1-1/2" iron pipe size	G		175	.091		4.22	5		9.22	12.30
5780	2" iron pipe size	G		170	.094		4.85	5.15		10	13.25
5790	2-1/2" iron pipe size	G		170	.094		5.15	5.15		10.30	13.55
5800	3" iron pipe size	G		160	.100		5.50	5.50		11	14.45
5810	4" iron pipe size	G		150	.107		6.90	5.85		12.75	16.55
5820	5" iron pipe size	G		145	.110		7.20	6.05		13.25	17.15
5830	6" iron pipe size	G		140	.114		7.55	6.25		13.80	17.90
5900	1-1/2" wall, 1/2" iron pipe size	G		160	.100		4.54	5.50		10.04	13.40
5920	3/4" iron pipe size	G		160	.100		4.61	5.50		10.11	13.45
5930	1" iron pipe size	G		160	.100		5	5.50		10.50	13.90
5940	1-1/4" iron pipe size	G		155	.103		5.40	5.65		11.05	14.60
5950	1-1/2" iron pipe size	G		155	.103		5.80	5.65		11.45	15.05
5960	2" iron pipe size	G		150	.107		6.45	5.85		12.30	16
5970	2-1/2" iron pipe size	G		150	.107		7	5.85		12.85	16.65
5980	3" iron pipe size	G		145	.110		7.30	6.05		13.35	17.30
5990	4" iron pipe size	G		135	.119		8.45	6.50		14.95	19.25
6000	5" iron pipe size	G		130	.123		9.55	6.75		16.30	21
6010	6" iron pipe size	G		120	.133		9.75	7.30		17.05	22
6020	7" iron pipe size	G		115	.139		11.45	7.65		19.10	24.50
6030	8" iron pipe size	G		105	.152		12.90	8.35		21.25	27
6040	9" iron pipe size	G		100	.160		15.50	8.80		24.30	30.50
6050	10" iron pipe size	G		95	.168		17.10	9.25		26.35	33
6060	12" iron pipe size	G		90	.178		20.50	9.75		30.25	37.50
6070	14" iron pipe size	G		85	.188		23	10.30		33.30	41
6080	16" iron pipe size	G		80	.200		25.50	10.95		36.45	45
6090	18" iron pipe size	G		75	.213		28.50	11.70		40.20	49
6120	2" wall, 1/2" iron pipe size	G		145	.110		7.15	6.05		13.20	17.15
6140	3/4" iron pipe size	G		145	.110		7.55	6.05		13.60	17.55
6150	1" iron pipe size	G		145	.110		7.95	6.05		14	18
6160	1-1/4" iron pipe size	G		140	.114		8.50	6.25		14.75	18.95
6170	1-1/2" iron pipe size	G		140	.114		8.90	6.25		15.15	19.40
6180	2" iron pipe size	G		135	.119		9.35	6.50		15.85	20
6190	2-1/2" iron pipe size	G		135	.119		11.25	6.50		17.75	22.50
6200	3" iron pipe size	G		130	.123		11.30	6.75		18.05	23
6210	4" iron pipe size	G		125	.128		13.05	7		20.05	25
6220	5" iron pipe size	G		120	.133		14.90	7.30		22.20	27.50
6230	6" iron pipe size	G		115	.139		16.25	7.65		23.90	29.50
6240	7" iron pipe size	G		110	.145		17.60	8		25.60	31.50

138

22 07 Plumbing Insulation

22 07 19 – Plumbing Piping Insulation

22 07 19.10 Piping Insulation		Crew	Daily Output	Labor-Hours	Unit	Material	2021 Bare Costs Labor	Equipment	Total	Total Incl O&P
6250	8" iron pipe size	G Q-14	105	.152	L.F.	19.80	8.35		28.15	35
6260	9" iron pipe size	G	100	.160		22	8.80		30.80	37.50
6270	10" iron pipe size	G	95	.168		24.50	9.25		33.75	41
6280	12" iron pipe size	G	90	.178		27	9.75		36.75	44.50
6290	14" iron pipe size	G	85	.188		29.50	10.30		39.80	48.50
6300	16" iron pipe size	G	80	.200		32.50	10.95		43.45	53
6310	18" iron pipe size	G	75	.213		35.50	11.70		47.20	57.50
6320	20" iron pipe size	G	65	.246		44	13.50		57.50	69
6330	22" iron pipe size	G	60	.267		48.50	14.65		63.15	76
6340	24" iron pipe size	G	55	.291		50	15.95		65.95	79.50
6360	3" wall, 1/2" iron pipe size	G	115	.139		13	7.65		20.65	26
6380	3/4" iron pipe size	G	115	.139		13.05	7.65		20.70	26
6390	1" iron pipe size	G	115	.139		13.15	7.65		20.80	26
6400	1-1/4" iron pipe size	G	110	.145		13.35	8		21.35	27
6410	1-1/2" iron pipe size	G	110	.145		13.40	8		21.40	27
6420	2" iron pipe size	G	105	.152		13.85	8.35		22.20	28
6430	2-1/2" iron pipe size	G	105	.152		16.30	8.35		24.65	31
6440	3" iron pipe size	G	100	.160		16.45	8.80		25.25	31.50
6450	4" iron pipe size	G	95	.168		21	9.25		30.25	37
6460	5" iron pipe size	G	90	.178		23	9.75		32.75	40
6470	6" iron pipe size	G	90	.178		25.50	9.75		35.25	43
6480	7" iron pipe size	G	85	.188		28.50	10.30		38.80	47.50
6490	8" iron pipe size	G	85	.188		30.50	10.30		40.80	49.50
6500	9" iron pipe size	G	80	.200		34.50	10.95		45.45	55
6510	10" iron pipe size	G	75	.213		37	11.70		48.70	58.50
6520	12" iron pipe size	G	70	.229		41	12.55		53.55	64
6530	14" iron pipe size	G	65	.246		45.50	13.50		59	71
6540	16" iron pipe size	G	60	.267		50.50	14.65		65.15	78
6550	18" iron pipe size	G	55	.291		55.50	15.95		71.45	85.50
6560	20" iron pipe size	G	50	.320		66	17.55		83.55	99.50
6570	22" iron pipe size	G	45	.356		71.50	19.50		91	109
6580	24" iron pipe size	G	40	.400		77	22		99	118
6600	Fiberglass, with all service jacket									
6640	1/2" wall, 1/2" iron pipe size	G Q-14	250	.064	L.F.	.79	3.51		4.30	6.20
6660	3/4" iron pipe size	G	240	.067		.90	3.66		4.56	6.60
6670	1" iron pipe size	G	230	.070		.94	3.82		4.76	6.90
6680	1-1/4" iron pipe size	G	220	.073		1	3.99		4.99	7.20
6690	1-1/2" iron pipe size	G	220	.073		1.14	3.99		5.13	7.35
6700	2" iron pipe size	G	210	.076		1.22	4.18		5.40	7.75
6710	2-1/2" iron pipe size	G	200	.080		1.27	4.39		5.66	8.10
6840	1" wall, 1/2" iron pipe size	G	240	.067		.96	3.66		4.62	6.65
6860	3/4" iron pipe size	G	230	.070		1.03	3.82		4.85	7
6870	1" iron pipe size	G	220	.073		1.14	3.99		5.13	7.35
6880	1-1/4" iron pipe size	G	210	.076		1.22	4.18		5.40	7.75
6890	1-1/2" iron pipe size	G	210	.076		1.32	4.18		5.50	7.85
6900	2" iron pipe size	G	200	.080		2.19	4.39		6.58	9.10
6910	2-1/2" iron pipe size	G	190	.084		2.19	4.62		6.81	9.45
6920	3" iron pipe size	G	180	.089		2.39	4.88		7.27	10.10
6930	3-1/2" iron pipe size	G	170	.094		2.85	5.15		8	11.05
6940	4" iron pipe size	G	150	.107		3.20	5.85		9.05	12.45
6950	5" iron pipe size	G	140	.114		3.45	6.25		9.70	13.40
6960	6" iron pipe size	G	120	.133		3.78	7.30		11.08	15.35
6970	7" iron pipe size	G	110	.145		4.50	8		12.50	17.15

For customer support on your Plumbing Costs with RSMeans Data, call 800.448.8182.

139

22 07 19 – Plumbing Piping Insulation

22 07 19.10 Piping Insulation		Crew	Daily Output	Labor-Hours	Unit	Material	2021 Bare Costs Labor	Equipment	Total	Total Incl O&P
6980	8" iron pipe size	G	Q-14	100	.160	L.F.	6.15	8.80		14.95	20
6990	9" iron pipe size	G		90	.178		6.50	9.75		16.25	22
7000	10" iron pipe size	G		90	.178		6.65	9.75		16.40	22.50
7010	12" iron pipe size	G		80	.200		7.25	10.95		18.20	25
7020	14" iron pipe size	G		80	.200		8.80	10.95		19.75	26.50
7030	16" iron pipe size	G		70	.229		11.25	12.55		23.80	31.50
7040	18" iron pipe size	G		70	.229		12.45	12.55		25	33
7050	20" iron pipe size	G		60	.267		13.90	14.65		28.55	38
7060	24" iron pipe size	G		60	.267		16.90	14.65		31.55	41
7080	1-1/2" wall, 1/2" iron pipe size	G		230	.070		2.55	3.82		6.37	8.65
7100	3/4" iron pipe size	G		220	.073		2.55	3.99		6.54	8.90
7110	1" iron pipe size	G		210	.076		2.74	4.18		6.92	9.40
7120	1-1/4" iron pipe size	G		200	.080		2.97	4.39		7.36	9.95
7130	1-1/2" iron pipe size	G		200	.080		3.48	4.39		7.87	10.55
7140	2" iron pipe size	G		190	.084		3.48	4.62		8.10	10.90
7150	2-1/2" iron pipe size	G		180	.089		3.69	4.88		8.57	11.50
7160	3" iron pipe size	G		170	.094		3.87	5.15		9.02	12.15
7170	3-1/2" iron pipe size	G		160	.100		4.24	5.50		9.74	13.05
7180	4" iron pipe size	G		140	.114		4.89	6.25		11.14	15
7190	5" iron pipe size	G		130	.123		4.92	6.75		11.67	15.75
7200	6" iron pipe size	G		110	.145		5.20	8		13.20	17.95
7210	7" iron pipe size	G		100	.160		5.80	8.80		14.60	19.75
7220	8" iron pipe size	G		90	.178		7.30	9.75		17.05	23
7230	9" iron pipe size	G		85	.188		7.60	10.30		17.90	24
7240	10" iron pipe size	G		80	.200		7.85	10.95		18.80	25.50
7250	12" iron pipe size	G		75	.213		8.90	11.70		20.60	27.50
7260	14" iron pipe size	G		70	.229		10.80	12.55		23.35	31
7270	16" iron pipe size	G		65	.246		14.05	13.50		27.55	36
7280	18" iron pipe size	G		60	.267		15.75	14.65		30.40	40
7290	20" iron pipe size	G		55	.291		16.10	15.95		32.05	42
7300	24" iron pipe size	G		50	.320		19.70	17.55		37.25	48.50
7320	2" wall, 1/2" iron pipe size	G		220	.073		3.97	3.99		7.96	10.45
7340	3/4" iron pipe size	G		210	.076		4.09	4.18		8.27	10.90
7350	1" iron pipe size	G		200	.080		4.35	4.39		8.74	11.50
7360	1-1/4" iron pipe size	G		190	.084		4.60	4.62		9.22	12.10
7370	1-1/2" iron pipe size	G		190	.084		4.48	4.62		9.10	12
7380	2" iron pipe size	G		180	.089		5.05	4.88		9.93	13
7390	2-1/2" iron pipe size	G		170	.094		5.45	5.15		10.60	13.90
7400	3" iron pipe size	G		160	.100		5.80	5.50		11.30	14.75
7410	3-1/2" iron pipe size	G		150	.107		6.25	5.85		12.10	15.85
7420	4" iron pipe size	G		130	.123		6.30	6.75		13.05	17.25
7430	5" iron pipe size	G		120	.133		7.70	7.30		15	19.70
7440	6" iron pipe size	G		100	.160		6.90	8.80		15.70	21
7450	7" iron pipe size	G		90	.178		9.05	9.75		18.80	25
7460	8" iron pipe size	G		80	.200		9.65	10.95		20.60	27.50
7470	9" iron pipe size	G		75	.213		10.60	11.70		22.30	29.50
7480	10" iron pipe size	G		70	.229		11.55	12.55		24.10	32
7490	12" iron pipe size	G		65	.246		12.95	13.50		26.45	35
7500	14" iron pipe size	G		60	.267		16.90	14.65		31.55	41
7510	16" iron pipe size	G		55	.291		18.55	15.95		34.50	45
7520	18" iron pipe size	G		50	.320		20.50	17.55		38.05	50
7530	20" iron pipe size	G		45	.356		23.50	19.50		43	56
7540	24" iron pipe size	G		40	.400		25	22		47	61

22 07 19 – Plumbing Piping Insulation

22 07 19.10 Piping Insulation		Crew	Daily Output	Labor-Hours	Unit	Material	2021 Bare Costs Labor	Equipment	Total	Total Incl O&P
7560	2-1/2" wall, 1/2" iron pipe size	G	Q-14	210	.076	L.F.	4.70	4.18		8.88	11.55
7562	3/4" iron pipe size	G		200	.080		4.91	4.39		9.30	12.10
7564	1" iron pipe size	G		190	.084		5.10	4.62		9.72	12.70
7566	1-1/4" iron pipe size	G		185	.086		5.30	4.74		10.04	13.10
7568	1-1/2" iron pipe size	G		180	.089		5.60	4.88		10.48	13.60
7570	2" iron pipe size	G		170	.094		5.85	5.15		11	14.35
7572	2-1/2" iron pipe size	G		160	.100		6.75	5.50		12.25	15.85
7574	3" iron pipe size	G		150	.107		7.10	5.85		12.95	16.75
7576	3-1/2" iron pipe size	G		140	.114		7.75	6.25		14	18.10
7578	4" iron pipe size	G		120	.133		8.15	7.30		15.45	20
7580	5" iron pipe size	G		110	.145		5.65	8		13.65	18.40
7582	6" iron pipe size	G		90	.178		11.70	9.75		21.45	28
7584	7" iron pipe size	G		80	.200		11.75	10.95		22.70	30
7586	8" iron pipe size	G		70	.229		12.45	12.55		25	33
7588	9" iron pipe size	G		65	.246		13.65	13.50		27.15	35.50
7590	10" iron pipe size	G		60	.267		14.85	14.65		29.50	39
7592	12" iron pipe size	G		55	.291		18.05	15.95		34	44.50
7594	14" iron pipe size	G		50	.320		21.50	17.55		39.05	50.50
7596	16" iron pipe size	G		45	.356		24.50	19.50		44	57
7598	18" iron pipe size	G		40	.400		26.50	22		48.50	62.50
7602	24" iron pipe size	G		30	.533		35	29.50		64.50	83.50
7620	3" wall, 1/2" iron pipe size	G		200	.080		6	4.39		10.39	13.30
7622	3/4" iron pipe size	G		190	.084		6.35	4.62		10.97	14.05
7624	1" iron pipe size	G		180	.089		6.75	4.88		11.63	14.90
7626	1-1/4" iron pipe size	G		175	.091		6.90	5		11.90	15.20
7628	1-1/2" iron pipe size	G		170	.094		7.20	5.15		12.35	15.80
7630	2" iron pipe size	G		160	.100		7.75	5.50		13.25	16.95
7632	2-1/2" iron pipe size	G		150	.107		8.10	5.85		13.95	17.85
7634	3" iron pipe size	G		140	.114		8.60	6.25		14.85	19.05
7636	3-1/2" iron pipe size	G		130	.123		9.50	6.75		16.25	21
7638	4" iron pipe size	G		110	.145		10.20	8		18.20	23.50
7640	5" iron pipe size	G		100	.160		11.55	8.80		20.35	26
7642	6" iron pipe size	G		80	.200		12.35	10.95		23.30	30.50
7644	7" iron pipe size	G		70	.229		14.10	12.55		26.65	35
7646	8" iron pipe size	G		60	.267		15.35	14.65		30	39.50
7648	9" iron pipe size	G		55	.291		16.55	15.95		32.50	42.50
7650	10" iron pipe size	G		50	.320		17.80	17.55		35.35	46.50
7652	12" iron pipe size	G		45	.356		22	19.50		41.50	54.50
7654	14" iron pipe size	G		40	.400		26	22		48	62
7656	16" iron pipe size	G		35	.457		29.50	25		54.50	71
7658	18" iron pipe size	G		32	.500		31.50	27.50		59	76.50
7660	20" iron pipe size	G		30	.533		34	29.50		63.50	82.50
7662	24" iron pipe size	G		28	.571		43.50	31.50		75	95.50
7664	26" iron pipe size	G		26	.615		47.50	34		81.50	104
7666	30" iron pipe size	G		24	.667		49.50	36.50		86	111
7800	For fiberglass with standard canvas jacket, deduct						5%				
7802	For fittings, add 3 L.F. for each fitting										
7804	plus 4 L.F. for each flange of the fitting										
7820	Polyethylene tubing flexible closed cell foam, UV resistant										
7828	Standard temperature (-90°F to +212°F)										
7830	3/8" wall, 1/8" iron pipe size	G	1 Asbe	130	.062	L.F.	.35	3.75		4.10	6.15
7831	1/4" iron pipe size	G		130	.062		.36	3.75		4.11	6.15
7832	3/8" iron pipe size	G		130	.062		.33	3.75		4.08	6.10

22 07 19 – Plumbing Piping Insulation

22 07 19.10 Piping Insulation		Crew	Daily Output	Labor-Hours	Unit	Material	2021 Bare Costs Labor	Equipment	Total	Total Incl O&P
7833	1/2" iron pipe size	G	1 Asbe	126	.063	L.F.	.46	3.87		4.33	6.40
7834	3/4" iron pipe size	G		122	.066		.45	4		4.45	6.60
7835	1" iron pipe size	G		120	.067		.58	4.06		4.64	6.85
7836	1-1/4" iron pipe size	G		118	.068		.74	4.13		4.87	7.10
7837	1-1/2" iron pipe size	G		118	.068		.77	4.13		4.90	7.15
7838	2" iron pipe size	G		116	.069		1.22	4.20		5.42	7.80
7839	2-1/2" iron pipe size	G		114	.070		1.17	4.28		5.45	7.85
7840	3" iron pipe size	G		112	.071		2.39	4.35		6.74	9.30
7842	1/2" wall, 1/8" iron pipe size	G		120	.067		.42	4.06		4.48	6.65
7843	1/4" iron pipe size	G		120	.067		.55	4.06		4.61	6.80
7844	3/8" iron pipe size	G		120	.067		.58	4.06		4.64	6.85
7845	1/2" iron pipe size	G		118	.068		.67	4.13		4.80	7.05
7846	3/4" iron pipe size	G		116	.069		.68	4.20		4.88	7.20
7847	1" iron pipe size	G		114	.070		.71	4.28		4.99	7.35
7848	1-1/4" iron pipe size	G		112	.071		.82	4.35		5.17	7.55
7849	1-1/2" iron pipe size	G		110	.073		.91	4.43		5.34	7.80
7850	2" iron pipe size	G		108	.074		1.71	4.51		6.22	8.80
7851	2-1/2" iron pipe size	G		106	.075		2.21	4.60		6.81	9.50
7852	3" iron pipe size	G		104	.077		2.88	4.69		7.57	10.30
7853	3-1/2" iron pipe size	G		102	.078		3.22	4.78		8	10.85
7854	4" iron pipe size	G		100	.080		3.84	4.88		8.72	11.65
7855	3/4" wall, 1/8" iron pipe size	G		110	.073		.63	4.43		5.06	7.50
7856	1/4" iron pipe size	G		110	.073		.68	4.43		5.11	7.55
7857	3/8" iron pipe size	G		108	.074		.95	4.51		5.46	7.95
7858	1/2" iron pipe size	G		106	.075		1.13	4.60		5.73	8.30
7859	3/4" iron pipe size	G		104	.077		1.34	4.69		6.03	8.60
7860	1" iron pipe size	G		102	.078		1.58	4.78		6.36	9.05
7861	1-1/4" iron pipe size	G		100	.080		1.75	4.88		6.63	9.40
7862	1-1/2" iron pipe size	G		100	.080		2.22	4.88		7.10	9.90
7863	2" iron pipe size	G		98	.082		3.04	4.98		8.02	10.95
7864	2-1/2" iron pipe size	G		96	.083		3.84	5.10		8.94	11.95
7865	3" iron pipe size	G		94	.085		3.79	5.20		8.99	12.10
7866	3-1/2" iron pipe size	G		92	.087		5.40	5.30		10.70	14.05
7867	4" iron pipe size	G		90	.089		6.05	5.40		11.45	14.95
7868	1" wall, 1/4" iron pipe size	G		100	.080		1.65	4.88		6.53	9.25
7869	3/8" iron pipe size	G		98	.082		1.56	4.98		6.54	9.30
7870	1/2" iron pipe size	G		96	.083		2.09	5.10		7.19	10.05
7871	3/4" iron pipe size	G		94	.085		2.39	5.20		7.59	10.60
7872	1" iron pipe size	G		92	.087		2.92	5.30		8.22	11.30
7873	1-1/4" iron pipe size	G		90	.089		3.31	5.40		8.71	11.95
7874	1-1/2" iron pipe size	G		90	.089		4.07	5.40		9.47	12.80
7875	2" iron pipe size	G		88	.091		4.90	5.55		10.45	13.90
7876	2-1/2" iron pipe size	G		86	.093		6.80	5.65		12.45	16.10
7877	3" iron pipe size	G		84	.095		7.35	5.80		13.15	16.95
7878	Contact cement, quart can	G				Ea.	13.95			13.95	15.30
7879	Rubber tubing, flexible closed cell foam										
7880	3/8" wall, 1/4" iron pipe size	G	1 Asbe	120	.067	L.F.	.70	4.06		4.76	6.95
7900	3/8" iron pipe size	G		120	.067		.75	4.06		4.81	7.05
7910	1/2" iron pipe size	G		115	.070		.87	4.24		5.11	7.45
7920	3/4" iron pipe size	G		115	.070		.96	4.24		5.20	7.55
7930	1" iron pipe size	G		110	.073		1.09	4.43		5.52	8
7940	1-1/4" iron pipe size	G		110	.073		1.24	4.43		5.67	8.15
7950	1-1/2" iron pipe size	G		110	.073		1.53	4.43		5.96	8.50

For customer support on your Plumbing Costs with RSMeans Data, call 800.448.8182.

22 07 Plumbing Insulation

22 07 19 – Plumbing Piping Insulation

22 07 19.10 Piping Insulation		Crew	Daily Output	Labor-Hours	Unit	Material	2021 Bare Costs Labor	Equipment	Total	Total Incl O&P
8100	1/2" wall, 1/4" iron pipe size	G 1 Asbe	90	.089	L.F.	1.14	5.40		6.54	9.55
8120	3/8" iron pipe size	G	90	.089		1.13	5.40		6.53	9.55
8130	1/2" iron pipe size	G	89	.090		1.25	5.50		6.75	9.80
8140	3/4" iron pipe size	G	89	.090		1.29	5.50		6.79	9.80
8150	1" iron pipe size	G	88	.091		1.56	5.55		7.11	10.20
8160	1-1/4" iron pipe size	G	87	.092		1.72	5.60		7.32	10.45
8170	1-1/2" iron pipe size	G	87	.092		1.97	5.60		7.57	10.70
8180	2" iron pipe size	G	86	.093		2.52	5.65		8.17	11.40
8190	2-1/2" iron pipe size	G	86	.093		2.99	5.65		8.64	11.95
8200	3" iron pipe size	G	85	.094		3.51	5.75		9.26	12.65
8210	3-1/2" iron pipe size	G	85	.094		4.84	5.75		10.59	14.10
8220	4" iron pipe size	G	80	.100		5.25	6.10		11.35	15.15
8230	5" iron pipe size	G	80	.100		7.10	6.10		13.20	17.15
8240	6" iron pipe size	G	75	.107		7.10	6.50		13.60	17.75
8300	3/4" wall, 1/4" iron pipe size	G	90	.089		1.60	5.40		7	10.05
8320	3/8" iron pipe size	G	90	.089		1.74	5.40		7.14	10.20
8330	1/2" iron pipe size	G	89	.090		1.90	5.50		7.40	10.50
8340	3/4" iron pipe size	G	89	.090		2.28	5.50		7.78	10.90
8350	1" iron pipe size	G	88	.091		2.77	5.55		8.32	11.55
8360	1-1/4" iron pipe size	G	87	.092		3.17	5.60		8.77	12.05
8370	1-1/2" iron pipe size	G	87	.092		4.21	5.60		9.81	13.20
8380	2" iron pipe size	G	86	.093		5.10	5.65		10.75	14.30
8390	2-1/2" iron pipe size	G	86	.093		6.30	5.65		11.95	15.55
8400	3" iron pipe size	G	85	.094		7	5.75		12.75	16.50
8410	3-1/2" iron pipe size	G	85	.094		5.25	5.75		11	14.55
8420	4" iron pipe size	G	80	.100		8.65	6.10		14.75	18.90
8430	5" iron pipe size	G	80	.100		10.30	6.10		16.40	20.50
8440	6" iron pipe size	G	80	.100		12.50	6.10		18.60	23
8444	1" wall, 1/2" iron pipe size	G	86	.093		3.40	5.65		9.05	12.40
8445	3/4" iron pipe size	G	84	.095		4.19	5.80		9.99	13.50
8446	1" iron pipe size	G	84	.095		5.10	5.80		10.90	14.50
8447	1-1/4" iron pipe size	G	82	.098		5.90	5.95		11.85	15.60
8448	1-1/2" iron pipe size	G	82	.098		6.55	5.95		12.50	16.30
8449	2" iron pipe size	G	80	.100		7.85	6.10		13.95	18
8450	2-1/2" iron pipe size	G	80	.100		10.65	6.10		16.75	21
8456	Rubber insulation tape, 1/8" x 2" x 30'	G			Ea.	25			25	27.50
8460	Polyolefin tubing, flexible closed cell foam, UV stabilized, work									
8462	temp. -165°F to +210°F, 0 water vapor transmission									
8464	3/8" wall, 1/8" iron pipe size	G 1 Asbe	140	.057	L.F.	.48	3.48		3.96	5.90
8466	1/4" iron pipe size	G	140	.057		.50	3.48		3.98	5.90
8468	3/8" iron pipe size	G	140	.057		.56	3.48		4.04	5.95
8470	1/2" iron pipe size	G	136	.059		.62	3.59		4.21	6.20
8472	3/4" iron pipe size	G	132	.061		.62	3.69		4.31	6.35
8474	1" iron pipe size	G	130	.062		.83	3.75		4.58	6.65
8476	1-1/4" iron pipe size	G	128	.063		1.02	3.81		4.83	6.95
8478	1-1/2" iron pipe size	G	128	.063		1.24	3.81		5.05	7.20
8480	2" iron pipe size	G	126	.063		1.50	3.87		5.37	7.55
8482	2-1/2" iron pipe size	G	123	.065		2.18	3.96		6.14	8.45
8484	3" iron pipe size	G	121	.066		2.42	4.03		6.45	8.80
8486	4" iron pipe size	G	118	.068		4.32	4.13		8.45	11.05
8500	1/2" wall, 1/8" iron pipe size	G	130	.062		.70	3.75		4.45	6.50
8502	1/4" iron pipe size	G	130	.062		.76	3.75		4.51	6.60
8504	3/8" iron pipe size	G	130	.062		.82	3.75		4.57	6.65

143

22 07 19.10 Piping Insulation

			Crew	Daily Output	Labor-Hours	Unit	Material	2021 Bare Costs Labor	Equipment	Total	Total Incl O&P
8506	1/2" iron pipe size	G	1 Asbe	128	.063	L.F.	.88	3.81		4.69	6.80
8508	3/4" iron pipe size	G		126	.063		1.03	3.87		4.90	7.05
8510	1" iron pipe size	G		123	.065		1.18	3.96		5.14	7.35
8512	1-1/4" iron pipe size	G		121	.066		1.39	4.03		5.42	7.70
8514	1-1/2" iron pipe size	G		119	.067		1.67	4.10		5.77	8.10
8516	2" iron pipe size	G		117	.068		2.11	4.17		6.28	8.70
8518	2-1/2" iron pipe size	G		114	.070		2.86	4.28		7.14	9.70
8520	3" iron pipe size	G		112	.071		3.72	4.35		8.07	10.75
8522	4" iron pipe size	G		110	.073		5.15	4.43		9.58	12.45
8534	3/4" wall, 1/8" iron pipe size	G		120	.067		1.08	4.06		5.14	7.40
8536	1/4" iron pipe size	G		120	.067		1.14	4.06		5.20	7.45
8538	3/8" iron pipe size	G		117	.068		1.31	4.17		5.48	7.85
8540	1/2" iron pipe size	G		114	.070		1.45	4.28		5.73	8.15
8542	3/4" iron pipe size	G		112	.071		1.57	4.35		5.92	8.40
8544	1" iron pipe size	G		110	.073		2.20	4.43		6.63	9.20
8546	1-1/4" iron pipe size	G		108	.074		2.94	4.51		7.45	10.15
8548	1-1/2" iron pipe size	G		108	.074		3.44	4.51		7.95	10.70
8550	2" iron pipe size	G		106	.075		4.35	4.60		8.95	11.85
8552	2-1/2" iron pipe size	G		104	.077		5.50	4.69		10.19	13.20
8554	3" iron pipe size	G		102	.078		6.85	4.78		11.63	14.80
8556	4" iron pipe size	G		100	.080		9	4.88		13.88	17.35
8570	1" wall, 1/8" iron pipe size	G		110	.073		1.95	4.43		6.38	8.95
8572	1/4" iron pipe size	G		108	.074		2.63	4.51		7.14	9.80
8574	3/8" iron pipe size	G		106	.075		2.04	4.60		6.64	9.30
8576	1/2" iron pipe size	G		104	.077		2.16	4.69		6.85	9.55
8578	3/4" iron pipe size	G		102	.078		2.75	4.78		7.53	10.35
8580	1" iron pipe size	G		100	.080		3.27	4.88		8.15	11.05
8582	1-1/4" iron pipe size	G		97	.082		3.72	5.05		8.77	11.80
8584	1-1/2" iron pipe size	G		97	.082		4.25	5.05		9.30	12.40
8586	2" iron pipe size	G		95	.084		6	5.15		11.15	14.45
8588	2-1/2" iron pipe size	G		93	.086		7.85	5.25		13.10	16.60
8590	3" iron pipe size	G		91	.088		9.55	5.35		14.90	18.70
8606	Contact adhesive (R-320)	G				Qt.	19.25			19.25	21
8608	Contact adhesive (R-320)	G				Gal.	77			77	84.50
8610	NOTE: Preslit/preglued vs. unslit, same price										

22 07 19.30 Piping Insulation Protective Jacketing, PVC

			Crew	Daily Output	Labor-Hours	Unit	Material	2021 Bare Costs Labor	Equipment	Total	Total Incl O&P
0010	**PIPING INSULATION PROTECTIVE JACKETING, PVC**										
0100	PVC, white, 48" lengths cut from roll goods										
0120	20 mil thick										
0140	Size based on OD of insulation										
0150	1-1/2" ID		Q-14	270	.059	L.F.	.36	3.25		3.61	5.35
0152	2" ID			260	.062		.41	3.38		3.79	5.60
0154	2-1/2" ID			250	.064		.55	3.51		4.06	5.95
0156	3" ID			240	.067		.56	3.66		4.22	6.20
0158	3-1/2" ID			230	.070		.74	3.82		4.56	6.65
0160	4" ID			220	.073		.72	3.99		4.71	6.90
0162	4-1/2" ID			210	.076		.82	4.18		5	7.30
0164	5" ID			200	.080		1.02	4.39		5.41	7.80
0166	5-1/2" ID			190	.084		1.11	4.62		5.73	8.25
0168	6" ID			180	.089		1.06	4.88		5.94	8.60
0170	6-1/2" ID			175	.091		1.15	5		6.15	8.90
0172	7" ID			170	.094		1.38	5.15		6.53	9.40

22 07 19.30 Piping Insulation Protective Jacketing, PVC	Crew	Daily Output	Labor-Hours	Unit	Material	2021 Bare Costs Labor	Equipment	Total	Total Incl O&P
0174	7-1/2" ID	Q-14	164	.098	L.F.	1.33	5.35		6.68	9.65
0176	8" ID		161	.099		1.61	5.45		7.06	10.10
0178	8-1/2" ID		158	.101		1.68	5.55		7.23	10.35
0180	9" ID		155	.103		1.78	5.65		7.43	10.60
0182	9-1/2" ID		152	.105		1.66	5.75		7.41	10.70
0184	10" ID		149	.107		1.96	5.90		7.86	11.15
0186	10-1/2" ID		146	.110		2.07	6		8.07	11.50
0188	11" ID		143	.112		2.16	6.15		8.31	11.80
0190	11-1/2" ID		140	.114		1.98	6.25		8.23	11.80
0192	12" ID		137	.117		2.35	6.40		8.75	12.40
0194	12-1/2" ID		134	.119		2.42	6.55		8.97	12.65
0195	13" ID		132	.121		2.54	6.65		9.19	12.95
0196	13-1/2" ID		132	.121		2.62	6.65		9.27	13.05
0198	14" ID		130	.123		2.72	6.75		9.47	13.35
0200	15" ID		128	.125		2.88	6.85		9.73	13.65
0202	16" ID		126	.127		3.09	6.95		10.04	14.05
0204	17" ID		124	.129		3.26	7.10		10.36	14.45
0206	18" ID		122	.131		3.46	7.20		10.66	14.80
0208	19" ID		120	.133		3.65	7.30		10.95	15.20
0210	20" ID		118	.136		3.90	7.45		11.35	15.70
0212	21" ID		116	.138		4.04	7.55		11.59	16
0214	22" ID		114	.140		4.26	7.70		11.96	16.50
0216	23" ID		112	.143		4.45	7.85		12.30	16.90
0218	24" ID		110	.145		4.64	8		12.64	17.30
0220	25" ID		108	.148		4.83	8.15		12.98	17.75
0222	26" ID		106	.151		5.05	8.30		13.35	18.20
0224	27" ID		104	.154		5.20	8.45		13.65	18.65
0226	28" ID		102	.157		5.40	8.60		14	19.10
0228	29" ID		100	.160		5.60	8.80		14.40	19.55
0230	30" ID		98	.163		5.80	8.95		14.75	20
0300	For colors, add				Ea.	10%				
1000	30 mil thick									
1010	Size based on OD of insulation									
1020	2" ID	Q-14	260	.062	L.F.	.67	3.38		4.05	5.90
1022	2-1/2" ID		250	.064		.82	3.51		4.33	6.25
1024	3" ID		240	.067		.96	3.66		4.62	6.65
1026	3-1/2" ID		230	.070		1.11	3.82		4.93	7.05
1028	4" ID		220	.073		1.24	3.99		5.23	7.45
1030	4-1/2" ID		210	.076		1.39	4.18		5.57	7.95
1032	5" ID		200	.080		1.52	4.39		5.91	8.35
1034	5-1/2" ID		190	.084		1.68	4.62		6.30	8.90
1036	6" ID		180	.089		1.80	4.88		6.68	9.45
1038	6-1/2" ID		175	.091		1.97	5		6.97	9.80
1040	7" ID		170	.094		2.10	5.15		7.25	10.20
1042	7-1/2" ID		164	.098		2.26	5.35		7.61	10.70
1044	8" ID		161	.099		2.38	5.45		7.83	10.95
1046	8-1/2" ID		158	.101		2.54	5.55		8.09	11.30
1048	9" ID		155	.103		2.67	5.65		8.32	11.60
1050	9-1/2" ID		152	.105		2.83	5.75		8.58	11.95
1052	10" ID		149	.107		2.97	5.90		8.87	12.25
1054	10-1/2" ID		146	.110		3.11	6		9.11	12.60
1056	11" ID		143	.112		3.27	6.15		9.42	13
1058	11-1/2" ID		140	.114		3.40	6.25		9.65	13.35

For customer support on your Plumbing Costs with RSMeans Data, call 800.448.8182.

145

22 07 19.30 Piping Insulation Protective Jacketing, PVC		Crew	Daily Output	Labor-Hours	Unit	Material	2021 Bare Costs Labor	Equipment	Total	Total Incl O&P
1060	12" ID	Q-14	137	.117	L.F.	3.54	6.40		9.94	13.70
1062	12-1/2" ID		134	.119		3.69	6.55		10.24	14.05
1063	13" ID		132	.121		3.82	6.65		10.47	14.35
1064	13-1/2" ID		132	.121		3.99	6.65		10.64	14.55
1066	14" ID		130	.123		4.10	6.75		10.85	14.85
1068	15" ID		128	.125		4.39	6.85		11.24	15.35
1070	16" ID		126	.127		4.70	6.95		11.65	15.80
1072	17" ID		124	.129		4.99	7.10		12.09	16.35
1074	18" ID		122	.131		5.25	7.20		12.45	16.80
1076	19" ID		120	.133		5.55	7.30		12.85	17.30
1078	20" ID		118	.136		5.80	7.45		13.25	17.80
1080	21" ID		116	.138		6.10	7.55		13.65	18.25
1082	22" ID		114	.140		6.40	7.70		14.10	18.80
1084	23" ID		112	.143		6.70	7.85		14.55	19.40
1086	24" ID		110	.145		7	8		15	19.90
1088	25" ID		108	.148		7.30	8.15		15.45	20.50
1090	26" ID		106	.151		7.60	8.30		15.90	21
1092	27" ID		104	.154		7.85	8.45		16.30	21.50
1094	28" ID		102	.157		8.15	8.60		16.75	22
1096	29" ID		100	.160		8.45	8.80		17.25	22.50
1098	30" ID		98	.163		8.75	8.95		17.70	23.50
1300	For colors, add				Ea.	10%				
2000	PVC, white, fitting covers									
2020	Fiberglass insulation inserts included with sizes 1-3/4" thru 9-3/4"									
2030	Size is based on OD of insulation									
2040	90° elbow fitting									
2060	1-3/4"	Q-14	135	.119	Ea.	.55	6.50		7.05	10.55
2062	2"		130	.123		.68	6.75		7.43	11.10
2064	2-1/4"		128	.125		.79	6.85		7.64	11.35
2068	2-1/2"		126	.127		.84	6.95		7.79	11.55
2070	2-3/4"		123	.130		.99	7.15		8.14	12
2072	3"		120	.133		.96	7.30		8.26	12.25
2074	3-3/8"		116	.138		1.13	7.55		8.68	12.80
2076	3-3/4"		113	.142		1.21	7.75		8.96	13.25
2078	4-1/8"		110	.145		1.60	8		9.60	13.95
2080	4-3/4"		105	.152		1.93	8.35		10.28	14.90
2082	5-1/4"		100	.160		2.25	8.80		11.05	15.90
2084	5-3/4"		95	.168		2.84	9.25		12.09	17.25
2086	6-1/4"		90	.178		4.73	9.75		14.48	20
2088	6-3/4"		87	.184		4.99	10.10		15.09	21
2090	7-1/4"		85	.188		6.25	10.30		16.55	22.50
2092	7-3/4"		83	.193		6.60	10.55		17.15	23.50
2094	8-3/4"		80	.200		8.45	10.95		19.40	26
2096	9-3/4"		77	.208		11.25	11.40		22.65	30
2098	10-7/8"		74	.216		12.60	11.85		24.45	32
2100	11-7/8"		71	.225		14.40	12.35		26.75	35
2102	12-7/8"		68	.235		19.95	12.90		32.85	42
2104	14-1/8"		66	.242		21	13.30		34.30	43.50
2106	15-1/8"		64	.250		22.50	13.70		36.20	45.50
2108	16-1/8"		63	.254		24.50	13.95		38.45	48.50
2110	17-1/8"		62	.258		27	14.15		41.15	51.50
2112	18-1/8"		61	.262		36.50	14.40		50.90	62
2114	19-1/8"		60	.267		47.50	14.65		62.15	74.50

22 07 19.30 Piping Insulation Protective Jacketing, PVC		Crew	Daily Output	Labor-Hours	Unit	Material	2021 Bare Costs Labor	Equipment	Total	Total Incl O&P
2116	20-1/8"	Q-14	59	.271	Ea.	61	14.85		75.85	90
2200	45° elbow fitting									
2220	1-3/4" thru 9-3/4" same price as 90° elbow fitting									
2320	10-7/8"	Q-14	74	.216	Ea.	12.55	11.85		24.40	32
2322	11-7/8"		71	.225		13.85	12.35		26.20	34
2324	12-7/8"		68	.235		15.45	12.90		28.35	37
2326	14-1/8"		66	.242		17.65	13.30		30.95	40
2328	15-1/8"		64	.250		18.95	13.70		32.65	42
2330	16-1/8"		63	.254		21.50	13.95		35.45	45.50
2332	17-1/8"		62	.258		24.50	14.15		38.65	48.50
2334	18-1/8"		61	.262		30	14.40		44.40	55
2336	19-1/8"		60	.267		40.50	14.65		55.15	67
2338	20-1/8"		59	.271		46	14.85		60.85	73.50
2400	Tee fitting									
2410	1-3/4"	Q-14	96	.167	Ea.	1.04	9.15		10.19	15.15
2412	2"		94	.170		1.18	9.35		10.53	15.60
2414	2-1/4"		91	.176		1.27	9.65		10.92	16.15
2416	2-1/2"		88	.182		1.39	9.95		11.34	16.80
2418	2-3/4"		85	.188		1.53	10.30		11.83	17.50
2420	3"		82	.195		1.66	10.70		12.36	18.20
2422	3-3/8"		79	.203		1.91	11.10		13.01	19.10
2424	3-3/4"		76	.211		2.37	11.55		13.92	20.50
2426	4-1/8"		73	.219		2.57	12		14.57	21
2428	4-3/4"		70	.229		3.19	12.55		15.74	22.50
2430	5-1/4"		67	.239		3.84	13.10		16.94	24
2432	5-3/4"		63	.254		5.10	13.95		19.05	27
2434	6-1/4"		60	.267		6.70	14.65		21.35	30
2436	6-3/4"		59	.271		8.30	14.85		23.15	32
2438	7-1/4"		57	.281		13.40	15.40		28.80	38.50
2440	7-3/4"		54	.296		14.70	16.25		30.95	41
2442	8-3/4"		52	.308		17.85	16.90		34.75	45.50
2444	9-3/4"		50	.320		21	17.55		38.55	50
2446	10-7/8"		48	.333		21.50	18.30		39.80	51.50
2448	11-7/8"		47	.340		23.50	18.65		42.15	54.50
2450	12-7/8"		46	.348		26	19.10		45.10	57.50
2452	14-1/8"		45	.356		28.50	19.50		48	61.50
2454	15-1/8"		44	.364		31	19.95		50.95	64.50
2456	16-1/8"		43	.372		33	20.50		53.50	67.50
2458	17-1/8"		42	.381		35.50	21		56.50	71
2460	18-1/8"		41	.390		39	21.50		60.50	75.50
2462	19-1/8"		40	.400		42.50	22		64.50	80.50
2464	20-1/8"		39	.410		47	22.50		69.50	86.50
4000	Mechanical grooved fitting cover, including insert									
4020	90° elbow fitting									
4030	3/4" & 1"	Q-14	140	.114	Ea.	5.35	6.25		11.60	15.50
4040	1-1/4" & 1-1/2"		135	.119		6.75	6.50		13.25	17.40
4042	2"		130	.123		8.60	6.75		15.35	19.85
4044	2-1/2"		125	.128		9.60	7		16.60	21.50
4046	3"		120	.133		10.75	7.30		18.05	23
4048	3-1/2"		115	.139		12.45	7.65		20.10	25.50
4050	4"		110	.145		13.85	8		21.85	27.50
4052	5"		100	.160		17.20	8.80		26	32.50
4054	6"		90	.178		25.50	9.75		35.25	43

For customer support on your Plumbing Costs with RSMeans Data, call 800.448.8182.

147

22 07 19.30 Piping Insulation Protective Jacketing, PVC

		Crew	Daily Output	Labor-Hours	Unit	Material	2021 Bare Costs Labor	Equipment	Total	Total Incl O&P
4056	8"	Q-14	80	.200	Ea.	27.50	10.95		38.45	47.50
4058	10"		75	.213		35.50	11.70		47.20	57
4060	12"		68	.235		51	12.90		63.90	76
4062	14"		65	.246		75.50	13.50		89	104
4064	16"		63	.254		103	13.95		116.95	135
4066	18"		61	.262		141	14.40		155.40	177
4100	45° elbow fitting									
4120	3/4" & 1"	Q-14	140	.114	Ea.	4.83	6.25		11.08	14.90
4130	1-1/4" & 1-1/2"		135	.119		6.15	6.50		12.65	16.70
4140	2"		130	.123		8.40	6.75		15.15	19.60
4142	2-1/2"		125	.128		9.35	7		16.35	21
4144	3"		120	.133		9.60	7.30		16.90	22
4146	3-1/2"		115	.139		12.85	7.65		20.50	26
4148	4"		110	.145		15.80	8		23.80	29.50
4150	5"		100	.160		16.75	8.80		25.55	32
4152	6"		90	.178		23.50	9.75		33.25	40.50
4154	8"		80	.200		24.50	10.95		35.45	44
4156	10"		75	.213		31.50	11.70		43.20	53
4158	12"		68	.235		47.50	12.90		60.40	72
4160	14"		65	.246		65	13.50		78.50	92
4162	16"		63	.254		78.50	13.95		92.45	108
4164	18"		61	.262		125	14.40		139.40	159
4200	Tee fitting									
4220	3/4" & 1"	Q-14	93	.172	Ea.	7.05	9.45		16.50	22
4230	1-1/4" & 1-1/2"		90	.178		8.95	9.75		18.70	25
4240	2"		87	.184		14.40	10.10		24.50	31.50
4242	2-1/2"		84	.190		16	10.45		26.45	33.50
4244	3"		80	.200		17.15	10.95		28.10	35.50
4246	3-1/2"		77	.208		18.40	11.40		29.80	38
4248	4"		73	.219		23	12		35	43.50
4250	5"		67	.239		28.50	13.10		41.60	51.50
4252	6"		60	.267		37.50	14.65		52.15	64
4254	8"		54	.296		41	16.25		57.25	70
4256	10"		50	.320		46.50	17.55		64.05	78
4258	12"		46	.348		71	19.10		90.10	107
4260	14"		43	.372		103	20.50		123.50	144
4262	16"		42	.381		105	21		126	148
4264	18"		41	.390		156	21.50		177.50	205

22 07 19.40 Pipe Insulation Protective Jacketing, Aluminum

		Crew	Daily Output	Labor-Hours	Unit	Material	2021 Bare Costs Labor	Equipment	Total	Total Incl O&P
0010	**PIPE INSULATION PROTECTIVE JACKETING, ALUMINUM**									
0100	Metal roll jacketing									
0120	Aluminum with polykraft moisture barrier									
0140	Smooth, based on OD of insulation, .016" thick									
0180	1/2" ID	Q-14	220	.073	L.F.	.29	3.99		4.28	6.40
0190	3/4" ID		215	.074		.37	4.08		4.45	6.65
0200	1" ID		210	.076		.45	4.18		4.63	6.90
0210	1-1/4" ID		205	.078		.54	4.28		4.82	7.15
0220	1-1/2" ID		202	.079		.62	4.34		4.96	7.35
0230	1-3/4" ID		199	.080		.79	4.41		5.20	7.60
0240	2" ID		195	.082		.88	4.50		5.38	7.85
0250	2-1/4" ID		191	.084		.98	4.59		5.57	8.15
0260	2-1/2" ID		187	.086		.97	4.69		5.66	8.25

For customer support on your Plumbing Costs with RSMeans Data, call 800.448.8182.

22 07 Plumbing Insulation

22 07 19 – Plumbing Piping Insulation

22 07 19.40 Pipe Insulation Protective Jacketing, Aluminum		Crew	Daily Output	Labor-Hours	Unit	Material	2021 Bare Costs Labor	Equipment	Total	Total Incl O&P
0270	2-3/4" ID	Q-14	184	.087	L.F.	1.06	4.77		5.83	8.45
0280	3" ID		180	.089		1.27	4.88		6.15	8.85
0290	3-1/4" ID		176	.091		1.37	4.99		6.36	9.15
0300	3-1/2" ID		172	.093		1.33	5.10		6.43	9.25
0310	3-3/4" ID		169	.095		1.41	5.20		6.61	9.50
0320	4" ID		165	.097		1.50	5.30		6.80	9.80
0330	4-1/4" ID		161	.099		1.74	5.45		7.19	10.25
0340	4-1/2" ID		157	.102		1.68	5.60		7.28	10.40
0350	4-3/4" ID		154	.104		1.93	5.70		7.63	10.80
0360	5" ID		150	.107		2.03	5.85		7.88	11.20
0370	5-1/4" ID		146	.110		2.12	6		8.12	11.55
0380	5-1/2" ID		143	.112		2.02	6.15		8.17	11.60
0390	5-3/4" ID		139	.115		2.33	6.30		8.63	12.20
0400	6" ID		135	.119		2.42	6.50		8.92	12.60
0410	6-1/4" ID		133	.120		2.51	6.60		9.11	12.85
0420	6-1/2" ID		131	.122		2.37	6.70		9.07	12.85
0430	7" ID		128	.125		2.52	6.85		9.37	13.25
0440	7-1/4" ID		125	.128		2.61	7		9.61	13.60
0450	7-1/2" ID		123	.130		2.69	7.15		9.84	13.85
0460	8" ID		121	.132		2.86	7.25		10.11	14.25
0470	8-1/2" ID		119	.134		3.03	7.35		10.38	14.65
0480	9" ID		116	.138		3.20	7.55		10.75	15.05
0490	9-1/2" ID		114	.140		3.38	7.70		11.08	15.50
0500	10" ID		112	.143		3.55	7.85		11.40	15.90
0510	10-1/2" ID		110	.145		4.09	8		12.09	16.70
0520	11" ID		107	.150		3.88	8.20		12.08	16.80
0530	11-1/2" ID		105	.152		4.06	8.35		12.41	17.25
0540	12" ID		103	.155		4.24	8.50		12.74	17.70
0550	12-1/2" ID		100	.160		4.41	8.80		13.21	18.25
0560	13" ID		99	.162		4.58	8.85		13.43	18.60
0570	14" ID		98	.163		4.93	8.95		13.88	19.10
0580	15" ID		96	.167		5.25	9.15		14.40	19.80
0590	16" ID		95	.168		5.60	9.25		14.85	20.50
0600	17" ID		93	.172		5.95	9.45		15.40	21
0610	18" ID		92	.174		6.20	9.55		15.75	21.50
0620	19" ID		90	.178		6.65	9.75		16.40	22
0630	20" ID		89	.180		7	9.85		16.85	23
0640	21" ID		87	.184		7.35	10.10		17.45	23.50
0650	22" ID		86	.186		7.65	10.20		17.85	24
0660	23" ID		84	.190		8	10.45		18.45	25
0670	24" ID		83	.193		8.35	10.55		18.90	25.50
0710	For smooth .020" thick, add					27%	10%			
0720	For smooth .024" thick, add					52%	20%			
0730	For smooth .032" thick, add					104%	33%			
0800	For stucco embossed, add					1%				
0820	For corrugated, add					2.50%				
0900	White aluminum with polysurlyn moisture barrier									
0910	Smooth, % is an add to polykraft lines of same thickness									
0940	For smooth .016" thick, add				L.F.	35%				
0960	For smooth .024" thick, add				"	22%				
1000	Aluminum fitting covers									
1010	Size is based on OD of insulation									
1020	90° LR elbow, 2 piece									

For customer support on your Plumbing Costs with RSMeans Data, call 800.448.8182.

149

22 07 Plumbing Insulation

22 07 19 – Plumbing Piping Insulation

22 07 19.40 Pipe Insulation Protective Jacketing, Aluminum		Crew	Daily Output	Labor-Hours	Unit	Material	2021 Bare Costs Labor	Equipment	Total	Total Incl O&P
1100	1-1/2"	Q-14	140	.114	Ea.	4.42	6.25		10.67	14.45
1110	1-3/4"		135	.119		4.42	6.50		10.92	14.80
1120	2"		130	.123		5.55	6.75		12.30	16.45
1130	2-1/4"		128	.125		5.55	6.85		12.40	16.60
1140	2-1/2"		126	.127		5.55	6.95		12.50	16.75
1150	2-3/4"		123	.130		5.55	7.15		12.70	17
1160	3"		120	.133		6.05	7.30		13.35	17.85
1170	3-1/4"		117	.137		6.05	7.50		13.55	18.10
1180	3-1/2"		115	.139		7.15	7.65		14.80	19.50
1190	3-3/4"		113	.142		7.30	7.75		15.05	19.95
1200	4"		110	.145		7.65	8		15.65	20.50
1210	4-1/4"		108	.148		7.85	8.15		16	21
1220	4-1/2"		106	.151		7.85	8.30		16.15	21.50
1230	4-3/4"		104	.154		7.85	8.45		16.30	21.50
1240	5"		102	.157		8.85	8.60		17.45	23
1250	5-1/4"		100	.160		10.25	8.80		19.05	24.50
1260	5-1/2"		97	.165		12.75	9.05		21.80	28
1270	5-3/4"		95	.168		12.75	9.25		22	28
1280	6"		92	.174		10.85	9.55		20.40	26.50
1290	6-1/4"		90	.178		10.85	9.75		20.60	27
1300	6-1/2"		87	.184		13.10	10.10		23.20	30
1310	7"		85	.188		15.90	10.30		26.20	33.50
1320	7-1/4"		84	.190		15.90	10.45		26.35	33.50
1330	7-1/2"		83	.193		23.50	10.55		34.05	41.50
1340	8"		82	.195		17.40	10.70		28.10	35.50
1350	8-1/2"		80	.200		32	10.95		42.95	52.50
1360	9"		78	.205		32	11.25		43.25	52.50
1370	9-1/2"		77	.208		23.50	11.40		34.90	43.50
1380	10"		76	.211		23.50	11.55		35.05	43.50
1390	10-1/2"		75	.213		24	11.70		35.70	44.50
1400	11"		74	.216		24	11.85		35.85	44.50
1410	11-1/2"		72	.222		27.50	12.20		39.70	49
1420	12"		71	.225		27.50	12.35		39.85	49.50
1430	12-1/2"		69	.232		49	12.70		61.70	73.50
1440	13"		68	.235		49	12.90		61.90	74
1450	14"		66	.242		65.50	13.30		78.80	92.50
1460	15"		64	.250		68.50	13.70		82.20	96.50
1470	16"		63	.254		75	13.95		88.95	104
2000	45° elbow, 2 piece									
2010	2-1/2"	Q-14	126	.127	Ea.	4.59	6.95		11.54	15.70
2020	2-3/4"		123	.130		4.59	7.15		11.74	15.95
2030	3"		120	.133		5.20	7.30		12.50	16.90
2040	3-1/4"		117	.137		5.20	7.50		12.70	17.15
2050	3-1/2"		115	.139		5.90	7.65		13.55	18.15
2060	3-3/4"		113	.142		5.90	7.75		13.65	18.40
2070	4"		110	.145		6	8		14	18.80
2080	4-1/4"		108	.148		6.85	8.15		15	20
2090	4-1/2"		106	.151		6.85	8.30		15.15	20
2100	4-3/4"		104	.154		6.85	8.45		15.30	20.50
2110	5"		102	.157		8.05	8.60		16.65	22
2120	5-1/4"		100	.160		8.05	8.80		16.85	22.50
2130	5-1/2"		97	.165		8.45	9.05		17.50	23
2140	6"		92	.174		8.45	9.55		18	24

150

22 07 Plumbing Insulation

22 07 19 – Plumbing Piping Insulation

22 07 19.40 Pipe Insulation Protective Jacketing, Aluminum		Crew	Daily Output	Labor-Hours	Unit	Material	2021 Bare Costs Labor	Equipment	Total	Total Incl O&P
2150	6-1/2"	Q-14	87	.184	Ea.	11.65	10.10		21.75	28.50
2160	7"		85	.188		11.65	10.30		21.95	28.50
2170	7-1/2"		83	.193		11.80	10.55		22.35	29
2180	8"		82	.195		11.80	10.70		22.50	29.50
2190	8-1/2"		80	.200		14.65	10.95		25.60	33
2200	9"		78	.205		14.65	11.25		25.90	33.50
2210	9-1/2"		77	.208		20.50	11.40		31.90	40
2220	10"		76	.211		20.50	11.55		32.05	40
2230	10-1/2"		75	.213		19.05	11.70		30.75	39
2240	11"		74	.216		19.05	11.85		30.90	39
2250	11-1/2"		72	.222		23	12.20		35.20	43.50
2260	12"		71	.225		23	12.35		35.35	44
2270	13"		68	.235		27	12.90		39.90	49.50
2280	14"		66	.242		33	13.30		46.30	56.50
2290	15"		64	.250		50.50	13.70		64.20	77
2300	16"		63	.254		55	13.95		68.95	82
2310	17"		62	.258		53.50	14.15		67.65	80.50
2320	18"		61	.262		61.50	14.40		75.90	89.50
2330	19"		60	.267		75	14.65		89.65	105
2340	20"		59	.271		72.50	14.85		87.35	103
2350	21"		58	.276		78	15.15		93.15	109
3000	Tee, 4 piece									
3010	2-1/2"	Q-14	88	.182	Ea.	29	9.95		38.95	47.50
3020	2-3/4"		86	.186		29	10.20		39.20	47.50
3030	3"		84	.190		32.50	10.45		42.95	52
3040	3-1/4"		82	.195		32.50	10.70		43.20	52.50
3050	3-1/2"		80	.200		34	10.95		44.95	54.50
3060	4"		78	.205		34.50	11.25		45.75	55
3070	4-1/4"		76	.211		36.50	11.55		48.05	57.50
3080	4-1/2"		74	.216		36.50	11.85		48.35	58
3090	4-3/4"		72	.222		36.50	12.20		48.70	58.50
3100	5"		70	.229		37.50	12.55		50.05	60.50
3110	5-1/4"		68	.235		37.50	12.90		50.40	61.50
3120	5-1/2"		66	.242		39.50	13.30		52.80	63.50
3130	6"		64	.250		39.50	13.70		53.20	64
3140	6-1/2"		60	.267		43	14.65		57.65	70
3150	7"		58	.276		43	15.15		58.15	70.50
3160	7-1/2"		56	.286		48	15.65		63.65	77
3170	8"		54	.296		48	16.25		64.25	78
3180	8-1/2"		52	.308		49.50	16.90		66.40	80
3190	9"		50	.320		49.50	17.55		67.05	81
3200	9-1/2"		49	.327		36.50	17.90		54.40	67.50
3210	10"		48	.333		36.50	18.30		54.80	68
3220	10-1/2"		47	.340		39	18.65		57.65	71.50
3230	11"		46	.348		39	19.10		58.10	72
3240	11-1/2"		45	.356		41	19.50		60.50	75
3250	12"		44	.364		41	19.95		60.95	75.50
3260	13"		43	.372		43	20.50		63.50	78.50
3270	14"		42	.381		45	21		66	81.50
3280	15"		41	.390		49	21.50		70.50	86.50
3290	16"		40	.400		50	22		72	88.50
3300	17"		39	.410		57.50	22.50		80	98
3310	18"		38	.421		60	23		83	102

For customer support on your Plumbing Costs with RSMeans Data, call 800.448.8182.

151

22 07 Plumbing Insulation

22 07 19 – Plumbing Piping Insulation

22 07 19.40 Pipe Insulation Protective Jacketing, Aluminum		Crew	Daily Output	Labor-Hours	Unit	Material	2021 Bare Costs Labor	Equipment	Total	Total Incl O&P
3320	19"	Q-14	37	.432	Ea.	66	23.50		89.50	110
3330	20"		36	.444		67.50	24.50		92	112
3340	22"		35	.457		86	25		111	133
3350	23"		34	.471		88.50	26		114.50	137
3360	24"		31	.516		91	28.50		119.50	144

22 07 19.50 Pipe Insulation Protective Jacketing, St. Stl.

		Crew	Daily Output	Labor-Hours	Unit	Material	2021 Bare Costs Labor	Equipment	Total	Total Incl O&P
0010	**PIPE INSULATION PROTECTIVE JACKETING, STAINLESS STEEL**									
0100	Metal roll jacketing									
0120	Type 304 with moisture barrier									
0140	Smooth, based on OD of insulation, .010" thick									
0260	2-1/2" ID	Q-14	250	.064	L.F.	3.09	3.51		6.60	8.75
0270	2-3/4" ID		245	.065		3.36	3.58		6.94	9.20
0280	3" ID		240	.067		3.63	3.66		7.29	9.60
0290	3-1/4" ID		235	.068		3.95	3.73		7.68	10.05
0300	3-1/2" ID		230	.070		4.18	3.82		8	10.45
0310	3-3/4" ID		225	.071		4.50	3.90		8.40	10.90
0320	4" ID		220	.073		4.74	3.99		8.73	11.30
0330	4-1/4" ID		215	.074		5	4.08		9.08	11.75
0340	4-1/2" ID		210	.076		5.30	4.18		9.48	12.20
0350	5" ID		200	.080		5.85	4.39		10.24	13.10
0360	5-1/2" ID		190	.084		6.40	4.62		11.02	14.05
0370	6" ID		180	.089		6.95	4.88		11.83	15.05
0380	6-1/2" ID		175	.091		7.50	5		12.50	15.90
0390	7" ID		170	.094		8.05	5.15		13.20	16.75
0400	7-1/2" ID		164	.098		8.60	5.35		13.95	17.65
0410	8" ID		161	.099		9.10	5.45		14.55	18.40
0420	8-1/2" ID		158	.101		9.65	5.55		15.20	19.15
0430	9" ID		155	.103		10.25	5.65		15.90	19.90
0440	9-1/2" ID		152	.105		10.75	5.75		16.50	20.50
0450	10" ID		149	.107		11.30	5.90		17.20	21.50
0460	10-1/2" ID		146	.110		11.90	6		17.90	22.50
0470	11" ID		143	.112		12.45	6.15		18.60	23
0480	12" ID		137	.117		13.50	6.40		19.90	24.50
0490	13" ID		132	.121		14.60	6.65		21.25	26.50
0500	14" ID		130	.123		15.70	6.75		22.45	27.50
0700	For smooth .016" thick, add					45%	33%			
1000	Stainless steel, Type 316, fitting covers									
1010	Size is based on OD of insulation									
1020	90° LR elbow, 2 piece									
1100	1-1/2"	Q-14	126	.127	Ea.	12.90	6.95		19.85	25
1110	2-3/4"		123	.130		13.45	7.15		20.60	25.50
1120	3"		120	.133		14.05	7.30		21.35	26.50
1130	3-1/4"		117	.137		14.05	7.50		21.55	27
1140	3-1/2"		115	.139		14.75	7.65		22.40	28
1150	3-3/4"		113	.142		15.60	7.75		23.35	29
1160	4"		110	.145		17.15	8		25.15	31
1170	4-1/4"		108	.148		23	8.15		31.15	37.50
1180	4-1/2"		106	.151		23	8.30		31.30	37.50
1190	5"		102	.157		23.50	8.60		32.10	38.50
1200	5-1/2"		97	.165		35	9.05		44.05	52.50
1210	6"		92	.174		38	9.55		47.55	56.50
1220	6-1/2"		87	.184		52.50	10.10		62.60	73.50

152

22 07 Plumbing Insulation

22 07 19 – Plumbing Piping Insulation

22 07 19.50 Pipe Insulation Protective Jacketing, St. Stl.		Crew	Daily Output	Labor-Hours	Unit	Material	2021 Bare Costs Labor	2021 Bare Costs Equipment	Total	Total Incl O&P
1230	7"	Q-14	85	.188	Ea.	52.50	10.30		62.80	74
1240	7-1/2"		83	.193		61.50	10.55		72.05	83.50
1250	8"		80	.200		61.50	10.95		72.45	84.50
1260	8-1/2"		80	.200		63.50	10.95		74.45	87
1270	9"		78	.205		96.50	11.25		107.75	123
1280	9-1/2"		77	.208		95	11.40		106.40	121
1290	10"		76	.211		95	11.55		106.55	122
1300	10-1/2"		75	.213		113	11.70		124.70	143
1310	11"		74	.216		108	11.85		119.85	137
1320	12"		71	.225		123	12.35		135.35	154
1330	13"		68	.235		170	12.90		182.90	207
1340	14"		66	.242		172	13.30		185.30	210
2000	45° elbow, 2 piece									
2010	2-1/2"	Q-14	126	.127	Ea.	11.20	6.95		18.15	23
2020	2-3/4"		123	.130		11.20	7.15		18.35	23.50
2030	3"		120	.133		12.05	7.30		19.35	24.50
2040	3-1/4"		117	.137		12.05	7.50		19.55	24.50
2050	3-1/2"		115	.139		12.20	7.65		19.85	25
2060	3-3/4"		113	.142		12.20	7.75		19.95	25.50
2070	4"		110	.145		15.70	8		23.70	29.50
2080	4-1/4"		108	.148		22	8.15		30.15	36.50
2090	4-1/2"		106	.151		22	8.30		30.30	36.50
2100	4-3/4"		104	.154		22	8.45		30.45	37
2110	5"		102	.157		22.50	8.60		31.10	37.50
2120	5-1/2"		97	.165		22.50	9.05		31.55	38.50
2130	6"		92	.174		26.50	9.55		36.05	44
2140	6-1/2"		87	.184		26.50	10.10		36.60	45
2150	7"		85	.188		45	10.30		55.30	65.50
2160	7-1/2"		83	.193		46	10.55		56.55	66.50
2170	8"		82	.195		46	10.70		56.70	67
2180	8-1/2"		80	.200		53.50	10.95		64.45	76
2190	9"		78	.205		53.50	11.25		64.75	76
2200	9-1/2"		77	.208		63	11.40		74.40	87
2210	10"		76	.211		63	11.55		74.55	87
2220	10-1/2"		75	.213		76	11.70		87.70	101
2230	11"		74	.216		76	11.85		87.85	102
2240	12"		71	.225		82.50	12.35		94.85	109
2250	13"		68	.235		97	12.90		109.90	127

22 11 Facility Water Distribution

22 11 13 – Facility Water Distribution Piping

22 11 13.14 Pipe, Brass

0010	**PIPE, BRASS**, Plain end									
0900	Field threaded, coupling & clevis hanger assembly 10' OC									
0920	Regular weight									
1120	1/2" diameter	1 Plum	48	.167	L.F.	8.30	11.30		19.60	26
1140	3/4" diameter		46	.174		10.85	11.75		22.60	29.50
1160	1" diameter		43	.186		5.30	12.60		17.90	24.50
1180	1-1/4" diameter	Q-1	72	.222		22	13.55		35.55	44
1200	1-1/2" diameter		65	.246		26.50	15		41.50	51.50
1220	2" diameter		53	.302		17.60	18.40		36	47

For customer support on your Plumbing Costs with RSMeans Data, call 800.448.8182.

153

22 11 13.14 Pipe, Brass

		Crew	Daily Output	Labor-Hours	Unit	Material	2021 Bare Costs Labor	2021 Bare Costs Equipment	Total	Total Incl O&P
1240	2-1/2" diameter	Q-1	41	.390	L.F.	29.50	24		53.50	68
1260	3" diameter	↓	31	.516		41	31.50		72.50	92
1300	4" diameter	Q-2	37	.649	↓	156	41		197	232
1930	To delete coupling & hanger, subtract									
1940	1/2" diam.					40%	46%			
1950	3/4" diam. to 1-1/2" diam.					39%	39%			
1960	2" diam. to 4" diam.					48%	35%			

22 11 13.16 Pipe Fittings, Brass

		Crew	Daily Output	Labor-Hours	Unit	Material	2021 Bare Costs Labor	2021 Bare Costs Equipment	Total	Total Incl O&P
0010	**PIPE FITTINGS, BRASS**, Rough bronze, threaded, lead free.									
1000	Standard wt., 90° elbow									
1040	1/8"	1 Plum	13	.615	Ea.	23.50	41.50		65	88
1060	1/4"		13	.615		23.50	41.50		65	88
1080	3/8"		13	.615		23.50	41.50		65	88
1100	1/2"		12	.667		23.50	45		68.50	93.50
1120	3/4"		11	.727		31.50	49		80.50	108
1140	1"	↓	10	.800		51	54		105	137
1160	1-1/4"	Q-1	17	.941		82.50	57.50		140	177
1180	1-1/2"		16	1		104	61		165	206
1200	2"		14	1.143		165	69.50		234.50	285
1220	2-1/2"		11	1.455		400	88.50		488.50	570
1240	3"	↓	8	2		610	122		732	850
1260	4"	Q-2	11	2.182		1,225	138		1,363	1,550
1280	5"		8	3		2,825	190		3,015	3,375
1300	6"	↓	7	3.429		5,025	217		5,242	5,850
1500	45° elbow, 1/8"	1 Plum	13	.615		29	41.50		70.50	94
1540	1/4"		13	.615		29	41.50		70.50	94
1560	3/8"		13	.615		29	41.50		70.50	94
1580	1/2"		12	.667		29	45		74	99.50
1600	3/4"		11	.727		41	49		90	119
1620	1"	↓	10	.800		70	54		124	158
1640	1-1/4"	Q-1	17	.941		111	57.50		168.50	208
1660	1-1/2"		16	1		140	61		201	245
1680	2"		14	1.143		226	69.50		295.50	355
1700	2-1/2"		11	1.455		430	88.50		518.50	605
1720	3"	↓	8	2		550	122		672	785
1740	4"	Q-2	11	2.182		1,225	138		1,363	1,550
1760	5"		8	3		2,300	190		2,490	2,800
1780	6"	↓	7	3.429		3,175	217		3,392	3,825
2000	Tee, 1/8"	1 Plum	9	.889		27.50	60		87.50	121
2040	1/4"		9	.889		27.50	60		87.50	121
2060	3/8"		9	.889		28.50	60		88.50	121
2080	1/2"		8	1		28	67.50		95.50	132
2100	3/4"		7	1.143		39	77.50		116.50	158
2120	1"	↓	6	1.333		75.50	90.50		166	219
2140	1-1/4"	Q-1	10	1.600		122	97.50		219.50	279
2160	1-1/2"		9	1.778		138	108		246	315
2180	2"		8	2		228	122		350	435
2200	2-1/2"		7	2.286		545	139		684	805
2220	3"	↓	5	3.200		830	195		1,025	1,200
2240	4"	Q-2	7	3.429		2,050	217		2,267	2,600
2260	5"		5	4.800		3,375	305		3,680	4,175
2280	6"	↓	4	6		5,550	380		5,930	6,700

22 11 13 – Facility Water Distribution Piping

22 11 13.16 Pipe Fittings, Brass

		Crew	Daily Output	Labor-Hours	Unit	Material	2021 Bare Costs Labor	Equipment	Total	Total Incl O&P
2500	Coupling, 1/8"	1 Plum	26	.308	Ea.	19.65	21		40.65	52.50
2540	1/4"		22	.364		19.65	24.50		44.15	58.50
2560	3/8"		18	.444		19.65	30		49.65	66.50
2580	1/2"		15	.533		19.65	36		55.65	75.50
2600	3/4"		14	.571		27.50	38.50		66	88
2620	1"		13	.615		47	41.50		88.50	114
2640	1-1/4"	Q-1	22	.727		78.50	44.50		123	153
2660	1-1/2"		20	.800		102	48.50		150.50	185
2680	2"		18	.889		169	54		223	267
2700	2-1/2"		14	1.143		288	69.50		357.50	420
2720	3"		10	1.600		400	97.50		497.50	585
2740	4"	Q-2	12	2		820	126		946	1,100
2760	5"		10	2.400		1,575	152		1,727	1,950
2780	6"		9	2.667		2,275	168		2,443	2,750
3000	Union, 125 lb.									
3020	1/8"	1 Plum	12	.667	Ea.	40.50	45		85.50	112
3040	1/4"		12	.667		43.50	45		88.50	116
3060	3/8"		12	.667		43.50	45		88.50	116
3080	1/2"		11	.727		43.50	49		92.50	122
3100	3/4"		10	.800		60	54		114	147
3120	1"		9	.889		90.50	60		150.50	190
3140	1-1/4"	Q-1	16	1		131	61		192	235
3160	1-1/2"		15	1.067		156	65		221	268
3180	2"		13	1.231		210	75		285	345
3200	2-1/2"		10	1.600		735	97.50		832.50	950
3220	3"		7	2.286		980	139		1,119	1,275
3240	4"	Q-2	10	2.400		2,725	152		2,877	3,225

22 11 13.23 Pipe/Tube, Copper

		Crew	Daily Output	Labor-Hours	Unit	Material	2021 Bare Costs Labor	Equipment	Total	Total Incl O&P
0010	**PIPE/TUBE, COPPER**, Solder joints	R221113-50								
0100	Solder									
0120	Solder, lead free, roll				Lb.	16.80			16.80	18.50
1000	Type K tubing, couplings & clevis hanger assemblies 10' OC									
1100	1/4" diameter	R221113-70	1 Plum	84	.095	L.F.	5	6.45		11.45	15.10
1120	3/8" diameter		82	.098		5.05	6.60		11.65	15.40
1140	1/2" diameter		78	.103		5.70	6.95		12.65	16.60
1160	5/8" diameter		77	.104		4.02	7.05		11.07	14.90
1180	3/4" diameter		74	.108		9.45	7.30		16.75	21.50
1200	1" diameter		66	.121		12.20	8.20		20.40	25.50
1220	1-1/4" diameter		56	.143		15.30	9.65		24.95	31.50
1240	1-1/2" diameter		50	.160		19.15	10.85		30	37
1260	2" diameter		40	.200		29.50	13.55		43.05	52.50
1280	2-1/2" diameter	Q-1	60	.267		46	16.25		62.25	75.50
1300	3" diameter		54	.296		63.50	18.05		81.55	96.50
1320	3-1/2" diameter		42	.381		86.50	23		109.50	130
1330	4" diameter		38	.421		109	25.50		134.50	159
1340	5" diameter		32	.500		87.50	30.50		118	142
1360	6" diameter	Q-2	38	.632		136	40		176	210
1380	8" diameter	"	34	.706		200	44.50		244.50	287
1390	For other than full hard temper, add					13%				
1440	For silver solder, add						15%			
1800	For medical clean (oxygen class), add					12%				
1950	To delete cplgs. & hngrs., 1/4"-1" pipe, subtract					27%	60%			

22 11 13.23 Pipe/Tube, Copper		Crew	Daily Output	Labor-Hours	Unit	Material	2021 Bare Costs Labor	Equipment	Total	Total Incl O&P
1960	1-1/4"-3" pipe, subtract					14%	52%			
1970	3-1/2"-5" pipe, subtract					10%	60%			
1980	6"-8" pipe, subtract					19%	53%			
2000	Type L tubing, couplings & clevis hanger assemblies 10' OC									
2100	1/4" diameter	1 Plum	88	.091	L.F.	1.83	6.15		7.98	11.20
2120	3/8" diameter		84	.095		2.13	6.45		8.58	11.95
2140	1/2" diameter		81	.099		2.51	6.70		9.21	12.75
2160	5/8" diameter		79	.101		6.05	6.85		12.90	16.90
2180	3/4" diameter		76	.105		3.04	7.15		10.19	14
2200	1" diameter		68	.118		3.55	7.95		11.50	15.80
2220	1-1/4" diameter		58	.138		10.70	9.35		20.05	26
2240	1-1/2" diameter		52	.154		6.35	10.40		16.75	22.50
2260	2" diameter		42	.190		11.50	12.90		24.40	32
2280	2-1/2" diameter	Q-1	62	.258		15.85	15.70		31.55	41
2300	3" diameter		56	.286		46	17.40		63.40	77
2320	3-1/2" diameter		43	.372		64	22.50		86.50	105
2340	4" diameter		39	.410		41	25		66	82.50
2360	5" diameter		34	.471		178	28.50		206.50	239
2380	6" diameter	Q-2	40	.600		114	38		152	182
2400	8" diameter	"	36	.667		161	42		203	240
2410	For other than full hard temper, add					21%				
2590	For silver solder, add						15%			
2900	For medical clean (oxygen class), add					12%				
2940	To delete cplgs. & hngrs., 1/4"-1" pipe, subtract					37%	63%			
2960	1-1/4"-3" pipe, subtract					12%	53%			
2970	3-1/2"-5" pipe, subtract					12%	63%			
2980	6"-8" pipe, subtract					24%	55%			
3000	Type M tubing, couplings & clevis hanger assemblies 10' OC									
3100	1/4" diameter	1 Plum	90	.089	L.F.	3.88	6		9.88	13.25
3120	3/8" diameter		87	.092		4.29	6.25		10.54	14
3140	1/2" diameter		84	.095		3.75	6.45		10.20	13.75
3160	5/8" diameter		81	.099		2.65	6.70		9.35	12.90
3180	3/4" diameter		78	.103		5.40	6.95		12.35	16.30
3200	1" diameter		70	.114		7.35	7.75		15.10	19.65
3220	1-1/4" diameter		60	.133		11.90	9.05		20.95	26.50
3240	1-1/2" diameter		54	.148		14.50	10.05		24.55	31
3260	2" diameter		44	.182		23	12.30		35.30	43.50
3280	2-1/2" diameter	Q-1	64	.250		34.50	15.25		49.75	60.50
3300	3" diameter		58	.276		45.50	16.80		62.30	75
3320	3-1/2" diameter		45	.356		64.50	21.50		86	104
3340	4" diameter		40	.400		84	24.50		108.50	129
3360	5" diameter		36	.444		168	27		195	225
3370	6" diameter	Q-2	42	.571		225	36		261	300
3380	8" diameter	"	38	.632		390	40		430	490
3440	For silver solder, add						15%			
3960	To delete cplgs. & hngrs., 1/4"-1" pipe, subtract					35%	65%			
3970	1-1/4"-3" pipe, subtract					19%	56%			
3980	3-1/2"-5" pipe, subtract					13%	65%			
3990	6"-8" pipe, subtract					28%	58%			
4000	Type DWV tubing, couplings & clevis hanger assemblies 10' OC									
4100	1-1/4" diameter	1 Plum	60	.133	L.F.	10.90	9.05		19.95	25.50
4120	1-1/2" diameter		54	.148		6.65	10.05		16.70	22.50
4140	2" diameter		44	.182		9.25	12.30		21.55	28.50

156

22 11 Facility Water Distribution

22 11 13 – Facility Water Distribution Piping

22 11 13.23 Pipe/Tube, Copper		Crew	Daily Output	Labor-Hours	Unit	Material	2021 Bare Costs Labor	Equipment	Total	Total Incl O&P
4160	3" diameter	Q-1	58	.276	L.F.	18.95	16.80		35.75	46
4180	4" diameter		40	.400		68	24.50		92.50	111
4200	5" diameter		36	.444		136	27		163	191
4220	6" diameter	Q-2	42	.571		203	36		239	277
4240	8" diameter	"	38	.632		640	40		680	765
4730	To delete cplgs. & hngrs., 1-1/4"-2" pipe, subtract					16%	53%			
4740	3"-4" pipe, subtract					13%	60%			
4750	5"-8" pipe, subtract					23%	58%			
5200	ACR tubing, type L, hard temper, cleaned and									
5220	capped, no couplings or hangers									
5240	3/8" OD				L.F.	.49			.49	.54
5250	1/2" OD					.75			.75	.83
5260	5/8" OD					.95			.95	1.05
5270	3/4" OD					1.29			1.29	1.42
5280	7/8" OD					1.52			1.52	1.67
5290	1-1/8" OD					2.19			2.19	2.41
5300	1-3/8" OD					2.92			2.92	3.21
5310	1-5/8" OD					3.80			3.80	4.18
5320	2-1/8" OD					5.80			5.80	6.35
5330	2-5/8" OD					8.35			8.35	9.20
5340	3-1/8" OD					11.10			11.10	12.25
5350	3-5/8" OD					72			72	79
5360	4-1/8" OD					20.50			20.50	22.50
5380	ACR tubing, type L, hard, cleaned and capped									
5381	No couplings or hangers									
5384	3/8"	1 Stpi	160	.050	L.F.	.49	3.42		3.91	5.65
5385	1/2"		160	.050		.75	3.42		4.17	5.95
5386	5/8"		160	.050		.95	3.42		4.37	6.15
5387	3/4"		130	.062		1.29	4.21		5.50	7.70
5388	7/8"		130	.062		1.52	4.21		5.73	7.95
5389	1-1/8"		115	.070		2.19	4.75		6.94	9.50
5390	1-3/8"		100	.080		2.92	5.45		8.37	11.35
5391	1-5/8"		90	.089		3.80	6.10		9.90	13.25
5392	2-1/8"		80	.100		5.80	6.85		12.65	16.55
5393	2-5/8"	Q-5	125	.128		8.35	7.90		16.25	21
5394	3-1/8"		105	.152		11.10	9.40		20.50	26.50
5395	4-1/8"		95	.168		20.50	10.35		30.85	38
5800	Refrigeration tubing, dryseal, 50' coils									
5840	1/8" OD				Coil	11.40			11.40	12.55
5850	3/16" OD					52.50			52.50	57.50
5860	1/4" OD					15.35			15.35	16.85
5870	5/16" OD					81			81	89
5880	3/8" OD					21.50			21.50	24
5890	1/2" OD					31			31	34
5900	5/8" OD					41.50			41.50	45.50
5910	3/4" OD					49			49	54
5920	7/8" OD					73			73	80.50
5930	1-1/8" OD					107			107	117
5940	1-3/8" OD					185			185	203
5950	1-5/8" OD					234			234	258

For customer support on your Plumbing Costs with RSMeans Data, call 800.448.8182.

157

22 11 13.25 Pipe/Tube Fittings, Copper	Crew	Daily Output	Labor-Hours	Unit	Material	2021 Bare Costs Labor	Equipment	Total	Total Incl O&P
0010 **PIPE/TUBE FITTINGS, COPPER**, Wrought unless otherwise noted									
0020 For silver solder, add						15%			
0040 Solder joints, copper x copper									
0070 90° elbow, 1/4"	1 Plum	22	.364	Ea.	5.05	24.50		29.55	42.50
0090 3/8"		22	.364		4.89	24.50		29.39	42.50
0100 1/2"		20	.400		1.77	27		28.77	42.50
0110 5/8"		19	.421		3.39	28.50		31.89	46
0120 3/4"		19	.421		3.89	28.50		32.39	47
0130 1"		16	.500		9.55	34		43.55	61
0140 1-1/4"		15	.533		15.80	36		51.80	71.50
0150 1-1/2"		13	.615		22	41.50		63.50	86.50
0160 2"		11	.727		40	49		89	118
0170 2-1/2"	Q-1	13	1.231		73.50	75		148.50	193
0180 3"		11	1.455		104	88.50		192.50	247
0190 3-1/2"		10	1.600		320	97.50		417.50	500
0200 4"		9	1.778		228	108		336	415
0210 5"		6	2.667		895	162		1,057	1,225
0220 6"	Q-2	9	2.667		1,200	168		1,368	1,575
0230 8"	"	8	3		4,425	190		4,615	5,125
0250 45° elbow, 1/4"	1 Plum	22	.364		9	24.50		33.50	47
0270 3/8"		22	.364		7.70	24.50		32.20	45.50
0280 1/2"		20	.400		3.18	27		30.18	44
0290 5/8"		19	.421		13.50	28.50		42	57.50
0300 3/4"		19	.421		5.40	28.50		33.90	48.50
0310 1"		16	.500		13.55	34		47.55	65.50
0320 1-1/4"		15	.533		18.20	36		54.20	74
0330 1-1/2"		13	.615		22.50	41.50		64	86.50
0340 2"		11	.727		37.50	49		86.50	115
0350 2-1/2"	Q-1	13	1.231		72.50	75		147.50	192
0360 3"		13	1.231		101	75		176	223
0370 3-1/2"		10	1.600		174	97.50		271.50	335
0380 4"		9	1.778		209	108		317	390
0390 5"		6	2.667		755	162		917	1,075
0400 6"	Q-2	9	2.667		1,200	168		1,368	1,550
0410 8"	"	8	3		5,225	190		5,415	6,000
0450 Tee, 1/4"	1 Plum	14	.571		10.10	38.50		48.60	68.50
0470 3/8"		14	.571		8.10	38.50		46.60	66.50
0480 1/2"		13	.615		2.96	41.50		44.46	65.50
0490 5/8"		12	.667		18.50	45		63.50	88
0500 3/4"		12	.667		7.15	45		52.15	75.50
0510 1"		10	.800		11.50	54		65.50	93.50
0520 1-1/4"		9	.889		29.50	60		89.50	123
0530 1-1/2"		8	1		44.50	67.50		112	150
0540 2"		7	1.143		71	77.50		148.50	193
0550 2-1/2"	Q-1	8	2		131	122		253	325
0560 3"		7	2.286		188	139		327	415
0570 3-1/2"		6	2.667		545	162		707	840
0580 4"		5	3.200		440	195		635	775
0590 5"		4	4		1,350	244		1,594	1,850
0600 6"	Q-2	6	4		1,850	253		2,103	2,400
0610 8"	"	5	4.800		7,100	305		7,405	8,250
0612 Tee, reducing on the outlet, 1/4"	1 Plum	15	.533		27.50	36		63.50	84.50

158

For customer support on your Plumbing Costs with RSMeans Data, call 800.448.8182.

22 11 13.25 Pipe/Tube Fittings, Copper		Crew	Daily Output	Labor-Hours	Unit	Material	2021 Bare Costs Labor	Equipment	Total	Total Incl O&P
0613	3/8"	1 Plum	15	.533	Ea.	23.50	36		59.50	80
0614	1/2"		14	.571		21.50	38.50		60	81.50
0615	5/8"		13	.615		43.50	41.50		85	110
0616	3/4"		12	.667		9.95	45		54.95	78.50
0617	1"		11	.727		47.50	49		96.50	126
0618	1-1/4"		10	.800		46.50	54		100.50	133
0619	1-1/2"		9	.889		49.50	60		109.50	145
0620	2"		8	1		81	67.50		148.50	190
0621	2-1/2"	Q-1	9	1.778		223	108		331	405
0622	3"		8	2		253	122		375	460
0623	4"		6	2.667		455	162		617	740
0624	5"		5	3.200		1,975	195		2,170	2,475
0625	6"	Q-2	7	3.429		3,475	217		3,692	4,125
0626	8"	"	6	4		13,400	253		13,653	15,100
0630	Tee, reducing on the run, 1/4"	1 Plum	15	.533		31	36		67	88.50
0631	3/8"		15	.533		35.50	36		71.50	93
0632	1/2"		14	.571		30	38.50		68.50	90.50
0633	5/8"		13	.615		12.25	41.50		53.75	75.50
0634	3/4"		12	.667		24	45		69	94
0635	1"		11	.727		38	49		87	116
0636	1-1/4"		10	.800		61.50	54		115.50	149
0637	1-1/2"		9	.889		106	60		166	206
0638	2"		8	1		135	67.50		202.50	250
0639	2-1/2"	Q-1	9	1.778		288	108		396	475
0640	3"		8	2		400	122		522	620
0641	4"		6	2.667		860	162		1,022	1,175
0642	5"		5	3.200		2,400	195		2,595	2,950
0643	6"	Q-2	7	3.429		3,650	217		3,867	4,350
0644	8"	"	6	4		12,600	253		12,853	14,200
0650	Coupling, 1/4"	1 Plum	24	.333		1.78	22.50		24.28	35.50
0670	3/8"		24	.333		2.31	22.50		24.81	36
0680	1/2"		22	.364		1.92	24.50		26.42	39
0690	5/8"		21	.381		5.05	26		31.05	44
0700	3/4"		21	.381		3.85	26		29.85	42.50
0710	1"		18	.444		7.65	30		37.65	53.50
0715	1-1/4"		17	.471		13.40	32		45.40	62.50
0716	1-1/2"		15	.533		17.70	36		53.70	73.50
0718	2"		13	.615		29.50	41.50		71	94.50
0721	2-1/2"	Q-1	15	1.067		60.50	65		125.50	164
0722	3"		13	1.231		82.50	75		157.50	203
0724	3-1/2"		8	2		155	122		277	355
0726	4"		7	2.286		182	139		321	410
0728	5"		6	2.667		380	162		542	655
0731	6"	Q-2	8	3		630	190		820	975
0732	8"	"	7	3.429		2,025	217		2,242	2,550
0741	Coupling, reducing, concentric									
0743	1/2"	1 Plum	23	.348	Ea.	4.43	23.50		27.93	40
0745	3/4"		21.50	.372		6.70	25		31.70	45
0747	1"		19.50	.410		10.45	28		38.45	53
0748	1-1/4"		18	.444		19.05	30		49.05	66
0749	1-1/2"		16	.500		24.50	34		58.50	77.50
0751	2"		14	.571		46	38.50		84.50	108
0752	2-1/2"		13	.615		84.50	41.50		126	155

22 11 13.25 Pipe/Tube Fittings, Copper		Crew	Daily Output	Labor-Hours	Unit	Material	2021 Bare Costs Labor	Equipment	Total	Total Incl O&P
0753	3"	Q-1	14	1.143	Ea.	102	69.50		171.50	217
0755	4"	"	8	2		208	122		330	410
0757	5"	Q-2	7.50	3.200		1,125	202		1,327	1,550
0759	6"		7	3.429		1,800	217		2,017	2,300
0761	8"		6.50	3.692		4,800	233		5,033	5,650
0771	Cap, sweat									
0773	1/2"	1 Plum	40	.200	Ea.	1.86	13.55		15.41	22
0775	3/4"		38	.211		3.46	14.25		17.71	25.50
0777	1"		32	.250		8.10	16.95		25.05	34.50
0778	1-1/4"		29	.276		10.70	18.70		29.40	40
0779	1-1/2"		26	.308		15.65	21		36.65	48
0781	2"		22	.364		29	24.50		53.50	68.50
0791	Flange, sweat									
0793	3"	Q-1	22	.727	Ea.	390	44.50		434.50	495
0795	4"		18	.889		545	54		599	680
0797	5"		12	1.333		1,025	81		1,106	1,250
0799	6"	Q-2	18	1.333		1,075	84		1,159	1,300
0801	8"	"	16	1.500		2,025	95		2,120	2,375
0850	Unions, 1/4"	1 Plum	21	.381		59	26		85	104
0870	3/8"		21	.381		59.50	26		85.50	104
0880	1/2"		19	.421		35	28.50		63.50	81
0890	5/8"		18	.444		135	30		165	193
0900	3/4"		18	.444		43.50	30		73.50	93
0910	1"		15	.533		68	36		104	129
0920	1-1/4"		14	.571		112	38.50		150.50	181
0930	1-1/2"		12	.667		148	45		193	230
0940	2"		10	.800		251	54		305	360
0950	2-1/2"	Q-1	12	1.333		550	81		631	725
0960	3"	"	10	1.600		1,425	97.50		1,522.50	1,725
0980	Adapter, copper x male IPS, 1/4"	1 Plum	20	.400		25	27		52	68
0990	3/8"		20	.400		12.35	27		39.35	54
1000	1/2"		18	.444		5.35	30		35.35	51
1010	3/4"		17	.471		8.90	32		40.90	57.50
1020	1"		15	.533		23	36		59	79
1030	1-1/4"		13	.615		32.50	41.50		74	98
1040	1-1/2"		12	.667		38.50	45		83.50	110
1050	2"		11	.727		65	49		114	145
1060	2-1/2"	Q-1	10.50	1.524		227	93		320	390
1070	3"		10	1.600		282	97.50		379.50	455
1080	3-1/2"		9	1.778		262	108		370	450
1090	4"		8	2		370	122		492	585
1200	5", cast		6	2.667		2,200	162		2,362	2,650
1210	6", cast	Q-2	8.50	2.824		2,475	178		2,653	3,000
1214	Adapter, copper x female IPS									
1216	1/2"	1 Plum	18	.444	Ea.	8.45	30		38.45	54.50
1218	3/4"		17	.471		11.55	32		43.55	60.50
1220	1"		15	.533		26.50	36		62.50	83
1221	1-1/4"		13	.615		38.50	41.50		80	104
1222	1-1/2"		12	.667		60	45		105	134
1224	2"		11	.727		81.50	49		130.50	163
1250	Cross, 1/2"		10	.800		40.50	54		94.50	126
1260	3/4"		9.50	.842		78.50	57		135.50	172
1270	1"		8	1		134	67.50		201.50	248

22 11 13.25 Pipe/Tube Fittings, Copper		Crew	Daily Output	Labor-Hours	Unit	Material	2021 Bare Costs Labor	Equipment	Total	Total Incl O&P
1280	1-1/4"	1 Plum	7.50	1.067	Ea.	192	72		264	320
1290	1-1/2"		6.50	1.231		274	83.50		357.50	425
1300	2"	↓	5.50	1.455		515	98.50		613.50	715
1310	2-1/2"	Q-1	6.50	2.462		1,200	150		1,350	1,550
1320	3"	"	5.50	2.909	↓	895	177		1,072	1,250
1500	Tee fitting, mechanically formed (Type 1, 'branch sizes up to 2 in.')									
1520	1/2" run size, 3/8" to 1/2" branch size	1 Plum	80	.100	Ea.		6.75		6.75	10.10
1530	3/4" run size, 3/8" to 3/4" branch size		60	.133			9.05		9.05	13.45
1540	1" run size, 3/8" to 1" branch size		54	.148			10.05		10.05	14.95
1550	1-1/4" run size, 3/8" to 1-1/4" branch size		48	.167			11.30		11.30	16.85
1560	1-1/2" run size, 3/8" to 1-1/2" branch size		40	.200			13.55		13.55	20
1570	2" run size, 3/8" to 2" branch size		35	.229			15.45		15.45	23
1580	2-1/2" run size, 1/2" to 2" branch size		32	.250			16.95		16.95	25.50
1590	3" run size, 1" to 2" branch size		26	.308			21		21	31
1600	4" run size, 1" to 2" branch size	↓	24	.333	↓		22.50		22.50	33.50
1640	Tee fitting, mechanically formed (Type 2, branches 2-1/2" thru 4")									
1650	2-1/2" run size, 2-1/2" branch size	1 Plum	12.50	.640	Ea.		43.50		43.50	64.50
1660	3" run size, 2-1/2" to 3" branch size		12	.667			45		45	67.50
1670	3-1/2" run size, 2-1/2" to 3-1/2" branch size		11	.727			49		49	73.50
1680	4" run size, 2-1/2" to 4" branch size		10.50	.762			51.50		51.50	77
1698	5" run size, 2" to 4" branch size		9.50	.842			57		57	85
1700	6" run size, 2" to 4" branch size		8.50	.941			63.50		63.50	95
1710	8" run size, 2" to 4" branch size	↓	7	1.143	↓		77.50		77.50	115
1800	ACR fittings, OD size									
1802	Tee, straight									
1808	5/8"	1 Stpi	12	.667	Ea.	4.34	45.50		49.84	73
1810	3/4"		12	.667		27	45.50		72.50	97.50
1812	7/8"		10	.800		10.45	54.50		64.95	93
1813	1"		10	.800		74.50	54.50		129	163
1814	1-1/8"		10	.800		31.50	54.50		86	116
1816	1-3/8"		9	.889		43	61		104	138
1818	1-5/8"		8	1		65.50	68.50		134	174
1820	2-1/8"	↓	7	1.143		104	78		182	231
1822	2-5/8"	Q-5	8	2		191	123		314	395
1824	3-1/8"		7	2.286		275	141		416	510
1826	4-1/8"	↓	5	3.200	↓	645	197		842	1,000
1830	90° elbow									
1836	5/8"	1 Stpi	19	.421	Ea.	10.05	29		39.05	54
1838	3/4"		19	.421		1.75	29		30.75	45
1840	7/8"		16	.500		7.30	34		41.30	59
1842	1-1/8"		16	.500		6.40	34		40.40	58
1844	1-3/8"		15	.533		32.50	36.50		69	90
1846	1-5/8"		13	.615		27.50	42		69.50	93
1848	2-1/8"		11	.727		59	49.50		108.50	139
1850	2-5/8"	Q-5	13	1.231		107	75.50		182.50	231
1852	3-1/8"		11	1.455		152	89.50		241.50	300
1854	4-1/8"	↓	9	1.778	↓	335	109		444	530
1860	Coupling									
1866	5/8"	1 Stpi	21	.381	Ea.	1.92	26		27.92	41
1868	3/4"		21	.381		5.05	26		31.05	44.50
1870	7/8"		18	.444		3.85	30.50		34.35	49.50
1871	1"		18	.444		13.15	30.50		43.65	60
1872	1-1/8"	↓	18	.444		7.65	30.50		38.15	54

For customer support on your Plumbing Costs with RSMeans Data, call 800.448.8182.

161

22 11 13.25 Pipe/Tube Fittings, Copper		Crew	Daily Output	Labor-Hours	Unit	Material	2021 Bare Costs Labor	Equipment	Total	Total Incl O&P
1874	1-3/8"	1 Stpi	17	.471	Ea.	13.40	32		45.40	63
1876	1-5/8"		15	.533		17.70	36.50		54.20	74
1878	2-1/8"	↓	13	.615		29.50	42		71.50	95.50
1880	2-5/8"	Q-5	15	1.067		60.50	65.50		126	165
1882	3-1/8"	↓	13	1.231		82.50	75.50		158	204
1884	4-1/8"	↓	7	2.286	↓	182	141		323	410
2000	DWV, solder joints, copper x copper									
2030	90° elbow, 1-1/4"	1 Plum	13	.615	Ea.	27	41.50		68.50	92
2050	1-1/2"		12	.667		36	45		81	107
2070	2"	↓	10	.800		57.50	54		111.50	145
2090	3"	Q-1	10	1.600		139	97.50		236.50	298
2100	4"	"	9	1.778		675	108		783	905
2150	45° elbow, 1-1/4"	1 Plum	13	.615		22	41.50		63.50	86.50
2170	1-1/2"		12	.667		20.50	45		65.50	90
2180	2"	↓	10	.800		42	54		96	128
2190	3"	Q-1	10	1.600		94.50	97.50		192	249
2200	4"	"	9	1.778		115	108		223	289
2250	Tee, sanitary, 1-1/4"	1 Plum	9	.889		47.50	60		107.50	142
2270	1-1/2"		8	1		59	67.50		126.50	166
2290	2"	↓	7	1.143		81	77.50		158.50	204
2310	3"	Q-1	7	2.286		340	139		479	585
2330	4"	"	6	2.667		750	162		912	1,075
2400	Coupling, 1-1/4"	1 Plum	14	.571		11.30	38.50		49.80	70
2420	1-1/2"		13	.615		13.95	41.50		55.45	77.50
2440	2"	↓	11	.727		19.40	49		68.40	95
2460	3"	Q-1	11	1.455		45	88.50		133.50	182
2480	4"	"	10	1.600	↓	99.50	97.50		197	254
2602	Traps, see Section 22 13 16.60									
3500	Compression joint fittings									
3510	As used for plumbing and oil burner work									
3520	Fitting price includes nuts and sleeves									
3540	Sleeve, 1/8"				Ea.	.24			.24	.26
3550	3/16"					.24			.24	.26
3560	1/4"					.08			.08	.09
3570	5/16"					.14			.14	.15
3580	3/8"					.50			.50	.55
3600	1/2"					.66			.66	.73
3620	Nut, 1/8"					.21			.21	.23
3630	3/16"					.37			.37	.41
3640	1/4"					.36			.36	.40
3650	5/16"					.36			.36	.40
3660	3/8"					.59			.59	.65
3670	1/2"					1.03			1.03	1.13
3710	Union, 1/8"	1 Plum	26	.308		1.47	21		22.47	32.50
3720	3/16"		24	.333		1.40	22.50		23.90	35
3730	1/4"		24	.333		2.83	22.50		25.33	36.50
3740	5/16"		23	.348		3.63	23.50		27.13	39
3750	3/8"		22	.364		3.20	24.50		27.70	40.50
3760	1/2"		22	.364		5.65	24.50		30.15	43
3780	5/8"		21	.381		5.45	26		31.45	44.50
3820	Union tee, 1/8"		17	.471		5.55	32		37.55	53.50
3830	3/16"		16	.500		3.21	34		37.21	54
3840	1/4"	↓	15	.533	↓	5.65	36		41.65	60

For customer support on your Plumbing Costs with RSMeans Data, call 800.448.8182.

22 11 13 – Facility Water Distribution Piping

22 11 13.25 Pipe/Tube Fittings, Copper		Crew	Daily Output	Labor-Hours	Unit	Material	2021 Bare Costs Labor	2021 Bare Costs Equipment	Total	Total Incl O&P
3850	5/16"	1 Plum	15	.533	Ea.	4.08	36		40.08	58.50
3860	3/8"		15	.533		8.05	36		44.05	63
3870	1/2"		15	.533		13.55	36		49.55	69
3910	Union elbow, 1/4"		24	.333		3.03	22.50		25.53	37
3920	5/16"		23	.348		2.92	23.50		26.42	38
3930	3/8"		22	.364		5.90	24.50		30.40	43.50
3940	1/2"		22	.364		9.60	24.50		34.10	47.50
3980	Female connector, 1/8"		26	.308		1.64	21		22.64	33
4000	3/16" x 1/8"		24	.333		1.44	22.50		23.94	35
4010	1/4" x 1/8"		24	.333		2.47	22.50		24.97	36
4020	1/4"		24	.333		3.32	22.50		25.82	37
4030	3/8" x 1/4"		22	.364		3.32	24.50		27.82	40.50
4040	1/2" x 3/8"		22	.364		5.25	24.50		29.75	43
4050	5/8" x 1/2"		21	.381		7.45	26		33.45	46.50
4090	Male connector, 1/8"		26	.308		1.46	21		22.46	32.50
4100	3/16" x 1/8"		24	.333		.94	22.50		23.44	34.50
4110	1/4" x 1/8"		24	.333		1.89	22.50		24.39	35.50
4120	1/4"		24	.333		1.89	22.50		24.39	35.50
4130	5/16" x 1/8"		23	.348		2.23	23.50		25.73	37.50
4140	5/16" x 1/4"		23	.348		1.30	23.50		24.80	36.50
4150	3/8" x 1/8"		22	.364		2.64	24.50		27.14	40
4160	3/8" x 1/4"		22	.364		3.03	24.50		27.53	40.50
4170	3/8"		22	.364		3.03	24.50		27.53	40.50
4180	3/8" x 1/2"		22	.364		4.15	24.50		28.65	41.50
4190	1/2" x 3/8"		22	.364		4.03	24.50		28.53	41.50
4200	1/2"		22	.364		4.51	24.50		29.01	42
4210	5/8" x 1/2"		21	.381		5.20	26		31.20	44.50
4240	Male elbow, 1/8"		26	.308		3.63	21		24.63	35
4250	3/16" x 1/8"		24	.333		2.07	22.50		24.57	36
4260	1/4" x 1/8"		24	.333		3.03	22.50		25.53	37
4270	1/4"		24	.333		3.43	22.50		25.93	37.50
4280	3/8" x 1/4"		22	.364		4.37	24.50		28.87	42
4290	3/8"		22	.364		7.50	24.50		32	45.50
4300	1/2" x 1/4"		22	.364		7.25	24.50		31.75	45
4310	1/2" x 3/8"		22	.364		11.40	24.50		35.90	49.50
4340	Female elbow, 1/8"		26	.308		4.05	21		25.05	35.50
4350	1/4" x 1/8"		24	.333		4.42	22.50		26.92	38.50
4360	1/4"		24	.333		5.80	22.50		28.30	40
4370	3/8" x 1/4"		22	.364		5.75	24.50		30.25	43.50
4380	1/2" x 3/8"		22	.364		5.10	24.50		29.60	42.50
4390	1/2"		22	.364		6.85	24.50		31.35	44.50
4420	Male run tee, 1/4" x 1/8"		15	.533		5.40	36		41.40	60
4430	5/16" x 1/8"		15	.533		5.40	36		41.40	60
4440	3/8" x 1/4"		15	.533		8.40	36		44.40	63.50
4480	Male branch tee, 1/4" x 1/8"		15	.533		6.30	36		42.30	61
4490	1/4"		15	.533		6.75	36		42.75	61.50
4500	3/8" x 1/4"		15	.533		4.09	36		40.09	58.50
4510	1/2" x 3/8"		15	.533		6.35	36		42.35	61
4520	1/2"		15	.533		10.20	36		46.20	65
4800	Flare joint fittings									
4810	Refrigeration fittings									
4820	Flare joint nuts and labor not incl. in price. Add 1 nut per jnt.									
4830	90° elbow, 1/4"				Ea.	2.51			2.51	2.76

22 11 13.25 Pipe/Tube Fittings, Copper	Crew	Daily Output	Labor-Hours	Unit	Material	2021 Bare Costs Labor	Equipment	Total	Total Incl O&P
4840	3/8"				Ea.	3.06			3.06	3.37
4850	1/2"					3.93			3.93	4.32
4860	5/8"					5.40			5.40	5.95
4870	3/4"					12.50			12.50	13.75
5030	Tee, 1/4"					3.24			3.24	3.56
5040	5/16"					3.87			3.87	4.26
5050	3/8"					4.30			4.30	4.73
5060	1/2"					6.80			6.80	7.45
5070	5/8"					7			7	7.70
5080	3/4"					7.20			7.20	7.90
5140	Union, 3/16"					1.48			1.48	1.63
5150	1/4"					1.02			1.02	1.12
5160	5/16"					1.38			1.38	1.52
5170	3/8"					1.53			1.53	1.68
5180	1/2"					2.45			2.45	2.70
5190	5/8"					3.56			3.56	3.92
5200	3/4"					11.40			11.40	12.55
5260	Long flare nut, 3/16"	1 Stpi	42	.190		.40	13		13.40	19.90
5270	1/4"		41	.195		1.04	13.35		14.39	21
5280	5/16"		40	.200		1.89	13.65		15.54	22.50
5290	3/8"		39	.205		2.06	14		16.06	23.50
5300	1/2"		38	.211		2.94	14.40		17.34	24.50
5310	5/8"		37	.216		6.60	14.80		21.40	29.50
5320	3/4"		34	.235		13.80	16.10		29.90	39
5380	Short flare nut, 3/16"		42	.190		4.56	13		17.56	24.50
5390	1/4"		41	.195		.73	13.35		14.08	20.50
5400	5/16"		40	.200		.88	13.65		14.53	21.50
5410	3/8"		39	.205		1.50	14		15.50	22.50
5420	1/2"		38	.211		2.11	14.40		16.51	24
5430	5/8"		36	.222		2.67	15.20		17.87	25.50
5440	3/4"		34	.235		7.95	16.10		24.05	33
5500	90° elbow flare by MIPS, 1/4"					3.05			3.05	3.36
5510	3/8"					3.37			3.37	3.71
5520	1/2"					5.40			5.40	5.90
5530	5/8"					6.75			6.75	7.45
5540	3/4"					12.35			12.35	13.55
5600	Flare by FIPS, 1/4"					5.10			5.10	5.65
5610	3/8"					5.55			5.55	6.10
5620	1/2"					7.95			7.95	8.75
5670	Flare by sweat, 1/4"					2.31			2.31	2.54
5680	3/8"					5.35			5.35	5.90
5690	1/2"					6.10			6.10	6.70
5700	5/8"					16.25			16.25	17.90
5760	Tee flare by IPS, 1/4"					2.87			2.87	3.16
5770	3/8"					7.15			7.15	7.90
5780	1/2"					12			12	13.20
5790	5/8"					8.05			8.05	8.85
5850	Connector, 1/4"					1.11			1.11	1.22
5860	3/8"					1.65			1.65	1.82
5870	1/2"					2.01			2.01	2.21
5880	5/8"					3.79			3.79	4.17
5890	3/4"					5.70			5.70	6.25
5950	Seal cap, 1/4"					.62			.62	.68

22 11 13.25 Pipe/Tube Fittings, Copper	Crew	Daily Output	Labor-Hours	Unit	Material	2021 Bare Costs Labor	Equipment	Total	Total Incl O&P
5960	3/8"				Ea.	.95			.95	1.05
5970	1/2"					1.34			1.34	1.47
5980	5/8"					2.47			2.47	2.72
5990	3/4"					6.25			6.25	6.90
6000	Water service fittings									
6010	Flare joints nut and labor are included in the fitting price.									
6020	90° elbow, C x C, 3/8"	1 Plum	19	.421	Ea.	70.50	28.50		99	121
6030	1/2"		18	.444		101	30		131	156
6040	3/4"		16	.500		128	34		162	192
6050	1"		15	.533		183	36		219	255
6080	2"		10	.800		885	54		939	1,050
6090	90° elbow, C x MPT, 3/8"		19	.421		53.50	28.50		82	102
6100	1/2"		18	.444		76.50	30		106.50	129
6110	3/4"		16	.500		88.50	34		122.50	148
6120	1"		15	.533		241	36		277	320
6130	1-1/4"		13	.615		450	41.50		491.50	555
6140	1-1/2"		12	.667		500	45		545	615
6150	2"		10	.800		385	54		439	500
6160	90° elbow, C x FPT, 3/8"		19	.421		57.50	28.50		86	106
6170	1/2"		18	.444		82	30		112	135
6180	3/4"		16	.500		102	34		136	163
6190	1"		15	.533		234	36		270	310
6200	1-1/4"		13	.615		155	41.50		196.50	233
6210	1-1/2"		12	.667		595	45		640	725
6220	2"		10	.800		860	54		914	1,025
6230	Tee, C x C x C, 3/8"		13	.615		114	41.50		155.50	187
6240	1/2"		12	.667		147	45		192	230
6250	3/4"		11	.727		154	49		203	244
6260	1"		10	.800		233	54		287	335
6330	Tube nut, C x nut seat, 3/8"		40	.200		20.50	13.55		34.05	43
6340	1/2"		38	.211		20.50	14.25		34.75	44.50
6350	3/4"		34	.235		30	15.95		45.95	57
6360	1"		32	.250		54	16.95		70.95	85
6380	Coupling, C x C, 3/8"		19	.421		69.50	28.50		98	119
6390	1/2"		18	.444		90	30		120	144
6400	3/4"		16	.500		114	34		148	177
6410	1"		15	.533		211	36		247	286
6420	1-1/4"		12	.667		340	45		385	445
6430	1-1/2"		12	.667		515	45		560	640
6440	2"		10	.800		780	54		834	935
6450	Adapter, C x FPT, 3/8"		19	.421		52.50	28.50		81	101
6460	1/2"		18	.444		68	30		98	120
6470	3/4"		16	.500		82.50	34		116.50	141
6480	1"		15	.533		176	36		212	247
6490	1-1/4"		13	.615		137	41.50		178.50	213
6500	1-1/2"		12	.667		375	45		420	480
6510	2"		10	.800		365	54		419	480
6520	Adapter, C x MPT, 3/8"		19	.421		43	28.50		71.50	89.50
6530	1/2"		18	.444		61	30		91	112
6540	3/4"		16	.500		84	34		118	143
6550	1"		15	.533		148	36		184	217
6560	1-1/4"		13	.615		287	41.50		328.50	375
6570	1-1/2"		12	.667		370	45		415	480

22 11 13.25 Pipe/Tube Fittings, Copper

		Crew	Daily Output	Labor-Hours	Unit	Material	2021 Bare Costs Labor	2021 Bare Costs Equipment	Total	Total Incl O&P
6580	2"	1 Plum	10	.800	Ea.	500	54		554	630
6992	Tube connector fittings, See Section 22 11 13.76 for plastic ftng.									
7000	Insert type brass/copper, 100 psi @ 180°F, CTS									
7010	Adapter MPT 3/8" x 1/2" CTS	1 Plum	29	.276	Ea.	3.03	18.70		21.73	31.50
7020	1/2" x 1/2"		26	.308		3.10	21		24.10	34.50
7030	3/4" x 1/2"		26	.308		4.69	21		25.69	36
7040	3/4" x 3/4"		25	.320		4.70	21.50		26.20	37.50
7050	Adapter CTS 1/2" x 1/2" sweat		24	.333		3.94	22.50		26.44	38
7060	3/4" x 3/4" sweat		22	.364		2.60	24.50		27.10	40
7070	Coupler center set 3/8" CTS		25	.320		1.47	21.50		22.97	34
7080	1/2" CTS		23	.348		1.15	23.50		24.65	36.50
7090	3/4" CTS		22	.364		1.45	24.50		25.95	38.50
7100	Elbow 90°, copper 3/8"		25	.320		2.87	21.50		24.37	35.50
7110	1/2" CTS		23	.348		2.01	23.50		25.51	37
7120	3/4" CTS		22	.364		2.58	24.50		27.08	40
7130	Tee copper 3/8" CTS		17	.471		3.71	32		35.71	51.50
7140	1/2" CTS		15	.533		2.64	36		38.64	57
7150	3/4" CTS		14	.571		4.06	38.50		42.56	62
7160	3/8" x 3/8" x 1/2"		16	.500		4.78	34		38.78	56
7170	1/2" x 3/8" x 1/2"		15	.533		3.08	36		39.08	57.50
7180	3/4" x 1/2" x 3/4"		14	.571		3.87	38.50		42.37	62

22 11 13.27 Pipe/Tube, Grooved Joint for Copper

		Crew	Daily Output	Labor-Hours	Unit	Material	2021 Bare Costs Labor	2021 Bare Costs Equipment	Total	Total Incl O&P
0010	**PIPE/TUBE, GROOVED JOINT FOR COPPER**									
4000	Fittings: coupling material required at joints not incl. in fitting price.									
4001	Add 1 selected coupling, material only, per joint for installed price.									
4010	Coupling, rigid style									
4018	2" diameter	1 Plum	50	.160	Ea.	35.50	10.85		46.35	55
4020	2-1/2" diameter	Q-1	80	.200		39.50	12.20		51.70	61.50
4022	3" diameter		67	.239		44.50	14.55		59.05	70.50
4024	4" diameter		50	.320		66.50	19.50		86	103
4026	5" diameter		40	.400		111	24.50		135.50	159
4028	6" diameter	Q-2	50	.480		147	30.50		177.50	207
4100	Elbow, 90° or 45°									
4108	2" diameter	1 Plum	25	.320	Ea.	60.50	21.50		82	99
4110	2-1/2" diameter	Q-1	40	.400		65.50	24.50		90	109
4112	3" diameter		33	.485		91.50	29.50		121	145
4114	4" diameter		25	.640		210	39		249	289
4116	5" diameter		20	.800		595	48.50		643.50	730
4118	6" diameter	Q-2	25	.960		945	60.50		1,005.50	1,150
4200	Tee									
4208	2" diameter	1 Plum	17	.471	Ea.	99	32		131	157
4210	2-1/2" diameter	Q-1	27	.593		105	36		141	170
4212	3" diameter		22	.727		156	44.50		200.50	238
4214	4" diameter		17	.941		340	57.50		397.50	460
4216	5" diameter		13	1.231		955	75		1,030	1,150
4218	6" diameter	Q-2	17	1.412		1,175	89		1,264	1,425
4300	Reducer, concentric									
4310	3" x 2-1/2" diameter	Q-1	35	.457	Ea.	87	28		115	137
4312	4" x 2-1/2" diameter		32	.500		177	30.50		207.50	240
4314	4" x 3" diameter		29	.552		177	33.50		210.50	244
4316	5" x 3" diameter		25	.640		495	39		534	605
4318	5" x 4" diameter		22	.727		495	44.50		539.50	610

22 11 13.27 Pipe/Tube, Grooved Joint for Copper

		Crew	Daily Output	Labor-Hours	Unit	Material	2021 Bare Costs Labor	Equipment	Total	Total Incl O&P
4320	6" x 3" diameter	Q-2	28	.857	Ea.	535	54		589	670
4322	6" x 4" diameter		26	.923		535	58.50		593.50	675
4324	6" x 5" diameter	↓	24	1	↓	535	63		598	685
4350	Flange, w/groove gasket									
4351	ANSI class 125 and 150									
4355	2" diameter	1 Plum	23	.348	Ea.	264	23.50		287.50	325
4356	2-1/2" diameter	Q-1	37	.432		275	26.50		301.50	345
4358	3" diameter		31	.516		288	31.50		319.50	360
4360	4" diameter		23	.696		315	42.50		357.50	410
4362	5" diameter	↓	19	.842		425	51.50		476.50	540
4364	6" diameter	Q-2	23	1.043	↓	460	66		526	610

22 11 13.29 Pipe, Fittings and Valves, Copper, Pressed-Joint

		Crew	Daily Output	Labor-Hours	Unit	Material	2021 Bare Costs Labor	Equipment	Total	Total Incl O&P
0010	**PIPE, FITTINGS AND VALVES, COPPER, PRESSED-JOINT**									
0040	Pipe/tube includes coupling & clevis type hanger assy's, 10' OC									
0120	Type K									
0130	1/2" diameter	1 Plum	78	.103	L.F.	5.80	6.95		12.75	16.75
0134	3/4" diameter		74	.108		9.55	7.30		16.85	21.50
0138	1" diameter		66	.121		12.40	8.20		20.60	26
0142	1-1/4" diameter		56	.143		15.25	9.65		24.90	31
0146	1-1/2" diameter		50	.160		19.70	10.85		30.55	37.50
0150	2" diameter	↓	40	.200		29.50	13.55		43.05	52.50
0154	2-1/2" diameter	Q-1	60	.267		49	16.25		65.25	78.50
0158	3" diameter		54	.296		66.50	18.05		84.55	100
0162	4" diameter	↓	38	.421	↓	107	25.50		132.50	157
0180	To delete cplgs. & hngrs., 1/2" pipe, subtract					19%	48%			
0184	3/4"-2" pipe, subtract					14%	46%			
0186	2-1/2"-4" pipe, subtract					24%	34%			
0220	Type L									
0230	1/2" diameter	1 Plum	81	.099	L.F.	2.64	6.70		9.34	12.90
0234	3/4" diameter		76	.105		3.15	7.15		10.30	14.10
0238	1" diameter		68	.118		3.77	7.95		11.72	16.05
0242	1-1/4" diameter		58	.138		10.65	9.35		20	25.50
0246	1-1/2" diameter		52	.154		6.90	10.40		17.30	23
0250	2" diameter	↓	42	.190		11.50	12.90		24.40	32
0254	2-1/2" diameter	Q-1	62	.258		18.80	15.70		34.50	44
0258	3" diameter		56	.286		49.50	17.40		66.90	80.50
0262	4" diameter	↓	39	.410	↓	39	25		64	80.50
0280	To delete cplgs. & hngrs., 1/2" pipe, subtract					21%	52%			
0284	3/4"-2" pipe, subtract					17%	46%			
0286	2-1/2"-4" pipe, subtract					23%	35%			
0320	Type M									
0330	1/2" diameter	1 Plum	84	.095	L.F.	3.88	6.45		10.33	13.85
0334	3/4" diameter		78	.103		5.55	6.95		12.50	16.45
0338	1" diameter		70	.114		7.60	7.75		15.35	19.90
0342	1-1/4" diameter		60	.133		11.80	9.05		20.85	26.50
0346	1-1/2" diameter		54	.148		15.05	10.05		25.10	31.50
0350	2" diameter	↓	44	.182		23	12.30		35.30	43.50
0354	2-1/2" diameter	Q-1	64	.250		37.50	15.25		52.75	64
0358	3" diameter		58	.276		48.50	16.80		65.30	78.50
0362	4" diameter	↓	40	.400	↓	82	24.50		106.50	127
0380	To delete cplgs. & hngrs., 1/2" pipe, subtract					32%	49%			
0384	3/4"-2" pipe, subtract					21%	46%			

22 11 13.29 Pipe, Fittings and Valves, Copper, Pressed-Joint	Crew	Daily Output	Labor-Hours	Unit	Material	2021 Bare Costs Labor	Equipment	Total	Total Incl O&P
0386 2-1/2"-4" pipe, subtract					25%	36%			
1600 Fittings									
1610 Press joints, copper x copper									
1620 Note: Reducing fittings show most expensive size combination.									
1800 90° elbow, 1/2"	1 Plum	36.60	.219	Ea.	3.60	14.80		18.40	26
1810 3/4"		27.50	.291		5.80	19.70		25.50	36
1820 1"		25.90	.309		11.60	21		32.60	44
1830 1-1/4"		20.90	.383		23.50	26		49.50	64.50
1840 1-1/2"		18.30	.437		44.50	29.50		74	92.50
1850 2"		15.70	.510		62	34.50		96.50	120
1860 2-1/2"	Q-1	25.90	.618		181	37.50		218.50	255
1870 3"		22	.727		227	44.50		271.50	315
1880 4"		16.30	.982		281	60		341	400
2000 45° elbow, 1/2"	1 Plum	36.60	.219		4.35	14.80		19.15	27
2010 3/4"		27.50	.291		5.10	19.70		24.80	35
2020 1"		25.90	.309		16.30	21		37.30	49
2030 1-1/4"		20.90	.383		23.50	26		49.50	64.50
2040 1-1/2"		18.30	.437		38	29.50		67.50	85.50
2050 2"		15.70	.510		52.50	34.50		87	110
2060 2-1/2"	Q-1	25.90	.618		122	37.50		159.50	190
2070 3"		22	.727		171	44.50		215.50	254
2080 4"		16.30	.982		242	60		302	355
2200 Tee, 1/2"	1 Plum	27.50	.291		5.50	19.70		25.20	35.50
2210 3/4"		20.70	.386		9.65	26		35.65	49.50
2220 1"		19.40	.412		17.65	28		45.65	61
2230 1-1/4"		15.70	.510		30.50	34.50		65	85
2240 1-1/2"		13.80	.580		58	39.50		97.50	122
2250 2"		11.80	.678		71.50	46		117.50	147
2260 2-1/2"	Q-1	19.40	.825		228	50.50		278.50	325
2270 3"		16.50	.970		279	59		338	395
2280 4"		12.20	1.311		400	80		480	560
2400 Tee, reducing on the outlet									
2410 3/4"	1 Plum	20.70	.386	Ea.	8.40	26		34.40	48.50
2420 1"		19.40	.412		20.50	28		48.50	64
2430 1-1/4"		15.70	.510		30	34.50		64.50	84.50
2440 1-1/2"		13.80	.580		63	39.50		102.50	128
2450 2"		11.80	.678		98.50	46		144.50	177
2460 2-1/2"	Q-1	19.40	.825		294	50.50		344.50	400
2470 3"		16.50	.970		365	59		424	490
2480 4"		12.20	1.311		440	80		520	605
2600 Tee, reducing on the run									
2610 3/4"	1 Plum	20.70	.386	Ea.	16.60	26		42.60	57.50
2620 1"		19.40	.412		32	28		60	77
2630 1-1/4"		15.70	.510		63.50	34.50		98	121
2640 1-1/2"		13.80	.580		97	39.50		136.50	165
2650 2"		11.80	.678		107	46		153	186
2660 2-1/2"	Q-1	19.40	.825		345	50.50		395.50	450
2670 3"		16.50	.970		410	59		469	545
2680 4"		12.20	1.311		515	80		595	685
2800 Coupling, 1/2"	1 Plum	36.60	.219		3.25	14.80		18.05	25.50
2810 3/4"		27.50	.291		4.91	19.70		24.61	35
2820 1"		25.90	.309		9.85	21		30.85	42
2830 1-1/4"		20.90	.383		12.70	26		38.70	52.50

22 11 13.29 Pipe, Fittings and Valves, Copper, Pressed-Joint

		Crew	Daily Output	Labor-Hours	Unit	Material	2021 Bare Costs Labor	Equipment	Total	Total Incl O&P
4050	2"	1 Plum	15.70	.510	Ea.	67	34.50		101.50	126
4060	2-1/2"	Q-1	25.90	.618		182	37.50		219.50	256
4070	3"		22	.727		248	44.50		292.50	340
4080	4"		16.30	.982		320	60		380	440
4100	Stub out, 1/2"	1 Plum	50	.160		10.05	10.85		20.90	27.50
4110	3/4"		35	.229		19.40	15.45		34.85	44.50
4120	1"		30	.267		21	18.05		39.05	50
6000	Valves									
6200	Ball valve									
6210	1/2"	1 Plum	25.60	.313	Ea.	39.50	21		60.50	75
6220	3/4"		19.20	.417		52.50	28		80.50	100
6230	1"		18.10	.442		66	30		96	118
6240	1-1/4"		14.70	.544		108	37		145	173
6250	1-1/2"		12.80	.625		266	42.50		308.50	355
6260	2"		11	.727		385	49		434	495
6400	Check valve									
6410	1/2"	1 Plum	30	.267	Ea.	33.50	18.05		51.55	64
6420	3/4"		22.50	.356		43	24		67	83
6430	1"		21.20	.377		49.50	25.50		75	92
6440	1-1/4"		17.20	.465		71.50	31.50		103	126
6450	1-1/2"		15	.533		101	36		137	165
6460	2"		12.90	.620		184	42		226	266
6600	Butterfly valve, lug type									
6660	2-1/2"	Q-1	9	1.778	Ea.	157	108		265	335
6670	3"		8	2		193	122		315	395
6680	4"		5	3.200		241	195		436	555

22 11 13.44 Pipe, Steel

		Crew	Daily Output	Labor-Hours	Unit	Material	2021 Bare Costs Labor	Equipment	Total	Total Incl O&P
0010	**PIPE, STEEL**	R221113-50								
0020	All pipe sizes are to Spec. A-53 unless noted otherwise	R221113-70								
0032	Schedule 10, see Line 22 11 13.48 0500									
0050	Schedule 40, threaded, with couplings, and clevis hanger									
0060	assemblies sized for covering, 10' OC									
0540	Black, 1/4" diameter	1 Plum	66	.121	L.F.	7.85	8.20		16.05	21
0550	3/8" diameter		65	.123		10.25	8.35		18.60	24
0560	1/2" diameter		63	.127		2.69	8.60		11.29	15.80
0570	3/4" diameter		61	.131		3.27	8.90		12.17	16.85
0580	1" diameter		53	.151		3.73	10.20		13.93	19.35
0590	1-1/4" diameter	Q-1	89	.180		4.82	10.95		15.77	21.50
0600	1-1/2" diameter		80	.200		5.65	12.20		17.85	24.50
0610	2" diameter		64	.250		7.60	15.25		22.85	31
0620	2-1/2" diameter		50	.320		14.30	19.50		33.80	44.50
0630	3" diameter		43	.372		18.65	22.50		41.15	54.50
0640	3-1/2" diameter		40	.400		28	24.50		52.50	67
0650	4" diameter		36	.444		28	27		55	71
0809	A-106, gr. A/B, seamless w/cplgs. & clevis hanger assemblies									
0811	1/4" diameter	1 Plum	66	.121	L.F.	4.69	8.20		12.89	17.40
0812	3/8" diameter		65	.123		4.67	8.35		13.02	17.60
0813	1/2" diameter		63	.127		4.83	8.60		13.43	18.15
0814	3/4" diameter		61	.131		5.70	8.90		14.60	19.50
0815	1" diameter		53	.151		6.65	10.20		16.85	22.50
0816	1-1/4" diameter	Q-1	89	.180		8.60	10.95		19.55	26
0817	1-1/2" diameter		80	.200		9.90	12.20		22.10	29

22 11 13.29 Pipe, Fittings and Valves, Copper, Pressed-Joint		Crew	Daily Output	Labor-Hours	Unit	Material	2021 Bare Costs Labor	Equipment	Total	Total Incl O&P
2840	1-1/2"	1 Plum	18.30	.437	Ea.	23	29.50		52.50	69.50
2850	2"	↓	15.70	.510		29.50	34.50		64	84
2860	2-1/2"	Q-1	25.90	.618		90	37.50		127.50	155
2870	3"		22	.727		115	44.50		159.50	192
2880	4"	↓	16.30	.982		162	60		222	269
3000	Union, 1/2"	1 Plum	36.60	.219		29	14.80		43.80	54
3010	3/4"		27.50	.291		36	19.70		55.70	69
3020	1"		25.90	.309		58.50	21		79.50	95
3030	1-1/4"		20.90	.383		85	26		111	132
3040	1-1/2"		18.30	.437		111	29.50		140.50	166
3050	2"	↓	15.70	.510	↓	178	34.50		212.50	247
3200	Adapter, tube to MPT									
3210	1/2"	1 Plum	15.10	.530	Ea.	4.24	36		40.24	58
3220	3/4"		13.60	.588		7.60	40		47.60	68
3230	1"		11.60	.690		14.20	46.50		60.70	85
3240	1-1/4"		9.90	.808		30.50	54.50		85	115
3250	1-1/2"		9	.889		43	60		103	137
3260	2"	↓	7.90	1.013		82.50	68.50		151	193
3270	2-1/2"	Q-1	13.40	1.194		184	73		257	310
3280	3"		10.60	1.509		232	92		324	390
3290	4"	↓	7.70	2.078	↓	284	127		411	500
3400	Adapter, tube to FPT									
3410	1/2"	1 Plum	15.10	.530	Ea.	5.30	36		41.30	59.50
3420	3/4"		13.60	.588		8.25	40		48.25	68.50
3430	1"		11.60	.690		15.80	46.50		62.30	87
3440	1-1/4"		9.90	.808		36.50	54.50		91	122
3450	1-1/2"		9	.889		51	60		111	146
3460	2"	↓	7.90	1.013		84.50	68.50		153	195
3470	2-1/2"	Q-1	13.40	1.194		208	73		281	340
3480	3"		10.60	1.509		340	92		432	510
3490	4"	↓	7.70	2.078	↓	440	127		567	675
3600	Flange									
3620	1"	1 Plum	36.20	.221	Ea.	156	14.95		170.95	194
3630	1-1/4"	↓	29.30	.273		222	18.50		240.50	272
3640	1-1/2"		25.60	.313		245	21		266	300
3650	2"		22	.364		275	24.50		299.50	335
3660	2-1/2"	Q-1	36.20	.442		277	27		304	345
3670	3"		30.80	.519		335	31.50		366.50	415
3680	4"	↓	22.80	.702		375	43		418	475
3800	Cap, 1/2"	1 Plum	53.10	.151		6.90	10.20		17.10	23
3810	3/4"		39.80	.201		11.85	13.60		25.45	33.50
3820	1"		37.50	.213		18.25	14.45		32.70	41.50
3830	1-1/4"		30.40	.263		22	17.80		39.80	50.50
3840	1-1/2"		26.60	.301		34	20.50		54.50	67.50
3850	2"	↓	22.80	.351		41	24		65	80.50
3860	2-1/2"	Q-1	37.50	.427		127	26		153	179
3870	3"		31.90	.502		161	30.50		191.50	223
3880	4"	↓	23.60	.678	↓	188	41.50		229.50	269
4000	Reducer									
4010	3/4"	1 Plum	27.50	.291	Ea.	14.90	19.70		34.60	46
4020	1"		25.90	.309		27.50	21		48.50	61
4030	1-1/4"		20.90	.383		40.50	26		66.50	83
4040	1-1/2"	↓	18.30	.437		48.50	29.50		78	97.50

For customer support on your Plumbing Costs with RSMeans Data, call 800.448.8182.

169

22 11 Facility Water Distribution

22 11 13 – Facility Water Distribution Piping

22 11 13.44 Pipe, Steel		Crew	Daily Output	Labor-Hours	Unit	Material	2021 Bare Costs Labor	Equipment	Total	Total Incl O&P
0819	2" diameter	Q-1	64	.250	L.F.	11.20	15.25		26.45	35
0821	2-1/2" diameter		50	.320		16.65	19.50		36.15	47.50
0822	3" diameter		43	.372		22	22.50		44.50	58
0823	4" diameter	↓	36	.444	↓	36	27		63	80.50
1220	To delete coupling & hanger, subtract									
1230	1/4" diam. to 3/4" diam.					31%	56%			
1240	1" diam. to 1-1/2" diam.					23%	51%			
1250	2" diam. to 4" diam.					23%	41%			
1280	All pipe sizes are to Spec. A-53 unless noted otherwise									
1281	Schedule 40, threaded, with couplings and clevis hanger									
1282	assemblies sized for covering, 10' OC									
1290	Galvanized, 1/4" diameter	1 Plum	66	.121	L.F.	10.85	8.20		19.05	24
1300	3/8" diameter		65	.123		14.30	8.35		22.65	28
1310	1/2" diameter		63	.127		3.07	8.60		11.67	16.20
1320	3/4" diameter		61	.131		3.76	8.90		12.66	17.40
1330	1" diameter	↓	53	.151		4.45	10.20		14.65	20
1340	1-1/4" diameter	Q-1	89	.180		5.80	10.95		16.75	22.50
1350	1-1/2" diameter		80	.200		6.80	12.20		19	25.50
1360	2" diameter		64	.250		9.20	15.25		24.45	32.50
1370	2-1/2" diameter		50	.320		16.10	19.50		35.60	46.50
1380	3" diameter		43	.372		21	22.50		43.50	57
1390	3-1/2" diameter		40	.400		32.50	24.50		57	72.50
1400	4" diameter	↓	36	.444	↓	32.50	27		59.50	76.50
1750	To delete coupling & hanger, subtract									
1760	1/4" diam. to 3/4" diam.					31%	56%			
1770	1" diam. to 1-1/2" diam.					23%	51%			
1780	2" diam. to 4" diam.					23%	41%			
1900	Pipe nipple std black 2" long, 1/2" diameter	1 Plum	19	.421	Ea.	1.25	28.50		29.75	44
1910	Pipe nipple std black 2" long, 1" diameter	"	15	.533		2.32	36		38.32	56.50
1920	3" long, 2-1/2" diameter	Q-1	16	1	↓	12.20	61		73.20	104
2000	Welded, sch. 40, on yoke & roll hanger assy's, sized for covering, 10' OC									
2040	Black, 1" diameter	Q-15	93	.172	L.F.	4.27	10.50	1.16	15.93	21.50
2050	1-1/4" diameter		84	.190		5.70	11.60	1.28	18.58	25
2060	1-1/2" diameter		76	.211		6.20	12.85	1.41	20.46	27.50
2070	2" diameter		61	.262		7.40	16	1.76	25.16	34
2080	2-1/2" diameter		47	.340		12.65	20.50	2.29	35.44	47.50
2090	3" diameter		43	.372		15.50	22.50	2.50	40.50	54
2100	3-1/2" diameter		39	.410		19.05	25	2.76	46.81	61.50
2110	4" diameter		37	.432		22	26.50	2.91	51.41	66.50
2120	5" diameter	↓	32	.500		24.50	30.50	3.36	58.36	76
2130	6" diameter	Q-16	36	.667		26	42	2.99	70.99	95
2140	8" diameter		29	.828		71.50	52.50	3.71	127.71	161
2150	10" diameter		24	1		91	63	4.48	158.48	199
2160	12" diameter		19	1.263		115	80	5.65	200.65	252
2170	14" diameter (two rod roll type hanger for 14" diam. and up)		15	1.600		114	101	7.15	222.15	284
2180	16" diameter (two rod roll type hanger)		13	1.846		167	117	8.25	292.25	365
2190	18" diameter (two rod roll type hanger)		11	2.182		158	138	9.75	305.75	390
2200	20" diameter (two rod roll type hanger)		9	2.667		175	168	11.95	354.95	455
2220	24" diameter (two rod roll type hanger)	↓	8	3	↓	233	190	13.45	436.45	555
2560	To delete hanger, subtract									
2570	1" diam. to 1-1/2" diam.					15%	34%			
2580	2" diam. to 3-1/2" diam.					9%	21%			
2590	4" diam. to 12" diam.					5%	12%			

171

22 11 13.44 Pipe, Steel		Crew	Daily Output	Labor-Hours	Unit	Material	2021 Bare Costs Labor	Equipment	Total	Total Incl O&P
2596	14" diam. to 24" diam.					3%	10%			
3250	Flanged, 150 lb. weld neck, on yoke & roll hangers									
3260	sized for covering, 10' OC									
3290	Black, 1" diameter	Q-15	70	.229	L.F.	9.65	13.95	1.54	25.14	33.50
3300	1-1/4" diameter		64	.250		11.05	15.25	1.68	27.98	36.50
3310	1-1/2" diameter		58	.276		11.60	16.80	1.85	30.25	40
3320	2" diameter		45	.356		14.85	21.50	2.39	38.74	51.50
3330	2-1/2" diameter		36	.444		20.50	27	2.99	50.49	66.50
3340	3" diameter		32	.500		24	30.50	3.36	57.86	75.50
3350	3-1/2" diameter		29	.552		28.50	33.50	3.71	65.71	85.50
3360	4" diameter		26	.615		31.50	37.50	4.14	73.14	95.50
3370	5" diameter		21	.762		40	46.50	5.10	91.60	119
3380	6" diameter	Q-16	25	.960		43	60.50	4.30	107.80	143
3390	8" diameter		19	1.263		99.50	80	5.65	185.15	235
3400	10" diameter		16	1.500		149	95	6.70	250.70	310
3410	12" diameter		14	1.714		185	108	7.70	300.70	375
3470	For 300 lb. flanges, add					63%				
3480	For 600 lb. flanges, add					310%				
3960	To delete flanges & hanger, subtract									
3970	1" diam. to 2" diam.					76%	65%			
3980	2-1/2" diam. to 4" diam.					62%	59%			
3990	5" diam. to 12" diam.					60%	46%			
4750	Schedule 80, threaded, with couplings, and clevis hanger assemblies									
4760	sized for covering, 10' OC									
4790	Black, 1/4" diameter	1 Plum	54	.148	L.F.	7.35	10.05		17.40	23
4800	3/8" diameter		53	.151		9.55	10.20		19.75	26
4810	1/2" diameter		52	.154		3.90	10.40		14.30	19.85
4820	3/4" diameter		50	.160		4.66	10.85		15.51	21.50
4830	1" diameter		45	.178		5.40	12.05		17.45	24
4840	1-1/4" diameter	Q-1	75	.213		7.45	13		20.45	27.50
4850	1-1/2" diameter		69	.232		8.45	14.15		22.60	30.50
4860	2" diameter		56	.286		11.75	17.40		29.15	39
4870	2-1/2" diameter		44	.364		18.85	22		40.85	53.50
4880	3" diameter		38	.421		24.50	25.50		50	65
4890	3-1/2" diameter		35	.457		30.50	28		58.50	75
4900	4" diameter		32	.500		36	30.50		66.50	85
5430	To delete coupling & hanger, subtract									
5440	1/4" diam. to 1/2" diam.					31%	54%			
5450	3/4" diam. to 1-1/2" diam.					28%	49%			
5460	2" diam. to 4" diam.					21%	40%			
5510	Galvanized, 1/4" diameter	1 Plum	54	.148	L.F.	10.40	10.05		20.45	26.50
5520	3/8" diameter		53	.151		11.25	10.20		21.45	27.50
5530	1/2" diameter		52	.154		4.32	10.40		14.72	20.50
5540	3/4" diameter		50	.160		5.20	10.85		16.05	22
5550	1" diameter		45	.178		6.25	12.05		18.30	25
5560	1-1/4" diameter	Q-1	75	.213		8.60	13		21.60	29
5570	1-1/2" diameter		69	.232		9.85	14.15		24	32
5580	2" diameter		56	.286		13.65	17.40		31.05	41
5590	2-1/2" diameter		44	.364		21.50	22		43.50	56.50
5600	3" diameter		38	.421		28	25.50		53.50	69
5610	3-1/2" diameter		35	.457		32.50	28		60.50	77.50
5620	4" diameter		32	.500		41	30.50		71.50	91
5930	To delete coupling & hanger, subtract									

22 11 13.44 Pipe, Steel

	Crew	Daily Output	Labor-Hours	Unit	Material	2021 Bare Costs Labor	Equipment	Total	Total Incl O&P
5940	1/4" diam. to 1/2" diam.				31%	54%			
5950	3/4" diam. to 1-1/2" diam.				28%	49%			
5960	2" diam. to 4" diam.				21%	40%			
6000	Welded, on yoke & roller hangers								
6010	sized for covering, 10' OC								
6040	Black, 1" diameter	Q-15	85	.188	L.F.	5.85	11.45	1.26	18.56	25
6050	1-1/4" diameter		79	.203		7.90	12.35	1.36	21.61	28.50
6060	1-1/2" diameter		72	.222		8.90	13.55	1.49	23.94	31.50
6070	2" diameter		57	.281		11.10	17.10	1.89	30.09	40
6080	2-1/2" diameter		44	.364		17.10	22	2.44	41.54	54.50
6090	3" diameter		40	.400		22	24.50	2.69	49.19	63.50
6100	3-1/2" diameter		34	.471		27	28.50	3.16	58.66	76
6110	4" diameter		33	.485		31.50	29.50	3.26	64.26	82
6120	5" diameter, A-106B		26	.615		56	37.50	4.14	97.64	122
6130	6" diameter, A-106B	Q-16	30	.800		70	50.50	3.58	124.08	156
6140	8" diameter, A-106B		25	.960		107	60.50	4.30	171.80	213
6150	10" diameter, A-106B		20	1.200		163	76	5.40	244.40	298
6160	12" diameter, A-106B		15	1.600		355	101	7.15	463.15	550
6540	To delete hanger, subtract								
6550	1" diam. to 1-1/2" diam.				30%	14%			
6560	2" diam. to 3" diam.				23%	9%			
6570	3-1/2" diam. to 5" diam.				12%	6%			
6580	6" diam. to 12" diam.				10%	4%			
7250	Flanged, 300 lb. weld neck, on yoke & roll hangers								
7260	sized for covering, 10' OC								
7290	Black, 1" diameter	Q-15	66	.242	L.F.	12.90	14.75	1.63	29.28	38
7300	1-1/4" diameter		61	.262		14.90	16	1.76	32.66	42.50
7310	1-1/2" diameter		54	.296		15.95	18.05	1.99	35.99	46.50
7320	2" diameter		42	.381		20.50	23	2.56	46.06	60
7330	2-1/2" diameter		33	.485		28	29.50	3.26	60.76	78
7340	3" diameter		29	.552		31.50	33.50	3.71	68.71	88.50
7350	3-1/2" diameter		24	.667		43.50	40.50	4.48	88.48	113
7360	4" diameter		23	.696		48	42.50	4.67	95.17	121
7370	5" diameter		19	.842		80.50	51.50	5.65	137.65	171
7380	6" diameter	Q-16	23	1.043		92.50	66	4.67	163.17	206
7390	8" diameter		17	1.412		150	89	6.30	245.30	305
7400	10" diameter		14	1.714		248	108	7.70	363.70	440
7410	12" diameter		12	2		445	126	8.95	579.95	690
7470	For 600 lb. flanges, add				100%				
7940	To delete flanges & hanger, subtract								
7950	1" diam. to 1-1/2" diam.				75%	66%			
7960	2" diam. to 3" diam.				62%	60%			
7970	3-1/2" diam. to 5" diam.				54%	66%			
7980	6" diam. to 12" diam.				55%	62%			
9000	Threading pipe labor, one end, all schedules through 80								
9010	1/4" through 3/4" pipe size	1 Plum	80	.100	Ea.		6.75		6.75	10.10
9020	1" through 2" pipe size		73	.110			7.40		7.40	11.05
9030	2-1/2" pipe size		53	.151			10.20		10.20	15.25
9040	3" pipe size		50	.160			10.85		10.85	16.15
9050	3-1/2" pipe size	Q-1	89	.180			10.95		10.95	16.35
9060	4" pipe size		73	.219			13.35		13.35	19.95
9070	5" pipe size		53	.302			18.40		18.40	27.50
9080	6" pipe size		46	.348			21		21	31.50

22 11 13.44 Pipe, Steel	Crew	Daily Output	Labor-Hours	Unit	Material	2021 Bare Costs Labor	2021 Bare Costs Equipment	Total	Total Incl O&P
9090	8" pipe size	Q-1	29	.552	Ea.		33.50		33.50	50
9100	10" pipe size		21	.762			46.50		46.50	69.50
9110	12" pipe size		13	1.231			75		75	112
9120	Cutting pipe labor, one cut									
9124	Shop fabrication, machine cut									
9126	Schedule 40, straight pipe									
9128	2" pipe size or less	1 Stpi	62	.129	Ea.		8.80		8.80	13.15
9130	2-1/2" pipe size		56	.143			9.75		9.75	14.55
9132	3" pipe size		42	.190			13		13	19.45
9134	4" pipe size		31	.258			17.65		17.65	26.50
9136	5" pipe size		26	.308			21		21	31.50
9138	6" pipe size		19	.421			29		29	43
9140	8" pipe size		14	.571			39		39	58.50
9142	10" pipe size		10	.800			54.50		54.50	81.50
9144	12" pipe size		7	1.143			78		78	117
9146	14" pipe size	Q-5	10.50	1.524			94		94	140
9148	16" pipe size		8.60	1.860			114		114	171
9150	18" pipe size		7	2.286			141		141	210
9152	20" pipe size		5.80	2.759			170		170	253
9154	24" pipe size		4	4			246		246	365
9160	Schedule 80, straight pipe									
9164	2" pipe size or less	1 Stpi	42	.190	Ea.		13		13	19.45
9166	2-1/2" pipe size		37	.216			14.80		14.80	22
9168	3" pipe size		31	.258			17.65		17.65	26.50
9170	4" pipe size		23	.348			24		24	35.50
9172	5" pipe size		18	.444			30.50		30.50	45.50
9174	6" pipe size		14.60	.548			37.50		37.50	56
9176	8" pipe size		10	.800			54.50		54.50	81.50
9178	10" pipe size	Q-5	14	1.143			70.50		70.50	105
9180	12" pipe size	"	10	1.600			98.50		98.50	147
9200	Welding labor per joint									
9210	Schedule 40									
9230	1/2" pipe size	Q-15	32	.500	Ea.		30.50	3.36	33.86	49
9240	3/4" pipe size		27	.593			36	3.98	39.98	58.50
9250	1" pipe size		23	.696			42.50	4.67	47.17	68.50
9260	1-1/4" pipe size		20	.800			48.50	5.40	53.90	78.50
9270	1-1/2" pipe size		19	.842			51.50	5.65	57.15	82.50
9280	2" pipe size		16	1			61	6.70	67.70	98.50
9290	2-1/2" pipe size		13	1.231			75	8.25	83.25	121
9300	3" pipe size		12	1.333			81	8.95	89.95	131
9310	4" pipe size		10	1.600			97.50	10.75	108.25	157
9320	5" pipe size		9	1.778			108	11.95	119.95	175
9330	6" pipe size		8	2			122	13.45	135.45	197
9340	8" pipe size		5	3.200			195	21.50	216.50	315
9350	10" pipe size		4	4			244	27	271	395
9360	12" pipe size		3	5.333			325	36	361	525
9370	14" pipe size		2.60	6.154			375	41.50	416.50	605
9380	16" pipe size		2.20	7.273			445	49	494	715
9390	18" pipe size		2	8			485	54	539	785
9400	20" pipe size		1.80	8.889			540	59.50	599.50	875
9410	22" pipe size		1.70	9.412			575	63.50	638.50	925
9420	24" pipe size		1.50	10.667			650	71.50	721.50	1,050
9450	Schedule 80									

22 11 13 – Facility Water Distribution Piping

22 11 13.44 Pipe, Steel

		Crew	Daily Output	Labor-Hours	Unit	Material	2021 Bare Costs Labor	2021 Bare Costs Equipment	Total	Total Incl O&P
9460	1/2" pipe size	Q-15	27	.593	Ea.		36	3.98	39.98	58.50
9470	3/4" pipe size		23	.696			42.50	4.67	47.17	68.50
9480	1" pipe size		20	.800			48.50	5.40	53.90	78.50
9490	1-1/4" pipe size		19	.842			51.50	5.65	57.15	82.50
9500	1-1/2" pipe size		18	.889			54	5.95	59.95	87.50
9510	2" pipe size		15	1.067			65	7.15	72.15	105
9520	2-1/2" pipe size		12	1.333			81	8.95	89.95	131
9530	3" pipe size		11	1.455			88.50	9.75	98.25	143
9540	4" pipe size		8	2			122	13.45	135.45	197
9550	5" pipe size		6	2.667			162	17.90	179.90	262
9560	6" pipe size		5	3.200			195	21.50	216.50	315
9570	8" pipe size		4	4			244	27	271	395
9580	10" pipe size		3	5.333			325	36	361	525
9590	12" pipe size		2	8			485	54	539	785
9600	14" pipe size	Q-16	2.60	9.231			585	41.50	626.50	915
9610	16" pipe size		2.30	10.435			660	47	707	1,025
9620	18" pipe size		2	12			760	54	814	1,175
9630	20" pipe size		1.80	13.333			840	59.50	899.50	1,325
9640	22" pipe size		1.60	15			950	67	1,017	1,500
9650	24" pipe size		1.50	16			1,000	71.50	1,071.50	1,575

22 11 13.45 Pipe Fittings, Steel, Threaded

		Crew	Daily Output	Labor-Hours	Unit	Material	2021 Bare Costs Labor	2021 Bare Costs Equipment	Total	Total Incl O&P
0010	**PIPE FITTINGS, STEEL, THREADED**									
0020	Cast iron									
0040	Standard weight, black									
0060	90° elbow, straight									
0070	1/4"	1 Plum	16	.500	Ea.	12.65	34		46.65	64.50
0080	3/8"		16	.500		18.35	34		52.35	70.50
0090	1/2"		15	.533		8.05	36		44.05	63
0100	3/4"		14	.571		8.35	38.50		46.85	66.50
0110	1"		13	.615		9.95	41.50		51.45	73
0120	1-1/4"	Q-1	22	.727		14.10	44.50		58.60	81.50
0130	1-1/2"		20	.800		19.50	48.50		68	94
0140	2"		18	.889		30.50	54		84.50	115
0150	2-1/2"		14	1.143		73	69.50		142.50	184
0160	3"		10	1.600		119	97.50		216.50	276
0170	3-1/2"		8	2		325	122		447	535
0180	4"		6	2.667		221	162		383	485
0250	45° elbow, straight									
0260	1/4"	1 Plum	16	.500	Ea.	16.40	34		50.40	68.50
0270	3/8"		16	.500		17.60	34		51.60	70
0280	1/2"		15	.533		12.30	36		48.30	67.50
0300	3/4"		14	.571		12.40	38.50		50.90	71
0320	1"		13	.615		14.50	41.50		56	78
0330	1-1/4"	Q-1	22	.727		19.50	44.50		64	87.50
0340	1-1/2"		20	.800		32.50	48.50		81	108
0350	2"		18	.889		37.50	54		91.50	122
0360	2-1/2"		14	1.143		97	69.50		166.50	211
0370	3"		10	1.600		154	97.50		251.50	315
0380	3-1/2"		8	2		380	122		502	595
0400	4"		6	2.667		320	162		482	590
0500	Tee, straight									
0510	1/4"	1 Plum	10	.800	Ea.	21	54		75	104

For customer support on your Plumbing Costs with RSMeans Data, call 800.448.8182.

175

22 11 13.45 Pipe Fittings, Steel, Threaded		Crew	Daily Output	Labor-Hours	Unit	Material	2021 Bare Costs Labor	Equipment	Total	Total Incl O&P
0520	3/8"	1 Plum	10	.800	Ea.	19.30	54		73.30	102
0530	1/2"		9	.889		12.50	60		72.50	104
0540	3/4"		9	.889		14.60	60		74.60	106
0550	1"		8	1		13	67.50		80.50	115
0560	1-1/4"	Q-1	14	1.143		23.50	69.50		93	130
0570	1-1/2"		13	1.231		31	75		106	146
0580	2"		11	1.455		43	88.50		131.50	179
0590	2-1/2"		9	1.778		112	108		220	285
0600	3"		6	2.667		171	162		333	430
0610	3-1/2"		5	3.200		345	195		540	670
0620	4"		4	4		335	244		579	730
0660	Tee, reducing, run or outlet									
0661	1/2"	1 Plum	9	.889	Ea.	32.50	60		92.50	126
0662	3/4"		9	.889		20.50	60		80.50	113
0663	1"		8	1		17.60	67.50		85.10	120
0664	1-1/4"	Q-1	14	1.143		23.50	69.50		93	130
0665	1-1/2"		13	1.231		43.50	75		118.50	160
0666	2"		11	1.455		51.50	88.50		140	189
0667	2-1/2"		9	1.778		127	108		235	300
0668	3"		6	2.667		220	162		382	485
0669	3-1/2"		5	3.200		560	195		755	905
0670	4"		4	4		430	244		674	840
0674	Reducer, concentric									
0675	3/4"	1 Plum	18	.444	Ea.	22.50	30		52.50	69.50
0676	1"	"	15	.533		15.25	36		51.25	71
0677	1-1/4"	Q-1	26	.615		39.50	37.50		77	99.50
0678	1-1/2"		24	.667		56	40.50		96.50	122
0679	2"		21	.762		73	46.50		119.50	150
0680	2-1/2"		18	.889		104	54		158	195
0681	3"		14	1.143		173	69.50		242.50	294
0682	3-1/2"		12	1.333		340	81		421	495
0683	4"		10	1.600		355	97.50		452.50	535
0687	Reducer, eccentric									
0688	3/4"	1 Plum	16	.500	Ea.	47.50	34		81.50	103
0689	1"	"	14	.571		51	38.50		89.50	114
0690	1-1/4"	Q-1	25	.640		76	39		115	142
0691	1-1/2"		22	.727		103	44.50		147.50	179
0692	2"		20	.800		137	48.50		185.50	223
0693	2-1/2"		16	1		209	61		270	320
0694	3"		12	1.333		325	81		406	480
0695	3-1/2"		10	1.600		505	97.50		602.50	700
0696	4"		9	1.778		655	108		763	880
0700	Standard weight, galvanized cast iron									
0720	90° elbow, straight									
0730	1/4"	1 Plum	16	.500	Ea.	16.80	34		50.80	69
0740	3/8"		16	.500		16.80	34		50.80	69
0750	1/2"		15	.533		21	36		57	77.50
0760	3/4"		14	.571		16.35	38.50		54.85	75.50
0770	1"		13	.615		21.50	41.50		63	86
0780	1-1/4"	Q-1	22	.727		29	44.50		73.50	98
0790	1-1/2"		20	.800		46.50	48.50		95	124
0800	2"		18	.889		59	54		113	146
0810	2-1/2"		14	1.143		122	69.50		191.50	238

22 11 Facility Water Distribution

22 11 13 – Facility Water Distribution Piping

22 11 13.45 Pipe Fittings, Steel, Threaded	Crew	Daily Output	Labor-Hours	Unit	Material	2021 Bare Costs Labor	Equipment	Total	Total Incl O&P
0820	3"	Q-1	10	1.600	Ea.	185	97.50		282.50	350
0830	3-1/2"		8	2		295	122		417	505
0840	4"		6	2.667		340	162		502	615
0900	45° elbow, straight									
0910	1/4"	1 Plum	16	.500	Ea.	23	34		57	76
0920	3/8"		16	.500		20	34		54	72.50
0930	1/2"		15	.533		16.55	36		52.55	72
0940	3/4"		14	.571		18.85	38.50		57.35	78.50
0950	1"		13	.615		37	41.50		78.50	103
0960	1-1/4"	Q-1	22	.727		55	44.50		99.50	127
0970	1-1/2"		20	.800		64.50	48.50		113	144
0980	2"		18	.889		90.50	54		144.50	181
0990	2-1/2"		14	1.143		136	69.50		205.50	254
1000	3"		10	1.600		284	97.50		381.50	455
1010	3-1/2"		8	2		525	122		647	760
1020	4"		6	2.667		495	162		657	785
1100	Tee, straight									
1110	1/4"	1 Plum	10	.800	Ea.	20.50	54		74.50	104
1120	3/8"		10	.800		17.70	54		71.70	100
1130	1/2"		9	.889		28.50	60		88.50	122
1140	3/4"		9	.889		27	60		87	120
1150	1"		8	1		25.50	67.50		93	129
1160	1-1/4"	Q-1	14	1.143		44.50	69.50		114	153
1170	1-1/2"		13	1.231		58.50	75		133.50	177
1180	2"		11	1.455		73	88.50		161.50	213
1190	2-1/2"		9	1.778		150	108		258	325
1200	3"		6	2.667		490	162		652	780
1210	3-1/2"		5	3.200		420	195		615	750
1220	4"		4	4		580	244		824	1,000
1300	Extra heavy weight, black									
1310	Couplings, steel straight									
1320	1/4"	1 Plum	19	.421	Ea.	5.70	28.50		34.20	49
1330	3/8"		19	.421		6.20	28.50		34.70	49.50
1340	1/2"		19	.421		8.40	28.50		36.90	51.50
1350	3/4"		18	.444		8.95	30		38.95	55
1360	1"		15	.533		11.40	36		47.40	66.50
1370	1-1/4"	Q-1	26	.615		18.30	37.50		55.80	76
1380	1-1/2"		24	.667		18.30	40.50		58.80	80.50
1390	2"		21	.762		28	46.50		74.50	100
1400	2-1/2"		18	.889		41.50	54		95.50	127
1410	3"		14	1.143		49	69.50		118.50	158
1420	3-1/2"		12	1.333		66	81		147	194
1430	4"		10	1.600		77.50	97.50		175	231
1510	90° elbow, straight									
1520	1/2"	1 Plum	15	.533	Ea.	43.50	36		79.50	102
1530	3/4"		14	.571		44	38.50		82.50	106
1540	1"		13	.615		53.50	41.50		95	121
1550	1-1/4"	Q-1	22	.727		79.50	44.50		124	154
1560	1-1/2"		20	.800		98.50	48.50		147	181
1580	2"		18	.889		122	54		176	215
1590	2-1/2"		14	1.143		282	69.50		351.50	415
1600	3"		10	1.600		395	97.50		492.50	580
1610	4"		6	2.667		835	162		997	1,150

For customer support on your Plumbing Costs with RSMeans Data, call 800.448.8182.

177

22 11 13 – Facility Water Distribution Piping

22 11 13.45 Pipe Fittings, Steel, Threaded	Crew	Daily Output	Labor-Hours	Unit	Material	2021 Bare Costs Labor	Equipment	Total	Total Incl O&P
1650	45° elbow, straight									
1660	1/2"	1 Plum	15	.533	Ea.	59	36		95	119
1670	3/4"		14	.571		59.50	38.50		98	123
1680	1"		13	.615		71.50	41.50		113	141
1690	1-1/4"	Q-1	22	.727		118	44.50		162.50	196
1700	1-1/2"		20	.800		130	48.50		178.50	216
1710	2"		18	.889		185	54		239	284
1720	2-1/2"		14	1.143		305	69.50		374.50	440
1800	Tee, straight									
1810	1/2"	1 Plum	9	.889	Ea.	67.50	60		127.50	165
1820	3/4"		9	.889		67.50	60		127.50	165
1830	1"		8	1		81.50	67.50		149	191
1840	1-1/4"	Q-1	14	1.143		122	69.50		191.50	238
1850	1-1/2"		13	1.231		157	75		232	284
1860	2"		11	1.455		195	88.50		283.50	345
1870	2-1/2"		9	1.778		415	108		523	615
1880	3"		6	2.667		580	162		742	880
1890	4"		4	4		1,125	244		1,369	1,625
4000	Standard weight, black									
4010	Couplings, steel straight, merchants									
4030	1/4"	1 Plum	19	.421	Ea.	1.53	28.50		30.03	44
4040	3/8"		19	.421		1.83	28.50		30.33	44.50
4050	1/2"		19	.421		1.95	28.50		30.45	44.50
4060	3/4"		18	.444		2.49	30		32.49	47.50
4070	1"		15	.533		3.49	36		39.49	58
4080	1-1/4"	Q-1	26	.615		4.45	37.50		41.95	61
4090	1-1/2"		24	.667		5.65	40.50		46.15	66.50
4100	2"		21	.762		8.05	46.50		54.55	78.50
4110	2-1/2"		18	.889		26.50	54		80.50	111
4120	3"		14	1.143		37.50	69.50		107	145
4130	3-1/2"		12	1.333		66.50	81		147.50	194
4140	4"		10	1.600		66.50	97.50		164	218
4166	Plug, 1/4"	1 Plum	38	.211		3.59	14.25		17.84	25.50
4167	3/8"		38	.211		2.98	14.25		17.23	25
4168	1/2"		38	.211		3.58	14.25		17.83	25.50
4169	3/4"		32	.250		9.60	16.95		26.55	36
4170	1"		30	.267		10.20	18.05		28.25	38
4171	1-1/4"	Q-1	52	.308		11.75	18.75		30.50	41
4172	1-1/2"		48	.333		16.65	20.50		37.15	49
4173	2"		42	.381		21.50	23		44.50	58
4176	2-1/2"		36	.444		32	27		59	75.50
4180	4"		20	.800		66.50	48.50		115	146
4200	Standard weight, galvanized									
4210	Couplings, steel straight, merchants									
4230	1/4"	1 Plum	19	.421	Ea.	1.76	28.50		30.26	44.50
4240	3/8"		19	.421		2.25	28.50		30.75	45
4250	1/2"		19	.421		2.39	28.50		30.89	45
4260	3/4"		18	.444		3	30		33	48.50
4270	1"		15	.533		4.19	36		40.19	58.50
4280	1-1/4"	Q-1	26	.615		5.35	37.50		42.85	62
4290	1-1/2"		24	.667		6.65	40.50		47.15	68
4300	2"		21	.762		9.95	46.50		56.45	80.50
4310	2-1/2"		18	.889		25	54		79	108

22 11 13 – Facility Water Distribution Piping

22 11 13.45 Pipe Fittings, Steel, Threaded	Crew	Daily Output	Labor-Hours	Unit	Material	2021 Bare Costs Labor	Equipment	Total	Total Incl O&P
4320	3"	Q-1	14	1.143	Ea.	35.50	69.50		105	143
4330	3-1/2"		12	1.333		76.50	81		157.50	205
4340	4"		10	1.600		76.50	97.50		174	229
4370	Plug, galvanized, square head									
4374	1/2"	1 Plum	38	.211	Ea.	9.30	14.25		23.55	32
4375	3/4"		32	.250		8.65	16.95		25.60	35
4376	1"		30	.267		8.65	18.05		26.70	36.50
4377	1-1/4"	Q-1	52	.308		14.40	18.75		33.15	44
4378	1-1/2"		48	.333		19.45	20.50		39.95	52
4379	2"		42	.381		24	23		47	61
4380	2-1/2"		36	.444		51	27		78	96.50
4381	3"		28	.571		65.50	35		100.50	124
4382	4"		20	.800		179	48.50		227.50	270
4385	8"		20	.800		127	48.50		175.50	213
4700	Nipple, black									
4710	1/2" x 4" long	1 Plum	19	.421	Ea.	3.49	28.50		31.99	46.50
4712	3/4" x 4" long		18	.444		4.20	30		34.20	49.50
4714	1" x 4" long		15	.533		5.85	36		41.85	60.50
4716	1-1/4" x 4" long	Q-1	26	.615		7.30	37.50		44.80	64
4718	1-1/2" x 4" long		24	.667		8.60	40.50		49.10	70
4720	2" x 4" long		21	.762		12	46.50		58.50	82.50
4722	2-1/2" x 4" long		18	.889		32.50	54		86.50	117
4724	3" x 4" long		14	1.143		41.50	69.50		111	150
4726	4" x 4" long		10	1.600		56	97.50		153.50	207
4800	Nipple, galvanized									
4810	1/2" x 4" long	1 Plum	19	.421	Ea.	4.26	28.50		32.76	47
4812	3/4" x 4" long		18	.444		5.30	30		35.30	51
4814	1" x 4" long		15	.533		7.05	36		43.05	62
4816	1-1/4" x 4" long	Q-1	26	.615		8.70	37.50		46.20	65.50
4818	1-1/2" x 4" long		24	.667		11	40.50		51.50	72.50
4820	2" x 4" long		21	.762		13.95	46.50		60.45	85
4822	2-1/2" x 4" long		18	.889		36.50	54		90.50	122
4824	3" x 4" long		14	1.143		48.50	69.50		118	158
4826	4" x 4" long		10	1.600		65.50	97.50		163	217
5000	Malleable iron, 150 lb.									
5020	Black									
5040	90° elbow, straight									
5060	1/4"	1 Plum	16	.500	Ea.	6.25	34		40.25	57.50
5070	3/8"		16	.500		6.25	34		40.25	57.50
5080	1/2"		15	.533		4.32	36		40.32	59
5090	3/4"		14	.571		5.20	38.50		43.70	63.50
5100	1"		13	.615		9.05	41.50		50.55	72
5110	1-1/4"	Q-1	22	.727		14.95	44.50		59.45	82.50
5120	1-1/2"		20	.800		19.65	48.50		68.15	94
5130	2"		18	.889		34	54		88	119
5140	2-1/2"		14	1.143		75.50	69.50		145	187
5150	3"		10	1.600		110	97.50		207.50	267
5160	3-1/2"		8	2		320	122		442	530
5170	4"		6	2.667		211	162		373	475
5250	45° elbow, straight									
5270	1/4"	1 Plum	16	.500	Ea.	9.40	34		43.40	61
5280	3/8"		16	.500		9.40	34		43.40	61
5290	1/2"		15	.533		7.15	36		43.15	62

For customer support on your Plumbing Costs with RSMeans Data, call 800.448.8182.

179

22 11 13 – Facility Water Distribution Piping

22 11 13.45 Pipe Fittings, Steel, Threaded	Crew	Daily Output	Labor-Hours	Unit	Material	2021 Bare Costs Labor	Equipment	Total	Total Incl O&P
5300	3/4"	1 Plum	14	.571	Ea.	8.85	38.50		47.35	67
5310	1"		13	.615		11.10	41.50		52.60	74.50
5320	1-1/4"	Q-1	22	.727		19.65	44.50		64.15	87.50
5330	1-1/2"		20	.800		24.50	48.50		73	99.50
5340	2"		18	.889		37	54		91	122
5350	2-1/2"		14	1.143		106	69.50		175.50	221
5360	3"		10	1.600		138	97.50		235.50	297
5370	3-1/2"		8	2		244	122		366	450
5380	4"		6	2.667		271	162		433	540
5450	Tee, straight									
5470	1/4"	1 Plum	10	.800	Ea.	9.05	54		63.05	91
5480	3/8"		10	.800		9.05	54		63.05	91
5490	1/2"		9	.889		5.80	60		65.80	96.50
5500	3/4"		9	.889		8.30	60		68.30	99
5510	1"		8	1		14.20	67.50		81.70	117
5520	1-1/4"	Q-1	14	1.143		23	69.50		92.50	130
5530	1-1/2"		13	1.231		28.50	75		103.50	144
5540	2"		11	1.455		49	88.50		137.50	186
5550	2-1/2"		9	1.778		105	108		213	278
5560	3"		6	2.667		155	162		317	410
5570	3-1/2"		5	3.200		375	195		570	705
5580	4"		4	4		375	244		619	775
5601	Tee, reducing, on outlet									
5602	1/2"	1 Plum	9	.889	Ea.	9.85	60		69.85	101
5603	3/4"		9	.889		12.85	60		72.85	104
5604	1"		8	1		16.20	67.50		83.70	119
5605	1-1/4"	Q-1	14	1.143		28	69.50		97.50	135
5606	1-1/2"		13	1.231		40.50	75		115.50	157
5607	2"		11	1.455		69	88.50		157.50	208
5608	2-1/2"		9	1.778		137	108		245	310
5609	3"		6	2.667		217	162		379	480
5610	3-1/2"		5	3.200		465	195		660	805
5611	4"		4	4		445	244		689	850
5650	Coupling									
5670	1/4"	1 Plum	19	.421	Ea.	7.75	28.50		36.25	51
5680	3/8"		19	.421		7.75	28.50		36.25	51
5690	1/2"		19	.421		6	28.50		34.50	49
5700	3/4"		18	.444		7	30		37	52.50
5710	1"		15	.533		10.50	36		46.50	65.50
5720	1-1/4"	Q-1	26	.615		12.05	37.50		49.55	69.50
5730	1-1/2"		24	.667		16.30	40.50		56.80	78.50
5740	2"		21	.762		24	46.50		70.50	96
5750	2-1/2"		18	.889		66.50	54		120.50	154
5760	3"		14	1.143		90	69.50		159.50	203
5770	3-1/2"		12	1.333		171	81		252	310
5780	4"		10	1.600		181	97.50		278.50	345
5840	Reducer, concentric, 1/4"	1 Plum	19	.421		8.15	28.50		36.65	51.50
5850	3/8"		19	.421		8.15	28.50		36.65	51.50
5860	1/2"		19	.421		7	28.50		35.50	50
5870	3/4"		16	.500		7.85	34		41.85	59
5880	1"		15	.533		12.20	36		48.20	67.50
5890	1-1/4"	Q-1	26	.615		15.75	37.50		53.25	73.50
5900	1-1/2"		24	.667		19.90	40.50		60.40	82.50

22 11 13.45 Pipe Fittings, Steel, Threaded

		Crew	Daily Output	Labor-Hours	Unit	Material	2021 Bare Costs Labor	Equipment	Total	Total Incl O&P
5910	2"	Q-1	21	.762	Ea.	29	46.50		75.50	101
5911	2-1/2"		18	.889		66	54		120	154
5912	3"		14	1.143		100	69.50		169.50	214
5913	3-1/2"		12	1.333		345	81		426	495
5914	4"		10	1.600		325	97.50		422.50	505
5981	Bushing, 1/4"	1 Plum	19	.421		1.19	28.50		29.69	44
5982	3/8"		19	.421		1.41	28.50		29.91	44
5983	1/2"		19	.421		6.30	28.50		34.80	49.50
5984	3/4"		16	.500		7	34		41	58
5985	1"		15	.533		7.40	36		43.40	62
5986	1-1/4"	Q-1	26	.615		9.65	37.50		47.15	66.50
5987	1-1/2"		24	.667		12.20	40.50		52.70	74
5988	2"		21	.762		15.40	46.50		61.90	86.50
5989	Cap, 1/4"	1 Plum	38	.211		6.95	14.25		21.20	29
5991	3/8"		38	.211		6.65	14.25		20.90	29
5992	1/2"		38	.211		4.36	14.25		18.61	26.50
5993	3/4"		32	.250		5.90	16.95		22.85	32
5994	1"		30	.267		7.15	18.05		25.20	35
5995	1-1/4"	Q-1	52	.308		9.45	18.75		28.20	38.50
5996	1-1/2"		48	.333		12.95	20.50		33.45	45
5997	2"		42	.381		18.90	23		41.90	55.50
6000	For galvanized elbows, tees, and couplings, add					20%				
6058	For galvanized reducers, caps and bushings, add					20%				
6100	90° elbow, galvanized, 150 lb., reducing									
6110	3/4" x 1/2"	1 Plum	15.40	.519	Ea.	12.45	35		47.45	66
6112	1" x 3/4"		14	.571		15.95	38.50		54.45	75
6114	1" x 1/2"		14.50	.552		16.95	37.50		54.45	74.50
6116	1-1/4" x 1"	Q-1	24.20	.661		26.50	40.50		67	89.50
6118	1-1/4" x 3/4"		25.40	.630		32	38.50		70.50	93
6120	1-1/4" x 1/2"		26.20	.611		34	37		71	93
6122	1-1/2" x 1-1/4"		21.60	.741		42.50	45		87.50	115
6124	1-1/2" x 1"		23.50	.681		42.50	41.50		84	109
6126	1-1/2" x 3/4"		24.60	.650		42.50	39.50		82	106
6128	2" x 1-1/2"		20.50	.780		50	47.50		97.50	126
6130	2" x 1-1/4"		21	.762		57	46.50		103.50	133
6132	2" x 1"		22.80	.702		59.50	43		102.50	130
6134	2" x 3/4"		23.90	.669		60.50	41		101.50	128
6136	2-1/2" x 2"		12.30	1.301		169	79.50		248.50	305
6138	2-1/2" x 1-1/2"		12.50	1.280		189	78		267	325
6140	3" x 2-1/2"		8.60	1.860		310	113		423	510
6142	3" x 2"		11.80	1.356		267	82.50		349.50	415
6144	4" x 3"		8.20	1.951		695	119		814	940
6160	90° elbow, black, 150 lb., reducing									
6170	1" x 3/4"	1 Plum	14	.571	Ea.	10.90	38.50		49.40	69.50
6174	1-1/2" x 1"	Q-1	23.50	.681		26	41.50		67.50	90.50
6178	1-1/2" x 3/4"		24.60	.650		29.50	39.50		69	91.50
6182	2" x 1-1/2"		20.50	.780		37.50	47.50		85	112
6186	2" x 1"		22.80	.702		42.50	43		85.50	111
6190	2" x 3/4"		23.90	.669		45	41		86	111
6194	2-1/2" x 2"		12.30	1.301		103	79.50		182.50	231
7000	Union, with brass seat									
7010	1/4"	1 Plum	15	.533	Ea.	26	36		62	83
7020	3/8"		15	.533		22.50	36		58.50	78.50

For customer support on your Plumbing Costs with RSMeans Data, call 800.448.8182.

181

22 11 13.45 Pipe Fittings, Steel, Threaded

		Crew	Daily Output	Labor-Hours	Unit	Material	2021 Bare Costs Labor	Equipment	Total	Total Incl O&P
7030	1/2"	1 Plum	14	.571	Ea.	20	38.50		58.50	79.50
7040	3/4"		13	.615		18.75	41.50		60.25	82.50
7050	1"	↓	12	.667		24.50	45		69.50	94.50
7060	1-1/4"	Q-1	21	.762		43.50	46.50		90	117
7070	1-1/2"		19	.842		54	51.50		105.50	136
7080	2"		17	.941		63	57.50		120.50	155
7090	2-1/2"		13	1.231		187	75		262	320
7100	3"	↓	9	1.778	↓	182	108		290	365
7120	Union, galvanized									
7124	1/2"	1 Plum	14	.571	Ea.	27	38.50		65.50	87
7125	3/4"		13	.615		31	41.50		72.50	96
7126	1"	↓	12	.667		40.50	45		85.50	112
7127	1-1/4"	Q-1	21	.762		58.50	46.50		105	134
7128	1-1/2"		19	.842		70.50	51.50		122	154
7129	2"		17	.941		81	57.50		138.50	175
7130	2-1/2"		13	1.231		282	75		357	420
7131	3"	↓	9	1.778	↓	395	108		503	595
7500	Malleable iron, 300 lb.									
7520	Black									
7540	90° elbow, straight, 1/4"	1 Plum	16	.500	Ea.	22	34		56	74.50
7560	3/8"		16	.500		19.45	34		53.45	72
7570	1/2"		15	.533		26.50	36		62.50	83
7580	3/4"		14	.571		28.50	38.50		67	89
7590	1"	↓	13	.615		36.50	41.50		78	102
7600	1-1/4"	Q-1	22	.727		53	44.50		97.50	125
7610	1-1/2"		20	.800		62.50	48.50		111	142
7620	2"		18	.889		90	54		144	180
7630	2-1/2"		14	1.143		234	69.50		303.50	360
7640	3"		10	1.600		268	97.50		365.50	440
7650	4"	↓	6	2.667		660	162		822	965
7700	45° elbow, straight, 1/4"	1 Plum	16	.500		32.50	34		66.50	86.50
7720	3/8"		16	.500		32.50	34		66.50	86.50
7730	1/2"		15	.533		36.50	36		72.50	94
7740	3/4"		14	.571		40.50	38.50		79	102
7750	1"	↓	13	.615		44.50	41.50		86	111
7760	1-1/4"	Q-1	22	.727		71.50	44.50		116	145
7770	1-1/2"		20	.800		93	48.50		141.50	176
7780	2"		18	.889		141	54		195	236
7790	2-1/2"		14	1.143		330	69.50		399.50	470
7800	3"		10	1.600		435	97.50		532.50	620
7810	4"	↓	6	2.667		985	162		1,147	1,325
7850	Tee, straight, 1/4"	1 Plum	10	.800		27.50	54		81.50	112
7870	3/8"		10	.800		29	54		83	113
7880	1/2"		9	.889		37	60		97	131
7890	3/4"		9	.889		40	60		100	134
7900	1"	↓	8	1		48.50	67.50		116	154
7910	1-1/4"	Q-1	14	1.143		69	69.50		138.50	180
7920	1-1/2"		13	1.231		80.50	75		155.50	201
7930	2"		11	1.455		119	88.50		207.50	263
7940	2-1/2"		9	1.778		305	108		413	495
7950	3"		6	2.667		415	162		577	695
7960	4"	↓	4	4		1,225	244		1,469	1,725
8050	Couplings, straight, 1/4"	1 Plum	19	.421		19.75	28.50		48.25	64

For customer support on your Plumbing Costs with RSMeans Data, call 800.448.8182.

22 11 13.45 Pipe Fittings, Steel, Threaded	Crew	Daily Output	Labor-Hours	Unit	Material	2021 Bare Costs Labor	Equipment	Total	Total Incl O&P
8070	3/8"	1 Plum	19	.421	Ea.	19.75	28.50		48.25	64
8080	1/2"		19	.421		22	28.50		50.50	66.50
8090	3/4"		18	.444		25.50	30		55.50	73
8100	1"		15	.533		29	36		65	86
8110	1-1/4"	Q-1	26	.615		34.50	37.50		72	94
8120	1-1/2"		24	.667		51.50	40.50		92	118
8130	2"		21	.762		74.50	46.50		121	152
8140	2-1/2"		18	.889		143	54		197	239
8150	3"		14	1.143		207	69.50		276.50	330
8160	4"		10	1.600		375	97.50		472.50	560
8162	6"		10	1.600		375	97.50		472.50	560
8200	Galvanized									
8220	90° elbow, straight, 1/4"	1 Plum	16	.500	Ea.	39	34		73	93.50
8222	3/8"		16	.500		40	34		74	94.50
8224	1/2"		15	.533		47.50	36		83.50	107
8226	3/4"		14	.571		53.50	38.50		92	117
8228	1"		13	.615		68.50	41.50		110	138
8230	1-1/4"	Q-1	22	.727		108	44.50		152.50	184
8232	1-1/2"		20	.800		117	48.50		165.50	201
8234	2"		18	.889		196	54		250	297
8236	2-1/2"		14	1.143		435	69.50		504.50	585
8238	3"		10	1.600		515	97.50		612.50	710
8240	4"		6	2.667		1,600	162		1,762	2,025
8280	45° elbow, straight									
8282	1/2"	1 Plum	15	.533	Ea.	74	36		110	136
8284	3/4"		14	.571		82	38.50		120.50	148
8286	1"		13	.615		89.50	41.50		131	161
8288	1-1/4"	Q-1	22	.727		139	44.50		183.50	219
8290	1-1/2"		20	.800		179	48.50		227.50	270
8292	2"		18	.889		249	54		303	355
8310	Tee, straight, 1/4"	1 Plum	10	.800		56	54		110	143
8312	3/8"		10	.800		57	54		111	144
8314	1/2"		9	.889		71	60		131	169
8316	3/4"		9	.889		78.50	60		138.50	176
8318	1"		8	1		96.50	67.50		164	207
8320	1-1/4"	Q-1	14	1.143		138	69.50		207.50	256
8322	1-1/2"		13	1.231		143	75		218	270
8324	2"		11	1.455		229	88.50		317.50	385
8326	2-1/2"		9	1.778		775	108		883	1,000
8328	3"		6	2.667		860	162		1,022	1,175
8330	4"		4	4		2,200	244		2,444	2,800
8380	Couplings, straight, 1/4"	1 Plum	19	.421		25.50	28.50		54	71
8382	3/8"		19	.421		37	28.50		65.50	83
8384	1/2"		19	.421		38	28.50		66.50	84.50
8386	3/4"		18	.444		43	30		73	92
8388	1"		15	.533		56.50	36		92.50	117
8390	1-1/4"	Q-1	26	.615		74	37.50		111.50	138
8392	1-1/2"		24	.667		104	40.50		144.50	176
8394	2"		21	.762		126	46.50		172.50	209
8396	2-1/2"		18	.889		350	54		404	465
8398	3"		14	1.143		415	69.50		484.50	560
8399	4"		10	1.600		660	97.50		757.50	870
8529	Black									

For customer support on your Plumbing Costs with RSMeans Data, call 800.448.8182.

183

22 11 13.45 Pipe Fittings, Steel, Threaded	Crew	Daily Output	Labor-Hours	Unit	Material	2021 Bare Costs Labor	Equipment	Total	Total Incl O&P
8530	Reducer, concentric, 1/4"	1 Plum	19	.421	Ea.	26.50	28.50		55	71.50
8531	3/8"		19	.421		26	28.50		54.50	71
8532	1/2"		17	.471		31.50	32		63.50	82
8533	3/4"		16	.500		32	34		66	86
8534	1"		15	.533		38.50	36		74.50	96
8535	1-1/4"	Q-1	26	.615		50.50	37.50		88	112
8536	1-1/2"		24	.667		79	40.50		119.50	147
8537	2"		21	.762		99.50	46.50		146	180
8550	Cap, 1/4"	1 Plum	38	.211		20.50	14.25		34.75	44
8551	3/8"		38	.211		20.50	14.25		34.75	44
8552	1/2"		34	.235		20	15.95		35.95	46
8553	3/4"		32	.250		25.50	16.95		42.45	53.50
8554	1"		30	.267		33.50	18.05		51.55	63.50
8555	1-1/4"	Q-1	52	.308		37	18.75		55.75	68.50
8556	1-1/2"		48	.333		54	20.50		74.50	89.50
8557	2"		42	.381		75	23		98	117
8570	Plug, 1/4"	1 Plum	38	.211		3.96	14.25		18.21	26
8571	3/8"		38	.211		2.76	14.25		17.01	24.50
8572	1/2"		34	.235		3.52	15.95		19.47	28
8573	3/4"		32	.250		4.40	16.95		21.35	30.50
8574	1"		30	.267		6.75	18.05		24.80	34.50
8575	1-1/4"	Q-1	52	.308		13.85	18.75		32.60	43
8576	1-1/2"		48	.333		15.85	20.50		36.35	48
8577	2"		42	.381		25	23		48	62
9500	Union with brass seat, 1/4"	1 Plum	15	.533		47.50	36		83.50	106
9530	3/8"		15	.533		48	36		84	107
9540	1/2"		14	.571		38.50	38.50		77	100
9550	3/4"		13	.615		43	41.50		84.50	109
9560	1"		12	.667		56	45		101	129
9570	1-1/4"	Q-1	21	.762		91	46.50		137.50	170
9580	1-1/2"		19	.842		95.50	51.50		147	182
9590	2"		17	.941		118	57.50		175.50	216
9600	2-1/2"		13	1.231		380	75		455	530
9610	3"		9	1.778		495	108		603	705
9620	4"		5	3.200		1,650	195		1,845	2,100
9630	Union, all iron, 1/4"	1 Plum	15	.533		73	36		109	135
9650	3/8"		15	.533		73	36		109	135
9660	1/2"		14	.571		77	38.50		115.50	143
9670	3/4"		13	.615		83	41.50		124.50	153
9680	1"		12	.667		102	45		147	180
9690	1-1/4"	Q-1	21	.762		156	46.50		202.50	241
9700	1-1/2"		19	.842		95	51.50		146.50	182
9710	2"		17	.941		80	57.50		137.50	174
9720	2-1/2"		13	1.231		107	75		182	229
9730	3"		9	1.778		162	108		270	340
9750	For galvanized unions, add					15%				
9757	Forged steel, 3000 lb.									
9758	Black									
9760	90° elbow, 1/4"	1 Plum	16	.500	Ea.	19.05	34		53.05	71.50
9761	3/8"		16	.500		19.05	34		53.05	71.50
9762	1/2"		15	.533		16.45	36		52.45	72
9763	3/4"		14	.571		18.25	38.50		56.75	77.50
9764	1"		13	.615		30	41.50		71.50	94.50

22 11 13 – Facility Water Distribution Piping

22 11 13.45 Pipe Fittings, Steel, Threaded		Crew	Daily Output	Labor-Hours	Unit	Material	2021 Bare Costs Labor	Equipment	Total	Total Incl O&P
9765	1-1/4"	Q-1	22	.727	Ea.	52	44.50		96.50	123
9766	1-1/2"		20	.800		67	48.50		115.50	146
9767	2"	↓	18	.889		81	54		135	170
9780	45° elbow, 1/4"	1 Plum	16	.500		24	34		58	77
9781	3/8"		16	.500		24	34		58	77
9782	1/2"		15	.533		24	36		60	80
9783	3/4"		14	.571		27.50	38.50		66	88
9784	1"	↓	13	.615		37.50	41.50		79	104
9785	1-1/4"	Q-1	22	.727		52	44.50		96.50	123
9786	1-1/2"		20	.800		74.50	48.50		123	155
9787	2"	↓	18	.889		103	54		157	194
9800	Tee, 1/4"	1 Plum	10	.800		23	54		77	107
9801	3/8"		10	.800		23	54		77	107
9802	1/2"		9	.889		21	60		81	113
9803	3/4"		9	.889		28	60		88	121
9804	1"	↓	8	1		36.50	67.50		104	142
9805	1-1/4"	Q-1	14	1.143		72	69.50		141.50	183
9806	1-1/2"		13	1.231		84	75		159	204
9807	2"	↓	11	1.455		91	88.50		179.50	232
9820	Reducer, concentric, 1/4"	1 Plum	19	.421		11.55	28.50		40.05	55
9821	3/8"		19	.421		12.05	28.50		40.55	56
9822	1/2"		17	.471		12.05	32		44.05	61
9823	3/4"		16	.500		14.30	34		48.30	66
9824	1"	↓	15	.533		18.60	36		54.60	74.50
9825	1-1/4"	Q-1	26	.615		38	37.50		75.50	97.50
9826	1-1/2"		24	.667		41	40.50		81.50	106
9827	2"	↓	21	.762		59.50	46.50		106	135
9840	Cap, 1/4"	1 Plum	38	.211		8.45	14.25		22.70	31
9841	3/8"		38	.211		8.75	14.25		23	31
9842	1/2"		34	.235		8.50	15.95		24.45	33.50
9843	3/4"		32	.250		11.85	16.95		28.80	38.50
9844	1"	↓	30	.267		18.20	18.05		36.25	47
9845	1-1/4"	Q-1	52	.308		24	18.75		42.75	54
9846	1-1/2"		48	.333		28.50	20.50		49	61.50
9847	2"	↓	42	.381		41	23		64	79.50
9860	Plug, 1/4"	1 Plum	38	.211		3.25	14.25		17.50	25
9861	3/8"		38	.211		4.23	14.25		18.48	26
9862	1/2"		34	.235		3.51	15.95		19.46	28
9863	3/4"		32	.250		4.43	16.95		21.38	30.50
9864	1"	↓	30	.267		6.80	18.05		24.85	34.50
9865	1-1/4"	Q-1	52	.308		15.95	18.75		34.70	45.50
9866	1-1/2"		48	.333		21	20.50		41.50	54
9867	2"	↓	42	.381		33.50	23		56.50	71.50
9880	Union, bronze seat, 1/4"	1 Plum	15	.533		74	36		110	136
9881	3/8"		15	.533		74	36		110	136
9882	1/2"		14	.571		71	38.50		109.50	136
9883	3/4"		13	.615		96	41.50		137.50	168
9884	1"	↓	12	.667		107	45		152	186
9885	1-1/4"	Q-1	21	.762		186	46.50		232.50	275
9886	1-1/2"		19	.842		215	51.50		266.50	315
9887	2"	↓	17	.941		255	57.50		312.50	365
9900	Coupling, 1/4"	1 Plum	19	.421		7.60	28.50		36.10	51
9901	3/8"	↓	19	.421		7.60	28.50		36.10	51

185

22 11 13 – Facility Water Distribution Piping

22 11 13.45 Pipe Fittings, Steel, Threaded

		Crew	Daily Output	Labor-Hours	Unit	Material	2021 Bare Costs Labor	Equipment	Total	Total Incl O&P
9902	1/2"	1 Plum	17	.471	Ea.	6.10	32		38.10	54
9903	3/4"		16	.500		7.90	34		41.90	59
9904	1"		15	.533		13.75	36		49.75	69
9905	1-1/4"	Q-1	26	.615		23	37.50		60.50	81
9906	1-1/2"		24	.667		29.50	40.50		70	93
9907	2"		21	.762		40.50	46.50		87	114

22 11 13.47 Pipe Fittings, Steel

		Crew	Daily Output	Labor-Hours	Unit	Material	2021 Bare Costs Labor	Equipment	Total	Total Incl O&P
0010	**PIPE FITTINGS, STEEL**, flanged, welded & special									
0020	Flanged joints, CI, standard weight, black. One gasket & bolt									
0040	set, mat'l only, required at each joint, not included (see line 0620)									
0060	90° elbow, straight, 1-1/2" pipe size	Q-1	14	1.143	Ea.	770	69.50		839.50	950
0080	2" pipe size		13	1.231		425	75		500	575
0090	2-1/2" pipe size		12	1.333		455	81		536	625
0100	3" pipe size		11	1.455		380	88.50		468.50	550
0110	4" pipe size		8	2		470	122		592	700
0120	5" pipe size		7	2.286		1,125	139		1,264	1,425
0130	6" pipe size	Q-2	9	2.667		740	168		908	1,075
0140	8" pipe size		8	3		1,275	190		1,465	1,675
0150	10" pipe size		7	3.429		2,800	217		3,017	3,400
0160	12" pipe size		6	4		5,625	253		5,878	6,575
0171	90° elbow, reducing									
0172	2-1/2" by 2" pipe size	Q-1	12	1.333	Ea.	1,500	81		1,581	1,775
0173	3" by 2-1/2" pipe size		11	1.455		1,425	88.50		1,513.50	1,675
0174	4" by 3" pipe size		8	2		1,050	122		1,172	1,325
0175	5" by 3" pipe size		7	2.286		2,425	139		2,564	2,850
0176	6" by 4" pipe size	Q-2	9	2.667		1,300	168		1,468	1,675
0177	8" by 6" pipe size		8	3		1,900	190		2,090	2,350
0178	10" by 8" pipe size		7	3.429		3,600	217		3,817	4,300
0179	12" by 10" pipe size		6	4		6,925	253		7,178	7,975
0200	45° elbow, straight, 1-1/2" pipe size	Q-1	14	1.143		930	69.50		999.50	1,125
0220	2" pipe size		13	1.231		615	75		690	785
0230	2-1/2" pipe size		12	1.333		655	81		736	840
0240	3" pipe size		11	1.455		635	88.50		723.50	830
0250	4" pipe size		8	2		715	122		837	965
0260	5" pipe size		7	2.286		1,725	139		1,864	2,075
0270	6" pipe size	Q-2	9	2.667		1,175	168		1,343	1,525
0280	8" pipe size		8	3		1,700	190		1,890	2,150
0290	10" pipe size		7	3.429		3,600	217		3,817	4,300
0300	12" pipe size		6	4		5,525	253		5,778	6,450
0310	Cross, straight									
0311	2-1/2" pipe size	Q-1	6	2.667	Ea.	1,425	162		1,587	1,800
0312	3" pipe size		5	3.200		1,400	195		1,595	1,850
0313	4" pipe size		4	4		1,850	244		2,094	2,400
0314	5" pipe size		3	5.333		4,200	325		4,525	5,100
0315	6" pipe size	Q-2	5	4.800		3,850	305		4,155	4,700
0316	8" pipe size		4	6		6,550	380		6,930	7,800
0317	10" pipe size		3	8		7,775	505		8,280	9,300
0318	12" pipe size		2	12		12,200	760		12,960	14,500
0350	Tee, straight, 1-1/2" pipe size	Q-1	10	1.600		885	97.50		982.50	1,125
0370	2" pipe size		9	1.778		465	108		573	670
0380	2-1/2" pipe size		8	2		675	122		797	920
0390	3" pipe size		7	2.286		470	139		609	730

22 11 13.47 Pipe Fittings, Steel

22 11 13.47 Pipe Fittings, Steel		Crew	Daily Output	Labor-Hours	Unit	Material	2021 Bare Costs Labor	2021 Bare Costs Equipment	Total	Total Incl O&P
0400	4" pipe size	Q-1	5	3.200	Ea.	720	195		915	1,075
0410	5" pipe size	↓	4	4		1,925	244		2,169	2,500
0420	6" pipe size	Q-2	6	4		1,050	253		1,303	1,525
0430	8" pipe size		5	4.800		1,775	305		2,080	2,400
0440	10" pipe size		4	6		4,800	380		5,180	5,850
0450	12" pipe size	↓	3	8	↓	7,525	505		8,030	9,025
0459	Tee, reducing on outlet									
0460	2-1/2" by 2" pipe size	Q-1	8	2	Ea.	1,300	122		1,422	1,600
0461	3" by 2-1/2" pipe size		7	2.286		1,475	139		1,614	1,825
0462	4" by 3" pipe size		5	3.200		1,475	195		1,670	1,925
0463	5" by 4" pipe size	↓	4	4		3,475	244		3,719	4,200
0464	6" by 4" pipe size	Q-2	6	4		1,425	253		1,678	1,950
0465	8" by 6" pipe size		5	4.800		2,225	305		2,530	2,900
0466	10" by 8" pipe size		4	6		5,075	380		5,455	6,175
0467	12" by 10" pipe size	↓	3	8	↓	9,300	505		9,805	11,000
0476	Reducer, concentric									
0477	3" by 2-1/2"	Q-1	12	1.333	Ea.	845	81		926	1,050
0478	4" by 3"		9	1.778		950	108		1,058	1,200
0479	5" by 4"	↓	8	2		1,475	122		1,597	1,800
0480	6" by 4"	Q-2	10	2.400		1,275	152		1,427	1,625
0481	8" by 6"		9	2.667		1,600	168		1,768	2,025
0482	10" by 8"		8	3		3,300	190		3,490	3,925
0483	12" by 10"	↓	7	3.429	↓	5,700	217		5,917	6,600
0492	Reducer, eccentric									
0493	4" by 3"	Q-1	8	2	Ea.	1,550	122		1,672	1,875
0494	5" by 4"	"	7	2.286		2,525	139		2,664	2,975
0495	6" by 4"	Q-2	9	2.667		1,450	168		1,618	1,850
0496	8" by 6"		8	3		1,850	190		2,040	2,300
0497	10" by 8"		7	3.429		4,700	217		4,917	5,500
0498	12" by 10"	↓	6	4		6,425	253		6,678	7,450
0500	For galvanized elbows and tees, add					100%				
0520	For extra heavy weight elbows and tees, add					140%				
0620	Gasket and bolt set, 150 lb., 1/2" pipe size	1 Plum	20	.400		3.57	27		30.57	44.50
0622	3/4" pipe size		19	.421		3.73	28.50		32.23	46.50
0624	1" pipe size		18	.444		4.59	30		34.59	50
0626	1-1/4" pipe size		17	.471		4.53	32		36.53	52.50
0628	1-1/2" pipe size		15	.533		4.29	36		40.29	58.50
0630	2" pipe size		13	.615		9.05	41.50		50.55	72
0640	2-1/2" pipe size		12	.667		8.40	45		53.40	77
0650	3" pipe size		11	.727		9.70	49		58.70	84
0660	3-1/2" pipe size		9	.889		15.75	60		75.75	107
0670	4" pipe size		8	1		16.80	67.50		84.30	120
0680	5" pipe size		7	1.143		28.50	77.50		106	147
0690	6" pipe size		6	1.333		30	90.50		120.50	168
0700	8" pipe size		5	1.600		31	108		139	196
0710	10" pipe size		4.50	1.778		58.50	120		178.50	244
0720	12" pipe size		4.20	1.905		62.50	129		191.50	261
0730	14" pipe size		4	2		52.50	135		187.50	260
0740	16" pipe size		3	2.667		69.50	181		250.50	345
0750	18" pipe size		2.70	2.963		135	201		336	445
0760	20" pipe size		2.30	3.478		219	235		454	590
0780	24" pipe size		1.90	4.211		273	285		558	725
0790	26" pipe size		1.60	5		370	340		710	915

22 11 13.47 Pipe Fittings, Steel		Crew	Daily Output	Labor-Hours	Unit	Material	2021 Bare Costs Labor	Equipment	Total	Total Incl O&P
0810	30" pipe size	1 Plum	1.40	5.714	Ea.	720	385		1,105	1,375
0830	36" pipe size	↓	1.10	7.273	↓	1,350	490		1,840	2,200
0850	For 300 lb. gasket set, add					40%				
2000	Flanged unions, 125 lb., black, 1/2" pipe size	1 Plum	17	.471	Ea.	108	32		140	167
2040	3/4" pipe size		17	.471		150	32		182	213
2050	1" pipe size	↓	16	.500		146	34		180	211
2060	1-1/4" pipe size	Q-1	28	.571		172	35		207	241
2070	1-1/2" pipe size		27	.593		159	36		195	229
2080	2" pipe size		26	.615		187	37.50		224.50	262
2090	2-1/2" pipe size		24	.667		253	40.50		293.50	340
2100	3" pipe size		22	.727		286	44.50		330.50	380
2110	3-1/2" pipe size		18	.889		485	54		539	615
2120	4" pipe size		16	1		390	61		451	520
2130	5" pipe size	↓	14	1.143		910	69.50		979.50	1,100
2140	6" pipe size	Q-2	19	1.263		850	80		930	1,050
2150	8" pipe size	"	16	1.500		2,050	95		2,145	2,425
2200	For galvanized unions, add				↓	150%				
2290	Threaded flange									
2300	Cast iron									
2310	Black, 125 lb., per flange									
2320	1" pipe size	1 Plum	27	.296	Ea.	54.50	20		74.50	90
2330	1-1/4" pipe size	Q-1	44	.364		74	22		96	114
2340	1-1/2" pipe size		40	.400		60.50	24.50		85	103
2350	2" pipe size		36	.444		68	27		95	116
2360	2-1/2" pipe size		28	.571		79.50	35		114.50	140
2370	3" pipe size		20	.800		103	48.50		151.50	186
2380	3-1/2" pipe size		16	1		153	61		214	260
2390	4" pipe size		12	1.333		123	81		204	256
2400	5" pipe size	↓	10	1.600		195	97.50		292.50	360
2410	6" pipe size	Q-2	14	1.714		221	108		329	405
2420	8" pipe size		12	2		350	126		476	575
2430	10" pipe size		10	2.400		620	152		772	905
2440	12" pipe size	↓	8	3	↓	1,475	190		1,665	1,875
2460	For galvanized flanges, add					95%				
2490	Blind flange									
2492	Cast iron									
2494	Black, 125 lb., per flange									
2496	1" pipe size	1 Plum	27	.296	Ea.	98	20		118	138
2500	1-1/2" pipe size	Q-1	40	.400		110	24.50		134.50	158
2502	2" pipe size		36	.444		124	27		151	178
2504	2-1/2" pipe size		28	.571		137	35		172	202
2506	3" pipe size		20	.800		164	48.50		212.50	253
2508	4" pipe size		12	1.333		210	81		291	350
2510	5" pipe size	↓	10	1.600		340	97.50		437.50	520
2512	6" pipe size	Q-2	14	1.714		370	108		478	570
2514	8" pipe size		12	2		580	126		706	830
2516	10" pipe size		10	2.400		860	152		1,012	1,175
2518	12" pipe size	↓	8	3	↓	1,750	190		1,940	2,200
2520	For galvanized flanges, add					80%				
2570	Threaded flange									
2580	Forged steel									
2590	Black, 150 lb., per flange									
2600	1/2" pipe size	1 Plum	30	.267	Ea.	26.50	18.05		44.55	56.50

22 11 13 – Facility Water Distribution Piping

22 11 13.47 Pipe Fittings, Steel

		Crew	Daily Output	Labor-Hours	Unit	Material	2021 Bare Costs Labor	Equipment	Total	Total Incl O&P
2610	3/4" pipe size	1 Plum	28	.286	Ea.	26.50	19.35		45.85	58.50
2620	1" pipe size		27	.296		26.50	20		46.50	59.50
2630	1-1/4" pipe size	Q-1	44	.364		26.50	22		48.50	62.50
2640	1-1/2" pipe size		40	.400		26.50	24.50		51	66
2650	2" pipe size		36	.444		29.50	27		56.50	73
2660	2-1/2" pipe size		28	.571		40.50	35		75.50	96.50
2670	3" pipe size		20	.800		41	48.50		89.50	118
2690	4" pipe size		12	1.333		46.50	81		127.50	172
2700	5" pipe size		10	1.600		75.50	97.50		173	228
2710	6" pipe size	Q-2	14	1.714		81.50	108		189.50	252
2720	8" pipe size		12	2		142	126		268	345
2730	10" pipe size		10	2.400		255	152		407	505
2860	Black, 300 lb., per flange									
2870	1/2" pipe size	1 Plum	30	.267	Ea.	29.50	18.05		47.55	59.50
2880	3/4" pipe size		28	.286		29.50	19.35		48.85	61.50
2890	1" pipe size		27	.296		29.50	20		49.50	62.50
2900	1-1/4" pipe size	Q-1	44	.364		29.50	22		51.50	65.50
2910	1-1/2" pipe size		40	.400		29.50	24.50		54	69
2920	2" pipe size		36	.444		38	27		65	82.50
2930	2-1/2" pipe size		28	.571		47.50	35		82.50	104
2940	3" pipe size		20	.800		48.50	48.50		97	126
2960	4" pipe size		12	1.333		69	81		150	197
2970	6" pipe size	Q-2	14	1.714		151	108		259	330
3000	Weld joint, butt, carbon steel, standard weight									
3040	90° elbow, long radius									
3050	1/2" pipe size	Q-15	16	1	Ea.	41	61	6.70	108.70	143
3060	3/4" pipe size		16	1		41	61	6.70	108.70	143
3070	1" pipe size		16	1		24.50	61	6.70	92.20	125
3080	1-1/4" pipe size		14	1.143		24.50	69.50	7.70	101.70	139
3090	1-1/2" pipe size		13	1.231		24.50	75	8.25	107.75	148
3100	2" pipe size		10	1.600		21	97.50	10.75	129.25	180
3110	2-1/2" pipe size		8	2		32.50	122	13.45	167.95	233
3120	3" pipe size		7	2.286		23	139	15.35	177.35	250
3130	4" pipe size		5	3.200		54	195	21.50	270.50	375
3136	5" pipe size		4	4		110	244	27	381	515
3140	6" pipe size	Q-16	5	4.800		119	305	21.50	445.50	610
3150	8" pipe size		3.75	6.400		226	405	28.50	659.50	885
3160	10" pipe size		3	8		445	505	36	986	1,275
3170	12" pipe size		2.50	9.600		520	605	43	1,168	1,525
3180	14" pipe size		2	12		950	760	54	1,764	2,225
3190	16" pipe size		1.50	16		1,300	1,000	71.50	2,371.50	3,000
3191	18" pipe size		1.25	19.200		1,375	1,225	86	2,686	3,425
3192	20" pipe size		1.15	20.870		2,250	1,325	93.50	3,668.50	4,550
3194	24" pipe size		1.02	23.529		3,400	1,475	105	4,980	6,075
3200	45° elbow, long									
3210	1/2" pipe size	Q-15	16	1	Ea.	61	61	6.70	128.70	165
3220	3/4" pipe size		16	1		61	61	6.70	128.70	165
3230	1" pipe size		16	1		21.50	61	6.70	89.20	122
3240	1-1/4" pipe size		14	1.143		21.50	69.50	7.70	98.70	136
3250	1-1/2" pipe size		13	1.231		21.50	75	8.25	104.75	145
3260	2" pipe size		10	1.600		21.50	97.50	10.75	129.75	180
3270	2-1/2" pipe size		8	2		26	122	13.45	161.45	225
3280	3" pipe size		7	2.286		27	139	15.35	181.35	254

22 11 Facility Water Distribution

22 11 13 – Facility Water Distribution Piping

22 11 13.47 Pipe Fittings, Steel		Crew	Daily Output	Labor-Hours	Unit	Material	2021 Bare Costs Labor	Equipment	Total	Total Incl O&P
3290	4" pipe size	Q-15	5	3.200	Ea.	48.50	195	21.50	265	370
3296	5" pipe size	↓	4	4		73.50	244	27	344.50	475
3300	6" pipe size	Q-16	5	4.800		95	305	21.50	421.50	585
3310	8" pipe size		3.75	6.400		158	405	28.50	591.50	810
3320	10" pipe size		3	8		315	505	36	856	1,150
3330	12" pipe size		2.50	9.600		445	605	43	1,093	1,450
3340	14" pipe size		2	12		600	760	54	1,414	1,850
3341	16" pipe size		1.50	16		1,075	1,000	71.50	2,146.50	2,750
3342	18" pipe size		1.25	19.200		1,525	1,225	86	2,836	3,575
3343	20" pipe size		1.15	20.870		1,575	1,325	93.50	2,993.50	3,825
3345	24" pipe size		1.05	22.857		2,525	1,450	102	4,077	5,050
3346	26" pipe size		.85	28.235		2,625	1,775	126	4,526	5,725
3347	30" pipe size		.45	53.333		2,900	3,375	239	6,514	8,500
3349	36" pipe size	↓	.38	63.158	↓	3,200	4,000	283	7,483	9,750
3350	Tee, straight									
3352	For reducing tees and concentrics see starting line 4600									
3360	1/2" pipe size	Q-15	10	1.600	Ea.	108	97.50	10.75	216.25	275
3370	3/4" pipe size		10	1.600		108	97.50	10.75	216.25	275
3380	1" pipe size		10	1.600		53.50	97.50	10.75	161.75	216
3390	1-1/4" pipe size		9	1.778		67	108	11.95	186.95	249
3400	1-1/2" pipe size		8	2		67	122	13.45	202.45	270
3410	2" pipe size		6	2.667		53.50	162	17.90	233.40	320
3420	2-1/2" pipe size		5	3.200		73.50	195	21.50	290	395
3430	3" pipe size		4	4		82	244	27	353	485
3440	4" pipe size		3	5.333		115	325	36	476	650
3446	5" pipe size	↓	2.50	6.400		190	390	43	623	835
3450	6" pipe size	Q-16	3	8		199	505	36	740	1,025
3460	8" pipe size		2.50	9.600		350	605	43	998	1,325
3470	10" pipe size		2	12		685	760	54	1,499	1,925
3480	12" pipe size		1.60	15		955	950	67	1,972	2,550
3481	14" pipe size		1.30	18.462		1,675	1,175	82.50	2,932.50	3,700
3482	16" pipe size		1	24		1,875	1,525	108	3,508	4,475
3483	18" pipe size		.80	30		2,950	1,900	134	4,984	6,225
3484	20" pipe size		.75	32		4,675	2,025	143	6,843	8,300
3486	24" pipe size		.70	34.286		6,025	2,175	154	8,354	10,000
3487	26" pipe size		.55	43.636		6,600	2,750	195	9,545	11,600
3488	30" pipe size		.30	80		7,225	5,050	360	12,635	15,900
3490	36" pipe size	↓	.25	96		7,975	6,075	430	14,480	18,300
3491	Eccentric reducer, 1-1/2" pipe size	Q-15	14	1.143		44.50	69.50	7.70	121.70	161
3492	2" pipe size		11	1.455		51.50	88.50	9.75	149.75	200
3493	2-1/2" pipe size		9	1.778		56.50	108	11.95	176.45	238
3494	3" pipe size		8	2		74	122	13.45	209.45	278
3495	4" pipe size	↓	6	2.667		150	162	17.90	329.90	425
3496	6" pipe size	Q-16	5	4.800		223	305	21.50	549.50	725
3497	8" pipe size		4	6		258	380	27	665	880
3498	10" pipe size		3	8		445	505	36	986	1,275
3499	12" pipe size	↓	2.50	9.600		655	605	43	1,303	1,675
3501	Cap, 1-1/2" pipe size	Q-15	28	.571		19.05	35	3.84	57.89	77
3502	2" pipe size		22	.727		22	44.50	4.89	71.39	96
3503	2-1/2" pipe size		18	.889		24	54	5.95	83.95	114
3504	3" pipe size		16	1		23.50	61	6.70	91.20	124
3505	4" pipe size	↓	12	1.333		33.50	81	8.95	123.45	167
3506	6" pipe size	Q-16	10	2.400	↓	57.50	152	10.75	220.25	300

For customer support on your Plumbing Costs with RSMeans Data, call 800.448.8182.

22 11 13.47 Pipe Fittings, Steel

		Crew	Daily Output	Labor-Hours	Unit	Material	2021 Bare Costs Labor	Equipment	Total	Total Incl O&P
3507	8" pipe size	Q-16	8	3	Ea.	87	190	13.45	290.45	395
3508	10" pipe size		6	4		156	253	17.90	426.90	565
3509	12" pipe size		5	4.800		223	305	21.50	549.50	725
3511	14" pipe size		4	6		465	380	27	872	1,100
3512	16" pipe size		4	6		600	380	27	1,007	1,250
3513	18" pipe size	↓	3	8	↓	525	505	36	1,066	1,375
3517	Weld joint, butt, carbon steel, extra strong									
3519	90° elbow, long									
3520	1/2" pipe size	Q-15	13	1.231	Ea.	54.50	75	8.25	137.75	181
3530	3/4" pipe size		12	1.333		54.50	81	8.95	144.45	191
3540	1" pipe size		11	1.455		26.50	88.50	9.75	124.75	172
3550	1-1/4" pipe size		10	1.600		26.50	97.50	10.75	134.75	186
3560	1-1/2" pipe size		9	1.778		26.50	108	11.95	146.45	204
3570	2" pipe size		8	2		27	122	13.45	162.45	226
3580	2-1/2" pipe size		7	2.286		38	139	15.35	192.35	266
3590	3" pipe size		6	2.667		48.50	162	17.90	228.40	315
3600	4" pipe size		4	4		79.50	244	27	350.50	480
3606	5" pipe size	↓	3.50	4.571		186	279	30.50	495.50	655
3610	6" pipe size	Q-16	4.50	5.333		201	335	24	560	755
3620	8" pipe size		3.50	6.857		385	435	30.50	850.50	1,100
3630	10" pipe size		2.50	9.600		815	605	43	1,463	1,850
3640	12" pipe size	↓	2.25	10.667	↓	990	675	48	1,713	2,150
3650	45° elbow, long									
3660	1/2" pipe size	Q-15	13	1.231	Ea.	60.50	75	8.25	143.75	188
3670	3/4" pipe size		12	1.333		60.50	81	8.95	150.45	197
3680	1" pipe size		11	1.455		28	88.50	9.75	126.25	174
3690	1-1/4" pipe size		10	1.600		28	97.50	10.75	136.25	188
3700	1-1/2" pipe size		9	1.778		28	108	11.95	147.95	206
3710	2" pipe size		8	2		28	122	13.45	163.45	228
3720	2-1/2" pipe size		7	2.286		62.50	139	15.35	216.85	293
3730	3" pipe size		6	2.667		36	162	17.90	215.90	300
3740	4" pipe size		4	4		57	244	27	328	455
3746	5" pipe size	↓	3.50	4.571		136	279	30.50	445.50	600
3750	6" pipe size	Q-16	4.50	5.333		156	335	24	515	705
3760	8" pipe size		3.50	6.857		271	435	30.50	736.50	975
3770	10" pipe size		2.50	9.600		520	605	43	1,168	1,525
3780	12" pipe size	↓	2.25	10.667	↓	770	675	48	1,493	1,900
3800	Tee, straight									
3810	1/2" pipe size	Q-15	9	1.778	Ea.	133	108	11.95	252.95	320
3820	3/4" pipe size		8.50	1.882		127	115	12.65	254.65	325
3830	1" pipe size		8	2		67	122	13.45	202.45	271
3840	1-1/4" pipe size		7	2.286		54	139	15.35	208.35	284
3850	1-1/2" pipe size		6	2.667		68	162	17.90	247.90	335
3860	2" pipe size		5	3.200		61	195	21.50	277.50	380
3870	2-1/2" pipe size		4	4		101	244	27	372	505
3880	3" pipe size		3.50	4.571		125	279	30.50	434.50	585
3890	4" pipe size		2.50	6.400		153	390	43	586	795
3896	5" pipe size	↓	2.25	7.111		365	435	48	848	1,100
3900	6" pipe size	Q-16	2.25	10.667		281	675	48	1,004	1,375
3910	8" pipe size		2	12		535	760	54	1,349	1,775
3920	10" pipe size		1.75	13.714		865	865	61.50	1,791.50	2,325
3930	12" pipe size	↓	1.50	16		1,275	1,000	71.50	2,346.50	2,975
4000	Eccentric reducer, 1-1/2" pipe size	Q-15	10	1.600	↓	22.50	97.50	10.75	130.75	181

22 11 13.47 Pipe Fittings, Steel		Crew	Daily Output	Labor-Hours	Unit	Material	2021 Bare Costs Labor	Equipment	Total	Total Incl O&P
4010	2" pipe size	Q-15	9	1.778	Ea.	61.50	108	11.95	181.45	243
4020	2-1/2" pipe size		8	2		82.50	122	13.45	217.95	288
4030	3" pipe size		7	2.286		61	139	15.35	215.35	292
4040	4" pipe size		5	3.200		99	195	21.50	315.50	425
4046	5" pipe size		4.70	3.404		320	207	23	550	685
4050	6" pipe size	Q-16	4.50	5.333		305	335	24	664	865
4060	8" pipe size		3.50	6.857		400	435	30.50	865.50	1,125
4070	10" pipe size		2.50	9.600		665	605	43	1,313	1,675
4080	12" pipe size		2.25	10.667		845	675	48	1,568	1,975
4090	14" pipe size		2.10	11.429		1,600	720	51	2,371	2,875
4100	16" pipe size		1.90	12.632		1,975	800	56.50	2,831.50	3,450
4151	Cap, 1-1/2" pipe size	Q-15	24	.667		25	40.50	4.48	69.98	93
4152	2" pipe size		18	.889		22	54	5.95	81.95	112
4153	2-1/2" pipe size		16	1		30.50	61	6.70	98.20	132
4154	3" pipe size		14	1.143		33	69.50	7.70	110.20	149
4155	4" pipe size		10	1.600		45	97.50	10.75	153.25	206
4156	6" pipe size	Q-16	9	2.667		95.50	168	11.95	275.45	370
4157	8" pipe size		7	3.429		143	217	15.35	375.35	500
4158	10" pipe size		5	4.800		217	305	21.50	543.50	715
4159	12" pipe size		4	6		293	380	27	700	920
4190	Weld fittings, reducing, standard weight									
4200	Welding ring w/spacer pins, 2" pipe size				Ea.	1.69			1.69	1.86
4210	2-1/2" pipe size					2.32			2.32	2.55
4220	3" pipe size					2.31			2.31	2.54
4230	4" pipe size					2.88			2.88	3.17
4236	5" pipe size					3.54			3.54	3.89
4240	6" pipe size					2.54			2.54	2.79
4250	8" pipe size					2.76			2.76	3.04
4260	10" pipe size					4.32			4.32	4.75
4270	12" pipe size					5.05			5.05	5.55
4280	14" pipe size					5.85			5.85	6.45
4290	16" pipe size					4.99			4.99	5.50
4300	18" pipe size					7.60			7.60	8.35
4310	20" pipe size					9.50			9.50	10.45
4330	24" pipe size					13.55			13.55	14.95
4340	26" pipe size					16.20			16.20	17.85
4350	30" pipe size					18.55			18.55	20.50
4370	36" pipe size					23			23	25
4600	Tee, reducing on outlet									
4601	2-1/2" x 2" pipe size	Q-15	5	3.200	Ea.	97	195	21.50	313.50	420
4602	3" x 2-1/2" pipe size		4	4		140	244	27	411	550
4604	4" x 3" pipe size		3	5.333		113	325	36	474	650
4605	5" x 4" pipe size		2.50	6.400		315	390	43	748	980
4606	6" x 5" pipe size	Q-16	3	8		440	505	36	981	1,275
4607	8" x 6" pipe size		2.50	9.600		545	605	43	1,193	1,550
4608	10" x 8" pipe size		2	12		780	760	54	1,594	2,050
4609	12" x 10" pipe size		1.60	15		1,225	950	67	2,242	2,850
4610	16" x 12" pipe size		1.50	16		2,325	1,000	71.50	3,396.50	4,125
4611	14" x 12" pipe size		1.52	15.789		2,225	1,000	70.50	3,295.50	4,025
4618	Reducer, concentric									
4619	2-1/2" by 2" pipe size	Q-15	10	1.600	Ea.	39	97.50	10.75	147.25	200
4620	3" by 2-1/2" pipe size		9	1.778		32	108	11.95	151.95	211
4621	3-1/2" by 3" pipe size		8	2		104	122	13.45	239.45	310

22 11 13.47 Pipe Fittings, Steel

		Crew	Daily Output	Labor-Hours	Unit	Material	2021 Bare Costs Labor	2021 Bare Costs Equipment	Total	Total Incl O&P
4622	4" by 2-1/2" pipe size	Q-15	7	2.286	Ea.	53.50	139	15.35	207.85	284
4623	5" by 3" pipe size		7	2.286		134	139	15.35	288.35	370
4624	6" by 4" pipe size	Q-16	6	4		111	253	17.90	381.90	515
4625	8" by 6" pipe size		5	4.800		113	305	21.50	439.50	605
4626	10" by 8" pipe size		4	6		246	380	27	653	865
4627	12" by 10" pipe size		3	8		272	505	36	813	1,100
4660	Reducer, eccentric									
4662	3" x 2" pipe size	Q-15	8	2	Ea.	69.50	122	13.45	204.95	273
4664	4" x 3" pipe size		6	2.667		84.50	162	17.90	264.40	355
4666	4" x 2" pipe size		6	2.667		103	162	17.90	282.90	375
4670	6" x 4" pipe size	Q-16	5	4.800		174	305	21.50	500.50	670
4672	6" x 3" pipe size		5	4.800		273	305	21.50	599.50	780
4676	8" x 6" pipe size		4	6		222	380	27	629	840
4678	8" x 4" pipe size		4	6		405	380	27	812	1,050
4682	10" x 8" pipe size		3	8		291	505	36	832	1,125
4684	10" x 6" pipe size		3	8		400	505	36	941	1,225
4688	12" x 10" pipe size		2.50	9.600		455	605	43	1,103	1,450
4690	12" x 8" pipe size		2.50	9.600		715	605	43	1,363	1,750
4691	14" x 12" pipe size		2.20	10.909		795	690	49	1,534	1,950
4693	16" x 14" pipe size		1.80	13.333		945	840	59.50	1,844.50	2,375
4694	16" x 12" pipe size		2	12		1,625	760	54	2,439	2,950
4696	18" x 16" pipe size		1.60	15		1,375	950	67	2,392	3,000
5000	Weld joint, socket, forged steel, 3000 lb., schedule 40 pipe									
5010	90° elbow, straight									
5020	1/4" pipe size	Q-15	22	.727	Ea.	29.50	44.50	4.89	78.89	104
5030	3/8" pipe size		22	.727		29.50	44.50	4.89	78.89	104
5040	1/2" pipe size		20	.800		16.50	48.50	5.40	70.40	96.50
5050	3/4" pipe size		20	.800		13.85	48.50	5.40	67.75	93.50
5060	1" pipe size		20	.800		22	48.50	5.40	75.90	103
5070	1-1/4" pipe size		18	.889		42.50	54	5.95	102.45	134
5080	1-1/2" pipe size		16	1		43	61	6.70	110.70	146
5090	2" pipe size		12	1.333		62	81	8.95	151.95	199
5100	2-1/2" pipe size		10	1.600		206	97.50	10.75	314.25	385
5110	3" pipe size		8	2		380	122	13.45	515.45	615
5120	4" pipe size		6	2.667		925	162	17.90	1,104.90	1,275
5130	45° elbow, straight									
5134	1/4" pipe size	Q-15	22	.727	Ea.	29.50	44.50	4.89	78.89	104
5135	3/8" pipe size		22	.727		29.50	44.50	4.89	78.89	104
5136	1/2" pipe size		20	.800		22	48.50	5.40	75.90	103
5137	3/4" pipe size		20	.800		25.50	48.50	5.40	79.40	106
5140	1" pipe size		20	.800		33.50	48.50	5.40	87.40	115
5150	1-1/4" pipe size		18	.889		46	54	5.95	105.95	138
5160	1-1/2" pipe size		16	1		56	61	6.70	123.70	160
5170	2" pipe size		12	1.333		90.50	81	8.95	180.45	230
5180	2-1/2" pipe size		10	1.600		238	97.50	10.75	346.25	420
5190	3" pipe size		8	2		395	122	13.45	530.45	630
5200	4" pipe size		6	2.667		775	162	17.90	954.90	1,125
5250	Tee, straight									
5254	1/4" pipe size	Q-15	15	1.067	Ea.	33	65	7.15	105.15	141
5255	3/8" pipe size		15	1.067		33	65	7.15	105.15	141
5256	1/2" pipe size		13	1.231		20.50	75	8.25	103.75	144
5257	3/4" pipe size		13	1.231		25	75	8.25	108.25	149
5260	1" pipe size		13	1.231		34	75	8.25	117.25	159

For customer support on your Plumbing Costs with RSMeans Data, call 800.448.8182.

193

22 11 13.47 Pipe Fittings, Steel		Crew	Daily Output	Labor-Hours	Unit	Material	2021 Bare Costs Labor	Equipment	Total	Total Incl O&P
5270	1-1/4" pipe size	Q-15	12	1.333	Ea.	51.50	81	8.95	141.45	188
5280	1-1/2" pipe size		11	1.455		68.50	88.50	9.75	166.75	218
5290	2" pipe size		8	2		100	122	13.45	235.45	305
5300	2-1/2" pipe size		6	2.667		246	162	17.90	425.90	535
5310	3" pipe size		5	3.200		590	195	21.50	806.50	960
5320	4" pipe size		4	4		1,100	244	27	1,371	1,625
5350	For reducing sizes, add					60%				
5450	Couplings									
5451	1/4" pipe size	Q-15	23	.696	Ea.	15.20	42.50	4.67	62.37	85.50
5452	3/8" pipe size		23	.696		15.20	42.50	4.67	62.37	85.50
5453	1/2" pipe size		21	.762		6.90	46.50	5.10	58.50	83
5454	3/4" pipe size		21	.762		8.95	46.50	5.10	60.55	85
5460	1" pipe size		20	.800		9.90	48.50	5.40	63.80	89.50
5470	1-1/4" pipe size		20	.800		17.55	48.50	5.40	71.45	97.50
5480	1-1/2" pipe size		18	.889		19.35	54	5.95	79.30	109
5490	2" pipe size		14	1.143		31	69.50	7.70	108.20	146
5500	2-1/2" pipe size		12	1.333		69	81	8.95	158.95	207
5510	3" pipe size		9	1.778		156	108	11.95	275.95	345
5520	4" pipe size		7	2.286		234	139	15.35	388.35	480
5570	Union, 1/4" pipe size		21	.762		32	46.50	5.10	83.60	111
5571	3/8" pipe size		21	.762		32	46.50	5.10	83.60	111
5572	1/2" pipe size		19	.842		38	51.50	5.65	95.15	124
5573	3/4" pipe size		19	.842		31.50	51.50	5.65	88.65	117
5574	1" pipe size		19	.842		40.50	51.50	5.65	97.65	127
5575	1-1/4" pipe size		17	.941		82	57.50	6.30	145.80	183
5576	1-1/2" pipe size		15	1.067		71	65	7.15	143.15	183
5577	2" pipe size		11	1.455		108	88.50	9.75	206.25	261
5600	Reducer, 1/4" pipe size		23	.696		40	42.50	4.67	87.17	113
5601	3/8" pipe size		23	.696		42	42.50	4.67	89.17	115
5602	1/2" pipe size		21	.762		24.50	46.50	5.10	76.10	102
5603	3/4" pipe size		21	.762		24.50	46.50	5.10	76.10	102
5604	1" pipe size		21	.762		36.50	46.50	5.10	88.10	115
5605	1-1/4" pipe size		19	.842		55	51.50	5.65	112.15	143
5607	1-1/2" pipe size		17	.941		61	57.50	6.30	124.80	159
5608	2" pipe size		13	1.231		67	75	8.25	150.25	195
5612	Cap, 1/4" pipe size		46	.348		18.15	21	2.34	41.49	54
5613	3/8" pipe size		46	.348		18.15	21	2.34	41.49	54
5614	1/2" pipe size		42	.381		10.95	23	2.56	36.51	49.50
5615	3/4" pipe size		42	.381		12.80	23	2.56	38.36	51.50
5616	1" pipe size		42	.381		19.40	23	2.56	44.96	59
5617	1-1/4" pipe size		38	.421		18.25	25.50	2.83	46.58	61.50
5618	1-1/2" pipe size		34	.471		25	28.50	3.16	56.66	74.50
5619	2" pipe size		26	.615		39	37.50	4.14	80.64	104
5630	T-O-L, 1/4" pipe size, nozzle		23	.696		8.15	42.50	4.67	55.32	77.50
5631	3/8" pipe size, nozzle		23	.696		8.40	42.50	4.67	55.57	78
5632	1/2" pipe size, nozzle		22	.727		8.20	44.50	4.89	57.59	80.50
5633	3/4" pipe size, nozzle		21	.762		9.40	46.50	5.10	61	85.50
5634	1" pipe size, nozzle		20	.800		10.95	48.50	5.40	64.85	90.50
5635	1-1/4" pipe size, nozzle		18	.889		13.90	54	5.95	73.85	103
5636	1-1/2" pipe size, nozzle		16	1		17.40	61	6.70	85.10	118
5637	2" pipe size, nozzle		12	1.333		19.15	81	8.95	109.10	152
5638	2-1/2" pipe size, nozzle		10	1.600		66.50	97.50	10.75	174.75	230
5639	4" pipe size, nozzle		6	2.667		150	162	17.90	329.90	425

22 11 Facility Water Distribution

22 11 13 – Facility Water Distribution Piping

22 11 13.47 Pipe Fittings, Steel

		Crew	Daily Output	Labor-Hours	Unit	Material	2021 Bare Costs Labor	Equipment	Total	Total Incl O&P
5640	W-O-L, 1/4" pipe size, nozzle	Q-15	23	.696	Ea.	20.50	42.50	4.67	67.67	91
5641	3/8" pipe size, nozzle		23	.696		16.85	42.50	4.67	64.02	87
5642	1/2" pipe size, nozzle		22	.727		16.90	44.50	4.89	66.29	90
5643	3/4" pipe size, nozzle		21	.762		17.85	46.50	5.10	69.45	95
5644	1" pipe size, nozzle		20	.800		18.65	48.50	5.40	72.55	99
5645	1-1/4" pipe size, nozzle		18	.889		21.50	54	5.95	81.45	111
5646	1-1/2" pipe size, nozzle		16	1		22	61	6.70	89.70	123
5647	2" pipe size, nozzle		12	1.333		22.50	81	8.95	112.45	155
5648	2-1/2" pipe size, nozzle		10	1.600		51	97.50	10.75	159.25	213
5649	3" pipe size, nozzle		8	2		56	122	13.45	191.45	258
5650	4" pipe size, nozzle		6	2.667		71	162	17.90	250.90	340
5651	5" pipe size, nozzle		5	3.200		175	195	21.50	391.50	505
5652	6" pipe size, nozzle		4	4		196	244	27	467	610
5653	8" pipe size, nozzle		3	5.333		375	325	36	736	940
5654	10" pipe size, nozzle		2.60	6.154		530	375	41.50	946.50	1,175
5655	12" pipe size, nozzle		2.20	7.273		1,000	445	49	1,494	1,825
5674	S-O-L, 1/4" pipe size, outlet		23	.696		10.10	42.50	4.67	57.27	80
5675	3/8" pipe size, outlet		23	.696		10.10	42.50	4.67	57.27	80
5676	1/2" pipe size, outlet		22	.727		10.15	44.50	4.89	59.54	82.50
5677	3/4" pipe size, outlet		21	.762		10.25	46.50	5.10	61.85	86.50
5678	1" pipe size, outlet		20	.800		11.40	48.50	5.40	65.30	91
5679	1-1/4" pipe size, outlet		18	.889		19.05	54	5.95	79	109
5680	1-1/2" pipe size, outlet		16	1		19.05	61	6.70	86.75	119
5681	2" pipe size, outlet	▼	12	1.333	▼	19.70	81	8.95	109.65	152
6000	Weld-on flange, forged steel									
6020	Slip-on, 150 lb. flange (welded front and back)									
6050	1/2" pipe size	Q-15	18	.889	Ea.	22	54	5.95	81.95	112
6060	3/4" pipe size		18	.889		22	54	5.95	81.95	112
6070	1" pipe size		17	.941		22	57.50	6.30	85.80	116
6080	1-1/4" pipe size		16	1		22	61	6.70	89.70	122
6090	1-1/2" pipe size		15	1.067		22	65	7.15	94.15	129
6100	2" pipe size		12	1.333		26.50	81	8.95	116.45	160
6110	2-1/2" pipe size		10	1.600		40.50	97.50	10.75	148.75	201
6120	3" pipe size		9	1.778		30.50	108	11.95	150.45	209
6130	3-1/2" pipe size		7	2.286		33.50	139	15.35	187.85	262
6140	4" pipe size		6	2.667		31.50	162	17.90	211.40	297
6150	5" pipe size		5	3.200		64.50	195	21.50	281	385
6160	6" pipe size	Q-16	6	4		51.50	253	17.90	322.40	450
6170	8" pipe size		5	4.800		91	305	21.50	417.50	580
6180	10" pipe size		4	6		159	380	27	566	770
6190	12" pipe size		3	8		237	505	36	778	1,050
6191	14" pipe size		2.50	9.600		268	605	43	916	1,250
6192	16" pipe size	▼	1.80	13.333	▼	480	840	59.50	1,379.50	1,850
6200	300 lb. flange									
6210	1/2" pipe size	Q-15	17	.941	Ea.	23.50	57.50	6.30	87.30	118
6220	3/4" pipe size		17	.941		23.50	57.50	6.30	87.30	118
6230	1" pipe size		16	1		23.50	61	6.70	91.20	124
6240	1-1/4" pipe size		13	1.231		23.50	75	8.25	106.75	147
6250	1-1/2" pipe size		12	1.333		23.50	81	8.95	113.45	157
6260	2" pipe size		11	1.455		35	88.50	9.75	133.25	181
6270	2-1/2" pipe size		9	1.778		35.50	108	11.95	155.45	214
6280	3" pipe size		7	2.286		37.50	139	15.35	191.85	266
6290	4" pipe size		6	2.667		57	162	17.90	236.90	325

For customer support on your Plumbing Costs with RSMeans Data, call 800.448.8182.

22 11 13.47 Pipe Fittings, Steel		Crew	Daily Output	Labor-Hours	Unit	Material	2021 Bare Costs Labor	Equipment	Total	Total Incl O&P
6300	5" pipe size	Q-15	4	4	Ea.	106	244	27	377	510
6310	6" pipe size	Q-16	5	4.800		104	305	21.50	430.50	595
6320	8" pipe size		4	6		178	380	27	585	790
6330	10" pipe size		3.40	7.059		310	445	31.50	786.50	1,050
6340	12" pipe size		2.80	8.571		330	540	38.50	908.50	1,225
6400	Welding neck, 150 lb. flange									
6410	1/2" pipe size	Q-15	40	.400	Ea.	24.50	24.50	2.69	51.69	66.50
6420	3/4" pipe size		36	.444		24.50	27	2.99	54.49	71
6430	1" pipe size		32	.500		24.50	30.50	3.36	58.36	76
6440	1-1/4" pipe size		29	.552		24.50	33.50	3.71	61.71	81
6450	1-1/2" pipe size		26	.615		24.50	37.50	4.14	66.14	87.50
6460	2" pipe size		20	.800		32.50	48.50	5.40	86.40	114
6470	2-1/2" pipe size		16	1		35.50	61	6.70	103.20	137
6480	3" pipe size		14	1.143		39	69.50	7.70	116.20	155
6500	4" pipe size		10	1.600		41	97.50	10.75	149.25	202
6510	5" pipe size		8	2		63.50	122	13.45	198.95	266
6520	6" pipe size	Q-16	10	2.400		62.50	152	10.75	225.25	305
6530	8" pipe size		7	3.429		113	217	15.35	345.35	465
6540	10" pipe size		6	4		183	253	17.90	453.90	595
6550	12" pipe size		5	4.800		286	305	21.50	612.50	795
6551	14" pipe size		4.50	5.333		420	335	24	779	995
6552	16" pipe size		3	8		570	505	36	1,111	1,425
6553	18" pipe size		2.50	9.600		770	605	43	1,418	1,800
6554	20" pipe size		2.30	10.435		930	660	47	1,637	2,050
6556	24" pipe size		2	12		1,275	760	54	2,089	2,575
6557	26" pipe size		1.70	14.118		1,375	890	63.50	2,328.50	2,925
6558	30" pipe size		.90	26.667		1,575	1,675	119	3,369	4,400
6559	36" pipe size		.75	32		1,825	2,025	143	3,993	5,175
6560	300 lb. flange									
6570	1/2" pipe size	Q-15	36	.444	Ea.	33.50	27	2.99	63.49	80.50
6580	3/4" pipe size		34	.471		33.50	28.50	3.16	65.16	83
6590	1" pipe size		30	.533		33.50	32.50	3.58	69.58	89
6600	1-1/4" pipe size		28	.571		33.50	35	3.84	72.34	92.50
6610	1-1/2" pipe size		24	.667		33.50	40.50	4.48	78.48	102
6620	2" pipe size		18	.889		43.50	54	5.95	103.45	135
6630	2-1/2" pipe size		14	1.143		49.50	69.50	7.70	126.70	167
6640	3" pipe size		12	1.333		43	81	8.95	132.95	178
6650	4" pipe size		8	2		73.50	122	13.45	208.95	278
6660	5" pipe size		7	2.286		107	139	15.35	261.35	340
6670	6" pipe size	Q-16	9	2.667		110	168	11.95	289.95	385
6680	8" pipe size		6	4		210	253	17.90	480.90	625
6690	10" pipe size		5	4.800		395	305	21.50	721.50	915
6700	12" pipe size		4	6		430	380	27	837	1,075
7740	Plain ends for plain end pipe, mechanically coupled									
7750	Cplg. & labor required at joints not included, add 1 per									
7760	joint for installed price, see line 9180									
7770	Malleable iron, painted, unless noted otherwise									
7800	90° elbow, 1"				Ea.	177			177	194
7810	1-1/2"					209			209	230
7820	2"					310			310	345
7830	2-1/2"					365			365	400
7840	3"					380			380	415
7860	4"					430			430	470

For customer support on your Plumbing Costs with RSMeans Data, call 800.448.8182.

22 11 13.47 Pipe Fittings, Steel		Crew	Daily Output	Labor-Hours	Unit	Material	2021 Bare Costs Labor	Equipment	Total	Total Incl O&P
7870	5" welded steel				Ea.	490			490	540
7880	6"					620			620	685
7890	8" welded steel					1,175			1,175	1,275
7900	10" welded steel					1,475			1,475	1,625
7910	12" welded steel					1,650			1,650	1,825
7970	45° elbow, 1"					106			106	117
7980	1-1/2"					158			158	173
7990	2"					330			330	365
8000	2-1/2"					330			330	365
8010	3"					380			380	415
8030	4"					395			395	435
8040	5" welded steel					490			490	540
8050	6"					565			565	620
8060	8"					660			660	725
8070	10" welded steel					680			680	750
8080	12" welded steel					1,225			1,225	1,350
8140	Tee, straight 1"					198			198	217
8150	1-1/2"					254			254	280
8160	2"					254			254	280
8170	2-1/2"					330			330	365
8180	3"					510			510	560
8200	4"					730			730	800
8210	5" welded steel					995			995	1,100
8220	6"					865			865	950
8230	8" welded steel					1,250			1,250	1,375
8240	10" welded steel					1,950			1,950	2,150
8250	12" welded steel					2,275			2,275	2,525
8340	Segmentally welded steel, painted									
8390	Wye 2"				Ea.	288			288	315
8400	2-1/2"					288			288	315
8410	3"					345			345	380
8430	4"					375			375	410
8440	5"					655			655	720
8450	6"					795			795	875
8460	8"					1,175			1,175	1,275
8470	10"					1,800			1,800	1,975
8480	12"					1,650			1,650	1,800
8540	Wye, lateral 2"					345			345	375
8550	2-1/2"					395			395	435
8560	3"					470			470	515
8580	4"					645			645	710
8590	5"					1,100			1,100	1,200
8600	6"					1,125			1,125	1,250
8610	8"					1,900			1,900	2,075
8620	10"					1,900			1,900	2,075
8630	12"					3,475			3,475	3,825
8690	Cross, 2"					175			175	192
8700	2-1/2"					175			175	192
8710	3"					208			208	229
8730	4"					440			440	485
8740	5"					655			655	720
8750	6"					685			685	755
8760	8"					995			995	1,100

22 11 13.47 Pipe Fittings, Steel		Crew	Daily Output	Labor-Hours	Unit	Material	2021 Bare Costs Labor	Equipment	Total	Total Incl O&P
8770	10"				Ea.	1,700			1,700	1,875
8780	12"					1,775			1,775	1,950
8800	Tees, reducing 2" x 1"					235			235	259
8810	2" x 1-1/2"					235			235	259
8820	3" x 1"					234			234	257
8830	3" x 1-1/2"					218			218	239
8840	3" x 2"					235			235	258
8850	4" x 1"					345			345	380
8860	4" x 1-1/2"					345			345	380
8870	4" x 2"					315			315	345
8880	4" x 2-1/2"					196			196	216
8890	4" x 3"					325			325	355
8900	6" x 2"					360			360	395
8910	6" x 3"					550			550	605
8920	6" x 4"					460			460	505
8930	8" x 2"					640			640	705
8940	8" x 3"					715			715	785
8950	8" x 4"					780			780	855
8960	8" x 5"					450			450	500
8970	8" x 6"					735			735	805
8980	10" x 4"					510			510	560
8990	10" x 6"					525			525	575
9000	10" x 8"					820			820	900
9010	12" x 6"					1,350			1,350	1,500
9020	12" x 8"					1,025			1,025	1,125
9030	12" x 10"				↓	1,000			1,000	1,100
9080	Adapter nipples 3" long									
9090	1"				Ea.	24			24	26
9100	1-1/2"					31			31	34.50
9110	2"					31			31	34.50
9120	2-1/2"					36			36	39.50
9130	3"					43.50			43.50	48
9140	4"					70.50			70.50	77.50
9150	6"				↓	185			185	204
9180	Coupling, mechanical, plain end pipe to plain end pipe or fitting									
9190	1"	Q-1	29	.552	Ea.	107	33.50		140.50	168
9200	1-1/2"		28	.571		107	35		142	170
9210	2"		27	.593		107	36		143	172
9220	2-1/2"		26	.615		107	37.50		144.50	174
9230	3"		25	.640		156	39		195	230
9240	3-1/2"		24	.667		181	40.50		221.50	260
9250	4"	↓	22	.727		181	44.50		225.50	265
9260	5"	Q-2	28	.857		256	54		310	365
9270	6"		24	1		315	63		378	440
9280	8"		19	1.263		550	80		630	725
9290	10"		16	1.500		715	95		810	925
9300	12"	↓	12	2	↓	895	126		1,021	1,175
9310	Outlets for precut holes through pipe wall									
9331	Strapless type, with gasket									
9332	4" to 8" pipe x 1/2"	1 Plum	13	.615	Ea.	117	41.50		158.50	191
9333	4" to 8" pipe x 3/4"		13	.615		125	41.50		166.50	200
9334	10" pipe and larger x 1/2"		11	.727		117	49		166	203
9335	10" pipe and larger x 3/4"	↓	11	.727	↓	125	49		174	212

For customer support on your Plumbing Costs with RSMeans Data, call 800.448.8182.

22 11 13 – Facility Water Distribution Piping

22 11 13.47 Pipe Fittings, Steel

		Crew	Daily Output	Labor-Hours	Unit	Material	2021 Bare Costs Labor	Equipment	Total	Total Incl O&P
9341	Thermometer wells with gasket									
9342	4" to 8" pipe, 6" stem	1 Plum	14	.571	Ea.	175	38.50		213.50	251
9343	8" pipe and larger, 6" stem	"	13	.615	"	175	41.50		216.50	255
9800	Mech. tee, cast iron, grooved or threaded, w/nuts and bolts									
9805	2" x 1"	1 Plum	50	.160	Ea.	106	10.85		116.85	133
9810	2" x 1-1/4"		50	.160		124	10.85		134.85	153
9815	2" x 1-1/2"	↓	50	.160		129	10.85		139.85	158
9820	2-1/2" x 1"	Q-1	80	.200		111	12.20		123.20	140
9825	2-1/2" x 1-1/4"		80	.200		163	12.20		175.20	197
9830	2-1/2" x 1-1/2"		80	.200		163	12.20		175.20	197
9835	2-1/2" x 2"		80	.200		111	12.20		123.20	140
9840	3" x 1"		80	.200		120	12.20		132.20	150
9845	3" x 1-1/4"		80	.200		156	12.20		168.20	190
9850	3" x 1-1/2"		80	.200		156	12.20		168.20	190
9852	3" x 2"		80	.200		190	12.20		202.20	228
9856	4" x 1"		67	.239		149	14.55		163.55	186
9858	4" x 1-1/4"		67	.239		195	14.55		209.55	237
9860	4" x 1-1/2"		67	.239		197	14.55		211.55	239
9862	4" x 2"		67	.239		204	14.55		218.55	247
9864	4" x 2-1/2"		67	.239		211	14.55		225.55	254
9866	4" x 3"	↓	67	.239		211	14.55		225.55	254
9870	6" x 1-1/2"	Q-2	50	.480		247	30.50		277.50	320
9872	6" x 2"		50	.480		251	30.50		281.50	320
9874	6" x 2-1/2"		50	.480		249	30.50		279.50	320
9876	6" x 3"		50	.480		284	30.50		314.50	355
9878	6" x 4"		50	.480		315	30.50		345.50	395
9880	8" x 2"		42	.571		425	36		461	525
9882	8" x 2-1/2"		42	.571		425	36		461	525
9884	8" x 3"		42	.571		450	36		486	550
9886	8" x 4"	↓	42	.571	↓	915	36		951	1,050
9940	For galvanized fittings for plain end pipe, add					20%				

22 11 13.48 Pipe, Fittings and Valves, Steel, Grooved-Joint

		Crew	Daily Output	Labor-Hours	Unit	Material	2021 Bare Costs Labor	Equipment	Total	Total Incl O&P
0010	**PIPE, FITTINGS AND VALVES, STEEL, GROOVED-JOINT** R221113-70									
0012	Fittings are ductile iron. Steel fittings noted.									
0020	Pipe includes coupling & clevis type hanger assemblies, 10' OC									
0500	Schedule 10, black									
0550	2" diameter	1 Plum	43	.186	L.F.	7.90	12.60		20.50	27.50
0560	2-1/2" diameter	Q-1	61	.262		9.65	16		25.65	34.50
0570	3" diameter		55	.291		11.15	17.75		28.90	39
0580	3-1/2" diameter		53	.302		14.65	18.40		33.05	43.50
0590	4" diameter		49	.327		15.20	19.90		35.10	46.50
0600	5" diameter	↓	40	.400		19.40	24.50		43.90	58
0610	6" diameter	Q-2	46	.522		24.50	33		57.50	76
0620	8" diameter	"	41	.585	↓	35.50	37		72.50	94
0700	To delete couplings & hangers, subtract									
0710	2" diam. to 5" diam.					25%	20%			
0720	6" diam. to 8" diam.					27%	15%			
1000	Schedule 40, black									
1040	3/4" diameter	1 Plum	71	.113	L.F.	5.85	7.65		13.50	17.85
1050	1" diameter		63	.127		5.40	8.60		14	18.80
1060	1-1/4" diameter		58	.138		6.95	9.35		16.30	21.50
1070	1-1/2" diameter	↓	51	.157		7.85	10.60		18.45	24.50

For customer support on your Plumbing Costs with RSMeans Data, call 800.448.8182.

199

22 11 13.48 Pipe, Fittings and Valves, Steel, Grooved-Joint	Crew	Daily Output	Labor-Hours	Unit	Material	2021 Bare Costs Labor	Equipment	Total	Total Incl O&P
1080	2" diameter	1 Plum	40	.200	L.F.	9.30	13.55		22.85	30
1090	2-1/2" diameter	Q-1	57	.281		14.90	17.10		32	42
1100	3" diameter		50	.320		18.35	19.50		37.85	49
1110	4" diameter		45	.356		26	21.50		47.50	61.50
1120	5" diameter		37	.432		31.50	26.50		58	74
1130	6" diameter	Q-2	42	.571		35	36		71	92.50
1140	8" diameter		37	.649		87.50	41		128.50	157
1150	10" diameter		31	.774		120	49		169	205
1160	12" diameter		27	.889		135	56		191	232
1170	14" diameter		20	1.200		150	76		226	278
1180	16" diameter		17	1.412		219	89		308	375
1190	18" diameter		14	1.714		221	108		329	405
1200	20" diameter		12	2		282	126		408	500
1210	24" diameter		10	2.400		305	152		457	565
1740	To delete coupling & hanger, subtract									
1750	3/4" diam. to 2" diam.					65%	27%			
1760	2-1/2" diam. to 5" diam.					41%	18%			
1770	6" diam. to 12" diam.					31%	13%			
1780	14" diam. to 24" diam.					35%	10%			
1800	Galvanized									
1840	3/4" diameter	1 Plum	71	.113	L.F.	6.30	7.65		13.95	18.35
1850	1" diameter		63	.127		6.30	8.60		14.90	19.75
1860	1-1/4" diameter		58	.138		8.15	9.35		17.50	23
1870	1-1/2" diameter		51	.157		9.25	10.60		19.85	26
1880	2" diameter		40	.200		11.40	13.55		24.95	32.50
1890	2-1/2" diameter	Q-1	57	.281		17.05	17.10		34.15	44.50
1900	3" diameter		50	.320		21	19.50		40.50	52.50
1910	4" diameter		45	.356		30	21.50		51.50	65.50
1920	5" diameter		37	.432		28	26.50		54.50	70
1930	6" diameter	Q-2	42	.571		36.50	36		72.50	94
1940	8" diameter		37	.649		55	41		96	122
1950	10" diameter		31	.774		119	49		168	204
1960	12" diameter		27	.889		144	56		200	242
2540	To delete coupling & hanger, subtract									
2550	3/4" diam. to 2" diam.					36%	27%			
2560	2-1/2" diam. to 5" diam.					19%	18%			
2570	6" diam. to 12" diam.					14%	13%			
2600	Schedule 80, black									
2610	3/4" diameter	1 Plum	65	.123	L.F.	6.60	8.35		14.95	19.70
2650	1" diameter		61	.131		6.90	8.90		15.80	21
2660	1-1/4" diameter		55	.145		9.10	9.85		18.95	24.50
2670	1-1/2" diameter		49	.163		10.45	11.05		21.50	28
2680	2" diameter		38	.211		13	14.25		27.25	36
2690	2-1/2" diameter	Q-1	54	.296		19.45	18.05		37.50	48.50
2700	3" diameter		48	.333		24.50	20.50		45	57.50
2710	4" diameter		44	.364		36	22		58	72.50
2720	5" diameter		35	.457		63	28		91	111
2730	6" diameter	Q-2	40	.600		75	38		113	139
2740	8" diameter		35	.686		118	43.50		161.50	195
2750	10" diameter		29	.828		177	52.50		229.50	272
2760	12" diameter		24	1		365	63		428	500
3240	To delete coupling & hanger, subtract									
3250	3/4" diam. to 2" diam.					30%	25%			

For customer support on your Plumbing Costs with RSMeans Data, call 800.448.8182.

22 11 13.48 Pipe, Fittings and Valves, Steel, Grooved-Joint	Crew	Daily Output	Labor-Hours	Unit	Material	2021 Bare Costs Labor	Equipment	Total	Total Incl O&P
3260	2-1/2" diam. to 5" diam.					14%	17%			
3270	6" diam. to 12" diam.					12%	12%			
3300	Galvanized									
3310	3/4" diameter	1 Plum	65	.123	L.F.	7.15	8.35		15.50	20.50
3350	1" diameter		61	.131		7.75	8.90		16.65	22
3360	1-1/4" diameter		55	.145		10.25	9.85		20.10	26
3370	1-1/2" diameter		46	.174		11.85	11.75		23.60	30.50
3380	2" diameter		38	.211		15.60	14.25		29.85	38.50
3390	2-1/2" diameter	Q-1	54	.296		23	18.05		41.05	52
3400	3" diameter		48	.333		29.50	20.50		50	63
3410	4" diameter		44	.364		42.50	22		64.50	79.50
3420	5" diameter		35	.457		47	28		75	93.50
3430	6" diameter	Q-2	40	.600		64	38		102	127
3440	8" diameter		35	.686		140	43.50		183.50	219
3450	10" diameter		29	.828		245	52.50		297.50	345
3460	12" diameter		24	1		655	63		718	815
3920	To delete coupling & hanger, subtract									
3930	3/4" diam. to 2" diam.					30%	25%			
3940	2-1/2" diam. to 5" diam.					15%	17%			
3950	6" diam. to 12" diam.					11%	12%			
3990	Fittings: coupling material required at joints not incl. in fitting price.									
3994	Add 1 selected coupling, material only, per joint for installed price.									
4000	Elbow, 90° or 45°, painted									
4030	3/4" diameter	1 Plum	50	.160	Ea.	95	10.85		105.85	121
4040	1" diameter		50	.160		51	10.85		61.85	72
4050	1-1/4" diameter		40	.200		51	13.55		64.55	76
4060	1-1/2" diameter		33	.242		51	16.40		67.40	80.50
4070	2" diameter		25	.320		51	21.50		72.50	88.50
4080	2-1/2" diameter	Q-1	40	.400		51	24.50		75.50	92.50
4090	3" diameter		33	.485		89.50	29.50		119	143
4100	4" diameter		25	.640		97	39		136	165
4110	5" diameter		20	.800		230	48.50		278.50	325
4120	6" diameter	Q-2	25	.960		271	60.50		331.50	390
4130	8" diameter		21	1.143		560	72		632	730
4140	10" diameter		18	1.333		1,050	84		1,134	1,275
4150	12" diameter		15	1.600		1,150	101		1,251	1,400
4170	14" diameter		12	2		1,125	126		1,251	1,425
4180	16" diameter		11	2.182		1,450	138		1,588	1,800
4190	18" diameter	Q-3	14	2.286		1,825	147		1,972	2,250
4200	20" diameter		12	2.667		2,425	172		2,597	2,900
4210	24" diameter		10	3.200		3,500	206		3,706	4,150
4250	For galvanized elbows, add					26%				
4690	Tee, painted									
4700	3/4" diameter	1 Plum	38	.211	Ea.	102	14.25		116.25	134
4740	1" diameter		33	.242		78.50	16.40		94.90	111
4750	1-1/4" diameter		27	.296		78.50	20		98.50	117
4760	1-1/2" diameter		22	.364		78.50	24.50		103	124
4770	2" diameter		17	.471		78.50	32		110.50	134
4780	2-1/2" diameter	Q-1	27	.593		78.50	36		114.50	141
4790	3" diameter		22	.727		107	44.50		151.50	184
4800	4" diameter		17	.941		163	57.50		220.50	265
4810	5" diameter		13	1.231		380	75		455	525
4820	6" diameter	Q-2	17	1.412		435	89		524	615

For customer support on your Plumbing Costs with RSMeans Data, call 800.448.8182.

201

22 11 Facility Water Distribution

22 11 13 – Facility Water Distribution Piping

22 11 13.48 Pipe, Fittings and Valves, Steel, Grooved-Joint	Crew	Daily Output	Labor-Hours	Unit	Material	2021 Bare Costs Labor	Equipment	Total	Total Incl O&P
4830	8" diameter	Q-2	14	1.714	Ea.	955	108		1,063	1,200
4840	10" diameter		12	2		1,250	126		1,376	1,575
4850	12" diameter		10	2.400		1,625	152		1,777	2,000
4851	14" diameter		9	2.667		1,500	168		1,668	1,900
4852	16" diameter		8	3		1,675	190		1,865	2,125
4853	18" diameter	Q-3	10	3.200		2,100	206		2,306	2,600
4854	20" diameter		9	3.556		3,000	229		3,229	3,675
4855	24" diameter		8	4		4,600	258		4,858	5,425
4900	For galvanized tees, add					24%				
4906	Couplings, rigid style, painted									
4908	1" diameter	1 Plum	100	.080	Ea.	39.50	5.40		44.90	51.50
4909	1-1/4" diameter		100	.080		37	5.40		42.40	48.50
4910	1-1/2" diameter		67	.119		39.50	8.10		47.60	55.50
4912	2" diameter		50	.160		50.50	10.85		61.35	71.50
4914	2-1/2" diameter	Q-1	80	.200		57	12.20		69.20	80.50
4916	3" diameter		67	.239		66.50	14.55		81.05	94.50
4918	4" diameter		50	.320		92	19.50		111.50	130
4920	5" diameter		40	.400		118	24.50		142.50	167
4922	6" diameter	Q-2	50	.480		156	30.50		186.50	218
4924	8" diameter		42	.571		245	36		281	325
4926	10" diameter		35	.686		475	43.50		518.50	590
4928	12" diameter		32	.750		530	47.50		577.50	655
4930	14" diameter		24	1		495	63		558	640
4931	16" diameter		20	1.200		530	76		606	695
4932	18" diameter		18	1.333		610	84		694	795
4933	20" diameter		16	1.500		835	95		930	1,050
4934	24" diameter		13	1.846		1,075	117		1,192	1,350
4940	Flexible, standard, painted									
4950	3/4" diameter	1 Plum	100	.080	Ea.	29	5.40		34.40	40
4960	1" diameter		100	.080		29	5.40		34.40	40
4970	1-1/4" diameter		80	.100		37.50	6.75		44.25	51
4980	1-1/2" diameter		67	.119		40.50	8.10		48.60	56.50
4990	2" diameter		50	.160		43.50	10.85		54.35	64
5000	2-1/2" diameter	Q-1	80	.200		50.50	12.20		62.70	73.50
5010	3" diameter		67	.239		56	14.55		70.55	83
5020	3-1/2" diameter		57	.281		80	17.10		97.10	114
5030	4" diameter		50	.320		80.50	19.50		100	118
5040	5" diameter		40	.400		121	24.50		145.50	170
5050	6" diameter	Q-2	50	.480		143	30.50		173.50	203
5070	8" diameter		42	.571		231	36		267	310
5090	10" diameter		35	.686		375	43.50		418.50	480
5110	12" diameter		32	.750		430	47.50		477.50	540
5120	14" diameter		24	1		610	63		673	770
5130	16" diameter		20	1.200		800	76		876	995
5140	18" diameter		18	1.333		940	84		1,024	1,150
5150	20" diameter		16	1.500		1,475	95		1,570	1,775
5160	24" diameter		13	1.846		1,625	117		1,742	1,950
5176	Lightweight style, painted									
5178	1-1/2" diameter	1 Plum	67	.119	Ea.	36	8.10		44.10	51.50
5180	2" diameter	"	50	.160		43.50	10.85		54.35	64
5182	2-1/2" diameter	Q-1	80	.200		50.50	12.20		62.70	73.50
5184	3" diameter		67	.239		56	14.55		70.55	83
5186	3-1/2" diameter		57	.281		67	17.10		84.10	99

22 11 Facility Water Distribution

22 11 13 – Facility Water Distribution Piping

22 11 13.48 Pipe, Fittings and Valves, Steel, Grooved-Joint		Crew	Daily Output	Labor-Hours	Unit	Material	2021 Bare Costs Labor	Equipment	Total	Total Incl O&P
5188	4" diameter	Q-1	50	.320	Ea.	80.50	19.50		100	118
5190	5" diameter	↓	40	.400		121	24.50		145.50	170
5192	6" diameter	Q-2	50	.480		143	30.50		173.50	203
5194	8" diameter		42	.571		231	36		267	310
5196	10" diameter		35	.686		470	43.50		513.50	580
5198	12" diameter	↓	32	.750		520	47.50		567.50	640
5200	For galvanized couplings, add				↓	33%				
5220	Tee, reducing, painted									
5225	2" x 1-1/2" diameter	Q-1	38	.421	Ea.	165	25.50		190.50	221
5226	2-1/2" x 2" diameter		28	.571		165	35		200	234
5227	3" x 2-1/2" diameter		23	.696		146	42.50		188.50	224
5228	4" x 3" diameter		18	.889		196	54		250	297
5229	5" x 4" diameter	↓	15	1.067		415	65		480	555
5230	6" x 4" diameter	Q-2	18	1.333		460	84		544	630
5231	8" x 6" diameter		15	1.600		955	101		1,056	1,200
5232	10" x 8" diameter		13	1.846		1,075	117		1,192	1,350
5233	12" x 10" diameter		11	2.182		1,325	138		1,463	1,675
5234	14" x 12" diameter		10	2.400		1,550	152		1,702	1,950
5235	16" x 12" diameter	↓	9	2.667		1,350	168		1,518	1,725
5236	18" x 12" diameter	Q-3	12	2.667		1,600	172		1,772	2,025
5237	18" x 16" diameter		11	2.909		2,025	187		2,212	2,525
5238	20" x 16" diameter		10	3.200		2,650	206		2,856	3,225
5239	24" x 20" diameter	↓	9	3.556	↓	4,200	229		4,429	4,975
5240	Reducer, concentric, painted									
5241	2-1/2" x 2" diameter	Q-1	43	.372	Ea.	59.50	22.50		82	99.50
5242	3" x 2-1/2" diameter		35	.457		71.50	28		99.50	121
5243	4" x 3" diameter		29	.552		86	33.50		119.50	145
5244	5" x 4" diameter	↓	22	.727		118	44.50		162.50	196
5245	6" x 4" diameter	Q-2	26	.923		137	58.50		195.50	238
5246	8" x 6" diameter		23	1.043		355	66		421	490
5247	10" x 8" diameter		20	1.200		690	76		766	875
5248	12" x 10" diameter	↓	16	1.500	↓	1,225	95		1,320	1,500
5255	Eccentric, painted									
5256	2-1/2" x 2" diameter	Q-1	42	.381	Ea.	123	23		146	171
5257	3" x 2-1/2" diameter		34	.471		141	28.50		169.50	198
5258	4" x 3" diameter		28	.571		171	35		206	240
5259	5" x 4" diameter	↓	21	.762		232	46.50		278.50	325
5260	6" x 4" diameter	Q-2	25	.960		270	60.50		330.50	390
5261	8" x 6" diameter		22	1.091		545	69		614	705
5262	10" x 8" diameter		19	1.263		1,425	80		1,505	1,700
5263	12" x 10" diameter	↓	15	1.600	↓	1,975	101		2,076	2,325
5270	Coupling, reducing, painted									
5272	2" x 1-1/2" diameter	1 Plum	52	.154	Ea.	57.50	10.40		67.90	78.50
5274	2-1/2" x 2" diameter	Q-1	82	.195		74	11.90		85.90	99.50
5276	3" x 2" diameter		69	.232		84	14.15		98.15	113
5278	4" x 2" diameter		52	.308		132	18.75		150.75	174
5280	5" x 4" diameter	↓	42	.381		149	23		172	199
5282	6" x 4" diameter	Q-2	52	.462		224	29		253	291
5284	8" x 6" diameter	"	44	.545	↓	335	34.50		369.50	420
5290	Outlet coupling, painted									
5294	1-1/2" x 1" pipe size	1 Plum	65	.123	Ea.	76	8.35		84.35	96
5296	2" x 1" pipe size	"	48	.167		77	11.30		88.30	101
5298	2-1/2" x 1" pipe size	Q-1	78	.205		118	12.50		130.50	149

For customer support on your Plumbing Costs with RSMeans Data, call 800.448.8182.

203

22 11 13.48 Pipe, Fittings and Valves, Steel, Grooved-Joint		Crew	Daily Output	Labor-Hours	Unit	Material	2021 Bare Costs Labor	Equipment	Total	Total Incl O&P
5300	2-1/2" x 1-1/4" pipe size	Q-1	70	.229	Ea.	133	13.95		146.95	167
5302	3" x 1" pipe size		65	.246		151	15		166	189
5304	4" x 3/4" pipe size		48	.333		166	20.50		186.50	214
5306	4" x 1-1/2" pipe size		46	.348		235	21		256	291
5308	6" x 1-1/2" pipe size	Q-2	44	.545		330	34.50		364.50	415
5320	Outlet, strap-on T, painted									
5324	Threaded female branch									
5330	2" pipe x 1-1/2" branch size	1 Plum	50	.160	Ea.	54.50	10.85		65.35	76
5334	2-1/2" pipe x 1-1/2" branch size	Q-1	80	.200		66.50	12.20		78.70	91
5338	3" pipe x 1-1/2" branch size		67	.239		68.50	14.55		83.05	96.50
5342	3" pipe x 2" branch size		59	.271		79	16.50		95.50	112
5346	4" pipe x 2" branch size		50	.320		83	19.50		102.50	121
5350	4" pipe x 2-1/2" branch size		47	.340		86.50	20.50		107	127
5354	4" pipe x 3" branch size		43	.372		92	22.50		114.50	135
5358	5" pipe x 3" branch size		40	.400		113	24.50		137.50	162
5362	6" pipe x 2" branch size	Q-2	50	.480		104	30.50		134.50	160
5366	6" pipe x 3" branch size		47	.511		118	32.50		150.50	178
5370	6" pipe x 4" branch size		44	.545		132	34.50		166.50	197
5374	8" pipe x 2-1/2" branch size		42	.571		181	36		217	253
5378	8" pipe x 4" branch size		38	.632		196	40		236	276
5390	Grooved branch									
5394	2" pipe x 1-1/4" branch size	1 Plum	78	.103	Ea.	54.50	6.95		61.45	70.50
5398	2" pipe x 1-1/2" branch size	"	50	.160		54.50	10.85		65.35	76
5402	2-1/2" pipe x 1-1/2" branch size	Q-1	80	.200		66.50	12.20		78.70	91
5406	3" pipe x 1-1/2" branch size		67	.239		68.50	14.55		83.05	96.50
5410	3" pipe x 2" branch size		59	.271		79	16.50		95.50	112
5414	4" pipe x 2" branch size		50	.320		83	19.50		102.50	121
5418	4" pipe x 2-1/2" branch size		47	.340		86.50	20.50		107	127
5422	4" pipe x 3" branch size		43	.372		92	22.50		114.50	135
5426	5" pipe x 3" branch size		40	.400		113	24.50		137.50	162
5430	6" pipe x 2" branch size	Q-2	50	.480		104	30.50		134.50	160
5434	6" pipe x 3" branch size		47	.511		118	32.50		150.50	178
5438	6" pipe x 4" branch size		44	.545		132	34.50		166.50	197
5442	8" pipe x 2-1/2" branch size		42	.571		181	36		217	253
5446	8" pipe x 4" branch size		38	.632		196	40		236	276
5750	Flange, w/groove gasket, black steel									
5754	See Line 22 11 13.47 0620 for gasket & bolt set									
5760	ANSI class 125 and 150, painted									
5780	2" pipe size	1 Plum	23	.348	Ea.	167	23.50		190.50	218
5790	2-1/2" pipe size	Q-1	37	.432		207	26.50		233.50	268
5800	3" pipe size		31	.516		223	31.50		254.50	293
5820	4" pipe size		23	.696		298	42.50		340.50	390
5830	5" pipe size		19	.842		345	51.50		396.50	455
5840	6" pipe size	Q-2	23	1.043		375	66		441	515
5850	8" pipe size		17	1.412		425	89		514	605
5860	10" pipe size		14	1.714		670	108		778	900
5870	12" pipe size		12	2		880	126		1,006	1,150
5880	14" pipe size		10	2.400		1,625	152		1,777	2,000
5890	16" pipe size		9	2.667		1,875	168		2,043	2,325
5900	18" pipe size		6	4		2,325	253		2,578	2,925
5910	20" pipe size		5	4.800		2,775	305		3,080	3,500
5920	24" pipe size		4.50	5.333		3,550	335		3,885	4,425
5940	ANSI class 350, painted									

22 11 Facility Water Distribution

22 11 13 – Facility Water Distribution Piping

22 11 13.48 Pipe, Fittings and Valves, Steel, Grooved-Joint	Crew	Daily Output	Labor-Hours	Unit	Material	2021 Bare Costs Labor	Equipment	Total	Total Incl O&P
5946 2" pipe size	1 Plum	23	.348	Ea.	209	23.50		232.50	264
5948 2-1/2" pipe size	Q-1	37	.432		241	26.50		267.50	305
5950 3" pipe size		31	.516		330	31.50		361.50	405
5952 4" pipe size		23	.696		435	42.50		477.50	545
5954 5" pipe size		19	.842		500	51.50		551.50	620
5956 6" pipe size	Q-2	23	1.043		580	66		646	735
5958 8" pipe size		17	1.412		665	89		754	865
5960 10" pipe size		14	1.714		1,050	108		1,158	1,300
5962 12" pipe size		12	2		1,125	126		1,251	1,450
6100 Cross, painted									
6110 2" diameter	1 Plum	12.50	.640	Ea.	139	43.50		182.50	218
6112 2-1/2" diameter	Q-1	20	.800		139	48.50		187.50	226
6114 3" diameter		15.50	1.032		248	63		311	365
6116 4" diameter		12.50	1.280		410	78		488	565
6118 6" diameter	Q-2	12.50	1.920		1,075	121		1,196	1,350
6120 8" diameter		10.50	2.286		1,400	144		1,544	1,775
6122 10" diameter		9	2.667		2,300	168		2,468	2,800
6124 12" diameter		7.50	3.200		3,325	202		3,527	3,975
7400 Suction diffuser									
7402 Grooved end inlet x flanged outlet									
7410 3" x 3"	Q-1	27	.593	Ea.	1,350	36		1,386	1,525
7412 4" x 4"		19	.842		1,825	51.50		1,876.50	2,075
7414 5" x 5"		14	1.143		2,125	69.50		2,194.50	2,450
7416 6" x 6"	Q-2	20	1.200		2,700	76		2,776	3,075
7418 8" x 8"		15	1.600		5,000	101		5,101	5,650
7420 10" x 10"		12	2		6,800	126		6,926	7,675
7422 12" x 12"		9	2.667		11,200	168		11,368	12,600
7424 14" x 14"		7	3.429		10,400	217		10,617	11,700
7426 16" x 14"		6	4		10,700	253		10,953	12,200
7500 Strainer, tee type, painted									
7506 2" pipe size	1 Plum	21	.381	Ea.	900	26		926	1,025
7508 2-1/2" pipe size	Q-1	30	.533		945	32.50		977.50	1,100
7510 3" pipe size		28	.571		1,050	35		1,085	1,225
7512 4" pipe size		20	.800		1,200	48.50		1,248.50	1,400
7514 5" pipe size		15	1.067		1,725	65		1,790	2,000
7516 6" pipe size	Q-2	23	1.043		1,875	66		1,941	2,175
7518 8" pipe size		16	1.500		2,875	95		2,970	3,325
7520 10" pipe size		13	1.846		4,250	117		4,367	4,850
7522 12" pipe size		10	2.400		5,450	152		5,602	6,225
7524 14" pipe size		8	3		14,000	190		14,190	15,700
7526 16" pipe size		7	3.429		17,400	217		17,617	19,500
7570 Expansion joint, max. 3" travel									
7572 2" diameter	1 Plum	24	.333	Ea.	1,125	22.50		1,147.50	1,275
7574 3" diameter	Q-1	31	.516		1,300	31.50		1,331.50	1,475
7576 4" diameter	"	23	.696		1,725	42.50		1,767.50	1,950
7578 6" diameter	Q-2	38	.632		2,675	40		2,715	2,975
7790 Valves: coupling material required at joints not incl. in valve price.									
7794 Add 1 selected coupling, material only, per joint for installed price.									
7800 Ball valve w/handle, carbon steel trim									
7810 1-1/2" pipe size	1 Plum	31	.258	Ea.	190	17.45		207.45	235
7812 2" pipe size	"	24	.333		211	22.50		233.50	266
7814 2-1/2" pipe size	Q-1	38	.421		460	25.50		485.50	545
7816 3" pipe size		31	.516		740	31.50		771.50	860

For customer support on your Plumbing Costs with RSMeans Data, call 800.448.8182.

205

22 11 13.48 Pipe, Fittings and Valves, Steel, Grooved-Joint	Crew	Daily Output	Labor-Hours	Unit	Material	2021 Bare Costs Labor	Equipment	Total	Total Incl O&P
7818	4" pipe size	Q-1	23	.696	Ea.	1,150	42.50		1,192.50	1,325
7820	6" pipe size	Q-2	24	1		3,525	63		3,588	3,975
7830	With gear operator									
7834	2-1/2" pipe size	Q-1	38	.421	Ea.	930	25.50		955.50	1,075
7836	3" pipe size		31	.516		1,325	31.50		1,356.50	1,500
7838	4" pipe size		23	.696		1,675	42.50		1,717.50	1,925
7840	6" pipe size	Q-2	24	1		3,950	63		4,013	4,450
7870	Check valve									
7874	2-1/2" pipe size	Q-1	38	.421	Ea.	385	25.50		410.50	460
7876	3" pipe size		31	.516		450	31.50		481.50	540
7878	4" pipe size		23	.696		475	42.50		517.50	590
7880	5" pipe size		18	.889		790	54		844	950
7882	6" pipe size	Q-2	24	1		940	63		1,003	1,125
7884	8" pipe size		19	1.263		1,275	80		1,355	1,525
7886	10" pipe size		16	1.500		3,425	95		3,520	3,925
7888	12" pipe size		13	1.846		4,075	117		4,192	4,650
7900	Plug valve, balancing, w/lever operator									
7906	3" pipe size	Q-1	31	.516	Ea.	830	31.50		861.50	960
7908	4" pipe size	"	23	.696		1,025	42.50		1,067.50	1,200
7909	6" pipe size	Q-2	24	1		1,575	63		1,638	1,825
7916	With gear operator									
7920	3" pipe size	Q-1	31	.516	Ea.	1,575	31.50		1,606.50	1,775
7922	4" pipe size	"	23	.696		1,625	42.50		1,667.50	1,875
7924	6" pipe size	Q-2	24	1		2,200	63		2,263	2,525
7926	8" pipe size		19	1.263		3,375	80		3,455	3,850
7928	10" pipe size		16	1.500		4,275	95		4,370	4,850
7930	12" pipe size		13	1.846		6,500	117		6,617	7,325
8000	Butterfly valve, 2 position handle, with standard trim									
8010	1-1/2" pipe size	1 Plum	30	.267	Ea.	360	18.05		378.05	420
8020	2" pipe size	"	23	.348		360	23.50		383.50	430
8030	3" pipe size	Q-1	30	.533		515	32.50		547.50	620
8050	4" pipe size	"	22	.727		565	44.50		609.50	690
8070	6" pipe size	Q-2	23	1.043		1,150	66		1,216	1,350
8080	8" pipe size		18	1.333		1,550	84		1,634	1,825
8090	10" pipe size		15	1.600		2,525	101		2,626	2,925
8200	With stainless steel trim									
8240	1-1/2" pipe size	1 Plum	30	.267	Ea.	455	18.05		473.05	525
8250	2" pipe size	"	23	.348		455	23.50		478.50	535
8270	3" pipe size	Q-1	30	.533		610	32.50		642.50	725
8280	4" pipe size	"	22	.727		665	44.50		709.50	795
8300	6" pipe size	Q-2	23	1.043		1,250	66		1,316	1,475
8310	8" pipe size		18	1.333		3,075	84		3,159	3,500
8320	10" pipe size		15	1.600		5,125	101		5,226	5,775
8322	12" pipe size		12	2		5,875	126		6,001	6,650
8324	14" pipe size		10	2.400		7,525	152		7,677	8,500
8326	16" pipe size		9	2.667		12,500	168		12,668	14,100
8328	18" pipe size	Q-3	12	2.667		13,400	172		13,572	15,000
8330	20" pipe size		10	3.200		16,500	206		16,706	18,400
8332	24" pipe size		9	3.556		22,000	229		22,229	24,500
8336	Note: sizes 8" up w/manual gear operator									
9000	Cut one groove, labor									
9010	3/4" pipe size	Q-1	152	.105	Ea.		6.40		6.40	9.55
9020	1" pipe size		140	.114			6.95		6.95	10.40

For customer support on your Plumbing Costs with RSMeans Data, call 800.448.8182.

22 11 13.48 Pipe, Fittings and Valves, Steel, Grooved-Joint

		Crew	Daily Output	Labor-Hours	Unit	Material	2021 Bare Costs Labor	2021 Bare Costs Equipment	Total	Total Incl O&P
9030	1-1/4" pipe size	Q-1	124	.129	Ea.		7.85		7.85	11.75
9040	1-1/2" pipe size		114	.140			8.55		8.55	12.75
9050	2" pipe size		104	.154			9.35		9.35	14
9060	2-1/2" pipe size		96	.167			10.15		10.15	15.15
9070	3" pipe size		88	.182			11.10		11.10	16.55
9080	3-1/2" pipe size		83	.193			11.75		11.75	17.55
9090	4" pipe size		78	.205			12.50		12.50	18.65
9100	5" pipe size		72	.222			13.55		13.55	20
9110	6" pipe size		70	.229			13.95		13.95	21
9120	8" pipe size		54	.296			18.05		18.05	27
9130	10" pipe size		38	.421			25.50		25.50	38.50
9140	12" pipe size		30	.533			32.50		32.50	48.50
9150	14" pipe size		20	.800			48.50		48.50	72.50
9160	16" pipe size		19	.842			51.50		51.50	76.50
9170	18" pipe size		18	.889			54		54	81
9180	20" pipe size		17	.941			57.50		57.50	85.50
9190	24" pipe size	▼	15	1.067	▼		65		65	97
9210	Roll one groove									
9220	3/4" pipe size	Q-1	266	.060	Ea.		3.66		3.66	5.45
9230	1" pipe size		228	.070			4.28		4.28	6.40
9240	1-1/4" pipe size		200	.080			4.87		4.87	7.25
9250	1-1/2" pipe size		178	.090			5.50		5.50	8.15
9260	2" pipe size		116	.138			8.40		8.40	12.55
9270	2-1/2" pipe size		110	.145			8.85		8.85	13.25
9280	3" pipe size		100	.160			9.75		9.75	14.55
9290	3-1/2" pipe size		94	.170			10.35		10.35	15.50
9300	4" pipe size		86	.186			11.35		11.35	16.90
9310	5" pipe size		84	.190			11.60		11.60	17.30
9320	6" pipe size		80	.200			12.20		12.20	18.20
9330	8" pipe size		66	.242			14.75		14.75	22
9340	10" pipe size		58	.276			16.80		16.80	25
9350	12" pipe size		46	.348			21		21	31.50
9360	14" pipe size		30	.533			32.50		32.50	48.50
9370	16" pipe size		28	.571			35		35	52
9380	18" pipe size		27	.593			36		36	54
9390	20" pipe size		25	.640			39		39	58
9400	24" pipe size	▼	23	.696	▼		42.50		42.50	63.50

22 11 13.60 Tubing, Stainless Steel

		Crew	Daily Output	Labor-Hours	Unit	Material	2021 Bare Costs Labor	2021 Bare Costs Equipment	Total	Total Incl O&P
0010	**TUBING, STAINLESS STEEL**									
5000	Tubing									
5010	Type 304, no joints, no hangers									
5020	.035 wall									
5021	1/4"	1 Plum	160	.050	L.F.	12.35	3.39		15.74	18.60
5022	3/8"		160	.050		5.10	3.39		8.49	10.70
5023	1/2"		160	.050		9.35	3.39		12.74	15.30
5024	5/8"		160	.050		8.10	3.39		11.49	13.95
5025	3/4"		133	.060		9.75	4.07		13.82	16.80
5026	7/8"		133	.060		10.30	4.07		14.37	17.40
5027	1"	▼	114	.070	▼	11.30	4.75		16.05	19.55
5040	.049 wall									
5041	1/4"	1 Plum	160	.050	L.F.	4.77	3.39		8.16	10.30
5042	3/8"	▼	160	.050		6.60	3.39		9.99	12.30

For customer support on your Plumbing Costs with RSMeans Data, call 800.448.8182.

207

22 11 13.60 Tubing, Stainless Steel

		Crew	Daily Output	Labor-Hours	Unit	Material	2021 Bare Costs Labor	Equipment	Total	Total Incl O&P
5043	1/2"	1 Plum	160	.050	L.F.	7.30	3.39		10.69	13.10
5044	5/8"		160	.050		9.90	3.39		13.29	15.95
5045	3/4"		133	.060		10.55	4.07		14.62	17.70
5046	7/8"		133	.060		12.85	4.07		16.92	20
5047	1"		114	.070		11.65	4.75		16.40	19.90
5060	.065 wall									
5061	1/4"	1 Plum	160	.050	L.F.	5.35	3.39		8.74	10.90
5062	3/8"		160	.050		7.35	3.39		10.74	13.15
5063	1/2"		160	.050		8.90	3.39		12.29	14.85
5064	5/8"		160	.050		10.85	3.39		14.24	17
5065	3/4"		133	.060		11.50	4.07		15.57	18.75
5066	7/8"		133	.060		13.20	4.07		17.27	20.50
5067	1"		114	.070		12.90	4.75		17.65	21.50
5210	Type 316									
5220	.035 wall									
5221	1/4"	1 Plum	160	.050	L.F.	3.96	3.39		7.35	9.40
5222	3/8"		160	.050		5.70	3.39		9.09	11.35
5223	1/2"		160	.050		7	3.39		10.39	12.75
5224	5/8"		160	.050		9.10	3.39		12.49	15.05
5225	3/4"		133	.060		15	4.07		19.07	22.50
5226	7/8"		133	.060		19.50	4.07		23.57	27.50
5227	1"		114	.070		15.45	4.75		20.20	24
5240	.049 wall									
5241	1/4"	1 Plum	160	.050	L.F.	6.80	3.39		10.19	12.55
5242	3/8"		160	.050		8.85	3.39		12.24	14.75
5243	1/2"		160	.050		7.90	3.39		11.29	13.70
5244	5/8"		160	.050		10.90	3.39		14.29	17.05
5245	3/4"		133	.060		10.50	4.07		14.57	17.65
5246	7/8"		133	.060		17.55	4.07		21.62	25.50
5247	1"		114	.070		15.90	4.75		20.65	24.50
5260	.065 wall									
5261	1/4"	1 Plum	160	.050	L.F.	5.20	3.39		8.59	10.75
5262	3/8"		160	.050		8.30	3.39		11.69	14.15
5263	1/2"		160	.050		8.90	3.39		12.29	14.85
5264	5/8"		160	.050		12.40	3.39		15.79	18.70
5265	3/4"		133	.060		13.35	4.07		17.42	21
5266	7/8"		133	.060		20.50	4.07		24.57	28.50
5267	1"		114	.070		19.55	4.75		24.30	28.50

22 11 13.61 Tubing Fittings, Stainless Steel

		Crew	Daily Output	Labor-Hours	Unit	Material	2021 Bare Costs Labor	Equipment	Total	Total Incl O&P
0010	**TUBING FITTINGS, STAINLESS STEEL**									
8200	Tube fittings, compression type									
8202	Type 316									
8204	90° elbow									
8206	1/4"	1 Plum	24	.333	Ea.	17.65	22.50		40.15	53
8207	3/8"		22	.364		21.50	24.50		46	61
8208	1/2"		22	.364		35.50	24.50		60	76
8209	5/8"		21	.381		40	26		66	82.50
8210	3/4"		21	.381		63	26		89	108
8211	7/8"		20	.400		97	27		124	148
8212	1"		20	.400		121	27		148	174
8220	Union tee									
8222	1/4"	1 Plum	15	.533	Ea.	25	36		61	81.50

22 11 13 – Facility Water Distribution Piping

22 11 13.61 Tubing Fittings, Stainless Steel

		Crew	Daily Output	Labor-Hours	Unit	Material	2021 Bare Costs Labor	Equipment	Total	Total Incl O&P
8224	3/8"	1 Plum	15	.533	Ea.	32	36		68	89.50
8225	1/2"		15	.533		49.50	36		85.50	109
8226	5/8"		14	.571		56	38.50		94.50	119
8227	3/4"		14	.571		74.50	38.50		113	140
8228	7/8"		13	.615		146	41.50		187.50	222
8229	1"		13	.615		159	41.50		200.50	237
8234	Union									
8236	1/4"	1 Plum	24	.333	Ea.	12.25	22.50		34.75	47
8237	3/8"		22	.364		17.40	24.50		41.90	56
8238	1/2"		22	.364		26	24.50		50.50	65.50
8239	5/8"		21	.381		34	26		60	75.50
8240	3/4"		21	.381		42	26		68	85
8241	7/8"		20	.400		69	27		96	117
8242	1"		20	.400		73	27		100	121
8250	Male connector									
8252	1/4" x 1/4"	1 Plum	24	.333	Ea.	7.85	22.50		30.35	42
8253	3/8" x 3/8"		22	.364		12.25	24.50		36.75	50.50
8254	1/2" x 1/2"		22	.364		18.10	24.50		42.60	57
8256	3/4" x 3/4"		21	.381		28	26		54	69
8258	1" x 1"		20	.400		48.50	27		75.50	94

22 11 13.64 Pipe, Stainless Steel

		Crew	Daily Output	Labor-Hours	Unit	Material	2021 Bare Costs Labor	Equipment	Total	Total Incl O&P
0010	**PIPE, STAINLESS STEEL** R221113-70									
0020	Welded, with clevis type hanger assemblies, 10' OC									
0500	Schedule 5, type 304									
0540	1/2" diameter	Q-15	128	.125	L.F.	9.45	7.60	.84	17.89	22.50
0550	3/4" diameter		116	.138		12	8.40	.93	21.33	27
0560	1" diameter		103	.155		13.40	9.45	1.04	23.89	30
0570	1-1/4" diameter		93	.172		18.75	10.50	1.16	30.41	37.50
0580	1-1/2" diameter		85	.188		23	11.45	1.26	35.71	43.50
0590	2" diameter		69	.232		39	14.15	1.56	54.71	65.50
0600	2-1/2" diameter		53	.302		41	18.40	2.03	61.43	74.50
0610	3" diameter		48	.333		76	20.50	2.24	98.74	117
0620	4" diameter		44	.364		131	22	2.44	155.44	180
0630	5" diameter		36	.444		159	27	2.99	188.99	219
0640	6" diameter	Q-16	42	.571		345	36	2.56	383.56	435
0650	8" diameter		34	.706		530	44.50	3.16	577.66	655
0660	10" diameter		26	.923		315	58.50	4.14	377.64	435
0670	12" diameter		21	1.143		410	72	5.10	487.10	570
0700	To delete hangers, subtract									
0710	1/2" diam. to 1-1/2" diam.					8%	19%			
0720	2" diam. to 5" diam.					4%	9%			
0730	6" diam. to 12" diam.					3%	4%			
0750	For small quantities, add				L.F.	10%				
1250	Schedule 5, type 316									
1290	1/2" diameter	Q-15	128	.125	L.F.	9.40	7.60	.84	17.84	22.50
1300	3/4" diameter		116	.138		13.85	8.40	.93	23.18	29
1310	1" diameter		103	.155		20	9.45	1.04	30.49	37.50
1320	1-1/4" diameter		93	.172		25	10.50	1.16	36.66	44.50
1330	1-1/2" diameter		85	.188		34.50	11.45	1.26	47.21	56.50
1340	2" diameter		69	.232		54	14.15	1.56	69.71	81.50
1350	2-1/2" diameter		53	.302		90	18.40	2.03	110.43	129
1360	3" diameter		48	.333		114	20.50	2.24	136.74	159

For customer support on your Plumbing Costs with RSMeans Data, call 800.448.8182.

209

22 11 13.64 Pipe, Stainless Steel	Crew	Daily Output	Labor-Hours	Unit	Material	2021 Bare Costs Labor	Equipment	Total	Total Incl O&P
1370	4" diameter	Q-15	44	.364	L.F.	142	22	2.44	166.44	192
1380	5" diameter		36	.444		163	27	2.99	192.99	223
1390	6" diameter	Q-16	42	.571		194	36	2.56	232.56	270
1400	8" diameter		34	.706		292	44.50	3.16	339.66	390
1410	10" diameter		26	.923		345	58.50	4.14	407.64	465
1420	12" diameter		21	1.143		445	72	5.10	522.10	600
1490	For small quantities, add					10%				
1940	To delete hanger, subtract									
1950	1/2" diam. to 1-1/2" diam.					5%	19%			
1960	2" diam. to 5" diam.					3%	9%			
1970	6" diam. to 12" diam.					2%	4%			
2000	Schedule 10, type 304									
2040	1/4" diameter	Q-15	131	.122	L.F.	9.70	7.45	.82	17.97	22.50
2050	3/8" diameter		128	.125		7.95	7.60	.84	16.39	21
2060	1/2" diameter		125	.128		8.65	7.80	.86	17.31	22
2070	3/4" diameter		113	.142		8.85	8.65	.95	18.45	23.50
2080	1" diameter		100	.160		13.20	9.75	1.08	24.03	30.50
2090	1-1/4" diameter		91	.176		15.20	10.70	1.18	27.08	34
2100	1-1/2" diameter		83	.193		17.05	11.75	1.30	30.10	37.50
2110	2" diameter		67	.239		21	14.55	1.60	37.15	46.50
2120	2-1/2" diameter		51	.314		27	19.10	2.11	48.21	60.50
2130	3" diameter		46	.348		33	21	2.34	56.34	70
2140	4" diameter		42	.381		40.50	23	2.56	66.06	82
2150	5" diameter		35	.457		72	28	3.07	103.07	124
2160	6" diameter	Q-16	40	.600		64	38	2.69	104.69	130
2170	8" diameter		33	.727		51.50	46	3.26	100.76	129
2180	10" diameter		25	.960		70.50	60.50	4.30	135.30	173
2190	12" diameter		21	1.143		98	72	5.10	175.10	222
2250	For small quantities, add					10%				
2650	To delete hanger, subtract									
2660	1/4" diam. to 3/4" diam.					9%	22%			
2670	1" diam. to 2" diam.					4%	15%			
2680	2-1/2" diam. to 5" diam.					3%	8%			
2690	6" diam. to 12" diam.					3%	4%			
2750	Schedule 10, type 316									
2790	1/4" diameter	Q-15	131	.122	L.F.	6.95	7.45	.82	15.22	19.60
2800	3/8" diameter		128	.125		8.50	7.60	.84	16.94	21.50
2810	1/2" diameter		125	.128		9.70	7.80	.86	18.36	23.50
2820	3/4" diameter		113	.142		11.65	8.65	.95	21.25	27
2830	1" diameter		100	.160		17.05	9.75	1.08	27.88	34.50
2840	1-1/4" diameter		91	.176		22	10.70	1.18	33.88	41.50
2850	1-1/2" diameter		83	.193		24.50	11.75	1.30	37.55	46
2860	2" diameter		67	.239		30	14.55	1.60	46.15	56.50
2870	2-1/2" diameter		51	.314		38.50	19.10	2.11	59.71	73.50
2880	3" diameter		46	.348		46	21	2.34	69.34	84.50
2890	4" diameter		42	.381		58.50	23	2.56	84.06	102
2900	5" diameter		35	.457		56.50	28	3.07	87.57	107
2910	6" diameter	Q-16	40	.600		87	38	2.69	127.69	155
2920	8" diameter		33	.727		73	46	3.26	122.26	152
2930	10" diameter		25	.960		85.50	60.50	4.30	150.30	189
2940	12" diameter		21	1.143		119	72	5.10	196.10	245
2990	For small quantities, add					10%				
3430	To delete hanger, subtract									

210

22 11 13.64 Pipe, Stainless Steel	Crew	Daily Output	Labor-Hours	Unit	Material	2021 Bare Costs Labor	Equipment	Total	Total Incl O&P
3440	1/4" diam. to 3/4" diam.					6%	22%			
3450	1" diam. to 2" diam.					3%	15%			
3460	2-1/2" diam. to 5" diam.					2%	8%			
3470	6" diam. to 12" diam.					2%	4%			
3500	Threaded, couplings and clevis hanger assemblies, 10' OC									
3520	Schedule 40, type 304									
3540	1/4" diameter	1 Plum	54	.148	L.F.	5.30	10.05		15.35	21
3550	3/8" diameter		53	.151		6.10	10.20		16.30	22
3560	1/2" diameter		52	.154		5.80	10.40		16.20	22
3570	3/4" diameter		51	.157		8.90	10.60		19.50	25.50
3580	1" diameter		45	.178		13.85	12.05		25.90	33
3590	1-1/4" diameter	Q-1	76	.211		17.35	12.85		30.20	38.50
3600	1-1/2" diameter		69	.232		24.50	14.15		38.65	48
3610	2" diameter		57	.281		35	17.10		52.10	64
3620	2-1/2" diameter		44	.364		47	22		69	85
3630	3" diameter		38	.421		60	25.50		85.50	105
3640	4" diameter	Q-2	51	.471		85	29.50		114.50	138
3740	For small quantities, add					10%				
4200	To delete couplings & hangers, subtract									
4210	1/4" diam. to 3/4" diam.					15%	56%			
4220	1" diam. to 2" diam.					18%	49%			
4230	2-1/2" diam. to 4" diam.					34%	40%			
4250	Schedule 40, type 316									
4290	1/4" diameter	1 Plum	54	.148	L.F.	12.25	10.05		22.30	28.50
4300	3/8" diameter		53	.151		13.35	10.20		23.55	30
4310	1/2" diameter		52	.154		15.55	10.40		25.95	32.50
4320	3/4" diameter		51	.157		18.05	10.60		28.65	35.50
4330	1" diameter		45	.178		25.50	12.05		37.55	46
4340	1-1/4" diameter	Q-1	76	.211		34	12.85		46.85	56.50
4350	1-1/2" diameter		69	.232		39	14.15		53.15	64
4360	2" diameter		57	.281		52.50	17.10		69.60	83.50
4370	2-1/2" diameter		44	.364		86	22		108	128
4380	3" diameter		38	.421		109	25.50		134.50	159
4390	4" diameter	Q-2	51	.471		144	29.50		173.50	204
4490	For small quantities, add					10%				
4900	To delete couplings & hangers, subtract									
4910	1/4" diam. to 3/4" diam.					12%	56%			
4920	1" diam. to 2" diam.					14%	49%			
4930	2-1/2" diam. to 4" diam.					27%	40%			
5000	Schedule 80, type 304									
5040	1/4" diameter	1 Plum	53	.151	L.F.	21	10.20		31.20	38.50
5050	3/8" diameter		52	.154		26.50	10.40		36.90	44.50
5060	1/2" diameter		51	.157		31	10.60		41.60	50
5070	3/4" diameter		48	.167		32	11.30		43.30	52.50
5080	1" diameter		43	.186		40.50	12.60		53.10	63.50
5090	1-1/4" diameter	Q-1	73	.219		57	13.35		70.35	83
5100	1-1/2" diameter		67	.239		65	14.55		79.55	92.50
5110	2" diameter		54	.296		81.50	18.05		99.55	117
5190	For small quantities, add					10%				
5700	To delete couplings & hangers, subtract									
5710	1/4" diam. to 3/4" diam.					10%	53%			
5720	1" diam. to 2" diam.					14%	47%			
5750	Schedule 80, type 316									

22 11 13.64 Pipe, Stainless Steel		Crew	Daily Output	Labor-Hours	Unit	Material	2021 Bare Costs Labor	Equipment	Total	Total Incl O&P
5790	1/4" diameter	1 Plum	53	.151	L.F.	26.50	10.20		36.70	45
5800	3/8" diameter		52	.154		33	10.40		43.40	51.50
5810	1/2" diameter		51	.157		37.50	10.60		48.10	57.50
5820	3/4" diameter		48	.167		41	11.30		52.30	62
5830	1" diameter		43	.186		52.50	12.60		65.10	76.50
5840	1-1/4" diameter	Q-1	73	.219		67	13.35		80.35	93.50
5850	1-1/2" diameter		67	.239		75	14.55		89.55	104
5860	2" diameter		54	.296		93	18.05		111.05	129
5950	For small quantities, add					10%				
7000	To delete couplings & hangers, subtract									
7010	1/4" diam. to 3/4" diam.					9%	53%			
7020	1" diam. to 2" diam.					14%	47%			
8000	Weld joints with clevis type hanger assemblies, 10' OC									
8010	Schedule 40, type 304									
8050	1/8" pipe size	Q-15	126	.127	L.F.	4.01	7.75	.85	12.61	16.90
8060	1/4" pipe size		125	.128		4.43	7.80	.86	13.09	17.45
8070	3/8" pipe size		122	.131		5.05	8	.88	13.93	18.50
8080	1/2" pipe size		118	.136		4.39	8.25	.91	13.55	18.20
8090	3/4" pipe size		109	.147		7.05	8.95	.99	16.99	22
8100	1" pipe size		95	.168		10.90	10.25	1.13	22.28	28.50
8110	1-1/4" pipe size		86	.186		12.55	11.35	1.25	25.15	32
8120	1-1/2" pipe size		78	.205		19.05	12.50	1.38	32.93	41
8130	2" pipe size		62	.258		26.50	15.70	1.73	43.93	54.50
8140	2-1/2" pipe size		49	.327		27.50	19.90	2.19	49.59	62.50
8150	3" pipe size		44	.364		33.50	22	2.44	57.94	72
8160	3-1/2" pipe size		44	.364		86	22	2.44	110.44	131
8170	4" pipe size		39	.410		46.50	25	2.76	74.26	92
8180	5" pipe size		32	.500		132	30.50	3.36	165.86	194
8190	6" pipe size	Q-16	37	.649		69.50	41	2.91	113.41	141
8191	6" pipe size		37	.649		68	41	2.91	111.91	139
8200	8" pipe size		29	.828		102	52.50	3.71	158.21	194
8210	10" pipe size		24	1		286	63	4.48	353.48	415
8220	12" pipe size		20	1.200		530	76	5.40	611.40	700
8300	Schedule 40, type 316									
8310	1/8" pipe size	Q-15	126	.127	L.F.	10.15	7.75	.85	18.75	23.50
8320	1/4" pipe size		125	.128		11.20	7.80	.86	19.86	25
8330	3/8" pipe size		122	.131		12.10	8	.88	20.98	26
8340	1/2" pipe size		118	.136		13.90	8.25	.91	23.06	28.50
8350	3/4" pipe size		109	.147		15.80	8.95	.99	25.74	32
8360	1" pipe size		95	.168		22	10.25	1.13	33.38	40.50
8370	1-1/4" pipe size		86	.186		28	11.35	1.25	40.60	49.50
8380	1-1/2" pipe size		78	.205		32.50	12.50	1.38	46.38	56
8390	2" pipe size		62	.258		42.50	15.70	1.73	59.93	72
8400	2-1/2" pipe size		49	.327		62.50	19.90	2.19	84.59	100
8410	3" pipe size		44	.364		77	22	2.44	101.44	121
8420	3-1/2" pipe size		44	.364		114	22	2.44	138.44	162
8430	4" pipe size		39	.410		98.50	25	2.76	126.26	149
8440	5" pipe size		32	.500		189	30.50	3.36	222.86	257
8450	6" pipe size	Q-16	37	.649		155	41	2.91	198.91	235
8460	8" pipe size		29	.828		219	52.50	3.71	275.21	325
8470	10" pipe size		24	1		395	63	4.48	462.48	535
8480	12" pipe size		20	1.200		535	76	5.40	616.40	710
8500	Schedule 80, type 304									

22 11 Facility Water Distribution

22 11 13 – Facility Water Distribution Piping

22 11 13.64 Pipe, Stainless Steel

		Crew	Daily Output	Labor-Hours	Unit	Material	2021 Bare Costs Labor	Equipment	Total	Total Incl O&P
8510	1/4" pipe size	Q-15	110	.145	L.F.	18.95	8.85	.98	28.78	35.50
8520	3/8" pipe size		109	.147		24	8.95	.99	33.94	41
8530	1/2" pipe size		106	.151		29	9.20	1.01	39.21	46.50
8540	3/4" pipe size		96	.167		29	10.15	1.12	40.27	48.50
8550	1" pipe size		87	.184		36	11.20	1.24	48.44	57.50
8560	1-1/4" pipe size		81	.198		48.50	12.05	1.33	61.88	73
8570	1-1/2" pipe size		74	.216		55.50	13.15	1.45	70.10	82.50
8580	2" pipe size		58	.276		68	16.80	1.85	86.65	102
8590	2-1/2" pipe size		46	.348		102	21	2.34	125.34	146
8600	3" pipe size		41	.390		101	24	2.62	127.62	149
8610	4" pipe size	▼	33	.485		124	29.50	3.26	156.76	184
8630	6" pipe size	Q-16	30	.800	▼	220	50.50	3.58	274.08	320
8640	Schedule 80, type 316									
8650	1/4" pipe size	Q-15	110	.145	L.F.	25.50	8.85	.98	35.33	42.50
8660	3/8" pipe size		109	.147		31.50	8.95	.99	41.44	49
8670	1/2" pipe size		106	.151		36	9.20	1.01	46.21	54.50
8680	3/4" pipe size		96	.167		38.50	10.15	1.12	49.77	59
8690	1" pipe size		87	.184		49	11.20	1.24	61.44	72
8700	1-1/4" pipe size		81	.198		60.50	12.05	1.33	73.88	86
8710	1-1/2" pipe size		74	.216		67.50	13.15	1.45	82.10	95.50
8720	2" pipe size		58	.276		82	16.80	1.85	100.65	118
8730	2-1/2" pipe size		46	.348		70	21	2.34	93.34	111
8740	3" pipe size		41	.390		118	24	2.62	144.62	168
8760	4" pipe size	▼	33	.485		156	29.50	3.26	188.76	220
8770	6" pipe size	Q-16	30	.800	▼	283	50.50	3.58	337.08	390
9100	Threading pipe labor, sst, one end, schedules 40 & 80									
9110	1/4" through 3/4" pipe size	1 Plum	61.50	.130	Ea.		8.80		8.80	13.15
9120	1" through 2" pipe size		55.90	.143			9.70		9.70	14.45
9130	2-1/2" pipe size		41.50	.193			13.05		13.05	19.50
9140	3" pipe size	▼	38.50	.208			14.05		14.05	21
9150	3-1/2" pipe size	Q-1	68.40	.234			14.25		14.25	21.50
9160	4" pipe size		73	.219			13.35		13.35	19.95
9170	5" pipe size		40.70	.393			24		24	36
9180	6" pipe size		35.40	.452			27.50		27.50	41
9190	8" pipe size		22.30	.717			43.50		43.50	65
9200	10" pipe size		16.10	.994			60.50		60.50	90.50
9210	12" pipe size	▼	12.30	1.301	▼		79.50		79.50	118
9250	Welding labor per joint for stainless steel									
9260	Schedule 5 and 10									
9270	1/4" pipe size	Q-15	36	.444	Ea.		27	2.99	29.99	44
9280	3/8" pipe size		35	.457			28	3.07	31.07	45
9290	1/2" pipe size		35	.457			28	3.07	31.07	45
9300	3/4" pipe size		28	.571			35	3.84	38.84	56
9310	1" pipe size		25	.640			39	4.30	43.30	62.50
9320	1-1/4" pipe size		22	.727			44.50	4.89	49.39	71.50
9330	1-1/2" pipe size		21	.762			46.50	5.10	51.60	75
9340	2" pipe size		18	.889			54	5.95	59.95	87.50
9350	2-1/2" pipe size		12	1.333			81	8.95	89.95	131
9360	3" pipe size		9.73	1.644			100	11.05	111.05	162
9370	4" pipe size		7.37	2.171			132	14.60	146.60	213
9380	5" pipe size		6.15	2.602			159	17.50	176.50	256
9390	6" pipe size		5.71	2.802			171	18.85	189.85	276
9400	8" pipe size	▼	3.69	4.336			264	29	293	425

For customer support on your Plumbing Costs with RSMeans Data, call 800.448.8182.

213

22 11 13.64 Pipe, Stainless Steel

		Crew	Daily Output	Labor-Hours	Unit	Material	2021 Bare Costs Labor	2021 Bare Costs Equipment	Total	Total Incl O&P
9410	10" pipe size	Q-15	2.91	5.498	Ea.		335	37	372	540
9420	12" pipe size	↓	2.31	6.926	↓		420	46.50	466.50	680
9500	Schedule 40									
9510	1/4" pipe size	Q-15	28	.571	Ea.		35	3.84	38.84	56
9520	3/8" pipe size		27	.593			36	3.98	39.98	58.50
9530	1/2" pipe size		25.40	.630			38.50	4.23	42.73	62
9540	3/4" pipe size		22.22	.720			44	4.84	48.84	71
9550	1" pipe size		20.25	.790			48	5.30	53.30	78
9560	1-1/4" pipe size		18.82	.850			52	5.70	57.70	84
9570	1-1/2" pipe size		17.78	.900			55	6.05	61.05	88.50
9580	2" pipe size		15.09	1.060			64.50	7.15	71.65	104
9590	2-1/2" pipe size		7.96	2.010			122	13.50	135.50	198
9600	3" pipe size		6.43	2.488			152	16.70	168.70	244
9610	4" pipe size		4.88	3.279			200	22	222	320
9620	5" pipe size		4.26	3.756			229	25	254	370
9630	6" pipe size		3.77	4.244			259	28.50	287.50	415
9640	8" pipe size		2.44	6.557			400	44	444	645
9650	10" pipe size		1.92	8.333			510	56	566	820
9660	12" pipe size	↓	1.52	10.526	↓		640	70.50	710.50	1,025
9750	Schedule 80									
9760	1/4" pipe size	Q-15	21.55	.742	Ea.		45	4.99	49.99	73
9770	3/8" pipe size		20.75	.771			47	5.20	52.20	75.50
9780	1/2" pipe size		19.54	.819			50	5.50	55.50	80.50
9790	3/4" pipe size		17.09	.936			57	6.30	63.30	92
9800	1" pipe size		15.58	1.027			62.50	6.90	69.40	101
9810	1-1/4" pipe size		14.48	1.105			67.50	7.45	74.95	108
9820	1-1/2" pipe size		13.68	1.170			71.50	7.85	79.35	115
9830	2" pipe size		11.61	1.378			84	9.25	93.25	135
9840	2-1/2" pipe size		6.12	2.614			159	17.55	176.55	257
9850	3" pipe size		4.94	3.239			197	22	219	320
9860	4" pipe size		3.75	4.267			260	28.50	288.50	420
9870	5" pipe size		3.27	4.893			298	33	331	480
9880	6" pipe size		2.90	5.517			335	37	372	540
9890	8" pipe size		1.87	8.556			520	57.50	577.50	845
9900	10" pipe size		1.48	10.811			660	72.50	732.50	1,075
9910	12" pipe size		1.17	13.675			835	92	927	1,350
9920	Schedule 160, 1/2" pipe size		17	.941			57.50	6.30	63.80	92.50
9930	3/4" pipe size		14.81	1.080			66	7.25	73.25	106
9940	1" pipe size		13.50	1.185			72	7.95	79.95	117
9950	1-1/4" pipe size		12.55	1.275			77.50	8.55	86.05	125
9960	1-1/2" pipe size		11.85	1.350			82.50	9.05	91.55	133
9970	2" pipe size		10	1.600			97.50	10.75	108.25	157
9980	3" pipe size		4.28	3.738			228	25	253	370
9990	4" pipe size	↓	3.25	4.923	↓		300	33	333	485

22 11 13.66 Pipe Fittings, Stainless Steel

		Crew	Daily Output	Labor-Hours	Unit	Material	2021 Bare Costs Labor	2021 Bare Costs Equipment	Total	Total Incl O&P
0010	**PIPE FITTINGS, STAINLESS STEEL**									
0100	Butt weld joint, schedule 5, type 304									
0120	90° elbow, long									
0140	1/2"	Q-15	17.50	.914	Ea.	17.85	55.50	6.15	79.50	109
0150	3/4"		14	1.143		17.85	69.50	7.70	95.05	132
0160	1"		12.50	1.280		18.85	78	8.60	105.45	146
0170	1-1/4"	↓	11	1.455		26	88.50	9.75	124.25	171

22 11 Facility Water Distribution

22 11 13 – Facility Water Distribution Piping

22 11 13.66 Pipe Fittings, Stainless Steel	Crew	Daily Output	Labor-Hours	Unit	Material	2021 Bare Costs Labor	Equipment	Total	Total Incl O&P
0180	1-1/2"	Q-15	10.50	1.524	Ea.	22	93	10.25	125.25	174
0190	2"		9	1.778		26	108	11.95	145.95	204
0200	2-1/2"		6	2.667		59.50	162	17.90	239.40	325
0210	3"		4.86	3.292		55.50	201	22	278.50	385
0220	3-1/2"		4.27	3.747		163	228	25	416	550
0230	4"		3.69	4.336		93	264	29	386	530
0240	5"		3.08	5.195		345	315	35	695	890
0250	6"	Q-16	4.29	5.594		268	355	25	648	850
0260	8"		2.76	8.696		565	550	39	1,154	1,475
0270	10"		2.18	11.009		870	695	49.50	1,614.50	2,075
0280	12"		1.73	13.873		1,250	875	62	2,187	2,725
0320	For schedule 5, type 316, add					30%				
0600	45° elbow, long									
0620	1/2"	Q-15	17.50	.914	Ea.	17.85	55.50	6.15	79.50	109
0630	3/4"		14	1.143		17.85	69.50	7.70	95.05	132
0640	1"		12.50	1.280		18.85	78	8.60	105.45	146
0650	1-1/4"		11	1.455		26	88.50	9.75	124.25	171
0660	1-1/2"		10.50	1.524		22	93	10.25	125.25	174
0670	2"		9	1.778		26	108	11.95	145.95	204
0680	2-1/2"		6	2.667		59.50	162	17.90	239.40	325
0690	3"		4.86	3.292		44.50	201	22	267.50	375
0700	3-1/2"		4.27	3.747		163	228	25	416	550
0710	4"		3.69	4.336		75.50	264	29	368.50	510
0720	5"		3.08	5.195		277	315	35	627	815
0730	6"	Q-16	4.29	5.594		188	355	25	568	765
0740	8"		2.76	8.696		395	550	39	984	1,300
0750	10"		2.18	11.009		695	695	49.50	1,439.50	1,875
0760	12"		1.73	13.873		865	875	62	1,802	2,325
0800	For schedule 5, type 316, add					25%				
1100	Tee, straight									
1130	1/2"	Q-15	11.66	1.372	Ea.	53.50	83.50	9.20	146.20	194
1140	3/4"		9.33	1.715		53.50	104	11.50	169	228
1150	1"		8.33	1.921		56.50	117	12.90	186.40	251
1160	1-1/4"		7.33	2.183		45.50	133	14.65	193.15	264
1170	1-1/2"		7	2.286		44.50	139	15.35	198.85	274
1180	2"		6	2.667		46.50	162	17.90	226.40	315
1190	2-1/2"		4	4		112	244	27	383	520
1200	3"		3.24	4.938		81.50	300	33	414.50	575
1210	3-1/2"		2.85	5.614		234	340	37.50	611.50	810
1220	4"		2.46	6.504		127	395	43.50	565.50	775
1230	5"		2	8		400	485	54	939	1,225
1240	6"	Q-16	2.85	8.421		315	530	37.50	882.50	1,175
1250	8"		1.84	13.043		675	825	58.50	1,558.50	2,025
1260	10"		1.45	16.552		1,100	1,050	74	2,224	2,825
1270	12"		1.15	20.870		1,525	1,325	93.50	2,943.50	3,750
1320	For schedule 5, type 316, add					25%				
2000	Butt weld joint, schedule 10, type 304									
2020	90° elbow, long									
2040	1/2"	Q-15	17	.941	Ea.	6.65	57.50	6.30	70.45	100
2050	3/4"		14	1.143		6.80	69.50	7.70	84	120
2060	1"		12.50	1.280		7.60	78	8.60	94.20	134
2070	1-1/4"		11	1.455		11.65	88.50	9.75	109.90	156
2080	1-1/2"		10.50	1.524		9.50	93	10.25	112.75	161

For customer support on your Plumbing Costs with RSMeans Data, call 800.448.8182.

215

22 11 13.66 Pipe Fittings, Stainless Steel		Crew	Daily Output	Labor-Hours	Unit	Material	2021 Bare Costs Labor	Equipment	Total	Total Incl O&P
2090	2"	Q-15	9	1.778	Ea.	11.45	108	11.95	131.40	188
2100	2-1/2"		6	2.667		20.50	162	17.90	200.40	284
2110	3"		4.86	3.292		24.50	201	22	247.50	350
2120	3-1/2"		4.27	3.747		162	228	25	415	545
2130	4"		3.69	4.336		41	264	29	334	470
2140	5"		3.08	5.195		345	315	35	695	890
2150	6"	Q-16	4.29	5.594		109	355	25	489	680
2160	8"		2.76	8.696		180	550	39	769	1,050
2170	10"		2.18	11.009		450	695	49.50	1,194.50	1,600
2180	12"		1.73	13.873		690	875	62	1,627	2,125
2500	45° elbow, long									
2520	1/2"	Q-15	17.50	.914	Ea.	8.50	55.50	6.15	70.15	99
2530	3/4"		14	1.143		8.65	69.50	7.70	85.85	122
2540	1"		12.50	1.280		9.55	78	8.60	96.15	136
2550	1-1/4"		11	1.455		16.05	88.50	9.75	114.30	160
2560	1-1/2"		10.50	1.524		9.75	93	10.25	113	161
2570	2"		9	1.778		11.65	108	11.95	131.60	188
2580	2-1/2"		6	2.667		18.45	162	17.90	198.35	282
2590	3"		4.86	3.292		21	201	22	244	345
2600	3-1/2"		4.27	3.747		162	228	25	415	545
2610	4"		3.69	4.336		34	264	29	327	465
2620	5"		3.08	5.195		275	315	35	625	810
2630	6"	Q-16	4.29	5.594		73.50	355	25	453.50	640
2640	8"		2.76	8.696		130	550	39	719	1,000
2650	10"		2.18	11.009		320	695	49.50	1,064.50	1,450
2660	12"		1.73	13.873		490	875	62	1,427	1,900
2670	Reducer, concentric									
2674	1" x 3/4"	Q-15	13.25	1.208	Ea.	17.90	73.50	8.10	99.50	139
2676	2" x 1-1/2"	"	9.75	1.641		13.05	100	11.05	124.10	176
2678	6" x 4"	Q-16	4.91	4.888		32.50	310	22	364.50	520
2680	Caps									
2682	1"	Q-15	25	.640	Ea.	20	39	4.30	63.30	84.50
2684	1-1/2"		21	.762		21.50	46.50	5.10	73.10	99
2685	2"		18	.889		19	54	5.95	78.95	109
2686	4"		7.38	2.168		30	132	14.55	176.55	246
2687	6"	Q-16	8.58	2.797		49	177	12.55	238.55	330
3000	Tee, straight									
3030	1/2"	Q-15	11.66	1.372	Ea.	23.50	83.50	9.20	116.20	161
3040	3/4"		9.33	1.715		24	104	11.50	139.50	195
3050	1"		8.33	1.921		24	117	12.90	153.90	216
3060	1-1/4"		7.33	2.183		37	133	14.65	184.65	255
3070	1-1/2"		7	2.286		24	139	15.35	178.35	251
3080	2"		6	2.667		26	162	17.90	205.90	290
3090	2-1/2"		4	4		68	244	27	339	470
3100	3"		3.24	4.938		48	300	33	381	540
3110	3-1/2"		2.85	5.614		129	340	37.50	506.50	695
3120	4"		2.46	6.504		72	395	43.50	510.50	715
3130	5"		2	8		400	485	54	939	1,225
3140	6"	Q-16	2.85	8.421		164	530	37.50	731.50	1,025
3150	8"		1.84	13.043		335	825	58.50	1,218.50	1,650
3151	10"		1.45	16.552		420	1,050	74	1,544	2,100
3152	12"		1.15	20.870		960	1,325	93.50	2,378.50	3,125
3154	For schedule 10, type 316, add					25%				

216

22 11 Facility Water Distribution

22 11 13 – Facility Water Distribution Piping

22 11 13.66 Pipe Fittings, Stainless Steel

		Crew	Daily Output	Labor-Hours	Unit	Material	2021 Bare Costs Labor	Equipment	Total	Total Incl O&P
3281	Butt weld joint, schedule 40, type 304									
3284	90° elbow, long, 1/2"	Q-15	12.70	1.260	Ea.	7.45	77	8.45	92.90	132
3288	3/4"		11.10	1.441		7.60	88	9.70	105.30	150
3289	1"		10.13	1.579		8.60	96	10.60	115.20	165
3290	1-1/4"		9.40	1.702		12.80	104	11.45	128.25	182
3300	1-1/2"		8.89	1.800		7.65	110	12.10	129.75	186
3310	2"		7.55	2.119		10.55	129	14.25	153.80	220
3320	2-1/2"		3.98	4.020		22.50	245	27	294.50	420
3330	3"		3.21	4.984		28.50	305	33.50	367	525
3340	3-1/2"		2.83	5.654		294	345	38	677	880
3350	4"		2.44	6.557		53	400	44	497	700
3360	5"		2.13	7.512		460	460	50.50	970.50	1,250
3370	6"	Q-16	2.83	8.481		144	535	38	717	1,000
3380	8"		1.83	13.115		262	830	59	1,151	1,575
3390	10"		1.44	16.667		1,375	1,050	74.50	2,499.50	3,175
3400	12"		1.14	21.053		1,750	1,325	94.50	3,169.50	4,000
3410	For schedule 40, type 316, add					25%				
3460	45° elbow, long, 1/2"	Q-15	12.70	1.260	Ea.	7.70	77	8.45	93.15	133
3470	3/4"		11.10	1.441		7.75	88	9.70	105.45	150
3480	1"		10.13	1.579		7.85	96	10.60	114.45	164
3490	1-1/4"		9.40	1.702		11.35	104	11.45	126.80	180
3500	1-1/2"		8.89	1.800		8.80	110	12.10	130.90	187
3510	2"		7.55	2.119		10.95	129	14.25	154.20	221
3520	2-1/2"		3.98	4.020		22.50	245	27	294.50	420
3530	3"		3.21	4.984		20.50	305	33.50	359	515
3540	3-1/2"		2.83	5.654		294	345	38	677	880
3550	4"		2.44	6.557		37	400	44	481	685
3560	5"		2.13	7.512		325	460	50.50	835.50	1,100
3570	6"	Q-16	2.83	8.481		92.50	535	38	665.50	945
3580	8"		1.83	13.115		183	830	59	1,072	1,500
3590	10"		1.44	16.667		965	1,050	74.50	2,089.50	2,725
3600	12"		1.14	21.053		1,225	1,325	94.50	2,644.50	3,425
3610	For schedule 40, type 316, add					25%				
3660	Tee, straight 1/2"	Q-15	8.46	1.891	Ea.	18.45	115	12.70	146.15	207
3670	3/4"		7.40	2.162		18.65	132	14.55	165.20	234
3680	1"		6.74	2.374		18.80	145	15.95	179.75	254
3690	1-1/4"		6.27	2.552		31.50	155	17.15	203.65	285
3700	1-1/2"		5.92	2.703		19	165	18.15	202.15	287
3710	2"		5.03	3.181		20.50	194	21.50	236	335
3720	2-1/2"		2.65	6.038		51	370	40.50	461.50	650
3730	3"		2.14	7.477		41	455	50	546	780
3740	3-1/2"		1.88	8.511		215	520	57	792	1,075
3750	4"		1.62	9.877		59	600	66.50	725.50	1,050
3760	5"		1.42	11.268		495	685	75.50	1,255.50	1,650
3770	6"	Q-16	1.88	12.766		133	805	57	995	1,400
3780	8"		1.22	19.672		385	1,250	88	1,723	2,375
3790	10"		.96	25		1,525	1,575	112	3,212	4,150
3800	12"		.76	31.579		2,025	2,000	141	4,166	5,350
3810	For schedule 40, type 316, add					25%				
3820	Tee, reducing on outlet, 3/4" x 1/2"	Q-15	7.73	2.070	Ea.	61.50	126	13.90	201.40	271
3822	1" x 1/2"		7.24	2.210		76.50	135	14.85	226.35	300
3824	1" x 3/4"		6.96	2.299		70.50	140	15.45	225.95	305
3826	1-1/4" x 1"		6.43	2.488		173	152	16.70	341.70	435

22 11 13.66 Pipe Fittings, Stainless Steel		Crew	Daily Output	Labor-Hours	Unit	Material	2021 Bare Costs Labor	Equipment	Total	Total Incl O&P
3828	1-1/2" x 1/2"	Q-15	6.58	2.432	Ea.	104	148	16.35	268.35	355
3830	1-1/2" x 3/4"		6.35	2.520		97	154	16.95	267.95	355
3832	1-1/2" x 1"		6.18	2.589		74.50	158	17.40	249.90	335
3834	2" x 1"		5.50	2.909		133	177	19.55	329.55	435
3836	2" x 1-1/2"		5.30	3.019		112	184	20.50	316.50	420
3838	2-1/2" x 2"		3.15	5.079		169	310	34	513	685
3840	3" x 1-1/2"		2.72	5.882		171	360	39.50	570.50	765
3842	3" x 2"		2.65	6.038		142	370	40.50	552.50	750
3844	4" x 2"		2.10	7.619		355	465	51	871	1,150
3846	4" x 3"		1.77	9.040		258	550	61	869	1,175
3848	5" x 4"		1.48	10.811		595	660	72.50	1,327.50	1,725
3850	6" x 3"	Q-16	2.19	10.959		655	690	49	1,394	1,800
3852	6" x 4"		2.04	11.765		570	745	52.50	1,367.50	1,775
3854	8" x 4"		1.46	16.438		1,325	1,050	73.50	2,448.50	3,075
3856	10" x 8"		.69	34.783		2,250	2,200	156	4,606	5,925
3858	12" x 10"		.55	43.636		3,000	2,750	195	5,945	7,650
3950	Reducer, concentric, 3/4" x 1/2"	Q-15	11.85	1.350		12.55	82.50	9.05	104.10	147
3952	1" x 3/4"		10.60	1.509		14.50	92	10.15	116.65	164
3954	1-1/4" x 3/4"		10.19	1.570		43	95.50	10.55	149.05	202
3956	1-1/4" x 1"		9.76	1.639		21	100	11	132	184
3958	1-1/2" x 3/4"		9.88	1.619		36.50	98.50	10.90	145.90	199
3960	1-1/2" x 1"		9.47	1.690		68.50	103	11.35	182.85	242
3962	2" x 1"		8.65	1.850		13.70	113	12.45	139.15	197
3964	2" x 1-1/2"		8.16	1.961		10.95	119	13.20	143.15	205
3966	2-1/2" x 1"		5.71	2.802		62.50	171	18.85	252.35	345
3968	2-1/2" x 2"		5.21	3.071		49	187	20.50	256.50	355
3970	3" x 1"		4.88	3.279		43	200	22	265	370
3972	3" x 1-1/2"		4.72	3.390		19.60	207	23	249.60	355
3974	3" x 2"		4.51	3.548		17.65	216	24	257.65	370
3976	4" x 2"		3.69	4.336		25.50	264	29	318.50	455
3978	4" x 3"		2.77	5.776		19.60	350	39	408.60	590
3980	5" x 3"		2.56	6.250		310	380	42	732	955
3982	5" x 4"		2.27	7.048		250	430	47.50	727.50	965
3984	6" x 3"	Q-16	3.57	6.723		160	425	30	615	845
3986	6" x 4"		3.19	7.524		51	475	33.50	559.50	805
3988	8" x 4"		2.44	9.836		165	620	44	829	1,150
3990	8" x 6"		2.22	10.811		115	685	48.50	848.50	1,200
3992	10" x 6"		1.91	12.565		465	795	56.50	1,316.50	1,750
3994	10" x 8"		1.61	14.907		390	940	67	1,397	1,900
3995	12" x 6"		1.63	14.724		885	930	66	1,881	2,450
3996	12" x 8"		1.41	17.021		660	1,075	76.50	1,811.50	2,400
3997	12" x 10"		1.27	18.898		465	1,200	84.50	1,749.50	2,375
4000	Socket weld joint, 3,000 lb., type 304									
4100	90° elbow									
4140	1/4"	Q-15	13.47	1.188	Ea.	26	72.50	8	106.50	146
4150	3/8"		12.97	1.234		34	75	8.30	117.30	159
4160	1/2"		12.21	1.310		36	80	8.80	124.80	168
4170	3/4"		10.68	1.498		43.50	91.50	10.05	145.05	195
4180	1"		9.74	1.643		65.50	100	11.05	176.55	233
4190	1-1/4"		9.05	1.768		194	108	11.90	313.90	385
4200	1-1/2"		8.55	1.871		140	114	12.60	266.60	340
4210	2"		7.26	2.204		222	134	14.80	370.80	460
4300	45° elbow									

22 11 13.66 Pipe Fittings, Stainless Steel		Crew	Daily Output	Labor-Hours	Unit	Material	2021 Bare Costs Labor	Equipment	Total	Total Incl O&P
4340	1/4"	Q-15	13.47	1.188	Ea.	47.50	72.50	8	128	169
4350	3/8"		12.97	1.234		47.50	75	8.30	130.80	174
4360	1/2"		12.21	1.310		47.50	80	8.80	136.30	181
4370	3/4"		10.68	1.498		54	91.50	10.05	155.55	207
4380	1"		9.74	1.643		78.50	100	11.05	189.55	247
4390	1-1/4"		9.05	1.768		230	108	11.90	349.90	425
4400	1-1/2"		8.55	1.871		132	114	12.60	258.60	330
4410	2"		7.26	2.204		237	134	14.80	385.80	475
4500	Tee									
4540	1/4"	Q-15	8.97	1.784	Ea.	59.50	109	12	180.50	241
4550	3/8"		8.64	1.852		71	113	12.45	196.45	260
4560	1/2"		8.13	1.968		70.50	120	13.25	203.75	271
4570	3/4"		7.12	2.247		81	137	15.10	233.10	310
4580	1"		6.48	2.469		109	150	16.60	275.60	365
4590	1-1/4"		6.03	2.653		240	162	17.85	419.85	525
4600	1-1/2"		5.69	2.812		266	171	18.90	455.90	570
4610	2"		4.83	3.313		380	202	22.50	604.50	745
5000	Socket weld joint, 3,000 lb., type 316									
5100	90° elbow									
5140	1/4"	Q-15	13.47	1.188	Ea.	55.50	72.50	8	136	178
5150	3/8"		12.97	1.234		64.50	75	8.30	147.80	192
5160	1/2"		12.21	1.310		62.50	80	8.80	151.30	198
5170	3/4"		10.68	1.498		82.50	91.50	10.05	184.05	238
5180	1"		9.74	1.643		117	100	11.05	228.05	290
5190	1-1/4"		9.05	1.768		259	108	11.90	378.90	460
5200	1-1/2"		8.55	1.871		227	114	12.60	353.60	435
5210	2"		7.26	2.204		365	134	14.80	513.80	615
5300	45° elbow									
5340	1/4"	Q-15	13.47	1.188	Ea.	112	72.50	8	192.50	240
5350	3/8"		12.97	1.234		112	75	8.30	195.30	244
5360	1/2"		12.21	1.310		90	80	8.80	178.80	228
5370	3/4"		10.68	1.498		100	91.50	10.05	201.55	257
5380	1"		9.74	1.643		147	100	11.05	258.05	325
5390	1-1/4"		9.05	1.768		267	108	11.90	386.90	470
5400	1-1/2"		8.55	1.871		232	114	12.60	358.60	440
5410	2"		7.26	2.204		345	134	14.80	493.80	595
5500	Tee									
5540	1/4"	Q-15	8.97	1.784	Ea.	75	109	12	196	258
5550	3/8"		8.64	1.852		91.50	113	12.45	216.95	283
5560	1/2"		8.13	1.968		82.50	120	13.25	215.75	284
5570	3/4"		7.12	2.247		102	137	15.10	254.10	335
5580	1"		6.48	2.469		152	150	16.60	318.60	410
5590	1-1/4"		6.03	2.653		305	162	17.85	484.85	595
5600	1-1/2"		5.69	2.812		325	171	18.90	514.90	630
5610	2"		4.83	3.313		505	202	22.50	729.50	880
5700	For socket weld joint, 6,000 lb., type 304 and 316, add					100%				
6000	Threaded companion flange									
6010	Stainless steel, 150 lb., type 304									
6020	1/2" diam.	1 Plum	30	.267	Ea.	33.50	18.05		51.55	64
6030	3/4" diam.		28	.286		38.50	19.35		57.85	71.50
6040	1" diam.		27	.296		42	20		62	76.50
6050	1-1/4" diam.	Q-1	44	.364		54.50	22		76.50	92.50
6060	1-1/2" diam.		40	.400		55	24.50		79.50	96.50

22 11 13.66 Pipe Fittings, Stainless Steel		Crew	Daily Output	Labor-Hours	Unit	Material	2021 Bare Costs Labor	Equipment	Total	Total Incl O&P
6070	2" diam.	Q-1	36	.444	Ea.	72	27		99	120
6080	2-1/2" diam.		28	.571		106	35		141	168
6090	3" diam.		20	.800		107	48.50		155.50	191
6110	4" diam.		12	1.333		147	81		228	283
6130	6" diam.	Q-2	14	1.714		259	108		367	445
6140	8" diam.	"	12	2		485	126		611	725
6150	For type 316, add					40%				
6260	Weld flanges, stainless steel, type 304									
6270	Slip on, 150 lb. (welded, front and back)									
6280	1/2" diam.	Q-15	12.70	1.260	Ea.	31	77	8.45	116.45	158
6290	3/4" diam.		11.11	1.440		32.50	88	9.70	130.20	178
6300	1" diam.		10.13	1.579		35.50	96	10.60	142.10	195
6310	1-1/4" diam.		9.41	1.700		48.50	104	11.45	163.95	221
6320	1-1/2" diam.		8.89	1.800		48.50	110	12.10	170.60	231
6330	2" diam.		7.55	2.119		62	129	14.25	205.25	277
6340	2-1/2" diam.		3.98	4.020		88.50	245	27	360.50	490
6350	3" diam.		3.21	4.984		93.50	305	33.50	432	595
6370	4" diam.		2.44	6.557		127	400	44	571	785
6390	6" diam.	Q-16	1.89	12.698		195	800	57	1,052	1,475
6400	8" diam.	"	1.22	19.672		355	1,250	88	1,693	2,325
6410	For type 316, add					40%				
6530	Weld neck 150 lb.									
6540	1/2" diam.	Q-15	25.40	.630	Ea.	26	38.50	4.23	68.73	90.50
6550	3/4" diam.		22.22	.720		31	44	4.84	79.84	105
6560	1" diam.		20.25	.790		34	48	5.30	87.30	115
6570	1-1/4" diam.		18.82	.850		54.50	52	5.70	112.20	144
6580	1-1/2" diam.		17.78	.900		47.50	55	6.05	108.55	141
6590	2" diam.		15.09	1.060		51	64.50	7.15	122.65	160
6600	2-1/2" diam.		7.96	2.010		81.50	122	13.50	217	287
6610	3" diam.		6.43	2.488		83.50	152	16.70	252.20	335
6630	4" diam.		4.88	3.279		127	200	22	349	460
6640	5" diam.		4.26	3.756		133	229	25	387	515
6650	6" diam.	Q-16	5.66	4.240		182	268	19	469	620
6652	8" diam.		3.65	6.575		325	415	29.50	769.50	1,025
6654	10" diam.		2.88	8.333		360	525	37.50	922.50	1,225
6656	12" diam.		2.28	10.526		455	665	47	1,167	1,550
6670	For type 316, add					23%				
7000	Threaded joint, 150 lb., type 304									
7030	90° elbow									
7040	1/8"	1 Plum	13	.615	Ea.	17.65	41.50		59.15	81.50
7050	1/4"		13	.615		17.65	41.50		59.15	81.50
7070	3/8"		13	.615		21	41.50		62.50	85
7080	1/2"		12	.667		19.15	45		64.15	88.50
7090	3/4"		11	.727		23.50	49		72.50	99.50
7100	1"		10	.800		33	54		87	117
7110	1-1/4"	Q-1	17	.941		53	57.50		110.50	144
7120	1-1/2"		16	1		63	61		124	160
7130	2"		14	1.143		87.50	69.50		157	201
7140	2-1/2"		11	1.455		205	88.50		293.50	355
7150	3"		8	2		289	122		411	500
7160	4"	Q-2	11	2.182		490	138		628	745
7180	45° elbow									
7190	1/8"	1 Plum	13	.615	Ea.	25.50	41.50		67	90

22 11 13.66 Pipe Fittings, Stainless Steel

		Crew	Daily Output	Labor-Hours	Unit	Material	2021 Bare Costs Labor	2021 Bare Costs Equipment	Total	Total Incl O&P
7200	1/4"	1 Plum	13	.615	Ea.	25.50	41.50		67	90
7210	3/8"		13	.615		26.50	41.50		68	91
7220	1/2"		12	.667		27	45		72	97
7230	3/4"		11	.727		31	49		80	108
7240	1"		10	.800		36.50	54		90.50	121
7250	1-1/4"	Q-1	17	.941		52.50	57.50		110	143
7260	1-1/2"		16	1		66.50	61		127.50	164
7270	2"		14	1.143		94	69.50		163.50	208
7280	2-1/2"		11	1.455		279	88.50		367.50	435
7290	3"		8	2		405	122		527	625
7300	4"	Q-2	11	2.182		695	138		833	970
7320	Tee, straight									
7330	1/8"	1 Plum	9	.889	Ea.	27	60		87	120
7340	1/4"		9	.889		27	60		87	120
7350	3/8"		9	.889		29.50	60		89.50	123
7360	1/2"		8	1		28.50	67.50		96	133
7370	3/4"		7	1.143		33	77.50		110.50	152
7380	1"		6.50	1.231		42.50	83.50		126	171
7390	1-1/4"	Q-1	11	1.455		74.50	88.50		163	214
7400	1-1/2"		10	1.600		95	97.50		192.50	249
7410	2"		9	1.778		119	108		227	293
7420	2-1/2"		7	2.286		291	139		430	530
7430	3"		5	3.200		440	195		635	770
7440	4"	Q-2	7	3.429		980	217		1,197	1,400
7460	Coupling, straight									
7470	1/8"	1 Plum	19	.421	Ea.	7.35	28.50		35.85	50.50
7480	1/4"		19	.421		8.70	28.50		37.20	52
7490	3/8"		19	.421		10.35	28.50		38.85	54
7500	1/2"		19	.421		14	28.50		42.50	58
7510	3/4"		18	.444		18.85	30		48.85	65.50
7520	1"		15	.533		29.50	36		65.50	86.50
7530	1-1/4"	Q-1	26	.615		48	37.50		85.50	109
7540	1-1/2"		24	.667		53	40.50		93.50	119
7550	2"		21	.762		85	46.50		131.50	163
7560	2-1/2"		18	.889		196	54		250	297
7570	3"		14	1.143		267	69.50		336.50	400
7580	4"	Q-2	16	1.500		380	95		475	560
7600	Reducer, concentric, 1/2"	1 Plum	12	.667		23	45		68	93
7610	3/4"		11	.727		30	49		79	107
7612	1"		10	.800		50	54		104	136
7614	1-1/4"	Q-1	17	.941		106	57.50		163.50	202
7616	1-1/2"		16	1		116	61		177	218
7618	2"		14	1.143		180	69.50		249.50	300
7620	2-1/2"		11	1.455		475	88.50		563.50	655
7622	3"		8	2		530	122		652	760
7624	4"	Q-2	11	2.182		875	138		1,013	1,175
7710	Union									
7720	1/8"	1 Plum	12	.667	Ea.	35	45		80	106
7730	1/4"		12	.667		36.50	45		81.50	108
7740	3/8"		12	.667		40.50	45		85.50	112
7750	1/2"		11	.727		48	49		97	127
7760	3/4"		10	.800		65	54		119	153
7770	1"		9	.889		95	60		155	195

22 11 13.66 Pipe Fittings, Stainless Steel

		Crew	Daily Output	Labor-Hours	Unit	Material	2021 Bare Costs Labor	Equipment	Total	Total Incl O&P
7780	1-1/4"	Q-1	16	1	Ea.	223	61		284	335
7790	1-1/2"		15	1.067		244	65		309	365
7800	2"		13	1.231		310	75		385	450
7810	2-1/2"		10	1.600		595	97.50		692.50	800
7820	3"		7	2.286		790	139		929	1,075
7830	4"	Q-2	10	2.400		1,075	152		1,227	1,425
7838	Caps									
7840	1/2"	1 Plum	24	.333	Ea.	12.05	22.50		34.55	47
7841	3/4"		22	.364		16.50	24.50		41	55
7842	1"		20	.400		28	27		55	71.50
7843	1-1/2"	Q-1	32	.500		62	30.50		92.50	114
7844	2"	"	28	.571		73.50	35		108.50	133
7845	4"	Q-2	22	1.091		425	69		494	570
7850	For 150 lb., type 316, add					25%				

22 11 13.74 Pipe, Plastic

		Crew	Daily Output	Labor-Hours	Unit	Material	2021 Bare Costs Labor	Equipment	Total	Total Incl O&P
0010	**PIPE, PLASTIC** R221113-70									
0020	Fiberglass reinforced, couplings 10' OC, clevis hanger assy's, 3 per 10'									
0080	General service									
0120	2" diameter	Q-1	59	.271	L.F.	11.65	16.50		28.15	37.50
0140	3" diameter		52	.308		15.40	18.75		34.15	45
0150	4" diameter		48	.333		21	20.50		41.50	53.50
0160	6" diameter		39	.410		39	25		64	80
0170	8" diameter	Q-2	49	.490		66	31		97	119
0180	10" diameter		41	.585		98	37		135	163
0190	12" diameter		36	.667		130	42		172	206
0600	PVC, high impact/pressure, cplgs. 10' OC, clevis hanger assy's, 3 per 10'									
1020	Schedule 80									
1070	1/2" diameter	1 Plum	50	.160	L.F.	6.45	10.85		17.30	23.50
1080	3/4" diameter		47	.170		7.55	11.50		19.05	25.50
1090	1" diameter		43	.186		7.45	12.60		20.05	27
1100	1-1/4" diameter		39	.205		9.70	13.90		23.60	31
1110	1-1/2" diameter		34	.235		11.80	15.95		27.75	37
1120	2" diameter	Q-1	55	.291		15.35	17.75		33.10	43.50
1140	3" diameter		50	.320		29.50	19.50		49	61.50
1150	4" diameter		46	.348		45.50	21		66.50	81.50
1170	6" diameter		38	.421		64	25.50		89.50	109
1730	To delete coupling & hangers, subtract									
1740	1/2" diam.					62%	80%			
1750	3/4" diam. to 1-1/4" diam.					58%	73%			
1760	1-1/2" diam. to 6" diam.					40%	57%			
1800	PVC, couplings 10' OC, clevis hanger assemblies, 3 per 10'									
1820	Schedule 40									
1860	1/2" diameter	1 Plum	54	.148	L.F.	4.17	10.05		14.22	19.55
1870	3/4" diameter		51	.157		4.77	10.60		15.37	21
1880	1" diameter		46	.174		3.35	11.75		15.10	21
1890	1-1/4" diameter		42	.190		3.36	12.90		16.26	23
1900	1-1/2" diameter		36	.222		4.90	15.05		19.95	28
1910	2" diameter	Q-1	59	.271		4.85	16.50		21.35	30
1920	2-1/2" diameter		56	.286		7.85	17.40		25.25	34.50
1930	3" diameter		53	.302		10.05	18.40		28.45	38.50
1940	4" diameter		48	.333		13	20.50		33.50	45
1950	5" diameter		43	.372		25	22.50		47.50	61.50

22 11 Facility Water Distribution

22 11 13 – Facility Water Distribution Piping

22 11 13.74 Pipe, Plastic		Crew	Daily Output	Labor-Hours	Unit	Material	2021 Bare Costs Labor	Equipment	Total	Total Incl O&P
1960	6" diameter	Q-1	39	.410	L.F.	25	25		50	65
1970	8" diameter	Q-2	48	.500		28	31.50		59.50	77.50
1980	10" diameter		43	.558		66	35.50		101.50	125
1990	12" diameter		42	.571		80.50	36		116.50	143
2000	14" diameter		31	.774		131	49		180	217
2010	16" diameter		23	1.043		190	66		256	310
2340	To delete coupling & hangers, subtract									
2360	1/2" diam. to 1-1/4" diam.					65%	74%			
2370	1-1/2" diam. to 6" diam.					44%	57%			
2380	8" diam. to 12" diam.					41%	53%			
2390	14" diam. to 16" diam.					48%	45%			
2420	Schedule 80									
2440	1/4" diameter	1 Plum	58	.138	L.F.	3.90	9.35		13.25	18.25
2450	3/8" diameter		55	.145		3.90	9.85		13.75	19
2460	1/2" diameter		50	.160		4.54	10.85		15.39	21
2470	3/4" diameter		47	.170		4.93	11.50		16.43	22.50
2480	1" diameter		43	.186		3.58	12.60		16.18	22.50
2490	1-1/4" diameter		39	.205		3.38	13.90		17.28	24
2500	1-1/2" diameter		34	.235		5.35	15.95		21.30	30
2510	2" diameter	Q-1	55	.291		4.95	17.75		22.70	32
2520	2-1/2" diameter		52	.308		6.40	18.75		25.15	35
2530	3" diameter		50	.320		11.05	19.50		30.55	41
2540	4" diameter		46	.348		18.40	21		39.40	51.50
2550	5" diameter		42	.381		32.50	23		55.50	70
2560	6" diameter		38	.421		29	25.50		54.50	70.50
2570	8" diameter	Q-2	47	.511		47	32.50		79.50	99.50
2580	10" diameter		42	.571		115	36		151	181
2590	12" diameter		38	.632		135	40		175	209
2830	To delete coupling & hangers, subtract									
2840	1/4" diam. to 1/2" diam.					66%	80%			
2850	3/4" diam. to 1-1/4" diam.					61%	73%			
2860	1-1/2" diam. to 6" diam.					41%	57%			
2870	8" diam. to 12" diam.					31%	50%			
2900	Schedule 120									
2910	1/2" diameter	1 Plum	50	.160	L.F.	5.15	10.85		16	22
2950	3/4" diameter		47	.170		5.80	11.50		17.30	23.50
2960	1" diameter		43	.186		4.81	12.60		17.41	24
2970	1-1/4" diameter		39	.205		6.25	13.90		20.15	27.50
2980	1-1/2" diameter		33	.242		7.30	16.40		23.70	32.50
2990	2" diameter	Q-1	54	.296		9.50	18.05		27.55	37.50
3000	2-1/2" diameter		52	.308		13.85	18.75		32.60	43.50
3010	3" diameter		49	.327		27	19.90		46.90	59.50
3020	4" diameter		45	.356		28	21.50		49.50	63.50
3030	6" diameter		37	.432		66.50	26.50		93	113
3240	To delete coupling & hangers, subtract									
3250	1/2" diam. to 1-1/4" diam.					52%	74%			
3260	1-1/2" diam. to 4" diam.					30%	57%			
3270	6" diam.					17%	50%			
3300	PVC, pressure, couplings 10' OC, clevis hanger assy's, 3 per 10'									
3310	SDR 26, 160 psi									
3350	1-1/4" diameter	1 Plum	42	.190	L.F.	3.64	12.90		16.54	23.50
3360	1-1/2" diameter	"	36	.222		5.05	15.05		20.10	28
3370	2" diameter	Q-1	59	.271		5.25	16.50		21.75	30.50

For customer support on your Plumbing Costs with RSMeans Data, call 800.448.8182.

22 11 13 – Facility Water Distribution Piping

22 11 13.74 Pipe, Plastic	Crew	Daily Output	Labor-Hours	Unit	Material	2021 Bare Costs Labor	Equipment	Total	Total Incl O&P
3380	2-1/2" diameter	Q-1	56	.286	L.F.	7.15	17.40		24.55	34
3390	3" diameter		53	.302		9.25	18.40		27.65	37.50
3400	4" diameter		48	.333		16.05	20.50		36.55	48
3420	6" diameter		39	.410		27	25		52	67
3430	8" diameter	Q-2	48	.500		44	31.50		75.50	95
3660	To delete coupling & clevis hanger assy's, subtract									
3670	1-1/4" diam.					63%	68%			
3680	1-1/2" diam. to 4" diam.					48%	57%			
3690	6" diam. to 8" diam.					60%	54%			
3720	SDR 21, 200 psi, 1/2" diameter	1 Plum	54	.148	L.F.	4.24	10.05		14.29	19.60
3740	3/4" diameter		51	.157		4.33	10.60		14.93	20.50
3750	1" diameter		46	.174		2.65	11.75		14.40	20.50
3760	1-1/4" diameter		42	.190		3.79	12.90		16.69	23.50
3770	1-1/2" diameter		36	.222		3.93	15.05		18.98	27
3780	2" diameter	Q-1	59	.271		4.49	16.50		20.99	29.50
3790	2-1/2" diameter		56	.286		7.15	17.40		24.55	34
3800	3" diameter		53	.302		8.20	18.40		26.60	36.50
3810	4" diameter		48	.333		14.35	20.50		34.85	46.50
3830	6" diameter		39	.410		23	25		48	63
3840	8" diameter	Q-2	48	.500		37.50	31.50		69	88.50
4000	To delete coupling & hangers, subtract									
4010	1/2" diam. to 3/4" diam.					71%	77%			
4020	1" diam. to 1-1/4" diam.					63%	70%			
4030	1-1/2" diam. to 6" diam.					44%	57%			
4040	8" diam.					46%	54%			
4100	DWV type, schedule 40, couplings 10' OC, clevis hanger assy's, 3 per 10'									
4210	ABS, schedule 40, foam core type									
4212	Plain end black									
4214	1-1/2" diameter	1 Plum	39	.205	L.F.	2.79	13.90		16.69	23.50
4216	2" diameter	Q-1	62	.258		3.27	15.70		18.97	27
4218	3" diameter		56	.286		5.65	17.40		23.05	32
4220	4" diameter		51	.314		8.05	19.10		27.15	37.50
4222	6" diameter		42	.381		19.75	23		42.75	56
4240	To delete coupling & hangers, subtract									
4244	1-1/2" diam. to 6" diam.					43%	48%			
4400	PVC									
4410	1-1/4" diameter	1 Plum	42	.190	L.F.	3.42	12.90		16.32	23
4420	1-1/2" diameter	"	36	.222		2.35	15.05		17.40	25
4460	2" diameter	Q-1	59	.271		3.13	16.50		19.63	28
4470	3" diameter		53	.302		5.35	18.40		23.75	33.50
4480	4" diameter		48	.333		7.40	20.50		27.90	38.50
4490	6" diameter		39	.410		15.75	25		40.75	55
4500	8" diameter	Q-2	48	.500		24	31.50		55.50	73
4510	To delete coupling & hangers, subtract									
4520	1-1/4" diam. to 1-1/2" diam.					48%	60%			
4530	2" diam. to 8" diam.					42%	54%			
4532	to delete hangers, 2" diam. to 8" diam.	Q-1	50	.320	L.F.	2.65	19.50		22.15	32
4550	PVC, schedule 40, foam core type									
4552	Plain end, white									
4554	1-1/2" diameter	1 Plum	39	.205	L.F.	2.50	13.90		16.40	23.50
4556	2" diameter	Q-1	62	.258		2.93	15.70		18.63	26.50
4558	3" diameter		56	.286		4.95	17.40		22.35	31.50
4560	4" diameter		51	.314		6.80	19.10		25.90	36

22 11 13 – Facility Water Distribution Piping

22 11 13.74 Pipe, Plastic		Crew	Daily Output	Labor-Hours	Unit	Material	2021 Bare Costs Labor	Equipment	Total	Total Incl O&P
4562	6" diameter	Q-1	42	.381	L.F.	14.45	23		37.45	50.50
4564	8" diameter	Q-2	51	.471		22	29.50		51.50	68.50
4568	10" diameter		48	.500		26	31.50		57.50	75.50
4570	12" diameter		46	.522		31	33		64	83.50
4580	To delete coupling & hangers, subtract									
4582	1-1/2" diam. to 2" diam.					58%	54%			
4584	3" diam. to 12" diam.					46%	42%			
4800	PVC, clear pipe, cplgs. 10' OC, clevis hanger assy's 3 per 10', Sched. 40									
4840	1/4" diameter	1 Plum	59	.136	L.F.	4.40	9.20		13.60	18.55
4850	3/8" diameter		56	.143		4.40	9.65		14.05	19.30
4860	1/2" diameter		54	.148		5.45	10.05		15.50	21
4870	3/4" diameter		51	.157		6.10	10.60		16.70	22.50
4880	1" diameter		46	.174		5.65	11.75		17.40	24
4890	1-1/4" diameter		42	.190		7.20	12.90		20.10	27
4900	1-1/2" diameter		36	.222		8.35	15.05		23.40	31.50
4910	2" diameter	Q-1	59	.271		10.80	16.50		27.30	36.50
4920	2-1/2" diameter		56	.286		16.25	17.40		33.65	44
4930	3" diameter		53	.302		21	18.40		39.40	50.50
4940	3-1/2" diameter		50	.320		27.50	19.50		47	59.50
4950	4" diameter		48	.333		28.50	20.50		49	61.50
5250	To delete coupling & hangers, subtract									
5260	1/4" diam. to 3/8" diam.					60%	81%			
5270	1/2" diam. to 3/4" diam.					41%	77%			
5280	1" diam. to 1-1/2" diam.					26%	67%			
5290	2" diam. to 4" diam.					16%	58%			
5300	CPVC, socket joint, couplings 10' OC, clevis hanger assemblies, 3 per 10'									
5302	Schedule 40									
5304	1/2" diameter	1 Plum	54	.148	L.F.	5.30	10.05		15.35	21
5305	3/4" diameter		51	.157		6.45	10.60		17.05	23
5306	1" diameter		46	.174		5.30	11.75		17.05	23.50
5307	1-1/4" diameter		42	.190		6.50	12.90		19.40	26.50
5308	1-1/2" diameter		36	.222		6.35	15.05		21.40	29.50
5309	2" diameter	Q-1	59	.271		9.35	16.50		25.85	35
5310	2-1/2" diameter		56	.286		15.20	17.40		32.60	42.50
5311	3" diameter		53	.302		19.80	18.40		38.20	49.50
5312	4" diameter		48	.333		31	20.50		51.50	64.50
5314	6" diameter		43	.372		49	22.50		71.50	87.50
5318	To delete coupling & hangers, subtract									
5319	1/2" diam. to 3/4" diam.					37%	77%			
5320	1" diam. to 1-1/4" diam.					27%	70%			
5321	1-1/2" diam. to 3" diam.					21%	57%			
5322	4" diam. to 6" diam.					16%	57%			
5324	Schedule 80									
5325	1/2" diameter	1 Plum	50	.160	L.F.	5.50	10.85		16.35	22
5326	3/4" diameter		47	.170		6.30	11.50		17.80	24
5327	1" diameter		43	.186		5.85	12.60		18.45	25.50
5328	1-1/4" diameter		39	.205		7.25	13.90		21.15	28.50
5329	1-1/2" diameter		34	.235		8.55	15.95		24.50	33.50
5330	2" diameter	Q-1	55	.291		10.95	17.75		28.70	38.50
5331	2-1/2" diameter		52	.308		17.30	18.75		36.05	47
5332	3" diameter		50	.320		19.45	19.50		38.95	50.50
5333	4" diameter		46	.348		29.50	21		50.50	64
5334	6" diameter		38	.421		55	25.50		80.50	99

225

For customer support on your Plumbing Costs with RSMeans Data, call 800.448.8182.

22 11 13.74 Pipe, Plastic	Crew	Daily Output	Labor-Hours	Unit	Material	2021 Bare Costs Labor	Equipment	Total	Total Incl O&P
5335 8" diameter	Q-2	47	.511	L.F.	193	32.50		225.50	261
5339 To delete couplings & hangers, subtract									
5340 1/2" diam. to 3/4" diam.					44%	77%			
5341 1" diam. to 1-1/4" diam.					32%	71%			
5342 1-1/2" diam. to 4" diam.					25%	58%			
5343 6" diam. to 8" diam.					20%	53%			
5360 CPVC, threaded, couplings 10' OC, clevis hanger assemblies, 3 per 10'									
5380 Schedule 40									
5460 1/2" diameter	1 Plum	54	.148	L.F.	6.15	10.05		16.20	21.50
5470 3/4" diameter		51	.157		7.90	10.60		18.50	24.50
5480 1" diameter		46	.174		6.85	11.75		18.60	25
5490 1-1/4" diameter		42	.190		7.65	12.90		20.55	27.50
5500 1-1/2" diameter		36	.222		7.35	15.05		22.40	30.50
5510 2" diameter	Q-1	59	.271		10.55	16.50		27.05	36
5520 2-1/2" diameter		56	.286		16.50	17.40		33.90	44
5530 3" diameter		53	.302		21.50	18.40		39.90	51.50
5540 4" diameter		48	.333		38.50	20.50		59	73
5550 6" diameter		43	.372		54.50	22.50		77	94
5730 To delete coupling & hangers, subtract									
5740 1/2" diam. to 3/4" diam.					37%	77%			
5750 1" diam. to 1-1/4" diam.					27%	70%			
5760 1-1/2" diam. to 3" diam.					21%	57%			
5770 4" diam. to 6" diam.					16%	57%			
5800 Schedule 80									
5860 1/2" diameter	1 Plum	50	.160	L.F.	6.35	10.85		17.20	23
5870 3/4" diameter		47	.170		7.75	11.50		19.25	26
5880 1" diameter		43	.186		7.35	12.60		19.95	27
5890 1-1/4" diameter		39	.205		8.45	13.90		22.35	30
5900 1-1/2" diameter		34	.235		9.55	15.95		25.50	34.50
5910 2" diameter	Q-1	55	.291		12.15	17.75		29.90	40
5920 2-1/2" diameter		52	.308		18.60	18.75		37.35	48.50
5930 3" diameter		50	.320		21.50	19.50		41	52.50
5940 4" diameter		46	.348		37	21		58	72
5950 6" diameter		38	.421		60.50	25.50		86	105
5960 8" diameter	Q-2	47	.511		196	32.50		228.50	264
6060 To delete couplings & hangers, subtract									
6070 1/2" diam. to 3/4" diam.					44%	77%			
6080 1" diam. to 1-1/4" diam.					32%	71%			
6090 1-1/2" diam. to 4" diam.					25%	58%			
6100 6" diam. to 8" diam.					20%	53%			
6240 CTS, 1/2" diameter	1 Plum	54	.148	L.F.	4.51	10.05		14.56	19.90
6250 3/4" diameter		51	.157		4.26	10.60		14.86	20.50
6260 1" diameter		46	.174		7.95	11.75		19.70	26.50
6270 1-1/4" diameter		42	.190		12.70	12.90		25.60	33.50
6280 1-1/2" diameter		36	.222		16.45	15.05		31.50	40.50
6290 2" diameter	Q-1	59	.271		27	16.50		43.50	54
6370 To delete coupling & hangers, subtract									
6380 1/2" diam.					51%	79%			
6390 3/4" diam.					40%	76%			
6392 1" thru 2" diam.					72%	68%			
6500 Residential installation, plastic pipe									
6510 Couplings 10' OC, strap hangers 3 per 10'									
6520 PVC, Schedule 40									

22 11 13.74 Pipe, Plastic

		Crew	Daily Output	Labor-Hours	Unit	Material	2021 Bare Costs Labor	Equipment	Total	Total Incl O&P
6530	1/2" diameter	1 Plum	138	.058	L.F.	1.19	3.92		5.11	7.15
6540	3/4" diameter		128	.063		1.83	4.23		6.06	8.30
6550	1" diameter		119	.067		2.47	4.55		7.02	9.50
6560	1-1/4" diameter		111	.072		2.17	4.88		7.05	9.70
6570	1-1/2" diameter		104	.077		3.21	5.20		8.41	11.30
6580	2" diameter	Q-1	197	.081		3.05	4.95		8	10.75
6590	2-1/2" diameter		162	.099		6.30	6		12.30	15.95
6600	4" diameter		123	.130		7.20	7.95		15.15	19.75
6700	PVC, DWV, Schedule 40									
6720	1-1/4" diameter	1 Plum	100	.080	L.F.	3.21	5.40		8.61	11.65
6730	1-1/2" diameter	"	94	.085		2.19	5.75		7.94	11
6740	2" diameter	Q-1	178	.090		2.99	5.50		8.49	11.45
6760	4" diameter	"	110	.145		8.20	8.85		17.05	22.50
7280	PEX, flexible, no couplings or hangers									
7282	Note: For labor costs add 25% to the couplings and fittings labor total.									
7285	For fittings see section 23 83 16.10 7000									
7300	Non-barrier type, hot/cold tubing rolls									
7310	1/4" diameter x 100'				L.F.	.56			.56	.62
7350	3/8" diameter x 100'					.57			.57	.63
7360	1/2" diameter x 100'					.84			.84	.92
7370	1/2" diameter x 500'					.74			.74	.81
7380	1/2" diameter x 1000'					.64			.64	.70
7400	3/4" diameter x 100'					1.05			1.05	1.16
7410	3/4" diameter x 500'					1.21			1.21	1.33
7420	3/4" diameter x 1000'					1.21			1.21	1.33
7460	1" diameter x 100'					2.09			2.09	2.30
7470	1" diameter x 300'					2.09			2.09	2.30
7480	1" diameter x 500'					2.11			2.11	2.32
7500	1-1/4" diameter x 100'					3.56			3.56	3.92
7510	1-1/4" diameter x 300'					3.56			3.56	3.92
7540	1-1/2" diameter x 100'					4.88			4.88	5.35
7550	1-1/2" diameter x 300'					4.95			4.95	5.45
7596	Most sizes available in red or blue									
7700	Non-barrier type, hot/cold tubing straight lengths									
7710	1/2" diameter x 20'				L.F.	.70			.70	.77
7750	3/4" diameter x 20'					1.11			1.11	1.22
7760	1" diameter x 20'					1.84			1.84	2.02
7770	1-1/4" diameter x 20'					3.32			3.32	3.65
7780	1-1/2" diameter x 20'					5.20			5.20	5.70
7790	2" diameter					9.45			9.45	10.40
7796	Most sizes available in red or blue									
9000	Polypropylene pipe									
9002	For fusion weld fittings and accessories see line 22 11 13.76 9400									
9004	Note: sizes 1/2" thru 4" use socket fusion									
9005	Sizes 6" thru 10" use butt fusion									
9010	SDR 7.4 (domestic hot water piping)									
9011	Enhanced to minimize thermal expansion and high temperature life									
9016	13' lengths, size is ID, includes joints 13' OC and hangers 3 per 10'									
9020	3/8" diameter	1 Plum	53	.151	L.F.	2.51	10.20		12.71	18
9022	1/2" diameter		52	.154		3.49	10.40		13.89	19.40
9024	3/4" diameter		50	.160		4.24	10.85		15.09	21
9026	1" diameter		45	.178		5.80	12.05		17.85	24.50
9028	1-1/4" diameter		40	.200		8.25	13.55		21.80	29

22 11 13 – Facility Water Distribution Piping

22 11 13.74 Pipe, Plastic

		Crew	Daily Output	Labor-Hours	Unit	Material	2021 Bare Costs Labor	2021 Bare Costs Equipment	Total	Total Incl O&P
9030	1-1/2" diameter	1 Plum	35	.229	L.F.	12.20	15.45		27.65	36.50
9032	2" diameter	Q-1	58	.276		17.75	16.80		34.55	44.50
9034	2-1/2" diameter		55	.291		25	17.75		42.75	54
9036	3" diameter		52	.308		33.50	18.75		52.25	64.50
9038	3-1/2" diameter		49	.327		46.50	19.90		66.40	81
9040	4" diameter		46	.348		51.50	21		72.50	88
9042	6" diameter	↓	39	.410		69.50	25		94.50	114
9044	8" diameter	Q-2	48	.500		106	31.50		137.50	164
9046	10" diameter	"	43	.558	↓	168	35.50		203.50	238
9050	To delete joint & hangers, subtract									
9052	3/8" diam. to 1" diam.					45%	65%			
9054	1-1/4" diam. to 4" diam.					15%	45%			
9056	6" diam. to 10" diam.					5%	24%			
9060	SDR 11 (domestic cold water piping)									
9062	13' lengths, size is ID, includes joints 13' OC and hangers 3 per 10'									
9064	1/2" diameter	1 Plum	57	.140	L.F.	3	9.50		12.50	17.50
9066	3/4" diameter		54	.148		3.45	10.05		13.50	18.75
9068	1" diameter		49	.163		4.37	11.05		15.42	21.50
9070	1-1/4" diameter		45	.178		6.15	12.05		18.20	24.50
9072	1-1/2" diameter	↓	40	.200		8.55	13.55		22.10	29.50
9074	2" diameter	Q-1	62	.258		12.55	15.70		28.25	37.50
9076	2-1/2" diameter		59	.271		16.85	16.50		33.35	43
9078	3" diameter		56	.286		23.50	17.40		40.90	51.50
9080	3-1/2" diameter		53	.302		34	18.40		52.40	65
9082	4" diameter		50	.320		37	19.50		56.50	69.50
9084	6" diameter	↓	47	.340		47	20.50		67.50	83
9086	8" diameter	Q-2	51	.471		72.50	29.50		102	125
9088	10" diameter	"	46	.522	↓	110	33		143	170
9090	To delete joint & hangers, subtract									
9092	1/2" diam. to 1" diam.					45%	65%			
9094	1-1/4" diam. to 4" diam.					15%	45%			
9096	6" diam. to 10" diam.					5%	24%			

22 11 13.76 Pipe Fittings, Plastic

		Crew	Daily Output	Labor-Hours	Unit	Material	2021 Bare Costs Labor	2021 Bare Costs Equipment	Total	Total Incl O&P
0010	**PIPE FITTINGS, PLASTIC**									
0030	Epoxy resin, fiberglass reinforced, general service									
0100	3"	Q-1	20.80	.769	Ea.	114	47		161	196
0110	4"		16.50	.970		123	59		182	223
0120	6"	↓	10.10	1.584		228	96.50		324.50	395
0130	8"	Q-2	9.30	2.581		420	163		583	705
0140	10"		8.50	2.824		525	178		703	845
0150	12"	↓	7.60	3.158	↓	755	200		955	1,125
0170	Elbow, 90°, flanged									
0172	2"	Q-1	23	.696	Ea.	261	42.50		303.50	350
0173	3"		16	1		299	61		360	420
0174	4"		13	1.231		390	75		465	540
0176	6"	↓	8	2		705	122		827	960
0177	8"	Q-2	9	2.667		1,275	168		1,443	1,650
0178	10"		7	3.429		1,750	217		1,967	2,250
0179	12"	↓	5	4.800	↓	2,350	305		2,655	3,050
0186	Elbow, 45°, flanged									
0188	2"	Q-1	23	.696	Ea.	261	42.50		303.50	350
0189	3"	↓	16	1		299	61		360	420

22 11 13.76 Pipe Fittings, Plastic

		Crew	Daily Output	Labor-Hours	Unit	Material	2021 Bare Costs Labor	Equipment	Total	Total Incl O&P
0190	4"	Q-1	13	1.231	Ea.	390	75		465	540
0192	6"		8	2		705	122		827	960
0193	8"	Q-2	9	2.667		1,300	168		1,468	1,675
0194	10"		7	3.429		1,800	217		2,017	2,300
0195	12"		5	4.800		2,450	305		2,755	3,150
0352	Tee, flanged									
0354	2"	Q-1	17	.941	Ea.	355	57.50		412.50	475
0355	3"		10	1.600		470	97.50		567.50	660
0356	4"		8	2		525	122		647	755
0358	6"		5	3.200		900	195		1,095	1,275
0359	8"	Q-2	6	4		1,700	253		1,953	2,250
0360	10"		5	4.800		2,475	305		2,780	3,175
0361	12"		4	6		3,425	380		3,805	4,350
0365	Wye, flanged									
0367	2"	Q-1	17	.941	Ea.	700	57.50		757.50	855
0368	3"		10	1.600		965	97.50		1,062.50	1,200
0369	4"		8	2		1,275	122		1,397	1,600
0371	6"		5	3.200		1,825	195		2,020	2,300
0372	8"	Q-2	6	4		2,975	253		3,228	3,650
0373	10"		5	4.800		4,600	305		4,905	5,525
0374	12"		4	6		6,725	380		7,105	7,950
0380	Couplings									
0410	2"	Q-1	33.10	.483	Ea.	28	29.50		57.50	75
0420	3"		20.80	.769		32.50	47		79.50	106
0430	4"		16.50	.970		45	59		104	138
0440	6"		10.10	1.584		104	96.50		200.50	258
0450	8"	Q-2	9.30	2.581		176	163		339	435
0460	10"		8.50	2.824		242	178		420	535
0470	12"		7.60	3.158		345	200		545	680
0473	High corrosion resistant couplings, add					30%				
0474	Reducer, concentric, flanged									
0475	2" x 1-1/2"	Q-1	30	.533	Ea.	680	32.50		712.50	795
0476	3" x 2"		24	.667		740	40.50		780.50	875
0477	4" x 3"		19	.842		830	51.50		881.50	990
0479	6" x 4"		15	1.067		830	65		895	1,000
0480	8" x 6"	Q-2	16	1.500		1,375	95		1,470	1,650
0481	10" x 8"		13	1.846		1,275	117		1,392	1,575
0482	12" x 10"		11	2.182		1,950	138		2,088	2,350
0486	Adapter, bell x male or female									
0488	2"	Q-1	28	.571	Ea.	34.50	35		69.50	90
0489	3"		20	.800		52	48.50		100.50	130
0491	4"		17	.941		62	57.50		119.50	154
0492	6"		12	1.333		142	81		223	277
0493	8"	Q-2	15	1.600		205	101		306	375
0494	10"	"	11	2.182		294	138		432	530
0528	Flange									
0532	2"	Q-1	46	.348	Ea.	45.50	21		66.50	81.50
0533	3"		32	.500		63	30.50		93.50	115
0534	4"		26	.615		80	37.50		117.50	144
0536	6"		16	1		152	61		213	258
0537	8"	Q-2	18	1.333		233	84		317	380
0538	10"		14	1.714		330	108		438	520
0539	12"		10	2.400		415	152		567	680

22 11 13.76 Pipe Fittings, Plastic		Crew	Daily Output	Labor-Hours	Unit	Material	2021 Bare Costs Labor	Equipment	Total	Total Incl O&P
2100	PVC schedule 80, socket joint									
2110	90° elbow, 1/2"	1 Plum	30.30	.264	Ea.	2.90	17.85		20.75	29.50
2130	3/4"		26	.308		3.71	21		24.71	35
2140	1"		22.70	.352		6	24		30	42
2150	1-1/4"		20.20	.396		7.80	27		34.80	48.50
2160	1-1/2"		18.20	.440		8.55	30		38.55	54
2170	2"	Q-1	33.10	.483		10.35	29.50		39.85	55.50
2180	3"		20.80	.769		26.50	47		73.50	99
2190	4"		16.50	.970		41.50	59		100.50	134
2200	6"		10.10	1.584		118	96.50		214.50	274
2210	8"	Q-2	9.30	2.581		325	163		488	600
2250	45° elbow, 1/2"	1 Plum	30.30	.264		5.50	17.85		23.35	32.50
2270	3/4"		26	.308		8.35	21		29.35	40
2280	1"		22.70	.352		12.50	24		36.50	49.50
2290	1-1/4"		20.20	.396		15.55	27		42.55	57
2300	1-1/2"		18.20	.440		18.85	30		48.85	65
2310	2"	Q-1	33.10	.483		24.50	29.50		54	71
2320	3"		20.80	.769		61	47		108	137
2330	4"		16.50	.970		112	59		171	212
2340	6"		10.10	1.584		142	96.50		238.50	300
2350	8"	Q-2	9.30	2.581		305	163		468	585
2400	Tee, 1/2"	1 Plum	20.20	.396		8.20	27		35.20	49
2420	3/4"		17.30	.462		8.60	31.50		40.10	56
2430	1"		15.20	.526		10.75	35.50		46.25	65
2440	1-1/4"		13.50	.593		29	40		69	91.50
2450	1-1/2"		12.10	.661		29.50	45		74.50	99.50
2460	2"	Q-1	20	.800		37	48.50		85.50	113
2470	3"		13.90	1.151		49	70		119	159
2480	4"		11	1.455		58	88.50		146.50	196
2490	6"		6.70	2.388		198	146		344	435
2500	8"	Q-2	6.20	3.871		460	245		705	870
2510	Flange, socket, 150 lb., 1/2"	1 Plum	55.60	.144		15.50	9.75		25.25	31.50
2514	3/4"		47.60	.168		16.60	11.40		28	35.50
2518	1"		41.70	.192		18.45	13		31.45	40
2522	1-1/2"		33.30	.240		19.45	16.25		35.70	46
2526	2"	Q-1	60.60	.264		26	16.10		42.10	52.50
2530	4"		30.30	.528		56	32		88	110
2534	6"		18.50	.865		87.50	52.50		140	175
2538	8"	Q-2	17.10	1.404		157	88.50		245.50	305
2550	Coupling, 1/2"	1 Plum	30.30	.264		5.45	17.85		23.30	32.50
2570	3/4"		26	.308		7.55	21		28.55	39.50
2580	1"		22.70	.352		7.90	24		31.90	44
2590	1-1/4"		20.20	.396		10.85	27		37.85	52
2600	1-1/2"		18.20	.440		16.45	30		46.45	62.50
2610	2"	Q-1	33.10	.483		17.45	29.50		46.95	63
2620	3"		20.80	.769		35.50	47		82.50	109
2630	4"		16.50	.970		73.50	59		132.50	169
2640	6"		10.10	1.584		95.50	96.50		192	249
2650	8"	Q-2	9.30	2.581		130	163		293	385
2660	10"		8.50	2.824		410	178		588	720
2670	12"		7.60	3.158		475	200		675	820
2700	PVC (white), schedule 40, socket joints									
2760	90° elbow, 1/2"	1 Plum	33.30	.240	Ea.	.55	16.25		16.80	25

22 11 Facility Water Distribution

22 11 13 – Facility Water Distribution Piping

22 11 13.76 Pipe Fittings, Plastic		Crew	Daily Output	Labor-Hours	Unit	Material	2021 Bare Costs Labor	Equipment	Total	Total Incl O&P
2770	3/4"	1 Plum	28.60	.280	Ea.	.62	18.95		19.57	29
2780	1"		25	.320		1.11	21.50		22.61	33.50
2790	1-1/4"		22.20	.360		1.80	24.50		26.30	38.50
2800	1-1/2"	↓	20	.400		2.10	27		29.10	43
2810	2"	Q-1	36.40	.440		3.29	27		30.29	43.50
2820	2-1/2"		26.70	.599		10.05	36.50		46.55	65.50
2830	3"		22.90	.699		12	42.50		54.50	76.50
2840	4"		18.20	.879		21.50	53.50		75	104
2850	5"		12.10	1.322		51	80.50		131.50	177
2860	6"	↓	11.10	1.441		68	88		156	206
2870	8"	Q-2	10.30	2.330		176	147		323	415
2980	45° elbow, 1/2"	1 Plum	33.30	.240		.91	16.25		17.16	25.50
2990	3/4"		28.60	.280		1.41	18.95		20.36	30
3000	1"		25	.320		1.63	21.50		23.13	34.50
3010	1-1/4"		22.20	.360		2.18	24.50		26.68	39
3020	1-1/2"	↓	20	.400		2.85	27		29.85	43.50
3030	2"	Q-1	36.40	.440		3.72	27		30.72	44
3040	2-1/2"		26.70	.599		9.70	36.50		46.20	65
3050	3"		22.90	.699		14.40	42.50		56.90	79.50
3060	4"		18.20	.879		27	53.50		80.50	110
3070	5"		12.10	1.322		51	80.50		131.50	177
3080	6"	↓	11.10	1.441		66.50	88		154.50	204
3090	8"	Q-2	10.30	2.330		160	147		307	395
3180	Tee, 1/2"	1 Plum	22.20	.360		.66	24.50		25.16	37
3190	3/4"		19	.421		.76	28.50		29.26	43.50
3200	1"		16.70	.479		1.19	32.50		33.69	50
3210	1-1/4"		14.80	.541		2.12	36.50		38.62	57
3220	1-1/2"	↓	13.30	.602		2.69	40.50		43.19	64
3230	2"	Q-1	24.20	.661		3.92	40.50		44.42	64.50
3240	2-1/2"		17.80	.899		12.35	55		67.35	95
3250	3"		15.20	1.053		15.10	64		79.10	112
3260	4"		12.10	1.322		30.50	80.50		111	154
3270	5"		8.10	1.975		71	120		191	258
3280	6"	↓	7.40	2.162		103	132		235	310
3290	8"	Q-2	6.80	3.529		240	223		463	600
3380	Coupling, 1/2"	1 Plum	33.30	.240		.34	16.25		16.59	25
3390	3/4"		28.60	.280		.48	18.95		19.43	29
3400	1"		25	.320		.84	21.50		22.34	33.50
3410	1-1/4"		22.20	.360		1.11	24.50		25.61	37.50
3420	1-1/2"	↓	20	.400		1.24	27		28.24	42
3430	2"	Q-1	36.40	.440		1.89	27		28.89	42
3440	2-1/2"		26.70	.599		4.19	36.50		40.69	59
3450	3"		22.90	.699		6.55	42.50		49.05	70.50
3460	4"		18.20	.879		9.50	53.50		63	90.50
3470	5"		12.10	1.322		16.65	80.50		97.15	138
3480	6"	↓	11.10	1.441		30	88		118	164
3490	8"	Q-2	10.30	2.330		53.50	147		200.50	279
3600	Cap, schedule 40, PVC socket 1/2"	1 Plum	60.60	.132		.48	8.95		9.43	13.90
3610	3/4"		51.90	.154		.56	10.45		11.01	16.20
3620	1"		45.50	.176		.88	11.90		12.78	18.70
3630	1-1/4"		40.40	.198		1.19	13.40		14.59	21.50
3640	1-1/2"	↓	36.40	.220		1.36	14.90		16.26	23.50
3650	2"	Q-1	66.10	.242		1.63	14.75		16.38	24

22 11 13 – Facility Water Distribution Piping

22 11 13.76 Pipe Fittings, Plastic		Crew	Daily Output	Labor-Hours	Unit	Material	2021 Bare Costs Labor	Equipment	Total	Total Incl O&P
3660	2-1/2"	Q-1	48.50	.330	Ea.	4.98	20		24.98	35.50
3670	3"		41.60	.385		5.45	23.50		28.95	41
3680	4"		33.10	.483		12.90	29.50		42.40	58
3690	6"	↓	20.20	.792		31	48.50		79.50	106
3700	8"	Q-2	18.60	1.290	↓	78	81.50		159.50	208
3710	Reducing insert, schedule 40, socket weld									
3712	3/4"	1 Plum	31.50	.254	Ea.	.54	17.20		17.74	26
3713	1"		27.50	.291		.99	19.70		20.69	30.50
3715	1-1/2"	↓	22	.364		1.40	24.50		25.90	38.50
3716	2"	Q-1	40	.400		2.31	24.50		26.81	39
3717	4"		20	.800		12.25	48.50		60.75	86
3718	6"	↓	12.20	1.311		30.50	80		110.50	153
3719	8"	Q-2	11.30	2.124	↓	110	134		244	320
3730	Reducing insert, socket weld x female/male thread									
3732	1/2"	1 Plum	38.30	.209	Ea.	2.39	14.15		16.54	23.50
3733	3/4"		32.90	.243		1.48	16.45		17.93	26
3734	1"		28.80	.278		2.07	18.80		20.87	30.50
3736	1-1/2"	↓	23	.348		3.72	23.50		27.22	39
3737	2"	Q-1	41.90	.382		3.98	23.50		27.48	39
3738	4"	"	20.90	.766	↓	38.50	46.50		85	112
3742	Male adapter, socket weld x male thread									
3744	1/2"	1 Plum	38.30	.209	Ea.	.47	14.15		14.62	21.50
3745	3/4"		32.90	.243		.52	16.45		16.97	25
3746	1"		28.80	.278		.93	18.80		19.73	29
3748	1-1/2"	↓	23	.348		1.68	23.50		25.18	37
3749	2"	Q-1	41.90	.382		1.90	23.50		25.40	36.50
3750	4"	"	20.90	.766	↓	10.95	46.50		57.45	81.50
3754	Female adapter, socket weld x female thread									
3756	1/2"	1 Plum	38.30	.209	Ea.	.58	14.15		14.73	21.50
3757	3/4"		32.90	.243		.74	16.45		17.19	25.50
3758	1"		28.80	.278		.86	18.80		19.66	29
3760	1-1/2"	↓	23	.348		1.52	23.50		25.02	36.50
3761	2"	Q-1	41.90	.382		2.04	23.50		25.54	36.50
3762	4"	"	20.90	.766		11.45	46.50		57.95	82
3800	PVC, schedule 80, socket joints									
3810	Reducing insert									
3812	3/4"	1 Plum	28.60	.280	Ea.	1.46	18.95		20.41	30
3813	1"		25	.320		3.78	21.50		25.28	36.50
3815	1-1/2"	↓	20	.400		8.05	27		35.05	49.50
3816	2"	Q-1	36.40	.440		11.50	27		38.50	52.50
3817	4"		18.20	.879		44	53.50		97.50	129
3818	6"	↓	11.10	1.441		61	88		149	198
3819	8"	Q-2	10.20	2.353	↓	282	149		431	530
3830	Reducing insert, socket weld x female/male thread									
3832	1/2"	1 Plum	34.80	.230	Ea.	11	15.55		26.55	35
3833	3/4"		29.90	.268		6.40	18.10		24.50	34
3834	1"		26.10	.307		10.05	21		31.05	42
3836	1-1/2"	↓	20.90	.383		12.65	26		38.65	52.50
3837	2"	Q-1	38	.421		18.55	25.50		44.05	59
3838	4"	"	19	.842	↓	90	51.50		141.50	176
3844	Adapter, male socket x male thread									
3846	1/2"	1 Plum	34.80	.230	Ea.	4.05	15.55		19.60	27.50
3847	3/4"	↓	29.90	.268	↓	4.47	18.10		22.57	32

22 11 13 – Facility Water Distribution Piping

22 11 13.76 Pipe Fittings, Plastic	Crew	Daily Output	Labor-Hours	Unit	Material	2021 Bare Costs Labor	Equipment	Total	Total Incl O&P
3848	1"	1 Plum	26.10	.307	Ea.	7.70	21		28.70	39.50
3850	1-1/2"	↓	20.90	.383		12.95	26		38.95	53
3851	2"	Q-1	38	.421		18.75	25.50		44.25	59
3852	4"	"	19	.842	↓	42	51.50		93.50	123
3860	Adapter, female socket x female thread									
3862	1/2"	1 Plum	34.80	.230	Ea.	3.83	15.55		19.38	27
3863	3/4"		29.90	.268		5.70	18.10		23.80	33.50
3864	1"		26.10	.307		8.40	21		29.40	40.50
3866	1-1/2"	↓	20.90	.383		16.65	26		42.65	57
3867	2"	Q-1	38	.421		29	25.50		54.50	70.50
3868	4"	"	19	.842	↓	89	51.50		140.50	174
3872	Union, socket joints									
3874	1/2"	1 Plum	25.80	.310	Ea.	8.40	21		29.40	41
3875	3/4"		22.10	.362		10.65	24.50		35.15	48.50
3876	1"		19.30	.415		12.15	28		40.15	55.50
3878	1-1/2"	↓	15.50	.516		27.50	35		62.50	82
3879	2"	Q-1	28.10	.569	↓	37	34.50		71.50	93
3888	Cap									
3890	1/2"	1 Plum	54.50	.147	Ea.	4	9.95		13.95	19.25
3891	3/4"		46.70	.171		4.22	11.60		15.82	22
3892	1"		41	.195		7.50	13.20		20.70	28
3894	1-1/2"	↓	32.80	.244		9.05	16.50		25.55	34.50
3895	2"	Q-1	59.50	.269		20.50	16.40		36.90	47
3896	4"		30	.533		61	32.50		93.50	116
3897	6"	↓	18.20	.879		227	53.50		280.50	330
3898	8"	Q-2	16.70	1.437	↓	292	91		383	455
4500	DWV, ABS, non pressure, socket joints									
4540	1/4 bend, 1-1/4"	1 Plum	20.20	.396	Ea.	8.15	27		35.15	49
4560	1-1/2"	"	18.20	.440		6.35	30		36.35	51.50
4570	2"	Q-1	33.10	.483		6.05	29.50		35.55	50.50
4580	3"		20.80	.769		24.50	47		71.50	97
4590	4"		16.50	.970		50.50	59		109.50	144
4600	6"	↓	10.10	1.584	↓	216	96.50		312.50	380
4650	1/8 bend, same as 1/4 bend									
4800	Tee, sanitary									
4820	1-1/4"	1 Plum	13.50	.593	Ea.	10.85	40		50.85	72
4830	1-1/2"	"	12.10	.661		9.40	45		54.40	77.50
4840	2"	Q-1	20	.800		14.45	48.50		62.95	88.50
4850	3"		13.90	1.151		39.50	70		109.50	149
4860	4"		11	1.455		70	88.50		158.50	209
4862	Tee, sanitary, reducing, 2" x 1-1/2"		22	.727		12.60	44.50		57.10	80
4864	3" x 2"		15.30	1.046		31.50	63.50		95	130
4868	4" x 3"	↓	12.10	1.322	↓	70.50	80.50		151	198
4870	Combination Y and 1/8 bend									
4872	1-1/2"	1 Plum	12.10	.661	Ea.	24.50	45		69.50	94
4874	2"	Q-1	20	.800		25	48.50		73.50	100
4876	3"		13.90	1.151		59	70		129	170
4878	4"		11	1.455		123	88.50		211.50	268
4880	3" x 1-1/2"		15.50	1.032		66.50	63		129.50	168
4882	4" x 3"	↓	12.10	1.322		88.50	80.50		169	218
4900	Wye, 1-1/4"	1 Plum	13.50	.593		12.50	40		52.50	74
4902	1-1/2"	"	12.10	.661		15.55	45		60.55	84
4904	2"	Q-1	20	.800		20.50	48.50		69	95

For customer support on your Plumbing Costs with RSMeans Data, call 800.448.8182.

233

22 11 13.76 Pipe Fittings, Plastic		Crew	Daily Output	Labor-Hours	Unit	Material	2021 Bare Costs Labor	Equipment	Total	Total Incl O&P
4906	3"	Q-1	13.90	1.151	Ea.	43.50	70		113.50	153
4908	4"		11	1.455		89	88.50		177.50	230
4910	6"		6.70	2.388		268	146		414	510
4918	3" x 1-1/2"		15.50	1.032		38.50	63		101.50	137
4920	4" x 3"		12.10	1.322		77	80.50		157.50	205
4922	6" x 4"		6.90	2.319		242	141		383	475
4930	Double wye, 1-1/2"	1 Plum	9.10	.879		48	59.50		107.50	142
4932	2"	Q-1	16.60	.964		57	58.50		115.50	151
4934	3"		10.40	1.538		128	93.50		221.50	281
4936	4"		8.25	1.939		251	118		369	450
4940	2" x 1-1/2"		16.80	.952		54.50	58		112.50	147
4942	3" x 2"		10.60	1.509		89.50	92		181.50	236
4944	4" x 3"		8.45	1.893		208	115		323	400
4946	6" x 4"		7.25	2.207		285	134		419	515
4950	Reducer bushing, 2" x 1-1/2"		36.40	.440		4.99	27		31.99	45.50
4952	3" x 1-1/2"		27.30	.586		22	35.50		57.50	77.50
4954	4" x 2"		18.20	.879		40.50	53.50		94	125
4956	6" x 4"		11.10	1.441		114	88		202	256
4960	Couplings, 1-1/2"	1 Plum	18.20	.440		3.26	30		33.26	48
4962	2"	Q-1	33.10	.483		4.09	29.50		33.59	48.50
4963	3"		20.80	.769		12.30	47		59.30	83.50
4964	4"		16.50	.970		20.50	59		79.50	111
4966	6"		10.10	1.584		86	96.50		182.50	239
4970	2" x 1-1/2"		33.30	.480		8.75	29.50		38.25	53
4972	3" x 1-1/2"		21	.762		26	46.50		72.50	98
4974	4" x 3"		16.70	.958		38.50	58.50		97	130
4978	Closet flange, 4"	1 Plum	32	.250		23.50	16.95		40.45	51
4980	4" x 3"	"	34	.235		25.50	15.95		41.45	52
5000	DWV, PVC, schedule 40, socket joints									
5040	1/4 bend, 1-1/4"	1 Plum	20.20	.396	Ea.	16.55	27		43.55	58
5060	1-1/2"	"	18.20	.440		4.75	30		34.75	50
5070	2"	Q-1	33.10	.483		7.45	29.50		36.95	52
5080	3"		20.80	.769		22	47		69	94
5090	4"		16.50	.970		43.50	59		102.50	136
5100	6"		10.10	1.584		152	96.50		248.50	310
5105	8"	Q-2	9.30	2.581		197	163		360	460
5106	10"	"	8.50	2.824		259	178		437	550
5110	1/4 bend, long sweep, 1-1/2"	1 Plum	18.20	.440		11	30		41	56.50
5112	2"	Q-1	33.10	.483		12.25	29.50		41.75	57.50
5114	3"		20.80	.769		28.50	47		75.50	101
5116	4"		16.50	.970		53.50	59		112.50	147
5150	1/8 bend, 1-1/4"	1 Plum	20.20	.396		11.30	27		38.30	52.50
5170	1-1/2"	"	18.20	.440		4.64	30		34.64	49.50
5180	2"	Q-1	33.10	.483		6.95	29.50		36.45	51.50
5190	3"		20.80	.769		19.65	47		66.65	91.50
5200	4"		16.50	.970		36	59		95	128
5210	6"		10.10	1.584		133	96.50		229.50	291
5215	8"	Q-2	9.30	2.581		87	163		250	340
5216	10"		8.50	2.824		247	178		425	540
5217	12"		7.60	3.158		299	200		499	630
5250	Tee, sanitary 1-1/4"	1 Plum	13.50	.593		17.75	40		57.75	79.50
5254	1-1/2"	"	12.10	.661		8.30	45		53.30	76
5255	2"	Q-1	20	.800		12.20	48.50		60.70	86

22 11 13.76 Pipe Fittings, Plastic

		Crew	Daily Output	Labor-Hours	Unit	Material	2021 Bare Costs Labor	Equipment	Total	Total Incl O&P
5256	3"	Q-1	13.90	1.151	Ea.	32	70		102	141
5257	4"		11	1.455		58.50	88.50		147	197
5259	6"	↓	6.70	2.388		235	146		381	475
5261	8"	Q-2	6.20	3.871	↓	520	245		765	940
5276	Tee, sanitary, reducing									
5281	2" x 1-1/2" x 1-1/2"	Q-1	23	.696	Ea.	10.70	42.50		53.20	75.50
5282	2" x 1-1/2" x 2"		22	.727		12.95	44.50		57.45	80.50
5283	2" x 2" x 1-1/2"		22	.727		10.75	44.50		55.25	78
5284	3" x 3" x 1-1/2"		15.50	1.032		23.50	63		86.50	120
5285	3" x 3" x 2"		15.30	1.046		24	63.50		87.50	122
5286	4" x 4" x 1-1/2"		12.30	1.301		61.50	79.50		141	186
5287	4" x 4" x 2"		12.20	1.311		50.50	80		130.50	175
5288	4" x 4" x 3"		12.10	1.322		68.50	80.50		149	196
5291	6" x 6" x 4"	↓	6.90	2.319	↓	228	141		369	460
5294	Tee, double sanitary									
5295	1-1/2"	1 Plum	9.10	.879	Ea.	18.25	59.50		77.75	109
5296	2"	Q-1	16.60	.964		24.50	58.50		83	115
5297	3"		10.40	1.538		69	93.50		162.50	216
5298	4"	↓	8.25	1.939	↓	111	118		229	298
5303	Wye, reducing									
5304	2" x 1-1/2" x 1-1/2"	Q-1	23	.696	Ea.	21	42.50		63.50	86.50
5305	2" x 2" x 1-1/2"		22	.727		18.15	44.50		62.65	86
5306	3" x 3" x 2"		15.30	1.046		59.50	63.50		123	160
5307	4" x 4" x 2"		12.20	1.311		43.50	80		123.50	167
5309	4" x 4" x 3"	↓	12.10	1.322		59	80.50		139.50	185
5314	Combination Y & 1/8 bend, 1-1/2"	1 Plum	12.10	.661		20	45		65	89
5315	2"	Q-1	20	.800		25	48.50		73.50	100
5317	3"		13.90	1.151		55	70		125	166
5318	4"		11	1.455		109	88.50		197.50	252
5319	6"	↓	6.70	2.388		420	146		566	675
5320	8"	Q-2	6.20	3.871		630	245		875	1,050
5321	10"		5.70	4.211		1,125	266		1,391	1,625
5322	12"	↓	5.10	4.706	↓	1,325	297		1,622	1,900
5324	Combination Y & 1/8 bend, reducing									
5325	2" x 2" x 1-1/2"	Q-1	22	.727	Ea.	28	44.50		72.50	97
5327	3" x 3" x 1-1/2"		15.50	1.032		50.50	63		113.50	150
5328	3" x 3" x 2"		15.30	1.046		37.50	63.50		101	137
5329	4" x 4" x 2"	↓	12.20	1.311		57	80		137	182
5331	Wye, 1-1/4"	1 Plum	13.50	.593		22.50	40		62.50	85
5332	1-1/2"	"	12.10	.661		15.10	45		60.10	83.50
5333	2"	Q-1	20	.800		14.85	48.50		63.35	89
5334	3"		13.90	1.151		40	70		110	149
5335	4"		11	1.455		73	88.50		161.50	212
5336	6"	↓	6.70	2.388		210	146		356	450
5337	8"	Q-2	6.20	3.871		370	245		615	775
5338	10"		5.70	4.211		525	266		791	970
5339	12"		5.10	4.706		850	297		1,147	1,375
5341	2" x 1-1/2"	Q-1	22	.727		10.75	44.50		55.25	78
5342	3" x 1-1/2"		15.50	1.032		15.90	63		78.90	112
5343	4" x 3"		12.10	1.322		35	80.50		115.50	159
5344	6" x 4"	↓	6.90	2.319		94.50	141		235.50	315
5345	8" x 6"	Q-2	6.40	3.750		205	237		442	580
5347	Double wye, 1-1/2"	1 Plum	9.10	.879	↓	34	59.50		93.50	127

22 11 13.76 Pipe Fittings, Plastic		Crew	Daily Output	Labor-Hours	Unit	Material	2021 Bare Costs Labor	Equipment	Total	Total Incl O&P
5348	2"	Q-1	16.60	.964	Ea.	40.50	58.50		99	132
5349	3"		10.40	1.538		79	93.50		172.50	227
5350	4"	↓	8.25	1.939	↓	170	118		288	365
5353	Double wye, reducing									
5354	2" x 2" x 1-1/2" x 1-1/2"	Q-1	16.80	.952	Ea.	34.50	58		92.50	125
5355	3" x 3" x 2" x 2"		10.60	1.509		58.50	92		150.50	202
5356	4" x 4" x 3" x 3"		8.45	1.893		127	115		242	310
5357	6" x 6" x 4" x 4"	↓	7.25	2.207		445	134		579	690
5374	Coupling, 1-1/4"	1 Plum	20.20	.396		10.70	27		37.70	52
5376	1-1/2"	"	18.20	.440		2.23	30		32.23	47
5378	2"	Q-1	33.10	.483		3.06	29.50		32.56	47.50
5380	3"		20.80	.769		10.65	47		57.65	82
5390	4"		16.50	.970		18.10	59		77.10	108
5400	6"	↓	10.10	1.584		59.50	96.50		156	210
5402	8"	Q-2	9.30	2.581		98.50	163		261.50	350
5404	2" x 1-1/2"	Q-1	33.30	.480		6.80	29.50		36.30	51
5406	3" x 1-1/2"		21	.762		19.90	46.50		66.40	91.50
5408	4" x 3"		16.70	.958		32.50	58.50		91	123
5410	Reducer bushing, 2" x 1-1/4"		36.50	.438		3.73	26.50		30.23	44
5411	2" x 1-1/2"		36.40	.440		3.87	27		30.87	44.50
5412	3" x 1-1/2"		27.30	.586		11	35.50		46.50	65.50
5413	3" x 2"		27.10	.590		9.80	36		45.80	64.50
5414	4" x 2"		18.20	.879		32.50	53.50		86	116
5415	4" x 3"		16.70	.958		16.70	58.50		75.20	105
5416	6" x 4"	↓	11.10	1.441		83	88		171	223
5418	8" x 6"	Q-2	10.20	2.353		163	149		312	400
5425	Closet flange 4"	Q-1	32	.500		28	30.50		58.50	76
5426	4" x 3"	"	34	.471	↓	29.50	28.50		58	75.50
5450	Solvent cement for PVC, industrial grade, per quart				Qt.	32			32	35
5500	CPVC, Schedule 80, threaded joints									
5540	90° elbow, 1/4"	1 Plum	32	.250	Ea.	12.80	16.95		29.75	39.50
5560	1/2"		30.30	.264		7.40	17.85		25.25	34.50
5570	3/4"		26	.308		11.10	21		32.10	43
5580	1"		22.70	.352		15.60	24		39.60	52.50
5590	1-1/4"		20.20	.396		30	27		57	73
5600	1-1/2"	↓	18.20	.440		32.50	30		62.50	80
5610	2"	Q-1	33.10	.483		43.50	29.50		73	91.50
5620	2-1/2"		24.20	.661		134	40.50		174.50	208
5630	3"		20.80	.769		144	47		191	228
5640	4"		16.50	.970		225	59		284	335
5650	6"	↓	10.10	1.584	↓	291	96.50		387.50	465
5660	45° elbow same as 90° elbow									
5700	Tee, 1/4"	1 Plum	22	.364	Ea.	25	24.50		49.50	64.50
5702	1/2"		20.20	.396		25	27		52	67.50
5704	3/4"		17.30	.462		36	31.50		67.50	86
5706	1"		15.20	.526		38.50	35.50		74	95.50
5708	1-1/4"		13.50	.593		39	40		79	103
5710	1-1/2"	↓	12.10	.661		40.50	45		85.50	112
5712	2"	Q-1	20	.800		45	48.50		93.50	122
5714	2-1/2"		16.20	.988		222	60		282	335
5716	3"		13.90	1.151		258	70		328	390
5718	4"		11	1.455		610	88.50		698.50	800
5720	6"	↓	6.70	2.388		765	146		911	1,050

22 11 13.76 Pipe Fittings, Plastic

		Crew	Daily Output	Labor-Hours	Unit	Material	2021 Bare Costs Labor	Equipment	Total	Total Incl O&P
5730	Coupling, 1/4"	1 Plum	32	.250	Ea.	16.30	16.95		33.25	43.50
5732	1/2"		30.30	.264		13.40	17.85		31.25	41.50
5734	3/4"		26	.308		21.50	21		42.50	55
5736	1"		22.70	.352		24.50	24		48.50	62.50
5738	1-1/4"		20.20	.396		26	27		53	68.50
5740	1-1/2"		18.20	.440		28	30		58	75.50
5742	2"	Q-1	33.10	.483		33	29.50		62.50	80.50
5744	2-1/2"		24.20	.661		59	40.50		99.50	125
5746	3"		20.80	.769		69	47		116	146
5748	4"		16.50	.970		141	59		200	243
5750	6"		10.10	1.584		214	96.50		310.50	380
5752	8"	Q-2	9.30	2.581		445	163		608	735
5900	CPVC, Schedule 80, socket joints									
5904	90° elbow, 1/4"	1 Plum	32	.250	Ea.	12.15	16.95		29.10	39
5906	1/2"		30.30	.264		4.77	17.85		22.62	32
5908	3/4"		26	.308		6.10	21		27.10	37.50
5910	1"		22.70	.352		9.65	24		33.65	46
5912	1-1/4"		20.20	.396		21	27		48	63
5914	1-1/2"		18.20	.440		23.50	30		53.50	70
5916	2"	Q-1	33.10	.483		28	29.50		57.50	75
5918	2-1/2"		24.20	.661		65	40.50		105.50	132
5920	3"		20.80	.769		73.50	47		120.50	151
5922	4"		16.50	.970		132	59		191	233
5924	6"		10.10	1.584		266	96.50		362.50	435
5926	8"		9.30	1.720		650	105		755	870
5930	45° elbow, 1/4"	1 Plum	32	.250		18.10	16.95		35.05	45.50
5932	1/2"		30.30	.264		5.85	17.85		23.70	33
5934	3/4"		26	.308		8.40	21		29.40	40.50
5936	1"		22.70	.352		13.40	24		37.40	50.50
5938	1-1/4"		20.20	.396		26.50	27		53.50	69
5940	1-1/2"		18.20	.440		27	30		57	74
5942	2"	Q-1	33.10	.483		30.50	29.50		60	77.50
5944	2-1/2"		24.20	.661		62	40.50		102.50	128
5946	3"		20.80	.769		79.50	47		126.50	158
5948	4"		16.50	.970		109	59		168	208
5950	6"		10.10	1.584		335	96.50		431.50	515
5952	8"		9.30	1.720		700	105		805	925
5960	Tee, 1/4"	1 Plum	22	.364		11.20	24.50		35.70	49.50
5962	1/2"		20.20	.396		11.20	27		38.20	52.50
5964	3/4"		17.30	.462		11.40	31.50		42.90	59
5966	1"		15.20	.526		13.95	35.50		49.45	68.50
5968	1-1/4"		13.50	.593		29.50	40		69.50	92.50
5970	1-1/2"		12.10	.661		33.50	45		78.50	104
5972	2"	Q-1	20	.800		37.50	48.50		86	114
5974	2-1/2"		16.20	.988		95.50	60		155.50	195
5976	3"		13.90	1.151		95.50	70		165.50	210
5978	4"		11	1.455		127	88.50		215.50	272
5980	6"		6.70	2.388		330	146		476	580
5982	8"	Q-2	6.20	3.871		945	245		1,190	1,425
5990	Coupling, 1/4"	1 Plum	32	.250		12.95	16.95		29.90	40
5992	1/2"		30.30	.264		5.05	17.85		22.90	32
5994	3/4"		26	.308		7.05	21		28.05	39
5996	1"		22.70	.352		9.50	24		33.50	46

For customer support on your Plumbing Costs with RSMeans Data, call 800.448.8182.

237

22 11 Facility Water Distribution

22 11 13 – Facility Water Distribution Piping

22 11 13.76 Pipe Fittings, Plastic	Crew	Daily Output	Labor-Hours	Unit	Material	2021 Bare Costs Labor	Equipment	Total	Total Incl O&P
5998	1-1/4"	1 Plum	20.20	.396	Ea.	14.20	27		41.20	55.50
6000	1-1/2"	↓	18.20	.440		17.85	30		47.85	64
6002	2"	Q-1	33.10	.483		21	29.50		50.50	67
6004	2-1/2"		24.20	.661		46	40.50		86.50	111
6006	3"		20.80	.769		50	47		97	126
6008	4"		16.50	.970		66	59		125	161
6010	6"	↓	10.10	1.584		155	96.50		251.50	315
6012	8"	Q-2	9.30	2.581	↓	420	163		583	705
6200	CTS, 100 psi at 180°F, hot and cold water									
6230	90° elbow, 1/2"	1 Plum	20	.400	Ea.	.32	27		27.32	41
6250	3/4"		19	.421		.52	28.50		29.02	43
6251	1"		16	.500		1.52	34		35.52	52
6252	1-1/4"		15	.533		3.04	36		39.04	57.50
6253	1-1/2"	↓	14	.571		5.45	38.50		43.95	63.50
6254	2"	Q-1	23	.696		10.50	42.50		53	75
6260	45° elbow, 1/2"	1 Plum	20	.400		.40	27		27.40	41
6280	3/4"		19	.421		.70	28.50		29.20	43.50
6281	1"		16	.500		1.76	34		35.76	52.50
6282	1-1/4"		15	.533		3.74	36		39.74	58
6283	1-1/2"	↓	14	.571		5.40	38.50		43.90	63.50
6284	2"	Q-1	23	.696		11.35	42.50		53.85	76
6290	Tee, 1/2"	1 Plum	13	.615		.41	41.50		41.91	62.50
6310	3/4"		12	.667		.76	45		45.76	68.50
6311	1"		11	.727		3.81	49		52.81	77.50
6312	1-1/4"		10	.800		5.90	54		59.90	87.50
6313	1-1/2"	↓	10	.800		7.65	54		61.65	89.50
6314	2"	Q-1	17	.941		12.35	57.50		69.85	99
6320	Coupling, 1/2"	1 Plum	22	.364		.26	24.50		24.76	37.50
6340	3/4"		21	.381		.34	26		26.34	39
6341	1"		18	.444		1.55	30		31.55	46.50
6342	1-1/4"		17	.471		2	32		34	49.50
6343	1-1/2"	↓	16	.500		2.83	34		36.83	53.50
6344	2"	Q-1	28	.571	↓	5.40	35		40.40	58
6360	Solvent cement for CPVC, commercial grade, per quart				Qt.	55			55	60.50
7340	PVC flange, slip-on, Sch 80 std., 1/2"	1 Plum	22	.364	Ea.	16	24.50		40.50	54.50
7350	3/4"		21	.381		17.05	26		43.05	57.50
7360	1"		18	.444		19	30		49	66
7370	1-1/4"		17	.471		19.60	32		51.60	69
7380	1-1/2"	↓	16	.500		20	34		54	72.50
7390	2"	Q-1	26	.615		26.50	37.50		64	85.50
7400	2-1/2"		24	.667		41	40.50		81.50	106
7410	3"		18	.889		45.50	54		99.50	131
7420	4"		15	1.067		57.50	65		122.50	160
7430	6"	↓	10	1.600		90.50	97.50		188	245
7440	8"	Q-2	11	2.182		162	138		300	385
7550	Union, schedule 40, socket joints, 1/2"	1 Plum	19	.421		5.25	28.50		33.75	48.50
7560	3/4"		18	.444		5.95	30		35.95	51.50
7570	1"		15	.533		6.10	36		42.10	61
7580	1-1/4"		14	.571		18.80	38.50		57.30	78
7590	1-1/2"		13	.615		20.50	41.50		62	84.50
7600	2"	Q-1	20	.800	↓	27.50	48.50		76	103
7992	Polybutyl/polyethyl pipe, for copper fittings see Line 22 11 13.25 7000									
8000	Compression type, PVC, 160 psi cold water									

238

22 11 13.76 Pipe Fittings, Plastic

		Crew	Daily Output	Labor-Hours	Unit	Material	2021 Bare Costs Labor	Equipment	Total	Total Incl O&P
8010	Coupling, 3/4" CTS	1 Plum	21	.381	Ea.	4.06	26		30.06	43
8020	1" CTS		18	.444		5.05	30		35.05	50.50
8030	1-1/4" CTS		17	.471		7.10	32		39.10	55.50
8040	1-1/2" CTS		16	.500		9.75	34		43.75	61
8050	2" CTS		15	.533		13.50	36		49.50	69
8060	Female adapter, 3/4" FPT x 3/4" CTS		23	.348		7.15	23.50		30.65	43
8070	3/4" FPT x 1" CTS		21	.381		8.15	26		34.15	47.50
8080	1" FPT x 1" CTS		20	.400		8.30	27		35.30	49.50
8090	1-1/4" FPT x 1-1/4" CTS		18	.444		10.85	30		40.85	57
8100	1-1/2" FPT x 1-1/2" CTS		16	.500		12.40	34		46.40	64
8110	2" FPT x 2" CTS		13	.615		18.05	41.50		59.55	82
8130	Male adapter, 3/4" MPT x 3/4" CTS		23	.348		5.95	23.50		29.45	41.50
8140	3/4" MPT x 1" CTS		21	.381		6.95	26		32.95	46
8150	1" MPT x 1" CTS		20	.400		7.10	27		34.10	48.50
8160	1-1/4" MPT x 1-1/4" CTS		18	.444		9.55	30		39.55	55.50
8170	1-1/2" MPT x 1-1/2" CTS		16	.500		11.50	34		45.50	63
8180	2" MPT x 2" CTS		13	.615		14.90	41.50		56.40	78.50
8200	Spigot adapter, 3/4" IPS x 3/4" CTS		23	.348		2.17	23.50		25.67	37.50
8210	3/4" IPS x 1" CTS		21	.381		2.67	26		28.67	41.50
8220	1" IPS x 1" CTS		20	.400		2.66	27		29.66	43.50
8230	1-1/4" IPS x 1-1/4" CTS		18	.444		4.01	30		34.01	49.50
8240	1-1/2" IPS x 1-1/2" CTS		16	.500		4.20	34		38.20	55
8250	2" IPS x 2" CTS		13	.615		5.20	41.50		46.70	67.50
8270	Price includes insert stiffeners									
8280	250 psi is same price as 160 psi									
8300	Insert type, nylon, 160 & 250 psi, cold water									
8310	Clamp ring stainless steel, 3/4" IPS	1 Plum	115	.070	Ea.	2.86	4.71		7.57	10.20
8320	1" IPS		107	.075		3.45	5.05		8.50	11.35
8330	1-1/4" IPS		101	.079		2.94	5.35		8.29	11.25
8340	1-1/2" IPS		95	.084		4.09	5.70		9.79	13
8350	2" IPS		85	.094		5.35	6.35		11.70	15.35
8370	Coupling, 3/4" IPS		22	.364		.93	24.50		25.43	38
8380	1" IPS		19	.421		.97	28.50		29.47	43.50
8390	1-1/4" IPS		18	.444		1.44	30		31.44	46.50
8400	1-1/2" IPS		17	.471		1.70	32		33.70	49.50
8410	2" IPS		16	.500		3.22	34		37.22	54
8430	Elbow, 90°, 3/4" IPS		22	.364		1.86	24.50		26.36	39
8440	1" IPS		19	.421		2.06	28.50		30.56	45
8450	1-1/4" IPS		18	.444		2.30	30		32.30	47.50
8460	1-1/2" IPS		17	.471		2.72	32		34.72	50.50
8470	2" IPS		16	.500		3.79	34		37.79	54.50
8490	Male adapter, 3/4" IPS x 3/4" MPT		25	.320		.93	21.50		22.43	33.50
8500	1" IPS x 1" MPT		21	.381		.96	26		26.96	39.50
8510	1-1/4" IPS x 1-1/4" MPT		20	.400		1.52	27		28.52	42
8520	1-1/2" IPS x 1-1/2" MPT		18	.444		1.70	30		31.70	47
8530	2" IPS x 2" MPT		15	.533		3.27	36		39.27	57.50
8550	Tee, 3/4" IPS		14	.571		1.80	38.50		40.30	59.50
8560	1" IPS		13	.615		2.35	41.50		43.85	64.50
8570	1-1/4" IPS		12	.667		3.66	45		48.66	71.50
8580	1-1/2" IPS		11	.727		4.16	49		53.16	78
8590	2" IPS		10	.800		8.20	54		62.20	90
8610	Insert type, PVC, 100 psi @ 180°F, hot & cold water									
8620	Coupler, male, 3/8" CTS x 3/8" MPT	1 Plum	29	.276	Ea.	.77	18.70		19.47	29

22 11 Facility Water Distribution

22 11 13 – Facility Water Distribution Piping

22 11 13.76 Pipe Fittings, Plastic		Crew	Daily Output	Labor-Hours	Unit	Material	2021 Bare Costs Labor	Equipment	Total	Total Incl O&P
8630	3/8" CTS x 1/2" MPT	1 Plum	28	.286	Ea.	.77	19.35		20.12	30
8640	1/2" CTS x 1/2" MPT		27	.296		.79	20		20.79	31
8650	1/2" CTS x 3/4" MPT		26	.308		2.41	21		23.41	33.50
8660	3/4" CTS x 1/2" MPT		25	.320		2.18	21.50		23.68	35
8670	3/4" CTS x 3/4" MPT		25	.320		1.05	21.50		22.55	33.50
8700	Coupling, 3/8" CTS x 1/2" CTS		25	.320		4.21	21.50		25.71	37
8710	1/2" CTS		23	.348		5.60	23.50		29.10	41
8730	3/4" CTS		22	.364		12.90	24.50		37.40	51
8750	Elbow 90°, 3/8" CTS		25	.320		4.12	21.50		25.62	37
8760	1/2" CTS		23	.348		5.55	23.50		29.05	41
8770	3/4" CTS		22	.364		7.60	24.50		32.10	45.50
8800	Rings, crimp, copper, 3/8" CTS		120	.067		.21	4.51		4.72	7
8810	1/2" CTS		117	.068		.22	4.63		4.85	7.15
8820	3/4" CTS		115	.070		.29	4.71		5	7.35
8850	Reducer tee, bronze, 3/8" x 3/8" x 1/2" CTS		17	.471		4.86	32		36.86	53
8860	1/2" x 1/2" x 3/4" CTS		15	.533		10.60	36		46.60	65.50
8870	3/4" x 1/2" x 1/2" CTS		14	.571		10.45	38.50		48.95	69
8890	3/4" x 3/4" x 1/2" CTS		14	.571		10.45	38.50		48.95	69
8900	1" x 1/2" x 1/2" CTS		14	.571		17.70	38.50		56.20	77
8930	Tee, 3/8" CTS		17	.471		3.08	32		35.08	51
8940	1/2" CTS		15	.533		2.85	36		38.85	57
8950	3/4" CTS	↓	14	.571	↓	3.59	38.50		42.09	61.50
8960	Copper rings included in fitting price									
9000	Flare type, assembled, acetal, hot & cold water									
9010	Coupling, 1/4" & 3/8" CTS	1 Plum	24	.333	Ea.	3.52	22.50		26.02	37.50
9020	1/2" CTS		22	.364		4.03	24.50		28.53	41.50
9030	3/4" CTS		21	.381		5.95	26		31.95	45
9040	1" CTS		18	.444		7.60	30		37.60	53.50
9050	Elbow 90°, 1/4" CTS		26	.308		3.90	21		24.90	35.50
9060	3/8" CTS		24	.333		4.19	22.50		26.69	38
9070	1/2" CTS		22	.364		4.95	24.50		29.45	42.50
9080	3/4" CTS		21	.381		7.60	26		33.60	47
9090	1" CTS		18	.444		10.95	30		40.95	57
9110	Tee, 1/4" CTS		16	.500		4.14	34		38.14	55
9114	3/8" CTS		15	.533		4.63	36		40.63	59
9120	1/2" CTS		14	.571		5.55	38.50		44.05	63.50
9130	3/4" CTS		13	.615		8.65	41.50		50.15	71.50
9140	1" CTS	↓	12	.667	↓	11.60	45		56.60	80.50
9400	Polypropylene, fittings and accessories									
9404	Fittings fusion welded, sizes are ID									
9408	Note: sizes 1/2" thru 4" use socket fusion									
9410	Sizes 6" thru 10" use butt fusion									
9416	Coupling									
9420	3/8"	1 Plum	39	.205	Ea.	.95	13.90		14.85	21.50
9422	1/2"		37.40	.214		1.77	14.50		16.27	23.50
9424	3/4"		35.40	.226		1.97	15.30		17.27	25
9426	1"		29.70	.269		2.60	18.25		20.85	30
9428	1-1/4"		27.60	.290		3.11	19.60		22.71	33
9430	1-1/2"	↓	24.80	.323		6.55	22		28.55	39.50
9432	2"	Q-1	43	.372		13.10	22.50		35.60	48.50
9434	2-1/2"		35.60	.449		14.65	27.50		42.15	57
9436	3"		30.90	.518		32	31.50		63.50	82.50
9438	3-1/2"	↓	27.80	.576		52.50	35		87.50	110

240

22 11 13.76 Pipe Fittings, Plastic	Crew	Daily Output	Labor-Hours	Unit	Material	2021 Bare Costs Labor	Equipment	Total	Total Incl O&P
9440	4"	Q-1	25	.640	Ea.	68.50	39		107.50	134
9442	Reducing coupling, female to female									
9446	2" to 1-1/2"	Q-1	49	.327	Ea.	18.85	19.90		38.75	50
9448	2-1/2" to 2"		41.20	.388		20.50	23.50		44	58.50
9450	3" to 2-1/2"		33.10	.483		28	29.50		57.50	74.50
9470	Reducing bushing, female to female									
9472	1/2" to 3/8"	1 Plum	38.20	.209	Ea.	1.26	14.20		15.46	22.50
9474	3/4" to 3/8" or 1/2"		36.50	.219		1.97	14.85		16.82	24
9476	1" to 3/4" or 1/2"		33.10	.242		2.63	16.35		18.98	27.50
9478	1-1/4" to 3/4" or 1"		28.70	.279		4.05	18.85		22.90	32.50
9480	1-1/2" to 1/2" thru 1-1/4"		26.20	.305		6.70	20.50		27.20	38.50
9482	2" to 1/2" thru 1-1/2"	Q-1	43	.372		13.40	22.50		35.90	49
9484	2-1/2" to 1/2" thru 2"		41.20	.388		15	23.50		38.50	52
9486	3" to 1-1/2" thru 2-1/2"		33.10	.483		33.50	29.50		63	80.50
9488	3-1/2" to 2" thru 3"		29.20	.548		53.50	33.50		87	109
9490	4" to 2-1/2" thru 3-1/2"		26.20	.611		83.50	37		120.50	148
9491	6" to 4" SDR 7.4		16.50	.970		100	59		159	198
9492	6" to 4" SDR 11		16.50	.970		83.50	59		142.50	180
9493	8" to 6" SDR 7.4		10.10	1.584		147	96.50		243.50	305
9494	8" to 6" SDR 11		10.10	1.584		94	96.50		190.50	247
9495	10" to 8" SDR 7.4		7.80	2.051		202	125		327	410
9496	10" to 8" SDR 11		7.80	2.051		139	125		264	340
9500	90° elbow									
9504	3/8"	1 Plum	39	.205	Ea.	1.27	13.90		15.17	22
9506	1/2"		37.40	.214		1.89	14.50		16.39	23.50
9508	3/4"		35.40	.226		2.43	15.30		17.73	25.50
9510	1"		29.70	.269		3.51	18.25		21.76	31
9512	1-1/4"		27.60	.290		5.40	19.60		25	35.50
9514	1-1/2"		24.80	.323		11.65	22		33.65	45.50
9516	2"	Q-1	43	.372		17.85	22.50		40.35	53.50
9518	2-1/2"		35.60	.449		39.50	27.50		67	84.50
9520	3"		30.90	.518		66	31.50		97.50	120
9522	3-1/2"		27.80	.576		93.50	35		128.50	156
9524	4"		25	.640		144	39		183	216
9526	6" SDR 7.4		5.55	2.883		208	176		384	490
9528	6" SDR 11		5.55	2.883		173	176		349	450
9530	8" SDR 7.4	Q-2	8.10	2.963		320	187		507	630
9532	8" SDR 11		8.10	2.963		288	187		475	595
9534	10" SDR 7.4		7.50	3.200		430	202		632	770
9536	10" SDR 11		7.50	3.200		390	202		592	725
9551	45° elbow									
9554	3/8"	1 Plum	39	.205	Ea.	1.27	13.90		15.17	22
9556	1/2"		37.40	.214		1.89	14.50		16.39	23.50
9558	3/4"		35.40	.226		2.43	15.30		17.73	25.50
9564	1"		29.70	.269		3.51	18.25		21.76	31
9566	1-1/4"		27.60	.290		5.40	19.60		25	35.50
9568	1-1/2"		24.80	.323		11.65	22		33.65	45.50
9570	2"	Q-1	43	.372		17.65	22.50		40.15	53.50
9572	2-1/2"		35.60	.449		39	27.50		66.50	84
9574	3"		30.90	.518		72	31.50		103.50	127
9576	3-1/2"		27.80	.576		103	35		138	166
9578	4"		25	.640		158	39		197	232
9580	6" SDR 7.4		5.55	2.883		196	176		372	480

For customer support on your Plumbing Costs with RSMeans Data, call 800.448.8182.

241

22 11 13.76 Pipe Fittings, Plastic	Crew	Daily Output	Labor-Hours	Unit	Material	2021 Bare Costs Labor	Equipment	Total	Total Incl O&P
9582	6" SDR 11	Q-1	5.55	2.883	Ea.	169	176		345	450
9584	8" SDR 7.4	Q-2	8.10	2.963		289	187		476	600
9586	8" SDR 11		8.10	2.963		272	187		459	580
9588	10" SDR 7.4		7.50	3.200		465	202		667	810
9590	10" SDR 11		7.50	3.200		360	202		562	695
9600	Tee									
9604	3/8"	1 Plum	26	.308	Ea.	1.69	21		22.69	33
9606	1/2"		24.90	.321		2.55	22		24.55	35.50
9608	3/4"		23.70	.338		3.51	23		26.51	38
9610	1"		19.90	.402		4.46	27		31.46	45.50
9612	1-1/4"		18.50	.432		6.80	29.50		36.30	51
9614	1-1/2"		16.60	.482		19.45	32.50		51.95	70
9616	2"	Q-1	26.80	.597		28	36.50		64.50	85
9618	2-1/2"		23.80	.672		46.50	41		87.50	113
9620	3"		20.60	.777		77	47.50		124.50	155
9622	3-1/2"		18.40	.870		120	53		173	211
9624	4"		16.70	.958		160	58.50		218.50	263
9626	6" SDR 7.4		3.70	4.324		220	263		483	635
9628	6" SDR 11		3.70	4.324		176	263		439	590
9630	8" SDR 7.4	Q-2	5.40	4.444		385	281		666	845
9632	8" SDR 11		5.40	4.444		355	281		636	815
9634	10" SDR 7.4		5	4.800		665	305		970	1,175
9636	10" SDR 11		5	4.800		560	305		865	1,075
9638	For reducing tee use same tee price									
9660	End cap									
9662	3/8"	1 Plum	78	.103	Ea.	1.80	6.95		8.75	12.35
9664	1/2"		74.60	.107		2.77	7.25		10.02	13.90
9666	3/4"		71.40	.112		3.51	7.60		11.11	15.15
9668	1"		59.50	.134		4.25	9.10		13.35	18.30
9670	1-1/4"		55.60	.144		6.70	9.75		16.45	22
9672	1-1/2"		49.50	.162		9.25	10.95		20.20	26.50
9674	2"	Q-1	80	.200		15.45	12.20		27.65	35
9676	2-1/2"		71.40	.224		22.50	13.65		36.15	45
9678	3"		61.70	.259		50.50	15.80		66.30	79
9680	3-1/2"		55.20	.290		61	17.65		78.65	93.50
9682	4"		50	.320		93	19.50		112.50	131
9684	6" SDR 7.4		26.50	.604		127	37		164	195
9686	6" SDR 11		26.50	.604		98.50	37		135.50	163
9688	8" SDR 7.4	Q-2	16.30	1.472		128	93		221	280
9690	8" SDR 11		16.30	1.472		106	93		199	256
9692	10" SDR 7.4		14.90	1.611		191	102		293	360
9694	10" SDR 11		14.90	1.611		131	102		233	296
9800	Accessories and tools									
9802	Pipe clamps for suspension, not including rod or beam clamp									
9804	3/8"	1 Plum	74	.108	Ea.	2.32	7.30		9.62	13.45
9805	1/2"		70	.114		2.91	7.75		10.66	14.75
9806	3/4"		68	.118		2.95	7.95		10.90	15.15
9807	1"		66	.121		3.66	8.20		11.86	16.30
9808	1-1/4"		64	.125		3.75	8.45		12.20	16.80
9809	1-1/2"		62	.129		4.10	8.75		12.85	17.55
9810	2"	Q-1	110	.145		5.10	8.85		13.95	18.85
9811	2-1/2"		104	.154		6.50	9.35		15.85	21
9812	3"		98	.163		7.05	9.95		17	22.50

For customer support on your Plumbing Costs with RSMeans Data, call 800.448.8182.

22 11 13.76 Pipe Fittings, Plastic

		Crew	Daily Output	Labor-Hours	Unit	Material	2021 Bare Costs Labor	Equipment	Total	Total Incl O&P
9813	3-1/2"	Q-1	92	.174	Ea.	7.70	10.60		18.30	24.50
9814	4"		86	.186		8.65	11.35		20	26.50
9815	6"		70	.229		10.55	13.95		24.50	32.50
9816	8"	Q-2	100	.240		37.50	15.15		52.65	64
9817	10"	"	94	.255		43	16.15		59.15	71.50
9820	Pipe cutter									
9822	For 3/8" thru 1-1/4"				Ea.	116			116	128
9824	For 1-1/2" thru 4"				"	305			305	335
9826	Note: Pipes may be cut with standard									
9827	iron saw with blades for plastic.									
9982	For plastic hangers see Line 22 05 29.10 8000									
9986	For copper/brass fittings see Line 22 11 13.25 7000									

22 11 13.78 Pipe, High Density Polyethylene Plastic (HDPE)

		Crew	Daily Output	Labor-Hours	Unit	Material	2021 Bare Costs Labor	Equipment	Total	Total Incl O&P
0010	**PIPE, HIGH DENSITY POLYETHYLENE PLASTIC (HDPE)**									
0020	Not incl. hangers, trenching, backfill, hoisting or digging equipment.									
0030	Standard length is 40', add a weld for each joint									
0035	For HDPE weld machine see 015433401685 in equipment rental									
0040	Single wall									
0050	Straight									
0054	1" diameter DR 11				L.F.	.91			.91	1
0058	1-1/2" diameter DR 11					1.17			1.17	1.29
0062	2" diameter DR 11					1.93			1.93	2.12
0066	3" diameter DR 11					2.34			2.34	2.57
0070	3" diameter DR 17					1.87			1.87	2.06
0074	4" diameter DR 11					3.92			3.92	4.31
0078	4" diameter DR 17					3.92			3.92	4.31
0082	6" diameter DR 11					9.75			9.75	10.70
0086	6" diameter DR 17					6.40			6.40	7.05
0090	8" diameter DR 11					16.15			16.15	17.80
0094	8" diameter DR 26					7.60			7.60	8.35
0098	10" diameter DR 11					25.50			25.50	28
0102	10" diameter DR 26					11.70			11.70	12.85
0106	12" diameter DR 11					37			37	40.50
0110	12" diameter DR 26					17.55			17.55	19.30
0114	16" diameter DR 11					56.50			56.50	62
0118	16" diameter DR 26					25.50			25.50	28
0122	18" diameter DR 11					72			72	79.50
0126	18" diameter DR 26					33			33	36.50
0130	20" diameter DR 11					87.50			87.50	96.50
0134	20" diameter DR 26					39			39	43
0138	22" diameter DR 11					107			107	118
0142	22" diameter DR 26					48.50			48.50	53.50
0146	24" diameter DR 11					127			127	139
0150	24" diameter DR 26					56.50			56.50	62
0154	28" diameter DR 17					117			117	129
0158	28" diameter DR 26					78			78	86
0162	30" diameter DR 21					109			109	120
0166	30" diameter DR 26					89.50			89.50	98.50
0170	36" diameter DR 26					89.50			89.50	98.50
0174	42" diameter DR 26					173			173	191
0178	48" diameter DR 26					228			228	251
0182	54" diameter DR 26					286			286	315

For customer support on your Plumbing Costs with RSMeans Data, call 800.448.8182.

243

22 11 13.78 Pipe, High Density Polyethylene Plastic (HDPE)	Crew	Daily Output	Labor-Hours	Unit	Material	2021 Bare Costs Labor	Equipment	Total	Total Incl O&P
0300	90° elbow									
0304	1" diameter DR 11				Ea.	5.65			5.65	6.20
0308	1-1/2" diameter DR 11					7.05			7.05	7.75
0312	2" diameter DR 11					7.05			7.05	7.75
0316	3" diameter DR 11					14.10			14.10	15.55
0320	3" diameter DR 17					14.10			14.10	15.55
0324	4" diameter DR 11					19.75			19.75	22
0328	4" diameter DR 17					19.75			19.75	22
0332	6" diameter DR 11					45			45	49.50
0336	6" diameter DR 17					45			45	49.50
0340	8" diameter DR 11					112			112	123
0344	8" diameter DR 26					99			99	109
0348	10" diameter DR 11					415			415	460
0352	10" diameter DR 26					385			385	425
0356	12" diameter DR 11					440			440	485
0360	12" diameter DR 26					400			400	440
0364	16" diameter DR 11					520			520	575
0368	16" diameter DR 26					515			515	565
0372	18" diameter DR 11					655			655	720
0376	18" diameter DR 26					615			615	680
0380	20" diameter DR 11					770			770	845
0384	20" diameter DR 26					745			745	820
0388	22" diameter DR 11					795			795	875
0392	22" diameter DR 26					770			770	845
0396	24" diameter DR 11					900			900	990
0400	24" diameter DR 26					875			875	960
0404	28" diameter DR 17					1,100			1,100	1,200
0408	28" diameter DR 26					1,025			1,025	1,125
0412	30" diameter DR 17					1,550			1,550	1,700
0416	30" diameter DR 26					1,400			1,400	1,550
0420	36" diameter DR 26					1,800			1,800	1,975
0424	42" diameter DR 26					2,300			2,300	2,550
0428	48" diameter DR 26					2,700			2,700	2,975
0432	54" diameter DR 26					6,425			6,425	7,050
0500	45° elbow									
0512	2" diameter DR 11				Ea.	5.65			5.65	6.25
0516	3" diameter DR 11					14.10			14.10	15.55
0520	3" diameter DR 17					14.10			14.10	15.55
0524	4" diameter DR 11					19.75			19.75	22
0528	4" diameter DR 17					19.75			19.75	22
0532	6" diameter DR 11					45			45	49.50
0536	6" diameter DR 17					45			45	49.50
0540	8" diameter DR 11					112			112	123
0544	8" diameter DR 26					65			65	71.50
0548	10" diameter DR 11					415			415	460
0552	10" diameter DR 26					385			385	425
0556	12" diameter DR 11					440			440	485
0560	12" diameter DR 26					400			400	440
0564	16" diameter DR 11					231			231	254
0568	16" diameter DR 26					218			218	240
0572	18" diameter DR 11					245			245	270
0576	18" diameter DR 26					225			225	247
0580	20" diameter DR 11					370			370	410

22 11 Facility Water Distribution

22 11 13 – Facility Water Distribution Piping

22 11 13.78 Pipe, High Density Polyethylene Plastic (HDPE)	Crew	Daily Output	Labor-Hours	Unit	Material	2021 Bare Costs Labor	Equipment	Total	Total Incl O&P
0584	20" diameter DR 26				Ea.	355			355	390
0588	22" diameter DR 11					515			515	565
0592	22" diameter DR 26					490			490	535
0596	24" diameter DR 11					630			630	690
0600	24" diameter DR 26					605			605	665
0604	28" diameter DR 17					715			715	785
0608	28" diameter DR 26					695			695	765
0612	30" diameter DR 17					885			885	975
0616	30" diameter DR 26					860			860	945
0620	36" diameter DR 26					1,100			1,100	1,200
0624	42" diameter DR 26					1,400			1,400	1,550
0628	48" diameter DR 26					1,525			1,525	1,675
0632	54" diameter DR 26					2,050			2,050	2,250
0700	Tee									
0704	1" diameter DR 11				Ea.	7.35			7.35	8.10
0708	1-1/2" diameter DR 11					10.35			10.35	11.35
0712	2" diameter DR 11					8.85			8.85	9.75
0716	3" diameter DR 11					16.25			16.25	17.85
0720	3" diameter DR 17					16.25			16.25	17.85
0724	4" diameter DR 11					23.50			23.50	26
0728	4" diameter DR 17					23.50			23.50	26
0732	6" diameter DR 11					59			59	65
0736	6" diameter DR 17					59			59	65
0740	8" diameter DR 11					146			146	161
0744	8" diameter DR 17					146			146	161
0748	10" diameter DR 11					435			435	480
0752	10" diameter DR 17					435			435	480
0756	12" diameter DR 11					580			580	640
0760	12" diameter DR 17					580			580	640
0764	16" diameter DR 11					305			305	335
0768	16" diameter DR 17					253			253	278
0772	18" diameter DR 11					435			435	475
0776	18" diameter DR 17					355			355	390
0780	20" diameter DR 11					530			530	580
0784	20" diameter DR 17					435			435	475
0788	22" diameter DR 11					680			680	745
0792	22" diameter DR 17					535			535	590
0796	24" diameter DR 11					840			840	925
0800	24" diameter DR 17					710			710	780
0804	28" diameter DR 17					1,375			1,375	1,500
0812	30" diameter DR 17					1,575			1,575	1,725
0820	36" diameter DR 17					2,600			2,600	2,850
0824	42" diameter DR 26					2,900			2,900	3,175
0828	48" diameter DR 26					3,100			3,100	3,425
1000	Flange adptr, w/back-up ring and 1/2 cost of plated bolt set									
1004	1" diameter DR 11				Ea.	28			28	31
1008	1-1/2" diameter DR 11					28			28	31
1012	2" diameter DR 11					17.70			17.70	19.50
1016	3" diameter DR 11					20.50			20.50	23
1020	3" diameter DR 17					20.50			20.50	23
1024	4" diameter DR 11					28			28	31
1028	4" diameter DR 17					28			28	31
1032	6" diameter DR 11					40			40	44

For customer support on your Plumbing Costs with RSMeans Data, call 800.448.8182.

245

22 11 13.78 Pipe, High Density Polyethylene Plastic (HDPE)	Crew	Daily Output	Labor-Hours	Unit	Material	2021 Bare Costs Labor	Equipment	Total	Total Incl O&P
1036	6" diameter DR 17				Ea.	40			40	44
1040	8" diameter DR 11					57.50			57.50	63.50
1044	8" diameter DR 26					57.50			57.50	63.50
1048	10" diameter DR 11					91.50			91.50	101
1052	10" diameter DR 26					91.50			91.50	101
1056	12" diameter DR 11					134			134	148
1060	12" diameter DR 26					134			134	148
1064	16" diameter DR 11					288			288	315
1068	16" diameter DR 26					288			288	315
1072	18" diameter DR 11					375			375	410
1076	18" diameter DR 26					375			375	410
1080	20" diameter DR 11					520			520	570
1084	20" diameter DR 26					520			520	570
1088	22" diameter DR 11					560			560	620
1092	22" diameter DR 26					560			560	615
1096	24" diameter DR 17					610			610	670
1100	24" diameter DR 32.5					610			610	670
1104	28" diameter DR 15.5					825			825	910
1108	28" diameter DR 32.5					825			825	910
1112	30" diameter DR 11					960			960	1,050
1116	30" diameter DR 21					960			960	1,050
1120	36" diameter DR 26					1,050			1,050	1,150
1124	42" diameter DR 26					1,200			1,200	1,325
1128	48" diameter DR 26					1,450			1,450	1,600
1132	54" diameter DR 26				▼	1,775			1,775	1,950
1200	Reducer									
1208	2" x 1-1/2" diameter DR 11				Ea.	8.50			8.50	9.35
1212	3" x 2" diameter DR 11					8.50			8.50	9.35
1216	4" x 2" diameter DR 11					9.90			9.90	10.85
1220	4" x 3" diameter DR 11					12.70			12.70	14
1224	6" x 4" diameter DR 11					29.50			29.50	32.50
1228	8" x 6" diameter DR 11					45			45	49.50
1232	10" x 8" diameter DR 11					77.50			77.50	85.50
1236	12" x 8" diameter DR 11					127			127	140
1240	12" x 10" diameter DR 11					102			102	112
1244	14" x 12" diameter DR 11					113			113	124
1248	16" x 14" diameter DR 11					144			144	158
1252	18" x 16" diameter DR 11					175			175	193
1256	20" x 18" diameter DR 11					345			345	380
1260	22" x 20" diameter DR 11					425			425	470
1264	24" x 22" diameter DR 11					480			480	530
1268	26" x 24" diameter DR 11					565			565	620
1272	28" x 24" diameter DR 11					720			720	790
1276	32" x 28" diameter DR 17					930			930	1,025
1280	36" x 32" diameter DR 17				▼	1,275			1,275	1,400
4000	Welding labor per joint, not including welding machine									
4010	Pipe joint size (cost based on thickest wall for each diam.)									
4030	1" pipe size	4 Skwk	273	.117	Ea.		6.70		6.70	10.05
4040	1-1/2" pipe size		175	.183			10.45		10.45	15.70
4050	2" pipe size		128	.250			14.30		14.30	21.50
4060	3" pipe size		100	.320			18.25		18.25	27.50
4070	4" pipe size	▼	77	.416			23.50		23.50	35.50
4080	6" pipe size	5 Skwk	63	.635	▼		36.50		36.50	54.50

22 11 13.78 Pipe, High Density Polyethylene Plastic (HDPE)		Crew	Daily Output	Labor-Hours	Unit	Material	2021 Bare Costs Labor	Equipment	Total	Total Incl O&P
4090	8" pipe size	5 Skwk	48	.833	Ea.		47.50		47.50	71.50
4100	10" pipe size	↓	40	1			57		57	86
4110	12" pipe size	6 Skwk	41	1.171			67		67	101
4120	16" pipe size		34	1.412			80.50		80.50	121
4130	18" pipe size	↓	32	1.500			85.50		85.50	129
4140	20" pipe size	8 Skwk	37	1.730			99		99	149
4150	22" pipe size		35	1.829			104		104	157
4160	24" pipe size		34	1.882			107		107	162
4170	28" pipe size		33	1.939			111		111	167
4180	30" pipe size		32	2			114		114	172
4190	36" pipe size		31	2.065			118		118	177
4200	42" pipe size		30	2.133			122		122	183
4210	48" pipe size	9 Skwk	33	2.182			125		125	187
4220	54" pipe size	"	31	2.323			133		133	200
5000	Dual wall contained pipe									
5040	Straight									
5054	1" DR 11 x 3" DR 11				L.F.	5.35			5.35	5.85
5058	1" DR 11 x 4" DR 11					5.60			5.60	6.20
5062	1-1/2" DR 11 x 4" DR 17					6.20			6.20	6.80
5066	2" DR 11 x 4" DR 17					6.55			6.55	7.20
5070	2" DR 11 x 6" DR 17					9.90			9.90	10.90
5074	3" DR 11 x 6" DR 17					11.10			11.10	12.20
5078	3" DR 11 x 6" DR 26					9.45			9.45	10.40
5086	3" DR 17 x 8" DR 17					12.05			12.05	13.30
5090	4" DR 11 x 8" DR 17					16.75			16.75	18.40
5094	4" DR 17 x 8" DR 26					13.10			13.10	14.40
5098	6" DR 11 x 10" DR 17					26			26	28.50
5102	6" DR 17 x 10" DR 26					17.65			17.65	19.45
5106	6" DR 26 x 10" DR 26					16.55			16.55	18.20
5110	8" DR 17 x 12" DR 26					25.50			25.50	28
5114	8" DR 26 x 12" DR 32.5					23.50			23.50	25.50
5118	10" DR 17 x 14" DR 26					38			38	42
5122	10" DR 17 x 16" DR 26					27.50			27.50	30
5126	10" DR 26 x 16" DR 26					36.50			36.50	40
5130	12" DR 26 x 16" DR 26					38			38	42
5134	12" DR 17 x 18" DR 26					53			53	58.50
5138	12" DR 26 x 18" DR 26					50.50			50.50	55.50
5142	14" DR 26 x 20" DR 32.5					51			51	56.50
5146	16" DR 26 x 22" DR 32.5					56.50			56.50	62.50
5150	18" DR 26 x 24" DR 32.5					70.50			70.50	77.50
5154	20" DR 32.5 x 28" DR 32.5					79.50			79.50	87.50
5158	22" DR 32.5 x 30" DR 32.5					79.50			79.50	87.50
5162	24" DR 32.5 x 32" DR 32.5					83			83	91
5166	36" DR 32.5 x 42" DR 32.5					97			97	107
5300	Force transfer coupling									
5354	1" DR 11 x 3" DR 11				Ea.	185			185	203
5358	1" DR 11 x 4" DR 17					217			217	239
5362	1-1/2" DR 11 x 4" DR 17					208			208	228
5366	2" DR 11 x 4" DR 17					208			208	228
5370	2" DR 11 x 6" DR 17					355			355	390
5374	3" DR 11 x 6" DR 17					380			380	420
5378	3" DR 11 x 6" DR 26					380			380	420
5382	3" DR 11 x 8" DR 11					380			380	420

22 11 13.78 Pipe, High Density Polyethylene Plastic (HDPE)	Crew	Daily Output	Labor-Hours	Unit	Material	2021 Bare Costs Labor	Equipment	Total	Total Incl O&P
5386	3" DR 11 x 8" DR 17				Ea.	415			415	460
5390	4" DR 11 x 8" DR 17					425			425	470
5394	4" DR 17 x 8" DR 26					430			430	475
5398	6" DR 11 x 10" DR 17					510			510	560
5402	6" DR 17 x 10" DR 26					485			485	535
5406	6" DR 26 x 10" DR 26					665			665	730
5410	8" DR 17 x 12" DR 26					705			705	775
5414	8" DR 26 x 12" DR 32.5					815			815	895
5418	10" DR 17 x 14" DR 26					960			960	1,050
5422	10" DR 17 x 16" DR 26					975			975	1,075
5426	10" DR 26 x 16" DR 26					1,075			1,075	1,200
5430	12" DR 26 x 16" DR 26					1,325			1,325	1,450
5434	12" DR 17 x 18" DR 26					1,325			1,325	1,450
5438	12" DR 26 x 18" DR 26					1,450			1,450	1,600
5442	14" DR 26 x 20" DR 32.5					1,600			1,600	1,775
5446	16" DR 26 x 22" DR 32.5					1,775			1,775	1,950
5450	18" DR 26 x 24" DR 32.5					1,975			1,975	2,175
5454	20" DR 32.5 x 28" DR 32.5					2,150			2,150	2,375
5458	22" DR 32.5 x 30" DR 32.5					2,375			2,375	2,625
5462	24" DR 32.5 x 32" DR 32.5					2,625			2,625	2,900
5466	36" DR 32.5 x 42" DR 32.5					2,825			2,825	3,100
5600	90° elbow									
5654	1" DR 11 x 3" DR 11				Ea.	202			202	222
5658	1" DR 11 x 4" DR 17					238			238	262
5662	1-1/2" DR 11 x 4" DR 17					245			245	270
5666	2" DR 11 x 4" DR 17					240			240	264
5670	2" DR 11 x 6" DR 17					345			345	380
5674	3" DR 11 x 6" DR 17					375			375	415
5678	3" DR 17 x 6" DR 26					380			380	420
5682	3" DR 17 x 8" DR 11					510			510	560
5686	3" DR 17 x 8" DR 17					545			545	600
5690	4" DR 11 x 8" DR 17					460			460	510
5694	4" DR 17 x 8" DR 26					445			445	490
5698	6" DR 11 x 10" DR 17					805			805	890
5702	6" DR 17 x 10" DR 26					820			820	900
5706	6" DR 26 x 10" DR 26					800			800	880
5710	8" DR 17 x 12" DR 26					1,075			1,075	1,175
5714	8" DR 26 x 12" DR 32.5					885			885	975
5718	10" DR 17 x 14" DR 26					1,050			1,050	1,150
5722	10" DR 17 x 16" DR 26					1,025			1,025	1,125
5726	10" DR 26 x 16" DR 26					1,025			1,025	1,125
5730	12" DR 26 x 16" DR 26					1,050			1,050	1,150
5734	12" DR 17 x 18" DR 26					1,150			1,150	1,250
5738	12" DR 26 x 18" DR 26					1,150			1,150	1,275
5742	14" DR 26 x 20" DR 32.5					1,250			1,250	1,375
5746	16" DR 26 x 22" DR 32.5					1,300			1,300	1,425
5750	18" DR 26 x 24" DR 32.5					1,350			1,350	1,475
5754	20" DR 32.5 x 28" DR 32.5					1,400			1,400	1,525
5758	22" DR 32.5 x 30" DR 32.5					1,425			1,425	1,575
5762	24" DR 32.5 x 32" DR 32.5					1,475			1,475	1,625
5766	36" DR 32.5 x 42" DR 32.5					1,525			1,525	1,700
5800	45° elbow									
5804	1" DR 11 x 3" DR 11				Ea.	234			234	258

22 11 Facility Water Distribution

22 11 13 – Facility Water Distribution Piping

22 11 13.78 Pipe, High Density Polyethylene Plastic (HDPE)	Crew	Daily Output	Labor-Hours	Unit	Material	2021 Bare Costs Labor	Equipment	Total	Total Incl O&P
5808	1" DR 11 x 4" DR 17				Ea.	192			192	211
5812	1-1/2" DR 11 x 4" DR 17					195			195	215
5816	2" DR 11 x 4" DR 17					186			186	205
5820	2" DR 11 x 6" DR 17					267			267	294
5824	3" DR 11 x 6" DR 17					287			287	315
5828	3" DR 17 x 6" DR 26					289			289	320
5832	3" DR 17 x 8" DR 11					370			370	405
5836	3" DR 17 x 8" DR 17					415			415	455
5840	4" DR 11 x 8" DR 17					340			340	375
5844	4" DR 17 x 8" DR 26					345			345	380
5848	6" DR 11 x 10" DR 17					535			535	590
5852	6" DR 17 x 10" DR 26					545			545	600
5856	6" DR 26 x 10" DR 26					610			610	675
5860	8" DR 17 x 12" DR 26					725			725	800
5864	8" DR 26 x 12" DR 32.5					680			680	750
5868	10" DR 17 x 14" DR 26					770			770	845
5872	10" DR 17 x 16" DR 26					740			740	815
5876	10" DR 26 x 16" DR 26					785			785	865
5880	12" DR 26 x 16" DR 26					710			710	780
5884	12" DR 17 x 18" DR 26					1,000			1,000	1,100
5888	12" DR 26 x 18" DR 26					895			895	985
5892	14" DR 26 x 20" DR 32.5					970			970	1,075
5896	16" DR 26 x 22" DR 32.5					1,000			1,000	1,100
5900	18" DR 26 x 24" DR 32.5					1,075			1,075	1,175
5904	20" DR 32.5 x 28" DR 32.5					1,050			1,050	1,175
5908	22" DR 32.5 x 30" DR 32.5					1,100			1,100	1,200
5912	24" DR 32.5 x 32" DR 32.5					1,150			1,150	1,250
5916	36" DR 32.5 x 42" DR 32.5					1,175			1,175	1,300
6000	Access port with 4" riser									
6050	1" DR 11 x 4" DR 17				Ea.	279			279	305
6054	1-1/2" DR 11 x 4" DR 17					281			281	310
6058	2" DR 11 x 6" DR 17					340			340	375
6062	3" DR 11 x 6" DR 17					350			350	380
6066	3" DR 17 x 6" DR 26					335			335	370
6070	3" DR 17 x 8" DR 11					350			350	380
6074	3" DR 17 x 8" DR 17					360			360	395
6078	4" DR 11 x 8" DR 17					370			370	410
6082	4" DR 17 x 8" DR 26					355			355	390
6086	6" DR 11 x 10" DR 17					435			435	475
6090	6" DR 17 x 10" DR 26					395			395	435
6094	6" DR 26 x 10" DR 26					415			415	455
6098	8" DR 17 x 12" DR 26					445			445	490
6102	8" DR 26 x 12" DR 32.5					440			440	480
6200	End termination with vent plug									
6204	1" DR 11 x 3" DR 11				Ea.	94.50			94.50	104
6208	1" DR 11 x 4" DR 17					138			138	152
6212	1-1/2" DR 11 x 4" DR 17					138			138	152
6216	2" DR 11 x 4" DR 17					138			138	152
6220	2" DR 11 x 6" DR 17					280			280	310
6224	3" DR 11 x 6" DR 17					280			280	310
6228	3" DR 17 x 6" DR 26					280			280	310
6232	3" DR 17 x 8" DR 11					271			271	298
6236	3" DR 17 x 8" DR 17					305			305	335

22 11 13.78 Pipe, High Density Polyethylene Plastic (HDPE)	Crew	Daily Output	Labor-Hours	Unit	Material	2021 Bare Costs Labor	Equipment	Total	Total Incl O&P
6240	4" DR 11 x 8" DR 17				Ea.	360			360	400
6244	4" DR 17 x 8" DR 26					340			340	375
6248	6" DR 11 x 10" DR 17					395			395	435
6252	6" DR 17 x 10" DR 26					365			365	400
6256	6" DR 26 x 10" DR 26					530			530	585
6260	8" DR 17 x 12" DR 26					555			555	610
6264	8" DR 26 x 12" DR 32.5					665			665	735
6268	10" DR 17 x 14" DR 26					885			885	970
6272	10" DR 17 x 16" DR 26					835			835	915
6276	10" DR 26 x 16" DR 26					940			940	1,025
6280	12" DR 26 x 16" DR 26					1,050			1,050	1,150
6284	12" DR 17 x 18" DR 26					1,225			1,225	1,325
6288	12" DR 26 x 18" DR 26					1,325			1,325	1,450
6292	14" DR 26 x 20" DR 32.5					1,475			1,475	1,625
6296	16" DR 26 x 22" DR 32.5					1,650			1,650	1,825
6300	18" DR 26 x 24" DR 32.5					1,875			1,875	2,050
6304	20" DR 32.5 x 28" DR 32.5					2,075			2,075	2,275
6308	22" DR 32.5 x 30" DR 32.5					2,325			2,325	2,550
6312	24" DR 32.5 x 32" DR 32.5					2,600			2,600	2,875
6316	36" DR 32.5 x 42" DR 32.5				▼	2,875			2,875	3,150
6600	Tee									
6604	1" DR 11 x 3" DR 11				Ea.	235			235	259
6608	1" DR 11 x 4" DR 17					255			255	281
6612	1-1/2" DR 11 x 4" DR 17					264			264	290
6616	2" DR 11 x 4" DR 17					240			240	263
6620	2" DR 11 x 6" DR 17					385			385	420
6624	3" DR 11 x 6" DR 17					450			450	500
6628	3" DR 17 x 6" DR 26					430			430	470
6632	3" DR 17 x 8" DR 11					525			525	580
6636	3" DR 17 x 8" DR 17					570			570	625
6640	4" DR 11 x 8" DR 17					685			685	755
6644	4" DR 17 x 8" DR 26					690			690	755
6648	6" DR 11 x 10" DR 17					895			895	985
6652	6" DR 17 x 10" DR 26					895			895	985
6656	6" DR 26 x 10" DR 26					1,100			1,100	1,200
6660	8" DR 17 x 12" DR 26					1,100			1,100	1,200
6664	8" DR 26 x 12" DR 32.5					1,425			1,425	1,550
6668	10" DR 17 x 14" DR 26					1,650			1,650	1,825
6672	10" DR 17 x 16" DR 26					1,825			1,825	2,025
6676	10" DR 26 x 16" DR 26					1,825			1,825	2,025
6680	12" DR 26 x 16" DR 26					2,100			2,100	2,300
6684	12" DR 17 x 18" DR 26					2,100			2,100	2,300
6688	12" DR 26 x 18" DR 26					2,375			2,375	2,600
6692	14" DR 26 x 20" DR 32.5					2,675			2,675	2,950
6696	16" DR 26 x 22" DR 32.5					3,025			3,025	3,325
6700	18" DR 26 x 24" DR 32.5					3,450			3,450	3,775
6704	20" DR 32.5 x 28" DR 32.5					3,875			3,875	4,275
6708	22" DR 32.5 x 30" DR 32.5					4,400			4,400	4,850
6712	24" DR 32.5 x 32" DR 32.5					4,975			4,975	5,475
6716	36" DR 32.5 x 42" DR 32.5				▼	5,725			5,725	6,275
6800	Wye									
6816	2" DR 11 x 4" DR 17				Ea.	585			585	640
6820	2" DR 11 x 6" DR 17				▼	720			720	795

22 11 13 – Facility Water Distribution Piping

22 11 13.78 Pipe, High Density Polyethylene Plastic (HDPE)	Crew	Daily Output	Labor-Hours	Unit	Material	2021 Bare Costs Labor	Equipment	Total	Total Incl O&P
6824	3" DR 11 x 6" DR 17				Ea.	765			765	840
6828	3" DR 17 x 6" DR 26					765			765	840
6832	3" DR 17 x 8" DR 11					815			815	895
6836	3" DR 17 x 8" DR 17					885			885	970
6840	4" DR 11 x 8" DR 17					970			970	1,075
6844	4" DR 17 x 8" DR 26					610			610	670
6848	6" DR 11 x 10" DR 17					1,500			1,500	1,650
6852	6" DR 17 x 10" DR 26					935			935	1,025
6856	6" DR 26 x 10" DR 26					1,300			1,300	1,425
6860	8" DR 17 x 12" DR 26					1,825			1,825	2,000
6864	8" DR 26 x 12" DR 32.5					1,500			1,500	1,650
6868	10" DR 17 x 14" DR 26					1,600			1,600	1,750
6872	10" DR 17 x 16" DR 26					1,725			1,725	1,900
6876	10" DR 26 x 16" DR 26					1,875			1,875	2,050
6880	12" DR 26 x 16" DR 26					2,025			2,025	2,225
6884	12" DR 17 x 18" DR 26					2,175			2,175	2,375
6888	12" DR 26 x 18" DR 26					2,350			2,350	2,575
6892	14" DR 26 x 20" DR 32.5					2,525			2,525	2,775
6896	16" DR 26 x 22" DR 32.5					2,650			2,650	2,925
6900	18" DR 26 x 24" DR 32.5					3,025			3,025	3,350
6904	20" DR 32.5 x 28" DR 32.5					3,175			3,175	3,500
6908	22" DR 32.5 x 30" DR 32.5					3,425			3,425	3,775
6912	24" DR 32.5 x 32" DR 32.5					3,700			3,700	4,075
9000	Welding labor per joint, not including welding machine									
9010	Pipe joint size, outer pipe (cost based on the thickest walls)									
9020	Straight pipe									
9050	3" pipe size	4 Skwk	96	.333	Ea.		19.05		19.05	28.50
9060	4" pipe size	"	77	.416			23.50		23.50	35.50
9070	6" pipe size	5 Skwk	60	.667			38		38	57.50
9080	8" pipe size	"	40	1			57		57	86
9090	10" pipe size	6 Skwk	41	1.171			67		67	101
9100	12" pipe size		39	1.231			70.50		70.50	106
9110	14" pipe size		38	1.263			72		72	109
9120	16" pipe size		35	1.371			78.50		78.50	118
9130	18" pipe size	8 Skwk	45	1.422			81		81	122
9140	20" pipe size		42	1.524			87		87	131
9150	22" pipe size		40	1.600			91.50		91.50	137
9160	24" pipe size		38	1.684			96		96	145
9170	28" pipe size		37	1.730			99		99	149
9180	30" pipe size		36	1.778			102		102	153
9190	32" pipe size		35	1.829			104		104	157
9200	42" pipe size		32	2			114		114	172

22 11 19 – Domestic Water Piping Specialties

22 11 19.10 Flexible Connectors

		Crew	Daily Output	Labor-Hours	Unit	Material	Labor	Equipment	Total	Total Incl O&P
0010	**FLEXIBLE CONNECTORS**, Corrugated, 5/8" OD, 3/4" ID									
0050	Gas, seamless brass, steel fittings									
0200	12" long	1 Plum	36	.222	Ea.	20.50	15.05		35.55	45.50
0220	18" long		36	.222		25.50	15.05		40.55	51
0240	24" long		34	.235		30.50	15.95		46.45	57.50
0260	30" long		34	.235		33	15.95		48.95	60
0280	36" long		32	.250		36.50	16.95		53.45	65.50
0320	48" long		30	.267		46	18.05		64.05	77.50

For customer support on your Plumbing Costs with RSMeans Data, call 800.448.8182.

251

22 11 Facility Water Distribution

22 11 19 – Domestic Water Piping Specialties

22 11 19.10 Flexible Connectors		Crew	Daily Output	Labor-Hours	Unit	Material	2021 Bare Costs Labor	Equipment	Total	Total Incl O&P
0340	60" long	1 Plum	30	.267	Ea.	55	18.05		73.05	87
0360	72" long	↓	30	.267	↓	63.50	18.05		81.55	96.50
2000	Water, copper tubing, dielectric separators									
2100	12" long	1 Plum	36	.222	Ea.	8.30	15.05		23.35	31.50
2220	15" long		36	.222		9.20	15.05		24.25	32.50
2240	18" long		36	.222		8.50	15.05		23.55	32
2260	24" long	↓	34	.235	↓	12.30	15.95		28.25	37.50

22 11 19.14 Flexible Metal Hose

		Crew	Daily Output	Labor-Hours	Unit	Material	2021 Bare Costs Labor	Equipment	Total	Total Incl O&P
0010	**FLEXIBLE METAL HOSE**, Connectors, standard lengths									
0100	Bronze braided, bronze ends									
0120	3/8" diameter x 12"	1 Stpi	26	.308	Ea.	28.50	21		49.50	62.50
0140	1/2" diameter x 12"		24	.333		28.50	23		51.50	65.50
0160	3/4" diameter x 12"		20	.400		36.50	27.50		64	81.50
0180	1" diameter x 18"		19	.421		51	29		80	99
0200	1-1/2" diameter x 18"		13	.615		57	42		99	126
0220	2" diameter x 18"	↓	11	.727	↓	96.50	49.50		146	180
1000	Carbon steel ends									
1020	1/4" diameter x 12"	1 Stpi	28	.286	Ea.	38.50	19.55		58.05	71.50
1040	3/8" diameter x 12"		26	.308		39.50	21		60.50	75
1060	1/2" diameter x 12"		24	.333		34.50	23		57.50	72
1080	1/2" diameter x 24"		24	.333		59.50	23		82.50	99.50
1120	3/4" diameter x 12"		20	.400		47	27.50		74.50	93
1140	3/4" diameter x 24"		20	.400		70.50	27.50		98	119
1160	3/4" diameter x 36"		20	.400		83	27.50		110.50	133
1180	1" diameter x 18"		19	.421		90	29		119	143
1200	1" diameter x 30"		19	.421		83.50	29		112.50	135
1220	1" diameter x 36"		19	.421		106	29		135	159
1240	1-1/4" diameter x 18"		15	.533		305	36.50		341.50	390
1260	1-1/4" diameter x 36"		15	.533		122	36.50		158.50	189
1280	1-1/2" diameter x 18"		13	.615		130	42		172	206
1300	1-1/2" diameter x 36"		13	.615		136	42		178	212
1320	2" diameter x 24"		11	.727		178	49.50		227.50	270
1340	2" diameter x 36"		11	.727		182	49.50		231.50	274
1360	2-1/2" diameter x 24"		9	.889		340	61		401	465
1380	2-1/2" diameter x 36"		9	.889		128	61		189	232
1400	3" diameter x 24"		7	1.143		455	78		533	615
1420	3" diameter x 36"	↓	7	1.143	↓	1,025	78		1,103	1,275
2000	Carbon steel braid, carbon steel solid ends									
2100	1/2" diameter x 12"	1 Stpi	24	.333	Ea.	54	23		77	93.50
2120	3/4" diameter x 12"		20	.400		65.50	27.50		93	113
2140	1" diameter x 12"		19	.421		105	29		134	159
2160	1-1/4" diameter x 12"		15	.533		59	36.50		95.50	119
2180	1-1/2" diameter x 12"	↓	13	.615	↓	65	42		107	135
3000	Stainless steel braid, welded on carbon steel ends									
3100	1/2" diameter x 12"	1 Stpi	24	.333	Ea.	67.50	23		90.50	109
3120	3/4" diameter x 12"		20	.400		81.50	27.50		109	131
3140	3/4" diameter x 24"		20	.400		104	27.50		131.50	155
3160	3/4" diameter x 36"		20	.400		105	27.50		132.50	157
3180	1" diameter x 12"		19	.421		119	29		148	174
3200	1" diameter x 24"		19	.421		137	29		166	193
3220	1" diameter x 36"		19	.421		127	29		156	183
3240	1-1/4" diameter x 12"	↓	15	.533		146	36.50		182.50	215

22 11 Facility Water Distribution

22 11 19 – Domestic Water Piping Specialties

22 11 19.14 Flexible Metal Hose		Crew	Daily Output	Labor-Hours	Unit	Material	2021 Bare Costs Labor	Equipment	Total	Total Incl O&P
3260	1-1/4" diameter x 24"	1 Stpi	15	.533	Ea.	157	36.50		193.50	228
3280	1-1/4" diameter x 36"		15	.533		177	36.50		213.50	250
3300	1-1/2" diameter x 12"		13	.615		67.50	42		109.50	138
3320	1-1/2" diameter x 24"		13	.615		172	42		214	252
3340	1-1/2" diameter x 36"		13	.615		200	42		242	283
3400	Metal stainless steel braid, over corrugated stainless steel, flanged ends									
3410	150 psi									
3420	1/2" diameter x 12"	1 Stpi	24	.333	Ea.	95.50	23		118.50	139
3430	1" diameter x 12"		20	.400		140	27.50		167.50	195
3440	1-1/2" diameter x 12"		15	.533		153	36.50		189.50	223
3450	2-1/2" diameter x 9"		12	.667		63.50	45.50		109	138
3460	3" diameter x 9"		9	.889		32	61		93	126
3470	4" diameter x 9"		7	1.143		40.50	78		118.50	162
3480	4" diameter x 30"		5	1.600		375	109		484	575
3490	4" diameter x 36"		4.80	1.667		395	114		509	605
3500	6" diameter x 11"		5	1.600		77	109		186	248
3510	6" diameter x 36"		3.80	2.105		515	144		659	780
3520	8" diameter x 12"		4	2		291	137		428	525
3530	10" diameter x 13"		3	2.667		206	182		388	500
3540	12" diameter x 14"	Q-5	4	4		330	246		576	725
6000	Molded rubber with helical wire reinforcement									
6010	150 psi									
6020	1-1/2" diameter x 12"	1 Stpi	15	.533	Ea.	31	36.50		67.50	88.50
6030	2" diameter x 12"		12	.667		101	45.50		146.50	179
6040	3" diameter x 12"		8	1		105	68.50		173.50	218
6050	4" diameter x 12"		6	1.333		133	91		224	283
6060	6" diameter x 18"		4	2		199	137		336	425
6070	8" diameter x 24"		3	2.667		279	182		461	575
6080	10" diameter x 24"		2	4		340	273		613	785
6090	12" diameter x 24"	Q-5	3	5.333		385	330		715	915
7000	Molded teflon with stainless steel flanges									
7010	150 psi									
7020	2-1/2" diameter x 3-3/16"	Q-1	7.80	2.051	Ea.	595	125		720	840
7030	3" diameter x 3-5/8"		6.50	2.462		555	150		705	835
7040	4" diameter x 3-5/8"		5	3.200		705	195		900	1,075
7050	6" diameter x 4"		4.30	3.721		980	227		1,207	1,425
7060	8" diameter x 6"		3.80	4.211		1,550	257		1,807	2,100

22 11 19.18 Mixing Valve

		Crew	Daily Output	Labor-Hours	Unit	Material	2021 Bare Costs Labor	Equipment	Total	Total Incl O&P
0010	**MIXING VALVE**, Automatic, water tempering.									
0040	1/2" size	1 Stpi	19	.421	Ea.	655	29		684	765
0050	3/4" size		18	.444		565	30.50		595.50	665
0100	1" size		16	.500		930	34		964	1,075
0120	1-1/4" size		13	.615		1,250	42		1,292	1,450
0140	1-1/2" size		10	.800		1,525	54.50		1,579.50	1,750
0160	2" size		8	1		1,900	68.50		1,968.50	2,200
0170	2-1/2" size		6	1.333		1,900	91		1,991	2,225
0180	3" size		4	2		4,550	137		4,687	5,200
0190	4" size		3	2.667		4,550	182		4,732	5,275

22 11 19.22 Pressure Reducing Valve

	Crew	Daily Output	Labor-Hours	Unit	Material	2021 Bare Costs Labor	Equipment	Total	Total Incl O&P
0010 **PRESSURE REDUCING VALVE**, Steam, pilot operated.									
0100 Threaded, iron body									
0200 1-1/2" size	1 Stpi	8	1	Ea.	1,700	68.50		1,768.50	1,975
0220 2" size	"	5	1.600	"	2,450	109		2,559	2,875
1000 Flanged, iron body, 125 lb. flanges									
1020 2" size	1 Stpi	8	1	Ea.	3,350	68.50		3,418.50	3,775
1040 2-1/2" size	"	4	2		2,625	137		2,762	3,075
1060 3" size	Q-5	4.50	3.556		2,925	219		3,144	3,525
1080 4" size	"	3	5.333		3,500	330		3,830	4,350
1500 For 250 lb. flanges, add					5%				

22 11 19.26 Pressure Regulators

	Crew	Daily Output	Labor-Hours	Unit	Material	2021 Bare Costs Labor	Equipment	Total	Total Incl O&P
0010 **PRESSURE REGULATORS**									
0100 Gas appliance regulators									
0106 Main burner and pilot applications									
0108 Rubber seat poppet type									
0109 1/8" pipe size	1 Stpi	24	.333	Ea.	19	23		42	55
0110 1/4" pipe size		24	.333		22.50	23		45.50	59
0112 3/8" pipe size		24	.333		23	23		46	59.50
0113 1/2" pipe size		24	.333		25.50	23		48.50	62
0114 3/4" pipe size		20	.400		30	27.50		57.50	74
0122 Lever action type									
0123 3/8" pipe size	1 Stpi	24	.333	Ea.	29.50	23		52.50	66.50
0124 1/2" pipe size		24	.333		29.50	23		52.50	66.50
0125 3/4" pipe size		20	.400		58	27.50		85.50	105
0126 1" pipe size		19	.421		58	29		87	107
0132 Double diaphragm type									
0133 3/8" pipe size	1 Stpi	24	.333	Ea.	46	23		69	84.50
0134 1/2" pipe size		24	.333		57.50	23		80.50	97.50
0135 3/4" pipe size		20	.400		95	27.50		122.50	146
0136 1" pipe size		19	.421		109	29		138	163
0137 1-1/4" pipe size		15	.533		380	36.50		416.50	475
0138 1-1/2" pipe size		13	.615		700	42		742	835
0139 2" pipe size		11	.727		700	49.50		749.50	845
0140 2-1/2" pipe size	Q-5	15	1.067		1,375	65.50		1,440.50	1,600
0141 3" pipe size		13	1.231		1,375	75.50		1,450.50	1,625
0142 4" pipe size (flanged)		8	2		2,425	123		2,548	2,850
0160 Main burner only									
0162 Straight-thru-flow design									
0163 1/2" pipe size	1 Stpi	24	.333	Ea.	47	23		70	85.50
0164 3/4" pipe size		20	.400		64	27.50		91.50	112
0165 1" pipe size		19	.421		85	29		114	137
0166 1-1/4" pipe size		15	.533		85	36.50		121.50	148
0200 Oil, light, hot water, ordinary steam, threaded									
0220 Bronze body, 1/4" size	1 Stpi	24	.333	Ea.	211	23		234	266
0230 3/8" size		24	.333		194	23		217	247
0240 1/2" size		24	.333		254	23		277	315
0250 3/4" size		20	.400		300	27.50		327.50	370
0260 1" size		19	.421		435	29		464	525
0270 1-1/4" size		15	.533		585	36.50		621.50	700
0320 Iron body, 1/4" size		24	.333		130	23		153	177
0330 3/8" size		24	.333		174	23		197	226
0340 1/2" size		24	.333		159	23		182	209

254

22 11 Facility Water Distribution

22 11 19 – Domestic Water Piping Specialties

22 11 19.26 Pressure Regulators

		Crew	Daily Output	Labor-Hours	Unit	Material	2021 Bare Costs Labor	Equipment	Total	Total Incl O&P
0350	3/4" size	1 Stpi	20	.400	Ea.	193	27.50		220.50	253
0360	1" size		19	.421		247	29		276	315
0370	1-1/4" size		15	.533		345	36.50		381.50	430
9002	For water pressure regulators, see Section 22 05 23.20									

22 11 19.30 Pressure and Temperature Safety Plug

		Crew	Daily Output	Labor-Hours	Unit	Material	2021 Bare Costs Labor	Equipment	Total	Total Incl O&P
0010	**PRESSURE & TEMPERATURE SAFETY PLUG**									
1000	3/4" external thread, 3/8" diam. element									
1020	Carbon steel									
1050	7-1/2" insertion	1 Stpi	32	.250	Ea.	50.50	17.10		67.60	81
1120	304 stainless steel									
1150	7-1/2" insertion	1 Stpi	32	.250	Ea.	39.50	17.10		56.60	69
1220	316 stainless steel									
1250	7-1/2" insertion	1 Stpi	32	.250	Ea.	71	17.10		88.10	104

22 11 19.32 Pressure and Temperature Measurement Plug

		Crew	Daily Output	Labor-Hours	Unit	Material	2021 Bare Costs Labor	Equipment	Total	Total Incl O&P
0010	**PRESSURE & TEMPERATURE MEASUREMENT PLUG**									
0020	A permanent access port for insertion of									
0030	a pressure or temperature measuring probe									
0100	Plug, brass									
0110	1/4" MNPT, 1-1/2" long	1 Stpi	32	.250	Ea.	6.80	17.10		23.90	33
0120	3" long		31	.258		12.60	17.65		30.25	40.50
0140	1/2" MNPT, 1-1/2" long		30	.267		9.30	18.25		27.55	37
0150	3" long		29	.276		15.30	18.85		34.15	45
0200	Pressure gauge probe adapter									
0210	1/8" diameter, 1-1/2" probe				Ea.	22			22	24
0220	3" probe				"	35.50			35.50	39
0300	Temperature gauge, 5" stem									
0310	Analog				Ea.	11.25			11.25	12.40
0330	Digital				"	30.50			30.50	34
0400	Pressure gauge, compound									
0410	1/4" MNPT				Ea.	34			34	37.50
0500	Pressure and temperature test kit									
0510	Contains 2 thermometers and 2 pressure gauges									
0520	Kit				Ea.	365			365	405

22 11 19.34 Sleeves and Escutcheons

		Crew	Daily Output	Labor-Hours	Unit	Material	2021 Bare Costs Labor	Equipment	Total	Total Incl O&P
0010	**SLEEVES & ESCUTCHEONS**									
0100	Pipe sleeve									
0110	Steel, w/water stop, 12" long, with link seal									
0120	2" diam. for 1/2" carrier pipe	1 Plum	8.40	.952	Ea.	74.50	64.50		139	178
0130	2-1/2" diam. for 3/4" carrier pipe		8	1		83.50	67.50		151	193
0140	2-1/2" diam. for 1" carrier pipe		8	1		79.50	67.50		147	189
0150	3" diam. for 1-1/4" carrier pipe		7.20	1.111		101	75		176	224
0160	3-1/2" diam. for 1-1/2" carrier pipe		6.80	1.176		120	79.50		199.50	250
0170	4" diam. for 2" carrier pipe		6	1.333		114	90.50		204.50	260
0180	4" diam. for 2-1/2" carrier pipe		6	1.333		112	90.50		202.50	258
0190	5" diam. for 3" carrier pipe		5.40	1.481		130	100		230	293
0200	6" diam. for 4" carrier pipe		4.80	1.667		142	113		255	325
0210	10" diam. for 6" carrier pipe	Q-1	8	2		280	122		402	490
0220	12" diam. for 8" carrier pipe		7.20	2.222		390	135		525	625
0230	14" diam. for 10" carrier pipe		6.40	2.500		395	152		547	660
0240	16" diam. for 12" carrier pipe		5.80	2.759		440	168		608	735
0250	18" diam. for 14" carrier pipe		5.20	3.077		775	187		962	1,125
0260	24" diam. for 18" carrier pipe		4	4		1,150	244		1,394	1,650

For customer support on your Plumbing Costs with RSMeans Data, call 800.448.8182.

22 11 Facility Water Distribution

22 11 19 – Domestic Water Piping Specialties

22 11 19.34 Sleeves and Escutcheons

		Crew	Daily Output	Labor-Hours	Unit	Material	2021 Bare Costs Labor	Equipment	Total	Total Incl O&P
0270	24" diam. for 20" carrier pipe	Q-1	4	4	Ea.	1,000	244		1,244	1,475
0280	30" diam. for 24" carrier pipe	↓	3.20	5	↓	1,325	305		1,630	1,900
0500	Wall sleeve									
0510	Ductile iron with rubber gasket seal									
0520	3"	1 Plum	8.40	.952	Ea.	212	64.50		276.50	330
0530	4"		7.20	1.111		226	75		301	360
0540	6"		6	1.333		275	90.50		365.50	440
0550	8"		4	2		1,225	135		1,360	1,550
0560	10"		3	2.667		1,475	181		1,656	1,900
0570	12"	↓	2.40	3.333	↓	1,750	226		1,976	2,250
5000	Escutcheon									
5100	Split ring, pipe									
5110	Chrome plated									
5120	1/2"	1 Plum	160	.050	Ea.	1.18	3.39		4.57	6.35
5130	3/4"		160	.050		1.31	3.39		4.70	6.50
5140	1"		135	.059		1.41	4.01		5.42	7.55
5150	1-1/2"		115	.070		1.97	4.71		6.68	9.20
5160	2"		100	.080		2.19	5.40		7.59	10.50
5170	4"		80	.100		5.20	6.75		11.95	15.85
5180	6"	↓	68	.118	↓	8.20	7.95		16.15	21
5400	Shallow flange type									
5410	Chrome plated steel									
5420	1/2" CTS	1 Plum	180	.044	Ea.	.36	3.01		3.37	4.89
5430	3/4" CTS		180	.044		.39	3.01		3.40	4.92
5440	1/2" IPS		180	.044		.45	3.01		3.46	4.99
5450	3/4" IPS		180	.044		.45	3.01		3.46	4.99
5460	1" IPS		175	.046		.52	3.09		3.61	5.20
5470	1-1/2" IPS		170	.047		.86	3.19		4.05	5.70
5480	2" IPS	↓	160	.050	↓	1.03	3.39		4.42	6.20

22 11 19.38 Water Supply Meters

		Crew	Daily Output	Labor-Hours	Unit	Material	2021 Bare Costs Labor	Equipment	Total	Total Incl O&P
0010	**WATER SUPPLY METERS**									
1000	Detector, serves dual systems such as fire and domestic or									
1020	process water, wide range cap., UL and FM approved									
1100	3" mainline x 2" by-pass, 400 GPM	Q-1	3.60	4.444	Ea.	8,425	271		8,696	9,675
1140	4" mainline x 2" by-pass, 700 GPM	"	2.50	6.400		8,425	390		8,815	9,850
1180	6" mainline x 3" by-pass, 1,600 GPM	Q-2	2.60	9.231		12,900	585		13,485	15,100
1220	8" mainline x 4" by-pass, 2,800 GPM		2.10	11.429		19,100	720		19,820	22,100
1260	10" mainline x 6" by-pass, 4,400 GPM		2	12		27,300	760		28,060	31,200
1300	10" x 12" mainlines x 6" by-pass, 5,400 GPM	↓	1.70	14.118	↓	37,000	890		37,890	42,000
2000	Domestic/commercial, bronze									
2020	Threaded									
2060	5/8" diameter, to 20 GPM	1 Plum	16	.500	Ea.	56.50	34		90.50	113
2080	3/4" diameter, to 30 GPM		14	.571		103	38.50		141.50	172
2100	1" diameter, to 50 GPM	↓	12	.667	↓	157	45		202	241
2300	Threaded/flanged									
2340	1-1/2" diameter, to 100 GPM	1 Plum	8	1	Ea.	385	67.50		452.50	520
2360	2" diameter, to 160 GPM	"	6	1.333	"	520	90.50		610.50	705
2600	Flanged, compound									
2640	3" diameter, 320 GPM	Q-1	3	5.333	Ea.	2,650	325		2,975	3,400
2660	4" diameter, to 500 GPM		1.50	10.667		4,250	650		4,900	5,650
2680	6" diameter, to 1,000 GPM		1	16		6,875	975		7,850	9,025
2700	8" diameter, to 1,800 GPM	↓	.80	20		10,700	1,225		11,925	13,600

22 11 19 – Domestic Water Piping Specialties

22 11 19.38 Water Supply Meters	Crew	Daily Output	Labor-Hours	Unit	Material	2021 Bare Costs Labor	Equipment	Total	Total Incl O&P
7000	Turbine									
7260	Flanged									
7300	2" diameter, to 160 GPM	1 Plum	7	1.143	Ea.	645	77.50		722.50	825
7320	3" diameter, to 450 GPM	Q-1	3.60	4.444		1,100	271		1,371	1,625
7340	4" diameter, to 650 GPM	"	2.50	6.400		1,825	390		2,215	2,575
7360	6" diameter, to 1,800 GPM	Q-2	2.60	9.231		3,025	585		3,610	4,200
7380	8" diameter, to 2,500 GPM		2.10	11.429		5,175	720		5,895	6,775
7400	10" diameter, to 5,500 GPM		1.70	14.118		6,950	890		7,840	8,975

22 11 19.42 Backflow Preventers

		Crew	Daily Output	Labor-Hours	Unit	Material	2021 Bare Costs Labor	Equipment	Total	Total Incl O&P
0010	**BACKFLOW PREVENTERS**, Includes valves									
0020	and four test cocks, corrosion resistant, automatic operation									
1000	Double check principle									
1010	Threaded, with ball valves									
1020	3/4" pipe size	1 Plum	16	.500	Ea.	240	34		274	315
1030	1" pipe size		14	.571		335	38.50		373.50	430
1040	1-1/2" pipe size		10	.800		640	54		694	785
1050	2" pipe size		7	1.143		690	77.50		767.50	875
1080	Threaded, with gate valves									
1100	3/4" pipe size	1 Plum	16	.500	Ea.	1,275	34		1,309	1,450
1120	1" pipe size		14	.571		1,300	38.50		1,338.50	1,475
1140	1-1/2" pipe size		10	.800		1,150	54		1,204	1,350
1160	2" pipe size		7	1.143		2,050	77.50		2,127.50	2,375
1200	Flanged, valves are gate									
1210	3" pipe size	Q-1	4.50	3.556	Ea.	3,250	217		3,467	3,900
1220	4" pipe size	"	3	5.333		2,900	325		3,225	3,675
1230	6" pipe size	Q-2	3	8		4,525	505		5,030	5,725
1240	8" pipe size		2	12		7,725	760		8,485	9,625
1250	10" pipe size		1	24		11,600	1,525		13,125	15,100
1300	Flanged, valves are OS&Y									
1370	1" pipe size	1 Plum	5	1.600	Ea.	1,400	108		1,508	1,700
1374	1-1/2" pipe size		5	1.600		1,275	108		1,383	1,550
1378	2" pipe size		4.80	1.667		2,175	113		2,288	2,575
1380	3" pipe size	Q-1	4.50	3.556		2,975	217		3,192	3,600
1400	4" pipe size	"	3	5.333		3,700	325		4,025	4,525
1420	6" pipe size	Q-2	3	8		5,275	505		5,780	6,550
1430	8" pipe size	"	2	12		11,000	760		11,760	13,200
4000	Reduced pressure principle									
4100	Threaded, bronze, valves are ball									
4120	3/4" pipe size	1 Plum	16	.500	Ea.	450	34		484	545
4140	1" pipe size		14	.571		480	38.50		518.50	590
4150	1-1/4" pipe size		12	.667		1,175	45		1,220	1,375
4160	1-1/2" pipe size		10	.800		970	54		1,024	1,150
4180	2" pipe size		7	1.143		1,550	77.50		1,627.50	1,825
5000	Flanged, bronze, valves are OS&Y									
5060	2-1/2" pipe size	Q-1	5	3.200	Ea.	5,225	195		5,420	6,050
5080	3" pipe size		4.50	3.556		5,725	217		5,942	6,625
5100	4" pipe size		3	5.333		7,350	325		7,675	8,575
5120	6" pipe size	Q-2	3	8		10,400	505		10,905	12,200
5200	Flanged, iron, valves are gate									
5210	2-1/2" pipe size	Q-1	5	3.200	Ea.	3,200	195		3,395	3,800
5220	3" pipe size		4.50	3.556		2,475	217		2,692	3,050
5230	4" pipe size		3	5.333		3,675	325		4,000	4,500

For customer support on your Plumbing Costs with RSMeans Data, call 800.448.8182.

257

22 11 19.42 Backflow Preventers

		Crew	Daily Output	Labor-Hours	Unit	Material	2021 Bare Costs Labor	Equipment	Total	Total Incl O&P
5240	6" pipe size	Q-2	3	8	Ea.	5,025	505		5,530	6,275
5250	8" pipe size		2	12		9,275	760		10,035	11,300
5260	10" pipe size	↓	1	24	↓	14,100	1,525		15,625	17,800
5600	Flanged, iron, valves are OS&Y									
5660	2-1/2" pipe size	Q-1	5	3.200	Ea.	3,825	195		4,020	4,525
5680	3" pipe size		4.50	3.556		3,250	217		3,467	3,900
5700	4" pipe size		3	5.333		4,725	325		5,050	5,675
5720	6" pipe size	Q-2	3	8		5,650	505		6,155	6,950
5740	8" pipe size		2	12		9,950	760		10,710	12,000
5760	10" pipe size	↓	1	24	↓	19,200	1,525		20,725	23,400

22 11 19.50 Vacuum Breakers

		Crew	Daily Output	Labor-Hours	Unit	Material	2021 Bare Costs Labor	Equipment	Total	Total Incl O&P
0010	**VACUUM BREAKERS** R221113-40									
0013	See also backflow preventers Section 22 11 19.42									
1000	Anti-siphon continuous pressure type									
1010	Max. 150 psi - 210°F									
1020	Bronze body									
1030	1/2" size	1 Stpi	24	.333	Ea.	221	23		244	277
1040	3/4" size		20	.400		221	27.50		248.50	284
1050	1" size		19	.421		226	29		255	291
1060	1-1/4" size		15	.533		450	36.50		486.50	550
1070	1-1/2" size		13	.615		545	42		587	665
1080	2" size	↓	11	.727	↓	560	49.50		609.50	690
1200	Max. 125 psi with atmospheric vent									
1210	Brass, in-line construction									
1220	1/4" size	1 Stpi	24	.333	Ea.	151	23		174	200
1230	3/8" size	"	24	.333		151	23		174	200
1260	For polished chrome finish, add				↓	13%				
2000	Anti-siphon, non-continuous pressure type									
2010	Hot or cold water 125 psi - 210°F									
2020	Bronze body									
2030	1/4" size	1 Stpi	24	.333	Ea.	92	23		115	135
2040	3/8" size		24	.333		92	23		115	135
2050	1/2" size		24	.333		103	23		126	148
2060	3/4" size		20	.400		124	27.50		151.50	177
2070	1" size		19	.421		191	29		220	253
2080	1-1/4" size		15	.533		335	36.50		371.50	420
2090	1-1/2" size		13	.615		370	42		412	470
2100	2" size		11	.727		610	49.50		659.50	745
2110	2-1/2" size		8	1		1,750	68.50		1,818.50	2,025
2120	3" size	↓	6	1.333	↓	2,325	91		2,416	2,675
2150	For polished chrome finish, add					50%				
3000	Air gap fitting									
3020	1/2" NPT size	1 Plum	19	.421	Ea.	62	28.50		90.50	111
3030	1" NPT size	"	15	.533		71	36		107	133
3040	2" NPT size	Q-1	21	.762		139	46.50		185.50	223
3050	3" NPT size		14	1.143		275	69.50		344.50	410
3060	4" NPT size	↓	10	1.600	↓	275	97.50		372.50	450

22 11 19.54 Water Hammer Arresters/Shock Absorbers

		Crew	Daily Output	Labor-Hours	Unit	Material	2021 Bare Costs Labor	Equipment	Total	Total Incl O&P
0010	**WATER HAMMER ARRESTERS/SHOCK ABSORBERS**									
0490	Copper									
0500	3/4" male IPS for 1 to 11 fixtures	1 Plum	12	.667	Ea.	31	45		76	102
0600	1" male IPS for 12 to 32 fixtures	↓	8	1	↓	52	67.50		119.50	158

22 11 19.54 Water Hammer Arresters/Shock Absorbers		Crew	Daily Output	Labor- Hours	Unit	Material	2021 Bare Costs Labor	Equipment	Total	Total Incl O&P
0700	1-1/4" male IPS for 33 to 60 fixtures	1 Plum	8	1	Ea.	52.50	67.50		120	159
0800	1-1/2" male IPS for 61 to 113 fixtures		8	1		77	67.50		144.50	186
0900	2" male IPS for 114 to 154 fixtures		8	1		118	67.50		185.50	231
1000	2-1/2" male IPS for 155 to 330 fixtures		4	2		330	135		465	565
4000	Bellows type									
4010	3/4" FNPT, to 11 fixture units	1 Plum	10	.800	Ea.	415	54		469	535
4020	1" FNPT, to 32 fixture units		8.80	.909		785	61.50		846.50	950
4030	1" FNPT, to 60 fixture units		8.80	.909		1,175	61.50		1,236.50	1,375
4040	1" FNPT, to 113 fixture units		8.80	.909		2,925	61.50		2,986.50	3,325
4050	1" FNPT, to 154 fixture units		6.60	1.212		3,300	82		3,382	3,775
4060	1-1/2" FNPT, to 300 fixture units		5	1.600		4,075	108		4,183	4,625

22 11 19.64 Hydrants

		Crew	Daily Output	Labor- Hours	Unit	Material	2021 Bare Costs Labor	Equipment	Total	Total Incl O&P
0010	**HYDRANTS**									
0050	Wall type, moderate climate, bronze, encased									
0200	3/4" IPS connection	1 Plum	16	.500	Ea.	930	34		964	1,075
0300	1" IPS connection		14	.571		2,475	38.50		2,513.50	2,750
0500	Anti-siphon type, 3/4" connection		16	.500		865	34		899	1,000
1000	Non-freeze, bronze, exposed									
1100	3/4" IPS connection, 4" to 9" thick wall	1 Plum	14	.571	Ea.	530	38.50		568.50	640
1120	10" to 14" thick wall		12	.667		465	45		510	580
1140	15" to 19" thick wall		12	.667		590	45		635	720
1160	20" to 24" thick wall		10	.800		805	54		859	965
1200	For 1" IPS connection, add					15%	10%			
1240	For 3/4" adapter type vacuum breaker, add				Ea.	93.50			93.50	103
1280	For anti-siphon type, add				"	198			198	218
2000	Non-freeze bronze, encased, anti-siphon type									
2100	3/4" IPS connection, 5" to 9" thick wall	1 Plum	14	.571	Ea.	1,325	38.50		1,363.50	1,500
2120	10" to 14" thick wall		12	.667		1,725	45		1,770	1,975
2140	15" to 19" thick wall		12	.667		1,850	45		1,895	2,100
2160	20" to 24" thick wall		10	.800		1,900	54		1,954	2,175
2200	For 1" IPS connection, add					10%	10%			
3000	Ground box type, bronze frame, 3/4" IPS connection									
3080	Non-freeze, all bronze, polished face, set flush									
3100	2' depth of bury	1 Plum	8	1	Ea.	1,200	67.50		1,267.50	1,425
3120	3' depth of bury		8	1		1,300	67.50		1,367.50	1,525
3140	4' depth of bury		8	1		1,400	67.50		1,467.50	1,625
3160	5' depth of bury		7	1.143		1,500	77.50		1,577.50	1,775
3180	6' depth of bury		7	1.143		1,575	77.50		1,652.50	1,850
3200	7' depth of bury		6	1.333		1,800	90.50		1,890.50	2,100
3220	8' depth of bury		5	1.600		1,900	108		2,008	2,225
3240	9' depth of bury		4	2		2,000	135		2,135	2,400
3260	10' depth of bury		4	2		2,075	135		2,210	2,500
3400	For 1" IPS connection, add					15%	10%			
3450	For 1-1/4" IPS connection, add					325%	14%			
3500	For 1-1/2" IPS connection, add					370%	18%			
3550	For 2" IPS connection, add					445%	24%			
3600	For tapped drain port in box, add					120			120	132
4000	Non-freeze, CI body, bronze frame & scoriated cover									
4010	with hose storage									
4100	2' depth of bury	1 Plum	7	1.143	Ea.	2,350	77.50		2,427.50	2,700
4120	3' depth of bury		7	1.143		2,450	77.50		2,527.50	2,825
4140	4' depth of bury		7	1.143		2,525	77.50		2,602.50	2,900

For customer support on your Plumbing Costs with RSMeans Data, call 800.448.8182.

259

22 11 19 – Domestic Water Piping Specialties

22 11 19.64 Hydrants		Crew	Daily Output	Labor-Hours	Unit	Material	2021 Bare Costs Labor	Equipment	Total	Total Incl O&P
4160	5' depth of bury	1 Plum	6.50	1.231	Ea.	2,575	83.50		2,658.50	2,950
4180	6' depth of bury		6	1.333		2,625	90.50		2,715.50	3,000
4200	7' depth of bury		5.50	1.455		2,700	98.50		2,798.50	3,125
4220	8' depth of bury		5	1.600		2,775	108		2,883	3,200
4240	9' depth of bury		4.50	1.778		2,600	120		2,720	3,025
4260	10' depth of bury	↓	4	2		2,975	135		3,110	3,475
4280	For 1" IPS connection, add					570			570	625
4300	For tapped drain port in box, add				↓	126			126	139
5000	Moderate climate, all bronze, polished face									
5020	and scoriated cover, set flush									
5100	3/4" IPS connection	1 Plum	16	.500	Ea.	880	34		914	1,025
5120	1" IPS connection	"	14	.571		2,000	38.50		2,038.50	2,250
5200	For tapped drain port in box, add				↓	126			126	139
6000	Ground post type, all non-freeze, all bronze, aluminum casing									
6010	guard, exposed head, 3/4" IPS connection									
6100	2' depth of bury	1 Plum	8	1	Ea.	1,400	67.50		1,467.50	1,650
6120	3' depth of bury		8	1		1,475	67.50		1,542.50	1,725
6140	4' depth of bury		8	1		1,650	67.50		1,717.50	1,925
6160	5' depth of bury		7	1.143		1,550	77.50		1,627.50	1,825
6180	6' depth of bury		7	1.143		1,900	77.50		1,977.50	2,200
6200	7' depth of bury		6	1.333		1,775	90.50		1,865.50	2,075
6220	8' depth of bury		5	1.600		1,875	108		1,983	2,225
6240	9' depth of bury		4	2		2,000	135		2,135	2,400
6260	10' depth of bury	↓	4	2	↓	2,100	135		2,235	2,525
6300	For 1" IPS connection, add					40%	10%			
6350	For 1-1/4" IPS connection, add					140%	14%			
6400	For 1-1/2" IPS connection, add					225%	18%			
6450	For 2" IPS connection, add					315%	24%			

22 11 23 – Domestic Water Pumps

22 11 23.10 General Utility Pumps

		Crew	Daily Output	Labor-Hours	Unit	Material	2021 Bare Costs Labor	Equipment	Total	Total Incl O&P
0010	**GENERAL UTILITY PUMPS**									
2000	Single stage									
3000	Double suction,									
3140	50 HP, 5" D x 6" S	Q-2	.33	72.727	Ea.	8,650	4,600		13,250	16,400
3180	60 HP, 6" D x 8" S	Q-3	.30	107		17,600	6,875		24,475	29,700
3190	75 HP, to 2,500 GPM		.28	114		16,900	7,375		24,275	29,600
3220	100 HP, to 3,000 GPM		.26	123		21,000	7,925		28,925	34,900
3240	150 HP, to 4,000 GPM	↓	.24	133	↓	24,700	8,600		33,300	40,000
4000	Centrifugal, end suction, mounted on base									
4010	Horizontal mounted, with drip proof motor, rated @ 100' head									
4020	Vertical split case, single stage									
4040	100 GPM, 5 HP, 1-1/2" discharge	Q-1	1.70	9.412	Ea.	4,675	575		5,250	6,000
4050	200 GPM, 10 HP, 2" discharge		1.30	12.308		5,100	750		5,850	6,750
4060	250 GPM, 10 HP, 3" discharge	↓	1.28	12.500		5,600	760		6,360	7,300
4070	300 GPM, 15 HP, 2" discharge	Q-2	1.56	15.385		6,300	970		7,270	8,400
4080	500 GPM, 20 HP, 4" discharge		1.44	16.667		7,900	1,050		8,950	10,300
4090	750 GPM, 30 HP, 4" discharge		1.20	20		8,350	1,275		9,625	11,100
4100	1,050 GPM, 40 HP, 5" discharge		1	24		8,650	1,525		10,175	11,800
4110	1,500 GPM, 60 HP, 6" discharge		.60	40		17,600	2,525		20,125	23,200
4120	2,000 GPM, 75 HP, 6" discharge		.50	48		18,600	3,025		21,625	25,000
4130	3,000 GPM, 100 HP, 8" discharge	↓	.40	60	↓	22,900	3,800		26,700	30,900
4200	Horizontal split case, single stage									

22 11 Facility Water Distribution

22 11 23 – Domestic Water Pumps

22 11 23.10 General Utility Pumps

		Crew	Daily Output	Labor-Hours	Unit	Material	2021 Bare Costs Labor	Equipment	Total	Total Incl O&P
4210	100 GPM, 7.5 HP, 1-1/2" discharge	Q-1	1.70	9.412	Ea.	5,000	575		5,575	6,350
4220	250 GPM, 15 HP, 2-1/2" discharge	"	1.30	12.308		8,000	750		8,750	9,925
4230	500 GPM, 20 HP, 4" discharge	Q-2	1.60	15		9,600	950		10,550	12,000
4240	750 GPM, 25 HP, 5" discharge		1.54	15.584		9,625	985		10,610	12,100
4250	1,000 GPM, 40 HP, 5" discharge		1.20	20		12,000	1,275		13,275	15,100
4260	1,500 GPM, 50 HP, 6" discharge	Q-3	1.42	22.535		15,300	1,450		16,750	19,000
4270	2,000 GPM, 75 HP, 8" discharge		1.14	28.070		19,800	1,800		21,600	24,500
4280	3,000 GPM, 100 HP, 10" discharge		.96	33.333		20,300	2,150		22,450	25,500
4290	3,500 GPM, 150 HP, 10" discharge		.86	37.209		20,700	2,400		23,100	26,300
4300	4,000 GPM, 200 HP, 10" discharge		.66	48.485		24,700	3,125		27,825	31,900
4330	Horizontal split case, two stage, 500' head									
4340	100 GPM, 40 HP, 1-1/2" discharge	Q-2	1.70	14.118	Ea.	18,900	890		19,790	22,100
4350	200 GPM, 50 HP, 1-1/2" discharge	"	1.44	16.667		20,400	1,050		21,450	24,000
4360	300 GPM, 75 HP, 2" discharge	Q-3	1.57	20.382		27,200	1,325		28,525	31,900
4370	400 GPM, 100 HP, 3" discharge		1.14	28.070		31,500	1,800		33,300	37,400
4380	800 GPM, 200 HP, 4" discharge		.86	37.209		43,800	2,400		46,200	52,000
5000	Centrifugal, in-line									
5006	Vertical mount, iron body, 125 lb. flgd, 1,800 RPM TEFC mtr									
5010	Single stage									
5012	.5 HP, 1-1/2" suction & discharge	Q-1	3.20	5	Ea.	1,575	305		1,880	2,175
5014	.75 HP, 2" suction & discharge		2.80	5.714		3,100	350		3,450	3,925
5015	1 HP, 3" suction & discharge		2.50	6.400		3,350	390		3,740	4,250
5016	1.5 HP, 3" suction & discharge		2.30	6.957		3,500	425		3,925	4,475
5018	2 HP, 4" suction & discharge		2.10	7.619		4,725	465		5,190	5,900
5020	3 HP, 5" suction & discharge		1.90	8.421		4,800	515		5,315	6,050
5030	5 HP, 6" suction & discharge		1.60	10		5,450	610		6,060	6,875
5040	7.5 HP, 6" suction & discharge		1.30	12.308		6,275	750		7,025	8,025
5050	10 HP, 8" suction & discharge	Q-2	1.80	13.333		8,325	840		9,165	10,400
5060	15 HP, 8" suction & discharge		1.70	14.118		9,875	890		10,765	12,200
5064	20 HP, 8" suction & discharge		1.60	15		13,000	950		13,950	15,700
5070	30 HP, 8" suction & discharge		1.50	16		14,500	1,000		15,500	17,400
5080	40 HP, 8" suction & discharge		1.40	17.143		17,400	1,075		18,475	20,700
5090	50 HP, 8" suction & discharge		1.30	18.462		17,700	1,175		18,875	21,300
5094	60 HP, 8" suction & discharge		.90	26.667		21,300	1,675		22,975	25,900
5100	75 HP, 8" suction & discharge	Q-3	.60	53.333		21,300	3,425		24,725	28,500
5110	100 HP, 8" suction & discharge	"	.50	64		24,200	4,125		28,325	32,800

22 11 23.11 Miscellaneous Pumps

		Crew	Daily Output	Labor-Hours	Unit	Material	2021 Bare Costs Labor	Equipment	Total	Total Incl O&P
0010	**MISCELLANEOUS PUMPS**									
0020	Water pump, portable, gasoline powered									
0100	170 GPH, 2" discharge	Q-1	11	1.455	Ea.	790	88.50		878.50	1,000
0110	343 GPH, 3" discharge		10.50	1.524		735	93		828	945
0120	608 GPH, 4" discharge		10	1.600		1,625	97.50		1,722.50	1,950
0500	Pump, propylene body, housing and impeller									
0510	22 GPM, 1/3 HP, 40' HD	1 Plum	5	1.600	Ea.	750	108		858	985
0520	33 GPM, 1/2 HP, 40' HD	Q-1	5	3.200		490	195		685	830
0530	53 GPM, 3/4 HP, 40' HD	"	4	4		965	244		1,209	1,425
0600	Rotary pump, CI									
0614	282 GPH, 3/4 HP, 1" discharge	Q-1	4.50	3.556	Ea.	1,200	217		1,417	1,650
0618	277 GPH, 1 HP, 1" discharge		4	4		1,200	244		1,444	1,675
0624	1,100 GPH, 1.5 HP, 1-1/4" discharge		3.60	4.444		3,275	271		3,546	4,025
0628	1,900 GPH, 2 HP, 1-1/4" discharge		3.20	5		4,425	305		4,730	5,300
1000	Turbine pump, CI									

For customer support on your Plumbing Costs with RSMeans Data, call 800.448.8182.

261

22 11 23 – Domestic Water Pumps

22 11 23.11 Miscellaneous Pumps

		Crew	Daily Output	Labor-Hours	Unit	Material	2021 Bare Costs Labor	Equipment	Total	Total Incl O&P
1010	50 GPM, 2 HP, 3" discharge	Q-1	.80	20	Ea.	3,800	1,225		5,025	6,000
1020	100 GPM, 3 HP, 4" discharge	Q-2	.96	25		7,175	1,575		8,750	10,300
1030	250 GPM, 15 HP, 6" discharge	"	.94	25.532		8,900	1,625		10,525	12,200
1040	500 GPM, 25 HP, 6" discharge	Q-3	1.22	26.230		8,975	1,700		10,675	12,400
1050	1,000 GPM, 50 HP, 8" discharge		1.14	28.070		12,400	1,800		14,200	16,300
1060	2,000 GPM, 100 HP, 10" discharge		1	32		14,700	2,050		16,750	19,200
1070	3,000 GPM, 150 HP, 10" discharge		.80	40		23,600	2,575		26,175	29,900
1080	4,000 GPM, 200 HP, 12" discharge		.70	45.714		25,800	2,950		28,750	32,800
1090	6,000 GPM, 300 HP, 14" discharge		.60	53.333		39,400	3,425		42,825	48,400
1100	10,000 GPM, 300 HP, 18" discharge	▼	.58	55.172	▼	45,500	3,550		49,050	55,500
2000	Centrifugal stainless steel pumps									
2100	100 GPM, 100' TDH, 3 HP	Q-1	1.80	8.889	Ea.	12,800	540		13,340	14,900
2130	250 GPM, 100' TDH, 10 HP	"	1.28	12.500		18,000	760		18,760	20,900
2160	500 GPM, 100' TDH, 20 HP	Q-2	1.44	16.667	▼	20,600	1,050		21,650	24,200
2200	Vertical turbine stainless steel pumps									
2220	100 GPM, 100' TDH, 7.5 HP	Q-2	.95	25.263	Ea.	72,000	1,600		73,600	81,500
2240	250 GPM, 100' TDH, 15 HP		.94	25.532		79,500	1,625		81,125	90,000
2260	500 GPM, 100' TDH, 20 HP	▼	.93	25.806		87,500	1,625		89,125	98,500
2280	750 GPM, 100' TDH, 30 HP	Q-3	1.19	26.891		95,000	1,725		96,725	107,000
2300	1,000 GPM, 100' TDH, 40 HP	"	1.16	27.586		103,000	1,775		104,775	115,500
2320	100 GPM, 200' TDH, 15 HP	Q-2	.94	25.532		110,500	1,625		112,125	124,000
2340	250 GPM, 200' TDH, 25 HP	Q-3	1.22	26.230		118,000	1,700		119,700	132,500
2360	500 GPM, 200' TDH, 40 HP		1.16	27.586		126,000	1,775		127,775	141,000
2380	750 GPM, 200' TDH, 60 HP		1.13	28.319		119,500	1,825		121,325	133,500
2400	1,000 GPM, 200' TDH, 75 HP	▼	1.12	28.571	▼	131,000	1,850		132,850	147,000

22 11 23.13 Domestic-Water Packaged Booster Pumps

		Crew	Daily Output	Labor-Hours	Unit	Material	2021 Bare Costs Labor	Equipment	Total	Total Incl O&P
0010	**DOMESTIC-WATER PACKAGED BOOSTER PUMPS**									
0200	Pump system, with diaphragm tank, control, press. switch									
0300	1 HP pump	Q-1	1.30	12.308	Ea.	7,600	750		8,350	9,475
0400	1-1/2 HP pump		1.25	12.800		7,675	780		8,455	9,600
0420	2 HP pump		1.20	13.333		7,225	810		8,035	9,150
0440	3 HP pump	▼	1.10	14.545		7,975	885		8,860	10,100
0460	5 HP pump	Q-2	1.50	16		8,125	1,000		9,125	10,400
0480	7-1/2 HP pump		1.42	16.901		9,825	1,075		10,900	12,400
0500	10 HP pump	▼	1.34	17.910	▼	10,300	1,125		11,425	13,000
2000	Pump system, variable speed, base, controls, starter									
2010	Duplex, 100' head									
2020	400 GPM, 7-1/2 HP, 4" discharge	Q-2	.70	34.286	Ea.	45,900	2,175		48,075	53,500
2025	Triplex, 100' head									
2030	1,000 GPM, 15 HP, 6" discharge	Q-2	.50	48	Ea.	49,200	3,025		52,225	58,500
2040	1,700 GPM, 30 HP, 6" discharge	"	.30	80	"	73,500	5,050		78,550	88,500

262

For customer support on your Plumbing Costs with RSMeans Data, call 800.448.8182.

22 12 Facility Potable-Water Storage Tanks

22 12 21 – Facility Underground Potable-Water Storage Tanks

22 12 21.13 Fiberglass, Undrgrnd Pot.-Water Storage Tanks	Crew	Daily Output	Labor-Hours	Unit	Material	2021 Bare Costs Labor	Equipment	Total	Total Incl O&P
0010 **FIBERGLASS, UNDERGROUND POTABLE-WATER STORAGE TANKS**									
0020 Excludes excavation, backfill & piping									
0030 Single wall									
2000 600 gallon capacity	B-21B	3.75	10.667	Ea.	4,325	515	127	4,967	5,650
2010 1,000 gallon capacity		3.50	11.429		5,450	550	136	6,136	6,975
2020 2,000 gallon capacity		3.25	12.308		7,675	595	146	8,416	9,475
2030 4,000 gallon capacity		3	13.333		10,100	645	159	10,904	12,200
2040 6,000 gallon capacity		2.65	15.094		11,500	730	180	12,410	13,900
2050 8,000 gallon capacity		2.30	17.391		14,100	840	207	15,147	17,000
2060 10,000 gallon capacity		2	20		15,400	965	238	16,603	18,700
2070 12,000 gallon capacity		1.50	26.667		22,100	1,275	315	23,690	26,700
2080 15,000 gallon capacity		1	40		24,200	1,925	475	26,600	30,100
2090 20,000 gallon capacity		.75	53.333		31,600	2,575	635	34,810	39,300
2100 25,000 gallon capacity		.50	80		48,000	3,850	950	52,800	59,500
2110 30,000 gallon capacity		.35	114		58,000	5,500	1,350	64,850	73,500
2120 40,000 gallon capacity	▼	.30	133	▼	79,000	6,425	1,575	87,000	98,000

22 12 23 – Facility Indoor Potable-Water Storage Tanks

22 12 23.13 Facility Steel, Indoor Pot.-Water Storage Tanks

	Crew	Daily Output	Labor-Hours	Unit	Material	2021 Bare Costs Labor	Equipment	Total	Total Incl O&P
0010 **FACILITY STEEL, INDOOR POT.-WATER STORAGE TANKS**									
2000 Galvanized steel, 15 gal., 14" diam. x 26" LOA	1 Plum	12	.667	Ea.	1,075	45		1,120	1,250
2060 30 gal., 14" diam. x 49" LOA		11	.727		1,200	49		1,249	1,375
2080 80 gal., 20" diam. x 64" LOA		9	.889		1,775	60		1,835	2,050
2100 135 gal., 24" diam. x 75" LOA		6	1.333		2,625	90.50		2,715.50	3,025
2120 240 gal., 30" diam. x 86" LOA		4	2		4,900	135		5,035	5,575
2140 300 gal., 36" diam. x 76" LOA	▼	3	2.667		6,875	181		7,056	7,825
2160 400 gal., 36" diam. x 100" LOA	Q-1	4	4		8,350	244		8,594	9,575
2180 500 gal., 36" diam. x 126" LOA	"	3	5.333		10,400	325		10,725	12,000
3000 Glass lined, P.E., 80 gal., 20" diam. x 60" LOA	1 Plum	9	.889		4,175	60		4,235	4,700
3060 140 gal., 24" diam. x 75" LOA		6	1.333		5,750	90.50		5,840.50	6,450
3080 200 gal., 30" diam. x 71" LOA		4	2		7,825	135		7,960	8,800
3100 350 gal., 36" diam. x 86" LOA	▼	3	2.667		9,450	181		9,631	10,700
3120 450 gal., 42" diam. x 79" LOA	Q-1	4	4		12,900	244		13,144	14,500
3140 600 gal., 48" diam. x 81" LOA		3	5.333		15,700	325		16,025	17,800
3160 750 gal., 48" diam. x 105" LOA		3	5.333		17,600	325		17,925	19,900
3180 900 gal., 54" diam. x 95" LOA		2.50	6.400		24,500	390		24,890	27,600
3200 1,500 gal., 54" diam. x 153" LOA		2	8		31,500	485		31,985	35,400
3220 1,800 gal., 54" diam. x 181" LOA		1.50	10.667		35,300	650		35,950	39,800
3240 2,000 gal., 60" diam. x 165" LOA	▼	1	16		40,400	975		41,375	46,000
3260 3,500 gal., 72" diam. x 201" LOA	Q-2	1.50	16	▼	51,000	1,000		52,000	58,000

22 13 Facility Sanitary Sewerage

22 13 16 – Sanitary Waste and Vent Piping

22 13 16.20 Pipe, Cast Iron

		Crew	Daily Output	Labor-Hours	Unit	Material	2021 Bare Costs Labor	Equipment	Total	Total Incl O&P
0010 **PIPE, CAST IRON**, Soil, on clevis hanger assemblies, 5' OC	R221113-70									
0020 Single hub, service wt., lead & oakum joints 10' OC										
2120 2" diameter	R221316-10	Q-1	63	.254	L.F.	13.90	15.45		29.35	38.50
2140 3" diameter			60	.267		18.65	16.25		34.90	45
2160 4" diameter	R221316-20	▼	55	.291		24.50	17.75		42.25	53
2180 5" diameter		Q-2	76	.316		29	19.95		48.95	62
2200 6" diameter		"	73	.329	▼	41.50	21		62.50	76.50

For customer support on your Plumbing Costs with RSMeans Data, call 800.448.8182.

263

22 13 16.20 Pipe, Cast Iron

		Crew	Daily Output	Labor-Hours	Unit	Material	2021 Bare Costs Labor	2021 Bare Costs Equipment	Total	Total Incl O&P
2220	8" diameter	Q-3	59	.542	L.F.	66	35		101	125
2240	10" diameter		54	.593		108	38		146	176
2260	12" diameter		48	.667		156	43		199	235
2261	15" diameter		40	.800		230	51.50		281.50	330
2320	For service weight, double hub, add					10%				
2340	For extra heavy, single hub, add					48%	4%			
2360	For extra heavy, double hub, add					71%	4%			
2400	Lead for caulking (1#/diam. in.)	Q-1	160	.100	Lb.	1.02	6.10		7.12	10.20
2420	Oakum for caulking (1/8#/diam. in.)	"	40	.400	"	4.15	24.50		28.65	41
2960	To delete hangers, subtract									
2970	2" diam. to 4" diam.					16%	19%			
2980	5" diam. to 8" diam.					14%	14%			
2990	10" diam. to 15" diam.					13%	19%			
3000	Single hub, service wt., push-on gasket joints 10' OC									
3010	2" diameter	Q-1	66	.242	L.F.	15.15	14.75		29.90	38.50
3020	3" diameter		63	.254		20.50	15.45		35.95	45.50
3030	4" diameter		57	.281		26.50	17.10		43.60	54.50
3040	5" diameter	Q-2	79	.304		32.50	19.20		51.70	64
3050	6" diameter	"	75	.320		45	20		65	79.50
3060	8" diameter	Q-3	62	.516		73.50	33.50		107	130
3070	10" diameter		56	.571		121	37		158	188
3080	12" diameter		49	.653		172	42		214	252
3082	15" diameter		40	.800		250	51.50		301.50	350
3100	For service weight, double hub, add					65%				
3110	For extra heavy, single hub, add					48%	4%			
3120	For extra heavy, double hub, add					29%	4%			
3130	To delete hangers, subtract									
3140	2" diam. to 4" diam.					12%	21%			
3150	5" diam. to 8" diam.					10%	16%			
3160	10" diam. to 15" diam.					9%	21%			
4000	No hub, couplings 10' OC									
4100	1-1/2" diameter	Q-1	71	.225	L.F.	9.95	13.75		23.70	31.50
4120	2" diameter		67	.239		10.35	14.55		24.90	33
4140	3" diameter		64	.250		13.55	15.25		28.80	37.50
4160	4" diameter		58	.276		17.85	16.80		34.65	44.50
4180	5" diameter	Q-2	83	.289		21.50	18.25		39.75	51
4200	6" diameter	"	79	.304		30.50	19.20		49.70	62
4220	8" diameter	Q-3	69	.464		55	30		85	105
4240	10" diameter		61	.525		90	34		124	150
4244	12" diameter		58	.552		136	35.50		171.50	202
4248	15" diameter		52	.615		204	39.50		243.50	284
4280	To delete hangers, subtract									
4290	1-1/2" diam. to 6" diam.					22%	47%			
4300	8" diam. to 10" diam.					21%	44%			
4310	12" diam. to 15" diam.					19%	40%			

22 13 16.30 Pipe Fittings, Cast Iron

		Crew	Daily Output	Labor-Hours	Unit	Material	2021 Bare Costs Labor	2021 Bare Costs Equipment	Total	Total Incl O&P
0010	**PIPE FITTINGS, CAST IRON**, Soil									
0040	Hub and spigot, service weight, lead & oakum joints									
0080	1/4 bend, 2"	Q-1	16	1	Ea.	25.50	61		86.50	119
0120	3"		14	1.143		34	69.50		103.50	142
0140	4"		13	1.231		53.50	75		128.50	171
0160	5"	Q-2	18	1.333		74.50	84		158.50	208

22 13 16.30 Pipe Fittings, Cast Iron		Crew	Daily Output	Labor-Hours	Unit	Material	2021 Bare Costs Labor	Equipment	Total	Total Incl O&P
0180	6"	Q-2	17	1.412	Ea.	93	89		182	235
0200	8"	Q-3	11	2.909		280	187		467	590
0220	10"		10	3.200		410	206		616	760
0224	12"		9	3.556		555	229		784	950
0226	15"		7	4.571		1,950	295		2,245	2,575
0242	Short sweep, CI, 90°, 2"	Q-1	16	1		25	61		86	119
0243	3"		14	1.143		44.50	69.50		114	153
0244	4"		13	1.231		68.50	75		143.50	187
0245	6"	Q-2	17	1.412		139	89		228	285
0246	8"	Q-3	11	2.909		300	187		487	610
0247	10"		10	3.200		615	206		821	990
0248	12"		9	3.556		1,425	229		1,654	1,925
0251	Long sweep elbow									
0252	2"	Q-1	16	1	Ea.	39	61		100	134
0253	3"		14	1.143		55	69.50		124.50	165
0254	4"		13	1.231		81	75		156	202
0255	6"	Q-2	17	1.412		163	89		252	310
0256	8"	Q-3	11	2.909		370	187		557	685
0257	10"		10	3.200		625	206		831	995
0258	12"		9	3.556		1,000	229		1,229	1,475
0259	15"		7	4.571		2,675	295		2,970	3,400
0266	Closet bend, 3" diameter with flange 10" x 16"	Q-1	14	1.143		141	69.50		210.50	259
0268	16" x 16"		12	1.333		157	81		238	294
0270	Closet bend, 4" diameter, 2-1/2" x 4" ring, 6" x 16"		13	1.231		123	75		198	247
0280	8" x 16"		13	1.231		107	75		182	230
0290	10" x 12"		12	1.333		100	81		181	231
0300	10" x 18"		11	1.455		147	88.50		235.50	294
0310	12" x 16"		11	1.455		124	88.50		212.50	268
0330	16" x 16"		10	1.600		161	97.50		258.50	325
0340	1/8 bend, 2"		16	1		18.15	61		79.15	111
0350	3"		14	1.143		28.50	69.50		98	136
0360	4"		13	1.231		41.50	75		116.50	158
0380	5"	Q-2	18	1.333		58.50	84		142.50	191
0400	6"	"	17	1.412		70.50	89		159.50	211
0420	8"	Q-3	11	2.909		211	187		398	510
0440	10"		10	3.200		305	206		511	645
0460	12"		9	3.556		575	229		804	970
0461	15"		7	4.571		1,350	295		1,645	1,950
0500	Sanitary tee, 2"	Q-1	10	1.600		35.50	97.50		133	184
0540	3"		9	1.778		57.50	108		165.50	226
0620	4"		8	2		70.50	122		192.50	260
0700	5"	Q-2	12	2		141	126		267	345
0800	6"	"	11	2.182		159	138		297	380
0880	8"	Q-3	7	4.571		420	295		715	905
0881	10"		7	4.571		775	295		1,070	1,300
0882	12"		6	5.333		1,375	345		1,720	2,025
0883	15"		4	8		2,800	515		3,315	3,850
0900	Sanitary tee, tapped									
0901	2" x 2"	Q-1	9	1.778	Ea.	48	108		156	215
0902	3" x 2"		8	2		53.50	122		175.50	241
0903	4" x 2"		7	2.286		87.50	139		226.50	305
0910	Sanitary cross, tapped [double tapped sanitary tee]									
0911	2" x 2"	Q-1	7	2.286	Ea.	74	139		213	290

22 13 16.30 Pipe Fittings, Cast Iron		Crew	Daily Output	Labor-Hours	Unit	Material	2021 Bare Costs Labor	Equipment	Total	Total Incl O&P
0912	3" x 2"	Q-1	6	2.667	Ea.	94	162		256	345
0913	4" x 2"	▼	5	3.200	▼	92	195		287	390
0940	Sanitary tee, reducing									
0942	3" x 2"	Q-1	10	1.600	Ea.	49.50	97.50		147	200
0943	4" x 3"		9	1.778		65	108		173	234
0944	4" x 2"		9	1.778		60	108		168	228
0945	5" x 3"	Q-2	12.50	1.920		123	121		244	315
0946	6" x 4"		10	2.400		180	152		332	425
0947	6" x 3"		11	2.182		129	138		267	350
0948	6" x 2"	▼	12.50	1.920		125	121		246	320
0949	8" x 6"	Q-3	9	3.556		355	229		584	730
0950	8" x 5"		9	3.556		360	229		589	735
0951	8" x 4"		10	3.200		270	206		476	605
0954	10" x 6"		8	4		610	258		868	1,050
0958	12" x 8"		7	4.571		1,100	295		1,395	1,675
0962	15" x 10"		5	6.400		1,775	410		2,185	2,575
1000	Tee, 2"	Q-1	10	1.600		51.50	97.50		149	202
1060	3"		9	1.778		76.50	108		184.50	247
1120	4"	▼	8	2		98.50	122		220.50	290
1200	5"	Q-2	12	2		209	126		335	420
1300	6"	"	11	2.182		206	138		344	430
1380	8"	Q-3	7	4.571	▼	320	295		615	790
1400	Combination Y and 1/8 bend									
1420	2"	Q-1	10	1.600	Ea.	44.50	97.50		142	194
1460	3"		9	1.778		68	108		176	237
1520	4"	▼	8	2		93.50	122		215.50	285
1540	5"	Q-2	12	2		177	126		303	385
1560	6"		11	2.182		224	138		362	455
1580	8"	▼	7	3.429		555	217		772	935
1582	12"	Q-3	6	5.333	▼	1,150	345		1,495	1,800
1584	Combination Y & 1/8 bend, reducing									
1586	3" x 2"	Q-1	10	1.600	Ea.	51	97.50		148.50	202
1587	4" x 2"		9.50	1.684		69	103		172	229
1588	4" x 3"	▼	9	1.778		81.50	108		189.50	252
1589	6" x 2"	Q-2	12.50	1.920		145	121		266	340
1590	6" x 3"		12	2		154	126		280	360
1591	6" x 4"	▼	11	2.182		162	138		300	385
1592	8" x 2"	Q-3	11	2.909		305	187		492	615
1593	8" x 4"	▼	10	3.200		269	206		475	605
1594	8" x 6"		9	3.556		370	229		599	745
1600	Double Y, 2"	Q-1	8	2		79	122		201	269
1610	3"		7	2.286		99	139		238	315
1620	4"	▼	6.50	2.462		129	150		279	365
1630	5"	Q-2	9	2.667		236	168		404	510
1640	6"	"	8	3		340	190		530	655
1650	8"	Q-3	5.50	5.818		815	375		1,190	1,450
1660	10"		5	6.400		1,900	410		2,310	2,725
1670	12"	▼	4.50	7.111	▼	2,175	460		2,635	3,075
1676	Combination double Y & 1/8 bend									
1678	2"	Q-1	8	2	Ea.	91.50	122		213.50	283
1680	3"		7	2.286		117	139		256	335
1682	4"	▼	6.50	2.462		192	150		342	435
1684	6"	Q-2	8	3	▼	665	190		855	1,025

22 13 Facility Sanitary Sewerage

22 13 16 – Sanitary Waste and Vent Piping

22 13 16.30 Pipe Fittings, Cast Iron	Crew	Daily Output	Labor-Hours	Unit	Material	2021 Bare Costs Labor	Equipment	Total	Total Incl O&P
1690 Combination double Y & 1/8 bend, CI, reducing									
1692 3" x 2"	Q-1	8	2	Ea.	102	122		224	294
1694 4" x 2"		7	2.286		119	139		258	340
1696 4" x 3"		7	2.286		135	139		274	355
1698 6" x 4"	Q-2	9	2.667		435	168		603	730
1700 Double Y, CI, reducing									
1702 3" x 2"	Q-1	8	2	Ea.	84.50	122		206.50	275
1703 4" x 2"		7	2.286		101	139		240	320
1704 4" x 3"		7	2.286		107	139		246	325
1706 6" x 3"	Q-2	9.50	2.526		258	160		418	520
1707 6" x 4"	"	9	2.667		240	168		408	515
1708 8" x 4"	Q-3	8.50	3.765		565	243		808	980
1709 8" x 6"		8	4		580	258		838	1,025
1711 10" x 6"		8	4		1,225	258		1,483	1,725
1712 10" x 8"		6.50	4.923		1,825	315		2,140	2,500
1713 12" x 6"		7.50	4.267		2,300	275		2,575	2,925
1714 12" x 8"		6	5.333		2,250	345		2,595	3,000
1740 Reducer, 3" x 2"	Q-1	15	1.067		25	65		90	125
1750 4" x 2"		14.50	1.103		28.50	67		95.50	132
1760 4" x 3"		14	1.143		32.50	69.50		102	140
1770 5" x 2"		14	1.143		69	69.50		138.50	180
1780 5" x 3"		13.50	1.185		73	72		145	188
1790 5" x 4"		13	1.231		42	75		117	158
1800 6" x 2"		13.50	1.185		68	72		140	183
1810 6" x 3"		13	1.231		66.50	75		141.50	185
1830 6" x 4"		12.50	1.280		65.50	78		143.50	188
1840 6" x 5"		11	1.455		70.50	88.50		159	210
1880 8" x 3"	Q-2	13.50	1.778		127	112		239	310
1900 8" x 4"		13	1.846		109	117		226	294
1920 8" x 5"		12	2		115	126		241	315
1940 8" x 6"		12	2		112	126		238	315
1942 10" x 4"		12.50	1.920		165	121		286	360
1943 10" x 6"		11.50	2.087		184	132		316	400
1944 10" x 8"		9.50	2.526		185	160		345	440
1945 12" x 4"		11.50	2.087		268	132		400	490
1946 12" x 6"		11	2.182		288	138		426	520
1947 12" x 8"		9	2.667		296	168		464	575
1948 12" x 10"		8.50	2.824		300	178		478	595
1949 15" x 6"	Q-3	12	2.667		560	172		732	870
1950 15" x 8"		10	3.200		610	206		816	980
1951 15" x 10"		9.50	3.368		630	217		847	1,025
1952 15" x 12"		9	3.556		640	229		869	1,050
1960 Increaser, 2" x 3"	Q-1	15	1.067		63	65		128	166
1980 2" x 4"		14	1.143		60	69.50		129.50	170
2000 2" x 5"		13	1.231		66	75		141	185
2020 3" x 4"		13	1.231		65.50	75		140.50	185
2040 3" x 5"		13	1.231		66	75		141	185
2060 3" x 6"		12	1.333		88	81		169	218
2070 4" x 5"		13	1.231		78	75		153	198
2080 4" x 6"		12	1.333		89	81		170	219
2090 4" x 8"	Q-2	13	1.846		184	117		301	375
2100 5" x 6"	Q-1	11	1.455		132	88.50		220.50	277
2110 5" x 8"	Q-2	12	2		208	126		334	420

22 13 16.30 Pipe Fittings, Cast Iron		Crew	Daily Output	Labor-Hours	Unit	Material	2021 Bare Costs Labor	Equipment	Total	Total Incl O&P
2120	6" x 8"	Q-2	12	2	Ea.	214	126		340	425
2130	6" x 10"		8	3		380	190		570	705
2140	8" x 10"		6.50	3.692		390	233		623	780
2150	10" x 12"		5.50	4.364		695	276		971	1,175
2500	Y, 2"	Q-1	10	1.600		32.50	97.50		130	181
2510	3"		9	1.778		60	108		168	228
2520	4"		8	2		80.50	122		202.50	271
2530	5"	Q-2	12	2		143	126		269	345
2540	6"	"	11	2.182		186	138		324	410
2550	8"	Q-3	7	4.571		455	295		750	940
2560	10"		6	5.333		735	345		1,080	1,325
2570	12"		5	6.400		1,750	410		2,160	2,550
2580	15"		4	8		3,700	515		4,215	4,850
2581	Y, reducing									
2582	3" x 2"	Q-1	10	1.600	Ea.	47	97.50		144.50	197
2584	4" x 2"		9.50	1.684		62.50	103		165.50	222
2586	4" x 3"		9	1.778		69	108		177	238
2588	6" x 2"	Q-2	12.50	1.920		124	121		245	315
2590	6" x 3"		12.25	1.959		126	124		250	325
2592	6" x 4"		12	2		125	126		251	325
2594	8" x 2"	Q-3	11	2.909		278	187		465	585
2596	8" x 3"		10.50	3.048		277	196		473	600
2598	8" x 4"		10	3.200		239	206		445	575
2600	8" x 6"		9	3.556		297	229		526	665
2602	10" x 3"		10	3.200		415	206		621	765
2604	10" x 4"		9.50	3.368		410	217		627	775
2606	10" x 6"		8	4		445	258		703	875
2608	10" x 8"		8	4		610	258		868	1,050
2610	12" x 4"		9	3.556		690	229		919	1,100
2612	12" x 6"		8.50	3.765		710	243		953	1,150
2614	12" x 8"		8	4		890	258		1,148	1,375
2616	12" x 10"		7.50	4.267		1,550	275		1,825	2,100
2618	15" x 4"		7	4.571		2,300	295		2,595	3,000
2620	15" x 6"		6.50	4.923		2,425	315		2,740	3,125
2622	15" x 8"		6.50	4.923		2,450	315		2,765	3,175
2624	15" x 10"		5	6.400		2,425	410		2,835	3,275
2626	15" x 12"		4.50	7.111		2,600	460		3,060	3,525
3000	For extra heavy, add					44%	4%			
3600	Hub and spigot, service weight gasket joint									
3605	Note: gaskets and joint labor have									
3606	been included with all listed fittings.									
3610	1/4 bend, 2"	Q-1	20	.800	Ea.	38.50	48.50		87	115
3620	3"		17	.941		50.50	57.50		108	141
3630	4"		15	1.067		74	65		139	179
3640	5"	Q-2	21	1.143		107	72		179	225
3650	6"	"	19	1.263		126	80		206	258
3660	8"	Q-3	12	2.667		355	172		527	645
3670	10"		11	2.909		535	187		722	870
3680	12"		10	3.200		715	206		921	1,100
3690	15"		8	4		2,125	258		2,383	2,725
3692	Short sweep, CI, 90°, 2"	Q-1	20	.800		37.50	48.50		86	114
3693	3"		17	.941		61	57.50		118.50	153
3694	4"		15	1.067		89	65		154	195

22 13 16.30 Pipe Fittings, Cast Iron		Crew	Daily Output	Labor-Hours	Unit	Material	2021 Bare Costs Labor	2021 Bare Costs Equipment	Total	Total Incl O&P
3695	6"	Q-2	19	1.263	Ea.	172	80		252	310
3696	8"	Q-3	12	2.667		375	172		547	665
3697	10"		11	2.909		745	187		932	1,100
3698	12"		10	3.200		1,600	206		1,806	2,050
3700	Closet bend, 3" diameter with ring 10" x 16"	Q-1	17	.941		157	57.50		214.50	259
3710	16" x 16"		15	1.067		173	65		238	288
3730	Closet bend, 4" diameter, 1" x 4" ring, 6" x 16"		15	1.067		143	65		208	254
3740	8" x 16"		15	1.067		128	65		193	237
3750	10" x 12"		14	1.143		120	69.50		189.50	236
3760	10" x 18"		13	1.231		168	75		243	296
3770	12" x 16"		13	1.231		144	75		219	271
3780	16" x 16"		12	1.333		182	81		263	320
3786	Long sweep elbow									
3787	2"	Q-1	20	.800	Ea.	51.50	48.50		100	129
3788	3"		17	.941		71.50	57.50		129	164
3789	4"		15	1.067		102	65		167	209
3790	6"	Q-2	19	1.263		196	80		276	335
3791	8"	Q-3	12	2.667		440	172		612	740
3792	10"		11	2.909		750	187		937	1,100
3793	12"		10	3.200		1,175	206		1,381	1,600
3794	15"		8	4		2,875	258		3,133	3,525
3800	1/8 bend, 2"	Q-1	20	.800		31	48.50		79.50	107
3810	3"		17	.941		45	57.50		102.50	135
3820	4"		15	1.067		62	65		127	166
3830	5"	Q-2	21	1.143		91	72		163	208
3840	6"	"	19	1.263		104	80		184	233
3850	8"	Q-3	12	2.667		284	172		456	565
3860	10"		11	2.909		430	187		617	755
3870	12"		10	3.200		735	206		941	1,125
3880	15"		8	4		1,550	258		1,808	2,075
3882	Sanitary tee, tapped									
3884	2" x 2"	Q-1	11	1.455	Ea.	73.50	88.50		162	213
3886	3" x 2"		10	1.600		82.50	97.50		180	236
3888	4" x 2"		9	1.778		120	108		228	295
3890	Sanitary cross, tapped [double tapped sanitary tee]									
3892	2" x 2"	Q-1	9	1.778	Ea.	99.50	108		207.50	271
3894	3" x 2"		8	2		123	122		245	320
3896	4" x 2"		7	2.286		125	139		264	345
3900	Sanitary tee, 2"		12	1.333		61	81		142	188
3910	3"		10	1.600		90	97.50		187.50	244
3920	4"		9	1.778		112	108		220	285
3930	5"	Q-2	13	1.846		205	117		322	400
3940	6"	"	11	2.182		226	138		364	455
3950	8"	Q-3	8.50	3.765		570	243		813	985
3952	10"		8	4		1,025	258		1,283	1,500
3954	12"		7	4.571		1,700	295		1,995	2,300
3956	15"		6	5.333		3,175	345		3,520	4,025
3960	Sanitary tee, reducing									
3961	3" x 2"	Q-1	10.50	1.524	Ea.	78.50	93		171.50	226
3962	4" x 2"		10	1.600		93.50	97.50		191	248
3963	4" x 3"		9.50	1.684		102	103		205	265
3964	5" x 3"	Q-2	13.50	1.778		171	112		283	355
3965	6" x 2"		13	1.846		171	117		288	360

For customer support on your Plumbing Costs with RSMeans Data, call 800.448.8182.

269

22 13 16.30 Pipe Fittings, Cast Iron		Crew	Daily Output	Labor-Hours	Unit	Material	2021 Bare Costs Labor	Equipment	Total	Total Incl O&P
3966	6" x 3"	Q-2	12.50	1.920	Ea.	179	121		300	380
3967	6" x 4"		12	2		234	126		360	445
3968	8" x 4"	Q-3	10.50	3.048		365	196		561	695
3969	8" x 5"		10	3.200		465	206		671	825
3970	8" x 6"		9.50	3.368		460	217		677	835
3971	10" x 6"		9	3.556		770	229		999	1,175
3972	12" x 8"		8.50	3.765		1,350	243		1,593	1,825
3973	15" x 10"		6	5.333		2,075	345		2,420	2,825
3980	Tee, 2"	Q-1	12	1.333		77	81		158	206
3990	3"		10	1.600		109	97.50		206.50	265
4000	4"		9	1.778		139	108		247	315
4010	5"	Q-2	13	1.846		273	117		390	475
4020	6"	"	11	2.182		272	138		410	505
4030	8"	Q-3	8	4		465	258		723	895
4060	Combination Y and 1/8 bend									
4070	2"	Q-1	12	1.333	Ea.	70	81		151	198
4080	3"		10	1.600		100	97.50		197.50	255
4090	4"		9	1.778		135	108		243	310
4100	5"	Q-2	13	1.846		242	117		359	440
4110	6"	"	11	2.182		291	138		429	525
4120	8"	Q-3	8	4		700	258		958	1,150
4121	12"	"	7	4.571		1,475	295		1,770	2,075
4130	Combination Y & 1/8 bend, reducing									
4132	3" x 2"	Q-1	10.50	1.524	Ea.	80	93		173	228
4134	4" x 2"		10	1.600		102	97.50		199.50	258
4136	4" x 3"		9.50	1.684		118	103		221	283
4138	6" x 2"	Q-2	13	1.846		191	117		308	385
4140	6" x 3"		12.50	1.920		204	121		325	405
4142	6" x 4"		12	2		216	126		342	425
4144	8" x 2"	Q-3	11	2.909		410	187		597	735
4146	8" x 4"		10.50	3.048		360	196		556	695
4148	8" x 6"		9.50	3.368		475	217		692	850
4160	Double Y, 2"	Q-1	10	1.600		117	97.50		214.50	274
4170	3"		8	2		148	122		270	345
4180	4"		7	2.286		191	139		330	420
4190	5"	Q-2	10	2.400		330	152		482	590
4200	6"	"	9	2.667		435	168		603	730
4210	8"	Q-3	6	5.333		1,025	345		1,370	1,650
4220	10"		5	6.400		2,275	410		2,685	3,125
4230	12"		4.50	7.111		2,650	460		3,110	3,600
4234	Combination double Y & 1/8 bend									
4235	2"	Q-1	10	1.600	Ea.	130	97.50		227.50	287
4236	3"		8	2		166	122		288	365
4237	4"		7	2.286		254	139		393	485
4238	6"	Q-2	9	2.667		765	168		933	1,100
4242	Combination double Y & 1/8 bend, CI, reducing									
4243	3" x 2"	Q-1	9	1.778	Ea.	143	108		251	320
4244	4" x 2"		8	2		165	122		287	365
4245	4" x 3"		7.50	2.133		188	130		318	400
4246	6" x 4"	Q-2	10	2.400		510	152		662	785
4248	Double Y, CI, reducing									
4249	3" x 2"	Q-1	9	1.778	Ea.	126	108		234	300
4250	4" x 2"		8	2		147	122		269	345

22 13 16.30 Pipe Fittings, Cast Iron		Crew	Daily Output	Labor-Hours	Unit	Material	2021 Bare Costs Labor	Equipment	Total	Total Incl O&P
4251	4" x 3"	Q-1	7.50	2.133	Ea.	144	130		274	350
4252	6" x 3"	Q-2	10.50	2.286		325	144		469	570
4253	6" x 4"	"	10	2.400		315	152		467	570
4254	8" x 4"	Q-3	9.50	3.368		680	217		897	1,075
4255	8" x 6"		9	3.556		720	229		949	1,125
4256	10" x 6"		8.50	3.765		1,425	243		1,668	1,925
4257	10" x 8"		7.50	4.267		2,100	275		2,375	2,725
4258	12" x 6"		8	4		2,525	258		2,783	3,150
4259	12" x 8"	↓	7	4.571		2,550	295		2,845	3,275
4260	Reducer, 3" x 2"	Q-1	17	.941		54	57.50		111.50	145
4270	4" x 2"		16.50	.970		62	59		121	156
4280	4" x 3"		16	1		69.50	61		130.50	167
4290	5" x 2"		16	1		114	61		175	216
4300	5" x 3"		15.50	1.032		121	63		184	227
4310	5" x 4"		15	1.067		94.50	65		159.50	201
4320	6" x 2"		15.50	1.032		114	63		177	219
4330	6" x 3"		15	1.067		116	65		181	225
4336	6" x 4"		14	1.143		119	69.50		188.50	235
4340	6" x 5"	↓	13	1.231		136	75		211	262
4360	8" x 3"	Q-2	15	1.600		217	101		318	390
4370	8" x 4"		15	1.600		203	101		304	375
4380	8" x 5"		14	1.714		220	108		328	405
4390	8" x 6"		14	1.714		218	108		326	400
4394	10" x 4"		13.50	1.778		310	112		422	515
4395	10" x 6"		13	1.846		345	117		462	555
4396	10" x 8"		12.50	1.920		385	121		506	605
4397	12" x 4"		12	2		450	126		576	685
4398	12" x 6"		11.50	2.087		480	132		612	725
4399	12" x 8"		11	2.182		530	138		668	790
4400	12" x 10"	↓	10.50	2.286		590	144		734	860
4401	15" x 6"	Q-3	13	2.462		790	159		949	1,100
4402	15" x 8"		12	2.667		875	172		1,047	1,225
4403	15" x 10"		11	2.909		950	187		1,137	1,325
4404	15" x 12"	↓	10	3.200		995	206		1,201	1,400
4430	Increaser, 2" x 3"	Q-1	17	.941		79	57.50		136.50	173
4440	2" x 4"		16	1		80.50	61		141.50	180
4450	2" x 5"		15	1.067		98.50	65		163.50	205
4460	3" x 4"		15	1.067		86	65		151	192
4470	3" x 5"		15	1.067		98.50	65		163.50	205
4480	3" x 6"		14	1.143		121	69.50		190.50	237
4490	4" x 5"		15	1.067		110	65		175	218
4500	4" x 6"	↓	14	1.143		122	69.50		191.50	239
4510	4" x 8"	Q-2	15	1.600		257	101		358	435
4520	5" x 6"	Q-1	13	1.231		165	75		240	294
4530	5" x 8"	Q-2	14	1.714		281	108		389	470
4540	6" x 8"		14	1.714		286	108		394	475
4550	6" x 10"		10	2.400		510	152		662	785
4560	8" x 10"		8.50	2.824		520	178		698	835
4570	10" x 12"	↓	7.50	3.200		860	202		1,062	1,250
4600	Y, 2"	Q-1	12	1.333		58	81		139	185
4610	3"		10	1.600		93	97.50		190.50	247
4620	4"	↓	9	1.778		122	108		230	296
4630	5"	Q-2	13	1.846		207	117		324	400

22 13 16.30 Pipe Fittings, Cast Iron		Crew	Daily Output	Labor-Hours	Unit	Material	2021 Bare Costs Labor	Equipment	Total	Total Incl O&P
4640	6"	Q-2	11	2.182	Ea.	252	138		390	485
4650	8"	Q-3	8	4		600	258		858	1,050
4660	10"		7	4.571		985	295		1,280	1,525
4670	12"		6	5.333		2,075	345		2,420	2,800
4672	15"	↓	5	6.400	↓	4,075	410		4,485	5,125
4680	Y, reducing									
4681	3" x 2"	Q-1	10.50	1.524	Ea.	76	93		169	223
4682	4" x 2"		10	1.600		96	97.50		193.50	251
4683	4" x 3"	↓	9.50	1.684		106	103		209	270
4684	6" x 2"	Q-2	13	1.846		170	117		287	360
4685	6" x 3"		12.50	1.920		176	121		297	375
4686	6" x 4"	↓	12	2		179	126		305	385
4687	8" x 2"	Q-3	11	2.909		365	187		552	680
4688	8" x 3"		10.75	2.977		365	192		557	685
4689	8" x 4"		10.50	3.048		330	196		526	660
4690	8" x 6"		9.50	3.368		405	217		622	770
4691	10" x 3"		10	3.200		560	206		766	925
4692	10" x 4"		9.50	3.368		555	217		772	935
4693	10" x 6"		9	3.556		605	229		834	1,000
4694	10" x 8"		8.80	3.636		810	234		1,044	1,250
4695	12" x 4"		9.50	3.368		875	217		1,092	1,275
4696	12" x 6"		9	3.556		905	229		1,134	1,350
4697	12" x 8"		8.50	3.765		1,125	243		1,368	1,575
4698	12" x 10"		8	4		1,825	258		2,083	2,400
4699	15" x 4"		8	4		2,525	258		2,783	3,150
4700	15" x 6"		7.50	4.267		2,650	275		2,925	3,300
4701	15" x 8"		7	4.571		2,725	295		3,020	3,450
4702	15" x 10"		6	5.333		2,725	345		3,070	3,525
4703	15" x 12"	↓	5	6.400	↓	2,950	410		3,360	3,875
4900	For extra heavy, add					44%	4%			
4940	Gasket and making push-on joint									
4950	2"	Q-1	40	.400	Ea.	12.70	24.50		37.20	50.50
4960	3"		35	.457		16.30	28		44.30	59.50
4970	4"	↓	32	.500		20.50	30.50		51	68
4980	5"	Q-2	43	.558		32	35.50		67.50	88
4990	6"	"	40	.600		33	38		71	93
5000	8"	Q-3	32	1		73	64.50		137.50	176
5010	10"		29	1.103		127	71		198	245
5020	12"		25	1.280		162	82.50		244.50	300
5022	15"	↓	21	1.524	↓	193	98		291	360
5030	Note: gaskets and joint labor have									
5040	been included with all listed fittings.									
5990	No hub									
6000	Cplg. & labor required at joints not incl. in fitting									
6010	price. Add 1 coupling per joint for installed price									
6020	1/4 bend, 1-1/2"				Ea.	13.15			13.15	14.50
6060	2"					14.30			14.30	15.70
6080	3"					19.95			19.95	22
6120	4"					29.50			29.50	32.50
6140	5"					71.50			71.50	78.50
6160	6"					71.50			71.50	78.50
6180	8"					201			201	221
6181	10"				↓	430			430	475

22 13 16.30 Pipe Fittings, Cast Iron		Crew	Daily Output	Labor-Hours	Unit	Material	2021 Bare Costs Labor	Equipment	Total	Total Incl O&P
6182	12"				Ea.	1,250			1,250	1,375
6183	15"					1,650			1,650	1,800
6184	1/4 bend, long sweep, 1-1/2"					33.50			33.50	37
6186	2"					31.50			31.50	34.50
6188	3"					38			38	41.50
6189	4"					60			60	66
6190	5"					117			117	128
6191	6"					133			133	146
6192	8"					345			345	380
6193	10"					670			670	735
6200	1/8 bend, 1-1/2"					11			11	12.10
6210	2"					12.30			12.30	13.55
6212	3"					16.50			16.50	18.15
6214	4"					21.50			21.50	24
6216	5"					45			45	49.50
6218	6"					48			48	53
6220	8"					138			138	152
6222	10"					263			263	289
6364	Closet flange									
6366	4"				Ea.	19.95			19.95	22
6370	Closet bend, no hub									
6376	4" x 16"				Ea.	106			106	117
6380	Sanitary tee, tapped, 1-1/2"					26			26	28.50
6382	2" x 1-1/2"					23			23	25.50
6384	2"					25			25	27.50
6386	3" x 2"					37			37	40.50
6388	3"					63.50			63.50	70
6390	4" x 1-1/2"					32.50			32.50	36
6392	4" x 2"					37			37	41
6393	4"					37			37	41
6394	6" x 1-1/2"					85.50			85.50	94
6396	6" x 2"					90			90	99.50
6459	Sanitary tee, 1-1/2"					18.40			18.40	20.50
6460	2"					19.75			19.75	21.50
6470	3"					24.50			24.50	27
6472	4"					46			46	50.50
6474	5"					108			108	119
6476	6"					110			110	121
6478	8"					445			445	490
6480	10"					515			515	565
6724	Sanitary tee, reducing									
6725	3" x 2"				Ea.	21.50			21.50	23.50
6726	4" x 3"					35.50			35.50	39
6727	5" x 4"					83.50			83.50	91.50
6728	6" x 3"					81			81	89
6729	8" x 4"					235			235	258
6730	Y, 1-1/2"					18.70			18.70	20.50
6740	2"					18.25			18.25	20
6750	3"					26.50			26.50	29.50
6760	4"					42.50			42.50	47
6762	5"					101			101	111
6764	6"					113			113	125
6768	8"					267			267	293

For customer support on your Plumbing Costs with RSMeans Data, call 800.448.8182.

273

22 13 16.30 Pipe Fittings, Cast Iron	Crew	Daily Output	Labor-Hours	Unit	Material	2021 Bare Costs Labor	Equipment	Total	Total Incl O&P
6769	10"				Ea.	590			590	650
6770	12"					1,175			1,175	1,275
6771	15"					2,725			2,725	2,975
6791	Y, reducing, 3" x 2"					19.75			19.75	21.50
6792	4" x 2"					28.50			28.50	31
6793	5" x 2"					62.50			62.50	69
6794	6" x 2"					69.50			69.50	76.50
6795	6" x 4"					90.50			90.50	99.50
6796	8" x 4"					155			155	171
6797	8" x 6"					191			191	210
6798	10" x 6"					430			430	475
6799	10" x 8"					515			515	565
6800	Double Y, 2"					29			29	32
6920	3"					53.50			53.50	59
7000	4"					109			109	120
7100	6"					192			192	211
7120	8"					545			545	600
7200	Combination Y and 1/8 bend									
7220	1-1/2"				Ea.	19.90			19.90	22
7260	2"					21			21	23
7320	3"					33			33	36
7400	4"					63			63	69.50
7480	5"					129			129	142
7500	6"					174			174	191
7520	8"					405			405	445
7800	Reducer, 3" x 2"					10.20			10.20	11.25
7820	4" x 2"					15.55			15.55	17.10
7840	4" x 3"					15.55			15.55	17.10
7842	6" x 3"					42			42	46
7844	6" x 4"					42			42	46
7846	6" x 5"					43			43	47
7848	8" x 2"					66.50			66.50	73
7850	8" x 3"					61.50			61.50	67.50
7852	8" x 4"					64.50			64.50	71
7854	8" x 5"					73			73	80.50
7856	8" x 6"					72			72	79
7858	10" x 4"					127			127	140
7860	10" x 6"					134			134	148
7862	10" x 8"					158			158	173
7864	12" x 4"					263			263	289
7866	12" x 6"					282			282	310
7868	12" x 8"					290			290	320
7870	12" x 10"					295			295	325
7872	15" x 4"					550			550	605
7874	15" x 6"					520			520	570
7876	15" x 8"					600			600	655
7878	15" x 10"					620			620	680
7880	15" x 12"					625			625	690
8000	Coupling, standard (by CISPI Mfrs.)									
8020	1-1/2"	Q-1	48	.333	Ea.	13.90	20.50		34.40	46
8040	2"		44	.364		15.50	22		37.50	50
8080	3"		38	.421		18.20	25.50		43.70	58.50
8120	4"		33	.485		24	29.50		53.50	70.50

22 13 16 – Sanitary Waste and Vent Piping

22 13 16.30 Pipe Fittings, Cast Iron		Crew	Daily Output	Labor-Hours	Unit	Material	2021 Bare Costs Labor	Equipment	Total	Total Incl O&P
8160	5"	Q-2	44	.545	Ea.	60.50	34.50		95	118
8180	6"	"	40	.600		61.50	38		99.50	124
8200	8"	Q-3	33	.970		100	62.50		162.50	204
8220	10"	"	26	1.231	↓	124	79.50		203.50	254
8300	Coupling, cast iron clamp & neoprene gasket (by MG)									
8310	1-1/2"	Q-1	48	.333	Ea.	9.70	20.50		30.20	41
8320	2"		44	.364		13.05	22		35.05	47.50
8330	3"		38	.421		12.80	25.50		38.30	52.50
8340	4"		33	.485		19.45	29.50		48.95	65.50
8350	5"	Q-2	44	.545		36	34.50		70.50	91
8360	6"	"	40	.600		34.50	38		72.50	94.50
8380	8"	Q-3	33	.970		112	62.50		174.50	217
8400	10"	"	26	1.231	↓	175	79.50		254.50	310
8410	Reducing, no hub									
8416	2" x 1-1/2"	Q-1	44	.364	Ea.	10.80	22		32.80	45
8600	Coupling, stainless steel, heavy duty									
8620	1-1/2"	Q-1	48	.333	Ea.	5.35	20.50		25.85	36.50
8630	2"		44	.364		5.80	22		27.80	39.50
8640	2" x 1-1/2"		44	.364		13.05	22		35.05	47.50
8650	3"		38	.421		6.05	25.50		31.55	45
8660	4"		33	.485		6.80	29.50		36.30	51.50
8670	4" x 3"		33	.485		18.75	29.50		48.25	64.50
8680	5"	Q-2	44	.545		14.90	34.50		49.40	68
8690	6"	"	40	.600		16.40	38		54.40	74.50
8700	8"	Q-3	33	.970		28	62.50		90.50	124
8710	10"		26	1.231		35	79.50		114.50	157
8712	12"		22	1.455		72	93.50		165.50	219
8715	15"	↓	18	1.778	↓	112	115		227	294

22 13 16.40 Pipe Fittings, Cast Iron for Drainage

		Crew	Daily Output	Labor-Hours	Unit	Material	2021 Bare Costs Labor	Equipment	Total	Total Incl O&P
0010	**PIPE FITTINGS, CAST IRON FOR DRAINAGE**, Special									
0020	Cast iron, drainage, threaded, black									
0030	90° elbow, straight									
0031	1-1/2" pipe size	Q-1	20	.800	Ea.	48.50	48.50		97	126
0032	2" pipe size		18	.889		72.50	54		126.50	161
0033	3" pipe size		10	1.600		273	97.50		370.50	445
0034	4" pipe size	↓	6	2.667	↓	435	162		597	715
0040	90° long turn elbow, straight									
0041	1-1/2" pipe size	Q-1	20	.800	Ea.	67.50	48.50		116	147
0042	2" pipe size		18	.889		101	54		155	192
0043	3" pipe size		10	1.600		365	97.50		462.50	550
0044	4" pipe size	↓	6	2.667	↓	675	162		837	985
0050	90° street elbow, straight									
0051	1-1/2" pipe size	Q-1	20	.800	Ea.	69	48.50		117.50	148
0052	2" pipe size	"	18	.889	"	91	54		145	181
0060	45° elbow									
0061	1-1/2" pipe size	Q-1	20	.800	Ea.	47	48.50		95.50	125
0062	2" pipe size		18	.889		68	54		122	156
0063	3" pipe size		10	1.600		267	97.50		364.50	440
0064	4" pipe size	↓	6	2.667	↓	415	162		577	700
0070	45° street elbow									
0071	1-1/2" pipe size	Q-1	20	.800	Ea.	64.50	48.50		113	143
0072	2" pipe size	"	18	.889	"	104	54		158	195

For customer support on your Plumbing Costs with RSMeans Data, call 800.448.8182.

275

22 13 16.40 Pipe Fittings, Cast Iron for Drainage		Crew	Daily Output	Labor-Hours	Unit	Material	2021 Bare Costs Labor	Equipment	Total	Total Incl O&P
0092	Tees, straight									
0093	1-1/2" pipe size	Q-1	13	1.231	Ea.	79.50	75		154.50	200
0094	2" pipe size	"	11	1.455	"	131	88.50		219.50	276
0100	TY's, straight									
0101	1-1/2" pipe size	Q-1	13	1.231	Ea.	78	75		153	198
0102	2" pipe size		11	1.455		128	88.50		216.50	273
0103	3" pipe size		6	2.667		510	162		672	800
0104	4" pipe size		4	4		700	244		944	1,125
0120	45° Y branch, straight									
0121	1-1/2" pipe size	Q-1	13	1.231	Ea.	95.50	75		170.50	217
0122	2" pipe size		11	1.455		178	88.50		266.50	330
0123	3" pipe size		6	2.667		615	162		777	915
0124	4" pipe size		4	4		915	244		1,159	1,375
0147	Double Y branch, straight									
0148	1-1/2" pipe size	Q-1	10	1.600	Ea.	256	97.50		353.50	425
0149	2" pipe size	"	7	2.286	"	216	139		355	445
0160	P trap									
0161	1-1/2" pipe size	Q-1	15	1.067	Ea.	142	65		207	253
0162	2" pipe size		13	1.231		243	75		318	380
0163	3" pipe size		7	2.286		790	139		929	1,075
0164	4" pipe size		5	3.200		1,775	195		1,970	2,250
0180	Tucker connection									
0181	1-1/2" pipe size	Q-1	24	.667	Ea.	147	40.50		187.50	223
0182	2" pipe size	"	21	.762	"	181	46.50		227.50	270
0205	Cast iron, drainage, threaded, galvanized									
0206	90° elbow, straight									
0207	1-1/2" pipe size	Q-1	20	.800	Ea.	71.50	48.50		120	151
0208	2" pipe size		18	.889		108	54		162	200
0209	3" pipe size		10	1.600		410	97.50		507.50	595
0210	4" pipe size		6	2.667		795	162		957	1,125
0216	90° long turn elbow, straight									
0217	1-1/2" pipe size	Q-1	20	.800	Ea.	81.50	48.50		130	162
0218	2" pipe size		18	.889		140	54		194	235
0219	3" pipe size		10	1.600		380	97.50		477.50	560
0220	4" pipe size		6	2.667		535	162		697	830
0226	90° street elbow, straight									
0227	1-1/2" pipe size	Q-1	20	.800	Ea.	90.50	48.50		139	172
0228	2" pipe size	"	18	.889	"	149	54		203	245
0236	45° elbow									
0237	1-1/2" pipe size	Q-1	20	.800	Ea.	70.50	48.50		119	151
0238	2" pipe size		18	.889		101	54		155	192
0239	3" pipe size		10	1.600		350	97.50		447.50	530
0240	4" pipe size		6	2.667		575	162		737	875
0246	45° street elbow									
0247	1-1/2" pipe size	Q-1	20	.800	Ea.	94	48.50		142.50	176
0248	2" pipe size	"	18	.889	"	144	54		198	240
0268	Tees, straight									
0269	1-1/2" pipe size	Q-1	13	1.231	Ea.	108	75		183	230
0270	2" pipe size	"	11	1.455	"	180	88.50		268.50	330
0276	TY's, straight									
0277	1-1/2" pipe size	Q-1	13	1.231	Ea.	109	75		184	232
0278	2" pipe size		11	1.455		178	88.50		266.50	330
0279	3" pipe size		6	2.667		765	162		927	1,075

22 13 Facility Sanitary Sewerage

22 13 16 – Sanitary Waste and Vent Piping

22 13 16.40 Pipe Fittings, Cast Iron for Drainage

		Crew	Daily Output	Labor-Hours	Unit	Material	2021 Bare Costs Labor	2021 Bare Costs Equipment	Total	Total Incl O&P
0280	4" pipe size	Q-1	4	4	Ea.	860	244		1,104	1,300
0296	45° Y branch, straight									
0297	1-1/2" pipe size	Q-1	13	1.231	Ea.	121	75		196	245
0298	2" pipe size		11	1.455		242	88.50		330.50	400
0299	3" pipe size		6	2.667		900	162		1,062	1,225
0300	4" pipe size	↓	4	4	↓	1,325	244		1,569	1,825
0323	Double Y branch, straight									
0324	1-1/2" pipe size	Q-1	10	1.600	Ea.	161	97.50		258.50	320
0325	2" pipe size	"	7	2.286	"	410	139		549	660
0336	P trap									
0337	1-1/2" pipe size	Q-1	15	1.067	Ea.	195	65		260	310
0338	2" pipe size		13	1.231		244	75		319	380
0339	3" pipe size		7	2.286		1,175	139		1,314	1,475
0340	4" pipe size	↓	5	3.200	↓	1,725	195		1,920	2,200
0356	Tucker connection									
0357	1-1/2" pipe size	Q-1	24	.667	Ea.	223	40.50		263.50	305
0358	2" pipe size	"	21	.762	"	335	46.50		381.50	440
1000	Drip pan elbow (safety valve discharge elbow)									
1010	Cast iron, threaded inlet									
1014	2-1/2"	Q-1	8	2	Ea.	1,825	122		1,947	2,175
1015	3"		6.40	2.500		1,950	152		2,102	2,375
1017	4"	↓	4.80	3.333	↓	2,625	203		2,828	3,200
1018	Cast iron, flanged inlet									
1019	6"	Q-2	3.60	6.667	Ea.	2,225	420		2,645	3,075
1020	8"	"	2.60	9.231	"	2,750	585		3,335	3,900

22 13 16.50 Shower Drains

		Crew	Daily Output	Labor-Hours	Unit	Material	2021 Bare Costs Labor	2021 Bare Costs Equipment	Total	Total Incl O&P
0010	**SHOWER DRAINS**									
2780	Shower, with strainer, uniform diam. trap, bronze top									
2800	2" and 3" pipe size	Q-1	8	2	Ea.	460	122		582	690
2820	4" pipe size	"	7	2.286		525	139		664	790
2840	For galvanized body, add				↓	278			278	305
2860	With strainer, backwater valve, drum trap									
2880	1-1/2", 2" & 3" pipe size	Q-1	8	2	Ea.	460	122		582	685
2890	4" pipe size	"	7	2.286		640	139		779	910
2900	For galvanized body, add				↓	220			220	242

22 13 16.60 Traps

		Crew	Daily Output	Labor-Hours	Unit	Material	2021 Bare Costs Labor	2021 Bare Costs Equipment	Total	Total Incl O&P
0010	**TRAPS**									
0030	Cast iron, service weight									
0050	Running P trap, without vent									
1100	2"	Q-1	16	1	Ea.	184	61		245	294
1140	3"		14	1.143		184	69.50		253.50	305
1150	4"	↓	13	1.231		184	75		259	315
1160	6"	Q-2	17	1.412	↓	815	89		904	1,025
1180	Running trap, single hub, with vent									
2080	3" pipe size, 3" vent	Q-1	14	1.143	Ea.	146	69.50		215.50	264
2120	4" pipe size, 4" vent	"	13	1.231		198	75		273	330
2140	5" pipe size, 4" vent	Q-2	11	2.182		315	138		453	550
2160	6" pipe size, 4" vent		10	2.400		850	152		1,002	1,150
2180	6" pipe size, 6" vent	↓	8	3		980	190		1,170	1,350
2200	8" pipe size, 4" vent	Q-3	10	3.200		3,925	206		4,131	4,625
2220	8" pipe size, 6" vent	"	8	4		3,025	258		3,283	3,725
2300	For double hub, vent, add				↓	10%	20%			

22 13 16.60 Traps	Crew	Daily Output	Labor-Hours	Unit	Material	2021 Bare Costs Labor	Equipment	Total	Total Incl O&P
2800	S trap,								
2850	4" pipe size	Q-1	13	1.231	Ea.	94	75		169	216
3000	P trap, B&S, 2" pipe size		16	1		38.50	61		99.50	133
3040	3" pipe size		14	1.143		57.50	69.50		127	167
3060	4" pipe size	↓	13	1.231		82.50	75		157.50	203
3080	5" pipe size	Q-2	18	1.333		208	84		292	355
3100	6" pipe size	"	17	1.412		255	89		344	415
3120	8" pipe size	Q-3	11	2.909		770	187		957	1,125
3130	10" pipe size	"	10	3.200		1,800	206		2,006	2,275
3150	P trap, no hub, 1-1/2" pipe size	Q-1	17	.941		21	57.50		78.50	109
3160	2" pipe size		16	1		19.60	61		80.60	113
3170	3" pipe size		14	1.143		43	69.50		112.50	152
3180	4" pipe size	↓	13	1.231		76.50	75		151.50	196
3190	6" pipe size	Q-2	17	1.412		185	89		274	335
3350	Deep seal trap, B&S								
3400	1-1/4" pipe size	Q-1	14	1.143	Ea.	77.50	69.50		147	190
3410	1-1/2" pipe size		14	1.143		77.50	69.50		147	190
3420	2" pipe size		14	1.143		66	69.50		135.50	177
3440	3" pipe size		12	1.333		84	81		165	214
3460	4" pipe size		11	1.455		133	88.50		221.50	279
3500	For trap primer connection, add				↓	178			178	196
3540	For trap with floor cleanout, add					70%	5%			
3580	For trap with adjustable cleanout, add	Q-1	10	1.600	Ea.	181	97.50		278.50	345
4700	Copper, drainage, drum trap									
4800	3" x 5" solid, 1-1/2" pipe size	1 Plum	16	.500	Ea.	355	34		389	440
4840	3" x 6" swivel, 1-1/2" pipe size	"	16	.500	"	790	34		824	915
5100	P trap, standard pattern									
5200	1-1/4" pipe size	1 Plum	18	.444	Ea.	234	30		264	300
5240	1-1/2" pipe size		17	.471		178	32		210	244
5260	2" pipe size		15	.533		275	36		311	355
5280	3" pipe size	↓	11	.727	↓	785	49		834	940
5340	With cleanout, swivel joint and slip joint									
5360	1-1/4" pipe size	1 Plum	18	.444	Ea.	151	30		181	211
5400	1-1/2" pipe size		17	.471		465	32		497	560
5420	2" pipe size	↓	15	.533	↓	400	36		436	495
5750	Chromed brass, tubular, P trap, without cleanout, 20 ga.									
5800	1-1/4" pipe size	1 Plum	18	.444	Ea.	18.60	30		48.60	65.50
5840	1-1/2" pipe size	"	17	.471	"	21	32		53	71
5900	With cleanout, 20 ga.									
5940	1-1/4" pipe size	1 Plum	18	.444	Ea.	30.50	30		60.50	79
6000	1-1/2" pipe size	"	17	.471	"	33	32		65	84
6350	S trap, without cleanout, 20 ga.									
6400	1-1/4" pipe size	1 Plum	18	.444	Ea.	48.50	30		78.50	98
6440	1-1/2" pipe size	"	17	.471	"	55.50	32		87.50	109
6550	With cleanout, 20 ga.									
6600	1-1/4" pipe size	1 Plum	18	.444	Ea.	75.50	30		105.50	128
6640	1-1/2" pipe size	"	17	.471		61.50	32		93.50	116
6660	Corrosion resistant, glass, P trap, 1-1/2" pipe size	Q-1	17	.941		80.50	57.50		138	174
6670	2" pipe size		16	1		105	61		166	207
6680	3" pipe size		14	1.143		216	69.50		285.50	340
6690	4" pipe size	↓	13	1.231		320	75		395	465
6700	6" pipe size	Q-2	17	1.412	↓	1,250	89		1,339	1,500
6710	ABS DWV P trap, solvent weld joint									

22 13 Facility Sanitary Sewerage

22 13 16 – Sanitary Waste and Vent Piping

22 13 16.60 Traps

		Crew	Daily Output	Labor-Hours	Unit	Material	2021 Bare Costs Labor	Equipment	Total	Total Incl O&P
6720	1-1/2" pipe size	1 Plum	18	.444	Ea.	22	30		52	69
6722	2" pipe size		17	.471		26.50	32		58.50	76.50
6724	3" pipe size		15	.533		114	36		150	179
6726	4" pipe size		14	.571		209	38.50		247.50	288
6732	PVC DWV P trap, solvent weld joint									
6733	1-1/2" pipe size	1 Plum	18	.444	Ea.	17.60	30		47.60	64.50
6734	2" pipe size		17	.471		21.50	32		53.50	71
6735	3" pipe size		15	.533		72.50	36		108.50	134
6736	4" pipe size		14	.571		174	38.50		212.50	250
6760	PP DWV, dilution trap, 1-1/2" pipe size		16	.500		281	34		315	360
6770	P trap, 1-1/2" pipe size		17	.471		70	32		102	125
6780	2" pipe size		16	.500		103	34		137	165
6790	3" pipe size		14	.571		191	38.50		229.50	268
6800	4" pipe size		13	.615		320	41.50		361.50	410
6830	S trap, 1-1/2" pipe size		16	.500		58	34		92	114
6840	2" pipe size		15	.533		86.50	36		122.50	150
6850	Universal trap, 1-1/2" pipe size		14	.571		115	38.50		153.50	184
6860	PVC DWV hub x hub, basin trap, 1-1/4" pipe size		18	.444		43.50	30		73.50	93
6870	Sink P trap, 1-1/2" pipe size		18	.444		12.15	30		42.15	58.50
6880	Tubular S trap, 1-1/2" pipe size		17	.471		31.50	32		63.50	82
6890	PVC sch. 40 DWV, drum trap									
6900	1-1/2" pipe size	1 Plum	16	.500	Ea.	44	34		78	99
6910	P trap, 1-1/2" pipe size		18	.444		10.20	30		40.20	56.50
6920	2" pipe size		17	.471		13.75	32		45.75	62.50
6930	3" pipe size		15	.533		36	36		72	93.50
6940	4" pipe size		14	.571		106	38.50		144.50	175
6950	P trap w/clean out, 1-1/2" pipe size		18	.444		17.25	30		47.25	64
6960	2" pipe size		17	.471		20.50	32		52.50	70
6970	P trap adjustable, 1-1/2" pipe size		17	.471		11.70	32		43.70	60.50
6980	P trap adj. w/union & cleanout, 1-1/2" pipe size		16	.500		44.50	34		78.50	99.50
7000	Trap primer, flow through type, 1/2" diameter		24	.333		47.50	22.50		70	86
7100	With sediment strainer		22	.364		51	24.50		75.50	93
7450	Trap primer distribution unit									
7500	2 openings	1 Plum	18	.444	Ea.	30.50	30		60.50	79
7540	3 openings		17	.471		33	32		65	84
7560	4 openings		16	.500		37.50	34		71.50	91.50
7850	Trap primer manifold									
7900	2 outlet	1 Plum	18	.444	Ea.	63.50	30		93.50	115
7940	4 outlet		16	.500		101	34		135	162
7960	6 outlet		15	.533		140	36		176	208
7980	8 outlet		13	.615		180	41.50		221.50	260

22 13 16.80 Vent Flashing and Caps

		Crew	Daily Output	Labor-Hours	Unit	Material	2021 Bare Costs Labor	Equipment	Total	Total Incl O&P
0010	**VENT FLASHING AND CAPS**									
0120	Vent caps									
0140	Cast iron									
0160	1-1/4" to 1-1/2" pipe	1 Plum	23	.348	Ea.	40.50	23.50		64	79.50
0170	2" to 2-1/8" pipe		22	.364		47	24.50		71.50	88.50
0180	2-1/2" to 3-5/8" pipe		21	.381		51.50	26		77.50	95
0190	4" to 4-1/8" pipe		19	.421		74	28.50		102.50	124
0200	5" to 6" pipe		17	.471		105	32		137	164
0300	PVC									
0320	1-1/4" to 1-1/2" pipe	1 Plum	24	.333	Ea.	13.70	22.50		36.20	48.50

279

For customer support on your Plumbing Costs with RSMeans Data, call 800.448.8182.

22 13 16 – Sanitary Waste and Vent Piping

22 13 16.80 Vent Flashing and Caps	Crew	Daily Output	Labor-Hours	Unit	Material	2021 Bare Costs Labor	Equipment	Total	Total Incl O&P
0330	2" to 2-1/8" pipe	1 Plum	23	.348	Ea.	15.50	23.50		39	52
0900	Vent flashing									
1000	Aluminum with lead ring									
1020	1-1/4" pipe	1 Plum	20	.400	Ea.	5.55	27		32.55	46.50
1030	1-1/2" pipe		20	.400		5.30	27		32.30	46.50
1040	2" pipe		18	.444		5.30	30		35.30	51
1050	3" pipe		17	.471		5.85	32		37.85	54
1060	4" pipe		16	.500		7.05	34		41.05	58.50
1350	Copper with neoprene ring									
1400	1-1/4" pipe	1 Plum	20	.400	Ea.	78.50	27		105.50	127
1430	1-1/2" pipe		20	.400		78.50	27		105.50	127
1440	2" pipe		18	.444		78.50	30		108.50	131
1450	3" pipe		17	.471		94.50	32		126.50	152
1460	4" pipe		16	.500		94.50	34		128.50	155
2000	Galvanized with neoprene ring									
2020	1-1/4" pipe	1 Plum	20	.400	Ea.	14.10	27		41.10	56
2030	1-1/2" pipe		20	.400		18.70	27		45.70	61
2040	2" pipe		18	.444		20.50	30		50.50	67.50
2050	3" pipe		17	.471		22.50	32		54.50	72.50
2060	4" pipe		16	.500		26	34		60	79.50
2980	Neoprene, one piece									
3000	1-1/4" pipe	1 Plum	24	.333	Ea.	5.65	22.50		28.15	39.50
3030	1-1/2" pipe		24	.333		3.14	22.50		25.64	37
3040	2" pipe		23	.348		4.48	23.50		27.98	40
3050	3" pipe		21	.381		6.55	26		32.55	45.50
3060	4" pipe		20	.400		9.80	27		36.80	51.50
4000	Lead, 4#, 8" skirt, vent through roof									
4100	2" pipe	1 Plum	18	.444	Ea.	43	30		73	92
4110	3" pipe		17	.471		49	32		81	102
4120	4" pipe		16	.500		55.50	34		89.50	112
4130	6" pipe		14	.571		83.50	38.50		122	150

22 13 19 – Sanitary Waste Piping Specialties

22 13 19.13 Sanitary Drains

		Crew	Daily Output	Labor-Hours	Unit	Material	2021 Bare Costs Labor	Equipment	Total	Total Incl O&P
0010	**SANITARY DRAINS**									
0400	Deck, auto park, CI, 13" top									
0440	3", 4", 5", and 6" pipe size	Q-1	8	2	Ea.	2,100	122		2,222	2,500
0480	For galvanized body, add				"	1,300			1,300	1,425
0800	Promenade, heelproof grate, CI, 14" top									
0840	2", 3", and 4" pipe size	Q-1	10	1.600	Ea.	845	97.50		942.50	1,075
0860	5" and 6" pipe size		9	1.778		1,050	108		1,158	1,300
0880	8" pipe size		8	2		1,200	122		1,322	1,475
0940	For galvanized body, add					675			675	740
0960	With polished bronze top, 2"-3"-4" diam.					1,550			1,550	1,700
1200	Promenade, heelproof grate, CI, lateral, 14" top									
1240	2", 3" and 4" pipe size	Q-1	10	1.600	Ea.	1,100	97.50		1,197.50	1,350
1260	5" and 6" pipe size		9	1.778		1,250	108		1,358	1,525
1280	8" pipe size		8	2		1,425	122		1,547	1,750
1340	For galvanized body, add					550			550	605
1360	For polished bronze top, add					1,075			1,075	1,175
1500	Promenade, slotted grate, CI, 11" top									
1540	2", 3", 4", 5", and 6" pipe size	Q-1	12	1.333	Ea.	555	81		636	730
1600	For galvanized body, add					385			385	425

22 13 19.13 Sanitary Drains

		Crew	Daily Output	Labor-Hours	Unit	Material	2021 Bare Costs Labor	Equipment	Total	Total Incl O&P
1640	With polished bronze top				Ea.	1,125			1,125	1,250
2000	Floor, medium duty, CI, deep flange, 7" diam. top									
2040	2" and 3" pipe size	Q-1	12	1.333	Ea.	355	81		436	510
2080	For galvanized body, add					162			162	178
2120	With polished bronze top					440			440	485
2160	Heavy duty, CI, 12" diam. anti-tilt grate									
2180	2", 3", 4", 5" and 6" pipe size	Q-1	10	1.600	Ea.	730	97.50		827.50	950
2220	For galvanized body, add					495			495	545
2240	With polished bronze top					1,250			1,250	1,375
2300	Extra-heavy duty, CI, 15" anti-tilt grate									
2320	4", 5", 6", and 8" pipe size	Q-1	8	2	Ea.	1,825	122		1,947	2,175
2360	For galvanized body, add					810			810	890
2380	With polished bronze top					2,550			2,550	2,800
2400	Heavy duty, with sediment bucket, CI, 12" diam. loose grate									
2420	2", 3", 4", 5", and 6" pipe size	Q-1	9	1.778	Ea.	1,050	108		1,158	1,300
2440	For galvanized body, add					735			735	810
2460	With polished bronze top					1,575			1,575	1,725
2500	Heavy duty, cleanout & trap w/bucket, CI, 15" top									
2540	2", 3", and 4" pipe size	Q-1	6	2.667	Ea.	6,950	162		7,112	7,875
2560	For galvanized body, add					2,800			2,800	3,100
2580	With polished bronze top					11,200			11,200	12,300
2600	Medium duty, with perforated SS basket, CI, body,									
2610	18" top for refuse container washing area									
2620	2" thru 6" pipe size	Q-1	4	4	Ea.	4,725	244		4,969	5,575
2630	Acid resistant									
2638	PVC									
2640	2", 3" and 4" pipe size	Q-1	16	1	Ea.	395	61		456	525
2644	Cast iron, epoxy coated									
2646	2", 3" and 4" pipe size	Q-1	14	1.143	Ea.	650	69.50		719.50	825
2650	PVC or ABS thermoplastic									
2660	3" and 4" pipe size	Q-1	16	1	Ea.	425	61		486	560
2680	Extra heavy duty, oil intercepting, gas seal cone,									
2690	with cleanout, loose grate, CI, body 16" top									
2700	3" and 4" diameter outlet, 4" slab depth	Q-1	4	4	Ea.	10,400	244		10,644	11,900
2720	4" diameter outlet, 8" slab depth		3	5.333		8,625	325		8,950	9,950
2740	4" diam. outlet, 10"-12" slab depth, 16" top		2	8		13,400	485		13,885	15,500
2910	Prison cell, vandal-proof, 1-1/2", and 2" diam. pipe		12	1.333		565	81		646	740
2920	3" pipe size		10	1.600		540	97.50		637.50	735
2930	Trap drain, light duty, backwater valve CI top									
2950	8" diameter top, 2" pipe size	Q-1	12	1.333	Ea.	510	81		591	680
2960	10" diameter top, 3" pipe size		10	1.600		700	97.50		797.50	915
2970	12" diameter top, 4" pipe size		8	2		980	122		1,102	1,250

22 13 19.14 Floor Receptors

		Crew	Daily Output	Labor-Hours	Unit	Material	2021 Bare Costs Labor	Equipment	Total	Total Incl O&P
0010	**FLOOR RECEPTORS**, For connection to 2", 3" & 4" diameter pipe									
0200	12-1/2" square top, 25 sq. in. open area	Q-1	10	1.600	Ea.	1,750	97.50		1,847.50	2,075
0300	For grate with 4" diameter x 3-3/4" high funnel, add					1,875			1,875	2,050
0400	For grate with 6" diameter x 6" high funnel, add					245			245	269
0500	For full hinged grate with open center, add					108			108	119
0600	For aluminum bucket, add					213			213	235
0700	For acid-resisting bucket, add					310			310	345
0900	For stainless steel mesh bucket liner, add					315			315	350
1000	For bronze antisplash dome strainer, add					152			152	167

For customer support on your Plumbing Costs with RSMeans Data, call 800.448.8182.

281

22 13 Facility Sanitary Sewerage

22 13 19 - Sanitary Waste Piping Specialties

22 13 19.14 Floor Receptors

		Crew	Daily Output	Labor-Hours	Unit	Material	2021 Bare Costs Labor	2021 Bare Costs Equipment	Total	Total Incl O&P
1100	For partial solid cover, add				Ea.	77.50			77.50	85.50
1200	For trap primer connection, add					122			122	134
2000	12-5/8" diameter top, 40 sq. in. open area	Q-1	10	1.600		815	97.50		912.50	1,050
2100	For options, add same prices as square top									
3000	8" x 4" rectangular top, 7.5 sq. in. open area	Q-1	14	1.143	Ea.	990	69.50		1,059.50	1,200
3100	For trap primer connections, add					98.50			98.50	108
4000	24" x 16" rectangular top, 70 sq. in. open area	Q-1	4	4		5,725	244		5,969	6,675
4100	For trap primer connection, add					254			254	280

22 13 19.15 Sink Waste Treatment

		Crew	Daily Output	Labor-Hours	Unit	Material	2021 Bare Costs Labor	2021 Bare Costs Equipment	Total	Total Incl O&P
0010	**SINK WASTE TREATMENT**, System for commercial kitchens									
0100	includes clock timer & fittings									
0200	System less chemical, wall mounted cabinet	1 Plum	16	.500	Ea.	480	34		514	580
2000	Chemical, 1 gallon, add					42.50			42.50	46.50
2100	6 gallons, add					191			191	210
2200	15 gallons, add					520			520	570
2300	30 gallons, add					975			975	1,075
2400	55 gallons, add					1,650			1,650	1,825

22 13 19.39 Floor Drain Trap Seal

		Crew	Daily Output	Labor-Hours	Unit	Material	2021 Bare Costs Labor	2021 Bare Costs Equipment	Total	Total Incl O&P
0010	**FLOOR DRAIN TRAP SEAL**									
0100	Inline									
0110	2"	1 Plum	20	.400	Ea.	27.50	27		54.50	70.50
0120	3"		16	.500		43.50	34		77.50	98.50
0130	3.5"		14	.571		32	38.50		70.50	92.50
0140	4"		10	.800		53	54		107	140

22 13 23 - Sanitary Waste Interceptors

22 13 23.10 Interceptors

		Crew	Daily Output	Labor-Hours	Unit	Material	2021 Bare Costs Labor	2021 Bare Costs Equipment	Total	Total Incl O&P
0010	**INTERCEPTORS**									
0150	Grease, fabricated steel, 4 GPM, 8 lb. fat capacity	1 Plum	4	2	Ea.	1,750	135		1,885	2,125
0200	7 GPM, 14 lb. fat capacity		4	2		2,300	135		2,435	2,725
1000	10 GPM, 20 lb. fat capacity		4	2		2,875	135		3,010	3,350
1040	15 GPM, 30 lb. fat capacity		4	2		4,025	135		4,160	4,625
1060	20 GPM, 40 lb. fat capacity		3	2.667		5,200	181		5,381	6,000
1080	25 GPM, 50 lb. fat capacity	Q-1	3.50	4.571		5,525	279		5,804	6,500
1100	35 GPM, 70 lb. fat capacity		3	5.333		7,225	325		7,550	8,425
1120	50 GPM, 100 lb. fat capacity		2	8		9,050	485		9,535	10,700
1140	75 GPM, 150 lb. fat capacity		2	8		19,300	485		19,785	21,900
1160	100 GPM, 200 lb. fat capacity		2	8		21,700	485		22,185	24,600
1180	150 GPM, 300 lb. fat capacity		2	8		24,200	485		24,685	27,300
1200	200 GPM, 400 lb. fat capacity		1.50	10.667		35,000	650		35,650	39,500
1220	250 GPM, 500 lb. fat capacity		1.30	12.308		38,400	750		39,150	43,400
1240	300 GPM, 600 lb. fat capacity		1	16		45,300	975		46,275	51,500
1260	400 GPM, 800 lb. fat capacity	Q-2	1.20	20		39,800	1,275		41,075	45,600
1280	500 GPM, 1,000 lb. fat capacity	"	1	24		65,500	1,525		67,025	74,500
1580	For seepage pan, add					7%				
3000	Hair, cast iron, 1-1/4" and 1-1/2" pipe connection	1 Plum	8	1	Ea.	625	67.50		692.50	785
3100	For chrome-plated cast iron, add					470			470	520
3200	For polished bronze, add					1,100			1,100	1,200
3400	Lint interceptor, fabricated steel									
3410	Size based on 10 GPM per machine									
3420	30 GPM, 2" pipe size	Q-1	3	5.333	Ea.	7,300	325		7,625	8,500
3430	70 GPM, 3" pipe size		2.50	6.400		8,500	390		8,890	9,925
3440	100 GPM, 4" pipe size		2	8		10,000	485		10,485	11,700

For customer support on your Plumbing Costs with RSMeans Data, call 800.448.8182.

22 13 Facility Sanitary Sewerage

22 13 23 – Sanitary Waste Interceptors

22 13 23.10 Interceptors

		Crew	Daily Output	Labor-Hours	Unit	Material	2021 Bare Costs Labor	2021 Bare Costs Equipment	Total	Total Incl O&P
3450	200 GPM, 4" pipe size	Q-1	1.50	10.667	Ea.	10,700	650		11,350	12,700
3460	300 GPM, 6" pipe size	↓	1	16		14,100	975		15,075	17,000
3470	400 GPM, 6" pipe size	Q-2	1.20	20		16,000	1,275		17,275	19,500
3480	500 GPM, 6" pipe size	"	1	24		16,300	1,525		17,825	20,300
4000	Oil, fabricated steel, 10 GPM, 2" pipe size	1 Plum	4	2		3,700	135		3,835	4,275
4100	15 GPM, 2" or 3" pipe size		4	2		5,075	135		5,210	5,800
4120	20 GPM, 2" or 3" pipe size	↓	3	2.667		6,700	181		6,881	7,650
4140	25 GPM, 2" or 3" pipe size	Q-1	3.50	4.571		6,675	279		6,954	7,775
4160	35 GPM, 2", 3", or 4" pipe size		3	5.333		8,125	325		8,450	9,400
4180	50 GPM, 2", 3", or 4" pipe size		2	8		10,900	485		11,385	12,700
4200	75 GPM, 3" pipe size		2	8		20,600	485		21,085	23,400
4220	100 GPM, 3" pipe size		2	8		20,500	485		20,985	23,200
4240	150 GPM, 4" pipe size		2	8		25,400	485		25,885	28,600
4260	200 GPM, 4" pipe size		1.50	10.667		35,900	650		36,550	40,400
4280	250 GPM, 5" pipe size		1.30	12.308		41,500	750		42,250	46,800
4300	300 GPM, 5" pipe size	↓	1	16		47,000	975		47,975	53,000
4320	400 GPM, 6" pipe size	Q-2	1.20	20		60,500	1,275		61,775	68,500
4340	500 GPM, 6" pipe size	"	1	24	↓	76,000	1,525		77,525	86,500
5000	Sand interceptor, fabricated steel									
5020	20 GPM, 4" pipe size	Q-1	3	5.333	Ea.	11,400	325		11,725	13,000
5030	50 GPM, 4" pipe size		2.50	6.400		12,100	390		12,490	13,900
5040	150 GPM, 4" pipe size		2	8		36,500	485		36,985	40,900
5050	250 GPM, 6" pipe size	↓	1.30	12.308		38,000	750		38,750	42,900
5060	500 GPM, 6" pipe size	Q-2	1	24		48,800	1,525		50,325	56,000
6000	Solids, precious metals recovery, CI, 1-1/4" to 2" pipe	1 Plum	4	2		730	135		865	1,000
6100	Dental lab., large, CI, 1-1/2" to 2" pipe	"	3	2.667	↓	2,550	181		2,731	3,100

22 13 26 – Sanitary Waste Separators

22 13 26.10 Separators

		Crew	Daily Output	Labor-Hours	Unit	Material	2021 Bare Costs Labor	2021 Bare Costs Equipment	Total	Total Incl O&P
0010	**SEPARATORS**, Entrainment eliminator, steel body, 150 PSIG									
0100	1/4" size	1 Stpi	24	.333	Ea.	293	23		316	355
0120	1/2" size		24	.333		240	23		263	298
0140	3/4" size		20	.400		245	27.50		272.50	310
0160	1" size		19	.421		315	29		344	390
0180	1-1/4" size		15	.533		330	36.50		366.50	420
0200	1-1/2" size		13	.615		365	42		407	470
0220	2" size	↓	11	.727	↓	410	49.50		459.50	525

22 13 29 – Sanitary Sewerage Pumps

22 13 29.13 Wet-Pit-Mounted, Vertical Sewerage Pumps

		Crew	Daily Output	Labor-Hours	Unit	Material	2021 Bare Costs Labor	2021 Bare Costs Equipment	Total	Total Incl O&P
0010	**WET-PIT-MOUNTED, VERTICAL SEWERAGE PUMPS**									
0020	Controls incl. alarm/disconnect panel w/wire. Excavation not included									
0260	Simplex, 9 GPM at 60 PSIG, 91 gal. tank				Ea.	3,750			3,750	4,100
0300	Unit with manway, 26" ID, 18" high					4,150			4,150	4,550
0340	26" ID, 36" high					4,075			4,075	4,475
0380	43" ID, 4' high					4,325			4,325	4,750
0600	Simplex, 9 GPM at 60 PSIG, 150 gal. tank, indoor					4,350			4,350	4,775
0700	Unit with manway, 26" ID, 36" high					4,975			4,975	5,475
0740	26" ID, 4' high					5,125			5,125	5,625
2000	Duplex, 18 GPM at 60 PSIG, 150 gal. tank, indoor					8,050			8,050	8,850
2060	Unit with manway, 43" ID, 4' high					9,575			9,575	10,500
2400	For core only				↓	2,025			2,025	2,225
3000	Indoor residential type installation									
3020	Simplex, 9 GPM at 60 PSIG, 91 gal. HDPE tank				Ea.	3,750			3,750	4,125

22 13 29.14 Sewage Ejector Pumps	Crew	Daily Output	Labor-Hours	Unit	Material	2021 Bare Costs Labor	Equipment	Total	Total Incl O&P
0010 **SEWAGE EJECTOR PUMPS**, With operating and level controls									
0100 Simplex system incl. tank, cover, pump 15' head									
0500 37 gal. PE tank, 12 GPM, 1/2 HP, 2" discharge	Q-1	3.20	5	Ea.	515	305		820	1,025
0510 3" discharge		3.10	5.161		550	315		865	1,075
0530 87 GPM, .7 HP, 2" discharge		3.20	5		795	305		1,100	1,325
0540 3" discharge		3.10	5.161		860	315		1,175	1,425
0600 45 gal. coated stl. tank, 12 GPM, 1/2 HP, 2" discharge		3	5.333		925	325		1,250	1,500
0610 3" discharge		2.90	5.517		960	335		1,295	1,550
0630 87 GPM, .7 HP, 2" discharge		3	5.333		1,175	325		1,500	1,775
0640 3" discharge		2.90	5.517		1,250	335		1,585	1,875
0660 134 GPM, 1 HP, 2" discharge		2.80	5.714		1,275	350		1,625	1,925
0680 3" discharge		2.70	5.926		1,350	360		1,710	2,025
0700 70 gal. PE tank, 12 GPM, 1/2 HP, 2" discharge		2.60	6.154		995	375		1,370	1,650
0710 3" discharge		2.40	6.667		1,075	405		1,480	1,775
0730 87 GPM, .7 HP, 2" discharge		2.50	6.400		1,325	390		1,715	2,025
0740 3" discharge		2.30	6.957		1,375	425		1,800	2,125
0760 134 GPM, 1 HP, 2" discharge		2.20	7.273		1,400	445		1,845	2,200
0770 3" discharge		2	8		1,500	485		1,985	2,375
0800 75 gal. coated stl. tank, 12 GPM, 1/2 HP, 2" discharge		2.40	6.667		1,100	405		1,505	1,825
0810 3" discharge		2.20	7.273		1,150	445		1,595	1,925
0830 87 GPM, .7 HP, 2" discharge		2.30	6.957		1,400	425		1,825	2,150
0840 3" discharge		2.10	7.619		1,475	465		1,940	2,300
0860 134 GPM, 1 HP, 2" discharge		2	8		1,500	485		1,985	2,375
0880 3" discharge	▼	1.80	8.889	▼	1,575	540		2,115	2,525
1040 Duplex system incl. tank, covers, pumps									
1060 110 gal. fiberglass tank, 24 GPM, 1/2 HP, 2" discharge	Q-1	1.60	10	Ea.	2,000	610		2,610	3,100
1080 3" discharge		1.40	11.429		2,100	695		2,795	3,350
1100 174 GPM, .7 HP, 2" discharge		1.50	10.667		2,575	650		3,225	3,825
1120 3" discharge		1.30	12.308		2,675	750		3,425	4,075
1140 268 GPM, 1 HP, 2" discharge		1.20	13.333		2,800	810		3,610	4,300
1160 3" discharge	▼	1	16		2,900	975		3,875	4,650
1260 135 gal. coated stl. tank, 24 GPM, 1/2 HP, 2" discharge	Q-2	1.70	14.118		2,050	890		2,940	3,600
2000 3" discharge		1.60	15		2,175	950		3,125	3,825
2640 174 GPM, .7 HP, 2" discharge		1.60	15		2,700	950		3,650	4,400
2660 3" discharge		1.50	16		2,850	1,000		3,850	4,625
2700 268 GPM, 1 HP, 2" discharge		1.30	18.462		2,925	1,175		4,100	4,975
3040 3" discharge		1.10	21.818		3,125	1,375		4,500	5,475
3060 275 gal. coated stl. tank, 24 GPM, 1/2 HP, 2" discharge		1.50	16		2,575	1,000		3,575	4,350
3080 3" discharge		1.40	17.143		2,600	1,075		3,675	4,500
3100 174 GPM, .7 HP, 2" discharge		1.40	17.143		3,325	1,075		4,400	5,275
3120 3" discharge		1.30	18.462		3,575	1,175		4,750	5,700
3140 268 GPM, 1 HP, 2" discharge		1.10	21.818		3,650	1,375		5,025	6,050
3160 3" discharge	▼	.90	26.667	▼	3,850	1,675		5,525	6,750
3260 Pump system accessories, add									
3300 Alarm horn and lights, 115 V mercury switch	Q-1	8	2	Ea.	102	122		224	294
3340 Switch, mag. contactor, alarm bell, light, 3 level control		5	3.200		520	195		715	860
3380 Alternator, mercury switch activated	▼	4	4	▼	935	244		1,179	1,400

22 14 23 – Storm Drainage Piping Specialties

22 14 23.33 Backwater Valves	Crew	Daily Output	Labor-Hours	Unit	Material	2021 Bare Costs Labor	Equipment	Total	Total Incl O&P
0010 **BACKWATER VALVES**, CI Body									
6980 Bronze gate and automatic flapper valves									
7000 3" and 4" pipe size	Q-1	13	1.231	Ea.	2,675	75		2,750	3,050
7100 5" and 6" pipe size	"	13	1.231	"	4,100	75		4,175	4,625
7240 Bronze flapper valve, bolted cover									
7260 2" pipe size	Q-1	16	1	Ea.	785	61		846	950
7280 3" pipe size		14.50	1.103		1,225	67		1,292	1,450
7300 4" pipe size		13	1.231		1,150	75		1,225	1,375
7320 5" pipe size	Q-2	18	1.333		1,800	84		1,884	2,125
7340 6" pipe size	"	17	1.412		2,550	89		2,639	2,950
7360 8" pipe size	Q-3	10	3.200		3,650	206		3,856	4,325
7380 10" pipe size	"	9	3.556		5,975	229		6,204	6,925
7500 For threaded cover, same cost									
7540 Revolving disk type, same cost as flapper type									

22 14 26 – Facility Storm Drains

22 14 26.13 Roof Drains

	Crew	Daily Output	Labor-Hours	Unit	Material	2021 Bare Costs Labor	Equipment	Total	Total Incl O&P
0010 **ROOF DRAINS**									
0140 Cornice, CI, 45° or 90° outlet									
0200 3" and 4" pipe size	Q-1	12	1.333	Ea.	400	81		481	555
0260 For galvanized body, add					128			128	141
0280 For polished bronze dome, add					95.50			95.50	105
3860 Roof, flat metal deck, CI body, 12" CI dome									
3880 2" pipe size	Q-1	15	1.067	Ea.	335	65		400	460
3890 3" pipe size		14	1.143		470	69.50		539.50	620
3900 4" pipe size		13	1.231		565	75		640	730
3910 5" pipe size		12	1.333		730	81		811	925
3920 6" pipe size		10	1.600		930	97.50		1,027.50	1,175
4280 Integral expansion joint, CI body, 12" CI dome									
4300 2" pipe size	Q-1	8	2	Ea.	440	122		562	665
4320 3" pipe size		7	2.286		440	139		579	695
4340 4" pipe size		6	2.667		440	162		602	725
4360 5" pipe size		4	4		960	244		1,204	1,425
4380 6" pipe size		3	5.333		440	325		765	970
4400 8" pipe size		3	5.333		1,675	325		2,000	2,300
4440 For galvanized body, add					425			425	470
4620 Main, all aluminum, 12" low profile dome									
4640 2", 3" and 4" pipe size	Q-1	14	1.143	Ea.	665	69.50		734.50	835
4660 5" and 6" pipe size		13	1.231		880	75		955	1,075
4680 8" pipe size		10	1.600		825	97.50		922.50	1,050
4690 Main, CI body, 12" poly. dome, 2", 3" & 4" pipe		8	2		495	122		617	725
4710 5" and 6" pipe size		6	2.667		705	162		867	1,025
4720 8" pipe size		4	4		925	244		1,169	1,400
4730 For underdeck clamp, add		22	.727		345	44.50		389.50	445
4740 For vandalproof dome, add					86.50			86.50	95
4750 For galvanized body, add					725			725	800
4760 Main, ABS body and dome, 2" pipe size	Q-1	14	1.143		140	69.50		209.50	258
4780 3" pipe size		14	1.143		140	69.50		209.50	258
4800 4" pipe size		14	1.143		140	69.50		209.50	258
4820 For underdeck clamp, add		24	.667		46.50	40.50		87	112
4900 Terrace planting area, with perforated overflow, CI									
4920 2", 3" and 4" pipe size	Q-1	8	2	Ea.	815	122		937	1,075

285

22 14 26 – Facility Storm Drains

22 14 26.16 Facility Area Drains	Crew	Daily Output	Labor-Hours	Unit	Material	2021 Bare Costs Labor	2021 Bare Costs Equipment	Total	Total Incl O&P
0010 **FACILITY AREA DRAINS**									
4980 Scupper floor, oblique strainer, CI									
5000 6" x 7" top, 2", 3" and 4" pipe size	Q-1	16	1	Ea.	435	61		496	565
5100 8" x 12" top, 5" and 6" pipe size	"	14	1.143		840	69.50		909.50	1,025
5160 For galvanized body, add					40%				
5200 For polished bronze strainer, add					85%				

22 14 26.19 Facility Trench Drains

22 14 26.19 Facility Trench Drains	Crew	Daily Output	Labor-Hours	Unit	Material	2021 Bare Costs Labor	2021 Bare Costs Equipment	Total	Total Incl O&P
0010 **FACILITY TRENCH DRAINS**									
5980 Trench, floor, heavy duty, modular, CI, 12" x 12" top									
6000 2", 3", 4", 5" & 6" pipe size	Q-1	8	2	Ea.	1,175	122		1,297	1,475
6100 For unit with polished bronze top		8	2		1,725	122		1,847	2,075
6200 For 12" extension section, CI top		8	2		1,175	122		1,297	1,475
6240 For 12" extension section, polished bronze top		8	2		1,900	122		2,022	2,250
6600 Trench, floor, for cement concrete encasement									
6610 Not including trenching or concrete									
6640 Polyester polymer concrete									
6650 4" internal width, with grate									
6660 Light duty steel grate	Q-1	120	.133	L.F.	69.50	8.10		77.60	88.50
6670 Medium duty steel grate		115	.139		125	8.50		133.50	150
6680 Heavy duty iron grate		110	.145		123	8.85		131.85	149
6700 12" internal width, with grate									
6770 Heavy duty galvanized grate	Q-1	80	.200	L.F.	168	12.20		180.20	203
6800 Fiberglass									
6810 8" internal width, with grate									
6820 Medium duty galvanized grate	Q-1	115	.139	L.F.	113	8.50		121.50	137
6830 Heavy duty iron grate	"	110	.145	"	199	8.85		207.85	232

22 14 29 – Sump Pumps

22 14 29.13 Wet-Pit-Mounted, Vertical Sump Pumps

22 14 29.13 Wet-Pit-Mounted, Vertical Sump Pumps	Crew	Daily Output	Labor-Hours	Unit	Material	2021 Bare Costs Labor	2021 Bare Costs Equipment	Total	Total Incl O&P
0010 **WET-PIT-MOUNTED, VERTICAL SUMP PUMPS**									
0400 Molded PVC base, 21 GPM at 15' head, 1/3 HP	1 Plum	5	1.600	Ea.	138	108		246	315
0800 Iron base, 21 GPM at 15' head, 1/3 HP		5	1.600		155	108		263	330
1200 Solid brass, 21 GPM at 15' head, 1/3 HP		5	1.600		250	108		358	435
2000 Sump pump, single stage									
2010 25 GPM, 1 HP, 1-1/2" discharge	Q-1	1.80	8.889	Ea.	4,000	540		4,540	5,200
2020 75 GPM, 1-1/2 HP, 2" discharge		1.50	10.667		4,225	650		4,875	5,625
2030 100 GPM, 2 HP, 2-1/2" discharge		1.30	12.308		4,300	750		5,050	5,850
2040 150 GPM, 3 HP, 3" discharge		1.10	14.545		4,300	885		5,185	6,050
2050 200 GPM, 3 HP, 3" discharge		1	16		4,550	975		5,525	6,450
2060 300 GPM, 10 HP, 4" discharge	Q-2	1.20	20		4,925	1,275		6,200	7,275
2070 500 GPM, 15 HP, 5" discharge		1.10	21.818		5,600	1,375		6,975	8,200
2080 800 GPM, 20 HP, 6" discharge		1	24		6,600	1,525		8,125	9,550
2090 1,000 GPM, 30 HP, 6" discharge		.85	28.235		7,250	1,775		9,025	10,700
2100 1,600 GPM, 50 HP, 8" discharge		.72	33.333		11,300	2,100		13,400	15,600
2110 2,000 GPM, 60 HP, 8" discharge	Q-3	.85	37.647		11,600	2,425		14,025	16,300
2202 For general purpose float switch, copper coated float, add	Q-1	5	3.200		109	195		304	410

22 14 Facility Storm Drainage

22 14 29 – Sump Pumps

22 14 29.16 Submersible Sump Pumps

22 14 29.16 Submersible Sump Pumps	Crew	Daily Output	Labor-Hours	Unit	Material	2021 Bare Costs Labor	Equipment	Total	Total Incl O&P
0010 **SUBMERSIBLE SUMP PUMPS**									
1000 Elevator sump pumps, automatic									
1010 Complete systems, pump, oil detector, controls and alarm									
1020 1-1/2" discharge, does not include the sump pit/tank									
1040 1/3 HP, 115 V	1 Plum	4.40	1.818	Ea.	1,900	123		2,023	2,275
1050 1/2 HP, 115 V		4	2		2,025	135		2,160	2,425
1060 1/2 HP, 230 V		4	2		2,000	135		2,135	2,400
1070 3/4 HP, 115 V		3.60	2.222		2,125	150		2,275	2,550
1080 3/4 HP, 230 V	▼	3.60	2.222	▼	2,100	150		2,250	2,525
1100 Sump pump only									
1110 1/3 HP, 115 V	1 Plum	6.40	1.250	Ea.	177	84.50		261.50	320
1120 1/2 HP, 115 V		5.80	1.379		253	93.50		346.50	420
1130 1/2 HP, 230 V		5.80	1.379		294	93.50		387.50	465
1140 3/4 HP, 115 V		5.40	1.481		335	100		435	520
1150 3/4 HP, 230 V	▼	5.40	1.481	▼	380	100		480	570
1200 Oil detector, control and alarm only									
1210 115 V	1 Plum	8	1	Ea.	1,825	67.50		1,892.50	2,125
1220 230 V	"	8	1	"	1,825	67.50		1,892.50	2,125
7000 Sump pump, automatic									
7100 Plastic, 1-1/4" discharge, 1/4 HP	1 Plum	6.40	1.250	Ea.	173	84.50		257.50	315
7140 1/3 HP		6	1.333		229	90.50		319.50	385
7160 1/2 HP		5.40	1.481		250	100		350	425
7180 1-1/2" discharge, 1/2 HP		5.20	1.538		350	104		454	540
7500 Cast iron, 1-1/4" discharge, 1/4 HP		6	1.333		243	90.50		333.50	400
7540 1/3 HP		6	1.333		285	90.50		375.50	450
7560 1/2 HP	▼	5	1.600	▼	345	108		453	540

22 14 53 – Rainwater Storage Tanks

22 14 53.13 Fiberglass, Rainwater Storage Tank

	Crew	Daily Output	Labor-Hours	Unit	Material	2021 Bare Costs Labor	Equipment	Total	Total Incl O&P
0010 **FIBERGLASS, RAINWATER STORAGE TANK**									
2000 600 gallon	B-21B	3.75	10.667	Ea.	4,325	515	127	4,967	5,650
2010 1,000 gallon		3.50	11.429		5,450	550	136	6,136	6,975
2020 2,000 gallon		3.25	12.308		7,675	595	146	8,416	9,475
2030 4,000 gallon		3	13.333		10,100	645	159	10,904	12,200
2040 6,000 gallon		2.65	15.094		11,500	730	180	12,410	13,900
2050 8,000 gallon		2.30	17.391		14,100	840	207	15,147	17,000
2060 10,000 gallon		2	20		15,400	965	238	16,603	18,700
2070 12,000 gallon		1.50	26.667		22,100	1,275	315	23,690	26,700
2080 15,000 gallon		1	40		24,200	1,925	475	26,600	30,100
2090 20,000 gallon		.75	53.333		31,600	2,575	635	34,810	39,300
2100 25,000 gallon		.50	80		48,000	3,850	950	52,800	59,500
2110 30,000 gallon		.35	114		58,000	5,500	1,350	64,850	73,500
2120 40,000 gallon	▼	.30	133	▼	79,000	6,425	1,575	87,000	98,000

22 15 13 – General Service Compressed-Air Piping

22 15 13.10 Compressor Accessories	Crew	Daily Output	Labor-Hours	Unit	Material	2021 Bare Costs Labor	Equipment	Total	Total Incl O&P
0010 **COMPRESSOR ACCESSORIES**									
3460 Air filter, regulator, lubricator combination									
3470 Flush mount									
3480 Adjustable range 0-140 psi									
3500 1/8" NPT, 34 SCFM	1 Stpi	17	.471	Ea.	109	32		141	168
3510 1/4" NPT, 61 SCFM		17	.471		196	32		228	264
3520 3/8" NPT, 85 SCFM		16	.500		200	34		234	271
3530 1/2" NPT, 150 SCFM		15	.533		215	36.50		251.50	291
3540 3/4" NPT, 171 SCFM		14	.571		264	39		303	350
3550 1" NPT, 150 SCFM	↓	13	.615	↓	425	42		467	530
4000 Couplers, air line, sleeve type									
4010 Female, connection size NPT									
4020 1/4"	1 Stpi	38	.211	Ea.	8.50	14.40		22.90	31
4030 3/8"		36	.222		7.70	15.20		22.90	31
4040 1/2"		35	.229		22	15.60		37.60	47.50
4050 3/4"	↓	34	.235	↓	22	16.10		38.10	48
4100 Male									
4110 1/4"	1 Stpi	38	.211	Ea.	7.80	14.40		22.20	30
4120 3/8"		36	.222		9.55	15.20		24.75	33
4130 1/2"		35	.229		16.90	15.60		32.50	42
4140 3/4"	↓	34	.235	↓	15.55	16.10		31.65	41
4150 Coupler, combined male and female halves									
4160 1/2"	1 Stpi	17	.471	Ea.	38.50	32		70.50	90.50
4170 3/4"	"	15	.533	"	37.50	36.50		74	95.50

22 15 19 – General Service Packaged Air Compressors and Receivers

22 15 19.10 Air Compressors

	Crew	Daily Output	Labor-Hours	Unit	Material	2021 Bare Costs Labor	Equipment	Total	Total Incl O&P
0010 **AIR COMPRESSORS**									
5250 Air, reciprocating air cooled, splash lubricated, tank mounted									
5300 Single stage, 1 phase, 140 psi									
5303 1/2 HP, 17 gal. tank	1 Stpi	3	2.667	Ea.	1,975	182		2,157	2,425
5305 3/4 HP, 30 gal. tank		2.60	3.077		1,625	210		1,835	2,100
5307 1 HP, 30 gal. tank	↓	2.20	3.636		2,375	249		2,624	3,000
5309 2 HP, 30 gal. tank	Q-5	4	4		2,850	246		3,096	3,500
5310 3 HP, 30 gal. tank		3.60	4.444		3,150	273		3,423	3,850
5314 3 HP, 60 gal. tank		3.50	4.571		3,275	281		3,556	4,025
5320 5 HP, 60 gal. tank		3.20	5		3,475	310		3,785	4,250
5330 5 HP, 80 gal. tank		3	5.333		3,475	330		3,805	4,325
5340 7.5 HP, 80 gal. tank	↓	2.60	6.154		5,200	380		5,580	6,300
5600 2 stage pkg., 3 phase									
5650 6 CFM at 125 psi, 1-1/2 HP, 60 gal. tank	Q-5	3	5.333	Ea.	2,775	330		3,105	3,550
5670 10.9 CFM at 125 psi, 3 HP, 80 gal. tank		1.50	10.667		3,225	655		3,880	4,525
5680 38.7 CFM at 125 psi, 10 HP, 120 gal. tank	↓	.60	26.667		6,250	1,650		7,900	9,325
5690 105 CFM at 125 psi, 25 HP, 250 gal. tank	Q-6	.60	40	↓	17,600	2,550		20,150	23,100
5800 With single stage pump									
5850 8.3 CFM at 125 psi, 2 HP, 80 gal. tank	Q-6	3.50	6.857	Ea.	2,975	435		3,410	3,925
5860 38.7 CFM at 125 psi, 10 HP, 120 gal. tank	"	.90	26.667	"	5,600	1,700		7,300	8,700
6000 Reciprocating, 2 stage, tank mtd, 3 ph, cap rated @ 175 PSIG									
6050 Pressure lubricated, hvy. duty, 9.7 CFM, 3 HP, 120 gal. tank	Q-5	1.30	12.308	Ea.	5,025	755		5,780	6,650
6054 5 CFM, 1-1/2 HP, 80 gal. tank		2.80	5.714		2,850	350		3,200	3,650
6056 6.4 CFM, 2 HP, 80 gal. tank		2	8		2,975	490		3,465	4,000
6058 8.1 CFM, 3 HP, 80 gal. tank		1.70	9.412		3,475	580		4,055	4,700
6059 14.8 CFM, 5 HP, 80 gal. tank	↓	1	16	↓	3,450	985		4,435	5,275

22 15 General Service Compressed-Air Systems

22 15 19 – General Service Packaged Air Compressors and Receivers

22 15 19.10 Air Compressors		Crew	Daily Output	Labor-Hours	Unit	Material	2021 Bare Costs Labor	Equipment	Total	Total Incl O&P
6060	16.5 CFM, 5 HP, 120 gal. tank	Q-5	1	16	Ea.	5,600	985		6,585	7,625
6063	13 CFM, 6 HP, 80 gal. tank		.90	17.778		5,275	1,100		6,375	7,425
6066	19.8 CFM, 7.5 HP, 80 gal. tank		.80	20		6,200	1,225		7,425	8,650
6070	25.8 CFM, 7-1/2 HP, 120 gal. tank		.80	20		6,050	1,225		7,275	8,475
6078	34.8 CFM, 10 HP, 80 gal. tank		.70	22.857		7,225	1,400		8,625	10,100
6080	34.8 CFM, 10 HP, 120 gal. tank		.60	26.667		8,100	1,650		9,750	11,400
6090	53.7 CFM, 15 HP, 120 gal. tank	Q-6	.80	30		8,550	1,925		10,475	12,300
6100	76.7 CFM, 20 HP, 120 gal. tank		.70	34.286		12,900	2,175		15,075	17,400
6104	76.7 CFM, 20 HP, 240 gal. tank		.68	35.294		12,500	2,250		14,750	17,100
6110	90.1 CFM, 25 HP, 120 gal. tank		.63	38.095		12,400	2,425		14,825	17,300
6120	101 CFM, 30 HP, 120 gal. tank		.57	42.105		10,800	2,675		13,475	15,900
6130	101 CFM, 30 HP, 250 gal. tank		.52	46.154		12,900	2,950		15,850	18,600
6200	Oil-less, 13.6 CFM, 5 HP, 120 gal. tank	Q-5	.88	18.182		13,700	1,125		14,825	16,700
6210	13.6 CFM, 5 HP, 250 gal. tank		.80	20		14,600	1,225		15,825	17,900
6220	18.2 CFM, 7.5 HP, 120 gal. tank		.73	21.918		13,600	1,350		14,950	17,000
6230	18.2 CFM, 7.5 HP, 250 gal. tank		.67	23.881		14,600	1,475		16,075	18,300
6250	30.5 CFM, 10 HP, 120 gal. tank		.57	28.070		16,100	1,725		17,825	20,300
6260	30.5 CFM, 10 HP, 250 gal. tank		.53	30.189		17,100	1,850		18,950	21,600
6270	41.3 CFM, 15 HP, 120 gal. tank	Q-6	.70	34.286		17,500	2,175		19,675	22,500
6280	41.3 CFM, 15 HP, 250 gal. tank	"	.67	35.821		18,500	2,275		20,775	23,700

22 31 Domestic Water Softeners

22 31 13 – Residential Domestic Water Softeners

22 31 13.10 Residential Water Softeners

0010	**RESIDENTIAL WATER SOFTENERS**									
7350	Water softener, automatic, to 30 grains per gallon	2 Plum	5	3.200	Ea.	405	217		622	770
7400	To 100 grains per gallon	"	4	4	"	865	271		1,136	1,350

22 31 16 – Commercial Domestic Water Softeners

22 31 16.10 Water Softeners

0010	**WATER SOFTENERS**									
5800	Softener systems, automatic, intermediate sizes									
5820	available, may be used in multiples.									
6000	Hardness capacity between regenerations and flow									
6060	40,000 grains, 14 GPM	Q-1	4	4	Ea.	2,700	244		2,944	3,325
6070	50,000 grains, 17 GPM		3.60	4.444		2,350	271		2,621	3,000
6080	90,000 grains, 25 GPM		2.80	5.714		6,650	350		7,000	7,850
6100	150,000 grains, 37 GPM cont., 51 GPM peak		1.20	13.333		4,825	810		5,635	6,500
6200	300,000 grains, 81 GPM cont., 113 GPM peak		1	16		8,850	975		9,825	11,200
6300	750,000 grains, 160 GPM cont., 230 GPM peak		.80	20		12,800	1,225		14,025	15,900
6400	900,000 grains, 185 GPM cont., 270 GPM peak		.70	22.857		20,600	1,400		22,000	24,800
8000	Water treatment, salts, 50 lb. bag									
8020	Salt, water softener, bag, pelletized				Lb.	4.11			4.11	4.52
8030	Salt, water softener, bag, crystal rock salt				"	.23			.23	.25

22 32 Domestic Water Filtration Equipment

22 32 19 – Domestic-Water Off-Floor Cartridge Filters

22 32 19.10 Water Filters		Crew	Daily Output	Labor-Hours	Unit	Material	2021 Bare Costs Labor	Equipment	Total	Total Incl O&P
0010	**WATER FILTERS**, Purification and treatment									
1000	Cartridge style, dirt and rust type	1 Plum	12	.667	Ea.	79.50	45		124.50	155
1200	Replacement cartridge		32	.250		16.25	16.95		33.20	43.50
1600	Taste and odor type		12	.667		109	45		154	188
1700	Replacement cartridge		32	.250		39	16.95		55.95	68.50
3000	Central unit, dirt/rust/odor/taste/scale		4	2		245	135		380	470
3100	Replacement cartridge, standard		20	.400		109	27		136	161
3600	Replacement cartridge, heavy duty		20	.400		88	27		115	137
8000	Commercial, fully automatic or push button automatic									
8200	Iron removal, 660 GPH, 1" pipe size	Q-1	1.50	10.667	Ea.	1,875	650		2,525	3,050
8240	1,500 GPH, 1-1/4" pipe size		1	16		3,175	975		4,150	4,950
8280	2,340 GPH, 1-1/2" pipe size		.80	20		3,500	1,225		4,725	5,675
8320	3,420 GPH, 2" pipe size		.60	26.667		6,450	1,625		8,075	9,500
8360	4,620 GPH, 2-1/2" pipe size		.50	32		10,200	1,950		12,150	14,200
8500	Neutralizer for acid water, 780 GPH, 1" pipe size		1.50	10.667		1,825	650		2,475	2,975
8540	1,140 GPH, 1-1/4" pipe size		1	16		2,050	975		3,025	3,700
8580	1,740 GPH, 1-1/2" pipe size		.80	20		2,975	1,225		4,200	5,100
8620	2,520 GPH, 2" pipe size		.60	26.667		3,875	1,625		5,500	6,675
8660	3,480 GPH, 2-1/2" pipe size		.50	32		6,450	1,950		8,400	9,975
8800	Sediment removal, 780 GPH, 1" pipe size		1.50	10.667		1,750	650		2,400	2,900
8840	1,140 GPH, 1-1/4" pipe size		1	16		2,075	975		3,050	3,725
8880	1,740 GPH, 1-1/2" pipe size		.80	20		2,750	1,225		3,975	4,850
8920	2,520 GPH, 2" pipe size		.60	26.667		3,975	1,625		5,600	6,800
8960	3,480 GPH, 2-1/2" pipe size		.50	32		6,250	1,950		8,200	9,775
9200	Taste and odor removal, 660 GPH, 1" pipe size		1.50	10.667		2,425	650		3,075	3,650
9240	1,500 GPH, 1-1/4" pipe size		1	16		4,175	975		5,150	6,025
9280	2,340 GPH, 1-1/2" pipe size		.80	20		4,750	1,225		5,975	7,050
9320	3,420 GPH, 2" pipe size		.60	26.667		7,275	1,625		8,900	10,400
9360	4,620 GPH, 2-1/2" pipe size		.50	32		9,750	1,950		11,700	13,600

22 33 Electric Domestic Water Heaters

22 33 13 – Instantaneous Electric Domestic Water Heaters

22 33 13.10 Hot Water Dispensers

		Crew	Daily Output	Labor-Hours	Unit	Material	2021 Bare Costs Labor	Equipment	Total	Total Incl O&P
0010	**HOT WATER DISPENSERS**									
0160	Commercial, 100 cup, 11.3 amp	1 Plum	14	.571	Ea.	455	38.50		493.50	560
3180	Household, 60 cup	"	14	.571	"	239	38.50		277.50	320

22 33 13.20 Instantaneous Elec. Point-Of-Use Water Heaters

			Crew	Daily Output	Labor-Hours	Unit	Material	2021 Bare Costs Labor	Equipment	Total	Total Incl O&P
0010	**INSTANTANEOUS ELECTRIC POINT-OF-USE WATER HEATERS**										
8965	Point of use, electric, glass lined										
8969	Energy saver										
8970	2.5 gal. single element	G	1 Plum	2.80	2.857	Ea.	315	193		508	635
8971	4 gal. single element	G		2.80	2.857		360	193		553	685
8974	6 gal. single element	G		2.50	3.200		465	217		682	840
8975	10 gal. single element	G		2.50	3.200		465	217		682	835
8976	15 gal. single element	G		2.40	3.333		545	226		771	930
8977	20 gal. single element	G		2.40	3.333		685	226		911	1,100
8978	30 gal. single element	G		2.30	3.478		700	235		935	1,125
8979	40 gal. single element	G		2.20	3.636		1,375	246		1,621	1,900
8988	Commercial (ASHRAE energy std. 90)										
8989	6 gallon	G	1 Plum	2.50	3.200	Ea.	1,025	217		1,242	1,475
8990	10 gallon	G		2.50	3.200		1,200	217		1,417	1,650

290

22 33 13 – Instantaneous Electric Domestic Water Heaters

22 33 13.20 Instantaneous Elec. Point-Of-Use Water Heaters		Crew	Daily Output	Labor-Hours	Unit	Material	2021 Bare Costs Labor	Equipment	Total	Total Incl O&P
8991	15 gallon	G	1 Plum	2.40	3.333	Ea.	1,200	226		1,426	1,650
8992	20 gallon	G		2.40	3.333		1,375	226		1,601	1,850
8993	30 gallon	G		2.30	3.478		2,925	235		3,160	3,550
8995	Under the sink, copper, w/bracket										
8996	2.5 gallon	G	1 Plum	4	2	Ea.	410	135		545	655

22 33 30 – Residential, Electric Domestic Water Heaters

22 33 30.13 Residential, Small-Capacity Elec. Water Heaters

		Crew	Daily Output	Labor-Hours	Unit	Material	2021 Bare Costs Labor	Equipment	Total	Total Incl O&P
0010	**RESIDENTIAL, SMALL-CAPACITY ELECTRIC DOMESTIC WATER HEATERS**									
1000	Residential, electric, glass lined tank, 5 yr., 10 gal., single element D2020–210	1 Plum	2.30	3.478	Ea.	465	235		700	860
1040	20 gallon, single element		2.20	3.636		685	246		931	1,125
1060	30 gallon, double element		2.20	3.636		1,225	246		1,471	1,725
1080	40 gallon, double element		2	4		1,400	271		1,671	1,925
1100	52 gallon, double element		2	4		1,575	271		1,846	2,150
1120	66 gallon, double element		1.80	4.444		2,150	300		2,450	2,800
1140	80 gallon, double element		1.60	5		2,425	340		2,765	3,175
1180	120 gallon, double element		1.40	5.714		3,350	385		3,735	4,275

22 33 33 – Light-Commercial Electric Domestic Water Heaters

22 33 33.10 Commercial Electric Water Heaters

		Crew	Daily Output	Labor-Hours	Unit	Material	2021 Bare Costs Labor	Equipment	Total	Total Incl O&P
0010	**COMMERCIAL ELECTRIC WATER HEATERS**									
4000	Commercial, 100° rise. NOTE: for each size tank, a range of									
4010	heaters between the ones shown is available									
4020	Electric									
4100	5 gal., 3 kW, 12 GPH, 208 volt	1 Plum	2	4	Ea.	4,200	271		4,471	5,000
4120	10 gal., 6 kW, 25 GPH, 208 volt		2	4		4,575	271		4,846	5,425
4130	30 gal., 24 kW, 98 GPH, 208 volt		1.92	4.167		7,325	282		7,607	8,475
4136	40 gal., 36 kW, 148 GPH, 208 volt		1.88	4.255		11,100	288		11,388	12,600
4140	50 gal., 9 kW, 37 GPH, 208 volt		1.80	4.444		6,225	300		6,525	7,300
4160	50 gal., 36 kW, 148 GPH, 208 volt		1.80	4.444		9,700	300		10,000	11,200
4180	80 gal., 12 kW, 49 GPH, 208 volt		1.50	5.333		7,700	360		8,060	9,025
4200	80 gal., 36 kW, 148 GPH, 208 volt		1.50	5.333		11,000	360		11,360	12,600
4220	100 gal., 36 kW, 148 GPH, 208 volt		1.20	6.667		11,800	450		12,250	13,700
4240	120 gal., 36 kW, 148 GPH, 208 volt		1.20	6.667		12,300	450		12,750	14,200
4260	150 gal., 15 kW, 61 GPH, 480 volt		1	8		34,400	540		34,940	38,600
4280	150 gal., 120 kW, 490 GPH, 480 volt		1	8		46,800	540		47,340	52,500
4300	200 gal., 15 kW, 61 GPH, 480 volt	Q-1	1.70	9.412		36,500	575		37,075	41,000
4320	200 gal., 120 kW, 490 GPH, 480 volt		1.70	9.412		49,200	575		49,775	55,000
4340	250 gal., 15 kW, 61 GPH, 480 volt		1.50	10.667		37,700	650		38,350	42,500
4360	250 gal., 150 kW, 615 GPH, 480 volt		1.50	10.667		46,200	650		46,850	52,000
4380	300 gal., 30 kW, 123 GPH, 480 volt		1.30	12.308		41,500	750		42,250	46,800
4400	300 gal., 180 kW, 738 GPH, 480 volt		1.30	12.308		71,500	750		72,250	80,000
4420	350 gal., 30 kW, 123 GPH, 480 volt		1.10	14.545		36,900	885		37,785	41,900
4440	350 gal., 180 kW, 738 GPH, 480 volt		1.10	14.545		52,000	885		52,885	58,500
4460	400 gal., 30 kW, 123 GPH, 480 volt		1	16		50,500	975		51,475	57,000
4480	400 gal., 210 kW, 860 GPH, 480 volt		1	16		84,000	975		84,975	94,000
4500	500 gal., 30 kW, 123 GPH, 480 volt		.80	20		59,000	1,225		60,225	67,000
4520	500 gal., 240 kW, 984 GPH, 480 volt		.80	20		98,500	1,225		99,725	110,000
4540	600 gal., 30 kW, 123 GPH, 480 volt	Q-2	1.20	20		66,500	1,275		67,775	75,000
4560	600 gal., 300 kW, 1,230 GPH, 480 volt		1.20	20		116,000	1,275		117,275	129,500
4580	700 gal., 30 kW, 123 GPH, 480 volt		1	24		58,500	1,525		60,025	67,000
4600	700 gal., 300 kW, 1,230 GPH, 480 volt		1	24		88,000	1,525		89,525	99,500
4620	800 gal., 60 kW, 245 GPH, 480 volt		.90	26.667		79,500	1,675		81,175	90,000
4640	800 gal., 300 kW, 1,230 GPH, 480 volt		.90	26.667		90,000	1,675		91,675	101,500

For customer support on your Plumbing Costs with RSMeans Data, call 800.448.8182.

291

22 33 Electric Domestic Water Heaters

22 33 33 – Light-Commercial Electric Domestic Water Heaters

22 33 33.10 Commercial Electric Water Heaters

22 33 33.10 Commercial Electric Water Heaters		Crew	Daily Output	Labor-Hours	Unit	Material	2021 Bare Costs Labor	Equipment	Total	Total Incl O&P
4660	1,000 gal., 60 kW, 245 GPH, 480 volt	Q-2	.70	34.286	Ea.	69,500	2,175		71,675	79,500
4680	1,000 gal., 480 kW, 1,970 GPH, 480 volt		.70	34.286		115,500	2,175		117,675	130,000
4700	1,200 gal., 60 kW, 245 GPH, 480 volt		.60	40		78,500	2,525		81,025	90,500
4720	1,200 gal., 480 kW, 1,970 GPH, 480 volt		.60	40		119,500	2,525		122,025	135,000
4740	1,500 gal., 60 kW, 245 GPH, 480 volt		.50	48		103,500	3,025		106,525	118,500
4760	1,500 gal., 480 kW, 1,970 GPH, 480 volt		.50	48		143,500	3,025		146,525	162,500
5400	Modulating step control for under 90 kW, 2-5 steps	1 Elec	5.30	1.509		1,000	96		1,096	1,250
5440	For above 90 kW, 1 through 5 steps beyond standard, add		3.20	2.500		268	159		427	530
5460	For above 90 kW, 6 through 10 steps beyond standard, add		2.70	2.963		595	189		784	935
5480	For above 90 kW, 11 through 18 steps beyond standard, add		1.60	5		885	320		1,205	1,450

22 34 Fuel-Fired Domestic Water Heaters

22 34 13 – Instantaneous, Tankless, Gas Domestic Water Heaters

22 34 13.10 Instantaneous, Tankless, Gas Water Heaters

22 34 13.10 Instantaneous, Tankless, Gas Water Heaters			Crew	Daily Output	Labor-Hours	Unit	Material	2021 Bare Costs Labor	Equipment	Total	Total Incl O&P
0010	**INSTANTANEOUS, TANKLESS, GAS WATER HEATERS**										
9410	Natural gas/propane, 3.2 GPM	G	1 Plum	2	4	Ea.	605	271		876	1,075
9420	6.4 GPM	G		1.90	4.211		770	285		1,055	1,275
9430	8.4 GPM	G		1.80	4.444		865	300		1,165	1,400
9440	9.5 GPM	G		1.60	5		1,025	340		1,365	1,625

22 34 30 – Residential Gas Domestic Water Heaters

22 34 30.13 Residential, Atmos, Gas Domestic Wtr Heaters

22 34 30.13		Crew	Daily Output	Labor-Hours	Unit	Material	2021 Bare Costs Labor	Equipment	Total	Total Incl O&P
0010	**RESIDENTIAL, ATMOSPHERIC, GAS DOMESTIC WATER HEATERS**									
2000	Gas fired, foam lined tank, 10 yr., vent not incl.									
2040	30 gallon	1 Plum	2	4	Ea.	2,225	271		2,496	2,825
2060	40 gallon		1.90	4.211		1,850	285		2,135	2,450
2080	50 gallon		1.80	4.444		1,975	300		2,275	2,625
2090	60 gallon		1.70	4.706		1,875	320		2,195	2,550
2100	75 gallon		1.50	5.333		2,575	360		2,935	3,375
2120	100 gallon		1.30	6.154		3,475	415		3,890	4,425
2900	Water heater, safety-drain pan, 26" round		20	.400		21	27		48	63.50
3000	Tank leak safety, water & gas shut off see 22 05 23.20 8800									

22 34 36 – Commercial Gas Domestic Water Heaters

22 34 36.13 Commercial, Atmos., Gas Domestic Water Htrs.

22 34 36.13		Crew	Daily Output	Labor-Hours	Unit	Material	2021 Bare Costs Labor	Equipment	Total	Total Incl O&P
0010	**COMMERCIAL, ATMOSPHERIC, GAS DOMESTIC WATER HEATERS**									
6000	Gas fired, flush jacket, std. controls, vent not incl.									
6040	75 MBH input, 73 GPH	1 Plum	1.40	5.714	Ea.	3,700	385		4,085	4,650
6060	98 MBH input, 95 GPH		1.40	5.714		9,075	385		9,460	10,600
6080	120 MBH input, 110 GPH		1.20	6.667		9,250	450		9,700	10,900
6100	120 MBH input, 115 GPH		1.10	7.273		8,750	490		9,240	10,400
6120	140 MBH input, 130 GPH		1	8		10,600	540		11,140	12,500
6140	155 MBH input, 150 GPH		.80	10		12,100	675		12,775	14,300
6160	180 MBH input, 170 GPH		.70	11.429		11,800	775		12,575	14,100
6180	200 MBH input, 192 GPH		.60	13.333		11,700	905		12,605	14,300
6200	250 MBH input, 245 GPH		.50	16		12,700	1,075		13,775	15,600
6220	260 MBH input, 250 GPH	Q-1	.80	20		13,100	1,225		14,325	16,200
6240	360 MBH input, 360 GPH		.80	20		15,400	1,225		16,625	18,700
6260	500 MBH input, 480 GPH		.70	22.857		21,700	1,400		23,100	25,900
6280	725 MBH input, 690 GPH		.60	26.667		28,500	1,625		30,125	33,700
6900	For low water cutoff, add	1 Plum	8	1		405	67.50		472.50	545
6960	For bronze body hot water circulator, add	"	4	2		2,300	135		2,435	2,725

22 34 Fuel-Fired Domestic Water Heaters

22 34 46 – Oil-Fired Domestic Water Heaters

22 34 46.10 Residential Oil-Fired Water Heaters		Crew	Daily Output	Labor-Hours	Unit	Material	2021 Bare Costs Labor	Equipment	Total	Total Incl O&P
0010	**RESIDENTIAL OIL-FIRED WATER HEATERS**									
3000	Oil fired, glass lined tank, 5 yr., vent not included, 30 gallon D2020-260	1 Plum	2	4	Ea.	1,675	271		1,946	2,250
3040	50 gallon		1.80	4.444		1,525	300		1,825	2,125
3060	70 gallon		1.50	5.333		2,350	360		2,710	3,125

22 34 46.20 Commercial Oil-Fired Water Heaters

		Crew	Daily Output	Labor-Hours	Unit	Material	2021 Bare Costs Labor	Equipment	Total	Total Incl O&P
0010	**COMMERCIAL OIL-FIRED WATER HEATERS**									
	Oil fired, glass lined, UL listed, std. controls, vent not incl.									
8060	140 gal., 140 MBH input, 134 GPH	Q-1	2.13	7.512	Ea.	24,900	460		25,360	28,100
8080	140 gal., 199 MBH input, 191 GPH		2	8		25,700	485		26,185	29,000
8100	140 gal., 255 MBH input, 247 GPH		1.60	10		26,500	610		27,110	30,000
8120	140 gal., 270 MBH input, 259 GPH		1.20	13.333		32,700	810		33,510	37,200
8140	140 gal., 400 MBH input, 384 GPH		1	16		33,600	975		34,575	38,500
8160	140 gal., 540 MBH input, 519 GPH		.96	16.667		35,100	1,025		36,125	40,100
8180	140 gal., 720 MBH input, 691 GPH		.92	17.391		35,800	1,050		36,850	40,900
8200	221 gal., 300 MBH input, 288 GPH		.88	18.182		47,300	1,100		48,400	53,500
8220	221 gal., 600 MBH input, 576 GPH		.86	18.605		61,500	1,125		62,625	69,000
8240	221 gal., 800 MBH input, 768 GPH		.82	19.512		53,000	1,200		54,200	60,500
8260	201 gal., 1,000 MBH input, 960 GPH	Q-2	1.26	19.048		54,500	1,200		55,700	62,000
8280	201 gal., 1,250 MBH input, 1,200 GPH		1.22	19.672		55,000	1,250		56,250	62,500
8300	201 gal., 1,500 MBH input, 1,441 GPH		1.16	20.690		59,500	1,300		60,800	67,500
8320	411 gal., 600 MBH input, 576 GPH		1.12	21.429		60,500	1,350		61,850	68,500
8340	411 gal., 800 MBH input, 768 GPH		1.08	22.222		62,500	1,400		63,900	71,000
8360	411 gal., 1,000 MBH input, 960 GPH		1.04	23.077		64,000	1,450		65,450	72,000
8380	411 gal., 1,250 MBH input, 1,200 GPH		.98	24.490		65,500	1,550		67,050	74,500
8400	397 gal., 1,500 MBH input, 1,441 GPH		.92	26.087		69,000	1,650		70,650	78,500
8420	397 gal., 1,750 MBH input, 1,681 GPH		.86	27.907		71,500	1,775		73,275	81,000
8430	397 gal., 2,000 MBH input, 1,921 GPH		.82	29.268		77,000	1,850		78,850	88,000
8440	375 gal., 2,250 MBH input, 2,161 GPH		.76	31.579		79,500	2,000		81,500	90,000
8450	375 gal., 2,500 MBH input, 2,401 GPH		.82	29.268		82,500	1,850		84,350	93,500
8500	Oil fired, polymer lined									
8510	400 MBH, 125 gallon	Q-2	1	24	Ea.	39,800	1,525		41,325	46,000
8520	400 MBH, 600 gallon		.80	30		73,500	1,900		75,400	84,000
8530	800 MBH, 400 gallon		.67	35.821		66,500	2,275		68,775	76,500
8540	800 MBH, 600 gallon		.60	40		100,500	2,525		103,025	114,500
8550	1,000 MBH, 600 gallon		.50	48		89,500	3,025		92,525	102,500
8560	1,200 MBH, 900 gallon		.40	60		98,500	3,800		102,300	114,000
8900	For low water cutoff, add	1 Plum	8	1		380	67.50		447.50	520
8960	For bronze body hot water circulator, add	"	4	2		1,050	135		1,185	1,350

22 35 Domestic Water Heat Exchangers

22 35 30 – Water Heating by Steam

22 35 30.10 Water Heating Transfer Package

		Crew	Daily Output	Labor-Hours	Unit	Material	2021 Bare Costs Labor	Equipment	Total	Total Incl O&P
0010	**WATER HEATING TRANSFER PACKAGE**, Complete controls,									
0020	expansion tank, converter, air separator									
1000	Hot water, 180°F enter, 200°F exit, 15# steam									
1010	One pump system, 28 GPM	Q-6	.75	32	Ea.	22,300	2,050		24,350	27,600
1020	35 GPM		.70	34.286		25,500	2,175		27,675	31,300
1040	55 GPM		.65	36.923		28,600	2,350		30,950	35,000
1060	130 GPM		.55	43.636		36,300	2,775		39,075	44,200
1080	255 GPM		.40	60		47,600	3,825		51,425	58,000

For customer support on your Plumbing Costs with RSMeans Data, call 800.448.8182.

293

22 35 Domestic Water Heat Exchangers

22 35 30 – Water Heating by Steam

22 35 30.10 Water Heating Transfer Package	Crew	Daily Output	Labor-Hours	Unit	Material	2021 Bare Costs Labor	Equipment	Total	Total Incl O&P
1100	550 GPM	Q-6	.30	80	Ea.	73,000	5,100		78,100	87,500
1120	800 GPM		.25	96		74,000	6,125		80,125	90,500
1220	Two pump system, 28 GPM		.70	34.286		30,300	2,175		32,475	36,600
1240	35 GPM		.65	36.923		37,300	2,350		39,650	44,500
1260	55 GPM		.60	40		38,400	2,550		40,950	46,000
1280	130 GPM		.50	48		50,500	3,050		53,550	60,000
1300	255 GPM		.35	68.571		66,500	4,375		70,875	79,500
1320	550 GPM		.25	96		81,000	6,125		87,125	98,000
1340	800 GPM		.20	120		97,500	7,650		105,150	119,000

22 35 43 – Domestic Water Heat Reclaimers

22 35 43.10 Drainwater Heat Recovery

		Crew	Daily Output	Labor-Hours	Unit	Material	Labor	Equipment	Total	Total Incl O&P
0010	**DRAINWATER HEAT RECOVERY**									
9005	Drainwater heat recov unit, copp coil type, for 1/2" supply, 3" waste	1 Plum	3	2.667	Ea.	505	181		686	825
9010	For 1/2" supply, 4" waste		3	2.667		795	181		976	1,150
9020	For 3/4" supply, 3" waste		3	2.667		585	181		766	915
9030	For 3/4" supply, 4" waste		3	2.667		745	181		926	1,100
9040	For 1" supply, 4" waste, double manifold	1 Plum	3	2.667	Ea.	1,375	181		1,556	1,800

22 41 Residential Plumbing Fixtures

22 41 06 – Plumbing Fixtures General

22 41 06.10 Plumbing Fixture Notes

		Crew	Daily Output	Labor-Hours	Unit	Material	Labor	Equipment	Total	Total Incl O&P
0010	**PLUMBING FIXTURE NOTES**, Incl. trim fittings unless otherwise noted R224000-30									
0080	For rough-in, supply, waste, and vent, see add for each type									
0122	For electric water coolers, see Section 22 47 16.10									
0160	For color, unless otherwise noted, add				Ea.	20%				

22 41 13 – Residential Water Closets, Urinals, and Bidets

22 41 13.13 Water Closets

		Crew	Daily Output	Labor-Hours	Unit	Material	Labor	Equipment	Total	Total Incl O&P
0010	**WATER CLOSETS** D2010-110									
0022	For seats, see Section 22 41 13.44									
0032	For automatic flush, see Line 22 42 39.10 0972									
0150	Tank type, vitreous china, incl. seat, supply pipe w/stop, 1.6 gpf or noted									
0200	Wall hung R224000-30									
0400	Two piece, close coupled	Q-1	5.30	3.019	Ea.	440	184		624	755
0960	For rough-in, supply, waste, vent and carrier	"	2.73	5.861	"	1,425	355		1,780	2,100
0999	Floor mounted									
1020	One piece, low profile	Q-1	5.30	3.019	Ea.	865	184		1,049	1,225
1050	One piece		5.30	3.019		905	184		1,089	1,275
1100	Two piece, close coupled		5.30	3.019		191	184		375	485
1102	Economy		5.30	3.019		106	184		290	390
1110	Two piece, close coupled, dual flush		5.30	3.019		284	184		468	585
1140	Two piece, close coupled, 1.28 gpf, ADA [G]		5.30	3.019		268	184		452	570
1960	For color, add					30%				
1961	For designer colors and trim, add					55%				
1980	For rough-in, supply, waste and vent	Q-1	3.05	5.246	Ea.	345	320		665	855

22 41 13.19 Bidets

		Crew	Daily Output	Labor-Hours	Unit	Material	Labor	Equipment	Total	Total Incl O&P
0010	**BIDETS**									
0180	Vitreous china, with trim on fixture	Q-1	5	3.200	Ea.	645	195		840	1,000
0200	With trim for wall mounting	"	5	3.200		775	195		970	1,150
9590	For color, add					40%				
9591	For designer colors and trim, add					50%				

For customer support on your Plumbing Costs with RSMeans Data, call 800.448.8182.

22 41 Residential Plumbing Fixtures

22 41 13 – Residential Water Closets, Urinals, and Bidets

22 41 13.19 Bidets

		Crew	Daily Output	Labor-Hours	Unit	Material	2021 Bare Costs Labor	2021 Bare Costs Equipment	Total	Total Incl O&P
9600	For rough-in, supply, waste and vent, add	Q-1	1.78	8.989	Ea.	390	550		940	1,250

22 41 13.44 Toilet Seats

		Crew	Daily Output	Labor-Hours	Unit	Material	2021 Bare Costs Labor	2021 Bare Costs Equipment	Total	Total Incl O&P
0010	**TOILET SEATS**									
0100	Molded composition, white									
0150	Industrial, w/o cover, open front, regular bowl	1 Plum	24	.333	Ea.	22	22.50		44.50	57.50
0200	With self-sustaining hinge		24	.333		23	22.50		45.50	59
0220	With self-sustaining check hinge		24	.333		22.50	22.50		45	58
0240	Extra heavy, with check hinge	↓	24	.333	↓	21.50	22.50		44	57
0260	Elongated bowl, same price									
0300	Junior size, w/o cover, open front	1 Plum	24	.333	Ea.	43.50	22.50		66	81.50
0320	Regular primary bowl, open front		24	.333		43.50	22.50		66	81.50
0340	Regular baby bowl, open front, check hinge		24	.333		37	22.50		59.50	74
0380	Open back & front, w/o cover, reg. or elongated bowl	↓	24	.333	↓	17.90	22.50		40.40	53
0400	Residential									
0420	Regular bowl, w/cover, closed front	1 Plum	24	.333	Ea.	29	22.50		51.50	65.50
0440	Open front	"	24	.333	"	26	22.50		48.50	62
0460	Elongated bowl, add					25%				
0500	Self-raising hinge, w/o cover, open front									
0520	Regular bowl	1 Plum	24	.333	Ea.	103	22.50		125.50	147
0540	Elongated bowl	"	24	.333	"	25	22.50		47.50	61
0700	Molded wood, white, with cover									
0720	Closed front, regular bowl, square back	1 Plum	24	.333	Ea.	10.05	22.50		32.55	44.50
0740	Extended back		24	.333		14.05	22.50		36.55	49
0780	Elongated bowl, square back		24	.333		13.40	22.50		35.90	48
0800	Open front	↓	24	.333	↓	14.15	22.50		36.65	49
0850	Decorator styles									
0890	Vinyl top, patterned	1 Plum	24	.333	Ea.	21.50	22.50		44	57
0900	Vinyl padded, plain colors, regular bowl		24	.333		21.50	22.50		44	57
0930	Elongated bowl	↓	24	.333	↓	25	22.50		47.50	61
1000	Solid plastic, white									
1030	Industrial, w/o cover, open front, regular bowl	1 Plum	24	.333	Ea.	26.50	22.50		49	63
1080	Extra heavy, concealed check hinge		24	.333		22	22.50		44.50	57.50
1100	Self-sustaining hinge		24	.333		25	22.50		47.50	61
1150	Elongated bowl		24	.333		23	22.50		45.50	58.50
1170	Concealed check		24	.333		17.55	22.50		40.05	53
1190	Self-sustaining hinge, concealed check		24	.333		55.50	22.50		78	94.50
1220	Residential, with cover, closed front, regular bowl		24	.333		41	22.50		63.50	78.50
1240	Elongated bowl		24	.333		49.50	22.50		72	88
1260	Open front, regular bowl		24	.333		43	22.50		65.50	81
1280	Elongated bowl	↓	24	.333	↓	52.50	22.50		75	91

22 41 16 – Residential Lavatories and Sinks

22 41 16.13 Lavatories

			Crew	Daily Output	Labor-Hours	Unit	Material	2021 Bare Costs Labor	2021 Bare Costs Equipment	Total	Total Incl O&P
0010	**LAVATORIES**, With trim, white unless noted otherwise	D2010-310									
0500	Vanity top, porcelain enamel on cast iron										
0600	20" x 18"	R224000-30	Q-1	6.40	2.500	Ea.	400	152		552	665
0640	33" x 19" oval			6.40	2.500		560	152		712	840
0680	20" x 17" oval			6.40	2.500		146	152		298	385
0720	19" round			6.40	2.500		490	152		642	765
0760	20" x 12" triangular bowl		↓	6.40	2.500	↓	278	152		430	530
0860	For color, add						25%				
0861	For designer colors and trim, add						70%				
1000	Cultured marble, 19" x 17", single bowl		Q-1	6.40	2.500	Ea.	159	152		311	400

22 41 Residential Plumbing Fixtures

22 41 16 – Residential Lavatories and Sinks

22 41 16.13 Lavatories		Crew	Daily Output	Labor-Hours	Unit	Material	2021 Bare Costs Labor	Equipment	Total	Total Incl O&P
1040	25" x 19", single bowl	Q-1	6.40	2.500	Ea.	159	152		311	400
1080	31" x 19", single bowl		6.40	2.500		185	152		337	430
1120	25" x 22", single bowl		6.40	2.500		179	152		331	425
1160	37" x 22", single bowl		6.40	2.500		247	152		399	500
1200	49" x 22", single bowl		6.40	2.500		280	152		432	535
1580	For color, same price									
1900	Stainless steel, self-rimming, 25" x 22", single bowl, ledge	Q-1	6.40	2.500	Ea.	345	152		497	605
1960	17" x 22", single bowl		6.40	2.500		335	152		487	595
2040	18-3/4" round		6.40	2.500		830	152		982	1,150
2600	Steel, enameled, 20" x 17", single bowl		5.80	2.759		169	168		337	435
2660	19" round		5.80	2.759		194	168		362	465
2720	18" round		5.80	2.759		122	168		290	385
2860	For color, add					10%				
2861	For designer colors and trim, add					20%				
2900	Vitreous china, 20" x 16", single bowl	Q-1	5.40	2.963	Ea.	242	181		423	535
2960	20" x 17", single bowl		5.40	2.963		145	181		326	430
3020	19" round, single bowl		5.40	2.963		144	181		325	425
3080	19" x 16", single bowl		5.40	2.963		252	181		433	545
3140	17" x 14", single bowl		5.40	2.963		178	181		359	465
3200	22" x 13", single bowl		5.40	2.963		248	181		429	540
3560	For color, add					50%				
3561	For designer colors and trim, add					100%				
3580	Rough-in, supply, waste and vent for all above lavatories	Q-1	2.30	6.957	Ea.	405	425		830	1,075
4000	Wall hung									
4040	Porcelain enamel on cast iron, 16" x 14", single bowl	Q-1	8	2	Ea.	510	122		632	740
4060	18" x 15", single bowl		8	2		340	122		462	555
4120	19" x 17", single bowl		8	2		440	122		562	665
4180	20" x 18", single bowl		8	2		305	122		427	515
4240	22" x 19", single bowl		8	2		795	122		917	1,050
4580	For color, add					30%				
4581	For designer colors and trim, add					75%				
6000	Vitreous china, 18" x 15", single bowl with backsplash	Q-1	7	2.286	Ea.	192	139		331	420
6060	19" x 17", single bowl		7	2.286		173	139		312	400
6120	20" x 18", single bowl		7	2.286		285	139		424	525
6210	27" x 20", ADA compliant		7	2.286		915	139		1,054	1,200
6500	For color, add					30%				
6501	For designer colors and trim, add					50%				
6960	Rough-in, supply, waste and vent for above lavatories	Q-1	1.66	9.639	Ea.	665	585		1,250	1,600
7000	Pedestal type									
7600	Vitreous china, 27" x 21", white	Q-1	6.60	2.424	Ea.	730	148		878	1,025
7610	27" x 21", colored		6.60	2.424		920	148		1,068	1,250
7620	27" x 21", premium color		6.60	2.424		1,050	148		1,198	1,375
7660	26" x 20", white		6.60	2.424		710	148		858	1,000
7670	26" x 20", colored		6.60	2.424		895	148		1,043	1,200
7680	26" x 20", premium color		6.60	2.424		1,075	148		1,223	1,425
7700	24" x 20", white		6.60	2.424		490	148		638	755
7710	24" x 20", colored		6.60	2.424		590	148		738	870
7720	24" x 20", premium color		6.60	2.424		650	148		798	935
7760	21" x 18", white		6.60	2.424		293	148		441	540
7770	21" x 18", colored		6.60	2.424		335	148		483	585
7990	Rough-in, supply, waste and vent for pedestal lavatories		1.66	9.639		665	585		1,250	1,600

22 41 16 – Residential Lavatories and Sinks

22 41 16.16 Sinks		Crew	Daily Output	Labor-Hours	Unit	Material	2021 Bare Costs Labor	2021 Bare Costs Equipment	Total	Total Incl O&P
0010	**SINKS**, With faucets and drain	D2010-410								
2000	Kitchen, counter top style, PE on CI, 24" x 21" single bowl	Q-1	5.60	2.857	Ea.	305	174		479	595
2100	31" x 22" single bowl		5.60	2.857		910	174		1,084	1,250
2200	32" x 21" double bowl		4.80	3.333		390	203		593	735
2310	For color, add					20%				
2311	For designer colors and trim, add					50%				
3000	Stainless steel, self rimming, 19" x 18" single bowl	Q-1	5.60	2.857	Ea.	630	174		804	950
3100	25" x 22" single bowl		5.60	2.857		695	174		869	1,025
3200	33" x 22" double bowl		4.80	3.333		1,000	203		1,203	1,400
3300	43" x 22" double bowl		4.80	3.333		1,175	203		1,378	1,575
3400	22" x 43" triple bowl		4.40	3.636		1,075	222		1,297	1,525
3500	Corner double bowl each 14" x 16"		4.80	3.333		780	203		983	1,150
4000	Steel, enameled, with ledge, 24" x 21" single bowl		5.60	2.857		480	174		654	790
4100	32" x 21" double bowl		4.80	3.333		480	203		683	835
4960	For color sinks except stainless steel, add					10%				
4961	For designer colors and trim, add					20%				
4980	For rough-in, supply, waste and vent, counter top sinks	Q-1	2.14	7.477		355	455		810	1,075
5000	Kitchen, raised deck, PE on CI									
5100	32" x 21", dual level, double bowl	Q-1	2.60	6.154	Ea.	485	375		860	1,100
5200	42" x 21", double bowl & disposer well	"	2.20	7.273		995	445		1,440	1,750
5700	For color, add					20%				
5701	For designer colors and trim, add					50%				
5790	For rough-in, supply, waste & vent, sinks	Q-1	1.85	8.649		355	525		880	1,175

22 41 19 – Residential Bathtubs

22 41 19.10 Baths		Crew	Daily Output	Labor-Hours	Unit	Material	2021 Bare Costs Labor	2021 Bare Costs Equipment	Total	Total Incl O&P
0010	**BATHS**	D2010-510								
0100	Tubs, recessed porcelain enamel on cast iron, with trim									
0180	48" x 42"	Q-1	4	4	Ea.	3,225	244		3,469	3,925
0220	72" x 36"	"	3	5.333	"	3,150	325		3,475	3,950
0300	Mat bottom									
0340	4'-6" long	Q-1	5	3.200	Ea.	1,650	195		1,845	2,100
0380	5' long		4.40	3.636		1,125	222		1,347	1,575
0420	5'-6" long		4	4		1,800	244		2,044	2,350
0480	Above floor drain, 5' long		4	4		830	244		1,074	1,275
0560	Corner 48" x 44"		4.40	3.636		2,925	222		3,147	3,550
0750	For color, add					30%				
0760	For designer colors & trim, add					60%				
2000	Enameled formed steel, 4'-6" long	Q-1	5.80	2.759	Ea.	540	168		708	845
2300	Above floor drain, 5' long	"	5.50	2.909	"	575	177		752	900
2350	For color, add					10%				
4000	Soaking, acrylic, w/pop-up drain 66" x 36" x 20" deep	Q-1	5.50	2.909	Ea.	2,125	177		2,302	2,600
4100	60" x 42" x 20" deep		5	3.200		1,300	195		1,495	1,725
4200	72" x 42" x 23" deep		4.80	3.333		2,200	203		2,403	2,725
4310	For color, add					5%				
4311	For designer colors & trim, add					20%				
4600	Module tub & showerwall surround, molded fiberglass									
4610	5' long x 34" wide x 76" high	Q-1	4	4	Ea.	830	244		1,074	1,275
4620	For color, add					10%				
4621	For designer colors and trim, add					25%				
4750	ADA compliant with 1-1/2" OD grab bar, antiskid bottom ♿									
4760	60" x 32-3/4" x 72" high	Q-1	4	4	Ea.	595	244		839	1,025
4770	60" x 30" x 71" high with molded seat		3.50	4.571		750	279		1,029	1,250

For customer support on your Plumbing Costs with RSMeans Data, call 800.448.8182.

297

22 41 19 – Residential Bathtubs

22 41 19.10 Baths		Crew	Daily Output	Labor-Hours	Unit	Material	2021 Bare Costs Labor	Equipment	Total	Total Incl O&P
9600	Rough-in, supply, waste and vent, for all above tubs, add	Q-1	2.07	7.729	Ea.	460	470		930	1,200

22 41 23 – Residential Showers

22 41 23.20 Showers

		Crew	Daily Output	Labor-Hours	Unit	Material	2021 Bare Costs Labor	Equipment	Total	Total Incl O&P
0010	**SHOWERS** D2010-710									
1500	Stall, with drain only. Add for valve and door/curtain									
1510	Baked enamel, molded stone receptor, 30" square	Q-1	5.20	3.077	Ea.	1,350	187		1,537	1,775
1520	32" square		5	3.200		1,150	195		1,345	1,575
1530	36" square		4.80	3.333		3,075	203		3,278	3,700
1540	Terrazzo receptor, 32" square		5	3.200		1,350	195		1,545	1,775
1560	36" square		4.80	3.333		1,775	203		1,978	2,250
1580	36" corner angle		4.80	3.333		2,200	203		2,403	2,725
1600	For color, add					10%				
1601	For designer colors and trim, add					15%				
1604	For thermostatic valve, add				Ea.	1,200			1,200	1,300
3000	Fiberglass, one piece, with 3 walls, 32" x 32" square	Q-1	5.50	2.909		370	177		547	675
3100	36" x 36" square	"	5.50	2.909		415	177		592	725
3200	ADA compliant, 1-1/2" OD grab bars, nonskid floor									
3210	48" x 34-1/2" x 72" corner seat	Q-1	5	3.200	Ea.	640	195		835	995
3220	60" x 34-1/2" x 72" corner seat		4	4		770	244		1,014	1,200
3230	48" x 34-1/2" x 72" fold up seat		5	3.200		1,150	195		1,345	1,550
3250	64" x 65-3/4" x 81-1/2" fold. seat, ADA		3.80	4.211		1,425	257		1,682	1,925
3260	For thermostatic valve, add					1,200			1,200	1,300
4000	Polypropylene, stall only, w/molded-stone floor, 30" x 30"	Q-1	2	8		765	485		1,250	1,575
4100	32" x 32"	"	2	8		775	485		1,260	1,575
4110	For thermostatic valve, add					1,200			1,200	1,300
4200	Rough-in, supply, waste and vent for above showers	Q-1	2.05	7.805		445	475		920	1,200

22 41 23.40 Shower System Components

		Crew	Daily Output	Labor-Hours	Unit	Material	2021 Bare Costs Labor	Equipment	Total	Total Incl O&P
0010	**SHOWER SYSTEM COMPONENTS**									
4500	Receptor only									
4510	For tile, 36" x 36"	1 Plum	4	2	Ea.	405	135		540	645
4520	Fiberglass receptor only, 32" x 32"		8	1		112	67.50		179.50	224
4530	34" x 34"		7.80	1.026		208	69.50		277.50	335
4540	36" x 36"		7.60	1.053		130	71.50		201.50	250
4600	Rectangular									
4620	32" x 48"	1 Plum	7.40	1.081	Ea.	155	73		228	280
4630	34" x 54"		7.20	1.111		186	75		261	315
4640	34" x 60"		7	1.143		197	77.50		274.50	330
5000	Built-in, head, arm, 2.5 GPM valve		4	2		93	135		228	305
5200	Head, arm, by-pass, integral stops, handles		3.60	2.222		305	150		455	560
5500	Head, water economizer, 1.6 GPM G		24	.333		53.50	22.50		76	92.50
5800	Mixing valve, built-in		6	1.333		74.50	90.50		165	217
5900	Exposed		6	1.333		820	90.50		910.50	1,050

22 41 36 – Residential Laundry Trays

22 41 36.10 Laundry Sinks

		Crew	Daily Output	Labor-Hours	Unit	Material	2021 Bare Costs Labor	Equipment	Total	Total Incl O&P
0010	**LAUNDRY SINKS**, With trim D2010-420									
0020	Porcelain enamel on cast iron, black iron frame									
0050	24" x 21", single compartment	Q-1	6	2.667	Ea.	620	162		782	920
0100	26" x 21", single compartment		6	2.667		680	162		842	985
0200	48" x 20", double compartment		5	3.200		795	195		990	1,150
2000	Molded stone, on wall hanger or legs									
2020	22" x 23", single compartment	Q-1	6	2.667	Ea.	179	162		341	440

22 41 36 – Residential Laundry Trays

22 41 36.10 Laundry Sinks	Crew	Daily Output	Labor-Hours	Unit	Material	2021 Bare Costs Labor	Equipment	Total	Total Incl O&P
2100	45" x 21", double compartment	Q-1	5	3.200	Ea.	365	195		560	690
3000	Plastic, on wall hanger or legs									
3020	18" x 23", single compartment	Q-1	6.50	2.462	Ea.	146	150		296	385
3100	20" x 24", single compartment		6.50	2.462		163	150		313	405
3200	36" x 23", double compartment		5.50	2.909		216	177		393	505
3300	40" x 24", double compartment		5.50	2.909		284	177		461	575
5000	Stainless steel, counter top, 22" x 17" single compartment		6	2.667		77	162		239	325
5200	33" x 22", double compartment		5	3.200		92.50	195		287.50	395
9600	Rough-in, supply, waste and vent, for all laundry sinks		2.14	7.477		355	455		810	1,075

22 41 39 – Residential Faucets, Supplies and Trim

22 41 39.10 Faucets and Fittings

22 41 39.10 Faucets and Fittings	Crew	Daily Output	Labor-Hours	Unit	Material	2021 Bare Costs Labor	Equipment	Total	Total Incl O&P
0010	**FAUCETS AND FITTINGS**									
0150	Bath, faucets, diverter spout combination, sweat	1 Plum	8	1	Ea.	86	67.50		153.50	196
0200	For integral stops, IPS unions, add					111			111	123
0300	Three valve combinations, spout, head, arm, flange, sweat	1 Plum	6	1.333		96	90.50		186.50	241
0400	For integral stops, IPS unions, add				Pr.	67			67	73.50
0420	Bath, press-bal mix valve w/diverter, spout, shower head, arm/flange	1 Plum	8	1	Ea.	200	67.50		267.50	320
0500	Drain, central lift, 1-1/2" IPS male		20	.400		53	27		80	99
0600	Trip lever, 1-1/2" IPS male		20	.400		60	27		87	107
0700	Pop up, 1-1/2" IPS male		18	.444		70.50	30		100.50	123
0800	Chain and stopper, 1-1/2" IPS male		24	.333		33	22.50		55.50	70
0810	Bidet									
0812	Fitting, over the rim, swivel spray/pop-up drain	1 Plum	8	1	Ea.	231	67.50		298.50	355
1000	Kitchen sink faucets, top mount, cast spout		10	.800		83.50	54		137.50	173
1100	For spray, add		24	.333		17.20	22.50		39.70	52.50
1110	For basket strainer w/tail piece, add		24	.333		15.60	22.50		38.10	50.50
1200	Wall type, swing tube spout		10	.800		73.50	54		127.50	162
1240	For soap dish, add					3.67			3.67	4.04
1250	For basket strainer w/tail piece, add					46.50			46.50	51
1300	Single control lever handle									
1310	With pull out spray									
1320	Polished chrome	1 Plum	10	.800	Ea.	225	54		279	330
2000	Laundry faucets, shelf type, IPS or copper unions		12	.667		61.50	45		106.50	135
2100	Lavatory faucet, centerset, without drain		10	.800		67	54		121	155
2120	With pop-up drain		6.66	1.201		86.50	81.50		168	216
2130	For acrylic handles, add					5.25			5.25	5.80
2150	Concealed, 12" centers	1 Plum	10	.800		89.50	54		143.50	180
2160	With pop-up drain	"	6.66	1.201		107	81.50		188.50	239
2210	Porcelain cross handles and pop-up drain									
2220	Polished chrome	1 Plum	6.66	1.201	Ea.	214	81.50		295.50	355
2230	Polished brass	"	6.66	1.201	"	283	81.50		364.50	430
2260	Single lever handle and pop-up drain									
2280	Satin nickel	1 Plum	6.66	1.201	Ea.	300	81.50		381.50	450
2290	Polished chrome		6.66	1.201		215	81.50		296.50	360
2600	Shelfback, 4" to 6" centers, 17 ga. tailpiece		10	.800		80.50	54		134.50	170
2650	With pop-up drain		6.66	1.201		100	81.50		181.50	231
2700	Shampoo faucet with supply tube		24	.333		47	22.50		69.50	85.50
2800	Self-closing, center set		10	.800		149	54		203	245
2810	Automatic sensor and operator, with faucet head		6.15	1.301		590	88		678	780
4000	Shower by-pass valve with union		18	.444		57	30		87	108
4100	Shower arm with flange and head		22	.364		19.10	24.50		43.60	58
4140	Shower, hand held, pin mount, massage action, chrome		22	.364		73.50	24.50		98	118

For customer support on your Plumbing Costs with RSMeans Data, call 800.448.8182.

299

22 41 39.10 Faucets and Fittings	Crew	Daily Output	Labor-Hours	Unit	Material	2021 Bare Costs Labor	Equipment	Total	Total Incl O&P
4142	Polished brass	1 Plum	22	.364	Ea.	160	24.50		184.50	213
4144	Shower, hand held, wall mtd, adj. spray, 2 wall mounts, chrome		20	.400		115	27		142	167
4146	Polished brass		20	.400		245	27		272	310
4148	Shower, hand held head, bar mounted 24", adj. spray, chrome		20	.400		184	27		211	243
4150	Polished brass		20	.400		380	27		407	455
4200	Shower thermostatic mixing valve, concealed, with shower head trim kit		8	1		365	67.50		432.50	500
4220	Shower pressure balancing mixing valve									
4230	With shower head, arm, flange and diverter tub spout									
4240	Chrome	1 Plum	6.14	1.303	Ea.	420	88		508	590
4250	Satin nickel		6.14	1.303		550	88		638	735
4260	Polished graphite		6.14	1.303		535	88		623	720
5000	Sillcock, compact, brass, IPS or copper to hose		24	.333		12.90	22.50		35.40	47.50
6000	Stop and waste valves, bronze									
6100	Angle, solder end 1/2"	1 Plum	24	.333	Ea.	42.50	22.50		65	80
6110	3/4"		20	.400		49	27		76	94.50
6300	Straightway, solder end 3/8"		24	.333		19.30	22.50		41.80	55
6310	1/2"		24	.333		21.50	22.50		44	57
6320	3/4"		20	.400		28	27		55	71
6410	Straightway, threaded 1/2"		24	.333		29	22.50		51.50	65.50
6420	3/4"		20	.400		32.50	27		59.50	76
6430	1"		19	.421		22	28.50		50.50	66.50
7800	Water closet, wax gasket		96	.083		1.25	5.65		6.90	9.80
7820	Gasket toilet tank to bowl		32	.250		2.89	16.95		19.84	28.50
7830	Replacement diaphragm washer assy for ballcock valve		12	.667		2.89	45		47.89	70.50
7850	Dual flush valve		12	.667		111	45		156	190
8000	Water supply stops, polished chrome plate									
8200	Angle, 3/8"	1 Plum	24	.333	Ea.	11.30	22.50		33.80	46
8300	1/2"		22	.364		12.30	24.50		36.80	50.50
8400	Straight, 3/8"		26	.308		14.65	21		35.65	47
8500	1/2"		24	.333		11.85	22.50		34.35	46.50
8600	Water closet, angle, w/flex riser, 3/8"		24	.333		40	22.50		62.50	77.50
9100	Miscellaneous									
9720	Teflon tape, 1/2" x 520" roll				Ea.	.77			.77	.85

		Crew	Daily Output	Labor-Hours	Unit	Material	2021 Bare Costs Labor	Equipment	Total	Total Incl O&P
0010	**WASHER/DRYER ACCESSORIES**									
1020	Valves ball type single lever									
1030	1/2" diam., IPS	1 Plum	21	.381	Ea.	66	26		92	111
1040	1/2" diam., solder	"	21	.381	"	66	26		92	111
1050	Recessed box, 16 ga., two hose valves and drain									
1060	1/2" size, 1-1/2" drain	1 Plum	18	.444	Ea.	161	30		191	222
1070	1/2" size, 2" drain	"	17	.471	"	144	32		176	207
1080	With grounding electric receptacle									
1090	1/2" size, 1-1/2" drain	1 Plum	18	.444	Ea.	178	30		208	241
1100	1/2" size, 2" drain	"	17	.471	"	191	32		223	258
1110	With grounding and dryer receptacle									
1120	1/2" size, 1-1/2" drain	1 Plum	18	.444	Ea.	163	30		193	225
1130	1/2" size, 2" drain	"	17	.471	"	220	32		252	290
1140	Recessed box, 16 ga., ball valves with single lever and drain									
1150	1/2" size, 1-1/2" drain	1 Plum	19	.421	Ea.	305	28.50		333.50	380
1160	1/2" size, 2" drain	"	18	.444	"	263	30		293	335
1170	With grounding electric receptacle									
1180	1/2" size, 1-1/2" drain	1 Plum	19	.421	Ea.	330	28.50		358.50	405

22 41 Residential Plumbing Fixtures

22 41 39 – Residential Faucets, Supplies and Trim

22 41 39.70 Washer/Dryer Accessories

		Crew	Daily Output	Labor-Hours	Unit	Material	2021 Bare Costs Labor	2021 Bare Costs Equipment	Total	Total Incl O&P
1190	1/2" size, 2" drain	1 Plum	18	.444	Ea.	292	30		322	365
1200	With grounding and dryer receptacles									
1210	1/2" size, 1-1/2" drain	1 Plum	19	.421	Ea.	290	28.50		318.50	365
1220	1/2" size, 2" drain	"	18	.444	"	320	30		350	395
1300	Recessed box, 20 ga., two hose valves and drain (economy type)									
1310	1/2" size, 1-1/2" drain	1 Plum	19	.421	Ea.	126	28.50		154.50	181
1320	1/2" size, 2" drain		18	.444		117	30		147	174
1330	Box with drain only		24	.333		72.50	22.50		95	114
1340	1/2" size, 1-1/2" ABS/PVC drain		19	.421		143	28.50		171.50	200
1350	1/2" size, 2" ABS/PVC drain		18	.444		143	30		173	202
1352	Box with drain and 15 A receptacle		24	.333		74	22.50		96.50	115
1360	1/2" size, 2" drain ABS/PVC, 15 A receptacle		24	.333		103	22.50		125.50	148
1400	Wall mounted									
1410	1/2" size, 1-1/2" plastic drain	1 Plum	19	.421	Ea.	35	28.50		63.50	81.50
1420	1/2" size, 2" plastic drain	"	18	.444	"	21	30		51	68
1500	Dryer vent kit									
1510	8' flex duct, clamps and outside hood	1 Plum	20	.400	Ea.	14.15	27		41.15	56
1980	Rough-in, supply, waste, and vent for washer boxes	"	3.46	2.310	"	410	156		566	685

22 42 Commercial Plumbing Fixtures

22 42 13 – Commercial Water Closets, Urinals, and Bidets

22 42 13.13 Water Closets

		Crew	Daily Output	Labor-Hours	Unit	Material	2021 Bare Costs Labor	2021 Bare Costs Equipment	Total	Total Incl O&P
0010	**WATER CLOSETS**									
3000	Bowl only, with flush valve, seat, 1.6 gpf unless noted									
3100	Wall hung	Q-1	5.80	2.759	Ea.	945	168		1,113	1,300
3200	For rough-in, supply, waste and vent, single WC		2.56	6.250		1,450	380		1,830	2,175
3300	Floor mounted		5.80	2.759		335	168		503	620
3350	With wall outlet		5.80	2.759		550	168		718	855
3360	With floor outlet, 1.28 gpf ▣G		5.80	2.759		550	168		718	855
3362	With floor outlet, 1.28 gpf, ADA ▣G		5.80	2.759		575	168		743	880
3370	For rough-in, supply, waste and vent, single WC		2.84	5.634		375	345		720	920
3390	Floor mounted children's size, 10-3/4" high									
3392	With automatic flush sensor, 1.6 gpf	Q-1	6.20	2.581	Ea.	660	157		817	960
3396	With automatic flush sensor, 1.28 gpf		6.20	2.581		610	157		767	905
3400	For rough-in, supply, waste and vent, single WC		2.84	5.634		375	345		720	920
3500	Gang side by side carrier system, rough-in, supply, waste & vent									
3510	For single hook-up	Q-1	1.97	8.122	Ea.	1,675	495		2,170	2,575
3520	For each additional hook-up, add	"	2.14	7.477	"	1,550	455		2,005	2,375
3550	Gang back to back carrier system, rough-in, supply, waste & vent									
3560	For pair hook-up	Q-1	1.76	9.091	Pr.	2,825	555		3,380	3,925
3570	For each additional pair hook-up, add	"	1.81	8.840	"	2,700	540		3,240	3,750

22 42 13.16 Urinals

		Crew	Daily Output	Labor-Hours	Unit	Material	2021 Bare Costs Labor	2021 Bare Costs Equipment	Total	Total Incl O&P
0010	**URINALS** D2010-210									
0102	For automatic flush see Line 22 42 39.10 0972									
3000	Wall hung, vitreous china, with self-closing valve R224000-30									
3100	Siphon jet type	Q-1	3	5.333	Ea.	335	325		660	855
3120	Blowout type		3	5.333		495	325		820	1,025
3140	Water saving .5 gpf ▣G		3	5.333		575	325		900	1,125
3300	Rough-in, supply, waste & vent		2.83	5.654		835	345		1,180	1,425
5000	Stall type, vitreous china, includes valve		2.50	6.400		765	390		1,155	1,425
6980	Rough-in, supply, waste and vent		1.99	8.040		500	490		990	1,275

22 42 Commercial Plumbing Fixtures

22 42 13 – Commercial Water Closets, Urinals, and Bidets

22 42 13.16 Urinals

		Crew	Daily Output	Labor-Hours	Unit	Material	2021 Bare Costs Labor	Equipment	Total	Total Incl O&P
8000	Waterless (no flush) urinal									
8010	Wall hung									
8014	Fiberglass reinforced polyester									
8020	Standard unit	Q-1	21.30	.751	Ea.	445	46		491	560
8030	ADA compliant unit	Q-1	21.30	.751	Ea.	410	46		456	525
8070	For solid color, add					69			69	75.50
8080	For 2" brass flange (new const.), add	Q-1	96	.167		25.50	10.15		35.65	43
8200	Vitreous china									
8220	ADA compliant unit, 14"	Q-1	21.30	.751	Ea.	234	46		280	325
8250	ADA compliant unit, 15.5"	"	21.30	.751		283	46		329	380
8270	For solid color, add					69			69	75.50
8290	Rough-in, supply, waste & vent	Q-1	2.92	5.479		805	335		1,140	1,400
8400	Trap liquid									
8410	1 quart				Ea.	19			19	21
8420	1 gallon				"	67			67	74

22 42 16 – Commercial Lavatories and Sinks

22 42 16.13 Lavatories

0010	**LAVATORIES**, With trim, white unless noted otherwise									
0020	Commercial lavatories same as residential. See Section 22 41 16									

22 42 16.16 Commercial Sinks

		Crew	Daily Output	Labor-Hours	Unit	Material	Labor	Equipment	Total	Total Incl O&P
0010	**COMMERCIAL SINKS**									
5900	Scullery sink, stainless steel									
5910	1 bowl and drain board, 43" x 22" OD	Q-1	5.40	2.963	Ea.	1,650	181		1,831	2,100
5920	2 bowls and drain board, 49" x 22" OD		4.60	3.478		2,700	212		2,912	3,300
5930	3 bowls and drain board, 43" x 22" OD		4.20	3.810		4,700	232		4,932	5,500
5940	1 bowl and drain board, 50" x 28" OD, with legs		5.40	2.963		1,750	181		1,931	2,200

22 42 16.30 Classroom Sinks

		Crew	Daily Output	Labor-Hours	Unit	Material	Labor	Equipment	Total	Total Incl O&P
0010	**CLASSROOM SINKS**									
6020	Countertop, stainless steel									
6024	with faucet, bubbler and strainer, ADA compliant									
6036	25" x 17" single bowl	Q-1	5.20	3.077	Ea.	1,200	187		1,387	1,575
6040	28" x 22" single bowl		5.20	3.077		1,275	187		1,462	1,700
6044	31" x 19" single bowl		5.20	3.077		1,350	187		1,537	1,750
6070	37" x 17" double bowl		4.40	3.636		1,950	222		2,172	2,475
6100	For rough-in, supply, waste and vent, counter top classroom sinks		2.14	7.477		355	455		810	1,075

22 42 16.34 Laboratory Countertops and Sinks

		Crew	Daily Output	Labor-Hours	Unit	Material	Labor	Equipment	Total	Total Incl O&P
0010	**LABORATORY COUNTERTOPS AND SINKS**									
0050	Laboratory sinks, corrosion resistant									
1000	Stainless steel sink, bench mounted, with									
1020	plug & waste fitting with 1-1/2" straight threads									
1030	Single bowl, 2 drainboards, backnut & strainer									
1050	18-1/2" x 15-1/2" x 12-1/2" sink, 54" x 24" OD	Q-1	3	5.333	Ea.	1,350	325		1,675	1,950
1100	Single bowl, single drainboard, backnut & strainer									
1130	18-1/2" x 15-1/2" x 12-1/2" sink, 47" x 24" OD	Q-1	3	5.333	Ea.	980	325		1,305	1,550
1146	Double bowl, single drainboard, backnut & strainer									
1150	18-1/2" x 15-1/2" x 12-1/2" sink, 70" x 24" OD	Q-1	3	5.333	Ea.	1,500	325		1,825	2,100
1280	Polypropylene									
1290	Flanged 1-1/4" wide, rectangular with strainer									
1300	plug & waste fitting, 1-1/2" straight threads									
1320	12" x 12" x 8" sink, 14-1/2" x 14-1/2" OD	Q-1	4	4	Ea.	264	244		508	655
1340	16" x 16" x 8" sink, 18-1/2" x 18-1/2" OD		4	4		375	244		619	780

22 42 Commercial Plumbing Fixtures

22 42 16 – Commercial Lavatories and Sinks

22 42 16.34 Laboratory Countertops and Sinks

		Crew	Daily Output	Labor-Hours	Unit	Material	2021 Bare Costs Labor	2021 Bare Costs Equipment	Total	Total Incl O&P
1360	21" x 18" x 10" sink, 23-1/2" x 20-1/2" OD	Q-1	4	4	Ea.	385	244		629	790
1490	For rough-in, supply, waste & vent, add	↓	2.02	7.921	↓	227	485		712	970
1600	Polypropylene									
1620	Cup sink, oval, integral strainers									
1640	6" x 3" I.D., 7" x 4" OD	Q-1	6	2.667	Ea.	152	162		314	410
1660	9" x 3" I.D., 10" x 4-1/2" OD	"	6	2.667		179	162		341	440
1740	1-1/2" diam. x 11" long					45			45	49.50
1980	For rough-in, supply, waste & vent, add	Q-1	1.70	9.412	↓	249	575		824	1,125

22 42 16.40 Service Sinks

		Crew	Daily Output	Labor-Hours	Unit	Material	2021 Bare Costs Labor	2021 Bare Costs Equipment	Total	Total Incl O&P
0010	**SERVICE SINKS**									
6650	Service, floor, corner, PE on CI, 28" x 28"	Q-1	4.40	3.636	Ea.	1,350	222		1,572	1,825
6750	Vinyl coated rim guard, add					65			65	71.50
6755	Mop sink, molded stone, 22" x 18"	1 Plum	3.33	2.402		575	163		738	875
6760	Mop sink, molded stone, 24" x 36"		3.33	2.402		283	163		446	555
6770	Mop sink, molded stone, 24" x 36", w/rim 3 sides	↓	3.33	2.402		277	163		440	550
6790	For rough-in, supply, waste & vent, floor service sinks	Q-1	1.64	9.756		1,350	595		1,945	2,375
7000	Service, wall, PE on CI, roll rim, 22" x 18"		4	4		1,000	244		1,244	1,475
7100	24" x 20"	↓	4	4		1,075	244		1,319	1,550
7600	For stainless steel rim guard, two sides only, add					51.50			51.50	56.50
7800	For stainless steel rim guard, front only, add					57			57	62.50
8600	Vitreous china, 22" x 20"	Q-1	4	4		850	244		1,094	1,300
8960	For stainless steel rim guard, front or one side, add					80.50			80.50	89
8980	For rough-in, supply, waste & vent, wall service sinks	Q-1	1.30	12.308	↓	2,100	750		2,850	3,425

22 42 23 – Commercial Showers

22 42 23.30 Group Showers

		Crew	Daily Output	Labor-Hours	Unit	Material	2021 Bare Costs Labor	2021 Bare Costs Equipment	Total	Total Incl O&P
0010	**GROUP SHOWERS**									
6000	Group, w/pressure balancing valve, rough-in and rigging not included									
6800	Column, 6 heads, no receptors, less partitions	Q-1	3	5.333	Ea.	9,675	325		10,000	11,100
6900	With stainless steel partitions		1	16		12,800	975		13,775	15,600
7600	5 heads, no receptors, less partitions		3	5.333		6,650	325		6,975	7,775
7620	4 heads (1 ADA compliant) no receptors, less partitions		3	5.333		6,175	325		6,500	7,250
7700	With stainless steel partitions	↓	1	16	↓	6,650	975		7,625	8,775
8000	Wall, 2 heads, no receptors, less partitions	Q-1	4	4	Ea.	2,825	244		3,069	3,475
8100	With stainless steel partitions	"	2	8	"	6,200	485		6,685	7,525

22 42 33 – Wash Fountains

22 42 33.20 Commercial Wash Fountains

		Crew	Daily Output	Labor-Hours	Unit	Material	2021 Bare Costs Labor	2021 Bare Costs Equipment	Total	Total Incl O&P
0010	**COMMERCIAL WASH FOUNTAINS** D2010-610									
1900	Group, foot control									
2000	Precast terrazzo, circular, 36" diam., 5 or 6 persons	Q-2	3	8	Ea.	7,225	505		7,730	8,700
2100	54" diam. for 8 or 10 persons		2.50	9.600		10,900	605		11,505	12,900
2400	Semi-circular, 36" diam. for 3 persons		3	8		6,100	505		6,605	7,450
2500	54" diam. for 4 or 5 persons		2.50	9.600		11,200	605		11,805	13,200
2700	Quarter circle (corner), 54" diam. for 3 persons		3.50	6.857		7,325	435		7,760	8,725
3000	Stainless steel, circular, 36" diameter		3.50	6.857		6,900	435		7,335	8,250
3100	54" diameter		2.80	8.571		8,325	540		8,865	9,950
3400	Semi-circular, 36" diameter		3.50	6.857		4,350	435		4,785	5,425
3500	54" diameter		2.80	8.571		7,275	540		7,815	8,800
5000	Thermoplastic, pre-assembled, circular, 36" diameter		6	4		4,575	253		4,828	5,400
5100	54" diameter		4	6		4,375	380		4,755	5,400
5400	Semi-circular, 36" diameter		6	4		4,950	253		5,203	5,825
5600	54" diameter	↓	4	6	↓	5,025	380		5,405	6,100

22 42 33 – Wash Fountains

22 42 33.20 Commercial Wash Fountains	Crew	Daily Output	Labor-Hours	Unit	Material	2021 Bare Costs Labor	Equipment	Total	Total Incl O&P
5610	Group, infrared control, barrier free									
5614	Precast terrazzo								
5620	Semi-circular 36" diam. for 3 persons	Q-2	3	8	Ea.	7,500	505		8,005	9,000
5630	46" diam. for 4 persons		2.80	8.571		8,125	540		8,665	9,725
5640	Circular, 54" diam. for 8 persons, button control		2.50	9.600		9,750	605		10,355	11,600
5700	Rough-in, supply, waste and vent for above wash fountains	Q-1	1.82	8.791		510	535		1,045	1,350
6200	Duo for small washrooms, stainless steel		2	8		3,300	485		3,785	4,350
6400	Bowl with backsplash		2	8		2,400	485		2,885	3,350
6500	Rough-in, supply, waste & vent for duo fountains		2.02	7.921		315	485		800	1,075

22 42 39 – Commercial Faucets, Supplies, and Trim

22 42 39.10 Faucets and Fittings

		Crew	Daily Output	Labor-Hours	Unit	Material	2021 Bare Costs Labor	Equipment	Total	Total Incl O&P
0010	**FAUCETS AND FITTINGS**									
0840	Flush valves, with vacuum breaker									
0850	Water closet									
0860	Exposed, rear spud	1 Plum	8	1	Ea.	152	67.50		219.50	268
0870	Top spud		8	1		198	67.50		265.50	320
0880	Concealed, rear spud		8	1		212	67.50		279.50	335
0890	Top spud		8	1		172	67.50		239.50	290
0900	Wall hung		8	1		223	67.50		290.50	345
0910	Dual flush flushometer		12	.667		237	45		282	330
0912	Flushometer retrofit kit		18	.444		17.70	30		47.70	64.50
0920	Urinal									
0930	Exposed, stall	1 Plum	8	1	Ea.	198	67.50		265.50	320
0940	Wall (washout)		8	1		163	67.50		230.50	281
0950	Pedestal, top spud		8	1		142	67.50		209.50	258
0960	Concealed, stall		8	1		169	67.50		236.50	287
0970	Wall (washout)		8	1		182	67.50		249.50	300
0971	Automatic flush sensor and operator for									
0972	urinals or water closets, standard [G]	1 Plum	8	1	Ea.	525	67.50		592.50	675
0980	High efficiency water saving									
0984	Water closets, 1.28 gpf [G]	1 Plum	8	1	Ea.	435	67.50		502.50	580
0988	Urinals, .5 gpf [G]	"	8	1	"	435	67.50		502.50	580
2790	Faucets for lavatories									
2800	Self-closing, center set	1 Plum	10	.800	Ea.	149	54		203	245
2810	Automatic sensor and operator, with faucet head		6.15	1.301		590	88		678	780
3000	Service sink faucet, cast spout, pail hook, hose end		14	.571		76	38.50		114.50	141

22 42 39.30 Carriers and Supports

		Crew	Daily Output	Labor-Hours	Unit	Material	2021 Bare Costs Labor	Equipment	Total	Total Incl O&P
0010	**CARRIERS AND SUPPORTS,** For plumbing fixtures									
0500	Drinking fountain, wall mounted									
0600	Plate type with studs, top back plate	1 Plum	7	1.143	Ea.	73.50	77.50		151	196
0700	Top front and back plate		7	1.143		161	77.50		238.50	292
0800	Top & bottom, front & back plates, w/bearing jacks		7	1.143		187	77.50		264.50	320
3000	Lavatory, concealed arm									
3050	Floor mounted, single									
3100	High back fixture	1 Plum	6	1.333	Ea.	800	90.50		890.50	1,025
3200	Flat slab fixture		6	1.333		935	90.50		1,025.50	1,150
3220	ADA compliant		6	1.333		770	90.50		860.50	980
3250	Floor mounted, back to back									
3300	High back fixtures	1 Plum	5	1.600	Ea.	1,150	108		1,258	1,400
3400	Flat slab fixtures		5	1.600		1,400	108		1,508	1,700
3430	ADA compliant		5	1.600		895	108		1,003	1,150
3500	Wall mounted, in stud or masonry									

22 42 Commercial Plumbing Fixtures

22 42 39 – Commercial Faucets, Supplies, and Trim

22 42 39.30 Carriers and Supports		Crew	Daily Output	Labor-Hours	Unit	Material	2021 Bare Costs Labor	Equipment	Total	Total Incl O&P
3600	High back fixture	1 Plum	6	1.333	Ea.	475	90.50		565.50	655
3700	Flat slab fixture	"	6	1.333	"	460	90.50		550.50	645
4000	Exposed arm type, floor mounted									
4100	Single high back or flat slab fixture	1 Plum	6	1.333	Ea.	1,150	90.50		1,240.50	1,375
4200	Back to back, high back or flat slab fixtures		5	1.600		1,800	108		1,908	2,125
4300	Wall mounted, high back or flat slab lavatory	↓	6	1.333	↓	790	90.50		880.50	1,000
4600	Sink, floor mounted									
4650	Exposed arm system									
4700	Single heavy fixture	1 Plum	5	1.600	Ea.	530	108		638	745
4750	Single heavy sink with slab		5	1.600		1,800	108		1,908	2,125
4800	Back to back, standard fixtures		5	1.600		855	108		963	1,100
4850	Back to back, heavy fixtures		5	1.600		1,475	108		1,583	1,775
4900	Back to back, heavy sink with slab	↓	5	1.600	↓	2,125	108		2,233	2,475
4950	Exposed offset arm system									
5000	Single heavy deep fixture	1 Plum	5	1.600	Ea.	1,275	108		1,383	1,550
5100	Plate type system									
5200	With bearing jacks, single fixture	1 Plum	5	1.600	Ea.	1,700	108		1,808	2,000
5300	With exposed arms, single heavy fixture		5	1.600		1,275	108		1,383	1,550
5400	Wall mounted, exposed arms, single heavy fixture		5	1.600		790	108		898	1,025
6000	Urinal, floor mounted, 2" or 3" coupling, blowout type		6	1.333		850	90.50		940.50	1,075
6100	With fixture or hanger bolts, blowout or washout		6	1.333		610	90.50		700.50	805
6200	With bearing plate		6	1.333		680	90.50		770.50	880
6300	Wall mounted, plate type system	↓	6	1.333	↓	610	90.50		700.50	805
6980	Water closet, siphon jet									
7000	Horizontal, adjustable, caulk									
7040	Single, 4" pipe size	1 Plum	5.33	1.501	Ea.	1,275	102		1,377	1,575
7050	4" pipe size, ADA compliant		5.33	1.501		660	102		762	875
7060	5" pipe size		5.33	1.501		1,125	102		1,227	1,400
7100	Double, 4" pipe size		5	1.600		2,400	108		2,508	2,800
7110	4" pipe size, ADA compliant		5	1.600		1,700	108		1,808	2,000
7120	5" pipe size	↓	5	1.600	↓	1,625	108		1,733	1,950
7160	Horizontal, adjustable, extended, caulk									
7180	Single, 4" pipe size	1 Plum	5.33	1.501	Ea.	1,700	102		1,802	2,000
7200	5" pipe size		5.33	1.501		2,100	102		2,202	2,475
7240	Double, 4" pipe size		5	1.600		2,775	108		2,883	3,200
7260	5" pipe size	↓	5	1.600	↓	2,925	108		3,033	3,375
7400	Vertical, adjustable, caulk or thread									
7440	Single, 4" pipe size	1 Plum	5.33	1.501	Ea.	1,450	102		1,552	1,750
7460	5" pipe size		5.33	1.501		1,825	102		1,927	2,150
7480	6" pipe size		5	1.600		1,875	108		1,983	2,225
7520	Double, 4" pipe size		5	1.600		2,575	108		2,683	2,975
7540	5" pipe size		5	1.600		2,350	108		2,458	2,725
7560	6" pipe size	↓	4	2	↓	2,600	135		2,735	3,050
7600	Vertical, adjustable, extended, caulk									
7620	Single, 4" pipe size	1 Plum	5.33	1.501	Ea.	1,225	102		1,327	1,500
7640	5" pipe size		5.33	1.501		1,075	102		1,177	1,325
7680	6" pipe size		5	1.600		2,050	108		2,158	2,425
7720	Double, 4" pipe size	↓	5	1.600	↓	2,350	108		2,458	2,725
7740	5" pipe size	1 Plum	5	1.600	Ea.	1,250	108		1,358	1,525
7760	6" pipe size	"	4	2	"	1,375	135		1,510	1,725
7780	Water closet, blow out									
7800	Vertical offset, caulk or thread									
7820	Single, 4" pipe size	1 Plum	5.33	1.501	Ea.	1,150	102		1,252	1,425

For customer support on your Plumbing Costs with RSMeans Data, call 800.448.8182.

22 42 Commercial Plumbing Fixtures

22 42 39 – Commercial Faucets, Supplies, and Trim

22 42 39.30 Carriers and Supports	Crew	Daily Output	Labor-Hours	Unit	Material	2021 Bare Costs Labor	2021 Bare Costs Equipment	Total	Total Incl O&P
7840	Double, 4" pipe size	1 Plum	5	1.600	Ea.	1,975	108		2,083	2,325
7880	Vertical offset, extended, caulk									
7900	Single, 4" pipe size	1 Plum	5.33	1.501	Ea.	1,450	102		1,552	1,750
7920	Double, 4" pipe size	"	5	1.600	"	2,275	108		2,383	2,650
7960	Vertical, for floor mounted back-outlet									
7980	Single, 4" thread, 2" vent	1 Plum	5.33	1.501	Ea.	945	102		1,047	1,200
8000	Double, 4" thread, 2" vent	"	6	1.333	"	2,775	90.50		2,865.50	3,175
8040	Vertical, for floor mounted back-outlet, extended									
8060	Single, 4" caulk, 2" vent	1 Plum	6	1.333	Ea.	945	90.50		1,035.50	1,175
8080	Double, 4" caulk, 2" vent	"	6	1.333	"	2,650	90.50		2,740.50	3,025
8200	Water closet, residential									
8220	Vertical centerline, floor mount									
8240	Single, 3" caulk, 2" or 3" vent	1 Plum	6	1.333	Ea.	925	90.50		1,015.50	1,150
8260	4" caulk, 2" or 4" vent		6	1.333		1,200	90.50		1,290.50	1,425
8280	3" copper sweat, 3" vent		6	1.333		830	90.50		920.50	1,050
8300	4" copper sweat, 4" vent		6	1.333		1,000	90.50		1,090.50	1,225
8400	Vertical offset, floor mount									
8420	Single, 3" or 4" caulk, vent	1 Plum	4	2	Ea.	1,150	135		1,285	1,475
8440	3" or 4" copper sweat, vent		5	1.600		1,150	108		1,258	1,425
8460	Double, 3" or 4" caulk, vent		4	2		1,975	135		2,110	2,375
8480	3" or 4" copper sweat, vent		5	1.600		1,975	108		2,083	2,325
9000	Water cooler (electric), floor mounted									
9100	Plate type with bearing plate, single	1 Plum	6	1.333	Ea.	680	90.50		770.50	880
9140	Plate type with bearing plate, back to back	"	4	2	"	760	135		895	1,025

22 43 Healthcare Plumbing Fixtures

22 43 13 – Healthcare Water Closets

22 43 13.40 Water Closets

		Crew	Daily Output	Labor-Hours	Unit	Material	2021 Bare Costs Labor	2021 Bare Costs Equipment	Total	Total Incl O&P
0010	**WATER CLOSETS**									
1000	Bowl only, 1 piece, w/seat and flush valve, ADA compliant, 18" high									
1030	Floor mounted									
1150	With wall outlet	Q-1	5.30	3.019	Ea.	335	184		519	645
1180	For rough-in, supply, waste and vent		2.84	5.634		375	345		720	920
1200	With floor outlet		5.30	3.019		375	184		559	685
1800	For rough-in, supply, waste and vent		3.05	5.246		345	320		665	855
3100	Wall hung		5.80	2.759		945	168		1,113	1,300
3150	Hospital type, slotted rim for bed pan									
3156	Elongated bowl, top spud	Q-1	5.80	2.759	Ea.	590	168		758	900
3160	Elongated bowl, rear spud		5.80	2.759		375	168		543	660
3200	For rough-in, supply, waste and vent, single WC		2.56	6.250		1,450	380		1,830	2,175
3300	Floor mounted									
3320	Bariatric (1,200 lb. capacity), elongated bowl, ADA compliant	Q-1	4.60	3.478	Ea.	2,900	212		3,112	3,525
3360	Hospital type, slotted rim for bed pan									
3370	Elongated bowl, top spud	Q-1	5	3.200	Ea.	315	195		510	635
3380	Elongated bowl, rear spud		5	3.200		415	195		610	745
3500	For rough-in, supply, waste and vent		3.05	5.246		345	320		665	855

22 43 Healthcare Plumbing Fixtures

22 43 16 – Healthcare Sinks

22 43 16.10 Sinks		Crew	Daily Output	Labor-Hours	Unit	Material	2021 Bare Costs Labor	Equipment	Total	Total Incl O&P
0010	**SINKS**									
0020	Vitreous china									
6702	Hospital type, without trim (see Section 22 41 39.10)									
6710	20" x 18", contoured splash shield	Q-1	8	2	Ea.	98	122		220	290
6730	28" x 20", surgeon, side decks		8	2		710	122		832	960
6740	28" x 22", surgeon scrub-up, deep bowl		8	2		865	122		987	1,125
6750	20" x 27", patient, ADA compliant		7	2.286		440	139		579	690
6760	30" x 22", all purpose		7	2.286		715	139		854	1,000
6770	30" x 22", plaster work		7	2.286		715	139		854	1,000
6820	20" x 24" clinic service, liquid/solid waste		6	2.667		990	162		1,152	1,350

22 43 19 – Healthcare Bathtubs

22 43 19.10 Bathtubs		Crew	Daily Output	Labor-Hours	Unit	Material	2021 Bare Costs Labor	Equipment	Total	Total Incl O&P
0010	**BATHTUBS**									
5002	Hospital type, with trim, see Section 22 41 39.10									
5050	Bathing pool, porcelain enamel on cast iron, grab bars									
5060	pop-up drain, 72" x 36"	Q-1	3	5.333	Ea.	4,275	325		4,600	5,175
5100	Perineal (sitz), vitreous china		3	5.333		1,325	325		1,650	1,950
5120	For pedestal, vitreous china, add		8	2		268	122		390	475
5300	Whirlpool, porcelain enamel on cast iron, 72" x 36"		1	16		4,825	975		5,800	6,775
5310	For color, add					5%				
5311	For designer colors and trim, add					15%				

22 43 23 – Healthcare Showers

22 43 23.10 Showers		Crew	Daily Output	Labor-Hours	Unit	Material	2021 Bare Costs Labor	Equipment	Total	Total Incl O&P
0010	**SHOWERS**									
5950	Module, ADA compl, SS panel, fixed & hand held head, control									
5960	valves, grab bar, curtain & rod, folding seat	1 Plum	4	2	Ea.	1,575	135		1,710	1,925

22 43 39 – Healthcare Faucets

22 43 39.10 Faucets and Fittings		Crew	Daily Output	Labor-Hours	Unit	Material	2021 Bare Costs Labor	Equipment	Total	Total Incl O&P
0010	**FAUCETS AND FITTINGS**									
2850	Medical, bedpan cleanser, with pedal valve,	1 Plum	12	.667	Ea.	870	45		915	1,025
2860	With screwdriver stop valve	1 Plum	12	.667	Ea.	450	45		495	565
2870	With self-closing spray valve		12	.667		278	45		323	375
2900	Faucet, gooseneck spout, wrist handles, grid drain		10	.800		145	54		199	240
2940	Mixing valve, knee action, screwdriver stops		4	2		485	135		620	735

22 45 Emergency Plumbing Fixtures

22 45 13 – Emergency Showers

22 45 13.10 Emergency Showers		Crew	Daily Output	Labor-Hours	Unit	Material	2021 Bare Costs Labor	Equipment	Total	Total Incl O&P
0010	**EMERGENCY SHOWERS**, Rough-in not included									
5000	Shower, single head, drench, ball valve, pull, freestanding	Q-1	4	4	Ea.	385	244		629	785
5200	Horizontal or vertical supply		4	4		630	244		874	1,050
6000	Multi-nozzle, eye/face wash combination		4	4		840	244		1,084	1,300
6400	Multi-nozzle, 12 spray, shower only		4	4		2,050	244		2,294	2,625
6600	For freeze-proof, add		6	2.667		520	162		682	810
8000	Walk-thru decontamination with eye-face wash		2	8		4,275	485		4,760	5,425
8200	For freeze proof, add		4	4		605	244		849	1,025

22 45 16.10 Eyewash Safety Equipment	Crew	Daily Output	Labor-Hours	Unit	Material	2021 Bare Costs Labor	2021 Bare Costs Equipment	Total	Total Incl O&P
0010 **EYEWASH SAFETY EQUIPMENT**, Rough-in not included									
1000 Eye wash fountain									
1400 Plastic bowl, pedestal mounted	Q-1	4	4	Ea.	340	244		584	740
1600 Unmounted		4	4		315	244		559	710
1800 Wall mounted		4	4		540	244		784	955
2000 Stainless steel, pedestal mounted		4	4		365	244		609	765
2200 Unmounted		4	4		305	244		549	700
2400 Wall mounted		4	4		283	244		527	675

22 45 19.10 Self-Contained Eyewash Safety Equipment

	Crew	Daily Output	Labor-Hours	Unit	Material	Labor	Equipment	Total	Total Incl O&P
0010 **SELF-CONTAINED EYEWASH SAFETY EQUIPMENT**									
3000 Eye wash, portable, self-contained				Ea.	2,050			2,050	2,250

22 45 26.10 Eye/Face Wash Safety Equipment

	Crew	Daily Output	Labor-Hours	Unit	Material	Labor	Equipment	Total	Total Incl O&P
0010 **EYE/FACE WASH SAFETY EQUIPMENT**, Rough-in not included									
4000 Eye and face wash, combination fountain									
4200 Stainless steel, pedestal mounted	Q-1	4	4	Ea.	1,075	244		1,319	1,550
4400 Unmounted		4	4		300	244		544	695
4600 Wall mounted		4	4		325	244		569	725

22 46 Security Plumbing Fixtures

22 46 13 – Security Water Closets and Urinals

22 46 13.10 Security Water Closets and Urinals

	Crew	Daily Output	Labor-Hours	Unit	Material	Labor	Equipment	Total	Total Incl O&P
0010 **SECURITY WATER CLOSETS AND URINALS**, Stainless steel									
2000 Urinal, back supply and flush									
2200 Wall hung	Q-1	4	4	Ea.	3,400	244		3,644	4,125
2240 Stall		2.50	6.400		4,500	390		4,890	5,525
2300 For urinal rough-in, supply, waste and vent		1.49	10.738		260	655		915	1,250
3000 Water closet, integral seat, back supply and flush									
3300 Wall hung, wall outlet	Q-1	5.80	2.759	Ea.	1,950	168		2,118	2,375
3400 Floor mount, wall outlet		5.80	2.759		2,525	168		2,693	3,025
3440 Floor mount, floor outlet		5.80	2.759		2,550	168		2,718	3,075
3480 For recessed tissue holder, add					176			176	193
3500 For water closet rough-in, supply, waste and vent	Q-1	1.19	13.445		350	820		1,170	1,600
5000 Water closet and lavatory units, push button filler valves,									
5010 soap & paper holders, seat									
5300 Wall hung	Q-1	5	3.200	Ea.	3,850	195		4,045	4,550
5400 Floor mount		5	3.200		3,850	195		4,045	4,550
6300 For unit rough-in, supply, waste and vent		1	16		415	975		1,390	1,900

22 46 16 – Security Lavatories and Sinks

22 46 16.13 Security Lavatories

	Crew	Daily Output	Labor-Hours	Unit	Material	Labor	Equipment	Total	Total Incl O&P
0010 **SECURITY LAVATORIES**, Stainless steel									
1000 Lavatory, wall hung, push button filler valve									
1100 Rectangular bowl	Q-1	8	2	Ea.	1,375	122		1,497	1,700
1200 Oval bowl		8	2		1,400	122		1,522	1,725
1240 Oval bowl, corner mount		8	2		1,650	122		1,772	1,975
1300 For lavatory rough-in, supply, waste and vent		1.50	10.667		495	650		1,145	1,525

22 46 Security Plumbing Fixtures

22 46 63 – Security Service Sink

22 46 63.10 Security Service Sink	Crew	Daily Output	Labor-Hours	Unit	Material	2021 Bare Costs Labor	Equipment	Total	Total Incl O&P
0010 **SECURITY SERVICE SINK**, Stainless steel									
1700 Service sink, with soap dish									
1740 24" x 19" size	Q-1	3	5.333	Ea.	4,200	325		4,525	5,075
1790 For sink rough-in, supply, waste and vent	"	.89	17.978	"	1,000	1,100		2,100	2,725

22 46 73 – Security Shower

22 46 73.10 Security Shower	Crew	Daily Output	Labor-Hours	Unit	Material	2021 Bare Costs Labor	Equipment	Total	Total Incl O&P
0010 **SECURITY SHOWER**, Stainless steel									
1800 Shower cabinet, unitized									
1840 36" x 36" x 88"	Q-1	2.20	7.273	Ea.	10,400	445		10,845	12,100
1900 Shower package for built-in									
1940 Hot & cold valves, recessed soap dish	Q-1	6	2.667	Ea.	450	162		612	735

22 47 Drinking Fountains and Water Coolers

22 47 13 – Drinking Fountains

22 47 13.10 Drinking Water Fountains	Crew	Daily Output	Labor-Hours	Unit	Material	2021 Bare Costs Labor	Equipment	Total	Total Incl O&P
0010 **DRINKING WATER FOUNTAINS**, For connection to cold water supply									
0802 For remote water chiller, see Section 22 47 23.10									
1000 Wall mounted, non-recessed R224000-30									
1200 Aluminum,									
1280 Dual bubbler type	1 Plum	3.20	2.500	Ea.	2,675	169		2,844	3,200
1400 Bronze, with no back	1 Plum	4	2	Ea.	1,125	135		1,260	1,450
1600 Cast iron, enameled, low back, single bubbler D2010-810		4	2		1,250	135		1,385	1,575
1640 Dual bubbler type		3.20	2.500		2,000	169		2,169	2,450
1680 Triple bubbler type		3.20	2.500		1,900	169		2,069	2,350
1800 Cast aluminum, enameled, for correctional institutions		4	2		1,300	135		1,435	1,650
2000 Fiberglass, 12" back, single bubbler unit		4	2		1,800	135		1,935	2,175
2040 Dual bubbler		3.20	2.500		2,775	169		2,944	3,300
2080 Triple bubbler		3.20	2.500		3,050	169		3,219	3,600
2200 Polymarble, no back, single bubbler		4	2		945	135		1,080	1,250
2240 Dual bubbler		3.20	2.500		2,525	169		2,694	3,025
2280 Triple bubbler		3.20	2.500		2,075	169		2,244	2,525
2400 Precast stone, no back		4	2		1,125	135		1,260	1,450
2700 Stainless steel, single bubbler, no back		4	2		870	135		1,005	1,150
2740 With back		4	2		1,125	135		1,260	1,450
2780 Dual handle, ADA compliant		4	2		785	135		920	1,075
2820 Dual level, ADA compliant		3.20	2.500		1,450	169		1,619	1,850
2840 Vandal resistant type		4	2		625	135		760	890
3300 Vitreous china									
3340 7" back	1 Plum	4	2	Ea.	670	135		805	935
3940 For vandal-resistant bottom plate, add					72.50			72.50	80
3960 For freeze-proof valve system, add	1 Plum	2	4		830	271		1,101	1,325
3980 For rough-in, supply and waste, add	"	2.21	3.620		370	245		615	770
4000 Wall mounted, semi-recessed									
4200 Poly-marble, single bubbler	1 Plum	4	2	Ea.	950	135		1,085	1,250
4600 Stainless steel, satin finish, single bubbler		4	2		1,700	135		1,835	2,075
4900 Vitreous china, single bubbler		4	2		1,075	135		1,210	1,375
5980 For rough-in, supply and waste, add		1.83	4.372		370	296		666	845
6000 Wall mounted, fully recessed									
6400 Poly-marble, single bubbler	1 Plum	4	2	Ea.	1,750	135		1,885	2,125
6440 For water glass filler, add					102			102	112

For customer support on your Plumbing Costs with RSMeans Data, call 800.448.8182.

309

22 47 Drinking Fountains and Water Coolers

22 47 13 – Drinking Fountains

22 47 13.10 Drinking Water Fountains

		Crew	Daily Output	Labor-Hours	Unit	Material	2021 Bare Costs Labor	Equipment	Total	Total Incl O&P
6800	Stainless steel, single bubbler	1 Plum	4	2	Ea.	1,350	135		1,485	1,675
6900	Fountain and cuspidor combination		2	4		2,625	271		2,896	3,275
7560	For freeze-proof valve system, add		2	4		1,000	271		1,271	1,500
7580	For rough-in, supply and waste, add	↓	1.83	4.372	↓	370	296		666	845
7600	Floor mounted, pedestal type									
7700	Aluminum, architectural style, CI base	1 Plum	2	4	Ea.	2,350	271		2,621	2,975
7780	ADA compliant unit		2	4		1,500	271		1,771	2,050
8000	Bronze, architectural style		2	4		2,800	271		3,071	3,475
8040	Enameled steel cylindrical column style		2	4		2,250	271		2,521	2,875
8200	Precast stone/concrete, cylindrical column		1	8		1,650	540		2,190	2,600
8240	ADA compliant unit		1	8		3,075	540		3,615	4,175
8400	Stainless steel, architectural style		2	4		2,400	271		2,671	3,025
8600	Enameled iron, heavy duty service, 2 bubblers		2	4		3,475	271		3,746	4,225
8660	4 bubblers		2	4		4,175	271		4,446	4,975
8880	For freeze-proof valve system, add		2	4		680	271		951	1,150
8900	For rough-in, supply and waste, add	↓	1.83	4.372	↓	370	296		666	845
9100	Deck mounted									
9500	Stainless steel, circular receptor	1 Plum	4	2	Ea.	455	135		590	700
9540	14" x 9" receptor		4	2		375	135		510	610
9580	25" x 17" deep receptor, with water glass filler		3	2.667		295	181		476	595
9760	White enameled steel, 14" x 9" receptor		4	2		395	135		530	635
9860	White enameled cast iron, 24" x 16" receptor		3	2.667		515	181		696	835
9980	For rough-in, supply and waste, add	↓	1.83	4.372		370	296		666	845

22 47 16 – Pressure Water Coolers

22 47 16.10 Electric Water Coolers

		Crew	Daily Output	Labor-Hours	Unit	Material	2021 Bare Costs Labor	Equipment	Total	Total Incl O&P
0010	**ELECTRIC WATER COOLERS** D2010-820									
0100	Wall mounted, non-recessed									
0140	4 GPH	Q-1	4	4	Ea.	690	244		934	1,125
0160	8 GPH, barrier free, sensor operated		4	4		1,750	244		1,994	2,300
0180	8.2 GPH		4	4		1,100	244		1,344	1,600
0220	14.3 GPH		4	4		1,175	244		1,419	1,675
0600	8 GPH hot and cold water	↓	4	4		1,175	244		1,419	1,675
0640	For stainless steel cabinet, add					93			93	102
1000	Dual height, 8.2 GPH	Q-1	3.80	4.211		1,775	257		2,032	2,325
1040	14.3 GPH	"	3.80	4.211		2,025	257		2,282	2,600
1240	For stainless steel cabinet, add					211			211	232
2600	ADA compliant, 8 GPH	Q-1	4	4		1,150	244		1,394	1,625
3000	Simulated recessed, 8 GPH		4	4		890	244		1,134	1,350
3040	11.5 GPH	↓	4	4		1,025	244		1,269	1,500
3200	For glass filler, add					101			101	111
3240	For stainless steel cabinet, add					87			87	96
3300	Semi-recessed, 8.1 GPH	Q-1	4	4		1,025	244		1,269	1,500
3320	12 GPH	"	4	4		1,175	244		1,419	1,650
3340	For glass filler, add					164			164	180
3360	For stainless steel cabinet, add					146			146	160
3400	Full recessed, stainless steel, 8 GPH	Q-1	3.50	4.571		2,475	279		2,754	3,150
3420	11.5 GPH	"	3.50	4.571		1,825	279		2,104	2,425
3460	For glass filler, add					199			199	219
3600	For mounting can only					219			219	241
4600	Floor mounted, flush-to-wall									
4640	4 GPH	1 Plum	3	2.667	Ea.	980	181		1,161	1,350
4680	8.2 GPH	↓	3	2.667		1,025	181		1,206	1,400

For customer support on your Plumbing Costs with RSMeans Data, call 800.448.8182.

22 47 Drinking Fountains and Water Coolers

22 47 16 – Pressure Water Coolers

22 47 16.10 Electric Water Coolers

		Crew	Daily Output	Labor-Hours	Unit	Material	2021 Bare Costs Labor	Equipment	Total	Total Incl O&P
4720	14.3 GPH	1 Plum	3	2.667	Ea.	1,175	181		1,356	1,575
4960	14 GPH hot and cold water	↓	3	2.667		1,200	181		1,381	1,600
4980	For stainless steel cabinet, add					141			141	155
5000	Dual height, 8.2 GPH	1 Plum	2	4		1,200	271		1,471	1,725
5040	14.3 GPH	"	2	4		1,250	271		1,521	1,775
5120	For stainless steel cabinet, add					207			207	227
5600	Explosion proof, 16 GPH	1 Plum	3	2.667		2,775	181		2,956	3,325
6000	Refrigerator compartment type, 4.5 GPH		3	2.667		1,525	181		1,706	1,950
6600	Bottle supply type, 1.0 GPH		4	2		405	135		540	645
6640	Hot and cold, 1.0 GPH		4	2		565	135		700	825
9800	For supply, waste & vent, all coolers	↓	2.21	3.620	↓	370	245		615	770

22 47 23 – Remote Water Coolers

22 47 23.10 Remote Water Coolers

		Crew	Daily Output	Labor-Hours	Unit	Material	2021 Bare Costs Labor	Equipment	Total	Total Incl O&P
0010	**REMOTE WATER COOLERS**, 80°F inlet									
0100	Air cooled, 50°F outlet, 115 V, 4.1 GPH	1 Plum	6	1.333	Ea.	665	90.50		755.50	865
0200	5.7 GPH		5.50	1.455		1,150	98.50		1,248.50	1,400
0300	8.0 GPH		5	1.600		735	108		843	965
0400	10.0 GPH		4.50	1.778		1,200	120		1,320	1,500
0500	13.4 GPH	↓	4	2		1,925	135		2,060	2,325
0700	29 GPH	Q-1	5	3.200		2,250	195		2,445	2,775
1000	230 V, 32 GPH	"	5	3.200	↓	2,225	195		2,420	2,750

22 51 Swimming Pool Plumbing Systems

22 51 19 – Swimming Pool Water Treatment Equipment

22 51 19.50 Swimming Pool Filtration Equipment

		Crew	Daily Output	Labor-Hours	Unit	Material	2021 Bare Costs Labor	Equipment	Total	Total Incl O&P
0010	**SWIMMING POOL FILTRATION EQUIPMENT**									
0900	Filter system, sand or diatomite type, incl. pump, 6,000 gal./hr.	2 Plum	1.80	8.889	Total	2,325	600		2,925	3,450
1020	Add for chlorination system, 800 S.F. pool		3	5.333	Ea.	257	360		617	825
1040	5,000 S.F. pool	↓	3	5.333	"	1,950	360		2,310	2,700

22 52 Fountain Plumbing Systems

22 52 16 – Fountain Pumps

22 52 16.10 Fountain Water Pumps

		Crew	Daily Output	Labor-Hours	Unit	Material	2021 Bare Costs Labor	Equipment	Total	Total Incl O&P
0010	**FOUNTAIN WATER PUMPS**									
0100	Pump w/controls									
0200	Single phase, 100' cord, 1/2 HP pump	2 Skwk	4.40	3.636	Ea.	1,325	208		1,533	1,750
0300	3/4 HP pump		4.30	3.721		1,325	212		1,537	1,800
0400	1 HP pump		4.20	3.810		1,775	218		1,993	2,275
0500	1-1/2 HP pump		4.10	3.902		3,250	223		3,473	3,900
0600	2 HP pump		4	4		4,325	228		4,553	5,125
0700	Three phase, 200' cord, 5 HP pump	↓	3.90	4.103	↓	6,175	234		6,409	7,150
0800	7-1/2 HP pump	2 Skwk	3.80	4.211	Ea.	12,900	240		13,140	14,600
0900	10 HP pump		3.70	4.324		14,500	247		14,747	16,300
1000	15 HP pump	↓	3.60	4.444	↓	20,200	254		20,454	22,600
2000	DESIGN NOTE: Use two horsepower per surface acre.									

For customer support on your Plumbing Costs with RSMeans Data, call 800.448.8182.

311

22 52 Fountain Plumbing Systems

22 52 33 – Fountain Ancillary

22 52 33.10 Fountain Miscellaneous	Crew	Daily Output	Labor-Hours	Unit	Material	2021 Bare Costs Labor	Equipment	Total	Total Incl O&P
0010 **FOUNTAIN MISCELLANEOUS**									
1300 Lights w/mounting kits, 200 watt	2 Skwk	18	.889	Ea.	1,200	51		1,251	1,375
1400 300 watt		18	.889		1,375	51		1,426	1,575
1500 500 watt		18	.889		1,600	51		1,651	1,825
1600 Color blender	↓	12	1.333	↓	605	76		681	780

22 62 Vacuum Systems for Laboratory and Healthcare Facilities

22 62 19 – Vacuum Equipment for Laboratory and Healthcare Facilities

22 62 19.70 Healthcare Vacuum Equipment

22 62 19.70 Healthcare Vacuum Equipment	Crew	Daily Output	Labor-Hours	Unit	Material	2021 Bare Costs Labor	Equipment	Total	Total Incl O&P
0010 **HEALTHCARE VACUUM EQUIPMENT**									
0300 Dental oral									
0310 Duplex									
0330 165 SCFM with 77 gal. separator	Q-2	1.30	18.462	Ea.	52,000	1,175		53,175	59,500
1100 Vacuum system									
1110 Vacuum outlet alarm panel	1 Plum	3.20	2.500	Ea.	1,050	169		1,219	1,400
2000 Medical, with receiver									
2100 Rotary vane type, lubricated, with controls									
2110 Simplex									
2120 1.5 HP, 80 gal. tank	Q-1	8	2	Ea.	6,700	122		6,822	7,525
2130 2 HP, 80 gal. tank		7	2.286		6,925	139		7,064	7,825
2140 3 HP, 80 gal. tank		6.60	2.424		7,300	148		7,448	8,250
2150 5 HP, 80 gal. tank	↓	6	2.667	↓	7,950	162		8,112	9,000
2200 Duplex									
2210 1 HP, 80 gal. tank	Q-1	8.40	1.905	Ea.	10,700	116		10,816	12,000
2220 1.5 HP, 80 gal. tank		7.80	2.051		11,000	125		11,125	12,300
2230 2 HP, 80 gal. tank		6.80	2.353		11,800	143		11,943	13,200
2240 3 HP, 120 gal. tank		6	2.667		12,900	162		13,062	14,400
2250 5 HP, 120 gal. tank	↓	5.40	2.963		14,600	181		14,781	16,400
2260 7.5 HP, 200 gal. tank	Q-2	7	3.429		20,900	217		21,117	23,300
2270 10 HP, 200 gal. tank		6.40	3.750		24,200	237		24,437	27,000
2280 15 HP, 200 gal. tank		5.80	4.138		43,300	261		43,561	48,000
2290 20 HP, 200 gal. tank		5	4.800		49,300	305		49,605	54,500
2300 25 HP, 200 gal. tank	↓	4	6	↓	57,500	380		57,880	63,500
2400 Triplex									
2410 7.5 HP, 200 gal. tank	Q-2	6.40	3.750	Ea.	33,200	237		33,437	36,900
2420 10 HP, 200 gal. tank		5.70	4.211		38,700	266		38,966	43,000
2430 15 HP, 200 gal. tank	↓	4.90	4.898	↓	65,500	310		65,810	72,500
2440 20 HP, 200 gal. tank	Q-2	4	6	Ea.	74,500	380		74,880	82,500
2450 25 HP, 200 gal. tank	"	3.70	6.486	"	86,500	410		86,910	95,500
2500 Quadruplex									
2510 7.5 HP, 200 gal. tank	Q-2	5.30	4.528	Ea.	41,400	286		41,686	45,900
2520 10 HP, 200 gal. tank		4.70	5.106		47,700	325		48,025	53,000
2530 15 HP, 200 gal. tank		4	6		86,000	380		86,380	95,500
2540 20 HP, 200 gal. tank		3.60	6.667		98,000	420		98,420	108,500
2550 25 HP, 200 gal. tank	↓	3.20	7.500	↓	114,000	475		114,475	126,000
4000 Liquid ring type, water sealed, with controls									
4200 Duplex									
4210 1.5 HP, 120 gal. tank	Q-1	6	2.667	Ea.	21,100	162		21,262	23,400
4220 3 HP, 120 gal. tank	"	5.50	2.909		22,000	177		22,177	24,500
4230 4 HP, 120 gal. tank	Q-2	6.80	3.529		23,500	223		23,723	26,100

22 62 19 – Vacuum Equipment for Laboratory and Healthcare Facilities

22 62 19.70 Healthcare Vacuum Equipment		Crew	Daily Output	Labor-Hours	Unit	Material	2021 Bare Costs Labor	Equipment	Total	Total Incl O&P
4240	5 HP, 120 gal. tank	Q-2	6.50	3.692	Ea.	24,800	233		25,033	27,700
4250	7.5 HP, 200 gal. tank		6.20	3.871		30,200	245		30,445	33,600
4260	10 HP, 200 gal. tank		5.80	4.138		37,600	261		37,861	41,800
4270	15 HP, 200 gal. tank		5.10	4.706		45,600	297		45,897	50,500
4280	20 HP, 200 gal. tank		4.60	5.217		59,000	330		59,330	65,500
4290	30 HP, 200 gal. tank	↓	4	6	↓	70,500	380		70,880	78,000

22 63 13 – Gas Piping for Laboratory and Healthcare Facilities

22 63 13.70 Healthcare Gas Piping

		Crew	Daily Output	Labor-Hours	Unit	Material	2021 Bare Costs Labor	Equipment	Total	Total Incl O&P
0010	**HEALTHCARE GAS PIPING**									
0030	Air compressor intake filter									
0034	Rooftop									
0036	Filter/silencer									
0040	1"	1 Stpi	10	.800	Ea.	258	54.50		312.50	365
0044	1-1/4"		9.60	.833		265	57		322	375
0048	1-1/2"		9.20	.870		284	59.50		343.50	405
0052	2"		8.80	.909		284	62		346	410
0056	2-1/2"		8.40	.952		380	65		445	510
0060	3"		8	1		455	68.50		523.50	600
0064	4"		7.60	1.053		475	72		547	625
0068	5"	↓	7.40	1.081	↓	665	74		739	840
0076	Inline									
0080	Filter/silencer									
0084	2"	1 Stpi	8.20	.976	Ea.	860	66.50		926.50	1,050
0088	2-1/2"		8	1		885	68.50		953.50	1,075
0090	3"		7.60	1.053		980	72		1,052	1,175
0092	4"		7.20	1.111		1,150	76		1,226	1,375
0094	5"		6.80	1.176		1,225	80.50		1,305.50	1,475
0096	6"	↓	6.40	1.250	↓	1,400	85.50		1,485.50	1,650
1000	Nitrogen or oxygen system									
1010	Cylinder manifold									
1020	5 cylinder	1 Plum	.80	10	Ea.	5,650	675		6,325	7,225
1026	10 cylinder	"	.40	20		6,800	1,350		8,150	9,525
1050	Nitrogen generator, 30 LPM	Q-5	.40	40	↓	30,400	2,450		32,850	37,200
1900	Vaporizers									
1910	LOX vaporizers									
1920	Nominal capacity									
1930	1410 SCFM	Q-1	2	8	Ea.	1,925	485		2,410	2,850
1940	5650 SCFM		1.40	11.429		4,475	695		5,170	5,950
1950	12,703 SCFM	↓	.80	20	↓	6,850	1,225		8,075	9,375
1980	Removal of LOX vaporizers									
1982	Nominal capacity									
1986	1410 SCFM	Q-1	4	4	Ea.		244		244	365
1990	5650 SCFM		2.80	5.714			350		350	520
1994	12,703 SCFM	↓	1.60	10	↓		610		610	910
3000	Outlets and valves									
3010	Recessed, wall mounted									
3012	Single outlet	1 Plum	3.20	2.500	Ea.	67.50	169		236.50	330
3100	Ceiling outlet									
3190	Zone valve with box									

For customer support on your Plumbing Costs with RSMeans Data, call 800.448.8182.

313

22 63 13.70 Healthcare Gas Piping	Crew	Daily Output	Labor-Hours	Unit	Material	2021 Bare Costs Labor	Equipment	Total	Total Incl O&P
3192	Cleaned for oxygen service, not including gauges									
3194	1/2" valve size	1 Plum	4.60	1.739	Ea.	225	118		343	425
3196	3/4" valve size		4.30	1.860		248	126		374	460
3198	1" valve size		4	2		275	135		410	500
3202	1-1/4" valve size		3.80	2.105		305	143		448	555
3206	1-1/2" valve size		3.60	2.222		345	150		495	600
3210	2" valve size		3.20	2.500		395	169		564	690
3214	2-1/2" valve size		3.10	2.581		1,025	175		1,200	1,375
3218	3" valve size		3	2.667		1,425	181		1,606	1,850
3224	Gauges for zone valve box									
3226	0-100 psi (O2, air, N2, CO2)	1 Plum	16	.500	Ea.	18	34		52	70.50
3228	Vacuum, WAGD (Waste Anesthesia Gas)		16	.500		18	34		52	70.50
3230	0-300 psi (Nitrogen)		16	.500		18	34		52	70.50
4000	Alarm panel, medical gases and vacuum									
4010	Alarm panel	1 Plum	3.20	2.500	Ea.	1,050	169		1,219	1,400
4030	Master alarm panel									
4034	Can also monitor area alarms									
4038	and communicate with PC-based alarm monitor.									
4040	10 signal	1 Elec	3.20	2.500	Ea.	1,025	159		1,184	1,350
4044	20 signal	1 Elec	3	2.667	Ea.	1,200	170		1,370	1,550
4048	30 signal		2.80	2.857		1,575	182		1,757	2,025
4052	40 signal		2.60	3.077		1,800	196		1,996	2,275
4056	50 signal		2.40	3.333		1,925	212		2,137	2,450
4060	60 signal		2.20	3.636		2,250	232		2,482	2,825
4100	Area alarm panel									
4104	Does not include specific gas transducers.									
4108	3 module alarm panel									
4112	P-P-P	1 Elec	2.80	2.857	Ea.	1,400	182		1,582	1,800
4116	P-P-V		2.80	2.857		945	182		1,127	1,300
4120	D-D-D		2.80	2.857		1,025	182		1,207	1,425
4130	6 module alarm panel									
4132	4-P, 3-V	1 Elec	2.20	3.636	Ea.	1,475	232		1,707	1,975
4136	3-P, 2-V, B		2.20	3.636		1,150	232		1,382	1,600
4140	5-D, B		2.20	3.636		1,450	232		1,682	1,950
4144	6-D		2.20	3.636		1,925	232		2,157	2,475
4148	3-P, 3-V		2.20	3.636		1,800	232		2,032	2,325
4170	Note: P=pressure, V=vacuum, B=blank, D=dual display									
4180	Alarm transducers, gas specific									
4182	Oxygen	1 Elec	24	.333	Ea.	159	21		180	207
4184	Vacuum		24	.333		156	21		177	203
4186	Nitrous oxide		24	.333		156	21		177	203
4188	Medical air		24	.333		159	21		180	207
4190	Carbon dioxide		24	.333		159	21		180	207
4192	Nitrogen		24	.333		159	21		180	207
4194	WAGD		24	.333		170	21		191	219
4300	Ball valves cleaned for oxygen service									
4310	with copper extensions and gauge port									
4320	1/4" diam.	1 Plum	24	.333	Ea.	64.50	22.50		87	105
4330	1/2" diam.		22	.364		63.50	24.50		88	107
4334	3/4" diam.		20	.400		81.50	27		108.50	131
4338	1" diam.		19	.421		113	28.50		141.50	168
4342	1-1/4" diam.		15	.533		142	36		178	211
4346	1-1/2" diam.		13	.615		188	41.50		229.50	269

22 63 13.70 Healthcare Gas Piping	Crew	Daily Output	Labor-Hours	Unit	Material	2021 Bare Costs Labor	Equipment	Total	Total Incl O&P
4350	2" diam.	1 Plum	11	.727	Ea.	310	49		359	415
4354	2-1/2" diam.	Q-1	15	1.067		855	65		920	1,025
4358	3" diam.		13	1.231		1,275	75		1,350	1,500
4362	4" diam.		10	1.600		2,425	97.50		2,522.50	2,825
5000	Manifold									
5010	Automatic switchover type									
5020	Note: Both a control panel and header assembly are required									
5030	Control panel									
5040	Oxygen	Q-1	4	4	Ea.	4,900	244		5,144	5,775
5060	Header assembly									
5066	Oxygen									
5070	2 x 2	Q-1	6	2.667	Ea.	805	162		967	1,125
5074	3 x 3		5.50	2.909		945	177		1,122	1,300
5078	4 x 4		5	3.200		1,225	195		1,420	1,650
5082	5 x 5		4.50	3.556		1,375	217		1,592	1,825
5086	6 x 6		4	4		1,725	244		1,969	2,275
5090	7 x 7		3.50	4.571		1,875	279		2,154	2,475
7000	Medical air compressors									
7020	Oil-less with inlet filter, duplexed dryers and aftercoolers									
7030	Duplex systems, horizontal tank, 208/230/460/575 V, 3 Ph.									
7040	1 HP, 80 gal. tank, 3.8 ACFM @50PSIG, 2.7 ACFM @100PSIG	Q-5	4	4	Ea.	30,100	246		30,346	33,600
7050	1.5 HP, 80 gal. tank, 6.8 ACFM @50PSIG, 4.5 ACFM @100PSIG		3.80	4.211		37,500	259		37,759	41,700
7060	2 HP, 80 gal. tank, 8.2 ACFM @50PSIG, 6.3 ACFM @100PSIG		3.60	4.444		37,700	273		37,973	41,800
7070	3 HP, 120 gal. tank, 11.2 ACFM @50PSIG, 9.4 ACFM @100PSIG		3.20	5		29,200	310		29,510	32,600
7080	5 HP, 120 gal. tank, 17.4 ACFM @50PSIG, 15.6 ACFM @100PSIG		2.80	5.714		48,700	350		49,050	54,000
7090	7.5 HP, 250 gal. tank, 31.6 ACFM @50PSIG, 25.9 ACFM @100PSIG		2.40	6.667		64,500	410		64,910	71,500
7100	10 HP, 250 gal. tank, 43 ACFM @50PSIG, 35.2 ACFM @100PSIG		2	8		70,000	490		70,490	77,500
7110	15 HP, 250 gal. tank, 69 ACFM @50PSIG, 56.5 ACFM @100PSIG		1.80	8.889		84,000	545		84,545	93,500
7200	Aftercooler, air-cooled									
7210	Steel manifold, copper tube, aluminum fins									
7220	35 SCFM @ 100PSIG	Q-5	4	4	Ea.	1,050	246		1,296	1,525
7300	Air dryer system									
7310	Refrigerated type									
7320	Flow @ 125 psi									
7330	20 SCFM	Q-5	6	2.667	Ea.	1,550	164		1,714	1,950
7332	25 SCFM		5.80	2.759		1,625	170		1,795	2,025
7334	35 SCFM		5.40	2.963		1,975	182		2,157	2,450
7336	50 SCFM		5	3.200		2,450	197		2,647	3,000
7338	75 SCFM		4.60	3.478		3,075	214		3,289	3,700
7340	100 SCFM		4.30	3.721		3,700	229		3,929	4,400
7342	125 SCFM		4	4		4,750	246		4,996	5,600
7360	Desiccant type									
7362	Flow with energy saving controls									
7366	40 SCFM	Q-5	3.60	4.444	Ea.	6,275	273		6,548	7,300
7368	60 SCFM		3.40	4.706		6,850	290		7,140	7,950
7370	90 SCFM		3.20	5		8,125	310		8,435	9,400
7372	115 SCFM		3	5.333		8,750	330		9,080	10,100
7374	165 SCFM	Q-5	2.90	5.517	Ea.	9,400	340		9,740	10,800
7378	260 SCFM	Q-6	3.60	6.667		11,100	425		11,525	12,800
7382	370 SCFM		3.40	7.059		12,900	450		13,350	14,900
7386	590 SCFM		3.20	7.500		15,900	480		16,380	18,200
7390	1130 SCFM		3	8		24,700	510		25,210	28,000
7460	Dew point monitor									

22 63 Gas Systems for Laboratory and Healthcare Facilities

22 63 13 – Gas Piping for Laboratory and Healthcare Facilities

22 63 13.70 Healthcare Gas Piping	Crew	Daily Output	Labor-Hours	Unit	Material	2021 Bare Costs Labor	Equipment	Total	Total Incl O&P
7466	LCD readout, high dew point alarm, probe included									
7470	Monitor	1 Stpi	2	4	Ea.	3,300	273		3,573	4,025

22 66 Chemical-Waste Systems for Lab. and Healthcare Facilities

22 66 53 – Laboratory Chemical-Waste and Vent Piping

22 66 53.30 Glass Pipe

		Crew	Daily Output	Labor-Hours	Unit	Material	2021 Bare Costs Labor	Equipment	Total	Total Incl O&P
0010	GLASS PIPE, Borosilicate, couplings & clevis hanger assemblies, 10' OC R221113-70									
0020	Drainage									
1100	1-1/2" diameter	Q-1	52	.308	L.F.	13.60	18.75		32.35	43
1120	2" diameter		44	.364		17.90	22		39.90	52.50
1140	3" diameter		39	.410		22.50	25		47.50	62.50
1160	4" diameter		30	.533		43	32.50		75.50	95.50
1180	6" diameter		26	.615		77	37.50		114.50	141
1870	To delete coupling & hanger, subtract									
1880	1-1/2" diam. to 2" diam.					19%	22%			
1890	3" diam. to 6" diam.					20%	17%			
2000	Process supply (pressure), beaded joints									
2040	1/2" diameter	1 Plum	36	.222	L.F.	7.45	15.05		22.50	30.50
2060	3/4" diameter		31	.258		8.45	17.45		25.90	35.50
2080	1" diameter		27	.296		22.50	20		42.50	54.50
2100	1-1/2" diameter	Q-1	47	.340		15.45	20.50		35.95	48
2120	2" diameter		39	.410		20	25		45	59.50
2140	3" diameter		34	.471		27	28.50		55.50	73
2160	4" diameter		25	.640		44.50	39		83.50	107
2180	6" diameter		21	.762		115	46.50		161.50	197
2860	To delete coupling & hanger, subtract									
2870	1/2" diam. to 1" diam.					25%	33%			
2880	1-1/2" diam. to 3" diam.					22%	21%			
2890	4" diam. to 6" diam.					23%	15%			
3800	Conical joint, transparent									
3980	6" diameter	Q-1	21	.762	L.F.	175	46.50		221.50	262
4500	To delete couplings & hangers, subtract									
4530	6" diam.					22%	26%			

22 66 53.40 Pipe Fittings, Glass

		Crew	Daily Output	Labor-Hours	Unit	Material	2021 Bare Costs Labor	Equipment	Total	Total Incl O&P
0010	PIPE FITTINGS, GLASS									
0020	Drainage, beaded ends									
0040	Coupling & labor required at joints not incl. in fitting									
0050	price. Add 1 per joint for installed price									
0070	90° bend or sweep, 1-1/2"				Ea.	35.50			35.50	39
0090	2"					45.50			45.50	50
0100	3"					74			74	81.50
0110	4"					118			118	130
0120	6" (sweep only)					320			320	350
0200	45° bend or sweep same as 90°									
0350	Tee, single sanitary, 1-1/2"				Ea.	57.50			57.50	63.50
0370	2"					57.50			57.50	63.50
0380	3"					83			83	91
0390	4"					153			153	168
0400	6"					410			410	450
0410	Tee, straight, 1-1/2"					71			71	78
0430	2"					71			71	78

For customer support on your Plumbing Costs with RSMeans Data, call 800.448.8182.

22 66 53.40 Pipe Fittings, Glass		Crew	Daily Output	Labor-Hours	Unit	Material	2021 Bare Costs Labor	Equipment	Total	Total Incl O&P
0440	3"				Ea.	103			103	113
0450	4"					142			142	156
0460	6"					445			445	485
0500	Coupling, stainless steel, TFE seal ring									
0520	1-1/2"	Q-1	32	.500	Ea.	27	30.50		57.50	75
0530	2"		30	.533		34	32.50		66.50	86
0540	3"		25	.640		32.50	39		71.50	94
0550	4"		23	.696		78.50	42.50		121	150
0560	6"		20	.800		177	48.50		225.50	267
0600	Coupling, stainless steel, bead to plain end									
0610	1-1/2"	Q-1	36	.444	Ea.	33	27		60	77
0620	2"		34	.471		43	28.50		71.50	90.50
0630	3"		29	.552		73.50	33.50		107	131
0640	4"		27	.593		109	36		145	174
0650	6"		24	.667		345	40.50		385.50	440
2350	Coupling, Viton liner, for temperatures to 400°F									
2370	1/2"	Q-1	40	.400	Ea.	70	24.50		94.50	114
2380	3/4"		37	.432		79.50	26.50		106	127
2390	1"		35	.457		138	28		166	194
2400	1-1/2"		32	.500		33.50	30.50		64	82.50
2410	2"		30	.533		37.50	32.50		70	89.50
2420	3"		25	.640		49	39		88	112
2430	4"		23	.696		84.50	42.50		127	157
2440	6"		20	.800		242	48.50		290.50	340
2550	For beaded joint armored fittings, add					200%				
2600	Conical ends. Flange set, gasket & labor not incl. in fitting									
2620	price. Add 1 per joint for installed price.									
2650	90° sweep elbow, 1"				Ea.	119			119	131
2670	1-1/2"					283			283	310
2680	2"					248			248	273
2690	3"					455			455	500
2700	4"					815			815	900
2710	6"					1,200			1,200	1,300
2750	Cross (straight), add					55%				
2850	Tee, add					20%				

22 66 53.60 Corrosion Resistant Pipe		Crew	Daily Output	Labor-Hours	Unit	Material	2021 Bare Costs Labor	Equipment	Total	Total Incl O&P
0010	**CORROSION RESISTANT PIPE**, No couplings or hangers R221113-70									
0020	Iron alloy, drain, mechanical joint									
1000	1-1/2" diameter	Q-1	70	.229	L.F.	75.50	13.95		89.45	104
1100	2" diameter		66	.242		77.50	14.75		92.25	108
1120	3" diameter		60	.267		85	16.25		101.25	118
1140	4" diameter		52	.308		108	18.75		126.75	147
1980	Iron alloy, drain, B&S joint									
2000	2" diameter	Q-1	54	.296	L.F.	90.50	18.05		108.55	127
2100	3" diameter		52	.308		88	18.75		106.75	125
2120	4" diameter		48	.333		107	20.50		127.50	149
2140	6" diameter	Q-2	59	.407		162	25.50		187.50	217
2160	8" diameter	"	54	.444		330	28		358	405
2980	Plastic, epoxy, fiberglass filament wound, B&S joint									
3000	2" diameter	Q-1	62	.258	L.F.	12	15.70		27.70	36.50
3100	3" diameter		51	.314		14.05	19.10		33.15	44
3120	4" diameter		45	.356		20	21.50		41.50	54.50

22 66 53.60 Corrosion Resistant Pipe

		Crew	Daily Output	Labor-Hours	Unit	Material	2021 Bare Costs Labor	Equipment	Total	Total Incl O&P
3140	6" diameter	Q-1	32	.500	L.F.	28	30.50		58.50	76.50
3160	8" diameter	Q-2	38	.632		44	40		84	108
3180	10" diameter		32	.750		60.50	47.50		108	137
3200	12" diameter		28	.857		73.50	54		127.50	162
3980	Polyester, fiberglass filament wound, B&S joint									
4000	2" diameter	Q-1	62	.258	L.F.	13.05	15.70		28.75	38
4100	3" diameter		51	.314		17.15	19.10		36.25	47.50
4120	4" diameter		45	.356		25	21.50		46.50	60
4140	6" diameter		32	.500		36.50	30.50		67	86
4160	8" diameter	Q-2	38	.632		87	40		127	155
4180	10" diameter		32	.750		116	47.50		163.50	199
4200	12" diameter		28	.857		127	54		181	221
4980	Polypropylene, acid resistant, fire retardant, Schedule 40									
5000	1-1/2" diameter	Q-1	68	.235	L.F.	7.40	14.35		21.75	29.50
5100	2" diameter		62	.258		15.40	15.70		31.10	40.50
5120	3" diameter		51	.314		20.50	19.10		39.60	51
5140	4" diameter		45	.356		26	21.50		47.50	61
5160	6" diameter	Q-1	32	.500	L.F.	52	30.50		82.50	103
5980	Proxylene, fire retardant, Schedule 40									
6000	1-1/2" diameter	Q-1	68	.235	L.F.	11.40	14.35		25.75	34
6100	2" diameter		62	.258		15.60	15.70		31.30	40.50
6120	3" diameter		51	.314		28.50	19.10		47.60	59.50
6140	4" diameter		45	.356		40	21.50		61.50	76.50
6160	6" diameter		32	.500		67.50	30.50		98	120
6820	For Schedule 80, add					35%	2%			

22 66 53.70 Pipe Fittings, Corrosion Resistant

		Crew	Daily Output	Labor-Hours	Unit	Material	2021 Bare Costs Labor	Equipment	Total	Total Incl O&P
0010	**PIPE FITTINGS, CORROSION RESISTANT**									
0030	Iron alloy									
0050	Mechanical joint									
0060	1/4 bend, 1-1/2"	Q-1	12	1.333	Ea.	128	81		209	262
0080	2"		10	1.600		209	97.50		306.50	375
0090	3"		9	1.778		251	108		359	440
0100	4"		8	2		288	122		410	495
0110	1/8 bend, 1-1/2"		12	1.333		81.50	81		162.50	211
0130	2"		10	1.600		139	97.50		236.50	298
0140	3"		9	1.778		186	108		294	365
0150	4"		8	2		249	122		371	455
0160	Tee and Y, sanitary, straight									
0170	1-1/2"	Q-1	8	2	Ea.	139	122		261	335
0180	2"		7	2.286		186	139		325	415
0190	3"		6	2.667		288	162		450	555
0200	4"		5	3.200		530	195		725	875
0360	Coupling, 1-1/2"		14	1.143		67	69.50		136.50	178
0380	2"		12	1.333		76	81		157	205
0390	3"		11	1.455		79.50	88.50		168	220
0400	4"		10	1.600		90	97.50		187.50	244
0500	Bell & Spigot									
0510	1/4 and 1/16 bend, 2"	Q-1	16	1	Ea.	121	61		182	224
0520	3"		14	1.143		282	69.50		351.50	415
0530	4"		13	1.231		289	75		364	430
0540	6"	Q-2	17	1.412		575	89		664	765
0550	8"	"	12	2		2,350	126		2,476	2,800

22 66 Chemical-Waste Systems for Lab. and Healthcare Facilities

22 66 53 – Laboratory Chemical-Waste and Vent Piping

22 66 53.70 Pipe Fittings, Corrosion Resistant		Crew	Daily Output	Labor-Hours	Unit	Material	2021 Bare Costs Labor	Equipment	Total	Total Incl O&P
0620	1/8 bend, 2"	Q-1	16	1	Ea.	136	61		197	241
0640	3"		14	1.143		252	69.50		321.50	380
0650	4"	↓	13	1.231		250	75		325	385
0660	6"	Q-2	17	1.412		480	89		569	660
0680	8"	"	12	2		1,875	126		2,001	2,250
0700	Tee, sanitary, 2"	Q-1	10	1.600		258	97.50		355.50	430
0710	3"		9	1.778		835	108		943	1,075
0720	4"	↓	8	2	↓	695	122		817	945
0730	6"	Q-2	11	2.182	Ea.	870	138		1,008	1,150
0740	8"	"	8	3		2,375	190		2,565	2,900
1800	Y, sanitary, 2"	Q-1	10	1.600		272	97.50		369.50	445
1820	3"		9	1.778		485	108		593	695
1830	4"	↓	8	2		435	122		557	660
1840	6"	Q-2	11	2.182		1,450	138		1,588	1,800
1850	8"	"	8	3	↓	4,000	190		4,190	4,675
3000	Epoxy, filament wound									
3030	Quick-lock joint									
3040	90° elbow, 2"	Q-1	28	.571	Ea.	99	35		134	161
3060	3"		16	1		113	61		174	215
3070	4"		13	1.231		154	75		229	282
3080	6"	↓	8	2		225	122		347	430
3090	8"	Q-2	9	2.667		415	168		583	705
3100	10"		7	3.429		520	217		737	900
3110	12"	↓	6	4		745	253		998	1,200
3120	45° elbow, 2"	Q-1	28	.571		75.50	35		110.50	135
3130	3"		16	1		117	61		178	220
3140	4"		13	1.231		125	75		200	249
3150	6"	↓	8	2		225	122		347	430
3160	8"	Q-2	9	2.667		415	168		583	705
3170	10"		7	3.429		520	217		737	900
3180	12"	↓	6	4		745	253		998	1,200
3190	Tee, 2"	Q-1	19	.842		234	51.50		285.50	335
3200	3"		11	1.455		282	88.50		370.50	440
3210	4"		9	1.778		340	108		448	535
3220	6"	↓	5	3.200		560	195		755	910
3230	8"	Q-2	6	4		650	253		903	1,100
3240	10"		5	4.800		905	305		1,210	1,450
3250	12"	↓	4	6	↓	1,425	380		1,805	2,125
4000	Polypropylene, acid resistant									
4020	Non-pressure, electrofusion joints									
4050	1/4 bend, 1-1/2"	1 Plum	16	.500	Ea.	29	34		63	82.50
4060	2"	Q-1	28	.571		56.50	35		91.50	114
4080	3"		17	.941		61	57.50		118.50	153
4090	4"		14	1.143		99	69.50		168.50	213
4110	6"	↓	8	2	↓	237	122		359	440
4150	1/4 bend, long sweep									
4170	1-1/2"	1 Plum	16	.500	Ea.	32	34		66	85.50
4180	2"	Q-1	28	.571		56.50	35		91.50	114
4200	3"		17	.941		70	57.50		127.50	163
4210	4"		14	1.143		102	69.50		171.50	216
4250	1/8 bend, 1-1/2"	1 Plum	16	.500	↓	27.50	34		61.50	80.50
4260	2"	Q-1	28	.571	Ea.	35	35		70	90.50
4280	3"	↓	17	.941	↓	64	57.50		121.50	156

22 66 53.70 Pipe Fittings, Corrosion Resistant	Crew	Daily Output	Labor-Hours	Unit	Material	2021 Bare Costs Labor	Equipment	Total	Total Incl O&P
4290	4"	Q-1	14	1.143	Ea.	71.50	69.50		141	183
4310	6"	▼	8	2	▼	198	122		320	400
4400	Tee, sanitary									
4420	1-1/2"	1 Plum	10	.800	Ea.	36.50	54		90.50	121
4430	2"	Q-1	17	.941		42.50	57.50		100	133
4450	3"		11	1.455		85	88.50		173.50	226
4460	4"		9	1.778		128	108		236	305
4480	6"		5	3.200		850	195		1,045	1,225
4490	Tee, sanitary reducing, 2" x 2" x 1-1/2"		17	.941		42.50	57.50		100	133
4491	3" x 3" x 1-1/2"		11	1.455		76	88.50		164.50	216
4492	3" x 3" x 2"		11	1.455		85	88.50		173.50	226
4493	4" x 4" x 2"		10	1.600		115	97.50		212.50	271
4494	4" x 4" x 3"		9	1.778		125	108		233	299
4496	6" x 6" x 4"	▼	5	3.200		420	195		615	750
4650	Wye 45°, 1-1/2"	1 Plum	10	.800		39.50	54		93.50	125
4652	2"	Q-1	17	.941		55	57.50		112.50	146
4653	3"		11	1.455		91	88.50		179.50	232
4654	4"		9	1.778		133	108		241	310
4656	6"	▼	5	3.200	▼	345	195		540	665
4660	Wye, reducing									
4662	2" x 2" x 1-1/2"	Q-1	17	.941	Ea.	51.50	57.50		109	143
4666	3" x 3" x 2"		11	1.455		103	88.50		191.50	246
4668	4" x 4" x 2"		10	1.600		123	97.50		220.50	280
4669	4" x 4" x 3"		9	1.778		129	108		237	305
4671	6" x 6" x 2"		6	2.667		207	162		369	470
4673	6" x 6" x 3"		5.50	2.909		229	177		406	515
4675	6" x 6" x 4"	▼	5	3.200	▼	238	195		433	550
4678	Combination Y & 1/8 bend									
4681	1-1/2"	1 Plum	10	.800	Ea.	48	54		102	134
4683	2"	Q-1	17	.941		61	57.50		118.50	153
4684	3"		11	1.455		103	88.50		191.50	246
4685	4"	▼	9	1.778	▼	142	108		250	320
4689	Combination Y & 1/8 bend, reducing									
4692	2" x 2" x 1-1/2"	Q-1	17	.941	Ea.	58	57.50		115.50	149
4694	3" x 3" x 1-1/2"		12	1.333		86.50	81		167.50	217
4695	3" x 3" x 2"		11	1.455		91	88.50		179.50	232
4697	4" x 4" x 2"		10	1.600		129	97.50		226.50	287
4699	4" x 4" x 3"	▼	9	1.778	▼	135	108		243	310
4710	Hub adapter									
4712	1-1/2"	1 Plum	16	.500	Ea.	67	34		101	124
4713	2"	Q-1	28	.571	"	76	35		111	136
4714	3"	Q-1	17	.941	Ea.	96	57.50		153.50	191
4715	4"	"	14	1.143	"	125	69.50		194.50	241
4719	Mechanical joint adapter									
4721	1-1/2"	1 Plum	16	.500	Ea.	41	34		75	95.50
4722	2"	Q-1	28	.571		42.50	35		77.50	99
4723	3"		17	.941		61	57.50		118.50	153
4724	4"	▼	14	1.143	▼	91	69.50		160.50	204
4728	Couplings									
4731	1-1/2"	1 Plum	16	.500	Ea.	23	34		57	75.50
4732	2"	Q-1	28	.571		29	35		64	84
4733	3"		17	.941		36.50	57.50		94	126
4734	4"	▼	14	1.143		51.50	69.50		121	161

22 66 Chemical-Waste Systems for Lab. and Healthcare Facilities

22 66 53 – Laboratory Chemical-Waste and Vent Piping

22 66 53.70 Pipe Fittings, Corrosion Resistant	Crew	Daily Output	Labor-Hours	Unit	Material	2021 Bare Costs Labor	Equipment	Total	Total Incl O&P
4736	6"	Q-1	8	2	Ea.	82	122		204	273

22 66 83 – Chemical-Waste Tanks

22 66 83.13 Chemical-Waste Dilution Tanks

		Crew	Daily Output	Labor-Hours	Unit	Material	Labor	Equipment	Total	Total Incl O&P
0010	**CHEMICAL-WASTE DILUTION TANKS**									
7000	Tanks, covers included									
7800	Polypropylene									
7810	Continuous service to 200°F									
7830	2 gallon, 8" x 8" x 8"	Q-1	20	.800	Ea.	123	48.50		171.50	208
7850	7 gallon, 12" x 12" x 12"		20	.800		181	48.50		229.50	272
7870	16 gallon, 18" x 12" x 18"		17	.941		201	57.50		258.50	305
8010	33 gallon, 24" x 18" x 18"		12	1.333		268	81		349	415
8070	44 gallon, 24" x 18" x 24"		10	1.600		335	97.50		432.50	515
8080	89 gallon, 36" x 24" x 24"	↓	8	2	↓	545	122		667	780
8150	Polyethylene, heavy duty walls									
8160	Continuous service to 180°F									
8180	5 gallon, 12" x 6" x 18"	Q-1	20	.800	Ea.	36	48.50		84.50	112
8210	15 gallon, 14" I.D. x 27" deep		17	.941		61	57.50		118.50	153
8230	55 gallon, 22" I.D. x 36" deep		10	1.600		141	97.50		238.50	300
8250	100 gallon, 28" I.D. x 42" deep		8	2		340	122		462	555
8270	200 gallon, 36" I.D. x 48" deep		6	2.667		525	162		687	815
8290	360 gallon, 48" I.D. x 48" deep	↓	5	3.200	↓	640	195		835	990

For customer support on your Plumbing Costs with RSMeans Data, call 800.448.8182.

321

Division Notes

	CREW	DAILY OUTPUT	LABOR-HOURS	UNIT	BARE COSTS				TOTAL INCL O&P
					MAT.	LABOR	EQUIP.	TOTAL	

Estimating Tips

The labor adjustment factors listed in Subdivision 22 01 02.20 also apply to Division 23.

23 10 00 Facility Fuel Systems

- The prices in this subdivision for above- and below-ground storage tanks do not include foundations or hold-down slabs, unless noted. The estimator should refer to Divisions 3 and 31 for foundation system pricing. In addition to the foundations, required tank accessories, such as tank gauges, leak detection devices, and additional manholes and piping, must be added to the tank prices.

23 50 00 Central Heating Equipment

- When estimating the cost of an HVAC system, check to see who is responsible for providing and installing the temperature control system. It is possible to overlook controls, assuming that they would be included in the electrical estimate.
- When looking up a boiler, be careful on specified capacity. Some manufacturers rate their products on output while others use input.
- Include HVAC insulation for pipe, boiler, and duct (wrap and liner).
- Be careful when looking up mechanical items to get the correct pressure rating and connection type (thread, weld, flange).

23 70 00 Central HVAC Equipment

- Combination heating and cooling units are sized by the air conditioning requirements. (See Reference No. R236000-20 for the preliminary sizing guide.)
- A ton of air conditioning is nominally 400 CFM.
- Rectangular duct is taken off by the linear foot for each size, but its cost is usually estimated by the pound. Remember that SMACNA standards now base duct on internal pressure.
- Prefabricated duct is estimated and purchased like pipe: straight sections and fittings.
- Note that cranes or other lifting equipment are not included on any lines in Division 23. For example, if a crane is required to lift a heavy piece of pipe into place high above a gym floor, or to put a rooftop unit on the roof of a four-story building, etc., it must be added. Due to the potential for extreme variation—from nothing additional required to a major crane or helicopter—we feel that including a nominal amount for "lifting contingency" would be useless and detract from the accuracy of the estimate. When using equipment rental cost data from RSMeans, do not forget to include the cost of the operator(s).

Reference Numbers

Reference numbers are shown at the beginning of some major classifications. These numbers refer to related items in the Reference Section. The reference information may be an estimating procedure, an alternate pricing method, or technical information.

Note: Not all subdivisions listed here necessarily appear. ■

Same Data. Simplified.

Enjoy the convenience and efficiency of accessing your costs anywhere:

- **Skip the multiplier** by setting your location
- **Quickly search,** edit, favorite and share costs
- **Stay on top of price changes** with automatic updates

Discover more at rsmeans.com/online

Note: "Powered in part by CINX™, based on liscensed proprietary information of Harrison Publishing House, Inc."

Note: Trade Service, in part, has been used as a reference source for some of the material prices used in Division 23.

23 05 02 – HVAC General

23 05 02.10 Air Conditioning, General	Crew	Daily Output	Labor-Hours	Unit	Material	2021 Bare Costs Labor	Equipment	Total	Total Incl O&P
0010 **AIR CONDITIONING, GENERAL** Prices are for standard efficiencies (SEER 13)									
0020 for upgrade to SEER 14 add					10%				

23 05 05 – Selective Demolition for HVAC

23 05 05.10 HVAC Demolition

		Crew	Daily Output	Labor-Hours	Unit	Material	2021 Bare Costs Labor	Equipment	Total	Total Incl O&P
0010	**HVAC DEMOLITION** R220105-10									
0100	Air conditioner, split unit, 3 ton	Q-5	2	8	Ea.		490		490	735
0150	Package unit, 3 ton R024119-10	Q-6	3	8			510		510	760
0190	Rooftop, self contained, up to 5 ton	1 Plum	1.20	6.667			450		450	675
0250	Air curtain	Q-9	20	.800	L.F.		47		47	71
0254	Air filters, up thru 16,000 CFM		20	.800	Ea.		47		47	71
0256	20,000 thru 60,000 CFM		16	1			59		59	89
0297	Boiler blowdown	Q-5	8	2			123		123	184
0298	Boilers									
0300	Electric, up thru 148 kW	Q-19	2	12	Ea.		745		745	1,125
0310	150 thru 518 kW	"	1	24			1,500		1,500	2,225
0320	550 thru 2,000 kW	Q-21	.40	80			5,100		5,100	7,600
0330	2,070 kW and up	"	.30	107			6,800		6,800	10,100
0340	Gas and/or oil, up thru 150 MBH	Q-7	2.20	14.545			945		945	1,400
0350	160 thru 2,000 MBH		.80	40			2,600		2,600	3,875
0360	2,100 thru 4,500 MBH		.50	64			4,175		4,175	6,225
0370	4,600 thru 7,000 MBH		.30	107			6,950		6,950	10,400
0380	7,100 thru 12,000 MBH		.16	200			13,000		13,000	19,400
0390	12,200 thru 25,000 MBH		.12	267			17,300		17,300	25,900
0400	Central station air handler unit, up thru 15 ton	Q-5	1.60	10			615		615	920
0410	17.5 thru 30 ton	"	.80	20			1,225		1,225	1,825
0430	Computer room unit									
0434	Air cooled split, up thru 10 ton	Q-5	.67	23.881	Ea.		1,475		1,475	2,200
0436	12 thru 23 ton		.53	30.189			1,850		1,850	2,775
0440	Chilled water, up thru 10 ton		1.30	12.308			755		755	1,125
0444	12 thru 23 ton		1	16			985		985	1,475
0450	Glycol system, up thru 10 ton		.53	30.189			1,850		1,850	2,775
0454	12 thru 23 ton		.40	40			2,450		2,450	3,675
0460	Water cooled, not including condenser, up thru 10 ton		.80	20			1,225		1,225	1,825
0464	12 thru 23 ton		.60	26.667			1,650		1,650	2,450
0600	Condenser, up thru 50 ton		1	16			985		985	1,475
0610	51 thru 100 ton	Q-6	.90	26.667			1,700		1,700	2,550
0620	101 thru 1,000 ton	"	.70	34.286			2,175		2,175	3,275
0660	Condensing unit, up thru 10 ton	Q-5	1.25	12.800			790		790	1,175
0670	11 thru 50 ton	"	.40	40			2,450		2,450	3,675
0680	60 thru 100 ton	Q-6	.30	80			5,100		5,100	7,625
0700	Cooling tower, up thru 400 ton		.80	30			1,925		1,925	2,850
0710	450 thru 600 ton		.53	45.283			2,900		2,900	4,300
0720	700 thru 1,300 ton		.40	60			3,825		3,825	5,725
0780	Dehumidifier, up thru 155 lb./hr.	Q-1	8	2			122		122	182
0790	240 lb./hr. and up	"	2	8			485		485	725
1560	Ductwork									
1570	Metal, steel, sst, fabricated	Q-9	1000	.016	Lb.		.94		.94	1.42
1580	Aluminum, fabricated		485	.033	"		1.94		1.94	2.93
1590	Spiral, prefabricated		400	.040	L.F.		2.36		2.36	3.55
1600	Fiberglass, prefabricated		400	.040			2.36		2.36	3.55
1610	Flex, prefabricated		500	.032			1.88		1.88	2.84
1620	Glass fiber reinforced plastic, prefabricated		280	.057			3.37		3.37	5.10

23 05 05 – Selective Demolition for HVAC

23 05 05.10 HVAC Demolition	Crew	Daily Output	Labor-Hours	Unit	Material	2021 Bare Costs Labor	Equipment	Total	Total Incl O&P
1630	Diffusers, registers or grills, up thru 20" max dimension	1 Shee	50	.160	Ea.		10.45		10.45	15.80
1640	21 thru 36" max dimension		36	.222			14.55		14.55	22
1650	Above 36" max dimension	↓	30	.267			17.45		17.45	26.50
1700	Evaporator, up thru 12,000 BTUH	Q-5	5.30	3.019			186		186	277
1710	12,500 thru 30,000 BTUH	"	2.70	5.926			365		365	545
1720	31,000 BTUH and up	Q-6	1.50	16			1,025		1,025	1,525
1730	Evaporative cooler, up thru 5 HP	Q-9	2.70	5.926			350		350	525
1740	10 thru 30 HP	"	.67	23.881	↓		1,400		1,400	2,125
1750	Exhaust systems									
1760	Exhaust components	1 Shee	8	1	System		65.50		65.50	98.50
1770	Weld fume hoods	"	20	.400	Ea.		26		26	39.50
2120	Fans, up thru 1 HP or 2,000 CFM	Q-9	8	2			118		118	178
2124	1-1/2 thru 10 HP or 20,000 CFM	↓	5.30	3.019			178		178	268
2128	15 thru 30 HP or above 20,000 CFM	↓	4	4			236		236	355
2150	Fan coil air conditioner, chilled water, up thru 7.5 ton	Q-5	14	1.143			70.50		70.50	105
2154	Direct expansion, up thru 10 ton		8	2			123		123	184
2158	11 thru 30 ton	↓	2	8			490		490	735
2170	Flue shutter damper	Q-9	8	2			118		118	178
2200	Furnace, electric	Q-20	2	10			600		600	900
2300	Gas or oil, under 120 MBH	Q-9	4	4			236		236	355
2340	Over 120 MBH	"	3	5.333			315		315	475
2730	Heating and ventilating unit	Q-5	2.70	5.926			365		365	545
2740	Heater, electric, wall, baseboard and quartz	1 Elec	10	.800			51		51	75.50
2750	Heater, electric, unit, cabinet, fan and convector	"	8	1			63.50		63.50	94.50
2760	Heat exchanger, shell and tube type	Q-5	1.60	10			615		615	920
2770	Plate type	Q-6	.60	40	↓		2,550		2,550	3,800
2810	Heat pump									
2820	Air source, split, 4 thru 10 ton	Q-5	.90	17.778	Ea.		1,100		1,100	1,625
2830	15 thru 25 ton	Q-6	.80	30			1,925		1,925	2,850
2850	Single package, up thru 12 ton	Q-5	1	16			985		985	1,475
2860	Water source, up thru 15 ton	"	.90	17.778	↓		1,100		1,100	1,625
2870	20 thru 50 ton	Q-6	.80	30	Ea.		1,925		1,925	2,850
2910	Heat recovery package, up thru 20,000 CFM	Q-5	2	8			490		490	735
2920	25,000 CFM and up		1.20	13.333			820		820	1,225
2930	Heat transfer package, up thru 130 GPM		.80	20			1,225		1,225	1,825
2934	255 thru 800 GPM		.42	38.095			2,350		2,350	3,500
2940	Humidifier		10.60	1.509			93		93	139
2961	Hydronic unit heaters, up thru 200 MBH		14	1.143			70.50		70.50	105
2962	Above 200 MBH		8	2			123		123	184
2964	Valance units	↓	32	.500	↓		31		31	46
2966	Radiant floor heating									
2967	System valves, controls, manifolds	Q-5	16	1	Ea.		61.50		61.50	92
2968	Per room distribution		8	2			123		123	184
2970	Hydronic heating, baseboard radiation		16	1			61.50		61.50	92
2976	Convectors and free standing radiators	↓	18	.889			54.50		54.50	81.50
2980	Induced draft fan, up thru 1 HP	Q-9	4.60	3.478			205		205	310
2984	1-1/2 HP thru 7-1/2 HP	"	2.20	7.273			430		430	645
2988	Infrared unit	Q-5	16	1			61.50		61.50	92
2992	Louvers	1 Shee	46	.174	S.F.		11.40		11.40	17.15
3000	Mechanical equipment, light items. Unit is weight, not cooling.	Q-5	.90	17.778	Ton		1,100		1,100	1,625
3600	Heavy items		1.10	14.545	"		895		895	1,325
3720	Make-up air unit, up thru 6,000 CFM		3	5.333	Ea.		330		330	490
3730	6,500 thru 30,000 CFM	↓	1.60	10			615		615	920

23 05 05 – Selective Demolition for HVAC

23 05 05.10 HVAC Demolition	Crew	Daily Output	Labor-Hours	Unit	Material	2021 Bare Costs Labor	Equipment	Total	Total Incl O&P
3740	35,000 thru 75,000 CFM	Q-6	1	24	Ea.		1,525		1,525	2,275
3800	Mixing boxes, constant and VAV	Q-9	18	.889			52.50		52.50	79
4000	Packaged terminal air conditioner, up thru 18,000 BTUH	Q-5	8	2			123		123	184
4010	24,000 thru 48,000 BTUH	"	2.80	5.714			350		350	525
5000	Refrigerant compressor, reciprocating or scroll									
5010	Up thru 5 ton	1 Stpi	6	1.333	Ea.		91		91	136
5020	5.08 thru 10 ton	Q-5	6	2.667			164		164	245
5030	15 thru 50 ton	"	3	5.333			330		330	490
5040	60 thru 130 ton	Q-6	2.80	8.571			545		545	815
5090	Remove refrigerant from system	1 Stpi	40	.200	Lb.		13.65		13.65	20.50
5100	Rooftop air conditioner, up thru 10 ton	Q-5	1.40	11.429	Ea.		705		705	1,050
5110	12 thru 40 ton	Q-6	1	24			1,525		1,525	2,275
5120	50 thru 140 ton		.50	48			3,050		3,050	4,575
5130	150 thru 300 ton		.30	80			5,100		5,100	7,625
6000	Self contained single package air conditioner, up thru 10 ton	Q-5	1.60	10			615		615	920
6010	15 thru 60 ton	Q-6	1.20	20	Ea.		1,275		1,275	1,900
6100	Space heaters, up thru 200 MBH	Q-5	10	1.600			98.50		98.50	147
6110	Over 200 MBH		5	3.200			197		197	294
6200	Split ductless, both sections		8	2			123		123	184
6300	Steam condensate meter	1 Stpi	11	.727			49.50		49.50	74
6600	Thru-the-wall air conditioner	L-2	8	2			97		97	145
7000	Vent chimney, prefabricated, up thru 12" diameter	Q-9	94	.170	V.L.F.		10.05		10.05	15.10
7010	14" thru 36" diameter		40	.400			23.50		23.50	35.50
7020	38" thru 48" diameter		32	.500			29.50		29.50	44.50
7030	54" thru 60" diameter	Q-10	14	1.714			105		105	158
7400	Ventilators, up thru 14" neck diameter	Q-9	58	.276	Ea.		16.25		16.25	24.50
7410	16" thru 50" neck diameter		40	.400			23.50		23.50	35.50
7450	Relief vent, up thru 24" x 96"		22	.727			43		43	64.50
7460	48" x 60" thru 96" x 144"		10	1.600			94		94	142
8000	Water chiller up thru 10 ton	Q-5	2.50	6.400			395		395	590
8010	15 thru 100 ton	Q-6	.48	50			3,200		3,200	4,750
8020	110 thru 500 ton	Q-7	.29	110			7,175		7,175	10,700
8030	600 thru 1000 ton		.23	139			9,050		9,050	13,500
8040	1100 ton and up		.20	160			10,400		10,400	15,500
8400	Window air conditioner	1 Carp	16	.500			27.50		27.50	41
8401	HVAC demo, water chiller, 1100 ton and up	"	12	.667	Ton		36.50		36.50	54.50

23 05 23 – General-Duty Valves for HVAC Piping

23 05 23.20 Valves, Bronze/Brass

		Crew	Daily Output	Labor-Hours	Unit	Material	2021 Bare Costs Labor	Equipment	Total	Total Incl O&P
0010	**VALVES, BRONZE/BRASS**									
0020	Brass									
1300	Ball combination valves, shut-off and union									
1310	Solder, with strainer, drain and PT ports									
1320	1/2"	1 Stpi	17	.471	Ea.	74	32		106	130
1330	3/4"		16	.500		81	34		115	140
1340	1"		14	.571		109	39		148	179
1350	1-1/4"		12	.667		141	45.50		186.50	223
1360	1-1/2"		10	.800		206	54.50		260.50	310
1370	2"		8	1		259	68.50		327.50	385
1410	Threaded, with strainer, drain and PT ports									
1420	1/2"	1 Stpi	20	.400	Ea.	74	27.50		101.50	123
1430	3/4"		18	.444		81	30.50		111.50	135
1440	1"		17	.471		109	32		141	168

326

23 05 Common Work Results for HVAC

23 05 23 – General-Duty Valves for HVAC Piping

23 05 23.20 Valves, Bronze/Brass		Crew	Daily Output	Labor-Hours	Unit	Material	2021 Bare Costs Labor	Equipment	Total	Total Incl O&P
1450	1-1/4"	1 Stpi	12	.667	Ea.	141	45.50		186.50	223
1460	1-1/2"		11	.727		206	49.50		255.50	300
1470	2"		9	.889		259	61		320	375

23 05 23.30 Valves, Iron Body		Crew	Daily Output	Labor-Hours	Unit	Material	2021 Bare Costs Labor	Equipment	Total	Total Incl O&P
0010	**VALVES, IRON BODY** R220523-90									
0022	For grooved joint, see Section 22 11 13.48									
0100	Angle, 125 lb.									
0110	Flanged									
0116	2"	1 Plum	5	1.600	Ea.	1,600	108		1,708	1,900
0118	4"	Q-1	3	5.333		2,650	325		2,975	3,375
0120	6"	Q-2	3	8		5,150	505		5,655	6,425
0122	8"	"	2.50	9.600		9,450	605		10,055	11,300
1020	Butterfly, wafer type, gear actuator, 200 lb.									
1030	2"	1 Plum	14	.571	Ea.	118	38.50		156.50	188
1040	2-1/2"	Q-1	9	1.778		120	108		228	294
1050	3"		8	2		124	122		246	320
1060	4"		5	3.200		138	195		333	445
1070	5"	Q-2	5	4.800		155	305		460	625
1080	6"		5	4.800		176	305		481	650
1090	8"		4.50	5.333		219	335		554	745
1100	10"		4	6		294	380		674	890
1110	12"		3	8		725	505		1,230	1,550
1200	Wafer type, lever actuator, 200 lb.									
1220	2"	1 Plum	14	.571	Ea.	203	38.50		241.50	281
1230	2-1/2"	Q-1	9	1.778		214	108		322	395
1240	3"		8	2		226	122		348	430
1250	4"		5	3.200		268	195		463	585
1260	5"	Q-2	5	4.800		420	305		725	915
1270	6"		5	4.800		450	305		755	950
1280	8"		4.50	5.333		695	335		1,030	1,275
1290	10"		4	6		715	380		1,095	1,350
1300	12"		3	8		1,250	505		1,755	2,125
1650	Gate, 125 lb., N.R.S.									
2150	Flanged									
2200	2"	1 Plum	5	1.600	Ea.	915	108		1,023	1,150
2240	2-1/2"	Q-1	5	3.200		935	195		1,130	1,325
2260	3"		4.50	3.556		1,050	217		1,267	1,475
2280	4"		3	5.333		1,500	325		1,825	2,125
2290	5"	Q-2	3.40	7.059		2,550	445		2,995	3,500
2300	6"		3	8		2,300	505		2,805	3,275
2320	8"		2.50	9.600		4,400	605		5,005	5,725
2340	10"		2.20	10.909		7,725	690		8,415	9,525
2360	12"		1.70	14.118		10,600	890		11,490	13,000
2420	For 250 lb. flanged, add					200%	10%			
3550	OS&Y, 125 lb., flanged									
3600	2"	1 Plum	5	1.600	Ea.	625	108		733	845
3640	2-1/2"	Q-1	5	3.200	"	630	195		825	980
3660	3"	Q-1	4.50	3.556	Ea.	690	217		907	1,075
3670	3-1/2"		3	5.333		800	325		1,125	1,375
3680	4"		3	5.333		1,000	325		1,325	1,575
3690	5"	Q-2	3.40	7.059		1,650	445		2,095	2,500
3700	6"		3	8		1,650	505		2,155	2,575

23 05 23.30 Valves, Iron Body		Crew	Daily Output	Labor-Hours	Unit	Material	2021 Bare Costs Labor	Equipment	Total	Total Incl O&P
3720	8"	Q-2	2.50	9.600	Ea.	2,950	605		3,555	4,150
3740	10"		2.20	10.909		5,375	690		6,065	6,950
3760	12"	↓	1.70	14.118	↓	7,350	890		8,240	9,400
3900	For 175 lb., flanged, add					200%	10%			
4350	Globe, OS&Y									
4540	Class 125, flanged									
4550	2"	1 Plum	5	1.600	Ea.	1,250	108		1,358	1,525
4560	2-1/2"	Q-1	5	3.200		1,250	195		1,445	1,675
4570	3"		4.50	3.556		1,525	217		1,742	2,000
4580	4"	↓	3	5.333		2,175	325		2,500	2,875
4590	5"	Q-2	3.40	7.059		3,975	445		4,420	5,050
4600	6"		3	8		3,975	505		4,480	5,125
4610	8"		2.50	9.600		7,775	605		8,380	9,450
4612	10"		2.20	10.909		12,200	690		12,890	14,400
4614	12"	↓	1.70	14.118	↓	11,900	890		12,790	14,400
5040	Class 250, flanged									
5050	2"	1 Plum	4.50	1.778	Ea.	2,000	120		2,120	2,375
5060	2-1/2"	Q-1	4.50	3.556		2,600	217		2,817	3,175
5070	3"		4	4		2,675	244		2,919	3,325
5080	4"	↓	2.70	5.926		3,925	360		4,285	4,875
5090	5"	Q-2	3	8		7,275	505		7,780	8,750
5100	6"		2.70	8.889		7,050	560		7,610	8,600
5110	8"		2.20	10.909		11,900	690		12,590	14,100
5120	10"		2	12		19,500	760		20,260	22,500
5130	12"	↓	1.60	15		29,200	950		30,150	33,600
5240	Valve sprocket rim w/chain, for 2" valve	1 Stpi	30	.267		269	18.25		287.25	325
5250	2-1/2" valve		27	.296		175	20.50		195.50	223
5260	3-1/2" valve		25	.320		520	22		542	610
5270	6" valve		20	.400		186	27.50		213.50	246
5280	8" valve		18	.444		600	30.50		630.50	705
5290	12" valve		16	.500		258	34		292	335
5300	16" valve		12	.667		291	45.50		336.50	390
5310	20" valve		10	.800		291	54.50		345.50	400
5320	36" valve	↓	8	1	↓	410	68.50		478.50	550
5450	Swing check, 125 lb., threaded									
5500	2"	1 Plum	11	.727	Ea.	615	49		664	750
5540	2-1/2"	Q-1	15	1.067		790	65		855	965
5550	3"	"	13	1.231	↓	855	75		930	1,050
5560	4"	Q-1	10	1.600	Ea.	1,350	97.50		1,447.50	1,650
5950	Flanged									
5994	1"	1 Plum	7	1.143	Ea.	330	77.50		407.50	480
5998	1-1/2"		6	1.333		550	90.50		640.50	740
6000	2"	↓	5	1.600		535	108		643	745
6040	2-1/2"	Q-1	5	3.200		490	195		685	825
6050	3"		4.50	3.556		520	217		737	900
6060	4"	↓	3	5.333		820	325		1,145	1,375
6070	6"	Q-2	3	8		1,400	505		1,905	2,275
6080	8"		2.50	9.600		2,625	605		3,230	3,800
6090	10"		2.20	10.909		4,475	690		5,165	5,950
6100	12"		1.70	14.118		7,450	890		8,340	9,525
6102	14"		1.55	15.484		18,100	980		19,080	21,400
6104	16"		1.40	17.143		24,100	1,075		25,175	28,100
6110	18"	↓	1.30	18.462		36,600	1,175		37,775	42,100

23 05 23 – General-Duty Valves for HVAC Piping

23 05 23.30 Valves, Iron Body

		Crew	Daily Output	Labor-Hours	Unit	Material	2021 Bare Costs Labor	Equipment	Total	Total Incl O&P
6112	20"	Q-2	1	24	Ea.	58,500	1,525		60,025	66,500
6114	24"	↓	.75	32	↓	73,500	2,025		75,525	84,000
6160	For 250 lb. flanged, add					200%	20%			
6600	Silent check, bronze trim									
6610	Compact wafer type, for 125 or 150 lb. flanges									
6630	1-1/2"	1 Plum	11	.727	Ea.	163	49		212	254
6640	2"	"	9	.889		520	60		580	660
6650	2-1/2"	Q-1	9	1.778		217	108		325	400
6660	3"		8	2		231	122		353	435
6670	4"	↓	5	3.200		295	195		490	615
6680	5"	Q-2	6	4		390	253		643	805
6690	6"		6	4		520	253		773	945
6700	8"		4.50	5.333		880	335		1,215	1,475
6710	10"		4	6		1,525	380		1,905	2,250
6720	12"	↓	3	8	↓	2,900	505		3,405	3,950
6740	For 250 or 300 lb. flanges, thru 6" no change									
6741	For 8" and 10", add				Ea.	11%	10%			
6750	Twin disc									
6752	2"	1 Plum	9	.889	Ea.	480	60		540	615
6754	4"	Q-1	5	3.200		780	195		975	1,150
6756	6"	Q-2	5	4.800		1,150	305		1,455	1,725
6758	8"		4.50	5.333		1,775	335		2,110	2,450
6760	10"		4	6		2,825	380		3,205	3,700
6762	12"		3	8		3,675	505		4,180	4,800
6764	18"		1.50	16		15,900	1,000		16,900	19,000
6766	24"	↓	.75	32	↓	22,400	2,025		24,425	27,600
6800	Full flange type, 150 lb.									
6900	Globe type, 125 lb.									
6911	2-1/2"	Q-1	9	1.778	Ea.	485	108		593	695
6912	3"	↓	8	2		430	122		552	655
6913	4"	↓	5	3.200		770	195		965	1,125
6914	5"	Q-2	6	4		1,075	253		1,328	1,550
6915	6"		5	4.800		1,350	305		1,655	1,925
6916	8"		4.50	5.333		2,450	335		2,785	3,200
6917	10"		4	6		3,100	380		3,480	3,975
6918	12"	↓	3	8		5,350	505		5,855	6,625
6940	For 250 lb., add					40%	10%			
6980	Screwed end type, 125 lb.									
6981	1"	1 Plum	19	.421	Ea.	126	28.50		154.50	182
6982	1-1/4"		15	.533		167	36		203	237
6983	1-1/2"		13	.615		202	41.50		243.50	285
6984	2"	↓	11	.727	↓	277	49		326	380

23 05 23.70 Valves, Semi-Steel

		Crew	Daily Output	Labor-Hours	Unit	Material	2021 Bare Costs Labor	Equipment	Total	Total Incl O&P
0010	**VALVES, SEMI-STEEL** R220523-90									
1020	Lubricated plug valve, threaded, 200 psi									
1030	1/2"	1 Plum	18	.444	Ea.	118	30		148	174
1040	3/4"		16	.500		124	34		158	187
1050	1"		14	.571		130	38.50		168.50	201
1060	1-1/4"		12	.667		154	45		199	237
1070	1-1/2"		11	.727		149	49		198	238
1080	2"	↓	8	1		615	67.50		682.50	780
1090	2-1/2"	Q-1	5	3.200		675	195		870	1,025

For customer support on your Plumbing Costs with RSMeans Data, call 800.448.8182.

329

23 05 23.70 Valves, Semi-Steel		Crew	Daily Output	Labor-Hours	Unit	Material	2021 Bare Costs Labor	Equipment	Total	Total Incl O&P
1100	3"	Q-1	4.50	3.556	Ea.	415	217		632	785
6990	Flanged, 200 psi									
7000	2"	1 Plum	8	1	Ea.	605	67.50		672.50	765
7010	2-1/2"	Q-1	5	3.200		905	195		1,100	1,275
7020	3"		4.50	3.556		1,125	217		1,342	1,575
7030	4"		3	5.333		1,550	325		1,875	2,175
7036	5"		2.50	6.400		970	390		1,360	1,650
7040	6"	Q-2	3	8		1,275	505		1,780	2,150
7050	8"		2.50	9.600		2,275	605		2,880	3,400
7060	10"		2.20	10.909		3,800	690		4,490	5,200
7070	12"		1.70	14.118		5,400	890		6,290	7,275

23 05 23.90 Valves, Stainless Steel

		Crew	Daily Output	Labor-Hours	Unit	Material	2021 Bare Costs Labor	Equipment	Total	Total Incl O&P
0010	**VALVES, STAINLESS STEEL** R220523-90									
1700	Check, 200 lb., threaded									
1710	1/4"	1 Plum	24	.333	Ea.	144	22.50		166.50	193
1720	1/2"		22	.364		144	24.50		168.50	195
1730	3/4"		20	.400		148	27		175	204
1750	1"		19	.421		193	28.50		221.50	255
1760	1-1/2"		13	.615		360	41.50		401.50	460
1770	2"		11	.727		615	49		664	750
1800	150 lb., flanged									
1810	2-1/2"	Q-1	5	3.200	Ea.	1,375	195		1,570	1,800
1820	3"		4.50	3.556		2,000	217		2,217	2,525
1830	4"		3	5.333		3,000	325		3,325	3,775
1840	6"	Q-2	3	8		5,300	505		5,805	6,575
1850	8"	"	2.50	9.600		10,800	605		11,405	12,800
2100	Gate, OS&Y, 150 lb., flanged									
2120	1/2"	1 Plum	18	.444	Ea.	510	30		540	605
2140	3/4"		16	.500		490	34		524	590
2150	1"		14	.571		615	38.50		653.50	740
2160	1-1/2"		11	.727		1,200	49		1,249	1,375
2170	2"		8	1		1,425	67.50		1,492.50	1,650
2180	2-1/2"	Q-1	5	3.200		1,825	195		2,020	2,300
2190	3"		4.50	3.556		1,825	217		2,042	2,325
2200	4"		3	5.333		2,700	325		3,025	3,450
2205	5"		2.80	5.714		5,225	350		5,575	6,275
2210	6"	Q-2	3	8		5,050	505		5,555	6,300
2220	8"		2.50	9.600		8,775	605		9,380	10,600
2230	10"		2.30	10.435		15,200	660		15,860	17,700
2240	12"		1.90	12.632		20,300	800		21,100	23,600
2260	For 300 lb., flanged, add					120%	15%			
2600	600 lb., flanged									
2620	1/2"	1 Plum	16	.500	Ea.	181	34		215	250
2640	3/4"		14	.571		195	38.50		233.50	273
2650	1"		12	.667		235	45		280	325
2660	1-1/2"		10	.800		375	54		429	495
2670	2"		7	1.143		515	77.50		592.50	685
2680	2-1/2"	Q-1	4	4		6,675	244		6,919	7,700
2690	3"	"	3.60	4.444		6,675	271		6,946	7,725
3100	Globe, OS&Y, 150 lb., flanged									
3120	1/2"	1 Plum	18	.444	Ea.	485	30		515	580
3140	3/4"		16	.500		525	34		559	630

23 05 23 – General-Duty Valves for HVAC Piping

23 05 23.90 Valves, Stainless Steel

		Crew	Daily Output	Labor-Hours	Unit	Material	2021 Bare Costs Labor	Equipment	Total	Total Incl O&P
3150	1"	1 Plum	14	.571	Ea.	690	38.50		728.50	815
3160	1-1/2"		11	.727		1,075	49		1,124	1,250
3170	2"		8	1		1,425	67.50		1,492.50	1,650
3180	2-1/2"	Q-1	5	3.200		3,100	195		3,295	3,700
3190	3"		4.50	3.556		3,100	217		3,317	3,725
3200	4"		3	5.333		4,950	325		5,275	5,925
3210	6"	Q-2	3	8		8,325	505		8,830	9,925

23 05 23.94 Hospital Type Valves

		Crew	Daily Output	Labor-Hours	Unit	Material	2021 Bare Costs Labor	Equipment	Total	Total Incl O&P
0010	**HOSPITAL TYPE VALVES**									
0300	Chiller valves									
0330	Manual operation balancing valve									
0340	2-1/2" line size	Q-1	9	1.778	Ea.	650	108		758	875
0350	3" line size	"	8	2	"	665	122		787	910
0360	4" line size	Q-1	5	3.200	Ea.	960	195		1,155	1,350
0370	5" line size	Q-2	5	4.800		1,200	305		1,505	1,775
0380	6" line size		5	4.800		1,500	305		1,805	2,100
0390	8" line size		4.50	5.333		3,350	335		3,685	4,175
0400	10" line size		4	6		5,450	380		5,830	6,575
0410	12" line size		3	8		7,200	505		7,705	8,675
0600	Automatic flow limiting valve									
0620	2-1/2" line size	Q-1	8	2	Ea.	370	122		492	585
0630	3" line size		7	2.286		580	139		719	850
0640	4" line size		6	2.667		830	162		992	1,150
0645	5" line size	Q-2	5	4.800		1,125	305		1,430	1,675
0650	6" line size		5	4.800		1,375	305		1,680	1,975
0660	8" line size		4.50	5.333		2,100	335		2,435	2,800
0670	10" line size		4	6		3,350	380		3,730	4,275
0680	12" line size		3	8		5,575	505		6,080	6,900

23 05 93 – Testing, Adjusting, and Balancing for HVAC

23 05 93.10 Balancing, Air

		Crew	Daily Output	Labor-Hours	Unit	Material	2021 Bare Costs Labor	Equipment	Total	Total Incl O&P
0010	**BALANCING, AIR** (Subcontractor's quote incl. material and labor)									
0900	Heating and ventilating equipment									
1000	Centrifugal fans, utility sets				Ea.				420	420
1100	Heating and ventilating unit								630	630
1200	In-line fan								630	630
1300	Propeller and wall fan								119	119
1400	Roof exhaust fan								280	280
2000	Air conditioning equipment, central station								910	910
2100	Built-up low pressure unit								840	840
2200	Built-up high pressure unit								980	980
2300	Built-up high pressure dual duct								1,550	1,550
2400	Built-up variable volume								1,825	1,825
2500	Multi-zone A.C. and heating unit								630	630
2600	For each zone over one, add								140	140
2700	Package A.C. unit								350	350
2800	Rooftop heating and cooling unit								490	490
3000	Supply, return, exhaust, registers & diffusers, avg. height ceiling								84	84
3100	High ceiling								126	126
3200	Floor height								70	70
3300	Off mixing box								56	56
3500	Induction unit								91	91
3600	Lab fume hood								420	420

23 05 93.10 Balancing, Air

		Crew	Daily Output	Labor-Hours	Unit	Material	2021 Bare Costs Labor	Equipment	Total	Total Incl O&P
3700	Linear supply				Ea.				210	210
3800	Linear supply high								245	245
4000	Linear return				Ea.				70	70
4100	Light troffers								84	84
4200	Moduline - master								84	84
4300	Moduline - slaves								42	42
4400	Regenerators								560	560
4500	Taps into ceiling plenums								105	105
4600	Variable volume boxes								84	84

23 05 93.20 Balancing, Water

		Crew	Daily Output	Labor-Hours	Unit	Material	2021 Bare Costs Labor	Equipment	Total	Total Incl O&P
0010	**BALANCING, WATER** (Subcontractor's quote incl. material and labor)									
0050	Air cooled condenser				Ea.				256	256
0080	Boiler								515	515
0100	Cabinet unit heater								88	88
0200	Chiller								620	620
0300	Convector								73	73
0400	Converter								365	365
0500	Cooling tower								475	475
0600	Fan coil unit, unit ventilator								132	132
0700	Fin tube and radiant panels								146	146
0800	Main and duct re-heat coils								135	135
0810	Heat exchanger								135	135
0900	Main balancing cocks								110	110
1000	Pumps								320	320
1100	Unit heater								102	102

23 05 93.50 Piping, Testing

		Crew	Daily Output	Labor-Hours	Unit	Material	2021 Bare Costs Labor	Equipment	Total	Total Incl O&P
0010	**PIPING, TESTING**									
0100	Nondestructive testing									
0110	Nondestructive hydraulic pressure test, isolate & 1 hr. hold									
0120	1" - 4" pipe									
0140	0-250 L.F.	1 Stpi	1.33	6.015	Ea.		410		410	615
0160	250-500 L.F.	"	.80	10			685		685	1,025
0180	500-1000 L.F.	Q-5	1.14	14.035			865		865	1,300
0200	1000-2000 L.F.		.80	20			1,225		1,225	1,825
0320	0-250 L.F.		1	16			985		985	1,475
0340	250-500 L.F.		.73	21.918			1,350		1,350	2,025
0360	500-1000 L.F.		.53	30.189			1,850		1,850	2,775
0380	1000-2000 L.F.		.38	42.105			2,600		2,600	3,875
1000	Pneumatic pressure test, includes soaping joints									
1120	1" - 4" pipe									
1140	0-250 L.F.	Q-5	2.67	5.993	Ea.	13.05	370		383.05	565
1160	250-500 L.F.		1.33	12.030		26	740		766	1,125
1180	500-1000 L.F.		.80	20		39	1,225		1,264	1,875
1200	1000-2000 L.F.		.50	32		52	1,975		2,027	3,000
1300	6" - 10" pipe									
1320	0-250 L.F.	Q-5	1.33	12.030	Ea.	13.05	740		753.05	1,125
1340	250-500 L.F.		.67	23.881		26	1,475		1,501	2,225
1360	500-1000 L.F.		.40	40		52	2,450		2,502	3,725
1380	1000-2000 L.F.		.25	64		65.50	3,950		4,015.50	5,950
2110	2" diam.	1 Stpi	8	1		16.50	68.50		85	120
2120	3" diam.		8	1		16.50	68.50		85	120
2130	4" diam.		8	1		25	68.50		93.50	129

23 05 Common Work Results for HVAC

23 05 93 – Testing, Adjusting, and Balancing for HVAC

23 05 93.50 Piping, Testing		Crew	Daily Output	Labor-Hours	Unit	Material	2021 Bare Costs Labor	2021 Bare Costs Equipment	Total	Total Incl O&P
2140	6" diam.	1 Stpi	8	1	Ea.	25	68.50		93.50	129
2150	8" diam.		6.60	1.212		25	83		108	151
2160	10" diam.	↓	6	1.333	↓	33	91		124	173
3000	Liquid penetration of welds									
3110	2" diam.	1 Stpi	14	.571	Ea.	9.95	39		48.95	69.50
3120	3" diam.		13.60	.588		9.95	40		49.95	71
3130	4" diam.		13.40	.597		9.95	41		50.95	72
3140	6" diam.		13.20	.606		9.95	41.50		51.45	73
3150	8" diam.		13	.615		14.95	42		56.95	79.50
3160	10" diam.	↓	12.80	.625	↓	14.95	42.50		57.45	80.50

23 07 HVAC Insulation

23 07 13 – Duct Insulation

23 07 13.10 Duct Thermal Insulation

			Crew	Daily Output	Labor-Hours	Unit	Material	2021 Bare Costs Labor	2021 Bare Costs Equipment	Total	Total Incl O&P
0010	**DUCT THERMAL INSULATION**										
0110	Insulation req'd. is based on the surface size/area to be covered										
3730	Sheet insulation										
3760	Polyethylene foam, closed cell, UV resistant										
3770	Standard temperature (-90°F to +212°F)										
3771	1/4" thick	G	Q-14	450	.036	S.F.	1.83	1.95		3.78	4.99
3772	3/8" thick	G		440	.036		2.88	1.99		4.87	6.20
3773	1/2" thick	G		420	.038		3.52	2.09		5.61	7.05
3774	3/4" thick	G		400	.040		5.05	2.19		7.24	8.90
3775	1" thick	G		380	.042		6.80	2.31		9.11	11.05
3776	1-1/2" thick	G		360	.044		10.75	2.44		13.19	15.60
3777	2" thick	G		340	.047		14.20	2.58		16.78	19.60
3778	2-1/2" thick	G	↓	320	.050	↓	18.25	2.74		20.99	24
3779	Adhesive (see line 7878)										
3780	Foam, rubber										
3782	1" thick	G	1 Stpi	50	.160	S.F.	5.90	10.95		16.85	23
7000	Board insulation										
7020	Mineral wool, 1200° F										
7022	6 lb. density, plain										
7024	1" thick	G	Q-14	370	.043	S.F.	1	2.37		3.37	4.73
7026	1-1/2" thick	G		350	.046		.50	2.51		3.01	4.39
7028	2" thick	G	↓	330	.048	↓	.67	2.66		3.33	4.81
7030	3" thick	G	Q-14	300	.053	S.F.	.83	2.93		3.76	5.40
7032	4" thick	G	"	280	.057	"	1.35	3.13		4.48	6.30
7038	8 lb. density, plain										
7040	1" thick	G	Q-14	360	.044	S.F.	.41	2.44		2.85	4.18
7042	1-1/2" thick	G		340	.047		.60	2.58		3.18	4.61
7044	2" thick	G		320	.050		.82	2.74		3.56	5.10
7046	3" thick	G		290	.055		1.23	3.03		4.26	6
7048	4" thick	G	↓	270	.059	↓	1.60	3.25		4.85	6.75
7060	10 lb. density, plain										
7062	1" thick	G	Q-14	350	.046	S.F.	.57	2.51		3.08	4.47
7064	1-1/2" thick	G		330	.048		.86	2.66		3.52	5
7066	2" thick	G		310	.052		1.14	2.83		3.97	5.60
7068	3" thick	G		280	.057		1.72	3.13		4.85	6.70
7070	4" thick	G	↓	260	.062	↓	2.19	3.38		5.57	7.55
7878	Contact cement, quart can						Ea.	13.95			13.95	15.30

23 07 HVAC Insulation

23 07 16 – HVAC Equipment Insulation

23 07 16.10 HVAC Equipment Thermal Insulation		Crew	Daily Output	Labor-Hours	Unit	Material	2021 Bare Costs Labor	Equipment	Total	Total Incl O&P
0010	**HVAC EQUIPMENT THERMAL INSULATION**									
0110	Insulation req'd. is based on the surface size/area to be covered									
1000	Boiler, 1-1/2" calcium silicate only	G	Q-14	110	.145	S.F.	4.98	8		12.98	17.70
1020	Plus 2" fiberglass	G	"	80	.200	"	6.50	10.95		17.45	24
2000	Breeching, 2" calcium silicate									
2020	Rectangular	G	Q-14	42	.381	S.F.	9.60	21		30.60	42.50
2040	Round	G	"	38.70	.413	"	10.05	22.50		32.55	45.50
2300	Calcium silicate block, +200°F to +1,200°F									
2310	On irregular surfaces, valves and fittings									
2340	1" thick	G	Q-14	30	.533	S.F.	4.35	29.50		33.85	50
2360	1-1/2" thick	G		25	.640		4.91	35		39.91	59
2380	2" thick	G		22	.727		4.26	40		44.26	65.50
2400	3" thick	↓		18	.889	↓	6.55	49		55.55	81.50
2410	On plane surfaces									
2420	1" thick	G	Q-14	126	.127	S.F.	4.35	6.95		11.30	15.45
2430	1-1/2" thick	G		120	.133		4.91	7.30		12.21	16.60
2440	2" thick	G		100	.160		4.26	8.80		13.06	18.10
2450	3" thick	G	↓	70	.229	↓	6.55	12.55		19.10	26.50

23 09 Instrumentation and Control for HVAC

23 09 13 – Instrumentation and Control Devices for HVAC

23 09 13.60 Water Level Controls

		Crew	Daily Output	Labor-Hours	Unit	Material	Labor	Equipment	Total	Total Incl O&P
0010	**WATER LEVEL CONTROLS**									
3000	Low water cut-off for hot water boiler, 50 psi maximum									
3100	1" top & bottom equalizing pipes, manual reset	1 Stpi	14	.571	Ea.	490	39		529	600
3200	1" top & bottom equalizing pipes	"	14	.571	"	445	39		484	550
3300	2-1/2" side connection for nipple-to-boiler	1 Stpi	14	.571	Ea.	440	39		479	545

23 09 23.10 Control Components/DDC Systems

		Crew	Daily Output	Labor-Hours	Unit	Material	Labor	Equipment	Total	Total Incl O&P
0010	**CONTROL COMPONENTS/DDC SYSTEMS** (Sub's quote incl. M & L)									
0100	Analog inputs									
0110	Sensors (avg. 50' run in 1/2" EMT)									
0120	Duct temperature				Ea.				465	465
0130	Space temperature								665	665
0140	Duct humidity, +/- 3%								595	595
0150	Space humidity, +/- 2%								1,150	1,150
0160	Duct static pressure								680	680
0170	CFM/transducer								920	920
0172	Water temperature								940	940
0174	Water flow								3,450	3,450
0176	Water pressure differential								975	975
0177	Steam flow								2,400	2,400
0178	Steam pressure								1,025	1,025
0180	KW/transducer								1,350	1,350
0182	KWH totalization (not incl. elec. meter pulse xmtr.)								625	625
0190	Space static pressure				↓				1,075	1,075
1000	Analog outputs (avg. 50' run in 1/2" EMT)									
1010	P/I transducer				Ea.				635	635
1020	Analog output, matl. in MUX								305	305
1030	Pneumatic (not incl. control device)								645	645
1040	Electric (not incl. control device)				↓				380	380

23 09 23.10 Control Components/DDC Systems	Crew	Daily Output	Labor-Hours	Unit	Material	2021 Bare Costs Labor	Equipment	Total	Total Incl O&P
2000 Status (alarms)									
2100 Digital inputs (avg. 50' run in 1/2" EMT)									
2110 Freeze				Ea.				435	435
2120 Fire								395	395
2130 Differential pressure (air)	2 Elec	3	5.333		283	340		623	815
2140 Differential pressure (water)								975	975
2150 Current sensor								435	435
2160 Duct high temperature thermostat								570	570
2170 Duct smoke detector				↓				705	705
2200 Digital output (avg. 50' run in 1/2" EMT)									
2210 Start/stop				Ea.				340	340
2220 On/off (maintained contact)				"				585	585
3000 Controller MUX panel, incl. function boards									
3100 48 point				Ea.				5,275	5,275
3110 128 point				"				7,225	7,225
3200 DDC controller (avg. 50' run in conduit)									
3210 Mechanical room									
3214 16 point controller (incl. 120 volt/1 phase power supply)				Ea.				3,275	3,275
3229 32 point controller (incl. 120 volt/1 phase power supply)				"				5,425	5,425
3230 Includes software programming and checkout									
3260 Space									
3266 VAV terminal box (incl. space temp. sensor)				Ea.				840	840
3280 Host computer (avg. 50' run in conduit)									
3281 Package complete with PC, keyboard,									
3282 printer, monitor, basic software				Ea.				3,150	3,150
4000 Front end costs									
4100 Computer (P.C.) with software program				Ea.				6,350	6,350
4200 Color graphics software								3,925	3,925
4300 Color graphics slides								490	490
4350 Additional printer				↓				980	980
4400 Communications trunk cable				L.F.				3.80	3.80
4500 Engineering labor (not incl. dftg.)				Point				94	94
4600 Calibration labor								120	120
4700 Start-up, checkout labor				↓				120	120
4800 Programming labor, as req'd									
5000 Communications bus (data transmission cable)									
5010 #18 twisted shielded pair in 1/2" EMT conduit				C.L.F.				380	380
8000 Applications software									
8050 Basic maintenance manager software (not incl. data base entry)				Ea.				1,950	1,950
8100 Time program				Point				6.85	6.85
8120 Duty cycle								13.65	13.65
8140 Optimum start/stop								41.50	41.50
8160 Demand limiting								20.50	20.50
8180 Enthalpy program				↓				41.50	41.50
8200 Boiler optimization				Ea.				1,225	1,225
8220 Chiller optimization				"				1,625	1,625
8240 Custom applications									
8260 Cost varies with complexity									

23 09 53 – Pneumatic and Electric Control System for HVAC

23 09 53.10 Control Components	Crew	Daily Output	Labor-Hours	Unit	Material	2021 Bare Costs Labor	2021 Bare Costs Equipment	Total	Total Incl O&P
0010 **CONTROL COMPONENTS** R230500-10									
0680 Controller for VAV box, includes actuator	1 Stpi	7.30	1.096	Ea.	360	75		435	505
2000 Gauges, pressure or vacuum									
2100 2" diameter dial	1 Stpi	32	.250	Ea.	10.90	17.10		28	37.50
2200 2-1/2" diameter dial	"	32	.250	"	13.45	17.10		30.55	40.50
2300 3-1/2" diameter dial	1 Stpi	32	.250	Ea.	22	17.10		39.10	49.50
2400 4-1/2" diameter dial	"	32	.250	"	25	17.10		42.10	53
2700 Flanged iron case, black ring									
2800 3-1/2" diameter dial	1 Stpi	32	.250	Ea.	175	17.10		192.10	219
2900 4-1/2" diameter dial		32	.250		105	17.10		122.10	142
3000 6" diameter dial		32	.250		167	17.10		184.10	210
3010 Steel case, 0-300 psi									
3012 2" diameter dial	1 Stpi	16	.500	Ea.	8.80	34		42.80	60.50
3014 4" diameter dial	"	16	.500	"	24	34		58	77.50
3020 Aluminum case, 0-300 psi									
3022 3-1/2" diameter dial	1 Stpi	16	.500	Ea.	12.55	34		46.55	65
3024 4-1/2" diameter dial		16	.500		125	34		159	189
3026 6" diameter dial		16	.500		222	34		256	295
3028 8-1/2" diameter dial		16	.500		330	34		364	415
3030 Brass case, 0-300 psi									
3032 2" diameter dial	1 Stpi	16	.500	Ea.	48	34		82	104
3034 4-1/2" diameter dial	"	16	.500	"	99.50	34		133.50	161
3040 Steel case, high pressure, 0-10,000 psi									
3042 4-1/2" diameter dial	1 Stpi	16	.500	Ea.	293	34		327	375
3044 6-1/2" diameter dial		16	.500		299	34		333	380
3046 8-1/2" diameter dial		16	.500		395	34		429	485
3080 Pressure gauge, differential, magnehelic									
3084 0-2" W.C., with air filter kit	1 Stpi	6	1.333	Ea.	36.50	91		127.50	177
3300 For compound pressure-vacuum, add					18%				
4000 Thermometers									
4100 Dial type, 3-1/2" diameter, vapor type, union connection	1 Stpi	32	.250	Ea.	226	17.10		243.10	275
4120 Liquid type, union connection		32	.250		435	17.10		452.10	505
4500 Stem type, 6-1/2" case, 2" stem, 1/2" NPT		32	.250		60	17.10		77.10	91.50
4520 4" stem, 1/2" NPT		32	.250		73.50	17.10		90.60	107
4600 9" case, 3-1/2" stem, 3/4" NPT		28	.286		48	19.55		67.55	82
4620 6" stem, 3/4" NPT		28	.286		57.50	19.55		77.05	92.50
4640 8" stem, 3/4" NPT		28	.286		192	19.55		211.55	240
4660 12" stem, 1" NPT		26	.308		188	21		209	239
4670 Bi-metal, dial type, steel case brass stem									
4672 2" dial, 4" - 9" stem	1 Stpi	16	.500	Ea.	41.50	34		75.50	96.50
4673 2-1/2" dial, 4" - 9" stem		16	.500		41.50	34		75.50	96.50
4674 3-1/2" dial, 4" - 9" stem		16	.500		51	34		85	107
4680 Mercury filled, industrial, union connection type									
4682 Angle stem, 7" scale	1 Stpi	16	.500	Ea.	61.50	34		95.50	119
4683 9" scale	"	16	.500	"	170	34		204	238
4684 12" scale	1 Stpi	16	.500	Ea.	218	34		252	290
4686 Straight stem, 7" scale		16	.500		166	34		200	234
4687 9" scale		16	.500		145	34		179	210
4688 12" scale		16	.500		175	34		209	243
4690 Mercury filled, industrial, separable socket type, with well									
4692 Angle stem, with socket, 7" scale	1 Stpi	16	.500	Ea.	200	34		234	271
4693 9" scale		16	.500		200	34		234	271

23 09 Instrumentation and Control for HVAC

23 09 53 – Pneumatic and Electric Control System for HVAC

	23 09 53.10 Control Components	Crew	Daily Output	Labor-Hours	Unit	Material	2021 Bare Costs Labor	Equipment	Total	Total Incl O&P
4694	12" scale	1 Stpi	16	.500	Ea.	237	34		271	310
4696	Straight stem, with socket, 7" scale		16	.500		156	34		190	223
4697	9" scale		16	.500		156	34		190	223
4698	12" scale		16	.500		192	34		226	262
6000	Valves, motorized zone									
6100	Sweat connections, 1/2" C x C	1 Stpi	20	.400	Ea.	139	27.50		166.50	194
6110	3/4" C x C		20	.400		78.50	27.50		106	127
6120	1" C x C		19	.421		183	29		212	244
6140	1/2" C x C, with end switch, 2 wire		20	.400		138	27.50		165.50	192
6150	3/4" C x C, with end switch, 2 wire		20	.400		85.50	27.50		113	135
6160	1" C x C, with end switch, 2 wire		19	.421		180	29		209	241

23 11 Facility Fuel Piping

23 11 13 – Facility Fuel-Oil Piping

23 11 13.10 Fuel Oil Specialties

		Crew	Daily Output	Labor-Hours	Unit	Material	2021 Bare Costs Labor	Equipment	Total	Total Incl O&P
0010	**FUEL OIL SPECIALTIES**									
0020	Foot valve, single poppet, metal to metal construction									
0040	Bevel seat, 1/2" diameter	1 Stpi	20	.400	Ea.	92.50	27.50		120	143
0060	3/4" diameter		18	.444		120	30.50		150.50	178
1000	Oil filters, 3/8" IPT, 20 gal. per hour		20	.400		21	27.50		48.50	64
2000	Remote tank gauging system, self contained									
3000	Valve, ball check, globe type, 3/8" diameter	1 Stpi	24	.333	Ea.	13.30	23		36.30	48.50
3500	Fusible, 3/8" diameter		24	.333		29	23		52	65.50
3600	1/2" diameter		24	.333		69.50	23		92.50	111
3610	3/4" diameter		20	.400		170	27.50		197.50	228
3620	1" diameter		19	.421		460	29		489	550
4000	Nonfusible, 3/8" diameter		24	.333		27	23		50	64
4500	Shutoff, gate type, lever handle, spring-fusible kit									
4520	1/4" diameter	1 Stpi	14	.571	Ea.	32	39		71	93.50
4540	3/8" diameter		12	.667		77.50	45.50		123	154
4560	1/2" diameter		10	.800		114	54.50		168.50	207
4570	3/4" diameter		8	1		116	68.50		184.50	230
5000	Vent alarm, whistling signal					37.50			37.50	41
5500	Vent protector/breather, 1-1/4" diameter	1 Stpi	32	.250	Ea.	22.50	17.10		39.60	50.50

23 11 23 – Facility Natural-Gas Piping

23 11 23.10 Gas Meters

		Crew	Daily Output	Labor-Hours	Unit	Material	2021 Bare Costs Labor	Equipment	Total	Total Incl O&P
0010	**GAS METERS**									
4000	Residential									
4010	Gas meter, residential, 3/4" pipe size	1 Plum	14	.571	Ea.	291	38.50		329.50	380
4020	Gas meter, residential, 1" pipe size		12	.667		296	45		341	395
4030	Gas meter, residential, 1-1/4" pipe size		10	.800		267	54		321	375

23 11 23.20 Gas Piping, Flexible (Csst)

		Crew	Daily Output	Labor-Hours	Unit	Material	2021 Bare Costs Labor	Equipment	Total	Total Incl O&P
0010	**GAS PIPING, FLEXIBLE (CSST)**									
0100	Tubing with lightning protection									
0110	3/8"	1 Stpi	65	.123	L.F.	3.32	8.40		11.72	16.20
0120	1/2"		62	.129		2.90	8.80		11.70	16.35
0130	3/4"		60	.133		4.50	9.10		13.60	18.55
0140	1"		55	.145		6.50	9.95		16.45	22
0150	1-1/4"		50	.160		8.65	10.95		19.60	26
0160	1-1/2"		45	.178		15.20	12.15		27.35	35

23 11 23.20 Gas Piping, Flexible (Csst)	Crew	Daily Output	Labor-Hours	Unit	Material	2021 Bare Costs Labor	Equipment	Total	Total Incl O&P
0170	2"	1 Stpi	40	.200	L.F.	22	13.65		35.65	44.50
0200	Tubing for underground/underslab burial									
0210	3/8"	1 Stpi	65	.123	L.F.	4.62	8.40		13.02	17.65
0220	1/2"		62	.129		5.30	8.80		14.10	18.95
0230	3/4"		60	.133		6.70	9.10		15.80	21
0240	1"		55	.145		9	9.95		18.95	25
0250	1-1/4"		50	.160		12.10	10.95		23.05	29.50
0260	1-1/2"		45	.178		22.50	12.15		34.65	43
0270	2"		40	.200		27	13.65		40.65	50
3000	Fittings									
3010	Straight									
3100	Tube to NPT									
3110	3/8"	1 Stpi	29	.276	Ea.	13.90	18.85		32.75	43.50
3120	1/2"		27	.296		14.80	20.50		35.30	46.50
3130	3/4"		25	.320		22	22		44	57
3140	1"		23	.348		34.50	24		58.50	73.50
3150	1-1/4"		20	.400		69	27.50		96.50	117
3160	1-1/2"		17	.471		142	32		174	204
3170	2"		15	.533		240	36.50		276.50	320
3200	Coupling									
3210	3/8"	1 Stpi	29	.276	Ea.	23.50	18.85		42.35	54
3220	1/2"		27	.296		27.50	20.50		48	60
3230	3/4"		25	.320		41.50	22		63.50	78
3240	1"		23	.348		58	24		82	99.50
3250	1-1/4"		20	.400		128	27.50		155.50	181
3260	1-1/2"		17	.471		277	32		309	355
3270	2"		15	.533		465	36.50		501.50	565
3300	Flange fitting									
3310	3/8"	1 Stpi	25	.320	Ea.	21	22		43	56
3320	1/2"		22	.364		22	25		47	61.50
3330	3/4"		19	.421		29	29		58	74.50
3340	1"		16	.500		39.50	34		73.50	94.50
3350	1-1/4"		12	.667		83.50	45.50		129	160
3400	90° flange valve									
3410	3/8"	1 Stpi	25	.320	Ea.	40	22		62	76.50
3420	1/2"		22	.364		40.50	25		65.50	81.50
3430	3/4"		19	.421		49.50	29		78.50	97.50
4000	Tee									
4120	1/2"	1 Stpi	20.50	.390	Ea.	42.50	26.50		69	87
4130	3/4"		19	.421		56.50	29		85.50	105
4140	1"		17.50	.457		102	31.50		133.50	159
5000	Reducing									
5110	Tube to NPT									
5120	3/4" to 1/2" NPT	1 Stpi	26	.308	Ea.	21.50	21		42.50	55
5130	1" to 3/4" NPT	"	24	.333	"	33	23		56	70.50
5200	Reducing tee									
5210	1/2" x 3/8" x 3/8"	1 Stpi	21	.381	Ea.	56	26		82	101
5220	3/4" x 1/2" x 1/2"		20	.400		60	27.50		87.50	107
5230	1" x 3/4" x 1/2"		18	.444		175	30.50		205.50	238
5240	1-1/4" x 1-1/4" x 1"		15.60	.513		233	35		268	310
5250	1-1/2" x 1-1/2" x 1-1/4"		13.30	.602		355	41		396	450
5260	2" x 2" x 1-1/2"		11.80	.678		535	46.50		581.50	660
5300	Manifold with four ports and mounting bracket									

23 11 Facility Fuel Piping

23 11 23 – Facility Natural-Gas Piping

23 11 23.20 Gas Piping, Flexible (Csst)	Crew	Daily Output	Labor-Hours	Unit	Material	2021 Bare Costs Labor	Equipment	Total	Total Incl O&P
5302	Labor to mount manifold does not include making pipe								
5304	connections which are included in fitting labor.								
5310	3/4" x 1/2" x 1/2" (4)	1 Stpi	76	.105	Ea.	43	7.20		50.20	58.50
5330	1-1/4" x 1" x 3/4" (4)		72	.111		62	7.60		69.60	79.50
5350	2" x 1-1/2" x 1" (4)		68	.118		77.50	8.05		85.55	97.50
5600	Protective striker plate								
5610	Quarter plate, 3" x 2"	1 Stpi	88	.091	Ea.	.97	6.20		7.17	10.30
5620	Half plate, 3" x 7"		82	.098		3.32	6.65		9.97	13.60
5630	Full plate, 3" x 12"		78	.103		7	7		14	18.15

23 12 Facility Fuel Pumps

23 12 13 – Facility Fuel-Oil Pumps

23 12 13.10 Pump and Motor Sets

		Crew	Daily Output	Labor-Hours	Unit	Material	2021 Bare Costs Labor	Equipment	Total	Total Incl O&P
0010	**PUMP AND MOTOR SETS**									
1810	Light fuel and diesel oils									
1820	20 GPH, 1/3 HP	Q-5	6	2.667	Ea.	1,625	164		1,789	2,050
1850	145 GPH, 1/2 HP	"	4	4	"	1,725	246		1,971	2,275

23 13 Facility Fuel-Storage Tanks

23 13 13 – Facility Underground Fuel-Oil, Storage Tanks

23 13 13.09 Single-Wall Steel Fuel-Oil Tanks

		Crew	Daily Output	Labor-Hours	Unit	Material	2021 Bare Costs Labor	Equipment	Total	Total Incl O&P
0010	**SINGLE-WALL STEEL FUEL-OIL TANKS**									
5000	Tanks, steel ugnd., sti-p3, not incl. hold-down bars									
5500	Excavation, pad, pumps and piping not included									
5510	Single wall, 500 gallon capacity, 7 ga. shell	Q-5	2.70	5.926	Ea.	2,000	365		2,365	2,750
5520	1,000 gallon capacity, 7 ga. shell	"	2.50	6.400		2,900	395		3,295	3,775
5530	2,000 gallon capacity, 1/4" thick shell	Q-7	4.60	6.957		4,175	455		4,630	5,275
5535	2,500 gallon capacity, 7 ga. shell	Q-5	3	5.333		5,625	330		5,955	6,675
5540	5,000 gallon capacity, 1/4" thick shell	Q-7	3.20	10		13,200	650		13,850	15,500
5560	10,000 gallon capacity, 1/4" thick shell		2	16		10,800	1,050		11,850	13,400
5580	15,000 gallon capacity, 5/16" thick shell		1.70	18.824		11,100	1,225		12,325	14,000
5600	20,000 gallon capacity, 5/16" thick shell		1.50	21.333		27,000	1,400		28,400	31,800
5610	25,000 gallon capacity, 3/8" thick shell		1.30	24.615		30,900	1,600		32,500	36,400
5620	30,000 gallon capacity, 3/8" thick shell		1.10	29.091		34,800	1,900		36,700	41,000
5630	40,000 gallon capacity, 3/8" thick shell		.90	35.556		45,200	2,325		47,525	53,000
5640	50,000 gallon capacity, 3/8" thick shell		.80	40		51,500	2,600		54,100	60,500

23 13 13.13 Dbl-Wall Steel, Undrgrnd Fuel-Oil, Stor. Tanks

		Crew	Daily Output	Labor-Hours	Unit	Material	2021 Bare Costs Labor	Equipment	Total	Total Incl O&P
0010	**DOUBLE-WALL STEEL, UNDERGROUND FUEL-OIL, STORAGE TANKS**									
6200	Steel, underground, 360°, double wall, UL listed,									
6210	with sti-P3 corrosion protection,									
6220	(dielectric coating, cathodic protection, electrical									
6230	isolation) 30 year warranty,									
6240	not incl. manholes or hold-downs.									
6250	500 gallon capacity	Q-5	2.40	6.667	Ea.	3,550	410		3,960	4,525
6260	1,000 gallon capacity	"	2.25	7.111		3,875	440		4,315	4,900
6270	2,000 gallon capacity	Q-7	4.16	7.692		5,125	500		5,625	6,375
6280	3,000 gallon capacity		3.90	8.205		7,375	535		7,910	8,900
6290	4,000 gallon capacity		3.64	8.791		8,700	570		9,270	10,400
6300	5,000 gallon capacity		2.91	10.997		9,325	715		10,040	11,400

23 13 13 — Facility Underground Fuel-Oil, Storage Tanks

23 13 13.13 Dbl-Wall Steel, Undrgrnd Fuel-Oil, Stor. Tanks		Crew	Daily Output	Labor-Hours	Unit	Material	2021 Bare Costs Labor	Equipment	Total	Total Incl O&P
6310	6,000 gallon capacity	Q-7	2.42	13.223	Ea.	12,000	860		12,860	14,500
6320	8,000 gallon capacity		2.08	15.385		11,800	1,000		12,800	14,500
6330	10,000 gallon capacity		1.82	17.582		14,700	1,150		15,850	17,900
6340	12,000 gallon capacity		1.70	18.824		16,400	1,225		17,625	19,800
6350	15,000 gallon capacity		1.33	24.060		22,000	1,575		23,575	26,500
6360	20,000 gallon capacity		1.33	24.060		36,300	1,575		37,875	42,300
6370	25,000 gallon capacity		1.16	27.586		43,400	1,800		45,200	50,500
6380	30,000 gallon capacity		1.03	31.068		40,100	2,025		42,125	47,200
6390	40,000 gallon capacity		.80	40		135,000	2,600		137,600	152,500
6395	50,000 gallon capacity		.73	43.836		164,000	2,850		166,850	185,000
6400	For hold-downs 500-2,000 gal., add		16	2	Set	215	130		345	430
6410	For hold-downs 3,000-6,000 gal., add		12	2.667		370	173		543	670
6420	For hold-downs 8,000-12,000 gal., add		11	2.909		530	189		719	865
6430	For hold-downs 15,000 gal., add		9	3.556		725	231		956	1,150
6440	For hold-downs 20,000 gal., add		8	4		835	260		1,095	1,300
6450	For hold-downs 20,000 gal. plus, add		6	5.333		1,000	345		1,345	1,625
6500	For manways, add				Ea.	1,200			1,200	1,325
6600	In place with hold-downs									
6652	550 gallon capacity	Q-5	1.84	8.696	Ea.	3,775	535		4,310	4,950

23 13 13.23 Glass-Fiber-Reinfcd-Plastic, Fuel-Oil, Storage

		Crew	Daily Output	Labor-Hours	Unit	Material	2021 Bare Costs Labor	Equipment	Total	Total Incl O&P
0010	**GLASS-FIBER-REINFCD-PLASTIC, UNDERGRND. FUEL-OIL, STORAGE**									
0210	Fiberglass, underground, single wall, UL listed, not including									
0220	manway or hold-down strap									
0225	550 gallon capacity	Q-5	2.67	5.993	Ea.	4,225	370		4,595	5,200
0230	1,000 gallon capacity	"	2.46	6.504		5,450	400		5,850	6,575
0240	2,000 gallon capacity	Q-7	4.57	7.002		8,725	455		9,180	10,300
0245	3,000 gallon capacity		3.90	8.205		8,850	535		9,385	10,500
0250	4,000 gallon capacity		3.55	9.014		11,800	585		12,385	13,900
0255	5,000 gallon capacity		3.20	10		10,800	650		11,450	12,800
0260	6,000 gallon capacity		2.67	11.985		10,100	780		10,880	12,300
0270	8,000 gallon capacity		2.29	13.974		15,800	910		16,710	18,700
0280	10,000 gallon capacity		2	16		15,900	1,050		16,950	19,100
0282	12,000 gallon capacity		1.88	17.021		16,800	1,100		17,900	20,200
0284	15,000 gallon capacity		1.68	19.048		24,300	1,250		25,550	28,600
0290	20,000 gallon capacity		1.45	22.069		33,800	1,425		35,225	39,400
0300	25,000 gallon capacity		1.28	25		58,000	1,625		59,625	66,500
0320	30,000 gallon capacity		1.14	28.070		87,500	1,825		89,325	99,000
0340	40,000 gallon capacity		.89	35.955		125,000	2,350		127,350	140,500
0360	48,000 gallon capacity		.81	39.506		185,000	2,575		187,575	207,500
0500	For manway, fittings and hold-downs, add					20%	15%			
0600	For manways, add					2,850			2,850	3,125
1000	For helical heating coil, add	Q-5	2.50	6.400		5,625	395		6,020	6,800
1020	Fiberglass, underground, double wall, UL listed									
1030	includes manways, not incl. hold-down straps									
1040	600 gallon capacity	Q-5	2.42	6.612	Ea.	9,275	405		9,680	10,800
1050	1,000 gallon capacity	"	2.25	7.111		12,700	440		13,140	14,700
1060	2,500 gallon capacity	Q-7	4.16	7.692		17,600	500		18,100	20,000
1070	3,000 gallon capacity		3.90	8.205		19,700	535		20,235	22,400
1080	4,000 gallon capacity		3.64	8.791		20,600	570		21,170	23,500
1090	6,000 gallon capacity		2.42	13.223		25,500	860		26,360	29,400
1100	8,000 gallon capacity		2.08	15.385		32,500	1,000		33,500	37,200
1110	10,000 gallon capacity		1.82	17.582		39,300	1,150		40,450	44,900

340

23 13 Facility Fuel-Storage Tanks

23 13 13 – Facility Underground Fuel-Oil, Storage Tanks

23 13 13.23 Glass-Fiber-Reinfcd-Plastic, Fuel-Oil, Storage

		Crew	Daily Output	Labor-Hours	Unit	Material	2021 Bare Costs Labor	2021 Bare Costs Equipment	Total	Total Incl O&P
1120	12,000 gallon capacity	Q-7	1.70	18.824	Ea.	40,800	1,225		42,025	46,700
1122	15,000 gallon capacity		1.52	21.053		73,000	1,375		74,375	82,000
1124	20,000 gallon capacity		1.33	24.060		78,500	1,575		80,075	88,500
1126	25,000 gallon capacity		1.16	27.586		91,000	1,800		92,800	102,500
1128	30,000 gallon capacity		1.03	31.068		109,000	2,025		111,025	123,000
1140	For hold-down straps, add					2%	10%			
1150	For hold-downs 500-4,000 gal., add	Q-7	16	2	Set	450	130		580	695
1160	For hold-downs 5,000-15,000 gal., add		8	4		905	260		1,165	1,375
1170	For hold-downs 20,000 gal., add		5.33	6.004		1,350	390		1,740	2,075
1180	For hold-downs 25,000 gal., add		4	8		1,800	520		2,320	2,775
1190	For hold-downs 30,000 gal., add		2.60	12.308		2,725	800		3,525	4,175
2210	Fiberglass, underground, single wall, UL listed, including									
2220	hold-down straps, no manways									
2225	550 gallon capacity	Q-5	2	8	Ea.	4,675	490		5,165	5,875
2230	1,000 gallon capacity	"	1.88	8.511		5,900	525		6,425	7,250
2240	2,000 gallon capacity	Q-7	3.55	9.014		9,175	585		9,760	11,000
2250	4,000 gallon capacity		2.90	11.034		12,200	720		12,920	14,600
2260	6,000 gallon capacity		2	16		11,000	1,050		12,050	13,700
2270	8,000 gallon capacity		1.78	17.978		16,700	1,175		17,875	20,100
2280	10,000 gallon capacity		1.60	20		16,800	1,300		18,100	20,500
2282	12,000 gallon capacity		1.52	21.053		17,700	1,375		19,075	21,600
2284	15,000 gallon capacity		1.39	23.022		25,200	1,500		26,700	29,900
2290	20,000 gallon capacity		1.14	28.070		35,200	1,825		37,025	41,400
2300	25,000 gallon capacity		.96	33.333		60,000	2,175		62,175	69,000
2320	30,000 gallon capacity		.80	40		90,500	2,600		93,100	103,500
3020	Fiberglass, underground, double wall, UL listed									
3030	includes manways and hold-down straps									
3040	600 gallon capacity	Q-5	1.86	8.602	Ea.	9,725	530		10,255	11,500
3050	1,000 gallon capacity	"	1.70	9.412		13,200	580		13,780	15,400
3060	2,500 gallon capacity	Q-7	3.29	9.726		18,000	635		18,635	20,700
3070	3,000 gallon capacity		3.13	10.224		20,100	665		20,765	23,100
3080	4,000 gallon capacity		2.93	10.922		21,000	710		21,710	24,200
3090	6,000 gallon capacity		1.86	17.204		26,400	1,125		27,525	30,800
3100	8,000 gallon capacity		1.65	19.394		33,400	1,250		34,650	38,600
3110	10,000 gallon capacity		1.48	21.622		40,200	1,400		41,600	46,300
3120	12,000 gallon capacity		1.40	22.857		41,700	1,475		43,175	48,100
3122	15,000 gallon capacity		1.28	25		74,000	1,625		75,625	83,500
3124	20,000 gallon capacity		1.06	30.189		80,000	1,975		81,975	90,500
3126	25,000 gallon capacity		.90	35.556		92,500	2,325		94,825	105,500
3128	30,000 gallon capacity		.74	43.243		112,000	2,825		114,825	127,000

23 13 23 – Facility Aboveground Fuel-Oil, Storage Tanks

23 13 23.16 Horizontal, Stl, Abvgrd Fuel-Oil, Storage Tanks

		Crew	Daily Output	Labor-Hours	Unit	Material	2021 Bare Costs Labor	2021 Bare Costs Equipment	Total	Total Incl O&P
0010	**HORIZONTAL, STEEL, ABOVEGROUND FUEL-OIL, STORAGE TANKS**									
3000	Steel, storage, aboveground, including cradles, coating,									
3020	fittings, not including foundation, pumps or piping									
3040	Single wall, 275 gallon	Q-5	5	3.200	Ea.	475	197		672	820
3060	550 gallon	"	2.70	5.926		4,725	365		5,090	5,750
3080	1,000 gallon	Q-7	5	6.400		7,950	415		8,365	9,375
3100	1,500 gallon		4.75	6.737		12,100	440		12,540	14,000
3120	2,000 gallon		4.60	6.957		11,700	455		12,155	13,600
3320	Double wall, 500 gallon capacity	Q-5	2.40	6.667		1,975	410		2,385	2,775
3330	2,000 gallon capacity	Q-7	4.15	7.711		6,575	500		7,075	8,000

For customer support on your Plumbing Costs with RSMeans Data, call 800.448.8182.

341

23 13 Facility Fuel-Storage Tanks

23 13 23 – Facility Aboveground Fuel-Oil, Storage Tanks

23 13 23.16 Horizontal, Stl, Abvgrd Fuel-Oil, Storage Tanks		Crew	Daily Output	Labor-Hours	Unit	Material	2021 Bare Costs Labor	Equipment	Total	Total Incl O&P
3340	4,000 gallon capacity	Q-7	3.60	8.889	Ea.	14,300	580		14,880	16,700
3350	6,000 gallon capacity		2.40	13.333		16,300	865		17,165	19,200
3360	8,000 gallon capacity		2	16		19,300	1,050		20,350	22,900
3370	10,000 gallon capacity		1.80	17.778		30,500	1,150		31,650	35,300
3380	15,000 gallon capacity		1.50	21.333		40,900	1,400		42,300	47,100
3390	20,000 gallon capacity		1.30	24.615		48,100	1,600		49,700	55,500
3400	25,000 gallon capacity		1.15	27.826		59,500	1,800		61,300	68,000
3410	30,000 gallon capacity		1	32		67,000	2,075		69,075	76,500

23 13 23.26 Horizontal, Conc., Abvgrd Fuel-Oil, Stor. Tanks

		Crew	Daily Output	Labor-Hours	Unit	Material	2021 Bare Costs Labor	Equipment	Total	Total Incl O&P
0010	**HORIZONTAL, CONCRETE, ABOVEGROUND FUEL-OIL, STORAGE TANKS**									
0050	Concrete, storage, aboveground, including pad & pump									
0100	500 gallon	F-3	2	20	Ea.	11,100	1,125	238	12,463	14,100
0200	1,000 gallon	"	2	20		15,600	1,125	238	16,963	19,000
0300	2,000 gallon	F-4	2	24		20,000	1,325	495	21,820	24,500

23 21 Hydronic Piping and Pumps

23 21 20 – Hydronic HVAC Piping Specialties

23 21 20.10 Air Control

		Crew	Daily Output	Labor-Hours	Unit	Material	2021 Bare Costs Labor	Equipment	Total	Total Incl O&P
0010	**AIR CONTROL**									
0030	Air separator, with strainer									
0040	2" diameter	Q-5	6	2.667	Ea.	1,625	164		1,789	2,025
0080	2-1/2" diameter		5	3.200		1,800	197		1,997	2,300
0100	3" diameter		4	4		2,800	246		3,046	3,450
1000	Micro-bubble separator for total air removal, closed loop system									
1010	Requires bladder type tank in system.									
1020	Water (hot or chilled) or glycol system									
1030	Threaded									
1040	3/4" diameter	1 Stpi	20	.400	Ea.	83.50	27.50		111	133
1050	1" diameter	1 Stpi	19	.421	Ea.	97	29		126	149
1060	1-1/4" diameter		16	.500		128	34		162	192
1070	1-1/2" diameter		13	.615		169	42		211	249
1080	2" diameter		11	.727		1,150	49.50		1,199.50	1,350
1090	2-1/2" diameter	Q-5	15	1.067		1,375	65.50		1,440.50	1,600
1100	3" diameter		13	1.231		1,850	75.50		1,925.50	2,150
1110	4" diameter		10	1.600		2,050	98.50		2,148.50	2,400
1800	With drain/dismantle/cleaning access flange									
1810	Flanged									
1820	2" diameter	Q-5	4.60	3.478	Ea.	3,250	214		3,464	3,900
1830	2-1/2" diameter		4	4		3,525	246		3,771	4,250
1840	3" diameter		3.60	4.444		4,300	273		4,573	5,150
1850	4" diameter		2.50	6.400		4,925	395		5,320	6,025
1860	5" diameter	Q-6	2.80	8.571		9,050	545		9,595	10,800
1870	6" diameter	"	2.60	9.231		11,600	590		12,190	13,700

23 21 20.14 Air Purging Scoop

		Crew	Daily Output	Labor-Hours	Unit	Material	2021 Bare Costs Labor	Equipment	Total	Total Incl O&P
0010	**AIR PURGING SCOOP**, With tappings.									
0020	For air vent and expansion tank connection									
0100	1" pipe size, threaded	1 Stpi	19	.421	Ea.	29.50	29		58.50	75.50
0110	1-1/4" pipe size, threaded		15	.533		29.50	36.50		66	87
0120	1-1/2" pipe size, threaded		13	.615		65	42		107	135
0130	2" pipe size, threaded		11	.727		74	49.50		123.50	156

23 21 Hydronic Piping and Pumps

23 21 20 – Hydronic HVAC Piping Specialties

23 21 20.18 Automatic Air Vent

	Crew	Daily Output	Labor-Hours	Unit	Material	2021 Bare Costs Labor	Equipment	Total	Total Incl O&P
0010	**AUTOMATIC AIR VENT**								
0020	Cast iron body, stainless steel internals, float type								
0180	1/2" NPT inlet, 250 psi	1 Stpi	10	.800	Ea.	370	54.50		424.50	485
0220	3/4" NPT inlet, 250 psi	"	10	.800	"	370	54.50		424.50	485
0600	Forged steel body, stainless steel internals, float type								
0640	1/2" NPT inlet, 750 psi	1 Stpi	12	.667	Ea.	1,225	45.50		1,270.50	1,425
0680	3/4" NPT inlet, 750 psi	"	12	.667	"	1,225	45.50		1,270.50	1,425
1100	Formed steel body, noncorrosive								
1110	1/8" NPT inlet, 150 psi	1 Stpi	32	.250	Ea.	16.10	17.10		33.20	43
1120	1/4" NPT inlet, 150 psi		32	.250		54	17.10		71.10	85
1130	3/4" NPT inlet, 150 psi		32	.250		54	17.10		71.10	85
1300	Chrome plated brass, automatic/manual, for radiators								
1310	1/8" NPT inlet, nickel plated brass	1 Stpi	32	.250	Ea.	9.35	17.10		26.45	36

23 21 20.26 Circuit Setter

	Crew	Daily Output	Labor-Hours	Unit	Material	2021 Bare Costs Labor	Equipment	Total	Total Incl O&P
0010	**CIRCUIT SETTER**, Balance valve								
0018	Threaded								
0019	1/2" pipe size	1 Stpi	22	.364	Ea.	85	25		110	131
0020	3/4" pipe size	"	20	.400	"	90.50	27.50		118	141

23 21 20.30 Cocks, Drains and Specialties

	Crew	Daily Output	Labor-Hours	Unit	Material	2021 Bare Costs Labor	Equipment	Total	Total Incl O&P
0010	**COCKS, DRAINS AND SPECIALTIES**								
1000	Boiler drain								
1010	Pipe thread to hose								
1020	Bronze								
1030	1/2" size	1 Stpi	36	.222	Ea.	16.10	15.20		31.30	40
1040	3/4" size	"	34	.235	"	17.50	16.10		33.60	43.50
1100	Solder to hose								
1110	Bronze								
1120	1/2" size	1 Stpi	46	.174	Ea.	16.10	11.90		28	35.50
1130	3/4" size	"	44	.182	"	17.50	12.45		29.95	38
1600	With built-in vacuum breaker								
1610	1/2" I.P. or solder	1 Stpi	36	.222	Ea.	34.50	15.20		49.70	60.50
1630	With tamper proof vacuum breaker								
1640	1/2" I.P. or solder	1 Stpi	36	.222	Ea.	53	15.20		68.20	81
1650	3/4" I.P. or solder	"	34	.235	"	63	16.10		79.10	93
3000	Cocks								
3010	Air, lever or tee handle								
3020	Bronze, single thread								
3030	1/8" size	1 Stpi	52	.154	Ea.	13.65	10.50		24.15	31
3040	1/4" size		46	.174		14.45	11.90		26.35	33.50
3050	3/8" size		40	.200		14.90	13.65		28.55	37
3060	1/2" size		36	.222		17.60	15.20		32.80	42
3100	Bronze, double thread								
3110	1/8" size	1 Stpi	26	.308	Ea.	17.45	21		38.45	50.50
3120	1/4" size		22	.364		18.30	25		43.30	57
3130	3/8" size		18	.444		19	30.50		49.50	66.50
3140	1/2" size		15	.533		23.50	36.50		60	80
4500	Gauge cock, brass								
4510	1/4" FPT	1 Stpi	24	.333	Ea.	14.45	23		37.45	50
4512	1/4" MPT	"	24	.333	"	18.55	23		41.55	54.50
4600	Pigtail, steam syphon								
4604	1/4"	1 Stpi	24	.333	Ea.	26.50	23		49.50	63
4650	Snubber valve								

For customer support on your Plumbing Costs with RSMeans Data, call 800.448.8182.

343

23 21 Hydronic Piping and Pumps

23 21 20 – Hydronic HVAC Piping Specialties

23 21 20.30 Cocks, Drains and Specialties

		Crew	Daily Output	Labor-Hours	Unit	Material	2021 Bare Costs Labor	Equipment	Total	Total Incl O&P
4654	1/4"	1 Stpi	22	.364	Ea.	4.91	25		29.91	42.50
4660	Nipple, black steel									
4664	1/4" x 3"	1 Stpi	37	.216	Ea.	3.59	14.80		18.39	26

23 21 20.34 Dielectric Unions

		Crew	Daily Output	Labor-Hours	Unit	Material	2021 Bare Costs Labor	Equipment	Total	Total Incl O&P
0010	**DIELECTRIC UNIONS**, Standard gaskets for water and air									
0020	250 psi maximum pressure									
0280	Female IPT to sweat, straight									
0300	1/2" pipe size	1 Plum	24	.333	Ea.	8.80	22.50		31.30	43
0340	3/4" pipe size		20	.400		9.80	27		36.80	51.50
0360	1" pipe size		19	.421		13.25	28.50		41.75	57
0380	1-1/4" pipe size	1 Plum	15	.533	Ea.	17.95	36		53.95	73.50
0400	1-1/2" pipe size		13	.615		31	41.50		72.50	96
0420	2" pipe size		11	.727		42	49		91	120
0580	Female IPT to brass pipe thread, straight									
0600	1/2" pipe size	1 Plum	24	.333	Ea.	18.90	22.50		41.40	54.50
0640	3/4" pipe size		20	.400		20.50	27		47.50	63.50
0660	1" pipe size		19	.421		38.50	28.50		67	85
0680	1-1/4" pipe size		15	.533		38	36		74	96
0700	1-1/2" pipe size		13	.615		55.50	41.50		97	123
0720	2" pipe size		11	.727		113	49		162	198
0780	Female IPT to female IPT, straight									
0800	1/2" pipe size	1 Plum	24	.333	Ea.	16.25	22.50		38.75	51.50
0840	3/4" pipe size		20	.400		18.55	27		45.55	61
0860	1" pipe size		19	.421		24.50	28.50		53	69.50
0880	1-1/4" pipe size		15	.533		32.50	36		68.50	89.50
0900	1-1/2" pipe size		13	.615		65.50	41.50		107	134
0920	2" pipe size		11	.727		97	49		146	181
2000	175 psi maximum pressure									
2180	Female IPT to sweat									
2240	2" pipe size	1 Plum	9	.889	Ea.	198	60		258	310
2260	2-1/2" pipe size	Q-1	15	1.067		236	65		301	355
2280	3" pipe size		14	1.143		325	69.50		394.50	460
2300	4" pipe size		11	1.455		855	88.50		943.50	1,075
2480	Female IPT to brass pipe									
2500	1-1/2" pipe size	1 Plum	11	.727	Ea.	264	49		313	365
2540	2" pipe size	"	9	.889		277	60		337	395
2560	2-1/2" pipe size	Q-1	15	1.067		320	65		385	445
2580	3" pipe size		14	1.143		320	69.50		389.50	455
2600	4" pipe size		11	1.455		455	88.50		543.50	630

23 21 20.42 Expansion Joints

		Crew	Daily Output	Labor-Hours	Unit	Material	2021 Bare Costs Labor	Equipment	Total	Total Incl O&P
0010	**EXPANSION JOINTS**									
0100	Bellows type, neoprene cover, flanged spool									
0140	6" face to face, 1-1/4" diameter	1 Stpi	11	.727	Ea.	340	49.50		389.50	450
0160	1-1/2" diameter	"	10.60	.755		340	51.50		391.50	450
0180	2" diameter	Q-5	13.30	1.203		345	74		419	490
0190	2-1/2" diameter		12.40	1.290		355	79.50		434.50	510
0200	3" diameter		11.40	1.404		400	86.50		486.50	570
0210	4" diameter		8.40	1.905		430	117		547	650
0220	5" diameter		7.60	2.105		525	130		655	775
0230	6" diameter		6.80	2.353		545	145		690	810
0240	8" diameter		5.40	2.963		630	182		812	965
0250	10" diameter		5	3.200		870	197		1,067	1,250

344

23 21 20.42 Expansion Joints

		Crew	Daily Output	Labor-Hours	Unit	Material	2021 Bare Costs Labor	Equipment	Total	Total Incl O&P
0260	12" diameter	Q-5	4.60	3.478	Ea.	995	214		1,209	1,425
0480	10" face to face, 2" diameter		13	1.231		495	75.50		570.50	660
0500	2-1/2" diameter		12	1.333		520	82		602	695
0520	3" diameter		11	1.455		530	89.50		619.50	720
0540	4" diameter		8	2		605	123		728	850
0560	5" diameter		7	2.286		720	141		861	1,000
0580	6" diameter		6	2.667		745	164		909	1,075
0600	8" diameter		5	3.200		890	197		1,087	1,275
0620	10" diameter		4.60	3.478		980	214		1,194	1,400
0640	12" diameter		4	4		1,225	246		1,471	1,725
0660	14" diameter		3.80	4.211		1,500	259		1,759	2,025
0680	16" diameter		2.90	5.517		1,750	340		2,090	2,425
0700	18" diameter		2.50	6.400		1,950	395		2,345	2,750
0720	20" diameter		2.10	7.619		2,050	470		2,520	2,950
0740	24" diameter		1.80	8.889		2,400	545		2,945	3,475
0760	26" diameter		1.40	11.429		2,675	705		3,380	4,000
0780	30" diameter		1.20	13.333		3,000	820		3,820	4,525
0800	36" diameter		1	16		3,450	985		4,435	5,250

23 21 20.46 Expansion Tanks

		Crew	Daily Output	Labor-Hours	Unit	Material	2021 Bare Costs Labor	Equipment	Total	Total Incl O&P
0010	**EXPANSION TANKS**									
1502	Plastic, corrosion resistant, see Section 22 66 83.13									
1505	Aboveground fuel-oil, storage tanks, see Section 23 13 23									
1507	Underground fuel-oil storage tanks, see Section 23 13 13									
1512	Tank leak detection systems, see Section 28 33 33.50									
2000	Steel, liquid expansion, ASME, painted, 15 gallon capacity	Q-5	17	.941	Ea.	880	58		938	1,050
2020	24 gallon capacity		14	1.143		725	70.50		795.50	905
2040	30 gallon capacity		12	1.333		1,025	82		1,107	1,250
2060	40 gallon capacity		10	1.600		1,200	98.50		1,298.50	1,475
2360	Galvanized									
2370	15 gallon capacity	Q-5	17	.941	Ea.	1,500	58		1,558	1,725
2380	24 gallon capacity		14	1.143		1,525	70.50		1,595.50	1,775
2390	30 gallon capacity		12	1.333		1,675	82		1,757	1,975
3000	Steel ASME expansion, rubber diaphragm, 19 gal. cap. accept.		12	1.333		3,350	82		3,432	3,825
3020	31 gallon capacity		8	2		3,725	123		3,848	4,275
3040	61 gallon capacity		6	2.667		5,450	164		5,614	6,225

23 21 20.50 Float Valves

		Crew	Daily Output	Labor-Hours	Unit	Material	2021 Bare Costs Labor	Equipment	Total	Total Incl O&P
0010	**FLOAT VALVES**									
0020	With ball and bracket									
0030	Single seat, threaded									
0040	Brass body									
0050	1/2"	1 Stpi	11	.727	Ea.	88	49.50		137.50	171
0060	3/4"		9	.889		99.50	61		160.50	200
0070	1"		7	1.143		177	78		255	310
0080	1-1/2"		4.50	1.778		242	122		364	445
0090	2"		3.60	2.222		258	152		410	510
0300	For condensate receivers, CI, in-line mount									
0320	1" inlet	1 Stpi	7	1.143	Ea.	161	78		239	294
0360	For condensate receiver, CI, external float, flanged tank mount									
0370	3/4" inlet	1 Stpi	5	1.600	Ea.	99.50	109		208.50	272

For customer support on your Plumbing Costs with RSMeans Data, call 800.448.8182.

345

23 21 20.54 Flow Check Control

		Crew	Daily Output	Labor-Hours	Unit	Material	2021 Bare Costs Labor	Equipment	Total	Total Incl O&P
0010	**FLOW CHECK CONTROL**									
0100	Bronze body, soldered									
0110	3/4" size	1 Stpi	20	.400	Ea.	77	27.50		104.50	126
0120	1" size	"	19	.421	"	93	29		122	145
0200	Cast iron body, threaded									
0210	3/4" size	1 Stpi	20	.400	Ea.	59.50	27.50		87	107
0220	1" size		19	.421		67.50	29		96.50	117
0230	1-1/4" size		15	.533		81.50	36.50		118	144
0240	1-1/2" size		13	.615		125	42		167	200
0250	2" size		11	.727		179	49.50		228.50	271

23 21 20.58 Hydronic Heating Control Valves

		Crew	Daily Output	Labor-Hours	Unit	Material	2021 Bare Costs Labor	Equipment	Total	Total Incl O&P
0010	**HYDRONIC HEATING CONTROL VALVES**									
0050	Hot water, nonelectric, thermostatic									
0100	Radiator supply, 1/2" diameter	1 Stpi	24	.333	Ea.	37	23		60	74.50
0120	3/4" diameter	"	20	.400	"	41.50	27.50		69	86.50
1000	Manual, radiator supply									
1010	1/2" pipe size, angle union	1 Stpi	24	.333	Ea.	76.50	23		99.50	119
1020	3/4" pipe size, angle union	"	20	.400	"	96.50	27.50		124	147
1100	Radiator, balancing, straight, sweat connections									
1110	1/2" pipe size	1 Stpi	24	.333	Ea.	28.50	23		51.50	65.50
1120	3/4" pipe size	"	20	.400	"	40	27.50		67.50	85
1200	Steam, radiator, supply									
1210	1/2" pipe size, angle union	1 Stpi	24	.333	Ea.	74.50	23		97.50	116
1220	3/4" pipe size, angle union	"	20	.400	"	81	27.50		108.50	130
8000	System balancing and shut-off									
8020	Butterfly, quarter turn, calibrated, threaded or solder									
8040	Bronze, -30°F to +350°F, pressure to 175 psi									
8060	1/2" size	1 Stpi	22	.364	Ea.	27.50	25		52.50	67.50
8070	3/4" size	"	20	.400	"	44.50	27.50		72	90

23 21 20.66 Monoflow Tee Fitting

		Crew	Daily Output	Labor-Hours	Unit	Material	2021 Bare Costs Labor	Equipment	Total	Total Incl O&P
0010	**MONOFLOW TEE FITTING**									
1100	For one pipe hydronic, supply and return									
1110	Copper, soldered									
1120	3/4" x 1/2" size	1 Stpi	13	.615	Ea.	20.50	42		62.50	85.50

23 21 20.74 Strainers, Basket Type

		Crew	Daily Output	Labor-Hours	Unit	Material	2021 Bare Costs Labor	Equipment	Total	Total Incl O&P
0010	**STRAINERS, BASKET TYPE**, Perforated stainless steel basket									
0100	Brass or monel available									
2000	Simplex style									
2300	Bronze body									
2320	Screwed, 3/8" pipe size	1 Stpi	22	.364	Ea.	185	25		210	240
2340	1/2" pipe size		20	.400		192	27.50		219.50	252
2360	3/4" pipe size		17	.471		254	32		286	325
2380	1" pipe size		15	.533		365	36.50		401.50	460
2400	1-1/4" pipe size		13	.615		405	42		447	510
2420	1-1/2" pipe size		12	.667		410	45.50		455.50	520
2440	2" pipe size		10	.800		605	54.50		659.50	745
2460	2-1/2" pipe size	Q-5	15	1.067		1,000	65.50		1,065.50	1,200
2480	3" pipe size	"	14	1.143		1,500	70.50		1,570.50	1,750
2600	Flanged, 2" pipe size	1 Stpi	6	1.333		890	91		981	1,100
2620	2-1/2" pipe size	Q-5	4.50	3.556		1,525	219		1,744	2,000
2640	3" pipe size		3.50	4.571		1,725	281		2,006	2,325

23 21 20 – Hydronic HVAC Piping Specialties

23 21 20.74 Strainers, Basket Type		Crew	Daily Output	Labor-Hours	Unit	Material	2021 Bare Costs Labor	Equipment	Total	Total Incl O&P
2660	4" pipe size	Q-5	3	5.333	Ea.	2,650	330		2,980	3,425
2680	5" pipe size	Q-6	3.40	7.059		4,225	450		4,675	5,325
2700	6" pipe size		3	8		5,150	510		5,660	6,425
2710	8" pipe size	↓	2.50	9.600	↓	8,400	610		9,010	10,100
3600	Iron body									
3700	Screwed, 3/8" pipe size	1 Stpi	22	.364	Ea.	168	25		193	221
3720	1/2" pipe size		20	.400		173	27.50		200.50	231
3740	3/4" pipe size		17	.471		220	32		252	290
3760	1" pipe size		15	.533		224	36.50		260.50	300
3780	1-1/4" pipe size		13	.615		293	42		335	385
3800	1-1/2" pipe size		12	.667		325	45.50		370.50	425
3820	2" pipe size	↓	10	.800		385	54.50		439.50	505
3840	2-1/2" pipe size	Q-5	15	1.067		515	65.50		580.50	665
3860	3" pipe size	"	14	1.143		625	70.50		695.50	795
4000	Flanged, 2" pipe size	1 Stpi	6	1.333		595	91		686	790
4020	2-1/2" pipe size	Q-5	4.50	3.556		805	219		1,024	1,200
4040	3" pipe size		3.50	4.571		840	281		1,121	1,350
4060	4" pipe size	↓	3	5.333		1,300	330		1,630	1,925
4080	5" pipe size	Q-6	3.40	7.059		1,950	450		2,400	2,825
4100	6" pipe size		3	8		2,500	510		3,010	3,500
4120	8" pipe size		2.50	9.600		4,525	610		5,135	5,900
4140	10" pipe size	↓	2.20	10.909	↓	9,925	695		10,620	12,000
7000	Stainless steel body									
7200	Screwed, 1" pipe size	1 Stpi	15	.533	Ea.	480	36.50		516.50	585
7210	1-1/4" pipe size		13	.615		745	42		787	885
7220	1-1/2" pipe size		12	.667		745	45.50		790.50	890
7240	2" pipe size	↓	10	.800		1,100	54.50		1,154.50	1,300
7260	2-1/2" pipe size	Q-5	15	1.067		1,575	65.50		1,640.50	1,825
7280	3" pipe size	"	14	1.143		2,175	70.50		2,245.50	2,475
7400	Flanged, 2" pipe size	1 Stpi	6	1.333		1,725	91		1,816	2,025
7420	2-1/2" pipe size	Q-5	4.50	3.556		3,450	219		3,669	4,100
7440	3" pipe size		3.50	4.571		3,525	281		3,806	4,300
7460	4" pipe size	↓	3	5.333		5,525	330		5,855	6,575
7480	6" pipe size	Q-6	3	8		11,100	510		11,610	13,000
7500	8" pipe size	"	2.50	9.600	↓	16,100	610		16,710	18,600
8100	Duplex style									
8200	Bronze body									
8240	Screwed, 3/4" pipe size	1 Stpi	16	.500	Ea.	1,375	34		1,409	1,550
8260	1" pipe size		14	.571		1,375	39		1,414	1,550
8280	1-1/4" pipe size		12	.667		2,325	45.50		2,370.50	2,625
8300	1-1/2" pipe size	↓	11	.727		2,325	49.50		2,374.50	2,625
8320	2" pipe size		9	.889		4,350	61		4,411	4,875
8340	2-1/2" pipe size	Q-5	14	1.143		5,625	70.50		5,695.50	6,275
8420	Flanged, 2" pipe size	1 Stpi	6	1.333		5,350	91		5,441	6,000
8440	2-1/2" pipe size	Q-5	4.50	3.556		7,400	219		7,619	8,450
8460	3" pipe size		3.50	4.571		8,050	281		8,331	9,300
8480	4" pipe size	↓	3	5.333		10,900	330		11,230	12,500
8500	5" pipe size	Q-6	3.40	7.059		26,500	450		26,950	29,900
8520	6" pipe size	"	3	8	↓	25,700	510		26,210	29,100
8700	Iron body									
8740	Screwed, 3/4" pipe size	1 Stpi	16	.500	Ea.	1,775	34		1,809	2,000
8760	1" pipe size		14	.571		1,775	39		1,814	2,000
8780	1-1/4" pipe size	↓	12	.667	↓	1,975	45.50		2,020.50	2,250

For customer support on your Plumbing Costs with RSMeans Data, call 800.448.8182.

347

23 21 20.74 Strainers, Basket Type

		Crew	Daily Output	Labor-Hours	Unit	Material	2021 Bare Costs Labor	Equipment	Total	Total Incl O&P
8800	1-1/2" pipe size	1 Stpi	11	.727	Ea.	1,975	49.50		2,024.50	2,250
8820	2" pipe size	↓	9	.889		3,325	61		3,386	3,775
8840	2-1/2" pipe size	Q-5	14	1.143		3,675	70.50		3,745.50	4,150
9000	Flanged, 2" pipe size	1 Stpi	6	1.333		3,550	91		3,641	4,050
9020	2-1/2" pipe size	Q-5	4.50	3.556		3,800	219		4,019	4,500
9040	3" pipe size		3.50	4.571		4,150	281		4,431	4,975
9060	4" pipe size		3	5.333		7,000	330		7,330	8,200
9080	5" pipe size	Q-6	3.40	7.059		15,200	450		15,650	17,500
9100	6" pipe size		3	8		15,200	510		15,710	17,600
9120	8" pipe size		2.50	9.600		26,800	610		27,410	30,400
9140	10" pipe size		2.20	10.909		43,100	695		43,795	48,500
9160	12" pipe size		1.70	14.118		47,600	900		48,500	54,000
9170	14" pipe size		1.40	17.143		56,000	1,100		57,100	63,000
9180	16" pipe size	↓	1	24	↓	69,000	1,525		70,525	78,500
9700	Stainless steel body									
9740	Screwed, 1" pipe size	1 Stpi	14	.571	Ea.	3,225	39		3,264	3,600
9760	1-1/2" pipe size		11	.727		4,825	49.50		4,874.50	5,400
9780	2" pipe size		9	.889		7,350	61		7,411	8,200
9860	Flanged, 2" pipe size	↓	6	1.333		7,350	91		7,441	8,225
9880	2-1/2" pipe size	Q-5	4.50	3.556		12,300	219		12,519	13,900
9900	3" pipe size		3.50	4.571		13,500	281		13,781	15,200
9920	4" pipe size		3	5.333		17,500	330		17,830	19,800
9940	6" pipe size	Q-6	3	8		26,700	510		27,210	30,200
9960	8" pipe size	"	2.50	9.600	↓	69,000	610		69,610	76,500

23 21 20.76 Strainers, Y Type, Bronze Body

		Crew	Daily Output	Labor-Hours	Unit	Material	2021 Bare Costs Labor	Equipment	Total	Total Incl O&P
0010	**STRAINERS, Y TYPE, BRONZE BODY**									
0050	Screwed, 125 lb., 1/4" pipe size	1 Stpi	24	.333	Ea.	45.50	23		68.50	84
0070	3/8" pipe size		24	.333		45.50	23		68.50	84
0100	1/2" pipe size		20	.400		45.50	27.50		73	91
0120	3/4" pipe size		19	.421		56.50	29		85.50	105
0140	1" pipe size		17	.471		83	32		115	140
0150	1-1/4" pipe size		15	.533		132	36.50		168.50	200
0160	1-1/2" pipe size		14	.571		178	39		217	255
0180	2" pipe size		13	.615		234	42		276	320
0182	3" pipe size		12	.667		1,575	45.50		1,620.50	1,800
0200	300 lb., 2-1/2" pipe size	Q-5	17	.941		890	58		948	1,050
0220	3" pipe size		16	1		1,325	61.50		1,386.50	1,550
0240	4" pipe size	↓	15	1.067	↓	2,425	65.50		2,490.50	2,775
0500	For 300 lb. rating 1/4" thru 2", add					15%				
1000	Flanged, 150 lb., 1-1/2" pipe size	1 Stpi	11	.727	Ea.	565	49.50		614.50	695
1020	2" pipe size	"	8	1		690	68.50		758.50	860
1030	2-1/2" pipe size	Q-5	5	3.200		1,150	197		1,347	1,575
1040	3" pipe size		4.50	3.556		1,425	219		1,644	1,900
1060	4" pipe size	↓	3	5.333		2,200	330		2,530	2,900
1080	5" pipe size	Q-6	3.40	7.059		3,150	450		3,600	4,125
1100	6" pipe size		3	8		4,225	510		4,735	5,400
1106	8" pipe size	↓	2.60	9.231	↓	4,600	590		5,190	5,925
1500	For 300 lb. rating, add					40%				

23 21 20.78 Strainers, Y Type, Iron Body

		Crew	Daily Output	Labor-Hours	Unit	Material	2021 Bare Costs Labor	Equipment	Total	Total Incl O&P
0010	**STRAINERS, Y TYPE, IRON BODY**									
0050	Screwed, 250 lb., 1/4" pipe size	1 Stpi	20	.400	Ea.	18.80	27.50		46.30	61.50
0070	3/8" pipe size	↓	20	.400	↓	18.80	27.50		46.30	61.50

23 21 Hydronic Piping and Pumps

23 21 20 – Hydronic HVAC Piping Specialties

23 21 20.78 Strainers, Y Type, Iron Body

		Crew	Daily Output	Labor-Hours	Unit	Material	2021 Bare Costs Labor	Equipment	Total	Total Incl O&P
0100	1/2" pipe size	1 Stpi	20	.400	Ea.	18.80	27.50		46.30	61.50
0120	3/4" pipe size		18	.444		22	30.50		52.50	69.50
0140	1" pipe size		16	.500		29	34		63	83
0150	1-1/4" pipe size		15	.533		40.50	36.50		77	99
0160	1-1/2" pipe size		12	.667		55.50	45.50		101	129
0180	2" pipe size		8	1		83	68.50		151.50	194
0200	2-1/2" pipe size	Q-5	12	1.333		340	82		422	495
0220	3" pipe size		11	1.455		370	89.50		459.50	540
0240	4" pipe size		5	3.200		630	197		827	985
0500	For galvanized body, add					50%				
1000	Flanged, 125 lb., 1-1/2" pipe size	1 Stpi	11	.727	Ea.	260	49.50		309.50	360
1020	2" pipe size	"	8	1		185	68.50		253.50	305
1030	2-1/2" pipe size	Q-5	5	3.200		165	197		362	475
1040	3" pipe size		4.50	3.556		242	219		461	590
1060	4" pipe size		3	5.333		400	330		730	930
1080	5" pipe size	Q-6	3.40	7.059		475	450		925	1,200
1100	6" pipe size		3	8		805	510		1,315	1,650
1120	8" pipe size		2.50	9.600		1,350	610		1,960	2,400
1140	10" pipe size		2	12		2,450	765		3,215	3,850
1160	12" pipe size		1.70	14.118		3,700	900		4,600	5,400
1170	14" pipe size		1.30	18.462		6,825	1,175		8,000	9,250
1180	16" pipe size		1	24		9,675	1,525		11,200	12,900
1500	For 250 lb. rating, add					20%				
2000	For galvanized body, add					50%				
2500	For steel body, add					40%				

23 21 20.80 Suction Diffusers

		Crew	Daily Output	Labor-Hours	Unit	Material	2021 Bare Costs Labor	Equipment	Total	Total Incl O&P
0010	**SUCTION DIFFUSERS**									
0100	Cast iron body with integral straightening vanes, strainer									
1000	Flanged									
1010	2" inlet, 1-1/2" pump side	1 Stpi	6	1.333	Ea.	410	91		501	585
1020	2" pump side	"	5	1.600		480	109		589	690
1030	3" inlet, 2" pump side	Q-5	6.50	2.462		510	151		661	785

23 21 20.84 Thermoflo Indicator

		Crew	Daily Output	Labor-Hours	Unit	Material	2021 Bare Costs Labor	Equipment	Total	Total Incl O&P
0010	**THERMOFLO INDICATOR**, For balancing									
1000	Sweat connections, 1-1/4" pipe size	1 Stpi	12	.667	Ea.	830	45.50		875.50	985
1020	1-1/2" pipe size	"	10	.800	"	845	54.50		899.50	1,000

23 21 20.88 Venturi Flow

		Crew	Daily Output	Labor-Hours	Unit	Material	2021 Bare Costs Labor	Equipment	Total	Total Incl O&P
0010	**VENTURI FLOW**, Measuring device									
0050	1/2" diameter	1 Stpi	24	.333	Ea.	292	23		315	355
0100	3/4" diameter		20	.400		268	27.50		295.50	335
0120	1" diameter		19	.421		286	29		315	360
0140	1-1/4" diameter		15	.533		355	36.50		391.50	445
0160	1-1/2" diameter		13	.615		355	42		397	460
0180	2" diameter		11	.727		380	49.50		429.50	495
0200	2-1/2" diameter	Q-5	16	1		520	61.50		581.50	665
0220	3" diameter		14	1.143		540	70.50		610.50	695
0240	4" diameter		11	1.455		805	89.50		894.50	1,025
0260	5" diameter	Q-6	4	6		1,050	385		1,435	1,725
0280	6" diameter		3.50	6.857		1,175	435		1,610	1,950
0300	8" diameter		3	8		1,500	510		2,010	2,400
0320	10" diameter		2	12		3,550	765		4,315	5,075
0330	12" diameter		1.80	13.333		5,050	850		5,900	6,825

For customer support on your Plumbing Costs with RSMeans Data, call 800.448.8182.

349

23 21 20 – Hydronic HVAC Piping Specialties

23 21 20.88 Venturi Flow		Crew	Daily Output	Labor-Hours	Unit	Material	2021 Bare Costs		Total	Total Incl O&P
							Labor	Equipment		
0340	14" diameter	Q-6	1.60	15	Ea.	5,725	955		6,680	7,725
0350	16" diameter	↓	1.40	17.143		6,575	1,100		7,675	8,850
0500	For meter, add				↓	2,450			2,450	2,675

23 21 23 – Hydronic Pumps

23 21 23.13 In-Line Centrifugal Hydronic Pumps

		Crew	Daily Output	Labor-Hours	Unit	Material	Labor	Equipment	Total	Total Incl O&P
0010	**IN-LINE CENTRIFUGAL HYDRONIC PUMPS**									
0600	Bronze, sweat connections, 1/40 HP, in line									
0640	3/4" size	Q-1	16	1	Ea.	273	61		334	390
1000	Flange connection, 3/4" to 1-1/2" size									
1040	1/12 HP	Q-1	6	2.667	Ea.	680	162		842	985
1060	1/8 HP		6	2.667		1,250	162		1,412	1,600
1100	1/3 HP		6	2.667		1,375	162		1,537	1,775
1140	2" size, 1/6 HP		5	3.200		1,775	195		1,970	2,250
1180	2-1/2" size, 1/4 HP		5	3.200		2,275	195		2,470	2,800
1220	3" size, 1/4 HP		4	4		2,450	244		2,694	3,075
1260	1/3 HP		4	4		2,900	244		3,144	3,575
1300	1/2 HP		4	4		2,975	244		3,219	3,650
1340	3/4 HP		4	4		2,975	244		3,219	3,650
1380	1 HP	↓	4	4	↓	4,725	244		4,969	5,575
2000	Cast iron, flange connection									
2040	3/4" to 1-1/2" size, in line, 1/12 HP	Q-1	6	2.667	Ea.	480	162		642	765
2060	1/8 HP		6	2.667		775	162		937	1,100
2100	1/3 HP		6	2.667		865	162		1,027	1,200
2140	2" size, 1/6 HP		5	3.200		950	195		1,145	1,350
2180	2-1/2" size, 1/4 HP		5	3.200		1,150	195		1,345	1,550
2220	3" size, 1/4 HP		4	4		1,150	244		1,394	1,625
2260	1/3 HP		4	4		1,500	244		1,744	2,025
2300	1/2 HP		4	4		1,550	244		1,794	2,100
2340	3/4 HP		4	4		1,800	244		2,044	2,350
2380	1 HP	↓	4	4	↓	2,600	244		2,844	3,225
2600	For nonferrous impeller, add					3%				
3000	High head, bronze impeller									
3030	1-1/2" size, 1/2 HP	Q-1	5	3.200	Ea.	1,400	195		1,595	1,850
3040	1-1/2" size, 3/4 HP		5	3.200		1,700	195		1,895	2,175
3050	2" size, 1 HP		4	4		2,050	244		2,294	2,625
3090	2" size, 1-1/2 HP	↓	4	4	↓	2,375	244		2,619	3,000
4000	Close coupled, end suction, bronze impeller									
4040	1-1/2" size, 1-1/2 HP, to 40 GPM	Q-1	3	5.333	Ea.	2,700	325		3,025	3,450
4090	2" size, 2 HP, to 50 GPM		3	5.333		2,950	325		3,275	3,725
4100	2" size, 3 HP, to 90 GPM		2.30	6.957		3,000	425		3,425	3,925
4190	2-1/2" size, 3 HP, to 150 GPM		2	8		3,800	485		4,285	4,900
4300	3" size, 5 HP, to 225 GPM		1.80	8.889		4,000	540		4,540	5,200
4410	3" size, 10 HP, to 350 GPM		1.60	10		6,900	610		7,510	8,475
4420	4" size, 7-1/2 HP, to 350 GPM	↓	1.60	10		4,750	610		5,360	6,125
4520	4" size, 10 HP, to 600 GPM	Q-2	1.70	14.118		7,350	890		8,240	9,400
4530	5" size, 15 HP, to 1,000 GPM		1.70	14.118		7,375	890		8,265	9,425
4610	5" size, 20 HP, to 1,350 GPM		1.50	16		7,825	1,000		8,825	10,100
4620	5" size, 25 HP, to 1,550 GPM	↓	1.50	16	↓	9,425	1,000		10,425	11,900
5000	Base mounted, bronze impeller, coupling guard									
5040	1-1/2" size, 1-1/2 HP, to 40 GPM	Q-1	2.30	6.957	Ea.	7,525	425		7,950	8,900
5090	2" size, 2 HP, to 50 GPM		2.30	6.957		8,450	425		8,875	9,925
5100	2" size, 3 HP, to 90 GPM	↓	2	8		8,350	485		8,835	9,925

23 21 Hydronic Piping and Pumps

23 21 23 – Hydronic Pumps

23 21 23.13 In-Line Centrifugal Hydronic Pumps

		Crew	Daily Output	Labor-Hours	Unit	Material	2021 Bare Costs Labor	Equipment	Total	Total Incl O&P
5190	2-1/2" size, 3 HP, to 150 GPM	Q-1	1.80	8.889	Ea.	9,925	540		10,465	11,700
5300	3" size, 5 HP, to 225 GPM		1.60	10		12,200	610		12,810	14,300
5410	4" size, 5 HP, to 350 GPM		1.50	10.667		12,500	650		13,150	14,700
5420	4" size, 7-1/2 HP, to 350 GPM		1.50	10.667		11,800	650		12,450	13,900
5520	5" size, 10 HP, to 600 GPM	Q-2	1.60	15		14,900	950		15,850	17,800
5530	5" size, 15 HP, to 1,000 GPM		1.60	15		16,000	950		16,950	18,900
5610	6" size, 20 HP, to 1,350 GPM		1.40	17.143		25,300	1,075		26,375	29,500
5620	6" size, 25 HP, to 1,550 GPM		1.40	17.143		20,400	1,075		21,475	24,000
5800	The above pump capacities are based on 1,800 RPM,									
5810	at a 60' head. Increasing the RPM									
5820	or decreasing the head will increase the GPM.									

23 21 29 – Automatic Condensate Pump Units

23 21 29.10 Condensate Removal Pump System

			Crew	Daily Output	Labor-Hours	Unit	Material	2021 Bare Costs Labor	Equipment	Total	Total Incl O&P
0010	**CONDENSATE REMOVAL PUMP SYSTEM**										
0020	Pump with 1 gal. ABS tank										
0100	115 V										
0120	1/50 HP, 200 GPH	G	1 Stpi	12	.667	Ea.	196	45.50		241.50	283
0140	1/18 HP, 270 GPH	G		10	.800		195	54.50		249.50	297
0160	1/5 HP, 450 GPH	G		8	1		455	68.50		523.50	600
0200	230 V										
0240	1/18 HP, 270 GPH		1 Stpi	10	.800	Ea.	217	54.50		271.50	320
0260	1/5 HP, 450 GPH	G	"	8	1	"	525	68.50		593.50	675

23 23 Refrigerant Piping

23 23 23 – Refrigerants

23 23 23.10 Anti-Freeze

					Unit	Material	2021 Bare Costs Labor	Equipment	Total	Total Incl O&P
0010	**ANTI-FREEZE**, Inhibited									
0900	Ethylene glycol concentrated									
1000	55 gallon drums, small quantities				Gal.	13.70			13.70	15.10
1200	Large quantities				Gal.	10.70			10.70	11.75
2000	Propylene glycol, for solar heat, small quantities					23			23	25
2100	Large quantities					11.90			11.90	13.10

23 34 HVAC Fans

23 34 14 – Blower HVAC Fans

23 34 14.10 Blower Type HVAC Fans

		Crew	Daily Output	Labor-Hours	Unit	Material	2021 Bare Costs Labor	Equipment	Total	Total Incl O&P
0010	**BLOWER TYPE HVAC FANS**									
2500	Ceiling fan, right angle, extra quiet, 0.10" S.P.									
2520	95 CFM	Q-20	20	1	Ea.	276	60		336	395
2540	210 CFM		19	1.053		325	63		388	455
2560	385 CFM		18	1.111		415	66.50		481.50	555
2640	For wall or roof cap, add	1 Shee	16	.500		276	32.50		308.50	355
2660	For straight thru fan, add					10%				
2680	For speed control switch, add	1 Elec	16	.500		150	32		182	213

For customer support on your Plumbing Costs with RSMeans Data, call 800.448.8182.

351

23 34 HVAC Fans

23 34 23 – HVAC Power Ventilators

23 34 23.10 HVAC Power Circulators and Ventilators	Crew	Daily Output	Labor-Hours	Unit	Material	2021 Bare Costs Labor	Equipment	Total	Total Incl O&P
0010 **HVAC POWER CIRCULATORS AND VENTILATORS**									
6650 Residential, bath exhaust, grille, back draft damper									
6660 50 CFM	Q-20	24	.833	Ea.	48	50		98	128
6670 110 CFM		22	.909		96	54.50		150.50	188
6680 Light combination, squirrel cage, 100 watt, 70 CFM	↓	24	.833	↓	112	50		162	198
6700 Light/heater combination, ceiling mounted									
6710 70 CFM, 1,450 watt	Q-20	24	.833	Ea.	239	50		289	340
6800 Heater combination, recessed, 70 CFM		24	.833	↓	62.50	50		112.50	144
6820 With 2 infrared bulbs	↓	23	.870	↓	94	52		146	182
6940 Residential roof jacks and wall caps									
6944 Wall cap with back draft damper									
6946 3" & 4" diam. round duct	1 Shee	11	.727	Ea.	18.75	47.50		66.25	92.50
6948 6" diam. round duct	"	11	.727	"	54.50	47.50		102	132
6958 Roof jack with bird screen and back draft damper									
6960 3" & 4" diam. round duct	1 Shee	11	.727	Ea.	24	47.50		71.50	98.50
6962 3-1/4" x 10" rectangular duct	"	10	.800	"	38.50	52.50		91	122
6980 Transition									
6982 3-1/4" x 10" to 6" diam. round	1 Shee	20	.400	Ea.	38	26		64	81.50

23 35 Special Exhaust Systems

23 35 16 – Engine Exhaust Systems

23 35 16.10 Engine Exhaust Removal Systems

	Crew	Daily Output	Labor-Hours	Unit	Material	2021 Bare Costs Labor	Equipment	Total	Total Incl O&P
0010 **ENGINE EXHAUST REMOVAL SYSTEMS** D3090–320									
0500 Engine exhaust, garage, in-floor system									
0510 Single tube outlet assemblies									
0520 For transite pipe ducting, self-storing tube									
0530 3" tubing adapter plate	1 Shee	16	.500	Ea.	276	32.50		308.50	355
0540 4" tubing adapter plate		16	.500		280	32.50		312.50	360
0550 5" tubing adapter plate	↓	16	.500	↓	277	32.50		309.50	355
0600 For vitrified tile ducting									
0610 3" tubing adapter plate, self-storing tube	1 Shee	16	.500	Ea.	273	32.50		305.50	350
0620 4" tubing adapter plate, self-storing tube		16	.500		279	32.50		311.50	355
0660 5" tubing adapter plate, self-storing tube	↓	16	.500	↓	279	32.50		311.50	355
0800 Two tube outlet assemblies									
0810 For transite pipe ducting, self-storing tube									
0820 3" tubing, dual exhaust adapter plate	1 Shee	16	.500	Ea.	281	32.50		313.50	360
0850 For vitrified tile ducting									
0860 3" tubing, dual exhaust, self-storing tube	1 Shee	16	.500	Ea.	310	32.50		342.50	390
0870 3" tubing, double outlet, non-storing tubes	"	16	.500	"	310	32.50		342.50	390
0900 Accessories for metal tubing (overhead systems also)									
0910 Adapters, for metal tubing end									
0920 3" tail pipe type				Ea.	53.50			53.50	58.50
0930 4" tail pipe type					59			59	65
0940 5" tail pipe type					60			60	66
0990 5" diesel stack type					315			315	345
1000 6" diesel stack type				↓	375			375	410
1100 Bullnose (guide) required for in-floor assemblies									
1110 3" tubing size				Ea.	32			32	35
1120 4" tubing size					33			33	36.50
1130 5" tubing size				↓	35			35	38.50

23 35 16.10 Engine Exhaust Removal Systems	Crew	Daily Output	Labor-Hours	Unit	Material	2021 Bare Costs Labor	Equipment	Total	Total Incl O&P
1150	Plain rings, for tubing end									
1160	3" tubing size				Ea.	23			23	25
1170	4" tubing size				"	38			38	42
1200	Tubing, galvanized, flexible (for overhead systems also)									
1210	3" ID				L.F.	10.75			10.75	11.85
1220	4" ID					13.20			13.20	14.50
1230	5" ID					15.55			15.55	17.10
1240	6" ID					18			18	19.80
1250	Stainless steel, flexible (for overhead system also)									
1260	3" ID				L.F.	22			22	24
1270	4" ID					30			30	33
1280	5" ID					34			34	37
1290	6" ID					39.50			39.50	43
1500	Engine exhaust, garage, overhead components, for neoprene tubing									
1510	Alternate metal tubing & accessories see above									
1550	Adapters, for neoprene tubing end									
1560	3" tail pipe, adjustable, neoprene				Ea.	63			63	69.50
1570	3" tail pipe, heavy wall neoprene					71			71	78.50
1580	4" tail pipe, heavy wall neoprene					105			105	115
1590	5" tail pipe, heavy wall neoprene					124			124	136
1650	Connectors, tubing									
1660	3" interior, aluminum				Ea.	23			23	25
1670	4" interior, aluminum					38			38	42
1710	5" interior, neoprene					59.50			59.50	65.50
1750	3" spiralock, neoprene					23			23	25
1760	4" spiralock, neoprene					38			38	42
1780	Y for 3" ID tubing, neoprene, dual exhaust					213			213	234
1790	Y for 4" ID tubing, aluminum, dual exhaust					168			168	185
1850	Elbows, aluminum, splice into tubing for strap									
1860	3" neoprene tubing size				Ea.	57.50			57.50	63
1870	4" neoprene tubing size					40.50			40.50	44.50
1900	Flange assemblies, connect tubing to overhead duct					72			72	79
2000	Hardware and accessories									
2020	Cable, galvanized, 1/8" diameter				L.F.	.53			.53	.58
2040	Cleat, tie down cable or rope				Ea.	5.50			5.50	6.05
2060	Pulley					7.60			7.60	8.35
2080	Pulley hook, universal					5.50			5.50	6.05
2100	Rope, nylon, 1/4" diameter				L.F.	.38			.38	.42
2120	Winch, 1" diameter				Ea.	111			111	122
2150	Lifting strap, mounts on neoprene									
2160	3" tubing size				Ea.	27.50			27.50	30.50
2170	4" tubing size					27.50			27.50	30.50
2180	5" tubing size					27.50			27.50	30.50
2190	6" tubing size					27.50			27.50	30.50
2200	Tubing, neoprene, 11' lengths									
2210	3" ID				L.F.	12.90			12.90	14.20
2220	4" ID					16.85			16.85	18.55
2230	5" ID					30.50			30.50	33.50
2500	Engine exhaust, thru-door outlet									
2510	3" tube size	1 Carp	16	.500	Ea.	67	27.50		94.50	115
2530	4" tube size	"	16	.500	"	58.50	27.50		86	106
3000	Tubing, exhaust, flex hose, with									
3010	coupler, damper and tail pipe adapter									

23 35 Special Exhaust Systems

23 35 16 – Engine Exhaust Systems

23 35 16.10 Engine Exhaust Removal Systems

23 35 16.10 Engine Exhaust Removal Systems	Crew	Daily Output	Labor-Hours	Unit	Material	2021 Bare Costs Labor	Equipment	Total	Total Incl O&P
3020	Neoprene									
3040	3" x 20'	1 Shee	6	1.333	Ea.	380	87.50		467.50	550
3050	4" x 15'	1 Shee	5.40	1.481	Ea.	425	97		522	615
3060	4" x 20'		5	1.600		510	105		615	720
3070	5" x 15'		4.40	1.818		675	119		794	925
3100	Galvanized									
3110	3" x 20'	1 Shee	6	1.333	Ea.	320	87.50		407.50	480
3120	4" x 17'		5.60	1.429		350	93.50		443.50	525
3130	4" x 20'		5	1.600		390	105		495	590
3140	5" x 17'		4.60	1.739		415	114		529	630

23 35 43 – Welding Fume Elimination Systems

23 35 43.10 Welding Fume Elimination System Components

	23 35 43.10 Welding Fume Elimination System Components	Crew	Daily Output	Labor-Hours	Unit	Material	2021 Bare Costs Labor	Equipment	Total	Total Incl O&P
0010	**WELDING FUME ELIMINATION SYSTEM COMPONENTS**									
7500	Welding fume elimination accessories for garage exhaust systems									
7600	Cut off (blast gate)									
7610	3" tubing size, 3" x 6" opening	1 Shee	24	.333	Ea.	29	22		51	65
7620	4" tubing size, 4" x 8" opening		24	.333		29.50	22		51.50	65.50
7630	5" tubing size, 5" x 10" opening		24	.333		33.50	22		55.50	69.50
7640	6" tubing size		24	.333		37	22		59	73.50
7650	8" tubing size		24	.333		48	22		70	85.50
7700	Hoods, magnetic, with handle & screen									
7710	3" tubing size, 3" x 6" opening	1 Shee	24	.333	Ea.	107	22		129	151
7720	4" tubing size, 4" x 8" opening		24	.333		95	22		117	137
7730	5" tubing size, 5" x 10" opening		24	.333		95	22		117	137

23 38 Ventilation Hoods

23 38 13 – Commercial-Kitchen Hoods

23 38 13.10 Hood and Ventilation Equipment

	23 38 13.10 Hood and Ventilation Equipment	Crew	Daily Output	Labor-Hours	Unit	Material	2021 Bare Costs Labor	Equipment	Total	Total Incl O&P
0010	**HOOD AND VENTILATION EQUIPMENT**									
2970	Exhaust hood, sst, gutter on all sides, 4' x 4' x 2'	1 Carp	1.80	4.444	Ea.	1,675	243		1,918	2,225
2980	4' x 4' x 7'	"	1.60	5		6,850	274		7,124	7,925
7800	Vent hood, wall canopy with fire protection, 30"	L-3A	9	1.333		110	79.50		189.50	240
7810	Without fire protection, 36"		10	1.200		94	71.50		165.50	210
7820	Island canopy with fire protection, 30"		7	1.714		730	102		832	960
7830	Without fire protection, 36"		8	1.500		790	89.50		879.50	1,000
7840	Back shelf with fire protection, 30"		11	1.091		410	65		475	550
7850	Without fire protection, black, 36"		12	1		470	59.50		529.50	605
7852	Without fire protection, stainless steel, 36"		12	1		475	59.50		534.50	610
7860	Range hood & CO$_2$ system, 30"	1 Carp	2.50	3.200		4,800	175		4,975	5,525
7950	Hood fire protection system, electric stove	Q-1	3	5.333		1,975	325		2,300	2,650
7952	Hood fire protection system, gas stove	"	3	5.333		2,375	325		2,700	3,075

23 51 Breechings, Chimneys, and Stacks

23 51 13 – Draft Control Devices

23 51 13.16 Vent Dampers

		Crew	Daily Output	Labor-Hours	Unit	Material	2021 Bare Costs Labor	Equipment	Total	Total Incl O&P
0010	**VENT DAMPERS**									
5000	Vent damper, bi-metal, gas, 3" diameter	Q-9	24	.667	Ea.	74.50	39.50		114	141
5010	4" diameter	"	24	.667	"	73.50	39.50		113	140

23 51 13.19 Barometric Dampers

		Crew	Daily Output	Labor-Hours	Unit	Material	2021 Bare Costs Labor	Equipment	Total	Total Incl O&P
0010	**BAROMETRIC DAMPERS**									
1000	Barometric, gas fired system only, 6" size for 5" and 6" pipes	1 Shee	20	.400	Ea.	103	26		129	153
1020	7" size, for 6" and 7" pipes		19	.421		112	27.50		139.50	165
1040	8" size, for 7" and 8" pipes		18	.444		144	29		173	202
1060	9" size, for 8" and 9" pipes	↓	16	.500	↓	160	32.50		192.50	226
2000	All fuel, oil, oil/gas, coal									
2020	10" for 9" and 10" pipes	1 Shee	15	.533	Ea.	249	35		284	325
2040	12" for 11" and 12" pipes		15	.533		325	35		360	410
2060	14" for 13" and 14" pipes		14	.571		420	37.50		457.50	520
2080	16" for 15" and 16" pipes		13	.615		590	40.50		630.50	710
2100	18" for 17" and 18" pipes		12	.667		785	43.50		828.50	930
2120	20" for 19" and 21" pipes		10	.800		935	52.50		987.50	1,100
2140	24" for 22" and 25" pipes	Q-9	12	1.333		1,150	78.50		1,228.50	1,400
2160	28" for 26" and 30" pipes		10	1.600		1,425	94		1,519	1,725
2180	32" for 31" and 34" pipes	↓	8	2		1,825	118		1,943	2,175
3260	For thermal switch for above, add	1 Shee	24	.333	↓	102	22		124	146

23 51 23 – Gas Vents

23 51 23.10 Gas Chimney Vents

		Crew	Daily Output	Labor-Hours	Unit	Material	2021 Bare Costs Labor	Equipment	Total	Total Incl O&P
0010	**GAS CHIMNEY VENTS**, Prefab metal, UL listed									
0020	Gas, double wall, galvanized steel									
0080	3" diameter	Q-9	72	.222	V.L.F.	7.95	13.10		21.05	28.50
0100	4" diameter		68	.235		9.65	13.85		23.50	31.50
0120	5" diameter		64	.250		10.40	14.75		25.15	33.50
0140	6" diameter		60	.267		13.85	15.70		29.55	39
0160	7" diameter		56	.286		25.50	16.85		42.35	53.50
0180	8" diameter		52	.308		27	18.10		45.10	57.50
0200	10" diameter		48	.333		52	19.65		71.65	86.50
0220	12" diameter		44	.364		62	21.50		83.50	101
0240	14" diameter		42	.381		103	22.50		125.50	147
0260	16" diameter		40	.400		148	23.50		171.50	199
0280	18" diameter	↓	38	.421		183	25		208	239
0300	20" diameter	Q-10	36	.667		215	40.50		255.50	299
0320	22" diameter		34	.706		273	43		316	365
0340	24" diameter	↓	32	.750	↓	335	46		381	440
0600	For 4", 5" and 6" oval, add					50%				
0650	Gas, double wall, galvanized steel, fittings									
0660	Elbow 45°, 3" diameter	Q-9	36	.444	Ea.	14	26		40	55
0670	4" diameter	"	34	.471	"	16.95	27.50		44.45	60.50
0680	5" diameter	Q-9	32	.500	Ea.	19.80	29.50		49.30	66.50
0690	6" diameter		30	.533		24.50	31.50		56	74.50
0700	7" diameter		28	.571		39	33.50		72.50	94
0710	8" diameter		26	.615		52	36.50		88.50	112
0720	10" diameter		24	.667		111	39.50		150.50	181
0730	12" diameter		22	.727		118	43		161	195
0740	14" diameter		21	.762		185	45		230	272
0750	16" diameter		20	.800		241	47		288	335
0760	18" diameter	↓	19	.842		315	49.50		364.50	425
0770	20" diameter	Q-10	18	1.333		355	81.50		436.50	515

23 51 23.10 Gas Chimney Vents		Crew	Daily Output	Labor-Hours	Unit	Material	2021 Bare Costs Labor	Equipment	Total	Total Incl O&P
0780	22" diameter	Q-10	17	1.412	Ea.	565	86		651	750
0790	24" diameter	↓	16	1.500	↓	720	91.50		811.50	930
0916	Adjustable length									
0918	3" diameter, to 12"	Q-9	36	.444	Ea.	17.80	26		43.80	59
0920	4" diameter, to 12"		34	.471		20.50	27.50		48	65
0924	6" diameter, to 12"		30	.533		26.50	31.50		58	77
0928	8" diameter, to 12"		26	.615		50	36.50		86.50	110
0930	10" diameter, to 18"		24	.667		154	39.50		193.50	229
0932	12" diameter, to 18"		22	.727		157	43		200	238
0936	16" diameter, to 18"		20	.800		315	47		362	420
0938	18" diameter, to 18"	↓	19	.842		365	49.50		414.50	475
0944	24" diameter, to 18"	Q-10	16	1.500		690	91.50		781.50	900
0950	Elbow 90°, adjustable, 3" diameter	Q-9	36	.444		24	26		50	65.50
0960	4" diameter		34	.471		28	27.50		55.50	73
0970	5" diameter		32	.500		33.50	29.50		63	81
0980	6" diameter		30	.533		40.50	31.50		72	92
0990	7" diameter		28	.571		69.50	33.50		103	128
1010	8" diameter		26	.615		70.50	36.50		107	133
1020	Wall thimble, 4 to 7" adjustable, 3" diameter		36	.444		14.95	26		40.95	56
1022	4" diameter		34	.471		16.85	27.50		44.35	60.50
1024	5" diameter		32	.500		20.50	29.50		50	67
1026	6" diameter		30	.533		21	31.50		52.50	70.50
1028	7" diameter		28	.571		43.50	33.50		77	99
1030	8" diameter		26	.615		53.50	36.50		90	114
1040	Roof flashing, 3" diameter		36	.444		8.20	26		34.20	48.50
1050	4" diameter		34	.471		9.55	27.50		37.05	52.50
1060	5" diameter		32	.500		29	29.50		58.50	76.50
1070	6" diameter		30	.533		23	31.50		54.50	72.50
1080	7" diameter		28	.571		30	33.50		63.50	84
1090	8" diameter		26	.615		32	36.50		68.50	90
1100	10" diameter		24	.667		43.50	39.50		83	107
1110	12" diameter	↓	22	.727	↓	60.50	43		103.50	132
1120	14" diameter	Q-9	20	.800	Ea.	138	47		185	223
1130	16" diameter		18	.889		160	52.50		212.50	255
1140	18" diameter	↓	16	1		228	59		287	340
1150	20" diameter	Q-10	18	1.333		300	81.50		381.50	455
1160	22" diameter		14	1.714		375	105		480	570
1170	24" diameter	↓	12	2		435	122		557	665
1200	Tee, 3" diameter	Q-9	27	.593		36.50	35		71.50	92.50
1210	4" diameter		26	.615		39.50	36.50		76	98
1220	5" diameter		25	.640		41.50	37.50		79	103
1230	6" diameter		24	.667		47.50	39.50		87	111
1240	7" diameter		23	.696		68	41		109	137
1250	8" diameter		22	.727		74	43		117	146
1260	10" diameter		21	.762		198	45		243	285
1270	12" diameter		20	.800		203	47		250	294
1280	14" diameter		18	.889		360	52.50		412.50	475
1290	16" diameter		16	1		535	59		594	680
1300	18" diameter	↓	14	1.143		645	67.50		712.50	810
1310	20" diameter	Q-10	17	1.412		885	86		971	1,100
1320	22" diameter		13	1.846		1,125	113		1,238	1,425
1330	24" diameter	↓	12	2		1,300	122		1,422	1,600
1460	Tee cap, 3" diameter	Q-9	45	.356	↓	2.45	21		23.45	34

23 51 23.10 Gas Chimney Vents		Crew	Daily Output	Labor-Hours	Unit	Material	2021 Bare Costs Labor	Equipment	Total	Total Incl O&P
1470	4" diameter	Q-9	42	.381	Ea.	2.63	22.50		25.13	37
1480	5" diameter		40	.400		3.38	23.50		26.88	39
1490	6" diameter		37	.432		4.91	25.50		30.41	44
1500	7" diameter		35	.457		8.25	27		35.25	49.50
1510	8" diameter		34	.471		9	27.50		36.50	52
1520	10" diameter		32	.500		72.50	29.50		102	125
1530	12" diameter		30	.533		72.50	31.50		104	128
1540	14" diameter		28	.571		75.50	33.50		109	134
1550	16" diameter		25	.640		77	37.50		114.50	142
1560	18" diameter	↓	24	.667		91	39.50		130.50	159
1570	20" diameter	Q-10	27	.889		103	54.50		157.50	195
1580	22" diameter		22	1.091		99	66.50		165.50	209
1590	24" diameter	↓	21	1.143		117	70		187	233
1750	Top, 3" diameter	Q-9	46	.348		18.65	20.50		39.15	51.50
1760	4" diameter		44	.364		19.65	21.50		41.15	54
1770	5" diameter		42	.381		18.15	22.50		40.65	54
1780	6" diameter		40	.400		24.50	23.50		48	62
1790	7" diameter		38	.421		54	25		79	96.50
1800	8" diameter		36	.444		66.50	26		92.50	113
1810	10" diameter		34	.471		109	27.50		136.50	162
1820	12" diameter	↓	32	.500	↓	132	29.50		161.50	191
1830	14" diameter	Q-9	30	.533	Ea.	212	31.50		243.50	281
1840	16" diameter		28	.571		273	33.50		306.50	350
1850	18" diameter	↓	26	.615		375	36.50		411.50	470
1860	20" diameter	Q-10	28	.857		570	52.50		622.50	705
1870	22" diameter		22	1.091		870	66.50		936.50	1,050
1880	24" diameter	↓	20	1.200	↓	1,075	73.50		1,148.50	1,300
1900	Gas, double wall, galvanized steel, oval									
1904	4" x 1'	Q-9	68	.235	V.L.F.	19.45	13.85		33.30	42.50
1906	5" x 1'		64	.250		97	14.75		111.75	129
1908	5"/6" x 1'	↓	60	.267	↓	48	15.70		63.70	76
1910	Oval fittings									
1912	Adjustable length									
1914	4" diameter to 12" long	Q-9	34	.471	Ea.	20	27.50		47.50	64
1916	5" diameter to 12" long		32	.500		21.50	29.50		51	68
1918	5"/6" diameter to 12" long	↓	30	.533	↓	29	31.50		60.50	79.50
1920	Elbow 45°									
1922	4"	Q-9	34	.471	Ea.	32.50	27.50		60	78
1924	5"		32	.500		65.50	29.50		95	117
1926	5"/6"	↓	30	.533	↓	67	31.50		98.50	121
1930	Elbow 45°, flat									
1932	4"	Q-9	34	.471	Ea.	32.50	27.50		60	78
1934	5"		32	.500		65.50	29.50		95	117
1936	5"/6"	↓	30	.533	↓	67	31.50		98.50	121
1940	Top									
1942	4"	Q-9	44	.364	Ea.	36.50	21.50		58	72.50
1944	5"		42	.381		35	22.50		57.50	72.50
1946	5"/6"	↓	40	.400	↓	45.50	23.50		69	85.50
1950	Adjustable flashing									
1952	4"	Q-9	34	.471	Ea.	13.75	27.50		41.25	57
1954	5"		32	.500		40	29.50		69.50	88.50
1956	5"/6"	↓	30	.533	↓	40.50	31.50		72	92
1960	Tee									

For customer support on your Plumbing Costs with RSMeans Data, call 800.448.8182.

357

23 51 Breechings, Chimneys, and Stacks

23 51 23 – Gas Vents

23 51 23.10 Gas Chimney Vents		Crew	Daily Output	Labor-Hours	Unit	Material	2021 Bare Costs Labor	Equipment	Total	Total Incl O&P
1962	4"	Q-9	26	.615	Ea.	48.50	36.50		85	108
1964	5"		25	.640		98.50	37.50		136	165
1966	5"/6"		24	.667		101	39.50		140.50	170
1970	Tee with short snout									
1972	4"	Q-9	26	.615	Ea.	51.50	36.50		88	111

23 51 26 – All-Fuel Vent Chimneys

23 51 26.10 All-Fuel Vent Chimneys, Press. Tight, Dbl. Wall

		Crew	Daily Output	Labor-Hours	Unit	Material	2021 Bare Costs Labor	Equipment	Total	Total Incl O&P
0010	**ALL-FUEL VENT CHIMNEYS, PRESSURE TIGHT, DOUBLE WALL**									
3200	All fuel, pressure tight, double wall, 1" insulation, UL listed, 1,400°F.									
3210	304 stainless steel liner, aluminized steel outer jacket									
3220	6" diameter	Q-9	60	.267	L.F.	64.50	15.70		80.20	94.50
3221	8" diameter	Q-9	52	.308	L.F.	63	18.10		81.10	97
3222	10" diameter		48	.333		70.50	19.65		90.15	108
3223	12" diameter		44	.364		67.50	21.50		89	107
3224	14" diameter		42	.381		89.50	22.50		112	133
3225	16" diameter		40	.400		109	23.50		132.50	156
3226	18" diameter		38	.421		97	25		122	144
3227	20" diameter	Q-10	36	.667		140	40.50		180.50	216
3228	24" diameter	"	32	.750		175	46		221	261
3260	For 316 stainless steel liner, add					30%				
3280	All fuel, pressure tight, double wall fittings									
3284	304 stainless steel inner, aluminized steel jacket									
3288	Adjustable 20"/29" section									
3292	6" diameter	Q-9	30	.533	Ea.	200	31.50		231.50	268
3293	8" diameter		26	.615		226	36.50		262.50	305
3294	10" diameter		24	.667		258	39.50		297.50	345
3295	12" diameter		22	.727		289	43		332	385
3296	14" diameter		21	.762		249	45		294	340
3297	16" diameter		20	.800		335	47		382	435
3298	18" diameter		19	.842		405	49.50		454.50	520
3299	20" diameter	Q-10	18	1.333		425	81.50		506.50	595
3300	24" diameter	"	16	1.500		550	91.50		641.50	740
3350	Elbow 90° fixed									
3354	6" diameter	Q-9	30	.533	Ea.	410	31.50		441.50	500
3355	8" diameter		26	.615		460	36.50		496.50	560
3356	10" diameter		24	.667		525	39.50		564.50	635
3357	12" diameter		22	.727		595	43		638	720
3358	14" diameter		21	.762		675	45		720	810
3359	16" diameter		20	.800		760	47		807	905
3360	18" diameter		19	.842		860	49.50		909.50	1,025
3361	20" diameter	Q-10	18	1.333		975	81.50		1,056.50	1,200
3362	24" diameter	"	16	1.500		1,250	91.50		1,341.50	1,525
3380	For 316 stainless steel liner, add					30%				
3400	Elbow 45°									
3404	6" diameter	Q-9	30	.533	Ea.	206	31.50		237.50	274
3405	8" diameter		26	.615		246	36.50		282.50	325
3406	10" diameter		24	.667		263	39.50		302.50	350
3407	12" diameter		22	.727		300	43		343	395
3408	14" diameter		21	.762		360	45		405	465
3409	16" diameter		20	.800		405	47		452	515
3410	18" diameter		19	.842		460	49.50		509.50	580
3411	20" diameter	Q-10	18	1.333		540	81.50		621.50	715

For customer support on your Plumbing Costs with RSMeans Data, call 800.448.8182.

23 51 Breechings, Chimneys, and Stacks

23 51 26 – All-Fuel Vent Chimneys

23 51 26.10 All-Fuel Vent Chimneys, Press. Tight, Dbl. Wall	Crew	Daily Output	Labor-Hours	Unit	Material	2021 Bare Costs Labor	Equipment	Total	Total Incl O&P
3412	24" diameter	Q-10	16	1.500	Ea.	690	91.50		781.50	900
3430	For 316 stainless steel liner, add				Ea.	30%				
3450	Tee 90°									
3454	6" diameter	Q-9	24	.667	Ea.	248	39.50		287.50	330
3455	8" diameter		22	.727		269	43		312	360
3456	10" diameter		21	.762		299	45		344	400
3457	12" diameter		20	.800		345	47		392	450
3458	14" diameter		18	.889		420	52.50		472.50	540
3459	16" diameter		16	1		475	59		534	610
3460	18" diameter	↓	14	1.143		560	67.50		627.50	715
3461	20" diameter	Q-10	17	1.412		625	86		711	820
3462	24" diameter	"	12	2		740	122		862	995
3480	For tee cap, add					35%	20%			
3500	For 316 stainless steel liner, add				↓	30%				
3520	Plate support, galvanized									
3524	6" diameter	Q-9	26	.615	Ea.	123	36.50		159.50	190
3525	8" diameter		22	.727		144	43		187	224
3526	10" diameter		20	.800		157	47		204	244
3527	12" diameter		18	.889		165	52.50		217.50	261
3528	14" diameter		17	.941		196	55.50		251.50	299
3529	16" diameter		16	1		206	59		265	315
3530	18" diameter	↓	15	1.067		217	63		280	335
3531	20" diameter	Q-10	16	1.500		226	91.50		317.50	385
3532	24" diameter	"	14	1.714	↓	235	105		340	415
3570	Bellows, lined									
3574	6" diameter	Q-9	30	.533	Ea.	1,500	31.50		1,531.50	1,700
3575	8" diameter		26	.615		1,575	36.50		1,611.50	1,775
3576	10" diameter		24	.667		1,575	39.50		1,614.50	1,775
3577	12" diameter		22	.727		1,600	43		1,643	1,850
3578	14" diameter		21	.762		1,300	45		1,345	1,500
3579	16" diameter		20	.800		1,675	47		1,722	1,925
3580	18" diameter	↓	19	.842		1,725	49.50		1,774.50	1,975
3581	20" diameter	Q-10	18	1.333		1,750	81.50		1,831.50	2,075
3590	For all 316 stainless steel construction, add				↓	55%				
3600	Ventilated roof thimble, 304 stainless steel									
3620	6" diameter	Q-9	26	.615	Ea.	265	36.50		301.50	345
3624	8" diameter		22	.727		257	43		300	350
3625	10" diameter		20	.800		256	47		303	355
3626	12" diameter		18	.889		292	52.50		344.50	400
3627	14" diameter		17	.941		305	55.50		360.50	420
3628	16" diameter		16	1		340	59		399	465
3629	18" diameter	↓	15	1.067		365	63		428	500
3630	20" diameter	Q-10	16	1.500	↓	375	91.50		466.50	550
3631	24" diameter	Q-10	14	1.714	Ea.	415	105		520	615
3650	For 316 stainless steel, add					30%				
3670	Exit cone, 316 stainless steel only									
3674	6" diameter	Q-9	46	.348	Ea.	194	20.50		214.50	244
3675	8" diameter		42	.381		199	22.50		221.50	252
3676	10" diameter		40	.400		197	23.50		220.50	252
3677	12" diameter		38	.421		224	25		249	284
3678	14" diameter		37	.432		221	25.50		246.50	282
3679	16" diameter		36	.444		245	26		271	310
3680	18" diameter	↓	35	.457	↓	297	27		324	365

For customer support on your Plumbing Costs with RSMeans Data, call 800.448.8182.

359

23 51 Breechings, Chimneys, and Stacks

23 51 26 – All-Fuel Vent Chimneys

23 51 26.10 All-Fuel Vent Chimneys, Press. Tight, Dbl. Wall		Crew	Daily Output	Labor-Hours	Unit	Material	2021 Bare Costs Labor	Equipment	Total	Total Incl O&P
3681	20" diameter	Q-10	28	.857	Ea.	355	52.50		407.50	470
3682	24" diameter	"	26	.923	↓	475	56.50		531.50	610
3720	Roof guide, 304 stainless steel									
3724	6" diameter	Q-9	25	.640	Ea.	97	37.50		134.50	164
3725	8" diameter		21	.762		113	45		158	192
3726	10" diameter		19	.842		123	49.50		172.50	211
3727	12" diameter		17	.941		128	55.50		183.50	225
3728	14" diameter		16	1		149	59		208	253
3729	16" diameter		15	1.067		157	63		220	268
3730	18" diameter		14	1.143		167	67.50		234.50	286
3731	20" diameter	Q-10	15	1.600		176	97.50		273.50	340
3732	24" diameter	"	13	1.846		184	113		297	370
3750	For 316 stainless steel, add				↓	30%				
3770	Rain cap with bird screen									
3774	6" diameter	Q-9	46	.348	Ea.	296	20.50		316.50	355
3775	8" diameter		42	.381		340	22.50		362.50	410
3776	10" diameter		40	.400		400	23.50		423.50	475
3777	12" diameter		38	.421		460	25		485	550
3778	14" diameter		37	.432		535	25.50		560.50	630
3779	16" diameter		36	.444		610	26		636	710
3780	18" diameter	↓	35	.457		690	27		717	800
3781	20" diameter	Q-10	28	.857		785	52.50		837.50	945
3782	24" diameter	"	26	.923	↓	945	56.50		1,001.50	1,125

23 51 26.30 All-Fuel Vent Chimneys, Double Wall, St. Stl.

		Crew	Daily Output	Labor-Hours	Unit	Material	2021 Bare Costs Labor	Equipment	Total	Total Incl O&P
0010	**ALL-FUEL VENT CHIMNEYS, DOUBLE WALL, STAINLESS STEEL**									
7780	All fuel, pressure tight, double wall, 4" insulation, UL listed, 1,400°F.									
7790	304 stainless steel liner, aluminized steel outer jacket									
7800	6" diameter	Q-9	60	.267	V.L.F.	54	15.70		69.70	82.50
7804	8" diameter		52	.308		78	18.10		96.10	114
7806	10" diameter		48	.333		89.50	19.65		109.15	128
7808	12" diameter		44	.364		104	21.50		125.50	148
7810	14" diameter	↓	42	.381	↓	117	22.50		139.50	162
7880	For 316 stainless steel liner, add				L.F.	30%				
8000	All fuel, double wall, stainless steel fittings									
8010	Roof support, 6" diameter	Q-9	30	.533	Ea.	121	31.50		152.50	181
8030	8" diameter		26	.615		127	36.50		163.50	195
8040	10" diameter		24	.667		151	39.50		190.50	225
8050	12" diameter		22	.727		159	43		202	240
8060	14" diameter		21	.762		171	45		216	256
8100	Elbow 45°, 6" diameter		30	.533		246	31.50		277.50	320
8140	8" diameter		26	.615		281	36.50		317.50	365
8160	10" diameter		24	.667		340	39.50		379.50	435
8180	12" diameter		22	.727		365	43		408	465
8200	14" diameter		21	.762		420	45		465	535
8300	Insulated tee, 6" diameter		30	.533		310	31.50		341.50	390
8360	8" diameter		26	.615		325	36.50		361.50	415
8380	10" diameter		24	.667		350	39.50		389.50	445
8400	12" diameter		22	.727		425	43		468	530
8420	14" diameter		20	.800		500	47		547	620
8500	Boot tee, 6" diameter		28	.571		705	33.50		738.50	825
8520	8" diameter		24	.667		750	39.50		789.50	885
8530	10" diameter		22	.727		850	43		893	1,000

23 51 Breechings, Chimneys, and Stacks

23 51 26 – All-Fuel Vent Chimneys

23 51 26.30 All-Fuel Vent Chimneys, Double Wall, St. Stl.		Crew	Daily Output	Labor-Hours	Unit	Material	2021 Bare Costs Labor	Equipment	Total	Total Incl O&P
8540	12" diameter	Q-9	20	.800	Ea.	990	47		1,037	1,175
8550	14" diameter		18	.889		1,025	52.50		1,077.50	1,200
8600	Rain cap with bird screen, 6" diameter		30	.533		315	31.50		346.50	400
8640	8" diameter		26	.615		345	36.50		381.50	435
8660	10" diameter		24	.667		350	39.50		389.50	445
8680	12" diameter		22	.727		495	43		538	610
8700	14" diameter		21	.762		520	45		565	640
8800	Flat roof flashing, 6" diameter		30	.533		116	31.50		147.50	175
8840	8" diameter		26	.615		118	36.50		154.50	185
8860	10" diameter		24	.667		127	39.50		166.50	198
8880	12" diameter		22	.727		141	43		184	220
8900	14" diameter		21	.762		149	45		194	232

23 52 Heating Boilers

23 52 13 – Electric Boilers

23 52 13.10 Electric Boilers, ASME

			Crew	Daily Output	Labor-Hours	Unit	Material	2021 Bare Costs Labor	Equipment	Total	Total Incl O&P
0010	**ELECTRIC BOILERS, ASME**, Standard controls and trim	D3020-102									
1000	Steam, 6 KW, 20.5 MBH		Q-19	1.20	20	Ea.	4,500	1,250		5,750	6,800
1040	9 KW, 30.7 MBH			1.20	20		4,325	1,250		5,575	6,600
1060	18 KW, 61.4 MBH			1.20	20		4,450	1,250		5,700	6,750
1080	24 KW, 81.8 MBH			1.10	21.818		5,150	1,350		6,500	7,700
1120	36 KW, 123 MBH			1.10	21.818		5,700	1,350		7,050	8,300
2000	Hot water, 7.5 KW, 25.6 MBH		Q-19	1.30	18.462	Ea.	5,150	1,150		6,300	7,375
2020	15 KW, 51.2 MBH			1.30	18.462		5,425	1,150		6,575	7,700
2040	30 KW, 102 MBH			1.20	20		5,525	1,250		6,775	7,925
2060	45 KW, 164 MBH			1.20	20		5,400	1,250		6,650	7,775
2070	60 KW, 205 MBH			1.20	20		5,775	1,250		7,025	8,200
2080	75 KW, 256 MBH			1.10	21.818		6,100	1,350		7,450	8,725

23 52 16 – Condensing Boilers

23 52 16.24 Condensing Boilers

			Crew	Daily Output	Labor-Hours	Unit	Material	2021 Bare Costs Labor	Equipment	Total	Total Incl O&P
0010	**CONDENSING BOILERS**, Cast iron, high efficiency										
0020	Packaged with standard controls, circulator and trim										
0030	Intermittent (spark) pilot, natural or LP gas										
0040	Hot water, DOE MBH output (AFUE %)										
0100	42 MBH (84.0%)	G	Q-5	1.80	8.889	Ea.	1,700	545		2,245	2,700
0120	57 MBH (84.3%)	G		1.60	10		1,875	615		2,490	2,975
0140	85 MBH (84.0%)	G		1.40	11.429		2,075	705		2,780	3,325
0160	112 MBH (83.7%)	G		1.20	13.333		2,350	820		3,170	3,800
0180	140 MBH (83.3%)	G	Q-6	1.60	15		2,625	955		3,580	4,325
0200	167 MBH (83.0%)	G		1.40	17.143		3,225	1,100		4,325	5,175
0220	194 MBH (82.7%)	G		1.20	20		3,275	1,275		4,550	5,500

23 52 19 – Pulse Combustion Boilers

23 52 19.20 Pulse Type Combustion Boilers

			Crew	Daily Output	Labor-Hours	Unit	Material	2021 Bare Costs Labor	Equipment	Total	Total Incl O&P
0010	**PULSE TYPE COMBUSTION BOILERS**, High efficiency										
7990	Special feature gas fired boilers										
8000	Pulse combustion, standard controls/trim										
8010	Hot water, DOE MBH output (AFUE %)										
8030	71 MBH (95.2%)		Q-5	1.60	10	Ea.	4,025	615		4,640	5,350
8050	94 MBH (95.3%)	G		1.40	11.429		4,450	705		5,155	5,950
8080	139 MBH (95.6%)	G		1.20	13.333		5,000	820		5,820	6,725

23 52 Heating Boilers

23 52 19 – Pulse Combustion Boilers

23 52 19.20 Pulse Type Combustion Boilers	Crew	Daily Output	Labor-Hours	Unit	Material	2021 Bare Costs Labor	Equipment	Total	Total Incl O&P
8090 207 MBH (95.4%)	Q-5	1.16	13.793	Ea.	5,650	850		6,500	7,500
8120 270 MBH (96.4%)		1.12	14.286		7,825	880		8,705	9,925
8130 365 MBH (91.7%)		1.07	14.953		8,850	920		9,770	11,100

23 52 23 – Cast-Iron Boilers

23 52 23.20 Gas-Fired Boilers

0010 **GAS-FIRED BOILERS**, Natural or propane, standard controls, packaged									
1000 Cast iron, with insulated jacket									
2000 Steam, gross output, 81 MBH	Q-7	1.40	22.857	Ea.	3,125	1,475		4,600	5,675
2020 102 MBH		1.30	24.615		2,600	1,600		4,200	5,250
2040 122 MBH		1	32		3,425	2,075		5,500	6,875
2060 163 MBH		.90	35.556		3,425	2,325		5,750	7,225
2080 203 MBH		.90	35.556		4,125	2,325		6,450	7,975
2100 240 MBH		.85	37.647		4,225	2,450		6,675	8,275
2120 280 MBH		.80	40		4,475	2,600		7,075	8,800
2140 320 MBH		.70	45.714		4,800	2,975		7,775	9,750
3000 Hot water, gross output, 80 MBH	Q-7	1.46	21.918	Ea.	2,850	1,425		4,275	5,275
3020 100 MBH		1.35	23.704		2,800	1,550		4,350	5,375
3040 122 MBH		1.10	29.091		3,025	1,900		4,925	6,150
3060 163 MBH		1	32		3,550	2,075		5,625	7,025
3080 203 MBH		1	32		4,150	2,075		6,225	7,675
3100 240 MBH		.95	33.684		4,275	2,200		6,475	7,975
3120 280 MBH		.90	35.556		3,850	2,325		6,175	7,700
3140 320 MBH		.80	40		4,225	2,600		6,825	8,525
7000 For tankless water heater, add					10%				
7050 For additional zone valves up to 312 MBH, add					199			199	219

23 52 23.40 Oil-Fired Boilers

0010 **OIL-FIRED BOILERS**, Standard controls, flame retention burner, packaged									
1000 Cast iron, with insulated flush jacket									
2000 Steam, gross output, 109 MBH	Q-7	1.20	26.667	Ea.	2,475	1,725		4,200	5,325
2020 144 MBH		1.10	29.091		2,800	1,900		4,700	5,900
2040 173 MBH		1	32		3,150	2,075		5,225	6,575
2060 207 MBH		.90	35.556		3,375	2,325		5,700	7,175
3000 Hot water, same price as steam									
4000 For tankless coil in smaller sizes, add				Ea.	15%				

23 52 26 – Steel Boilers

23 52 26.40 Oil-Fired Boilers

0010 **OIL-FIRED BOILERS**, Standard controls, flame retention burner									
5000 Steel, with insulated flush jacket									
7000 Hot water, gross output, 103 MBH	Q-6	1.60	15	Ea.	2,225	955		3,180	3,875
7020 122 MBH		1.45	16.506		2,100	1,050		3,150	3,900
7040 137 MBH		1.36	17.595		2,850	1,125		3,975	4,800
7060 168 MBH		1.30	18.405		2,800	1,175		3,975	4,825
7080 225 MBH		1.22	19.704		4,300	1,250		5,550	6,600
7340 For tankless coil in steam or hot water, add					7%				

23 52 28 – Swimming Pool Boilers

23 52 28.10 Swimming Pool Heaters

0010 **SWIMMING POOL HEATERS**, Not including wiring, external									
0020 piping, base or pad									
0160 Gas fired, input, 155 MBH	Q-6	1.50	16	Ea.	1,850	1,025		2,875	3,575
0200 199 MBH		1	24		2,000	1,525		3,525	4,475
0220 250 MBH		.70	34.286		2,175	2,175		4,350	5,675

23 52 Heating Boilers

23 52 28 - Swimming Pool Boilers

23 52 28.10 Swimming Pool Heaters

		Crew	Daily Output	Labor-Hours	Unit	Material	2021 Bare Costs Labor	Equipment	Total	Total Incl O&P
0240	300 MBH	Q-6	.60	40	Ea.	2,275	2,550		4,825	6,325
0260	399 MBH		.50	48		2,575	3,050		5,625	7,400
0280	500 MBH		.40	60		7,350	3,825		11,175	13,800
0300	650 MBH		.35	68.571		7,800	4,375		12,175	15,100
0320	750 MBH		.33	72.727		8,525	4,650		13,175	16,300
0360	990 MBH		.22	109		11,500	6,950		18,450	23,000
0370	1,260 MBH		.21	114		17,600	7,300		24,900	30,200
0380	1,440 MBH	Q-6	.19	126	Ea.	19,400	8,050		27,450	33,400
0400	1,800 MBH		.14	171		20,800	10,900		31,700	39,100
0410	2,070 MBH		.13	185		24,500	11,800		36,300	44,600
2000	Electric, 12 KW, 4,800 gallon pool	Q-19	3	8		2,225	500		2,725	3,200
2020	15 KW, 7,200 gallon pool		2.80	8.571		2,350	535		2,885	3,375
2040	24 KW, 9,600 gallon pool		2.40	10		2,675	625		3,300	3,875
2060	30 KW, 12,000 gallon pool		2	12		2,725	745		3,470	4,125
2080	36 KW, 14,400 gallon pool		1.60	15		3,175	935		4,110	4,900
2100	57 KW, 24,000 gallon pool		1.20	20		3,975	1,250		5,225	6,225
9000	To select pool heater: 12 BTUH x S.F. pool area									
9010	X temperature differential = required output									
9050	For electric, KW = gallons x 2.5 divided by 1,000									
9100	For family home type pool, double the									
9110	Rated gallon capacity = 1/2°F rise per hour									

23 52 88 - Burners

23 52 88.10 Replacement Type Burners

		Crew	Daily Output	Labor-Hours	Unit	Material	2021 Bare Costs Labor	Equipment	Total	Total Incl O&P
0010	**REPLACEMENT TYPE BURNERS**									
0990	Residential, conversion, gas fired, LP or natural									
1000	Gun type, atmospheric input 50 to 225 MBH	Q-1	2.50	6.400	Ea.	1,725	390		2,115	2,475
1020	100 to 400 MBH	"	2	8	"	2,875	485		3,360	3,875
3000	Flame retention oil fired assembly, input									
3020	.50 to 2.25 GPH	Q-1	2.40	6.667	Ea.	360	405		765	1,000
3040	2.0 to 5.0 GPH	"	2	8		335	485		820	1,100
4600	Gas safety, shut off valve, 3/4" threaded	1 Stpi	20	.400		192	27.50		219.50	252
4610	1" threaded		19	.421		185	29		214	247
4620	1-1/4" threaded		15	.533		209	36.50		245.50	285
4630	1-1/2" threaded		13	.615		226	42		268	310
4640	2" threaded		11	.727		253	49.50		302.50	355
4650	2-1/2" threaded	Q-1	15	1.067		290	65		355	415
4660	3" threaded		13	1.231		575	75		650	745
4670	4" flanged		3	5.333		2,850	325		3,175	3,600
4680	6" flanged	Q-2	3	8		6,325	505		6,830	7,725

23 54 Furnaces

23 54 16 - Fuel-Fired Furnaces

23 54 16.14 Condensing Furnaces

		Crew	Daily Output	Labor-Hours	Unit	Material	2021 Bare Costs Labor	Equipment	Total	Total Incl O&P
0010	**CONDENSING FURNACES**, High efficiency									
0020	Oil fired, packaged, complete									
0030	Upflow									
0040	Output @ 95% A.F.U.E.									
0100	49 MBH @ 1,000 CFM	Q-9	3.70	4.324	Ea.	7,875	255		8,130	9,025
0110	73.5 MBH @ 2,000 CFM	"	3.60	4.444	"	8,225	262		8,487	9,450
0120	96 MBH @ 2,000 CFM	Q-9	3.40	4.706	Ea.	8,225	277		8,502	9,475

363

23 54 Furnaces

23 54 16 – Fuel-Fired Furnaces

23 54 16.14 Condensing Furnaces	Crew	Daily Output	Labor-Hours	Unit	Material	2021 Bare Costs Labor	Equipment	Total	Total Incl O&P
0130	115.6 MBH @ 2,000 CFM	Q-9	3.40	4.706	Ea.	8,225	277		8,502	9,475
0140	147 MBH @ 2,000 CFM		3.30	4.848		18,800	286		19,086	21,100
0150	192 MBH @ 4,000 CFM		2.60	6.154		19,500	360		19,860	21,900
0170	231.5 MBH @ 4,000 CFM		2.30	6.957		19,500	410		19,910	22,000
0260	For variable speed motor, add					765			765	840
0270	Note: Also available in horizontal, counterflow and lowboy configurations.									

23 55 Fuel-Fired Heaters

23 55 23 – Gas-Fired Radiant Heaters

23 55 23.10 Infrared Type Heating Units

		Crew	Daily Output	Labor-Hours	Unit	Material	2021 Bare Costs Labor	Equipment	Total	Total Incl O&P
0010	**INFRARED TYPE HEATING UNITS**									
0020	Gas fired, unvented, electric ignition, 100% shutoff.									
0030	Piping and wiring not included									
0100	Input, 30 MBH	Q-5	6	2.667	Ea.	515	164		679	810
0120	45 MBH		5	3.200		1,225	197		1,422	1,650
0140	50 MBH		4.50	3.556		1,250	219		1,469	1,700
0160	60 MBH		4	4		1,225	246		1,471	1,725
0180	75 MBH		3	5.333		1,250	330		1,580	1,875
0200	90 MBH		2.50	6.400		1,450	395		1,845	2,200
0220	105 MBH		2	8		1,625	490		2,115	2,525
0240	120 MBH		2	8		1,625	490		2,115	2,525
1000	Gas fired, vented, electric ignition, tubular									
1020	Piping and wiring not included, 20' to 80' lengths									
1030	Single stage, input, 60 MBH	Q-6	4.50	5.333	Ea.	1,575	340		1,915	2,225
1040	80 MBH		3.90	6.154		1,575	395		1,970	2,300
1050	100 MBH		3.40	7.059		1,575	450		2,025	2,400
1060	125 MBH		2.90	8.276		1,525	530		2,055	2,475
1070	150 MBH		2.70	8.889		1,525	565		2,090	2,525
1080	170 MBH		2.50	9.600		1,525	610		2,135	2,600
1090	200 MBH		2.20	10.909		1,750	695		2,445	2,975
1100	Note: Final pricing may vary due to									
1110	tube length and configuration package selected									
1130	Two stage, input, 60 MBH high, 45 MBH low	Q-6	4.50	5.333	Ea.	1,925	340		2,265	2,600
1140	80 MBH high, 60 MBH low		3.90	6.154		1,925	395		2,320	2,675
1150	100 MBH high, 65 MBH low		3.40	7.059		1,925	450		2,375	2,775
1160	125 MBH high, 95 MBH low		2.90	8.276		1,925	530		2,455	2,925
1170	150 MBH high, 100 MBH low		2.70	8.889		1,925	565		2,490	2,975
1180	170 MBH high, 125 MBH low		2.50	9.600		2,150	610		2,760	3,300
1190	200 MBH high, 150 MBH low		2.20	10.909		2,150	695		2,845	3,425
1220	Note: Final pricing may vary due to									
1230	tube length and configuration package selected									

23 56 Solar Energy Heating Equipment

23 56 16 – Packaged Solar Heating Equipment

23 56 16.40 Solar Heating Systems

		Crew	Daily Output	Labor-Hours	Unit	Material	2021 Bare Costs Labor	Equipment	Total	Total Incl O&P
0010	**SOLAR HEATING SYSTEMS**	D2020-265								
0020	System/package prices, not including connecting									
0030	pipe, insulation, or special heating/plumbing fixtures	D2020-270								
0152	For solar ultraviolet pipe insulation see Section 22 07 19.10									
0500	Hot water, standard package, low temperature	D2020-275								
0540	1 collector, circulator, fittings, 65 gal. tank	G	Q-1	.50	32	Ea.	5,500	1,950		7,450	8,925
0580	2 collectors, circulator, fittings, 120 gal. tank	D2020-280 G		.40	40		5,475	2,425		7,900	9,650
0620	3 collectors, circulator, fittings, 120 gal. tank	G		.34	47.059		8,800	2,875		11,675	14,000
0700	Medium temperature package	D2020-285								
0720	1 collector, circulator, fittings, 80 gal. tank	G	Q-1	.50	32	Ea.	5,150	1,950		7,100	8,575
0740	2 collectors, circulator, fittings, 120 gal. tank	D2020-290 G		.40	40		8,200	2,425		10,625	12,700
0780	3 collectors, circulator, fittings, 120 gal. tank	G		.30	53.333		8,925	3,250		12,175	14,700
0980	For each additional 120 gal. tank, add	D2020-295 G					1,800			1,800	1,975

23 56 19 – Solar Heating Components

23 56 19.50 Solar Heating Ancillary

		Crew	Daily Output	Labor-Hours	Unit	Material	2021 Bare Costs Labor	Equipment	Total	Total Incl O&P
0010	**SOLAR HEATING ANCILLARY**									
2300	Circulators, air	D3010-650								
2310	Blowers									
2330	100-300 S.F. system, 1/10 HP	D3010-660 G	Q-9	16	1	Ea.	271	59		330	385
2340	300-500 S.F. system, 1/5 HP	G		15	1.067		360	63		423	490
2350	Two speed, 100-300 S.F., 1/10 HP	D3010-675 G		14	1.143		148	67.50		215.50	265
2400	Reversible fan, 20" diameter, 2 speed	G		18	.889		117	52.50		169.50	207
2550	Booster fan 6" diameter, 120 CFM	G		16	1		38	59		97	131
2570	6" diameter, 225 CFM	G		16	1		46.50	59		105.50	141
2580	8" diameter, 150 CFM	G		16	1		42.50	59		101.50	136
2590	8" diameter, 310 CFM	G		14	1.143		65.50	67.50		133	174
2600	8" diameter, 425 CFM	G		14	1.143		74	67.50		141.50	183
2650	Rheostat	G		32	.500		16	29.50		45.50	62
2660	Shutter/damper	G		12	1.333		59	78.50		137.50	183
2670	Shutter motor	G		16	1		147	59		206	251
2800	Circulators, liquid, 1/25 HP, 5.3 GPM	G	Q-1	14	1.143		208	69.50		277.50	335
2820	1/20 HP, 17 GPM	G		12	1.333		310	81		391	460
2850	1/20 HP, 17 GPM, stainless steel	G		12	1.333		263	81		344	410
2870	1/12 HP, 30 GPM	G		10	1.600		360	97.50		457.50	540
3000	Collector panels, air with aluminum absorber plate									
3010	Wall or roof mount									
3040	Flat black, plastic glazing									
3080	4' x 8'	G	Q-9	6	2.667	Ea.	685	157		842	990
3100	4' x 10'	G		5	3.200	"	845	188		1,033	1,225
3200	Flush roof mount, 10' to 16' x 22" wide	G		96	.167	L.F.	182	9.80		191.80	215
3210	Manifold, by L.F. width of collectors	G		160	.100	"	151	5.90		156.90	175
3300	Collector panels, liquid with copper absorber plate									
3320	Black chrome, tempered glass glazing									
3330	Alum. frame, 4' x 8', 5/32" single glazing	G	Q-1	9.50	1.684	Ea.	1,050	103		1,153	1,300
3390	Alum. frame, 4' x 10', 5/32" single glazing	G		6	2.667		1,200	162		1,362	1,575
3450	Flat black, alum. frame, 3.5' x 7.5'	G		9	1.778		765	108		873	1,000
3500	4' x 8'	G		5.50	2.909		930	177		1,107	1,300
3520	4' x 10'	G		10	1.600		1,150	97.50		1,247.50	1,425
3540	4' x 12.5'	G		5	3.200		1,250	195		1,445	1,675
3550	Liquid with fin tube absorber plate									
3560	Alum. frame 4' x 8' tempered glass	G	Q-1	10	1.600	Ea.	580	97.50		677.50	785
3580	Liquid with vacuum tubes, 4' x 6'-10"	G		9	1.778		950	108		1,058	1,200

For customer support on your Plumbing Costs with RSMeans Data, call 800.448.8182.

365

23 56 19.50 Solar Heating Ancillary		Crew	Daily Output	Labor-Hours	Unit	Material	2021 Bare Costs Labor	Equipment	Total	Total Incl O&P
3600	Liquid, full wetted, plastic, alum. frame, 4' x 10'	G	Q-1	5	3.200	Ea.	330	195		525	650
3650	Collector panel mounting, flat roof or ground rack	G		7	2.286		267	139		406	500
3670	Roof clamps	G		70	.229	Set	3.57	13.95		17.52	25
3700	Roof strap, teflon	G	1 Plum	205	.039	L.F.	25	2.64		27.64	31.50
3900	Differential controller with two sensors										
3930	Thermostat, hard wired	G	1 Plum	8	1	Ea.	115	67.50		182.50	228
3950	Line cord and receptacle	G		12	.667		158	45		203	242
4050	Pool valve system	G		2.50	3.200		149	217		366	490
4070	With 12 VAC actuator	G		2	4		330	271		601	770
4080	Pool pump system, 2" pipe size	G		6	1.333		200	90.50		290.50	355
4100	Five station with digital read-out	G		3	2.667		278	181		459	575
4150	Sensors										
4200	Brass plug, 1/2" MPT	G	1 Plum	32	.250	Ea.	15.40	16.95		32.35	42.50
4210	Brass plug, reversed	G		32	.250		30	16.95		46.95	58.50
4220	Freeze prevention	G		32	.250		26	16.95		42.95	54
4240	Screw attached	G		32	.250		15.20	16.95		32.15	42
4250	Brass, immersion	G		32	.250		15.20	16.95		32.15	42
4300	Heat exchanger										
4315	includes coil, blower, circulator										
4316	and controller for DHW and space hot air										
4330	Fluid to air coil, up flow, 45 MBH	G	Q-1	4	4	Ea.	335	244		579	735
4380	70 MBH	G		3.50	4.571		380	279		659	830
4400	80 MBH	G		3	5.333		500	325		825	1,025
4580	Fluid to fluid package includes two circulating pumps										
4590	expansion tank, check valve, relief valve										
4600	controller, high temperature cutoff and sensors	G	Q-1	2.50	6.400	Ea.	785	390		1,175	1,450
4650	Heat transfer fluid										
4700	Propylene glycol, inhibited anti-freeze	G	1 Plum	28	.286	Gal.	23	19.35		42.35	54
4800	Solar storage tanks, knocked down										
4810	Air, galvanized steel clad, double wall, 4" fiberglass insulation										
5120	45 mil reinforced polypropylene lining,										
5140	4' high, 4' x 4' = 64 C.F./450 gallons	G	Q-9	2	8	Ea.	2,975	470		3,445	3,975
5150	4' x 8' = 128 C.F./900 gallons	G		1.50	10.667		4,750	630		5,380	6,200
5160	4' x 12' = 190 C.F./1,300 gallons	G		1.30	12.308		5,975	725		6,700	7,675
5170	8' x 8' = 250 C.F./1,700 gallons	G		1	16		7,575	940		8,515	9,750
5190	6'-3" high, 7' x 7' = 306 C.F./2,000 gallons	G	Q-10	1.20	20		14,000	1,225		15,225	17,300
5200	7' x 10'-6" = 459 C.F./3,000 gallons	G		.80	30		17,400	1,825		19,225	21,900
5210	7' x 14' = 613 C.F./4,000 gallons	G		.60	40		22,100	2,450		24,550	28,000
5220	10'-6" x 10'-6" = 689 C.F./4,500 gallons	G		.50	48		20,800	2,925		23,725	27,300
5230	10'-6" x 14' = 919 C.F./6,000 gallons	G		.40	60		24,300	3,675		27,975	32,200
5240	14' x 14' = 1,225 C.F./8,000 gallons	G	Q-11	.40	80		29,100	4,975		34,075	39,500
5250	14' x 17'-6" = 1,531 C.F./10,000 gallons	G		.30	107		31,500	6,650		38,150	44,600
5260	17'-6" x 17'-6" = 1,914 C.F./12,500 gallons	G		.25	128		43,200	7,975		51,175	59,500
5270	17'-6" x 21' = 2,297 C.F./15,000 gallons	G		.20	160		40,500	9,975		50,475	59,500
5280	21' x 21' = 2,756 C.F./18,000 gallons	G		.18	178		42,800	11,100		53,900	64,000
5290	30 mil reinforced Hypalon lining, add						.02%				
7000	Solar control valves and vents										
7050	Air purger, 1" pipe size	G	1 Plum	12	.667	Ea.	52.50	45		97.50	126
7070	Air eliminator, automatic 3/4" size	G		32	.250		31	16.95		47.95	59.50
7090	Air vent, automatic, 1/8" fitting	G		32	.250		15.95	16.95		32.90	43
7100	Manual, 1/8" NPT	G		32	.250		3.31	16.95		20.26	29
7120	Backflow preventer, 1/2" pipe size	G		16	.500		89	34		123	149
7130	3/4" pipe size	G		16	.500		74	34		108	132

23 56 Solar Energy Heating Equipment

23 56 19 – Solar Heating Components

23 56 19.50 Solar Heating Ancillary

		Crew	Daily Output	Labor-Hours	Unit	Material	2021 Bare Costs Labor	Equipment	Total	Total Incl O&P
7150	Balancing valve, 3/4" pipe size	G 1 Plum	20	.400	Ea.	56	27		83	102
7180	Draindown valve, 1/2" copper tube	G	9	.889		220	60		280	330
7200	Flow control valve, 1/2" pipe size	G	22	.364		143	24.50		167.50	194
7220	Expansion tank, up to 5 gal.	G	32	.250		65.50	16.95		82.45	97.50
7250	Hydronic controller (aquastat)	G	8	1		217	67.50		284.50	340
7400	Pressure gauge, 2" dial	G	32	.250		25.50	16.95		42.45	53.50
7450	Relief valve, temp. and pressure 3/4" pipe size	G 1 Plum	30	.267	Ea.	26.50	18.05		44.55	56
7500	Solenoid valve, normally closed									
7520	Brass, 3/4" NPT, 24V	G 1 Plum	9	.889	Ea.	161	60		221	267
7530	1" NPT, 24V	G	9	.889		257	60		317	375
7750	Vacuum relief valve, 3/4" pipe size	G	32	.250		31	16.95		47.95	59.50
7800	Thermometers									
7820	Digital temperature monitoring, 4 locations	G 1 Plum	2.50	3.200	Ea.	167	217		384	510
7900	Upright, 1/2" NPT	G	8	1		59.50	67.50		127	166
7970	Remote probe, 2" dial	G	8	1		35	67.50		102.50	140
7990	Stem, 2" dial, 9" stem	G	16	.500		23	34		57	75.50
8250	Water storage tank with heat exchanger and electric element									
8270	66 gal. with 2" x 2 lb. density insulation	G 1 Plum	1.60	5	Ea.	2,075	340		2,415	2,800
8300	80 gal. with 2" x 2 lb. density insulation	G	1.60	5		1,800	340		2,140	2,475
8380	120 gal. with 2" x 2 lb. density insulation	G	1.40	5.714		1,750	385		2,135	2,500
8400	120 gal. with 2" x 2 lb. density insul., 40 S.F. heat coil	G	1.40	5.714		2,600	385		2,985	3,425
8500	Water storage module, plastic									
8600	Tubular, 12" diameter, 4' high	G 1 Carp	48	.167	Ea.	101	9.10		110.10	125
8610	12" diameter, 8' high	G	40	.200		159	10.95		169.95	191
8620	18" diameter, 5' high	G	38	.211		179	11.50		190.50	214
8630	18" diameter, 10' high	G	32	.250		262	13.70		275.70	310
8640	58" diameter, 5' high	G 2 Carp	32	.500		855	27.50		882.50	980
8650	Cap, 12" diameter	G				20			20	22
8660	18" diameter	G				25.50			25.50	28.50

23 57 Heat Exchangers for HVAC

23 57 16 – Steam-to-Water Heat Exchangers

23 57 16.10 Shell/Tube Type Steam-to-Water Heat Exch.

		Crew	Daily Output	Labor-Hours	Unit	Material	2021 Bare Costs Labor	Equipment	Total	Total Incl O&P
0010	**SHELL AND TUBE TYPE STEAM-TO-WATER HEAT EXCHANGERS**									
0016	Shell & tube type, 2 or 4 pass, 3/4" OD copper tubes,									
0020	C.I. heads, C.I. tube sheet, steel shell									
0100	Hot water 40°F to 180°F, by steam at 10 psi									
0120	8 GPM	Q-5	6	2.667	Ea.	2,825	164		2,989	3,350
0140	10 GPM		5	3.200		4,225	197		4,422	4,950
0160	40 GPM		4	4		6,550	246		6,796	7,575
0500	For bronze head and tube sheet, add					50%				

23 57 19 – Liquid-to-Liquid Heat Exchangers

23 57 19.13 Plate-Type, Liquid-to-Liquid Heat Exchangers

		Crew	Daily Output	Labor-Hours	Unit	Material	2021 Bare Costs Labor	Equipment	Total	Total Incl O&P
0010	**PLATE-TYPE, LIQUID-TO-LIQUID HEAT EXCHANGERS**									
3000	Plate type,									
3100	400 GPM	Q-6	.80	30	Ea.	48,300	1,925		50,225	56,000
3120	800 GPM	"	.50	48	"	83,000	3,050		86,050	96,000
3140	1,200 GPM	Q-7	.34	94.118	Ea.	123,500	6,125		129,625	145,000
3160	1,800 GPM	"	.24	133	"	164,000	8,675		172,675	193,500

23 57 Heat Exchangers for HVAC

23 57 19 – Liquid-to-Liquid Heat Exchangers

23 57 19.16 Shell-Type, Liquid-to-Liquid Heat Exchangers	Crew	Daily Output	Labor-Hours	Unit	Material	2021 Bare Costs Labor	Equipment	Total	Total Incl O&P
0010 **SHELL-TYPE, LIQUID-TO-LIQUID HEAT EXCHANGERS**									
1000 Hot water 40°F to 140°F, by water at 200°F									
1020 7 GPM	Q-5	6	2.667	Ea.	12,600	164		12,764	14,000
1040 16 GPM		5	3.200		18,500	197		18,697	20,600
1060 34 GPM		4	4		27,300	246		27,546	30,500
1080 55 GPM		3	5.333		38,800	330		39,130	43,200
1100 74 GPM		1.50	10.667		50,000	655		50,655	56,000
1120 86 GPM		1.40	11.429		65,500	705		66,205	73,000
1140 112 GPM	Q-6	2	12		81,500	765		82,265	91,000
1160 126 GPM		1.80	13.333		102,000	850		102,850	113,500
1180 152 GPM		1	24		80,500	1,525		82,025	91,000

23 72 Air-to-Air Energy Recovery Equipment

23 72 16 – Heat-Pipe Air-To-Air Energy-Recovery Equipment

23 72 16.10 Heat Pipes

	Crew	Daily Output	Labor-Hours	Unit	Material	2021 Bare Costs Labor	Equipment	Total	Total Incl O&P
0010 **HEAT PIPES**									
8000 Heat pipe type, glycol, 50% efficient									
8010 100 MBH, 1,700 CFM	1 Stpi	.80	10	Ea.	4,900	685		5,585	6,400
8020 160 MBH, 2,700 CFM		.60	13.333		6,575	910		7,485	8,575
8030 620 MBH, 4,000 CFM		.40	20		9,925	1,375		11,300	13,000

23 81 Decentralized Unitary HVAC Equipment

23 81 13 – Packaged Terminal Air-Conditioners

23 81 13.10 Packaged Cabinet Type Air-Conditioners

	Crew	Daily Output	Labor-Hours	Unit	Material	2021 Bare Costs Labor	Equipment	Total	Total Incl O&P
0010 **PACKAGED CABINET TYPE AIR-CONDITIONERS**, Cabinet, wall sleeve,									
0100 louver, electric heat, thermostat, manual changeover, 208 V									
0200 6,000 BTUH cooling, 8,800 BTU heat	Q-5	6	2.667	Ea.	930	164		1,094	1,275
0220 9,000 BTUH cooling, 13,900 BTU heat		5	3.200		980	197		1,177	1,375
0240 12,000 BTUH cooling, 13,900 BTU heat		4	4		1,225	246		1,471	1,725
0260 15,000 BTUH cooling, 13,900 BTU heat		3	5.333		1,550	330		1,880	2,200

23 81 19 – Self-Contained Air-Conditioners

23 81 19.10 Window Unit Air Conditioners

	Crew	Daily Output	Labor-Hours	Unit	Material	2021 Bare Costs Labor	Equipment	Total	Total Incl O&P
0010 **WINDOW UNIT AIR CONDITIONERS**									
4000 Portable/window, 15 amp, 125 V grounded receptacle required									
4060 5,000 BTUH	1 Carp	8	1	Ea.	236	54.50		290.50	340
4340 6,000 BTUH		8	1		241	54.50		295.50	345
4480 8,000 BTUH		6	1.333		305	73		378	450
4500 10,000 BTUH		6	1.333		365	73		438	515
4520 12,000 BTUH	L-2	8	2		1,875	97		1,972	2,200
4600 Window/thru-the-wall, 15 amp, 230 V grounded receptacle required									
4780 18,000 BTUH	L-2	6	2.667	Ea.	600	129		729	855
4940 25,000 BTUH	L-2	4	4	Ea.	910	194		1,104	1,300
4960 29,000 BTUH	"	4	4	"	865	194		1,059	1,250

23 81 Decentralized Unitary HVAC Equipment

23 81 43 – Air-Source Unitary Heat Pumps

23 81 43.10 Air-Source Heat Pumps

		Crew	Daily Output	Labor-Hours	Unit	Material	2021 Bare Costs Labor	Equipment	Total	Total Incl O&P
0010	**AIR-SOURCE HEAT PUMPS**, Not including interconnecting tubing									
1000	Air to air, split system, not including curbs, pads, fan coil and ductwork									
1010	For curbs/pads see Section 23 91 10									
1012	Outside condensing unit only, for fan coil see Section 23 82 19.10									
1015	1.5 ton cooling, 7 MBH heat @ 0°F	Q-5	2.40	6.667	Ea.	1,475	410		1,885	2,225
1020	2 ton cooling, 8.5 MBH heat @ 0°F		2	8		1,600	490		2,090	2,475
1030	2.5 ton cooling, 10 MBH heat @ 0°F		1.60	10		1,775	615		2,390	2,875
1040	3 ton cooling, 13 MBH heat @ 0°F		1.20	13.333		1,950	820		2,770	3,375
1050	3.5 ton cooling, 18 MBH heat @ 0°F		1	16		2,075	985		3,060	3,775
1054	4 ton cooling, 24 MBH heat @ 0°F		.80	20		2,250	1,225		3,475	4,300
1060	5 ton cooling, 27 MBH heat @ 0°F		.50	32		2,450	1,975		4,425	5,650
1500	Single package, not including curbs, pads, or plenums									
1502	0.5 ton cooling, supplementary heat included	Q-5	8	2	Ea.	2,700	123		2,823	3,150
1504	0.75 ton cooling, supplementary heat included		6	2.667		2,850	164		3,014	3,375
1506	1 ton cooling, supplementary heat included		4	4		3,750	246		3,996	4,500
1510	1.5 ton cooling, 5 MBH heat @ 0°F		1.55	10.323		3,300	635		3,935	4,575
1520	2 ton cooling, 6.5 MBH heat @ 0°F		1.50	10.667		3,325	655		3,980	4,650
1540	2.5 ton cooling, 8 MBH heat @ 0°F		1.40	11.429		3,225	705		3,930	4,600
1560	3 ton cooling, 10 MBH heat @ 0°F		1.20	13.333		3,725	820		4,545	5,325
1570	3.5 ton cooling, 11 MBH heat @ 0°F		1	16		4,125	985		5,110	6,025
1580	4 ton cooling, 13 MBH heat @ 0°F		.96	16.667		4,425	1,025		5,450	6,400
1620	5 ton cooling, 27 MBH heat @ 0°F		.65	24.615		5,025	1,525		6,550	7,775
1640	7.5 ton cooling, 35 MBH heat @ 0°F		.40	40		7,725	2,450		10,175	12,200
6000	Air to water, single package, excluding storage tank and ductwork									
6010	Includes circulating water pump, air duct connections, digital temperature									
6020	controller with remote tank temp. probe and sensor for storage tank.									
6040	Water heating - air cooling capacity									
6110	35.5 MBH heat water, 2.3 ton cool air	Q-5	1.60	10	Ea.	17,100	615		17,715	19,700
6120	58 MBH heat water, 3.8 ton cool air		1.10	14.545		19,500	895		20,395	22,800
6130	76 MBH heat water, 4.9 ton cool air		.87	18.391		23,300	1,125		24,425	27,300
6140	98 MBH heat water, 6.5 ton cool air		.62	25.806		30,100	1,600		31,700	35,500
6150	113 MBH heat water, 7.4 ton cool air		.59	27.119		33,000	1,675		34,675	38,800
6160	142 MBH heat water, 9.2 ton cool air		.52	30.769		40,000	1,900		41,900	46,800
6170	171 MBH heat water, 11.1 ton cool air	Q-5	.49	32.653	Ea.	47,100	2,000		49,100	55,000

23 81 46 – Water-Source Unitary Heat Pumps

23 81 46.10 Water Source Heat Pumps

		Crew	Daily Output	Labor-Hours	Unit	Material	2021 Bare Costs Labor	Equipment	Total	Total Incl O&P
0010	**WATER SOURCE HEAT PUMPS**, Not incl. connecting tubing or water source									
2000	Water source to air, single package									
2100	1 ton cooling, 13 MBH heat @ 75°F	Q-5	2	8	Ea.	2,075	490		2,565	3,025
2120	1.5 ton cooling, 17 MBH heat @ 75°F		1.80	8.889		1,925	545		2,470	2,950
2140	2 ton cooling, 19 MBH heat @ 75°F		1.70	9.412		2,300	580		2,880	3,400
2160	2.5 ton cooling, 25 MBH heat @ 75°F		1.60	10		2,500	615		3,115	3,675
2180	3 ton cooling, 27 MBH heat @ 75°F		1.40	11.429		2,825	705		3,530	4,150
2190	3.5 ton cooling, 29 MBH heat @ 75°F		1.30	12.308		2,775	755		3,530	4,200
2200	4 ton cooling, 31 MBH heat @ 75°F		1.20	13.333		3,200	820		4,020	4,750
2220	5 ton cooling, 29 MBH heat @ 75°F		.90	17.778		3,350	1,100		4,450	5,325
3960	For supplementary heat coil, add					10%				

For customer support on your Plumbing Costs with RSMeans Data, call 800.448.8182.

369

23 82 Convection Heating and Cooling Units

23 82 19 – Fan Coil Units

23 82 19.10 Fan Coil Air Conditioning

		Crew	Daily Output	Labor-Hours	Unit	Material	2021 Bare Costs Labor	Equipment	Total	Total Incl O&P
0010	**FAN COIL AIR CONDITIONING**									
0030	Fan coil AC, cabinet mounted, filters and controls									
0320	1 ton cooling	Q-5	6	2.667	Ea.	1,825	164		1,989	2,250
0940	Direct expansion, for use w/air cooled condensing unit, 1-1/2 ton cooling		5	3.200		700	197		897	1,075
0950	2 ton cooling		4.80	3.333		685	205		890	1,050
0960	2-1/2 ton cooling		4.40	3.636		800	224		1,024	1,225
0970	3 ton cooling		3.80	4.211		980	259		1,239	1,450
0980	3-1/2 ton cooling		3.60	4.444		1,000	273		1,273	1,500
0990	4 ton cooling		3.40	4.706		970	290		1,260	1,500
1000	5 ton cooling		3	5.333		1,250	330		1,580	1,900
1500	For hot water coil, add					40%	10%			

23 82 29 – Radiators

23 82 29.10 Hydronic Heating

		Crew	Daily Output	Labor-Hours	Unit	Material	2021 Bare Costs Labor	Equipment	Total	Total Incl O&P
0010	**HYDRONIC HEATING**, Terminal units, not incl. main supply pipe									
1000	Radiation									
1100	Panel, baseboard, C.I., including supports, no covers	Q-5	46	.348	L.F.	53.50	21.50		75	91
3000	Radiators, cast iron									
3100	Free standing or wall hung, 6 tube, 25" high	Q-5	96	.167	Section	70.50	10.25		80.75	93
3150	4 tube, 25" high		96	.167		52.50	10.25		62.75	73.50
3200	4 tube, 19" high		96	.167		49	10.25		59.25	69.50
3250	Adj. brackets, 2 per wall radiator up to 30 sections	1 Stpi	32	.250	Ea.	51	17.10		68.10	82
3500	Recessed, 20" high x 5" deep, without grille	Q-5	60	.267	Section	58	16.40		74.40	88.50
3525	Free standing or wall hung, 30" high	"	60	.267		46.50	16.40		62.90	76
3600	For inlet grille, add					6.50			6.50	7.15
9500	To convert SFR to BTU rating: Hot water, 150 x SFR									
9510	Forced hot water, 180 x SFR; steam, 240 x SFR									

23 82 33 – Convectors

23 82 33.10 Convector Units

		Crew	Daily Output	Labor-Hours	Unit	Material	2021 Bare Costs Labor	Equipment	Total	Total Incl O&P
0010	**CONVECTOR UNITS**, Terminal units, not incl. main supply pipe									
2204	Convector, multifin, 2 pipe w/cabinet									
2210	17" H x 24" L	Q-5	10	1.600	Ea.	111	98.50		209.50	269
2214	17" H x 36" L		8.60	1.860		166	114		280	355
2218	17" H x 48" L		7.40	2.162		221	133		354	440
2222	21" H x 24" L		9	1.778		124	109		233	299
2228	21" H x 48" L		6.80	2.353		247	145		392	490
2240	For knob operated damper, add					140%				
2241	For metal trim strips, add	Q-5	64	.250	Ea.	6.55	15.40		21.95	30
2243	For snap-on inlet grille, add					10%	10%			
2245	For hinged access door, add	Q-5	64	.250	Ea.	39.50	15.40		54.90	66.50
2246	For air chamber, auto-venting, add	"	58	.276	"	8.60	16.95		25.55	35

23 82 36 – Finned-Tube Radiation Heaters

23 82 36.10 Finned Tube Radiation

		Crew	Daily Output	Labor-Hours	Unit	Material	2021 Bare Costs Labor	Equipment	Total	Total Incl O&P
0010	**FINNED TUBE RADIATION**, Terminal units, not incl. main supply pipe									
1310	Baseboard, pkgd, 1/2" copper tube, alum. fin, 7" high	Q-5	60	.267	L.F.	11.75	16.40		28.15	37.50
1320	3/4" copper tube, alum. fin, 7" high	"	58	.276	"	7.85	16.95		24.80	34
1381	Rough in baseboard panel & fin tube, supply & balance valves	1 Stpi	1.06	7.547	Ea.	296	515		811	1,100
1500	Note: fin tube may also require corners, caps, etc.									

For customer support on your Plumbing Costs with RSMeans Data, call 800.448.8182.

23 83 16 – Radiant-Heating Hydronic Piping

23 83 16.10 Radiant Floor Heating	Crew	Daily Output	Labor-Hours	Unit	Material	2021 Bare Costs Labor	Equipment	Total	Total Incl O&P
0010 **RADIANT FLOOR HEATING**									
0100 Tubing, PEX (cross-linked polyethylene)									
0110 Oxygen barrier type for systems with ferrous materials									
0120 1/2"	Q-5	800	.020	L.F.	1.07	1.23		2.30	3.02
0130 3/4"		535	.030		1.50	1.84		3.34	4.40
0140 1"		400	.040		2.35	2.46		4.81	6.25
0200 Non barrier type for ferrous free systems									
0210 1/2"	Q-5	800	.020	L.F.	.59	1.23		1.82	2.49
0220 3/4"		535	.030		1.07	1.84		2.91	3.93
0230 1"		400	.040		1.84	2.46		4.30	5.70
1000 Manifolds									
1110 Brass									
1120 With supply and return valves, flow meter, thermometer,									
1122 auto air vent and drain/fill valve.									
1130 1", 2 circuit	Q-5	14	1.143	Ea.	355	70.50		425.50	495
1140 1", 3 circuit	"	13.50	1.185	"	405	73		478	555
1150 1", 4 circuit	Q-5	13	1.231	Ea.	440	75.50		515.50	595
1154 1", 5 circuit		12.50	1.280		520	79		599	695
1158 1", 6 circuit		12	1.333		570	82		652	745
1162 1", 7 circuit		11.50	1.391		620	85.50		705.50	815
1166 1", 8 circuit		11	1.455		690	89.50		779.50	890
1172 1", 9 circuit		10.50	1.524		740	94		834	955
1174 1", 10 circuit		10	1.600		795	98.50		893.50	1,025
1178 1", 11 circuit		9.50	1.684		830	104		934	1,075
1182 1", 12 circuit		9	1.778		915	109		1,024	1,175
1610 Copper manifold header (cut to size)									
1620 1" header, 12 circuit 1/2" sweat outlets	Q-5	3.33	4.805	Ea.	124	296		420	575
1630 1-1/4" header, 12 circuit 1/2" sweat outlets		3.20	5		143	310		453	620
1640 1-1/4" header, 12 circuit 3/4" sweat outlets		3	5.333		155	330		485	660
1650 1-1/2" header, 12 circuit 3/4" sweat outlets		3.10	5.161		186	320		506	680
1660 2" header, 12 circuit 3/4" sweat outlets		2.90	5.517		273	340		613	805
3000 Valves									
3110 Thermostatic zone valve actuator with end switch	Q-5	40	.400	Ea.	53	24.50		77.50	94.50
3114 Thermostatic zone valve actuator	"	36	.444	"	100	27.50		127.50	151
3120 Motorized straight zone valve with operator complete									
3130 3/4"	Q-5	35	.457	Ea.	164	28		192	222
3140 1"		32	.500		177	31		208	241
3150 1-1/4"		29.60	.541		226	33.50		259.50	298
3500 4 way mixing valve, manual, brass									
3530 1"	Q-5	13.30	1.203	Ea.	234	74		308	365
3540 1-1/4"		11.40	1.404		254	86.50		340.50	410
3550 1-1/2"		11	1.455		325	89.50		414.50	490
3560 2"		10.60	1.509		460	93		553	645
3800 Mixing valve motor, 4 way for valves, 1" and 1-1/4"		34	.471		385	29		414	470
3810 Mixing valve motor, 4 way for valves, 1-1/2" and 2"		30	.533		445	33		478	540
5000 Radiant floor heating, zone control panel									
5120 4 zone actuator valve control, expandable	Q-5	20	.800	Ea.	150	49		199	239
5130 6 zone actuator valve control, expandable		18	.889		279	54.50		333.50	385
6070 Thermal track, straight panel for long continuous runs, 5.333 S.F.		40	.400		36.50	24.50		61	77
6080 Thermal track, utility panel, for direction reverse at run end, 5.333 S.F.		40	.400		36.50	24.50		61	77
6090 Combination panel, for direction reverse plus straight run, 5.333 S.F.		40	.400		36.50	24.50		61	77
7000 PEX tubing fittings									
7100 Compression type									

23 83 Radiant Heating Units

23 83 16 – Radiant-Heating Hydronic Piping

23 83 16.10 Radiant Floor Heating		Crew	Daily Output	Labor-Hours	Unit	Material	2021 Bare Costs Labor	Equipment	Total	Total Incl O&P
7116	Coupling									
7120	1/2" x 1/2"	1 Stpi	27	.296	Ea.	7	20.50		27.50	37.50
7124	3/4" x 3/4"	1 Stpi	23	.348	Ea.	16.85	24		40.85	54
7130	Adapter									
7132	1/2" x female sweat 1/2"	1 Stpi	27	.296	Ea.	4.54	20.50		25.04	35
7134	1/2" x female sweat 3/4"		26	.308		5.10	21		26.10	37
7136	5/8" x female sweat 3/4"	↓	24	.333	↓	7.30	23		30.30	42
7140	Elbow									
7142	1/2" x female sweat 1/2"	1 Stpi	27	.296	Ea.	6.95	20.50		27.45	37.50
7144	1/2" x female sweat 3/4"		26	.308		8.10	21		29.10	40.50
7146	5/8" x female sweat 3/4"	↓	24	.333	↓	9.10	23		32.10	44
7200	Insert type									
7206	PEX x male NPT									
7210	1/2" x 1/2"	1 Stpi	29	.276	Ea.	3.12	18.85		21.97	31.50
7220	3/4" x 3/4"		27	.296		4.60	20.50		25.10	35
7230	1" x 1"	↓	26	.308		7.75	21		28.75	40
7300	PEX coupling									
7310	1/2" x 1/2"	1 Stpi	30	.267	Ea.	1.18	18.25		19.43	28.50
7320	3/4" x 3/4"		29	.276		1.78	18.85		20.63	30
7330	1" x 1"	↓	28	.286	↓	2.76	19.55		22.31	32
7400	PEX stainless crimp ring									
7410	1/2" x 1/2"	1 Stpi	86	.093	Ea.	.58	6.35		6.93	10.15
7420	3/4" x 3/4"		84	.095		.79	6.50		7.29	10.55
7430	1" x 1"	↓	82	.098	↓	1.13	6.65		7.78	11.20

23 83 33 – Electric Radiant Heaters

23 83 33.10 Electric Heating		Crew	Daily Output	Labor-Hours	Unit	Material	2021 Bare Costs Labor	Equipment	Total	Total Incl O&P
0010	**ELECTRIC HEATING**, not incl. conduit or feed wiring									
1100	Rule of thumb: Baseboard units, including control	1 Elec	4.40	1.818	kW	119	116		235	300
1300	Baseboard heaters, 2' long, 350 watt		8	1	Ea.	31.50	63.50		95	129
1400	3' long, 750 watt		8	1		36.50	63.50		100	135
1600	4' long, 1,000 watt		6.70	1.194		38.50	76		114.50	156
1800	5' long, 935 watt		5.70	1.404		52.50	89.50		142	191
2000	6' long, 1,500 watt		5	1.600		53	102		155	210
2200	7' long, 1,310 watt		4.40	1.818		63.50	116		179.50	242
2400	8' long, 2,000 watt		4	2		73	127		200	270
2600	9' long, 1,680 watt		3.60	2.222		85.50	142		227.50	305
2800	10' long, 1,875 watt	↓	3.30	2.424	↓	97.50	154		251.50	335
2950	Wall heaters with fan, 120 to 277 volt									
3160	Recessed, residential, 750 watt	1 Elec	6	1.333	Ea.	83	85		168	218
3170	1,000 watt		6	1.333		335	85		420	490
3180	1,250 watt		5	1.600		113	102		215	276
3190	1,500 watt		4	2		360	127		487	585
3600	Thermostats, integral	↓	16	.500	↓	29.50	32		61.50	79.50
3800	Line voltage, 1 pole	1 Elec	8	1	Ea.	16.40	63.50		79.90	113
3810	2 pole	"	8	1	"	23	63.50		86.50	120
4000	Heat trace system, 400 degree									
4020	115 V, 2.5 watts/L.F.	1 Elec	530	.015	L.F.	10.05	.96		11.01	12.50
4030	5 watts/L.F.		530	.015		10.05	.96		11.01	12.50
4050	10 watts/L.F.		530	.015		8.40	.96		9.36	10.70
4060	208 V, 5 watts/L.F.		530	.015		10.05	.96		11.01	12.50
4080	480 V, 8 watts/L.F.	↓	530	.015	↓	10.05	.96		11.01	12.50
4200	Heater raceway									

23 83 Radiant Heating Units

23 83 33 – Electric Radiant Heaters

23 83 33.10 Electric Heating	Crew	Daily Output	Labor-Hours	Unit	Material	2021 Bare Costs Labor	Equipment	Total	Total Incl O&P
4260 Heat transfer cement									
4280 1 gallon				Ea.	80.50			80.50	89
4300 5 gallon				"	244			244	268
4320 Cable tie									
4340 3/4" pipe size	1 Elec	470	.017	Ea.	.02	1.08		1.10	1.63
4360 1" pipe size		444	.018		.02	1.15		1.17	1.73
4380 1-1/4" pipe size		400	.020		.02	1.27		1.29	1.91
4400 1-1/2" pipe size		355	.023		.02	1.44		1.46	2.15
4420 2" pipe size		320	.025		.02	1.59		1.61	2.39
4440 3" pipe size		160	.050		.02	3.19		3.21	4.75
4460 4" pipe size		100	.080		.06	5.10		5.16	7.60
4480 Thermostat NEMA 3R, 22 amp, 0-150 degree, 10' cap.		8	1		190	63.50		253.50	305
4500 Thermostat NEMA 4X, 25 amp, 40 degree, 5-1/2' cap.		7	1.143		200	73		273	330
4520 Thermostat NEMA 4X, 22 amp, 25-325 degree, 10' cap.		7	1.143		525	73		598	685
4540 Thermostat NEMA 4X, 22 amp, 15-140 degree		6	1.333		440	85		525	610
4580 Thermostat NEMA 4, 7, 9, 22 amp, 25-325 degree, 10' cap.		3.60	2.222		640	142		782	915
4600 Thermostat NEMA 4, 7, 9, 22 amp, 15-140 degree		3	2.667		640	170		810	955
4720 Fiberglass application tape, 36 yard roll		11	.727		55.50	46.50		102	131
5000 Radiant heating ceiling panels, 2' x 4', 500 watt		16	.500		294	32		326	375
5050 750 watt		16	.500		257	32		289	330
5200 For recessed plaster frame, add		32	.250		70.50	15.95		86.45	101
5300 Infrared quartz heaters, 120 volts, 1,000 watt		6.70	1.194		236	76		312	370
5350 1,500 watt		5	1.600		233	102		335	405
5400 240 volts, 1,500 watt		5	1.600		239	102		341	415
5450 2,000 watt		4	2		277	127		404	495
5500 3,000 watt		3	2.667		275	170		445	550
5550 4,000 watt		2.60	3.077		315	196		511	640
5570 Modulating control		.80	10		124	635		759	1,075
5600 Unit heaters, heavy duty, with fan & mounting bracket									
5650 Single phase, 208-240-277 volt, 3 kW	1 Elec	6	1.333	Ea.	445	85		530	610
5750 5 kW	1 Elec	5.50	1.455	Ea.	279	92.50		371.50	445
5800 7 kW		5	1.600		670	102		772	885
5850 10 kW		4	2		1,025	127		1,152	1,350
5950 15 kW		3.80	2.105		1,750	134		1,884	2,125
6000 480 volt, 3 kW		6	1.333		435	85		520	600
6020 4 kW		5.80	1.379		365	88		453	535
6040 5 kW		5.50	1.455		445	92.50		537.50	630
6060 7 kW		5	1.600		580	102		682	790
6080 10 kW		4	2		1,050	127		1,177	1,350
6100 13 kW		3.80	2.105		1,850	134		1,984	2,250
6120 15 kW		3.70	2.162		1,075	138		1,213	1,375
6140 20 kW		3.50	2.286		1,475	146		1,621	1,850
6300 3 phase, 208-240 volt, 5 kW		5.50	1.455		425	92.50		517.50	605
6320 7 kW		5	1.600		705	102		807	925
6340 10 kW		4	2		740	127		867	1,000
6360 15 kW		3.70	2.162		1,125	138		1,263	1,425
6380 20 kW		3.50	2.286		1,475	146		1,621	1,850
6400 25 kW		3.30	2.424		1,700	154		1,854	2,075
6500 480 volt, 5 kW		5.50	1.455		615	92.50		707.50	820
6520 7 kW		5	1.600		795	102		897	1,025
6540 10 kW		4	2		825	127		952	1,100
6560 13 kW		3.80	2.105		1,525	134		1,659	1,900
6580 15 kW		3.70	2.162		1,375	138		1,513	1,700

373

23 83 Radiant Heating Units

23 83 33 – Electric Radiant Heaters

23 83 33.10 Electric Heating	Crew	Daily Output	Labor-Hours	Unit	Material	2021 Bare Costs Labor	Equipment	Total	Total Incl O&P
6600	20 kW	1 Elec	3.50	2.286	Ea.	2,125	146		2,271	2,575
6800	Vertical discharge heaters, with fan									
6820	Single phase, 208-240-277 volt, 10 kW	1 Elec	4	2	Ea.	780	127		907	1,050
6840	15 kW		3.70	2.162		1,050	138		1,188	1,350
6900	3 phase, 208-240 volt, 10 kW		4	2		970	127		1,097	1,275
6920	15 kW		3.70	2.162		1,500	138		1,638	1,850
6940	20 kW		3.50	2.286		2,775	146		2,921	3,275
7100	480 volt, 10 kW		4	2		1,150	127		1,277	1,450
7120	15 kW		3.70	2.162		1,850	138		1,988	2,250
7140	20 kW		3.50	2.286		2,550	146		2,696	3,025
7160	25 kW		3.30	2.424		3,325	154		3,479	3,875
7900	Cabinet convector heaters, 240 volt, three phase,									
7920	3' long, 2,000 watt	1 Elec	5.30	1.509	Ea.	2,300	96		2,396	2,675
7940	3,000 watt		5.30	1.509		2,400	96		2,496	2,800
7960	4,000 watt		5.30	1.509		2,475	96		2,571	2,850
7980	6,000 watt		4.60	1.739		2,675	111		2,786	3,125
8000	8,000 watt		4.60	1.739		2,750	111		2,861	3,200
8020	4' long, 4,000 watt		4.60	1.739		2,500	111		2,611	2,925
8040	6,000 watt		4	2		2,600	127		2,727	3,050
8060	8,000 watt		4	2		2,675	127		2,802	3,125
8080	10,000 watt		4	2		2,700	127		2,827	3,175
8100	Available also in 208 or 277 volt									
8200	Cabinet unit heaters, 120 to 277 volt, 1 pole,									
8220	wall mounted, 2 kW	1 Elec	4.60	1.739	Ea.	2,075	111		2,186	2,475
8230	3 kW		4.60	1.739		2,125	111		2,236	2,525
8240	4 kW		4.40	1.818		2,275	116		2,391	2,700
8250	5 kW		4.40	1.818		2,475	116		2,591	2,900
8260	6 kW		4.20	1.905		2,450	121		2,571	2,875
8270	8 kW		4	2		2,550	127		2,677	3,025
8280	10 kW		3.80	2.105		2,550	134		2,684	3,000
8290	12 kW		3.50	2.286		2,700	146		2,846	3,200
8300	13.5 kW		2.90	2.759		3,925	176		4,101	4,575
8310	16 kW		2.70	2.963		3,350	189		3,539	3,950
8320	20 kW		2.30	3.478		4,025	222		4,247	4,750
8330	24 kW		1.90	4.211		4,325	268		4,593	5,150
8350	Recessed, 2 kW		4.40	1.818		2,375	116		2,491	2,775
8370	3 kW		4.40	1.818		2,475	116		2,591	2,900
8380	4 kW		4.20	1.905		2,400	121		2,521	2,825
8390	5 kW		4.20	1.905		2,550	121		2,671	3,000
8400	6 kW		4	2		1,825	127		1,952	2,225
8410	8 kW		3.80	2.105		2,725	134		2,859	3,200
8420	10 kW		3.50	2.286		3,375	146		3,521	3,925
8430	12 kW		2.90	2.759		3,425	176		3,601	4,000
8440	13.5 kW		2.70	2.963		3,275	189		3,464	3,875
8450	16 kW		2.30	3.478		2,325	222		2,547	2,875
8460	20 kW		1.90	4.211		4,125	268		4,393	4,925
8470	24 kW		1.60	5		4,100	320		4,420	5,000
8490	Ceiling mounted, 2 kW		3.20	2.500		2,175	159		2,334	2,600
8510	3 kW		3.20	2.500		2,275	159		2,434	2,725
8520	4 kW		3	2.667		2,375	170		2,545	2,850
8530	5 kW		3	2.667		2,500	170		2,670	3,000
8540	6 kW		2.80	2.857		2,525	182		2,707	3,050
8550	8 kW		2.40	3.333		2,725	212		2,937	3,325

For customer support on your Plumbing Costs with RSMeans Data, call 800.448.8182.

23 83 Radiant Heating Units

23 83 33 – Electric Radiant Heaters

23 83 33.10 Electric Heating		Crew	Daily Output	Labor-Hours	Unit	Material	2021 Bare Costs Labor	Equipment	Total	Total Incl O&P
8560	10 kW	1 Elec	2.20	3.636	Ea.	3,250	232		3,482	3,925
8570	12 kW		2	4		2,875	255		3,130	3,525
8580	13.5 kW		1.50	5.333		2,850	340		3,190	3,650
8590	16 kW		1.30	6.154		2,975	390		3,365	3,850
8600	20 kW		.90	8.889		4,025	565		4,590	5,275
8610	24 kW	▼	.60	13.333	▼	4,100	850		4,950	5,775
8630	208 to 480 V, 3 pole									
8650	Wall mounted, 2 kW	1 Elec	4.60	1.739	Ea.	2,025	111		2,136	2,425
8670	3 kW		4.60	1.739		2,450	111		2,561	2,875
8680	4 kW		4.40	1.818		2,550	116		2,666	3,000
8690	5 kW		4.40	1.818		2,700	116		2,816	3,150
8700	6 kW		4.20	1.905		2,725	121		2,846	3,175
8710	8 kW		4	2		2,900	127		3,027	3,375
8720	10 kW		3.80	2.105		2,975	134		3,109	3,475
8730	12 kW		3.50	2.286		3,050	146		3,196	3,600
8740	13.5 kW		2.90	2.759		3,100	176		3,276	3,650
8750	16 kW		2.70	2.963		3,250	189		3,439	3,850
8760	20 kW		2.30	3.478		4,825	222		5,047	5,625
8770	24 kW		1.90	4.211		4,500	268		4,768	5,350
8790	Recessed, 2 kW		4.40	1.818		2,250	116		2,366	2,650
8810	3 kW		4.40	1.818		2,375	116		2,491	2,800
8820	4 kW		4.20	1.905		2,525	121		2,646	2,950
8830	5 kW		4.20	1.905		1,800	121		1,921	2,150
8840	6 kW		4	2		2,550	127		2,677	3,000
8850	8 kW		3.80	2.105		2,700	134		2,834	3,175
8860	10 kW		3.50	2.286		1,950	146		2,096	2,375
8870	12 kW		2.90	2.759		2,975	176		3,151	3,525
8880	13.5 kW		2.70	2.963		2,650	189		2,839	3,175
8890	16 kW		2.30	3.478		3,075	222		3,297	3,725
8900	20 kW		1.90	4.211		4,475	268		4,743	5,325
8920	24 kW		1.60	5		4,650	320		4,970	5,600
8940	Ceiling mount, 2 kW		3.20	2.500		2,325	159		2,484	2,775
8950	3 kW		3.20	2.500		2,225	159		2,384	2,650
8960	4 kW		3	2.667		2,600	170		2,770	3,100
8970	5 kW		3	2.667		2,350	170		2,520	2,850
8980	6 kW		2.80	2.857		2,675	182		2,857	3,225
8990	8 kW		2.40	3.333		1,875	212		2,087	2,400
9000	10 kW		2.20	3.636		2,550	232		2,782	3,150
9020	13.5 kW		1.50	5.333		3,425	340		3,765	4,275
9030	16 kW		1.30	6.154		4,525	390		4,915	5,550
9040	20 kW		.90	8.889		4,475	565		5,040	5,775
9060	24 kW		.60	13.333		3,025	850		3,875	4,575
9230	13.5 kW, 40,956 BTU		2.20	3.636		1,925	232		2,157	2,450
9250	24 kW, 81,912 BTU	▼	2	4	▼	2,100	255		2,355	2,700

23 91 Prefabricated Equipment Supports

23 91 10 – Prefabricated Curbs, Pads and Stands

23 91 10.10 Prefabricated Pads and Stands	Crew	Daily Output	Labor-Hours	Unit	Material	2021 Bare Costs Labor	Equipment	Total	Total Incl O&P
0010 **PREFABRICATED PADS AND STANDS**									
6000 Pad, fiberglass reinforced concrete with polystyrene foam core									
6220 30" x 36"	1 Shee	8	1	Ea.	89	65.50		154.50	197
6340 36" x 54"	Q-9	6	2.667	"	144	157		301	395

Estimating Tips
26 05 00 Common Work Results for Electrical

- Conduit should be taken off in three main categories—power distribution, branch power, and branch lighting—so the estimator can concentrate on systems and components, therefore making it easier to ensure all items have been accounted for.

- For cost modifications for elevated conduit installation, add the percentages to labor according to the height of installation and only to the quantities exceeding the different height levels, not to the total conduit quantities. Refer to subdivision 26 01 02.20 for labor adjustment factors.

- Remember that aluminum wiring of equal ampacity is larger in diameter than copper and may require larger conduit.

- If more than three wires at a time are being pulled, deduct percentages from the labor hours of that grouping of wires.

- When taking off grounding systems, identify separately the type and size of wire, and list each unique type of ground connection.

- The estimator should take the weights of materials into consideration when completing a takeoff. Topics to consider include: How will the materials be supported? What methods of support are available? How high will the support structure have to reach? Will the final support structure be able to withstand the total burden? Is the support material included or separate from the fixture, equipment, and material specified?

- Do not overlook the costs for equipment used in the installation. If scaffolding or highlifts are available in the field, contractors may use them in lieu of the proposed ladders and rolling staging.

26 20 00 Low-Voltage Electrical Transmission

- Supports and concrete pads may be shown on drawings for the larger equipment, or the support system may be only a piece of plywood for the back of a panelboard. In either case, they must be included in the costs.

26 40 00 Electrical and Cathodic Protection

- When taking off cathodic protection systems, identify the type and size of cable, and list each unique type of anode connection.

26 50 00 Lighting

- Fixtures should be taken off room by room using the fixture schedule, specifications, and the ceiling plan. For large concentrations of lighting fixtures in the same area, deduct the percentages from labor hours.

Reference Numbers

Reference numbers are shown at the beginning of some major classifications. These numbers refer to related items in the Reference Section. The reference information may be an estimating procedure, an alternate pricing method, or technical information.

Note: Not all subdivisions listed here necessarily appear. ∎

Same Data. Simplified.

Enjoy the convenience and efficiency of accessing your costs anywhere:

- **Skip the multiplier** by setting your location
- **Quickly search,** edit, favorite and share costs
- **Stay on top of price changes** with automatic updates

Discover more at rsmeans.com/online

26 05 33.95 Cutting and Drilling	Crew	Daily Output	Labor-Hours	Unit	Material	2021 Bare Costs Labor	Equipment	Total	Total Incl O&P
0010 **CUTTING AND DRILLING**									
0100 Hole drilling to 10' high, concrete wall									
0110 8" thick, 1/2" pipe size	R-31	12	.667	Ea.	.26	42.50	5.20	47.96	69
0120 3/4" pipe size		12	.667		.26	42.50	5.20	47.96	69
0130 1" pipe size		9.50	.842		.36	53.50	6.60	60.46	87
0140 1-1/4" pipe size		9.50	.842		.36	53.50	6.60	60.46	87
0150 1-1/2" pipe size		9.50	.842		.36	53.50	6.60	60.46	87
0160 2" pipe size		4.40	1.818		.55	116	14.25	130.80	188
0170 2-1/2" pipe size		4.40	1.818		.55	116	14.25	130.80	188
0180 3" pipe size		4.40	1.818		.55	116	14.25	130.80	188
0190 3-1/2" pipe size		3.30	2.424		.67	154	19	173.67	251
0200 4" pipe size		3.30	2.424		.67	154	19	173.67	251
0500 12" thick, 1/2" pipe size		9.40	.851		.39	54	6.65	61.04	88.50
0520 3/4" pipe size		9.40	.851		.39	54	6.65	61.04	88.50
0540 1" pipe size		7.30	1.096		.53	70	8.60	79.13	114
0560 1-1/4" pipe size		7.30	1.096		.53	70	8.60	79.13	114
0570 1-1/2" pipe size		7.30	1.096		.53	70	8.60	79.13	114
0580 2" pipe size		3.60	2.222		.83	142	17.40	160.23	230
0590 2-1/2" pipe size		3.60	2.222		.83	142	17.40	160.23	230
0600 3" pipe size		3.60	2.222		.83	142	17.40	160.23	230
0610 3-1/2" pipe size		2.80	2.857		1.01	182	22.50	205.51	296
0630 4" pipe size		2.50	3.200		1.01	204	25	230.01	335
0650 16" thick, 1/2" pipe size		7.60	1.053		.52	67	8.25	75.77	109
0670 3/4" pipe size		7	1.143		.52	73	8.95	82.47	118
0690 1" pipe size		6	1.333		.71	85	10.45	96.16	138
0710 1-1/4" pipe size		5.50	1.455		.71	92.50	11.40	104.61	151
0730 1-1/2" pipe size		5.50	1.455		.71	92.50	11.40	104.61	151
0750 2" pipe size		3	2.667		1.10	170	21	192.10	276
0770 2-1/2" pipe size		2.70	2.963		1.10	189	23	213.10	305
0790 3" pipe size		2.50	3.200		1.10	204	25	230.10	335
0810 3-1/2" pipe size		2.30	3.478		1.34	222	27	250.34	360
0830 4" pipe size		2	4		1.34	255	31.50	287.84	415
0850 20" thick, 1/2" pipe size		6.40	1.250		.65	79.50	9.80	89.95	129
0870 3/4" pipe size		6	1.333		.65	85	10.45	96.10	138
0890 1" pipe size		5	1.600		.89	102	12.55	115.44	166
0910 1-1/4" pipe size		4.80	1.667		.89	106	13.05	119.94	173
0930 1-1/2" pipe size		4.60	1.739		.89	111	13.60	125.49	181
0950 2" pipe size		2.70	2.963		1.38	189	23	213.38	305
0970 2-1/2" pipe size		2.40	3.333		1.38	212	26	239.38	345
0990 3" pipe size		2.20	3.636		1.38	232	28.50	261.88	380
1010 3-1/2" pipe size		2	4		1.68	255	31.50	288.18	415
1030 4" pipe size		1.70	4.706		1.68	300	37	338.68	485
1050 24" thick, 1/2" pipe size		5.50	1.455		.78	92.50	11.40	104.68	151
1070 3/4" pipe size		5.10	1.569		.78	100	12.30	113.08	162
1090 1" pipe size		4.30	1.860		1.07	119	14.55	134.62	193
1110 1-1/4" pipe size		4	2		1.07	127	15.65	143.72	207
1130 1-1/2" pipe size		4	2		1.07	127	15.65	143.72	207
1150 2" pipe size		2.40	3.333		1.65	212	26	239.65	345
1170 2-1/2" pipe size		2.20	3.636		1.65	232	28.50	262.15	380
1190 3" pipe size		2	4		1.65	255	31.50	288.15	415
1210 3-1/2" pipe size		1.80	4.444		2.02	283	35	320.02	460
1230 4" pipe size		1.50	5.333		2.02	340	42	384.02	555
1500 Brick wall, 8" thick, 1/2" pipe size		18	.444		.26	28.50	3.48	32.24	46

26 05 33 – Raceway and Boxes for Electrical Systems

26 05 33.95 Cutting and Drilling	Crew	Daily Output	Labor-Hours	Unit	Material	2021 Bare Costs Labor	Equipment	Total	Total Incl O&P
1520	3/4" pipe size	R-31	18	.444	Ea.	.26	28.50	3.48	32.24	46
1540	1" pipe size		13.30	.602		.36	38.50	4.71	43.57	62.50
1560	1-1/4" pipe size		13.30	.602		.36	38.50	4.71	43.57	62.50
1580	1-1/2" pipe size		13.30	.602		.36	38.50	4.71	43.57	62.50
1600	2" pipe size		5.70	1.404		.55	89.50	11	101.05	146
1620	2-1/2" pipe size		5.70	1.404		.55	89.50	11	101.05	146
1640	3" pipe size		5.70	1.404		.55	89.50	11	101.05	146
1660	3-1/2" pipe size		4.40	1.818		.67	116	14.25	130.92	188
1680	4" pipe size		4	2		.67	127	15.65	143.32	207
1700	12" thick, 1/2" pipe size		14.50	.552		.39	35	4.32	39.71	57
1720	3/4" pipe size		14.50	.552		.39	35	4.32	39.71	57
1740	1" pipe size		11	.727		.53	46.50	5.70	52.73	76
1760	1-1/4" pipe size		11	.727		.53	46.50	5.70	52.73	76
1780	1-1/2" pipe size		11	.727		.53	46.50	5.70	52.73	76
1800	2" pipe size		5	1.600		.83	102	12.55	115.38	166
1820	2-1/2" pipe size		5	1.600		.83	102	12.55	115.38	166
1840	3" pipe size		5	1.600		.83	102	12.55	115.38	166
1860	3-1/2" pipe size		3.80	2.105		1.01	134	16.50	151.51	218
1880	4" pipe size		3.30	2.424		1.01	154	19	174.01	251
1900	16" thick, 1/2" pipe size		12.30	.650		.52	41.50	5.10	47.12	67.50
1920	3/4" pipe size		12.30	.650		.52	41.50	5.10	47.12	67.50
1940	1" pipe size		9.30	.860		.71	55	6.75	62.46	89.50
1960	1-1/4" pipe size		9.30	.860		.71	55	6.75	62.46	89.50
1980	1-1/2" pipe size		9.30	.860		.71	55	6.75	62.46	89.50
2000	2" pipe size		4.40	1.818		1.10	116	14.25	131.35	189
2010	2-1/2" pipe size		4.40	1.818		1.10	116	14.25	131.35	189
2030	3" pipe size		4.40	1.818		1.10	116	14.25	131.35	189
2050	3-1/2" pipe size		3.30	2.424		1.34	154	19	174.34	251
2070	4" pipe size		3	2.667		1.34	170	21	192.34	276
2090	20" thick, 1/2" pipe size		10.70	.748		.65	47.50	5.85	54	78
2110	3/4" pipe size		10.70	.748		.65	47.50	5.85	54	78
2130	1" pipe size		8	1		.89	63.50	7.85	72.24	104
2150	1-1/4" pipe size		8	1		.89	63.50	7.85	72.24	104
2170	1-1/2" pipe size		8	1		.89	63.50	7.85	72.24	104
2190	2" pipe size		4	2		1.38	127	15.65	144.03	208
2210	2-1/2" pipe size		4	2		1.38	127	15.65	144.03	208
2230	3" pipe size		4	2		1.38	127	15.65	144.03	208
2250	3-1/2" pipe size		3	2.667		1.68	170	21	192.68	277
2270	4" pipe size		2.70	2.963		1.68	189	23	213.68	305
2290	24" thick, 1/2" pipe size		9.40	.851		.78	54	6.65	61.43	88.50
2310	3/4" pipe size		9.40	.851		.78	54	6.65	61.43	88.50
2330	1" pipe size		7.10	1.127		1.07	72	8.80	81.87	118
2350	1-1/4" pipe size		7.10	1.127		1.07	72	8.80	81.87	118
2370	1-1/2" pipe size		7.10	1.127		1.07	72	8.80	81.87	118
2390	2" pipe size		3.60	2.222		1.65	142	17.40	161.05	231
2410	2-1/2" pipe size		3.60	2.222		1.65	142	17.40	161.05	231
2430	3" pipe size		3.60	2.222		1.65	142	17.40	161.05	231
2450	3-1/2" pipe size		2.80	2.857		2.02	182	22.50	206.52	297
2470	4" pipe size		2.50	3.200		2.02	204	25	231.02	335
3000	Knockouts to 8' high, metal boxes & enclosures									
3020	With hole saw, 1/2" pipe size	1 Elec	53	.151	Ea.		9.60		9.60	14.30
3040	3/4" pipe size		47	.170			10.85		10.85	16.10
3050	1" pipe size		40	.200			12.75		12.75	18.95

26 05 33 – Raceway and Boxes for Electrical Systems

26 05 33.95 Cutting and Drilling		Crew	Daily Output	Labor-Hours	Unit	Material	2021 Bare Costs Labor	Equipment	Total	Total Incl O&P
3060	1-1/4" pipe size	1 Elec	36	.222	Ea.		14.15		14.15	21
3070	1-1/2" pipe size		32	.250			15.95		15.95	23.50
3080	2" pipe size		27	.296			18.85		18.85	28
3090	2-1/2" pipe size		20	.400			25.50		25.50	38
4010	3" pipe size		16	.500			32		32	47.50
4030	3-1/2" pipe size		13	.615			39		39	58.50
4050	4" pipe size		11	.727			46.50		46.50	69

26 05 83 – Wiring Connections

26 05 83.10 Motor Connections

		Crew	Daily Output	Labor-Hours	Unit	Material	2021 Bare Costs Labor	Equipment	Total	Total Incl O&P
0010	**MOTOR CONNECTIONS**									
0020	Flexible conduit and fittings, 115 volt, 1 phase, up to 1 HP motor	1 Elec	8	1	Ea.	6.20	63.50		69.70	101
0050	2 HP motor		6.50	1.231		10.30	78.50		88.80	127
0100	3 HP motor		5.50	1.455		8.95	92.50		101.45	148
0110	230 volt, 3 phase, 3 HP motor		6.78	1.180		7.15	75		82.15	120
0112	5 HP motor		5.47	1.463		6.35	93		99.35	145
0114	7-1/2 HP motor		4.61	1.735		10.35	111		121.35	175
0120	10 HP motor		4.20	1.905		16.80	121		137.80	198
0150	15 HP motor		3.30	2.424		16.80	154		170.80	247
0200	25 HP motor		2.70	2.963		25	189		214	310
0400	50 HP motor		2.20	3.636		51.50	232		283.50	400
0600	100 HP motor		1.50	5.333		119	340		459	635

26 05 90 – Residential Applications

26 05 90.10 Residential Wiring

		Crew	Daily Output	Labor-Hours	Unit	Material	2021 Bare Costs Labor	Equipment	Total	Total Incl O&P
0010	**RESIDENTIAL WIRING**									
0020	20' avg. runs and #14/2 wiring incl. unless otherwise noted									
1000	Service & panel, includes 24' SE-AL cable, service eye, meter,									
1010	Socket, panel board, main bkr., ground rod, 15 or 20 amp									
1020	1-pole circuit breakers, and misc. hardware									
1100	100 amp, with 10 branch breakers	1 Elec	1.19	6.723	Ea.	400	430		830	1,075
1110	With PVC conduit and wire		.92	8.696		425	555		980	1,300
1120	With RGS conduit and wire		.73	10.959		600	700		1,300	1,675
1150	150 amp, with 14 branch breakers		1.03	7.767		650	495		1,145	1,450
1170	With PVC conduit and wire		.82	9.756		690	620		1,310	1,675
1180	With RGS conduit and wire		.67	11.940		1,050	760		1,810	2,275
1200	200 amp, with 18 branch breakers	2 Elec	1.80	8.889		1,075	565		1,640	2,050
1220	With PVC conduit and wire		1.46	10.959		1,125	700		1,825	2,275
1230	With RGS conduit and wire		1.24	12.903		1,550	820		2,370	2,950
1800	Lightning surge suppressor	1 Elec	32	.250		50.50	15.95		66.45	79.50
2000	Switch devices									
2100	Single pole, 15 amp, ivory, with a 1-gang box, cover plate,									
2110	Type NM (Romex) cable	1 Elec	17.10	.468	Ea.	6.85	30		36.85	52
2120	Type MC cable		14.30	.559		20.50	35.50		56	75.50
2130	EMT & wire		5.71	1.401		29	89.50		118.50	165
2150	3-way, #14/3, type NM cable		14.55	.550		8.50	35		43.50	61.50
2170	Type MC cable		12.31	.650		26	41.50		67.50	90
2180	EMT & wire		5	1.600		31	102		133	185
2200	4-way, #14/3, type NM cable		14.55	.550		15.40	35		50.40	69
2220	Type MC cable		12.31	.650		32.50	41.50		74	97.50
2230	EMT & wire		5	1.600		38	102		140	193
2250	S.P., 20 amp, #12/2, type NM cable		13.33	.600		11.85	38		49.85	70
2270	Type MC cable		11.43	.700		24.50	44.50		69	93.50
2280	EMT & wire		4.85	1.649		35.50	105		140.50	195

380

For customer support on your Plumbing Costs with RSMeans Data, call 800.448.8182.

26 05 90.10 Residential Wiring	Crew	Daily Output	Labor-Hours	Unit	Material	2021 Bare Costs Labor	Equipment	Total	Total Incl O&P
2290	S.P. rotary dimmer, 600 W, no wiring	1 Elec	17	.471	Ea.	27.50	30		57.50	74.50
2300	S.P. rotary dimmer, 600 W, type NM cable		14.55	.550		31	35		66	86
2320	Type MC cable		12.31	.650		44.50	41.50		86	111
2330	EMT & wire		5	1.600		55	102		157	212
2350	3-way rotary dimmer, type NM cable		13.33	.600		22	38		60	81.50
2370	Type MC cable		11.43	.700		36	44.50		80.50	106
2380	EMT & wire		4.85	1.649		46	105		151	207
2400	Interval timer wall switch, 20 amp, 1-30 min., #12/2									
2410	Type NM cable	1 Elec	14.55	.550	Ea.	57.50	35		92.50	116
2420	Type MC cable		12.31	.650		64.50	41.50		106	133
2430	EMT & wire		5	1.600		81.50	102		183.50	241
2500	Decorator style									
2510	S.P., 15 amp, type NM cable	1 Elec	17.10	.468	Ea.	11.60	30		41.60	57.50
2520	Type MC cable		14.30	.559		25.50	35.50		61	81
2530	EMT & wire		5.71	1.401		34	89.50		123.50	170
2550	3-way, #14/3, type NM cable		14.55	.550		13.30	35		48.30	66.50
2570	Type MC cable		12.31	.650		30.50	41.50		72	95
2580	EMT & wire		5	1.600		36	102		138	191
2600	4-way, #14/3, type NM cable		14.55	.550		20	35		55	74
2620	Type MC cable		12.31	.650		37.50	41.50		79	103
2630	EMT & wire		5	1.600		42.50	102		144.50	198
2650	S.P., 20 amp, #12/2, type NM cable		13.33	.600		16.60	38		54.60	75.50
2670	Type MC cable		11.43	.700		29.50	44.50		74	99
2680	EMT & wire		4.85	1.649		40.50	105		145.50	201
2700	S.P., slide dimmer, type NM cable		17.10	.468		37	30		67	85
2720	Type MC cable		14.30	.559		50.50	35.50		86	109
2730	EMT & wire		5.71	1.401		60.50	89.50		150	200
2750	S.P., touch dimmer, type NM cable		17.10	.468		24.50	30		54.50	71.50
2770	Type MC cable		14.30	.559		38	35.50		73.50	95
2780	EMT & wire		5.71	1.401		48.50	89.50		138	186
2800	3-way touch dimmer, type NM cable		13.33	.600		48.50	38		86.50	111
2820	Type MC cable		11.43	.700		62.50	44.50		107	135
2830	EMT & wire		4.85	1.649		72.50	105		177.50	236
3000	Combination devices									
3100	S.P. switch/15 amp recpt., ivory, 1-gang box, plate									
3110	Type NM cable	1 Elec	11.43	.700	Ea.	19.50	44.50		64	88
3120	Type MC cable		10	.800		33	51		84	112
3130	EMT & wire		4.40	1.818		43.50	116		159.50	220
3150	S.P. switch/pilot light, type NM cable		11.43	.700		20.50	44.50		65	89
3170	Type MC cable		10	.800		34	51		85	113
3180	EMT & wire		4.43	1.806		44	115		159	220
3190	2-S.P. switches, 2-#14/2, no wiring		14	.571		5.20	36.50		41.70	59.50
3200	2-S.P. switches, 2-#14/2, type NM cables		10	.800		21	51		72	98.50
3220	Type MC cable		8.89	.900		39	57.50		96.50	128
3230	EMT & wire		4.10	1.951		46.50	124		170.50	236
3250	3-way switch/15 amp recpt., #14/3, type NM cable		10	.800		26	51		77	104
3270	Type MC cable	1 Elec	8.89	.900	Ea.	43.50	57.50		101	133
3280	EMT & wire		4.10	1.951		48.50	124		172.50	239
3300	2-3 way switches, 2-#14/3, type NM cables		8.89	.900		33	57.50		90.50	122
3320	Type MC cable		8	1		58	63.50		121.50	159
3330	EMT & wire		4	2		55.50	127		182.50	250
3350	S.P. switch/20 amp recpt., #12/2, type NM cable		10	.800		27	51		78	105
3370	Type MC cable		8.89	.900		34	57.50		91.50	122

26 05 90.10 Residential Wiring	Crew	Daily Output	Labor-Hours	Unit	Material	2021 Bare Costs Labor	Equipment	Total	Total Incl O&P
3380	EMT & wire	1 Elec	4.10	1.951	Ea.	50.50	124		174.50	241
3400	Decorator style									
3410	S.P. switch/15 amp recpt., type NM cable	1 Elec	11.43	.700	Ea.	24.50	44.50		69	93
3420	Type MC cable		10	.800		38	51		89	118
3430	EMT & wire		4.40	1.818		48	116		164	225
3450	S.P. switch/pilot light, type NM cable		11.43	.700		25	44.50		69.50	94
3470	Type MC cable		10	.800		39	51		90	118
3480	EMT & wire		4.40	1.818		49	116		165	226
3500	2-S.P. switches, 2-#14/2, type NM cables		10	.800		26	51		77	104
3520	Type MC cable		8.89	.900		44	57.50		101.50	133
3530	EMT & wire		4.10	1.951		51	124		175	241
3550	3-way/15 amp recpt., #14/3, type NM cable		10	.800		31	51		82	110
3570	Type MC cable		8.89	.900		48	57.50		105.50	138
3580	EMT & wire		4.10	1.951		53.50	124		177.50	244
3650	2-3 way switches, 2-#14/3, type NM cables		8.89	.900		38	57.50		95.50	127
3670	Type MC cable		8	1		63	63.50		126.50	164
3680	EMT & wire		4	2		60.50	127		187.50	256
3700	S.P. switch/20 amp recpt., #12/2, type NM cable		10	.800		31.50	51		82.50	111
3720	Type MC cable		8.89	.900		38.50	57.50		96	128
3730	EMT & wire		4.10	1.951		55.50	124		179.50	246
4000	Receptacle devices									
4010	Duplex outlet, 15 amp recpt., ivory, 1-gang box, plate									
4015	Type NM cable	1 Elec	14.55	.550	Ea.	7.30	35		42.30	60
4020	Type MC cable		12.31	.650		21	41.50		62.50	84.50
4030	EMT & wire		5.33	1.501		29.50	95.50		125	175
4050	With #12/2, type NM cable		12.31	.650		8.50	41.50		50	71
4070	Type MC cable		10.67	.750		21.50	48		69.50	94.50
4080	EMT & wire		4.71	1.699		32.50	108		140.50	197
4100	20 amp recpt., #12/2, type NM cable		12.31	.650		18.10	41.50		59.60	81.50
4120	Type MC cable		10.67	.750		31	48		79	105
4130	EMT & wire		4.71	1.699		42	108		150	207
4140	For GFI see Section 26 05 90.10 line 4300 below									
4150	Decorator style, 15 amp recpt., type NM cable	1 Elec	14.55	.550	Ea.	12.05	35		47.05	65.50
4170	Type MC cable	1 Elec	12.31	.650	Ea.	26	41.50		67.50	90
4180	EMT & wire		5.33	1.501		34	95.50		129.50	180
4200	With #12/2, type NM cable		12.31	.650		13.25	41.50		54.75	76
4220	Type MC cable		10.67	.750		26	48		74	99.50
4230	EMT & wire		4.71	1.699		37	108		145	202
4250	20 amp recpt., #12/2, type NM cable		12.31	.650		23	41.50		64.50	86.50
4270	Type MC cable		10.67	.750		35.50	48		83.50	111
4280	EMT & wire		4.71	1.699		46.50	108		154.50	213
4300	GFI, 15 amp recpt., type NM cable		12.31	.650		21	41.50		62.50	84.50
4320	Type MC cable		10.67	.750		34.50	48		82.50	109
4330	EMT & wire		4.71	1.699		43	108		151	209
4350	GFI with #12/2, type NM cable		10.67	.750		22	48		70	95.50
4370	Type MC cable		9.20	.870		35	55.50		90.50	121
4380	EMT & wire		4.21	1.900		46	121		167	231
4400	20 amp recpt., #12/2, type NM cable		10.67	.750		62.50	48		110.50	140
4420	Type MC cable		9.20	.870		75.50	55.50		131	166
4430	EMT & wire		4.21	1.900		86.50	121		207.50	275
4500	Weather-proof cover for above receptacles, add		32	.250		3.29	15.95		19.24	27
4550	Air conditioner outlet, 20 amp-240 volt recpt.									
4560	30' of #12/2, 2 pole circuit breaker									

382

26 05 90.10 Residential Wiring	Crew	Daily Output	Labor-Hours	Unit	Material	2021 Bare Costs Labor	Equipment	Total	Total Incl O&P
4570	Type NM cable	1 Elec	10	.800	Ea.	59.50	51		110.50	141
4580	Type MC cable		9	.889		74	56.50		130.50	166
4590	EMT & wire		4	2		84	127		211	282
4600	Decorator style, type NM cable		10	.800		64.50	51		115.50	147
4620	Type MC cable		9	.889		79	56.50		135.50	171
4630	EMT & wire	↓	4	2	↓	89	127		216	287
4650	Dryer outlet, 30 amp-240 volt recpt., 20' of #10/3									
4660	2 pole circuit breaker									
4670	Type NM cable	1 Elec	6.41	1.248	Ea.	54	79.50		133.50	178
4680	Type MC cable		5.71	1.401		64	89.50		153.50	203
4690	EMT & wire	↓	3.48	2.299	↓	75	146		221	300
4700	Range outlet, 50 amp-240 volt recpt., 30' of #8/3									
4710	Type NM cable	1 Elec	4.21	1.900	Ea.	85.50	121		206.50	274
4720	Type MC cable		4	2		108	127		235	310
4730	EMT & wire		2.96	2.703		104	172		276	370
4750	Central vacuum outlet, type NM cable		6.40	1.250		61.50	79.50		141	186
4770	Type MC cable		5.71	1.401		74	89.50		163.50	215
4780	EMT & wire	↓	3.48	2.299	↓	90.50	146		236.50	320
4800	30 amp-110 volt locking recpt., #10/2 circ. bkr.									
4810	Type NM cable	1 Elec	6.20	1.290	Ea.	74.50	82		156.50	204
4830	EMT & wire	"	3.20	2.500	"	106	159		265	355
4900	Low voltage outlets									
4910	Telephone recpt., 20' of 4/C phone wire	1 Elec	26	.308	Ea.	7.60	19.60		27.20	37.50
4920	TV recpt., 20' of RG59U coax wire, F type connector	"	16	.500	"	15.40	32		47.40	64.50
4950	Door bell chime, transformer, 2 buttons, 60' of bellwire									
4970	Economy model	1 Elec	11.50	.696	Ea.	55.50	44.50		100	127
4980	Custom model		11.50	.696		94	44.50		138.50	170
4990	Luxury model, 3 buttons	↓	9.50	.842	↓	193	53.50		246.50	292
6000	Lighting outlets									
6050	Wire only (for fixture), type NM cable	1 Elec	32	.250	Ea.	5.90	15.95		21.85	30
6070	Type MC cable		24	.333		12.25	21		33.25	45
6080	EMT & wire		10	.800		19.95	51		70.95	97.50
6100	Box (4") and wire (for fixture), type NM cable		25	.320		14.95	20.50		35.45	47
6120	Type MC cable		20	.400		21.50	25.50		47	61.50
6130	EMT & wire	↓	11	.727	↓	29	46.50		75.50	101
6200	Fixtures (use with line 6050 or 6100 above)									
6210	Canopy style, economy grade	1 Elec	40	.200	Ea.	22.50	12.75		35.25	44
6220	Custom grade		40	.200		53	12.75		65.75	77.50
6250	Dining room chandelier, economy grade		19	.421		80.50	27		107.50	129
6260	Custom grade		19	.421		340	27		367	415
6270	Luxury grade		15	.533		3,050	34		3,084	3,400
6310	Kitchen fixture (fluorescent), economy grade		30	.267		76	17		93	109
6320	Custom grade		25	.320		220	20.50		240.50	273
6350	Outdoor, wall mounted, economy grade		30	.267		38.50	17		55.50	67.50
6360	Custom grade		30	.267		116	17		133	153
6370	Luxury grade		25	.320		248	20.50		268.50	305
6410	Outdoor PAR floodlights, 1 lamp, 150 watt		20	.400		15.55	25.50		41.05	55
6420	2 lamp, 150 watt each		20	.400		28	25.50		53.50	69
6430	For infrared security sensor, add		32	.250		102	15.95		117.95	136
6450	Outdoor, quartz-halogen, 300 watt flood		20	.400		40	25.50		65.50	82
6600	Recessed downlight, round, pre-wired, 50 or 75 watt trim		30	.267		46.50	17		63.50	76
6610	With shower light trim		30	.267		71.50	17		88.50	104
6620	With wall washer trim	↓	28	.286		72	18.20		90.20	106

26 05 90.10 Residential Wiring	Crew	Daily Output	Labor-Hours	Unit	Material	2021 Bare Costs Labor	Equipment	Total	Total Incl O&P
6630	With eye-ball trim	1 Elec	28	.286	Ea.	31	18.20		49.20	61
6700	Porcelain lamp holder		40	.200		2.75	12.75		15.50	22
6710	With pull switch		40	.200		11.30	12.75		24.05	31.50
6750	Fluorescent strip, 2-20 watt tube, wrap around diffuser, 24"		24	.333		43.50	21		64.50	79.50
6770	2-34 watt tubes, 48"		20	.400		148	25.50		173.50	200
6800	Bathroom heat lamp, 1-250 watt		28	.286		33.50	18.20		51.70	64
6810	2-250 watt lamps	1 Elec	28	.286	Ea.	60	18.20		78.20	93
6820	For timer switch, see Section 26 05 90.10 line 2400									
6900	Outdoor post lamp, incl. post, fixture, 35' of #14/2									
6910	Type NM cable	1 Elec	3.50	2.286	Ea.	310	146		456	555
6920	Photo-eye, add		27	.296		15.20	18.85		34.05	44.50
6950	Clock dial time switch, 24 hr., w/enclosure, type NM cable		11.43	.700		71	44.50		115.50	145
6970	Type MC cable		11	.727		84.50	46.50		131	162
6980	EMT & wire		4.85	1.649		93	105		198	258
7000	Alarm systems									
7050	Smoke detectors, box, #14/3, type NM cable	1 Elec	14.55	.550	Ea.	31	35		66	86
7070	Type MC cable		12.31	.650		42.50	41.50		84	108
7080	EMT & wire		5	1.600		47.50	102		149.50	203
7090	For relay output to security system, add					9.80			9.80	10.80
8000	Residential equipment									
8050	Disposal hook-up, incl. switch, outlet box, 3' of flex									
8060	20 amp-1 pole circ. bkr., and 25' of #12/2									
8070	Type NM cable	1 Elec	10	.800	Ea.	25	51		76	103
8080	Type MC cable		8	1		38.50	63.50		102	137
8090	EMT & wire		5	1.600		52.50	102		154.50	209
8100	Trash compactor or dishwasher hook-up, incl. outlet box,									
8110	3' of flex, 15 amp-1 pole circ. bkr., and 25' of #14/2									
8120	Type NM cable	1 Elec	10	.800	Ea.	21.50	51		72.50	99
8130	Type MC cable		8	1		36.50	63.50		100	135
8140	EMT & wire		5	1.600		48	102		150	204
8150	Hot water sink dispenser hook-up, use line 8100									
8200	Vent/exhaust fan hook-up, type NM cable	1 Elec	32	.250	Ea.	5.90	15.95		21.85	30
8220	Type MC cable		24	.333		12.25	21		33.25	45
8230	EMT & wire		10	.800		19.95	51		70.95	97.50
8250	Bathroom vent fan, 50 CFM (use with above hook-up)									
8260	Economy model	1 Elec	15	.533	Ea.	18.60	34		52.60	71
8270	Low noise model		15	.533		50.50	34		84.50	107
8280	Custom model		12	.667		116	42.50		158.50	191
8300	Bathroom or kitchen vent fan, 110 CFM									
8310	Economy model	1 Elec	15	.533	Ea.	63	34		97	120
8320	Low noise model	"	15	.533	"	101	34		135	162
8350	Paddle fan, variable speed (w/o lights)									
8360	Economy model (AC motor)	1 Elec	10	.800	Ea.	132	51		183	221
8362	With light kit		10	.800		171	51		222	265
8370	Custom model (AC motor)		10	.800		320	51		371	430
8372	With light kit	1 Elec	10	.800	Ea.	360	51		411	470
8380	Luxury model (DC motor)		8	1		279	63.50		342.50	400
8382	With light kit		8	1		320	63.50		383.50	445
8390	Remote speed switch for above, add		12	.667		40.50	42.50		83	108
8500	Whole house exhaust fan, ceiling mount, 36", variable speed									
8510	Remote switch, incl. shutters, 20 amp-1 pole circ. bkr.									
8520	30' of #12/2, type NM cable	1 Elec	4	2	Ea.	1,500	127		1,627	1,875
8530	Type MC cable		3.50	2.286		1,525	146		1,671	1,900

26 05 Common Work Results for Electrical

26 05 90 – Residential Applications

26 05 90.10 Residential Wiring

		Crew	Daily Output	Labor-Hours	Unit	Material	2021 Bare Costs Labor	Equipment	Total	Total Incl O&P
8540	EMT & wire	1 Elec	3	2.667	Ea.	1,550	170		1,720	1,950
8600	Whirlpool tub hook-up, incl. timer switch, outlet box									
8610	3' of flex, 20 amp-1 pole GFI circ. bkr.									
8620	30' of #12/2, type NM cable	1 Elec	5	1.600	Ea.	241	102		343	415
8630	Type MC cable	↓	4.20	1.905		252	121		373	455
8640	EMT & wire	↓	3.40	2.353	↓	263	150		413	510
8650	Hot water heater hook-up, incl. 1-2 pole circ. bkr., box;									
8660	3' of flex, 20' of #10/2, type NM cable	1 Elec	5	1.600	Ea.	29	102		131	183
8670	Type MC cable		4.20	1.905		43	121		164	228
8680	EMT & wire	↓	3.40	2.353	↓	48.50	150		198.50	277
9000	Heating/air conditioning									
9050	Furnace/boiler hook-up, incl. firestat, local on-off switch									
9060	Emergency switch, and 40' of type NM cable	1 Elec	4	2	Ea.	51	127		178	245
9070	Type MC cable		3.50	2.286		69	146		215	292
9080	EMT & wire	↓	1.50	5.333	↓	88	340		428	600
9100	Air conditioner hook-up, incl. local 60 amp disc. switch									
9110	3' sealtite, 40 amp, 2 pole circuit breaker									
9130	40' of #8/2, type NM cable	1 Elec	3.50	2.286	Ea.	149	146		295	380
9140	Type MC cable		3	2.667		180	170		350	450
9150	EMT & wire	↓	1.30	6.154	↓	186	390		576	785
9200	Heat pump hook-up, 1-40 & 1-100 amp 2 pole circ. bkr.									
9210	Local disconnect switch, 3' sealtite									
9220	40' of #8/2 & 30' of #3/2									
9230	Type NM cable	1 Elec	1.30	6.154	Ea.	425	390		815	1,050
9240	Type MC cable		1.08	7.407		510	470		980	1,250
9250	EMT & wire	↓	.94	8.511	↓	555	540		1,095	1,425
9500	Thermostat hook-up, using low voltage wire									
9520	Heating only, 25' of #18-3	1 Elec	24	.333	Ea.	6.15	21		27.15	38.50
9530	Heating/cooling, 25' of #18-4	"	20	.400	"	8.75	25.50		34.25	47.50

26 24 Switchboards and Panelboards

26 24 19 – Motor-Control Centers

26 24 19.40 Motor Starters and Controls

		Crew	Daily Output	Labor-Hours	Unit	Material	2021 Bare Costs Labor	Equipment	Total	Total Incl O&P
0010	**MOTOR STARTERS AND CONTROLS**									
0050	Magnetic, FVNR, with enclosure and heaters, 480 volt									
0080	2 HP, size 00	1 Elec	3.50	2.286	Ea.	120	146		266	345
0100	5 HP, size 0	↓	2.30	3.478		385	222		607	750
0200	10 HP, size 1	↓	1.60	5		169	320		489	660
0300	25 HP, size 2	2 Elec	2.20	7.273		335	465		800	1,050
0400	50 HP, size 3		1.80	8.889		555	565		1,120	1,450
0500	100 HP, size 4		1.20	13.333		1,250	850		2,100	2,625
0600	200 HP, size 5		.90	17.778		2,975	1,125		4,100	4,925
0610	400 HP, size 6	↓	.80	20		8,350	1,275		9,625	11,100
0620	NEMA 7, 5 HP, size 0	1 Elec	1.60	5		2,550	320		2,870	3,275
0630	10 HP, size 1	"	1.10	7.273		2,650	465		3,115	3,625
0640	25 HP, size 2	2 Elec	1.80	8.889		4,300	565		4,865	5,575
0650	50 HP, size 3		1.20	13.333		6,450	850		7,300	8,350
0660	100 HP, size 4		.90	17.778		8,325	1,125		9,450	10,800
0670	200 HP, size 5	↓	.50	32		13,500	2,050		15,550	17,900
0700	Combination, with motor circuit protectors, 5 HP, size 0	1 Elec	1.80	4.444		1,450	283		1,733	2,000
0800	10 HP, size 1	"	1.30	6.154		1,500	390		1,890	2,225

26 24 19 – Motor-Control Centers

26 24 19.40 Motor Starters and Controls	Crew	Daily Output	Labor-Hours	Unit	Material	2021 Bare Costs Labor	Equipment	Total	Total Incl O&P
0900	25 HP, size 2	2 Elec	2	8	Ea.	2,075	510		2,585	3,050
1000	50 HP, size 3		1.32	12.121		3,025	770		3,795	4,475
1200	100 HP, size 4		.80	20		6,650	1,275		7,925	9,225
1220	NEMA 7, 5 HP, size 0	1 Elec	1.30	6.154		3,400	390		3,790	4,325
1230	10 HP, size 1	"	1	8		3,475	510		3,985	4,575
1240	25 HP, size 2	2 Elec	1.32	12.121		4,650	770		5,420	6,275
1250	50 HP, size 3		.80	20		7,700	1,275		8,975	10,400
1260	100 HP, size 4		.60	26.667		12,000	1,700		13,700	15,700
1270	200 HP, size 5		.40	40		26,000	2,550		28,550	32,400
1400	Combination, with fused switch, 5 HP, size 0	1 Elec	1.80	4.444		1,250	283		1,533	1,800
1600	10 HP, size 1	"	1.30	6.154		1,325	390		1,715	2,025
1800	25 HP, size 2	2 Elec	2	8		2,050	510		2,560	3,025
2000	50 HP, size 3		1.32	12.121		3,675	770		4,445	5,175
2200	100 HP, size 4		.80	20		7,300	1,275		8,575	9,925
3500	Magnetic FVNR with NEMA 12, enclosure & heaters, 480 volt									
3600	5 HP, size 0	1 Elec	2.20	3.636	Ea.	256	232		488	625
3700	10 HP, size 1	"	1.50	5.333		385	340		725	930
3800	25 HP, size 2	2 Elec	2	8		720	510		1,230	1,550
3900	50 HP, size 3		1.60	10		1,100	635		1,735	2,175
4000	100 HP, size 4		1	16		2,650	1,025		3,675	4,450
4100	200 HP, size 5		.80	20		6,375	1,275		7,650	8,925
4200	Combination, with motor circuit protectors, 5 HP, size 0	1 Elec	1.70	4.706		850	300		1,150	1,375
4300	10 HP, size 1	"	1.20	6.667		885	425		1,310	1,600
4400	25 HP, size 2	2 Elec	1.80	8.889		1,325	565		1,890	2,300
4500	50 HP, size 3		1.20	13.333		2,150	850		3,000	3,625
4600	100 HP, size 4		.74	21.622		4,850	1,375		6,225	7,400
4700	Combination, with fused switch, 5 HP, size 0	1 Elec	1.70	4.706		700	300		1,000	1,225
4800	10 HP, size 1	"	1.20	6.667		725	425		1,150	1,425
4900	25 HP, size 2	2 Elec	1.80	8.889		1,100	565		1,665	2,075
5000	50 HP, size 3		1.20	13.333		1,775	850		2,625	3,200
5100	100 HP, size 4		.74	21.622		3,600	1,375		4,975	6,000
5200	Factory installed controls, adders to size 0 thru 5									
5300	Start-stop push button	1 Elec	32	.250	Ea.	54	15.95		69.95	82.50
5400	Hand-off-auto-selector switch		32	.250		54	15.95		69.95	82.50
5500	Pilot light		32	.250		101	15.95		116.95	135
5600	Start-stop-pilot		32	.250		155	15.95		170.95	194
5700	Auxiliary contact, NO or NC		32	.250		74	15.95		89.95	105
5800	NO-NC		32	.250		148	15.95		163.95	187

26 29 Low-Voltage Controllers

26 29 13 – Enclosed Controllers

26 29 13.20 Control Stations

		Crew	Daily Output	Labor-Hours	Unit	Material	Labor	Equipment	Total	Total Incl O&P
0010	**CONTROL STATIONS**									
0050	NEMA 1, heavy duty, stop/start	1 Elec	8	1	Ea.	155	63.50		218.50	265
0100	Stop/start, pilot light		6.20	1.290		211	82		293	355
0200	Hand/off/automatic		6.20	1.290		114	82		196	248
0400	Stop/start/reverse		5.30	1.509		209	96		305	375
0500	NEMA 7, heavy duty, stop/start		6	1.333		575	85		660	755
0600	Stop/start, pilot light		4	2		700	127		827	960

Estimating Tips

- When estimating material costs for electronic safety and security systems, it is always prudent to obtain manufacturers' quotations for equipment prices and special installation requirements that may affect the total cost.

- Fire alarm systems consist of control panels, annunciator panels, batteries with rack, charger, and fire alarm actuating and indicating devices. Some fire alarm systems include speakers, telephone lines, door closer controls, and other components. Be careful not to overlook the costs related to installation for these items. Also be aware of costs for integrated automation instrumentation and terminal devices, control equipment, control wiring, and programming. Insurance underwriters may have specific requirements for the type of materials to be installed or design requirements based on the hazard to be protected. Local jurisdictions may have requirements not covered by code. It is advisable to be aware of any special conditions.

- Security equipment includes items such as CCTV, access control, and other detection and identification systems to perform alert and alarm functions. Be sure to consider the costs related to installation for this security equipment, such as for integrated automation instrumentation and terminal devices, control equipment, control wiring, and programming.

Reference Numbers

Reference numbers are shown at the beginning of some major classifications. These numbers refer to related items in the Reference Section. The reference information may be an estimating procedure, an alternate pricing method, or technical information.

Same Data. Simplified.

Enjoy the convenience and efficiency of accessing your costs anywhere:

- **Skip the multiplier** by setting your location
- **Quickly search,** edit, favorite and share costs
- **Stay on top of price changes** with automatic updates

Discover more at rsmeans.com/online

Note: Trade Service, in part, has been used as a reference source for some of the material prices used in Division 28.

28 31 Intrusion Detection

28 31 16 – Intrusion Detection Systems Infrastructure

28 31 16.50 Intrusion Detection	Crew	Daily Output	Labor-Hours	Unit	Material	2021 Bare Costs Labor	Equipment	Total	Total Incl O&P
0010 **INTRUSION DETECTION**, not including wires & conduits									
0100 Burglar alarm, battery operated, mechanical trigger	1 Elec	4	2	Ea.	289	127		416	510
0200 Electrical trigger		4	2		345	127		472	570
0400 For outside key control, add		8	1		89	63.50		152.50	193
0600 For remote signaling circuitry, add		8	1		147	63.50		210.50	257
0800 Card reader, flush type, standard		2.70	2.963		680	189		869	1,025
1000 Multi-code		2.70	2.963		1,250	189		1,439	1,650

28 42 Gas Detection and Alarm

28 42 15 – Gas Detection Sensors

28 42 15.50 Tank Leak Detection Systems

	Crew	Daily Output	Labor-Hours	Unit	Material	2021 Bare Costs Labor	Equipment	Total	Total Incl O&P
0010 **TANK LEAK DETECTION SYSTEMS** Liquid and vapor									
0100 For hydrocarbons and hazardous liquids/vapors									
0120 Controller, data acquisition, incl. printer, modem, RS232 port									
0140 24 channel, for use with all probes				Ea.	3,325			3,325	3,650
0160 9 channel, for external monitoring				"	1,250			1,250	1,375
0200 Probes									
0210 Well monitoring									
0220 Liquid phase detection				Ea.	680			680	745
0230 Hydrocarbon vapor, fixed position					935			935	1,025
0240 Hydrocarbon vapor, float mounted					625			625	690
0250 Both liquid and vapor hydrocarbon					685			685	750
0300 Secondary containment, liquid phase									
0310 Pipe trench/manway sump				Ea.	705			705	775
0320 Double wall pipe and manual sump					590			590	650
0330 Double wall fiberglass annular space					425			425	465
0340 Double wall steel tank annular space					256			256	281
0500 Accessories									
0510 Modem, non-dedicated phone line				Ea.	325			325	360
0600 Monitoring, internal									
0610 Automatic tank gauge, incl. overfill				Ea.	1,275			1,275	1,400
0620 Product line				"	1,325			1,325	1,450
0700 Monitoring, special									
0710 Cathodic protection				Ea.	720			720	790
0720 Annular space chemical monitor				"	980			980	1,075

28 46 Fire Detection and Alarm

28 46 11 – Fire Sensors and Detectors

28 46 11.27 Other Sensors

	Crew	Daily Output	Labor-Hours	Unit	Material	2021 Bare Costs Labor	Equipment	Total	Total Incl O&P
0010 **OTHER SENSORS**									
5200 Smoke detector, ceiling type	1 Elec	6.20	1.290	Ea.	126	82		208	260
5240 Smoke detector, addressable type		6	1.333		214	85		299	360
5400 Duct type		3.20	2.500		295	159		454	560
5420 Duct addressable type		3.20	2.500		180	159		339	435

28 46 11.50 Fire and Heat Detectors

	Crew	Daily Output	Labor-Hours	Unit	Material	2021 Bare Costs Labor	Equipment	Total	Total Incl O&P
0010 **FIRE & HEAT DETECTORS**									
5000 Detector, rate of rise	1 Elec	8	1	Ea.	29.50	63.50		93	127
5100 Fixed temp fire alarm	"	7	1.143	"	39	73		112	151

28 46 Fire Detection and Alarm

28 46 20 – Fire Alarm

28 46 20.50 Alarm Panels and Devices		Crew	Daily Output	Labor-Hours	Unit	Material	2021 Bare Costs Labor	Equipment	Total	Total Incl O&P
0010	**ALARM PANELS AND DEVICES**, not including wires & conduits									
3600	4 zone	2 Elec	2	8	Ea.	435	510		945	1,225
3700	5 zone		1.50	10.667		740	680		1,420	1,825
3800	8 zone		1	16		420	1,025		1,445	1,975
3900	10 zone		1.25	12.800		890	815		1,705	2,175
4000	12 zone		.67	23.988		2,550	1,525		4,075	5,100
4020	Alarm device, tamper, flow	1 Elec	8	1	Ea.	206	63.50		269.50	320
4025	Fire alarm, loop expander card		16	.500		810	32		842	940
4050	Actuating device		8	1		330	63.50		393.50	455
4200	Battery and rack		4	2		465	127		592	700
4400	Automatic charger		8	1		525	63.50		588.50	675
4600	Signal bell		8	1		68	63.50		131.50	170
4610	Fire alarm signal bell 10" red 20-24 V P		8	1		153	63.50		216.50	264
4800	Trouble buzzer or manual station		8	1		75.50	63.50		139	178
5425	Duct smoke and heat detector 2 wire		8	1		155	63.50		218.50	265
5430	Fire alarm duct detector controller		3	2.667		183	170		353	455
5435	Fire alarm duct detector sensor kit		8	1		77.50	63.50		141	180
5440	Remote test station for smoke detector duct type		5.30	1.509		54.50	96		150.50	203
5460	Remote fire alarm indicator light		5.30	1.509		28	96		124	174
5600	Strobe and horn		5.30	1.509		129	96		225	285
5610	Strobe and horn (ADA type)		5.30	1.509		155	96		251	315
5620	Visual alarm (ADA type)		6.70	1.194		103	76		179	226
5800	electric bell		6.70	1.194		50.50	76		126.50	169
6000	Door holder, electro-magnetic		4	2		92	127		219	290
6200	Combination holder and closer		3.20	2.500		157	159		316	410
6600	Drill switch		8	1		485	63.50		548.50	630
6800	Master box		2.70	2.963		5,850	189		6,039	6,725
7000	Break glass station		8	1		62.50	63.50		126	163
7800	Remote annunciator, 8 zone lamp		1.80	4.444		219	283		502	660
8000	12 zone lamp	2 Elec	2.60	6.154		430	390		820	1,050
8200	16 zone lamp	"	2.20	7.273		385	465		850	1,100

For customer support on your Plumbing Costs with RSMeans Data, call 800.448.8182.

389

Division Notes

	CREW	DAILY OUTPUT	LABOR-HOURS	UNIT	BARE COSTS				TOTAL INCL O&P
					MAT.	LABOR	EQUIP.	TOTAL	

Estimating Tips
31 05 00 Common Work Results for Earthwork

- Estimating the actual cost of performing earthwork requires careful consideration of the variables involved. This includes items such as type of soil, whether water will be encountered, dewatering, whether banks need bracing, disposal of excavated earth, and length of haul to fill or spoil sites, etc. If the project has large quantities of cut or fill, consider raising or lowering the site to reduce costs, while paying close attention to the effect on site drainage and utilities.

- If the project has large quantities of fill, creating a borrow pit on the site can significantly lower the costs.

- It is very important to consider what time of year the project is scheduled for completion. Bad weather can create large cost overruns from dewatering, site repair, and lost productivity from cold weather.

Reference Numbers

Reference numbers are shown at the beginning of some major classifications. These numbers refer to related items in the Reference Section. The reference information may be an estimating procedure, an alternate pricing method, or technical information.

Note: Not all subdivisions listed here necessarily appear. ■

Same Data. Simplified.

Enjoy the convenience and efficiency of accessing your costs anywhere:

- **Skip the multiplier** by setting your location
- **Quickly search,** edit, favorite and share costs
- **Stay on top of price changes** with automatic updates

Discover more at rsmeans.com/online

31 23 16.13 Excavating, Trench	Crew	Daily Output	Labor-Hours	Unit	Material	2021 Bare Costs Labor	2021 Bare Costs Equipment	Total	Total Incl O&P
0010 **EXCAVATING, TRENCH** G1030–805									
0011 Or continuous footing									
0020 Common earth with no sheeting or dewatering included									
0050 1' to 4' deep, 3/8 C.Y. excavator	B-11C	150	.107	B.C.Y.		5.50	1.44	6.94	9.80
0060 1/2 C.Y. excavator	B-11M	200	.080			4.14	1.18	5.32	7.45
0090 4' to 6' deep, 1/2 C.Y. excavator	"	200	.080			4.14	1.18	5.32	7.45
0100 5/8 C.Y. excavator	B-12Q	250	.064			3.39	2.42	5.81	7.70
0300 1/2 C.Y. excavator, truck mounted	B-12J	200	.080			4.23	4.25	8.48	11
0500 6' to 10' deep, 3/4 C.Y. excavator	B-12F	225	.071			3.76	3.12	6.88	9.05
0600 1 C.Y. excavator, truck mounted	B-12K	400	.040			2.12	2.46	4.58	5.85
0900 10' to 14' deep, 3/4 C.Y. excavator	B-12F	200	.080			4.23	3.51	7.74	10.15
1000 1-1/2 C.Y. excavator	B-12B	540	.030			1.57	1.29	2.86	3.76
1300 14' to 20' deep, 1 C.Y. excavator	B-12A	320	.050			2.65	2.60	5.25	6.80
1340 20' to 24' deep, 1 C.Y. excavator	"	288	.056			2.94	2.89	5.83	7.55
1352 4' to 6' deep, 1/2 C.Y. excavator w/trench box	B-13H	188	.085			4.50	5.15	9.65	12.35
1354 5/8 C.Y. excavator	"	235	.068			3.60	4.13	7.73	9.90
1362 6' to 10' deep, 3/4 C.Y. excavator w/trench box	B-13G	212	.075			3.99	3.87	7.86	10.20
1374 10' to 14' deep, 3/4 C.Y. excavator w/trench box	"	188	.085			4.50	4.37	8.87	11.50
1376 1-1/2 C.Y. excavator	B-13E	508	.032			1.67	1.60	3.27	4.25
1381 14' to 20' deep, 1 C.Y. excavator w/trench box	B-13D	301	.053			2.81	3.16	5.97	7.65
1386 20' to 24' deep, 1 C.Y. excavator w/trench box	"	271	.059			3.13	3.51	6.64	8.50
1400 By hand with pick and shovel 2' to 6' deep, light soil	1 Clab	8	1			44.50		44.50	66.50
1500 Heavy soil	"	4	2			89		89	133
1700 For tamping backfilled trenches, air tamp, add	A-1G	100	.080	E.C.Y.		3.55	.55	4.10	5.90
1900 Vibrating plate, add	B-18	180	.133	"		6	.92	6.92	9.95
2100 Trim sides and bottom for concrete pours, common earth		1500	.016	S.F.		.72	.11	.83	1.20
2300 Hardpan		600	.040	"		1.80	.28	2.08	2.99
5020 Loam & sandy clay with no sheeting or dewatering included									
5050 1' to 4' deep, 3/8 C.Y. tractor loader/backhoe	B-11C	162	.099	B.C.Y.		5.10	1.33	6.43	9.05
5060 1/2 C.Y. excavator	B-11M	216	.074			3.83	1.09	4.92	6.90
5080 4' to 6' deep, 1/2 C.Y. excavator	"	216	.074			3.83	1.09	4.92	6.90
5090 5/8 C.Y. excavator	B-12Q	276	.058			3.07	2.19	5.26	7
5130 1/2 C.Y. excavator, truck mounted	B-12J	216	.074			3.92	3.94	7.86	10.20
5140 6' to 10' deep, 3/4 C.Y. excavator	B-12F	243	.066			3.49	2.89	6.38	8.40
5160 1 C.Y. excavator, truck mounted	B-12K	432	.037			1.96	2.28	4.24	5.45
5190 10' to 14' deep, 3/4 C.Y. excavator	B-12F	216	.074			3.92	3.25	7.17	9.40
5210 1-1/2 C.Y. excavator	B-12B	583	.027			1.45	1.19	2.64	3.48
5250 14' to 20' deep, 1 C.Y. excavator	B-12A	346	.046			2.45	2.41	4.86	6.30
5300 20' to 24' deep, 1 C.Y. excavator	"	311	.051			2.72	2.68	5.40	7
5352 4' to 6' deep, 1/2 C.Y. excavator w/trench box	B-13H	205	.078			4.13	4.73	8.86	11.35
5354 5/8 C.Y. excavator	"	257	.062			3.30	3.77	7.07	9.05
5362 6' to 10' deep, 3/4 C.Y. excavator w/trench box	B-13G	231	.069			3.67	3.55	7.22	9.35
5370 10' to 14' deep, 3/4 C.Y. excavator w/trench box	"	205	.078			4.13	4	8.13	10.55
5374 1-1/2 C.Y. excavator	B-13E	554	.029			1.53	1.47	3	3.90
5382 14' to 20' deep, 1 C.Y. excavator w/trench box	B-13D	329	.049			2.57	2.89	5.46	7
5392 20' to 24' deep, 1 C.Y. excavator w/trench box	"	295	.054			2.87	3.23	6.10	7.85
6020 Sand & gravel with no sheeting or dewatering included									
6050 1' to 4' deep, 3/8 C.Y. excavator	B-11C	165	.097	B.C.Y.		5	1.31	6.31	8.90
6060 1/2 C.Y. excavator	B-11M	220	.073	"		3.76	1.07	4.83	6.80
6080 4' to 6' deep, 1/2 C.Y. excavator	B-11M	220	.073	B.C.Y.		3.76	1.07	4.83	6.80
6090 5/8 C.Y. excavator	B-12Q	275	.058			3.08	2.20	5.28	7
6130 1/2 C.Y. excavator, truck mounted	B-12J	220	.073			3.85	3.87	7.72	10
6140 6' to 10' deep, 3/4 C.Y. excavator	B-12F	248	.065			3.41	2.83	6.24	8.20

31 23 Excavation and Fill

31 23 16 – Excavation

31 23 16.13 Excavating, Trench

		Crew	Daily Output	Labor-Hours	Unit	Material	2021 Bare Costs Labor	2021 Bare Costs Equipment	Total	Total Incl O&P
6160	1 C.Y. excavator, truck mounted	B-12K	440	.036	B.C.Y.		1.92	2.24	4.16	5.35
6190	10' to 14' deep, 3/4 C.Y. excavator	B-12F	220	.073			3.85	3.19	7.04	9.25
6210	1-1/2 C.Y. excavator	B-12B	594	.027			1.43	1.17	2.60	3.42
6250	14' to 20' deep, 1 C.Y. excavator	B-12A	352	.045			2.41	2.37	4.78	6.20
6300	20' to 24' deep, 1 C.Y. excavator	"	317	.050			2.67	2.63	5.30	6.85
6352	4' to 6' deep, 1/2 C.Y. excavator w/trench box	B-13H	209	.077			4.05	4.64	8.69	11.15
6354	5/8 C.Y. excavator	"	261	.061			3.24	3.72	6.96	8.95
6362	6' to 10' deep, 3/4 C.Y. excavator w/trench box	B-13G	236	.068			3.59	3.48	7.07	9.20
6370	10' to 14' deep, 3/4 C.Y. excavator w/trench box	"	209	.077			4.05	3.93	7.98	10.35
6374	1-1/2 C.Y. excavator	B-13E	564	.028			1.50	1.44	2.94	3.83
6382	14' to 20' deep, 1 C.Y. excavator w/trench box	B-13D	334	.048			2.54	2.85	5.39	6.90
6392	20' to 24' deep, 1 C.Y. excavator w/trench box	"	301	.053	▼		2.81	3.16	5.97	7.65
7020	Dense hard clay with no sheeting or dewatering included									
7050	1' to 4' deep, 3/8 C.Y. excavator	B-11C	132	.121	B.C.Y.		6.25	1.64	7.89	11.15
7060	1/2 C.Y. excavator	B-11M	176	.091			4.70	1.34	6.04	8.45
7080	4' to 6' deep, 1/2 C.Y. excavator	"	176	.091			4.70	1.34	6.04	8.45
7090	5/8 C.Y. excavator	B-12Q	220	.073			3.85	2.75	6.60	8.75
7130	1/2 C.Y. excavator, truck mounted	B-12J	176	.091			4.81	4.83	9.64	12.45
7140	6' to 10' deep, 3/4 C.Y. excavator	B-12F	198	.081			4.28	3.54	7.82	10.30
7160	1 C.Y. excavator, truck mounted	B-12K	352	.045			2.41	2.80	5.21	6.65
7190	10' to 14' deep, 3/4 C.Y. excavator	B-12F	176	.091			4.81	3.99	8.80	11.55
7210	1-1/2 C.Y. excavator	B-12B	475	.034			1.78	1.46	3.24	4.27
7250	14' to 20' deep, 1 C.Y. excavator	B-12A	282	.057			3	2.95	5.95	7.75
7300	20' to 24' deep, 1 C.Y. excavator	"	254	.063	▼		3.33	3.28	6.61	8.60

31 23 16.14 Excavating, Utility Trench

		Crew	Daily Output	Labor-Hours	Unit	Material	2021 Bare Costs Labor	2021 Bare Costs Equipment	Total	Total Incl O&P
0010	**EXCAVATING, UTILITY TRENCH** G1030–805									
0011	Common earth									
0050	Trenching with chain trencher, 12 HP, operator walking									
0100	4" wide trench, 12" deep	B-53	800	.010	L.F.		.56	.20	.76	1.05
0150	18" deep		750	.011			.59	.21	.80	1.11
0200	24" deep		700	.011			.63	.23	.86	1.20
0300	6" wide trench, 12" deep		650	.012			.68	.24	.92	1.29
0350	18" deep		600	.013			.74	.26	1	1.39
0400	24" deep		550	.015			.81	.29	1.10	1.52
0450	36" deep		450	.018			.99	.35	1.34	1.86
0600	8" wide trench, 12" deep	▼	475	.017	▼		.93	.33	1.26	1.76
0650	18" deep	B-53	400	.020	L.F.		1.11	.40	1.51	2.09
0700	24" deep		350	.023			1.27	.45	1.72	2.39
0750	36" deep	▼	300	.027	▼		1.48	.53	2.01	2.79
1000	Backfill by hand including compaction, add									
1050	4" wide trench, 12" deep	A-1G	800	.010	L.F.		.44	.07	.51	.74
1100	18" deep		530	.015			.67	.10	.77	1.11
1150	24" deep		400	.020			.89	.14	1.03	1.48
1300	6" wide trench, 12" deep		540	.015			.66	.10	.76	1.09
1350	18" deep		405	.020			.88	.13	1.01	1.46
1400	24" deep		270	.030			1.32	.20	1.52	2.18
1450	36" deep		180	.044			1.97	.30	2.27	3.27
1600	8" wide trench, 12" deep		400	.020			.89	.14	1.03	1.48
1650	18" deep		265	.030			1.34	.21	1.55	2.23
1700	24" deep		200	.040			1.78	.27	2.05	2.95
1750	36" deep	▼	135	.059	▼		2.63	.40	3.03	4.38
2000	Chain trencher, 40 HP operator riding									

31 23 Excavation and Fill

31 23 16 – Excavation

31 23 16.14 Excavating, Utility Trench

		Crew	Daily Output	Labor-Hours	Unit	Material	2021 Bare Costs Labor	Equipment	Total	Total Incl O&P
2050	6" wide trench and backfill, 12" deep	B-54	1200	.007	L.F.		.37	.38	.75	.96
2100	18" deep		1000	.008			.44	.45	.89	1.16
2150	24" deep		975	.008			.46	.46	.92	1.19
2200	36" deep		900	.009			.49	.50	.99	1.29
2250	48" deep		750	.011			.59	.60	1.19	1.54
2300	60" deep		650	.012			.68	.69	1.37	1.78
2400	8" wide trench and backfill, 12" deep		1000	.008			.44	.45	.89	1.16
2450	18" deep		950	.008			.47	.47	.94	1.22
2500	24" deep		900	.009			.49	.50	.99	1.29
2550	36" deep		800	.010			.56	.56	1.12	1.45
2600	48" deep		650	.012			.68	.69	1.37	1.78
2700	12" wide trench and backfill, 12" deep		975	.008			.46	.46	.92	1.19
2750	18" deep		860	.009			.52	.52	1.04	1.35
2800	24" deep		800	.010			.56	.56	1.12	1.45
2850	36" deep		725	.011			.61	.62	1.23	1.59
3000	16" wide trench and backfill, 12" deep		835	.010			.53	.54	1.07	1.38
3050	18" deep		750	.011			.59	.60	1.19	1.54
3100	24" deep	▼	700	.011	▼		.63	.64	1.27	1.66
3200	Compaction with vibratory plate, add							35%	35%	
5100	Hand excavate and trim for pipe bells after trench excavation									
5200	8" pipe	1 Clab	155	.052	L.F.		2.29		2.29	3.42
5300	18" pipe	"	130	.062	"		2.73		2.73	4.08

31 23 19 – Dewatering

31 23 19.20 Dewatering Systems

		Crew	Daily Output	Labor-Hours	Unit	Material	2021 Bare Costs Labor	Equipment	Total	Total Incl O&P
0010	**DEWATERING SYSTEMS**									
0020	Excavate drainage trench, 2' wide, 2' deep	B-11C	90	.178	C.Y.		9.20	2.40	11.60	16.35
0100	2' wide, 3' deep, with backhoe loader	B-11C	135	.119	C.Y.		6.15	1.60	7.75	10.90
0200	Excavate sump pits by hand, light soil	1 Clab	7.10	1.127			50		50	74.50
0300	Heavy soil	"	3.50	2.286	▼		101		101	151
0500	Pumping 8 hrs., attended 2 hrs./day, incl. 20 L.F.									
0550	of suction hose & 100 L.F. discharge hose									
0600	2" diaphragm pump used for 8 hrs.	B-10H	4	3	Day		162	25	187	270
0650	4" diaphragm pump used for 8 hrs.	B-10I	4	3			162	37.50	199.50	283
0800	8 hrs. attended, 2" diaphragm pump	B-10H	1	12			650	99	749	1,075
0900	3" centrifugal pump	B-10J	1	12			650	92	742	1,075
1000	4" diaphragm pump	B-10I	1	12			650	149	799	1,125
1100	6" centrifugal pump	B-10K	1	12	▼		650	297	947	1,300
1300	CMP, incl. excavation 3' deep, 12" diameter	B-6	115	.209	L.F.	12	10.05	1.88	23.93	30
1400	18" diameter		100	.240	"	18.75	11.55	2.16	32.46	40
1600	Sump hole construction, incl. excavation and gravel, pit		1250	.019	C.F.	1.12	.92	.17	2.21	2.81
1700	With 12" gravel collar, 12" pipe, corrugated, 16 ga.		70	.343	L.F.	23.50	16.50	3.09	43.09	54
1800	15" pipe, corrugated, 16 ga.		55	.436		31	21	3.93	55.93	70
1900	18" pipe, corrugated, 16 ga.		50	.480		35.50	23	4.32	62.82	79
2000	24" pipe, corrugated, 14 ga.		40	.600		43	29	5.40	77.40	96
2200	Wood lining, up to 4' x 4', add	▼	300	.080	SFCA	15.95	3.85	.72	20.52	24
9950	See Section 31 23 19.40 for wellpoints									
9960	See Section 31 23 19.30 for deep well systems									

31 23 19.30 Wells

		Crew	Daily Output	Labor-Hours	Unit	Material	2021 Bare Costs Labor	Equipment	Total	Total Incl O&P
0010	**WELLS**									
0011	For dewatering 10' to 20' deep, 2' diameter									
0020	with steel casing, minimum	B-6	165	.145	V.L.F.	37.50	7	1.31	45.81	53.50
0050	Average	▼	98	.245	▼	46	11.80	2.21	60.01	70.50

394

31 23 Excavation and Fill

31 23 19 – Dewatering

31 23 19.30 Wells

		Crew	Daily Output	Labor-Hours	Unit	Material	2021 Bare Costs Labor	2021 Bare Costs Equipment	Total	Total Incl O&P
0100	Maximum	B-6	49	.490	V.L.F.	52	23.50	4.41	79.91	97.50
0300	For dewatering pumps see 01 54 33 in Reference Section									
0500	For domestic water wells, see Section 33 21 13.10									

31 23 19.40 Wellpoints

		Crew	Daily Output	Labor-Hours	Unit	Material	Labor	Equipment	Total	Total Incl O&P
0010	**WELLPOINTS** R312319-90									
0011	For equipment rental, see 01 54 33 in Reference Section									
0100	Installation and removal of single stage system									
0110	Labor only, 0.75 labor-hours per L.F.	1 Clab	10.70	.748	LF Hdr		33		33	49.50
0200	2.0 labor-hours per L.F.	"	4	2	"		89		89	133
0400	Pump operation, 4 @ 6 hr. shifts									
0410	Per 24 hr. day	4 Eqlt	1.27	25.197	Day		1,400		1,400	2,075
0500	Per 168 hr. week, 160 hr. straight, 8 hr. double time	↓	.18	178	Week		9,875		9,875	14,700
0550	Per 4.3 week month	↓	.04	800	Month		44,400		44,400	66,000
0600	Complete installation, operation, equipment rental, fuel &									
0610	removal of system with 2" wellpoints 5' OC									
0700	100' long header, 6" diameter, first month	4 Eqlt	3.23	9.907	LF Hdr	160	550		710	995
0800	Thereafter, per month		4.13	7.748		128	430		558	780
1000	200' long header, 8" diameter, first month		6	5.333		153	296		449	610
1100	Thereafter, per month		8.39	3.814		72	212		284	395
1300	500' long header, 8" diameter, first month		10.63	3.010		56	167		223	310
1400	Thereafter, per month		20.91	1.530		40	85		125	171
1600	1,000' long header, 10" diameter, first month		11.62	2.754		48	153		201	281
1700	Thereafter, per month	↓	41.81	.765	↓	24	42.50		66.50	90
1900	Note: above figures include pumping 168 hrs. per week,									
1910	the pump operator, and one stand-by pump.									

31 23 23 – Fill

31 23 23.15 Borrow, Loading and/or Spreading

		Crew	Daily Output	Labor-Hours	Unit	Material	Labor	Equipment	Total	Total Incl O&P
0010	**BORROW, LOADING AND/OR SPREADING**									
0020	Material only, bank run gravel				Ton	21			21	23
0500	Haul 2 mi. spread, 200 HP dozer, bank run gravel	B-15	1100	.025			1.34	2.44	3.78	4.68
1000	Hand spread, bank run gravel	A-5	33	.545			24.50	1.51	26.01	38
1800	Delivery charge, minimum 20 tons, 1 hr. round trip, add	B-34B	130	.062			3.16	4.46	7.62	9.60
1820	1-1/2 hr. round trip, add		93	.086			4.41	6.25	10.66	13.45
1840	2 hr. round trip, add	↓	65	.123	↓		6.30	8.90	15.20	19.25

31 23 23.16 Fill By Borrow and Utility Bedding

		Crew	Daily Output	Labor-Hours	Unit	Material	Labor	Equipment	Total	Total Incl O&P
0010	**FILL BY BORROW AND UTILITY BEDDING**									
0049	Utility bedding, for pipe & conduit, not incl. compaction G1030-805									
0050	Crushed or screened bank run gravel	B-6	150	.160	L.C.Y.	33.50	7.70	1.44	42.64	49.50
0100	Crushed stone 3/4" to 1/2"		150	.160		25.50	7.70	1.44	34.64	41.50
0200	Sand, dead or bank	↓	150	.160	↓	18.75	7.70	1.44	27.89	33.50
0500	Compacting bedding in trench	A-1D	90	.089	E.C.Y.		3.95	.35	4.30	6.30
0600	If material source exceeds 2 miles, add for extra mileage.									
0610	See Section 31 23 23.20 for hauling mileage add.									

31 23 23.17 General Fill

		Crew	Daily Output	Labor-Hours	Unit	Material	Labor	Equipment	Total	Total Incl O&P
0010	**GENERAL FILL**									
0011	Spread dumped material, no compaction									
0020	By dozer	B-10B	1000	.012	L.C.Y.		.65	1.52	2.17	2.64
0100	By hand	1 Clab	12	.667	"		29.50		29.50	44
0500	Gravel fill, compacted, under floor slabs, 4" deep	B-37	10000	.005	S.F.	.60	.22	.03	.85	1.01
0600	6" deep		8600	.006		.89	.26	.03	1.18	1.40
0700	9" deep	↓	7200	.007	↓	1.49	.31	.04	1.84	2.14

For customer support on your Plumbing Costs with RSMeans Data, call 800.448.8182.

395

31 23 23.17 General Fill		Crew	Daily Output	Labor-Hours	Unit	Material	2021 Bare Costs Labor	Equipment	Total	Total Incl O&P
0800	12" deep	B-37	6000	.008	S.F.	2.08	.37	.04	2.49	2.90
1000	Alternate pricing method, 4" deep	↓	120	.400	E.C.Y.	44.50	18.65	2.16	65.31	79.50
1100	6" deep		160	.300	"	44.50	13.95	1.62	60.07	72
1200	9" deep	B-37	200	.240	E.C.Y.	44.50	11.20	1.29	56.99	67
1300	12" deep	"	220	.218	"	44.50	10.15	1.18	55.83	65.50
1400	Granular fill				L.C.Y.	30			30	33

31 23 23.20 Hauling

		Crew	Daily Output	Labor-Hours	Unit	Material	2021 Bare Costs Labor	Equipment	Total	Total Incl O&P
0010	**HAULING**									
0011	Excavated or borrow, loose cubic yards									
0012	no loading equipment, including hauling, waiting, loading/dumping									
0013	time per cycle (wait, load, travel, unload or dump & return)									
0014	8 C.Y. truck, 15 MPH avg., cycle 0.5 miles, 10 min. wait/ld./uld.	B-34A	320	.025	L.C.Y.		1.28	1.27	2.55	3.32
0016	cycle 1 mile		272	.029			1.51	1.50	3.01	3.91
0018	cycle 2 miles		208	.038			1.97	1.96	3.93	5.10
0020	cycle 4 miles		144	.056			2.85	2.83	5.68	7.35
0022	cycle 6 miles		112	.071			3.66	3.64	7.30	9.50
0024	cycle 8 miles		88	.091			4.66	4.63	9.29	12.05
0026	20 MPH avg., cycle 0.5 mile		336	.024			1.22	1.21	2.43	3.16
0028	cycle 1 mile		296	.027			1.39	1.38	2.77	3.58
0030	cycle 2 miles		240	.033			1.71	1.70	3.41	4.43
0032	cycle 4 miles		176	.045			2.33	2.32	4.65	6.05
0034	cycle 6 miles		136	.059			3.02	3	6.02	7.80
0036	cycle 8 miles		112	.071			3.66	3.64	7.30	9.50
0044	25 MPH avg., cycle 4 miles		192	.042			2.14	2.12	4.26	5.55
0046	cycle 6 miles		160	.050			2.57	2.55	5.12	6.65
0048	cycle 8 miles		128	.063			3.21	3.18	6.39	8.30
0050	30 MPH avg., cycle 4 miles		216	.037			1.90	1.89	3.79	4.92
0052	cycle 6 miles		176	.045			2.33	2.32	4.65	6.05
0054	cycle 8 miles		144	.056			2.85	2.83	5.68	7.35
0114	15 MPH avg., cycle 0.5 mile, 15 min. wait/ld./uld.		224	.036			1.83	1.82	3.65	4.74
0116	cycle 1 mile		200	.040			2.05	2.04	4.09	5.30
0118	cycle 2 miles		168	.048			2.44	2.43	4.87	6.30
0120	cycle 4 miles		120	.067			3.42	3.40	6.82	8.85
0122	cycle 6 miles		96	.083			4.28	4.25	8.53	11.05
0124	cycle 8 miles		80	.100			5.15	5.10	10.25	13.25
0126	20 MPH avg., cycle 0.5 mile		232	.034			1.77	1.76	3.53	4.57
0128	cycle 1 mile		208	.038			1.97	1.96	3.93	5.10
0130	cycle 2 miles		184	.043			2.23	2.22	4.45	5.75
0132	cycle 4 miles		144	.056			2.85	2.83	5.68	7.35
0134	cycle 6 miles		112	.071			3.66	3.64	7.30	9.50
0136	cycle 8 miles		96	.083			4.28	4.25	8.53	11.05
0144	25 MPH avg., cycle 4 miles		152	.053			2.70	2.68	5.38	7
0146	cycle 6 miles		128	.063			3.21	3.18	6.39	8.30
0148	cycle 8 miles		112	.071			3.66	3.64	7.30	9.50
0150	30 MPH avg., cycle 4 miles	↓	168	.048	↓		2.44	2.43	4.87	6.30
0152	cycle 6 miles	B-34A	144	.056	L.C.Y.		2.85	2.83	5.68	7.35
0154	cycle 8 miles		120	.067			3.42	3.40	6.82	8.85
0214	15 MPH avg., cycle 0.5 mile, 20 min. wait/ld./uld.		176	.045			2.33	2.32	4.65	6.05
0216	cycle 1 mile		160	.050			2.57	2.55	5.12	6.65
0218	cycle 2 miles		136	.059			3.02	3	6.02	7.80
0220	cycle 4 miles		104	.077			3.95	3.92	7.87	10.20
0222	cycle 6 miles	↓	88	.091	↓		4.66	4.63	9.29	12.05

31 23 Excavation and Fill

31 23 23 – Fill

31 23 23.20 Hauling		Crew	Daily Output	Labor-Hours	Unit	Material	2021 Bare Costs Labor	Equipment	Total	Total Incl O&P
0224	cycle 8 miles	B-34A	72	.111	L.C.Y.		5.70	5.65	11.35	14.75
0226	20 MPH avg., cycle 0.5 mile		176	.045			2.33	2.32	4.65	6.05
0228	cycle 1 mile		168	.048			2.44	2.43	4.87	6.30
0230	cycle 2 miles		144	.056			2.85	2.83	5.68	7.35
0232	cycle 4 miles		120	.067			3.42	3.40	6.82	8.85
0234	cycle 6 miles		96	.083			4.28	4.25	8.53	11.05
0236	cycle 8 miles		88	.091			4.66	4.63	9.29	12.05
0244	25 MPH avg., cycle 4 miles		128	.063			3.21	3.18	6.39	8.30
0246	cycle 6 miles		112	.071			3.66	3.64	7.30	9.50
0248	cycle 8 miles		96	.083			4.28	4.25	8.53	11.05
0250	30 MPH avg., cycle 4 miles		136	.059			3.02	3	6.02	7.80
0252	cycle 6 miles		120	.067			3.42	3.40	6.82	8.85
0254	cycle 8 miles		104	.077			3.95	3.92	7.87	10.20
0314	15 MPH avg., cycle 0.5 mile, 25 min. wait/ld./uld.		144	.056			2.85	2.83	5.68	7.35
0316	cycle 1 mile		128	.063			3.21	3.18	6.39	8.30
0318	cycle 2 miles		112	.071			3.66	3.64	7.30	9.50
0320	cycle 4 miles		96	.083			4.28	4.25	8.53	11.05
0322	cycle 6 miles		80	.100			5.15	5.10	10.25	13.25
0324	cycle 8 miles		64	.125			6.40	6.35	12.75	16.60
0326	20 MPH avg., cycle 0.5 mile		144	.056			2.85	2.83	5.68	7.35
0328	cycle 1 mile		136	.059			3.02	3	6.02	7.80
0330	cycle 2 miles		120	.067			3.42	3.40	6.82	8.85
0332	cycle 4 miles		104	.077			3.95	3.92	7.87	10.20
0334	cycle 6 miles		88	.091			4.66	4.63	9.29	12.05
0336	cycle 8 miles		80	.100			5.15	5.10	10.25	13.25
0344	25 MPH avg., cycle 4 miles		112	.071			3.66	3.64	7.30	9.50
0346	cycle 6 miles		96	.083			4.28	4.25	8.53	11.05
0348	cycle 8 miles		88	.091			4.66	4.63	9.29	12.05
0350	30 MPH avg., cycle 4 miles		112	.071			3.66	3.64	7.30	9.50
0352	cycle 6 miles		104	.077			3.95	3.92	7.87	10.20
0354	cycle 8 miles		96	.083			4.28	4.25	8.53	11.05
0414	15 MPH avg., cycle 0.5 mile, 30 min. wait/ld./uld.		120	.067			3.42	3.40	6.82	8.85
0416	cycle 1 mile		112	.071			3.66	3.64	7.30	9.50
0418	cycle 2 miles		96	.083			4.28	4.25	8.53	11.05
0420	cycle 4 miles	B-34A	80	.100	L.C.Y.		5.15	5.10	10.25	13.25
0422	cycle 6 miles		72	.111			5.70	5.65	11.35	14.75
0424	cycle 8 miles		64	.125			6.40	6.35	12.75	16.60
0426	20 MPH avg., cycle 0.5 mile		120	.067			3.42	3.40	6.82	8.85
0428	cycle 1 mile		112	.071			3.66	3.64	7.30	9.50
0430	cycle 2 miles		104	.077			3.95	3.92	7.87	10.20
0432	cycle 4 miles		88	.091			4.66	4.63	9.29	12.05
0434	cycle 6 miles		80	.100			5.15	5.10	10.25	13.25
0436	cycle 8 miles		72	.111			5.70	5.65	11.35	14.75
0444	25 MPH avg., cycle 4 miles		96	.083			4.28	4.25	8.53	11.05
0446	cycle 6 miles		88	.091			4.66	4.63	9.29	12.05
0448	cycle 8 miles		80	.100			5.15	5.10	10.25	13.25
0450	30 MPH avg., cycle 4 miles		96	.083			4.28	4.25	8.53	11.05
0452	cycle 6 miles		88	.091			4.66	4.63	9.29	12.05
0454	cycle 8 miles		80	.100			5.15	5.10	10.25	13.25
0514	15 MPH avg., cycle 0.5 mile, 35 min. wait/ld./uld.		104	.077			3.95	3.92	7.87	10.20
0516	cycle 1 mile		96	.083			4.28	4.25	8.53	11.05
0518	cycle 2 miles		88	.091			4.66	4.63	9.29	12.05
0520	cycle 4 miles		72	.111			5.70	5.65	11.35	14.75

For customer support on your Plumbing Costs with RSMeans Data, call 800.448.8182.

397

31 23 23.20 Hauling		Crew	Daily Output	Labor-Hours	Unit	Material	2021 Bare Costs Labor	2021 Bare Costs Equipment	Total	Total Incl O&P
0522	cycle 6 miles	B-34A	64	.125	L.C.Y.		6.40	6.35	12.75	16.60
0524	cycle 8 miles		56	.143			7.35	7.30	14.65	18.95
0526	20 MPH avg., cycle 0.5 mile		104	.077			3.95	3.92	7.87	10.20
0528	cycle 1 mile		96	.083			4.28	4.25	8.53	11.05
0530	cycle 2 miles		96	.083			4.28	4.25	8.53	11.05
0532	cycle 4 miles		80	.100			5.15	5.10	10.25	13.25
0534	cycle 6 miles		72	.111			5.70	5.65	11.35	14.75
0536	cycle 8 miles		64	.125			6.40	6.35	12.75	16.60
0544	25 MPH avg., cycle 4 miles		88	.091			4.66	4.63	9.29	12.05
0546	cycle 6 miles		80	.100			5.15	5.10	10.25	13.25
0548	cycle 8 miles		72	.111			5.70	5.65	11.35	14.75
0550	30 MPH avg., cycle 4 miles		88	.091			4.66	4.63	9.29	12.05
0552	cycle 6 miles		80	.100			5.15	5.10	10.25	13.25
0554	cycle 8 miles	▼	72	.111			5.70	5.65	11.35	14.75
1014	12 C.Y. truck, cycle 0.5 mile, 15 MPH avg., 15 min. wait/ld./uld.	B-34B	336	.024			1.22	1.72	2.94	3.73
1016	cycle 1 mile		300	.027			1.37	1.93	3.30	4.17
1018	cycle 2 miles		252	.032			1.63	2.30	3.93	4.96
1020	cycle 4 miles		180	.044			2.28	3.22	5.50	6.95
1022	cycle 6 miles		144	.056			2.85	4.02	6.87	8.70
1024	cycle 8 miles		120	.067			3.42	4.83	8.25	10.40
1025	cycle 10 miles		96	.083			4.28	6.05	10.33	13.05
1026	20 MPH avg., cycle 0.5 mile	▼	348	.023	▼		1.18	1.66	2.84	3.59
1028	cycle 1 mile	B-34B	312	.026	L.C.Y.		1.32	1.86	3.18	4.01
1030	cycle 2 miles		276	.029			1.49	2.10	3.59	4.53
1032	cycle 4 miles		216	.037			1.90	2.68	4.58	5.80
1034	cycle 6 miles		168	.048			2.44	3.45	5.89	7.45
1036	cycle 8 miles		144	.056			2.85	4.02	6.87	8.70
1038	cycle 10 miles		120	.067			3.42	4.83	8.25	10.40
1040	25 MPH avg., cycle 4 miles		228	.035			1.80	2.54	4.34	5.50
1042	cycle 6 miles		192	.042			2.14	3.02	5.16	6.50
1044	cycle 8 miles		168	.048			2.44	3.45	5.89	7.45
1046	cycle 10 miles		144	.056			2.85	4.02	6.87	8.70
1050	30 MPH avg., cycle 4 miles		252	.032			1.63	2.30	3.93	4.96
1052	cycle 6 miles		216	.037			1.90	2.68	4.58	5.80
1054	cycle 8 miles		180	.044			2.28	3.22	5.50	6.95
1056	cycle 10 miles		156	.051			2.63	3.71	6.34	8
1060	35 MPH avg., cycle 4 miles		264	.030			1.55	2.19	3.74	4.73
1062	cycle 6 miles		228	.035			1.80	2.54	4.34	5.50
1064	cycle 8 miles		204	.039			2.01	2.84	4.85	6.15
1066	cycle 10 miles		180	.044			2.28	3.22	5.50	6.95
1068	cycle 20 miles		120	.067			3.42	4.83	8.25	10.40
1069	cycle 30 miles		84	.095			4.89	6.90	11.79	14.90
1070	cycle 40 miles		72	.111			5.70	8.05	13.75	17.35
1072	40 MPH avg., cycle 6 miles		240	.033			1.71	2.41	4.12	5.20
1074	cycle 8 miles		216	.037			1.90	2.68	4.58	5.80
1076	cycle 10 miles		192	.042			2.14	3.02	5.16	6.50
1078	cycle 20 miles		120	.067			3.42	4.83	8.25	10.40
1080	cycle 30 miles		96	.083			4.28	6.05	10.33	13.05
1082	cycle 40 miles		72	.111			5.70	8.05	13.75	17.35
1084	cycle 50 miles		60	.133			6.85	9.65	16.50	21
1094	45 MPH avg., cycle 8 miles		216	.037			1.90	2.68	4.58	5.80
1096	cycle 10 miles		204	.039			2.01	2.84	4.85	6.15
1098	cycle 20 miles	▼	132	.061	▼		3.11	4.39	7.50	9.50

31 23 23 – Fill

31 23 23.20 Hauling		Crew	Daily Output	Labor-Hours	Unit	Material	2021 Bare Costs Labor	Equipment	Total	Total Incl O&P
1100	cycle 30 miles	B-34B	108	.074	L.C.Y.		3.80	5.35	9.15	11.60
1102	cycle 40 miles		84	.095			4.89	6.90	11.79	14.90
1104	cycle 50 miles		72	.111			5.70	8.05	13.75	17.35
1106	50 MPH avg., cycle 10 miles		216	.037			1.90	2.68	4.58	5.80
1108	cycle 20 miles		144	.056			2.85	4.02	6.87	8.70
1110	cycle 30 miles		108	.074			3.80	5.35	9.15	11.60
1112	cycle 40 miles		84	.095			4.89	6.90	11.79	14.90
1114	cycle 50 miles		72	.111			5.70	8.05	13.75	17.35
1214	15 MPH avg., cycle 0.5 mile, 20 min. wait/ld./uld.		264	.030			1.55	2.19	3.74	4.73
1216	cycle 1 mile	↓	240	.033	↓		1.71	2.41	4.12	5.20
1218	cycle 2 miles	B-34B	204	.039	L.C.Y.		2.01	2.84	4.85	6.15
1220	cycle 4 miles		156	.051			2.63	3.71	6.34	8
1222	cycle 6 miles		132	.061			3.11	4.39	7.50	9.50
1224	cycle 8 miles		108	.074			3.80	5.35	9.15	11.60
1225	cycle 10 miles		96	.083			4.28	6.05	10.33	13.05
1226	20 MPH avg., cycle 0.5 mile		264	.030			1.55	2.19	3.74	4.73
1228	cycle 1 mile		252	.032			1.63	2.30	3.93	4.96
1230	cycle 2 miles		216	.037			1.90	2.68	4.58	5.80
1232	cycle 4 miles		180	.044			2.28	3.22	5.50	6.95
1234	cycle 6 miles		144	.056			2.85	4.02	6.87	8.70
1236	cycle 8 miles		132	.061			3.11	4.39	7.50	9.50
1238	cycle 10 miles		108	.074			3.80	5.35	9.15	11.60
1240	25 MPH avg., cycle 4 miles		192	.042			2.14	3.02	5.16	6.50
1242	cycle 6 miles		168	.048			2.44	3.45	5.89	7.45
1244	cycle 8 miles		144	.056			2.85	4.02	6.87	8.70
1246	cycle 10 miles		132	.061			3.11	4.39	7.50	9.50
1250	30 MPH avg., cycle 4 miles		204	.039			2.01	2.84	4.85	6.15
1252	cycle 6 miles		180	.044			2.28	3.22	5.50	6.95
1254	cycle 8 miles		156	.051			2.63	3.71	6.34	8
1256	cycle 10 miles		144	.056			2.85	4.02	6.87	8.70
1260	35 MPH avg., cycle 4 miles		216	.037			1.90	2.68	4.58	5.80
1262	cycle 6 miles		192	.042			2.14	3.02	5.16	6.50
1264	cycle 8 miles		168	.048			2.44	3.45	5.89	7.45
1266	cycle 10 miles		156	.051			2.63	3.71	6.34	8
1268	cycle 20 miles		108	.074			3.80	5.35	9.15	11.60
1269	cycle 30 miles		72	.111			5.70	8.05	13.75	17.35
1270	cycle 40 miles		60	.133			6.85	9.65	16.50	21
1272	40 MPH avg., cycle 6 miles		192	.042			2.14	3.02	5.16	6.50
1274	cycle 8 miles		180	.044			2.28	3.22	5.50	6.95
1276	cycle 10 miles		156	.051			2.63	3.71	6.34	8
1278	cycle 20 miles		108	.074			3.80	5.35	9.15	11.60
1280	cycle 30 miles		84	.095			4.89	6.90	11.79	14.90
1282	cycle 40 miles		72	.111			5.70	8.05	13.75	17.35
1284	cycle 50 miles		60	.133			6.85	9.65	16.50	21
1294	45 MPH avg., cycle 8 miles		180	.044			2.28	3.22	5.50	6.95
1296	cycle 10 miles		168	.048			2.44	3.45	5.89	7.45
1298	cycle 20 miles		120	.067			3.42	4.83	8.25	10.40
1300	cycle 30 miles		96	.083			4.28	6.05	10.33	13.05
1302	cycle 40 miles		72	.111			5.70	8.05	13.75	17.35
1304	cycle 50 miles		60	.133			6.85	9.65	16.50	21
1306	50 MPH avg., cycle 10 miles		180	.044			2.28	3.22	5.50	6.95
1308	cycle 20 miles	↓	132	.061	↓		3.11	4.39	7.50	9.50
1310	cycle 30 miles	B-34B	96	.083	L.C.Y.		4.28	6.05	10.33	13.05

31 23 23.20 Hauling		Crew	Daily Output	Labor-Hours	Unit	Material	2021 Bare Costs Labor	Equipment	Total	Total Incl O&P
1312	cycle 40 miles	B-34B	84	.095	L.C.Y.		4.89	6.90	11.79	14.90
1314	cycle 50 miles		72	.111			5.70	8.05	13.75	17.35
1414	15 MPH avg., cycle 0.5 mile, 25 min. wait/ld./uld.		204	.039			2.01	2.84	4.85	6.15
1416	cycle 1 mile		192	.042			2.14	3.02	5.16	6.50
1418	cycle 2 miles		168	.048			2.44	3.45	5.89	7.45
1420	cycle 4 miles		132	.061			3.11	4.39	7.50	9.50
1422	cycle 6 miles		120	.067			3.42	4.83	8.25	10.40
1424	cycle 8 miles		96	.083			4.28	6.05	10.33	13.05
1425	cycle 10 miles		84	.095			4.89	6.90	11.79	14.90
1426	20 MPH avg., cycle 0.5 mile		216	.037			1.90	2.68	4.58	5.80
1428	cycle 1 mile		204	.039			2.01	2.84	4.85	6.15
1430	cycle 2 miles		180	.044			2.28	3.22	5.50	6.95
1432	cycle 4 miles		156	.051			2.63	3.71	6.34	8
1434	cycle 6 miles		132	.061			3.11	4.39	7.50	9.50
1436	cycle 8 miles		120	.067			3.42	4.83	8.25	10.40
1438	cycle 10 miles		96	.083			4.28	6.05	10.33	13.05
1440	25 MPH avg., cycle 4 miles		168	.048			2.44	3.45	5.89	7.45
1442	cycle 6 miles		144	.056			2.85	4.02	6.87	8.70
1444	cycle 8 miles		132	.061			3.11	4.39	7.50	9.50
1446	cycle 10 miles		108	.074			3.80	5.35	9.15	11.60
1450	30 MPH avg., cycle 4 miles		168	.048			2.44	3.45	5.89	7.45
1452	cycle 6 miles		156	.051			2.63	3.71	6.34	8
1454	cycle 8 miles		132	.061			3.11	4.39	7.50	9.50
1456	cycle 10 miles		120	.067			3.42	4.83	8.25	10.40
1460	35 MPH avg., cycle 4 miles		180	.044			2.28	3.22	5.50	6.95
1462	cycle 6 miles		156	.051			2.63	3.71	6.34	8
1464	cycle 8 miles		144	.056			2.85	4.02	6.87	8.70
1466	cycle 10 miles		132	.061			3.11	4.39	7.50	9.50
1468	cycle 20 miles		96	.083			4.28	6.05	10.33	13.05
1469	cycle 30 miles		72	.111			5.70	8.05	13.75	17.35
1470	cycle 40 miles		60	.133			6.85	9.65	16.50	21
1472	40 MPH avg., cycle 6 miles		168	.048			2.44	3.45	5.89	7.45
1474	cycle 8 miles		156	.051			2.63	3.71	6.34	8
1476	cycle 10 miles		144	.056			2.85	4.02	6.87	8.70
1478	cycle 20 miles		96	.083			4.28	6.05	10.33	13.05
1480	cycle 30 miles		84	.095			4.89	6.90	11.79	14.90
1482	cycle 40 miles		60	.133			6.85	9.65	16.50	21
1484	cycle 50 miles		60	.133			6.85	9.65	16.50	21
1494	45 MPH avg., cycle 8 miles		156	.051			2.63	3.71	6.34	8
1496	cycle 10 miles		144	.056			2.85	4.02	6.87	8.70
1498	cycle 20 miles		108	.074			3.80	5.35	9.15	11.60
1500	cycle 30 miles		84	.095			4.89	6.90	11.79	14.90
1502	cycle 40 miles		72	.111			5.70	8.05	13.75	17.35
1504	cycle 50 miles		60	.133			6.85	9.65	16.50	21
1506	50 MPH avg., cycle 10 miles		156	.051			2.63	3.71	6.34	8
1508	cycle 20 miles		120	.067			3.42	4.83	8.25	10.40
1510	cycle 30 miles		96	.083			4.28	6.05	10.33	13.05
1512	cycle 40 miles		72	.111			5.70	8.05	13.75	17.35
1514	cycle 50 miles		60	.133			6.85	9.65	16.50	21
1614	15 MPH avg., cycle 0.5 mile, 30 min. wait/ld./uld.		180	.044			2.28	3.22	5.50	6.95
1616	cycle 1 mile		168	.048			2.44	3.45	5.89	7.45
1618	cycle 2 miles		144	.056			2.85	4.02	6.87	8.70
1620	cycle 4 miles		120	.067			3.42	4.83	8.25	10.40

31 23 Excavation and Fill

31 23 23 – Fill

31 23 23.20 Hauling		Crew	Daily Output	Labor-Hours	Unit	Material	2021 Bare Costs Labor	2021 Bare Costs Equipment	Total	Total Incl O&P
1622	cycle 6 miles	B-34B	108	.074	L.C.Y.		3.80	5.35	9.15	11.60
1624	cycle 8 miles		84	.095			4.89	6.90	11.79	14.90
1625	cycle 10 miles		84	.095			4.89	6.90	11.79	14.90
1626	20 MPH avg., cycle 0.5 mile		180	.044			2.28	3.22	5.50	6.95
1628	cycle 1 mile		168	.048			2.44	3.45	5.89	7.45
1630	cycle 2 miles		156	.051			2.63	3.71	6.34	8
1632	cycle 4 miles		132	.061			3.11	4.39	7.50	9.50
1634	cycle 6 miles		120	.067			3.42	4.83	8.25	10.40
1636	cycle 8 miles		108	.074			3.80	5.35	9.15	11.60
1638	cycle 10 miles		96	.083			4.28	6.05	10.33	13.05
1640	25 MPH avg., cycle 4 miles		144	.056			2.85	4.02	6.87	8.70
1642	cycle 6 miles		132	.061			3.11	4.39	7.50	9.50
1644	cycle 8 miles		108	.074			3.80	5.35	9.15	11.60
1646	cycle 10 miles		108	.074			3.80	5.35	9.15	11.60
1650	30 MPH avg., cycle 4 miles		144	.056			2.85	4.02	6.87	8.70
1652	cycle 6 miles		132	.061			3.11	4.39	7.50	9.50
1654	cycle 8 miles		120	.067			3.42	4.83	8.25	10.40
1656	cycle 10 miles		108	.074			3.80	5.35	9.15	11.60
1660	35 MPH avg., cycle 4 miles		156	.051			2.63	3.71	6.34	8
1662	cycle 6 miles		144	.056			2.85	4.02	6.87	8.70
1664	cycle 8 miles		132	.061			3.11	4.39	7.50	9.50
1666	cycle 10 miles		120	.067			3.42	4.83	8.25	10.40
1668	cycle 20 miles		84	.095			4.89	6.90	11.79	14.90
1669	cycle 30 miles		72	.111			5.70	8.05	13.75	17.35
1670	cycle 40 miles		60	.133			6.85	9.65	16.50	21
1672	40 MPH avg., cycle 6 miles		144	.056			2.85	4.02	6.87	8.70
1674	cycle 8 miles		132	.061			3.11	4.39	7.50	9.50
1676	cycle 10 miles		120	.067			3.42	4.83	8.25	10.40
1678	cycle 20 miles		96	.083			4.28	6.05	10.33	13.05
1680	cycle 30 miles		72	.111			5.70	8.05	13.75	17.35
1682	cycle 40 miles		60	.133			6.85	9.65	16.50	21
1684	cycle 50 miles		48	.167			8.55	12.05	20.60	26
1694	45 MPH avg., cycle 8 miles		144	.056			2.85	4.02	6.87	8.70
1696	cycle 10 miles		132	.061			3.11	4.39	7.50	9.50
1698	cycle 20 miles		96	.083			4.28	6.05	10.33	13.05
1700	cycle 30 miles		84	.095			4.89	6.90	11.79	14.90
1702	cycle 40 miles		60	.133			6.85	9.65	16.50	21
1704	cycle 50 miles		60	.133			6.85	9.65	16.50	21
1706	50 MPH avg., cycle 10 miles		132	.061			3.11	4.39	7.50	9.50
1708	cycle 20 miles		108	.074			3.80	5.35	9.15	11.60
1710	cycle 30 miles		84	.095			4.89	6.90	11.79	14.90
1712	cycle 40 miles		72	.111			5.70	8.05	13.75	17.35
1714	cycle 50 miles		60	.133			6.85	9.65	16.50	21
2000	Hauling, 8 C.Y. truck, small project cost per hour	B-34A	8	1	Hr.		51.50	51	102.50	133
2100	12 C.Y. truck	B-34B	8	1			51.50	72.50	124	156
2150	16.5 C.Y. truck	B-34C	8	1			51.50	79.50	131	164
2175	18 C.Y. 8 wheel truck	B-34I	8	1			51.50	94	145.50	181
2200	20 C.Y. truck	B-34D	8	1			51.50	81.50	133	167
2300	Grading at dump, or embankment if required, by dozer	B-10B	1000	.012	L.C.Y.		.65	1.52	2.17	2.64
2310	Spotter at fill or cut, if required	1 Clab	8	1	Hr.		44.50		44.50	66.50
9014	18 C.Y. truck, 8 wheels,15 min. wait/ld./uld.,15 MPH, cycle 0.5 mi.	B-34I	504	.016	L.C.Y.		.81	1.50	2.31	2.86
9016	cycle 1 mile		450	.018			.91	1.67	2.58	3.20
9018	cycle 2 miles		378	.021			1.09	1.99	3.08	3.81

For customer support on your Plumbing Costs with RSMeans Data, call 800.448.8182.

401

31 23 23.20 Hauling		Crew	Daily Output	Labor-Hours	Unit	Material	Labor	Equipment	Total	Total Incl O&P
							2021 Bare Costs			
9020	cycle 4 miles	B-34I	270	.030	L.C.Y.		1.52	2.79	4.31	5.35
9022	cycle 6 miles		216	.037			1.90	3.49	5.39	6.70
9024	cycle 8 miles		180	.044			2.28	4.19	6.47	8
9025	cycle 10 miles		144	.056			2.85	5.25	8.10	10
9026	20 MPH avg., cycle 0.5 mile		522	.015			.79	1.44	2.23	2.77
9028	cycle 1 mile		468	.017			.88	1.61	2.49	3.08
9030	cycle 2 miles		414	.019			.99	1.82	2.81	3.48
9032	cycle 4 miles		324	.025			1.27	2.33	3.60	4.45
9034	cycle 6 miles		252	.032			1.63	2.99	4.62	5.70
9036	cycle 8 miles		216	.037			1.90	3.49	5.39	6.70
9038	cycle 10 miles		180	.044			2.28	4.19	6.47	8
9040	25 MPH avg., cycle 4 miles		342	.023			1.20	2.20	3.40	4.21
9042	cycle 6 miles		288	.028			1.43	2.62	4.05	5
9044	cycle 8 miles		252	.032			1.63	2.99	4.62	5.70
9046	cycle 10 miles		216	.037			1.90	3.49	5.39	6.70
9050	30 MPH avg., cycle 4 miles		378	.021			1.09	1.99	3.08	3.81
9052	cycle 6 miles		324	.025			1.27	2.33	3.60	4.45
9054	cycle 8 miles		270	.030			1.52	2.79	4.31	5.35
9056	cycle 10 miles		234	.034			1.75	3.22	4.97	6.15
9060	35 MPH avg., cycle 4 miles		396	.020			1.04	1.90	2.94	3.64
9062	cycle 6 miles		342	.023			1.20	2.20	3.40	4.21
9064	cycle 8 miles		288	.028			1.43	2.62	4.05	5
9066	cycle 10 miles		270	.030			1.52	2.79	4.31	5.35
9068	cycle 20 miles		162	.049			2.53	4.65	7.18	8.90
9070	cycle 30 miles		126	.063			3.26	6	9.26	11.45
9072	cycle 40 miles		90	.089			4.56	8.35	12.91	16
9074	40 MPH avg., cycle 6 miles		360	.022			1.14	2.09	3.23	4
9076	cycle 8 miles		324	.025			1.27	2.33	3.60	4.45
9078	cycle 10 miles		288	.028			1.43	2.62	4.05	5
9080	cycle 20 miles		180	.044			2.28	4.19	6.47	8
9082	cycle 30 miles		144	.056			2.85	5.25	8.10	10
9084	cycle 40 miles		108	.074			3.80	7	10.80	13.40
9086	cycle 50 miles		90	.089			4.56	8.35	12.91	16
9094	45 MPH avg., cycle 8 miles		324	.025			1.27	2.33	3.60	4.45
9096	cycle 10 miles		306	.026			1.34	2.46	3.80	4.72
9098	cycle 20 miles		198	.040			2.07	3.81	5.88	7.30
9100	cycle 30 miles		144	.056			2.85	5.25	8.10	10
9102	cycle 40 miles		126	.063			3.26	6	9.26	11.45
9104	cycle 50 miles		108	.074			3.80	7	10.80	13.40
9106	50 MPH avg., cycle 10 miles		324	.025			1.27	2.33	3.60	4.45
9108	cycle 20 miles		216	.037			1.90	3.49	5.39	6.70
9110	cycle 30 miles		162	.049			2.53	4.65	7.18	8.90
9112	cycle 40 miles		126	.063			3.26	6	9.26	11.45
9114	cycle 50 miles		108	.074			3.80	7	10.80	13.40
9214	20 min. wait/ld./uld.,15 MPH, cycle 0.5 mi.		396	.020			1.04	1.90	2.94	3.64
9216	cycle 1 mile		360	.022			1.14	2.09	3.23	4
9218	cycle 2 miles		306	.026			1.34	2.46	3.80	4.72
9220	cycle 4 miles		234	.034			1.75	3.22	4.97	6.15
9222	cycle 6 miles		198	.040			2.07	3.81	5.88	7.30
9224	cycle 8 miles		162	.049			2.53	4.65	7.18	8.90
9225	cycle 10 miles		144	.056			2.85	5.25	8.10	10
9226	20 MPH avg., cycle 0.5 mile		396	.020			1.04	1.90	2.94	3.64
9228	cycle 1 mile		378	.021			1.09	1.99	3.08	3.81

31 23 23 – Fill

31 23 23.20 Hauling		Crew	Daily Output	Labor-Hours	Unit	Material	2021 Bare Costs Labor	Equipment	Total	Total Incl O&P
9230	cycle 2 miles	B-34I	324	.025	L.C.Y.		1.27	2.33	3.60	4.45
9232	cycle 4 miles		270	.030			1.52	2.79	4.31	5.35
9234	cycle 6 miles		216	.037			1.90	3.49	5.39	6.70
9236	cycle 8 miles		198	.040			2.07	3.81	5.88	7.30
9238	cycle 10 miles		162	.049			2.53	4.65	7.18	8.90
9240	25 MPH avg., cycle 4 miles		288	.028			1.43	2.62	4.05	5
9242	cycle 6 miles		252	.032			1.63	2.99	4.62	5.70
9244	cycle 8 miles		216	.037			1.90	3.49	5.39	6.70
9246	cycle 10 miles		198	.040			2.07	3.81	5.88	7.30
9250	30 MPH avg., cycle 4 miles		306	.026			1.34	2.46	3.80	4.72
9252	cycle 6 miles		270	.030			1.52	2.79	4.31	5.35
9254	cycle 8 miles		234	.034			1.75	3.22	4.97	6.15
9256	cycle 10 miles		216	.037			1.90	3.49	5.39	6.70
9260	35 MPH avg., cycle 4 miles		324	.025			1.27	2.33	3.60	4.45
9262	cycle 6 miles		288	.028			1.43	2.62	4.05	5
9264	cycle 8 miles		252	.032			1.63	2.99	4.62	5.70
9266	cycle 10 miles		234	.034			1.75	3.22	4.97	6.15
9268	cycle 20 miles		162	.049			2.53	4.65	7.18	8.90
9270	cycle 30 miles		108	.074			3.80	7	10.80	13.40
9272	cycle 40 miles		90	.089			4.56	8.35	12.91	16
9274	40 MPH avg., cycle 6 miles		288	.028			1.43	2.62	4.05	5
9276	cycle 8 miles		270	.030			1.52	2.79	4.31	5.35
9278	cycle 10 miles		234	.034			1.75	3.22	4.97	6.15
9280	cycle 20 miles		162	.049			2.53	4.65	7.18	8.90
9282	cycle 30 miles		126	.063			3.26	6	9.26	11.45
9284	cycle 40 miles		108	.074			3.80	7	10.80	13.40
9286	cycle 50 miles		90	.089			4.56	8.35	12.91	16
9294	45 MPH avg., cycle 8 miles		270	.030			1.52	2.79	4.31	5.35
9296	cycle 10 miles		252	.032			1.63	2.99	4.62	5.70
9298	cycle 20 miles		180	.044			2.28	4.19	6.47	8
9300	cycle 30 miles		144	.056			2.85	5.25	8.10	10
9302	cycle 40 miles		108	.074			3.80	7	10.80	13.40
9304	cycle 50 miles		90	.089			4.56	8.35	12.91	16
9306	50 MPH avg., cycle 10 miles		270	.030			1.52	2.79	4.31	5.35
9308	cycle 20 miles		198	.040			2.07	3.81	5.88	7.30
9310	cycle 30 miles		144	.056			2.85	5.25	8.10	10
9312	cycle 40 miles		126	.063			3.26	6	9.26	11.45
9314	cycle 50 miles		108	.074			3.80	7	10.80	13.40
9414	25 min. wait/ld./uld.,15 MPH, cycle 0.5 mi.		306	.026			1.34	2.46	3.80	4.72
9416	cycle 1 mile		288	.028			1.43	2.62	4.05	5
9418	cycle 2 miles		252	.032			1.63	2.99	4.62	5.70
9420	cycle 4 miles		198	.040			2.07	3.81	5.88	7.30
9422	cycle 6 miles		180	.044			2.28	4.19	6.47	8
9424	cycle 8 miles		144	.056			2.85	5.25	8.10	10
9425	cycle 10 miles		126	.063			3.26	6	9.26	11.45
9426	20 MPH avg., cycle 0.5 mile		324	.025			1.27	2.33	3.60	4.45
9428	cycle 1 mile	B-34I	306	.026	L.C.Y.		1.34	2.46	3.80	4.72
9430	cycle 2 miles		270	.030			1.52	2.79	4.31	5.35
9432	cycle 4 miles		234	.034			1.75	3.22	4.97	6.15
9434	cycle 6 miles		198	.040			2.07	3.81	5.88	7.30
9436	cycle 8 miles		180	.044			2.28	4.19	6.47	8
9438	cycle 10 miles		144	.056			2.85	5.25	8.10	10
9440	25 MPH avg., cycle 4 miles		252	.032			1.63	2.99	4.62	5.70

For customer support on your Plumbing Costs with RSMeans Data, call 800.448.8182.

403

31 23 23 – Fill

31 23 23.20 Hauling		Crew	Daily Output	Labor-Hours	Unit	Material	2021 Bare Costs Labor	2021 Bare Costs Equipment	Total	Total Incl O&P
9442	cycle 6 miles	B-34I	216	.037	L.C.Y.		1.90	3.49	5.39	6.70
9444	cycle 8 miles		198	.040			2.07	3.81	5.88	7.30
9446	cycle 10 miles		180	.044			2.28	4.19	6.47	8
9450	30 MPH avg., cycle 4 miles		252	.032			1.63	2.99	4.62	5.70
9452	cycle 6 miles		234	.034			1.75	3.22	4.97	6.15
9454	cycle 8 miles		198	.040			2.07	3.81	5.88	7.30
9456	cycle 10 miles		180	.044			2.28	4.19	6.47	8
9460	35 MPH avg., cycle 4 miles		270	.030			1.52	2.79	4.31	5.35
9462	cycle 6 miles		234	.034			1.75	3.22	4.97	6.15
9464	cycle 8 miles		216	.037			1.90	3.49	5.39	6.70
9466	cycle 10 miles		198	.040			2.07	3.81	5.88	7.30
9468	cycle 20 miles		144	.056			2.85	5.25	8.10	10
9470	cycle 30 miles		108	.074			3.80	7	10.80	13.40
9472	cycle 40 miles		90	.089			4.56	8.35	12.91	16
9474	40 MPH avg., cycle 6 miles		252	.032			1.63	2.99	4.62	5.70
9476	cycle 8 miles		234	.034			1.75	3.22	4.97	6.15
9478	cycle 10 miles		216	.037			1.90	3.49	5.39	6.70
9480	cycle 20 miles		144	.056			2.85	5.25	8.10	10
9482	cycle 30 miles		126	.063			3.26	6	9.26	11.45
9484	cycle 40 miles		90	.089			4.56	8.35	12.91	16
9486	cycle 50 miles		90	.089			4.56	8.35	12.91	16
9494	45 MPH avg., cycle 8 miles		234	.034			1.75	3.22	4.97	6.15
9496	cycle 10 miles		216	.037			1.90	3.49	5.39	6.70
9498	cycle 20 miles		162	.049			2.53	4.65	7.18	8.90
9500	cycle 30 miles		126	.063			3.26	6	9.26	11.45
9502	cycle 40 miles		108	.074			3.80	7	10.80	13.40
9504	cycle 50 miles		90	.089			4.56	8.35	12.91	16
9506	50 MPH avg., cycle 10 miles		234	.034			1.75	3.22	4.97	6.15
9508	cycle 20 miles		180	.044			2.28	4.19	6.47	8
9510	cycle 30 miles		144	.056			2.85	5.25	8.10	10
9512	cycle 40 miles		108	.074			3.80	7	10.80	13.40
9514	cycle 50 miles		90	.089			4.56	8.35	12.91	16
9614	30 min. wait/ld./uld.,15 MPH, cycle 0.5 mi.		270	.030			1.52	2.79	4.31	5.35
9616	cycle 1 mile		252	.032			1.63	2.99	4.62	5.70
9618	cycle 2 miles		216	.037			1.90	3.49	5.39	6.70
9620	cycle 4 miles		180	.044			2.28	4.19	6.47	8
9622	cycle 6 miles		162	.049			2.53	4.65	7.18	8.90
9624	cycle 8 miles		126	.063			3.26	6	9.26	11.45
9625	cycle 10 miles		126	.063			3.26	6	9.26	11.45
9626	20 MPH avg., cycle 0.5 mile		270	.030			1.52	2.79	4.31	5.35
9628	cycle 1 mile		252	.032			1.63	2.99	4.62	5.70
9630	cycle 2 miles		234	.034			1.75	3.22	4.97	6.15
9632	cycle 4 miles		198	.040			2.07	3.81	5.88	7.30
9634	cycle 6 miles		180	.044			2.28	4.19	6.47	8
9636	cycle 8 miles		162	.049			2.53	4.65	7.18	8.90
9638	cycle 10 miles		144	.056			2.85	5.25	8.10	10
9640	25 MPH avg., cycle 4 miles		216	.037			1.90	3.49	5.39	6.70
9642	cycle 6 miles		198	.040			2.07	3.81	5.88	7.30
9644	cycle 8 miles		180	.044			2.28	4.19	6.47	8
9646	cycle 10 miles		162	.049			2.53	4.65	7.18	8.90
9650	30 MPH avg., cycle 4 miles		216	.037			1.90	3.49	5.39	6.70
9652	cycle 6 miles		198	.040			2.07	3.81	5.88	7.30
9654	cycle 8 miles		180	.044			2.28	4.19	6.47	8

31 23 23.20 Hauling		Crew	Daily Output	Labor-Hours	Unit	Material	2021 Bare Costs Labor	Equipment	Total	Total Incl O&P
9656	cycle 10 miles	B-34I	162	.049	L.C.Y.		2.53	4.65	7.18	8.90
9660	35 MPH avg., cycle 4 miles		234	.034			1.75	3.22	4.97	6.15
9662	cycle 6 miles		216	.037			1.90	3.49	5.39	6.70
9664	cycle 8 miles		198	.040			2.07	3.81	5.88	7.30
9666	cycle 10 miles		180	.044			2.28	4.19	6.47	8
9668	cycle 20 miles		126	.063			3.26	6	9.26	11.45
9670	cycle 30 miles		108	.074			3.80	7	10.80	13.40
9672	cycle 40 miles		90	.089			4.56	8.35	12.91	16
9674	40 MPH avg., cycle 6 miles		216	.037			1.90	3.49	5.39	6.70
9676	cycle 8 miles		198	.040			2.07	3.81	5.88	7.30
9678	cycle 10 miles		180	.044			2.28	4.19	6.47	8
9680	cycle 20 miles		144	.056			2.85	5.25	8.10	10
9682	cycle 30 miles		108	.074			3.80	7	10.80	13.40
9684	cycle 40 miles		90	.089			4.56	8.35	12.91	16
9686	cycle 50 miles		72	.111			5.70	10.45	16.15	20
9694	45 MPH avg., cycle 8 miles		216	.037			1.90	3.49	5.39	6.70
9696	cycle 10 miles		198	.040			2.07	3.81	5.88	7.30
9698	cycle 20 miles		144	.056			2.85	5.25	8.10	10
9700	cycle 30 miles		126	.063			3.26	6	9.26	11.45
9702	cycle 40 miles		108	.074			3.80	7	10.80	13.40
9704	cycle 50 miles		90	.089			4.56	8.35	12.91	16
9706	50 MPH avg., cycle 10 miles		198	.040			2.07	3.81	5.88	7.30
9708	cycle 20 miles		162	.049			2.53	4.65	7.18	8.90
9710	cycle 30 miles	B-34I	126	.063	L.C.Y.		3.26	6	9.26	11.45
9712	cycle 40 miles		108	.074			3.80	7	10.80	13.40
9714	cycle 50 miles		90	.089			4.56	8.35	12.91	16

For customer support on your Plumbing Costs with RSMeans Data, call 800.448.8182.

405

Division Notes

		CREW	DAILY OUTPUT	LABOR-HOURS	UNIT	BARE COSTS				TOTAL INCL O&P
						MAT.	LABOR	EQUIP.	TOTAL	

Estimating Tips

32 01 00 Operations and Maintenance of Exterior Improvements

- Recycling of asphalt pavement is becoming very popular and is an alternative to removal and replacement. It can be a good value engineering proposal if removed pavement can be recycled, either at the project site or at another site that is reasonably close to the project site. Sections on repair of flexible and rigid pavement are included.

32 10 00 Bases, Ballasts, and Paving

- When estimating paving, keep in mind the project schedule. Also note that prices for asphalt and concrete are generally higher in the cold seasons. Lines for pavement markings, including tactile warning systems and fence lines, are included.

32 90 00 Planting

- The timing of planting and guarantee specifications often dictate the costs for establishing tree and shrub growth and a stand of grass or ground cover. Establish the work performance schedule to coincide with the local planting season. Maintenance and growth guarantees can add 20–100% to the total landscaping cost and can be contractually cumbersome. The cost to replace trees and shrubs can be as high as 5% of the total cost, depending on the planting zone, soil conditions, and time of year.

Reference Numbers

Reference numbers are shown at the beginning of some major classifications. These numbers refer to related items in the Reference Section. The reference information may be an estimating procedure, an alternate pricing method, or technical information.

Note: Not all subdivisions listed here necessarily appear. ■

Same Data. Simplified.

Enjoy the convenience and efficiency of accessing your costs anywhere:

- **Skip the multiplier** by setting your location
- **Quickly search,** edit, favorite and share costs
- **Stay on top of price changes** with automatic updates

Discover more at rsmeans.com/online

32 12 Flexible Paving

32 12 16 – Asphalt Paving

32 12 16.13 Plant-Mix Asphalt Paving

		Crew	Daily Output	Labor-Hours	Unit	Material	2021 Bare Costs Labor	Equipment	Total	Total Incl O&P
0010	**PLANT-MIX ASPHALT PAVING**									
0020	For highways and large paved areas, excludes hauling									
0025	See Section 31 23 23.20 for hauling costs									
0080	Binder course, 1-1/2" thick	B-25	7725	.011	S.Y.	5.60	.55	.35	6.50	7.35
0120	2" thick		6345	.014		7.45	.67	.43	8.55	9.65
0130	2-1/2" thick		5620	.016		9.30	.76	.49	10.55	11.85
0160	3" thick		4905	.018		11.15	.87	.56	12.58	14.15
0170	3-1/2" thick		4520	.019		13	.95	.61	14.56	16.40
0200	4" thick	▼	4140	.021		14.85	1.03	.66	16.54	18.60
0300	Wearing course, 1" thick	B-25B	10575	.009		3.68	.45	.28	4.41	5.05
0340	1-1/2" thick		7725	.012		6.20	.61	.39	7.20	8.15
0380	2" thick		6345	.015		8.30	.75	.47	9.52	10.80
0420	2-1/2" thick		5480	.018		10.25	.87	.54	11.66	13.20
0460	3" thick		4900	.020		12.25	.97	.61	13.83	15.55
0470	3-1/2" thick		4520	.021		14.35	1.05	.66	16.06	18.10
0480	4" thick	▼	4140	.023		16.40	1.15	.72	18.27	20.50
0600	Porous pavement, 1-1/2" open graded friction course	B-25	7725	.011	▼	15.30	.55	.35	16.20	18.05
0800	Alternate method of figuring paving costs									
0810	Binder course, 1-1/2" thick	B-25	630	.140	Ton	68.50	6.80	4.35	79.65	90
0811	2" thick		690	.128		68.50	6.20	3.97	78.67	88.50
0812	3" thick		800	.110		68.50	5.35	3.42	77.27	86.50
0813	4" thick	▼	900	.098		68.50	4.75	3.04	76.29	85.50
0850	Wearing course, 1" thick	B-25B	575	.167		72	8.25	5.20	85.45	97.50
0851	1-1/2" thick		630	.152		72	7.55	4.74	84.29	96
0852	2" thick		690	.139		72	6.90	4.33	83.23	94.50
0853	2-1/2" thick		765	.125		72	6.20	3.90	82.10	93
0854	3" thick		800	.120	▼	72	5.95	3.73	81.68	92.50
1000	Pavement replacement over trench, 2" thick	B-21	90	.311	S.Y.	7.65	15.90	2.12	25.67	35
1050	4" thick		70	.400		15.15	20.50	2.73	38.38	50
1080	6" thick	▼	55	.509	▼	24	26	3.47	53.47	69.50

32 12 16.14 Asphaltic Concrete Paving

		Crew	Daily Output	Labor-Hours	Unit	Material	2021 Bare Costs Labor	Equipment	Total	Total Incl O&P
0011	**ASPHALTIC CONCRETE PAVING**, parking lots & driveways									
0015	No asphalt hauling included									
0018	Use 6.05 C.Y. per inch per M.S.F. for hauling									
0020	6" stone base, 2" binder course, 1" topping	B-25C	9000	.005	S.F.	2.06	.26	.27	2.59	2.94
0025	2" binder course, 2" topping		9000	.005		2.49	.26	.27	3.02	3.42
0030	3" binder course, 2" topping		9000	.005		2.91	.26	.27	3.44	3.88
0035	4" binder course, 2" topping		9000	.005		3.32	.26	.27	3.85	4.33
0040	1-1/2" binder course, 1" topping		9000	.005		1.85	.26	.27	2.38	2.72
0042	3" binder course, 1" topping		9000	.005		2.47	.26	.27	3	3.40
0045	3" binder course, 3" topping		9000	.005		3.35	.26	.27	3.88	4.37
0050	4" binder course, 3" topping		9000	.005		3.76	.26	.27	4.29	4.82
0055	4" binder course, 4" topping		9000	.005		4.19	.26	.27	4.72	5.30
0300	Binder course, 1-1/2" thick		35000	.001		.62	.07	.07	.76	.86
0400	2" thick		25000	.002		.80	.10	.10	1	1.13
0500	3" thick		15000	.003		1.24	.16	.16	1.56	1.79
0600	4" thick		10800	.004		1.63	.22	.22	2.07	2.36
0800	Sand finish course, 3/4" thick		41000	.001		.31	.06	.06	.43	.49
0900	1" thick	▼	34000	.001		.38	.07	.07	.52	.60
1000	Fill pot holes, hot mix, 2" thick	B-16	4200	.008		.82	.36	.14	1.32	1.58
1100	4" thick		3500	.009		1.20	.43	.17	1.80	2.14
1120	6" thick	▼	3100	.010		1.61	.48	.19	2.28	2.70

32 12 Flexible Paving

32 12 16 – Asphalt Paving

32 12 16.14 Asphaltic Concrete Paving		Crew	Daily Output	Labor-Hours	Unit	Material	2021 Bare Costs Labor	2021 Bare Costs Equipment	Total	Total Incl O&P
1140	Cold patch, 2" thick	B-51	3000	.016	S.F.	1.24	.73	.07	2.04	2.53
1160	4" thick	↓	2700	.018	↓	2.37	.81	.07	3.25	3.90
1180	6" thick	↓	1900	.025	↓	3.69	1.15	.10	4.94	5.85

32 84 Planting Irrigation

32 84 13 – Drip Irrigation

32 84 13.10 Subsurface Drip Irrigation

			Crew	Daily Output	Labor-Hours	Unit	Material	2021 Bare Costs Labor	2021 Bare Costs Equipment	Total	Total Incl O&P
0010	**SUBSURFACE DRIP IRRIGATION**										
0011	Looped grid, pressure compensating										
0100	Preinserted PE emitter, line, hand bury, irregular area, small	G	3 Skwk	1200	.020	L.F.	.26	1.14		1.40	2.01
0150	Medium	G		1800	.013		.26	.76		1.02	1.44
0200	Large	G		2520	.010		.26	.54		.80	1.11
0250	Rectangular area, small	G		2040	.012		.26	.67		.93	1.30
0300	Medium	G		2640	.009		.26	.52		.78	1.07
0350	Large	G		3600	.007		.26	.38		.64	.86
0400	Install in trench, irregular area, small	G		4050	.006		.26	.34		.60	.80
0450	Medium	G		7488	.003		.26	.18		.44	.57
0500	Large	G		16560	.001		.26	.08		.34	.41
0550	Rectangular area, small	G		8100	.003		.26	.17		.43	.54
0600	Medium			21960	.001		.26	.06		.32	.38
0650	Large	G	↓	33264	.001		.26	.04		.30	.35
0700	Trenching and backfill	G	B-53	500	.016			.89	.32	1.21	1.67
0800	Vinyl tubing, 1/4", material only	G					.08			.08	.09
0850	Supply tubing, 1/2", material only, 100' coil	G					.12			.12	.13
0900	500' coil	G				↓	.11			.11	.12
0950	Compression fittings	G	1 Skwk	90	.089	Ea.	1.57	5.10		6.67	9.40
1000	Barbed fittings, 1/4"	G		360	.022		.13	1.27		1.40	2.05
1100	Flush risers	G		60	.133		4.21	7.60		11.81	16.10
1150	Flush ends, figure eight	G		180	.044		.61	2.54		3.15	4.49
1300	Auto flush, spring loaded	G		90	.089		2.28	5.10		7.38	10.15
1350	Volumetric	G		90	.089		8.05	5.10		13.15	16.50
1400	Air relief valve, inline with compensation tee, 1/2"	G		45	.178		16.70	10.15		26.85	33.50
1450	1"	G		30	.267		18.55	15.25		33.80	43.50
1500	Round box for flush ends, 6"	G		30	.267		7.95	15.25		23.20	32
1600	Screen filter, 3/4" screen	G		12	.667		9.10	38		47.10	67.50
1650	1" disk	G		8	1		28	57		85	117
1700	1-1/2" disk	G		4	2		112	114		226	295
1750	2" disk	G	↓	3	2.667	↓	151	152		303	395
1800	Typical installation 18" OC, small, minimum					S.F.				1.53	2.35
1850	Maximum									1.78	2.75
1900	Large, minimum									1.48	2.28
2000	Maximum					↓				1.71	2.64
2100	For non-pressure compensating systems, deduct									10%	10%

32 84 23 – Underground Sprinklers

32 84 23.10 Sprinkler Irrigation System

			Crew	Daily Output	Labor-Hours	Unit	Material	2021 Bare Costs Labor	2021 Bare Costs Equipment	Total	Total Incl O&P
0010	**SPRINKLER IRRIGATION SYSTEM**	G2050-710									
0011	For lawns										
0800	Residential system, custom, 1" supply		B-20	2000	.012	S.F.	.28	.59		.87	1.20
0900	1-1/2" supply		"	1800	.013	"	.42	.66		1.08	1.44
1020	Pop up spray head w/risers, hi-pop, full circle pattern, 4"		2 Skwk	76	.211	Ea.	4.24	12		16.24	23

For customer support on your Plumbing Costs with RSMeans Data, call 800.448.8182.

409

32 84 23.10 Sprinkler Irrigation System	Crew	Daily Output	Labor-Hours	Unit	Material	2021 Bare Costs Labor	2021 Bare Costs Equipment	Total	Total Incl O&P
1030	1/2 circle pattern, 4"	2 Skwk	76	.211	Ea.	7.10	12		19.10	26
1040	6", full circle pattern		76	.211		20.50	12		32.50	40.50
1050	1/2 circle pattern, 6"		76	.211		10.90	12		22.90	30
1060	12", full circle pattern		76	.211		15.65	12		27.65	35.50
1070	1/2 circle pattern, 12"		76	.211		17.50	12		29.50	37.50
1080	Pop up bubbler head w/risers, hi-pop bubbler head, 4"		76	.211		5.05	12		17.05	23.50
1110	Impact full/part circle sprinklers, 28'-54' @ 25-60 psi		37	.432		23	24.50		47.50	62.50
1120	Spaced 37'-49' @ 25-50 psi		37	.432		19.40	24.50		43.90	58.50
1130	Spaced 43'-61' @ 30-60 psi		37	.432		65	24.50		89.50	109
1140	Spaced 54'-78' @ 40-80 psi		37	.432		110	24.50		134.50	158
1145	Impact rotor pop-up full/part commercial circle sprinklers									
1150	Spaced 42'-65' @ 35-80 psi	2 Skwk	25	.640	Ea.	20.50	36.50		57	77.50
1160	Spaced 48'-76' @ 45-85 psi	"	25	.640	"	19.20	36.50		55.70	76
1165	Impact rotor pop-up part. circle comm., 53'-75', 55-100 psi, w/accessories									
1180	Sprinkler, premium, pop-up rotator, 50'-100'	2 Skwk	25	.640	Ea.	96.50	36.50		133	161
1250	Plastic case, 2 nozzle, metal cover		25	.640		117	36.50		153.50	184
1260	Rubber cover		25	.640		102	36.50		138.50	167
1270	Iron case, 2 nozzle, metal cover		22	.727		147	41.50		188.50	225
1280	Rubber cover		22	.727		146	41.50		187.50	224
1282	Impact rotor pop-up full circle commercial, 39'-99', 30-100 psi									
1284	Plastic case, metal cover	2 Skwk	25	.640	Ea.	93.50	36.50		130	158
1286	Rubber cover		25	.640		136	36.50		172.50	205
1288	Iron case, metal cover		22	.727		180	41.50		221.50	262
1290	Rubber cover		22	.727		183	41.50		224.50	265
1292	Plastic case, 2 nozzle, metal cover		22	.727		135	41.50		176.50	211
1294	Rubber cover		22	.727		138	41.50		179.50	214
1296	Iron case, 2 nozzle, metal cover		20	.800		158	45.50		203.50	243
1298	Rubber cover		20	.800		187	45.50		232.50	275
1305	Electric remote control valve, plastic, 3/4"		18	.889		20.50	51		71.50	99
1310	1"		18	.889		23.50	51		74.50	103
1320	1-1/2"		18	.889		98.50	51		149.50	185
1330	2"		18	.889		117	51		168	205
1335	Quick coupling valves, brass, locking cover									
1340	Inlet coupling valve, 3/4"	2 Skwk	18.75	.853	Ea.	23.50	48.50		72	99.50
1350	1"		18.75	.853		28	48.50		76.50	105
1360	Controller valve boxes, 6" round boxes		18.75	.853		5.30	48.50		53.80	79.50
1370	10" round boxes		14.25	1.123		11	64		75	109
1380	12" square box		9.75	1.641		42	93.50		135.50	188
1388	Electromech. control, 14 day 3-60 min., auto start to 23/day									
1390	4 station	2 Skwk	1.04	15.385	Ea.	68.50	880		948.50	1,400
1400	7 station		.64	25		178	1,425		1,603	2,350
1410	12 station		.40	40		202	2,275		2,477	3,650
1420	Dual programs, 18 station		.24	66.667		267	3,800		4,067	6,025
1430	23 station		.16	100		310	5,700		6,010	8,950
1435	Backflow preventer, bronze, 0-175 psi, w/valves, test cocks									
1440	3/4"	2 Skwk	6	2.667	Ea.	90	152		242	330
1450	1"		6	2.667		103	152		255	340
1460	1-1/2"		6	2.667		282	152		434	540
1470	2"		6	2.667		395	152		547	665
1475	Pressure vacuum breaker, brass, 15-150 psi									
1480	3/4"	2 Skwk	6	2.667	Ea.	49.50	152		201.50	284
1490	1"		6	2.667		57.50	152		209.50	292
1500	1-1/2"		6	2.667		85	152		237	325

32 84 Planting Irrigation

32 84 23 – Underground Sprinklers

32 84 23.10 Sprinkler Irrigation System	Crew	Daily Output	Labor-Hours	Unit	Material	2021 Bare Costs Labor	Equipment	Total	Total Incl O&P
1510	2"	2 Skwk	6	2.667	Ea.	159	152		311	405
6000	Riser mounted gear drive sprinkler, nozzle and case									
6200	Plastic, medium volume	1 Skwk	30	.267	Ea.	19.95	15.25		35.20	45
7000	Riser mounted impact sprinkler, body, part or full circle									
7200	Full circle plastic, low/medium volume	1 Skwk	25	.320	Ea.	19.80	18.25		38.05	49.50
7400	Brass, low/medium volume		25	.320		44.50	18.25		62.75	76.50
7600	Medium volume		25	.320		59.50	18.25		77.75	93
7800	Female thread		25	.320		106	18.25		124.25	145
8000	Riser mounted impact sprinkler, body, full circle only									
8200	Plastic, low/medium volume	1 Skwk	25	.320	Ea.	10.70	18.25		28.95	39.50
8400	Brass, low/medium volume		25	.320		51.50	18.25		69.75	84
8600	High volume		25	.320		91.50	18.25		109.75	128
8800	Very high volume		25	.320		169	18.25		187.25	214

Division Notes

	CREW	DAILY OUTPUT	LABOR-HOURS	UNIT	BARE COSTS				TOTAL INCL O&P
					MAT.	LABOR	EQUIP.	TOTAL	

Estimating Tips
33 10 00 Water Utilities
33 30 00 Sanitary Sewerage Utilities
33 40 00 Storm Drainage Utilities

- Never assume that the water, sewer, and drainage lines will go in at the early stages of the project. Consider the site access needs before dividing the site in half with open trenches, loose pipe, and machinery obstructions. Always inspect the site to establish that the site drawings are complete. Check off all existing utilities on your drawings as you locate them. Be especially careful with underground utilities because appurtenances are sometimes buried during regrading or repaving operations. If you find any discrepancies, mark up the site plan for further research. Differing site conditions can be very costly if discovered later in the project.

- See also Section 33 01 00 for restoration of pipe where removal/replacement may be undesirable. Use of new types of piping materials can reduce the overall project cost. Owners/design engineers should consider the installing contractor as a valuable source of current information on utility products and local conditions that could lead to significant cost savings.

Reference Numbers
Reference numbers are shown at the beginning of some major classifications. These numbers refer to related items in the Reference Section. The reference information may be an estimating procedure, an alternate pricing method, or technical information.

Note: Not all subdivisions listed here necessarily appear. ■

Same Data. Simplified.
Enjoy the convenience and efficiency of accessing your costs anywhere:
- **Skip the multiplier** by setting your location
- **Quickly search,** edit, favorite and share costs
- **Stay on top of price changes** with automatic updates

Discover more at rsmeans.com/online

33 01 Operation and Maintenance of Utilities

33 01 10 – Operation and Maintenance of Water Utilities

33 01 10.10 Corrosion Resistance		Crew	Daily Output	Labor-Hours	Unit	Material	2021 Bare Costs Labor	Equipment	Total	Total Incl O&P
0010	**CORROSION RESISTANCE**									
0012	Wrap & coat, add to pipe, 4" diameter				L.F.	2.50			2.50	2.75
0020	5" diameter					2.78			2.78	3.06
0040	6" diameter					3.71			3.71	4.08
0060	8" diameter					5.05			5.05	5.55
0080	10" diameter					6.20			6.20	6.80
0100	12" diameter					6.65			6.65	7.30
0120	14" diameter					8.80			8.80	9.70
0140	16" diameter					10.40			10.40	11.45
0160	18" diameter					10.95			10.95	12
0180	20" diameter					12.35			12.35	13.60
0200	24" diameter					14.85			14.85	16.30
0220	Small diameter pipe, 1" diameter, add					.80			.80	.88
0240	2" diameter					1.10			1.10	1.21
0260	2-1/2" diameter					1.52			1.52	1.67
0280	3" diameter					1.94			1.94	2.13
0300	Fittings, field covered, add				S.F.	9.85			9.85	10.85
0500	Coating, bituminous, per diameter inch, 1 coat, add				L.F.	.19			.19	.21
0540	3 coat					.62			.62	.68
0560	Coal tar epoxy, per diameter inch, 1 coat, add					.25			.25	.28
0600	3 coat					.69			.69	.76
1000	Polyethylene H.D. extruded, 0.025" thk., 1/2" diameter, add					.10			.10	.11
1020	3/4" diameter					.17			.17	.19
1040	1" diameter					.20			.20	.22
1060	1-1/4" diameter					.26			.26	.29
1080	1-1/2" diameter					.29			.29	.32
1100	0.030" thk., 2" diameter					.40			.40	.44
1120	2-1/2" diameter					.52			.52	.57
1140	0.035" thk., 3" diameter					.55			.55	.61
1160	3-1/2" diameter					.64			.64	.70
1180	4" diameter					.75			.75	.83
1200	0.040" thk., 5" diameter					.92			.92	1.01
1220	6" diameter					1.10			1.10	1.21
1240	8" diameter					1.40			1.40	1.54
1260	10" diameter					1.80			1.80	1.98
1280	12" diameter					2.16			2.16	2.38
1300	0.060" thk., 14" diameter					2.85			2.85	3.14
1320	16" diameter					3.27			3.27	3.60
1340	18" diameter					3.25			3.25	3.58
1360	20" diameter					3.62			3.62	3.98
1380	Fittings, field wrapped, add				S.F.	2.36			2.36	2.60

33 01 10.20 Pipe Repair

33 01 10.20 Pipe Repair		Crew	Daily Output	Labor-Hours	Unit	Material	2021 Bare Costs Labor	Equipment	Total	Total Incl O&P
0010	**PIPE REPAIR**									
0020	Not including excavation or backfill									
0100	Clamp, stainless steel, lightweight, for steel pipe									
0110	3" long, 1/2" diameter pipe	1 Plum	34	.235	Ea.	16.15	15.95		32.10	42
0120	3/4" diameter pipe		32	.250		16.70	16.95		33.65	44
0130	1" diameter pipe		30	.267		17.65	18.05		35.70	46.50
0140	1-1/4" diameter pipe		28	.286		18.65	19.35		38	49.50
0150	1-1/2" diameter pipe		26	.308		19.50	21		40.50	52.50
0160	2" diameter pipe		24	.333		21.50	22.50		44	57
0170	2-1/2" diameter pipe		23	.348		23	23.50		46.50	60

33 01 10.20 Pipe Repair	Crew	Daily Output	Labor-Hours	Unit	Material	2021 Bare Costs Labor	Equipment	Total	Total Incl O&P
0180	3" diameter pipe	1 Plum	22	.364	Ea.	25	24.50		49.50	64.50
0190	3-1/2" diameter pipe	↓	21	.381		26.50	26		52.50	67.50
0200	4" diameter pipe	B-20	44	.545		26.50	27		53.50	69.50
0210	5" diameter pipe		42	.571		36.50	28		64.50	82
0220	6" diameter pipe		38	.632		36.50	31		67.50	86.50
0230	8" diameter pipe		30	.800		43	39.50		82.50	107
0240	10" diameter pipe		28	.857		232	42.50		274.50	320
0250	12" diameter pipe		24	1		244	49.50		293.50	340
0260	14" diameter pipe		22	1.091		268	54		322	375
0270	16" diameter pipe		20	1.200		150	59		209	254
0280	18" diameter pipe		18	1.333		305	65.50		370.50	435
0290	20" diameter pipe		16	1.500		320	74		394	465
0300	24" diameter pipe	↓	14	1.714		355	84.50		439.50	520
0360	For 6" long, add					100%	40%			
0370	For 9" long, add					200%	100%			
0380	For 12" long, add					300%	150%			
0390	For 18" long, add				↓	500%	200%			
0400	Pipe freezing for live repairs of systems 3/8" to 6"									
0410	Note: Pipe freezing can also be used to install a valve into a live system									
0420	Pipe freezing each side 3/8"	2 Skwk	8	2	Ea.	590	114		704	820
0425	Pipe freezing each side 3/8", second location same kit		8	2		34	114		148	210
0430	Pipe freezing each side 3/4"		8	2		525	114		639	750
0435	Pipe freezing each side 3/4", second location same kit		8	2		34	114		148	210
0440	Pipe freezing each side 1-1/2"		6	2.667		525	152		677	810
0445	Pipe freezing each side 1-1/2" , second location same kit		6	2.667		34	152		186	267
0450	Pipe freezing each side 2"		6	2.667		830	152		982	1,150
0455	Pipe freezing each side 2", second location same kit		6	2.667		34	152		186	267
0460	Pipe freezing each side 2-1/2" to 3"		6	2.667		955	152		1,107	1,275
0465	Pipe freezing each side 2-1/2" to 3", second location same kit		6	2.667		68	152		220	305
0470	Pipe freezing each side 4"		4	4		1,575	228		1,803	2,075
0475	Pipe freezing each side 4", second location same kit		4	4		71	228		299	425
0480	Pipe freezing each side 5" to 6"		4	4		4,850	228		5,078	5,700
0485	Pipe freezing each side 5" to 6", second location same kit	↓	4	4		214	228		442	580
0490	Pipe freezing extra 20 lb. CO_2 cylinders (3/8" to 2" - 1 ea, 3" - 2 ea)					240			240	264
0500	Pipe freezing extra 50 lb. CO_2 cylinders (4" - 2 ea, 5"-6" - 6 ea)				↓	575			575	630
1000	Clamp, stainless steel, with threaded service tap									
1040	Full seal for iron, steel, PVC pipe									
1100	6" long, 2" diameter pipe	1 Plum	17	.471	Ea.	91	32		123	148
1110	2-1/2" diameter pipe		16	.500		94.50	34		128.50	155
1120	3" diameter pipe		15.60	.513		185	34.50		219.50	256
1130	3-1/2" diameter pipe	↓	15	.533		115	36		151	181
1140	4" diameter pipe	B-20	32	.750		123	37		160	191
1150	6" diameter pipe		28	.857		219	42.50		261.50	305
1160	8" diameter pipe		21	1.143		345	56.50		401.50	465
1170	10" diameter pipe		20	1.200		400	59		459	530
1180	12" diameter pipe		17	1.412		425	69.50		494.50	575
1200	8" long, 2" diameter pipe	1 Plum	11.72	.683		124	46		170	206
1210	2-1/2" diameter pipe		11	.727		129	49		178	216
1220	3" diameter pipe		10.75	.744		174	50.50		224.50	266
1230	3-1/2" diameter pipe		10.34	.774		136	52.50		188.50	227
1240	4" diameter pipe	B-20	22	1.091		605	54		659	745
1250	6" diameter pipe		19.31	1.243		213	61.50		274.50	325
1260	8" diameter pipe	↓	14.48	1.657	↓	241	81.50		322.50	385

For customer support on your Plumbing Costs with RSMeans Data, call 800.448.8182.

415

33 01 Operation and Maintenance of Utilities

33 01 10 – Operation and Maintenance of Water Utilities

33 01 10.20 Pipe Repair		Crew	Daily Output	Labor-Hours	Unit	Material	2021 Bare Costs Labor	2021 Bare Costs Equipment	Total	Total Incl O&P
1270	10" diameter pipe	B-20	13.80	1.739	Ea.	310	85.50		395.50	470
1280	12" diameter pipe		11.72	2.048		345	101		446	530
1300	12" long, 2" diameter pipe	1 Plum	9.44	.847		207	57.50		264.50	315
1310	2-1/2" diameter pipe	1 Plum	8.89	.900	Ea.	211	61		272	325
1320	3" diameter pipe		8.67	.923		273	62.50		335.50	395
1330	3-1/2" diameter pipe		8.33	.960		231	65		296	350
1340	4" diameter pipe	B-20	17.78	1.350		267	66.50		333.50	395
1350	6" diameter pipe		15.56	1.542		310	76		386	455
1360	8" diameter pipe		11.67	2.057		355	101		456	545
1370	10" diameter pipe		11.11	2.160		455	107		562	665
1380	12" diameter pipe		9.44	2.542		520	125		645	765
1400	20" long, 2" diameter pipe	1 Plum	8.10	.988		251	67		318	375
1410	2-1/2" diameter pipe		7.62	1.050		277	71		348	410
1420	3" diameter pipe		7.43	1.077		345	73		418	490
1430	3-1/2" diameter pipe		7.14	1.120		335	76		411	485
1440	4" diameter pipe	B-20	15.24	1.575		330	77.50		407.50	475
1450	6" diameter pipe		13.33	1.800		505	89		594	690
1460	8" diameter pipe		10	2.400		555	118		673	785
1470	10" diameter pipe		9.52	2.521		675	124		799	930
1480	12" diameter pipe		8.10	2.963		795	146		941	1,100
1600	Clamp, stainless steel, single section									
1640	Full seal for iron, steel, PVC pipe									
1700	6" long, 2" diameter pipe	1 Plum	17	.471	Ea.	94	32		126	152
1710	2-1/2" diameter pipe		16	.500		94.50	34		128.50	155
1720	3" diameter pipe		15.60	.513		110	34.50		144.50	173
1730	3-1/2" diameter pipe		15	.533		115	36		151	181
1740	4" diameter pipe	B-20	32	.750		122	37		159	190
1750	6" diameter pipe		27	.889		152	44		196	234
1760	8" diameter pipe		21	1.143		179	56.50		235.50	282
1770	10" diameter pipe		20	1.200		269	59		328	385
1780	12" diameter pipe		17	1.412		274	69.50		343.50	405
1800	8" long, 2" diameter pipe	1 Plum	11.72	.683		124	46		170	206
1805	2-1/2" diameter pipe		11.03	.725		129	49		178	216
1810	3" diameter pipe		10.76	.743		136	50.50		186.50	224
1815	3-1/2" diameter pipe		10.34	.774		136	52.50		188.50	227
1820	4" diameter pipe	B-20	22.07	1.087		153	53.50		206.50	249
1825	6" diameter pipe		19.31	1.243		183	61.50		244.50	293
1830	8" diameter pipe		14.48	1.657		214	81.50		295.50	360
1835	10" diameter pipe		13.79	1.740		286	86		372	445
1840	12" diameter pipe		11.72	2.048		395	101		496	580
1850	12" long, 2" diameter pipe	1 Plum	9.44	.847		207	57.50		264.50	315
1855	2-1/2" diameter pipe		8.89	.900		192	61		253	300
1860	3" diameter pipe		8.67	.923		219	62.50		281.50	335
1865	3-1/2" diameter pipe		8.33	.960		231	65		296	350
1870	4" diameter pipe	B-20	17.78	1.350		243	66.50		309.50	365
1875	6" diameter pipe		15.56	1.542		292	76		368	435
1880	8" diameter pipe		11.67	2.057		345	101		446	530
1885	10" diameter pipe		11.11	2.160		450	107		557	655
1890	12" diameter pipe		9.44	2.542		520	125		645	760
1900	20" long, 2" diameter pipe	1 Plum	8.10	.988		251	67		318	375
1905	2-1/2" diameter pipe		7.62	1.050		277	71		348	410
1910	3" diameter pipe		7.43	1.077		320	73		393	460
1915	3-1/2" diameter pipe		7.14	1.120		335	76		411	485

For customer support on your Plumbing Costs with RSMeans Data, call 800.448.8182.

33 01 Operation and Maintenance of Utilities

33 01 10 – Operation and Maintenance of Water Utilities

33 01 10.20 Pipe Repair

		Crew	Daily Output	Labor-Hours	Unit	Material	2021 Bare Costs Labor	Equipment	Total	Total Incl O&P
1920	4" diameter pipe	B-20	15.24	1.575	Ea.	400	77.50		477.50	555
1925	6" diameter pipe		13.33	1.800		485	89		574	670
1930	8" diameter pipe		10	2.400		555	118		673	790
1935	10" diameter pipe		9.52	2.521		690	124		814	940
1940	12" diameter pipe		8.10	2.963		750	146		896	1,050
2000	Clamp, stainless steel, two section									
2040	Full seal, for iron, steel, PVC pipe									
2100	8" long, 4" diameter pipe	B-20	24	1	Ea.	241	49.50		290.50	340
2110	6" diameter pipe		20	1.200		277	59		336	395
2120	8" diameter pipe		13	1.846		315	91		406	485
2130	10" diameter pipe		12	2		320	98.50		418.50	505
2140	12" diameter pipe		10	2.400		410	118		528	625
2200	10" long, 4" diameter pipe		16	1.500		315	74		389	460
2210	6" diameter pipe		13	1.846		355	91		446	525
2220	8" diameter pipe		9	2.667		395	131		526	630
2230	10" diameter pipe		8	3		520	148		668	790
2240	12" diameter pipe		7	3.429		595	169		764	910
2242	Clamp, stainless steel, three section									
2244	Full seal, for iron, steel, PVC pipe									
2250	10" long, 14" diameter pipe	B-20	6.40	3.750	Ea.	860	185		1,045	1,225
2260	16" diameter pipe		6	4		870	197		1,067	1,250
2270	18" diameter pipe		5	4.800		955	237		1,192	1,400
2280	20" diameter pipe		4.60	5.217		955	257		1,212	1,425
2290	24" diameter pipe		4	6		1,525	296		1,821	2,125
2320	For 12" long, add to 10"					15%	25%			
2330	For 20" long, add to 10"					70%	55%			
8000	For internal cleaning and inspection, see Section 33 01 30.11									
8100	For pipe testing, see Section 23 05 93.50									

33 01 30 – Operation and Maintenance of Sewer Utilities

33 01 30.11 Television Inspection of Sewers

		Crew	Daily Output	Labor-Hours	Unit	Material	2021 Bare Costs Labor	Equipment	Total	Total Incl O&P
0010	**TELEVISION INSPECTION OF SEWERS**									
0100	Pipe internal cleaning & inspection, cleaning, pressure pipe systems									
0120	Pig method, lengths 1000' to 10,000'									
0140	4" diameter thru 24" diameter, minimum				L.F.				3.60	4.14
0160	Maximum				"				18	21
6000	Sewage/sanitary systems									
6100	Power rodder with header & cutters									
6110	Mobilization charge, minimum				Total				695	800
6120	Mobilization charge, maximum				"				9,125	10,600
6140	Cleaning 4"-12" diameter				L.F.				6	6.60
6190	14"-24" diameter								8	8.80
6240	30" diameter								9.60	10.50
6250	36" diameter								12	13.20
6260	48" diameter								16	17.60
6270	60" diameter								24	26.50
6280	72" diameter								48	53
9000	Inspection, television camera with video									
9060	up to 500 linear feet				Total				715	820

417

For customer support on your Plumbing Costs with RSMeans Data, call 800.448.8182.

33 01 30.23 Pipe Bursting

		Crew	Daily Output	Labor-Hours	Unit	Material	2021 Bare Costs Labor	2021 Bare Costs Equipment	Total	Total Incl O&P
0010	**PIPE BURSTING**									
0011	300' runs, replace with HDPE pipe									
0020	Not including excavation, backfill, shoring, or dewatering									
0100	6" to 15" diameter, minimum				L.F.				200	220
0200	Maximum								550	605
0300	18" to 36" diameter, minimum								650	715
0400	Maximum								1,050	1,150
0500	Mobilize and demobilize, minimum				Job				3,000	3,300
0600	Maximum				"				32,100	35,400

33 01 30.72 Cured-In-Place Pipe Lining

		Crew	Daily Output	Labor-Hours	Unit	Material	Labor	Equipment	Total	Total Incl O&P
0010	**CURED-IN-PLACE PIPE LINING**									
0011	Not incl. bypass or cleaning									
0020	Less than 10,000 L.F., urban, 6" to 10"	C-17E	130	.615	L.F.	9.45	35.50	.28	45.23	63.50
0050	10" to 12"		125	.640		11.65	37	.29	48.94	68.50
0070	12" to 16"		115	.696		11.95	40	.32	52.27	73.50
0100	16" to 20"		95	.842		14	48.50	.39	62.89	89
0200	24" to 36"		90	.889		15.40	51	.41	66.81	94.50
0300	48" to 72"		80	1		24	57.50	.46	81.96	114

33 01 30.74 Sliplining, Excludes Cleaning

		Crew	Daily Output	Labor-Hours	Unit	Material	Labor	Equipment	Total	Total Incl O&P
0010	**SLIPLINING, excludes cleaning** and video inspection									
0020	Pipe relined with one pipe size smaller than original (4" for 6")									
0100	6" diameter, original size	B-6B	600	.080	L.F.	5.60	3.61	1.67	10.88	13.45
0150	8" diameter, original size		600	.080		10	3.61	1.67	15.28	18.25
0200	10" diameter, original size		600	.080		13.25	3.61	1.67	18.53	22
0250	12" diameter, original size		400	.120		17.15	5.40	2.50	25.05	29.50
0300	14" diameter, original size	B-6B	400	.120	L.F.	18.05	5.40	2.50	25.95	30.50
0350	16" diameter, original size	B-6C	300	.160		21.50	7.20	7.90	36.60	43
0400	18" diameter, original size	"	300	.160		27	7.20	7.90	42.10	49
1000	Pipe HDPE lining, make service line taps	B-6	4	6	Ea.	101	289	54	444	600

33 01 30.75 Cured-In-place Pipe Lining

		Crew	Daily Output	Labor-Hours	Unit	Material	Labor	Equipment	Total	Total Incl O&P
0010	**CURED-IN-PLACE PIPE LINING**									
0050	Bypass, Sewer	C-17E	500	.160	L.F.		9.20	.07	9.27	13.95
0090	Mobilize - (Any Size) 1 Mob per Job.		1	80	Job		4,600	37	4,637	6,975
0091	Relocate - (°12") (8hrs, 3 per day)		3	26.667	Ea.		1,525	12.25	1,537.25	2,325
0092	Relocate - (12" to 24") (8hrs, 2 per day)		2	40			2,300	18.40	2,318.40	3,475
0093	Relocate - (24" to 42") (8hrs, 1.5 per day)		1.50	53.333			3,075	24.50	3,099.50	4,650
0094	Relocate - (42") (8hrs, 1.5 per day)		1	80			4,600	37	4,637	6,975
0095	Additional Travel, $2.00 per Mile		3362.50	.024	Mile		1.37	.01	1.38	2.07
0096	Demobilize - (Any Size) 1 Demob per Job.		1	80	Job		4,600	37	4,637	6,975
0097	Refrigeration Truck (12", Cost per Each Location)		3	26.667	Ea.		1,525	12.25	1,537.25	2,325
0098	Refrigeration Truck (12"-24", Each Pipe / Location)		3	26.667			1,525	12.25	1,537.25	2,325
0099	Refrigeration Truck (24"-42", Each Pipe / Location)		1.50	53.333			3,075	24.50	3,099.50	4,650
0100	Refrigeration Truck (42", Each Pipe / Location)		1	80			4,600	37	4,637	6,975
1210	12", 6mm, Epoxy, 250k, 2ft		6	13.333	L.F.	48	765	6.15	819.15	1,200
1211	12", 6mm, Epoxy, 250k, 4ft		12	6.667		48	385	3.07	436.07	630
1212	12", 6mm, Epoxy, 250k, 10ft		30	2.667		48	153	1.23	202.23	285
1213	12", 6mm, Epoxy, 250k, 20ft		60	1.333		48	76.50	.61	125.11	169
1214	12", 6mm, Epoxy, 250k, 50ft		150	.533		48	30.50	.25	78.75	99.50
1215	12", 6mm, Epoxy, 250k, 100ft		300	.267		48	15.35	.12	63.47	76
1216	12", 6mm, Epoxy, 250k, 300ft		900	.089		48	5.10	.04	53.14	60.50
1217	12", 6mm, Epoxy, 250k, 328ft		1000	.080		48	4.60	.04	52.64	60

33 01 Operation and Maintenance of Utilities

33 01 30 – Operation and Maintenance of Sewer Utilities

33 01 30.75 Cured-In-place Pipe Lining	Crew	Daily Output	Labor-Hours	Unit	Material	2021 Bare Costs Labor	Equipment	Total	Total Incl O&P
1218	12", 6mm, Epoxy, 250k, Transitions, add				L.F.	20%				
1219	12", 6mm, Epoxy, 250k, 45 and 90 deg, add					50%				
1230	12", 5mm, Epoxy, 400k, 2ft	C-17E	6	13.333		40	765	6.15	811.15	1,200
1231	12", 5mm, Epoxy, 400k, 4ft		12	6.667		40	385	3.07	428.07	620
1232	12", 5mm, Epoxy, 400k, 10ft		30	2.667		40	153	1.23	194.23	276
1233	12", 5mm, Epoxy, 400k, 20ft		60	1.333		40	76.50	.61	117.11	160
1234	12", 5mm, Epoxy, 400k, 50ft		150	.533		40	30.50	.25	70.75	90.50
1235	12", 5mm, Epoxy, 400k, 100ft		300	.267		40	15.35	.12	55.47	67
1236	12", 5mm, Epoxy, 400k, 300ft		900	.089		40	5.10	.04	45.14	51.50
1237	12", 5mm, Epoxy, 400k, 328ft		1000	.080		40	4.60	.04	44.64	51
1238	12", 5mm, Epoxy, 400k, Transitions, add					20%				
1239	12", 5mm, Epoxy, 400k, 45 and 90 deg, add					50%				
2410	24", 12mm, Epoxy, 250k, 2ft	C-17E	4	20		192	1,150	9.20	1,351.20	1,950
2411	24", 12mm, Epoxy, 250k, 4ft		8	10		192	575	4.60	771.60	1,075
2412	24", 12mm, Epoxy, 250k, 10ft		20	4		192	230	1.84	423.84	560
2413	24", 12mm, Epoxy, 250k, 20ft		40	2		192	115	.92	307.92	385
2414	24", 12mm, Epoxy, 250k, 50ft		100	.800		192	46	.37	238.37	281
2415	24", 12mm, Epoxy, 250k, 100ft		200	.400		192	23	.18	215.18	247
2416	24", 12mm, Epoxy, 250k, 300ft		600	.133		192	7.65	.06	199.71	224
2417	24", 12mm, Epoxy, 250k, 328ft		656	.122		192	7	.06	199.06	223
2418	24", 12mm, Epoxy, 250k, Transitions, add					20%				
2419	24", 12mm, Epoxy, 250k, 45 and 90 deg, add					50%				
2430	24", 10.5mm, Epoxy, 400k, 2ft	C-17E	4	20		168	1,150	9.20	1,327.20	1,925
2431	24", 10.5mm, Epoxy, 400k, 4ft		8	10		168	575	4.60	747.60	1,050
2432	24", 10.5mm, Epoxy, 400k, 10ft		20	4		168	230	1.84	399.84	530
2433	24", 10.5mm, Epoxy, 400k, 20ft		40	2		168	115	.92	283.92	360
2434	24", 10.5mm, Epoxy, 400k, 50ft		100	.800		168	46	.37	214.37	254
2435	24", 10.5mm, Epoxy, 400k, 100ft		200	.400		168	23	.18	191.18	220
2436	24", 10.5mm, Epoxy, 400k, 300ft		600	.133		168	7.65	.06	175.71	197
2437	24", 10.5mm, Epoxy, 400k, 328ft		656	.122		168	7	.06	175.06	196
2438	24", 10.5mm, Epoxy, 400k, Transitions, add					20%				
2439	24", 10.5mm, Epoxy, 400k, 45 and 90 deg, add					50%				

33 05 Common Work Results for Utilities

33 05 07 – Trenchless Installation of Utility Piping

33 05 07.23 Utility Boring and Jacking

		Crew	Daily Output	Labor-Hours	Unit	Material	2021 Bare Costs Labor	Equipment	Total	Total Incl O&P
0010	**UTILITY BORING AND JACKING**									
0011	Casing only, 100' minimum,									
0020	not incl. jacking pits or dewatering									
0100	Roadwork, 1/2" thick wall, 24" diameter casing	B-42	20	3.200	L.F.	142	159	53.50	354.50	455
0200	36" diameter		16	4		219	199	66.50	484.50	615
0300	48" diameter		15	4.267		305	212	71	588	735
0500	Railroad work, 24" diameter		15	4.267		142	212	71	425	555
0600	36" diameter		14	4.571		219	228	76	523	665
0700	48" diameter		12	5.333		305	265	89	659	835
0900	For ledge, add								20%	20%
1000	Small diameter boring, 3", sandy soil	B-82	900	.018		23	.89	.20	24.09	27
1040	Rocky soil	"	500	.032		23	1.60	.37	24.97	28.50
1100	Prepare jacking pits, incl. mobilization & demobilization, minimum				Ea.				3,225	3,700
1101	Maximum				"				22,000	25,500

For customer support on your Plumbing Costs with RSMeans Data, call 800.448.8182.

419

33 05 07 – Trenchless Installation of Utility Piping

33 05 07.36 Microtunneling	Crew	Daily Output	Labor-Hours	Unit	Material	2021 Bare Costs Labor	Equipment	Total	Total Incl O&P
0010 **MICROTUNNELING**									
0011 Not including excavation, backfill, shoring,									
0020 or dewatering, average 50'/day, slurry method									
0100 24" to 48" outside diameter, minimum				L.F.				965	965
0110 Adverse conditions, add				"				500	500
1000 Rent microtunneling machine, average monthly lease				Month				97,500	107,000
1010 Operating technician				Day				630	705
1100 Mobilization and demobilization, minimum				Job				41,200	45,900
1110 Maximum				"				445,500	490,500

33 05 61 – Concrete Manholes

33 05 61.10 Storm Drainage Manholes, Frames and Covers

33 05 61.10 Storm Drainage Manholes, Frames and Covers	Crew	Daily Output	Labor-Hours	Unit	Material	2021 Bare Costs Labor	Equipment	Total	Total Incl O&P
0010 **STORM DRAINAGE MANHOLES, FRAMES & COVERS**									
0020 Excludes footing, excavation, backfill (See line items for frame & cover)									
0050 Brick, 4' inside diameter, 4' deep	D-1	1	16	Ea.	700	780		1,480	1,950
0100 6' deep		.70	22.857		1,000	1,125		2,125	2,775
0150 8' deep		.50	32		1,300	1,550		2,850	3,775
0200 Add for 1' depth increase		4	4	V.L.F.	176	195		371	485
0400 Concrete blocks (radial), 4' ID, 4' deep		1.50	10.667	Ea.	430	520		950	1,250
0500 6' deep		1	16		570	780		1,350	1,800
0600 8' deep		.70	22.857		710	1,125		1,835	2,450
0700 For depths over 8', add		5.50	2.909	V.L.F.	73	142		215	293
0800 Concrete, cast in place, 4' x 4', 8" thick, 4' deep	C-14H	2	24	Ea.	680	1,275	13.70	1,968.70	2,650
0900 6' deep		1.50	32		985	1,725	18.25	2,728.25	3,650
1000 8' deep		1	48		1,400	2,575	27.50	4,002.50	5,400
1100 For depths over 8', add		8	6	V.L.F.	166	320	3.42	489.42	665
1110 Precast, 4' ID, 4' deep	B-22	4.10	7.317	Ea.	905	380	70	1,355	1,625
1120 6' deep		3	10		1,075	515	95.50	1,685.50	2,050
1130 8' deep		2	15		1,900	775	143	2,818	3,400
1140 For depths over 8', add		16	1.875	V.L.F.	107	97	17.90	221.90	283
1150 5' ID, 4' deep	B-6	3	8	Ea.	1,875	385	72	2,332	2,725
1160 6' deep		2	12		3,250	575	108	3,933	4,550
1170 8' deep		1.50	16		2,750	770	144	3,664	4,325
1180 For depths over 8', add		12	2	V.L.F.	470	96	18	584	680
1190 6' ID, 4' deep		2	12	Ea.	2,100	575	108	2,783	3,275
1200 6' deep		1.50	16		4,600	770	144	5,514	6,350
1210 8' deep		1	24		5,375	1,150	216	6,741	7,900
1220 For depths over 8', add		8	3	V.L.F.	425	144	27	596	715
1250 Slab tops, precast, 8" thick									
1300 4' diameter manhole	B-6	8	3	Ea.	288	144	27	459	560
1400 5' diameter manhole	"	7.50	3.200	"	685	154	29	868	1,000
1500 6' diameter manhole	B-6	7	3.429	Ea.	715	165	31	911	1,075
3800 Steps, heavyweight cast iron, 7" x 9"	1 Bric	40	.200		17.25	10.75		28	35
3900 8" x 9"		40	.200		20.50	10.75		31.25	39
3928 12" x 10-1/2"		40	.200		31	10.75		41.75	50
4000 Standard sizes, galvanized steel		40	.200		24	10.75		34.75	42.50
4100 Aluminum		40	.200		31	10.75		41.75	50
4150 Polyethylene		40	.200		32.50	10.75		43.25	52

33 05 Common Work Results for Utilities

33 05 63 – Concrete Vaults and Chambers

33 05 63.13 Precast Concrete Utility Structures

	Crew	Daily Output	Labor-Hours	Unit	Material	2021 Bare Costs Labor	2021 Bare Costs Equipment	Total	Total Incl O&P
0010 **PRECAST CONCRETE UTILITY STRUCTURES**, 6" thick									
0050 5' x 10' x 6' high, ID	B-13	2	28	Ea.	1,825	1,350	293	3,468	4,350
0100 6' x 10' x 6' high, ID		2	28		1,900	1,350	293	3,543	4,450
0150 5' x 12' x 6' high, ID		2	28		2,000	1,350	293	3,643	4,550
0200 6' x 12' x 6' high, ID		1.80	31.111		2,250	1,500	325	4,075	5,075
0250 6' x 13' x 6' high, ID		1.50	37.333		2,950	1,800	390	5,140	6,375
0300 8' x 14' x 7' high, ID	↓	1	56	↓	3,200	2,700	585	6,485	8,175
0350 Hand hole, precast concrete, 1-1/2" thick									
0400 1'-0" x 2'-0" x 1'-9", ID, light duty	B-1	4	6	Ea.	525	270		795	985
0450 4'-6" x 3'-2" x 2'-0", OD, heavy duty	B-6	3	8	"	1,600	385	72	2,057	2,400

33 05 71 – Cleanouts

	Crew	Daily Output	Labor-Hours	Unit	Material	2021 Bare Costs Labor	2021 Bare Costs Equipment	Total	Total Incl O&P
0300 Cleanout Access Housing, X-Hvy Dty, Ductile Iron	B-20	10	2.400	Ea.	1,400	118		1,518	1,700
0306 Vandal Proof, Add					96.50			96.50	106
0313 Galvanized, Add				↓	775			775	850

33 05 97 – Identification and Signage for Utilities

33 05 97.05 Utility Connection

	Crew	Daily Output	Labor-Hours	Unit	Material	2021 Bare Costs Labor	2021 Bare Costs Equipment	Total	Total Incl O&P
0010 **UTILITY CONNECTION**									
0020 Water, sanitary, stormwater, gas, single connection	B-14	1	48	Ea.	4,925	2,225	216	7,366	9,000
0030 Telecommunication	"	3	16	"	420	745	72	1,237	1,650

33 05 97.10 Utility Accessories

	Crew	Daily Output	Labor-Hours	Unit	Material	2021 Bare Costs Labor	2021 Bare Costs Equipment	Total	Total Incl O&P
0010 **UTILITY ACCESSORIES**									
0400 Underground tape, detectable, reinforced, alum. foil core, 2"	1 Clab	150	.053	C.L.F.	3	2.37		5.37	6.85

33 11 Groundwater Sources

33 11 13 – Potable Water Supply Wells

33 11 13.10 Wells and Accessories

	Crew	Daily Output	Labor-Hours	Unit	Material	2021 Bare Costs Labor	2021 Bare Costs Equipment	Total	Total Incl O&P
0010 **WELLS & ACCESSORIES**									
0011 Domestic									
0100 Drilled, 4" to 6" diameter	B-23	120	.333	L.F.		14.95	13.50	28.45	37.50
0200 8" diameter	"	95.20	.420	"		18.80	17	35.80	46.50
0400 Gravel pack well, 40' deep, incl. gravel & casing, complete									
0500 24" diameter casing x 18" diameter screen	B-23	.13	308	Total	38,700	13,800	12,400	64,900	77,000
0600 36" diameter casing x 18" diameter screen		.12	333	"	39,800	14,900	13,500	68,200	81,000
0800 Observation wells, 1-1/4" riser pipe	↓	163	.245	V.L.F.	18.15	11	9.95	39.10	47.50
0900 For flush Buffalo roadway box, add	1 Skwk	16.60	.482	Ea.	52	27.50		79.50	99
1200 Test well, 2-1/2" diameter, up to 50' deep (15 to 50 GPM)	B-23	1.51	26.490	"	895	1,175	1,075	3,145	3,925
1300 Over 50' deep, add	"	121.80	.328	L.F.	24	14.70	13.30	52	63
1500 Pumps, installed in wells to 100' deep, 4" submersible									
1510 1/2 HP	Q-1	3.22	4.969	Ea.	580	305		885	1,100
1520 3/4 HP		2.66	6.015		980	365		1,345	1,625
1600 1 HP	↓	2.29	6.987		805	425		1,230	1,525
1700 1-1/2 HP	Q-22	1.60	10		2,250	610	297	3,157	3,700
1800 2 HP		1.33	12.030		1,800	735	360	2,895	3,500
1900 3 HP		1.14	14.035		2,050	855	415	3,320	4,000
2000 5 HP		1.14	14.035		2,575	855	415	3,845	4,550
3000 Pump, 6" submersible, 25' to 150' deep, 25 HP, 103 to 400 GPM		.89	17.978		9,325	1,100	535	10,960	12,500
3100 25' to 500' deep, 30 HP, 104 to 400 GPM	↓	.73	21.918	↓	11,000	1,325	650	12,975	14,800
8000 Steel well casing	B-23A	3020	.008	Lb.	1.29	.40	.29	1.98	2.33
8110 Well screen assembly, stainless steel, 2" diameter	↓	273	.088	L.F.	95.50	4.39	3.23	103.12	115

421

33 11 13 – Potable Water Supply Wells

33 11 13.10 Wells and Accessories		Crew	Daily Output	Labor-Hours	Unit	Material	2021 Bare Costs Labor	2021 Bare Costs Equipment	Total	Total Incl O&P
8120	3" diameter	B-23A	253	.095	L.F.	141	4.74	3.48	149.22	166
8130	4" diameter		200	.120		181	6	4.40	191.40	213
8140	5" diameter		168	.143		170	7.15	5.25	182.40	203
8150	6" diameter		126	.190		190	9.50	7	206.50	231
8160	8" diameter		98.50	.244		253	12.15	8.95	274.10	305
8170	10" diameter		73	.329		315	16.40	12.05	343.45	390
8180	12" diameter		62.50	.384		370	19.15	14.10	403.25	455
8190	14" diameter		54.30	.442		415	22	16.20	453.20	510
8200	16" diameter		48.30	.497		460	25	18.25	503.25	560
8210	18" diameter		39.20	.612		580	30.50	22.50	633	710
8220	20" diameter		31.20	.769		655	38.50	28	721.50	815
8230	24" diameter		23.80	1.008		810	50.50	37	897.50	1,000
8240	26" diameter		21	1.143		1,025	57	42	1,124	1,250
8244	Well casing or drop pipe, PVC, 1/2" diameter		550	.044		1.30	2.18	1.60	5.08	6.45
8245	3/4" diameter		550	.044		1.31	2.18	1.60	5.09	6.45
8246	1" diameter		550	.044		1.35	2.18	1.60	5.13	6.50
8247	1-1/4" diameter		520	.046		1.70	2.30	1.69	5.69	7.15
8248	1-1/2" diameter		490	.049		1.80	2.45	1.80	6.05	7.60
8249	1-3/4" diameter		380	.063		1.84	3.15	2.32	7.31	9.25
8250	2" diameter		280	.086		2.11	4.28	3.14	9.53	12.20
8252	3" diameter		260	.092		4.18	4.61	3.39	12.18	15.20
8254	4" diameter		205	.117		4.43	5.85	4.30	14.58	18.30
8255	5" diameter		170	.141		4.45	7.05	5.20	16.70	21
8256	6" diameter		130	.185		8.50	9.20	6.75	24.45	30.50
8258	8" diameter		100	.240		13	12	8.80	33.80	42
8260	10" diameter		73	.329		21	16.40	12.05	49.45	61
8262	12" diameter		62	.387		23.50	19.35	14.20	57.05	70.50
8300	Slotted PVC, 1-1/4" diameter		521	.046		2.88	2.30	1.69	6.87	8.45
8310	1-1/2" diameter		488	.049		2.98	2.46	1.80	7.24	8.90
8320	2" diameter		273	.088		3.86	4.39	3.23	11.48	14.35
8330	3" diameter		253	.095		6.80	4.74	3.48	15.02	18.35
8340	4" diameter		200	.120		7.25	6	4.40	17.65	21.50
8350	5" diameter		168	.143		15.65	7.15	5.25	28.05	33.50
8360	6" diameter		126	.190		16.80	9.50	7	33.30	40.50
8370	8" diameter		98.50	.244		24.50	12.15	8.95	45.60	55
8400	Artificial gravel pack, 2" screen, 6" casing	B-23B	174	.138		4.52	6.90	6.40	17.82	22.50
8405	8" casing		111	.216		6.20	10.80	10.05	27.05	34
8410	10" casing		74.50	.322		8.50	16.10	15	39.60	50
8415	12" casing		60	.400		13.60	19.95	18.60	52.15	65.50
8420	14" casing		50.20	.478		13.90	24	22	59.90	75.50
8425	16" casing		40.70	.590		18.20	29.50	27.50	75.20	94
8430	18" casing		36	.667		22.50	33.50	31	87	108
8435	20" casing		29.50	.814		24.50	40.50	38	103	129
8440	24" casing		25.70	.934		28.50	46.50	43.50	118.50	149
8445	26" casing		24.60	.976		31.50	48.50	45.50	125.50	157
8450	30" casing		20	1.200		36	60	56	152	191
8455	36" casing		16.40	1.463		39.50	73	68	180.50	228
8500	Develop well		8	3	Hr.	305	150	139	594	710
8550	Pump test well		8	3		96.50	150	139	385.50	480
8560	Standby well	B-23A	8	3		99	150	110	359	455
8570	Standby, drill rig		8	3			150	110	260	345
8580	Surface seal well, concrete filled		1	24	Ea.	1,075	1,200	880	3,155	3,925
8590	Well test pump, install & remove	B-23	1	40			1,800	1,625	3,425	4,450

33 11 Groundwater Sources

33 11 13 – Potable Water Supply Wells

33 11 13.10 Wells and Accessories

		Crew	Daily Output	Labor-Hours	Unit	Material	2021 Bare Costs Labor	Equipment	Total	Total Incl O&P
8600	Well sterilization, chlorine	2 Clab	1	16	Ea.	125	710		835	1,200
8610	Well water pressure switch	1 Clab	12	.667		97.50	29.50		127	151
8630	Well water pressure switch with manual reset	"	12	.667		130	29.50		159.50	186
9950	See Section 31 23 19.40 for wellpoints									
9960	See Section 31 23 19.30 for drainage wells									

33 11 13.20 Water Supply Wells, Pumps

		Crew	Daily Output	Labor-Hours	Unit	Material	2021 Bare Costs Labor	Equipment	Total	Total Incl O&P
0010	**WATER SUPPLY WELLS, PUMPS**									
0011	With pressure control									
1000	Deep well, jet, 42 gal. galvanized tank									
1040	3/4 HP	1 Plum	.80	10	Ea.	1,125	675		1,800	2,250
3000	Shallow well, jet, 30 gal. galvanized tank									
3040	1/2 HP	1 Plum	2	4	Ea.	905	271		1,176	1,400

33 14 Water Utility Transmission and Distribution

33 14 13 – Public Water Utility Distribution Piping

33 14 13.15 Water Supply, Ductile Iron Pipe

		Crew	Daily Output	Labor-Hours	Unit	Material	2021 Bare Costs Labor	Equipment	Total	Total Incl O&P
0010	**WATER SUPPLY, DUCTILE IRON PIPE** R331113-80									
0020	Not including excavation or backfill									
2000	Pipe, class 50 water piping, 18' lengths									
2020	Mechanical joint, 4" diameter	B-21	200	.140	L.F.	47	7.15	.95	55.10	63.50
2040	6" diameter		160	.175		55	8.95	1.19	65.14	75
2060	8" diameter		133.33	.210		50	10.70	1.43	62.13	72.50
2080	10" diameter		114.29	.245		65.50	12.50	1.67	79.67	92.50
2100	12" diameter		105.26	.266		88.50	13.60	1.81	103.91	120
2120	14" diameter		100	.280		104	14.30	1.91	120.21	138
2140	16" diameter		72.73	.385		106	19.65	2.63	128.28	148
2160	18" diameter		68.97	.406		141	20.50	2.77	164.27	189
2170	20" diameter		57.14	.490		142	25	3.34	170.34	198
2180	24" diameter		47.06	.595		157	30.50	4.06	191.56	223
3000	Push-on joint, 4" diameter		400	.070		24	3.57	.48	28.05	32.50
3020	6" diameter		333.33	.084		23.50	4.29	.57	28.36	33
3040	8" diameter		200	.140		33	7.15	.95	41.10	48.50
3060	10" diameter		181.82	.154		51	7.85	1.05	59.90	69.50
3080	12" diameter		160	.175		54	8.95	1.19	64.14	74
3100	14" diameter		133.33	.210		54	10.70	1.43	66.13	77
3120	16" diameter		114.29	.245		57.50	12.50	1.67	71.67	83.50
3140	18" diameter		100	.280		64	14.30	1.91	80.21	93.50
3160	20" diameter		88.89	.315		66.50	16.10	2.15	84.75	99.50
3180	24" diameter		76.92	.364		92.50	18.60	2.48	113.58	133
8000	Piping, fittings, mechanical joint, AWWA C110									
8006	90° bend, 4" diameter	B-20	16	1.500	Ea.	173	74		247	300
8020	6" diameter	"	12.80	1.875		256	92.50		348.50	420
8040	8" diameter	B-21	10.67	2.624		500	134	17.90	651.90	770
8060	10" diameter		11.43	2.450		695	125	16.70	836.70	965
8080	12" diameter		10.53	2.659		985	136	18.15	1,139.15	1,300
8100	14" diameter		10	2.800		1,325	143	19.10	1,487.10	1,700
8120	16" diameter		7.27	3.851		1,700	197	26.50	1,923.50	2,175
8140	18" diameter		6.90	4.058		2,375	207	27.50	2,609.50	2,975
8160	20" diameter		5.71	4.904		2,975	250	33.50	3,258.50	3,675
8180	24" diameter		4.70	5.957		4,675	305	40.50	5,020.50	5,650
8200	Wye or tee, 4" diameter	B-20	10.67	2.249		475	111		586	685

33 14 13.15 Water Supply, Ductile Iron Pipe

		Crew	Daily Output	Labor-Hours	Unit	Material	2021 Bare Costs Labor	2021 Bare Costs Equipment	Total	Total Incl O&P
8220	6" diameter	B-20	8.53	2.814	Ea.	645	139		784	920
8240	8" diameter	B-21	7.11	3.938		1,000	201	27	1,228	1,425
8260	10" diameter		7.62	3.675		975	188	25	1,188	1,375
8280	12" diameter		7.02	3.989		1,900	204	27	2,131	2,425
8300	14" diameter		6.67	4.198		2,150	214	28.50	2,392.50	2,725
8320	16" diameter		4.85	5.773		2,475	295	39.50	2,809.50	3,200
8340	18" diameter		4.60	6.087		4,600	310	41.50	4,951.50	5,550
8360	20" diameter		3.81	7.349		7,425	375	50	7,850	8,800
8380	24" diameter		3.14	8.917		10,100	455	61	10,616	11,800
8398	45° bend, 4" diameter	B-20	16	1.500		201	74		275	330
8400	6" diameter	"	12.80	1.875		279	92.50		371.50	445
8405	8" diameter	B-21	10.67	2.624		410	134	17.90	561.90	670
8410	12" diameter		10.53	2.659		870	136	18.15	1,024.15	1,175
8420	16" diameter		7.27	3.851		1,675	197	26.50	1,898.50	2,150
8430	20" diameter		5.71	4.904		2,400	250	33.50	2,683.50	3,050
8440	24" diameter		4.70	5.957		3,375	305	40.50	3,720.50	4,200
8450	Decreaser, 6" x 4" diameter	B-20	14.22	1.688		231	83		314	380
8460	8" x 6" diameter	B-21	11.64	2.406		345	123	16.40	484.40	580
8470	10" x 6" diameter		13.33	2.101		435	107	14.35	556.35	655
8480	12" x 6" diameter		12.70	2.205		615	113	15.05	743.05	860
8490	16" x 6" diameter		10	2.800		995	143	19.10	1,157.10	1,325
8500	20" x 6" diameter		8.42	3.325		1,800	170	22.50	1,992.50	2,250
8552	For water utility valves see Section 33 14 19									
8700	Joint restraint, ductile iron mechanical joints									
8710	4" diameter	B-20	32	.750	Ea.	26.50	37		63.50	85
8720	6" diameter	"	25.60	.938		34.50	46		80.50	107
8730	8" diameter	B-21	21.33	1.313		52	67	8.95	127.95	167
8740	10" diameter		18.28	1.532		89.50	78	10.45	177.95	227
8750	12" diameter		16.84	1.663		99	85	11.35	195.35	248
8760	14" diameter		16	1.750		141	89.50	11.95	242.45	300
8770	16" diameter		11.64	2.406		145	123	16.40	284.40	360
8780	18" diameter		11.03	2.539		202	130	17.30	349.30	435
8785	20" diameter		9.14	3.063		248	156	21	425	530
8790	24" diameter		7.53	3.718		340	190	25.50	555.50	685
9600	Steel sleeve with tap, 4" diameter	B-20	3	8		550	395		945	1,200
9620	6" diameter		2	12		590	590		1,180	1,525
9630	8" diameter		2	12		750	590		1,340	1,700

33 14 13.20 Water Supply, Polyethylene Pipe, C901

		Crew	Daily Output	Labor-Hours	Unit	Material	2021 Bare Costs Labor	2021 Bare Costs Equipment	Total	Total Incl O&P
0010	**WATER SUPPLY, POLYETHYLENE PIPE, C901**									
0020	Not including excavation or backfill									
1000	Piping, 160 psi, 3/4" diameter	Q-1A	525	.019	L.F.	.96	1.30		2.26	3
1120	1" diameter		485	.021		.66	1.40		2.06	2.83
1140	1-1/2" diameter		450	.022		2.06	1.51		3.57	4.53
1160	2" diameter		365	.027		2.72	1.87		4.59	5.75
2000	Fittings, insert type, nylon, 160 & 250 psi, cold water									
2220	Clamp ring, stainless steel, 3/4" diameter	Q-1A	345	.029	Ea.	5.55	1.97		7.52	9.05
2240	1" diameter		321	.031		5.55	2.12		7.67	9.25
2260	1-1/2" diameter		285	.035		5.55	2.39		7.94	9.65
2280	2" diameter		255	.039		5.55	2.67		8.22	10.10
2300	Coupling, 3/4" diameter		66	.152		1.40	10.30		11.70	16.95
2320	1" diameter		57	.175		1.88	11.95		13.83	19.90
2340	1-1/2" diameter		51	.196		4.56	13.35		17.91	25

33 14 Water Utility Transmission and Distribution

33 14 13 – Public Water Utility Distribution Piping

33 14 13.20 Water Supply, Polyethylene Pipe, C901	Crew	Daily Output	Labor-Hours	Unit	Material	2021 Bare Costs Labor	Equipment	Total	Total Incl O&P
2360	2" diameter	Q-1A	48	.208	Ea.	5.80	14.20		20	27.50
2400	Elbow, 90°, 3/4" diameter		66	.152		2.29	10.30		12.59	17.90
2420	1" diameter		57	.175		2.78	11.95		14.73	21
2440	1-1/2" diameter		51	.196		7.15	13.35		20.50	28
2460	2" diameter		48	.208		9.55	14.20		23.75	31.50

33 14 13.25 Water Supply, Polyvinyl Chloride Pipe

		Crew	Daily Output	Labor-Hours	Unit	Material	2021 Bare Costs Labor	Equipment	Total	Total Incl O&P
0010	**WATER SUPPLY, POLYVINYL CHLORIDE PIPE** R331113-80									
0020	Not including excavation or backfill, unless specified									
2100	PVC pipe, Class 150, 1-1/2" diameter	Q-1A	750	.013	L.F.	.64	.91		1.55	2.06
2120	2" diameter		686	.015		.81	.99		1.80	2.37
2140	2-1/2" diameter		500	.020		1.18	1.36		2.54	3.33
2160	3" diameter	B-20	430	.056		1.77	2.75		4.52	6.05
3010	AWWA C905, PR 100, DR 25									
3030	14" diameter	B-21	213	.131	L.F.	13.75	6.70	.90	21.35	26
3040	16" diameter		200	.140		18.80	7.15	.95	26.90	32.50
3050	18" diameter		160	.175		23	8.95	1.19	33.14	39.50
3060	20" diameter		133	.211		28.50	10.75	1.44	40.69	48.50
3070	24" diameter		107	.262		42	13.35	1.78	57.13	68.50
3080	30" diameter		80	.350		70.50	17.85	2.39	90.74	107
3090	36" diameter		80	.350		109	17.85	2.39	129.24	149
3100	42" diameter		60	.467		148	24	3.18	175.18	201
3200	48" diameter		60	.467		188	24	3.18	215.18	246
3960	Pressure pipe, class 200, ASTM 2241, SDR 21, 3/4" diameter	Q-1A	1000	.010		.21	.68		.89	1.25
3980	1" diameter		900	.011		.25	.76		1.01	1.41
4000	1-1/2" diameter		750	.013		1.16	.91		2.07	2.64
4010	2" diameter		686	.015		.86	.99		1.85	2.43
4020	2-1/2" diameter		500	.020		2.16	1.36		3.52	4.41
4030	3" diameter	B-20	430	.056		2.30	2.75		5.05	6.65
4040	4" diameter		375	.064		3.78	3.16		6.94	8.90
4050	6" diameter		316	.076		8	3.74		11.74	14.40
4060	8" diameter		260	.092		8.45	4.55		13	16.10
4090	Including trenching to 3' deep, 3/4" diameter	Q-1C	300	.080		.21	4.82	6.30	11.33	14.40
4100	1" diameter		280	.086		.25	5.15	6.75	12.15	15.45
4110	1-1/2" diameter		260	.092		1.16	5.55	7.30	14.01	17.60
4120	2" diameter		220	.109		.86	6.60	8.60	16.06	20
4130	2-1/2" diameter		200	.120		2.16	7.25	9.45	18.86	23.50
4140	3" diameter		175	.137		2.30	8.25	10.80	21.35	27
4150	4" diameter		150	.160		3.78	9.65	12.65	26.08	32.50
4160	6" diameter		125	.192		8	11.55	15.15	34.70	42.50
4165	Fittings									
4170	Elbow, 90°, 3/4"	Q-1A	114	.088	Ea.	1.22	5.95		7.17	10.25
4180	1"		100	.100		3.19	6.80		9.99	13.65
4190	1-1/2"		80	.125		19.80	8.50		28.30	34.50
4200	2"		72	.139		14.20	9.45		23.65	30
4210	3"	B-20	46	.522		20.50	25.50		46	61
4220	4"		36	.667		35	33		68	87.50
4230	6"		24	1		64.50	49.50		114	145
4240	8"		14	1.714		126	84.50		210.50	265
4250	Elbow, 45°, 3/4"	Q-1A	114	.088		1.19	5.95		7.14	10.20
4260	1"		100	.100		9.65	6.80		16.45	21
4270	1-1/2"		80	.125		16.25	8.50		24.75	30.50
4280	2"		72	.139		19.60	9.45		29.05	35.50

For customer support on your Plumbing Costs with RSMeans Data, call 800.448.8182.

425

33 14 13 – Public Water Utility Distribution Piping

33 14 13.25 Water Supply, Polyvinyl Chloride Pipe		Crew	Daily Output	Labor-Hours	Unit	Material	2021 Bare Costs Labor	Equipment	Total	Total Incl O&P
4290	2-1/2"	Q-1A	54	.185	Ea.	22.50	12.60		35.10	43.50
4300	3"	B-20	46	.522		28.50	25.50		54	69.50
4310	4"		36	.667		48.50	33		81.50	103
4320	6"		24	1		89	49.50		138.50	172
4330	8"		14	1.714		126	84.50		210.50	265
4340	Tee, 3/4"	Q-1A	76	.132		3.25	8.95		12.20	16.95
4350	1"		66	.152		12.50	10.30		22.80	29
4360	1-1/2"		54	.185		17.65	12.60		30.25	38
4370	2"		48	.208		21	14.20		35.20	44
4380	2-1/2"		36	.278		23	18.90		41.90	53.50
4390	3"	B-20	30	.800		24.50	39.50		64	86
4400	4"		24	1		38.50	49.50		88	117
4410	6"		14.80	1.622		84.50	80		164.50	213
4420	8"		9	2.667		149	131		280	360
4430	Coupling, 3/4"	Q-1A	114	.088		1.69	5.95		7.64	10.75
4440	1"		100	.100		7.35	6.80		14.15	18.25
4450	1-1/2"		80	.125		9	8.50		17.50	22.50
4460	2"		72	.139		9.15	9.45		18.60	24
4470	2-1/2"		54	.185		17.35	12.60		29.95	38
4480	3"	B-20	46	.522		12.50	25.50		38	52.50
4490	4"		36	.667		23	33		56	74
4500	6"		24	1		32.50	49.50		82	110
4510	8"		14	1.714		68.50	84.50		153	203
4520	Pressure pipe Class 150, SDR 18, AWWA C900, 4" diameter		380	.063	L.F.	3.79	3.11		6.90	8.85
4530	6" diameter		316	.076		5.50	3.74		9.24	11.65
4540	8" diameter	B-21	264	.106		7.05	5.40	.72	13.17	16.65
4550	10" diameter		220	.127		9.55	6.50	.87	16.92	21
4560	12" diameter		186	.151		13	7.70	1.03	21.73	27
8000	Fittings with rubber gasket									
8003	Class 150, DR 18									
8006	90° bend , 4" diameter	B-20	100	.240	Ea.	42	11.85		53.85	63.50
8020	6" diameter	"	90	.267		74.50	13.15		87.65	102
8040	8" diameter	B-21	80	.350		144	17.85	2.39	164.24	188
8060	10" diameter		50	.560		280	28.50	3.82	312.32	355
8080	12" diameter		30	.933		370	47.50	6.35	423.85	485
8100	Tee, 4" diameter		90	.311		57.50	15.90	2.12	75.52	90
8120	6" diameter		80	.350		129	17.85	2.39	149.24	171
8140	8" diameter		70	.400		184	20.50	2.73	207.23	236
8160	10" diameter		40	.700		257	35.50	4.77	297.27	340
8180	12" diameter		20	1.400		360	71.50	9.55	441.05	515
8200	45° bend, 4" diameter	B-20	100	.240		42	11.85		53.85	63.50
8220	6" diameter	"	90	.267		72.50	13.15		85.65	99.50
8240	8" diameter	B-21	50	.560		138	28.50	3.82	170.32	199
8260	10" diameter		50	.560		157	28.50	3.82	189.32	220
8280	12" diameter		30	.933		239	47.50	6.35	292.85	340
8300	Reducing tee 6" x 4"		100	.280		126	14.30	1.91	142.21	162
8320	8" x 6"		90	.311		223	15.90	2.12	241.02	271
8330	10" x 6"		90	.311		195	15.90	2.12	213.02	241
8340	10" x 8"		90	.311		209	15.90	2.12	227.02	256
8350	12" x 6"		90	.311		263	15.90	2.12	281.02	315
8360	12" x 8"		90	.311		291	15.90	2.12	309.02	345
8400	Tapped service tee (threaded type) 6" x 6" x 3/4"		100	.280		95	14.30	1.91	111.21	129
8430	6" x 6" x 1"		90	.311		95	15.90	2.12	113.02	131

33 14 13 – Public Water Utility Distribution Piping

33 14 13.25 Water Supply, Polyvinyl Chloride Pipe		Crew	Daily Output	Labor-Hours	Unit	Material	2021 Bare Costs Labor	2021 Bare Costs Equipment	Total	Total Incl O&P
8440	6" x 6" x 1-1/2"	B-21	90	.311	Ea.	95	15.90	2.12	113.02	131
8450	6" x 6" x 2"		90	.311		95	15.90	2.12	113.02	131
8460	8" x 8" x 3/4"		90	.311		140	15.90	2.12	158.02	180
8470	8" x 8" x 1"		90	.311		140	15.90	2.12	158.02	180
8480	8" x 8" x 1-1/2"		90	.311		140	15.90	2.12	158.02	180
8490	8" x 8" x 2"		90	.311		140	15.90	2.12	158.02	180
8500	Repair coupling 4"	B-20	100	.240		26	11.85		37.85	46
8520	6" diameter		90	.267		40	13.15		53.15	63.50
8540	8" diameter		50	.480		96	23.50		119.50	142
8560	10" diameter		50	.480		202	23.50		225.50	258
8580	12" diameter		50	.480		256	23.50		279.50	320
8600	Plug end 4"		100	.240		22.50	11.85		34.35	42.50
8620	6" diameter		90	.267		40.50	13.15		53.65	64
8640	8" diameter		50	.480		69	23.50		92.50	112
8660	10" diameter		50	.480		110	23.50		133.50	157
8680	12" diameter		50	.480		148	23.50		171.50	199
8700	PVC pipe, joint restraint									
8710	4" diameter	B-20	32	.750	Ea.	46.50	37		83.50	107
8720	6" diameter		25.60	.938		57.50	46		103.50	133
8730	8" diameter		21.33	1.125		83.50	55.50		139	175
8740	10" diameter		18.28	1.313		133	64.50		197.50	243
8750	12" diameter		16.84	1.425		140	70.50		210.50	259
8760	14" diameter		16	1.500		190	74		264	320
8770	16" diameter		11.64	2.062		256	102		358	435
8780	18" diameter		11.03	2.176		315	107		422	505
8785	20" diameter		9.14	2.626		445	129		574	685
8790	24" diameter		7.53	3.187		515	157		672	800

33 14 13.35 Water Supply, HDPE		Crew	Daily Output	Labor-Hours	Unit	Material	2021 Bare Costs Labor	2021 Bare Costs Equipment	Total	Total Incl O&P
0010	**WATER SUPPLY, HDPE**									
0011	Butt fusion joints, SDR 21 40' lengths not including excavation or backfill									
0100	4" diameter	B-22A	400	.100	L.F.	3.20	5.10	2.01	10.31	13.35
0200	6" diameter		380	.105		5.60	5.35	2.11	13.06	16.50
0300	8" diameter		320	.125		10	6.35	2.51	18.86	23.50
0400	10" diameter		300	.133		13.25	6.75	2.68	22.68	27.50
0500	12" diameter		260	.154		17.15	7.80	3.09	28.04	34
0600	14" diameter	B-22B	220	.182		18.05	9.25	6.65	33.95	41
0700	16" diameter		180	.222		21.50	11.30	8.15	40.95	49.50
0800	18" diameter		140	.286		27	14.50	10.50	52	62.50
0850	20" diameter		130	.308		34	15.60	11.30	60.90	73.50
0900	24" diameter		100	.400		70	20.50	14.70	105.20	124
1000	Fittings									
1100	Elbows, 90 degrees									
1200	4" diameter	B-22A	32	1.250	Ea.	16.55	63.50	25	105.05	141
1300	6" diameter		28	1.429		47	72.50	28.50	148	192
1400	8" diameter		24	1.667		94.50	84.50	33.50	212.50	267
1500	10" diameter		18	2.222		260	113	44.50	417.50	505
1600	12" diameter		12	3.333		283	169	67	519	635
1700	14" diameter	B-22B	9	4.444		625	226	163	1,014	1,200
1800	16" diameter		6	6.667		925	340	245	1,510	1,800
1900	18" diameter		4	10		1,250	510	365	2,125	2,550
2000	24" diameter		3	13.333		1,375	675	490	2,540	3,075
2100	Tees									

For customer support on your Plumbing Costs with RSMeans Data, call 800.448.8182.

427

33 14 13 – Public Water Utility Distribution Piping

33 14 13.35 Water Supply, HDPE

		Crew	Daily Output	Labor-Hours	Unit	Material	2021 Bare Costs Labor	Equipment	Total	Total Incl O&P
2200	4" diameter	B-22A	30	1.333	Ea.	22.50	67.50	27	117	156
2300	6" diameter		26	1.538		68.50	78	31	177.50	227
2400	8" diameter		22	1.818		131	92.50	36.50	260	320
2500	10" diameter		15	2.667		173	135	53.50	361.50	450
2600	12" diameter	↓	10	4		505	203	80.50	788.50	950
2700	14" diameter	B-22B	8	5		595	254	184	1,033	1,225
2800	16" diameter		6	6.667		700	340	245	1,285	1,550
2900	18" diameter		4	10		765	510	365	1,640	2,000
3000	24" diameter	↓	2	20	↓	935	1,025	735	2,695	3,350
4100	Caps									
4110	4" diameter	B-22A	34	1.176	Ea.	15.60	59.50	23.50	98.60	132
4120	6" diameter		30	1.333		35	67.50	27	129.50	169
4130	8" diameter		26	1.538		58.50	78	31	167.50	216
4150	10" diameter		20	2		146	102	40	288	355
4160	12" diameter	↓	14	2.857	↓	176	145	57.50	378.50	475

33 14 13.45 Water Supply, Copper Pipe

		Crew	Daily Output	Labor-Hours	Unit	Material	2021 Bare Costs Labor	Equipment	Total	Total Incl O&P
0010	**WATER SUPPLY, COPPER PIPE**									
0020	Not including excavation or backfill									
2000	Tubing, type K, 20' joints, 3/4" diameter	Q-1	400	.040	L.F.	7.95	2.44		10.39	12.40
2200	1" diameter		320	.050		10.90	3.05		13.95	16.55
3000	1-1/2" diameter		265	.060		16.85	3.68		20.53	24
3020	2" diameter		230	.070		26	4.24		30.24	35
3040	2-1/2" diameter		146	.110		39.50	6.70		46.20	53.50
3060	3" diameter		134	.119		54.50	7.30		61.80	71
4012	4" diameter	↓	95	.168		90	10.25		100.25	114
4016	6" diameter	Q-2	80	.300		57.50	18.95		76.45	91.50
4018	8" diameter	"	80	.300	↓	134	18.95		152.95	177
5000	Tubing, type L									
5108	2" diameter	Q-1	230	.070	L.F.	7.95	4.24		12.19	15.10
6010	3" diameter		134	.119		37	7.30		44.30	52
6012	4" diameter	↓	95	.168		22	10.25		32.25	39.50
6016	6" diameter	Q-2	80	.300	↓	49.50	18.95		68.45	83
7165	Fittings, brass, corporation stops, no lead, 3/4" diameter	1 Plum	19	.421	Ea.	76.50	28.50		105	127
7166	1" diameter		16	.500		100	34		134	161
7167	1-1/2" diameter		13	.615		213	41.50		254.50	296
7168	2" diameter		11	.727		335	49		384	445
7170	Curb stops, no lead, 3/4" diameter		19	.421		100	28.50		128.50	153
7171	1" diameter		16	.500		145	34		179	210
7172	1-1/2" diameter		13	.615		275	41.50		316.50	360
7173	2" diameter		11	.727		310	49		359	415

33 14 17 – Site Water Utility Service Laterals

33 14 17.15 Tapping, Crosses and Sleeves

		Crew	Daily Output	Labor-Hours	Unit	Material	2021 Bare Costs Labor	Equipment	Total	Total Incl O&P
0010	**TAPPING, CROSSES AND SLEEVES**									
4000	Drill and tap pressurized main (labor only)									
4100	6" main, 1" to 2" service	Q-1	3	5.333	Ea.		325		325	485
4150	8" main, 1" to 2" service	"	2.75	5.818	"		355		355	530
4500	Tap and insert gate valve									
4600	8" main, 4" branch	B-21	3.20	8.750	Ea.		445	59.50	504.50	735
4650	6" branch	↓	2.70	10.370			530	70.50	600.50	870
4700	10" main, 4" branch		2.70	10.370	↓		530	70.50	600.50	870
4750	6" branch	B-21	2.35	11.915	Ea.		610	81.50	691.50	1,000
4800	12" main, 6" branch	↓	2.35	11.915			610	81.50	691.50	1,000

33 14 17.15 Tapping, Crosses and Sleeves	Crew	Daily Output	Labor-Hours	Unit	Material	2021 Bare Costs Labor	Equipment	Total	Total Incl O&P
4850	8" branch	B-21	2.35	11.915	Ea.		610	81.50	691.50	1,000
7020	Crosses, 4" x 4"		37	.757		1,400	38.50	5.15	1,443.65	1,625
7030	6" x 4"		25	1.120		1,675	57	7.65	1,739.65	1,950
7040	6" x 6"		25	1.120		1,675	57	7.65	1,739.65	1,950
7060	8" x 6"		21	1.333		2,050	68	9.10	2,127.10	2,350
7080	8" x 8"		21	1.333		2,225	68	9.10	2,302.10	2,550
7100	10" x 6"		21	1.333		4,050	68	9.10	4,127.10	4,575
7120	10" x 10"		21	1.333		4,400	68	9.10	4,477.10	4,925
7140	12" x 6"		18	1.556		4,050	79.50	10.60	4,140.10	4,600
7160	12" x 12"		18	1.556		5,125	79.50	10.60	5,215.10	5,775
7180	14" x 6"		16	1.750		10,100	89.50	11.95	10,201.45	11,200
7200	14" x 14"		16	1.750		10,700	89.50	11.95	10,801.45	11,900
7220	16" x 6"		14	2		10,900	102	13.65	11,015.65	12,100
7240	16" x 10"		14	2		11,000	102	13.65	11,115.65	12,300
7260	16" x 16"		14	2		11,600	102	13.65	11,715.65	12,900
7280	18" x 6"		10	2.800		16,200	143	19.10	16,362.10	18,100
7300	18" x 12"		10	2.800		16,400	143	19.10	16,562.10	18,200
7320	18" x 18"		10	2.800		16,800	143	19.10	16,962.10	18,700
7340	20" x 6"		8	3.500		13,100	179	24	13,303	14,800
7360	20" x 12"		8	3.500		14,100	179	24	14,303	15,800
7380	20" x 20"		8	3.500		20,600	179	24	20,803	22,900
7400	24" x 6"		6	4.667		17,000	238	32	17,270	19,100
7420	24" x 12"		6	4.667		17,200	238	32	17,470	19,300
7440	24" x 18"		6	4.667		26,100	238	32	26,370	29,100
7460	24" x 24"		6	4.667		26,700	238	32	26,970	29,800
7600	Cut-in sleeves with rubber gaskets, 4"		18	1.556		640	79.50	10.60	730.10	835
7620	6"		12	2.333		745	119	15.90	879.90	1,025
7640	8"		10	2.800		950	143	19.10	1,112.10	1,275
7660	10"		10	2.800		1,325	143	19.10	1,487.10	1,675
7680	12"		9	3.111		1,800	159	21	1,980	2,225
7800	Cut-in valves with rubber gaskets, 4"		18	1.556		460	79.50	10.60	550.10	640
7820	6"		12	2.333		600	119	15.90	734.90	855
7840	8"		10	2.800		910	143	19.10	1,072.10	1,225
7860	10"		10	2.800		1,375	143	19.10	1,537.10	1,725
7880	12"		9	3.111		1,725	159	21	1,905	2,125
7900	Tapping valve 4", MJ, ductile iron		18	1.556		670	79.50	10.60	760.10	865
7920	6", MJ, ductile iron		12	2.333		885	119	15.90	1,019.90	1,175
7940	Tapping valve 8", MJ, ductile iron		10	2.800		850	143	19.10	1,012.10	1,175
7960	Tapping valve 10", MJ, ductile iron		10	2.800		1,175	143	19.10	1,337.10	1,525
7980	Tapping valve 12", MJ, ductile iron		8	3.500		2,125	179	24	2,328	2,625
8000	Sleeves with rubber gaskets, 4" x 4"		37	.757		1,125	38.50	5.15	1,168.65	1,300
8010	6" x 4"		25	1.120		1,050	57	7.65	1,114.65	1,250
8020	6" x 6"		25	1.120		1,325	57	7.65	1,389.65	1,575
8030	8" x 4"		21	1.333		950	68	9.10	1,027.10	1,150
8040	8" x 6"		21	1.333		1,650	68	9.10	1,727.10	1,900
8060	8" x 8"		21	1.333		1,450	68	9.10	1,527.10	1,675
8070	10" x 4"		21	1.333		885	68	9.10	962.10	1,075
8080	10" x 6"		21	1.333		3,200	68	9.10	3,277.10	3,600
8090	10" x 8"		21	1.333		1,400	68	9.10	1,477.10	1,625
8100	10" x 10"		21	1.333		2,500	68	9.10	2,577.10	2,850
8110	12" x 4"		18	1.556		3,175	79.50	10.60	3,265.10	3,600
8120	12" x 6"		18	1.556		3,175	79.50	10.60	3,265.10	3,600
8130	12" x 8"		18	1.556		3,375	79.50	10.60	3,465.10	3,825

For customer support on your Plumbing Costs with RSMeans Data, call 800.448.8182.

429

33 14 17 – Site Water Utility Service Laterals

33 14 17.15 Tapping, Crosses and Sleeves

		Crew	Daily Output	Labor-Hours	Unit	Material	2021 Bare Costs Labor	Equipment	Total	Total Incl O&P
8135	12" x 10"	B-21	18	1.556	Ea.	2,850	79.50	10.60	2,940.10	3,250
8140	12" x 12"		18	1.556		3,175	79.50	10.60	3,265.10	3,625
8160	14" x 6"		16	1.750		1,600	89.50	11.95	1,701.45	1,925
8180	14" x 14"		16	1.750		2,775	89.50	11.95	2,876.45	3,200
8200	16" x 6"		14	2		1,750	102	13.65	1,865.65	2,100
8220	16" x 10"		14	2		3,825	102	13.65	3,940.65	4,375
8240	16" x 16"		14	2		9,325	102	13.65	9,440.65	10,500
8260	18" x 6"		10	2.800		1,250	143	19.10	1,412.10	1,600
8280	18" x 12"		10	2.800		3,800	143	19.10	3,962.10	4,425
8300	18" x 18"		10	2.800		12,300	143	19.10	12,462.10	13,800
8320	20" x 6"		8	3.500		2,525	179	24	2,728	3,075
8340	20" x 12"		8	3.500		3,575	179	24	3,778	4,225
8360	20" x 20"		8	3.500		15,100	179	24	15,303	16,900
8380	24" x 6"		6	4.667		2,100	238	32	2,370	2,700
8400	24" x 12"		6	4.667		9,600	238	32	9,870	11,000
8420	24" x 18"		6	4.667		11,200	238	32	11,470	12,700
8440	24" x 24"		6	4.667		12,600	238	32	12,870	14,300
8800	Hydrant valve box, 6' long	B-20	20	1.200		232	59		291	345
8820	8' long		18	1.333		237	65.50		302.50	360
8830	Valve box w/lid 4' deep		14	1.714		98.50	84.50		183	235
8840	Valve box and large base w/lid		14	1.714		340	84.50		424.50	495

33 14 19 – Valves and Hydrants for Water Utility Service

33 14 19.10 Valves

		Crew	Daily Output	Labor-Hours	Unit	Material	2021 Bare Costs Labor	Equipment	Total	Total Incl O&P
0010	**VALVES**, water distribution									
0011	See Sections 22 05 23.20 and 22 05 23.60									
3000	Butterfly valves with boxes, cast iron, mechanical joint									
3100	4" diameter	B-6	6	4	Ea.	665	192	36	893	1,050
3140	6" diameter		6	4		875	192	36	1,103	1,300
3180	8" diameter		6	4		1,100	192	36	1,328	1,550
3300	10" diameter		6	4		1,500	192	36	1,728	1,975
3340	12" diameter		6	4		2,025	192	36	2,253	2,575
3400	14" diameter		4	6		4,325	289	54	4,668	5,275
3440	16" diameter		4	6		6,150	289	54	6,493	7,250
3460	18" diameter		4	6		6,550	289	54	6,893	7,700
3480	20" diameter		4	6		10,100	289	54	10,443	11,600
3500	24" diameter		4	6		14,400	289	54	14,743	16,400
3510	30" diameter		4	6		12,000	289	54	12,343	13,700
3520	36" diameter		4	6		14,900	289	54	15,243	16,900
3530	42" diameter		4	6		18,900	289	54	19,243	21,300
3540	48" diameter		4	6		24,300	289	54	24,643	27,300
3600	With lever operator									
3610	4" diameter	B-6	6	4	Ea.	565	192	36	793	950
3614	6" diameter		6	4		780	192	36	1,008	1,175
3616	8" diameter		6	4		1,000	192	36	1,228	1,425
3618	10" diameter		6	4		1,400	192	36	1,628	1,875
3620	12" diameter		6	4		1,950	192	36	2,178	2,450
3622	14" diameter		4	6		4,000	289	54	4,343	4,900
3624	16" diameter		4	6		5,800	289	54	6,143	6,875
3626	18" diameter		4	6		6,225	289	54	6,568	7,325
3628	20" diameter		4	6		9,725	289	54	10,068	11,200
3630	24" diameter		4	6		14,100	289	54	14,443	16,000
3700	Check valves, flanged									

33 14 Water Utility Transmission and Distribution

33 14 19 — Valves and Hydrants for Water Utility Service

33 14 19.10 Valves		Crew	Daily Output	Labor-Hours	Unit	Material	2021 Bare Costs Labor	Equipment	Total	Total Incl O&P
3710	4" diameter	B-6	6	4	Ea.	820	192	36	1,048	1,225
3714	6" diameter		6	4		1,400	192	36	1,628	1,850
3716	8" diameter		6	4		2,625	192	36	2,853	3,225
3718	10" diameter		6	4		4,475	192	36	4,703	5,250
3720	12" diameter		6	4		7,450	192	36	7,678	8,525
3726	18" diameter		4	6		36,600	289	54	36,943	40,800
3730	24" diameter		4	6		73,500	289	54	73,843	81,500
3800	Gate valves, C.I., 125 psi, mechanical joint, w/boxes									
3810	4" diameter	B-6	6	4	Ea.	1,100	192	36	1,328	1,550
3814	6" diameter		6	4		1,750	192	36	1,978	2,250
3816	8" diameter		6	4		3,050	192	36	3,278	3,675
3818	10" diameter		6	4		5,475	192	36	5,703	6,350
3820	12" diameter		6	4		7,450	192	36	7,678	8,500
3822	14" diameter		4	6		15,300	289	54	15,643	17,400
3824	16" diameter		4	6		21,600	289	54	21,943	24,300
3826	18" diameter		4	6		26,200	289	54	26,543	29,300
3828	20" diameter		4	6		36,400	289	54	36,743	40,500
3830	24" diameter		4	6		54,000	289	54	54,343	59,500
3831	30" diameter		4	6		58,500	289	54	58,843	65,000
3832	36" diameter		4	6		91,500	289	54	91,843	101,000
3880	Sleeve, for tapping mains, 8" x 4", add				Ea.	950			950	1,050
3884	10" x 6", add					3,200			3,200	3,500
3888	12" x 6", add					3,175			3,175	3,475
3892	12" x 8", add					3,375			3,375	3,700

33 14 19.20 Valves		Crew	Daily Output	Labor-Hours	Unit	Material	2021 Bare Costs Labor	Equipment	Total	Total Incl O&P
0010	**VALVES**									
0011	Special trim or use									
9000	Valves, gate valve, N.R.S. PIV with post, 4" diameter	B-6	6	4	Ea.	2,625	192	36	2,853	3,200
9020	6" diameter		6	4		3,275	192	36	3,503	3,925
9040	8" diameter		6	4		4,550	192	36	4,778	5,350
9060	10" diameter		6	4		7,000	192	36	7,228	8,025
9080	12" diameter		6	4		8,950	192	36	9,178	10,200
9100	14" diameter		6	4		16,600	192	36	16,828	18,600
9120	OS&Y, 4" diameter		6	4		925	192	36	1,153	1,350
9140	6" diameter		6	4		1,300	192	36	1,528	1,750
9160	8" diameter		6	4		1,900	192	36	2,128	2,400
9180	10" diameter		6	4		2,925	192	36	3,153	3,550
9200	12" diameter		6	4		4,800	192	36	5,028	5,600
9220	14" diameter		4	6		4,975	289	54	5,318	5,975
9400	Check valves, rubber disc, 2-1/2" diameter		6	4		430	192	36	658	795
9420	3" diameter		6	4		590	192	36	818	975
9440	4" diameter		6	4		820	192	36	1,048	1,225
9480	6" diameter		6	4		1,400	192	36	1,628	1,850
9500	8" diameter		6	4		2,625	192	36	2,853	3,225
9520	10" diameter		6	4		4,475	192	36	4,703	5,250
9540	12" diameter		6	4		7,450	192	36	7,678	8,525
9542	14" diameter		4	6		7,900	289	54	8,243	9,175
9700	Detector check valves, reducing, 4" diameter		6	4		1,500	192	36	1,728	1,975
9720	6" diameter		6	4		2,425	192	36	2,653	3,000
9740	8" diameter		6	4		3,525	192	36	3,753	4,200
9760	10" diameter		6	4		6,400	192	36	6,628	7,350
9800	Galvanized, 4" diameter		6	4		2,050	192	36	2,278	2,600

33 14 Water Utility Transmission and Distribution

33 14 19 – Valves and Hydrants for Water Utility Service

33 14 19.20 Valves		Crew	Daily Output	Labor-Hours	Unit	Material	2021 Bare Costs Labor	2021 Bare Costs Equipment	Total	Total Incl O&P
9820	6" diameter	B-6	6	4	Ea.	2,825	192	36	3,053	3,450
9840	8" diameter		6	4		3,800	192	36	4,028	4,500
9860	10" diameter	↓	6	4	↓	6,925	192	36	7,153	7,925

33 14 19.30 Fire Hydrants

		Crew	Daily Output	Labor-Hours	Unit	Material	2021 Bare Costs Labor	2021 Bare Costs Equipment	Total	Total Incl O&P
0010	**FIRE HYDRANTS** G3010-410									
0020	Mechanical joints unless otherwise noted									
1000	Fire hydrants, two way; excavation and backfill not incl.									
1100	4-1/2" valve size, depth 2'-0"	B-21	10	2.800	Ea.	3,075	143	19.10	3,237.10	3,625
1200	4'-6"		9	3.111		3,325	159	21	3,505	3,925
1260	6'-0"		7	4		3,575	204	27.50	3,806.50	4,275
1340	8'-0"	↓	6	4.667	↓	3,725	238	32	3,995	4,500
1420	10'-0"	B-21	5	5.600	Ea.	3,975	286	38	4,299	4,850
2000	5-1/4" valve size, depth 2'-0"		10	2.800		3,325	143	19.10	3,487.10	3,900
2080	4'-0"		9	3.111		3,650	159	21	3,830	4,275
2160	6'-0"		7	4		3,975	204	27.50	4,206.50	4,700
2240	8'-0"		6	4.667		4,350	238	32	4,620	5,200
2320	10'-0"	↓	5	5.600		4,675	286	38	4,999	5,625
2350	For threeway valves, add					7%				
2400	Lower barrel extensions with stems, 1'-0"	B-20	14	1.714		340	84.50		424.50	500
2440	2'-0"		13	1.846		415	91		506	590
2480	3'-0"		12	2		1,375	98.50		1,473.50	1,650
2520	4'-0"	↓	10	2.400	↓	785	118		903	1,025
5000	Indicator post									
5020	Adjustable, valve size 4" to 14", 4' bury	B-21	10	2.800	Ea.	1,400	143	19.10	1,562.10	1,775
5060	8' bury		7	4		1,625	204	27.50	1,856.50	2,100
5080	10' bury		6	4.667		2,125	238	32	2,395	2,750
5100	12' bury		5	5.600		1,750	286	38	2,074	2,400
5120	14' bury		4	7		1,675	355	47.50	2,077.50	2,450
5500	Non-adjustable, valve size 4" to 14", 3' bury		10	2.800		870	143	19.10	1,032.10	1,200
5520	3'-6" bury		10	2.800		870	143	19.10	1,032.10	1,200
5540	4' bury	↓	9	3.111	↓	870	159	21	1,050	1,225

33 16 Water Utility Storage Tanks

33 16 36 – Ground-Level Reinforced Concrete Water Storage Tanks

33 16 36.16 Prestressed Conc. Water Storage Tanks

		Crew	Daily Output	Labor-Hours	Unit	Material	2021 Bare Costs Labor	2021 Bare Costs Equipment	Total	Total Incl O&P
0010	**PRESTRESSED CONC. WATER STORAGE TANKS**									
0020	Not including fdn., pipe or pumps, 250,000 gallons				Ea.				299,000	329,500

33 31 Sanitary Sewerage Piping

33 31 11 – Public Sanitary Sewerage Gravity Piping

33 31 11.10 Sewage Collection, Valves

		Crew	Daily Output	Labor-Hours	Unit	Material	2021 Bare Costs Labor	2021 Bare Costs Equipment	Total	Total Incl O&P
0010	**SEWAGE COLLECTION, VALVES**									
1000	Backwater sewer line valve									
1010	Offset type, bronze swing check assy and bronze cover									
1020	B&S connections									
1040	2" size	2 Skwk	14	1.143	Ea.	775	65.50		840.50	955
1050	3" size		12.50	1.280		1,100	73		1,173	1,300
1060	4" size	↓	11	1.455		1,100	83		1,183	1,325
1070	6" size	3 Skwk	15	1.600	↓	1,775	91.50		1,866.50	2,100

33 31 Sanitary Sewerage Piping

33 31 11 – Public Sanitary Sewerage Gravity Piping

33 31 11.10 Sewage Collection, Valves

	Crew	Daily Output	Labor-Hours	Unit	Material	2021 Bare Costs Labor	2021 Bare Costs Equipment	Total	Total Incl O&P
1080	8" size	3 Skwk	7.50	3.200	Ea.	2,225	183		2,408	2,725
1110	Backwater drainage control									
1120	Offset type, w/bronze manually operated shear gate									
1130	B&S connections									
1160	4" size	2 Skwk	11	1.455	Ea.	3,225	83		3,308	3,675
1170	6" size	3 Skwk	15	1.600	"	4,375	91.50		4,466.50	4,950
3416	16" diameter	B-21	80	.350	L.F.	218	17.85	2.39	238.24	269
3418	18" diameter	B-13B	80	.700		380	34	12.40	426.40	485
3420	20" diameter		60	.933		415	45	16.50	476.50	540
3424	24" diameter		60	.933		565	45	16.50	626.50	705
3436	36" diameter		40	1.400		875	67.50	25	967.50	1,100

33 31 11.25 Sewage Collection, Polyvinyl Chloride Pipe

	Crew	Daily Output	Labor-Hours	Unit	Material	2021 Bare Costs Labor	2021 Bare Costs Equipment	Total	Total Incl O&P
0010	**SEWAGE COLLECTION, POLYVINYL CHLORIDE PIPE**									
0020	Not including excavation or backfill									
2000	20' lengths, SDR 35, B&S, 4" diameter	B-20	375	.064	L.F.	2.21	3.16		5.37	7.15
2040	6" diameter	"	350	.069		4.26	3.38		7.64	9.75
2120	10" diameter	B-21	330	.085		12.05	4.33	.58	16.96	20.50
2160	12" diameter		320	.088		17.05	4.47	.60	22.12	26
2170	14" diameter		280	.100		19.45	5.10	.68	25.23	30
2200	15" diameter		240	.117		22	5.95	.80	28.75	34
2250	16" diameter		220	.127		29	6.50	.87	36.37	42
4000	Piping, DWV PVC, no exc./bkfill., 10' L, Sch 40, 4" diameter	B-20	375	.064		4.78	3.16		7.94	9.95
4010	6" diameter		350	.069		11.20	3.38		14.58	17.35
4020	8" diameter		335	.072		17.85	3.53		21.38	25

33 31 11.37 Centrif. Cst. Fbgs-Reinf. Polymer Mort. Util. Pipe

	Crew	Daily Output	Labor-Hours	Unit	Material	2021 Bare Costs Labor	2021 Bare Costs Equipment	Total	Total Incl O&P
0010	**CENTRIF. CST. FBGS-REINF. POLYMER MORT. UTIL. PIPE**									
0020	Not including excavation or backfill									
3348	48" diameter	B-13B	40	1.400	L.F.	181	67.50	25	273.50	325
3354	54" diameter		40	1.400		295	67.50	25	387.50	455
3363	63" diameter		20	2.800		365	135	49.50	549.50	655
3418	18" diameter		60	.933		128	45	16.50	189.50	226
3424	24" diameter		60	.933		141	45	16.50	202.50	240
3436	36" diameter		40	1.400		159	67.50	25	251.50	305
3444	44" diameter		40	1.400		175	67.50	25	267.50	320
3448	48" diameter		40	1.400		195	67.50	25	287.50	345

33 34 Onsite Wastewater Disposal

33 34 13 – Septic Tanks

33 34 13.13 Concrete Septic Tanks

	Crew	Daily Output	Labor-Hours	Unit	Material	2021 Bare Costs Labor	2021 Bare Costs Equipment	Total	Total Incl O&P
0010	**CONCRETE SEPTIC TANKS** G3020-300									
0011	Not including excavation or piping									
0015	Septic tanks, precast, 1,000 gallon	B-21	8	3.500	Ea.	1,000	179	24	1,203	1,400
0060	1,500 gallon		7	4		1,675	204	27.50	1,906.50	2,175
0100	2,000 gallon		5	5.600		2,550	286	38	2,874	3,300
0200	5,000 gallon	B-13	3.50	16		7,350	770	168	8,288	9,425
0300	15,000 gallon, 4 piece	B-13B	1.70	32.941		25,700	1,600	580	27,880	31,300
0400	25,000 gallon, 4 piece		1.10	50.909		48,400	2,450	900	51,750	57,500
0500	40,000 gallon, 4 piece		.80	70		53,500	3,375	1,250	58,125	65,500
0520	50,000 gallon, 5 piece	B-13C	.60	93.333		61,500	4,500	3,850	69,850	79,000
0640	75,000 gallon, cast in place	C-14C	.25	448		75,000	23,400	108	98,508	117,500

33 34 13 – Septic Tanks

33 34 13.13 Concrete Septic Tanks

		Crew	Daily Output	Labor-Hours	Unit	Material	2021 Bare Costs Labor	Equipment	Total	Total Incl O&P
0660	100,000 gallon	C-14C	.15	747	Ea.	93,000	39,000	179	132,179	160,000
1150	Leaching field chambers, 13' x 3'-7" x 1'-4", standard	B-13	16	3.500		475	169	36.50	680.50	820
1200	Heavy duty, 8' x 4' x 1'-6"		14	4		325	193	42	560	690
1300	13' x 3'-9" x 1'-6"		12	4.667		1,400	225	49	1,674	1,925
1350	20' x 4' x 1'-6"		5	11.200		1,250	540	117	1,907	2,300
1400	Leaching pit, precast concrete, 3' diameter, 3' deep	B-21	8	3.500		710	179	24	913	1,075
1500	6' diameter, 3' section		4.70	5.957		1,100	305	40.50	1,445.50	1,725
2000	Velocity reducing pit, precast conc., 6' diameter, 3' deep		4.70	5.957		1,875	305	40.50	2,220.50	2,550

33 34 13.33 Polyethylene Septic Tanks

		Crew	Daily Output	Labor-Hours	Unit	Material	2021 Bare Costs Labor	Equipment	Total	Total Incl O&P
0010	**POLYETHYLENE SEPTIC TANKS**									
0015	High density polyethylene, 1,000 gallon	B-21	8	3.500	Ea.	1,600	179	24	1,803	2,050
0020	1,250 gallon		8	3.500		1,325	179	24	1,528	1,750
0025	1,500 gallon		7	4		1,400	204	27.50	1,631.50	1,850

33 34 16 – Septic Tank Effluent Filters

33 34 16.13 Septic Tank Gravity Effluent Filters

		Crew	Daily Output	Labor-Hours	Unit	Material	2021 Bare Costs Labor	Equipment	Total	Total Incl O&P
0010	**SEPTIC TANK GRAVITY EFFLUENT FILTERS**									
3000	Effluent filter, 4" diameter	1 Skwk	8	1	Ea.	36	57		93	126
3020	6" diameter		7	1.143		43.50	65.50		109	146
3030	8" diameter		7	1.143		254	65.50		319.50	375
3040	8" diameter, very fine		7	1.143		505	65.50		570.50	655
3050	10" diameter, very fine		6	1.333		240	76		316	380
3060	10" diameter		6	1.333		274	76		350	415
3080	12" diameter		6	1.333		660	76		736	840
3090	15" diameter		5	1.600		1,100	91.50		1,191.50	1,325

33 34 51 – Drainage Field Systems

33 34 51.10 Drainage Field Excavation and Fill

		Crew	Daily Output	Labor-Hours	Unit	Material	2021 Bare Costs Labor	Equipment	Total	Total Incl O&P
0010	**DRAINAGE FIELD EXCAVATION AND FILL**									
2200	Septic tank & drainage field excavation with 3/4 C.Y. backhoe	B-12F	145	.110	C.Y.		5.85	4.84	10.69	14
2400	4' trench for disposal field, 3/4 C.Y. backhoe	"	335	.048	L.F.		2.53	2.09	4.62	6.05
2600	Gravel fill, run of bank	B-6	150	.160	C.Y.	26.50	7.70	1.44	35.64	42.50
2800	Crushed stone, 3/4"	"	150	.160	"	41	7.70	1.44	50.14	58

33 34 51.13 Utility Septic Tank Tile Drainage Field

		Crew	Daily Output	Labor-Hours	Unit	Material	2021 Bare Costs Labor	Equipment	Total	Total Incl O&P
0010	**UTILITY SEPTIC TANK TILE DRAINAGE FIELD**									
0015	Distribution box, concrete, 5 outlets	2 Clab	20	.800	Ea.	75	35.50		110.50	136
0020	7 outlets		16	1		93.50	44.50		138	170
0025	9 outlets		8	2		530	89		619	720
0115	Distribution boxes, HDPE, 5 outlets		20	.800		75	35.50		110.50	136
0117	6 outlets		15	1.067		75	47.50		122.50	153
0118	7 outlets		15	1.067		74.50	47.50		122	153
0120	8 outlets		10	1.600		79	71		150	193
0240	Distribution boxes, outlet flow leveler	1 Clab	50	.160		2.99	7.10		10.09	13.90
0300	Precast concrete, galley, 4' x 4' x 4'	B-21	16	1.750		246	89.50	11.95	347.45	420
0350	HDPE infiltration chamber 12" H x 15" W	2 Clab	300	.053	L.F.	7.05	2.37		9.42	11.30
0351	12" H x 15" W end cap	1 Clab	32	.250	Ea.	19.05	11.10		30.15	37.50
0355	chamber 12" H x 22" W	2 Clab	300	.053	L.F.	6.70	2.37		9.07	10.90
0356	12" H x 22" W end cap	1 Clab	32	.250	Ea.	23.50	11.10		34.60	42
0360	chamber 13" H x 34" W	2 Clab	300	.053	L.F.	14.45	2.37		16.82	19.40
0361	13" H x 34" W end cap	1 Clab	32	.250	Ea.	51	11.10		62.10	72.50
0365	chamber 16" H x 34" W	2 Clab	300	.053	L.F.	19.40	2.37		21.77	25
0366	16" H x 34" W end cap	1 Clab	32	.250	Ea.	16.90	11.10		28	35
0370	chamber 8" H x 16" W	2 Clab	300	.053	L.F.	10	2.37		12.37	14.55

33 34 Onsite Wastewater Disposal

33 34 51 – Drainage Field Systems

33 34 51.13 Utility Septic Tank Tile Drainage Field	Crew	Daily Output	Labor-Hours	Unit	Material	2021 Bare Costs Labor	Equipment	Total	Total Incl O&P
0371	8" H x 16" W end cap	1 Clab	32	.250	Ea.	11.20	11.10		22.30	29

33 41 Subdrainage

33 41 16 – Subdrainage Piping

33 41 16.25 Piping, Subdrainage, Corrugated Metal

		Crew	Daily Output	Labor-Hours	Unit	Material	2021 Bare Costs Labor	Equipment	Total	Total Incl O&P
0010	**PIPING, SUBDRAINAGE, CORRUGATED METAL** G1030–805									
0021	Not including excavation and backfill									
2010	Aluminum, perforated									
2020	6" diameter, 18 ga.	B-20	380	.063	L.F.	6.35	3.11		9.46	11.65
2200	8" diameter, 16 ga.	"	370	.065		10.45	3.20		13.65	16.30
2220	10" diameter, 16 ga.	B-21	360	.078		13.05	3.97	.53	17.55	21
2240	12" diameter, 16 ga.		285	.098		14.65	5	.67	20.32	24.50
2260	18" diameter, 16 ga.		205	.137		22	6.95	.93	29.88	35.50
3000	Uncoated galvanized, perforated									
3020	6" diameter, 18 ga.	B-20	380	.063	L.F.	6.75	3.11		9.86	12.10
3200	8" diameter, 16 ga.	"	370	.065		8.10	3.20		11.30	13.75
3220	10" diameter, 16 ga.	B-21	360	.078		8.60	3.97	.53	13.10	16
3240	12" diameter, 16 ga.		285	.098		9.60	5	.67	15.27	18.80
3260	18" diameter, 16 ga.		205	.137		14.65	6.95	.93	22.53	27.50
4000	Steel, perforated, asphalt coated									
4020	6" diameter, 18 ga.	B-20	380	.063	L.F.	6	3.11		9.11	11.25
4030	8" diameter, 18 ga.	"	370	.065		9.30	3.20		12.50	15.05
4040	10" diameter, 16 ga.	B-21	360	.078		9.40	3.97	.53	13.90	16.90
4050	12" diameter, 16 ga.		285	.098		9.40	5	.67	15.07	18.60
4060	18" diameter, 16 ga.		205	.137		16.85	6.95	.93	24.73	30

33 41 16.30 Piping, Subdrainage, Plastic

		Crew	Daily Output	Labor-Hours	Unit	Material	2021 Bare Costs Labor	Equipment	Total	Total Incl O&P
0010	**PIPING, SUBDRAINAGE, PLASTIC**									
0020	Not including excavation and backfill									
2100	Perforated PVC, 4" diameter	B-14	314	.153	L.F.	2.21	7.10	.69	10	13.80
2110	6" diameter		300	.160		4.26	7.45	.72	12.43	16.60
2120	8" diameter		290	.166		8	7.70	.74	16.44	21
2130	10" diameter		280	.171		11.15	8	.77	19.92	25
2140	12" diameter		270	.178		15.75	8.30	.80	24.85	30.50

33 42 Stormwater Conveyance

33 42 11 – Stormwater Gravity Piping

33 42 11.40 Piping, Storm Drainage, Corrugated Metal

		Crew	Daily Output	Labor-Hours	Unit	Material	2021 Bare Costs Labor	Equipment	Total	Total Incl O&P
0010	**PIPING, STORM DRAINAGE, CORRUGATED METAL**									
0020	Not including excavation or backfill									
2000	Corrugated metal pipe, galvanized									
2020	Bituminous coated with paved invert, 20' lengths									
2040	8" diameter, 16 ga.	B-14	330	.145	L.F.	7.25	6.80	.65	14.70	18.80
2060	10" diameter, 16 ga.		260	.185		7.55	8.60	.83	16.98	22
2080	12" diameter, 16 ga.		210	.229		12	10.65	1.03	23.68	30
2100	15" diameter, 16 ga.		200	.240		12.70	11.20	1.08	24.98	32
2120	18" diameter, 16 ga.		190	.253		18.75	11.75	1.14	31.64	39.50
2140	24" diameter, 14 ga.		160	.300		20.50	13.95	1.35	35.80	45.50
2160	30" diameter, 14 ga.	B-13	120	.467		26.50	22.50	4.89	53.89	68
2180	36" diameter, 12 ga.		120	.467		29.50	22.50	4.89	56.89	71.50

For customer support on your Plumbing Costs with RSMeans Data, call 800.448.8182.

435

33 42 11.40 Piping, Storm Drainage, Corrugated Metal		Crew	Daily Output	Labor-Hours	Unit	Material	2021 Bare Costs Labor	Equipment	Total	Total Incl O&P
2200	48" diameter, 12 ga.	B-13	100	.560	L.F.	40	27	5.85	72.85	91
2220	60" diameter, 10 ga.	B-13B	75	.747		70	36	13.20	119.20	146
2240	72" diameter, 8 ga.	"	45	1.244		73	60	22	155	194
2250	End sections, 8" diameter, 16 ga.	B-14	20	2.400	Ea.	44	112	10.80	166.80	227
2255	10" diameter, 16 ga.		20	2.400		51.50	112	10.80	174.30	235
2260	12" diameter, 16 ga.		18	2.667		105	124	12	241	315
2265	15" diameter, 16 ga.		18	2.667		208	124	12	344	425
2270	18" diameter, 16 ga.		16	3		242	140	13.50	395.50	490
2275	24" diameter, 16 ga.	B-13	16	3.500		320	169	36.50	525.50	645
2280	30" diameter, 16 ga.		14	4		520	193	42	755	910
2285	36" diameter, 14 ga.		14	4		650	193	42	885	1,050
2290	48" diameter, 14 ga.		10	5.600		1,875	270	58.50	2,203.50	2,525
2292	60" diameter, 14 ga.		6	9.333		1,850	450	98	2,398	2,825
2294	72" diameter, 14 ga.	B-13B	5	11.200		2,850	540	198	3,588	4,150
2300	Bends or elbows, 8" diameter	B-14	28	1.714		103	80	7.70	190.70	242
2320	10" diameter		25	1.920		128	89.50	8.65	226.15	284
2340	12" diameter, 16 ga.		23	2.087		151	97	9.40	257.40	320
2342	18" diameter, 16 ga.		20	2.400		216	112	10.80	338.80	415
2344	24" diameter, 14 ga.		16	3		360	140	13.50	513.50	620
2346	30" diameter, 14 ga.		15	3.200		440	149	14.40	603.40	725
2348	36" diameter, 14 ga.	B-13	15	3.733		575	180	39	794	945
2350	48" diameter, 12 ga.	"	12	4.667		740	225	49	1,014	1,200
2352	60" diameter, 10 ga.	B-13B	10	5.600		995	270	99	1,364	1,625
2354	72" diameter, 10 ga.	"	6	9.333		1,225	450	165	1,840	2,200
2360	Wyes or tees, 8" diameter	B-14	25	1.920		146	89.50	8.65	244.15	305
2380	10" diameter		21	2.286		182	106	10.30	298.30	370
2400	12" diameter, 16 ga.		19	2.526		208	118	11.35	337.35	420
2410	18" diameter, 16 ga.		16	3		289	140	13.50	442.50	545
2412	24" diameter, 14 ga.		16	3		525	140	13.50	678.50	800
2414	30" diameter, 14 ga.	B-13	12	4.667		685	225	49	959	1,150
2416	36" diameter, 14 ga.		11	5.091		795	246	53.50	1,094.50	1,300
2418	48" diameter, 12 ga.		10	5.600		1,075	270	58.50	1,403.50	1,650
2420	60" diameter, 10 ga.	B-13B	8	7		1,575	340	124	2,039	2,375
2422	72" diameter, 10 ga.	"	5	11.200		1,900	540	198	2,638	3,100
2500	Galvanized, uncoated, 20' lengths									
2520	8" diameter, 16 ga.	B-14	355	.135	L.F.	8.20	6.30	.61	15.11	19.05
2540	10" diameter, 16 ga.		280	.171		8.15	8	.77	16.92	22
2560	12" diameter, 16 ga.		220	.218		8.90	10.15	.98	20.03	26
2580	15" diameter, 16 ga.		220	.218		11.60	10.15	.98	22.73	29
2600	18" diameter, 16 ga.		205	.234		12.90	10.90	1.05	24.85	31.50
2620	24" diameter, 14 ga.		175	.274		24.50	12.80	1.23	38.53	47
2640	30" diameter, 14 ga.	B-13	130	.431		28	21	4.51	53.51	66.50
2660	36" diameter, 12 ga.		130	.431		32	21	4.51	57.51	71
2680	48" diameter, 12 ga.		110	.509		47.50	24.50	5.35	77.35	94.50
2690	60" diameter, 10 ga.	B-13B	78	.718		78	34.50	12.70	125.20	151
2695	72" diameter, 10 ga.	"	60	.933		90	45	16.50	151.50	184
2711	Bends or elbows, 12" diameter, 16 ga.	B-14	30	1.600	Ea.	123	74.50	7.20	204.70	254
2712	15" diameter, 16 ga.		25.04	1.917		156	89.50	8.65	254.15	315
2714	18" diameter, 16 ga.		20	2.400		175	112	10.80	297.80	370
2716	24" diameter, 14 ga.		16	3		265	140	13.50	418.50	515
2718	30" diameter, 14 ga.		15	3.200		355	149	14.40	518.40	630
2720	36" diameter, 14 ga.	B-13	15	3.733		475	180	39	694	830
2722	48" diameter, 12 ga.		12	4.667		640	225	49	914	1,100

33 42 11.40 Piping, Storm Drainage, Corrugated Metal

		Crew	Daily Output	Labor-Hours	Unit	Material	2021 Bare Costs Labor	Equipment	Total	Total Incl O&P
2724	60" diameter, 10 ga.	B-13	10	5.600	Ea.	995	270	58.50	1,323.50	1,575
2726	72" diameter, 10 ga.	↓	6	9.333		1,250	450	98	1,798	2,150
2728	Wyes or tees, 12" diameter, 16 ga.	B-14	22.48	2.135		158	99.50	9.60	267.10	335
2730	18" diameter, 16 ga.		15	3.200		245	149	14.40	408.40	505
2732	24" diameter, 14 ga.		15	3.200		425	149	14.40	588.40	710
2734	30" diameter, 14 ga.	↓	14	3.429		550	160	15.45	725.45	860
2736	36" diameter, 14 ga.	B-13	14	4		700	193	42	935	1,100
2738	48" diameter, 12 ga.		12	4.667		1,025	225	49	1,299	1,525
2740	60" diameter, 10 ga.		10	5.600		1,475	270	58.50	1,803.50	2,100
2742	72" diameter, 10 ga.	↓	6	9.333		1,775	450	98	2,323	2,725
2780	End sections, 8" diameter	B-14	35	1.371		102	64	6.15	172.15	214
2785	10" diameter		35	1.371		80.50	64	6.15	150.65	191
2790	12" diameter		35	1.371		96	64	6.15	166.15	207
2800	18" diameter	↓	30	1.600		112	74.50	7.20	193.70	242
2810	24" diameter	B-13	25	2.240		189	108	23.50	320.50	395
2820	30" diameter		25	2.240		340	108	23.50	471.50	560
2825	36" diameter		20	2.800		455	135	29.50	619.50	735
2830	48" diameter	↓	10	5.600		990	270	58.50	1,318.50	1,575
2835	60" diameter	B-13B	5	11.200		1,950	540	198	2,688	3,150
2840	72" diameter	"	4	14		2,250	675	248	3,173	3,750
2850	Couplings, 12" diameter					14.40			14.40	15.85
2855	18" diameter					17.80			17.80	19.55
2860	24" diameter					22.50			22.50	25
2865	30" diameter					27			27	29.50
2870	36" diameter					40.50			40.50	45
2875	48" diameter					59.50			59.50	65.50
2880	60" diameter					56.50			56.50	62
2885	72" diameter			↓		64			64	70.50

33 42 11.60 Sewage/Drainage Collection, Concrete Pipe

		Crew	Daily Output	Labor-Hours	Unit	Material	2021 Bare Costs Labor	Equipment	Total	Total Incl O&P
0010	**SEWAGE/DRAINAGE COLLECTION, CONCRETE PIPE**									
0020	Not including excavation or backfill									
1000	Non-reinforced pipe, extra strength, B&S or T&G joints									
1010	6" diameter	B-14	265.04	.181	L.F.	8.05	8.45	.81	17.31	22.50
1020	8" diameter		224	.214		8.90	10	.96	19.86	25.50
1030	10" diameter		216	.222		9.85	10.35	1	21.20	27.50
1040	12" diameter		200	.240		10.25	11.20	1.08	22.53	29
1050	15" diameter		180	.267		14.70	12.40	1.20	28.30	36
1060	18" diameter		144	.333		18.45	15.55	1.50	35.50	45
1070	21" diameter	↓	112	.429		19.30	19.95	1.93	41.18	53.50
1080	24" diameter		100	.480	↓	22.50	22.50	2.16	47.16	60.50
2000	Reinforced culvert, class 3, no gaskets									
2010	12" diameter	B-14	150	.320	L.F.	16.50	14.90	1.44	32.84	41.50
2020	15" diameter		150	.320		20.50	14.90	1.44	36.84	46
2030	18" diameter		132	.364		25	16.95	1.64	43.59	55.50
2035	21" diameter		120	.400		28.50	18.65	1.80	48.95	61
2040	24" diameter		100	.480		27.50	22.50	2.16	52.16	66
2045	27" diameter	B-13	92	.609		46.50	29.50	6.40	82.40	103
2050	30" diameter		88	.636		44	30.50	6.65	81.15	102
2060	36" diameter	↓	72	.778		80	37.50	8.15	125.65	153
2070	42" diameter	B-13B	68	.824		104	40	14.55	158.55	190
2080	48" diameter		64	.875		108	42.50	15.45	165.95	199
2085	54" diameter	↓	56	1	↓	148	48.50	17.70	214.20	254

	33 42 11.60 Sewage/Drainage Collection, Concrete Pipe	Crew	Daily Output	Labor-Hours	Unit	Material	2021 Bare Costs Labor	Equipment	Total	Total Incl O&P
2090	60" diameter	B-13B	48	1.167	L.F.	229	56.50	20.50	306	360
2100	72" diameter		40	1.400		254	67.50	25	346.50	410
2120	84" diameter		32	1.750		410	84.50	31	525.50	615
2140	96" diameter		24	2.333		495	113	41.50	649.50	760
2200	With gaskets, class 3, 12" diameter	B-21	168	.167		18.15	8.50	1.14	27.79	34
2220	15" diameter		160	.175		22.50	8.95	1.19	32.64	39.50
2230	18" diameter		152	.184		28	9.40	1.26	38.66	46
2235	21" diameter		152	.184		31	9.40	1.26	41.66	50
2240	24" diameter		136	.206		34	10.50	1.40	45.90	54.50
2260	30" diameter	B-13	88	.636		52.50	30.50	6.65	89.65	111
2270	36" diameter	"	72	.778		90	37.50	8.15	135.65	163
2290	48" diameter	B-13B	64	.875		121	42.50	15.45	178.95	214
2310	72" diameter	"	40	1.400		275	67.50	25	367.50	430
2330	Flared ends, 12" diameter	B-21	31	.903	Ea.	340	46	6.15	392.15	450
2340	15" diameter		25	1.120		395	57	7.65	459.65	530
2400	18" diameter		20	1.400		445	71.50	9.55	526.05	610
2420	24" diameter		14	2		515	102	13.65	630.65	740
2440	36" diameter	B-13	10	5.600		1,125	270	58.50	1,453.50	1,725
3080	Radius pipe, add to pipe prices, 12" to 60" diameter				L.F.	50%				
3090	Over 60" diameter, add				"	20%				
3500	Reinforced elliptical, 8' lengths, C507 class 3									
3520	14" x 23" inside, round equivalent 18" diameter	B-21	82	.341	L.F.	41	17.45	2.33	60.78	73.50
3530	24" x 38" inside, round equivalent 30" diameter	B-13	58	.966		70	46.50	10.10	126.60	158
3540	29" x 45" inside, round equivalent 36" diameter		52	1.077		110	52	11.30	173.30	211
3550	38" x 60" inside, round equivalent 48" diameter		38	1.474		178	71	15.45	264.45	320
3560	48" x 76" inside, round equivalent 60" diameter		26	2.154		203	104	22.50	329.50	405
3570	58" x 91" inside, round equivalent 72" diameter		22	2.545		345	123	26.50	494.50	590
3780	Concrete slotted pipe, class 4 mortar joint									
3800	12" diameter	B-21	168	.167	L.F.	24.50	8.50	1.14	34.14	41
3840	18" diameter	"	152	.184	"	31	9.40	1.26	41.66	49.50
3900	Concrete slotted pipe, Class 4 O-ring joint									
3940	12" diameter	B-21	168	.167	L.F.	31.50	8.50	1.14	41.14	48.50
3960	18" diameter	"	152	.184	"	32	9.40	1.26	42.66	50.50

33 42 13 – Stormwater Culverts

33 42 13.15 Oval Arch Culverts

		Crew	Daily Output	Labor-Hours	Unit	Material	2021 Bare Costs Labor	Equipment	Total	Total Incl O&P
0010	**OVAL ARCH CULVERTS**									
3000	Corrugated galvanized or aluminum, coated & paved									
3020	17" x 13", 16 ga., 15" equivalent	B-14	200	.240	L.F.	16.90	11.20	1.08	29.18	36.50
3040	21" x 15", 16 ga., 18" equivalent		150	.320		17.30	14.90	1.44	33.64	42.50
3060	28" x 20", 14 ga., 24" equivalent		125	.384		26.50	17.90	1.73	46.13	57.50
3080	35" x 24", 14 ga., 30" equivalent		100	.480		32.50	22.50	2.16	57.16	72
3100	42" x 29", 12 ga., 36" equivalent	B-13	100	.560		30	27	5.85	62.85	80
3120	49" x 33", 12 ga., 42" equivalent	"	90	.622		39	30	6.50	75.50	94.50
3140	57" x 38", 12 ga., 48" equivalent	B-13	75	.747	L.F.	51.50	36	7.85	95.35	119
3160	Steel, plain oval arch culverts, plain									
3180	17" x 13", 16 ga., 15" equivalent	B-14	225	.213	L.F.	11.05	9.95	.96	21.96	28
3200	21" x 15", 16 ga., 18" equivalent		175	.274		14.40	12.80	1.23	28.43	36.50
3220	28" x 20", 14 ga., 24" equivalent		150	.320		20	14.90	1.44	36.34	45.50
3240	35" x 24", 14 ga., 30" equivalent	B-13	108	.519		26	25	5.45	56.45	72.50
3260	42" x 29", 12 ga., 36" equivalent		108	.519		46.50	25	5.45	76.95	94.50
3280	49" x 33", 12 ga., 42" equivalent		92	.609		56	29.50	6.40	91.90	113
3300	57" x 38", 12 ga., 48" equivalent		75	.747		56.50	36	7.85	100.35	125

33 42 Stormwater Conveyance

33 42 13 – Stormwater Culverts

33 42 13.15 Oval Arch Culverts

		Crew	Daily Output	Labor-Hours	Unit	Material	2021 Bare Costs Labor	Equipment	Total	Total Incl O&P
3320	End sections, 17" x 13"	B-13	22	2.545	Ea.	132	123	26.50	281.50	360
3340	42" x 29"	↓	17	3.294	"	545	159	34.50	738.50	875
3360	Multi-plate arch, steel	B-20	1690	.014	Lb.	1.29	.70		1.99	2.47

33 42 33 – Stormwater Curbside Drains and Inlets

33 42 33.13 Catch Basins

		Crew	Daily Output	Labor-Hours	Unit	Material	2021 Bare Costs Labor	Equipment	Total	Total Incl O&P
0010	**CATCH BASINS**									
0011	Not including footing & excavation									
1600	Frames & grates, C.I., 24" square, 500 lb.	B-6	7.80	3.077	Ea.	395	148	27.50	570.50	685
1700	26" D shape, 600 lb.		7	3.429		735	165	31	931	1,075
1800	Light traffic, 18" diameter, 100 lb.		10	2.400		216	115	21.50	352.50	435
1900	24" diameter, 300 lb.		8.70	2.759		209	133	25	367	455
2000	36" diameter, 900 lb.		5.80	4.138		720	199	37.50	956.50	1,125
2100	Heavy traffic, 24" diameter, 400 lb.		7.80	3.077		291	148	27.50	466.50	570
2200	36" diameter, 1,150 lb.		3	8		955	385	72	1,412	1,700
2300	Mass. State standard, 26" diameter, 475 lb.		7	3.429		695	165	31	891	1,050
2400	30" diameter, 620 lb.		7	3.429		380	165	31	576	695
2500	Watertight, 24" diameter, 350 lb.		7.80	3.077		430	148	27.50	605.50	725
2600	26" diameter, 500 lb.		7	3.429		445	165	31	641	770
2700	32" diameter, 575 lb.	↓	6	4	↓	915	192	36	1,143	1,325
2800	3 piece cover & frame, 10" deep,									
2900	1,200 lb., for heavy equipment	B-6	3	8	Ea.	1,075	385	72	1,532	1,850
3000	Raised for paving 1-1/4" to 2" high									
3100	4 piece expansion ring									
3200	20" to 26" diameter	1 Clab	3	2.667	Ea.	223	118		341	425
3300	30" to 36" diameter	"	3	2.667	"	310	118		428	515
3320	Frames and covers, existing, raised for paving, 2", including									
3340	row of brick, concrete collar, up to 12" wide frame	B-6	18	1.333	Ea.	52.50	64	12	128.50	166
3360	20" to 26" wide frame		11	2.182		79	105	19.65	203.65	266
3380	30" to 36" wide frame		9	2.667	↓	98	128	24	250	325
3400	Inverts, single channel brick	D-1	3	5.333	↓	113	260		373	515
3500	Concrete	D-1	5	3.200	Ea.	137	156		293	385
3600	Triple channel, brick		2	8		186	390		576	790
3700	Concrete	↓	3	5.333	↓	156	260		416	560

33 52 Hydrocarbon Transmission and Distribution

33 52 13 – Liquid Hydrocarbon Piping

33 52 13.16 Gasoline Piping

		Crew	Daily Output	Labor-Hours	Unit	Material	2021 Bare Costs Labor	Equipment	Total	Total Incl O&P
0010	**GASOLINE PIPING**									
0020	Primary containment pipe, fiberglass-reinforced									
0030	Plastic pipe 15' & 30' lengths									
0040	2" diameter	Q-6	425	.056	L.F.	7.25	3.60		10.85	13.40
0050	3" diameter		400	.060		9.85	3.83		13.68	16.50
0060	4" diameter	↓	375	.064	↓	12.75	4.08		16.83	20
0100	Fittings									
0110	Elbows, 90° & 45°, bell ends, 2"	Q-6	24	1	Ea.	43.50	64		107.50	143
0120	3" diameter		22	1.091		57	69.50		126.50	167
0130	4" diameter		20	1.200		71.50	76.50		148	193
0200	Tees, bell ends, 2"		21	1.143		57.50	73		130.50	172
0210	3" diameter		18	1.333		65.50	85		150.50	199
0230	Flanges bell ends, 2"	↓	24	1		31	64		95	130

33 52 13 – Liquid Hydrocarbon Piping

33 52 13.16 Gasoline Piping

	33 52 13.16 Gasoline Piping	Crew	Daily Output	Labor-Hours	Unit	Material	2021 Bare Costs Labor	Equipment	Total	Total Incl O&P
0240	3" diameter	Q-6	22	1.091	Ea.	38.50	69.50		108	147
0250	4" diameter		20	1.200		43	76.50		119.50	161
0260	Sleeve couplings, 2"		21	1.143		11.95	73		84.95	122
0270	3" diameter		18	1.333		18.80	85		103.80	148
0280	4" diameter		15	1.600		21	102		123	175
0290	Threaded adapters, 2"		21	1.143		19.55	73		92.55	131
0300	3" diameter		18	1.333		25.50	85		110.50	155
0310	4" diameter		15	1.600		36.50	102		138.50	192
0320	Reducers, 2"		27	.889		26.50	56.50		83	114
0330	3" diameter		22	1.091		26.50	69.50		96	133
0340	4" diameter		20	1.200		35.50	76.50		112	153
1010	Gas station product line for secondary containment (double wall)									
1100	Fiberglass reinforced plastic pipe 25' lengths									
1120	Pipe, plain end, 3" diameter	Q-6	375	.064	L.F.	36	4.08		40.08	45.50
1130	4" diameter		350	.069		29.50	4.37		33.87	38.50
1140	5" diameter		325	.074		33	4.71		37.71	43
1150	6" diameter		300	.080		36.50	5.10		41.60	47.50
1200	Fittings									
1230	Elbows, 90° & 45°, 3" diameter	Q-6	18	1.333	Ea.	152	85		237	294
1240	4" diameter		16	1.500		183	95.50		278.50	345
1250	5" diameter		14	1.714		175	109		284	355
1260	6" diameter		12	2		218	128		346	430
1270	Tees, 3" diameter		15	1.600		189	102		291	360
1280	4" diameter	Q-6	12	2	Ea.	180	128		308	390
1290	5" diameter		9	2.667		235	170		405	515
1300	6" diameter		6	4		330	255		585	745
1310	Couplings, 3" diameter		18	1.333		53.50	85		138.50	186
1320	4" diameter		16	1.500		122	95.50		217.50	277
1330	5" diameter		14	1.714		217	109		326	400
1340	6" diameter		12	2		340	128		468	565
1350	Cross-over nipples, 3" diameter		18	1.333		10.20	85		95.20	138
1360	4" diameter		16	1.500		12.40	95.50		107.90	157
1370	5" diameter		14	1.714		17.20	109		126.20	182
1380	6" diameter		12	2		18	128		146	210
1400	Telescoping, reducers, concentric 4" x 3"		18	1.333		43.50	85		128.50	175
1410	5" x 4"		17	1.412		93	90		183	236
1420	6" x 5"		16	1.500		261	95.50		356.50	430

33 52 16 – Gas Hydrocarbon Piping

33 52 16.20 Piping, Gas Service and Distribution, P.E.

	33 52 16.20 Piping, Gas Service and Distribution, P.E.	Crew	Daily Output	Labor-Hours	Unit	Material	2021 Bare Costs Labor	Equipment	Total	Total Incl O&P
0010	**PIPING, GAS SERVICE AND DISTRIBUTION, POLYETHYLENE**									
0020	Not including excavation or backfill									
1000	60 psi coils, compression coupling @ 100', 1/2" diameter, SDR 11	B-20A	608	.053	L.F.	.56	2.80		3.36	4.80
1010	1" diameter, SDR 11		544	.059		1.13	3.13		4.26	5.90
1040	1-1/4" diameter, SDR 11		544	.059		1.61	3.13		4.74	6.45
1100	2" diameter, SDR 11		488	.066		2.84	3.49		6.33	8.30
1160	3" diameter, SDR 11		408	.078		5.05	4.17		9.22	11.75
1500	60 psi 40' joints with coupling, 3" diameter, SDR 11	B-21A	408	.098		7.50	5.35	1.06	13.91	17.40
1540	4" diameter, SDR 11		352	.114		14.50	6.25	1.23	21.98	26.50
1600	6" diameter, SDR 11		328	.122		37	6.70	1.32	45.02	52
1640	8" diameter, SDR 11		272	.147		62	8.05	1.59	71.64	82

33 52 16.23 Medium Density Polyethylene Piping	Crew	Daily Output	Labor-Hours	Unit	Material	2021 Bare Costs Labor	Equipment	Total	Total Incl O&P
2010 **MEDIUM DENSITY POLYETHYLENE PIPING**									
2020 ASTM D2513, not including excavation or backfill									
2200 Butt fused pipe									
2205 80psi coils, butt fusion joint @ 100', 1/2" CTS diameter, SDR 7	B-22C	2050	.008	L.F.	.17	.40	.07	.64	.85
2210 1" CTS diameter, SDR 11.5		1900	.008		.24	.43	.07	.74	.98
2215 80psi coils, butt fusion joint @ 100', IPS 1/2" diameter, SDR 9.3		1950	.008		.25	.42	.07	.74	.98
2220 3/4" diameter, SDR 11		1950	.008		.32	.42	.07	.81	1.05
2225 1" diameter, SDR 11		1850	.009		.77	.44	.07	1.28	1.59
2230 1-1/4" diameter, SDR 11		1850	.009		.78	.44	.07	1.29	1.60
2235 1-1/2" diameter, SDR 11		1750	.009		.96	.46	.08	1.50	1.84
2240 2" diameter, SDR 11		1750	.009		1.29	.46	.08	1.83	2.20
2245 80psi 40' lengths, butt fusion joint, IPS 3" diameter, SDR 11.5	B-22A	660	.061		2.48	3.08	1.22	6.78	8.65
2250 4" diameter, SDR 11.5	"	500	.080		3.91	4.06	1.61	9.58	12.10
2400 Socket fused pipe									
2405 80psi coils, socket fusion coupling @ 100', 1/2" CTS diameter, SDR 7	B-20	1050	.023	L.F.	.19	1.13		1.32	1.90
2410 1" CTS diameter, SDR 11.5		1000	.024		.26	1.18		1.44	2.06
2415 80psi coils, socket fusion coupling @ 100', IPS 1/2" diameter, SDR 9.3		1000	.024		.27	1.18		1.45	2.06
2420 3/4" diameter, SDR 11		1000	.024		.34	1.18		1.52	2.14
2425 1" diameter, SDR 11		950	.025		.79	1.25		2.04	2.73
2430 1-1/4" diameter, SDR 11		950	.025		.80	1.25		2.05	2.74
2435 1-1/2" diameter, SDR 11		900	.027		.98	1.31		2.29	3.05
2440 2" diameter, SDR 11		850	.028		1.32	1.39		2.71	3.53
2445 80psi 40' lengths, socket fusion coupling, IPS 3" diameter, SDR 11.5	B-21A	340	.118		2.88	6.45	1.27	10.60	14.15
2450 4" diameter, SDR 11.5	"	260	.154		4.49	8.45	1.66	14.60	19.35
2600 Compression coupled pipe									
2605 80psi coils, compression coupling @ 100', 1/2" CTS diameter, SDR 7	B-20	2250	.011	L.F.	.30	.53		.83	1.12
2610 1" CTS diameter, SDR 11.5		2175	.011		.49	.54		1.03	1.34
2615 80psi coils, compression coupling @ 100', IPS 1/2" diameter, SDR 9.3		2175	.011		.48	.54		1.02	1.34
2620 3/4" diameter, SDR 11		2100	.011		.55	.56		1.11	1.45
2625 1" diameter, SDR 11		2025	.012		1.08	.58		1.66	2.06
3000 Fittings, butt fusion									
3010 SDR 11, IPS unless noted CTS									
3100 Caps, 3/4" diameter	B-22C	28.50	.561	Ea.	3.24	28.50	4.73	36.47	51.50
3105 1" diameter		27	.593		2.98	30	5	37.98	54
3110 1-1/4" diameter		27	.593		4.77	30	5	39.77	56
3115 1-1/2" diameter		25.50	.627		2.76	32	5.30	40.06	56.50
3120 2" diameter		24	.667		8.70	34	5.60	48.30	66.50
3125 3" diameter		24	.667		13.75	34	5.60	53.35	72
3130 4" diameter		18	.889		29	45	7.50	81.50	108
3200 Reducers, 1" x 3/4" diameters		13.50	1.185		13.60	60	10	83.60	116
3205 1-1/4" x 1" diameters		13.50	1.185		15.20	60	10	85.20	118
3210 1-1/2" x 3/4" diameters		12.75	1.255		19.85	63.50	10.60	93.95	129
3215 1-1/2" x 1" diameters		12.75	1.255		16.85	63.50	10.60	90.95	126
3220 1-1/2" x 1-1/4" diameters		12.75	1.255		16.85	63.50	10.60	90.95	126
3225 2" x 1" diameters		12	1.333		17.25	67.50	11.25	96	132
3230 2" x 1-1/4" diameters		12	1.333		17.25	67.50	11.25	96	132
3235 2" x 1-1/2" diameters		12	1.333		16	67.50	11.25	94.75	131
3240 3" x 2" diameters		12	1.333		20	67.50	11.25	98.75	135
3245 4" x 2" diameters		9	1.778		32	90	15	137	187
3250 4" x 3" diameters		9	1.778		29	90	15	134	183
3300 Elbows, 90°, 3/4" diameter		14.25	1.123		4.04	57	9.45	70.49	100
3302 1" diameter		13.50	1.185		4.94	60	10	74.94	106

For customer support on your Plumbing Costs with RSMeans Data, call 800.448.8182.

441

33 52 16.23 Medium Density Polyethylene Piping		Crew	Daily Output	Labor-Hours	Unit	Material	2021 Bare Costs Labor	2021 Bare Costs Equipment	Total	Total Incl O&P
3304	1-1/4" diameter	B-22C	13.50	1.185	Ea.	9.50	60	10	79.50	111
3306	1-1/2" diameter		12.75	1.255		8.25	63.50	10.60	82.35	116
3308	2" diameter		12	1.333		14.10	67.50	11.25	92.85	129
3310	3" diameter		12	1.333		28.50	67.50	11.25	107.25	144
3312	4" diameter		9	1.778		39	90	15	144	195
3350	45°, 3" diameter		12	1.333		31.50	67.50	11.25	110.25	148
3352	4" diameter		9	1.778		37	90	15	142	192
3400	Tees, 3/4" diameter		9.50	1.684		16.50	85.50	14.20	116.20	162
3405	1" diameter		9	1.778		4.97	90	15	109.97	157
3410	1-1/4" diameter		9	1.778		12.70	90	15	117.70	166
3415	1-1/2" diameter		8.50	1.882		18.40	95.50	15.85	129.75	181
3420	2" diameter		8	2		19.50	102	16.85	138.35	192
3425	3" diameter		8	2		35.50	102	16.85	154.35	210
3430	4" diameter	▼	6	2.667	▼	44.50	135	22.50	202	277
3500	Tapping tees, high volume, butt fusion outlets									
3505	1-1/2" punch, 2" x 2" outlet	B-22C	11	1.455	Ea.	110	74	12.25	196.25	246
3510	1-7/8" punch, 3" x 2" outlet		11	1.455		110	74	12.25	196.25	246
3515	4" x 2" outlet	▼	9	1.778		110	90	15	215	273
3550	For protective sleeves, add				▼	2.90			2.90	3.19
3600	Service saddles, saddle contour x outlet, butt fusion outlets									
3602	1-1/4" x 3/4" outlet	B-22C	17	.941	Ea.	14.15	48	7.95	70.10	96
3604	1-1/4" x 1" outlet		17	.941		6.65	48	7.95	62.60	87.50
3606	1-1/4" x 1-1/4" outlet		16	1		6.65	51	8.45	66.10	92.50
3608	1-1/2" x 3/4" outlet		16	1		6.65	51	8.45	66.10	92.50
3610	1-1/2" x 1-1/4" outlet		15	1.067		14.15	54	9	77.15	106
3612	2" x 3/4" outlet		12	1.333		6.65	67.50	11.25	85.40	121
3614	2" x 1" outlet		12	1.333		6.65	67.50	11.25	85.40	121
3616	2" x 1-1/4" outlet		11	1.455		6.65	74	12.25	92.90	132
3618	3" x 3/4" outlet		12	1.333		6.65	67.50	11.25	85.40	121
3620	3" x 1" outlet		12	1.333		6.65	67.50	11.25	85.40	121
3622	3" x 1-1/4" outlet		11	1.455		6.65	74	12.25	92.90	132
3624	4" x 3/4" outlet		10	1.600		6.65	81	13.50	101.15	144
3626	4" x 1" outlet		10	1.600		6.65	81	13.50	101.15	144
3628	4" x 1-1/4" outlet	▼	9	1.778		14.15	90	15	119.15	167
3685	For protective sleeves, 3/4" diameter outlets, add					.86			.86	.95
3690	1" diameter outlets, add					1.44			1.44	1.58
3695	1-1/4" diameter outlets, add				▼	2.48			2.48	2.73
3700	Branch saddles, contour x outlet, butt outlets, round base									
3702	2" x 2" outlet	B-22C	11	1.455	Ea.	17.30	74	12.25	103.55	144
3704	3" x 2" outlet		11	1.455		29	74	12.25	115.25	157
3706	3" x 3" outlet		11	1.455		27	74	12.25	113.25	154
3708	4" x 2" outlet		9	1.778		30	90	15	135	185
3710	4" x 3" outlet		9	1.778		50	90	15	155	207
3712	4" x 4" outlet		9	1.778		36	90	15	141	191
3750	Rectangular base, 2" x 2" outlet		11	1.455		17.30	74	12.25	103.55	144
3752	3" x 2" outlet		11	1.455		18.80	74	12.25	105.05	145
3754	3" x 3" outlet		11	1.455		27	74	12.25	113.25	154
3756	4" x 2" outlet		9	1.778		21.50	90	15	126.50	175
3758	4" x 3" outlet		9	1.778		25.50	90	15	130.50	180
3760	4" x 4" outlet	▼	9	1.778	▼	31	90	15	136	186
4000	Fittings, socket fusion									
4010	SDR 11, IPS unless noted CTS									
4100	Caps, 1/2" CTS diameter	B-20	30	.800	Ea.	2.37	39.50		41.87	61.50

33 52 16.23 Medium Density Polyethylene Piping		Crew	Daily Output	Labor-Hours	Unit	Material	2021 Bare Costs Labor	Equipment	Total	Total Incl O&P
4105	1/2" diameter	B-20	28.50	.842	Ea.	2.37	41.50		43.87	64.50
4110	3/4" diameter		28.50	.842		3.63	41.50		45.13	66
4115	1" CTS diameter		28	.857		3.92	42.50		46.42	68
4120	1" diameter		27	.889		4.66	44		48.66	70.50
4125	1-1/4" diameter		27	.889		5.65	44		49.65	71.50
4130	1-1/2" diameter		25.50	.941		8.50	46.50		55	79
4135	2" diameter		24	1		6	49.50		55.50	80.50
4140	3" diameter		24	1		13.85	49.50		63.35	89.50
4145	4" diameter		18	1.333		20.50	65.50		86	121
4200	Reducers, 1/2" x 1/2" CTS diameters		14.25	1.684		6.15	83		89.15	131
4202	3/4" x 1/2" CTS diameters		14.25	1.684		5.70	83		88.70	130
4204	3/4" x 1/2" diameters		14.25	1.684		6.10	83		89.10	131
4206	1" CTS x 1/2" diameters		14	1.714		5.95	84.50		90.45	134
4208	1" CTS x 3/4" diameters		13.50	1.778		6	87.50		93.50	138
4210	1" x 1/2" CTS diameters		13.50	1.778		5.65	87.50		93.15	137
4212	1" x 1/2" diameters		13.50	1.778		6.30	87.50		93.80	138
4214	1" x 3/4" diameters		13.50	1.778		8.75	87.50		96.25	141
4216	1" x 1" CTS diameters		13.50	1.778		5.15	87.50		92.65	137
4218	1-1/4" x 1/2" CTS diameters		13.50	1.778		5.95	87.50		93.45	138
4220	1-1/4" x 1/2" diameters		13.50	1.778		8.65	87.50		96.15	141
4222	1-1/4" x 3/4" diameters		13.50	1.778		12.45	87.50		99.95	145
4224	1-1/4" x 1" CTS diameters		13.50	1.778		8.65	87.50		96.15	141
4226	1-1/4" x 1" diameters		13.50	1.778		10.60	87.50		98.10	143
4228	1-1/2" x 3/4" diameters		12.75	1.882		9.50	93		102.50	149
4230	1-1/2" x 1" diameters		12.75	1.882		8.05	93		101.05	148
4232	1-1/2" x 1-1/4" diameters		12.75	1.882		9.10	93		102.10	149
4234	2" x 3/4" diameters		12	2		9.95	98.50		108.45	159
4236	2" x 1" CTS diameters		12	2		9.80	98.50		108.30	159
4238	2" x 1" diameters		12	2		10.70	98.50		109.20	160
4240	2" x 1-1/4" diameters		12	2		9.75	98.50		108.25	159
4242	2" x 1-1/2" diameters		12	2		10.35	98.50		108.85	159
4244	3" x 2" diameters		12	2		11.75	98.50		110.25	161
4246	4" x 2" diameters		9	2.667		24.50	131		155.50	224
4248	4" x 3" diameters		9	2.667		28	131		159	228
4300	Couplings, 1/2" CTS diameter		15	1.600		1.80	79		80.80	120
4305	1/2" diameter		14.25	1.684		1.50	83		84.50	126
4310	3/4" diameter		14.25	1.684		1.68	83		84.68	126
4315	1" CTS diameter		14	1.714		2.03	84.50		86.53	129
4320	1" diameter		13.50	1.778		1.71	87.50		89.21	133
4325	1-1/4" diameter		13.50	1.778		2.06	87.50		89.56	133
4330	1-1/2" diameter		12.75	1.882		2.43	93		95.43	142
4335	2" diameter		12	2		2.59	98.50		101.09	151
4340	3" diameter		12	2		16.20	98.50		114.70	166
4345	4" diameter		9	2.667		23	131		154	223
4400	Elbows, 90°, 1/2" CTS diameter		15	1.600		3.41	79		82.41	122
4405	1/2" diameter		14.25	1.684		3.35	83		86.35	128
4410	3/4" diameter		14.25	1.684		3.96	83		86.96	128
4415	1" CTS diameter		14	1.714		3.81	84.50		88.31	131
4420	1" diameter		13.50	1.778		4.24	87.50		91.74	136
4425	1-1/4" diameter		13.50	1.778		4.04	87.50		91.54	135
4430	1-1/2" diameter		12.75	1.882		5.10	93		98.10	145
4435	2" diameter		12	2		4.16	98.50		102.66	153
4440	3" diameter		12	2		22	98.50		120.50	172

For customer support on your Plumbing Costs with RSMeans Data, call 800.448.8182.

443

33 52 16.23 Medium Density Polyethylene Piping		Crew	Daily Output	Labor-Hours	Unit	Material	2021 Bare Costs Labor	Equipment	Total	Total Incl O&P
4445	4" diameter	B-20	9	2.667	Ea.	57	131		188	260
4460	45°, 1" diameter		13.50	1.778		6.80	87.50		94.30	138
4465	2" diameter		12	2		9.10	98.50		107.60	158
4500	Tees, 1/2" CTS diameter		10	2.400		3.26	118		121.26	181
4505	1/2" diameter		9.50	2.526		2.89	125		127.89	189
4510	3/4" diameter		9.50	2.526		3.89	125		128.89	190
4515	1" CTS diameter		9.33	2.571		3.81	127		130.81	194
4520	1" diameter		9	2.667		4.75	131		135.75	202
4525	1-1/4" diameter		9	2.667		5.25	131		136.25	203
4530	1-1/2" diameter		8.50	2.824		8.70	139		147.70	218
4535	2" diameter		8	3		10.45	148		158.45	232
4540	3" diameter		8	3		32	148		180	256
4545	4" diameter		6	4		59.50	197		256.50	360
4600	Tapping tees, type II, 3/4" punch, socket fusion outlets									
4601	Saddle contour x outlet diameter, IPS unless noted CTS									
4602	1-1/4" x 1/2" CTS outlet	B-20	17	1.412	Ea.	11.95	69.50		81.45	117
4604	1-1/4" x 1/2" outlet		17	1.412		11.95	69.50		81.45	117
4606	1-1/4" x 3/4" outlet		17	1.412		11.95	69.50		81.45	117
4608	1-1/4" x 1" CTS outlet		17	1.412		13.05	69.50		82.55	118
4610	1-1/4" x 1" outlet		17	1.412		13.05	69.50		82.55	118
4612	1-1/4" x 1-1/4" outlet		16	1.500		22	74		96	135
4614	1-1/2" x 1/2" CTS outlet		16	1.500		11.95	74		85.95	124
4616	1-1/2" x 1/2" outlet		16	1.500		11.95	74		85.95	124
4618	1-1/2" x 3/4" outlet		16	1.500		11.95	74		85.95	124
4620	1-1/2" x 1" CTS outlet		16	1.500		13.05	74		87.05	125
4622	1-1/2" x 1" outlet		16	1.500		13.05	74		87.05	125
4624	1-1/2" x 1-1/4" outlet		15	1.600		22	79		101	142
4626	2" x 1/2" CTS outlet		12	2		11.95	98.50		110.45	161
4628	2" x 1/2" outlet		12	2		11.95	98.50		110.45	161
4630	2" x 3/4" outlet		12	2		11.95	98.50		110.45	161
4632	2" x 1" CTS outlet		12	2		13.05	98.50		111.55	162
4634	2" x 1" outlet		12	2		13.05	98.50		111.55	162
4636	2" x 1-1/4" outlet		11	2.182		22	108		130	185
4638	3" x 1/2" CTS outlet		12	2		11.95	98.50		110.45	161
4640	3" x 1/2" outlet		12	2		11.95	98.50		110.45	161
4642	3" x 3/4" outlet		12	2		11.95	98.50		110.45	161
4644	3" x 1" CTS outlet		12	2		13.05	98.50		111.55	162
4646	3" x 1" outlet		12	2		13.05	98.50		111.55	162
4648	3" x 1-1/4" outlet		11	2.182		22	108		130	185
4650	4" x 1/2" CTS outlet		10	2.400		11.95	118		129.95	190
4652	4" x 1/2" outlet		10	2.400		11.95	118		129.95	190
4654	4" x 3/4" outlet		10	2.400		11.95	118		129.95	190
4656	4" x 1" CTS outlet		10	2.400		13.05	118		131.05	191
4658	4" x 1" outlet		10	2.400		13.05	118		131.05	191
4660	4" x 1-1/4" outlet		9	2.667		22	131		153	221
4685	For protective sleeves, 1/2" CTS to 3/4" diameter outlets, add					.86			.86	.95
4690	1" CTS & IPS diameter outlets, add					1.44			1.44	1.58
4695	1-1/4" diameter outlets, add					2.48			2.48	2.73
4700	Tapping tees, high volume, socket fusion outlets									
4705	1-1/2" punch, 2" x 1-1/4" outlet	B-20	11	2.182	Ea.	110	108		218	282
4710	1-7/8" punch, 3" x 1-1/4" outlet		11	2.182		110	108		218	282
4715	4" x 1-1/4" outlet		9	2.667		110	131		241	320
4750	For protective sleeves, add					2.90			2.90	3.19

33 52 16.23 Medium Density Polyethylene Piping	Crew	Daily Output	Labor-Hours	Unit	Material	2021 Bare Costs Labor	Equipment	Total	Total Incl O&P
4800	Service saddles, saddle contour x outlet, socket fusion outlets									
4801	IPS unless noted CTS									
4802	1-1/4" x 1/2" CTS outlet	B-20	17	1.412	Ea.	3.70	69.50		73.20	108
4804	1-1/4" x 1/2" outlet	"	17	1.412	"	3.70	69.50		73.20	108
4806	1-1/4" x 3/4" outlet	B-20	17	1.412	Ea.	3.70	69.50		73.20	108
4808	1-1/4" x 1" CTS outlet		17	1.412		6.10	69.50		75.60	111
4810	1-1/4" x 1" outlet		17	1.412		13.90	69.50		83.40	119
4812	1-1/4" x 1-1/4" outlet		16	1.500		9.80	74		83.80	122
4814	1-1/2" x 1/2" CTS outlet		16	1.500		3.70	74		77.70	115
4816	1-1/2" x 1/2" outlet		16	1.500		3.70	74		77.70	115
4818	1-1/2" x 3/4" outlet		16	1.500		12.65	74		86.65	125
4820	1-1/2" x 1-1/4" outlet		15	1.600		9.80	79		88.80	129
4822	2" x 1/2" CTS outlet		12	2		3.70	98.50		102.20	152
4824	2" x 1/2" outlet		12	2		3.70	98.50		102.20	152
4826	2" x 3/4" outlet		12	2		3.70	98.50		102.20	152
4828	2" x 1" CTS outlet		12	2		6.10	98.50		104.60	155
4830	2" x 1" outlet		12	2		6.10	98.50		104.60	155
4832	2" x 1-1/4" outlet		11	2.182		15.75	108		123.75	178
4834	3" x 1/2" CTS outlet		12	2		3.70	98.50		102.20	152
4836	3" x 1/2" outlet		12	2		3.70	98.50		102.20	152
4838	3" x 3/4" outlet		12	2		3.70	98.50		102.20	152
4840	3" x 1" CTS outlet		12	2		6.10	98.50		104.60	155
4842	3" x 1" outlet		12	2		6.10	98.50		104.60	155
4844	3" x 1-1/4" outlet		11	2.182		14.50	108		122.50	177
4846	4" x 1/2" CTS outlet		10	2.400		3.70	118		121.70	181
4848	4" x 1/2" outlet		10	2.400		3.70	118		121.70	181
4850	4" x 3/4" outlet		10	2.400		3.70	118		121.70	181
4852	4" x 1" CTS outlet		10	2.400		6.10	118		124.10	184
4854	4" x 1" outlet		10	2.400		6.10	118		124.10	184
4856	4" x 1-1/4" outlet	↓	9	2.667		15.75	131		146.75	214
4885	For protective sleeves, 1/2" CTS to 3/4" diameter outlets, add					.86			.86	.95
4890	1" CTS & IPS diameter outlets, add					1.44			1.44	1.58
4895	1-1/4" diameter outlets, add					2.48			2.48	2.73
4900	Spigot fittings, tees, SDR 7, 1/2" CTS diameter	B-20	10	2.400		25.50	118		143.50	205
4901	SDR 9.3, 1/2" diameter		9.50	2.526		11.30	125		136.30	198
4902	SDR 10, 1-1/4" diameter		9	2.667		11.30	131		142.30	209
4903	SDR 11, 3/4" diameter		9.50	2.526		23.50	125		148.50	212
4904	2" diameter		8	3		21.50	148		169.50	245
4906	SDR 11.5, 1" CTS diameter		9.33	2.571		60	127		187	256
4907	3" diameter		8	3		61	148		209	288
4908	4" diameter		6	4		157	197		354	470
4921	90° elbows, SDR 7, 1/2" CTS diameter		15	1.600		13.45	79		92.45	133
4922	SDR 9.3, 1/2" diameter		14.25	1.684		5.80	83		88.80	130
4923	SDR 10, 1-1/4" diameter		13.50	1.778		5.90	87.50		93.40	138
4924	SDR 11, 3/4" diameter		14.25	1.684		6.60	83		89.60	131
4925	2" diameter		12	2		25	98.50		123.50	176
4927	SDR 11.5, 1" CTS diameter		14	1.714		23	84.50		107.50	152
4928	3" diameter		12	2		56.50	98.50		155	210
4929	4" diameter		9	2.667		121	131		252	330
4935	Caps, SDR 7, 1/2" CTS diameter		30	.800		4.46	39.50		43.96	64
4936	SDR 9.3, 1/2" diameter		28.50	.842		4.46	41.50		45.96	67
4937	SDR 10, 1-1/4" diameter		27	.889		4.62	44		48.62	70.50
4938	SDR 11, 3/4" diameter	↓	28.50	.842	↓	4.77	41.50		46.27	67.50

For customer support on your Plumbing Costs with RSMeans Data, call 800.448.8182.

445

33 52 16.23 Medium Density Polyethylene Piping	Crew	Daily Output	Labor-Hours	Unit	Material	2021 Bare Costs Labor	Equipment	Total	Total Incl O&P
4939	1" diameter	B-20	27	.889	Ea.	5.30	44		49.30	71.50
4940	2" diameter		24	1		5.70	49.50		55.20	80.50
4942	SDR 11.5, 1" CTS diameter		28	.857		18.70	42.50		61.20	84
4943	3" diameter		24	1		30.50	49.50		80	108
4944	4" diameter		18	1.333		29	65.50		94.50	131
4946	SDR 13.5, 4" diameter		18	1.333		73.50	65.50		139	179
4950	Reducers, SDR 10 x SDR 11, 1-1/4" x 3/4" diameters		13.50	1.778		15.25	87.50		102.75	148
4951	1-1/4" x 1" diameters		13.50	1.778		15.25	87.50		102.75	148
4952	SDR 10 x SDR 11.5, 1-1/4" x 1" CTS diameters		13.50	1.778		15.25	87.50		102.75	148
4953	SDR 11 x SDR 7, 3/4" x 1/2" CTS diameters		14.25	1.684		6.80	83		89.80	132
4954	SDR 11 x SDR 9.3, 3/4" x 1/2" diameters		14.25	1.684		6.80	83		89.80	132
4955	SDR 11 x SDR 10, 2" x 1-1/4" diameters		12	2		15.80	98.50		114.30	165
4956	SDR 11 x SDR 11, 1" x 3/4" diameters		13.50	1.778		7.85	87.50		95.35	140
4957	2" x 3/4" diameters		12	2		15.60	98.50		114.10	165
4958	2" x 1" diameters		12	2		15.70	98.50		114.20	165
4959	SDR 11 x SDR 11.5, 1" x 1" CTS diameters		13.50	1.778		8	87.50		95.50	140
4960	2" x 1" CTS diameters		12	2		15.60	98.50		114.10	165
4962	SDR 11.5 x SDR 11, 3" x 2" diameters		12	2		21	98.50		119.50	171
4963	4" x 2" diameters		9	2.667		32	131		163	232
4964	SDR 11.5 x SDR 11.5, 4" x 3" diameters		9	2.667		34.50	131		165.50	235
6100	Fittings, compression									
6101	MDPE gas pipe, ASTM D2513/ASTM F1924-98									
6102	Caps, SDR 7, 1/2" CTS diameter	B-20	60	.400	Ea.	12.85	19.70		32.55	43.50
6104	SDR 9.3, 1/2" IPS diameter		58	.414		22.50	20.50		43	55
6106	SDR 10, 1/2" CTS diameter		60	.400		11.35	19.70		31.05	42
6108	1-1/4" IPS diameter		54	.444		74.50	22		96.50	115
6110	SDR 11, 3/4" IPS diameter		58	.414		20	20.50		40.50	52.50
6112	1" IPS diameter		54	.444		32.50	22		54.50	68.50
6114	1-1/4" IPS diameter		54	.444		78	22		100	119
6116	1-1/2" IPS diameter		52	.462		95	23		118	139
6118	2" IPS diameter		50	.480		83.50	23.50		107	128
6120	SDR 11.5, 1" CTS diameter		56	.429		20.50	21		41.50	54.50
6122	SDR 12.5, 1" CTS diameter		56	.429		24	21		45	58
6202	Reducers, SDR 7 x SDR 10, 1/2" CTS x 1/2" CTS diameters		30	.800		17.35	39.50		56.85	78
6204	SDR 9.3 x SDR 7, 1/2" IPS x 1/2" CTS diameters		29	.828		50	41		91	116
6206	SDR 11 x SDR 7, 3/4" IPS x 1/2" CTS diameters		29	.828		49.50	41		90.50	116
6208	1" IPS x 1/2" CTS diameters		27	.889		60	44		104	132
6210	SDR 11 x SDR 9.3, 3/4" IPS x 1/2" IPS diameters		29	.828		49	41		90	115
6212	1" IPS x 1/2" IPS diameters		27	.889		62	44		106	134
6214	SDR 11 x SDR 10, 2" IPS x 1-1/4" IPS diameters		25	.960		95.50	47.50		143	176
6216	SDR 11 x SDR 11, 1" IPS x 3/4" IPS diameters		27	.889		60	44		104	132
6218	1-1/4" IPS x 1" IPS diameters		27	.889		57	44		101	128
6220	2" IPS x 1-1/4" IPS diameters		25	.960		95.50	47.50		143	176
6222	SDR 11 x SDR 11.5, 1" IPS x 1" CTS diameters		27	.889		46	44		90	117
6224	SDR 11 x SDR 12.5, 1" IPS x 1" CTS diameters		27	.889		56	44		100	128
6226	SDR 11.5 x SDR 7, 1" CTS x 1/2" CTS diameters		28	.857		34	42.50		76.50	101
6228	SDR 11.5 x SDR 9.3, 1" CTS x 1/2" IPS diameters		28	.857		39.50	42.50		82	107
6230	SDR 11.5 x SDR 11, 1" CTS x 3/4" IPS diameters		28	.857		55	42.50		97.50	124
6232	SDR 12.5 x SDR 7, 1" CTS x 1/2" CTS diameters		28	.857		36	42.50		78.50	103
6234	SDR 12.5 x SDR 9.3, 1" CTS x 1/2" IPS diameters		28	.857		40.50	42.50		83	108
6236	SDR 12.5 x SDR 11, 1" CTS x 3/4" IPS diameters		28	.857		34.50	42.50		77	102
6302	Couplings, SDR 7, 1/2" CTS diameter		30	.800		13.10	39.50		52.60	73.50
6304	SDR 9.3, 1/2" IPS diameter		29	.828		23	41		64	86.50

33 52 16 – Gas Hydrocarbon Piping

33 52 16.23 Medium Density Polyethylene Piping	Crew	Daily Output	Labor-Hours	Unit	Material	2021 Bare Costs Labor	Equipment	Total	Total Incl O&P
6306	SDR 10, 1/2" CTS diameter	B-20	30	.800	Ea.	12.40	39.50		51.90	72.50
6308	SDR 11, 3/4" IPS diameter		29	.828		23.50	41		64.50	86.50
6310	1" IPS diameter		27	.889		31.50	44		75.50	100
6312	SDR 11.5, 1" CTS diameter		28	.857		24.50	42.50		67	90.50
6314	SDR 12.5, 1" CTS diameter		28	.857		27.50	42.50		70	94
6402	Repair couplings, SDR 10, 1-1/4" IPS diameter		27	.889		71.50	44		115.50	144
6404	SDR 11, 1-1/4" IPS diameter		27	.889		73	44		117	146
6406	1-1/2" IPS diameter		26	.923		78.50	45.50		124	155
6408	2" IPS diameter		25	.960		91.50	47.50		139	172
6502	Elbows, 90°, SDR 7, 1/2" CTS diameter		30	.800		23	39.50		62.50	84.50
6504	SDR 9.3, 1/2" IPS diameter		29	.828		40	41		81	105
6506	SDR 10, 1-1/4" IPS diameter		27	.889		98	44		142	174
6508	SDR 11, 3/4" IPS diameter		29	.828		36.50	41		77.50	101
6510	1" IPS diameter		27	.889		36.50	44		80.50	106
6512	1-1/4" IPS diameter		27	.889		97.50	44		141.50	173
6514	1-1/2" IPS diameter		26	.923		146	45.50		191.50	229
6516	2" IPS diameter	B-20	25	.960	Ea.	124	47.50		171.50	208
6518	SDR 11.5, 1" CTS diameter		28	.857		37	42.50		79.50	104
6520	SDR 12.5, 1" CTS diameter		28	.857		53.50	42.50		96	123
6602	Tees, SDR 7, 1/2" CTS diameter		20	1.200		40.50	59		99.50	133
6604	SDR 9.3, 1/2" IPS diameter		19.33	1.241		45	61		106	141
6606	SDR 10, 1-1/4" IPS diameter		18	1.333		127	65.50		192.50	239
6608	SDR 11, 3/4" IPS diameter		19.33	1.241		56.50	61		117.50	154
6610	1" IPS diameter		18	1.333		62	65.50		127.50	167
6612	1-1/4" IPS diameter		18	1.333		104	65.50		169.50	214
6614	1-1/2" IPS diameter		17.33	1.385		216	68.50		284.50	340
6616	2" IPS diameter		16.67	1.440		143	71		214	264
6618	SDR 11.5, 1" CTS diameter		18.67	1.286		57	63.50		120.50	158
6620	SDR 12.5, 1" CTS diameter		18.67	1.286		59	63.50		122.50	160
8100	Fittings, accessories									
8105	SDR 11, IPS unless noted CTS									
8110	Protective sleeves, for high volume tapping tees, butt fusion outlets	1 Skwk	16	.500	Ea.	5.45	28.50		33.95	49
8120	For tapping tees, socket outlets, CTS, 1/2" diameter		18	.444		1.42	25.50		26.92	39.50
8122	1" diameter		18	.444		2.51	25.50		28.01	41
8124	IPS, 1/2" diameter		18	.444		1.96	25.50		27.46	40
8126	3/4" diameter		18	.444		2.51	25.50		28.01	41
8128	1" diameter		18	.444		3.42	25.50		28.92	42
8130	1-1/4" diameter		18	.444		4.31	25.50		29.81	42.50
8205	Tapping tee test caps, type I, yellow PE cap, "aldyl style"		45	.178		44	10.15		54.15	64
8210	Type II, yellow polyethylene cap		45	.178		44	10.15		54.15	64
8215	High volume, yellow polyethylene cap		45	.178		56.50	10.15		66.65	78
8250	Quick connector, female x female inlets, 1/4" N.P.T.		50	.160		18.30	9.15		27.45	34
8255	Test hose, 24" length, 3/8" ID, male outlets, 1/4" N.P.T.		50	.160		25	9.15		34.15	41.50
8260	Quick connector and test hose assembly		50	.160		43.50	9.15		52.65	61.50
8305	Purge point caps, butt fusion, SDR 10, 1-1/4" diameter	B-22C	27	.593		32	30	5	67	85.50
8310	SDR 11, 1-1/4" diameter		27	.593		32	30	5	67	85.50
8315	2" diameter		24	.667		32.50	34	5.60	72.10	92
8320	3" diameter		24	.667		45	34	5.60	84.60	106
8325	4" diameter		18	.889		44	45	7.50	96.50	124
8340	Socket fusion, SDR 11, 1-1/4" diameter		27	.593		8	30	5	43	59.50
8345	2" diameter		24	.667		9.60	34	5.60	49.20	67.50
8350	3" diameter		24	.667		58.50	34	5.60	98.10	121
8355	4" diameter		18	.889		43	45	7.50	95.50	123

33 52 Hydrocarbon Transmission and Distribution

33 52 16 – Gas Hydrocarbon Piping

33 52 16.23 Medium Density Polyethylene Piping

		Crew	Daily Output	Labor-Hours	Unit	Material	2021 Bare Costs Labor	Equipment	Total	Total Incl O&P
8360	Purge test quick connector, female x female inlets, 1/4" N.P.T.	1 Skwk	50	.160	Ea.	18.30	9.15		27.45	34
8365	Purge test hose, 24" length, 3/8" ID, male outlets, 1/4" N.P.T.	"	50	.160		25	9.15		34.15	41.50
8370	Purge test quick connector and test hose assembly	1 Skwk	50	.160	Ea.	43.50	9.15		52.65	61.50
8405	Transition fittings, MDPE x zinc plated steel, SDR 7, 1/2" CTS x 1/2" MPT	B-22C	30	.533		24.50	27	4.50	56	72.50
8410	SDR 9.3, 1/2" IPS x 3/4" MPT		28.50	.561		38.50	28.50	4.73	71.73	89.50
8415	SDR 10, 1-1/4" IPS x 1-1/4" MPT		27	.593		36.50	30	5	71.50	91
8420	SDR 11, 3/4" IPS x 3/4" MPT		28.50	.561		19.40	28.50	4.73	52.63	69
8425	1" IPS x 1" MPT		27	.593		23.50	30	5	58.50	76.50
8430	1-1/4" IPS x 1-1/4" MPT		17	.941		44.50	48	7.95	100.45	129
8435	1-1/2" IPS x 1-1/2" MPT		25.50	.627		45.50	32	5.30	82.80	103
8440	2" IPS x 2" MPT		24	.667		52	34	5.60	91.60	114
8445	SDR 11.5, 1" CTS x 1" MPT		28	.571		32	29	4.82	65.82	84

33 52 16.26 High Density Polyethylene Piping

		Crew	Daily Output	Labor-Hours	Unit	Material	2021 Bare Costs Labor	Equipment	Total	Total Incl O&P
2010	**HIGH DENSITY POLYETHYLENE PIPING**									
2020	ASTM D2513, not including excavation or backfill									
2200	Butt fused pipe									
2205	125psi coils, butt fusion joint @ 100', 1/2" CTS diameter, SDR 7	B-22C	2050	.008	L.F.	.13	.40	.07	.60	.80
2210	160psi coils, butt fusion joint @ 100', IPS 1/2" diameter, SDR 9		1950	.008		.17	.42	.07	.66	.89
2215	3/4" diameter, SDR 11		1950	.008		.25	.42	.07	.74	.98
2220	1" diameter, SDR 11		1850	.009		.41	.44	.07	.92	1.19
2225	1-1/4" diameter, SDR 11		1850	.009		.60	.44	.07	1.11	1.40
2230	1-1/2" diameter, SDR 11		1750	.009		1.13	.46	.08	1.67	2.02
2235	2" diameter, SDR 11		1750	.009		.92	.46	.08	1.46	1.79
2240	3" diameter, SDR 11		1650	.010		1.51	.49	.08	2.08	2.49
2245	160psi 40' lengths, butt fusion joint, IPS 3" diameter, SDR 11	B-22A	660	.061		1.64	3.08	1.22	5.94	7.75
2250	4" diameter, SDR 11		500	.080		2.71	4.06	1.61	8.38	10.80
2255	6" diameter, SDR 11		500	.080		5.85	4.06	1.61	11.52	14.25
2260	8" diameter, SDR 11		420	.095		9.95	4.83	1.91	16.69	20.50
2265	10" diameter, SDR 11		340	.118		23	5.95	2.36	31.31	36.50
2270	12" diameter, SDR 11		300	.133		32	6.75	2.68	41.43	48
2400	Socket fused pipe									
2405	125psi coils, socket fusion coupling @ 100', 1/2" CTS diameter, SDR 7	B-20	1050	.023	L.F.	.15	1.13		1.28	1.85
2410	160psi coils, socket fusion coupling @ 100', IPS 1/2" diameter, SDR 9		1000	.024		.19	1.18		1.37	1.98
2415	3/4" diameter, SDR 11		1000	.024		.27	1.18		1.45	2.06
2420	1" diameter, SDR 11		950	.025		.43	1.25		1.68	2.33
2425	1-1/4" diameter, SDR 11		950	.025		.62	1.25		1.87	2.54
2430	1-1/2" diameter, SDR 11		900	.027		1.15	1.31		2.46	3.24
2435	2" diameter, SDR 11		850	.028		.95	1.39		2.34	3.12
2440	3" diameter, SDR 11		850	.028		1.62	1.39		3.01	3.87
2445	160psi 40' lengths, socket fusion coupling, IPS 3" diameter, SDR 11	B-21A	340	.118		1.93	6.45	1.27	9.65	13.10
2450	4" diameter, SDR 11	"	260	.154		3.25	8.45	1.66	13.36	18
2600	Compression coupled pipe									
2610	160psi coils, compression coupling @ 100', IPS 1/2" diameter, SDR 9	B-20	1000	.024	L.F.	.40	1.18		1.58	2.21
2615	3/4" diameter, SDR 11		1000	.024		.48	1.18		1.66	2.30
2620	1" diameter, SDR 11		950	.025		.72	1.25		1.97	2.66
2625	1-1/4" diameter, SDR 11		950	.025		1.33	1.25		2.58	3.32
2630	1-1/2" diameter, SDR 11		900	.027		1.92	1.31		3.23	4.08
2635	2" diameter, SDR 11		850	.028		1.84	1.39		3.23	4.10
3000	Fittings, butt fusion									
3010	SDR 11, IPS unless noted CTS									
3105	Caps, 1/2" diameter	B-22C	28.50	.561	Ea.	6.60	28.50	4.73	39.83	55
3110	3/4" diameter		28.50	.561		3.43	28.50	4.73	36.66	51.50

448

For customer support on your Plumbing Costs with RSMeans Data, call 800.448.8182.

33 52 16 – Gas Hydrocarbon Piping

33 52 16.26 **High Density Polyethylene Piping**		Crew	Daily Output	Labor-Hours	Unit	Material	2021 Bare Costs Labor	Equipment	Total	Total Incl O&P
3115	1" diameter	B-22C	27	.593	Ea.	3.74	30	5	38.74	54.50
3120	1-1/4" diameter		27	.593		6.10	30	5	41.10	57
3125	1-1/2" diameter		25.50	.627		4.49	32	5.30	41.79	58
3130	2" diameter		24	.667		8	34	5.60	47.60	65.50
3135	3" diameter		24	.667		16	34	5.60	55.60	74.50
3140	4" diameter	▼	18	.889		22	45	7.50	74.50	100
3145	6" diameter	B-22A	18	2.222		61.50	113	44.50	219	286
3150	8" diameter		15	2.667		98	135	53.50	286.50	370
3155	10" diameter		12	3.333		330	169	67	566	685
3160	12" diameter	▼	10.50	3.810		350	193	76.50	619.50	760
3205	Reducers, 1/2" x 1/2" CTS diameters	B-22C	14.25	1.123		12.25	57	9.45	78.70	109
3210	3/4" x 1/2" CTS diameters		14.25	1.123		8.80	57	9.45	75.25	106
3215	1" x 1/2" CTS diameters		13.50	1.185		13.65	60	10	83.65	116
3220	1" x 1/2" diameters		13.50	1.185		16.80	60	10	86.80	120
3225	1" x 3/4" diameters		13.50	1.185		13.60	60	10	83.60	116
3230	1-1/4" x 1" diameters		13.50	1.185		15.20	60	10	85.20	118
3232	1-1/2" x 3/4" diameters		12.75	1.255		7.90	63.50	10.60	82	116
3233	1-1/2" x 1" diameters		12.75	1.255		7.90	63.50	10.60	82	116
3234	1-1/2" x 1-1/4" diameters		12.75	1.255		7.90	63.50	10.60	82	116
3235	2" x 1" diameters		12	1.333		17.25	67.50	11.25	96	132
3240	2" x 1-1/4" diameters		12	1.333		17.25	67.50	11.25	96	132
3245	2" x 1-1/2" diameters		12	1.333		9.05	67.50	11.25	87.80	123
3250	3" x 2" diameters		12	1.333		10.55	67.50	11.25	89.30	125
3255	4" x 2" diameters		9	1.778		14.40	90	15	119.40	167
3260	4" x 3" diameters	▼	9	1.778		15.45	90	15	120.45	169
3262	6" x 3" diameters	B-22A	9	4.444		35.50	226	89	350.50	470
3265	6" x 4" diameters		9	4.444		47.50	226	89	362.50	485
3270	8" x 6" diameters		7.50	5.333		81	271	107	459	610
3275	10" x 8" diameters		6	6.667		170	340	134	644	840
3280	12" x 8" diameters		5.25	7.619		250	385	153	788	1,025
3285	12" x 10" diameters	▼	5.25	7.619		116	385	153	654	875
3305	Elbows, 90°, 3/4" diameter	B-22C	14.25	1.123		5.85	57	9.45	72.30	102
3310	1" diameter		13.50	1.185		5.75	60	10	75.75	107
3315	1-1/4" diameter		13.50	1.185		6.70	60	10	76.70	108
3320	1-1/2" diameter		12.75	1.255		7.60	63.50	10.60	81.70	116
3325	2" diameter		12	1.333		8.25	67.50	11.25	87	122
3330	3" diameter		12	1.333		16.05	67.50	11.25	94.80	131
3335	4" diameter	▼	9	1.778		21	90	15	126	175
3340	6" diameter	B-22A	9	4.444		55	226	89	370	495
3345	8" diameter		7.50	5.333		164	271	107	542	705
3350	10" diameter		6	6.667		395	340	134	869	1,075
3355	12" diameter		5.25	7.619		370	385	153	908	1,150
3380	45°, 3/4" diameter	B-22C	14.25	1.123		6.50	57	9.45	72.95	103
3385	1" diameter		13.50	1.185		6.50	60	10	76.50	108
3390	1-1/4" diameter		13.50	1.185		6.65	60	10	76.65	108
3395	1-1/2" diameter		12.75	1.255		8.65	63.50	10.60	82.75	117
3400	2" diameter		12	1.333		9.70	67.50	11.25	88.45	124
3405	3" diameter		12	1.333		16.45	67.50	11.25	95.20	131
3410	4" diameter	▼	9	1.778		20.50	90	15	125.50	174
3415	6" diameter	B-22A	9	4.444		55	226	89	370	495
3420	8" diameter		7.50	5.333		171	271	107	549	710
3425	10" diameter		6	6.667		375	340	134	849	1,075
3430	12" diameter	▼	5.25	7.619		735	385	153	1,273	1,550

For customer support on your Plumbing Costs with RSMeans Data, call 800.448.8182.

449

33 52 Hydrocarbon Transmission and Distribution

33 52 16 – Gas Hydrocarbon Piping

33 52 16.26 High Density Polyethylene Piping		Crew	Daily Output	Labor-Hours	Unit	Material	2021 Bare Costs Labor	Equipment	Total	Total Incl O&P
3505	Tees, 1/2" CTS diameter	B-22C	10	1.600	Ea.	8	81	13.50	102.50	146
3510	1/2" diameter		9.50	1.684		8.80	85.50	14.20	108.50	153
3515	3/4" diameter		9.50	1.684		6.55	85.50	14.20	106.25	151
3520	1" diameter		9	1.778		7.05	90	15	112.05	159
3525	1-1/4" diameter		9	1.778		7.50	90	15	112.50	160
3530	1-1/2" diameter		8.50	1.882		12.35	95.50	15.85	123.70	174
3535	2" diameter		8	2		9.75	102	16.85	128.60	181
3540	3" diameter		8	2		18.05	102	16.85	136.90	190
3545	4" diameter	▼	6	2.667		26.50	135	22.50	184	257
3550	6" diameter	B-22A	6	6.667		65.50	340	134	539.50	725
3555	8" diameter		5	8		264	405	161	830	1,075
3560	10" diameter		4	10		490	510	201	1,201	1,525
3565	12" diameter	▼	3.50	11.429	▼	650	580	229	1,459	1,825
3600	Tapping tees, type II, 3/4" punch, butt fusion outlets									
3601	Saddle contour x outlet diameter, IPS unless noted CTS									
3602	1-1/4" x 1/2" CTS outlet	B-22C	17	.941	Ea.	11.20	48	7.95	67.15	92.50
3604	1-1/4" x 1/2" outlet		17	.941		11.20	48	7.95	67.15	92.50
3606	1-1/4" x 3/4" outlet		17	.941		11.20	48	7.95	67.15	92.50
3608	1-1/4" x 1" outlet		17	.941		13.05	48	7.95	69	94.50
3610	1-1/4" x 1-1/4" outlet		16	1		23.50	51	8.45	82.95	111
3612	1-1/2" x 1/2" CTS outlet		16	1		11.20	51	8.45	70.65	97.50
3614	1-1/2" x 1/2" outlet		16	1		11.20	51	8.45	70.65	97.50
3616	1-1/2" x 3/4" outlet		16	1		11.20	51	8.45	70.65	97.50
3618	1-1/2" x 1" outlet		16	1		13.05	51	8.45	72.50	99.50
3620	1-1/2" x 1-1/4" outlet		15	1.067		23.50	54	9	86.50	116
3622	2" x 1/2" CTS outlet		12	1.333		11.20	67.50	11.25	89.95	126
3624	2" x 1/2" outlet		12	1.333		11.20	67.50	11.25	89.95	126
3626	2" x 3/4" outlet		12	1.333		11.20	67.50	11.25	89.95	126
3628	2" x 1" outlet		12	1.333		13.05	67.50	11.25	91.80	128
3630	2" x 1-1/4" outlet		11	1.455		23.50	74	12.25	109.75	150
3632	3" x 1/2" CTS outlet		12	1.333		11.20	67.50	11.25	89.95	126
3634	3" x 1/2" outlet		12	1.333		11.20	67.50	11.25	89.95	126
3636	3" x 3/4" outlet		12	1.333		11.20	67.50	11.25	89.95	126
3638	3" x 1" outlet		12	1.333		13.05	67.50	11.25	91.80	128
3640	3" x 1-1/4" outlet		11	1.455		13.05	74	12.25	99.30	139
3642	4" x 1/2" CTS outlet		10	1.600		11.20	81	13.50	105.70	149
3644	4" x 1/2" outlet		10	1.600		11.20	81	13.50	105.70	149
3646	4" x 3/4" outlet		10	1.600		11.20	81	13.50	105.70	149
3648	4" x 1" outlet		10	1.600		13.05	81	13.50	107.55	151
3650	4" x 1-1/4" outlet	▼	9	1.778		23.50	90	15	128.50	177
3652	6" x 1/2" CTS outlet	B-22A	10	4		11.20	203	80.50	294.70	405
3654	6" x 1/2" outlet		10	4		11.20	203	80.50	294.70	405
3656	6" x 3/4" outlet		10	4		11.20	203	80.50	294.70	405
3658	6" x 1" outlet		10	4		13.05	203	80.50	296.55	410
3660	6" x 1-1/4" outlet		9	4.444		23.50	226	89	338.50	460
3662	8" x 1/2" CTS outlet		8	5		11.20	254	100	365.20	500
3664	8" x 1/2" outlet		8	5		11.20	254	100	365.20	500
3666	8" x 3/4" outlet		8	5		11.20	254	100	365.20	500
3668	8" x 1" outlet		8	5		13.05	254	100	367.05	505
3670	8" x 1-1/4" outlet	▼	7	5.714		23.50	290	115	428.50	585
3675	For protective sleeves, 1/2" to 3/4" diameter outlets, add					.86			.86	.95
3680	1" diameter outlets, add					1.44			1.44	1.58
3685	1-1/4" diameter outlets, add				▼	2.48			2.48	2.73

450

For customer support on your Plumbing Costs with RSMeans Data, call 800.448.8182.

33 52 16 – Gas Hydrocarbon Piping

33 52 16.26 High Density Polyethylene Piping	Crew	Daily Output	Labor-Hours	Unit	Material	2021 Bare Costs Labor	Equipment	Total	Total Incl O&P
3700	Tapping tees, high volume, butt fusion outlets									
3705	1-1/2" punch, 2" x 2" outlet	B-22C	11	1.455	Ea.	110	74	12.25	196.25	246
3710	1-7/8" punch, 3" x 2" outlet		11	1.455		110	74	12.25	196.25	246
3715	4" x 2" outlet	↓	9	1.778		110	90	15	215	273
3720	6" x 2" outlet	B-22A	9	4.444		135	226	89	450	580
3725	8" x 2" outlet		7	5.714		110	290	115	515	680
3730	10" x 2" outlet		6	6.667		110	340	134	584	775
3735	12" x 2" outlet	↓	5	8		148	405	161	714	945
3750	For protective sleeves, add				↓	2.90			2.90	3.19
3800	Service saddles, saddle contour x outlet, butt fusion outlets									
3801	IPS unless noted CTS									
3802	1-1/4" x 3/4" outlet	B-22C	17	.941	Ea.	6.65	48	7.95	62.60	87.50
3804	1-1/4" x 1" outlet		16	1		6.65	51	8.45	66.10	92.50
3806	1-1/4" x 1-1/4" outlet		16	1		6.95	51	8.45	66.40	93
3808	1-1/2" x 3/4" outlet		16	1		14.15	51	8.45	73.60	101
3810	1-1/2" x 1-1/4" outlet		15	1.067		14.30	54	9	77.30	107
3812	2" x 3/4" outlet		12	1.333		14.15	67.50	11.25	92.90	129
3814	2" x 1" outlet		11	1.455		6.65	74	12.25	92.90	132
3816	2" x 1-1/4" outlet		11	1.455		6.95	74	12.25	93.20	132
3818	3" x 3/4" outlet		12	1.333		14.15	67.50	11.25	92.90	129
3820	3" x 1" outlet		11	1.455		14.15	74	12.25	100.40	140
3822	3" x 1-1/4" outlet		11	1.455		6.95	74	12.25	93.20	132
3824	4" x 3/4" outlet		10	1.600		14.15	81	13.50	108.65	152
3826	4" x 1" outlet		9	1.778		14.15	90	15	119.15	167
3828	4" x 1-1/4" outlet	↓	9	1.778		14.30	90	15	119.30	167
3830	6" x 3/4" outlet	B-22A	10	4		6.65	203	80.50	290.15	400
3832	6" x 1" outlet		9	4.444		6.65	226	89	321.65	440
3834	6" x 1-1/4" outlet		9	4.444		6.95	226	89	321.95	440
3836	8" x 3/4" outlet		8	5		14.15	254	100	368.15	505
3838	8" x 1" outlet		7	5.714		6.65	290	115	411.65	570
3840	8" x 1-1/4" outlet	↓	7	5.714		6.95	290	115	411.95	570
3880	For protective sleeves, 3/4" diameter outlets, add					.86			.86	.95
3885	1" diameter outlets, add					1.44			1.44	1.58
3890	1-1/4" diameter outlets, add					2.48			2.48	2.73
3905	Branch saddles, contour x outlet, 2" x 2" outlet	B-22C	11	1.455		18.65	74	12.25	104.90	145
3910	3" x 2" outlet		11	1.455		18.65	74	12.25	104.90	145
3915	3" x 3" outlet		11	1.455		25	74	12.25	111.25	152
3920	4" x 2" outlet		9	1.778		18.65	90	15	123.65	172
3925	4" x 3" outlet		9	1.778		25	90	15	130	179
3930	4" x 4" outlet	↓	9	1.778		34	90	15	139	189
3935	6" x 2" outlet	B-22A	9	4.444		18.65	226	89	333.65	455
3940	6" x 3" outlet		9	4.444		25	226	89	340	460
3945	6" x 4" outlet		9	4.444		34	226	89	349	470
3950	6" x 6" outlet		9	4.444		81.50	226	89	396.50	525
3955	8" x 2" outlet		7	5.714		18.65	290	115	423.65	580
3960	8" x 3" outlet		7	5.714		25	290	115	430	590
3965	8" x 4" outlet		7	5.714		34	290	115	439	600
3970	8" x 6" outlet	↓	7	5.714		81.50	290	115	486.50	650
3980	Ball valves, full port, 3/4" diameter	B-22C	14.25	1.123		57.50	57	9.45	123.95	159
3982	1" diameter		13.50	1.185		57.50	60	10	127.50	164
3984	1-1/4" diameter		13.50	1.185		57.50	60	10	127.50	165
3986	1-1/2" diameter		12.75	1.255		84.50	63.50	10.60	158.60	200
3988	2" diameter	↓	12	1.333	↓	132	67.50	11.25	210.75	258

33 52 16.26 High Density Polyethylene Piping		Crew	Daily Output	Labor-Hours	Unit	Material	2021 Bare Costs Labor	2021 Bare Costs Equipment	Total	Total Incl O&P
3990	3" diameter	B-22C	12	1.333	Ea.	310	67.50	11.25	388.75	455
3992	4" diameter		9	1.778		410	90	15	515	600
3994	6" diameter	B-22A	9	4.444		1,075	226	89	1,390	1,625
3996	8" diameter	"	7.50	5.333		1,750	271	107	2,128	2,450
4000	Fittings, socket fusion									
4010	SDR 11, IPS unless noted CTS									
4105	Caps, 1/2" CTS diameter	B-20	30	.800	Ea.	2.37	39.50		41.87	61.50
4110	1/2" diameter		28.50	.842		2.38	41.50		43.88	64.50
4115	3/4" diameter		28.50	.842		2.60	41.50		44.10	65
4120	1" diameter		27	.889		3.29	44		47.29	69
4125	1-1/4" diameter		27	.889		3.45	44		47.45	69.50
4130	1-1/2" diameter		25.50	.941		3.81	46.50		50.31	73.50
4135	2" diameter		24	1		4.01	49.50		53.51	78.50
4140	3" diameter		24	1		24.50	49.50		74	101
4145	4" diameter		18	1.333		40	65.50		105.50	143
4205	Reducers, 1/2" x 1/2" CTS diameters		14.25	1.684		6	83		89	131
4210	3/4" x 1/2" CTS diameters		14.25	1.684		6.15	83		89.15	131
4215	3/4" x 1/2" diameters		14.25	1.684		6.15	83		89.15	131
4220	1" x 1/2" CTS diameters		13.50	1.778		5.95	87.50		93.45	138
4225	1" x 1/2" diameters		13.50	1.778		6.30	87.50		93.80	138
4230	1" x 3/4" diameters		13.50	1.778		5.15	87.50		92.65	137
4235	1-1/4" x 1/2" CTS diameters		13.50	1.778		5.95	87.50		93.45	138
4240	1-1/4" x 3/4" diameters		13.50	1.778		6.90	87.50		94.40	139
4245	1-1/4" x 1" diameters		13.50	1.778		5.65	87.50		93.15	137
4250	1-1/2" x 3/4" diameters		12.75	1.882		8.05	93		101.05	148
4252	1-1/2" x 1" diameters		12.75	1.882		8.05	93		101.05	148
4255	1-1/2" x 1-1/4" diameters		12.75	1.882		7.95	93		100.95	148
4260	2" x 3/4" diameters		12	2		9.90	98.50		108.40	159
4265	2" x 1" diameters		12	2		9.55	98.50		108.05	159
4270	2" x 1-1/4" diameters		12	2		10.35	98.50		108.85	159
4275	3" x 2" diameters		12	2		11.85	98.50		110.35	161
4280	4" x 2" diameters		9	2.667		28	131		159	228
4285	4" x 3" diameters		9	2.667		26	131		157	226
4305	Couplings, 1/2" CTS diameter		15	1.600		1.95	79		80.95	120
4310	1/2" diameter		14.25	1.684		1.95	83		84.95	126
4315	3/4" diameter		14.25	1.684		1.54	83		84.54	126
4320	1" diameter		13.50	1.778		1.52	87.50		89.02	133
4325	1-1/4" diameter		13.50	1.778		2.22	87.50		89.72	133
4330	1-1/2" diameter		12.75	1.882		2.42	93		95.42	142
4335	2" diameter		12	2		2.51	98.50		101.01	151
4340	3" diameter		12	2		11.45	98.50		109.95	161
4345	4" diameter		9	2.667		21.50	131		152.50	221
4405	Elbows, 90°, 1/2" CTS diameter		15	1.600		3.35	79		82.35	122
4410	1/2" diameter		14.25	1.684		3.38	83		86.38	128
4415	3/4" diameter		14.25	1.684		2.78	83		85.78	127
4420	1" diameter		13.50	1.778		3.36	87.50		90.86	135
4425	1-1/4" diameter		13.50	1.778		4.62	87.50		92.12	136
4430	1-1/2" diameter		12.75	1.882		6.75	93		99.75	146
4435	2" diameter		12	2		6.85	98.50		105.35	156
4440	3" diameter		12	2		25	98.50		123.50	176
4445	4" diameter		9	2.667		100	131		231	305
4450	45°, 3/4" diameter		14.25	1.684		6.80	83		89.80	132
4455	1" diameter		13.50	1.778		7.65	87.50		95.15	139

33 52 16 – Gas Hydrocarbon Piping

33 52 16.26 **High Density Polyethylene Piping**	Crew	Daily Output	Labor-Hours	Unit	Material	2021 Bare Costs Labor	Equipment	Total	Total Incl O&P
4460	1-1/4" diameter	B-20	13.50	1.778	Ea.	6.45	87.50		93.95	138
4465	1-1/2" diameter		12.75	1.882		10.10	93		103.10	150
4470	2" diameter		12	2		7.50	98.50		106	156
4475	3" diameter		12	2		41	98.50		139.50	194
4480	4" diameter		9	2.667		53.50	131		184.50	256
4505	Tees, 1/2" CTS diameter		10	2.400		2.89	118		120.89	180
4510	1/2" diameter		9.50	2.526		2.87	125		127.87	189
4515	3/4" diameter		9.50	2.526		2.86	125		127.86	189
4520	1" diameter		9	2.667		3.53	131		134.53	201
4525	1-1/4" diameter		9	2.667		4.91	131		135.91	202
4530	1-1/2" diameter		8.50	2.824		8.45	139		147.45	217
4535	2" diameter		8	3		9.05	148		157.05	231
4540	3" diameter		8	3		37.50	148		185.50	262
4545	4" diameter	↓	6	4	↓	72.50	197		269.50	375
4600	Tapping tees, type II, 3/4" punch, socket fusion outlets									
4601	Saddle countour x outlet diameter, IPS unless noted CTS									
4602	1-1/4" x 1/2" CTS outlet	B-20	17	1.412	Ea.	11.20	69.50		80.70	116
4604	1-1/4" x 1/2" outlet		17	1.412		11.20	69.50		80.70	116
4606	1-1/4" x 3/4" outlet		17	1.412		11.20	69.50		80.70	116
4608	1-1/4" x 1" outlet		17	1.412		13.05	69.50		82.55	118
4610	1-1/4" x 1-1/4" outlet		16	1.500		23.50	74		97.50	137
4612	1-1/2" x 1/2" CTS outlet		16	1.500		11.20	74		85.20	123
4614	1-1/2" x 1/2" outlet		16	1.500		11.20	74		85.20	123
4616	1-1/2" x 3/4" outlet		16	1.500		11.20	74		85.20	123
4618	1-1/2" x 1" outlet		16	1.500		13.05	74		87.05	125
4620	1-1/2" x 1-1/4" outlet		15	1.600		23.50	79		102.50	144
4622	2" x 1/2" CTS outlet		12	2		11.20	98.50		109.70	160
4624	2" x 1/2" outlet		12	2		11.20	98.50		109.70	160
4626	2" x 3/4" outlet		12	2		11.20	98.50		109.70	160
4628	2" x 1" outlet		12	2		13.05	98.50		111.55	162
4630	2" x 1-1/4" outlet		11	2.182		23.50	108		131.50	187
4632	3" x 1/2" CTS outlet		12	2		11.20	98.50		109.70	160
4634	3" x 1/2" outlet		12	2		11.20	98.50		109.70	160
4636	3" x 3/4" outlet		12	2		11.20	98.50		109.70	160
4638	3" x 1" outlet		12	2		13.05	98.50		111.55	162
4640	3" x 1-1/4" outlet		11	2.182		23.50	108		131.50	187
4642	4" x 1/2" CTS outlet		10	2.400		11.20	118		129.20	189
4644	4" x 1/2" outlet		10	2.400		11.20	118		129.20	189
4646	4" x 3/4" outlet		10	2.400		11.20	118		129.20	189
4648	4" x 1" outlet		10	2.400		13.05	118		131.05	191
4650	4" x 1-1/4" outlet	↓	9	2.667		23.50	131		154.50	223
4652	6" x 1/2" CTS outlet	B-21A	10	4		11.20	219	43.50	273.70	385
4654	6" x 1/2" outlet	B-20	10	2.400		11.20	118		129.20	189
4656	6" x 3/4" outlet	B-21A	10	4		11.20	219	43.50	273.70	385
4658	6" x 1" outlet		10	4		13.05	219	43.50	275.55	385
4660	6" x 1-1/4" outlet		9	4.444		23.50	244	48	315.50	445
4662	8" x 1/2" CTS outlet		8	5		11.20	274	54	339.20	480
4664	8" x 1/2" outlet		8	5		11.20	274	54	339.20	480
4666	8" x 3/4" outlet		8	5		11.20	274	54	339.20	480
4668	8" x 1" outlet		8	5		13.05	274	54	341.05	485
4670	8" x 1-1/4" outlet	↓	7	5.714		23.50	315	62	400.50	560
4675	For protective sleeves, 1/2" to 3/4" diameter outlets, add					.86			.86	.95
4680	1" diameter outlets, add				↓	1.44			1.44	1.58

33 52 16.26 High Density Polyethylene Piping	Crew	Daily Output	Labor-Hours	Unit	Material	2021 Bare Costs Labor	Equipment	Total	Total Incl O&P
4685	1-1/4" diameter outlets, add				Ea.	2.48			2.48	2.73
4700	Tapping tees, high volume, socket fusion outlets									
4705	1-1/2" punch, 2" x 1-1/4" outlet	B-20	11	2.182	Ea.	110	108		218	282
4710	1-7/8" punch, 3" x 1-1/4" outlet		11	2.182		110	108		218	282
4715	4" x 1-1/4" outlet	↓	9	2.667		110	131		241	320
4720	6" x 1-1/4" outlet	B-21A	9	4.444		110	244	48	402	540
4725	8" x 1-1/4" outlet		7	5.714		110	315	62	487	655
4730	10" x 1-1/4" outlet		6	6.667		110	365	72	547	745
4735	12" x 1-1/4" outlet	↓	5	8		110	440	86.50	636.50	870
4750	For protective sleeves, add				↓	2.90			2.90	3.19
4800	Service saddles, saddle contour x outlet, socket fusion outlets									
4801	IPS unless noted CTS									
4802	1-1/4" x 1/2" CTS outlet	B-20	17	1.412	Ea.	5.90	69.50		75.40	111
4804	1-1/4" x 1/2" outlet		17	1.412		5.90	69.50		75.40	111
4806	1-1/4" x 3/4" outlet		17	1.412		5.90	69.50		75.40	111
4808	1-1/4" x 1" outlet		16	1.500		6.95	74		80.95	119
4810	1-1/4" x 1-1/4" outlet		16	1.500		9.80	74		83.80	122
4812	1-1/2" x 1/2" CTS outlet		16	1.500		5.90	74		79.90	118
4814	1-1/2" x 1/2" outlet		16	1.500		5.90	74		79.90	118
4816	1-1/2" x 3/4" outlet		16	1.500		5.90	74		79.90	118
4818	1-1/2" x 1-1/4" outlet		15	1.600		9.80	79		88.80	129
4820	2" x 1/2" CTS outlet		12	2		5.90	98.50		104.40	155
4822	2" x 1/2" outlet		12	2		5.90	98.50		104.40	155
4824	2" x 3/4" outlet		12	2		5.90	98.50		104.40	155
4826	2" x 1" outlet		11	2.182		6.95	108		114.95	169
4828	2" x 1-1/4" outlet		11	2.182		9.80	108		117.80	172
4830	3" x 1/2" CTS outlet		12	2		5.90	98.50		104.40	155
4832	3" x 1/2" outlet		12	2		5.90	98.50		104.40	155
4834	3" x 3/4" outlet		12	2		5.90	98.50		104.40	155
4836	3" x 1" outlet		11	2.182		6.95	108		114.95	169
4838	3" x 1-1/4" outlet		11	2.182		9.80	108		117.80	172
4840	4" x 1/2" CTS outlet		10	2.400		5.90	118		123.90	184
4842	4" x 1/2" outlet		10	2.400		5.90	118		123.90	184
4844	4" x 3/4" outlet		10	2.400		5.90	118		123.90	184
4846	4" x 1" outlet		9	2.667		6.95	131		137.95	205
4848	4" x 1-1/4" outlet	↓	9	2.667		9.80	131		140.80	208
4850	6" x 1/2" CTS outlet	B-21A	10	4		5.90	219	43.50	268.40	380
4852	6" x 1/2" outlet		10	4		5.90	219	43.50	268.40	380
4854	6" x 3/4" outlet		10	4		5.90	219	43.50	268.40	380
4856	6" x 1" outlet		9	4.444		6.95	244	48	298.95	425
4858	6" x 1-1/4" outlet		9	4.444		9.80	244	48	301.80	430
4860	8" x 1/2" CTS outlet		8	5		5.90	274	54	333.90	475
4862	8" x 1/2" outlet		8	5		5.90	274	54	333.90	475
4864	8" x 3/4" outlet		8	5		5.90	274	54	333.90	475
4866	8" x 1" outlet		7	5.714		7.05	315	62	384.05	540
4868	8" x 1-1/4" outlet	↓	7	5.714		10	315	62	387	545
4880	For protective sleeves, 1/2" to 3/4" diameter outlets, add					.86			.86	.95
4885	1" diameter outlets, add					1.44			1.44	1.58
4890	1-1/4" diameter outlets, add					2.48			2.48	2.73
4900	Reducer tees, 1-1/4" x 3/4" x 3/4" diameters	B-20	9	2.667	↓	7.50	131		138.50	205
4902	1-1/4" x 3/4" x 1" diameters	B-20	9	2.667	Ea.	9.70	131		140.70	208
4904	1-1/4" x 3/4" x 1-1/4" diameters		9	2.667		9.70	131		140.70	208
4906	1-1/4" x 1" x 3/4" diameters	↓	9	2.667		9.70	131		140.70	208

For customer support on your Plumbing Costs with RSMeans Data, call 800.448.8182.

33 52 16 – Gas Hydrocarbon Piping

33 52 16.26 High Density Polyethylene Piping	Crew	Daily Output	Labor-Hours	Unit	Material	2021 Bare Costs Labor	Equipment	Total	Total Incl O&P
4908	1-1/4" x 1" x 1" diameters	B-20	9	2.667	Ea.	7.40	131		138.40	205
4910	1-1/4" x 1" x 1-1/4" diameters		9	2.667		9.70	131		140.70	208
4912	1-1/4" x 1-1/4" x 3/4" diameters		9	2.667		7	131		138	205
4914	1-1/4" x 1-1/4" x 1" diameters		9	2.667		7.40	131		138.40	205
4916	1-1/2" x 3/4" x 3/4" diameters		8.50	2.824		12.30	139		151.30	222
4918	1-1/2" x 3/4" x 1" diameters		8.50	2.824		12.30	139		151.30	222
4920	1-1/2" x 3/4" x 1-1/4" diameters		8.50	2.824		12.30	139		151.30	222
4922	1-1/2" x 3/4" x 1-1/2" diameters		8.50	2.824		12.30	139		151.30	222
4924	1-1/2" x 1" x 3/4" diameters		8.50	2.824		12.30	139		151.30	222
4926	1-1/2" x 1" x 1" diameters		8.50	2.824		12.30	139		151.30	222
4928	1-1/2" x 1" x 1-1/4" diameters		8.50	2.824		12.30	139		151.30	222
4930	1-1/2" x 1" x 1-1/2" diameters		8.50	2.824		12.30	139		151.30	222
4932	1-1/2" x 1-1/4" x 3/4" diameters		8.50	2.824		10.35	139		149.35	219
4934	1-1/2" x 1-1/4" x 1" diameters		8.50	2.824		10.20	139		149.20	219
4936	1-1/2" x 1-1/4" x 1-1/4" diameters		8.50	2.824		10.35	139		149.35	219
4938	1-1/2" x 1-1/4" x 1-1/2" diameters		8.50	2.824		12.30	139		151.30	222
4940	1-1/2" x 1-1/2" x 3/4" diameters		8.50	2.824		9.65	139		148.65	219
4942	1-1/2" x 1-1/2" x 1" diameters		8.50	2.824		9.65	139		148.65	219
4944	1-1/2" x 1-1/2" x 1-1/4" diameters		8.50	2.824		9.65	139		148.65	219
4946	2" x 1-1/4" x 3/4" diameters		8	3		11.20	148		159.20	233
4948	2" x 1-1/4" x 1" diameters		8	3		10.85	148		158.85	233
4950	2" x 1-1/4" x 1-1/4" diameters		8	3		11.30	148		159.30	233
4952	2" x 1-1/2" x 3/4" diameters		8	3		11.20	148		159.20	233
4954	2" x 1-1/2" x 1" diameters		8	3		10.85	148		158.85	233
4956	2" x 1-1/2" x 1-1/4" diameters		8	3		10.85	148		158.85	233
4958	2" x 2" x 3/4" diameters		8	3		11.15	148		159.15	233
4960	2" x 2" x 1" diameters		8	3		10.95	148		158.95	233
4962	2" x 2" x 1-1/4" diameters	▼	8	3	▼	11.20	148		159.20	233
8100	Fittings, accessories									
8105	SDR 11, IPS unless noted CTS									
8110	Flange adapters, 2" x 6" long	1 Skwk	32	.250	Ea.	15.55	14.30		29.85	38.50
8115	3" x 6" long		32	.250		19.20	14.30		33.50	42.50
8120	4" x 6" long	▼	24	.333		24	19.05		43.05	55
8125	6" x 8" long	2 Skwk	24	.667		38	38		76	99.50
8130	8" x 9" long	"	20	.800		53.50	45.50		99	127
8135	Backup flanges, 2" diameter	1 Skwk	32	.250		19.80	14.30		34.10	43.50
8140	3" diameter		32	.250		26.50	14.30		40.80	51
8145	4" diameter	▼	24	.333		52.50	19.05		71.55	86
8150	6" diameter	2 Skwk	24	.667		54	38		92	117
8155	8" diameter	"	20	.800	▼	92	45.50		137.50	170
8200	Tapping tees, test caps									
8205	Type II, yellow polyethylene cap	1 Skwk	45	.178	Ea.	44	10.15		54.15	64
8210	High volume, black polyethylene cap		45	.178		56.50	10.15		66.65	78
8215	Quick connector, female x female inlets, 1/4" N.P.T.		50	.160		18.30	9.15		27.45	34
8220	Test hose, 24" length, 3/8" ID, male outlets, 1/4" N.P.T.		50	.160		25	9.15		34.15	41.50
8225	Quick connector and test hose assembly	▼	50	.160		43.50	9.15		52.65	61.50
8300	Threaded transition fittings									
8302	HDPE x MPT zinc plated steel, SDR 7, 1/2" CTS x 1/2" MPT	B-22C	30	.533	Ea.	30	27	4.50	61.50	79
8304	SDR 9.3, 1/2" IPS x 3/4" MPT		28.50	.561		40	28.50	4.73	73.23	91.50
8306	SDR 11, 3/4" IPS x 3/4" MPT		28.50	.561		19.70	28.50	4.73	52.93	69
8308	1" IPS x 1" MPT		27	.593		26	30	5	61	79
8310	1-1/4" IPS x 1-1/4" MPT		27	.593		41	30	5	76	95.50
8312	1-1/2" IPS x 1-1/2" MPT	▼	25.50	.627	▼	47.50	32	5.30	84.80	106

33 52 16.26 High Density Polyethylene Piping	Crew	Daily Output	Labor-Hours	Unit	Material	2021 Bare Costs Labor	Equipment	Total	Total Incl O&P
8314	2" IPS x 2" MPT	B-22C	24	.667	Ea.	54.50	34	5.60	94.10	117
8322	HDPE x MPT 316 stainless steel, SDR 11, 3/4" IPS x 3/4" MPT		28.50	.561		26	28.50	4.73	59.23	76.50
8324	1" IPS x 1" MPT		27	.593		27	30	5	62	80
8326	1-1/4" IPS x 1-1/4" MPT		27	.593		33	30	5	68	87
8328	2" IPS x 2" MPT		24	.667		40	34	5.60	79.60	101
8330	3" IPS x 3" MPT		24	.667		79.50	34	5.60	119.10	144
8332	4" IPS x 4" MPT		18	.889		109	45	7.50	161.50	196
8334	6" IPS x 6" MPT	B-22A	18	2.222		203	113	44.50	360.50	440
8342	HDPE x FPT 316 stainless steel, SDR 11, 3/4" IPS x 3/4" FPT	B-22C	28.50	.561		34	28.50	4.73	67.23	85
8344	1" IPS x 1" FPT		27	.593		45.50	30	5	80.50	101
8346	1-1/4" IPS x 1-1/4" FPT		27	.593		73.50	30	5	108.50	132
8348	1-1/2" IPS x 1-1/2" FPT		25.50	.627		70.50	32	5.30	107.80	131
8350	2" IPS x 2" FPT		24	.667		93	34	5.60	132.60	160
8352	3" IPS x 3" FPT		24	.667		180	34	5.60	219.60	255
8354	4" IPS x 4" FPT		18	.889		259	45	7.50	311.50	360
8362	HDPE x MPT epoxy carbon steel, SDR 11, 3/4" IPS x 3/4" MPT		28.50	.561		19.75	28.50	4.73	52.98	69
8364	1" IPS x 1" MPT		27	.593		22	30	5	57	74.50
8366	1-1/4" IPS x 1-1/4" MPT		27	.593		23	30	5	58	76
8368	1-1/2" IPS x 1-1/2" MPT		25.50	.627		25	32	5.30	62.30	81
8370	2" IPS x 2" MPT		24	.667		27.50	34	5.60	67.10	86.50
8372	3" IPS x 3" MPT		24	.667		42	34	5.60	81.60	103
8374	4" IPS x 4" MPT		18	.889		57	45	7.50	109.50	139
8382	HDPE x FPT epoxy carbon steel, SDR 11, 3/4" IPS x 3/4" FPT	B-22C	28.50	.561	Ea.	24	28.50	4.73	57.23	74
8384	1" IPS x 1" FPT		27	.593		26.50	30	5	61.50	79.50
8386	1-1/4" IPS x 1-1/4" FPT		27	.593		56.50	30	5	91.50	113
8388	1-1/2" IPS x 1-1/2" FPT		25.50	.627		57	32	5.30	94.30	116
8390	2" IPS x 2" FPT		24	.667		60.50	34	5.60	100.10	123
8392	3" IPS x 3" FPT		24	.667		103	34	5.60	142.60	170
8394	4" IPS x 4" FPT		18	.889		129	45	7.50	181.50	218
8402	HDPE x MPT poly coated carbon steel, SDR 11, 1" IPS x 1" MPT		27	.593		21.50	30	5	56.50	74
8404	1-1/4" IPS x 1-1/4" MPT		27	.593		28	30	5	63	81
8406	1-1/2" IPS x 1-1/2" MPT		25.50	.627		33	32	5.30	70.30	90
8408	2" IPS x 2" MPT		24	.667		37.50	34	5.60	77.10	97.50
8410	3" IPS x 3" MPT		24	.667		72.50	34	5.60	112.10	136
8412	4" IPS x 4" MPT		18	.889		113	45	7.50	165.50	200
8414	6" IPS x 6" MPT	B-22A	18	2.222		263	113	44.50	420.50	505
8602	Socket fused HDPE x MPT brass, SDR 11, 3/4" IPS x 3/4" MPT	B-20	28.50	.842		11.65	41.50		53.15	75
8604	1" IPS x 3/4" MPT		27	.889		14.10	44		58.10	81
8606	1" IPS x 1" MPT		27	.889		13.60	44		57.60	80.50
8608	1-1/4" IPS x 3/4" MPT		27	.889		14.60	44		58.60	81.50
8610	1-1/4" IPS x 1" MPT		27	.889		14.60	44		58.60	81.50
8612	1-1/4" IPS x 1-1/4" MPT		27	.889		21	44		65	88.50
8614	1-1/2" IPS x 1-1/2" MPT		25.50	.941		28.50	46.50		75	101
8616	2" IPS x 2" MPT		24	1		29	49.50		78.50	106
8618	Socket fused HDPE x FPT brass, SDR 11, 3/4" IPS x 3/4" FPT		28.50	.842		11.65	41.50		53.15	75
8620	1" IPS x 1/2" FPT		27	.889		11.65	44		55.65	78.50
8622	1" IPS x 3/4" FPT		27	.889		14.10	44		58.10	81
8624	1" IPS x 1" FPT		27	.889		13.60	44		57.60	80.50
8626	1-1/4" IPS x 3/4" FPT		27	.889		14.60	44		58.60	81.50
8628	1-1/4" IPS x 1" FPT		27	.889		14.10	44		58.10	81
8630	1-1/4" IPS x 1-1/4" FPT		27	.889		21	44		65	88.50
8632	1-1/2" IPS x 1-1/2" FPT		25.50	.941		29	46.50		75.50	102
8634	2" IPS x 1-1/2" FPT		24	1		29.50	49.50		79	107

33 52 Hydrocarbon Transmission and Distribution

33 52 16 – Gas Hydrocarbon Piping

33 52 16.26 High Density Polyethylene Piping	Crew	Daily Output	Labor-Hours	Unit	Material	2021 Bare Costs Labor	2021 Bare Costs Equipment	Total	Total Incl O&P
8636	2" IPS x 2" FPT	B-20	24	1	Ea.	31	49.50		80.50	109

33 61 Hydronic Energy Distribution

33 61 13 – Underground Hydronic Energy Distribution

33 61 13.20 Pipe Conduit, Prefabricated/Preinsulated

		Crew	Daily Output	Labor-Hours	Unit	Material	2021 Bare Costs Labor	2021 Bare Costs Equipment	Total	Total Incl O&P
0010	**PIPE CONDUIT, PREFABRICATED/PREINSULATED** R221113-70									
0020	Does not include trenching, fittings or crane.									
0300	For cathodic protection, add 12 to 14%									
0310	of total built-up price (casing plus service pipe)									
0580	Polyurethane insulated system, 250°F max. temp.									
0620	Black steel service pipe, standard wt., 1/2" insulation									
0660	3/4" diam. pipe size	Q-17	54	.296	L.F.	94.50	18.25	1.99	114.74	133
0670	1" diam. pipe size		50	.320		104	19.70	2.15	125.85	146
0680	1-1/4" diam. pipe size		47	.340		116	21	2.29	139.29	162
0690	1-1/2" diam. pipe size		45	.356		126	22	2.39	150.39	174
0700	2" diam. pipe size		42	.381		131	23.50	2.56	157.06	182
0710	2-1/2" diam. pipe size		34	.471		133	29	3.16	165.16	192
0720	3" diam. pipe size		28	.571		154	35	3.84	192.84	226
0730	4" diam. pipe size		22	.727		194	45	4.89	243.89	285
0740	5" diam. pipe size		18	.889		247	54.50	5.95	307.45	360
0750	6" diam. pipe size	Q-18	23	1.043		287	66.50	4.67	358.17	420
0760	8" diam. pipe size		19	1.263		420	80.50	5.65	506.15	590
0770	10" diam. pipe size		16	1.500		530	95.50	6.70	632.20	735
0780	12" diam. pipe size		13	1.846		660	118	8.25	786.25	910
0790	14" diam. pipe size		11	2.182		735	139	9.75	883.75	1,025
0800	16" diam. pipe size		10	2.400		845	153	10.75	1,008.75	1,175
0810	18" diam. pipe size		8	3		975	191	13.45	1,179.45	1,375
0820	20" diam. pipe size		7	3.429		1,075	219	15.35	1,309.35	1,550
0830	24" diam. pipe size		6	4		1,325	255	17.90	1,597.90	1,850
0900	For 1" thick insulation, add					10%				
0940	For 1-1/2" thick insulation, add					13%				
0980	For 2" thick insulation, add					20%				
1500	Gland seal for system, 3/4" diam. pipe size	Q-17	32	.500	Ea.	1,100	31	3.36	1,134.36	1,250
1510	1" diam. pipe size		32	.500		1,100	31	3.36	1,134.36	1,250
1540	1-1/4" diam. pipe size		30	.533		1,175	33	3.58	1,211.58	1,325
1550	1-1/2" diam. pipe size		30	.533		1,175	33	3.58	1,211.58	1,325
1560	2" diam. pipe size		28	.571		1,375	35	3.84	1,413.84	1,575
1570	2-1/2" diam. pipe size		26	.615		1,500	38	4.14	1,542.14	1,700
1580	3" diam. pipe size		24	.667		1,600	41	4.48	1,645.48	1,850
1590	4" diam. pipe size		22	.727		1,900	45	4.89	1,949.89	2,175
1600	5" diam. pipe size		19	.842		2,325	52	5.65	2,382.65	2,650
1610	6" diam. pipe size	Q-18	26	.923		2,475	59	4.14	2,538.14	2,825
1620	8" diam. pipe size		25	.960		2,875	61.50	4.30	2,940.80	3,275
1630	10" diam. pipe size		23	1.043		3,350	66.50	4.67	3,421.17	3,800
1640	12" diam. pipe size		21	1.143		3,700	73	5.10	3,778.10	4,200
1650	14" diam. pipe size		19	1.263		4,175	80.50	5.65	4,261.15	4,700
1660	16" diam. pipe size		18	1.333		4,925	85	5.95	5,015.95	5,525
1670	18" diam. pipe size		16	1.500		5,225	95.50	6.70	5,327.20	5,875
1680	20" diam. pipe size		14	1.714		5,925	109	7.70	6,041.70	6,700
1690	24" diam. pipe size		12	2		6,575	128	8.95	6,711.95	7,425
2000	Elbow, 45° for system									

457

33 61 13.20 Pipe Conduit, Prefabricated/Preinsulated	Crew	Daily Output	Labor-Hours	Unit	Material	2021 Bare Costs Labor	Equipment	Total	Total Incl O&P
2020	3/4" diam. pipe size	Q-17	14	1.143	Ea.	720	70.50	7.70	798.20	905
2040	1" diam. pipe size		13	1.231		740	75.50	8.25	823.75	930
2050	1-1/4" diam. pipe size		11	1.455		835	89.50	9.75	934.25	1,075
2060	1-1/2" diam. pipe size		9	1.778		840	109	11.95	960.95	1,100
2070	2" diam. pipe size		6	2.667		905	164	17.90	1,086.90	1,250
2080	2-1/2" diam. pipe size		4	4		990	246	27	1,263	1,500
2090	3" diam. pipe size		3.50	4.571		1,150	281	30.50	1,461.50	1,700
2100	4" diam. pipe size		3	5.333		1,325	330	36	1,691	2,000
2110	5" diam. pipe size		2.80	5.714		1,725	350	38.50	2,113.50	2,450
2120	6" diam. pipe size	Q-18	4	6		1,950	385	27	2,362	2,725
2130	8" diam. pipe size		3	8		2,825	510	36	3,371	3,900
2140	10" diam. pipe size		2.40	10		3,450	640	45	4,135	4,775
2150	12" diam. pipe size		2	12		4,525	765	54	5,344	6,200
2160	14" diam. pipe size		1.80	13.333		5,625	850	59.50	6,534.50	7,550
2170	16" diam. pipe size		1.60	15		6,700	955	67	7,722	8,875
2180	18" diam. pipe size		1.30	18.462		8,400	1,175	82.50	9,657.50	11,100
2190	20" diam. pipe size		1	24		10,600	1,525	108	12,233	14,000
2200	24" diam. pipe size		.70	34.286		13,300	2,175	154	15,629	18,000
2260	For elbow, 90°, add					25%				
2300	For tee, straight, add					85%	30%			
2340	For tee, reducing, add					170%	30%			
2380	For weldolet, straight, add					50%				
2400	Polyurethane insulation, 1"									
2410	FRP carrier and casing									
2420	4"	Q-5	18	.889	L.F.	55.50	54.50		110	143
2422	6"	Q-6	23	1.043		96	66.50		162.50	206
2424	8"		19	1.263		157	80.50		237.50	292
2426	10"		16	1.500		211	95.50		306.50	375
2428	12"		13	1.846		275	118		393	480
2430	FRP carrier and PVC casing									
2440	4"	Q-5	18	.889	L.F.	23	54.50		77.50	107
2444	8"	Q-6	19	1.263		49.50	80.50		130	175
2446	10"		16	1.500		71.50	95.50		167	222
2448	12"		13	1.846		97.50	118		215.50	283
2450	PVC carrier and casing									
2460	4"	Q-1	36	.444	L.F.	11.85	27		38.85	53.50
2462	6"	"	29	.552		16.95	33.50		50.45	68.50
2464	8"	Q-2	36	.667		24	42		66	89.50
2466	10"		32	.750		31.50	47.50		79	105
2468	12"		31	.774		37.50	49		86.50	114

33 63 Steam Energy Distribution

33 63 13 – Underground Steam and Condensate Distribution Piping

33 63 13.10 Calcium Silicate Insulated System	Crew	Daily Output	Labor-Hours	Unit	Material	2021 Bare Costs Labor	Equipment	Total	Total Incl O&P
0010 **CALCIUM SILICATE INSULATED SYSTEM**									
0011 High temp. (1200 degrees F)									
2840 Steel casing with protective exterior coating									
2850 6-5/8" diameter	Q-18	52	.462	L.F.	137	29.50	2.07	168.57	197
2860 8-5/8" diameter		50	.480		149	30.50	2.15	181.65	212
2870 10-3/4" diameter		47	.511		175	32.50	2.29	209.79	243
2880 12-3/4" diameter		44	.545		191	35	2.44	228.44	265
2890 14" diameter		41	.585		215	37.50	2.62	255.12	295
2900 16" diameter		39	.615		231	39.50	2.76	273.26	315
2910 18" diameter		36	.667		254	42.50	2.99	299.49	345
2920 20" diameter		34	.706		286	45	3.16	334.16	385
2930 22" diameter		32	.750		395	48	3.36	446.36	510
2940 24" diameter		29	.828		455	53	3.71	511.71	585
2950 26" diameter		26	.923		520	59	4.14	583.14	665
2960 28" diameter		23	1.043		640	66.50	4.67	711.17	810
2970 30" diameter		21	1.143		685	73	5.10	763.10	870
2980 32" diameter		19	1.263		770	80.50	5.65	856.15	975
2990 34" diameter		18	1.333		780	85	5.95	870.95	990
3000 36" diameter	↓	16	1.500		835	95.50	6.70	937.20	1,075
3040 For multi-pipe casings, add					10%				
3060 For oversize casings, add				↓	2%				
3400 Steel casing gland seal, single pipe									
3420 6-5/8" diameter	Q-18	25	.960	Ea.	2,100	61.50	4.30	2,165.80	2,400
3440 8-5/8" diameter		23	1.043		2,450	66.50	4.67	2,521.17	2,800
3450 10-3/4" diameter		21	1.143		2,725	73	5.10	2,803.10	3,125
3460 12-3/4" diameter		19	1.263		3,225	80.50	5.65	3,311.15	3,675
3470 14" diameter		17	1.412		3,575	90	6.30	3,671.30	4,100
3480 16" diameter		16	1.500		4,200	95.50	6.70	4,302.20	4,750
3490 18" diameter		15	1.600		4,650	102	7.15	4,759.15	5,250
3500 20" diameter		13	1.846		5,150	118	8.25	5,276.25	5,825
3510 22" diameter		12	2		5,775	128	8.95	5,911.95	6,550
3520 24" diameter		11	2.182		6,500	139	9.75	6,648.75	7,375
3530 26" diameter		10	2.400		7,425	153	10.75	7,588.75	8,400
3540 28" diameter		9.50	2.526		8,375	161	11.30	8,547.30	9,450
3550 30" diameter		9	2.667		8,525	170	11.95	8,706.95	9,650
3560 32" diameter		8.50	2.824		9,575	180	12.65	9,767.65	10,800
3570 34" diameter		8	3		10,400	191	13.45	10,604.45	11,800
3580 36" diameter	↓	7	3.429		11,100	219	15.35	11,334.35	12,500
3620 For multi-pipe casings, add				↓	5%				
4000 Steel casing anchors, single pipe									
4020 6-5/8" diameter	Q-18	8	3	Ea.	1,875	191	13.45	2,079.45	2,375
4040 8-5/8" diameter		7.50	3.200		1,950	204	14.35	2,168.35	2,475
4050 10-3/4" diameter		7	3.429		2,550	219	15.35	2,784.35	3,150
4060 12-3/4" diameter		6.50	3.692		2,725	236	16.55	2,977.55	3,375
4070 14" diameter		6	4		3,225	255	17.90	3,497.90	3,950
4080 16" diameter		5.50	4.364		3,725	278	19.55	4,022.55	4,525
4090 18" diameter		5	4.800		4,225	305	21.50	4,551.50	5,125
4100 20" diameter		4.50	5.333		4,650	340	24	5,014	5,625
4110 22" diameter		4	6		5,175	385	27	5,587	6,300
4120 24" diameter		3.50	6.857		5,650	435	30.50	6,115.50	6,900
4130 26" diameter		3	8		6,375	510	36	6,921	7,825
4140 28" diameter		2.50	9.600		7,025	610	43	7,678	8,700
4150 30" diameter	↓	2	12	↓	7,450	765	54	8,269	9,400

459

33 63 13 – Underground Steam and Condensate Distribution Piping

33 63 13.10 Calcium Silicate Insulated System		Crew	Daily Output	Labor-Hours	Unit	Material	2021 Bare Costs		Total	Total Incl O&P
							Labor	Equipment		
4160	32" diameter	Q-18	1.50	16	Ea.	8,900	1,025	71.50	9,996.50	11,400
4170	34" diameter		1	24		10,000	1,525	108	11,633	13,400
4180	36" diameter	↓	1	24		10,900	1,525	108	12,533	14,400
4220	For multi-pipe, add				↓	5%	20%			
4800	Steel casing elbow									
4820	6-5/8" diameter	Q-18	15	1.600	Ea.	2,675	102	7.15	2,784.15	3,075
4830	8-5/8" diameter		15	1.600		2,850	102	7.15	2,959.15	3,275
4850	10-3/4" diameter		14	1.714		3,450	109	7.70	3,566.70	3,950
4860	12-3/4" diameter		13	1.846		4,075	118	8.25	4,201.25	4,650
4870	14" diameter		12	2		4,375	128	8.95	4,511.95	5,000
4880	16" diameter		11	2.182		4,775	139	9.75	4,923.75	5,475
4890	18" diameter		10	2.400		5,450	153	10.75	5,613.75	6,225
4900	20" diameter		9	2.667		5,800	170	11.95	5,981.95	6,675
4910	22" diameter		8	3		6,250	191	13.45	6,454.45	7,175
4920	24" diameter		7	3.429		6,875	219	15.35	7,109.35	7,925
4930	26" diameter		6	4		7,550	255	17.90	7,822.90	8,700
4940	28" diameter		5	4.800		8,150	305	21.50	8,476.50	9,425
4950	30" diameter		4	6		8,175	385	27	8,587	9,600
4960	32" diameter		3	8		9,300	510	36	9,846	11,000
4970	34" diameter		2	12		10,200	765	54	11,019	12,400
4980	36" diameter	↓	2	12	↓	10,900	765	54	11,719	13,200
5500	Black steel service pipe, std. wt., 1" thick insulation									
5510	3/4" diameter pipe size	Q-17	54	.296	L.F.	85.50	18.25	1.99	105.74	123
5540	1" diameter pipe size		50	.320		88	19.70	2.15	109.85	129
5550	1-1/4" diameter pipe size		47	.340		111	21	2.29	134.29	157
5560	1-1/2" diameter pipe size		45	.356		116	22	2.39	140.39	163
5570	2" diameter pipe size	↓	42	.381	↓	127	23.50	2.56	153.06	177
5580	2-1/2" diameter pipe size	Q-17	34	.471	L.F.	132	29	3.16	164.16	192
5590	3" diameter pipe size		28	.571		116	35	3.84	154.84	184
5600	4" diameter pipe size		22	.727		121	45	4.89	170.89	205
5610	5" diameter pipe size	↓	18	.889		166	54.50	5.95	226.45	271
5620	6" diameter pipe size	Q-18	23	1.043	↓	179	66.50	4.67	250.17	300
6000	Black steel service pipe, std. wt., 1-1/2" thick insul.									
6010	3/4" diameter pipe size	Q-17	54	.296	L.F.	101	18.25	1.99	121.24	140
6040	1" diameter pipe size		50	.320		106	19.70	2.15	127.85	149
6050	1-1/4" diameter pipe size		47	.340		114	21	2.29	137.29	160
6060	1-1/2" diameter pipe size		45	.356		113	22	2.39	137.39	159
6070	2" diameter pipe size		42	.381		121	23.50	2.56	147.06	171
6080	2-1/2" diameter pipe size		34	.471		111	29	3.16	143.16	168
6090	3" diameter pipe size		28	.571		123	35	3.84	161.84	192
6100	4" diameter pipe size		22	.727		141	45	4.89	190.89	227
6110	5" diameter pipe size	↓	18	.889		189	54.50	5.95	249.45	296
6120	6" diameter pipe size	Q-18	23	1.043		197	66.50	4.67	268.17	320
6130	8" diameter pipe size		19	1.263		253	80.50	5.65	339.15	405
6140	10" diameter pipe size		16	1.500		330	95.50	6.70	432.20	510
6150	12" diameter pipe size	↓	13	1.846		375	118	8.25	501.25	595
6190	For 2" thick insulation, add					15%				
6220	For 2-1/2" thick insulation, add					25%				
6260	For 3" thick insulation, add				↓	30%				
6800	Black steel service pipe, ex. hvy. wt., 1" thick insul.									
6820	3/4" diameter pipe size	Q-17	50	.320	L.F.	102	19.70	2.15	123.85	145
6840	1" diameter pipe size		47	.340		109	21	2.29	132.29	154
6850	1-1/4" diameter pipe size	↓	44	.364		120	22.50	2.44	144.94	168

33 63 13 – Underground Steam and Condensate Distribution Piping

33 63 13.10 Calcium Silicate Insulated System		Crew	Daily Output	Labor-Hours	Unit	Material	2021 Bare Costs Labor	Equipment	Total	Total Incl O&P
6860	1-1/2" diameter pipe size	Q-17	42	.381	L.F.	121	23.50	2.56	147.06	172
6870	2" diameter pipe size		40	.400		134	24.50	2.69	161.19	186
6880	2-1/2" diameter pipe size		31	.516		109	32	3.47	144.47	171
6890	3" diameter pipe size		27	.593		126	36.50	3.98	166.48	198
6900	4" diameter pipe size		21	.762		155	47	5.10	207.10	247
6910	5" diameter pipe size		17	.941		146	58	6.30	210.30	254
6920	6" diameter pipe size	Q-18	22	1.091		141	69.50	4.89	215.39	264
7400	Black steel service pipe, ex. hvy. wt., 1-1/2" thick insul.									
7420	3/4" diameter pipe size	Q-17	50	.320	L.F.	75.50	19.70	2.15	97.35	115
7440	1" diameter pipe size		47	.340		95	21	2.29	118.29	139
7450	1-1/4" diameter pipe size		44	.364		105	22.50	2.44	129.94	151
7460	1-1/2" diameter pipe size		42	.381		103	23.50	2.56	129.06	151
7470	2" diameter pipe size		40	.400		109	24.50	2.69	136.19	159
7480	2-1/2" diameter pipe size	Q-17	31	.516	L.F.	102	32	3.47	137.47	163
7490	3" diameter pipe size		27	.593		112	36.50	3.98	152.48	183
7500	4" diameter pipe size		21	.762		143	47	5.10	195.10	233
7510	5" diameter pipe size		17	.941		197	58	6.30	261.30	310
7520	6" diameter pipe size	Q-18	22	1.091		198	69.50	4.89	272.39	325
7530	8" diameter pipe size		18	1.333		325	85	5.95	415.95	490
7540	10" diameter pipe size		15	1.600		305	102	7.15	414.15	495
7550	12" diameter pipe size		13	1.846		365	118	8.25	491.25	585
7590	For 2" thick insulation, add					13%				
7640	For 2-1/2" thick insulation, add					18%				
7680	For 3" thick insulation, add					24%				

33 63 13.20 Combined Steam Pipe With Condensate Return

		Crew	Daily Output	Labor-Hours	Unit	Material	2021 Bare Costs Labor	Equipment	Total	Total Incl O&P
0010	**COMBINED STEAM PIPE WITH CONDENSATE RETURN**									
9010	8" and 4" in 24" case	Q-18	100	.240	L.F.	465	15.30	1.08	481.38	535
9020	6" and 3" in 20" case		104	.231		390	14.70	1.03	405.73	455
9030	3" and 1-1/2" in 16" case		110	.218		293	13.90	.98	307.88	340
9040	2" and 1-1/4" in 12-3/4" case		114	.211		246	13.45	.94	260.39	292
9050	1-1/2" and 1-1/4" in 10-3/4" case		116	.207		291	13.20	.93	305.13	340
9100	Steam pipe only (no return)									
9110	6" in 18" case	Q-18	108	.222	L.F.	320	14.20	1	335.20	370
9120	4" in 14" case		112	.214		246	13.65	.96	260.61	293
9130	3" in 14" case		114	.211		234	13.45	.94	248.39	278
9140	2-1/2" in 12-3/4" case		118	.203		218	13	.91	231.91	259
9150	2" in 12-3/4" case		122	.197		205	12.55	.88	218.43	246

For customer support on your Plumbing Costs with RSMeans Data, call 800.448.8182.

461

Division Notes

	CREW	DAILY OUTPUT	LABOR-HOURS	UNIT	BARE COSTS				TOTAL INCL O&P
I					MAT.	LABOR	EQUIP.	TOTAL	

Estimating Tips

Products such as conveyors, material handling cranes and hoists, and other items specified in this division require trained installers. The general contractor may not have any choice as to who will perform the installation or when it will be performed. Long lead times are often required for these products, making early decisions in purchasing and scheduling necessary. The installation of this type of equipment may require the embedment of mounting hardware during the construction of floors, structural walls, or interior walls/partitions. Electrical connections will require coordination with the electrical contractor.

Reference Numbers

Reference numbers are shown at the beginning of some major classifications. These numbers refer to related items in the Reference Section. The reference information may be an estimating procedure, an alternate pricing method, or technical information.

41 22 23.10 Material Handling	Crew	Daily Output	Labor-Hours	Unit	Material	2021 Bare Costs Labor	Equipment	Total	Total Incl O&P
0010	**MATERIAL HANDLING**, cranes, hoists and lifts								
1500	Cranes, portable hydraulic, floor type, 2,000 lb. capacity				Ea.	4,150			4,150	4,550
1600	4,000 lb. capacity					5,375			5,375	5,900
1800	Movable gantry type, 12' to 15' range, 2,000 lb. capacity					3,200			3,200	3,525
1900	6,000 lb. capacity					5,775			5,775	6,350
2100	Hoists, electric overhead, chain, hook hung, 15' lift, 1 ton cap.					2,550			2,550	2,800
2200	3 ton capacity					3,950			3,950	4,350
2500	5 ton capacity					7,625			7,625	8,400
2600	For hand-pushed trolley, add					15%				
2700	For geared trolley, add					30%				
2800	For motor trolley, add					75%				
3000	For lifts over 15', 1 ton, add				L.F.	25			25	27.50
3100	5 ton, add				"	88			88	97
3300	Lifts, scissor type, portable, electric, 36" high, 2,000 lb.				Ea.	4,175			4,175	4,575
3400	48" high, 4,000 lb.				Ea.	4,700			4,700	5,175

Estimating Tips

This section involves equipment and construction costs for air, noise, and odor pollution control systems. These systems may be interrelated and care must be taken that the complete systems are estimated. For example, air pollution equipment may include dust and air-entrained particles that have to be collected. The vacuum systems could be noisy, requiring silencers to reduce noise pollution, and the collected solids have to be disposed of to prevent solid pollution.

Reference Numbers

Reference numbers are shown at the beginning of some major classifications. These numbers refer to related items in the Reference Section. The reference information may be an estimating procedure, an alternate pricing method, or technical information.

44 11 16.10 Dust Collection Systems	Crew	Daily Output	Labor-Hours	Unit	Material	2021 Bare Costs Labor	Equipment	Total	Total Incl O&P
0010 **DUST COLLECTION SYSTEMS** Commercial/Industrial									
0120 Central vacuum units									
0130 Includes stand, filters and motorized shaker									
0200 500 CFM, 10" inlet, 2 HP	Q-20	2.40	8.333	Ea.	4,650	500		5,150	5,875
0220 1,000 CFM, 10" inlet, 3 HP		2.20	9.091		4,900	545		5,445	6,200
0240 1,500 CFM, 10" inlet, 5 HP		2	10		6,100	600		6,700	7,600
0260 3,000 CFM, 13" inlet, 10 HP		1.50	13.333		16,100	800		16,900	18,900
0280 5,000 CFM, 16" inlet, 2 @ 10 HP	↓	1	20	↓	17,200	1,200		18,400	20,700
1000 Vacuum tubing, galvanized									
1100 2-1/8" OD, 16 ga.	Q-9	440	.036	L.F.	3.17	2.14		5.31	6.70
1110 2-1/2" OD, 16 ga.		420	.038		3.50	2.24		5.74	7.25
1120 3" OD, 16 ga.		400	.040		4.78	2.36		7.14	8.80
1130 3-1/2" OD, 16 ga.		380	.042		6.80	2.48		9.28	11.20
1140 4" OD, 16 ga.		360	.044		6.50	2.62		9.12	11.10
1150 5" OD, 14 ga.		320	.050		11.70	2.95		14.65	17.30
1160 6" OD, 14 ga.		280	.057		15.40	3.37		18.77	22
1170 8" OD, 14 ga.		200	.080		21.50	4.71		26.21	30.50
1180 10" OD, 12 ga.		160	.100		62.50	5.90		68.40	78
1190 12" OD, 12 ga.		120	.133		59	7.85		66.85	76.50
1200 14" OD, 12 ga.	↓	80	.200	↓	66.50	11.80		78.30	91
1940 Hose, flexible wire reinforced rubber									
1956 3" diam.	Q-9	400	.040	L.F.	7	2.36		9.36	11.25
1960 4" diam.		360	.044		8.25	2.62		10.87	13
1970 5" diam.		320	.050		10.25	2.95		13.20	15.70
1980 6" diam.	↓	280	.057	↓	10.90	3.37		14.27	17.10
2000 90° elbow, slip fit									
2110 2-1/8" diam.	Q-9	70	.229	Ea.	13.40	13.45		26.85	35
2120 2-1/2" diam.		65	.246		18.70	14.50		33.20	42.50
2130 3" diam.		60	.267		25.50	15.70		41.20	51.50
2140 3-1/2" diam.		55	.291		31.50	17.15		48.65	60.50
2150 4" diam.		50	.320		39	18.85		57.85	71.50
2160 5" diam.		45	.356		73.50	21		94.50	113
2170 6" diam.		40	.400		102	23.50		125.50	148
2180 8" diam.	↓	30	.533	↓	193	31.50		224.50	261
2400 45° elbow, slip fit									
2410 2-1/8" diam.	Q-9	70	.229	Ea.	11.70	13.45		25.15	33.50
2420 2-1/2" diam.		65	.246		17.25	14.50		31.75	41
2430 3" diam.	↓	60	.267	↓	21	15.70		36.70	46.50
2440 3-1/2" diam.	Q-9	55	.291	Ea.	26	17.15		43.15	55
2450 4" diam.		50	.320		34	18.85		52.85	66
2460 5" diam.		45	.356		57.50	21		78.50	94.50
2470 6" diam.		40	.400		77.50	23.50		101	121
2480 8" diam.	↓	35	.457	↓	152	27		179	208
2800 90° TY, slip fit thru 6" diam.									
2810 2-1/8" diam.	Q-9	42	.381	Ea.	24.50	22.50		47	61
2820 2-1/2" diam.		39	.410		30.50	24		54.50	70
2830 3" diam.		36	.444		37.50	26		63.50	81
2840 3-1/2" diam.		33	.485		54	28.50		82.50	103
2850 4" diam.		30	.533		48	31.50		79.50	101
2860 5" diam.		27	.593		79.50	35		114.50	140
2870 6" diam.	↓	24	.667		118	39.50		157.50	188
2880 8" diam., butt end					201			201	221
2890 10" diam., butt end	↓			↓	430			430	475

44 11 Particulate Control Equipment

44 11 16 – Fugitive Dust Barrier Systems

44 11 16.10 Dust Collection Systems		Crew	Daily Output	Labor-Hours	Unit	Material	2021 Bare Costs Labor	Equipment	Total	Total Incl O&P
2900	12" diam., butt end				Ea.	430			430	475
2910	14" diam., butt end					770			770	850
2920	6" x 4" diam., butt end					130			130	144
2930	8" x 4" diam., butt end					247			247	272
2940	10" x 4" diam., butt end					325			325	355
2950	12" x 4" diam., butt end					370			370	405
3100	90° elbow, butt end, segmented									
3110	8" diam., butt end, segmented				Ea.	166			166	182
3120	10" diam., butt end, segmented					480			480	530
3130	12" diam., butt end, segmented					610			610	670
3140	14" diam., butt end, segmented					760			760	840
3200	45° elbow, butt end, segmented									
3210	8" diam., butt end, segmented				Ea.	152			152	167
3220	10" diam., butt end, segmented					345			345	380
3230	12" diam., butt end, segmented					355			355	390
3240	14" diam., butt end, segmented					615			615	680
3400	All butt end fittings require one coupling per joint.									
3410	Labor for fitting included with couplings.									
3460	Compression coupling, galvanized, neoprene gasket									
3470	2-1/8" diam.	Q-9	44	.364	Ea.	13.55	21.50		35.05	47.50
3480	2-1/2" diam.		44	.364		13.55	21.50		35.05	47.50
3490	3" diam.		38	.421		19.45	25		44.45	59
3500	3-1/2" diam.		35	.457		22	27		49	64.50
3510	4" diam.		33	.485		24	28.50		52.50	69
3520	5" diam.		29	.552		27	32.50		59.50	79
3530	6" diam.		26	.615		30	36.50		66.50	88
3540	8" diam.	Q-9	22	.727	Ea.	57.50	43		100.50	128
3550	10" diam.		20	.800		83	47		130	162
3560	12" diam.		18	.889		103	52.50		155.50	192
3570	14" diam.		16	1		158	59		217	263
3800	Air gate valves, galvanized									
3810	2-1/8" diam.	Q-9	30	.533	Ea.	122	31.50		153.50	182
3820	2-1/2" diam.		28	.571		132	33.50		165.50	197
3830	3" diam.		26	.615		141	36.50		177.50	210
3840	4" diam.		23	.696		163	41		204	242
3850	6" diam.		18	.889		227	52.50		279.50	330

For customer support on your Plumbing Costs with RSMeans Data, call 800.448.8182.

467

Division Notes

		CREW	DAILY OUTPUT	LABOR-HOURS	UNIT	BARE COSTS				TOTAL INCL O&P
						MAT.	LABOR	EQUIP.	TOTAL	
						CREW	DAILY OUTPUT	LABOR-HOURS	UNIT	MAT.	LABOR	EQUIP.	TOTAL	TOTAL INCL O&P

Estimating Tips

This division contains information about water and wastewater equipment and systems, which was formerly located in Division 44. The main areas of focus are total wastewater treatment plants and components of wastewater treatment plants. Also included in this section are oil/water separators for wastewater treatment.

Reference Numbers

Reference numbers are shown at the beginning of some major classifications. These numbers refer to related items in the Reference Section. The reference information may be an estimating procedure, an alternate pricing method, or technical information.

Division 46

Water & Wastewater Equipment

Same Data. Simplified.

Enjoy the convenience and efficiency of accessing your costs anywhere:

- **Skip the multiplier** by setting your location
- **Quickly search,** edit, favorite and share costs
- **Stay on top of price changes** with automatic updates

Discover more at rsmeans.com/online

46 07 Packaged Water and Wastewater Treatment Equipment

46 07 53 – Packaged Wastewater Treatment Equipment

46 07 53.10 Biological Pkg. Wastewater Treatment Plants		Crew	Daily Output	Labor-Hours	Unit	Material	2021 Bare Costs Labor	Equipment	Total	Total Incl O&P
0010	**BIOLOGICAL PACKAGED WASTEWATER TREATMENT PLANTS**									
0011	Not including fencing or external piping									
0020	Steel packaged, blown air aeration plants									
0100	1,000 GPD				Gal.				55	60.50
0200	5,000 GPD								22	24
0300	15,000 GPD								22	24
0400	30,000 GPD								15.40	16.95
0500	50,000 GPD								11	12.10
0600	100,000 GPD								9.90	10.90
0700	200,000 GPD								8.80	9.70
0800	500,000 GPD				↓				7.70	8.45
1000	Concrete, extended aeration, primary and secondary treatment									
1010	10,000 GPD				Gal.				22	24
1100	30,000 GPD								15.40	16.95
1200	50,000 GPD								11	12.10
1400	100,000 GPD								9.90	10.90
1500	500,000 GPD				↓				7.70	8.45
1700	Municipal wastewater treatment facility									
1720	1.0 MGD				Gal.				11	12.10
1740	1.5 MGD								10.60	11.65
1760	2.0 MGD								10	11
1780	3.0 MGD								7.80	8.60
1800	5.0 MGD				↓				5.80	6.70
2000	Holding tank system, not incl. excavation or backfill									
2010	Recirculating chemical water closet	2 Plum	4	4	Ea.	575	271		846	1,025
2100	For voltage converter, add	"	16	1		288	67.50		355.50	415
2200	For high level alarm, add	1 Plum	7.80	1.026	↓	152	69.50		221.50	271

46 07 53.20 Wastewater Treatment System

		Crew	Daily Output	Labor-Hours	Unit	Material	2021 Bare Costs Labor	Equipment	Total	Total Incl O&P
0010	**WASTEWATER TREATMENT SYSTEM**									
0020	Fiberglass, 1,000 gallon	B-21	1.29	21.705	Ea.	4,475	1,100	148	5,723	6,750
0100	1,500 gallon	"	1.03	27.184	"	9,600	1,400	185	11,185	12,900

46 23 Grit Removal And Handling Equipment

46 23 23 – Vortex Grit Removal Equipment

46 23 23.10 Rainwater Filters

		Crew	Daily Output	Labor-Hours	Unit	Material	2021 Bare Costs Labor	Equipment	Total	Total Incl O&P
0010	**RAINWATER FILTERS**									
0100	42 gal./min	B-21	3.50	8	Ea.	41,400	410	54.50	41,864.50	46,300
0200	65 gal./min	↓	3.50	8		41,500	410	54.50	41,964.50	46,300
0300	208 gal./min	↓	3.50	8	↓	41,500	410	54.50	41,964.50	46,400

46 25 13 – Coalescing Oil-Water Separators

46 25 13.20 Oil/Water Separators	Crew	Daily Output	Labor-Hours	Unit	Material	2021 Bare Costs Labor	Equipment	Total	Total Incl O&P
0010 **OIL/WATER SEPARATORS**									
0020 Underground, tank only									
0030 Excludes excavation, backfill & piping									
0100 200 GPM	B-21	3.50	8	Ea.	41,200	410	54.50	41,664.50	46,000
0110 400 GPM		3.25	8.615		90,500	440	59	90,999	100,500
0120 600 GPM		2.75	10.182		91,000	520	69.50	91,589.50	101,000
0130 800 GPM		2.50	11.200		114,500	570	76.50	115,146.50	127,000
0140 1,000 GPM		2	14		127,500	715	95.50	128,310.50	141,500
0150 1,200 GPM		1.50	18.667		142,000	955	127	143,082	157,500
0160 1,500 GPM	B-21	1	28	Ea.	161,500	1,425	191	163,116	180,000

For customer support on your Plumbing Costs with RSMeans Data, call 800.448.8182.

471

Assemblies Section

Table of Contents

RSMeans data: Assemblies— How They Work

Assemblies estimating provides a fast and reasonably accurate way to develop construction costs. An assembly is the grouping of individual work items—with appropriate quantities— to provide a cost for a major construction component in a convenient unit of measure.

An assemblies estimate is often used during early stages of design development to compare the cost impact of various design alternatives on total building cost.

Assemblies estimates are also used as an efficient tool to verify construction estimates.

Assemblies estimates do not require a completed design or detailed drawings. Instead, they are based on the general size of the structure and other known parameters of the project. The degree of accuracy of an assemblies estimate is generally within +/- 15%.

Most assemblies consist of three major elements: a graphic, the system components, and the cost data itself. The **Graphic** is a visual representation showing the typical appearance of the assembly

1 Unique 12-character Identifier

Our assemblies are identified by a **unique 12-character identifier**. The assemblies are numbered using UNIFORMAT II, ASTM Standard E1557. The first 5 characters represent this system to Level 3. The last 7 characters represent further breakdown in order to arrange items in understandable groups of similar tasks. Line numbers are consistent across all of our publications, so a line number in any assemblies data set will always refer to the same item.

2 Reference Box

Information is available in the Reference Section to assist the estimator with estimating procedures, alternate pricing methods, and additional technical information.

The **Reference Box** indicates the exact location of this information in the Reference Section. The "R" stands for "reference," and the remaining characters are the line numbers.

3 Narrative Descriptions

Our assemblies descriptions appear in two formats: narrative and table. **Narrative descriptions** are shown in a hierarchical structure to make them readable. In order to read a complete description, read up through the indents to the top of the section. Include everything that is above and to the left that is not contradicted by information below.

4 System Components

System components are listed separately to detail what is included in the development of the total system price.

Narrative Format

D40 Fire Protection

D4020 Standpipes

System Components	QUANTITY	UNIT	COST PER FLOOR		
			MAT.	INST.	TOTAL
SYSTEM D4020 310 0560					
WET STANDPIPE RISER, CLASS I, STEEL, BLACK, SCH. 40 PIPE, 10' HEIGHT					
4" DIAMETER PIPE, ONE FLOOR					
Pipe, steel, black, schedule 40, threaded, 4" diam.	20.000	L.F.	610	810	1,420
Pipe, Tee, malleable iron, black, 150 lb. threaded, 4" pipe size	2.000	Ea.	820	730	1,550
Pipe, 90° elbow, malleable iron, black, 150 lb threaded, 4" pipe size	1.000	Ea.	232	242	474
Pipe, nipple, steel, black, schedule 40, 2-1/2" pipe size x 3" long	2.000	Ea.	26.90	182	208.90
Fire valve, gate, 300 lb., brass w/handwheel, 2-1/2" pipe size	1.000	Ea.	203	114	317
Fire valve, pressure reducing rgh brs, 2-1/2" pipe size	1.000	Ea.	1,170	228	1,398
Valve, swing check, w/ball drip, CI w/brs. ftngs., 4" pipe size	1.000	Ea.	480	475	955
Standpipe conn wall dble. flush brs. w/plugs & chains 2-1/2"x2-1/2"x4"	1.000	Ea.	645	286	931
Valve, swing check, bronze, 125 lb, regrinding disc, 2-1/2" pipe size	1.000	Ea.	1,075	97	1,172
Roof manifold, fire, w/valves & caps, horiz/vert brs 2-1/2"x2-1/2"x4"	1.000	Ea.	277	298	575
Fire, hydrolator, vent & drain, 2-1/2" pipe size	1.000	Ea.	123	66	189
Valve, gate, iron body 125 lb., OS&Y, threaded, 4" pipe size	1.000	Ea.	1,100	485	1,585
Tamper switch (valve supervisory switch)	1.000	Ea.	294	49.50	343.50
TOTAL			7,055.90	4,062.50	11,118.40

D4020 310	Wet Standpipe Risers, Class I		COST PER FLOOR		
			MAT.	INST.	TOTAL
0550	Wet standpipe risers, Class I, steel, black, sch. 40, 10' height				
0560	4" diameter pipe, one floor		7,050	4,075	11,125
0580	Additional floors		1,650	1,250	2,900
0600	6" diameter pipe, one floor		10,900	7,050	17,950
0620	Additional floors		2,625	1,975	4,600
0640	8" diameter pipe, one floor	R211226 -10	16,500	8,525	25,025
0660	Additional floors		4,050	2,425	6,475
0680		R211226 -20			

For supplemental customizable square foot estimating forms, visit: **RSMeans.com/2021books** in question. It is frequently accompanied by additional explanatory technical information describing the class of items. The **System Components** is a listing of the individual tasks that make up the assembly, including the quantity and unit of measure for each item, along with the cost of material and installation. The **Assemblies**

data below lists prices for other similar systems with dimensional and/or size variations.

All of our assemblies costs represent the cost for the installing contractor. An allowance for profit has been added to all material, labor, and equipment rental costs. A markup for labor burdens, including workers' compensation, fixed

overhead, and business overhead, is included with installation costs.

The information in RSMeans cost data represents a "national average" cost. This data should be modified to the project location using the **City Cost Indexes** or **Location Factors** tables found in the Reference Section.

Table Format

D20 Plumbing

D2010 Plumbing Fixtures

Example of Plumbing Cost Calculations:
The bathroom system includes the individual fixtures such as bathtub, lavatory, shower and water closet. These fixtures are listed below as separate items merely as a checklist.

D2010 951	Plumbing Systems 20 Unit, 2 Story Apartment Building							
						COST EACH		
	FIXTURE	SYSTEM	LINE	QUANTITY	UNIT	MAT.	INST.	TOTAL
0440	Bathroom	D2010 926	3640	20	Ea.	68,500	57,500	126,000
0480	Bathtub							
0520	Booster pump[1]	not req'd.						
0560	Drinking fountain							
0600	Garbage disposal[1]	not incl.						
0660								
0680	Grease interceptor							
0720	Water heater	D2020 250	2140	1	Ea.	17,100	3,925	21,025
0760	Kitchen sink	D2010 410	1960	20	Ea.	30,200	20,400	50,600
0800	Laundry sink	D2010 420	1840	4	Ea.	5,125	4,025	9,150
0840	Lavatory							
0900								
0920	Roof drain, 1 floor	D2040 210	4200	2	Ea.	2,725	2,100	4,825
0960	Roof drain, add'l floor	D2040 210	4240	20	L.F.	530	530	1,060
1000	Service sink	D2010 440	4300	1	Ea.	3,450	1,400	4,850
1040	Sewage ejector[1]	not req'd.						
1080	Shower							
1100								
1160	Sump pump							
1200	Urinal							
1240	Water closet							
1320								
1360	**SUB TOTAL**					127,500	90,000	217,500
1481	Water controls	R221113-40		10%[2]		12,800	9,000	21,800
1521	Pipe & fittings[3]	R221113-40		30%[2]		38,300	27,000	65,300
1560	Other							
1601	Quality/complexity	R221113-40		15%[2]		19,200	13,500	32,700
1680								
1720	**TOTAL**					198,000	140,000	338,000
1741								

[1]**Note:** Cost for items such as booster pumps, backflow preventers, sewage ejectors, water meters, etc., may be obtained from the unit price section in the front of this data set.

Water controls, pipe and fittings, and the Quality/Complexity factors come from Table R221113-40.

[2]Percentage of subtotal.

[3]Long, easily discernable runs of pipe would be more accurately priced from Unit Price Section 22 11 13. If this is done, reduce the miscellaneous percentage in proportion.

5 Unit of Measure

All RSMeans data: Assemblies include a typical **Unit of Measure** used for estimating that item. For instance, for continuous footings or foundation walls the unit is linear feet (L.F.). For spread footings the unit is each (Ea.). The estimator needs to take special care that the unit in the data matches the unit in the takeoff. Abbreviations and unit conversions can be found in the Reference Section.

6 Table Descriptions

Table descriptions work similar to Narrative Descriptions, except that if there is a blank in the column at a particular line number, read up to the description above in the same column.

RSMeans data: Assemblies— How They Work (Continued)

Sample Estimate

This sample demonstrates the elements of an estimate, including a tally of the RSMeans data lines. Published assemblies costs include all markups for labor burden and profit for the installing contractor. This estimate adds a summary of the markups applied by a general contractor on the installing contractor's work. These figures represent the total cost to the owner. The location factor with RSMeans data is applied at the bottom of the estimate to adjust the cost of the work to a specific location.

Project Name:	Interior Fit-out, ABC Office			
Location:	**Anywhere, USA**		Date: 1/1/2021	STD
Assembly Number	**Description**	**Qty.**	**Unit**	**Subtotal**
① C1010 124 1200	Wood partition, 2 x 4 @ 16" OC w/5/8" FR gypsum board	560	S.F.	$3,169.60
C1020 114 1800	Metal door & frame, flush hollow core, 3'-0" x 7'-0"	2	Ea.	$2,580.00
C3010 230 0080	Painting, brushwork, primer & 2 coats	1,120	S.F.	$1,512.00
C3020 410 0140	Carpet, tufted, nylon, roll goods, 12' wide, 26 oz	240	S.F.	$684.00
C3030 210 6000	Acoustic ceilings, 24" x 48" tile, tee grid suspension	200	S.F.	$1,812.00
D5020 125 0560	Receptacles incl plate, box, conduit, wire, 20 A duplex	8	Ea.	$2,448.00
D5020 125 0720	Light switch incl plate, box, conduit, wire, 20 A single pole	2	Ea.	$598.00
D5020 210 0560	Fluorescent fixtures, recess mounted, 20 per 1000 SF	200	S.F.	$2,414.00
	Assembly Subtotal			**$15,217.60**
	Sales Tax @ **②**		5 %	$ 380.44
	General Requirements @ **③**		7 %	$ 1,065.23
	Subtotal A			**$16,663.27**
	GC Overhead @ **④**		5 %	$ 833.16
	Subtotal B			**$17,496.44**
	GC Profit @ **⑤**		5 %	$ 874.82
	Subtotal C			**$18,371.26**
	Adjusted by Location Factor ⑥		114.2	$ 20,979.98
	Architects Fee @ **⑦**		8 %	$ 1,678.40
	Contingency @ **⑧**		15 %	$ 3,147.00
	Project Total Cost			**$ 25,805.38**

This estimate is based on an interactive spreadsheet. You are free to download it and adjust it to your methodology.
A copy of this spreadsheet is available at **RSMeans.com/2021books**.

① Work Performed

The body of the estimate shows the RSMeans data selected, including line numbers, a brief description of each item, its takeoff quantity and unit, and the total installed cost, including the installing contractor's overhead and profit.

② Sales Tax

If the work is subject to state or local sales taxes, the amount must be added to the estimate. In a conceptual estimate it can be assumed that one half of the total represents material costs. Therefore, apply the sales tax rate to 50% of the assembly subtotal.

③ General Requirements

This item covers project-wide needs provided by the general contractor. These items vary by project but may include temporary facilities and utilities, security, testing, project cleanup, etc. In assemblies estimates a percentage is used—typically between 5% and 15% of project cost.

④ General Contractor Overhead

This entry represents the general contractor's markup on all work to cover project administration costs.

⑤ General Contractor Profit

This entry represents the GC's profit on all work performed. The value included here can vary widely by project and is influenced by the GC's perception of the project's financial risk and market conditions.

⑥ Location Factor

RSMeans published data are based on national average costs. If necessary, adjust the total cost of the project using a location factor from the "Location Factor" table or the "City Cost Indexes" table found in the Reference Section. Use location factors if the work is general, covering the work of multiple trades. If the work is by a single trade (e.g., masonry) use the more specific data found in the City Cost Indexes.

To adjust costs by location factors, multiply the base cost by the factor and divide by 100.

⑦ Architect's Fee

If appropriate, add the design cost to the project estimate. These fees vary based on project complexity and size. Typical design and engineering fees can be found in the Reference Section.

⑧ Contingency

A factor for contingency may be added to any estimate to represent the cost of unknowns that may occur between the time that the estimate is performed and the time the project is constructed. The amount of the allowance will depend on the stage of design at which the estimate is done, as well as the contractor's assessment of the risk involved.

Same Data. Simplified.

Enjoy the convenience and efficiency of accessing your costs anywhere:

- **Skip the multiplier** by setting your location
- **Quickly search,** edit, favorite and share costs
- **Stay on top of price changes** with automatic updates

Discover more at rsmeans.com/online

D2010 Plumbing Fixtures

Systems are complete with trim seat and rough-in (supply, waste and vent) for connection to supply branches and waste mains.

One Piece Wall Hung   **Supply**   **Waste/Vent**   **Floor Mount**

System Components	QUANTITY	UNIT	COST EACH MAT.	COST EACH INST.	COST EACH TOTAL
SYSTEM D2010 110 1880					
WATER CLOSET, VITREOUS CHINA					
TANK TYPE, WALL HUNG, TWO PIECE					
Water closet, tank type vit china wall hung 2 pc. w/seat supply & stop	1.000	Ea.	480	275	755
Pipe Steel galvanized, schedule 40, threaded, 2" diam.	4.000	L.F.	40.60	90	130.60
Pipe, CI soil, no hub, cplg 10' OC, hanger 5' OC, 4" diam.	2.000	L.F.	39.30	50	89.30
Pipe, coupling, standard coupling, CI soil, no hub, 4" diam.	2.000	Ea.	53	88	141
Copper tubing type L solder joint, hangar 10' O.C., 1/2" diam.	6.000	L.F.	16.56	60	76.56
Wrought copper 90° elbow for solder joints 1/2" diam.	2.000	Ea.	3.90	81	84.90
Wrought copper Tee for solder joints 1/2" diam.	1.000	Ea.	3.26	62	65.26
Supports/carrier, water closet, siphon jet, horiz, single, 4" waste	1.000	Ea.	1,425	152	1,577
TOTAL			2,061.62	858	2,919.62

D2010 110	Water Closet Systems		COST EACH MAT.	COST EACH INST.	COST EACH TOTAL
1800	Water closet, vitreous china				
1840	Tank type, wall hung				
1880	Close coupled two piece	R221113 -40	2,050	860	2,910
1920	Floor mount, one piece		1,425	910	2,335
1960	One piece low profile	R224000 -30	1,375	910	2,285
2000	Two piece close coupled		640	910	1,550
2040	Bowl only with flush valve				
2080	Wall hung		2,750	975	3,725
2120	Floor mount		825	925	1,750
2160	Floor mount, ADA compliant with 18" high bowl		865	950	1,815

D2010 Plumbing Fixtures

Systems are complete with trim, seat, flush valve and rough-in (supply, waste and vent) for connection to supply branches and waste mains.

Side by Side

Back to Back

Supply   Waste/Vent   Supply   Waste/Vent

System Components	QUANTITY	UNIT	COST EACH		
			MAT.	INST.	TOTAL
SYSTEM D2010 120 1760					
WATER CLOSETS, BATTERY MOUNT, WALL HUNG, SIDE BY SIDE, FIRST CLOSET					
Water closet, bowl only w/flush valve, seat, wall hung	1.000	Ea.	1,050	251	1,301
Pipe, CI soil, no hub, cplg 10' OC, hanger 5' OC, 4" diam.	3.000	L.F.	58.95	75	133.95
Coupling, standard, CI, soil, no hub, 4" diam.	2.000	Ea.	53	88	141
Copper tubing, type L, solder joints, hangers 10' OC, 1" diam.	6.000	L.F.	23.40	71.40	94.80
Copper tubing, type DWV, solder joints, hangers 10'OC, 2" diam.	6.000	L.F.	61.20	110.10	171.30
Wrought copper 90° elbow for solder joints 1" diam.	1.000	Ea.	10.50	50.50	61
Wrought copper Tee for solder joints 1" diam.	1.000	Ea.	12.65	81	93.65
Support/carrier, siphon jet, horiz, adjustable single, 4" pipe	1.000	Ea.	1,425	152	1,577
Valve, gate, bronze, 125 lb, NRS, soldered 1" diam.	1.000	Ea.	134	42.50	176.50
Wrought copper, DWV, 90° elbow, 2" diam.	1.000	Ea.	63.50	81	144.50
TOTAL			2,892.20	1,002.50	3,894.70

D2010 120	Water Closets, Group		COST EACH		
			MAT.	INST.	TOTAL
1760	Water closets, battery mount, wall hung, side by side, first closet	R221113 -40	2,900	1,000	3,900
1800	Each additional water closet, add		2,775	945	3,720
3000	Back to back, first pair of closets	R224000 -30	5,200	1,325	6,525
3100	Each additional pair of closets, back to back		5,075	1,300	6,375
9000	Back to back, first pair of closets, auto sensor flush valve, 1.28 gpf		5,625	1,450	7,075
9100	Ea additional pair of cls, back to back, auto sensor flush valve, 1.28 gpf		5,450	1,350	6,800

481

For customer support on your Plumbing Costs with RSMeans Data, call 800.448.8182.

D2010 Plumbing Fixtures

Systems are complete with trim, flush valve and rough-in (supply, waste and vent) for connection to supply branches and waste mains.

Stall Type

Supply   **Waste/Vent**

Wall Hung

System Components	QUANTITY	UNIT	COST EACH		
			MAT.	INST.	TOTAL
SYSTEM D2010 210 2000					
URINAL, VITREOUS CHINA, WALL HUNG					
Urinal, wall hung, vitreous china, incl. hanger	1.000	Ea.	370	485	855
Pipe, steel, galvanized, schedule 40, threaded, 1-1/2" diam.	5.000	L.F.	37.50	91	128.50
Copper tubing type DWV, solder joint, hangers 10' OC, 2" diam.	3.000	L.F.	30.60	55.05	85.65
Combination Y & 1/8 bend for CI soil pipe, no hub, 3" diam.	1.000	Ea.	27		27
Pipe, CI, no hub, cplg. 10' OC, hanger 5' OC, 3" diam.	4.000	L.F.	59.60	90	149.60
Pipe coupling standard, CI soil, no hub, 3" diam.	2.000	Ea.	40	77	117
Copper tubing type L, solder joint, hanger 10' OC 3/4" diam.	5.000	L.F.	16.75	53.25	70
Wrought copper 90° elbow for solder joints 3/4" diam.	1.000	Ea.	4.28	42.50	46.78
Wrought copper Tee for solder joints, 3/4" diam.	1.000	Ea.	7.85	67.50	75.35
TOTAL			593.58	961.30	1,554.88

D2010 210	Urinal Systems		COST EACH		
			MAT.	INST.	TOTAL
2000	Urinal, vitreous china, wall hung	R224000 -30	595	960	1,555
2040	Stall type		1,400	1,150	2,550

D2010 Plumbing Fixtures

Systems are complete with trim, flush valve and rough-in (supply, waste and vent) for connection to supply branches and waste mains.

Side by Side

Back to Back

Waste/Vent

Supply

Supply

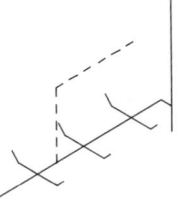

Waste/Vent

System Components	QUANTITY	UNIT	COST EACH		
			MAT.	INST.	TOTAL
SYSTEM D2010 220 1760					
URINALS, BATTERY MOUNT, WALL HUNG, SIDE BY SIDE, FIRST URINAL					
Urinal, wall hung, vitreous china, with hanger & trim	1.000	Ea.	370	485	855
No hub cast iron soil pipe, 3″ diameter	4.000	L.F.	59.60	90	149.60
No hub cast iron sanitary tee, 3″ diameter	1.000	Ea.	27		27
No hub coupling, 3″ diameter	2.000	Ea.	40	77	117
Copper tubing, type L, 3/4″ diameter	5.000	L.F.	16.75	53.25	70
Copper tubing, type DWV, 2″ diameter	2.000	L.F.	20.40	36.70	57.10
Copper 90° elbow, 3/4″ diameter	2.000	Ea.	8.56	85	93.56
Copper 90° elbow, type DWV, 2″ diameter	2.000	Ea.	127	162	289
Galvanized steel pipe, 1-1/2″ diameter	5.000	L.F.	37.50	91	128.50
Cast iron drainage elbow, 90°, 1-1/2″ diameter	1.000	Ea.	7.85	67.50	75.35
TOTAL			714.66	1,147.45	1,862.11

D2010 220	Urinal Systems, Battery Mount		COST EACH		
			MAT.	INST.	TOTAL
1760	Urinals, battery mount, side by side, first urinal	R221113 -40	715	1,150	1,865
1800	Each additional urinal, add		685	1,050	1,735
2000	Back to back, first pair of urinals	R224000 -30	1,325	1,750	3,075
2100	Each additional pair of urinals, back to back		955	1,375	2,330

D2010 Plumbing Fixtures

Systems are complete with trim and rough-in (supply, waste and vent) to connect to supply branches and waste mains.

Vanity Top

Supply   Waste/Vent

Wall Hung

System Components	QUANTITY	UNIT	COST EACH		
			MAT.	INST.	TOTAL
SYSTEM D2010 310 1560					
LAVATORY W/TRIM, VANITY TOP, P.E. ON C.I., 20" X 18"					
Lavatory w/trim, PE on CI, white, vanity top, 20" x 18" oval	1.000	Ea.	440	227	667
Pipe, steel, galvanized, schedule 40, threaded, 1-1/4" diam.	4.000	L.F.	25.40	65.40	90.80
Copper tubing type DWV, solder joint, hanger 10' OC 1-1/4" diam.	4.000	L.F.	48	53.80	101.80
Wrought copper DWV, Tee, sanitary, 1-1/4" diam.	1.000	Ea.	52	90	142
P trap w/cleanout, 20 ga., 1-1/4" diam.	1.000	Ea.	257	45	302
Copper tubing type L, solder joint, hanger 10' OC 1/2" diam.	10.000	L.F.	27.60	100	127.60
Wrought copper 90° elbow for solder joints 1/2" diam.	2.000	Ea.	3.90	81	84.90
Wrought copper Tee for solder joints, 1/2" diam.	2.000	Ea.	6.52	124	130.52
Stop, chrome, angle supply, 1/2" diam.	2.000	Ea.	27	74	101
TOTAL			887.42	860.20	1,747.62

D2010 310	Lavatory Systems		COST EACH		
			MAT.	INST.	TOTAL
1560	Lavatory w/trim, vanity top, PE on CI, 20" x 18", Vanity top by others.	R221113 -40	885	860	1,745
1600	19" x 16" oval		605	860	1,465
1640	18" round		985	860	1,845
1680	Cultured marble, 19" x 17"		620	860	1,480
1720	25" x 19"		620	860	1,480
1760	Stainless, self-rimming, 25" x 22"		825	860	1,685
1800	17" x 22"	R224000 -30	815	860	1,675
1840	Steel enameled, 20" x 17"		635	885	1,520
1880	19" round		660	885	1,545
1920	Vitreous china, 20" x 16"		715	900	1,615
1960	19" x 16"		725	900	1,625
2000	22" x 13"		720	900	1,620
2040	Wall hung, PE on CI, 18" x 15"		1,100	950	2,050
2080	19" x 17"		1,225	950	2,175
2120	20" x 18"		1,075	950	2,025
2160	Vitreous china, 18" x 15"		945	975	1,920
2200	19" x 17"		925	975	1,900
2240	24" x 20"		1,050	975	2,025
2300	20" x 27", handicap		1,800	1,050	2,850

D2010 Plumbing Fixtures

Systems are complete with trim, flush valve and rough-in (supply, waste and vent) for connection to supply branches and waste mains.

Side by Side

Back to Back

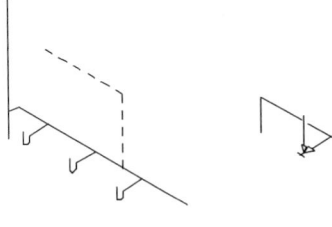

Waste/Vent

Supply
(Two supply systems required)

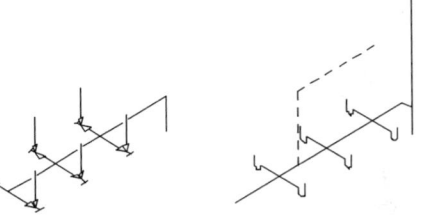

Supply
(Two supply systems required)

Waste/Vent

System Components	QUANTITY	UNIT	COST EACH		
			MAT.	INST.	TOTAL
SYSTEM D2010 320 1760					
LAVATORIES, BATTERY MOUNT, WALL HUNG, SIDE BY SIDE, FIRST LAVATORY					
Lavatory w/trim wall hung PE on CI 20″ x 18″	1.000	Ea.	335	182	517
Stop, chrome, angle supply, 3/8″ diameter	2.000	Ea.	24.90	67	91.90
Concealed arm support	1.000	Ea.	880	135	1,015
P trap w/cleanout, 20 ga. C.P., 1-1/4″ diameter	1.000	Ea.	34	45	79
Copper tubing, type L, 1/2″ diameter	10.000	L.F.	27.60	100	127.60
Copper tubing, type DWV, 1-1/4″ diameter	4.000	L.F.	48	53.80	101.80
Copper 90° elbow, 1/2″ diameter	2.000	Ea.	3.90	81	84.90
Copper tee, 1/2″ diameter	2.000	Ea.	6.52	124	130.52
DWV copper sanitary tee, 1-1/4″ diameter	2.000	Ea.	104	180	284
Galvanized steel pipe, 1-1/4″ diameter	4.000	L.F.	25.40	65.40	90.80
Black cast iron 90° elbow, 1-1/4″ diameter	1.000	Ea.	15.50	66	81.50
TOTAL			1,504.82	1,099.20	2,604.02

D2010 320	Lavatory Systems, Battery Mount		COST EACH		
			MAT.	INST.	TOTAL
1760	Lavatories, battery mount, side by side, first lavatory	R221113 -40	1,500	1,100	2,600
1800	Each additional lavatory, add		1,375	765	2,140
2000	Back to back, first pair of lavatories		2,350	1,750	4,100
2100	Each additional pair of lavatories, back to back		2,225	1,450	3,675

D2010 Plumbing Fixtures

Systems are complete with trim and rough-in (supply, waste and vent) to connect to supply branches and waste mains.

Countertop Single Bowl

Supply

Waste/Vent

Countertop Double Bowl

System Components	QUANTITY	UNIT	COST EACH		
			MAT.	INST.	TOTAL
SYSTEM D2010 410 1720					
KITCHEN SINK W/TRIM, COUNTERTOP, P.E. ON C.I., 24″ X 21″, SINGLE BOWL					
Kitchen sink, counter top style, PE on CI, 24″ x 21″ single bowl	1.000	Ea.	335	260	595
Pipe, steel, galvanized, schedule 40, threaded, 1-1/4″ diam.	4.000	L.F.	25.40	65.40	90.80
Copper tubing, type DWV, solder, hangers 10′ OC 1-1/2″ diam.	6.000	L.F.	43.80	89.70	133.50
Wrought copper, DWV, Tee, sanitary, 1-1/2″ diam.	1.000	Ea.	65	101	166
P trap, standard, copper, 1-1/2″ diam.	1.000	Ea.	196	47.50	243.50
Copper tubing, type L, solder joints, hangers 10′ OC 1/2″ diam.	10.000	L.F.	27.60	100	127.60
Wrought copper 90° elbow for solder joints 1/2″ diam.	2.000	Ea.	3.90	81	84.90
Wrought copper Tee for solder joints, 1/2″ diam.	2.000	Ea.	6.52	124	130.52
Stop, angle supply, chrome, 1/2″ CTS	2.000	Ea.	27	74	101
TOTAL			730.22	942.60	1,672.82

D2010 410	Kitchen Sink Systems		COST EACH		
			MAT.	INST.	TOTAL
1720	Kitchen sink w/trim, countertop, PE on CI, 24″x21″, single bowl	R221113 -40	730	945	1,675
1760	30″ x 21″ single bowl		1,400	945	2,345
1800	32″ x 21″ double bowl		840	1,025	1,865
1880	Stainless steel, 19″ x 18″ single bowl		1,075	945	2,020
1920	25″ x 22″ single bowl		1,150	945	2,095
1960	33″ x 22″ double bowl		1,500	1,025	2,525
2000	43″ x 22″ double bowl		1,700	1,025	2,725
2040	44″ x 22″ triple bowl		1,625	1,075	2,700
2080	44″ x 24″ corner double bowl		1,275	1,025	2,300
2120	Steel, enameled, 24″ x 21″ single bowl		925	945	1,870
2160	32″ x 21″ double bowl		940	1,025	1,965
2240	Raised deck, PE on CI, 32″ x 21″, dual level, double bowl		950	1,275	2,225
2280	42″ x 21″ dual level, triple bowl		1,525	1,400	2,925

D2010 Plumbing Fixtures

Systems are complete with trim and rough-in (supply, waste and vent) to connect to supply branches and waste mains.

Single Compartment Sink

Supply   Waste/Vent

Double Compartment Sink

System Components	QUANTITY	UNIT	COST EACH		
			MAT.	INST.	TOTAL
SYSTEM D2010 420 1760					
LAUNDRY SINK W/TRIM, PE ON CI, BLACK IRON FRAME					
24″ X 20″ OD, SINGLE COMPARTMENT					
Laundry sink PE on CI w/trim & frame, 24″ x 21″ OD, 1 compartment	1.000	Ea.	680	242	922
Pipe, steel, galvanized, schedule 40, threaded, 1-1/4″ diam	4.000	L.F.	25.40	65.40	90.80
Copper tubing, type DWV, solder joint, hanger 10′ OC 1-1/2″diam	6.000	L.F.	43.80	89.70	133.50
Wrought copper, DWV, Tee, sanitary, 1-1/2″ diam	1.000	Ea.	65	101	166
P trap, standard, copper, 1-1/2″ diam	1.000	Ea.	196	47.50	243.50
Copper tubing type L, solder joints, hangers 10′ OC, 1/2″ diam	10.000	L.F.	27.60	100	127.60
Wrought copper 90° elbow for solder joints 1/2″ diam	2.000	Ea.	3.90	81	84.90
Wrought copper Tee for solder joints, 1/2″ diam	2.000	Ea.	6.52	124	130.52
Stop, angle supply, 1/2″ diam	2.000	Ea.	27	74	101
TOTAL			1,075.22	924.60	1,999.82

D2010 420	Laundry Sink Systems		COST EACH		
			MAT.	INST.	TOTAL
1740	Laundry sink w/trim, PE on CI, black iron frame	R221113 -40			
1760	24″ x 20″, single compartment		1,075	925	2,000
1800	24″ x 23″ single compartment	R224000 -30	1,150	925	2,075
1840	48″ x 21″ double compartment		1,275	1,000	2,275
1920	Molded stone, on wall, 22″ x 21″ single compartment		590	925	1,515
1960	45″x 21″ double compartment		810	1,000	1,810
2040	Plastic, on wall or legs, 18″ x 23″ single compartment		555	905	1,460
2080	20″ x 24″ single compartment		575	905	1,480
2120	36″ x 23″ double compartment		650	980	1,630
2160	40″ x 24″ double compartment		720	980	1,700

487

For customer support on your Plumbing Costs with RSMeans Data, call 800.448.8182.

D2010 Plumbing Fixtures

Corrosion resistant laboratory sink systems are complete with trim and rough-in (supply, waste and vent) to connect to supply branches and waste mains.

Laboratory Sink

Supply   **Waste/Vent**

Polypropylene Cup Sink

System Components	QUANTITY	UNIT	COST EACH		
			MAT.	INST.	TOTAL
SYSTEM D2010 430 1600					
LABORATORY SINK W/TRIM, STAINLESS STEEL, SINGLE BOWL					
DOUBLE DRAINBOARD, 54" X 24" O.D.					
Sink w/trim, stainless steel, 1 bowl, 2 drainboards 54" x 24" OD	1.000	Ea.	1,475	485	1,960
Pipe, polypropylene, schedule 40, acid resistant 1-1/2" diam.	10.000	L.F.	81.50	215	296.50
Tee, sanitary, polypropylene, acid resistant, 1-1/2" diam.	1.000	Ea.	40	81	121
P trap, polypropylene, acid resistant, 1-1/2" diam.	1.000	Ea.	77	47.50	124.50
Copper tubing type L, solder joint, hanger 10' O.C. 1/2" diam.	10.000	L.F.	27.60	100	127.60
Wrought copper 90° elbow for solder joints 1/2" diam.	2.000	Ea.	3.90	81	84.90
Wrought copper Tee for solder joints, 1/2" diam.	2.000	Ea.	6.52	124	130.52
Stop, angle supply, chrome, 1/2" diam.	2.000	Ea.	27	74	101
TOTAL			1,738.52	1,207.50	2,946.02

D2010 430	Laboratory Sink Systems		COST EACH		
			MAT.	INST.	TOTAL
1580	Laboratory sink w/trim,				
1590	Stainless steel, single bowl,	R221113 -40			
1600	Double drainboard, 54" x 24" O.D.		1,750	1,200	2,950
1640	Single drainboard, 47" x 24"O.D.		1,350	1,200	2,550
1670	Stainless steel, double bowl,				
1680	70" x 24" O.D.		1,900	1,200	3,100
1750	Polyethylene, single bowl,				
1760	Flanged, 14-1/2" x 14-1/2" O.D.		555	1,100	1,655
1800	18-1/2" x 18-1/2" O.D.		680	1,100	1,780
1840	23-1/2" x 20-1/2" O.D.		690	1,100	1,790
1920	Polypropylene, cup sink, oval, 7" x 4" O.D.		450	960	1,410
1960	10" x 4-1/2" O.D.		480	960	1,440

D2010 Plumbing Fixtures

Corrosion resistant laboratory sink systems are complete with trim and rough–in (supply, waste and vent) to connect to supply branches and waste mains.

Wall Hung　　　　**Supply**　　**Waste/Vent**　　　　**Corner, Floor**

System Components	QUANTITY	UNIT	COST EACH		
			MAT.	INST.	TOTAL
SYSTEM D2010 440 4260					
SERVICE SINK, PE ON CI, CORNER FLOOR, 28"X28", W/RIM GUARD & TRIM					
Service sink, corner floor, PE on CI, 28" x 28", w/rim guard & trim	1.000	Ea.	1,500	330	1,830
Copper tubing type DWV, solder joint, hanger 10'OC 3" diam.	6.000	L.F.	126	150	276
Copper tubing type DWV, solder joint, hanger 10'OC 2" diam	4.000	L.F.	40.80	73.40	114.20
Wrought copper DWV, Tee, sanitary, 3" diam.	1.000	Ea.	375	208	583
P trap with cleanout & slip joint, copper 3" diam	1.000	Ea.	865	73.50	938.50
Copper tubing, type L, solder joints, hangers 10' OC, 1/2" diam	10.000	L.F.	27.60	100	127.60
Wrought copper 90° elbow for solder joints 1/2" diam	2.000	Ea.	3.90	81	84.90
Wrought copper Tee for solder joints, 1/2" diam	2.000	Ea.	6.52	124	130.52
Stop, angle supply, chrome, 1/2" diam	2.000	Ea.	27	74	101
TOTAL			2,971.82	1,213.90	4,185.72

D2010 440	Service Sink Systems		COST EACH		
			MAT.	INST.	TOTAL
4260	Service sink w/trim, PE on CI, corner floor, 28" x 28", w/rim guard	R221113 -40	2,975	1,225	4,200
4300	Wall hung w/rim guard, 22" x 18"		3,450	1,400	4,850
4340	24" x 20"	R224000 -30	3,525	1,400	4,925
4380	Vitreous china, wall hung 22" x 20"		3,275	1,400	4,675

D2010 Plumbing Fixtures

Systems are complete with trim and rough-in (supply, waste and vent) to connect to supply branches and waste mains.

Recessed Bathtub   **Supply**   **Waste/Vent**   **Corner Bathtub**

System Components	QUANTITY	UNIT	COST EACH		
			MAT.	INST.	TOTAL
SYSTEM D2010 510 2000					
BATHTUB, RECESSED, PORCELAIN ENAMEL ON CAST IRON,, 48" x 42"					
Bath tub, porcelain enamel on cast iron, w/fittings, 48" x 42"	1.000	Ea.	3,550	365	3,915
Pipe, steel, galvanized, schedule 40, threaded, 1-1/4" diam.	4.000	L.F.	25.40	65.40	90.80
Pipe, CI no hub soil w/couplings 10' OC, hangers 5' OC, 4" diam.	3.000	L.F.	58.95	75	133.95
Combination Y and 1/8 bend for C.I. soil pipe, no hub, 4" pipe size	1.000	Ea.	69.50		69.50
Drum trap, 3" x 5", copper, 1-1/2" diam.	1.000	Ea.	390	50.50	440.50
Copper tubing type L, solder joints, hangers 10' OC 1/2" diam.	10.000	L.F.	27.60	100	127.60
Wrought copper 90° elbow, solder joints, 1/2" diam.	2.000	Ea.	3.90	81	84.90
Wrought copper Tee, solder joints, 1/2" diam.	2.000	Ea.	6.52	124	130.52
Stop, angle supply, 1/2" diameter	2.000	Ea.	27	74	101
Copper tubing type DWV, solder joints, hanger 10' OC 1-1/2" diam.	3.000	L.F.	21.90	44.85	66.75
Pipe coupling, standard, C.I. soil no hub, 4" pipe size	2.000	Ea.	53	88	141
TOTAL			4,233.77	1,067.75	5,301.52

D2010 510	Bathtub Systems		COST EACH		
			MAT.	INST.	TOTAL
2000	Bathtub, recessed, P.E. on CI., 48" x 42"	R224000 -30	4,225	1,075	5,300
2040	72" x 36"		4,150	1,200	5,350
2080	Mat bottom, 5' long	R221113 -40	1,925	1,025	2,950
2120	5'-6" long		2,650	1,075	3,725
2160	Corner, 48" x 42"		3,900	1,025	4,925
2200	Formed steel, enameled, 4'-6" long		1,275	955	2,230

D2010 Plumbing Fixtures

Circular Fountain

Systems are complete with trim, flush valve and rough-in (supply, waste and vent) for connection to supply branches and waste mains.

Supply

Waste/Vent

Semi-Circular Fountain

System Components	QUANTITY	UNIT	COST EACH		
			MAT.	INST.	TOTAL
SYSTEM D2010 610 1760					
GROUP WASH FOUNTAIN, PRECAST TERRAZZO					
CIRCULAR, 36″ DIAMETER					
Wash fountain, group, precast terrazzo, foot control 36″ diam.	1.000	Ea.	7,950	755	8,705
Copper tubing type DWV, solder joint, hanger 10′ OC, 2″ diam.	10.000	L.F.	102	183.50	285.50
P trap, standard, copper, 2″ diam.	1.000	Ea.	300	54	354
Wrought copper, Tee, sanitary, 2″ diam.	1.000	Ea.	89	115	204
Copper tubing type L, solder joint, hanger 10′ OC 1/2″ diam.	20.000	L.F.	55.20	200	255.20
Wrought copper 90° elbow for solder joints 1/2″ diam.	3.000	Ea.	5.85	121.50	127.35
Wrought copper Tee for solder joints, 1/2″ diam.	2.000	Ea.	6.52	124	130.52
TOTAL			8,508.57	1,553	10,061.57

D2010 610	Group Wash Fountain Systems		COST EACH		
			MAT.	INST.	TOTAL
1740	Group wash fountain, precast terrazzo				
1760	Circular, 36″ diameter	R221113 -40	8,500	1,550	10,050
1800	54″ diameter		12,600	1,700	14,300
1840	Semi-circular, 36″ diameter	R224000 -30	7,250	1,550	8,800
1880	54″ diameter		12,900	1,700	14,600
1960	Stainless steel, circular, 36″ diameter		8,150	1,450	9,600
2000	54″ diameter		9,700	1,600	11,300
2040	Semi-circular, 36″ diameter		5,325	1,450	6,775
2080	54″ diameter		8,550	1,600	10,150
2160	Thermoplastic, circular, 36″ diameter		5,575	1,175	6,750
2200	54″ diameter		5,375	1,375	6,750
2240	Semi-circular, 36″ diameter		6,000	1,175	7,175
2280	54″ diameter		6,075	1,375	7,450

D2010 Plumbing Fixtures

Systems are complete with trim and rough-in (supply, waste and vent) for connection to supply branches and waste mains.

Three Wall

Supply   Waste/Vent

Corner Angle

System Components			COST EACH		
	QUANTITY	UNIT	MAT.	INST.	TOTAL
SYSTEM D2010 710 1560					
SHOWER, STALL, BAKED ENAMEL, MOLDED STONE RECEPTOR, 30" SQUARE					
Shower stall, enameled steel, molded stone receptor, 30" square	1.000	Ea.	1,500	280	1,780
Copper tubing type DWV, solder joints, hangers 10' OC, 2" diam.	6.000	L.F.	43.80	89.70	133.50
Wrought copper DWV, Tee, sanitary, 2" diam.	1.000	Ea.	65	101	166
Trap, standard, copper, 2" diam.	1.000	Ea.	196	47.50	243.50
Copper tubing type L, solder joint, hanger 10' OC 1/2" diam.	16.000	L.F.	44.16	160	204.16
Wrought copper 90° elbow for solder joints 1/2" diam.	3.000	Ea.	5.85	121.50	127.35
Wrought copper Tee for solder joints, 1/2" diam.	2.000	Ea.	6.52	124	130.52
Stop and waste, straightway, bronze, solder joint 1/2" diam.	2.000	Ea.	47	67	114
TOTAL			1,908.33	990.70	2,899.03

D2010 710	Shower Systems		COST EACH		
			MAT.	INST.	TOTAL
1560	Shower, stall, baked enamel, molded stone receptor, 30" square		1,900	990	2,890
1600	32" square	R221113 -40	1,675	1,000	2,675
1640	Terrazzo receptor, 32" square		1,875	1,000	2,875
1680	36" square	R224000 -30	2,350	1,025	3,375
1720	36" corner angle		2,675	420	3,095
1800	Fiberglass one piece, three walls, 32" square		820	975	1,795
1840	36" square		870	975	1,845
1880	Polypropylene, molded stone receptor, 30" square		1,250	1,425	2,675
1920	32" square		1,275	1,425	2,700
1960	Built-in head, arm, bypass, stops and handles		122	375	497
2050	Shower, stainless steel panels, handicap				
2100	w/fixed and handheld head, control valves, grab bar, and seat		4,675	4,450	9,125
2500	Shower, group with six heads, thermostatic mix valves & balancing valve		13,000	1,100	14,100
2520	Five heads		9,300	990	10,290

D2010 Plumbing Fixtures

Systems are complete with trim and rough-in (supply, waste and vent) to connect to supply branches and waste mains.

Wall Mounted, No Back   **Supply   Waste/Vent**   **Wall Mounted, Low Back**

System Components	QUANTITY	UNIT	COST EACH		
			MAT.	INST.	TOTAL
SYSTEM D2010 810 1800					
DRINKING FOUNTAIN, ONE BUBBLER, WALL MOUNTED					
NON RECESSED, BRONZE, NO BACK					
Drinking fountain, wall mount, bronze, 1 bubbler	1.000	Ea.	1,250	202	1,452
Copper tubing, type L, solder joint, hanger 10' OC 3/8″ diam.	5.000	L.F.	11.70	48	59.70
Stop, supply, straight, chrome, 3/8″ diam.	1.000	Ea.	21.50	33.50	55
Wrought copper 90° elbow for solder joints 3/8″ diam.	1.000	Ea.	5.40	37	42.40
Wrought copper Tee for solder joints, 3/8″ diam.	1.000	Ea.	8.90	57.50	66.40
Copper tubing, type DWV, solder joint, hanger 10' OC 1-1/4″ diam.	4.000	L.F.	48	53.80	101.80
P trap, standard, copper drainage, 1-1/4″ diam.	1.000	Ea.	257	45	302
Wrought copper, DWV, Tee, sanitary, 1-1/4″ diam.	1.000	Ea.	52	90	142
TOTAL			1,654.50	566.80	2,221.30

D2010 810	Drinking Fountain Systems		COST EACH		
			MAT.	INST.	TOTAL
1740	Drinking fountain, one bubbler, wall mounted	R221113 -40			
1760	Non recessed				
1800	Bronze, no back	R224000 -30	1,650	565	2,215
1840	Cast iron, enameled, low back		1,775	565	2,340
1880	Fiberglass, 12″ back		2,375	565	2,940
1920	Stainless steel, no back		1,375	565	1,940
1960	Semi-recessed, poly marble		1,450	565	2,015
2040	Stainless steel		2,275	565	2,840
2080	Vitreous china		1,575	565	2,140
2120	Full recessed, poly marble		2,325	565	2,890
2200	Stainless steel		1,875	565	2,440
2240	Floor mounted, pedestal type, aluminum		2,975	770	3,745
2320	Bronze		3,475	770	4,245
2360	Stainless steel		3,025	770	3,795

D2010 Plumbing Fixtures

Systems are complete with trim and rough-in (supply, waste and vent) for connection to supply branches and waste mains.

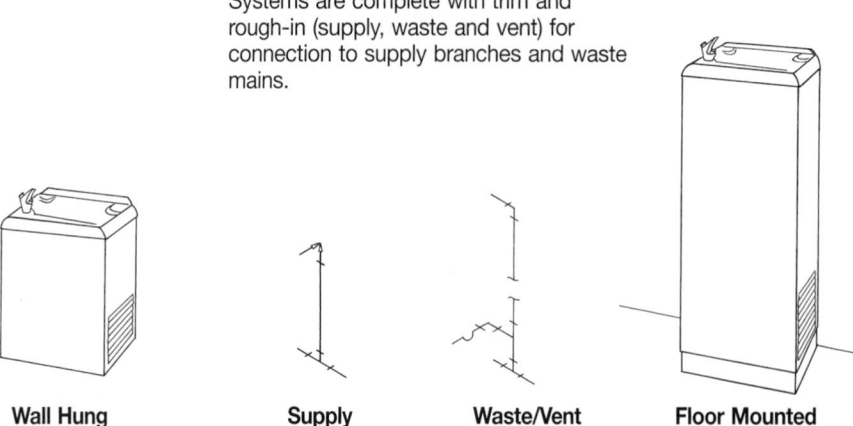

Wall Hung   Supply   Waste/Vent   Floor Mounted

System Components	QUANTITY	UNIT	COST EACH		
			MAT.	INST.	TOTAL
SYSTEM D2010 820 1840					
WATER COOLER, ELECTRIC, SELF CONTAINED, WALL HUNG, 8.2 G.P.H.					
Water cooler, wall mounted, 8.2 GPH	1.000	Ea.	1,225	365	1,590
Copper tubing type DWV, solder joint, hanger 10' OC 1-1/4" diam.	4.000	L.F.	48	53.80	101.80
Wrought copper DWV, Tee, sanitary 1-1/4" diam.	1.000	Ea.	52	90	142
P trap, copper drainage, 1-1/4" diam.	1.000	Ea.	257	45	302
Copper tubing type L, solder joint, hanger 10' OC 3/8" diam.	5.000	L.F.	11.70	48	59.70
Wrought copper 90° elbow for solder joints 3/8" diam.	1.000	Ea.	5.40	37	42.40
Wrought copper Tee for solder joints, 3/8" diam.	1.000	Ea.	8.90	57.50	66.40
Stop and waste, straightway, bronze, solder, 3/8" diam.	1.000	Ea.	21.50	33.50	55
TOTAL			1,629.50	729.80	2,359.30

D2010 820	Water Cooler Systems		COST EACH		
			MAT.	INST.	TOTAL
1840	Water cooler, electric, wall hung, 8.2 G.P.H.	R221113 -40	1,625	730	2,355
1880	Dual height, 14.3 G.P.H.		2,625	750	3,375
1920	Wheelchair type, 7.5 G.P.H.	R224000 -30	1,650	730	2,380
1960	Semi recessed, 8.1 G.P.H.		1,525	730	2,255
2000	Full recessed, 8 G.P.H.		3,125	780	3,905
2040	Floor mounted, 14.3 G.P.H.		1,700	635	2,335
2080	Dual height, 14.3 G.P.H.		1,775	770	2,545
2120	Refrigerated compartment type, 1.5 G.P.H.		2,075	635	2,710

D2010 Plumbing Fixtures

Two Fixture Bathroom Systems consisting of a lavatory, water closet, and rough-in service piping.

• Prices for plumbing and fixtures only.

*Common wall is with an adjacent bathroom.

System Components	QUANTITY	UNIT	COST EACH MAT.	COST EACH INST.	COST EACH TOTAL
SYSTEM D2010 920 1180					
BATHROOM, LAVATORY & WATER CLOSET, 2 WALL PLUMBING, STAND ALONE					
Water closet, two piece, close coupled	1.000	Ea.	211	275	486
Water closet, rough-in waste & vent	1.000	Set	380	475	855
Lavatory w/ftngs., wall hung, white, PE on CI, 20" x 18"	1.000	Ea.	335	182	517
Lavatory, rough-in waste & vent	1.000	Set	730	875	1,605
Copper tubing type L, solder joint, hanger 10' OC 1/2" diam.	10.000	L.F.	27.60	100	127.60
Pipe, steel, galvanized, schedule 40, threaded, 2" diam.	12.000	L.F.	121.80	270	391.80
Pipe, CI soil, no hub, coupling 10' OC, hanger 5' OC, 4" diam.	7.000	L.F.	185.50	185.50	371
TOTAL			1,990.90	2,362.50	4,353.40

D2010 920	Two Fixture Bathroom, Two Wall Plumbing		COST EACH MAT.	COST EACH INST.	COST EACH TOTAL
1180	Bathroom, lavatory & water closet, 2 wall plumbing, stand alone	R221113 -40	2,000	2,375	4,375
1200	Share common plumbing wall*		1,850	2,050	3,900

D2010 922	Two Fixture Bathroom, One Wall Plumbing		COST EACH MAT.	COST EACH INST.	COST EACH TOTAL
2220	Bathroom, lavatory & water closet, one wall plumbing, stand alone	R221113 -40	1,900	2,125	4,025
2240	Share common plumbing wall*		1,650	1,800	3,450
2260		R224000 -30			
2280					

D2010 Plumbing Fixtures

Three Fixture Bathroom Systems consisting of a lavatory, water closet, bathtub or shower and rough-in service piping.

- Prices for plumbing and fixtures only.

*Common wall is with an adjacent bathroom.

System Components	QUANTITY	UNIT	COST EACH		
			MAT.	INST.	TOTAL
SYSTEM D2010 924 1170					
BATHROOM, LAVATORY, WATER CLOSET & BATHTUB					
ONE WALL PLUMBING, STAND ALONE					
Wtr closet, rough-in, supply, waste and vent	1.000	Set	380	475	855
Wtr closet, 2 pc close cpld vit china flr mntd w/seat supply & stop	1.000	Ea.	211	275	486
Lavatory w/ftngs, wall hung, white, PE on CI, 20" x 18"	1.000	Ea.	335	182	517
Lavatory, rough-in waste & vent	1.000	Set	730	875	1,605
Bathtub, white PE on CI, w/ftgs, mat bottom, recessed, 5' long	1.000	Ea.	1,250	330	1,580
Baths, rough-in waste and vent	1.000	Set	505	705	1,210
TOTAL			3,411	2,842	6,253

D2010 924	Three Fixture Bathroom, One Wall Plumbing		COST EACH		
			MAT.	INST.	TOTAL
1150	Bathroom, three fixture, one wall plumbing	R221113 -40			
1160	Lavatory, water closet & bathtub				
1170	Stand alone	R224000 -30	3,400	2,850	6,250
1180	Share common plumbing wall *		2,725	2,000	4,725

D2010 926	Three Fixture Bathroom, Two Wall Plumbing		COST EACH		
			MAT.	INST.	TOTAL
2130	Bathroom, three fixture, two wall plumbing	R221113 -40			
2140	Lavatory, water closet & bathtub				
2160	Stand alone		3,375	2,800	6,175
2180	Long plumbing wall common *		2,900	2,250	5,150
3610	Lavatory, bathtub & water closet				
3620	Stand alone		3,600	3,200	6,800
3640	Long plumbing wall common *		3,425	2,875	6,300
4660	Water closet, corner bathtub & lavatory				
4680	Stand alone		5,375	2,850	8,225
4700	Long plumbing wall common *		4,875	2,125	7,000
6100	Water closet, stall shower & lavatory				
6120	Stand alone		3,725	3,175	6,900
6140	Long plumbing wall common *		3,550	2,950	6,500
7060	Lavatory, corner stall shower & water closet				
7080	Stand alone		4,575	2,825	7,400
7100	Short plumbing wall common *		3,900	1,900	5,800

D2010 Plumbing Fixtures

Four Fixture Bathroom Systems consisting of a lavatory, water closet, bathtub, shower and rough-in service piping.

- Prices for plumbing and fixtures only.

*Common wall is with an adjacent bathroom.

System Components	QUANTITY	UNIT	COST EACH		
			MAT.	INST.	TOTAL
SYSTEM D2010 928 1160					
BATHROOM, BATHTUB, WATER CLOSET, STALL SHOWER & LAVATORY					
TWO WALL PLUMBING, STAND ALONE					
Wtr closet, 2 pc close cpld vit china flr mntd w/seat supply & stop	1.000	Ea.	211	275	486
Water closet, rough-in waste & vent	1.000	Set	380	475	855
Lavatory w/ftngs, wall hung, white PE on CI, 20" x 18"	1.000	Ea.	335	182	517
Lavatory, rough-in waste & vent	1.000	Set	730	875	1,605
Bathtub, white PE on CI, w/ftgs, mat bottom, recessed, 5' long	1.000	Ea.	1,250	330	1,580
Baths, rough-in waste and vent	1.000	Set	505	705	1,210
Shower stall, bkd enam, molded stone receptor, door & trim 32" sq.	1.000	Ea.	1,275	291	1,566
Shower stall, rough-in supply, waste & vent	1.000	Set	490	710	1,200
TOTAL			5,176	3,843	9,019

D2010 928	Four Fixture Bathroom, Two Wall Plumbing		COST EACH		
			MAT.	INST.	TOTAL
1140	Bathroom, four fixture, two wall plumbing	R221113 -40			
1150	Bathtub, water closet, stall shower & lavatory				
1160	Stand alone	R224000 -30	5,175	3,850	9,025
1180	Long plumbing wall common *		4,025	2,350	6,375
2260	Bathtub, lavatory, corner stall shower & water closet				
2280	Stand alone		5,675	3,075	8,750
2320	Long plumbing wall common *		5,175	2,375	7,550
3620	Bathtub, stall shower, lavatory & water closet				
3640	Stand alone		5,175	3,850	9,025
3660	Long plumbing wall (opp. door) common *		4,675	3,150	7,825

D2010 930	Four Fixture Bathroom, Three Wall Plumbing		COST EACH		
			MAT.	INST.	TOTAL
4680	Bathroom, four fixture, three wall plumbing	R221113 -40			
4700	Bathtub, stall shower, lavatory & water closet				
4720	Stand alone		6,475	4,225	10,700
4760	Long plumbing wall (opposite door) common *		6,350	3,900	10,250

497

For customer support on your Plumbing Costs with RSMeans Data, call 800.448.8182.

D2010 Plumbing Fixtures

Five Fixture Bathroom Systems consisting of two lavatories, a water closet, bathtub, shower and rough-in service piping.

• Prices for plumbing and fixtures only.

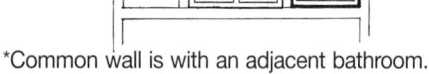

*Common wall is with an adjacent bathroom.

System Components	QUANTITY	UNIT	COST EACH		
			MAT.	INST.	TOTAL
SYSTEM D2010 932 1360					
BATHROOM, BATHTUB, WATER CLOSET, STALL SHOWER & TWO LAVATORIES					
TWO WALL PLUMBING, STAND ALONE					
Wtr closet, 2 pc close cpld vit china flr mntd incl seat, supply & stop	1.000	Ea.	211	275	486
Water closet, rough-in waste & vent	1.000	Set	380	475	855
Lavatory w/ftngs, wall hung, white PE on CI, 20" x 18"	2.000	Ea.	670	364	1,034
Lavatory, rough-in waste & vent	2.000	Set	1,460	1,750	3,210
Bathtub, white PE on CI, w/ftgs, mat bottom, recessed, 5' long	1.000	Ea.	1,250	330	1,580
Baths, rough-in waste and vent	1.000	Set	505	705	1,210
Shower stall, bkd enam molded stone receptor, door & ftng, 32" sq.	1.000	Ea.	1,275	291	1,566
Shower stall, rough-in supply, waste & vent	1.000	Set	490	710	1,200
TOTAL			6,241	4,900	11,141

D2010 932	Five Fixture Bathroom, Two Wall Plumbing		COST EACH		
			MAT.	INST.	TOTAL
1320	Bathroom, five fixture, two wall plumbing				
1340	Bathtub, water closet, stall shower & two lavatories	R221113 -40			
1360	Stand alone	R224000 -30	6,250	4,900	11,150
1400	One short plumbing wall common *		5,725	4,200	9,925
1500	Bathtub, two lavatories, corner stall shower & water closet				
1520	Stand alone		7,400	4,925	12,325
1540	Long plumbing wall common*		6,475	3,825	10,300

D2010 934	Five Fixture Bathroom, Three Wall Plumbing		COST EACH		
			MAT.	INST.	TOTAL
2360	Bathroom, five fixture, three wall plumbing				
2380	Water closet, bathtub, two lavatories & stall shower	R221113 -40			
2400	Stand alone		7,400	4,925	12,325
2440	One short plumbing wall common *		6,875	4,200	11,075

D2010 936	Five Fixture Bathroom, One Wall Plumbing		COST EACH		
			MAT.	INST.	TOTAL
4080	Bathroom, five fixture, one wall plumbing				
4100	Bathtub, two lavatories, corner stall shower & water closet	R221113 -40			
4120	Stand alone		6,950	4,400	11,350
4160	Share common wall *		5,675	2,850	8,525

D2010 Plumbing Fixtures

Example of Plumbing Cost Calculations: The bathroom system includes the individual fixtures such as bathtub, lavatory, shower and water closet. These fixtures are listed below as separate items merely as a checklist.

D2010 951		Plumbing Systems 20 Unit, 2 Story Apartment Building						
	FIXTURE	SYSTEM	LINE	QUANTITY	UNIT	COST EACH		
						MAT.	INST.	TOTAL
0440	Bathroom	D2010 926	3640	20	Ea.	68,500	57,500	126,000
0480	Bathtub							
0520	Booster pump[1]	not req'd.						
0560	Drinking fountain							
0600	Garbage disposal[1]	not incl.						
0660								
0680	Grease interceptor							
0720	Water heater	D2020 250	2140	1	Ea.	17,100	3,925	21,025
0760	Kitchen sink	D2010 410	1960	20	Ea.	30,200	20,400	50,600
0800	Laundry sink	D2010 420	1840	4	Ea.	5,125	4,025	9,150
0840	Lavatory							
0900								
0920	Roof drain, 1 floor	D2040 210	4200	2	Ea.	2,725	2,100	4,825
0960	Roof drain, add'l floor	D2040 210	4240	20	L.F.	530	530	1,060
1000	Service sink	D2010 440	4300	1	Ea.	3,450	1,400	4,850
1040	Sewage ejector[1]	not req'd.						
1080	Shower							
1100								
1160	Sump pump							
1200	Urinal							
1240	Water closet							
1320								
1360	**SUB TOTAL**					127,500	90,000	217,500
1481	Water controls	R221113-40		10%[2]		12,800	9,000	21,800
1521	Pipe & fittings[3]	R221113-40		30%[2]		38,300	27,000	65,300
1560	Other							
1601	Quality/complexity	R221113-40		15%[2]		19,200	13,500	32,700
1680								
1720	**TOTAL**					198,000	140,000	338,000
1741								

[1]**Note:** Cost for items such as booster pumps, backflow preventers, sewage ejectors, water meters, etc., may be obtained from the unit price section in the front of this data set.

Water controls, pipe and fittings, and the Quality/Complexity factors come from Table R221113-40.

[2]Percentage of subtotal.

[3]Long, easily discernable runs of pipe would be more accurately priced from Unit Price Section 22 11 13. If this is done, reduce the miscellaneous percentage in proportion.

D2020 Domestic Water Distribution

Installation includes piping and fittings within 10' of heater. Electric water heaters do not require venting.

1 Kilowatt hour will raise:			
Gallons of Water	Degrees F	Gallons of Water	Degrees F
4.1	100°	6.8	60°
4.5	90°	8.2	50°
5.1	80°	10.0	40°
5.9	70°		

System Components	QUANTITY	UNIT	COST EACH		
			MAT.	INST.	TOTAL
SYSTEM D2020 210 1780					
ELECTRIC WATER HEATER, RESIDENTIAL, 100°F RISE					
10 GALLON TANK, 7 GPH					
Water heater, residential electric, glass lined tank, 10 gal.	1.000	Ea.	510	350	860
Copper tubing, type L, solder joint, hanger 10' OC 1/2" diam.	30.000	L.F.	82.80	300	382.80
Wrought copper 90° elbow for solder joints 1/2" diam.	4.000	Ea.	7.80	162	169.80
Wrought copper Tee for solder joints, 1/2" diam.	2.000	Ea.	6.52	124	130.52
Union, wrought copper, 1/2" diam.	2.000	Ea.	77	85	162
Valve, gate, bronze, 125 lb, NRS, soldered 1/2" diam.	2.000	Ea.	168	67	235
Relief valve, bronze, press & temp, self-close, 3/4" IPS	1.000	Ea.	293	29	322
Wrought copper adapter, CTS to MPT 3/4" IPS	1.000	Ea.	9.80	47.50	57.30
Copper tubing, type L, solder joints, 3/4" diam.	1.000	L.F.	3.35	10.65	14
Wrought copper 90° elbow for solder joints 3/4" diam.	1.000	Ea.	4.28	42.50	46.78
TOTAL			1,162.55	1,217.65	2,380.20

D2020 210	Electric Water Heaters - Residential Systems		COST EACH		
			MAT.	INST.	TOTAL
1760	Electric water heater, residential, 100°F rise				
1780	10 gallon tank, 7 GPH	R224000 -10	1,175	1,225	2,400
1820	20 gallon tank, 7 GPH	R224000 -20	1,475	1,300	2,775
1860	30 gallon tank, 7 GPH		2,275	1,375	3,650
1900	40 gallon tank, 8 GPH		2,650	1,500	4,150
1940	52 gallon tank, 10 GPH		2,875	1,525	4,400
1980	66 gallon tank, 13 GPH		4,575	1,725	6,300
2020	80 gallon tank, 16 GPH		4,900	1,800	6,700
2060	120 gallon tank, 23 GPH		6,825	2,100	8,925

D2020 Domestic Water Distribution

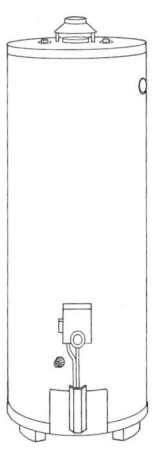

Installation includes piping and fittings within 10' of heater. Gas heaters require vent piping (not included with these units).

System Components	QUANTITY	UNIT	COST EACH MAT.	COST EACH INST.	COST EACH TOTAL
SYSTEM D2020 220 2260					
GAS FIRED WATER HEATER, RESIDENTIAL, 100°F RISE					
30 GALLON TANK, 32 GPH					
Water heater, residential, gas, glass lined tank, 30 gallon	1.000	Ea.	2,425	405	2,830
Copper tubing, type L, solder joint, hanger 10' OC, 3/4" diam	33.000	L.F.	110.55	351.45	462
Wrought copper 90° elbow for solder joints, 3/4" diam	5.000	Ea.	21.40	212.50	233.90
Wrought copper Tee for solder joints, 3/4" diam	2.000	Ea.	15.70	135	150.70
Wrought copper union for soldered joints, 3/4" diam.	2.000	Ea.	96	90	186
Valve bronze, 125 lb., NRS, soldered 3/4" diam	2.000	Ea.	186	81	267
Relief valve, press & temp, bronze, self-close, 3/4" diam	1.000	Ea.	293	29	322
Wrought copper, adapter, CTS to MPT 3/4" IPS	1.000	Ea.	9.80	47.50	57.30
Pipe steel black, schedule 40, threaded, 1/2" diam	10.000	L.F.	29.60	128.50	158.10
Pipe, 90° elbow, malleable iron black, 150 lb., threaded, 1/2" diam	2.000	Ea.	9.50	108	117.50
Pipe, union with brass seat, malleable iron black, 1/2" diam	1.000	Ea.	22	57.50	79.50
Valve, gas stop w/o check, brass, 1/2" IPS	1.000	Ea.	23	33.50	56.50
TOTAL			3,241.55	1,678.95	4,920.50

D2020 220	Gas Fired Water Heaters - Residential Systems		COST EACH MAT.	COST EACH INST.	COST EACH TOTAL
2200	Gas fired water heater, residential, 100°F rise				
2260	30 gallon tank, 32 GPH	R224000 -10	3,250	1,675	4,925
2300	40 gallon tank, 32 GPH	R224000 -20	3,050	1,875	4,925
2340	50 gallon tank, 63 GPH		3,200	1,875	5,075
2380	75 gallon tank, 63 GPH		4,400	2,100	6,500
2420	100 gallon tank, 63 GPH		5,375	2,200	7,575

D2020 Domestic Water Distribution

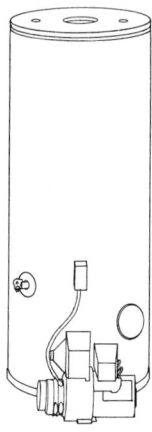

Installation includes piping and fittings within 10' of heater. Oil fired heaters require vent piping (not included in these prices).

System Components	QUANTITY	UNIT	COST EACH		
			MAT.	INST.	TOTAL
SYSTEM D2020 230 2220					
OIL FIRED WATER HEATER, RESIDENTIAL, 100°F RISE					
30 GALLON TANK, 103 GPH					
Water heater, residential, oil glass lined tank, 30 Gal	1.000	Ea.	1,850	405	2,255
Copper tubing, type L, solder joint, hanger 10' O.C. 3/4" diam.	33.000	L.F.	110.55	351.45	462
Wrought copper 90° elbow for solder joints 3/4" diam.	5.000	Ea.	21.40	212.50	233.90
Wrought copper Tee for solder joints, 3/4" diam.	2.000	Ea.	15.70	135	150.70
Wrought copper union for soldered joints, 3/4" diam.	2.000	Ea.	96	90	186
Valve, gate, bronze, 125 lb, NRS, soldered 3/4" diam.	2.000	Ea.	186	81	267
Relief valve, bronze, press & temp, self-close, 3/4" IPS	1.000	Ea.	293	29	322
Wrought copper adapter, CTS to MPT, 3/4" IPS	1.000	Ea.	9.80	47.50	57.30
Copper tubing, type L, solder joint, hanger 10' OC 3/8" diam.	10.000	L.F.	23.40	96	119.40
Wrought copper 90° elbow for solder joints 3/8" diam.	2.000	Ea.	10.80	74	84.80
Valve, globe, fusible, 3/8" diam.	1.000	Ea.	31.50	34	65.50
TOTAL			2,648.15	1,555.45	4,203.60

D2020 230	Oil Fired Water Heaters - Residential Systems		COST EACH		
			MAT.	INST.	TOTAL
2200	Oil fired water heater, residential, 100°F rise				
2220	30 gallon tank, 103 GPH		2,650	1,550	4,200
2260	50 gallon tank, 145 GPH		2,675	1,750	4,425
2300	70 gallon tank, 164 GPH	R224000 -20	4,100	1,975	6,075
2340	85 gallon tank, 181 GPH		11,200	2,000	13,200

Systems below include piping and fittings within 10' of heater. Electric water heaters do not require venting.

System Components	QUANTITY	UNIT	COST EACH		
			MAT.	INST.	TOTAL
SYSTEM D2020 240 1820					
ELECTRIC WATER HEATER, COMMERCIAL, 100°F RISE					
50 GALLON TANK, 9 KW, 37 GPH					
Water heater, commercial, electric, 50 Gal, 9 KW, 37 GPH	1.000	Ea.	6,850	450	7,300
Copper tubing, type L, solder joint, hanger 10' OC, 3/4" diam	34.000	L.F.	113.90	362.10	476
Wrought copper 90° elbow for solder joints 3/4" diam	5.000	Ea.	21.40	212.50	233.90
Wrought copper Tee for solder joints, 3/4" diam	2.000	Ea.	15.70	135	150.70
Wrought copper union for soldered joints, 3/4" diam.	2.000	Ea.	96	90	186
Valve, gate, bronze, 125 lb, NRS, soldered 3/4" diam	2.000	Ea.	186	81	267
Relief valve, bronze, press & temp, self-close, 3/4" IPS	1.000	Ea.	293	29	322
Wrought copper adapter, copper tubing to male, 3/4" IPS	1.000	Ea.	9.80	47.50	57.30
TOTAL			7,585.80	1,407.10	8,992.90

D2020 240	Electric Water Heaters - Commercial Systems		COST EACH		
			MAT.	INST.	TOTAL
1800	Electric water heater, commercial, 100°F rise				
1820	50 gallon tank, 9 KW 37 GPH		7,575	1,400	8,975
1860	80 gal, 12 KW 49 GPH	R224000 -10	9,950	1,750	11,700
1900	36 KW 147 GPH		13,600	1,875	15,475
1940	120 gal, 36 KW 147 GPH	R224000 -20	15,000	2,025	17,025
1980	150 gal, 120 KW 490 GPH		53,000	2,175	55,175
2020	200 gal, 120 KW 490 GPH		55,500	2,225	57,725
2060	250 gal, 150 KW 615 GPH		53,500	2,600	56,100
2100	300 gal, 180 KW 738 GPH		81,500	2,750	84,250
2140	350 gal, 30 KW 123 GPH		43,000	2,950	45,950
2180	180 KW 738 GPH		59,500	2,950	62,450
2220	500 gal, 30 KW 123 GPH		67,500	3,475	70,975
2260	240 KW 984 GPH		110,500	3,475	113,975
2300	700 gal, 30 KW 123 GPH		67,500	3,975	71,475
2340	300 KW 1230 GPH		100,000	3,975	103,975
2380	1000 gal, 60 KW 245 GPH		82,000	5,525	87,525
2420	480 KW 1970 GPH		132,500	5,550	138,050
2460	1500 gal, 60 KW 245 GPH		119,500	6,850	126,350
2500	480 KW 1970 GPH		163,500	6,850	170,350

503

D2020 Domestic Water Distribution

Units may be installed in multiples for increased capacity.

Included below is the heater with self-energizing gas controls, safety pilots, insulated jacket, hi-limit aquastat and pressure relief valve.

Installation includes piping and fittings within 10' of heater. Gas heaters require vent piping (not included in these prices).

System Components	QUANTITY	UNIT	COST EACH		
			MAT.	INST.	TOTAL
SYSTEM D2020 250 1780					
GAS FIRED WATER HEATER, COMMERCIAL, 100°F RISE					
75.5 MBH INPUT, 63 GPH					
Water heater, commercial, gas, 75.5 MBH, 63 GPH	1.000	Ea.	4,075	575	4,650
Copper tubing, type L, solder joint, hanger 10' OC, 1-1/4" diam	30.000	L.F.	354	418.50	772.50
Wrought copper 90° elbow for solder joints 1-1/4" diam	4.000	Ea.	69.60	216	285.60
Wrought copper tee for solder joints, 1-1/4" diam	2.000	Ea.	65	180	245
Wrought copper union for soldered joints, 1-1/4" diam	2.000	Ea.	246	115	361
Valve, gate, bronze, 125 lb, NRS, soldered 1-1/4" diam	2.000	Ea.	410	108	518
Relief valve, bronze, press & temp, self-close, 3/4" IPS	1.000	Ea.	293	29	322
Copper tubing, type L, solder joints, 3/4" diam	8.000	L.F.	26.80	85.20	112
Wrought copper 90° elbow for solder joints 3/4" diam	1.000	Ea.	4.28	42.50	46.78
Wrought copper, adapter, CTS to MPT, 3/4" IPS	1.000	Ea.	9.80	47.50	57.30
Pipe steel black, schedule 40, threaded, 3/4" diam	10.000	L.F.	36	132.50	168.50
Pipe, 90° elbow, malleable iron black, 150 lb threaded, 3/4" diam	2.000	Ea.	11.50	115	126.50
Pipe, union with brass seat, malleable iron black, 3/4" diam	1.000	Ea.	20.50	62	82.50
Valve, gas stop w/o check, brass, 3/4" IPS	1.000	Ea.	27	37	64
TOTAL			5,648.48	2,163.20	7,811.68

D2020 250	Gas Fired Water Heaters - Commercial Systems		COST EACH		
			MAT.	INST.	TOTAL
1760	Gas fired water heater, commercial, 100°F rise				
1780	75.5 MBH input, 63 GPH		5,650	2,175	7,825
1820	95 MBH input, 86 GPH	R224000 -10	11,500	2,175	13,675
1860	100 MBH input, 91 GPH		11,800	2,275	14,075
1900	115 MBH input, 110 GPH		11,200	2,325	13,525
1980	155 MBH input, 150 GPH		14,900	2,600	17,500
2020	175 MBH input, 168 GPH	R224000 -20	14,700	2,800	17,500
2060	200 MBH input, 192 GPH		14,800	3,175	17,975
2100	240 MBH input, 230 GPH		15,900	3,450	19,350
2140	300 MBH input, 278 GPH		17,100	3,925	21,025
2180	390 MBH input, 374 GPH		20,100	3,975	24,075
2220	500 MBH input, 480 GPH		27,100	4,275	31,375
2260	600 MBH input, 576 GPH		34,600	4,625	39,225

504

For customer support on your Plumbing Costs with RSMeans Data, call 800.448.8182.

D2020   Domestic Water Distribution

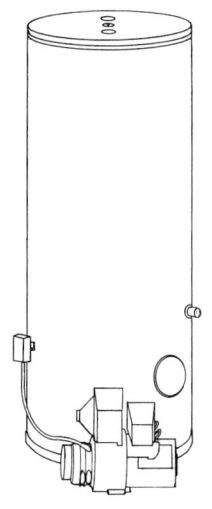

Units may be installed in multiples for increased capacity.

Included below is the heater, wired-in flame retention burners, cadmium cell primary controls, hi-limit controls, ASME pressure relief valves, draft controls, and insulated jacket.

Oil fired water heater systems include piping and fittings within 10′ of heater. Oil fired heaters require vent piping (not included in these systems).

System Components	QUANTITY	UNIT	COST EACH		
			MAT.	INST.	TOTAL
SYSTEM D2020 260 1820					
OIL FIRED WATER HEATER, COMMERCIAL, 100°F RISE					
140 GAL., 140 MBH INPUT, 134 GPH					
Water heater, commercial, oil, 140 gal., 140 MBH input, 134 GPH	1.000	Ea.	27,400	685	28,085
Copper tubing, type L, solder joint, hanger 10′ OC, 3/4″ diam.	34.000	L.F.	113.90	362.10	476
Wrought copper 90° elbow for solder joints 3/4″ diam.	5.000	Ea.	21.40	212.50	233.90
Wrought copper Tee for solder joints, 3/4″ diam.	2.000	Ea.	15.70	135	150.70
Wrought copper union for soldered joints, 3/4″ diam.	2.000	Ea.	96	90	186
Valve, bronze, 125 lb, NRS, soldered 3/4″ diam.	2.000	Ea.	186	81	267
Relief valve, bronze, press & temp, self-close, 3/4″ IPS	1.000	Ea.	293	29	322
Wrought copper adapter, copper tubing to male, 3/4″ IPS	1.000	Ea.	9.80	47.50	57.30
Copper tubing, type L, solder joint, hanger 10′ OC, 3/8″ diam.	10.000	L.F.	23.40	96	119.40
Wrought copper 90° elbow for solder joints 3/8″ diam.	2.000	Ea.	10.80	74	84.80
Valve, globe, fusible, 3/8″ IPS	1.000	Ea.	31.50	34	65.50
TOTAL			28,201.50	1,846.10	30,047.60

D2020 260	Oil Fired Water Heaters - Commercial Systems		COST EACH		
			MAT.	INST.	TOTAL
1800	Oil fired water heater, commercial, 100°F rise				
1820	140 gal., 140 MBH input, 134 GPH		28,200	1,850	30,050
1900	140 gal., 255 MBH input, 247 GPH	R224000 -10	30,600	2,325	32,925
1940	140 gal., 270 MBH input, 259 GPH		37,600	2,625	40,225
1980	140 gal., 400 MBH input, 384 GPH	R224000 -20	38,800	3,050	41,850
2060	140 gal., 720 MBH input, 691 GPH		41,100	3,175	44,275
2100	221 gal., 300 MBH input, 288 GPH		54,500	3,475	57,975
2140	221 gal., 600 MBH input, 576 GPH		70,000	3,525	73,525
2180	221 gal., 800 MBH input, 768 GPH		61,500	3,650	65,150
2220	201 gal., 1000 MBH input, 960 GPH		63,000	3,675	66,675
2260	201 gal., 1250 MBH input, 1200 GPH		64,500	3,775	68,275
2300	201 gal., 1500 MBH input, 1441 GPH		69,500	3,875	73,375
2340	411 gal., 600 MBH input, 576 GPH		70,500	3,975	74,475
2380	411 gal., 800 MBH input, 768 GPH		73,500	4,075	77,575
2420	411 gal., 1000 MBH input, 960 GPH		76,000	4,700	80,700
2460	411 gal., 1250 MBH input, 1200 GPH		78,000	4,825	82,825
2500	397 gal., 1500 MBH input, 1441 GPH		82,000	4,975	86,975

In this closed-loop indirect collection system, fluid with a low freezing temperature, such as propylene glycol, transports heat from the collectors to water storage. The transfer fluid is contained in a closed-loop consisting of collectors, supply and return piping, and a remote heat exchanger. The heat exchanger transfers heat energy from the

fluid in the collector loop to potable water circulated in a storage loop. A typical two-or-three panel system contains 5 to 6 gallons of heat transfer fluid.

When the collectors become approximately 20°F warmer than the storage temperature, a controller activates the circulator on the collector and storage

loops. The circulators will move the fluid and potable water through the heat exchanger until heat collection no longer occurs. At that point, the system shuts down. Since the heat transfer medium is a fluid with a very low freezing temperature, there is no need for it to be drained from the system between periods of collection.

D2020 Domestic Water Distribution

System Components	QUANTITY	UNIT	COST EACH		
			MAT.	INST.	TOTAL
SYSTEM D2020 265 2760					
SOLAR, CLOSED LOOP, ADD-ON HOT WATER SYS., EXTERNAL HEAT EXCHANGER					
3/4″ TUBING, TWO 3′X7′ BLACK CHROME COLLECTORS					
A,B,G,L,K,M Heat exchanger fluid-fluid pkg incl 2 circulators, expansion tank,					
Check valve, relief valve, controller, hi temp cutoff, & 2 sensors	1.000	Ea.	860	580	1,440
C Thermometer, 2″ dial	3.000	Ea.	75	151.50	226.50
D, T Fill & drain valve, brass, 3/4″ connection	1.000	Ea.	14.20	33.50	47.70
E Air vent, manual, 1/8″ fitting	2.000	Ea.	7.28	51	58.28
F Air purger	1.000	Ea.	58	67.50	125.50
H Strainer, Y type, bronze body, 3/4″ IPS	1.000	Ea.	62	43	105
I Valve, gate, bronze, NRS, soldered 3/4″ diam	6.000	Ea.	558	243	801
J Neoprene vent flashing	2.000	Ea.	21.50	81	102.50
N-1, N Relief valve temp & press, 150 psi 210°F self-closing 3/4″ IPS	1.000	Ea.	29	27	56
O Pipe covering, urethane, ultraviolet cover, 1″ wall 3/4″ diam	20.000	L.F.	60.60	146	206.60
P Pipe covering, fiberglass, all service jacket, 1″ wall, 3/4″ diam	50.000	L.F.	56.50	292.50	349
Q Collector panel solar energy blk chrome on copper, 1/8″ temp glass 3′x7′	2.000	Ea.	2,300	306	2,606
Roof clamps for solar energy collector panels	2.000	Set	7.86	42	49.86
R Valve, swing check, bronze, regrinding disc, 3/4″ diam	2.000	Ea.	272	81	353
S Pressure gauge, 60 psi, 2″ dial	1.000	Ea.	28	25.50	53.50
U Valve, water tempering, bronze, sweat connections, 3/4″ diam	1.000	Ea.	193	40.50	233.50
W-2, V Tank water storage w/heating element, drain, relief valve, existing	1.000	Ea.			
Copper tubing type L, solder joint, hanger 10′ OC 3/4″ diam	20.000	L.F.	67	213	280
Copper tubing, type M, solder joint, hanger 10′ OC 3/4″ diam	70.000	L.F.	416.50	724.50	1,141
Sensor wire, #22-2 conductor multistranded	.500	C.L.F.	7.83	37.75	45.58
Solar energy heat transfer fluid, propylene glycol anti-freeze	6.000	Gal.	150	174	324
Wrought copper fittings & solder, 3/4″ diam	76.000	Ea.	325.28	3,230	3,555.28
TOTAL			5,569.55	6,590.25	12,159.80

D2020 265	Solar, Closed Loop, Add-On Hot Water Systems		COST EACH		
			MAT.	INST.	TOTAL
2550	Solar, closed loop, add-on hot water system, external heat exchanger				
2570	3/8″ tubing, 3 ea. 4′ x 4′-4″ vacuum tube collectors		6,175	6,200	12,375
2580	1/2″ tubing, 4 ea. 4 x 4′-4″ vacuum tube collectors	R235616 -60	6,950	6,675	13,625
2600	2 ea. 3′x7′ black chrome collectors		5,025	6,300	11,325
2620	3 ea. 3′x7′ black chrome collectors		6,200	6,475	12,675
2640	2 ea. 3′x7′ flat black collectors		4,425	6,300	10,725
2660	3 ea. 3′x7′ flat black collectors		5,250	6,500	11,750
2700	3/4″ tubing, 3 ea. 3′x7′ black chrome collectors		6,725	6,775	13,500
2720	3 ea. 3′x7′ flat black absorber plate collectors		5,800	6,800	12,600
2740	2 ea. 4′x9′ flat black w/plastic glazing collectors		5,325	6,825	12,150
2760	2 ea. 3′x7′ black chrome collectors		5,575	6,600	12,175
2780	1″ tubing, 4 ea 2′x9′ plastic absorber & glazing collectors		6,875	7,650	14,525
2800	4 ea. 3′x7′ black chrome absorber collectors		8,925	7,700	16,625
2820	4 ea. 3′x7′ flat black absorber collectors		7,675	7,725	15,400

507

In the drainback indirect-collection system, the heat transfer fluid is distilled water contained in a loop consisting of collectors, supply and return piping, and an unpressurized holding tank. A large heat exchanger containing incoming potable water is immersed in the holding tank. When a controller activates solar collection, the distilled water is pumped through the collectors and heated and pumped back down to the holding tank. When the temperature differential between the water in the collectors and water in storage is such that collection no longer occurs, the pump turns off and gravity causes the distilled water in the collector loop to drain back to the holding tank. All the loop piping is pitched so that the water can drain out of the collectors and piping and not freeze there. As hot water is needed in the home, incoming water first flows through the holding tank with the immersed heat exchanger and is warmed and then flows through a conventional heater for any supplemental heating that is necessary.

508

For customer support on your Plumbing Costs with RSMeans Data, call 800.448.8182.

D2020 Domestic Water Distribution

System Components	QUANTITY	UNIT	COST EACH		
			MAT.	INST.	TOTAL
SYSTEM D2020 270 2760					
SOLAR, DRAINBACK, ADD ON, HOT WATER, IMMERSED HEAT EXCHANGER					
3/4" TUBING, THREE EA 3'X7' BLACK CHROME COLLECTOR					
A, B Differential controller 2 sensors, thermostat, solar energy system	1.000	Ea.	174	67.50	241.50
C Thermometer 2" dial	3.000	Ea.	75	151.50	226.50
D, T Fill & drain valve, brass, 3/4" connection	1.000	Ea.	14.20	33.50	47.70
E-1 Automatic air vent 1/8" fitting	1.000	Ea.	17.55	25.50	43.05
H Strainer, Y type, bronze body, 3/4" IPS	1.000	Ea.	62	43	105
I Valve, gate, bronze, NRS, soldered 3/4" diam	2.000	Ea.	186	81	267
J Neoprene vent flashing	2.000	Ea.	21.50	81	102.50
L Circulator, solar heated liquid, 1/20 HP	1.000	Ea.	340	121	461
N Relief valve temp. & press. 150 psi 210°F self-closing 3/4" IPS	1.000	Ea.	29	27	56
O Pipe covering, urethane, ultraviolet cover, 1" wall, 3/4" diam	20.000	L.F.	60.60	146	206.60
P Pipe covering, fiberglass, all service jacket, 1" wall, 3/4" diam	50.000	L.F.	56.50	292.50	349
Q Collector panel solar energy blk chrome on copper, 1/8" temp glas 3'x7'	3.000	Ea.	3,450	459	3,909
Roof clamps for solar energy collector panels	3.000	Set	11.79	63	74.79
R Valve, swing check, bronze, regrinding disc, 3/4" diam	1.000	Ea.	136	40.50	176.50
U Valve, water tempering, bronze sweat connections, 3/4" diam	1.000	Ea.	193	40.50	233.50
V Tank, water storage w/heating element, drain, relief valve, existing	1.000	Ea.			
W Tank, water storage immersed heat exchr elec 2"x1/2# insul 120 gal	1.000	Ea.	1,925	575	2,500
X Valve, globe, bronze, rising stem, 3/4" diam, soldered	3.000	Ea.	570	121.50	691.50
Y Flow control valve	1.000	Ea.	157	37	194
Z Valve, ball, bronze, solder 3/4" diam, solar loop flow control	1.000	Ea.	33	40.50	73.50
Copper tubing, type L, solder joint, hanger 10' OC 3/4" diam	20.000	L.F.	67	213	280
Copper tubing, type M, solder joint, hanger 10' OC 3/4" diam	70.000	L.F.	416.50	724.50	1,141
Sensor wire, #22-2 conductor, multistranded	.500	C.L.F.	7.83	37.75	45.58
Wrought copper fittings & solder, 3/4" diam	76.000	Ea.	325.28	3,230	3,555.28
TOTAL			8,328.75	6,651.75	14,980.50

D2020 270	Solar, Drainback, Hot Water Systems		COST EACH		
			MAT.	INST.	TOTAL
2550	Solar, drainback, hot water, immersed heat exchanger				
2560	3/8" tubing, 3 ea. 4' x 4'-4" vacuum tube collectors		7,750	6,025	13,775
2580	1/2" tubing, 4 ea 4' x 4'-4" vacuum tube collectors, 80 gal tank	R235616 -60	8,525	6,500	15,025
2600	120 gal tank		8,475	6,575	15,050
2640	2 ea. 3'x7' blk chrome collectors, 80 gal tank		6,600	6,125	12,725
2660	3 ea. 3'x7' blk chrome collectors, 120 gal tank		7,700	6,375	14,075
2700	2 ea. 3'x7' flat blk collectors, 120 gal tank		5,925	6,200	12,125
2720	3 ea. 3'x7' flat blk collectors, 120 gal tank		6,775	6,400	13,175
2760	3/4" tubing, 3 ea 3'x7' black chrome collectors, 120 gal tank		8,325	6,650	14,975
2780	3 ea. 3'x7' flat black absorber collectors, 120 gal tank		7,400	6,675	14,075
2800	2 ea. 4'x9' flat blk w/plastic glazing collectors 120 gal tank		6,925	6,700	13,625
2840	1" tubing, 4 ea. 2'x9' plastic absorber & glazing collectors, 120 gal tank		8,600	7,550	16,150
2860	4 ea. 3'x7' black chrome absorber collectors, 120 gal tank		10,600	7,575	18,175
2880	4 ea. 3'x7' flat black absorber collectors, 120 gal tank		9,400	7,625	17,025

D2020 Domestic Water Distribution

In the draindown direct-collection system, incoming domestic water is heated in the collectors. When the controller activates solar collection, domestic water is first heated as it flows through the collectors and is then pumped to storage. When conditions are no longer suitable for heat collection, the pump shuts off and the water in the loop drains down and out of the system by means of solenoid valves and properly pitched piping.

D2020 Domestic Water Distribution

System Components	QUANTITY	UNIT	COST EACH		
			MAT.	INST.	TOTAL
SYSTEM D2020 275 2760					
SOLAR, DRAINDOWN, HOT WATER, DIRECT COLLECTION					
3/4″ TUBING, THREE 3′X7′ BLACK CHROME COLLECTORS					
A, B Differential controller, 2 sensors, thermostat, solar energy system	1.000	Ea.	174	67.50	241.50
A-1 Solenoid valve, solar heating loop, brass, 3/4″ diam, 24 volts	3.000	Ea.	531	270	801
B-1 Solar energy sensor, freeze prevention	1.000	Ea.	28.50	25.50	54
C Thermometer, 2″ dial	3.000	Ea.	75	151.50	226.50
E-1 Vacuum relief valve, 3/4″ diam	1.000	Ea.	34	25.50	59.50
F-1 Air vent, automatic, 1/8″ fitting	1.000	Ea.	17.55	25.50	43.05
H Strainer, Y type, bronze body, 3/4″ IPS	1.000	Ea.	62	43	105
I Valve, gate, bronze, NRS, soldered, 3/4″ diam	2.000	Ea.	186	81	267
J Vent flashing neoprene	2.000	Ea.	21.50	81	102.50
K Circulator, solar heated liquid, 1/25 HP	1.000	Ea.	229	104	333
N Relief valve temp & press 150 psi 210°F self-closing 3/4″ IPS	1.000	Ea.	29	27	56
O Pipe covering, urethane, ultraviolet cover, 1″ wall, 3/4″ diam	20.000	L.F.	60.60	146	206.60
P Pipe covering, fiberglass, all service jacket, 1″ wall, 3/4″ diam	50.000	L.F.	56.50	292.50	349
Roof clamps for solar energy collector panels	3.000	Set	11.79	63	74.79
Q Collector panel solar energy blk chrome on copper, 1/8″ temp glass 3′x7′	3.000	Ea.	3,450	459	3,909
R Valve, swing check, bronze, regrinding disc, 3/4″ diam, soldered	2.000	Ea.	272	81	353
T Drain valve, brass, 3/4″ connection	2.000	Ea.	28.40	67	95.40
U Valve, water tempering, bronze, sweat connections, 3/4″ diam	1.000	Ea.	193	40.50	233.50
W-2, W Tank, water storage elec elem 2″x1/2# insul 120 gal	1.000	Ea.	1,925	575	2,500
X Valve, globe, bronze, rising stem, 3/4″ diam, soldered	1.000	Ea.	190	40.50	230.50
Copper tubing, type L, solder joints, hangers 10′ OC 3/4″ diam	20.000	L.F.	67	213	280
Copper tubing, type M, solder joints, hangers 10′ OC 3/4″ diam	70.000	L.F.	416.50	724.50	1,141
Sensor wire, #22-2 conductor, multistranded	.500	C.L.F.	7.83	37.75	45.58
Wrought copper fittings & solder, 3/4″ diam	76.000	Ea.	325.28	3,230	3,555.28
TOTAL			8,391.45	6,871.25	15,262.70

D2020 275	Solar, Draindown, Hot Water Systems	COST EACH		
		MAT.	INST.	TOTAL
2550	Solar, draindown, hot water			
2560	3/8″ tubing, 3 ea. 4′ x 4′-4″ vacuum tube collectors, 80 gal tank	7,600	6,550	14,150
2580	1/2″ tubing, 4 ea. 4′ x 4′-4″ vacuum tube collectors, 80 gal tank	8,650	6,725	15,375
2600	120 gal tank	8,600	6,800	15,400
2640	2 ea. 3′x7′ black chrome collectors, 80 gal tank	6,750	6,350	13,100
2660	3 ea. 3′x7′ black chrome collectors, 120 gal tank	7,850	6,600	14,450
2700	2 ea. 3′x7′ flat black collectors, 120 gal tank	6,075	6,450	12,525
2720	3 ea. 3′x7′ flat black collectors, 120 gal tank	6,825	6,600	13,425
2760	3/4″ tubing, 3 ea. 3′x7′ black chrome collectors, 120 gal tank	8,400	6,875	15,275
2780	3 ea. 3′x7′ flat collectors, 120 gal tank	7,450	6,900	14,350
2800	2 ea. 4′x9′ flat black & plastic glazing collectors, 120 gal tank	7,000	6,925	13,925
2840	1″ tubing, 4 ea. 2′x9′ plastic absorber & glazing collectors, 120 gal tank	12,400	9,125	21,525
2860	4 ea. 3′x7′ black chrome absorber collectors, 120 gal tank	12,000	8,625	20,625
2880	4 ea. 3′x7′ flat black absorber collectors, 120 gal tank	9,600	7,825	17,425

Note in row 2580: R235616 -60

In the recirculation system, (a direct-collection system), incoming domestic water is heated in the collectors. When the controller activates solar collection, domestic water is heated as it flows through the collectors and then it flows back to storage. When conditions are not suitable for heat collection, the pump shuts off and the flow of the water stops. In this type of system, water remains in the collector loop at all times. A "frost sensor" at the collector activates the circulation of warm water from storage through the collectors when protection from freezing is required.

D2020 Domestic Water Distribution

System Components	QUANTITY	UNIT	COST EACH		
			MAT.	INST.	TOTAL
SYSTEM D2020 280 2820					
SOLAR, RECIRCULATION, HOT WATER					
3/4″ TUBING, TWO 3′X7′ BLACK CHROME COLLECTORS					
A, B Differential controller 2 sensors, thermostat, for solar energy system	1.000	Ea.	174	67.50	241.50
A-1 Solenoid valve, solar heating loop, brass, 3/4″ IPS, 24 volts	2.000	Ea.	354	180	534
B-1 Solar energy sensor freeze prevention	1.000	Ea.	28.50	25.50	54
C Thermometer, 2″ dial	3.000	Ea.	75	151.50	226.50
D Drain valve, brass, 3/4″ connection	1.000	Ea.	14.20	33.50	47.70
F-1 Air vent, automatic, 1/8″ fitting	1.000	Ea.	17.55	25.50	43.05
H Strainer, Y type, bronze body, 3/4″ IPS	1.000	Ea.	62	43	105
I Valve, gate, bronze, 125 lb, soldered 3/4″ diam	3.000	Ea.	279	121.50	400.50
J Vent flashing, neoprene	2.000	Ea.	21.50	81	102.50
L Circulator, solar heated liquid, 1/20 HP	1.000	Ea.	340	121	461
N Relief valve, temp & press 150 psi 210° F self-closing 3/4″ IPS	2.000	Ea.	58	54	112
O Pipe covering, urethane, ultraviolet cover, 1″ wall, 3/4″ diam	20.000	L.F.	60.60	146	206.60
P Pipe covering, fiberglass, all service jacket, 1″ wall 3/4″ diam	50.000	L.F.	56.50	292.50	349
Q Collector panel solar energy blk chrome on copper, 1/8″ temp glass 3′x7′	2.000	Ea.	2,300	306	2,606
Roof clamps for solar energy collector panels	2.000	Set	7.86	42	49.86
R Valve, swing check, bronze, 125 lb, regrinding disc, soldered 3/4″ diam	2.000	Ea.	272	81	353
U Valve, water tempering, bronze, sweat connections, 3/4″ diam	1.000	Ea.	193	40.50	233.50
V Tank, water storage, w/heating element, drain, relief valve, existing	1.000	Ea.			
X Valve, globe, bronze, 125 lb, 3/4″ diam	2.000	Ea.	380	81	461
Copper tubing, type L, solder joints, hangers 10′ OC 3/4″ diam	20.000	L.F.	67	213	280
Copper tubing, type M, solder joints, hangers 10′ OC 3/4″ diam	70.000	L.F.	416.50	724.50	1,141
Wrought copper fittings & solder, 3/4″ diam	76.000	Ea.	325.28	3,230	3,555.28
Sensor wire, #22-2 conductor, multistranded	.500	C.L.F.	7.83	37.75	45.58
TOTAL			5,510.32	6,098.25	11,608.57

D2020 280	Solar, Recirculation, Domestic Hot Water Systems		COST EACH		
			MAT.	INST.	TOTAL
2550	Solar, recirculation, hot water				
2560	3/8″ tubing, 3 ea. 4′ x 4′-4″ vacuum tube collectors		6,050	5,700	11,750
2580	1/2″ tubing, 4 ea. 4′ x 4′-4″ vacuum tube collectors		6,825	6,200	13,025
2640	2 ea. 3′x7′ black chrome collectors	R235616 -60	4,900	5,800	10,700
2660	3 ea. 3′x7′ black chrome collectors		6,050	5,975	12,025
2700	2 ea. 3′x7′ flat black collectors		4,275	5,825	10,100
2720	3 ea. 3′x7′ flat black collectors		5,125	6,000	11,125
2760	3/4″ tubing, 3 ea. 3′x7′ black chrome collectors		6,675	6,275	12,950
2780	3 ea. 3′x7′ flat black absorber plate collectors		5,725	6,300	12,025
2800	2 ea. 4′x9′ flat black w/plastic glazing collectors		5,250	6,325	11,575
2820	2 ea. 3′x7′ black chrome collectors		5,500	6,100	11,600
2840	1″ tubing, 4 ea. 2′x9′ black plastic absorber & glazing collectors		7,125	7,175	14,300
2860	4 ea. 3′x7′ black chrome absorber collectors		9,175	7,200	16,375
2880	4 ea. 3′x7′ flat black absorber collectors		7,925	7,225	15,150

The thermosyphon domestic hot water system, a direct collection system, operates under city water pressure and does not require pumps for system operation. An insulated water storage tank is located above the collectors. As the sun heats the collectors, warm water in them rises by means of natural convection; the colder water in the storage tank flows into the collectors by means of gravity. As long as the sun is shining the water continues to flow through the collectors and to become warmer.

To prevent freezing, the system must be drained or the collectors covered with an insulated lid when the temperature drops below 32°F.

D2020 Domestic Water Distribution

System Components	QUANTITY	UNIT	COST EACH		
			MAT.	INST.	TOTAL
SYSTEM D2020 285 0960					
SOLAR, THERMOSYPHON, WATER HEATER					
3/4" TUBING, TWO 3'X7' BLACK CHROME COLLECTORS					
D-1 Framing lumber, fir, 2" x 6" x 8', tank cradle	.008	M.B.F.	9.40	8.40	17.80
F-1 Framing lumber, fir, 2" x 4" x 24', sleepers	.016	M.B.F.	18.40	25.20	43.60
I Valve, gate, bronze, 125 lb, soldered 1/2" diam	2.000	Ea.	168	67	235
J Vent flashing, neoprene	4.000	Ea.	43	162	205
O Pipe covering, urethane, ultraviolet cover, 1" wall, 1/2"diam	40.000	L.F.	95.20	286	381.20
P Pipe covering fiberglass all service jacket 1" wall 1/2" diam	160.000	L.F.	169.60	896	1,065.60
Q Collector panel solar, blk chrome on copper, 3/16" temp glass 3'-6" x 7.5'	2.000	Ea.	1,680	324	2,004
Y Flow control valve, globe, bronze, 125#, soldered, 1/2" diam	1.000	Ea.	142	33.50	175.50
U Valve, water tempering, bronze,sweat connections, 1/2" diam	1.000	Ea.	131	33.50	164.50
W Tank, water storage, solar energy system, 80 Gal, 2" x 1/2 lb insul	1.000	Ea.	1,975	505	2,480
Copper tubing type L, solder joints, hangers 10' OC 1/2" diam	150.000	L.F.	414	1,500	1,914
Copper tubing type M, solder joints, hangers 10' OC 1/2" diam	50.000	L.F.	206.50	480	686.50
Sensor wire, #22-2 gauge multistranded	.500	C.L.F.	7.83	37.75	45.58
Wrought copper fittings & solder, 1/2" diam	75.000	Ea.	146.25	3,037.50	3,183.75
TOTAL			5,206.18	7,395.85	12,602.03

D2020 285	Thermosyphon, Hot Water		COST EACH		
			MAT.	INST.	TOTAL
0960	Solar, thermosyphon, hot water, two collector system	R235616 -60	5,200	7,400	12,600
0970					

For customer support on your Plumbing Costs with RSMeans Data, call 800.448.8182.

515

ATTIC FLOOR

CWS

HWS

This domestic hot water pre-heat system
includes heat exchanger with a
circulating pump, blower, air-to-water,
coil and controls, mounted in the upper
collector manifold. Heat from the hot air
coming out of the collectors is transferred
through the heat exchanger. For each
degree of DHW preheating gained, one
degree less heating is needed from the
fuel fired water heater. The system is
simple, inexpensive to operate and can
provide a substantial portion of DHW
requirements for modest additional cost.

D2020 Domestic Water Distribution

System Components	QUANTITY	UNIT	COST EACH		
			MAT.	INST.	TOTAL
SYSTEM D2020 290 2560					
SOLAR, HOT WATER, AIR TO WATER HEAT EXCHANGE					
THREE COLLECTORS, OPTICAL BLACK ON ALUM., 7.5' x 3.5', 80 GAL TANK					
A, B Differential controller, 2 sensors, thermostat, solar energy system	1.000	Ea.	174	67.50	241.50
C Thermometer, 2" dial	2.000	Ea.	50	101	151
C-1 Heat exchanger, air to fluid, up flow 70 MBH	1.000	Ea.	415	415	830
E Air vent, manual, for solar energy system 1/8" fitting	1.000	Ea.	3.64	25.50	29.14
F Air purger	1.000	Ea.	58	67.50	125.50
G Expansion tank	1.000	Ea.	72	25.50	97.50
H Strainer, Y type, bronze body, 1/2" IPS	1.000	Ea.	50	41	91
I Valve, gate, bronze, 125 lb, NRS, soldered 1/2" diam	5.000	Ea.	420	167.50	587.50
N Relief valve, temp & pressure solar 150 psi 210°F self-closing	1.000	Ea.	29	27	56
P Pipe covering, fiberglass, all service jacket, 1" wall, 1/2" diam	60.000	L.F.	63.60	336	399.60
Q Collector panel solar energy, air, black on alum plate, 7.5' x 3.5'	3.000	Ea.	2,520	486	3,006
B-1, R-1 Shutter damper for solar heater circulator	1.000	Ea.	65	118	183
B-1, R-1 Shutter motor for solar heater circulator	1.000	Ea.	162	89	251
R Backflow preventer, 1/2" pipe size	2.000	Ea.	196	101	297
T Drain valve, brass, 3/4" connection	1.000	Ea.	14.20	33.50	47.70
U Valve, water tempering, bronze, sweat connections, 1/2 " diam	1.000	Ea.	131	33.50	164.50
W Tank, water storage, solar energy system, 80 Gal, 2" x 2 lb insul	1.000	Ea.	1,975	505	2,480
V Tank, water storage, w/heating element, drain, relief valve, existing	1.000	System			
X Valve, globe, bronze, 125 lb, rising stem, 1/2" diam	1.000	Ea.	142	33.50	175.50
Copper tubing type M, solder joints, hangers 10' OC, 1/2" diam	50.000	L.F.	206.50	480	686.50
Copper tubing type L, solder joints, hangers 10' OC 1/2" diam	10.000	L.F.	27.60	100	127.60
Wrought copper fittings & solder, 1/2" diam	10.000	Ea.	19.50	405	424.50
Sensor wire, #22-2 conductor multistranded	.500	C.L.F.	7.83	37.75	45.58
Q-1, Q-2 Ductwork, fiberglass, aluminized jacket, 1-1/2" thick, 8" diam	32.000	S.F.	236.80	252.80	489.60
Q-3 Manifold for flush mount solar energy collector panels, air	6.000	L.F.	996	53.40	1,049.40
TOTAL			8,034.67	4,001.95	12,036.62

D2020 290	Solar, Hot Water, Air To Water Heat Exchange		COST EACH		
			MAT.	INST.	TOTAL
2550	Solar, hot water, air to water heat exchange				
2560	Three collectors, optical black on aluminum, 7.5' x 3.5', 80 Gal tank		8,025	4,000	12,025
2580	Four collectors, optical black on aluminum, 7.5' x 3.5', 80 Gal tank	R235616 -60	9,275	4,250	13,525
2600	Four collectors, optical black on aluminum, 7.5' x 3.5', 120 Gal tank		9,225	4,300	13,525

517

For customer support on your Plumbing Costs with RSMeans Data, call 800.448.8182.

In this closed-loop indirect collection system, fluid with a low freezing temperature, such as propylene glycol, transports heat from the collectors to water storage. The transfer fluid is contained in a closed-loop consisting of collectors, supply and return piping, and a heat exchanger immersed in the storage tank. A typical two-or-three panel system contains 5 to 6 gallons of heat transfer fluid.

When the collectors become approximately 20°F warmer than the storage temperature, a controller activates the circulator. The circulator moves the fluid continuously through the collectors until the temperature difference between the collectors and storage is such that heat collection no longer occurs; at that point, the circulator shuts off. Since the heat transfer fluid has a very low freezing temperature, there is no need for it to be drained from the collectors between periods of collection.

518

For customer support on your Plumbing Costs with RSMeans Data, call 800.448.8182.

System Components	QUANTITY	UNIT	COST EACH		
			MAT.	INST.	TOTAL
SYSTEM D2020 295 2760					
SOLAR, CLOSED LOOP, HOT WATER SYSTEM, IMMERSED HEAT EXCHANGER					
3/4" TUBING, THREE 3' X 7' BLACK CHROME COLLECTORS					
A, B Differential controller, 2 sensors, thermostat, solar energy system	1.000	Ea.	174	67.50	241.50
C Thermometer 2" dial	3.000	Ea.	75	151.50	226.50
D, T Fill & drain valves, brass, 3/4" connection	3.000	Ea.	42.60	100.50	143.10
E Air vent, manual, 1/8" fitting	1.000	Ea.	3.64	25.50	29.14
F Air purger	1.000	Ea.	58	67.50	125.50
G Expansion tank	1.000	Ea.	72	25.50	97.50
I Valve, gate, bronze, NRS, soldered 3/4" diam	3.000	Ea.	279	121.50	400.50
J Neoprene vent flashing	2.000	Ea.	21.50	81	102.50
K Circulator, solar heated liquid, 1/25 HP	1.000	Ea.	229	104	333
N-1, N Relief valve, temp & press 150 psi 210°F self-closing 3/4" IPS	2.000	Ea.	58	54	112
O Pipe covering, urethane ultraviolet cover, 1" wall, 3/4" diam	20.000	L.F.	60.60	146	206.60
P Pipe covering, fiberglass, all service jacket, 1" wall, 3/4" diam	50.000	L.F.	56.50	292.50	349
Roof clamps for solar energy collector panel	3.000	Set	11.79	63	74.79
Q Collector panel solar blk chrome on copper, 1/8" temp glass, 3'x7'	3.000	Ea.	3,450	459	3,909
R-1 Valve, swing check, bronze, regrinding disc, 3/4" diam, soldered	1.000	Ea.	136	40.50	176.50
S Pressure gauge, 60 psi, 2-1/2" dial	1.000	Ea.	28	25.50	53.50
U Valve, water tempering, bronze, sweat connections, 3/4" diam	1.000	Ea.	193	40.50	233.50
W-2, W Tank, water storage immersed heat exchr elec elem 2"x2# insul 120 Gal	1.000	Ea.	1,925	575	2,500
X Valve, globe, bronze, rising stem, 3/4" diam, soldered	1.000	Ea.	190	40.50	230.50
Copper tubing type L, solder joint, hanger 10' OC 3/4" diam	20.000	L.F.	67	213	280
Copper tubing, type M, solder joint, hanger 10' OC 3/4" diam	70.000	L.F.	416.50	724.50	1,141
Sensor wire, #22-2 conductor multistranded	.500	C.L.F.	7.83	37.75	45.58
Solar energy heat transfer fluid, propylene glycol, anti-freeze	6.000	Gal.	150	174	324
Wrought copper fittings & solder, 3/4" diam	76.000	Ea.	325.28	3,230	3,555.28
TOTAL			8,030.24	6,860.25	14,890.49

D2020 295	Solar, Closed Loop, Hot Water Systems	COST EACH		
		MAT.	INST.	TOTAL
2550	Solar, closed loop, hot water system, immersed heat exchanger			
2560	3/8" tubing, 3 ea. 4' x 4'-4" vacuum tube collectors, 80 gal. tank	7,575	6,250	13,825
2580	1/2" tubing, 4 ea. 4' x 4'-4" vacuum tube collectors, 80 gal. tank	8,325	6,725	15,050
2600	120 gal. tank	8,275	6,800	15,075
2640	2 ea. 3'x7' black chrome collectors, 80 gal. tank	6,475	6,425	12,900
2660	120 gal. tank	6,375	6,400	12,775
2700	2 ea. 3'x7' flat black collectors, 120 gal. tank	5,750	6,425	12,175
2720	3 ea. 3'x7' flat black collectors, 120 gal. tank	6,600	6,600	13,200
2760	3/4" tubing, 3 ea. 3'x7' black chrome collectors, 120 gal. tank	8,025	6,850	14,875
2780	3 ea. 3'x7' flat black collectors, 120 gal. tank	7,100	6,875	13,975
2800	2 ea. 4'x9' flat black w/plastic glazing collectors 120 gal. tank	6,625	6,900	13,525
2840	1" tubing, 4 ea. 2'x9' plastic absorber & glazing collectors 120 gal. tank	8,075	7,725	15,800
2860	4 ea. 3'x7' black chrome collectors, 120 gal. tank	10,100	7,750	17,850
2880	4 ea. 3'x7' flat black absorber collectors, 120 gal. tank	8,875	7,800	16,675

(R235616-60)

For customer support on your Plumbing Costs with RSMeans Data, call 800.448.8182.

519

D2040 Rain Water Drainage

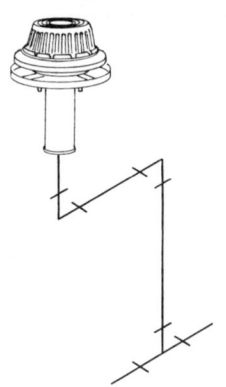

Design Assumptions: Vertical conductor size is based on a maximum rate of rainfall of 4″ per hour. To convert roof area to other rates multiply "Max. S.F. Roof Area" shown by four and divide the result by desired local rate. The answer is the local roof area that may be handled by the indicated pipe diameter.

Basic cost is for roof drain, 10′ of vertical leader and 10′ of horizontal, plus connection to the main.

Pipe Dia.	Max. S.F. Roof Area	Gallons per Min.
2″	544	23
3″	1610	67
4″	3460	144
5″	6280	261
6″	10,200	424
8″	22,000	913

System Components	QUANTITY	UNIT	COST EACH		
			MAT.	INST.	TOTAL
SYSTEM D2040 210 1880					
ROOF DRAIN, DWV PVC PIPE, 2″ DIAM., 10′ HIGH					
Drain, roof, main, ABS, dome type 2″ pipe size	1.000	Ea.	154	104	258
Clamp, roof drain, underdeck	1.000	Ea.	51	60.50	111.50
Pipe, Tee, PVC DWV, schedule 40, 2″ pipe size	1.000	Ea.	13.40	72.50	85.90
Pipe, PVC, DWV, schedule 40, 2″ diam.	20.000	L.F.	68.80	490	558.80
Pipe, elbow, PVC schedule 40, 2″ diam.	2.000	Ea.	7.24	80	87.24
TOTAL			294.44	807	1,101.44

D2040 210	Roof Drain Systems	COST EACH		
		MAT.	INST.	TOTAL
1880	Roof drain, DWV PVC, 2″ diam., piping, 10′ high	294	805	1,099
1920	For each additional foot add	3.44	24.50	27.94
1960	3″ diam., 10′ high	385	945	1,330
2000	For each additional foot add	5.85	27.50	33.35
2040	4″ diam., 10′ high	480	1,075	1,555
2080	For each additional foot add	8.15	30.50	38.65
2120	5″ diam., 10′ high	1,625	1,225	2,850
2160	For each additional foot add	27.50	34	61.50
2200	6″ diam., 10′ high	1,700	1,375	3,075
2240	For each additional foot add	17.30	37.50	54.80
2280	8″ diam., 10′ high	3,275	2,300	5,575
2320	For each additional foot add	30.50	47	77.50
3940	C.I., soil, single hub, service wt., 2″ diam. piping, 10′ high	760	885	1,645
3980	For each additional foot add	15.25	23	38.25
4120	3″ diam., 10′ high	1,075	965	2,040
4160	For each additional foot add	20.50	24.50	45
4200	4″ diam., 10′ high	1,350	1,050	2,400
4240	For each additional foot add	26.50	26.50	53
4280	5″ diam., 10′ high	1,775	1,150	2,925
4320	For each additional foot add	32	30	62
4360	6″ diam., 10′ high	2,350	1,225	3,575
4400	For each additional foot add	45.50	31	76.50
4440	8″ diam., 10′ high	4,400	2,525	6,925
4480	For each additional foot add	72.50	52	124.50
6040	Steel galv. sch 40 threaded, 2″ diam. piping, 10′ high	735	855	1,590
6080	For each additional foot add	10.15	22.50	32.65
6120	3″ diam., 10′ high	1,450	1,225	2,675
6160	For each additional foot add	23	34	57

D2040 Rain Water Drainage

D2040 210	Roof Drain Systems	COST EACH		
		MAT.	INST.	TOTAL
6200	4" diam., 10' high	2,175	1,600	3,775
6240	For each additional foot add	36	40.50	76.50
6280	5" diam., 10' high	2,350	1,325	3,675
6320	For each additional foot add	30.50	39.50	70
6360	6" diam, 10' high	2,975	1,700	4,675
6400	For each additional foot add	40	54	94
6440	8" diam., 10' high	5,325	2,475	7,800
6480	For each additional foot add	60.50	61	121.50

D3010 Energy Supply

Basis for Heat Loss Estimate, Apartment Type Structures:

1. Masonry walls and flat roof are insulated. U factor is assumed at .08.
2. Window glass area taken as BOCA minimum, 1/10th of floor area. Double insulating glass with 1/4″ air space, U = .65.
3. Infiltration = 0.3 C.F. per hour per S.F. of net wall.
4. Concrete floor loss is 2 BTUH per S.F.
5. Temperature difference taken as 70°F.
6. Ventilating or makeup air has not been included and must be added if desired. Air shafts are not used.

System Components	QUANTITY	UNIT	COST EACH MAT.	COST EACH INST.	COST EACH TOTAL
SYSTEM D3010 510 1760					
HEATING SYSTEM, FIN TUBE RADIATION, FORCED HOT WATER					
1,000 S.F. AREA, 10,000 C.F. VOLUME					
Boiler, oil fired, CI, burner, ctrls/insul/breech/pipe/ftng/valves, 109 MBH	1.000	Ea.	4,768.75	4,550	9,318.75
Circulating pump, CI flange connection, 1/12 HP	1.000	Ea.	525	242	767
Expansion tank, painted steel, ASME 18 Gal capacity	1.000	Ea.	800	105	905
Storage tank, steel, above ground, 275 Gal capacity w/supports	1.000	Ea.	525	294	819
Copper tubing type L, solder joint, hanger 10′ OC, 3/4″ diam	100.000	L.F.	335	1,065	1,400
Radiation, 3/4″ copper tube w/alum fin baseboard pkg, 7″ high	30.000	L.F.	258	765	1,023
Pipe covering, calcium silicate w/cover, 1′ wall, 3/4′ diam	100.000	L.F.	520	790	1,310
TOTAL			7,731.75	7,811	15,542.75
COST PER S.F.			7.73	7.81	15.54

D3010 510	Apartment Building Heating - Fin Tube Radiation	COST PER S.F. MAT.	COST PER S.F. INST.	COST PER S.F. TOTAL
1740	Heating systems, fin tube radiation, forced hot water			
1760	1,000 S.F. area, 10,000 C.F. volume	7.73	7.80	15.53
1800	10,000 S.F. area, 100,000 C.F. volume	3.48	4.69	8.17
1840	20,000 S.F. area, 200,000 C.F. volume	3.74	5.25	8.99
1880	30,000 S.F. area, 300,000 C.F. volume	3.64	5.10	8.74

Many styles of active solar energy systems exist. Those shown on the following page represent the majority of systems now being installed in different regions of the country.

The five active domestic hot water (DHW) systems which follow are typically specified as two or three panel systems with additional variations being the type of glazing and size of the storage tanks. Various combinations have been costed for the user's evaluation and comparison. The basic specifications from which the following systems were developed satisfy the construction detail requirements specified in the HUD Intermediate Minimum Property Standards (IMPS). If these standards are not complied with, the renewable energy system's costs could be significantly lower than shown.

To develop the system's specifications and costs it was necessary to make a number of assumptions about the systems. Certain systems are more appropriate to one climatic region than another or the systems may require modifications to be usable in particular locations. Specific instances in which the systems are not appropriate throughout the country as specified, or in which modification will be needed include the following:

- The freeze protection mechanisms provided in the DHW systems vary greatly. In harsh climates, where freeze is a major concern, a closed-loop indirect collection system may be more appropriate than a direct collection system.

- The thermosyphon water heater system described cannot be used when temperatures drop below 32°F.

- In warm climates it may be necessary to modify the systems installed to prevent overheating.

For each renewable resource (solar) system a schematic diagram and descriptive summary of the system is provided along with a list of all the components priced as part of the system. The costs were developed based on these specifications.

Considerations affecting costs which may increase or decrease beyond the estimates presented here include the following:

- Special structural qualities (allowance for earthquake, future expansion, high winds, and unusual spans or shapes);

- Isolated building site or rough terrain that would affect the transportation of personnel, material, or equipment;

- Unusual climatic conditions during the construction process;

- Substitution of other materials or system components for those used in the system specifications.

In this closed-loop indirect collection system, fluid with a low freezing temperature, propylene glycol, transports heat from the collectors to water storage. The transfer fluid is contained in a closed loop consisting of collectors, supply and return piping, and a heat exchanger immersed in the storage tank.

When the collectors become approximately 20°F warmer than the storage temperature, the controller activates the circulator. The circulator moves the fluid continuously until the temperature difference between fluid in the collectors and storage is such that the collection will no longer occur and then the circulator turns off. Since the heat transfer fluid has a very low freezing temperature, there is no need for it to be drained from the collectors between periods of collection.

D3010 Energy Supply

System Components	QUANTITY	UNIT	COST EACH		
			MAT.	INST.	TOTAL
SYSTEM D3010 650 2750					
SOLAR, CLOSED LOOP, SPACE/HOT WATER					
1" TUBING, TEN 3'X7' BLK CHROME ON COPPER ABSORBER COLLECTORS					
A, B Differential controller 2 sensors, thermostat, solar energy system	2.000	Ea.	348	135	483
C Thermometer, 2" dial	10.000	Ea.	250	505	755
C-1 Heat exchanger, solar energy system, fluid to air, up flow, 80 MBH	1.000	Ea.	550	485	1,035
D, T Fill & drain valves, brass, 3/4" connection	5.000	Ea.	71	167.50	238.50
D-1 Fan center	1.000	Ea.	133	168	301
E Air vent, manual, 1/8" fitting	1.000	Ea.	3.64	25.50	29.14
E-2 Thermostat, 2 stage for sensing room temperature	1.000	Ea.	272	102	374
F Air purger	2.000	Ea.	116	135	251
F-1 Controller, liquid temperature, solar energy system	1.000	Ea.	131	162	293
G Expansion tank	2.000	Ea.	144	51	195
I Valve, gate, bronze, 125 lb, soldered, 1" diam	5.000	Ea.	670	212.50	882.50
J Vent flashing, neoprene	2.000	Ea.	21.50	81	102.50
K Circulator, solar heated liquid, 1/25 HP	2.000	Ea.	458	208	666
L Circulator, solar heated liquid, 1/20 HP	1.000	Ea.	340	121	461
N Relief valve, temp & pressure 150 psi 210°F self-closing	4.000	Ea.	116	108	224
N-1 Relief valve, pressure poppet, bronze, 30 psi, 3/4" IPS	3.000	Ea.	366	87	453
O Pipe covering, urethane, ultraviolet cover, 1" wall, 1" diam	50.000	L.F.	151.50	365	516.50
P Pipe covering, fiberglass, all service jacket, 1" wall, 1" diam	60.000	L.F.	67.80	351	418.80
Q Collector panel solar energy blk chrome on copper 1/8" temp glass 3'x7'	10.000	Ea.	11,500	1,530	13,030
Roof clamps for solar energy collector panel	10.000	Set	39.30	210	249.30
R Valve, swing check, bronze, 125 lb, regrinding disc, 3/4" & 1" diam	4.000	Ea.	776	170	946
S Pressure gage, 0-60 psi, for solar energy system	2.000	Ea.	56	51	107
U Valve, water tempering, bronze, sweat connections, 3/4" diam	1.000	Ea.	193	40.50	233.50
W-2, W-1, W Tank, water storage immersed heat xchr elec elem 2"x1/2# ins 120 gal	4.000	Ea.	7,700	2,300	10,000
X Valve, globe, bronze, 125 lb, rising stem, 1" diam	3.000	Ea.	894	127.50	1,021.50
Y Valve, flow control	1.000	Ea.	157	37	194
Copper tubing, type M, solder joint, hanger 10' OC 1" diam	110.000	L.F.	891	1,270.50	2,161.50
Copper tubing, type L, solder joint, hanger 10' OC 3/4" diam	20.000	L.F.	67	213	280
Wrought copper fittings & solder, 3/4" & 1" diam	121.000	Ea.	1,270.50	6,110.50	7,381
Sensor, wire, #22-2 conductor, multistranded	.700	C.L.F.	10.96	52.85	63.81
Ductwork, galvanized steel, for heat exchanger	8.000	Lb.	5.20	75.20	80.40
Solar energy heat transfer fluid propylene glycol, anti-freeze	25.000	Gal.	625	725	1,350
TOTAL			28,394.40	16,382.55	44,776.95

D3010 650	Solar, Closed Loop, Space/Hot Water Systems		COST EACH		
			MAT.	INST.	TOTAL
2540	Solar, closed loop, space/hot water				
2550	1/2" tubing, 12 ea. 4'x4'4" vacuum tube collectors		26,900	15,300	42,200
2600	3/4" tubing, 12 ea. 4'x4'4" vacuum tube collectors	R235616 -60	28,400	15,700	44,100
2650	10 ea. 3' x 7' black chrome absorber collectors		27,300	15,300	42,600
2700	10 ea. 3' x 7' flat black absorber collectors		24,200	15,400	39,600
2750	1" tubing, 10 ea. 3' x 7' black chrome absorber collectors		28,400	16,400	44,800
2800	10 ea. 3' x 7' flat black absorber collectors		25,500	16,500	42,000
2850	6 ea. 4' x 9' flat black w/plastic glazing collectors		23,200	16,400	39,600
2900	12 ea. 2' x 9' plastic absorber and glazing collectors		24,700	16,700	41,400

This draindown pool system uses a differential thermostat similar to those used in solar domestic hot water and space heating applications. To heat the pool, the pool water passes through the conventional pump-filter loop and then flows through the collectors. When collection is not possible, or when the pool temperature is reached, all water drains from the solar loop back to the pool through the existing piping. The modes are controlled by solenoid valves or other automatic valves in conjunction with a vacuum breaker relief valve, which facilitates draindown.

D3010 Energy Supply

System Components	QUANTITY	UNIT	COST EACH		
			MAT.	INST.	TOTAL
SYSTEM D3010 660 2640					
SOLAR SWIMMING POOL HEATER, ROOF MOUNTED COLLECTORS					
TEN 4′ X 10′ FULLY WETTED UNGLAZED PLASTIC ABSORBERS					
A Differential thermostat/controller, 110V, adj pool pump system	1.000	Ea.	365	405	770
A-1 Solenoid valve, PVC, normally 1 open 1 closed (included)	2.000	Ea.			
B Sensor, thermistor type (included)	2.000	Ea.			
E-1 Valve, vacuum relief	1.000	Ea.	34	25.50	59.50
Q Collector panel, solar energy, plastic, liquid full wetted, 4′ x 10′	10.000	Ea.	3,600	2,910	6,510
R Valve, ball check, PVC, socket, 1-1/2″ diam	1.000	Ea.	120	40.50	160.50
Z Valve, ball, PVC, socket, 1-1/2″ diam	3.000	Ea.	280.50	121.50	402
Pipe, PVC, sch 40, 1-1/2″ diam	80.000	L.F.	432	1,800	2,232
Pipe fittings, PVC sch 40, socket joint, 1-1/2″ diam	10.000	Ea.	23.10	405	428.10
Sensor wire, #22-2 conductor, multistranded	.500	C.L.F.	7.83	37.75	45.58
Roof clamps for solar energy collector panels	10.000	Set	39.30	210	249.30
Roof strap, teflon for solar energy collector panels	26.000	L.F.	715	102.44	817.44
TOTAL			5,616.73	6,057.69	11,674.42

D3010 660	Solar Swimming Pool Heater Systems		COST EACH		
			MAT.	INST.	TOTAL
2530	Solar swimming pool heater systems, roof mounted collectors				
2540	10 ea. 3′x7′ black chrome absorber, 1/8″ temp. glass		13,500	4,675	18,175
2560	10 ea. 4′x8′ black chrome absorber, 3/16″ temp. glass	R235616 -60	15,300	5,575	20,875
2580	10 ea. 3′8″x6′ flat black absorber, 3/16″ temp. glass		10,400	4,775	15,175
2600	10 ea. 4′x9′ flat black absorber, plastic glazing		12,300	5,800	18,100
2620	10 ea. 2′x9′ rubber absorber, plastic glazing		14,900	6,250	21,150
2640	10 ea. 4′x10′ fully wetted unglazed plastic absorber		5,625	6,050	11,675
2660	Ground mounted collectors				
2680	10 ea. 3′x7′ black chrome absorber, 1/8″ temp. glass		13,700	5,200	18,900
2700	10 ea. 4′x8′ black chrome absorber, 3/16″ temp. glass		15,400	6,100	21,500
2720	10 ea. 3′8″x6′ flat blk absorber, 3/16″ temp. glass		10,600	5,300	15,900
2740	10 ea. 4′x9′ flat blk absorber, plastic glazing		12,400	6,325	18,725
2760	10 ea. 2′x9′ rubber absorber, plastic glazing		15,000	6,575	21,575
2780	10 ea. 4′x10′ fully wetted unglazed plastic absorber		5,800	6,575	12,375

The complete Solar Air Heating System provides maximum savings of conventional fuel with both space heating and year-round domestic hot water heating. It allows for the home air conditioning to operate simultaneously and independently from the solar domestic water heating in summer. The system's modes of operation are:

Mode 1: The building is heated directly from the collectors with air circulated by the Solar Air Mover.

Mode 2: When heat is not needed in the building, the dampers change within the air mover to circulate the air from the collectors to the rock storage bin.

Mode 3: When heat is not available from the collector array and is available in rock storage, the air mover draws heated air from rock storage and directs it into the building. When heat is not available from the collectors or the rock storage bin, the auxiliary heating unit will provide heat for the building. The size of the collector array is typically 25% the size of the main floor area.

D3010 Energy Supply

System Components		QUANTITY	UNIT	COST EACH		
				MAT.	INST.	TOTAL
	SYSTEM D3010 675 1210					
	SOLAR, SPACE/HOT WATER, AIR TO WATER HEAT EXCHANGE					
A, B	Differential controller 2 sensors thermos., solar energy sys liquid loop	1.000	Ea.	174	67.50	241.50
A-1, B	Differential controller 2 sensors 6 station solar energy sys air loop	1.000	Ea.	305	269	574
B-1	Solar energy sensor, freeze prevention	1.000	Ea.	28.50	25.50	54
C	Thermometer for solar energy system, 2" dial	2.000	Ea.	50	101	151
C-1	Heat exchanger, solar energy system, air to fluid, up flow, 70 MBH	1.000	Ea.	415	415	830
D	Drain valve, brass, 3/4" connection	2.000	Ea.	28.40	67	95.40
E	Air vent, manual, for solar energy system 1/8" fitting	1.000	Ea.	3.64	25.50	29.14
E-1	Thermostat, 2 stage for sensing room temperature	1.000	Ea.	272	102	374
F	Air purger	1.000	Ea.	58	67.50	125.50
G	Expansion tank, for solar energy system	1.000	Ea.	72	25.50	97.50
I	Valve, gate, bronze, 125 lb, NRS, soldered 3/4" diam	2.000	Ea.	186	81	267
K	Circulator, solar heated liquid, 1/25 HP	1.000	Ea.	229	104	333
N	Relief valve temp & press 150 psi 210°F self-closing, 3/4" IPS	1.000	Ea.	29	27	56
N-1	Relief valve, pressure, poppet, bronze, 30 psi, 3/4" IPS	1.000	Ea.	122	29	151
P	Pipe covering, fiberglass, all service jacket, 1" wall, 3/4" diam	60.000	L.F.	67.80	351	418.80
Q-3	Manifold for flush mount solar energy collector panels	20.000	L.F.	3,320	178	3,498
Q	Collector panel solar energy, air, black on alum. plate, 7.5' x 3.5'	10.000	Ea.	8,400	1,620	10,020
R	Valve, swing check, bronze, 125 lb, regrinding disc, 3/4" diam	2.000	Ea.	272	81	353
S	Pressure gage, 2" dial, for solar energy system	1.000	Ea.	28	25.50	53.50
U	Valve, water tempering, bronze, sweat connections, 3/4" diam	1.000	Ea.	193	40.50	233.50
W-2, W	Tank, water storage, solar, elec element 2"x1/2# insul, 80 Gal	1.000	Ea.	1,975	505	2,480
X	Valve, globe, bronze, 125 lb, soldered, 3/4" diam	1.000	Ea.	190	40.50	230.50
	Copper tubing type L, solder joints, hangers 10' OC 3/4" diam	10.000	L.F.	33.50	106.50	140
	Copper tubing type M, solder joints, hangers 10' OC 3/4" diam	60.000	L.F.	357	621	978
	Wrought copper fittings & solder, 3/4" diam	26.000	Ea.	111.28	1,105	1,216.28
	Sensor wire, #22-2 conductor multistranded	1.200	C.L.F.	18.78	90.60	109.38
Q-1	Duct work, rigid fiberglass, rectangular	400.000	S.F.	384	2,520	2,904
	Duct work, spiral preformed, steel, PVC coated both sides, 12" x 10"	8.000	Ea.	168	472	640
R-2	Shutter/damper for solar heater circulator	9.000	Ea.	585	1,062	1,647
K-1	Shutter motor for solar heater circulator blower	9.000	Ea.	1,458	801	2,259
K-1	Fan, solar energy heated air circulator, space & DHW system	1.000	Ea.	1,775	2,850	4,625
Z-1	Tank, solar energy air storage, 6'-3"H 7'x7' = 306 CF/2000 Gal	1.000	Ea.	15,400	1,850	17,250
Z-2	Crushed stone 1-1/2"	11.000	C.Y.	313.50	94.27	407.77
C-2	Thermometer, remote probe, 2" dial	1.000	Ea.	38.50	101	139.50
R-1	Solenoid valve	1.000	Ea.	177	90	267
V, R-4, R-3	Furnace, supply diffusers, return grilles, existing	1.000	Ea.	15,400	1,850	17,250
	TOTAL			52,637.90	17,861.37	70,499.27

D3010 675	Air To Water Heat Exchange		COST EACH		
			MAT.	INST.	TOTAL
1210 1220	Solar, air to water heat exchange, for space/hot water heating	R235616 -60	52,500	17,900	70,400

529

D3020 Heat Generating Systems

Boiler

Baseboard Radiation

Small Electric Boiler
System Considerations:
1. Terminal units are fin tube baseboard radiation rated at 720 BTU/hr with 200° water temperature or 820 BTU/hr steam.
2. Primary use being for residential or smaller supplementary areas, the floor levels are based on 7-1/2' ceiling heights.
3. All distribution piping is copper for boilers through 205 MBH. All piping for larger systems is steel pipe.

System Components	QUANTITY	UNIT	COST EACH		
			MAT.	INST.	TOTAL
SYSTEM D3020 102 1120					
SMALL HEATING SYSTEM, HYDRONIC, ELECTRIC BOILER					
1,480 S.F., 61 MBH, STEAM, 1 FLOOR					
Boiler, electric steam, std cntrls, trim, ftngs and valves, 18 KW, 61.4 MBH	1.000	Ea.	5,390	2,035	7,425
Copper tubing type L, solder joint, hanger 10'OC, 1-1/4" diam	160.000	L.F.	1,888	2,232	4,120
Radiation, 3/4" copper tube w/alum fin baseboard pkg 7" high	60.000	L.F.	516	1,530	2,046
Rough in baseboard panel or fin tube with valves & traps	10.000	Set	3,250	7,700	10,950
Pipe covering, calcium silicate w/cover, 1" wall 1-1/4" diam	160.000	L.F.	832	1,304	2,136
Low water cut-off, quick hookup, in gage glass tappings	1.000	Ea.	400	51	451
TOTAL			12,276	14,852	27,128
COST PER S.F.			8.29	10.04	18.33

D3020 102	Small Heating Systems, Hydronic, Electric Boilers	COST PER S.F.		
		MAT.	INST.	TOTAL
1100	Small heating systems, hydronic, electric boilers			
1120	Steam, 1 floor, 1480 S.F., 61 M.B.H.	8.31	10.07	18.38
1160	3,000 S.F., 123 M.B.H.	6.40	8.80	15.20
1200	5,000 S.F., 205 M.B.H.	5.05	8.10	13.15
1240	2 floors, 12,400 S.F., 512 M.B.H.	4.39	8.05	12.44
1280	3 floors, 24,800 S.F., 1023 M.B.H.	4.80	8	12.80
1360	Hot water, 1 floor, 1,000 S.F., 41 M.B.H.	13.75	5.60	19.35
1400	2,500 S.F., 103 M.B.H.	9.15	10.05	19.20
1440	2 floors, 4,850 S.F., 205 M.B.H.	8.25	12.05	20.30
1480	3 floors, 9,700 S.F., 410 M.B.H.	8.35	12.45	20.80

Cast Iron Garage Exhaust System

Dual Exhaust System

System Components	QUANTITY	UNIT	COST EACH		
			MAT.	INST.	TOTAL
SYSTEM D3090 320 1040					
GARAGE, EXHAUST, SINGLE 3″ EXHAUST OUTLET, CARS & LIGHT TRUCKS					
A Outlet top assy, for engine exhaust system with adapters and ftngs, 3″ diam	1.000	Ea.	300	49.50	349.50
F Bullnose (guide) for engine exhaust system, 3″ diam	1.000	Ea.	35		35
G Galvanized flexible tubing for engine exhaust system, 3″ diam	8.000	L.F.	94.80		94.80
H Adapter for metal tubing end of engine exhaust system, 3″ tail pipe	1.000	Ea.	58.50		58.50
J Pipe, sewer, cast iron, push-on joint, 8″ diam.	18.000	L.F.	1,449	891	2,340
Excavating utility trench, chain trencher, 8″ wide, 24″ deep	18.000	L.F.		43.02	43.02
Backfill utility trench by hand, incl. compaction, 8″ wide 24″ deep	18.000	L.F.		53.10	53.10
Stand for blower, concrete over polystyrene core, 6″ high	1.000	Ea.	6	24.50	30.50
AC&V duct spiral reducer 10″x8″	1.000	Ea.	25.50	44.50	70
AC&V duct spiral reducer 12″x10″	1.000	Ea.	24	59	83
AC&V duct spiral preformed 45° elbow, 8″ diam	1.000	Ea.	8.65	51	59.65
AC&V utility fan, belt drive, 3 phase, 2000 CFM, 1 HP	2.000	Ea.	3,150	780	3,930
Safety switch, heavy duty fused, 240V, 3 pole, 30 amp	1.000	Ea.	122	237	359
TOTAL			5,273.45	2,232.62	7,506.07

D3090 320	Garage Exhaust Systems	COST PER BAY		
		MAT.	INST.	TOTAL
1040	Garage, single 3″ exhaust outlet, cars & light trucks, one bay	5,275	2,225	7,500
1060	Additional bays up to seven bays	1,300	600	1,900
1500	4″ outlet, trucks, one bay	5,300	2,225	7,525
1520	Additional bays up to six bays	1,325	600	1,925
1600	5″ outlet, diesel trucks, one bay	5,600	2,225	7,825
1650	Additional single bays up to six bays	1,800	710	2,510
1700	Two adjoining bays	5,600	2,225	7,825
2000	Dual exhaust, 3″ outlets, pair of adjoining bays	6,575	2,775	9,350
2100	Additional pairs of adjoining bays	1,950	710	2,660

Dry Pipe System: A system employing automatic sprinklers attached to a piping system containing air under pressure, the release of which from the opening of sprinklers permits the water pressure to open a valve known as a "dry pipe valve". The water then flows into the piping system and out the opened sprinklers.

All areas are assumed to be open.

System Components	QUANTITY	UNIT	COST EACH		
			MAT.	INST.	TOTAL
SYSTEM D4010 310 0580					
DRY PIPE SPRINKLER, STEEL, BLACK, SCH. 40 PIPE					
LIGHT HAZARD, ONE FLOOR, 2000 S.F.					
Valve, gate, iron body 125 lb., OS&Y, flanged, 4" pipe size	1.000	Ea.	1,100	485	1,585
Backflow pvntr, auto, 2 OS&Y v, tst cocks, dbl chk, flgd, 4"	1.000	Ea.	4,050	485	4,535
Tamper switch (valve supervisory switch)	3.000	Ea.	882	148.50	1,030.50
Valve, swing check, bronze, 125 lb, regrinding disc, 2-1/2" pipe size	1.000	Ea.	1,075	97	1,172
Valve, angle, bronze, 150 lb., rising stem, threaded, 2" pipe size	1.000	Ea.	950	73.50	1,023.50
*Alarm valve, 2-1/2" pipe size	1.000	Ea.	1,800	475	2,275
Alarm, water motor, complete with gong	1.000	Ea.	355	199	554
Fire alarm horn, electric	1.000	Ea.	55.50	113	168.50
Valve swing check w/balldrip Cl with brass trim, 4" pipe size	1.000	Ea.	480	475	955
Pipe, steel, black, schedule 40, 4" diam.	8.000	L.F.	192	341.60	533.60
Dry pipe valve, trim & gauges, 4" pipe size	1.000	Ea.	3,775	1,425	5,200
Pipe, steel, black, schedule 40, threaded, cplg & hngr 10'OC 2-1/2" diam.	15.000	L.F.	235.50	435	670.50
Pipe, steel, black, schedule 40, threaded, cplg & hngr 10'OC 2" diam.	9.375	L.F.	78.28	210.94	289.22
Pipe, steel, black, schedule 40, threaded, cplg & hngr 10'OC 1-1/4" diam.	28.125	L.F.	149.06	459.84	608.90
Pipe, steel, black, schedule 40, threaded, cplg & hngr 10'OC 1" diam.	84.000	L.F.	344.40	1,281	1,625.40
Pipe Tee, malleable iron black, 150 lb. threaded, 4" pipe size	2.000	Ea.	820	730	1,550
Pipe Tee, malleable iron black, 150 lb. threaded, 2-1/2" pipe size	2.000	Ea.	232	324	556
Pipe Tee, malleable iron black, 150 lb. threaded, 2" pipe size	1.000	Ea.	53.50	132	185.50
Pipe Tee, malleable iron black, 150 lb. threaded, 1-1/4" pipe size	4.000	Ea.	102	416	518
Pipe Tee, malleable iron black, 150 lb. threaded, 1" pipe size	3.000	Ea.	46.80	303	349.80
Pipe 90° elbow malleable iron black, 150 lb. threaded, 1" pipe size	5.000	Ea.	50	310	360
Sprinkler head dry K5.6, 1" NPT, 3" to 6" length	12.000	Ea.	2,592	684	3,276
Air compressor, 200 Gal sprinkler system capacity, 1/3 HP	1.000	Ea.	1,550	610	2,160
*Standpipe connection, wall, flush, brs. w/plug & chain 2-1/2"x2-1/2"	1.000	Ea.	167	286	453
Valve gate bronze, 300 psi, NRS, class 150, threaded, 1" pipe size	1.000	Ea.	152	42.50	194.50
TOTAL			21,287.04	10,541.88	31,828.92
COST PER S.F.			7.98	3.95	11.93

*Not included in systems under 2000 S.F.

D4010 310	Dry Pipe Sprinkler Systems		COST PER S.F.		
			MAT.	INST.	TOTAL
0520	Dry pipe sprinkler systems, steel, black, sch. 40 pipe				
0530	Light hazard, one floor, 500 S.F.		15.40	7.50	22.90
0560	1000 S.F.	R211313 -10	9.50	4.42	13.92
0580	2000 S.F.		7.99	3.94	11.93

D4010 Sprinklers

D4010 310	Dry Pipe Sprinkler Systems		COST PER S.F.		
			MAT.	INST.	TOTAL
0600	5000 S.F.	R211313 -20	4.24	2.98	7.22
0620	10,000 S.F.		2.83	2.42	5.25
0640	50,000 S.F.		2	2.13	4.13
0660	Each additional floor, 500 S.F.		2.93	3.56	6.49
0680	1000 S.F.		2.54	2.91	5.45
0700	2000 S.F.		2.40	2.68	5.08
0720	5000 S.F.		1.95	2.30	4.25
0740	10,000 S.F.		1.73	2.11	3.84
0760	50,000 S.F.		1.62	1.88	3.50
1000	Ordinary hazard, one floor, 500 S.F.		15.70	7.55	23.25
1020	1000 S.F.		9.80	4.46	14.26
1040	2000 S.F.		8.60	4.66	13.26
1060	5000 S.F.		4.89	3.18	8.07
1080	10,000 S.F.		3.64	3.17	6.81
1100	50,000 S.F.		3.17	2.98	6.15
1140	Each additional floor, 500 S.F.		3.25	3.65	6.90
1160	1000 S.F.		2.95	3.25	6.20
1180	2000 S.F.		2.90	2.96	5.86
1200	5000 S.F.		2.70	2.54	5.24
1220	10,000 S.F.		2.42	2.47	4.89
1240	50,000 S.F.		2.38	2.14	4.52
1500	Extra hazard, one floor, 500 S.F.		19.30	9.40	28.70
1520	1000 S.F.		15.40	6.85	22.25
1540	2000 S.F.		9.50	5.90	15.40
1560	5000 S.F.		5.55	4.39	9.94
1580	10,000 S.F.		5.65	4.17	9.82
1600	50,000 S.F.		5.90	4	9.90
1660	Each additional floor, 500 S.F.		4.70	4.50	9.20
1680	1000 S.F.		4.21	4.22	8.43
1700	2000 S.F.		3.90	4.23	8.13
1720	5000 S.F.		3.35	3.70	7.05
1740	10,000 S.F.		4.11	3.38	7.49
1760	50,000 S.F.		4.16	3.25	7.41
2020	Grooved steel, black, sch. 40 pipe, light hazard, one floor, 2000 S.F.		8.35	3.87	12.22
2060	10,000 S.F.		2.99	2.11	5.10
2100	Each additional floor, 2000 S.F.		2.65	2.17	4.82
2150	10,000 S.F.		1.89	1.79	3.68
2200	Ordinary hazard, one floor, 2000 S.F.		8.80	4.09	12.89
2250	10,000 S.F.		3.67	2.69	6.36
2300	Each additional floor, 2000 S.F.		3.06	2.39	5.45
2350	10,000 S.F.		2.81	2.40	5.21
2400	Extra hazard, one floor, 2000 S.F.		9.80	5.10	14.90
2450	10,000 S.F.		5.25	3.48	8.73
2500	Each additional floor, 2000 S.F.		4.36	3.50	7.86
2550	10,000 S.F.		3.90	2.98	6.88
3050	Grooved steel, black, sch. 10 pipe, light hazard, one floor, 2000 S.F.		8.25	3.84	12.09
3100	10,000 S.F.		2.93	2.08	5.01
3150	Each additional floor, 2000 S.F.		2.55	2.14	4.69
3200	10,000 S.F.		1.84	1.76	3.60
3250	Ordinary hazard, one floor, 2000 S.F.		8.70	4.06	12.76
3300	10,000 S.F.		3.51	2.63	6.14
3350	Each additional floor, 2000 S.F.		2.97	2.36	5.33
3400	10,000 S.F.		2.49	2.33	4.82
3450	Extra hazard, one floor, 2000 S.F.		9.75	5.05	14.80
3500	10,000 S.F.		5.10	3.43	8.53
3550	Each additional floor, 2000 S.F.		4.29	3.47	7.76
3600	10,000 S.F.		3.80	2.94	6.74
4050	Copper tubing, type L, light hazard, one floor, 2000 S.F.		8.30	3.87	12.17

D4010 310	Dry Pipe Sprinkler Systems	COST PER S.F.		
		MAT.	INST.	TOTAL
4100	10,000 S.F.	3.05	2.15	5.20
4150	Each additional floor, 2000 S.F.	2.60	2.21	4.81
4200	10,000 S.F.	1.96	1.84	3.80
4250	Ordinary hazard, one floor, 2000 S.F.	8.75	4.28	13.03
4300	10,000 S.F.	4.37	2.48	6.85
4350	Each additional floor, 2000 S.F.	4.13	2.51	6.64
4400	10,000 S.F.	3.22	2.12	5.34
4450	Extra hazard, one floor, 2000 S.F.	9.90	5.20	15.10
4500	10,000 S.F.	7.60	3.74	11.34
4550	Each additional floor, 2000 S.F.	5.20	3.53	8.73
4600	10,000 S.F.	5.45	3.21	8.66
5050	Copper tubing, type L, T-drill system, light hazard, one floor			
5060	2000 S.F.	8.45	3.64	12.09
5100	10,000 S.F.	2.94	1.80	4.74
5150	Each additional floor, 2000 S.F.	2.78	1.99	4.77
5200	10,000 S.F.	1.86	1.50	3.36
5250	Ordinary hazard, one floor, 2000 S.F.	8.60	3.72	12.32
5300	10,000 S.F.	3.67	2.27	5.94
5350	Each additional floor, 2000 S.F.	2.89	2.02	4.91
5400	10,000 S.F.	2.52	1.85	4.37
5450	Extra hazard, one floor, 2000 S.F.	9.45	4.34	13.79
5500	10,000 S.F.	5.45	2.84	8.29
5550	Each additional floor, 2000 S.F.	3.99	2.75	6.74
5600	10,000 S.F.	3.71	2.30	6.01

For customer support on your Plumbing Costs with RSMeans Data, call 800.448.8182.

Pre-Action System: A system employing automatic sprinklers attached to a piping system containing air that may or may not be under pressure, with a supplemental heat responsive system of generally more sensitive characteristics than the automatic sprinklers themselves, installed in the same areas as the sprinklers. Actuation of the heat responsive system, as from a fire, opens a valve which permits water to flow into the sprinkler piping system and to be discharged from those sprinklers which were opened by heat from the fire.

All areas are assumed to be open.

System Components	QUANTITY	UNIT	COST EACH		
			MAT.	INST.	TOTAL
SYSTEM D4010 350 0580					
PREACTION SPRINKLER SYSTEM, STEEL, BLACK, SCH. 40 PIPE					
LIGHT HAZARD, 1 FLOOR, 2000 S.F.					
Valve, gate, iron body 125 lb., OS&Y, flanged, 4" pipe size	1.000	Ea.	1,100	485	1,585
Backflow pvntr, auto, 2 OS&Y v, tst cocks, dbl chk, flgd, 4"	1.000	Ea.	4,050	485	4,535
Tamper switch (valve supervisory switch)	3.000	Ea.	882	148.50	1,030.50
*Valve, swing check w/ball drip Cl with brass trim 4" pipe size	1.000	Ea.	480	475	955
Valve, swing check, bronze, 125 lb, regrinding disc, 2-1/2" pipe size	1.000	Ea.	1,075	97	1,172
Valve, angle, bronze, 150 lb., rising stem, threaded, 2" pipe size	1.000	Ea.	950	73.50	1,023.50
*Alarm valve, 2-1/2" pipe size	1.000	Ea.	1,800	475	2,275
Alarm, water motor, complete with gong	1.000	Ea.	355	199	554
Fire alarm horn, electric	1.000	Ea.	55.50	113	168.50
Thermostatic release for release line	2.000	Ea.	1,870	79	1,949
Pipe, steel, black, schedule 40, 4" diam.	8.000	L.F.	192	341.60	533.60
Dry pipe valve, trim & gauges, 4" pipe size	1.000	Ea.	3,775	1,425	5,200
Pipe, steel, black, schedule 40, threaded, cplg. & hngr. 10'OC 2-1/2" diam.	15.000	L.F.	235.50	435	670.50
Pipe steel black, schedule 40, threaded, cplg. & hngr. 10'OC 2" diam.	9.375	L.F.	78.28	210.94	289.22
Pipe, steel, black, schedule 40, threaded, cplg. & hngr. 10'OC 1-1/4" diam.	28.125	L.F.	149.06	459.84	608.90
Pipe, steel, black, schedule 40, threaded, cplg. & hngr. 10'OC 1" diam.	84.000	L.F.	344.40	1,281	1,625.40
Pipe, Tee, malleable iron, black, 150 lb. threaded, 4" diam.	2.000	Ea.	820	730	1,550
Pipe, Tee, malleable iron, black, 150 lb. threaded, 2-1/2" pipe size	2.000	Ea.	232	324	556
Pipe, Tee, malleable iron, black, 150 lb. threaded, 2" pipe size	1.000	Ea.	53.50	132	185.50
Pipe, Tee, malleable iron, black, 150 lb. threaded, 1-1/4" pipe size	4.000	Ea.	102	416	518
Pipe, Tee, malleable iron, black, 150 lb. threaded, 1" pipe size	3.000	Ea.	46.80	303	349.80
Pipe, 90° elbow, malleable iron, blk., 150 lb. threaded, 1" pipe size	5.000	Ea.	50	310	360
Sprinkler head, std. spray, brass 135°-286°F 1/2" NPT, 3/8" orifice	12.000	Ea.	258	594	852
Air compressor auto complete 200 Gal sprinkler sys. cap., 1/3 HP	1.000	Ea.	1,550	610	2,160
*Standpipe conn.,wall, flush, brass w/plug & chain 2-1/2" x 2-1/2"	1.000	Ea.	167	286	453
Valve, gate, bronze, 300 psi, NRS, class 150, threaded, 1" pipe size	1.000	Ea.	152	42.50	194.50
TOTAL			20,823.04	10,530.88	31,353.92
COST PER S.F.			7.81	3.95	11.76

*Not included in systems under 2000 S.F.

D4010 350	Preaction Sprinkler Systems	COST PER S.F.		
		MAT.	INST.	TOTAL
0520	Preaction sprinkler systems, steel, black, sch. 40 pipe			
0530	Light hazard, one floor, 500 S.F.	15.20	6.05	21.25

D4010 350	Preaction Sprinkler Systems		COST PER S.F.		
			MAT.	INST.	TOTAL
0560	1000 S.F.	R211313 -10	8.15	4.48	12.63
0580	2000 S.F.		7.80	3.94	11.74
0600	5000 S.F.	R211313 -20	3.87	2.96	6.83
0620	10,000 S.F.		2.50	2.41	4.91
0640	50,000 S.F.		1.70	2.12	3.82
0660	Each additional floor, 500 S.F.		3.92	3.18	7.10
0680	1000 S.F.		2.36	2.90	5.26
0700	2000 S.F.		2.23	2.68	4.91
0720	5000 S.F.		2.30	2.34	4.64
0740	10,000 S.F.		1.91	2.13	4.04
0760	50,000 S.F.		1.40	1.94	3.34
1000	Ordinary hazard, one floor, 500 S.F.		15.55	6.50	22.05
1020	1000 S.F.		9.35	4.44	13.79
1040	2000 S.F.		8.50	4.66	13.16
1060	5000 S.F.		4.28	3.16	7.44
1080	10,000 S.F.		2.97	3.14	6.11
1100	50,000 S.F.		2.52	2.95	5.47
1140	Each additional floor, 500 S.F.		3.49	3.66	7.15
1160	1000 S.F.		2.37	2.93	5.30
1180	2000 S.F.		2.08	2.93	5.01
1200	5000 S.F.		2.22	2.74	4.96
1220	10,000 S.F.		1.88	2.83	4.71
1240	50,000 S.F.		1.82	2.51	4.33
1500	Extra hazard, one floor, 500 S.F.		21	8.40	29.40
1520	1000 S.F.		14.20	6.25	20.45
1540	2000 S.F.		8.45	5.90	14.35
1560	5000 S.F.		4.80	4.73	9.53
1580	10,000 S.F.		4.52	4.60	9.12
1600	50,000 S.F.		4.67	4.44	9.11
1660	Each additional floor, 500 S.F.		4.35	4.50	8.85
1680	1000 S.F.		3.38	4.21	7.59
1700	2000 S.F.		2.96	4.21	7.17
1720	5000 S.F.		2.38	3.72	6.10
1740	10,000 S.F.		2.89	3.42	6.31
1760	50,000 S.F.		2.77	3.20	5.97
2020	Grooved steel, black, sch. 40 pipe, light hazard, one floor, 2000 S.F.		8.20	3.86	12.06
2060	10,000 S.F.		2.66	2.10	4.76
2100	Each additional floor of 2000 S.F.		2.48	2.17	4.65
2150	10,000 S.F.		1.57	1.78	3.35
2200	Ordinary hazard, one floor, 2000 S.F.		8.30	4.07	12.37
2250	10,000 S.F.		3	2.67	5.67
2300	Each additional floor, 2000 S.F.		2.60	2.37	4.97
2350	10,000 S.F.		1.91	2.35	4.26
2400	Extra hazard, one floor, 2000 S.F.		8.75	5.05	13.80
2450	10,000 S.F.		4.01	3.44	7.45
2500	Each additional floor, 2000 S.F.		3.31	3.46	6.77
2550	10,000 S.F.		2.63	2.94	5.57
3050	Grooved steel, black, sch. 10 pipe light hazard, one floor, 2000 S.F.		8.10	3.84	11.94
3100	10,000 S.F.		2.60	2.07	4.67
3150	Each additional floor, 2000 S.F.		2.38	2.13	4.51
3200	10,000 S.F.		1.51	1.75	3.26
3250	Ordinary hazard, one floor, 2000 S.F.		8.10	3.81	11.91
3300	10,000 S.F.		2.55	2.60	5.15
3350	Each additional floor, 2000 S.F.		2.50	2.35	4.85
3400	10,000 S.F.		1.81	2.30	4.11
3450	Extra hazard, one floor, 2000 S.F.		8.70	5.05	13.75
3500	10,000 S.F.		3.82	3.38	7.20
3550	Each additional floor, 2000 S.F.		3.24	3.44	6.68

D40 Fire Protection

D4010 Sprinklers

D4010 350	Preaction Sprinkler Systems	COST PER S.F.		
		MAT.	INST.	TOTAL
3600	10,000 S.F.	2.53	2.90	5.43
4050	Copper tubing, type L, light hazard, one floor, 2000 S.F.	8.15	3.87	12.02
4100	10,000 S.F.	2.73	2.14	4.87
4150	Each additional floor, 2000 S.F.	2.43	2.21	4.64
4200	10,000 S.F.	1.29	1.81	3.10
4250	Ordinary hazard, one floor, 2000 S.F.	8.90	4.23	13.13
4300	10,000 S.F.	3.70	2.45	6.15
4350	Each additional floor, 2000 S.F.	2.97	2.22	5.19
4400	10,000 S.F.	2.32	1.93	4.25
4450	Extra hazard, one floor, 2000 S.F.	9.65	5.10	14.75
4500	10,000 S.F.	6.30	3.69	9.99
4550	Each additional floor, 2000 S.F.	4.17	3.49	7.66
4600	10,000 S.F.	4.16	3.17	7.33
5050	Copper tubing, type L, T-drill system, light hazard, one floor			
5060	2000 S.F.	8.30	3.64	11.94
5100	10,000 S.F.	2.62	1.79	4.41
5150	Each additional floor, 2000 S.F.	2.60	1.98	4.58
5200	10,000 S.F.	1.53	1.48	3.01
5250	Ordinary hazard, one floor, 2000 S.F.	8.15	3.70	11.85
5300	10,000 S.F.	2.99	2.24	5.23
5350	Each additional floor, 2000 S.F.	2.44	2.01	4.45
5400	10,000 S.F.	1.91	1.93	3.84
5450	Extra hazard, one floor, 2000 S.F.	8.40	4.30	12.70
5500	10,000 S.F.	4.15	2.78	6.93
5550	Each additional floor, 2000 S.F.	2.94	2.71	5.65
5600	10,000 S.F.	2.44	2.25	4.69

Deluge System: A system employing open sprinklers attached to a piping system connected to a water supply through a valve which is opened by the operation of a heat responsive system installed in the same areas as the sprinklers. When this valve opens, water flows into the piping system and discharges from all sprinklers attached thereto.

All areas are assumed to be open.

System Components	QUANTITY	UNIT	COST EACH		
			MAT.	INST.	TOTAL
SYSTEM D4010 370 0580					
DELUGE SPRINKLER SYSTEM, STEEL BLACK SCH. 40 PIPE					
LIGHT HAZARD, 1 FLOOR, 2000 S.F.					
Valve, gate, iron body 125 lb., OS&Y, flanged, 4" pipe size	1.000	Ea.	1,100	485	1,585
Backflow pvntr, auto, 2 OS&Y v, tst cocks, dbl chk, flgd, 4"	1.000	Ea.	4,050	485	4,535
Tamper switch (valve supervisory switch)	3.000	Ea.	882	148.50	1,030.50
Valve, swing check w/ball drip, Cl w/brass ftngs., 4" pipe size	1.000	Ea.	480	475	955
Valve, swing check, bronze, 125 lb, regrinding disc, 2-1/2" pipe size	1.000	Ea.	1,075	97	1,172
Valve, angle, bronze, 150 lb., rising stem, threaded, 2" pipe size	1.000	Ea.	950	73.50	1,023.50
*Alarm valve, 2-1/2" pipe size	1.000	Ea.	1,800	475	2,275
Alarm, water motor, complete with gong	1.000	Ea.	355	199	554
Fire alarm horn, electric	1.000	Ea.	55.50	113	168.50
Thermostatic release for release line	2.000	Ea.	1,870	79	1,949
Pipe, steel, black, schedule 40, 4" diam.	8.000	L.F.	192	341.60	533.60
Deluge valve trim, pressure relief, emergency release, gauge, 4" pipe size	1.000	Ea.	6,625	1,425	8,050
Deluge system, monitoring panel w/deluge valve & trim	1.000	Ea.	8,425	44	8,469
Pipe, steel, black, schedule 40, threaded, cplg & hngr 10' OC 2-1/2" diam.	15.000	L.F.	235.50	435	670.50
Pipe, steel, black, schedule 40, threaded, cplg & hngr 10' OC 2" diam.	9.375	L.F.	78.28	210.94	289.22
Pipe, steel, black, schedule 40, threaded, cplg & hngr 10' OC 1-1/4" diam.	28.125	L.F.	149.06	459.84	608.90
Pipe, steel, black, schedule 40, threaded, cplg & hngr 10' OC 1" diam.	84.000	L.F.	344.40	1,281	1,625.40
Pipe, Tee, malleable iron, black, 150 lb. threaded, 4" pipe size	2.000	Ea.	820	730	1,550
Pipe, Tee, malleable iron, black, 150 lb. threaded, 2-1/2" pipe size	2.000	Ea.	232	324	556
Pipe, Tee, malleable iron, black, 150 lb. threaded, 2" pipe size	1.000	Ea.	53.50	132	185.50
Pipe, Tee, malleable iron, black, 150 lb. threaded, 1-1/4" pipe size	4.000	Ea.	102	416	518
Pipe, Tee, malleable iron, black, 150 lb. threaded, 1" pipe size	3.000	Ea.	46.80	303	349.80
Pipe, 90° elbow, malleable iron, black, 150 lb. threaded 1" pipe size	5.000	Ea.	50	310	360
Sprinkler head, std spray, brass 135°-286°F 1/2" NPT, 3/8" orifice	12.000	Ea.	258	594	852
Air compressor, auto, complete, 200 Gal sprinkler sys. cap., 1/3 HP	1.000	Ea.	1,550	610	2,160
*Standpipe connection, wall, flush w/plug & chain 2-1/2" x 2-1/2"	1.000	Ea.	167	286	453
Valve, gate, bronze, 300 psi, NRS, class 150, threaded, 1" pipe size	1.000	Ea.	152	42.50	194.50
TOTAL			32,098.04	10,574.88	42,672.92
COST PER S.F.			12.04	3.97	16.01

*Not included in systems under 2000 S.F.

D4010 370	Deluge Sprinkler Systems	COST PER S.F.		
		MAT.	INST.	TOTAL
0520	Deluge sprinkler systems, steel, black, sch. 40 pipe			
0530	Light hazard, one floor, 500 S.F.	30	6.10	36.10

D4010 Sprinklers

D4010 370	Deluge Sprinkler Systems		COST PER S.F.		
			MAT.	INST.	TOTAL
0560	1000 S.F.	R211313 -10	16.80	4.35	21.15
0580	2000 S.F.		12.04	3.98	16.02
0600	5000 S.F.	R211313 -20	5.55	2.97	8.52
0620	10,000 S.F.		3.35	2.42	5.77
0640	50,000 S.F.		1.86	2.12	3.98
0660	Each additional floor, 500 S.F.		3.10	3.18	6.28
0680	1000 S.F.		2.36	2.90	5.26
0700	2000 S.F.		2.23	2.68	4.91
0720	5000 S.F.		1.58	2.28	3.86
0740	10,000 S.F.		1.41	2.10	3.51
0760	50,000 S.F.		1.37	1.94	3.31
1000	Ordinary hazard, one floor, 500 S.F.		31.50	6.95	38.45
1020	1000 S.F.		16.85	4.48	21.33
1040	2000 S.F.		12.75	4.68	17.43
1060	5000 S.F.		5.95	3.16	9.11
1080	10,000 S.F.		3.81	3.15	6.96
1100	50,000 S.F.		2.73	3	5.73
1140	Each additional floor, 500 S.F.		3.49	3.66	7.15
1160	1000 S.F.		2.37	2.93	5.30
1180	2000 S.F.		2.08	2.93	5.01
1200	5000 S.F.		2.10	2.52	4.62
1220	10,000 S.F.		1.90	2.50	4.40
1240	50,000 S.F.		1.76	2.30	4.06
1500	Extra hazard, one floor, 500 S.F.		36	8.50	44.50
1520	1000 S.F.		22.50	6.50	29
1540	2000 S.F.		12.65	5.90	18.55
1560	5000 S.F.		6.20	4.36	10.56
1580	10,000 S.F.		5.25	4.22	9.47
1600	50,000 S.F.		5.25	4.08	9.33
1660	Each additional floor, 500 S.F.		4.35	4.50	8.85
1680	1000 S.F.		3.38	4.21	7.59
1700	2000 S.F.		2.96	4.21	7.17
1720	5000 S.F.		2.38	3.72	6.10
1740	10,000 S.F.		2.93	3.55	6.48
1760	50,000 S.F.		2.89	3.42	6.31
2000	Grooved steel, black, sch. 40 pipe, light hazard, one floor				
2020	2000 S.F.		12.40	3.88	16.28
2060	10,000 S.F.		3.54	2.12	5.66
2100	Each additional floor, 2,000 S.F.		2.48	2.17	4.65
2150	10,000 S.F.		1.57	1.78	3.35
2200	Ordinary hazard, one floor, 2000 S.F.		8.30	4.07	12.37
2250	10,000 S.F.		3.84	2.67	6.51
2300	Each additional floor, 2000 S.F.		2.60	2.37	4.97
2350	10,000 S.F.		1.91	2.35	4.26
2400	Extra hazard, one floor, 2000 S.F.		13	5.05	18.05
2450	10,000 S.F.		4.90	3.46	8.36
2500	Each additional floor, 2000 S.F.		3.31	3.46	6.77
2550	10,000 S.F.		2.63	2.94	5.57
3000	Grooved steel, black, sch. 10 pipe, light hazard, one floor				
3050	2000 S.F.		11.70	3.69	15.39
3100	10,000 S.F.		3.45	2.07	5.52
3150	Each additional floor, 2000 S.F.		2.38	2.13	4.51
3200	10,000 S.F.		1.51	1.75	3.26
3250	Ordinary hazard, one floor, 2000 S.F.		12.45	4.06	16.51
3300	10,000 S.F.		3.39	2.60	5.99
3350	Each additional floor, 2000 S.F.		2.50	2.35	4.85
3400	10,000 S.F.		1.81	2.30	4.11
3450	Extra hazard, one floor, 2000 S.F.		12.95	5.05	18

539

For customer support on your Plumbing Costs with RSMeans Data, call 800.448.8182.

D4010 370	Deluge Sprinkler Systems	COST PER S.F.		
		MAT.	INST.	TOTAL
3500	10,000 S.F.	4.67	3.38	8.05
3550	Each additional floor, 2000 S.F.	3.24	3.44	6.68
3600	10,000 S.F.	2.53	2.90	5.43
4000	Copper tubing, type L, light hazard, one floor			
4050	2000 S.F.	12.35	3.88	16.23
4100	10,000 S.F.	3.51	2.14	5.65
4150	Each additional floor, 2000 S.F.	2.42	2.21	4.63
4200	10,000 S.F.	1.29	1.81	3.10
4250	Ordinary hazard, one floor, 2000 S.F.	12.50	4.28	16.78
4300	10,000 S.F.	3.83	2.51	6.34
4350	Each additional floor, 2000 S.F.	2.46	2.25	4.71
4400	10,000 S.F.	1.77	1.97	3.74
4450	Extra hazard, one floor, 2000 S.F.	13.05	5.15	18.20
4500	10,000 S.F.	6.10	3.77	9.87
4550	Each additional floor, 2000 S.F.	3.37	3.55	6.92
4600	10,000 S.F.	3.50	3.22	6.72
5000	Copper tubing, type L, T-drill system, light hazard, one floor			
5050	2000 S.F.	12.55	3.66	16.21
5100	10,000 S.F.	3.46	1.80	5.26
5150	Each additional floor, 2000 S.F.	2.60	2.01	4.61
5200	10,000 S.F.	1.53	1.48	3.01
5250	Ordinary hazard, one floor, 2000 S.F.	10.55	3.48	14.03
5300	10,000 S.F.	3.84	2.24	6.08
5350	Each additional floor, 2000 S.F.	2.61	2	4.61
5400	10,000 S.F.	1.91	1.93	3.84
5450	Extra hazard, one floor, 2000 S.F.	10.80	4.08	14.88
5500	10,000 S.F.	4.99	2.78	7.77
5550	Each additional floor, 2000 S.F.	2.94	2.71	5.65
5600	10,000 S.F.	2.44	2.25	4.69

D4010 Sprinklers

On-off multicycle sprinkler system is a fixed fire protection system utilizing water as its extinguishing agent. It is a time delayed, recycling, preaction type which automatically shuts the water off when heat is reduced below the detector operating temperature and turns the water back on when that temperature is exceeded.

The system senses a fire condition through a closed circuit electrical detector system which controls water flow to the fire automatically. Batteries supply up to 90 hour emergency power supply for system operation. The piping system is dry (until water is required) and is monitored with pressurized air. Should any leak in the system piping occur, an alarm will sound, but water will not enter the system until heat is sensed by a Firecycle detector.

All areas are assumed to be open.

System Components	QUANTITY	UNIT	COST EACH		
			MAT.	INST.	TOTAL
SYSTEM D4010 390 0580					
ON-OFF MULTICYCLE SPRINKLER SYSTEM, STEEL, BLACK, SCH. 40 PIPE					
LIGHT HAZARD, ONE FLOOR, 2000 S.F.					
Valve, gate, iron body 125 lb., OS&Y, flanged, 4" pipe size	1.000	Ea.	1,100	485	1,585
Backflow pvntr, auto, 2 OS&Y v, tst cocks, dbl chk, flgd, 4"	1.000	Ea.	4,050	485	4,535
Tamper switch (valve supervisory switch)	3.000	Ea.	882	148.50	1,030.50
Valve, angle, bronze, 150 lb., rising stem, threaded, 2" pipe size	1.000	Ea.	950	73.50	1,023.50
Valve, swing check, bronze, 125 lb, regrinding disc, 2-1/2" pipe size	1.000	Ea.	1,075	97	1,172
*Alarm valve, 2-1/2" pipe size	1.000	Ea.	1,800	475	2,275
Alarm, water motor, complete with gong	1.000	Ea.	355	199	554
Pipe, steel, black, schedule 40, 4" diam.	8.000	L.F.	192	341.60	533.60
Fire alarm, horn, electric	1.000	Ea.	55.50	113	168.50
Pipe, steel, black, schedule 40, threaded, cplg & hngr 10' OC 2-1/2" diam.	15.000	L.F.	235.50	435	670.50
Pipe, steel, black, schedule 40, threaded, cplg & hngr 10' OC 2" diam.	9.375	L.F.	78.28	210.94	289.22
Pipe, steel, black, schedule 40, threaded, cplg & hngr 10' OC 1-1/4" diam.	28.125	L.F.	149.06	459.84	608.90
Pipe, steel, black, schedule 40, threaded, cplg & hngr 10' OC 1" diam.	84.000	L.F.	344.40	1,281	1,625.40
Pipe, Tee, malleable iron, black, 150 lb. threaded, 4" pipe size	2.000	Ea.	820	730	1,550
Pipe, Tee, malleable iron, black, 150 lb. threaded, 2-1/2" pipe size	2.000	Ea.	232	324	556
Pipe, Tee, malleable iron, black, 150 lb. threaded, 2" pipe size	1.000	Ea.	53.50	132	185.50
Pipe, Tee, malleable iron, black, 150 lb. threaded, 1-1/4" pipe size	4.000	Ea.	102	416	518
Pipe, Tee, malleable iron, black, 150 lb. threaded, 1" pipe size	3.000	Ea.	46.80	303	349.80
Pipe, 90° elbow, malleable iron, black, 150 lb. threaded, 1" pipe size	5.000	Ea.	50	310	360
Sprinkler head std spray, brass 135°-286°F 1/2" NPT, 3/8" orifice	12.000	Ea.	258	594	852
Firecycle controls, incls panel, battery, solenoid valves, press switches	1.000	Ea.	25,700	3,025	28,725
Detector, firecycle system	2.000	Ea.	1,920	99	2,019
Firecycle pkg, swing check & flow control valves w/trim 4" pipe size	1.000	Ea.	7,800	1,425	9,225
Air compressor, auto, complete, 200 Gal sprinkler sys. cap., 1/3 HP	1.000	Ea.	1,550	610	2,160
*Standpipe connection, wall, flush, brass w/plug & chain 2-1/2"x2-1/2"	1.000	Ea.	167	286	453
Valve, gate, bronze 300 psi, NRS, class 150, threaded, 1" diam.	1.000	Ea.	152	42.50	194.50
TOTAL			50,118.04	13,100.88	63,218.92
COST PER S.F.			18.79	4.91	23.70

*Not included in systems under 2000 S.F.

D4010 390	On-off multicycle Sprinkler Systems	COST PER S.F.		
		MAT.	INST.	TOTAL
0520	On-off multicycle sprinkler systems, steel, black, sch. 40 pipe			
0530	Light hazard, one floor, 500 S.F.	59.50	11.55	71.05

541

D4010 Sprinklers

D4010 390	On-off multicycle Sprinkler Systems		COST PER S.F.		
			MAT.	INST.	TOTAL
0560	1000 S.F.	R211313 -10	31.50	7.25	38.75
0580	2000 S.F.		18.75	4.91	23.66
0600	5000 S.F.	R211313 -20	8.25	3.35	11.60
0620	10,000 S.F.		4.78	2.61	7.39
0640	50,000 S.F.		2.20	2.16	4.36
0660	Each additional floor of 500 S.F.		3.14	3.20	6.34
0680	1000 S.F.		2.38	2.91	5.29
0700	2000 S.F.		1.88	2.67	4.55
0720	5000 S.F.		1.59	2.29	3.88
0740	10,000 S.F.		1.49	2.11	3.60
0760	50,000 S.F.		1.43	1.94	3.37
1000	Ordinary hazard, one floor, 500 S.F.		57	11.70	68.70
1020	1000 S.F.		31.50	7.20	38.70
1040	2000 S.F.		19.15	5.60	24.75
1060	5000 S.F.		8.70	3.55	12.25
1080	10,000 S.F.		5.25	3.35	8.60
1100	50,000 S.F.		3.23	3.36	6.59
1140	Each additional floor, 500 S.F.		3.53	3.68	7.21
1160	1000 S.F.		2.39	2.94	5.33
1180	2000 S.F.		2.38	2.70	5.08
1200	5000 S.F.		2.11	2.53	4.64
1220	10,000 S.F.		1.83	2.46	4.29
1240	50,000 S.F.		1.80	2.19	3.99
1500	Extra hazard, one floor, 500 S.F.		65.50	13.95	79.45
1520	1000 S.F.		36	9	45
1540	2000 S.F.		19.40	6.85	26.25
1560	5000 S.F.		8.90	4.74	13.64
1580	10,000 S.F.		6.70	4.78	11.48
1600	50,000 S.F.		5.60	5.25	10.85
1660	Each additional floor, 500 S.F.		4.39	4.51	8.90
1680	1000 S.F.		3.40	4.22	7.62
1700	2000 S.F.		2.98	4.22	7.20
1720	5000 S.F.		2.39	3.73	6.12
1740	10,000 S.F.		2.97	3.43	6.40
1760	50,000 S.F.		2.92	3.30	6.22
2020	Grooved steel, black, sch. 40 pipe, light hazard, one floor				
2030	2000 S.F.		17.35	4.59	21.94
2060	10,000 S.F.		5.35	3.16	8.51
2100	Each additional floor, 2000 S.F.		2.49	2.17	4.66
2150	10,000 S.F.		1.65	1.79	3.44
2200	Ordinary hazard, one floor, 2000 S.F.		19.30	5.05	24.35
2250	10,000 S.F.		5.75	3.02	8.77
2300	Each additional floor, 2000 S.F.		2.61	2.38	4.99
2350	10,000 S.F.		2	2.36	4.36
2400	Extra hazard, one floor, 2000 S.F.		17.90	5.80	23.70
2450	10,000 S.F.		6.20	3.62	9.82
2500	Each additional floor, 2000 S.F.		3.33	3.47	6.80
2550	10,000 S.F.		2.71	2.94	5.65
3050	Grooved steel, black, sch. 10 pipe, light hazard, one floor,				
3060	2000 S.F.		19.10	4.80	23.90
3100	10,000 S.F.		4.88	2.27	7.15
3150	Each additional floor, 2000 S.F.		2.40	2.14	4.54
3200	10,000 S.F.		1.59	1.76	3.35
3250	Ordinary hazard, one floor, 2000 S.F.		19.20	5	24.20
3300	10,000 S.F.		5.20	2.82	8.02
3350	Each additional floor, 2000 S.F.		2.52	2.35	4.87
3400	10,000 S.F.		1.90	2.31	4.21
3450	Extra hazard, one floor, 2000 S.F.		19.70	6	25.70

For customer support on your Plumbing Costs with RSMeans Data, call 800.448.8182.

D40 Fire Protection

D4010 Sprinklers

D4010 390	On-off multicycle Sprinkler Systems	COST PER S.F.		
		MAT.	INST.	TOTAL
3500	10,000 S.F.	6.05	3.56	9.61
3550	Each additional floor, 2000 S.F.	3.26	3.44	6.70
3600	10,000 S.F.	2.61	2.91	5.52
4060	Copper tubing, type L, light hazard, one floor, 2000 S.F.	19.10	4.83	23.93
4100	10,000 S.F.	5	2.34	7.34
4150	Each additional floor, 2000 S.F.	2.44	2.22	4.66
4200	10,000 S.F.	1.72	1.84	3.56
4250	Ordinary hazard, one floor, 2000 S.F.	19.90	5.20	25.10
4300	10,000 S.F.	5.95	2.65	8.60
4350	Each additional floor, 2000 S.F.	2.99	2.22	5.21
4400	10,000 S.F.	2.36	1.91	4.27
4450	Extra hazard, one floor, 2000 S.F.	20.50	6.05	26.55
4500	10,000 S.F.	8.60	3.92	12.52
4550	Each additional floor, 2000 S.F.	4.19	3.50	7.69
4600	10,000 S.F.	4.25	3.17	7.42
5060	Copper tubing, type L, T-drill system, light hazard, one floor 2000 S.F.	19.30	4.60	23.90
5100	10,000 S.F.	5.10	1.98	7.08
5150	Each additional floor, 2000 S.F.	2.95	2.07	5.02
5200	10,000 S.F.	1.61	1.49	3.10
5250	Ordinary hazard, one floor, 2000 S.F.	19.65	4.62	24.27
5300	10,000 S.F.	5.85	2.39	8.24
5350	Each additional floor, 2000 S.F.	2.98	1.96	4.94
5400	10,000 S.F.	2.59	1.88	4.47
5450	Extra hazard, one floor, 2000 S.F.	19.95	5.20	25.15
5500	10,000 S.F.	7.45	2.90	10.35
5550	Each additional floor, 2000 S.F.	3.51	2.67	6.18
5600	10,000 S.F.	3.14	2.21	5.35

Wet Pipe System. A system employing automatic sprinklers attached to a piping system containing water and connected to a water supply so that water discharges immediately from sprinklers opened by heat from a fire.

All areas are assumed to be open.

System Components	QUANTITY	UNIT	COST EACH		
			MAT.	INST.	TOTAL
SYSTEM D4010 410 0580					
WET PIPE SPRINKLER, STEEL, BLACK, SCH. 40 PIPE					
LIGHT HAZARD, ONE FLOOR, 2000 S.F.					
Valve, gate, iron body, 125 lb., OS&Y, flanged, 4" diam.	1.000	Ea.	1,100	485	1,585
Backflow pvntr, auto, 2 OS&Y v, tst cocks, dbl chk, flgd, 4"	1.000	Ea.	4,050	485	4,535
Tamper switch (valve supervisory switch)	3.000	Ea.	882	148.50	1,030.50
Valve, swing check, bronze, 125 lb, regrinding disc, 2-1/2" pipe size	1.000	Ea.	1,075	97	1,172
Valve, angle, bronze, 150 lb., rising stem, threaded, 2" diam.	1.000	Ea.	950	73.50	1,023.50
*Alarm valve, 2-1/2" pipe size	1.000	Ea.	1,800	475	2,275
Alarm, water motor, complete with gong	1.000	Ea.	355	199	554
Valve, swing check, w/balldrip CI with brass trim 4" pipe size	1.000	Ea.	480	475	955
Pipe, steel, black, schedule 40, 4" diam.	8.000	L.F.	192	341.60	533.60
Fire alarm horn, electric	1.000	Ea.	55.50	113	168.50
Pipe, steel, black, schedule 40, threaded, cplg & hngr 10' OC, 2-1/2" diam.	15.000	L.F.	235.50	435	670.50
Pipe, steel, black, schedule 40, threaded, cplg & hngr 10' OC, 2" diam.	9.375	L.F.	78.28	210.94	289.22
Pipe, steel, black, schedule 40, threaded, cplg & hngr 10' OC, 1-1/4" diam.	28.125	L.F.	149.06	459.84	608.90
Pipe, steel, black, schedule 40, threaded cplg & hngr 10' OC, 1" diam.	84.000	L.F.	344.40	1,281	1,625.40
Pipe Tee, malleable iron black, 150 lb. threaded, 4" pipe size	2.000	Ea.	820	730	1,550
Pipe Tee, malleable iron black, 150 lb. threaded, 2-1/2" pipe size	2.000	Ea.	232	324	556
Pipe Tee, malleable iron black, 150 lb. threaded, 2" pipe size	1.000	Ea.	53.50	132	185.50
Pipe Tee, malleable iron black, 150 lb. threaded, 1-1/4" pipe size	4.000	Ea.	102	416	518
Pipe Tee, malleable iron black, 150 lb. threaded, 1" pipe size	3.000	Ea.	46.80	303	349.80
Pipe 90° elbow, malleable iron black, 150 lb. threaded, 1" pipe size	5.000	Ea.	50	310	360
Sprinkler head, standard spray, brass 135°-286°F 1/2" NPT, 3/8" orifice	12.000	Ea.	258	594	852
Valve, gate, bronze, NRS, class 150, threaded, 1" pipe size	1.000	Ea.	152	42.50	194.50
*Standpipe connection, wall, single, flush w/plug & chain 2-1/2"x2-1/2"	1.000	Ea.	167	286	453
TOTAL			13,628.04	8,416.88	22,044.92
COST PER S.F.			5.11	3.16	8.27

*Not included in systems under 2000 S.F.

D4010 410	Wet Pipe Sprinkler Systems		COST PER S.F.		
			MAT.	INST.	TOTAL
0520	Wet pipe sprinkler systems, steel, black, sch. 40 pipe				
0530	Light hazard, one floor, 500 S.F.		6.15	4	10.15
0560	1000 S.F.	R211313 -10	4.87	3.49	8.36
0580	2000 S.F.		5.11	3.16	8.27
0600	5000 S.F.	R211313 -20	2.51	2.63	5.14
0620	10,000 S.F.		1.61	2.24	3.85

D4010 Sprinklers

D4010 410	Wet Pipe Sprinkler Systems	COST PER S.F.		
		MAT.	INST.	TOTAL
0640	50,000 S.F.	1.17	2.07	3.24
0660	Each additional floor, 500 S.F.	1.73	3.20	4.93
0680	1000 S.F. R211313 -40	1.69	2.95	4.64
0700	2000 S.F.	1.52	2.65	4.17
0720	5000 S.F.	1.02	2.26	3.28
0740	10,000 S.F.	.92	2.08	3
0760	50,000 S.F.	.79	1.63	2.42
1000	Ordinary hazard, one floor, 500 S.F.	6.50	4.25	10.75
1020	1000 S.F.	4.84	3.42	8.26
1040	2000 S.F.	5.45	3.85	9.30
1060	5000 S.F.	2.92	2.83	5.75
1080	10,000 S.F.	2.08	2.97	5.05
1100	50,000 S.F.	1.96	2.86	4.82
1140	Each additional floor, 500 S.F.	2.09	3.60	5.69
1160	1000 S.F.	1.67	2.90	4.57
1180	2000 S.F.	1.73	2.92	4.65
1200	5000 S.F.	1.69	2.76	4.45
1220	10,000 S.F.	1.39	2.81	4.20
1240	50,000 S.F.	1.38	2.49	3.87
1500	Extra hazard, one floor, 500 S.F.	9.30	5.50	14.80
1520	1000 S.F.	8.30	4.89	13.19
1540	2000 S.F.	5.75	5.25	11
1560	5000 S.F.	3.66	4.66	8.32
1580	10,000 S.F.	3.57	4.47	8.04
1600	50,000 S.F.	3.90	4.28	8.18
1660	Each additional floor, 500 S.F.	2.95	4.44	7.39
1680	1000 S.F.	2.68	4.18	6.86
1700	2000 S.F.	2.26	4.18	6.44
1720	5000 S.F.	1.81	3.70	5.51
1740	10,000 S.F.	2.40	3.40	5.80
1760	50,000 S.F.	2.36	3.24	5.60
2020	Grooved steel, black sch. 40 pipe, light hazard, one floor, 2000 S.F.	5.50	3.07	8.57
2060	10,000 S.F.	1.77	1.92	3.69
2100	Each additional floor, 2000 S.F.	1.77	2.14	3.91
2150	10,000 S.F.	1.08	1.76	2.84
2200	Ordinary hazard, one floor, 2000 S.F.	5.60	3.28	8.88
2250	10,000 S.F.	2.11	2.49	4.60
2300	Each additional floor, 2000 S.F.	1.89	2.35	4.24
2350	10,000 S.F.	1.42	2.33	3.75
2400	Extra hazard, one floor, 2000 S.F.	4.22	4.02	8.24
2450	10,000 S.F.	2.59	3.20	5.79
2500	Each additional floor, 2000 S.F.	2.61	3.43	6.04
2550	10,000 S.F.	2.13	2.91	5.04
3050	Grooved steel, black sch. 10 pipe, light hazard, one floor, 2000 S.F.	3.55	2.81	6.36
3100	10,000 S.F.	1.71	1.90	3.61
3150	Each additional floor, 2000 S.F.	1.68	2.10	3.78
3200	10,000 S.F.	1.02	1.73	2.75
3250	Ordinary hazard, one floor, 2000 S.F.	5.55	3.25	8.80
3300	10,000 S.F.	1.94	2.43	4.37
3350	Each additional floor, 2000 S.F.	1.80	2.32	4.12
3400	10,000 S.F.	1.32	2.28	3.60
3450	Extra hazard, one floor, 2000 S.F.	6	4.23	10.23
3500	10,000 S.F.	2.93	3.21	6.14
3550	Each additional floor, 2000 S.F.	2.54	3.41	5.95
3600	10,000 S.F.	2.03	2.88	4.91
4050	Copper tubing, type L, light hazard, one floor, 2000 S.F.	5.45	3.07	8.52
4100	10,000 S.F.	1.84	1.97	3.81
4150	Each additional floor, 2000 S.F.	1.72	2.18	3.90

545

D4010 Sprinklers

D4010 410	Wet Pipe Sprinkler Systems	COST PER S.F.		
		MAT.	INST.	TOTAL
4200	10,000 S.F.	1.15	1.81	2.96
4250	Ordinary hazard, one floor, 2000 S.F.	5.60	3.48	9.08
4300	10,000 S.F.	2.15	2.33	4.48
4350	Each additional floor, 2000 S.F.	1.91	2.41	4.32
4400	10,000 S.F.	1.40	2.13	3.53
4450	Extra hazard, one floor, 2000 S.F.	6.15	4.35	10.50
4500	10,000 S.F.	4.32	3.57	7.89
4550	Each additional floor, 2000 S.F.	2.67	3.52	6.19
4600	10,000 S.F.	3.01	3.20	6.21
5050	Copper tubing, type L, T-drill system, light hazard, one floor			
5060	2000 S.F.	5.60	2.85	8.45
5100	10,000 S.F.	1.73	1.62	3.35
5150	Each additional floor, 2000 S.F.	1.90	1.95	3.85
5200	10,000 S.F.	1.04	1.46	2.50
5250	Ordinary hazard, one floor, 2000 S.F.	3.60	2.67	6.27
5300	10,000 S.F.	2.10	2.07	4.17
5350	Each additional floor, 2000 S.F.	1.73	1.97	3.70
5400	10,000 S.F.	1.42	1.91	3.33
5450	Extra hazard, one floor, 2000 S.F.	5.70	3.51	9.21
5500	10,000 S.F.	3.26	2.61	5.87
5550	Each additional floor, 2000 S.F.	4.43	3	7.43
5600	10,000 S.F.	1.95	2.23	4.18

D4010 Sprinklers

Plan

Longitudinal Movement

Elevation

Horizontal Views

System Components			COST PER EACH		
	QUANTITY	UNIT	MAT.	INST.	TOTAL
SYSTEM D4010 412 1000					
WET SPRINKLER SYSTEM, SCHEDULE 10 BLACK STEEL GROOVED PIPE					
SEPARATION ASSEMBLY, 2 INCH DIAMETER					
Hanger support for 2 inch pipe	2.000	Ea.	112	88	200
Pipe roller support yoke	2.000	Ea.	24.90	20.80	45.70
Spool pieces 2 inch	3.000	Ea.	26.10	56.40	82.50
2 inch grooved elbows	6.000	Ea.	336	195	531
Coupling grooved joint 2 inch pipe	10.000	Ea.	555	161.50	716.50
Roll grooved joint labor only	8.000	Ea.		100.40	100.40
TOTAL			1,054	622.10	1,676.10

D4010 412	Wet Sprinkler Seismic Components	COST PER EACH		
		MAT.	INST.	TOTAL
1000	Wet sprinkler sys, Sch. 10, blk steel grooved, separation assembly, 2 inch	1,050	620	1,670
1100	2-1/2 inch	1,175	690	1,865
1200	3 inch	1,550	790	2,340
1300	4 inch	1,925	985	2,910
1400	6 inch	3,850	1,425	5,275
1500	8 inch	6,875	1,675	8,550
2000	Wet sprinkler sys, Sch. 10, blk steel grooved, flexible coupling, 2-1/2 inch	670	345	1,015
2100	3 inch	695	405	1,100
2200	4 inch	935	540	1,475
2300	6 inch	1,325	755	2,080

D4010 412	Wet Sprinkler Seismic Components	COST PER EACH		
		MAT.	INST.	TOTAL
2400	8 inch	1,875	945	2,820
2990	Allowance for seismic movement by providing annular space in concrete wall			
2995	If additional cores are performed for the same size and at the same time reduce			
2996	price of additional cores by 50 %. If different size cores reduce by 30 %.			
3000	Wet pipe sprinkler systems, 2 inch black steel, annular space 6 inch wall	4.07	720	724.07
3010	8 inch wall	4.30	725	729.30
3020	10 inch wall	4.54	735	739.54
3030	12 inch wall	4.77	740	744.77
3100	Wet pipe sprinkler systems, 2-1/2/3 inch black steel, annular space 6" wall	4.07	740	744.07
3110	8 inch wall	4.30	745	749.30
3120	10 inch wall	4.54	755	759.54
3130	12 inch wall	4.77	760	764.77
3200	Wet pipe sprinkler systems, 4 inch black steel, annular space 6" wall	4.07	820	824.07
3210	8 inch wall	4.30	825	829.30
3220	10 inch wall	4.54	835	839.54
3230	12 inch wall	4.77	840	844.77
3300	Wet pipe sprinkler systems, 6 inch black steel, annular space 6" wall	4.07	855	859.07
3310	8 inch wall	4.30	860	864.30
3320	10 inch wall	4.54	870	874.54
3330	12 inch wall	4.77	875	879.77
3400	Wet pipe sprinkler systems, 8 inch black steel, annular space 6" wall	4.07	885	889.07
3410	8 inch wall	4.30	895	899.30
3420	10 inch wall	4.54	900	904.54
3430	12 inch wall	4.77	910	914.77
3996	Lateral, longitudinal and 4-way seismic strut braces			
4000	Wet pipe sprinkler systems, 2 inch pipe double lateral strut brace	134	335	469
4010	2 -1/2 inch	143	335	478
4020	3 inch	150	335	485
4030	4 inch	164	340	504
4040	6 inch	238	340	578
4050	8 inch	291	340	631
4100	Wet pipe sprinkler systems, 2 inch pipe longitudinal strut brace	134	335	469
4110	2 -1/2 inch	143	335	478
4120	3 inch	150	335	485
4130	4 inch	164	340	504
4140	6 inch	238	340	578
4150	8 inch	291	340	631
4200	Wet pipe sprinkler systems, 2 inch pipe single lateral strut brace	95	221	316
4210	2 -1/2 inch	104	222	326
4220	3 inch	111	222	333
4230	4 inch	124	225	349
4240	6 inch	197	226	423
4250	8 inch	250	227	477
4300	Wet pipe sprinkler systems, 2 inch pipe longitudinal wire brace	79	232	311
4310	2 -1/2 inch	88	232	320
4320	3 inch	95	233	328
4330	4 inch	138	240	378
4340	6 inch	183	238	421
4350	8 inch	236	238	474
4400	Wet pipe sprinkler systems, 2 inch pipe 4-way wire brace	102	355	457
4410	2 -1/2 inch	111	355	466
4420	3 inch	118	360	478
4430	4 inch	194	370	564
4440	6 inch	209	365	574
4450	8 inch	262	365	627
4500	Wet pipe sprinkler systems, 2 inch pipe 4-way strut brace	211	565	776
4510	2 -1/2 inch	220	565	785
4520	3 inch	227	565	792

D4010 Sprinklers

D4010 412	Wet Sprinkler Seismic Components	COST PER EACH		
		MAT.	INST.	TOTAL
4530	4 inch	246	570	816
4540	6 inch	320	570	890
4550	8 inch	370	570	940

D4010 Sprinklers

Wet Pipe System. A system employing automatic sprinklers attached to a piping system containing water and connected to a water supply so that water discharges immediately from sprinklers opened by heat from a fire.

All areas are assumed to be open.

System Components	QUANTITY	UNIT	COST PER S.F.		
			MAT.	INST.	TOTAL
SYSTEM D4010 413 0580					
SEISMIC WET PIPE SPRINKLER, STEEL, BLACK, SCH. 40 PIPE					
LIGHT HAZARD, ONE FLOOR, 3' FLEXIBLE FEED, 2000 S.F.					
Valve, gate, iron body, 125 lb., OS&Y, flanged, 4" diam.	1.000	Ea.	.41	.18	.59
Backflow pvntr, auto, 2 OS&Y v, tst cocks, dbl chk, flgd, 4"	1.000	Ea.	1.52	.18	1.70
Tamper switch (valve supervisory switch)	3.000	Ea.	.33	.06	.39
Valve, swing check, bronze, 125 lb, regrinding disc, 2-1/2" pipe size	1.000	Ea.	.40	.04	.44
Valve, angle, bronze, 150 lb., rising stem, threaded, 2" diam.	1.000	Ea.	.36	.03	.39
*Alarm valve, 2-1/2" pipe size	1.000	Ea.	.67	.18	.85
Alarm, water motor, complete with gong	1.000	Ea.	.13	.07	.20
Valve, swing check, w/balldrip CI with brass trim 4" pipe size	1.000	Ea.	.18	.18	.36
Pipe, steel, black, schedule 40, 4" diam.	8.000	L.F.	.07	.13	.20
Fire alarm horn, electric	1.000	Ea.	.02	.04	.06
Pipe, steel, black, schedule 40, threaded, cplg & hngr 10' OC, 2-1/2" diam.	15.000	L.F.	.09	.16	.25
Pipe, steel, black, schedule 40, threaded, cplg & hngr 10' OC, 2" diam.	9.375	L.F.	.03	.08	.11
Pipe, steel, black, schedule 40, threaded, cplg & hngr 10' OC, 1-1/4" diam.	28.125	L.F.	.06	.17	.23
Pipe Tee, malleable iron black, 150 lb. threaded, 4" pipe size	2.000	Ea.	.31	.27	.58
Pipe Tee, malleable iron black, 150 lb. threaded, 2-1/2" pipe size	2.000	Ea.	.09	.12	.21
Pipe Tee, malleable iron black, 150 lb. threaded, 2" pipe size	1.000	Ea.	.02	.05	.07
Pipe Tee, malleable iron black, 150 lb. threaded, 1-1/4" pipe size	4.000	Ea.	.04	.16	.20
Pipe Tee, malleable iron black, 150 lb. threaded, 1" pipe size	3.000	Ea.	.02	.11	.13
Pipe 90° elbow, malleable iron black, 150 lb. threaded, 1" pipe size	5.000	Ea.	.02	.12	.14
Sprinkler head, standard spray, brass 135°-286°F 1/2" NPT, 3/8" orifice	12.000	Ea.	.10	.22	.32
Valve, gate, bronze, NRS, class 150, threaded, 1" pipe size	1.000	Ea.	.06	.02	.08
*Standpipe connection, wall, single, flush w/plug & chain 2-1/2"x2-1/2"	1.000	Ea.	.06	.11	.17
Flexible connector 36 inches long	12.000	Ea.	.31	.12	.43
TOTAL			5.30	2.80	8.10

*Not included in systems under 2000 S.F.

D4010 413	Seismic Wet Pipe Flexible Feed Sprinkler Systems	COST PER S.F.		
		MAT.	INST.	TOTAL
0520	Seismic wet pipe 3' flexible feed sprinkler systems, steel, black, sch. 40 pipe			
0525	Note: See D4010-412 for strut braces, flexible joints and separation assemblies			
0530	Light hazard, one floor, 3' flexible feed, 500 S.F.	6.25	3.30	9.55
0560	1000 S.F.	5	3.04	8.04
0580	2000 S.F.	5.30	2.79	8.09
0600	5000 S.F.	2.71	2.30	5.01

550

D4010 Sprinklers

D4010 413	Seismic Wet Pipe Flexible Feed Sprinkler Systems	COST PER S.F.		
		MAT.	INST.	TOTAL
0620	10,000 S.F.	1.83	2.09	3.92
0640	50,000 S.F.	1.33	1.81	3.14
0660	Each additional floor, 500 S.F.	1.92	2.54	4.46
0680	1000 S.F.	1.95	2.53	4.48
0700	2000 S.F.	1.66	2.13	3.79
0720	5000 S.F.	1.22	1.93	3.15
0740	10,000 S.F.	1.16	1.93	3.09
0760	50,000 S.F.	1.01	1.53	2.54
1000	Ordinary hazard, one floor, 500 S.F.	6.70	3.59	10.29
1020	1000 S.F.	5.05	2.93	7.98
1040	2000 S.F.	5.75	3.55	9.30
1060	5000 S.F.	3.23	2.62	5.85
1080	10,000 S.F.	2.40	2.83	5.23
1100	50,000 S.F.	2.24	2.59	4.83
1140	Each additional floor, 500 S.F.	5.10	3.19	8.29
1160	1000 S.F.	1.91	2.41	4.32
1180	2000 S.F.	2.02	2.62	4.64
1200	5000 S.F.	2.04	2.56	4.60
1220	10,000 S.F.	1.74	2.68	4.42
1240	50,000 S.F.	1.67	2.22	3.89
1500	Extra hazard, one floor, 500 S.F.	9.50	4.44	13.94
1520	1000 S.F.	8.55	3.88	12.43
1540	2000 S.F.	6.20	4.77	10.97
1560	5000 S.F.	4.07	4.36	8.43
1580	10,000 S.F.	4.05	4.25	8.30
1600	50,000 S.F.	4.41	4.03	8.44
1660	Each additional floor, 500 S.F.	2.74	3.33	6.07
1680	1000 S.F.	3.15	3.21	6.36
1700	2000 S.F.	2.77	3.73	6.50
1720	5000 S.F.	2.26	3.40	5.66
1740	10,000 S.F.	2.90	3.17	6.07
1760	50,000 S.F.	2.87	2.98	5.85
2020	Grooved steel, black sch. 40 pipe, light hazard, one floor, 2000 S.F.	3.70	2.41	6.11
2060	10,000 S.F.	1.75	1.73	3.48
2100	Each additional floor, 2000 S.F.	1.83	1.72	3.55
2150	10,000 S.F.	1.08	1.57	2.65
2200	Ordinary hazard, one floor, 2000 S.F.	5.85	3.05	8.90
2250	10,000 S.F.	2.40	2.40	4.80
2300	Each additional floor, 2000 S.F.	2.12	2.12	4.24
2350	10,000 S.F.	1.74	2.25	3.99
2400	Extra hazard, one floor, 2000 S.F.	6.40	3.89	10.29
2450	10,000 S.F.	3.49	3.11	6.60
2500	Each additional floor, 2000 S.F.	3.03	3.08	6.11
2550	10,000 S.F.	2.57	2.76	5.33
3050	Grooved steel black sch. 10 pipe, light hazard, one floor, 2000 S.F.	5.45	2.62	8.07
3100	10,000 S.F.	1.89	1.79	3.68
3150	Each additional floor, 2000 S.F.	1.73	1.68	3.41
3200	10,000 S.F.	1.02	1.55	2.57
3250	Ordinary hazard, one floor, 2000 S.F.	5.75	3.03	8.78
3300	10,000 S.F.	2.30	2.35	4.65
3350	Each additional floor, 2000 S.F.	1.92	2.07	3.99
3400	10,000 S.F.	1.59	2.19	3.78
3450	Extra hazard, one floor, 2000 S.F.	6.30	3.87	10.17
3500	10,000 S.F.	3.35	3.05	6.40
3550	Each additional floor, 2000 S.F.	2.95	3.06	6.01
3600	10,000 S.F.	2.47	2.72	5.19

Wet Pipe System. A system employing automatic sprinklers attached to a piping system containing water and connected to a water supply so that water discharges immediately from sprinklers opened by heat from a fire.

All areas are assumed to be open.

System Components	QUANTITY	UNIT	COST PER S.F.		
			MAT.	INST.	TOTAL
SYSTEM D4010 414 0580					
SEISMIC WET PIPE SPRINKLER, STEEL, BLACK, SCH. 40 PIPE					
LIGHT HAZARD, ONE FLOOR, 4' FLEXIBLE FEED, 2000 S.F.					
Valve, gate, iron body, 125 lb., OS&Y, flanged, 4" diam.	1.000	Ea.	.41	.18	.59
Backflow pvntr, auto, 2 OS&Y v, tst cocks, dbl chk, flgd, 4"	1.000	Ea.	1.52	.18	1.70
Tamper switch (valve supervisory switch)	3.000	Ea.	.33	.06	.39
Valve, swing check, bronze, 125 lb, regrinding disc, 2-1/2" pipe size	1.000	Ea.	.40	.04	.44
Valve, angle, bronze, 150 lb., rising stem, threaded, 2" diam.	1.000	Ea.	.36	.03	.39
*Alarm valve, 2-1/2" pipe size	1.000	Ea.	.67	.18	.85
Alarm, water motor, complete with gong	1.000	Ea.	.13	.07	.20
Valve, swing check, w/balldrip Cl with brass trim 4" pipe size	1.000	Ea.	.18	.18	.36
Pipe, steel, black, schedule 40, 4" diam.	8.000	L.F.	.07	.13	.20
Fire alarm horn, electric	1.000	Ea.	.02	.04	.06
Pipe, steel, black, schedule 40, threaded, cplg & hngr 10' OC, 2-1/2" diam.	15.000	L.F.	.09	.16	.25
Pipe, steel, black, schedule 40, threaded, cplg & hngr 10' OC, 2" diam.	9.375	L.F.	.03	.08	.11
Pipe, steel, black, schedule 40, threaded, cplg & hngr 10' OC, 1-1/4" diam.	28.125	L.F.	.06	.17	.23
Pipe Tee, malleable iron black, 150 lb. threaded, 4" pipe size	2.000	Ea.	.31	.27	.58
Pipe Tee, malleable iron black, 150 lb. threaded, 2-1/2" pipe size	2.000	Ea.	.09	.12	.21
Pipe Tee, malleable iron black, 150 lb. threaded, 2" pipe size	1.000	Ea.	.02	.05	.07
Pipe Tee, malleable iron black, 150 lb. threaded, 1-1/4" pipe size	4.000	Ea.	.04	.16	.20
Pipe Tee, malleable iron black, 150 lb. threaded, 1" pipe size	3.000	Ea.	.02	.11	.13
Pipe 90° elbow, malleable iron black, 150 lb. threaded, 1" pipe size	5.000	Ea.	.02	.12	.14
Sprinkler head, standard spray, brass 135°-286°F 1/2" NPT, 3/8" orifice	12.000	Ea.	.10	.22	.32
Valve, gate, bronze, NRS, class 150, threaded, 1" pipe size	1.000	Ea.	.06	.02	.08
*Standpipe connection, wall, single, flush w/plug & chain 2-1/2"x2-1/2"	1.000	Ea.	.06	.11	.17
Flexible connector 48 inches long	12.000	Ea.	.36	.16	.52
TOTAL			5.35	2.84	8.19

*Not included in systems under 2000 S.F.

D4010 414	Seismic Wet Pipe Flexible Feed Sprinkler Systems	COST PER S.F.		
		MAT.	INST.	TOTAL
0520	Seismic wet pipe 4' flexible feed sprinkler systems, steel, black, sch. 40 pipe			
0525	Note: See D4010-412 for strut braces, flexible joints and separation assemblies			
0530	Light hazard, one floor, 4' flexible feed, 500 S.F.	6.30	3.34	9.64
0560	1000 S.F.	5.05	3.08	8.13
0580	2000 S.F.	5.35	2.84	8.19
0600	5000 S.F.	2.77	2.35	5.12
0620	10,000 S.F.	1.87	2.13	4
0640	50,000 S.F.	1.38	1.84	3.22

D4010 Sprinklers

D4010 414	Seismic Wet Pipe Flexible Feed Sprinkler Systems	COST PER S.F.		
		MAT.	INST.	TOTAL
0660	Each additional floor, 500 S.F.	1.99	2.59	4.58
0680	1000 S.F.	2.02	2.59	4.61
0700	2000 S.F.	1.71	2.17	3.88
0720	5000 S.F.	1.28	1.98	3.26
0740	10,000 S.F.	1.20	1.97	3.17
0760	50,000 S.F.	1.06	1.57	2.63
1000	Ordinary hazard, one floor, 500 S.F.	6.75	3.64	10.39
1020	1000 S.F.	5.15	2.98	8.13
1040	2000 S.F.	5.80	3.61	9.41
1060	5000 S.F.	3.30	2.68	5.98
1080	10,000 S.F.	2.47	2.89	5.36
1100	50,000 S.F.	2.31	2.64	4.95
1140	Each additional floor, 500 S.F.	2.34	3	5.34
1160	1000 S.F.	1.97	2.46	4.43
1180	2000 S.F.	2.09	2.68	4.77
1200	5000 S.F.	2.11	2.62	4.73
1220	10,000 S.F.	1.81	2.73	4.54
1240	50,000 S.F.	1.74	2.28	4.02
1500	Extra hazard, one floor, 500 S.F.	9.60	4.53	14.13
1520	1000 S.F.	8.65	3.97	12.62
1540	2000 S.F.	6.30	4.86	11.16
1560	5000 S.F.	4.17	4.44	8.61
1580	10,000 S.F.	4.15	4.33	8.48
1600	50,000 S.F.	4.52	4.12	8.64
1660	Each additional floor, 500 S.F.	2.85	3.41	6.26
1680	1000 S.F.	3.25	3.30	6.55
1700	2000 S.F.	2.87	3.82	6.69
1720	5000 S.F.	2.36	3.48	5.84
1740	10,000 S.F.	3	3.26	6.26
1760	50,000 S.F.	2.98	3.07	6.05
2020	Grooved steel, black sch. 40 pipe, light hazard, one floor, 2000 S.F.	5.60	2.69	8.29
2060	10,000 S.F.	1.76	1.75	3.51
2100	Each additional floor, 2000 S.F.	1.88	1.76	3.64
2150	10,000 S.F.	1.09	1.58	2.67
2200	Ordinary hazard, one floor, 2000 S.F.	5.90	3.11	9.01
2250	10,000 S.F.	2.47	2.46	4.93
2300	Each additional floor, 2000 S.F.	2.19	2.18	4.37
2350	10,000 S.F.	1.81	2.30	4.11
2400	Extra hazard, one floor, 2000 S.F.	6.50	3.98	10.48
2450	10,000 S.F.	3.59	3.19	6.78
2500	Each additional floor, 2000 S.F.	3.13	3.17	6.30
2550	10,000 S.F.	2.68	2.85	5.53
3050	Grooved steel black sch. 10 pipe, light hazard, one floor, 2000 S.F.	5.50	2.67	8.17
3100	10,000 S.F.	1.94	1.83	3.77
3150	Each additional floor, 2000 S.F.	1.78	1.73	3.51
3200	10,000 S.F.	1.03	1.56	2.59
3250	Ordinary hazard, one floor, 2000 S.F.	5.80	3.08	8.88
3300	10,000 S.F.	2.37	2.41	4.78
3350	Each additional floor, 2000 S.F.	2.10	2.15	4.25
3400	10,000 S.F.	1.71	2.25	3.96
3450	Extra hazard, one floor, 2000 S.F.	6.40	3.95	10.35
3500	10,000 S.F.	3.45	3.14	6.59
3550	Each additional floor, 2000 S.F.	3.06	3.15	6.21
3600	10,000 S.F.	2.58	2.81	5.39

D40 Fire Protection

D4020 Standpipes

Roof ————→ Roof connections with hose gate valves (for combustible roof)

Hose connections on each floor (size based on class of service)

Check Valve ————→ Siamese inlet connections (for fire department use)

System Components	QUANTITY	UNIT	COST PER FLOOR		
			MAT.	INST.	TOTAL
SYSTEM D4020 310 0560					
WET STANDPIPE RISER, CLASS I, STEEL, BLACK, SCH. 40 PIPE, 10' HEIGHT					
4" DIAMETER PIPE, ONE FLOOR					
Pipe, steel, black, schedule 40, threaded, 4" diam.	20.000	L.F.	610	810	1,420
Pipe, Tee, malleable iron, black, 150 lb. threaded, 4" pipe size	2.000	Ea.	820	730	1,550
Pipe, 90° elbow, malleable iron, black, 150 lb threaded, 4" pipe size	1.000	Ea.	232	242	474
Pipe, nipple, steel, black, schedule 40, 2-1/2" pipe size x 3" long	2.000	Ea.	26.90	182	208.90
Fire valve, gate, 300 lb., brass w/handwheel, 2-1/2" pipe size	1.000	Ea.	203	114	317
Fire valve, pressure reducing rgh brs, 2-1/2" pipe size	1.000	Ea.	1,170	228	1,398
Valve, swing check, w/ball drip, CI w/brs. ftngs., 4" pipe size	1.000	Ea.	480	475	955
Standpipe conn wall dble. flush brs. w/plugs & chains 2-1/2"x2-1/2"x4"	1.000	Ea.	645	286	931
Valve, swing check, bronze, 125 lb, regrinding disc, 2-1/2" pipe size	1.000	Ea.	1,075	97	1,172
Roof manifold, fire, w/valves & caps, horiz/vert brs 2-1/2"x2-1/2"x4"	1.000	Ea.	277	298	575
Fire, hydrolator, vent & drain, 2-1/2" pipe size	1.000	Ea.	123	66	189
Valve, gate, iron body 125 lb., OS&Y, threaded, 4" pipe size	1.000	Ea.	1,100	485	1,585
Tamper switch (valve supervisory switch)	1.000	Ea.	294	49.50	343.50
TOTAL			7,055.90	4,062.50	11,118.40

D4020 310	Wet Standpipe Risers, Class I		COST PER FLOOR		
			MAT.	INST.	TOTAL
0550	Wet standpipe risers, Class I, steel, black, sch. 40, 10' height				
0560	4" diameter pipe, one floor		7,050	4,075	11,125
0580	Additional floors		1,650	1,250	2,900
0600	6" diameter pipe, one floor		10,900	7,050	17,950
0620	Additional floors	R211226 -10	2,625	1,975	4,600
0640	8" diameter pipe, one floor		16,500	8,525	25,025
0660	Additional floors	R211226 -20	4,050	2,425	6,475
0680					

D4020 310	Wet Standpipe Risers, Class II		COST PER FLOOR		
			MAT.	INST.	TOTAL
1030	Wet standpipe risers, Class II, steel, black sch. 40, 10' height				
1040	2" diameter pipe, one floor		3,100	1,500	4,600
1060	Additional floors		895	555	1,450
1080	2-1/2" diameter pipe, one floor		3,925	2,125	6,050
1100	Additional floors		1,000	650	1,650
1120					

554

For customer support on your Plumbing Costs with RSMeans Data, call 800.448.8182.

D40 Fire Protection

D4020 Standpipes

D4020 310	Wet Standpipe Risers, Class III	COST PER FLOOR		
		MAT.	INST.	TOTAL
1530	Wet standpipe risers, Class III, steel, black, sch. 40, 10' height			
1540	4" diameter pipe, one floor	7,200	4,075	11,275
1560	Additional floors	1,500	1,050	2,550
1580	6" diameter pipe, one floor	11,100	7,050	18,150
1600	Additional floors	2,675	1,975	4,650
1620	8" diameter pipe, one floor	16,700	8,525	25,225
1640	Additional floors	4,100	2,425	6,525

555

D4020 Standpipes

- Roof
- Roof connections with hose gate valves (for combustible roof)
- Hose connections on each floor (size based on class of service)
- Check Valve
- Siamese inlet connections (for fire department use)

System Components	QUANTITY	UNIT	COST PER FLOOR		
			MAT.	INST.	TOTAL
SYSTEM D4020 330 0540					
DRY STANDPIPE RISER, CLASS I, PIPE, STEEL, BLACK, SCH 40, 10' HEIGHT					
4" DIAMETER PIPE, ONE FLOOR					
Pipe, steel, black, schedule 40, threaded, 4" diam.	20.000	L.F.	610	810	1,420
Pipe, Tee, malleable iron, black, 150 lb. threaded, 4" pipe size	2.000	Ea.	820	730	1,550
Pipe, 90° elbow, malleable iron, black, 150 lb threaded, 4" pipe size	1.000	Ea.	232	242	474
Pipe, nipple, steel, black, schedule 40, 2-1/2" pipe size x 3" long	2.000	Ea.	26.90	182	208.90
Fire valve gate NRS 300 lb., brass w/handwheel, 2-1/2" pipe size	1.000	Ea.	203	114	317
Tamper switch (valve supervisory switch)	1.000	Ea.	294	49.50	343.50
Fire valve, pressure reducing rgh brs, 2-1/2" pipe size	1.000	Ea.	585	114	699
Standpipe conn wall dble. flush brs. w/plugs & chains 2-1/2"x2-1/2"x4"	1.000	Ea.	645	286	931
Valve swing check w/ball drip CI w/brs. ftngs., 4"pipe size	1.000	Ea.	480	475	955
Roof manifold, fire, w/valves & caps, horiz/vert brs 2-1/2"x2-1/2"x4"	1.000	Ea.	277	298	575
TOTAL			4,172.90	3,300.50	7,473.40

D4020 330	Dry Standpipe Risers, Class I		COST PER FLOOR		
			MAT.	INST.	TOTAL
0530	Dry standpipe riser, Class I, steel, black, sch. 40, 10' height				
0540	4" diameter pipe, one floor		4,175	3,300	7,475
0560	Additional floors		1,525	1,175	2,700
0580	6" diameter pipe, one floor		7,775	5,600	13,375
0600	Additional floors		2,500	1,925	4,425
0620	8" diameter pipe, one floor	R211226 -10	11,900	6,800	18,700
0640	Additional floors		3,925	2,350	6,275
0660		R211226 -20			

D4020 330	Dry Standpipe Risers, Class II		COST PER FLOOR		
			MAT.	INST.	TOTAL
1030	Dry standpipe risers, Class II, steel, black, sch. 40, 10' height				
1040	2" diameter pipe, one floor		2,250	1,500	3,750
1060	Additional floors		770	490	1,260
1080	2-1/2" diameter pipe, one floor		3,100	1,750	4,850
1100	Additional floors		885	585	1,470
1120					

D40 Fire Protection

D4020 Standpipes

D4020 330	Dry Standpipe Risers, Class III	COST PER FLOOR		
		MAT.	INST.	TOTAL
1530	Dry standpipe risers, Class III, steel, black, sch. 40, 10' height			
1540	4" diameter pipe, one floor	3,925	3,200	7,125
1560	Additional floors	1,400	1,075	2,475
1580	6" diameter pipe, one floor	7,850	5,600	13,450
1600	Additional floors	2,550	1,925	4,475
1620	8" diameter pipe, one floor	12,000	6,800	18,800
1640	Additional floors	4,000	2,350	6,350

D40 Fire Protection

D4020 Standpipes

D4020 410	Fire Hose Equipment	COST EACH		
		MAT.	INST.	TOTAL
0100	Adapters, reducing, 1 piece, FxM, hexagon, cast brass, 2-1/2" x 1-1/2"	60		60
0200	Pin lug, 1-1/2" x 1"	50.50		50.50
0250	3" x 2-1/2"	134		134
0300	For polished chrome, add 75% mat.			
0400	Cabinets, D.S. glass in door, recessed, steel box, not equipped			
0500	Single extinguisher, steel door & frame	202	179	381
0550	Stainless steel door & frame	262	179	441
0600	Valve, 2-1/2" angle, steel door & frame	230	119	349
0650	Aluminum door & frame	279	119	398
0700	Stainless steel door & frame	375	119	494
0750	Hose rack assy, 2-1/2" x 1-1/2" valve & 100' hose, steel door & frame	430	238	668
0800	Aluminum door & frame	610	238	848
0850	Stainless steel door & frame	845	238	1,083
0900	Hose rack assy & extinguisher,2-1/2"x1-1/2" valve & hose,steel door & frame	310	286	596
0950	Aluminum	805	286	1,091
1000	Stainless steel	705	286	991
1550	Compressor, air, dry pipe system, automatic, 200 gal., 3/4 H.P.	1,550	610	2,160
1600	520 gal., 1 H.P.	1,800	610	2,410
1650	Alarm, electric pressure switch (circuit closer)	128	30.50	158.50
2500	Couplings, hose, rocker lug, cast brass, 1-1/2"	70		70
2550	2-1/2"	54.50		54.50
3000	Escutcheon plate, for angle valves, polished brass, 1-1/2"	17.30		17.30
3050	2-1/2"	27		27
3500	Fire pump, electric, w/controller, fittings, relief valve			
3550	4" pump, 30 HP, 500 GPM	22,700	4,450	27,150
3600	5" pump, 40 H.P., 1000 G.P.M.	33,200	5,050	38,250
3650	5" pump, 100 H.P., 1000 G.P.M.	37,200	5,600	42,800
3700	For jockey pump system, add	4,300	715	5,015
5000	Hose, per linear foot, synthetic jacket, lined,			
5100	300 lb. test, 1-1/2" diameter	3.64	.55	4.19
5150	2-1/2" diameter	6.40	.65	7.05
5200	500 lb. test, 1-1/2" diameter	3.98	.55	4.53
5250	2-1/2" diameter	6.70	.65	7.35
5500	Nozzle, plain stream, polished brass, 1-1/2" x 10"	64.50		64.50
5550	2-1/2" x 15" x 13/16" or 1-1/2"	119		119
5600	Heavy duty combination adjustable fog and straight stream w/handle 1-1/2"	490		490
5650	2-1/2" direct connection	515		515
6000	Rack, for 1-1/2" diameter hose 100 ft. long, steel	99	71.50	170.50
6050	Brass	150	71.50	221.50
6500	Reel, steel, for 50 ft. long 1-1/2" diameter hose	173	102	275
6550	For 75 ft. long 2-1/2" diameter hose	293	102	395
7050	Siamese, w/plugs & chains, polished brass, sidewalk, 4" x 2-1/2" x 2-1/2"	385	570	955
7100	6" x 2-1/2" x 2-1/2"	925	715	1,640
7200	Wall type, flush, 4" x 2-1/2" x 2-1/2"	645	286	931
7250	6" x 2-1/2" x 2-1/2"	955	310	1,265
7300	Projecting, 4" x 2-1/2" x 2-1/2"	590	286	876
7350	6" x 2-1/2" x 2-1/2"	985	310	1,295
7400	For chrome plate, add 15% mat.			
8000	Valves, angle, wheel handle, 300 Lb., rough brass, 1-1/2"	111	66	177
8050	2-1/2"	214	114	328
8100	Combination pressure restricting, 1-1/2"	106	66	172
8150	2-1/2"	187	114	301
8200	Pressure restricting, adjustable, satin brass, 1-1/2"	385	66	451
8250	2-1/2"	585	114	699
8300	Hydrolator, vent and drain, rough brass, 1-1/2"	123	66	189
8350	2-1/2"	123	66	189
8400	Cabinet assy, incls. adapter, rack, hose, and nozzle	1,150	430	1,580

D4090 Other Fire Protection Systems

General: Automatic fire protection (suppression) systems other than water sprinklers may be desired for special environments, high risk areas, isolated locations or unusual hazards. Some typical applications would include:

Paint dip tanks
Securities vaults
Electronic data processing
Tape and data storage
Transformer rooms
Spray booths
Petroleum storage
High rack storage

Piping and wiring costs are dependent on the individual application and must be added to the component costs shown below.

All areas are assumed to be open.

D4090 910	Fire Suppression Unit Components	COST EACH		
		MAT.	INST.	TOTAL
0020	Detectors with brackets			
0040	Fixed temperature heat detector	43	108	151
0060	Rate of temperature rise detector	32.50	94.50	127
0080	Ion detector (smoke) detector	138	122	260
0200	Extinguisher agent			
0240	200 lb FM200, container	7,575	365	7,940
0280	75 lb carbon dioxide cylinder	1,600	242	1,842
0320	Dispersion nozzle			
0340	FM200 1-1/2" dispersion nozzle	285	57.50	342.50
0380	Carbon dioxide 3" x 5" dispersion nozzle	179	45	224
0420	Control station			
0440	Single zone control station with batteries	1,200	755	1,955
0470	Multizone (4) control station with batteries	3,475	1,525	5,000
0490				
0500	Electric mechanical release	1,200	390	1,590
0520				
0550	Manual pull station	100	135	235
0570				
0640	Battery standby power 10" x 10" x 17"	510	189	699
0700				
0740	Bell signalling device	75	94.50	169.50

D4090 920	FM200 Systems	COST PER C.F.		
		MAT.	INST.	TOTAL
0820	Average FM200 system, minimum			2.06
0840	Maximum			4.08

559

For customer support on your Plumbing Costs with RSMeans Data, call 800.448.8182.

G1030 Site Earthwork

Trenching Systems are shown on a cost per linear foot basis. The systems include: excavation; backfill and removal of spoil; and compaction for various depths and trench bottom widths. The backfill has been reduced to accommodate a pipe of suitable diameter and bedding.

The slope for trench sides varies from none to 1:1.

The Expanded System Listing shows Trenching Systems that range from 2' to 12' in width. Depths range from 2' to 25'.

System Components	QUANTITY	UNIT	COST PER L.F.		
			EQUIP.	LABOR	TOTAL
SYSTEM G1030 805 1310					
TRENCHING, COMMON EARTH, NO SLOPE, 2' WIDE, 2' DP, 3/8 C.Y. BUCKET					
Excavation, trench, hyd. backhoe, track mtd., 3/8 C.Y. bucket	.148	B.C.Y.	.24	1.21	1.45
Backfill and load spoil, from stockpile	.153	L.C.Y.	.13	.37	.50
Compaction by vibrating plate, 6" lifts, 4 passes	.118	E.C.Y.	.03	.45	.48
Remove excess spoil, 8 C.Y. dump truck, 2 mile roundtrip	.040	L.C.Y.	.15	.20	.35
TOTAL			.55	2.23	2.78

G1030 805	Trenching Common Earth	COST PER L.F.		
		EQUIP.	LABOR	TOTAL
1310	Trenching, common earth, no slope, 2' wide, 2' deep, 3/8 C.Y. bucket	.54	2.24	2.78
1320	3' deep, 3/8 C.Y. bucket	.76	3.36	4.12
1330	4' deep, 3/8 C.Y. bucket	.98	4.48	5.46
1340	6' deep, 3/8 C.Y. bucket	1.28	5.80	7.08
1350	8' deep, 1/2 C.Y. bucket	1.66	7.70	9.36
1360	10' deep, 1 C.Y. bucket	3.61	9.20	12.81
1400	4' wide, 2' deep, 3/8 C.Y. bucket	1.28	4.45	5.73
1410	3' deep, 3/8 C.Y. bucket	1.71	6.70	8.41
1420	4' deep, 1/2 C.Y. bucket	1.94	7.45	9.39
1430	6' deep, 1/2 C.Y. bucket	3.20	12	15.20
1440	8' deep, 1/2 C.Y. bucket	6.75	15.50	22.25
1450	10' deep, 1 C.Y. bucket	8.20	19.25	27.45
1460	12' deep, 1 C.Y. bucket	10.55	24.50	35.05
1470	15' deep, 1-1/2 C.Y. bucket	7.70	22	29.70
1480	18' deep, 2-1/2 C.Y. bucket	13	30.50	43.50
1520	6' wide, 6' deep, 5/8 C.Y. bucket w/trench box	9.05	17.95	27
1530	8' deep, 3/4 C.Y. bucket	11.75	23.50	35.25
1540	10' deep, 1 C.Y. bucket	11.70	24.50	36.20
1550	12' deep, 1-1/2 C.Y. bucket	10.40	26	36.40
1560	16' deep, 2-1/2 C.Y. bucket	16.60	32.50	49.10
1570	20' deep, 3-1/2 C.Y. bucket	20.50	39	59.50
1580	24' deep, 3-1/2 C.Y. bucket	24.50	47	71.50
1640	8' wide, 12' deep, 1-1/2 C.Y. bucket w/trench box	14.50	33	47.50
1650	15' deep, 1-1/2 C.Y. bucket	18.80	43.50	62.30
1660	18' deep, 2-1/2 C.Y. bucket	24	43.50	67.50
1680	24' deep, 3-1/2 C.Y. bucket	33	60.50	93.50
1730	10' wide, 20' deep, 3-1/2 C.Y. bucket w/trench box	27	57.50	84.50
1740	24' deep, 3-1/2 C.Y. bucket	39.50	69.50	109
1780	12' wide, 20' deep, 3-1/2 C.Y. bucket w/trench box	42	73	115
1790	25' deep, bucket	52	93	145
1800	1/2 to 1 slope, 2' wide, 2' deep, 3/8 C.Y. bucket	.76	3.36	4.12
1810	3' deep, 3/8 C.Y. bucket	1.25	5.90	7.15

G1030 Site Earthwork

G1030 805	Trenching Common Earth	COST PER L.F.		
		EQUIP.	LABOR	TOTAL
1820	4' deep, 3/8 C.Y. bucket	1.85	8.95	10.80
1840	6' deep, 3/8 C.Y. bucket	3.03	14.55	17.58
1860	8' deep, 1/2 C.Y. bucket	4.78	23	27.78
1880	10' deep, 1 C.Y. bucket	12.45	32.50	44.95
2300	4' wide, 2' deep, 3/8 C.Y. bucket	1.49	5.55	7.04
2310	3' deep, 3/8 C.Y. bucket	2.20	9.20	11.40
2320	4' deep, 1/2 C.Y. bucket	2.72	11.35	14.07
2340	6' deep, 1/2 C.Y. bucket	5.35	21.50	26.85
2360	8' deep, 1/2 C.Y. bucket	13.10	31.50	44.60
2380	10' deep, 1 C.Y. bucket	18.15	44	62.15
2400	12' deep, 1 C.Y. bucket	24	59.50	83.50
2430	15' deep, 1-1/2 C.Y. bucket	22	64	86
2460	18' deep, 2-1/2 C.Y. bucket	45	99.50	144.50
2840	6' wide, 6' deep, 5/8 C.Y. bucket w/trench box	13.15	26.50	39.65
2860	8' deep, 3/4 C.Y. bucket	19.15	39.50	58.65
2880	10' deep, 1 C.Y. bucket	15.60	40.50	56.10
2900	12' deep, 1-1/2 C.Y. bucket	19.75	53	72.75
2940	16' deep, 2-1/2 C.Y. bucket	38	77.50	115.50
2980	20' deep, 3-1/2 C.Y. bucket	51.50	105	156.50
3020	24' deep, 3-1/2 C.Y. bucket	73	143	216
3100	8' wide, 12' deep, 1-1/2 C.Y. bucket w/trench box	24.50	60.50	85
3120	15' deep, 1-1/2 C.Y. bucket	35.50	87.50	123
3140	18' deep, 2-1/2 C.Y. bucket	52.50	104	156.50
3180	24' deep, 3-1/2 C.Y. bucket	81.50	157	238.50
3270	10' wide, 20' deep, 3-1/2 C.Y. bucket w/trench box	52.50	121	173.50
3280	24' deep, 3-1/2 C.Y. bucket	90	171	261
3370	12' wide, 20' deep, 3-1/2 C.Y. bucket w/trench box	75.50	140	215.50
3380	25' deep, 3-1/2 C.Y. bucket	105	200	305
3500	1 to 1 slope, 2' wide, 2' deep, 3/8 C.Y. bucket	.98	4.48	5.46
3520	3' deep, 3/8 C.Y. bucket	3.09	10.25	13.34
3540	4' deep, 3/8 C.Y. bucket	2.72	13.45	16.17
3560	6' deep, 1/2 C.Y. bucket	3.03	14.55	17.58
3580	8' deep, 1/2 C.Y. bucket	5.95	29	34.95
3600	10' deep, 1 C.Y. bucket	21.50	55.50	77
3800	4' wide, 2' deep, 3/8 C.Y. bucket	1.71	6.70	8.41
3820	3' deep, 3/8 C.Y. bucket	2.69	11.75	14.44
3840	4' deep, 1/2 C.Y. bucket	3.50	15.20	18.70
3860	6' deep, 1/2 C.Y. bucket	7.55	30.50	38.05
3880	8' deep, 1/2 C.Y. bucket	19.50	47	66.50
3900	10' deep, 1 C.Y. bucket	28	68.50	96.50
3920	12' deep, 1 C.Y. bucket	41.50	99.50	141
3940	15' deep, 1-1/2 C.Y. bucket	36	106	142
3960	18' deep, 2-1/2 C.Y. bucket	62	137	199
4030	6' wide, 6' deep, 5/8 C.Y. bucket w/trench box	17.30	35.50	52.80
4040	8' deep, 3/4 C.Y. bucket	25	49	74
4050	10' deep, 1 C.Y. bucket	22.50	59.50	82
4060	12' deep, 1-1/2 C.Y. bucket	30	81	111
4070	16' deep, 2-1/2 C.Y. bucket	59.50	123	182.50
4080	20' deep, 3-1/2 C.Y. bucket	83	171	254
4090	24' deep, 3-1/2 C.Y. bucket	121	240	361
4500	8' wide, 12' deep, 1-1/2 C.Y. bucket w/trench box	34.50	88	122.50
4550	15' deep, 1-1/2 C.Y. bucket	52	132	184
4600	18' deep, 2-1/2 C.Y. bucket	79.50	161	240.50
4650	24' deep, 3-1/2 C.Y. bucket	130	253	383
4800	10' wide, 20' deep, 3-1/2 C.Y. bucket w/trench box	78	184	262
4850	24' deep, 3-1/2 C.Y. bucket	138	268	406
4950	12' wide, 20' deep, 3-1/2 C.Y. bucket w/trench box	109	207	316
4980	25' deep, 3-1/2 C.Y. bucket	157	305	462

G1030 Site Earthwork

Trenching Systems are shown on a cost per linear foot basis. The systems include: excavation; backfill and removal of spoil; and compaction for various depths and trench bottom widths. The backfill has been reduced to accommodate a pipe of suitable diameter and bedding.

The slope for trench sides varies from none to 1:1.

The Expanded System Listing shows Trenching Systems that range from 2' to 12' in width. Depths range from 2' to 25'.

System Components	QUANTITY	UNIT	COST PER L.F. EQUIP.	LABOR	TOTAL
SYSTEM G1030 806 1310					
TRENCHING, LOAM & SANDY CLAY, NO SLOPE, 2' WIDE, 2' DP, 3/8 C.Y. BUCKET					
Excavation, trench, hyd. backhoe, track mtd., 3/8 C.Y. bucket	.148	B.C.Y.	.22	1.12	1.34
Backfill and load spoil, from stockpile	.165	L.C.Y.	.14	.40	.54
Compaction by vibrating plate 18" wide, 6" lifts, 4 passes	.118	E.C.Y.	.03	.45	.48
Remove excess spoil, 8 C.Y. dump truck, 2 mile roundtrip	.042	L.C.Y.	.16	.21	.37
TOTAL			.55	2.18	2.73

G1030 806	Trenching Loam & Sandy Clay	COST PER L.F. EQUIP.	LABOR	TOTAL
1310	Trenching, loam & sandy clay, no slope, 2' wide, 2' deep, 3/8 C.Y. bucket	.54	2.19	2.73
1320	3' deep, 3/8 C.Y. bucket	.81	3.59	4.40
1330	4' deep, 3/8 C.Y. bucket	.97	4.37	5.35
1340	6' deep, 3/8 C.Y. bucket	1.82	5.20	7.05
1350	8' deep, 1/2 C.Y. bucket	2.38	6.90	9.25
1360	10' deep, 1 C.Y. bucket	2.70	7.35	10.05
1400	4' wide, 2' deep, 3/8 C.Y. bucket	1.30	4.37	5.65
1410	3' deep, 3/8 C.Y. bucket	1.73	6.55	8.30
1420	4' deep, 1/2 C.Y. bucket	1.96	7.35	9.30
1430	6' deep, 1/2 C.Y. bucket	4.30	10.80	15.10
1440	8' deep, 1/2 C.Y. bucket	6.55	15.30	22
1450	10' deep, 1 C.Y. bucket	6.40	15.65	22
1460	12' deep, 1 C.Y. bucket	8.05	19.50	27.50
1470	15' deep, 1-1/2 C.Y. bucket	8.05	22.50	30.50
1480	18' deep, 2-1/2 C.Y. bucket	11.20	25	36
1520	6' wide, 6' deep, 5/8 C.Y. bucket w/trench box	8.70	17.60	26.50
1530	8' deep, 3/4 C.Y. bucket	11.30	23	34.50
1540	10' deep, 1 C.Y. bucket	10.70	23.50	34
1550	12' deep, 1-1/2 C.Y. bucket	10.25	26	36.50
1560	16' deep, 2-1/2 C.Y. bucket	16.20	33	49
1570	20' deep, 3-1/2 C.Y. bucket	19	39	58
1580	24' deep, 3-1/2 C.Y. bucket	24	47.50	71.50
1640	8' wide, 12' deep, 1-1/4 C.Y. bucket w/trench box	14.45	33.50	47.50
1650	15' deep, 1-1/2 C.Y. bucket	17.70	42.50	60
1660	18' deep, 2-1/2 C.Y. bucket	24.50	47.50	72
1680	24' deep, 3-1/2 C.Y. bucket	32	61.50	93.50
1730	10' wide, 20' deep, 3-1/2 C.Y. bucket w/trench box	32.50	62	94.50
1740	24' deep, 3-1/2 C.Y. bucket	40.50	76	117
1780	12' wide, 20' deep, 3-1/2 C.Y. bucket w/trench box	39	73.50	112
1790	25' deep, 3-1/2 C.Y. bucket	50.50	94.50	145
1800	1/2:1 slope, 2' wide, 2' deep, 3/8 C.Y. bucket	.76	3.28	4.04
1810	3' deep, 3/8 C.Y. bucket	1.24	5.75	7
1820	4' deep, 3/8 C.Y. bucket	1.83	8.75	10.60
1840	6' deep, 3/8 C.Y. bucket	4.37	13.05	17.40

G1030 Site Earthwork

G1030 806	Trenching Loam & Sandy Clay	COST PER L.F.		
		EQUIP.	LABOR	TOTAL
1860	8' deep, 1/2 C.Y. bucket	6.95	21	28
1880	10' deep, 1 C.Y. bucket	9.25	26	35.50
2300	4' wide, 2' deep, 3/8 C.Y. bucket	1.51	5.45	7
2310	3' deep, 3/8 C.Y. bucket	2.21	9	11.25
2320	4' deep, 1/2 C.Y. bucket	2.74	11.15	13.90
2340	6' deep, 1/2 C.Y. bucket	7.25	19.20	26.50
2360	8' deep, 1/2 C.Y. bucket	12.75	31	43.50
2380	10' deep, 1 C.Y. bucket	14.10	36	50
2400	12' deep, 1 C.Y. bucket	23.50	59.50	83
2430	15' deep, 1-1/2 C.Y. bucket	23	66	89
2460	18' deep, 2-1/2 C.Y. bucket	44.50	101	145
2840	6' wide, 6' deep, 5/8 C.Y. bucket w/trench box	12.70	26.50	39.50
2860	8' deep, 3/4 C.Y. bucket	18.35	39	57.50
2880	10' deep, 1 C.Y. bucket	19.20	44	63
2900	12' deep, 1-1/2 C.Y. bucket	20	53.50	73.50
2940	16' deep, 2-1/2 C.Y. bucket	37	78.50	116
2980	20' deep, 3-1/2 C.Y. bucket	50	106	156
3020	24' deep, 3-1/2 C.Y. bucket	70.50	145	216
3100	8' wide, 12' deep, 1-1/2 C.Y. bucket w/trench box	24	60.50	85
3120	15' deep, 1-1/2 C.Y. bucket	33	85	118
3140	18' deep, 2-1/2 C.Y. bucket	51	105	156
3180	24' deep, 3-1/2 C.Y. bucket	79	159	238
3270	10' wide, 20' deep, 3-1/2 C.Y. bucket w/trench box	63.50	129	192
3280	24' deep, 3-1/2 C.Y. bucket	87.50	174	261
3320	12' wide, 20' deep, 3-1/2 C.Y. bucket w/trench box	70	140	211
3380	25' deep, 3-1/2 C.Y. bucket w/trench box	95.50	188	284
3500	1:1 slope, 2' wide, 2' deep, 3/8 C.Y. bucket	.97	4.38	5.35
3520	3' deep, 3/8 C.Y. bucket	1.72	8.20	9.95
3540	4' deep, 3/8 C.Y. bucket	2.69	13.15	15.80
3560	6' deep, 1/2 C.Y. bucket	4.37	13.05	17.40
3580	8' deep, 1/2 C.Y. bucket	11.45	34.50	46
3600	10' deep, 1 C.Y. bucket	15.85	45	60.50
3800	4' wide, 2' deep, 3/8 C.Y. bucket	1.73	6.55	8.30
3820	3' deep, 1/2 C.Y. bucket	2.69	11.50	14.20
3840	4' deep, 1/2 C.Y. bucket	3.52	15	18.50
3860	6' deep, 1/2 C.Y. bucket	10.25	27.50	38
3880	8' deep, 1/2 C.Y. bucket	18.95	46.50	65.50
3900	10' deep, 1 C.Y. bucket	22	56	78
3920	12' deep, 1 C.Y. bucket	31.50	79.50	111
3940	15' deep, 1-1/2 C.Y. bucket	37.50	110	147
3960	18' deep, 2-1/2 C.Y. bucket	61	139	200
4030	6' wide, 6' deep, 5/8 C.Y. bucket w/trench box	16.70	35.50	52.50
4040	8' deep, 3/4 C.Y. bucket	25.50	55	80.50
4050	10' deep, 1 C.Y. bucket	27.50	64.50	92
4060	12' deep, 1-1/2 C.Y. bucket	30	81	111
4070	16' deep, 2-1/2 C.Y. bucket	58	124	182
4080	20' deep, 3-1/2 C.Y. bucket	81	173	254
4090	24' deep, 3-1/2 C.Y. bucket	118	243	360
4500	8' wide, 12' deep, 1-1/4 C.Y. bucket w/trench box	34	88	122
4550	15' deep, 1-1/2 C.Y. bucket	48	128	176
4600	18' deep, 2-1/2 C.Y. bucket	77.50	163	241
4650	24' deep, 3-1/2 C.Y. bucket	126	257	385
4800	10' wide, 20' deep, 3-1/2 C.Y. bucket w/trench box	94.50	196	291
4850	24' deep, 3-1/2 C.Y. bucket	134	272	405
4950	12' wide, 20' deep, 3-1/2 C.Y. bucket w/trench box	101	208	310
4980	25' deep, 3-1/2 C.Y. bucket	152	305	460

G1030 Site Earthwork

Trenching Systems are shown on a cost per linear foot basis. The systems include: excavation; backfill and removal of spoil; and compaction for various depths and trench bottom widths. The backfill has been reduced to accommodate a pipe of suitable diameter and bedding.

The slope for trench sides varies from none to 1:1.

The Expanded System Listing shows Trenching Systems that range from 2' to 12' in width. Depths range from 2' to 25'.

System Components	QUANTITY	UNIT	COST PER L.F.		
			EQUIP.	LABOR	TOTAL
SYSTEM G1030 807 1310					
TRENCHING, SAND & GRAVEL, NO SLOPE, 2' WIDE, 2' DEEP, 3/8 C.Y. BUCKET					
Excavation, trench, hyd. backhoe, track mtd., 3/8 C.Y. bucket	.148	B.C.Y.	.21	1.10	1.31
Backfill and load spoil, from stockpile	.140	L.C.Y.	.12	.34	.46
Compaction by vibrating plate 18" wide, 6" lifts, 4 passes	.118	E.C.Y.	.03	.45	.48
Remove excess spoil, 8 C.Y. dump truck, 2 mile roundtrip	.035	L.C.Y.	.13	.18	.31
TOTAL			.49	2.07	2.56

G1030 807	Trenching Sand & Gravel	COST PER L.F.		
		EQUIP.	LABOR	TOTAL
1310	Trenching, sand & gravel, no slope, 2' wide, 2' deep, 3/8 C.Y. bucket	.49	2.07	2.56
1320	3' deep, 3/8 C.Y. bucket	.75	3.45	4.20
1330	4' deep, 3/8 C.Y. bucket	.89	4.15	5.05
1340	6' deep, 3/8 C.Y. bucket	1.72	4.97	6.70
1350	8' deep, 1/2 C.Y. bucket	2.26	6.60	8.85
1360	10' deep, 1 C.Y. bucket	2.52	6.95	9.45
1400	4' wide, 2' deep, 3/8 C.Y. bucket	1.16	4.10	5.25
1410	3' deep, 3/8 C.Y. bucket	1.56	6.20	7.75
1420	4' deep, 1/2 C.Y. bucket	1.78	6.95	8.75
1430	6' deep, 1/2 C.Y. bucket	4.07	10.30	14.40
1440	8' deep, 1/2 C.Y. bucket	6.20	14.55	20.50
1450	10' deep, 1 C.Y. bucket	6	14.75	21
1460	12' deep, 1 C.Y. bucket	7.60	18.45	26
1470	15' deep, 1-1/2 C.Y. bucket	7.50	21.50	29
1480	18' deep, 2-1/2 C.Y. bucket	10.55	23.50	34
1520	6' wide, 6' deep, 5/8 C.Y. bucket w/trench box	8.25	16.65	25
1530	8' deep, 3/4 C.Y. bucket	10.75	22	32.50
1540	10' deep, 1 C.Y. bucket	10.05	22	32
1550	12' deep, 1-1/2 C.Y. bucket	9.60	24.50	34.50
1560	16' deep, 2 C.Y. bucket	15.50	31	46.50
1570	20' deep, 3-1/2 C.Y. bucket	17.90	36.50	54.50
1580	24' deep, 3-1/2 C.Y. bucket	22.50	44.50	67
1640	8' wide, 12' deep, 1-1/2 C.Y. bucket w/trench box	13.45	31.50	45
1650	15' deep, 1-1/2 C.Y. bucket	16.40	40	56
1660	18' deep, 2-1/2 C.Y. bucket	23	44.50	68
1680	24' deep, 3-1/2 C.Y. bucket	30.50	57.50	88
1730	10' wide, 20' deep, 3-1/2 C.Y. bucket w/trench box	30.50	58	88
1740	24' deep, 3-1/2 C.Y. bucket	38	71.50	109
1780	12' wide, 20' deep, 3-1/2 C.Y. bucket w/trench box	36.50	68.50	105
1790	25' deep, 3-1/2 C.Y. bucket	47.50	88.50	136
1800	1/2:1 slope, 2' wide, 2' deep, 3/8 C.Y. bucket	.69	3.11	3.80
1810	3' deep, 3/8 C.Y. bucket	1.14	5.45	6.60
1820	4' deep, 3/8 C.Y. bucket	1.68	8.35	10
1840	6' deep, 3/8 C.Y. bucket	4.16	12.45	16.60

G1030 Site Earthwork

G1030 807	Trenching Sand & Gravel	COST PER L.F.		
		EQUIP.	LABOR	TOTAL
1860	8' deep, 1/2 C.Y. bucket	6.60	19.90	26.50
1880	10' deep, 1 C.Y. bucket	8.65	24.50	33
2300	4' wide, 2' deep, 3/8 C.Y. bucket	1.36	5.15	6.50
2310	3' deep, 3/8 C.Y. bucket	2.01	8.55	10.55
2320	4' deep, 1/2 C.Y. bucket	2.49	10.60	13.05
2340	6' deep, 1/2 C.Y. bucket	6.95	18.35	25.50
2360	8' deep, 1/2 C.Y. bucket	12.05	29.50	41.50
2380	10' deep, 1 C.Y. bucket	13.30	34	47
2400	12' deep, 1 C.Y. bucket	22.50	56.50	78.50
2430	15' deep, 1-1/2 C.Y. bucket	21.50	62.50	84
2460	18' deep, 2-1/2 C.Y. bucket	42	95	137
2840	6' wide, 6' deep, 5/8 C.Y. bucket w/trench box	12.05	25.50	37.50
2860	8' deep, 3/4 C.Y. bucket	17.45	37	54.50
2880	10' deep, 1 C.Y. bucket	18.15	41.50	59.50
2900	12' deep, 1-1/2 C.Y. bucket	18.80	50.50	69.50
2940	16' deep, 2 C.Y. bucket	35.50	74	110
2980	20' deep, 3-1/2 C.Y. bucket	47.50	99.50	147
3020	24' deep, 3-1/2 C.Y. bucket	67	137	204
3100	8' wide, 12' deep, 1-1/4 C.Y. bucket w/trench box	22.50	57.50	80
3120	15' deep, 1-1/2 C.Y. bucket	31	80.50	111
3140	18' deep, 2-1/2 C.Y. bucket	48.50	99	148
3180	24' deep, 3-1/2 C.Y. bucket	75	150	224
3270	10' wide, 20' deep, 3-1/2 C.Y. bucket w/trench box	60	121	181
3280	24' deep, 3-1/2 C.Y. bucket	82.50	163	246
3370	12' wide, 20' deep, 3-1/2 C.Y. bucket w/trench box	67	134	201
3380	25' deep, 3-1/2 C.Y. bucket	96	188	284
3500	1:1 slope, 2' wide, 2' deep, 3/8 C.Y. bucket	1.76	5.25	7
3520	3' deep, 3/8 C.Y. bucket	1.58	7.80	9.40
3540	4' deep, 3/8 C.Y. bucket	2.48	12.50	15
3560	6' deep, 3/8 C.Y. bucket	4.16	12.45	16.65
3580	8' deep, 1/2 C.Y. bucket	10.95	33	44
3600	10' deep, 1 C.Y. bucket	14.80	42	57
3800	4' wide, 2' deep, 3/8 C.Y. bucket	1.56	6.20	7.75
3820	3' deep, 3/8 C.Y. bucket	2.45	10.90	13.35
3840	4' wide, 2' deep, 1/2 C.Y. bucket	3.21	14.20	17.40
3860	6' deep, 1/2 C.Y. bucket	9.80	26.50	36
3880	8' deep, 1/2 C.Y. bucket	17.95	44.50	62
3900	10' deep, 1 C.Y. bucket	20.50	53	73.50
3920	12' deep, 1 C.Y. bucket	30	75	105
3940	15' deep, 1-1/2 C.Y. bucket	35	104	139
3960	18' deep, 2-1/2 C.Y. bucket	57.50	131	188
4030	6' wide, 6' deep, 5/8 C.Y. bucket w/trench box	15.85	34	49.50
4040	8' deep, 3/4 C.Y. bucket	24	52.50	76.50
4050	10' deep, 1 C.Y. bucket	26	61	87
4060	12' deep, 1-1/2 C.Y. bucket	28	76.50	104
4070	16' deep, 2 C.Y. bucket	55.50	117	173
4080	20' deep, 3-1/2 C.Y. bucket	76.50	163	239
4090	24' deep, 3-1/2 C.Y. bucket	111	228	340
4500	8' wide, 12' deep, 1-1/2 C.Y. bucket w/trench box	32	83.50	115
4550	15' deep, 1-1/2 C.Y. bucket	45	121	166
4600	18' deep, 2-1/2 C.Y. bucket	73.50	153	227
4650	24' deep, 3-1/2 C.Y. bucket	119	241	360
4800	10' wide, 20' deep, 3-1/2 C.Y. bucket w/trench box	89.50	184	273
4850	24' deep, 3-1/2 C.Y. bucket	127	255	380
4950	12' wide, 20' deep, 3-1/2 C.Y. bucket w/trench box	95.50	195	290
4980	25' deep, 3-1/2 C.Y. bucket	144	288	430

G1030   Site Earthwork

The Pipe Bedding System is shown for various pipe diameters. Compacted bank sand is used for pipe bedding and to fill 12" over the pipe. No backfill is included. Various side slopes are shown to accommodate different soil conditions. Pipe sizes vary from 6" to 84" diameter.

System Components	QUANTITY	UNIT	COST PER L.F.		
			MAT.	INST.	TOTAL
SYSTEM G1030 815 1440					
PIPE BEDDING, SIDE SLOPE 0 TO 1, 1' WIDE, PIPE SIZE 6" DIAMETER					
Borrow, bank sand, 2 mile haul, machine spread	.086	C.Y.	1.75	.73	2.48
Compaction, vibrating plate	.086	C.Y.		.25	.25
TOTAL			1.75	.98	2.73

G1030 815	Pipe Bedding	COST PER L.F.		
		MAT.	INST.	TOTAL
1440	Pipe bedding, side slope 0 to 1, 1' wide, pipe size 6" diameter	1.75	.97	2.72
1460	2' wide, pipe size 8" diameter	3.79	2.11	5.90
1480	Pipe size 10" diameter	3.87	2.15	6.02
1500	Pipe size 12" diameter	3.96	2.20	6.16
1520	3' wide, pipe size 14" diameter	6.40	3.57	9.97
1540	Pipe size 15" diameter	6.50	3.60	10.10
1560	Pipe size 16" diameter	6.55	3.64	10.19
1580	Pipe size 18" diameter	6.65	3.71	10.36
1600	4' wide, pipe size 20" diameter	9.45	5.25	14.70
1620	Pipe size 21" diameter	9.55	5.30	14.85
1640	Pipe size 24" diameter	9.75	5.45	15.20
1660	Pipe size 30" diameter	9.95	5.55	15.50
1680	6' wide, pipe size 32" diameter	17	9.45	26.45
1700	Pipe size 36" diameter	17.40	9.70	27.10
1720	7' wide, pipe size 48" diameter	27.50	15.40	42.90
1740	8' wide, pipe size 60" diameter	33.50	18.75	52.25
1760	10' wide, pipe size 72" diameter	47	26	73
1780	12' wide, pipe size 84" diameter	62	34.50	96.50
2140	Side slope 1/2 to 1, 1' wide, pipe size 6" diameter	3.27	1.82	5.09
2160	2' wide, pipe size 8" diameter	5.55	3.09	8.64
2180	Pipe size 10" diameter	5.95	3.32	9.27
2200	Pipe size 12" diameter	6.30	3.51	9.81
2220	3' wide, pipe size 14" diameter	9.10	5.05	14.15
2240	Pipe size 15" diameter	9.35	5.20	14.55
2260	Pipe size 16" diameter	9.55	5.30	14.85
2280	Pipe size 18" diameter	10.05	5.60	15.65
2300	4' wide, pipe size 20" diameter	13.30	7.40	20.70
2320	Pipe size 21" diameter	13.55	7.55	21.10
2340	Pipe size 24" diameter	14.45	8.05	22.50
2360	Pipe size 30" diameter	16	8.90	24.90
2380	6' wide, pipe size 32" diameter	23.50	13.15	36.65
2400	Pipe size 36" diameter	25	13.95	38.95
2420	7' wide, pipe size 48" diameter	39	22	61
2440	8' wide, pipe size 60" diameter	49.50	27.50	77
2460	10' wide, pipe size 72" diameter	68	38	106
2480	12' wide, pipe size 84" diameter	89.50	49.50	139
2620	Side slope 1 to 1, 1' wide, pipe size 6" diameter	4.79	2.66	7.45
2640	2' wide, pipe size 8" diameter	7.40	4.10	11.50

G1030 Site Earthwork

G1030 815	Pipe Bedding	COST PER L.F.		
		MAT.	INST.	TOTAL
2660	Pipe size 10" diameter	8	4.46	12.46
2680	Pipe size 12" diameter	8.75	4.86	13.61
2700	3' wide, pipe size 14" diameter	11.85	6.60	18.45
2720	Pipe size 15" diameter	12.20	6.80	19
2740	Pipe size 16" diameter	12.60	7	19.60
2760	Pipe size 18" diameter	13.45	7.50	20.95
2780	4' wide, pipe size 20" diameter	17.10	9.50	26.60
2800	Pipe size 21" diameter	17.55	9.75	27.30
2820	Pipe size 24" diameter	19.05	10.60	29.65
2840	Pipe size 30" diameter	22	12.25	34.25
2860	6' wide, pipe size 32" diameter	30	16.80	46.80
2880	Pipe size 36" diameter	33	18.25	51.25
2900	7' wide, pipe size 48" diameter	50.50	28	78.50
2920	8' wide, pipe size 60" diameter	66	36.50	102.50
2940	10' wide, pipe size 72" diameter	89.50	50	139.50
2960	12' wide, pipe size 84" diameter	117	65	182

G2040 Site Development

- Chain Link Fence
- Concrete Sidewalk
- Pool Equipment (Building not included in price)
- 12' x 30' Swimming Pool
- Fence Gate

The Swimming Pool System is a complete package. Everything from excavation to deck hardware is included in system costs. Below are three basic types of pool systems: residential, motel, and municipal. Systems elements include: excavation, pool materials, installation, deck hardware, pumps and filters, sidewalk, and fencing.

The Expanded System Listing shows three basic types of pools with a variety of finishes and basic materials. These systems are either vinyl lined with metal sides; gunite shell with a cement plaster finish or tile finish; or concrete sided with vinyl lining. Pool sizes listed here vary from 12' x 30' to 60' x 82.5'. All costs are on a per unit basis.

System Components	QUANTITY	UNIT	COST EACH		
			MAT.	INST.	TOTAL
SYSTEM G2040 920 1000					
SWIMMING POOL, RESIDENTIAL, CONC. SIDES, VINYL LINED, 12' X 30'					
Swimming pool, residential, in-ground including equipment	360.000	S.F.	11,700	5,832	17,532
4" thick reinforced concrete sidewalk, broom finish, no base	400.000	S.F.	980	1,192	2,172
Chain link fence, residential, 4' high	124.000	L.F.	762.60	596.44	1,359.04
Fence gate, chain link, 4' high	1.000	Ea.	104	192	296
TOTAL			13,546.60	7,812.44	21,359.04

G2040 920	Swimming Pools	COST EACH		
		MAT.	INST.	TOTAL
1000	Swimming pool, residential class, concrete sides, vinyl lined, 12' x 30'	13,500	7,800	21,300
1100	16' x 32'	15,400	8,825	24,225
1200	20' x 40'	19,800	11,300	31,100
1500	Tile finish, 12' x 30'	29,600	33,100	62,700
1600	16' x 32'	30,100	34,800	64,900
1700	20' x 40'	45,100	48,900	94,000
2000	Metal sides, vinyl lined, 12' x 30'	11,600	5,125	16,725
2100	16' x 32'	13,100	5,750	18,850
2200	20' x 40'	16,900	7,250	24,150
3000	Gunite shell, cement plaster finish, 12' x 30'	23,400	14,000	37,400
3100	16' x 32'	29,700	18,300	48,000
3200	20' x 40'	41,200	18,100	59,300
4000	Motel class, concrete sides, vinyl lined, 20' x 40'	28,400	15,500	43,900
4100	28' x 60'	39,900	21,700	61,600
4500	Tile finish, 20' x 40'	53,000	56,500	109,500
4600	28' x 60'	84,500	91,500	176,000
5000	Metal sides, vinyl lined, 20' x 40'	29,400	11,300	40,700
5100	28' x 60'	39,500	15,200	54,700
6000	Gunite shell, cement plaster finish, 20' x 40'	76,500	45,100	121,600
6100	28' x 60'	114,500	67,500	182,000
7000	Municipal class, gunite shell, cement plaster finish, 42' x 75'	305,000	158,000	463,000
7100	60' x 82.5'	392,000	203,000	595,000
7500	Concrete walls, tile finish, 42' x 75'	341,500	214,500	556,000
7600	60' x 82.5'	445,000	284,500	729,500
7700	Tile finish and concrete gutter, 42' x 75'	375,000	214,500	589,500
7800	60' x 82.5'	485,000	284,500	769,500
7900	Tile finish and stainless gutter, 42' x 75'	439,000	214,500	653,500
8000	60' x 82.5'	649,000	328,500	977,500

G2050 Landscaping

There are three basic types of Site Irrigation Systems: pop-up, riser mounted and quick coupling. Sprinkler heads are spray, impact or gear driven. Each system includes: the hardware for spraying the water; the pipe and fittings needed to deliver the water; and all other accessory equipment such as valves, couplings, nipples, and nozzles. Excavation heads and backfill costs are also included in the system.

The Expanded System Listing shows a wide variety of Site Irrigation Systems.

System Components	QUANTITY	UNIT	COST PER S.F.		
			MAT.	INST.	TOTAL
SYSTEM G2050 710 1000					
SITE IRRIGATION, POP UP SPRAY, PLASTIC, 10' RADIUS, 1000 S.F., PVC PIPE					
Excavation, chain trencher	54.000	L.F.		56.70	56.70
Pipe, fittings & nipples, PVC Schedule 40, 1" diameter	5.000	Ea.	4.60	127.50	132.10
Fittings, bends or elbows, 4" diameter	5.000	Ea.	47.25	287.50	334.75
Couplings PVC plastic, high pressure, 1" diameter	5.000	Ea.	4.60	162.50	167.10
Valves, bronze, globe, 125 lb. rising stem, threaded, 1" diameter	1.000	Ea.	298	42.50	340.50
Head & nozzle, pop-up spray, PVC plastic	5.000	Ea.	23.70	115	138.70
Backfill by hand with compaction	54.000	L.F.		39.96	39.96
Total cost per 1,000 S.F.			378.15	831.66	1,209.81
Total cost per S.F.			.38	.83	1.21

G2050 710	Site Irrigation	COST PER S.F.		
		MAT.	INST.	TOTAL
1000	Site irrigation, pop up spray, 10' radius, 1000 S.F., PVC pipe	.38	.83	1.21
1100	Polyethylene pipe	.42	.72	1.14
1200	14' radius, 8000 S.F., PVC pipe	.13	.36	.49
1300	Polyethylene pipe	.13	.33	.46
1400	18' square, 1000 S.F. PVC pipe	.39	.57	.96
1500	Polyethylene pipe	.39	.49	.88
1600	24' square, 8000 S.F., PVC pipe	.11	.23	.34
1700	Polyethylene pipe	.11	.22	.33
1800	4' x 30' strip, 200 S.F., PVC pipe	1.83	2.35	4.18
1900	Polyethylene pipe	1.82	2.18	4
2000	6' x 40' strip, 800 S.F., PVC pipe	.49	.78	1.27
2200	Economy brass, 11' radius, 1000 S.F., PVC pipe	.49	.80	1.29
2300	Polyethylene pipe	.48	.72	1.20
2400	14' radius, 8000 S.F., PVC pipe	.16	.36	.52
2500	Polyethylene pipe	.16	.33	.49
2600	3' x 28' strip, 200 S.F., PVC pipe	2.03	2.35	4.38
2700	Polyethylene pipe	2.02	2.18	4.20
2800	7' x 36' strip, 1000 S.F., PVC pipe	.45	.62	1.07
2900	Polyethylene pipe	.44	.57	1.01
3000	Hd brass, 11' radius, 1000 S.F., PVC pipe	.53	.80	1.33
3100	Polyethylene pipe	.53	.72	1.25
3200	14' radius, 8000 S.F., PVC pipe	.18	.36	.54
3300	Polyethylene pipe	.18	.33	.51
3400	Riser mounted spray, plastic, 10' radius, 1000 S.F., PVC pipe	.42	.80	1.22
3500	Polyethylene pipe	.42	.72	1.14
3600	12' radius, 5000 S.F., PVC pipe	.19	.47	.66

G20 Site Improvements

G2050 Landscaping

G2050 710	Site Irrigation	COST PER S.F.		
		MAT.	INST.	TOTAL
3700	Polyethylene pipe	.19	.44	.63
3800	19' square, 2000 S.F., PVC pipe	.19	.26	.45
3900	Polyethylene pipe	.19	.23	.42
4000	24' square, 8000 S.F., PVC pipe	.11	.23	.34
4100	Polyethylene pipe	.11	.22	.33
4200	5' x 32' strip, 300 S.F., PVC pipe	1.22	1.58	2.80
4300	Polyethylene pipe	1.22	1.46	2.68
4400	6' x 40' strip, 800 S.F., PVC pipe	.49	.78	1.27
4500	Polyethylene pipe	.49	.72	1.21
4600	Brass, 11' radius, 1000 S.F., PVC pipe	.46	.80	1.26
4700	Polyethylene pipe	.46	.72	1.18
4800	14' radius, 8000 S.F., PVC pipe	.15	.36	.51
4900	Polythylene pipe	.15	.33	.48
5000	Pop up gear drive stream type, plastic, 30' radius, 10,000 S.F., PVC pipe	.23	.69	.92
5020	Polyethylene pipe	.16	.21	.37
5040	40,000 S.F., PVC pipe	.22	.69	.91
5060	Polyethylene pipe	.12	.20	.32
5080	Riser mounted gear drive stream type, plas., 30' rad., 10,000 S.F.,PVC pipe	.22	.69	.91
5100	Polyethylene pipe	.15	.21	.36
5120	40,000 S.F., PVC pipe	.21	.69	.90
5140	Polyethylene pipe	.11	.20	.31
5200	Q.C. valve thread type w/impact head, brass, 75' rad, 20,000 S.F., PVC pipe	.09	.27	.36
5250	Polyethylene pipe	.06	.07	.13
5300	100,000 S.F., PVC pipe	.15	.43	.58
5350	Polyethylene pipe	.08	.10	.18
5400	Q.C. valve lug type w/impact head, brass, 75' rad, 20,000 S.F., PVC pipe	.09	.27	.36
5450	Polyethylene pipe	.06	.07	.13
5500	100,000 S.F. PVC pipe	.15	.43	.58
5550	Polyethylene pipe	.08	.10	.18
6000	Site irrigation, pop up impact type, plastic, 40' rad.,10000 S.F., PVC pipe	.13	.45	.58
6100	Polyethylene pipe	.09	.12	.21
6200	45,000 S.F., PVC pipe	.14	.49	.63
6300	Polyethylene pipe	.07	.13	.20
6400	High medium volume brass, 40' radius, 10000 S.F., PVC pipe	.14	.45	.59
6500	Polyethylene pipe	.09	.12	.21
6600	45,000 S.F. PVC pipe	.15	.49	.64
6700	Polyethylene pipe	.07	.13	.20
6800	High volume brass, 60' radius, 25,000 S.F., PVC pipe	.10	.35	.45
6900	Polyethylene pipe	.07	.09	.16
7000	100,000 S.F., PVC pipe	.13	.35	.48
7100	Polyethylene pipe	.07	.08	.15
7200	Riser mounted part/full impact type, plas., 40' rad., 10,000 S.F., PVC pipe	.13	.45	.58
7300	Polyethylene pipe	.08	.24	.32
7400	45,000 S.F., PVC pipe	.14	.49	.63
7500	Polyethylene pipe	.06	.13	.19
7600	Low medium volume brass, 40' radius, 10,000 S.F., PVC pipe	.14	.45	.59
7700	Polyethylene pipe	.09	.13	.22
7800	45,000 S.F., PVC pipe	.14	.49	.63
7900	Polyethylene pipe	.07	.13	.20
8000	Medium volume brass, 50' radius, 30,000 S.F., PVC pipe	.11	.37	.48
8100	Polyethylene pipe	.07	.10	.17
8200	70,000 S.F., PVC pipe	.18	.60	.78
8300	Polyethylene pipe	.09	.15	.24
8400	Riser mounted full only impact type, plas., 40' rad., 10,000 S.F., PVC pipe	.12	.45	.57
8500	Polyethylene pipe	.08	.13	.21
8600	45,000 S.F., PVC pipe	.13	.49	.62
8700	Polyethylene pipe	.06	.13	.19
8800	Low medium volume brass, 40' radius, 10,000 S.F., PVC pipe	.14	.45	.59

G20 Site Improvements

G2050 Landscaping

G2050 710	Site Irrigation	COST PER S.F.		
		MAT.	INST.	TOTAL
8900	Polyethylene pipe	.09	.13	.22
9000	45,000 S.F., PVC pipe	.15	.49	.64
9100	Polyethylene pipe	.07	.13	.20
9200	High volume brass, 80' radius, 75,000 S.F., PVC pipe	.09	.26	.35
9300	Polyethylene pipe	.01	.02	.03
9400	150,000 S.F., PVC pipe	.09	.24	.33
9500	Polyethylene pipe	.01	.02	.03
9600	Very high volume brass, 100' radius, 100,000 S.F., PVC pipe	.08	.20	.28
9700	Polyethylene pipe	.01	.02	.03
9800	200,000 S.F., PVC pipe	.08	.20	.28
9900	Polyethylene pipe	.01	.02	.03

For customer support on your Plumbing Costs with RSMeans Data, call 800.448.8182.

G2050 Landscaping

Valve
Backflow Preventer
Pressure Regulator
Filter
Tubing Adapter
Drip Tubing

Drip irrigation systems can be for either lawns or individual plants. Shown below are drip irrigation systems for lawns. These drip type systems are most cost effective where there are water restrictions on use or the cost of water is very high. The supply uses PVC piping with the laterals polyethylene drip lines.

System Components	QUANTITY	UNIT	COST PER S.F.		
			MAT.	INST.	TOTAL
SYSTEM G2050 720 1000					
SITE IRRIGATION, DRIP SYSTEM, 800 S.F. (20′ X 40′)					
Excavate supply and header lines 4″ x 18″ deep	50.000	L.F.		55.50	55.50
Excavate drip lines 4″ x 4″ deep	560.000	L.F.		935.20	935.20
Install supply manifold (stop valve, strainer, PRV, adapters)	1.000	Ea.	23.50	172	195.50
Install 3/4″ PVC supply line and risers to 4 ″ drip level	75.000	L.F.	24.75	171.75	196.50
Install 3/4″ PVC fittings for supply line	33.000	Ea.	60.39	567.60	627.99
Install PVC to drip line adapters	14.000	Ea.	26.18	240.80	266.98
Install 1/4 inch drip lines with hose clamps	560.000	L.F.	61.60	319.20	380.80
Backfill trenches	3.180	C.Y.		152.64	152.64
Cleanup area when completed	1.000	Ea.		133	133
Total cost per 800 S.F.			196.42	2,747.69	2,944.11
Total cost per S.F.			.25	3.43	3.68

G2050 720	Site Irrigation	COST PER S.F.		
		MAT.	INST.	TOTAL
1000	Site irrigation, drip lawn watering system, 800 S.F. (20′ x 40′ area)	.25	3.44	3.69
1100	1000 S.F. (20′ x 50′ area)	.21	3.09	3.30
1200	1600 S.F. (40′ x 40′ area), 2 zone	.26	3.30	3.56
1300	2000 S.F. (40′ x 50′ area), 2 zone	.23	2.98	3.21
1400	2400 S.F. (60′ x 40′ area), 3 zone	.25	3.26	3.51
1500	3000 S.F. (60′ x 50′ area), 3 zone	.22	2.93	3.15
1600	3200 S.F. (2 x 40′ x 40′), 4 zone	.26	3.24	3.50
1700	4000 S.F. (2 x 40′ x 50′), 4 zone	.22	2.93	3.15
1800	4800 S.F. (2 x 60′ x 40′), 6 zone with control	.27	3.22	3.49
1900	6000 S.F. (2 x 60′ x 50′), 6 zone with control	.23	2.90	3.13
2000	6400 S.F. (4 x 40′ x 40′), 8 zone with control	.27	3.26	3.53
2100	8000 S.F. (4 x 40′ x 50′), 8 zone with control	.23	2.95	3.18
2200	3200 S.F. (2 x 40′ x 40′), 4 zone with control	.28	3.31	3.59
2300	4000 S.F. (2 x 40′ x 50′), 4 zone with control	.24	2.99	3.23
2400	2400 S.F. (60′ x 40′), 3 zone with control	.28	3.36	3.64
2500	3000 S.F. (60′ x 50′), 3 zone with control	.24	3.01	3.25
2600	4800 S.F. (2 x 60′ x 40′), 6 zone, manual	.25	3.17	3.42
2700	6000 S.F. (2 x 60′ x 50′), 6 zone, manual	.21	2.86	3.07
2800	1000 S.F. (10′ x 100′ area)	.13	1.92	2.05
2900	2000 S.F. (2 x 10′ x 100′ area)	.15	1.85	2

G3010 Water Supply

The Water Service Systems are for copper service taps from 1″ to 2″ diameter into pressurized mains from 6″ to 8″ diameter. Costs are given for two offsets with depths varying from 2′ to 10′. Included in system components are excavation and backfill and required curb stops with boxes.

System Components	QUANTITY	UNIT	COST EACH MAT.	COST EACH INST.	COST EACH TOTAL
SYSTEM G3010 121 1000					
WATER SERVICE, LEAD FREE, 6″ MAIN, 1″ COPPER SERVICE, 10′ OFFSET, 2′ DEEP					
Trench excavation, 1/2 C.Y. backhoe, 1 laborer	2.220	C.Y.		16.51	16.51
Drill & tap pressurized main, 6″, 1″ to 2″ service	1.000	Ea.		485	485
Corporation stop, 1″ diameter	1.000	Ea.	110	50.50	160.50
Saddles, 3/4″ & 1″ diameter, add	1.000	Ea.	73		73
Copper tubing, type K, 1″ diameter	12.000	L.F.	144	54.60	198.60
Piping, curb stops, no lead, 1″ diameter	1.000	Ea.	159	50.50	209.50
Curb box, cast iron, 1/2″ to 1″ curb stops	1.000	Ea.	44.50	67.50	112
Backfill by hand, no compaction, heavy soil	2.890	C.Y.		8.50	8.50
Compaction, vibrating plate	2.220	C.Y.		22.11	22.11
TOTAL			530.50	755.22	1,285.72

G3010 121	Water Service, Lead Free	COST EACH MAT.	COST EACH INST.	COST EACH TOTAL
1000	Water service, Lead Free, 6″ main, 1″ copper service, 10′ offset, 2′ deep	530	755	1,285
1040	4′ deep	530	800	1,330
1060	6′ deep	530	1,025	1,555
1080	8′ deep	530	1,125	1,655
1100	10′ deep	530	1,250	1,780
1200	20′ offset, 2′ deep	530	800	1,330
1240	4′ deep	530	895	1,425
1260	6′ deep	530	1,350	1,880
1280	8′ deep	530	1,550	2,080
1300	10′ deep	530	1,775	2,305
1400	1-1/2″ copper service, 10′ offset, 2′ deep	985	825	1,810
1440	4′ deep	985	870	1,855
1460	6′ deep	985	1,100	2,085
1480	8′ deep	985	1,200	2,185
1500	10′ deep	985	1,300	2,285
1520	20′ offset, 2′ deep	985	870	1,855
1530	4′ deep	985	965	1,950
1540	6′ deep	985	1,425	2,410
1560	8′ deep	985	1,625	2,610
1580	10′ deep	985	1,850	2,835
1600	2″ copper service, 10′ offset, 2′ deep	1,300	855	2,155
1610	4′ deep	1,275	905	2,180
1620	6′ deep	1,275	1,125	2,400
1630	8′ deep	1,275	1,225	2,500
1650	10′ deep	1,275	1,350	2,625
1660	20′ offset, 2′ deep	1,275	905	2,180
1670	4′ deep	1,275	1,000	2,275
1680	6′ deep	1,275	1,450	2,725

575

For customer support on your Plumbing Costs with RSMeans Data, call 800.448.8182.

G3010 Water Supply

G3010 121	Water Service, Lead Free	COST EACH		
		MAT.	INST.	TOTAL
1690	8' deep	1,275	1,650	2,925
1695	10' deep	1,275	1,875	3,150
1700	8" main, 1" copper service, 10' offset, 2' deep	530	800	1,330
1710	4' deep	530	845	1,375
1720	6' deep	530	1,075	1,605
1730	8' deep	530	1,175	1,705
1740	10' deep	530	1,275	1,805
1750	20' offset, 2' deep	530	845	1,375
1760	4' deep	530	940	1,470
1770	6' deep	530	1,400	1,930
1780	8' deep	530	1,600	2,130
1790	10' deep	530	1,825	2,355
1800	1-1/2" copper service, 10' offset, 2' deep	985	870	1,855
1810	4' deep	985	915	1,900
1820	6' deep	985	1,150	2,135
1830	8' deep	985	1,675	2,660
1840	10' deep	985	1,350	2,335
1850	20' offset, 2' deep	985	915	1,900
1860	4' deep	985	915	1,900
1870	6' deep	985	1,450	2,435
1880	8' deep	985	1,675	2,660
1890	10' deep	985	1,875	2,860
1900	2" copper service, 10' offset, 2' deep	1,300	900	2,200
1910	4' deep	1,275	950	2,225
1920	6' deep	1,275	1,175	2,450
1930	8' deep	1,275	1,275	2,550
1940	10' deep	1,275	1,375	2,650
1950	20' offset, 2' deep	1,275	950	2,225
1960	4' deep	1,275	1,050	2,325
1970	6' deep	1,275	1,500	2,775
1980	8' deep	1,275	1,700	2,975
1990	10' deep	1,275	1,925	3,200

G3010 Water Supply

The Fire Hydrant Systems include: four different hydrants with three different lengths of offsets, at several depths of trenching. Excavation and backfill is included with each system as well as thrust blocks as necessary and pipe bedding. Finally spreading of excess material and fine grading complete the components.

System Components	QUANTITY	UNIT	COST PER EACH		
			MAT.	INST.	TOTAL
SYSTEM G3010 410 1500					
HYDRANT, 4-1/2" VALVE SIZE, TWO WAY, 10' OFFSET, 2' DEEP					
Excavation, trench, 1 C.Y. hydraulic backhoe	3.556	C.Y.		32.11	32.11
Sleeve, 12" x 6"	1.000	Ea.	3,475	130.65	3,605.65
Ductile iron pipe, 6" diameter	8.000	L.F.	484	117.28	601.28
Gate valve, 6"	1.000	Ea.	3,600	326.50	3,926.50
Hydrant valve box, 6' long	1.000	Ea.	256	88.50	344.50
4-1/2" valve size, depth 2'-0"	1.000	Ea.	3,400	235	3,635
Thrust blocks, at valve, shoe and sleeve	1.600	C.Y.	366.40	497.73	864.13
Borrow, crushed stone, 3/8"	4.089	C.Y.	149.25	35.05	184.30
Backfill and compact, dozer, air tamped	4.089	C.Y.		85.06	85.06
Spread excess excavated material	.204	C.Y.		9.79	9.79
Fine grade, hand	400.000	S.F.		308	308
TOTAL			11,730.65	1,865.67	13,596.32

G3010 410	Fire Hydrants	COST PER EACH		
		MAT.	INST.	TOTAL
1000	Hydrant, 4-1/2" valve size, two way, 0' offset, 2' deep	11,200	1,325	12,525
1100	4' deep	11,500	1,350	12,850
1200	6' deep	11,800	1,425	13,225
1300	8' deep	11,900	1,475	13,375
1400	10' deep	12,200	1,550	13,750
1500	10' offset, 2' deep	11,700	1,875	13,575
1600	4' deep	12,000	2,200	14,200
1700	6' deep	12,300	2,850	15,150
1800	8' deep	12,400	3,800	16,200
1900	10' deep	12,700	5,800	18,500
2000	20' offset, 2' deep	12,300	2,425	14,725
2100	4' deep	12,600	2,975	15,575
2200	6' deep	12,900	3,950	16,850
2300	8' deep	13,000	5,325	18,325
2400	10' deep	13,300	7,200	20,500
2500	Hydrant, 4-1/2" valve size, three way, 0' offset, 2' deep	11,500	1,325	12,825
2600	4' deep	11,800	1,350	13,150
2700	6' deep	12,100	1,425	13,525
2800	8' deep	12,200	1,500	13,700
2900	10' deep	12,500	1,600	14,100
3000	10' offset, 2' deep	12,000	1,875	13,875
3100	4' deep	12,300	2,225	14,525
3200	6' deep	12,600	2,875	15,475
3300	8' deep	12,700	3,825	16,525

G3010 Water Supply

G3010 410	Fire Hydrants	COST PER EACH		
		MAT.	INST.	TOTAL
3400	10' deep	13,000	5,825	18,825
3500	20' offset, 2' deep	12,600	2,450	15,050
3600	4' deep	12,900	3,000	15,900
3700	6' deep	13,200	3,975	17,175
3800	8' deep	13,300	5,350	18,650
3900	10' deep	13,600	7,225	20,825
5000	5-1/4" valve size, two way, 0' offset, 2' deep	11,500	1,325	12,825
5100	4' deep	11,900	1,350	13,250
5200	6' deep	12,200	1,425	13,625
5300	8' deep	12,600	1,475	14,075
5400	10' deep	13,000	1,550	14,550
5500	10' offset, 2' deep	12,000	1,875	13,875
5600	4' deep	12,400	2,200	14,600
5700	6' deep	12,700	2,850	15,550
5800	8' deep	13,100	3,800	16,900
5900	10' deep	13,500	5,800	19,300
6000	20' offset, 2' deep	12,600	2,425	15,025
6100	4' deep	13,000	2,975	15,975
6200	6' deep	13,300	3,950	17,250
6300	8' deep	13,700	5,325	19,025
6400	10' deep	14,100	7,200	21,300
6500	Three way, 0' offset, 2' deep	11,800	1,325	13,125
6600	4' deep	12,200	1,350	13,550
6700	6' deep	12,500	1,425	13,925
6800	8' deep	13,000	1,500	14,500
6900	10' deep	13,400	1,600	15,000
7000	10' offset, 2' deep	12,300	1,875	14,175
7100	4' deep	12,600	2,225	14,825
7200	6' deep	13,000	2,875	15,875
7300	8' deep	13,500	3,825	17,325
7400	10' deep	13,800	5,825	19,625
7500	20' offset, 2' deep	12,900	2,450	15,350
7600	4' deep	13,200	2,700	15,900
7700	6' deep	13,600	3,975	17,575

G3020 Sanitary Sewer

The Septic System includes: a septic tank; leaching field; concrete distribution boxes; plus excavation and gravel backfill.

The Expanded System Listing shows systems with tanks ranging from 1000 gallons to 2000 gallons. Tanks are either concrete or fiberglass. Cost is on a complete unit basis.

System Components	QUANTITY	UNIT	COST EACH		
			MAT.	INST.	TOTAL
SYSTEM G3020 302 0300					
1000 GALLON TANK, LEACHING FIELD, 600 S.F., 50' FROM BLDG.					
Mobilization	2.000	Ea.		1,772	1,772
Septic tank, precast, 1000 gallon	1.000	Ea.	1,100	293.50	1,393.50
Effluent filter	1.000	Ea.	39.50	86	125.50
Distribution box, precast, 5 outlets	1.000	Ea.	103	66.50	169.50
Flow leveler	3.000	Ea.	9.87	31.80	41.67
Sewer pipe, PVC, SDR 35, 4" diameter	160.000	L.F.	388.80	755.20	1,144
Tee	1.000	Ea.	21	86	107
Elbow	2.000	Ea.	18.90	115	133.90
Viewport cap	1.000	Ea.	7.80	28.50	36.30
Filter fabric	67.000	S.Y.	115.24	29.48	144.72
Detectable marking tape	1.600	C.L.F.	5.28	5.65	10.93
Excavation	160.000	C.Y.		2,240	2,240
Backfill	133.000	L.C.Y.		433.58	433.58
Spoil	55.000	L.C.Y.		405.35	405.35
Compaction	113.000	E.C.Y.		1,066.72	1,066.72
Stone fill, 3/4" to 1-1/2"	39.000	C.Y.	1,755	510.51	2,265.51
TOTAL			3,564.39	7,925.79	11,490.18

G3020 302	Septic Systems	COST EACH		
		MAT.	INST.	TOTAL
0300	1000 gal. concrete septic tank, 600 S.F. leaching field, 50 feet from bldg.	3,575	7,925	11,500
0301	200 feet from building	3,925	9,825	13,750
0305	750 S.F. leaching field, 50 feet from building	4,075	8,625	12,700
0306	200 feet from building	4,450	10,500	14,950
0400	1250 gal. concrete septic tank, 600 S.F. leaching field, 50 feet from bldg.	4,125	8,225	12,350
0401	200 feet from building	4,475	10,200	14,675
0405	750 S.F. leaching field, 50 feet from building	4,625	8,950	13,575
0406	200 feet from building	5,000	10,900	15,900
0410	1000 S.F. leaching field, 50 feet from building	5,525	10,200	15,725
0411	200 feet from building	5,900	12,200	18,100
0500	1500 gal. concrete septic tank, 600 S.F. leaching field, 50 feet from bldg.	4,325	8,525	12,850
0501	200 feet from building	4,675	10,400	15,075
0505	750 S.F. leaching field, 50 feet from building	4,850	9,250	14,100
0506	200 feet from building	5,225	12,200	17,425
0510	1000 S.F. leaching field, 50 feet from building	5,725	10,600	16,325
0511	200 feet from building	6,100	12,500	18,600
0515	1100 S.F. leaching field, 50 feet from building	6,075	11,000	17,075
0516	200 feet from building	6,450	13,100	19,550
0600	2000 gal concrete tank, 1200 S.F. leaching field, 50 feet from building	7,425	11,700	19,125
0601	200 feet from building	7,800	13,600	21,400

G3020 302	Septic Systems	COST EACH		
		MAT.	INST.	TOTAL
0605	1500 S.F. leaching field, 50 feet from building	8,475	13,200	21,675
0606	200 feet from building	8,850	15,100	23,950
0710	2500 gal concrete tank, 1500 S.F. leaching field, 50 feet from building	8,275	14,000	22,275
0711	200 feet from building	8,650	15,900	24,550
0715	1600 S.F. leaching field, 50 feet from building	8,650	14,500	23,150
0716	200 feet from building	9,025	16,400	25,425
0720	1800 S.F. leaching field, 50 feet from building	9,350	15,600	24,950
0721	200 feet from building	9,725	17,500	27,225
1300	1000 gal. 2 compartment tank, 600 S.F. leaching field, 50 feet from bldg.	3,250	7,925	11,175
1301	200 feet from building	3,600	9,825	13,425
1305	750 S.F. leaching field, 50 feet from building	3,750	8,625	12,375
1306	200 feet from building	4,125	10,500	14,625
1400	1250 gal. 2 compartment tank, 600 S.F. leaching field, 50 feet from bldg.	3,650	8,225	11,875
1401	200 feet from building	4,000	10,200	14,200
1405	750 S.F. leaching field, 50 feet from building	4,150	8,950	13,100
1406	200 feet from building	4,525	10,900	15,425
1410	1000 S.F. leaching field, 50 feet from building	5,050	10,200	15,250
1411	200 feet from building	5,425	12,200	17,625
1500	1500 gal. 2 compartment tank, 600 S.F. leaching field, 50 feet from bldg.	4,225	8,525	12,750
1501	200 feet from building	4,575	10,400	14,975
1505	750 S.F. leaching field, 50 feet from building	4,750	9,250	14,000
1506	200 feet from building	5,125	12,200	17,325
1510	1000 S.F. leaching field, 50 feet from building	5,625	10,600	16,225
1511	200 feet from building	6,000	12,500	18,500
1515	1100 S.F. leaching field, 50 feet from building	5,975	11,000	16,975
1516	200 feet from building	6,350	13,100	19,450
1600	2000 gal 2 compartment tank, 1200 S.F. leaching field, 50 feet from bldg.	7,450	11,700	19,150
1601	200 feet from building	7,825	13,600	21,425
1605	1500 S.F. leaching field, 50 feet from building	8,500	13,200	21,700
1606	200 feet from building	8,875	15,100	23,975
1710	2500 gal 2 compartment tank, 1500 S.F. leaching field, 50 feet from bldg.	9,225	14,000	23,225
1711	200 feet from building	9,600	15,900	25,500
1715	1600 S.F. leaching field, 50 feet from building	9,600	14,500	24,100
1716	200 feet from building	9,975	16,400	26,375
1720	1800 S.F. leaching field, 50 feet from building	10,300	15,600	25,900
1721	200 feet from building	10,700	17,500	28,200

G3030 Storm Sewer

Manhole   Catch Basin

The Manhole and Catch Basin System includes: excavation with a backhoe; a formed concrete footing; frame and cover; cast iron steps and compacted backfill.

The Expanded System Listing shows manholes that have a 4', 5' and 6' inside diameter riser. Depths range from 4' to 14'. Construction material shown is either concrete, concrete block, precast concrete, or brick.

System Components	QUANTITY	UNIT	COST PER EACH		
			MAT.	INST.	TOTAL
SYSTEM G3030 210 1920					
MANHOLE/CATCH BASIN, BRICK, 4' I.D. RISER, 4' DEEP					
Excavation, hydraulic backhoe, 3/8 C.Y. bucket	14.815	B.C.Y.		133.78	133.78
Trim sides and bottom of excavation	64.000	S.F.		76.80	76.80
Forms in place, manhole base, 4 uses	20.000	SFCA	23.40	120	143.40
Reinforcing in place footings, #4 to #7	.019	Ton	26.13	25.65	51.78
Concrete, 3000 psi	.925	C.Y.	126.73		126.73
Place and vibrate concrete, footing, direct chute	.925	C.Y.		56.04	56.04
Catch basin or MH, brick, 4' ID, 4' deep	1.000	Ea.	770	1,175	1,945
Catch basin or MH steps; heavy galvanized cast iron	1.000	Ea.	19	16.20	35.20
Catch basin or MH frame and cover	1.000	Ea.	320	251.50	571.50
Fill, granular	12.954	L.C.Y.	427.48		427.48
Backfill, spread with wheeled front end loader	12.954	L.C.Y.		38.08	38.08
Backfill compaction, 12" lifts, air tamp	12.954	E.C.Y.		139.52	139.52
TOTAL			1,712.74	2,032.57	3,745.31

G3030 210	Manholes & Catch Basins	COST PER EACH		
		MAT.	INST.	TOTAL
1920	Manhole/catch basin, brick, 4' I.D. riser, 4' deep	1,725	2,025	3,750
1940	6' deep	2,450	2,825	5,275
1960	8' deep	3,300	3,875	7,175
1980	10' deep	4,025	4,800	8,825
3000	12' deep	5,150	5,250	10,400
3020	14' deep	6,475	7,325	13,800
3200	Block, 4' I.D. riser, 4' deep	1,425	1,650	3,075
3220	6' deep	1,975	2,325	4,300
3240	8' deep	2,650	3,200	5,850
3260	10' deep	3,150	3,975	7,125
3280	12' deep	4,075	5,050	9,125
3300	14' deep	5,150	6,175	11,325
4620	Concrete, cast-in-place, 4' I.D. riser, 4' deep	1,700	2,775	4,475
4640	6' deep	2,425	3,725	6,150
4660	8' deep	3,425	5,375	8,800
4680	10' deep	4,125	6,700	10,825
4700	12' deep	5,250	8,300	13,550
4720	14' deep	6,525	9,975	16,500
5820	Concrete, precast, 4' I.D. riser, 4' deep	1,950	1,500	3,450
5840	6' deep	2,525	2,025	4,550

581

G3030 Storm Sewer

G3030 210	Manholes & Catch Basins	COST PER EACH		
		MAT.	INST.	TOTAL
5860	8' deep	3,975	2,850	6,825
5880	10' deep	4,550	3,500	8,050
5900	12' deep	5,525	4,250	9,775
5920	14' deep	6,700	5,500	12,200
6000	5' I.D. riser, 4' deep	3,100	1,625	4,725
6020	6' deep	5,025	2,275	7,300
6040	8' deep	5,050	3,025	8,075
6060	10' deep	6,775	3,850	10,625
6080	12' deep	8,675	4,925	13,600
6100	14' deep	10,700	6,025	16,725
6200	6' I.D. riser, 4' deep	3,525	2,125	5,650
6220	6' deep	6,775	2,825	9,600
6240	8' deep	8,300	3,950	12,250
6260	10' deep	10,000	5,000	15,000
6280	12' deep	12,000	6,325	18,325
6300	14' deep	14,100	7,675	21,775

Reference Section

All the reference information is in one section, making it easy to find what you need to know and easy to use the data set on a daily basis. This section is visually identified by a vertical black bar on the page edges.

In this Reference Section, we've included Equipment Rental Costs, a listing of rental and operating costs; Crew Listings, a full listing of all crews and equipment, and their costs; Historical Cost Indexes for cost comparisons over time; City Cost Indexes and Location Factors for adjusting costs to the region you are in; Reference Tables, where you will find explanations, estimating information and procedures, and technical data; Change Orders, information on pricing changes to contract documents; and an explanation of all the Abbreviations in the data set.

Table of Contents

Estimating Tips

- This section contains the average costs to rent and operate hundreds of pieces of construction equipment. This is useful information when one is estimating the time and material requirements of any particular operation in order to establish a unit or total cost. Bare equipment costs shown on a unit cost line include, not only rental, but also operating costs for equipment under normal use.

Rental Costs

- Equipment rental rates are obtained from the following industry sources throughout North America: contractors, suppliers, dealers, manufacturers, and distributors.

- Rental rates vary throughout the country, with larger cities generally having lower rates. Lease plans for new equipment are available for periods in excess of six months, with a percentage of payments applying toward purchase.

- Monthly rental rates vary from 2% to 5% of the purchase price of the equipment depending on the anticipated life of the equipment and its wearing parts.

- Weekly rental rates are about 1/3 of the monthly rates, and daily rental rates are about 1/3 of the weekly rate.

- Rental rates can also be treated as reimbursement costs for contractor-owned equipment. Owned equipment costs include depreciation, loan payments, interest, taxes, insurance, storage, and major repairs.

Operating Costs

- The operating costs include parts and labor for routine servicing, such as the repair and replacement of pumps, filters, and worn lines. Normal operating expendables, such as fuel, lubricants, tires, and electricity (where applicable), are also included.

- Extraordinary operating expendables with highly variable wear patterns, such as diamond bits and blades, are excluded. These costs can be found as material costs in the Unit Price section.

- The hourly operating costs listed do not include the operator's wages.

Equipment Cost/Day

- Any power equipment required by a crew is shown in the Crew Listings with a daily cost.

- This daily cost of equipment needed by a crew includes both the rental cost and the operating cost and is based on dividing the weekly rental rate by 5 (the number of working days in the week), then adding the hourly operating cost multiplied by 8 (the number of hours in a day). This "Equipment Cost/Day" is shown in the far right column of the Equipment Rental section.

- If equipment is needed for only one or two days, it is best to develop your own cost by including components for daily rent and hourly operating costs. This is important when the listed Crew for a task does not contain the equipment needed, such as a crane for lifting mechanical heating/cooling equipment up onto a roof.

- If the quantity of work is less than the crew's Daily Output shown for a Unit Price line item that includes a bare unit equipment cost, the recommendation is to estimate one day's rental cost and operating cost for equipment shown in the Crew Listing for that line item.

- Please note, in some cases the equipment description in the crew is followed by a time period in parenthesis. For example: (daily) or (monthly). In these cases the equipment cost/day is calculated by adding the rental cost per time period to the hourly operating cost multiplied by 8.

Mobilization, Demobilization Costs

- The cost to move construction equipment from an equipment yard or rental company to the job site and back again is not included in equipment rental costs listed in the Reference Section. It is also not included in the bare equipment cost of any Unit Price line item or in any equipment costs shown in the Crew Listings.

- Mobilization (to the site) and demobilization (from the site) costs can be found in the Unit Price section.

- If a piece of equipment is already at the job site, it is not appropriate to utilize mobilization or demobilization costs again in an estimate. ∎

Same Data. Simplified.

Enjoy the convenience and efficiency of accessing your costs anywhere:

- **Skip the multiplier** by setting your location
- **Quickly search,** edit, favorite and share costs
- **Stay on top of price changes** with automatic updates

Discover more at rsmeans.com/online

01 54 33 | Equipment Rental

		UNIT	HOURLY OPER. COST	RENT PER DAY	RENT PER WEEK	RENT PER MONTH	EQUIPMENT COST/DAY	
10	0010	**CONCRETE EQUIPMENT RENTAL** without operators R015433 -10							**10**
	0200	Bucket, concrete lightweight, 1/2 C.Y.	Ea.	.76	38.50	116.16	350	29.30	
	0300	1 C.Y.		.99	63.50	190	570	45.90	
	0400	1-1/2 C.Y.		1.29	60.50	181.82	545	46.70	
	0500	2 C.Y.		1.40	74	222.23	665	55.65	
	0580	8 C.Y.		7.01	94.50	282.83	850	112.65	
	0600	Cart, concrete, self-propelled, operator walking, 10 C.F.		2.88	157	469.50	1,400	116.95	
	0700	Operator riding, 18 C.F.		4.85	167	500.50	1,500	138.95	
	0800	Conveyer for concrete, portable, gas, 16" wide, 26' long		10.71	162	484.86	1,450	182.70	
	0900	46' long		11.09	177	530.31	1,600	194.80	
	1000	56' long		11.26	194	580.82	1,750	206.25	
	1100	Core drill, electric, 2-1/2 H.P., 1" to 8" bit diameter		1.58	83.50	250	750	62.65	
	1150	11 H.P., 8" to 18" cores		5.45	130	390	1,175	121.60	
	1200	Finisher, concrete floor, gas, riding trowel, 96" wide		9.75	155	465.11	1,400	171.05	
	1300	Gas, walk-behind, 3 blade, 36" trowel		2.06	101	302.50	910	76.95	
	1400	4 blade, 48" trowel		3.10	116	347.50	1,050	94.30	
	1500	Float, hand-operated (Bull float), 48" wide		.08	13	39	117	8.45	
	1570	Curb builder, 14 H.P., gas, single screw		14.14	256	767.69	2,300	266.65	
	1590	Double screw		12.50	256	767.69	2,300	253.55	
	1600	Floor grinder, concrete and terrazzo, electric, 22" path		3.07	119	357.50	1,075	96.05	
	1700	Edger, concrete, electric, 7" path		1.19	57.50	172.50	520	44.05	
	1750	Vacuum pick-up system for floor grinders, wet/dry		1.63	103	309.38	930	74.95	
	1800	Mixer, powered, mortar and concrete, gas, 6 C.F., 18 H.P.		7.48	89	267.50	805	113.35	
	1900	10 C.F., 25 H.P.		9.08	124	372.50	1,125	147.15	
	2000	16 C.F.		9.44	146	436.88	1,300	162.90	
	2100	Concrete, stationary, tilt drum, 2 C.Y.		7.98	81	242.43	725	112.35	
	2120	Pump, concrete, truck mounted, 4" line, 80' boom		31.24	290	868.70	2,600	423.65	
	2140	5" line, 110' boom		39.99	290	868.70	2,600	493.65	
	2160	Mud jack, 50 C.F. per hr.		6.50	231	691.93	2,075	190.35	
	2180	225 C.F. per hr.		8.61	296	888.91	2,675	246.65	
	2190	Shotcrete pump rig, 12 C.Y./hr.		16.73	226	676.78	2,025	269.20	
	2200	35 C.Y./hr.		15.91	290	868.70	2,600	301.05	
	2600	Saw, concrete, manual, gas, 18 H.P.		5.59	115	345	1,025	113.70	
	2650	Self-propelled, gas, 30 H.P.		7.97	82	245.62	735	112.85	
	2675	V-groove crack chaser, manual, gas, 6 H.P.		1.65	100	300	900	73.25	
	2700	Vibrators, concrete, electric, 60 cycle, 2 H.P.		.47	70	210	630	45.80	
	2800	3 H.P.		.57	81	242.50	730	53.05	
	2900	Gas engine, 5 H.P.		1.56	17.05	51.21	154	22.70	
	3000	8 H.P.		2.10	17.25	51.74	155	27.15	
	3050	Vibrating screed, gas engine, 8 H.P.		2.83	108	325	975	87.65	
	3120	Concrete transit mixer, 6 x 4, 250 H.P., 8 C.Y., rear discharge		51.08	70.50	212.13	635	451.10	
	3200	Front discharge		59.30	136	409.10	1,225	556.20	
	3300	6 x 6, 285 H.P., 12 C.Y., rear discharge		58.56	167	500	1,500	568.50	
	3400	Front discharge		61.02	167	500	1,500	588.15	
20	0010	**EARTHWORK EQUIPMENT RENTAL** without operators R015433 -10							**20**
	0040	Aggregate spreader, push type, 8' to 12' wide	Ea.	2.62	65	195	585	59.95	
	0045	Tailgate type, 8' wide		2.56	64	191.92	575	58.90	
	0055	Earth auger, truck mounted, for fence & sign posts, utility poles		13.95	152	454.55	1,375	202.55	
	0060	For borings and monitoring wells		42.95	84	252.53	760	394.15	
	0070	Portable, trailer mounted		2.32	92.50	277.50	835	74.05	
	0075	Truck mounted, for caissons, water wells		86.00	134	402	1,200	768.40	
	0080	Horizontal boring machine, 12" to 36" diameter, 45 H.P.		22.93	110	330	990	249.45	
	0090	12" to 48" diameter, 65 H.P.		31.47	130	390	1,175	329.75	
	0095	Auger, for fence posts, gas engine, hand held		.45	84.50	254.05	760	54.40	
	0100	Excavator, diesel hydraulic, crawler mounted, 1/2 C.Y. cap.		21.92	470	1,408.39	4,225	457	
	0120	5/8 C.Y. capacity		29.30	615	1,851.77	5,550	604.75	
	0140	3/4 C.Y. capacity		32.95	730	2,190.83	6,575	701.80	
	0150	1 C.Y. capacity		41.58	835	2,500	7,500	832.65	

For customer support on your Plumbing Costs with RSMeans Data, call 800.448.8182.

01 54 33 | Equipment Rental

		UNIT	HOURLY OPER. COST	RENT PER DAY	RENT PER WEEK	RENT PER MONTH	EQUIPMENT COST/DAY	
0200	1-1/2 C.Y. capacity	Ea.	49.02	505	1,518	4,550	695.80	20
0300	2 C.Y. capacity		57.09	810	2,430	7,300	942.70	
0320	2-1/2 C.Y. capacity		83.38	1,500	4,500	13,500	1,567	
0325	3-1/2 C.Y. capacity		121.20	2,025	6,072	18,200	2,184	
0330	4-1/2 C.Y. capacity		152.99	3,700	11,132	33,400	3,450	
0335	6 C.Y. capacity		194.11	3,250	9,765.80	29,300	3,506	
0340	7 C.Y. capacity		176.77	3,425	10,303.22	30,900	3,475	
0342	Excavator attachments, bucket thumbs		3.43	261	783.90	2,350	184.20	
0345	Grapples		3.17	225	674.15	2,025	160.20	
0346	Hydraulic hammer for boom mounting, 4000 ft lb.		13.60	900	2,702.04	8,100	649.20	
0347	5000 ft lb.		16.10	960	2,884.20	8,650	705.60	
0348	8000 ft lb.		23.75	1,225	3,643.20	10,900	918.65	
0349	12,000 ft lb.		25.95	1,125	3,373	10,100	882.20	
0350	Gradall type, truck mounted, 3 ton @ 15' radius, 5/8 C.Y.		43.83	835	2,500	7,500	850.65	
0370	1 C.Y. capacity		59.93	840	2,525.30	7,575	984.55	
0400	Backhoe-loader, 40 to 45 H.P., 5/8 C.Y. capacity		12.01	286	857.50	2,575	267.55	
0450	45 H.P. to 60 H.P., 3/4 C.Y. capacity		18.19	118	353.54	1,050	216.20	
0460	80 H.P., 1-1/4 C.Y. capacity		20.54	118	353.54	1,050	235.05	
0470	112 H.P., 1-1/2 C.Y. capacity		33.28	620	1,855.22	5,575	637.30	
0482	Backhoe-loader attachment, compactor, 20,000 lb.		6.50	157	470.34	1,400	146.05	
0485	Hydraulic hammer, 750 ft lb.		3.71	108	324.01	970	94.50	
0486	Hydraulic hammer, 1200 ft lb.		6.61	207	621.89	1,875	177.25	
0500	Brush chipper, gas engine, 6" cutter head, 35 H.P.		9.25	212	635	1,900	201	
0550	Diesel engine, 12" cutter head, 130 H.P.		23.88	292	875	2,625	366.05	
0600	15" cutter head, 165 H.P.		26.82	515	1,545	4,625	523.60	
0750	Bucket, clamshell, general purpose, 3/8 C.Y.		1.41	92.50	277.78	835	66.85	
0800	1/2 C.Y.		1.53	92.50	277.78	835	67.80	
0850	3/4 C.Y.		1.65	92.50	277.78	835	68.80	
0900	1 C.Y.		1.71	92.50	277.78	835	69.25	
0950	1-1/2 C.Y.		2.81	92.50	277.78	835	78.05	
1000	2 C.Y.		2.94	95	285	855	80.55	
1010	Bucket, dragline, medium duty, 1/2 C.Y.		.83	92.50	277.78	835	62.15	
1020	3/4 C.Y.		.79	92.50	277.78	835	61.85	
1030	1 C.Y.		.80	92.50	277.78	835	62	
1040	1-1/2 C.Y.		1.27	92.50	277.78	835	65.70	
1050	2 C.Y.		1.30	92.50	277.78	835	65.95	
1070	3 C.Y.		2.10	92.50	277.78	835	72.30	
1200	Compactor, manually guided 2-drum vibratory smooth roller, 7.5 H.P.		7.28	181	542.50	1,625	166.75	
1250	Rammer/tamper, gas, 8"		2.23	48.50	146.06	440	47	
1260	15"		2.65	55.50	167.23	500	54.65	
1300	Vibratory plate, gas, 18" plate, 3000 lb. blow		2.15	24.50	73.55	221	31.90	
1350	21" plate, 5000 lb. blow		2.64	241	722.50	2,175	165.60	
1370	Curb builder/extruder, 14 H.P., gas, single screw		12.50	256	767.69	2,300	253.55	
1390	Double screw		12.50	256	767.69	2,300	253.55	
1500	Disc harrow attachment, for tractor		.48	83.50	249.80	750	53.80	
1810	Feller buncher, shearing & accumulating trees, 100 H.P.		43.75	465	1,393.97	4,175	628.80	
1860	Grader, self-propelled, 25,000 lb.		33.65	1,125	3,366.73	10,100	942.50	
1910	30,000 lb.		33.15	1,350	4,040.48	12,100	1,073	
1920	40,000 lb.		52.35	1,325	4,000	12,000	1,219	
1930	55,000 lb.		67.53	1,675	5,000	15,000	1,540	
1950	Hammer, pavement breaker, self-propelled, diesel, 1000 to 1250 lb.		28.65	600	1,800	5,400	589.25	
2000	1300 to 1500 lb.		43.19	1,025	3,051.51	9,150	955.80	
2050	Pile driving hammer, steam or air, 4150 ft lb. @ 225 bpm		12.26	505	1,518	4,550	401.70	
2100	8750 ft lb. @ 145 bpm		14.47	710	2,125.20	6,375	540.85	
2150	15,000 ft lb. @ 60 bpm		14.81	845	2,530	7,600	624.45	
2200	24,450 ft lb. @ 111 bpm		15.83	980	2,934.80	8,800	713.60	
2250	Leads, 60' high for pile driving hammers up to 20,000 ft lb.		3.70	305	910.80	2,725	211.80	
2300	90' high for hammers over 20,000 ft lb.		5.50	545	1,639.44	4,925	371.85	

		UNIT	HOURLY OPER. COST	RENT PER DAY	RENT PER WEEK	RENT PER MONTH	EQUIPMENT COST/DAY	
20	2350	Diesel type hammer, 22,400 ft lb.	Ea.	17.98	495	1,489.40	4,475	441.70	20
	2400	41,300 ft lb.		25.91	625	1,881.35	5,650	583.55	
	2450	141,000 ft lb.		41.70	995	2,978.80	8,925	929.35	
	2500	Vib. elec. hammer/extractor, 200 kW diesel generator, 34 H.P.		41.74	725	2,168.78	6,500	767.70	
	2550	80 H.P.		73.68	1,050	3,135.58	9,400	1,217	
	2600	150 H.P.		136.35	2,000	6,035.99	18,100	2,298	
	2700	Hydro excavator w/ext boom 12 C.Y., 1200 gallons		38.15	1,625	4,857.60	14,600	1,277	
	2800	Log chipper, up to 22″ diameter, 600 H.P.		45.00	325	975	2,925	555	
	2850	Logger, for skidding & stacking logs, 150 H.P.		43.84	940	2,813.18	8,450	913.35	
	2860	Mulcher, diesel powered, trailer mounted		20.06	310	924.26	2,775	345.35	
	2900	Rake, spring tooth, with tractor		14.85	375	1,123.58	3,375	343.50	
	3000	Roller, vibratory, tandem, smooth drum, 20 H.P.		7.88	325	978.60	2,925	258.75	
	3050	35 H.P.		10.22	275	825	2,475	246.80	
	3100	Towed type vibratory compactor, smooth drum, 50 H.P.		25.50	525	1,581.85	4,750	520.35	
	3150	Sheepsfoot, 50 H.P.		25.87	365	1,100	3,300	426.95	
	3170	Landfill compactor, 220 H.P.		70.51	1,675	5,035.45	15,100	1,571	
	3200	Pneumatic tire roller, 80 H.P.		13.04	410	1,228.10	3,675	349.90	
	3250	120 H.P.		19.56	565	1,700	5,100	496.50	
	3300	Sheepsfoot vibratory roller, 240 H.P.		62.77	1,425	4,303.41	12,900	1,363	
	3320	340 H.P.		84.58	2,200	6,565.78	19,700	1,990	
	3350	Smooth drum vibratory roller, 75 H.P.		23.55	660	1,982.18	5,950	584.80	
	3400	125 H.P.		27.86	750	2,247.17	6,750	672.35	
	3410	Rotary mower, brush, 60″, with tractor		18.96	365	1,097.45	3,300	371.15	
	3420	Rototiller, walk-behind, gas, 5 H.P.		2.15	60	180	540	53.25	
	3422	8 H.P.		2.84	116	347.50	1,050	92.20	
	3440	Scrapers, towed type, 7 C.Y. capacity		6.50	129	386.72	1,150	129.30	
	3450	10 C.Y. capacity		7.27	172	517.37	1,550	161.65	
	3500	15 C.Y. capacity		7.46	199	595.76	1,775	178.85	
	3525	Self-propelled, single engine, 14 C.Y. capacity		134.48	2,250	6,739.92	20,200	2,424	
	3550	Dual engine, 21 C.Y. capacity		142.64	2,525	7,575.90	22,700	2,656	
	3600	31 C.Y. capacity		189.53	3,650	10,954.15	32,900	3,707	
	3640	44 C.Y. capacity		234.76	4,700	14,083.90	42,300	4,695	
	3650	Elevating type, single engine, 11 C.Y. capacity		62.42	935	2,800	8,400	1,059	
	3700	22 C.Y. capacity		115.65	1,625	4,850	14,600	1,895	
	3710	Screening plant, 110 H.P. w/5′ x 10′ screen		21.32	650	1,952.56	5,850	561.10	
	3720	5′ x 16′ screen		26.92	1,225	3,666.67	11,000	948.65	
	3850	Shovel, crawler-mounted, front-loading, 7 C.Y. capacity		220.61	3,975	11,915.21	35,700	4,148	
	3855	12 C.Y. capacity		339.92	5,500	16,514.06	49,500	6,022	
	3860	Shovel/backhoe bucket, 1/2 C.Y.		2.72	74	221.58	665	66.05	
	3870	3/4 C.Y.		2.69	83	248.76	745	71.25	
	3880	1 C.Y.		2.78	92	275.42	825	77.35	
	3890	1-1/2 C.Y.		2.98	108	324.01	970	88.65	
	3910	3 C.Y.		3.47	146	438.98	1,325	115.55	
	3950	Stump chipper, 18″ deep, 30 H.P.		6.95	223	668	2,000	189.20	
	4110	Dozer, crawler, torque converter, diesel 80 H.P.		25.48	335	1,010.12	3,025	405.85	
	4150	105 H.P.		34.64	605	1,818.22	5,450	640.80	
	4200	140 H.P.		41.64	660	1,980	5,950	729.15	
	4260	200 H.P.		63.72	1,675	5,050.60	15,200	1,520	
	4310	300 H.P.		81.45	1,900	5,667.20	17,000	1,785	
	4360	410 H.P.		107.70	3,250	9,727.46	29,200	2,807	
	4370	500 H.P.		134.58	3,925	11,788.10	35,400	3,434	
	4380	700 H.P.		232.23	5,525	16,587.71	49,800	5,175	
	4400	Loader, crawler, torque conv., diesel, 1-1/2 C.Y., 80 H.P.		29.80	555	1,667.71	5,000	572	
	4450	1-1/2 to 1-3/4 C.Y., 95 H.P.		30.54	705	2,120	6,350	668.35	
	4510	1-3/4 to 2-1/4 C.Y., 130 H.P.		48.19	900	2,700	8,100	925.50	
	4530	2-1/2 to 3-1/4 C.Y., 190 H.P.		58.30	1,125	3,400	10,200	1,146	
	4560	3-1/2 to 5 C.Y., 275 H.P.		72.05	1,475	4,400	13,200	1,456	
	4610	Front end loader, 4WD, articulated frame, diesel, 1 to 1-1/4 C.Y., 70 H.P.		16.77	286	857.06	2,575	305.60	

01 54 33 | Equipment Rental

			UNIT	HOURLY OPER. COST	RENT PER DAY	RENT PER WEEK	RENT PER MONTH	EQUIPMENT COST/DAY	
20	4620	1-1/2 to 1-3/4 C.Y., 95 H.P.		Ea.	20.17	465	1,400	4,200	441.40	**20**
	4650	1-3/4 to 2 C.Y., 130 H.P.			21.26	380	1,140	3,425	398.05	
	4710	2-1/2 to 3-1/2 C.Y., 145 H.P.			29.79	665	2,000	6,000	638.30	
	4730	3 to 4-1/2 C.Y., 185 H.P.			32.38	835	2,500	7,500	759	
	4760	5-1/4 to 5-3/4 C.Y., 270 H.P.			53.67	900	2,692.98	8,075	967.95	
	4810	7 to 9 C.Y., 475 H.P.			91.99	2,575	7,744.59	23,200	2,285	
	4870	9 to 11 C.Y., 620 H.P.			133.10	2,725	8,204.77	24,600	2,706	
	4880	Skid-steer loader, wheeled, 10 C.F., 30 H.P. gas			9.66	170	511.19	1,525	179.50	
	4890	1 C.Y., 78 H.P., diesel			18.60	495	1,487.50	4,475	446.30	
	4892	Skid-steer attachment, auger			.75	149	447.50	1,350	95.50	
	4893	Backhoe			.75	123	370.35	1,100	80.05	
	4894	Broom			.71	136	407.50	1,225	87.20	
	4895	Forks			.16	32	96	288	20.45	
	4896	Grapple			.73	89.50	267.93	805	59.45	
	4897	Concrete hammer			1.06	183	550	1,650	118.50	
	4898	Tree spade			.61	105	313.56	940	67.55	
	4899	Trencher			.66	102	305	915	66.25	
	4900	Trencher, chain, boom type, gas, operator walking, 12 H.P.			4.21	208	624.51	1,875	158.60	
	4910	Operator riding, 40 H.P.			16.83	525	1,580	4,750	450.70	
	5000	Wheel type, diesel, 4' deep, 12" wide			69.33	975	2,926.54	8,775	1,140	
	5100	6' deep, 20" wide			88.37	885	2,655	7,975	1,238	
	5150	Chain type, diesel, 5' deep, 8" wide			16.45	365	1,097.45	3,300	351.05	
	5200	Diesel, 8' deep, 16" wide			90.47	1,950	5,853.08	17,600	1,894	
	5202	Rock trencher, wheel type, 6" wide x 18" deep			47.46	91	272.73	820	434.20	
	5206	Chain type, 18" wide x 7' deep			105.44	286	858.60	2,575	1,015	
	5210	Tree spade, self-propelled			13.79	232	696.98	2,100	249.70	
	5250	Truck, dump, 2-axle, 12 ton, 8 C.Y. payload, 220 H.P.			24.16	355	1,071.50	3,225	407.60	
	5300	Three axle dump, 16 ton, 12 C.Y. payload, 400 H.P.			45.03	365	1,095.41	3,275	579.35	
	5310	Four axle dump, 25 ton, 18 C.Y. payload, 450 H.P.			50.45	585	1,750	5,250	753.60	
	5350	Dump trailer only, rear dump, 16-1/2 C.Y.			5.80	153	459.89	1,375	138.35	
	5400	20 C.Y.			6.26	172	517.37	1,550	153.55	
	5450	Flatbed, single axle, 1-1/2 ton rating			19.22	74.50	223.67	670	198.50	
	5500	3 ton rating			25.95	1,075	3,212.18	9,625	850.05	
	5550	Off highway rear dump, 25 ton capacity			63.42	1,525	4,600	13,800	1,427	
	5600	35 ton capacity			67.70	675	2,020.24	6,050	945.65	
	5610	50 ton capacity			84.88	2,175	6,500	19,500	1,979	
	5620	65 ton capacity			90.64	2,025	6,062.12	18,200	1,938	
	5630	100 ton capacity			122.70	2,975	8,936.41	26,800	2,769	
	6000	Vibratory plow, 25 H.P., walking			6.84	305	909.11	2,725	236.55	
40	0010	**GENERAL EQUIPMENT RENTAL** without operators	R015433 -10							**40**
	0020	Aerial lift, scissor type, to 20' high, 1200 lb. capacity, electric		Ea.	3.52	154	462.50	1,400	120.70	
	0030	To 30' high, 1200 lb. capacity			3.82	235	705	2,125	171.55	
	0040	Over 30' high, 1500 lb. capacity			5.20	288	865	2,600	214.55	
	0070	Articulating boom, to 45' high, 500 lb. capacity, diesel	R015433 -15		10.04	250	750	2,250	230.30	
	0075	To 60' high, 500 lb. capacity			13.82	305	909.11	2,725	292.40	
	0080	To 80' high, 500 lb. capacity			16.24	995	2,987.50	8,975	727.45	
	0085	To 125' high, 500 lb. capacity			18.56	1,675	4,997.50	15,000	1,148	
	0100	Telescoping boom to 40' high, 500 lb. capacity, diesel			11.37	320	954.56	2,875	281.90	
	0105	To 45' high, 500 lb. capacity			12.66	325	974.77	2,925	296.20	
	0110	To 60' high, 500 lb. capacity			16.56	267	800	2,400	292.45	
	0115	To 80' high, 500 lb. capacity			21.52	360	1,077.80	3,225	387.75	
	0120	To 100' high, 500 lb. capacity			29.06	985	2,950	8,850	822.45	
	0125	To 120' high, 500 lb. capacity			29.51	1,525	4,545	13,600	1,145	
	0195	Air compressor, portable, 6.5 CFM, electric			.91	53.50	160	480	39.30	
	0196	Gasoline			.66	56	167.50	505	38.80	
	0200	Towed type, gas engine, 60 CFM			9.55	129	387.50	1,175	153.85	
	0300	160 CFM			10.60	213	637.50	1,925	212.30	

For customer support on your Plumbing Costs with RSMeans Data, call 800.448.8182.

539

		UNIT	HOURLY OPER. COST	RENT PER DAY	RENT PER WEEK	RENT PER MONTH	EQUIPMENT COST/DAY	
40	0400	Diesel engine, rotary screw, 250 CFM	Ea.	12.23	175	525	1,575	202.85	**40**
	0500	365 CFM		16.19	355	1,070	3,200	343.55	
	0550	450 CFM		20.19	300	900	2,700	341.50	
	0600	600 CFM		34.51	251	752.54	2,250	426.55	
	0700	750 CFM		35.04	525	1,580	4,750	596.30	
	0930	Air tools, breaker, pavement, 60 lb.		.57	81.50	245	735	53.60	
	0940	80 lb.		.57	81	242.50	730	53.05	
	0950	Drills, hand (jackhammer), 65 lb.		.68	64	192.50	580	43.90	
	0960	Track or wagon, swing boom, 4" drifter		62.50	1,050	3,135.41	9,400	1,127	
	0970	5" drifter		62.50	1,050	3,135.41	9,400	1,127	
	0975	Track mounted quarry drill, 6" diameter drill		102.94	1,900	5,722.33	17,200	1,968	
	0980	Dust control per drill		1.05	25.50	76.26	229	23.65	
	0990	Hammer, chipping, 12 lb.		.61	46.50	140	420	32.85	
	1000	Hose, air with couplings, 50' long, 3/4" diameter		.07	11	33	99	7.15	
	1100	1" diameter		.08	12.35	37	111	8.05	
	1200	1-1/2" diameter		.22	35	105	315	22.80	
	1300	2" diameter		.24	45	135	405	28.95	
	1400	2-1/2" diameter		.36	57.50	172.50	520	37.40	
	1410	3" diameter		.42	58.50	175	525	38.35	
	1450	Drill, steel, 7/8" x 2'		.09	13.05	39.12	117	8.50	
	1460	7/8" x 6'		.12	19.85	59.58	179	12.85	
	1520	Moil points		.03	7	21	63	4.40	
	1525	Pneumatic nailer w/accessories		.48	39.50	119.19	360	27.70	
	1530	Sheeting driver for 60 lb. breaker		.04	7.85	23.52	70.50	5.05	
	1540	For 90 lb. breaker		.13	10.65	31.88	95.50	7.45	
	1550	Spade, 25 lb.		.51	7.50	22.47	67.50	8.55	
	1560	Tamper, single, 35 lb.		.60	58.50	175	525	39.75	
	1570	Triple, 140 lb.		.90	62.50	187.09	560	44.60	
	1580	Wrenches, impact, air powered, up to 3/4" bolt		.43	50	149.75	450	33.40	
	1590	Up to 1-1/4" bolt		.58	67.50	202.50	610	45.15	
	1600	Barricades, barrels, reflectorized, 1 to 99 barrels		.03	4	12	36	2.65	
	1610	100 to 200 barrels		.03	4.46	13.38	40	2.90	
	1620	Barrels with flashers, 1 to 99 barrels		.04	6.45	19.39	58	4.15	
	1630	100 to 200 barrels		.03	5.15	15.52	46.50	3.35	
	1640	Barrels with steady burn type C lights		.05	8.55	25.61	77	5.50	
	1650	Illuminated board, trailer mounted, with generator		3.32	141	423.30	1,275	111.20	
	1670	Portable barricade, stock, with flashers, 1 to 6 units		.04	6.45	19.34	58	4.15	
	1680	25 to 50 units		.03	6	18.03	54	3.85	
	1685	Butt fusion machine, wheeled, 1.5 HP electric, 2" - 8" diameter pipe		2.66	190	568.50	1,700	134.95	
	1690	Tracked, 20 HP diesel, 4" - 12" diameter pipe		10.04	565	1,702.05	5,100	420.75	
	1695	83 HP diesel, 8" - 24" diameter pipe		51.84	1,125	3,358.65	10,100	1,086	
	1700	Carts, brick, gas engine, 1000 lb. capacity		2.98	65.50	196.97	590	63.20	
	1800	1500 lb., 7-1/2' lift		2.95	70	210.50	630	65.70	
	1822	Dehumidifier, medium, 6 lb./hr., 150 CFM		1.20	77.50	232.03	695	56.05	
	1824	Large, 18 lb./hr., 600 CFM		2.22	585	1,750	5,250	367.75	
	1830	Distributor, asphalt, trailer mounted, 2000 gal., 38 H.P. diesel		11.13	355	1,071.32	3,225	303.25	
	1840	3000 gal., 38 H.P. diesel		13.02	385	1,149.71	3,450	334.10	
	1850	Drill, rotary hammer, electric		1.13	72	216.17	650	52.25	
	1860	Carbide bit, 1-1/2" diameter, add to electric rotary hammer		.03	41.50	125	375	25.25	
	1865	Rotary, crawler, 250 H.P.		137.40	2,450	7,375	22,100	2,574	
	1870	Emulsion sprayer, 65 gal., 5 H.P. gas engine		2.80	108	324.01	970	87.20	
	1880	200 gal., 5 H.P. engine		7.31	181	543.50	1,625	167.20	
	1900	Floor auto-scrubbing machine, walk-behind, 28" path		5.69	223	670	2,000	179.55	
	1930	Floodlight, mercury vapor, or quartz, on tripod, 1000 watt		.46	36.50	110	330	25.70	
	1940	2000 watt		.60	28.50	85.55	257	21.90	
	1950	Floodlights, trailer mounted with generator, 1 - 300 watt light		3.59	79.50	238.30	715	76.35	
	1960	2 - 1000 watt lights		4.54	88.50	265.48	795	89.40	
	2000	4 - 300 watt lights		4.29	101	303.11	910	95	

01 54 33 | Equipment Rental

		UNIT	HOURLY OPER. COST	RENT PER DAY	RENT PER WEEK	RENT PER MONTH	EQUIPMENT COST/DAY	
40	2005	Foam spray rig, incl. box trailer, compressor, generator, proportioner	Ea.	25.76	540	1,620.05	4,850	530.15	**40**
	2015	Forklift, pneumatic tire, rough terr, straight mast, 5000 lb, 12' lift, gas		18.82	221	663.70	2,000	283.30	
	2025	8000 lb, 12' lift		22.95	365	1,097.45	3,300	403.10	
	2030	5000 lb, 12' lift, diesel		15.59	247	740.71	2,225	272.85	
	2035	8000 lb, 12' lift, diesel		16.90	280	839.82	2,525	303.15	
	2045	All terrain, telescoping boom, diesel, 5000 lb, 10' reach, 19' lift		17.40	200	600	1,800	259.25	
	2055	6600 lb, 29' reach, 42' lift		21.29	236	707.08	2,125	311.75	
	2065	10,000 lb, 31' reach, 45' lift		23.30	267	800	2,400	346.45	
	2070	Cushion tire, smooth floor, gas, 5000 lb capacity		8.32	262	785	2,350	223.60	
	2075	8000 lb capacity		11.47	320	957.50	2,875	283.25	
	2085	Diesel, 5000 lb capacity		7.82	257	770	2,300	216.55	
	2090	12,000 lb capacity		12.16	400	1,206.59	3,625	338.60	
	2095	20,000 lb capacity		17.41	800	2,405	7,225	620.30	
	2100	Generator, electric, gas engine, 1.5 kW to 3 kW		2.60	46.50	140	420	48.80	
	2200	5 kW		3.25	91	272.50	820	80.45	
	2300	10 kW		5.98	108	322.50	970	112.35	
	2400	25 kW		7.47	405	1,222.25	3,675	304.25	
	2500	Diesel engine, 20 kW		9.29	222	665	2,000	207.35	
	2600	50 kW		16.09	360	1,085	3,250	345.70	
	2700	100 kW		28.85	485	1,455	4,375	521.85	
	2800	250 kW		54.84	825	2,472.50	7,425	933.25	
	2850	Hammer, hydraulic, for mounting on boom, to 500 ft lb.		2.93	94.50	283.25	850	80.05	
	2860	1000 ft lb.		4.65	141	423.30	1,275	121.85	
	2900	Heaters, space, oil or electric, 50 MBH		1.48	47	141.42	425	40.10	
	3000	100 MBH		2.74	47	141.42	425	50.25	
	3100	300 MBH		8.00	136	409.10	1,225	145.80	
	3150	500 MBH		13.28	208	625	1,875	231.20	
	3200	Hose, water, suction with coupling, 20' long, 2" diameter		.02	5.65	17	51	3.55	
	3210	3" diameter		.03	14.15	42.50	128	8.75	
	3220	4" diameter		.03	28.50	85.01	255	17.25	
	3230	6" diameter		.11	41	122.96	370	25.50	
	3240	8" diameter		.27	54	161.92	485	34.55	
	3250	Discharge hose with coupling, 50' long, 2" diameter		.01	6.50	19.50	58.50	4	
	3260	3" diameter		.01	7.35	22	66	4.50	
	3270	4" diameter		.02	21	63.25	190	12.80	
	3280	6" diameter		.06	29.50	88.04	264	18.10	
	3290	8" diameter		.24	38	113.85	340	24.70	
	3295	Insulation blower		.84	118	353.54	1,050	77.40	
	3300	Ladders, extension type, 16' to 36' long		.18	41.50	125	375	26.45	
	3400	40' to 60' long		.64	123	368	1,100	78.75	
	3405	Lance for cutting concrete		2.22	66.50	200	600	57.80	
	3407	Lawn mower, rotary, 22", 5 H.P.		1.06	38.50	115	345	31.50	
	3408	48" self-propelled		2.92	160	481	1,450	119.60	
	3410	Level, electronic, automatic, with tripod and leveling rod		1.06	43	129.50	390	34.40	
	3430	Laser type, for pipe and sewer line and grade		2.20	117	350	1,050	87.55	
	3440	Rotating beam for interior control		.91	64	192.50	580	45.80	
	3460	Builder's optical transit, with tripod and rod		.10	43	129.50	390	26.70	
	3500	Light towers, towable, with diesel generator, 2000 watt		4.31	102	307.29	920	95.90	
	3600	4000 watt		4.55	142	425	1,275	121.40	
	3700	Mixer, powered, plaster and mortar, 6 C.F., 7 H.P.		2.08	86.50	260	780	68.60	
	3800	10 C.F., 9 H.P.		2.26	124	372.50	1,125	92.60	
	3850	Nailer, pneumatic		.48	30.50	91	273	22.10	
	3900	Paint sprayers complete, 8 CFM		.86	62.50	186.74	560	44.20	
	4000	17 CFM		1.61	111	333.84	1,000	79.70	
	4020	Pavers, bituminous, rubber tires, 8' wide, 50 H.P., diesel		32.31	575	1,724.57	5,175	603.40	
	4030	10' wide, 150 H.P.		96.77	1,975	5,905.34	17,700	1,955	
	4050	Crawler, 8' wide, 100 H.P., diesel		88.64	2,075	6,245.03	18,700	1,958	
	4060	10' wide, 150 H.P.		105.21	2,375	7,133.45	21,400	2,268	

01 54 33 | Equipment Rental

		UNIT	HOURLY OPER. COST	RENT PER DAY	RENT PER WEEK	RENT PER MONTH	EQUIPMENT COST/DAY	
40	4070	Concrete paver, 12' to 24' wide, 250 H.P.	Ea.	88.67	1,700	5,121.45	15,400	1,734	**40**
	4080	Placer-spreader-trimmer, 24' wide, 300 H.P.		118.92	2,575	7,760.56	23,300	2,503	
	4100	Pump, centrifugal gas pump, 1-1/2" diam., 65 GPM		3.97	54.50	164.10	490	64.60	
	4200	2" diameter, 130 GPM		5.04	44	132.50	400	66.80	
	4300	3" diameter, 250 GPM		5.18	55	165	495	74.40	
	4400	6" diameter, 1500 GPM		22.46	92.50	277.78	835	235.25	
	4500	Submersible electric pump, 1-1/4" diameter, 55 GPM		.40	35	105	315	24.25	
	4600	1-1/2" diameter, 83 GPM		.45	43.50	130	390	29.60	
	4700	2" diameter, 120 GPM		1.66	55	165	495	46.30	
	4800	3" diameter, 300 GPM		3.07	109	327.50	985	90.05	
	4900	4" diameter, 560 GPM		14.90	76	227.50	685	164.70	
	5000	6" diameter, 1590 GPM		22.30	65.50	196.97	590	217.80	
	5100	Diaphragm pump, gas, single, 1-1/2" diameter		1.14	38.50	116.16	350	32.35	
	5200	2" diameter		4.02	92.50	277.78	835	87.70	
	5300	3" diameter		4.09	86	258	775	84.35	
	5400	Double, 4" diameter		6.10	96	287.88	865	106.35	
	5450	Pressure washer 5 GPM, 3000 psi		3.92	110	330	990	97.35	
	5460	7 GPM, 3000 psi		5.00	90	270	810	93.95	
	5470	High pressure water jet 10 KSI		40.02	730	2,185.92	6,550	757.40	
	5480	40 KSI		28.21	990	2,975.28	8,925	820.75	
	5500	Trash pump, self-priming, gas, 2" diameter		3.85	109	328.29	985	96.50	
	5600	Diesel, 4" diameter		6.74	163	489.91	1,475	151.95	
	5650	Diesel, 6" diameter		17.02	163	489.91	1,475	234.15	
	5655	Grout pump		18.92	284	851.83	2,550	321.75	
	5700	Salamanders, L.P. gas fired, 100,000 Btu		2.92	57.50	172.50	520	57.85	
	5705	50,000 Btu		1.68	23	69	207	27.25	
	5720	Sandblaster, portable, open top, 3 C.F. capacity		.61	132	395	1,175	83.85	
	5730	6 C.F. capacity		1.02	133	399	1,200	87.90	
	5740	Accessories for above		.14	24	72.12	216	15.55	
	5750	Sander, floor		.78	73.50	220	660	50.20	
	5760	Edger		.52	35	105	315	25.20	
	5800	Saw, chain, gas engine, 18" long		1.77	63.50	190	570	52.20	
	5900	Hydraulic powered, 36" long		.79	59	176.77	530	41.65	
	5950	60" long		.79	65.50	196.97	590	45.70	
	6000	Masonry, table mounted, 14" diameter, 5 H.P.		1.34	76.50	230	690	56.70	
	6050	Portable cut-off, 8 H.P.		1.83	72.50	217.50	655	58.15	
	6100	Circular, hand held, electric, 7-1/4" diameter		.23	13.85	41.50	125	10.15	
	6200	12" diameter		.24	41	122.50	370	26.40	
	6250	Wall saw, w/hydraulic power, 10 H.P.		3.32	99.50	299	895	86.40	
	6275	Shot blaster, walk-behind, 20" wide		4.79	284	851.83	2,550	208.70	
	6280	Sidewalk broom, walk-behind		2.27	83.50	250.85	755	68.30	
	6300	Steam cleaner, 100 gallons per hour		3.38	83.50	250.85	755	77.20	
	6310	200 gallons per hour		4.38	101	303.11	910	95.70	
	6340	Tar kettle/pot, 400 gallons		16.65	128	383.85	1,150	209.95	
	6350	Torch, cutting, acetylene-oxygen, 150' hose, excludes gases		.46	15.50	46.51	140	12.95	
	6360	Hourly operating cost includes tips and gas		21.17	7.10	21.23	63.50	173.60	
	6410	Toilet, portable chemical		.13	23.50	70.03	210	15.05	
	6420	Recycle flush type		.16	29	86.75	260	18.65	
	6430	Toilet, fresh water flush, garden hose,		.20	34.50	103.47	310	22.25	
	6440	Hoisted, non-flush, for high rise		.16	28	84.66	254	18.20	
	6465	Tractor, farm with attachment		17.57	395	1,184.37	3,550	377.45	
	6480	Trailers, platform, flush deck, 2 axle, 3 ton capacity		1.71	96	287.41	860	71.15	
	6500	25 ton capacity		6.31	145	433.76	1,300	137.20	
	6600	40 ton capacity		8.14	206	616.66	1,850	188.45	
	6700	3 axle, 50 ton capacity		8.82	228	683.33	2,050	207.25	
	6800	75 ton capacity		11.21	305	907.63	2,725	271.20	
	6810	Trailer mounted cable reel for high voltage line work		5.94	28.50	85.86	258	64.75	
	6820	Trailer mounted cable tensioning rig		11.79	28.50	85.86	258	111.50	

01 54 33 | Equipment Rental

		UNIT	HOURLY OPER. COST	RENT PER DAY	RENT PER WEEK	RENT PER MONTH	EQUIPMENT COST/DAY	
40	6830	Cable pulling rig	Ea.	74.51	28.50	85.86	258	613.30	**40**
	6850	Portable cable/wire puller, 8000 lb max pulling capacity		3.74	121	363.64	1,100	102.65	
	6900	Water tank trailer, engine driven discharge, 5000 gallons		7.24	160	480.79	1,450	154.10	
	6925	10,000 gallons		9.87	218	653.25	1,950	209.60	
	6950	Water truck, off highway, 6000 gallons		72.61	845	2,529.89	7,600	1,087	
	7010	Tram car for high voltage line work, powered, 2 conductor		6.95	29.50	88	264	73.20	
	7020	Transit (builder's level) with tripod		.10	17.75	53.30	160	11.45	
	7030	Trench box, 3000 lb., 6' x 8'		.57	97.50	293.15	880	63.15	
	7040	7200 lb., 6' x 20'		.73	189	566.72	1,700	119.15	
	7050	8000 lb., 8' x 16'		1.09	205	615	1,850	131.70	
	7060	9500 lb., 8' x 20'		1.22	235	705.51	2,125	150.85	
	7065	11,000 lb., 8' x 24'		1.28	221	663.70	2,000	142.95	
	7070	12,000 lb., 10' x 20'		1.51	267	799.57	2,400	172	
	7100	Truck, pickup, 3/4 ton, 2 wheel drive		9.35	62.50	187.09	560	112.20	
	7200	4 wheel drive		9.60	167	500	1,500	176.75	
	7250	Crew carrier, 9 passenger		12.81	109	328.29	985	168.15	
	7290	Flat bed truck, 20,000 lb. GVW		15.44	134	402.40	1,200	204.05	
	7300	Tractor, 4 x 2, 220 H.P.		22.52	218	653.25	1,950	310.80	
	7410	330 H.P.		32.72	298	893.64	2,675	440.45	
	7500	6 x 4, 380 H.P.		36.53	345	1,034.74	3,100	499.15	
	7600	450 H.P.		44.76	420	1,254.23	3,775	608.95	
	7610	Tractor, with A frame, boom and winch, 225 H.P.		25.04	296	886.76	2,650	377.65	
	7620	Vacuum truck, hazardous material, 2500 gallons		12.94	315	940.67	2,825	291.70	
	7625	5,000 gallons		13.18	445	1,332.62	4,000	371.95	
	7650	Vacuum, HEPA, 16 gallon, wet/dry		.86	122	365	1,100	79.90	
	7655	55 gallon, wet/dry		.79	25.50	76.50	230	21.60	
	7660	Water tank, portable		.74	162	485.11	1,450	102.90	
	7690	Sewer/catch basin vacuum, 14 C.Y., 1500 gallons		17.52	670	2,012	6,025	542.60	
	7700	Welder, electric, 200 amp		3.86	33.50	101.01	305	51.10	
	7800	300 amp		5.61	104	313.14	940	107.55	
	7900	Gas engine, 200 amp		9.05	59	176.77	530	107.80	
	8000	300 amp		10.25	111	333.96	1,000	148.75	
	8100	Wheelbarrow, any size		.07	11.15	33.50	101	7.20	
	8200	Wrecking ball, 4000 lb.		2.53	62	186	560	57.45	
50	0010	**HIGHWAY EQUIPMENT RENTAL** without operators							**50**
	0050	Asphalt batch plant, portable drum mixer, 100 ton/hr. [R015433-10]	Ea.	89.47	1,550	4,677.24	14,000	1,651	
	0060	200 ton/hr.		103.21	1,675	4,990.80	15,000	1,824	
	0070	300 ton/hr.		121.29	1,950	5,853.08	17,600	2,141	
	0100	Backhoe attachment, long stick, up to 185 H.P., 10-1/2' long		.37	26	77.34	232	18.45	
	0140	Up to 250 H.P., 12' long		.42	29	86.75	260	20.70	
	0180	Over 250 H.P., 15' long		.57	39.50	118.11	355	28.20	
	0200	Special dipper arm, up to 100 H.P., 32' long		1.17	80.50	241.44	725	57.65	
	0240	Over 100 H.P., 33' long		1.46	101	303.11	910	72.30	
	0280	Catch basin/sewer cleaning truck, 3 ton, 9 C.Y., 1000 gal.		35.81	425	1,280.36	3,850	542.60	
	0300	Concrete batch plant, portable, electric, 200 C.Y./hr.		24.47	565	1,698.44	5,100	535.50	
	0520	Grader/dozer attachment, ripper/scarifier, rear mounted, up to 135 H.P.		3.19	64	192.32	575	63.95	
	0540	Up to 180 H.P.		4.18	97	290.56	870	91.60	
	0580	Up to 250 H.P.		5.92	155	465.11	1,400	140.40	
	0700	Pvmt. removal bucket, for hyd. excavator, up to 90 H.P.		2.18	59	176.64	530	52.80	
	0740	Up to 200 H.P.		2.33	75.50	225.76	675	63.80	
	0780	Over 200 H.P.		2.55	92.50	276.98	830	75.80	
	0900	Aggregate spreader, self-propelled, 187 H.P.		51.21	750	2,247.17	6,750	859.10	
	1000	Chemical spreader, 3 C.Y.		3.20	99.50	299	895	85.40	
	1900	Hammermill, traveling, 250 H.P.		68.03	520	1,565.69	4,700	857.40	
	2000	Horizontal borer, 3" diameter, 13 H.P. gas driven		5.47	234	702.03	2,100	184.15	
	2150	Horizontal directional drill, 20,000 lb. thrust, 78 H.P. diesel		27.86	535	1,606.09	4,825	544.10	
	2160	30,000 lb. thrust, 115 H.P.		34.24	625	1,868.72	5,600	647.65	
	2170	50,000 lb. thrust, 170 H.P.		49.09	720	2,156.61	6,475	824.05	

01 54 33 | Equipment Rental

		UNIT	HOURLY OPER. COST	RENT PER DAY	RENT PER WEEK	RENT PER MONTH	EQUIPMENT COST/DAY	
50	2190	Mud trailer for HDD, 1500 gallons, 175 H.P., gas	Ea.	25.76	177	530.31	1,600	312.15	50
	2200	Hydromulcher, diesel, 3000 gallon, for truck mounting		17.61	193	580	1,750	256.85	
	2300	Gas, 600 gallon		7.57	96	287.88	865	118.15	
	2400	Joint & crack cleaner, walk behind, 25 H.P.		3.19	46	137.38	410	53	
	2500	Filler, trailer mounted, 400 gallons, 20 H.P.		8.43	173	517.50	1,550	170.95	
	3000	Paint striper, self-propelled, 40 gallon, 22 H.P.		6.83	123	368.69	1,100	128.35	
	3100	120 gallon, 120 H.P.		17.83	385	1,151.54	3,450	372.90	
	3200	Post drivers, 6" I-Beam frame, for truck mounting		12.54	325	969.72	2,900	294.25	
	3400	Road sweeper, self-propelled, 8' wide, 90 H.P.		36.34	720	2,164.75	6,500	723.65	
	3450	Road sweeper, vacuum assisted, 4 C.Y., 220 gallons		58.98	680	2,038.13	6,125	879.45	
	4000	Road mixer, self-propelled, 130 H.P.		46.79	835	2,508.46	7,525	876	
	4100	310 H.P.		75.91	2,175	6,558.59	19,700	1,919	
	4220	Cold mix paver, incl. pug mill and bitumen tank, 165 H.P.		96.11	2,350	7,015.87	21,000	2,172	
	4240	Pavement brush, towed		3.47	101	303.11	910	88.35	
	4250	Paver, asphalt, wheel or crawler, 130 H.P., diesel		95.37	2,300	6,898.28	20,700	2,143	
	4300	Paver, road widener, gas, 1' to 6', 67 H.P.		47.22	985	2,952.67	8,850	968.30	
	4400	Diesel, 2' to 14', 88 H.P.		57.06	1,175	3,501.40	10,500	1,157	
	4600	Slipform pavers, curb and gutter, 2 track, 75 H.P.		58.53	1,275	3,814.96	11,400	1,231	
	4700	4 track, 165 H.P.		36.11	855	2,560.72	7,675	801.05	
	4800	Median barrier, 215 H.P.		59.13	1,350	4,076.26	12,200	1,288	
	4901	Trailer, low bed, 75 ton capacity		10.84	286	857.06	2,575	258.10	
	5000	Road planer, walk behind, 10" cutting width, 10 H.P.		2.48	258	773	2,325	174.45	
	5100	Self-propelled, 12" cutting width, 64 H.P.		8.34	192	575.77	1,725	181.85	
	5120	Traffic line remover, metal ball blaster, truck mounted, 115 H.P.		47.03	1,025	3,100	9,300	996.25	
	5140	Grinder, truck mounted, 115 H.P.		51.41	1,025	3,100	9,300	1,031	
	5160	Walk-behind, 11 H.P.		3.59	143	430	1,300	114.75	
	5200	Pavement profiler, 4' to 6' wide, 450 H.P.		218.77	1,275	3,838.46	11,500	2,518	
	5300	8' to 10' wide, 750 H.P.		334.94	1,350	4,015.23	12,000	3,483	
	5400	Roadway plate, steel, 1" x 8' x 20'		.09	61.50	184.69	555	37.65	
	5600	Stabilizer, self-propelled, 150 H.P.		41.55	1,050	3,131.37	9,400	958.70	
	5700	310 H.P.		76.95	1,325	3,939.47	11,800	1,404	
	5800	Striper, truck mounted, 120 gallon paint, 460 H.P.		49.24	350	1,046	3,150	603.10	
	5900	Thermal paint heating kettle, 115 gallons		7.78	75	225	675	107.25	
	6000	Tar kettle, 330 gallon, trailer mounted		12.40	96	287.50	865	156.70	
	7000	Tunnel locomotive, diesel, 8 to 12 ton		30.11	625	1,881.35	5,650	617.20	
	7005	Electric, 10 ton		29.60	715	2,142.65	6,425	665.30	
	7010	Muck cars, 1/2 C.Y. capacity		2.33	27	81	243	34.80	
	7020	1 C.Y. capacity		2.54	35	105.56	315	41.45	
	7030	2 C.Y. capacity		2.69	39.50	118.11	355	45.10	
	7040	Side dump, 2 C.Y. capacity		2.90	49	146.33	440	52.50	
	7050	3 C.Y. capacity		3.90	53.50	160.96	485	63.40	
	7060	5 C.Y. capacity		5.69	69.50	207.99	625	87.10	
	7100	Ventilating blower for tunnel, 7-1/2 H.P.		2.16	53.50	159.91	480	49.30	
	7110	10 H.P.		2.45	55.50	167.23	500	53.05	
	7120	20 H.P.		3.58	72.50	217.40	650	72.15	
	7140	40 H.P.		6.21	96	287.43	860	107.20	
	7160	60 H.P.		8.79	103	308.33	925	132	
	7175	75 H.P.		10.49	160	480.79	1,450	180.10	
	7180	200 H.P.		21.03	315	945.90	2,850	357.45	
	7800	Windrow loader, elevating		54.49	1,700	5,125	15,400	1,461	
60	0010	**LIFTING AND HOISTING EQUIPMENT RENTAL** without operators R015433 -10							60
	0150	Crane, flatbed mounted, 3 ton capacity	Ea.	14.58	203	610.30	1,825	238.75	
	0200	Crane, climbing, 106' jib, 6000 lb. capacity, 410 fpm R312316 -45		40.13	2,675	8,034	24,100	1,928	
	0300	101' jib, 10,250 lb. capacity, 270 fpm		46.90	2,300	6,868.82	20,600	1,749	
	0500	Tower, static, 130' high, 106' jib, 6200 lb. capacity at 400 fpm		45.61	2,300	6,916	20,700	1,748	
	0520	Mini crawler spider crane, up to 24" wide, 1990 lb. lifting capacity		12.65	555	1,672.31	5,025	435.70	
	0525	Up to 30" wide, 6450 lb. lifting capacity		14.70	660	1,985.87	5,950	514.75	
	0530	Up to 52" wide, 6680 lb. lifting capacity		23.38	810	2,430.08	7,300	673.05	

01 54 33 | Equipment Rental

		UNIT	HOURLY OPER. COST	RENT PER DAY	RENT PER WEEK	RENT PER MONTH	EQUIPMENT COST/DAY	
60	0535	Up to 55" wide, 8920 lb. lifting capacity	Ea.	26.10	895	2,691.37	8,075	747.10	60
	0540	Up to 66" wide, 13,350 lb. lifting capacity		35.34	1,400	4,180.77	12,500	1,119	
	0600	Crawler mounted, lattice boom, 1/2 C.Y., 15 tons at 12' radius		37.34	855	2,558	7,675	810.30	
	0700	3/4 C.Y., 20 tons at 12' radius		54.72	960	2,874	8,625	1,013	
	0800	1 C.Y., 25 tons at 12' radius		68.10	1,025	3,038	9,125	1,152	
	0900	1-1/2 C.Y., 40 tons at 12' radius		66.98	1,150	3,476	10,400	1,231	
	1000	2 C.Y., 50 tons at 12' radius		89.66	1,375	4,120	12,400	1,541	
	1100	3 C.Y., 75 tons at 12' radius		65.65	2,400	7,210	21,600	1,967	
	1200	100 ton capacity, 60' boom		86.78	2,700	8,080.96	24,200	2,310	
	1300	165 ton capacity, 60' boom		107.18	3,025	9,091.08	27,300	2,676	
	1400	200 ton capacity, 70' boom		139.61	3,875	11,616.38	34,800	3,440	
	1500	350 ton capacity, 80' boom		184.04	4,200	12,626.50	37,900	3,998	
	1600	Truck mounted, lattice boom, 6 x 4, 20 tons at 10' radius		40.16	2,000	6,026	18,100	1,526	
	1700	25 tons at 10' radius		43.16	2,400	7,210	21,600	1,787	
	1800	8 x 4, 30 tons at 10' radius		54.71	2,575	7,725	23,200	1,983	
	1900	40 tons at 12' radius		54.71	2,825	8,446	25,300	2,127	
	2000	60 tons at 15' radius		54.23	1,675	5,000.09	15,000	1,434	
	2050	82 tons at 15' radius		60.03	1,800	5,404.14	16,200	1,561	
	2100	90 tons at 15' radius		67.06	1,950	5,883.95	17,700	1,713	
	2200	115 tons at 15' radius		75.66	2,200	6,591.03	19,800	1,923	
	2300	150 tons at 18' radius		81.91	2,775	8,343	25,000	2,324	
	2350	165 tons at 18' radius		87.91	2,825	8,498	25,500	2,403	
	2400	Truck mounted, hydraulic, 12 ton capacity		29.80	395	1,186.89	3,550	475.80	
	2500	25 ton capacity		36.72	490	1,464.67	4,400	586.70	
	2550	33 ton capacity		54.71	910	2,727.32	8,175	983.15	
	2560	40 ton capacity		54.71	910	2,727.32	8,175	983.15	
	2600	55 ton capacity		54.32	925	2,777.83	8,325	990.15	
	2700	80 ton capacity		71.13	1,475	4,444.53	13,300	1,458	
	2720	100 ton capacity		75.72	1,575	4,722.31	14,200	1,550	
	2740	120 ton capacity		106.34	1,850	5,555.66	16,700	1,962	
	2760	150 ton capacity		113.80	2,050	6,186.99	18,600	2,148	
	2800	Self-propelled, 4 x 4, with telescoping boom, 5 ton		15.29	435	1,298	3,900	381.95	
	2900	12-1/2 ton capacity		21.63	435	1,298	3,900	432.65	
	3000	15 ton capacity		34.77	455	1,363.66	4,100	550.85	
	3050	20 ton capacity		22.55	655	1,969.73	5,900	574.35	
	3100	25 ton capacity		37.06	1,425	4,293.01	12,900	1,155	
	3150	40 ton capacity		45.35	665	1,994.99	5,975	761.80	
	3200	Derricks, guy, 20 ton capacity, 60' boom, 75' mast		22.97	1,450	4,378	13,100	1,059	
	3300	100' boom, 115' mast		36.41	2,050	6,180	18,500	1,527	
	3400	Stiffleg, 20 ton capacity, 70' boom, 37' mast		25.66	635	1,906	5,725	586.50	
	3500	100' boom, 47' mast		39.71	685	2,060	6,175	729.70	
	3550	Helicopter, small, lift to 1250 lb. maximum, w/pilot		100.14	2,175	6,500.12	19,500	2,101	
	3600	Hoists, chain type, overhead, manual, 3/4 ton		.15	10.25	30.70	92	7.30	
	3900	10 ton		.80	6.25	18.78	56.50	10.10	
	4000	Hoist and tower, 5000 lb. cap., portable electric, 40' high		4.82	146	438	1,325	126.15	
	4100	For each added 10' section, add		.12	32.50	98	294	20.55	
	4200	Hoist and single tubular tower, 5000 lb. electric, 100' high		7.03	108	325	975	121.20	
	4300	For each added 6'-6" section, add		.21	39.50	118	355	25.25	
	4400	Hoist and double tubular tower, 5000 lb., 100' high		7.65	108	325	975	126.20	
	4500	For each added 6'-6" section, add		.23	42.50	128	385	27.45	
	4550	Hoist and tower, mast type, 6000 lb., 100' high		9.32	95.50	286.87	860	131.95	
	4570	For each added 10' section, add		.13	32.50	98	294	20.65	
	4600	Hoist and tower, personnel, electric, 2000 lb., 100' @ 125 fpm		17.67	25.50	75.76	227	156.55	
	4700	3000 lb., 100' @ 200 fpm		20.22	25.50	75.76	227	176.95	
	4800	3000 lb., 150' @ 300 fpm		22.44	25.50	75.76	227	194.70	
	4900	4000 lb., 100' @ 300 fpm		23.22	25.50	75.76	227	200.90	
	5000	6000 lb., 100' @ 275 fpm		24.95	25.50	75.76	227	214.75	
	5100	For added heights up to 500', add	L.F.	.01	3.33	10	30	2.10	

01 54 33 | Equipment Rental

			UNIT	HOURLY OPER. COST	RENT PER DAY	RENT PER WEEK	RENT PER MONTH	EQUIPMENT COST/DAY	
60	5200	Jacks, hydraulic, 20 ton	Ea.	.05	19.85	59.60	179	12.30	60
	5500	100 ton		.40	26	78.50	236	18.95	
	6100	Jacks, hydraulic, climbing w/50' jackrods, control console, 30 ton cap.		2.19	32	96	288	36.70	
	6150	For each added 10' jackrod section, add		.05	5	15	45	3.40	
	6300	50 ton capacity		3.52	34.50	103	310	48.75	
	6350	For each added 10' jackrod section, add		.06	5	15	45	3.50	
	6500	125 ton capacity		9.20	53.50	160	480	105.55	
	6550	For each added 10' jackrod section, add		.62	5	15	45	8	
	6600	Cable jack, 10 ton capacity with 200' cable		1.60	36.50	110	330	34.80	
	6650	For each added 50' of cable, add	▼	.22	15.35	46	138	11	
70	0010	**WELLPOINT EQUIPMENT RENTAL** without operators							70
	0020	Based on 2 months rental							
	0100	Combination jetting & wellpoint pump, 60 H.P. diesel	Ea.	15.83	305	922	2,775	311.05	
	0200	High pressure gas jet pump, 200 H.P., 300 psi	"	34.17	278	833.35	2,500	440.05	
	0300	Discharge pipe, 8" diameter	L.F.	.01	1.41	4.24	12.75	.90	
	0350	12" diameter		.01	2.09	6.26	18.80	1.35	
	0400	Header pipe, flows up to 150 GPM, 4" diameter		.01	.74	2.22	6.65	.50	
	0500	400 GPM, 6" diameter		.01	1.08	3.23	9.70	.70	
	0600	800 GPM, 8" diameter		.01	1.41	4.24	12.75	.95	
	0700	1500 GPM, 10" diameter		.01	1.75	5.25	15.75	1.15	
	0800	2500 GPM, 12" diameter		.03	2.09	6.26	18.80	1.45	
	0900	4500 GPM, 16" diameter		.03	2.42	7.27	22	1.70	
	0950	For quick coupling aluminum and plastic pipe, add	▼	.03	9.60	28.84	86.50	6.05	
	1100	Wellpoint, 25' long, with fittings & riser pipe, 1-1/2" or 2" diameter	Ea.	.07	133	399	1,200	80.35	
	1200	Wellpoint pump, diesel powered, 4" suction, 20 H.P.		7.07	152	454.55	1,375	147.50	
	1300	6" suction, 30 H.P.		9.49	140	420	1,250	159.90	
	1400	8" suction, 40 H.P.		12.85	253	757.59	2,275	254.35	
	1500	10" suction, 75 H.P.		18.75	268	803.05	2,400	310.60	
	1600	12" suction, 100 H.P.		27.51	300	904.06	2,700	400.90	
	1700	12" suction, 175 H.P.	▼	39.37	320	959.61	2,875	506.90	
80	0010	**MARINE EQUIPMENT RENTAL** without operators	Ea.						80
	0200	Barge, 400 Ton, 30' wide x 90' long		17.84	1,200	3,632.05	10,900	869.15	
	0240	800 Ton, 45' wide x 90' long		22.41	1,500	4,468.20	13,400	1,073	
	2000	Tugboat, diesel, 100 H.P.		29.92	240	721.18	2,175	383.65	
	2040	250 H.P.		58.10	435	1,306.49	3,925	726.10	
	2080	380 H.P.		126.49	1,300	3,919.48	11,800	1,796	
	3000	Small work boat, gas, 16-foot, 50 H.P.		11.48	48.50	145.01	435	120.85	
	4000	Large, diesel, 48-foot, 200 H.P.	▼	75.57	1,375	4,154.64	12,500	1,436	

Note: Rows 0010 for sections 70 and 80 reference R015433-10.

Crew A-1

Crew A-1	Hr.	Daily	Hr.	Daily	Bare Costs	Incl. O&P
1 Building Laborer	$44.40	$355.20	$66.25	$530.00	$44.40	$66.25
1 Concrete Saw, Gas Manual		113.70		125.07	14.21	15.63
8 L.H., Daily Totals		$468.90		$655.07	$58.61	$81.88

Crew A-1A

Crew A-1A	Hr.	Daily	Hr.	Daily	Bare Costs	Incl. O&P
1 Skilled Worker	$57.10	$456.80	$85.90	$687.20	$57.10	$85.90
1 Shot Blaster, 20"		208.70		229.57	26.09	28.70
8 L.H., Daily Totals		$665.50		$916.77	$83.19	$114.60

Crew A-1B

Crew A-1B	Hr.	Daily	Hr.	Daily	Bare Costs	Incl. O&P
1 Building Laborer	$44.40	$355.20	$66.25	$530.00	$44.40	$66.25
1 Concrete Saw		112.85		124.14	14.11	15.52
8 L.H., Daily Totals		$468.05		$654.13	$58.51	$81.77

Crew A-1C

Crew A-1C	Hr.	Daily	Hr.	Daily	Bare Costs	Incl. O&P
1 Building Laborer	$44.40	$355.20	$66.25	$530.00	$44.40	$66.25
1 Chain Saw, Gas, 18"		52.20		57.42	6.53	7.18
8 L.H., Daily Totals		$407.40		$587.42	$50.92	$73.43

Crew A-1D

Crew A-1D	Hr.	Daily	Hr.	Daily	Bare Costs	Incl. O&P
1 Building Laborer	$44.40	$355.20	$66.25	$530.00	$44.40	$66.25
1 Vibrating Plate, Gas, 18"		31.90		35.09	3.99	4.39
8 L.H., Daily Totals		$387.10		$565.09	$48.39	$70.64

Crew A-1E

Crew A-1E	Hr.	Daily	Hr.	Daily	Bare Costs	Incl. O&P
1 Building Laborer	$44.40	$355.20	$66.25	$530.00	$44.40	$66.25
1 Vibrating Plate, Gas, 21"		165.60		182.16	20.70	22.77
8 L.H., Daily Totals		$520.80		$712.16	$65.10	$89.02

Crew A-1F

Crew A-1F	Hr.	Daily	Hr.	Daily	Bare Costs	Incl. O&P
1 Building Laborer	$44.40	$355.20	$66.25	$530.00	$44.40	$66.25
1 Rammer/Tamper, Gas, 8"		47.00		51.70	5.88	6.46
8 L.H., Daily Totals		$402.20		$581.70	$50.27	$72.71

Crew A-1G

Crew A-1G	Hr.	Daily	Hr.	Daily	Bare Costs	Incl. O&P
1 Building Laborer	$44.40	$355.20	$66.25	$530.00	$44.40	$66.25
1 Rammer/Tamper, Gas, 15"		54.65		60.12	6.83	7.51
8 L.H., Daily Totals		$409.85		$590.12	$51.23	$73.76

Crew A-1H

Crew A-1H	Hr.	Daily	Hr.	Daily	Bare Costs	Incl. O&P
1 Building Laborer	$44.40	$355.20	$66.25	$530.00	$44.40	$66.25
1 Exterior Steam Cleaner		77.20		84.92	9.65	10.62
8 L.H., Daily Totals		$432.40		$614.92	$54.05	$76.86

Crew A-1J

Crew A-1J	Hr.	Daily	Hr.	Daily	Bare Costs	Incl. O&P
1 Building Laborer	$44.40	$355.20	$66.25	$530.00	$44.40	$66.25
1 Cultivator, Walk-Behind, 5 H.P.		53.25		58.58	6.66	7.32
8 L.H., Daily Totals		$408.45		$588.58	$51.06	$73.57

Crew A-1K

Crew A-1K	Hr.	Daily	Hr.	Daily	Bare Costs	Incl. O&P
1 Building Laborer	$44.40	$355.20	$66.25	$530.00	$44.40	$66.25
1 Cultivator, Walk-Behind, 8 H.P.		92.20		101.42	11.53	12.68
8 L.H., Daily Totals		$447.40		$631.42	$55.92	$78.93

Crew A-1M

Crew A-1M	Hr.	Daily	Hr.	Daily	Bare Costs	Incl. O&P
1 Building Laborer	$44.40	$355.20	$66.25	$530.00	$44.40	$66.25
1 Snow Blower, Walk-Behind		68.30		75.13	8.54	9.39
8 L.H., Daily Totals		$423.50		$605.13	$52.94	$75.64

Crew A-2

Crew A-2	Hr.	Daily	Hr.	Daily	Bare Costs	Incl. O&P
2 Laborers	$44.40	$710.40	$66.25	$1060.00	$45.87	$68.50
1 Truck Driver (light)	48.80	390.40	73.00	584.00		
1 Flatbed Truck, Gas, 1.5 Ton		198.50		218.35	8.27	9.10
24 L.H., Daily Totals		$1299.30		$1862.35	$54.14	$77.60

Crew A-2A

Crew A-2A	Hr.	Daily	Hr.	Daily	Bare Costs	Incl. O&P
2 Laborers	$44.40	$710.40	$66.25	$1060.00	$45.87	$68.50
1 Truck Driver (light)	48.80	390.40	73.00	584.00		
1 Flatbed Truck, Gas, 1.5 Ton		198.50		218.35		
1 Concrete Saw		112.85		124.14	12.97	14.27
24 L.H., Daily Totals		$1412.15		$1986.48	$58.84	$82.77

Crew A-2B

Crew A-2B	Hr.	Daily	Hr.	Daily	Bare Costs	Incl. O&P
1 Truck Driver (light)	$48.80	$390.40	$73.00	$584.00	$48.80	$73.00
1 Flatbed Truck, Gas, 1.5 Ton		198.50		218.35	24.81	27.29
8 L.H., Daily Totals		$588.90		$802.35	$73.61	$100.29

Crew A-3A

Crew A-3A	Hr.	Daily	Hr.	Daily	Bare Costs	Incl. O&P
1 Equip. Oper. (light)	$55.50	$444.00	$82.70	$661.60	$55.50	$82.70
1 Pickup Truck, 4x4, 3/4 Ton		176.75		194.43	22.09	24.30
8 L.H., Daily Totals		$620.75		$856.02	$77.59	$107.00

Crew A-3B

Crew A-3B	Hr.	Daily	Hr.	Daily	Bare Costs	Incl. O&P
1 Equip. Oper. (medium)	$59.00	$472.00	$87.90	$703.20	$55.15	$82.30
1 Truck Driver (heavy)	51.30	410.40	76.70	613.60		
1 Dump Truck, 12 C.Y., 400 H.P.		579.35		637.28		
1 F.E. Loader, W.M., 2.5 C.Y.		638.30		702.13	76.10	83.71
16 L.H., Daily Totals		$2100.05		$2656.22	$131.25	$166.01

Crew A-3C

Crew A-3C	Hr.	Daily	Hr.	Daily	Bare Costs	Incl. O&P
1 Equip. Oper. (light)	$55.50	$444.00	$82.70	$661.60	$55.50	$82.70
1 Loader, Skid Steer, 78 H.P.		446.30		490.93	55.79	61.37
8 L.H., Daily Totals		$890.30		$1152.53	$111.29	$144.07

Crew A-3D

Crew A-3D	Hr.	Daily	Hr.	Daily	Bare Costs	Incl. O&P
1 Truck Driver (light)	$48.80	$390.40	$73.00	$584.00	$48.80	$73.00
1 Pickup Truck, 4x4, 3/4 Ton		176.75		194.43		
1 Flatbed Trailer, 25 Ton		137.20		150.92	39.24	43.17
8 L.H., Daily Totals		$704.35		$929.35	$88.04	$116.17

Crew A-3E

Crew A-3E	Hr.	Daily	Hr.	Daily	Bare Costs	Incl. O&P
1 Equip. Oper. (crane)	$61.45	$491.60	$91.55	$732.40	$56.38	$84.13
1 Truck Driver (heavy)	51.30	410.40	76.70	613.60		
1 Pickup Truck, 4x4, 3/4 Ton		176.75		194.43	11.05	12.15
16 L.H., Daily Totals		$1078.75		$1540.43	$67.42	$96.28

Crew A-3F

Crew A-3F	Hr.	Daily	Hr.	Daily	Bare Costs	Incl. O&P
1 Equip. Oper. (crane)	$61.45	$491.60	$91.55	$732.40	$56.38	$84.13
1 Truck Driver (heavy)	51.30	410.40	76.70	613.60		
1 Pickup Truck, 4x4, 3/4 Ton		176.75		194.43		
1 Truck Tractor, 6x4, 380 H.P.		499.15		549.07		
1 Lowbed Trailer, 75 Ton		258.10		283.91	58.38	64.21
16 L.H., Daily Totals		$1836.00		$2373.40	$114.75	$148.34

Crew No.	Bare Costs Hr.	Bare Costs Daily	Incl. Subs O&P Hr.	Incl. Subs O&P Daily	Cost Per Labor-Hour Bare Costs	Cost Per Labor-Hour Incl. O&P
Crew A-3G						
1 Equip. Oper. (crane)	$61.45	$491.60	$91.55	$732.40	$56.38	$84.13
1 Truck Driver (heavy)	51.30	410.40	76.70	613.60		
1 Pickup Truck, 4x4, 3/4 Ton		176.75		194.43		
1 Truck Tractor, 6x4, 450 H.P.		608.95		669.85		
1 Lowbed Trailer, 75 Ton		258.10		283.91	65.24	71.76
16 L.H., Daily Totals		$1945.80		$2494.18	$121.61	$155.89
Crew A-3H						
1 Equip. Oper. (crane)	$61.45	$491.60	$91.55	$732.40	$61.45	$91.55
1 Hyd. Crane, 12 Ton (Daily)		733.15		806.47	91.64	100.81
8 L.H., Daily Totals		$1224.75		$1538.87	$153.09	$192.36
Crew A-3I						
1 Equip. Oper. (crane)	$61.45	$491.60	$91.55	$732.40	$61.45	$91.55
1 Hyd. Crane, 25 Ton (Daily)		810.50		891.55	101.31	111.44
8 L.H., Daily Totals		$1302.10		$1623.95	$162.76	$202.99
Crew A-3J						
1 Equip. Oper. (crane)	$61.45	$491.60	$91.55	$732.40	$61.45	$91.55
1 Hyd. Crane, 40 Ton (Daily)		1287.00		1415.70	160.88	176.96
8 L.H., Daily Totals		$1778.60		$2148.10	$222.32	$268.51
Crew A-3K						
1 Equip. Oper. (crane)	$61.45	$491.60	$91.55	$732.40	$56.98	$84.90
1 Equip. Oper. (oiler)	52.50	420.00	78.25	626.00		
1 Hyd. Crane, 55 Ton (Daily)		1377.00		1514.70		
1 P/U Truck, 3/4 Ton (Daily)		143.85		158.24	95.05	104.56
16 L.H., Daily Totals		$2432.45		$3031.34	$152.03	$189.46
Crew A-3L						
1 Equip. Oper. (crane)	$61.45	$491.60	$91.55	$732.40	$56.98	$84.90
1 Equip. Oper. (oiler)	52.50	420.00	78.25	626.00		
1 Hyd. Crane, 80 Ton (Daily)		2058.00		2263.80		
1 P/U Truck, 3/4 Ton (Daily)		143.85		158.24	137.62	151.38
16 L.H., Daily Totals		$3113.45		$3780.43	$194.59	$236.28
Crew A-3M						
1 Equip. Oper. (crane)	$61.45	$491.60	$91.55	$732.40	$56.98	$84.90
1 Equip. Oper. (oiler)	52.50	420.00	78.25	626.00		
1 Hyd. Crane, 100 Ton (Daily)		2253.00		2478.30		
1 P/U Truck, 3/4 Ton (Daily)		143.85		158.24	149.80	164.78
16 L.H., Daily Totals		$3308.45		$3994.93	$206.78	$249.68
Crew A-3N						
1 Equip. Oper. (crane)	$61.45	$491.60	$91.55	$732.40	$61.45	$91.55
1 Tower Crane (monthly)		1737.00		1910.70	217.13	238.84
8 L.H., Daily Totals		$2228.60		$2643.10	$278.57	$330.39
Crew A-3P						
1 Equip. Oper. (light)	$55.50	$444.00	$82.70	$661.60	$55.50	$82.70
1 A.T. Forklift, 31' reach, 45' lift		346.45		381.10	43.31	47.64
8 L.H., Daily Totals		$790.45		$1042.69	$98.81	$130.34
Crew A-3Q						
1 Equip. Oper. (light)	$55.50	$444.00	$82.70	$661.60	$55.50	$82.70
1 Pickup Truck, 4x4, 3/4 Ton		176.75		194.43		
1 Flatbed Trailer, 3 Ton		71.15		78.27	30.99	34.09
8 L.H., Daily Totals		$691.90		$934.29	$86.49	$116.79

Crew No.	Bare Costs Hr.	Bare Costs Daily	Incl. Subs O&P Hr.	Incl. Subs O&P Daily	Cost Per Labor-Hour Bare Costs	Cost Per Labor-Hour Incl. O&P
Crew A-3R						
1 Equip. Oper. (light)	$55.50	$444.00	$82.70	$661.60	$55.50	$82.70
1 Forklift, Smooth Floor, 8,000 Lb.		283.25		311.57	35.41	38.95
8 L.H., Daily Totals		$727.25		$973.17	$90.91	$121.65
Crew A-4						
2 Carpenters	$54.70	$875.20	$81.65	$1306.40	$51.95	$77.40
1 Painter, Ordinary	46.45	371.60	68.90	551.20		
24 L.H., Daily Totals		$1246.80		$1857.60	$51.95	$77.40
Crew A-5						
2 Laborers	$44.40	$710.40	$66.25	$1060.00	$44.89	$67.00
.25 Truck Driver (light)	48.80	97.60	73.00	146.00		
.25 Flatbed Truck, Gas, 1.5 Ton		49.63		54.59	2.76	3.03
18 L.H., Daily Totals		$857.63		$1260.59	$47.65	$70.03
Crew A-6						
1 Instrument Man	$57.10	$456.80	$85.90	$687.20	$54.88	$82.15
1 Rodman/Chainman	52.65	421.20	78.40	627.20		
1 Level, Electronic		34.40		37.84	2.15	2.37
16 L.H., Daily Totals		$912.40		$1352.24	$57.02	$84.52
Crew A-7						
1 Chief of Party	$68.50	$548.00	$102.35	$818.80	$59.42	$88.88
1 Instrument Man	57.10	456.80	85.90	687.20		
1 Rodman/Chainman	52.65	421.20	78.40	627.20		
1 Level, Electronic		34.40		37.84	1.43	1.58
24 L.H., Daily Totals		$1460.40		$2171.04	$60.85	$90.46
Crew A-8						
1 Chief of Party	$68.50	$548.00	$102.35	$818.80	$57.73	$86.26
1 Instrument Man	57.10	456.80	85.90	687.20		
2 Rodmen/Chainmen	52.65	842.40	78.40	1254.40		
1 Level, Electronic		34.40		37.84	1.08	1.18
32 L.H., Daily Totals		$1881.60		$2798.24	$58.80	$87.44
Crew A-9						
1 Asbestos Foreman	$61.45	$491.60	$94.00	$752.00	$61.01	$93.34
7 Asbestos Workers	60.95	3413.20	93.25	5222.00		
64 L.H., Daily Totals		$3904.80		$5974.00	$61.01	$93.34
Crew A-10A						
1 Asbestos Foreman	$61.45	$491.60	$94.00	$752.00	$61.12	$93.50
2 Asbestos Workers	60.95	975.20	93.25	1492.00		
24 L.H., Daily Totals		$1466.80		$2244.00	$61.12	$93.50
Crew A-10B						
1 Asbestos Foreman	$61.45	$491.60	$94.00	$752.00	$61.08	$93.44
3 Asbestos Workers	60.95	1462.80	93.25	2238.00		
32 L.H., Daily Totals		$1954.40		$2990.00	$61.08	$93.44
Crew A-10C						
3 Asbestos Workers	$60.95	$1462.80	$93.25	$2238.00	$60.95	$93.25
1 Flatbed Truck, Gas, 1.5 Ton		198.50		218.35	8.27	9.10
24 L.H., Daily Totals		$1661.30		$2456.35	$69.22	$102.35

For customer support on your Plumbing Costs with RSMeans Data, call 800.448.8182.

Crew A-10D

Crew No.	Bare Costs Hr.	Daily	Incl. Subs O&P Hr.	Daily	Cost Per Labor-Hour Bare Costs	Incl. O&P
2 Asbestos Workers	$60.95	$975.20	$93.25	$1492.00	$58.96	$89.08
1 Equip. Oper. (crane)	61.45	491.60	91.55	732.40		
1 Equip. Oper. (oiler)	52.50	420.00	78.25	626.00		
1 Hydraulic Crane, 33 Ton		983.15		1081.46	30.72	33.80
32 L.H., Daily Totals		$2869.95		$3931.86	$89.69	$122.87

Crew A-11

Crew No.	Bare Costs Hr.	Daily	Incl. Subs O&P Hr.	Daily	Cost Per Labor-Hour Bare Costs	Incl. O&P
1 Asbestos Foreman	$61.45	$491.60	$94.00	$752.00	$61.01	$93.34
7 Asbestos Workers	60.95	3413.20	93.25	5222.00		
2 Chip. Hammers, 12 Lb., Elec.		65.70		72.27	1.03	1.13
64 L.H., Daily Totals		$3970.50		$6046.27	$62.04	$94.47

Crew A-12

Crew No.	Bare Costs Hr.	Daily	Incl. Subs O&P Hr.	Daily	Cost Per Labor-Hour Bare Costs	Incl. O&P
1 Asbestos Foreman	$61.45	$491.60	$94.00	$752.00	$61.01	$93.34
7 Asbestos Workers	60.95	3413.20	93.25	5222.00		
1 Trk-Mtd Vac, 14 CY, 1500 Gal.		542.60		596.86		
1 Flatbed Truck, 20,000 GVW		204.05		224.46	11.67	12.83
64 L.H., Daily Totals		$4651.45		$6795.31	$72.68	$106.18

Crew A-13

Crew No.	Bare Costs Hr.	Daily	Incl. Subs O&P Hr.	Daily	Cost Per Labor-Hour Bare Costs	Incl. O&P
1 Equip. Oper. (light)	$55.50	$444.00	$82.70	$661.60	$55.50	$82.70
1 Trk-Mtd Vac, 14 CY, 1500 Gal.		542.60		596.86		
1 Flatbed Truck, 20,000 GVW		204.05		224.46	93.33	102.66
8 L.H., Daily Totals		$1190.65		$1482.92	$148.83	$185.36

Crew B-1

Crew No.	Bare Costs Hr.	Daily	Incl. Subs O&P Hr.	Daily	Cost Per Labor-Hour Bare Costs	Incl. O&P
1 Labor Foreman (outside)	$46.40	$371.20	$69.25	$554.00	$45.07	$67.25
2 Laborers	44.40	710.40	66.25	1060.00		
24 L.H., Daily Totals		$1081.60		$1614.00	$45.07	$67.25

Crew B-1A

Crew No.	Bare Costs Hr.	Daily	Incl. Subs O&P Hr.	Daily	Cost Per Labor-Hour Bare Costs	Incl. O&P
1 Labor Foreman (outside)	$46.40	$371.20	$69.25	$554.00	$45.07	$67.25
2 Laborers	44.40	710.40	66.25	1060.00		
2 Cutting Torches		25.90		28.49		
2 Sets of Gases		347.20		381.92	15.55	17.10
24 L.H., Daily Totals		$1454.70		$2024.41	$60.61	$84.35

Crew B-1B

Crew No.	Bare Costs Hr.	Daily	Incl. Subs O&P Hr.	Daily	Cost Per Labor-Hour Bare Costs	Incl. O&P
1 Labor Foreman (outside)	$46.40	$371.20	$69.25	$554.00	$49.16	$73.33
2 Laborers	44.40	710.40	66.25	1060.00		
1 Equip. Oper. (crane)	61.45	491.60	91.55	732.40		
2 Cutting Torches		25.90		28.49		
2 Sets of Gases		347.20		381.92		
1 Hyd. Crane, 12 Ton		475.80		523.38	26.53	29.18
32 L.H., Daily Totals		$2422.10		$3280.19	$75.69	$102.51

Crew B-1C

Crew No.	Bare Costs Hr.	Daily	Incl. Subs O&P Hr.	Daily	Cost Per Labor-Hour Bare Costs	Incl. O&P
1 Labor Foreman (outside)	$46.40	$371.20	$69.25	$554.00	$45.07	$67.25
2 Laborers	44.40	710.40	66.25	1060.00		
1 Telescoping Boom Lift, to 60'		292.45		321.69	12.19	13.40
24 L.H., Daily Totals		$1374.05		$1935.69	$57.25	$80.65

Crew B-1D

Crew No.	Bare Costs Hr.	Daily	Incl. Subs O&P Hr.	Daily	Cost Per Labor-Hour Bare Costs	Incl. O&P
2 Laborers	$44.40	$710.40	$66.25	$1060.00	$44.40	$66.25
1 Small Work Boat, Gas, 50 H.P.		120.85		132.94		
1 Pressure Washer, 7 GPM		93.95		103.35	13.43	14.77
16 L.H., Daily Totals		$925.20		$1296.28	$57.83	$81.02

Crew B-1E

Crew No.	Bare Costs Hr.	Daily	Incl. Subs O&P Hr.	Daily	Cost Per Labor-Hour Bare Costs	Incl. O&P
1 Labor Foreman (outside)	$46.40	$371.20	$69.25	$554.00	$44.90	$67.00
3 Laborers	44.40	1065.60	66.25	1590.00		
1 Work Boat, Diesel, 200 H.P.		1436.00		1579.60		
2 Pressure Washers, 7 GPM		187.90		206.69	50.75	55.82
32 L.H., Daily Totals		$3060.70		$3930.29	$95.65	$122.82

Crew B-1F

Crew No.	Bare Costs Hr.	Daily	Incl. Subs O&P Hr.	Daily	Cost Per Labor-Hour Bare Costs	Incl. O&P
2 Skilled Workers	$57.10	$913.60	$85.90	$1374.40	$52.87	$79.35
1 Laborer	44.40	355.20	66.25	530.00		
1 Small Work Boat, Gas, 50 H.P.		120.85		132.94		
1 Pressure Washer, 7 GPM		93.95		103.35	8.95	9.85
24 L.H., Daily Totals		$1483.60		$2140.68	$61.82	$89.19

Crew B-1G

Crew No.	Bare Costs Hr.	Daily	Incl. Subs O&P Hr.	Daily	Cost Per Labor-Hour Bare Costs	Incl. O&P
2 Laborers	$44.40	$710.40	$66.25	$1060.00	$44.40	$66.25
1 Small Work Boat, Gas, 50 H.P.		120.85		132.94	7.55	8.31
16 L.H., Daily Totals		$831.25		$1192.93	$51.95	$74.56

Crew B-1H

Crew No.	Bare Costs Hr.	Daily	Incl. Subs O&P Hr.	Daily	Cost Per Labor-Hour Bare Costs	Incl. O&P
2 Skilled Workers	$57.10	$913.60	$85.90	$1374.40	$52.87	$79.35
1 Laborer	44.40	355.20	66.25	530.00		
1 Small Work Boat, Gas, 50 H.P.		120.85		132.94	5.04	5.54
24 L.H., Daily Totals		$1389.65		$2037.34	$57.90	$84.89

Crew B-1J

Crew No.	Bare Costs Hr.	Daily	Incl. Subs O&P Hr.	Daily	Cost Per Labor-Hour Bare Costs	Incl. O&P
1 Labor Foreman (inside)	$44.90	$359.20	$67.00	$536.00	$44.65	$66.63
1 Laborer	44.40	355.20	66.25	530.00		
16 L.H., Daily Totals		$714.40		$1066.00	$44.65	$66.63

Crew B-1K

Crew No.	Bare Costs Hr.	Daily	Incl. Subs O&P Hr.	Daily	Cost Per Labor-Hour Bare Costs	Incl. O&P
1 Carpenter Foreman (inside)	$55.20	$441.60	$82.40	$659.20	$54.95	$82.03
1 Carpenter	54.70	437.60	81.65	653.20		
16 L.H., Daily Totals		$879.20		$1312.40	$54.95	$82.03

Crew B-2

Crew No.	Bare Costs Hr.	Daily	Incl. Subs O&P Hr.	Daily	Cost Per Labor-Hour Bare Costs	Incl. O&P
1 Labor Foreman (outside)	$46.40	$371.20	$69.25	$554.00	$44.80	$66.85
4 Laborers	44.40	1420.80	66.25	2120.00		
40 L.H., Daily Totals		$1792.00		$2674.00	$44.80	$66.85

Crew B-2A

Crew No.	Bare Costs Hr.	Daily	Incl. Subs O&P Hr.	Daily	Cost Per Labor-Hour Bare Costs	Incl. O&P
1 Labor Foreman (outside)	$46.40	$371.20	$69.25	$554.00	$45.07	$67.25
2 Laborers	44.40	710.40	66.25	1060.00		
1 Telescoping Boom Lift, to 60'		292.45		321.69	12.19	13.40
24 L.H., Daily Totals		$1374.05		$1935.69	$57.25	$80.65

Crew B-3

Crew No.	Bare Costs Hr.	Daily	Incl. Subs O&P Hr.	Daily	Cost Per Labor-Hour Bare Costs	Incl. O&P
1 Labor Foreman (outside)	$46.40	$371.20	$69.25	$554.00	$49.47	$73.84
2 Laborers	44.40	710.40	66.25	1060.00		
1 Equip. Oper. (medium)	59.00	472.00	87.90	703.20		
2 Truck Drivers (heavy)	51.30	820.80	76.70	1227.20		
1 Crawler Loader, 3 C.Y.		1146.00		1260.60		
2 Dump Trucks, 12 C.Y., 400 H.P.		1158.70		1274.57	48.01	52.82
48 L.H., Daily Totals		$4679.10		$6079.57	$97.48	$126.66

Crew B-3A

Crew No.	Bare Costs Hr.	Daily	Incl. Subs O&P Hr.	Daily	Cost Per Labor-Hour Bare Costs	Incl. O&P
4 Laborers	$44.40	$1420.80	$66.25	$2120.00	$47.32	$70.58
1 Equip. Oper. (medium)	59.00	472.00	87.90	703.20		
1 Hyd. Excavator, 1.5 C.Y.		695.80		765.38	17.40	19.13
40 L.H., Daily Totals		$2588.60		$3588.58	$64.72	$89.71

For customer support on your Plumbing Costs with RSMeans Data, call 800.448.8182.

599

Crew No.	Bare Costs		Incl. Subs O&P		Cost Per Labor-Hour	
Crew B-3B	Hr.	Daily	Hr.	Daily	Bare Costs	Incl. O&P
2 Laborers	$44.40	$710.40	$66.25	$1060.00	$49.77	$74.28
1 Equip. Oper. (medium)	59.00	472.00	87.90	703.20		
1 Truck Driver (heavy)	51.30	410.40	76.70	613.60		
1 Backhoe Loader, 80 H.P.		235.05		258.56		
1 Dump Truck, 12 C.Y., 400 H.P.		579.35		637.28	25.45	28.00
32 L.H., Daily Totals		$2407.20		$3272.64	$75.22	$102.27
Crew B-3C	Hr.	Daily	Hr.	Daily	Bare Costs	Incl. O&P
3 Laborers	$44.40	$1065.60	$66.25	$1590.00	$48.05	$71.66
1 Equip. Oper. (medium)	59.00	472.00	87.90	703.20		
1 Crawler Loader, 4 C.Y.		1456.00		1601.60	45.50	50.05
32 L.H., Daily Totals		$2993.60		$3894.80	$93.55	$121.71
Crew B-4	Hr.	Daily	Hr.	Daily	Bare Costs	Incl. O&P
1 Labor Foreman (outside)	$46.40	$371.20	$69.25	$554.00	$45.88	$68.49
4 Laborers	44.40	1420.80	66.25	2120.00		
1 Truck Driver (heavy)	51.30	410.40	76.70	613.60		
1 Truck Tractor, 220 H.P.		310.80		341.88		
1 Flatbed Trailer, 40 Ton		188.45		207.29	10.40	11.44
48 L.H., Daily Totals		$2701.65		$3836.78	$56.28	$79.93
Crew B-5	Hr.	Daily	Hr.	Daily	Bare Costs	Incl. O&P
1 Labor Foreman (outside)	$46.40	$371.20	$69.25	$554.00	$48.86	$72.86
4 Laborers	44.40	1420.80	66.25	2120.00		
2 Equip. Oper. (medium)	59.00	944.00	87.90	1406.40		
1 Air Compressor, 250 cfm		202.85		223.13		
2 Breakers, Pavement, 60 lb.		107.20		117.92		
2 -50' Air Hoses, 1.5"		45.60		50.16		
1 Crawler Loader, 3 C.Y.		1146.00		1260.60	26.82	29.50
56 L.H., Daily Totals		$4237.65		$5732.22	$75.67	$102.36
Crew B-5A	Hr.	Daily	Hr.	Daily	Bare Costs	Incl. O&P
1 Labor Foreman (outside)	$46.40	$371.20	$69.25	$554.00	$49.08	$73.22
6 Laborers	44.40	2131.20	66.25	3180.00		
2 Equip. Oper. (medium)	59.00	944.00	87.90	1406.40		
1 Equip. Oper. (light)	55.50	444.00	82.70	661.60		
2 Truck Drivers (heavy)	51.30	820.80	76.70	1227.20		
1 Air Compressor, 365 cfm		343.55		377.90		
2 Breakers, Pavement, 60 lb.		107.20		117.92		
8 -50' Air Hoses, 1"		64.40		70.84		
2 Dump Trucks, 8 C.Y., 220 H.P.		815.20		896.72	13.86	15.24
96 L.H., Daily Totals		$6041.55		$8492.58	$62.93	$88.46
Crew B-5B	Hr.	Daily	Hr.	Daily	Bare Costs	Incl. O&P
1 Powderman	$57.10	$456.80	$85.90	$687.20	$54.83	$81.97
2 Equip. Oper. (medium)	59.00	944.00	87.90	1406.40		
3 Truck Drivers (heavy)	51.30	1231.20	76.70	1840.80		
1 F.E. Loader, W.M., 2.5 C.Y.		638.30		702.13		
3 Dump Trucks, 12 C.Y., 400 H.P.		1738.05		1911.86		
1 Air Compressor, 365 cfm		343.55		377.90	56.66	62.33
48 L.H., Daily Totals		$5351.90		$6926.29	$111.50	$144.30

Crew No.	Bare Costs		Incl. Subs O&P		Cost Per Labor-Hour	
Crew B-5C	Hr.	Daily	Hr.	Daily	Bare Costs	Incl. O&P
3 Laborers	$44.40	$1065.60	$66.25	$1590.00	$51.09	$76.23
1 Equip. Oper. (medium)	59.00	472.00	87.90	703.20		
2 Truck Drivers (heavy)	51.30	820.80	76.70	1227.20		
1 Equip. Oper. (crane)	61.45	491.60	91.55	732.40		
1 Equip. Oper. (oiler)	52.50	420.00	78.25	626.00		
2 Dump Trucks, 12 C.Y., 400 H.P.		1158.70		1274.57		
1 Crawler Loader, 4 C.Y.		1456.00		1601.60		
1 S.P. Crane, 4x4, 25 Ton		1155.00		1270.50	58.90	64.79
64 L.H., Daily Totals		$7039.70		$9025.47	$110.00	$141.02
Crew B-5D	Hr.	Daily	Hr.	Daily	Bare Costs	Incl. O&P
1 Labor Foreman (outside)	$46.40	$371.20	$69.25	$554.00	$49.16	$73.34
4 Laborers	44.40	1420.80	66.25	2120.00		
2 Equip. Oper. (medium)	59.00	944.00	87.90	1406.40		
1 Truck Driver (heavy)	51.30	410.40	76.70	613.60		
1 Air Compressor, 250 cfm		202.85		223.13		
2 Breakers, Pavement, 60 lb.		107.20		117.92		
2 -50' Air Hoses, 1.5"		45.60		50.16		
1 Crawler Loader, 3 C.Y.		1146.00		1260.60		
1 Dump Truck, 12 C.Y., 400 H.P.		579.35		637.28	32.52	35.77
64 L.H., Daily Totals		$5227.40		$6983.10	$81.68	$109.11
Crew B-5E	Hr.	Daily	Hr.	Daily	Bare Costs	Incl. O&P
1 Labor Foreman (outside)	$46.40	$371.20	$69.25	$554.00	$49.16	$73.34
4 Laborers	44.40	1420.80	66.25	2120.00		
2 Equip. Oper. (medium)	59.00	944.00	87.90	1406.40		
1 Truck Driver (heavy)	51.30	410.40	76.70	613.60		
1 Water Tank Trailer, 5000 Gal.		154.10		169.51		
1 High Pressure Water Jet 40 KSI		820.75		902.83		
2 -50' Air Hoses, 1.5"		45.60		50.16		
1 Crawler Loader, 3 C.Y.		1146.00		1260.60		
1 Dump Truck, 12 C.Y., 400 H.P.		579.35		637.28	42.90	47.19
64 L.H., Daily Totals		$5892.20		$7714.38	$92.07	$120.54
Crew B-6	Hr.	Daily	Hr.	Daily	Bare Costs	Incl. O&P
2 Laborers	$44.40	$710.40	$66.25	$1060.00	$48.10	$71.73
1 Equip. Oper. (light)	55.50	444.00	82.70	661.60		
1 Backhoe Loader, 48 H.P.		216.20		237.82	9.01	9.91
24 L.H., Daily Totals		$1370.60		$1959.42	$57.11	$81.64
Crew B-6A	Hr.	Daily	Hr.	Daily	Bare Costs	Incl. O&P
.5 Labor Foreman (outside)	$46.40	$185.60	$69.25	$277.00	$50.64	$75.51
1 Laborer	44.40	355.20	66.25	530.00		
1 Equip. Oper. (medium)	59.00	472.00	87.90	703.20		
1 Vacuum Truck, 5000 Gal.		371.95		409.14	18.60	20.46
20 L.H., Daily Totals		$1384.75		$1919.35	$69.24	$95.97
Crew B-6B	Hr.	Daily	Hr.	Daily	Bare Costs	Incl. O&P
2 Labor Foremen (outside)	$46.40	$742.40	$69.25	$1108.00	$45.07	$67.25
4 Laborers	44.40	1420.80	66.25	2120.00		
1 S.P. Crane, 4x4, 5 Ton		381.95		420.14		
1 Flatbed Truck, Gas, 1.5 Ton		198.50		218.35		
1 Butt Fusion Mach., 4"-12" diam.		420.75		462.82	20.86	22.94
48 L.H., Daily Totals		$3164.40		$4329.32	$65.92	$90.19

Crew No.	Bare Costs		Incl. Subs O&P		Cost Per Labor-Hour	

Crew B-6C

Crew B-6C	Hr.	Daily	Hr.	Daily	Bare Costs	Incl. O&P
2 Labor Foremen (outside)	$46.40	$742.40	$69.25	$1108.00	$45.07	$67.25
4 Laborers	44.40	1420.80	66.25	2120.00		
1 S.P. Crane, 4x4, 12 Ton		432.65		475.92		
1 Flatbed Truck, Gas, 3 Ton		850.05		935.05		
1 Butt Fusion Mach., 8"-24" diam.		1086.00		1194.60	49.35	54.28
48 L.H., Daily Totals		$4531.90		$5833.57	$94.41	$121.53

Crew B-6D

Crew B-6D	Hr.	Daily	Hr.	Daily	Bare Costs	Incl. O&P
.5 Labor Foreman (outside)	$46.40	$185.60	$69.25	$277.00	$50.64	$75.51
1 Laborer	44.40	355.20	66.25	530.00		
1 Equip. Oper. (medium)	59.00	472.00	87.90	703.20		
1 Hydro Excavator, 12 C.Y.		1277.00		1404.70	63.85	70.23
20 L.H., Daily Totals		$2289.80		$2914.90	$114.49	$145.75

Crew B-7

Crew B-7	Hr.	Daily	Hr.	Daily	Bare Costs	Incl. O&P
1 Labor Foreman (outside)	$46.40	$371.20	$69.25	$554.00	$47.17	$70.36
4 Laborers	44.40	1420.80	66.25	2120.00		
1 Equip. Oper. (medium)	59.00	472.00	87.90	703.20		
1 Brush Chipper, 12", 130 H.P.		366.05		402.65		
1 Crawler Loader, 3 C.Y.		1146.00		1260.60		
2 Chain Saws, Gas, 36" Long		83.30		91.63	33.24	36.56
48 L.H., Daily Totals		$3859.35		$5132.09	$80.40	$106.92

Crew B-7A

Crew B-7A	Hr.	Daily	Hr.	Daily	Bare Costs	Incl. O&P
2 Laborers	$44.40	$710.40	$66.25	$1060.00	$48.10	$71.73
1 Equip. Oper. (light)	55.50	444.00	82.70	661.60		
1 Rake w/Tractor		343.50		377.85		
2 Chain Saws, Gas, 18"		104.40		114.84	18.66	20.53
24 L.H., Daily Totals		$1602.30		$2214.29	$66.76	$92.26

Crew B-7B

Crew B-7B	Hr.	Daily	Hr.	Daily	Bare Costs	Incl. O&P
1 Labor Foreman (outside)	$46.40	$371.20	$69.25	$554.00	$47.76	$71.26
4 Laborers	44.40	1420.80	66.25	2120.00		
1 Equip. Oper. (medium)	59.00	472.00	87.90	703.20		
1 Truck Driver (heavy)	51.30	410.40	76.70	613.60		
1 Brush Chipper, 12", 130 H.P.		366.05		402.65		
1 Crawler Loader, 3 C.Y.		1146.00		1260.60		
2 Chain Saws, Gas, 36" Long		83.30		91.63		
1 Dump Truck, 8 C.Y., 220 H.P.		407.60		448.36	35.77	39.34
56 L.H., Daily Totals		$4677.35		$6194.05	$83.52	$110.61

Crew B-7C

Crew B-7C	Hr.	Daily	Hr.	Daily	Bare Costs	Incl. O&P
1 Labor Foreman (outside)	$46.40	$371.20	$69.25	$554.00	$47.76	$71.26
4 Laborers	44.40	1420.80	66.25	2120.00		
1 Equip. Oper. (medium)	59.00	472.00	87.90	703.20		
1 Truck Driver (heavy)	51.30	410.40	76.70	613.60		
1 Brush Chipper, 12", 130 H.P.		366.05		402.65		
1 Crawler Loader, 3 C.Y.		1146.00		1260.60		
2 Chain Saws, Gas, 36" Long		83.30		91.63		
1 Dump Truck, 12 C.Y., 400 H.P.		579.35		637.28	38.83	42.72
56 L.H., Daily Totals		$4849.10		$6382.97	$86.59	$113.98

Crew B-8

Crew B-8	Hr.	Daily	Hr.	Daily	Bare Costs	Incl. O&P
1 Labor Foreman (outside)	$46.40	$371.20	$69.25	$554.00	$51.04	$76.15
2 Laborers	44.40	710.40	66.25	1060.00		
2 Equip. Oper. (medium)	59.00	944.00	87.90	1406.40		
1 Equip. Oper. (oiler)	52.50	420.00	78.25	626.00		
2 Truck Drivers (heavy)	51.30	820.80	76.70	1227.20		
1 Hyd. Crane, 25 Ton		586.70		645.37		
1 Crawler Loader, 3 C.Y.		1146.00		1260.60		
2 Dump Trucks, 12 C.Y., 400 H.P.		1158.70		1274.57	45.18	49.70
64 L.H., Daily Totals		$6157.80		$8054.14	$96.22	$125.85

Crew B-9

Crew B-9	Hr.	Daily	Hr.	Daily	Bare Costs	Incl. O&P
1 Labor Foreman (outside)	$46.40	$371.20	$69.25	$554.00	$44.80	$66.85
4 Laborers	44.40	1420.80	66.25	2120.00		
1 Air Compressor, 250 cfm		202.85		223.13		
2 Breakers, Pavement, 60 lb.		107.20		117.92		
2 -50' Air Hoses, 1.5"		45.60		50.16	8.89	9.78
40 L.H., Daily Totals		$2147.65		$3065.22	$53.69	$76.63

Crew B-9A

Crew B-9A	Hr.	Daily	Hr.	Daily	Bare Costs	Incl. O&P
2 Laborers	$44.40	$710.40	$66.25	$1060.00	$46.70	$69.73
1 Truck Driver (heavy)	51.30	410.40	76.70	613.60		
1 Water Tank Trailer, 5000 Gal.		154.10		169.51		
1 Truck Tractor, 220 H.P.		310.80		341.88		
2 -50' Discharge Hoses, 3"		9.00		9.90	19.75	21.72
24 L.H., Daily Totals		$1594.70		$2194.89	$66.45	$91.45

Crew B-9B

Crew B-9B	Hr.	Daily	Hr.	Daily	Bare Costs	Incl. O&P
2 Laborers	$44.40	$710.40	$66.25	$1060.00	$46.70	$69.73
1 Truck Driver (heavy)	51.30	410.40	76.70	613.60		
2 -50' Discharge Hoses, 3"		9.00		9.90		
1 Water Tank Trailer, 5000 Gal.		154.10		169.51		
1 Truck Tractor, 220 H.P.		310.80		341.88		
1 Pressure Washer		97.35		107.08	23.80	26.18
24 L.H., Daily Totals		$1692.05		$2301.97	$70.50	$95.92

Crew B-9D

Crew B-9D	Hr.	Daily	Hr.	Daily	Bare Costs	Incl. O&P
1 Labor Foreman (outside)	$46.40	$371.20	$69.25	$554.00	$44.80	$66.85
4 Common Laborers	44.40	1420.80	66.25	2120.00		
1 Air Compressor, 250 cfm		202.85		223.13		
2 -50' Air Hoses, 1.5"		45.60		50.16		
2 Air Powered Tampers		79.50		87.45	8.20	9.02
40 L.H., Daily Totals		$2119.95		$3034.74	$53.00	$75.87

Crew B-9E

Crew B-9E	Hr.	Daily	Hr.	Daily	Bare Costs	Incl. O&P
1 Cement Finisher	$51.80	$414.40	$75.90	$607.20	$48.10	$71.08
1 Laborer	44.40	355.20	66.25	530.00		
1 Chip. Hammers, 12 Lb., Elec.		32.85		36.13	2.05	2.26
16 L.H., Daily Totals		$802.45		$1173.34	$50.15	$73.33

Crew B-10

Crew B-10	Hr.	Daily	Hr.	Daily	Bare Costs	Incl. O&P
1 Equip. Oper. (medium)	$59.00	$472.00	$87.90	$703.20	$54.13	$80.68
.5 Laborer	44.40	177.60	66.25	265.00		
12 L.H., Daily Totals		$649.60		$968.20	$54.13	$80.68

Crew B-10A

Crew B-10A	Hr.	Daily	Hr.	Daily	Bare Costs	Incl. O&P
1 Equip. Oper. (medium)	$59.00	$472.00	$87.90	$703.20	$54.13	$80.68
.5 Laborer	44.40	177.60	66.25	265.00		
1 Roller, 2-Drum, W.B., 7.5 H.P.		166.75		183.43	13.90	15.29
12 L.H., Daily Totals		$816.35		$1151.63	$68.03	$95.97

Crew B-10B

Crew No.	Bare Costs Hr.	Daily	Incl. Subs O&P Hr.	Daily	Cost Per Labor-Hour Bare Costs	Incl. O&P
1 Equip. Oper. (medium)	$59.00	$472.00	$87.90	$703.20	$54.13	$80.68
.5 Laborer	44.40	177.60	66.25	265.00		
1 Dozer, 200 H.P.		1520.00		1672.00	126.67	139.33
12 L.H., Daily Totals		$2169.60		$2640.20	$180.80	$220.02

Crew B-10C

Crew No.	Bare Costs Hr.	Daily	Incl. Subs O&P Hr.	Daily	Cost Per Labor-Hour Bare Costs	Incl. O&P
1 Equip. Oper. (medium)	$59.00	$472.00	$87.90	$703.20	$54.13	$80.68
.5 Laborer	44.40	177.60	66.25	265.00		
1 Dozer, 200 H.P.		1520.00		1672.00		
1 Vibratory Roller, Towed, 23 Ton		520.35		572.38	170.03	187.03
12 L.H., Daily Totals		$2689.95		$3212.59	$224.16	$267.72

Crew B-10D

Crew No.	Bare Costs Hr.	Daily	Incl. Subs O&P Hr.	Daily	Cost Per Labor-Hour Bare Costs	Incl. O&P
1 Equip. Oper. (medium)	$59.00	$472.00	$87.90	$703.20	$54.13	$80.68
.5 Laborer	44.40	177.60	66.25	265.00		
1 Dozer, 200 H.P.		1520.00		1672.00		
1 Sheepsft. Roller, Towed		426.95		469.64	162.25	178.47
12 L.H., Daily Totals		$2596.55		$3109.84	$216.38	$259.15

Crew B-10E

Crew No.	Bare Costs Hr.	Daily	Incl. Subs O&P Hr.	Daily	Cost Per Labor-Hour Bare Costs	Incl. O&P
1 Equip. Oper. (medium)	$59.00	$472.00	$87.90	$703.20	$54.13	$80.68
.5 Laborer	44.40	177.60	66.25	265.00		
1 Tandem Roller, 5 Ton		258.75		284.63	21.56	23.72
12 L.H., Daily Totals		$908.35		$1252.83	$75.70	$104.40

Crew B-10F

Crew No.	Bare Costs Hr.	Daily	Incl. Subs O&P Hr.	Daily	Cost Per Labor-Hour Bare Costs	Incl. O&P
1 Equip. Oper. (medium)	$59.00	$472.00	$87.90	$703.20	$54.13	$80.68
.5 Laborer	44.40	177.60	66.25	265.00		
1 Tandem Roller, 10 Ton		246.80		271.48	20.57	22.62
12 L.H., Daily Totals		$896.40		$1239.68	$74.70	$103.31

Crew B-10G

Crew No.	Bare Costs Hr.	Daily	Incl. Subs O&P Hr.	Daily	Cost Per Labor-Hour Bare Costs	Incl. O&P
1 Equip. Oper. (medium)	$59.00	$472.00	$87.90	$703.20	$54.13	$80.68
.5 Laborer	44.40	177.60	66.25	265.00		
1 Sheepsfoot Roller, 240 H.P.		1363.00		1499.30	113.58	124.94
12 L.H., Daily Totals		$2012.60		$2467.50	$167.72	$205.63

Crew B-10H

Crew No.	Bare Costs Hr.	Daily	Incl. Subs O&P Hr.	Daily	Cost Per Labor-Hour Bare Costs	Incl. O&P
1 Equip. Oper. (medium)	$59.00	$472.00	$87.90	$703.20	$54.13	$80.68
.5 Laborer	44.40	177.60	66.25	265.00		
1 Diaphragm Water Pump, 2"		87.70		96.47		
1 -20' Suction Hose, 2"		3.55		3.90		
2 -50' Discharge Hoses, 2"		8.00		8.80	8.27	9.10
12 L.H., Daily Totals		$748.85		$1077.38	$62.40	$89.78

Crew B-10I

Crew No.	Bare Costs Hr.	Daily	Incl. Subs O&P Hr.	Daily	Cost Per Labor-Hour Bare Costs	Incl. O&P
1 Equip. Oper. (medium)	$59.00	$472.00	$87.90	$703.20	$54.13	$80.68
.5 Laborer	44.40	177.60	66.25	265.00		
1 Diaphragm Water Pump, 4"		106.35		116.99		
1 -20' Suction Hose, 4"		17.25		18.98		
2 -50' Discharge Hoses, 4"		25.60		28.16	12.43	13.68
12 L.H., Daily Totals		$798.80		$1132.32	$66.57	$94.36

Crew B-10J

Crew No.	Bare Costs Hr.	Daily	Incl. Subs O&P Hr.	Daily	Cost Per Labor-Hour Bare Costs	Incl. O&P
1 Equip. Oper. (medium)	$59.00	$472.00	$87.90	$703.20	$54.13	$80.68
.5 Laborer	44.40	177.60	66.25	265.00		
1 Centrifugal Water Pump, 3"		74.40		81.84		
1 -20' Suction Hose, 3"		8.75		9.63		
2 -50' Discharge Hoses, 3"		9.00		9.90	7.68	8.45
12 L.H., Daily Totals		$741.75		$1069.57	$61.81	$89.13

Crew B-10K

Crew No.	Bare Costs Hr.	Daily	Incl. Subs O&P Hr.	Daily	Cost Per Labor-Hour Bare Costs	Incl. O&P
1 Equip. Oper. (medium)	$59.00	$472.00	$87.90	$703.20	$54.13	$80.68
.5 Laborer	44.40	177.60	66.25	265.00		
1 Centr. Water Pump, 6"		235.25		258.77		
1 -20' Suction Hose, 6"		25.50		28.05		
2 -50' Discharge Hoses, 6"		36.20		39.82	24.75	27.22
12 L.H., Daily Totals		$946.55		$1294.85	$78.88	$107.90

Crew B-10L

Crew No.	Bare Costs Hr.	Daily	Incl. Subs O&P Hr.	Daily	Cost Per Labor-Hour Bare Costs	Incl. O&P
1 Equip. Oper. (medium)	$59.00	$472.00	$87.90	$703.20	$54.13	$80.68
.5 Laborer	44.40	177.60	66.25	265.00		
1 Dozer, 80 H.P.		405.85		446.44	33.82	37.20
12 L.H., Daily Totals		$1055.45		$1414.64	$87.95	$117.89

Crew B-10M

Crew No.	Bare Costs Hr.	Daily	Incl. Subs O&P Hr.	Daily	Cost Per Labor-Hour Bare Costs	Incl. O&P
1 Equip. Oper. (medium)	$59.00	$472.00	$87.90	$703.20	$54.13	$80.68
.5 Laborer	44.40	177.60	66.25	265.00		
1 Dozer, 300 H.P.		1785.00		1963.50	148.75	163.63
12 L.H., Daily Totals		$2434.60		$2931.70	$202.88	$244.31

Crew B-10N

Crew No.	Bare Costs Hr.	Daily	Incl. Subs O&P Hr.	Daily	Cost Per Labor-Hour Bare Costs	Incl. O&P
1 Equip. Oper. (medium)	$59.00	$472.00	$87.90	$703.20	$54.13	$80.68
.5 Laborer	44.40	177.60	66.25	265.00		
1 F.E. Loader, T.M., 1.5 C.Y.		572.00		629.20	47.67	52.43
12 L.H., Daily Totals		$1221.60		$1597.40	$101.80	$133.12

Crew B-10O

Crew No.	Bare Costs Hr.	Daily	Incl. Subs O&P Hr.	Daily	Cost Per Labor-Hour Bare Costs	Incl. O&P
1 Equip. Oper. (medium)	$59.00	$472.00	$87.90	$703.20	$54.13	$80.68
.5 Laborer	44.40	177.60	66.25	265.00		
1 F.E. Loader, T.M., 2.25 C.Y.		925.50		1018.05	77.13	84.84
12 L.H., Daily Totals		$1575.10		$1986.25	$131.26	$165.52

Crew B-10P

Crew No.	Bare Costs Hr.	Daily	Incl. Subs O&P Hr.	Daily	Cost Per Labor-Hour Bare Costs	Incl. O&P
1 Equip. Oper. (medium)	$59.00	$472.00	$87.90	$703.20	$54.13	$80.68
.5 Laborer	44.40	177.60	66.25	265.00		
1 Crawler Loader, 3 C.Y.		1146.00		1260.60	95.50	105.05
12 L.H., Daily Totals		$1795.60		$2228.80	$149.63	$185.73

Crew B-10Q

Crew No.	Bare Costs Hr.	Daily	Incl. Subs O&P Hr.	Daily	Cost Per Labor-Hour Bare Costs	Incl. O&P
1 Equip. Oper. (medium)	$59.00	$472.00	$87.90	$703.20	$54.13	$80.68
.5 Laborer	44.40	177.60	66.25	265.00		
1 Crawler Loader, 4 C.Y.		1456.00		1601.60	121.33	133.47
12 L.H., Daily Totals		$2105.60		$2569.80	$175.47	$214.15

Crew B-10R

Crew No.	Bare Costs Hr.	Daily	Incl. Subs O&P Hr.	Daily	Cost Per Labor-Hour Bare Costs	Incl. O&P
1 Equip. Oper. (medium)	$59.00	$472.00	$87.90	$703.20	$54.13	$80.68
.5 Laborer	44.40	177.60	66.25	265.00		
1 F.E. Loader, W.M., 1 C.Y.		305.60		336.16	25.47	28.01
12 L.H., Daily Totals		$955.20		$1304.36	$79.60	$108.70

Crew B-10S

Crew No.	Bare Costs Hr.	Daily	Incl. Subs O&P Hr.	Daily	Cost Per Labor-Hour Bare Costs	Incl. O&P
1 Equip. Oper. (medium)	$59.00	$472.00	$87.90	$703.20	$54.13	$80.68
.5 Laborer	44.40	177.60	66.25	265.00		
1 F.E. Loader, W.M., 1.5 C.Y.		441.40		485.54	36.78	40.46
12 L.H., Daily Totals		$1091.00		$1453.74	$90.92	$121.15

Crew B-10T

Crew No.	Bare Costs Hr.	Daily	Incl. Subs O&P Hr.	Daily	Cost Per Labor-Hour Bare Costs	Incl. O&P
1 Equip. Oper. (medium)	$59.00	$472.00	$87.90	$703.20	$54.13	$80.68
.5 Laborer	44.40	177.60	66.25	265.00		
1 F.E. Loader, W.M., 2.5 C.Y.		638.30		702.13	53.19	58.51
12 L.H., Daily Totals		$1287.90		$1670.33	$107.33	$139.19

For customer support on your Plumbing Costs with RSMeans Data, call 800.448.8182.

Crew No.	Bare Costs			Incl. Subs O&P		Cost Per Labor-Hour	
Crew B-10U	Hr.	Daily	Hr.	Daily	Bare Costs	Incl. O&P
1 Equip. Oper. (medium)	$59.00	$472.00	$87.90	$703.20	$54.13	$80.68
.5 Laborer	44.40	177.60	66.25	265.00		
1 F.E. Loader, W.M., 5.5 C.Y.		967.95		1064.74	80.66	88.73
12 L.H., Daily Totals		$1617.55		$2032.94	$134.80	$169.41
Crew B-10V	Hr.	Daily	Hr.	Daily	Bare Costs	Incl. O&P
1 Equip. Oper. (medium)	$59.00	$472.00	$87.90	$703.20	$54.13	$80.68
.5 Laborer	44.40	177.60	66.25	265.00		
1 Dozer, 700 H.P.		5175.00		5692.50	431.25	474.38
12 L.H., Daily Totals		$5824.60		$6660.70	$485.38	$555.06
Crew B-10W	Hr.	Daily	Hr.	Daily	Bare Costs	Incl. O&P
1 Equip. Oper. (medium)	$59.00	$472.00	$87.90	$703.20	$54.13	$80.68
.5 Laborer	44.40	177.60	66.25	265.00		
1 Dozer, 105 H.P.		640.80		704.88	53.40	58.74
12 L.H., Daily Totals		$1290.40		$1673.08	$107.53	$139.42
Crew B-10X	Hr.	Daily	Hr.	Daily	Bare Costs	Incl. O&P
1 Equip. Oper. (medium)	$59.00	$472.00	$87.90	$703.20	$54.13	$80.68
.5 Laborer	44.40	177.60	66.25	265.00		
1 Dozer, 410 H.P.		2807.00		3087.70	233.92	257.31
12 L.H., Daily Totals		$3456.60		$4055.90	$288.05	$337.99
Crew B-10Y	Hr.	Daily	Hr.	Daily	Bare Costs	Incl. O&P
1 Equip. Oper. (medium)	$59.00	$472.00	$87.90	$703.20	$54.13	$80.68
.5 Laborer	44.40	177.60	66.25	265.00		
1 Vibr. Roller, Towed, 12 Ton		584.80		643.28	48.73	53.61
12 L.H., Daily Totals		$1234.40		$1611.48	$102.87	$134.29
Crew B-11A	Hr.	Daily	Hr.	Daily	Bare Costs	Incl. O&P
1 Equipment Oper. (med.)	$59.00	$472.00	$87.90	$703.20	$51.70	$77.08
1 Laborer	44.40	355.20	66.25	530.00		
1 Dozer, 200 H.P.		1520.00		1672.00	95.00	104.50
16 L.H., Daily Totals		$2347.20		$2905.20	$146.70	$181.57
Crew B-11B	Hr.	Daily	Hr.	Daily	Bare Costs	Incl. O&P
1 Equipment Oper. (light)	$55.50	$444.00	$82.70	$661.60	$49.95	$74.47
1 Laborer	44.40	355.20	66.25	530.00		
1 Air Powered Tamper		39.75		43.73		
1 Air Compressor, 365 cfm		343.55		377.90		
2 -50' Air Hoses, 1.5"		45.60		50.16	26.81	29.49
16 L.H., Daily Totals		$1228.10		$1663.39	$76.76	$103.96
Crew B-11C	Hr.	Daily	Hr.	Daily	Bare Costs	Incl. O&P
1 Equipment Oper. (med.)	$59.00	$472.00	$87.90	$703.20	$51.70	$77.08
1 Laborer	44.40	355.20	66.25	530.00		
1 Backhoe Loader, 48 H.P.		216.20		237.82	13.51	14.86
16 L.H., Daily Totals		$1043.40		$1471.02	$65.21	$91.94
Crew B-11J	Hr.	Daily	Hr.	Daily	Bare Costs	Incl. O&P
1 Equipment Oper. (med.)	$59.00	$472.00	$87.90	$703.20	$51.70	$77.08
1 Laborer	44.40	355.20	66.25	530.00		
1 Grader, 30,000 Lbs.		1073.00		1180.30		
1 Ripper, Beam & 1 Shank		91.60		100.76	72.79	80.07
16 L.H., Daily Totals		$1991.80		$2514.26	$124.49	$157.14

Crew No.	Bare Costs			Incl. Subs O&P		Cost Per Labor-Hour	
Crew B-11K	Hr.	Daily	Hr.	Daily	Bare Costs	Incl. O&P
1 Equipment Oper. (med.)	$59.00	$472.00	$87.90	$703.20	$51.70	$77.08
1 Laborer	44.40	355.20	66.25	530.00		
1 Trencher, Chain Type, 8' D		1894.00		2083.40	118.38	130.21
16 L.H., Daily Totals		$2721.20		$3316.60	$170.07	$207.29
Crew B-11L	Hr.	Daily	Hr.	Daily	Bare Costs	Incl. O&P
1 Equipment Oper. (med.)	$59.00	$472.00	$87.90	$703.20	$51.70	$77.08
1 Laborer	44.40	355.20	66.25	530.00		
1 Grader, 30,000 Lbs.		1073.00		1180.30	67.06	73.77
16 L.H., Daily Totals		$1900.20		$2413.50	$118.76	$150.84
Crew B-11M	Hr.	Daily	Hr.	Daily	Bare Costs	Incl. O&P
1 Equipment Oper. (med.)	$59.00	$472.00	$87.90	$703.20	$51.70	$77.08
1 Laborer	44.40	355.20	66.25	530.00		
1 Backhoe Loader, 80 H.P.		235.05		258.56	14.69	16.16
16 L.H., Daily Totals		$1062.25		$1491.76	$66.39	$93.23
Crew B-11N	Hr.	Daily	Hr.	Daily	Bare Costs	Incl. O&P
1 Labor Foreman (outside)	$46.40	$371.20	$69.25	$554.00	$52.47	$78.36
2 Equipment Operators (med.)	59.00	944.00	87.90	1406.40		
6 Truck Drivers (heavy)	51.30	2462.40	76.70	3681.60		
1 F.E. Loader, W.M., 5.5 C.Y.		967.95		1064.74		
1 Dozer, 410 H.P.		2807.00		3087.70		
6 Dump Trucks, Off Hwy., 50 Ton		11874.00		13061.40	217.35	239.08
72 L.H., Daily Totals		$19426.55		$22855.85	$269.81	$317.44
Crew B-11Q	Hr.	Daily	Hr.	Daily	Bare Costs	Incl. O&P
1 Equipment Operator (med.)	$59.00	$472.00	$87.90	$703.20	$54.13	$80.68
.5 Laborer	44.40	177.60	66.25	265.00		
1 Dozer, 140 H.P.		729.15		802.07	60.76	66.84
12 L.H., Daily Totals		$1378.75		$1770.27	$114.90	$147.52
Crew B-11R	Hr.	Daily	Hr.	Daily	Bare Costs	Incl. O&P
1 Equipment Operator (med.)	$59.00	$472.00	$87.90	$703.20	$54.13	$80.68
.5 Laborer	44.40	177.60	66.25	265.00		
1 Dozer, 200 H.P.		1520.00		1672.00	126.67	139.33
12 L.H., Daily Totals		$2169.60		$2640.20	$180.80	$220.02
Crew B-11S	Hr.	Daily	Hr.	Daily	Bare Costs	Incl. O&P
1 Equipment Operator (med.)	$59.00	$472.00	$87.90	$703.20	$54.13	$80.68
.5 Laborer	44.40	177.60	66.25	265.00		
1 Dozer, 300 H.P.		1785.00		1963.50		
1 Ripper, Beam & 1 Shank		91.60		100.76	156.38	172.02
12 L.H., Daily Totals		$2526.20		$3032.46	$210.52	$252.71
Crew B-11T	Hr.	Daily	Hr.	Daily	Bare Costs	Incl. O&P
1 Equipment Operator (med.)	$59.00	$472.00	$87.90	$703.20	$54.13	$80.68
.5 Laborer	44.40	177.60	66.25	265.00		
1 Dozer, 410 H.P.		2807.00		3087.70		
1 Ripper, Beam & 2 Shanks		140.40		154.44	245.62	270.18
12 L.H., Daily Totals		$3597.00		$4210.34	$299.75	$350.86
Crew B-11U	Hr.	Daily	Hr.	Daily	Bare Costs	Incl. O&P
1 Equipment Operator (med.)	$59.00	$472.00	$87.90	$703.20	$54.13	$80.68
.5 Laborer	44.40	177.60	66.25	265.00		
1 Dozer, 520 H.P.		3434.00		3777.40	286.17	314.78
12 L.H., Daily Totals		$4083.60		$4745.60	$340.30	$395.47

For customer support on your Plumbing Costs with RSMeans Data, call 800.448.8182.

603

Crew B-11V	Hr.	Daily	Hr.	Daily	Bare Costs	Incl. O&P
3 Laborers	$44.40	$1065.60	$66.25	$1590.00	$44.40	$66.25
1 Roller, 2-Drum, W.B., 7.5 H.P.		166.75		183.43	6.95	7.64
24 L.H., Daily Totals		$1232.35		$1773.43	$51.35	$73.89

Crew B-11W	Hr.	Daily	Hr.	Daily	Bare Costs	Incl. O&P
1 Equipment Operator (med.)	$59.00	$472.00	$87.90	$703.20	$51.37	$76.76
1 Common Laborer	44.40	355.20	66.25	530.00		
10 Truck Drivers (heavy)	51.30	4104.00	76.70	6136.00		
1 Dozer, 200 H.P.		1520.00		1672.00		
1 Vibratory Roller, Towed, 23 Ton		520.35		572.38		
10 Dump Trucks, 8 C.Y., 220 H.P.		4076.00		4483.60	63.71	70.08
96 L.H., Daily Totals		$11047.55		$14097.18	$115.08	$146.85

Crew B-11Y	Hr.	Daily	Hr.	Daily	Bare Costs	Incl. O&P
1 Labor Foreman (outside)	$46.40	$371.20	$69.25	$554.00	$49.49	$73.80
5 Common Laborers	44.40	1776.00	66.25	2650.00		
3 Equipment Operators (med.)	59.00	1416.00	87.90	2109.60		
1 Dozer, 80 H.P.		405.85		446.44		
2 Rollers, 2-Drum, W.B., 7.5 H.P.		333.50		366.85		
4 Vibrating Plates, Gas, 21"		662.40		728.64	19.47	21.42
72 L.H., Daily Totals		$4964.95		$6855.52	$68.96	$95.22

Crew B-12A	Hr.	Daily	Hr.	Daily	Bare Costs	Incl. O&P
1 Equip. Oper. (crane)	$61.45	$491.60	$91.55	$732.40	$52.92	$78.90
1 Laborer	44.40	355.20	66.25	530.00		
1 Hyd. Excavator, 1 C.Y.		832.65		915.91	52.04	57.24
16 L.H., Daily Totals		$1679.45		$2178.32	$104.97	$136.14

Crew B-12B	Hr.	Daily	Hr.	Daily	Bare Costs	Incl. O&P
1 Equip. Oper. (crane)	$61.45	$491.60	$91.55	$732.40	$52.92	$78.90
1 Laborer	44.40	355.20	66.25	530.00		
1 Hyd. Excavator, 1.5 C.Y.		695.80		765.38	43.49	47.84
16 L.H., Daily Totals		$1542.60		$2027.78	$96.41	$126.74

Crew B-12C	Hr.	Daily	Hr.	Daily	Bare Costs	Incl. O&P
1 Equip. Oper. (crane)	$61.45	$491.60	$91.55	$732.40	$52.92	$78.90
1 Laborer	44.40	355.20	66.25	530.00		
1 Hyd. Excavator, 2 C.Y.		942.70		1036.97	58.92	64.81
16 L.H., Daily Totals		$1789.50		$2299.37	$111.84	$143.71

Crew B-12D	Hr.	Daily	Hr.	Daily	Bare Costs	Incl. O&P
1 Equip. Oper. (crane)	$61.45	$491.60	$91.55	$732.40	$52.92	$78.90
1 Laborer	44.40	355.20	66.25	530.00		
1 Hyd. Excavator, 3.5 C.Y.		2184.00		2402.40	136.50	150.15
16 L.H., Daily Totals		$3030.80		$3664.80	$189.43	$229.05

Crew B-12E	Hr.	Daily	Hr.	Daily	Bare Costs	Incl. O&P
1 Equip. Oper. (crane)	$61.45	$491.60	$91.55	$732.40	$52.92	$78.90
1 Laborer	44.40	355.20	66.25	530.00		
1 Hyd. Excavator, .5 C.Y.		457.00		502.70	28.56	31.42
16 L.H., Daily Totals		$1303.80		$1765.10	$81.49	$110.32

Crew B-12F	Hr.	Daily	Hr.	Daily	Bare Costs	Incl. O&P
1 Equip. Oper. (crane)	$61.45	$491.60	$91.55	$732.40	$52.92	$78.90
1 Laborer	44.40	355.20	66.25	530.00		
1 Hyd. Excavator, .75 C.Y.		701.80		771.98	43.86	48.25
16 L.H., Daily Totals		$1548.60		$2034.38	$96.79	$127.15

Crew B-12G	Hr.	Daily	Hr.	Daily	Bare Costs	Incl. O&P
1 Equip. Oper. (crane)	$61.45	$491.60	$91.55	$732.40	$52.92	$78.90
1 Laborer	44.40	355.20	66.25	530.00		
1 Crawler Crane, 15 Ton		810.30		891.33		
1 Clamshell Bucket, .5 C.Y.		67.80		74.58	54.88	60.37
16 L.H., Daily Totals		$1724.90		$2228.31	$107.81	$139.27

Crew B-12H	Hr.	Daily	Hr.	Daily	Bare Costs	Incl. O&P
1 Equip. Oper. (crane)	$61.45	$491.60	$91.55	$732.40	$52.92	$78.90
1 Laborer	44.40	355.20	66.25	530.00		
1 Crawler Crane, 25 Ton		1152.00		1267.20		
1 Clamshell Bucket, 1 C.Y.		69.25		76.17	76.33	83.96
16 L.H., Daily Totals		$2068.05		$2605.78	$129.25	$162.86

Crew B-12I	Hr.	Daily	Hr.	Daily	Bare Costs	Incl. O&P
1 Equip. Oper. (crane)	$61.45	$491.60	$91.55	$732.40	$52.92	$78.90
1 Laborer	44.40	355.20	66.25	530.00		
1 Crawler Crane, 20 Ton		1013.00		1114.30		
1 Dragline Bucket, .75 C.Y.		61.85		68.03	67.18	73.90
16 L.H., Daily Totals		$1921.65		$2444.74	$120.10	$152.80

Crew B-12J	Hr.	Daily	Hr.	Daily	Bare Costs	Incl. O&P
1 Equip. Oper. (crane)	$61.45	$491.60	$91.55	$732.40	$52.92	$78.90
1 Laborer	44.40	355.20	66.25	530.00		
1 Gradall, 5/8 C.Y.		850.65		935.72	53.17	58.48
16 L.H., Daily Totals		$1697.45		$2198.11	$106.09	$137.38

Crew B-12K	Hr.	Daily	Hr.	Daily	Bare Costs	Incl. O&P
1 Equip. Oper. (crane)	$61.45	$491.60	$91.55	$732.40	$52.92	$78.90
1 Laborer	44.40	355.20	66.25	530.00		
1 Gradall, 3 Ton, 1 C.Y.		984.55		1083.01	61.53	67.69
16 L.H., Daily Totals		$1831.35		$2345.41	$114.46	$146.59

Crew B-12L	Hr.	Daily	Hr.	Daily	Bare Costs	Incl. O&P
1 Equip. Oper. (crane)	$61.45	$491.60	$91.55	$732.40	$52.92	$78.90
1 Laborer	44.40	355.20	66.25	530.00		
1 Crawler Crane, 15 Ton		810.30		891.33		
1 F.E. Attachment, .5 C.Y.		66.05		72.66	54.77	60.25
16 L.H., Daily Totals		$1723.15		$2226.39	$107.70	$139.15

Crew B-12M	Hr.	Daily	Hr.	Daily	Bare Costs	Incl. O&P
1 Equip. Oper. (crane)	$61.45	$491.60	$91.55	$732.40	$52.92	$78.90
1 Laborer	44.40	355.20	66.25	530.00		
1 Crawler Crane, 20 Ton		1013.00		1114.30		
1 F.E. Attachment, .75 C.Y.		71.25		78.38	67.77	74.54
16 L.H., Daily Totals		$1931.05		$2455.07	$120.69	$153.44

Crew B-12N	Hr.	Daily	Hr.	Daily	Bare Costs	Incl. O&P
1 Equip. Oper. (crane)	$61.45	$491.60	$91.55	$732.40	$52.92	$78.90
1 Laborer	44.40	355.20	66.25	530.00		
1 Crawler Crane, 25 Ton		1152.00		1267.20		
1 F.E. Attachment, 1 C.Y.		77.35		85.08	76.83	84.52
16 L.H., Daily Totals		$2076.15		$2614.68	$129.76	$163.42

Crew B-12O	Hr.	Daily	Hr.	Daily	Bare Costs	Incl. O&P
1 Equip. Oper. (crane)	$61.45	$491.60	$91.55	$732.40	$52.92	$78.90
1 Laborer	44.40	355.20	66.25	530.00		
1 Crawler Crane, 40 Ton		1231.00		1354.10		
1 F.E. Attachment, 1.5 C.Y.		88.65		97.52	82.48	90.73
16 L.H., Daily Totals		$2166.45		$2714.01	$135.40	$169.63

Crews - Standard

Crew B-12P	Hr.	Daily	Hr.	Daily	Bare Costs	Incl. O&P
1 Equip. Oper. (crane)	$61.45	$491.60	$91.55	$732.40	$52.92	$78.90
1 Laborer	44.40	355.20	66.25	530.00		
1 Crawler Crane, 40 Ton		1231.00		1354.10		
1 Dragline Bucket, 1.5 C.Y.		65.70		72.27	81.04	89.15
16 L.H., Daily Totals		$2143.50		$2688.77	$133.97	$168.05

Crew B-12Q	Hr.	Daily	Hr.	Daily	Bare Costs	Incl. O&P
1 Equip. Oper. (crane)	$61.45	$491.60	$91.55	$732.40	$52.92	$78.90
1 Laborer	44.40	355.20	66.25	530.00		
1 Hyd. Excavator, 5/8 C.Y.		604.75		665.23	37.80	41.58
16 L.H., Daily Totals		$1451.55		$1927.63	$90.72	$120.48

Crew B-12S	Hr.	Daily	Hr.	Daily	Bare Costs	Incl. O&P
1 Equip. Oper. (crane)	$61.45	$491.60	$91.55	$732.40	$52.92	$78.90
1 Laborer	44.40	355.20	66.25	530.00		
1 Hyd. Excavator, 2.5 C.Y.		1567.00		1723.70	97.94	107.73
16 L.H., Daily Totals		$2413.80		$2986.10	$150.86	$186.63

Crew B-12T	Hr.	Daily	Hr.	Daily	Bare Costs	Incl. O&P
1 Equip. Oper. (crane)	$61.45	$491.60	$91.55	$732.40	$52.92	$78.90
1 Laborer	44.40	355.20	66.25	530.00		
1 Crawler Crane, 75 Ton		1967.00		2163.70		
1 F.E. Attachment, 3 C.Y.		115.55		127.11	130.16	143.18
16 L.H., Daily Totals		$2929.35		$3553.20	$183.08	$222.08

Crew B-12V	Hr.	Daily	Hr.	Daily	Bare Costs	Incl. O&P
1 Equip. Oper. (crane)	$61.45	$491.60	$91.55	$732.40	$52.92	$78.90
1 Laborer	44.40	355.20	66.25	530.00		
1 Crawler Crane, 75 Ton		1967.00		2163.70		
1 Dragline Bucket, 3 C.Y.		72.30		79.53	127.46	140.20
16 L.H., Daily Totals		$2886.10		$3505.63	$180.38	$219.10

Crew B-12Y	Hr.	Daily	Hr.	Daily	Bare Costs	Incl. O&P
1 Equip. Oper. (crane)	$61.45	$491.60	$91.55	$732.40	$50.08	$74.68
2 Laborers	44.40	710.40	66.25	1060.00		
1 Hyd. Excavator, 3.5 C.Y.		2184.00		2402.40	91.00	100.10
24 L.H., Daily Totals		$3386.00		$4194.80	$141.08	$174.78

Crew B-12Z	Hr.	Daily	Hr.	Daily	Bare Costs	Incl. O&P
1 Equip. Oper. (crane)	$61.45	$491.60	$91.55	$732.40	$50.08	$74.68
2 Laborers	44.40	710.40	66.25	1060.00		
1 Hyd. Excavator, 2.5 C.Y.		1567.00		1723.70	65.29	71.82
24 L.H., Daily Totals		$2769.00		$3516.10	$115.38	$146.50

Crew B-13	Hr.	Daily	Hr.	Daily	Bare Costs	Incl. O&P
1 Labor Foreman (outside)	$46.40	$371.20	$69.25	$554.00	$48.28	$72.01
4 Laborers	44.40	1420.80	66.25	2120.00		
1 Equip. Oper. (crane)	61.45	491.60	91.55	732.40		
1 Equip. Oper. (oiler)	52.50	420.00	78.25	626.00		
1 Hyd. Crane, 25 Ton		586.70		645.37	10.48	11.52
56 L.H., Daily Totals		$3290.30		$4677.77	$58.76	$83.53

Crew B-13A	Hr.	Daily	Hr.	Daily	Bare Costs	Incl. O&P
1 Labor Foreman (outside)	$46.40	$371.20	$69.25	$554.00	$50.83	$75.85
2 Laborers	44.40	710.40	66.25	1060.00		
2 Equipment Operators (med.)	59.00	944.00	87.90	1406.40		
2 Truck Drivers (heavy)	51.30	820.80	76.70	1227.20		
1 Crawler Crane, 75 Ton		1967.00		2163.70		
1 Crawler Loader, 4 C.Y.		1456.00		1601.60		
2 Dump Trucks, 8 C.Y., 220 H.P.		815.20		896.72	75.68	83.25
56 L.H., Daily Totals		$7084.60		$8909.62	$126.51	$159.10

Crew B-13B	Hr.	Daily	Hr.	Daily	Bare Costs	Incl. O&P
1 Labor Foreman (outside)	$46.40	$371.20	$69.25	$554.00	$48.28	$72.01
4 Laborers	44.40	1420.80	66.25	2120.00		
1 Equip. Oper. (crane)	61.45	491.60	91.55	732.40		
1 Equip. Oper. (oiler)	52.50	420.00	78.25	626.00		
1 Hyd. Crane, 55 Ton		990.15		1089.17	17.68	19.45
56 L.H., Daily Totals		$3693.75		$5121.56	$65.96	$91.46

Crew B-13C	Hr.	Daily	Hr.	Daily	Bare Costs	Incl. O&P
1 Labor Foreman (outside)	$46.40	$371.20	$69.25	$554.00	$48.28	$72.01
4 Laborers	44.40	1420.80	66.25	2120.00		
1 Equip. Oper. (crane)	61.45	491.60	91.55	732.40		
1 Equip. Oper. (oiler)	52.50	420.00	78.25	626.00		
1 Crawler Crane, 100 Ton		2310.00		2541.00	41.25	45.38
56 L.H., Daily Totals		$5013.60		$6573.40	$89.53	$117.38

Crew B-13D	Hr.	Daily	Hr.	Daily	Bare Costs	Incl. O&P
1 Laborer	$44.40	$355.20	$66.25	$530.00	$52.92	$78.90
1 Equip. Oper. (crane)	61.45	491.60	91.55	732.40		
1 Hyd. Excavator, 1 C.Y.		832.65		915.91		
1 Trench Box		119.15		131.07	59.49	65.44
16 L.H., Daily Totals		$1798.60		$2309.38	$112.41	$144.34

Crew B-13E	Hr.	Daily	Hr.	Daily	Bare Costs	Incl. O&P
1 Laborer	$44.40	$355.20	$66.25	$530.00	$52.92	$78.90
1 Equip. Oper. (crane)	61.45	491.60	91.55	732.40		
1 Hyd. Excavator, 1.5 C.Y.		695.80		765.38		
1 Trench Box		119.15		131.07	50.93	56.03
16 L.H., Daily Totals		$1661.75		$2158.84	$103.86	$134.93

Crew B-13F	Hr.	Daily	Hr.	Daily	Bare Costs	Incl. O&P
1 Laborer	$44.40	$355.20	$66.25	$530.00	$52.92	$78.90
1 Equip. Oper. (crane)	61.45	491.60	91.55	732.40		
1 Hyd. Excavator, 3.5 C.Y.		2184.00		2402.40		
1 Trench Box		119.15		131.07	143.95	158.34
16 L.H., Daily Totals		$3149.95		$3795.86	$196.87	$237.24

Crew B-13G	Hr.	Daily	Hr.	Daily	Bare Costs	Incl. O&P
1 Laborer	$44.40	$355.20	$66.25	$530.00	$52.92	$78.90
1 Equip. Oper. (crane)	61.45	491.60	91.55	732.40		
1 Hyd. Excavator, .75 C.Y.		701.80		771.98		
1 Trench Box		119.15		131.07	51.31	56.44
16 L.H., Daily Totals		$1667.75		$2165.45	$104.23	$135.34

Crew B-13H	Hr.	Daily	Hr.	Daily	Bare Costs	Incl. O&P
1 Laborer	$44.40	$355.20	$66.25	$530.00	$52.92	$78.90
1 Equip. Oper. (crane)	61.45	491.60	91.55	732.40		
1 Gradall, 5/8 C.Y.		850.65		935.72		
1 Trench Box		119.15		131.07	60.61	66.67
16 L.H., Daily Totals		$1816.60		$2329.18	$113.54	$145.57

For customer support on your Plumbing Costs with RSMeans Data, call 800.448.8182.

605

Crews - Standard

Crew No.	Bare Costs Hr.	Daily	Incl. Subs O&P Hr.	Daily	Cost Per Labor-Hour Bare Costs	Incl. O&P
Crew B-13I	Hr.	Daily	Hr.	Daily	Bare Costs	Incl. O&P
1 Laborer	$44.40	$355.20	$66.25	$530.00	$52.92	$78.90
1 Equip. Oper. (crane)	61.45	491.60	91.55	732.40		
1 Gradall, 3 Ton, 1 C.Y.		984.55		1083.01		
1 Trench Box		119.15		131.07	68.98	75.88
16 L.H., Daily Totals		$1950.50		$2476.47	$121.91	$154.78
Crew B-13J	Hr.	Daily	Hr.	Daily	Bare Costs	Incl. O&P
1 Laborer	$44.40	$355.20	$66.25	$530.00	$52.92	$78.90
1 Equip. Oper. (crane)	61.45	491.60	91.55	732.40		
1 Hyd. Excavator, 2.5 C.Y.		1567.00		1723.70		
1 Trench Box		119.15		131.07	105.38	115.92
16 L.H., Daily Totals		$2532.95		$3117.17	$158.31	$194.82
Crew B-13K	Hr.	Daily	Hr.	Daily	Bare Costs	Incl. O&P
2 Equip. Opers. (crane)	$61.45	$983.20	$91.55	$1464.80	$61.45	$91.55
1 Hyd. Excavator, .75 C.Y.		701.80		771.98		
1 Hyd. Hammer, 4000 ft-lb		649.20		714.12		
1 Hyd. Excavator, .75 C.Y.		701.80		771.98	128.30	141.13
16 L.H., Daily Totals		$3036.00		$3722.88	$189.75	$232.68
Crew B-13L	Hr.	Daily	Hr.	Daily	Bare Costs	Incl. O&P
2 Equip. Opers. (crane)	$61.45	$983.20	$91.55	$1464.80	$61.45	$91.55
1 Hyd. Excavator, 1.5 C.Y.		695.80		765.38		
1 Hyd. Hammer, 5000 ft-lb		705.60		776.16		
1 Hyd. Excavator, .75 C.Y.		701.80		771.98	131.45	144.60
16 L.H., Daily Totals		$3086.40		$3778.32	$192.90	$236.15
Crew B-13M	Hr.	Daily	Hr.	Daily	Bare Costs	Incl. O&P
2 Equip. Opers. (crane)	$61.45	$983.20	$91.55	$1464.80	$61.45	$91.55
1 Hyd. Excavator, 2.5 C.Y.		1567.00		1723.70		
1 Hyd. Hammer, 8000 ft-lb		918.65		1010.52		
1 Hyd. Excavator, 1.5 C.Y.		695.80		765.38	198.84	218.72
16 L.H., Daily Totals		$4164.65		$4964.40	$260.29	$310.27
Crew B-13N	Hr.	Daily	Hr.	Daily	Bare Costs	Incl. O&P
2 Equip. Opers. (crane)	$61.45	$983.20	$91.55	$1464.80	$61.45	$91.55
1 Hyd. Excavator, 3.5 C.Y.		2184.00		2402.40		
1 Hyd. Hammer, 12,000 ft-lb		882.20		970.42		
1 Hyd. Excavator, 1.5 C.Y.		695.80		765.38	235.13	258.64
16 L.H., Daily Totals		$4745.20		$5603.00	$296.57	$350.19
Crew B-14	Hr.	Daily	Hr.	Daily	Bare Costs	Incl. O&P
1 Labor Foreman (outside)	$46.40	$371.20	$69.25	$554.00	$46.58	$69.49
4 Laborers	44.40	1420.80	66.25	2120.00		
1 Equip. Oper. (light)	55.50	444.00	82.70	661.60		
1 Backhoe Loader, 48 H.P.		216.20		237.82	4.50	4.95
48 L.H., Daily Totals		$2452.20		$3573.42	$51.09	$74.45
Crew B-14A	Hr.	Daily	Hr.	Daily	Bare Costs	Incl. O&P
1 Equip. Oper. (crane)	$61.45	$491.60	$91.55	$732.40	$55.77	$83.12
.5 Laborer	44.40	177.60	66.25	265.00		
1 Hyd. Excavator, 4.5 C.Y.		3450.00		3795.00	287.50	316.25
12 L.H., Daily Totals		$4119.20		$4792.40	$343.27	$399.37
Crew B-14B	Hr.	Daily	Hr.	Daily	Bare Costs	Incl. O&P
1 Equip. Oper. (crane)	$61.45	$491.60	$91.55	$732.40	$55.77	$83.12
.5 Laborer	44.40	177.60	66.25	265.00		
1 Hyd. Excavator, 6 C.Y.		3506.00		3856.60	292.17	321.38
12 L.H., Daily Totals		$4175.20		$4854.00	$347.93	$404.50

Crew No.	Bare Costs Hr.	Daily	Incl. Subs O&P Hr.	Daily	Cost Per Labor-Hour Bare Costs	Incl. O&P
Crew B-14C	Hr.	Daily	Hr.	Daily	Bare Costs	Incl. O&P
1 Equip. Oper. (crane)	$61.45	$491.60	$91.55	$732.40	$55.77	$83.12
.5 Laborer	44.40	177.60	66.25	265.00		
1 Hyd. Excavator, 7 C.Y.		3475.00		3822.50	289.58	318.54
12 L.H., Daily Totals		$4144.20		$4819.90	$345.35	$401.66
Crew B-14F	Hr.	Daily	Hr.	Daily	Bare Costs	Incl. O&P
1 Equip. Oper. (crane)	$61.45	$491.60	$91.55	$732.40	$55.77	$83.12
.5 Laborer	44.40	177.60	66.25	265.00		
1 Hyd. Shovel, 7 C.Y.		4148.00		4562.80	345.67	380.23
12 L.H., Daily Totals		$4817.20		$5560.20	$401.43	$463.35
Crew B-14G	Hr.	Daily	Hr.	Daily	Bare Costs	Incl. O&P
1 Equip. Oper. (crane)	$61.45	$491.60	$91.55	$732.40	$55.77	$83.12
.5 Laborer	44.40	177.60	66.25	265.00		
1 Hyd. Shovel, 12 C.Y.		6022.00		6624.20	501.83	552.02
12 L.H., Daily Totals		$6691.20		$7621.60	$557.60	$635.13
Crew B-14J	Hr.	Daily	Hr.	Daily	Bare Costs	Incl. O&P
1 Equip. Oper. (medium)	$59.00	$472.00	$87.90	$703.20	$54.13	$80.68
.5 Laborer	44.40	177.60	66.25	265.00		
1 F.E. Loader, 8 C.Y.		2285.00		2513.50	190.42	209.46
12 L.H., Daily Totals		$2934.60		$3481.70	$244.55	$290.14
Crew B-14K	Hr.	Daily	Hr.	Daily	Bare Costs	Incl. O&P
1 Equip. Oper. (medium)	$59.00	$472.00	$87.90	$703.20	$54.13	$80.68
.5 Laborer	44.40	177.60	66.25	265.00		
1 F.E. Loader, 10 C.Y.		2706.00		2976.60	225.50	248.05
12 L.H., Daily Totals		$3355.60		$3944.80	$279.63	$328.73
Crew B-15	Hr.	Daily	Hr.	Daily	Bare Costs	Incl. O&P
1 Equipment Oper. (med.)	$59.00	$472.00	$87.90	$703.20	$52.51	$78.41
.5 Laborer	44.40	177.60	66.25	265.00		
2 Truck Drivers (heavy)	51.30	820.80	76.70	1227.20		
2 Dump Trucks, 12 C.Y., 400 H.P.		1158.70		1274.57		
1 Dozer, 200 H.P.		1520.00		1672.00	95.67	105.23
28 L.H., Daily Totals		$4149.10		$5141.97	$148.18	$183.64
Crew B-16	Hr.	Daily	Hr.	Daily	Bare Costs	Incl. O&P
1 Labor Foreman (outside)	$46.40	$371.20	$69.25	$554.00	$46.63	$69.61
2 Laborers	44.40	710.40	66.25	1060.00		
1 Truck Driver (heavy)	51.30	410.40	76.70	613.60		
1 Dump Truck, 12 C.Y., 400 H.P.		579.35		637.28	18.10	19.92
32 L.H., Daily Totals		$2071.35		$2864.89	$64.73	$89.53
Crew B-17	Hr.	Daily	Hr.	Daily	Bare Costs	Incl. O&P
2 Laborers	$44.40	$710.40	$66.25	$1060.00	$48.90	$72.97
1 Equip. Oper. (light)	55.50	444.00	82.70	661.60		
1 Truck Driver (heavy)	51.30	410.40	76.70	613.60		
1 Backhoe Loader, 48 H.P.		216.20		237.82		
1 Dump Truck, 8 C.Y., 220 H.P.		407.60		448.36	19.49	21.44
32 L.H., Daily Totals		$2188.60		$3021.38	$68.39	$94.42
Crew B-17A	Hr.	Daily	Hr.	Daily	Bare Costs	Incl. O&P
2 Labor Foremen (outside)	$46.40	$742.40	$69.25	$1108.00	$47.54	$71.08
6 Laborers	44.40	2131.20	66.25	3180.00		
1 Skilled Worker Foreman (out)	59.10	472.80	88.90	711.20		
1 Skilled Worker	57.10	456.80	85.90	687.20		
80 L.H., Daily Totals		$3803.20		$5686.40	$47.54	$71.08

Crews - Standard

Crew No.	Bare Costs Hr.	Daily	Incl. Subs O&P Hr.	Daily	Cost Per Labor-Hour Bare Costs	Incl. O&P
Crew B-17B	Hr.	Daily	Hr.	Daily	Bare Costs	Incl. O&P
2 Laborers	$44.40	$710.40	$66.25	$1060.00	$48.90	$72.97
1 Equip. Oper. (light)	55.50	444.00	82.70	661.60		
1 Truck Driver (heavy)	51.30	410.40	76.70	613.60		
1 Backhoe Loader, 48 H.P.		216.20		237.82		
1 Dump Truck, 12 C.Y., 400 H.P.		579.35		637.28	24.86	27.35
32 L.H., Daily Totals		$2360.35		$3210.30	$73.76	$100.32
Crew B-18	Hr.	Daily	Hr.	Daily	Bare Costs	Incl. O&P
1 Labor Foreman (outside)	$46.40	$371.20	$69.25	$554.00	$45.07	$67.25
2 Laborers	44.40	710.40	66.25	1060.00		
1 Vibrating Plate, Gas, 21"		165.60		182.16	6.90	7.59
24 L.H., Daily Totals		$1247.20		$1796.16	$51.97	$74.84
Crew B-19	Hr.	Daily	Hr.	Daily	Bare Costs	Incl. O&P
1 Pile Driver Foreman (outside)	$57.90	$463.20	$89.50	$716.00	$57.11	$87.06
4 Pile Drivers	55.90	1788.80	86.40	2764.80		
2 Equip. Oper. (crane)	61.45	983.20	91.55	1464.80		
1 Equip. Oper. (oiler)	52.50	420.00	78.25	626.00		
1 Crawler Crane, 40 Ton		1231.00		1354.10		
1 Lead, 90' High		371.85		409.04		
1 Hammer, Diesel, 22k ft-lb		441.70		485.87	31.95	35.14
64 L.H., Daily Totals		$5699.75		$7820.60	$89.06	$122.20
Crew B-19A	Hr.	Daily	Hr.	Daily	Bare Costs	Incl. O&P
1 Pile Driver Foreman (outside)	$57.90	$463.20	$89.50	$716.00	$57.11	$87.06
4 Pile Drivers	55.90	1788.80	86.40	2764.80		
2 Equip. Oper. (crane)	61.45	983.20	91.55	1464.80		
1 Equip. Oper. (oiler)	52.50	420.00	78.25	626.00		
1 Crawler Crane, 75 Ton		1967.00		2163.70		
1 Lead, 90' High		371.85		409.04		
1 Hammer, Diesel, 41k ft-lb		583.55		641.90	45.66	50.23
64 L.H., Daily Totals		$6577.60		$8786.24	$102.78	$137.29
Crew B-19B	Hr.	Daily	Hr.	Daily	Bare Costs	Incl. O&P
1 Pile Driver Foreman (outside)	$57.90	$463.20	$89.50	$716.00	$57.11	$87.06
4 Pile Drivers	55.90	1788.80	86.40	2764.80		
2 Equip. Oper. (crane)	61.45	983.20	91.55	1464.80		
1 Equip. Oper. (oiler)	52.50	420.00	78.25	626.00		
1 Crawler Crane, 40 Ton		1231.00		1354.10		
1 Lead, 90' High		371.85		409.04		
1 Hammer, Diesel, 22k ft-lb		441.70		485.87		
1 Barge, 400 Ton		869.15		956.07	45.53	50.08
64 L.H., Daily Totals		$6568.90		$8776.67	$102.64	$137.14
Crew B-19C	Hr.	Daily	Hr.	Daily	Bare Costs	Incl. O&P
1 Pile Driver Foreman (outside)	$57.90	$463.20	$89.50	$716.00	$57.11	$87.06
4 Pile Drivers	55.90	1788.80	86.40	2764.80		
2 Equip. Oper. (crane)	61.45	983.20	91.55	1464.80		
1 Equip. Oper. (oiler)	52.50	420.00	78.25	626.00		
1 Crawler Crane, 75 Ton		1967.00		2163.70		
1 Lead, 90' High		371.85		409.04		
1 Hammer, Diesel, 41k ft-lb		583.55		641.90		
1 Barge, 400 Ton		869.15		956.07	59.24	65.17
64 L.H., Daily Totals		$7446.75		$9742.31	$116.36	$152.22
Crew B-20	Hr.	Daily	Hr.	Daily	Bare Costs	Incl. O&P
1 Labor Foreman (outside)	$46.40	$371.20	$69.25	$554.00	$49.30	$73.80
1 Skilled Worker	57.10	456.80	85.90	687.20		
1 Laborer	44.40	355.20	66.25	530.00		
24 L.H., Daily Totals		$1183.20		$1771.20	$49.30	$73.80

Crew No.	Bare Costs Hr.	Daily	Incl. Subs O&P Hr.	Daily	Cost Per Labor-Hour Bare Costs	Incl. O&P
Crew B-20A	Hr.	Daily	Hr.	Daily	Bare Costs	Incl. O&P
1 Labor Foreman (outside)	$46.40	$371.20	$69.25	$554.00	$53.16	$79.34
1 Laborer	44.40	355.20	66.25	530.00		
1 Plumber	67.70	541.60	101.05	808.40		
1 Plumber Apprentice	54.15	433.20	80.80	646.40		
32 L.H., Daily Totals		$1701.20		$2538.80	$53.16	$79.34
Crew B-21	Hr.	Daily	Hr.	Daily	Bare Costs	Incl. O&P
1 Labor Foreman (outside)	$46.40	$371.20	$69.25	$554.00	$51.04	$76.34
1 Skilled Worker	57.10	456.80	85.90	687.20		
1 Laborer	44.40	355.20	66.25	530.00		
.5 Equip. Oper. (crane)	61.45	245.80	91.55	366.20		
.5 S.P. Crane, 4x4, 5 Ton		190.97		210.07	6.82	7.50
28 L.H., Daily Totals		$1619.97		$2347.47	$57.86	$83.84
Crew B-21A	Hr.	Daily	Hr.	Daily	Bare Costs	Incl. O&P
1 Labor Foreman (outside)	$46.40	$371.20	$69.25	$554.00	$54.82	$81.78
1 Laborer	44.40	355.20	66.25	530.00		
1 Plumber	67.70	541.60	101.05	808.40		
1 Plumber Apprentice	54.15	433.20	80.80	646.40		
1 Equip. Oper. (crane)	61.45	491.60	91.55	732.40		
1 S.P. Crane, 4x4, 12 Ton		432.65		475.92	10.82	11.90
40 L.H., Daily Totals		$2625.45		$3747.11	$65.64	$93.68
Crew B-21B	Hr.	Daily	Hr.	Daily	Bare Costs	Incl. O&P
1 Labor Foreman (outside)	$46.40	$371.20	$69.25	$554.00	$48.21	$71.91
3 Laborers	44.40	1065.60	66.25	1590.00		
1 Equip. Oper. (crane)	61.45	491.60	91.55	732.40		
1 Hyd. Crane, 12 Ton		475.80		523.38	11.90	13.08
40 L.H., Daily Totals		$2404.20		$3399.78	$60.10	$84.99
Crew B-21C	Hr.	Daily	Hr.	Daily	Bare Costs	Incl. O&P
1 Labor Foreman (outside)	$46.40	$371.20	$69.25	$554.00	$48.28	$72.01
4 Laborers	44.40	1420.80	66.25	2120.00		
1 Equip. Oper. (crane)	61.45	491.60	91.55	732.40		
1 Equip. Oper. (oiler)	52.50	420.00	78.25	626.00		
2 Cutting Torches		25.90		28.49		
2 Sets of Gases		347.20		381.92		
1 Lattice Boom Crane, 90 Ton		1713.00		1884.30	37.25	40.98
56 L.H., Daily Totals		$4789.70		$6327.11	$85.53	$112.98
Crew B-22	Hr.	Daily	Hr.	Daily	Bare Costs	Incl. O&P
1 Labor Foreman (outside)	$46.40	$371.20	$69.25	$554.00	$51.73	$77.35
1 Skilled Worker	57.10	456.80	85.90	687.20		
1 Laborer	44.40	355.20	66.25	530.00		
.75 Equip. Oper. (crane)	61.45	368.70	91.55	549.30		
.75 S.P. Crane, 4x4, 5 Ton		286.46		315.11	9.55	10.50
30 L.H., Daily Totals		$1838.36		$2635.61	$61.28	$87.85
Crew B-22A	Hr.	Daily	Hr.	Daily	Bare Costs	Incl. O&P
1 Labor Foreman (outside)	$46.40	$371.20	$69.25	$554.00	$50.75	$75.84
1 Skilled Worker	57.10	456.80	85.90	687.20		
2 Laborers	44.40	710.40	66.25	1060.00		
1 Equipment Operator, Crane	61.45	491.60	91.55	732.40		
1 S.P. Crane, 4x4, 5 Ton		381.95		420.14		
1 Butt Fusion Mach., 4"-12" diam.		420.75		462.82	20.07	22.07
40 L.H., Daily Totals		$2832.70		$3916.57	$70.82	$97.91

For customer support on your Plumbing Costs with RSMeans Data, call 800.448.8182.

607

Crew No.	Bare Costs		Incl. Subs O&P		Cost Per Labor-Hour	

Crew B-22B	Hr.	Daily	Hr.	Daily	Bare Costs	Incl. O&P
1 Labor Foreman (outside)	$46.40	$371.20	$69.25	$554.00	$50.75	$75.84
1 Skilled Worker	57.10	456.80	85.90	687.20		
2 Laborers	44.40	710.40	66.25	1060.00		
1 Equip. Oper. (crane)	61.45	491.60	91.55	732.40		
1 S.P. Crane, 4x4, 5 Ton		381.95		420.14		
1 Butt Fusion Mach., 8"-24" diam.		1086.00		1194.60	36.70	40.37
40 L.H., Daily Totals		$3497.95		$4648.35	$87.45	$116.21

Crew B-22C	Hr.	Daily	Hr.	Daily	Bare Costs	Incl. O&P
1 Skilled Worker	$57.10	$456.80	$85.90	$687.20	$50.75	$76.08
1 Laborer	44.40	355.20	66.25	530.00		
1 Butt Fusion Mach., 2"-8" diam.		134.95		148.44	8.43	9.28
16 L.H., Daily Totals		$946.95		$1365.65	$59.18	$85.35

Crew B-23	Hr.	Daily	Hr.	Daily	Bare Costs	Incl. O&P
1 Labor Foreman (outside)	$46.40	$371.20	$69.25	$554.00	$44.80	$66.85
4 Laborers	44.40	1420.80	66.25	2120.00		
1 Drill Rig, Truck-Mounted		768.40		845.24		
1 Flatbed Truck, Gas, 3 Ton		850.05		935.05	40.46	44.51
40 L.H., Daily Totals		$3410.45		$4454.30	$85.26	$111.36

Crew B-23A	Hr.	Daily	Hr.	Daily	Bare Costs	Incl. O&P
1 Labor Foreman (outside)	$46.40	$371.20	$69.25	$554.00	$49.93	$74.47
1 Laborer	44.40	355.20	66.25	530.00		
1 Equip. Oper. (medium)	59.00	472.00	87.90	703.20		
1 Drill Rig, Truck-Mounted		768.40		845.24		
1 Pickup Truck, 3/4 Ton		112.20		123.42	36.69	40.36
24 L.H., Daily Totals		$2079.00		$2755.86	$86.63	$114.83

Crew B-23B	Hr.	Daily	Hr.	Daily	Bare Costs	Incl. O&P
1 Labor Foreman (outside)	$46.40	$371.20	$69.25	$554.00	$49.93	$74.47
1 Laborer	44.40	355.20	66.25	530.00		
1 Equip. Oper. (medium)	59.00	472.00	87.90	703.20		
1 Drill Rig, Truck-Mounted		768.40		845.24		
1 Pickup Truck, 3/4 Ton		112.20		123.42		
1 Centr. Water Pump, 6"		235.25		258.77	46.49	51.14
24 L.H., Daily Totals		$2314.25		$3014.64	$96.43	$125.61

Crew B-24	Hr.	Daily	Hr.	Daily	Bare Costs	Incl. O&P
1 Cement Finisher	$51.80	$414.40	$75.90	$607.20	$50.30	$74.60
1 Laborer	44.40	355.20	66.25	530.00		
1 Carpenter	54.70	437.60	81.65	653.20		
24 L.H., Daily Totals		$1207.20		$1790.40	$50.30	$74.60

Crew B-25	Hr.	Daily	Hr.	Daily	Bare Costs	Incl. O&P
1 Labor Foreman (outside)	$46.40	$371.20	$69.25	$554.00	$48.56	$72.43
7 Laborers	44.40	2486.40	66.25	3710.00		
3 Equip. Oper. (medium)	59.00	1416.00	87.90	2109.60		
1 Asphalt Paver, 130 H.P.		2143.00		2357.30		
1 Tandem Roller, 10 Ton		246.80		271.48		
1 Roller, Pneum. Whl., 12 Ton		349.90		384.89	31.13	34.25
88 L.H., Daily Totals		$7013.30		$9387.27	$79.70	$106.67

Crew B-25B	Hr.	Daily	Hr.	Daily	Bare Costs	Incl. O&P
1 Labor Foreman (outside)	$46.40	$371.20	$69.25	$554.00	$49.43	$73.72
7 Laborers	44.40	2486.40	66.25	3710.00		
4 Equip. Oper. (medium)	59.00	1888.00	87.90	2812.80		
1 Asphalt Paver, 130 H.P.		2143.00		2357.30		
2 Tandem Rollers, 10 Ton		493.60		542.96		
1 Roller, Pneum. Whl., 12 Ton		349.90		384.89	31.11	34.22
96 L.H., Daily Totals		$7732.10		$10361.95	$80.54	$107.94

Crew B-25C	Hr.	Daily	Hr.	Daily	Bare Costs	Incl. O&P
1 Labor Foreman (outside)	$46.40	$371.20	$69.25	$554.00	$49.60	$73.97
3 Laborers	44.40	1065.60	66.25	1590.00		
2 Equip. Oper. (medium)	59.00	944.00	87.90	1406.40		
1 Asphalt Paver, 130 H.P.		2143.00		2357.30		
1 Tandem Roller, 10 Ton		246.80		271.48	49.79	54.77
48 L.H., Daily Totals		$4770.60		$6179.18	$99.39	$128.73

Crew B-25D	Hr.	Daily	Hr.	Daily	Bare Costs	Incl. O&P
1 Labor Foreman (outside)	$46.40	$371.20	$69.25	$554.00	$49.82	$74.30
3 Laborers	44.40	1065.60	66.25	1590.00		
2.125 Equip. Oper. (medium)	59.00	1003.00	87.90	1494.30		
.125 Truck Driver (heavy)	51.30	51.30	76.70	76.70		
.125 Truck Tractor, 6x4, 380 H.P.		62.39		68.63		
.125 Dist. Tanker, 3000 Gallon		41.76		45.94		
1 Asphalt Paver, 130 H.P.		2143.00		2357.30		
1 Tandem Roller, 10 Ton		246.80		271.48	49.88	54.87
50 L.H., Daily Totals		$4985.06		$6458.35	$99.70	$129.17

Crew B-25E	Hr.	Daily	Hr.	Daily	Bare Costs	Incl. O&P
1 Labor Foreman (outside)	$46.40	$371.20	$69.25	$554.00	$50.03	$74.61
3 Laborers	44.40	1065.60	66.25	1590.00		
2.250 Equip. Oper. (medium)	59.00	1062.00	87.90	1582.20		
.25 Truck Driver (heavy)	51.30	102.60	76.70	153.40		
.25 Truck Tractor, 6x4, 380 H.P.		124.79		137.27		
.25 Dist. Tanker, 3000 Gallon		83.53		91.88		
1 Asphalt Paver, 130 H.P.		2143.00		2357.30		
1 Tandem Roller, 10 Ton		246.80		271.48	49.96	54.96
52 L.H., Daily Totals		$5199.51		$6737.52	$99.99	$129.57

Crew B-26	Hr.	Daily	Hr.	Daily	Bare Costs	Incl. O&P
1 Labor Foreman (outside)	$46.40	$371.20	$69.25	$554.00	$49.23	$73.32
6 Laborers	44.40	2131.20	66.25	3180.00		
2 Equip. Oper. (medium)	59.00	944.00	87.90	1406.40		
1 Rodman (reinf.)	58.90	471.20	88.05	704.40		
1 Cement Finisher	51.80	414.40	75.90	607.20		
1 Grader, 30,000 Lbs.		1073.00		1180.30		
1 Paving Mach. & Equip.		2503.00		2753.30	40.64	44.70
88 L.H., Daily Totals		$7908.00		$10385.60	$89.86	$118.02

Crew B-26A	Hr.	Daily	Hr.	Daily	Bare Costs	Incl. O&P
1 Labor Foreman (outside)	$46.40	$371.20	$69.25	$554.00	$49.23	$73.32
6 Laborers	44.40	2131.20	66.25	3180.00		
2 Equip. Oper. (medium)	59.00	944.00	87.90	1406.40		
1 Rodman (reinf.)	58.90	471.20	88.05	704.40		
1 Cement Finisher	51.80	414.40	75.90	607.20		
1 Grader, 30,000 Lbs.		1073.00		1180.30		
1 Paving Mach. & Equip.		2503.00		2753.30		
1 Concrete Saw		112.85		124.14	41.92	46.11
88 L.H., Daily Totals		$8020.85		$10509.74	$91.15	$119.43

Crew B-26B	Hr.	Daily	Hr.	Daily	Bare Costs	Incl. O&P
1 Labor Foreman (outside)	$46.40	$371.20	$69.25	$554.00	$50.04	$74.53
6 Laborers	44.40	2131.20	66.25	3180.00		
3 Equip. Oper. (medium)	59.00	1416.00	87.90	2109.60		
1 Rodman (reinf.)	58.90	471.20	88.05	704.40		
1 Cement Finisher	51.80	414.40	75.90	607.20		
1 Grader, 30,000 Lbs.		1073.00		1180.30		
1 Paving Mach. & Equip.		2503.00		2753.30		
1 Concrete Pump, 110' Boom		493.65		543.01	42.39	46.63
96 L.H., Daily Totals		$8873.65		$11631.82	$92.43	$121.16

For customer support on your Plumbing Costs with RSMeans Data, call 800.448.8182.

Crews - Standard

Crew No.	Bare Costs Hr.	Daily	Incl. Subs O&P Hr.	Daily	Cost Per Labor-Hour Bare Costs	Incl. O&P
Crew B-26C						
1 Labor Foreman (outside)	$46.40	$371.20	$69.25	$554.00	$48.25	$71.86
6 Laborers	44.40	2131.20	66.25	3180.00		
1 Equip. Oper. (medium)	59.00	472.00	87.90	703.20		
1 Rodman (reinf.)	58.90	471.20	88.05	704.40		
1 Cement Finisher	51.80	414.40	75.90	607.20		
1 Paving Mach. & Equip.		2503.00		2753.30		
1 Concrete Saw		112.85		124.14	32.70	35.97
80 L.H., Daily Totals		$6475.85		$8626.24	$80.95	$107.83
Crew B-27						
1 Labor Foreman (outside)	$46.40	$371.20	$69.25	$554.00	$44.90	$67.00
3 Laborers	44.40	1065.60	66.25	1590.00		
1 Berm Machine		253.55		278.90	7.92	8.72
32 L.H., Daily Totals		$1690.35		$2422.91	$52.82	$75.72
Crew B-28						
2 Carpenters	$54.70	$875.20	$81.65	$1306.40	$51.27	$76.52
1 Laborer	44.40	355.20	66.25	530.00		
24 L.H., Daily Totals		$1230.40		$1836.40	$51.27	$76.52
Crew B-29						
1 Labor Foreman (outside)	$46.40	$371.20	$69.25	$554.00	$48.28	$72.01
4 Laborers	44.40	1420.80	66.25	2120.00		
1 Equip. Oper. (crane)	61.45	491.60	91.55	732.40		
1 Equip. Oper. (oiler)	52.50	420.00	78.25	626.00		
1 Gradall, 5/8 C.Y.		850.65		935.72	15.19	16.71
56 L.H., Daily Totals		$3554.25		$4968.11	$63.47	$88.72
Crew B-30						
1 Equip. Oper. (medium)	$59.00	$472.00	$87.90	$703.20	$53.87	$80.43
2 Truck Drivers (heavy)	51.30	820.80	76.70	1227.20		
1 Hyd. Excavator, 1.5 C.Y.		695.80		765.38		
2 Dump Trucks, 12 C.Y., 400 H.P.		1158.70		1274.57	77.27	85.00
24 L.H., Daily Totals		$3147.30		$3970.35	$131.14	$165.43
Crew B-31						
1 Labor Foreman (outside)	$46.40	$371.20	$69.25	$554.00	$46.86	$69.93
3 Laborers	44.40	1065.60	66.25	1590.00		
1 Carpenter	54.70	437.60	81.65	653.20		
1 Air Compressor, 250 cfm		202.85		223.13		
1 Sheeting Driver		7.45		8.20		
2 -50' Air Hoses, 1.5"		45.60		50.16	6.40	7.04
40 L.H., Daily Totals		$2130.30		$3078.69	$53.26	$76.97
Crew B-32						
1 Laborer	$44.40	$355.20	$66.25	$530.00	$55.35	$82.49
3 Equip. Oper. (medium)	59.00	1416.00	87.90	2109.60		
1 Grader, 30,000 Lbs.		1073.00		1180.30		
1 Tandem Roller, 10 Ton		246.80		271.48		
1 Dozer, 200 H.P.		1520.00		1672.00	88.74	97.62
32 L.H., Daily Totals		$4611.00		$5763.38	$144.09	$180.11
Crew B-32A						
1 Laborer	$44.40	$355.20	$66.25	$530.00	$54.13	$80.68
2 Equip. Oper. (medium)	59.00	944.00	87.90	1406.40		
1 Grader, 30,000 Lbs.		1073.00		1180.30		
1 Roller, Vibratory, 25 Ton		672.35		739.59	72.72	80.00
24 L.H., Daily Totals		$3044.55		$3856.28	$126.86	$160.68

Crew No.	Bare Costs Hr.	Daily	Incl. Subs O&P Hr.	Daily	Cost Per Labor-Hour Bare Costs	Incl. O&P
Crew B-32B						
1 Laborer	$44.40	$355.20	$66.25	$530.00	$54.13	$80.68
2 Equip. Oper. (medium)	59.00	944.00	87.90	1406.40		
1 Dozer, 200 H.P.		1520.00		1672.00		
1 Roller, Vibratory, 25 Ton		672.35		739.59	91.35	100.48
24 L.H., Daily Totals		$3491.55		$4347.98	$145.48	$181.17
Crew B-32C						
1 Labor Foreman (outside)	$46.40	$371.20	$69.25	$554.00	$52.03	$77.58
2 Laborers	44.40	710.40	66.25	1060.00		
3 Equip. Oper. (medium)	59.00	1416.00	87.90	2109.60		
1 Grader, 30,000 Lbs.		1073.00		1180.30		
1 Tandem Roller, 10 Ton		246.80		271.48		
1 Dozer, 200 H.P.		1520.00		1672.00	59.16	65.08
48 L.H., Daily Totals		$5337.40		$6847.38	$111.20	$142.65
Crew B-33A						
1 Equip. Oper. (medium)	$59.00	$472.00	$87.90	$703.20	$54.83	$81.71
.5 Laborer	44.40	177.60	66.25	265.00		
.25 Equip. Oper. (medium)	59.00	118.00	87.90	175.80		
1 Scraper, Towed, 7 C.Y.		129.30		142.23		
1.25 Dozers, 300 H.P.		2231.25		2454.38	168.61	185.47
14 L.H., Daily Totals		$3128.15		$3740.61	$223.44	$267.19
Crew B-33B						
1 Equip. Oper. (medium)	$59.00	$472.00	$87.90	$703.20	$54.83	$81.71
.5 Laborer	44.40	177.60	66.25	265.00		
.25 Equip. Oper. (medium)	59.00	118.00	87.90	175.80		
1 Scraper, Towed, 10 C.Y.		161.65		177.82		
1.25 Dozers, 300 H.P.		2231.25		2454.38	170.92	188.01
14 L.H., Daily Totals		$3160.50		$3776.19	$225.75	$269.73
Crew B-33C						
1 Equip. Oper. (medium)	$59.00	$472.00	$87.90	$703.20	$54.83	$81.71
.5 Laborer	44.40	177.60	66.25	265.00		
.25 Equip. Oper. (medium)	59.00	118.00	87.90	175.80		
1 Scraper, Towed, 15 C.Y.		178.85		196.74		
1.25 Dozers, 300 H.P.		2231.25		2454.38	172.15	189.37
14 L.H., Daily Totals		$3177.70		$3795.11	$226.98	$271.08
Crew B-33D						
1 Equip. Oper. (medium)	$59.00	$472.00	$87.90	$703.20	$54.83	$81.71
.5 Laborer	44.40	177.60	66.25	265.00		
.25 Equip. Oper. (medium)	59.00	118.00	87.90	175.80		
1 S.P. Scraper, 14 C.Y.		2424.00		2666.40		
.25 Dozer, 300 H.P.		446.25		490.88	205.02	225.52
14 L.H., Daily Totals		$3637.85		$4301.27	$259.85	$307.23
Crew B-33E						
1 Equip. Oper. (medium)	$59.00	$472.00	$87.90	$703.20	$54.83	$81.71
.5 Laborer	44.40	177.60	66.25	265.00		
.25 Equip. Oper. (medium)	59.00	118.00	87.90	175.80		
1 S.P. Scraper, 21 C.Y.		2656.00		2921.60		
.25 Dozer, 300 H.P.		446.25		490.88	221.59	243.75
14 L.H., Daily Totals		$3869.85		$4556.48	$276.42	$325.46

Crews - Standard

Crew B-33F

Crew No.	Bare Costs Hr.	Bare Costs Daily	Incl. Subs O&P Hr.	Incl. Subs O&P Daily	Cost Per Labor-Hour Bare Costs	Cost Per Labor-Hour Incl. O&P
1 Equip. Oper. (medium)	$59.00	$472.00	$87.90	$703.20	$54.83	$81.71
.5 Laborer	44.40	177.60	66.25	265.00		
.25 Equip. Oper. (medium)	59.00	118.00	87.90	175.80		
1 Elev. Scraper, 11 C.Y.		1059.00		1164.90		
.25 Dozer, 300 H.P.		446.25		490.88	107.52	118.27
14 L.H., Daily Totals		$2272.85		$2799.78	$162.35	$199.98

Crew B-33G

Crew No.	Hr.	Daily	Hr.	Daily	Bare Costs	Incl. O&P
1 Equip. Oper. (medium)	$59.00	$472.00	$87.90	$703.20	$54.83	$81.71
.5 Laborer	44.40	177.60	66.25	265.00		
.25 Equip. Oper. (medium)	59.00	118.00	87.90	175.80		
1 Elev. Scraper, 22 C.Y.		1895.00		2084.50		
.25 Dozer, 300 H.P.		446.25		490.88	167.23	183.96
14 L.H., Daily Totals		$3108.85		$3719.38	$222.06	$265.67

Crew B-33H

Crew No.	Hr.	Daily	Hr.	Daily	Bare Costs	Incl. O&P
.5 Laborer	$44.40	$177.60	$66.25	$265.00	$54.83	$81.71
1 Equipment Operator (med.)	59.00	472.00	87.90	703.20		
.25 Equipment Operator (med.)	59.00	118.00	87.90	175.80		
1 S.P. Scraper, 44 C.Y.		4695.00		5164.50		
.25 Dozer, 410 H.P.		701.75		771.92	385.48	424.03
14 L.H., Daily Totals		$6164.35		$7080.43	$440.31	$505.74

Crew B-33J

Crew No.	Hr.	Daily	Hr.	Daily	Bare Costs	Incl. O&P
1 Equipment Operator (med.)	$59.00	$472.00	$87.90	$703.20	$59.00	$87.90
1 S.P. Scraper, 14 C.Y.		2424.00		2666.40	303.00	333.30
8 L.H., Daily Totals		$2896.00		$3369.60	$362.00	$421.20

Crew B-33K

Crew No.	Hr.	Daily	Hr.	Daily	Bare Costs	Incl. O&P
1 Equipment Operator (med.)	$59.00	$472.00	$87.90	$703.20	$54.83	$81.71
.25 Equipment Operator (med.)	59.00	118.00	87.90	175.80		
.5 Laborer	44.40	177.60	66.25	265.00		
1 S.P. Scraper, 31 C.Y.		3707.00		4077.70		
.25 Dozer, 410 H.P.		701.75		771.92	314.91	346.40
14 L.H., Daily Totals		$5176.35		$5993.63	$369.74	$428.12

Crew B-34A

Crew No.	Hr.	Daily	Hr.	Daily	Bare Costs	Incl. O&P
1 Truck Driver (heavy)	$51.30	$410.40	$76.70	$613.60	$51.30	$76.70
1 Dump Truck, 8 C.Y., 220 H.P.		407.60		448.36	50.95	56.05
8 L.H., Daily Totals		$818.00		$1061.96	$102.25	$132.75

Crew B-34B

Crew No.	Hr.	Daily	Hr.	Daily	Bare Costs	Incl. O&P
1 Truck Driver (heavy)	$51.30	$410.40	$76.70	$613.60	$51.30	$76.70
1 Dump Truck, 12 C.Y., 400 H.P.		579.35		637.28	72.42	79.66
8 L.H., Daily Totals		$989.75		$1250.89	$123.72	$156.36

Crew B-34C

Crew No.	Hr.	Daily	Hr.	Daily	Bare Costs	Incl. O&P
1 Truck Driver (heavy)	$51.30	$410.40	$76.70	$613.60	$51.30	$76.70
1 Truck Tractor, 6x4, 380 H.P.		499.15		549.07		
1 Dump Trailer, 16.5 C.Y.		138.35		152.19	79.69	87.66
8 L.H., Daily Totals		$1047.90		$1314.85	$130.99	$164.36

Crew B-34D

Crew No.	Hr.	Daily	Hr.	Daily	Bare Costs	Incl. O&P
1 Truck Driver (heavy)	$51.30	$410.40	$76.70	$613.60	$51.30	$76.70
1 Truck Tractor, 6x4, 380 H.P.		499.15		549.07		
1 Dump Trailer, 20 C.Y.		153.55		168.91	81.59	89.75
8 L.H., Daily Totals		$1063.10		$1331.57	$132.89	$166.45

Crew B-34E

Crew No.	Hr.	Daily	Hr.	Daily	Bare Costs	Incl. O&P
1 Truck Driver (heavy)	$51.30	$410.40	$76.70	$613.60	$51.30	$76.70
1 Dump Truck, Off Hwy., 25 Ton		1427.00		1569.70	178.38	196.21
8 L.H., Daily Totals		$1837.40		$2183.30	$229.68	$272.91

Crew B-34F

Crew No.	Hr.	Daily	Hr.	Daily	Bare Costs	Incl. O&P
1 Truck Driver (heavy)	$51.30	$410.40	$76.70	$613.60	$51.30	$76.70
1 Dump Truck, Off Hwy., 35 Ton		945.65		1040.21	118.21	130.03
8 L.H., Daily Totals		$1356.05		$1653.82	$169.51	$206.73

Crew B-34G

Crew No.	Hr.	Daily	Hr.	Daily	Bare Costs	Incl. O&P
1 Truck Driver (heavy)	$51.30	$410.40	$76.70	$613.60	$51.30	$76.70
1 Dump Truck, Off Hwy., 50 Ton		1979.00		2176.90	247.38	272.11
8 L.H., Daily Totals		$2389.40		$2790.50	$298.68	$348.81

Crew B-34H

Crew No.	Hr.	Daily	Hr.	Daily	Bare Costs	Incl. O&P
1 Truck Driver (heavy)	$51.30	$410.40	$76.70	$613.60	$51.30	$76.70
1 Dump Truck, Off Hwy., 65 Ton		1938.00		2131.80	242.25	266.48
8 L.H., Daily Totals		$2348.40		$2745.40	$293.55	$343.18

Crew B-34I

Crew No.	Hr.	Daily	Hr.	Daily	Bare Costs	Incl. O&P
1 Truck Driver (heavy)	$51.30	$410.40	$76.70	$613.60	$51.30	$76.70
1 Dump Truck, 18 C.Y., 450 H.P.		753.60		828.96	94.20	103.62
8 L.H., Daily Totals		$1164.00		$1442.56	$145.50	$180.32

Crew B-34J

Crew No.	Hr.	Daily	Hr.	Daily	Bare Costs	Incl. O&P
1 Truck Driver (heavy)	$51.30	$410.40	$76.70	$613.60	$51.30	$76.70
1 Dump Truck, Off Hwy., 100 Ton		2769.00		3045.90	346.13	380.74
8 L.H., Daily Totals		$3179.40		$3659.50	$397.43	$457.44

Crew B-34K

Crew No.	Hr.	Daily	Hr.	Daily	Bare Costs	Incl. O&P
1 Truck Driver (heavy)	$51.30	$410.40	$76.70	$613.60	$51.30	$76.70
1 Truck Tractor, 6x4, 450 H.P.		608.95		669.85		
1 Lowbed Trailer, 75 Ton		258.10		283.91	108.38	119.22
8 L.H., Daily Totals		$1277.45		$1567.36	$159.68	$195.92

Crew B-34L

Crew No.	Hr.	Daily	Hr.	Daily	Bare Costs	Incl. O&P
1 Equip. Oper. (light)	$55.50	$444.00	$82.70	$661.60	$55.50	$82.70
1 Flatbed Truck, Gas, 1.5 Ton		198.50		218.35	24.81	27.29
8 L.H., Daily Totals		$642.50		$879.95	$80.31	$109.99

Crew B-34M

Crew No.	Hr.	Daily	Hr.	Daily	Bare Costs	Incl. O&P
1 Equip. Oper. (light)	$55.50	$444.00	$82.70	$661.60	$55.50	$82.70
1 Flatbed Truck, Gas, 3 Ton		850.05		935.05	106.26	116.88
8 L.H., Daily Totals		$1294.05		$1596.66	$161.76	$199.58

Crew B-34N

Crew No.	Hr.	Daily	Hr.	Daily	Bare Costs	Incl. O&P
1 Truck Driver (heavy)	$51.30	$410.40	$76.70	$613.60	$55.15	$82.30
1 Equip. Oper. (medium)	59.00	472.00	87.90	703.20		
1 Truck Tractor, 6x4, 380 H.P.		499.15		549.07		
1 Flatbed Trailer, 40 Ton		188.45		207.29	42.98	47.27
16 L.H., Daily Totals		$1570.00		$2073.16	$98.13	$129.57

Crew B-34P

Crew No.	Hr.	Daily	Hr.	Daily	Bare Costs	Incl. O&P
1 Pipe Fitter	$68.35	$546.80	$102.00	$816.00	$58.72	$87.63
1 Truck Driver (light)	48.80	390.40	73.00	584.00		
1 Equip. Oper. (medium)	59.00	472.00	87.90	703.20		
1 Flatbed Truck, Gas, 3 Ton		850.05		935.05		
1 Backhoe Loader, 48 H.P.		216.20		237.82	44.43	48.87
24 L.H., Daily Totals		$2475.45		$3276.07	$103.14	$136.50

Crew No.	Bare Costs		Incl. Subs O&P		Cost Per Labor-Hour	
Crew B-34Q	Hr.	Daily	Hr.	Daily	Bare Costs	Incl. O&P
1 Pipe Fitter	$68.35	$546.80	$102.00	$816.00	$59.53	$88.85
1 Truck Driver (light)	48.80	390.40	73.00	584.00		
1 Equip. Oper. (crane)	61.45	491.60	91.55	732.40		
1 Flatbed Trailer, 25 Ton		137.20		150.92		
1 Dump Truck, 8 C.Y., 220 H.P.		407.60		448.36		
1 Hyd. Crane, 25 Ton		586.70		645.37	47.15	51.86
24 L.H., Daily Totals		$2560.30		$3377.05	$106.68	$140.71
Crew B-34R	Hr.	Daily	Hr.	Daily	Bare Costs	Incl. O&P
1 Pipe Fitter	$68.35	$546.80	$102.00	$816.00	$59.53	$88.85
1 Truck Driver (light)	48.80	390.40	73.00	584.00		
1 Equip. Oper. (crane)	61.45	491.60	91.55	732.40		
1 Flatbed Trailer, 25 Ton		137.20		150.92		
1 Dump Truck, 8 C.Y., 220 H.P.		407.60		448.36		
1 Hyd. Crane, 25 Ton		586.70		645.37		
1 Hyd. Excavator, 1 C.Y.		832.65		915.91	81.84	90.02
24 L.H., Daily Totals		$3392.95		$4292.97	$141.37	$178.87
Crew B-34S	Hr.	Daily	Hr.	Daily	Bare Costs	Incl. O&P
2 Pipe Fitters	$68.35	$1093.60	$102.00	$1632.00	$62.36	$93.06
1 Truck Driver (heavy)	51.30	410.40	76.70	613.60		
1 Equip. Oper. (crane)	61.45	491.60	91.55	732.40		
1 Flatbed Trailer, 40 Ton		188.45		207.29		
1 Truck Tractor, 6x4, 380 H.P.		499.15		549.07		
1 Hyd. Crane, 80 Ton		1458.00		1603.80		
1 Hyd. Excavator, 2 C.Y.		942.70		1036.97	96.51	106.16
32 L.H., Daily Totals		$5083.90		$6375.13	$158.87	$199.22
Crew B-34T	Hr.	Daily	Hr.	Daily	Bare Costs	Incl. O&P
2 Pipe Fitters	$68.35	$1093.60	$102.00	$1632.00	$62.36	$93.06
1 Truck Driver (heavy)	51.30	410.40	76.70	613.60		
1 Equip. Oper. (crane)	61.45	491.60	91.55	732.40		
1 Flatbed Trailer, 40 Ton		188.45		207.29		
1 Truck Tractor, 6x4, 380 H.P.		499.15		549.07		
1 Hyd. Crane, 80 Ton		1458.00		1603.80	67.05	73.75
32 L.H., Daily Totals		$4141.20		$5338.16	$129.41	$166.82
Crew B-34U	Hr.	Daily	Hr.	Daily	Bare Costs	Incl. O&P
1 Truck Driver (heavy)	$51.30	$410.40	$76.70	$613.60	$53.40	$79.70
1 Equip. Oper. (light)	55.50	444.00	82.70	661.60		
1 Truck Tractor, 220 H.P.		310.80		341.88		
1 Flatbed Trailer, 25 Ton		137.20		150.92	28.00	30.80
16 L.H., Daily Totals		$1302.40		$1768.00	$81.40	$110.50
Crew B-34V	Hr.	Daily	Hr.	Daily	Bare Costs	Incl. O&P
1 Truck Driver (heavy)	$51.30	$410.40	$76.70	$613.60	$56.08	$83.65
1 Equip. Oper. (crane)	61.45	491.60	91.55	732.40		
1 Equip. Oper. (light)	55.50	444.00	82.70	661.60		
1 Truck Tractor, 6x4, 450 H.P.		608.95		669.85		
1 Equipment Trailer, 50 Ton		207.25		227.97		
1 Pickup Truck, 4x4, 3/4 Ton		176.75		194.43	41.37	45.51
24 L.H., Daily Totals		$2338.95		$3099.84	$97.46	$129.16

Crew No.	Bare Costs		Incl. Subs O&P		Cost Per Labor-Hour	
Crew B-34W	Hr.	Daily	Hr.	Daily	Bare Costs	Incl. O&P
5 Truck Drivers (heavy)	$51.30	$2052.00	$76.70	$3068.00	$53.92	$80.50
2 Equip. Opers. (crane)	61.45	983.20	91.55	1464.80		
1 Equip. Oper. (mechanic)	61.50	492.00	91.65	733.20		
1 Laborer	44.40	355.20	66.25	530.00		
4 Truck Tractors, 6x4, 380 H.P.		1996.60		2196.26		
2 Equipment Trailers, 50 Ton		414.50		455.95		
2 Flatbed Trailers, 40 Ton		376.90		414.59		
1 Pickup Truck, 4x4, 3/4 Ton		176.75		194.43		
1 S.P. Crane, 4x4, 20 Ton		574.35		631.78	49.15	54.07
72 L.H., Daily Totals		$7421.50		$9689.01	$103.08	$134.57
Crew B-35	Hr.	Daily	Hr.	Daily	Bare Costs	Incl. O&P
1 Labor Foreman (outside)	$46.40	$371.20	$69.25	$554.00	$56.75	$84.76
1 Skilled Worker	57.10	456.80	85.90	687.20		
2 Welders	67.70	1083.20	101.05	1616.80		
1 Laborer	44.40	355.20	66.25	530.00		
1 Equip. Oper. (crane)	61.45	491.60	91.55	732.40		
1 Equip. Oper. (oiler)	52.50	420.00	78.25	626.00		
2 Welder, Electric, 300 amp		215.10		236.61		
1 Hyd. Excavator, .75 C.Y.		701.80		771.98	16.37	18.01
56 L.H., Daily Totals		$4094.90		$5754.99	$73.12	$102.77
Crew B-35A	Hr.	Daily	Hr.	Daily	Bare Costs	Incl. O&P
1 Labor Foreman (outside)	$46.40	$371.20	$69.25	$554.00	$53.42	$79.79
2 Laborers	44.40	710.40	66.25	1060.00		
1 Skilled Worker	57.10	456.80	85.90	687.20		
1 Welder (plumber)	67.70	541.60	101.05	808.40		
1 Equip. Oper. (crane)	61.45	491.60	91.55	732.40		
1 Equip. Oper. (oiler)	52.50	420.00	78.25	626.00		
1 Welder, Gas Engine, 300 amp		148.75		163.63		
1 Crawler Crane, 75 Ton		1967.00		2163.70	37.78	41.56
56 L.H., Daily Totals		$5107.35		$6795.32	$91.20	$121.35
Crew B-36	Hr.	Daily	Hr.	Daily	Bare Costs	Incl. O&P
1 Labor Foreman (outside)	$46.40	$371.20	$69.25	$554.00	$50.64	$75.51
2 Laborers	44.40	710.40	66.25	1060.00		
2 Equip. Oper. (medium)	59.00	944.00	87.90	1406.40		
1 Dozer, 200 H.P.		1520.00		1672.00		
1 Aggregate Spreader		59.95		65.94		
1 Tandem Roller, 10 Ton		246.80		271.48	45.67	50.24
40 L.H., Daily Totals		$3852.35		$5029.82	$96.31	$125.75
Crew B-36A	Hr.	Daily	Hr.	Daily	Bare Costs	Incl. O&P
1 Labor Foreman (outside)	$46.40	$371.20	$69.25	$554.00	$53.03	$79.05
2 Laborers	44.40	710.40	66.25	1060.00		
4 Equip. Oper. (medium)	59.00	1888.00	87.90	2812.80		
1 Dozer, 200 H.P.		1520.00		1672.00		
1 Aggregate Spreader		59.95		65.94		
1 Tandem Roller, 10 Ton		246.80		271.48		
1 Roller, Pneum. Whl., 12 Ton		349.90		384.89	38.87	42.76
56 L.H., Daily Totals		$5146.25		$6821.11	$91.90	$121.81

For customer support on your Plumbing Costs with RSMeans Data, call 800.448.8182.

611

Crew No.	Bare Costs		Incl. Subs O&P		Cost Per Labor-Hour	

Crew B-36B

Crew B-36B	Hr.	Daily	Hr.	Daily	Bare Costs	Incl. O&P
1 Labor Foreman (outside)	$46.40	$371.20	$69.25	$554.00	$52.81	$78.76
2 Laborers	44.40	710.40	66.25	1060.00		
4 Equip. Oper. (medium)	59.00	1888.00	87.90	2812.80		
1 Truck Driver (heavy)	51.30	410.40	76.70	613.60		
1 Grader, 30,000 Lbs.		1073.00		1180.30		
1 F.E. Loader, Crl, 1.5 C.Y.		668.35		735.18		
1 Dozer, 300 H.P.		1785.00		1963.50		
1 Roller, Vibratory, 25 Ton		672.35		739.59		
1 Truck Tractor, 6x4, 450 H.P.		608.95		669.85		
1 Water Tank Trailer, 5000 Gal.		154.10		169.51	77.53	85.28
64 L.H., Daily Totals		$8341.75		$10498.33	$130.34	$164.04

Crew B-36C	Hr.	Daily	Hr.	Daily	Bare Costs	Incl. O&P
1 Labor Foreman (outside)	$46.40	$371.20	$69.25	$554.00	$54.94	$81.93
3 Equip. Oper. (medium)	59.00	1416.00	87.90	2109.60		
1 Truck Driver (heavy)	51.30	410.40	76.70	613.60		
1 Grader, 30,000 Lbs.		1073.00		1180.30		
1 Dozer, 300 H.P.		1785.00		1963.50		
1 Roller, Vibratory, 25 Ton		672.35		739.59		
1 Truck Tractor, 6x4, 450 H.P.		608.95		669.85		
1 Water Tank Trailer, 5000 Gal.		154.10		169.51	107.33	118.07
40 L.H., Daily Totals		$6491.00		$7999.94	$162.28	$200.00

Crew B-36D	Hr.	Daily	Hr.	Daily	Bare Costs	Incl. O&P
1 Labor Foreman (outside)	$46.40	$371.20	$69.25	$554.00	$55.85	$83.24
3 Equip. Oper. (medium)	59.00	1416.00	87.90	2109.60		
1 Grader, 30,000 Lbs.		1073.00		1180.30		
1 Dozer, 300 H.P.		1785.00		1963.50		
1 Roller, Vibratory, 25 Ton		672.35		739.59	110.32	121.36
32 L.H., Daily Totals		$5317.55		$6546.98	$166.17	$204.59

Crew B-37	Hr.	Daily	Hr.	Daily	Bare Costs	Incl. O&P
1 Labor Foreman (outside)	$46.40	$371.20	$69.25	$554.00	$46.58	$69.49
4 Laborers	44.40	1420.80	66.25	2120.00		
1 Equip. Oper. (light)	55.50	444.00	82.70	661.60		
1 Tandem Roller, 5 Ton		258.75		284.63	5.39	5.93
48 L.H., Daily Totals		$2494.75		$3620.22	$51.97	$75.42

Crew B-37A	Hr.	Daily	Hr.	Daily	Bare Costs	Incl. O&P
2 Laborers	$44.40	$710.40	$66.25	$1060.00	$45.87	$68.50
1 Truck Driver (light)	48.80	390.40	73.00	584.00		
1 Flatbed Truck, Gas, 1.5 Ton		198.50		218.35		
1 Tar Kettle, T.M.		156.70		172.37	14.80	16.28
24 L.H., Daily Totals		$1456.00		$2034.72	$60.67	$84.78

Crew B-37B	Hr.	Daily	Hr.	Daily	Bare Costs	Incl. O&P
3 Laborers	$44.40	$1065.60	$66.25	$1590.00	$45.50	$67.94
1 Truck Driver (light)	48.80	390.40	73.00	584.00		
1 Flatbed Truck, Gas, 1.5 Ton		198.50		218.35		
1 Tar Kettle, T.M.		156.70		172.37	11.10	12.21
32 L.H., Daily Totals		$1811.20		$2564.72	$56.60	$80.15

Crew B-37C	Hr.	Daily	Hr.	Daily	Bare Costs	Incl. O&P
2 Laborers	$44.40	$710.40	$66.25	$1060.00	$46.60	$69.63
2 Truck Drivers (light)	48.80	780.80	73.00	1168.00		
2 Flatbed Trucks, Gas, 1.5 Ton		397.00		436.70		
1 Tar Kettle, T.M.		156.70		172.37	17.30	19.03
32 L.H., Daily Totals		$2044.90		$2837.07	$63.90	$88.66

Crew B-37D	Hr.	Daily	Hr.	Daily	Bare Costs	Incl. O&P
1 Laborer	$44.40	$355.20	$66.25	$530.00	$46.60	$69.63
1 Truck Driver (light)	48.80	390.40	73.00	584.00		
1 Pickup Truck, 3/4 Ton		112.20		123.42	7.01	7.71
16 L.H., Daily Totals		$857.80		$1237.42	$53.61	$77.34

Crew B-37E	Hr.	Daily	Hr.	Daily	Bare Costs	Incl. O&P
3 Laborers	$44.40	$1065.60	$66.25	$1590.00	$49.33	$73.62
1 Equip. Oper. (light)	55.50	444.00	82.70	661.60		
1 Equip. Oper. (medium)	59.00	472.00	87.90	703.20		
2 Truck Drivers (light)	48.80	780.80	73.00	1168.00		
4 Barrels w/ Flasher		16.60		18.26		
1 Concrete Saw		112.85		124.14		
1 Rotary Hammer Drill		52.25		57.48		
1 Hammer Drill Bit		25.25		27.77		
1 Loader, Skid Steer, 30 H.P.		179.50		197.45		
1 Conc. Hammer Attach.		118.50		130.35		
1 Vibrating Plate, Gas, 18"		31.90		35.09		
2 Flatbed Trucks, Gas, 1.5 Ton		397.00		436.70	16.68	18.34
56 L.H., Daily Totals		$3696.25		$5150.03	$66.00	$91.96

Crew B-37F	Hr.	Daily	Hr.	Daily	Bare Costs	Incl. O&P
3 Laborers	$44.40	$1065.60	$66.25	$1590.00	$45.50	$67.94
1 Truck Driver (light)	48.80	390.40	73.00	584.00		
4 Barrels w/ Flasher		16.60		18.26		
1 Concrete Mixer, 10 C.F.		147.15		161.87		
1 Air Compressor, 60 cfm		153.85		169.24		
1 -50' Air Hose, 3/4"		7.15		7.87		
1 Spade (Chipper)		8.55		9.40		
1 Flatbed Truck, Gas, 1.5 Ton		198.50		218.35	16.62	18.28
32 L.H., Daily Totals		$1987.80		$2758.98	$62.12	$86.22

Crew B-37G	Hr.	Daily	Hr.	Daily	Bare Costs	Incl. O&P
1 Labor Foreman (outside)	$46.40	$371.20	$69.25	$554.00	$46.58	$69.49
4 Laborers	44.40	1420.80	66.25	2120.00		
1 Equip. Oper. (light)	55.50	444.00	82.70	661.60		
1 Berm Machine		253.55		278.90		
1 Tandem Roller, 5 Ton		258.75		284.63	10.67	11.74
48 L.H., Daily Totals		$2748.30		$3899.13	$57.26	$81.23

Crew B-37H	Hr.	Daily	Hr.	Daily	Bare Costs	Incl. O&P
1 Labor Foreman (outside)	$46.40	$371.20	$69.25	$554.00	$46.58	$69.49
4 Laborers	44.40	1420.80	66.25	2120.00		
1 Equip. Oper. (light)	55.50	444.00	82.70	661.60		
1 Tandem Roller, 5 Ton		258.75		284.63		
1 Flatbed Truck, Gas, 1.5 Ton		198.50		218.35		
1 Tar Kettle, T.M.		156.70		172.37	12.79	14.07
48 L.H., Daily Totals		$2849.95		$4010.95	$59.37	$83.56

Crew No.	Bare Costs		Incl. Subs O&P		Cost Per Labor-Hour	

Crew B-37I

Crew B-37I	Hr.	Daily	Hr.	Daily	Bare Costs	Incl. O&P
3 Laborers	$44.40	$1065.60	$66.25	$1590.00	$49.33	$73.62
1 Equip. Oper. (light)	55.50	444.00	82.70	661.60		
1 Equip. Oper. (medium)	59.00	472.00	87.90	703.20		
2 Truck Drivers (light)	48.80	780.80	73.00	1168.00		
4 Barrels w/ Flasher		16.60		18.26		
1 Concrete Saw		112.85		124.14		
1 Rotary Hammer Drill		52.25		57.48		
1 Hammer Drill Bit		25.25		27.77		
1 Air Compressor, 60 cfm		153.85		169.24		
1 -50' Air Hose, 3/4"		7.15		7.87		
1 Spade (Chipper)		8.55		9.40		
1 Loader, Skid Steer, 30 H.P.		179.50		197.45		
1 Conc. Hammer Attach.		118.50		130.35		
1 Concrete Mixer, 10 C.F.		147.15		161.87		
1 Vibrating Plate, Gas, 18"		31.90		35.09		
2 Flatbed Trucks, Gas, 1.5 Ton		397.00		436.70	22.33	24.56
56 L.H., Daily Totals		$4012.95		$5498.40	$71.66	$98.19

Crew B-37J

Crew B-37J	Hr.	Daily	Hr.	Daily	Bare Costs	Incl. O&P
1 Labor Foreman (outside)	$46.40	$371.20	$69.25	$554.00	$46.58	$69.49
4 Laborers	44.40	1420.80	66.25	2120.00		
1 Equip. Oper. (light)	55.50	444.00	82.70	661.60		
1 Air Compressor, 60 cfm		153.85		169.24		
1 -50' Air Hose, 3/4"		7.15		7.87		
2 Concrete Mixers, 10 C.F.		294.30		323.73		
2 Flatbed Trucks, Gas, 1.5 Ton		397.00		436.70		
1 Shot Blaster, 20"		208.70		229.57	22.10	24.31
48 L.H., Daily Totals		$3297.00		$4502.70	$68.69	$93.81

Crew B-37K

Crew B-37K	Hr.	Daily	Hr.	Daily	Bare Costs	Incl. O&P
1 Labor Foreman (outside)	$46.40	$371.20	$69.25	$554.00	$46.58	$69.49
4 Laborers	44.40	1420.80	66.25	2120.00		
1 Equip. Oper. (light)	55.50	444.00	82.70	661.60		
1 Air Compressor, 60 cfm		153.85		169.24		
1 -50' Air Hose, 3/4"		7.15		7.87		
2 Flatbed Trucks, Gas, 1.5 Ton		397.00		436.70		
1 Shot Blaster, 20"		208.70		229.57	15.97	17.57
48 L.H., Daily Totals		$3002.70		$4178.97	$62.56	$87.06

Crew B-38

Crew B-38	Hr.	Daily	Hr.	Daily	Bare Costs	Incl. O&P
1 Labor Foreman (outside)	$46.40	$371.20	$69.25	$554.00	$49.94	$74.47
2 Laborers	44.40	710.40	66.25	1060.00		
1 Equip. Oper. (light)	55.50	444.00	82.70	661.60		
1 Equip. Oper. (medium)	59.00	472.00	87.90	703.20		
1 Backhoe Loader, 48 H.P.		216.20		237.82		
1 Hyd. Hammer (1200 lb.)		177.25		194.97		
1 F.E. Loader, W.M., 4 C.Y.		759.00		834.90		
1 Pvmt. Rem. Bucket		63.80		70.18	30.41	33.45
40 L.H., Daily Totals		$3213.85		$4316.68	$80.35	$107.92

Crew B-39

Crew B-39	Hr.	Daily	Hr.	Daily	Bare Costs	Incl. O&P
1 Labor Foreman (outside)	$46.40	$371.20	$69.25	$554.00	$46.58	$69.49
4 Laborers	44.40	1420.80	66.25	2120.00		
1 Equip. Oper. (light)	55.50	444.00	82.70	661.60		
1 Air Compressor, 250 cfm		202.85		223.13		
2 Breakers, Pavement, 60 lb.		107.20		117.92		
2 -50' Air Hoses, 1.5"		45.60		50.16	7.41	8.15
48 L.H., Daily Totals		$2591.65		$3726.82	$53.99	$77.64

Crew B-40

Crew B-40	Hr.	Daily	Hr.	Daily	Bare Costs	Incl. O&P
1 Pile Driver Foreman (outside)	$57.90	$463.20	$89.50	$716.00	$57.11	$87.06
4 Pile Drivers	55.90	1788.80	86.40	2764.80		
2 Equip. Oper. (crane)	61.45	983.20	91.55	1464.80		
1 Equip. Oper. (oiler)	52.50	420.00	78.25	626.00		
1 Crawler Crane, 40 Ton		1231.00		1354.10		
1 Vibratory Hammer & Gen.		2298.00		2527.80	55.14	60.65
64 L.H., Daily Totals		$7184.20		$9453.50	$112.25	$147.71

Crew B-40B

Crew B-40B	Hr.	Daily	Hr.	Daily	Bare Costs	Incl. O&P
1 Labor Foreman (outside)	$46.40	$371.20	$69.25	$554.00	$48.92	$72.97
3 Laborers	44.40	1065.60	66.25	1590.00		
1 Equip. Oper. (crane)	61.45	491.60	91.55	732.40		
1 Equip. Oper. (oiler)	52.50	420.00	78.25	626.00		
1 Lattice Boom Crane, 40 Ton		2127.00		2339.70	44.31	48.74
48 L.H., Daily Totals		$4475.40		$5842.10	$93.24	$121.71

Crew B-41

Crew B-41	Hr.	Daily	Hr.	Daily	Bare Costs	Incl. O&P
1 Labor Foreman (outside)	$46.40	$371.20	$69.25	$554.00	$45.91	$68.49
4 Laborers	44.40	1420.80	66.25	2120.00		
.25 Equip. Oper. (crane)	61.45	122.90	91.55	183.10		
.25 Equip. Oper. (oiler)	52.50	105.00	78.25	156.50		
.25 Crawler Crane, 40 Ton		307.75		338.52	6.99	7.69
44 L.H., Daily Totals		$2327.65		$3352.13	$52.90	$76.18

Crew B-42

Crew B-42	Hr.	Daily	Hr.	Daily	Bare Costs	Incl. O&P
1 Labor Foreman (outside)	$46.40	$371.20	$69.25	$554.00	$49.78	$74.67
4 Laborers	44.40	1420.80	66.25	2120.00		
1 Equip. Oper. (crane)	61.45	491.60	91.55	732.40		
1 Equip. Oper. (oiler)	52.50	420.00	78.25	626.00		
1 Welder	60.30	482.40	93.30	746.40		
1 Hyd. Crane, 25 Ton		586.70		645.37		
1 Welder, Gas Engine, 300 amp		148.75		163.63		
1 Horz. Boring Csg. Mch.		329.75		362.73	16.64	18.31
64 L.H., Daily Totals		$4251.20		$5950.52	$66.43	$92.98

Crew B-43

Crew B-43	Hr.	Daily	Hr.	Daily	Bare Costs	Incl. O&P
1 Labor Foreman (outside)	$46.40	$371.20	$69.25	$554.00	$48.92	$72.97
3 Laborers	44.40	1065.60	66.25	1590.00		
1 Equip. Oper. (crane)	61.45	491.60	91.55	732.40		
1 Equip. Oper. (oiler)	52.50	420.00	78.25	626.00		
1 Drill Rig, Truck-Mounted		768.40		845.24	16.01	17.61
48 L.H., Daily Totals		$3116.80		$4347.64	$64.93	$90.58

Crew B-44

Crew B-44	Hr.	Daily	Hr.	Daily	Bare Costs	Incl. O&P
1 Pile Driver Foreman (outside)	$57.90	$463.20	$89.50	$716.00	$56.10	$85.56
4 Pile Drivers	55.90	1788.80	86.40	2764.80		
2 Equip. Oper. (crane)	61.45	983.20	91.55	1464.80		
1 Laborer	44.40	355.20	66.25	530.00		
1 Crawler Crane, 40 Ton		1231.00		1354.10		
1 Lead, 60' High		211.80		232.98		
1 Hammer, Diesel, 15K ft.-lbs.		624.45		686.89	32.30	35.53
64 L.H., Daily Totals		$5657.65		$7749.57	$88.40	$121.09

Crew B-45

Crew B-45	Hr.	Daily	Hr.	Daily	Bare Costs	Incl. O&P
1 Equip. Oper. (medium)	$59.00	$472.00	$87.90	$703.20	$55.15	$82.30
1 Truck Driver (heavy)	51.30	410.40	76.70	613.60		
1 Dist. Tanker, 3000 Gallon		334.10		367.51		
1 Truck Tractor, 6x4, 380 H.P.		499.15		549.07	52.08	57.29
16 L.H., Daily Totals		$1715.65		$2233.38	$107.23	$139.59

For customer support on your Plumbing Costs with RSMeans Data, call 800.448.8182.

613

Crew No.	Bare Costs		Incl. Subs O&P		Cost Per Labor-Hour	
Crew B-46	Hr.	Daily	Hr.	Daily	Bare Costs	Incl. O&P
1 Pile Driver Foreman (outside)	$57.90	$463.20	$89.50	$716.00	$50.48	$76.84
2 Pile Drivers	55.90	894.40	86.40	1382.40		
3 Laborers	44.40	1065.60	66.25	1590.00		
1 Chain Saw, Gas, 36" Long		41.65		45.81	.87	.95
48 L.H., Daily Totals		$2464.85		$3734.22	$51.35	$77.80
Crew B-47	Hr.	Daily	Hr.	Daily	Bare Costs	Incl. O&P
1 Blast Foreman (outside)	$46.40	$371.20	$69.25	$554.00	$48.77	$72.73
1 Driller	44.40	355.20	66.25	530.00		
1 Equip. Oper. (light)	55.50	444.00	82.70	661.60		
1 Air Track Drill, 4"		1127.00		1239.70		
1 Air Compressor, 600 cfm		426.55		469.20		
2 -50' Air Hoses, 3"		76.70		84.37	67.93	74.72
24 L.H., Daily Totals		$2800.65		$3538.88	$116.69	$147.45
Crew B-47A	Hr.	Daily	Hr.	Daily	Bare Costs	Incl. O&P
1 Drilling Foreman (outside)	$46.40	$371.20	$69.25	$554.00	$53.45	$79.68
1 Equip. Oper. (heavy)	61.45	491.60	91.55	732.40		
1 Equip. Oper. (oiler)	52.50	420.00	78.25	626.00		
1 Air Track Drill, 5"		1127.00		1239.70	46.96	51.65
24 L.H., Daily Totals		$2409.80		$3152.10	$100.41	$131.34
Crew B-47C	Hr.	Daily	Hr.	Daily	Bare Costs	Incl. O&P
1 Laborer	$44.40	$355.20	$66.25	$530.00	$49.95	$74.47
1 Equip. Oper. (light)	55.50	444.00	82.70	661.60		
1 Air Compressor, 750 cfm		596.30		655.93		
2 -50' Air Hoses, 3"		76.70		84.37		
1 Air Track Drill, 4"		1127.00		1239.70	112.50	123.75
16 L.H., Daily Totals		$2599.20		$3171.60	$162.45	$198.22
Crew B-47E	Hr.	Daily	Hr.	Daily	Bare Costs	Incl. O&P
1 Labor Foreman (outside)	$46.40	$371.20	$69.25	$554.00	$44.90	$67.00
3 Laborers	44.40	1065.60	66.25	1590.00		
1 Flatbed Truck, Gas, 3 Ton		850.05		935.05	26.56	29.22
32 L.H., Daily Totals		$2286.85		$3079.05	$71.46	$96.22
Crew B-47G	Hr.	Daily	Hr.	Daily	Bare Costs	Incl. O&P
1 Labor Foreman (outside)	$46.40	$371.20	$69.25	$554.00	$47.67	$71.11
2 Laborers	44.40	710.40	66.25	1060.00		
1 Equip. Oper. (light)	55.50	444.00	82.70	661.60		
1 Air Track Drill, 4"		1127.00		1239.70		
1 Air Compressor, 600 cfm		426.55		469.20		
2 -50' Air Hoses, 3"		76.70		84.37		
1 Gunite Pump Rig		321.75		353.93	61.00	67.10
32 L.H., Daily Totals		$3477.60		$4422.80	$108.68	$138.21
Crew B-47H	Hr.	Daily	Hr.	Daily	Bare Costs	Incl. O&P
1 Skilled Worker Foreman (out)	$59.10	$472.80	$88.90	$711.20	$57.60	$86.65
3 Skilled Workers	57.10	1370.40	85.90	2061.60		
1 Flatbed Truck, Gas, 3 Ton		850.05		935.05	26.56	29.22
32 L.H., Daily Totals		$2693.25		$3707.86	$84.16	$115.87

Crew No.	Bare Costs		Incl. Subs O&P		Cost Per Labor-Hour	
Crew B-48	Hr.	Daily	Hr.	Daily	Bare Costs	Incl. O&P
1 Labor Foreman (outside)	$46.40	$371.20	$69.25	$554.00	$49.86	$74.36
3 Laborers	44.40	1065.60	66.25	1590.00		
1 Equip. Oper. (crane)	61.45	491.60	91.55	732.40		
1 Equip. Oper. (oiler)	52.50	420.00	78.25	626.00		
1 Equip. Oper. (light)	55.50	444.00	82.70	661.60		
1 Centr. Water Pump, 6"		235.25		258.77		
1 -20' Suction Hose, 6"		25.50		28.05		
1 -50' Discharge Hose, 6"		18.10		19.91		
1 Drill Rig, Truck-Mounted		768.40		845.24	18.70	20.57
56 L.H., Daily Totals		$3839.65		$5315.98	$68.57	$94.93
Crew B-49	Hr.	Daily	Hr.	Daily	Bare Costs	Incl. O&P
1 Labor Foreman (outside)	$46.40	$371.20	$69.25	$554.00	$52.25	$78.46
3 Laborers	44.40	1065.60	66.25	1590.00		
2 Equip. Oper. (crane)	61.45	983.20	91.55	1464.80		
2 Equip. Oper. (oilers)	52.50	840.00	78.25	1252.00		
1 Equip. Oper. (light)	55.50	444.00	82.70	661.60		
2 Pile Drivers	55.90	894.40	86.40	1382.40		
1 Hyd. Crane, 25 Ton		586.70		645.37		
1 Centr. Water Pump, 6"		235.25		258.77		
1 -20' Suction Hose, 6"		25.50		28.05		
1 -50' Discharge Hose, 6"		18.10		19.91		
1 Drill Rig, Truck-Mounted		768.40		845.24	18.57	20.42
88 L.H., Daily Totals		$6232.35		$8702.15	$70.82	$98.89
Crew B-50	Hr.	Daily	Hr.	Daily	Bare Costs	Incl. O&P
2 Pile Driver Foremen (outside)	$57.90	$926.40	$89.50	$1432.00	$54.27	$82.68
6 Pile Drivers	55.90	2683.20	86.40	4147.20		
2 Equip. Oper. (crane)	61.45	983.20	91.55	1464.80		
1 Equip. Oper. (oiler)	52.50	420.00	78.25	626.00		
3 Laborers	44.40	1065.60	66.25	1590.00		
1 Crawler Crane, 40 Ton		1231.00		1354.10		
1 Lead, 60' High		211.80		232.98		
1 Hammer, Diesel, 15K ft.-lbs.		624.45		686.89		
1 Air Compressor, 600 cfm		426.55		469.20		
2 -50' Air Hoses, 3"		76.70		84.37		
1 Chain Saw, Gas, 36" Long		41.65		45.81	23.32	25.66
112 L.H., Daily Totals		$8690.55		$12133.37	$77.59	$108.33
Crew B-51	Hr.	Daily	Hr.	Daily	Bare Costs	Incl. O&P
1 Labor Foreman (outside)	$46.40	$371.20	$69.25	$554.00	$45.47	$67.88
4 Laborers	44.40	1420.80	66.25	2120.00		
1 Truck Driver (light)	48.80	390.40	73.00	584.00		
1 Flatbed Truck, Gas, 1.5 Ton		198.50		218.35	4.14	4.55
48 L.H., Daily Totals		$2380.90		$3476.35	$49.60	$72.42
Crew B-52	Hr.	Daily	Hr.	Daily	Bare Costs	Incl. O&P
1 Carpenter Foreman (outside)	$56.70	$453.60	$84.60	$676.80	$50.76	$75.55
1 Carpenter	54.70	437.60	81.65	653.20		
3 Laborers	44.40	1065.60	66.25	1590.00		
1 Cement Finisher	51.80	414.40	75.90	607.20		
.5 Rodman (reinf.)	58.90	235.60	88.05	352.20		
.5 Equip. Oper. (medium)	59.00	236.00	87.90	351.60		
.5 Crawler Loader, 3 C.Y.		573.00		630.30	10.23	11.26
56 L.H., Daily Totals		$3415.80		$4861.30	$61.00	$86.81
Crew B-53	Hr.	Daily	Hr.	Daily	Bare Costs	Incl. O&P
1 Equip. Oper. (light)	$55.50	$444.00	$82.70	$661.60	$55.50	$82.70
1 Trencher, Chain, 12 H.P.		158.60		174.46	19.82	21.81
8 L.H., Daily Totals		$602.60		$836.06	$75.33	$104.51

Crews - Standard

Crew No.	Bare Costs		Incl. Subs O&P		Cost Per Labor-Hour	
Crew B-54	Hr.	Daily	Hr.	Daily	Bare Costs	Incl. O&P
1 Equip. Oper. (light)	$55.50	$444.00	$82.70	$661.60	$55.50	$82.70
1 Trencher, Chain, 40 H.P.		450.70		495.77	56.34	61.97
8 L.H., Daily Totals		$894.70		$1157.37	$111.84	$144.67

Crew B-54A	Hr.	Daily	Hr.	Daily	Bare Costs	Incl. O&P
.17 Labor Foreman (outside)	$46.40	$63.10	$69.25	$94.18	$57.17	$85.19
1 Equipment Operator (med.)	59.00	472.00	87.90	703.20		
1 Wheel Trencher, 67 H.P.		1140.00		1254.00	121.79	133.97
9.36 L.H., Daily Totals		$1675.10		$2051.38	$178.96	$219.16

Crew B-54B	Hr.	Daily	Hr.	Daily	Bare Costs	Incl. O&P
.25 Labor Foreman (outside)	$46.40	$92.80	$69.25	$138.50	$56.48	$84.17
1 Equipment Operator (med.)	59.00	472.00	87.90	703.20		
1 Wheel Trencher, 150 H.P.		1238.00		1361.80	123.80	136.18
10 L.H., Daily Totals		$1802.80		$2203.50	$180.28	$220.35

Crew B-54C	Hr.	Daily	Hr.	Daily	Bare Costs	Incl. O&P
1 Laborer	$44.40	$355.20	$66.25	$530.00	$51.70	$77.08
1 Equipment Operator (med.)	59.00	472.00	87.90	703.20		
1 Wheel Trencher, 67 H.P.		1140.00		1254.00	71.25	78.38
16 L.H., Daily Totals		$1967.20		$2487.20	$122.95	$155.45

Crew B-54D	Hr.	Daily	Hr.	Daily	Bare Costs	Incl. O&P
1 Laborer	$44.40	$355.20	$66.25	$530.00	$51.70	$77.08
1 Equipment Operator (med.)	59.00	472.00	87.90	703.20		
1 Rock Trencher, 6" Width		434.20		477.62	27.14	29.85
16 L.H., Daily Totals		$1261.40		$1710.82	$78.84	$106.93

Crew B-54E	Hr.	Daily	Hr.	Daily	Bare Costs	Incl. O&P
1 Laborer	$44.40	$355.20	$66.25	$530.00	$51.70	$77.08
1 Equipment Operator (med.)	59.00	472.00	87.90	703.20		
1 Rock Trencher, 18" Width		1015.00		1116.50	63.44	69.78
16 L.H., Daily Totals		$1842.20		$2349.70	$115.14	$146.86

Crew B-55	Hr.	Daily	Hr.	Daily	Bare Costs	Incl. O&P
2 Laborers	$44.40	$710.40	$66.25	$1060.00	$45.87	$68.50
1 Truck Driver (light)	48.80	390.40	73.00	584.00		
1 Truck-Mounted Earth Auger		394.15		433.57		
1 Flatbed Truck, Gas, 3 Ton		850.05		935.05	51.84	57.03
24 L.H., Daily Totals		$2345.00		$3012.62	$97.71	$125.53

Crew B-56	Hr.	Daily	Hr.	Daily	Bare Costs	Incl. O&P
1 Laborer	$44.40	$355.20	$66.25	$530.00	$49.95	$74.47
1 Equip. Oper. (light)	55.50	444.00	82.70	661.60		
1 Air Track Drill, 4"		1127.00		1239.70		
1 Air Compressor, 600 cfm		426.55		469.20		
1 -50' Air Hose, 3"		38.35		42.19	99.49	109.44
16 L.H., Daily Totals		$2391.10		$2942.69	$149.44	$183.92

Crew B-57	Hr.	Daily	Hr.	Daily	Bare Costs	Incl. O&P
1 Labor Foreman (outside)	$46.40	$371.20	$69.25	$554.00	$50.77	$75.71
2 Laborers	44.40	710.40	66.25	1060.00		
1 Equip. Oper. (crane)	61.45	491.60	91.55	732.40		
1 Equip. Oper. (light)	55.50	444.00	82.70	661.60		
1 Equip. Oper. (oiler)	52.50	420.00	78.25	626.00		
1 Crawler Crane, 25 Ton		1152.00		1267.20		
1 Clamshell Bucket, 1 C.Y.		69.25		76.17		
1 Centr. Water Pump, 6"		235.25		258.77		
1 -20' Suction Hose, 6"		25.50		28.05		
20 -50' Discharge Hoses, 6"		362.00		398.20	38.42	42.26
48 L.H., Daily Totals		$4281.20		$5662.40	$89.19	$117.97

Crew B-58	Hr.	Daily	Hr.	Daily	Bare Costs	Incl. O&P
2 Laborers	$44.40	$710.40	$66.25	$1060.00	$48.10	$71.73
1 Equip. Oper. (light)	55.50	444.00	82.70	661.60		
1 Backhoe Loader, 48 H.P.		216.20		237.82		
1 Small Helicopter, w/ Pilot		2101.00		2311.10	96.55	106.21
24 L.H., Daily Totals		$3471.60		$4270.52	$144.65	$177.94

Crew B-59	Hr.	Daily	Hr.	Daily	Bare Costs	Incl. O&P
1 Truck Driver (heavy)	$51.30	$410.40	$76.70	$613.60	$51.30	$76.70
1 Truck Tractor, 220 H.P.		310.80		341.88		
1 Water Tank Trailer, 5000 Gal.		154.10		169.51	58.11	63.92
8 L.H., Daily Totals		$875.30		$1124.99	$109.41	$140.62

Crew B-59A	Hr.	Daily	Hr.	Daily	Bare Costs	Incl. O&P
2 Laborers	$44.40	$710.40	$66.25	$1060.00	$46.70	$69.73
1 Truck Driver (heavy)	51.30	410.40	76.70	613.60		
1 Water Tank Trailer, 5000 Gal.		154.10		169.51		
1 Truck Tractor, 220 H.P.		310.80		341.88	19.37	21.31
24 L.H., Daily Totals		$1585.70		$2184.99	$66.07	$91.04

Crew B-60	Hr.	Daily	Hr.	Daily	Bare Costs	Incl. O&P
1 Labor Foreman (outside)	$46.40	$371.20	$69.25	$554.00	$51.45	$76.71
2 Laborers	44.40	710.40	66.25	1060.00		
1 Equip. Oper. (crane)	61.45	491.60	91.55	732.40		
2 Equip. Oper. (light)	55.50	888.00	82.70	1323.20		
1 Equip. Oper. (oiler)	52.50	420.00	78.25	626.00		
1 Crawler Crane, 40 Ton		1231.00		1354.10		
1 Lead, 60' High		211.80		232.98		
1 Hammer, Diesel, 15K ft.-lbs.		624.45		686.89		
1 Backhoe Loader, 48 H.P.		216.20		237.82	40.78	44.85
56 L.H., Daily Totals		$5164.65		$6807.40	$92.23	$121.56

Crew B-61	Hr.	Daily	Hr.	Daily	Bare Costs	Incl. O&P
1 Labor Foreman (outside)	$46.40	$371.20	$69.25	$554.00	$47.02	$70.14
3 Laborers	44.40	1065.60	66.25	1590.00		
1 Equip. Oper. (light)	55.50	444.00	82.70	661.60		
1 Cement Mixer, 2 C.Y.		112.35		123.58		
1 Air Compressor, 160 cfm		212.30		233.53	8.12	8.93
40 L.H., Daily Totals		$2205.45		$3162.72	$55.14	$79.07

Crew B-62	Hr.	Daily	Hr.	Daily	Bare Costs	Incl. O&P
2 Laborers	$44.40	$710.40	$66.25	$1060.00	$48.10	$71.73
1 Equip. Oper. (light)	55.50	444.00	82.70	661.60		
1 Loader, Skid Steer, 30 H.P.		179.50		197.45	7.48	8.23
24 L.H., Daily Totals		$1333.90		$1919.05	$55.58	$79.96

Crew B-62A

Crew No.	Bare Costs Hr.	Daily	Incl. Subs O&P Hr.	Daily	Cost Per Labor-Hour Bare Costs	Incl. O&P
2 Laborers	$44.40	$710.40	$66.25	$1060.00	$48.10	$71.73
1 Equip. Oper. (light)	55.50	444.00	82.70	661.60		
1 Loader, Skid Steer, 30 H.P.		179.50		197.45		
1 Trencher Attachment		66.25		72.88	10.24	11.26
24 L.H., Daily Totals		$1400.15		$1991.93	$58.34	$83.00

Crew B-63

Crew No.	Bare Costs Hr.	Daily	Incl. Subs O&P Hr.	Daily	Bare Costs	Incl. O&P
4 Laborers	$44.40	$1420.80	$66.25	$2120.00	$46.62	$69.54
1 Equip. Oper. (light)	55.50	444.00	82.70	661.60		
1 Loader, Skid Steer, 30 H.P.		179.50		197.45	4.49	4.94
40 L.H., Daily Totals		$2044.30		$2979.05	$51.11	$74.48

Crew B-63B

Crew No.	Bare Costs Hr.	Daily	Incl. Subs O&P Hr.	Daily	Bare Costs	Incl. O&P
1 Labor Foreman (inside)	$44.90	$359.20	$67.00	$536.00	$47.30	$70.55
2 Laborers	44.40	710.40	66.25	1060.00		
1 Equip. Oper. (light)	55.50	444.00	82.70	661.60		
1 Loader, Skid Steer, 78 H.P.		446.30		490.93	13.95	15.34
32 L.H., Daily Totals		$1959.90		$2748.53	$61.25	$85.89

Crew B-64

Crew No.	Bare Costs Hr.	Daily	Incl. Subs O&P Hr.	Daily	Bare Costs	Incl. O&P
1 Laborer	$44.40	$355.20	$66.25	$530.00	$46.60	$69.63
1 Truck Driver (light)	48.80	390.40	73.00	584.00		
1 Power Mulcher (small)		201.00		221.10		
1 Flatbed Truck, Gas, 1.5 Ton		198.50		218.35	24.97	27.47
16 L.H., Daily Totals		$1145.10		$1553.45	$71.57	$97.09

Crew B-65

Crew No.	Bare Costs Hr.	Daily	Incl. Subs O&P Hr.	Daily	Bare Costs	Incl. O&P
1 Laborer	$44.40	$355.20	$66.25	$530.00	$46.60	$69.63
1 Truck Driver (light)	48.80	390.40	73.00	584.00		
1 Power Mulcher (Large)		345.35		379.88		
1 Flatbed Truck, Gas, 1.5 Ton		198.50		218.35	33.99	37.39
16 L.H., Daily Totals		$1289.45		$1712.23	$80.59	$107.01

Crew B-66

Crew No.	Bare Costs Hr.	Daily	Incl. Subs O&P Hr.	Daily	Bare Costs	Incl. O&P
1 Equip. Oper. (light)	$55.50	$444.00	$82.70	$661.60	$55.50	$82.70
1 Loader-Backhoe, 40 H.P.		267.55		294.31	33.44	36.79
8 L.H., Daily Totals		$711.55		$955.90	$88.94	$119.49

Crew B-67

Crew No.	Bare Costs Hr.	Daily	Incl. Subs O&P Hr.	Daily	Bare Costs	Incl. O&P
1 Millwright	$58.75	$470.00	$84.90	$679.20	$57.13	$83.80
1 Equip. Oper. (light)	55.50	444.00	82.70	661.60		
1 R.T. Forklift, 5,000 Lb., diesel		272.85		300.13	17.05	18.76
16 L.H., Daily Totals		$1186.85		$1640.93	$74.18	$102.56

Crew B-67B

Crew No.	Bare Costs Hr.	Daily	Incl. Subs O&P Hr.	Daily	Bare Costs	Incl. O&P
1 Millwright Foreman (inside)	$59.25	$474.00	$85.60	$684.80	$59.00	$85.25
1 Millwright	58.75	470.00	84.90	679.20		
16 L.H., Daily Totals		$944.00		$1364.00	$59.00	$85.25

Crew B-68

Crew No.	Bare Costs Hr.	Daily	Incl. Subs O&P Hr.	Daily	Bare Costs	Incl. O&P
2 Millwrights	$58.75	$940.00	$84.90	$1358.40	$57.67	$84.17
1 Equip. Oper. (light)	55.50	444.00	82.70	661.60		
1 R.T. Forklift, 5,000 Lb., diesel		272.85		300.13	11.37	12.51
24 L.H., Daily Totals		$1656.85		$2320.14	$69.04	$96.67

Crew B-68A

Crew No.	Bare Costs Hr.	Daily	Incl. Subs O&P Hr.	Daily	Bare Costs	Incl. O&P
1 Millwright Foreman (inside)	$59.25	$474.00	$85.60	$684.80	$58.92	$85.13
2 Millwrights	58.75	940.00	84.90	1358.40		
1 Forklift, Smooth Floor, 8,000 Lb.		283.25		311.57	11.80	12.98
24 L.H., Daily Totals		$1697.25		$2354.78	$70.72	$98.12

Crew B-68B

Crew No.	Bare Costs Hr.	Daily	Incl. Subs O&P Hr.	Daily	Bare Costs	Incl. O&P
1 Millwright Foreman (inside)	$59.25	$474.00	$85.60	$684.80	$62.79	$92.40
2 Millwrights	58.75	940.00	84.90	1358.40		
2 Electricians	63.70	1019.20	94.65	1514.40		
2 Plumbers	67.70	1083.20	101.05	1616.80		
1 R.T. Forklift, 5,000 Lb., gas		283.30		311.63	5.06	5.56
56 L.H., Daily Totals		$3799.70		$5486.03	$67.85	$97.96

Crew B-68C

Crew No.	Bare Costs Hr.	Daily	Incl. Subs O&P Hr.	Daily	Bare Costs	Incl. O&P
1 Millwright Foreman (inside)	$59.25	$474.00	$85.60	$684.80	$62.35	$91.55
1 Millwright	58.75	470.00	84.90	679.20		
1 Electrician	63.70	509.60	94.65	757.20		
1 Plumber	67.70	541.60	101.05	808.40		
1 R.T. Forklift, 5,000 Lb., gas		283.30		311.63	8.85	9.74
32 L.H., Daily Totals		$2278.50		$3241.23	$71.20	$101.29

Crew B-68D

Crew No.	Bare Costs Hr.	Daily	Incl. Subs O&P Hr.	Daily	Bare Costs	Incl. O&P
1 Labor Foreman (inside)	$44.90	$359.20	$67.00	$536.00	$48.27	$71.98
1 Laborer	44.40	355.20	66.25	530.00		
1 Equip. Oper. (light)	55.50	444.00	82.70	661.60		
1 R.T. Forklift, 5,000 Lb., gas		283.30		311.63	11.80	12.98
24 L.H., Daily Totals		$1441.70		$2039.23	$60.07	$84.97

Crew B-68E

Crew No.	Bare Costs Hr.	Daily	Incl. Subs O&P Hr.	Daily	Bare Costs	Incl. O&P
1 Struc. Steel Foreman (inside)	$60.80	$486.40	$94.10	$752.80	$60.40	$93.46
3 Struc. Steel Workers	60.30	1447.20	93.30	2239.20		
1 Welder	60.30	482.40	93.30	746.40		
1 Forklift, Smooth Floor, 8,000 Lb.		283.25		311.57	7.08	7.79
40 L.H., Daily Totals		$2699.25		$4049.97	$67.48	$101.25

Crew B-68F

Crew No.	Bare Costs Hr.	Daily	Incl. Subs O&P Hr.	Daily	Bare Costs	Incl. O&P
1 Skilled Worker Foreman (out)	$59.10	$472.80	$88.90	$711.20	$57.77	$86.90
2 Skilled Workers	57.10	913.60	85.90	1374.40		
1 R.T. Forklift, 5,000 Lb., gas		283.30		311.63	11.80	12.98
24 L.H., Daily Totals		$1669.70		$2397.23	$69.57	$99.88

Crew B-68G

Crew No.	Bare Costs Hr.	Daily	Incl. Subs O&P Hr.	Daily	Bare Costs	Incl. O&P
2 Structural Steel Workers	$60.30	$964.80	$93.30	$1492.80	$60.30	$93.30
1 R.T. Forklift, 5,000 Lb., gas		283.30		311.63	17.71	19.48
16 L.H., Daily Totals		$1248.10		$1804.43	$78.01	$112.78

Crew B-69

Crew No.	Bare Costs Hr.	Daily	Incl. Subs O&P Hr.	Daily	Bare Costs	Incl. O&P
1 Labor Foreman (outside)	$46.40	$371.20	$69.25	$554.00	$48.92	$72.97
3 Laborers	44.40	1065.60	66.25	1590.00		
1 Equip. Oper. (crane)	61.45	491.60	91.55	732.40		
1 Equip. Oper. (oiler)	52.50	420.00	78.25	626.00		
1 Hyd. Crane, 80 Ton		1458.00		1603.80	30.38	33.41
48 L.H., Daily Totals		$3806.40		$5106.20	$79.30	$106.38

Crew B-69A

Crew No.	Bare Costs Hr.	Daily	Incl. Subs O&P Hr.	Daily	Bare Costs	Incl. O&P
1 Labor Foreman (outside)	$46.40	$371.20	$69.25	$554.00	$48.40	$71.97
3 Laborers	44.40	1065.60	66.25	1590.00		
1 Equip. Oper. (medium)	59.00	472.00	87.90	703.20		
1 Concrete Finisher	51.80	414.40	75.90	607.20		
1 Curb/Gutter Paver, 2-Track		1231.00		1354.10	25.65	28.21
48 L.H., Daily Totals		$3554.20		$4808.50	$74.05	$100.18

Crew B-69B

Crew No.	Bare Costs Hr.	Bare Costs Daily	Incl. Subs O&P Hr.	Incl. Subs O&P Daily	Cost Per Labor-Hour Bare Costs	Cost Per Labor-Hour Incl. O&P
1 Labor Foreman (outside)	$46.40	$371.20	$69.25	$554.00	$48.40	$71.97
3 Laborers	44.40	1065.60	66.25	1590.00		
1 Equip. Oper. (medium)	59.00	472.00	87.90	703.20		
1 Cement Finisher	51.80	414.40	75.90	607.20		
1 Curb/Gutter Paver, 4-Track		801.05		881.15	16.69	18.36
48 L.H., Daily Totals		$3124.25		$4335.56	$65.09	$90.32

Crew B-70

Crew No.	Bare Costs Hr.	Bare Costs Daily	Incl. Subs O&P Hr.	Incl. Subs O&P Daily	Cost Per Labor-Hour Bare Costs	Cost Per Labor-Hour Incl. O&P
1 Labor Foreman (outside)	$46.40	$371.20	$69.25	$554.00	$50.94	$75.96
3 Laborers	44.40	1065.60	66.25	1590.00		
3 Equip. Oper. (medium)	59.00	1416.00	87.90	2109.60		
1 Grader, 30,000 Lbs.		1073.00		1180.30		
1 Ripper, Beam & 1 Shank		91.60		100.76		
1 Road Sweeper, S.P., 8' wide		723.65		796.01		
1 F.E. Loader, W.M., 1.5 C.Y.		441.40		485.54	41.60	45.76
56 L.H., Daily Totals		$5182.45		$6816.22	$92.54	$121.72

Crew B-70A

Crew No.	Bare Costs Hr.	Bare Costs Daily	Incl. Subs O&P Hr.	Incl. Subs O&P Daily	Cost Per Labor-Hour Bare Costs	Cost Per Labor-Hour Incl. O&P
1 Laborer	$44.40	$355.20	$66.25	$530.00	$56.08	$83.57
4 Equip. Oper. (medium)	59.00	1888.00	87.90	2812.80		
1 Grader, 40,000 Lbs.		1219.00		1340.90		
1 F.E. Loader, W.M., 2.5 C.Y.		638.30		702.13		
1 Dozer, 80 H.P.		405.85		446.44		
1 Roller, Pneum. Whl., 12 Ton		349.90		384.89	65.33	71.86
40 L.H., Daily Totals		$4856.25		$6217.15	$121.41	$155.43

Crew B-71

Crew No.	Bare Costs Hr.	Bare Costs Daily	Incl. Subs O&P Hr.	Incl. Subs O&P Daily	Cost Per Labor-Hour Bare Costs	Cost Per Labor-Hour Incl. O&P
1 Labor Foreman (outside)	$46.40	$371.20	$69.25	$554.00	$50.94	$75.96
3 Laborers	44.40	1065.60	66.25	1590.00		
3 Equip. Oper. (medium)	59.00	1416.00	87.90	2109.60		
1 Pvmt. Profiler, 750 H.P.		3483.00		3831.30		
1 Road Sweeper, S.P., 8' wide		723.65		796.01		
1 F.E. Loader, W.M., 1.5 C.Y.		441.40		485.54	83.00	91.30
56 L.H., Daily Totals		$7500.85		$9366.45	$133.94	$167.26

Crew B-72

Crew No.	Bare Costs Hr.	Bare Costs Daily	Incl. Subs O&P Hr.	Incl. Subs O&P Daily	Cost Per Labor-Hour Bare Costs	Cost Per Labor-Hour Incl. O&P
1 Labor Foreman (outside)	$46.40	$371.20	$69.25	$554.00	$51.95	$77.45
3 Laborers	44.40	1065.60	66.25	1590.00		
4 Equip. Oper. (medium)	59.00	1888.00	87.90	2812.80		
1 Pvmt. Profiler, 750 H.P.		3483.00		3831.30		
1 Hammermill, 250 H.P.		857.40		943.14		
1 Windrow Loader		1461.00		1607.10		
1 Mix Paver, 165 H.P.		2172.00		2389.20		
1 Roller, Pneum. Whl., 12 Ton		349.90		384.89	130.05	143.06
64 L.H., Daily Totals		$11648.10		$14112.43	$182.00	$220.51

Crew B-73

Crew No.	Bare Costs Hr.	Bare Costs Daily	Incl. Subs O&P Hr.	Incl. Subs O&P Daily	Cost Per Labor-Hour Bare Costs	Cost Per Labor-Hour Incl. O&P
1 Labor Foreman (outside)	$46.40	$371.20	$69.25	$554.00	$53.77	$80.16
2 Laborers	44.40	710.40	66.25	1060.00		
5 Equip. Oper. (medium)	59.00	2360.00	87.90	3516.00		
1 Road Mixer, 310 H.P.		1919.00		2110.90		
1 Tandem Roller, 10 Ton		246.80		271.48		
1 Hammermill, 250 H.P.		857.40		943.14		
1 Grader, 30,000 Lbs.		1073.00		1180.30		
.5 F.E. Loader, W.M., 1.5 C.Y.		220.70		242.77		
.5 Truck Tractor, 220 H.P.		155.40		170.94		
.5 Water Tank Trailer, 5000 Gal.		77.05		84.75	71.08	78.19
64 L.H., Daily Totals		$7990.95		$10134.29	$124.86	$158.35

Crew B-74

Crew No.	Bare Costs Hr.	Bare Costs Daily	Incl. Subs O&P Hr.	Incl. Subs O&P Daily	Cost Per Labor-Hour Bare Costs	Cost Per Labor-Hour Incl. O&P
1 Labor Foreman (outside)	$46.40	$371.20	$69.25	$554.00	$53.67	$80.06
1 Laborer	44.40	355.20	66.25	530.00		
4 Equip. Oper. (medium)	59.00	1888.00	87.90	2812.80		
2 Truck Drivers (heavy)	51.30	820.80	76.70	1227.20		
1 Grader, 30,000 Lbs.		1073.00		1180.30		
1 Ripper, Beam & 1 Shank		91.60		100.76		
2 Stabilizers, 310 H.P.		2808.00		3088.80		
1 Flatbed Truck, Gas, 3 Ton		850.05		935.05		
1 Chem. Spreader, Towed		85.40		93.94		
1 Roller, Vibratory, 25 Ton		672.35		739.59		
1 Water Tank Trailer, 5000 Gal.		154.10		169.51		
1 Truck Tractor, 220 H.P.		310.80		341.88	94.46	103.90
64 L.H., Daily Totals		$9480.50		$11773.83	$148.13	$183.97

Crew B-75

Crew No.	Bare Costs Hr.	Bare Costs Daily	Incl. Subs O&P Hr.	Incl. Subs O&P Daily	Cost Per Labor-Hour Bare Costs	Cost Per Labor-Hour Incl. O&P
1 Labor Foreman (outside)	$46.40	$371.20	$69.25	$554.00	$54.01	$80.54
1 Laborer	44.40	355.20	66.25	530.00		
4 Equip. Oper. (medium)	59.00	1888.00	87.90	2812.80		
1 Truck Driver (heavy)	51.30	410.40	76.70	613.60		
1 Grader, 30,000 Lbs.		1073.00		1180.30		
1 Ripper, Beam & 1 Shank		91.60		100.76		
2 Stabilizers, 310 H.P.		2808.00		3088.80		
1 Dist. Tanker, 3000 Gallon		334.10		367.51		
1 Truck Tractor, 6x4, 380 H.P.		499.15		549.07		
1 Roller, Vibratory, 25 Ton		672.35		739.59	97.83	107.61
56 L.H., Daily Totals		$8503.00		$10536.42	$151.84	$188.15

Crew B-76

Crew No.	Bare Costs Hr.	Bare Costs Daily	Incl. Subs O&P Hr.	Incl. Subs O&P Daily	Cost Per Labor-Hour Bare Costs	Cost Per Labor-Hour Incl. O&P
1 Dock Builder Foreman (outside)	$57.90	$463.20	$89.50	$716.00	$56.98	$86.98
5 Dock Builders	55.90	2236.00	86.40	3456.00		
2 Equip. Oper. (crane)	61.45	983.20	91.55	1464.80		
1 Equip. Oper. (oiler)	52.50	420.00	78.25	626.00		
1 Crawler Crane, 50 Ton		1541.00		1695.10		
1 Barge, 400 Ton		869.15		956.07		
1 Hammer, Diesel, 15K ft.-lbs.		624.45		686.89		
1 Lead, 60' High		211.80		232.98		
1 Air Compressor, 600 cfm		426.55		469.20		
2 -50' Air Hoses, 3"		76.70		84.37	52.08	57.29
72 L.H., Daily Totals		$7852.05		$10387.42	$109.06	$144.27

Crew B-76A

Crew No.	Bare Costs Hr.	Bare Costs Daily	Incl. Subs O&P Hr.	Incl. Subs O&P Daily	Cost Per Labor-Hour Bare Costs	Cost Per Labor-Hour Incl. O&P
1 Labor Foreman (outside)	$46.40	$371.20	$69.25	$554.00	$47.79	$71.29
5 Laborers	44.40	1776.00	66.25	2650.00		
1 Equip. Oper. (crane)	61.45	491.60	91.55	732.40		
1 Equip. Oper. (oiler)	52.50	420.00	78.25	626.00		
1 Crawler Crane, 50 Ton		1541.00		1695.10		
1 Barge, 400 Ton		869.15		956.07	37.66	41.42
64 L.H., Daily Totals		$5468.95		$7213.56	$85.45	$112.71

Crew B-77

Crew No.	Bare Costs Hr.	Bare Costs Daily	Incl. Subs O&P Hr.	Incl. Subs O&P Daily	Cost Per Labor-Hour Bare Costs	Cost Per Labor-Hour Incl. O&P
1 Labor Foreman (outside)	$46.40	$371.20	$69.25	$554.00	$45.68	$68.20
3 Laborers	44.40	1065.60	66.25	1590.00		
1 Truck Driver (light)	48.80	390.40	73.00	584.00		
1 Crack Cleaner, 25 H.P.		53.00		58.30		
1 Crack Filler, Trailer Mtd.		170.95		188.04		
1 Flatbed Truck, Gas, 3 Ton		850.05		935.05	26.85	29.54
40 L.H., Daily Totals		$2901.20		$3909.40	$72.53	$97.73

Crew No.	Bare Costs		Incl. Subs O&P		Cost Per Labor-Hour	
Crew B-78	Hr.	Daily	Hr.	Daily	Bare Costs	Incl. O&P
1 Labor Foreman (outside)	$46.40	$371.20	$69.25	$554.00	$45.47	$67.88
4 Laborers	44.40	1420.80	66.25	2120.00		
1 Truck Driver (light)	48.80	390.40	73.00	584.00		
1 Paint Striper, S.P., 40 Gallon		128.35		141.19		
1 Flatbed Truck, Gas, 3 Ton		850.05		935.05		
1 Pickup Truck, 3/4 Ton		112.20		123.42	22.72	24.99
48 L.H., Daily Totals		$3273.00		$4457.66	$68.19	$92.87
Crew B-78A	Hr.	Daily	Hr.	Daily	Bare Costs	Incl. O&P
1 Equip. Oper. (light)	$55.50	$444.00	$82.70	$661.60	$55.50	$82.70
1 Line Rem. (Metal Balls) 115 H.P.		996.25		1095.88	124.53	136.98
8 L.H., Daily Totals		$1440.25		$1757.47	$180.03	$219.68
Crew B-78B	Hr.	Daily	Hr.	Daily	Bare Costs	Incl. O&P
2 Laborers	$44.40	$710.40	$66.25	$1060.00	$45.63	$68.08
.25 Equip. Oper. (light)	55.50	111.00	82.70	165.40		
1 Pickup Truck, 3/4 Ton		112.20		123.42		
1 Line Rem.,11 H.P.,Walk Behind		114.75		126.22		
.25 Road Sweeper, S.P., 8' wide		180.91		199.00	22.66	24.92
18 L.H., Daily Totals		$1229.26		$1674.05	$68.29	$93.00
Crew B-78C	Hr.	Daily	Hr.	Daily	Bare Costs	Incl. O&P
1 Labor Foreman (outside)	$46.40	$371.20	$69.25	$554.00	$45.47	$67.88
4 Laborers	44.40	1420.80	66.25	2120.00		
1 Truck Driver (light)	48.80	390.40	73.00	584.00		
1 Paint Striper, T.M., 120 Gal.		603.10		663.41		
1 Flatbed Truck, Gas, 3 Ton		850.05		935.05		
1 Pickup Truck, 3/4 Ton		112.20		123.42	32.61	35.87
48 L.H., Daily Totals		$3747.75		$4979.89	$78.08	$103.75
Crew B-78D	Hr.	Daily	Hr.	Daily	Bare Costs	Incl. O&P
2 Labor Foremen (outside)	$46.40	$742.40	$69.25	$1108.00	$45.24	$67.53
7 Laborers	44.40	2486.40	66.25	3710.00		
1 Truck Driver (light)	48.80	390.40	73.00	584.00		
1 Paint Striper, T.M., 120 Gal.		603.10		663.41		
1 Flatbed Truck, Gas, 3 Ton		850.05		935.05		
3 Pickup Trucks, 3/4 Ton		336.60		370.26		
1 Air Compressor, 60 cfm		153.85		169.24		
1 -50' Air Hose, 3/4"		7.15		7.87		
1 Breaker, Pavement, 60 lb.		53.60		58.96	25.05	27.56
80 L.H., Daily Totals		$5623.55		$7606.78	$70.29	$95.08
Crew B-78E	Hr.	Daily	Hr.	Daily	Bare Costs	Incl. O&P
2 Labor Foremen (outside)	$46.40	$742.40	$69.25	$1108.00	$45.10	$67.31
9 Laborers	44.40	3196.80	66.25	4770.00		
1 Truck Driver (light)	48.80	390.40	73.00	584.00		
1 Paint Striper, T.M., 120 Gal.		603.10		663.41		
1 Flatbed Truck, Gas, 3 Ton		850.05		935.05		
4 Pickup Trucks, 3/4 Ton		448.80		493.68		
2 Air Compressors, 60 cfm		307.70		338.47		
2 -50' Air Hoses, 3/4"		14.30		15.73		
2 Breakers, Pavement, 60 lb.		107.20		117.92	24.28	26.71
96 L.H., Daily Totals		$6660.75		$9026.26	$69.38	$94.02

Crew No.	Bare Costs		Incl. Subs O&P		Cost Per Labor-Hour	
Crew B-78F	Hr.	Daily	Hr.	Daily	Bare Costs	Incl. O&P
2 Labor Foremen (outside)	$46.40	$742.40	$69.25	$1108.00	$45.00	$67.16
11 Laborers	44.40	3907.20	66.25	5830.00		
1 Truck Driver (light)	48.80	390.40	73.00	584.00		
1 Paint Striper, T.M., 120 Gal.		603.10		663.41		
1 Flatbed Truck, Gas, 3 Ton		850.05		935.05		
7 Pickup Trucks, 3/4 Ton		785.40		863.94		
3 Air Compressors, 60 cfm		461.55		507.70		
3 -50' Air Hoses, 3/4"		21.45		23.59		
3 Breakers, Pavement, 60 lb.		160.80		176.88	25.74	28.31
112 L.H., Daily Totals		$7922.35		$10692.58	$70.74	$95.47
Crew B-79	Hr.	Daily	Hr.	Daily	Bare Costs	Incl. O&P
1 Labor Foreman (outside)	$46.40	$371.20	$69.25	$554.00	$45.68	$68.20
3 Laborers	44.40	1065.60	66.25	1590.00		
1 Truck Driver (light)	48.80	390.40	73.00	584.00		
1 Paint Striper, T.M., 120 Gal.		603.10		663.41		
1 Heating Kettle, 115 Gallon		107.25		117.97		
1 Flatbed Truck, Gas, 3 Ton		850.05		935.05		
2 Pickup Trucks, 3/4 Ton		224.40		246.84	44.62	49.08
40 L.H., Daily Totals		$3612.00		$4691.28	$90.30	$117.28
Crew B-79A	Hr.	Daily	Hr.	Daily	Bare Costs	Incl. O&P
1.5 Equip. Oper. (light)	$55.50	$666.00	$82.70	$992.40	$55.50	$82.70
.5 Line Remov. (Grinder) 115 H.P.		515.50		567.05		
1 Line Rem. (Metal Balls) 115 H.P.		996.25		1095.88	125.98	138.58
12 L.H., Daily Totals		$2177.75		$2655.32	$181.48	$221.28
Crew B-79B	Hr.	Daily	Hr.	Daily	Bare Costs	Incl. O&P
1 Laborer	$44.40	$355.20	$66.25	$530.00	$44.40	$66.25
1 Set of Gases		173.60		190.96	21.70	23.87
8 L.H., Daily Totals		$528.80		$720.96	$66.10	$90.12
Crew B-79C	Hr.	Daily	Hr.	Daily	Bare Costs	Incl. O&P
1 Labor Foreman (outside)	$46.40	$371.20	$69.25	$554.00	$45.31	$67.64
5 Laborers	44.40	1776.00	66.25	2650.00		
1 Truck Driver (light)	48.80	390.40	73.00	584.00		
1 Paint Striper, T.M., 120 Gal.		603.10		663.41		
1 Heating Kettle, 115 Gallon		107.25		117.97		
1 Flatbed Truck, Gas, 3 Ton		850.05		935.05		
3 Pickup Trucks, 3/4 Ton		336.60		370.26		
1 Air Compressor, 60 cfm		153.85		169.24		
1 -50' Air Hose, 3/4"		7.15		7.87		
1 Breaker, Pavement, 60 lb.		53.60		58.96	37.71	41.48
56 L.H., Daily Totals		$4649.20		$6110.76	$83.02	$109.12
Crew B-79D	Hr.	Daily	Hr.	Daily	Bare Costs	Incl. O&P
2 Labor Foremen (outside)	$46.40	$742.40	$69.25	$1108.00	$45.45	$67.84
5 Laborers	44.40	1776.00	66.25	2650.00		
1 Truck Driver (light)	48.80	390.40	73.00	584.00		
1 Paint Striper, T.M., 120 Gal.		603.10		663.41		
1 Heating Kettle, 115 Gallon		107.25		117.97		
1 Flatbed Truck, Gas, 3 Ton		850.05		935.05		
4 Pickup Trucks, 3/4 Ton		448.80		493.68		
1 Air Compressor, 60 cfm		153.85		169.24		
1 -50' Air Hose, 3/4"		7.15		7.87		
1 Breaker, Pavement, 60 lb.		53.60		58.96	34.75	38.22
64 L.H., Daily Totals		$5132.60		$6788.18	$80.20	$106.07

Crew No.	Bare Costs		Incl. Subs O&P		Cost Per Labor-Hour	
Crew B-79E	Hr.	Daily	Hr.	Daily	Bare Costs	Incl. O&P
2 Labor Foremen (outside)	$46.40	$742.40	$69.25	$1108.00	$45.24	$67.53
7 Laborers	44.40	2486.40	66.25	3710.00		
1 Truck Driver (light)	48.80	390.40	73.00	584.00		
1 Paint Striper, T.M., 120 Gal.		603.10		663.41		
1 Heating Kettle, 115 Gallon		107.25		117.97		
1 Flatbed Truck, Gas, 3 Ton		850.05		935.05		
5 Pickup Trucks, 3/4 Ton		561.00		617.10		
2 Air Compressors, 60 cfm		307.70		338.47		
2 -50' Air Hoses, 3/4"		14.30		15.73		
2 Breakers, Pavement, 60 lb.		107.20		117.92	31.88	35.07
80 L.H., Daily Totals		$6169.80		$8207.66	$77.12	$102.60
Crew B-80	Hr.	Daily	Hr.	Daily	Bare Costs	Incl. O&P
1 Labor Foreman (outside)	$46.40	$371.20	$69.25	$554.00	$48.77	$72.80
1 Laborer	44.40	355.20	66.25	530.00		
1 Truck Driver (light)	48.80	390.40	73.00	584.00		
1 Equip. Oper. (light)	55.50	444.00	82.70	661.60		
1 Flatbed Truck, Gas, 3 Ton		850.05		935.05		
1 Earth Auger, Truck-Mtd.		202.55		222.81	32.89	36.18
32 L.H., Daily Totals		$2613.40		$3487.46	$81.67	$108.98
Crew B-80A	Hr.	Daily	Hr.	Daily	Bare Costs	Incl. O&P
3 Laborers	$44.40	$1065.60	$66.25	$1590.00	$44.40	$66.25
1 Flatbed Truck, Gas, 3 Ton		850.05		935.05	35.42	38.96
24 L.H., Daily Totals		$1915.65		$2525.05	$79.82	$105.21
Crew B-80B	Hr.	Daily	Hr.	Daily	Bare Costs	Incl. O&P
3 Laborers	$44.40	$1065.60	$66.25	$1590.00	$47.17	$70.36
1 Equip. Oper. (light)	55.50	444.00	82.70	661.60		
1 Crane, Flatbed Mounted, 3 Ton		238.75		262.63	7.46	8.21
32 L.H., Daily Totals		$1748.35		$2514.22	$54.64	$78.57
Crew B-80C	Hr.	Daily	Hr.	Daily	Bare Costs	Incl. O&P
2 Laborers	$44.40	$710.40	$66.25	$1060.00	$45.87	$68.50
1 Truck Driver (light)	48.80	390.40	73.00	584.00		
1 Flatbed Truck, Gas, 1.5 Ton		198.50		218.35		
1 Manual Fence Post Auger, Gas		54.40		59.84	10.54	11.59
24 L.H., Daily Totals		$1353.70		$1922.19	$56.40	$80.09
Crew B-81	Hr.	Daily	Hr.	Daily	Bare Costs	Incl. O&P
1 Laborer	$44.40	$355.20	$66.25	$530.00	$51.57	$76.95
1 Equip. Oper. (medium)	59.00	472.00	87.90	703.20		
1 Truck Driver (heavy)	51.30	410.40	76.70	613.60		
1 Hydromulcher, T.M., 3000 Gal.		256.85		282.54		
1 Truck Tractor, 220 H.P.		310.80		341.88	23.65	26.02
24 L.H., Daily Totals		$1805.25		$2471.22	$75.22	$102.97
Crew B-81A	Hr.	Daily	Hr.	Daily	Bare Costs	Incl. O&P
1 Laborer	$44.40	$355.20	$66.25	$530.00	$46.60	$69.63
1 Truck Driver (light)	48.80	390.40	73.00	584.00		
1 Hydromulcher, T.M., 600 Gal.		118.15		129.97		
1 Flatbed Truck, Gas, 3 Ton		850.05		935.05	60.51	66.56
16 L.H., Daily Totals		$1713.80		$2179.02	$107.11	$136.19
Crew B-82	Hr.	Daily	Hr.	Daily	Bare Costs	Incl. O&P
1 Laborer	$44.40	$355.20	$66.25	$530.00	$49.95	$74.47
1 Equip. Oper. (light)	55.50	444.00	82.70	661.60		
1 Horiz. Borer, 6 H.P.		184.15		202.57	11.51	12.66
16 L.H., Daily Totals		$983.35		$1394.17	$61.46	$87.14

Crew No.	Bare Costs		Incl. Subs O&P		Cost Per Labor-Hour	
Crew B-82A	Hr.	Daily	Hr.	Daily	Bare Costs	Incl. O&P
2 Laborers	$44.40	$710.40	$66.25	$1060.00	$49.95	$74.47
2 Equip. Opers. (light)	55.50	888.00	82.70	1323.20		
2 Dump Truck, 8 C.Y., 220 H.P.		815.20		896.72		
1 Flatbed Trailer, 25 Ton		137.20		150.92		
1 Horiz. Dir. Drill, 20k lb. Thrust		544.10		598.51		
1 Mud Trailer for HDD, 1500 Gal.		312.15		343.37		
1 Pickup Truck, 4x4, 3/4 Ton		176.75		194.43		
1 Flatbed Trailer, 3 Ton		71.15		78.27		
1 Loader, Skid Steer, 78 H.P.		446.30		490.93	78.21	86.04
32 L.H., Daily Totals		$4101.25		$5136.34	$128.16	$160.51
Crew B-82B	Hr.	Daily	Hr.	Daily	Bare Costs	Incl. O&P
2 Laborers	$44.40	$710.40	$66.25	$1060.00	$49.95	$74.47
2 Equip. Opers. (light)	55.50	888.00	82.70	1323.20		
2 Dump Truck, 8 C.Y., 220 H.P.		815.20		896.72		
1 Flatbed Trailer, 25 Ton		137.20		150.92		
1 Horiz. Dir. Drill, 30k lb. Thrust		647.65		712.41		
1 Mud Trailer for HDD, 1500 Gal.		312.15		343.37		
1 Pickup Truck, 4x4, 3/4 Ton		176.75		194.43		
1 Flatbed Trailer, 3 Ton		71.15		78.27		
1 Loader, Skid Steer, 78 H.P.		446.30		490.93	81.45	89.59
32 L.H., Daily Totals		$4204.80		$5250.24	$131.40	$164.07
Crew B-82C	Hr.	Daily	Hr.	Daily	Bare Costs	Incl. O&P
2 Laborers	$44.40	$710.40	$66.25	$1060.00	$49.95	$74.47
2 Equip. Opers. (light)	55.50	888.00	82.70	1323.20		
2 Dump Truck, 8 C.Y., 220 H.P.		815.20		896.72		
1 Flatbed Trailer, 25 Ton		137.20		150.92		
1 Horiz. Dir. Drill, 50k lb. Thrust		824.05		906.46		
1 Mud Trailer for HDD, 1500 Gal.		312.15		343.37		
1 Pickup Truck, 4x4, 3/4 Ton		176.75		194.43		
1 Flatbed Trailer, 3 Ton		71.15		78.27		
1 Loader, Skid Steer, 78 H.P.		446.30		490.93	86.96	95.66
32 L.H., Daily Totals		$4381.20		$5444.28	$136.91	$170.13
Crew B-82D	Hr.	Daily	Hr.	Daily	Bare Costs	Incl. O&P
1 Equip. Oper. (light)	$55.50	$444.00	$82.70	$661.60	$55.50	$82.70
1 Mud Trailer for HDD, 1500 Gal.		312.15		343.37	39.02	42.92
8 L.H., Daily Totals		$756.15		$1004.97	$94.52	$125.62
Crew B-83	Hr.	Daily	Hr.	Daily	Bare Costs	Incl. O&P
1 Tugboat Captain	$59.00	$472.00	$87.90	$703.20	$51.70	$77.08
1 Tugboat Hand	44.40	355.20	66.25	530.00		
1 Tugboat, 250 H.P.		726.10		798.71	45.38	49.92
16 L.H., Daily Totals		$1553.30		$2031.91	$97.08	$126.99
Crew B-84	Hr.	Daily	Hr.	Daily	Bare Costs	Incl. O&P
1 Equip. Oper. (medium)	$59.00	$472.00	$87.90	$703.20	$59.00	$87.90
1 Rotary Mower/Tractor		371.15		408.26	46.39	51.03
8 L.H., Daily Totals		$843.15		$1111.46	$105.39	$138.93
Crew B-85	Hr.	Daily	Hr.	Daily	Bare Costs	Incl. O&P
3 Laborers	$44.40	$1065.60	$66.25	$1590.00	$48.70	$72.67
1 Equip. Oper. (medium)	59.00	472.00	87.90	703.20		
1 Truck Driver (heavy)	51.30	410.40	76.70	613.60		
1 Telescoping Boom Lift, to 80'		387.75		426.52		
1 Brush Chipper, 12", 130 H.P.		366.05		402.65		
1 Pruning Saw, Rotary		26.40		29.04	19.50	21.46
40 L.H., Daily Totals		$2728.20		$3765.02	$68.20	$94.13

Crew No.	Bare Costs		Incl. Subs O&P		Cost Per Labor-Hour	
	Hr.	Daily	Hr.	Daily	Bare Costs	Incl. O&P
Crew B-86	Hr.	Daily	Hr.	Daily	Bare Costs	Incl. O&P
1 Equip. Oper. (medium)	$59.00	$472.00	$87.90	$703.20	$59.00	$87.90
1 Stump Chipper, S.P.		189.20		208.12	23.65	26.02
8 L.H., Daily Totals		$661.20		$911.32	$82.65	$113.92
Crew B-86A	Hr.	Daily	Hr.	Daily	Bare Costs	Incl. O&P
1 Equip. Oper. (medium)	$59.00	$472.00	$87.90	$703.20	$59.00	$87.90
1 Grader, 30,000 Lbs.		1073.00		1180.30	134.13	147.54
8 L.H., Daily Totals		$1545.00		$1883.50	$193.13	$235.44
Crew B-86B	Hr.	Daily	Hr.	Daily	Bare Costs	Incl. O&P
1 Equip. Oper. (medium)	$59.00	$472.00	$87.90	$703.20	$59.00	$87.90
1 Dozer, 200 H.P.		1520.00		1672.00	190.00	209.00
8 L.H., Daily Totals		$1992.00		$2375.20	$249.00	$296.90
Crew B-87	Hr.	Daily	Hr.	Daily	Bare Costs	Incl. O&P
1 Laborer	$44.40	$355.20	$66.25	$530.00	$56.08	$83.57
4 Equip. Oper. (medium)	59.00	1888.00	87.90	2812.80		
2 Feller Bunchers, 100 H.P.		1257.60		1383.36		
1 Log Chipper, 22" Tree		555.00		610.50		
1 Dozer, 105 H.P.		640.80		704.88		
1 Chain Saw, Gas, 36" Long		41.65		45.81	62.38	68.61
40 L.H., Daily Totals		$4738.25		$6087.35	$118.46	$152.18
Crew B-88	Hr.	Daily	Hr.	Daily	Bare Costs	Incl. O&P
1 Laborer	$44.40	$355.20	$66.25	$530.00	$56.91	$84.81
6 Equip. Oper. (medium)	59.00	2832.00	87.90	4219.20		
2 Feller Bunchers, 100 H.P.		1257.60		1383.36		
1 Log Chipper, 22" Tree		555.00		610.50		
2 Log Skidders, 50 H.P.		1826.70		2009.37		
1 Dozer, 105 H.P.		640.80		704.88		
1 Chain Saw, Gas, 36" Long		41.65		45.81	77.17	84.89
56 L.H., Daily Totals		$7508.95		$9503.13	$134.09	$169.70
Crew B-89	Hr.	Daily	Hr.	Daily	Bare Costs	Incl. O&P
1 Equip. Oper. (light)	$55.50	$444.00	$82.70	$661.60	$52.15	$77.85
1 Truck Driver (light)	48.80	390.40	73.00	584.00		
1 Flatbed Truck, Gas, 3 Ton		850.05		935.05		
1 Concrete Saw		112.85		124.14		
1 Water Tank, 65 Gal.		102.90		113.19	66.61	73.27
16 L.H., Daily Totals		$1900.20		$2417.98	$118.76	$151.12
Crew B-89A	Hr.	Daily	Hr.	Daily	Bare Costs	Incl. O&P
1 Skilled Worker	$57.10	$456.80	$85.90	$687.20	$50.75	$76.08
1 Laborer	44.40	355.20	66.25	530.00		
1 Core Drill (Large)		121.60		133.76	7.60	8.36
16 L.H., Daily Totals		$933.60		$1350.96	$58.35	$84.44
Crew B-89B	Hr.	Daily	Hr.	Daily	Bare Costs	Incl. O&P
1 Equip. Oper. (light)	$55.50	$444.00	$82.70	$661.60	$52.15	$77.85
1 Truck Driver (light)	48.80	390.40	73.00	584.00		
1 Wall Saw, Hydraulic, 10 H.P.		86.40		95.04		
1 Generator, Diesel, 100 kW		521.85		574.03		
1 Water Tank, 65 Gal.		102.90		113.19		
1 Flatbed Truck, Gas, 3 Ton		850.05		935.05	97.58	107.33
16 L.H., Daily Totals		$2395.60		$2962.92	$149.72	$185.18
Crew B-89C	Hr.	Daily	Hr.	Daily	Bare Costs	Incl. O&P
1 Cement Finisher	$51.80	$414.40	$75.90	$607.20	$51.80	$75.90
1 Masonry cut-off saw, gas		58.15		63.97	7.27	8.00
8 L.H., Daily Totals		$472.55		$671.16	$59.07	$83.90

Crew No.	Bare Costs		Incl. Subs O&P		Cost Per Labor-Hour	
	Hr.	Daily	Hr.	Daily	Bare Costs	Incl. O&P
Crew B-90	Hr.	Daily	Hr.	Daily	Bare Costs	Incl. O&P
1 Labor Foreman (outside)	$46.40	$371.20	$69.25	$554.00	$49.15	$73.35
3 Laborers	44.40	1065.60	66.25	1590.00		
2 Equip. Oper. (light)	55.50	888.00	82.70	1323.20		
2 Truck Drivers (heavy)	51.30	820.80	76.70	1227.20		
1 Road Mixer, 310 H.P.		1919.00		2110.90		
1 Dist. Truck, 2000 Gal.		303.25		333.57	34.72	38.19
64 L.H., Daily Totals		$5367.85		$7138.88	$83.87	$111.54
Crew B-90A	Hr.	Daily	Hr.	Daily	Bare Costs	Incl. O&P
1 Labor Foreman (outside)	$46.40	$371.20	$69.25	$554.00	$53.03	$79.05
2 Laborers	44.40	710.40	66.25	1060.00		
4 Equip. Oper. (medium)	59.00	1888.00	87.90	2812.80		
2 Graders, 30,000 Lbs.		2146.00		2360.60		
1 Tandem Roller, 10 Ton		246.80		271.48		
1 Roller, Pneum. Whl., 12 Ton		349.90		384.89	48.98	53.87
56 L.H., Daily Totals		$5712.30		$7443.77	$102.01	$132.92
Crew B-90B	Hr.	Daily	Hr.	Daily	Bare Costs	Incl. O&P
1 Labor Foreman (outside)	$46.40	$371.20	$69.25	$554.00	$52.03	$77.58
2 Laborers	44.40	710.40	66.25	1060.00		
3 Equip. Oper. (medium)	59.00	1416.00	87.90	2109.60		
1 Roller, Pneum. Whl., 12 Ton		349.90		384.89		
1 Road Mixer, 310 H.P.		1919.00		2110.90	47.27	52.00
48 L.H., Daily Totals		$4766.50		$6219.39	$99.30	$129.57
Crew B-90C	Hr.	Daily	Hr.	Daily	Bare Costs	Incl. O&P
1 Labor Foreman (outside)	$46.40	$371.20	$69.25	$554.00	$50.45	$75.28
4 Laborers	44.40	1420.80	66.25	2120.00		
3 Equip. Oper. (medium)	59.00	1416.00	87.90	2109.60		
3 Truck Drivers (heavy)	51.30	1231.20	76.70	1840.80		
3 Road Mixers, 310 H.P.		5757.00		6332.70	65.42	71.96
88 L.H., Daily Totals		$10196.20		$12957.10	$115.87	$147.24
Crew B-90D	Hr.	Daily	Hr.	Daily	Bare Costs	Incl. O&P
1 Labor Foreman (outside)	$46.40	$371.20	$69.25	$554.00	$49.52	$73.89
6 Laborers	44.40	2131.20	66.25	3180.00		
3 Equip. Oper. (medium)	59.00	1416.00	87.90	2109.60		
3 Truck Drivers (heavy)	51.30	1231.20	76.70	1840.80		
3 Road Mixers, 310 H.P.		5757.00		6332.70	55.36	60.89
104 L.H., Daily Totals		$10906.60		$14017.10	$104.87	$134.78
Crew B-90E	Hr.	Daily	Hr.	Daily	Bare Costs	Incl. O&P
1 Labor Foreman (outside)	$46.40	$371.20	$69.25	$554.00	$50.26	$74.96
4 Laborers	44.40	1420.80	66.25	2120.00		
3 Equip. Oper. (medium)	59.00	1416.00	87.90	2109.60		
1 Truck Driver (heavy)	51.30	410.40	76.70	613.60		
1 Road Mixer, 310 H.P.		1919.00		2110.90	26.65	29.32
72 L.H., Daily Totals		$5537.40		$7508.10	$76.91	$104.28
Crew B-91	Hr.	Daily	Hr.	Daily	Bare Costs	Incl. O&P
1 Labor Foreman (outside)	$46.40	$371.20	$69.25	$554.00	$52.81	$78.76
2 Laborers	44.40	710.40	66.25	1060.00		
4 Equip. Oper. (medium)	59.00	1888.00	87.90	2812.80		
1 Truck Driver (heavy)	51.30	410.40	76.70	613.60		
1 Dist. Tanker, 3000 Gallon		334.10		367.51		
1 Truck Tractor, 6x4, 380 H.P.		499.15		549.07		
1 Aggreg. Spreader, S.P.		859.10		945.01		
1 Roller, Pneum. Whl., 12 Ton		349.90		384.89		
1 Tandem Roller, 10 Ton		246.80		271.48	35.77	39.34
64 L.H., Daily Totals		$5669.05		$7558.35	$88.58	$118.10

For customer support on your Plumbing Costs with RSMeans Data, call 800.448.8182.

Crew No.	Bare Costs		Incl. Subs O&P		Cost Per Labor-Hour	
Crew B-91B	Hr.	Daily	Hr.	Daily	Bare Costs	Incl. O&P
1 Laborer	$44.40	$355.20	$66.25	$530.00	$51.70	$77.08
1 Equipment Oper. (med.)	59.00	472.00	87.90	703.20		
1 Road Sweeper, Vac. Assist.		879.45		967.39	54.97	60.46
16 L.H., Daily Totals		$1706.65		$2200.59	$106.67	$137.54
Crew B-91C	Hr.	Daily	Hr.	Daily	Bare Costs	Incl. O&P
1 Laborer	$44.40	$355.20	$66.25	$530.00	$46.60	$69.63
1 Truck Driver (light)	48.80	390.40	73.00	584.00		
1 Catch Basin Cleaning Truck		542.60		596.86	33.91	37.30
16 L.H., Daily Totals		$1288.20		$1710.86	$80.51	$106.93
Crew B-91D	Hr.	Daily	Hr.	Daily	Bare Costs	Incl. O&P
1 Labor Foreman (outside)	$46.40	$371.20	$69.25	$554.00	$51.23	$76.42
5 Laborers	44.40	1776.00	66.25	2650.00		
5 Equip. Oper. (medium)	59.00	2360.00	87.90	3516.00		
2 Truck Drivers (heavy)	51.30	820.80	76.70	1227.20		
1 Aggreg. Spreader, S.P.		859.10		945.01		
2 Truck Tractors, 6x4, 380 H.P.		998.30		1098.13		
2 Dist. Tankers, 3000 Gallon		668.20		735.02		
2 Pavement Brushes, Towed		176.70		194.37		
2 Rollers Pneum. Whl., 12 Ton		699.80		769.78	32.71	35.98
104 L.H., Daily Totals		$8730.10		$11689.51	$83.94	$112.40
Crew B-92	Hr.	Daily	Hr.	Daily	Bare Costs	Incl. O&P
1 Labor Foreman (outside)	$46.40	$371.20	$69.25	$554.00	$44.90	$67.00
3 Laborers	44.40	1065.60	66.25	1590.00		
1 Crack Cleaner, 25 H.P.		53.00		58.30		
1 Air Compressor, 60 cfm		153.85		169.24		
1 Tar Kettle, T.M.		156.70		172.37		
1 Flatbed Truck, Gas, 3 Ton		850.05		935.05	37.92	41.72
32 L.H., Daily Totals		$2650.40		$3478.96	$82.83	$108.72
Crew B-93	Hr.	Daily	Hr.	Daily	Bare Costs	Incl. O&P
1 Equip. Oper. (medium)	$59.00	$472.00	$87.90	$703.20	$59.00	$87.90
1 Feller Buncher, 100 H.P.		628.80		691.68	78.60	86.46
8 L.H., Daily Totals		$1100.80		$1394.88	$137.60	$174.36
Crew B-94A	Hr.	Daily	Hr.	Daily	Bare Costs	Incl. O&P
1 Laborer	$44.40	$355.20	$66.25	$530.00	$44.40	$66.25
1 Diaphragm Water Pump, 2"		87.70		96.47		
1 -20' Suction Hose, 2"		3.55		3.90		
2 -50' Discharge Hoses, 2"		8.00		8.80	12.41	13.65
8 L.H., Daily Totals		$454.45		$639.17	$56.81	$79.90
Crew B-94B	Hr.	Daily	Hr.	Daily	Bare Costs	Incl. O&P
1 Laborer	$44.40	$355.20	$66.25	$530.00	$44.40	$66.25
1 Diaphragm Water Pump, 4"		106.35		116.99		
1 -20' Suction Hose, 4"		17.25		18.98		
2 -50' Discharge Hoses, 4"		25.60		28.16	18.65	20.52
8 L.H., Daily Totals		$504.40		$694.12	$63.05	$86.77
Crew B-94C	Hr.	Daily	Hr.	Daily	Bare Costs	Incl. O&P
1 Laborer	$44.40	$355.20	$66.25	$530.00	$44.40	$66.25
1 Centrifugal Water Pump, 3"		74.40		81.84		
1 -20' Suction Hose, 3"		8.75		9.63		
2 -50' Discharge Hoses, 3"		9.00		9.90	11.52	12.67
8 L.H., Daily Totals		$447.35		$631.37	$55.92	$78.92

Crew No.	Bare Costs		Incl. Subs O&P		Cost Per Labor-Hour	
Crew B-94D	Hr.	Daily	Hr.	Daily	Bare Costs	Incl. O&P
1 Laborer	$44.40	$355.20	$66.25	$530.00	$44.40	$66.25
1 Centr. Water Pump, 6"		235.25		258.77		
1 -20' Suction Hose, 6"		25.50		28.05		
2 -50' Discharge Hoses, 6"		36.20		39.82	37.12	40.83
8 L.H., Daily Totals		$652.15		$856.64	$81.52	$107.08
Crew C-1	Hr.	Daily	Hr.	Daily	Bare Costs	Incl. O&P
3 Carpenters	$54.70	$1312.80	$81.65	$1959.60	$52.13	$77.80
1 Laborer	44.40	355.20	66.25	530.00		
32 L.H., Daily Totals		$1668.00		$2489.60	$52.13	$77.80
Crew C-2	Hr.	Daily	Hr.	Daily	Bare Costs	Incl. O&P
1 Carpenter Foreman (outside)	$56.70	$453.60	$84.60	$676.80	$53.32	$79.58
4 Carpenters	54.70	1750.40	81.65	2612.80		
1 Laborer	44.40	355.20	66.25	530.00		
48 L.H., Daily Totals		$2559.20		$3819.60	$53.32	$79.58
Crew C-2A	Hr.	Daily	Hr.	Daily	Bare Costs	Incl. O&P
1 Carpenter Foreman (outside)	$56.70	$453.60	$84.60	$676.80	$52.83	$78.62
3 Carpenters	54.70	1312.80	81.65	1959.60		
1 Cement Finisher	51.80	414.40	75.90	607.20		
1 Laborer	44.40	355.20	66.25	530.00		
48 L.H., Daily Totals		$2536.00		$3773.60	$52.83	$78.62
Crew C-3	Hr.	Daily	Hr.	Daily	Bare Costs	Incl. O&P
1 Rodman Foreman (outside)	$60.90	$487.20	$91.05	$728.40	$55.10	$82.31
4 Rodmen (reinf.)	58.90	1884.80	88.05	2817.60		
1 Equip. Oper. (light)	55.50	444.00	82.70	661.60		
2 Laborers	44.40	710.40	66.25	1060.00		
3 Stressing Equipment		56.85		62.53		
.5 Grouting Equipment		123.33		135.66	2.82	3.10
64 L.H., Daily Totals		$3706.57		$5465.79	$57.92	$85.40
Crew C-4	Hr.	Daily	Hr.	Daily	Bare Costs	Incl. O&P
1 Rodman Foreman (outside)	$60.90	$487.20	$91.05	$728.40	$59.40	$88.80
3 Rodmen (reinf.)	58.90	1413.60	88.05	2113.20		
3 Stressing Equipment		56.85		62.53	1.78	1.95
32 L.H., Daily Totals		$1957.65		$2904.14	$61.18	$90.75
Crew C-4A	Hr.	Daily	Hr.	Daily	Bare Costs	Incl. O&P
2 Rodmen (reinf.)	$58.90	$942.40	$88.05	$1408.80	$58.90	$88.05
4 Stressing Equipment		75.80		83.38	4.74	5.21
16 L.H., Daily Totals		$1018.20		$1492.18	$63.64	$93.26
Crew C-5	Hr.	Daily	Hr.	Daily	Bare Costs	Incl. O&P
1 Rodman Foreman (outside)	$60.90	$487.20	$91.05	$728.40	$58.64	$87.58
4 Rodmen (reinf.)	58.90	1884.80	88.05	2817.60		
1 Equip. Oper. (crane)	61.45	491.60	91.55	732.40		
1 Equip. Oper. (oiler)	52.50	420.00	78.25	626.00		
1 Hyd. Crane, 25 Ton		586.70		645.37	10.48	11.52
56 L.H., Daily Totals		$3870.30		$5549.77	$69.11	$99.10
Crew C-6	Hr.	Daily	Hr.	Daily	Bare Costs	Incl. O&P
1 Labor Foreman (outside)	$46.40	$371.20	$69.25	$554.00	$45.97	$68.36
4 Laborers	44.40	1420.80	66.25	2120.00		
1 Cement Finisher	51.80	414.40	75.90	607.20		
2 Gas Engine Vibrators		54.30		59.73	1.13	1.24
48 L.H., Daily Totals		$2260.70		$3340.93	$47.10	$69.60

Crews - Standard

Crew C-6A

Crew No.	Bare Costs Hr.	Bare Costs Daily	Incl. Subs O&P Hr.	Incl. Subs O&P Daily	Cost Per Labor-Hour Bare Costs	Cost Per Labor-Hour Incl. O&P
2 Cement Finishers	$51.80	$828.80	$75.90	$1214.40	$51.80	$75.90
1 Concrete Vibrator, Elec, 2 HP		45.80		50.38	2.86	3.15
16 L.H., Daily Totals		$874.60		$1264.78	$54.66	$79.05

Crew C-7

Crew No.	Hr.	Daily	Hr.	Daily	Bare Costs	Incl. O&P
1 Labor Foreman (outside)	$46.40	$371.20	$69.25	$554.00	$47.97	$71.39
5 Laborers	44.40	1776.00	66.25	2650.00		
1 Cement Finisher	51.80	414.40	75.90	607.20		
1 Equip. Oper. (medium)	59.00	472.00	87.90	703.20		
1 Equip. Oper. (oiler)	52.50	420.00	78.25	626.00		
2 Gas Engine Vibrators		54.30		59.73		
1 Concrete Bucket, 1 C.Y.		45.90		50.49		
1 Hyd. Crane, 55 Ton		990.15		1089.17	15.14	16.66
72 L.H., Daily Totals		$4543.95		$6339.78	$63.11	$88.05

Crew C-7A

Crew No.	Hr.	Daily	Hr.	Daily	Bare Costs	Incl. O&P
1 Labor Foreman (outside)	$46.40	$371.20	$69.25	$554.00	$46.38	$69.24
5 Laborers	44.40	1776.00	66.25	2650.00		
2 Truck Drivers (heavy)	51.30	820.80	76.70	1227.20		
2 Conc. Transit Mixers		1176.30		1293.93	18.38	20.22
64 L.H., Daily Totals		$4144.30		$5725.13	$64.75	$89.46

Crew C-7B

Crew No.	Hr.	Daily	Hr.	Daily	Bare Costs	Incl. O&P
1 Labor Foreman (outside)	$46.40	$371.20	$69.25	$554.00	$47.79	$71.29
5 Laborers	44.40	1776.00	66.25	2650.00		
1 Equipment Operator, Crane	61.45	491.60	91.55	732.40		
1 Equipment Oiler	52.50	420.00	78.25	626.00		
1 Conc. Bucket, 2 C.Y.		55.65		61.22		
1 Lattice Boom Crane, 165 Ton		2403.00		2643.30	38.42	42.26
64 L.H., Daily Totals		$5517.45		$7266.92	$86.21	$113.55

Crew C-7C

Crew No.	Hr.	Daily	Hr.	Daily	Bare Costs	Incl. O&P
1 Labor Foreman (outside)	$46.40	$371.20	$69.25	$554.00	$48.30	$72.04
5 Laborers	44.40	1776.00	66.25	2650.00		
2 Equipment Operators (med.)	59.00	944.00	87.90	1406.40		
2 F.E. Loaders, W.M., 4 C.Y.		1518.00		1669.80	23.72	26.09
64 L.H., Daily Totals		$4609.20		$6280.20	$72.02	$98.13

Crew C-7D

Crew No.	Hr.	Daily	Hr.	Daily	Bare Costs	Incl. O&P
1 Labor Foreman (outside)	$46.40	$371.20	$69.25	$554.00	$46.77	$69.77
5 Laborers	44.40	1776.00	66.25	2650.00		
1 Equip. Oper. (medium)	59.00	472.00	87.90	703.20		
1 Concrete Conveyer		206.25		226.88	3.68	4.05
56 L.H., Daily Totals		$2825.45		$4134.07	$50.45	$73.82

Crew C-8

Crew No.	Hr.	Daily	Hr.	Daily	Bare Costs	Incl. O&P
1 Labor Foreman (outside)	$46.40	$371.20	$69.25	$554.00	$48.89	$72.53
3 Laborers	44.40	1065.60	66.25	1590.00		
2 Cement Finishers	51.80	828.80	75.90	1214.40		
1 Equip. Oper. (medium)	59.00	472.00	87.90	703.20		
1 Concrete Pump (Small)		423.65		466.01	7.57	8.32
56 L.H., Daily Totals		$3161.25		$4527.61	$56.45	$80.85

Crew C-8A

Crew No.	Hr.	Daily	Hr.	Daily	Bare Costs	Incl. O&P
1 Labor Foreman (outside)	$46.40	$371.20	$69.25	$554.00	$47.20	$69.97
3 Laborers	44.40	1065.60	66.25	1590.00		
2 Cement Finishers	51.80	828.80	75.90	1214.40		
48 L.H., Daily Totals		$2265.60		$3358.40	$47.20	$69.97

Crew C-8B

Crew No.	Hr.	Daily	Hr.	Daily	Bare Costs	Incl. O&P
1 Labor Foreman (outside)	$46.40	$371.20	$69.25	$554.00	$47.72	$71.18
3 Laborers	44.40	1065.60	66.25	1590.00		
1 Equip. Oper. (medium)	59.00	472.00	87.90	703.20		
1 Vibrating Power Screed		87.65		96.42		
1 Roller, Vibratory, 25 Ton		672.35		739.59		
1 Dozer, 200 H.P.		1520.00		1672.00	57.00	62.70
40 L.H., Daily Totals		$4188.80		$5355.20	$104.72	$133.88

Crew C-8C

Crew No.	Hr.	Daily	Hr.	Daily	Bare Costs	Incl. O&P
1 Labor Foreman (outside)	$46.40	$371.20	$69.25	$554.00	$48.40	$71.97
3 Laborers	44.40	1065.60	66.25	1590.00		
1 Cement Finisher	51.80	414.40	75.90	607.20		
1 Equip. Oper. (medium)	59.00	472.00	87.90	703.20		
1 Shotcrete Rig, 12 C.Y./hr		269.20		296.12		
1 Air Compressor, 160 cfm		212.30		233.53		
4 -50' Air Hoses, 1"		32.20		35.42		
4 -50' Air Hoses, 2"		115.80		127.38	13.11	14.43
48 L.H., Daily Totals		$2952.70		$4146.85	$61.51	$86.39

Crew C-8D

Crew No.	Hr.	Daily	Hr.	Daily	Bare Costs	Incl. O&P
1 Labor Foreman (outside)	$46.40	$371.20	$69.25	$554.00	$49.52	$73.53
1 Laborer	44.40	355.20	66.25	530.00		
1 Cement Finisher	51.80	414.40	75.90	607.20		
1 Equipment Oper. (light)	55.50	444.00	82.70	661.60		
1 Air Compressor, 250 cfm		202.85		223.13		
2 -50' Air Hoses, 1"		16.10		17.71	6.84	7.53
32 L.H., Daily Totals		$1803.75		$2593.65	$56.37	$81.05

Crew C-8E

Crew No.	Hr.	Daily	Hr.	Daily	Bare Costs	Incl. O&P
1 Labor Foreman (outside)	$46.40	$371.20	$69.25	$554.00	$47.82	$71.10
3 Laborers	44.40	1065.60	66.25	1590.00		
1 Cement Finisher	51.80	414.40	75.90	607.20		
1 Equipment Oper. (light)	55.50	444.00	82.70	661.60		
1 Shotcrete Rig, 35 C.Y./hr.		301.05		331.15		
1 Air Compressor, 250 cfm		202.85		223.13		
4 -50' Air Hoses, 1"		32.20		35.42		
4 -50' Air Hoses, 2"		115.80		127.38	13.58	14.94
48 L.H., Daily Totals		$2947.10		$4129.89	$61.40	$86.04

Crew C-9

Crew No.	Hr.	Daily	Hr.	Daily	Bare Costs	Incl. O&P
1 Cement Finisher	$51.80	$414.40	$75.90	$607.20	$49.02	$72.78
2 Laborers	44.40	710.40	66.25	1060.00		
1 Equipment Oper. (light)	55.50	444.00	82.70	661.60		
1 Grout Pump, 50 C.F./hr.		190.35		209.38		
1 Air Compressor, 160 cfm		212.30		233.53		
2 -50' Air Hoses, 1"		16.10		17.71		
2 -50' Air Hoses, 2"		57.90		63.69	14.90	16.38
32 L.H., Daily Totals		$2045.45		$2853.11	$63.92	$89.16

Crew C-10

Crew No.	Hr.	Daily	Hr.	Daily	Bare Costs	Incl. O&P
1 Laborer	$44.40	$355.20	$66.25	$530.00	$49.33	$72.68
2 Cement Finishers	51.80	828.80	75.90	1214.40		
24 L.H., Daily Totals		$1184.00		$1744.40	$49.33	$72.68

Crew C-10B

Crew No.	Hr.	Daily	Hr.	Daily	Bare Costs	Incl. O&P
3 Laborers	$44.40	$1065.60	$66.25	$1590.00	$47.36	$70.11
2 Cement Finishers	51.80	828.80	75.90	1214.40		
1 Concrete Mixer, 10 C.F.		147.15		161.87		
2 Trowels, 48" Walk-Behind		188.60		207.46	8.39	9.23
40 L.H., Daily Totals		$2230.15		$3173.72	$55.75	$79.34

For customer support on your Plumbing Costs with RSMeans Data, call 800.448.8182.

Crew No.	Bare Costs		Incl. Subs O&P		Cost Per Labor-Hour	
Crew C-10C	Hr.	Daily	Hr.	Daily	Bare Costs	Incl. O&P
1 Laborer	$44.40	$355.20	$66.25	$530.00	$49.33	$72.68
2 Cement Finishers	51.80	828.80	75.90	1214.40		
1 Trowel, 48" Walk-Behind		94.30		103.73	3.93	4.32
24 L.H., Daily Totals		$1278.30		$1848.13	$53.26	$77.01
Crew C-10D	Hr.	Daily	Hr.	Daily	Bare Costs	Incl. O&P
1 Laborer	$44.40	$355.20	$66.25	$530.00	$49.33	$72.68
2 Cement Finishers	51.80	828.80	75.90	1214.40		
1 Vibrating Power Screed		87.65		96.42		
1 Trowel, 48" Walk-Behind		94.30		103.73	7.58	8.34
24 L.H., Daily Totals		$1365.95		$1944.55	$56.91	$81.02
Crew C-10E	Hr.	Daily	Hr.	Daily	Bare Costs	Incl. O&P
1 Laborer	$44.40	$355.20	$66.25	$530.00	$49.33	$72.68
2 Cement Finishers	51.80	828.80	75.90	1214.40		
1 Vibrating Power Screed		87.65		96.42		
1 Cement Trowel, 96" Ride-On		171.05		188.16	10.78	11.86
24 L.H., Daily Totals		$1442.70		$2028.97	$60.11	$84.54
Crew C-10F	Hr.	Daily	Hr.	Daily	Bare Costs	Incl. O&P
1 Laborer	$44.40	$355.20	$66.25	$530.00	$49.33	$72.68
2 Cement Finishers	51.80	828.80	75.90	1214.40		
1 Telescoping Boom Lift, to 60'		292.45		321.69	12.19	13.40
24 L.H., Daily Totals		$1476.45		$2066.09	$61.52	$86.09
Crew C-11	Hr.	Daily	Hr.	Daily	Bare Costs	Incl. O&P
1 Struc. Steel Foreman (outside)	$62.30	$498.40	$96.40	$771.20	$59.78	$91.78
6 Struc. Steel Workers	60.30	2894.40	93.30	4478.40		
1 Equip. Oper. (crane)	61.45	491.60	91.55	732.40		
1 Equip. Oper. (oiler)	52.50	420.00	78.25	626.00		
1 Lattice Boom Crane, 150 Ton		2324.00		2556.40	32.28	35.51
72 L.H., Daily Totals		$6628.40		$9164.40	$92.06	$127.28
Crew C-12	Hr.	Daily	Hr.	Daily	Bare Costs	Incl. O&P
1 Carpenter Foreman (outside)	$56.70	$453.60	$84.60	$676.80	$54.44	$81.22
3 Carpenters	54.70	1312.80	81.65	1959.60		
1 Laborer	44.40	355.20	66.25	530.00		
1 Equip. Oper. (crane)	61.45	491.60	91.55	732.40		
1 Hyd. Crane, 12 Ton		475.80		523.38	9.91	10.90
48 L.H., Daily Totals		$3089.00		$4422.18	$64.35	$92.13
Crew C-13	Hr.	Daily	Hr.	Daily	Bare Costs	Incl. O&P
1 Struc. Steel Worker	$60.30	$482.40	$93.30	$746.40	$58.43	$89.42
1 Welder	60.30	482.40	93.30	746.40		
1 Carpenter	54.70	437.60	81.65	653.20		
1 Welder, Gas Engine, 300 amp		148.75		163.63	6.20	6.82
24 L.H., Daily Totals		$1551.15		$2309.63	$64.63	$96.23
Crew C-14	Hr.	Daily	Hr.	Daily	Bare Costs	Incl. O&P
1 Carpenter Foreman (outside)	$56.70	$453.60	$84.60	$676.80	$53.39	$79.54
5 Carpenters	54.70	2188.00	81.65	3266.00		
4 Laborers	44.40	1420.80	66.25	2120.00		
4 Rodmen (reinf.)	58.90	1884.80	88.05	2817.60		
2 Cement Finishers	51.80	828.80	75.90	1214.40		
1 Equip. Oper. (crane)	61.45	491.60	91.55	732.40		
1 Equip. Oper. (oiler)	52.50	420.00	78.25	626.00		
1 Hyd. Crane, 80 Ton		1458.00		1603.80	10.13	11.14
144 L.H., Daily Totals		$9145.60		$13057.00	$63.51	$90.67

Crew No.	Bare Costs		Incl. Subs O&P		Cost Per Labor-Hour	
Crew C-14A	Hr.	Daily	Hr.	Daily	Bare Costs	Incl. O&P
1 Carpenter Foreman (outside)	$56.70	$453.60	$84.60	$676.80	$54.68	$81.58
16 Carpenters	54.70	7001.60	81.65	10451.20		
4 Rodmen (reinf.)	58.90	1884.80	88.05	2817.60		
2 Laborers	44.40	710.40	66.25	1060.00		
1 Cement Finisher	51.80	414.40	75.90	607.20		
1 Equip. Oper. (medium)	59.00	472.00	87.90	703.20		
1 Gas Engine Vibrator		27.15		29.86		
1 Concrete Pump (Small)		423.65		466.01	2.25	2.48
200 L.H., Daily Totals		$11387.60		$16811.88	$56.94	$84.06
Crew C-14B	Hr.	Daily	Hr.	Daily	Bare Costs	Incl. O&P
1 Carpenter Foreman (outside)	$56.70	$453.60	$84.60	$676.80	$54.57	$81.36
16 Carpenters	54.70	7001.60	81.65	10451.20		
4 Rodmen (reinf.)	58.90	1884.80	88.05	2817.60		
2 Laborers	44.40	710.40	66.25	1060.00		
2 Cement Finishers	51.80	828.80	75.90	1214.40		
1 Equip. Oper. (medium)	59.00	472.00	87.90	703.20		
1 Gas Engine Vibrator		27.15		29.86		
1 Concrete Pump (Small)		423.65		466.01	2.17	2.38
208 L.H., Daily Totals		$11802.00		$17419.08	$56.74	$83.75
Crew C-14C	Hr.	Daily	Hr.	Daily	Bare Costs	Incl. O&P
1 Carpenter Foreman (outside)	$56.70	$453.60	$84.60	$676.80	$52.29	$77.96
6 Carpenters	54.70	2625.60	81.65	3919.20		
2 Rodmen (reinf.)	58.90	942.40	88.05	1408.80		
4 Laborers	44.40	1420.80	66.25	2120.00		
1 Cement Finisher	51.80	414.40	75.90	607.20		
1 Gas Engine Vibrator		27.15		29.86	.24	.27
112 L.H., Daily Totals		$5883.95		$8761.86	$52.54	$78.23
Crew C-14D	Hr.	Daily	Hr.	Daily	Bare Costs	Incl. O&P
1 Carpenter Foreman (outside)	$56.70	$453.60	$84.60	$676.80	$54.35	$81.07
18 Carpenters	54.70	7876.80	81.65	11757.60		
2 Rodmen (reinf.)	58.90	942.40	88.05	1408.80		
2 Laborers	44.40	710.40	66.25	1060.00		
1 Cement Finisher	51.80	414.40	75.90	607.20		
1 Equip. Oper. (medium)	59.00	472.00	87.90	703.20		
1 Gas Engine Vibrator		27.15		29.86		
1 Concrete Pump (Small)		423.65		466.01	2.25	2.48
200 L.H., Daily Totals		$11320.40		$16709.48	$56.60	$83.55
Crew C-14E	Hr.	Daily	Hr.	Daily	Bare Costs	Incl. O&P
1 Carpenter Foreman (outside)	$56.70	$453.60	$84.60	$676.80	$53.34	$79.52
2 Carpenters	54.70	875.20	81.65	1306.40		
4 Rodmen (reinf.)	58.90	1884.80	88.05	2817.60		
3 Laborers	44.40	1065.60	66.25	1590.00		
1 Cement Finisher	51.80	414.40	75.90	607.20		
1 Gas Engine Vibrator		27.15		29.86	.31	.34
88 L.H., Daily Totals		$4720.75		$7027.86	$53.64	$79.86
Crew C-14F	Hr.	Daily	Hr.	Daily	Bare Costs	Incl. O&P
1 Labor Foreman (outside)	$46.40	$371.20	$69.25	$554.00	$49.56	$73.02
2 Laborers	44.40	710.40	66.25	1060.00		
6 Cement Finishers	51.80	2486.40	75.90	3643.20		
1 Gas Engine Vibrator		27.15		29.86	.38	.41
72 L.H., Daily Totals		$3595.15		$5287.06	$49.93	$73.43

For customer support on your Plumbing Costs with RSMeans Data, call 800.448.8182.

623

Crew C-14G

	Bare Costs		Incl. Subs O&P		Cost Per Labor-Hour	
Crew C-14G	Hr.	Daily	Hr.	Daily	Bare Costs	Incl. O&P
1 Labor Foreman (outside)	$46.40	$371.20	$69.25	$554.00	$48.91	$72.19
2 Laborers	44.40	710.40	66.25	1060.00		
4 Cement Finishers	51.80	1657.60	75.90	2428.80		
1 Gas Engine Vibrator		27.15		29.86	.48	.53
56 L.H., Daily Totals		$2766.35		$4072.67	$49.40	$72.73

Crew C-14H

	Bare Costs		Incl. Subs O&P		Cost Per Labor-Hour	
Crew C-14H	Hr.	Daily	Hr.	Daily	Bare Costs	Incl. O&P
1 Carpenter Foreman (outside)	$56.70	$453.60	$84.60	$676.80	$53.53	$79.68
2 Carpenters	54.70	875.20	81.65	1306.40		
1 Rodman (reinf.)	58.90	471.20	88.05	704.40		
1 Laborer	44.40	355.20	66.25	530.00		
1 Cement Finisher	51.80	414.40	75.90	607.20		
1 Gas Engine Vibrator		27.15		29.86	.57	.62
48 L.H., Daily Totals		$2596.75		$3854.67	$54.10	$80.31

Crew C-14L

	Bare Costs		Incl. Subs O&P		Cost Per Labor-Hour	
Crew C-14L	Hr.	Daily	Hr.	Daily	Bare Costs	Incl. O&P
1 Carpenter Foreman (outside)	$56.70	$453.60	$84.60	$676.80	$51.19	$76.28
6 Carpenters	54.70	2625.60	81.65	3919.20		
4 Laborers	44.40	1420.80	66.25	2120.00		
1 Cement Finisher	51.80	414.40	75.90	607.20		
1 Gas Engine Vibrator		27.15		29.86	.28	.31
96 L.H., Daily Totals		$4941.55		$7353.06	$51.47	$76.59

Crew C-14M

	Bare Costs		Incl. Subs O&P		Cost Per Labor-Hour	
Crew C-14M	Hr.	Daily	Hr.	Daily	Bare Costs	Incl. O&P
1 Carpenter Foreman (outside)	$56.70	$453.60	$84.60	$676.80	$53.08	$79.03
2 Carpenters	54.70	875.20	81.65	1306.40		
1 Rodman (reinf.)	58.90	471.20	88.05	704.40		
2 Laborers	44.40	710.40	66.25	1060.00		
1 Cement Finisher	51.80	414.40	75.90	607.20		
1 Equip. Oper. (medium)	59.00	472.00	87.90	703.20		
1 Gas Engine Vibrator		27.15		29.86		
1 Concrete Pump (Small)		423.65		466.01	7.04	7.75
64 L.H., Daily Totals		$3847.60		$5553.88	$60.12	$86.78

Crew C-15

	Bare Costs		Incl. Subs O&P		Cost Per Labor-Hour	
Crew C-15	Hr.	Daily	Hr.	Daily	Bare Costs	Incl. O&P
1 Carpenter Foreman (outside)	$56.70	$453.60	$84.60	$676.80	$51.31	$76.28
2 Carpenters	54.70	875.20	81.65	1306.40		
3 Laborers	44.40	1065.60	66.25	1590.00		
2 Cement Finishers	51.80	828.80	75.90	1214.40		
1 Rodman (reinf.)	58.90	471.20	88.05	704.40		
72 L.H., Daily Totals		$3694.40		$5492.00	$51.31	$76.28

Crew C-16

	Bare Costs		Incl. Subs O&P		Cost Per Labor-Hour	
Crew C-16	Hr.	Daily	Hr.	Daily	Bare Costs	Incl. O&P
1 Labor Foreman (outside)	$46.40	$371.20	$69.25	$554.00	$48.89	$72.53
3 Laborers	44.40	1065.60	66.25	1590.00		
2 Cement Finishers	51.80	828.80	75.90	1214.40		
1 Equip. Oper. (medium)	59.00	472.00	87.90	703.20		
1 Gunite Pump Rig		321.75		353.93		
2 -50' Air Hoses, 3/4"		14.30		15.73		
2 -50' Air Hoses, 2"		57.90		63.69	7.03	7.74
56 L.H., Daily Totals		$3131.55		$4494.94	$55.92	$80.27

Crew C-16A

	Bare Costs		Incl. Subs O&P		Cost Per Labor-Hour	
Crew C-16A	Hr.	Daily	Hr.	Daily	Bare Costs	Incl. O&P
1 Laborer	$44.40	$355.20	$66.25	$530.00	$51.75	$76.49
2 Cement Finishers	51.80	828.80	75.90	1214.40		
1 Equip. Oper. (medium)	59.00	472.00	87.90	703.20		
1 Gunite Pump Rig		321.75		353.93		
2 -50' Air Hoses, 3/4"		14.30		15.73		
2 -50' Air Hoses, 2"		57.90		63.69		
1 Telescoping Boom Lift, to 60'		292.45		321.69	21.45	23.59
32 L.H., Daily Totals		$2342.40		$3202.64	$73.20	$100.08

Crew C-17

	Bare Costs		Incl. Subs O&P		Cost Per Labor-Hour	
Crew C-17	Hr.	Daily	Hr.	Daily	Bare Costs	Incl. O&P
2 Skilled Worker Foremen (out)	$59.10	$945.60	$88.90	$1422.40	$57.50	$86.50
8 Skilled Workers	57.10	3654.40	85.90	5497.60		
80 L.H., Daily Totals		$4600.00		$6920.00	$57.50	$86.50

Crew C-17A

	Bare Costs		Incl. Subs O&P		Cost Per Labor-Hour	
Crew C-17A	Hr.	Daily	Hr.	Daily	Bare Costs	Incl. O&P
2 Skilled Worker Foremen (out)	$59.10	$945.60	$88.90	$1422.40	$57.55	$86.56
8 Skilled Workers	57.10	3654.40	85.90	5497.60		
.125 Equip. Oper. (crane)	61.45	61.45	91.55	91.55		
.125 Hyd. Crane, 80 Ton		182.25		200.47	2.25	2.48
81 L.H., Daily Totals		$4843.70		$7212.02	$59.80	$89.04

Crew C-17B

	Bare Costs		Incl. Subs O&P		Cost Per Labor-Hour	
Crew C-17B	Hr.	Daily	Hr.	Daily	Bare Costs	Incl. O&P
2 Skilled Worker Foremen (out)	$59.10	$945.60	$88.90	$1422.40	$57.60	$86.62
8 Skilled Workers	57.10	3654.40	85.90	5497.60		
.25 Equip. Oper. (crane)	61.45	122.90	91.55	183.10		
.25 Hyd. Crane, 80 Ton		364.50		400.95		
.25 Trowel, 48" Walk-Behind		23.57		25.93	4.73	5.21
82 L.H., Daily Totals		$5110.98		$7529.98	$62.33	$91.83

Crew C-17C

	Bare Costs		Incl. Subs O&P		Cost Per Labor-Hour	
Crew C-17C	Hr.	Daily	Hr.	Daily	Bare Costs	Incl. O&P
2 Skilled Worker Foremen (out)	$59.10	$945.60	$88.90	$1422.40	$57.64	$86.68
8 Skilled Workers	57.10	3654.40	85.90	5497.60		
.375 Equip. Oper. (crane)	61.45	184.35	91.55	274.65		
.375 Hyd. Crane, 80 Ton		546.75		601.42	6.59	7.25
83 L.H., Daily Totals		$5331.10		$7796.07	$64.23	$93.93

Crew C-17D

	Bare Costs		Incl. Subs O&P		Cost Per Labor-Hour	
Crew C-17D	Hr.	Daily	Hr.	Daily	Bare Costs	Incl. O&P
2 Skilled Worker Foremen (out)	$59.10	$945.60	$88.90	$1422.40	$57.69	$86.74
8 Skilled Workers	57.10	3654.40	85.90	5497.60		
.5 Equip. Oper. (crane)	61.45	245.80	91.55	366.20		
.5 Hyd. Crane, 80 Ton		729.00		801.90	8.68	9.55
84 L.H., Daily Totals		$5574.80		$8088.10	$66.37	$96.29

Crew C-17E

	Bare Costs		Incl. Subs O&P		Cost Per Labor-Hour	
Crew C-17E	Hr.	Daily	Hr.	Daily	Bare Costs	Incl. O&P
2 Skilled Worker Foremen (out)	$59.10	$945.60	$88.90	$1422.40	$57.50	$86.50
8 Skilled Workers	57.10	3654.40	85.90	5497.60		
1 Hyd. Jack with Rods		36.70		40.37	.46	.50
80 L.H., Daily Totals		$4636.70		$6960.37	$57.96	$87.00

Crew C-18

	Bare Costs		Incl. Subs O&P		Cost Per Labor-Hour	
Crew C-18	Hr.	Daily	Hr.	Daily	Bare Costs	Incl. O&P
.125 Labor Foreman (outside)	$46.40	$46.40	$69.25	$69.25	$44.62	$66.58
1 Laborer	44.40	355.20	66.25	530.00		
1 Concrete Cart, 10 C.F.		116.95		128.65	12.99	14.29
9 L.H., Daily Totals		$518.55		$727.89	$57.62	$80.88

Crew C-19

	Bare Costs		Incl. Subs O&P		Cost Per Labor-Hour	
Crew C-19	Hr.	Daily	Hr.	Daily	Bare Costs	Incl. O&P
.125 Labor Foreman (outside)	$46.40	$46.40	$69.25	$69.25	$44.62	$66.58
1 Laborer	44.40	355.20	66.25	530.00		
1 Concrete Cart, 18 C.F.		138.95		152.85	15.44	16.98
9 L.H., Daily Totals		$540.55		$752.10	$60.06	$83.57

Crew C-20

	Bare Costs		Incl. Subs O&P		Cost Per Labor-Hour	
Crew C-20	Hr.	Daily	Hr.	Daily	Bare Costs	Incl. O&P
1 Labor Foreman (outside)	$46.40	$371.20	$69.25	$554.00	$47.40	$70.54
5 Laborers	44.40	1776.00	66.25	2650.00		
1 Cement Finisher	51.80	414.40	75.90	607.20		
1 Equip. Oper. (medium)	59.00	472.00	87.90	703.20		
2 Gas Engine Vibrators		54.30		59.73		
1 Concrete Pump (Small)		423.65		466.01	7.47	8.21
64 L.H., Daily Totals		$3511.55		$5040.15	$54.87	$78.75

Crews - Standard

Crew No.	Bare Costs		Incl. Subs O&P		Cost Per Labor-Hour	

Crew C-21

Crew C-21	Hr.	Daily	Hr.	Daily	Bare Costs	Incl. O&P
1 Labor Foreman (outside)	$46.40	$371.20	$69.25	$554.00	$47.40	$70.54
5 Laborers	44.40	1776.00	66.25	2650.00		
1 Cement Finisher	51.80	414.40	75.90	607.20		
1 Equip. Oper. (medium)	59.00	472.00	87.90	703.20		
2 Gas Engine Vibrators		54.30		59.73		
1 Concrete Conveyer		206.25		226.88	4.07	4.48
64 L.H., Daily Totals		$3294.15		$4801.01	$51.47	$75.02

Crew C-22

Crew C-22	Hr.	Daily	Hr.	Daily	Bare Costs	Incl. O&P
1 Rodman Foreman (outside)	$60.90	$487.20	$91.05	$728.40	$59.19	$88.47
4 Rodmen (reinf.)	58.90	1884.80	88.05	2817.60		
.125 Equip. Oper. (crane)	61.45	61.45	91.55	91.55		
.125 Equip. Oper. (oiler)	52.50	52.50	78.25	78.25		
.125 Hyd. Crane, 25 Ton		73.34		80.67	1.75	1.92
42 L.H., Daily Totals		$2559.29		$3796.47	$60.94	$90.39

Crew C-23

Crew C-23	Hr.	Daily	Hr.	Daily	Bare Costs	Incl. O&P
2 Skilled Worker Foremen (out)	$59.10	$945.60	$88.90	$1422.40	$57.48	$86.30
6 Skilled Workers	57.10	2740.80	85.90	4123.20		
1 Equip. Oper. (crane)	61.45	491.60	91.55	732.40		
1 Equip. Oper. (oiler)	52.50	420.00	78.25	626.00		
1 Lattice Boom Crane, 90 Ton		1713.00		1884.30	21.41	23.55
80 L.H., Daily Totals		$6311.00		$8788.30	$78.89	$109.85

Crew C-23A

Crew C-23A	Hr.	Daily	Hr.	Daily	Bare Costs	Incl. O&P
1 Labor Foreman (outside)	$46.40	$371.20	$69.25	$554.00	$49.83	$74.31
2 Laborers	44.40	710.40	66.25	1060.00		
1 Equip. Oper. (crane)	61.45	491.60	91.55	732.40		
1 Equip. Oper. (oiler)	52.50	420.00	78.25	626.00		
1 Crawler Crane, 100 Ton		2310.00		2541.00		
3 Conc. Buckets, 8 C.Y.		337.95		371.75	66.20	72.82
40 L.H., Daily Totals		$4641.15		$5885.15	$116.03	$147.13

Crew C-24

Crew C-24	Hr.	Daily	Hr.	Daily	Bare Costs	Incl. O&P
2 Skilled Worker Foremen (out)	$59.10	$945.60	$88.90	$1422.40	$57.48	$86.30
6 Skilled Workers	57.10	2740.80	85.90	4123.20		
1 Equip. Oper. (crane)	61.45	491.60	91.55	732.40		
1 Equip. Oper. (oiler)	52.50	420.00	78.25	626.00		
1 Lattice Boom Crane, 150 Ton		2324.00		2556.40	29.05	31.95
80 L.H., Daily Totals		$6922.00		$9460.40	$86.53	$118.26

Crew C-25

Crew C-25	Hr.	Daily	Hr.	Daily	Bare Costs	Incl. O&P
2 Rodmen (reinf.)	$58.90	$942.40	$88.05	$1408.80	$47.42	$73.15
2 Rodmen Helpers	35.95	575.20	58.25	932.00		
32 L.H., Daily Totals		$1517.60		$2340.80	$47.42	$73.15

Crew C-27

Crew C-27	Hr.	Daily	Hr.	Daily	Bare Costs	Incl. O&P
2 Cement Finishers	$51.80	$828.80	$75.90	$1214.40	$51.80	$75.90
1 Concrete Saw		112.85		124.14	7.05	7.76
16 L.H., Daily Totals		$941.65		$1338.54	$58.85	$83.66

Crew C-28

Crew C-28	Hr.	Daily	Hr.	Daily	Bare Costs	Incl. O&P
1 Cement Finisher	$51.80	$414.40	$75.90	$607.20	$51.80	$75.90
1 Portable Air Compressor, Gas		38.80		42.68	4.85	5.34
8 L.H., Daily Totals		$453.20		$649.88	$56.65	$81.23

Crew C-29

Crew C-29	Hr.	Daily	Hr.	Daily	Bare Costs	Incl. O&P
1 Laborer	$44.40	$355.20	$66.25	$530.00	$44.40	$66.25
1 Pressure Washer		97.35		107.08	12.17	13.39
8 L.H., Daily Totals		$452.55		$637.09	$56.57	$79.64

Crew C-30

Crew C-30	Hr.	Daily	Hr.	Daily	Bare Costs	Incl. O&P
1 Laborer	$44.40	$355.20	$66.25	$530.00	$44.40	$66.25
1 Concrete Mixer, 10 C.F.		147.15		161.87	18.39	20.23
8 L.H., Daily Totals		$502.35		$691.87	$62.79	$86.48

Crew C-31

Crew C-31	Hr.	Daily	Hr.	Daily	Bare Costs	Incl. O&P
1 Cement Finisher	$51.80	$414.40	$75.90	$607.20	$51.80	$75.90
1 Grout Pump		321.75		353.93	40.22	44.24
8 L.H., Daily Totals		$736.15		$961.13	$92.02	$120.14

Crew C-32

Crew C-32	Hr.	Daily	Hr.	Daily	Bare Costs	Incl. O&P
1 Cement Finisher	$51.80	$414.40	$75.90	$607.20	$48.10	$71.08
1 Laborer	44.40	355.20	66.25	530.00		
1 Crack Chaser Saw, Gas, 6 H.P.		73.25		80.58		
1 Vacuum Pick-Up System		74.95		82.44	9.26	10.19
16 L.H., Daily Totals		$917.80		$1300.22	$57.36	$81.26

Crew D-1

Crew D-1	Hr.	Daily	Hr.	Daily	Bare Costs	Incl. O&P
1 Bricklayer	$53.70	$429.60	$80.90	$647.20	$48.70	$73.38
1 Bricklayer Helper	43.70	349.60	65.85	526.80		
16 L.H., Daily Totals		$779.20		$1174.00	$48.70	$73.38

Crew D-2

Crew D-2	Hr.	Daily	Hr.	Daily	Bare Costs	Incl. O&P
3 Bricklayers	$53.70	$1288.80	$80.90	$1941.60	$50.15	$75.50
2 Bricklayer Helpers	43.70	699.20	65.85	1053.60		
.5 Carpenter	54.70	218.80	81.65	326.60		
44 L.H., Daily Totals		$2206.80		$3321.80	$50.15	$75.50

Crew D-3

Crew D-3	Hr.	Daily	Hr.	Daily	Bare Costs	Incl. O&P
3 Bricklayers	$53.70	$1288.80	$80.90	$1941.60	$49.94	$75.20
2 Bricklayer Helpers	43.70	699.20	65.85	1053.60		
.25 Carpenter	54.70	109.40	81.65	163.30		
42 L.H., Daily Totals		$2097.40		$3158.50	$49.94	$75.20

Crew D-4

Crew D-4	Hr.	Daily	Hr.	Daily	Bare Costs	Incl. O&P
1 Bricklayer	$53.70	$429.60	$80.90	$647.20	$49.15	$73.83
2 Bricklayer Helpers	43.70	699.20	65.85	1053.60		
1 Equip. Oper. (light)	55.50	444.00	82.70	661.60		
1 Grout Pump, 50 C.F./hr.		190.35		209.38	5.95	6.54
32 L.H., Daily Totals		$1763.15		$2571.78	$55.10	$80.37

Crew D-5

Crew D-5	Hr.	Daily	Hr.	Daily	Bare Costs	Incl. O&P
1 Bricklayer	53.70	429.60	80.90	647.20	53.70	80.90
8 L.H., Daily Totals		$429.60		$647.20	$53.70	$80.90

Crew D-6

Crew D-6	Hr.	Daily	Hr.	Daily	Bare Costs	Incl. O&P
3 Bricklayers	$53.70	$1288.80	$80.90	$1941.60	$48.94	$73.71
3 Bricklayer Helpers	43.70	1048.80	65.85	1580.40		
.25 Carpenter	54.70	109.40	81.65	163.30		
50 L.H., Daily Totals		$2447.00		$3685.30	$48.94	$73.71

Crew D-7

Crew D-7	Hr.	Daily	Hr.	Daily	Bare Costs	Incl. O&P
1 Tile Layer	$51.70	$413.60	$75.55	$604.40	$46.65	$68.17
1 Tile Layer Helper	41.60	332.80	60.80	486.40		
16 L.H., Daily Totals		$746.40		$1090.80	$46.65	$68.17

Crew D-8

Crew D-8	Hr.	Daily	Hr.	Daily	Bare Costs	Incl. O&P
3 Bricklayers	$53.70	$1288.80	$80.90	$1941.60	$49.70	$74.88
2 Bricklayer Helpers	43.70	699.20	65.85	1053.60		
40 L.H., Daily Totals		$1988.00		$2995.20	$49.70	$74.88

For customer support on your Plumbing Costs with RSMeans Data, call 800.448.8182.

625

Crew No.	Bare Costs		Incl. Subs O&P		Cost Per Labor-Hour	
Crew D-9	Hr.	Daily	Hr.	Daily	Bare Costs	Incl. O&P
3 Bricklayers	$53.70	$1288.80	$80.90	$1941.60	$48.70	$73.38
3 Bricklayer Helpers	43.70	1048.80	65.85	1580.40		
48 L.H., Daily Totals		$2337.60		$3522.00	$48.70	$73.38
Crew D-10	Hr.	Daily	Hr.	Daily	Bare Costs	Incl. O&P
1 Bricklayer Foreman (outside)	$55.70	$445.60	$83.90	$671.20	$53.64	$80.55
1 Bricklayer	53.70	429.60	80.90	647.20		
1 Bricklayer Helper	43.70	349.60	65.85	526.80		
1 Equip. Oper. (crane)	61.45	491.60	91.55	732.40		
1 S.P. Crane, 4x4, 12 Ton		432.65		475.92	13.52	14.87
32 L.H., Daily Totals		$2149.05		$3053.51	$67.16	$95.42
Crew D-11	Hr.	Daily	Hr.	Daily	Bare Costs	Incl. O&P
1 Bricklayer Foreman (outside)	$55.70	$445.60	$83.90	$671.20	$51.03	$76.88
1 Bricklayer	53.70	429.60	80.90	647.20		
1 Bricklayer Helper	43.70	349.60	65.85	526.80		
24 L.H., Daily Totals		$1224.80		$1845.20	$51.03	$76.88
Crew D-12	Hr.	Daily	Hr.	Daily	Bare Costs	Incl. O&P
1 Bricklayer Foreman (outside)	$55.70	$445.60	$83.90	$671.20	$49.20	$74.13
1 Bricklayer	53.70	429.60	80.90	647.20		
2 Bricklayer Helpers	43.70	699.20	65.85	1053.60		
32 L.H., Daily Totals		$1574.40		$2372.00	$49.20	$74.13
Crew D-13	Hr.	Daily	Hr.	Daily	Bare Costs	Incl. O&P
1 Bricklayer Foreman (outside)	$55.70	$445.60	$83.90	$671.20	$52.16	$78.28
1 Bricklayer	53.70	429.60	80.90	647.20		
2 Bricklayer Helpers	43.70	699.20	65.85	1053.60		
1 Carpenter	54.70	437.60	81.65	653.20		
1 Equip. Oper. (crane)	61.45	491.60	91.55	732.40		
1 S.P. Crane, 4x4, 12 Ton		432.65		475.92	9.01	9.91
48 L.H., Daily Totals		$2936.25		$4233.52	$61.17	$88.20
Crew D-14	Hr.	Daily	Hr.	Daily	Bare Costs	Incl. O&P
3 Bricklayers	$53.70	$1288.80	$80.90	$1941.60	$51.20	$77.14
1 Bricklayer Helper	43.70	349.60	65.85	526.80		
32 L.H., Daily Totals		$1638.40		$2468.40	$51.20	$77.14
Crew E-1	Hr.	Daily	Hr.	Daily	Bare Costs	Incl. O&P
1 Welder Foreman (outside)	$62.30	$498.40	$96.40	$771.20	$59.37	$90.80
1 Welder	60.30	482.40	93.30	746.40		
1 Equip. Oper. (light)	55.50	444.00	82.70	661.60		
1 Welder, Gas Engine, 300 amp		148.75		163.63	6.20	6.82
24 L.H., Daily Totals		$1573.55		$2342.82	$65.56	$97.62
Crew E-2	Hr.	Daily	Hr.	Daily	Bare Costs	Incl. O&P
1 Struc. Steel Foreman (outside)	$62.30	$498.40	$96.40	$771.20	$59.64	$91.34
4 Struc. Steel Workers	60.30	1929.60	93.30	2985.60		
1 Equip. Oper. (crane)	61.45	491.60	91.55	732.40		
1 Equip. Oper. (oiler)	52.50	420.00	78.25	626.00		
1 Lattice Boom Crane, 90 Ton		1713.00		1884.30	30.59	33.65
56 L.H., Daily Totals		$5052.60		$6999.50	$90.22	$124.99
Crew E-3	Hr.	Daily	Hr.	Daily	Bare Costs	Incl. O&P
1 Struc. Steel Foreman (outside)	$62.30	$498.40	$96.40	$771.20	$60.97	$94.33
1 Struc. Steel Worker	60.30	482.40	93.30	746.40		
1 Welder	60.30	482.40	93.30	746.40		
1 Welder, Gas Engine, 300 amp		148.75		163.63	6.20	6.82
24 L.H., Daily Totals		$1611.95		$2427.63	$67.16	$101.15

Crew No.	Bare Costs		Incl. Subs O&P		Cost Per Labor-Hour	
Crew E-3A	Hr.	Daily	Hr.	Daily	Bare Costs	Incl. O&P
1 Struc. Steel Foreman (outside)	$62.30	$498.40	$96.40	$771.20	$60.97	$94.33
1 Struc. Steel Worker	60.30	482.40	93.30	746.40		
1 Welder	60.30	482.40	93.30	746.40		
1 Welder, Gas Engine, 300 amp		148.75		163.63		
1 Telescoping Boom Lift, to 40'		281.90		310.09	17.94	19.74
24 L.H., Daily Totals		$1893.85		$2737.72	$78.91	$114.07
Crew E-4	Hr.	Daily	Hr.	Daily	Bare Costs	Incl. O&P
1 Struc. Steel Foreman (outside)	$62.30	$498.40	$96.40	$771.20	$60.80	$94.08
3 Struc. Steel Workers	60.30	1447.20	93.30	2239.20		
1 Welder, Gas Engine, 300 amp		148.75		163.63	4.65	5.11
32 L.H., Daily Totals		$2094.35		$3174.03	$65.45	$99.19
Crew E-5	Hr.	Daily	Hr.	Daily	Bare Costs	Incl. O&P
2 Struc. Steel Foremen (outside)	$62.30	$996.80	$96.40	$1542.40	$60.03	$92.24
5 Struc. Steel Workers	60.30	2412.00	93.30	3732.00		
1 Equip. Oper. (crane)	61.45	491.60	91.55	732.40		
1 Welder	60.30	482.40	93.30	746.40		
1 Equip. Oper. (oiler)	52.50	420.00	78.25	626.00		
1 Lattice Boom Crane, 90 Ton		1713.00		1884.30		
1 Welder, Gas Engine, 300 amp		148.75		163.63	23.27	25.60
80 L.H., Daily Totals		$6664.55		$9427.13	$83.31	$117.84
Crew E-6	Hr.	Daily	Hr.	Daily	Bare Costs	Incl. O&P
3 Struc. Steel Foremen (outside)	$62.30	$1495.20	$96.40	$2313.60	$59.96	$92.17
9 Struc. Steel Workers	60.30	4341.60	93.30	6717.60		
1 Equip. Oper. (crane)	61.45	491.60	91.55	732.40		
1 Welder	60.30	482.40	93.30	746.40		
1 Equip. Oper. (oiler)	52.50	420.00	78.25	626.00		
1 Equip. Oper. (light)	55.50	444.00	82.70	661.60		
1 Lattice Boom Crane, 90 Ton		1713.00		1884.30		
1 Welder, Gas Engine, 300 amp		148.75		163.63		
1 Air Compressor, 160 cfm		212.30		233.53		
2 Impact Wrenches		90.30		99.33	16.91	18.60
128 L.H., Daily Totals		$9839.15		$14178.39	$76.87	$110.77
Crew E-7	Hr.	Daily	Hr.	Daily	Bare Costs	Incl. O&P
1 Struc. Steel Foreman (outside)	$62.30	$498.40	$96.40	$771.20	$60.03	$92.24
4 Struc. Steel Workers	60.30	1929.60	93.30	2985.60		
1 Equip. Oper. (crane)	61.45	491.60	91.55	732.40		
1 Equip. Oper. (oiler)	52.50	420.00	78.25	626.00		
1 Welder Foreman (outside)	62.30	498.40	96.40	771.20		
2 Welders	60.30	964.80	93.30	1492.80		
1 Lattice Boom Crane, 90 Ton		1713.00		1884.30		
2 Welder, Gas Engine, 300 amp		297.50		327.25	25.13	27.64
80 L.H., Daily Totals		$6813.30		$9590.75	$85.17	$119.88
Crew E-8	Hr.	Daily	Hr.	Daily	Bare Costs	Incl. O&P
1 Struc. Steel Foreman (outside)	$62.30	$498.40	$96.40	$771.20	$59.73	$91.67
4 Struc. Steel Workers	60.30	1929.60	93.30	2985.60		
1 Welder Foreman (outside)	62.30	498.40	96.40	771.20		
4 Welders	60.30	1929.60	93.30	2985.60		
1 Equip. Oper. (crane)	61.45	491.60	91.55	732.40		
1 Equip. Oper. (oiler)	52.50	420.00	78.25	626.00		
1 Equip. Oper. (light)	55.50	444.00	82.70	661.60		
1 Lattice Boom Crane, 90 Ton		1713.00		1884.30		
4 Welder, Gas Engine, 300 amp		595.00		654.50	22.19	24.41
104 L.H., Daily Totals		$8519.60		$12072.40	$81.92	$116.08

For customer support on your Plumbing Costs with RSMeans Data, call 800.448.8182.

Crews - Standard

Crew E-9	Hr.	Daily	Hr.	Daily	Bare Costs	Incl. O&P
2 Struc. Steel Foremen (outside)	$62.30	$996.80	$96.40	$1542.40	$59.96	$92.17
5 Struc. Steel Workers	60.30	2412.00	93.30	3732.00		
1 Welder Foreman (outside)	62.30	498.40	96.40	771.20		
5 Welders	60.30	2412.00	93.30	3732.00		
1 Equip. Oper. (crane)	61.45	491.60	91.55	732.40		
1 Equip. Oper. (oiler)	52.50	420.00	78.25	626.00		
1 Equip. Oper. (light)	55.50	444.00	82.70	661.60		
1 Lattice Boom Crane, 90 Ton		1713.00		1884.30		
5 Welder, Gas Engine, 300 amp		743.75		818.13	19.19	21.11
128 L.H., Daily Totals		$10131.55		$14500.03	$79.15	$113.28

Crew E-10	Hr.	Daily	Hr.	Daily	Bare Costs	Incl. O&P
1 Welder Foreman (outside)	$62.30	$498.40	$96.40	$771.20	$61.30	$94.85
1 Welder	60.30	482.40	93.30	746.40		
1 Welder, Gas Engine, 300 amp		148.75		163.63		
1 Flatbed Truck, Gas, 3 Ton		850.05		935.05	62.42	68.67
16 L.H., Daily Totals		$1979.60		$2616.28	$123.72	$163.52

Crew E-11	Hr.	Daily	Hr.	Daily	Bare Costs	Incl. O&P
2 Painters, Struc. Steel	$47.20	$755.20	$75.80	$1212.80	$48.58	$75.14
1 Building Laborer	44.40	355.20	66.25	530.00		
1 Equip. Oper. (light)	55.50	444.00	82.70	661.60		
1 Air Compressor, 250 cfm		202.85		223.13		
1 Sandblaster, Portable, 3 C.F.		83.85		92.23		
1 Set Sand Blasting Accessories		15.55		17.11	9.45	10.39
32 L.H., Daily Totals		$1856.65		$2736.88	$58.02	$85.53

Crew E-11A	Hr.	Daily	Hr.	Daily	Bare Costs	Incl. O&P
2 Painters, Struc. Steel	$47.20	$755.20	$75.80	$1212.80	$48.58	$75.14
1 Building Laborer	44.40	355.20	66.25	530.00		
1 Equip. Oper. (light)	55.50	444.00	82.70	661.60		
1 Air Compressor, 250 cfm		202.85		223.13		
1 Sandblaster, Portable, 3 C.F.		83.85		92.23		
1 Set Sand Blasting Accessories		15.55		17.11		
1 Telescoping Boom Lift, to 60'		292.45		321.69	18.58	20.44
32 L.H., Daily Totals		$2149.10		$3058.57	$67.16	$95.58

Crew E-11B	Hr.	Daily	Hr.	Daily	Bare Costs	Incl. O&P
2 Painters, Struc. Steel	$47.20	$755.20	$75.80	$1212.80	$46.27	$72.62
1 Building Laborer	44.40	355.20	66.25	530.00		
2 Paint Sprayer, 8 C.F.M.		88.40		97.24		
1 Telescoping Boom Lift, to 60'		292.45		321.69	15.87	17.46
24 L.H., Daily Totals		$1491.25		$2161.74	$62.14	$90.07

Crew E-12	Hr.	Daily	Hr.	Daily	Bare Costs	Incl. O&P
1 Welder Foreman (outside)	$62.30	$498.40	$96.40	$771.20	$58.90	$89.55
1 Equip. Oper. (light)	55.50	444.00	82.70	661.60		
1 Welder, Gas Engine, 300 amp		148.75		163.63	9.30	10.23
16 L.H., Daily Totals		$1091.15		$1596.43	$68.20	$99.78

Crew E-13	Hr.	Daily	Hr.	Daily	Bare Costs	Incl. O&P
1 Welder Foreman (outside)	$62.30	$498.40	$96.40	$771.20	$60.03	$91.83
.5 Equip. Oper. (light)	55.50	222.00	82.70	330.80		
1 Welder, Gas Engine, 300 amp		148.75		163.63	12.40	13.64
12 L.H., Daily Totals		$869.15		$1265.63	$72.43	$105.47

Crew E-14	Hr.	Daily	Hr.	Daily	Bare Costs	Incl. O&P
1 Welder Foreman (outside)	$62.30	$498.40	$96.40	$771.20	$62.30	$96.40
1 Welder, Gas Engine, 300 amp		148.75		163.63	18.59	20.45
8 L.H., Daily Totals		$647.15		$934.83	$80.89	$116.85

Crew E-16	Hr.	Daily	Hr.	Daily	Bare Costs	Incl. O&P
1 Welder Foreman (outside)	$62.30	$498.40	$96.40	$771.20	$61.30	$94.85
1 Welder	60.30	482.40	93.30	746.40		
1 Welder, Gas Engine, 300 amp		148.75		163.63	9.30	10.23
16 L.H., Daily Totals		$1129.55		$1681.22	$70.60	$105.08

Crew E-17	Hr.	Daily	Hr.	Daily	Bare Costs	Incl. O&P
1 Struc. Steel Foreman (outside)	$62.30	$498.40	$96.40	$771.20	$61.30	$94.85
1 Structural Steel Worker	60.30	482.40	93.30	746.40		
16 L.H., Daily Totals		$980.80		$1517.60	$61.30	$94.85

Crew E-18	Hr.	Daily	Hr.	Daily	Bare Costs	Incl. O&P
1 Struc. Steel Foreman (outside)	$62.30	$498.40	$96.40	$771.20	$60.44	$92.84
3 Structural Steel Workers	60.30	1447.20	93.30	2239.20		
1 Equipment Operator (med.)	59.00	472.00	87.90	703.20		
1 Lattice Boom Crane, 20 Ton		1526.00		1678.60	38.15	41.97
40 L.H., Daily Totals		$3943.60		$5392.20	$98.59	$134.81

Crew E-19	Hr.	Daily	Hr.	Daily	Bare Costs	Incl. O&P
1 Struc. Steel Foreman (outside)	$62.30	$498.40	$96.40	$771.20	$59.37	$90.80
1 Structural Steel Worker	60.30	482.40	93.30	746.40		
1 Equip. Oper. (light)	55.50	444.00	82.70	661.60		
1 Lattice Boom Crane, 20 Ton		1526.00		1678.60	63.58	69.94
24 L.H., Daily Totals		$2950.80		$3857.80	$122.95	$160.74

Crew E-20	Hr.	Daily	Hr.	Daily	Bare Costs	Incl. O&P
1 Struc. Steel Foreman (outside)	$62.30	$498.40	$96.40	$771.20	$59.72	$91.59
5 Structural Steel Workers	60.30	2412.00	93.30	3732.00		
1 Equip. Oper. (crane)	61.45	491.60	91.55	732.40		
1 Equip. Oper. (oiler)	52.50	420.00	78.25	626.00		
1 Lattice Boom Crane, 40 Ton		2127.00		2339.70	33.23	36.56
64 L.H., Daily Totals		$5949.00		$8201.30	$92.95	$128.15

Crew E-22	Hr.	Daily	Hr.	Daily	Bare Costs	Incl. O&P
1 Skilled Worker Foreman (out)	$59.10	$472.80	$88.90	$711.20	$57.77	$86.90
2 Skilled Workers	57.10	913.60	85.90	1374.40		
24 L.H., Daily Totals		$1386.40		$2085.60	$57.77	$86.90

Crew E-24	Hr.	Daily	Hr.	Daily	Bare Costs	Incl. O&P
3 Structural Steel Workers	$60.30	$1447.20	$93.30	$2239.20	$59.98	$91.95
1 Equipment Operator (med.)	59.00	472.00	87.90	703.20		
1 Hyd. Crane, 25 Ton		586.70		645.37	18.33	20.17
32 L.H., Daily Totals		$2505.90		$3587.77	$78.31	$112.12

Crew E-25	Hr.	Daily	Hr.	Daily	Bare Costs	Incl. O&P
1 Welder Foreman (outside)	$62.30	$498.40	$96.40	$771.20	$62.30	$96.40
1 Cutting Torch		12.95		14.24	1.62	1.78
8 L.H., Daily Totals		$511.35		$785.45	$63.92	$98.18

Crew E-26	Hr.	Daily	Hr.	Daily	Bare Costs	Incl. O&P
1 Struc. Steel Foreman (outside)	$62.30	$498.40	$96.40	$771.20	$61.64	$94.84
1 Struc. Steel Worker	60.30	482.40	93.30	746.40		
1 Welder	60.30	482.40	93.30	746.40		
.25 Electrician	63.70	127.40	94.65	189.30		
.25 Plumber	67.70	135.40	101.05	202.10		
1 Welder, Gas Engine, 300 amp		148.75		163.63	5.31	5.84
28 L.H., Daily Totals		$1874.75		$2819.03	$66.96	$100.68

Crews - Standard

Crew E-27

Crew No.	Bare Costs Hr.	Daily	Incl. Subs O&P Hr.	Daily	Cost Per L.H. Bare Costs	Incl. O&P
1 Struc. Steel Foreman (outside)	$62.30	$498.40	$96.40	$771.20	$59.72	$91.59
5 Struc. Steel Workers	60.30	2412.00	93.30	3732.00		
1 Equip. Oper. (crane)	61.45	491.60	91.55	732.40		
1 Equip. Oper. (oiler)	52.50	420.00	78.25	626.00		
1 Hyd. Crane, 12 Ton		475.80		523.38		
1 Hyd. Crane, 80 Ton		1458.00		1603.80	30.22	33.24
64 L.H., Daily Totals		$5755.80		$7988.78	$89.93	$124.82

Crew F-3

Crew No.	Bare Costs Hr.	Daily	Incl. Subs O&P Hr.	Daily	Cost Per L.H. Bare Costs	Incl. O&P
4 Carpenters	$54.70	$1750.40	$81.65	$2612.80	$56.05	$83.63
1 Equip. Oper. (crane)	61.45	491.60	91.55	732.40		
1 Hyd. Crane, 12 Ton		475.80		523.38	11.90	13.08
40 L.H., Daily Totals		$2717.80		$3868.58	$67.94	$96.71

Crew F-4

Crew No.	Bare Costs Hr.	Daily	Incl. Subs O&P Hr.	Daily	Cost Per L.H. Bare Costs	Incl. O&P
4 Carpenters	$54.70	$1750.40	$81.65	$2612.80	$55.46	$82.73
1 Equip. Oper. (crane)	61.45	491.60	91.55	732.40		
1 Equip. Oper. (oiler)	52.50	420.00	78.25	626.00		
1 Hyd. Crane, 55 Ton		990.15		1089.17	20.63	22.69
48 L.H., Daily Totals		$3652.15		$5060.36	$76.09	$105.42

Crew F-5

Crew No.	Bare Costs Hr.	Daily	Incl. Subs O&P Hr.	Daily	Cost Per L.H. Bare Costs	Incl. O&P
1 Carpenter Foreman (outside)	$56.70	$453.60	$84.60	$676.80	$55.20	$82.39
3 Carpenters	54.70	1312.80	81.65	1959.60		
32 L.H., Daily Totals		$1766.40		$2636.40	$55.20	$82.39

Crew F-6

Crew No.	Bare Costs Hr.	Daily	Incl. Subs O&P Hr.	Daily	Cost Per L.H. Bare Costs	Incl. O&P
2 Carpenters	$54.70	$875.20	$81.65	$1306.40	$51.93	$77.47
2 Building Laborers	44.40	710.40	66.25	1060.00		
1 Equip. Oper. (crane)	61.45	491.60	91.55	732.40		
1 Hyd. Crane, 12 Ton		475.80		523.38	11.90	13.08
40 L.H., Daily Totals		$2553.00		$3622.18	$63.83	$90.55

Crew F-7

Crew No.	Bare Costs Hr.	Daily	Incl. Subs O&P Hr.	Daily	Cost Per L.H. Bare Costs	Incl. O&P
2 Carpenters	$54.70	$875.20	$81.65	$1306.40	$49.55	$73.95
2 Building Laborers	44.40	710.40	66.25	1060.00		
32 L.H., Daily Totals		$1585.60		$2366.40	$49.55	$73.95

Crew G-1

Crew No.	Bare Costs Hr.	Daily	Incl. Subs O&P Hr.	Daily	Cost Per L.H. Bare Costs	Incl. O&P
1 Roofer Foreman (outside)	$50.20	$401.60	$81.35	$650.80	$44.99	$72.92
4 Roofers Composition	48.20	1542.40	78.15	2500.80		
2 Roofer Helpers	35.95	575.20	58.25	932.00		
1 Application Equipment		194.80		214.28		
1 Tar Kettle/Pot		209.95		230.94		
1 Crew Truck		168.15		184.97	10.23	11.25
56 L.H., Daily Totals		$3092.10		$4713.79	$55.22	$84.17

Crew G-2

Crew No.	Bare Costs Hr.	Daily	Incl. Subs O&P Hr.	Daily	Cost Per L.H. Bare Costs	Incl. O&P
1 Plasterer	$49.85	$398.80	$74.25	$594.00	$46.27	$68.95
1 Plasterer Helper	44.55	356.40	66.35	530.80		
1 Building Laborer	44.40	355.20	66.25	530.00		
1 Grout Pump, 50 C.F./hr.		190.35		209.38	7.93	8.72
24 L.H., Daily Totals		$1300.75		$1864.18	$54.20	$77.67

Crew G-2A

Crew No.	Bare Costs Hr.	Daily	Incl. Subs O&P Hr.	Daily	Cost Per L.H. Bare Costs	Incl. O&P
1 Roofer Composition	$48.20	$385.60	$78.15	$625.20	$42.85	$67.55
1 Roofer Helper	35.95	287.60	58.25	466.00		
1 Building Laborer	44.40	355.20	66.25	530.00		
1 Foam Spray Rig, Trailer-Mtd.		530.15		583.16		
1 Pickup Truck, 3/4 Ton		112.20		123.42	26.76	29.44
24 L.H., Daily Totals		$1670.75		$2327.78	$69.61	$96.99

Crew G-3

Crew No.	Bare Costs Hr.	Daily	Incl. Subs O&P Hr.	Daily	Cost Per L.H. Bare Costs	Incl. O&P
2 Sheet Metal Workers	$65.45	$1047.20	$98.70	$1579.20	$54.92	$82.47
2 Building Laborers	44.40	710.40	66.25	1060.00		
32 L.H., Daily Totals		$1757.60		$2639.20	$54.92	$82.47

Crew G-4

Crew No.	Bare Costs Hr.	Daily	Incl. Subs O&P Hr.	Daily	Cost Per L.H. Bare Costs	Incl. O&P
1 Labor Foreman (outside)	$46.40	$371.20	$69.25	$554.00	$45.07	$67.25
2 Building Laborers	44.40	710.40	66.25	1060.00		
1 Flatbed Truck, Gas, 1.5 Ton		198.50		218.35		
1 Air Compressor, 160 cfm		212.30		233.53	17.12	18.83
24 L.H., Daily Totals		$1492.40		$2065.88	$62.18	$86.08

Crew G-5

Crew No.	Bare Costs Hr.	Daily	Incl. Subs O&P Hr.	Daily	Cost Per L.H. Bare Costs	Incl. O&P
1 Roofer Foreman (outside)	$50.20	$401.60	$81.35	$650.80	$43.70	$70.83
2 Roofers Composition	48.20	771.20	78.15	1250.40		
2 Roofer Helpers	35.95	575.20	58.25	932.00		
1 Application Equipment		194.80		214.28	4.87	5.36
40 L.H., Daily Totals		$1942.80		$3047.48	$48.57	$76.19

Crew G-6A

Crew No.	Bare Costs Hr.	Daily	Incl. Subs O&P Hr.	Daily	Cost Per L.H. Bare Costs	Incl. O&P
2 Roofers Composition	$48.20	$771.20	$78.15	$1250.40	$48.20	$78.15
1 Small Compressor, Electric		39.30		43.23		
2 Pneumatic Nailers		55.40		60.94	5.92	6.51
16 L.H., Daily Totals		$865.90		$1354.57	$54.12	$84.66

Crew G-7

Crew No.	Bare Costs Hr.	Daily	Incl. Subs O&P Hr.	Daily	Cost Per L.H. Bare Costs	Incl. O&P
1 Carpenter	$54.70	$437.60	$81.65	$653.20	$54.70	$81.65
1 Small Compressor, Electric		39.30		43.23		
1 Pneumatic Nailer		27.70		30.47	8.38	9.21
8 L.H., Daily Totals		$504.60		$726.90	$63.08	$90.86

Crew H-1

Crew No.	Bare Costs Hr.	Daily	Incl. Subs O&P Hr.	Daily	Cost Per L.H. Bare Costs	Incl. O&P
2 Glaziers	$52.65	$842.40	$78.40	$1254.40	$56.48	$85.85
2 Struc. Steel Workers	60.30	964.80	93.30	1492.80		
32 L.H., Daily Totals		$1807.20		$2747.20	$56.48	$85.85

Crew H-2

Crew No.	Bare Costs Hr.	Daily	Incl. Subs O&P Hr.	Daily	Cost Per L.H. Bare Costs	Incl. O&P
2 Glaziers	$52.65	$842.40	$78.40	$1254.40	$49.90	$74.35
1 Building Laborer	44.40	355.20	66.25	530.00		
24 L.H., Daily Totals		$1197.60		$1784.40	$49.90	$74.35

Crew H-3

Crew No.	Bare Costs Hr.	Daily	Incl. Subs O&P Hr.	Daily	Cost Per L.H. Bare Costs	Incl. O&P
1 Glazier	$52.65	$421.20	$78.40	$627.20	$47.35	$71.03
1 Helper	42.05	336.40	63.65	509.20		
16 L.H., Daily Totals		$757.60		$1136.40	$47.35	$71.03

Crew H-4

Crew No.	Bare Costs Hr.	Daily	Incl. Subs O&P Hr.	Daily	Cost Per L.H. Bare Costs	Incl. O&P
1 Carpenter	$54.70	$437.60	$81.65	$653.20	$51.44	$77.05
1 Carpenter Helper	42.05	336.40	63.65	509.20		
.5 Electrician	63.70	254.80	94.65	378.60		
20 L.H., Daily Totals		$1028.80		$1541.00	$51.44	$77.05

Crew No.	Bare Costs		Incl. Subs O&P		Cost Per Labor-Hour	
Crew J-1	Hr.	Daily	Hr.	Daily	Bare Costs	Incl. O&P
3 Plasterers	$49.85	$1196.40	$74.25	$1782.00	$47.73	$71.09
2 Plasterer Helpers	44.55	712.80	66.35	1061.60		
1 Mixing Machine, 6 C.F.		113.35		124.69	2.83	3.12
40 L.H., Daily Totals		$2022.55		$2968.28	$50.56	$74.21
Crew J-2	Hr.	Daily	Hr.	Daily	Bare Costs	Incl. O&P
3 Plasterers	$49.85	$1196.40	$74.25	$1782.00	$48.79	$72.41
2 Plasterer Helpers	44.55	712.80	66.35	1061.60		
1 Lather	54.10	432.80	79.00	632.00		
1 Mixing Machine, 6 C.F.		113.35		124.69	2.36	2.60
48 L.H., Daily Totals		$2455.35		$3600.28	$51.15	$75.01
Crew J-3	Hr.	Daily	Hr.	Daily	Bare Costs	Incl. O&P
1 Terrazzo Worker	$51.75	$414.00	$75.60	$604.80	$47.65	$69.63
1 Terrazzo Helper	43.55	348.40	63.65	509.20		
1 Floor Grinder, 22" Path		96.05		105.66		
1 Terrazzo Mixer		162.90		179.19	16.18	17.80
16 L.H., Daily Totals		$1021.35		$1398.85	$63.83	$87.43
Crew J-4	Hr.	Daily	Hr.	Daily	Bare Costs	Incl. O&P
2 Cement Finishers	$51.80	$828.80	$75.90	$1214.40	$49.33	$72.68
1 Laborer	44.40	355.20	66.25	530.00		
1 Floor Grinder, 22" Path		96.05		105.66		
1 Floor Edger, 7" Path		44.05		48.45		
1 Vacuum Pick-Up System		74.95		82.44	8.96	9.86
24 L.H., Daily Totals		$1399.05		$1980.95	$58.29	$82.54
Crew J-4A	Hr.	Daily	Hr.	Daily	Bare Costs	Incl. O&P
2 Cement Finishers	$51.80	$828.80	$75.90	$1214.40	$48.10	$71.08
2 Laborers	44.40	710.40	66.25	1060.00		
1 Floor Grinder, 22" Path		96.05		105.66		
1 Floor Edger, 7" Path		44.05		48.45		
1 Vacuum Pick-Up System		74.95		82.44		
1 Floor Auto Scrubber		179.55		197.51	12.33	13.56
32 L.H., Daily Totals		$1933.80		$2708.46	$60.43	$84.64
Crew J-4B	Hr.	Daily	Hr.	Daily	Bare Costs	Incl. O&P
1 Laborer	$44.40	$355.20	$66.25	$530.00	$44.40	$66.25
1 Floor Auto Scrubber		179.55		197.51	22.44	24.69
8 L.H., Daily Totals		$534.75		$727.51	$66.84	$90.94
Crew J-6	Hr.	Daily	Hr.	Daily	Bare Costs	Incl. O&P
2 Painters	$46.45	$743.20	$68.90	$1102.40	$48.20	$71.69
1 Building Laborer	44.40	355.20	66.25	530.00		
1 Equip. Oper. (light)	55.50	444.00	82.70	661.60		
1 Air Compressor, 250 cfm		202.85		223.13		
1 Sandblaster, Portable, 3 C.F.		83.85		92.23		
1 Set Sand Blasting Accessories		15.55		17.11	9.45	10.39
32 L.H., Daily Totals		$1844.65		$2626.47	$57.65	$82.08
Crew J-7	Hr.	Daily	Hr.	Daily	Bare Costs	Incl. O&P
2 Painters	$46.45	$743.20	$68.90	$1102.40	$46.45	$68.90
1 Floor Belt Sander		50.20		55.22		
1 Floor Sanding Edger		25.20		27.72	4.71	5.18
16 L.H., Daily Totals		$818.60		$1185.34	$51.16	$74.08

Crew No.	Bare Costs		Incl. Subs O&P		Cost Per Labor-Hour	
Crew K-1	Hr.	Daily	Hr.	Daily	Bare Costs	Incl. O&P
1 Carpenter	$54.70	$437.60	$81.65	$653.20	$51.75	$77.33
1 Truck Driver (light)	48.80	390.40	73.00	584.00		
1 Flatbed Truck, Gas, 3 Ton		850.05		935.05	53.13	58.44
16 L.H., Daily Totals		$1678.05		$2172.26	$104.88	$135.77
Crew K-2	Hr.	Daily	Hr.	Daily	Bare Costs	Incl. O&P
1 Struc. Steel Foreman (outside)	$62.30	$498.40	$96.40	$771.20	$57.13	$87.57
1 Struc. Steel Worker	60.30	482.40	93.30	746.40		
1 Truck Driver (light)	48.80	390.40	73.00	584.00		
1 Flatbed Truck, Gas, 3 Ton		850.05		935.05	35.42	38.96
24 L.H., Daily Totals		$2221.25		$3036.66	$92.55	$126.53
Crew L-1	Hr.	Daily	Hr.	Daily	Bare Costs	Incl. O&P
1 Electrician	$63.70	$509.60	$94.65	$757.20	$65.70	$97.85
1 Plumber	67.70	541.60	101.05	808.40		
16 L.H., Daily Totals		$1051.20		$1565.60	$65.70	$97.85
Crew L-2	Hr.	Daily	Hr.	Daily	Bare Costs	Incl. O&P
1 Carpenter	$54.70	$437.60	$81.65	$653.20	$48.38	$72.65
1 Carpenter Helper	42.05	336.40	63.65	509.20		
16 L.H., Daily Totals		$774.00		$1162.40	$48.38	$72.65
Crew L-3	Hr.	Daily	Hr.	Daily	Bare Costs	Incl. O&P
1 Carpenter	$54.70	$437.60	$81.65	$653.20	$59.64	$89.16
.5 Electrician	63.70	254.80	94.65	378.60		
.5 Sheet Metal Worker	65.45	261.80	98.70	394.80		
16 L.H., Daily Totals		$954.20		$1426.60	$59.64	$89.16
Crew L-3A	Hr.	Daily	Hr.	Daily	Bare Costs	Incl. O&P
1 Carpenter Foreman (outside)	$56.70	$453.60	$84.60	$676.80	$59.62	$89.30
.5 Sheet Metal Worker	65.45	261.80	98.70	394.80		
12 L.H., Daily Totals		$715.40		$1071.60	$59.62	$89.30
Crew L-4	Hr.	Daily	Hr.	Daily	Bare Costs	Incl. O&P
2 Skilled Workers	$57.10	$913.60	$85.90	$1374.40	$52.08	$78.48
1 Helper	42.05	336.40	63.65	509.20		
24 L.H., Daily Totals		$1250.00		$1883.60	$52.08	$78.48
Crew L-5	Hr.	Daily	Hr.	Daily	Bare Costs	Incl. O&P
1 Struc. Steel Foreman (outside)	$62.30	$498.40	$96.40	$771.20	$60.75	$93.49
5 Struc. Steel Workers	60.30	2412.00	93.30	3732.00		
1 Equip. Oper. (crane)	61.45	491.60	91.55	732.40		
1 Hyd. Crane, 25 Ton		586.70		645.37	10.48	11.52
56 L.H., Daily Totals		$3988.70		$5880.97	$71.23	$105.02
Crew L-5A	Hr.	Daily	Hr.	Daily	Bare Costs	Incl. O&P
1 Struc. Steel Foreman (outside)	$62.30	$498.40	$96.40	$771.20	$61.09	$93.64
2 Structural Steel Workers	60.30	964.80	93.30	1492.80		
1 Equip. Oper. (crane)	61.45	491.60	91.55	732.40		
1 S.P. Crane, 4x4, 25 Ton		1155.00		1270.50	36.09	39.70
32 L.H., Daily Totals		$3109.80		$4266.90	$97.18	$133.34

For customer support on your Plumbing Costs with RSMeans Data, call 800.448.8182.

629

Crew No.	Bare Costs		Incl. Subs O&P		Cost Per Labor-Hour	
Crew L-5B	Hr.	Daily	Hr.	Daily	Bare Costs	Incl. O&P
1 Struc. Steel Foreman (outside)	$62.30	$498.40	$96.40	$771.20	$62.33	$94.01
2 Structural Steel Workers	60.30	964.80	93.30	1492.80		
2 Electricians	63.70	1019.20	94.65	1514.40		
2 Steamfitters/Pipefitters	68.35	1093.60	102.00	1632.00		
1 Equip. Oper. (crane)	61.45	491.60	91.55	732.40		
1 Equip. Oper. (oiler)	52.50	420.00	78.25	626.00		
1 Hyd. Crane, 80 Ton		1458.00		1603.80	20.25	22.27
72 L.H., Daily Totals		$5945.60		$8372.60	$82.58	$116.29
Crew L-6	Hr.	Daily	Hr.	Daily	Bare Costs	Incl. O&P
1 Plumber	$67.70	$541.60	$101.05	$808.40	$66.37	$98.92
.5 Electrician	63.70	254.80	94.65	378.60		
12 L.H., Daily Totals		$796.40		$1187.00	$66.37	$98.92
Crew L-7	Hr.	Daily	Hr.	Daily	Bare Costs	Incl. O&P
2 Carpenters	$54.70	$875.20	$81.65	$1306.40	$53.04	$79.11
1 Building Laborer	44.40	355.20	66.25	530.00		
.5 Electrician	63.70	254.80	94.65	378.60		
28 L.H., Daily Totals		$1485.20		$2215.00	$53.04	$79.11
Crew L-8	Hr.	Daily	Hr.	Daily	Bare Costs	Incl. O&P
2 Carpenters	$54.70	$875.20	$81.65	$1306.40	$57.30	$85.53
.5 Plumber	67.70	270.80	101.05	404.20		
20 L.H., Daily Totals		$1146.00		$1710.60	$57.30	$85.53
Crew L-9	Hr.	Daily	Hr.	Daily	Bare Costs	Incl. O&P
1 Labor Foreman (inside)	$44.90	$359.20	$67.00	$536.00	$50.19	$75.58
2 Building Laborers	44.40	710.40	66.25	1060.00		
1 Struc. Steel Worker	60.30	482.40	93.30	746.40		
.5 Electrician	63.70	254.80	94.65	378.60		
36 L.H., Daily Totals		$1806.80		$2721.00	$50.19	$75.58
Crew L-10	Hr.	Daily	Hr.	Daily	Bare Costs	Incl. O&P
1 Struc. Steel Foreman (outside)	$62.30	$498.40	$96.40	$771.20	$61.35	$93.75
1 Structural Steel Worker	60.30	482.40	93.30	746.40		
1 Equip. Oper. (crane)	61.45	491.60	91.55	732.40		
1 Hyd. Crane, 12 Ton		475.80		523.38	19.82	21.81
24 L.H., Daily Totals		$1948.20		$2773.38	$81.17	$115.56
Crew L-11	Hr.	Daily	Hr.	Daily	Bare Costs	Incl. O&P
2 Wreckers	$44.40	$710.40	$67.40	$1078.40	$51.44	$77.26
1 Equip. Oper. (crane)	61.45	491.60	91.55	732.40		
1 Equip. Oper. (light)	55.50	444.00	82.70	661.60		
1 Hyd. Excavator, 2.5 C.Y.		1567.00		1723.70		
1 Loader, Skid Steer, 78 H.P.		446.30		490.93	62.92	69.21
32 L.H., Daily Totals		$3659.30		$4687.03	$114.35	$146.47
Crew M-1	Hr.	Daily	Hr.	Daily	Bare Costs	Incl. O&P
3 Elevator Constructors	$90.30	$2167.20	$133.80	$3211.20	$85.79	$127.11
1 Elevator Apprentice	72.25	578.00	107.05	856.40		
5 Hand Tools		50.50		55.55	1.58	1.74
32 L.H., Daily Totals		$2795.70		$4123.15	$87.37	$128.85

Crew No.	Bare Costs		Incl. Subs O&P		Cost Per Labor-Hour	
Crew M-3	Hr.	Daily	Hr.	Daily	Bare Costs	Incl. O&P
1 Electrician Foreman (outside)	$65.70	$525.60	$97.65	$781.20	$67.62	$100.41
1 Common Laborer	44.40	355.20	66.25	530.00		
.25 Equipment Operator (med.)	59.00	118.00	87.90	175.80		
1 Elevator Constructor	90.30	722.40	133.80	1070.40		
1 Elevator Apprentice	72.25	578.00	107.05	856.40		
.25 S.P. Crane, 4x4, 20 Ton		143.59		157.95	4.22	4.65
34 L.H., Daily Totals		$2442.79		$3571.75	$71.85	$105.05
Crew M-4	Hr.	Daily	Hr.	Daily	Bare Costs	Incl. O&P
1 Electrician Foreman (outside)	$65.70	$525.60	$97.65	$781.20	$66.92	$99.38
1 Common Laborer	44.40	355.20	66.25	530.00		
.25 Equipment Operator, Crane	61.45	122.90	91.55	183.10		
.25 Equip. Oper. (oiler)	52.50	105.00	78.25	156.50		
1 Elevator Constructor	90.30	722.40	133.80	1070.40		
1 Elevator Apprentice	72.25	578.00	107.05	856.40		
.25 S.P. Crane, 4x4, 40 Ton		190.45		209.50	5.29	5.82
36 L.H., Daily Totals		$2599.55		$3787.09	$72.21	$105.20
Crew Q-1	Hr.	Daily	Hr.	Daily	Bare Costs	Incl. O&P
1 Plumber	$67.70	$541.60	$101.05	$808.40	$60.92	$90.92
1 Plumber Apprentice	54.15	433.20	80.80	646.40		
16 L.H., Daily Totals		$974.80		$1454.80	$60.92	$90.92
Crew Q-1A	Hr.	Daily	Hr.	Daily	Bare Costs	Incl. O&P
.25 Plumber Foreman (outside)	$69.70	$139.40	$104.00	$208.00	$68.10	$101.64
1 Plumber	67.70	541.60	101.05	808.40		
10 L.H., Daily Totals		$681.00		$1016.40	$68.10	$101.64
Crew Q-1C	Hr.	Daily	Hr.	Daily	Bare Costs	Incl. O&P
1 Plumber	$67.70	$541.60	$101.05	$808.40	$60.28	$89.92
1 Plumber Apprentice	54.15	433.20	80.80	646.40		
1 Equip. Oper. (medium)	59.00	472.00	87.90	703.20		
1 Trencher, Chain Type, 8' D		1894.00		2083.40	78.92	86.81
24 L.H., Daily Totals		$3340.80		$4241.40	$139.20	$176.72
Crew Q-2	Hr.	Daily	Hr.	Daily	Bare Costs	Incl. O&P
2 Plumbers	$67.70	$1083.20	$101.05	$1616.80	$63.18	$94.30
1 Plumber Apprentice	54.15	433.20	80.80	646.40		
24 L.H., Daily Totals		$1516.40		$2263.20	$63.18	$94.30
Crew Q-3	Hr.	Daily	Hr.	Daily	Bare Costs	Incl. O&P
1 Plumber Foreman (inside)	$68.20	$545.60	$101.80	$814.40	$64.44	$96.17
2 Plumbers	67.70	1083.20	101.05	1616.80		
1 Plumber Apprentice	54.15	433.20	80.80	646.40		
32 L.H., Daily Totals		$2062.00		$3077.60	$64.44	$96.17
Crew Q-4	Hr.	Daily	Hr.	Daily	Bare Costs	Incl. O&P
1 Plumber Foreman (inside)	$68.20	$545.60	$101.80	$814.40	$64.44	$96.17
1 Plumber	67.70	541.60	101.05	808.40		
1 Welder (plumber)	67.70	541.60	101.05	808.40		
1 Plumber Apprentice	54.15	433.20	80.80	646.40		
1 Welder, Electric, 300 amp		107.55		118.31	3.36	3.70
32 L.H., Daily Totals		$2169.55		$3195.91	$67.80	$99.87
Crew Q-5	Hr.	Daily	Hr.	Daily	Bare Costs	Incl. O&P
1 Steamfitter	$68.35	$546.80	$102.00	$816.00	$61.52	$91.83
1 Steamfitter Apprentice	54.70	437.60	81.65	653.20		
16 L.H., Daily Totals		$984.40		$1469.20	$61.52	$91.83

Crews - Standard

Crew No.	Bare Costs Hr.	Daily	Incl. Subs O&P Hr.	Daily	Cost Per Labor-Hour Bare Costs	Incl. O&P
Crew Q-6	Hr.	Daily	Hr.	Daily	Bare Costs	Incl. O&P
2 Steamfitters	$68.35	$1093.60	$102.00	$1632.00	$63.80	$95.22
1 Steamfitter Apprentice	54.70	437.60	81.65	653.20		
24 L.H., Daily Totals		$1531.20		$2285.20	$63.80	$95.22
Crew Q-7	Hr.	Daily	Hr.	Daily	Bare Costs	Incl. O&P
1 Steamfitter Foreman (inside)	$68.85	$550.80	$102.75	$822.00	$65.06	$97.10
2 Steamfitters	68.35	1093.60	102.00	1632.00		
1 Steamfitter Apprentice	54.70	437.60	81.65	653.20		
32 L.H., Daily Totals		$2082.00		$3107.20	$65.06	$97.10
Crew Q-8	Hr.	Daily	Hr.	Daily	Bare Costs	Incl. O&P
1 Steamfitter Foreman (inside)	$68.85	$550.80	$102.75	$822.00	$65.06	$97.10
1 Steamfitter	68.35	546.80	102.00	816.00		
1 Welder (steamfitter)	68.35	546.80	102.00	816.00		
1 Steamfitter Apprentice	54.70	437.60	81.65	653.20		
1 Welder, Electric, 300 amp		107.55		118.31	3.36	3.70
32 L.H., Daily Totals		$2189.55		$3225.51	$68.42	$100.80
Crew Q-9	Hr.	Daily	Hr.	Daily	Bare Costs	Incl. O&P
1 Sheet Metal Worker	$65.45	$523.60	$98.70	$789.60	$58.90	$88.83
1 Sheet Metal Apprentice	52.35	418.80	78.95	631.60		
16 L.H., Daily Totals		$942.40		$1421.20	$58.90	$88.83
Crew Q-10	Hr.	Daily	Hr.	Daily	Bare Costs	Incl. O&P
2 Sheet Metal Workers	$65.45	$1047.20	$98.70	$1579.20	$61.08	$92.12
1 Sheet Metal Apprentice	52.35	418.80	78.95	631.60		
24 L.H., Daily Totals		$1466.00		$2210.80	$61.08	$92.12
Crew Q-11	Hr.	Daily	Hr.	Daily	Bare Costs	Incl. O&P
1 Sheet Metal Foreman (inside)	$65.95	$527.60	$99.50	$796.00	$62.30	$93.96
2 Sheet Metal Workers	65.45	1047.20	98.70	1579.20		
1 Sheet Metal Apprentice	52.35	418.80	78.95	631.60		
32 L.H., Daily Totals		$1993.60		$3006.80	$62.30	$93.96
Crew Q-12	Hr.	Daily	Hr.	Daily	Bare Costs	Incl. O&P
1 Sprinkler Installer	$66.50	$532.00	$99.35	$794.80	$59.85	$89.42
1 Sprinkler Apprentice	53.20	425.60	79.50	636.00		
16 L.H., Daily Totals		$957.60		$1430.80	$59.85	$89.42
Crew Q-13	Hr.	Daily	Hr.	Daily	Bare Costs	Incl. O&P
1 Sprinkler Foreman (inside)	$67.00	$536.00	$100.10	$800.80	$63.30	$94.58
2 Sprinkler Installers	66.50	1064.00	99.35	1589.60		
1 Sprinkler Apprentice	53.20	425.60	79.50	636.00		
32 L.H., Daily Totals		$2025.60		$3026.40	$63.30	$94.58
Crew Q-14	Hr.	Daily	Hr.	Daily	Bare Costs	Incl. O&P
1 Asbestos Worker	$60.95	$487.60	$93.25	$746.00	$54.85	$83.90
1 Asbestos Apprentice	48.75	390.00	74.55	596.40		
16 L.H., Daily Totals		$877.60		$1342.40	$54.85	$83.90
Crew Q-15	Hr.	Daily	Hr.	Daily	Bare Costs	Incl. O&P
1 Plumber	$67.70	$541.60	$101.05	$808.40	$60.92	$90.92
1 Plumber Apprentice	54.15	433.20	80.80	646.40		
1 Welder, Electric, 300 amp		107.55		118.31	6.72	7.39
16 L.H., Daily Totals		$1082.35		$1573.11	$67.65	$98.32
Crew Q-16	Hr.	Daily	Hr.	Daily	Bare Costs	Incl. O&P
2 Plumbers	$67.70	$1083.20	$101.05	$1616.80	$63.18	$94.30
1 Plumber Apprentice	54.15	433.20	80.80	646.40		
1 Welder, Electric, 300 amp		107.55		118.31	4.48	4.93
24 L.H., Daily Totals		$1623.95		$2381.51	$67.66	$99.23
Crew Q-17	Hr.	Daily	Hr.	Daily	Bare Costs	Incl. O&P
1 Steamfitter	$68.35	$546.80	$102.00	$816.00	$61.52	$91.83
1 Steamfitter Apprentice	54.70	437.60	81.65	653.20		
1 Welder, Electric, 300 amp		107.55		118.31	6.72	7.39
16 L.H., Daily Totals		$1091.95		$1587.51	$68.25	$99.22
Crew Q-17A	Hr.	Daily	Hr.	Daily	Bare Costs	Incl. O&P
1 Steamfitter	$68.35	$546.80	$102.00	$816.00	$61.50	$91.73
1 Steamfitter Apprentice	54.70	437.60	81.65	653.20		
1 Equip. Oper. (crane)	61.45	491.60	91.55	732.40		
1 Hyd. Crane, 12 Ton		475.80		523.38		
1 Welder, Electric, 300 amp		107.55		118.31	24.31	26.74
24 L.H., Daily Totals		$2059.35		$2843.28	$85.81	$118.47
Crew Q-18	Hr.	Daily	Hr.	Daily	Bare Costs	Incl. O&P
2 Steamfitters	$68.35	$1093.60	$102.00	$1632.00	$63.80	$95.22
1 Steamfitter Apprentice	54.70	437.60	81.65	653.20		
1 Welder, Electric, 300 amp		107.55		118.31	4.48	4.93
24 L.H., Daily Totals		$1638.75		$2403.51	$68.28	$100.15
Crew Q-19	Hr.	Daily	Hr.	Daily	Bare Costs	Incl. O&P
1 Steamfitter	$68.35	$546.80	$102.00	$816.00	$62.25	$92.77
1 Steamfitter Apprentice	54.70	437.60	81.65	653.20		
1 Electrician	63.70	509.60	94.65	757.20		
24 L.H., Daily Totals		$1494.00		$2226.40	$62.25	$92.77
Crew Q-20	Hr.	Daily	Hr.	Daily	Bare Costs	Incl. O&P
1 Sheet Metal Worker	$65.45	$523.60	$98.70	$789.60	$59.86	$89.99
1 Sheet Metal Apprentice	52.35	418.80	78.95	631.60		
.5 Electrician	63.70	254.80	94.65	378.60		
20 L.H., Daily Totals		$1197.20		$1799.80	$59.86	$89.99
Crew Q-21	Hr.	Daily	Hr.	Daily	Bare Costs	Incl. O&P
2 Steamfitters	$68.35	$1093.60	$102.00	$1632.00	$63.77	$95.08
1 Steamfitter Apprentice	54.70	437.60	81.65	653.20		
1 Electrician	63.70	509.60	94.65	757.20		
32 L.H., Daily Totals		$2040.80		$3042.40	$63.77	$95.08
Crew Q-22	Hr.	Daily	Hr.	Daily	Bare Costs	Incl. O&P
1 Plumber	$67.70	$541.60	$101.05	$808.40	$60.92	$90.92
1 Plumber Apprentice	54.15	433.20	80.80	646.40		
1 Hyd. Crane, 12 Ton		475.80		523.38	29.74	32.71
16 L.H., Daily Totals		$1450.60		$1978.18	$90.66	$123.64
Crew Q-22A	Hr.	Daily	Hr.	Daily	Bare Costs	Incl. O&P
1 Plumber	$67.70	$541.60	$101.05	$808.40	$56.92	$84.91
1 Plumber Apprentice	54.15	433.20	80.80	646.40		
1 Laborer	44.40	355.20	66.25	530.00		
1 Equip. Oper. (crane)	61.45	491.60	91.55	732.40		
1 Hyd. Crane, 12 Ton		475.80		523.38	14.87	16.36
32 L.H., Daily Totals		$2297.40		$3240.58	$71.79	$101.27

For customer support on your Plumbing Costs with RSMeans Data, call 800.448.8182.

631

Crew No.	Bare Costs		Incl. Subs O&P		Cost Per Labor-Hour	

Crew Q-23	Hr.	Daily	Hr.	Daily	Bare Costs	Incl. O&P
1 Plumber Foreman (outside)	$69.70	$557.60	$104.00	$832.00	$65.47	$97.65
1 Plumber	67.70	541.60	101.05	808.40		
1 Equip. Oper. (medium)	59.00	472.00	87.90	703.20		
1 Lattice Boom Crane, 20 Ton		1526.00		1678.60	63.58	69.94
24 L.H., Daily Totals		$3097.20		$4022.20	$129.05	$167.59

Crew R-1	Hr.	Daily	Hr.	Daily	Bare Costs	Incl. O&P
1 Electrician Foreman	$64.20	$513.60	$95.40	$763.20	$59.53	$88.46
3 Electricians	63.70	1528.80	94.65	2271.60		
2 Electrician Apprentices	50.95	815.20	75.70	1211.20		
48 L.H., Daily Totals		$2857.60		$4246.00	$59.53	$88.46

Crew R-1A	Hr.	Daily	Hr.	Daily	Bare Costs	Incl. O&P
1 Electrician	$63.70	$509.60	$94.65	$757.20	$57.33	$85.17
1 Electrician Apprentice	50.95	407.60	75.70	605.60		
16 L.H., Daily Totals		$917.20		$1362.80	$57.33	$85.17

Crew R-1B	Hr.	Daily	Hr.	Daily	Bare Costs	Incl. O&P
1 Electrician	$63.70	$509.60	$94.65	$757.20	$55.20	$82.02
2 Electrician Apprentices	50.95	815.20	75.70	1211.20		
24 L.H., Daily Totals		$1324.80		$1968.40	$55.20	$82.02

Crew R-1C	Hr.	Daily	Hr.	Daily	Bare Costs	Incl. O&P
2 Electricians	$63.70	$1019.20	$94.65	$1514.40	$57.33	$85.17
2 Electrician Apprentices	50.95	815.20	75.70	1211.20		
1 Portable cable puller, 8000 lb.		102.65		112.92	3.21	3.53
32 L.H., Daily Totals		$1937.05		$2838.51	$60.53	$88.70

Crew R-2	Hr.	Daily	Hr.	Daily	Bare Costs	Incl. O&P
1 Electrician Foreman	$64.20	$513.60	$95.40	$763.20	$59.81	$88.90
3 Electricians	63.70	1528.80	94.65	2271.60		
2 Electrician Apprentices	50.95	815.20	75.70	1211.20		
1 Equip. Oper. (crane)	61.45	491.60	91.55	732.40		
1 S.P. Crane, 4x4, 5 Ton		381.95		420.14	6.82	7.50
56 L.H., Daily Totals		$3731.15		$5398.55	$66.63	$96.40

Crew R-3	Hr.	Daily	Hr.	Daily	Bare Costs	Incl. O&P
1 Electrician Foreman	$64.20	$513.60	$95.40	$763.20	$63.45	$94.33
1 Electrician	63.70	509.60	94.65	757.20		
.5 Equip. Oper. (crane)	61.45	245.80	91.55	366.20		
.5 S.P. Crane, 4x4, 5 Ton		190.97		210.07	9.55	10.50
20 L.H., Daily Totals		$1459.97		$2096.67	$73.00	$104.83

Crew R-4	Hr.	Daily	Hr.	Daily	Bare Costs	Incl. O&P
1 Struc. Steel Foreman (outside)	$62.30	$498.40	$96.40	$771.20	$61.38	$94.19
3 Struc. Steel Workers	60.30	1447.20	93.30	2239.20		
1 Electrician	63.70	509.60	94.65	757.20		
1 Welder, Gas Engine, 300 amp		148.75		163.63	3.72	4.09
40 L.H., Daily Totals		$2603.95		$3931.22	$65.10	$98.28

Crew R-5	Hr.	Daily	Hr.	Daily	Bare Costs	Incl. O&P
1 Electrician Foreman	$64.20	$513.60	$95.40	$763.20	$55.87	$83.45
4 Electrician Linemen	63.70	2038.40	94.65	3028.80		
2 Electrician Operators	63.70	1019.20	94.65	1514.40		
4 Electrician Groundmen	42.05	1345.60	63.65	2036.80		
1 Crew Truck		168.15		184.97		
1 Flatbed Truck, 20,000 GVW		204.05		224.46		
1 Pickup Truck, 3/4 Ton		112.20		123.42		
.2 Hyd. Crane, 55 Ton		198.03		217.83		
.2 Hyd. Crane, 12 Ton		95.16		104.68		
.2 Earth Auger, Truck-Mtd.		40.51		44.56		
1 Tractor w/Winch		377.65		415.42	13.59	14.95
88 L.H., Daily Totals		$6112.55		$8658.52	$69.46	$98.39

Crew R-6	Hr.	Daily	Hr.	Daily	Bare Costs	Incl. O&P
1 Electrician Foreman	$64.20	$513.60	$95.40	$763.20	$55.87	$83.45
4 Electrician Linemen	63.70	2038.40	94.65	3028.80		
2 Electrician Operators	63.70	1019.20	94.65	1514.40		
4 Electrician Groundmen	42.05	1345.60	63.65	2036.80		
1 Crew Truck		168.15		184.97		
1 Flatbed Truck, 20,000 GVW		204.05		224.46		
1 Pickup Truck, 3/4 Ton		112.20		123.42		
.2 Hyd. Crane, 55 Ton		198.03		217.83		
.2 Hyd. Crane, 12 Ton		95.16		104.68		
.2 Earth Auger, Truck-Mtd.		40.51		44.56		
1 Tractor w/Winch		377.65		415.42		
3 Cable Trailers		194.25		213.68		
.5 Tensioning Rig		55.75		61.33		
.5 Cable Pulling Rig		306.65		337.32	19.91	21.91
88 L.H., Daily Totals		$6669.20		$9270.84	$75.79	$105.35

Crew R-7	Hr.	Daily	Hr.	Daily	Bare Costs	Incl. O&P
1 Electrician Foreman	$64.20	$513.60	$95.40	$763.20	$45.74	$68.94
5 Electrician Groundmen	42.05	1682.00	63.65	2546.00		
1 Crew Truck		168.15		184.97	3.50	3.85
48 L.H., Daily Totals		$2363.75		$3494.17	$49.24	$72.80

Crew R-8	Hr.	Daily	Hr.	Daily	Bare Costs	Incl. O&P
1 Electrician Foreman	$64.20	$513.60	$95.40	$763.20	$56.57	$84.44
3 Electrician Linemen	63.70	1528.80	94.65	2271.60		
2 Electrician Groundmen	42.05	672.80	63.65	1018.40		
1 Pickup Truck, 3/4 Ton		112.20		123.42		
1 Crew Truck		168.15		184.97	5.84	6.42
48 L.H., Daily Totals		$2995.55		$4361.59	$62.41	$90.87

Crew R-9	Hr.	Daily	Hr.	Daily	Bare Costs	Incl. O&P
1 Electrician Foreman	$64.20	$513.60	$95.40	$763.20	$52.94	$79.24
1 Electrician Lineman	63.70	509.60	94.65	757.20		
2 Electrician Operators	63.70	1019.20	94.65	1514.40		
4 Electrician Groundmen	42.05	1345.60	63.65	2036.80		
1 Pickup Truck, 3/4 Ton		112.20		123.42		
1 Crew Truck		168.15		184.97	4.38	4.82
64 L.H., Daily Totals		$3668.35		$5379.98	$57.32	$84.06

Crew R-10	Hr.	Daily	Hr.	Daily	Bare Costs	Incl. O&P
1 Electrician Foreman	$64.20	$513.60	$95.40	$763.20	$60.17	$89.61
4 Electrician Linemen	63.70	2038.40	94.65	3028.80		
1 Electrician Groundman	42.05	336.40	63.65	509.20		
1 Crew Truck		168.15		184.97		
3 Tram Cars		219.60		241.56	8.08	8.89
48 L.H., Daily Totals		$3276.15		$4727.73	$68.25	$98.49

Crew No.		Bare Costs		Incl. Subs O&P		Cost Per Labor-Hour	
Crew R-11	Hr.	Daily	Hr.	Daily	Bare Costs	Incl. O&P
1 Electrician Foreman	$64.20	$513.60	$95.40	$763.20	$60.69	$90.26
4 Electricians	63.70	2038.40	94.65	3028.80		
1 Equip. Oper. (crane)	61.45	491.60	91.55	732.40		
1 Common Laborer	44.40	355.20	66.25	530.00		
1 Crew Truck		168.15		184.97		
1 Hyd. Crane, 12 Ton		475.80		523.38	11.50	12.65
56 L.H., Daily Totals		$4042.75		$5762.74	$72.19	$102.91

Crew R-12	Hr.	Daily	Hr.	Daily	Bare Costs	Incl. O&P
1 Carpenter Foreman (inside)	$55.20	$441.60	$82.40	$659.20	$51.90	$77.75
4 Carpenters	54.70	1750.40	81.65	2612.80		
4 Common Laborers	44.40	1420.80	66.25	2120.00		
1 Equip. Oper. (medium)	59.00	472.00	87.90	703.20		
1 Steel Worker	60.30	482.40	93.30	746.40		
1 Dozer, 200 H.P.		1520.00		1672.00		
1 Pickup Truck, 3/4 Ton		112.20		123.42	18.55	20.40
88 L.H., Daily Totals		$6199.40		$8637.02	$70.45	$98.15

Crew R-13	Hr.	Daily	Hr.	Daily	Bare Costs	Incl. O&P
1 Electrician Foreman	$64.20	$513.60	$95.40	$763.20	$61.55	$91.52
3 Electricians	63.70	1528.80	94.65	2271.60		
.25 Equip. Oper. (crane)	61.45	122.90	91.55	183.10		
1 Equipment Oiler	52.50	420.00	78.25	626.00		
.25 Hydraulic Crane, 33 Ton		245.79		270.37	5.85	6.44
42 L.H., Daily Totals		$2831.09		$4114.27	$67.41	$97.96

Crew R-15	Hr.	Daily	Hr.	Daily	Bare Costs	Incl. O&P
1 Electrician Foreman	$64.20	$513.60	$95.40	$763.20	$62.42	$92.78
4 Electricians	63.70	2038.40	94.65	3028.80		
1 Equipment Oper. (light)	55.50	444.00	82.70	661.60		
1 Telescoping Boom Lift, to 40'		281.90		310.09	5.87	6.46
48 L.H., Daily Totals		$3277.90		$4763.69	$68.29	$99.24

Crew R-15A	Hr.	Daily	Hr.	Daily	Bare Costs	Incl. O&P
1 Electrician Foreman	$64.20	$513.60	$95.40	$763.20	$55.98	$83.32
2 Electricians	63.70	1019.20	94.65	1514.40		
2 Common Laborers	44.40	710.40	66.25	1060.00		
1 Equip. Oper. (light)	55.50	444.00	82.70	661.60		
1 Telescoping Boom Lift, to 40'		281.90		310.09	5.87	6.46
48 L.H., Daily Totals		$2969.10		$4309.29	$61.86	$89.78

Crew R-18	Hr.	Daily	Hr.	Daily	Bare Costs	Incl. O&P
.25 Electrician Foreman	$64.20	$128.40	$95.40	$190.80	$55.89	$83.05
1 Electrician	63.70	509.60	94.65	757.20		
2 Electrician Apprentices	50.95	815.20	75.70	1211.20		
26 L.H., Daily Totals		$1453.20		$2159.20	$55.89	$83.05

Crew R-19	Hr.	Daily	Hr.	Daily	Bare Costs	Incl. O&P
.5 Electrician Foreman	$64.20	$256.80	$95.40	$381.60	$63.80	$94.80
2 Electricians	63.70	1019.20	94.65	1514.40		
20 L.H., Daily Totals		$1276.00		$1896.00	$63.80	$94.80

Crew R-21	Hr.	Daily	Hr.	Daily	Bare Costs	Incl. O&P
1 Electrician Foreman	$64.20	$513.60	$95.40	$763.20	$63.71	$94.67
3 Electricians	63.70	1528.80	94.65	2271.60		
.1 Equip. Oper. (medium)	59.00	47.20	87.90	70.32		
.1 S.P. Crane, 4x4, 25 Ton		115.50		127.05	3.52	3.87
32.8 L.H., Daily Totals		$2205.10		$3232.17	$67.23	$98.54

Crew No.		Bare Costs		Incl. Subs O&P		Cost Per Labor-Hour	
Crew R-22	Hr.	Daily	Hr.	Daily	Bare Costs	Incl. O&P
.66 Electrician Foreman	$64.20	$338.98	$95.40	$503.71	$58.30	$86.62
2 Electricians	63.70	1019.20	94.65	1514.40		
2 Electrician Apprentices	50.95	815.20	75.70	1211.20		
37.28 L.H., Daily Totals		$2173.38		$3229.31	$58.30	$86.62

Crew R-30	Hr.	Daily	Hr.	Daily	Bare Costs	Incl. O&P
.25 Electrician Foreman (outside)	$65.70	$131.40	$97.65	$195.30	$51.98	$77.40
1 Electrician	63.70	509.60	94.65	757.20		
2 Laborers (Semi-Skilled)	44.40	710.40	66.25	1060.00		
26 L.H., Daily Totals		$1351.40		$2012.50	$51.98	$77.40

Crew R-31	Hr.	Daily	Hr.	Daily	Bare Costs	Incl. O&P
1 Electrician	$63.70	$509.60	$94.65	$757.20	$63.70	$94.65
1 Core Drill, Electric, 2.5 H.P.		62.65		68.92	7.83	8.61
8 L.H., Daily Totals		$572.25		$826.12	$71.53	$103.26

Crew W-41E	Hr.	Daily	Hr.	Daily	Bare Costs	Incl. O&P
.5 Plumber Foreman (outside)	$69.70	$278.80	$104.00	$416.00	$58.78	$87.72
1 Plumber	67.70	541.60	101.05	808.40		
1 Laborer	44.40	355.20	66.25	530.00		
20 L.H., Daily Totals		$1175.60		$1754.40	$58.78	$87.72

For customer support on your Plumbing Costs with RSMeans Data, call 800.448.8182.

633

Historical Cost Indexes

The table below lists both the RSMeans® historical cost index based on Jan. 1, 1993 = 100 as well as the computed value of an index based on Jan. 1, 2021 costs. Since the Jan. 1, 2021 figure is estimated, space is left to write in the actual index figures as they become available through the quarterly *RSMeans Construction Cost Indexes.*

To compute the actual index based on Jan. 1, 2021 = 100, divide the historical cost index for a particular year by the actual Jan. 1, 2021 construction cost index. Space has been left to advance the index figures as the year progresses.

Year	Historical Cost Index Jan. 1, 1993 = 100		Current Index Based on Jan. 1, 2021 = 100		Year	Historical Cost Index Jan. 1, 1993 = 100	Current Index Based on Jan. 1, 2021 = 100		Year	Historical Cost Index Jan. 1, 1993 = 100	Current Index Based on Jan. 1, 2021 = 100	
	Est.	Actual	Est.	Actual		Actual	Est.	Actual		Actual	Est.	Actual
Oct 2021					July 2004	143.7	60.7		1984	82.0	34.6	
July 2021					2003	132.0	55.8		1983	80.2	33.9	
April 2021					2002	128.7	54.4		1982	76.1	32.2	
Jan 2021	236.7		100.0		2001	125.1	52.9		1981	70.0	29.6	
2020		234.6	99.1		2000	120.9	51.1		1980	62.9	26.6	
2019		232.2	98.1		1999	117.6	49.7		1979	57.8	24.4	
2018		222.9	94.2		1998	115.1	48.6		1978	53.5	22.6	
2017		213.6	90.2		1997	112.8	47.7		1977	49.5	20.9	
2016		207.3	87.6		1996	110.2	46.6		1976	46.9	19.8	
2015		206.2	87.1		1995	107.6	45.5		1975	44.8	18.9	
2014		204.9	86.6		1994	104.4	44.1		1974	41.4	17.5	
2013		201.2	85.0		1993	101.7	43.0		1973	37.7	15.9	
2012		194.6	82.2		1992	99.4	42.0		1972	34.8	14.7	
2011		191.2	80.8		1991	96.8	40.9		1971	32.1	13.6	
2010		183.5	77.5		1990	94.3	39.8		1970	28.7	12.1	
2009		180.1	76.1		1989	92.1	38.9		1969	26.9	11.4	
2008		180.4	76.2		1988	89.9	38.0					
2007		169.4	71.6		1987	87.7	37.0					
2006		162.0	68.4		1986	84.2	35.6					
2005		151.6	64.0		1985	82.6	34.9					

Adjustments to Costs

The "Historical Cost Index" can be used to convert national average building costs at a particular time to the approximate building costs for some other time.

Example:

Estimate and compare construction costs for different years in the same city.

To estimate the national average construction cost of a building in 1970, knowing that it cost $900,000 in 2021:

INDEX in 1970 = 28.7

INDEX in 2021 = 236.7

Note: The city cost indexes for Canada can be used to convert U.S. national averages to local costs in Canadian dollars.

Example:

To estimate and compare the cost of a building in Toronto, ON in 2021 with the known cost of $600,000 (US$) in New York, NY in 2021:

INDEX Toronto = 112.5

INDEX New York = 134.1

$$\frac{\text{INDEX Toronto}}{\text{INDEX New York}} \times \text{Cost New York} = \text{Cost Toronto}$$

$$\frac{112.5}{134.1} \times \$600,000 = .839 \times \$600,000 = \$503,400$$

The construction cost of the building in Toronto is $503,400 (CN$).

Time Adjustment Using the Historical Cost Indexes:

$$\frac{\text{Index for Year A}}{\text{Index for Year B}} \times \text{Cost in Year B} = \text{Cost in Year A}$$

$$\frac{\text{INDEX 1970}}{\text{INDEX 2021}} \times \text{Cost 2021} = \text{Cost 1970}$$

$$\frac{28.7}{236.7} \times \$900,000 = .121 \times \$900,000 = \$108,900$$

The construction cost of the building in 1970 was $108,900.

*Historical Cost Index updates and other resources are provided on the following website: **rsmeans.com/2021books**

For customer support on your Plumbing Costs with RSMeans Data, call 800.448.8182.

How to Use the City Cost Indexes

What you should know before you begin

RSMeans City Cost Indexes (CCI) are an extremely useful tool for when you want to compare costs from city to city and region to region.

This publication contains average construction cost indexes for 731 U.S. and Canadian cities covering over 930 three-digit zip code locations, as listed directly under each city.

Keep in mind that a City Cost Index number is a percentage ratio of a specific city's cost to the national average cost of the same item at a stated time period.

In other words, these index figures represent relative construction factors (or, if you prefer, multipliers) for material and installation costs, as well as the weighted average for Total In Place costs for each CSI MasterFormat division. Installation costs include both labor and equipment rental costs. When estimating equipment rental rates only for a specific location, use 01 54 33 EQUIPMENT RENTAL COSTS in the Reference Section.

The 30 City Average Index is the average of 30 major U.S. cities and serves as a national average. This national average represents the baseline from which all locations are calculated.

Index figures for both material and installation are based on the 30 major city average of 100 and represent the cost relationship as of July 1, 2020. The index for each division is computed from representative material and labor quantities for that division. The weighted average for each city is a weighted total of the components listed above it. It does not include relative productivity between trades or cities.

As changes occur in local material prices, labor rates, and equipment rental rates (including fuel costs), the impact of these changes should be accurately measured by the change in the City Cost Index for each particular city (as compared to the 30 city average).

Therefore, if you know (or have estimated) building costs in one city today, you can easily convert those costs to expected building costs in another city.

In addition, by using the Historical Cost Index, you can easily convert national average building costs at a particular time to the approximate building costs for some other time. The City Cost Indexes can then be applied to calculate the costs for a particular city.

Quick calculations

Location Adjustment Using the City Cost Indexes:

$$\frac{\text{Index for City A}}{\text{Index for City B}} \times \text{Cost in City B} = \text{Cost in City A}$$

Time Adjustment for the National Average Using the Historical Cost Index:

$$\frac{\text{Index for Year A}}{\text{Index for Year B}} \times \text{Cost in Year B} = \text{Cost in Year A}$$

Adjustment from the National Average:

$$\frac{\text{Index for City A}}{100} \times \text{National Average Cost} = \text{Cost in City A}$$

Since each of the other RSMeans data sets contains many different items, any *one* item multiplied by the particular city index may give incorrect results. However, the larger the number of items compiled, the closer the results should be to actual costs for that particular city.

The City Cost Indexes for Canadian cities are calculated using Canadian material and equipment prices and labor rates in Canadian dollars. When compared to the baseline mentioned previously, the resulting index enables the user to convert baseline costs to local costs. Therefore, indexes for Canadian cities can be used to convert U.S. national average prices to local costs in Canadian dollars.

How to use this section

1. Compare costs from city to city.

In using the RSMeans Indexes, remember that an index number is not a fixed number but a ratio: It's a percentage ratio of a building component's cost at any stated time to the national average cost of that same component at the same time period. Put in the form of an equation:

$$\frac{\text{Specific City Cost}}{\text{National Average Cost}} \times 100 = \text{City Index Number}$$

Therefore, when making cost comparisons between cities, do not subtract one city's index number from the index number of another city and read the result as a percentage difference. Instead, divide one city's index number by that of the other city. The resulting number may then be used as a multiplier to calculate cost differences from city to city.

The formula used to find cost differences between cities for the purpose of comparison is as follows:

$$\frac{\text{City A Index}}{\text{City B Index}} \times \text{City B Cost (Known)} = \text{City A Cost (Unknown)}$$

In addition, you can use RSMeans CCI to calculate and compare costs division by division between cities using the same basic formula. (Just be sure that you're comparing similar divisions.)

2. Compare a specific city's construction costs with the national average.

When you're studying construction location feasibility, it's advisable to compare a prospective project's cost index with an index of the national average cost.

For example, divide the weighted average index of construction costs of a specific city by that of the 30 City Average, which = 100.

$$\frac{\text{City Index}}{100} = \text{\% of National Average}$$

As a result, you get a ratio that indicates the relative cost of construction in that city in comparison with the national average.

3. Convert U.S. national average to actual costs in Canadian City.

$$\frac{\text{Index for Canadian City}}{100} \times \text{National Average Cost} = \text{Cost in Canadian City in \$ CAN}$$

For customer support on your Plumbing Costs with RSMeans Data, call 800.448.8182.

635

4. Adjust construction cost data based on a national average.

When you use a source of construction cost data which is based on a national average (such as RSMeans cost data), it is necessary to adjust those costs to a specific location.

$$\frac{\text{City Index}}{100} \times \frac{\text{Cost Based on}}{\text{National Average Costs}} = \frac{\text{City Cost}}{(\text{Unknown})}$$

5. When applying the City Cost Indexes to demolition projects, use the appropriate division installation index. For example, for removal of existing doors and windows, use the Division 8 (Openings) index.

What you might like to know about how we developed the Indexes

The information presented in the CCI is organized according to the Construction Specifications Institute (CSI) MasterFormat 2018 classification system.

To create a reliable index, RSMeans researched the building type most often constructed in the United States and Canada. Because it was concluded that no one type of building completely represented the building construction industry, nine different types of buildings were combined to create a composite model.

The exact material, labor, and equipment quantities are based on detailed analyses of these nine building types, and then each quantity is weighted in proportion to expected usage. These various material items, labor hours, and equipment rental rates are thus combined to form a composite building representing as closely as possible the actual usage of materials, labor, and equipment in the North American building construction industry.

The following structures were chosen to make up that composite model:

1. Factory, 1 story
2. Office, 2–4 stories
3. Store, Retail
4. Town Hall, 2–3 stories
5. High School, 2–3 stories
6. Hospital, 4–8 stories
7. Garage, Parking
8. Apartment, 1–3 stories
9. Hotel/Motel, 2–3 stories

For the purposes of ensuring the timeliness of the data, the components of the index for the composite model have been streamlined. They currently consist of:

- specific quantities of 66 commonly used construction materials;
- specific labor-hours for 21 building construction trades; and
- specific days of equipment rental for 6 types of construction equipment (normally used to install the 66 material items by the 21 trades.) Fuel costs and routine maintenance costs are included in the equipment cost.

Material and equipment price quotations are gathered quarterly from cities in the United States and Canada. These prices and the latest negotiated labor wage rates for 21 different building trades are used to compile the quarterly update of the City Cost Index.

The 30 major U.S. cities used to calculate the national average are:

Atlanta, GA	Memphis, TN
Baltimore, MD	Milwaukee, WI
Boston, MA	Minneapolis, MN
Buffalo, NY	Nashville, TN
Chicago, IL	New Orleans, LA
Cincinnati, OH	New York, NY
Cleveland, OH	Philadelphia, PA
Columbus, OH	Phoenix, AZ
Dallas, TX	Pittsburgh, PA
Denver, CO	St. Louis, MO
Detroit, MI	San Antonio, TX
Houston, TX	San Diego, CA
Indianapolis, IN	San Francisco, CA
Kansas City, MO	Seattle, WA
Los Angeles, CA	Washington, DC

What the CCI does not indicate

The weighted average for each city is a total of the divisional components weighted to reflect typical usage. It does not include the productivity variations between trades or cities.

In addition, the CCI does not take into consideration factors such as the following:

- managerial efficiency
- competitive conditions
- automation
- restrictive union practices
- unique local requirements
- regional variations due to specific building codes

For customer support on your Plumbing Costs with RSMeans Data, call 800.448.8182.

Section 1

DIVISION		UNITED STATES 30 CITY AVERAGE			ALABAMA ANNISTON 362			BIRMINGHAM 350 - 352			BUTLER 369			DECATUR 356			DOTHAN 363		
		MAT.	INST.	TOTAL	MAT.	INST.	TOTAL	MAT.	INST.	TOTAL	MAT.	INST.	TOTAL	MAT.	INST.	TOTAL	MAT.	INST.	TOTAL
015433	CONTRACTOR EQUIPMENT		100.0	100.0		104.5	104.5		104.7	104.7		102.1	102.1		104.5	104.5		102.1	102.1
0241, 31 - 34	SITE & INFRASTRUCTURE, DEMOLITION	100.0	100.0	100.0	90.9	92.7	92.1	92.0	93.1	92.7	105.4	88.1	93.3	85.9	92.0	90.2	102.9	88.7	93.1
0310	Concrete Forming & Accessories	100.0	100.0	100.0	86.5	65.3	68.5	92.0	66.5	70.3	83.0	65.9	68.5	93.0	61.4	66.2	91.4	68.2	71.7
0320	Concrete Reinforcing	100.0	100.0	100.0	86.9	72.3	79.5	86.9	72.3	83.9	91.9	72.9	82.2	89.8	68.7	79.1	91.9	72.3	82.0
0330	Cast-in-Place Concrete	100.0	100.0	100.0	87.8	66.3	79.8	102.6	68.4	89.8	85.6	66.6	78.5	100.7	66.1	87.8	85.6	68.0	79.0
03	CONCRETE	100.0	100.0	100.0	92.3	68.6	81.6	94.4	69.9	83.4	93.3	69.1	82.4	94.3	66.1	81.6	92.5	70.6	82.6
04	MASONRY	100.0	100.0	100.0	85.1	69.1	75.2	86.0	67.7	74.7	89.3	69.1	76.8	84.1	68.6	74.5	90.6	60.2	71.8
05	METALS	100.0	100.0	100.0	98.1	94.5	97.0	98.1	95.0	97.2	97.0	95.5	96.6	100.2	93.2	98.1	97.1	95.4	96.6
06	WOOD, PLASTICS & COMPOSITES	100.0	100.0	100.0	84.9	65.9	74.6	94.0	65.9	78.7	79.4	65.9	72.1	98.1	60.4	77.6	91.4	68.7	79.1
07	THERMAL & MOISTURE PROTECTION	100.0	100.0	100.0	95.5	63.9	81.9	96.0	67.3	83.6	95.5	67.3	83.4	95.1	66.4	82.8	95.5	65.0	82.4
08	OPENINGS	100.0	100.0	100.0	96.8	67.7	90.0	103.3	67.7	95.1	96.8	67.8	90.1	109.4	63.7	98.8	96.9	69.3	90.5
0920	Plaster & Gypsum Board	100.0	100.0	100.0	90.5	65.5	74.0	92.0	65.5	74.5	87.1	65.5	72.9	100.4	59.8	73.6	97.8	68.3	78.4
0950, 0980	Ceilings & Acoustic Treatment	100.0	100.0	100.0	80.8	65.5	70.4	82.1	65.5	70.8	80.8	65.5	70.4	82.4	59.8	67.1	80.8	68.3	72.3
0960	Flooring	100.0	100.0	100.0	83.8	80.6	82.9	94.8	76.3	89.4	88.6	80.6	86.3	85.6	80.6	84.2	94.0	82.2	90.6
0970, 0990	Wall Finishes & Painting/Coating	100.0	100.0	100.0	90.3	58.9	71.7	88.5	60.1	71.7	90.3	48.0	65.3	80.4	61.4	69.2	90.3	80.9	84.8
09	FINISHES	100.0	100.0	100.0	80.0	67.8	73.4	86.5	67.4	76.1	83.0	66.7	74.1	82.3	64.7	72.7	85.9	72.3	78.4
COVERS	DIVS. 10 - 14, 25, 28, 41, 43, 44, 46	100.0	100.0	100.0	100.0	70.2	93.4	100.0	83.4	96.3	100.0	77.0	94.9	100.0	69.5	93.3	100.0	72.0	93.8
21, 22, 23	FIRE SUPPRESSION, PLUMBING & HVAC	100.0	100.0	100.0	101.2	50.4	80.8	100.0	66.4	86.6	98.7	65.7	85.5	100.1	64.7	85.9	98.7	65.4	85.3
26, 27, 3370	ELECTRICAL, COMMUNICATIONS & UTIL.	100.0	100.0	100.0	96.4	58.4	77.0	97.2	61.2	78.8	98.4	65.4	81.5	93.0	64.0	78.2	97.2	75.3	86.0
MF2018	WEIGHTED AVERAGE	100.0	100.0	100.0	95.6	67.4	83.5	97.0	71.7	86.2	95.9	71.6	85.5	97.0	70.1	85.4	96.0	72.9	86.1

Section 2 — ALABAMA

DIVISION		EVERGREEN 364			GADSDEN 359			HUNTSVILLE 357 - 358			JASPER 355			MOBILE 365 - 366			MONTGOMERY 360 - 361		
		MAT.	INST.	TOTAL	MAT.	INST.	TOTAL	MAT.	INST.	TOTAL	MAT.	INST.	TOTAL	MAT.	INST.	TOTAL	MAT.	INST.	TOTAL
015433	CONTRACTOR EQUIPMENT		102.1	102.1		104.5	104.5		104.5	104.5		104.5	104.5		102.1	102.1		102.1	102.1
0241, 31 - 34	SITE & INFRASTRUCTURE, DEMOLITION	105.8	88.0	93.4	91.3	92.7	92.3	85.6	92.6	90.5	91.3	92.8	92.4	98.0	89.1	91.8	96.0	88.5	90.8
0310	Concrete Forming & Accessories	80.0	66.7	68.7	85.2	65.8	68.7	93.0	65.2	69.4	90.1	66.1	69.7	90.5	66.7	70.3	93.7	65.9	70.1
0320	Concrete Reinforcing	92.0	72.8	82.2	95.4	72.3	83.6	89.8	73.7	81.6	89.8	72.4	81.0	89.6	72.2	80.8	97.6	71.8	84.5
0330	Cast-in-Place Concrete	85.6	66.3	78.4	100.7	67.3	88.3	98.0	67.0	86.5	111.6	67.1	95.1	89.7	66.3	81.0	86.8	67.6	79.7
03	CONCRETE	93.6	69.3	82.7	98.8	69.2	85.4	93.0	69.0	82.2	102.5	69.2	87.5	88.8	69.2	80.0	87.2	69.2	79.1
04	MASONRY	89.3	68.7	76.5	82.6	68.6	72.3	85.2	65.4	73.0	80.4	70.3	74.1	87.9	58.8	69.9	85.2	59.2	69.2
05	METALS	97.0	95.4	96.5	98.1	95.0	97.2	100.2	95.3	98.7	98.1	94.9	97.1	99.1	94.7	97.7	98.1	94.4	97.0
06	WOOD, PLASTICS & COMPOSITES	76.0	67.4	71.3	88.5	65.9	76.3	98.1	65.3	80.3	95.2	65.9	79.3	89.9	67.4	77.7	93.3	65.9	78.5
07	THERMAL & MOISTURE PROTECTION	95.5	66.5	83.0	95.4	66.8	83.1	95.1	66.6	82.8	95.4	62.4	81.2	95.1	64.9	82.1	93.5	65.1	81.3
08	OPENINGS	96.8	68.6	90.3	105.9	67.7	97.0	109.1	67.7	99.5	105.9	67.7	97.0	99.5	68.6	92.3	98.8	67.7	91.6
0920	Plaster & Gypsum Board	86.5	67.0	73.7	92.6	65.5	74.7	100.4	64.8	77.0	97.2	65.5	76.3	94.2	67.0	76.3	89.9	65.5	73.8
0950, 0980	Ceilings & Acoustic Treatment	80.8	67.0	71.4	79.0	65.5	69.8	84.3	64.8	71.1	79.0	65.5	69.8	86.1	67.0	73.1	84.8	65.5	71.7
0960	Flooring	86.4	80.6	84.7	82.2	77.3	80.8	85.6	77.3	83.2	84.1	80.6	83.1	93.5	63.7	84.8	94.0	63.7	85.2
0970, 0990	Wall Finishes & Painting/Coating	90.3	48.0	65.3	80.4	58.9	67.7	80.4	67.3	72.6	80.4	58.9	67.7	93.6	49.9	67.7	91.3	58.9	72.1
09	FINISHES	82.3	67.4	74.2	80.0	67.1	72.9	82.7	67.5	74.4	81.1	68.1	74.0	85.8	64.2	74.0	86.1	64.4	74.2
COVERS	DIVS. 10 - 14, 25, 28, 41, 43, 44, 46	100.0	70.1	93.4	100.0	82.8	96.2	100.0	82.6	96.2	100.0	71.4	93.7	100.0	83.7	96.4	100.0	83.0	96.2
21, 22, 23	FIRE SUPPRESSION, PLUMBING & HVAC	98.7	65.5	85.4	102.0	62.8	86.3	100.1	63.5	85.4	102.0	66.2	87.6	100.0	62.8	85.1	100.0	65.0	85.9
26, 27, 3370	ELECTRICAL, COMMUNICATIONS & UTIL.	95.7	65.4	80.2	93.0	61.2	76.7	93.9	65.7	79.5	92.7	58.4	75.2	99.1	58.1	78.1	99.7	73.9	86.5
MF2018	WEIGHTED AVERAGE	95.6	71.4	85.2	97.1	70.6	85.7	96.9	71.4	86.0	97.5	71.0	86.2	96.4	68.8	84.5	95.8	71.4	85.3

Section 3 — ALABAMA / ALASKA

DIVISION		ALABAMA PHENIX CITY 368			SELMA 367			TUSCALOOSA 354			ALASKA ANCHORAGE 995 - 996			FAIRBANKS 997			JUNEAU 998		
		MAT.	INST.	TOTAL	MAT.	INST.	TOTAL	MAT.	INST.	TOTAL	MAT.	INST.	TOTAL	MAT.	INST.	TOTAL	MAT.	INST.	TOTAL
015433	CONTRACTOR EQUIPMENT		102.1	102.1		102.1	102.1		104.5	104.5		115.9	115.9		115.9	115.9		115.9	115.9
0241, 31 - 34	SITE & INFRASTRUCTURE, DEMOLITION	109.7	88.5	95.0	102.8	88.5	92.9	86.2	92.7	90.8	122.1	130.7	128.1	121.2	130.7	127.8	140.2	130.7	133.6
0310	Concrete Forming & Accessories	86.5	66.1	69.1	84.2	65.9	68.7	93.0	66.0	70.1	115.1	117.7	117.3	119.9	117.9	118.2	123.8	117.7	118.6
0320	Concrete Reinforcing	91.9	69.2	80.4	91.9	72.4	82.0	89.8	72.3	80.9	152.4	121.3	136.6	141.8	121.3	131.4	127.6	121.3	124.4
0330	Cast-in-Place Concrete	85.6	67.7	78.9	85.6	66.6	78.5	102.1	67.6	89.3	97.6	117.2	104.9	108.1	117.6	111.7	120.4	117.2	119.2
03	CONCRETE	96.5	68.9	84.1	92.0	69.0	81.7	94.9	69.4	83.4	116.9	117.3	117.1	108.4	117.6	112.5	124.8	117.3	121.4
04	MASONRY	89.2	69.1	76.8	92.5	69.1	78.1	84.3	66.3	73.2	183.5	119.5	144.0	185.0	119.8	144.7	157.6	119.5	134.1
05	METALS	97.0	94.2	96.1	97.0	95.1	96.4	99.4	95.0	98.1	125.0	104.8	118.9	125.2	104.9	119.1	119.7	104.8	115.2
06	WOOD, PLASTICS & COMPOSITES	84.6	65.8	74.4	81.6	65.9	73.1	98.1	65.9	80.6	114.0	115.8	115.0	125.8	115.8	120.4	123.4	115.8	119.3
07	THERMAL & MOISTURE PROTECTION	95.9	67.8	83.8	95.4	67.3	83.3	95.2	67.0	83.1	169.5	117.6	147.2	181.1	118.7	154.2	185.1	117.6	156.0
08	OPENINGS	96.8	66.8	89.8	96.8	67.7	90.0	109.1	67.7	99.5	130.6	116.5	127.3	135.8	116.5	131.4	130.4	116.5	127.2
0920	Plaster & Gypsum Board	91.7	65.3	74.3	89.9	65.5	73.8	100.4	65.5	77.4	142.8	116.1	125.2	172.8	116.1	135.4	154.4	116.1	129.5
0950, 0980	Ceilings & Acoustic Treatment	80.8	65.3	70.3	80.8	65.5	70.4	84.3	65.5	71.5	132.6	116.1	121.4	127.9	116.1	119.9	138.9	116.1	123.4
0960	Flooring	90.6	80.6	87.7	89.1	80.6	86.6	85.6	77.3	83.2	125.1	122.6	124.4	122.1	122.6	122.2	129.4	122.6	127.4
0970, 0990	Wall Finishes & Painting/Coating	90.3	81.0	84.8	90.3	58.9	71.7	80.4	58.9	67.7	108.7	113.2	111.3	113.7	117.7	116.1	110.8	113.2	112.2
09	FINISHES	84.7	70.2	76.8	83.2	67.8	74.8	82.6	67.2	74.2	125.7	118.3	121.7	128.6	118.9	123.3	129.3	118.3	123.3
COVERS	DIVS. 10 - 14, 25, 28, 41, 43, 44, 46	100.0	82.9	96.2	100.0	70.7	93.6	100.0	82.9	96.2	100.0	109.3	102.0	100.0	109.4	102.1	100.0	109.3	102.0
21, 22, 23	FIRE SUPPRESSION, PLUMBING & HVAC	98.7	65.0	85.2	98.7	64.3	84.9	100.1	66.5	86.6	100.5	106.1	102.8	100.4	110.3	104.3	100.5	106.1	102.8
26, 27, 3370	ELECTRICAL, COMMUNICATIONS & UTIL.	97.8	65.7	81.4	96.9	73.9	85.1	93.5	61.2	77.0	112.5	108.8	110.7	120.6	108.8	114.6	108.9	108.8	108.9
MF2018	WEIGHTED AVERAGE	96.5	72.0	86.0	95.7	72.4	85.7	97.0	71.5	86.0	119.3	113.6	116.9	120.3	114.7	117.9	119.0	113.6	116.7

For customer support on your Plumbing Costs with RSMeans Data, call 800.448.8182.

637

ALASKA / ARIZONA

DIVISION		ALASKA KETCHIKAN 999			CHAMBERS 865			FLAGSTAFF 860			GLOBE 855			KINGMAN 864			MESA/TEMPE 852		
		MAT.	INST.	TOTAL	MAT.	INST.	TOTAL	MAT.	INST.	TOTAL	MAT.	INST.	TOTAL	MAT.	INST.	TOTAL	MAT.	INST.	TOTAL
015433	CONTRACTOR EQUIPMENT		115.9	115.9		91.1	91.1		91.1	91.1		92.2	92.2		91.1	91.1		92.2	92.2
0241, 31 - 34	SITE & INFRASTRUCTURE, DEMOLITION	176.2	130.7	144.5	71.0	95.0	87.7	89.5	95.0	93.3	102.0	95.9	97.8	70.9	95.0	87.7	92.1	95.9	94.8
0310	Concrete Forming & Accessories	111.6	117.7	116.7	98.2	68.0	72.5	103.7	69.2	74.4	93.9	68.0	71.9	96.3	67.9	72.2	97.1	68.1	72.5
0320	Concrete Reinforcing	108.4	121.3	114.9	103.2	79.1	91.0	103.0	79.0	90.8	109.0	79.1	93.8	103.4	79.1	91.0	109.7	79.1	94.2
0330	Cast-in-Place Concrete	218.8	117.2	181.0	89.7	69.4	82.2	89.8	69.4	82.2	84.1	69.0	78.5	89.5	69.4	82.0	84.8	69.1	79.0
03	CONCRETE	186.7	117.3	155.4	92.7	70.3	82.7	112.5	70.9	93.8	104.5	70.3	89.1	92.4	70.3	82.5	95.5	70.4	84.2
04	MASONRY	192.7	119.5	147.5	95.2	60.7	73.9	95.3	59.0	72.9	100.4	60.6	75.8	95.2	60.7	73.9	100.6	60.6	75.9
05	METALS	125.4	104.8	119.1	101.9	72.4	92.9	102.4	72.3	93.3	97.4	73.2	90.0	102.6	72.3	93.4	97.7	73.3	90.3
06	WOOD, PLASTICS & COMPOSITES	116.0	115.8	115.9	102.8	69.3	84.6	109.3	70.9	88.4	94.2	69.4	80.8	97.5	69.3	82.2	98.1	69.4	82.5
07	THERMAL & MOISTURE PROTECTION	186.0	117.6	156.5	94.1	70.1	83.8	96.1	69.6	84.7	100.2	68.7	86.6	94.1	69.9	83.7	100.0	68.4	86.4
08	OPENINGS	131.5	116.5	128.0	108.2	66.4	98.5	108.3	72.2	99.9	97.2	66.7	90.2	108.4	68.7	99.2	97.3	68.8	90.7
0920	Plaster & Gypsum Board	158.3	116.1	130.5	92.6	68.6	76.8	96.2	70.2	79.1	88.0	68.6	75.2	84.6	68.6	74.1	90.8	68.6	76.2
0950, 0980	Ceilings & Acoustic Treatment	121.3	116.1	117.8	107.2	68.6	81.1	108.1	70.2	82.4	93.5	68.6	76.6	108.1	68.6	81.3	93.5	68.6	76.6
0960	Flooring	122.1	122.6	122.2	89.9	58.8	80.9	92.1	58.8	82.4	104.0	58.8	90.9	88.7	58.8	80.0	105.6	58.8	92.0
0970, 0990	Wall Finishes & Painting/Coating	113.7	113.2	113.4	88.5	68.3	76.6	88.5	68.3	76.6	92.8	68.3	78.3	88.5	68.3	76.6	92.8	68.3	78.3
09	FINISHES	129.1	118.3	123.2	90.1	66.1	77.0	93.2	67.0	78.9	96.5	66.2	80.0	88.7	66.1	76.4	96.3	66.2	79.9
COVERS	DIVS. 10 - 14, 25, 28, 41, 43, 44, 46	100.0	109.3	102.0	100.0	83.0	96.3	100.0	83.1	96.3	100.0	83.2	96.3	100.0	82.9	96.2	100.0	83.2	96.3
21, 22, 23	FIRE SUPPRESSION, PLUMBING & HVAC	98.6	106.1	101.6	98.3	77.0	89.7	100.2	77.2	91.0	97.6	77.0	89.3	98.3	77.0	89.7	100.0	77.2	90.9
26, 27, 3370	ELECTRICAL, COMMUNICATIONS & UTIL.	120.6	108.8	114.6	102.8	83.6	93.0	101.8	62.2	81.5	96.9	64.4	80.3	102.9	64.4	83.2	94.0	64.4	78.9
MF2018	WEIGHTED AVERAGE	131.0	113.6	123.5	98.3	74.4	88.0	102.0	71.7	89.0	98.7	71.8	87.2	98.2	71.8	86.9	97.7	72.0	86.7

ARIZONA / ARKANSAS

DIVISION		PHOENIX 850,853			PRESCOTT 863			SHOW LOW 859			TUCSON 856 - 857			BATESVILLE 725			CAMDEN 717		
		MAT.	INST.	TOTAL	MAT.	INST.	TOTAL	MAT.	INST.	TOTAL	MAT.	INST.	TOTAL	MAT.	INST.	TOTAL	MAT.	INST.	TOTAL
015433	CONTRACTOR EQUIPMENT		93.2	93.2		91.1	91.1		92.2	92.2		92.2	92.2		91.9	91.9		91.9	91.9
0241, 31 - 34	SITE & INFRASTRUCTURE, DEMOLITION	92.6	94.5	93.9	77.5	95.0	89.7	104.2	95.9	98.4	87.7	95.9	93.4	76.6	89.7	85.7	79.6	89.7	86.6
0310	Concrete Forming & Accessories	101.8	71.6	76.1	99.8	68.0	72.8	101.3	68.1	73.1	97.6	69.3	73.6	79.5	56.4	59.9	79.8	56.3	59.8
0320	Concrete Reinforcing	107.8	79.2	93.2	103.0	79.1	90.9	109.7	79.1	94.2	91.0	79.0	84.9	93.3	65.3	79.1	98.6	65.0	81.5
0330	Cast-in-Place Concrete	84.5	69.3	78.8	89.8	69.4	82.2	84.2	69.1	78.5	87.3	69.1	80.5	74.8	76.4	75.4	78.1	76.4	77.5
03	CONCRETE	98.1	72.1	86.4	98.1	70.3	85.6	107.0	70.3	90.5	94.2	70.9	83.7	76.3	65.7	71.5	79.1	65.7	73.0
04	MASONRY	93.5	62.7	74.5	95.3	60.7	73.9	100.4	60.6	75.8	87.5	58.9	69.8	92.1	42.3	61.4	104.9	36.9	62.9
05	METALS	99.1	75.2	91.8	102.4	72.3	93.3	97.2	73.2	89.9	98.4	73.2	90.7	100.2	72.8	91.9	96.2	72.4	89.0
06	WOOD, PLASTICS & COMPOSITES	101.1	72.9	85.7	104.4	69.5	85.4	102.7	69.4	84.6	98.3	71.0	83.5	92.9	57.2	73.5	91.0	57.2	72.6
07	THERMAL & MOISTURE PROTECTION	101.3	69.2	87.5	94.7	70.2	84.2	100.4	69.0	86.9	101.1	68.1	86.9	102.7	57.4	83.2	93.5	55.8	77.3
08	OPENINGS	98.6	72.8	92.6	108.3	66.7	98.7	96.5	68.8	90.1	93.7	72.3	88.7	102.0	53.1	90.7	105.6	54.9	94.3
0920	Plaster & Gypsum Board	98.0	71.9	80.8	93.0	68.8	77.1	92.6	68.6	76.8	95.3	70.2	78.8	78.6	56.4	64.0	87.9	56.4	67.1
0950, 0980	Ceilings & Acoustic Treatment	104.1	71.9	82.3	106.4	68.8	80.9	93.5	68.6	76.6	94.3	70.2	78.0	85.7	56.4	65.8	80.0	56.4	64.0
0960	Flooring	107.7	58.8	93.5	90.7	58.8	81.5	107.3	58.8	93.2	96.1	58.8	85.3	86.3	53.0	76.6	92.4	35.5	75.9
0970, 0990	Wall Finishes & Painting/Coating	97.5	65.0	78.3	88.5	68.3	76.6	92.8	68.3	78.3	93.3	68.3	78.5	90.7	53.5	68.7	94.5	53.5	70.2
09	FINISHES	100.9	68.5	83.2	90.7	66.2	77.3	98.3	66.2	80.8	94.3	67.2	79.5	76.9	54.9	64.9	80.4	51.7	64.7
COVERS	DIVS. 10 - 14, 25, 28, 41, 43, 44, 46	100.0	84.7	96.6	100.0	82.9	96.2	100.0	83.2	96.3	100.0	83.4	96.3	100.0	65.9	92.5	100.0	65.9	92.5
21, 22, 23	FIRE SUPPRESSION, PLUMBING & HVAC	99.9	78.6	91.3	100.2	77.0	90.9	97.6	77.0	89.3	100.0	77.1	90.8	97.6	52.2	79.4	97.5	56.9	81.2
26, 27, 3370	ELECTRICAL, COMMUNICATIONS & UTIL.	99.6	62.2	80.5	101.5	64.4	82.5	94.3	64.4	79.0	96.1	59.5	77.4	96.5	60.7	78.2	94.4	57.2	75.4
MF2018	WEIGHTED AVERAGE	99.0	73.0	87.9	99.6	71.7	87.6	98.9	71.9	87.3	96.6	71.4	85.8	93.6	60.0	79.2	94.1	59.6	79.3

ARKANSAS

DIVISION		FAYETTEVILLE 727			FORT SMITH 729			HARRISON 726			HOT SPRINGS 719			JONESBORO 724			LITTLE ROCK 720 - 722		
		MAT.	INST.	TOTAL	MAT.	INST.	TOTAL	MAT.	INST.	TOTAL	MAT.	INST.	TOTAL	MAT.	INST.	TOTAL	MAT.	INST.	TOTAL
015433	CONTRACTOR EQUIPMENT		91.9	91.9		91.9	91.9		91.9	91.9		91.9	91.9		115.7	115.7		91.9	91.9
0241, 31 - 34	SITE & INFRASTRUCTURE, DEMOLITION	76.0	89.7	85.6	81.2	89.8	87.2	81.5	89.7	87.2	82.7	89.7	87.6	100.5	107.5	105.4	89.1	90.1	89.8
0310	Concrete Forming & Accessories	75.3	56.5	59.3	93.6	61.7	66.5	83.8	56.3	60.4	77.4	56.5	59.6	82.7	56.5	60.5	90.3	62.0	66.3
0320	Concrete Reinforcing	93.4	62.6	77.7	94.5	65.9	80.0	92.9	65.4	78.9	96.7	65.3	80.8	90.1	65.5	77.6	100.0	67.9	83.7
0330	Cast-in-Place Concrete	74.8	76.5	75.4	85.6	76.7	82.2	83.0	76.4	80.5	79.9	76.5	78.6	81.5	77.7	80.1	82.9	76.8	80.6
03	CONCRETE	76.0	65.3	71.2	83.7	68.3	76.7	83.1	65.7	75.2	82.4	65.8	74.9	80.6	67.3	74.6	84.5	68.9	77.4
04	MASONRY	83.5	41.8	57.8	90.1	62.1	72.8	92.6	43.3	62.1	78.8	42.0	56.1	84.8	39.9	57.1	86.0	62.1	71.2
05	METALS	100.2	72.0	91.6	102.5	74.6	94.0	101.4	72.8	92.7	96.2	72.9	89.1	96.7	87.8	94.0	102.8	75.6	94.5
06	WOOD, PLASTICS & COMPOSITES	89.2	57.2	71.8	110.0	63.6	84.8	98.9	57.2	76.2	88.4	57.2	71.4	97.1	57.5	75.6	96.5	63.6	78.6
07	THERMAL & MOISTURE PROTECTION	103.5	57.2	83.6	103.9	63.6	86.5	103.0	57.7	83.5	93.7	57.3	78.0	108.4	56.6	86.1	98.0	63.9	83.3
08	OPENINGS	102.0	54.9	91.1	104.1	60.1	93.8	102.8	52.0	91.0	105.5	54.8	93.8	107.8	56.0	95.8	97.2	59.7	88.5
0920	Plaster & Gypsum Board	78.0	56.4	63.8	84.8	63.0	70.4	83.9	56.4	65.8	86.6	56.4	66.7	89.9	56.4	67.8	94.5	63.0	73.8
0950, 0980	Ceilings & Acoustic Treatment	85.7	56.4	65.8	87.4	63.0	70.9	87.4	56.4	66.4	80.0	56.4	64.0	90.7	56.4	67.4	89.5	63.0	71.5
0960	Flooring	83.4	51.9	74.3	92.9	73.3	87.2	88.6	53.0	78.3	91.3	51.9	79.8	62.6	47.0	58.1	94.0	82.1	90.5
0970, 0990	Wall Finishes & Painting/Coating	90.7	53.5	68.7	90.7	53.5	68.7	90.7	53.5	68.7	94.5	53.5	70.2	80.6	53.5	64.6	91.2	53.5	68.9
09	FINISHES	76.1	54.5	64.3	80.3	63.1	70.9	79.0	54.7	65.8	80.2	54.7	66.3	74.9	53.8	63.4	86.3	64.9	74.6
COVERS	DIVS. 10 - 14, 25, 28, 41, 43, 44, 46	100.0	65.9	92.5	100.0	78.4	95.2	100.0	67.0	92.7	100.0	65.9	92.5	100.0	64.9	92.3	100.0	78.4	95.3
21, 22, 23	FIRE SUPPRESSION, PLUMBING & HVAC	97.6	61.2	83.0	100.1	49.1	79.7	97.6	48.5	77.9	97.5	52.5	79.5	100.4	52.2	81.1	99.8	52.6	80.9
26, 27, 3370	ELECTRICAL, COMMUNICATIONS & UTIL.	90.8	53.1	71.5	94.0	58.4	75.8	95.2	58.9	76.6	96.3	66.8	81.2	100.2	60.9	80.1	101.1	66.8	83.6
MF2018	WEIGHTED AVERAGE	92.5	60.7	78.9	95.9	63.6	82.0	95.0	59.1	79.6	93.5	61.0	79.6	95.5	62.7	81.4	96.2	65.9	83.2

For customer support on your Plumbing Costs with RSMeans Data, call 800.448.8182.

City Cost Indexes

ARKANSAS / CALIFORNIA

		PINE BLUFF 716			RUSSELLVILLE 728			TEXARKANA 718			WEST MEMPHIS 723			ALHAMBRA 917 - 918			ANAHEIM 928		
DIVISION		MAT.	INST.	TOTAL	MAT.	INST.	TOTAL	MAT.	INST.	TOTAL	MAT.	INST.	TOTAL	MAT.	INST.	TOTAL	MAT.	INST.	TOTAL
015433	CONTRACTOR EQUIPMENT		91.9	91.9		91.9	91.9		93.2	93.2		115.7	115.7		97.7	97.7		101.8	101.8
0241, 31 - 34	SITE & INFRASTRUCTURE, DEMOLITION	84.7	89.8	88.3	77.8	89.7	86.1	95.5	91.9	93.0	107.3	107.7	107.6	97.1	110.3	106.3	100.2	110.6	107.4
0310	Concrete Forming & Accessories	77.1	61.9	64.2	80.2	56.3	59.9	82.8	56.4	60.4	87.9	56.8	61.5	116.4	140.0	136.4	101.6	140.1	134.3
0320	Concrete Reinforcing	98.4	67.9	82.9	94.0	65.2	79.4	98.0	65.3	81.4	90.1	63.7	76.7	102.5	130.1	116.5	95.4	129.9	112.9
0330	Cast-in-Place Concrete	79.9	76.7	78.7	78.4	76.4	77.6	87.1	76.4	83.2	85.5	77.8	82.6	83.4	129.8	100.7	85.2	133.2	103.1
03	CONCRETE	83.1	68.8	76.6	79.3	65.6	73.1	81.4	65.7	74.4	87.3	67.1	78.2	95.2	133.3	112.3	96.3	134.6	113.5
04	MASONRY	110.4	62.1	80.5	89.4	40.9	59.4	92.9	41.3	61.0	73.6	39.8	52.7	111.7	141.9	130.4	78.5	140.2	116.6
05	METALS	97.0	75.4	90.4	100.2	72.6	91.8	89.7	72.8	84.6	95.7	87.4	93.2	90.3	110.5	96.5	107.1	111.7	108.5
06	WOOD, PLASTICS & COMPOSITES	88.0	63.6	74.8	94.2	57.2	74.1	95.6	57.2	74.7	103.0	57.5	78.3	102.8	137.1	121.4	97.9	137.4	119.3
07	THERMAL & MOISTURE PROTECTION	93.8	64.0	81.0	103.7	57.0	83.6	94.4	57.1	78.4	108.9	56.6	86.4	96.2	133.2	112.1	105.0	136.3	118.5
08	OPENINGS	106.8	60.5	96.0	102.0	54.1	90.9	111.8	54.1	98.4	105.3	54.8	93.5	89.8	135.7	100.4	106.2	135.9	113.1
0920	Plaster & Gypsum Board	86.3	63.0	71.0	78.6	56.4	64.0	89.2	56.4	67.6	92.1	56.4	68.6	96.3	138.3	124.0	111.2	138.3	129.1
0950, 0980	Ceilings & Acoustic Treatment	80.0	63.0	68.5	85.7	56.4	65.8	83.4	56.4	65.1	88.8	56.4	66.8	112.6	138.3	130.0	112.6	138.3	130.0
0960	Flooring	91.1	73.3	85.9	85.9	53.0	76.3	93.2	44.8	79.1	64.9	47.0	59.7	100.0	120.4	105.9	97.3	120.4	104.0
0970, 0990	Wall Finishes & Painting/Coating	94.5	53.5	70.2	90.7	53.5	68.7	94.5	53.5	70.2	80.6	53.5	64.6	105.4	121.1	114.7	90.7	121.1	108.7
09	FINISHES	80.2	63.1	70.9	77.1	54.8	64.9	82.4	53.4	66.6	76.2	54.0	64.1	100.8	134.3	119.1	98.1	134.5	118.0
COVERS	DIVS. 10 - 14, 25, 28, 41, 43, 44, 46	100.0	78.4	95.2	100.0	65.9	92.5	100.0	65.9	92.5	100.0	66.7	92.7	100.0	119.6	104.3	100.0	120.1	104.4
21, 22, 23	FIRE SUPPRESSION, PLUMBING & HVAC	100.0	52.6	81.0	97.6	51.9	79.3	100.0	55.6	82.2	97.9	64.2	84.4	97.5	131.7	111.2	100.0	131.8	112.7
26, 27, 3370	ELECTRICAL, COMMUNICATIONS & UTIL.	94.5	58.9	76.3	94.0	58.9	76.1	96.2	58.9	77.1	101.7	62.7	81.7	115.3	130.5	123.1	90.3	111.0	100.9
MF2018	WEIGHTED AVERAGE	95.9	64.6	82.4	93.7	59.6	79.1	94.9	60.4	80.1	95.3	65.5	82.5	98.2	129.4	111.6	99.3	126.9	111.2

CALIFORNIA

		BAKERSFIELD 932 - 933			BERKELEY 947			EUREKA 955			FRESNO 936 - 938			INGLEWOOD 903 - 905			LONG BEACH 906 - 908		
DIVISION		MAT.	INST.	TOTAL	MAT.	INST.	TOTAL	MAT.	INST.	TOTAL	MAT.	INST.	TOTAL	MAT.	INST.	TOTAL	MAT.	INST.	TOTAL
015433	CONTRACTOR EQUIPMENT		99.8	99.8		102.8	102.8		99.6	99.6		99.8	99.8		99.0	99.0		99.0	99.0
0241, 31 - 34	SITE & INFRASTRUCTURE, DEMOLITION	97.1	108.4	105.0	111.1	110.7	110.8	111.5	107.4	108.7	101.1	107.8	105.8	91.4	106.7	102.0	98.7	106.7	104.3
0310	Concrete Forming & Accessories	101.0	138.9	133.2	112.1	167.2	158.9	109.5	152.9	146.4	98.2	152.1	144.0	106.9	140.3	135.3	101.6	140.3	134.5
0320	Concrete Reinforcing	103.5	129.9	116.9	89.1	131.8	110.8	103.9	131.3	117.8	86.5	130.8	109.0	103.0	130.1	116.8	102.1	130.1	116.4
0330	Cast-in-Place Concrete	89.3	132.3	105.3	111.5	133.6	119.7	92.5	129.8	106.3	96.8	129.4	108.9	79.5	132.3	99.2	90.5	132.3	106.1
03	CONCRETE	89.8	133.7	109.5	101.4	147.1	122.0	106.8	139.3	121.5	93.5	138.7	113.9	89.7	134.4	109.8	99.2	134.4	115.0
04	MASONRY	92.5	139.6	121.6	124.9	156.3	144.3	103.7	152.4	133.8	97.2	146.3	127.5	75.6	142.1	116.6	84.7	142.1	120.1
05	METALS	105.1	111.1	106.9	108.5	114.0	110.2	106.9	113.6	108.9	105.3	112.1	107.4	99.7	112.1	103.5	99.6	112.1	103.4
06	WOOD, PLASTICS & COMPOSITES	92.8	136.2	116.4	106.7	172.6	142.5	111.8	156.0	135.8	98.8	156.0	129.9	96.9	137.5	119.0	90.7	137.5	116.1
07	THERMAL & MOISTURE PROTECTION	100.7	128.2	112.5	99.9	157.2	124.5	108.6	150.6	126.7	92.1	134.4	110.3	100.3	134.9	115.2	100.5	134.9	115.3
08	OPENINGS	96.1	133.1	104.7	92.7	159.5	108.2	105.5	141.6	113.9	97.0	144.0	108.0	89.4	135.9	100.2	89.4	135.9	100.2
0920	Plaster & Gypsum Board	97.4	137.0	123.5	113.1	174.1	153.3	116.2	157.4	143.3	94.3	157.4	135.8	105.2	138.3	127.0	101.3	138.3	125.7
0950, 0980	Ceilings & Acoustic Treatment	104.7	137.0	126.6	98.5	174.1	149.8	116.0	157.4	144.0	103.9	157.4	140.1	110.8	138.3	129.5	110.8	138.3	129.5
0960	Flooring	102.9	120.4	108.0	115.9	142.0	123.5	101.1	142.0	113.0	102.1	118.0	106.7	101.8	120.4	107.2	98.9	120.4	105.1
0970, 0990	Wall Finishes & Painting/Coating	92.1	109.3	102.3	113.3	162.7	142.5	92.1	137.6	119.0	104.1	123.9	115.8	105.2	121.1	114.6	105.2	121.1	114.6
09	FINISHES	94.0	132.5	115.0	107.6	163.7	138.3	103.1	150.9	129.2	93.5	145.3	121.8	104.5	134.6	120.9	103.5	134.6	120.4
COVERS	DIVS. 10 - 14, 25, 28, 41, 43, 44, 46	100.0	120.2	104.4	100.0	129.9	106.6	100.0	126.7	105.9	100.0	126.7	105.9	100.0	120.4	104.5	100.0	120.4	104.5
21, 22, 23	FIRE SUPPRESSION, PLUMBING & HVAC	100.1	129.0	111.7	97.7	168.0	125.9	97.5	129.1	110.2	100.2	130.3	112.2	97.1	131.8	111.0	97.1	131.8	111.0
26, 27, 3370	ELECTRICAL, COMMUNICATIONS & UTIL.	105.1	110.7	108.0	103.0	161.0	132.7	97.0	124.3	111.0	94.9	108.0	101.6	101.0	130.5	116.1	100.7	130.5	116.0
MF2018	WEIGHTED AVERAGE	98.7	125.3	110.1	102.7	151.2	123.5	102.7	133.2	115.8	98.4	129.3	111.7	96.0	129.6	110.4	97.6	129.6	111.3

CALIFORNIA

		LOS ANGELES 900 - 902			MARYSVILLE 959			MODESTO 953			MOJAVE 935			OAKLAND 946			OXNARD 930		
DIVISION		MAT.	INST.	TOTAL	MAT.	INST.	TOTAL	MAT.	INST.	TOTAL	MAT.	INST.	TOTAL	MAT.	INST.	TOTAL	MAT.	INST.	TOTAL
015433	CONTRACTOR EQUIPMENT		105.2	105.2		99.6	99.6		99.6	99.6		99.8	99.8		102.8	102.8		98.7	98.7
0241, 31 - 34	SITE & INFRASTRUCTURE, DEMOLITION	97.2	112.4	107.8	108.0	107.3	107.5	102.7	107.3	105.9	94.2	108.4	104.1	116.0	110.7	112.3	101.4	106.5	104.9
0310	Concrete Forming & Accessories	104.8	140.6	135.3	99.7	152.5	144.5	96.2	152.3	143.9	108.4	138.8	134.3	101.3	167.2	157.3	100.4	140.2	134.2
0320	Concrete Reinforcing	103.8	130.2	117.2	103.9	130.6	117.5	107.6	130.5	119.3	104.1	129.9	117.2	91.1	131.8	111.8	102.3	129.8	116.3
0330	Cast-in-Place Concrete	82.1	133.4	101.2	103.3	129.5	113.1	92.5	129.4	106.2	83.7	132.3	101.8	101.7	133.6	113.6	97.5	132.5	110.6
03	CONCRETE	96.8	135.0	114.0	107.3	138.9	121.5	98.3	138.8	116.5	85.6	133.6	107.2	101.7	147.1	122.1	93.8	134.3	112.0
04	MASONRY	91.0	142.1	122.5	104.7	141.7	127.5	102.1	141.7	126.6	95.4	139.6	122.7	132.8	156.3	147.3	98.3	138.0	122.8
05	METALS	106.1	114.5	108.6	106.4	112.3	108.1	103.4	112.1	106.0	102.3	111.0	104.9	104.1	114.0	107.1	100.2	111.3	103.6
06	WOOD, PLASTICS & COMPOSITES	99.9	137.7	120.4	98.8	156.0	129.9	94.3	156.0	127.8	98.4	136.2	118.9	94.6	172.6	137.0	93.3	137.4	117.3
07	THERMAL & MOISTURE PROTECTION	96.5	134.8	113.0	108.0	139.1	121.4	107.6	135.5	119.6	97.7	125.6	109.7	98.6	157.2	123.8	101.1	134.3	115.4
08	OPENINGS	98.9	136.1	107.5	104.7	144.9	114.1	103.5	144.9	113.1	92.0	133.1	101.6	92.8	159.5	108.3	94.6	135.9	104.2
0920	Plaster & Gypsum Board	100.5	138.3	125.4	108.7	157.4	140.8	111.2	157.4	141.6	103.8	137.0	125.7	107.2	174.1	151.3	97.8	138.3	124.5
0950, 0980	Ceilings & Acoustic Treatment	117.7	138.3	131.7	115.1	157.4	143.8	112.6	157.4	143.0	106.1	137.0	127.1	101.3	174.1	150.6	107.7	138.3	128.4
0960	Flooring	100.0	120.4	105.9	96.8	123.5	104.6	97.7	129.9	106.7	102.3	120.4	107.5	109.6	142.0	119.0	94.9	120.4	102.3
0970, 0990	Wall Finishes & Painting/Coating	101.9	121.1	113.2	92.1	137.6	119.0	92.1	137.6	119.0	88.9	109.3	100.9	113.3	162.7	142.5	88.9	115.6	104.7
09	FINISHES	105.5	134.8	121.5	100.1	147.7	126.1	99.6	148.8	126.5	93.8	132.5	114.9	105.9	163.7	137.5	91.3	133.9	114.6
COVERS	DIVS. 10 - 14, 25, 28, 41, 43, 44, 46	100.0	121.1	104.6	100.0	126.7	105.9	100.0	126.7	105.9	100.0	120.2	104.4	100.0	129.9	106.6	100.0	120.4	104.5
21, 22, 23	FIRE SUPPRESSION, PLUMBING & HVAC	99.9	131.9	112.7	97.5	130.3	110.7	100.0	130.3	112.1	97.6	129.0	110.2	100.2	168.0	127.4	100.1	131.8	112.8
26, 27, 3370	ELECTRICAL, COMMUNICATIONS & UTIL.	99.4	130.5	115.3	93.8	122.4	108.4	96.1	108.4	102.4	93.4	110.7	102.3	102.2	161.0	132.3	99.2	117.5	108.5
MF2018	WEIGHTED AVERAGE	100.2	130.4	113.2	101.9	131.3	114.5	100.6	129.4	112.9	95.6	125.2	108.3	102.8	151.2	123.5	97.9	127.1	110.4

For customer support on your Plumbing Costs with RSMeans Data, call 800.448.8182.

639

DIVISION		CALIFORNIA																
		PALM SPRINGS			PALO ALTO			PASADENA			REDDING			RICHMOND			RIVERSIDE		
		922			943			910 - 912			960			948			925		
		MAT.	INST.	TOTAL	MAT.	INST.	TOTAL	MAT.	INST.	TOTAL	MAT.	INST.	TOTAL	MAT.	INST.	TOTAL	MAT.	INST.	TOTAL
015433	CONTRACTOR EQUIPMENT		100.7	100.7		102.8	102.8		97.7	97.7		99.6	99.6		102.8	102.8		100.7	100.7
0241, 31 - 34	SITE & INFRASTRUCTURE, DEMOLITION	91.8	108.6	103.5	107.4	110.7	109.7	93.9	110.3	105.3	126.7	107.3	113.2	115.2	110.6	112.0	98.7	108.6	105.6
0310	Concrete Forming & Accessories	98.4	140.1	133.8	99.5	167.2	157.0	104.4	140.0	134.6	103.2	152.7	145.3	114.9	166.8	159.0	102.3	140.1	134.4
0320	Concrete Reinforcing	109.3	129.9	119.8	89.1	131.1	110.5	103.4	130.1	116.9	132.1	130.6	131.3	89.1	131.3	110.5	106.2	129.9	118.2
0330	Cast-in-Place Concrete	81.2	133.2	106.0	94.4	133.6	109.0	79.1	129.8	98.0	101.6	129.5	112.0	108.5	133.5	117.8	88.4	133.2	105.0
03	CONCRETE	90.5	134.6	110.3	91.4	147.0	116.4	90.7	133.3	109.9	114.8	139.0	125.7	103.6	146.8	123.1	96.5	134.6	113.6
04	MASONRY	76.7	139.8	115.7	105.9	146.3	130.8	97.9	141.9	125.1	132.2	141.7	138.1	124.6	150.4	140.6	77.7	139.8	116.1
05	METALS	107.8	111.7	109.0	101.7	113.3	105.2	90.4	110.5	96.5	107.4	112.3	108.8	101.7	113.2	105.2	107.3	111.7	108.6
06	WOOD, PLASTICS & COMPOSITES	92.6	137.4	116.9	92.2	172.6	135.9	88.1	137.1	114.7	110.1	156.3	135.2	110.0	172.6	144.0	97.9	137.4	119.3
07	THERMAL & MOISTURE PROTECTION	104.2	136.2	118.0	97.8	154.4	122.2	95.8	133.2	111.9	131.6	139.2	134.8	98.5	154.5	122.6	105.1	136.2	118.5
08	OPENINGS	102.1	135.9	109.9	92.8	159.5	108.3	89.8	135.7	100.4	118.1	143.8	124.0	92.8	159.5	108.3	104.8	135.9	112.0
0920	Plaster & Gypsum Board	106.2	138.3	127.4	105.7	174.1	150.7	90.3	138.3	121.9	108.9	157.8	141.1	114.2	174.1	153.7	110.5	138.3	128.8
0950, 0980	Ceilings & Acoustic Treatment	109.2	138.3	128.9	99.4	174.1	150.0	112.6	138.3	130.0	148.5	157.8	154.8	99.4	174.1	150.0	118.7	138.3	132.0
0960	Flooring	99.4	120.4	105.5	108.2	142.0	118.0	94.2	120.4	101.8	93.5	132.9	105.0	118.2	142.0	125.1	100.9	120.4	106.6
0970, 0990	Wall Finishes & Painting/Coating	89.1	121.1	108.1	113.3	162.7	142.5	105.4	121.1	114.7	100.0	137.6	122.3	113.3	162.7	142.5	89.1	121.1	108.1
09	FINISHES	96.6	134.5	117.3	104.3	163.7	136.8	98.0	134.3	117.8	106.4	149.4	129.8	109.0	163.7	138.9	99.8	134.5	118.7
COVERS	DIVS. 10 - 14, 25, 28, 41, 43, 44, 46	100.0	120.2	104.4	100.0	129.9	106.6	100.0	119.6	104.3	100.0	126.8	105.9	100.0	129.8	106.6	100.0	120.2	104.4
21, 22, 23	FIRE SUPPRESSION, PLUMBING & HVAC	97.5	131.8	111.2	97.7	168.7	126.2	97.5	131.7	111.2	100.4	130.3	112.4	97.7	159.7	122.6	100.0	131.8	112.7
26, 27, 3370	ELECTRICAL, COMMUNICATIONS & UTIL.	93.3	113.5	103.6	102.1	169.9	136.8	112.3	130.5	121.6	101.9	122.4	112.4	102.7	136.5	120.0	90.1	113.5	102.1
MF2018	WEIGHTED AVERAGE	97.5	127.1	110.2	98.8	151.4	121.4	96.3	129.4	110.5	108.8	131.5	118.6	102.0	145.2	120.5	99.3	127.1	111.2

DIVISION		CALIFORNIA																
		SACRAMENTO			SALINAS			SAN BERNARDINO			SAN DIEGO			SAN FRANCISCO			SAN JOSE		
		942,956 - 958			939			923 - 924			919 - 921			940 - 941			951		
		MAT.	INST.	TOTAL	MAT.	INST.	TOTAL	MAT.	INST.	TOTAL	MAT.	INST.	TOTAL	MAT.	INST.	TOTAL	MAT.	INST.	TOTAL
015433	CONTRACTOR EQUIPMENT		101.6	101.6		99.8	99.8		100.7	100.7		102.5	102.5		110.9	110.9		101.3	101.3
0241, 31 - 34	SITE & INFRASTRUCTURE, DEMOLITION	97.4	116.3	110.6	114.7	107.9	110.0	78.2	108.6	99.4	101.9	108.0	106.0	116.9	114.2	115.0	133.8	102.3	111.9
0310	Concrete Forming & Accessories	99.6	155.0	146.7	103.6	155.3	147.5	106.0	140.1	135.0	105.9	128.0	124.6	105.2	167.7	158.3	102.1	166.9	157.2
0320	Concrete Reinforcing	84.6	130.9	108.2	102.8	131.0	117.2	106.2	129.9	118.2	100.1	129.8	115.2	103.7	132.6	118.4	94.9	131.4	113.5
0330	Cast-in-Place Concrete	89.5	130.5	104.7	96.3	129.8	108.8	61.0	133.2	87.9	90.6	123.5	102.9	118.1	133.8	123.9	107.5	132.7	116.9
03	CONCRETE	93.0	140.1	114.2	102.7	140.4	119.7	71.1	134.6	99.7	101.2	125.8	112.3	112.7	148.1	128.6	104.3	147.0	123.5
04	MASONRY	106.8	147.2	131.8	95.4	147.3	127.5	84.0	139.8	118.5	89.8	132.3	116.0	142.9	153.6	149.5	132.3	150.4	143.5
05	METALS	99.8	107.5	102.1	105.0	113.2	107.5	107.3	111.7	108.6	107.4	113.7	109.3	109.4	122.6	113.4	101.4	119.0	106.8
06	WOOD, PLASTICS & COMPOSITES	88.9	159.3	127.1	98.1	159.0	131.2	101.6	137.4	121.0	99.4	124.9	113.2	97.0	172.6	138.1	106.5	172.4	142.3
07	THERMAL & MOISTURE PROTECTION	108.4	141.0	122.4	98.3	145.2	118.5	103.3	136.2	117.5	100.6	120.3	109.1	103.3	155.7	125.8	104.1	155.6	126.3
08	OPENINGS	105.8	146.7	115.3	95.7	152.0	108.8	102.1	135.9	110.0	98.3	126.1	104.7	98.5	159.8	112.7	94.6	159.4	109.7
0920	Plaster & Gypsum Board	103.2	160.5	140.9	98.9	160.5	139.5	111.8	138.3	129.3	96.3	125.3	115.4	105.3	174.1	150.6	107.3	174.1	151.3
0950, 0980	Ceilings & Acoustic Treatment	99.4	160.5	140.8	106.1	160.5	143.0	112.6	138.3	130.0	123.1	125.3	124.6	108.6	174.1	153.0	113.6	174.1	154.6
0960	Flooring	108.2	127.6	113.9	96.9	140.6	109.6	102.8	120.4	108.0	103.6	120.4	108.4	110.0	142.0	119.3	90.8	142.0	105.7
0970, 0990	Wall Finishes & Painting/Coating	109.9	137.6	126.3	89.7	162.7	132.9	89.1	121.1	108.1	103.0	121.1	113.7	111.2	179.4	151.6	92.4	162.7	134.0
09	FINISHES	103.2	150.4	128.9	93.4	155.2	127.1	98.0	134.5	117.9	105.5	125.5	116.4	107.5	165.6	139.2	98.9	163.6	134.2
COVERS	DIVS. 10 - 14, 25, 28, 41, 43, 44, 46	100.0	127.6	106.1	100.0	127.1	106.0	100.0	120.2	104.4	100.0	117.0	103.7	100.0	130.0	106.6	100.0	129.4	106.5
21, 22, 23	FIRE SUPPRESSION, PLUMBING & HVAC	100.1	129.9	112.1	97.6	136.3	113.1	97.5	131.8	111.2	99.9	128.4	111.3	100.1	185.2	134.2	100.0	168.6	127.5
26, 27, 3370	ELECTRICAL, COMMUNICATIONS & UTIL.	97.7	122.4	110.3	94.4	127.6	111.4	93.3	111.8	102.7	98.7	102.7	100.8	102.3	176.3	140.1	99.1	169.9	135.3
MF2018	WEIGHTED AVERAGE	100.1	132.7	114.1	99.1	135.7	114.8	95.2	126.8	108.8	101.0	120.8	109.5	106.3	158.1	128.5	102.6	151.7	123.6

DIVISION		CALIFORNIA																
		SAN LUIS OBISPO			SAN MATEO			SAN RAFAEL			SANTA ANA			SANTA BARBARA			SANTA CRUZ		
		934			944			949			926 - 927			931			950		
		MAT.	INST.	TOTAL	MAT.	INST.	TOTAL	MAT.	INST.	TOTAL	MAT.	INST.	TOTAL	MAT.	INST.	TOTAL	MAT.	INST.	TOTAL
015433	CONTRACTOR EQUIPMENT		99.8	99.8		102.8	102.8		102.0	102.0		100.7	100.7		99.8	99.8		101.3	101.3
0241, 31 - 34	SITE & INFRASTRUCTURE, DEMOLITION	106.9	108.4	108.0	113.1	110.7	111.4	110.0	116.2	114.3	90.3	108.6	103.0	101.3	108.4	106.3	133.4	102.2	111.6
0310	Concrete Forming & Accessories	110.1	140.2	135.7	105.4	167.1	157.8	106.1	167.0	158.6	106.1	140.1	135.0	101.0	140.1	134.3	102.1	155.6	147.6
0320	Concrete Reinforcing	104.1	129.9	117.2	89.1	131.6	110.7	89.7	131.6	111.0	109.8	129.9	120.1	102.3	129.8	116.3	117.4	131.1	124.3
0330	Cast-in-Place Concrete	103.6	132.4	114.3	105.0	133.6	115.7	122.2	132.6	126.1	77.9	133.2	98.5	97.2	132.4	110.3	106.7	131.6	116.0
03	CONCRETE	101.2	134.3	116.1	100.3	147.0	121.3	121.8	146.4	132.9	88.0	134.6	109.0	93.7	134.3	111.9	106.6	141.4	122.3
04	MASONRY	97.1	137.3	121.9	124.3	153.6	142.4	101.5	153.6	133.7	73.6	140.2	114.7	95.7	137.3	121.4	136.1	147.5	143.1
05	METALS	103.0	111.2	105.5	101.6	113.7	105.3	102.7	110.0	104.9	107.4	111.7	108.7	100.7	111.2	103.9	109.0	117.6	111.6
06	WOOD, PLASTICS & COMPOSITES	100.5	137.4	120.6	99.7	172.6	139.3	97.4	172.4	138.1	103.5	137.4	121.9	93.3	137.4	117.3	106.5	159.2	135.1
07	THERMAL & MOISTURE PROTECTION	98.5	133.7	113.6	98.2	155.9	123.1	102.1	156.2	125.4	104.6	136.3	118.2	97.9	133.7	113.3	103.6	147.5	122.5
08	OPENINGS	93.9	133.8	103.2	92.8	159.5	108.3	103.4	159.4	116.4	101.4	135.9	109.4	95.5	135.9	104.9	95.9	152.1	108.9
0920	Plaster & Gypsum Board	104.7	138.3	126.8	110.8	174.1	152.5	112.7	174.1	153.2	113.3	138.3	129.8	97.8	138.3	124.5	115.3	160.5	145.1
0950, 0980	Ceilings & Acoustic Treatment	106.1	138.3	127.9	99.4	174.1	150.0	106.2	174.1	152.2	112.6	138.3	130.0	107.7	138.3	128.4	113.4	160.5	145.3
0960	Flooring	103.1	120.4	108.1	111.8	142.0	120.6	123.4	140.6	128.4	103.4	120.4	108.3	96.0	120.4	103.1	94.8	140.6	108.1
0970, 0990	Wall Finishes & Painting/Coating	88.9	115.6	104.7	113.3	162.7	142.5	109.3	169.9	145.2	89.1	121.1	108.1	88.9	115.6	104.7	92.5	162.7	134.0
09	FINISHES	95.1	133.9	116.3	106.5	163.7	137.8	109.1	164.1	139.2	99.5	134.5	118.6	91.8	133.9	114.8	101.2	155.2	130.7
COVERS	DIVS. 10 - 14, 25, 28, 41, 43, 44, 46	100.0	126.4	105.8	100.0	129.9	106.6	100.0	129.2	106.4	100.0	120.2	104.2	100.0	119.5	104.3	100.0	127.4	106.0
21, 22, 23	FIRE SUPPRESSION, PLUMBING & HVAC	97.6	131.8	111.3	97.7	163.6	124.1	97.7	185.1	132.7	97.5	131.8	111.2	100.1	131.8	112.8	100.0	136.7	114.7
26, 27, 3370	ELECTRICAL, COMMUNICATIONS & UTIL.	93.4	113.1	103.5	102.1	163.0	133.3	98.9	124.3	111.9	93.3	112.8	103.3	92.4	114.2	103.5	98.3	127.6	113.3
MF2018	WEIGHTED AVERAGE	98.4	126.6	110.5	101.2	150.2	122.2	103.9	149.3	123.4	97.2	127.0	110.0	97.2	126.7	109.8	104.5	136.0	118.0

City Cost Indexes

CALIFORNIA / COLORADO

DIVISION		SANTA ROSA 954 MAT.	INST.	TOTAL	STOCKTON 952 MAT.	INST.	TOTAL	SUSANVILLE 961 MAT.	INST.	TOTAL	VALLEJO 945 MAT.	INST.	TOTAL	VAN NUYS 913-916 MAT.	INST.	TOTAL	ALAMOSA 811 MAT.	INST.	TOTAL
015433	CONTRACTOR EQUIPMENT		100.1	100.1		99.6	99.6		99.6	99.6		102.0	102.0		97.7	97.7		93.1	93.1
0241, 31 - 34	SITE & INFRASTRUCTURE, DEMOLITION	104.2	107.3	106.4	102.5	107.3	105.8	134.4	107.3	115.5	98.7	116.1	110.8	111.4	110.3	110.6	139.7	88.5	104.1
0310	Concrete Forming & Accessories	99.3	166.2	156.2	100.1	154.6	146.4	104.4	152.5	145.3	101.3	165.9	156.2	111.5	140.0	135.7	102.0	62.7	68.6
0320	Concrete Reinforcing	104.8	131.5	118.4	107.6	130.5	119.3	132.1	130.4	131.2	90.8	131.4	111.5	103.4	130.1	116.9	111.5	69.8	90.3
0330	Cast-in-Place Concrete	101.5	131.1	112.5	90.1	129.9	104.8	92.5	129.5	106.3	97.4	131.7	110.2	83.5	129.8	100.7	100.1	73.5	90.2
03	CONCRETE	107.3	145.9	124.7	97.4	139.9	116.5	118.4	138.9	127.6	97.6	145.6	119.2	105.2	133.3	117.9	111.4	68.4	92.0
04	MASONRY	102.3	152.5	133.3	102.0	141.7	126.5	130.7	139.2	136.0	77.4	152.5	123.8	111.7	141.9	130.4	129.7	60.6	87.0
05	METALS	107.7	115.1	109.9	103.6	112.2	106.2	106.3	112.1	108.0	102.7	109.3	104.7	89.5	110.5	95.9	104.3	78.4	96.4
06	WOOD, PLASTICS & COMPOSITES	94.1	172.2	136.5	99.9	159.0	132.0	111.9	156.0	135.9	86.6	172.4	133.2	97.2	137.1	118.9	99.6	63.0	79.7
07	THERMAL & MOISTURE PROTECTION	105.0	152.9	125.6	108.0	138.1	121.0	133.4	138.4	135.6	100.3	153.6	123.2	96.9	133.2	112.5	104.9	68.2	89.1
08	OPENINGS	103.0	159.3	116.1	103.5	146.6	113.5	119.0	144.9	125.0	105.1	159.4	117.7	89.6	135.7	100.3	98.5	66.4	91.0
0920	Plaster & Gypsum Board	108.2	174.1	151.6	111.2	160.5	143.7	109.6	157.4	141.1	106.9	174.1	151.2	94.4	138.1	123.3	81.5	61.9	68.6
0950, 0980	Ceilings & Acoustic Treatment	112.6	174.1	154.3	122.1	160.5	148.1	140.0	157.4	151.8	108.1	174.1	152.8	109.2	138.3	128.9	103.8	61.9	75.4
0960	Flooring	99.9	133.5	109.7	97.2	129.9	106.7	94.0	132.9	105.3	117.5	142.0	124.6	97.3	120.4	104.0	108.4	71.3	97.6
0970, 0990	Wall Finishes & Painting/Coating	89.1	169.9	136.9	92.1	137.6	119.0	100.0	137.6	122.3	110.3	169.9	145.6	105.4	121.1	114.7	102.7	60.8	77.9
09	FINISHES	98.6	162.5	133.5	101.5	150.6	128.3	106.0	149.3	129.6	105.8	164.1	137.6	100.1	134.3	118.8	98.4	63.5	79.4
COVERS	DIVS. 10 - 14, 25, 28, 41, 43, 44, 46	100.0	128.3	106.2	100.0	127.0	106.0	100.0	126.7	105.9	100.0	128.9	106.4	100.0	119.6	104.3	100.0	84.8	96.6
21, 22, 23	FIRE SUPPRESSION, PLUMBING & HVAC	97.5	184.5	132.4	100.0	130.3	112.1	97.9	130.3	110.9	100.2	147.9	119.3	97.5	131.7	111.2	97.5	66.0	84.9
26, 27, 3370	ELECTRICAL, COMMUNICATIONS & UTIL.	93.6	124.3	109.3	96.1	114.3	105.4	102.2	122.4	112.6	95.1	127.9	111.9	112.3	130.5	121.6	97.0	64.8	80.5
MF2018	WEIGHTED AVERAGE	101.4	148.5	121.6	100.7	130.8	113.6	108.8	131.2	118.4	99.5	141.5	117.5	99.3	129.4	112.2	103.5	68.8	88.6

COLORADO

DIVISION		BOULDER 803 MAT.	INST.	TOTAL	COLORADO SPRINGS 808-809 MAT.	INST.	TOTAL	DENVER 800-802 MAT.	INST.	TOTAL	DURANGO 813 MAT.	INST.	TOTAL	FORT COLLINS 805 MAT.	INST.	TOTAL	FORT MORGAN 807 MAT.	INST.	TOTAL
015433	CONTRACTOR EQUIPMENT		95.2	95.2		93.1	93.1		99.4	99.4		93.1	93.1		95.2	95.2		95.2	95.2
0241, 31 - 34	SITE & INFRASTRUCTURE, DEMOLITION	98.0	95.8	96.5	99.9	91.3	93.9	104.9	102.3	103.1	133.3	88.5	102.1	110.8	95.7	100.2	100.5	95.5	97.0
0310	Concrete Forming & Accessories	101.7	73.6	77.8	92.3	65.1	69.2	101.0	64.8	70.3	108.4	62.5	69.4	99.3	68.6	73.3	102.3	62.4	68.4
0320	Concrete Reinforcing	98.3	69.8	83.8	97.5	69.9	83.5	97.5	69.9	83.5	111.5	69.7	90.3	98.3	69.7	83.8	98.5	69.7	83.9
0330	Cast-in-Place Concrete	112.5	72.8	97.7	115.5	73.9	100.0	120.5	71.3	102.2	115.1	73.4	99.6	127.0	71.4	106.3	110.3	72.7	96.3
03	CONCRETE	104.6	73.0	90.4	108.2	69.6	90.8	111.0	68.6	91.9	113.6	68.2	93.2	115.9	70.3	95.4	103.1	67.9	87.3
04	MASONRY	101.3	63.3	77.8	103.2	63.6	78.7	104.4	63.8	79.3	116.7	60.6	82.1	119.2	60.9	83.2	116.7	60.9	82.2
05	METALS	97.0	77.4	91.1	100.0	78.1	93.4	102.4	78.6	95.2	104.3	78.2	96.4	98.2	77.3	91.9	96.8	77.3	90.8
06	WOOD, PLASTICS & COMPOSITES	106.7	76.3	90.2	96.1	65.5	79.5	104.3	65.3	83.1	109.5	63.0	84.2	104.2	70.9	86.1	106.7	62.5	82.2
07	THERMAL & MOISTURE PROTECTION	96.2	72.7	86.1	96.9	71.4	85.9	94.7	70.9	84.5	104.9	68.2	89.1	96.5	70.4	85.3	96.1	69.6	84.7
08	OPENINGS	97.5	73.4	91.9	101.7	67.8	93.8	104.5	67.7	96.0	105.4	66.4	96.4	97.4	70.4	91.2	97.4	66.1	90.1
0920	Plaster & Gypsum Board	116.5	76.0	89.8	100.0	64.8	76.8	111.7	64.8	80.8	95.3	61.9	73.3	110.7	70.5	84.2	116.5	61.9	80.5
0950, 0980	Ceilings & Acoustic Treatment	86.2	76.0	79.3	93.0	64.8	73.9	97.5	64.8	75.3	103.8	61.9	75.4	86.2	70.5	75.5	86.2	61.9	69.7
0960	Flooring	108.3	71.3	97.6	99.7	71.3	91.5	105.9	71.3	95.9	113.4	71.3	101.2	104.6	71.3	95.0	108.7	71.3	97.9
0970, 0990	Wall Finishes & Painting/Coating	99.4	63.0	77.8	99.1	63.0	77.7	106.2	63.0	80.6	102.7	60.8	77.9	99.4	60.8	76.5	99.4	60.8	76.5
09	FINISHES	100.5	72.2	85.0	97.0	65.5	79.8	102.3	65.3	82.1	101.0	63.5	80.6	99.1	68.2	82.2	100.6	63.2	80.2
COVERS	DIVS. 10 - 14, 25, 28, 41, 43, 44, 46	100.0	86.1	96.9	100.0	84.7	96.6	100.0	84.2	96.5	100.0	84.7	96.6	100.0	84.7	96.6	100.0	83.8	96.4
21, 22, 23	FIRE SUPPRESSION, PLUMBING & HVAC	97.6	74.6	88.4	100.2	73.8	89.6	100.0	73.6	89.4	97.5	57.2	81.3	100.1	73.3	89.3	97.6	73.3	87.9
26, 27, 3370	ELECTRICAL, COMMUNICATIONS & UTIL.	96.5	80.0	88.1	99.6	74.9	87.0	101.3	80.2	90.5	96.5	51.5	73.5	96.5	77.2	86.6	96.8	80.0	88.2
MF2018	WEIGHTED AVERAGE	98.9	75.9	89.0	101.0	73.0	89.0	102.7	74.4	90.6	103.9	65.0	87.2	102.1	73.7	89.9	99.5	72.9	88.1

COLORADO

DIVISION		GLENWOOD SPRINGS 816 MAT.	INST.	TOTAL	GOLDEN 804 MAT.	INST.	TOTAL	GRAND JUNCTION 815 MAT.	INST.	TOTAL	GREELEY 806 MAT.	INST.	TOTAL	MONTROSE 814 MAT.	INST.	TOTAL	PUEBLO 810 MAT.	INST.	TOTAL
015433	CONTRACTOR EQUIPMENT		96.3	96.3		95.2	95.2		96.3	96.3		95.2	95.2		94.7	94.7		93.1	93.1
0241, 31 - 34	SITE & INFRASTRUCTURE, DEMOLITION	148.9	96.4	112.3	111.2	95.8	100.5	132.8	96.6	107.6	96.7	95.7	96.0	142.3	92.5	107.6	124.4	88.9	99.7
0310	Concrete Forming & Accessories	98.7	52.4	59.4	94.6	63.4	68.1	107.2	73.0	78.2	97.3	72.6	76.3	98.2	62.7	68.0	104.4	65.4	71.2
0320	Concrete Reinforcing	110.3	69.7	89.7	98.5	69.7	83.9	110.7	69.8	89.9	98.3	69.7	83.8	110.2	69.8	89.6	106.6	69.8	87.9
0330	Cast-in-Place Concrete	100.1	72.3	89.7	110.4	74.1	96.9	110.8	73.7	97.0	106.1	72.7	93.7	100.1	72.9	90.0	99.4	74.2	90.0
03	CONCRETE	116.4	63.3	92.5	113.4	68.9	93.4	110.1	73.2	93.5	99.6	72.6	87.4	107.4	68.2	89.8	100.3	69.9	86.6
04	MASONRY	101.6	60.8	76.4	119.7	63.3	84.9	136.8	64.0	91.8	112.5	60.9	80.7	109.5	60.6	79.3	97.8	61.6	75.4
05	METALS	104.0	77.9	96.0	96.9	77.3	91.0	105.6	78.2	97.3	98.2	77.4	91.9	103.0	78.3	95.5	107.2	79.0	98.7
06	WOOD, PLASTICS & COMPOSITES	94.6	49.2	69.9	98.5	62.5	79.0	107.2	75.7	90.1	101.3	76.2	87.7	95.6	62.8	77.8	102.3	65.8	82.5
07	THERMAL & MOISTURE PROTECTION	104.8	66.7	88.4	97.3	71.3	86.1	103.8	71.6	89.9	95.8	71.0	85.1	104.9	68.2	89.1	103.3	69.5	88.7
08	OPENINGS	104.4	58.8	93.8	97.5	66.1	90.2	105.1	73.1	97.7	97.4	73.7	91.9	105.6	66.3	96.5	100.3	67.9	92.8
0920	Plaster & Gypsum Board	122.5	48.0	73.4	108.0	61.9	77.6	135.9	75.2	96.0	109.2	76.0	87.3	80.6	61.9	68.3	84.9	64.8	71.7
0950, 0980	Ceilings & Acoustic Treatment	103.0	48.0	65.7	86.2	61.9	69.7	103.0	75.2	84.2	86.2	76.0	79.3	103.8	61.9	75.4	86.2	64.8	69.7
0960	Flooring	107.5	71.3	97.0	102.3	71.3	93.3	112.8	71.3	100.7	103.5	71.3	94.2	110.8	71.3	99.4	109.5	75.5	99.7
0970, 0990	Wall Finishes & Painting/Coating	102.7	60.8	77.9	99.4	63.0	77.8	102.7	63.0	79.2	99.4	60.8	76.5	102.7	60.8	77.9	102.7	60.8	77.9
09	FINISHES	104.3	55.4	77.6	98.9	64.1	79.9	106.0	71.8	87.3	97.8	71.3	83.3	99.0	63.4	79.6	98.6	66.3	80.9
COVERS	DIVS. 10 - 14, 25, 28, 41, 43, 44, 46	100.0	82.6	96.2	100.0	84.6	96.6	100.0	86.3	97.0	100.0	85.3	96.8	100.0	84.4	96.6	100.0	85.4	96.8
21, 22, 23	FIRE SUPPRESSION, PLUMBING & HVAC	97.5	67.0	85.3	97.6	74.6	88.4	99.9	75.8	90.2	100.1	73.3	89.3	97.5	75.8	88.8	99.9	73.8	89.5
26, 27, 3370	ELECTRICAL, COMMUNICATIONS & UTIL.	94.0	51.5	72.3	96.8	80.0	88.2	96.2	51.5	73.3	96.5	77.2	86.6	96.2	51.5	73.3	97.0	64.8	80.5
MF2018	WEIGHTED AVERAGE	103.7	65.4	87.2	101.0	73.7	89.3	105.5	72.2	91.2	99.3	74.7	88.7	102.5	69.3	88.3	101.4	71.4	88.5

For customer support on your Plumbing Costs with RSMeans Data, call 800.448.8182.

641

DIVISION		COLORADO			CONNECTICUT													
		SALIDA			BRIDGEPORT			BRISTOL			HARTFORD			MERIDEN			NEW BRITAIN		
		812			066			060			061			064			060		
		MAT.	INST.	TOTAL	MAT.	INST.	TOTAL	MAT.	INST.	TOTAL	MAT.	INST.	TOTAL	MAT.	INST.	TOTAL	MAT.	INST.	TOTAL
015433	CONTRACTOR EQUIPMENT		94.7	94.7		97.9	97.9		97.9	97.9		97.9	97.9		98.3	98.3		97.9	97.9
0241, 31 - 34	SITE & INFRASTRUCTURE, DEMOLITION	133.0	92.4	104.8	103.3	102.3	102.6	102.4	102.3	102.3	97.8	102.3	100.9	100.2	103.1	102.2	102.6	102.3	102.4
0310	Concrete Forming & Accessories	107.1	52.5	60.7	98.6	119.5	116.3	98.6	119.3	116.2	99.9	119.3	116.4	98.3	119.3	116.2	99.0	119.3	116.3
0320	Concrete Reinforcing	109.9	69.7	89.5	125.5	145.1	135.4	125.5	145.1	135.4	120.3	145.1	132.9	125.5	145.1	135.4	125.5	145.1	135.4
0330	Cast-in-Place Concrete	114.7	72.8	99.1	102.1	128.8	112.1	95.7	128.7	108.0	100.2	128.7	110.8	92.1	128.7	105.7	97.2	128.7	111.3
03	CONCRETE	108.7	63.5	88.4	98.6	126.2	111.0	95.5	126.1	109.3	95.9	126.1	109.5	93.8	126.1	108.3	96.3	126.1	109.7
04	MASONRY	138.1	60.6	90.2	116.6	130.9	125.4	107.4	130.9	121.9	106.8	130.9	121.7	107.0	130.9	121.8	109.7	130.9	122.8
05	METALS	102.7	78.1	95.2	91.2	115.7	98.6	91.2	115.6	98.6	95.7	115.6	101.7	88.5	115.6	96.7	87.7	115.6	96.1
06	WOOD, PLASTICS & COMPOSITES	103.6	49.3	74.1	100.8	117.5	109.9	100.8	117.5	109.9	91.0	117.5	105.4	100.8	117.5	109.9	100.8	117.5	109.9
07	THERMAL & MOISTURE PROTECTION	103.8	66.7	87.8	98.0	125.8	110.0	98.1	121.9	108.4	102.8	121.9	111.0	98.1	121.9	108.4	98.1	122.2	108.5
08	OPENINGS	98.6	58.8	89.3	98.4	122.7	104.0	98.4	122.7	104.0	96.5	122.7	102.5	100.6	122.7	105.7	98.4	122.7	104.0
0920	Plaster & Gypsum Board	80.9	48.0	59.2	116.7	117.7	117.4	116.7	117.7	117.4	104.7	117.7	113.3	117.8	117.7	117.7	116.7	117.7	117.4
0950, 0980	Ceilings & Acoustic Treatment	103.8	48.0	66.0	97.9	117.7	111.3	97.9	117.7	111.3	94.1	117.7	110.1	101.1	117.7	112.4	97.9	117.7	111.3
0960	Flooring	116.1	71.3	103.1	90.6	131.2	102.4	90.6	128.7	101.6	94.9	131.2	105.4	90.6	128.7	101.6	90.6	128.7	101.6
0970, 0990	Wall Finishes & Painting/Coating	102.7	60.8	77.9	89.8	128.8	112.9	89.8	128.8	112.9	94.3	128.8	114.8	89.8	128.8	112.9	89.8	128.8	112.9
09	FINISHES	99.5	55.5	75.5	90.9	122.2	108.0	90.9	121.8	107.8	92.0	122.2	108.5	91.7	121.8	108.1	90.9	121.8	107.8
COVERS	DIVS. 10 - 14, 25, 28, 41, 43, 44, 46	100.0	83.0	96.3	100.0	113.4	102.9	100.0	113.4	102.9	100.0	113.4	102.9	100.0	113.4	102.9	100.0	113.4	102.9
21, 22, 23	FIRE SUPPRESSION, PLUMBING & HVAC	97.5	65.0	84.5	100.1	118.5	107.5	100.1	118.5	107.5	100.0	118.5	107.4	97.6	118.5	106.0	100.1	118.5	107.5
26, 27, 3370	ELECTRICAL, COMMUNICATIONS & UTIL.	96.4	64.8	80.2	97.4	107.6	102.6	97.4	106.2	101.9	97.0	109.3	103.3	97.3	108.4	103.0	97.4	106.2	101.9
MF2018	WEIGHTED AVERAGE	103.1	66.6	87.5	98.1	118.4	106.8	97.2	118.1	106.2	97.8	118.5	106.7	96.2	118.4	105.7	96.9	118.1	106.0

DIVISION		CONNECTICUT																
		NEW HAVEN			NEW LONDON			NORWALK			STAMFORD			WATERBURY			WILLIMANTIC		
		065			063			068			069			067			062		
		MAT.	INST.	TOTAL	MAT.	INST.	TOTAL	MAT.	INST.	TOTAL	MAT.	INST.	TOTAL	MAT.	INST.	TOTAL	MAT.	INST.	TOTAL
015433	CONTRACTOR EQUIPMENT		98.3	98.3		98.3	98.3		97.9	97.9		97.9	97.9		97.9	97.9		97.9	97.9
0241, 31 - 34	SITE & INFRASTRUCTURE, DEMOLITION	102.5	103.1	102.9	94.9	102.8	100.4	103.1	102.3	102.5	103.7	102.3	102.8	103.1	102.3	102.6	103.0	102.1	102.3
0310	Concrete Forming & Accessories	98.3	119.4	116.2	98.3	119.3	116.1	98.6	119.5	116.4	98.6	119.8	116.6	98.6	119.4	116.2	98.6	119.2	116.1
0320	Concrete Reinforcing	125.5	145.1	135.4	98.3	145.1	122.1	125.5	145.2	135.5	125.5	145.3	135.5	125.5	145.1	135.4	125.5	145.1	135.4
0330	Cast-in-Place Concrete	98.9	128.7	110.0	84.3	128.7	100.8	100.5	130.0	111.5	102.1	130.1	112.6	102.1	128.7	112.0	95.4	127.0	107.1
03	CONCRETE	108.9	126.1	116.6	84.3	126.0	103.1	97.8	126.6	110.8	98.6	126.8	111.3	98.6	126.1	111.0	95.4	125.4	108.9
04	MASONRY	107.7	130.9	122.0	106.0	130.9	121.4	107.1	132.2	122.6	108.0	132.2	122.9	108.0	130.9	122.1	107.3	130.9	121.9
05	METALS	87.9	115.6	96.3	87.6	115.6	96.1	91.2	115.9	98.7	91.2	116.2	98.8	91.2	115.6	98.6	90.9	115.4	98.4
06	WOOD, PLASTICS & COMPOSITES	100.8	117.5	109.9	100.8	117.5	109.9	100.8	117.5	109.9	100.8	117.5	109.9	100.8	117.5	109.9	100.8	117.5	109.9
07	THERMAL & MOISTURE PROTECTION	98.2	122.5	108.7	98.0	121.9	108.3	98.2	126.4	110.3	98.1	126.4	110.3	98.1	122.5	108.6	98.3	121.3	108.2
08	OPENINGS	98.4	122.7	104.0	100.8	122.7	105.9	98.4	122.7	104.0	98.4	122.7	104.0	98.4	122.7	104.0	100.8	122.7	105.9
0920	Plaster & Gypsum Board	116.7	117.7	117.4	116.7	117.7	117.4	116.7	117.7	117.4	116.7	117.7	117.4	116.7	117.7	117.4	116.7	117.7	117.4
0950, 0980	Ceilings & Acoustic Treatment	97.9	117.7	111.3	96.0	117.7	110.7	97.9	117.7	111.3	97.9	117.7	111.3	97.9	117.7	111.3	96.0	117.7	110.7
0960	Flooring	90.6	131.2	102.4	90.6	128.7	101.6	90.6	128.7	101.6	90.6	131.2	102.4	90.6	131.2	102.4	90.6	125.2	100.6
0970, 0990	Wall Finishes & Painting/Coating	89.8	128.8	112.9	89.8	128.8	112.9	89.8	128.8	112.9	89.8	128.8	112.9	89.8	128.8	112.9	89.8	128.8	112.9
09	FINISHES	91.0	122.2	108.0	90.1	121.8	107.4	90.9	121.8	107.8	91.0	122.2	108.0	90.8	122.2	107.9	90.6	121.2	107.3
COVERS	DIVS. 10 - 14, 25, 28, 41, 43, 44, 46	100.0	113.4	102.9	100.0	113.4	102.9	100.0	113.4	102.9	100.0	113.6	103.0	100.0	113.4	103.0	100.0	113.4	102.9
21, 22, 23	FIRE SUPPRESSION, PLUMBING & HVAC	100.1	118.5	107.5	97.6	118.5	106.0	100.1	118.5	107.5	100.1	118.6	107.5	100.1	118.5	107.5	100.1	118.4	107.4
26, 27, 3370	ELECTRICAL, COMMUNICATIONS & UTIL.	97.3	108.6	103.1	94.5	108.4	101.6	97.4	107.4	102.5	97.4	153.8	126.2	96.9	109.7	103.5	97.4	109.3	103.5
MF2018	WEIGHTED AVERAGE	98.4	118.5	107.0	94.3	118.4	104.7	97.5	118.6	106.6	97.7	125.2	109.5	97.6	118.6	106.6	97.4	118.2	106.4

DIVISION		D.C.			DELAWARE									FLORIDA				
		WASHINGTON			DOVER			NEWARK			WILMINGTON			DAYTONA BEACH			FORT LAUDERDALE		
		200 - 205			199			197			198			321			333		
		MAT.	INST.	TOTAL	MAT.	INST.	TOTAL	MAT.	INST.	TOTAL	MAT.	INST.	TOTAL	MAT.	INST.	TOTAL	MAT.	INST.	TOTAL
015433	CONTRACTOR EQUIPMENT		105.6	105.6		121.3	121.3		121.3	121.3		121.5	121.5		102.1	102.1		95.2	95.2
0241, 31 - 34	SITE & INFRASTRUCTURE, DEMOLITION	105.2	95.9	98.7	100.5	113.6	109.7	102.7	113.6	110.3	105.0	114.0	111.2	113.2	88.9	96.3	92.1	76.8	81.4
0310	Concrete Forming & Accessories	97.0	74.0	77.4	96.8	101.9	101.2	96.5	101.9	101.1	95.0	101.9	100.9	97.1	62.7	67.9	92.0	63.3	67.6
0320	Concrete Reinforcing	100.9	88.7	94.7	109.1	118.4	113.8	103.3	118.4	111.0	107.4	118.4	113.0	93.3	62.6	77.7	94.5	61.4	77.7
0330	Cast-in-Place Concrete	112.6	80.6	100.7	100.6	107.5	103.1	90.0	107.5	96.5	102.3	107.5	104.3	88.1	65.9	79.8	90.1	62.0	79.6
03	CONCRETE	109.9	79.8	96.4	96.3	107.7	101.4	92.4	107.7	99.3	96.8	107.7	101.7	87.5	65.6	77.6	90.0	64.2	78.4
04	MASONRY	100.4	80.9	88.3	97.5	100.1	99.1	101.3	100.1	100.5	97.1	100.1	98.9	88.4	59.5	70.6	93.3	62.3	74.1
05	METALS	99.1	95.8	98.1	92.8	123.7	102.2	97.5	123.7	105.5	97.4	123.7	105.4	101.5	89.0	97.7	99.3	89.2	96.3
06	WOOD, PLASTICS & COMPOSITES	95.0	72.7	82.9	88.7	100.2	95.0	91.1	100.2	96.0	81.7	100.2	91.8	96.4	61.8	77.6	80.2	65.2	72.1
07	THERMAL & MOISTURE PROTECTION	103.7	84.9	95.6	104.3	112.7	107.9	107.0	112.7	109.5	103.9	112.7	107.7	99.8	65.7	85.1	98.5	62.5	83.0
08	OPENINGS	99.3	75.9	93.9	90.4	110.4	95.1	93.4	110.4	97.4	88.8	110.4	93.9	95.1	60.9	87.2	96.4	62.4	88.5
0920	Plaster & Gypsum Board	105.7	71.9	83.5	97.8	100.2	99.4	95.9	100.2	98.7	96.8	100.2	99.1	95.0	61.2	72.8	109.2	64.7	79.9
0950, 0980	Ceilings & Acoustic Treatment	105.6	71.9	82.8	99.1	100.2	99.9	96.6	100.2	99.1	90.3	100.2	97.0	83.7	61.2	68.5	86.1	64.7	71.6
0960	Flooring	96.4	78.1	91.1	90.4	109.0	95.8	87.7	109.0	93.9	91.5	109.0	96.6	101.0	56.7	88.2	96.8	76.7	91.0
0970, 0990	Wall Finishes & Painting/Coating	99.1	75.0	84.8	88.6	122.2	108.5	84.6	122.2	106.8	86.8	122.2	107.7	97.1	63.1	77.0	88.4	58.6	70.8
09	FINISHES	97.9	74.0	84.9	91.5	104.5	98.6	87.7	104.5	96.9	90.1	104.5	97.9	90.4	61.0	74.4	90.5	65.1	76.7
COVERS	DIVS. 10 - 14, 25, 28, 41, 43, 44, 46	100.0	96.3	99.2	100.0	105.4	101.2	100.0	105.4	101.2	100.0	105.4	101.2	100.0	82.7	96.2	100.0	82.7	96.2
21, 22, 23	FIRE SUPPRESSION, PLUMBING & HVAC	100.1	90.3	96.2	100.0	122.0	108.8	100.2	122.0	108.9	100.1	122.0	108.9	99.9	76.7	90.6	100.0	68.2	87.2
26, 27, 3370	ELECTRICAL, COMMUNICATIONS & UTIL.	101.3	102.4	101.9	96.7	112.4	104.7	99.2	112.4	106.0	98.0	112.4	105.4	94.4	60.3	77.0	91.7	69.1	80.1
MF2018	WEIGHTED AVERAGE	101.2	87.7	95.4	96.3	112.3	103.1	97.2	112.3	103.7	97.0	112.3	103.6	96.6	70.3	85.3	96.0	69.5	84.6

City Cost Indexes

FLORIDA

DIVISION		FORT MYERS 339, 341			GAINESVILLE 326, 344			JACKSONVILLE 320, 322			LAKELAND 338			MELBOURNE 329			MIAMI 330 - 332, 340		
		MAT.	INST.	TOTAL	MAT.	INST.	TOTAL	MAT.	INST.	TOTAL	MAT.	INST.	TOTAL	MAT.	INST.	TOTAL	MAT.	INST.	TOTAL
015433	CONTRACTOR EQUIPMENT		102.1	102.1		102.1	102.1		102.1	102.1		102.1	102.1		102.1	102.1		95.2	95.2
0241, 31 - 34	SITE & INFRASTRUCTURE, DEMOLITION	103.8	89.0	93.5	122.4	88.7	99.0	113.1	88.9	96.3	105.7	88.9	94.0	120.9	89.1	98.8	93.5	77.1	82.1
0310	Concrete Forming & Accessories	88.1	62.1	66.0	92.3	56.9	62.3	96.8	60.1	65.6	84.8	61.5	65.0	93.5	62.9	67.5	97.3	63.3	68.4
0320	Concrete Reinforcing	95.5	78.9	87.1	99.0	60.7	79.5	93.3	60.8	76.8	97.9	78.3	87.9	94.4	67.4	80.7	101.4	60.9	80.8
0330	Cast-in-Place Concrete	93.9	63.4	82.6	101.2	63.0	87.0	89.0	64.0	79.7	96.1	65.2	84.6	106.2	66.0	91.3	87.0	63.2	78.2
03	CONCRETE	90.5	67.1	80.0	98.3	61.6	81.8	87.9	63.4	76.9	92.2	67.4	81.0	98.8	66.5	84.2	88.3	64.6	77.6
04	MASONRY	87.1	64.8	73.3	100.7	55.9	73.0	88.1	56.5	68.6	101.9	58.3	74.9	85.0	59.6	69.3	92.6	54.7	69.2
05	METALS	101.5	95.2	99.6	100.5	87.9	96.7	100.1	88.1	96.4	101.4	94.2	99.3	110.2	90.8	104.3	99.7	88.3	96.2
06	WOOD, PLASTICS & COMPOSITES	77.1	63.2	69.6	90.0	56.2	71.6	96.4	60.3	76.8	72.4	61.0	66.2	91.8	61.8	75.5	90.3	65.2	76.7
07	THERMAL & MOISTURE PROTECTION	98.3	64.4	83.7	100.2	60.7	83.2	100.1	61.5	83.5	98.2	63.0	83.1	100.3	64.3	84.8	98.4	61.3	82.4
08	OPENINGS	97.7	65.2	90.2	94.8	57.4	86.1	95.1	59.7	86.9	97.7	64.0	89.8	94.4	62.0	86.9	99.9	62.4	91.2
0920	Plaster & Gypsum Board	105.3	62.7	77.2	91.8	55.4	67.8	95.0	59.7	71.8	101.6	60.4	74.5	91.8	61.2	71.7	95.5	64.7	75.2
0950, 0980	Ceilings & Acoustic Treatment	81.6	62.7	68.8	78.1	55.4	62.7	83.7	59.7	67.4	81.6	60.4	67.3	82.9	61.2	68.2	84.9	64.7	71.2
0960	Flooring	93.7	79.8	89.7	98.4	56.7	86.3	101.0	58.5	88.7	91.6	58.4	82.0	98.6	57.8	86.8	98.8	58.4	87.1
0970, 0990	Wall Finishes & Painting/Coating	92.6	63.6	75.4	97.1	63.6	77.3	97.1	63.6	77.3	92.6	63.6	75.4	97.1	83.0	88.8	88.9	58.6	70.9
09	FINISHES	90.3	65.3	76.6	89.1	56.9	71.6	90.5	59.7	73.7	89.2	60.6	73.6	89.8	63.4	75.4	89.1	61.6	74.1
COVERS	DIVS. 10 - 14, 25, 28, 41, 43, 44, 46	100.0	79.4	95.5	100.0	81.1	95.8	100.0	78.2	95.2	100.0	80.4	95.7	100.0	82.8	96.2	100.0	83.0	96.2
21, 22, 23	FIRE SUPPRESSION, PLUMBING & HVAC	98.7	57.5	82.2	98.8	62.3	84.2	99.9	62.3	84.9	98.7	59.4	82.9	99.9	75.3	90.1	100.0	62.5	85.0
26, 27, 3370	ELECTRICAL, COMMUNICATIONS & UTIL.	93.5	61.5	77.1	94.7	57.2	75.5	94.2	63.4	78.5	91.9	60.0	75.6	95.5	64.3	79.5	95.1	72.3	83.5
MF2018	WEIGHTED AVERAGE	96.3	68.3	84.3	98.2	64.9	83.9	96.4	66.5	83.6	97.0	67.1	84.2	99.4	71.2	87.3	96.5	67.4	84.0

FLORIDA

DIVISION		ORLANDO 327 - 328, 347			PANAMA CITY 324			PENSACOLA 325			SARASOTA 342			ST. PETERSBURG 337			TALLAHASSEE 323		
		MAT.	INST.	TOTAL	MAT.	INST.	TOTAL	MAT.	INST.	TOTAL	MAT.	INST.	TOTAL	MAT.	INST.	TOTAL	MAT.	INST.	TOTAL
015433	CONTRACTOR EQUIPMENT		102.1	102.1		102.1	102.1		102.1	102.1		102.1	102.1		102.1	102.1		102.1	102.1
0241, 31 - 34	SITE & INFRASTRUCTURE, DEMOLITION	113.6	89.2	96.6	126.3	88.2	99.8	126.4	88.6	100.1	116.9	89.0	97.4	107.3	88.5	94.2	106.7	88.7	94.1
0310	Concrete Forming & Accessories	101.5	62.5	68.4	96.1	60.3	65.7	94.2	63.6	68.2	93.8	61.6	66.5	91.4	59.3	64.1	98.6	56.9	63.1
0320	Concrete Reinforcing	98.5	67.4	82.7	97.5	69.8	83.4	100.0	69.2	84.3	93.7	78.3	85.8	97.9	78.2	87.9	99.1	60.7	79.6
0330	Cast-in-Place Concrete	102.6	65.8	88.9	93.5	58.8	80.6	115.6	63.9	96.4	103.7	65.4	89.5	97.1	61.0	83.7	93.1	62.8	81.9
03	CONCRETE	94.0	66.2	81.5	96.4	63.2	81.5	106.2	66.4	88.2	96.0	67.5	83.2	93.7	64.9	80.8	90.7	61.5	77.6
04	MASONRY	92.3	59.5	72.1	92.4	57.4	70.8	113.1	57.4	78.7	90.1	58.8	70.7	138.6	52.1	85.2	90.1	55.4	68.7
05	METALS	99.2	90.4	96.5	101.3	91.4	98.3	102.4	90.8	98.9	102.8	93.6	100.0	102.4	93.9	99.8	99.0	88.0	95.7
06	WOOD, PLASTICS & COMPOSITES	93.5	61.8	76.3	95.2	64.2	78.4	93.1	64.2	77.4	94.4	61.0	76.3	81.5	61.0	70.4	95.5	56.2	74.1
07	THERMAL & MOISTURE PROTECTION	101.9	65.7	86.3	100.5	58.8	82.5	100.4	61.8	83.8	98.5	63.2	83.3	98.5	58.8	81.4	100.6	61.2	83.6
08	OPENINGS	99.4	62.0	90.7	93.2	63.7	86.4	93.2	63.7	86.4	100.0	63.6	91.5	96.5	64.0	88.9	100.3	57.4	90.3
0920	Plaster & Gypsum Board	96.0	61.2	73.1	94.3	63.7	74.1	102.8	63.7	77.0	100.9	60.4	74.2	107.7	60.4	76.6	99.3	55.4	70.4
0950, 0980	Ceilings & Acoustic Treatment	90.0	61.2	70.5	82.9	63.7	69.9	82.9	63.7	69.9	88.7	60.4	69.6	83.5	60.4	67.9	88.3	55.4	66.0
0960	Flooring	97.2	58.4	85.9	100.6	75.0	93.2	96.4	56.7	84.9	103.5	56.7	89.9	95.7	56.7	84.4	99.0	56.7	86.7
0970, 0990	Wall Finishes & Painting/Coating	93.5	61.9	74.8	97.1	63.6	77.3	97.1	63.6	77.3	98.2	63.6	77.7	92.6	63.6	75.4	92.3	63.6	75.3
09	FINISHES	92.1	61.2	75.2	91.5	63.5	76.2	91.1	62.1	75.3	95.8	60.4	76.5	91.7	58.9	73.8	91.2	56.8	72.4
COVERS	DIVS. 10 - 14, 25, 28, 41, 43, 44, 46	100.0	82.7	96.2	100.0	76.5	94.8	100.0	81.0	95.8	100.0	80.6	95.7	100.0	77.1	95.0	100.0	78.7	95.3
21, 22, 23	FIRE SUPPRESSION, PLUMBING & HVAC	100.1	57.1	82.8	99.9	51.5	80.5	99.9	63.2	85.2	99.9	58.8	83.4	100.0	55.7	82.2	100.0	66.6	86.6
26, 27, 3370	ELECTRICAL, COMMUNICATIONS & UTIL.	98.9	64.7	81.4	93.3	57.2	74.8	96.7	51.9	73.8	96.1	60.0	77.6	91.9	62.1	76.7	98.4	57.2	77.3
MF2018	WEIGHTED AVERAGE	98.3	67.1	84.9	98.0	64.2	83.5	100.6	66.4	86.0	99.1	67.0	85.3	99.6	65.2	84.8	97.5	65.7	83.9

FLORIDA / GEORGIA

DIVISION		TAMPA 335 - 336, 346			WEST PALM BEACH 334, 349			ALBANY 317, 398			ATHENS 306			ATLANTA 300 - 303, 399			AUGUSTA 308 - 309		
		MAT.	INST.	TOTAL	MAT.	INST.	TOTAL	MAT.	INST.	TOTAL	MAT.	INST.	TOTAL	MAT.	INST.	TOTAL	MAT.	INST.	TOTAL
015433	CONTRACTOR EQUIPMENT		102.1	102.1		95.2	95.2		96.1	96.1		94.6	94.6		96.5	96.5		94.6	94.6
0241, 31 - 34	SITE & INFRASTRUCTURE, DEMOLITION	107.8	89.0	94.7	88.9	76.8	80.4	102.4	79.4	86.4	101.3	94.9	96.9	98.3	95.3	96.2	94.6	95.3	95.0
0310	Concrete Forming & Accessories	94.1	61.9	66.7	95.2	62.9	67.7	87.8	67.7	70.7	92.3	44.3	51.5	96.9	71.9	75.7	93.4	72.7	75.8
0320	Concrete Reinforcing	94.5	78.3	86.2	97.1	59.2	77.8	90.4	73.3	81.7	98.9	65.4	81.9	98.2	73.4	85.6	99.3	68.0	83.4
0330	Cast-in-Place Concrete	94.9	65.8	84.1	85.6	61.8	76.8	85.7	67.7	79.0	109.6	68.5	94.3	113.1	71.6	97.6	103.5	70.1	91.1
03	CONCRETE	92.4	67.8	81.3	86.7	63.6	76.3	86.8	70.4	79.4	101.4	57.7	81.7	103.6	72.7	89.7	94.2	71.6	84.0
04	MASONRY	94.2	59.3	72.6	92.8	53.3	68.4	91.9	67.7	77.0	76.6	77.8	77.3	89.9	69.7	77.4	90.1	69.2	77.2
05	METALS	101.4	94.3	99.2	98.4	87.7	95.1	101.3	97.5	100.1	94.9	79.5	90.2	95.7	84.6	92.4	94.6	80.5	90.3
06	WOOD, PLASTICS & COMPOSITES	85.3	61.0	72.1	85.0	65.2	74.3	79.5	68.4	73.5	95.0	36.8	63.4	97.8	73.1	84.4	96.5	75.4	85.0
07	THERMAL & MOISTURE PROTECTION	98.7	63.5	83.6	98.2	61.2	82.3	99.6	68.7	86.3	97.9	69.5	85.7	99.4	72.4	87.8	97.7	71.3	86.3
08	OPENINGS	97.7	64.0	89.8	96.0	62.0	88.1	87.6	70.5	83.6	93.0	50.9	83.3	97.6	73.4	92.0	93.1	73.3	88.5
0920	Plaster & Gypsum Board	110.1	60.4	77.4	113.8	64.7	81.5	103.8	68.0	80.2	97.4	35.5	56.6	94.0	72.6	81.8	98.7	75.1	83.1
0950, 0980	Ceilings & Acoustic Treatment	86.1	60.4	68.7	81.6	64.7	70.2	81.9	68.0	72.5	101.2	35.5	56.7	94.0	72.6	79.5	102.2	75.1	83.8
0960	Flooring	96.8	58.4	85.6	98.6	56.7	86.5	102.3	65.9	91.8	94.0	84.7	91.3	97.0	68.0	88.6	94.1	65.9	85.9
0970, 0990	Wall Finishes & Painting/Coating	92.6	63.6	75.4	88.4	58.6	70.8	94.5	96.5	95.6	101.8	96.5	98.7	106.2	96.5	100.5	101.8	80.6	89.2
09	FINISHES	93.0	60.8	75.4	90.6	61.1	74.5	92.4	70.0	80.2	97.7	55.2	74.5	97.8	73.5	84.5	97.4	72.5	83.8
COVERS	DIVS. 10 - 14, 25, 28, 41, 43, 44, 46	100.0	80.8	95.8	100.0	82.7	96.2	100.0	83.3	96.3	100.0	79.6	95.5	100.0	86.4	97.0	100.0	83.9	96.5
21, 22, 23	FIRE SUPPRESSION, PLUMBING & HVAC	100.0	60.0	84.0	98.7	59.9	83.1	99.9	69.9	87.9	97.6	67.3	85.5	100.0	72.0	88.8	100.1	63.6	85.5
26, 27, 3370	ELECTRICAL, COMMUNICATIONS & UTIL.	91.6	64.7	77.8	92.7	69.1	80.6	96.1	63.4	79.4	100.0	65.4	82.3	99.5	72.6	85.7	100.6	69.8	84.8
MF2018	WEIGHTED AVERAGE	97.4	68.1	84.8	95.1	66.0	82.6	95.8	72.6	85.8	96.7	67.9	84.4	98.7	75.6	88.8	96.9	72.7	86.5

City Cost Indexes

GEORGIA

DIVISION		COLUMBUS 318 - 319			DALTON 307			GAINESVILLE 305			MACON 310 - 312			SAVANNAH 313 - 314			STATESBORO 304		
		MAT.	INST.	TOTAL	MAT.	INST.	TOTAL	MAT.	INST.	TOTAL	MAT.	INST.	TOTAL	MAT.	INST.	TOTAL	MAT.	INST.	TOTAL
015433	CONTRACTOR EQUIPMENT		96.1	96.1		110.1	110.1		94.6	94.6		105.9	105.9		97.0	97.0		97.7	97.7
0241, 31 - 34	SITE & INFRASTRUCTURE, DEMOLITION	102.3	79.5	86.5	101.0	100.3	100.5	101.2	94.9	96.8	104.0	94.6	97.5	102.2	80.9	87.4	102.1	81.4	87.7
0310	Concrete Forming & Accessories	87.7	67.8	70.8	84.9	65.5	68.4	96.0	41.8	49.9	87.6	67.8	70.8	91.3	70.4	73.6	79.7	52.1	56.2
0320	Concrete Reinforcing	90.3	73.3	81.7	98.4	62.5	80.2	98.7	65.3	81.7	91.5	73.3	82.3	97.3	68.0	82.4	98.0	36.6	66.8
0330	Cast-in-Place Concrete	85.4	67.8	78.8	106.5	67.2	91.8	115.2	68.3	97.7	84.2	69.7	78.8	89.0	67.6	81.0	109.4	67.2	93.7
03	CONCRETE	86.6	70.5	79.4	100.5	67.4	85.6	103.4	56.4	82.2	86.2	71.2	79.5	88.0	70.7	80.2	100.9	57.0	81.1
04	MASONRY	91.9	67.8	77.0	77.4	77.1	77.2	85.4	77.8	80.7	104.2	67.9	81.8	86.8	67.7	75.0	79.4	77.8	78.4
05	METALS	100.9	97.9	100.0	96.1	93.8	95.4	94.2	78.9	89.6	96.5	98.0	96.9	97.5	95.3	96.9	99.7	84.0	94.9
06	WOOD, PLASTICS & COMPOSITES	79.5	68.4	73.5	78.2	67.1	72.1	98.9	34.1	63.7	85.0	68.4	76.0	88.2	72.3	79.6	71.5	48.5	59.0
07	THERMAL & MOISTURE PROTECTION	99.5	69.4	86.5	100.0	71.7	87.8	97.9	69.2	85.5	98.0	71.5	86.6	97.5	68.9	85.2	98.5	68.1	85.5
08	OPENINGS	87.6	70.5	83.6	94.2	66.9	87.9	93.0	49.4	82.9	87.4	70.5	83.5	95.1	71.3	89.6	95.1	49.5	84.5
0920	Plaster & Gypsum Board	103.8	68.0	80.2	85.0	66.6	72.9	99.9	32.7	55.6	108.4	68.0	81.8	102.7	72.0	82.5	85.8	47.5	60.6
0950, 0980	Ceilings & Acoustic Treatment	81.9	68.0	72.5	117.4	66.6	83.0	101.2	32.7	54.8	77.1	68.0	70.9	88.3	72.0	77.3	111.8	47.5	68.2
0960	Flooring	102.3	65.9	91.8	94.5	84.7	91.6	95.5	84.7	92.3	81.0	65.9	76.6	98.3	65.9	88.9	111.2	84.7	103.5
0970, 0990	Wall Finishes & Painting/Coating	94.5	96.5	95.6	91.6	71.9	80.0	101.8	96.5	98.7	96.5	96.5	96.5	91.5	86.3	88.4	100.0	71.9	83.4
09	FINISHES	92.3	70.0	80.1	107.5	70.2	87.2	98.3	53.6	73.9	82.3	70.0	75.6	92.3	71.2	80.8	110.2	59.4	82.5
COVERS	DIVS. 10 - 14, 25, 28, 41, 43, 44, 46	100.0	83.3	96.3	100.0	28.4	84.2	100.0	36.2	85.9	100.0	83.4	96.3	100.0	83.3	96.3	100.0	42.2	87.3
21, 22, 23	FIRE SUPPRESSION, PLUMBING & HVAC	100.0	66.2	86.5	97.7	62.0	83.4	97.6	67.1	85.4	100.0	69.3	87.7	100.1	67.1	86.8	98.1	67.8	86.0
26, 27, 3370	ELECTRICAL, COMMUNICATIONS & UTIL.	96.2	69.4	82.5	109.3	63.7	86.0	100.0	69.3	84.3	94.8	64.2	79.1	99.8	70.2	84.7	100.3	58.6	79.0
MF2018	WEIGHTED AVERAGE	95.7	72.7	85.8	98.6	71.0	86.7	97.3	66.5	84.1	94.6	74.0	85.8	96.3	73.1	86.3	98.7	65.8	84.6

GEORGIA · HAWAII · IDAHO

DIVISION		GEORGIA VALDOSTA 316			WAYCROSS 315			HAWAII HILO 967			HONOLULU 968			STATES & POSS., GUAM 969			IDAHO BOISE 836 - 837		
		MAT.	INST.	TOTAL	MAT.	INST.	TOTAL	MAT.	INST.	TOTAL	MAT.	INST.	TOTAL	MAT.	INST.	TOTAL	MAT.	INST.	TOTAL
015433	CONTRACTOR EQUIPMENT		96.1	96.1		96.1	96.1		100.4	100.4		100.4	100.4		165.0	165.0		97.6	97.6
0241, 31 - 34	SITE & INFRASTRUCTURE, DEMOLITION	111.6	79.4	89.2	108.5	80.6	89.1	147.0	107.5	119.5	155.4	107.5	122.0	191.2	103.5	130.1	86.5	97.0	93.8
0310	Concrete Forming & Accessories	79.0	42.0	47.6	80.9	64.7	67.1	106.5	123.7	121.1	119.6	123.7	123.1	108.9	53.0	61.4	98.7	79.8	82.7
0320	Concrete Reinforcing	92.4	61.2	76.5	92.4	61.4	76.7	137.5	126.6	131.9	159.9	126.6	142.9	237.2	28.2	131.0	105.1	78.8	91.8
0330	Cast-in-Place Concrete	83.9	67.5	77.8	94.8	67.4	84.6	170.9	125.6	154.1	137.5	125.6	133.1	148.2	98.5	129.7	90.6	83.1	87.8
03	CONCRETE	91.8	56.7	76.0	94.6	66.7	82.0	144.8	123.7	135.3	138.4	123.7	131.8	146.1	65.7	109.9	96.8	81.0	89.7
04	MASONRY	97.2	77.8	85.2	98.0	77.8	85.5	148.6	120.0	130.9	135.5	120.0	125.9	205.8	35.5	100.6	125.2	87.2	101.7
05	METALS	100.4	93.1	98.2	99.5	88.8	96.3	114.4	106.6	112.0	128.1	106.6	121.6	147.7	76.6	126.1	108.5	82.1	100.5
06	WOOD, PLASTICS & COMPOSITES	68.1	33.9	49.5	69.6	65.2	67.2	113.9	124.5	119.6	131.3	124.5	127.6	126.6	55.4	87.9	94.6	78.5	85.9
07	THERMAL & MOISTURE PROTECTION	99.8	67.5	85.9	99.6	70.3	87.0	131.8	120.2	126.8	149.6	120.2	136.9	154.9	59.8	114.0	94.9	85.5	90.9
08	OPENINGS	84.2	48.2	75.9	84.6	62.9	79.5	116.3	123.0	117.9	130.9	123.0	129.1	120.7	45.0	103.1	99.2	73.1	93.1
0920	Plaster & Gypsum Board	95.8	32.5	54.1	95.8	64.7	75.4	112.1	124.9	120.6	155.7	124.9	135.5	228.9	43.7	106.9	94.2	77.9	83.5
0950, 0980	Ceilings & Acoustic Treatment	80.2	32.5	47.9	78.3	64.7	69.1	137.5	124.9	129.0	144.3	124.9	131.2	254.9	43.7	111.8	101.2	77.9	85.4
0960	Flooring	96.1	86.3	93.2	97.3	84.7	93.6	110.8	140.7	119.5	130.2	140.7	133.3	130.1	40.8	104.2	93.2	79.6	89.3
0970, 0990	Wall Finishes & Painting/Coating	94.5	96.5	95.6	94.5	71.9	81.1	94.7	140.6	121.9	105.1	140.6	126.1	100.0	32.4	60.0	93.4	40.5	62.1
09	FINISHES	90.0	53.8	70.2	89.6	69.3	78.5	110.5	129.0	120.6	125.7	129.0	127.5	183.1	49.4	110.1	91.4	75.7	82.8
COVERS	DIVS. 10 - 14, 25, 28, 41, 43, 44, 46	100.0	79.1	95.4	100.0	49.8	89.0	100.0	112.6	102.8	100.0	112.6	102.8	100.0	67.3	92.8	100.0	88.6	97.5
21, 22, 23	FIRE SUPPRESSION, PLUMBING & HVAC	100.0	69.3	87.7	98.5	64.2	84.7	100.4	112.0	105.1	100.4	112.0	105.1	103.0	34.3	75.5	100.0	74.8	89.9
26, 27, 3370	ELECTRICAL, COMMUNICATIONS & UTIL.	94.6	52.9	73.3	98.3	54.0	75.6	109.0	124.3	116.9	110.7	124.3	117.7	158.3	36.8	96.1	97.6	69.8	83.4
MF2018	WEIGHTED AVERAGE	96.0	66.2	83.2	96.2	68.4	84.2	115.9	118.3	116.9	120.4	118.3	119.5	137.8	52.6	101.3	100.7	79.4	91.5

IDAHO · ILLINOIS

DIVISION		IDAHO COEUR D'ALENE 838			IDAHO FALLS 834			LEWISTON 835			POCATELLO 832			TWIN FALLS 833			ILLINOIS BLOOMINGTON 617		
		MAT.	INST.	TOTAL	MAT.	INST.	TOTAL	MAT.	INST.	TOTAL	MAT.	INST.	TOTAL	MAT.	INST.	TOTAL	MAT.	INST.	TOTAL
015433	CONTRACTOR EQUIPMENT		92.5	92.5		97.6	97.6		92.5	92.5		97.6	97.6		97.6	97.6		104.7	104.7
0241, 31 - 34	SITE & INFRASTRUCTURE, DEMOLITION	84.5	91.0	89.0	84.5	96.8	93.1	91.2	91.9	91.7	87.7	97.0	94.2	94.5	96.7	96.0	97.0	100.8	99.6
0310	Concrete Forming & Accessories	106.0	82.0	85.6	92.4	78.6	80.7	111.3	83.4	87.6	98.9	79.5	82.4	100.1	78.5	81.7	83.5	119.2	113.8
0320	Concrete Reinforcing	113.2	99.0	106.0	107.2	77.4	92.0	113.2	99.0	106.0	105.6	78.8	92.0	107.6	77.3	92.2	96.1	105.3	100.8
0330	Cast-in-Place Concrete	98.0	84.1	92.8	86.2	83.3	85.2	101.8	84.5	95.4	93.1	83.0	89.4	95.6	83.3	91.0	98.1	117.0	105.2
03	CONCRETE	103.0	85.7	95.2	88.8	80.2	85.0	106.6	86.5	97.5	96.0	80.8	89.1	103.6	80.2	93.1	93.9	116.7	104.1
04	MASONRY	126.8	85.0	100.9	120.2	84.6	98.2	127.2	86.5	102.0	122.7	87.2	100.8	125.4	84.6	100.2	110.9	124.2	119.2
05	METALS	101.8	89.0	97.9	117.0	81.4	106.2	101.3	89.2	97.6	117.1	81.7	106.3	117.1	81.2	106.2	97.0	123.8	105.1
06	WOOD, PLASTICS & COMPOSITES	97.6	82.1	89.2	88.1	78.5	82.9	103.7	82.1	92.0	94.6	78.5	85.9	95.7	78.5	86.3	83.4	116.3	101.3
07	THERMAL & MOISTURE PROTECTION	147.9	84.1	120.4	94.1	75.4	86.1	148.3	85.5	121.3	94.7	76.4	86.8	95.5	82.7	90.0	98.7	115.2	105.8
08	OPENINGS	116.4	78.1	107.5	102.1	72.8	95.3	110.1	82.0	103.6	99.9	68.1	92.5	102.9	63.6	93.8	93.3	122.1	100.0
0920	Plaster & Gypsum Board	170.4	81.8	112.0	80.8	77.9	78.9	172.2	81.8	112.7	82.2	77.9	79.4	84.4	77.9	80.1	88.8	116.9	107.3
0950, 0980	Ceilings & Acoustic Treatment	136.3	81.8	99.4	104.7	77.9	86.6	136.3	81.8	99.4	109.8	77.9	88.2	107.2	77.9	87.4	89.7	116.9	108.1
0960	Flooring	131.9	79.6	116.7	93.1	79.6	89.2	135.1	79.6	119.0	96.5	79.6	91.6	97.7	79.6	92.5	87.3	124.9	98.2
0970, 0990	Wall Finishes & Painting/Coating	112.5	74.5	90.0	93.4	40.5	62.1	112.5	73.2	89.3	93.3	40.5	62.1	93.4	40.5	62.1	88.7	138.3	118.0
09	FINISHES	155.0	80.7	114.4	89.7	75.0	81.7	156.3	81.2	115.3	92.3	75.7	83.2	93.1	75.0	83.2	86.9	122.5	106.3
COVERS	DIVS. 10 - 14, 25, 28, 41, 43, 44, 46	100.0	93.6	98.6	100.0	87.7	97.3	100.0	94.6	98.8	100.0	88.6	97.5	100.0	87.7	97.3	100.0	106.7	101.5
21, 22, 23	FIRE SUPPRESSION, PLUMBING & HVAC	99.5	81.7	92.4	100.8	73.4	89.8	100.7	86.9	95.2	99.9	74.8	89.8	99.9	69.6	87.7	97.5	107.7	101.6
26, 27, 3370	ELECTRICAL, COMMUNICATIONS & UTIL.	89.3	83.5	86.4	89.0	69.4	79.0	87.5	80.1	83.7	94.9	67.6	81.0	90.3	69.4	79.6	94.4	92.5	93.4
MF2018	WEIGHTED AVERAGE	107.7	84.4	97.7	100.2	78.2	90.8	107.8	85.7	98.3	101.7	78.6	91.8	102.9	77.2	91.9	96.1	112.2	103.0

ILLINOIS

DIVISION		CARBONDALE 629			CENTRALIA 628			CHAMPAIGN 618 - 619			CHICAGO 606 - 608			DECATUR 625			EAST ST. LOUIS 620 - 622		
		MAT.	INST.	TOTAL	MAT.	INST.	TOTAL	MAT.	INST.	TOTAL	MAT.	INST.	TOTAL	MAT.	INST.	TOTAL	MAT.	INST.	TOTAL
015433	CONTRACTOR EQUIPMENT		113.3	113.3		113.3	113.3		105.6	105.6		98.9	98.9		105.6	105.6		113.3	113.3
0241, 31 - 34	SITE & INFRASTRUCTURE, DEMOLITION	98.6	101.5	100.7	99.0	103.4	102.0	106.1	101.6	103.0	103.2	103.7	103.5	93.6	101.2	98.9	101.0	102.0	101.7
0310	Concrete Forming & Accessories	88.3	108.2	105.2	89.8	114.7	110.9	89.9	116.7	112.6	98.1	162.0	152.4	90.4	115.9	112.1	85.7	111.5	107.6
0320	Concrete Reinforcing	90.2	106.3	98.4	90.2	107.0	98.8	96.1	100.6	98.4	100.2	154.7	128.0	88.5	100.9	94.8	90.1	107.0	98.7
0330	Cast-in-Place Concrete	88.9	102.3	93.8	89.3	121.6	101.3	113.8	111.9	113.1	126.4	154.6	136.9	97.0	113.7	103.2	90.8	121.5	102.3
03	CONCRETE	81.1	107.3	92.9	81.6	117.1	97.6	106.4	112.8	109.3	120.9	157.3	130.8	91.6	113.1	101.3	82.6	115.7	97.5
04	MASONRY	76.0	112.4	98.5	76.0	125.5	106.6	133.9	121.8	126.4	104.5	166.2	142.6	71.8	120.2	101.7	76.3	125.5	106.7
05	METALS	96.6	130.7	107.0	96.6	133.8	107.9	97.0	119.7	103.9	99.0	146.6	113.5	100.3	119.2	106.0	97.8	133.5	108.6
06	WOOD, PLASTICS & COMPOSITES	88.7	105.4	97.8	91.2	110.9	101.9	90.4	114.9	103.7	101.1	161.7	134.0	89.8	114.9	103.4	85.8	106.9	97.3
07	THERMAL & MOISTURE PROTECTION	95.1	102.3	98.2	95.1	113.2	102.9	99.5	114.7	106.0	99.0	152.0	121.8	100.8	113.0	106.1	95.1	112.5	102.6
08	OPENINGS	89.2	116.9	95.6	89.2	120.0	96.3	93.9	116.8	99.2	104.4	170.9	119.9	100.3	116.9	104.2	89.3	117.7	95.9
0920	Plaster & Gypsum Board	95.3	105.7	102.2	96.6	111.3	106.3	90.6	115.4	107.0	94.8	163.5	140.1	97.8	115.4	109.4	93.8	107.2	102.6
0950, 0980	Ceilings & Acoustic Treatment	84.6	105.7	98.9	84.6	111.3	102.7	89.7	115.4	107.1	99.8	163.5	143.0	90.5	115.4	107.4	84.6	107.2	99.9
0960	Flooring	111.2	113.8	112.0	112.0	127.7	116.5	90.4	114.7	97.4	98.7	164.8	117.9	99.5	115.2	104.1	110.2	127.7	115.3
0970, 0990	Wall Finishes & Painting/Coating	109.1	107.7	108.3	109.1	114.6	112.3	88.7	115.4	104.5	104.9	162.3	138.9	100.8	114.4	108.9	109.1	114.6	112.3
09	FINISHES	92.8	108.1	101.1	93.2	117.1	106.3	88.8	116.6	104.0	98.4	163.5	133.9	92.8	116.2	105.6	92.4	114.4	104.4
COVERS	DIVS. 10 - 14, 25, 28, 41, 43, 44, 46	100.0	102.3	100.5	100.0	104.3	100.9	100.0	105.4	101.2	100.0	128.1	106.2	100.0	104.8	101.1	100.0	103.8	100.8
21, 22, 23	FIRE SUPPRESSION, PLUMBING & HVAC	97.4	106.5	101.1	97.4	98.8	98.0	97.5	106.3	101.0	99.8	138.0	115.1	99.9	96.4	98.5	99.9	99.3	99.7
26, 27, 3370	ELECTRICAL, COMMUNICATIONS & UTIL.	91.7	108.8	100.5	93.0	108.9	101.1	97.5	92.6	95.0	98.1	134.1	116.5	95.4	89.8	92.5	92.6	102.1	97.5
MF2018	WEIGHTED AVERAGE	92.6	109.8	100.0	92.8	113.0	101.5	99.6	109.8	104.0	101.4	146.1	120.5	96.4	107.0	101.0	93.7	111.3	101.2

ILLINOIS

DIVISION		EFFINGHAM 624			GALESBURG 614			JOLIET 604			KANKAKEE 609			LA SALLE 613			NORTH SUBURBAN 600 - 603		
		MAT.	INST.	TOTAL	MAT.	INST.	TOTAL	MAT.	INST.	TOTAL	MAT.	INST.	TOTAL	MAT.	INST.	TOTAL	MAT.	INST.	TOTAL
015433	CONTRACTOR EQUIPMENT		105.6	105.6		104.7	104.7		95.4	95.4		95.4	95.4		104.7	104.7		95.4	95.4
0241, 31 - 34	SITE & INFRASTRUCTURE, DEMOLITION	97.9	100.4	99.7	99.4	99.8	99.7	98.9	101.7	100.8	92.7	101.3	98.7	98.8	101.3	100.6	98.2	102.0	100.8
0310	Concrete Forming & Accessories	94.9	116.6	113.3	89.8	119.5	115.0	97.4	161.1	151.5	90.8	146.0	137.7	103.9	126.0	122.6	96.7	158.8	149.5
0320	Concrete Reinforcing	91.4	96.2	93.9	95.5	108.6	102.1	100.2	140.6	120.7	101.0	138.9	120.3	95.7	137.5	116.9	100.2	153.3	127.2
0330	Cast-in-Place Concrete	96.7	109.2	101.3	101.0	111.5	104.9	115.0	148.1	127.3	107.1	138.1	118.6	100.9	125.0	109.9	115.0	154.5	129.7
03	CONCRETE	92.3	111.0	100.7	96.8	115.4	105.2	102.0	151.9	124.5	95.8	141.3	116.3	97.6	128.2	111.4	101.9	155.4	126.0
04	MASONRY	80.2	114.1	101.2	111.1	123.7	118.9	103.7	160.0	138.5	100.0	148.4	129.9	111.1	145.6	132.4	100.7	162.1	138.6
05	METALS	97.4	114.5	102.6	97.0	124.1	105.2	96.9	135.6	108.7	96.9	133.8	108.1	97.1	142.8	110.9	98.1	142.0	111.4
06	WOOD, PLASTICS & COMPOSITES	92.1	117.3	105.8	90.3	118.7	105.7	99.2	162.3	133.5	92.1	144.6	120.6	106.5	123.3	115.6	98.1	157.7	130.5
07	THERMAL & MOISTURE PROTECTION	100.2	109.8	104.4	98.9	112.1	104.6	103.5	148.2	122.7	102.5	141.1	119.1	99.1	128.9	111.9	103.8	148.9	123.2
08	OPENINGS	95.3	116.2	100.2	93.3	118.6	99.2	101.9	166.5	116.9	94.8	156.8	109.2	93.3	135.2	103.0	102.0	168.4	117.4
0920	Plaster & Gypsum Board	97.8	117.9	111.0	90.6	119.3	109.5	87.3	164.1	137.9	85.5	146.0	125.3	98.3	124.1	115.2	90.9	159.5	136.1
0950, 0980	Ceilings & Acoustic Treatment	84.6	117.9	107.1	89.7	119.3	109.8	99.4	164.1	143.3	99.4	146.0	131.0	89.7	124.1	113.0	99.4	159.5	140.1
0960	Flooring	100.5	116.9	105.3	90.2	124.9	100.3	96.7	154.8	113.5	93.7	155.7	111.7	96.7	122.9	104.3	97.2	164.8	116.8
0970, 0990	Wall Finishes & Painting/Coating	100.8	112.2	107.6	88.7	101.1	96.0	97.0	167.3	138.6	97.0	138.3	121.4	88.7	136.5	117.0	99.0	168.2	139.9
09	FINISHES	91.9	117.4	105.8	88.1	119.4	105.2	93.7	162.0	131.0	92.2	147.5	122.4	91.1	125.3	109.8	94.5	161.8	131.2
COVERS	DIVS. 10 - 14, 25, 28, 41, 43, 44, 46	100.0	77.2	95.0	100.0	101.4	100.3	100.0	125.0	105.5	100.0	117.8	103.9	100.0	102.7	100.6	100.0	121.3	104.7
21, 22, 23	FIRE SUPPRESSION, PLUMBING & HVAC	97.5	104.7	100.4	97.5	105.7	100.8	99.9	133.0	113.1	97.4	132.4	111.4	97.5	127.0	109.3	99.8	136.4	114.5
26, 27, 3370	ELECTRICAL, COMMUNICATIONS & UTIL.	93.4	108.8	101.3	95.3	83.8	89.4	97.2	136.5	117.3	92.5	133.8	113.7	92.5	133.9	113.7	97.0	135.7	116.8
MF2018	WEIGHTED AVERAGE	95.2	109.2	101.2	96.8	109.5	102.3	99.4	142.2	117.8	96.3	136.1	113.3	97.0	128.9	110.7	99.4	144.1	118.6

ILLINOIS

DIVISION		PEORIA 615 - 616			QUINCY 623			ROCK ISLAND 612			ROCKFORD 610 - 611			SOUTH SUBURBAN 605			SPRINGFIELD 626 - 627		
		MAT.	INST.	TOTAL	MAT.	INST.	TOTAL	MAT.	INST.	TOTAL	MAT.	INST.	TOTAL	MAT.	INST.	TOTAL	MAT.	INST.	TOTAL
015433	CONTRACTOR EQUIPMENT		104.7	104.7		105.6	105.6		104.7	104.7		104.7	104.7		95.4	95.4		105.6	105.6
0241, 31 - 34	SITE & INFRASTRUCTURE, DEMOLITION	99.8	100.7	100.4	96.7	101.2	99.9	97.5	99.6	99.0	99.2	101.8	101.0	98.2	101.7	100.6	98.4	101.6	100.6
0310	Concrete Forming & Accessories	92.8	119.2	115.2	92.8	116.3	112.8	91.4	101.9	100.3	97.2	131.3	126.2	96.7	158.8	149.4	91.6	116.6	112.8
0320	Concrete Reinforcing	93.1	107.7	100.5	91.0	88.5	89.7	95.5	101.8	98.7	88.1	138.0	113.5	100.2	153.3	127.2	91.3	100.9	96.2
0330	Cast-in-Place Concrete	98.0	119.0	105.8	96.8	109.6	101.6	98.8	101.3	99.8	100.3	130.5	111.5	115.0	154.4	129.6	92.1	111.1	99.2
03	CONCRETE	94.0	117.7	104.7	91.9	109.7	99.9	94.7	102.7	98.3	94.8	132.6	111.8	101.9	155.3	126.0	89.7	112.5	100.0
04	MASONRY	110.8	124.0	119.0	100.2	114.1	108.8	110.9	102.4	105.6	86.3	145.3	122.8	100.7	162.1	138.6	81.7	121.6	106.4
05	METALS	99.7	125.0	107.3	97.5	113.0	102.2	97.0	119.6	103.9	99.7	142.8	112.8	98.1	142.0	111.4	97.8	119.3	104.4
06	WOOD, PLASTICS & COMPOSITES	98.8	116.5	108.4	89.6	117.3	104.6	92.0	100.4	96.5	98.8	127.2	114.2	98.1	157.7	130.5	91.0	114.9	104.0
07	THERMAL & MOISTURE PROTECTION	99.7	115.6	106.5	100.2	108.5	103.8	98.9	101.2	99.9	102.0	132.6	115.2	103.8	148.9	123.2	103.1	114.6	108.1
08	OPENINGS	98.8	123.0	104.4	96.0	114.9	100.4	93.3	106.3	96.3	98.8	140.5	108.5	102.0	168.4	117.4	100.8	117.4	104.6
0920	Plaster & Gypsum Board	95.1	117.0	109.5	96.6	117.9	110.6	90.6	100.5	97.1	95.1	128.0	116.8	90.9	159.5	136.1	99.5	115.4	110.0
0950, 0980	Ceilings & Acoustic Treatment	94.8	117.0	109.8	84.6	117.9	107.1	89.7	100.5	97.0	94.8	128.0	117.3	99.4	159.5	140.1	91.0	115.4	107.5
0960	Flooring	93.7	123.5	102.3	99.5	111.5	103.0	91.3	97.8	93.2	93.7	129.1	103.9	97.2	164.8	116.8	102.0	115.2	105.8
0970, 0990	Wall Finishes & Painting/Coating	88.7	138.3	118.0	100.8	114.4	108.9	88.7	98.0	94.2	88.7	146.5	122.9	99.0	168.2	139.9	102.9	113.1	108.9
09	FINISHES	90.8	122.2	107.9	91.4	116.5	105.1	88.4	100.4	94.9	90.8	132.2	113.4	94.5	161.8	131.2	97.3	116.4	107.7
COVERS	DIVS. 10 - 14, 25, 28, 41, 43, 44, 46	100.0	106.6	101.4	100.0	79.8	95.5	100.0	99.5	99.9	100.0	116.5	103.6	100.0	121.3	104.7	100.0	105.3	101.2
21, 22, 23	FIRE SUPPRESSION, PLUMBING & HVAC	99.9	102.4	100.9	97.5	103.5	99.9	97.5	99.7	98.4	100.0	118.7	107.5	99.8	136.4	114.5	99.9	102.4	100.9
26, 27, 3370	ELECTRICAL, COMMUNICATIONS & UTIL.	96.3	93.7	95.0	91.1	84.2	87.6	87.8	95.4	91.7	96.5	128.7	113.0	97.0	135.7	116.8	97.9	84.2	90.9
MF2018	WEIGHTED AVERAGE	98.5	111.5	104.0	95.8	105.1	99.8	95.8	102.1	98.5	97.5	128.7	110.9	99.4	144.0	118.6	97.1	107.7	101.7

For customer support on your Plumbing Costs with RSMeans Data, call 800.448.8182.

645

City Cost Indexes

INDIANA

DIVISION		ANDERSON 460 MAT.	INST.	TOTAL	BLOOMINGTON 474 MAT.	INST.	TOTAL	COLUMBUS 472 MAT.	INST.	TOTAL	EVANSVILLE 476-477 MAT.	INST.	TOTAL	FORT WAYNE 467-468 MAT.	INST.	TOTAL	GARY 463-464 MAT.	INST.	TOTAL
015433	CONTRACTOR EQUIPMENT		97.4	97.4		83.9	83.9		83.9	83.9		113.7	113.7		97.4	97.4		97.4	97.4
0241, 31-34	SITE & INFRASTRUCTURE, DEMOLITION	99.1	93.7	95.4	87.0	92.1	90.6	83.7	91.9	89.4	92.4	121.2	112.5	100.2	93.6	95.6	99.8	97.3	98.0
0310	Concrete Forming & Accessories	95.8	79.6	82.0	100.6	80.1	83.2	94.5	79.9	82.1	93.8	80.8	82.7	93.7	75.7	78.4	95.8	112.6	110.0
0320	Concrete Reinforcing	95.6	85.4	90.4	89.4	85.1	87.2	89.8	87.2	88.5	97.9	83.4	90.5	95.6	80.2	87.8	95.6	112.7	104.3
0330	Cast-in-Place Concrete	103.7	77.1	93.8	99.3	76.2	90.7	98.8	74.1	89.6	94.9	85.7	91.5	110.2	76.2	97.6	108.4	113.7	110.4
03	CONCRETE	92.9	80.3	87.2	98.1	79.2	89.6	97.3	78.7	89.0	98.1	83.1	91.3	95.9	77.4	87.5	95.1	112.7	103.0
04	MASONRY	87.3	75.6	80.1	88.1	72.0	78.1	87.9	73.1	78.7	83.7	78.9	80.7	90.4	72.9	79.6	88.7	109.2	101.3
05	METALS	99.1	89.1	96.1	99.5	74.2	91.8	99.5	74.4	91.9	92.3	84.0	89.8	99.1	87.2	95.5	99.1	105.1	100.9
06	WOOD, PLASTICS & COMPOSITES	97.5	79.8	87.9	112.3	80.6	95.1	106.6	80.4	92.4	93.2	79.9	86.0	97.2	76.0	85.6	94.7	111.0	103.5
07	THERMAL & MOISTURE PROTECTION	113.0	77.2	97.6	98.6	77.7	89.6	98.0	78.1	89.4	103.0	84.0	94.8	112.8	78.9	98.2	111.4	106.1	109.1
08	OPENINGS	95.3	78.0	91.3	99.7	78.6	94.8	95.9	79.0	92.0	93.8	78.2	90.2	95.3	73.2	90.1	95.3	114.2	99.7
0920	Plaster & Gypsum Board	104.6	79.5	88.1	98.6	80.7	86.8	95.5	80.6	85.7	94.0	79.0	84.1	103.7	75.6	85.2	97.3	111.6	106.7
0950, 0980	Ceilings & Acoustic Treatment	91.9	79.5	83.5	79.6	80.7	80.4	79.6	80.6	80.3	83.4	79.0	80.4	91.9	75.6	80.8	91.9	111.6	105.2
0960	Flooring	95.1	77.5	90.0	99.1	83.1	94.4	94.0	83.1	90.9	94.0	73.8	88.1	95.1	73.5	88.9	95.1	114.3	100.7
0970, 0990	Wall Finishes & Painting/Coating	94.0	67.6	78.4	85.4	79.5	81.9	85.4	79.5	81.9	91.0	86.6	88.4	94.0	72.1	81.0	94.0	121.4	110.2
09	FINISHES	91.8	77.9	84.2	90.1	80.7	85.0	88.2	80.6	84.0	88.8	79.9	84.0	91.5	75.1	82.5	90.6	113.9	103.3
COVERS	DIVS. 10-14, 25, 28, 41, 43, 44, 46	100.0	89.2	97.6	100.0	85.8	96.9	100.0	85.8	96.9	100.0	93.1	98.5	100.0	89.1	97.6	100.0	105.1	101.1
21, 22, 23	FIRE SUPPRESSION, PLUMBING & HVAC	99.9	78.7	91.4	99.8	78.9	91.4	97.3	78.6	89.8	100.0	80.4	92.1	99.9	73.8	89.5	99.9	105.1	102.0
26, 27, 3370	ELECTRICAL, COMMUNICATIONS & UTIL.	86.1	85.2	85.6	98.7	86.6	92.5	97.9	87.0	92.3	94.7	83.4	88.9	86.8	76.7	81.6	97.5	110.7	104.2
MF2018	WEIGHTED AVERAGE	96.1	81.8	90.0	97.8	80.3	90.3	96.4	80.4	89.5	95.5	84.8	90.9	96.7	78.2	88.8	97.5	108.3	102.1

INDIANA

DIVISION		INDIANAPOLIS 461-462 MAT.	INST.	TOTAL	KOKOMO 469 MAT.	INST.	TOTAL	LAFAYETTE 479 MAT.	INST.	TOTAL	LAWRENCEBURG 470 MAT.	INST.	TOTAL	MUNCIE 473 MAT.	INST.	TOTAL	NEW ALBANY 471 MAT.	INST.	TOTAL
015433	CONTRACTOR EQUIPMENT		85.3	85.3		97.4	97.4		83.9	83.9		103.6	103.6		95.5	95.5		93.3	93.3
0241, 31-34	SITE & INFRASTRUCTURE, DEMOLITION	100.0	90.9	93.6	95.5	93.6	94.2	84.5	92.0	89.8	82.0	107.3	99.7	86.9	92.6	90.9	79.2	94.5	89.8
0310	Concrete Forming & Accessories	100.2	85.1	87.3	99.1	78.0	81.2	91.9	80.9	82.5	90.9	76.6	78.8	91.6	79.1	81.0	89.8	77.4	79.2
0320	Concrete Reinforcing	95.6	87.6	91.5	86.4	87.3	86.9	89.4	85.3	87.3	88.7	77.2	82.9	98.9	85.3	92.0	90.0	83.0	86.4
0330	Cast-in-Place Concrete	100.6	84.7	94.7	102.7	80.9	94.6	99.4	80.0	92.2	92.9	74.4	86.0	104.3	76.2	93.9	95.9	74.3	87.9
03	CONCRETE	97.9	84.7	92.0	90.0	81.2	86.0	97.6	80.9	90.1	90.7	76.6	84.4	96.4	79.7	88.9	96.1	77.5	87.7
04	MASONRY	88.0	79.1	82.5	87.0	73.6	78.7	93.3	75.5	82.3	73.2	71.8	72.3	89.8	75.7	81.1	79.8	68.1	72.6
05	METALS	97.0	75.7	90.5	95.6	89.6	93.8	97.9	74.4	90.8	94.4	84.4	91.4	101.3	88.9	97.5	96.5	82.0	92.1
06	WOOD, PLASTICS & COMPOSITES	98.6	86.0	91.8	100.8	77.4	88.1	103.6	81.5	91.6	91.3	76.5	83.3	104.9	79.4	91.0	93.7	78.6	85.5
07	THERMAL & MOISTURE PROTECTION	103.3	81.4	93.9	111.9	76.5	96.6	98.0	79.4	90.0	104.0	76.2	92.0	101.2	78.5	91.4	99.6	72.2	82.1
08	OPENINGS	104.1	82.2	99.0	90.3	77.3	87.3	94.4	78.9	90.8	95.8	73.8	90.7	93.0	77.8	89.5	93.4	77.5	89.7
0920	Plaster & Gypsum Board	96.4	85.8	89.4	109.5	77.1	88.1	93.3	81.7	85.7	71.9	76.5	74.9	94.0	79.5	84.5	91.9	78.5	83.0
0950, 0980	Ceilings & Acoustic Treatment	94.9	85.8	88.7	91.9	77.1	81.9	76.2	81.7	80.0	86.0	76.5	79.5	79.6	79.5	79.6	83.4	78.5	80.1
0960	Flooring	96.9	81.8	92.5	99.0	91.3	96.8	93.0	81.8	89.8	69.0	82.9	73.1	93.4	77.5	88.8	91.4	55.8	81.1
0970, 0990	Wall Finishes & Painting/Coating	99.3	79.5	87.6	94.0	69.7	79.6	85.4	83.5	84.3	86.2	74.5	79.3	85.4	67.6	74.8	91.0	68.0	77.4
09	FINISHES	96.0	84.2	89.6	93.4	79.6	85.8	86.9	81.5	83.9	78.6	77.7	78.1	84.7	77.6	82.0	88.0	72.3	79.5
COVERS	DIVS. 10-14, 25, 28, 41, 43, 44, 46	100.0	90.6	97.9	100.0	86.4	97.0	100.0	88.4	97.5	100.0	84.8	96.6	100.0	88.1	97.4	100.0	83.5	96.4
21, 22, 23	FIRE SUPPRESSION, PLUMBING & HVAC	99.9	79.7	91.8	97.5	78.8	90.0	97.3	77.9	89.5	98.2	74.5	88.7	99.8	78.6	91.3	97.5	76.0	88.9
26, 27, 3370	ELECTRICAL, COMMUNICATIONS & UTIL.	102.6	87.0	94.6	90.3	79.1	84.6	97.4	80.5	88.8	92.7	72.9	82.6	90.5	75.9	83.0	93.3	76.4	84.6
MF2018	WEIGHTED AVERAGE	99.1	83.0	92.2	94.5	81.0	88.7	96.1	80.1	89.2	92.9	78.5	86.8	96.3	80.3	89.4	94.2	77.2	86.9

INDIANA / IOWA

DIVISION		SOUTH BEND 465-466 MAT.	INST.	TOTAL	TERRE HAUTE 478 MAT.	INST.	TOTAL	WASHINGTON 475 MAT.	INST.	TOTAL	BURLINGTON 526 MAT.	INST.	TOTAL	CARROLL 514 MAT.	INST.	TOTAL	CEDAR RAPIDS 522-524 MAT.	INST.	TOTAL
015433	CONTRACTOR EQUIPMENT		108.9	108.9		113.7	113.7		113.7	113.7		101.4	101.4		101.4	101.4		98.2	98.2
0241, 31-34	SITE & INFRASTRUCTURE, DEMOLITION	98.8	94.3	95.7	94.0	121.6	113.3	94.0	121.8	113.4	98.9	96.5	97.2	88.2	97.3	94.5	100.5	96.0	97.4
0310	Concrete Forming & Accessories	95.2	78.1	80.6	94.5	79.2	81.5	95.6	82.1	84.1	94.3	96.8	96.4	81.6	79.8	80.1	100.4	86.3	88.4
0320	Concrete Reinforcing	95.8	86.0	90.8	97.9	85.4	91.6	90.5	84.9	87.7	94.4	100.6	97.6	95.1	88.3	91.6	95.1	82.5	88.7
0330	Cast-in-Place Concrete	103.6	79.9	94.8	91.8	80.6	87.7	99.9	86.0	94.7	107.9	56.4	88.7	107.9	84.4	99.2	108.2	85.9	99.9
03	CONCRETE	92.5	81.7	87.6	100.9	81.0	91.9	106.7	84.1	96.5	96.0	83.7	90.5	94.8	83.7	89.8	96.1	86.1	91.6
04	MASONRY	92.9	75.6	82.2	91.0	74.8	81.0	83.9	78.6	80.7	99.7	73.4	83.5	101.3	74.0	84.4	105.3	81.7	90.7
05	METALS	102.1	104.8	102.9	93.1	85.2	90.7	87.7	85.2	86.9	88.3	99.9	91.8	88.3	95.8	90.6	90.6	93.9	91.6
06	WOOD, PLASTICS & COMPOSITES	95.8	77.2	85.7	95.5	78.9	86.5	95.9	81.5	88.1	93.0	101.0	97.4	78.7	82.9	81.0	100.4	86.1	92.6
07	THERMAL & MOISTURE PROTECTION	104.5	80.4	94.1	103.0	81.4	93.7	102.9	84.2	94.9	104.1	79.2	93.4	104.4	76.3	92.3	105.2	82.3	95.3
08	OPENINGS	95.8	77.7	91.6	94.3	77.5	90.4	91.2	79.4	88.4	96.5	98.6	97.0	101.0	82.2	96.6	101.5	83.6	97.3
0920	Plaster & Gypsum Board	94.5	76.9	82.9	94.0	78.0	83.4	94.3	80.6	85.3	105.0	101.2	102.5	100.5	82.6	88.7	110.1	86.0	94.3
0950, 0980	Ceilings & Acoustic Treatment	90.3	76.9	81.2	83.4	78.0	79.7	79.2	80.6	80.1	95.0	101.2	99.2	95.0	82.6	86.6	97.5	86.0	89.7
0960	Flooring	93.6	89.1	92.3	94.0	78.2	89.4	94.9	79.3	90.3	95.3	71.4	88.4	89.6	82.2	87.5	109.6	87.1	103.0
0970, 0990	Wall Finishes & Painting/Coating	88.9	84.7	86.4	91.0	81.9	85.6	91.0	86.6	88.4	92.3	87.5	89.5	92.3	85.8	88.4	93.9	72.6	81.3
09	FINISHES	90.0	80.5	84.9	88.8	79.2	83.6	88.6	81.9	85.0	92.4	92.4	92.4	88.7	81.2	84.6	97.8	85.1	90.9
COVERS	DIVS. 10-14, 25, 28, 41, 43, 44, 46	100.0	90.3	97.9	100.0	90.5	97.9	100.0	93.3	98.5	100.0	92.1	98.3	100.0	85.8	96.9	100.0	94.0	98.7
21, 22, 23	FIRE SUPPRESSION, PLUMBING & HVAC	99.9	76.7	90.6	100.0	78.2	91.2	97.5	80.4	90.7	97.7	85.7	92.9	97.7	72.6	87.6	100.1	83.0	93.3
26, 27, 3370	ELECTRICAL, COMMUNICATIONS & UTIL.	95.9	86.0	90.8	93.1	85.5	89.2	93.5	83.5	88.4	98.7	74.0	86.0	99.3	78.8	88.8	96.4	82.0	89.1
MF2018	WEIGHTED AVERAGE	97.5	83.6	91.5	96.3	83.8	90.9	94.9	85.4	90.8	96.0	86.3	91.8	95.7	81.3	89.6	98.1	85.8	92.8

646

For customer support on your Plumbing Costs with RSMeans Data, call 800.448.8182.

IOWA

DIVISION		COUNCIL BLUFFS 515			CRESTON 508			DAVENPORT 527 - 528			DECORAH 521			DES MOINES 500 - 503, 509			DUBUQUE 520		
		MAT.	INST.	TOTAL	MAT.	INST.	TOTAL	MAT.	INST.	TOTAL	MAT.	INST.	TOTAL	MAT.	INST.	TOTAL	MAT.	INST.	TOTAL
015433	CONTRACTOR EQUIPMENT		97.5	97.5		101.4	101.4		101.4	101.4		101.4	101.4		103.2	103.2		97.0	97.0
0241, 31 - 34	SITE & INFRASTRUCTURE, DEMOLITION	104.3	93.0	96.4	95.3	98.4	97.4	99.1	99.4	99.3	97.5	96.3	96.6	101.2	99.9	100.3	98.3	93.2	94.8
0310	Concrete Forming & Accessories	80.9	75.6	76.4	75.8	82.6	81.6	99.9	98.6	98.8	91.8	73.7	76.5	92.7	86.1	87.1	82.4	83.8	83.6
0320	Concrete Reinforcing	97.0	82.1	89.4	94.5	88.3	91.3	95.1	101.9	98.6	94.4	88.0	91.1	99.0	85.7	92.3	93.7	82.3	87.9
0330	Cast-in-Place Concrete	112.5	81.4	100.9	114.7	62.5	95.3	104.2	98.0	101.9	105.0	78.2	95.0	97.6	90.5	95.0	105.9	85.4	98.3
03	CONCRETE	98.3	79.7	89.9	98.3	77.3	88.8	94.2	99.4	96.5	93.9	78.7	87.1	91.3	88.3	90.0	92.9	84.8	89.3
04	MASONRY	106.7	77.8	88.8	99.3	81.6	88.3	102.1	94.1	97.2	120.9	76.2	93.3	87.3	85.5	86.2	106.3	70.3	84.1
05	METALS	95.7	93.5	95.1	92.1	96.3	93.4	90.6	106.4	95.4	88.4	95.4	90.5	98.1	97.2	97.8	89.3	93.1	90.4
06	WOOD, PLASTICS & COMPOSITES	77.5	74.2	75.7	70.6	82.9	77.3	100.4	98.1	99.1	89.8	72.9	80.7	87.7	85.3	86.4	79.3	84.6	82.2
07	THERMAL & MOISTURE PROTECTION	104.5	76.9	92.6	105.4	81.5	95.1	104.6	94.2	100.1	104.3	73.4	91.0	97.0	85.9	92.2	104.8	78.6	93.5
08	OPENINGS	100.5	77.6	95.2	111.0	84.0	104.7	101.5	99.5	101.0	99.5	80.2	95.0	98.9	88.2	96.4	100.5	84.8	96.9
0920	Plaster & Gypsum Board	100.5	73.9	82.9	96.1	82.6	87.2	110.1	98.2	102.3	103.8	72.3	83.1	93.5	85.0	87.9	100.5	84.5	90.0
0950, 0980	Ceilings & Acoustic Treatment	95.0	73.9	80.7	88.5	82.6	84.5	97.5	98.2	98.0	95.0	72.3	79.6	91.9	85.0	87.3	95.0	84.5	87.9
0960	Flooring	88.5	87.1	88.1	83.9	71.4	80.3	97.7	92.4	96.2	94.9	71.4	88.1	97.9	94.8	97.0	100.8	71.4	92.3
0970, 0990	Wall Finishes & Painting/Coating	89.1	72.6	79.3	83.2	76.2	79.1	92.3	92.9	92.6	92.3	85.8	88.4	88.8	85.8	87.0	93.1	80.5	85.7
09	FINISHES	89.4	77.1	82.7	82.5	79.9	81.1	94.3	96.8	95.7	92.0	74.4	82.4	90.2	87.7	88.8	93.1	81.2	86.6
COVERS	DIVS. 10 - 14, 25, 28, 41, 43, 44, 46	100.0	89.4	97.7	100.0	88.4	97.4	100.0	97.6	99.5	100.0	86.2	97.0	100.0	94.4	98.8	100.0	92.9	98.4
21, 22, 23	FIRE SUPPRESSION, PLUMBING & HVAC	100.1	74.4	89.8	97.6	80.2	90.6	100.1	96.0	98.5	97.7	75.3	88.7	99.8	84.7	93.8	100.1	77.3	91.0
26, 27, 3370	ELECTRICAL, COMMUNICATIONS & UTIL.	101.3	82.3	91.6	92.5	78.8	85.5	94.4	89.7	92.0	96.4	48.4	71.9	103.6	84.4	93.8	100.0	78.8	89.1
MF2018	WEIGHTED AVERAGE	98.8	80.8	91.1	96.7	83.1	90.9	97.1	96.9	97.0	96.7	75.9	87.8	97.3	88.4	93.5	97.1	81.9	90.6

IOWA

DIVISION		FORT DODGE 505			MASON CITY 504			OTTUMWA 525			SHENANDOAH 516			SIBLEY 512			SIOUX CITY 510 - 511		
		MAT.	INST.	TOTAL	MAT.	INST.	TOTAL	MAT.	INST.	TOTAL	MAT.	INST.	TOTAL	MAT.	INST.	TOTAL	MAT.	INST.	TOTAL
015433	CONTRACTOR EQUIPMENT		101.4	101.4		101.4	101.4		97.0	97.0		97.5	97.5		101.4	101.4		101.4	101.4
0241, 31 - 34	SITE & INFRASTRUCTURE, DEMOLITION	103.6	95.1	97.7	103.6	96.2	98.4	98.8	91.0	93.4	102.6	93.1	96.0	108.6	96.2	100.0	110.2	97.3	101.2
0310	Concrete Forming & Accessories	76.4	74.1	74.4	80.2	73.3	74.3	89.6	73.7	76.1	82.8	79.2	79.7	83.2	38.7	45.4	100.4	74.4	78.3
0320	Concrete Reinforcing	94.5	88.0	91.2	94.3	88.0	91.1	94.4	100.6	97.5	97.0	88.3	92.6	97.0	87.9	92.4	95.1	79.3	87.1
0330	Cast-in-Place Concrete	107.6	43.0	83.6	107.6	73.3	94.9	108.6	65.6	92.6	108.8	85.0	99.9	106.6	57.1	88.1	107.2	71.9	94.1
03	CONCRETE	93.6	66.7	81.5	93.8	76.8	86.2	95.5	76.6	87.0	95.8	83.6	90.3	94.7	55.5	77.1	95.5	75.3	86.4
04	MASONRY	98.1	58.6	73.7	110.9	75.0	88.7	102.6	54.0	72.6	106.2	76.0	87.6	124.9	58.8	84.1	99.4	68.4	80.2
05	METALS	92.2	95.0	93.1	92.3	95.4	93.2	88.2	100.1	91.8	94.7	96.3	95.2	88.5	94.3	90.3	90.6	91.4	90.9
06	WOOD, PLASTICS & COMPOSITES	71.1	82.9	77.5	75.0	72.9	73.9	86.9	79.7	83.0	79.3	80.8	80.1	80.1	35.1	55.7	100.4	74.3	86.2
07	THERMAL & MOISTURE PROTECTION	104.7	67.9	88.9	104.1	74.0	91.2	105.0	68.7	89.4	103.7	74.4	91.1	104.0	55.8	83.2	104.6	72.6	90.8
08	OPENINGS	104.8	73.7	97.6	96.8	80.2	93.0	101.0	83.9	97.0	92.1	78.7	89.0	97.7	47.3	86.0	101.5	75.5	95.4
0920	Plaster & Gypsum Board	96.1	82.6	87.2	96.1	72.3	80.5	101.4	79.5	87.0	100.5	80.6	87.4	100.5	33.5	56.3	110.1	73.8	86.2
0950, 0980	Ceilings & Acoustic Treatment	88.5	82.6	84.5	88.5	72.3	77.6	95.0	79.5	84.5	95.0	80.6	85.2	95.0	33.5	53.3	97.5	73.8	81.4
0960	Flooring	85.2	71.4	81.2	87.2	71.4	82.7	103.8	71.4	94.4	89.2	76.4	85.5	90.5	71.4	85.0	97.7	74.0	90.9
0970, 0990	Wall Finishes & Painting/Coating	83.2	83.4	83.3	83.2	85.8	84.7	93.1	85.8	88.8	89.1	85.8	87.1	92.3	85.8	88.4	92.3	68.2	78.0
09	FINISHES	84.3	75.5	79.5	84.9	74.1	79.0	94.3	74.8	83.6	89.5	79.3	83.9	91.7	47.7	67.7	95.7	73.5	83.6
COVERS	DIVS. 10 - 14, 25, 28, 41, 43, 44, 46	100.0	82.4	96.1	100.0	85.8	96.9	100.0	82.8	96.2	100.0	85.8	96.9	100.0	76.3	94.8	100.0	91.1	98.0
21, 22, 23	FIRE SUPPRESSION, PLUMBING & HVAC	97.6	69.5	86.3	97.6	78.6	89.9	97.7	72.6	87.6	97.7	85.7	92.9	97.7	67.7	85.6	100.1	78.1	91.3
26, 27, 3370	ELECTRICAL, COMMUNICATIONS & UTIL.	98.4	71.1	84.4	97.6	48.4	72.4	98.5	72.7	85.3	96.4	80.4	88.2	96.4	48.4	71.9	96.4	75.8	85.9
MF2018	WEIGHTED AVERAGE	96.4	74.1	86.8	96.2	76.2	87.6	96.6	76.3	87.9	96.4	83.8	91.0	97.0	63.3	82.5	97.8	78.7	89.6

DIVISION		IOWA						KANSAS											
		SPENCER 513			WATERLOO 506 - 507			BELLEVILLE 669			COLBY 677			DODGE CITY 678			EMPORIA 668		
		MAT.	INST.	TOTAL	MAT.	INST.	TOTAL	MAT.	INST.	TOTAL	MAT.	INST.	TOTAL	MAT.	INST.	TOTAL	MAT.	INST.	TOTAL
015433	CONTRACTOR EQUIPMENT		101.4	101.4		101.4	101.4		106.5	106.5		106.5	106.5		106.5	106.5		104.7	104.7
0241, 31 - 34	SITE & INFRASTRUCTURE, DEMOLITION	108.6	94.9	99.1	108.9	97.4	100.9	109.2	96.7	100.5	107.5	97.4	100.4	109.9	96.4	100.5	101.7	94.2	96.5
0310	Concrete Forming & Accessories	89.5	38.5	46.2	91.4	70.6	73.7	97.0	55.7	61.9	99.3	62.9	68.4	92.5	62.7	67.2	87.5	67.4	70.4
0320	Concrete Reinforcing	97.0	87.9	92.3	95.1	82.6	88.7	105.6	103.6	104.6	105.2	103.6	104.4	102.5	103.6	103.1	104.1	103.8	103.9
0330	Cast-in-Place Concrete	106.6	67.4	92.0	115.3	86.5	104.6	119.6	86.0	107.1	116.6	89.8	106.6	118.7	89.5	107.8	115.7	89.9	106.1
03	CONCRETE	95.2	59.0	78.9	99.7	79.3	90.5	112.6	75.9	96.1	109.9	80.6	96.7	111.2	80.3	97.3	105.1	82.6	95.0
04	MASONRY	124.9	58.8	84.1	98.9	76.7	85.2	90.4	59.7	71.5	99.5	66.0	78.8	108.0	65.9	82.0	96.0	67.4	78.3
05	METALS	88.4	94.1	90.2	94.5	93.8	94.3	96.0	99.3	97.0	93.2	100.2	95.3	94.7	99.1	96.0	95.7	100.3	97.1
06	WOOD, PLASTICS & COMPOSITES	86.8	35.1	58.7	88.5	66.2	76.5	97.0	52.6	72.9	101.1	59.0	78.2	92.5	59.0	74.3	87.3	64.8	75.1
07	THERMAL & MOISTURE PROTECTION	105.0	58.9	85.2	104.4	79.3	93.6	93.8	63.9	80.9	101.0	67.2	86.5	101.0	69.9	87.6	91.9	76.8	85.4
08	OPENINGS	109.0	47.3	94.7	97.3	75.1	92.1	96.2	63.7	88.7	100.2	71.0	93.4	100.1	67.2	92.5	94.2	72.8	89.2
0920	Plaster & Gypsum Board	101.4	33.5	56.7	104.9	65.4	78.9	89.2	51.4	64.3	99.3	58.0	72.1	92.9	58.0	69.9	86.2	63.9	71.5
0950, 0980	Ceilings & Acoustic Treatment	95.0	33.5	53.3	91.1	65.4	73.7	85.4	51.4	62.4	82.9	58.0	66.0	82.9	58.0	66.0	85.4	63.9	70.8
0960	Flooring	93.3	71.4	86.9	92.4	81.6	89.3	90.5	69.3	84.4	86.2	69.3	81.3	82.4	69.3	78.6	86.1	69.3	81.2
0970, 0990	Wall Finishes & Painting/Coating	92.3	58.7	72.4	83.2	83.4	83.3	92.0	56.5	71.0	92.8	56.5	71.3	92.8	56.5	71.3	92.0	56.5	71.0
09	FINISHES	92.6	43.5	65.8	88.4	73.0	80.0	85.9	56.7	70.0	86.0	62.0	72.9	84.1	62.0	72.0	83.2	65.4	73.5
COVERS	DIVS. 10 - 14, 25, 28, 41, 43, 44, 46	100.0	76.2	94.8	100.0	91.5	98.1	100.0	83.6	96.4	100.0	86.4	97.0	100.0	86.4	97.0	100.0	87.0	97.1
21, 22, 23	FIRE SUPPRESSION, PLUMBING & HVAC	97.7	69.7	86.4	100.0	81.9	92.8	97.4	71.3	87.0	97.6	72.1	87.4	100.1	72.1	88.9	97.4	74.9	88.4
26, 27, 3370	ELECTRICAL, COMMUNICATIONS & UTIL.	98.0	48.4	72.6	94.3	63.1	78.3	97.5	64.2	80.5	97.4	69.5	83.2	94.5	69.5	81.7	94.9	68.5	81.4
MF2018	WEIGHTED AVERAGE	98.5	63.6	83.6	97.5	79.3	89.7	98.1	72.1	87.0	98.4	75.7	88.7	99.3	75.4	89.1	96.4	77.2	88.2

For customer support on your Plumbing Costs with RSMeans Data, call 800.448.8182.

647

KANSAS

DIVISION		FORT SCOTT 667 MAT.	INST.	TOTAL	HAYS 676 MAT.	INST.	TOTAL	HUTCHINSON 675 MAT.	INST.	TOTAL	INDEPENDENCE 673 MAT.	INST.	TOTAL	KANSAS CITY 660-662 MAT.	INST.	TOTAL	LIBERAL 679 MAT.	INST.	TOTAL
015433	CONTRACTOR EQUIPMENT		105.6	105.6		106.5	106.5		106.5	106.5		106.5	106.5		103.1	103.1		106.5	106.5
0241, 31 - 34	SITE & INFRASTRUCTURE, DEMOLITION	98.5	94.2	95.5	111.8	96.9	101.5	91.3	97.4	95.5	110.8	97.4	101.5	93.3	94.5	94.1	111.7	97.0	101.5
0310	Concrete Forming & Accessories	105.6	82.1	85.6	96.7	60.5	65.9	87.1	58.1	62.5	107.9	69.6	75.4	101.7	97.2	97.8	92.9	60.4	65.3
0320	Concrete Reinforcing	103.4	103.7	103.6	102.5	103.6	103.1	102.5	103.6	103.1	101.9	103.7	102.8	100.3	105.3	102.9	104.0	103.6	103.8
0330	Cast-in-Place Concrete	107.2	84.8	98.9	92.0	86.2	89.8	85.2	89.4	86.8	119.2	89.6	108.2	91.8	98.6	94.4	92.0	86.0	89.7
03	CONCRETE	100.5	87.5	94.6	101.1	78.2	90.8	84.3	78.2	81.5	112.5	83.5	99.4	92.5	99.5	95.6	103.0	78.1	91.8
04	MASONRY	97.0	57.3	72.5	107.5	59.8	78.0	99.5	65.9	78.7	96.7	65.9	77.7	98.1	98.0	98.1	105.6	59.7	77.3
05	METALS	95.6	99.7	96.9	92.8	99.9	94.9	92.6	99.0	94.6	92.5	99.6	94.7	102.7	105.2	103.4	93.1	99.2	94.9
06	WOOD, PLASTICS & COMPOSITES	108.2	88.9	97.7	97.8	59.0	76.7	87.2	52.9	68.5	112.3	67.8	88.1	103.8	96.8	100.0	93.2	59.0	74.6
07	THERMAL & MOISTURE PROTECTION	92.9	74.4	85.0	101.4	64.6	85.5	99.7	66.2	85.3	101.1	77.0	90.7	92.6	99.8	95.7	101.5	64.0	85.4
08	OPENINGS	94.2	86.1	92.3	100.1	67.2	92.4	100.1	63.8	91.6	98.1	71.9	92.0	95.4	96.4	95.6	100.1	67.2	92.5
0920	Plaster & Gypsum Board	91.6	88.7	89.7	96.5	58.0	71.1	91.6	51.6	65.3	107.8	67.0	80.9	85.5	96.8	92.9	93.8	58.0	70.2
0950, 0980	Ceilings & Acoustic Treatment	85.4	88.7	87.7	82.9	58.0	66.0	82.9	51.6	61.7	82.9	67.0	72.1	85.4	96.8	93.1	82.9	58.0	66.0
0960	Flooring	99.7	69.1	90.9	85.0	69.3	80.4	79.7	69.3	76.6	90.4	69.1	84.2	80.9	97.5	85.7	82.7	68.1	78.5
0970, 0990	Wall Finishes & Painting/Coating	93.6	79.5	85.3	92.8	56.5	71.3	92.8	56.5	71.3	92.8	56.5	71.3	99.6	100.9	100.3	92.8	56.5	71.3
09	FINISHES	88.2	80.5	84.0	85.8	60.5	72.0	81.6	58.4	68.9	88.6	67.7	77.2	83.7	97.5	91.3	85.0	60.2	71.5
COVERS	DIVS. 10 - 14, 25, 28, 41, 43, 44, 46	100.0	86.7	97.1	100.0	84.3	96.5	100.0	85.7	96.8	100.0	87.4	97.2	100.0	94.5	98.8	100.0	84.3	96.5
21, 22, 23	FIRE SUPPRESSION, PLUMBING & HVAC	97.4	68.3	85.8	97.6	68.9	86.1	97.6	70.7	86.8	97.6	72.0	87.3	99.8	98.8	99.4	97.6	70.3	86.7
26, 27, 3370	ELECTRICAL, COMMUNICATIONS & UTIL.	94.2	69.5	81.6	96.4	69.5	82.7	91.8	62.4	76.8	93.8	74.6	84.0	99.5	97.9	98.7	94.5	69.5	81.7
MF2018	WEIGHTED AVERAGE	96.3	78.2	88.5	97.6	73.5	87.3	93.7	73.1	84.8	98.3	77.9	89.5	97.2	98.6	97.8	97.5	73.7	87.3

KANSAS / KENTUCKY

DIVISION		SALINA 674 MAT.	INST.	TOTAL	TOPEKA 664-666 MAT.	INST.	TOTAL	WICHITA 670-672 MAT.	INST.	TOTAL	ASHLAND 411-412 MAT.	INST.	TOTAL	BOWLING GREEN 421-422 MAT.	INST.	TOTAL	CAMPTON 413-414 MAT.	INST.	TOTAL
015433	CONTRACTOR EQUIPMENT		106.5	106.5		104.7	104.7		106.5	106.5		100.4	100.4		93.3	93.3		99.7	99.7
0241, 31 - 34	SITE & INFRASTRUCTURE, DEMOLITION	100.4	96.9	98.0	96.4	94.0	94.7	96.4	98.1	97.6	114.6	83.0	92.6	79.2	94.2	89.6	88.2	95.5	93.3
0310	Concrete Forming & Accessories	88.9	55.8	60.8	98.7	67.8	72.4	94.6	55.0	61.0	86.3	92.2	91.3	86.0	77.8	79.0	88.6	83.2	84.0
0320	Concrete Reinforcing	101.9	103.4	102.7	99.9	102.7	101.3	100.0	101.0	100.5	91.4	96.9	94.2	88.7	79.0	83.8	89.6	95.7	92.7
0330	Cast-in-Place Concrete	103.0	86.2	96.8	96.5	87.9	93.3	96.3	74.6	88.2	87.3	97.5	91.1	86.6	71.5	81.0	96.6	69.7	86.6
03	CONCRETE	98.2	76.0	88.2	94.2	82.0	88.7	93.8	71.2	83.7	91.7	95.9	93.6	90.3	76.2	84.0	93.9	80.9	88.1
04	MASONRY	122.3	56.0	81.4	91.8	65.6	75.6	96.6	50.5	68.1	89.6	94.0	92.3	91.6	70.6	78.6	89.0	55.9	68.5
05	METALS	94.5	99.6	96.0	99.7	100.4	99.9	96.5	97.8	96.9	95.5	108.8	99.6	97.2	83.8	93.1	96.5	90.7	94.7
06	WOOD, PLASTICS & COMPOSITES	88.7	52.6	69.1	102.0	66.9	82.9	98.6	52.9	73.8	75.3	90.2	83.4	87.9	78.4	82.7	86.3	90.9	88.8
07	THERMAL & MOISTURE PROTECTION	100.4	62.8	84.2	96.7	76.1	87.8	99.3	59.4	82.1	94.0	90.9	92.6	89.6	79.6	85.3	103.1	69.7	88.7
08	OPENINGS	100.1	63.7	91.6	103.6	75.2	97.0	104.5	63.3	94.9	92.6	91.6	92.4	93.4	80.5	90.4	94.6	89.9	93.5
0920	Plaster & Gypsum Board	91.6	51.4	65.1	97.9	66.0	76.9	94.8	51.6	66.4	60.0	90.2	79.9	87.3	78.2	81.3	87.3	90.2	89.2
0950, 0980	Ceilings & Acoustic Treatment	82.9	51.4	61.6	91.7	66.0	74.3	86.7	51.6	63.0	78.7	90.2	86.5	83.4	78.2	79.9	83.4	90.2	88.0
0960	Flooring	81.0	69.3	77.6	95.7	69.3	88.0	92.8	69.3	86.0	74.3	84.0	77.1	89.4	64.0	82.0	91.5	66.5	84.3
0970, 0990	Wall Finishes & Painting/Coating	92.8	56.5	71.3	96.1	69.5	80.4	91.5	56.5	70.8	92.9	92.8	92.8	91.0	90.4	90.7	91.0	55.3	69.9
09	FINISHES	82.8	56.7	68.6	92.5	67.2	78.7	89.6	56.3	71.5	75.9	90.7	84.0	86.7	76.3	81.0	87.4	78.1	82.4
COVERS	DIVS. 10 - 14, 25, 28, 41, 43, 44, 46	100.0	84.3	96.5	100.0	81.6	95.9	100.0	83.7	96.4	100.0	90.0	97.8	100.0	85.6	96.8	100.0	48.8	88.7
21, 22, 23	FIRE SUPPRESSION, PLUMBING & HVAC	100.1	70.9	88.4	100.0	74.7	89.9	99.8	68.5	87.3	97.3	86.9	93.2	100.0	78.8	91.5	97.5	78.1	89.8
26, 27, 3370	ELECTRICAL, COMMUNICATIONS & UTIL.	94.3	71.9	82.8	98.4	73.6	85.7	98.2	71.8	84.7	91.1	90.4	90.7	93.6	76.5	84.9	91.1	90.3	90.7
MF2018	WEIGHTED AVERAGE	98.0	72.8	87.2	98.3	77.7	89.5	97.8	70.8	86.2	93.6	92.0	92.9	94.7	78.9	87.9	94.8	79.9	88.4

KENTUCKY

DIVISION		CORBIN 407-409 MAT.	INST.	TOTAL	COVINGTON 410 MAT.	INST.	TOTAL	ELIZABETHTOWN 427 MAT.	INST.	TOTAL	FRANKFORT 406 MAT.	INST.	TOTAL	HAZARD 417-418 MAT.	INST.	TOTAL	HENDERSON 424 MAT.	INST.	TOTAL
015433	CONTRACTOR EQUIPMENT		99.7	99.7		103.6	103.6		93.3	93.3		99.7	99.7		99.7	99.7		113.7	113.7
0241, 31 - 34	SITE & INFRASTRUCTURE, DEMOLITION	90.8	96.0	94.4	83.5	107.3	100.1	73.6	93.9	87.7	89.1	96.5	94.2	85.9	96.6	93.4	82.2	120.8	109.1
0310	Concrete Forming & Accessories	84.8	78.1	79.1	84.0	72.7	74.4	80.6	74.0	75.0	101.4	79.3	82.7	85.0	84.1	84.3	91.6	79.1	81.0
0320	Concrete Reinforcing	92.1	92.3	92.2	88.3	80.9	84.5	89.2	84.3	86.7	96.9	82.2	89.5	90.0	92.5	91.3	88.8	81.1	84.9
0330	Cast-in-Place Concrete	92.8	74.3	85.9	92.4	82.1	88.6	78.3	69.0	74.8	94.1	76.4	87.6	92.9	71.7	85.0	76.5	87.0	80.4
03	CONCRETE	88.1	79.6	84.3	92.2	78.2	85.9	82.3	74.5	78.8	91.7	79.2	86.1	90.7	81.5	86.6	87.3	82.5	85.2
04	MASONRY	81.0	61.3	68.8	103.0	73.5	84.8	76.5	63.0	68.2	79.5	71.8	74.7	87.5	58.0	69.3	94.8	80.4	85.9
05	METALS	94.1	89.3	92.6	94.3	88.9	92.7	96.3	84.9	92.9	94.5	86.0	91.9	96.5	89.7	94.4	87.4	85.8	86.9
06	WOOD, PLASTICS & COMPOSITES	75.0	79.8	77.6	83.7	70.2	76.4	82.5	76.1	79.0	102.1	79.8	90.0	82.9	90.9	87.2	90.8	78.0	83.8
07	THERMAL & MOISTURE PROTECTION	107.0	71.5	91.7	104.3	73.7	91.1	89.0	70.5	81.1	105.6	75.7	92.7	103.0	71.5	89.4	102.2	85.1	94.9
08	OPENINGS	89.2	71.9	85.2	96.6	73.9	91.4	93.4	74.7	89.0	96.7	79.6	92.7	95.0	89.1	93.6	91.6	79.4	88.7
0920	Plaster & Gypsum Board	93.1	78.9	83.7	68.6	70.0	69.5	86.1	75.8	79.3	95.2	78.9	84.4	86.1	90.2	88.8	91.0	77.0	81.8
0950, 0980	Ceilings & Acoustic Treatment	79.8	78.9	79.2	86.0	70.0	75.2	83.4	75.8	78.3	90.1	78.9	82.5	83.4	90.2	88.0	79.2	77.0	77.7
0960	Flooring	86.0	66.5	80.3	66.6	86.3	72.3	86.8	76.3	83.8	95.1	66.5	86.8	89.8	66.5	83.0	93.0	78.6	88.9
0970, 0990	Wall Finishes & Painting/Coating	90.7	63.6	74.7	86.2	73.0	78.4	91.0	69.9	78.6	97.5	94.1	95.5	91.0	55.3	69.9	91.0	89.0	89.9
09	FINISHES	81.3	74.6	77.6	77.6	74.8	76.1	85.3	73.7	79.0	89.1	78.9	83.5	86.6	78.7	82.3	86.8	80.1	83.2
COVERS	DIVS. 10 - 14, 25, 28, 41, 43, 44, 46	100.0	91.7	98.2	100.0	87.7	97.3	100.0	71.5	93.7	100.0	58.6	90.9	100.0	49.6	88.9	100.0	56.9	90.5
21, 22, 23	FIRE SUPPRESSION, PLUMBING & HVAC	97.7	74.9	88.5	98.2	77.7	90.0	97.6	77.8	89.7	100.0	78.7	91.5	97.5	79.2	90.2	97.6	78.1	89.8
26, 27, 3370	ELECTRICAL, COMMUNICATIONS & UTIL.	90.9	90.3	90.6	94.6	74.2	84.2	90.9	78.2	84.4	102.9	73.2	87.7	91.1	90.4	90.7	93.1	75.5	84.1
MF2018	WEIGHTED AVERAGE	92.4	79.6	86.9	94.7	79.8	88.3	91.7	76.7	85.3	96.1	78.7	88.6	94.2	80.6	88.4	92.5	82.4	88.2

648

KENTUCKY

DIVISION		LEXINGTON 403 - 405			LOUISVILLE 400 - 402			OWENSBORO 423			PADUCAH 420			PIKEVILLE 415 - 416			SOMERSET 425 - 426		
		MAT.	INST.	TOTAL	MAT.	INST.	TOTAL	MAT.	INST.	TOTAL	MAT.	INST.	TOTAL	MAT.	INST.	TOTAL	MAT.	INST.	TOTAL
015433	CONTRACTOR EQUIPMENT		99.7	99.7		93.3	93.3		113.7	113.7		113.7	113.7		100.4	100.4		99.7	99.7
0241, 31 - 34	SITE & INFRASTRUCTURE, DEMOLITION	92.9	97.8	96.3	88.0	94.1	92.3	92.3	121.4	112.6	84.9	120.9	110.0	125.9	82.1	95.4	78.6	96.0	90.7
0310	Concrete Forming & Accessories	97.0	75.6	78.8	94.0	76.9	79.5	90.1	79.0	80.6	88.0	80.8	81.9	95.5	88.0	89.2	86.0	73.4	75.3
0320	Concrete Reinforcing	101.2	82.8	91.8	96.9	82.6	89.6	88.8	79.5	84.1	89.4	82.9	86.1	91.9	96.8	94.4	89.2	62.1	75.4
0330	Cast-in-Place Concrete	95.0	86.9	92.0	89.9	70.6	82.7	89.3	86.4	88.2	81.6	83.1	82.1	96.0	92.1	94.6	76.5	90.1	81.6
03	CONCRETE	91.0	81.2	86.6	89.2	76.1	83.3	99.1	82.0	91.4	91.8	82.3	87.5	105.1	92.2	99.3	77.6	77.8	77.7
04	MASONRY	79.6	72.6	75.3	75.7	68.0	70.9	87.8	80.3	83.1	90.6	80.1	84.1	87.9	85.0	86.1	82.4	64.0	71.1
05	METALS	96.5	86.1	93.4	100.5	84.6	95.7	88.9	85.6	87.9	85.9	87.6	86.4	95.4	108.6	99.4	96.4	78.3	90.9
06	WOOD, PLASTICS & COMPOSITES	91.3	73.6	81.7	87.9	78.4	82.7	88.5	78.0	82.8	86.0	80.7	83.1	85.7	90.2	88.2	83.4	73.6	78.1
07	THERMAL & MOISTURE PROTECTION	107.4	75.5	93.7	103.4	72.8	90.2	103.0	78.8	92.6	102.3	84.3	94.6	94.8	82.8	89.6	102.2	71.8	89.1
08	OPENINGS	89.4	76.0	86.3	87.4	75.5	84.6	91.6	79.0	88.7	90.9	81.3	88.7	93.2	88.2	92.1	94.1	68.0	88.0
0920	Plaster & Gypsum Board	102.7	72.5	82.8	95.9	78.2	84.3	89.5	77.0	81.2	88.5	79.8	82.8	63.7	90.2	81.2	86.1	72.5	77.1
0950, 0980	Ceilings & Acoustic Treatment	83.2	72.5	76.0	88.3	78.2	81.5	79.2	77.0	77.7	79.2	79.8	79.6	78.7	90.2	86.5	83.4	72.5	76.0
0960	Flooring	90.8	66.5	83.7	93.5	64.0	84.9	92.5	64.0	84.2	91.4	78.6	87.7	78.4	66.5	74.9	90.1	66.5	83.2
0970, 0990	Wall Finishes & Painting/Coating	90.7	81.8	85.4	96.9	69.9	80.9	91.0	89.0	89.9	91.0	75.1	81.6	92.9	94.1	93.6	91.0	69.9	78.6
09	FINISHES	84.7	74.1	78.9	88.2	73.8	80.4	87.0	76.9	81.5	86.2	79.7	82.7	78.4	85.0	82.0	86.0	71.6	78.1
COVERS	DIVS. 10 - 14, 25, 41, 43, 44, 46	100.0	90.3	97.9	100.0	87.5	97.2	100.0	94.9	98.9	100.0	86.3	97.0	100.0	49.2	88.8	100.0	88.9	97.6
21, 22, 23	FIRE SUPPRESSION, PLUMBING & HVAC	100.1	78.0	91.3	100.0	77.2	90.9	100.0	78.8	91.5	97.6	78.4	89.9	97.3	84.1	92.0	97.6	74.5	88.4
26, 27, 3370	ELECTRICAL, COMMUNICATIONS & UTIL.	93.4	76.5	84.8	96.0	76.5	86.0	93.1	75.5	84.1	95.3	77.5	86.2	93.9	90.4	92.1	91.5	90.3	90.9
MF2018	WEIGHTED AVERAGE	94.4	79.6	88.0	94.6	77.7	87.4	94.7	83.1	89.7	92.7	83.8	88.9	96.0	87.6	92.4	92.1	77.8	86.0

LOUISIANA

DIVISION		ALEXANDRIA 713 - 714			BATON ROUGE 707 - 708			HAMMOND 704			LAFAYETTE 705			LAKE CHARLES 706			MONROE 712		
		MAT.	INST.	TOTAL	MAT.	INST.	TOTAL	MAT.	INST.	TOTAL	MAT.	INST.	TOTAL	MAT.	INST.	TOTAL	MAT.	INST.	TOTAL
015433	CONTRACTOR EQUIPMENT		93.2	93.2		90.4	90.4		90.9	90.9		90.9	90.9		90.4	90.4		93.2	93.2
0241, 31 - 34	SITE & INFRASTRUCTURE, DEMOLITION	101.6	92.1	95.0	103.5	90.4	94.4	100.0	89.6	92.7	101.0	91.8	94.6	101.8	91.0	94.3	101.6	92.0	94.9
0310	Concrete Forming & Accessories	78.9	61.6	64.2	95.1	73.4	76.7	77.4	56.2	59.4	94.6	69.5	73.3	95.2	69.6	73.4	78.5	60.9	63.6
0320	Concrete Reinforcing	99.9	55.6	77.4	90.5	55.6	72.8	91.7	55.3	73.2	93.0	55.6	74.0	93.0	55.9	74.1	98.8	55.6	76.8
0330	Cast-in-Place Concrete	91.1	68.1	82.5	97.0	75.6	89.1	92.8	64.2	82.1	92.3	67.8	83.2	97.2	67.9	86.3	91.1	65.9	81.7
03	CONCRETE	86.5	63.5	76.2	91.0	71.7	82.3	89.0	59.6	75.8	90.0	66.9	79.6	92.4	67.0	81.0	86.3	62.5	75.6
04	MASONRY	109.4	61.6	79.9	91.6	64.0	74.5	94.4	53.6	69.2	94.4	64.0	75.6	93.9	68.2	78.0	104.1	60.3	77.1
05	METALS	88.5	71.0	83.2	99.8	74.3	92.1	90.0	68.7	83.5	89.3	69.3	83.2	89.3	69.7	83.3	88.4	70.8	83.1
06	WOOD, PLASTICS & COMPOSITES	91.0	60.5	74.5	99.7	75.1	86.3	83.4	55.8	68.4	104.2	70.0	85.6	102.4	70.0	84.8	90.4	60.5	74.2
07	THERMAL & MOISTURE PROTECTION	94.9	66.3	82.6	97.7	68.7	85.2	96.1	62.4	81.6	96.7	68.1	84.4	96.4	69.1	84.6	94.9	65.5	82.3
08	OPENINGS	113.4	60.3	101.1	101.2	72.3	94.5	98.2	56.5	88.5	101.9	63.9	93.1	101.9	63.9	93.1	113.4	58.7	100.7
0920	Plaster & Gypsum Board	86.8	59.8	69.0	97.1	74.9	82.4	97.9	55.0	69.6	106.1	69.6	82.0	106.1	69.6	82.0	86.5	59.8	68.9
0950, 0980	Ceilings & Acoustic Treatment	80.9	59.8	66.6	95.3	74.9	81.5	98.8	55.0	69.1	97.1	69.6	78.4	98.0	69.6	78.7	80.9	59.8	66.6
0960	Flooring	91.3	63.5	83.2	92.1	59.0	82.5	88.7	57.1	79.6	97.2	62.9	87.3	97.2	65.9	88.1	90.9	56.9	81.0
0970, 0990	Wall Finishes & Painting/Coating	94.5	61.3	74.9	91.5	61.3	73.7	95.5	61.4	75.3	95.5	61.3	75.3	95.5	61.3	75.3	94.5	61.3	74.9
09	FINISHES	81.5	61.6	70.6	87.8	69.9	78.0	89.0	56.4	71.2	92.3	67.6	78.8	92.5	68.2	79.2	81.4	59.9	69.7
COVERS	DIVS. 10 - 14, 25, 28, 41, 43, 44, 46	100.0	80.7	95.8	100.0	85.5	96.8	100.0	80.1	95.6	100.0	85.0	96.7	100.0	85.0	96.7	100.0	80.6	95.7
21, 22, 23	FIRE SUPPRESSION, PLUMBING & HVAC	100.0	64.7	85.9	100.0	65.5	86.2	97.7	62.3	83.5	100.1	65.6	86.3	100.1	65.6	86.4	100.0	63.3	85.3
26, 27, 3370	ELECTRICAL, COMMUNICATIONS & UTIL.	93.1	63.0	77.7	99.7	59.5	79.1	95.5	71.3	83.1	96.5	65.3	80.6	96.1	67.6	81.5	94.7	58.7	76.3
MF2018	WEIGHTED AVERAGE	96.0	66.6	83.4	97.6	69.7	85.7	94.4	64.5	81.6	96.0	68.9	84.4	96.2	69.8	84.9	95.8	65.1	82.6

LOUISIANA / MAINE

DIVISION		NEW ORLEANS 700 - 701			SHREVEPORT 710 - 711			THIBODAUX 703			AUGUSTA 043			BANGOR 044			BATH 045		
		MAT.	INST.	TOTAL	MAT.	INST.	TOTAL	MAT.	INST.	TOTAL	MAT.	INST.	TOTAL	MAT.	INST.	TOTAL	MAT.	INST.	TOTAL
015433	CONTRACTOR EQUIPMENT		86.7	86.7		93.2	93.2		90.9	90.9		97.9	97.9		97.9	97.9		97.9	97.9
0241, 31 - 34	SITE & INFRASTRUCTURE, DEMOLITION	100.9	92.9	95.4	103.9	92.0	95.6	102.3	91.6	94.8	87.8	98.5	95.2	90.7	99.8	97.0	88.6	98.5	95.5
0310	Concrete Forming & Accessories	96.8	70.6	74.5	94.6	61.6	66.6	88.6	65.6	69.2	101.1	78.7	82.1	95.2	78.0	80.6	90.6	78.7	80.5
0320	Concrete Reinforcing	94.7	56.0	75.0	100.1	55.4	77.4	91.7	55.1	73.1	105.3	82.9	93.9	94.5	82.1	88.7	93.5	82.6	88.0
0330	Cast-in-Place Concrete	93.1	70.7	84.7	94.0	66.1	83.6	99.7	65.8	87.1	90.5	110.3	97.8	73.2	111.1	87.3	73.2	110.2	87.0
03	CONCRETE	95.6	68.0	83.2	90.1	62.9	77.8	94.0	64.5	80.7	90.3	90.9	90.5	83.7	90.9	86.9	83.7	90.8	86.9
04	MASONRY	98.4	65.6	78.2	94.7	57.7	71.8	118.4	52.8	77.9	102.9	90.3	95.2	115.5	93.5	101.9	122.6	85.9	99.9
05	METALS	99.4	63.1	88.4	92.3	70.7	85.7	90.0	68.4	83.4	99.9	91.1	97.2	89.1	92.2	90.0	87.5	90.8	88.5
06	WOOD, PLASTICS & COMPOSITES	99.4	73.0	85.0	98.0	61.3	78.0	90.9	67.4	78.1	95.3	77.6	85.7	94.4	75.1	83.9	88.4	77.6	82.6
07	THERMAL & MOISTURE PROTECTION	93.9	69.7	83.5	93.2	64.9	81.0	96.2	63.8	82.2	104.8	76.0	92.4	102.5	82.9	94.0	102.4	75.4	90.8
08	OPENINGS	100.8	66.0	92.7	107.2	58.2	95.8	102.8	58.6	92.5	100.3	80.7	95.8	98.8	79.3	94.3	98.8	80.7	94.6
0920	Plaster & Gypsum Board	96.9	72.6	80.9	95.4	60.6	72.5	99.1	66.9	77.9	114.0	76.6	89.4	117.1	74.0	88.7	112.3	76.6	88.8
0950, 0980	Ceilings & Acoustic Treatment	102.1	72.6	82.1	87.6	60.6	69.3	98.8	66.9	77.2	105.6	76.6	86.0	90.9	74.0	79.5	90.0	76.6	80.9
0960	Flooring	102.3	65.7	91.7	97.6	56.9	85.8	94.6	39.4	78.6	88.2	47.3	76.3	80.9	102.1	87.0	79.4	47.3	70.0
0970, 0990	Wall Finishes & Painting/Coating	100.2	62.2	77.7	89.8	61.3	72.9	96.7	61.4	75.8	95.0	91.3	92.8	90.0	91.3	90.7	90.0	91.3	90.7
09	FINISHES	98.7	69.0	82.5	88.2	60.4	73.1	91.1	60.4	74.4	92.1	72.9	81.6	86.4	83.1	84.6	85.1	72.9	78.4
COVERS	DIVS. 10 - 14, 25, 28, 41, 43, 44, 46	100.0	85.8	96.9	100.0	84.0	96.5	100.0	83.4	96.4	100.0	99.3	99.9	100.0	101.8	100.4	100.0	99.3	99.9
21, 22, 23	FIRE SUPPRESSION, PLUMBING & HVAC	100.1	63.8	85.5	99.9	64.1	85.6	97.7	63.8	84.1	100.0	74.1	89.6	100.1	71.6	88.7	97.7	74.1	88.2
26, 27, 3370	ELECTRICAL, COMMUNICATIONS & UTIL.	100.7	71.3	85.6	100.4	67.1	83.4	94.2	71.3	82.5	98.8	78.6	88.5	97.9	71.1	84.2	96.0	78.6	87.1
MF2018	WEIGHTED AVERAGE	99.2	69.5	86.5	97.0	66.4	83.9	96.8	66.4	83.8	98.0	83.1	91.6	95.4	83.5	90.3	94.4	82.6	89.4

For customer support on your Plumbing Costs with RSMeans Data, call 800.448.8182.

649

		MAINE																
		HOULTON			KITTERY			LEWISTON			MACHIAS			PORTLAND			ROCKLAND		
DIVISION		047			039			042			046			040 - 041			048		
		MAT.	INST.	TOTAL	MAT.	INST.	TOTAL	MAT.	INST.	TOTAL	MAT.	INST.	TOTAL	MAT.	INST.	TOTAL	MAT.	INST.	TOTAL
015433	CONTRACTOR EQUIPMENT		97.9	97.9		97.9	97.9		97.9	97.9		97.9	97.9		97.9	97.9		97.9	97.9
0241, 31 - 34	SITE & INFRASTRUCTURE, DEMOLITION	90.4	98.5	96.0	81.5	98.5	93.3	88.4	99.8	96.3	89.8	98.5	95.8	88.6	99.8	96.4	86.4	98.5	94.8
0310	Concrete Forming & Accessories	99.4	78.5	81.6	90.8	79.0	80.7	101.2	78.0	81.5	96.2	78.4	81.1	102.8	78.0	81.7	97.3	78.7	81.5
0320	Concrete Reinforcing	94.5	81.1	87.7	89.3	82.6	85.9	116.1	83.1	99.3	94.5	81.1	87.7	104.7	83.1	93.7	94.5	82.9	88.6
0330	Cast-in-Place Concrete	73.2	110.2	86.9	71.9	110.4	86.2	74.7	111.1	88.2	73.2	110.1	86.9	90.8	111.1	98.4	74.7	110.3	87.9
03	CONCRETE	84.7	90.4	87.3	78.8	91.0	84.3	84.1	90.9	87.2	84.2	90.4	87.0	90.1	90.9	90.4	81.9	90.8	85.9
04	MASONRY	98.2	61.9	75.8	113.2	85.9	96.4	98.8	93.5	95.5	98.2	61.9	75.8	102.5	93.5	96.9	92.3	90.3	91.1
05	METALS	87.7	88.8	88.1	78.8	91.1	82.5	92.3	92.2	92.2	87.8	88.8	88.1	98.2	92.2	96.4	87.6	91.1	88.7
06	WOOD, PLASTICS & COMPOSITES	98.6	77.6	87.2	91.1	77.6	83.8	100.5	75.1	86.7	95.3	77.6	85.7	97.5	75.1	85.3	96.2	77.6	86.1
07	THERMAL & MOISTURE PROTECTION	102.6	69.1	88.2	105.4	74.5	92.1	102.3	82.9	93.9	102.5	68.4	87.8	106.5	82.9	96.3	102.2	76.4	91.1
08	OPENINGS	98.8	80.7	94.6	98.4	84.3	95.2	101.8	79.3	96.6	98.8	80.7	94.6	100.9	79.3	95.9	98.8	80.7	94.6
0920	Plaster & Gypsum Board	119.6	76.6	91.3	108.0	76.6	87.4	122.2	74.0	90.5	117.8	76.6	90.7	111.0	74.0	86.6	117.8	76.6	90.7
0950, 0980	Ceilings & Acoustic Treatment	90.0	76.6	80.9	101.6	76.6	84.7	100.3	74.0	82.5	90.0	76.6	80.9	104.7	74.0	83.9	90.0	76.6	80.9
0960	Flooring	82.1	44.2	71.1	85.2	49.3	74.8	83.5	102.1	88.9	81.3	44.2	70.6	87.7	102.1	91.9	81.7	44.2	70.8
0970, 0990	Wall Finishes & Painting/Coating	90.0	91.3	90.7	83.3	104.8	96.1	90.0	91.3	90.7	90.0	91.3	90.7	93.6	91.3	92.2	90.0	91.3	90.7
09	FINISHES	87.0	72.4	79.0	87.9	74.7	80.7	89.4	83.1	85.9	86.4	72.4	78.8	91.8	83.1	87.0	86.2	72.4	78.7
COVERS	DIVS. 10 - 14, 25, 28, 41, 43, 44, 46	100.0	99.3	99.8	100.0	99.3	99.9	100.0	101.8	100.4	100.0	99.3	99.9	100.0	101.8	100.4	100.0	99.3	99.9
21, 22, 23	FIRE SUPPRESSION, PLUMBING & HVAC	97.7	74.0	88.2	97.6	81.6	91.2	100.1	71.6	88.7	97.7	74.0	88.2	100.0	71.6	88.6	97.7	74.1	88.2
26, 27, 3370	ELECTRICAL, COMMUNICATIONS & UTIL.	99.8	78.6	89.0	92.5	78.6	85.4	99.8	75.3	87.3	99.8	78.6	89.0	102.8	75.3	88.7	99.7	78.6	88.9
MF2018	WEIGHTED AVERAGE	94.1	79.7	87.9	91.7	84.6	88.7	95.9	84.1	90.8	94.0	79.7	87.8	98.2	84.1	92.2	93.3	83.1	88.9

		MAINE			MARYLAND														
		WATERVILLE			ANNAPOLIS			BALTIMORE			COLLEGE PARK			CUMBERLAND			EASTON		
DIVISION		049			214			210 - 212			207 - 208			215			216		
		MAT.	INST.	TOTAL	MAT.	INST.	TOTAL	MAT.	INST.	TOTAL	MAT.	INST.	TOTAL	MAT.	INST.	TOTAL	MAT.	INST.	TOTAL
015433	CONTRACTOR EQUIPMENT		97.9	97.9		104.4	104.4		104.4	104.4		110.1	110.1		104.4	104.4		104.4	104.4
0241, 31 - 34	SITE & INFRASTRUCTURE, DEMOLITION	90.3	98.5	96.0	104.6	93.4	96.8	100.2	97.5	98.3	102.5	98.0	99.3	94.2	94.3	94.3	101.5	91.5	94.6
0310	Concrete Forming & Accessories	90.1	78.7	80.4	99.0	74.3	78.0	99.9	78.4	81.6	83.3	77.1	78.0	91.7	84.2	85.3	89.8	74.9	77.1
0320	Concrete Reinforcing	94.5	82.9	88.6	113.0	89.7	101.2	100.7	89.8	95.2	100.9	96.7	98.7	95.4	90.0	92.6	94.6	89.6	92.0
0330	Cast-in-Place Concrete	73.2	110.3	87.0	124.7	73.7	105.7	114.1	79.0	101.0	119.0	77.4	103.5	102.7	85.2	96.2	114.1	66.5	96.4
03	CONCRETE	85.0	90.8	87.6	105.7	78.2	93.3	110.9	81.5	97.7	109.5	82.1	97.2	94.4	86.7	90.9	102.9	75.8	90.7
04	MASONRY	109.3	90.3	97.6	96.6	69.1	79.6	99.4	74.6	84.1	113.9	75.1	89.9	97.9	87.3	91.3	111.7	59.8	79.7
05	METALS	87.7	91.1	88.7	102.7	103.1	102.8	99.0	98.3	98.8	86.0	108.8	92.9	97.0	104.7	99.4	97.3	99.9	98.1
06	WOOD, PLASTICS & COMPOSITES	87.8	77.6	82.3	96.2	75.9	85.2	102.9	79.0	89.9	78.1	76.3	77.2	88.4	83.3	85.6	86.2	82.3	84.1
07	THERMAL & MOISTURE PROTECTION	102.6	76.0	91.1	102.1	78.5	92.0	103.1	82.0	94.0	104.5	83.0	95.3	100.4	84.1	93.4	100.6	75.0	89.6
08	OPENINGS	98.8	80.7	94.6	100.1	81.4	95.7	98.8	83.1	95.1	94.7	83.4	92.1	99.7	87.1	96.8	99.1	84.9	95.0
0920	Plaster & Gypsum Board	112.3	76.6	88.8	103.0	75.6	84.9	104.8	78.4	87.4	96.7	75.8	83.0	105.0	83.1	90.6	105.0	82.2	89.9
0950, 0980	Ceilings & Acoustic Treatment	90.0	76.6	80.9	87.9	75.6	79.6	105.5	78.4	87.2	109.0	75.8	86.5	103.9	83.1	89.9	103.9	82.2	89.2
0960	Flooring	79.0	47.3	69.8	89.3	78.2	86.1	101.2	78.2	94.5	88.6	79.1	85.8	89.1	96.2	91.2	88.2	78.2	85.3
0970, 0990	Wall Finishes & Painting/Coating	90.0	91.3	90.7	91.7	76.6	82.7	104.1	77.3	88.2	100.0	74.4	84.9	96.9	86.9	91.0	96.9	74.4	83.6
09	FINISHES	85.1	72.9	78.5	86.5	74.8	80.1	101.7	77.9	88.7	94.6	76.5	84.8	94.9	86.7	90.4	95.1	76.1	84.7
COVERS	DIVS. 10 - 14, 25, 28, 41, 43, 44, 46	100.0	99.2	99.8	100.0	86.1	96.9	100.0	88.6	97.5	100.0	89.0	97.6	100.0	90.5	97.9	100.0	69.8	93.3
21, 22, 23	FIRE SUPPRESSION, PLUMBING & HVAC	97.7	74.1	88.2	100.0	82.7	93.1	100.0	83.8	93.5	97.7	88.1	93.8	97.4	73.6	87.9	97.4	74.4	88.2
26, 27, 3370	ELECTRICAL, COMMUNICATIONS & UTIL.	99.7	78.6	88.9	100.9	88.2	94.4	101.1	89.0	94.9	100.8	101.0	100.9	98.6	81.0	89.6	98.1	63.8	80.6
MF2018	WEIGHTED AVERAGE	94.4	83.1	89.6	100.2	83.1	92.8	101.4	85.0	94.3	98.1	88.6	94.0	97.3	85.5	92.2	99.0	76.0	89.1

		MARYLAND														MASSACHUSETTS		
		ELKTON			HAGERSTOWN			SALISBURY			SILVER SPRING			WALDORF			BOSTON		
DIVISION		219			217			218			209			206			020 - 022, 024		
		MAT.	INST.	TOTAL	MAT.	INST.	TOTAL	MAT.	INST.	TOTAL	MAT.	INST.	TOTAL	MAT.	INST.	TOTAL	MAT.	INST.	TOTAL
015433	CONTRACTOR EQUIPMENT		104.4	104.4		104.4	104.4		104.4	104.4		101.8	101.8		101.8	101.8		102.0	102.0
0241, 31 - 34	SITE & INFRASTRUCTURE, DEMOLITION	88.1	93.0	91.5	92.4	94.4	93.8	101.4	91.5	94.5	90.5	89.6	89.9	97.1	89.6	91.9	94.9	101.3	99.4
0310	Concrete Forming & Accessories	95.7	94.0	94.2	90.8	80.9	82.4	104.3	51.6	59.5	91.5	76.3	78.6	98.6	76.3	79.7	102.5	137.0	131.8
0320	Concrete Reinforcing	94.6	115.3	105.1	95.4	90.0	92.6	94.6	67.0	80.6	99.7	96.5	98.1	100.3	96.5	98.4	105.8	152.0	129.3
0330	Cast-in-Place Concrete	92.3	73.1	85.2	97.8	85.2	93.1	114.1	64.6	95.7	121.8	77.8	105.4	136.5	77.8	114.6	91.1	141.0	109.7
03	CONCRETE	86.7	91.2	88.7	90.5	85.2	88.2	103.9	60.7	84.4	107.5	81.6	95.8	118.7	81.7	102.0	98.4	140.3	117.3
04	MASONRY	96.6	68.4	79.2	103.4	87.3	93.4	111.4	56.3	77.4	113.4	75.4	89.9	97.4	75.4	83.8	108.3	148.1	132.9
05	METALS	97.3	112.4	101.9	97.2	104.7	99.5	97.3	91.3	95.5	90.3	104.6	94.7	90.3	104.6	94.7	97.0	133.0	107.9
06	WOOD, PLASTICS & COMPOSITES	94.1	103.0	98.9	87.5	78.9	82.8	105.2	52.8	76.8	84.9	75.6	79.9	92.1	75.6	83.2	103.2	136.5	121.3
07	THERMAL & MOISTURE PROTECTION	100.1	81.1	91.9	100.8	85.2	94.1	100.9	70.1	87.6	108.1	88.0	99.5	108.7	88.0	99.8	110.5	138.6	122.6
08	OPENINGS	98.1	103.0	99.2	98.0	83.1	94.6	98.3	62.5	90.0	86.7	83.0	85.9	87.4	83.0	86.4	97.8	143.1	108.4
0920	Plaster & Gypsum Board	108.9	103.5	105.3	105.0	78.6	87.6	116.2	51.8	73.8	103.1	75.8	85.1	106.4	75.8	86.3	107.2	137.4	127.1
0950, 0980	Ceilings & Acoustic Treatment	103.9	103.5	103.6	104.9	78.6	87.1	103.9	51.8	68.6	117.5	75.8	89.3	117.5	75.8	89.3	106.3	137.4	127.4
0960	Flooring	90.5	78.2	87.0	88.7	96.2	90.9	94.1	78.2	89.5	94.6	79.1	90.1	98.2	79.1	92.6	93.4	169.1	115.6
0970, 0990	Wall Finishes & Painting/Coating	96.9	74.4	83.6	96.9	74.4	83.6	96.9	74.4	83.6	107.2	74.4	87.8	107.2	74.4	87.8	97.4	153.2	130.4
09	FINISHES	95.5	90.4	92.7	94.8	82.7	88.2	98.5	57.8	76.2	95.1	75.9	84.6	96.9	76.0	85.5	101.6	145.2	125.4
COVERS	DIVS. 10 - 14, 25, 28, 41, 43, 44, 46	100.0	57.7	90.7	100.0	90.0	97.8	100.0	78.0	95.2	100.0	84.6	96.6	100.0	84.6	96.6	100.0	117.3	103.8
21, 22, 23	FIRE SUPPRESSION, PLUMBING & HVAC	97.4	80.4	90.6	99.9	84.9	93.9	97.4	72.3	87.3	97.7	87.9	93.7	97.7	87.9	93.7	100.1	127.5	111.1
26, 27, 3370	ELECTRICAL, COMMUNICATIONS & UTIL.	99.8	89.0	94.3	98.4	81.0	89.5	96.9	61.6	78.8	97.9	101.0	99.4	95.5	101.0	98.3	101.5	131.1	116.6
MF2018	WEIGHTED AVERAGE	96.2	87.6	92.5	97.5	87.0	93.0	99.5	68.5	86.2	97.3	87.4	93.0	98.1	87.4	93.5	100.0	133.3	114.3

MASSACHUSETTS

DIVISION		BROCKTON 023			BUZZARDS BAY 025			FALL RIVER 027			FITCHBURG 014			FRAMINGHAM 017			GREENFIELD 013		
		MAT.	INST.	TOTAL	MAT.	INST.	TOTAL	MAT.	INST.	TOTAL	MAT.	INST.	TOTAL	MAT.	INST.	TOTAL	MAT.	INST.	TOTAL
015433	CONTRACTOR EQUIPMENT		100.3	100.3		100.3	100.3		101.2	101.2		97.9	97.9		99.7	99.7		97.9	97.9
0241, 31 - 34	SITE & INFRASTRUCTURE, DEMOLITION	91.8	102.7	99.4	82.4	102.9	96.6	90.8	102.8	99.2	84.3	102.6	97.0	81.2	102.5	96.0	87.9	101.9	97.7
0310	Concrete Forming & Accessories	99.2	124.3	120.5	96.9	123.9	119.8	99.2	124.0	120.3	94.2	121.6	117.5	102.3	124.5	121.2	92.4	120.8	116.5
0320	Concrete Reinforcing	104.9	146.8	126.2	84.1	125.8	105.3	104.9	125.8	115.5	88.3	146.6	118.0	88.3	146.9	118.1	92.0	123.8	108.1
0330	Cast-in-Place Concrete	84.6	133.5	102.8	70.3	133.3	93.7	81.8	133.7	101.2	78.7	133.3	99.0	78.7	135.3	99.7	80.9	119.4	95.2
03	CONCRETE	89.7	130.8	108.2	76.1	126.9	98.9	88.4	127.1	105.8	77.1	129.2	100.6	79.7	131.5	103.0	80.4	120.1	98.3
04	MASONRY	101.9	138.0	124.2	94.8	138.0	121.5	102.5	137.9	124.4	100.3	134.7	121.6	106.8	139.0	126.7	105.0	121.6	115.2
05	METALS	91.7	126.4	102.2	86.7	117.7	96.1	91.7	118.0	99.7	88.3	122.8	98.8	88.3	126.5	99.9	90.6	111.6	97.0
06	WOOD, PLASTICS & COMPOSITES	99.2	123.1	112.2	95.9	123.1	110.7	99.2	123.4	112.3	96.4	119.9	109.2	103.2	122.9	113.9	94.4	123.1	110.0
07	THERMAL & MOISTURE PROTECTION	102.4	129.8	114.2	101.2	127.0	112.3	102.2	126.2	112.5	101.7	122.7	110.7	101.7	129.9	113.9	101.7	112.1	106.2
08	OPENINGS	99.5	132.1	107.0	95.4	122.4	101.7	99.5	121.8	104.6	101.5	130.3	108.2	91.8	132.0	101.1	101.7	122.1	106.4
0920	Plaster & Gypsum Board	90.2	123.4	112.1	86.0	123.4	110.7	90.2	123.4	112.1	110.9	120.1	117.0	114.0	123.4	120.2	111.8	123.4	119.5
0950, 0980	Ceilings & Acoustic Treatment	106.5	123.4	118.0	90.2	123.4	112.7	106.5	123.4	118.0	94.5	120.1	111.9	94.5	123.4	114.1	103.0	123.4	116.9
0960	Flooring	88.9	167.4	111.6	86.7	167.4	110.1	87.9	167.4	110.9	85.8	167.4	109.5	87.2	167.4	110.5	85.0	139.6	100.9
0970, 0990	Wall Finishes & Painting/Coating	91.0	142.9	121.7	91.0	142.9	121.7	91.0	142.9	121.7	87.2	142.9	120.5	88.1	142.9	120.5	87.2	112.2	102.0
09	FINISHES	90.8	134.6	114.7	85.8	134.6	112.4	90.5	134.8	114.7	86.5	132.7	111.7	87.2	134.4	113.0	88.4	124.4	108.0
COVERS	DIVS. 10 - 14, 25, 28, 41, 43, 44, 46	100.0	111.9	102.6	100.0	112.0	102.6	100.0	112.5	102.7	100.0	106.4	101.4	100.0	111.6	102.5	100.0	105.0	101.1
21, 22, 23	FIRE SUPPRESSION, PLUMBING & HVAC	100.3	106.1	102.6	97.8	105.8	101.0	100.3	105.8	102.5	97.8	106.8	101.4	97.8	123.7	108.2	97.8	100.6	98.9
26, 27, 3370	ELECTRICAL, COMMUNICATIONS & UTIL.	98.7	100.1	99.4	96.0	102.6	99.4	98.6	100.1	99.4	99.3	106.6	103.1	96.1	127.4	112.1	99.3	98.5	98.9
MF2018	WEIGHTED AVERAGE	96.5	119.2	106.2	91.7	117.7	102.8	96.3	117.4	105.3	93.4	118.7	104.3	92.8	127.0	107.4	94.7	110.9	101.6

MASSACHUSETTS

DIVISION		HYANNIS 026			LAWRENCE 019			LOWELL 018			NEW BEDFORD 027			PITTSFIELD 012			SPRINGFIELD 010 - 011		
		MAT.	INST.	TOTAL	MAT.	INST.	TOTAL	MAT.	INST.	TOTAL	MAT.	INST.	TOTAL	MAT.	INST.	TOTAL	MAT.	INST.	TOTAL
015433	CONTRACTOR EQUIPMENT		100.3	100.3		100.3	100.3		97.9	97.9		101.2	101.2		97.9	97.9		97.9	97.9
0241, 31 - 34	SITE & INFRASTRUCTURE, DEMOLITION	88.3	102.9	98.5	93.1	102.7	99.8	92.0	102.8	99.5	89.4	102.8	98.7	93.0	101.3	98.8	92.5	101.7	98.8
0310	Concrete Forming & Accessories	90.9	123.9	118.9	103.5	124.6	121.4	99.7	125.7	121.8	99.2	124.1	120.3	99.7	105.7	104.8	99.9	120.7	117.6
0320	Concrete Reinforcing	84.1	125.8	105.3	109.3	138.5	124.2	110.2	138.3	124.5	104.9	125.9	115.6	91.3	117.4	104.6	110.2	123.8	117.1
0330	Cast-in-Place Concrete	77.2	133.3	98.1	91.0	133.6	106.9	82.8	135.3	102.3	72.1	133.8	95.1	90.3	116.6	100.1	86.2	119.4	98.5
03	CONCRETE	81.7	126.9	102.0	92.9	129.5	109.4	85.1	130.4	105.5	83.7	127.2	103.3	86.5	111.2	97.6	86.8	120.1	101.8
04	MASONRY	100.9	138.0	123.8	112.5	138.8	128.7	99.3	137.8	123.1	100.8	138.6	124.1	99.9	113.8	108.5	99.6	121.6	113.2
05	METALS	88.2	117.7	97.1	91.0	123.4	100.8	91.0	120.0	99.8	91.7	118.1	99.7	90.8	108.6	96.2	93.5	111.5	99.0
06	WOOD, PLASTICS & COMPOSITES	89.0	123.1	107.5	103.7	123.1	114.2	102.9	123.1	113.9	99.2	123.4	112.3	102.9	105.5	104.3	102.9	123.1	113.9
07	THERMAL & MOISTURE PROTECTION	101.7	127.0	112.6	102.4	130.0	114.3	102.2	130.3	114.3	102.2	126.6	112.6	102.2	107.2	104.4	102.2	112.1	106.5
08	OPENINGS	96.0	122.4	102.1	95.5	129.9	103.5	102.8	129.9	109.0	99.5	126.6	105.8	102.8	110.7	104.6	102.8	122.1	107.2
0920	Plaster & Gypsum Board	81.4	123.4	109.1	116.4	123.4	121.0	116.4	123.4	121.0	90.2	123.4	112.1	116.4	105.3	109.1	116.4	123.4	121.0
0950, 0980	Ceilings & Acoustic Treatment	99.7	123.4	115.8	104.9	123.4	117.5	104.9	123.4	117.5	106.5	123.4	118.0	104.9	105.3	105.2	104.9	123.4	117.5
0960	Flooring	84.2	167.4	108.3	87.8	167.4	110.9	87.8	167.4	110.9	87.9	167.4	110.9	88.2	134.2	101.5	87.2	139.6	102.4
0970, 0990	Wall Finishes & Painting/Coating	91.0	142.9	121.7	87.3	142.9	120.2	87.2	142.9	120.1	91.0	142.9	121.7	87.2	112.2	102.0	88.0	112.2	102.3
09	FINISHES	86.6	134.6	112.8	90.4	134.6	114.6	90.4	135.4	115.0	90.4	134.8	114.6	90.5	111.7	102.1	90.3	124.4	108.9
COVERS	DIVS. 10 - 14, 25, 28, 41, 43, 44, 46	100.0	112.0	102.6	100.0	112.1	102.7	100.0	113.1	102.9	100.0	112.5	102.7	100.0	101.5	100.3	100.0	105.0	101.1
21, 22, 23	FIRE SUPPRESSION, PLUMBING & HVAC	100.3	106.5	102.8	100.1	121.9	108.8	100.1	123.4	109.4	100.3	105.7	102.4	100.1	96.3	98.5	100.1	100.9	100.3
26, 27, 3370	ELECTRICAL, COMMUNICATIONS & UTIL.	96.4	100.1	98.3	98.4	125.9	112.5	98.9	122.2	110.8	99.5	100.1	99.8	98.9	98.5	98.7	98.9	95.2	97.0
MF2018	WEIGHTED AVERAGE	93.8	117.5	103.9	96.8	125.8	109.2	96.0	125.4	108.6	95.7	117.6	105.1	96.2	105.1	100.0	96.6	110.4	102.5

DIVISION		MASSACHUSETTS WORCESTER 015 - 016			MICHIGAN ANN ARBOR 481			BATTLE CREEK 490			BAY CITY 487			DEARBORN 481			DETROIT 482		
		MAT.	INST.	TOTAL	MAT.	INST.	TOTAL	MAT.	INST.	TOTAL	MAT.	INST.	TOTAL	MAT.	INST.	TOTAL	MAT.	INST.	TOTAL
015433	CONTRACTOR EQUIPMENT		97.9	97.9		111.9	111.9		99.0	99.0		111.9	111.9		111.9	111.9		96.4	96.4
0241, 31 - 34	SITE & INFRASTRUCTURE, DEMOLITION	92.4	102.6	99.5	78.5	96.7	91.2	93.4	86.0	88.2	70.2	96.2	88.3	78.2	96.8	91.1	93.1	100.7	98.4
0310	Concrete Forming & Accessories	100.3	121.5	118.3	96.5	104.3	103.1	96.3	79.5	82.0	96.5	83.7	85.7	96.4	104.9	103.6	100.8	108.2	107.1
0320	Concrete Reinforcing	110.2	146.6	128.7	102.3	106.5	104.5	94.8	83.3	88.9	102.3	105.7	104.0	102.3	106.6	104.5	103.1	108.3	105.7
0330	Cast-in-Place Concrete	85.7	133.3	103.4	81.0	99.2	87.8	82.0	93.8	86.4	77.5	85.9	80.6	79.2	100.0	87.0	86.1	101.5	91.8
03	CONCRETE	86.6	129.2	105.7	85.3	103.7	93.6	84.2	84.9	84.5	83.6	89.7	86.4	84.4	104.3	93.4	91.5	105.1	97.6
04	MASONRY	99.1	134.7	121.1	99.9	98.4	99.0	97.3	80.1	86.7	99.5	79.3	87.0	99.7	100.3	100.1	102.0	101.9	102.0
05	METALS	93.6	122.7	102.4	100.0	116.6	105.0	100.7	84.5	95.8	100.6	114.9	104.9	100.1	116.7	105.1	102.3	95.7	100.3
06	WOOD, PLASTICS & COMPOSITES	103.4	119.9	112.3	92.4	106.0	99.8	92.1	77.8	84.3	92.4	84.3	88.0	92.4	106.0	99.8	99.2	109.8	105.0
07	THERMAL & MOISTURE PROTECTION	102.2	122.7	111.0	110.4	100.6	106.1	100.2	81.3	92.1	107.5	83.2	97.1	108.5	103.1	106.2	106.7	106.2	106.5
08	OPENINGS	102.8	130.3	109.2	96.3	101.6	97.5	88.4	76.3	85.6	96.3	85.9	93.9	96.3	101.9	97.6	98.7	104.1	99.9
0920	Plaster & Gypsum Board	116.4	120.1	118.8	100.4	105.8	104.0	90.2	73.8	79.4	100.4	83.5	89.3	100.4	105.8	104.0	97.3	109.9	105.6
0950, 0980	Ceilings & Acoustic Treatment	104.9	120.1	115.2	85.1	105.8	99.1	77.8	73.8	75.1	86.1	83.5	84.3	85.1	105.8	99.1	95.9	109.9	105.4
0960	Flooring	87.8	160.1	108.8	89.4	110.8	95.7	89.7	72.8	84.8	89.4	79.5	86.6	89.0	105.0	93.6	93.7	106.0	97.3
0970, 0990	Wall Finishes & Painting/Coating	87.2	142.9	120.1	83.5	98.0	92.1	86.8	78.3	81.7	83.5	81.2	82.1	83.5	96.3	91.1	91.4	97.9	95.3
09	FINISHES	90.4	131.2	112.7	89.4	105.1	98.0	83.3	77.4	80.1	89.2	82.2	85.4	89.2	104.3	97.5	96.3	107.1	102.2
COVERS	DIVS. 10 - 14, 25, 28, 41, 43, 44, 46	100.0	106.4	101.4	100.0	95.2	98.9	100.0	95.7	99.0	100.0	89.8	97.8	100.0	95.6	99.0	100.0	102.9	100.6
21, 22, 23	FIRE SUPPRESSION, PLUMBING & HVAC	100.1	106.8	102.7	100.1	93.4	97.4	100.1	84.5	93.8	100.1	80.4	92.2	100.1	102.3	101.0	100.0	103.7	101.5
26, 27, 3370	ELECTRICAL, COMMUNICATIONS & UTIL.	98.9	106.6	102.9	96.9	105.5	101.3	93.4	79.6	86.3	96.0	84.5	90.1	96.9	98.4	97.7	100.9	101.3	101.1
MF2018	WEIGHTED AVERAGE	96.6	118.5	106.0	96.4	101.7	98.6	94.6	82.5	89.4	95.8	87.5	92.2	96.2	102.9	99.0	99.1	103.0	100.7

For customer support on your Plumbing Costs with RSMeans Data, call 800.448.8182.

651

MICHIGAN

DIVISION		FLINT 484 - 485			GAYLORD 497			GRAND RAPIDS 493, 495			IRON MOUNTAIN 498 - 499			JACKSON 492			KALAMAZOO 491		
		MAT.	INST.	TOTAL	MAT.	INST.	TOTAL	MAT.	INST.	TOTAL	MAT.	INST.	TOTAL	MAT.	INST.	TOTAL	MAT.	INST.	TOTAL
015433	CONTRACTOR EQUIPMENT		111.9	111.9		106.8	106.8		99.0	99.0		93.8	93.8		106.8	106.8		99.0	99.0
0241, 31 - 34	SITE & INFRASTRUCTURE, DEMOLITION	68.3	96.3	87.8	88.0	82.8	84.4	92.3	85.9	87.9	96.6	93.2	94.2	110.9	85.5	93.2	93.7	86.0	88.3
0310	Concrete Forming & Accessories	99.5	84.1	86.4	95.2	74.3	77.4	95.1	78.0	80.5	87.1	79.7	80.8	91.7	84.6	85.7	96.3	79.3	81.9
0320	Concrete Reinforcing	102.3	106.1	104.2	88.4	89.7	89.0	99.5	83.0	91.1	88.1	87.3	87.7	85.8	106.1	96.1	94.8	78.4	86.5
0330	Cast-in-Place Concrete	81.6	88.3	84.1	81.8	80.6	81.3	86.1	92.0	88.3	96.8	69.7	86.7	81.6	92.5	85.7	83.6	93.7	87.4
03	CONCRETE	85.7	90.8	88.0	81.1	80.9	81.0	88.0	83.6	86.0	89.3	78.2	84.3	76.4	92.3	83.6	87.3	84.0	85.8
04	MASONRY	99.9	87.7	92.4	107.9	74.2	87.1	90.6	75.7	81.4	93.7	79.4	84.8	87.9	88.8	88.4	95.9	80.1	86.1
05	METALS	100.1	115.5	104.8	102.3	111.6	105.1	97.8	84.0	93.6	101.6	91.8	98.6	102.4	113.9	105.9	100.7	82.7	95.3
06	WOOD, PLASTICS & COMPOSITES	96.2	83.0	89.1	85.5	73.8	79.1	95.8	76.3	85.2	81.7	79.8	80.7	84.2	82.1	83.1	92.1	77.8	84.3
07	THERMAL & MOISTURE PROTECTION	107.8	86.9	98.8	98.8	70.6	86.6	102.1	72.7	89.4	102.7	76.2	91.3	98.0	90.8	94.9	100.2	81.3	92.1
08	OPENINGS	96.3	85.0	93.7	87.9	80.1	86.1	99.0	76.0	93.6	94.6	71.1	89.2	87.1	88.0	87.3	88.4	75.6	85.4
0920	Plaster & Gypsum Board	102.6	82.2	89.1	90.0	72.1	78.2	97.1	72.2	80.7	48.2	79.8	69.0	88.5	80.6	83.3	90.2	73.8	79.4
0950, 0980	Ceilings & Acoustic Treatment	85.1	82.2	83.1	77.0	72.1	73.6	89.6	72.2	77.8	74.0	79.8	77.9	77.0	80.6	79.4	77.8	73.8	75.1
0960	Flooring	89.4	88.9	89.3	83.0	88.5	84.6	93.6	78.7	89.3	101.9	90.3	98.5	81.7	81.3	81.6	89.7	72.8	84.8
0970, 0990	Wall Finishes & Painting/Coating	83.5	80.4	81.7	83.1	79.5	81.0	93.6	78.9	84.9	100.5	71.5	83.3	83.1	96.3	90.9	86.8	78.3	81.7
09	FINISHES	88.9	84.0	86.2	82.7	76.9	79.6	89.5	77.6	83.0	82.7	80.7	81.6	83.8	84.3	84.1	83.3	77.4	80.1
COVERS	DIVS. 10 - 14, 25, 28, 41, 43, 44, 46	100.0	90.6	97.9	100.0	79.2	95.4	100.0	95.4	99.0	100.0	87.5	97.3	100.0	93.5	98.6	100.0	95.7	99.0
21, 22, 23	FIRE SUPPRESSION, PLUMBING & HVAC	100.1	85.1	94.1	97.8	80.6	90.9	100.1	81.4	92.6	97.7	84.8	92.5	97.8	87.6	93.7	100.1	80.0	92.0
26, 27, 3370	ELECTRICAL, COMMUNICATIONS & UTIL.	96.9	91.3	94.0	91.4	74.3	82.7	98.6	85.0	91.6	97.4	82.6	89.8	95.1	105.5	100.4	93.2	75.9	84.4
MF2018	WEIGHTED AVERAGE	96.1	90.8	93.8	94.0	81.3	88.5	96.5	81.6	90.1	95.8	83.0	90.3	93.4	93.0	93.2	95.0	80.7	88.8

MICHIGAN / MINNESOTA

DIVISION		LANSING 488 - 489			MUSKEGON 494			ROYAL OAK 480, 483			SAGINAW 486			TRAVERSE CITY 496			BEMIDJI 566		
		MAT.	INST.	TOTAL	MAT.	INST.	TOTAL	MAT.	INST.	TOTAL	MAT.	INST.	TOTAL	MAT.	INST.	TOTAL	MAT.	INST.	TOTAL
015433	CONTRACTOR EQUIPMENT		111.9	111.9		99.0	99.0		91.9	91.9		111.9	111.9		93.8	93.8		100.3	100.3
0241, 31 - 34	SITE & INFRASTRUCTURE, DEMOLITION	89.5	96.5	94.4	91.4	85.9	87.6	82.3	97.2	92.7	71.1	96.2	88.6	81.9	91.1	88.3	95.9	99.8	98.6
0310	Concrete Forming & Accessories	93.5	77.0	79.5	96.6	79.2	81.8	92.0	105.1	103.1	96.5	81.9	84.1	87.1	72.3	74.5	85.8	84.3	84.5
0320	Concrete Reinforcing	105.3	105.9	105.6	95.4	83.0	89.1	92.7	89.8	91.2	102.3	105.7	104.0	89.5	82.5	85.9	97.1	101.4	99.3
0330	Cast-in-Place Concrete	99.5	87.2	95.0	81.7	91.9	85.5	70.9	100.3	81.8	80.0	85.8	82.2	75.5	76.9	76.0	102.4	105.7	103.6
03	CONCRETE	93.2	87.2	90.5	82.7	84.1	83.3	73.4	100.2	85.5	84.8	88.8	86.6	73.4	76.5	74.8	91.2	96.0	93.3
04	MASONRY	94.7	86.3	89.5	94.5	75.7	82.9	93.1	99.4	97.0	101.4	79.3	87.7	91.9	75.3	81.6	99.1	102.4	101.2
05	METALS	98.2	115.0	103.3	98.4	83.8	94.0	103.5	91.0	99.7	100.1	114.7	104.5	101.6	89.4	97.9	91.6	116.8	99.2
06	WOOD, PLASTICS & COMPOSITES	90.1	73.3	81.0	89.2	78.1	83.2	88.1	106.0	97.8	89.1	82.0	85.2	81.7	72.2	76.6	69.9	79.2	75.0
07	THERMAL & MOISTURE PROTECTION	106.8	85.5	97.6	99.1	73.2	87.9	106.0	102.7	104.6	108.8	83.1	97.8	101.7	67.9	87.1	106.0	92.2	100.1
08	OPENINGS	100.5	79.4	95.6	87.7	77.6	85.3	96.1	101.3	97.3	94.4	84.6	92.1	94.6	66.4	88.0	101.9	102.3	102.0
0920	Plaster & Gypsum Board	96.3	72.2	80.4	72.6	74.1	73.6	97.9	105.8	103.1	100.4	81.1	87.7	48.2	72.1	63.9	104.3	79.0	87.7
0950, 0980	Ceilings & Acoustic Treatment	84.9	72.2	76.3	77.8	74.1	75.3	84.4	105.8	98.9	85.1	81.1	82.4	74.0	72.1	72.7	121.3	79.0	92.6
0960	Flooring	96.2	81.3	91.9	88.4	74.7	84.4	86.9	105.0	92.1	89.4	79.5	86.6	101.9	88.5	98.0	92.3	88.4	91.2
0970, 0990	Wall Finishes & Painting/Coating	95.0	78.9	85.5	85.0	78.9	81.4	85.1	96.3	91.7	83.5	81.2	82.1	100.5	40.5	65.0	89.1	106.5	99.4
09	FINISHES	88.8	76.8	82.2	79.8	77.2	78.4	88.3	104.2	97.0	89.1	80.9	84.6	81.6	71.8	76.2	96.0	86.9	91.1
COVERS	DIVS. 10 - 14, 25, 28, 41, 43, 44, 46	100.0	90.7	97.9	100.0	95.5	99.0	100.0	100.6	100.1	100.0	89.5	97.7	100.0	85.6	96.8	100.0	93.9	98.6
21, 22, 23	FIRE SUPPRESSION, PLUMBING & HVAC	100.0	85.5	94.2	99.9	81.4	92.5	97.8	103.6	100.1	100.1	80.0	92.1	97.7	80.0	90.6	97.8	82.0	91.5
26, 27, 3370	ELECTRICAL, COMMUNICATIONS & UTIL.	97.5	89.8	93.5	93.6	74.7	84.0	98.9	101.3	100.1	95.1	86.6	90.7	93.0	75.8	84.2	102.3	101.1	101.7
MF2018	WEIGHTED AVERAGE	97.4	88.7	93.6	93.5	80.3	87.8	94.6	100.6	97.2	95.7	87.3	92.1	92.8	78.3	86.6	96.9	95.4	96.3

MINNESOTA

DIVISION		BRAINERD 564			DETROIT LAKES 565			DULUTH 556 - 558			MANKATO 560			MINNEAPOLIS 553 - 555			ROCHESTER 559		
		MAT.	INST.	TOTAL	MAT.	INST.	TOTAL	MAT.	INST.	TOTAL	MAT.	INST.	TOTAL	MAT.	INST.	TOTAL	MAT.	INST.	TOTAL
015433	CONTRACTOR EQUIPMENT		102.9	102.9		100.3	100.3		103.2	103.2		102.9	102.9		107.6	107.6		103.2	103.2
0241, 31 - 34	SITE & INFRASTRUCTURE, DEMOLITION	97.5	104.8	102.6	94.1	100.0	98.2	97.8	102.9	101.4	94.1	104.4	101.3	95.8	108.4	104.6	97.1	102.5	100.8
0310	Concrete Forming & Accessories	86.1	85.0	85.1	82.5	84.2	83.9	97.9	97.0	97.1	94.5	93.7	93.8	99.6	117.1	114.4	98.4	97.2	97.4
0320	Concrete Reinforcing	95.9	101.8	98.9	97.1	101.4	99.3	96.3	101.6	99.0	95.8	109.0	102.5	97.5	112.5	105.1	94.1	108.7	101.5
0330	Cast-in-Place Concrete	111.4	110.3	111.0	99.4	108.6	102.9	99.3	99.6	99.4	102.5	99.3	101.3	103.3	117.7	108.7	99.1	98.0	98.7
03	CONCRETE	95.1	98.0	96.4	88.7	96.9	92.4	94.4	99.8	96.8	90.7	94.4	94.6	100.8	117.1	108.1	92.0	100.6	95.9
04	MASONRY	123.4	115.3	118.4	123.3	108.0	113.8	101.6	107.9	105.5	111.9	109.3	110.3	112.2	122.0	118.3	103.3	105.7	104.8
05	METALS	92.7	117.1	100.1	91.5	116.2	99.0	97.5	119.0	104.1	92.5	120.6	101.1	99.7	126.4	107.8	98.0	122.2	105.3
06	WOOD, PLASTICS & COMPOSITES	86.0	76.4	80.8	66.8	76.6	72.1	93.8	94.4	94.1	95.9	91.6	93.6	96.7	113.9	106.0	97.8	95.6	96.6
07	THERMAL & MOISTURE PROTECTION	104.2	103.9	104.1	105.8	101.2	103.8	105.6	104.4	105.0	104.6	93.6	99.9	104.2	118.5	110.3	109.9	92.5	102.4
08	OPENINGS	88.3	100.7	91.2	101.8	100.8	101.6	106.3	106.5	106.3	92.9	110.8	97.1	101.6	125.0	107.0	100.6	113.8	103.6
0920	Plaster & Gypsum Board	91.4	76.3	81.5	103.4	76.3	85.5	94.4	94.8	94.6	96.0	92.0	93.3	97.4	114.4	108.6	103.5	95.9	98.5
0950, 0980	Ceilings & Acoustic Treatment	56.8	76.3	70.0	121.3	76.3	90.8	93.2	94.8	94.2	56.8	92.0	80.6	98.2	114.4	109.2	93.3	95.9	95.1
0960	Flooring	91.1	88.4	90.3	91.0	88.4	90.3	96.5	119.7	103.3	92.9	84.6	90.5	99.7	118.5	105.2	93.2	84.6	90.7
0970, 0990	Wall Finishes & Painting/Coating	83.6	106.5	97.1	89.1	85.6	87.0	89.5	107.2	100.0	94.4	102.5	99.2	97.4	127.1	115.0	85.3	100.8	94.5
09	FINISHES	80.8	87.0	84.2	95.4	84.5	89.4	91.5	101.6	97.0	82.5	93.0	88.2	100.5	118.2	110.2	91.2	95.8	93.7
COVERS	DIVS. 10 - 14, 25, 28, 41, 43, 44, 46	100.0	95.5	99.0	100.0	95.3	99.0	100.0	96.2	99.2	100.0	95.1	98.9	100.0	107.3	101.6	100.0	100.3	100.1
21, 22, 23	FIRE SUPPRESSION, PLUMBING & HVAC	97.1	85.7	92.5	97.8	84.7	92.6	99.8	96.5	98.5	97.1	84.8	92.2	99.9	113.0	105.2	99.9	93.5	97.4
26, 27, 3370	ELECTRICAL, COMMUNICATIONS & UTIL.	100.1	101.8	101.0	102.1	70.4	85.9	95.8	101.8	98.9	106.2	92.4	99.1	100.2	113.0	106.8	99.1	92.4	95.6
MF2018	WEIGHTED AVERAGE	95.9	98.6	97.0	97.6	92.2	95.3	98.5	102.7	100.3	96.0	98.0	96.8	100.8	116.5	107.7	98.3	100.3	99.1

For customer support on your Plumbing Costs with RSMeans Data, call 800.448.8182.

City Cost Indexes

		MINNESOTA																	MISSISSIPPI		
	DIVISION	SAINT PAUL			ST. CLOUD			THIEF RIVER FALLS			WILLMAR			WINDOM			BILOXI		
		550 - 551			563			567			562			561			395		
		MAT.	INST.	TOTAL	MAT.	INST.	TOTAL	MAT.	INST.	TOTAL	MAT.	INST.	TOTAL	MAT.	INST.	TOTAL	MAT.	INST.	TOTAL
015433	CONTRACTOR EQUIPMENT		103.2	103.2		102.9	102.9		100.3	100.3		102.9	102.9		102.9	102.9		102.7	102.7
0241, 31 - 34	SITE & INFRASTRUCTURE, DEMOLITION	95.0	103.8	101.1	92.7	105.8	101.8	94.8	99.6	98.1	92.1	104.7	100.9	86.0	103.8	98.4	107.1	89.9	95.1
0310	Concrete Forming & Accessories	97.7	116.8	113.9	83.4	112.9	108.5	86.4	83.4	83.8	83.2	89.1	88.2	87.6	84.3	84.8	92.0	65.7	69.6
0320	Concrete Reinforcing	95.5	109.2	102.5	95.9	109.2	102.7	97.5	101.2	99.4	95.5	109.0	102.4	95.5	107.9	101.8	86.0	50.0	67.7
0330	Cast-in-Place Concrete	100.5	116.4	106.4	98.2	115.1	104.5	101.5	82.7	94.5	99.8	81.7	93.0	86.3	85.6	86.0	120.0	66.8	100.2
03	CONCRETE	94.1	115.9	103.9	86.7	113.6	98.8	89.9	87.5	88.8	86.6	91.2	88.7	77.7	90.2	83.3	102.9	65.1	85.9
04	MASONRY	102.5	122.0	114.6	107.8	121.0	115.9	99.1	102.4	101.1	112.2	115.3	114.1	122.7	88.3	101.4	97.8	60.2	74.6
05	METALS	97.4	124.1	105.5	93.3	122.0	102.0	91.7	115.5	98.9	92.5	120.4	101.0	92.4	118.4	100.3	92.8	84.2	90.2
06	WOOD, PLASTICS & COMPOSITES	95.4	113.6	105.3	83.1	109.7	97.5	71.0	79.2	75.5	82.8	82.2	82.5	87.4	82.2	84.6	95.3	66.4	79.6
07	THERMAL & MOISTURE PROTECTION	104.1	119.3	110.7	104.4	113.1	108.1	106.8	90.3	99.7	104.1	103.4	103.8	104.1	84.5	95.7	99.6	63.2	83.9
08	OPENINGS	98.8	123.7	104.6	93.4	121.6	99.9	101.9	102.3	102.0	90.3	104.7	93.6	93.9	104.7	96.4	98.6	56.6	88.9
0920	Plaster & Gypsum Board	99.0	114.4	109.2	91.4	110.5	104.0	104.0	79.0	87.6	91.4	82.3	85.4	91.4	82.3	85.4	105.2	65.9	79.3
0950, 0980	Ceilings & Acoustic Treatment	96.2	114.4	108.5	56.8	110.5	93.2	121.3	79.0	92.6	56.8	82.3	74.0	56.8	82.3	74.0	94.2	65.9	75.0
0960	Flooring	93.1	124.6	102.2	87.9	118.5	96.8	91.9	88.4	90.9	89.3	88.4	89.1	91.8	88.4	90.8	93.5	56.2	82.7
0970, 0990	Wall Finishes & Painting/Coating	91.3	127.0	112.5	94.4	127.0	113.7	89.1	85.6	87.0	89.1	85.6	87.0	89.1	102.5	97.0	83.0	48.1	62.4
09	FINISHES	92.0	119.3	106.9	80.3	115.5	99.5	95.9	84.6	89.7	80.3	87.7	84.4	80.5	86.8	83.9	88.0	62.4	74.0
COVERS	DIVS. 10 - 14, 25, 28, 41, 43, 44, 46	100.0	106.6	101.5	100.0	101.1	100.2	100.0	93.7	98.6	100.0	96.1	99.1	100.0	92.4	98.3	100.0	73.4	94.1
21, 22, 23	FIRE SUPPRESSION, PLUMBING & HVAC	99.9	112.7	105.0	99.5	110.9	104.1	97.8	81.7	91.4	97.1	103.7	99.7	97.1	82.0	91.0	100.0	56.5	82.6
26, 27, 3370	ELECTRICAL, COMMUNICATIONS & UTIL.	102.2	113.0	107.7	100.1	113.0	106.7	99.6	70.4	84.6	100.1	85.8	92.8	106.2	92.3	99.1	101.7	55.0	77.8
MF2018	WEIGHTED AVERAGE	98.4	115.8	105.9	95.1	114.0	103.2	96.5	89.4	93.4	94.2	99.9	96.6	94.5	92.3	93.5	98.3	64.6	83.8

		MISSISSIPPI																			
	DIVISION	CLARKSDALE			COLUMBUS			GREENVILLE			GREENWOOD			JACKSON			LAUREL		
		386			397			387			389			390 - 392			394		
		MAT.	INST.	TOTAL	MAT.	INST.	TOTAL	MAT.	INST.	TOTAL	MAT.	INST.	TOTAL	MAT.	INST.	TOTAL	MAT.	INST.	TOTAL
015433	CONTRACTOR EQUIPMENT		102.7	102.7		102.7	102.7		102.7	102.7		102.7	102.7		102.7	102.7		102.7	102.7
0241, 31 - 34	SITE & INFRASTRUCTURE, DEMOLITION	99.7	88.5	91.9	106.0	89.4	94.4	105.6	89.9	94.7	102.8	88.2	92.7	101.9	89.9	93.6	111.3	88.6	95.5
0310	Concrete Forming & Accessories	83.4	44.8	50.6	80.8	47.2	52.3	80.2	63.6	66.2	92.5	45.0	52.2	91.0	66.1	69.9	81.0	60.7	63.7
0320	Concrete Reinforcing	101.0	66.0	83.2	92.1	66.4	79.0	101.6	67.3	84.2	101.0	66.1	83.3	95.6	51.9	73.4	92.7	32.8	62.3
0330	Cast-in-Place Concrete	102.8	50.0	83.1	122.4	55.3	97.4	105.9	56.6	87.5	110.5	49.7	87.9	104.5	65.9	90.2	119.7	51.2	94.2
03	CONCRETE	94.4	52.4	75.5	104.3	55.5	82.3	99.9	63.6	83.6	100.8	52.4	79.0	96.8	65.3	82.6	106.9	54.4	83.3
04	MASONRY	90.6	40.2	59.4	123.7	46.2	75.8	134.1	68.2	93.4	91.3	40.0	59.6	103.4	59.1	76.1	119.8	42.9	72.3
05	METALS	93.0	86.8	91.1	89.9	89.6	89.8	94.0	90.8	93.0	93.0	86.5	91.0	99.6	85.0	95.2	90.0	75.5	85.6
06	WOOD, PLASTICS & COMPOSITES	80.8	45.7	61.7	80.6	46.5	62.1	77.4	64.6	70.5	93.8	45.7	67.7	94.9	67.8	80.2	81.8	66.4	73.4
07	THERMAL & MOISTURE PROTECTION	96.2	48.6	75.7	99.5	53.5	79.7	96.7	63.4	82.3	96.6	50.8	76.9	97.8	62.6	82.6	99.7	54.4	80.2
08	OPENINGS	97.2	49.2	86.0	98.2	49.6	87.0	96.9	60.0	88.3	97.2	49.2	86.0	99.9	58.2	90.2	95.4	53.3	85.6
0920	Plaster & Gypsum Board	93.4	44.6	61.3	96.6	45.5	63.0	93.1	64.1	74.0	104.6	44.6	65.1	90.0	67.3	75.1	96.6	65.9	76.4
0950, 0980	Ceilings & Acoustic Treatment	86.2	44.6	58.0	88.1	45.5	59.2	90.0	64.1	72.5	86.2	44.6	58.0	94.8	67.3	76.2	88.1	65.9	73.1
0960	Flooring	97.7	46.5	82.9	87.1	52.2	76.9	96.1	46.5	81.7	103.5	46.5	87.0	91.7	57.9	81.9	85.7	46.5	74.3
0970, 0990	Wall Finishes & Painting/Coating	93.8	48.1	66.8	83.0	48.1	62.4	93.8	49.1	67.3	93.8	48.1	66.8	85.0	49.1	63.8	83.0	48.1	62.4
09	FINISHES	89.9	45.2	65.5	83.8	47.5	63.9	90.7	59.0	73.4	93.5	45.2	67.1	87.0	63.2	74.0	83.9	58.1	69.8
COVERS	DIVS. 10 - 14, 25, 28, 41, 43, 44, 46	100.0	48.9	88.8	100.0	50.0	89.0	100.0	72.6	94.0	100.0	48.9	88.8	100.0	72.9	94.0	100.0	34.9	85.7
21, 22, 23	FIRE SUPPRESSION, PLUMBING & HVAC	98.8	52.4	80.2	98.7	54.2	80.9	100.0	58.2	83.3	98.8	52.8	80.4	100.0	59.0	83.6	98.7	48.0	78.4
26, 27, 3370	ELECTRICAL, COMMUNICATIONS & UTIL.	96.0	42.6	68.7	99.2	55.0	76.6	96.0	56.1	75.6	96.0	39.9	67.3	103.3	56.1	79.2	100.7	57.7	78.7
MF2018	WEIGHTED AVERAGE	95.7	54.4	78.0	98.1	58.4	81.1	99.1	66.0	84.9	97.0	54.2	78.6	99.0	65.4	84.6	98.3	56.9	80.5

		MISSISSIPPI									MISSOURI								
	DIVISION	MCCOMB			MERIDIAN			TUPELO			BOWLING GREEN			CAPE GIRARDEAU			CHILLICOTHE		
		396			393			388			633			637			646		
		MAT.	INST.	TOTAL	MAT.	INST.	TOTAL	MAT.	INST.	TOTAL	MAT.	INST.	TOTAL	MAT.	INST.	TOTAL	MAT.	INST.	TOTAL
015433	CONTRACTOR EQUIPMENT		102.7	102.7		102.7	102.7		102.7	102.7		110.5	110.5		110.5	110.5		104.5	104.5
0241, 31 - 34	SITE & INFRASTRUCTURE, DEMOLITION	98.4	88.4	91.4	102.7	90.0	93.9	97.3	88.4	91.1	90.5	94.7	93.4	92.0	94.6	93.8	104.1	92.8	96.2
0310	Concrete Forming & Accessories	80.8	46.3	51.5	78.2	64.2	66.3	80.8	46.9	52.0	96.9	90.2	91.2	89.2	80.7	82.0	83.3	96.1	94.2
0320	Concrete Reinforcing	93.3	34.5	63.4	92.1	51.9	71.7	98.9	66.1	82.2	96.5	99.3	98.0	97.7	78.2	87.8	100.7	105.6	103.2
0330	Cast-in-Place Concrete	106.2	49.5	85.1	113.7	66.5	96.1	102.8	69.7	90.5	90.0	97.0	92.6	89.1	86.8	88.2	97.2	87.8	93.7
03	CONCRETE	93.1	47.5	72.6	97.9	64.6	82.9	94.0	60.2	78.8	91.9	95.7	93.6	90.9	84.3	87.9	102.0	95.6	99.1
04	MASONRY	125.1	39.6	72.3	97.3	59.6	74.0	123.0	42.7	73.3	110.9	97.4	102.5	107.5	80.2	90.6	99.2	92.9	95.3
05	METALS	90.1	74.8	85.5	91.0	84.9	89.2	92.9	86.8	91.0	96.4	118.9	103.2	97.4	109.0	101.0	84.0	111.2	92.3
06	WOOD, PLASTICS & COMPOSITES	80.6	48.1	62.9	77.8	64.6	70.6	78.0	46.5	60.9	101.3	89.8	95.1	93.1	78.7	85.3	93.9	97.5	95.9
07	THERMAL & MOISTURE PROTECTION	99.0	50.9	78.3	99.2	62.6	83.4	96.1	52.1	77.2	103.2	99.5	101.6	102.6	85.3	95.2	93.8	94.2	94.0
08	OPENINGS	98.3	42.5	85.3	98.0	56.1	88.3	97.2	52.5	86.8	100.7	97.0	99.8	100.7	75.4	94.8	87.9	97.7	90.2
0920	Plaster & Gypsum Board	96.6	47.1	64.0	96.6	64.1	75.2	93.1	45.5	61.7	97.7	89.9	92.6	97.1	78.4	84.8	98.8	97.3	97.8
0950, 0980	Ceilings & Acoustic Treatment	88.1	47.1	60.3	90.0	64.1	72.5	86.2	45.5	58.6	91.9	89.9	90.5	91.9	78.4	82.8	88.9	97.3	94.6
0960	Flooring	87.1	46.5	75.3	85.7	56.2	77.1	96.3	46.5	81.9	96.9	100.1	97.8	93.6	87.4	91.8	99.1	103.0	100.2
0970, 0990	Wall Finishes & Painting/Coating	83.0	48.1	62.4	83.0	48.1	62.4	93.8	46.4	65.8	97.0	109.1	104.1	97.0	68.8	80.3	93.8	105.9	101.0
09	FINISHES	83.1	46.5	63.1	83.4	61.2	71.3	89.4	46.1	65.8	97.4	93.5	95.3	96.3	79.5	87.1	98.1	98.8	98.5
COVERS	DIVS. 10 - 14, 25, 28, 41, 43, 44, 46	100.0	51.9	89.4	100.0	73.0	94.0	100.0	50.0	89.0	100.0	81.4	95.9	100.0	91.8	98.2	100.0	82.4	96.1
21, 22, 23	FIRE SUPPRESSION, PLUMBING & HVAC	98.7	51.8	79.9	100.0	58.7	83.5	98.9	53.8	80.8	97.6	100.4	98.7	100.0	99.8	99.9	97.7	101.5	99.2
26, 27, 3370	ELECTRICAL, COMMUNICATIONS & UTIL.	97.7	56.7	76.7	100.7	57.0	78.3	95.8	54.9	74.9	100.4	78.7	89.3	100.4	100.9	100.6	92.7	77.7	85.0
MF2018	WEIGHTED AVERAGE	96.5	54.5	78.5	96.6	65.1	83.0	97.0	58.2	80.4	98.1	96.0	97.2	98.5	91.8	95.6	94.8	95.4	95.0

For customer support on your Plumbing Costs with RSMeans Data, call 800.448.8182.

653

MISSOURI

DIVISION		COLUMBIA 652			FLAT RIVER 636			HANNIBAL 634			HARRISONVILLE 647			JEFFERSON CITY 650 - 651			JOPLIN 648		
		MAT.	INST.	TOTAL	MAT.	INST.	TOTAL	MAT.	INST.	TOTAL	MAT.	INST.	TOTAL	MAT.	INST.	TOTAL	MAT.	INST.	TOTAL
015433	CONTRACTOR EQUIPMENT		113.3	113.3		110.5	110.5		110.5	110.5		104.5	104.5		113.3	113.3		108.0	108.0
0241, 31 - 34	SITE & INFRASTRUCTURE, DEMOLITION	103.2	98.4	99.9	93.2	94.5	94.1	88.2	94.6	92.7	95.6	94.1	94.6	102.6	98.4	99.7	105.0	97.6	99.8
0310	Concrete Forming & Accessories	81.6	80.1	80.3	103.7	89.2	91.4	95.0	84.8	86.4	80.6	101.7	98.5	93.1	80.0	82.0	94.1	76.3	79.0
0320	Concrete Reinforcing	90.7	96.0	93.4	97.7	105.9	101.9	96.0	99.3	97.7	100.3	111.1	105.8	98.0	96.0	97.0	103.9	86.8	95.2
0330	Cast-in-Place Concrete	87.0	84.6	86.1	93.0	93.3	93.1	85.2	96.3	89.4	99.6	102.4	100.6	92.2	84.6	89.4	105.3	77.8	95.0
03	CONCRETE	83.8	86.3	84.9	94.9	95.1	95.0	88.2	93.1	90.4	97.8	104.2	100.8	90.8	86.2	88.7	100.7	79.8	91.3
04	MASONRY	139.0	86.5	106.6	107.9	78.0	89.4	103.0	96.6	99.0	93.9	101.5	98.6	97.7	86.5	90.8	93.0	81.9	86.1
05	METALS	100.6	116.3	105.4	96.3	120.4	103.6	96.4	118.5	103.1	84.4	114.7	93.6	100.2	116.2	105.0	86.9	99.8	90.8
06	WOOD, PLASTICS & COMPOSITES	80.9	77.6	79.1	110.7	90.0	99.4	99.3	83.5	90.7	90.4	101.6	96.5	95.8	77.6	85.9	106.0	75.7	89.5
07	THERMAL & MOISTURE PROTECTION	93.1	86.1	90.1	103.5	93.3	99.1	103.1	96.1	100.1	93.0	103.2	97.4	100.2	86.1	94.2	92.8	83.5	88.8
08	OPENINGS	96.5	82.3	93.2	100.7	99.1	100.3	100.7	86.2	97.3	87.6	104.8	91.6	96.9	82.3	93.5	88.9	78.5	86.5
0920	Plaster & Gypsum Board	84.6	77.1	79.7	104.1	90.0	94.8	97.4	83.4	88.2	94.8	101.5	99.2	93.9	77.1	82.8	105.7	74.9	85.4
0950, 0980	Ceilings & Acoustic Treatment	97.8	77.1	83.8	91.9	90.0	90.6	91.9	83.4	86.1	88.9	101.5	97.5	97.3	77.1	83.6	89.8	74.9	79.7
0960	Flooring	90.6	96.3	92.2	100.2	87.4	96.5	96.3	100.1	97.4	94.3	104.2	97.7	98.0	73.9	91.0	125.7	74.6	110.8
0970, 0990	Wall Finishes & Painting/Coating	91.6	80.2	84.9	97.0	73.6	83.2	97.0	96.6	96.7	98.1	105.6	102.5	88.4	80.2	83.6	93.4	77.9	84.2
09	FINISHES	84.9	82.2	83.4	99.4	86.3	92.3	96.9	87.8	92.0	95.7	102.6	99.4	90.8	78.3	84.0	105.1	76.6	89.6
COVERS	DIVS. 10 - 14, 25, 28, 41, 43, 44, 46	100.0	94.6	98.8	100.0	92.3	98.3	100.0	80.5	95.7	100.0	84.5	96.6	100.0	94.6	98.8	100.0	81.2	95.9
21, 22, 23	FIRE SUPPRESSION, PLUMBING & HVAC	99.9	98.1	99.2	97.6	99.0	98.1	97.6	99.9	98.5	97.6	103.0	99.8	99.9	98.1	99.2	100.1	72.7	89.1
26, 27, 3370	ELECTRICAL, COMMUNICATIONS & UTIL.	94.4	81.3	87.7	105.1	100.9	103.0	99.1	78.7	88.7	98.9	101.0	100.0	99.6	81.3	90.3	90.7	67.7	79.0
MF2018	WEIGHTED AVERAGE	97.6	91.3	94.9	99.1	96.2	97.9	97.0	94.0	95.7	94.2	102.6	97.8	97.7	90.8	94.8	95.9	79.7	89.0

MISSOURI

DIVISION		KANSAS CITY 640 - 641			KIRKSVILLE 635			POPLAR BLUFF 639			ROLLA 654 - 655			SEDALIA 653			SIKESTON 638		
		MAT.	INST.	TOTAL	MAT.	INST.	TOTAL	MAT.	INST.	TOTAL	MAT.	INST.	TOTAL	MAT.	INST.	TOTAL	MAT.	INST.	TOTAL
015433	CONTRACTOR EQUIPMENT		105.5	105.5		100.4	100.4		102.7	102.7		113.3	113.3		103.1	103.1		102.7	102.7
0241, 31 - 34	SITE & INFRASTRUCTURE, DEMOLITION	96.9	98.2	97.8	93.2	89.5	90.6	79.8	93.3	89.2	102.3	98.9	99.9	100.6	93.6	95.8	83.4	93.9	90.7
0310	Concrete Forming & Accessories	93.9	102.2	100.9	87.2	80.6	81.6	87.4	78.8	80.1	88.7	96.6	95.4	86.6	80.4	81.3	88.5	78.9	80.3
0320	Concrete Reinforcing	98.7	111.2	105.0	96.7	85.7	91.1	99.7	78.2	88.7	91.2	96.1	93.7	89.9	110.5	100.3	99.0	78.2	88.4
0330	Cast-in-Place Concrete	103.2	103.9	103.5	92.9	84.8	89.9	71.5	86.0	76.9	89.1	97.5	92.2	93.0	83.5	89.5	76.4	86.0	80.0
03	CONCRETE	100.7	104.8	102.5	107.0	84.1	96.7	81.8	82.4	82.1	85.5	98.2	91.2	99.8	87.8	94.4	85.6	82.5	84.2
04	MASONRY	99.7	103.7	102.2	114.4	86.6	97.3	105.8	75.7	87.2	112.7	87.2	96.9	119.0	83.3	96.9	105.6	75.7	87.1
05	METALS	93.5	113.4	99.5	96.0	102.1	97.8	96.6	98.7	97.2	100.0	116.9	105.2	98.7	112.2	102.8	96.9	98.8	97.5
06	WOOD, PLASTICS & COMPOSITES	101.3	101.5	101.4	85.8	78.8	82.0	84.8	78.7	81.5	88.9	98.4	94.0	82.3	79.0	80.5	86.5	78.7	82.3
07	THERMAL & MOISTURE PROTECTION	93.2	104.9	98.3	109.9	92.9	102.6	108.1	84.1	97.8	93.4	94.5	93.9	99.0	89.7	95.0	108.2	84.1	97.8
08	OPENINGS	94.4	104.8	96.8	106.2	81.1	100.4	107.2	75.4	99.8	96.5	93.7	95.9	101.7	88.8	98.7	107.2	75.4	99.8
0920	Plaster & Gypsum Board	101.3	101.5	101.5	92.2	78.5	83.2	92.6	78.4	83.3	87.0	98.4	94.5	80.6	78.5	79.2	94.4	78.4	83.9
0950, 0980	Ceilings & Acoustic Treatment	92.1	101.5	98.5	90.2	78.5	82.3	91.9	78.4	82.8	97.8	98.4	98.2	97.8	78.5	84.7	91.9	78.4	82.8
0960	Flooring	100.3	104.2	101.4	75.2	99.7	82.3	88.5	87.4	88.1	94.1	99.7	95.7	72.9	99.7	80.7	89.0	87.4	88.6
0970, 0990	Wall Finishes & Painting/Coating	96.8	105.9	102.2	92.8	80.4	85.4	92.0	68.8	78.3	91.6	92.6	92.2	91.6	105.9	100.1	92.0	68.8	78.3
09	FINISHES	99.0	102.8	101.1	96.4	83.4	89.3	95.9	78.6	86.4	86.4	96.7	92.0	83.6	85.5	84.7	96.6	78.9	86.9
COVERS	DIVS. 10 - 14, 25, 28, 41, 43, 44, 46	100.0	99.0	99.8	100.0	80.0	95.6	100.0	90.1	97.8	100.0	98.7	99.7	100.0	89.7	97.7	100.0	90.3	97.9
21, 22, 23	FIRE SUPPRESSION, PLUMBING & HVAC	100.0	103.0	101.2	97.6	99.3	98.3	97.6	97.0	97.4	97.5	100.8	98.8	97.4	97.0	97.3	97.6	97.0	97.3
26, 27, 3370	ELECTRICAL, COMMUNICATIONS & UTIL.	100.5	101.0	100.8	99.3	78.6	88.8	99.6	100.8	100.2	93.0	81.3	87.0	94.0	101.0	97.6	98.6	100.8	99.7
MF2018	WEIGHTED AVERAGE	98.2	103.6	100.5	100.6	88.8	95.5	96.9	89.2	93.6	95.9	96.6	96.2	98.2	93.7	96.3	97.4	89.4	94.0

MISSOURI / MONTANA

DIVISION		SPRINGFIELD 656 - 658			ST. JOSEPH 644 - 645			ST. LOUIS 630 - 631			BILLINGS 590 - 591			BUTTE 597			GREAT FALLS 594		
		MAT.	INST.	TOTAL	MAT.	INST.	TOTAL	MAT.	INST.	TOTAL	MAT.	INST.	TOTAL	MAT.	INST.	TOTAL	MAT.	INST.	TOTAL
015433	CONTRACTOR EQUIPMENT		105.6	105.6		104.5	104.5		111.0	111.0		100.5	100.5		100.3	100.3		100.3	100.3
0241, 31 - 34	SITE & INFRASTRUCTURE, DEMOLITION	102.8	96.1	98.1	99.0	91.9	94.1	97.6	99.2	98.7	96.6	97.5	97.2	103.4	96.3	98.4	107.2	97.2	100.2
0310	Concrete Forming & Accessories	94.3	77.5	80.0	93.0	91.5	91.8	100.6	102.6	102.3	98.3	66.4	71.2	84.9	66.5	69.3	98.2	65.7	70.6
0320	Concrete Reinforcing	87.4	95.6	91.6	97.5	110.6	104.1	89.3	107.2	98.4	90.7	82.6	85.6	98.4	81.1	89.6	90.8	80.6	85.6
0330	Cast-in-Place Concrete	94.6	75.9	87.6	97.8	98.2	97.9	100.5	103.5	101.6	115.6	69.8	98.6	127.2	69.8	105.9	134.5	69.7	110.4
03	CONCRETE	95.8	81.2	89.2	96.8	98.0	97.3	97.9	104.8	101.0	95.8	71.0	84.7	99.5	71.1	86.7	104.7	70.6	89.4
04	MASONRY	90.6	81.6	85.0	95.7	90.5	92.5	92.8	109.7	103.2	121.7	73.8	92.1	117.6	82.2	95.7	121.8	73.8	92.1
05	METALS	105.1	102.5	104.3	90.1	113.4	97.2	102.2	121.0	107.9	105.3	87.1	99.8	99.5	87.8	95.9	102.9	87.1	98.1
06	WOOD, PLASTICS & COMPOSITES	90.3	77.2	83.2	106.0	91.2	98.0	102.7	100.8	101.6	92.6	63.3	76.7	78.5	63.3	70.3	93.9	62.5	76.8
07	THERMAL & MOISTURE PROTECTION	97.4	77.7	88.9	93.4	90.5	92.1	101.0	106.0	103.1	105.6	69.7	90.2	105.4	73.3	91.6	106.1	71.0	91.0
08	OPENINGS	104.0	86.7	100.0	91.9	98.8	93.5	99.3	105.8	100.8	100.8	65.0	92.5	99.0	65.0	91.1	101.9	64.6	93.3
0920	Plaster & Gypsum Board	88.0	76.6	80.5	107.1	90.9	96.4	102.7	101.1	101.6	114.5	62.7	80.4	112.1	62.7	79.6	122.5	61.8	82.5
0950, 0980	Ceilings & Acoustic Treatment	97.8	76.6	83.5	96.6	90.9	92.7	93.6	101.1	98.7	90.6	62.7	71.7	96.5	62.7	73.6	98.2	61.8	73.5
0960	Flooring	93.0	73.9	87.5	104.0	102.4	103.5	99.3	100.1	99.5	92.9	78.0	88.6	90.8	94.0	91.7	97.5	78.0	91.8
0970, 0990	Wall Finishes & Painting/Coating	86.1	102.0	95.5	93.8	105.6	100.8	98.0	109.1	104.5	88.8	69.5	77.4	88.0	69.5	77.1	88.0	69.5	77.1
09	FINISHES	88.5	79.9	83.8	101.4	95.0	97.9	99.9	102.4	101.3	90.5	68.3	78.4	90.4	71.5	80.1	94.5	67.7	79.9
COVERS	DIVS. 10 - 14, 25, 28, 41, 43, 44, 46	100.0	92.4	98.3	100.0	96.0	99.1	100.0	102.3	100.5	100.0	90.3	97.9	100.0	90.3	97.9	100.0	90.2	97.9
21, 22, 23	FIRE SUPPRESSION, PLUMBING & HVAC	100.0	71.0	88.4	100.1	89.4	95.8	100.0	106.1	102.5	100.0	73.2	89.3	100.1	71.5	88.6	100.1	69.6	87.9
26, 27, 3370	ELECTRICAL, COMMUNICATIONS & UTIL.	98.1	70.5	84.0	99.0	77.7	88.1	103.0	100.9	101.9	96.9	69.9	83.1	103.3	70.9	86.7	96.2	69.1	82.4
MF2018	WEIGHTED AVERAGE	99.1	81.0	91.3	96.8	92.8	95.1	100.0	105.8	102.5	100.4	75.0	89.5	100.3	76.2	89.9	101.8	74.0	89.9

For customer support on your Plumbing Costs with RSMeans Data, call 800.448.8182.

MONTANA

DIVISION		HAVRE 595			HELENA 596			KALISPELL 599			MILES CITY 593			MISSOULA 598			WOLF POINT 592		
		MAT.	INST.	TOTAL	MAT.	INST.	TOTAL	MAT.	INST.	TOTAL	MAT.	INST.	TOTAL	MAT.	INST.	TOTAL	MAT.	INST.	TOTAL
015433	CONTRACTOR EQUIPMENT		100.3	100.3		100.3	100.3		100.3	100.3		100.3	100.3		100.3	100.3		100.3	100.3
0241, 31 - 34	SITE & INFRASTRUCTURE, DEMOLITION	111.0	96.4	100.8	96.8	96.3	96.4	93.1	96.2	95.3	99.7	96.3	97.3	85.4	96.0	92.8	117.7	96.4	102.8
0310	Concrete Forming & Accessories	77.6	66.4	68.1	99.4	66.5	71.5	88.2	67.6	70.7	96.7	66.6	71.1	88.2	66.5	69.7	89.0	65.9	69.4
0320	Concrete Reinforcing	99.2	81.1	90.0	104.7	79.9	92.1	101.0	85.9	93.3	98.8	81.2	89.8	100.1	85.9	92.9	100.2	80.4	90.2
0330	Cast-in-Place Concrete	137.1	69.7	112.0	99.5	69.8	88.4	110.5	69.8	95.3	120.9	69.8	101.9	93.7	69.7	84.8	135.5	69.8	111.1
03	CONCRETE	107.1	71.0	90.9	91.9	70.9	82.4	89.4	72.4	81.8	96.5	71.1	85.1	78.4	71.9	75.4	110.4	70.7	92.5
04	MASONRY	118.6	82.2	96.1	110.3	82.2	92.9	116.4	82.2	95.2	123.5	82.2	98.0	140.6	82.2	104.5	124.7	82.2	98.4
05	METALS	95.8	87.7	93.4	101.7	87.4	97.3	95.7	89.4	93.8	95.0	87.9	92.9	96.2	89.3	94.1	95.1	87.6	92.9
06	WOOD, PLASTICS & COMPOSITES	69.7	63.3	66.2	93.3	63.3	77.0	82.1	64.7	72.6	90.8	63.3	75.9	82.1	63.3	71.9	81.5	62.5	71.2
07	THERMAL & MOISTURE PROTECTION	105.8	67.3	89.2	100.4	73.3	88.8	104.9	74.0	91.6	105.3	68.8	89.6	104.5	72.8	90.8	106.4	68.7	90.1
08	OPENINGS	99.5	65.0	91.5	99.8	64.8	91.7	99.5	66.9	91.9	99.0	65.0	91.1	99.0	66.1	91.4	99.0	64.4	91.0
0920	Plaster & Gypsum Board	107.5	62.7	78.0	108.5	62.7	78.3	112.1	64.0	80.5	121.8	62.7	82.9	112.1	62.7	79.6	115.7	61.8	80.2
0950, 0980	Ceilings & Acoustic Treatment	96.5	62.7	73.6	100.6	62.7	74.9	96.5	64.0	74.5	94.8	62.7	73.0	96.5	62.7	73.6	94.8	61.8	72.4
0960	Flooring	88.4	94.0	90.0	101.6	94.0	99.4	92.5	94.0	92.9	97.4	94.0	96.4	92.5	94.0	92.9	93.9	94.0	93.9
0970, 0990	Wall Finishes & Painting/Coating	88.0	69.5	77.1	93.5	69.5	79.3	88.0	69.5	77.1	88.0	69.5	77.1	88.0	55.0	68.5	88.0	69.5	77.1
09	FINISHES	89.7	71.5	79.8	97.5	71.4	83.2	90.3	72.3	80.5	93.2	71.5	81.4	89.8	69.9	78.9	92.7	71.0	80.8
COVERS	DIVS. 10 - 14, 25, 28, 41, 43, 44, 46	100.0	90.3	97.9	100.0	90.3	97.9	100.0	90.5	97.9	100.0	90.3	97.9	100.0	90.3	97.9	100.0	90.3	97.9
21, 22, 23	FIRE SUPPRESSION, PLUMBING & HVAC	97.6	70.4	86.7	100.1	71.5	88.6	97.6	69.7	86.4	97.6	76.9	89.3	100.1	69.7	87.9	97.6	76.9	89.3
26, 27, 3370	ELECTRICAL, COMMUNICATIONS & UTIL.	96.2	70.9	83.3	102.8	70.9	86.5	100.3	69.8	84.7	96.2	76.3	86.0	101.3	69.7	85.1	96.2	76.3	86.0
MF2018	WEIGHTED AVERAGE	99.5	75.7	89.3	99.7	76.1	89.6	97.2	76.2	88.2	98.4	77.9	89.6	97.5	75.7	88.1	100.5	77.8	90.8

NEBRASKA

DIVISION		ALLIANCE 693			COLUMBUS 686			GRAND ISLAND 688			HASTINGS 689			LINCOLN 683 - 685			MCCOOK 690		
		MAT.	INST.	TOTAL	MAT.	INST.	TOTAL	MAT.	INST.	TOTAL	MAT.	INST.	TOTAL	MAT.	INST.	TOTAL	MAT.	INST.	TOTAL
015433	CONTRACTOR EQUIPMENT		97.8	97.8		104.7	104.7		104.7	104.7		104.7	104.7		104.7	104.7		104.7	104.7
0241, 31 - 34	SITE & INFRASTRUCTURE, DEMOLITION	100.9	100.8	100.8	105.6	95.0	98.2	110.6	95.2	99.9	109.3	95.0	99.3	99.0	95.2	96.4	103.9	94.9	97.7
0310	Concrete Forming & Accessories	86.6	56.6	61.1	97.5	77.0	80.1	97.1	70.7	74.6	100.5	74.5	78.4	92.5	77.4	79.7	92.1	57.3	62.6
0320	Concrete Reinforcing	110.9	89.9	100.2	100.7	88.4	94.5	100.1	78.3	89.0	100.1	78.3	89.1	101.2	78.6	89.7	103.3	78.5	90.7
0330	Cast-in-Place Concrete	108.2	83.9	99.1	109.4	84.0	99.9	115.8	79.7	102.4	115.8	75.9	101.0	89.3	84.3	87.4	117.1	75.9	101.7
03	CONCRETE	116.4	72.5	96.7	101.3	82.5	92.8	106.1	76.4	92.7	106.3	76.8	93.0	88.5	81.1	85.1	106.0	69.1	89.3
04	MASONRY	107.1	79.0	89.7	112.2	78.9	91.7	105.0	75.3	86.7	113.4	75.4	89.9	92.5	79.8	84.6	102.9	78.9	88.1
05	METALS	101.0	85.1	96.1	95.9	98.2	96.6	97.7	94.1	96.6	98.6	93.9	97.2	99.6	94.6	98.1	95.7	94.0	95.2
06	WOOD, PLASTICS & COMPOSITES	84.6	50.1	65.9	96.1	77.0	85.7	95.2	68.2	80.5	99.1	74.2	85.6	96.3	77.0	85.8	92.8	51.1	70.1
07	THERMAL & MOISTURE PROTECTION	104.3	67.7	88.5	103.4	81.7	94.1	103.6	79.6	93.3	103.6	79.0	93.0	100.4	81.8	92.4	99.3	77.4	89.9
08	OPENINGS	92.8	59.8	85.1	93.2	75.9	89.2	93.2	68.7	87.5	93.2	71.8	88.2	102.5	70.8	95.1	93.3	57.4	84.9
0920	Plaster & Gypsum Board	80.1	48.9	59.5	94.5	76.5	82.6	93.6	67.4	76.3	95.4	73.5	81.0	103.7	76.5	85.8	90.6	49.8	63.8
0950, 0980	Ceilings & Acoustic Treatment	93.7	48.9	63.3	88.3	76.5	80.3	88.3	67.4	74.1	88.3	73.5	78.3	92.2	76.5	81.6	89.7	49.8	62.7
0960	Flooring	92.6	90.3	92.0	85.8	90.3	87.1	85.5	81.5	84.4	86.8	84.0	86.0	99.5	89.5	96.6	91.1	90.3	90.9
0970, 0990	Wall Finishes & Painting/Coating	157.3	53.0	95.6	75.6	61.7	67.3	75.6	64.1	68.8	75.6	61.7	67.3	97.7	81.4	88.1	88.7	46.2	63.5
09	FINISHES	91.7	60.9	74.9	84.8	77.8	81.0	84.9	71.4	77.5	85.5	74.8	79.7	94.5	80.1	86.6	89.1	60.8	73.6
COVERS	DIVS. 10 - 14, 25, 28, 41, 43, 44, 46	100.0	85.6	96.8	100.0	85.2	96.7	100.0	87.8	97.3	100.0	84.9	96.7	100.0	88.7	97.5	100.0	85.8	96.9
21, 22, 23	FIRE SUPPRESSION, PLUMBING & HVAC	97.6	76.7	89.2	97.6	77.0	89.3	100.0	81.3	92.5	97.6	76.2	89.0	99.9	81.3	92.5	97.5	76.9	89.2
26, 27, 3370	ELECTRICAL, COMMUNICATIONS & UTIL.	92.4	67.2	79.5	95.1	83.4	89.1	93.7	67.3	80.2	93.1	81.4	87.1	104.4	67.3	85.4	94.4	67.3	80.5
MF2018	WEIGHTED AVERAGE	99.8	74.7	89.0	97.3	82.7	91.0	98.4	78.5	89.9	98.4	80.0	90.5	98.3	81.1	90.9	97.5	74.8	87.8

NEBRASKA / NEVADA

DIVISION		NORFOLK 687			NORTH PLATTE 691			OMAHA 680 - 681			VALENTINE 692			CARSON CITY 897			ELKO 898		
		MAT.	INST.	TOTAL	MAT.	INST.	TOTAL	MAT.	INST.	TOTAL	MAT.	INST.	TOTAL	MAT.	INST.	TOTAL	MAT.	INST.	TOTAL
015433	CONTRACTOR EQUIPMENT		94.1	94.1		104.7	104.7		94.1	94.1		97.6	97.6		97.6	97.6		97.6	97.6
0241, 31 - 34	SITE & INFRASTRUCTURE, DEMOLITION	86.5	94.1	91.8	105.1	94.7	97.9	90.3	94.3	93.1	88.1	99.7	96.2	86.4	97.7	94.3	69.7	96.3	88.2
0310	Concrete Forming & Accessories	83.4	76.1	77.2	94.6	77.2	79.8	97.1	75.5	78.7	83.0	54.4	58.7	105.7	78.3	82.4	110.8	99.1	100.8
0320	Concrete Reinforcing	100.8	68.4	84.3	102.8	78.4	90.4	97.7	78.5	88.0	103.3	68.2	85.5	108.1	119.5	113.9	116.1	117.9	117.0
0330	Cast-in-Place Concrete	109.9	73.1	96.2	117.1	62.0	96.6	92.6	79.8	87.9	103.2	55.3	85.4	97.0	83.6	92.0	93.9	74.7	86.7
03	CONCRETE	99.8	74.2	88.3	106.1	73.2	91.3	90.2	78.0	84.7	103.2	58.2	82.9	96.0	87.3	92.1	93.7	93.4	93.6
04	MASONRY	117.8	78.9	93.8	91.1	78.4	83.2	95.6	80.2	86.1	102.5	78.3	87.6	118.0	70.2	88.5	125.0	69.1	90.5
05	METALS	99.4	80.4	93.7	94.8	93.2	94.4	99.6	84.6	95.1	107.1	79.2	98.6	109.6	95.7	105.4	113.1	94.7	107.5
06	WOOD, PLASTICS & COMPOSITES	80.0	76.6	78.1	94.8	79.1	86.3	94.5	74.5	83.6	79.4	48.8	62.8	92.3	76.4	83.7	105.4	105.3	105.3
07	THERMAL & MOISTURE PROTECTION	103.2	79.7	93.1	99.2	79.5	90.7	99.3	80.5	91.2	100.0	74.9	89.2	108.6	79.7	96.2	104.8	75.7	92.3
08	OPENINGS	94.6	70.7	89.1	92.6	74.2	88.3	100.6	75.3	94.7	94.8	55.0	85.6	99.4	79.7	94.8	104.5	95.3	102.4
0920	Plaster & Gypsum Board	94.3	76.5	82.6	90.6	78.6	82.7	101.6	74.3	83.6	91.9	47.9	62.9	103.2	75.8	85.2	109.5	105.5	106.9
0950, 0980	Ceilings & Acoustic Treatment	103.1	76.5	85.1	89.7	78.6	82.2	104.1	74.3	83.9	106.1	47.9	66.7	102.3	75.8	84.4	102.7	105.5	104.6
0960	Flooring	106.9	90.3	102.1	92.2	81.5	89.1	101.8	90.3	98.4	118.5	90.3	110.3	97.8	69.4	89.6	101.1	69.4	91.9
0970, 0990	Wall Finishes & Painting/Coating	132.0	61.7	90.4	88.7	60.4	72.0	101.1	74.0	85.1	157.4	62.6	101.3	96.7	81.4	87.7	93.9	81.4	86.5
09	FINISHES	101.4	77.4	88.3	89.4	76.9	82.6	98.0	77.8	87.0	109.7	61.0	83.1	95.2	76.1	84.8	93.2	93.1	93.1
COVERS	DIVS. 10 - 14, 25, 28, 41, 43, 44, 46	100.0	84.8	96.7	100.0	62.4	91.7	100.0	87.5	97.2	100.0	57.2	90.6	100.0	102.0	100.4	100.0	92.9	98.4
21, 22, 23	FIRE SUPPRESSION, PLUMBING & HVAC	97.3	76.0	88.7	99.9	75.3	90.1	99.9	76.8	90.6	97.1	75.2	88.3	100.1	78.7	91.5	98.8	78.7	90.7
26, 27, 3370	ELECTRICAL, COMMUNICATIONS & UTIL.	94.0	83.4	88.6	92.6	67.3	79.7	103.3	83.4	93.1	90.0	67.3	78.4	99.6	90.4	94.9	96.9	90.4	93.6
MF2018	WEIGHTED AVERAGE	98.6	79.2	90.3	97.2	77.3	88.7	98.4	80.8	90.9	99.7	70.8	87.3	101.3	84.2	93.9	101.3	87.4	95.3

For customer support on your Plumbing Costs with RSMeans Data, call 800.448.8182.

655

City Cost Indexes

Table 1

DIVISION		NEVADA — ELY 893 MAT.	INST.	TOTAL	LAS VEGAS 889-891 MAT.	INST.	TOTAL	RENO 894-895 MAT.	INST.	TOTAL	NEW HAMPSHIRE — CHARLESTON 036 MAT.	INST.	TOTAL	CLAREMONT 037 MAT.	INST.	TOTAL	CONCORD 032-033 MAT.	INST.	TOTAL
015433	CONTRACTOR EQUIPMENT		97.6	97.6		97.6	97.6		97.6	97.6		97.9	97.9		97.9	97.9		97.9	97.9
0241, 31 - 34	SITE & INFRASTRUCTURE, DEMOLITION	75.5	97.5	90.8	78.9	100.1	93.7	75.3	97.7	90.9	83.5	100.4	95.3	77.6	100.4	93.5	90.8	100.6	97.6
0310	Concrete Forming & Accessories	103.7	104.5	104.3	104.4	107.0	106.6	99.8	78.3	81.6	88.1	85.0	85.4	94.4	85.0	86.4	99.5	94.5	95.3
0320	Concrete Reinforcing	114.8	118.1	116.5	105.8	122.0	114.0	108.5	120.9	114.8	89.3	86.2	87.7	89.3	86.2	87.7	104.0	86.6	95.2
0330	Cast-in-Place Concrete	100.8	98.1	99.8	97.6	107.9	101.4	106.6	83.6	98.0	86.1	113.0	96.1	79.1	113.0	91.7	105.0	113.8	108.3
03	CONCRETE	101.9	104.1	102.9	97.2	109.3	102.7	101.6	87.6	95.3	87.2	95.0	90.7	80.2	95.0	86.9	96.1	99.7	97.7
04	MASONRY	130.4	76.4	97.0	117.2	97.1	104.8	124.1	70.2	90.8	93.8	94.8	94.4	94.2	94.8	94.6	101.1	101.3	101.2
05	METALS	113.1	97.6	108.4	122.0	101.3	115.7	114.8	96.2	109.2	85.1	89.5	86.5	85.1	89.5	86.5	91.5	90.4	91.2
06	WOOD, PLASTICS & COMPOSITES	95.2	107.5	101.9	93.0	105.3	99.7	88.9	76.4	82.1	89.4	81.7	85.2	95.8	81.7	88.1	93.8	93.6	93.7
07	THERMAL & MOISTURE PROTECTION	105.4	94.8	100.8	118.9	98.9	110.3	104.8	79.7	94.0	104.9	104.5	104.8	104.7	104.5	104.6	111.4	109.4	110.5
08	OPENINGS	104.5	96.5	102.6	103.4	110.0	105.0	102.2	80.0	97.1	98.4	81.8	94.5	99.5	81.8	95.4	94.4	88.4	93.0
0920	Plaster & Gypsum Board	104.1	107.7	106.5	97.1	105.5	102.6	92.6	75.8	81.5	108.0	80.8	90.1	109.2	80.8	90.5	107.7	93.1	98.1
0950, 0980	Ceilings & Acoustic Treatment	102.7	107.7	106.1	108.6	105.5	106.5	106.1	75.8	85.6	101.6	80.8	87.5	101.6	80.8	87.5	101.7	93.1	95.8
0960	Flooring	98.8	69.4	90.2	90.5	99.9	93.2	95.5	69.4	88.0	83.7	104.3	89.7	86.0	104.3	91.3	93.2	104.3	96.5
0970, 0990	Wall Finishes & Painting/Coating	93.9	115.0	106.4	96.4	115.0	107.4	93.9	81.4	86.5	83.3	92.4	88.7	83.3	92.4	88.7	91.1	92.4	91.8
09	FINISHES	92.2	99.8	96.4	90.1	106.5	99.1	90.1	76.1	82.5	86.6	88.5	87.6	87.0	88.5	87.8	90.1	95.8	93.2
COVERS	DIVS. 10 - 14, 25, 28, 41, 43, 44, 46	100.0	72.3	93.9	100.0	101.8	100.4	100.0	102.0	100.4	100.0	90.3	97.9	100.0	90.3	97.9	100.0	104.6	101.0
21, 22, 23	FIRE SUPPRESSION, PLUMBING & HVAC	98.8	98.3	98.6	100.1	101.3	100.6	100.0	78.7	91.5	97.6	83.9	92.1	97.6	83.9	92.1	100.0	86.3	94.5
26, 27, 3370	ELECTRICAL, COMMUNICATIONS & UTIL.	97.1	95.2	96.2	101.3	110.4	106.0	97.5	90.4	93.9	93.7	77.1	85.2	93.7	77.1	85.2	95.4	77.1	86.1
MF2018	WEIGHTED AVERAGE	102.6	95.6	99.6	103.7	104.1	103.9	102.4	84.2	94.6	92.9	88.6	91.1	92.1	88.6	90.6	96.4	92.4	94.7

Table 2

DIVISION		NEW HAMPSHIRE — KEENE 034 MAT.	INST.	TOTAL	LITTLETON 035 MAT.	INST.	TOTAL	MANCHESTER 031 MAT.	INST.	TOTAL	NASHUA 030 MAT.	INST.	TOTAL	PORTSMOUTH 038 MAT.	INST.	TOTAL	NEW JERSEY — ATLANTIC CITY 082, 084 MAT.	INST.	TOTAL
015433	CONTRACTOR EQUIPMENT		97.9	97.9		97.9	97.9		97.9	97.9		97.9	97.9		97.9	97.9		95.4	95.4
0241, 31 - 34	SITE & INFRASTRUCTURE, DEMOLITION	90.5	100.4	97.4	77.8	99.3	92.8	88.6	100.6	96.9	92.0	100.5	97.9	85.7	101.0	96.3	88.7	104.0	99.3
0310	Concrete Forming & Accessories	92.9	85.2	86.3	107.0	79.0	83.2	98.9	94.7	95.4	101.8	94.7	95.7	89.7	94.2	93.5	113.8	141.1	137.0
0320	Concrete Reinforcing	89.3	86.2	87.7	90.0	86.2	88.1	104.4	86.6	95.4	111.5	86.6	98.9	89.3	86.6	87.9	80.2	141.7	111.4
0330	Cast-in-Place Concrete	86.5	113.1	96.4	77.6	104.0	87.4	98.0	116.0	104.7	81.8	115.9	94.5	77.6	115.8	91.8	81.4	139.4	103.0
03	CONCRETE	87.0	95.1	90.6	79.9	89.2	84.0	92.8	100.5	96.3	87.0	100.4	93.0	79.7	100.3	89.0	84.3	139.0	108.9
04	MASONRY	96.4	94.8	95.4	106.4	79.0	89.5	99.9	101.3	100.8	98.9	101.3	100.4	94.4	100.5	98.1	105.2	137.8	125.3
05	METALS	85.8	89.7	87.0	85.8	89.5	86.9	92.3	90.7	91.8	91.1	90.5	90.9	87.3	91.4	88.6	92.9	114.5	99.5
06	WOOD, PLASTICS & COMPOSITES	94.0	81.7	87.3	105.9	81.7	92.8	93.9	93.6	93.7	104.1	93.6	98.4	90.7	93.6	92.2	122.6	141.9	133.1
07	THERMAL & MOISTURE PROTECTION	105.5	104.5	105.1	104.9	97.3	101.6	107.1	109.7	108.2	106.0	109.7	107.6	105.5	108.5	106.8	100.7	134.5	115.2
08	OPENINGS	97.0	85.4	94.3	100.4	81.8	96.1	99.3	92.0	97.6	101.4	91.3	99.1	102.0	81.7	97.3	98.7	137.3	107.7
0920	Plaster & Gypsum Board	108.3	80.8	90.2	123.2	80.8	95.3	105.3	93.1	97.2	117.7	93.1	101.5	108.0	93.1	98.2	117.5	142.7	134.1
0950, 0980	Ceilings & Acoustic Treatment	101.6	80.8	87.5	101.6	80.8	87.5	102.6	93.1	96.1	113.6	93.1	99.7	102.6	93.1	96.1	90.9	142.7	126.0
0960	Flooring	85.7	104.3	91.1	96.0	103.2	98.1	94.1	108.6	98.3	89.3	104.3	93.7	83.8	104.3	89.8	92.5	164.3	113.4
0970, 0990	Wall Finishes & Painting/Coating	83.3	106.1	96.8	83.3	92.4	88.7	94.3	106.1	101.3	83.3	104.9	96.1	83.3	104.9	96.1	81.1	145.5	119.2
09	FINISHES	88.4	90.0	89.3	91.9	84.3	87.8	91.9	98.2	95.3	93.1	97.1	95.3	87.5	96.9	92.6	87.7	147.1	120.1
COVERS	DIVS. 10 - 14, 25, 28, 41, 43, 44, 46	100.0	96.6	99.3	100.0	97.5	99.5	100.0	104.6	101.0	100.0	104.6	101.0	100.0	104.3	100.9	100.0	116.6	103.7
21, 22, 23	FIRE SUPPRESSION, PLUMBING & HVAC	97.6	84.0	92.2	97.6	75.7	88.8	100.0	86.3	94.5	100.1	86.3	94.6	100.1	85.7	94.3	99.8	129.2	111.6
26, 27, 3370	ELECTRICAL, COMMUNICATIONS & UTIL.	93.7	77.1	85.2	94.6	50.3	72.0	93.5	77.1	85.1	95.8	77.1	86.2	94.1	77.1	85.4	93.4	143.6	119.1
MF2018	WEIGHTED AVERAGE	93.4	89.2	91.6	93.4	80.1	87.7	96.4	93.0	94.9	96.1	92.8	94.7	93.5	92.2	93.0	95.2	132.6	111.3

Table 3

DIVISION		NEW JERSEY — CAMDEN 081 MAT.	INST.	TOTAL	DOVER 078 MAT.	INST.	TOTAL	ELIZABETH 072 MAT.	INST.	TOTAL	HACKENSACK 076 MAT.	INST.	TOTAL	JERSEY CITY 073 MAT.	INST.	TOTAL	LONG BRANCH 077 MAT.	INST.	TOTAL
015433	CONTRACTOR EQUIPMENT		95.4	95.4		97.9	97.9		97.9	97.9		97.9	97.9		95.4	95.4		95.1	95.1
0241, 31 - 34	SITE & INFRASTRUCTURE, DEMOLITION	89.8	104.0	99.7	105.9	105.9	105.9	110.8	105.9	107.4	106.9	106.1	106.3	97.3	106.0	103.4	102.2	105.2	104.3
0310	Concrete Forming & Accessories	103.9	141.0	135.4	100.9	150.2	142.7	114.1	150.2	144.8	100.9	150.1	142.7	105.9	150.2	143.6	106.6	142.7	137.3
0320	Concrete Reinforcing	105.2	132.2	118.9	81.2	151.8	117.1	81.2	151.8	117.1	81.2	151.8	117.1	105.3	151.8	129.0	81.2	151.4	116.9
0330	Cast-in-Place Concrete	78.9	137.5	100.7	82.4	134.8	101.9	70.8	138.0	95.8	80.6	137.9	101.9	64.3	134.9	90.6	71.5	137.5	96.1
03	CONCRETE	84.6	136.6	108.0	83.2	143.5	110.3	80.0	144.6	109.0	81.7	144.5	110.0	77.8	143.3	107.3	81.4	140.7	108.1
04	MASONRY	96.5	137.8	122.0	90.5	148.2	126.1	107.5	148.2	132.6	94.4	148.2	127.6	86.7	148.2	124.7	100.0	138.3	123.7
05	METALS	98.3	111.1	102.2	85.8	123.3	97.2	87.3	123.3	98.2	85.9	123.2	97.2	91.2	120.3	100.1	85.9	119.2	96.0
06	WOOD, PLASTICS & COMPOSITES	109.9	141.9	127.2	98.6	150.2	126.6	114.4	150.2	133.9	98.6	150.2	126.6	99.5	150.2	127.1	100.6	143.3	123.8
07	THERMAL & MOISTURE PROTECTION	100.5	135.5	115.6	111.2	145.5	126.0	111.5	146.0	126.3	111.0	138.4	122.8	110.7	145.5	125.7	110.9	132.4	120.1
08	OPENINGS	100.8	135.1	108.8	104.1	145.7	113.8	102.5	145.7	112.5	101.9	145.7	112.1	100.5	145.7	111.0	96.5	141.2	106.9
0920	Plaster & Gypsum Board	112.5	142.7	132.4	108.5	151.3	136.7	116.1	151.3	139.3	108.5	151.3	136.7	111.3	151.3	137.6	110.3	144.3	132.7
0950, 0980	Ceilings & Acoustic Treatment	99.5	142.7	128.8	87.3	151.3	130.7	89.2	151.3	131.3	87.3	151.3	130.7	96.9	151.3	133.8	87.3	144.3	125.9
0960	Flooring	88.9	164.3	110.8	81.5	190.7	113.2	86.3	190.7	116.6	81.5	190.7	113.2	82.5	190.7	113.9	82.7	182.9	111.8
0970, 0990	Wall Finishes & Painting/Coating	81.1	145.5	119.2	83.0	145.5	120.0	83.0	145.5	120.0	83.0	145.5	120.0	83.1	145.5	120.1	83.1	145.5	120.1
09	FINISHES	87.4	147.1	120.0	84.4	157.8	124.4	87.8	157.8	126.0	84.2	157.8	124.4	86.7	158.1	125.7	85.3	151.3	121.3
COVERS	DIVS. 10 - 14, 25, 28, 41, 43, 44, 46	100.0	116.6	103.7	100.0	116.6	103.7	100.0	133.7	107.4	100.0	133.8	107.4	100.0	133.8	107.4	100.0	116.6	103.7
21, 22, 23	FIRE SUPPRESSION, PLUMBING & HVAC	99.9	129.1	111.6	99.9	138.3	115.3	100.1	137.9	115.2	100.1	138.3	115.3	100.1	138.3	115.4	99.9	134.0	113.6
26, 27, 3370	ELECTRICAL, COMMUNICATIONS & UTIL.	97.7	135.3	116.9	98.5	141.9	120.7	99.1	141.9	121.0	98.5	142.5	121.0	103.4	142.5	123.4	98.1	133.1	116.0
MF2018	WEIGHTED AVERAGE	96.3	130.7	111.1	94.6	139.6	113.9	95.7	139.7	114.6	94.4	139.6	113.8	94.8	139.4	113.9	94.0	133.6	111.0

		NEW JERSEY																
		NEW BRUNSWICK			NEWARK			PATERSON			POINT PLEASANT			SUMMIT			TRENTON		
DIVISION		088 - 089			070 - 071			074 - 075			087			079			085 - 086		
		MAT.	INST.	TOTAL	MAT.	INST.	TOTAL	MAT.	INST.	TOTAL	MAT.	INST.	TOTAL	MAT.	INST.	TOTAL	MAT.	INST.	TOTAL
015433	CONTRACTOR EQUIPMENT		95.1	95.1		97.9	97.9		97.9	97.9		95.1	95.1		97.9	97.9		95.1	95.1
0241, 31 - 34	SITE & INFRASTRUCTURE, DEMOLITION	101.3	105.6	104.3	110.5	105.9	107.3	108.8	106.1	106.9	102.8	105.2	104.5	108.3	105.9	106.6	88.2	105.2	100.0
0310	Concrete Forming & Accessories	107.4	149.9	143.5	104.9	150.2	143.4	102.8	150.1	143.0	101.2	142.7	136.4	103.8	150.2	143.2	102.8	142.5	136.5
0320	Concrete Reinforcing	81.1	151.5	116.9	103.6	151.8	128.1	105.3	151.8	129.0	81.1	151.4	116.8	81.2	151.8	117.1	106.5	119.6	113.1
0330	Cast-in-Place Concrete	100.7	140.4	115.4	91.0	138.0	108.5	82.0	137.9	102.8	100.7	139.6	115.1	68.5	138.0	94.3	99.1	137.4	113.3
03	CONCRETE	100.0	145.0	120.3	90.4	144.6	114.8	86.0	144.5	112.3	99.6	141.4	118.4	77.3	144.6	107.6	94.4	135.2	112.8
04	MASONRY	104.4	143.1	128.3	93.1	148.2	127.2	90.9	148.2	126.3	93.9	138.3	121.3	93.5	148.2	127.3	97.1	138.3	122.6
05	METALS	93.0	119.6	101.1	92.9	123.3	102.1	87.1	123.2	98.0	93.0	119.2	100.9	85.8	123.3	97.2	98.0	108.1	101.1
06	WOOD, PLASTICS & COMPOSITES	115.3	150.2	134.3	99.4	150.2	127.0	101.3	150.2	127.9	107.0	143.3	126.7	102.5	150.2	128.4	99.0	143.3	123.1
07	THERMAL & MOISTURE PROTECTION	101.0	141.6	118.5	112.8	146.0	127.1	111.3	138.4	123.0	101.0	134.8	115.5	111.7	146.0	126.4	101.5	136.5	116.6
08	OPENINGS	93.8	145.7	105.9	100.3	145.7	110.8	107.2	145.7	116.1	95.6	141.2	106.2	108.3	145.7	117.0	100.0	133.8	107.8
0920	Plaster & Gypsum Board	114.8	151.3	138.8	104.9	151.3	135.4	111.3	151.3	137.6	109.6	144.3	132.4	110.3	151.3	137.3	103.6	144.3	130.4
0950, 0980	Ceilings & Acoustic Treatment	90.9	151.3	131.8	95.8	151.3	133.4	96.9	151.3	133.8	90.9	144.3	127.1	87.3	151.3	130.7	94.9	144.3	128.4
0960	Flooring	90.2	190.7	119.4	94.0	190.7	122.1	82.5	190.7	113.9	87.8	164.3	110.0	82.7	190.7	114.0	94.9	182.9	120.4
0970, 0990	Wall Finishes & Painting/Coating	81.1	145.5	119.2	88.0	145.5	122.0	83.0	145.5	120.0	81.1	145.5	119.2	83.0	145.5	120.0	88.5	145.5	122.2
09	FINISHES	87.7	157.7	125.9	90.4	157.8	127.2	86.8	157.8	125.5	86.2	148.1	120.0	85.3	157.8	124.9	89.7	151.3	123.3
COVERS	DIVS. 10 - 14, 25, 28, 41, 43, 44, 46	100.0	133.6	107.4	100.0	133.8	107.4	100.0	133.8	107.4	100.0	113.9	103.1	100.0	133.7	107.4	100.0	116.8	103.7
21, 22, 23	FIRE SUPPRESSION, PLUMBING & HVAC	99.8	136.5	114.5	100.1	137.9	115.2	100.1	138.3	115.4	99.8	134.0	113.5	99.9	137.9	115.1	100.1	133.7	113.6
26, 27, 3370	ELECTRICAL, COMMUNICATIONS & UTIL.	94.0	137.0	116.0	107.7	142.5	125.5	103.4	141.9	123.1	93.4	133.1	113.7	99.1	141.9	121.0	101.6	131.8	117.0
MF2018	WEIGHTED AVERAGE	97.0	137.7	114.5	98.0	139.7	115.9	96.3	139.5	114.9	96.4	133.3	112.2	94.7	139.7	114.0	97.9	131.4	112.3

		NEW JERSEY			NEW MEXICO														
		VINELAND			ALBUQUERQUE			CARRIZOZO			CLOVIS			FARMINGTON			GALLUP		
DIVISION		080, 083			870 - 872			883			881			874			873		
		MAT.	INST.	TOTAL	MAT.	INST.	TOTAL	MAT.	INST.	TOTAL	MAT.	INST.	TOTAL	MAT.	INST.	TOTAL	MAT.	INST.	TOTAL
015433	CONTRACTOR EQUIPMENT		95.4	95.4		110.9	110.9		110.9	110.9		110.9	110.9		110.9	110.9		110.9	110.9
0241, 31 - 34	SITE & INFRASTRUCTURE, DEMOLITION	92.9	104.0	100.6	90.3	102.3	98.6	108.6	102.3	104.2	96.1	102.3	100.4	96.7	102.3	100.6	105.5	102.3	103.3
0310	Concrete Forming & Accessories	98.1	141.2	134.7	100.2	66.9	71.9	97.0	66.9	71.4	96.9	66.8	71.3	100.2	66.9	71.9	100.3	66.9	71.9
0320	Concrete Reinforcing	80.2	135.0	108.0	108.9	71.9	90.1	113.5	71.9	92.3	114.7	71.9	92.9	119.1	71.9	95.1	113.9	71.9	92.6
0330	Cast-in-Place Concrete	87.8	137.6	106.3	91.4	69.1	83.1	94.8	69.1	85.2	94.7	69.1	85.2	92.3	69.1	83.7	86.8	69.1	80.2
03	CONCRETE	88.5	137.3	110.4	93.9	69.8	83.1	115.0	69.8	94.7	103.3	69.8	88.2	97.4	69.8	85.0	103.6	69.8	88.4
04	MASONRY	93.5	138.3	121.2	105.7	58.7	76.7	106.9	58.7	77.1	106.9	58.7	77.1	114.3	58.7	80.0	100.9	58.7	74.9
05	METALS	92.9	112.7	98.9	110.1	90.6	104.2	105.7	90.6	101.1	105.3	90.5	100.8	107.8	90.6	102.5	106.9	90.6	101.9
06	WOOD, PLASTICS & COMPOSITES	103.5	141.9	124.3	104.0	68.8	84.9	94.6	68.8	80.6	94.6	68.8	80.6	104.1	68.8	84.9	104.1	68.8	84.9
07	THERMAL & MOISTURE PROTECTION	100.4	134.7	115.2	101.3	70.0	87.8	100.1	70.0	87.1	98.8	70.0	86.4	101.6	70.0	88.0	102.8	70.0	88.7
08	OPENINGS	95.2	136.1	104.7	100.9	67.7	93.2	98.7	67.7	91.5	98.8	67.7	91.6	103.3	67.7	95.1	103.4	67.7	95.1
0920	Plaster & Gypsum Board	108.1	142.7	130.9	107.8	67.7	81.4	80.9	67.7	72.2	80.9	67.7	72.2	98.8	67.7	78.3	98.8	67.7	78.3
0950, 0980	Ceilings & Acoustic Treatment	90.9	142.7	126.0	103.0	67.7	79.1	103.8	67.7	79.4	103.8	67.7	79.4	102.6	67.7	79.0	102.6	67.7	79.0
0960	Flooring	86.9	164.3	109.4	87.8	68.0	82.0	97.5	68.0	88.9	97.5	68.0	88.9	89.2	68.0	83.1	89.2	68.0	83.1
0970, 0990	Wall Finishes & Painting/Coating	81.1	145.5	119.2	96.1	51.9	69.9	93.4	51.9	68.8	93.4	51.9	68.8	90.8	51.9	67.7	90.8	51.9	67.7
09	FINISHES	85.0	147.2	119.0	89.8	65.5	76.5	92.9	65.5	78.0	91.7	65.5	77.4	88.8	65.5	76.1	90.0	65.5	76.6
COVERS	DIVS. 10 - 14, 25, 28, 41, 43, 44, 46	100.0	116.8	103.7	100.0	85.1	96.7	100.0	85.1	96.7	100.0	85.1	96.7	100.0	85.1	96.7	100.0	85.1	96.7
21, 22, 23	FIRE SUPPRESSION, PLUMBING & HVAC	99.8	129.4	111.7	100.1	68.7	87.5	98.4	68.7	86.5	98.4	68.4	86.4	99.9	68.7	87.4	98.4	68.7	86.5
26, 27, 3370	ELECTRICAL, COMMUNICATIONS & UTIL.	93.4	143.6	119.1	89.5	85.8	87.6	90.8	85.8	88.3	88.6	85.8	87.2	87.5	85.8	86.7	86.8	85.8	86.3
MF2018	WEIGHTED AVERAGE	94.6	132.3	110.8	99.2	75.0	88.8	101.3	75.0	90.0	99.1	74.9	88.7	99.8	75.0	89.2	99.7	75.0	89.1

		NEW MEXICO																	
		LAS CRUCES			LAS VEGAS			ROSWELL			SANTA FE			SOCORRO			TRUTH/CONSEQUENCES		
DIVISION		880			877			882			875			878			879		
		MAT.	INST.	TOTAL	MAT.	INST.	TOTAL	MAT.	INST.	TOTAL	MAT.	INST.	TOTAL	MAT.	INST.	TOTAL	MAT.	INST.	TOTAL
015433	CONTRACTOR EQUIPMENT		86.2	86.2		110.9	110.9		110.9	110.9		110.9	110.9		110.9	110.9		86.2	86.2
0241, 31 - 34	SITE & INFRASTRUCTURE, DEMOLITION	96.8	81.1	85.9	95.9	102.3	100.4	98.3	102.3	101.1	100.0	102.3	101.6	92.3	102.3	99.3	111.7	81.2	90.4
0310	Concrete Forming & Accessories	93.8	65.8	70.0	100.2	66.9	71.9	97.0	66.9	71.4	99.1	66.9	71.8	100.3	66.9	71.9	98.2	65.8	70.7
0320	Concrete Reinforcing	110.5	71.7	90.8	115.9	71.9	93.5	114.7	71.9	92.9	108.3	71.9	89.8	118.2	71.9	94.6	111.1	71.7	91.1
0330	Cast-in-Place Concrete	89.5	61.9	79.2	89.7	69.1	82.1	94.7	69.1	85.2	95.3	69.1	85.5	87.9	69.1	80.9	96.4	61.9	83.5
03	CONCRETE	82.0	66.4	75.0	95.0	69.8	83.7	104.1	69.8	88.7	94.2	69.8	83.2	93.9	69.8	83.1	87.8	66.4	78.2
04	MASONRY	102.3	58.4	75.2	101.2	58.7	75.0	118.1	58.7	81.4	94.3	58.7	72.3	101.1	58.7	74.9	106.4	58.4	73.6
05	METALS	104.1	82.8	97.6	106.6	90.6	101.7	106.6	90.6	101.7	103.6	90.6	99.7	106.9	90.6	101.9	106.4	82.8	99.3
06	WOOD, PLASTICS & COMPOSITES	83.9	67.8	75.1	104.1	68.8	84.9	94.6	68.8	80.6	99.1	68.8	82.7	104.1	68.8	84.9	95.4	67.8	80.4
07	THERMAL & MOISTURE PROTECTION	86.2	65.3	77.2	101.1	70.0	87.7	99.0	70.0	86.5	104.0	70.0	89.3	101.1	70.0	87.7	89.0	65.3	78.8
08	OPENINGS	94.1	67.1	87.9	99.8	67.7	92.3	98.6	67.7	91.5	100.2	67.7	92.6	99.6	67.7	92.2	92.9	67.1	86.9
0920	Plaster & Gypsum Board	79.0	67.7	71.6	98.8	67.7	78.3	80.9	67.7	72.2	111.2	67.7	82.6	98.8	67.7	78.3	99.7	67.7	78.7
0950, 0980	Ceilings & Acoustic Treatment	87.0	67.7	73.9	102.6	67.7	79.0	103.8	67.7	79.4	101.4	67.7	78.6	102.6	67.7	79.0	86.9	67.7	73.9
0960	Flooring	127.8	68.0	110.4	89.2	68.0	83.1	97.5	68.0	88.9	99.4	68.0	90.3	89.2	68.0	83.1	117.2	68.0	103.0
0970, 0990	Wall Finishes & Painting/Coating	82.6	51.9	64.4	90.8	51.9	67.7	93.4	51.9	68.8	98.2	51.9	70.8	90.8	51.9	67.7	83.3	51.9	64.7
09	FINISHES	100.5	64.7	81.0	88.6	65.5	76.0	91.8	65.5	77.4	96.1	65.5	79.4	88.6	65.5	76.0	98.7	64.7	80.2
COVERS	DIVS. 10 - 14, 25, 28, 41, 43, 44, 46	100.0	82.6	96.2	100.0	85.1	96.7	100.0	85.1	96.7	100.0	85.1	96.7	100.0	85.1	96.7	100.0	82.6	96.2
21, 22, 23	FIRE SUPPRESSION, PLUMBING & HVAC	100.3	68.5	87.5	98.4	68.7	86.5	99.9	68.7	87.4	100.0	68.7	87.5	98.4	68.7	86.5	98.3	68.5	86.4
26, 27, 3370	ELECTRICAL, COMMUNICATIONS & UTIL.	90.7	85.8	88.2	89.0	85.8	87.4	89.9	85.8	87.8	102.4	85.8	93.9	87.3	85.8	86.5	91.1	85.8	88.4
MF2018	WEIGHTED AVERAGE	96.5	71.7	85.9	98.1	75.0	88.2	100.5	75.0	89.6	99.7	75.0	89.1	97.7	75.0	87.9	97.3	71.7	86.3

For customer support on your Plumbing Costs with RSMeans Data, call 800.448.8182.

657

NEW MEXICO / NEW YORK

DIVISION		TUCUMCARI 884 MAT.	INST.	TOTAL	ALBANY 120-122 MAT.	INST.	TOTAL	BINGHAMTON 137-139 MAT.	INST.	TOTAL	BRONX 104 MAT.	INST.	TOTAL	BROOKLYN 112 MAT.	INST.	TOTAL	BUFFALO 140-142 MAT.	INST.	TOTAL
015433	CONTRACTOR EQUIPMENT		110.9	110.9		115.2	115.2		120.3	120.3		106.8	106.8		111.8	111.8		99.5	99.5
0241, 31 - 34	SITE & INFRASTRUCTURE, DEMOLITION	95.7	102.3	100.3	82.5	104.6	97.9	95.7	93.4	94.1	96.2	114.2	108.7	120.0	125.6	123.9	106.0	101.8	103.1
0310	Concrete Forming & Accessories	96.9	66.8	71.3	100.5	107.8	106.7	99.9	94.0	94.9	100.5	193.1	179.2	105.9	193.1	180.0	100.1	118.9	116.1
0320	Concrete Reinforcing	112.5	71.9	91.8	106.2	116.1	111.3	102.8	109.4	106.1	99.4	179.6	140.2	104.1	246.0	176.2	109.1	115.0	112.1
0330	Cast-in-Place Concrete	94.7	69.1	85.2	80.0	116.1	93.4	104.1	108.3	105.7	77.3	175.4	113.8	106.1	173.8	131.3	115.0	124.3	118.5
03	CONCRETE	102.6	69.8	87.8	84.6	113.1	97.4	93.4	103.8	98.1	85.9	183.7	129.9	104.5	193.3	144.4	108.1	119.5	113.2
04	MASONRY	117.9	58.7	81.3	87.4	116.9	105.6	106.7	106.2	106.4	87.5	189.3	150.4	117.1	189.2	161.7	123.5	123.6	123.5
05	METALS	105.3	90.5	100.8	98.3	125.9	106.7	90.7	134.4	104.0	85.9	173.7	112.6	98.2	174.0	121.2	98.9	107.1	101.4
06	WOOD, PLASTICS & COMPOSITES	94.6	68.8	80.6	95.6	104.7	100.5	103.7	90.7	96.6	103.5	194.1	152.7	105.9	193.8	153.6	99.6	118.2	109.7
07	THERMAL & MOISTURE PROTECTION	98.7	70.0	86.4	107.8	111.5	109.4	108.3	96.2	103.1	104.9	171.4	133.5	108.0	170.6	135.0	104.0	113.6	108.1
08	OPENINGS	98.6	67.7	91.4	97.2	105.0	99.0	92.4	95.3	93.1	94.0	202.1	119.1	89.8	201.9	115.8	99.1	112.0	102.1
0920	Plaster & Gypsum Board	80.9	67.7	72.2	103.7	104.7	104.4	105.9	90.2	95.5	95.9	196.6	162.2	102.3	196.6	164.4	103.6	118.6	113.5
0950, 0980	Ceilings & Acoustic Treatment	103.8	67.7	79.4	99.5	104.7	103.1	96.5	90.2	92.2	94.3	196.6	163.6	92.3	196.6	162.9	93.3	118.6	110.4
0960	Flooring	97.5	68.0	88.9	93.5	111.3	98.7	101.8	103.9	102.4	96.1	188.7	123.0	108.8	188.7	132.0	98.3	120.6	104.8
0970, 0990	Wall Finishes & Painting/Coating	93.4	51.9	68.8	93.4	103.9	99.6	87.8	106.9	99.1	104.8	166.6	141.3	114.1	166.6	145.1	98.9	118.6	110.6
09	FINISHES	91.6	65.5	77.4	90.4	107.6	99.8	91.7	96.5	94.3	97.5	190.5	148.2	104.7	190.2	151.4	98.7	119.9	110.3
COVERS	DIVS. 10 - 14, 25, 28, 41, 43, 44, 46	100.0	85.1	96.7	100.0	102.5	100.6	100.0	98.0	99.6	100.0	138.0	108.4	100.0	137.2	108.2	100.0	106.4	101.4
21, 22, 23	FIRE SUPPRESSION, PLUMBING & HVAC	98.4	68.4	86.4	100.1	111.8	104.8	100.5	98.7	99.8	100.2	176.4	130.7	99.8	176.3	130.4	100.0	99.7	99.9
26, 27, 3370	ELECTRICAL, COMMUNICATIONS & UTIL.	90.8	85.8	88.3	98.1	107.6	102.9	101.1	100.9	101.0	94.1	186.7	141.4	101.0	186.7	144.8	102.4	102.8	102.6
MF2018	WEIGHTED AVERAGE	99.8	74.9	89.1	95.7	111.5	102.5	96.9	102.8	99.5	94.1	176.8	129.6	101.2	178.9	134.6	102.2	110.0	105.6

NEW YORK

DIVISION		ELMIRA 148-149 MAT.	INST.	TOTAL	FAR ROCKAWAY 116 MAT.	INST.	TOTAL	FLUSHING 113 MAT.	INST.	TOTAL	GLENS FALLS 128 MAT.	INST.	TOTAL	HICKSVILLE 115,117,118 MAT.	INST.	TOTAL	JAMAICA 114 MAT.	INST.	TOTAL
015433	CONTRACTOR EQUIPMENT		123.0	123.0		111.8	111.8		111.8	111.8		115.2	115.2		111.8	111.8		111.8	111.8
0241, 31 - 34	SITE & INFRASTRUCTURE, DEMOLITION	104.9	93.8	97.2	123.3	125.6	124.9	123.3	125.6	124.9	73.1	104.2	94.8	112.9	124.4	120.9	117.5	125.6	123.1
0310	Concrete Forming & Accessories	83.9	97.4	95.3	92.2	193.1	177.9	96.1	193.1	178.5	84.1	100.7	98.2	88.5	161.5	150.5	96.1	193.1	178.5
0320	Concrete Reinforcing	108.5	108.2	108.4	104.1	246.0	176.2	105.9	246.0	177.1	101.8	116.0	109.0	104.1	179.4	142.4	104.1	246.0	176.2
0330	Cast-in-Place Concrete	103.2	107.8	104.9	114.8	173.8	136.7	114.8	173.8	136.7	77.1	114.0	90.9	97.4	166.2	123.0	106.1	173.8	131.3
03	CONCRETE	91.7	105.2	97.8	110.6	193.3	147.8	111.1	193.3	148.1	78.5	109.1	92.3	96.4	166.2	127.8	103.9	193.3	144.1
04	MASONRY	110.3	106.8	108.2	121.8	189.2	163.4	115.8	189.2	161.2	93.3	113.6	105.8	111.6	176.9	152.0	119.9	189.2	162.7
05	METALS	94.6	136.7	107.4	98.2	174.0	121.2	98.2	174.0	121.2	91.9	125.3	102.0	99.7	173.6	122.2	98.2	174.0	121.2
06	WOOD, PLASTICS & COMPOSITES	84.8	95.6	90.7	88.8	193.8	145.8	93.6	193.8	148.0	85.0	97.0	91.5	85.3	158.4	125.0	93.6	193.8	148.0
07	THERMAL & MOISTURE PROTECTION	103.7	96.4	100.6	107.9	170.6	134.9	108.0	170.6	135.0	100.6	109.2	104.3	107.6	161.9	131.0	107.8	170.6	134.8
08	OPENINGS	98.6	97.7	98.4	88.5	201.9	114.8	88.5	201.9	114.8	92.0	100.9	94.0	88.9	183.2	110.8	88.5	201.9	114.8
0920	Plaster & Gypsum Board	98.0	95.4	96.3	91.1	196.6	160.6	93.6	196.6	161.4	94.8	96.8	96.1	90.8	160.2	136.5	93.6	196.6	161.4
0950, 0980	Ceilings & Acoustic Treatment	99.9	95.4	96.9	81.3	196.6	159.4	81.3	196.6	159.4	86.5	196.8	160.2	80.3	160.2	134.5	81.3	196.6	159.4
0960	Flooring	88.8	103.9	93.2	103.7	188.7	128.3	105.2	188.7	129.4	84.9	111.3	92.6	102.6	185.7	126.7	105.2	188.7	129.4
0970, 0990	Wall Finishes & Painting/Coating	98.0	94.0	95.6	114.1	166.6	145.1	114.1	166.6	145.1	86.9	102.0	95.9	114.1	166.6	145.1	114.1	166.6	145.1
09	FINISHES	91.7	98.0	95.1	99.9	190.2	149.2	100.7	190.2	149.6	83.0	102.0	93.4	98.5	165.8	135.2	100.3	190.2	149.4
COVERS	DIVS. 10 - 14, 25, 28, 41, 43, 44, 46	100.0	99.1	99.8	100.0	137.2	108.2	100.0	137.2	108.2	100.0	98.1	99.6	100.0	134.8	107.7	100.0	137.2	108.2
21, 22, 23	FIRE SUPPRESSION, PLUMBING & HVAC	97.7	97.2	97.5	97.3	176.3	128.9	97.3	176.3	128.9	97.7	108.9	102.2	99.8	163.2	125.2	97.3	176.3	128.9
26, 27, 3370	ELECTRICAL, COMMUNICATIONS & UTIL.	100.4	104.3	102.4	107.9	186.7	148.2	107.9	186.7	148.2	92.8	107.6	100.4	100.4	144.4	122.9	99.4	186.7	144.1
MF2018	WEIGHTED AVERAGE	97.4	103.9	100.2	101.7	178.9	134.8	101.6	178.9	134.8	91.5	108.7	98.9	99.2	160.5	125.5	99.8	178.9	133.8

NEW YORK

DIVISION		JAMESTOWN 147 MAT.	INST.	TOTAL	KINGSTON 124 MAT.	INST.	TOTAL	LONG ISLAND CITY 111 MAT.	INST.	TOTAL	MONTICELLO 127 MAT.	INST.	TOTAL	MOUNT VERNON 105 MAT.	INST.	TOTAL	NEW ROCHELLE 108 MAT.	INST.	TOTAL
015433	CONTRACTOR EQUIPMENT		92.7	92.7		111.8	111.8		111.8	111.8		111.8	111.8		106.8	106.8		106.8	106.8
0241, 31 - 34	SITE & INFRASTRUCTURE, DEMOLITION	106.5	94.2	97.9	143.9	121.2	128.1	121.1	125.6	124.2	139.1	121.1	126.6	101.1	110.1	107.3	100.7	110.0	107.2
0310	Concrete Forming & Accessories	83.9	92.2	90.9	85.3	133.6	126.3	100.5	193.1	179.2	92.7	133.6	127.5	90.0	141.8	134.0	106.2	138.2	133.4
0320	Concrete Reinforcing	108.7	114.4	111.6	102.3	164.9	134.1	104.1	246.0	176.2	101.4	164.9	133.7	98.1	182.1	140.8	98.3	181.9	140.8
0330	Cast-in-Place Concrete	107.0	106.6	106.8	104.1	147.5	120.2	109.5	173.8	133.4	97.5	147.5	116.1	86.3	151.4	110.5	86.3	151.1	110.4
03	CONCRETE	94.7	101.0	97.5	98.7	143.2	118.7	107.0	193.3	145.8	94.0	143.2	116.1	94.6	152.2	120.5	94.0	150.5	119.4
04	MASONRY	120.0	104.7	110.6	107.8	154.7	136.8	113.9	189.2	160.5	100.1	154.7	133.8	92.8	156.2	132.0	92.8	156.2	132.0
05	METALS	92.0	102.5	95.2	100.0	136.5	111.1	98.2	174.0	121.2	99.9	136.7	111.1	85.7	165.6	110.0	86.0	164.9	109.9
06	WOOD, PLASTICS & COMPOSITES	83.4	88.7	86.3	86.2	127.6	108.7	99.9	193.8	150.9	93.1	127.6	111.8	92.4	136.3	116.2	111.5	132.4	122.9
07	THERMAL & MOISTURE PROTECTION	103.2	96.7	100.4	124.7	146.8	134.2	107.9	170.6	134.9	124.3	146.8	134.0	105.8	148.9	124.4	105.9	146.3	123.3
08	OPENINGS	98.4	95.7	97.8	93.7	146.1	105.8	88.5	201.9	114.8	89.5	146.0	102.6	94.0	169.4	111.5	94.1	167.2	111.1
0920	Plaster & Gypsum Board	85.7	88.3	87.4	96.9	128.5	117.7	98.5	196.6	163.1	97.5	128.5	117.9	90.9	137.1	121.3	104.3	133.1	123.3
0950, 0980	Ceilings & Acoustic Treatment	96.5	88.3	90.9	77.1	128.5	112.0	81.3	196.6	159.4	77.1	128.5	111.9	92.6	137.1	122.7	92.6	133.1	120.0
0960	Flooring	91.4	103.9	95.0	102.2	166.0	120.7	106.8	188.7	130.6	104.5	166.0	122.4	87.9	185.7	116.3	95.3	169.5	116.9
0970, 0990	Wall Finishes & Painting/Coating	99.6	101.0	100.4	113.6	130.5	123.6	114.1	166.6	145.1	113.6	126.6	121.3	103.2	166.6	140.7	103.2	166.6	140.7
09	FINISHES	90.2	94.3	92.5	96.5	138.5	119.4	101.6	190.2	150.0	96.9	138.0	119.3	94.5	151.4	125.5	98.4	145.8	124.3
COVERS	DIVS. 10 - 14, 25, 28, 41, 43, 44, 46	100.0	98.5	99.7	100.0	119.7	104.3	100.0	137.2	108.2	100.0	120.6	104.5	100.0	131.4	106.9	100.0	114.8	103.3
21, 22, 23	FIRE SUPPRESSION, PLUMBING & HVAC	97.6	95.3	96.6	97.6	136.7	113.3	99.8	176.3	130.4	97.6	141.0	115.0	97.8	144.9	116.7	97.8	144.6	116.6
26, 27, 3370	ELECTRICAL, COMMUNICATIONS & UTIL.	99.2	97.2	98.2	94.2	119.2	107.0	99.9	186.7	144.3	94.2	119.2	107.0	92.2	170.1	132.1	92.2	145.2	119.3
MF2018	WEIGHTED AVERAGE	97.5	97.8	97.7	99.8	136.0	115.3	100.8	178.9	134.4	98.4	136.9	114.9	94.4	151.3	118.8	94.9	146.0	116.8

City Cost Indexes

NEW YORK

DIVISION		NEW YORK 100 - 102			NIAGARA FALLS 143			PLATTSBURGH 129			POUGHKEEPSIE 125 - 126			QUEENS 110			RIVERHEAD 119		
		MAT.	INST.	TOTAL	MAT.	INST.	TOTAL	MAT.	INST.	TOTAL	MAT.	INST.	TOTAL	MAT.	INST.	TOTAL	MAT.	INST.	TOTAL
015433	CONTRACTOR EQUIPMENT		104.8	104.8		92.7	92.7		96.4	96.4		111.8	111.8		111.8	111.8		111.8	111.8
0241, 31 - 34	SITE & INFRASTRUCTURE, DEMOLITION	105.0	111.5	109.5	108.9	95.2	99.4	112.6	101.3	104.7	140.2	120.3	126.3	116.2	125.6	122.7	114.0	123.9	120.9
0310	Concrete Forming & Accessories	110.3	192.3	180.0	83.9	116.3	111.4	90.5	93.9	93.4	85.3	172.3	159.2	88.8	193.1	177.4	93.3	160.5	150.4
0320	Concrete Reinforcing	105.2	183.4	144.9	107.3	115.2	111.3	106.3	115.3	110.8	102.3	164.7	134.0	105.9	246.0	177.1	106.1	221.8	164.9
0330	Cast-in-Place Concrete	89.3	174.6	121.0	110.7	127.7	117.0	94.3	103.5	97.7	100.8	139.3	115.2	100.8	173.8	128.0	99.1	164.4	123.4
03	CONCRETE	99.1	183.5	137.1	97.0	119.3	107.0	91.3	100.6	95.5	96.2	157.8	123.9	99.6	193.3	141.7	97.1	171.0	130.3
04	MASONRY	101.2	187.1	154.2	128.3	127.8	128.0	87.7	99.4	94.9	100.1	142.5	126.3	108.1	189.2	158.2	117.3	175.8	153.5
05	METALS	96.9	173.1	120.0	94.6	103.1	97.2	96.0	100.0	97.2	100.0	136.3	111.0	98.2	174.0	121.2	100.1	161.6	118.8
06	WOOD, PLASTICS & COMPOSITES	109.2	194.2	155.4	83.4	111.4	98.6	91.7	90.7	91.2	86.2	184.5	139.6	85.4	193.8	144.3	90.5	158.4	127.4
07	THERMAL & MOISTURE PROTECTION	107.9	170.7	134.9	103.3	113.1	107.5	119.5	99.9	111.0	124.6	147.6	134.5	107.5	170.6	134.7	108.6	158.8	130.2
08	OPENINGS	97.1	203.0	121.7	98.5	107.7	100.6	99.4	97.4	98.9	93.7	175.5	112.7	88.5	201.9	114.8	88.9	176.6	109.3
0920	Plaster & Gypsum Board	101.2	196.6	164.0	85.7	111.6	102.8	114.7	90.0	98.4	96.9	187.0	156.2	90.8	196.6	160.5	91.9	160.2	136.9
0950, 0980	Ceilings & Acoustic Treatment	109.6	196.6	168.5	96.5	111.6	106.7	106.5	90.0	95.3	77.1	187.0	151.6	81.3	196.6	159.4	81.2	160.2	134.7
0960	Flooring	97.3	188.7	123.8	91.4	115.8	98.5	107.4	108.6	107.8	102.2	164.0	120.1	102.6	188.7	127.6	103.7	172.5	123.6
0970, 0990	Wall Finishes & Painting/Coating	101.4	166.6	140.0	99.6	114.4	108.4	108.3	97.6	102.0	113.6	126.6	121.3	114.1	166.6	145.1	114.1	166.6	145.1
09	FINISHES	102.3	190.0	150.2	90.4	116.2	104.5	94.9	96.3	95.7	96.3	169.0	136.0	98.9	190.2	148.7	99.1	162.7	133.8
COVERS	DIVS. 10 - 14, 25, 28, 41, 43, 44, 46	100.0	143.1	109.5	100.0	104.8	101.1	100.0	94.6	98.8	100.0	118.8	104.1	100.0	137.2	108.2	100.0	122.6	105.0
21, 22, 23	FIRE SUPPRESSION, PLUMBING & HVAC	100.1	175.3	130.2	97.6	107.5	101.6	97.6	102.6	99.6	97.6	123.0	107.8	99.8	176.3	130.4	99.9	157.9	123.2
26, 27, 3370	ELECTRICAL, COMMUNICATIONS & UTIL.	100.9	186.7	144.8	97.9	101.3	99.6	90.5	94.0	92.3	94.2	124.2	109.6	100.4	186.7	144.6	102.0	138.2	120.5
MF2018	WEIGHTED AVERAGE	99.8	176.2	132.6	98.6	110.2	103.5	96.5	99.0	97.5	99.0	140.2	116.7	99.2	178.9	133.4	100.0	156.8	124.3

NEW YORK

DIVISION		ROCHESTER 144 - 146			SCHENECTADY 123			STATEN ISLAND 103			SUFFERN 109			SYRACUSE 130 - 132			UTICA 133 - 135		
		MAT.	INST.	TOTAL	MAT.	INST.	TOTAL	MAT.	INST.	TOTAL	MAT.	INST.	TOTAL	MAT.	INST.	TOTAL	MAT.	INST.	TOTAL
015433	CONTRACTOR EQUIPMENT		119.9	119.9		115.2	115.2		106.8	106.8		106.8	106.8		115.2	115.2		115.2	115.2
0241, 31 - 34	SITE & INFRASTRUCTURE, DEMOLITION	92.9	108.7	103.9	83.2	104.6	98.1	105.1	114.2	111.5	98.0	108.8	105.5	94.3	103.2	100.5	72.8	102.7	93.6
0310	Concrete Forming & Accessories	106.0	101.9	102.5	100.9	107.7	106.7	89.4	193.1	177.7	99.0	144.7	137.8	99.0	91.6	92.7	100.0	88.7	90.4
0320	Concrete Reinforcing	110.9	108.0	109.4	100.8	116.1	108.6	99.4	222.2	161.8	98.3	156.4	127.8	103.8	109.2	106.5	103.8	103.2	103.5
0330	Cast-in-Place Concrete	101.1	105.3	102.7	93.1	116.0	101.7	86.3	175.6	119.5	83.7	143.0	105.8	96.6	105.3	99.9	88.4	103.9	94.2
03	CONCRETE	96.3	105.4	100.4	91.8	113.0	101.3	96.4	190.3	138.6	91.4	145.1	115.6	96.0	100.8	98.1	93.9	98.0	95.7
04	MASONRY	101.4	105.6	104.0	90.2	116.9	106.7	99.5	189.3	155.0	92.7	145.2	125.1	99.4	103.2	101.7	90.5	101.5	97.3
05	METALS	100.4	121.5	106.8	95.9	125.8	105.0	84.2	173.9	111.4	84.2	132.9	98.9	94.0	120.6	102.1	92.1	118.3	100.1
06	WOOD, PLASTICS & COMPOSITES	108.4	101.4	104.6	103.5	104.7	104.1	90.9	194.1	147.0	103.0	146.6	126.7	100.3	88.5	93.9	100.3	85.1	92.0
07	THERMAL & MOISTURE PROTECTION	104.8	101.8	103.5	102.3	111.5	106.2	105.2	171.4	133.7	105.8	144.3	122.4	102.7	96.8	100.2	91.1	96.7	93.5
08	OPENINGS	102.1	101.3	101.9	97.4	105.0	99.1	94.0	201.1	119.1	94.1	154.5	108.1	94.5	92.1	94.0	97.4	88.8	95.4
0920	Plaster & Gypsum Board	103.7	101.6	102.3	104.1	104.7	104.5	91.0	196.6	160.5	95.2	147.7	129.8	95.9	88.1	90.8	95.9	84.6	88.5
0950, 0980	Ceilings & Acoustic Treatment	102.4	101.6	101.8	93.2	104.7	101.0	94.3	196.6	163.6	92.6	147.7	129.9	96.5	88.1	90.8	96.5	84.6	88.4
0960	Flooring	90.0	108.5	95.4	92.1	111.3	97.7	91.5	188.7	119.7	91.4	177.3	116.3	90.7	95.4	92.0	88.5	95.5	90.6
0970, 0990	Wall Finishes & Painting/Coating	97.1	101.0	99.4	86.9	103.9	97.0	104.8	166.6	141.3	103.2	130.3	119.2	90.4	101.3	96.9	84.2	101.3	94.3
09	FINISHES	94.1	103.4	99.2	88.4	107.6	98.8	96.2	190.5	147.7	95.8	150.1	125.4	90.5	92.6	91.6	88.9	90.4	89.7
COVERS	DIVS. 10 - 14, 25, 28, 41, 43, 44, 46	100.0	100.4	100.1	100.0	102.5	100.6	100.0	138.0	108.4	100.0	117.0	103.8	100.0	97.0	99.3	100.0	91.6	98.1
21, 22, 23	FIRE SUPPRESSION, PLUMBING & HVAC	100.0	90.1	96.0	100.1	107.7	103.2	100.2	176.4	130.7	97.8	126.5	109.3	100.2	96.6	98.8	100.2	94.6	98.0
26, 27, 3370	ELECTRICAL, COMMUNICATIONS & UTIL.	104.7	92.1	98.3	96.9	107.6	102.4	94.1	186.7	141.5	98.9	117.4	108.4	101.0	103.5	102.3	98.9	103.5	101.3
MF2018	WEIGHTED AVERAGE	99.9	101.3	100.5	96.0	110.6	102.3	95.7	177.8	130.9	94.6	133.5	111.3	97.3	100.8	98.8	95.4	98.9	96.9

NEW YORK / NORTH CAROLINA

DIVISION		WATERTOWN 136			WHITE PLAINS 106			YONKERS 107			ASHEVILLE 287 - 288			CHARLOTTE 281 - 282			DURHAM 277		
		MAT.	INST.	TOTAL	MAT.	INST.	TOTAL	MAT.	INST.	TOTAL	MAT.	INST.	TOTAL	MAT.	INST.	TOTAL	MAT.	INST.	TOTAL
015433	CONTRACTOR EQUIPMENT		115.2	115.2		106.8	106.8		106.8	106.8		100.9	100.9		100.9	100.9		106.2	106.2
0241, 31 - 34	SITE & INFRASTRUCTURE, DEMOLITION	80.5	103.2	96.3	95.9	110.1	105.8	102.7	110.1	107.8	95.9	81.1	85.6	98.1	81.3	86.4	99.7	90.1	93.0
0310	Concrete Forming & Accessories	84.8	95.6	94.0	104.4	152.6	145.4	104.7	149.2	142.5	88.7	62.1	66.1	96.0	61.7	66.9	94.0	62.1	66.9
0320	Concrete Reinforcing	104.5	109.2	106.9	98.3	182.1	140.9	102.2	182.0	142.8	92.3	67.8	79.9	96.5	68.6	82.3	95.8	68.6	82.0
0330	Cast-in-Place Concrete	103.0	107.7	104.7	76.7	151.4	104.5	85.8	151.4	110.2	104.8	69.7	91.7	106.8	69.0	92.7	104.2	69.7	91.3
03	CONCRETE	106.3	103.4	105.0	85.6	157.1	117.8	94.1	155.6	121.8	96.8	67.4	83.6	96.5	67.1	83.3	96.6	67.5	83.5
04	MASONRY	91.7	107.0	101.2	92.2	156.2	131.7	96.3	156.2	133.3	86.5	62.4	71.6	90.2	61.0	72.2	85.6	62.4	71.3
05	METALS	92.2	120.6	100.8	85.7	165.6	110.0	93.5	165.5	115.4	98.9	90.5	96.3	98.9	90.5	96.3	114.7	90.0	107.3
06	WOOD, PLASTICS & COMPOSITES	81.7	92.5	87.6	109.1	151.0	131.9	109.0	146.6	129.4	87.4	60.7	72.9	92.3	60.7	75.1	90.3	60.7	74.2
07	THERMAL & MOISTURE PROTECTION	91.3	99.9	95.0	105.6	150.4	124.9	105.9	150.0	124.9	100.9	63.9	85.0	95.0	63.3	81.4	105.4	63.9	87.6
08	OPENINGS	97.4	96.5	97.2	94.1	177.5	113.5	97.8	173.0	115.2	90.9	61.6	84.1	97.8	61.7	89.4	97.4	61.7	89.1
0920	Plaster & Gypsum Board	86.7	92.2	90.4	98.2	152.2	133.7	101.4	147.7	131.9	107.3	59.5	75.8	104.4	59.5	74.8	94.1	59.5	71.3
0950, 0980	Ceilings & Acoustic Treatment	96.5	92.2	93.6	92.6	152.2	132.9	108.7	147.7	135.1	81.6	59.5	66.6	87.3	59.5	68.5	84.3	59.5	67.5
0960	Flooring	81.8	95.5	85.8	93.7	185.7	120.4	93.2	188.7	120.9	94.7	62.1	85.2	95.2	62.1	85.6	98.3	62.1	87.8
0970, 0990	Wall Finishes & Painting/Coating	84.2	95.4	90.9	103.2	166.6	140.7	103.2	166.6	140.7	103.7	58.6	77.0	97.4	58.6	74.4	98.9	58.6	75.1
09	FINISHES	86.2	95.1	91.1	96.6	160.0	131.2	96.5	158.0	131.9	88.6	61.5	73.8	88.7	61.1	73.6	87.0	61.5	73.1
COVERS	DIVS. 10 - 14, 25, 28, 41, 43, 44, 46	100.0	98.4	99.6	100.0	129.6	106.5	100.0	126.8	105.9	100.0	83.4	96.3	100.0	83.0	96.3	100.0	83.4	96.3
21, 22, 23	FIRE SUPPRESSION, PLUMBING & HVAC	100.2	89.8	96.0	100.3	144.9	118.2	100.3	144.8	118.1	100.5	61.5	84.9	99.9	61.9	84.7	100.5	61.5	84.9
26, 27, 3370	ELECTRICAL, COMMUNICATIONS & UTIL.	101.0	92.4	96.6	92.2	170.1	132.1	99.0	159.8	130.1	101.8	59.3	80.0	100.9	61.8	80.9	95.9	58.8	76.9
MF2018	WEIGHTED AVERAGE	97.0	99.2	98.0	94.1	153.5	119.6	98.1	151.2	120.9	96.9	67.0	84.1	97.6	67.3	84.6	99.8	67.7	86.0

For customer support on your Plumbing Costs with RSMeans Data, call 800.448.8182.

659

NORTH CAROLINA

DIVISION		ELIZABETH CITY 279			FAYETTEVILLE 283			GASTONIA 280			GREENSBORO 270, 272 - 274			HICKORY 286			KINSTON 285		
		MAT.	INST.	TOTAL	MAT.	INST.	TOTAL	MAT.	INST.	TOTAL	MAT.	INST.	TOTAL	MAT.	INST.	TOTAL	MAT.	INST.	TOTAL
015433	CONTRACTOR EQUIPMENT		110.3	110.3		106.2	106.2		100.9	100.9		106.2	106.2		106.2	106.2		106.2	106.2
0241, 31 - 34	SITE & INFRASTRUCTURE, DEMOLITION	104.1	91.8	95.5	95.2	90.0	91.6	95.7	81.4	85.8	99.6	90.1	93.0	94.8	88.8	90.6	93.7	88.7	90.2
0310	Concrete Forming & Accessories	80.3	62.5	65.2	88.3	61.6	65.6	94.8	62.4	67.2	93.8	62.1	66.9	85.3	62.1	65.6	81.8	61.6	64.6
0320	Concrete Reinforcing	93.9	71.7	82.6	95.9	68.6	82.0	92.7	68.6	80.5	94.7	68.6	81.4	92.3	68.6	80.3	91.9	68.6	80.0
0330	Cast-in-Place Concrete	104.4	71.4	92.1	109.9	68.9	94.6	102.4	69.8	90.3	103.4	69.7	90.8	104.7	69.7	91.7	101.1	68.9	89.1
03	CONCRETE	96.8	68.8	84.2	98.6	67.0	84.4	95.3	67.7	82.9	96.1	67.5	83.2	96.5	67.5	83.5	93.5	67.0	81.6
04	MASONRY	96.8	61.0	74.7	89.6	61.0	72.0	90.5	62.4	73.1	82.9	62.4	70.2	75.9	62.4	67.6	82.6	61.0	69.3
05	METALS	101.1	92.6	98.5	118.2	90.4	109.7	98.7	90.6	96.2	107.4	90.3	102.2	98.1	90.4	95.8	96.9	90.3	94.9
06	WOOD, PLASTICS & COMPOSITES	75.0	61.4	67.6	86.6	60.7	72.5	95.3	60.7	76.5	90.1	60.7	74.1	79.3	60.7	70.6	79.3	60.7	69.2
07	THERMAL & MOISTURE PROTECTION	104.6	63.2	86.8	100.4	63.3	84.4	101.1	63.9	85.1	105.2	63.9	87.4	101.3	63.9	85.2	101.1	63.9	84.8
08	OPENINGS	94.7	62.7	87.3	91.0	61.7	84.2	94.3	61.7	86.7	97.4	61.7	89.1	90.9	61.7	84.2	91.0	61.7	84.2
0920	Plaster & Gypsum Board	88.8	59.5	69.5	111.2	59.5	77.2	113.2	59.5	77.9	95.5	59.5	71.8	107.3	59.5	75.8	107.1	59.5	75.8
0950, 0980	Ceilings & Acoustic Treatment	84.3	59.5	67.5	83.3	59.5	67.2	85.0	59.5	67.7	84.3	59.5	67.5	81.6	59.5	66.6	85.0	59.5	67.7
0960	Flooring	90.4	79.8	87.3	94.8	62.1	85.3	98.0	79.8	92.7	98.3	62.1	87.8	94.6	79.8	90.3	91.7	79.8	88.3
0970, 0990	Wall Finishes & Painting/Coating	98.9	58.6	75.1	103.7	58.6	77.0	103.7	58.6	77.0	98.9	58.6	75.1	103.7	58.6	77.0	103.7	58.6	77.0
09	FINISHES	84.3	65.2	73.9	89.5	61.1	74.0	91.0	65.1	76.8	87.2	61.5	73.2	88.8	65.1	75.8	88.4	64.7	75.5
COVERS	DIVS. 10 - 14, 25, 28, 41, 43, 44, 46	100.0	81.7	96.0	100.0	83.0	96.2	100.0	83.4	96.4	100.0	80.7	95.7	100.0	83.4	96.3	100.0	82.9	96.2
21, 22, 23	FIRE SUPPRESSION, PLUMBING & HVAC	98.0	59.9	82.7	100.3	60.8	84.5	100.5	60.4	84.4	100.5	61.5	84.9	100.5	60.4	83.0	98.0	59.7	82.7
26, 27, 3370	ELECTRICAL, COMMUNICATIONS & UTIL.	95.6	66.2	80.6	101.4	58.8	79.6	101.2	61.9	81.1	95.1	59.3	76.7	99.4	61.9	80.2	99.2	57.2	77.7
MF2018	WEIGHTED AVERAGE	96.9	69.2	85.0	100.5	67.3	86.2	97.6	67.7	84.8	98.3	67.7	85.2	95.5	68.3	83.8	95.2	67.1	83.1

NORTH CAROLINA / NORTH DAKOTA

DIVISION		MURPHY 289			RALEIGH 275 - 276			ROCKY MOUNT 278			WILMINGTON 284			WINSTON-SALEM 271			BISMARCK 585		
		MAT.	INST.	TOTAL	MAT.	INST.	TOTAL	MAT.	INST.	TOTAL	MAT.	INST.	TOTAL	MAT.	INST.	TOTAL	MAT.	INST.	TOTAL
015433	CONTRACTOR EQUIPMENT		100.9	100.9		106.2	106.2		106.2	106.2		100.9	100.9		106.2	106.2		100.3	100.3
0241, 31 - 34	SITE & INFRASTRUCTURE, DEMOLITION	97.1	79.6	84.9	99.6	90.1	93.0	102.1	90.1	93.7	97.0	81.0	85.8	99.9	90.1	93.1	101.1	98.4	99.2
0310	Concrete Forming & Accessories	95.4	61.5	66.6	95.2	61.6	66.7	86.3	61.8	65.5	90.1	61.6	65.9	95.7	62.1	67.2	103.0	73.4	77.9
0320	Concrete Reinforcing	91.9	66.1	78.8	96.8	68.6	82.5	93.9	68.6	81.1	93.1	68.6	80.6	94.7	68.6	81.4	103.0	94.5	99.1
0330	Cast-in-Place Concrete	108.4	68.9	93.7	107.1	68.9	92.9	102.1	69.0	89.8	104.3	68.9	91.2	106.0	69.7	92.5	102.2	83.9	94.3
03	CONCRETE	99.7	66.5	84.8	95.1	67.0	82.5	97.5	67.1	83.8	96.7	67.0	83.4	97.5	67.5	84.0	94.4	81.4	88.5
04	MASONRY	78.9	61.0	67.9	81.6	61.0	68.9	75.3	61.0	66.5	76.5	61.0	66.9	83.2	62.4	70.3	98.4	82.4	88.5
05	METALS	95.9	89.5	94.0	99.8	90.4	96.9	100.3	90.5	97.3	97.6	90.4	95.4	104.6	90.4	100.3	98.7	92.6	96.8
06	WOOD, PLASTICS & COMPOSITES	96.0	60.6	76.7	90.4	60.7	74.3	81.8	60.7	70.4	89.2	60.7	73.7	90.1	60.7	74.1	102.1	69.5	84.4
07	THERMAL & MOISTURE PROTECTION	101.1	63.3	84.8	99.5	63.3	83.9	105.2	63.3	87.1	100.9	63.3	84.7	105.2	63.9	87.4	105.1	83.2	95.7
08	OPENINGS	90.9	61.1	84.0	98.7	61.7	90.1	93.9	61.7	86.5	91.0	61.7	84.2	97.4	61.7	89.1	106.3	79.3	100.0
0920	Plaster & Gypsum Board	112.5	59.3	77.5	91.1	59.5	70.3	90.3	59.5	70.0	109.2	59.5	76.5	95.5	59.5	71.8	100.0	69.0	79.6
0950, 0980	Ceilings & Acoustic Treatment	81.6	59.3	66.5	82.2	59.5	66.8	82.6	59.5	67.0	83.3	59.5	67.2	84.3	59.5	67.5	109.7	69.0	82.1
0960	Flooring	98.2	79.8	92.9	93.3	62.1	84.3	93.9	79.8	89.8	95.4	62.1	85.7	98.3	62.1	87.8	91.3	51.4	79.8
0970, 0990	Wall Finishes & Painting/Coating	103.7	58.6	77.0	92.1	58.6	72.3	98.9	58.6	75.1	103.7	58.6	77.0	98.9	58.6	75.1	93.7	59.0	73.2
09	FINISHES	90.6	64.6	76.4	85.6	61.1	72.3	85.3	64.7	74.1	89.4	61.1	74.0	87.2	61.5	73.2	95.6	68.3	80.7
COVERS	DIVS. 10 - 14, 25, 28, 41, 43, 44, 46	100.0	82.9	96.2	100.0	83.0	96.2	100.0	83.0	96.2	100.0	83.0	96.2	100.0	83.4	96.3	100.0	89.1	97.6
21, 22, 23	FIRE SUPPRESSION, PLUMBING & HVAC	98.0	59.6	82.6	100.0	60.8	84.3	98.0	59.7	82.6	100.5	60.8	84.6	100.5	61.5	84.9	99.9	75.5	90.1
26, 27, 3370	ELECTRICAL, COMMUNICATIONS & UTIL.	102.6	57.2	79.4	98.3	57.7	77.5	97.5	58.8	77.7	102.0	57.2	79.1	95.0	59.3	76.7	96.2	74.4	85.1
MF2018	WEIGHTED AVERAGE	96.3	66.2	83.4	96.9	67.1	84.1	96.1	67.5	83.8	96.5	66.3	83.6	98.0	67.8	85.0	99.1	80.0	90.9

NORTH DAKOTA

DIVISION		DEVILS LAKE 583			DICKINSON 586			FARGO 580 - 581			GRAND FORKS 582			JAMESTOWN 584			MINOT 587		
		MAT.	INST.	TOTAL	MAT.	INST.	TOTAL	MAT.	INST.	TOTAL	MAT.	INST.	TOTAL	MAT.	INST.	TOTAL	MAT.	INST.	TOTAL
015433	CONTRACTOR EQUIPMENT		100.3	100.3		100.3	100.3		100.3	100.3		100.3	100.3		100.3	100.3		100.3	100.3
0241, 31 - 34	SITE & INFRASTRUCTURE, DEMOLITION	106.5	98.5	100.9	114.1	98.4	103.2	97.7	98.4	98.2	110.0	98.4	101.9	105.5	98.5	100.6	107.6	98.4	101.2
0310	Concrete Forming & Accessories	101.9	73.5	77.7	90.9	73.4	76.0	97.2	73.5	77.1	94.9	73.4	76.6	92.6	73.4	76.3	90.6	73.4	76.0
0320	Concrete Reinforcing	100.2	94.4	97.3	101.1	94.5	97.8	94.6	94.8	94.7	98.7	94.4	96.5	100.7	94.8	97.7	102.1	94.5	98.2
0330	Cast-in-Place Concrete	122.1	84.0	107.9	110.4	83.9	100.5	99.9	83.9	93.9	110.4	83.9	100.5	120.5	84.0	106.9	110.4	83.9	100.5
03	CONCRETE	103.8	81.4	93.8	102.8	81.4	93.1	95.9	81.5	89.4	100.2	81.4	91.7	102.4	81.5	93.0	98.8	81.4	91.0
04	MASONRY	114.6	80.9	93.8	116.6	82.4	95.5	103.2	87.1	93.3	108.9	80.9	91.6	127.0	87.2	102.4	107.2	82.4	91.9
05	METALS	95.2	92.7	94.4	95.1	92.5	94.3	97.9	93.0	96.4	95.1	92.4	94.3	95.2	93.1	94.5	95.4	92.5	94.6
06	WOOD, PLASTICS & COMPOSITES	99.9	69.5	83.4	86.2	69.5	77.1	94.9	69.5	81.1	91.0	69.5	79.3	88.3	69.5	78.0	85.9	69.5	77.0
07	THERMAL & MOISTURE PROTECTION	106.5	82.8	96.3	107.1	83.2	96.8	106.5	85.0	97.2	106.8	82.8	96.4	106.3	85.0	97.2	106.5	83.2	96.5
08	OPENINGS	103.1	79.3	97.6	103.1	79.3	97.6	100.2	79.3	95.4	101.7	79.3	96.5	103.1	79.3	97.6	101.9	79.3	96.6
0920	Plaster & Gypsum Board	114.9	69.0	84.7	105.2	69.0	81.3	98.2	69.0	78.9	106.4	69.0	81.7	106.1	69.0	81.6	105.2	69.0	81.3
0950, 0980	Ceilings & Acoustic Treatment	101.8	69.0	79.6	101.8	69.0	79.6	95.2	69.0	77.4	101.8	69.0	79.6	101.8	69.0	79.6	101.8	69.0	79.6
0960	Flooring	95.3	51.4	82.6	88.8	51.4	77.9	100.0	51.4	85.9	90.6	51.4	79.3	89.5	51.4	78.4	88.5	51.4	77.8
0970, 0990	Wall Finishes & Painting/Coating	88.1	57.8	70.2	88.1	57.8	70.2	92.8	69.1	78.8	88.1	67.8	76.1	88.1	57.8	70.2	88.1	59.0	70.9
09	FINISHES	93.8	68.0	79.7	91.4	68.0	78.6	94.9	69.2	80.9	91.6	69.1	79.3	90.8	68.0	78.4	90.6	68.1	78.3
COVERS	DIVS. 10 - 14, 25, 28, 41, 43, 44, 46	100.0	89.1	97.6	100.0	89.1	97.6	100.0	89.1	97.6	100.0	89.1	97.6	100.0	89.1	97.6	100.0	89.1	97.6
21, 22, 23	FIRE SUPPRESSION, PLUMBING & HVAC	97.7	80.0	90.6	97.7	74.9	88.5	99.9	75.6	90.2	100.2	73.8	89.6	97.7	73.8	88.1	100.2	73.6	89.5
26, 27, 3370	ELECTRICAL, COMMUNICATIONS & UTIL.	94.4	69.9	81.9	102.2	69.5	85.5	101.2	71.1	85.8	97.5	69.9	83.4	94.4	69.9	81.9	100.4	74.4	87.1
MF2018	WEIGHTED AVERAGE	99.5	80.2	91.2	100.1	79.1	91.1	99.1	80.3	91.0	99.3	79.0	90.6	99.5	79.6	91.0	99.2	79.6	90.8

For customer support on your Plumbing Costs with RSMeans Data, call 800.448.8182.

		NORTH DAKOTA			OHIO														
	DIVISION	WILLISTON			AKRON			ATHENS			CANTON			CHILLICOTHE			CINCINNATI		
		588			442 - 443			457			446 - 447			456			451 - 452		
		MAT.	INST.	TOTAL	MAT.	INST.	TOTAL	MAT.	INST.	TOTAL	MAT.	INST.	TOTAL	MAT.	INST.	TOTAL	MAT.	INST.	TOTAL
015433	CONTRACTOR EQUIPMENT		100.3	100.3		91.4	91.4		87.5	87.5		91.4	91.4		98.4	98.4		96.0	96.0
0241, 31 - 34	SITE & INFRASTRUCTURE, DEMOLITION	108.0	96.2	99.8	99.8	99.1	99.4	116.0	89.8	97.8	100.0	98.8	99.2	102.4	100.0	100.8	97.9	99.0	98.7
0310	Concrete Forming & Accessories	96.8	73.1	76.7	100.8	85.2	87.6	92.9	79.2	81.3	100.8	75.7	79.5	95.7	83.9	85.6	99.1	82.1	84.7
0320	Concrete Reinforcing	103.1	94.5	98.7	95.4	93.7	94.6	90.7	91.8	91.2	95.4	77.2	86.2	88.0	91.4	89.7	93.3	80.0	86.6
0330	Cast-in-Place Concrete	110.4	83.8	100.5	107.0	91.7	101.3	111.5	98.3	106.6	108.0	89.0	100.9	101.2	95.2	99.0	96.6	77.5	89.5
03	CONCRETE	100.2	81.2	91.6	102.2	88.4	96.0	100.6	87.7	94.8	102.7	80.3	92.6	95.1	89.4	92.5	94.0	80.3	87.8
04	MASONRY	102.1	82.4	89.9	94.2	91.3	92.4	72.3	98.5	88.5	94.9	80.4	86.2	78.7	92.4	87.2	81.3	81.0	81.1
05	METALS	95.3	92.2	94.4	97.9	81.1	92.8	101.6	81.4	95.5	97.9	74.4	90.8	93.6	90.8	92.7	95.8	82.4	91.7
06	WOOD, PLASTICS & COMPOSITES	92.7	69.5	80.1	108.4	83.5	94.8	87.7	73.9	80.2	108.7	73.9	89.8	100.7	80.3	89.6	100.7	82.3	90.7
07	THERMAL & MOISTURE PROTECTION	106.7	83.2	96.6	105.3	93.7	100.3	101.9	93.5	98.3	106.4	89.0	98.9	104.0	90.9	98.4	101.9	82.0	93.3
08	OPENINGS	103.2	79.3	97.6	109.5	85.1	103.8	99.1	76.8	93.9	103.1	71.3	95.8	90.3	80.3	88.0	98.4	78.1	93.7
0920	Plaster & Gypsum Board	106.4	69.0	81.7	101.8	82.9	89.3	96.9	73.0	81.2	102.7	73.0	83.2	100.3	80.1	87.0	99.9	82.2	88.2
0950, 0980	Ceilings & Acoustic Treatment	101.8	69.0	79.6	94.9	82.9	86.8	110.6	73.0	85.1	94.9	73.0	80.1	103.2	80.1	87.6	97.0	82.2	87.0
0960	Flooring	91.6	51.4	79.9	98.1	86.1	94.6	121.9	77.8	109.1	98.3	76.6	92.0	99.5	77.8	93.2	100.8	81.2	95.1
0970, 0990	Wall Finishes & Painting/Coating	88.1	57.8	70.2	95.1	93.3	94.0	103.6	91.5	96.4	95.1	70.9	80.8	100.8	91.5	95.3	100.4	73.2	84.3
09	FINISHES	91.8	68.0	78.8	99.0	86.3	92.1	101.9	79.4	89.7	99.2	74.6	85.7	99.3	82.7	90.3	98.5	81.2	89.1
COVERS	DIVS. 10 - 14, 25, 28, 41, 43, 44, 46	100.0	89.0	97.6	100.0	92.2	98.3	100.0	89.2	97.6	100.0	89.8	97.7	100.0	88.7	97.5	100.0	89.8	97.6
21, 22, 23	FIRE SUPPRESSION, PLUMBING & HVAC	97.7	74.8	88.5	100.0	89.8	95.9	97.6	82.6	91.6	100.0	79.6	91.8	98.2	94.5	96.7	100.0	77.2	90.9
26, 27, 3370	ELECTRICAL, COMMUNICATIONS & UTIL.	97.9	70.4	83.8	94.9	83.9	89.3	97.7	91.9	94.7	94.2	87.1	90.6	97.2	91.9	94.5	96.1	72.9	84.3
MF2018	WEIGHTED AVERAGE	98.6	79.0	90.2	100.3	88.3	95.1	98.6	86.4	93.4	99.7	81.4	91.8	95.7	90.8	93.6	97.0	80.7	90.0

		OHIO																	
	DIVISION	CLEVELAND			COLUMBUS			DAYTON			HAMILTON			LIMA			LORAIN		
		441			430 - 432			453 - 454			450			458			440		
		MAT.	INST.	TOTAL	MAT.	INST.	TOTAL	MAT.	INST.	TOTAL	MAT.	INST.	TOTAL	MAT.	INST.	TOTAL	MAT.	INST.	TOTAL
015433	CONTRACTOR EQUIPMENT		92.5	92.5		92.0	92.0		91.6	91.6		98.4	98.4		90.9	90.9		91.4	91.4
0241, 31 - 34	SITE & INFRASTRUCTURE, DEMOLITION	98.4	97.4	97.7	101.3	92.5	95.2	98.0	99.3	98.9	97.9	99.5	99.1	109.5	89.5	95.5	99.1	99.2	99.2
0310	Concrete Forming & Accessories	100.2	89.2	90.9	98.7	77.9	81.0	97.5	79.4	82.2	97.6	79.6	82.3	92.8	78.5	80.7	100.6	76.6	80.2
0320	Concrete Reinforcing	95.9	91.8	93.8	99.2	81.0	90.0	93.3	81.4	87.3	93.3	79.5	86.3	90.7	81.6	86.1	95.4	94.0	94.7
0330	Cast-in-Place Concrete	103.7	98.7	101.8	100.6	82.4	93.8	86.7	83.0	85.3	92.9	83.4	89.4	102.4	92.2	98.6	101.8	94.2	99.0
03	CONCRETE	102.3	92.4	97.9	96.3	80.0	89.0	86.3	80.9	83.9	89.3	81.4	85.7	93.5	83.8	89.2	99.7	85.4	93.3
04	MASONRY	100.7	97.9	99.0	84.3	86.9	85.9	77.6	78.0	77.9	78.0	82.9	81.0	97.4	79.8	86.5	90.8	95.0	93.4
05	METALS	99.4	83.7	94.6	96.8	80.8	92.0	95.1	77.6	89.7	95.1	86.5	92.5	101.6	81.0	95.4	98.5	81.8	93.4
06	WOOD, PLASTICS & COMPOSITES	102.3	86.7	93.8	97.8	76.7	86.3	104.6	79.4	90.9	103.4	79.4	90.3	87.6	77.3	82.0	108.4	72.1	88.7
07	THERMAL & MOISTURE PROTECTION	102.4	97.8	100.4	96.7	84.9	91.6	108.3	81.7	96.8	104.1	82.2	94.7	101.5	86.0	94.8	106.3	93.6	100.9
08	OPENINGS	97.7	86.3	95.0	99.3	74.4	93.5	97.5	76.8	92.7	95.2	77.0	91.0	99.1	75.9	93.7	103.1	77.4	97.2
0920	Plaster & Gypsum Board	101.4	86.1	91.3	97.3	76.1	83.4	102.0	79.2	87.0	102.0	79.2	87.0	96.9	76.5	83.5	101.8	71.2	81.7
0950, 0980	Ceilings & Acoustic Treatment	90.4	86.1	87.5	91.4	76.1	81.1	105.0	79.2	87.5	104.1	79.2	87.2	109.7	76.5	87.2	94.9	71.2	78.9
0960	Flooring	99.5	90.5	96.9	94.2	77.8	89.5	103.2	74.8	95.0	100.6	81.2	95.0	120.8	79.8	108.9	98.3	92.8	96.7
0970, 0990	Wall Finishes & Painting/Coating	97.7	93.3	95.1	95.8	81.7	87.5	100.8	72.6	84.1	100.8	73.2	84.5	103.7	78.5	88.8	95.1	93.3	94.0
09	FINISHES	99.3	89.6	94.0	94.7	77.9	85.5	100.7	77.6	88.1	99.7	79.2	88.5	101.0	78.2	88.5	99.0	80.3	88.8
COVERS	DIVS. 10 - 14, 25, 28, 41, 43, 44, 46	100.0	98.0	99.6	100.0	89.5	97.7	100.0	86.9	97.1	100.0	87.1	97.2	100.0	86.2	97.0	100.0	92.8	98.4
21, 22, 23	FIRE SUPPRESSION, PLUMBING & HVAC	100.0	92.1	96.8	100.0	85.9	94.4	100.8	83.0	93.7	100.6	76.6	91.0	97.6	93.8	96.1	100.0	90.4	96.2
26, 27, 3370	ELECTRICAL, COMMUNICATIONS & UTIL.	94.6	93.0	93.8	99.0	82.2	90.4	94.8	77.5	86.0	95.1	77.3	86.0	98.0	77.5	87.5	94.4	80.1	87.1
MF2018	WEIGHTED AVERAGE	99.4	92.2	96.3	97.6	82.9	91.4	96.1	81.3	89.7	96.1	81.5	89.8	98.7	83.8	92.3	99.2	86.8	93.9

		OHIO																	
	DIVISION	MANSFIELD			MARION			SPRINGFIELD			STEUBENVILLE			TOLEDO			YOUNGSTOWN		
		448 - 449			433			455			439			434 - 436			444 - 445		
		MAT.	INST.	TOTAL	MAT.	INST.	TOTAL	MAT.	INST.	TOTAL	MAT.	INST.	TOTAL	MAT.	INST.	TOTAL	MAT.	INST.	TOTAL
015433	CONTRACTOR EQUIPMENT		91.4	91.4		91.8	91.8		91.6	91.6		95.6	95.6		95.6	95.6		91.4	91.4
0241, 31 - 34	SITE & INFRASTRUCTURE, DEMOLITION	95.4	99.0	97.9	95.5	95.8	95.7	98.3	99.2	99.0	140.6	103.8	115.0	99.3	95.9	96.9	99.7	98.9	99.2
0310	Concrete Forming & Accessories	89.7	75.6	77.7	96.6	79.4	82.0	97.5	79.3	82.0	97.6	81.2	83.7	100.4	86.7	88.8	100.8	78.8	82.1
0320	Concrete Reinforcing	86.9	81.3	84.1	91.4	81.3	86.3	93.3	81.4	87.3	89.2	97.6	93.4	99.2	85.1	92.0	95.4	87.7	91.5
0330	Cast-in-Place Concrete	99.0	90.4	95.8	87.2	90.9	88.6	89.0	82.8	86.7	94.7	92.2	93.7	95.5	92.3	94.3	105.9	89.9	100.0
03	CONCRETE	94.2	81.5	88.5	83.6	83.6	83.6	87.5	80.7	84.4	87.6	87.3	87.5	91.2	88.4	89.9	101.7	83.8	93.6
04	MASONRY	93.3	90.8	91.7	86.5	91.8	89.7	77.8	77.5	77.6	76.8	91.4	85.8	91.7	90.2	90.8	94.5	88.5	90.8
05	METALS	98.7	76.7	92.0	95.8	79.8	91.0	95.0	77.6	89.7	92.3	82.7	89.4	96.6	86.2	93.5	97.9	78.6	92.1
06	WOOD, PLASTICS & COMPOSITES	94.8	72.1	82.5	95.0	76.8	85.1	106.0	79.4	91.5	90.2	78.6	83.9	99.3	86.4	92.3	108.4	76.8	91.2
07	THERMAL & MOISTURE PROTECTION	104.4	90.9	98.6	93.4	91.0	92.4	108.1	81.5	96.7	105.0	89.5	98.3	93.6	92.2	93.0	106.6	90.7	99.8
08	OPENINGS	104.2	73.1	97.0	93.1	75.6	89.0	95.6	76.8	91.3	93.3	80.7	90.4	95.7	83.7	92.9	103.1	77.4	97.2
0920	Plaster & Gypsum Board	94.6	71.2	79.2	98.5	76.1	83.8	102.0	79.2	87.0	96.7	77.6	84.1	100.3	86.0	90.9	101.8	76.1	84.8
0950, 0980	Ceilings & Acoustic Treatment	95.7	71.2	79.1	97.6	76.1	83.1	105.0	79.2	87.5	94.7	77.6	83.1	97.6	86.0	89.8	94.9	76.1	82.1
0960	Flooring	93.3	95.8	94.0	93.6	95.8	94.3	103.2	74.8	95.0	121.5	93.3	113.3	94.0	93.8	93.9	98.3	93.0	96.8
0970, 0990	Wall Finishes & Painting/Coating	95.1	79.9	86.1	100.2	79.9	88.2	100.8	72.6	84.1	113.6	89.4	99.3	100.2	89.3	93.8	95.1	82.1	87.4
09	FINISHES	96.4	78.8	86.8	96.1	81.8	88.3	100.7	77.5	88.0	113.9	83.5	97.3	96.8	88.0	92.0	99.1	81.1	89.3
COVERS	DIVS. 10 - 14, 25, 28, 41, 43, 44, 46	100.0	90.2	97.8	100.0	87.4	97.2	100.0	86.7	97.1	100.0	88.0	97.4	100.0	92.9	98.4	100.0	90.4	97.9
21, 22, 23	FIRE SUPPRESSION, PLUMBING & HVAC	97.6	89.4	94.3	97.6	95.2	96.6	100.8	82.8	93.6	98.0	93.4	96.1	100.1	95.1	98.1	100.0	85.8	94.3
26, 27, 3370	ELECTRICAL, COMMUNICATIONS & UTIL.	92.2	93.0	92.6	93.1	93.0	93.1	94.8	82.2	88.3	88.0	109.1	98.8	99.2	106.0	102.7	94.4	76.2	85.0
MF2018	WEIGHTED AVERAGE	97.5	86.4	92.7	94.0	88.5	91.6	96.1	81.8	89.9	95.9	92.1	94.3	97.0	92.8	95.2	99.5	84.0	92.9

For customer support on your Plumbing Costs with RSMeans Data, call 800.448.8182.

661

DIVISION		OHIO			OKLAHOMA													
		ZANESVILLE			ARDMORE			CLINTON			DURANT			ENID			GUYMON		
		437 - 438			734			736			747			737			739		
		MAT.	INST.	TOTAL	MAT.	INST.	TOTAL	MAT.	INST.	TOTAL	MAT.	INST.	TOTAL	MAT.	INST.	TOTAL	MAT.	INST.	TOTAL
015433	CONTRACTOR EQUIPMENT		91.8	91.8		82.7	82.7		81.8	81.8		81.8	81.8		81.8	81.8		81.8	81.8
0241, 31 - 34	SITE & INFRASTRUCTURE, DEMOLITION	98.6	95.5	96.4	97.9	96.0	96.6	99.2	94.5	95.9	93.8	93.0	93.2	100.9	94.5	96.4	103.5	95.0	97.6
0310	Concrete Forming & Accessories	93.6	78.0	80.4	85.9	56.6	61.0	84.6	56.6	60.8	82.4	56.4	60.3	88.0	56.8	61.5	90.9	56.7	61.8
0320	Concrete Reinforcing	90.9	95.2	93.1	87.7	67.2	77.3	88.2	67.2	77.5	93.2	65.5	79.1	87.6	67.3	77.3	88.2	65.1	76.4
0330	Cast-in-Place Concrete	91.9	88.8	90.7	87.0	72.4	81.6	84.1	72.4	79.7	90.2	72.3	83.5	84.1	72.4	79.8	84.1	72.1	79.6
03	CONCRETE	87.2	84.6	86.0	82.8	63.8	74.3	82.4	63.8	74.0	85.5	63.4	75.5	82.8	63.9	74.3	85.4	63.3	75.4
04	MASONRY	83.5	86.2	85.2	94.6	57.5	71.7	118.1	57.5	80.6	87.3	63.4	72.5	100.2	57.5	73.8	97.1	56.6	72.1
05	METALS	97.3	85.1	93.6	100.5	61.2	88.5	100.6	61.2	88.6	92.6	60.7	82.9	102.0	61.3	89.6	101.0	59.7	88.5
06	WOOD, PLASTICS & COMPOSITES	90.3	76.8	82.9	90.9	56.0	75.9	98.7	56.0	75.5	89.3	56.0	71.2	102.0	56.0	77.0	105.4	56.0	78.5
07	THERMAL & MOISTURE PROTECTION	93.5	85.8	90.2	101.7	64.5	85.7	101.9	64.5	85.8	97.9	66.1	84.2	102.0	64.5	85.8	102.3	64.0	85.8
08	OPENINGS	93.1	79.7	90.0	105.9	56.0	94.3	105.9	56.0	94.3	98.1	55.5	88.2	107.1	56.7	95.4	106.0	55.5	94.3
0920	Plaster & Gypsum Board	94.5	76.1	82.4	93.7	55.3	68.4	93.4	55.3	68.3	79.6	55.3	63.6	94.3	55.3	68.6	94.3	55.3	68.6
0950, 0980	Ceilings & Acoustic Treatment	97.6	76.1	83.1	83.2	55.3	64.3	83.2	55.3	64.3	79.4	55.3	63.1	83.2	55.3	64.3	83.2	55.3	64.3
0960	Flooring	91.8	77.8	87.8	87.3	59.2	79.1	86.2	59.2	78.3	91.4	53.4	80.4	87.8	73.1	83.6	89.3	59.2	80.6
0970, 0990	Wall Finishes & Painting/Coating	100.2	91.5	95.0	84.6	45.1	61.2	84.6	45.1	61.2	92.9	45.1	64.6	84.6	45.1	61.2	84.6	43.4	60.2
09	FINISHES	95.2	78.5	86.1	81.0	54.9	66.8	80.8	54.9	66.7	81.4	53.8	66.3	81.5	57.8	68.6	82.4	56.1	68.0
COVERS	DIVS. 10 - 14, 25, 28, 41, 43, 44, 46	100.0	84.5	96.6	100.0	79.0	95.4	100.0	79.0	95.4	100.0	79.0	95.4	100.0	79.0	95.4	100.0	79.0	95.4
21, 22, 23	FIRE SUPPRESSION, PLUMBING & HVAC	97.6	91.5	95.1	97.7	67.8	85.7	97.7	67.8	85.7	97.8	67.7	85.7	100.2	67.8	87.2	97.7	65.0	84.6
26, 27, 3370	ELECTRICAL, COMMUNICATIONS & UTIL.	93.3	91.9	92.6	94.0	70.9	82.2	95.1	70.9	82.7	96.3	70.9	83.3	95.1	70.9	82.7	96.7	67.2	81.6
MF2018	WEIGHTED AVERAGE	94.6	87.1	91.4	95.7	66.2	83.0	96.9	66.1	83.6	93.5	66.3	81.8	97.1	66.5	84.0	96.8	64.8	83.1

DIVISION		OKLAHOMA																
		LAWTON			MCALESTER			MIAMI			MUSKOGEE			OKLAHOMA CITY			PONCA CITY		
		735			745			743			744			730 - 731			746		
		MAT.	INST.	TOTAL	MAT.	INST.	TOTAL	MAT.	INST.	TOTAL	MAT.	INST.	TOTAL	MAT.	INST.	TOTAL	MAT.	INST.	TOTAL
015433	CONTRACTOR EQUIPMENT		82.7	82.7		81.8	81.8		93.2	93.2		93.2	93.2		83.0	83.0		81.8	81.8
0241, 31 - 34	SITE & INFRASTRUCTURE, DEMOLITION	97.1	96.0	96.4	87.2	95.1	92.7	88.4	92.4	91.2	88.8	92.2	91.2	95.3	96.5	96.2	94.4	95.4	95.1
0310	Concrete Forming & Accessories	90.8	56.8	61.9	80.5	42.5	48.2	93.4	56.6	62.2	97.8	56.3	62.6	91.0	63.5	67.6	89.0	56.8	61.6
0320	Concrete Reinforcing	87.8	67.3	77.4	92.9	65.1	78.7	91.4	67.3	79.1	92.3	64.7	78.2	97.1	67.3	81.9	92.3	67.2	79.5
0330	Cast-in-Place Concrete	81.3	72.4	78.0	79.1	72.0	76.4	82.9	73.4	79.3	83.8	73.2	79.9	81.3	74.3	78.7	92.6	72.4	85.1
03	CONCRETE	79.5	63.9	72.5	76.5	56.9	67.7	81.1	65.2	73.9	82.6	64.5	74.5	82.5	67.6	75.8	87.6	63.9	76.9
04	MASONRY	97.0	57.5	72.5	104.4	57.4	75.3	90.1	57.5	70.0	106.1	50.2	71.6	99.6	57.5	73.5	82.9	57.5	67.2
05	METALS	105.8	61.3	92.3	92.5	59.6	82.5	92.5	76.4	87.6	94.0	74.4	88.0	98.7	61.4	87.4	92.5	61.3	83.0
06	WOOD, PLASTICS & COMPOSITES	104.5	56.0	78.1	86.8	37.2	59.9	102.1	56.1	77.1	106.6	56.1	79.2	92.9	65.0	77.7	97.5	56.0	75.0
07	THERMAL & MOISTURE PROTECTION	101.7	64.5	85.7	97.4	62.4	82.4	97.9	65.4	83.9	98.2	62.8	83.0	93.4	65.8	81.5	98.1	64.7	83.7
08	OPENINGS	108.9	56.7	96.8	98.1	45.2	85.8	98.1	56.1	88.3	99.3	55.6	89.1	103.4	61.6	93.7	98.1	56.0	88.3
0920	Plaster & Gypsum Board	95.8	55.3	69.1	78.7	36.0	50.6	85.7	55.3	65.7	87.5	55.3	66.3	100.2	64.5	76.7	84.5	55.3	65.3
0950, 0980	Ceilings & Acoustic Treatment	90.0	55.3	66.5	79.4	36.0	50.0	79.4	55.3	63.1	87.9	55.3	65.8	91.5	64.5	73.2	79.4	55.3	63.1
0960	Flooring	89.6	73.1	84.8	90.3	59.2	81.3	97.4	53.4	84.6	99.9	34.3	80.8	91.1	73.1	85.9	94.5	59.2	84.2
0970, 0990	Wall Finishes & Painting/Coating	84.6	45.1	61.2	92.9	43.4	63.6	92.9	43.4	63.6	92.9	43.4	63.6	88.6	45.1	62.9	92.9	45.1	64.6
09	FINISHES	83.1	57.8	69.3	80.4	43.7	60.3	83.3	53.7	67.1	86.0	50.3	66.5	86.5	63.1	73.7	83.1	55.9	68.2
COVERS	DIVS. 10 - 14, 25, 28, 41, 43, 44, 46	100.0	79.0	95.4	100.0	77.0	94.9	100.0	77.9	95.1	100.0	77.9	95.1	100.0	80.0	95.6	100.0	79.0	95.4
21, 22, 23	FIRE SUPPRESSION, PLUMBING & HVAC	100.2	67.8	87.2	97.8	65.0	84.6	97.8	65.0	84.7	100.2	62.7	85.2	100.1	67.8	87.1	97.8	65.0	84.6
26, 27, 3370	ELECTRICAL, COMMUNICATIONS & UTIL.	96.7	70.9	83.5	94.8	68.6	81.4	96.2	68.7	82.1	94.4	68.8	81.3	102.2	70.9	86.2	94.4	67.2	80.5
MF2018	WEIGHTED AVERAGE	97.6	66.6	84.3	92.8	61.9	79.5	93.2	66.5	81.7	95.2	64.4	82.0	96.8	68.2	84.5	93.6	65.2	81.4

DIVISION		OKLAHOMA												OREGON					
		POTEAU			SHAWNEE			TULSA			WOODWARD			BEND			EUGENE		
		749			748			740 - 741			738			977			974		
		MAT.	INST.	TOTAL	MAT.	INST.	TOTAL	MAT.	INST.	TOTAL	MAT.	INST.	TOTAL	MAT.	INST.	TOTAL	MAT.	INST.	TOTAL
015433	CONTRACTOR EQUIPMENT		91.9	91.9		81.8	81.8		93.2	93.2		81.8	81.8		99.8	99.8		99.8	99.8
0241, 31 - 34	SITE & INFRASTRUCTURE, DEMOLITION	75.7	87.9	84.2	97.4	95.4	96.0	95.0	91.4	92.5	99.6	94.5	96.0	114.1	103.2	106.5	103.7	103.2	103.4
0310	Concrete Forming & Accessories	86.9	56.4	61.0	82.2	56.7	60.6	98.0	57.5	63.6	84.7	44.6	50.6	106.6	96.8	98.3	103.0	96.8	97.7
0320	Concrete Reinforcing	93.3	67.2	80.1	92.3	67.2	79.5	92.5	67.3	79.7	87.6	67.2	77.3	100.4	111.7	106.1	104.8	111.7	108.3
0330	Cast-in-Place Concrete	82.9	73.3	79.3	95.6	72.4	87.0	91.3	75.3	85.4	84.1	72.3	79.7	115.1	99.9	109.4	111.4	99.9	107.1
03	CONCRETE	83.1	65.0	75.0	89.2	63.9	77.8	87.8	66.3	78.1	82.6	58.3	71.7	109.4	99.9	105.2	100.9	99.9	100.4
04	MASONRY	90.4	57.5	70.1	105.8	57.5	75.9	91.0	57.5	70.3	90.0	57.5	69.9	101.6	101.6	101.6	98.5	101.6	100.4
05	METALS	92.5	76.1	87.6	92.4	61.3	83.0	97.0	76.5	90.8	100.6	61.1	88.6	96.0	94.9	95.7	96.7	94.9	96.2
06	WOOD, PLASTICS & COMPOSITES	94.1	56.1	73.5	89.1	56.0	71.1	105.8	57.2	79.4	98.8	39.8	66.8	101.1	96.1	98.4	96.9	96.1	96.5
07	THERMAL & MOISTURE PROTECTION	98.0	65.4	84.0	98.1	64.0	83.4	98.3	65.6	84.2	101.9	62.9	85.1	110.2	98.7	105.3	109.3	101.7	106.0
08	OPENINGS	98.1	56.1	88.3	98.1	56.0	88.3	101.0	57.3	90.9	105.9	47.1	92.2	99.2	99.3	99.2	99.5	99.3	99.5
0920	Plaster & Gypsum Board	82.6	55.3	64.6	79.6	55.3	63.6	87.5	56.4	67.0	93.4	38.7	57.3	119.9	95.8	104.0	118.1	95.8	103.4
0950, 0980	Ceilings & Acoustic Treatment	79.4	55.3	63.1	79.4	55.3	63.1	87.9	56.4	66.6	83.2	38.7	53.0	95.9	95.8	95.9	96.9	95.8	96.2
0960	Flooring	93.9	53.4	82.1	91.4	59.2	82.0	98.7	63.7	88.5	86.2	56.5	77.5	97.8	104.2	99.7	96.3	104.2	98.6
0970, 0990	Wall Finishes & Painting/Coating	92.9	45.1	64.6	92.9	43.4	63.6	92.9	52.6	69.1	84.6	45.1	61.2	96.1	71.4	81.5	96.1	71.4	81.5
09	FINISHES	81.1	53.9	66.3	81.6	54.7	66.9	85.9	57.4	70.3	80.9	44.8	61.2	96.5	95.2	95.8	94.9	95.2	95.1
COVERS	DIVS. 10 - 14, 25, 28, 41, 43, 44, 46	100.0	79.3	95.4	100.0	79.0	95.4	100.0	78.0	95.2	100.0	77.3	95.0	100.0	98.3	99.6	100.0	98.3	99.6
21, 22, 23	FIRE SUPPRESSION, PLUMBING & HVAC	97.8	65.0	84.6	97.8	67.8	85.8	100.2	65.1	86.2	97.7	67.8	85.7	97.6	107.5	101.5	100.0	100.8	100.3
26, 27, 3370	ELECTRICAL, COMMUNICATIONS & UTIL.	94.5	68.7	81.3	96.4	70.9	83.4	96.3	68.7	82.2	96.5	70.9	83.4	100.6	93.7	97.1	99.3	93.7	96.4
MF2018	WEIGHTED AVERAGE	92.8	66.1	81.3	95.0	66.1	82.6	96.1	67.1	83.7	95.7	63.4	81.8	100.4	99.8	100.1	99.3	98.5	99.0

OREGON

DIVISION		KLAMATH FALLS 976 MAT.	INST.	TOTAL	MEDFORD 975 MAT.	INST.	TOTAL	PENDLETON 978 MAT.	INST.	TOTAL	PORTLAND 970-972 MAT.	INST.	TOTAL	SALEM 973 MAT.	INST.	TOTAL	VALE 979 MAT.	INST.	TOTAL
015433	CONTRACTOR EQUIPMENT		99.8	99.8		99.8	99.8		97.3	97.3		99.8	99.8		99.8	99.8		97.3	97.3
0241, 31 - 34	SITE & INFRASTRUCTURE, DEMOLITION	118.4	103.2	107.8	111.8	103.2	105.8	111.9	96.3	101.0	106.6	103.2	104.2	97.8	103.2	101.6	98.5	96.2	96.9
0310	Concrete Forming & Accessories	99.4	96.6	97.0	98.5	96.6	96.9	100.2	97.0	97.5	104.2	96.8	97.9	105.3	96.8	98.1	106.8	95.7	97.4
0320	Concrete Reinforcing	100.4	111.6	106.1	102.1	111.6	106.9	99.5	111.7	105.7	105.5	111.7	108.7	113.5	111.7	112.6	97.1	111.5	104.4
0330	Cast-in-Place Concrete	115.1	99.8	109.4	115.0	99.8	109.4	115.9	100.3	110.1	114.5	99.9	109.0	105.2	99.9	103.2	91.2	100.6	94.7
03	CONCRETE	112.1	99.8	106.6	106.9	99.8	103.7	94.0	100.2	96.8	102.5	99.9	101.4	98.3	99.9	99.0	78.2	99.6	87.9
04	MASONRY	114.5	101.6	106.5	95.4	101.6	99.2	105.6	101.7	103.2	100.0	101.6	101.0	104.3	101.6	102.6	103.4	101.7	102.3
05	METALS	96.0	94.6	95.6	96.3	94.6	95.8	103.1	95.5	100.8	97.8	94.9	97.0	103.8	94.9	101.1	103.0	94.2	100.3
06	WOOD, PLASTICS & COMPOSITES	91.9	96.1	94.2	90.8	96.1	93.7	94.0	96.2	95.2	98.1	96.1	97.0	93.1	96.1	94.7	102.6	96.2	99.1
07	THERMAL & MOISTURE PROTECTION	110.5	95.3	103.9	110.1	95.3	103.7	102.5	97.0	100.2	109.3	98.8	104.8	105.9	99.1	103.0	101.9	92.8	98.0
08	OPENINGS	99.2	99.3	99.3	102.0	99.3	101.4	95.5	99.4	96.4	97.4	99.3	97.9	101.2	99.3	100.7	95.4	89.2	94.0
0920	Plaster & Gypsum Board	114.8	95.8	102.3	114.2	95.8	102.1	100.3	95.8	97.4	118.0	95.8	103.4	113.0	95.8	101.7	106.4	95.8	99.4
0950, 0980	Ceilings & Acoustic Treatment	103.6	95.8	98.3	110.2	95.8	100.5	68.4	95.8	87.0	98.8	95.8	96.8	103.9	95.8	98.4	68.4	95.8	87.0
0960	Flooring	95.0	104.2	97.6	94.4	104.2	97.2	65.5	104.2	76.7	94.2	104.2	97.1	99.1	104.2	100.6	67.5	104.2	78.1
0970, 0990	Wall Finishes & Painting/Coating	96.1	67.0	78.9	96.1	67.0	78.9	87.6	76.2	80.9	95.9	76.2	84.3	93.4	75.1	82.6	87.6	71.4	78.0
09	FINISHES	96.9	94.7	95.7	97.1	94.7	95.8	82.6	95.8	83.5	94.7	95.7	95.2	95.1	95.6	95.4	69.3	95.3	83.5
COVERS	DIVS. 10 - 14, 25, 28, 41, 43, 44, 46	100.0	98.2	99.6	100.0	98.2	99.6	100.0	98.2	99.6	100.0	98.3	99.6	100.0	98.3	99.6	100.0	98.5	99.7
21, 22, 23	FIRE SUPPRESSION, PLUMBING & HVAC	97.6	107.5	101.5	100.0	107.5	103.0	99.3	114.5	105.4	100.0	104.7	101.9	100.0	107.5	103.0	99.3	72.6	88.6
26, 27, 3370	ELECTRICAL, COMMUNICATIONS & UTIL.	99.4	80.5	89.7	102.8	80.5	91.4	91.7	99.2	95.5	99.5	101.5	100.5	107.2	93.7	100.3	91.7	69.7	80.4
MF2018	WEIGHTED AVERAGE	101.3	97.8	99.8	100.8	97.8	99.5	96.4	101.7	98.7	99.6	100.4	100.0	101.1	99.9	100.6	94.1	87.9	91.4

PENNSYLVANIA

DIVISION		ALLENTOWN 181 MAT.	INST.	TOTAL	ALTOONA 166 MAT.	INST.	TOTAL	BEDFORD 155 MAT.	INST.	TOTAL	BRADFORD 167 MAT.	INST.	TOTAL	BUTLER 160 MAT.	INST.	TOTAL	CHAMBERSBURG 172 MAT.	INST.	TOTAL
015433	CONTRACTOR EQUIPMENT		115.2	115.2		115.2	115.2		113.2	113.2		115.2	115.2		115.2	115.2		114.4	114.4
0241, 31 - 34	SITE & INFRASTRUCTURE, DEMOLITION	93.0	101.8	99.1	95.9	101.6	99.9	100.4	99.3	99.6	92.0	100.5	97.9	87.2	102.4	97.8	88.0	99.8	96.2
0310	Concrete Forming & Accessories	98.3	111.2	109.3	83.2	84.9	84.6	81.9	83.5	83.2	85.4	99.2	97.1	84.7	96.7	94.9	88.5	78.9	79.8
0320	Concrete Reinforcing	103.8	116.6	110.3	100.6	108.8	104.8	99.5	108.8	104.2	102.8	108.8	105.9	101.3	122.8	112.2	103.4	115.5	109.5
0330	Cast-in-Place Concrete	87.6	102.8	93.2	97.6	89.3	94.5	94.0	88.3	91.9	93.2	93.2	93.2	86.2	98.9	90.9	95.1	96.5	95.6
03	CONCRETE	90.7	109.9	99.3	86.6	91.9	89.0	95.9	90.9	93.6	92.5	99.6	95.7	78.8	103.0	89.7	96.1	92.9	94.7
04	MASONRY	94.2	95.1	94.8	97.9	85.1	90.0	109.3	81.9	92.4	95.3	83.7	88.1	99.9	94.9	96.8	95.1	80.8	86.3
05	METALS	94.2	121.3	102.5	88.5	115.6	96.8	97.7	114.1	102.7	92.1	114.0	98.8	88.3	122.5	98.7	88.1	119.6	97.7
06	WOOD, PLASTICS & COMPOSITES	99.6	114.8	107.8	78.0	84.1	81.3	80.6	84.1	82.5	84.2	104.2	95.0	79.5	96.8	88.9	85.2	77.8	81.1
07	THERMAL & MOISTURE PROTECTION	102.7	110.8	106.2	101.5	92.5	97.6	101.5	89.8	96.5	102.5	91.0	97.6	101.0	97.1	99.3	93.2	85.4	89.8
08	OPENINGS	94.5	110.4	98.2	88.2	88.1	88.2	95.5	88.1	93.7	94.5	98.0	95.3	88.2	102.2	91.4	91.2	83.8	89.5
0920	Plaster & Gypsum Board	94.2	115.1	108.0	85.6	83.6	84.3	100.1	83.6	89.2	86.7	104.2	98.2	85.6	96.6	92.9	102.2	77.1	85.7
0950, 0980	Ceilings & Acoustic Treatment	88.8	115.1	106.6	91.4	83.6	86.1	101.3	83.6	89.3	91.5	104.2	100.1	92.3	96.6	95.2	92.1	77.1	81.9
0960	Flooring	90.7	98.3	92.9	84.2	102.3	89.4	96.7	107.2	99.7	85.0	107.2	91.5	85.1	108.9	92.0	88.1	80.6	85.9
0970, 0990	Wall Finishes & Painting/Coating	90.4	104.5	98.8	86.0	109.0	99.6	94.8	108.5	102.9	90.4	108.5	101.1	86.0	109.0	99.6	91.5	103.4	98.5
09	FINISHES	88.8	108.9	99.8	86.4	90.0	88.4	99.0	89.9	94.1	86.6	102.2	95.2	86.4	100.0	93.8	85.9	81.1	83.3
COVERS	DIVS. 10 - 14, 25, 28, 41, 43, 44, 46	100.0	101.6	100.4	100.0	96.4	99.2	100.0	95.1	98.9	100.0	98.1	99.6	100.0	99.3	99.8	100.0	94.6	98.8
21, 22, 23	FIRE SUPPRESSION, PLUMBING & HVAC	100.2	117.1	107.0	99.8	85.5	94.0	97.5	86.3	93.0	97.8	92.6	95.7	97.3	97.6	97.4	97.7	91.7	95.3
26, 27, 3370	ELECTRICAL, COMMUNICATIONS & UTIL.	100.3	98.1	99.2	90.7	110.3	100.8	98.7	110.3	104.7	94.3	110.3	102.5	91.3	110.4	101.0	91.7	87.9	89.8
MF2018	WEIGHTED AVERAGE	96.2	108.4	101.4	92.9	95.1	93.8	98.2	94.3	96.3	94.6	99.5	96.7	91.2	103.1	96.3	93.3	91.6	92.6

PENNSYLVANIA

DIVISION		DOYLESTOWN 189 MAT.	INST.	TOTAL	DUBOIS 158 MAT.	INST.	TOTAL	ERIE 164 - 165 MAT.	INST.	TOTAL	GREENSBURG 156 MAT.	INST.	TOTAL	HARRISBURG 170 - 171 MAT.	INST.	TOTAL	HAZLETON 182 MAT.	INST.	TOTAL
015433	CONTRACTOR EQUIPMENT		93.4	93.4		113.2	113.2		115.2	115.2		113.2	113.2		114.4	114.4		115.2	115.2
0241, 31 - 34	SITE & INFRASTRUCTURE, DEMOLITION	106.8	89.8	95.0	105.1	99.7	101.3	92.8	101.9	99.2	96.6	101.2	99.8	88.1	99.9	96.3	86.3	101.6	97.0
0310	Concrete Forming & Accessories	82.4	130.2	123.0	81.4	86.0	85.3	97.6	88.2	89.6	88.2	91.0	90.6	95.8	87.9	89.1	80.0	91.3	89.6
0320	Concrete Reinforcing	100.4	154.7	128.0	98.8	122.8	111.0	102.8	110.4	106.7	98.8	122.5	110.9	113.1	114.2	113.7	100.8	116.3	108.7
0330	Cast-in-Place Concrete	82.7	133.4	101.6	90.6	96.8	92.9	95.9	91.2	94.2	87.2	98.4	91.4	93.7	98.7	95.5	82.7	98.3	88.5
03	CONCRETE	86.3	134.6	108.0	98.0	97.3	97.7	94.9	94.6	89.6	91.4	100.2	95.3	91.7	97.5	94.3	83.5	99.3	90.6
04	MASONRY	97.6	128.4	116.6	109.8	95.4	100.9	86.9	88.7	88.0	120.0	90.9	102.1	89.9	85.6	87.3	107.2	91.4	97.4
05	METALS	91.9	123.9	101.6	97.7	119.6	104.3	88.7	116.3	97.1	97.6	120.5	104.6	94.2	119.3	101.9	94.0	120.5	102.0
06	WOOD, PLASTICS & COMPOSITES	79.8	129.4	106.7	79.5	84.1	82.0	96.0	86.8	91.0	86.8	89.6	88.3	98.0	87.6	92.4	78.3	89.5	84.4
07	THERMAL & MOISTURE PROTECTION	99.9	132.7	114.0	101.8	95.5	99.1	102.1	91.0	97.4	101.4	94.8	98.5	96.1	102.6	98.9	102.0	103.5	102.7
08	OPENINGS	96.9	137.0	106.2	95.5	91.2	94.5	88.3	90.5	88.8	95.4	98.3	96.1	98.0	88.9	95.9	95.0	96.0	95.3
0920	Plaster & Gypsum Board	84.7	130.2	114.7	98.9	83.6	88.8	94.2	86.4	89.1	101.1	89.3	93.4	108.1	87.2	94.3	85.1	89.2	87.8
0950, 0980	Ceilings & Acoustic Treatment	88.1	130.2	116.6	101.3	83.6	89.3	88.8	86.4	87.2	100.5	89.3	92.9	99.7	87.2	91.2	89.8	89.2	89.4
0960	Flooring	75.2	144.0	95.2	96.4	107.2	99.5	91.1	102.3	94.3	100.3	80.6	94.6	92.0	93.5	92.4	82.3	92.3	85.2
0970, 0990	Wall Finishes & Painting/Coating	90.0	147.5	124.0	94.8	109.0	103.2	96.5	94.8	95.5	94.8	109.0	103.2	96.6	87.0	90.9	90.4	106.8	100.1
09	FINISHES	80.5	134.3	109.9	99.2	91.4	95.0	89.8	91.0	90.5	99.7	90.6	94.8	91.1	88.2	89.5	84.9	92.8	89.2
COVERS	DIVS. 10 - 14, 25, 28, 41, 43, 44, 46	100.0	115.3	103.4	100.0	97.1	99.4	100.0	97.9	99.5	100.0	98.4	99.6	100.0	96.9	99.3	100.0	97.4	99.4
21, 22, 23	FIRE SUPPRESSION, PLUMBING & HVAC	97.3	135.6	112.6	97.5	89.5	94.3	99.8	95.7	98.1	97.5	89.3	94.2	100.1	93.7	97.5	97.8	97.7	97.8
26, 27, 3370	ELECTRICAL, COMMUNICATIONS & UTIL.	93.7	135.6	115.1	99.3	110.3	105.0	92.4	95.4	93.9	99.3	110.3	105.0	98.9	87.9	93.3	95.1	92.2	93.6
MF2018	WEIGHTED AVERAGE	93.8	129.2	109.0	98.7	98.3	98.5	92.8	96.2	94.3	98.2	98.7	98.4	96.1	94.9	95.6	94.1	98.3	95.9

For customer support on your Plumbing Costs with RSMeans Data, call 800.448.8182.

663

City Cost Indexes

PENNSYLVANIA

DIVISION		INDIANA 157			JOHNSTOWN 159			KITTANNING 162			LANCASTER 175 - 176			LEHIGH VALLEY 180			MONTROSE 188		
		MAT.	INST.	TOTAL	MAT.	INST.	TOTAL	MAT.	INST.	TOTAL	MAT.	INST.	TOTAL	MAT.	INST.	TOTAL	MAT.	INST.	TOTAL
015433	CONTRACTOR EQUIPMENT		113.2	113.2		113.2	113.2		115.2	115.2		114.4	114.4		115.2	115.2		115.2	115.2
0241, 31 - 34	SITE & INFRASTRUCTURE, DEMOLITION	94.7	100.3	98.6	100.8	100.5	100.6	89.8	102.0	98.3	80.0	100.1	94.0	90.3	102.0	98.5	89.0	101.7	97.9
0310	Concrete Forming & Accessories	82.4	95.5	93.5	81.4	84.8	84.3	84.7	92.2	91.0	86.5	88.8	88.4	91.8	111.6	108.6	80.9	91.8	90.2
0320	Concrete Reinforcing	98.1	122.9	110.7	99.5	122.3	111.1	101.3	122.6	112.2	103.0	114.2	108.7	100.8	111.1	106.0	105.6	119.6	112.7
0330	Cast-in-Place Concrete	85.5	97.1	89.8	94.8	89.1	92.7	89.6	97.4	92.5	80.8	99.5	87.8	89.6	102.0	94.2	87.8	95.7	90.7
03	CONCRETE	88.9	101.7	94.7	96.8	94.0	95.5	81.3	100.4	89.9	84.4	98.2	90.6	89.8	108.9	98.4	88.5	99.2	93.3
04	MASONRY	105.6	95.5	99.3	106.4	85.5	93.5	102.5	90.7	95.2	100.9	88.6	93.3	94.3	98.5	96.9	94.2	93.5	93.7
05	METALS	97.7	120.7	104.7	97.7	118.7	104.1	88.3	121.4	98.4	88.1	119.5	97.7	93.9	119.7	101.7	92.2	121.2	101.0
06	WOOD, PLASTICS & COMPOSITES	81.3	96.7	89.7	79.5	84.1	82.0	79.5	91.9	86.2	88.1	87.6	87.8	90.9	113.5	103.2	79.1	89.5	84.8
07	THERMAL & MOISTURE PROTECTION	101.2	96.9	99.4	101.5	91.0	97.0	101.2	95.0	98.5	92.6	103.8	97.4	102.4	112.8	106.9	102.0	93.5	98.4
08	OPENINGS	95.5	98.2	96.1	95.5	91.2	94.5	88.2	99.5	90.8	91.2	88.9	90.7	95.0	108.5	98.1	91.8	94.5	92.4
0920	Plaster & Gypsum Board	100.4	96.6	97.9	98.7	83.6	88.8	85.6	91.6	89.6	104.6	87.2	93.1	87.5	113.8	104.8	85.5	89.2	87.9
0950, 0980	Ceilings & Acoustic Treatment	101.3	96.6	98.1	100.5	83.6	89.1	92.3	91.6	91.8	92.1	87.2	88.8	89.8	113.8	106.1	91.5	89.2	89.9
0960	Flooring	97.2	107.2	100.1	96.4	80.6	91.8	85.1	107.2	91.5	89.1	98.9	91.9	87.7	97.3	90.5	83.0	107.2	90.0
0970, 0990	Wall Finishes & Painting/Coating	94.8	109.0	103.2	94.8	109.0	103.2	86.0	109.0	99.6	91.5	88.9	90.0	90.4	103.6	98.2	90.4	106.8	100.1
09	FINISHES	98.8	98.9	98.9	98.6	86.3	91.9	86.5	96.1	91.7	86.0	89.9	88.1	87.0	108.3	98.6	85.6	95.4	91.0
COVERS	DIVS. 10 - 14, 25, 28, 41, 43, 44, 46	100.0	98.4	99.7	100.0	96.3	99.2	100.0	98.1	99.6	100.0	97.6	99.5	100.0	104.3	101.0	100.0	98.1	99.6
21, 22, 23	FIRE SUPPRESSION, PLUMBING & HVAC	97.5	87.3	93.4	97.5	80.7	90.8	97.3	92.7	95.5	97.7	94.8	96.5	97.8	118.6	106.2	97.8	96.9	97.4
26, 27, 3370	ELECTRICAL, COMMUNICATIONS & UTIL.	99.3	110.3	105.0	99.3	110.3	105.0	90.7	110.3	100.8	93.0	95.6	94.3	95.1	132.5	114.3	94.3	97.6	96.0
MF2018	WEIGHTED AVERAGE	97.1	100.1	98.4	98.2	94.2	96.5	91.7	100.4	95.4	92.1	96.8	94.1	94.7	113.6	102.8	93.6	99.2	96.0

PENNSYLVANIA

DIVISION		NEW CASTLE 161			NORRISTOWN 194			OIL CITY 163			PHILADELPHIA 190 - 191			PITTSBURGH 150 - 152			POTTSVILLE 179		
		MAT.	INST.	TOTAL	MAT.	INST.	TOTAL	MAT.	INST.	TOTAL	MAT.	INST.	TOTAL	MAT.	INST.	TOTAL	MAT.	INST.	TOTAL
015433	CONTRACTOR EQUIPMENT		115.2	115.2		99.0	99.0		115.2	115.2		99.3	99.3		102.0	102.0		114.4	114.4
0241, 31 - 34	SITE & INFRASTRUCTURE, DEMOLITION	87.6	102.3	97.8	95.7	99.3	98.2	86.1	100.8	96.4	98.6	100.0	99.6	100.5	99.9	100.1	82.9	100.3	95.0
0310	Concrete Forming & Accessories	84.7	96.2	94.4	82.6	128.7	121.7	84.7	95.3	93.7	100.4	140.0	134.0	98.3	97.2	97.3	78.3	90.1	88.3
0320	Concrete Reinforcing	100.1	103.1	101.6	101.9	154.7	128.7	101.3	97.7	99.5	105.2	126.5	116.0	100.0	123.1	111.8	102.2	117.4	109.9
0330	Cast-in-Place Concrete	87.0	96.9	90.7	85.0	131.0	102.1	84.5	97.1	89.2	93.3	135.3	108.9	93.5	99.2	95.6	86.1	99.1	91.0
03	CONCRETE	79.1	98.7	88.0	86.9	133.0	107.7	77.6	97.5	86.6	101.3	134.8	116.4	97.5	102.3	99.7	87.9	99.3	93.0
04	MASONRY	99.6	94.6	96.5	108.7	124.6	118.5	98.8	90.6	93.7	98.9	131.3	118.9	101.3	99.6	100.2	94.6	89.4	91.4
05	METALS	88.3	115.0	96.4	93.5	123.8	102.7	88.3	113.9	96.1	98.1	113.5	102.8	99.1	107.4	101.6	88.3	121.2	98.3
06	WOOD, PLASTICS & COMPOSITES	79.5	96.8	88.9	76.1	129.3	105.0	79.5	96.8	88.9	96.8	141.8	121.2	99.7	97.2	98.3	77.8	87.6	83.1
07	THERMAL & MOISTURE PROTECTION	101.1	94.6	98.3	106.5	131.1	117.1	100.9	93.3	97.6	99.7	136.4	115.5	101.8	98.4	100.4	92.7	102.4	96.9
08	OPENINGS	88.2	94.1	89.5	88.7	137.0	99.9	88.2	97.7	90.4	101.3	137.3	109.6	97.8	102.5	98.9	91.3	95.7	92.3
0920	Plaster & Gypsum Board	85.6	96.6	92.9	83.8	130.2	114.3	85.6	96.6	92.9	97.8	142.8	127.4	98.7	96.9	97.5	99.1	87.2	91.3
0950, 0980	Ceilings & Acoustic Treatment	92.3	96.6	95.2	91.5	130.2	117.7	92.3	96.6	95.2	101.6	142.8	129.5	95.7	96.9	96.3	92.1	87.2	88.8
0960	Flooring	85.1	108.9	92.0	83.2	144.0	100.8	85.1	107.2	91.5	97.5	145.7	111.5	104.9	107.5	105.7	85.2	107.2	91.6
0970, 0990	Wall Finishes & Painting/Coating	86.0	109.0	99.6	85.8	147.5	122.3	86.0	109.0	99.6	91.8	152.1	127.5	97.5	111.4	105.7	91.5	106.8	100.6
09	FINISHES	86.4	99.8	93.7	83.3	133.3	110.6	86.3	98.9	93.2	98.4	142.8	122.7	100.4	100.2	100.3	84.3	93.9	89.6
COVERS	DIVS. 10 - 14, 25, 28, 41, 43, 44, 46	100.0	99.3	99.8	100.0	111.3	102.5	100.0	98.6	99.7	100.0	120.0	104.4	100.0	101.5	100.3	100.0	99.5	99.9
21, 22, 23	FIRE SUPPRESSION, PLUMBING & HVAC	97.3	99.0	98.0	97.6	133.6	112.0	97.3	96.3	96.9	100.2	140.0	116.1	99.9	98.3	99.2	97.7	97.9	97.8
26, 27, 3370	ELECTRICAL, COMMUNICATIONS & UTIL.	91.3	100.6	96.1	94.1	145.5	120.4	93.0	110.3	101.9	101.2	160.5	131.5	101.7	110.8	106.4	91.3	93.5	92.4
MF2018	WEIGHTED AVERAGE	91.2	100.3	95.1	94.0	130.0	109.4	91.1	100.3	95.1	99.9	135.3	115.1	99.6	102.2	100.7	92.0	98.4	94.8

PENNSYLVANIA

DIVISION		READING 195 - 196			SCRANTON 184 - 185			STATE COLLEGE 168			STROUDSBURG 183			SUNBURY 178			UNIONTOWN 154		
		MAT.	INST.	TOTAL	MAT.	INST.	TOTAL	MAT.	INST.	TOTAL	MAT.	INST.	TOTAL	MAT.	INST.	TOTAL	MAT.	INST.	TOTAL
015433	CONTRACTOR EQUIPMENT		121.3	121.3		115.2	115.2		114.4	114.4		115.2	115.2		115.2	115.2		113.2	113.2
0241, 31 - 34	SITE & INFRASTRUCTURE, DEMOLITION	99.9	112.5	108.7	93.5	101.8	99.3	84.2	100.2	95.4	88.0	101.9	97.7	94.7	101.2	99.3	95.3	101.1	99.3
0310	Concrete Forming & Accessories	98.2	88.5	90.0	98.5	89.0	90.5	84.4	85.1	85.0	86.2	93.1	92.0	89.0	87.1	87.4	75.9	96.5	93.4
0320	Concrete Reinforcing	103.3	152.9	128.5	103.8	119.7	111.9	102.0	108.9	105.5	104.1	119.8	112.1	105.0	115.5	110.4	98.8	122.9	111.1
0330	Cast-in-Place Concrete	76.2	99.7	84.9	91.4	95.8	93.1	88.2	89.4	88.6	86.2	97.2	90.3	94.2	97.6	95.4	85.5	98.4	90.3
03	CONCRETE	85.9	104.7	94.4	92.5	98.0	95.0	92.7	92.1	92.4	87.3	100.3	93.1	91.6	97.1	94.1	88.6	102.7	94.9
04	MASONRY	97.7	93.4	95.0	94.6	95.8	95.4	100.2	85.7	91.2	92.4	98.3	96.1	94.9	82.7	87.4	122.0	97.4	106.8
05	METALS	93.8	133.7	105.9	96.2	121.3	103.8	92.0	115.8	99.2	94.0	121.7	102.4	88.1	120.2	97.8	97.4	121.0	104.6
06	WOOD, PLASTICS & COMPOSITES	93.7	84.8	88.8	99.6	85.7	92.1	86.1	84.1	85.0	85.1	89.5	87.5	86.0	87.6	86.9	73.7	96.7	86.2
07	THERMAL & MOISTURE PROTECTION	106.9	106.2	106.6	102.6	93.9	98.8	101.7	100.4	101.1	102.2	93.9	98.6	93.6	95.3	94.4	101.1	97.8	99.7
08	OPENINGS	93.2	102.3	95.3	94.5	92.4	94.0	91.6	88.1	90.8	95.1	97.3	95.6	91.3	92.5	91.6	95.4	102.2	97.0
0920	Plaster & Gypsum Board	93.2	84.3	87.4	95.9	85.3	88.9	87.4	83.6	84.9	86.3	89.2	88.2	98.4	87.2	91.0	96.9	96.6	96.7
0950, 0980	Ceilings & Acoustic Treatment	82.9	84.3	83.9	96.5	85.3	88.9	88.9	83.6	85.3	88.1	89.2	88.8	88.7	87.2	87.7	100.5	96.6	97.9
0960	Flooring	87.7	101.2	91.6	90.7	96.0	92.2	88.1	101.2	91.9	85.5	97.2	88.9	86.0	107.2	92.1	93.5	107.2	97.5
0970, 0990	Wall Finishes & Painting/Coating	84.6	104.5	96.4	90.4	111.1	102.7	90.4	109.0	101.4	90.4	106.8	100.1	91.5	106.8	100.6	94.8	112.9	105.5
09	FINISHES	84.5	91.0	88.0	90.5	91.9	91.3	86.2	89.8	88.2	85.8	94.5	90.5	85.1	92.2	89.0	97.1	99.8	98.6
COVERS	DIVS. 10 - 14, 25, 28, 41, 43, 44, 46	100.0	97.6	99.5	100.0	96.5	99.2	100.0	96.5	99.2	100.0	99.2	99.8	100.0	94.9	98.9	100.0	99.1	99.8
21, 22, 23	FIRE SUPPRESSION, PLUMBING & HVAC	100.2	109.2	103.8	100.2	98.2	99.4	97.8	93.4	96.0	97.8	100.1	98.7	97.7	89.3	94.3	97.5	93.7	96.0
26, 27, 3370	ELECTRICAL, COMMUNICATIONS & UTIL.	100.2	93.6	96.8	100.3	97.6	98.9	93.5	110.3	102.1	95.1	145.2	120.8	91.7	94.1	92.9	96.3	110.3	103.5
MF2018	WEIGHTED AVERAGE	95.4	104.0	99.1	96.9	98.9	97.8	94.2	97.0	95.4	94.1	107.3	99.8	92.9	95.0	93.8	97.2	102.2	99.4

For customer support on your Plumbing Costs with RSMeans Data, call 800.448.8182.

PENNSYLVANIA

DIVISION		WASHINGTON 153			WELLSBORO 169			WESTCHESTER 193			WILKES-BARRE 186 - 187			WILLIAMSPORT 177			YORK 173 - 174		
		MAT.	INST.	TOTAL	MAT.	INST.	TOTAL	MAT.	INST.	TOTAL	MAT.	INST.	TOTAL	MAT.	INST.	TOTAL	MAT.	INST.	TOTAL
015433	CONTRACTOR EQUIPMENT		113.2	113.2		115.2	115.2		99.0	99.0		115.2	115.2		115.2	115.2		114.4	114.4
0241, 31 - 34	SITE & INFRASTRUCTURE, DEMOLITION	95.3	101.4	99.6	95.5	101.2	99.4	101.8	100.3	100.8	86.0	101.8	97.0	85.7	101.4	96.6	83.8	100.1	95.1
0310	Concrete Forming & Accessories	82.6	96.6	94.5	84.9	86.5	86.3	89.2	130.0	123.9	89.1	87.6	87.9	85.7	88.0	87.6	81.6	88.7	87.6
0320	Concrete Reinforcing	98.8	123.0	111.1	102.0	119.5	110.9	101.0	154.6	128.3	102.8	119.7	111.4	104.2	114.1	109.3	105.0	114.2	109.7
0330	Cast-in-Place Concrete	85.5	98.6	90.4	92.4	90.7	91.8	94.1	133.1	108.7	82.7	95.6	87.5	79.7	92.5	84.5	86.5	99.5	91.3
03	CONCRETE	89.0	102.8	95.2	94.9	95.0	94.9	94.7	134.4	112.5	84.4	97.3	90.2	79.9	95.4	86.9	89.0	98.2	93.1
04	MASONRY	105.3	97.5	100.5	101.4	83.1	90.1	103.2	128.4	118.8	107.5	95.5	100.1	86.9	88.0	87.6	95.9	88.6	91.3
05	METALS	97.4	121.4	104.7	92.0	120.5	100.7	93.5	123.9	102.7	92.2	121.2	101.0	88.1	118.9	97.5	89.8	119.4	98.8
06	WOOD, PLASTICS & COMPOSITES	81.4	96.7	89.7	83.6	86.8	85.4	82.7	129.3	108.0	87.6	84.1	85.7	82.5	87.6	85.3	81.3	87.6	84.7
07	THERMAL & MOISTURE PROTECTION	101.2	97.8	99.7	102.8	88.4	96.6	107.0	128.0	116.0	102.0	93.4	98.3	93.0	90.7	92.0	92.8	103.8	97.5
08	OPENINGS	95.4	102.2	96.9	94.5	92.6	94.0	88.7	137.0	99.9	91.8	91.5	91.7	91.3	92.2	91.5	91.2	88.9	90.7
0920	Plaster & Gypsum Board	100.2	96.6	97.8	86.1	86.4	86.3	84.4	130.2	114.5	87.0	83.6	84.7	99.1	87.2	91.3	99.5	87.2	91.4
0950, 0980	Ceilings & Acoustic Treatment	100.5	96.6	97.9	88.9	86.4	87.2	91.5	130.2	117.7	91.5	83.6	86.1	92.1	87.2	88.8	91.1	87.2	88.5
0960	Flooring	97.3	107.2	100.2	84.7	107.2	91.2	86.0	144.0	102.8	86.4	96.0	89.2	84.9	89.7	86.3	86.5	98.9	90.1
0970, 0990	Wall Finishes & Painting/Coating	94.8	112.9	105.5	90.4	106.8	100.1	85.8	147.5	122.3	90.4	104.5	98.8	91.5	104.5	99.2	91.5	87.0	88.8
09	FINISHES	98.7	99.8	99.3	86.3	92.2	89.5	84.6	134.3	111.7	86.6	90.1	88.5	84.8	89.2	87.2	84.6	89.6	87.3
COVERS	DIVS. 10 - 14, 25, 28, 41, 43, 44, 46	100.0	99.1	99.8	100.0	98.2	99.6	100.0	117.6	103.9	100.0	96.2	99.2	100.0	97.4	99.4	100.0	97.6	99.5
21, 22, 23	FIRE SUPPRESSION, PLUMBING & HVAC	97.5	95.3	96.6	97.8	89.3	94.4	97.6	135.5	112.8	97.8	98.8	98.2	97.7	94.3	96.3	100.1	94.8	98.0
26, 27, 3370	ELECTRICAL, COMMUNICATIONS & UTIL.	98.7	110.4	104.7	94.3	97.6	96.0	94.0	135.6	115.3	95.1	92.2	93.6	92.2	85.4	88.7	93.0	84.9	88.9
MF2018	WEIGHTED AVERAGE	96.9	102.7	99.4	95.2	95.2	95.2	95.0	129.9	110.0	93.8	97.9	95.6	90.8	94.6	92.4	93.2	95.3	94.1

		PUERTO RICO			RHODE ISLAND						SOUTH CAROLINA								
DIVISION		SAN JUAN 009			NEWPORT 028			PROVIDENCE 029			AIKEN 298			BEAUFORT 299			CHARLESTON 294		
		MAT.	INST.	TOTAL	MAT.	INST.	TOTAL	MAT.	INST.	TOTAL	MAT.	INST.	TOTAL	MAT.	INST.	TOTAL	MAT.	INST.	TOTAL
015433	CONTRACTOR EQUIPMENT		86.5	86.5		100.1	100.1		100.1	100.1		105.8	105.8		105.8	105.8		105.8	105.8
0241, 31 - 34	SITE & INFRASTRUCTURE, DEMOLITION	129.2	89.2	101.4	86.7	102.2	97.5	86.8	102.2	97.5	126.7	89.0	100.4	121.9	88.4	98.5	107.3	88.7	94.3
0310	Concrete Forming & Accessories	91.5	21.3	31.9	99.1	121.2	117.9	99.5	121.2	117.9	91.3	63.7	67.9	90.5	39.1	46.8	89.7	63.7	67.6
0320	Concrete Reinforcing	198.3	18.3	106.8	104.9	118.9	112.0	105.8	118.9	112.5	92.7	62.4	77.3	91.8	62.5	76.6	91.6	67.7	79.4
0330	Cast-in-Place Concrete	94.9	32.2	71.6	68.9	120.7	88.1	88.1	120.7	100.2	91.9	66.1	82.3	91.8	65.7	82.1	107.8	66.1	92.3
03	CONCRETE	96.2	25.6	64.4	82.1	120.1	99.2	87.7	120.1	102.3	104.5	66.0	87.2	101.7	48.8	77.9	97.6	66.9	83.3
04	MASONRY	87.7	20.4	46.1	96.0	124.0	113.3	100.4	124.0	115.0	78.2	65.3	70.3	92.4	65.3	75.7	93.8	65.3	76.2
05	METALS	109.8	38.4	88.1	91.6	112.3	97.9	94.1	112.3	99.6	99.1	89.4	96.2	99.1	79.7	93.3	101.0	91.1	98.0
06	WOOD, PLASTICS & COMPOSITES	97.6	20.7	55.8	99.2	120.3	110.7	97.7	120.3	110.0	87.9	65.6	75.8	86.9	33.3	57.8	85.7	65.6	74.8
07	THERMAL & MOISTURE PROTECTION	129.8	25.4	84.9	101.8	119.5	109.4	102.9	119.5	110.0	98.6	64.9	84.1	98.3	56.9	80.5	97.3	64.5	83.2
08	OPENINGS	155.2	19.0	123.6	99.5	120.2	104.3	100.8	120.2	105.3	95.8	62.6	88.1	95.8	38.8	82.6	99.4	63.8	91.1
0920	Plaster & Gypsum Board	170.9	18.6	70.6	89.5	120.6	109.9	99.7	120.6	113.4	92.0	64.6	73.9	95.6	31.4	53.3	96.7	64.6	75.5
0950, 0980	Ceilings & Acoustic Treatment	254.2	18.6	94.5	93.6	120.6	111.9	98.6	120.6	113.5	82.6	64.6	70.4	85.1	31.4	48.7	85.1	64.6	71.2
0960	Flooring	197.7	20.8	146.4	87.9	130.5	100.2	91.6	130.5	102.9	98.2	92.8	96.6	99.7	78.4	93.5	99.3	79.8	93.3
0970, 0990	Wall Finishes & Painting/Coating	186.1	22.2	89.1	91.0	118.7	107.4	95.3	118.7	109.1	97.0	67.8	79.7	97.0	58.4	74.1	97.0	67.8	79.7
09	FINISHES	191.9	21.4	98.8	87.9	123.1	107.1	89.2	123.1	107.7	89.6	69.7	78.7	90.6	47.1	66.9	88.7	67.5	77.1
COVERS	DIVS. 10 - 14, 25, 28, 41, 43, 44, 46	100.0	19.5	82.3	100.0	106.1	101.3	100.0	106.1	101.3	100.0	70.6	93.5	100.0	81.4	95.9	100.0	70.6	93.5
21, 22, 23	FIRE SUPPRESSION, PLUMBING & HVAC	103.5	17.9	69.2	100.3	113.2	105.5	100.1	113.2	105.3	98.0	56.0	81.2	98.0	56.0	81.2	100.5	58.3	83.6
26, 27, 3370	ELECTRICAL, COMMUNICATIONS & UTIL.	122.6	16.9	68.6	99.5	97.9	98.7	98.1	97.9	98.0	95.1	63.4	78.9	98.5	31.6	64.3	97.0	58.3	77.2
MF2018	WEIGHTED AVERAGE	118.1	27.2	79.1	95.0	113.7	103.0	96.3	113.7	103.8	97.7	67.9	84.9	98.3	56.0	80.1	98.4	67.6	85.2

		SOUTH CAROLINA															SOUTH DAKOTA		
DIVISION		COLUMBIA 290 - 292			FLORENCE 295			GREENVILLE 296			ROCK HILL 297			SPARTANBURG 293			ABERDEEN 574		
		MAT.	INST.	TOTAL	MAT.	INST.	TOTAL	MAT.	INST.	TOTAL	MAT.	INST.	TOTAL	MAT.	INST.	TOTAL	MAT.	INST.	TOTAL
015433	CONTRACTOR EQUIPMENT		105.8	105.8		105.8	105.8		105.8	105.8		105.8	105.8		105.8	105.8		100.3	100.3
0241, 31 - 34	SITE & INFRASTRUCTURE, DEMOLITION	106.5	88.7	94.1	116.2	88.5	96.9	111.3	88.8	95.6	109.1	88.0	94.4	111.0	88.8	95.5	100.8	98.2	99.0
0310	Concrete Forming & Accessories	91.7	63.7	67.9	78.7	63.6	65.9	89.2	63.7	67.5	87.4	63.4	67.0	91.7	63.7	67.9	96.6	77.8	80.6
0320	Concrete Reinforcing	96.7	67.7	81.9	91.2	67.7	79.3	91.1	66.0	78.4	91.9	67.7	79.6	91.1	67.7	79.2	105.3	72.4	88.5
0330	Cast-in-Place Concrete	109.0	66.1	93.1	91.8	65.9	82.2	91.8	66.1	82.2	91.8	66.0	82.2	91.8	66.1	82.2	112.8	81.2	101.1
03	CONCRETE	95.8	66.9	82.8	96.3	66.7	83.0	94.9	66.6	82.2	92.9	66.7	81.1	95.1	66.9	82.4	101.9	78.6	91.4
04	MASONRY	87.6	65.3	73.8	78.3	65.3	70.3	76.0	65.3	69.4	99.5	65.3	78.4	78.3	65.3	70.3	114.9	74.9	90.2
05	METALS	98.1	91.1	96.0	99.9	90.5	97.1	99.9	91.1	97.3	99.1	90.8	96.6	99.9	91.1	97.3	97.6	82.4	93.0
06	WOOD, PLASTICS & COMPOSITES	90.0	65.6	76.7	73.6	65.6	69.3	85.3	65.6	74.6	84.0	65.6	74.0	89.0	65.6	76.3	97.6	76.4	86.1
07	THERMAL & MOISTURE PROTECTION	93.4	64.5	80.9	97.7	64.5	83.4	97.6	64.5	83.4	97.4	60.2	81.4	97.6	64.5	83.4	103.9	79.4	93.4
08	OPENINGS	98.1	63.8	90.1	95.9	63.8	88.4	95.8	63.8	88.4	95.8	63.8	88.4	95.8	63.8	88.4	99.5	63.8	91.2
0920	Plaster & Gypsum Board	97.3	64.6	75.7	86.4	64.6	72.0	91.1	64.6	73.6	90.8	64.6	73.5	93.9	64.6	74.6	108.3	76.1	87.1
0950, 0980	Ceilings & Acoustic Treatment	87.6	64.6	72.0	83.4	64.6	70.6	82.6	64.6	70.4	82.6	64.6	70.4	82.6	64.6	70.4	93.9	76.1	81.8
0960	Flooring	91.4	78.5	87.6	90.7	78.4	87.1	97.0	78.4	91.6	95.9	78.4	90.8	98.4	78.4	92.6	91.9	45.3	78.4
0970, 0990	Wall Finishes & Painting/Coating	94.9	67.8	78.9	97.0	67.8	79.7	97.0	67.8	79.7	97.0	67.8	79.7	97.0	67.8	79.7	87.8	34.3	56.1
09	FINISHES	86.7	67.2	76.1	85.6	67.2	75.6	87.5	67.2	76.4	86.9	67.2	76.1	88.3	67.7	76.8	89.1	67.9	77.5
COVERS	DIVS. 10 - 14, 25, 28, 41, 43, 44, 46	100.0	70.6	93.5	100.0	70.6	93.5	100.0	70.6	93.5	100.0	70.6	93.5	100.0	70.6	93.5	100.0	86.5	97.0
21, 22, 23	FIRE SUPPRESSION, PLUMBING & HVAC	100.0	57.7	83.0	100.5	57.7	83.3	100.5	57.7	83.3	98.0	56.1	81.2	100.5	57.7	83.3	100.1	55.4	82.2
26, 27, 3370	ELECTRICAL, COMMUNICATIONS & UTIL.	97.3	61.1	78.8	95.0	61.1	77.7	97.0	61.3	78.8	97.0	61.4	78.8	97.0	61.3	78.8	98.3	65.0	81.2
MF2018	WEIGHTED AVERAGE	97.0	67.9	84.5	96.7	67.8	84.3	96.7	67.9	84.3	96.8	67.3	84.1	96.9	67.9	84.5	99.6	71.5	87.6

For customer support on your Plumbing Costs with RSMeans Data, call 800.448.8182.

665

City Cost Indexes

SOUTH DAKOTA

DIVISION		MITCHELL 573			MOBRIDGE 576			PIERRE 575			RAPID CITY 577			SIOUX FALLS 570 - 571			WATERTOWN 572		
		MAT.	INST.	TOTAL	MAT.	INST.	TOTAL	MAT.	INST.	TOTAL	MAT.	INST.	TOTAL	MAT.	INST.	TOTAL	MAT.	INST.	TOTAL
015433	CONTRACTOR EQUIPMENT		100.3	100.3		100.3	100.3		100.3	100.3		100.3	100.3		101.2	101.2		100.3	100.3
0241, 31 - 34	SITE & INFRASTRUCTURE, DEMOLITION	97.6	97.9	97.8	97.5	98.0	97.8	100.5	97.4	98.3	99.1	97.4	97.9	91.3	99.7	97.2	97.4	98.0	97.8
0310	Concrete Forming & Accessories	95.7	45.4	53.0	85.8	45.8	51.8	97.2	46.6	54.3	104.9	58.1	65.1	100.7	78.4	81.7	82.2	76.6	77.4
0320	Concrete Reinforcing	104.7	72.6	88.4	107.4	72.5	89.7	107.6	96.7	102.0	98.4	96.9	97.6	105.2	97.1	101.1	101.8	72.5	86.9
0330	Cast-in-Place Concrete	109.7	53.8	88.9	109.7	79.5	98.4	110.2	79.1	98.6	108.8	79.4	97.9	87.1	80.3	84.6	109.7	79.4	98.4
03	CONCRETE	99.5	54.5	79.3	99.2	63.6	83.2	98.3	67.8	84.6	98.9	73.1	87.3	87.5	82.8	85.4	98.3	77.5	88.9
04	MASONRY	102.5	77.1	86.8	111.2	74.4	88.4	113.0	74.1	89.0	110.9	75.6	89.1	93.2	77.1	83.2	138.5	75.3	99.4
05	METALS	96.6	82.2	92.3	96.7	82.7	92.4	100.3	88.9	96.8	99.3	89.8	96.4	99.0	92.0	96.9	96.6	82.3	92.4
06	WOOD, PLASTICS & COMPOSITES	96.5	34.6	53.0	84.5	34.4	57.3	102.2	34.9	65.6	102.1	49.3	73.4	98.8	76.4	86.6	80.4	76.4	78.2
07	THERMAL & MOISTURE PROTECTION	103.7	74.9	91.3	103.7	77.2	92.3	104.9	74.2	91.7	104.4	80.1	93.9	103.2	84.2	95.0	103.5	78.8	92.9
08	OPENINGS	98.6	40.8	85.2	101.2	40.1	87.0	100.7	58.1	90.8	103.4	66.1	94.7	102.3	81.0	97.3	98.6	63.3	90.4
0920	Plaster & Gypsum Board	106.8	33.2	58.3	100.6	32.9	56.0	102.8	33.4	57.1	107.5	48.3	68.5	98.5	76.1	83.7	98.5	76.1	83.7
0950, 0980	Ceilings & Acoustic Treatment	89.7	33.2	51.4	93.9	32.9	52.6	91.4	33.4	52.1	94.8	48.3	63.3	92.8	76.1	81.5	89.7	76.1	80.5
0960	Flooring	91.6	45.3	78.1	87.3	48.0	75.9	97.5	32.3	78.6	91.3	75.1	86.6	97.0	77.5	91.4	86.0	45.3	74.2
0970, 0990	Wall Finishes & Painting/Coating	87.8	37.5	58.0	87.8	38.5	58.6	95.5	96.2	95.9	87.8	96.2	92.8	99.9	96.2	97.7	87.8	34.3	56.1
09	FINISHES	87.8	42.3	63.0	86.6	42.8	62.7	92.3	46.2	67.2	89.0	63.4	75.0	93.6	79.8	86.1	85.0	66.6	74.9
COVERS	DIVS. 10 - 14, 25, 28, 41, 43, 44, 46	100.0	83.7	96.4	100.0	83.6	96.4	100.0	83.7	96.4	100.0	85.3	96.8	100.0	88.3	97.4	100.0	87.6	97.3
21, 22, 23	FIRE SUPPRESSION, PLUMBING & HVAC	97.6	50.1	78.6	97.6	70.6	86.8	100.0	78.4	91.3	100.1	78.5	91.4	100.0	71.6	88.6	97.6	53.2	79.8
26, 27, 3370	ELECTRICAL, COMMUNICATIONS & UTIL.	96.7	64.9	80.5	98.3	40.7	68.8	101.4	48.9	74.5	95.0	48.9	71.4	99.4	65.0	81.8	95.9	65.0	80.1
MF2018	WEIGHTED AVERAGE	97.5	62.4	82.5	98.1	64.4	83.7	100.2	69.4	87.0	99.4	73.3	88.2	97.5	79.2	89.6	98.6	70.8	86.7

TENNESSEE

DIVISION		CHATTANOOGA 373 - 374			COLUMBIA 384			COOKEVILLE 385			JACKSON 383			JOHNSON CITY 376			KNOXVILLE 377 - 379		
		MAT.	INST.	TOTAL	MAT.	INST.	TOTAL	MAT.	INST.	TOTAL	MAT.	INST.	TOTAL	MAT.	INST.	TOTAL	MAT.	INST.	TOTAL
015433	CONTRACTOR EQUIPMENT		107.6	107.6		102.1	102.1		102.1	102.1		108.3	108.3		101.4	101.4		101.4	101.4
0241, 31 - 34	SITE & INFRASTRUCTURE, DEMOLITION	104.2	98.6	100.3	89.9	88.8	89.1	95.6	85.6	88.6	99.1	98.7	98.8	110.4	85.2	92.9	90.8	88.0	88.9
0310	Concrete Forming & Accessories	96.5	58.9	64.5	80.4	64.0	66.5	80.5	32.5	39.7	87.2	41.9	48.7	82.3	59.0	62.5	94.6	63.4	68.1
0320	Concrete Reinforcing	95.2	67.0	80.9	88.7	66.8	77.6	88.7	66.4	77.4	88.7	67.3	77.8	95.9	63.6	79.5	95.2	63.6	79.1
0330	Cast-in-Place Concrete	98.5	63.3	85.4	91.6	62.0	80.6	103.9	40.5	80.3	101.4	67.8	88.9	79.2	58.7	71.6	92.5	64.8	82.2
03	CONCRETE	94.0	63.6	80.3	92.9	65.5	80.6	103.1	43.7	76.4	93.9	57.6	77.5	102.8	61.4	84.2	91.4	65.5	79.8
04	MASONRY	100.1	57.5	73.8	112.7	56.7	78.1	107.9	40.3	66.1	113.1	42.5	69.5	112.9	44.1	70.4	77.9	51.7	61.7
05	METALS	98.2	88.8	95.3	94.0	89.5	92.6	94.1	87.9	92.2	96.4	89.2	94.2	95.4	86.9	92.8	98.7	87.1	95.2
06	WOOD, PLASTICS & COMPOSITES	106.5	57.9	80.1	69.1	65.2	67.0	69.3	30.3	48.1	84.4	40.8	60.7	78.8	64.4	71.0	93.3	64.4	77.6
07	THERMAL & MOISTURE PROTECTION	96.1	62.1	81.5	94.2	62.9	80.7	94.7	50.4	75.6	96.6	55.3	78.8	92.3	54.6	76.1	90.0	61.2	77.6
08	OPENINGS	104.1	59.2	93.6	93.1	54.9	84.2	93.1	36.5	79.9	99.8	45.2	87.2	100.3	61.3	91.3	97.6	56.8	88.1
0920	Plaster & Gypsum Board	83.1	57.2	66.1	91.2	64.7	73.8	91.2	28.8	50.1	93.6	39.6	58.1	101.7	63.9	76.8	109.5	63.9	79.5
0950, 0980	Ceilings & Acoustic Treatment	107.2	57.2	73.3	80.2	64.7	69.7	80.2	28.8	45.4	89.6	39.6	55.8	102.6	63.9	76.4	103.4	63.9	76.7
0960	Flooring	98.0	57.1	86.1	82.0	55.8	74.4	82.0	51.7	73.2	81.3	56.1	74.0	93.3	45.4	79.4	98.3	51.3	84.7
0970, 0990	Wall Finishes & Painting/Coating	97.7	61.1	76.0	84.9	57.9	69.0	84.9	57.9	69.0	86.6	57.9	69.6	94.4	60.3	74.2	94.4	59.1	73.5
09	FINISHES	96.6	58.2	75.6	85.9	61.9	72.8	86.4	37.2	59.6	86.2	44.6	63.5	100.5	56.9	76.7	93.5	60.7	75.6
COVERS	DIVS. 10 - 14, 25, 28, 41, 43, 44, 46	100.0	69.2	93.2	100.0	68.7	93.1	100.0	36.6	86.0	100.0	62.5	91.7	100.0	77.3	95.0	100.0	81.0	95.8
21, 22, 23	FIRE SUPPRESSION, PLUMBING & HVAC	100.1	60.6	84.3	98.3	75.3	89.1	98.3	68.5	86.4	100.1	60.0	84.1	99.9	56.8	82.6	99.9	61.8	84.7
26, 27, 3370	ELECTRICAL, COMMUNICATIONS & UTIL.	100.7	84.9	92.6	93.3	51.8	72.1	94.8	62.8	78.4	99.7	52.8	75.7	91.4	41.8	66.0	96.7	56.3	76.0
MF2018	WEIGHTED AVERAGE	99.3	69.6	86.5	95.2	67.9	83.5	96.6	57.3	79.8	97.9	59.7	81.5	99.3	59.9	82.4	96.0	65.2	82.8

TENNESSEE / TEXAS

DIVISION		TENNESSEE — MCKENZIE 382			MEMPHIS 375, 380 - 381			NASHVILLE 370 - 372			TEXAS — ABILENE 795 - 796			AMARILLO 790 - 791			AUSTIN 786 - 787		
		MAT.	INST.	TOTAL	MAT.	INST.	TOTAL	MAT.	INST.	TOTAL	MAT.	INST.	TOTAL	MAT.	INST.	TOTAL	MAT.	INST.	TOTAL
015433	CONTRACTOR EQUIPMENT		102.1	102.1		101.5	101.5		105.2	105.2		93.2	93.2		93.2	93.2		92.7	92.7
0241, 31 - 34	SITE & INFRASTRUCTURE, DEMOLITION	95.3	85.8	88.7	88.1	94.9	92.8	101.6	97.5	98.7	93.7	91.8	92.4	92.8	91.1	91.6	95.9	90.8	92.4
0310	Concrete Forming & Accessories	88.2	34.8	42.9	95.1	66.3	70.7	96.3	64.9	69.6	95.6	60.8	66.0	96.9	53.0	59.6	94.4	54.4	60.4
0320	Concrete Reinforcing	88.8	67.4	77.9	102.8	67.9	85.1	101.3	67.0	83.9	95.6	52.7	73.8	106.3	51.0	78.2	90.0	48.5	68.9
0330	Cast-in-Place Concrete	101.6	53.2	83.6	95.5	78.0	89.0	91.1	65.6	81.6	84.8	66.8	78.1	81.6	66.7	76.1	94.6	66.9	84.3
03	CONCRETE	101.7	49.3	78.1	96.5	71.8	85.4	94.1	66.8	81.8	82.8	62.3	73.6	84.6	58.4	72.8	89.5	58.7	75.6
04	MASONRY	111.8	46.1	71.2	95.5	57.5	72.0	87.7	57.0	68.8	94.3	61.9	74.3	93.1	61.4	73.5	92.9	61.9	73.8
05	METALS	94.1	88.5	92.4	99.9	84.1	95.1	99.4	85.8	95.3	108.2	70.7	96.8	102.3	69.8	92.4	100.1	68.8	90.6
06	WOOD, PLASTICS & COMPOSITES	78.2	32.6	53.4	97.0	67.1	80.7	97.6	65.6	80.2	102.6	62.5	80.8	100.8	52.3	74.5	93.2	54.0	71.9
07	THERMAL & MOISTURE PROTECTION	94.7	52.6	76.6	89.1	68.6	80.3	94.8	63.9	81.5	98.9	65.2	84.4	98.7	63.4	83.5	94.8	65.0	81.9
08	OPENINGS	93.1	38.2	80.3	100.5	63.0	91.8	98.9	64.4	90.9	102.9	58.2	92.5	103.7	52.2	91.8	104.0	50.9	91.7
0920	Plaster & Gypsum Board	94.5	31.2	52.8	94.1	66.4	75.8	93.8	64.7	74.7	88.6	61.9	71.0	95.9	51.4	66.6	90.4	53.1	65.8
0950, 0980	Ceilings & Acoustic Treatment	80.2	31.2	47.0	98.9	66.4	76.9	98.7	64.7	75.7	90.4	61.9	71.1	97.5	51.4	66.3	89.4	53.1	64.8
0960	Flooring	84.9	55.8	76.4	97.2	57.1	85.6	97.4	57.1	85.7	92.5	69.9	85.9	96.5	65.9	87.6	92.3	65.9	84.6
0970, 0990	Wall Finishes & Painting/Coating	84.9	45.3	61.5	94.0	60.9	74.4	99.5	63.1	81.5	94.4	53.0	69.9	89.4	53.0	67.9	94.7	45.4	65.5
09	FINISHES	87.6	38.6	60.8	98.1	63.6	79.3	98.3	63.5	79.3	82.7	61.8	71.3	89.7	55.0	70.8	87.8	55.1	70.0
COVERS	DIVS. 10 - 14, 25, 28, 41, 43, 44, 46	100.0	24.5	83.4	100.0	81.9	96.0	100.0	81.7	96.0	100.0	80.5	95.7	100.0	73.8	94.2	100.0	77.8	95.1
21, 22, 23	FIRE SUPPRESSION, PLUMBING & HVAC	98.3	59.9	82.9	100.0	71.9	88.7	100.0	75.5	90.2	100.3	53.3	81.5	100.0	52.2	80.8	100.0	60.0	84.0
26, 27, 3370	ELECTRICAL, COMMUNICATIONS & UTIL.	94.6	56.7	75.2	101.4	65.1	82.9	94.8	62.8	78.4	97.9	55.3	76.1	100.6	60.3	80.0	96.1	58.7	76.9
MF2018	WEIGHTED AVERAGE	96.7	56.0	79.3	98.8	71.1	86.9	97.7	71.1	86.3	97.6	62.8	82.7	97.6	61.1	81.9	97.1	62.7	82.3

For customer support on your Plumbing Costs with RSMeans Data, call 800.448.8182.

TEXAS

DIVISION		BEAUMONT 776 - 777			BROWNWOOD 768			BRYAN 778			CHILDRESS 792			CORPUS CHRISTI 783 - 784			DALLAS 752 - 753		
		MAT.	INST.	TOTAL	MAT.	INST.	TOTAL	MAT.	INST.	TOTAL	MAT.	INST.	TOTAL	MAT.	INST.	TOTAL	MAT.	INST.	TOTAL
015433	CONTRACTOR EQUIPMENT		97.3	97.3		93.2	93.2		97.3	97.3		93.2	93.2		103.7	103.7		106.9	106.9
0241, 31 - 34	SITE & INFRASTRUCTURE, DEMOLITION	87.7	96.0	93.5	101.1	91.7	94.5	78.9	96.2	90.9	104.3	90.1	94.4	141.5	87.3	103.7	105.7	96.8	99.5
0310	Concrete Forming & Accessories	99.5	56.1	62.7	98.5	60.4	66.1	79.5	59.8	62.8	94.0	60.2	65.3	99.2	53.6	60.5	97.8	62.9	68.2
0320	Concrete Reinforcing	90.3	63.8	76.8	86.5	52.6	69.3	92.5	52.6	72.2	95.8	52.5	73.8	82.4	52.4	67.1	92.0	53.8	72.6
0330	Cast-in-Place Concrete	97.8	67.0	86.3	93.0	60.3	80.8	78.3	61.2	71.9	87.0	60.2	77.0	119.7	69.1	100.9	93.0	72.5	85.3
03	CONCRETE	95.0	62.2	80.2	89.3	59.8	76.0	78.9	60.0	70.4	90.6	59.7	76.7	99.6	60.8	82.1	93.8	66.2	81.4
04	MASONRY	94.3	63.0	75.0	124.4	59.7	85.8	123.8	61.9	85.6	98.3	61.4	75.5	84.5	62.1	70.6	98.2	60.7	75.0
05	METALS	96.9	76.5	90.7	97.2	70.2	89.0	96.8	72.3	89.4	105.7	70.1	94.9	95.7	85.3	92.5	101.8	84.0	96.4
06	WOOD, PLASTICS & COMPOSITES	111.1	55.6	81.0	109.6	62.5	84.0	76.8	61.0	68.2	101.9	62.5	80.5	118.0	53.0	82.7	98.8	64.7	80.2
07	THERMAL & MOISTURE PROTECTION	103.3	66.1	87.2	93.2	63.4	80.4	94.9	65.6	82.3	99.6	61.9	83.3	101.7	63.9	85.4	88.6	67.1	79.4
08	OPENINGS	97.1	56.6	87.7	102.4	58.2	92.1	97.9	57.6	88.5	98.2	58.2	88.9	103.7	51.5	91.6	97.5	60.4	88.9
0920	Plaster & Gypsum Board	104.4	54.8	71.7	86.6	61.9	70.3	91.4	60.4	71.0	88.2	61.9	70.9	99.4	51.9	68.1	94.2	63.7	74.1
0950, 0980	Ceilings & Acoustic Treatment	99.7	54.8	69.3	79.3	61.9	67.5	92.7	60.4	70.8	88.7	61.9	70.5	92.4	51.9	65.0	89.6	63.7	72.1
0960	Flooring	116.6	77.2	105.2	83.6	56.4	75.7	87.4	69.2	82.1	90.9	56.4	80.9	104.8	70.0	94.7	98.1	70.4	90.0
0970, 0990	Wall Finishes & Painting/Coating	92.7	55.3	70.6	91.7	53.0	68.8	90.0	58.9	71.6	94.4	53.0	69.9	112.2	45.4	72.7	101.4	55.2	74.0
09	FINISHES	94.3	59.4	75.3	78.1	59.1	67.7	82.4	61.1	70.7	83.1	59.1	70.0	97.7	54.5	74.6	95.8	63.5	78.2
COVERS	DIVS. 10 - 14, 25, 28, 41, 43, 44, 46	100.0	79.3	95.4	100.0	74.9	94.5	100.0	78.5	95.3	100.0	74.9	94.5	100.0	79.4	95.5	100.0	81.5	95.9
21, 22, 23	FIRE SUPPRESSION, PLUMBING & HVAC	100.1	64.4	85.8	97.6	50.6	78.8	97.7	63.8	84.1	97.8	52.9	79.8	100.1	49.8	80.0	100.0	60.7	84.2
26, 27, 3370	ELECTRICAL, COMMUNICATIONS & UTIL.	97.3	64.9	80.7	93.8	47.3	70.0	95.6	64.9	79.9	97.9	57.1	77.0	92.7	65.0	78.5	98.2	61.3	79.3
MF2018	WEIGHTED AVERAGE	97.5	67.1	84.4	96.6	60.2	81.0	94.6	66.4	82.5	97.5	61.8	82.2	99.2	63.0	83.7	98.5	67.7	85.3

TEXAS

DIVISION		DEL RIO 788			DENTON 762			EASTLAND 764			EL PASO 798 - 799, 885			FORT WORTH 760 - 761			GALVESTON 775		
		MAT.	INST.	TOTAL	MAT.	INST.	TOTAL	MAT.	INST.	TOTAL	MAT.	INST.	TOTAL	MAT.	INST.	TOTAL	MAT.	INST.	TOTAL
015433	CONTRACTOR EQUIPMENT		92.7	92.7		102.8	102.8		93.2	93.2		93.2	93.2		93.2	93.2		110.3	110.3
0241, 31 - 34	SITE & INFRASTRUCTURE, DEMOLITION	122.0	90.1	99.8	101.1	83.7	89.0	104.0	89.8	94.1	92.4	92.5	92.5	97.0	92.5	93.9	103.6	94.8	97.5
0310	Concrete Forming & Accessories	95.8	52.5	59.0	106.5	60.5	67.4	99.4	60.2	66.1	96.5	58.2	63.9	98.3	61.0	66.6	88.7	57.3	62.0
0320	Concrete Reinforcing	82.9	48.9	65.6	87.9	52.5	69.9	86.7	52.5	69.4	98.3	52.5	75.1	93.1	52.3	72.4	92.0	61.4	76.5
0330	Cast-in-Place Concrete	128.9	60.2	103.3	71.9	61.4	68.0	98.3	60.2	84.1	75.9	69.0	73.3	85.7	66.9	78.7	104.0	62.2	88.4
03	CONCRETE	120.1	55.5	91.1	70.3	61.3	66.2	93.7	59.7	78.4	80.8	61.8	72.3	85.2	62.3	74.9	96.5	61.8	80.9
04	MASONRY	99.5	61.8	76.2	133.1	60.6	88.3	94.2	61.9	74.2	85.5	64.0	72.2	93.8	60.5	73.2	89.5	62.0	72.5
05	METALS	95.6	68.0	87.2	96.8	85.5	93.4	97.0	70.0	88.8	103.7	70.5	93.6	98.8	70.7	90.2	98.5	91.8	96.5
06	WOOD, PLASTICS & COMPOSITES	98.2	51.9	73.0	122.3	62.7	89.9	117.0	62.5	87.4	90.5	58.6	73.2	101.2	62.5	80.2	93.5	57.5	73.9
07	THERMAL & MOISTURE PROTECTION	98.2	63.8	83.4	91.0	63.9	79.4	93.5	63.4	80.6	93.3	65.6	81.3	90.1	65.5	79.5	93.9	66.2	82.0
08	OPENINGS	98.9	49.2	87.3	120.9	57.9	106.2	71.7	58.2	68.6	97.8	54.3	87.7	102.0	58.1	91.8	101.9	57.7	91.6
0920	Plaster & Gypsum Board	96.0	50.9	66.3	91.3	61.9	71.9	86.6	61.9	70.3	97.2	57.8	71.2	93.6	61.9	72.7	98.4	56.6	70.9
0950, 0980	Ceilings & Acoustic Treatment	88.8	50.9	63.2	84.4	61.9	69.1	79.3	61.9	67.5	91.1	57.8	68.5	89.3	61.9	70.7	97.0	56.6	69.6
0960	Flooring	89.9	56.4	80.2	79.2	56.4	72.6	106.4	56.4	91.9	93.1	72.1	87.0	104.6	66.6	93.6	102.7	69.2	93.0
0970, 0990	Wall Finishes & Painting/Coating	99.5	43.7	66.5	101.9	50.2	71.3	93.1	53.0	69.4	99.7	50.9	70.8	97.2	53.2	71.1	101.6	55.1	74.1
09	FINISHES	90.3	51.8	69.3	77.0	58.9	67.1	85.1	59.1	70.9	88.4	59.7	72.7	90.7	61.2	74.6	91.9	58.6	73.7
COVERS	DIVS. 10 - 14, 25, 28, 41, 43, 44, 46	100.0	75.2	94.5	100.0	80.8	95.8	100.0	73.2	94.1	100.0	77.9	95.1	100.0	80.5	95.7	100.0	79.8	95.6
21, 22, 23	FIRE SUPPRESSION, PLUMBING & HVAC	97.6	60.7	82.8	97.6	57.1	81.4	97.6	50.6	78.8	99.9	55.6	82.6	99.9	56.6	82.6	97.6	63.9	84.1
26, 27, 3370	ELECTRICAL, COMMUNICATIONS & UTIL.	94.6	61.3	77.6	96.1	60.1	77.7	93.7	60.0	76.5	97.6	51.7	74.1	95.3	60.0	77.2	97.1	65.0	80.7
MF2018	WEIGHTED AVERAGE	100.2	62.0	83.8	96.7	64.4	82.9	93.2	61.8	79.7	95.7	64.7	82.4	96.3	64.1	82.5	97.4	68.1	84.8

TEXAS

DIVISION		GIDDINGS 789			GREENVILLE 754			HOUSTON 770 - 772			HUNTSVILLE 773			LAREDO 780			LONGVIEW 756		
		MAT.	INST.	TOTAL	MAT.	INST.	TOTAL	MAT.	INST.	TOTAL	MAT.	INST.	TOTAL	MAT.	INST.	TOTAL	MAT.	INST.	TOTAL
015433	CONTRACTOR EQUIPMENT		92.7	92.7		103.6	103.6		102.2	102.2		97.3	97.3		92.7	92.7		94.6	94.6
0241, 31 - 34	SITE & INFRASTRUCTURE, DEMOLITION	106.8	90.4	95.4	96.6	86.8	89.8	102.5	94.6	97.0	94.2	95.9	95.4	100.1	90.9	93.7	94.6	93.3	93.7
0310	Concrete Forming & Accessories	93.0	52.6	58.7	88.9	58.7	63.2	93.9	58.6	63.9	86.5	55.6	60.2	95.8	53.6	60.0	84.7	60.4	64.1
0320	Concrete Reinforcing	83.4	49.8	66.4	92.4	52.6	72.1	91.8	52.8	72.0	92.7	52.6	72.3	82.9	52.4	67.4	91.5	52.2	71.5
0330	Cast-in-Place Concrete	109.1	60.5	91.0	92.5	61.4	80.9	96.5	63.3	84.2	107.7	61.1	90.4	92.7	60.7	82.7	107.6	60.3	90.0
03	CONCRETE	96.4	55.9	78.2	87.5	60.4	75.3	95.3	60.6	79.7	103.0	58.0	82.8	90.4	59.0	76.3	103.4	59.7	83.7
04	MASONRY	106.7	61.9	79.1	154.1	60.5	96.3	88.7	62.9	72.8	124.6	61.9	85.9	92.5	61.9	73.6	150.1	60.5	94.7
05	METALS	95.1	68.9	87.1	99.3	84.1	94.7	101.2	77.8	94.1	96.8	72.1	89.3	98.1	70.1	89.6	92.1	69.1	85.1
06	WOOD, PLASTICS & COMPOSITES	97.4	51.9	72.7	87.7	60.2	72.7	98.9	59.0	77.2	86.1	55.6	69.5	98.2	52.9	73.6	81.6	62.6	71.3
07	THERMAL & MOISTURE PROTECTION	98.7	64.5	84.0	88.8	63.0	77.7	95.8	67.3	83.6	96.0	65.0	82.7	97.5	64.8	83.4	90.4	62.5	78.4
08	OPENINGS	98.0	50.1	86.9	94.1	56.9	85.4	105.1	56.5	93.8	97.8	54.0	87.7	99.0	51.2	87.9	85.1	57.8	78.7
0920	Plaster & Gypsum Board	95.1	50.9	66.0	84.9	59.2	68.0	97.7	58.0	71.5	96.0	54.8	68.9	96.7	51.9	67.2	84.0	61.9	69.4
0950, 0980	Ceilings & Acoustic Treatment	88.8	50.9	63.2	85.0	59.2	67.5	93.9	58.0	69.6	92.7	54.8	67.0	92.2	51.9	64.9	80.8	61.9	68.0
0960	Flooring	89.5	56.4	79.9	91.8	56.4	81.5	103.8	72.1	94.6	91.7	56.4	81.5	89.8	65.9	82.8	96.9	56.4	85.1
0970, 0990	Wall Finishes & Painting/Coating	99.5	45.4	67.5	95.8	53.0	70.5	99.4	58.9	75.5	90.0	58.9	71.6	99.5	45.4	67.5	86.0	50.2	64.8
09	FINISHES	89.0	52.0	68.8	90.3	57.7	72.5	97.9	60.9	77.7	85.1	55.6	69.0	89.4	54.4	70.3	95.5	58.8	75.5
COVERS	DIVS. 10 - 14, 25, 28, 41, 43, 44, 46	100.0	71.8	93.8	100.0	78.1	95.2	100.0	81.9	96.0	100.0	71.2	93.7	100.0	77.4	95.0	100.0	80.7	95.7
21, 22, 23	FIRE SUPPRESSION, PLUMBING & HVAC	97.7	62.8	83.7	97.7	59.3	82.3	100.1	65.5	86.2	97.7	63.8	84.1	100.1	60.2	84.1	97.6	58.8	82.1
26, 27, 3370	ELECTRICAL, COMMUNICATIONS & UTIL.	91.9	57.0	74.0	95.0	60.1	77.1	99.1	68.1	83.2	95.6	60.6	77.7	94.7	58.5	76.2	95.3	52.1	73.2
MF2018	WEIGHTED AVERAGE	96.7	61.9	81.8	98.0	64.5	83.6	99.3	67.9	85.8	98.3	64.3	83.7	96.6	62.7	82.1	98.0	62.6	82.8

For customer support on your Plumbing Costs with RSMeans Data, call 800.448.8182.

667

TEXAS

DIVISION		LUBBOCK 793-794			LUFKIN 759			MCALLEN 785			MCKINNEY 750			MIDLAND 797			ODESSA 797		
		MAT.	INST.	TOTAL	MAT.	INST.	TOTAL	MAT.	INST.	TOTAL	MAT.	INST.	TOTAL	MAT.	INST.	TOTAL	MAT.	INST.	TOTAL
015433	CONTRACTOR EQUIPMENT		105.6	105.6		94.6	94.6		103.8	103.8		103.6	103.6		105.6	105.6		93.2	93.2
0241, 31 - 34	SITE & INFRASTRUCTURE, DEMOLITION	118.2	91.0	99.3	89.6	94.8	93.2	146.3	87.3	105.2	93.1	86.8	88.7	121.2	90.0	99.5	94.0	92.5	92.9
0310	Concrete Forming & Accessories	94.8	54.5	60.5	87.9	55.5	60.5	99.7	52.6	59.7	88.0	58.7	63.2	98.6	60.7	66.4	95.6	60.8	66.0
0320	Concrete Reinforcing	96.9	52.8	74.4	93.1	68.7	80.7	82.5	52.3	67.2	92.4	52.6	72.2	97.9	52.7	74.9	95.6	52.6	73.7
0330	Cast-in-Place Concrete	85.0	70.2	79.5	96.2	60.1	82.8	130.0	61.6	104.6	86.7	61.4	77.3	90.3	67.8	81.9	84.8	66.8	78.1
03	CONCRETE	81.6	61.6	72.6	95.6	60.2	79.7	107.6	57.7	85.1	82.8	60.4	72.7	85.6	63.6	75.7	82.8	62.3	73.6
04	MASONRY	93.7	62.3	74.3	115.0	61.8	82.1	99.2	62.0	76.2	165.6	60.5	100.7	110.1	61.5	80.1	94.3	61.4	74.0
05	METALS	112.1	86.3	104.3	99.4	74.0	91.7	95.5	85.1	92.3	99.2	84.2	94.6	110.4	85.8	102.9	107.5	70.6	96.3
06	WOOD, PLASTICS & COMPOSITES	102.2	53.6	75.8	95.7	55.7	70.9	116.6	52.0	81.5	86.6	60.2	72.2	106.9	62.7	82.9	102.6	62.5	80.8
07	THERMAL & MOISTURE PROTECTION	88.1	65.3	78.3	90.1	61.6	77.8	101.4	64.7	85.6	88.5	63.0	77.6	88.4	65.1	78.4	98.9	64.2	84.0
08	OPENINGS	110.9	53.8	97.6	64.5	58.2	63.1	102.7	50.0	90.5	94.1	56.9	85.4	112.4	58.3	99.8	102.9	58.2	92.5
0920	Plaster & Gypsum Board	88.8	52.6	64.9	83.1	54.8	64.5	100.4	50.9	67.8	84.9	59.2	68.0	90.6	61.9	71.7	88.6	61.9	71.0
0950, 0980	Ceilings & Acoustic Treatment	91.3	52.6	65.0	76.5	54.8	61.8	92.2	50.9	64.3	85.0	59.2	67.5	89.6	61.9	70.8	90.4	61.9	71.1
0960	Flooring	86.4	71.3	82.0	129.5	56.4	108.3	104.4	75.2	95.9	91.4	56.4	81.2	87.6	65.9	81.3	92.5	65.9	84.8
0970, 0990	Wall Finishes & Painting/Coating	104.9	55.2	75.5	86.0	53.0	66.5	112.2	43.7	71.7	95.8	53.0	70.5	104.9	53.0	74.2	94.4	53.0	69.9
09	FINISHES	85.1	57.1	69.8	103.7	55.5	77.4	98.1	55.7	74.9	89.9	57.7	72.3	85.8	61.1	72.3	82.7	61.0	70.9
COVERS	DIVS. 10 - 14, 25, 28, 41, 43, 44, 46	100.0	79.9	95.6	100.0	71.5	93.7	100.0	77.5	95.0	100.0	80.8	95.8	100.0	75.3	94.6	100.0	75.0	94.5
21, 22, 23	FIRE SUPPRESSION, PLUMBING & HVAC	99.7	54.3	81.5	97.6	59.4	82.3	97.7	49.8	78.5	97.7	61.6	83.2	97.3	48.5	77.7	100.3	53.7	81.6
26, 27, 3370	ELECTRICAL, COMMUNICATIONS & UTIL.	96.6	61.6	78.7	96.5	62.9	79.3	92.5	32.4	61.8	95.1	60.1	77.2	96.6	60.4	78.1	98.0	60.3	78.7
MF2018	WEIGHTED AVERAGE	99.1	64.4	84.2	95.1	64.2	81.9	100.3	57.9	82.1	97.8	65.1	83.8	99.8	63.7	84.3	97.5	63.3	82.8

TEXAS

DIVISION		PALESTINE 758			SAN ANGELO 769			SAN ANTONIO 781-782			TEMPLE 765			TEXARKANA 755			TYLER 757		
		MAT.	INST.	TOTAL	MAT.	INST.	TOTAL	MAT.	INST.	TOTAL	MAT.	INST.	TOTAL	MAT.	INST.	TOTAL	MAT.	INST.	TOTAL
015433	CONTRACTOR EQUIPMENT		94.6	94.6		93.2	93.2		93.5	93.5		93.2	93.2		94.6	94.6		94.6	94.6
0241, 31 - 34	SITE & INFRASTRUCTURE, DEMOLITION	94.9	93.6	94.0	97.5	92.1	93.7	100.0	93.5	95.5	86.0	92.0	90.2	84.3	96.0	92.5	93.9	93.4	93.6
0310	Concrete Forming & Accessories	78.7	60.4	63.2	98.9	52.9	59.8	97.1	53.9	60.4	102.4	52.6	60.1	95.5	61.0	66.2	89.7	60.5	64.9
0320	Concrete Reinforcing	90.8	52.5	71.3	86.4	52.6	69.2	92.1	49.1	70.2	86.6	52.3	69.2	90.6	52.3	71.2	91.5	52.2	71.5
0330	Cast-in-Place Concrete	87.9	60.3	77.7	87.8	66.8	80.0	93.7	69.4	84.7	71.9	60.3	67.6	88.6	66.9	80.5	105.7	60.3	88.8
03	CONCRETE	97.0	59.8	80.3	85.2	58.7	73.3	94.3	59.3	78.5	72.8	56.2	65.3	88.6	62.3	76.8	102.9	59.8	83.5
04	MASONRY	109.4	60.5	79.2	121.0	61.9	84.5	95.5	62.0	74.8	132.2	61.9	88.8	169.9	60.5	102.3	159.7	60.5	98.4
05	METALS	99.2	69.2	90.1	97.3	70.6	89.2	99.5	66.7	89.6	97.0	69.7	88.7	92.0	69.7	85.3	99.0	69.2	90.0
06	WOOD, PLASTICS & COMPOSITES	79.5	62.6	70.3	109.9	51.9	78.4	99.5	53.0	74.3	119.9	51.9	83.0	94.1	62.6	77.0	90.8	62.6	75.5
07	THERMAL & MOISTURE PROTECTION	90.7	63.5	79.0	93.0	64.4	80.6	93.8	66.7	82.1	92.5	63.4	80.0	89.9	64.8	79.1	90.5	63.5	78.9
08	OPENINGS	64.5	58.3	63.0	102.4	52.3	90.8	101.7	50.8	89.9	68.3	52.3	64.5	85.0	58.2	78.8	64.4	58.2	63.0
0920	Plaster & Gypsum Board	81.3	61.9	68.5	86.6	61.9	70.3	98.5	51.9	67.9	86.6	50.9	63.1	88.6	61.9	71.0	83.1	61.9	69.1
0950, 0980	Ceilings & Acoustic Treatment	76.5	61.9	66.6	79.3	50.9	60.1	91.3	51.9	64.6	79.3	50.9	60.1	80.8	61.9	68.0	76.5	61.9	66.6
0960	Flooring	120.4	56.4	101.8	83.7	56.4	75.8	98.5	70.0	90.2	108.0	56.4	93.0	105.3	65.9	93.8	131.8	56.4	110.0
0970, 0990	Wall Finishes & Painting/Coating	86.0	53.0	66.5	91.7	53.0	68.8	99.4	45.4	67.5	93.1	45.4	64.9	86.0	53.0	66.5	86.0	53.0	66.5
09	FINISHES	101.4	59.2	78.3	77.8	52.8	64.2	99.0	55.4	75.2	84.3	52.0	66.6	98.0	61.1	77.8	104.7	59.2	79.8
COVERS	DIVS. 10 - 14, 25, 28, 41, 43, 44, 46	100.0	75.1	94.5	100.0	77.6	95.1	100.0	79.3	95.4	100.0	73.8	94.2	100.0	80.7	95.8	100.0	80.7	95.7
21, 22, 23	FIRE SUPPRESSION, PLUMBING & HVAC	97.6	58.0	81.7	97.6	53.4	79.9	100.1	61.1	84.5	97.6	54.4	80.3	97.6	56.8	81.2	97.6	59.5	82.3
26, 27, 3370	ELECTRICAL, COMMUNICATIONS & UTIL.	92.8	49.6	70.7	97.8	55.9	76.4	96.2	61.2	78.3	94.9	53.3	73.6	96.4	55.9	75.7	95.3	55.0	74.7
MF2018	WEIGHTED AVERAGE	94.5	62.0	80.6	96.3	60.8	81.1	98.6	63.5	83.5	91.7	60.0	78.1	97.2	63.6	82.8	98.2	63.0	83.2

TEXAS / UTAH

DIVISION		VICTORIA 779			WACO 766-767			WAXAHACHIE 751			WHARTON 774			WICHITA FALLS 763			LOGAN 843		
		MAT.	INST.	TOTAL	MAT.	INST.	TOTAL	MAT.	INST.	TOTAL	MAT.	INST.	TOTAL	MAT.	INST.	TOTAL	MAT.	INST.	TOTAL
015433	CONTRACTOR EQUIPMENT		109.0	109.0		93.2	93.2		103.6	103.6		110.3	110.3		93.2	93.2		96.9	96.9
0241, 31 - 34	SITE & INFRASTRUCTURE, DEMOLITION	108.6	92.0	97.1	94.4	92.4	93.0	94.9	87.0	89.4	113.6	94.2	100.1	95.2	92.5	93.3	98.5	94.1	95.4
0310	Concrete Forming & Accessories	87.9	52.7	58.0	100.9	60.9	66.9	88.0	60.7	64.8	83.4	54.5	58.9	100.9	60.9	66.9	101.8	67.5	72.7
0320	Concrete Reinforcing	88.4	52.4	70.1	86.3	48.6	67.1	92.4	52.6	72.2	91.9	52.4	71.8	86.3	51.1	68.4	105.6	86.4	95.8
0330	Cast-in-Place Concrete	116.8	62.1	96.5	77.8	66.8	73.7	91.5	61.6	80.4	120.0	61.1	98.1	83.1	66.8	77.1	86.4	75.2	82.2
03	CONCRETE	104.6	58.2	83.7	78.9	61.4	71.1	86.5	61.4	75.2	108.7	58.6	86.2	81.4	62.1	72.7	104.0	74.0	90.5
04	MASONRY	104.0	62.0	78.0	92.5	61.9	73.6	154.7	60.6	96.5	90.9	61.9	73.0	92.9	61.4	73.5	108.1	67.8	83.2
05	METALS	96.9	88.2	94.3	99.3	68.8	90.0	99.3	84.4	94.8	98.5	87.8	95.2	99.2	70.7	90.6	108.4	83.6	100.8
06	WOOD, PLASTICS & COMPOSITES	96.5	52.0	72.3	118.0	62.5	87.8	86.6	62.8	73.6	86.6	54.2	69.0	118.0	62.5	87.8	87.0	67.8	76.6
07	THERMAL & MOISTURE PROTECTION	97.8	62.8	82.8	93.4	65.5	81.4	88.7	63.2	77.7	94.3	65.5	81.9	93.4	64.5	81.0	97.0	71.7	86.1
08	OPENINGS	101.7	52.5	90.3	79.7	57.2	74.5	94.1	58.3	85.8	101.9	53.8	90.7	79.7	58.2	74.7	94.4	66.8	88.0
0920	Plaster & Gypsum Board	94.6	50.9	65.9	87.0	61.9	70.4	85.3	61.9	69.9	93.5	53.3	67.0	87.0	61.9	70.4	81.1	66.9	71.8
0950, 0980	Ceilings & Acoustic Treatment	97.8	50.9	66.1	81.0	61.9	68.0	86.7	61.9	69.9	97.0	53.3	67.4	81.0	61.9	68.0	104.7	66.9	79.1
0960	Flooring	101.4	56.4	88.3	107.3	65.9	95.3	91.4	56.4	81.2	100.0	68.8	90.9	108.0	74.3	98.3	97.5	59.7	86.5
0970, 0990	Wall Finishes & Painting/Coating	101.9	58.9	76.5	93.1	53.0	69.4	95.8	53.0	70.5	101.6	57.2	75.3	95.1	53.0	70.2	93.4	55.7	71.1
09	FINISHES	89.4	53.5	69.8	84.9	61.0	71.9	90.5	59.3	73.4	91.0	56.8	72.3	85.3	62.7	73.0	92.4	64.4	77.1
COVERS	DIVS. 10 - 14, 25, 28, 41, 43, 44, 46	100.0	71.1	93.6	100.0	80.5	95.7	100.0	81.0	95.8	100.0	71.3	93.7	100.0	74.9	94.5	100.0	83.7	96.4
21, 22, 23	FIRE SUPPRESSION, PLUMBING & HVAC	97.7	63.7	84.1	100.1	60.1	84.0	97.7	59.3	82.3	97.6	63.3	83.9	100.1	53.3	81.3	99.9	69.9	87.9
26, 27, 3370	ELECTRICAL, COMMUNICATIONS & UTIL.	101.9	55.3	78.1	98.0	55.4	76.2	95.0	60.1	77.2	100.8	60.6	80.2	99.7	55.3	77.0	95.8	70.5	82.8
MF2018	WEIGHTED AVERAGE	99.4	64.4	84.4	93.3	63.9	80.7	97.9	65.1	83.8	99.5	65.8	85.0	93.8	62.7	80.5	100.4	73.1	88.7

City Cost Indexes

DIVISION		UTAH											VERMONT					
		OGDEN 842,844			PRICE 845			PROVO 846-847			SALT LAKE CITY 840-841			BELLOWS FALLS 051			BENNINGTON 052		
		MAT.	INST.	TOTAL	MAT.	INST.	TOTAL	MAT.	INST.	TOTAL	MAT.	INST.	TOTAL	MAT.	INST.	TOTAL	MAT.	INST.	TOTAL
015433	CONTRACTOR EQUIPMENT		96.9	96.9		95.9	95.9		95.9	95.9		96.8	96.8		97.9	97.9		97.9	97.9
0241, 31 - 34	SITE & INFRASTRUCTURE, DEMOLITION	86.7	94.1	91.8	95.9	92.2	93.3	94.7	92.5	93.2	86.4	94.0	91.7	89.8	100.7	97.4	89.2	100.7	97.2
0310	Concrete Forming & Accessories	101.8	67.5	72.7	104.4	66.0	71.8	103.6	67.5	73.0	104.1	67.5	73.0	97.0	104.2	103.1	94.4	104.1	102.6
0320	Concrete Reinforcing	105.2	86.4	95.7	113.2	86.4	99.6	114.2	86.4	100.1	107.5	86.4	96.8	85.4	84.3	84.9	85.4	84.3	84.8
0330	Cast-in-Place Concrete	87.7	75.2	83.1	86.5	73.8	81.8	86.5	75.2	82.3	95.8	75.2	88.2	91.3	112.8	99.3	91.3	112.7	99.3
03	CONCRETE	93.6	74.0	84.7	105.1	72.8	90.5	103.5	73.9	90.2	112.7	74.0	95.3	86.3	103.4	94.0	86.2	103.3	93.9
04	MASONRY	103.0	67.8	81.3	113.9	66.5	84.6	114.0	67.8	85.5	116.7	67.8	86.5	101.6	88.0	93.2	112.4	88.0	97.3
05	METALS	108.9	83.6	101.2	105.5	83.5	98.8	106.5	83.6	99.5	112.4	83.6	103.6	85.9	89.6	87.0	85.9	89.4	86.9
06	WOOD, PLASTICS & COMPOSITES	87.0	67.8	76.6	90.3	65.9	77.1	88.6	67.8	77.3	89.0	67.8	77.5	106.9	109.8	108.5	103.8	109.8	107.1
07	THERMAL & MOISTURE PROTECTION	95.8	71.7	85.5	98.9	70.2	86.5	98.9	71.7	87.2	103.4	71.7	89.8	98.9	90.7	95.4	98.8	90.7	95.3
08	OPENINGS	94.4	66.8	88.0	98.3	78.5	93.7	98.3	66.8	91.0	96.2	66.8	89.4	102.4	99.3	101.7	102.4	99.3	101.7
0920	Plaster & Gypsum Board	81.1	66.9	71.8	83.9	65.0	71.4	81.7	66.9	72.0	93.1	66.9	75.9	104.9	109.8	108.1	103.4	109.8	107.6
0950, 0980	Ceilings & Acoustic Treatment	104.7	66.9	79.1	104.7	65.0	77.8	104.7	66.9	79.1	96.1	66.9	76.3	97.0	109.8	105.7	97.0	109.8	105.7
0960	Flooring	95.4	59.7	85.0	98.6	58.0	86.8	98.4	59.7	87.1	99.3	59.7	87.8	91.2	102.2	94.4	90.4	102.2	93.8
0970, 0990	Wall Finishes & Painting/Coating	93.4	55.7	71.1	93.4	55.7	71.1	93.4	55.7	71.1	96.5	56.8	73.0	87.1	104.2	97.2	87.1	104.2	97.2
09	FINISHES	90.6	64.4	76.3	93.6	62.8	76.8	93.1	64.4	77.4	92.5	64.5	77.2	86.3	104.8	96.4	85.9	104.8	96.2
COVERS	DIVS. 10 - 14, 25, 28, 41, 43, 44, 46	100.0	83.7	96.4	100.0	83.4	96.3	100.0	83.7	96.4	100.0	83.7	96.4	100.0	93.7	98.6	100.0	93.6	98.6
21, 22, 23	FIRE SUPPRESSION, PLUMBING & HVAC	99.9	69.9	87.9	98.6	64.6	85.0	99.9	69.9	87.9	100.1	69.9	88.0	97.7	89.2	94.3	97.7	89.2	94.3
26, 27, 3370	ELECTRICAL, COMMUNICATIONS & UTIL.	96.1	70.5	83.0	101.0	70.5	85.4	96.4	70.3	83.0	98.7	70.5	84.3	99.2	81.5	90.1	99.2	55.8	77.0
MF2018	WEIGHTED AVERAGE	98.5	73.1	87.6	101.1	71.7	88.5	100.8	72.9	88.8	102.9	73.1	90.1	94.4	93.6	94.1	94.8	90.0	92.8

DIVISION		VERMONT																
		BRATTLEBORO 053			BURLINGTON 054			GUILDHALL 059			MONTPELIER 056			RUTLAND 057			ST. JOHNSBURY 058		
		MAT.	INST.	TOTAL	MAT.	INST.	TOTAL	MAT.	INST.	TOTAL	MAT.	INST.	TOTAL	MAT.	INST.	TOTAL	MAT.	INST.	TOTAL
015433	CONTRACTOR EQUIPMENT		97.9	97.9		97.9	97.9		97.9	97.9		97.9	97.9		97.9	97.9		97.9	97.9
0241, 31 - 34	SITE & INFRASTRUCTURE, DEMOLITION	90.7	100.7	97.6	92.7	100.9	98.4	88.7	96.2	93.9	94.8	100.6	98.9	93.1	100.6	98.3	88.8	99.6	96.4
0310	Concrete Forming & Accessories	97.0	104.2	103.2	99.8	81.9	84.6	95.0	97.5	97.1	97.7	103.6	102.7	97.5	81.9	84.3	93.6	97.9	97.2
0320	Concrete Reinforcing	84.5	84.3	84.4	104.8	84.2	94.3	86.2	84.1	85.1	104.9	84.2	94.4	106.8	84.2	95.3	84.5	84.2	84.3
0330	Cast-in-Place Concrete	94.2	112.8	101.1	110.1	112.5	111.0	88.5	103.8	94.2	107.8	112.6	109.6	89.5	112.5	98.0	88.5	104.0	94.3
03	CONCRETE	88.3	103.4	95.1	98.6	93.2	96.1	83.9	97.1	89.9	97.4	103.0	99.9	88.7	93.1	90.7	83.7	97.4	89.8
04	MASONRY	112.0	88.0	97.2	113.8	88.1	97.9	112.3	73.0	88.0	104.3	88.1	94.3	92.5	88.1	89.8	140.6	73.0	98.8
05	METALS	85.9	89.6	87.0	93.5	88.8	92.1	85.9	88.5	86.7	91.0	88.9	90.3	91.5	88.7	90.7	85.9	88.9	86.8
06	WOOD, PLASTICS & COMPOSITES	107.2	109.8	108.6	100.8	80.6	89.8	102.8	109.8	106.6	96.5	109.8	103.8	107.4	80.6	92.8	97.5	109.8	104.2
07	THERMAL & MOISTURE PROTECTION	99.0	90.7	95.4	106.0	87.6	98.0	98.7	83.8	92.3	105.9	90.8	99.4	99.1	87.6	94.1	98.6	83.8	92.2
08	OPENINGS	102.4	99.6	101.8	104.1	79.6	98.4	102.4	95.7	100.9	105.6	95.7	103.6	105.6	79.6	99.5	102.4	95.7	100.9
0920	Plaster & Gypsum Board	104.9	109.8	108.1	106.4	79.7	88.8	110.7	109.8	110.1	106.7	109.8	108.7	105.3	79.7	88.4	112.2	109.8	110.6
0950, 0980	Ceilings & Acoustic Treatment	97.0	109.8	105.7	101.0	79.7	86.6	97.0	109.8	105.7	106.9	109.8	108.9	101.5	79.7	86.7	97.0	109.8	105.7
0960	Flooring	91.3	102.2	94.5	98.1	102.2	99.3	94.1	102.2	96.5	100.4	102.2	100.9	91.2	102.2	94.4	97.4	102.2	98.8
0970, 0990	Wall Finishes & Painting/Coating	87.1	104.2	97.2	96.4	90.7	93.0	87.1	90.7	89.2	95.3	90.7	92.5	87.1	90.7	89.2	87.1	90.7	89.2
09	FINISHES	86.5	104.8	96.5	92.5	86.1	89.0	87.9	99.5	94.3	94.2	103.4	99.2	87.4	86.1	86.7	89.1	99.5	94.8
COVERS	DIVS. 10 - 14, 25, 28, 41, 43, 44, 46	100.0	93.7	98.6	100.0	90.4	97.9	100.0	88.5	97.5	100.0	93.6	98.6	100.0	90.4	97.9	100.0	88.5	97.5
21, 22, 23	FIRE SUPPRESSION, PLUMBING & HVAC	97.7	89.2	94.3	100.1	69.4	87.8	97.7	61.5	83.2	97.6	69.4	86.3	100.2	69.4	87.9	97.7	61.5	83.2
26, 27, 3370	ELECTRICAL, COMMUNICATIONS & UTIL.	99.2	81.5	90.1	99.5	55.0	76.7	99.2	55.8	77.0	98.1	55.0	76.1	99.3	55.0	76.6	99.2	55.8	77.0
MF2018	WEIGHTED AVERAGE	95.2	93.6	94.5	99.2	80.6	91.2	94.7	80.2	88.5	97.8	85.3	92.4	96.3	80.5	89.5	96.1	80.5	89.4

DIVISION		VERMONT			VIRGINIA													
		WHITE RIVER JCT. 050			ALEXANDRIA 223			ARLINGTON 222			BRISTOL 242			CHARLOTTESVILLE 229			CULPEPER 227		
		MAT.	INST.	TOTAL	MAT.	INST.	TOTAL	MAT.	INST.	TOTAL	MAT.	INST.	TOTAL	MAT.	INST.	TOTAL	MAT.	INST.	TOTAL
015433	CONTRACTOR EQUIPMENT		97.9	97.9		107.5	107.5		106.2	106.2		106.2	106.2		110.4	110.4		106.2	106.2
0241, 31 - 34	SITE & INFRASTRUCTURE, DEMOLITION	93.1	99.7	97.7	114.2	92.2	98.9	124.3	89.9	100.4	108.9	89.2	95.2	113.4	91.7	98.3	112.0	89.9	96.6
0310	Concrete Forming & Accessories	91.6	98.4	97.4	90.4	70.2	73.2	89.1	70.1	73.0	85.2	65.6	68.5	83.6	47.6	53.0	80.7	72.9	74.0
0320	Concrete Reinforcing	85.4	84.2	84.8	83.9	87.4	85.7	94.7	87.4	91.0	94.7	69.8	82.0	94.1	70.1	81.9	94.7	87.4	91.0
0330	Cast-in-Place Concrete	94.2	104.8	98.2	110.1	76.5	97.6	114.7	76.5	95.7	106.7	45.4	83.9	111.0	76.9	98.3	109.7	76.4	97.3
03	CONCRETE	89.8	97.9	93.4	101.3	76.8	90.3	105.6	76.8	92.6	102.0	61.0	83.5	102.7	63.8	85.2	99.6	78.0	89.9
04	MASONRY	126.0	74.4	94.1	89.4	72.3	78.9	103.1	72.3	84.1	92.7	46.3	64.1	116.9	55.7	79.1	104.6	72.3	84.7
05	METALS	85.9	88.9	86.8	102.9	100.4	102.1	101.5	100.4	101.2	100.3	92.2	97.9	100.6	94.1	98.6	100.7	100.0	100.5
06	WOOD, PLASTICS & COMPOSITES	100.6	109.8	105.6	91.8	68.4	79.1	88.3	68.4	77.5	80.9	71.8	76.0	79.4	40.8	58.5	78.1	72.1	74.8
07	THERMAL & MOISTURE PROTECTION	99.0	84.5	92.8	103.0	79.8	93.0	105.1	79.8	94.2	104.4	59.5	85.0	103.9	68.0	88.5	104.1	79.7	93.6
08	OPENINGS	102.4	95.7	100.9	97.4	73.6	91.9	95.6	73.6	90.5	98.4	65.7	90.8	96.7	54.6	86.9	97.0	75.9	92.1
0920	Plaster & Gypsum Board	102.2	109.8	107.2	101.7	67.4	79.1	98.3	67.4	77.9	94.8	70.9	79.1	94.8	38.4	57.7	94.9	71.2	79.3
0950, 0980	Ceilings & Acoustic Treatment	97.0	109.8	105.7	93.4	67.4	75.8	91.7	67.4	75.2	90.8	70.9	77.4	90.8	38.4	55.3	91.7	71.2	77.8
0960	Flooring	89.3	102.2	93.0	97.0	76.9	91.2	95.5	76.9	90.1	92.1	57.6	82.1	90.6	57.6	81.0	90.6	76.9	86.7
0970, 0990	Wall Finishes & Painting/Coating	87.1	90.7	89.2	115.2	73.6	90.6	115.2	75.1	91.4	101.5	55.6	74.4	101.5	58.9	76.3	115.2	75.1	91.4
09	FINISHES	85.7	99.9	93.5	93.1	70.9	81.0	93.1	71.1	81.1	89.8	64.0	75.7	89.4	48.6	67.1	90.1	73.3	80.9
COVERS	DIVS. 10 - 14, 25, 28, 41, 43, 44, 46	100.0	89.0	97.6	100.0	87.3	97.2	100.0	84.9	96.7	100.0	76.0	94.7	100.0	80.4	95.7	100.0	85.3	96.8
21, 22, 23	FIRE SUPPRESSION, PLUMBING & HVAC	97.7	62.2	83.5	100.4	87.2	95.1	100.4	87.2	95.1	98.0	46.8	77.5	98.0	69.1	86.4	98.0	87.2	93.6
26, 27, 3370	ELECTRICAL, COMMUNICATIONS & UTIL.	99.2	55.7	76.9	96.9	96.7	96.8	94.6	99.3	97.0	96.5	37.2	66.2	96.5	70.0	83.0	98.9	99.3	99.1
MF2018	WEIGHTED AVERAGE	96.0	81.0	89.5	99.5	84.2	92.9	100.3	84.3	93.4	98.3	59.4	81.6	99.5	68.0	86.0	98.8	84.9	92.8

For customer support on your Plumbing Costs with RSMeans Data, call 800.448.8182.

669

VIRGINIA

DIVISION		FAIRFAX 220-221			FARMVILLE 239			FREDERICKSBURG 224-225			GRUNDY 246			HARRISONBURG 228			LYNCHBURG 245		
		MAT.	INST.	TOTAL	MAT.	INST.	TOTAL	MAT.	INST.	TOTAL	MAT.	INST.	TOTAL	MAT.	INST.	TOTAL	MAT.	INST.	TOTAL
015433	CONTRACTOR EQUIPMENT		106.2	106.2		110.4	110.4		106.2	106.2		106.2	106.2		106.2	106.2		106.2	106.2
0241, 31 - 34	SITE & INFRASTRUCTURE, DEMOLITION	123.0	89.9	100.0	109.6	90.7	96.4	111.5	89.7	96.3	106.7	88.5	94.1	120.0	88.3	97.9	107.5	90.0	95.3
0310	Concrete Forming & Accessories	83.6	70.1	72.1	98.5	45.2	53.2	83.6	67.1	69.6	88.3	38.1	45.7	79.7	38.7	44.8	85.2	71.5	73.6
0320	Concrete Reinforcing	94.7	87.4	91.0	90.0	69.7	79.7	95.4	82.5	88.8	93.4	45.3	69.0	94.7	58.9	76.5	94.1	70.6	82.1
0330	Cast-in-Place Concrete	107.1	76.5	95.7	106.2	86.0	98.7	108.7	76.2	96.6	106.7	51.0	86.0	107.1	53.8	87.3	106.7	75.7	95.2
03	CONCRETE	105.3	76.7	92.4	96.4	65.7	82.6	99.3	74.5	88.2	100.6	46.0	76.0	103.0	49.8	79.1	100.4	74.3	88.7
04	MASONRY	103.0	72.3	84.0	101.3	51.0	70.2	103.7	68.2	81.7	94.2	54.6	69.7	101.5	59.1	75.3	109.0	64.4	81.5
05	METALS	100.8	100.3	100.6	97.0	90.6	95.1	100.3	75.1	92.7	100.3	75.1	92.7	100.6	87.2	96.6	100.5	94.2	98.6
06	WOOD, PLASTICS & COMPOSITES	80.9	68.4	74.1	94.1	39.5	64.4	80.9	66.5	73.1	83.8	31.4	55.3	77.1	34.9	54.2	80.9	73.2	76.7
07	THERMAL & MOISTURE PROTECTION	104.8	75.6	92.2	104.3	54.9	83.0	104.1	77.1	92.5	104.3	47.4	79.8	104.6	62.9	86.6	104.1	71.1	89.9
08	OPENINGS	95.6	73.9	90.6	96.2	44.4	84.2	96.7	70.5	90.6	98.4	30.9	82.7	97.0	44.3	84.8	97.0	66.4	89.9
0920	Plaster & Gypsum Board	94.9	67.4	76.8	103.8	37.0	59.8	94.9	65.4	75.5	94.8	29.4	51.7	94.8	33.0	54.1	94.8	72.3	80.0
0950, 0980	Ceilings & Acoustic Treatment	91.7	67.4	75.2	86.3	37.0	52.9	91.7	65.4	73.9	90.8	29.4	49.2	90.8	33.0	51.6	90.8	72.3	78.3
0960	Flooring	92.4	77.7	88.1	92.5	52.2	80.8	92.4	74.8	87.3	93.4	29.2	74.8	90.5	76.9	86.5	92.1	67.6	85.0
0970, 0990	Wall Finishes & Painting/Coating	115.2	78.0	93.2	101.8	58.7	76.3	115.2	58.5	81.6	101.5	32.1	60.4	115.2	49.6	76.4	101.5	55.6	74.4
09	FINISHES	91.7	71.5	80.7	88.7	46.3	65.6	90.6	66.9	77.7	90.0	34.4	59.6	90.6	45.3	65.8	89.6	69.2	78.5
COVERS	DIVS. 10 - 14, 25, 28, 41, 43, 44, 46	100.0	84.9	96.7	100.0	76.2	94.8	100.0	79.8	95.6	100.0	74.3	94.4	100.0	61.2	91.5	100.0	79.5	95.5
21, 22, 23	FIRE SUPPRESSION, PLUMBING & HVAC	98.0	87.2	93.6	98.0	53.1	80.0	98.0	81.7	91.4	98.0	66.1	85.2	98.0	66.8	85.5	98.0	69.4	86.5
26, 27, 3370	ELECTRICAL, COMMUNICATIONS & UTIL.	97.6	99.3	98.5	90.5	70.0	80.0	94.8	94.0	94.4	96.5	70.0	82.9	96.7	88.5	92.5	97.6	69.4	83.2
MF2018	WEIGHTED AVERAGE	99.6	84.2	93.0	96.7	62.8	82.2	98.4	80.6	90.7	98.2	59.2	81.5	99.1	66.0	84.9	98.9	73.7	88.1

VIRGINIA

DIVISION		NEWPORT NEWS 236			NORFOLK 233-235			PETERSBURG 238			PORTSMOUTH 237			PULASKI 243			RICHMOND 230-232		
		MAT.	INST.	TOTAL	MAT.	INST.	TOTAL	MAT.	INST.	TOTAL	MAT.	INST.	TOTAL	MAT.	INST.	TOTAL	MAT.	INST.	TOTAL
015433	CONTRACTOR EQUIPMENT		110.4	110.4		111.0	111.0		110.4	110.4		110.3	110.3		106.2	106.2		110.4	110.4
0241, 31 - 34	SITE & INFRASTRUCTURE, DEMOLITION	108.4	91.7	96.8	105.1	92.7	96.5	112.1	91.7	97.9	107.0	91.0	95.8	106.0	89.0	94.2	106.0	91.7	96.0
0310	Concrete Forming & Accessories	97.7	61.3	66.8	103.7	61.4	67.8	90.9	62.5	66.8	86.7	48.0	53.8	88.3	41.2	48.3	95.3	80.7	82.9
0320	Concrete Reinforcing	89.8	67.3	78.3	98.0	67.3	82.4	89.4	70.7	79.9	89.4	66.6	77.8	93.4	86.2	89.7	98.8	70.7	84.5
0330	Cast-in-Place Concrete	103.2	76.2	93.1	108.0	77.2	96.6	109.6	77.4	97.6	102.2	61.2	86.9	106.7	84.6	98.5	97.1	77.5	89.8
03	CONCRETE	93.7	69.3	82.7	96.1	69.7	84.2	98.8	70.8	86.2	92.4	58.0	76.9	100.6	66.2	85.1	90.5	79.1	85.3
04	MASONRY	96.2	63.2	75.8	98.4	63.2	76.6	109.6	63.2	80.9	101.9	48.4	68.9	88.1	55.4	67.9	90.5	63.2	73.6
05	METALS	99.3	95.5	97.5	99.6	93.6	97.8	97.1	95.1	96.5	98.2	91.9	96.3	100.4	96.2	99.1	102.4	95.2	100.2
06	WOOD, PLASTICS & COMPOSITES	92.8	60.3	75.1	101.3	60.3	79.0	84.0	61.2	71.6	80.2	46.1	61.7	83.8	32.9	56.2	93.1	85.7	89.1
07	THERMAL & MOISTURE PROTECTION	104.3	70.9	89.9	102.3	71.1	88.8	104.2	72.1	90.4	104.2	60.8	85.6	104.3	57.9	84.3	101.5	74.8	90.0
08	OPENINGS	96.6	59.0	87.9	98.3	60.3	89.5	96.0	65.8	89.0	96.7	48.8	85.6	98.4	44.1	85.8	100.1	79.4	95.3
0920	Plaster & Gypsum Board	104.7	58.4	74.2	99.9	58.4	72.5	98.2	59.4	72.6	98.6	43.8	62.5	94.8	30.9	52.7	101.7	84.6	90.4
0950, 0980	Ceilings & Acoustic Treatment	90.5	58.4	68.8	94.8	58.4	70.1	87.1	59.4	68.3	90.5	43.8	58.9	90.8	30.9	50.2	92.7	84.6	87.2
0960	Flooring	92.5	67.6	85.3	90.6	67.6	83.9	88.7	72.7	84.0	85.8	57.6	77.6	93.4	57.6	83.0	93.6	72.7	87.5
0970, 0990	Wall Finishes & Painting/Coating	101.8	58.9	76.4	101.5	58.9	76.3	101.8	58.9	76.4	101.8	58.9	76.4	101.5	49.6	70.8	99.6	58.9	75.5
09	FINISHES	89.5	61.7	74.3	89.0	61.7	74.1	87.1	63.3	74.1	86.7	49.5	66.4	90.0	42.0	63.8	91.6	77.8	84.1
COVERS	DIVS. 10 - 14, 25, 28, 41, 43, 44, 46	100.0	79.5	95.5	100.0	79.5	95.5	100.0	82.2	96.1	100.0	65.2	92.3	100.0	74.6	94.4	100.0	84.8	96.7
21, 22, 23	FIRE SUPPRESSION, PLUMBING & HVAC	100.5	63.5	85.6	100.1	66.0	86.4	98.0	68.5	86.2	100.5	63.2	85.5	98.0	67.5	85.8	100.0	68.5	87.4
26, 27, 3370	ELECTRICAL, COMMUNICATIONS & UTIL.	93.1	64.7	78.6	95.5	61.9	78.3	93.2	70.0	81.3	91.5	61.9	76.4	96.5	85.4	90.8	97.6	70.0	83.5
MF2018	WEIGHTED AVERAGE	97.4	69.7	85.5	98.1	70.0	86.1	97.5	72.4	86.8	96.9	63.1	82.4	97.9	68.3	85.2	98.0	76.4	88.7

VIRGINIA / WASHINGTON

DIVISION		ROANOKE 240-241			STAUNTON 244			WINCHESTER 226			CLARKSTON 994			EVERETT 982			OLYMPIA 985		
		MAT.	INST.	TOTAL	MAT.	INST.	TOTAL	MAT.	INST.	TOTAL	MAT.	INST.	TOTAL	MAT.	INST.	TOTAL	MAT.	INST.	TOTAL
015433	CONTRACTOR EQUIPMENT		106.2	106.2		110.4	110.4		106.2	106.2		92.5	92.5		102.5	102.5		102.5	102.5
0241, 31 - 34	SITE & INFRASTRUCTURE, DEMOLITION	106.8	89.9	95.1	110.0	90.6	96.5	118.7	89.8	98.6	102.2	90.4	94.0	93.8	110.8	105.7	93.3	110.8	105.4
0310	Concrete Forming & Accessories	94.6	71.6	75.1	88.0	49.4	55.2	82.0	68.1	70.2	102.2	66.0	71.5	111.2	101.2	102.7	100.2	101.1	100.9
0320	Concrete Reinforcing	94.5	70.6	82.3	94.1	86.1	90.0	94.1	82.5	88.2	104.8	98.7	101.7	111.1	111.6	111.4	118.0	111.6	114.8
0330	Cast-in-Place Concrete	121.2	85.7	108.0	111.0	86.5	101.9	107.1	65.6	91.7	81.1	85.3	82.7	99.4	108.9	103.0	87.2	108.8	95.2
03	CONCRETE	105.2	77.8	92.9	102.0	70.5	87.8	102.6	71.3	88.5	96.1	78.7	88.3	90.4	105.1	97.0	86.2	105.0	94.7
04	MASONRY	95.6	64.4	76.3	104.5	54.4	73.6	99.0	71.2	81.8	99.6	93.6	95.9	117.9	103.8	109.2	110.3	103.8	106.3
05	METALS	102.6	94.3	100.1	100.6	95.9	99.1	100.7	98.3	100.0	93.7	88.0	92.0	104.4	96.0	101.8	103.1	95.6	100.8
06	WOOD, PLASTICS & COMPOSITES	92.7	73.2	82.1	83.8	44.7	62.5	79.4	66.7	72.5	103.6	60.2	80.0	107.5	99.8	103.3	93.6	99.8	97.0
07	THERMAL & MOISTURE PROTECTION	104.1	74.5	91.4	103.9	56.3	83.4	104.7	76.7	92.6	157.7	85.2	126.5	111.4	106.8	109.4	110.1	104.3	107.6
08	OPENINGS	97.4	66.4	90.2	97.0	50.9	86.3	98.5	70.3	91.9	119.0	66.9	106.9	108.2	103.6	107.2	110.7	102.1	108.7
0920	Plaster & Gypsum Board	101.7	72.3	82.4	94.8	42.4	60.3	94.9	65.6	75.6	151.9	58.9	90.7	109.4	99.9	103.1	103.2	99.9	101.0
0950, 0980	Ceilings & Acoustic Treatment	93.4	72.3	79.1	90.8	42.4	58.0	91.7	65.6	74.0	106.8	58.9	74.4	109.1	99.9	102.8	103.6	99.9	101.1
0960	Flooring	97.0	67.6	88.5	93.0	34.2	75.9	91.9	76.9	87.5	88.3	80.1	85.9	105.2	99.3	103.5	94.0	99.3	95.6
0970, 0990	Wall Finishes & Painting/Coating	101.5	55.6	74.4	101.5	31.5	60.1	115.2	79.6	94.1	83.8	74.0	78.0	90.2	95.6	93.4			
09	FINISHES	92.1	69.2	79.6	89.8	43.7	64.6	91.1	70.3	79.7	107.3	67.7	85.7	103.8	99.7	101.5	92.3	99.7	96.3
COVERS	DIVS. 10 - 14, 25, 28, 41, 43, 44, 46	100.0	79.4	95.5	100.0	77.2	95.0	100.0	84.3	96.6	100.0	91.2	98.1	100.0	100.4	100.1	100.0	100.8	100.2
21, 22, 23	FIRE SUPPRESSION, PLUMBING & HVAC	100.4	65.2	86.3	98.0	59.2	82.4	98.0	83.2	92.0	98.0	82.4	91.8	100.1	102.8	101.2	100.0	102.8	101.1
26, 27, 3370	ELECTRICAL, COMMUNICATIONS & UTIL.	96.6	56.1	75.9	95.5	70.1	82.5	95.2	94.0	94.6	87.4	97.9	92.8	103.5	101.5	102.5	100.9	100.1	100.5
MF2018	WEIGHTED AVERAGE	100.0	71.6	87.8	98.7	65.6	84.5	99.0	81.3	91.4	100.9	83.9	93.6	102.1	102.6	102.3	99.9	102.3	100.9

City Cost Indexes

WASHINGTON

DIVISION		RICHLAND 993 MAT.	INST.	TOTAL	SEATTLE 980-981,987 MAT.	INST.	TOTAL	SPOKANE 990-992 MAT.	INST.	TOTAL	TACOMA 983-984 MAT.	INST.	TOTAL	VANCOUVER 986 MAT.	INST.	TOTAL	WENATCHEE 988 MAT.	INST.	TOTAL
015433	CONTRACTOR EQUIPMENT		92.5	92.5		104.0	104.0		92.5	92.5		102.5	102.5		98.8	98.8		102.5	102.5
0241, 31 - 34	SITE & INFRASTRUCTURE, DEMOLITION	104.2	90.6	94.7	100.9	110.0	107.3	103.7	90.5	94.5	97.1	110.8	106.6	107.3	98.1	100.9	106.6	108.5	107.9
0310	Concrete Forming & Accessories	102.3	82.2	85.2	107.0	103.6	104.1	107.5	81.7	85.5	101.4	101.1	101.1	101.6	93.2	94.5	103.3	78.6	82.3
0320	Concrete Reinforcing	100.5	98.1	99.3	114.5	111.7	113.0	101.2	98.0	99.6	109.7	111.5	110.6	110.7	110.8	110.7	110.6	98.7	104.6
0330	Cast-in-Place Concrete	81.3	86.9	83.4	107.8	108.7	108.1	84.5	86.7	85.3	102.2	108.8	104.7	114.2	99.0	108.6	104.4	93.6	100.4
03	CONCRETE	95.8	86.5	91.6	99.2	106.3	102.4	97.8	86.1	92.6	92.0	105.0	97.9	101.5	98.1	99.9	99.0	87.3	93.8
04	MASONRY	100.6	85.5	91.3	125.2	103.8	112.0	101.2	85.5	91.5	116.6	102.4	107.8	116.6	90.6	100.6	120.0	96.1	105.2
05	METALS	94.1	87.8	92.2	105.8	99.1	103.8	96.5	87.3	93.7	106.2	95.5	102.9	103.6	96.0	101.3	103.6	88.2	98.3
06	WOOD, PLASTICS & COMPOSITES	103.8	80.2	91.0	101.3	102.7	102.1	112.7	80.2	95.0	97.0	99.8	98.5	89.5	92.2	91.0	98.8	75.5	86.1
07	THERMAL & MOISTURE PROTECTION	158.9	86.7	127.8	107.7	106.0	106.9	155.2	87.0	125.8	111.1	104.1	108.1	111.1	96.9	105.0	110.6	90.8	102.1
08	OPENINGS	117.0	76.9	107.7	108.4	105.2	107.7	117.6	76.8	108.1	109.0	102.1	107.4	105.0	96.6	103.0	108.5	75.1	100.7
0920	Plaster & Gypsum Board	151.9	79.5	104.2	107.8	102.9	104.5	144.5	79.5	101.7	109.2	99.9	103.0	107.4	92.4	97.5	112.5	74.9	87.7
0950, 0980	Ceilings & Acoustic Treatment	113.5	79.5	90.4	120.3	102.9	108.5	108.7	79.5	88.9	112.5	99.9	103.9	109.6	92.4	97.9	104.7	74.9	84.5
0960	Flooring	88.6	80.1	86.1	105.3	99.3	103.5	87.7	80.1	85.5	98.2	99.3	98.5	103.9	81.7	97.5	101.0	80.1	94.9
0970, 0990	Wall Finishes & Painting/Coating	83.8	77.0	79.8	101.9	95.6	98.2	83.9	78.4	80.7	91.1	95.6	93.8	93.8	73.8	82.0	91.1	72.8	80.3
09	FINISHES	108.9	80.4	93.3	109.6	101.4	105.1	106.8	80.5	92.4	102.5	99.7	101.0	100.5	88.5	93.9	103.2	77.1	89.0
COVERS	DIVS. 10 - 14, 25, 28, 41, 43, 44, 46	100.0	94.1	98.7	100.0	101.2	100.3	100.0	94.1	98.7	100.0	100.0	100.2	100.0	96.3	99.2	100.0	93.1	98.5
21, 22, 23	FIRE SUPPRESSION, PLUMBING & HVAC	100.5	110.9	104.7	100.1	112.5	105.0	100.5	85.5	94.5	100.1	102.8	101.2	100.2	100.0	100.1	97.7	94.1	96.2
26, 27, 3370	ELECTRICAL, COMMUNICATIONS & UTIL.	85.2	95.1	90.3	103.7	115.4	109.7	83.5	78.9	81.2	103.3	100.1	101.7	108.6	98.5	103.4	104.2	96.8	100.4
MF2018	WEIGHTED AVERAGE	101.4	92.2	97.5	104.3	107.4	105.6	101.7	84.5	94.3	102.5	102.1	102.3	103.4	96.2	100.3	102.9	91.0	97.8

WASHINGTON / WEST VIRGINIA

DIVISION		YAKIMA 989 MAT.	INST.	TOTAL	BECKLEY 258-259 MAT.	INST.	TOTAL	BLUEFIELD 247-248 MAT.	INST.	TOTAL	BUCKHANNON 262 MAT.	INST.	TOTAL	CHARLESTON 250-253 MAT.	INST.	TOTAL	CLARKSBURG 263-264 MAT.	INST.	TOTAL
015433	CONTRACTOR EQUIPMENT		102.5	102.5		106.2	106.2		106.2	106.2		106.2	106.2		106.2	106.2		106.2	106.2
0241, 31 - 34	SITE & INFRASTRUCTURE, DEMOLITION	99.6	109.3	106.4	97.9	91.3	93.3	100.6	91.3	94.1	106.8	91.2	95.9	97.5	92.3	93.9	107.6	91.2	96.2
0310	Concrete Forming & Accessories	102.0	96.1	97.0	84.4	89.7	88.9	85.2	89.5	88.8	84.6	85.9	85.7	96.5	89.6	90.7	82.0	86.0	85.4
0320	Concrete Reinforcing	110.2	98.2	104.1	88.8	89.4	89.1	93.0	84.9	88.9	93.7	84.7	89.1	96.0	89.4	92.7	93.7	98.0	95.9
0330	Cast-in-Place Concrete	109.2	85.6	100.4	103.1	95.1	100.1	104.4	95.0	100.9	104.0	94.4	100.5	99.6	96.1	98.3	114.0	90.0	105.1
03	CONCRETE	96.5	92.4	94.7	94.6	92.4	93.7	96.4	91.6	94.2	99.4	89.7	95.0	93.3	92.8	93.1	103.6	90.5	97.7
04	MASONRY	109.2	82.8	92.9	88.7	90.9	90.1	89.7	90.9	90.5	101.4	86.5	92.2	85.8	92.3	89.8	105.0	86.5	93.6
05	METALS	104.3	88.2	99.4	101.8	103.3	102.3	100.6	101.7	100.9	100.8	101.6	101.0	100.3	103.5	101.3	100.8	106.1	102.4
06	WOOD, PLASTICS & COMPOSITES	97.4	99.8	98.7	81.7	89.9	86.2	82.8	89.9	86.7	82.1	85.4	83.9	93.6	88.8	91.0	78.5	85.4	82.2
07	THERMAL & MOISTURE PROTECTION	111.3	87.7	101.1	106.5	88.4	98.7	104.1	88.4	97.3	104.4	88.6	97.6	102.8	88.9	96.8	104.3	88.3	97.4
08	OPENINGS	108.5	99.1	106.3	96.5	84.9	93.8	98.9	83.9	95.4	98.9	81.5	94.9	96.9	84.3	93.9	98.9	84.4	95.6
0920	Plaster & Gypsum Board	108.8	99.9	102.9	97.3	89.5	92.2	94.1	89.5	91.1	94.5	84.9	88.2	101.7	88.4	92.9	91.7	84.9	87.2
0950, 0980	Ceilings & Acoustic Treatment	106.6	99.9	102.0	81.9	89.5	87.1	89.1	89.5	89.4	90.8	84.9	86.8	93.8	88.4	90.1	90.8	84.9	86.8
0960	Flooring	99.0	80.1	93.5	91.3	101.6	94.3	89.9	101.6	93.3	89.6	96.4	91.6	97.3	101.6	98.5	88.6	96.4	90.9
0970, 0990	Wall Finishes & Painting/Coating	91.1	78.4	83.6	92.3	94.6	93.7	101.5	92.2	96.0	101.5	90.3	94.9	90.7	94.6	93.0	101.5	90.3	94.9
09	FINISHES	101.7	91.3	96.0	86.2	92.7	89.7	88.1	92.4	90.5	89.0	88.2	88.5	92.4	92.4	92.4	88.2	88.2	88.2
COVERS	DIVS. 10 - 14, 25, 28, 41, 43, 44, 46	100.0	97.3	99.4	100.0	92.1	98.3	100.0	92.1	98.3	100.0	90.8	98.0	100.0	92.1	98.3	100.0	90.8	98.0
21, 22, 23	FIRE SUPPRESSION, PLUMBING & HVAC	100.1	109.8	104.0	98.2	93.2	96.2	98.0	85.9	93.1	98.0	91.7	95.4	100.1	94.3	97.8	98.0	91.8	95.5
26, 27, 3370	ELECTRICAL, COMMUNICATIONS & UTIL.	106.3	95.1	100.6	93.7	86.2	89.9	95.5	86.2	90.7	96.9	91.5	94.1	98.8	86.2	92.3	96.9	91.5	94.1
MF2018	WEIGHTED AVERAGE	102.7	96.6	100.1	96.6	92.1	94.6	97.2	90.2	94.2	98.5	90.7	95.1	97.5	92.5	95.4	99.1	91.4	95.8

WEST VIRGINIA

DIVISION		GASSAWAY 266 MAT.	INST.	TOTAL	HUNTINGTON 255-257 MAT.	INST.	TOTAL	LEWISBURG 249 MAT.	INST.	TOTAL	MARTINSBURG 254 MAT.	INST.	TOTAL	MORGANTOWN 265 MAT.	INST.	TOTAL	PARKERSBURG 261 MAT.	INST.	TOTAL
015433	CONTRACTOR EQUIPMENT		106.2	106.2		106.2	106.2		106.2	106.2		106.2	106.2		106.2	106.2		106.2	106.2
0241, 31 - 34	SITE & INFRASTRUCTURE, DEMOLITION	104.3	91.2	95.2	102.6	92.3	95.5	116.2	91.2	98.8	101.7	92.0	94.9	101.6	92.0	94.9	109.8	92.2	97.5
0310	Concrete Forming & Accessories	83.9	89.2	88.4	96.3	91.7	92.4	82.3	89.2	88.1	84.4	81.7	82.1	82.3	86.1	85.5	86.7	88.3	88.1
0320	Concrete Reinforcing	93.7	94.2	94.0	90.1	95.8	93.0	93.7	84.9	89.2	88.8	97.9	93.4	93.7	98.0	95.9	93.0	84.7	88.8
0330	Cast-in-Place Concrete	108.9	94.4	103.5	112.4	96.5	106.5	104.4	94.9	100.9	108.0	90.8	101.6	104.0	94.5	100.5	106.3	91.4	100.8
03	CONCRETE	99.9	92.8	96.7	100.0	95.0	97.8	106.0	91.4	99.4	98.3	88.8	94.0	96.1	92.1	94.3	101.5	89.8	96.2
04	MASONRY	105.9	89.7	95.9	89.0	95.4	92.9	93.6	90.9	91.9	90.5	85.7	87.5	122.8	86.5	100.4	80.0	87.0	84.3
05	METALS	100.7	104.9	102.0	104.1	105.9	104.7	100.7	101.4	100.9	102.2	105.0	103.0	100.8	106.2	102.4	101.4	101.6	101.5
06	WOOD, PLASTICS & COMPOSITES	81.0	89.9	85.8	92.8	91.2	91.9	78.8	89.9	84.8	81.7	80.9	81.3	78.8	85.4	82.4	82.2	88.1	85.4
07	THERMAL & MOISTURE PROTECTION	104.1	90.0	98.0	106.6	90.4	99.7	105.0	88.4	97.9	106.8	83.0	96.6	104.1	88.6	97.4	104.2	89.1	97.7
08	OPENINGS	97.2	86.0	94.6	95.7	87.0	93.7	98.9	83.9	95.4	98.2	76.0	93.1	100.1	84.4	96.5	97.8	82.7	94.3
0920	Plaster & Gypsum Board	93.2	89.5	90.8	104.1	90.8	95.4	91.7	89.5	90.3	97.4	80.2	86.1	91.7	84.9	87.2	94.8	87.6	90.1
0950, 0980	Ceilings & Acoustic Treatment	90.8	89.5	89.9	82.8	90.8	88.2	90.8	89.5	89.9	82.8	80.2	81.0	90.8	84.9	86.8	90.8	87.6	88.7
0960	Flooring	89.4	101.6	93.0	98.6	106.7	100.9	88.7	101.6	92.4	91.3	102.3	94.5	88.7	96.4	90.9	92.5	95.0	93.2
0970, 0990	Wall Finishes & Painting/Coating	101.5	94.6	97.4	92.3	92.2	92.3	101.5	66.2	80.6	92.3	85.9	88.5	101.5	90.3	94.9	101.5	85.9	92.3
09	FINISHES	88.4	92.4	90.6	89.4	94.7	92.3	89.2	89.6	89.4	86.6	85.4	86.0	87.8	88.2	88.0	89.9	89.3	89.6
COVERS	DIVS. 10 - 14, 25, 28, 41, 43, 44, 46	100.0	91.3	98.1	100.0	92.5	98.3	100.0	66.7	92.7	100.0	83.0	96.3	100.0	85.2	96.7	100.0	91.5	98.1
21, 22, 23	FIRE SUPPRESSION, PLUMBING & HVAC	98.0	85.3	92.9	100.7	92.6	97.5	98.0	93.1	96.0	98.2	82.7	92.0	98.0	91.8	95.5	100.4	90.3	96.3
26, 27, 3370	ELECTRICAL, COMMUNICATIONS & UTIL.	96.9	86.2	91.4	97.4	91.8	94.6	93.1	86.2	89.5	99.5	78.7	88.9	97.1	91.5	94.2	97.0	90.4	93.6
MF2018	WEIGHTED AVERAGE	98.4	90.5	95.0	99.0	94.3	97.0	98.8	90.5	95.2	98.1	86.1	92.9	99.0	91.5	95.8	98.4	90.7	95.1

For customer support on your Plumbing Costs with RSMeans Data, call 800.448.8182.

671

City Cost Indexes

WEST VIRGINIA / WISCONSIN

DIVISION		PETERSBURG 268 MAT.	INST.	TOTAL	ROMNEY 267 MAT.	INST.	TOTAL	WHEELING 260 MAT.	INST.	TOTAL	BELOIT 535 MAT.	INST.	TOTAL	EAU CLAIRE 547 MAT.	INST.	TOTAL	GREEN BAY 541-543 MAT.	INST.	TOTAL
015433	CONTRACTOR EQUIPMENT		106.2	106.2		106.2	106.2		106.2	106.2		101.8	101.8		102.6	102.6		100.3	100.3
0241, 31 - 34	SITE & INFRASTRUCTURE, DEMOLITION	100.9	92.2	94.8	103.8	92.2	95.7	110.5	92.0	97.6	96.0	107.0	103.6	96.7	104.0	101.8	100.1	100.2	100.2
0310	Concrete Forming & Accessories	85.6	88.8	88.3	81.6	89.1	88.0	88.3	86.1	86.4	97.5	98.9	98.7	96.1	97.4	97.2	106.4	100.3	101.2
0320	Concrete Reinforcing	93.0	94.0	93.5	93.7	98.2	96.0	92.5	98.0	95.3	96.4	141.1	119.2	93.9	117.5	105.9	91.9	112.6	102.4
0330	Cast-in-Place Concrete	104.0	94.3	100.4	108.9	85.7	100.3	106.3	94.5	101.9	103.9	103.2	103.6	100.5	97.6	99.4	103.9	102.9	103.5
03	CONCRETE	96.3	92.5	94.6	99.7	90.4	95.5	101.5	92.1	97.3	94.9	107.6	100.6	93.1	101.1	96.7	96.3	103.4	99.5
04	MASONRY	96.7	86.5	90.4	94.0	89.7	91.4	104.2	86.4	93.2	98.5	102.4	100.9	91.3	98.1	95.5	122.4	99.9	108.5
05	METALS	100.9	104.2	101.9	100.9	106.2	102.5	101.6	106.2	103.0	97.9	113.0	102.5	94.4	106.7	98.2	96.8	105.1	99.3
06	WOOD, PLASTICS & COMPOSITES	83.1	89.9	86.8	77.9	89.9	84.4	83.8	85.4	84.7	98.6	96.9	97.6	102.0	97.0	99.3	109.6	100.0	104.4
07	THERMAL & MOISTURE PROTECTION	104.2	84.5	95.7	104.3	86.1	96.4	104.6	88.8	97.8	98.1	96.1	97.2	103.6	97.2	100.9	105.8	100.2	103.4
08	OPENINGS	100.1	86.0	96.8	100.0	86.9	97.0	98.6	84.4	95.3	100.4	111.4	103.0	105.4	102.2	104.7	101.3	104.2	102.0
0920	Plaster & Gypsum Board	94.5	89.5	91.2	91.4	89.5	90.2	94.8	84.9	88.3	93.7	97.2	96.0	105.7	97.2	100.1	104.2	100.4	101.7
0950, 0980	Ceilings & Acoustic Treatment	90.8	89.5	89.9	90.8	89.5	89.9	90.8	84.9	86.8	82.5	97.2	92.5	87.3	97.2	94.0	80.7	100.4	94.0
0960	Flooring	90.5	96.4	92.2	88.5	102.3	92.5	93.4	96.4	94.3	94.1	122.4	102.3	81.9	110.1	90.1	97.6	118.7	103.7
0970, 0990	Wall Finishes & Painting/Coating	101.5	90.3	94.9	101.5	90.3	94.9	101.5	90.3	94.9	94.3	106.3	101.4	83.0	78.4	80.3	91.5	82.1	85.9
09	FINISHES	88.7	90.9	89.9	87.9	92.1	90.2	90.1	88.2	89.1	90.0	103.7	97.5	86.0	97.7	92.4	90.2	102.0	96.6
COVERS	DIVS. 10 - 14, 25, 28, 41, 43, 44, 46	100.0	56.0	90.3	100.0	91.2	98.1	100.0	86.0	96.9	100.0	100.7	100.2	100.0	96.1	99.1	100.0	97.7	99.5
21, 22, 23	FIRE SUPPRESSION, PLUMBING & HVAC	98.0	85.6	93.0	98.0	85.4	92.9	100.4	91.8	97.0	100.1	97.6	99.1	100.1	90.2	96.1	100.3	85.3	94.3
26, 27, 3370	ELECTRICAL, COMMUNICATIONS & UTIL.	100.0	78.8	89.2	99.3	78.8	88.8	94.3	91.5	92.9	99.2	86.3	92.6	101.6	86.5	93.9	96.6	83.2	89.7
MF2018	WEIGHTED AVERAGE	98.2	87.7	93.7	98.4	89.2	94.4	99.5	91.5	96.1	98.0	101.4	99.5	97.5	96.5	97.1	99.4	95.8	97.9

WISCONSIN

DIVISION		KENOSHA 531 MAT.	INST.	TOTAL	LA CROSSE 546 MAT.	INST.	TOTAL	LANCASTER 538 MAT.	INST.	TOTAL	MADISON 537 MAT.	INST.	TOTAL	MILWAUKEE 530, 532 MAT.	INST.	TOTAL	NEW RICHMOND 540 MAT.	INST.	TOTAL
015433	CONTRACTOR EQUIPMENT		99.8	99.8		102.6	102.6		101.8	101.8		101.8	101.8		89.6	89.6		102.9	102.9
0241, 31 - 34	SITE & INFRASTRUCTURE, DEMOLITION	101.8	103.6	103.0	90.4	104.0	99.9	95.1	106.3	102.9	93.7	106.9	102.9	94.0	95.2	94.9	94.8	104.7	101.7
0310	Concrete Forming & Accessories	107.7	109.1	108.9	82.6	104.0	99.9	96.8	97.8	97.7	101.4	98.3	98.8	100.6	110.5	109.0	91.2	95.1	94.5
0320	Concrete Reinforcing	96.2	113.7	105.1	93.5	109.3	101.5	97.8	109.3	103.7	100.6	109.5	105.1	99.9	113.9	107.0	91.0	117.5	104.5
0330	Cast-in-Place Concrete	113.0	105.8	110.4	90.3	99.4	93.7	103.3	102.7	103.1	101.3	102.5	101.7	90.5	110.4	97.9	104.6	105.8	105.0
03	CONCRETE	99.9	108.5	103.8	84.5	100.2	91.6	94.6	101.5	97.7	94.8	101.7	97.9	92.1	110.1	100.2	91.1	102.9	96.4
04	MASONRY	96.2	110.3	104.9	90.4	98.1	95.2	98.6	102.3	100.9	97.1	101.2	99.6	100.6	113.0	108.3	117.5	100.5	107.0
05	METALS	98.7	104.2	100.4	94.3	103.7	97.2	95.4	101.2	97.2	99.9	102.1	100.6	99.2	96.5	98.4	94.6	106.3	98.2
06	WOOD, PLASTICS & COMPOSITES	106.0	108.8	107.5	86.2	97.0	92.1	97.7	96.9	97.3	98.4	96.9	97.5	101.3	109.2	105.6	90.8	92.8	91.9
07	THERMAL & MOISTURE PROTECTION	98.6	106.9	102.2	103.0	96.1	100.0	97.9	96.3	97.2	94.3	102.8	97.9	101.5	109.0	104.7	104.4	103.1	103.8
08	OPENINGS	94.7	108.5	97.9	105.4	94.7	102.9	96.2	94.6	95.8	103.5	101.7	103.0	101.4	109.5	103.3	90.6	93.0	91.1
0920	Plaster & Gypsum Board	82.7	109.5	100.4	100.7	97.2	98.4	93.0	97.2	95.8	105.1	97.2	99.9	98.0	109.5	105.6	92.1	93.2	92.8
0950, 0980	Ceilings & Acoustic Treatment	82.5	109.5	100.8	86.5	97.2	93.8	79.1	97.2	91.4	89.5	97.2	94.7	89.7	109.5	103.1	55.9	93.2	81.2
0960	Flooring	111.8	118.7	113.8	76.0	115.4	87.4	93.7	114.7	99.8	93.4	114.7	99.6	100.9	114.7	104.9	91.2	118.4	99.1
0970, 0990	Wall Finishes & Painting/Coating	104.4	121.9	114.8	83.0	83.9	83.5	94.3	100.3	97.8	93.6	106.3	101.1	106.1	122.7	115.9	94.4	83.9	88.1
09	FINISHES	94.4	112.6	104.3	83.0	99.3	91.9	89.1	101.3	95.7	90.2	102.1	96.7	99.4	112.8	106.7	81.4	98.0	90.5
COVERS	DIVS. 10 - 14, 25, 28, 41, 43, 44, 46	100.0	99.4	99.9	100.0	94.7	98.8	100.0	86.0	96.9	100.0	98.6	99.7	100.0	102.7	100.6	100.0	97.1	99.4
21, 22, 23	FIRE SUPPRESSION, PLUMBING & HVAC	100.3	98.1	99.4	100.1	90.1	96.1	97.6	91.2	95.1	100.0	97.4	98.9	100.0	105.6	102.2	97.1	91.0	94.6
26, 27, 3370	ELECTRICAL, COMMUNICATIONS & UTIL.	99.6	99.5	99.5	101.9	86.5	94.0	98.9	86.9	92.8	99.0	96.3	97.6	100.0	101.6	100.8	99.9	86.9	93.3
MF2018	WEIGHTED AVERAGE	98.7	104.6	101.2	95.9	95.9	95.9	96.4	96.7	96.5	98.4	100.3	99.2	98.9	105.9	101.9	95.6	97.0	96.2

WISCONSIN

DIVISION		OSHKOSH 549 MAT.	INST.	TOTAL	PORTAGE 539 MAT.	INST.	TOTAL	RACINE 534 MAT.	INST.	TOTAL	RHINELANDER 545 MAT.	INST.	TOTAL	SUPERIOR 548 MAT.	INST.	TOTAL	WAUSAU 544 MAT.	INST.	TOTAL
015433	CONTRACTOR EQUIPMENT		100.3	100.3		101.8	101.8		101.8	101.8		100.3	100.3		102.9	102.9		100.3	100.3
0241, 31 - 34	SITE & INFRASTRUCTURE, DEMOLITION	91.8	100.1	97.6	86.0	106.2	100.0	95.8	107.4	103.9	104.0	100.1	101.3	91.6	104.4	100.5	87.7	100.7	96.7
0310	Concrete Forming & Accessories	88.6	97.4	96.1	88.7	98.4	97.0	97.8	109.1	107.4	86.2	97.3	95.6	89.0	91.0	90.7	88.0	100.1	98.3
0320	Concrete Reinforcing	92.1	112.5	102.4	97.9	109.5	103.8	96.4	113.7	105.2	92.3	109.5	101.0	91.0	109.9	100.6	92.3	109.3	100.9
0330	Cast-in-Place Concrete	96.3	100.9	98.0	88.7	101.3	93.4	101.9	105.6	103.3	109.3	100.9	106.1	98.5	101.2	99.5	89.8	95.2	91.8
03	CONCRETE	86.3	101.4	93.1	82.7	101.5	91.1	98.1	106.4	100.4	98.1	100.8	99.3	88.5	98.2	91.4	81.5	100.1	89.9
04	MASONRY	104.9	99.8	101.8	97.4	102.3	100.4	98.5	110.3	105.8	122.2	99.7	108.3	116.7	101.5	107.9	104.3	99.7	101.5
05	METALS	94.8	104.4	97.8	96.1	105.3	98.9	99.6	104.2	101.0	94.7	103.3	97.3	95.5	104.2	98.2	94.6	103.8	97.4
06	WOOD, PLASTICS & COMPOSITES	87.9	97.0	92.8	87.1	96.9	92.4	98.9	108.8	104.3	85.3	97.0	91.6	89.0	88.2	88.6	87.3	100.0	94.2
07	THERMAL & MOISTURE PROTECTION	104.8	85.6	96.5	97.2	103.0	99.7	98.1	105.6	101.3	105.7	85.4	97.0	104.0	98.6	101.7	104.6	97.8	101.7
08	OPENINGS	97.7	99.0	98.0	96.4	102.6	97.8	100.4	108.5	102.3	97.7	94.6	97.0	90.0	93.8	90.9	97.9	99.9	98.4
0920	Plaster & Gypsum Board	90.8	97.2	95.0	85.6	97.2	93.3	93.7	109.5	104.1	90.8	97.2	95.0	92.0	88.5	89.7	90.8	100.4	97.1
0950, 0980	Ceilings & Acoustic Treatment	80.7	97.2	91.9	81.6	97.2	92.2	82.5	109.5	100.8	80.7	97.2	91.9	56.8	88.5	78.3	80.7	100.4	94.0
0960	Flooring	89.3	118.7	97.8	90.0	118.4	98.3	94.1	118.7	101.2	88.7	118.4	97.3	92.3	127.2	102.4	89.2	118.4	97.7
0970, 0990	Wall Finishes & Painting/Coating	89.1	107.0	99.7	94.3	100.3	97.8	94.3	119.0	108.9	89.1	82.1	84.9	83.6	108.6	98.3	89.1	85.4	86.9
09	FINISHES	85.1	102.9	94.8	86.9	102.3	95.3	90.0	112.3	102.1	86.0	100.1	93.7	80.8	99.2	90.8	84.8	102.3	94.4
COVERS	DIVS. 10 - 14, 25, 28, 41, 43, 44, 46	100.0	87.3	97.2	100.0	87.4	97.2	100.0	99.4	99.9	100.0	87.6	97.3	100.0	95.8	99.1	100.0	97.7	99.5
21, 22, 23	FIRE SUPPRESSION, PLUMBING & HVAC	97.8	84.9	92.7	97.6	98.0	97.8	100.1	98.2	99.3	97.8	90.6	94.9	97.1	93.0	95.4	97.8	90.8	95.0
26, 27, 3370	ELECTRICAL, COMMUNICATIONS & UTIL.	100.4	81.1	90.5	102.5	96.3	99.4	99.1	100.4	99.8	99.8	80.0	89.7	104.5	101.0	102.7	101.5	80.0	90.5
MF2018	WEIGHTED AVERAGE	95.6	94.2	95.0	94.8	100.5	97.3	98.1	105.0	101.1	98.2	94.6	96.6	95.3	98.7	96.8	95.0	95.8	95.3

City Cost Indexes

WYOMING

DIVISION		CASPER 826 MAT.	INST.	TOTAL	CHEYENNE 820 MAT.	INST.	TOTAL	NEWCASTLE 827 MAT.	INST.	TOTAL	RAWLINS 823 MAT.	INST.	TOTAL	RIVERTON 825 MAT.	INST.	TOTAL	ROCK SPRINGS 829 - 831 MAT.	INST.	TOTAL
015433	CONTRACTOR EQUIPMENT		97.6	97.6		97.6	97.6		97.6	97.6		97.6	97.6		97.6	97.6		97.6	97.6
0241, 31 - 34	SITE & INFRASTRUCTURE, DEMOLITION	95.6	94.7	94.9	90.9	94.7	93.5	84.1	94.2	91.1	97.2	94.2	95.1	91.0	94.1	93.2	88.1	94.2	92.3
0310	Concrete Forming & Accessories	99.8	54.3	61.1	101.2	67.2	72.3	91.8	72.5	75.4	96.1	72.7	76.2	90.8	61.9	66.2	98.3	62.8	68.2
0320	Concrete Reinforcing	109.6	84.9	97.0	105.0	84.9	94.8	112.9	84.7	98.6	112.6	84.8	98.4	113.6	84.7	98.9	113.6	83.0	98.0
0330	Cast-in-Place Concrete	104.4	76.0	93.8	98.6	76.0	90.2	99.6	75.0	90.4	99.7	75.1	90.5	99.6	75.0	90.5	99.6	75.0	90.4
03	CONCRETE	101.9	67.9	86.6	99.5	73.7	87.9	99.6	75.7	88.9	113.3	75.9	96.5	107.8	71.0	91.2	100.1	71.1	87.1
04	MASONRY	99.3	64.7	77.9	102.2	63.8	78.5	98.9	60.5	75.2	98.9	60.5	75.2	98.9	60.5	75.2	159.3	56.3	95.7
05	METALS	103.4	80.9	96.6	105.9	80.9	98.3	102.2	80.5	95.6	102.2	80.7	95.7	102.3	80.5	95.7	103.1	79.7	96.0
06	WOOD, PLASTICS & COMPOSITES	94.5	49.0	69.8	93.4	66.5	78.8	83.7	74.8	78.9	87.7	74.8	80.7	82.6	60.4	70.6	92.9	61.7	75.9
07	THERMAL & MOISTURE PROTECTION	100.4	66.2	85.7	95.4	67.9	83.6	97.0	65.5	83.4	98.5	76.8	89.2	97.9	69.4	85.6	97.1	68.3	84.7
08	OPENINGS	106.0	60.0	95.3	109.6	69.7	100.3	114.0	74.3	104.7	113.6	74.3	104.5	113.8	66.3	102.8	114.4	66.7	103.3
0920	Plaster & Gypsum Board	103.3	47.6	66.6	92.9	65.6	74.9	89.9	74.1	79.5	90.2	74.1	79.6	89.9	59.4	69.8	102.4	60.7	74.9
0950, 0980	Ceilings & Acoustic Treatment	105.4	47.6	66.2	99.3	65.6	76.5	101.2	74.1	82.9	101.2	74.1	82.9	101.2	59.4	72.9	101.2	60.7	73.7
0960	Flooring	104.2	72.0	94.8	102.2	72.0	93.5	96.1	46.0	81.6	98.9	46.0	83.6	95.6	62.3	85.9	101.2	46.0	85.2
0970, 0990	Wall Finishes & Painting/Coating	94.7	59.9	74.1	98.9	59.9	75.8	95.5	62.1	75.7	95.5	62.1	75.7	95.5	62.1	75.7	95.5	62.1	75.7
09	FINISHES	96.7	56.6	74.8	93.7	67.0	79.1	89.2	66.5	76.8	91.3	66.5	77.8	89.9	61.3	74.3	92.5	58.8	74.1
COVERS	DIVS. 10 - 14, 25, 28, 41, 43, 44, 46	100.0	93.7	98.6	100.0	89.0	97.6	100.0	93.8	98.6	100.0	93.8	98.6	100.0	79.0	95.4	100.0	84.3	96.5
21, 22, 23	FIRE SUPPRESSION, PLUMBING & HVAC	100.0	71.3	88.5	99.9	71.3	88.5	98.6	70.5	87.3	98.6	70.5	87.3	98.6	70.5	87.3	99.9	70.5	88.1
26, 27, 3370	ELECTRICAL, COMMUNICATIONS & UTIL.	101.2	61.8	81.0	103.1	62.7	82.4	101.9	63.9	82.4	101.9	63.9	82.4	101.9	60.6	80.7	100.0	78.5	89.0
MF2018	WEIGHTED AVERAGE	101.1	69.5	87.5	101.4	72.1	88.8	100.1	72.3	88.2	102.3	72.7	89.6	101.4	69.7	87.8	103.8	71.6	89.9

WYOMING / CANADA

DIVISION		SHERIDAN 828 MAT.	INST.	TOTAL	WHEATLAND 822 MAT.	INST.	TOTAL	WORLAND 824 MAT.	INST.	TOTAL	YELLOWSTONE NAT'L PA 821 MAT.	INST.	TOTAL	BARRIE, ONTARIO MAT.	INST.	TOTAL	BATHURST, NEW BRUNSWICK MAT.	INST.	TOTAL
015433	CONTRACTOR EQUIPMENT		97.6	97.6		97.6	97.6		97.6	97.6		97.6	97.6		103.0	103.0		102.7	102.7
0241, 31 - 34	SITE & INFRASTRUCTURE, DEMOLITION	91.2	94.7	93.6	88.6	94.2	92.5	86.1	94.2	91.7	86.2	94.2	91.7	118.0	99.7	105.3	105.8	95.5	98.6
0310	Concrete Forming & Accessories	98.9	62.7	68.1	93.7	49.7	56.3	93.8	62.7	67.4	93.8	62.8	67.5	124.4	84.6	90.6	102.5	59.4	65.9
0320	Concrete Reinforcing	113.6	84.8	98.9	112.9	84.2	98.3	113.6	84.7	98.9	115.6	84.7	99.9	177.7	89.4	132.8	144.5	59.2	101.1
0330	Cast-in-Place Concrete	103.0	76.0	92.9	104.0	74.9	93.2	99.6	75.0	90.4	99.6	75.0	90.4	156.6	84.6	129.8	111.1	57.9	91.3
03	CONCRETE	108.0	71.6	91.7	104.7	65.3	86.9	99.8	71.3	87.0	100.1	71.3	87.1	139.3	85.8	115.2	110.5	59.9	87.7
04	MASONRY	99.2	62.0	76.2	99.2	52.2	70.2	98.9	60.5	75.2	98.9	60.5	75.2	165.9	91.2	119.7	160.0	58.7	97.4
05	METALS	105.7	80.7	98.1	102.1	79.6	95.3	102.3	80.4	95.7	102.9	79.9	95.9	110.9	92.4	105.3	120.3	74.3	106.3
06	WOOD, PLASTICS & COMPOSITES	95.0	60.4	76.2	85.5	44.2	63.1	85.6	61.7	72.6	85.6	61.7	72.6	118.5	83.2	99.3	99.7	59.6	77.9
07	THERMAL & MOISTURE PROTECTION	98.2	66.7	84.7	97.2	59.5	81.0	97.0	66.4	83.9	96.4	66.4	83.5	114.4	88.5	103.2	109.9	60.1	88.5
08	OPENINGS	114.5	66.3	103.3	112.5	57.4	99.7	114.2	67.0	103.2	106.8	66.7	97.5	92.6	83.2	90.4	85.8	52.7	78.1
0920	Plaster & Gypsum Board	112.7	59.4	77.6	89.9	42.7	58.8	89.9	60.7	70.7	90.1	60.7	70.7	151.9	82.7	106.3	121.5	58.4	79.9
0950, 0980	Ceilings & Acoustic Treatment	104.0	59.4	73.8	101.2	42.7	61.6	101.2	60.7	73.7	102.0	60.7	74.0	95.1	82.7	86.7	115.9	58.4	76.9
0960	Flooring	99.5	62.3	88.7	97.6	45.2	82.4	97.6	46.0	82.6	97.6	46.0	82.6	118.9	90.2	110.6	100.1	42.9	83.5
0970, 0990	Wall Finishes & Painting/Coating	97.7	59.9	75.3	95.5	59.9	74.5	95.5	62.1	75.7	95.5	62.1	75.7	105.0	86.0	93.7	110.7	49.2	74.3
09	FINISHES	96.5	61.7	77.5	89.9	48.1	67.1	89.7	58.8	72.8	89.8	58.9	72.9	109.7	85.9	96.7	105.7	55.4	78.2
COVERS	DIVS. 10 - 14, 25, 28, 41, 43, 44, 46	100.0	94.9	98.9	100.0	83.2	96.3	100.0	77.8	95.1	100.0	77.8	95.1	139.2	66.4	123.2	131.1	59.1	115.3
21, 22, 23	FIRE SUPPRESSION, PLUMBING & HVAC	98.6	71.3	87.7	98.6	70.5	87.3	98.6	70.5	87.3	98.6	70.5	87.3	103.8	95.6	100.5	103.7	66.2	88.7
26, 27, 3370	ELECTRICAL, COMMUNICATIONS & UTIL.	104.4	60.6	82.0	101.9	78.5	89.9	101.9	78.5	89.9	100.7	88.2	94.3	115.3	84.9	99.8	108.7	57.2	82.4
MF2018	WEIGHTED AVERAGE	102.9	70.6	89.1	100.8	68.1	86.8	100.3	71.8	88.1	99.5	73.2	88.2	116.5	89.3	104.8	111.2	63.9	90.9

CANADA

DIVISION		BRANDON, MANITOBA MAT.	INST.	TOTAL	BRANTFORD, ONTARIO MAT.	INST.	TOTAL	BRIDGEWATER, NOVA SCOTIA MAT.	INST.	TOTAL	CALGARY, ALBERTA MAT.	INST.	TOTAL	CAP-DE-LA-MADELEINE, QUEBEC MAT.	INST.	TOTAL	CHARLESBOURG, QUEBEC MAT.	INST.	TOTAL
015433	CONTRACTOR EQUIPMENT		104.7	104.7		103.1	103.1		102.7	102.7		132.6	132.6		103.5	103.5		103.5	103.5
0241, 31 - 34	SITE & INFRASTRUCTURE, DEMOLITION	128.1	98.0	107.1	117.5	100.1	105.4	102.0	97.4	98.8	126.7	125.7	126.0	97.8	98.9	98.6	97.8	98.9	98.6
0310	Concrete Forming & Accessories	140.8	68.0	78.9	122.8	91.7	96.3	95.7	69.8	73.7	117.7	97.4	100.4	128.7	81.4	88.5	128.7	81.4	88.5
0320	Concrete Reinforcing	184.3	56.1	119.1	171.2	88.1	128.9	145.8	49.0	96.6	136.9	84.2	110.1	145.8	74.6	109.6	145.8	74.6	109.6
0330	Cast-in-Place Concrete	104.8	72.4	92.7	127.3	105.0	119.0	130.8	69.0	107.8	137.0	107.0	125.8	102.7	90.2	98.1	102.7	90.2	98.1
03	CONCRETE	118.4	68.4	95.9	122.4	95.9	110.5	120.6	66.9	96.5	125.2	99.3	113.5	108.0	83.8	97.1	108.0	83.8	97.1
04	MASONRY	213.4	62.0	119.8	166.5	95.4	122.6	162.1	67.0	103.3	204.0	88.8	132.8	162.8	77.9	110.4	162.8	77.9	110.4
05	METALS	135.2	79.1	118.2	118.0	93.1	110.4	117.3	76.9	105.0	129.6	104.4	122.0	116.4	86.0	107.2	116.4	86.0	107.2
06	WOOD, PLASTICS & COMPOSITES	154.7	68.8	108.0	121.8	90.3	104.7	90.4	69.4	79.0	99.9	97.2	98.4	133.5	81.3	105.2	133.5	81.3	105.2
07	THERMAL & MOISTURE PROTECTION	127.7	70.6	103.1	119.2	94.0	108.4	113.7	69.6	94.7	125.7	96.5	113.1	112.5	86.5	101.3	112.5	86.5	101.3
08	OPENINGS	102.4	61.6	92.9	89.7	89.2	89.6	84.2	63.1	79.3	82.9	86.0	83.7	91.1	74.4	87.2	91.1	74.4	87.2
0920	Plaster & Gypsum Board	113.7	67.5	83.3	113.1	90.0	97.9	118.4	68.4	85.5	125.1	96.0	106.0	142.5	80.7	101.8	142.5	80.7	101.8
0950, 0980	Ceilings & Acoustic Treatment	122.4	67.5	85.2	104.7	90.0	94.7	104.7	68.4	80.1	150.3	96.0	113.5	104.7	80.7	88.4	104.7	80.7	88.4
0960	Flooring	131.3	64.0	111.8	113.6	90.2	106.8	96.1	60.9	85.9	121.3	86.7	111.2	113.6	88.8	106.5	113.6	88.8	106.5
0970, 0990	Wall Finishes & Painting/Coating	117.6	55.5	80.8	109.8	94.4	100.7	109.8	60.9	80.9	115.0	109.2	111.6	109.8	85.8	95.6	109.8	85.8	95.6
09	FINISHES	119.9	66.2	90.6	106.6	92.0	98.6	101.7	67.5	83.0	124.6	97.2	109.7	110.2	83.6	95.7	110.2	83.6	95.7
COVERS	DIVS. 10 - 14, 25, 28, 41, 43, 44, 46	131.1	61.4	115.8	131.1	68.4	117.3	131.1	62.1	115.9	131.1	94.0	123.0	131.1	77.2	119.3	131.1	77.2	119.3
21, 22, 23	FIRE SUPPRESSION, PLUMBING & HVAC	103.9	80.7	94.6	103.8	98.5	101.7	103.8	81.3	94.8	104.4	91.6	99.3	104.1	86.1	96.9	104.1	86.1	96.9
26, 27, 3370	ELECTRICAL, COMMUNICATIONS & UTIL.	110.6	64.7	87.1	107.6	84.5	95.8	111.9	60.1	85.4	102.1	96.2	99.1	106.6	66.8	86.2	106.6	66.8	86.2
MF2018	WEIGHTED AVERAGE	121.8	72.4	100.6	113.7	93.1	104.9	111.9	72.2	94.8	118.3	97.6	109.4	111.3	82.1	98.8	111.3	82.1	98.8

For customer support on your Plumbing Costs with RSMeans Data, call 800.448.8182.

673

City Cost Indexes

CANADA

DIVISION		CHARLOTTETOWN, PRINCE EDWARD ISLAND			CHICOUTIMI, QUEBEC			CORNER BROOK, NEWFOUNDLAND			CORNWALL, ONTARIO			DALHOUSIE, NEW BRUNSWICK			DARTMOUTH, NOVA SCOTIA		
		MAT.	INST.	TOTAL	MAT.	INST.	TOTAL	MAT.	INST.	TOTAL	MAT.	INST.	TOTAL	MAT.	INST.	TOTAL	MAT.	INST.	TOTAL
015433	CONTRACTOR EQUIPMENT		120.9	120.9		103.3	103.3		103.2	103.2		103.1	103.1		103.0	103.0		102.0	102.0
0241, 31 - 34	SITE & INFRASTRUCTURE, DEMOLITION	132.8	108.0	115.5	102.7	98.4	99.7	132.3	95.4	106.6	115.7	99.6	104.5	101.1	95.7	97.3	119.5	97.0	103.8
0310	Concrete Forming & Accessories	115.9	54.4	63.6	131.2	89.2	95.5	117.6	77.7	83.7	120.5	84.7	90.1	101.7	59.7	66.0	107.4	69.8	75.4
0320	Concrete Reinforcing	164.4	47.9	105.2	109.0	93.9	101.3	169.0	50.2	108.6	171.2	87.8	128.8	146.3	59.2	102.1	176.7	49.0	111.8
0330	Cast-in-Place Concrete	147.6	58.3	114.4	104.7	94.9	101.0	122.0	65.5	101.0	114.5	95.7	107.5	106.9	58.1	88.7	117.7	68.8	99.5
03	CONCRETE	134.2	56.5	99.3	102.0	92.1	97.6	149.1	69.4	113.3	116.1	89.5	104.1	112.7	60.0	89.0	132.0	66.8	102.7
04	MASONRY	176.6	56.0	102.1	162.3	88.4	116.7	209.3	75.9	126.9	165.5	87.4	117.2	163.7	58.7	98.8	223.5	67.0	126.8
05	METALS	144.2	81.5	125.2	120.1	91.2	111.3	135.7	76.0	117.5	117.8	91.8	109.9	110.2	74.6	99.4	136.4	76.6	118.2
06	WOOD, PLASTICS & COMPOSITES	103.7	53.8	76.6	135.4	89.8	110.6	131.2	83.8	105.5	120.2	84.0	100.5	97.6	59.6	76.9	118.4	69.3	91.7
07	THERMAL & MOISTURE PROTECTION	126.0	58.4	96.9	110.4	96.0	104.2	132.2	67.8	104.5	119.0	88.6	105.9	118.1	60.1	93.2	129.2	69.5	103.5
08	OPENINGS	85.3	47.0	76.4	89.7	76.1	86.6	108.6	68.5	99.3	91.1	82.9	89.2	87.3	52.7	79.3	93.2	63.0	86.2
0920	Plaster & Gypsum Board	118.6	52.0	74.8	141.4	89.3	107.1	145.6	83.2	104.6	168.5	83.5	112.5	126.5	58.4	81.6	140.7	68.4	93.1
0950, 0980	Ceilings & Acoustic Treatment	128.3	52.0	76.6	115.1	89.3	97.6	123.2	83.2	96.1	107.2	83.5	91.2	107.0	58.4	74.1	130.0	68.4	88.3
0960	Flooring	116.7	57.1	99.4	115.8	88.8	108.0	113.2	51.2	95.2	113.6	88.9	106.5	102.3	65.6	91.7	108.1	60.9	94.4
0970, 0990	Wall Finishes & Painting/Coating	116.4	40.9	71.7	110.7	103.2	106.3	117.4	57.9	82.2	109.8	87.9	96.8	113.4	49.2	75.4	117.4	60.9	84.0
09	FINISHES	118.8	53.8	83.3	112.6	91.6	101.1	119.8	72.3	93.8	115.1	86.0	99.2	106.5	60.0	81.1	117.6	67.5	90.2
COVERS	DIVS. 10 - 14, 25, 28, 41, 43, 44, 46	131.1	59.3	115.3	131.1	80.9	120.1	131.1	61.9	115.9	131.1	66.0	116.8	131.1	59.1	115.3	131.1	62.1	115.9
21, 22, 23	FIRE SUPPRESSION, PLUMBING & HVAC	104.0	60.3	86.5	103.7	80.0	94.2	103.9	67.4	89.3	104.1	96.4	101.1	103.8	66.2	88.7	103.9	81.3	94.8
26, 27, 3370	ELECTRICAL, COMMUNICATIONS & UTIL.	108.1	48.6	77.7	105.4	83.1	94.0	108.0	52.8	79.8	107.8	85.4	96.4	108.6	54.1	80.7	112.1	60.1	85.5
MF2018	WEIGHTED AVERAGE	121.0	61.8	95.6	111.1	87.3	100.9	125.8	70.2	101.9	113.7	89.6	103.4	110.4	64.1	90.5	122.6	72.1	100.9

CANADA

DIVISION		EDMONTON, ALBERTA			FORT MCMURRAY, ALBERTA			FREDERICTON, NEW BRUNSWICK			GATINEAU, QUEBEC			GRANBY, QUEBEC			HALIFAX, NOVA SCOTIA		
		MAT.	INST.	TOTAL	MAT.	INST.	TOTAL	MAT.	INST.	TOTAL	MAT.	INST.	TOTAL	MAT.	INST.	TOTAL	MAT.	INST.	TOTAL
015433	CONTRACTOR EQUIPMENT		132.2	132.2		105.3	105.3		120.6	120.6		103.5	103.5		103.5	103.5		119.0	119.0
0241, 31 - 34	SITE & INFRASTRUCTURE, DEMOLITION	123.9	125.5	125.0	122.4	100.7	107.3	111.8	109.3	110.1	97.6	98.9	98.5	98.1	98.9	98.7	103.7	109.4	107.7
0310	Concrete Forming & Accessories	119.5	97.4	100.7	120.5	89.8	94.4	120.7	60.2	69.3	128.7	81.3	88.4	128.7	81.3	88.4	116.2	83.3	88.3
0320	Concrete Reinforcing	137.3	84.2	110.3	158.5	84.1	120.6	161.2	59.5	109.5	154.2	74.5	113.7	154.2	74.5	113.7	155.1	72.1	112.9
0330	Cast-in-Place Concrete	131.0	106.9	122.0	169.0	101.1	143.7	108.4	58.7	89.9	101.3	90.2	97.2	104.7	90.2	99.3	102.5	84.6	95.8
03	CONCRETE	122.5	99.2	112.0	140.6	93.0	119.2	115.4	61.3	91.1	108.3	83.7	97.3	109.9	83.7	98.1	110.9	83.1	98.4
04	MASONRY	199.0	88.8	130.9	206.9	85.5	131.9	183.6	60.0	107.2	162.7	77.9	110.3	163.0	77.9	110.4	181.0	85.3	121.9
05	METALS	136.5	104.2	126.7	147.2	91.6	130.3	144.5	86.5	126.9	116.4	85.9	107.1	116.6	85.8	107.3	144.4	97.6	130.2
06	WOOD, PLASTICS & COMPOSITES	98.8	97.2	97.9	116.0	89.3	101.5	109.0	59.8	82.3	133.5	81.3	105.2	133.5	81.3	105.2	104.1	82.5	92.4
07	THERMAL & MOISTURE PROTECTION	133.4	96.5	117.5	127.6	92.0	112.3	127.9	60.5	98.9	112.5	86.5	101.3	112.5	85.0	100.6	130.5	84.9	110.9
08	OPENINGS	82.0	86.0	82.9	91.1	81.6	88.9	86.7	51.7	78.6	91.1	69.9	86.2	91.1	69.9	86.2	87.1	75.2	84.4
0920	Plaster & Gypsum Board	122.4	96.0	105.0	113.6	88.6	97.1	121.7	58.4	80.0	112.4	80.7	91.5	112.4	80.7	91.5	115.5	81.7	91.9
0950, 0980	Ceilings & Acoustic Treatment	145.4	96.0	112.0	112.3	88.6	96.3	132.9	58.4	82.4	104.7	80.7	88.4	104.7	80.7	88.4	142.1	81.7	101.2
0960	Flooring	122.0	86.7	111.7	113.6	86.7	105.8	118.0	68.5	103.6	113.6	88.8	106.5	113.6	88.8	106.5	111.2	81.9	102.7
0970, 0990	Wall Finishes & Painting/Coating	114.8	109.2	111.5	109.9	90.6	98.5	116.0	62.7	84.4	109.8	85.8	95.6	109.8	85.8	95.6	119.2	90.0	101.9
09	FINISHES	119.6	97.2	107.4	109.5	89.6	98.6	120.1	62.1	88.5	105.8	83.6	93.7	105.8	83.6	93.7	115.1	84.2	98.2
COVERS	DIVS. 10 - 14, 25, 28, 41, 43, 44, 46	131.1	94.0	123.0	131.1	90.4	122.2	131.1	59.7	115.4	131.1	77.2	119.3	131.1	77.2	119.3	131.1	67.4	117.1
21, 22, 23	FIRE SUPPRESSION, PLUMBING & HVAC	104.3	91.6	99.2	104.2	95.3	100.6	104.1	75.3	92.6	104.1	86.1	96.9	103.8	86.1	96.7	104.2	80.9	94.9
26, 27, 3370	ELECTRICAL, COMMUNICATIONS & UTIL.	105.8	96.2	100.9	102.0	79.4	90.4	106.6	71.4	88.6	106.6	66.8	86.2	107.2	66.8	86.5	108.6	84.4	96.2
MF2018	WEIGHTED AVERAGE	118.9	97.5	109.7	122.9	90.2	108.9	118.7	71.2	98.4	111.0	81.9	98.5	111.3	81.8	98.6	117.7	85.8	104.0

CANADA

DIVISION		HAMILTON, ONTARIO			HULL, QUEBEC			JOLIETTE, QUEBEC			KAMLOOPS, BRITISH COLUMBIA			KINGSTON, ONTARIO			KITCHENER, ONTARIO		
		MAT.	INST.	TOTAL	MAT.	INST.	TOTAL	MAT.	INST.	TOTAL	MAT.	INST.	TOTAL	MAT.	INST.	TOTAL	MAT.	INST.	TOTAL
015433	CONTRACTOR EQUIPMENT		120.9	120.9		103.5	103.5		103.5	103.5		106.7	106.7		105.3	105.3		105.5	105.5
0241, 31 - 34	SITE & INFRASTRUCTURE, DEMOLITION	109.2	114.9	113.2	97.6	98.9	98.5	98.2	98.9	98.7	120.3	102.6	108.0	115.7	103.4	107.2	95.9	104.9	102.2
0310	Concrete Forming & Accessories	120.5	84.0	89.4	128.7	81.3	88.4	128.7	81.4	88.5	119.4	84.6	89.8	120.6	84.8	90.2	112.3	86.8	90.7
0320	Concrete Reinforcing	151.4	98.6	124.5	154.2	74.5	113.7	145.8	74.6	109.6	114.3	78.0	95.9	171.2	87.8	128.8	105.9	98.4	102.1
0330	Cast-in-Place Concrete	113.9	102.2	109.5	101.3	90.2	97.2	105.5	90.2	99.9	91.4	94.7	92.6	114.5	95.7	107.5	111.0	97.3	105.9
03	CONCRETE	113.1	98.3	106.4	108.3	83.7	97.3	109.3	83.8	97.8	117.4	87.4	103.9	118.0	89.5	105.1	99.2	92.7	96.3
04	MASONRY	167.0	99.7	125.4	162.7	77.9	110.3	163.1	77.9	110.5	170.0	85.7	117.9	172.4	87.5	119.9	143.0	96.9	114.5
05	METALS	143.5	105.7	132.0	116.6	85.9	107.3	116.6	86.0	107.3	118.4	88.2	109.3	119.1	91.8	110.8	131.7	96.3	120.9
06	WOOD, PLASTICS & COMPOSITES	102.6	93.1	97.4	133.5	81.3	105.2	133.5	81.3	105.2	103.4	83.0	92.3	120.2	84.1	100.6	108.5	84.8	95.6
07	THERMAL & MOISTURE PROTECTION	121.5	100.0	112.2	112.5	86.5	101.3	112.5	86.5	101.3	128.8	84.1	109.6	119.0	89.8	106.4	113.6	96.8	106.4
08	OPENINGS	83.5	91.9	85.4	91.1	69.9	86.2	91.1	74.4	87.2	87.9	80.8	86.3	91.1	82.6	89.1	81.8	85.5	82.7
0920	Plaster & Gypsum Board	118.6	92.3	101.3	112.4	80.7	91.5	142.5	80.7	101.8	98.6	82.1	87.8	171.0	83.6	113.5	102.6	84.4	90.6
0950, 0980	Ceilings & Acoustic Treatment	134.8	92.3	106.0	104.7	80.7	88.4	104.7	80.7	88.4	104.7	82.1	89.4	119.1	83.6	95.1	107.6	84.4	91.8
0960	Flooring	116.4	95.6	110.4	113.6	88.8	106.5	113.6	88.8	106.5	112.8	50.8	94.8	113.6	88.9	106.5	101.4	95.6	99.7
0970, 0990	Wall Finishes & Painting/Coating	110.4	103.5	106.3	109.8	85.8	95.6	109.8	85.8	95.6	109.8	78.4	91.2	109.8	81.3	92.9	103.8	93.3	97.6
09	FINISHES	115.6	95.3	104.5	105.8	83.6	93.7	110.2	83.6	95.7	106.1	78.0	90.8	117.8	85.4	100.1	99.6	88.7	93.6
COVERS	DIVS. 10 - 14, 25, 28, 41, 43, 44, 46	131.1	91.8	122.5	131.1	77.2	119.3	131.1	77.2	119.3	131.1	86.2	121.2	131.1	66.0	116.8	131.1	89.0	121.9
21, 22, 23	FIRE SUPPRESSION, PLUMBING & HVAC	104.4	90.5	98.8	103.8	86.1	96.7	103.8	86.1	96.7	103.8	89.5	98.1	104.1	96.6	101.1	103.7	89.2	97.9
26, 27, 3370	ELECTRICAL, COMMUNICATIONS & UTIL.	106.4	99.6	102.9	108.5	66.8	87.2	107.2	66.8	86.5	110.4	76.0	92.8	107.9	84.2	95.7	108.2	97.1	102.5
MF2018	WEIGHTED AVERAGE	116.6	98.1	108.6	111.2	81.9	98.6	111.5	82.1	98.9	113.6	85.7	101.6	114.7	89.7	104.0	109.9	93.4	102.8

For customer support on your Plumbing Costs with RSMeans Data, call 800.448.8182.

City Cost Indexes

CANADA

DIVISION		LAVAL, QUEBEC			LETHBRIDGE, ALBERTA			LLOYDMINSTER, ALBERTA			LONDON, ONTARIO			MEDICINE HAT, ALBERTA			MONCTON, NEW BRUNSWICK		
		MAT.	INST.	TOTAL	MAT.	INST.	TOTAL	MAT.	INST.	TOTAL	MAT.	INST.	TOTAL	MAT.	INST.	TOTAL	MAT.	INST.	TOTAL
015433	CONTRACTOR EQUIPMENT		103.5	103.5		105.3	105.3		105.3	105.3		122.4	122.4		105.3	105.3		102.7	102.7
0241, 31 - 34	SITE & INFRASTRUCTURE, DEMOLITION	98.0	98.9	98.6	115.1	101.3	105.5	115.0	100.7	105.0	103.8	114.7	111.4	113.8	100.8	104.7	105.2	95.8	98.7
0310	Concrete Forming & Accessories	128.9	81.3	88.5	122.0	89.9	94.7	120.1	80.3	86.3	120.7	88.4	93.2	121.9	80.3	86.6	102.5	62.0	68.1
0320	Concrete Reinforcing	154.2	74.5	113.7	158.5	84.1	120.6	158.5	84.1	120.6	151.4	97.3	123.9	158.5	84.0	120.6	144.5	70.8	107.0
0330	Cast-in-Place Concrete	104.7	90.2	99.3	126.8	101.1	117.2	117.6	97.5	110.1	120.4	100.1	112.8	117.6	97.5	110.1	107.0	68.2	92.6
03	CONCRETE	109.9	83.7	98.1	120.6	93.1	108.2	116.1	87.5	103.2	116.2	94.9	106.6	116.2	87.5	103.3	108.6	66.9	89.8
04	MASONRY	162.9	77.9	110.4	180.7	85.5	121.9	162.0	79.2	110.9	182.7	97.9	130.3	162.0	79.2	110.9	159.7	60.1	98.2
05	METALS	116.5	85.9	107.2	141.9	91.6	126.6	118.4	91.5	110.2	143.5	107.0	132.4	118.6	91.4	110.4	120.3	84.9	109.6
06	WOOD, PLASTICS & COMPOSITES	133.6	81.3	105.2	119.5	89.3	103.1	116.0	79.7	96.3	109.2	86.3	96.7	119.5	79.7	97.9	99.7	61.5	78.9
07	THERMAL & MOISTURE PROTECTION	113.0	86.5	101.6	124.8	92.0	110.7	121.3	87.5	106.7	122.7	97.6	111.9	128.1	87.5	110.6	114.6	64.2	92.9
08	OPENINGS	91.1	69.9	86.2	91.1	81.6	88.9	91.1	76.4	87.7	80.2	87.0	81.8	91.1	76.4	87.7	85.8	60.2	79.8
0920	Plaster & Gypsum Board	112.7	80.7	91.6	103.7	88.6	93.7	99.9	78.8	86.0	123.9	85.3	98.5	102.1	78.8	86.7	121.5	60.3	81.2
0950, 0980	Ceilings & Acoustic Treatment	104.7	80.7	88.4	112.3	88.6	96.3	104.7	78.8	87.1	142.7	85.3	103.8	104.7	78.8	87.1	115.9	60.3	78.2
0960	Flooring	113.6	88.8	106.5	113.6	86.7	105.8	113.6	86.7	105.8	112.8	95.6	107.8	113.6	86.7	105.8	100.1	67.2	90.5
0970, 0990	Wall Finishes & Painting/Coating	109.8	85.8	95.6	109.7	99.0	103.4	109.9	77.1	90.5	113.8	100.1	105.7	109.7	77.1	90.4	110.7	50.4	75.1
09	FINISHES	105.9	83.6	93.7	107.2	90.5	98.1	105.2	80.9	92.0	118.5	90.6	103.3	105.4	80.9	92.0	105.7	61.7	81.6
COVERS	DIVS. 10 - 14, 25, 28, 41, 43, 44, 46	131.1	77.2	119.3	131.1	90.4	122.2	131.1	87.3	121.5	131.1	90.7	122.2	131.1	87.3	121.5	131.1	60.7	115.6
21, 22, 23	FIRE SUPPRESSION, PLUMBING & HVAC	103.8	86.1	96.7	104.0	92.1	99.2	104.1	92.0	99.3	104.4	87.8	97.7	103.8	88.8	97.8	103.7	68.6	89.6
26, 27, 3370	ELECTRICAL, COMMUNICATIONS & UTIL.	105.6	66.8	85.8	103.4	79.4	91.1	101.2	79.4	90.0	103.6	97.1	100.3	101.2	79.4	90.0	112.3	59.0	85.0
MF2018	WEIGHTED AVERAGE	111.1	81.9	98.6	118.0	89.7	105.8	112.2	86.5	101.2	112.7	95.6	108.0	112.4	85.8	101.0	111.4	68.1	92.8

CANADA

DIVISION		MONTREAL, QUEBEC			MOOSE JAW, SASKATCHEWAN			NEW GLASGOW, NOVA SCOTIA			NEWCASTLE, NEW BRUNSWICK			NORTH BAY, ONTARIO			OSHAWA, ONTARIO		
		MAT.	INST.	TOTAL	MAT.	INST.	TOTAL	MAT.	INST.	TOTAL	MAT.	INST.	TOTAL	MAT.	INST.	TOTAL	MAT.	INST.	TOTAL
015433	CONTRACTOR EQUIPMENT		121.9	121.9		101.7	101.7		102.0	102.0		102.7	102.7		102.4	102.4		105.5	105.5
0241, 31 - 34	SITE & INFRASTRUCTURE, DEMOLITION	113.4	112.2	112.6	114.9	95.0	101.0	113.7	97.0	102.1	105.8	95.5	98.6	128.5	98.7	107.8	106.6	103.9	104.7
0310	Concrete Forming & Accessories	124.5	89.7	94.9	105.4	56.1	63.5	107.4	69.8	75.4	102.5	59.7	66.1	141.7	82.3	91.2	117.6	87.0	91.6
0320	Concrete Reinforcing	134.8	94.0	114.0	111.8	63.2	87.1	169.0	49.0	108.0	144.5	59.2	101.2	199.7	87.3	142.6	167.8	93.3	129.9
0330	Cast-in-Place Concrete	123.6	96.6	113.5	114.1	65.7	96.1	117.7	68.8	99.5	111.1	58.0	91.4	113.4	82.6	101.9	128.3	85.4	112.3
03	CONCRETE	116.0	93.6	105.9	101.9	61.7	83.8	131.0	66.8	102.2	110.5	60.0	87.8	133.0	83.7	110.8	121.0	87.9	106.1
04	MASONRY	168.0	88.4	118.9	160.7	57.6	97.0	208.9	67.0	121.2	160.0	58.7	97.4	215.0	83.6	133.8	146.0	89.9	111.4
05	METALS	151.6	102.6	136.7	115.4	75.9	103.4	133.8	76.6	116.4	120.3	74.5	106.4	134.5	91.2	121.3	121.1	95.0	113.2
06	WOOD, PLASTICS & COMPOSITES	108.8	90.2	98.7	100.7	54.7	75.7	118.4	69.3	91.7	99.7	59.6	77.9	157.8	82.6	117.0	115.5	85.9	99.4
07	THERMAL & MOISTURE PROTECTION	119.6	96.2	109.5	111.8	61.8	90.3	129.2	69.5	103.5	114.6	60.1	91.1	135.7	85.0	113.9	114.5	89.3	103.7
08	OPENINGS	85.7	78.2	83.9	87.1	52.9	79.2	93.2	63.0	86.2	85.8	52.7	78.1	100.8	80.5	96.1	86.9	87.1	86.9
0920	Plaster & Gypsum Board	117.7	89.3	99.0	96.1	53.4	68.0	139.1	68.4	92.6	121.5	58.4	79.9	136.4	82.1	100.6	105.7	85.4	92.4
0950, 0980	Ceilings & Acoustic Treatment	140.2	89.3	105.7	104.7	53.4	70.0	122.4	68.4	85.8	115.9	58.4	76.9	122.4	82.1	95.1	103.3	85.4	91.2
0960	Flooring	114.0	91.0	107.3	103.9	55.5	89.8	108.1	60.9	94.4	100.1	65.6	90.0	131.3	88.9	119.0	104.3	98.1	102.5
0970, 0990	Wall Finishes & Painting/Coating	112.2	103.2	106.9	109.8	62.6	81.9	117.4	60.9	84.0	110.7	49.2	74.3	117.4	87.3	99.6	103.8	107.5	106.0
09	FINISHES	116.5	92.2	103.2	101.7	56.4	76.9	115.9	67.5	89.4	105.7	59.9	80.7	122.9	84.3	101.9	100.7	91.0	95.4
COVERS	DIVS. 10 - 14, 25, 28, 41, 43, 44, 46	131.1	81.8	120.3	131.1	58.7	115.2	131.1	62.1	115.9	131.1	59.1	115.3	131.1	64.7	116.5	131.1	88.6	121.8
21, 22, 23	FIRE SUPPRESSION, PLUMBING & HVAC	104.6	80.1	94.8	104.1	71.9	91.2	103.9	81.3	94.8	103.7	66.2	88.7	103.9	94.5	100.1	103.7	100.0	102.2
26, 27, 3370	ELECTRICAL, COMMUNICATIONS & UTIL.	106.0	83.1	94.3	109.5	57.7	83.0	108.3	60.1	83.7	108.1	57.2	82.1	109.0	85.3	96.9	109.2	87.6	98.2
MF2018	WEIGHTED AVERAGE	118.7	89.9	106.3	109.7	65.6	90.8	120.7	72.1	94.5	111.3	64.5	91.2	123.7	87.4	108.1	112.0	92.9	103.8

CANADA

DIVISION		OTTAWA, ONTARIO			OWEN SOUND, ONTARIO			PETERBOROUGH, ONTARIO			PORTAGE LA PRAIRIE, MANITOBA			PRINCE ALBERT, SASKATCHEWAN			PRINCE GEORGE, BRITISH COLUMBIA		
		MAT.	INST.	TOTAL	MAT.	INST.	TOTAL	MAT.	INST.	TOTAL	MAT.	INST.	TOTAL	MAT.	INST.	TOTAL	MAT.	INST.	TOTAL
015433	CONTRACTOR EQUIPMENT		123.1	123.1		103.0	103.0		103.1	103.1		105.3	105.3		101.7	101.7		106.7	106.7
0241, 31 - 34	SITE & INFRASTRUCTURE, DEMOLITION	108.2	114.7	112.7	118.0	99.6	105.2	117.5	99.5	105.0	116.2	98.4	103.8	110.6	95.2	99.9	123.7	102.6	109.0
0310	Concrete Forming & Accessories	121.9	88.4	93.4	124.4	80.9	87.4	122.8	83.3	89.2	122.1	67.6	75.8	105.4	55.9	63.4	110.7	79.7	84.3
0320	Concrete Reinforcing	135.1	97.3	115.9	177.7	89.4	132.8	171.2	87.8	128.8	158.5	56.1	106.4	116.8	63.1	89.5	114.3	78.0	95.9
0330	Cast-in-Place Concrete	121.6	102.9	114.6	156.6	78.9	127.7	127.3	84.3	111.3	117.6	71.9	100.6	103.4	65.6	89.3	114.5	94.7	107.1
03	CONCRETE	114.9	96.0	106.4	139.3	82.2	113.6	122.4	84.9	105.5	110.1	68.1	91.2	97.4	61.6	81.3	127.9	85.2	108.7
04	MASONRY	164.0	100.4	124.7	165.9	88.9	118.3	166.5	89.9	119.2	165.2	61.0	100.8	159.9	57.6	96.7	172.0	85.7	118.7
05	METALS	144.3	108.4	133.4	110.9	92.2	105.2	118.0	91.9	110.1	118.6	79.3	106.7	115.4	75.7	103.4	118.4	88.2	109.3
06	WOOD, PLASTICS & COMPOSITES	107.8	85.2	95.5	118.5	79.5	97.3	121.8	81.4	99.8	119.5	68.8	91.9	100.7	54.7	75.7	103.4	76.3	88.7
07	THERMAL & MOISTURE PROTECTION	131.9	98.8	117.7	114.4	85.8	102.1	119.2	90.6	106.9	112.3	70.2	94.2	111.7	60.7	89.7	122.5	83.3	105.6
08	OPENINGS	88.5	86.3	88.0	92.6	79.9	89.6	89.7	82.3	88.0	91.1	61.7	84.2	86.1	52.9	78.4	87.9	77.1	85.4
0920	Plaster & Gypsum Board	120.1	84.1	96.4	151.9	78.9	103.8	113.1	80.8	91.8	101.7	67.5	79.2	96.1	53.4	68.0	98.6	75.2	83.2
0950, 0980	Ceilings & Acoustic Treatment	136.4	84.1	101.0	95.1	78.9	84.1	104.7	80.8	88.5	104.7	67.5	79.5	104.7	53.4	70.0	104.7	75.2	84.7
0960	Flooring	110.7	91.1	105.0	118.9	90.2	110.6	113.6	88.9	106.5	113.6	64.0	99.2	103.9	55.5	89.8	109.2	69.7	97.7
0970, 0990	Wall Finishes & Painting/Coating	110.6	94.9	101.3	105.0	86.0	93.7	109.8	89.4	97.7	109.9	55.5	77.7	109.8	53.4	76.4	109.8	78.4	91.2
09	FINISHES	117.0	89.1	101.8	109.7	83.2	95.2	106.6	85.0	94.8	105.3	66.0	83.8	101.7	55.4	76.4	105.0	77.3	89.9
COVERS	DIVS. 10 - 14, 25, 28, 41, 43, 44, 46	131.1	89.3	121.9	139.2	65.2	122.9	131.1	66.2	116.8	131.1	61.1	115.7	131.1	58.7	115.2	131.1	85.5	121.1
21, 22, 23	FIRE SUPPRESSION, PLUMBING & HVAC	104.4	89.5	98.4	103.8	94.4	100.0	103.8	97.9	101.4	103.8	80.1	94.3	104.1	64.8	88.4	103.8	89.5	98.1
26, 27, 3370	ELECTRICAL, COMMUNICATIONS & UTIL.	103.5	97.9	100.6	116.7	84.2	100.1	107.6	84.9	96.0	109.2	55.8	81.8	109.5	57.7	83.0	107.6	76.0	91.4
MF2018	WEIGHTED AVERAGE	117.4	96.4	108.4	116.6	87.6	104.2	113.7	89.4	103.3	112.2	70.9	94.5	108.9	63.9	89.6	114.5	85.0	101.9

City Cost Indexes

CANADA

DIVISION		QUEBEC CITY, QUEBEC			RED DEER, ALBERTA			REGINA, SASKATCHEWAN			RIMOUSKI, QUEBEC			ROUYN-NORANDA, QUEBEC			SAINT HYACINTHE, QUEBEC		
		MAT.	INST.	TOTAL	MAT.	INST.	TOTAL	MAT.	INST.	TOTAL	MAT.	INST.	TOTAL	MAT.	INST.	TOTAL	MAT.	INST.	TOTAL
015433	CONTRACTOR EQUIPMENT		122.8	122.8		105.3	105.3		127.7	127.7		103.5	103.5		103.5	103.5		103.5	103.5
0241, 31 - 34	SITE & INFRASTRUCTURE, DEMOLITION	115.8	112.1	113.2	113.8	100.8	104.7	126.6	119.3	121.5	97.8	98.5	98.3	97.6	98.9	98.5	98.0	98.9	98.6
0310	Concrete Forming & Accessories	121.8	89.8	94.6	134.6	80.3	88.5	123.8	93.8	98.3	128.7	89.3	95.2	128.7	81.3	88.4	128.7	81.3	88.4
0320	Concrete Reinforcing	135.0	94.0	114.2	158.5	84.0	120.6	137.5	91.4	114.1	110.0	93.9	101.8	154.2	74.5	113.7	154.2	74.5	113.7
0330	Cast-in-Place Concrete	118.2	96.3	110.1	117.6	97.5	110.1	138.1	99.2	123.6	106.6	94.9	102.2	101.3	90.2	97.2	104.7	90.2	99.3
03	CONCRETE	113.3	93.7	104.5	117.0	87.5	103.7	124.6	96.1	111.7	105.5	92.1	99.5	108.3	83.7	97.3	109.9	83.7	98.1
04	MASONRY	162.6	88.4	116.8	162.0	79.2	110.9	192.6	89.2	128.7	162.5	88.4	116.7	162.7	77.9	110.3	162.9	77.9	110.4
05	METALS	144.6	104.4	132.4	118.6	91.4	110.4	153.6	103.1	138.2	116.1	91.3	108.6	116.6	85.9	107.3	116.6	85.9	107.3
06	WOOD, PLASTICS & COMPOSITES	113.7	90.1	100.9	119.5	79.7	97.9	110.3	94.6	101.8	133.5	89.8	109.7	133.5	81.3	105.2	133.5	81.3	105.2
07	THERMAL & MOISTURE PROTECTION	119.2	96.2	109.3	138.5	87.5	116.5	139.3	88.5	117.4	112.5	96.0	105.4	112.5	86.5	101.3	112.7	86.5	101.4
08	OPENINGS	87.4	85.6	87.0	91.1	76.4	87.7	86.3	81.8	85.3	90.7	76.1	87.3	91.1	69.9	86.2	91.1	69.9	86.2
0920	Plaster & Gypsum Board	121.7	89.3	100.4	102.1	78.8	86.7	131.2	93.8	106.6	142.3	89.3	107.4	112.2	80.7	91.4	112.2	80.7	91.4
0950, 0980	Ceilings & Acoustic Treatment	139.5	89.3	105.5	104.7	78.8	87.1	147.1	93.8	111.0	103.8	89.3	94.0	103.8	80.7	88.1	103.8	80.7	88.1
0960	Flooring	112.3	91.0	106.1	116.3	86.7	107.7	123.9	95.6	115.7	115.0	88.8	107.4	113.6	88.8	106.5	113.6	88.8	106.5
0970, 0990	Wall Finishes & Painting/Coating	122.0	103.2	110.9	109.7	77.1	90.4	115.3	91.4	101.2	113.0	103.2	107.2	109.8	85.8	95.6	109.8	85.8	95.6
09	FINISHES	116.5	92.2	103.2	106.1	80.9	92.4	128.0	94.9	109.9	110.7	91.6	100.2	105.7	83.6	93.6	105.7	83.6	93.6
COVERS	DIVS. 10 - 14, 25, 28, 41, 43, 44, 46	131.1	81.6	120.2	131.1	87.3	121.5	131.1	70.8	117.8	131.1	80.9	120.1	131.1	77.2	119.3	131.1	77.2	119.3
21, 22, 23	FIRE SUPPRESSION, PLUMBING & HVAC	104.4	80.1	94.6	103.8	88.8	97.8	104.1	88.8	98.0	103.8	80.0	94.2	103.8	86.1	96.7	100.4	86.1	94.6
26, 27, 3370	ELECTRICAL, COMMUNICATIONS & UTIL.	105.7	83.1	94.2	101.2	79.4	90.0	104.3	93.7	98.9	107.2	83.1	94.9	107.2	66.8	86.5	107.8	66.8	86.8
MF2018	WEIGHTED AVERAGE	117.1	90.3	105.6	112.8	85.8	101.2	122.8	94.2	110.5	111.0	87.3	100.8	111.0	81.9	98.5	110.5	81.9	98.2

CANADA

DIVISION		SAINT JOHN, NEW BRUNSWICK			SARNIA, ONTARIO			SASKATOON, SASKATCHEWAN			SAULT STE MARIE, ONTARIO			SHERBROOKE, QUEBEC			SOREL, QUEBEC		
		MAT.	INST.	TOTAL	MAT.	INST.	TOTAL	MAT.	INST.	TOTAL	MAT.	INST.	TOTAL	MAT.	INST.	TOTAL	MAT.	INST.	TOTAL
015433	CONTRACTOR EQUIPMENT		102.7	102.7		103.1	103.1		101.7	101.7		103.1	103.1		103.5	103.5		103.5	103.5
0241, 31 - 34	SITE & INFRASTRUCTURE, DEMOLITION	105.9	97.2	99.8	116.1	99.6	104.6	112.3	97.9	102.3	106.1	99.2	101.3	98.1	98.9	98.7	98.2	98.9	98.7
0310	Concrete Forming & Accessories	122.6	65.0	73.6	121.5	90.1	94.8	105.3	93.2	95.0	110.8	86.2	89.9	128.7	81.3	88.4	128.7	81.4	88.5
0320	Concrete Reinforcing	144.5	71.4	107.3	122.2	89.2	105.4	117.1	91.3	104.0	109.8	87.9	98.7	154.2	74.5	113.7	145.8	74.6	109.6
0330	Cast-in-Place Concrete	109.5	70.0	94.8	117.4	97.2	110.9	109.7	96.0	104.6	105.5	83.0	97.1	104.7	90.2	99.3	105.5	90.2	99.9
03	CONCRETE	111.2	68.9	92.2	111.7	92.6	103.1	101.2	93.8	97.9	97.9	85.8	92.5	109.9	83.7	98.1	109.3	83.8	97.8
04	MASONRY	181.3	70.0	112.6	178.2	92.5	125.2	172.6	89.2	121.1	163.4	91.9	119.2	163.0	77.9	110.4	163.1	77.9	110.5
05	METALS	120.2	86.2	109.9	117.9	92.3	110.1	111.2	89.6	104.6	117.1	94.0	110.1	116.4	85.9	107.1	116.6	86.0	107.3
06	WOOD, PLASTICS & COMPOSITES	123.8	63.2	90.9	121.0	89.3	103.8	98.0	94.0	95.8	107.8	87.1	96.6	133.5	81.3	105.2	133.5	81.3	105.2
07	THERMAL & MOISTURE PROTECTION	114.9	71.0	96.0	119.3	94.1	108.4	114.4	87.4	102.8	118.0	89.4	105.7	112.5	86.5	101.3	112.5	86.5	101.3
08	OPENINGS	85.7	60.0	79.7	92.4	86.0	90.9	87.1	81.4	85.8	84.2	85.3	84.4	91.1	69.9	86.2	91.1	74.4	87.2
0920	Plaster & Gypsum Board	135.3	62.1	87.1	137.6	89.0	105.6	113.0	93.8	100.4	104.4	86.8	92.8	112.2	80.7	91.4	142.3	80.7	101.7
0950, 0980	Ceilings & Acoustic Treatment	121.0	62.1	81.1	108.9	89.0	95.4	124.0	93.8	103.6	104.7	86.8	92.5	103.8	80.7	88.1	103.8	80.7	88.1
0960	Flooring	111.6	67.2	98.7	113.6	97.2	108.9	106.8	95.6	103.6	106.7	94.2	103.1	113.6	88.8	106.5	113.6	88.8	106.5
0970, 0990	Wall Finishes & Painting/Coating	110.7	82.7	94.1	109.8	101.1	104.7	113.4	91.4	100.4	109.3	93.7	100.3	109.8	85.8	95.6	109.8	85.8	95.6
09	FINISHES	112.1	66.8	87.4	111.0	92.9	101.1	109.9	94.4	101.4	102.7	88.7	95.0	105.7	83.6	93.6	110.0	83.6	95.6
COVERS	DIVS. 10 - 14, 25, 28, 41, 43, 44, 46	131.1	61.8	115.9	131.1	67.5	117.1	131.1	69.6	117.6	131.1	87.7	121.6	131.1	77.2	119.3	131.1	77.2	119.3
21, 22, 23	FIRE SUPPRESSION, PLUMBING & HVAC	103.7	76.5	92.8	103.8	103.8	103.8	104.0	88.7	97.8	103.8	92.3	99.1	104.1	86.1	96.9	103.8	86.1	96.7
26, 27, 3370	ELECTRICAL, COMMUNICATIONS & UTIL.	114.8	83.7	98.9	110.5	87.1	98.5	107.5	93.7	100.4	109.2	85.3	97.0	107.2	66.8	86.5	107.2	66.8	86.5
MF2018	WEIGHTED AVERAGE	113.8	75.6	97.4	113.8	93.7	105.2	109.9	90.8	101.7	109.3	90.0	101.0	111.3	81.9	98.7	111.5	82.1	98.9

CANADA

DIVISION		ST. CATHARINES, ONTARIO			ST JEROME, QUEBEC			ST. JOHN'S, NEWFOUNDLAND			SUDBURY, ONTARIO			SUMMERSIDE, PRINCE EDWARD ISLAND			SYDNEY, NOVA SCOTIA		
		MAT.	INST.	TOTAL	MAT.	INST.	TOTAL	MAT.	INST.	TOTAL	MAT.	INST.	TOTAL	MAT.	INST.	TOTAL	MAT.	INST.	TOTAL
015433	CONTRACTOR EQUIPMENT		103.4	103.4		103.5	103.5		122.7	122.7		103.4	103.4		102.0	102.0		102.0	102.0
0241, 31 - 34	SITE & INFRASTRUCTURE, DEMOLITION	96.0	101.4	99.8	97.6	98.9	98.5	118.7	111.9	113.9	96.1	101.0	99.5	123.5	94.3	103.2	109.8	97.0	100.9
0310	Concrete Forming & Accessories	110.3	93.0	95.6	128.7	81.3	88.4	119.4	86.4	91.3	106.3	88.1	90.9	107.7	54.0	62.0	107.4	69.8	75.4
0320	Concrete Reinforcing	106.8	98.4	102.5	154.2	74.5	113.7	168.2	84.7	125.8	107.7	102.2	104.9	166.8	47.8	106.3	169.0	49.0	108.0
0330	Cast-in-Place Concrete	106.0	99.1	103.4	101.3	90.2	97.2	132.0	99.6	119.9	106.9	95.8	102.8	111.8	56.4	91.2	90.9	68.8	82.7
03	CONCRETE	96.8	96.2	96.5	108.3	83.7	97.3	128.0	91.7	111.7	97.1	93.4	95.4	139.1	54.9	101.2	118.2	66.8	95.1
04	MASONRY	142.6	99.2	115.8	162.7	77.9	110.3	187.8	91.7	128.5	142.7	95.6	113.6	208.2	56.0	114.2	206.6	67.0	120.3
05	METALS	121.0	96.3	113.5	116.6	85.9	107.3	147.1	100.2	132.9	120.5	95.8	113.0	133.8	70.0	114.4	133.8	76.6	116.4
06	WOOD, PLASTICS & COMPOSITES	106.0	92.5	98.7	133.5	81.3	105.2	105.1	84.3	93.8	101.9	87.2	93.9	119.0	53.4	83.3	118.4	69.3	91.7
07	THERMAL & MOISTURE PROTECTION	113.5	100.9	108.1	112.5	86.5	101.3	141.8	96.5	122.3	113.0	95.6	105.5	128.6	58.9	98.6	129.2	69.5	103.5
08	OPENINGS	81.3	90.2	83.4	91.1	69.9	86.2	85.0	75.4	82.8	82.0	85.7	82.8	104.7	46.8	91.2	93.2	63.0	86.2
0920	Plaster & Gypsum Board	94.2	92.3	92.9	112.2	80.7	91.4	138.3	83.2	102.0	95.6	86.8	89.8	140.0	52.0	82.1	139.1	68.4	92.6
0950, 0980	Ceilings & Acoustic Treatment	103.3	92.3	95.8	103.8	80.7	88.1	136.9	83.2	100.6	98.2	86.8	90.5	122.4	52.0	74.7	122.4	68.4	85.8
0960	Flooring	100.2	92.4	98.0	113.6	88.8	106.5	117.5	53.0	98.8	98.4	94.2	97.2	108.1	57.1	93.3	108.1	60.9	94.4
0970, 0990	Wall Finishes & Painting/Coating	103.8	103.5	103.6	109.8	85.8	95.6	118.0	101.2	108.0	103.8	94.4	98.2	117.4	40.9	72.1	117.4	60.9	84.0
09	FINISHES	97.2	94.1	95.5	105.7	83.6	93.6	123.2	81.7	100.5	95.9	89.7	92.5	117.0	53.4	82.3	115.9	67.5	89.4
COVERS	DIVS. 10 - 14, 25, 28, 41, 43, 44, 46	131.1	69.2	117.5	131.1	77.2	119.3	131.1	68.6	117.4	131.1	89.0	121.9	131.1	58.2	115.1	131.1	62.1	115.9
21, 22, 23	FIRE SUPPRESSION, PLUMBING & HVAC	103.7	89.1	97.9	103.8	86.1	96.7	104.2	84.8	96.4	103.6	88.6	97.6	103.9	60.2	86.4	103.9	81.3	94.8
26, 27, 3370	ELECTRICAL, COMMUNICATIONS & UTIL.	109.8	97.8	103.7	107.8	66.8	86.8	106.0	83.4	94.4	109.7	98.3	103.9	107.4	48.6	77.3	108.3	60.1	83.7
MF2018	WEIGHTED AVERAGE	107.7	94.3	102.0	111.1	81.9	98.6	121.5	88.8	107.4	107.6	93.1	101.4	123.1	59.4	95.8	119.0	72.1	98.8

For customer support on your Plumbing Costs with RSMeans Data, call 800.448.8182.

CANADA

DIVISION		THUNDER BAY, ONTARIO			TIMMINS, ONTARIO			TORONTO, ONTARIO			TROIS RIVIERES, QUEBEC			TRURO, NOVA SCOTIA			VANCOUVER, BRITISH COLUMBIA		
		MAT.	INST.	TOTAL	MAT.	INST.	TOTAL	MAT.	INST.	TOTAL	MAT.	INST.	TOTAL	MAT.	INST.	TOTAL	MAT.	INST.	TOTAL
015433	CONTRACTOR EQUIPMENT		103.4	103.4		103.1	103.1		122.5	122.5		102.9	102.9		102.7	102.7		142.3	142.3
0241, 31 - 34	SITE & INFRASTRUCTURE, DEMOLITION	100.6	101.3	101.1	117.5	99.1	104.7	115.6	114.5	114.8	110.1	98.4	102.0	102.2	97.4	98.9	114.4	130.8	125.8
0310	Concrete Forming & Accessories	117.5	91.3	95.3	122.8	82.3	88.4	119.4	99.1	102.2	149.5	81.4	91.6	95.7	69.8	73.7	120.6	92.8	96.9
0320	Concrete Reinforcing	95.5	97.9	96.7	171.2	87.3	128.5	151.4	100.3	125.4	169.0	74.5	121.0	145.8	49.0	96.6	122.5	93.4	107.7
0330	Cast-in-Place Concrete	116.7	98.4	109.9	127.3	82.8	110.7	113.9	111.9	113.1	94.1	90.1	92.6	132.2	69.0	108.7	110.9	92.3	104.0
03	CONCRETE	104.1	95.0	100.0	122.4	83.8	105.0	113.0	104.4	109.1	120.6	83.7	104.0	121.3	66.9	96.8	110.7	94.0	103.2
04	MASONRY	143.2	99.4	116.1	166.5	83.6	115.3	166.8	106.1	129.3	210.6	77.9	128.6	162.2	67.0	103.3	161.7	86.3	115.1
05	METALS	121.0	95.3	113.2	118.0	91.5	109.9	144.1	108.9	133.4	133.1	85.6	118.7	117.3	76.9	105.0	138.6	111.8	130.4
06	WOOD, PLASTICS & COMPOSITES	115.5	89.9	101.6	121.8	82.6	100.5	105.7	96.9	100.9	173.4	81.3	123.4	90.4	69.4	79.0	103.3	92.8	97.6
07	THERMAL & MOISTURE PROTECTION	113.8	98.2	107.1	119.2	85.0	104.5	129.3	106.4	119.4	127.9	86.5	110.1	113.7	69.6	94.7	134.4	88.0	114.4
08	OPENINGS	80.6	88.1	82.4	89.7	80.5	87.6	82.1	96.0	85.4	102.4	74.4	95.9	84.2	63.1	79.3	85.0	89.7	86.1
0920	Plaster & Gypsum Board	120.6	89.6	100.2	113.1	82.1	92.7	116.8	96.2	103.2	169.8	80.7	111.1	118.4	68.4	85.5	118.0	91.5	100.5
0950, 0980	Ceilings & Acoustic Treatment	98.2	89.6	92.4	104.7	82.1	89.4	133.9	96.2	108.4	120.7	80.7	93.6	104.7	68.4	80.1	143.7	91.5	108.3
0960	Flooring	104.3	99.0	102.8	113.6	88.9	106.5	109.1	101.4	106.9	131.3	88.8	119.0	96.1	60.9	85.9	127.9	90.7	117.1
0970, 0990	Wall Finishes & Painting/Coating	103.8	95.3	98.8	109.8	87.3	96.5	107.0	107.5	107.3	117.4	85.8	98.7	109.8	60.9	80.8	115.0	87.6	98.8
09	FINISHES	101.6	93.2	97.0	106.6	84.4	94.5	112.2	100.4	105.7	126.7	83.5	103.1	101.7	67.5	83.0	122.8	92.2	106.1
COVERS	DIVS. 10 - 14, 25, 28, 41, 43, 44, 46	131.1	69.3	117.5	131.1	64.8	116.5	131.1	93.9	122.9	131.1	77.2	119.3	131.1	62.1	115.9	131.1	91.7	122.5
21, 22, 23	FIRE SUPPRESSION, PLUMBING & HVAC	103.7	89.3	98.0	103.8	94.5	100.0	104.3	97.2	101.5	103.9	86.1	96.8	103.8	81.3	94.8	104.4	77.1	93.5
26, 27, 3370	ELECTRICAL, COMMUNICATIONS & UTIL.	108.2	96.5	102.2	109.2	85.3	97.0	104.8	100.0	102.3	108.6	66.8	87.2	106.9	60.1	83.0	101.8	80.5	90.9
MF2018	WEIGHTED AVERAGE	109.0	93.7	102.4	113.9	87.5	102.6	116.4	102.4	110.4	121.6	82.0	104.6	111.4	72.2	94.6	115.9	91.6	105.5

CANADA

DIVISION		VICTORIA, BRITISH COLUMBIA			WHITEHORSE, YUKON			WINDSOR, ONTARIO			WINNIPEG, MANITOBA			YARMOUTH, NOVA SCOTIA			YELLOWKNIFE, NWT		
		MAT.	INST.	TOTAL	MAT.	INST.	TOTAL	MAT.	INST.	TOTAL	MAT.	INST.	TOTAL	MAT.	INST.	TOTAL	MAT.	INST.	TOTAL
015433	CONTRACTOR EQUIPMENT		109.9	109.9		132.7	132.7		103.4	103.4		126.9	126.9		102.0	102.0		131.8	131.8
0241, 31 - 34	SITE & INFRASTRUCTURE, DEMOLITION	123.6	106.8	111.9	139.0	120.4	126.0	92.3	101.3	98.5	115.6	115.2	115.3	113.5	97.0	102.0	147.7	125.6	132.3
0310	Concrete Forming & Accessories	110.4	85.9	89.6	128.7	58.1	68.7	117.6	89.6	93.8	124.9	65.6	74.5	107.4	69.8	75.4	130.8	77.7	85.7
0320	Concrete Reinforcing	114.5	93.2	103.7	153.0	65.1	108.3	104.7	97.2	100.9	140.0	61.1	99.9	169.0	49.0	108.0	150.2	67.1	108.2
0330	Cast-in-Place Concrete	113.7	88.9	104.5	159.0	71.9	126.6	108.6	99.9	105.3	138.6	72.6	114.0	116.5	68.8	98.8	157.1	89.1	131.8
03	CONCRETE	129.1	88.5	110.8	143.6	66.0	108.7	98.3	94.7	96.7	125.2	68.8	99.8	130.5	66.8	101.8	142.5	81.3	114.9
04	MASONRY	175.6	86.2	120.4	240.3	59.2	128.4	142.7	98.4	115.3	188.9	64.9	112.3	208.8	67.0	121.2	235.0	70.3	133.2
05	METALS	112.2	91.4	105.9	151.7	91.8	133.5	121.0	95.8	113.4	153.6	89.3	134.1	133.8	76.6	116.4	149.9	94.7	133.1
06	WOOD, PLASTICS & COMPOSITES	101.8	84.6	92.5	123.9	56.7	87.4	115.5	88.2	100.7	110.1	66.2	86.3	118.4	69.3	91.7	132.9	79.2	103.7
07	THERMAL & MOISTURE PROTECTION	124.5	87.6	108.6	140.0	64.2	107.4	113.6	97.6	106.7	132.5	69.9	105.5	129.2	69.5	103.5	140.7	80.3	114.7
08	OPENINGS	88.6	80.9	86.8	102.8	54.4	91.5	80.4	87.6	82.1	85.6	59.8	79.6	93.2	63.0	86.2	97.1	67.4	90.2
0920	Plaster & Gypsum Board	106.4	83.8	91.5	172.9	54.6	95.0	106.6	87.8	94.2	124.1	64.2	84.6	139.1	68.4	92.6	179.7	77.7	112.5
0950, 0980	Ceilings & Acoustic Treatment	106.2	83.8	91.0	168.6	54.6	91.4	98.2	87.8	91.2	147.1	64.2	90.9	122.4	68.4	85.8	164.1	77.7	105.5
0960	Flooring	112.2	69.7	99.9	130.8	57.0	109.4	104.3	96.4	102.0	118.6	68.8	104.2	108.1	60.9	94.4	124.1	85.7	112.9
0970, 0990	Wall Finishes & Painting/Coating	113.4	87.6	98.1	121.7	54.5	82.0	103.8	96.3	99.4	111.5	53.4	77.1	117.4	60.9	84.0	124.4	77.9	96.9
09	FINISHES	108.4	83.7	94.9	145.3	57.2	97.2	99.2	91.5	95.0	123.1	65.2	91.5	115.9	67.5	89.4	142.1	78.7	107.5
COVERS	DIVS. 10 - 14, 25, 28, 41, 43, 44, 46	131.1	66.1	116.8	131.1	61.3	115.8	131.1	68.8	117.4	131.1	63.8	116.3	131.1	62.1	115.9	131.1	64.8	116.5
21, 22, 23	FIRE SUPPRESSION, PLUMBING & HVAC	103.8	75.2	92.3	104.2	73.2	91.8	103.7	89.2	97.9	104.4	64.3	88.3	103.9	81.3	94.8	104.7	90.4	99.0
26, 27, 3370	ELECTRICAL, COMMUNICATIONS & UTIL.	107.4	80.0	93.4	127.3	58.5	92.1	113.1	97.9	105.3	107.1	63.2	84.6	108.3	60.1	83.7	121.9	79.3	100.1
MF2018	WEIGHTED AVERAGE	114.2	84.2	101.3	133.0	70.5	106.2	108.3	93.5	101.9	122.1	71.2	100.3	120.7	72.1	99.8	131.3	85.0	111.4

For customer support on your Plumbing Costs with RSMeans Data, call 800.448.8182.

677

Costs shown in RSMeans cost data publications are based on national averages for materials and installation. To adjust these costs to a specific location, simply multiply the base cost by the factor and divide by 100 for that city. The data is arranged alphabetically by state and postal zip code numbers. For a city not listed, use the factor for a nearby city with similar economic characteristics.

STATE/ZIP	CITY	MAT.	INST.	TOTAL
ALABAMA				
350-352	Birmingham	98.2	70.9	86.4
354	Tuscaloosa	96.9	70.5	85.5
355	Jasper	97.6	69.6	85.5
356	Decatur	96.9	70.2	85.4
357-358	Huntsville	96.9	69.9	85.2
359	Gadsden	97.2	70.8	85.8
360-361	Montgomery	98.0	71.8	86.6
362	Anniston	97.1	66.6	83.9
363	Dothan	97.2	71.6	86.1
364	Evergreen	96.8	67.8	84.3
365-366	Mobile	97.8	68.7	85.2
367	Selma	96.9	71.2	85.8
368	Phenix City	97.7	70.8	86.1
369	Butler	97.1	69.7	85.2
ALASKA				
995-996	Anchorage	119.7	110.4	115.7
997	Fairbanks	120.2	111.0	116.2
998	Juneau	119.7	110.4	115.7
999	Ketchikan	130.1	110.9	121.8
ARIZONA				
850,853	Phoenix	98.4	72.1	87.0
851,852	Mesa/Tempe	98.0	71.1	86.4
855	Globe	98.2	70.7	86.3
856-857	Tucson	96.9	71.8	86.1
859	Show Low	98.4	70.8	86.5
860	Flagstaff	102.4	70.4	88.6
863	Prescott	99.9	71.1	87.5
864	Kingman	98.2	69.8	85.9
865	Chambers	98.2	70.4	86.2
ARKANSAS				
716	Pine Bluff	96.2	62.8	81.8
717	Camden	94.1	64.1	81.1
718	Texarkana	95.1	63.7	81.5
719	Hot Springs	93.4	62.5	80.1
720-722	Little Rock	96.0	63.5	81.9
723	West Memphis	94.2	67.5	82.7
724	Jonesboro	94.8	65.7	82.2
725	Batesville	92.4	62.9	79.6
726	Harrison	93.7	62.3	80.1
727	Fayetteville	91.3	64.2	79.6
728	Russellville	92.5	61.3	79.0
729	Fort Smith	95.1	61.9	80.8
CALIFORNIA				
900-902	Los Angeles	98.4	129.3	111.8
903-905	Inglewood	93.6	127.4	108.2
906-908	Long Beach	95.2	127.2	109.0
910-912	Pasadena	95.4	127.0	109.0
913-916	Van Nuys	98.5	127.0	110.8
917-918	Alhambra	97.4	127.0	110.2
919-921	San Diego	100.3	121.3	109.4
922	Palm Springs	97.3	121.9	108.0
923-924	San Bernardino	95.0	125.4	108.1
925	Riverside	99.4	126.0	110.9
926-927	Santa Ana	97.1	125.7	109.5
928	Anaheim	99.5	125.9	110.9
930	Oxnard	98.2	125.8	110.1
931	Santa Barbara	97.4	125.0	109.4
932-933	Bakersfield	99.0	124.8	110.1
934	San Luis Obispo	98.4	125.5	110.1
935	Mojave	95.6	123.6	107.7
936-938	Fresno	98.6	131.2	112.7
939	Salinas	99.2	137.8	115.9
940-941	San Francisco	107.6	159.0	129.8
942,956-958	Sacramento	101.0	133.1	114.9
943	Palo Alto	99.4	150.8	121.6
944	San Mateo	101.9	156.6	125.5
945	Vallejo	100.5	143.4	119.0
946	Oakland	103.9	153.2	125.2
947	Berkeley	103.3	151.9	124.3
948	Richmond	102.8	148.2	122.4
949	San Rafael	105.0	150.3	124.6
950	Santa Cruz	104.9	138.1	119.2
CALIFORNIA (CONT'D)				
951	San Jose	103.0	158.4	126.9
952	Stockton	101.0	131.9	114.3
953	Modesto	100.8	132.1	114.3
954	Santa Rosa	101.2	149.8	122.2
955	Eureka	102.6	137.6	117.7
959	Marysville	101.8	130.7	114.3
960	Redding	107.8	132.6	118.6
961	Susanville	107.3	131.0	117.5
COLORADO				
800-802	Denver	104.0	75.1	91.5
803	Boulder	99.2	73.7	88.2
804	Golden	101.5	73.0	89.2
805	Fort Collins	102.8	73.2	90.0
806	Greeley	99.9	73.3	88.4
807	Fort Morgan	99.9	72.5	88.0
808-809	Colorado Springs	101.8	72.3	89.1
810	Pueblo	101.6	70.7	88.2
811	Alamosa	103.3	69.2	88.6
812	Salida	102.9	69.8	88.6
813	Durango	103.6	65.5	87.1
814	Montrose	102.3	66.3	86.7
815	Grand Junction	105.6	70.0	90.2
816	Glenwood Springs	103.6	66.6	87.6
CONNECTICUT				
060	New Britain	99.7	116.9	107.1
061	Hartford	101.7	118.7	109.0
062	Willimantic	100.3	116.9	107.5
063	New London	96.7	116.6	105.3
064	Meriden	98.8	116.4	106.4
065	New Haven	101.5	117.0	108.2
066	Bridgeport	100.9	116.7	107.7
067	Waterbury	100.5	117.2	107.7
068	Norwalk	100.4	116.9	107.5
069	Stamford	100.5	123.9	110.6
D.C.				
200-205	Washington	101.3	88.0	95.5
DELAWARE				
197	Newark	98.7	110.6	103.8
198	Wilmington	99.1	110.7	104.1
199	Dover	99.5	110.4	104.2
FLORIDA				
320,322	Jacksonville	97.0	66.9	84.0
321	Daytona Beach	97.2	69.0	85.0
323	Tallahassee	98.1	66.7	84.6
324	Panama City	98.7	67.2	85.1
325	Pensacola	101.2	65.3	85.7
326,344	Gainesville	98.3	63.8	83.4
327-328,347	Orlando	97.9	66.3	84.3
329	Melbourne	99.9	70.6	87.2
330-332,340	Miami	96.2	70.4	85.1
333	Fort Lauderdale	96.5	72.2	86.0
334,349	West Palm Beach	95.1	69.0	83.8
335-336,346	Tampa	98.0	67.4	84.8
337	St. Petersburg	100.4	65.8	85.4
338	Lakeland	97.1	66.3	83.8
339,341	Fort Myers	96.4	66.7	83.6
342	Sarasota	99.1	65.3	84.5
GEORGIA				
300-303,399	Atlanta	98.7	76.7	89.2
304	Statesboro	97.9	67.1	84.6
305	Gainesville	96.5	64.3	82.6
306	Athens	95.9	63.8	82.0
307	Dalton	97.7	68.0	84.9
308-309	Augusta	96.4	73.5	86.5
310-312	Macon	95.3	74.4	86.3
313-314	Savannah	97.3	72.8	86.7
315	Waycross	96.6	66.9	83.8
316	Valdosta	96.7	65.4	83.2
317,398	Albany	96.5	72.2	86.0
318-319	Columbus	96.4	73.4	86.4

Location Factors - Commercial

STATE/ZIP	CITY	MAT.	INST.	TOTAL
HAWAII				
967	Hilo	115.6	117.2	116.3
968	Honolulu	120.3	117.8	119.2
STATES & POSS.				
969	Guam	138.0	63.4	105.7
IDAHO				
832	Pocatello	101.8	77.8	91.4
833	Twin Falls	102.9	76.9	91.7
834	Idaho Falls	100.2	76.9	90.1
835	Lewiston	108.4	84.9	98.2
836-837	Boise	100.9	78.2	91.1
838	Coeur d'Alene	108.1	81.5	96.6
ILLINOIS				
600-603	North Suburban	98.5	137.3	115.3
604	Joliet	98.5	140.3	116.5
605	South Suburban	98.5	137.2	115.3
606-608	Chicago	100.5	144.4	119.5
609	Kankakee	95.1	129.5	110.0
610-611	Rockford	96.3	126.3	109.3
612	Rock Island	94.5	95.5	94.9
613	La Salle	95.8	122.2	107.2
614	Galesburg	95.5	104.8	99.5
615-616	Peoria	97.4	109.9	102.8
617	Bloomington	94.8	105.4	99.4
618-619	Champaign	98.4	106.3	101.8
620-622	East St. Louis	94.1	109.8	100.9
623	Quincy	96.1	101.1	98.3
624	Effingham	95.3	104.9	99.4
625	Decatur	96.7	107.6	101.4
626-627	Springfield	97.8	110.4	103.2
628	Centralia	92.9	106.7	98.9
629	Carbondale	92.7	105.2	98.1
INDIANA				
460	Anderson	96.8	80.4	89.7
461-462	Indianapolis	99.5	82.9	92.3
463-464	Gary	98.3	105.8	101.5
465-466	South Bend	99.6	83.4	92.6
467-468	Fort Wayne	97.5	75.9	88.2
469	Kokomo	94.9	79.8	88.4
470	Lawrenceburg	93.0	78.3	86.6
471	New Albany	94.3	76.3	86.5
472	Columbus	96.5	79.4	89.1
473	Muncie	96.6	78.9	89.0
474	Bloomington	98.2	79.7	90.2
475	Washington	95.1	84.8	90.7
476-477	Evansville	96.0	83.2	90.5
478	Terre Haute	96.9	82.9	90.8
479	Lafayette	96.2	79.9	89.1
IOWA				
500-503,509	Des Moines	97.3	90.0	94.2
504	Mason City	95.0	71.2	84.7
505	Fort Dodge	95.0	71.0	84.6
506-507	Waterloo	96.5	78.4	88.7
508	Creston	95.2	81.1	89.1
510-511	Sioux City	98.0	80.3	90.3
512	Sibley	96.9	60.1	81.0
513	Spencer	98.3	59.9	81.7
514	Carroll	95.5	79.3	88.5
515	Council Bluffs	98.8	81.9	91.5
516	Shenandoah	96.1	78.7	88.6
520	Dubuque	97.3	80.6	90.1
521	Decorah	96.7	71.0	85.6
522-524	Cedar Rapids	98.3	86.6	93.2
525	Ottumwa	96.4	75.7	87.5
526	Burlington	95.8	81.1	89.5
527-528	Davenport	97.3	96.6	97.0
KANSAS				
660-662	Kansas City	96.4	99.2	97.6
664-666	Topeka	98.1	78.0	89.4
667	Fort Scott	95.2	73.9	86.0
668	Emporia	95.2	74.0	86.0
669	Belleville	96.9	69.1	84.9
670-672	Wichita	96.5	72.2	86.0
673	Independence	96.4	73.6	86.6
674	Salina	96.8	73.3	86.6
675	Hutchinson	91.9	69.7	82.3
676	Hays	96.0	70.4	84.9
677	Colby	96.5	71.7	85.7

STATE/ZIP	CITY	MAT.	INST.	TOTAL
KANSAS (CONT'D)				
678	Dodge City	98.0	74.8	88.0
679	Liberal	96.0	69.5	84.6
KENTUCKY				
400-402	Louisville	94.9	79.6	88.3
403-405	Lexington	94.2	78.8	87.5
406	Frankfort	96.8	78.6	88.9
407-409	Corbin	91.8	75.0	84.6
410	Covington	94.9	75.0	86.3
411-412	Ashland	93.6	86.3	90.5
413-414	Campton	94.9	76.3	86.9
415-416	Pikeville	96.1	82.1	90.1
417-418	Hazard	94.3	77.1	86.9
420	Paducah	93.0	80.0	87.4
421-422	Bowling Green	95.2	78.4	87.9
423	Owensboro	95.3	81.8	89.4
424	Henderson	92.7	78.7	86.7
425-426	Somerset	92.1	76.6	85.4
427	Elizabethtown	91.8	73.4	83.8
LOUISIANA				
700-701	New Orleans	97.7	68.3	85.0
703	Thibodaux	95.5	65.0	82.3
704	Hammond	93.2	63.4	80.3
705	Lafayette	95.1	65.9	82.5
706	Lake Charles	95.3	68.4	83.7
707-708	Baton Rouge	97.0	68.3	84.6
710-711	Shreveport	97.7	66.8	84.4
712	Monroe	96.0	64.6	82.4
713-714	Alexandria	96.1	64.7	82.5
MAINE				
039	Kittery	94.4	84.9	90.3
040-041	Portland	100.6	85.6	94.1
042	Lewiston	98.0	84.9	92.4
043	Augusta	101.6	86.0	94.9
044	Bangor	97.5	84.5	91.9
045	Bath	96.2	84.6	91.2
046	Machias	95.8	84.5	90.9
047	Houlton	95.9	84.4	91.0
048	Rockland	95.0	84.6	90.5
049	Waterville	96.2	84.5	91.2
MARYLAND				
206	Waldorf	97.4	84.8	92.0
207-208	College Park	97.5	85.8	92.5
209	Silver Spring	96.7	84.9	91.6
210-212	Baltimore	102.3	83.1	94.0
214	Annapolis	101.4	83.4	93.6
215	Cumberland	97.1	83.1	91.0
216	Easton	98.9	72.6	87.5
217	Hagerstown	97.7	86.2	92.8
218	Salisbury	99.3	65.0	84.5
219	Elkton	96.1	83.0	90.5
MASSACHUSETTS				
010-011	Springfield	99.1	110.8	104.2
012	Pittsfield	98.6	105.1	101.4
013	Greenfield	96.7	109.5	102.3
014	Fitchburg	95.4	113.2	103.1
015-016	Worcester	99.1	118.6	107.5
017	Framingham	94.9	122.4	106.8
018	Lowell	98.4	126.4	110.5
019	Lawrence	99.4	124.9	110.4
020-022, 024	Boston	100.2	132.9	114.3
023	Brockton	98.9	117.8	107.1
025	Buzzards Bay	93.3	114.1	102.3
026	Hyannis	96.0	116.9	105.1
027	New Bedford	98.1	116.3	106.0
MICHIGAN				
480,483	Royal Oak	94.1	96.9	95.3
481	Ann Arbor	96.3	102.0	98.7
482	Detroit	99.5	100.9	100.1
484-485	Flint	95.9	90.9	93.8
486	Saginaw	95.6	86.7	91.8
487	Bay City	95.7	85.3	91.2
488-489	Lansing	98.1	90.5	94.8
490	Battle Creek	95.7	79.4	88.6
491	Kalamazoo	96.0	80.9	89.5
492	Jackson	94.1	88.0	91.4
493,495	Grand Rapids	98.9	82.2	91.6
494	Muskegon	94.5	79.7	88.1

For customer support on your Plumbing Costs with RSMeans Data, call 800.448.8182.

679

Location Factors - Commercial

STATE/ZIP	CITY	MAT.	INST.	TOTAL	STATE/ZIP	CITY	MAT.	INST.	TOTAL
MICHIGAN (CONT'D)					**NEW HAMPSHIRE (CONT'D)**				
496	Traverse City	93.5	75.2	85.6	032-033	Concord	100.8	91.3	96.7
497	Gaylord	94.6	78.1	87.5	034	Keene	96.1	86.2	91.8
498-499	Iron Mountain	96.5	78.9	88.9	035	Littleton	96.1	77.7	88.1
					036	Charleston	95.6	85.6	91.3
MINNESOTA					037	Claremont	94.8	85.7	90.8
550-551	Saint Paul	99.3	119.2	107.9	038	Portsmouth	96.5	91.5	94.3
553-555	Minneapolis	99.7	116.6	107.0					
556-558	Duluth	99.9	103.9	101.6	**NEW JERSEY**				
559	Rochester	98.4	106.1	101.7	070-071	Newark	101.8	137.9	117.4
560	Mankato	95.5	105.0	99.6	072	Elizabeth	98.9	137.6	115.6
561	Windom	94.0	91.7	93.0	073	Jersey City	97.7	137.1	114.7
562	Willmar	93.7	96.6	94.9	074-075	Paterson	99.3	137.3	115.7
563	St. Cloud	94.9	115.9	104.0	076	Hackensack	97.4	137.1	114.6
564	Brainerd	95.5	97.9	96.5	077	Long Branch	97.1	126.6	109.9
565	Detroit Lakes	97.6	90.3	94.4	078	Dover	97.6	137.5	114.8
566	Bemidji	96.8	94.6	95.8	079	Summit	97.7	136.9	114.6
567	Thief River Falls	96.4	89.4	93.4	080,083	Vineland	97.2	126.6	109.9
					081	Camden	98.9	133.7	114.0
MISSISSIPPI					082,084	Atlantic City	97.8	132.6	112.8
386	Clarksdale	95.7	53.8	77.6	085-086	Trenton	100.5	124.4	110.8
387	Greenville	99.4	65.2	84.6	087	Point Pleasant	99.1	126.5	110.9
388	Tupelo	97.2	56.4	79.6	088-089	New Brunswick	99.7	132.2	113.8
389	Greenwood	97.0	53.5	78.2					
390-392	Jackson	98.6	67.9	85.3	**NEW MEXICO**				
393	Meridian	95.4	67.8	83.5	870-872	Albuquerque	98.9	71.6	87.1
394	Laurel	96.8	56.4	79.4	873	Gallup	99.2	71.6	87.3
395	Biloxi	97.1	66.6	83.9	874	Farmington	99.3	71.6	87.3
396	McComb	95.1	53.8	77.3	875	Santa Fe	100.5	72.2	88.3
397	Columbus	96.7	57.1	79.6	877	Las Vegas	97.4	71.6	86.3
					878	Socorro	97.1	71.6	86.1
MISSOURI					879	Truth/Consequences	96.8	68.6	84.6
630-631	St. Louis	97.5	104.6	100.6	880	Las Cruces	96.8	70.0	85.2
633	Bowling Green	95.6	92.3	94.2	881	Clovis	99.1	71.5	87.2
634	Hannibal	94.5	88.5	91.9	882	Roswell	100.7	71.6	88.1
635	Kirksville	98.0	83.6	91.8	883	Carrizozo	101.4	71.6	88.5
636	Flat River	96.6	90.6	94.0	884	Tucumcari	99.7	71.5	87.6
637	Cape Girardeau	96.4	90.6	93.9					
638	Sikeston	94.9	84.4	90.4	**NEW YORK**				
639	Poplar Bluff	94.4	84.3	90.0	100-102	New York	100.3	174.2	132.2
640-641	Kansas City	97.5	101.8	99.3	103	Staten Island	96.5	171.9	129.1
644-645	St. Joseph	96.0	93.3	94.9	104	Bronx	94.8	171.2	127.8
646	Chillicothe	93.7	90.7	92.4	105	Mount Vernon	94.9	147.0	117.4
647	Harrisonville	93.2	97.4	95.0	106	White Plains	94.9	149.8	118.6
648	Joplin	95.2	80.1	88.7	107	Yonkers	99.1	150.1	121.1
650-651	Jefferson City	95.7	90.0	93.2	108	New Rochelle	95.3	141.4	115.2
652	Columbia	95.7	94.5	95.2	109	Suffern	94.9	125.4	108.1
653	Sedalia	95.6	88.1	92.4	110	Queens	100.5	173.1	131.8
654-655	Rolla	93.4	90.6	92.2	111	Long Island City	102.2	173.1	132.8
656-658	Springfield	96.5	81.4	90.0	112	Brooklyn	102.7	173.1	133.1
					113	Flushing	102.5	173.1	133.0
MONTANA					114	Jamaica	100.9	173.1	132.1
590-591	Billings	101.4	77.2	90.9	115,117,118	Hicksville	100.5	151.7	122.6
592	Wolf Point	101.3	74.5	89.7	116	Far Rockaway	102.7	173.1	133.1
593	Miles City	99.1	74.6	88.5	119	Riverhead	101.3	156.9	125.3
594	Great Falls	102.8	77.1	91.7	120-122	Albany	98.3	109.4	103.1
595	Havre	100.2	72.5	88.3	123	Schenectady	97.9	108.6	102.5
596	Helena	101.2	74.7	89.7	124	Kingston	101.2	128.8	113.1
597	Butte	101.3	74.4	89.7	125-126	Poughkeepsie	100.3	131.7	113.9
598	Missoula	98.8	74.8	88.5	127	Monticello	99.8	129.5	112.7
599	Kalispell	98.0	74.1	87.7	128	Glens Falls	92.8	103.2	97.3
					129	Plattsburgh	97.7	92.8	95.6
NEBRASKA					130-132	Syracuse	98.7	101.6	99.9
680-681	Omaha	98.6	83.0	91.9	133-135	Utica	96.8	102.5	99.3
683-685	Lincoln	99.8	79.3	90.9	136	Watertown	98.4	98.9	98.6
686	Columbus	96.0	77.7	88.1	137-139	Binghamton	98.5	100.5	99.3
687	Norfolk	97.4	74.5	87.5	140-142	Buffalo	100.9	108.8	104.8
688	Grand Island	97.4	76.2	88.2	143	Niagara Falls	98.0	103.4	100.3
689	Hastings	97.1	75.2	87.7	144-146	Rochester	100.8	101.2	101.0
690	McCook	95.9	70.2	84.8	147	Jamestown	97.1	92.3	95.0
691	North Platte	96.1	74.9	86.9	148-149	Elmira	96.9	100.4	98.4
692	Valentine	98.0	67.4	84.8					
693	Alliance	98.4	70.4	86.3	**NORTH CAROLINA**				
					270,272-274	Greensboro	99.7	65.3	84.8
NEVADA					271	Winston-Salem	99.5	65.3	84.7
889-891	Las Vegas	104.2	107.0	105.4	275-276	Raleigh	99.0	65.3	84.4
893	Ely	102.9	89.5	97.1	277	Durham	101.2	65.2	85.6
894-895	Reno	102.8	85.6	95.4	278	Rocky Mount	97.1	64.5	83.0
897	Carson City	103.2	85.6	95.6	279	Elizabeth City	98.1	66.6	84.5
898	Elko	101.5	81.8	93.0	280	Gastonia	97.7	68.5	85.1
					281-282	Charlotte	98.4	72.1	87.0
NEW HAMPSHIRE					283	Fayetteville	100.5	64.0	84.7
030	Nashua	99.3	92.3	96.3	284	Wilmington	96.6	63.1	82.1
031	Manchester	100.5	93.0	97.3	285	Kinston	94.9	63.5	81.3

For customer support on your Plumbing Costs with RSMeans Data, call 800.448.8182.

STATE/ZIP	CITY	MAT.	INST.	TOTAL
NORTH CAROLINA (CONT'D)				
286	Hickory	95.2	69.0	83.9
287-288	Asheville	97.0	64.6	83.0
289	Murphy	96.1	62.6	81.6
NORTH DAKOTA				
580-581	Fargo	99.6	78.9	90.6
582	Grand Forks	98.8	81.0	91.1
583	Devils Lake	98.6	77.8	89.6
584	Jamestown	98.7	76.9	89.3
585	Bismarck	99.1	83.1	92.2
586	Dickinson	99.2	75.7	89.0
587	Minot	98.7	76.5	89.1
588	Williston	97.8	80.7	90.4
OHIO				
430-432	Columbus	99.2	83.2	92.3
433	Marion	95.5	83.6	90.3
434-436	Toledo	98.9	87.7	94.1
437-438	Zanesville	96.1	83.0	90.4
439	Steubenville	97.1	87.4	92.9
440	Lorain	98.7	81.8	91.4
441	Cleveland	99.4	91.5	96.0
442-443	Akron	99.7	83.0	92.5
444-445	Youngstown	99.1	79.7	90.7
446-447	Canton	99.2	77.5	89.8
448-449	Mansfield	96.6	81.1	89.9
450	Hamilton	96.9	76.7	88.2
451-452	Cincinnati	98.0	79.2	89.9
453-454	Dayton	96.9	76.6	88.1
455	Springfield	96.9	77.2	88.4
456	Chillicothe	96.3	85.3	91.5
457	Athens	99.2	81.8	91.7
458	Lima	99.6	79.0	90.7
OKLAHOMA				
730-731	Oklahoma City	97.6	67.8	84.8
734	Ardmore	95.9	63.8	82.0
735	Lawton	98.1	63.9	83.3
736	Clinton	97.3	63.7	82.8
737	Enid	97.8	64.2	83.3
738	Woodward	95.9	61.2	80.9
739	Guymon	97.0	61.7	81.7
740-741	Tulsa	96.9	65.5	83.3
743	Miami	93.5	63.0	80.3
744	Muskogee	96.1	65.7	83.0
745	McAlester	93.2	58.3	78.1
746	Ponca City	93.9	61.8	80.0
747	Durant	93.8	62.7	80.4
748	Shawnee	95.3	63.3	81.5
749	Poteau	93.0	62.6	79.9
OREGON				
970-972	Portland	102.0	104.7	103.2
973	Salem	103.3	102.8	103.0
974	Eugene	101.6	103.7	102.5
975	Medford	103.1	99.3	101.4
976	Klamath Falls	103.3	97.8	100.9
977	Bend	102.3	100.5	101.5
978	Pendleton	98.3	101.0	99.5
979	Vale	96.0	87.8	92.4
PENNSYLVANIA				
150-152	Pittsburgh	99.3	102.6	100.7
153	Washington	96.6	101.9	98.9
154	Uniontown	97.1	99.2	98.0
155	Bedford	98.0	90.1	94.6
156	Greensburg	98.0	95.8	97.1
157	Indiana	96.8	96.2	96.5
158	Dubois	98.5	94.0	96.6
159	Johnstown	98.0	97.2	97.7
160	Butler	92.1	98.7	95.0
161	New Castle	92.2	98.9	95.1
162	Kittanning	92.6	96.6	94.3
163	Oil City	92.1	94.4	93.1
164-165	Erie	94.0	94.6	94.3
166	Altoona	94.2	97.5	95.6
167	Bradford	95.6	95.0	95.3
168	State College	95.2	99.1	96.9
169	Wellsboro	96.2	89.3	93.2
170-171	Harrisburg	98.5	97.0	97.9
172	Chambersburg	95.4	86.0	91.3
173-174	York	96.2	94.6	95.5
175-176	Lancaster	94.2	96.5	95.2

STATE/ZIP	CITY	MAT.	INST.	TOTAL
PENNSYLVANIA (CONT'D)				
177	Williamsport	92.8	94.2	93.4
178	Sunbury	95.0	89.3	92.5
179	Pottsville	94.0	92.5	93.3
180	Lehigh Valley	95.6	107.5	100.8
181	Allentown	97.5	108.1	102.1
182	Hazleton	95.1	92.5	94.0
183	Stroudsburg	95.0	100.9	97.6
184-185	Scranton	98.3	96.8	97.7
186-187	Wilkes-Barre	94.9	96.8	95.7
188	Montrose	94.6	93.3	94.0
189	Doylestown	94.7	122.6	106.7
190-191	Philadelphia	99.7	136.9	115.8
193	Westchester	96.1	123.2	107.8
194	Norristown	95.1	123.4	107.3
195-196	Reading	97.3	104.7	100.5
PUERTO RICO				
009	San Juan	123.0	28.3	82.1
RHODE ISLAND				
028	Newport	97.0	112.6	103.8
029	Providence	101.0	113.3	106.3
SOUTH CAROLINA				
290-292	Columbia	98.8	67.6	85.3
293	Spartanburg	97.7	70.2	85.8
294	Charleston	99.4	67.0	85.4
295	Florence	97.4	67.0	84.2
296	Greenville	97.5	70.0	85.6
297	Rock Hill	97.3	68.2	84.7
298	Aiken	98.1	65.2	83.9
299	Beaufort	98.9	61.5	82.7
SOUTH DAKOTA				
570-571	Sioux Falls	99.3	82.1	91.8
572	Watertown	97.7	66.6	84.3
573	Mitchell	96.6	59.6	80.6
574	Aberdeen	99.1	66.7	85.1
575	Pierre	100.2	68.9	86.7
576	Mobridge	97.1	60.7	81.4
577	Rapid City	98.7	71.0	86.7
TENNESSEE				
370-372	Nashville	100.0	74.5	89.0
373-374	Chattanooga	99.0	68.7	85.9
375,380-381	Memphis	96.5	72.5	86.1
376	Johnson City	99.2	60.5	82.5
377-379	Knoxville	95.7	66.0	82.9
382	McKenzie	96.7	54.7	78.6
383	Jackson	98.1	60.0	81.6
384	Columbia	95.2	64.6	82.0
385	Cookeville	96.6	56.1	79.1
TEXAS				
750	McKinney	99.5	61.8	83.2
751	Waxahachie	99.5	62.0	83.3
752-753	Dallas	100.1	67.4	86.0
754	Greenville	99.6	61.3	83.1
755	Texarkana	99.1	60.4	82.4
756	Longview	99.8	59.2	82.3
757	Tyler	100.1	60.2	82.8
758	Palestine	95.9	58.9	80.0
759	Lufkin	96.6	60.5	81.0
760-761	Fort Worth	99.0	63.9	83.9
762	Denton	98.9	61.6	82.8
763	Wichita Falls	96.5	62.5	81.8
764	Eastland	95.7	58.9	79.8
765	Temple	94.1	56.4	77.8
766-767	Waco	95.8	63.2	81.7
768	Brownwood	99.1	57.2	81.0
769	San Angelo	98.7	57.3	80.8
770-772	Houston	101.3	67.3	86.6
773	Huntsville	99.9	60.7	83.0
774	Wharton	101.0	62.1	84.2
775	Galveston	98.8	64.0	83.8
776-777	Beaumont	99.2	64.9	84.4
778	Bryan	96.3	62.4	81.7
779	Victoria	101.0	60.8	83.6
780	Laredo	97.6	60.9	81.8
781-782	San Antonio	98.9	63.8	83.7
783-784	Corpus Christi	100.2	64.6	84.8
785	McAllen	100.9	55.2	81.2
786-787	Austin	98.6	62.3	82.9

For customer support on your Plumbing Costs with RSMeans Data, call 800.448.8182.

681

Location Factors - Commercial

STATE/ZIP	CITY	MAT.	INST.	TOTAL
TEXAS (CONT'D)				
788	Del Rio	100.7	58.2	82.3
789	Giddings	97.3	58.3	80.5
790-791	Amarillo	99.9	61.8	83.5
792	Childress	98.7	59.4	81.7
793-794	Lubbock	100.4	60.7	83.3
795-796	Abilene	99.0	61.6	82.9
797	Midland	100.9	61.7	84.0
798-799,885	El Paso	98.3	63.3	83.2
UTAH				
840-841	Salt Lake City	103.3	73.7	90.5
842,844	Ogden	98.6	74.0	88.0
843	Logan	100.7	74.0	89.2
845	Price	101.1	68.6	87.1
846-847	Provo	101.1	74.0	89.4
VERMONT				
050	White River Jct.	98.7	79.6	90.4
051	Bellows Falls	97.1	90.6	94.3
052	Bennington	97.5	87.2	93.0
053	Brattleboro	97.8	90.7	94.7
054	Burlington	103.1	82.3	94.1
056	Montpelier	101.2	86.0	94.6
057	Rutland	99.4	81.6	91.7
058	St. Johnsbury	98.7	79.2	90.3
059	Guildhall	97.3	79.2	89.5
VIRGINIA				
220-221	Fairfax	99.9	81.7	92.0
222	Arlington	101.0	81.6	92.6
223	Alexandria	100.0	84.8	93.4
224-225	Fredericksburg	98.6	79.9	90.5
226	Winchester	99.2	76.4	89.3
227	Culpeper	99.1	82.0	91.7
228	Harrisonburg	99.4	75.6	89.1
229	Charlottesville	99.8	74.2	88.7
230-232	Richmond	99.6	74.6	88.8
233-235	Norfolk	100.7	68.3	86.7
236	Newport News	99.5	69.5	86.5
237	Portsmouth	99.0	67.4	85.3
238	Petersburg	99.4	74.4	88.6
239	Farmville	98.5	66.0	84.5
240-241	Roanoke	100.5	69.1	87.0
242	Bristol	98.5	59.4	81.6
243	Pulaski	98.1	69.7	85.8
244	Staunton	99.0	67.8	85.5
245	Lynchburg	99.1	73.1	87.9
246	Grundy	98.4	63.2	83.2
WASHINGTON				
980-981,987	Seattle	105.4	108.4	106.7
982	Everett	104.5	104.2	104.4
983-984	Tacoma	104.9	105.7	105.2
985	Olympia	103.9	105.2	104.5
986	Vancouver	106.2	102.2	104.4
988	Wenatchee	105.0	85.6	96.6
989	Yakima	105.0	101.6	103.5
990-992	Spokane	101.0	86.8	94.9
993	Richland	100.7	93.0	97.4
994	Clarkston	99.9	79.5	91.1
WEST VIRGINIA				
247-248	Bluefield	97.3	85.2	92.1
249	Lewisburg	99.0	86.8	93.8
250-253	Charleston	96.6	91.1	94.2
254	Martinsburg	96.5	80.8	89.7
255-257	Huntington	97.8	91.7	95.1
258-259	Beckley	95.2	87.7	91.9
260	Wheeling	100.1	90.9	96.1
261	Parkersburg	98.9	90.0	95.1
262	Buckhannon	98.7	90.5	95.1
263-264	Clarksburg	99.3	91.0	95.7
265	Morgantown	99.3	91.4	95.9
266	Gassaway	98.7	87.3	93.7
267	Romney	98.5	85.3	92.8
268	Petersburg	98.3	85.6	92.8
WISCONSIN				
530,532	Milwaukee	97.9	111.7	103.9
531	Kenosha	99.2	106.2	102.2
534	Racine	98.5	106.2	101.8
535	Beloit	97.8	97.3	97.6
537	Madison	98.6	106.4	101.9

STATE/ZIP	CITY	MAT.	INST.	TOTAL
WISCONSIN (CONT'D)				
538	Lancaster	95.8	92.9	94.6
539	Portage	94.2	96.5	95.2
540	New Richmond	94.9	92.9	94.0
541-543	Green Bay	99.4	103.1	101.0
544	Wausau	94.5	92.9	93.8
545	Rhinelander	98.0	91.1	95.0
546	La Crosse	95.7	100.3	97.7
547	Eau Claire	97.4	100.4	98.7
548	Superior	94.6	94.6	94.6
549	Oshkosh	95.2	92.7	94.1
WYOMING				
820	Cheyenne	101.0	72.4	88.7
821	Yellowstone Nat'l Park	98.9	71.4	87.0
822	Wheatland	100.1	69.9	87.0
823	Rawlins	101.9	71.3	88.7
824	Worland	99.5	71.9	87.6
825	Riverton	100.8	71.4	88.1
826	Casper	102.5	72.3	89.4
827	Newcastle	99.4	71.3	87.3
828	Sheridan	102.4	71.1	88.9
829-831	Rock Springs	103.3	70.6	89.1
CANADIAN FACTORS (reflect Canadian currency)				
ALBERTA				
	Calgary	123.3	91.7	109.7
	Edmonton	125.3	91.9	110.9
	Fort McMurray	123.9	86.2	107.6
	Lethbridge	118.9	85.7	104.6
	Lloydminster	113.6	82.6	100.2
	Medicine Hat	113.7	81.9	100.0
	Red Deer	114.2	81.9	100.2
BRITISH COLUMBIA				
	Kamloops	115.6	81.6	100.9
	Prince George	116.5	81.0	101.2
	Vancouver	121.7	91.7	108.7
	Victoria	116.8	86.4	103.7
MANITOBA				
	Brandon	122.8	68.9	99.5
	Portage la Prairie	113.6	67.5	93.6
	Winnipeg	125.8	68.4	101.0
NEW BRUNSWICK				
	Bathurst	112.7	60.8	90.3
	Dalhousie	112.4	61.0	90.2
	Fredericton	121.9	67.8	98.5
	Moncton	113.0	73.2	95.8
	Newcastle	112.8	61.4	90.5
	St. John	115.6	72.9	97.1
NEWFOUNDLAND				
	Corner Brook	127.3	66.7	101.1
	St. Johns	126.2	85.2	108.5
NORTHWEST TERRITORIES				
	Yellowknife	135.0	81.1	111.7
NOVA SCOTIA				
	Bridgewater	113.3	68.7	94.0
	Dartmouth	124.0	68.6	100.1
	Halifax	120.5	84.9	105.1
	New Glasgow	121.9	68.6	98.9
	Sydney	120.1	68.6	97.9
	Truro	112.8	68.7	93.7
	Yarmouth	121.9	68.6	98.8
ONTARIO				
	Barrie	118.0	85.2	103.8
	Brantford	115.2	88.8	103.8
	Cornwall	115.3	85.5	102.4
	Hamilton	119.2	94.5	108.6
	Kingston	116.5	85.5	103.1
	Kitchener	109.9	91.2	101.8
	London	119.2	92.9	107.8
	North Bay	125.0	83.4	107.0
	Oshawa	112.4	93.6	104.3
	Ottawa	120.6	93.0	108.7
	Owen Sound	118.1	83.5	103.2
	Peterborough	115.2	85.2	102.3
	Sarnia	115.5	89.4	104.2

Location Factors - Commercial

STATE/ZIP	CITY	MAT.	INST.	TOTAL
ONTARIO (CONT'D)				
	Sault Ste. Marie	110.6	86.4	100.1
	St. Catharines	108.0	92.1	101.1
	Sudbury	107.7	90.3	100.2
	Thunder Bay	109.3	91.6	101.6
	Timmins	115.4	83.4	101.6
	Toronto	119.3	99.9	111.0
	Windsor	108.6	91.2	101.1
PRINCE EDWARD ISLAND				
	Charlottetown	124.1	58.7	95.8
	Summerside	124.4	56.4	95.0
QUEBEC				
	Cap-de-la-Madeleine	112.8	78.2	97.8
	Charlesbourg	112.8	78.2	97.8
	Chicoutimi	112.5	86.9	101.4
	Gatineau	112.5	78.0	97.6
	Granby	112.7	77.9	97.7
	Hull	112.6	78.0	97.7
	Joliette	113.0	78.2	97.9
	Laval	112.6	78.6	97.9
	Montreal	122.1	89.4	108.0
	Quebec City	122.2	89.9	108.2
	Rimouski	112.3	86.9	101.4
	Rouyn-Noranda	112.4	78.0	97.6
	Saint-Hyacinthe	111.8	78.0	97.2
	Sherbrooke	112.8	78.0	97.7
	Sorel	112.9	78.2	97.9
	Saint-Jerome	112.5	78.0	97.6
	Trois-Rivieres	122.8	78.1	103.5
SASKATCHEWAN				
	Moose Jaw	111.0	62.4	90.0
	Prince Albert	110.2	60.7	88.8
	Regina	127.0	88.5	110.3
	Saskatoon	112.0	85.1	100.4
YUKON				
	Whitehorse	136.0	68.6	106.9

R011105-05   Tips for Accurate Estimating

1. Use pre-printed or columnar forms for orderly sequence of dimensions and locations and for recording telephone quotations.

2. Use only the front side of each paper or form except for certain pre-printed summary forms.

3. Be consistent in listing dimensions: For example, length x width x height. This helps in rechecking to ensure that, the total length of partitions is appropriate for the building area.

4. Use printed (rather than measured) dimensions where given.

5. Add up multiple printed dimensions for a single entry where possible.

6. Measure all other dimensions carefully.

7. Use each set of dimensions to calculate multiple related quantities.

8. Convert foot and inch measurements to decimal feet when listing. Memorize decimal equivalents to .01 parts of a foot (1/8″ equals approximately .01″).

9. Do not "round off" quantities until the final summary.

10. Mark drawings with different colors as items are taken off.

11. Keep similar items together, different items separate.

12. Identify location and drawing numbers to aid in future checking for completeness.

13. Measure or list everything on the drawings or mentioned in the specifications.

14. It may be necessary to list items not called for to make the job complete.

15. Be alert for: Notes on plans such as N.T.S. (not to scale); changes in scale throughout the drawings; reduced size drawings; discrepancies between the specifications and the drawings.

16. Develop a consistent pattern of performing an estimate. For example:
    a. Start the quantity takeoff at the lower floor and move to the next higher floor.
    b. Proceed from the main section of the building to the wings.
    c. Proceed from south to north or vice versa, clockwise or counterclockwise.
    d. Take off floor plan quantities first, elevations next, then detail drawings.

17. List all gross dimensions that can be either used again for different quantities, or used as a rough check of other quantities for verification (exterior perimeter, gross floor area, individual floor areas, etc.).

18. Utilize design symmetry or repetition (repetitive floors, repetitive wings, symmetrical design around a center line, similar room layouts, etc.). Note: Extreme caution is needed here so as not to omit or duplicate an area.

19. Do not convert units until the final total is obtained. For instance, when estimating concrete work, keep all units to the nearest cubic foot, then summarize and convert to cubic yards.

20. When figuring alternatives, it is best to total all items involved in the basic system, then total all items involved in the alternates. Therefore you work with positive numbers in all cases. When adds and deducts are used, it is often confusing whether to add or subtract a portion of an item; especially on a complicated or involved alternate.

R011105-10   Unit Gross Area Requirements

The figures in the table below indicate typical ranges in square feet as a function of the "occupant" unit. This table is best used in the preliminary design stages to help determine the probable size requirement for the total project.

Building Type	Unit	Gross Area in S.F.		
		1/4	Median	3/4
Apartments	Unit	660	860	1,100
Auditorium & Play Theaters	Seat	18	25	38
Bowling Alleys	Lane		940	
Churches & Synagogues	Seat	20	28	39
Dormitories	Bed	200	230	275
Fraternity & Sorority Houses	Bed	220	315	370
Garages, Parking	Car	325	355	385
Hospitals	Bed	685	850	1,075
Hotels	Rental Unit	475	600	710
Housing for the elderly	Unit	515	635	755
Housing, Public	Unit	700	875	1,030
Ice Skating Rinks	Total	27,000	30,000	36,000
Motels	Rental Unit	360	465	620
Nursing Homes	Bed	290	350	450
Restaurants	Seat	23	29	39
Schools, Elementary	Pupil	65	77	90
Junior High & Middle		85	110	129
Senior High		102	130	145
Vocational		110	135	195
Shooting Ranges	Point		450	
Theaters & Movies	Seat		15	

R011105-20 Floor Area Ratios

Table below lists commonly used gross to net area and net to gross area ratios expressed in % for various building types.

Building Type	Gross to Net Ratio	Net to Gross Ratio	Building Type	Gross to Net Ratio	Net to Gross Ratio
Apartment	156	64	School Buildings (campus type)		
Bank	140	72	Administrative	150	67
Church	142	70	Auditorium	142	70
Courthouse	162	61	Biology	161	62
Department Store	123	81	Chemistry	170	59
Garage	118	85	Classroom	152	66
Hospital	183	55	Dining Hall	138	72
Hotel	158	63	Dormitory	154	65
Laboratory	171	58	Engineering	164	61
Library	132	76	Fraternity	160	63
Office	135	75	Gymnasium	142	70
Restaurant	141	70	Science	167	60
Warehouse	108	93	Service	120	83
			Student Union	172	59

The gross area of a building is the total floor area based on outside dimensions.

The net area of a building is the usable floor area for the function intended and excludes such items as stairways, corridors and mechanical rooms. In the case of a commercial building, it might be considered as the "leasable area."

R011105-30   Occupancy Determinations

Function of Space	SF/Person Required
Accessory storage areas, mechanical equipment rooms	300
Agriculture Building	300
Aircraft Hangars	500
Airport Terminal	
Baggage claim	20
Baggage handling	300
Concourse	100
Waiting areas	15
Assembly	
Gaming floors (keno, slots, etc.)	11
Exhibit Gallery and Museum	30
Assembly w/ fixed seats	load determined by seat number
Assembly w/o fixed seats	
Concentrated (chairs only-not fixed)	7
Standing space	5
Unconcentrated (tables and chairs)	15
Bowling centers, allow 5 persons for each lane including 15 feet of runway, and for additional areas	7
Business areas	100
Courtrooms-other than fixed seating areas	40
Day care	35
Dormitories	50
Educational	
Classroom areas	20
Shops and other vocational room areas	50
Exercise rooms	50
Fabrication and Manufacturing areas where hazardous materials are used	200
Industrial areas	100
Institutional areas	
Inpatient treatment areas	240
Outpatient areas	100
Sleeping areas	120
Kitchens commercial	200
Library	
Reading rooms	50
Stack area	100
Mercantile	
Areas on other floors	60
Basement and grade floor areas	30
Storage, stock, shipping areas	300
Parking garages	200
Residential	200
Skating rinks, swimming pools	
Rink and pool	50
Decks	15
Stages and platforms	15
Warehouses	500

Excerpted from the 2012 *International Building Code*, Copyright 2011. Washington, D.C.: International Code Council. Reproduced with permission. All rights reserved. www.ICCSAFE.org

R011105-40   Weather Data and Design Conditions

City	Latitude (1)		Winter Temperatures (1)			Winter Degree Days (2)	Summer (Design Dry Bulb) Temperatures and Relative Humidity		
	0	1'	Med. of Annual Extremes	99%	97½%		1%	2½%	5%
UNITED STATES									
Albuquerque, NM	35	0	5.1	12	16	4,400	96/61	94/61	92/61
Atlanta, GA	33	4	11.9	17	22	3,000	94/74	92/74	90/73
Baltimore, MD	39	2	7	14	17	4,600	94/75	91/75	89/74
Birmingham, AL	33	3	13	17	21	2,600	96/74	94/75	92/74
Bismarck, ND	46	5	-32	-23	-19	8,800	95/68	91/68	88/67
Boise, ID	43	3	1	3	10	5,800	96/65	94/64	91/64
Boston, MA	42	2	-1	6	9	5,600	91/73	88/71	85/70
Burlington, VT	44	3	-17	-12	-7	8,200	88/72	85/70	82/69
Charleston, WV	38	2	3	7	11	4,400	92/74	90/73	87/72
Charlotte, NC	35	1	13	18	22	3,200	95/74	93/74	91/74
Casper, WY	42	5	-21	-11	-5	7,400	92/58	90/57	87/57
Chicago, IL	41	5	-8	-3	2	6,600	94/75	91/74	88/73
Cincinnati, OH	39	1	0	1	6	4,400	92/73	90/72	88/72
Cleveland, OH	41	2	-3	1	5	6,400	91/73	88/72	86/71
Columbia, SC	34	0	16	20	24	2,400	97/76	95/75	93/75
Dallas, TX	32	5	14	18	22	2,400	102/75	100/75	97/75
Denver, CO	39	5	-10	-5	1	6,200	93/59	91/59	89/59
Des Moines, IA	41	3	-14	-10	-5	6,600	94/75	91/74	88/73
Detroit, MI	42	2	-3	3	6	6,200	91/73	88/72	86/71
Great Falls, MT	47	3	-25	-21	-15	7,800	91/60	88/60	85/59
Hartford, CT	41	5	-4	3	7	6,200	91/74	88/73	85/72
Houston, TX	29	5	24	28	33	1,400	97/77	95/77	93/77
Indianapolis, IN	39	4	-7	-2	2	5,600	92/74	90/74	87/73
Jackson, MS	32	2	16	21	25	2,200	97/76	95/76	93/76
Kansas City, MO	39	1	-4	2	6	4,800	99/75	96/74	93/74
Las Vegas, NV	36	1	18	25	28	2,800	108/66	106/65	104/65
Lexington, KY	38	0	-1	3	8	4,600	93/73	91/73	88/72
Little Rock, AR	34	4	11	15	20	3,200	99/76	96/77	94/77
Los Angeles, CA	34	0	36	41	43	2,000	93/70	89/70	86/69
Memphis, TN	35	0	10	13	18	3,200	98/77	95/76	93/76
Miami, FL	25	5	39	44	47	200	91/77	90/77	89/77
Milwaukee, WI	43	0	-11	-8	-4	7,600	90/74	87/73	84/71
Minneapolis, MN	44	5	-22	-16	-12	8,400	92/75	89/73	86/71
New Orleans, LA	30	0	28	29	33	1,400	93/78	92/77	90/77
New York, NY	40	5	6	11	15	5,000	92/74	89/73	87/72
Norfolk, VA	36	5	15	20	22	3,400	93/77	91/76	89/76
Oklahoma City, OK	35	2	4	9	13	3,200	100/74	97/74	95/73
Omaha, NE	41	2	-13	-8	-3	6,600	94/76	91/75	88/74
Philadelphia, PA	39	5	6	10	14	4,400	93/75	90/74	87/72
Phoenix, AZ	33	3	27	31	34	1,800	109/71	107/71	105/71
Pittsburgh, PA	40	3	-1	3	7	6,000	91/72	88/71	86/70
Portland, ME	43	4	-10	-6	-1	7,600	87/72	84/71	81/69
Portland, OR	45	4	18	17	23	4,600	89/68	85/67	81/65
Portsmouth, NH	43	1	-8	-2	2	7,200	89/73	85/71	83/70
Providence, RI	41	4	-1	5	9	6,000	89/73	86/72	83/70
Rochester, NY	43	1	-5	1	5	6,800	91/73	88/71	85/70
Salt Lake City, UT	40	5	0	3	8	6,000	97/62	95/62	92/61
San Francisco, CA	37	5	36	38	40	3,000	74/63	71/62	69/61
Seattle, WA	47	4	22	22	27	5,200	85/68	82/66	78/65
Sioux Falls, SD	43	4	-21	-15	-11	7,800	94/73	91/72	88/71
St. Louis, MO	38	4	-3	3	8	5,000	98/75	94/75	91/75
Tampa, FL	28	0	32	36	40	680	92/77	91/77	90/76
Trenton, NJ	40	1	4	11	14	5,000	91/75	88/74	85/73
Washington, DC	38	5	7	14	17	4,200	93/75	91/74	89/74
Wichita, KS	37	4	-3	3	7	4,600	101/72	98/73	96/73
Wilmington, DE	39	4	5	10	14	5,000	92/74	89/74	87/73
ALASKA									
Anchorage	61	1	-29	-23	-18	10,800	71/59	68/58	66/56
Fairbanks	64	5	-59	-51	-47	14,280	82/62	78/60	75/59
CANADA									
Edmonton, Alta.	53	3	-30	-29	-25	11,000	85/66	82/65	79/63
Halifax, N.S.	44	4	-4	1	5	8,000	79/66	76/65	74/64
Montreal, Que.	45	3	-20	-16	-10	9,000	88/73	85/72	83/71
Saskatoon, Sask.	52	1	-35	-35	-31	11,000	89/68	86/66	83/65
St. John's, N.F.	47	4	1	3	7	8,600	77/66	75/65	73/64
Saint John, N.B.	45	2	-15	-12	-8	8,200	80/67	77/65	75/64
Toronto, Ont.	43	4	-10	-5	-1	7,000	90/73	87/72	85/71
Vancouver, B.C.	49	1	13	15	19	6,000	79/67	77/66	74/65
Winnipeg, Man.	49	5	-31	-30	-27	10,800	89/73	86/71	84/70

(1) Handbook of Fundamentals, ASHRAE, Inc., NY 1989
(2) Local Climatological Annual Survey, USDC Env. Science Services
Administration, Asheville, NC

For customer support on your Plumbing Costs with RSMeans Data, call 800.448.8182.

687

R011105-50   Metric Conversion Factors

Description: This table is primarily for converting customary U.S. units in the left hand column to SI metric units in the right hand column. In addition, conversion factors for some commonly encountered Canadian and non-SI metric units are included.

	If You Know		Multiply By		To Find
Length	Inches	x	25.4[a]	=	Millimeters
	Feet	x	0.3048[a]	=	Meters
	Yards	x	0.9144[a]	=	Meters
	Miles (statute)	x	1.609	=	Kilometers
Area	Square inches	x	645.2	=	Square millimeters
	Square feet	x	0.0929	=	Square meters
	Square yards	x	0.8361	=	Square meters
Volume (Capacity)	Cubic inches	x	16,387	=	Cubic millimeters
	Cubic feet	x	0.02832	=	Cubic meters
	Cubic yards	x	0.7646	=	Cubic meters
	Gallons (U.S. liquids)[b]	x	0.003785	=	Cubic meters[c]
	Gallons (Canadian liquid)[b]	x	0.004546	=	Cubic meters[c]
	Ounces (U.S. liquid)[b]	x	29.57	=	Milliliters[c, d]
	Quarts (U.S. liquid)[b]	x	0.9464	=	Liters[c, d]
	Gallons (U.S. liquid)[b]	x	3.785	=	Liters[c, d]
Force	Kilograms force[d]	x	9.807	=	Newtons
	Pounds force	x	4.448	=	Newtons
	Pounds force	x	0.4536	=	Kilograms force[d]
	Kips	x	4448	=	Newtons
	Kips	x	453.6	=	Kilograms force[d]
Pressure, Stress, Strength (Force per unit area)	Kilograms force per square centimeter[d]	x	0.09807	=	Megapascals
	Pounds force per square inch (psi)	x	0.006895	=	Megapascals
	Kips per square inch	x	6.895	=	Megapascals
	Pounds force per square inch (psi)	x	0.07031	=	Kilograms force per square centimeter[d]
	Pounds force per square foot	x	47.88	=	Pascals
	Pounds force per square foot	x	4.882	=	Kilograms force per square meter[d]
Flow	Cubic feet per minute	x	0.4719	=	Liters per second
	Gallons per minute	x	0.0631	=	Liters per second
	Gallons per hour	x	1.05	=	Milliliters per second
Bending Moment Or Torque	Inch-pounds force	x	0.01152	=	Meter-kilograms force[d]
	Inch-pounds force	x	0.1130	=	Newton-meters
	Foot-pounds force	x	0.1383	=	Meter-kilograms force[d]
	Foot-pounds force	x	1.356	=	Newton-meters
	Meter-kilograms force[d]	x	9.807	=	Newton-meters
Mass	Ounces (avoirdupois)	x	28.35	=	Grams
	Pounds (avoirdupois)	x	0.4536	=	Kilograms
	Tons (metric)	x	1000	=	Kilograms
	Tons, short (2000 pounds)	x	907.2	=	Kilograms
	Tons, short (2000 pounds)	x	0.9072	=	Megagrams[e]
Mass per Unit Volume	Pounds mass per cubic foot	x	16.02	=	Kilograms per cubic meter
	Pounds mass per cubic yard	x	0.5933	=	Kilograms per cubic meter
	Pounds mass per gallon (U.S. liquid)[b]	x	119.8	=	Kilograms per cubic meter
	Pounds mass per gallon (Canadian liquid)[b]	x	99.78	=	Kilograms per cubic meter
Temperature	Degrees Fahrenheit	(F-32)/1.8		=	Degrees Celsius
	Degrees Fahrenheit	(F+459.67)/1.8		=	Degrees Kelvin
	Degrees Celsius	C+273.15		=	Degrees Kelvin

[a]The factor given is exact
[b]One U.S. gallon = 0.8327 Canadian gallon
[c]1 liter = 1000 milliliters = 1000 cubic centimeters
 1 cubic decimeter = 0.001 cubic meter

[d]Metric but not SI unit
[e]Called "tonne" in England and
 "metric ton" in other metric countries

R011105-60   Weights and Measures

Measures of Length
1 Mile = 1760 Yards = 5280 Feet
1 Yard = 3 Feet = 36 inches
1 Foot = 12 Inches
1 Mil = 0.001 Inch
1 Fathom = 2 Yards = 6 Feet
1 Rod = 5.5 Yards = 16.5 Feet
1 Hand = 4 Inches
1 Span = 9 Inches
1 Micro-inch = One Millionth Inch or 0.000001 Inch
1 Micron = One Millionth Meter + 0.00003937 Inch

Surveyor's Measure
1 Mile = 8 Furlongs = 80 Chains
1 Furlong = 10 Chains = 220 Yards
1 Chain = 4 Rods = 22 Yards = 66 Feet = 100 Links
1 Link = 7.92 Inches

Square Measure
1 Square Mile = 640 Acres = 6400 Square Chains
1 Acre = 10 Square Chains = 4840 Square Yards =
  43,560 Sq. Ft.
1 Square Chain = 16 Square Rods = 484 Square Yards =
  4356 Sq. Ft.
1 Square Rod = 30.25 Square Yards = 272.25 Square Feet = 625 Square
  Lines
1 Square Yard = 9 Square Feet
1 Square Foot = 144 Square Inches
An Acre equals a Square 208.7 Feet per Side

Cubic Measure
1 Cubic Yard = 27 Cubic Feet
1 Cubic Foot = 1728 Cubic Inches
1 Cord of Wood = 4 x 4 x 8 Feet = 128 Cubic Feet
1 Perch of Masonry = 16½ x 1½ x 1 Foot = 24.75 Cubic Feet

Avoirdupois or Commercial Weight
1 Gross or Long Ton = 2240 Pounds
1 Net or Short Ton = 2000 Pounds
1 Pound = 16 Ounces = 7000 Grains
1 Ounce = 16 Drachms = 437.5 Grains
1 Stone = 14 Pounds

Power
1 British Thermal Unit per Hour = 0.2931 Watts
1 Ton (Refrigeration) = 3.517 Kilowatts
1 Horsepower (Boiler) = 9.81 Kilowatts
1 Horsepower (550 ft-lb/s) = 0.746 Kilowatts

Shipping Measure
For Measuring Internal Capacity of a Vessel:
  1 Register Ton = 100 Cubic Feet

For Measurement of Cargo:
  Approximately 40 Cubic Feet of Merchandise is considered a Shipping
  Ton, unless that bulk would weigh more than 2000 Pounds, in which case
  Freight Charge may be based upon weight.

40 Cubic Feet = 32.143 U.S. Bushels = 31.16 Imp. Bushels

Liquid Measure
1 Imperial Gallon = 1.2009 U.S. Gallon = 277.42 Cu. In.
1 Cubic Foot = 7.48 U.S. Gallons

R011110-30 Engineering Fees

Typical **Structural Engineering Fees** based on type of construction and total project size. These fees are included in Architectural Fees.

Type of Construction	Total Project Size (in thousands of dollars)			
	$500	$500-$1,000	$1,000-$5,000	Over $5000
Industrial buildings, factories & warehouses	Technical payroll times 2.0 to 2.5	1.60%	1.25%	1.00%
Hotels, apartments, offices, dormitories, hospitals, public buildings, food stores		2.00%	1.70%	1.20%
Museums, banks, churches and cathedrals		2.00%	1.75%	1.25%
Thin shells, prestressed concrete, earthquake resistive		2.00%	1.75%	1.50%
Parking ramps, auditoriums, stadiums, convention halls, hangars & boiler houses		2.50%	2.00%	1.75%
Special buildings, major alterations, underpinning & future expansion	↓	Add to above 0.5%	Add to above 0.5%	Add to above 0.5%

For complex reinforced concrete or unusually complicated structures, add 20% to 50%.

Typical **Mechanical and Electrical Engineering Fees** are based on the size of the subcontract. The fee structure for both is shown below. These fees are included in Architectural Fees.

Type of Construction	Subcontract Size							
	$25,000	$50,000	$100,000	$225,000	$350,000	$500,000	$750,000	$1,000,000
Simple structures	6.4%	5.7%	4.8%	4.5%	4.4%	4.3%	4.2%	4.1%
Intermediate structures	8.0	7.3	6.5	5.6	5.1	5.0	4.9	4.8
Complex structures	10.1	9.0	9.0	8.0	7.5	7.5	7.0	7.0

For renovations, add 15% to 25% to applicable fee.

R012153-10 Repair and Remodeling

Cost figures are based on new construction utilizing the most cost-effective combination of labor, equipment and material with the work scheduled in proper sequence to allow the various trades to accomplish their work in an efficient manner.

The costs for repair and remodeling work must be modified due to the following factors that may be present in any given repair and remodeling project.

1. Equipment usage curtailment due to the physical limitations of the project, with only hand-operated equipment being used.

2. Increased requirement for shoring and bracing to hold up the building while structural changes are being made and to allow for temporary storage of construction materials on above-grade floors.

3. Material handling becomes more costly due to having to move within the confines of an enclosed building. For multi-story construction, low capacity elevators and stairwells may be the only access to the upper floors.

4. Large amount of cutting and patching and attempting to match the existing construction is required. It is often more economical to remove entire walls rather than create many new door and window openings. This sort of trade-off has to be carefully analyzed.

5. Cost of protection of completed work is increased since the usual sequence of construction usually cannot be accomplished.

6. Economies of scale usually associated with new construction may not be present. If small quantities of components must be custom fabricated due to job requirements, unit costs will naturally increase. Also, if only small work areas are available at a given time, job scheduling between trades becomes difficult and subcontractor quotations may reflect the excessive start-up and shut-down phases of the job.

7. Work may have to be done on other than normal shifts and may have to be done around an existing production facility which has to stay in production during the course of the repair and remodeling.

8. Dust and noise protection of adjoining non-construction areas can involve substantial special protection and alter usual construction methods.

9. Job may be delayed due to unexpected conditions discovered during demolition or removal. These delays ultimately increase construction costs.

10. Piping and ductwork runs may not be as simple as for new construction. Wiring may have to be snaked through walls and floors.

11. Matching "existing construction" may be impossible because materials may no longer be manufactured. Substitutions may be expensive.

12. Weather protection of existing structure requires additional temporary structures to protect building at openings.

13. On small projects, because of local conditions, it may be necessary to pay a tradesman for a minimum of four hours for a task that is completed in one hour.

All of the above areas can contribute to increased costs for a repair and remodeling project. Each of the above factors should be considered in the planning, bidding and construction stage in order to minimize the increased costs associated with repair and remodeling jobs.

R012153-60 Security Factors

Contractors entering, working in, and exiting secure facilities often lose productive time during a normal workday. The recommended allowances in this section are intended to provide for the loss of productivity by increasing labor costs. Note that different costs are associated with searches upon entry only and searches upon entry and exit. Time spent in a queue is unpredictable and not part of these allowances. Contractors should plan ahead for this situation.

Security checkpoints are designed to reflect the level of security required to gain access or egress. An extreme example is when contractors, along with any materials, tools, equipment, and vehicles, must be physically searched and have all materials, tools, equipment, and vehicles inventoried and documented prior to both entry and exit.

Physical searches without going through the documentation process represent the next level and take up less time.

Electronic searches—passing through a detector or x-ray machine with no documentation of materials, tools, equipment, and vehicles—take less time than physical searches.

Visual searches of materials, tools, equipment, and vehicles represent the next level of security.

Finally, access by means of an ID card or displayed sticker takes the least amount of time.

Another consideration is if the searches described above are performed each and every day, or if they are performed only on the first day with access granted by ID card or displayed sticker for the remainder of the project. The figures for this situation have been calculated to represent the initial check-in as described and subsequent entry by ID card or displayed sticker for up to 20 days on site. For the situation described above, where the time period is beyond 20 days, the impact on labor cost is negligible.

There are situations where tradespeople must be accompanied by an escort and observed during the work day. The loss of freedom of movement will slow down productivity for the tradesperson. Costs for the observer have not been included. Those costs are normally born by the owner.

R012153-65 Infectious Disease Precautions

Contractors entering and working on job sites may be required to take precautions to prevent the spread of infectious diseases. Those precautions will reduce the amount of productive time during a normal workday. The recommended allowances in this section are intended to provide for the loss of productive time by increasing labor costs.

The estimator should be aware that one, many or none of these precautions apply to the line items in an estimate. Job site requirements and sound judgement must be applied.

Labor cost implications are based upon:

Temperature checks	Once per day
Donning/doffing masks and gloves	Four times per day (2 times each)
Washing hands (additional over normal)	Eight times per day
Informational meetings	Once per day
Maintaining social distance	Throughout the day
Disinfecting tools or equipment	Throughout the day

R012909-80   Sales Tax by State

State sales tax on materials is tabulated below (5 states have no sales tax). Many states allow local jurisdictions, such as a county or city, to levy additional sales tax.

Some projects may be sales tax exempt, particularly those constructed with public funds.

State	Tax (%)	State	Tax (%)	State	Tax (%)	State	Tax (%)
Alabama	4	Illinois	6.25	Montana	0	Rhode Island	7
Alaska	0	Indiana	7	Nebraska	5.5	South Carolina	6
Arizona	5.6	Iowa	6	Nevada	6.85	South Dakota	4.5
Arkansas	6.5	Kansas	6.5	New Hampshire	0	Tennessee	7
California	7.25	Kentucky	6	New Jersey	6.625	Texas	6.25
Colorado	2.9	Louisiana	4.45	New Mexico	5.125	Utah	6.10
Connecticut	6.35	Maine	5.5	New York	4	Vermont	6
Delaware	0	Maryland	6	North Carolina	4.75	Virginia	5.3
District of Columbia	6	Massachusetts	6.25	North Dakota	5	Washington	6.5
Florida	6	Michigan	6	Ohio	5.75	West Virginia	6
Georgia	4	Minnesota	6.875	Oklahoma	4.5	Wisconsin	5
Hawaii	4	Mississippi	7	Oregon	0	Wyoming	4
Idaho	6	Missouri	4.225	Pennsylvania	6	Average	5.11 %

Sales Tax by Province (Canada)

GST - a value-added tax, which the government imposes on most goods and services provided in or imported into Canada. PST - a retail sales tax, which five of the provinces impose on the prices of most goods and some services. QST - a value-added tax, similar to the federal GST, which Quebec imposes. HST - Three provinces have combined their retail sales taxes with the federal GST into one harmonized tax.

Province	PST (%)	QST(%)	GST(%)	HST(%)
Alberta	7	0	5	0
British Columbia	7	0	5	0
Manitoba	7	0	5	0
New Brunswick	0	0	0	15
Newfoundland	0	0	0	15
Northwest Territories	0	0	5	0
Nova Scotia	0	0	0	15
Ontario	0	0	0	13
Prince Edward Island	0	0	0	15
Quebec	9.975	0	5	0
Saskatchewan	6	0	5	0
Yukon	0	0	5	0

R012909-85   Unemployment Taxes and Social Security Taxes

State unemployment tax rates vary not only from state to state, but also with the experience rating of the contractor. The federal unemployment tax rate is 6.2% of the first $7,000 of wages. This is reduced by a credit of up to 5.4% for timely payment to the state. The minimum federal unemployment tax is 0.6% after all credits.

Social security (FICA) for 2021 is estimated at time of publication to be 7.65% of wages up to $137,700.

R012909-86   Unemployment Tax by State

Information is from the U.S. Department of Labor, state unemployment tax rates.

State	Tax (%)	State	Tax (%)	State	Tax (%)	State	Tax (%)
Alabama	6.80	Illinois	6.93	Montana	6.30	Rhode Island	9.49
Alaska	5.4	Indiana	7.4	Nebraska	5.4	South Carolina	5.46
Arizona	12.76	Iowa	7.5	Nevada	5.4	South Dakota	9.35
Arkansas	14.3	Kansas	7.6	New Hampshire	7.5	Tennessee	10.0
California	6.2	Kentucky	9.3	New Jersey	5.8	Texas	6.5
Colorado	8.15	Louisiana	6.2	New Mexico	5.4	Utah	7.1
Connecticut	6.8	Maine	5.46	New York	9.1	Vermont	7.7
Delaware	8.20	Maryland	7.50	North Carolina	5.76	Virginia	6.21
District of Columbia	7	Massachusetts	12.65	North Dakota	10.74	Washington	7.73
Florida	5.4	Michigan	10.3	Ohio	9	West Virginia	8.5
Georgia	5.4	Minnesota	9.1	Oklahoma	5.5	Wisconsin	12.0
Hawaii	5.6	Mississippi	5.6	Oregon	5.4	Wyoming	8.8
Idaho	5.4	Missouri	8.37	Pennsylvania	11.03	Median	7.40%

R012909-90   Overtime

One way to improve the completion date of a project or eliminate negative float from a schedule is to compress activity duration times. This can be achieved by increasing the crew size or working overtime with the proposed crew.

To determine the costs of working overtime to compress activity duration times, consider the following examples. Below is an overtime efficiency and cost chart based on a five, six, or seven day week with an eight through twelve hour day. Payroll percentage increases for time and one half and double times are shown for the various working days.

Days per Week	Hours per Day	Production Efficiency					Payroll Cost Factors	
		1st Week	2nd Week	3rd Week	4th Week	Average 4 Weeks	@ 1-1/2 Times	@ 2 Times
5	8	100%	100%	100%	100%	100%	1.000	1.000
	9	100	100	95	90	96	1.056	1.111
	10	100	95	90	85	93	1.100	1.200
	11	95	90	75	65	81	1.136	1.273
	12	90	85	70	60	76	1.167	1.333
6	8	100	100	95	90	96	1.083	1.167
	9	100	95	90	85	93	1.130	1.259
	10	95	90	85	80	88	1.167	1.333
	11	95	85	70	65	79	1.197	1.394
	12	90	80	65	60	74	1.222	1.444
7	8	100	95	85	75	89	1.143	1.286
	9	95	90	80	70	84	1.183	1.365
	10	90	85	75	65	79	1.214	1.429
	11	85	80	65	60	73	1.240	1.481
	12	85	75	60	55	69	1.262	1.524

R013113-40   Builder's Risk Insurance

Builder's risk insurance is insurance on a building during construction. Premiums are paid by the owner or the contractor. Blasting, collapse and underground insurance would raise total insurance costs.

R013113-50   General Contractor's Overhead

There are two distinct types of overhead on a construction project: Project overhead and main office overhead. Project overhead includes those costs at a construction site not directly associated with the installation of construction materials. Examples of project overhead costs include the following:

1. Superintendent
2. Construction office and storage trailers
3. Temporary sanitary facilities
4. Temporary utilities
5. Security fencing
6. Photographs
7. Cleanup
8. Performance and payment bonds

The above project overhead items are also referred to as general requirements and therefore are estimated in Division 1. Division 1 is the first division listed in the CSI MasterFormat but it is usually the last division estimated. The sum of the costs in Divisions 1 through 49 is referred to as the sum of the direct costs.

All construction projects also include indirect costs. The primary components of indirect costs are the contractor's main office overhead and profit. The amount of the main office overhead expense varies depending on the following:

1. Owner's compensation
2. Project managers' and estimators' wages
3. Clerical support wages
4. Office rent and utilities
5. Corporate legal and accounting costs
6. Advertising
7. Automobile expenses
8. Association dues
9. Travel and entertainment expenses

These costs are usually calculated as a percentage of annual sales volume. This percentage can range from 35% for a small contractor doing less than $500,000 to 5% for a large contractor with sales in excess of $100 million.

R013113-55   Installing Contractor's Overhead

Installing contractors (subcontractors) also incur costs for general requirements and main office overhead.

Included within the total incl. overhead and profit costs is a percent mark-up for overhead that includes:

1. Compensation and benefits for office staff and project managers
2. Office rent, utilities, business equipment, and maintenance
3. Corporate legal and accounting costs

4. Advertising
5. Vehicle expenses (for office staff and project managers)
6. Association dues
7. Travel, entertainment
8. Insurance
9. Small tools and equipment

R013113-60   Workers' Compensation Insurance Rates by Trade

The table below tabulates the national averages for workers' compensation insurance rates by trade and type of building. The average "Insurance Rate" is multiplied by the "% of Building Cost" for each trade. This produces the "Workers' Compensation" cost by % of total labor cost, to be added for each trade by building type to determine the weighted average workers' compensation rate for the building types analyzed.

Trade	Insurance Rate (% Labor Cost) Range		Average	% of Building Cost Office Bldgs.	Schools & Apts.	Mfg.	Workers' Compensation Office Bldgs.	Schools & Apts.	Mfg.
Excavation, Grading, etc.	2.1 % to	16.5%	9.3%	4.8%	4.9%	4.5%	0.45%	0.46%	0.42%
Piles & Foundations	3.1 to	28.0	15.6	7.1	5.2	8.7	1.11	0.81	1.36
Concrete	2.5 to	24.9	13.7	5.0	14.8	3.7	0.69	2.03	0.51
Masonry	3.2 to	53.5	28.4	6.9	7.5	1.9	1.96	2.13	0.54
Structural Steel	3.3 to	30.8	17.1	10.7	3.9	17.6	1.83	0.67	3.01
Miscellaneous & Ornamental Metals	2.3 to	22.8	12.5	2.8	4.0	3.6	0.35	0.50	0.45
Carpentry & Millwork	2.6 to	29.1	15.8	3.7	4.0	0.5	0.58	0.63	0.08
Metal or Composition Siding	4.2 to	135.3	69.7	2.3	0.3	4.3	1.60	0.21	3.00
Roofing	4.2 to	105.1	54.7	2.3	2.6	3.1	1.26	1.42	1.70
Doors & Hardware	3.1 to	29.1	16.1	0.9	1.4	0.4	0.14	0.23	0.06
Sash & Glazing	2.6 to	20.5	11.5	3.5	4.0	1.0	0.40	0.46	0.12
Lath & Plaster	2.1 to	33.7	17.9	3.3	6.9	0.8	0.59	1.24	0.14
Tile, Marble & Floors	1.9 to	17.4	9.7	2.6	3.0	0.5	0.25	0.29	0.05
Acoustical Ceilings	1.7 to	21.5	11.6	2.4	0.2	0.3	0.28	0.02	0.03
Painting	2.8 to	37.7	20.3	1.5	1.6	1.6	0.30	0.32	0.32
Interior Partitions	2.6 to	29.1	15.8	3.9	4.3	4.4	0.62	0.68	0.70
Miscellaneous Items	1.6 to	97.7	49.7	5.2	3.7	9.7	2.58	1.84	4.82
Elevators	1.0 to	8.9	5.0	2.1	1.1	2.2	0.11	0.06	0.11
Sprinklers	1.4 to	14.2	7.8	0.5	—	2.0	0.04	—	0.16
Plumbing	1.4 to	13.2	7.3	4.9	7.2	5.2	0.36	0.53	0.38
Heat., Vent., Air Conditioning	2.8 to	15.8	9.3	13.5	11.0	12.9	1.26	1.02	1.20
Electrical	1.2 to	10.6	5.9	10.1	8.4	11.1	0.60	0.50	0.65
Total	1.0 % to	135.3%	68.1	100.0%	100.0%	100.0%	17.36%	16.05%	19.81%
	Overall Weighted Average		17.74%						

Workers' Compensation Insurance Rates by States

The table below lists the weighted average Workers' Compensation base rate for each state with a factor comparing this with the national average of 15.1%.

State	Weighted Average	Factor	State	Weighted Average	Factor	State	Weighted Average	Factor
Alabama	17.2%	159	Kentucky	17.5%	162	North Dakota	8.3%	77
Alaska	11.9	110	Louisiana	25.5	236	Ohio	6.4	59
Arizona	11.1	103	Maine	13.5	125	Oklahoma	13.3	123
Arkansas	7.3	68	Maryland	15.2	141	Oregon	10.8	100
California	32.6	302	Massachusetts	12.0	111	Pennsylvania	26.1	242
Colorado	8.1	75	Michigan	5.8	54	Rhode Island	11.0	102
Connecticut	19.8	183	Minnesota	20.7	192	South Carolina	24.6	228
Delaware	18.2	169	Mississippi	13.7	127	South Dakota	11.3	105
District of Columbia	11.3	105	Missouri	21.4	198	Tennessee	8.8	81
Florida	12.5	116	Montana	12.2	113	Texas	8.0	74
Georgia	50.0	463	Nebraska	16.6	154	Utah	9.2	85
Hawaii	14.0	130	Nevada	12.0	111	Vermont	13.6	126
Idaho	12.9	119	New Hampshire	12.7	118	Virginia	9.9	92
Illinois	27.9	258	New Jersey	21.5	199	Washington	10.4	96
Indiana	4.9	45	New Mexico	20.7	192	West Virginia	6.7	62
Iowa	15.1	140	New York	25.9	240	Wisconsin	16.0	148
Kansas	8.8	81	North Carolina	17.4	161	Wyoming	7.0	65
			Weighted Average for U.S. is	15.1% of payroll = 100%				

The weighted average skilled worker rate for 35 trades is 17.74%. For bidding purposes, apply the full value of Workers' Compensation directly to total labor costs, or if labor is 38%, materials 42% and overhead and profit 20% of total cost, carry 38/80 x 17.74% = 8.43% of cost (before overhead and profit) into overhead. Rates vary not only from state to state but also with the experience rating of the contractor.

Rates are the most current available at the time of publication.

R013113-80   Performance Bond

This table shows the cost of a Performance Bond for a construction job scheduled to be completed in 12 months. Add 1% of the premium cost per month for jobs requiring more than 12 months to complete. The rates are "standard" rates offered to contractors that the bonding company considers financially sound and capable of doing the work. Preferred rates are offered by some bonding companies based upon financial strength of the contractor. Actual rates vary from contractor to contractor and from bonding company to bonding company. Contractors should prequalify through a bonding agency before submitting a bid on a contract that requires a bond.

Contract Amount		Building Construction Class B Projects			Highways & Bridges					
					Class A New Construction			Class A-1 Highway Resurfacing		
First	$ 100,000 bid	$25.00 per M			$15.00 per M			$9.40 per M		
Next	400,000 bid	$ 2,500	plus $15.00	per M	$ 1,500	plus $10.00	per M	$ 940	plus $7.20	per M
Next	2,000,000 bid	8,500	plus 10.00	per M	5,500	plus 7.00	per M	3,820	plus 5.00	per M
Next	2,500,000 bid	28,500	plus 7.50	per M	19,500	plus 5.50	per M	15,820	plus 4.50	per M
Next	2,500,000 bid	47,250	plus 7.00	per M	33,250	plus 5.00	per M	28,320	plus 4.50	per M
Over	7,500,000 bid	64,750	plus 6.00	per M	45,750	plus 4.50	per M	39,570	plus 4.00	per M

R015423-10   Steel Tubular Scaffolding

On new construction, tubular scaffolding is efficient up to 60' high or five stories. Above this it is usually better to use a hung scaffolding if construction permits. Swing scaffolding operations may interfere with tenants. In this case, the tubular is more practical at all heights.

In repairing or cleaning the front of an existing building the cost of tubular scaffolding per S.F. of building front increases as the height increases above the first tier. The first tier cost is relatively high due to leveling and alignment.

The minimum efficient crew for erecting and dismantling is three workers. They can set up and remove 18 frame sections per day up to 5 stories high. For 6 to 12 stories high, a crew of four is most efficient. Use two or more on top and two on the bottom for handing up or hoisting. They can also set up and remove 18 frame sections per day. At 7' horizontal spacing, this will run about 800 S.F. per day of erecting and dismantling. Time for placing and removing planks must be added to the above. A crew of three can place and remove 72 planks per day up to 5 stories. For over 5 stories, a crew of four can place and remove 80 planks per day.

The table below shows the number of pieces required to erect tubular steel scaffolding for 1000 S.F. of building frontage. This area is made up of a scaffolding system that is 12 frames (11 bays) long by 2 frames high.

For jobs under twenty-five frames, add 50% to rental cost. Rental rates will be lower for jobs over three months duration. Large quantities for long periods can reduce rental rates by 20%.

Description of Component	Number of Pieces for 1000 S.F. of Building Front	Unit
5' Wide Standard Frame, 6'-4" High	24	Ea.
Leveling Jack & Plate	24	
Cross Brace	44	
Side Arm Bracket, 21"	12	
Guardrail Post	12	
Guardrail, 7' section	22	
Stairway Section	2	
Stairway Starter Bar	1	
Stairway Inside Handrail	2	
Stairway Outside Handrail	2	
Walk-Thru Frame Guardrail	2	

Scaffolding is often used as falsework over 15' high during construction of cast-in-place concrete beams and slabs. Two foot wide scaffolding is generally used for heavy beam construction. The span between frames depends upon the load to be carried with a maximum span of 5'.

Heavy duty shoring frames with a capacity of 10,000#/leg can be spaced up to 10' O.C. depending upon form support design and loading.

Scaffolding used as horizontal shoring requires less than half the material required with conventional shoring.

On new construction, erection is done by carpenters.

Rolling towers supporting horizontal shores can reduce labor and speed the job. For maintenance work, catwalks with spans up to 70' can be supported by the rolling towers.

R015433-10 Contractor Equipment

Rental Rates shown elsewhere in the data set pertain to late model high quality machines in excellent working condition, rented from equipment dealers. Rental rates from contractors may be substantially lower than the rental rates from equipment dealers depending upon economic conditions; for older, less productive machines, reduce rates by a maximum of 15%. Any overtime must be added to the base rates. For shift work, rates are lower. Usual rule of thumb is 150% of one shift rate for two shifts; 200% for three shifts.

For periods of less than one week, operated equipment is usually more economical to rent than renting bare equipment and hiring an operator.

Costs to move equipment to a job site (mobilization) or from a job site (demobilization) are not included in rental rates, nor in any Equipment costs on any Unit Price line items or crew listings. These costs can be found elsewhere. If a piece of equipment is already at a job site, it is not appropriate to utilize mob/demob costs in an estimate again.

Rental rates vary throughout the country with larger cities generally having lower rates. Lease plans for new equipment are available for periods in excess of six months with a percentage of payments applying toward purchase.

Rental rates can also be treated as reimbursement costs for contractor-owned equipment. Owned equipment costs include depreciation, loan payments, interest, taxes, insurance, storage, and major repairs.

Monthly rental rates vary from 2% to 5% of the cost of the equipment depending on the anticipated life of the equipment and its wearing parts. Weekly rates are about 1/3 the monthly rates and daily rental rates are about 1/3 the weekly rates.

The hourly operating costs for each piece of equipment include costs to the user such as fuel, oil, lubrication, normal expendables for the equipment, and a percentage of the mechanic's wages chargeable to maintenance. The hourly operating costs listed do not include the operator's wages.

The daily cost for equipment used in the standard crews is figured by dividing the weekly rate by five, then adding eight times the hourly operating cost to give the total daily equipment cost, not including the operator. This figure is in the right hand column of the Equipment listings under Equipment Cost/Day.

Pile Driving rates shown for the pile hammer and extractor do not include leads, cranes, boilers or compressors. Vibratory pile driving requires an added field specialist during set-up and pile driving operation for the electric model. The hydraulic model requires a field specialist for set-up only. Up to 125 reuses of sheet piling are possible using vibratory drivers. For normal conditions, crane capacity for hammer type and size is as follows.

Crane Capacity	Hammer Type and Size		
	Air or Steam	Diesel	Vibratory
25 ton	to 8,750 ft.-lb.		70 H.P.
40 ton	15,000 ft.-lb.	to 32,000 ft.-lb.	170 H.P.
60 ton	25,000 ft.-lb.		300 H.P.
100 ton		112,000 ft.-lb.	

Cranes should be specified for the job by size, building and site characteristics, availability, performance characteristics, and duration of time required.

Backhoes & Shovels rent for about the same as equivalent size cranes but maintenance and operating expenses are higher. The crane operator's rate must be adjusted for high boom heights. Average adjustments: for 150' boom add 2% per hour; over 185', add 4% per hour; over 210', add 6% per hour; over 250', add 8% per hour and over 295', add 12% per hour.

Tower Cranes of the climbing or static type have jibs from 50' to 200' and capacities at maximum reach range from 4,000 to 14,000 pounds. Lifting capacities increase up to maximum load as the hook radius decreases.

Typical rental rates, based on purchase price, are about 2% to 3% per month.

Erection and dismantling run between 500 and 2000 labor hours. Climbing operation takes 10 labor hours per 20' climb. Crane dead time is about 5 hours per 40' climb. If crane is bolted to side of the building add cost of ties and extra mast sections. Climbing cranes have from 80' to 180' of mast while static cranes have 80' to 800' of mast.

Truck Cranes can be converted to tower cranes by using tower attachments. Mast heights over 400' have been used.

A single 100' high material **Hoist and Tower** can be erected and dismantled in about 400 labor hours; a double 100' high hoist and tower in about 600 labor hours. Erection times for additional heights are 3 and 4 labor hours per vertical foot respectively up to 150', and 4 to 5 labor hours per vertical foot over 150' high. A 40' high portable Buck hoist takes about 160 labor hours to erect and dismantle. Additional heights take 2 labor hours per vertical foot to 80' and 3 labor hours per vertical foot for the next 100'. Most material hoists do not meet local code requirements for carrying personnel.

A 150' high **Personnel Hoist** requires about 500 to 800 labor hours to erect and dismantle. Budget erection time at 5 labor hours per vertical foot for all trades. Local code requirements or labor scarcity requiring overtime can add up to 50% to any of the above erection costs.

Earthmoving Equipment: The selection of earthmoving equipment depends upon the type and quantity of material, moisture content, haul distance, haul road, time available, and equipment available. Short haul cut and fill operations may require dozers only, while another operation may require excavators, a fleet of trucks, and spreading and compaction equipment. Stockpiled material and granular material are easily excavated with front end loaders. Scrapers are most economically used with hauls between 300' and 1-1/2 miles if adequate haul roads can be maintained. Shovels are often used for blasted rock and any material where a vertical face of 8' or more can be excavated. Special conditions may dictate the use of draglines, clamshells, or backhoes. Spreading and compaction equipment must be matched to the soil characteristics, the compaction required and the rate the fill is being supplied.

R015433-15 Heavy Lifting

Hydraulic Climbing Jacks

The use of hydraulic heavy lift systems is an alternative to conventional type crane equipment. The lifting, lowering, pushing, or pulling mechanism is a hydraulic climbing jack moving on a square steel jackrod from 1-5/8" to 4" square, or a steel cable. The jackrod or cable can be vertical or horizontal, stationary or movable, depending on the individual application. When the jackrod is stationary, the climbing jack will climb the rod and push or pull the load along with itself. When the climbing jack is stationary, the jackrod is movable with the load attached to the end and the climbing jack will lift or lower the jackrod with the attached load. The heavy lift system is normally operated by a single control lever located at the hydraulic pump.

The system is flexible in that one or more climbing jacks can be applied wherever a load support point is required, and the rate of lift synchronized.

Economic benefits have been demonstrated on projects such as: erection of ground assembled roofs and floors, complete bridge spans, girders and trusses, towers, chimney liners and steel vessels, storage tanks, and heavy machinery. Other uses are raising and lowering offshore work platforms, caissons, tunnel sections and pipelines.

For customer support on your Plumbing Costs with RSMeans Data, call 800.448.8182.

697

R015436-50   Mobilization

Costs to move rented construction equipment to a job site from an equipment dealer's or contractor's yard (mobilization) or off the job site (demobilization) are not included in the rental or operating rates, nor in the equipment cost on a unit price line or in a crew listing. These costs can be found consolidated in the Mobilization section of the data and elsewhere in particular site work sections. If a piece of equipment is already on the job site, it is not appropriate to include mob/demob costs in a new estimate that requires use of that equipment. The following table identifies approximate sizes of rented construction equipment that would be hauled on a towed trailer. Because this listing is not all-encompassing, the user can infer as to what size trailer might be required for a piece of equipment not listed.

3-ton Trailer	20-ton Trailer	40-ton Trailer	50-ton Trailer
20 H.P. Excavator	110 H.P. Excavator	200 H.P. Excavator	270 H.P. Excavator
50 H.P. Skid Steer	165 H.P. Dozer	300 H.P. Dozer	Small Crawler Crane
35 H.P. Roller	150 H.P. Roller	400 H.P. Scraper	500 H.P. Scraper
40 H.P. Trencher	Backhoe	450 H.P. Art. Dump Truck	500 H.P. Art. Dump Truck

R024119-10   Demolition Defined

Whole Building Demolition - Demolition of the whole building with no concern for any particular building element, component, or material type being demolished. This type of demolition is accomplished with large pieces of construction equipment that break up the structure, load it into trucks and haul it to a disposal site, but disposal or dump fees are not included. Demolition of below-grade foundation elements, such as footings, foundation walls, grade beams, slabs on grade, etc., is not included. Certain mechanical equipment containing flammable liquids or ozone-depleting refrigerants, electric lighting elements, communication equipment components, and other building elements may contain hazardous waste, and must be removed, either selectively or carefully, as hazardous waste before the building can be demolished.

Foundation Demolition - Demolition of below-grade foundation footings, foundation walls, grade beams, and slabs on grade. This type of demolition is accomplished by hand or pneumatic hand tools, and does not include saw cutting, or handling, loading, hauling, or disposal of the debris.

Gutting - Removal of building interior finishes and electrical/mechanical systems down to the load-bearing and sub-floor elements of the rough building frame, with no concern for any particular building element, component, or material type being demolished. This type of demolition is accomplished by hand or pneumatic hand tools, and includes loading into trucks, but not hauling, disposal or dump fees, scaffolding, or shoring. Certain mechanical equipment containing flammable liquids or ozone-depleting refrigerants, electric lighting elements, communication equipment components, and other building elements may contain hazardous waste, and must be removed, either selectively or carefully, as hazardous waste, before the building is gutted.

Selective Demolition - Demolition of a selected building element, component, or finish, with some concern for surrounding or adjacent elements, components, or finishes (see the first Subdivision (s) at the beginning of appropriate Divisions). This type of demolition is accomplished by hand or pneumatic hand tools, and does not include handling, loading,

storing, hauling, or disposal of the debris, scaffolding, or shoring. "Gutting" methods may be used in order to save time, but damage that is caused to surrounding or adjacent elements, components, or finishes may have to be repaired at a later time.

Careful Removal - Removal of a piece of service equipment, building element or component, or material type, with great concern for both the removed item and surrounding or adjacent elements, components or finishes. The purpose of careful removal may be to protect the removed item for later re-use, preserve a higher salvage value of the removed item, or replace an item while taking care to protect surrounding or adjacent elements, components, connections, or finishes from cosmetic and/or structural damage. An approximation of the time required to perform this type of removal is 1/3 to 1/2 the time it would take to install a new item of like kind (see Reference Number R220105-10). This type of removal is accomplished by hand or pneumatic hand tools, and does not include loading, hauling, or storing the removed item, scaffolding, shoring, or lifting equipment.

Cutout Demolition - Demolition of a small quantity of floor, wall, roof, or other assembly, with concern for the appearance and structural integrity of the surrounding materials. This type of demolition is accomplished by hand or pneumatic hand tools, and does not include saw cutting, handling, loading, hauling, or disposal of debris, scaffolding, or shoring.

Rubbish Handling - Work activities that involve handling, loading or hauling of debris. Generally, the cost of rubbish handling must be added to the cost of all types of demolition, with the exception of whole building demolition.

Minor Site Demolition - Demolition of site elements outside the footprint of a building. This type of demolition is accomplished by hand or pneumatic hand tools, or with larger pieces of construction equipment, and may include loading a removed item onto a truck (check the Crew for equipment used). It does not include saw cutting, hauling or disposal of debris, and, sometimes, handling or loading.

R024119-20   Dumpsters

Dumpster rental costs on construction sites are presented in two ways.

The cost per week rental includes the delivery of the dumpster; its pulling or emptying once per week, and its final removal. The assumption is made that the dumpster contractor could choose to empty a dumpster by simply bringing in an empty unit and removing the full one. These costs also include the disposal of the materials in the dumpster.

The Alternate Pricing can be used when actual planned conditions are not approximated by the weekly numbers. For example, these lines can be used when a dumpster is needed for 4 weeks and will need to be emptied 2 or 3 times per week. Conversely the Alternate Pricing lines can be used when a dumpster will be rented for several weeks or months but needs to be emptied only a few times over this period.

R026510-20   Underground Storage Tank Removal

Underground Storage Tank Removal can be divided into two categories: Non-Leaking and Leaking. Prior to removing an underground storage tank, tests should be made, with the proper authorities present, to determine whether a tank has been leaking or the surrounding soil has been contaminated.

To safely remove Liquid Underground Storage Tanks:
1. Excavate to the top of the tank.
2. Disconnect all piping.
3. Open all tank vents and access ports.

4. Remove all liquids and/or sludge.
5. Purge the tank with an inert gas.
6. Provide access to the inside of the tank and clean out the interior using proper personal protective equipment (PPE).
7. Excavate soil surrounding the tank using proper PPE for on-site personnel.
8. Pull and properly dispose of the tank.
9. Clean up the site of all contaminated material.
10. Install new tanks or close the excavation.

R028213-20   Asbestos Removal Process

Asbestos removal is accomplished by a specialty contractor who understands the federal and state regulations regarding the handling and disposal of the material. The process of asbestos removal is divided into many individual steps. An accurate estimate can be calculated only after all the steps have been priced.

The steps are generally as follows:

1. Obtain an asbestos abatement plan from an industrial hygienist.
2. Monitor the air quality in and around the removal area and along the path of travel between the removal area and transport area. This establishes the background contamination.
3. Construct a two part decontamination chamber at entrance to removal area.
4. Install a HEPA filter to create a negative pressure in the removal area.
5. Install wall, floor and ceiling protection as required by the plan, usually 2 layers of fireproof 6 mil polyethylene.

6. Industrial hygienist visually inspects work area to verify compliance with plan.
7. Provide temporary supports for conduit and piping affected by the removal process.
8. Proceed with asbestos removal and bagging process. Monitor air quality as described in Step #2. Discontinue operations when contaminate levels exceed applicable standards.
9. Document the legal disposal of materials in accordance with EPA standards.
10. Thoroughly clean removal area including all ledges, crevices and surfaces.
11. Post abatement inspection by industrial hygienist to verify plan compliance.
12. Provide a certificate from a licensed industrial hygienist attesting that contaminate levels are within acceptable standards before returning area to regular use.

Concrete   R0330 Cast-In-Place Concrete

R033053-60   Maximum Depth of Frost Penetration in Inches

THIS MAP IS REASONABLY ACCURATE FOR MOST PARTS OF THE UNITED STATES BUT IS NECESSARILY HIGHLY GENERALIZED, AND CONSEQUENTLY NOT TOO ACCURATE IN MOUNTAINOUS REGIONS, PARTICULARLY IN THE ROCKIES.

R050521-20   Welded Structural Steel

Usual weight reductions with welded design run 10% to 20% compared with bolted or riveted connections. This amounts to about the same total cost compared with bolted structures since field welding is more expensive than bolts. For normal spans of 18' to 24' figure 6 to 7 connections per ton.

Trusses — For welded trusses add 4% to weight of main members for connections. Up to 15% less steel can be expected in a welded truss compared to one that is shop bolted. Cost of erection is the same whether shop bolted or welded.

General — Typical electrodes for structural steel welding are E6010, E6011, E60T and E70T. Typical buildings vary between 2# to 8# of weld rod per

ton of steel. Buildings utilizing continuous design require about three times as much welding as conventional welded structures. In estimating field erection by welding, it is best to use the average linear feet of weld per ton to arrive at the welding cost per ton. The type, size and position of the weld will have a direct bearing on the cost per linear foot. A typical field welder will deposit 1.8# to 2# of weld rod per hour manually. Using semiautomatic methods can increase production by as much as 50% to 75%.

R131113-20   Swimming Pools

Pool prices given per square foot of surface area include pool structure, filter and chlorination equipment, pumps, related piping, ladders/steps, maintenance kit, skimmer and vacuum system. Decks and electrical service to equipment are not included.

Residential in-ground pool construction can be divided into two categories: vinyl lined and gunite. Vinyl lined pool walls are constructed of different materials including wood, concrete, plastic or metal. The bottom is often graded with sand over which the vinyl liner is installed. Vermiculite or soil cement bottoms may be substituted for an added cost.

Gunite pool construction is used both in residential and municipal installations. These structures are steel reinforced for strength and finished with a white cement limestone plaster.

Municipal pools will have a higher cost because plumbing codes require more expensive materials, chlorination equipment and higher filtration rates.

Municipal pools greater than 1,800 S.F. require gutter systems to control waves. This gutter may be formed into the concrete wall. Often a vinyl/stainless steel gutter or gutter/wall system is specified, which will raise the pool cost.

Competition pools usually require tile bottoms and sides with contrasting lane striping, which will also raise the pool cost.

R211226-10   Standpipe Systems

The basis for standpipe system design is National Fire Protection Association NFPA 14. However, the authority having jurisdiction should be consulted for special conditions, local requirements, and approval.

Standpipe systems, properly designed and maintained, are an effective and valuable time saving aid for extinguishing fires, especially in the upper stories of tall buildings, the interior of large commercial or industrial malls, or other areas where construction features or access make the laying of temporary hose lines time consuming and/or hazardous. Standpipes are frequently installed with automatic sprinkler systems for maximum protection.

There are three general classes of service for standpipe systems:
Class I A system with 2-1/2" (65 mm) hose connections for Fire Department use and those trained in handling heavy fire streams.
Class II A system with either 1-1/2" (40 mm) hose stations for use by trained personnel or a hose connection for the fire department's use.
Class III A system with 1-1/2" (40 mm) hose stations for trained personnel and 2-1/2" (65 mm) hose connections for fire department use.

Standpipe systems are also classified by the way water is supplied to the system. The following are the types of systems:
Automatic Dry Standpipe A system containing air or nitrogen under pressure, with a water supply capable of meeting water demands at all times. Opening of a valve allows the air/nitrogen to dissipate, releasing the pressurized dry valve's clapper to open and fill the system with water.
Automatic Wet Standpipe A system containing water with a water supply capable of meeting system demands at all times. Open a valve or connection and water will immediately flow.
Combination Standpipe A system that supplies both sprinkler systems and fire department/hose connections. Manual operation of approved remote control devices located at each hose station.
Manual Dry Standpipe A system not containing water. Water is manually pumped into the system by means of a fire department connection attached to a water supply.
Semiautomatic Dry Standpipe A system with an attached water supply held back at a valve which requires remote activation to open and allow flow of water into the system.
Wet Standpipe A system containing water at all times.

Reprinted with permission from NFPA 14-2013, *Installation of Standpipe and Hose Systems*, Copyright © 2013, National Fire Protection Association, Quincy, MA. This reprinted material is not the complete and official position of the NFPA on the referenced subject, which is represented only by the standard in its entirety.

For customer support on your Plumbing Costs with RSMeans Data, call 800.448.8182.

701

R211226-20   NFPA 14 Basic Standpipe Design

Class	Design-Use	Pipe Size Minimums	Water Supply Minimums
Class I	2 1/2" hose connection on each floor All areas within 150' of an exit in every exit stairway Fire Department Trained Personnel	Height to 100', 4" dia. Heights above 100', 6" dia. (275' max. except with pressure regulators 400' max.)	For each standpipe riser 500 GPM flow For common supply pipe allow 500 GPM for first standpipe plus 250 GPM for each additional standpipe (2500 GPM max. total) 30 min. duration 65 PSI at 500 GPM
Class II	1 1/2" hose connection with hose on each floor All areas within 130' of hose connection measured along path of hose travel Occupant personnel	Height to 50', 2" dia. Height above 50', 2 1/2" dia.	For each standpipe riser 100 GPM flow For multiple riser common supply pipe 100 GPM 300 min. duration, 65 PSI at 100 GPM
Class III	Both of above. Class I valved connections will meet Class III with additional 2 1/2" by 1 1/2" adapter and 1 1/2" hose.	Same as Class I	Same as Class I

*Note: Where 2 or more standpipes are installed in the same building or section of building they shall be interconnected at the bottom.

Combined Systems

Combined systems are systems where the risers supply both automatic sprinklers and 2-1/2" hose connection outlets for fire department use. In such a system the sprinkler spacing pattern shall be in accordance with NFPA 13 while the risers and supply piping will be sized in accordance with NFPA 14. When the building is completely sprinklered the risers may be sized by hydraulic calculation. The minimum size riser for buildings not completely sprinklered is 6".

The minimum water supply of a completely sprinklered, light hazard, high-rise occupancy building will be 500 GPM while the supply required for other types of completely sprinklered high-rise buildings is 1000 GPM.

General System Requirements

1. Approved valves will be provided at the riser for controlling branch lines to hose outlets.
2. A hose valve will be provided at each outlet for attachment of hose.
3. Where pressure at any standpipe outlet exceeds 100 PSI a pressure reducer must be installed to limit the pressure to 100 PSI. Note that the pressure head due to gravity in 100' of riser is 43.4 PSI. This must be overcome by city pressure, fire pumps, or gravity tanks to provide adequate pressure at the top of the riser.
4. Each hose valve on a wet system having linen hose shall have an automatic drip connection to prevent valve leakage from entering the hose.
5. Each riser will have a valve to isolate it from the rest of the system.
6. One or more fire department connections as an auxiliary supply shall be provided for each Class I or Class III standpipe system. In buildings having two or more zones, a connection will be provided for each zone.
7. There will be no shutoff valve in the fire department connection, but a check valve will be located in the line before it joins the system.
8. All hose connections street side will be identified on a cast plate or fitting as to purpose.

R211313-10 Sprinkler Systems (Automatic)

Sprinkler systems may be classified by type as follows:

1. **Wet Pipe System.** A system employing automatic sprinklers attached to a piping system containing water and connected to a water supply so that water discharges immediately from sprinklers opened by a fire.

2. **Dry Pipe System.** A system employing automatic sprinklers attached to a piping system containing air under pressure, the release of which as from the opening of sprinklers permits the water pressure to open a valve known as a "dry pipe valve". The water then flows into the piping system and out the opened sprinklers.

3. **Pre-Action System.** A system employing automatic sprinklers attached to a piping system containing air that may or may not be under pressure, with a supplemental heat responsive system of generally more sensitive characteristics than the automatic sprinklers themselves, installed in the same areas as the sprinklers; actuation of the heat responsive system, as from a fire, opens a valve which permits water to flow into the sprinkler piping system and to be discharged from any sprinklers which may be open.

4. **Deluge System.** A system employing open sprinklers attached to a piping system connected to a water supply through a valve which is opened by the operation of a heat responsive system installed in the same areas as the sprinklers. When this valve opens, water flows into the piping system and discharges from all sprinklers attached thereto.

5. **Combined Dry Pipe and Pre-Action Sprinkler System.** A system employing automatic sprinklers attached to a piping system containing air under pressure with a supplemental heat responsive system of generally more sensitive characteristics than the automatic sprinklers themselves, installed in the same areas as the sprinklers; operation of the heat responsive system, as from a fire, actuates tripping devices which open dry pipe valves simultaneously and without loss of air pressure in the system. Operation of the heat responsive system also opens approved air exhaust valves at the end of the feed main which facilitates the filling of the system with water which usually precedes the opening of sprinklers. The heat responsive system also serves as an automatic fire alarm system.

6. **Limited Water Supply System.** A system employing automatic sprinklers and conforming to these standards but supplied by a pressure tank of limited capacity.

7. **Chemical Systems.** Systems using halon, carbon dioxide, dry chemical or high expansion foam as selected for special requirements. Agent may extinguish flames by chemically inhibiting flame propagation, suffocate flames by excluding oxygen, interrupting chemical action of oxygen uniting with fuel or sealing and cooling the combustion center.

8. **Firecycle System.** Firecycle is a fixed fire protection sprinkler system utilizing water as its extinguishing agent. It is a time delayed, recycling, preaction type which automatically shuts the water off when heat is reduced below the detector operating temperature and turns the water back on when that temperature is exceeded. The system senses a fire condition through a closed circuit electrical detector system which controls water flow to the fire automatically. Batteries supply up to 90 hour emergency power supply for system operation. The piping system is dry (until water is required) and is monitored with pressurized air. Should any leak in the system piping occur, an alarm will sound, but water will not enter the system until heat is sensed by a firecycle detector.

Area coverage sprinkler systems may be laid out and fed from the supply in any one of several patterns as shown below. It is desirable, if possible, to utilize a central feed and achieve a shorter flow path from the riser to the furthest sprinkler. This permits use of the smallest sizes of pipe possible with resulting savings.

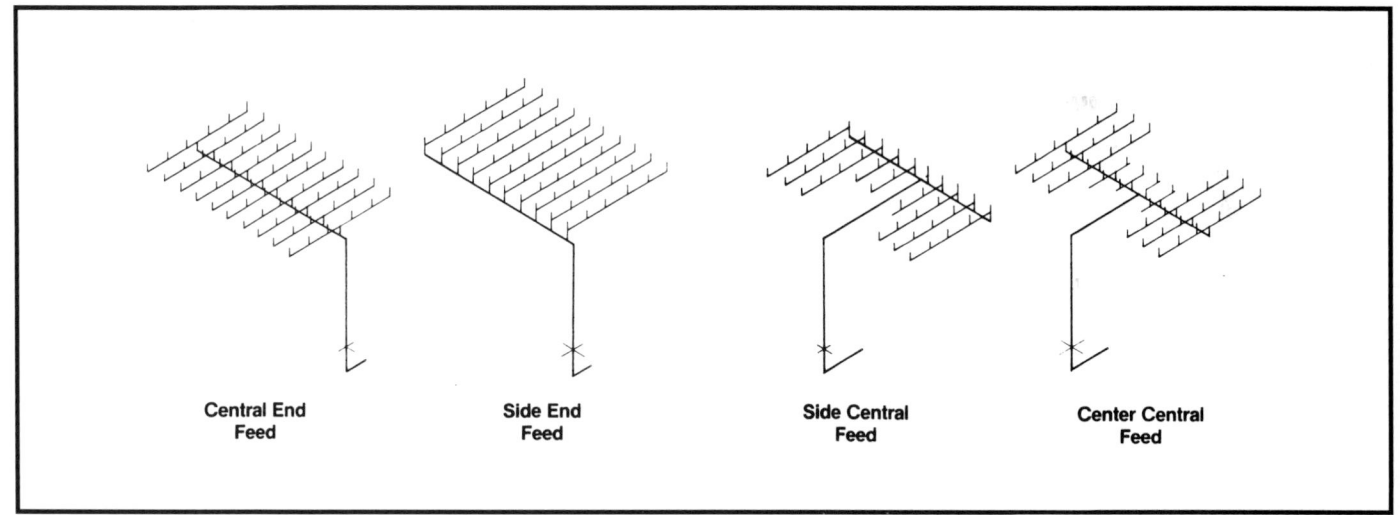

Central End Feed   Side End Feed   Side Central Feed   Center Central Feed

For customer support on your Plumbing Costs with RSMeans Data, call 800.448.8182.

703

R211313-20   System Classification

System Classification

Rules for installation of sprinkler systems vary depending on the classification of occupancy falling into one of three categories as follows:

Light Hazard Occupancy

The protection area allotted per sprinkler should not exceed 225 S.F., with the maximum distance between lines and sprinklers on lines being 15'. The sprinklers do not need to be staggered. Branch lines should not exceed eight sprinklers on either side of a cross main. Each large area requiring more than 100 sprinklers and without a sub-dividing partition should be supplied by feed mains or risers sized for ordinary hazard occupancy.
Maximum system area = 52,000 S.F.

Included in this group are:

Churches	Nursing Homes
Clubs	Offices
Educational	Residential
Hospitals	Restaurants
Institutional	Theaters and Auditoriums
Libraries	(except stages and prosceniums)
(except large stack rooms)	Unused Attics
Museums	

Ordinary Hazard Occupancy

The protection area allotted per sprinkler shall not exceed 130 S.F. of noncombustible ceiling and 130 S.F. of combustible ceiling. The maximum allowable distance between sprinkler lines and sprinklers on line is 15'. Sprinklers shall be staggered if the distance between heads exceeds 12'. Branch lines should not exceed eight sprinklers on either side of a cross main.
Maximum system area = 52,000 S.F.

Included in this group are:

Group 1	Group 2
Automotive Parking and Showrooms	Cereal Mills
Bakeries	Chemical Plants—Ordinary
Beverage manufacturing	Confectionery Products
Canneries	Distilleries
Dairy Products Manufacturing/Processing	Dry Cleaners
Electronic Plans	Feed Mills
Glass and Glass Products Manufacturing	Horse Stables
Laundries	Leather Goods Manufacturing
Restaurant Service Areas	Libraries—Large Stack Room Areas
	Machine Shops
	Metal Working
	Mercantile
	Paper and Pulp Mills
	Paper Process Plants
	Piers and Wharves
	Post Offices
	Printing and Publishing
	Repair Garages
	Stages
	Textile Manufacturing
	Tire Manufacturing
	Tobacco Products Manufacturing
	Wood Machining
	Wood Product Assembly

Extra Hazard Occupancy

The protection area allotted per sprinkler shall not exceed 100 S.F. of noncombustible ceiling and 100 S.F. of combustible ceiling. The maximum allowable distance between lines and between sprinklers on lines is 12'. Sprinklers on alternate lines shall be staggered if the distance between sprinklers on lines exceeds 8'. Branch lines should not exceed six sprinklers on either side of a cross main.
Maximum system area:
    Design by pipe schedule = 25,000 S.F.
    Design by hydraulic calculation = 40,000 S.F.

Included in this group are:

Group 1	Group 2
Aircraft hangars	Asphalt Saturating
Combustible Hydraulic Fluid Use Area	Flammable Liquids Spraying
Die Casting	Flow Coating
Metal Extruding	Manufactured/Modular Home
Plywood/Particle Board Manufacturing	Building Assemblies (where
Printing (inks with flash points < 100	finished enclosure is present
degrees F	and has combustible interiors)
Rubber Reclaiming, Compounding,	Open Oil Quenching
Drying, Milling, Vulcanizing	Plastics Processing
Saw Mills	Solvent Cleaning
Textile Picking, Opening, Blending,	Varnish and Paint Dipping
Garnetting, Carding, Combing of	
Cotton, Synthetics, Wood Shoddy,	
or Burlap	
Upholstering with Plastic Foams	

R211313-40 Adjustment for Sprinkler/Standpipe Installations

Quality/Complexity Multiplier (For all installations)

Economy installation, add	0 to 5%
Good quality, medium complexity, add	5 to 15%
Above average quality and complexity, add	15 to 25%

R220102-20 Labor Adjustment Factors

Labor Adjustment Factors are provided for Divisions 21, 22, and 23 to assist the mechanical estimator account for the various complexities and special conditions of any particular project. While a single percentage has been entered on each line of Division 22 01 02.20, it should be understood that these are just suggested midpoints of ranges of values commonly used by mechanical estimators. They may be increased or decreased depending on the severity of the special conditions.

The group for "existing occupied buildings" has been the subject of requests for explanation. Actually there are two stages to this group: buildings that are existing and "finished" but unoccupied, and those that also are occupied. Buildings that are "finished" may result in higher labor costs due to the workers having to be more careful not to damage finished walls, ceilings, floors, etc. and may necessitate special protective coverings and barriers. Also corridor bends and doorways may not accommodate long pieces of pipe or larger pieces of equipment. Work above an already hung ceiling can be very time consuming. The addition of occupants may force the work to be done on premium time (nights and/or weekends), eliminate the possible use of some preferred tools such as pneumatic drivers, powder charged drivers etc. The estimator should evaluate the access to the work area and just how the work is going to be accomplished to arrive at an increase in labor costs over "normal" new construction productivity.

R220105-10 Demolition (Selective vs. Removal for Replacement)

Demolition can be divided into two basic categories.

One type of demolition involves the removal of material with no concern for its replacement. The labor-hours to estimate this work are found under "Selective Demolition" in the Fire Protection, Plumbing and HVAC Divisions. It is selective in that individual items or all the material installed as a system or trade grouping such as plumbing or heating systems are removed. This may be accomplished by the easiest way possible, such as sawing, torch cutting, or sledge hammering as well as simple unbolting.

The second type of demolition is the removal of some items for repair or replacement. This removal may involve careful draining, opening of unions, disconnecting and tagging electrical connections, capping pipes/ducts to prevent entry of debris or leakage of the material contained as well as transporting the item away from its in-place location to a truck/dumpster. An approximation of the time required to accomplish this type of demolition is to use half of the time indicated as necessary to install a new unit. For example: installation of a new pump might be listed as requiring 6 labor-hours so if we had to estimate the removal of the old pump we would allow an additional 3 hours for a total of 9 hours. That is, the complete replacement of a defective pump with a new pump would be estimated to take 9 labor-hours.

For customer support on your Plumbing Costs with RSMeans Data, call 800.448.8182.

705

R220523-80   Valve Materials

VALVE MATERIALS

Bronze:
Bronze is one of the oldest materials used to make valves. It is most commonly used in hot and cold water systems and other non-corrosive services. It is often used as a seating surface in larger iron body valves to ensure tight closure.

Carbon Steel:
Carbon steel is a high strength material. Therefore, valves made from this metal are used in higher pressure services, such as steam lines up to 600 psi at 850°F. Many steel valves are available with butt-weld ends for economy and are generally used in high pressure steam service as well as other higher pressure non-corrosive services.

Forged Steel:
Valves from tough carbon steel are used in service up to 2000 psi and temperatures up to 1000°F in Gate, Globe and Check valves.

Iron:
Valves are normally used in medium to large pipe lines to control non-corrosive fluid and gases, where pressures do not exceed 250 psi at 450° or 500 psi cold water, oil or gas.

Stainless Steel:
Developed steel alloys can be used in over 90% corrosive services.

Plastic PVC:
This is used in a great variety of valves generally in high corrosive service with lower temperatures and pressures.

VALVE SERVICE PRESSURES

Pressure ratings on valves provide an indication of the safe operating pressure for a valve at some elevated temperature. This temperature is dependent upon the materials used and the fabrication of the valve. When specific data is not available, a good "rule-of-thumb" to follow is the temperature of saturated steam on the primary rating indicated on the valve body. Example: The valve has the number 150S printed on the side indicating 150 psi and hence, a maximum operating temperature of 367°F (temperature of saturated steam and 150 psi).

DEFINITIONS

1. "WOG" – Water, oil, gas (cold working pressures).
2. "SWP" – Steam working pressure.
3. 100% area (full port) – means the area through the valve is equal to or greater than the area of standard pipe.
4. "Standard Opening" – means that the area through the valve is less than the area of standard pipe and therefore these valves should be used only where restriction of flow is unimportant.
5. "Round Port" – means the valve has a full round opening through the plug and body, of the same size and area as standard pipe.
6. "Rectangular Port" – valves have rectangular shaped ports through the plug body. The area of the port is either equal to 100% of the area of standard pipe, or restricted (standard opening). In either case it is clearly marked.
7. "ANSI" – American National Standards Institute.

R220523-90   Valve Selection Considerations

INTRODUCTION: In any piping application, valve performance is critical. Valves should be selected to give the best performance at the lowest cost.

The following is a list of performance characteristics generally expected of valves.

1. Stopping flow or starting it.
2. Throttling flow (Modulation).
3. Flow direction changing.
4. Checking backflow (Permitting flow in only one direction).
5. Relieving or regulating pressure.

In order to properly select the right valve, some facts must be determined.

A. What liquid or gas will flow through the valve?
B. Does the fluid contain suspended particles?
C. Does the fluid remain in liquid form at all times?
D. Which metals does fluid corrode?
E. What are the pressure and temperature limits? (As temperature and pressure rise, so will the price of the valve.)
F. Is there constant line pressure?
G. Is the valve merely an on-off valve?
H. Will checking of backflow be required?
I. Will the valve operate frequently or infrequently?

Valves are classified by design type into such classifications as Gate, Globe, Angle, Check, Ball, Butterfly and Plug. They are also classified by end connection, stem, pressure restrictions and material such as bronze, cast iron, etc. Each valve has a specific use. A quality valve used correctly will provide a lifetime of trouble- free service, but a high quality valve installed in the wrong service may require frequent attention.

STEM TYPES
(OS & Y)—Rising Stem-Outside Screw and Yoke

Offers a visual indication of whether the valve is open or closed. Recommended where high temperatures, corrosives, and solids in the line might cause damage to inside-valve stem threads. The stem threads are engaged by the yoke bushing so the stem rises through the hand wheel as it is turned.

(R.S.)—Rising Stem-Inside Screw

Adequate clearance for operation must be provided because both the hand wheel and the stem rise.
The valve wedge position is indicated by the position of the stem and hand wheel.

(N.R.S.)—Non-Rising Stem-Inside Screw

A minimum clearance is required for operating this type of valve. Excessive wear or damage to stem threads inside the valve may be caused by heat, corrosion, and solids. Because the hand wheel and stem do not rise, wedge position cannot be visually determined.

VALVE TYPES
Gate Valves

Provide full flow, minute pressure drop, minimum turbulence and minimum fluid trapped in the line.
They are normally used where operation is infrequent.

Globe Valves

Globe valves are designed for throttling and/or frequent operation with positive shut-off. Particular attention must be paid to the several types of seating materials available to avoid unnecessary wear. The seats must be compatible with the fluid in service and may be composition or metal. The configuration of the Globe valve opening causes turbulence which results in increased resistance. Most bronze Globe valves are rising stem-inside screw, but they are also available on O.S. & Y.

Angle Valves

The fundamental difference between the Angle valve and the Globe valve is the fluid flow through the Angle valve. It makes a 90° turn and offers less resistance to flow than the Globe valve while replacing an elbow. An Angle valve thus reduces the number of joints and installation time.

Check Valves

Check valves are designed to prevent backflow by automatically seating when the direction of fluid is reversed.
Swing Check valves are generally installed with Gate-valves, as they provide comparable full flow. Usually recommended for lines where flow velocities are low and should not be used on lines with pulsating flow. Recommended for horizontal installation, or in vertical lines only where flow is upward.

Lift Check Valves

These are commonly used with Globe and Angle valves since they have similar diaphragm seating arrangements and are recommended for preventing backflow of steam, air, gas and water, and on vapor lines with high flow velocities. For horizontal lines, horizontal lift checks should be used and vertical lift checks for vertical lines.

Ball Valves

Ball valves are light and easily installed, yet because of modern elastomeric seats, provide tight closure. Flow is controlled by rotating up to 90° a drilled ball which fits tightly against resilient seals. This ball seats with flow in either direction, and valve handle indicates the degree of opening. Recommended for frequent operation readily adaptable to automation, ideal for installation where space is limited.

Butterfly Valves

Butterfly valves provide bubble-tight closure with excellent throttling characteristics. They can be used for full-open, closed and for throttling applications.

The Butterfly valve consists of a disc within the valve body which is controlled by a shaft. In its closed position, the valve disc seals against a resilient seat. The disc position throughout the full 90° rotation is visually indicated by the position of the operator.

A Butterfly valve is only a fraction of the weight of a Gate valve and requires no gaskets between flanges in most cases. Recommended for frequent operation and adaptable to automation where space is limited.

Wafer and Lug type bodies when installed between two pipe flanges, can be easily removed from the line. The pressure of the bolted flanges holds the valve in place.
Locating lugs makes installation easier.

Plug Valves

Lubricated plug valves, because of the wide range of service to which they are adapted, may be classified as all purpose valves. They can be safely used at all pressure and vacuums, and at all temperatures up to the limits of available lubricants. They are the most satisfactory valves for the handling of gritty suspensions and many other destructive, erosive, corrosive and chemical solutions.

R221113-40   Plumbing Approximations for Quick Estimating

Water Control

Water Meter; Backflow Preventer, ... 10 to 15% of Fixtures
Shock Absorbers; Vacuum Breakers;
Mixer.

Pipe And Fittings .... 30 to 60% of Fixtures

Note: Lower percentage for compact buildings or larger buildings with plumbing in one area.
Larger percentage for large buildings with plumbing spread out.
In extreme cases pipe may be more than 100% of fixtures.
Percentages **do not** include special purpose or process piping.

Plumbing Labor

1 & 2 Story Residential .... Rough-in Labor = 80% of Materials
Apartment Buildings .... Rough-in Labor = 90 to 100% of Materials
Labor for handling and placing fixtures is approximately 25 to 30% of fixtures

Quality/Complexity Multiplier (for all installations)
Economy installation, add. .... 0 to 5%
Good quality, medium complexity, add .... 5 to 15%
Above average quality and complexity, add .... 15 to 25%

R221113-50   Pipe Material Considerations

1. Malleable fittings should be used for gas service.
2. Malleable fittings are used where there are stresses/strains due to expansion and vibration.
3. Cast fittings may be broken as an aid to disassembling heating lines frozen by long use, temperature and minerals.
4. A cast iron pipe is extensively used for underground and submerged service.
5. Type M (light wall) copper tubing is available in hard temper only and is used for nonpressure and less severe applications than K and L.

6. Type L (medium wall) copper tubing, available hard or soft for interior service.
7. Type K (heavy wall) copper tubing, available in hard or soft temper for use where conditions are severe. For underground and interior service.
8. Hard drawn tubing requires fewer hangers or supports but should not be bent. Silver brazed fittings are recommended, but soft solder is normally used.
9. Type DMV (very light wall) copper tubing designed for drainage, waste and vent plus other non-critical pressure services.

Domestic/Imported Pipe and Fittings Costs

The prices shown in this publication for steel/cast iron pipe and steel, cast iron, and malleable iron fittings are based on domestic production sold at the normal trade discounts. The above listed items of foreign manufacture may be available at prices 1/3 to 1/2 of those shown. Some imported items after minor machining or finishing operations are being sold as domestic to further complicate the system.

Caution: Most pipe prices in this data set also include a coupling and pipe hangers which for the larger sizes can add significantly to the per foot cost and should be taken into account when comparing "book cost" with the quoted supplier's cost.

R221113-70 Piping to 10' High

When taking off pipe, it is important to identify the different material types and joining procedures, as well as distances between supports and components required for proper support.

During the takeoff, measure through all fittings. Do not subtract the lengths of the fittings, valves, or strainers, etc. This added length plus the final rounding of the totals will compensate for nipples and waste.

When rounding off totals always increase the actual amount to correspond with manufacturer's shipping lengths.

A. Both red brass and yellow brass pipe are normally furnished in 12' lengths, plain end. The Unit Price section includes in the linear foot costs two field threads and one coupling per 10' length. A carbon steel clevis type hanger assembly every 10' is also prorated into the linear foot costs, including both material and labor.

B. Cast iron soil pipe is furnished in either 5' or 10' lengths. For pricing purposes, the Unit Price section features 10' lengths with a joint and a carbon steel clevis hanger assembly every 5' prorated into the per foot costs of both material and labor.

Three methods of joining are considered: lead and oakum poured joints or push-on gasket type joints for the bell and spigot pipe and a joint clamp for the no-hub soil pipe. The labor and material costs for each of these individual joining procedures are also prorated into the linear costs per foot.

C. Copper tubing covers types K, L, M, and DWV which are furnished in 20' lengths. Means pricing data is based on a tubing cut each length and a coupling and two soft soldered joints every 10'. A carbon steel, clevis type hanger assembly every 10' is also prorated into the per foot costs. The prices for refrigeration tubing are for materials only. Labor for full lengths may be based on the type L labor but short cut measures in tight areas can increase the installation labor-hours from 20 to 40%.

D. Corrosion-resistant piping does not lend itself to one particular standard of hanging or support assembly due to its diversity of application and placement. The several varieties of corrosion-resistant piping do not include any material or labor costs for hanger assemblies (See the Unit Price section for appropriate selection).

E. Glass pipe is furnished in standard lengths either 5' or 10' long, beaded on one end. Special orders for diverse lengths beaded on both ends are also available. For pricing purposes, R.S. Means features 10' lengths with a coupling and a carbon steel band hanger assembly every 10' prorated into the per foot linear costs.

Glass pipe is also available with conical ends and standard lengths ranging from 6" through 3' in 6" increments, then up to 10' in 12" increments. Special lengths can be customized for particular installation requirements.

For pricing purposes, Means has based the labor and material pricing on 10' lengths. Included in these costs per linear foot are the prorated costs for a flanged assembly every 10' consisting of two flanges, a gasket, two insertable seals, and the required number of bolts and nuts. A carbon steel band hanger assembly based on 10' center lines has also been prorated into the costs per foot for labor and materials.

F. Plastic pipe of several compositions and joining methods are considered. Fiberglass reinforced pipe (FRP) is priced, based on 10' lengths (20' lengths are also available), with coupling and epoxy joints every 10'. FRP is furnished in both "General Service" and "High Strength." A carbon steel clevis hanger assembly, 3 for every 10', is built into the prorated labor and material costs on a per foot basis.

The PVC and CPVC pipe schedules 40, 80 and 120, plus SDR ratings are all based on 20' lengths with a coupling installed every 10', as well as a carbon steel clevis hanger assembly every 3'. The PVC and

ABS type DWV piping is based on 10' lengths with solvent weld couplings every 10', and with carbon steel clevis hanger assemblies, 3 for every 10'. The rest of the plastic piping in this section is based on flexible 100' coils and does not include any coupling or supports.

This section ends with PVC drain and sewer piping based on 10' lengths with bell and spigot ends and O-ring type, push-on joints.

G. Stainless steel piping includes both weld end and threaded piping, both in the type 304 and 316 specification and in the following schedules, 5, 10, 40, 80, and 160. Although this piping is usually furnished in 20' lengths, this cost grouping has a joint (either heli-arc butt-welded or threads and coupling) every 10'. A carbon steel clevis type hanger assembly is also included at 10' intervals and prorated into the linear foot costs.

H. Carbon steel pipe includes both black and galvanized. This section encompasses schedules 40 (standard) and 80 (extra heavy).

Several common methods of joining steel pipe — such as thread and coupled, butt welded, and flanged (150 lb. weld neck flanges) are also included.

For estimating purposes, it is assumed that the piping is purchased in 20' lengths and that a compatible joint is made up every 10'. These joints are prorated into the labor and material costs per linear foot. The following hanger and support assemblies every 10' are also included: carbon steel clevis for the T & C pipe, and single rod roll type for both the welded and flanged piping. All of these hangers are oversized to accommodate pipe insulation 3/4" thick through 5" pipe size and 1-1/2" thick from 6" through 12" pipe size.

I. Grooved joint steel pipe is priced both black and galvanized, in schedules 10, 40, and 80, furnished in 20' lengths. This section describes two joining methods: cut groove and roll groove. The schedule 10 piping is roll-grooved, while the heavier schedules are cut-grooved. The labor and material costs are prorated into per linear foot prices, including a coupled joint every 10', as well as a carbon steel clevis hanger assembly.

Notes:

The pipe hanger assemblies mentioned in the preceding paragraphs include the described hanger; appropriately sized steel, box-type insert and nut; plus 18" of threaded hanger rod.

C clamps are used when the pipe is to be supported from steel shapes rather than anchored in the slab. C clamps are slightly less costly than inserts. However, to save time in estimating, it is advisable to use the given line number cost, rather than substituting a C clamp for the insert.

Add to piping labor for elevated installation:

10' to 14.5' high	10%	30' to 34.5' high	40%
15' to 19.5' high	20%	35' to 39.5' high	50%
20' to 24.5' high	25%	40' and higher	55%
25' to 29.5' high	35%		

When using the percentage adds for elevated piping installations as shown above, bear in mind that the given heights are for the pipe supports, even though the insert, anchor, or clamp may be several feet higher than the pipe itself.

An allowance has been included in the piping installation time for testing and minor tightening of leaking joints, fittings, stuffing boxes, packing glands, etc. For extraordinary test requirements such as x-rays, prolonged pressure or demonstration tests, a percentage of the piping labor, based on the estimator's experience, must be added to the labor total. A testing service specializing in weld x-rays should be consulted for pricing if it is an estimate requirement. Equipment installation time includes start-up with associated adjustments.

R221316-10   Drainage Fixture Units for Fixtures and Groups

Fixture Type	Drainage Fixture Unit Value as Load Factors	Minimum Size of Trap (inches)
Automatic Clothes Washer, commercial (Note A)	3.0	2.0
Automatic Clothes Washer, residential	2.0	2.0
Full Bathroom Group w/bathtub or shower stall (1.6 gpf WC)	5.0	-
Full Bathroom Group w/bathtub or shower stall (WC flushing greater than 1.6 gpf) (Note D)	6.0	-
Bathtub (with or without overhead shower)	2.0	1.5
Bidet	1.0	1.25
Drinking Fountain	0.5	1025
Dishwasher, domestic	2.0	1.5
Emergency floor drains	0.0	2
Floor drains	2.0	2
Kitchen sinks, domestic	2.0	1.5
Lavatory	1.0	1.25
Shower (5.7 gpm or less flow rate)	2.0	1.5
Shower (5.7 gpm - 12.3 gpm flow rate)	3.0	2
Shower (12.3 gpm - 25.8 gpm flow rate)	5.0	3
Shower (25.8 gpm to 55.6 gpm flow rate)	6.0	4
Service Sink	2.0	1.5
Urinal	4.0	Note C
Urinal (1.0 gpf or less)	2.0	Note C
Urinal (nonwater supplied)	0.5	Note C
Water closet, flushometer tank, public or private	4.0	Note C
Water closet, private (1.6 gpf)	3.0	Note C
Water closet, private (flushing greater than 1.6 gpf)	4.0	Note C
Water closet, public (1.6 gpf)	4.0	Note C
Water closet, public (flushing greater than 1.6 gpf)	6.0	Note C

Notes:
A. A showerhead over a bathtub or whirlpool tub attachment does not increase the drainage fixture unit value.
B. Trap size shall be consistent with the fixture outlet size.
C. For fixtures added to a bathroom group, add the DFU value to those additional fixtures to the bathroom group fixture count.

Excerpted from the 2012 *International Plumbing Code*, Copyright 2011.
Washington, D.C.: International Code Council. Reproduced with permission.
All rights reserved. www.ICCSAFE.org

R221316-15   Drainage Fixture Units for Fixture Drains or Traps

Fixture Drain or Trap Size (inches)	Drainage Fixture Unit Value
1.25	1.0
1.5	2.0
2.0	3.0
2.5	4.0
3.0	5.0
4.0	6.0

Notes:
Drainage Fixture Unit values designate the relative load weight or different kinds of fixtures that shall be emplored in estimating the total load carried by a soil or waste pipe.

Excerpted from the 2012 *International Plumbing Code*, Copyright 2011.
Washington, D.C.: International Code Council. Reproduced with permission.
All rights reserved. www.ICCSAFE.org

R221316-20 Allowable Fixture Units (d.f.u.) for Branches and Stacks

Pipe Diam.	Horiz. Branch (not incl. drains)	Stack Size for 3 Stories or 3 Levels	Stack size for Over 3 levels	Maximum for 1 Story building Stack
1-1/2"	3	4	8	2
2"	6	10	24	6
2-1/2"	12	20	42	9
3"	20*	48*	72*	20*
4"	160	240	500	90
5"	360	540	1100	200
6"	620	960	1900	350
8"	1400	2200	3600	600
10"	2500	3800	5600	1000
12"	3900	6000	8400	1500
15"	7000			

*Not more than two water closets or bathroom groups within each branch interval nor more than six water closets or bathroom groups on the stack.

Stacks sized for the total may be reduced as load decreases at each story to a minimum diameter of 1/2 the maximum diameter.

R224000-10 Water Consumption Rates

Fixture Type	Water Supply Fixture Unit Value		
	Hot Water	Cold Water	Combined
Bathtub	1.0	1.0	1.4
Clothes Washer	1.0	1.0	1.4
Dishwasher	1.4	0.0	1.4
Kitchen Sink	1.0	1.0	1.4
Laundry Tub	1.0	1.0	1.4
Lavatory	0.5	0.5	0.7
Shower Stall	1.0	1.0	1.4
Water Closet (tank type)	0.0	2.2	2.2
Full Bath Group			
w/bathtub or shower stall	1.5	2.7	3.6
Half Bath Group			
w/W.C. and Lavatory	0.5	2.5	2.6
Kitchen Group			
w/Dishwasher and Sink	1.9	1.0	2.5
Laundry Group			
w/Clothes Washer and			
Laundry Tub	1.8	1.8	2.5
Hose bibb (sillcock)	0.0	2.5	2.5

Notes:
Typically, WSFU = 1GPM

Supply loads in the building water-distribution system shall be determined by total load on the pipe being sized, in terms of water supply fixture units (WSFU) and gallons per minute (GPM) flow rates. For fixtures not listed, choose a WSFU value of a fixture with similar flow characteristics. Water Fixture Supply Units determined the required water supply to fixtures and their service systems. Fixture units are equal to (1) cubic foot of water drained in a 1-1/4" pipe per minute. It is not a flow rate unit but a design factor.

Excerpted from the 2012 *International Plumbing Code*, Copyright 2011. Washington, D.C.: International Code Council. Reproduced with permission. All rights reserved. www.ICCSAFE.org

R224000-20   Fixture Demands in Gallons per Fixture per Hour

Table below is based on 140°F final temperature except for dishwashers in public places (*) where 180°F water is mandatory.

Supply Systems for Flush Tanks			Supply Systems for Flushometer Valves		
Load	Demand		Load	Demand	
WSFU	GPM	CU. FT.	WSFU	GPM	CU. FT.
1.0	3.0	0.041040	–	–	–
2.0	5.0	0.068400	–	–	–
3.0	6.5	0.868920	–	–	–
4.0	8.0	1.069440	–	–	–
5.0	9.4	1.256592	5.0	15	2.0052
6.0	10.7	1.430376	6.0	17.4	2.326032
7.0	11.8	1.577424	7.0	19.8	2.646364
8.0	12.8	1.711104	8.0	22.2	2.967696
9.0	13.7	1.831416	9.0	24.6	3.288528
10.0	14.6	1.951728	10.0	27	3.60936
11.0	15.4	2.058672	11.0	27.8	3.716304
12.0	16.0	2.138880	12.0	28.6	3.823248
13.0	16.5	2.205720	13.0	29.4	3.930192
14.0	17.0	2.272560	14.0	30.2	4.037136
15.0	17.5	2.339400	15.0	31	4.14408
16.0	18.0	2.906240	16.0	31.8	4.241024
17.0	18.4	2.459712	17.0	32.6	4.357968
18.0	18.8	2.513184	18.0	33.4	4.464912
19.0	19.2	2.566656	19.0	34.2	4.571856
20.0	19.6	2.620218	20.0	35.0	4.678800
25.0	21.5	2.874120	25.0	38.0	5.079840
30.0	23.3	3.114744	30.0	42.0	5.611356
35.0	24.9	3.328632	35.0	44.0	5.881920
40.0	26.3	3.515784	40.0	46.0	6.149280
45.0	27.7	3.702936	45.0	48.0	6.416640
50.0	29.1	3.890088	50.0	50.0	6.684000

Notes:

When designing a plumbing system that utilizes fixtures other than, or in addition to, water closets, use the data provided in the Supply Systems for Flush Tanks section of the above table.

To obtain the probable maximum demand, multiply the total demands for the fixtures (gal./fixture/hour) by the demand factor. The heater should have a heating capacity in gallons per hour equal to this maximum. The storage tank should have a capacity in gallons equal to the probable maximum demand multiplied by the storage capacity factor.

R224000-30   Minimum Plumbing Fixture Requirements

Classification	Occupancy	Description	Water Closet Male	Water Closet Female	Lavatories Male	Lavatories Female	Bathtubs/Showers	Drinking Fountains	Other
Assembly	A-1	Theaters and other buildings for the performing arts and motion pictures	1:125	1:65	1:200			1:500	1 Service Sink
	A-2	Nightclubs, bars, taverns dance halls	1:40		1:75			1:500	1 Service Sink
		Restaurants, banquet halls, food courts	1:75		1:200			1:500	1 Service Sink
	A-3	Auditorium w/o permanent seating, art galleries, exhibition halls, museums, lecture halls, libraries, arcades & gymnasiums	1:125	1:65	1:200			1:500	1 Service Sink
		Passenger terminals and transportation facilities	1:500		1:750			1:1000	1 Service Sink
		Places of worship and other religious services	1:150	1:75	1:200			1:1000	1 Service Sink
	A-4	Indoor sporting events and activities, coliseums, arenas, skating rinks, pools, and tennis courts	1:75 for the first 1500, then 1:120 for the remainder	1:40 for the first 1520, then 1:60 for the remainder	1:200	1:150		1:1000	1 Service Sink
	A-5	Outdoor sporting events and activities, stadiums, amusement parks, bleachers, grandstands	1:75 for the first 1500, then 1:120 for the remainder	1:40 for the first 1520, then 1:60 for the remainder	1:200	1:150		1:1000	1 Service Sink
Business	B	Buildings for the transaction of business, professional services, other services involving merchandise, office buildings, banks, light industrial	1:25 for the first 50, then 1:50 for the remainder		1:40 for the first 80, then 1:80 for the remainder			1:100	1 Service Sink
Educational	E	Educational Facilities	1:50		1:50			1:100	1 Service Sink
Factory and industrial	F-1 and F-2	Structures in which occupants are engaged in work fabricating, assembly or processing of products or materials	1:100		1:100		See International Plumbing Code	1:400	1 Service Sink
Institutional	I-1	Residential Care	1:10		1:10		1:8	1:100	1 Service Sink
	I-2	Hospitals, ambulatory nursing home care recipient	1 per room		1 per room		1:15	1:100	1 Service Sink
		Employees, other than residential care	1:25		1:35			1:100	
		Visitors, other than residential care	1:75		1:100			1:500	
	I-3	Prisons	1 per cell		1 per cell		1:15	1:100	1 Service Sink
		Reformatories, detention and correction centers	1:15		1:15		1:15	1:100	1 Service Sink
		Employees	1:25		1:35			1:100	
	I-4	Adult and child day care	1:15		1:15		1	1:100	1 Service Sink
Mercantile	M	Retail stores, service stations, shops, salesrooms, markets and shopping centers	1:500		1:750			1:1000	1 Service Sink
Residential	R-1	Hotels, Motels, boarding houses (transient)	1 per sleeping unit		1 per sleeping unit		1 per sleeping unit		1 Service Sink
	R-2	Dormitories, fraternities, sororities and boarding houses (not transient)	1:10		1:10		1:8	1:100	1 Service Sink
		Apartment House	1 per dwelling unit		1 per dwelling unit		1 per dwelling unit		1 Kitchen sink per dwelling; 1 clothes washer connection per 20 dwellings
	R-3	1 and 2 Family Dwellings	1 per dwelling unit		1:10		1 per dwelling unit		1 Kitchen sink per dwelling; 1 clothes washer connection per dwelling
	R-3	Congregate living facilities w/<16 people	1:10		1:10		1:8	1:100	1 Service Sink
	R-4	Congregate living facilities w/<16 people	1:10		1:10		1:8	1:100	1 Service Sink
Storage	S-1 and S-2	Structures for the storage of good, warehouses, storehouses and freight depots, low and moderate hazard	1:100		1:100		See International Plumbing Code	1;1000	1 Service Sink

Table 2902.1

Fig. R238313-11

R230500-10   Subcontractors

On the unit cost pages of the R.S. Means Cost Data data set, the last column is entitled "Total Incl. O&P". This is normally the cost of the installing contractor. In the HVAC Division, this is the cost of the mechanical contractor. If the particular work being estimated is to be performed by a sub to the mechanical contractor, the mechanical's profit and handling charge (usually 10%) is added to the total of the last column.

R235616-60   Solar Heating (Space and Hot Water)

Collectors should face as close to due South as possible, but variations of up to 20 degrees on either side of true South are acceptable. Local climate and collector type may influence the choice between east or west deviations. Obviously they should be located so they are not shaded from the sun's rays. Incline collectors at a slope of latitude minus 5 degrees for domestic hot water and latitude plus 15 degrees for space heating.

Flat plate collectors consist of a number of components as follows: Insulation to reduce heat loss through the bottom and sides of the collector. The enclosure which contains all the components in this assembly is usually weatherproof and prevents dust, wind and water from coming in contact with the absorber plate. The cover plate usually consists of one or more layers of a variety of glass or plastic and reduces the reradiation by creating an air space which traps the heat between the cover and the absorber plates.

The absorber plate must have a good thermal bond with the fluid passages. The absorber plate is usually metallic and treated with a surface coating which improves absorptivity. Black or dark paints or selective coatings are used for this purpose, and the design of this passage and plate combination helps determine a solar system's effectiveness.

Heat transfer fluid passage tubes are attached above and below or integral with an absorber plate for the purpose of transferring thermal energy from the absorber plate to a heat transfer medium. The heat exchanger is a device for transferring thermal energy from one fluid to another.

Piping and storage tanks should be well insulated to minimize heat losses.

Size domestic water heating storage tanks to hold 20 gallons of water per user, minimum, plus 10 gallons per dishwasher or washing machine. For domestic water heating an optimum collector size is approximately 3/4 square foot of area per gallon of water storage. For space heating of residences and small commercial applications the collector is commonly sized between 30% and 50% of the internal floor area. For space heating of large commercial applications, collector areas less than 30% of the internal floor area can still provide significant heat reductions.

A supplementary heat source is recommended for Northern states for December through February.

The solar energy transmission per square foot of collector surface varies greatly with the material used. Initial cost, heat transmittance and useful life are obviously interrelated.

R238313-10   Heat Trace Systems

Before you can determine the cost of a HEAT TRACE installation, the method of attachment must be established. There are (4) common methods:

1. Cable is simply attached to the pipe with polyester tape every 12'.
2. Cable is attached with a continuous cover of 2" wide aluminum tape.
3. Cable is attached with factory extruded heat transfer cement and covered with metallic raceway with clips every 10'.
4. Cable is attached between layers of pipe insulation using either clips or polyester tape.

In all of the above methods each component of the system must be priced individually.

Example: Components for method 3 must include:

A. Heat trace cable by voltage and watts per linear foot.
B. Heat transfer cement, 1 gallon per 60 linear feet of cover.
C. Metallic raceway by size and type.
D. Raceway clips by size of pipe.

When taking off linear foot lengths of cable add the following for each valve in the system. (E)

SCREWED OR WELDED VALVE:		
1/2"	=	6"
3/4"	=	9"
1"	=	1' -0"
1-1/2"	=	1' -6"
2"	=	2'
2-1/2"	=	2' -6"
3"	=	2' -6"
4"	=	4' -0"
6"	=	7' -0"
8"	=	9' -6"
10"	=	12' -6"
12"	=	15' -0"
14"	=	18' -0"
16"	=	21' -6"
18"	=	25' -6"
20"	=	28' -6"
24"	=	34' -0"
30"	=	40' -0"

FLANGED VALVE:		
1/2"	=	1' -0"
3/4"	=	1' -6"
1"	=	2' -0"
1-1/2"	=	2' -6"
2"	=	2' -6"
2-1/2"	=	3' -0"
3"	=	3' -6"
4"	=	4' -0"
6"	=	8' -0"
8"	=	11' -0"
10"	=	14' -0"
12"	=	16' -6"
14"	=	19' -6"
16"	=	23' -0"
18"	=	27' -0"
20"	=	30' -0"
24"	=	36' -0"
30"	=	42' -0"

BUTTERFLY VALVES:		
1/2"	=	0'
3/4"	=	0'
1"	=	1' -0"
1-1/2"	=	1' -6"
2"	=	2' -0"
2-1/2"	=	2' -6"
3"	=	2' -6"
4"	=	3' -0"
6"	=	3' -6"
8"	=	4' -0"
10"	=	4' -0"
12"	=	5' -0"
14"	=	5' -6"
16"	=	6' -0"
18"	=	6' -6"
20"	=	7' -0"
24"	=	8' -0"
30"	=	10' -0"

R238313-10 Heat Trace Systems (cont.)

Add the following quantities of heat transfer cement to linear foot totals for each valve:

Nominal Valve Size	Gallons of Cement per Valve
1/2"	0.14
3/4"	0.21
1"	0.29
1-1/2"	0.36
2"	0.43
2-1/2"	0.70
3"	0.71
4"	1.00
6"	1.43
8"	1.48
10"	1.50
12"	1.60
14"	1.75
16"	2.00
18"	2.25
20"	2.50
24"	3.00
30"	3.75

The following must be added to the list of components to accurately price HEAT TRACE systems:

1. Expediter fitting and clamp fasteners (F)
2. Junction box and nipple connected to expediter fitting (G)
3. Field installed terminal blocks within junction box
4. Ground lugs
5. Piping from power source to expediter fitting
6. Controls
7. Thermostats
8. Branch wiring
9. Cable splices
10. End of cable terminations
11. Branch piping fittings and boxes

Deduct the following percentages from labor if cable lengths in the same area exceed:

150' to 250'	10%	351' to 500'	20%
251' to 350'	15%	Over 500'	25%

Add the following percentages to labor for elevated installations:

15' to 20' high	10%	31' to 35' high	40%
21' to 25' high	20%	36' to 40' high	50%
26' to 30' high	30%	Over 40' high	60%

R238313-20 Spiral-Wrapped Heat Trace Cable (Pitch Table)

In order to increase the amount of heat, occasionally heat trace cable is wrapped in a spiral fashion around a pipe; increasing the number of feet of heater cable per linear foot of pipe.

Engineers first determine the heat loss per foot of pipe (based on the insulating material, its thickness, and the temperature differential across it). A ratio is then calculated by the formula:

$$\text{Feet of Heat Trace per Foot of Pipe} = \frac{\text{Watts/Foot of Heat Loss}}{\text{Watts/Foot of the Cable}}$$

The linear distance between wraps (pitch) is then taken from a chart or table. Generally, the pitch is listed on a drawing leaving the estimator to calculate the total length of heat tape required. An approximation may be taken from this table.

Feet of Heat Trace Per Foot of Pipe

Pitch In Inches	1	1¼	1½	2	2½	3	4	6	8	10	12	14	16	18	20	24
3.5	1.80															
4	1.65															
5	1.46	1.60	1.80													
6	1.34	1.45	1.55	1.75												
7	1.25	1.35	1.43	1.57	1.75											
8	1.20	1.28	1.34	1.45	1.60	1.80										
9	1.16	1.23	1.28	1.37	1.51	1.68										
10	1.13	1.19	1.24	1.32	1.44	1.57	1.82									
15	1.06	1.08	1.10	1.15	1.21	1.29	1.42	1.78								
20	1.04	1.05	1.06	1.08	1.13	1.17	1.25	1.49	1.73							
25		1.04	1.04	1.06	1.08	1.11	1.17	1.33	1.51	1.72						
30				1.04	1.05	1.07	1.12	1.24	1.37	1.54	1.70	1.80				
35					1.06	1.06	1.09	1.17	1.28	1.42	1.54	1.64	1.78			
40						1.05	1.07	1.14	1.22	1.33	1.44	1.52	1.64	1.75		
50							1.05	1.09	1.15	1.22	1.29	1.35	1.44	1.53	1.64	1.83
60								1.06	1.11	1.16	1.21	1.25	1.31	1.39	1.46	1.62
70								1.05	1.08	1.12	1.17	1.19	1.24	1.30	1.35	1.47
80									1.06	1.09	1.13	1.15	1.19	1.24	1.28	1.38
90									1.04	1.06	1.10	1.13	1.16	1.19	1.23	1.32
100										1.05	1.08	1.10	1.13	1.15	1.19	1.23

Note: Common practice would normally limit the lower end of the table to 5% of additional heat and above 80% an engineer would likely opt for two (2) parallel cables.

For customer support on your Plumbing Costs with RSMeans Data, call 800.448.8182.

717

R260519-90 Wire

Wire quantities are taken off by either measuring each cable run or by extending the conduit and raceway quantities times the number of conductors in the raceway. Ten percent should be added for waste and tie-ins. Keep in mind that the unit of measure of wire is C.L.F. not L.F. as in raceways so the formula would read:

$$\frac{\text{(L.F. Raceway x No. of Conductors) x 1.10}}{100} = \text{C.L.F.}$$

Price per C.L.F. of wire includes:
1. Setting up wire coils or spools on racks
2. Attaching wire to pull in means
3. Measuring and cutting wire
4. Pulling wire into a raceway
5. Identifying and tagging

Price does not include:
1. Connections to breakers, panelboards, or equipment
2. Splices

Job Conditions: Productivity is based on new construction to a height of 15' using rolling staging in an unobstructed area. Material staging is assumed to be within 100' of work being performed.

Economy of Scale: If more than three wires at a time are being pulled, deduct the following percentages from the labor of that grouping:

4-5 wires	25%
6-10 wires	30%
11-15 wires	35%
over 15	40%

If a wire pull is less than 100' in length and is interrupted several times by boxes, lighting outlets, etc., it may be necessary to add the following lengths to each wire being pulled:

Junction box to junction box	2 L.F.
Lighting panel to junction box	6 L.F.
Distribution panel to sub panel	8 L.F.
Switchboard to distribution panel	12 L.F.
Switchboard to motor control center	20 L.F.
Switchboard to cable tray	40 L.F.

Measure of Drops and Riser: It is important when taking off wire quantities to include the wire for drops to electrical equipment. If heights of electrical equipment are not clearly stated, use the following guide:

	Bottom A.F.F.	Top A.F.F.	Inside Cabinet
Safety switch to 100A	5'	6'	2'
Safety switch 400 to 600A	4'	6'	3'
100A panel 12 to 30 circuit	4'	6'	3'
42 circuit panel	3'	6'	4'
Switch box	3'	3'6"	1'
Switchgear	0'	8'	8'
Motor control centers	0'	8'	8'
Transformers - wall mount	4'	8'	2'
Transformers - floor mount	0'	12'	4'

R312316-45   Excavating Equipment

The table below lists theoretical hourly production in C.Y./hr. bank measure for some typical excavation equipment. Figures assume 50 minute hours, 83% job efficiency, 100% operator efficiency, 90° swing and properly sized hauling units, which must be modified for adverse digging and loading conditions. Actual production costs in the front of the data set average about 50% of the theoretical values listed here.

Equipment	Soil Type	B.C.Y. Weight	% Swell	1 C.Y.	1-1/2 C.Y.	2 C.Y.	2-1/2 C.Y.	3 C.Y.	3-1/2 C.Y.	4 C.Y.
Hydraulic Excavator "Backhoe" 15' Deep Cut	Moist loam, sandy clay	3400 lb.	40%	165	195	200	275	330	385	440
	Sand and gravel	3100	18	140	170	225	240	285	330	380
	Common earth	2800	30	150	180	230	250	300	350	400
	Clay, hard, dense	3000	33	120	140	190	200	240	260	320
Power Shovel Optimum Cut (Ft.)	Moist loam, sandy clay	3400	40	170 (6.0)	245 (7.0)	295 (7.8)	335 (8.4)	385 (8.8)	435 (9.1)	475 (9.4)
	Sand and gravel	3100	18	165 (6.0)	225 (7.0)	275 (7.8)	325 (8.4)	375 (8.8)	420 (9.1)	460 (9.4)
	Common earth	2800	30	145 (7.8)	200 (9.2)	250 (10.2)	295 (11.2)	335 (12.1)	375 (13.0)	425 (13.8)
	Clay, hard, dense	3000	33	120 (9.0)	175 (10.7)	220 (12.2)	255 (13.3)	300 (14.2)	335 (15.1)	375 (16.0)
Drag Line Optimum Cut (Ft.)	Moist loam, sandy clay	3400	40	130 (6.6)	180 (7.4)	220 (8.0)	250 (8.5)	290 (9.0)	325 (9.5)	385 (10.0)
	Sand and gravel	3100	18	130 (6.6)	175 (7.4)	210 (8.0)	245 (8.5)	280 (9.0)	315 (9.5)	375 (10.0)
	Common earth	2800	30	110 (8.0)	160 (9.0)	190 (9.9)	220 (10.5)	250 (11.0)	280 (11.5)	310 (12.0)
	Clay, hard, dense	3000	33	90 (9.3)	130 (10.7)	160 (11.8)	190 (12.3)	225 (12.8)	250 (13.3)	280 (12.0)

				Wheel Loaders				Track Loaders		
				3 C.Y.	4 C.Y.	6 C.Y.	8 C.Y.	2-1/4 C.Y.	3 C.Y.	4 C.Y.
Loading Tractors	Moist loam, sandy clay	3400	40	260	340	510	690	135	180	250
	Sand and gravel	3100	18	245	320	480	650	130	170	235
	Common earth	2800	30	230	300	460	620	120	155	220
	Clay, hard, dense	3000	33	200	270	415	560	110	145	200
	Rock, well-blasted	4000	50	180	245	380	520	100	130	180

R312319-90   Wellpoints

A single stage wellpoint system is usually limited to dewatering an average 15' depth below normal ground water level. Multi-stage systems are employed for greater depth with the pumping equipment installed only at the lowest header level. Ejectors with unlimited lift capacity can be economical when two or more stages of wellpoints can be replaced or when horizontal clearance is restricted, such as in deep trenches or tunneling projects, and where low water flows are expected. Wellpoints are usually spaced on 2-1/2' to 10' centers along a header pipe. Wellpoint spacing, header size, and pump size are all determined by the expected flow as dictated by soil conditions.

In almost all soils encountered in wellpoint dewatering, the wellpoints may be jetted into place. Cemented soils and stiff clays may require sand wicks about 12" in diameter around each wellpoint to increase efficiency and eliminate weeping into the excavation. These sand wicks require 1/2 to 3 C.Y. of washed filter sand and are installed by using a 12" diameter steel casing and hole puncher jetted into the ground 2' deeper than the wellpoint. Rock may require predrilled holes.

Labor required for the complete installation and removal of a single stage wellpoint system is in the range of 3/4 to 2 labor-hours per linear foot of header, depending upon jetting conditions, wellpoint spacing, etc.

Continuous pumping is necessary except in some free draining soil where temporary flooding is permissible (as in trenches which are backfilled after each day's work). Good practice requires provision of a stand-by pump during the continuous pumping operation.

Systems for continuous trenching below the water table should be installed three to four times the length of expected daily progress to ensure uninterrupted digging, and header pipe size should not be changed during the job.

For pervious free draining soils, deep wells in place of wellpoints may be economical because of lower installation and maintenance costs. Daily production ranges between two to three wells per day, for 25' to 40' depths, to one well per day for depths over 50'.

Detailed analysis and estimating for any dewatering problem is available at no cost from wellpoint manufacturers. Major firms will quote "sufficient equipment" quotes or their affiliates will offer lump sum proposals to cover complete dewatering responsibility.

Description for 200' System with 8" Header		Quantities
Equipment & Material	Wellpoints 25' long, 2" diameter @ 5' O.C.	40 Each
	Header pipe, 8" diameter	200 L.F.
	Discharge pipe, 8" diameter	100 L.F.
	8" valves	3 Each
	Combination jetting & wellpoint pump (standby)	1 Each
	Wellpoint pump, 8" diameter	1 Each
	Transportation to and from site	1 Day
	Fuel for 30 days x 60 gal./day	1800 Gallons
	Lubricants for 30 days x 16 lbs./day	480 Lbs.
	Sand for points	40 C.Y.
Labor	Technician to supervise installation	1 Week
	Labor for installation and removal of system	300 Labor-hours
	4 Operators straight time 40 hrs./wk. for 4.33 wks.	693 Hrs.
	4 Operators overtime 2 hrs./wk. for 4.33 wks.	35 Hrs.

Earthwork

R3141 Shoring

Reference Tables

R314116-40 Wood Sheet Piling

Wood sheet piling may be used for depths to 20' where there is no ground water. If moderate ground water is encountered Tongue & Groove

sheeting will help to keep it out. When considerable ground water is present, steel sheeting must be used.

For estimating purposes on trench excavation, sizes are as follows:

Depth	Sheeting	Wales	Braces	B.F. per S.F.
To 8'	3 x 12's	6 x 8's, 2 line	6 x 8's, @ 10'	4.0 @ 8'
8' x 12'	3 x 12's	10 x 10's, 2 line	10 x 10's, @ 9'	5.0 average
12' to 20'	3 x 12's	12 x 12's, 3 line	12 x 12's, @ 8'	7.0 average

Sheeting to be toed in at least 2' depending upon soil conditions. A five person crew with an air compressor and sheeting driver can drive and brace 440 SF/day at 8' deep, 360 SF/day at 12' deep, and 320 SF/day at 16' deep.

For normal soils, piling can be pulled in 1/3 the time to install. Pulling difficulty increases with the time in the ground. Production can be increased by high pressure jetting.

R314116-45 Steel Sheet Piling

Limiting weights are 22 to 38#/S.F. of wall surface with 27#/S.F. average for usual types and sizes. (Weights of piles themselves are from 30.7#/L.F. to 57#/L.F. but they are 15" to 21" wide.) Lightweight sections 12" to 28" wide from 3 ga. to 12 ga. thick are also available for shallow excavations. Piles may be driven two at a time with an impact or vibratory hammer (use vibratory to pull) hung from a crane without leads. A reasonable estimate of the life of steel sheet piling is 10 uses with up to 125 uses possible if a vibratory hammer is used. Used piling costs from 50% to 80% of new piling depending on location and market conditions. Sheet piling and H piles

can be rented for about 30% of the delivered mill price for the first month and 5% per month thereafter. Allow 1 labor-hour per pile for cleaning and trimming after driving. These costs increase with depth and hydrostatic head. Vibratory drivers are faster in wet granular soils and are excellent for pile extraction. Pulling difficulty increases with the time in the ground and may cost more than driving. It is often economical to abandon the sheet piling, especially if it can be used as the outer wall form. Allow about 1/3 additional length or more for toeing into ground. Add bracing, waler and strut costs. Waler costs can equal the cost per ton of sheeting.

Utilities

R3311 Water Utility Distribution Piping

R331113-80 Piping Designations

There are several systems currently in use to describe pipe and fittings. The following paragraphs will help to identify and clarify classifications of piping systems used for water distribution.

Piping may be classified by schedule. Piping schedules include 5S, 10S, 10, 20, 30, Standard, 40, 60, Extra Strong, 80, 100, 120, 140, 160 and Double Extra Strong. These schedules are dependent upon the pipe wall thickness. The wall thickness of a particular schedule may vary with pipe size.

Ductile iron pipe for water distribution is classified by Pressure Classes such as Class 150, 200, 250, 300 and 350. These classes are actually the rated water working pressure of the pipe in pounds per square inch (psi). The pipe in these pressure classes is designed to withstand the rated water working pressure plus a surge allowance of 100 psi.

The American Water Works Association (AWWA) provides standards for various types of **plastic pipe.** C-900 is the specification for polyvinyl chloride (PVC) piping used for water distribution in sizes ranging from 4" through 12". C-901 is the specification for polyethylene (PE) pressure pipe, tubing and fittings used for water distribution in sizes ranging from 1/2" through 3". C-905 is the specification for PVC piping sizes 14" and greater.

PVC pressure-rated pipe is identified using the standard dimensional ratio (SDR) method. This method is defined by the American Society for Testing and Materials (ASTM) Standard D 2241. This pipe is available in SDR numbers 64, 41, 32.5, 26, 21, 17, and 13.5. Pipe with an SDR of 64 will have the thinnest wall while pipe with an SDR of 13.5 will have the thickest wall. When the pressure rating (PR) of a pipe is given in psi, it is based on a line supplying water at 73 degrees F.

The National Sanitation Foundation (NSF) seal of approval is applied to products that can be used with potable water. These products have been tested to ANSI/NSF Standard 14.

Valves and strainers are classified by American National Standards Institute (ANSI) Classes. These Classes are 125, 150, 200, 250, 300, 400, 600, 900, 1500 and 2500. Within each class there is an operating pressure range dependent upon temperature. Design parameters should be compared to the appropriate material dependent, pressure-temperature rating chart for accurate valve selection.

For customer support on your Plumbing Costs with RSMeans Data, call 800.448.8182.

721

Change Orders

Change Order Considerations

A change order is a written document usually prepared by the design professional and signed by the owner, the architect/engineer, and the contractor. A change order states the agreement of the parties to: an addition, deletion, or revision in the work; an adjustment in the contract sum, if any; or an adjustment in the contract time, if any. Change orders, or "extras", in the construction process occur after execution of the construction contract and impact architects/engineers, contractors, and owners.

Change orders that are properly recognized and managed can ensure orderly, professional, and profitable progress for everyone involved in the project. There are many causes for change orders and change order requests. In all cases, change orders or change order requests should be addressed promptly and in a precise and prescribed manner. The following paragraphs include information regarding change order pricing and procedures.

The Causes of Change Orders

Reasons for issuing change orders include:

- Unforeseen field conditions that require a change in the work
- Correction of design discrepancies, errors, or omissions in the contract documents
- Owner-requested changes, either by design criteria, scope of work, or project objectives
- Completion date changes for reasons unrelated to the construction process
- Changes in building code interpretations, or other public authority requirements that require a change in the work
- Changes in availability of existing or new materials and products

Procedures

Properly written contract documents must include the correct change order procedures for all parties—owners, design professionals, and contractors—to follow in order to avoid costly delays and litigation.

Being "in the right" is not always a sufficient or acceptable defense. The contract provisions requiring notification and documentation must be adhered to within a defined or reasonable time frame.

The appropriate method of handling change orders is by a written proposal and acceptance by all parties involved. Prior to starting work on a project, all parties should identify their authorized agents who may sign and accept change orders, as well as any limits placed on their authority.

Time may be a critical factor when the need for a change arises. For such cases, the contractor might be directed to proceed on a "time and materials" basis, rather than wait for all paperwork to be processed—a delay that could impede progress. In this situation, the contractor must still follow the prescribed change order procedures including, but not limited to, notification and documentation.

Lack of documentation can be very costly, especially if legal judgments are to be made, and if certain field personnel are no longer available. For time and material change orders, the contractor should keep accurate daily records of all labor and material allocated to the change.

Owners or awarding authorities who do considerable and continual building construction (such as the federal government) realize the inevitability of change orders for numerous reasons, both predictable and unpredictable. As a result, the federal government, the American Institute of Architects (AIA), the Engineers Joint Contract Documents Committee (EJCDC), and other contractor, legal, and technical organizations have developed standards and procedures to be followed by all parties to achieve contract continuance and timely completion, while being financially fair to all concerned.

Pricing Change Orders

When pricing change orders, regardless of their cause, the most significant factor is when the change occurs. The need for a change may be perceived in the field or requested by the architect/engineer *before* any of the actual installation has begun, or may evolve or appear *during* construction when the item of work in question is partially installed. In the latter cases, the original sequence of construction is disrupted, along with all contiguous and supporting systems. Change orders cause the greatest impact when they occur *after* the installation has been completed and must be uncovered, or even replaced. Post-completion changes may be caused by necessary design changes, product failure, or changes in the owner's requirements that are not discovered until the building or the systems begin to function.

Specified procedures of notification and record keeping must be adhered to and enforced regardless of the stage of construction: *before, during,* or *after* installation. Some bidding documents anticipate change orders by requiring that unit prices including overhead and profit percentages—for additional as well as deductible changes—be listed. Generally these unit prices do not fully take into account the ripple effect, or impact on other trades, and should be used for general guidance only.

When pricing change orders, it is important to classify the time frame in which the change occurs. There are two basic time frames for change orders: *pre-installation change orders*, which occur before the start of construction, and *post-installation change orders*, which involve reworking after the original installation. Change orders that occur between these stages may be priced according to the extent of work completed using a combination of techniques developed for pricing *pre-* and *post-installation* changes.

Factors To Consider When Pricing Change Orders

As an estimator begins to prepare a change order, the following questions should be reviewed to determine their impact on the final price.

General

- *Is the change order work* pre-installation *or* post-installation?

  Change order work costs vary according to how much of the installation has been completed. Once workers have the project scoped in their minds, even though they have not started, it can be difficult to refocus. Consequently they may spend more than the normal amount of time understanding the change. Also, modifications to work in place, such as trimming or refitting, usually take more time than was initially estimated. The greater the amount of work in place, the more reluctant workers are to change it. Psychologically they may resent the change and as a result the rework takes longer than normal. Post-installation change order estimates must include demolition of existing work as required to accomplish the change. If the work is performed at a later time, additional obstacles, such as building finishes, may be present which must be protected. Regardless of whether the change occurs

722

pre-installation or post-installation, attempt to isolate the identifiable factors and price them separately. For example, add shipping costs that may be required pre-installation or any demolition required post-installation. Then analyze the potential impact on productivity of psychological and/or learning curve factors and adjust the output rates accordingly. One approach is to break down the typical workday into segments and quantify the impact on each segment.

Change Order Installation Efficiency

The labor-hours expressed (for new construction) are based on average installation time, using an efficiency level. For change order situations, adjustments to this efficiency level should reflect the daily labor-hour allocation for that particular occurrence.

- *Will the change substantially delay the original completion date?*

  A significant change in the project may cause the original completion date to be extended. The extended schedule may subject the contractor to new wage rates dictated by relevant labor contracts. Project supervision and other project overhead must also be extended beyond the original completion date. The schedule extension may also put installation into a new weather season. For example, underground piping scheduled for October installation was delayed until January. As a result, frost penetrated the trench area, thereby changing the degree of difficulty of the task. Changes and delays may have a ripple effect throughout the project. This effect must be analyzed and negotiated with the owner.

- *What is the net effect of a deduct change order?*

  In most cases, change orders resulting in a deduction or credit reflect only bare costs. The contractor may retain the overhead and profit based on the original bid.

Materials

- *Will you have to pay more or less for the new material, required by the change order, than you paid for the original purchase?*

  The same material prices or discounts will usually apply to materials purchased for change orders as new construction. In some

instances, however, the contractor may forfeit the advantages of competitive pricing for change orders. Consider the following example:

A contractor purchased over $20,000 worth of fan coil units for an installation and obtained the maximum discount. Some time later it was determined the project required an additional matching unit. The contractor has to purchase this unit from the original supplier to ensure a match. The supplier at this time may not discount the unit because of the small quantity, and he is no longer in a competitive situation. The impact of quantity on purchase can add between 0% and 25% to material prices and/or subcontractor quotes.

- *If materials have been ordered or delivered to the job site, will they be subject to a cancellation charge or restocking fee?*

  Check with the supplier to determine if ordered materials are subject to a cancellation charge. Delivered materials not used as a result of a change order may be subject to a restocking fee if returned to the supplier. Common restocking charges run between 20% and 40%. Also, delivery charges to return the goods to the supplier must be added.

Labor

- *How efficient is the existing crew at the actual installation?*

  Is the same crew that performed the initial work going to do the change order? Possibly the change consists of the installation of a unit identical to one already installed; therefore, the change should take less time. Be sure to consider this potential productivity increase and modify the productivity rates accordingly.

- *If the crew size is increased, what impact will that have on supervision requirements?*

  Under most bargaining agreements or management practices, there is a point at which a working foreman is replaced by a nonworking foreman. This replacement increases project overhead by adding a nonproductive worker. If additional workers are added to accelerate the project or to perform changes while maintaining the schedule, be sure to add additional supervision time if warranted. Calculate the

hours involved and the additional cost directly if possible.

- *What are the other impacts of increased crew size?*

  The larger the crew, the greater the potential for productivity to decrease. Some of the factors that cause this productivity loss are: overcrowding (producing restrictive conditions in the working space) and possibly a shortage of any special tools and equipment required. Such factors affect not only the crew working on the elements directly involved in the change order, but other crews whose movements may also be hampered. As the crew increases, check its basic composition for changes by the addition or deletion of apprentices or nonworking foreman, and quantify the potential effects of equipment shortages or other logistical factors.

- *As new crews, unfamiliar with the project, are brought onto the site, how long will it take them to become oriented to the project requirements?*

  The orientation time for a new crew to become 100% effective varies with the site and type of project. Orientation is easiest at a new construction site and most difficult at existing, very restrictive renovation sites. The type of work also affects orientation time. When all elements of the work are exposed, such as concrete or masonry work, orientation is decreased. When the work is concealed or less visible, such as existing electrical systems, orientation takes longer. Usually orientation can be accomplished in one day or less. Costs for added orientation should be itemized and added to the total estimated cost.

- *How much actual production can be gained by working overtime?*

  Short term overtime can be used effectively to accomplish more work in a day. However, as overtime is scheduled to run beyond several weeks, studies have shown marked decreases in output. The following chart shows the effect of long term overtime on worker efficiency. If the anticipated change requires extended overtime to keep the job on schedule, these factors can be used as a guide to predict the impact on time and cost. Add project overhead, particularly supervision, that may also be incurred.

For customer support on your Plumbing Costs with RSMeans Data, call 800.448.8182.

723

Days per Week	Hours per Day	Production Efficiency					Payroll Cost Factors	
		1st Week	2nd Week	3rd Week	4th Week	Average 4 Weeks	@ 1-1/2 Times	@ 2 Times
5	8	100%	100%	100%	100%	100%	100%	100%
	9	100	100	95	90	96	1.056	1.111
	10	100	95	90	85	93	1.100	1.200
	11	95	90	75	65	81	1.136	1.273
	12	90	85	70	60	76	1.167	1.333
6	8	100	100	95	90	96	1.083	1.167
	9	100	95	90	85	93	1.130	1.259
	10	95	90	85	80	88	1.167	1.333
	11	95	85	70	65	79	1.197	1.394
	12	90	80	65	60	74	1.222	1.444
7	8	100	95	85	75	89	1.143	1.286
	9	95	90	80	70	84	1.183	1.365
	10	90	85	75	65	79	1.214	1.429
	11	85	80	65	60	73	1.240	1.481
	12	85	75	60	55	69	1.262	1.524

Effects of Overtime

Caution: Under many labor agreements, Sundays and holidays are paid at a higher premium than the normal overtime rate.

The use of long-term overtime is counterproductive on almost any construction job; that is, the longer the period of overtime, the lower the actual production rate. Numerous studies have been conducted, and while they have resulted in slightly different numbers, all reach the same conclusion. The figure above tabulates the effects of overtime work on efficiency.

As illustrated, there can be a difference between the *actual* payroll cost per hour and the *effective* cost per hour for overtime work. This is due to the reduced production efficiency with the increase in weekly hours beyond 40. This difference between actual and effective cost results from overtime work over a prolonged period. Short-term overtime work does not result in as great a reduction in efficiency and, in such cases, effective cost may not vary significantly from the actual payroll cost. As the total hours per week are increased on a regular basis, more time is lost due to fatigue, lowered morale, and an increased accident rate.

As an example, assume a project where workers are working 6 days a week, 10 hours per day. From the figure above (based on productivity studies), the average effective productive hours over a 4-week period are:

$$0.875 \times 60 = 52.5$$

Depending upon the locale and day of week, overtime hours may be paid at time and a half or double time. For time and a half, the overall (average) *actual* payroll cost (including regular and overtime hours) is determined as follows:

$$\frac{40 \text{ reg. hrs.} + (20 \text{ overtime hrs.} \times 1.5)}{60 \text{ hrs.}} = 1.167$$

Based on 60 hours, the payroll cost per hour will be 116.7% of the normal rate at 40 hours per week. However, because the effective production (efficiency) for 60 hours is reduced to the equivalent of 52.5 hours, the effective cost of overtime is calculated as follows:

For time and a half:

$$\frac{40 \text{ reg. hrs.} + (20 \text{ overtime hrs.} \times 1.5)}{52.5 \text{ hrs.}} = 1.33$$

The installed cost will be 133% of the normal rate (for labor).

Thus, when figuring overtime, the actual cost per unit of work will be higher than the apparent overtime payroll dollar increase, due to the reduced productivity of the longer work week. These efficiency calculations are true only for those cost factors determined by hours worked. Costs that are applied weekly or monthly, such as equipment rentals, will not be similarly affected.

Equipment

- *What equipment is required to complete the change order?*

Change orders may require extending the rental period of equipment already on the job site, or the addition of special equipment brought in to accomplish the change work. In either case, the additional rental charges and operator labor charges must be added.

Summary

The preceding considerations and others you deem appropriate should be analyzed and applied to a change order estimate. The impact of each should be quantified and listed on the estimate to form an audit trail.

Change orders that are properly identified, documented, and managed help to ensure the orderly, professional, and profitable progress of the work. They also minimize potential claims or disputes at the end of the project.

Back by customer demand!

You asked and we listened. For customer convenience and estimating ease, we have made the 2021 Project Costs available for download at **RSMeans.com/2021books**. You will also find sample estimates, an RSMeans data overview video, and a book registration form to receive quarterly data updates throughout 2021.

Estimating Tips

- The cost figures available in the download were derived from hundreds of projects contained in the RSMeans database of completed construction projects. They include the contractor's overhead and profit. The figures have been adjusted to January of the current year.

- These projects were located throughout the U.S. and reflect a tremendous variation in square foot (S.F.) costs. This is due to differences, not only in labor and material costs, but also in individual owners' requirements. For instance, a bank in a large city would have different features than one in a rural area. This is true of all the different types of buildings analyzed. Therefore, caution should be exercised when using these Project Costs. For example, for courthouses, costs in the database are local courthouse costs and will not apply to the larger, more elaborate federal courthouses.

- None of the figures "go with" any others. All individual cost items were computed and tabulated separately. Thus, the sum of the median figures for plumbing, HVAC, and electrical will not normally total up to the total mechanical and electrical costs arrived at by separate analysis and tabulation of the projects.

- Each building was analyzed as to total and component costs and percentages. The figures were arranged in ascending order with the results tabulated as shown. The 1/4 column shows that 25% of the projects had lower costs and 75% had higher. The 3/4 column shows that 75% of the projects had lower costs and 25% had higher. The median column shows that 50% of the projects had lower costs and 50% had higher.

- Project Costs are useful in the conceptual stage when no details are available. As soon as details become available in the project design, the square foot approach should be discontinued and the project should be priced as to its particular components. When more precision is required, or for estimating the replacement cost of specific buildings, the current edition of *Square Foot Costs with RSMeans data* should be used.

- In using the figures in this section, it is recommended that the median column be used for preliminary figures if no additional information is available. The median figures, when multiplied by the total city construction cost index figures (see City Cost Indexes) and then multiplied by the project size modifier at the end of this section, should present a fairly accurate base figure, which would then have to be adjusted in view of the estimator's experience, local economic conditions, code requirements, and the owner's particular requirements. There is no need to factor in the percentage figures, as these should remain constant from city to city.

- The editors of this data would greatly appreciate receiving cost figures on one or more of your recent projects, which would then be included in the averages for next year. All cost figures received will be kept confidential, except that they will be averaged with other similar projects to arrive at square foot cost figures for next year.

See the website above for details and the discount available for submitting one or more of your projects.

Same Data. Simplified.

Enjoy the convenience and efficiency of accessing your costs anywhere:

- **Skip the multiplier** by setting your location
- **Quickly search,** edit, favorite and share costs
- **Stay on top of price changes** with automatic updates

Discover more at rsmeans.com/online

50 17 00 | Project Costs

			UNIT	UNIT COSTS			% OF TOTAL			
				1/4	MEDIAN	3/4	1/4	MEDIAN	3/4	
01	0000	**Auto Sales with Repair**	S.F.							01
	0100	Architectural		107	120	130	58%	64%	67%	
	0200	Plumbing		8.95	9.40	12.55	4.84%	5.20%	6.80%	
	0300	Mechanical		12	16.10	17.75	6.40%	8.70%	10.15%	
	0400	Electrical		18.45	23	28.50	9.05%	11.70%	15.90%	
	0500	Total Project Costs		180	187	193				
02	0000	**Banking Institutions**	S.F.							02
	0100	Architectural		161	198	241	59%	65%	69%	
	0200	Plumbing		6.50	9.10	12.60	2.12%	3.39%	4.19%	
	0300	Mechanical		12.95	17.90	21	4.41%	5.10%	10.75%	
	0400	Electrical		31.50	38	58.50	10.45%	13.05%	15.90%	
	0500	Total Project Costs		268	300	370				
03	0000	**Court House**	S.F.							03
	0100	Architectural		85	167	167	54.50%	58.50%	58.50%	
	0200	Plumbing		3.22	3.22	3.22	2.07%	2.07%	2.07%	
	0300	Mechanical		20	20	20	12.95%	12.95%	12.95%	
	0400	Electrical		26	26	26	16.60%	16.60%	16.60%	
	0500	Total Project Costs		155	286	286				
04	0000	**Data Centers**	S.F.							04
	0100	Architectural		192	192	192	68%	68%	68%	
	0200	Plumbing		10.55	10.55	10.55	3.71%	3.71%	3.71%	
	0300	Mechanical		27	27	27	9.45%	9.45%	9.45%	
	0400	Electrical		25.50	25.50	25.50	9%	9%	9%	
	0500	Total Project Costs		284	284	284				
05	0000	**Detention Centers**	S.F.							05
	0100	Architectural		178	188	200	52%	53%	60.50%	
	0200	Plumbing		18.80	23	27.50	5.15%	7.10%	7.25%	
	0300	Mechanical		24	34	41	7.55%	9.50%	13.80%	
	0400	Electrical		39	46.50	60.50	10.90%	14.85%	17.95%	
	0500	Total Project Costs		300	320	375				
06	0000	**Fire Stations**	S.F.							06
	0100	Architectural		101	130	190	46%	54.50%	61.50%	
	0200	Plumbing		10.25	14.05	18.35	4.59%	5.60%	6.30%	
	0300	Mechanical		15.10	22	29.50	6.05%	8.25%	10.20%	
	0400	Electrical		23.50	30	40	10.75%	12.55%	14.95%	
	0500	Total Project Costs		211	241	335				
07	0000	**Gymnasium**	S.F.							07
	0100	Architectural		89	118	118	57%	64.50%	64.50%	
	0200	Plumbing		2.19	7.15	7.15	1.58%	3.48%	3.48%	
	0300	Mechanical		3.35	30	30	2.42%	14.65%	14.65%	
	0400	Electrical		11	21.50	21.50	7.95%	10.35%	10.35%	
	0500	Total Project Costs		139	206	206				
08	0000	**Hospitals**	S.F.							08
	0100	Architectural		108	178	193	43%	47.50%	48%	
	0200	Plumbing		7.95	15.15	33	6%	7.45%	7.65%	
	0300	Mechanical		52.50	59	77	14.20%	17.95%	23.50%	
	0400	Electrical		24	48	62	10.95%	13.75%	16.85%	
	0500	Total Project Costs		253	375	405				
09	0000	**Industrial Buildings**	S.F.							09
	0100	Architectural		59	116	235	52%	56%	82%	
	0200	Plumbing		1.70	4.62	13.35	1.57%	2.11%	6.30%	
	0300	Mechanical		4.86	9.25	44	4.77%	5.55%	14.80%	
	0400	Electrical		7.40	8.45	70.50	7.85%	13.55%	16.20%	
	0500	Total Project Costs		88	133	435				
10	0000	**Medical Clinics & Offices**	S.F.							10
	0100	Architectural		90.50	122	163	48.50%	55.50%	62.50%	
	0200	Plumbing		9	13.50	21.50	4.47%	6.65%	8.75%	
	0300	Mechanical		14.60	22.50	41.50	7.80%	10.90%	16.05%	
	0400	Electrical		19.90	24.50	38	8.90%	11.40%	14.05%	
	0500	Total Project Costs		168	224	295				

		50 17 00 | Project Costs	UNIT	UNIT COSTS			% OF TOTAL			
				1/4	MEDIAN	3/4	1/4	MEDIAN	3/4	
11	0000	**Mixed Use**	S.F.							11
	0100	Architectural		92	130	198	45.50%	52.50%	61.50%	
	0200	Plumbing		6.25	11.45	12.15	3.31%	3.47%	4.18%	
	0300	Mechanical		15.25	25	46	4.68%	13.60%	17.05%	
	0400	Electrical		16.15	36	53.50	8.30%	11.40%	15.65%	
	0500	Total Project Costs		190	335	340				
12	0000	**Multi-Family Housing**	S.F.							12
	0100	Architectural		77.50	105	155	54.50%	61.50%	66.50%	
	0200	Plumbing		6.90	11.10	15.10	5.30%	6.85%	8%	
	0300	Mechanical		7.15	9.55	27.50	4.92%	6.90%	10.40%	
	0400	Electrical		10	15.70	22.50	6.20%	8%	10.25%	
	0500	Total Project Costs		128	210	253				
13	0000	**Nursing Home & Assisted Living**	S.F.							13
	0100	Architectural		72.50	94.50	119	51.50%	55.50%	63.50%	
	0200	Plumbing		7.80	11.75	12.90	6.25%	7.40%	8.80%	
	0300	Mechanical		6.40	9.45	18.50	4.04%	6.70%	9.55%	
	0400	Electrical		10.60	16.70	23.50	7%	10.75%	13.10%	
	0500	Total Project Costs		123	161	194				
14	0000	**Office Buildings**	S.F.							14
	0100	Architectural		93	130	179	54.50%	61%	69%	
	0200	Plumbing		5.15	8.10	15.15	2.70%	3.78%	5.85%	
	0300	Mechanical		10.10	17.15	26.50	5.60%	8.20%	11.10%	
	0400	Electrical		12.90	22	34	7.75%	10%	12.70%	
	0500	Total Project Costs		159	202	285				
15	0000	**Parking Garage**	S.F.							15
	0100	Architectural		32	39	40.50	70%	79%	88%	
	0200	Plumbing		1.05	1.10	2.06	2.05%	2.70%	2.83%	
	0300	Mechanical		.82	1.25	4.76	2.11%	3.62%	3.81%	
	0400	Electrical		2.80	3.08	6.45	5.30%	6.35%	7.95%	
	0500	Total Project Costs		39	47.50	51.50				
16	0000	**Parking Garage/Mixed Use**	S.F.							16
	0100	Architectural		103	113	115	61%	62%	65.50%	
	0200	Plumbing		3.33	4.36	6.65	2.47%	2.72%	3.66%	
	0300	Mechanical		14.20	16	23	7.80%	13.10%	13.60%	
	0400	Electrical		14.90	21.50	22	8.20%	12.65%	18.15%	
	0500	Total Project Costs		169	176	182				
17	0000	**Police Stations**	S.F.							17
	0100	Architectural		117	131	165	49%	56.50%	61%	
	0200	Plumbing		15.45	18.55	18.65	5.05%	5.55%	9.05%	
	0300	Mechanical		35	48.50	50.50	13%	14.55%	16.55%	
	0400	Electrical		26.50	29	30.50	9.15%	12.10%	14%	
	0500	Total Project Costs		219	270	305				
18	0000	**Police/Fire**	S.F.							18
	0100	Architectural		114	114	350	55.50%	66%	68%	
	0200	Plumbing		9.15	9.45	35	5.45%	5.50%	5.55%	
	0300	Mechanical		13.95	22	80	8.35%	12.70%	12.80%	
	0400	Electrical		15.90	20.50	91.50	9.50%	11.75%	14.55%	
	0500	Total Project Costs		167	173	630				
19	0000	**Public Assembly Buildings**	S.F.							19
	0100	Architectural		116	160	240	57.50%	61.50%	66%	
	0200	Plumbing		6.15	9	13.50	2.60%	3.36%	4.79%	
	0300	Mechanical		12.95	23	35.50	6.55%	8.95%	12.45%	
	0400	Electrical		19.15	26	41.50	8.60%	10.75%	13%	
	0500	Total Project Costs		186	255	375				
20	0000	**Recreational**	S.F.							20
	0100	Architectural		111	176	238	49.50%	59%	65.50%	
	0200	Plumbing		8.60	14.85	25.50	3.12%	4.76%	7.40%	
	0300	Mechanical		13.25	20	32	4.98%	6.60%	11.55%	
	0400	Electrical		19.15	27.50	40	7.15%	8.95%	10.75%	
	0500	Total Project Costs		198	296	445				

50 17 00 | Project Costs		UNIT	UNIT COSTS			% OF TOTAL		
			1/4	MEDIAN	3/4	1/4	MEDIAN	3/4
21	**0000 Restaurants**	S.F.						
	0100 Architectural		127	195	250	59%	60%	63.50%
	0200 Plumbing		13.90	32	40.50	7.35%	7.75%	8.95%
	0300 Mechanical		15	17.70	37.50	6.50%	8.15%	11.15%
	0400 Electrical		14.90	24.50	48.50	7.10%	10.30%	11.60%
	0500 Total Project Costs		210	305	420			
22	**0000 Retail**	S.F.						
	0100 Architectural		56.50	88.50	182	54.50%	60%	64.50%
	0200 Plumbing		7.05	9.85	12.25	4.18%	5.45%	8.45%
	0300 Mechanical		6.60	9.35	17.20	4.98%	6.15%	7.05%
	0400 Electrical		10.50	21.50	31.50	7.90%	11.25%	12.45%
	0500 Total Project Costs		86	153	292			
23	**0000 Schools**	S.F.						
	0100 Architectural		98	127	173	51%	56%	60.50%
	0200 Plumbing		7.75	10.70	15.65	3.66%	4.59%	7%
	0300 Mechanical		18.70	26.50	38.50	8.95%	12%	14.55%
	0400 Electrical		18.45	25.50	33	9.45%	11.25%	13.30%
	0500 Total Project Costs		178	227	310			
24	**0000 University, College & Private School Classroom & Admin Buildings**	S.F.						
	0100 Architectural		126	160	194	50.50%	55%	59.50%
	0200 Plumbing		8.50	11.50	19.90	3.02%	4.46%	7.35%
	0300 Mechanical		25	37.50	46.50	9.95%	11.95%	14.70%
	0400 Electrical		20	28.50	35.50	7.70%	9.90%	11.55%
	0500 Total Project Costs		219	293	380			
25	**0000 University, College & Private School Dormitories**	S.F.						
	0100 Architectural		81.50	143	152	53%	61.50%	68%
	0200 Plumbing		10.80	15.25	17.50	6.35%	6.65%	8.95%
	0300 Mechanical		4.84	20.50	32.50	4.13%	9%	11.80%
	0400 Electrical		5.75	19.95	30.50	4.75%	7.35%	10.60%
	0500 Total Project Costs		121	229	271			
26	**0000 University, College & Private School Science, Eng. & Lab Buildings**	S.F.						
	0100 Architectural		140	166	195	48.50%	56.50%	58%
	0200 Plumbing		9.65	14.65	24.50	3.29%	3.77%	5%
	0300 Mechanical		37	69	71	11.70%	19.40%	23.50%
	0400 Electrical		29	38	39	9%	12.05%	13.15%
	0500 Total Project Costs		293	315	370			
27	**0000 University, College & Private School Student Union Buildings**	S.F.						
	0100 Architectural		111	292	292	54.50%	54.50%	59.50%
	0200 Plumbing		16.80	16.80	25	3.13%	4.27%	11.45%
	0300 Mechanical		32	51.50	51.50	9.60%	9.60%	14.55%
	0400 Electrical		28	48.50	48.50	9.05%	12.80%	13.15%
	0500 Total Project Costs		219	535	535			
28	**0000 Warehouses**	S.F.						
	0100 Architectural		47.50	72.50	132	60.50%	67%	71.50%
	0200 Plumbing		2.48	5.30	10.20	2.82%	3.72%	5%
	0300 Mechanical		2.93	16.70	26	4.56%	8.15%	10.70%
	0400 Electrical		6.15	20	33.50	7.75%	10.10%	18.30%
	0500 Total Project Costs		71	113	228			

Square Foot Project Size Modifier

One factor that affects the S.F. cost of a particular building is the size. In general, for buildings built to the same specifications in the same locality, the larger building will have the lower S.F. cost. This is due mainly to the decreasing contribution of the exterior walls plus the economy of scale usually achievable in larger buildings. The Area Conversion Scale shown below will give a factor to convert costs for the typical size building to an adjusted cost for the particular project.

The Square Foot Base Size lists the median costs, most typical project size in our accumulated data, and the range in size of the projects.

The Size Factor for your project is determined by dividing your project area in S.F. by the typical project size for the particular Building Type. With this factor, enter the Area Conversion Scale at the appropriate Size Factor and determine the appropriate Cost Multiplier for your building size.

Example: Determine the cost per S.F. for a 107,200 S.F. Multi-family housing.

$$\frac{\text{Proposed building area} = 107,200 \text{ S.F.}}{\text{Typical size from below} = 53,600 \text{ S.F.}} = 2.00$$

Enter Area Conversion Scale at 2.0, intersect curve, read horizontally the appropriate cost multiplier of .94. Size adjusted cost becomes .94 x $210.00 = $197.40 based on national average costs.

Note: For Size Factors less than .50, the Cost Multiplier is 1.1
For Size Factors greater than 3.5, the Cost Multiplier is .90

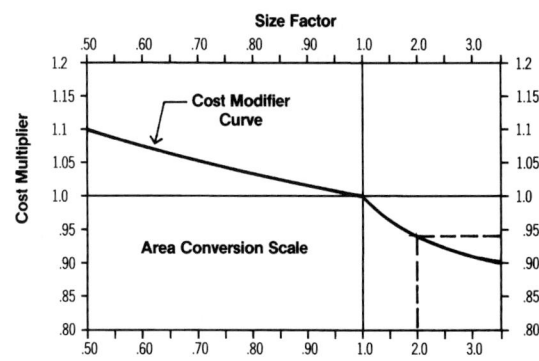

System	Median Cost (Total Project Costs)	Typical Size Gross S.F. (Median of Projects)	Typical Range (Low – High) (Projects)
Auto Sales with Repair	$187.00	24,900	4,680 – 29,253
Banking Institutions	300.00	9,300	3,267 – 38,148
Court House	286.00	47,600	24,680 – 70,500
Data Centers	284.00	14,400	14,369 – 14,369
Detention Centers	320.00	37,800	12,257 – 183,339
Fire Stations	241.00	12,500	6,300 – 49,577
Gymnasium	206.00	52,400	22,844 – 81,992
Hospitals	375.00	87,100	22,428 – 410,273
Industrial Buildings	133.00	21,100	5,146 – 200,625
Medical Clinics & Offices	224.00	22,500	2,273 – 327,000
Mixed Use	335.00	28,500	7,200 – 188,944
Multi-Family Housing	210.00	53,600	2,472 – 1,161,450
Nursing Home & Assisted Living	161.00	38,200	1,515 – 242,555
Office Buildings	202.00	20,600	1,115 – 930,000
Parking Garage	47.5	151,800	99,884 – 287,040
Parking Garage/Mixed Use	176.00	254,200	5,302 – 318,033
Police Stations	270.00	28,500	15,377 – 88,625
Police/Fire	173.00	44,300	8,600 – 50,279
Public Assembly Buildings	255.00	21,000	2,232 – 235,301
Recreational	296.00	27,700	1,000 – 223,787
Restaurants	305.00	6,000	3,975 – 42,000
Retail	153.00	26,000	4,035 – 84,270
Schools	227.00	69,900	1,344 – 410,848
University, College & Private School Classroom & Admin Buildings	293.00	48,200	8,806 – 196,187
University, College & Private School Dormitories	229.00	39,600	1,500 – 126,889
University, College & Private School Science, Eng. & Lab Buildings	315.00	60,000	5,311 – 117,643
University, College & Private School Student Union Buildings	535.00	47,300	42,075 – 50,000
Warehouses	113.00	11,100	640 – 303,750

Abbreviations

A	Area Square Feet; Ampere	Brk., brk	Brick	Csc	Cosecant
AAFES	Army and Air Force Exchange Service	brkt	Bracket	C.S.F.	Hundred Square Feet
		Brs.	Brass	CSI	Construction Specifications Institute
ABS	Acrylonitrile Butadiene Stryrene; Asbestos Bonded Steel	Brz.	Bronze		
		Bsn.	Basin	CT	Current Transformer
A.C., AC	Alternating Current;	Btr.	Better	CTS	Copper Tube Size
	Air-Conditioning;	BTU	British Thermal Unit	Cu	Copper, Cubic
	Asbestos Cement;	BTUH	BTU per Hour	Cu. Ft.	Cubic Foot
	Plywood Grade A & C	Bu.	Bushels	cw	Continuous Wave
ACI	American Concrete Institute	BUR	Built-up Roofing	C.W.	Cool White; Cold Water
ACR	Air Conditioning Refrigeration	BX	Interlocked Armored Cable	Cwt.	100 Pounds
ADA	Americans with Disabilities Act	°C	Degree Centigrade	C.W.X.	Cool White Deluxe
AD	Plywood, Grade A & D	c	Conductivity, Copper Sweat	C.Y.	Cubic Yard (27 cubic feet)
Addit.	Additional	C	Hundred; Centigrade	C.Y./Hr.	Cubic Yard per Hour
Adh.	Adhesive	C/C	Center to Center, Cedar on Cedar	Cyl.	Cylinder
Adj.	Adjustable	C-C	Center to Center	d	Penny (nail size)
af	Audio-frequency	Cab	Cabinet	D	Deep; Depth; Discharge
AFFF	Aqueous Film Forming Foam	Cair.	Air Tool Laborer	Dis., Disch.	Discharge
AFUE	Annual Fuel Utilization Efficiency	Cal.	Caliper	Db	Decibel
AGA	American Gas Association	Calc	Calculated	Dbl.	Double
Agg.	Aggregate	Cap.	Capacity	DC	Direct Current
A.H., Ah	Ampere Hours	Carp.	Carpenter	DDC	Direct Digital Control
A hr.	Ampere-hour	C.B.	Circuit Breaker	Demob.	Demobilization
A.H.U., AHU	Air Handling Unit	C.C.A.	Chromate Copper Arsenate	d.f.t.	Dry Film Thickness
A.I.A.	American Institute of Architects	C.C.F.	Hundred Cubic Feet	d.f.u.	Drainage Fixture Units
AIC	Ampere Interrupting Capacity	cd	Candela	D.H.	Double Hung
Allow.	Allowance	cd/sf	Candela per Square Foot	DHW	Domestic Hot Water
alt., alt	Alternate	CD	Grade of Plywood Face & Back	DI	Ductile Iron
Alum.	Aluminum	CDX	Plywood, Grade C & D, exterior glue	Diag.	Diagonal
a.m.	Ante Meridiem			Diam., Dia	Diameter
Amp.	Ampere	Cefi.	Cement Finisher	Distrib.	Distribution
Anod.	Anodized	Cem.	Cement	Div.	Division
ANSI	American National Standards Institute	CF	Hundred Feet	Dk.	Deck
		C.F.	Cubic Feet	D.L.	Dead Load; Diesel
APA	American Plywood Association	CFM	Cubic Feet per Minute	DLH	Deep Long Span Bar Joist
Approx.	Approximate	CFRP	Carbon Fiber Reinforced Plastic	dlx	Deluxe
Apt.	Apartment	c.g.	Center of Gravity	Do.	Ditto
Asb.	Asbestos	CHW	Chilled Water; Commercial Hot Water	DOP	Dioctyl Phthalate Penetration Test (Air Filters)
A.S.B.C.	American Standard Building Code				
Asbe.	Asbestos Worker	C.I., CI	Cast Iron	Dp., dp	Depth
ASCE	American Society of Civil Engineers	C.I.P., CIP	Cast in Place	D.P.S.T.	Double Pole, Single Throw
A.S.H.R.A.E.	American Society of Heating, Refrig. & AC Engineers	Circ.	Circuit	Dr.	Drive
		C.L.	Carload Lot	DR	Dimension Ratio
ASME	American Society of Mechanical Engineers	CL	Chain Link	Drink.	Drinking
		Clab.	Common Laborer	D.S.	Double Strength
ASTM	American Society for Testing and Materials	Clam	Common Maintenance Laborer	D.S.A.	Double Strength A Grade
		C.L.F.	Hundred Linear Feet	D.S.B.	Double Strength B Grade
Attchmt.	Attachment	CLF	Current Limiting Fuse	Dty.	Duty
Avg., Ave.	Average	CLP	Cross Linked Polyethylene	DWV	Drain Waste Vent
AWG	American Wire Gauge	cm	Centimeter	DX	Deluxe White, Direct Expansion
AWWA	American Water Works Assoc.	CMP	Corr. Metal Pipe	dyn	Dyne
Bbl.	Barrel	CMU	Concrete Masonry Unit	e	Eccentricity
B&B, BB	Grade B and Better; Balled & Burlapped	CN	Change Notice	E	Equipment Only; East; Emissivity
		Col.	Column	Ea.	Each
B&S	Bell and Spigot	CO₂	Carbon Dioxide	EB	Encased Burial
B.&W.	Black and White	Comb.	Combination	Econ.	Economy
b.c.c.	Body-centered Cubic	comm.	Commercial, Communication	E.C.Y	Embankment Cubic Yards
B.C.Y.	Bank Cubic Yards	Compr.	Compressor	EDP	Electronic Data Processing
BE	Bevel End	Conc.	Concrete	EIFS	Exterior Insulation Finish System
B.F.	Board Feet	Cont., cont	Continuous; Continued, Container	E.D.R.	Equiv. Direct Radiation
Bg. cem.	Bag of Cement	Corkbd.	Cork Board	Eq.	Equation
BHP	Boiler Horsepower; Brake Horsepower	Corr.	Corrugated	EL	Elevation
		Cos	Cosine	Elec.	Electrician; Electrical
B.I.	Black Iron	Cot	Cotangent	Elev.	Elevator; Elevating
bidir.	bidirectional	Cov.	Cover	EMT	Electrical Metallic Conduit; Thin Wall Conduit
Bit., Bitum.	Bituminous	C/P	Cedar on Paneling		
Bit., Conc.	Bituminous Concrete	CPA	Control Point Adjustment	Eng.	Engine, Engineered
Bk.	Backed	Cplg.	Coupling	EPDM	Ethylene Propylene Diene Monomer
Bkrs.	Breakers	CPM	Critical Path Method		
Bldg., bldg	Building	CPVC	Chlorinated Polyvinyl Chloride	EPS	Expanded Polystyrene
Blk.	Block	C.Pr.	Hundred Pair	Eqhv.	Equip. Oper., Heavy
Bm.	Beam	CRC	Cold Rolled Channel	Eqlt.	Equip. Oper., Light
Boil.	Boilermaker	Creos.	Creosote	Eqmd.	Equip. Oper., Medium
bpm	Blows per Minute	Crpt.	Carpet & Linoleum Layer	Eqmm.	Equip. Oper., Master Mechanic
BR	Bedroom	CRT	Cathode-ray Tube	Eqol.	Equip. Oper., Oilers
Brg., brng.	Bearing	CS	Carbon Steel, Constant Shear Bar Joist	Equip.	Equipment
Brhe.	Bricklayer Helper			ERW	Electric Resistance Welded
Bric.	Bricklayer				

E.S.	Energy Saver	H	High Henry	Lath.	Lather
Est.	Estimated	HC	High Capacity	Lav.	Lavatory
esu	Electrostatic Units	H.D., HD	Heavy Duty; High Density	lb.; #	Pound
E.W.	Each Way	H.D.O.	High Density Overlaid	L.B., LB	Load Bearing; L Conduit Body
EWT	Entering Water Temperature	HDPE	High Density Polyethylene Plastic	L. & E.	Labor & Equipment
Excav.	Excavation	Hdr.	Header	lb./hr.	Pounds per Hour
excl	Excluding	Hdwe.	Hardware	lb./L.F.	Pounds per Linear Foot
Exp., exp	Expansion, Exposure	H.I.D., HID	High Intensity Discharge	lbf/sq.in.	Pound-force per Square Inch
Ext., ext	Exterior; Extension	Help.	Helper Average	L.C.L.	Less than Carload Lot
Extru.	Extrusion	HEPA	High Efficiency Particulate Air	L.C.Y.	Loose Cubic Yard
f.	Fiber Stress		Filter	Ld.	Load
F	Fahrenheit; Female; Fill	Hg	Mercury	LE	Lead Equivalent
Fab., fab	Fabricated; Fabric	HIC	High Interrupting Capacity	LED	Light Emitting Diode
FBGS	Fiberglass	HM	Hollow Metal	L.F.	Linear Foot
F.C.	Footcandles	HMWPE	High Molecular Weight	L.F. Hdr	Linear Feet of Header
f.c.c.	Face-centered Cubic		Polyethylene	L.F. Nose	Linear Foot of Stair Nosing
f'c.	Compressive Stress in Concrete;	HO	High Output	L.F. Rsr	Linear Foot of Stair Riser
	Extreme Compressive Stress	Horiz.	Horizontal	Lg.	Long; Length; Large
F.E.	Front End	H.P., HP	Horsepower; High Pressure	L & H	Light and Heat
FEP	Fluorinated Ethylene Propylene	H.P.F.	High Power Factor	LH	Long Span Bar Joist
	(Teflon)	Hr.	Hour	L.H.	Labor Hours
F.G.	Flat Grain	Hrs./Day	Hours per Day	L.L., LL	Live Load
F.H.A.	Federal Housing Administration	HSC	High Short Circuit	L.L.D.	Lamp Lumen Depreciation
Fig.	Figure	Ht.	Height	lm	Lumen
Fin.	Finished	Htg.	Heating	lm/sf	Lumen per Square Foot
FIPS	Female Iron Pipe Size	Htrs.	Heaters	lm/W	Lumen per Watt
Fixt.	Fixture	HVAC	Heating, Ventilation & Air-	LOA	Length Over All
FJP	Finger jointed and primed		Conditioning	log	Logarithm
Fl. Oz.	Fluid Ounces	Hvy.	Heavy	L-O-L	Lateralolet
Flr.	Floor	HW	Hot Water	long.	Longitude
Flrs.	Floors	Hyd.; Hydr.	Hydraulic	L.P., LP	Liquefied Petroleum; Low Pressure
FM	Frequency Modulation;	Hz	Hertz (cycles)	L.P.F.	Low Power Factor
	Factory Mutual	I.	Moment of Inertia	LR	Long Radius
Fmg.	Framing	IBC	International Building Code	L.S.	Lump Sum
FM/UL	Factory Mutual/Underwriters Labs	I.C.	Interrupting Capacity	Lt.	Light
Fdn.	Foundation	ID	Inside Diameter	Lt. Ga.	Light Gauge
FNPT	Female National Pipe Thread	I.D.	Inside Dimension; Identification	L.T.L.	Less than Truckload Lot
Fori.	Foreman, Inside	I.F.	Inside Frosted	Lt. Wt.	Lightweight
Foro.	Foreman, Outside	I.M.C.	Intermediate Metal Conduit	L.V.	Low Voltage
Fount.	Fountain	In.	Inch	M	Thousand; Material; Male;
fpm	Feet per Minute	Incan.	Incandescent		Light Wall Copper Tubing
FPT	Female Pipe Thread	Incl.	Included; Including	M²CA	Meters Squared Contact Area
Fr	Frame	Int.	Interior	m/hr.; M.H.	Man-hour
F.R.	Fire Rating	Inst.	Installation	mA	Milliampere
FRK	Foil Reinforced Kraft	Insul., insul	Insulation/Insulated	Mach.	Machine
FSK	Foil/Scrim/Kraft	I.P.	Iron Pipe	Mag. Str.	Magnetic Starter
FRP	Fiberglass Reinforced Plastic	I.P.S., IPS	Iron Pipe Size	Maint.	Maintenance
FS	Forged Steel	IPT	Iron Pipe Threaded	Marb.	Marble Setter
FSC	Cast Body; Cast Switch Box	I.W.	Indirect Waste	Mat; Mat'l.	Material
Ft., ft	Foot; Feet	J	Joule	Max.	Maximum
Ftng.	Fitting	J.I.C.	Joint Industrial Council	MBF	Thousand Board Feet
Ftg.	Footing	K	Thousand; Thousand Pounds;	MBH	Thousand BTU's per hr.
Ft lb.	Foot Pound		Heavy Wall Copper Tubing, Kelvin	MC	Metal Clad Cable
Furn.	Furniture	K.A.H.	Thousand Amp. Hours	MCC	Motor Control Center
FVNR	Full Voltage Non-Reversing	kcmil	Thousand Circular Mils	M.C.F.	Thousand Cubic Feet
FVR	Full Voltage Reversing	KD	Knock Down	MCFM	Thousand Cubic Feet per Minute
FXM	Female by Male	K.D.A.T.	Kiln Dried After Treatment	M.C.M.	Thousand Circular Mils
Fy.	Minimum Yield Stress of Steel	kg	Kilogram	MCP	Motor Circuit Protector
g	Gram	kG	Kilogauss	MD	Medium Duty
G	Gauss	kgf	Kilogram Force	MDF	Medium-density fibreboard
Ga.	Gauge	kHz	Kilohertz	M.D.O.	Medium Density Overlaid
Gal., gal.	Gallon	Kip	1000 Pounds	Med.	Medium
Galv., galv	Galvanized	KJ	Kilojoule	MF	Thousand Feet
GC/MS	Gas Chromatograph/Mass	K.L.	Effective Length Factor	M.F.B.M.	Thousand Feet Board Measure
	Spectrometer	K.L.F.	Kips per Linear Foot	Mfg.	Manufacturing
Gen.	General	Km	Kilometer	Mfrs.	Manufacturers
GFI	Ground Fault Interrupter	KO	Knock Out	mg	Milligram
GFRC	Glass Fiber Reinforced Concrete	K.S.F.	Kips per Square Foot	MGD	Million Gallons per Day
Glaz.	Glazier	K.S.I.	Kips per Square Inch	MGPH	Million Gallons per Hour
GPD	Gallons per Day	kV	Kilovolt	MH, M.H.	Manhole; Metal Halide; Man-Hour
gpf	Gallon per Flush	kVA	Kilovolt Ampere	MHz	Megahertz
GPH	Gallons per Hour	kVAR	Kilovar (Reactance)	Mi.	Mile
gpm, GPM	Gallons per Minute	KW	Kilowatt	MI	Malleable Iron; Mineral Insulated
GR	Grade	KWh	Kilowatt-hour	MIPS	Male Iron Pipe Size
Gran.	Granular	L	Labor Only; Length; Long;	mj	Mechanical Joint
Grnd.	Ground		Medium Wall Copper Tubing	m	Meter
GVW	Gross Vehicle Weight	Lab.	Labor	mm	Millimeter
GWB	Gypsum Wall Board	lat	Latitude	Mill.	Millwright
				Min., min.	Minimum, Minute

For customer support on your Plumbing Costs with RSMeans Data, call 800.448.8182.

731

Misc.	Miscellaneous	PCM	Phase Contrast Microscopy	SBS	Styrene Butadiere Styrene
ml	Milliliter, Mainline	PDCA	Painting and Decorating	SC	Screw Cover
M.L.F.	Thousand Linear Feet		Contractors of America	SCFM	Standard Cubic Feet per Minute
Mo.	Month	P.E., PE	Professional Engineer;	Scaf.	Scaffold
Mobil.	Mobilization		Porcelain Enamel;	Sch., Sched.	Schedule
Mog.	Mogul Base		Polyethylene; Plain End	S.C.R.	Modular Brick
MPH	Miles per Hour	P.E.C.I.	Porcelain Enamel on Cast Iron	S.D.	Sound Deadening
MPT	Male Pipe Thread	Perf.	Perforated	SDR	Standard Dimension Ratio
MRGWB	Moisture Resistant Gypsum	PEX	Cross Linked Polyethylene	S.E.	Surfaced Edge
	Wallboard	Ph.	Phase	Sel.	Select
MRT	Mile Round Trip	P.I.	Pressure Injected	SER, SEU	Service Entrance Cable
ms	Millisecond	Pile.	Pile Driver	S.F.	Square Foot
M.S.F.	Thousand Square Feet	Pkg.	Package	S.F.C.A.	Square Foot Contact Area
Mstz.	Mosaic & Terrazzo Worker	Pl.	Plate	S.F. Flr.	Square Foot of Floor
M.S.Y.	Thousand Square Yards	Plah.	Plasterer Helper	S.F.G.	Square Foot of Ground
Mtd., mtd., mtd	Mounted	Plas.	Plasterer	S.F. Hor.	Square Foot Horizontal
Mthe.	Mosaic & Terrazzo Helper	plf	Pounds Per Linear Foot	SFR	Square Feet of Radiation
Mtng.	Mounting	Pluh.	Plumber Helper	S.F. Shlf.	Square Foot of Shelf
Mult.	Multi; Multiply	Plum.	Plumber	S4S	Surface 4 Sides
MUTCD	Manual on Uniform Traffic Control	Ply.	Plywood	Shee.	Sheet Metal Worker
	Devices	p.m.	Post Meridiem	Sin.	Sine
M.V.A.	Million Volt Amperes	Pntd.	Painted	Skwk.	Skilled Worker
M.V.A.R.	Million Volt Amperes Reactance	Pord.	Painter, Ordinary	SL	Saran Lined
MV	Megavolt	pp	Pages	S.L.	Slimline
MW	Megawatt	PP, PPL	Polypropylene	Sldr.	Solder
MXM	Male by Male	P.P.M.	Parts per Million	SLH	Super Long Span Bar Joist
MYD	Thousand Yards	Pr.	Pair	S.N.	Solid Neutral
N	Natural; North	P.E.S.B.	Pre-engineered Steel Building	SO	Stranded with oil resistant inside
nA	Nanoampere	Prefab.	Prefabricated		insulation
NA	Not Available; Not Applicable	Prefin.	Prefinished	S-O-L	Socketolet
N.B.C.	National Building Code	Prop.	Propelled	sp	Standpipe
NC	Normally Closed	PSF, psf	Pounds per Square Foot	S.P.	Static Pressure; Single Pole; Self-
NEMA	National Electrical Manufacturers	PSI, psi	Pounds per Square Inch		Propelled
	Assoc.	PSIG	Pounds per Square Inch Gauge	Spri.	Sprinkler Installer
NEHB	Bolted Circuit Breaker to 600V.	PSP	Plastic Sewer Pipe	spwg	Static Pressure Water Gauge
NFPA	National Fire Protection Association	Pspr.	Painter, Spray	S.P.D.T.	Single Pole, Double Throw
NLB	Non-Load-Bearing	Psst.	Painter, Structural Steel	SPF	Spruce Pine Fir; Sprayed
NM	Non-Metallic Cable	P.T.	Potential Transformer		Polyurethane Foam
nm	Nanometer	P. & T.	Pressure & Temperature	S.P.S.T.	Single Pole, Single Throw
No.	Number	Ptd.	Painted	SPT	Standard Pipe Thread
NO	Normally Open	Ptns.	Partitions	Sq.	Square; 100 Square Feet
N.O.C.	Not Otherwise Classified	Pu	Ultimate Load	Sq. Hd.	Square Head
Nose.	Nosing	PVC	Polyvinyl Chloride	Sq. In.	Square Inch
NPT	National Pipe Thread	Pvmt.	Pavement	S.S.	Single Strength; Stainless Steel
NQOD	Combination Plug-on/Bolt on	PRV	Pressure Relief Valve	S.S.B.	Single Strength B Grade
	Circuit Breaker to 240V.	Pwr.	Power	sst, ss	Stainless Steel
N.R.C., NRC	Noise Reduction Coefficient/	Q	Quantity Heat Flow	Sswk.	Structural Steel Worker
	Nuclear Regulator Commission	Qt.	Quart	Sswl.	Structural Steel Welder
N.R.S.	Non Rising Stem	Quan., Qty.	Quantity	St.; Stl.	Steel
ns	Nanosecond	Q.C.	Quick Coupling	STC	Sound Transmission Coefficient
NTP	Notice to Proceed	r	Radius of Gyration	Std.	Standard
nW	Nanowatt	R	Resistance	Stg.	Staging
OB	Opposing Blade	R.C.P.	Reinforced Concrete Pipe	STK	Select Tight Knot
OC	On Center	Rect.	Rectangle	STP	Standard Temperature & Pressure
OD	Outside Diameter	recpt.	Receptacle	Stpi.	Steamfitter, Pipefitter
O.D.	Outside Dimension	Reg.	Regular	Str.	Strength; Starter; Straight
ODS	Overhead Distribution System	Reinf.	Reinforced	Strd.	Stranded
O.G.	Ogee	Req'd.	Required	Struct.	Structural
O.H.	Overhead	Res.	Resistant	Sty.	Story
O&P	Overhead and Profit	Resi.	Residential	Subj.	Subject
Oper.	Operator	RF	Radio Frequency	Subs.	Subcontractors
Opng.	Opening	RFID	Radio-frequency Identification	Surf.	Surface
Orna.	Ornamental	Rgh.	Rough	Sw.	Switch
OSB	Oriented Strand Board	RGS	Rigid Galvanized Steel	Swbd.	Switchboard
OS&Y	Outside Screw and Yoke	RHW	Rubber, Heat & Water Resistant;	S.Y.	Square Yard
OSHA	Occupational Safety and Health		Residential Hot Water	Syn.	Synthetic
	Act	rms	Root Mean Square	S.Y.P.	Southern Yellow Pine
Ovhd.	Overhead	Rnd.	Round	Sys.	System
OWG	Oil, Water or Gas	Rodm.	Rodman	t.	Thickness
Oz.	Ounce	Rofc.	Roofer, Composition	T	Temperature; Ton
P.	Pole; Applied Load; Projection	Rofp.	Roofer, Precast	Tan	Tangent
p.	Page	Rohe.	Roofer Helpers (Composition)	T.C.	Terra Cotta
Pape.	Paperhanger	Rots.	Roofer, Tile & Slate	T & C	Threaded and Coupled
P.A.P.R.	Powered Air Purifying Respirator	R.O.W.	Right of Way	T.D.	Temperature Difference
PAR	Parabolic Reflector	RPM	Revolutions per Minute	TDD	Telecommunications Device for
P.B., PB	Push Button	R.S.	Rapid Start		the Deaf
Pc., Pcs.	Piece, Pieces	Rsr	Riser	T.E.M.	Transmission Electron Microscopy
P.C.	Portland Cement; Power Connector	RT	Round Trip	temp	Temperature, Tempered, Temporary
P.C.F.	Pounds per Cubic Foot	S.	Suction; Single Entrance; South	TFFN	Nylon Jacketed Wire

Abbreviations

TFE	Tetrafluoroethylene (Teflon)	U.L., UL	Underwriters Laboratory	w/	With
T. & G.	Tongue & Groove; Tar & Gravel	Uld.	Unloading	W.C., WC	Water Column; Water Closet
		Unfin.	Unfinished	W.F.	Wide Flange
Th., Thk.	Thick	UPS	Uninterruptible Power Supply	W.G.	Water Gauge
Thn.	Thin	URD	Underground Residential	Wldg.	Welding
Thrded	Threaded		Distribution	W. Mile	Wire Mile
Tilf.	Tile Layer, Floor	US	United States	W-O-L	Weldolet
Tilh.	Tile Layer, Helper	USGBC	U.S. Green Building Council	W.R.	Water Resistant
THHN	Nylon Jacketed Wire	USP	United States Primed	Wrck.	Wrecker
THW.	Insulated Strand Wire	UTMCD	Uniform Traffic Manual For Control	WSFU	Water Supply Fixture Unit
THWN	Nylon Jacketed Wire		Devices	W.S.P.	Water, Steam, Petroleum
T.L., TL	Truckload	UTP	Unshielded Twisted Pair	WT., Wt.	Weight
T.M.	Track Mounted	V	Volt	WWF	Welded Wire Fabric
Tot.	Total	VA	Volt Amperes	XFER	Transfer
T-O-L	Threadolet	VAT	Vinyl Asbestos Tile	XFMR	Transformer
tmpd	Tempered	V.C.T.	Vinyl Composition Tile	XHD	Extra Heavy Duty
TPO	Thermoplastic Polyolefin	VAV	Variable Air Volume	XHHW	Cross-Linked Polyethylene Wire
T.S.	Trigger Start	VC	Veneer Core	XLPE	Insulation
Tr.	Trade	VDC	Volts Direct Current	XLP	Cross-linked Polyethylene
Transf.	Transformer	Vent.	Ventilation	Xport	Transport
Trhv.	Truck Driver, Heavy	Vert.	Vertical	Y	Wye
Trlr	Trailer	V.F.	Vinyl Faced	yd	Yard
Trlt.	Truck Driver, Light	V.G.	Vertical Grain	yr	Year
TTY	Teletypewriter	VHF	Very High Frequency	Δ	Delta
TV	Television	VHO	Very High Output	%	Percent
T.W.	Thermoplastic Water Resistant Wire	Vib.	Vibrating	~	Approximately
		VLF	Vertical Linear Foot	Ø	Phase; diameter
UCI	Uniform Construction Index	VOC	Volatile Organic Compound	@	At
UF	Underground Feeder	Vol.	Volume	#	Pound; Number
UGND	Underground Feeder	VRP	Vinyl Reinforced Polyester	<	Less Than
UHF	Ultra High Frequency	W	Wire; Watt; Wide; West	>	Greater Than
U.I.	United Inch			Z	Zone

Index

Index

Index

For customer support on your Plumbing Costs with RSMeans Data, call 800.448.8182.

Index

For customer support on your Plumbing Costs with RSMeans Data, call 800.448.8182.

745

Index

For customer support on your Plumbing Costs with RSMeans Data, call 800.448.8182.

747

Index

Division Notes

	CREW	DAILY OUTPUT	LABOR-HOURS	UNIT	BARE COSTS				TOTAL INCL O&P
					MAT.	LABOR	EQUIP.	TOTAL	

Division Notes

	CREW	DAILY OUTPUT	LABOR-HOURS	UNIT	BARE COSTS				TOTAL INCL O&P
					MAT.	LABOR	EQUIP.	TOTAL	

Other Data & Services
A tradition of excellence in construction cost information and services since 1942

Table of Contents
Cost Data Selection Guide
RSMeans Data Online
Training

For more information visit our website at RSMeans.com

Unit prices according to the latest MasterFormat®

Cost Data Selection Guide

The following table provides definitive information on the content of each cost data publication. The number of lines of data provided in each unit price or assemblies division, as well as the number of crews, is listed for each data set. The presence of other elements such as reference tables, square foot models, equipment rental costs, historical cost indexes, and city cost indexes, is also indicated. You can use the table to help select the RSMeans data set that has the quantity and type of information you most need in your work.

Unit Cost Divisions	Building Construction	Mechanical	Electrical	Commercial Renovation	Square Foot	Site Work Landsc.	Green Building	Interior	Concrete Masonry	Open Shop	Heavy Construction	Light Commercial	Facilities Construction	Plumbing	Residential
1	621	454	475	576	0	543	200	377	505	620	560	322	1092	462	219
2	825	347	157	781	0	972	181	466	221	824	739	550	1268	355	340
3	1745	341	232	1265	0	1537	1043	355	2274	1745	1930	538	2028	317	445
4	960	22	0	920	0	724	180	613	1158	928	614	532	1175	0	446
5	1895	158	155	1099	0	858	1793	1106	735	1895	1031	985	1912	204	752
6	2468	18	18	2127	0	110	589	1544	281	2464	123	2157	2141	22	2677
7	1593	215	128	1633	0	580	761	532	523	1590	26	1326	1693	227	1046
8	2140	80	3	2733	0	255	1138	1813	105	2142	0	2328	2966	0	1552
9	2125	86	45	1943	0	313	464	2216	424	2062	15	1779	2379	54	1544
10	1088	17	10	684	0	232	32	898	136	1088	34	588	1179	237	224
11	1095	199	166	539	0	135	56	923	29	1062	0	229	1115	162	108
12	539	0	2	297	0	219	147	1552	14	506	0	272	1571	23	216
13	745	149	157	252	0	370	124	250	77	721	271	109	761	115	103
14	273	36	0	223	0	0	0	257	0	273	0	12	293	16	6
21	127	0	41	37	0	0	0	293	0	127	0	121	665	685	259
22	1165	7543	160	1226	0	2010	1061	849	20	1154	2109	875	7505	9400	719
23	1263	7010	639	1033	0	250	864	775	38	1246	191	980	5244	2011	579
25	0	0	14	14	0	0	0	0	0	0	0	0	0	0	0
26	1513	491	10473	1293	0	859	645	1160	55	1439	648	1360	10254	399	636
27	95	0	467	105	0	0	0	71	0	95	45	67	389	0	56
28	143	79	223	124	0	0	28	97	0	127	0	70	209	57	41
31	1511	733	610	807	0	3261	284	7	1216	1456	3280	607	1568	660	616
32	906	49	8	944	0	4523	419	418	367	877	1942	496	1803	140	544
33	1319	1100	563	310	0	3142	33	0	253	596	3277	135	1790	2165	161
34	112	0	47	9	0	195	0	0	36	67	226	0	141	0	0
35	18	0	0	0	0	327	0	0	0	18	442	0	84	0	0
41	63	0	0	34	0	8	0	22	0	62	31	0	69	14	0
44	75	79	0	0	0	0	0	0	0	0	0	0	75	75	0
46	23	16	0	0	0	274	261	0	0	23	264	0	33	33	0
48	8	0	36	2	0	0	21	0	0	8	15	8	21	0	8
Totals	26453	19222	14829	21010	0	21697	10324	16594	8467	25215	17813	16446	51423	17833	13297

Assem Div	Building Construction	Mechanical	Elec-trical	Commercial Renovation	Square Foot	Site Work Landscape	Assemblies	Green Building	Interior	Concrete Masonry	Heavy Construction	Light Commercial	Facilities Construction	Plumbing	Asm Div	Residential
A		15	0	190	165	579	599	0	0	537	573	155	24	0	1	378
B		0	0	848	2567	0	5673	56	329	1979	368	2108	174	0	2	211
C		0	0	647	956	0	1338	0	1645	146	0	846	251	0	3	591
D		1060	945	712	1866	72	2544	330	827	0	0	1353	1108	1090	4	851
E		0	0	85	262	0	302	0	5	0	0	259	5	0	5	391
F		0	0	0	114	0	143	0	0	0	0	114	0	0	6	357
G		527	447	318	312	3378	792	0	0	535	1349	205	293	677	7	307
															8	760
															9	80
															10	0
															11	0
															12	0
Totals		1602	1392	2800	6242	4029	11391	386	2806	3197	2290	5040	1855	1767		3926

Reference Section	Building Construction Costs	Mechanical	Electrical	Commercial Renovation	Square Foot	Site Work Landscape	Assem.	Green Building	Interior	Concrete Masonry	Open Shop	Heavy Construction	Light Commercial	Facilities Construction	Plumbing	Resi.
Reference Tables	yes	yes	yes	yes	no	yes	yes	yes	yes	yes	yes	yes	yes	yes	yes	yes
Models					111			25					50			28
Crews	584	584	584	564		584		584	584	584	562	584	562	564	584	560
Equipment Rental Costs	yes	yes	yes	yes		yes		yes	yes	yes	yes	yes	yes	yes	yes	yes
Historical Cost Indexes	yes	yes	yes	yes	yes	yes	yes	yes	yes	yes	yes	yes	yes	yes	yes	no
City Cost Indexes	yes	yes	yes	yes	yes	yes	yes	yes	yes	yes	yes	yes	yes	yes	yes	yes

Virtual classes offered monthly, call for the schedule—see note for details

2021 Training Class Options   📞 877.620.6245

RSMeans offers training classes in a variety of formats—eLearning training modules, instructor-led classes, on-site training, and virtual training classes. Due to our current Covid-19 situation, RSMeans has adjusted our current training offerings and instructor-led classes for 2021 have not been determined. Please visit our website at https://www.rsmeans.com/products/training/seminars for our current schedule of class offerings.

Training classes that are offered in one or more of our formats are: Building Systems and the Construction Process; Construction Cost Estimating: Concepts & Practice; Introduction to Estimating; Scope of Work for Facilities Estimating; Facilities Construction Estimating; Maintenance & Repair Estimating for Facilities; Mechanical & Electrical Estimating; RSMeans CostWorks CD; and RSMeans Online Training.

If you have any questions or you would like to register for any class or purchase a training module, call us at 877-620.6245.

Facilities Construction Estimating

In this two-day course, professionals working in facilities management can get help with their daily challenges to establish budgets for all phases of a project.

Some of what you'll learn:
- Determining the full scope of a project
- Understanding of RSMeans data and what is included in prices
- Identifying appropriate factors to be included in your estimate
- Creative solutions to estimating issues
- Organizing estimates for presentation and discussion
- Special estimating techniques for repair/remodel and maintenance projects
- Appropriate use of contingency, city cost indexes, and reference notes
- Techniques to get to the correct estimate quickly

Who should attend: facility managers, engineers, contractors, facility tradespeople, planners, and project managers.

Construction Cost Estimating: Concepts and Practice

This one- or two-day introductory course to improve estimating skills and effectiveness starts with the details of interpreting bid documents and ends with the summary of the estimate and bid submission.

Some of what you'll learn:
- Using the plans and specifications to create estimates
- The takeoff process—deriving all tasks with correct quantities
- Developing pricing using various sources; how subcontractor pricing fits in
- Summarizing the estimate to arrive at the final number
- Formulas for area and cubic measure, adding waste and adjusting productivity to specific projects
- Evaluating subcontractors' proposals and prices
- Adding insurance and bonds
- Understanding how labor costs are calculated
- Submitting bids and proposals

Who should attend: project managers, architects, engineers, owners' representatives, contractors, and anyone who's responsible for budgeting or estimating construction projects.

RSMeans data Training

Assessing Scope of Work for Facilities Construction Estimating

This two-day practical training program addresses the vital importance of understanding the scope of projects in order to produce accurate cost estimates for facilities repair and remodeling.

Some of what you'll learn:
- Discussions of site visits, plans/specs, record drawings of facilities, and site-specific lists
- Review of CSI divisions, including means, methods, materials, and the challenges of scoping each topic
- Exercises in scope identification and scope writing for accurate estimating of projects
- Hands-on exercises that require scope, takeoff, and pricing

Who should attend: corporate and government estimators, planners, facility managers, and others who need to produce accurate project estimates.

Maintenance & Repair Estimating for Facilities

This two-day course teaches attendees how to plan, budget, and estimate the cost of ongoing and preventive maintenance and repair for existing buildings and grounds.

Some of what you'll learn:
- The most financially favorable maintenance, repair, and replacement scheduling and estimating
- Auditing and value engineering facilities
- Preventive planning and facilities upgrading
- Determining both in-house and contract-out service costs
- Annual, asset-protecting M&R plan

Who should attend: facility managers, maintenance supervisors, buildings and grounds superintendents, plant managers, planners, estimators, and others involved in facilities planning and budgeting.

Mechanical & Electrical Estimating

This two-day course teaches attendees how to prepare more accurate and complete mechanical/electrical estimates, avoid the pitfalls of omission and double-counting, and understand the composition and rationale within the RSMeans mechanical/electrical database.

Some of what you'll learn:
- The unique way mechanical and electrical systems are interrelated
- M&E estimates—conceptual, planning, budgeting, and bidding stages
- Order of magnitude, square foot, assemblies, and unit price estimating
- Comparative cost analysis of equipment and design alternatives

Who should attend: architects, engineers, facilities managers, mechanical and electrical contractors, and others who need a highly reliable method for developing, understanding, and evaluating mechanical and electrical contracts.

Building Systems and the Construction Process

This one-day course was written to assist novices and those outside the industry in obtaining a solid understanding of the construction process – from both a building systems and construction administration approach.

Some of what you'll learn:
- Various systems used and how components come together to create a building
- Start with foundation and end with the physical systems of the structure such as HVAC and Electrical
- Focus on the process from start of design through project closeout

This training session requires you to bring a laptop computer to class.

Who should attend: building professionals or novices to help make the crossover to the construction industry; suited for anyone responsible for providing high level oversight on construction projects.

Practical Project Management for Construction Professionals

In this two-day course you will acquire the essential knowledge and develop the skills to effectively and efficiently execute the day-to-day responsibilities of the construction project manager.

Some of what you'll learn:
- General conditions of the construction contract
- Contract modifications: change orders and construction change directives
- Negotiations with subcontractors and vendors
- Effective writing: notification and communications
- Dispute resolution: claims and liens

Who should attend: architects, engineers, owners' representatives, and project managers.

RSMeans data Training

755

Training for our Online Estimating Solution

Construction estimating is vital to the decision-making process at each state of every project. Our online solution works the way you do. It's systematic, flexible, and intuitive. In this one-day class you will see how you can estimate any phase of any project faster and better.

Some of what you'll learn:
- Customizing our online estimating solution
- Making the most of RSMeans "Circle Reference" numbers
- How to integrate your cost data
- Generating reports, exporting estimates to MS Excel, sharing, collaborating, and more

Also offered as a self-paced or on-site training program!

Training for our CD Estimating Solution

This one-day course helps users become more familiar with the functionality of the CD. Each menu, icon, screen, and function found in the program is explained in depth. Time is devoted to hands-on estimating exercises.

Some of what you'll learn:
- Searching the database using all navigation methods
- Exporting RSMeans data to your preferred spreadsheet format
- Viewing crews, assembly components, and much more
- Automatically regionalizing the database

This training session requires you to bring a laptop computer to class.

When you register for this course you will receive an outline for your laptop requirements.

Also offered as a self-paced or on-site training program!

Site Work Estimating with RSMeans data

This one-day program focuses directly on site work costs. Accurately scoping, quantifying, and pricing site preparation, underground utility work, and improvements to exterior site elements are often the most difficult estimating tasks on any project. Some of what you'll learn:
- Evaluation of site work and understanding site scope including: site clearing, grading, excavation, disposal and trucking of materials, backfill and compaction, underground utilities, paving, sidewalks, and seeding & planting.
- Unit price site work estimates—Correct use of RSMeans site work cost data to develop a cost estimate.
- Using and modifying assemblies—Save valuable time when estimating site work activities using custom assemblies.

Who should attend: engineers, contractors, estimators, project managers, owners' representatives, and others who are concerned with the proper preparation and/or evaluation of site work estimates.

Please bring a laptop with capability to access the internet.

Facilities Estimating Using the CD

This two-day class combines hands-on skill-building with best estimating practices and real-life problems. You will learn key concepts, tips, pointers, and guidelines to save time and avoid cost oversights and errors.

Some of what you'll learn:
- Estimating process concepts
- Customizing and adapting RSMeans cost data
- Establishing scope of work to account for all known variables
- Budget estimating: when, why, and how
- Site visits: what to look for and what you can't afford to overlook
- How to estimate repair and remodeling variables

This training session requires you to bring a laptop computer to class.

Who should attend: facility managers, architects, engineers, contractors, facility tradespeople, planners, project managers, and anyone involved with JOC, SABRE, or IDIQ.

Registration Information

How to register
By Phone
Register by phone at 877.620.6245

Online
Register online at
RSMeans.com/products/services/training
Note: Purchase Orders or Credits Cards are required to register.

Two-day seminar registration fee - $1,300*.

One-Day Construction Cost Estimating or Building Systems and the Construction Process - $825*.

Two-day virtual training classes - $825*.

Three-day virtual training classes - $995*.

Instructor-led Government pricing
All federal government employees save off the regular seminar price. Other promotional discounts cannot be combined with the government discount. Call 781.422.5115 for government pricing.

CANCELLATION POLICY FOR INSTRUCTOR-LED CLASSES:
If you are unable to attend a seminar, substitutions may be made at any time before the session starts by notifying the seminar registrar at 781.422.5115 or your sales representative.
If you cancel twenty-one (21) days or more prior to the seminar, there will be no penalty and your registration fees will be refunded. These cancellations must be received by the seminar registrar or your sales representative and will be confirmed to be eligible for cancellation.
If you cancel fewer than twenty-one (21) days prior to the seminar, you will forfeit the registration fee.
In the unfortunate event of an RSMeans cancellation, RSMeans will work with you to reschedule your attendance in the same seminar at a later date or will fully refund your registration fee. RSMeans cannot be responsible for any non-refundable travel expenses incurred by you or another as a result of your registration, attendance at, or cancellation of an RSMeans seminar.
Any on-demand training modules are not eligible for cancellation, substitution, transfer, return, or refund.

AACE approved courses
Many seminars described and offered here have been approved for 14 hours (1.4 recertification credits) of credit by the AACE International Certification Board toward meeting the continuing education requirements for recertification as a Certified Cost Engineer/Certified Cost Consultant.

AIA Continuing Education
We are registered with the AIA Continuing Education System (AIA/CES) and are committed to developing quality learning activities in accordance with the CES criteria. Many seminars meet the AIA/CES criteria for Quality Level 2. AIA members may receive 14 learning units (LUs) for each two-day RSMeans course.

Daily course schedule
The first day of each seminar session begins at 8:30 a.m. and ends at 4:30 p.m. The second day begins at 8:00 a.m. and ends at 4:00 p.m. Participants are urged to bring a hand-held calculator since many actual problems will be worked out in each session.

Continental breakfast
Your registration includes the cost of a continental breakfast and a morning and afternoon refreshment break. These informal segments allow you to discuss topics of mutual interest with other seminar attendees. (You are free to make your own lunch and dinner arrangements.)

Hotel/transportation arrangements
We arrange to hold a block of rooms at most host hotels. To take advantage of special group rates when making your reservation, be sure to mention that you are attending the RSMeans Institute data seminar. You are, of course, free to stay at the lodging place of your choice. (Hotel reservations and transportation arrangements should be made directly by seminar attendees.)

Important
Class sizes are limited, so please register as soon as possible.

*Note: Pricing subject to change.